GRAPHS

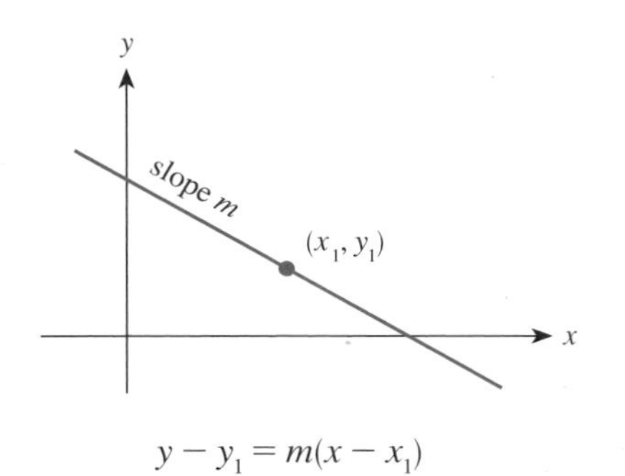

$y - y_1 = m(x - x_1)$

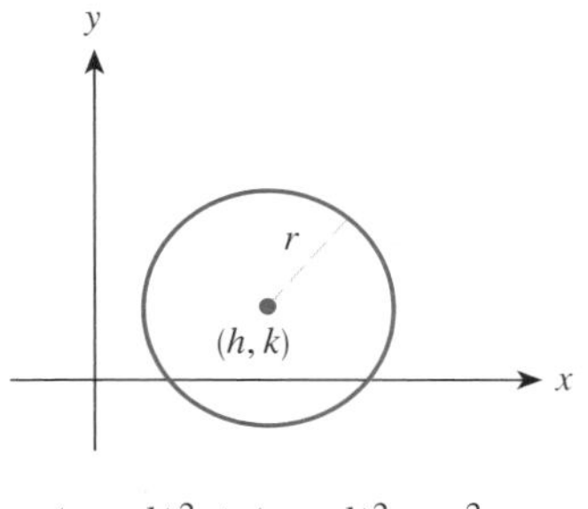

$(x - h)^2 + (y - k)^2 = r^2$

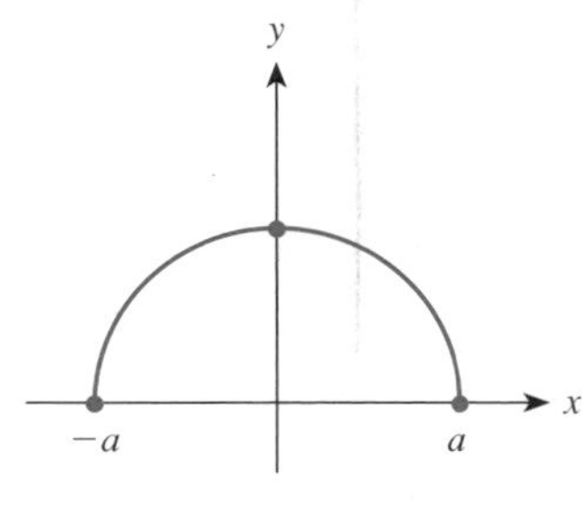

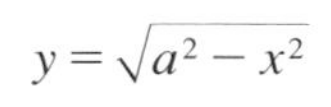

$y = \sqrt{a^2 - x^2}$

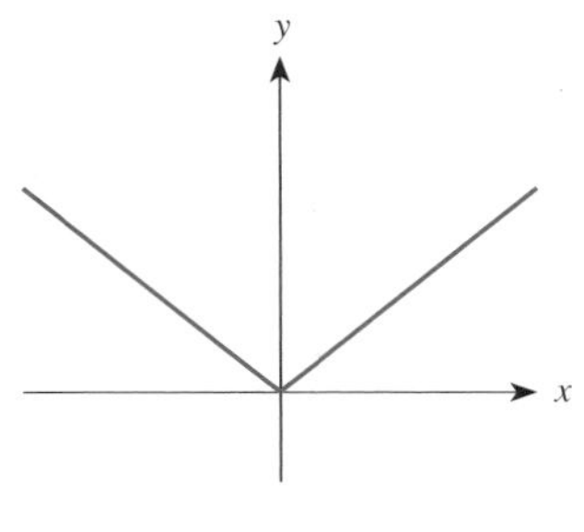

$y = |x|$

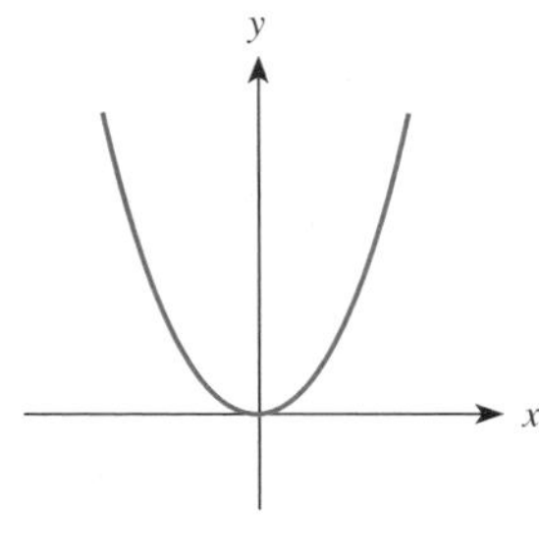

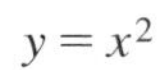

$y = x^2$

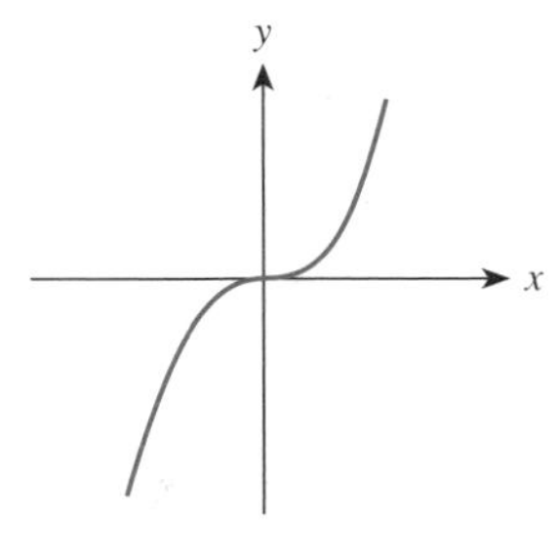

$y = x^3$

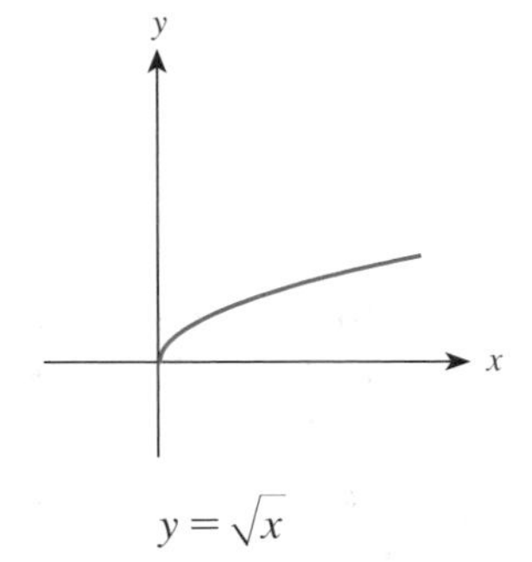

$y = \sqrt{x}$

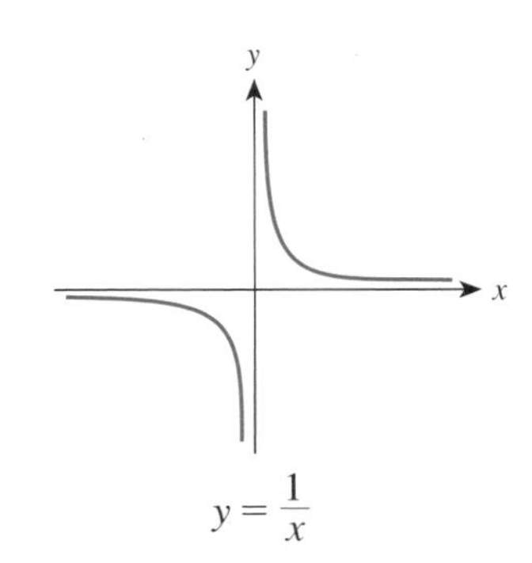

$y = \frac{1}{x}$

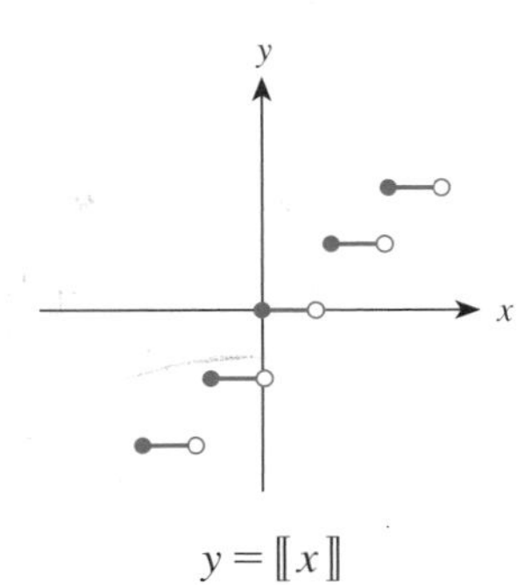

$y = [\![x]\!]$

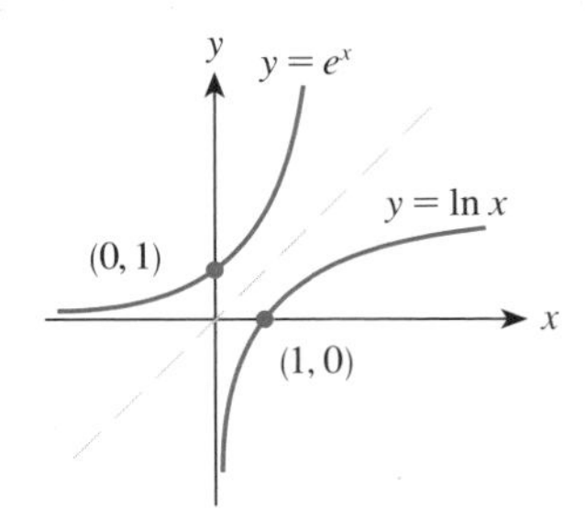

$y = e^x$ and $y = \ln x$

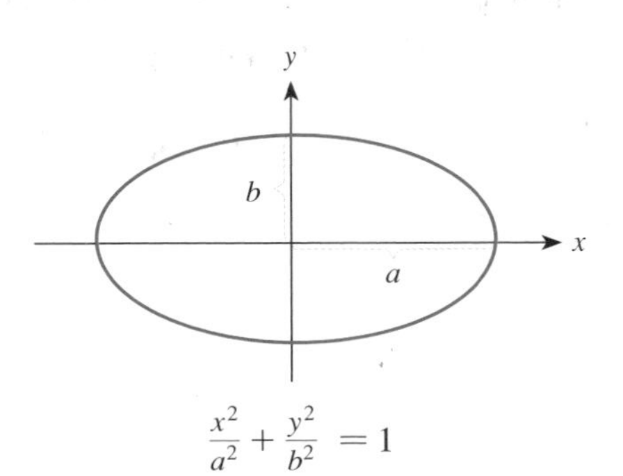

$\frac{x^2}{a^2} + \frac{y^2}{b^2} = 1$

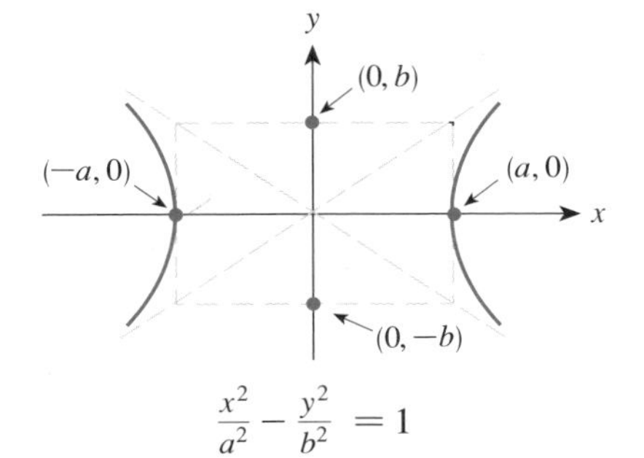

$\frac{x^2}{a^2} - \frac{y^2}{b^2} = 1$

Precalculus

with Unit-Circle Trigonometry

Fourth Edition

David Cohen
Late of University of California
Los Angeles

With

Theodore B. Lee
City College of San Francisco

David Sklar
San Francisco State University

Australia • Brazil • Japan • Korea • Mexico • Singapore • Spain • United Kingdom • United States

Precalculus: With Unit-Circle Trigonometry, Fourth Edition
David Cohen, Theodore B. Lee, David Sklar

Acquisitions Editor: John-Paul Ramin
Assistant Editor: Katherine Brayton
Editorial Assistant: Leata Holloway
Technology Project Manager: Earl Perry
Senior Marketing Manager: Karin Sandberg
Marketing Assistant: Erin Mitchell
Managing Marketing Communications Manager: Bryan Vann
Senior Project Manager, Editorial Production: Janet Hill
Senior Art Director: Vernon Boes
Senior Print/Media Buyer: Karen Hunt
Permissions Editor: Kiely Sisk
Production Service: Martha Emry
Text Designer: Rokusek Design
Art Editor: Martha Emry
Photo Researcher: Sue Howard
Copy Editor: Barbara Willette
Illustrator: Jade Myers
Cover Designer: Cheryl Carrington
Cover Image: Yva Momatiuk/John Eastcott
Compositor: G&S Typesetters, Inc.

Library of Congress Control Number: 2005921577
Student Edition:
ISBN-13: 978-0-534-40230-3
ISBN-10: 0-534-40230-5
Instructor's Edition:
ISBN-13: 978-0-534-40231-0
ISBN-10: 0-534-40231-3

Brooks/Cole
10 Davis Drive
Belmont, CA 94002-3098
USA

Printed in China by China Translation & Printing Services Limited
6 7 8 9 11 10 09

To our parents:
Ruth and Ernest Cohen
Lorraine and Robert Lee
Helen and Rubin Sklar

Tribute to David Cohen

Because he loved the part in "The Myth of Sisyphus," where Camus re-envisions Sisyphus eternally condemned to the task of pushing a giant boulder to the summit of a mountain only to watch it roll back down again—descending the hill and smiling, I like to remember my dad in similar moments of inbetweenness. When the unknown and hope would combine to make anything seem attainable. When becoming conscious of a situation did not necessarily signify limitation by it, but instead, liberation from it.

And so I imagine my dad, after hours of working at the computer, between a thoughtful sip of coffee and suddenly realizing the clearest way to word a problem. After years of early morning commutes to avoid 405 traffic, solitary interludes that began with waking to a still-dark sky and still-dreaming family, and ended with walking, transformed into a professor, into a UCLA lecture hall. After a lifetime of doing his best for those he cared about, intervals of good and poor health simply challenged him to give a little more. My dad endured his task bravely, from his decision 15 years ago to fight leukemia until May 2002, when the illness reminded us that even fathers and teachers are mortal, regardless of scheduled office hours or books left unwritten.

My dad once told me the five most important things in life to him, and teaching and writing these math books were both on that list. I was never a "registered" student of his, but to anyone using this book, especially students, I pass on what I know he would have told me: Your best is always good enough. Enjoy.

Emily Cohen

• • •

My dad was gifted at teaching. He had an innate ability to explain things simply and relevantly. He excelled at combining his explanations with a patient ear and his self-defined nerdy sense of humor.

I remember when my dad taught me to read. I had become upset one night while we were out to dinner because I couldn't read the menu. Frustrated, I expressed my dismay from my booster chair. My dad responded with the first of my reading lessons. He began by pointing out something I already knew, the letter "a." He carefully explained that this letter was actually a word . . . every day after that he would write new words on one of his ubiquitous lined yellow notepads and teach them to me. Eventually we began to form simple sentences. He always kept my attention because he made up sentences that made me laugh. The frogs go jog, the cat loves the dog, so on and so forth. My dad's sense of humor was quite captivating.

I was lucky in high school to have someone to assist me with all of my mathematical questions. My math teachers seemed to have a special aptitude for making formulas, theories, and problems both complicated and boring. My dad was good at simplifying these matters for me. He was so good at explaining math to me that I would often remark afterward, with the hindsight of an enlightened one, "Oh, that's all? Why didn't they just say that in the first place?" I took comfort in knowing that math didn't have to be complicated when it was explained well.

Hemingway said that the key to immortality was to write a book. I feel lucky because my dad left behind several. Try reading one—you might find that you actually like it!

Jennifer Cohen

• • •

I first met Dave Cohen in the spring of 1981, when I signed up for an algebra course he was teaching as part of my graduate studies at UCLA. I quickly took a liking to his conversational style of teaching and the clear explanations he would use to illustrate an idea. Dave and I began an informal, weekly meeting to discuss solutions to various problems he would pose. Usually he would buy me a cup of coffee and present me with some new trigonometric identity he had discovered, or some conic section property he had come across in an old 1886 algebra book. We would discuss the problem ("Did it have a solution? Was it really an identity? Could we prove it?"), then go about our business.

In the summer of 1981 I had the opportunity to TA a course called Precalculus, which used Dave's notes (sometimes written by hand), rather than a traditional textbook. I had no idea what the term "precalculus" meant, and soon learned that these notes Dave had prepared were beginning to define, or at the least greatly expand, the subject. It was the beginning of a textbook.

Two years later, after graduating UCLA, I was pleased to receive a copy of *Precalculus,* by David Cohen. Dave had scribbled a note to me saying: "Thanks for the great coffee breaks." I also was pleased to see that some of our weekly problems appeared in that book (others showed up years later). More importantly, that book, and every text Dave has ever written, talked to you. Even now, when I read this book, I can hear Dave talking, and sometimes even sense him listening. Lecturing was never a part of Dave Cohen's vocabulary.

Since that time Dave has written books entitled *College Algebra, Trigonometry,* and *Algebra and Trigonometry.* Most have gone on to multiple editions, including this one. In each edition Dave sought the better explanation, the "cooler" problem, the more interesting data set. I've had the honor of working with him on all of these books, and we had developed a mathematical friendship. Perhaps some of you reading this have a great study partner; one who intuitively knows what the other is thinking. That was the relationship Dave and I had.

I last saw Dave in November 2001, when we met at UCLA to (naturally) have a cup of coffee and discuss his latest textbook. He was excited about the quality of problems and interesting data sets he had found to use in the book and felt it was going to be the best book he had ever written. This present text is strongly based on that material.

In May 2002, just after completing much of this manuscript, Dave Cohen passed away from complications caused by his leukemia. His legacy of fine textbooks and great teaching has influenced a generation of students and teachers. Everyone who had contact with Dave feels a bit richer from the experience. His total lack of ego, curiosity about the world, and respect for what others think, made him one of the finest human beings I have ever met.

Ross Rueger

About the Co-Authors

David Sklar was a longtime friend of David Cohen, whom he met while they were both attending graduate school at San Francisco State University. Because of his relationship with the author, he has followed the progress of this book since work started on the first edition. He brings a unique blend of teaching and professional experience to the table, having taught at San Francisco State University, Sonoma State University, Menlo College, and City College of San Francisco. At the same time, David is a researcher and consultant in the field of optics—you'll notice that interest in some of the new group projects in the text. David is an active member of the American Mathematical Society, Mathematical Association of America, is past chairman of the Northern California section of the MAA, and serves on the mathematics department advisory board at San Francisco State University.

David is joined by Theodore Lee, professor of mathematics at City College of San Francisco. Ted is a highly respected teacher who brings nearly 30 years of teaching experience to this project and provides a valuable perspective on teaching precalculus mathematics. On three separate occasions he has been honored with distinguished teaching awards from his colleagues at CCSF. Ted has also been honored by Alpha Sigma Gamma, CCSF's student honor society, as a favorite teacher. A fourth-generation Californian, Ted received his bachelor's degree from the University of California, Berkeley, and his master's degree from the University of California, Los Angeles.

Both David and Ted understand the factors that make the book so special to the people who use it—the clear writing, the conversational style, the variety of problems (including many challenging ones), and the thoughtful use of technology. They have made every effort to maintain the standard and quality of David Cohen's work.

Contents

LIST OF PROJECTS

Preface

This text develops the elements of college algebra and trigonometry in a straightforward manner. As in the earlier editions, our goal has been to create a book that is *accessible* to the student. The presentation is student-oriented in three specific ways. First, we've tried to talk to, rather than lecture at, the student. Second, examples are consistently used to introduce, to explain, and to motivate concepts. And third, many of the initial exercises for each section are carefully coordinated with the worked examples in that section.

AUDIENCE

In writing *Precalculus with Unit Circle Trigonometry,* we have assumed that the students have been exposed to intermediate algebra, but that they have not necessarily mastered that subject. Also, for many college algebra students, there may be a gap of several years between their last mathematics course and the present one. Appendix B consists of review sections for such students, reviewing topics on integer exponents, *n*th roots, rational exponents, factoring, and fractional expressions. In Chapter 1, many references refer the reader to Appendix B for further practice.

CURRICULUM REFORM

This new edition of *Precalculus* reflects several of the major themes that have developed in the curriculum reform movement of the past decade. Graphs, visualization of data, and functions are now introduced much earlier and receive greater emphasis. Many sections now contain more examples and exercises involving applications and real-life data. In addition to the Writing Mathematics sections from the previous edition, there are now mini projects or projects at the ends of many sections. These give the students additional opportunities to discuss, explore, learn, and write mathematics, often using real-life data.

TECHNOLOGY

In the following discussion and throughout this text, the term *graphing utility* refers to either a graphing calculator or a computer with software for graphing and analyzing functions. Over the past decade, all of us in the mathematics teaching community have become increasingly aware of the graphing utility and its potential for making a positive impact on our students' learning. We are also aware of the limitations of the graphing utility as a *sole* analysis device.

The role of the graphing utility is expanded in this new edition: The existence of the graphing utility is taken for granted and some examples do make use of this technology. However, just as with the previous edition, this remains a text in which the central focus is on mathematics and its applications. If the instructor chooses, the text can be used without reference to the graphing utility, but a scientific calculator will be required since the text no longer includes logarithmic or trigonometric tables. Students already familiar with a graphing utility will, at a minimum, need to read the page in Section 1.5 explaining how to specify the dimensions of a viewing rectangle, since that notation accompanies some figures in the text. Overall, the quality and relevance of the graphing utility exercises is vastly improved over the previous edition. Graphing utility exercises (identified by the symbol Ⓖ) are now integrated into the regular exercise sets.

CHANGES IN THIS EDITION

Comments and suggestions from students, instructors, and reviewers have helped us to revise this text in a number of ways that we believe will make the book more useful to the instructor and more accessible to the student. As previously mentioned, graphing utility exercises now appear in virtually every exercise set as well as in the text examples. Some of the major changes occur in the following areas.

Chapter 1 Some of the previous edition's Chapter 1 review material has been shifted to two appendices in the new edition. Notes to students at appropriate places remind them to consult one of these appendices if the use of one of these topics seems unclear. Section 1.4 introduces graphs and data visualization and incorporates much of the material that appeared in Section 2.1 of the previous edition. Section 1.5 includes an introduction to graphing utilities and material formerly covered in Section 2.2; Sections 1.6 and 1.7 correspond to Sections 2.3 and 2.4 of earlier editions.

Chapter 2 Section 2.1 incorporates much of the material on quadratic equations that appeared in Section 1.4 of the former text, although graphing utilities are used more extensively in the examples and exercises. Section 2.2 covers equations that can be solved using some of the techniques used in solving linear and quadratic equations and includes a discussion of extraneous solutions. Sections 2.3 and 2.4 correspond to Sections 2.5 and 2.6 of the previous edition.

Chapter 3 Section 3.1 includes an expanded introduction to the function concept. Functions are introduced at length using algebraic, verbal, tabular, and graphical forms. Functions as models are introduced and used in examples and exercises. Implicit functions are introduced in the section project. In Section 3.3, material on the average rate of change of a function has been expanded, and there is an increased emphasis on applications. Sections 3.4, 3.5, and 3.6 correspond to Sections 3.3, 3.4, and 3.5 of the previous edition.

Chapter 4 Section 4.1 includes additional examples on applications of the regression line. There is also a new discussion on the use of spreadsheets in generating scatter plots and regression lines. In Section 4.2, there is new material on modeling real-life data with quadratic functions. Also, first and second differences are introduced as a tool for determining whether a set of data points may be generated by a linear or quadratic function. Section 4.3 on iteration and population dynamics is now marked "optional," and some exercises that don't refer directly to the text exposition have been deleted. In Section 4.5, in order to better focus on problem solving rather than algebraic manipulation in the applied max/min problems, the vertex formula $x = -b/2a$ is introduced as a simple tool for determining the vertex of a parabola. Instructors who require the use of completing the square rather than the vertex formula should find their students well prepared in view of the previous work on that topic in Sections 1.7, 2.1, and 4.2.

Chapter 5 Section 5.2 contains new material explaining in a careful, but student-friendly, manner exactly why the use of e as the base for an exponential function indeed simplifies matters. In Section 5.3 the Richter magnitude scale is discussed as an example of a logarithmic scale. Historical background motivating the use of the logarithmic scale is provided. The Section 5.6 project explains how loan payments in an amortization schedule are calculated under compound interest. In Section 5.7 the concept of an average relative growth rate is explained in the text, and a follow-up exercise gives an empirical introduction to the instantaneous relative growth rate. These ideas are used to supply insight into the interpretation and use of the exponential growth constant k.

Chapter 6 In Section 6.3 we have clarified the explanations in the use of the now four-step procedure for evaluating trigonometric functions. Section 6.3 ends with a project on using a basic trigonometric inequality to construct a linear approximation for the sine fuction. A mini project at the end of Section 6.4 uses the graphing utility to help understand the nature of identities and provide a way to visualize them. We now present both right-triangle and trigonometric identity approaches in the solutions of several examples (see Examples 4, 5, and 6 in Section 6.5), and encourage students to use whichever method they prefer. Two projects follow Section 6.5. The first uses data gathered from observations of transits of Venus to calculate the distance from Earth to the Sun. The second project discusses the construction of regular polygons.

Chapter 7 The exercises in this chapter make more extensive use of the graphing utility. A project on constructing square waves and other nontrigonometric wave forms follows Section 7.2. This material is used in a project on an introduction to Fourier series that follows Section 7.3. Section 7.3 includes a worked example using a trigonometric function to model weather data. Section 7.4 ends with a project developing a model for the motion of a piston in an internal combustion engine.

Chapter 8 The Section 8.1 project uses an addition formula for the sine function to design a Fresnel lens, and the project concluding Section 8.2 uses an addition formula for the cosine to derive a superposition formula useful in differential equations. The derivation leads to a more general formula for combining waveforms. The project at the end of Section 8.4 applies trigonometric identities and techniques for solving trigonometric equations to a measurement problem in the manufacture of eyeglass lenses. The last project in the chapter provides a careful development of the two versions of the inverse secant function most commonly encountered in first-year calculus.

Chapter 9 The project at the end of Section 9.1 uses Snell's Law to analyze the optical phenomenon of "lifting." The project at the end of Section 9.3 uses the geometric approach to vectors to develop vector algebra, providing an alternative to the component approach to vector algebra presented in Section 9.4. In Exercise Set

9.4, the exercises on the dot product, defined using components, have been rewritten to emphasize the algebra, as opposed to the arithmetic, of the dot product. These exercises are extended to reveal an expression for the dot product in purely geometric terms. The algebraic and geometric approaches to vectors are combined in the Section 9.4 project to develop vector and scalar descriptions of lines and circles in the plane and to use these constructions to analyze an optics problem on ray tracing. This project leads naturally to the project after Section 9.5 that deals with using parametric equations for lines and circles in the plane to develop formulas for an important trigonometric substitution used in calculus.

Chapter 10 In Section 10.1 the work on simple linear systems is applied to supply and demand models in determining market equilibrium. The Section 10.2 project provides an introduction to the Leontief input-output model in economics. Section 10.3 discusses the use of graphing calculators and spreadsheets in computing matrix products. Students can apply this technology in an extended project on the use of matrices in the study of communication networks. Section 10.4 discusses the use of graphing calculators and spreadsheets in computing matrix inverses. Section 10.5 does likewise for determinants. Section 10.4 also includes new examples and exercises on using matrices to code and decode messages. In an extended project, students will solve a Leontief input-output problem involving a seven-sector model for the U.S. economy.

Chapter 11 Section 11.2 defines focal length and focal ratio for a parabola in preparation for an example and exercises on radio telescopes and parabolic reflectors. Mini Project 2 at the end of the section discusses the classic string and T-square construction of the parabola. Section 11.4 includes new material on the perihelion and aphelion for an elliptical orbit. The use of hyperbolas in determining a location appears in the Section 11.5 project.

Chapter 12 In Section 12.2 long division for polynomials is discussed just prior to synthetic division. Two methods for solving certain types of cubic equations are explained and applied in a project for Section 12.4. This complements the discussion in the text on the history of polynomial equations.

Chapter 13 Section 13.3 now has an example using a recursive sequence to model population growth and a project to explore some variations to the model. Section 13.4 concludes with a project on sigma notation that develops some additional algebra for simplifying sums.

The Accompanying CD ***Interactive Video Skillbuilder CD* (0-534-40239-9)** The Interactive Video Skillbuilder CD-ROM contains video instruction covering each chapter of the text. The problems worked during each video lesson are shown first, so that the students can try working them before watching the solution. To help students evaluate their progress, each section contains a 10-question Web quiz (the quiz results can be e-mailed to the instructor) and each chapter contains a chapter test, with the answer to each problem on each test. This CD-ROM also includes MathCue tutorial and answers with step-by-step explanations, a Quiz function that enables students to generate quiz problems keyed to problem types from each section of the book, a Chapter Test that provides many problems keyed to problem types from each chapter, and a Solution Finder that allows students to enter their own basic problems and receive step-by-step graphing calculator tutorial for precalculus and college algebra, featuring examples, exercises, and video tutorials. Also new, English/Spanish closed-caption translations can be selected to display along with the video instruction.

SUPPLEMENTARY MATERIALS

For the Instructor

Instructor's Edition This special version of the complete student text contains a Resource Integration Guide, an easy-to-use tool that helps you quickly compile a teaching and learning program that complements both the text and your personal teaching style. A complete set of answers is printed in the back of the text. **ISBN 0-534-40231-3**

Test Bank The Test Bank includes multiple tests per chapter as well as final exams. The tests are made up of a combination of multiple-choice, free-response, true/false, and fill-in-the-blank questions. **ISBN 0-534-40238-0**

Complete Solutions Manual The Complete Solutions Manual provides worked-out solutions to all of the problems in the text. **ISBN 0-534-40234-8**

Text-Specific Videotapes These text-specific videotape sets, available at no charge to qualified adopters of the text, feature 10- to 20-minute problem-solving lessons that cover each section of every chapter. **ISBN 0-534-40219-4**

Instructor's Resource CD-ROM This CD-Rom provides the instructor with dynamic media tools for teaching precalculus. PowerPoint lecture slides, combined with all of the instructor supplements in electronic format, are available on this CD-Rom. **ISBN 0-534-40227-5**

iLrn™ Instructor Version Efficient and versatile, **iLrn** gives you the power to transform the teaching and learning experience. **iLrn Instructor Version** is made up of two components, **iLrn Testing** and **iLrn Tutorial. iLrn Testing** is an internet-ready, text-specific testing suite that allows instructors to customize exams and track student progress in an accessible, browser-based format. **iLrn** offers full algorithmic generation of problems and free-response mathematics. **iLrn Tutorial** is a text-specific, interactive tutorial software program that is delivered via the web (at **http://www.iLrn.com**) and is offered in both student and instructor versions. Like **iLrn Testing,** it is browser-based, making it an intuitive mathematical guide even for students with little technological proficiency. **iLrn Tutorial** allows students to work with real math notation in real time, providing instant analysis and feedback. The tracking program built into the instructor version of the software enables instructors to carefully monitor student progress. The complete integration of the testing, tutorial, and course management components simplifies your routine tasks. Results flow automatically to your grade book, and you can easily communicate with individuals, sections, or entire courses. **ISBN 0-534-28038-2**

A personalized study plan can be generated by course-specific diagnostic built into **iLrn.** With a personalized study plan, students can focus their time where they need it the most, creating a positive learning environment and paving a pathway to success in their mathematics course. Additional features you'll find when using **iLrn:**

- Manipulate and format your tests with the rich text format (RTF) conversion tool. RTF conversion allows instructors to open tests in most word processors, such as Microsoft Word, for further formatting and customization.

- iLrn offers more problem types to provide greater variety and more diverse challenges in your tests.

The iLrn Tutorial interface effectively engages students and helps them learn math concepts faster.

WebTutor ToolBox for WebCT and Blackboard
ISBN 0-534-27488-9 WebCT; ISBN 0-534-27489-7 Blackboard
Preloaded with content and available free via access code when packaged with this text, **WebTutor ToolBox for WebCT and Blackboard** pairs all the content of this text's rich Book Companion Website with all the sophisticated course management functionality of a **WebCT** or **Blackboard** product. You can assign materials (including online quizzes) and have the results flow automatically to your grade book. **ToolBox** is ready to use as soon as you log on—or, you can customize its preloaded content by uploading images and other resources, adding web links, or creating your own practice materials. Students only have access to student resources on the website. Instructors can enter an access code to reach password-protected Instructor Resources.

For the Student

Student Solutions Manual The Student Solutions Manual provides worked-out solutions to the odd-numbered problems in the text. **ISBN 0-534-40232-1**

Cengage Learning Mathematics Website
academic.cengage.com/math
When a Brooks/Cole mathematics text is adopted, the instructor and students have access to everything from book-specific resources to newsgroups. It's a great way to make teaching and learning an interactive and intriguing experience.

iLrn™ Tutorial Student Version Free access to this text-specific, interactive, web-based tutorial system is included with the text. **iLrn Tutorial Student Version** is browser-based, making it an intuitive mathematical guide even for students with little technological proficiency. Simple to use, **iLrn Tutorial** allows students to work with real math notation in real time, providing instant analysis and feedback. The entire textbook is available in PDF format through **iLrn Tutorial,** as are section-specific video tutorials, unlimited practice problems, and additional student resources such as a glossary, web links, and more. And, when students get stuck on a particular problem or concept, they need only log on to **vMentor™,** accessed through **iLrn Tutorial,** where they can talk (using their own computer microphones) to **vMentor** tutors who will skillfully guide them through the problem using an interactive whiteboard for illustration.

REVIEW BOARD

We have worked diligently with our editor to ensure that we're providing an updated version of the same Cohen textbook you know and trust. Brooks/Cole asked some of the longest-standing and most loyal users of the text to participate in a review board that compared the parts of this book we wrote to the parts David Cohen wrote. Their feedback has been very positive, and we're confident that we've maintained the quality and approach of David Cohen's work. We are grateful to the following review board participants for their contributions and would like to acknowledge them.

Donna J. Bailey, Truman State University
Satish Bhatnagar, University of Nevada, Las Vegas
M. Hilary Davies, University of Alaska, Anchorage
Greg Dietrich, Florida Community College at Jacksonville
John Gosselin, University of Georgia
Johnny A. Johnson, University of Houston
Richard Riggs, New Jersey City University
Ross Rueger, College of the Sequoias
Fred Schifando, Pennsylvania State University
Jeffrey S. Snapp, Harvard–Westlake School
Thomas J. Walters, University of California, Los Angeles (retired)
Sandra Wray-McAfee, University of Michigan, Dearborn
Loris I. Zucca, Kingwood College

ACKNOWLEDGMENTS

A wonderful team of editors, accuracy checkers, and proofreaders has helped to eliminate many errors from the original manuscript. The remaining errors are those of the authors. Inspired by Donald Knuth, we would like to offer a reward of $5.00 to the first person to inform us of each remaining error. We can be reached through our editor whose e-mail address is John-Paul.Ramin@Cengage.com.

Many students and teachers from both colleges and high schools have made constructive suggestions about the text and exercises, and we thank them for that. We would also like to thank David Cohen's cousin, Bruce Cohen, for helpful discussions on using technology in the classroom, and Tom Walters for suggesting a project on identities and graphs. We are particularly indebted to Eric Barkan for numerous discussions on the material and his detailed comments on each of the seemingly endless revisions of the applications-oriented projects as well as his help in preparing the manuscript.

Special thanks to Ross Rueger who wrote the supplementary manuals and prepared the answer section for the text. Ross worked with David Cohen on many of David's textbooks and we very much appreciate that he agreed to work with us on this new edition.

Thanks to Charles Heuer for his careful work in checking the text and the exercise answers for accuracy. It has been a rare pleasure to work with Martha Emry on the production of the text, and we thank her for her patience and extraordinary ability to create order out of chaos and to keep us on track at all times. To John-Paul Ramin, Janet Hill, Katherine Brayton, Karin Sandberg, Vernon Boes, Leata Holloway, and the staff at Brooks/Cole, thank you for all your work and help in bringing this manuscript into print. Finally we want to thank David Cohen's sister, Susan Cohen, and David's wife, Annie Cohen, for their warm encouragement.

Theodore B. Lee
David Sklar

1 CHAPTER Fundamentals

Real numbers, equations, graphs—these topics set the stage for our work in precalculus. How much from previous courses should you remember about solving equations? Section 1.3 provides a review of the fundamentals. The rest of the chapter reviews and begins to extend what you've learned in previous courses about graphs and graphing. For example, we use graphs to visualize trends in

- Spending by the television networks to broadcast the Olympic Games (Exercise 21, page 27)
- Internet usage (Exercise 23, pages 27–28)
- Carbon dioxide levels in the atmosphere (Example 5, page 25)
- U.S. population growth (Exercises 7 and 8 on page 53)

1.1 SETS OF REAL NUMBERS

Natural numbers have been used since time immemorial; fractions were employed by the ancient Egyptians as early as 1700 B.C.; and the Pythagoreans, in ancient Greece, about 400 B.C., discovered numbers, like $\sqrt{2}$, which cannot be fractions.—Stefan Drobot in *Real Numbers* (Englewood Cliffs, N.J.: Prentice-Hall, Inc., 1964)

What secrets lie hidden in decimals?—Stephan P. Richards in *A Number for Your Thoughts* (New Providence, N.J.: S. P. Richards, 1982)

Here, as in your previous mathematics courses, most of the numbers we deal with are real numbers. These are the numbers used in everyday life, in the sciences, in industry, and in business. Perhaps the simplest way to define a real number is this: A **real number** is any number that can be expressed in decimal form. Some examples of real numbers are

$$7\ (= 7.000\ldots)$$
$$\sqrt{2}\ (= 1.4142\ldots)$$
$$-2/3\ (= -0.\overline{6})$$

(Recall that the bar above the 6 in the decimal $-0.\overline{6}$ indicates that the 6 repeats indefinitely.)

Certain sets of real numbers are referred to often enough to be given special names. These are summarized in the box that follows.

As you've seen in previous courses, the real numbers can be represented as points on a *number line,* as shown in Figure 1. As indicated in Figure 1, the point associated with the number zero is referred to as the **origin.**

The fundamental fact here is that there is a **one-to-one correspondence** between the set of real numbers and the set of points on the line. This means that each real number is identified with exactly one point on the line; conversely, with each point on the line we identify exactly one real number. The real number associated with a given point is called the **coordinate** of the point. As a practical matter, we're

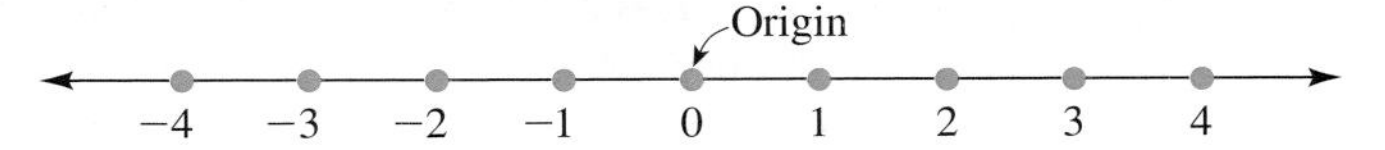

Figure 1

PROPERTY SUMMARY Sets of Real Numbers

Name	Definition and Comments	Examples
Natural numbers	These are the ordinary counting numbers: 1, 2, 3, and so on.	1, 4, 29, 1066
Integers	These are the natural numbers along with their negatives and zero.	−26, 0, 1, 1776
Rational numbers	As the name suggests, these are the real numbers that are *ratios* of two integers (with nonzero denominators). It can be proved that a real number is rational if and only if its decimal expansion *terminates* (e.g., 3.15) or *repeats* (e.g., $2.\overline{43}$).	$4\ (= \frac{4}{1}), -\frac{2}{3}$, $1.7\ (= \frac{17}{10}), 4.\overline{3}$, $4.1\overline{73}$
Irrational numbers	These are the real numbers that are not rational. Section A.3 of the Appendix contains a proof of the fact that the number $\sqrt{2}$ is irrational. The proof that π is irrational is more difficult. The first person to prove that π is irrational was the Swiss mathematician J. H. Lambert (1728–1777).	$\sqrt{2}, 3 + \sqrt{2}$, $3\sqrt{2}, \pi, 4 + \pi$, 4π

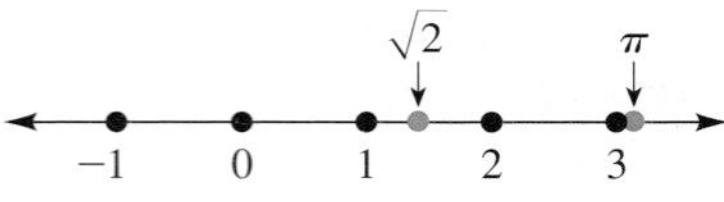

Figure 2

usually more interested in relative locations than precise locations on a number line. For instance, since π is approximately 3.1, we show π slightly to the right of 3 in Figure 2. Similarly, since $\sqrt{2}$ is approximately 1.4, we show $\sqrt{2}$ slightly less than halfway from 1 to 2 in Figure 2.

It is often convenient to use number lines that show reference points other than the integers used in Figure 2. For instance, Figure 3(a) displays a number line with reference points that are multiples of π. In this case it is the integers that we then locate approximately. For example, in Figure 3(b) we show the approximate location of the number 1 on such a line.

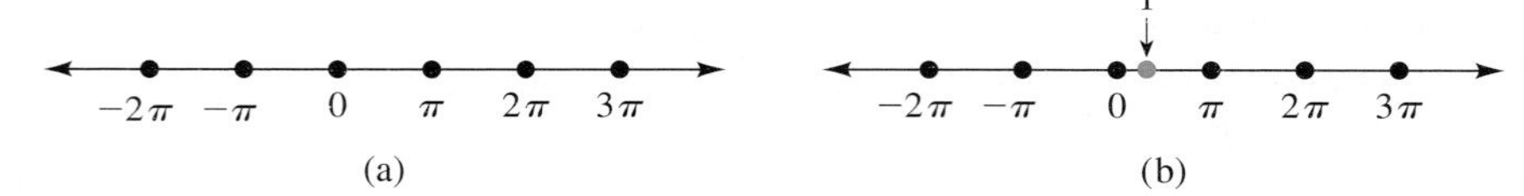

Figure 3

Two of the most basic relations for real numbers are **less than** and **greater than,** symbolized by $<$ and $>$, respectively. For ease of reference, we review these and two related symbols in the box on page 3.

In general, relationships involving real numbers and any of the four symbols $<$, $\leq$, $>$, and $\geq$ are called **inequalities.** One of the simplest uses of inequalities is in defining certain sets of real numbers called *intervals.* Roughly speaking, any uninterrupted portion of the number line is referred to as an **interval.** In the definitions that follow, you'll see notations such as $a < x < b$. This means that *both* of the inequalities $a < x$ and $x < b$ hold; in other words, the number x is between a and b.

(a) The open interval (a, b) contains all real numbers from a to b, excluding a and b.

(b) The closed interval $[a, b]$ contains all real numbers from a to b, including a and b.

Figure 4

DEFINITION Open Intervals and Closed Intervals

The **open interval** (a, b) consists of all real numbers x such that $a < x < b$. See Figure 4(a).

The **closed interval** $[a, b]$ consists of all real numbers x such that $a \leq x \leq b$. See Figure 4(b).

PROPERTY SUMMARY Notation for Less Than and Greater Than

Notation	Definition	Examples
$a < b$	a is less than b. On a number line, oriented as in Figure 1, 2, or 3, the point a lies to the left of b.	$2 < 3$; $-4 < 1$
$a \le b$	a is less than or equal to b.	$2 \le 3$; $3 \le 3$
$b > a$	b is greater than a. On a number line oriented as in Figure 1, 2, or 3, the point b lies to the right of a. ($b > a$ is equivalent to $a < b$.)	$3 > 2$; $0 > -1$
$b \ge a$	b is greater than or equal to a.	$3 \ge 2$; $3 \ge 3$

Note that the brackets in Figure 4(b) are used to indicate that the numbers a and b are included in the interval $[a, b]$, whereas the parentheses in Figure 4(a) indicate that a and b are excluded from the interval (a, b). At times you'll see notation such as $[a, b)$. This stands for the set of all real numbers x such that $a \le x < b$. Similarly, $(a, b]$ denotes the set of all numbers x such that $a < x \le b$.

EXAMPLE 1 Understanding interval notation

Show each interval on a number line, and specify inequalities describing the numbers x in each interval.

$$[-1, 2] \quad (-1, 2) \quad (-1, 2] \quad [-1, 2)$$

SOLUTION

See Figure 5.

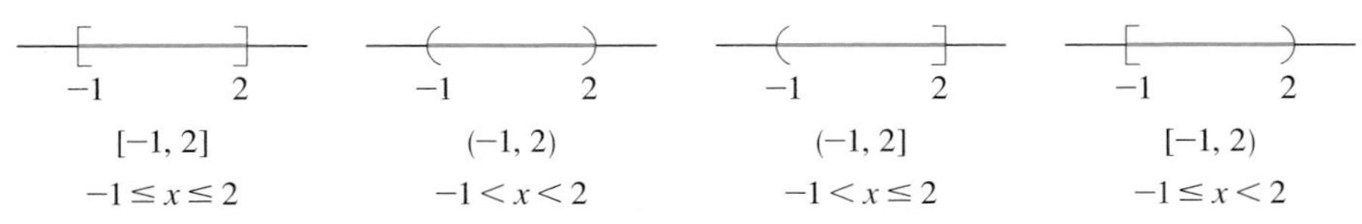

Figure 5

In addition to the four types of intervals shown in Figure 5, we can also consider **unbounded intervals.** These are intervals that extend indefinitely in one direction or the other, as shown, for example, in Figure 6. We also have a convenient notation for unbounded intervals; for example, we indicate the unbounded interval in Figure 6 with the notation $(2, \infty)$.

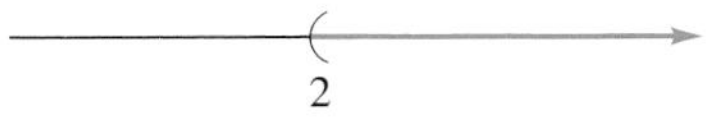

Figure 6

2

COMMENT AND CAUTION The symbol ∞ is read *infinity.* It is not a real number, and its use in the context $(2, \infty)$ is only to indicate that the interval has no right-hand boundary. In the box that follows we define the five types of unbounded intervals. Note that the last interval, $(-\infty, \infty)$, is actually the entire real-number line.

PROPERTY SUMMARY Unbounded Intervals

For a real number a the notations for unbounded intervals are:

Notation	Defining Inequality	Example
(a, ∞)	$x > a$	$(2, \infty)$; 2
$[a, \infty)$	$x \geq a$	$[2, \infty)$; 2
$(-\infty, a)$	$x < a$	$(-\infty, 2)$; 2
$(-\infty, a]$	$x \leq a$	$(-\infty, 2]$; 2
$(-\infty, \infty)$		$(-\infty, \infty)$; 2

EXAMPLE 2 Understanding notation for unbounded intervals

Indicate each set of real numbers on a number line:
(a) $(-\infty, 4]$; **(b)** $(-3, \infty)$.

0 1 2 3 4 5
Figure 7

−4 −3 −2 −1 0 1
Figure 8

SOLUTION

(a) The interval $(-\infty, 4]$ consists of all real numbers that are less than or equal to 4. See Figure 7.

(b) The interval $(-3, \infty)$ consists of all real numbers that are greater than -3. See Figure 8.

We conclude this section by mentioning that our treatment of the real-number system has been rather informal, and we have not derived any of the rules of arithmetic and algebra using the most basic properties of the real numbers. However, we do list those basic properties and derive some of their consequences in Section A.2 of the Appendix.

EXERCISE SET 1.1

A

In Exercises 1–10, determine whether the number is a natural number, an integer, a rational number, or an irrational number. (Some numbers fit in more than one category.) The following facts will be helpful in some cases: Any number of the form $\sqrt{n}$, where n is a natural number that is not a perfect square, is irrational. Also, the sum, difference, product, and quotient of an irrational number and a nonzero rational are all irrational. (For example, the following four numbers are irrational: $\sqrt{6}$, $\sqrt{10} - 2$, $3\sqrt{15}$, and $-5\sqrt{3}/2$.)

1. (a) -203
(b) $203/2$

2. (a) $27/4$
(b) $\sqrt{27/4}$

3. (a) 10^6
(b) $10^6/10^7$

4. (a) 8.7
(b) $8.\overline{7}$

5. (a) 8.74
(b) $8.\overline{74}$

6. (a) $\sqrt{99}$
(b) $\sqrt{99} + 1$

7. $3\sqrt{101} + 1$ **8.** $(3 - \sqrt{2}) + (3 + \sqrt{2})$
9. $(\sqrt{5} + 1)/4$ **10.** $(0.1234)/(0.5677)$

In each of Exercises 11–20, draw a number line similar to the one shown in Figure 1. Then indicate the approximate location of the given number. Where necessary, make use of the approximations $\sqrt{2} \approx 1.4$ *and* $\sqrt{3} \approx 1.7$. *(The symbol* $\approx$ *means* is approximately equal to.*)*

11. $11/4$ **12.** $-7/8$ **13.** $1 + \sqrt{2}$
14. $1 - \sqrt{2}$ **15.** $\sqrt{2} - 1$ **16.** $-\sqrt{2} - 1$
17. $\sqrt{2} + \sqrt{3}$ **18.** $\sqrt{2} - \sqrt{3}$ **19.** $(1 + \sqrt{2})/2$
20. $(2\sqrt{3} + 1)/2$

In Exercises 21–30, draw a number line similar to the one shown in Figure 3(a). Then indicate the approximate location of the given number.

21. $\pi/2$ **22.** $3\pi/2$ **23.** $\pi/6$ **24.** $7\pi/4$
25. -1 **26.** 3 **27.** $\pi/3$ **28.** $3/2$
29. $2\pi + 1$ **30.** $2\pi - 1$

In Exercises 31–40, say whether the statement is TRUE *or* FALSE. *(In Exercises 37–40, do not use a calculator or table; use instead the approximations* $\sqrt{2} \approx 1.4$ *and* $\pi \approx 3.1$.*)*

31. $-5 < -50$ **32.** $0 < -1$ **33.** $-2 \le -2$
34. $\sqrt{7} - 2 \ge 0$ **35.** $\frac{13}{14} > \frac{15}{16}$ **36.** $0.\overline{7} > 0.7$
37. $2\pi < 6$ **38.** $2 \le (\pi + 1)/2$ **39.** $2\sqrt{2} \ge 2$
40. $\pi^2 < 12$

In Exercises 41–54, express each interval using inequality notation and show the given interval on a number line.

41. $(2, 5)$ **42.** $(-2, 2)$ **43.** $[1, 4]$
44. $[-\frac{3}{2}, \frac{1}{2}]$ **45.** $[0, 3)$ **46.** $(-4, 0]$
47. $(-3, \infty)$ **48.** $(\sqrt{2}, \infty)$ **49.** $[-1, \infty)$
50. $[0, \infty)$ **51.** $(-\infty, 1)$ **52.** $(-\infty, -2)$
53. $(-\infty, \pi]$ **54.** $(-\infty, \infty)$

B

55. The value of the irrational number π, correct to ten decimal places (without rounding off), is 3.1415926535. By using a calculator, determine to how many decimal places each of the following quantities agrees with π.

(a) $(4/3)^4$: This is the value used for π in the Rhind papyrus, an ancient Babylonian text written about 1650 B.C.

(b) $22/7$: Archimedes (287–212 B.C.) showed that $223/71 < \pi < 22/7$. The use of the approximation $22/7$ for π was introduced to the Western world through the writings of Boethius (ca. 480–520), a Roman philosopher, mathematician, and statesman. Among all fractions with numerators and denominators less than 100, the fraction $22/7$ is the best approximation to π.

(c) $355/113$: This approximation of π was obtained in fifth-century China by Zu Chong-Zhi (430–501) and his son. According to David Wells in *The Penguin Dictionary of Curious and Interesting Numbers* (Harmondsworth, Middlesex, England: Viking Penguin, Ltd., 1986), "This is the best approximation of any fraction below 103993/33102."

(d) $\dfrac{63}{25}\left(\dfrac{17 + 15\sqrt{5}}{7 + 15\sqrt{5}}\right)$: This approximation for π was obtained by the Indian mathematician Scrinivasa Ramanujan (1887–1920).

Remark: A simple approximation that agrees with π through the first 14 decimal places is $\dfrac{355}{113}\left(1 - \dfrac{0.0003}{3533}\right)$. This approximation was also discovered by Ramanujan. For a fascinating account of the history of π, see the book by Petr Beckmann, *A History of* π, 16th ed. (New York: Barnes & Noble Books, 1989), and for a more modern look at π, see Richard Preston's article, "The Mountains of Pi," in *The New Yorker* (March 2, 1992, pp. 36–67).

C

In Exercises 56–58, give an example of irrational numbers a and b such that the indicated expression is **(a)** *rational and* **(b)** *irrational.*

56. $a + b$ **57.** ab **58.** a/b

59. (a) Give an example in which the result of raising a rational number to a rational power is an irrational number.

(b) Give an example in which the result of raising an irrational number to a rational power is a rational number.

60. Can an irrational number raised to an irrational power yield an answer that is rational? This problem shows that the answer is "yes." (However, if you study the following solution very carefully, you'll see that even though we've answered the question in the affirmative, we've not pinpointed the specific case in which an irrational number raised to an irrational power is rational.)

(a) Let $A = (\sqrt{2})^{\sqrt{2}}$. Now, either A is rational or A is irrational. If A is rational, we are done. Why?

(b) If A is irrational, we are done. Why?
Hint: Consider $A^{\sqrt{2}}$.

Remark: For more about this problem and related questions, see the article "Irrational Numbers," by J. P. Jones and S. Toporowski in *American Mathematical Monthly,* vol. 80 (1973), pp. 423–424.

1.2 ABSOLUTE VALUE

There has been a real need in analysis for a convenient symbolism for "absolute value" . . . and the two vertical bars introduced in 1841 by Weierstrass, as in $|z|$, have met with wide adoption; . . .—Florian Cajori in *A History of Mathematical Notations,* vol. 1 (La Salle, Ill.: The Open Court Publishing Co., 1928)

As an aid in measuring distances on the number line, we review the concept of *absolute value.* We begin with a definition of absolute value that is geometric in nature. Then, after you have developed some familiarity with the concept, we explain a more algebraic approach that is often useful in analytical work.

DEFINITION Absolute Value (geometric version)

The **absolute value** of a real number x, denoted by $|x|$, is the distance from x to the origin.

For instance, because the numbers 5 and -5 are both five units from the origin, we have $|5| = 5$ and $|-5| = 5$. Here are three more examples:

$$|17| = 17 \qquad |-2/3| = 2/3 \qquad |0| = 0$$

EXAMPLE 1 Evaluating expressions containing absolute values

Evaluate each expression:
(a) $5 - |6 - 7|$; **(b)** $||-2| - |-3||$.

SOLUTION

(a) $5 - |6 - 7| = 5 - |-1|$
$= 5 - 1 = 4$

(b) $||-2| - |-3|| = |2 - 3|$
$= |-1| = 1$

As we said at the beginning of this section, there is an equivalent, more algebraic way to define absolute value. According to this equivalent definition, the value of $|x|$ is x itself when $x \geq 0$, and the value of $|x|$ is $-x$ when $x < 0$. We can write this symbolically as follows:

DEFINITION Absolute Value (algebraic version)

$$|x| = \begin{cases} x & \text{when } x \geq 0 \\ -x & \text{when } x < 0 \end{cases}$$

EXAMPLE
$|-7| = -(-7) = 7$

By looking at examples with specific numbers, you should be able to convince yourself that both definitions yield the same result. We use the algebraic definition of absolute value in Examples 2 and 3.

EXAMPLE 2 Rewriting expressions to eliminate absolute value

Rewrite each expression in a form that does not contain absolute values:
(a) $|\pi - 4| + 1$; **(b)** $|x - 5|$, given that $x \geq 5$; **(c)** $|t - 5|$, given that $t < 5$.

SOLUTION

(a) The quantity $\pi - 4$ is negative (since $\pi \approx 3.14$), and therefore its absolute value is equal to $-(\pi - 4)$. In view of this, we have

$$|\pi - 4| + 1 = -(\pi - 4) + 1 = -\pi + 5$$

(b) Since $x \geq 5$, the quantity $x - 5$ is nonnegative, and therefore its absolute value is equal to $x - 5$ itself. Thus we have

$$|x - 5| = x - 5 \qquad \text{when } x \geq 5$$

(c) Since $t < 5$, the quantity $t - 5$ is negative. Therefore its absolute value is equal to $-(t - 5)$, which in turn is equal to $5 - t$. In view of this, we have

$$|t - 5| = 5 - t \qquad \text{when } t < 5$$

EXAMPLE 3 Simplifying an expression containing absolute values

Simplify the expression $|x - 1| + |x - 2|$, given that x is in the open interval $(1, 2)$.

SOLUTION

Since x is greater than 1, the quantity $x - 1$ is positive, and consequently,

$$|x - 1| = x - 1$$

On the other hand, we are also given that x is less than 2. Therefore the quantity $x - 2$ is negative, and we have

$$|x - 2| = -(x - 2) = -x + 2$$

Putting things together now, we can write

$$\begin{aligned} |x - 1| + |x - 2| &= (x - 1) + (-x + 2) \\ &= -1 + 2 = 1 \end{aligned}$$

In the box that follows, we list several basic properties of the absolute value. Each of these properties can be derived from the definitions. (With the exception of the *triangle inequality,* we shall omit the derivations. For a proof of the triangle inequality, see Exercise 67.)

PROPERTY SUMMARY Properties of Absolute Value

1. For all real numbers x, we have
   (a) $|x| \geq 0$;
   (b) $x \leq |x|$ and $-x \leq |x|$.
   (c) $|x|^2 = x^2$.
2. For all real numbers a and b, we have
   (a) $|ab| = |a||b|$ and $|a/b| = |a|/|b|$ $(b \neq 0)$;
   (b) $|a + b| \leq |a| + |b|$ (the triangle inequality).

EXAMPLE 4 Rewriting an expression to eliminate absolute value

Write the expression $|-2 - x^2|$ in an equivalent form that does not contain absolute values.

SOLUTION
Note that x^2 is nonnegative for any real number x, so $2 + x^2$ is positive. Then $-2 - x^2 = -(2 + x^2)$ is negative. Thus

$$\begin{aligned} |-2 - x^2| &= -(-2 - x^2) && \text{using the algebraic definition of absolute value} \\ &= 2 + x^2 \end{aligned}$$

Alternatively,

$$\begin{aligned} |-2 - x^2| &= |-1(2 + x^2)| \\ &= |-1||2 + x^2| && \text{using Property 2(a)} \\ &= 2 + x^2 \end{aligned}$$

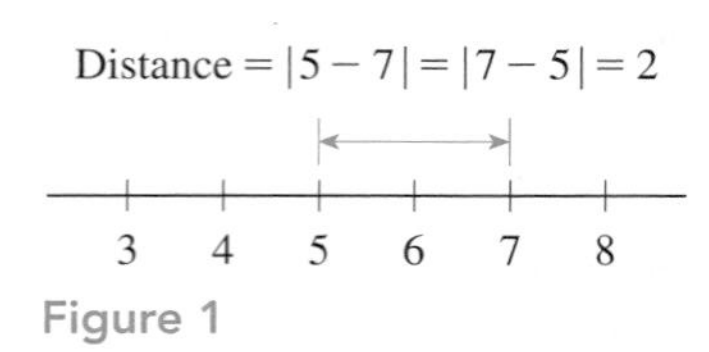

Figure 1

If we think of the real numbers as points on a number line, the distance between two numbers a and b is given by the absolute value of their difference. For instance, as indicated in Figure 1, the distance between 5 and 7, namely, 2, is given by either $|5 - 7|$ or $|7 - 5|$. For reference, we summarize this simple but important fact as follows.

PROPERTY SUMMARY Distance on a Number Line

For real numbers a and b, the **distance** between a and b is $|a - b| = |b - a|$.

EXAMPLE 5 Using absolute value to rewrite statements regarding distance

Rewrite each of the following statements using absolute value notation:
(a) The distance between 12 and −5 is 17.
(b) The distance between x and 2 is 4.
(c) The distance between x and 2 is less than 4.
(d) The number t is more than five units from the origin.

SOLUTION
(a) $|12 - (-5)| = 17$ or $|-5 - 12| = 17$
(b) $|x - 2| = 4$ or $|2 - x| = 4$
(c) $|x - 2| < 4$ or $|2 - x| < 4$
(d) $|t| > 5$

EXAMPLE 6 Displaying intervals defined by absolute value inequalities

In each case, the set of real numbers satisfying the given inequality is one or more intervals on the number line. Show the interval(s) on a number line.
(a) $|x| < 2$ **(b)** $|x| > 2$ **(c)** $|x - 3| < 1$ **(d)** $|x - 3| \geq 1$

Figure 2
$|x| < 2$

Figure 3
$|x| > 2$

Figure 4
$|x - 3| < 1$

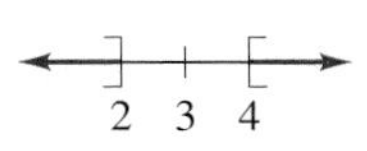

Figure 5
$|x - 3| \geq 1$

SOLUTION

(a) The given inequality tells us that the distance from x to the origin is less than two units. So, as indicated in Figure 2, the number x must lie in the open interval $(-2, 2)$.

(b) The condition $|x| > 2$ means that x is more than two units from the origin. Thus, as indicated in Figure 3, the number x lies either to the right of 2 or to the left of -2.

(c) The given inequality tells us that x must be less than one unit away from 3 on the number line. Looking one unit to either side of 3, then, we see that x must lie between 2 and 4 and x cannot equal 2 or 4. See Figure 4.

(d) The inequality $|x - 3| \geq 1$ says that x is at least one unit away from 3 on the number line. This means that either $x \geq 4$ or $x \leq 2$, as shown in Figure 5. [Here's an alternative way of thinking about this: The numbers satisfying the given inequality are precisely those numbers that do *not* satisfy the inequality in part (c). So for part (d), we need to shade that portion of the number line that was not shaded in part (c).]

EXERCISE SET 1.2 

A

In Exercises 1–16, evaluate each expression.

1. $|3|$
2. $3 + |-3|$
3. $|-6|$
4. $-6 - |-6|$
5. $|-1 + 3|$
6. $|-6 + 3|$
7. $|-\frac{4}{5}| - \frac{4}{5}$
8. $|\frac{4}{5}| - \frac{4}{5}$
9. $|-6 + 2| - |4|$
10. $|-3 - 4| - |-4|$
11. $||-8| + |-9||$
12. $||-8| - |-9||$
13. $\left|\dfrac{27 - 5}{5 - 27}\right|$
14. $\dfrac{|27 - 5|}{|5 - 27|}$
15. $|7(-8)| - |7| \cdot |-8|$
16. $|(-7)^2| + |-7|^2 - (-|-3|)^3$

In Exercises 17–24, evaluate each expression, given that $a = -2$, $b = 3$, and $c = -4$.

17. $|a - b|^2$
18. $a^2 - |bc|$
19. $|c| - |b| - |a|$
20. $|b + c| - |b| - |c|$
21. $|a + b|^2 - |b + c|^2$
22. $\dfrac{|a| + |b| + |c|}{|a + b + c|}$
23. $\dfrac{a + b + |a - b|}{2}$
24. $\dfrac{a + b - |a - b|}{2}$

In Exercises 25–38, rewrite each expression without using absolute value notation.

25. $|\sqrt{2} - 1| - 1$
26. $|1 - \sqrt{2}| + 1$
27. $|x - 3|$ given that $x \geq 3$
28. $|x - 3|$ given that $x < 3$
29. $|t^2 + 1|$
30. $|x^4 + 1|$
31. $|-\sqrt{3} - 4|$
32. $|-\sqrt{3} - \sqrt{5}|$
33. $|x - 3| + |x - 4|$ given that $x < 3$
34. $|x - 3| + |x - 4|$ given that $x > 4$
35. $|x - 3| + |x - 4|$ given that $3 < x < 4$
36. $|x - 3| + |x - 4|$ given that $x = 4$
37. $|x + 1| + 4|x + 3|$ given that $-\frac{5}{2} < x < -\frac{3}{2}$
38. $|x + 1| + 4|x + 3|$ given that $x < -3$

In Exercises 39–48, rewrite each statement using absolute value notation, as in Example 5.

39. The distance between x and 1 is 1/2.
40. The distance between x and 1 is less than 1/2.
41. The distance between x and 1 is at least 1/2.
42. The distance between x and 1 exceeds 1/2.
43. The distance between y and -4 is less than 1.
44. The distance between x^3 and -1 is at most 0.001.
45. The number y is less than three units from the origin.
46. The number y is less than one unit from the number t.
47. The distance between x^2 and a^2 is less than M.
48. The sum of the distances of a and b from the origin is greater than or equal to the distance of $a + b$ from the origin.

In Exercises 49–60, the set of real numbers satisfying the given inequality is one or more intervals on the number line. Show the interval(s) on a number line.

49. $|x| < 4$
50. $|x| < 2$
51. $|x| > 1$
52. $|x| > 0$
53. $|x - 5| < 3$
54. $|x - 4| < 4$
55. $|x - 3| \leq 4$
56. $|x - 1| \leq \frac{1}{2}$
57. $|x + \frac{1}{3}| < \frac{3}{2}$
58. $|x + \frac{\pi}{2}| > 1$
59. $|x - 5| \geq 2$
60. $|x + 5| \geq 2$

B

61. In parts (a) and (b), sketch the interval or intervals corresponding to the given inequality:
(a) $|x - 2| < 1$;
(b) $0 < |x - 2| < 1$.
(c) In what way do your answers in (a) and (b) differ? (The distinction is important in the study of *limits* in calculus.)

62. Show that for all real numbers a and b, we have

$$|a| - |b| \le |a - b|$$

Hint: Beginning with the identity $a = (a - b) + b$, take the absolute value of each side and then use the triangle inequality.

63. Show that

$$|a + b + c| \le |a| + |b| + |c|$$

for all real numbers a, b, and c. *Hint*: The left-hand side can be written $|a + (b + c)|$. Now use the triangle inequality.

64. Explain why there are no real numbers that satisfy the equation $|x^2 + 4x| = -12$.

C

65. (As background for this exercise, you might want to work Exercise 23.) Prove that

$$\max(a, b) = \frac{a + b + |a - b|}{2}$$

Hint: Consider three separate cases: $a = b$; $a > b$; and $b > a$.

66. (As background for this exercise, you might want to work Exercise 24.) Prove that

$$\min(a, b) = \frac{a + b - |a - b|}{2}$$

67. Complete the following steps to prove the triangle inequality.
(a) Let a and b be real numbers. Which property in the summary box on page 7 tells us that $a \le |a|$ and $b \le |b|$?
(b) Add the two inequalities in part (a) to obtain $a + b \le |a| + |b|$.
(c) In a similar fashion, add the two inequalities $-a \le |a|$ and $-b \le |b|$ and deduce that $-(a + b) \le |a| + |b|$.
(d) Why do the results in parts (b) and (c) imply that $|a + b| \le |a| + |b|$?

1.3 SOLVING EQUATIONS (REVIEW AND PREVIEW)

I learned algebra fortunately by not learning it at school, and knowing that the whole idea was to find out what x was, and it didn't make any difference how you did it.—Physicist Richard Feynman (1918–1988) in Jagdish Mehra's *The Beat of a Different Drum* (New York: Oxford University Press, 1994)

The title of al-Khwarizmi's second and most important book, Hisab al-jabr w'al muqabala [830] . . . *has given us the word* algebra. Al-jabr *means transposing a quantity from one side of an equation to the other, while* muqabala *signifies the simplification of the resulting equation.* —Stuart Hollingdale in *Makers of Mathematics* (Harmondsworth, Middlesex, England: Penguin Books, Ltd., 1989)

"Algebra is a merry science," Uncle Jakob would say. "We go hunting for a little animal whose name we don't know, so we call it x. When we bag our game we pounce on it and give it its right name."—Physicist Albert Einstein (1879–1955)

Consider the familiar expression for the area of a circle of radius r, namely, πr^2. Here π is a constant; its value never changes throughout the discussion. On the other hand, r is a variable; we can substitute any positive number for r to obtain the area of a particular circle. More generally, by a **constant** we mean either a particular number (such as π, or -17, or $\sqrt{2}$) or a letter with a value that remains fixed (although perhaps unspecified) throughout a given discussion. In contrast, a **variable** is a letter for which we can substitute any number selected from a given set of numbers. The given set of numbers is called the **domain** of the variable.

Some expressions will make sense only for certain values of the variable. For instance, $1/(x - 3)$ will be undefined when x is 3 (for then the denominator is zero). So in this case we would agree that the domain of the variable x consists of all real numbers except $x = 3$. Similarly, throughout this chapter we adopt the following convention.

The Domain Convention

The domain of a variable in a given expression is the set of all real-number values of the variable for which the expression is defined.

It's customary to use the letters near the end of the alphabet for variables; letters from the beginning of the alphabet are used for constants. For example, in the expression $ax + b$, the letter x is the variable and a and b are constants.

EXAMPLE 1 Specifying variables, constants, and the domain in an expression

Specify the variable, the constants, and the domain of the variable for each of the following expressions:

(a) $3x + 4$; **(b)** $\dfrac{1}{(t-1)(t+3)}$; **(c)** $ay^2 + by + c$; **(d)** $4x + 3x^{-1}$.

SOLUTION

	VARIABLE	CONSTANTS	DOMAIN
(a) $3x + 4$	x	3, 4	The set of all real numbers.
(b) $\dfrac{1}{(t-1)(t+3)}$	t	$1, -1, 3$	The set of all real numbers except $t = 1$ and $t = -3$.
(c) $ay^2 + by + c$	y	a, b, c	The set of all real numbers.
(d) $4x + 3x^{-1}$	x	4, 3	The set of all real numbers except $x = 0$.

Note: The number 2 that appears in part (c) above is an exponent; y^2 is shorthand notation for the product $y \times y$. Similarly, in part (d) x^{-1} is shorthand notation for $1/x$.

Now let's review the terminology and skills used in solving two basic types of equations: *linear equations* and *quadratic equations*.

DEFINITION Linear Equation in One Variable

A **linear** or **first-degree equation in one variable** is an equation that can be written in the form

$$ax + b = 0 \qquad \text{with } a \text{ and } b \text{ real numbers and } a \neq 0$$

Here are three examples of linear equations in one variable:

$$2x = 10, \qquad 3m + 1 = 2, \qquad \text{and} \qquad \frac{y}{2} = \frac{y}{3} + 1$$

As with any equation involving a variable, each of these equations is neither true nor false *until* we replace the variable with a number. By a **solution** or a **root** of an equation in one variable, we mean a value for the variable that makes the equation a true statement. For example, the value $x = 5$ is a solution of the equation $2x = 10$, since, with $x = 5$, the equation becomes $2(5) = 10$, which is certainly true. We also say in this case that the value $x = 5$ **satisfies** the equation. To check an equation means to verify that the original equation with the solution substituted for the variable is a true statement.

Equations that become true statements for *all* values in the domain of the variable are called **identities.** Two examples of identities are

$$x^2 - 9 = (x - 3)(x + 3) \qquad \text{and} \qquad \frac{4x^2}{x} = 4x$$

The first is true for all real numbers; the second is true for all real numbers except 0. In contrast to this, a **conditional equation** is true only for some (or perhaps none) of the values of the variable. Two examples of conditional equations are $2x = 10$ and $x = x + 1$. The first of these is true only when $x = 5$. The second equation has no solution (because, intuitively at least, no number can be one more than itself).

We say that two equations are **equivalent** when they have exactly the same solutions. In this section, and throughout the text, the basic method for solving an equation in one variable involves writing a sequence of equivalent equations until we finally reach an equivalent equation of the form

$$\text{variable} = \text{a number}$$

which explicitly displays a solution of the original equation. In generating equivalent equations, we rely on the following three principles. (These can be justified by using the properties of real numbers discussed in Appendix A.2.)

Procedures That Yield Equivalent Equations

1. Adding or subtracting the same quantity on both sides of an equation produces an equivalent equation.
2. Multiplying or dividing both sides of an equation by the same nonzero quantity produces an equivalent equation.
3. Simplifying an expression on either side of an equation produces an equivalent equation.

The examples that follow show how these principles are applied in solving various equations. *Note*: Beginning in Example 2, we use some basic factoring techniques from elementary or intermediate algebra. If you find that you need a quick reference for factoring formulas, see the inside back cover of this book. For a detailed review (with many examples and practice exercises) see Appendix B.4.

EXAMPLE 2 Solving equations equivalent to linear equations

(a) Solve: $3[1 - 2(x + 1)] = 2 - x$.

(b) Solve: $ax + b = c$; where a, b, and c are constants, $a \neq 0$.

(c) Solve: $\dfrac{1}{x + 5} = \dfrac{2}{x - 3} + \dfrac{2x + 2}{x^2 + 2x - 15}$.

SOLUTION

(a)
$$\begin{aligned}
3[1 - 2(x + 1)] &= 2 - x \\
3[1 - 2x - 2)] &= 2 - x && \text{simplifying the left-hand side} \\
3(-1 - 2x) &= 2 - x \\
-3 - 6x &= 2 - x \\
-3 - 5x &= 2 && \text{adding } x \text{ to both sides} \\
-5x &= 5 && \text{adding 3 to both sides} \\
x &= -1 && \text{dividing both sides by } -5
\end{aligned}$$

CHECK Replacing x with -1 in the original equation yields

$$\begin{aligned}
3[1 - 2(0)] &\stackrel{?}{=} 2 - (-1) \\
3(1) &\stackrel{?}{=} 2 + 1 \qquad \text{True}
\end{aligned}$$

(b)
$$\begin{aligned}
ax + b &= c \\
ax &= c - b && \text{subtracting } b \text{ from both sides} \\
x &= \frac{c - b}{a} && \text{dividing both sides by } a \text{ (recall that } a \neq 0)
\end{aligned}$$

CHECK Replacing x with $\frac{c - b}{a}$ in the original equation yields

$$\begin{aligned}
a\left(\frac{c - b}{a}\right) + b &\stackrel{?}{=} c \\
c - b + b &\stackrel{?}{=} c \\
c &\stackrel{?}{=} c \qquad \text{True}
\end{aligned}$$

(c) A common strategy in solving equations with fractions is to multiply through by the least common denominator. This eliminates the need to work with fractions. By factoring the denominator $x^2 + 2x - 15$, we obtain

$$\frac{1}{x + 5} = \frac{2}{x - 3} + \frac{2x + 2}{(x + 5)(x - 3)}$$

From this we see that the least common denominator for the three fractions is $(x + 5)(x - 3)$. Now, multiplying both sides by this least common denominator, we have

$$\frac{(x + 5)(x - 3)}{x + 5} = \frac{2(x + 5)(x - 3)}{x - 3} + \frac{(2x + 2)(x + 5)(x - 3)}{(x + 5)(x - 3)}$$

$$\begin{aligned}
x - 3 &= 2(x + 5) + 2x + 2 \\
x - 3 &= 2x + 10 + 2x + 2 && \text{simplifying} \\
x - 3 &= 4x + 12 && \text{simplifying} \\
-3x - 3 &= 12 && \text{subtracting } 4x \text{ from both sides} \\
-3x &= 15 && \text{adding 3 to both sides} \\
x &= -5 && \text{dividing both sides by } -3
\end{aligned}$$

CHECK The preceding steps show that *if* the equation has a solution, then the solution is $x = -5$. With $x = -5$, however, the left-hand side of the original equation becomes $1/(-5 + 5)$, or $1/0$, which is undefined. We conclude therefore that the given equation has no solution.

In Example 2(c) the value $x = -5$, which does not check in the original equation, is called an **extraneous root** or **extraneous solution.** How is it that an extraneous solution was generated in Example 2(c)? We multiplied both sides by $(x + 5)(x - 3)$. Since we didn't know at that stage whether the quantity $(x + 5)(x - 3)$ was nonzero, we could not be certain that the resulting equation was actually an equivalent equation. [Indeed, as it turns out with $x = -5$, the quantity $(x + 5)(x - 3)$ *is* equal to zero.] For this reason, it is always necessary to check in the original equation any solutions you obtain as a result of multiplying both sides of an equation by an expression involving the variable. We restate this advice in the box that follows.

PROPERTY SUMMARY Extraneous Solutions

Multiplying both sides of an equation by an expression involving the variable may introduce extraneous solutions that do not check in the original equation. Therefore, it is always necessary to check any candidates for solutions that you obtain in this manner.

EXAMPLE 3 Solving equations where the unknown is the denominator

Solve the given equation for x:

$$y = \frac{ax + b}{cx + d} \qquad \text{where } cx + d \neq 0,\ yc - a \neq 0$$

SOLUTION

Multiplying both sides of the given equation by the nonzero quantity $cx + d$ yields

$$\begin{aligned} y(cx + d) &= ax + b \\ ycx + yd &= ax + b && \text{simplifying} \\ ycx - ax &= b - yd && \text{gathering terms involving } x \\ x(yc - a) &= b - yd && \text{factoring} \\ x &= \frac{b - yd}{yc - a} && \text{dividing both sides by } yc - a \neq 0 \end{aligned}$$

You should check for yourself that the expression for x on the right-hand side of this last equation indeed satisfies the original equation.

In the example just concluded, we used a basic factoring technique from elementary algebra to solve the equations. Factoring is also useful in solving *quadratic equations.*

DEFINITION Quadratic Equation

A **quadratic equation** is an equation in one variable that can be written in the form

$$ax^2 + bx + c = 0 \qquad \text{with } a, b, \text{ and } c \text{ real numbers and } a \neq 0$$

To solve a quadratic equation by factoring, we rely on the following familiar and important property of the real-number system.

PROPERTY SUMMARY
Zero-Product Property of Real Numbers

$$pq = 0 \quad \text{if and only if} \quad p = 0 \text{ or } q = 0 \quad \text{(or both)}$$

EXAMPLE 4 **Applying the zero-product property to solve quadratic equations**

Solve:
(a) $8x^2 - 3 = 10x$; **(b)** $4x^2 - 9 = 0$.

SOLUTION

(a) In preparation for using the zero-product property, we first rewrite the equation so the right-hand side is zero. Then we have

$$8x^2 - 10x - 3 = 0$$
$$(2x - 3)(4x + 1) = 0 \qquad \text{Check the factoring.}$$
$$2x - 3 = 0 \quad \Big| \quad 4x + 1 = 0$$
$$x = \frac{3}{2} \quad \Big| \quad x = -\frac{1}{4}$$

You can check that the values $x = 3/2$ and $x = -1/4$ both satisfy the given equation.

(b) Using difference-of-squares factoring, we have

$$(2x - 3)(2x + 3) = 0$$
$$2x - 3 = 0 \quad \Big| \quad 2x + 3 = 0$$
$$x = \frac{3}{2} \quad \Big| \quad x = -\frac{3}{2}$$

You can check that the values $x = 3/2$ and $x = -3/2$ both satisfy the given equation.

Here's another perspective on Example 4(b). Instead of using factoring to solve the equation $4x^2 - 9 = 0$, we can instead rewrite it as $x^2 = 9/4$. Taking the principal square root of both sides then yields

$$\sqrt{x^2} = \sqrt{9/4}$$

and therefore

$$|x| = \frac{3}{2} \quad \text{(Why?)}$$

By looking at this last equation, we can see that there are two solutions, $x = 3/2$ and $x = -3/2$. (Those are the only two numbers with absolute values of $3/2$.) We abbreviate these two solutions by writing $x = \pm 3/2$. In practice, we usually omit showing the step involving the absolute value. For example, to solve the equation $9x^2 - 2 = 0$, just rewrite it as $x^2 = 2/9$. Then "taking square roots" immediately yields the two solutions $x = \pm\sqrt{2/9} = \pm\sqrt{2}/3$.

Not all quadratic equations can be solved by factoring. Consider, for example, the equation $x^2 - 2x - 4 = 0$. The only three possible factorizations with integer coefficients are

$$(x - 4)(x + 1) \qquad (x + 4)(x - 1) \qquad (x - 2)(x + 2)$$

but none yields the appropriate middle term, $-2x$, when multiplied out. In cases such as these, we can use the *quadratic formula,* given in the box that follows. (In Section 2.1, we'll derive this formula and look at some of its implications. For now, though, the focus is simply on using this formula to calculate solutions.)

The Quadratic Formula

The solutions of the quadratic equation $ax^2 + bx + c = 0$, where $a \neq 0$, are given by

$$x = \frac{-b \pm \sqrt{b^2 - 4ac}}{2a}$$

EXAMPLE 5 Using the quadratic formula to solve a quadratic equation

Use the quadratic formula to solve the equation $2x^2 = 3 - 4x$.

SOLUTION

We first rewrite the given equation as $2x^2 + 4x - 3 = 0$, so that it has the form $ax^2 + bx + c = 0$. By comparing these last two equations, we see that $a = 2$, $b = 4$, and $c = -3$. Therefore

$$x = \frac{-b \pm \sqrt{b^2 - 4ac}}{2a} = \frac{-4 \pm \sqrt{4^2 - 4(2)(-3)}}{2(2)}$$

$$= \frac{-4 \pm \sqrt{40}}{4} = \frac{-4 \pm 2\sqrt{10}}{4} = \frac{-2 \pm \sqrt{10}}{2}$$

Thus, the two solutions are $\dfrac{-2 + \sqrt{10}}{2}$ and $\dfrac{-2 - \sqrt{10}}{2}$.

The techniques that we've reviewed in this section for solving linear and quadratic equations will be used throughout this book; you'll see applications in analyzing graphs and functions and in solving many types of applied problems. Linear and quadratic equations both fall under the general heading of *polynomial equations.*

DEFINITION Polynomial Equation

A **polynomial equation** in one variable is an equation of the form

$$a_n x^n + a_{n-1} x^{n-1} + \cdots + a_1 x + a_0 = 0$$

where the subscripted letter *a*'s represent constants and the exponents on the variable are nonnegative integers.

EXAMPLES

(a) $4x^2 - 5x - 1 = 0$

(b) $x^3 - 2x^2 - 3x = 0$

(c) $2x^4 - \frac{4}{3}x^3 + x^2 - 3x + \sqrt{2} = 0$

If a_n is not zero, the **degree** of the polynomial equation is the largest exponent of the variable that appears in the equation. For example, the degrees of equations (a), (b), and (c) in the box above are 2, 3, and 4, respectively.

As we've seen in this section, polynomial equations of degree 1 (linear equations) and polynomial equations of degree 2 (quadratic equations) can be solved by using fairly basic algebra. So too can some higher-degree equations. For instance, we can use factoring and the zero-product property to solve equation (b) in the box above. We have

$$\begin{aligned} x^3 - 2x^2 - 3x &= 0 \\ x(x^2 - 2x - 3) &= 0 && \text{factoring out the common factor } x \\ x(x - 3)(x + 1) &= 0 && \text{factoring the quadratic} \end{aligned}$$

Therefore

$$x = 0 \quad \text{or} \quad x - 3 = 0 \quad \text{or} \quad x + 1 = 0 \quad \text{using the zero-product property}$$

From these last three equations we conclude that the solutions of the third-degree polynomial equation $x^3 - 2x^2 - 3x = 0$ are $x = 0, 3$, and -1. (You should check for yourself that each of these numbers indeed satisfies the equation.)

Unfortunately, not all polynomial equations are as easy to solve as this last one.

Chapter 12 contains a more complete discussion of polynomial equations and an answer to the following question: Is there a general formula, similar to the quadratic formula, for solving *any* polynomial equation?

EXERCISE SET 1.3

A

In Exercises 1–5, determine whether the given value is a solution of the equation.

1. $4x - 5 = -13; x = -2$

2. $\frac{1}{x} = \frac{3}{x} - 1; x = 2$

3. $\frac{2}{y-1} - \frac{3}{y} = \frac{7}{y^2 - y}; y = -3$

4. $(y - 1)(y + 5) = 0; y = 5$

5. $m^2 + m - \frac{5}{16} = 0; m = \frac{1}{4}$

6. Verify that the numbers $1 + \sqrt{5}$ and $1 - \sqrt{5}$ both satisfy the equation $x^2 - 2x - 4 = 0$.

Solve each equation in Exercises 7–19.

7. $2x - 3 = -5$

8. $2m - 1 + 3m + 5 = 6m - 8$

9. $1 - (2m + 5) = -3m$

10. $(x + 2)(x + 1) = x^2 + 11$

11. $t - \{4 - [t - (4 + t)]\} = 6$

12. $\frac{x}{3} + \frac{2x}{5} = \frac{-11}{5}$

13. $1 - \frac{y}{3} = 6$

14. $\frac{x-1}{4} + \frac{2x+3}{-1} = 0$

15. $\frac{1}{x} = \frac{4}{x} - 1$

16. $\frac{1}{y} + 1 = \frac{3}{y} - \frac{1}{2y}$

17. $\frac{1}{x-3} - \frac{2}{x+3} = \frac{1}{x^2 - 9}$

18. $\frac{1}{x-5} + \frac{1}{x+5} = \frac{2x+1}{x^2 - 25}$

19. $\frac{4}{x+2} + \frac{1}{x-2} = \frac{4}{x^2 - 4}$

20. $\frac{3}{2x+1} - \frac{4}{x+1} = \frac{2}{2x^2 + 3x + 1}$

21. $\frac{5}{x-4} - \frac{3}{2x^2 - 5x - 12} = \frac{1}{2x+3}$

22. (a) $\frac{2}{3x} = \frac{3}{x}$ (b) $\frac{2}{3x} = \frac{3}{x+1}$ (c) $\frac{2}{3x} = \frac{3}{x} + 1$

23. (a) $\frac{3}{x-2} = \frac{5}{9x}$ (b) $\frac{3}{x-2} = \frac{5}{9x-2}$ (c) $\frac{3}{x-2} = \frac{5}{\frac{5}{3}x - 2}$

In Exercises 24–33, solve each equation by factoring.

24. $x^2 - 5x - 6 = 0$

25. $x^2 - 5x = -6$

26. $10z^2 - 13z - 3 = 0$

27. $3t^2 - t - 4 = 0$

28. $(x + 1)^2 - 4 = 0$

29. $x^2 + 3x - 40 = 0$

30. $x(2x - 13) = -6$

31. $x(3x - 23) = 8$

32. $x(x + 1) = 156$

33. $x^2 + (2\sqrt{5})x + 5 = 0$

In Exercises 34–41, use the quadratic formula to solve each equation. In Exercises 34–39, give two forms for each solution: an expression containing a radical and a calculator approximation rounded off to two decimal places.

34. $2x^2 + 3x - 4 = 0$

35. $4x^2 - 3x - 9 = 0$

36. $x(x + 6) = -2$

37. $x(3x + 8) = -2$

38. $2x^2 - 10 = -\sqrt{2}x$

39. $\sqrt{3}x^2 + \sqrt{3} = 6x$

40. $12x^2 - 25x = -12$

41. $24x^2 + 23x = -5$

In Exercises 42–47, solve the equations using any method you choose.

42. $x^2 = 24$

43. $2y^2 - 50 = 0$

44. $\frac{1}{8} - t^2 = 0$

45. $x^2 - \sqrt{5} = 0$

46. **(a)** $u(u + 18) = -81$
(b) $u(u + 18) = 81$

47. **(a)** $x^2 + 156x + 5963 = 0$
(b) $144y^2 - 54y = 13$

48. Solve each of the following equations for x. *Hint:* As in the text, begin by factoring out a common factor.
(a) $x^3 - 13x^2 + 42x = 0$
(b) $x^3 - 6x^2 + x = 0$

For Exercises 49–58, solve each equation for x in terms of the other letters.

49. $3ax - 2b = b + 3$

50. $ax + b = bx - a$

51. $ax + b = bx + a$

52. $\frac{x}{a} + \frac{x}{b} = 1$

53. $\frac{1}{x} = a + b$

54. $\frac{1}{ax} = \frac{1}{bx} - \frac{1}{c}$

55. $\frac{1}{a} - \frac{1}{x} = \frac{1}{x} - \frac{1}{b}$

56. **(a)** $y = mx + b$, where $m \neq 0$
(b) $y - y_1 = m(x - x_1)$, where $m \neq 0$
(c) $\frac{x}{a} + \frac{y}{b} = 1$
(d) $Ax + By + C = 0$, where $A \neq 0$

57. $(ax + b)^2 - (bx + a)^2 = 0$, where $a \neq \pm b$

58. $(x - p)^2 + (x - q)^2 = p^2 + q^2$

B

In Exercises 59–64, solve each equation for x in terms of the other letters.

59. $a^2(a - x) = b^2(b + x) - 2abx$, where $a \neq b$

60. $\frac{b}{ax - 1} - \frac{a}{bx - 1} = 0$, where $a \neq b$

61. $\frac{a - x}{a - b} - 2 = \frac{c - x}{b - c}$

62. $\frac{x + 2p}{2q - x} + \frac{x - 2p}{2q + x} - \frac{4pq}{4q^2 - x^2} = 0$

63. $\frac{x - a}{x - b} = \frac{b - x}{a - x}$, where $a \neq b$

64. $1 - \frac{a}{b}\left(1 - \frac{a}{x}\right) - \frac{b}{a}\left(1 - \frac{b}{x}\right) = 0$

In Exercises 65–68, solve each equation for the indicated variable.

65. $S = 2\pi r^2 + 2\pi rh$; for h

66. $\frac{x_1 x}{a^2} + \frac{y_1 y}{b^2} = 1$; for y

67. $d = \frac{r}{1 + rt}$; for r

68. $S = \frac{rl - a}{r - 1}$; for r

Solve the equations in Exercises 69–74. (In these exercises, you'll need to multiply both sides of the equations by expressions involving the variable. Remember to check your answers in these cases.)

69. $\frac{3}{x + 5} + \frac{4}{x} = 2$

70. $\frac{5}{x + 2} - \frac{2x - 1}{5} = 0$

71. $1 - x - \frac{2}{6x + 1} = 0$

72. $\frac{x^2 - 3x}{x + 1} = \frac{4}{x + 1}$

73. $\frac{x}{x - 2} + \frac{x}{x + 2} = \frac{8}{x^2 - 4}$

74. $\frac{2x}{x^2 - 1} - \frac{1}{x + 3} = 0$

75. Given the equation $\frac{1}{x} = \frac{1}{a} + \frac{1}{b}$:
(a) Solve to show $x = \frac{ab}{a + b}$, provided $a + b \neq 0$.
(b) Check the solution.

1.4 RECTANGULAR COORDINATES. VISUALIZING DATA

The name coordinate does not appear in the work of Descartes. This term is due to Leibniz and so are abscissa *and* ordinate *(1692).*—David M. Burton in *The History of Mathematics: An Introduction,* 2nd ed. (Dubuque, Iowa: Wm. C. Brown Publishers, 1991)

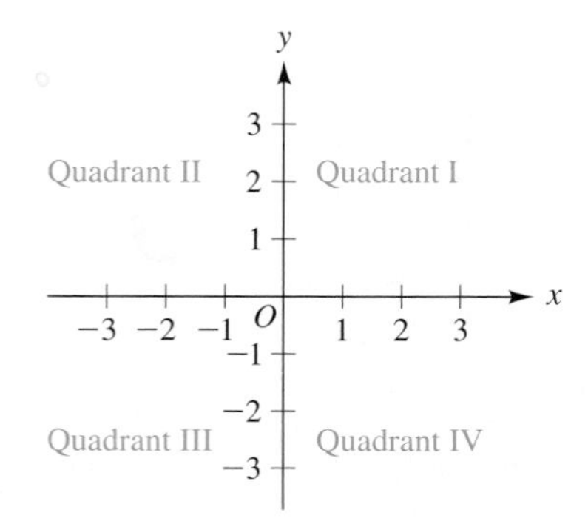

Figure 1

In previous courses you learned to work with a rectangular coordinate system such as that shown in Figure 1. In this section we review some of the most basic formulas and techniques that are useful here.

The point of intersection of the two perpendicular number lines, or **axes,** is called the **origin** and is denoted by the letter O. The horizontal and vertical axes are often labeled the ***x*-axis** and the ***y*-axis,** respectively, but any other variables will

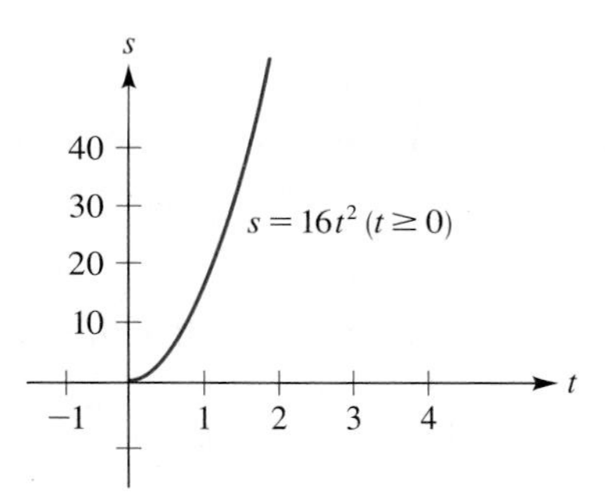

(a) A graph of the formula $s = 16t^2$ in a t-s coordinate system. [The formula relates the distance s (in feet) and the time t (in seconds) for an object falling in a vacuum.]

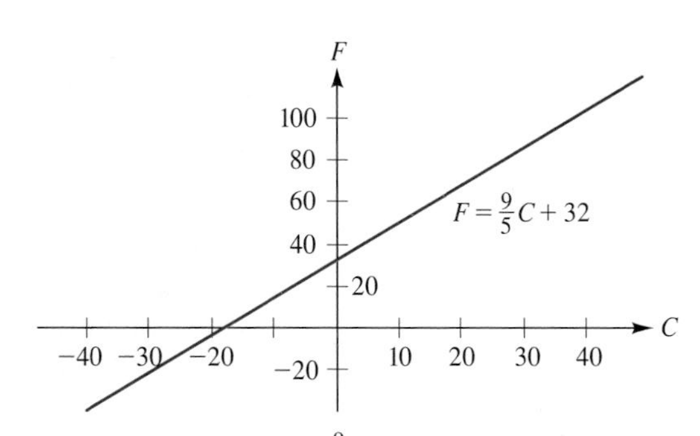

(b) A graph of the equation $F = \frac{9}{5}C + 32$ in a C-F coordinate system. (The equation gives the relationship between the temperature C on the Celsius scale and F on the Fahrenheit scale.)

Figure 2

do just as well for labeling the axes. See Figure 2 for examples of this. (We'll discuss curves or graphs like the ones in Figure 2 in later sections.)

Note that in Figures 1 and 2 the axes divide the plane into four regions, or **quadrants,** labeled I through IV, as shown in Figure 1. Unless indicated otherwise, we assume that the same unit of length is used on both axes. In Figure 1, the same scales are used on both axes; not so in Figure 2.

Now look at the point P in Figure 3(a). Starting from the origin O, one way to reach P is to move three units in the positive x-direction and then two units in the positive y-direction. That is, the location of P relative to the origin and the axes is "right 3, up 2." We say that the **coordinates** of P are (3, 2). The first number within the parentheses conveys the information "right 3," and the second number conveys the information "up 2." We say that the ***x*-coordinate** of P is 3 and the ***y*-coordinate** of P is 2. Likewise, the coordinates of point Q in Figure 3(a) are $(-2, 4)$. With this coordinate notation in mind, observe in Figure 3(b) that (3, 2) and (2, 3) represent different points; that is, the order in which the two numbers appear within the parentheses affects the location of the point. Figure 3(c) displays various points with given coordinates; you should check for yourself that the coordinates correspond correctly to the location of each point.

Some terminology and notation: The x-y coordinate system that we have described is often called a **Cartesian coordinate system.** The term *Cartesian* is used in honor of René Descartes, the seventeenth-century French philosopher and mathematician. The coordinates (x, y) of a point P are referred to as an **ordered pair.**

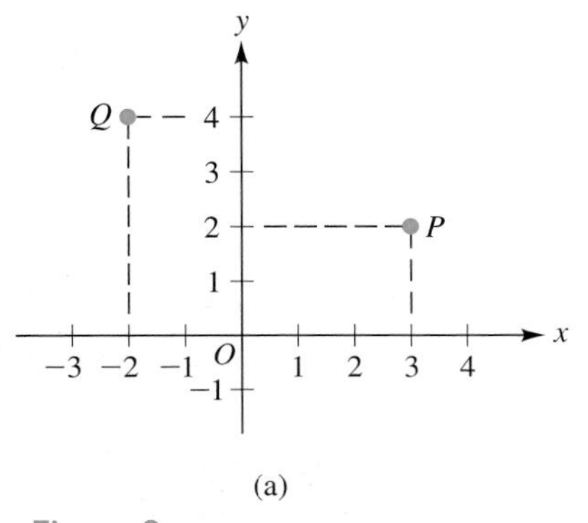

(a)

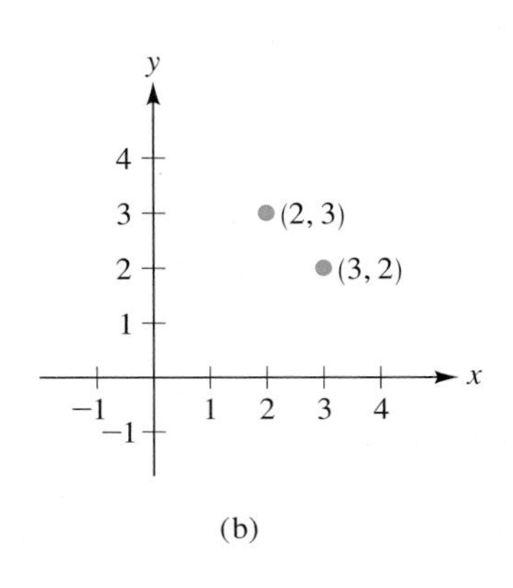

(b)

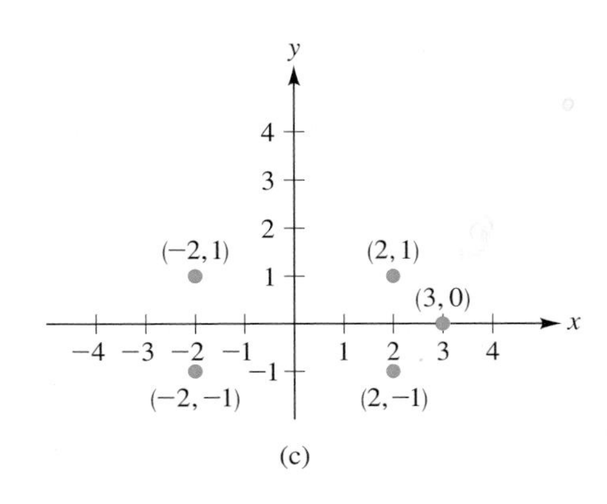

(c)

Figure 3

Recall, for example, that (3, 2) and (2, 3) represent different points; that is, the order of the numbers matters. The x-coordinate of a point is sometimes referred to as the **abscissa** of the point; the y-coordinate is the **ordinate.** The notation $P(x, y)$ means that P is a point that has coordinates (x, y). At times, we abbreviate the phrase *the point whose coordinates are* (x, y) to simply *the point* (x, y).

The next part of our work in this section depends on a key result from elementary geometry, the Pythagorean theorem. For reference, we state this theorem and its converse in the box that follows. (For proofs of the Pythagorean theorem, see Exercises 32 and 33 at the end of this section or Exercise 100 in the Chapter Review Exercises.)

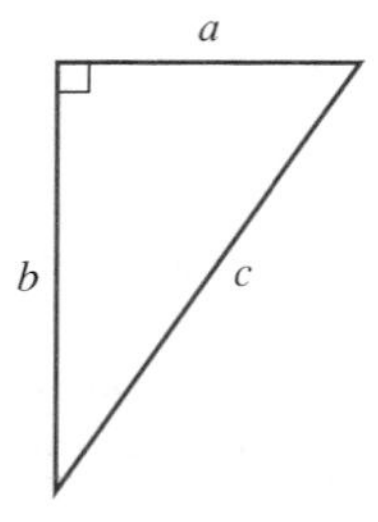

Figure 4

The Pythagorean Theorem and Its Converse

1. Pythagorean Theorem
   (See Figure 4.) In a right triangle the lengths of the sides are related by the equation
   $$a^2 + b^2 = c^2$$
   where a and b are the lengths of the sides forming the right angle and c is the length of the hypotenuse (the side opposite the right angle).
2. Converse
   If the lengths a, b, and c of the sides of a triangle are related by an equation of the form $a^2 + b^2 = c^2$, then the triangle is a right triangle, and c is the length of the hypotenuse.

EXAMPLE 1 Using the Pythagorean theorem to find a distance

Use the Pythagorean theorem to calculate the distance d between the points (2, 1) and (6, 3).

SOLUTION

We plot the two given points and draw a line connecting them, as shown in Figure 5. Then we draw the broken lines as shown, parallel to the axes, and apply the Pythagorean theorem to the right triangle that is formed. The base of the triangle is four units long. You can see this by simply counting spaces or by using absolute value, as discussed in Section 1.2: $|6 - 2| = 4$. The height of the triangle is found to be two units, either by counting spaces or by computing the absolute value: $|3 - 1| = 2$. Thus we have

$$d^2 = 4^2 + 2^2 = 20$$
$$d = \sqrt{20} = \sqrt{4}\sqrt{5} = 2\sqrt{5}$$

Note: Since d is a distance, we disregard the solution $-\sqrt{20}$ of the equation $d^2 = 20$.

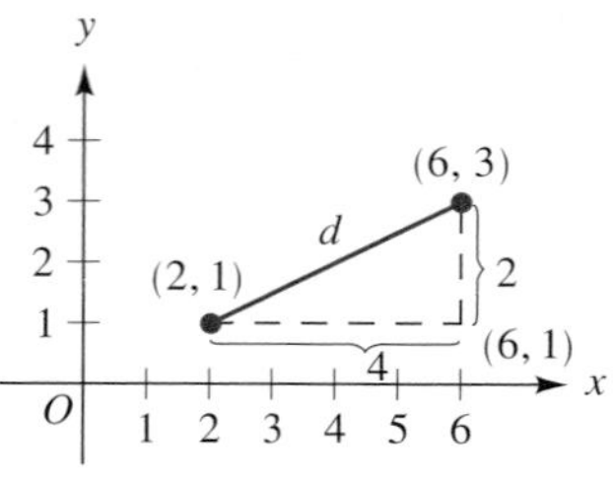

Figure 5

The method we used in Example 1 can be applied to derive a general formula for the distance d between any two points (x_1, y_1) and (x_2, y_2) (see Figure 6).

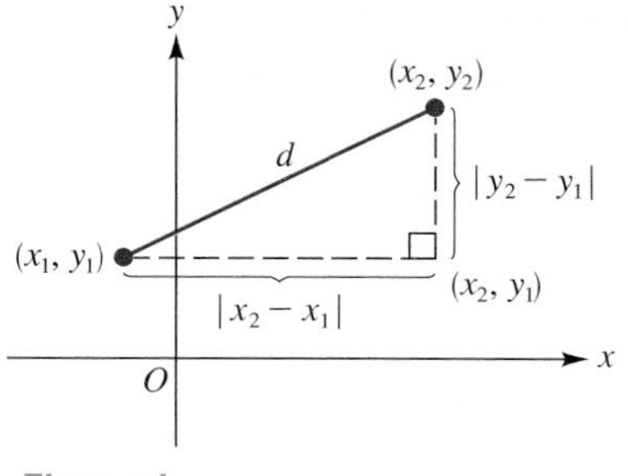

Figure 6

Just as before, we draw in the right triangle and apply the Pythagorean theorem. We have

$$\begin{aligned} d^2 &= |x_2 - x_1|^2 + |y_2 - y_1|^2 \\ &= (x_2 - x_1)^2 + (y_2 - y_1)^2 \quad \text{(Why?)} \end{aligned}$$

and therefore

$$d = \sqrt{(x_2 - x_1)^2 + (y_2 - y_1)^2}$$

This last equation is referred to as the **distance formula.** For reference, we restate it in the box that follows.

The Distance Formula

The distance d between the points (x_1, y_1) and (x_2, y_2) is given by

$$d = \sqrt{(x_2 - x_1)^2 + (y_2 - y_1)^2}$$

Examples 2–4 demonstrate some simple calculations involving the distance formula.

NOTE In computing the distance between two given points, it does not matter which one you treat as (x_1, y_1) and which as (x_2, y_2), because quantities such as $x_2 - x_1$ and $x_1 - x_2$ are negatives of each other and so have equal squares.

EXAMPLE 2 Using the distance formula

Calculate the distance between the points $(2, -6)$ and $(5, 3)$.

SOLUTION

Substituting $(2, -6)$ for (x_1, y_1) and $(5, 3)$ for (x_2, y_2) in the distance formula, we have

$$\begin{aligned} d &= \sqrt{(5 - 2)^2 + [3 - (-6)]^2} \\ &= \sqrt{3^2 + 9^2} = \sqrt{90} \\ &= \sqrt{9}\sqrt{10} = 3\sqrt{10} \end{aligned}$$

You should check for yourself that the same answer is obtained using $(2, -6)$ as (x_2, y_2) and $(5, 3)$ as (x_1, y_1).

EXAMPLE 3 Using the distance formula and the converse of the Pythagorean theorem

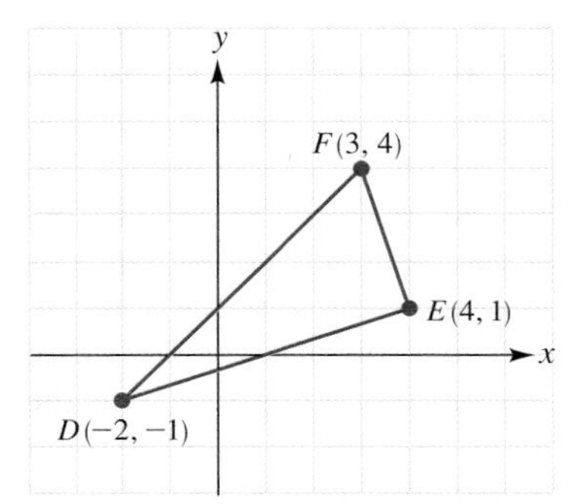

Figure 7

Is the triangle with vertices $D(-2, -1)$, $E(4, 1)$, and $F(3, 4)$ a right triangle?

SOLUTION

First we sketch the triangle in question (see Figure 7). From the sketch it appears that angle E could be a right angle, but certainly this is not a proof. We need to use the distance formula to calculate the lengths of the three sides and then check whether any relation of the form $a^2 + b^2 = c^2$ holds. The calculations are as follows:

$$DE = \sqrt{[4 - (-2)]^2 + [1 - (-1)]^2} = \sqrt{36 + 4} = \sqrt{40}$$
$$EF = \sqrt{(4 - 3)^2 + (1 - 4)^2} = \sqrt{1 + 9} = \sqrt{10}$$
$$DF = \sqrt{[3 - (-2)]^2 + [4 - (-1)]^2} = \sqrt{25 + 25} = \sqrt{50}$$

Because $(\sqrt{40})^2 + (\sqrt{10})^2 = (\sqrt{50})^2$, D, E, and F are vertices of a right triangle with hypotenuse $\overline{DF}$ and right angle at vertex E, and so $\triangle DEF$ is a right triangle. (In Section 1.6 you'll see that this result can be obtained more easily by using the concept of slope.)

EXAMPLE 4 Using the distance formula to find a radius

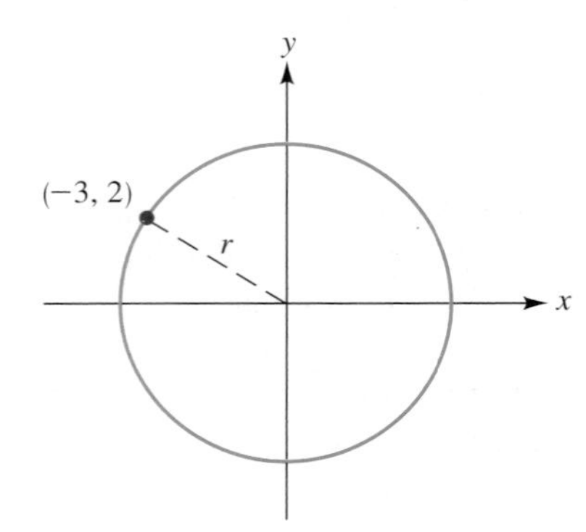

Figure 8

(a) Find the radius r of the circle in Figure 8. (Assume that the center of the circle is located at the origin.)

(b) Compute the area and the circumference of the circle. For each answer, give exact expressions and also calculator approximations rounded to one decimal place.

SOLUTION

(a) The radius r is the distance from center $(0, 0)$ to the given point $(-3, 2)$ on the circle. Using the distance formula, we have

$$r = \sqrt{(-3 - 0)^2 + (2 - 0)^2}$$
$$= \sqrt{9 + 4} = \sqrt{13} \text{ units}$$

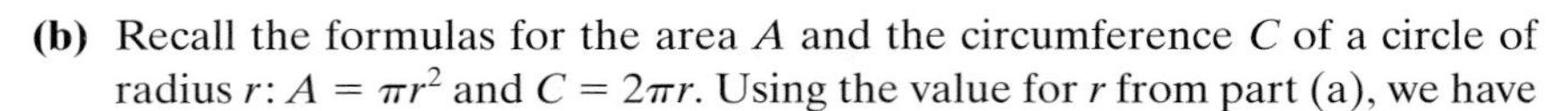

(b) Recall the formulas for the area A and the circumference C of a circle of radius r: $A = \pi r^2$ and $C = 2\pi r$. Using the value for r from part (a), we have

$$A = \pi r^2 = \pi(\sqrt{13})^2 \qquad C = 2\pi r = 2\pi\sqrt{13} \text{ units}$$
$$= 13\pi \text{ square units} \qquad \approx 22.7 \text{ units}$$
$$\approx 40.8 \text{ square units}$$

One of the important applications of rectangular coordinates is in displaying quantitative data. You see instances of this every day in newspapers, in magazines, and in textbooks as diverse as astronomy to zoology. We show some examples in the figures and discussion that follow.

Table 1 provides world population data for the period 1965–1995. In Figure 9(a) the familiar *bar graph* (or *column chart*) format is used to display the data from the table. In Figure 9(b) we've plotted the data in a rectangular coordinate system. On the horizontal axis the variable t represents years; the variable P represents population in units of one billion. The data in the first row of the table (which state that the population in 1965 was 3.345 billion) are plotted in Figure 9(b) as the point (1965, 3.345). Likewise, the second row of data in the table gives us the point (1975, 4.086), and so on. Sometimes it is more convenient to work with smaller numbers on the horizontal axis than those used in Figure 9(b). One very common way to do this is indicated in Figure 9(c), where we are now letting the variable t

TABLE 1 World Population 1965–1995

Year	Population (billions)
1965	3.345
1975	4.086
1985	4.850
1995	5.687

Source: U.S. Census Bureau (International Data Base)

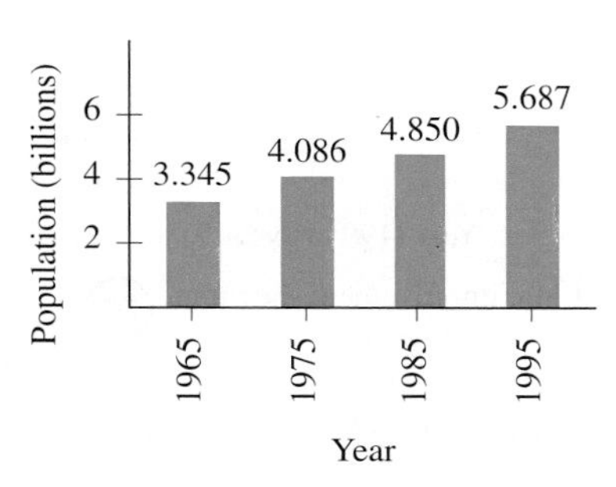

(a) Bar graph or column chart

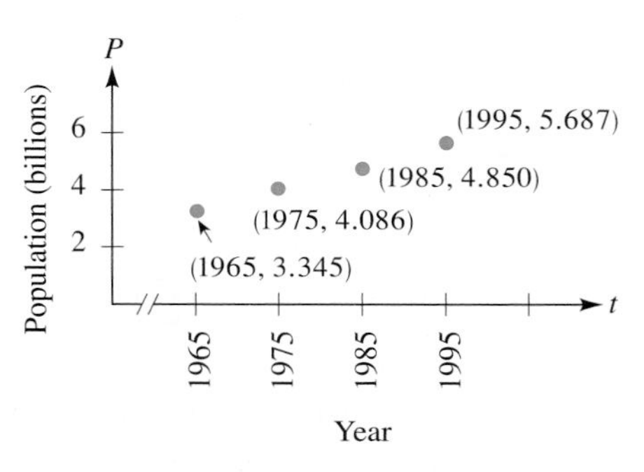

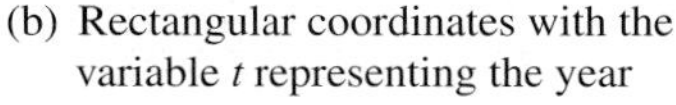
(b) Rectangular coordinates with the variable t representing the year

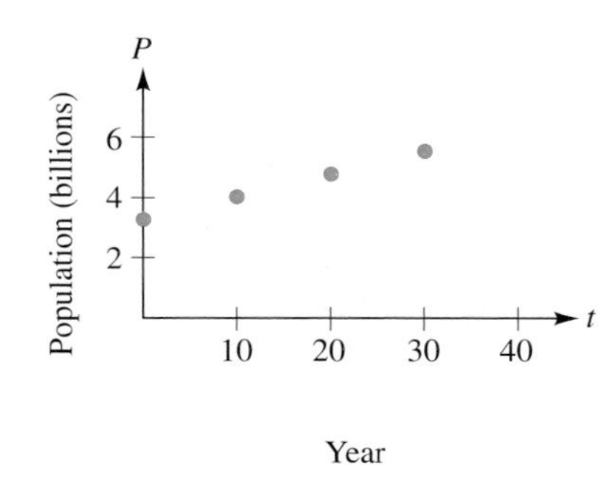

(c) Rectangular coordinates with the variable t representing the number of years since 1965

Figure 9
World population 1965–1995

represent *years since 1965*. In other words, the year 1965 is $t = 0$, 1966 is $t = 1$, 1967 is $t = 2$, and so on. The data in the first row of Table 1 are then plotted in Figure 9(c) as the point (0, 3.345), rather than (1965, 3.345). Likewise, the second row of data in Table 1 is plotted in Figure 9(c) as (10, 4.086), and so on.

Both the bar graph and the rectangular plots in Figure 9 make it immediately clear that the world population is increasing. Is it increasing at a steady rate? Is it increasing rapidly? In fact, one needs to exercise caution in using graphs to draw conclusions about *how fast* the quantity being graphed (in this case, population) is increasing or decreasing. For instance, Figure 10 shows another graph of world population, this time covering the period 1800–1995. Figure 9(b) and Figure 10 may lead to different interpretations about the nature of world population growth. [In Figure 10, the four blue dots are the data points that appear in Figure 9(b).]

As another example about the need for caution in interpreting graphs, look at Figure 11, which shows two very different interpretations of the data for SAT

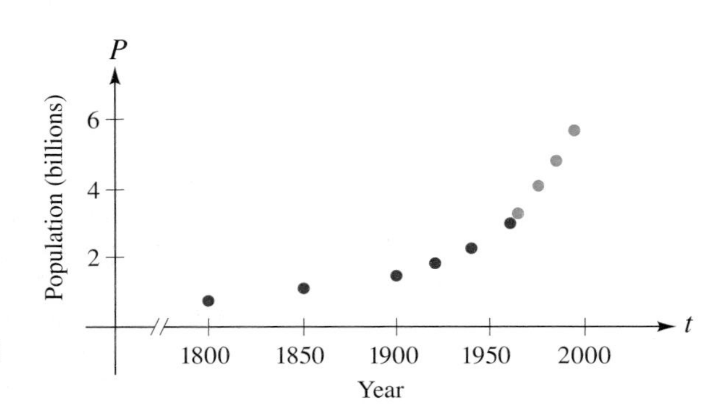

Figure 10
World population 1800–1995
Source: U.S. Census Bureau

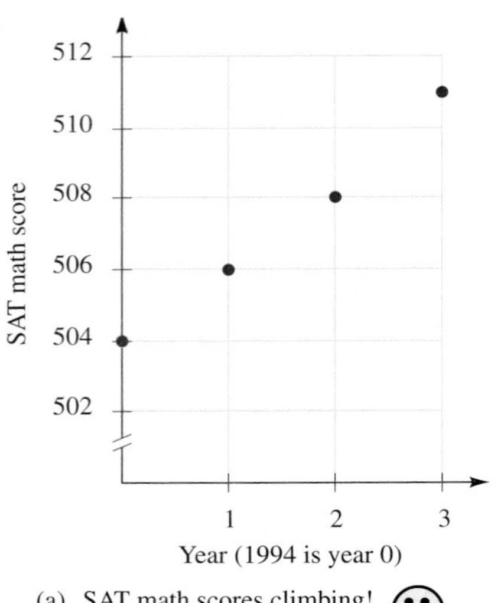

(a) SAT math scores climbing!

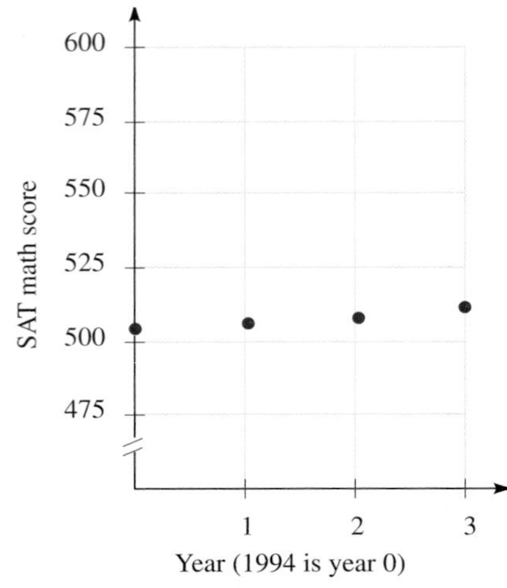

(b) Little improvement shown in SAT math scores!

Figure 11
Two visualizations and interpretations of the data in Table 2

TABLE 2 SAT Math Scores: National Averages 1994–1997

Year	SAT math score
0 (1994)	504
1 (1995)	506
2 (1996)	508
3 (1997)	511

Source: The College Board

mathematics scores in Table 2. The bottom line is that graphs are useful, even indispensible, in giving us an easy way to see general trends in data, but we must exercise care in drawing further conclusions, especially regarding rates of increase or decrease. A complete analysis of how fast a quantity is increasing or decreasing may require topics from calculus or the field of statistics. For straight-line graphs, however, the concept of *slope* (reviewed in Section 1.6) tells us definitively about rates of increase or decrease. Also, when we study functions in Chapter 3, we'll make a first step toward answering general questions about rates of change.

In Example 5 we make use of a simple result that you may recall from previous courses: the *midpoint formula.* This result is summarized in the box that follows. (For a proof of the formula, see Exercise 31 at the end of this section.)

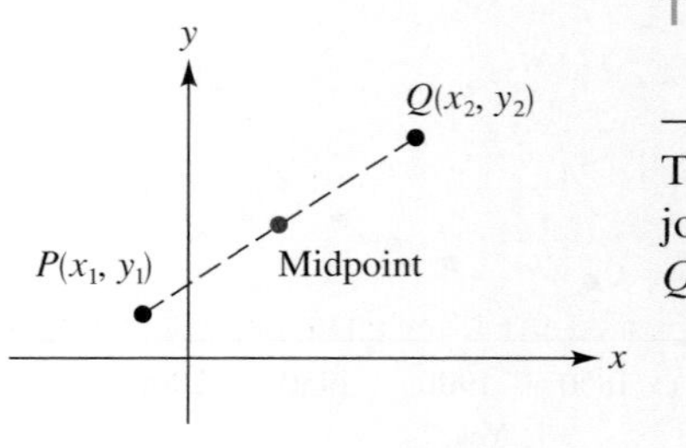

THE MIDPOINT FORMULA

The midpoint of the line segment joining the points $P(x_1, y_1)$ and $Q(x_2, y_2)$ is

$$\left(\frac{x_1 + x_2}{2}, \frac{y_1 + y_2}{2}\right)$$

Example

The midpoint of the line segment joining $(2, -15)$ and $(4, 5)$ is

$$\left(\frac{2 + 4}{2}, \frac{-15 + 5}{2}\right) = (3, -5)$$

EXAMPLE 5 An application of the midpoint formula

Data concerning the amount of carbon dioxide in the atmosphere (measured in *parts per million* or *ppm*) is used by environmental scientists in the study of global warming. Table 3 provides some figures for the period 1990–1996.

TABLE 3 Atmospheric Carbon Dioxide

Year	Carbon dioxide in atmosphere (ppm)
1990	354.0
1992	356.3
1994	358.9
1996	362.6

Source: C. D. Keeling and T. P. Whorf, Scripps Institution of Oceanography

(a) Plot the data in a rectangular coordinate system. Use the variable t on the horizontal axis, with $t = 0$ corresponding to the year 1990. Use the variable c (to denote carbon dioxide levels, in ppm) on the vertical axis.
(b) Use the midpoint formula and the data for 1992 and 1994 to estimate the amount of carbon dioxide for the year 1993.
(c) Compute the *percentage error* in the estimation in part (b), given that the actual 1993 value is 357.0 ppm. The general formula for percentage error in an estimation or approximation is

$$\text{percentage error} = \frac{|(\text{actual value}) - (\text{approximate value})|}{\text{actual value}} \times 100$$

SOLUTION

(a) See Figure 12. Note that if $t = 0$ corresponds to 1990, then $t = 2$ corresponds to 1992, $t = 4$ corresponds to 1994, and $t = 6$ corresponds to 1996.

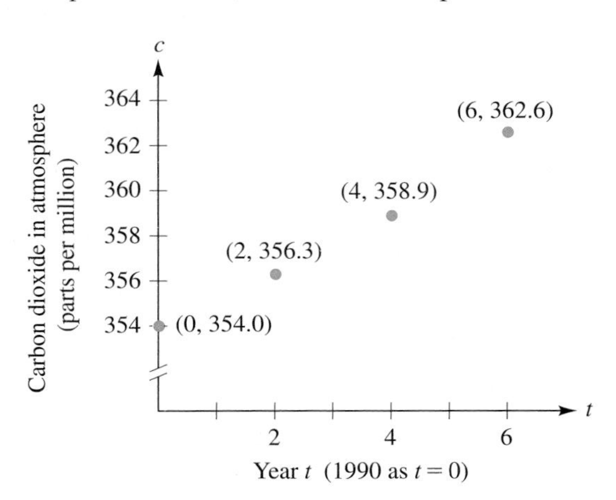

Figure 12

(b) The midpoint of the line segment joining the points (2, 356.3) and (4, 358.9) is

$$\left(\frac{2+4}{2}, \frac{356.3+358.9}{2}\right)$$

which, as you can verify, works out to

$$(3, 357.6)$$

Thus, our approximation for the carbon dioxide level in 1993 ($t = 3$) is 357.6 ppm.

(c) We have

$$\begin{aligned}\text{percentage error} &= \frac{|(\text{actual value}) - (\text{approximate value})|}{\text{actual value}} \times 100 \\ &= \frac{|357.0 - 357.6|}{357.0} \times 100 \\ &\approx 0.2\% \quad \text{using a calculator and rounding to one decimal place}\end{aligned}$$

In Figure 13 we show the given data for 1992 and 1994 along with the estimated and the actual value for 1993.

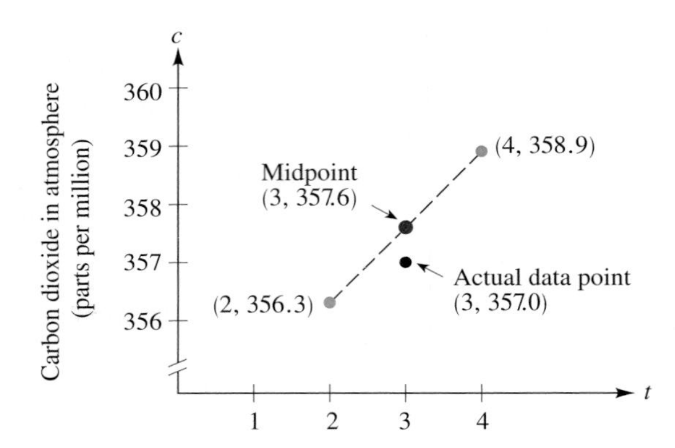

Figure 13
The midpoint of the line segment is close to the actual data point.

CAUTION Do not assume on the basis of this one example that the midpoint approximation always works as well as it does here. In this regard, be sure to work Exercise 23 at the end of this section.

EXERCISE SET 1.4

A

1. Plot the points $(5, 2)$, $(-4, 5)$, $(-4, 0)$, $(-1, -1)$, and $(5, -2)$.
2. Draw the square $ABCD$ with vertices (corners) $A(1, 0)$, $B(0, 1)$, $C(-1, 0)$, and $D(0, -1)$.
3. **(a)** Draw the right triangle PQR with vertices $P(1, 0)$, $Q(5, 0)$, and $R(5, 3)$.
   (b) Use the formula for the area of a triangle, $A = \frac{1}{2}bh$, to find the area of triangle PQR in part (a).
4. **(a)** Draw the trapezoid $ABCD$ with vertices $A(0, 0)$, $B(7, 0)$, $C(6, 4)$, and $D(4, 4)$.
   (b) Compute the area of the trapezoid. (See the inside front cover of this book for the appropriate formula.)

In Exercises 5–10, calculate the distance between the given points.

5. **(a)** $(0, 0)$ and $(-3, 4)$
   (b) $(2, 1)$ and $(7, 13)$
6. **(a)** $(-1, -3)$ and $(-5, 4)$
   (b) $(6, -2)$ and $(-1, 1)$
7. **(a)** $(-5, 0)$ and $(5, 0)$
   (b) $(0, -8)$ and $(0, 1)$
8. **(a)** $(-5, -3)$ and $(-9, -6)$
   (b) $(\frac{9}{2}, 3)$ and $(-2\frac{1}{2}, -1)$
9. $(1, \sqrt{3})$ and $(-1, -\sqrt{3})$
10. $(-3, 1)$ and $(374, -335)$
11. Which point is farther from the origin?
    (a) $(3, -2)$ or $(4, \frac{1}{2})$
    (b) $(-6, 7)$ or $(9, 0)$

12. Use the distance formula to show that, in each case, the triangle with given vertices is an isosceles triangle.
(a) $(0, 2), (7, 4), (2, -5)$
(b) $(-1, -8), (0, -1), (-4, -4)$
(c) $(-7, 4), (-3, 10), (1, 3)$

13. In each case, determine whether the triangle with the given vertices is a right triangle.
(a) $(7, -1), (-3, 5), (-12, -10)$
(b) $(4, 5), (-3, 9), (1, 3)$
(c) $(-8, -2), (1, -1), (10, 19)$

14. (a) Two of the three triangles specified in Exercise 13 are right triangles. Find their areas.
(b) Calculate the area of the remaining triangle in Exercise 13 by using the following formula for the area A of a triangle with vertices (x_1, y_1), (x_2, y_2), and (x_3, y_3):

$$A = \frac{1}{2}|x_1y_2 - x_2y_1 + x_2y_3 - x_3y_2 + x_3y_1 - x_1y_3|$$

The derivation of this formula is given in Exercise 34.
(c) Use the formula given in part (b) to check your answers in part (a).

15. Use the formula given in Exercise 14(b) to calculate the area of the triangle with vertices $(1, -4)$, $(5, 3)$, and $(13, 17)$. What do you conclude?

16. The coordinates of points A, B, and C are $A(-4, 6)$, $B(-1, 2)$, and $C(2, -2)$.
(a) Show that $AB = BC$ by using the distance formula.
(b) Show that $AB + BC = AC$ by using the distance formula.
(c) What can you conclude from parts (a) and (b)?

In Exercises 17 and 18, find the midpoint of the line segment joining points P and Q.

17. (a) $P(3, 2)$ and $Q(9, 8)$
(b) $P(-4, 0)$ and $Q(5, -3)$
(c) $P(3, -6)$ and $Q(-1, -2)$

18. (a) $P(12, 0)$ and $Q(12, 8)$
(b) $P(\frac{3}{5}, -\frac{2}{3})$ and $Q(0, 0)$
(c) $P(1, \pi)$ and $Q(3, 3\pi)$

In Exercises 19 and 20, the given points P and Q are the endpoints of a diameter of a circle. Find **(a)** *the center of the circle;* **(b)** *the radius of the circle.*

19. $P(-4, -2)$ and $Q(6, 4)$

20. $P(1, -3)$ and $Q(-5, -5)$

21. (a) Using a coordinate system similar to the one shown in the following figure (or a photocopy), plot the two points from Table A corresponding to the data for the years 1984 and 1992. Ignore the information about the name of the network.
(b) Use the midpoint formula with the two points that you plotted in part (a) to obtain an approximation for the dollar amount paid per TV hour for 1988. (Round the answer to one decimal place.)
(c) Compute the percentage error in the approximation in part (b). The actual 1988 value is given in the table.

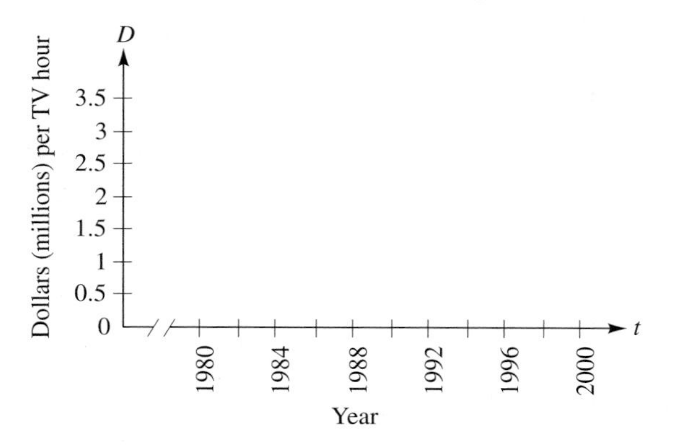

TABLE A How Much the Networks Paid (per TV hour) to Televise the Winter Olympics, 1980–1998

Year (network)	1980 (ABC)	1984 (ABC)	1988 (ABC)	1992 (CBS)	1994 (CBS)	1998 (CBS)
Millions of dollars per TV hour	0.3	1.5	3.3	2.1	2.5	2.9

Source: World Features Syndicate

22. (a) Set up a coordinate system with the horizontal t-axis (running from 0 to at least 6) representing years af-ter 1990 and the vertical P-axis (running from 14 to at least 27) representing percentage of sales due to imports; then use it to plot the data in Table B below. *Note*: you should use a broken vertical axis as in Figures 11 through 13.
(b) What are the coordinates of the point in your graph that corresponds to the data for 1990? For 1992?
(c) Use the midpoint formula with the two points that you listed in part (b) to estimate the percentage of import sales for the year 1991. (Round the answer to one decimal place.)
(d) Compute the percentage error in the estimate in part (c), given that the actual 1991 value is 24.9%. [The formula for percentage error is given in Example 5(c).]

TABLE B Percentage of Retail New Car Sales in United States due to Imports, 1990–1996

Year	1990	1992	1994	1996
Percentage due to imports	25.8	23.6	19.3	14.9

Source: American Automobile Manufacturers Assn.

23. Over the past two decades the Internet has grown very rapidly. Figures A and B provide estimates for the number

n of Internet host computers, worldwide, for the years $t = 1995–1997$ and 1985–1987. Source: *Network Wizards* (http//www.nw.com)

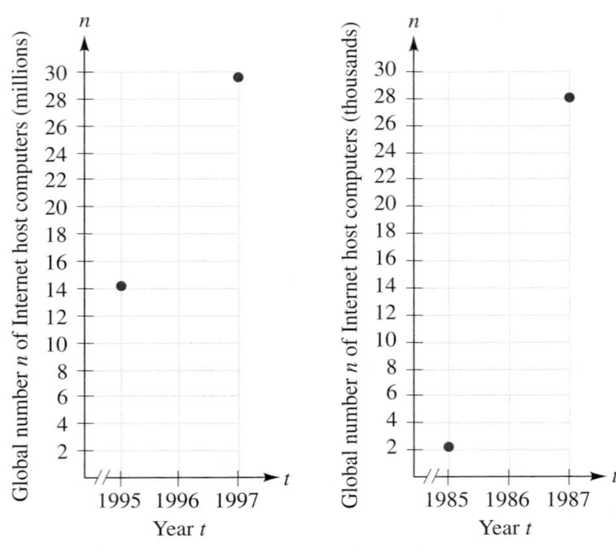

Figure A Figure B

(a) Use Figure A to complete the following table. Round the values of n to the nearest two million. Then use the midpoint formula and the numbers in your table to estimate the global number of Internet host computers for the year 1996.

t	1995	1997
n		

(b) Use Figure B to complete the following table. Round the values of n to the nearest two thousand. Then use the midpoint formula and the numbers in your table to estimate the global number of Internet host computers for the year 1986.

t	1985	1987
n		

(c) Compute the percentage errors to determine which estimate, the one for 1986 or the one for 1996, is more accurate. Use the following data from *Network Wizards* (http//www.nw.com) in computing the percentage errors: The number of host computers for 1986 and 1996 were 5089 and 21,819,000, respectively. (Round each answer to the nearest one percent.)

Note: You'll find out in part (c) that one estimate is very good, the other is way off. The point here is that without more initial information, it's hard to say whether the midpoint formula will produce a useful estimate. In subsequent chapters, we'll use *functions* and larger data sets to obtain more reliable estimates.

24. Have you or a friend ever run in a 10K (10,000 meter) race? When the author polled his precalculus class at UCLA in Fall 1997, he found that there were five students in the class (of 160) who said they had run a 10K in under 50 minutes. Of those five, two (one male, one female) said they had run a 10K in under 40 minutes. The world record for this event is well under 30 minutes. In this exercise you'll look at some of the world records in this event over the past decade.

(a) The table that follows lists the world records in the (men's) 10,000 meter race as of the end of the years 1993, 1995, and 1997. After converting the times into seconds, plot the three points corresponding to these records in a coordinate system similar to the one shown.

Year	Time	Runner
1993	26:58.38	Yobes Ondieki (Kenya)
1995	26:43.53	Haile Gebrselassie (Kenya)
1997	26:27.85	Paul Tergat (Kenya)

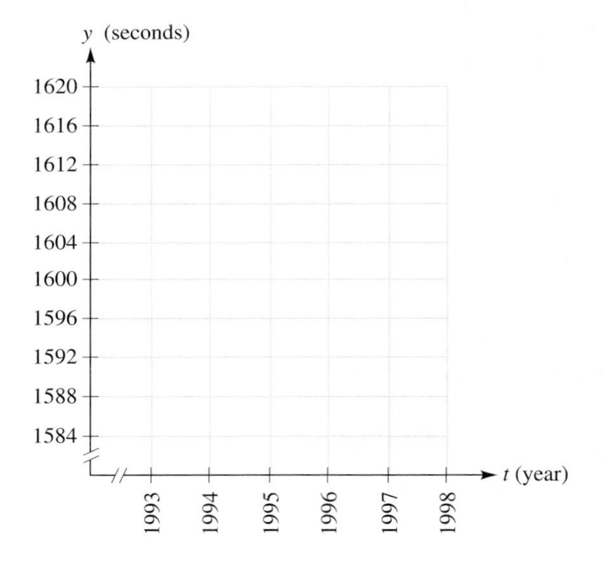

(b) Use the midpoint formula and the data for 1993 and 1995 to compute an estimate for what the world record might have been by the end of 1994. Then compute the percentage error (rounded to two decimal places), given that the record at the end of 1994 was 26:52.23 (set by William Seigei of Kenya). Was your estimate too high or too low?

(c) Use the midpoint formula and the data for 1995 and 1997 to compute an estimate for what the world record might have been by the end of 1996. Then compute the percentage error given that the record at the end of 1996 was 26:38.08 (set by Salah Hissou of Morocco). Was your estimate too high or too low? Is the percentage error more or less than that obtained in part (b)?

(d) Using a coordinate system similar to the one shown in part (a), or using a photocopy, plot the points corresponding to the (actual, not estimated) world records for the years 1993, 1994, 1995, 1996, 1997, and 1998. Except for 1998, all the records have been given above. The world record at the end of 1998 was 26:22.75 (set by Haile Gebrselassie of Kenya). Use the picture you obtain to say whether or not the record times seem to be leveling off.

25. **(a)** Sketch the parallelogram with vertices $A(-7, -1)$, $B(4, 3)$, $C(7, 8)$, and $D(-4, 4)$.
(b) Compute the midpoints of the diagonals $\overline{AC}$ and $\overline{BD}$.
(c) What conclusion can you draw from part (b)?

26. The vertices of $\triangle ABC$ are $A(1, 1)$, $B(9, 3)$, and $C(3, 5)$.
(a) Find the perimeter of $\triangle ABC$.
(b) Find the perimeter of the triangle that is formed by joining the midpoints of the three sides of $\triangle ABC$.
(c) Compute the ratio of the perimeter in part (a) to the perimeter in part (b).
(d) What theorem from geometry provides the answer for part (c) without using the results in (a) and (b)?

27. Use the Pythagorean theorem to find the length a in the figure. Then find b, c, d, e, f, and g.

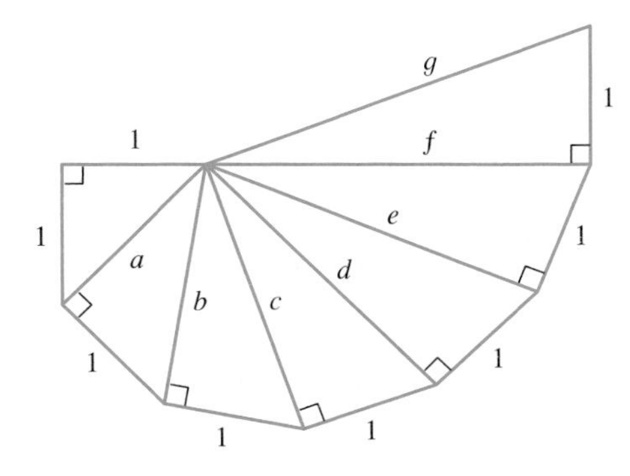

Note: This figure provides a geometric construction for the irrational numbers $\sqrt{2}, \sqrt{3}, \ldots, \sqrt{n}$, where n is a nonsquare natural number. According to Boyer's *A History of Mathematics,* 2nd ed. (New York: John Wiley & Sons, Inc., 1991), "Plato . . . says that his teacher Theodorus of Cyrene . . . was the first to prove the irrationality of the square roots of the nonsquare integers from 3 to 17, inclusive. It is not known how he did this, nor why he stopped with $\sqrt{17}$." One plausible reason for Theodorus's stopping with $\sqrt{17}$ may have to do with the figure shown here. Theodorus may have known that the figure begins to overlap itself at the stage where $\sqrt{18}$ would be constructed.

28. (A numerologist's delight) Using the Pythagorean theorem and your calculator, compute the area of a right triangle in which the lengths of the hypotenuse and one leg are 2045 and 693, respectively.

B

29. The diagonals of a parallelogram bisect each other. Steps (a), (b), and (c) outline a proof of this theorem. (See Exercise 25 for a particular instance of this theorem.)
(a) In the parallelogram $OABC$ shown in the figure, check that the coordinates of B must be $(a + b, c)$.
(b) Use the midpoint formula to calculate the midpoints of diagonals $\overline{OB}$ and $\overline{AC}$.
(c) The two answers in part (b) are identical. This shows that the two diagonals do indeed bisect each other, as we wished to prove.

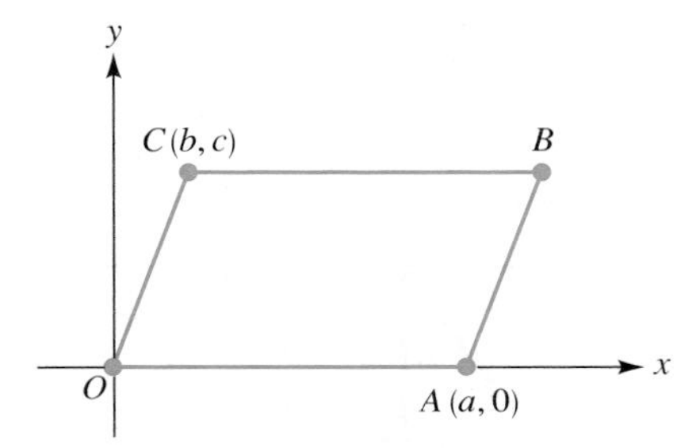

30. Prove that in a parallelogram, the sum of the squares of the lengths of the diagonals equals the sum of the squares of the lengths of the four sides. (Use the figure in Exercise 29.)

31. Suppose that the coordinates of points P, Q, and M are

$$P(x_1, y_1) \qquad Q(x_2, y_2)$$
$$M\left(\frac{x_1 + x_2}{2}, \frac{y_1 + y_2}{2}\right)$$

Follow steps (a) and (b) to prove that M is the midpoint of the line segment from P to Q.
(a) By computing both of the distances PM and MQ, show that $PM = MQ$. (This shows that M lies on the perpendicular bisector of line segment $\overline{PQ}$, but it does not show that M actually lies on $\overline{PQ}$.)
(b) Show that $PM + MQ = PQ$. (This shows that M does lie on $\overline{PQ}$.)

32. This problem outlines one of the shortest proofs of the Pythagorean theorem. The proof was discovered by the Hindu mathematician Bhāskara (1114–ca. 1185). (For other proofs, see the next exercise and also Exercise 100 on page 83.) In the figure we are given a right triangle ACB with the right angle at C, and we want to prove that $a^2 + b^2 = c^2$. In the figure, $\overline{CD}$ is drawn perpendicular to $\overline{AB}$.
(a) Check that $\angle CAD = \angle DCB$ and that $\triangle BCD$ and $\triangle BAC$ are similar.
(b) Use the result in part (a) to obtain the equation $a/y = c/a$, and conclude that $a^2 = cy$.
(c) Show that $\triangle ACD$ is similar to $\triangle ABC$, and use this to deduce that $b^2 = c^2 - cy$.

(d) Combine the two equations deduced in parts (b) and (c) to arrive at $a^2 + b^2 = c^2$.

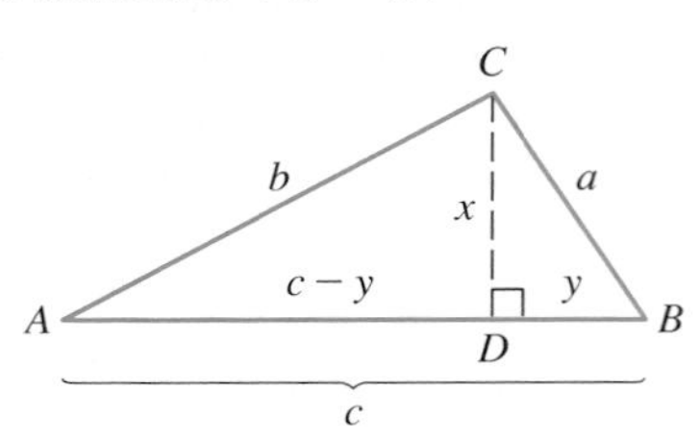

33. One of the oldest and simplest proofs of the Pythagorean theorem is found in the ancient Chinese text *Chou Pei Suan Ching.* This text was written during the Han period (206 B.C.–A.D. 222), but portions of it may date back to 600 B.C. The proof in *Chou Pei Suan Ching* is based on this diagram from the text. In this exercise we explain the details of the proof.

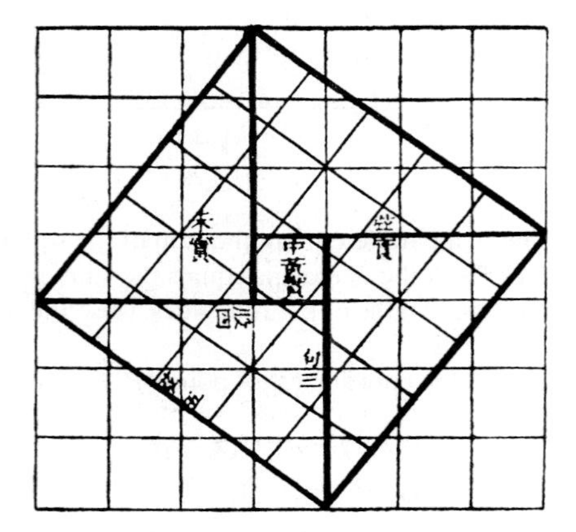

A diagram accompanying a proof of the "Pythagorean" theorem in the ancient Chinese text *Chou Pei Suan Ching* [from *Science and Civilisation in China,* vol. 3, by Joseph Needham (Cambridge, England: Cambridge University Press, 1959)].

(a) Starting with the right triangle in Figure A, we make four replicas of this triangle and arrange them to form the pattern shown in Figure B. Explain why the outer quadrilateral in Figure B is a square.

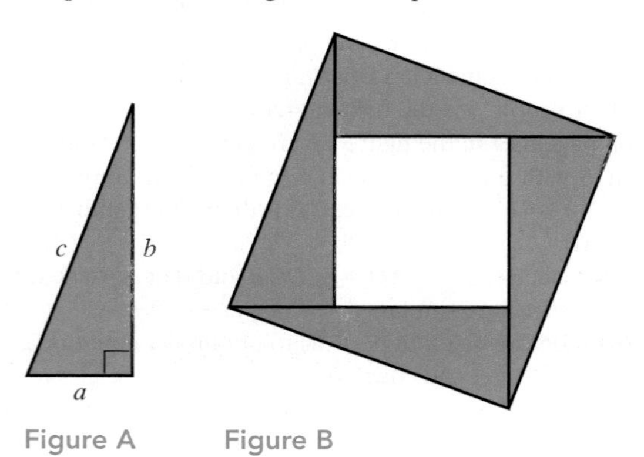

Figure A Figure B

(b) The unshaded region in the center of Figure B is a square. What is the length of each side?

(c) The area of the outer square in Figure B is $(\text{side})^2 = c^2$. This area can also be computed by adding up the areas of the four right triangles and the inner square. Compute the area in this fashion. After simplifying, you should obtain $a^2 + b^2$. Now conclude that $a^2 + b^2 = c^2$, since both expressions represent the same area.

C

34. This problem indicates a method for calculating the area of a triangle when the coordinates of the three vertices are given.

(a) Calculate the area of $\triangle ABC$ in the figure.

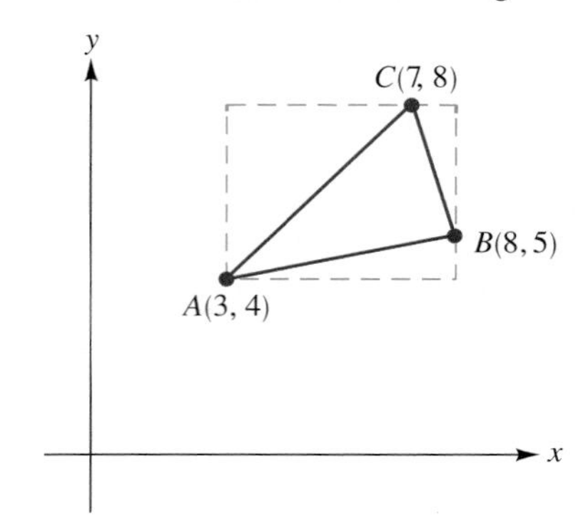

Hint: First calculate the area of the rectangle enclosing $\triangle ABC$, and then subtract the areas of the three right triangles.

(b) Calculate the area of the triangle with vertices (1, 3), (4, 1), and (10, 4). *Hint*: Work with an enclosing rectangle and three right triangles, as in part (a).

(c) Using the same technique that you used in parts (a) and (b), show that the area of the triangle in the following figure is given by

$$A = \tfrac{1}{2}(x_1y_2 - x_2y_1 + x_2y_3 - x_3y_2 + x_3y_1 - x_1y_3)$$

Remark: If we use absolute value signs instead of the parentheses, then the formula will hold regardless of the relative positions or quadrants of the three vertices. Thus the area of a triangle with vertices (x_1, y_1), (x_2, y_2), (x_3, y_3) is given by

$$A = \tfrac{1}{2}|x_1y_2 - x_2y_1 + x_2y_3 - x_3y_2 + x_3y_1 - x_1y_3|$$

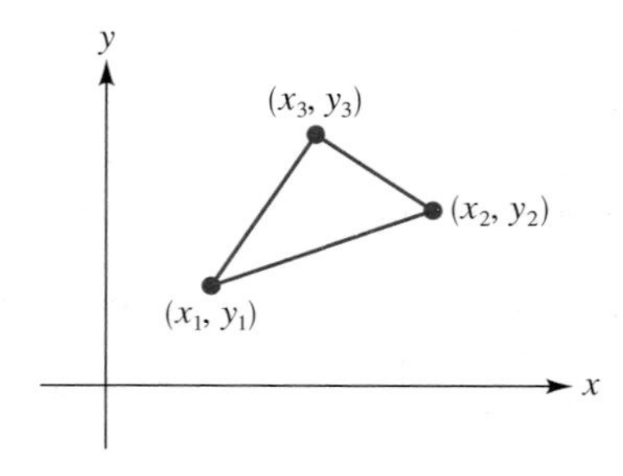

MINI PROJECT

DISCUSS, COMPUTE, REASSESS

As background for this discussion you should review Example 5. In that example, we used the midpoint formula to estimate the amount of carbon dioxide in the atmosphere for the year 1993 using data from the surrounding years 1992 and 1994. Consider now the following three figures (independent of Example 5). Each figure shows four data points, and there is a question mark regarding the location of a fifth. Note that the data points in each figure appear to be rising in a straight-line or near straight-line fashion.

(a) In your opinion, which of the following scenarios might lead to the most reliable estimate for the actual location of the fifth data point? The least reliable? What are the reasons or intuitions behind your replies?

(i) Figure A: Using the midpoint formula and the data for 1994 and 1996 to obtain a possible value for the global average temperature in 1995;

(ii) Figure B: Using the midpoint formula with the points (10, 50) and (30, 86) to obtain a possible value for the Fahrenheit temperature corresponding to 20 degrees on the Celsius scale;

(iii) Figure C: Using the midpoint formula and the data for 1965 and 1985 to obtain a possible value for the percentage of the earth's population living in urban areas in 1975.

(b) Carry out the three calculations described in part (a). Then use the following information to compute the percentage error for each of your estimates: **(i)** The average global temperature for 1995 was 57.9°F; **(ii)** The temperature 20°C corresponds to 68°F; **(iii)** In 1975, 37.8% of the world's population lived in urban areas.

(c) Use the percentage errors you've computed as a check on your part (a) answers. Were you correct about which estimate was the most reliable and which was the least? If you missed either one, how might you adjust the reasons or intuitions that you expressed in part (a)?

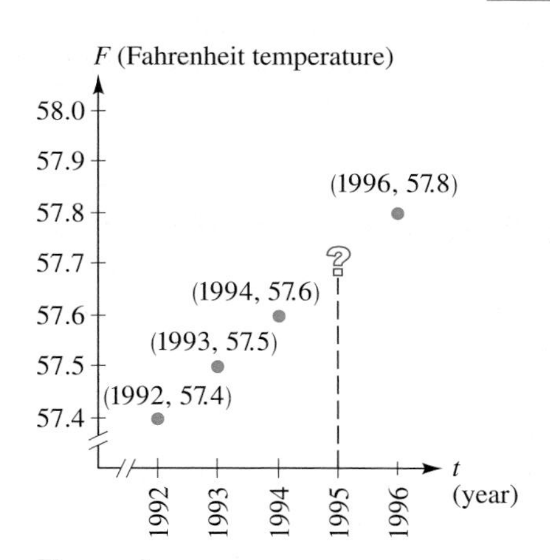

Figure A
Global average temperature at Earth's surface

Source: Adapted from data compiled by Worldwatch Institute from Goddard Institute for Space Studies

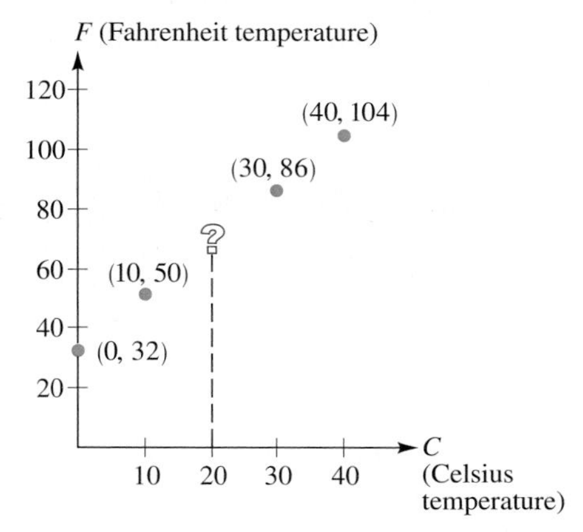

Figure B
Some correspondences between temperatures on the Celsius and Fahrenheit scales. [For example, the rightmost point (40, 104) indicates that a temperature of 40 degrees on the Celsius scale corresponds to 104 degrees on the Fahrenheit scale.]

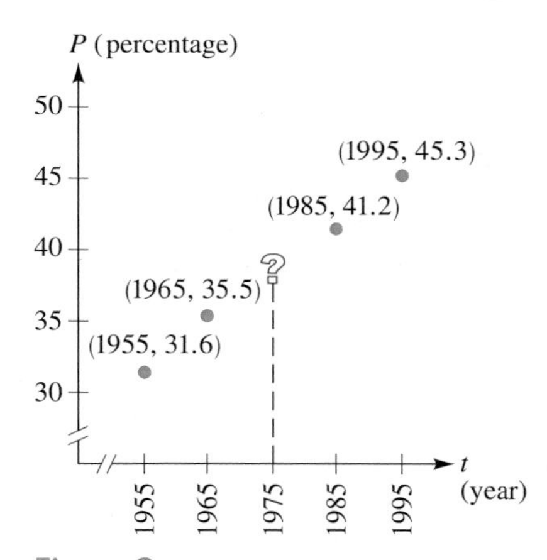

Figure C
Percentage of Earth's human population living in urban areas

Source: Compiled by Worldwatch Institute from United Nations, World Population Prospects: The 1996 Revision

1.5 GRAPHS AND GRAPHING UTILITIES

In this section we look at the connection between two basic skills that you've worked on in previous courses:

- graphing equations
- solving equations

We also look at these topics in terms of *graphing utilities.* By a **graphing utility** we mean either a graphing calculator or a computer with software for graphing and analyzing equations. Whether we are working "by hand" or with technology, the following definition underlies all.

DEFINITION The Graph of an Equation

The **graph** of an equation in two variables is the set of all points with coordinates satisfying the equation.

TABLE 1 $y = 3x - 2$

x	y
0	−2
1	1
2	4
3	7
−1	−5
−2	−8

Suppose, for example, that we want to graph the equation $y = 3x - 2$. We begin by noting that the domain of the variable x in the expression $3x - 2$ is the set of all real numbers. Now we choose values for x and in each case compute the corresponding y-value from the equation $y = 3x - 2$. For example, if x is zero, then $y = 3(0) - 2 = -2$. Table 1 summarizes the results of some of these calculations. The first line in Table 1 tells us that the point with coordinates $(0, -2)$ is on the graph of $y = 3x - 2$. Reading down the table, we see that some other points on the graph are $(1, 1)$, $(2, 4)$, $(3, 7)$, $(-1, -5)$, and $(-2, -8)$.

We now plot (locate) the points we have determined and we note, in this example, that they all appear to lie on a straight line. We draw the line indicated; this is the graph of $y = 3x - 2$ (see Figure 1). In Example 1 we ask a simple question that will test your understanding of the process we've just described and of graphing in general.

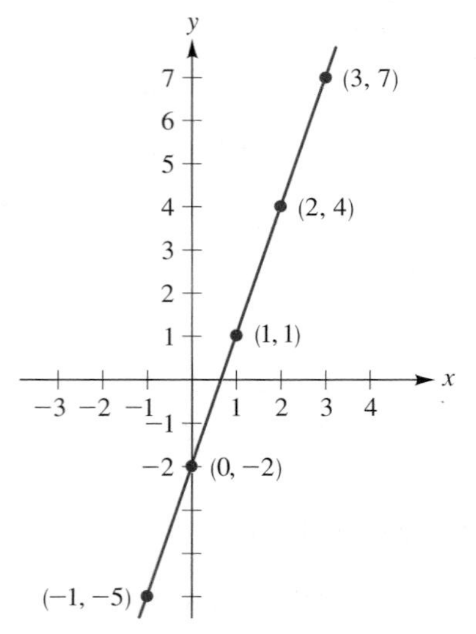

Figure 1
$y = 3x - 2$

EXAMPLE 1 Applying the definition of the "graph of an equation"

Does the point $(1\frac{2}{3}, 2\frac{2}{3})$ lie on the graph of $y = 3x - 2$?

SOLUTION

Looking at the graph in Figure 1, we certainly see that the point $(1\frac{2}{3}, 2\frac{2}{3})$ is very close to the line, but from this visual inspection alone, we cannot be certain that this point actually lies on the line. To settle the question, then, we check to see whether the values $x = 1\frac{2}{3} = \frac{5}{3}$ and $y = 2\frac{2}{3} = \frac{8}{3}$ together satisfy the equation $y = 3x - 2$. We have

$$\frac{8}{3} \stackrel{?}{=} 3\left(\frac{5}{3}\right) - 2$$

$$\frac{8}{3} \stackrel{?}{=} 3 \qquad \text{No}$$

Since the coordinates $(1\frac{2}{3}, 2\frac{2}{3})$ evidently do not satisfy the equation $y = 3x - 2$, we conclude that the point $(1\frac{2}{3}, 2\frac{2}{3})$ is not on the graph. [The calculation further tells us that the point $(1\frac{2}{3}, 3)$ *is* on the graph.]

A graphing utility produces graphs of equations in essentially the same way that we did in Table 1 and Figure 1. The graphing utility uses the given equation to compute pairs of points (x, y), and it plots these points one at a time on the viewing screen. Of course, the machine is able to quickly compute and plot many more points than we used in Table 1—that's an advantage. On the other hand, the graphing utility is not "smart." For example, it does not know ahead of time that the equation $y = 3x - 2$ will be a straight line. (*You'll* know this by the end of the next section, or maybe you already remember from a previous course that the graph of any equation of the form $y = mx + b$ is always a straight line.) Later in this section and in the exercises we'll say more about some of the limitations of a graphing utility and how some thought or experimentation on your part is often necessary.

GRAPHICAL PERSPECTIVE

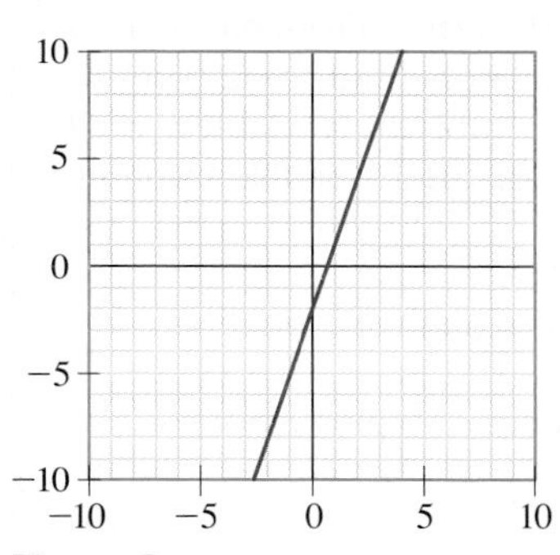

Figure 2
A graph of $y = 3x - 2$

In Figure 2 we show a graph of the equation $y = 3x - 2$ obtained with a graphing utility. Actually, since there are so many different types of graphing utilities, the picture we've chosen to show in this text is a kind of amalgam containing features from several types of utilities. If you have a graphing utility, you should now produce a graph of $y = 3x - 2$ for yourself and compare it to Figure 2. (If necessary refer to the user's manual that came with your calculator or software.) In this text, the label GRAPHICAL PERSPECTIVE (as you see above Figure 2) is a signal for those students with graphing utilities: Use your graphing utility, making adjustments as necessary, to obtain a view that is similar to the one shown here in the textbook. For many graphing utilities, the default view or **standard viewing rectangle** is $-10 \leq x \leq 10$ and $-10 \leq y \leq 10$. This is the view shown in Figure 2. In the next example we graph an equation first by hand, then with a graphing utility. In the graphing utility portion, you'll see a case in which the standard viewing rectangle is inappropriate and needs to be modified.

EXAMPLE 2 An application of linear equations and graphs to temperature scales

The equation $5F - 9C = 160$ relates temperature F on the Fahrenheit scale to temperature C on the Celsius scale.

(a) Solve the given equation for F in terms of C. Then set up a table and graph the equation in a C-F coordinate system (C on the horizontal axis, F on the vertical).

(b) Use a graphing utility to obtain a graph of $5F - 9C = 160$.

SOLUTION

(a) We have

$$\begin{aligned} 5F - 9C &= 160 \\ 5F &= 9C + 160 && \text{adding } 9C \text{ to both sides} \\ F &= \frac{9}{5}C + 32 && \text{dividing both sides by 5} \qquad (1) \end{aligned}$$

Using equation (1) now, we pick values for C and compute the corresponding values for F. To avoid working with fractions, we take values for C that

TABLE 2 $F = \frac{9}{5}C + 32$

C	F
0	32
5	41
10	50
15	59
−5	23
−10	14

are multiples of 5, as indicated in Table 2. (Check for yourself that the entries in the right-hand column of the table are correct.)

In view of the data in Table 2, we'll mark off the axes as indicated in Figure 3(a), with tick marks five units apart on the C-axis and ten units apart on the F-axis. (Other markings are feasible. For example, it would be reasonable to use markings of 20 instead of 10 on the vertical axis, but it would be very clumsy to use markings at one-unit intervals.) In Figure 3(b) we've plotted the data pairs (C, F) given by Table 2. These points appear to lie on a straight line, and we've drawn the line in Figure 3(c). This is the required graph.

Note: In elementary graphing, there is the question of how many points must be plotted before the essential features of a graph are clear. As you'll see throughout this text, there are a number of techniques and concepts that make it unnecessary to plot a large number of points. For instance, if you knew ahead of time that the graph of $F = \frac{9}{5}C + 32$ were a straight line, then you would need to plot only two points in order to draw the line.

(b) With a graphing utility, just as in graphing by hand, we first need to solve the given equation for one variable in terms of the other. From part (a) we have $F = \frac{9}{5}C + 32$. Because most graphing calculators (and some types of computer software) require that we use x and y for the variables, we'll rewrite the equation as $y = \frac{9}{5}x + 32$. Then we can enter it in the graphing utility as

$$y = \left(\frac{9}{5}\right) * x + 32 \qquad \text{The symbol * denotes multiplication.}$$

or better yet (with fewer keystrokes and less chance of typing error) as

$$y = \frac{9x}{5} + 32 \tag{2}$$

Figure 4(a) shows the "graph" of equation (2) in the standard viewing rectangle. Yes, nothing's there; evidently, no point of the required graph passes through the standard viewing rectangle. We need to find a more appropriate viewing rectangle. After some experimenting with the dimensions of the viewing rectangle, we obtain the views shown in Figure 4(b), then Figure 4(c). Figure 4(c) is an appropriate viewing rectangle; the essential features of the graph are clear, and there is not a lot of wasted space in the viewing screen, as there is in Figure 4(b).

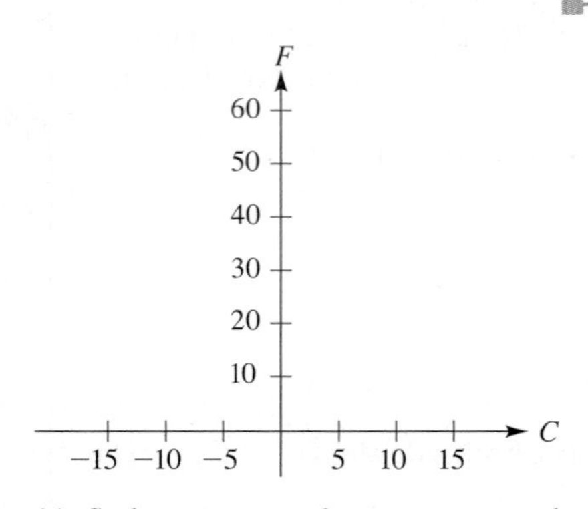

(a) Scales on axes are chosen to accommodate the range of numbers in Table 2.

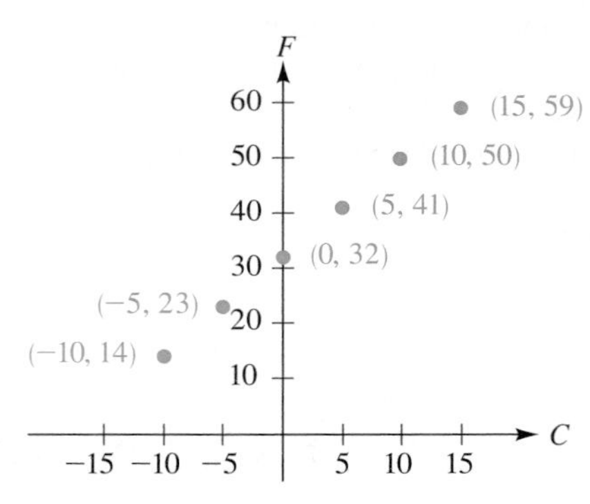

(b) Plot of data points (C, F) from Table 2.

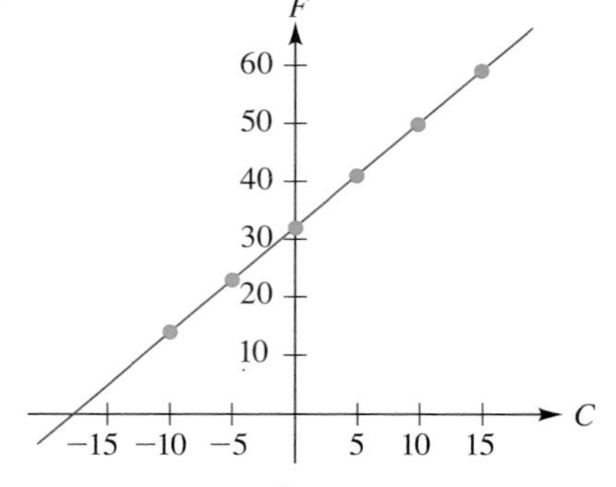

(c) Graph of $F = \frac{9}{5}C + 32$ (or $5F - 9C = 160$)

Figure 3

GRAPHICAL PERSPECTIVE

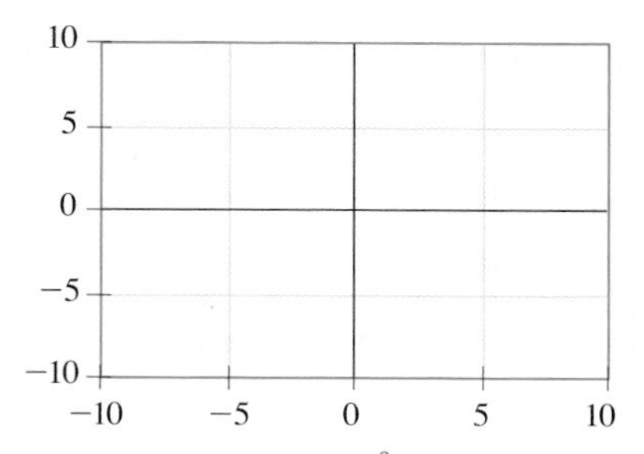

(a) The graph of $y = \frac{9}{5}x + 32$ does not pass through the standard viewing rectangle.

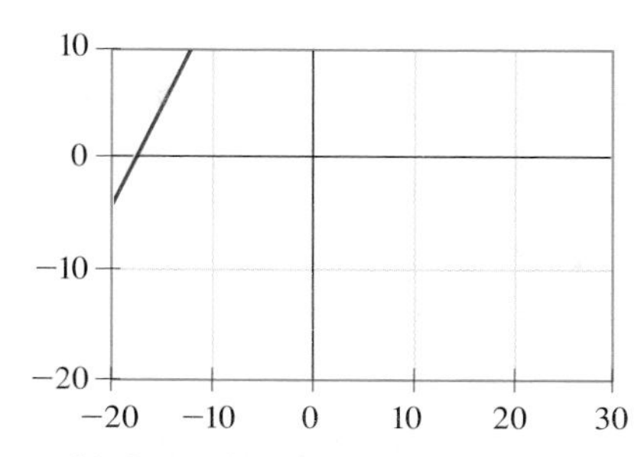

(b) Better, but still not an appropriate viewing rectangle.

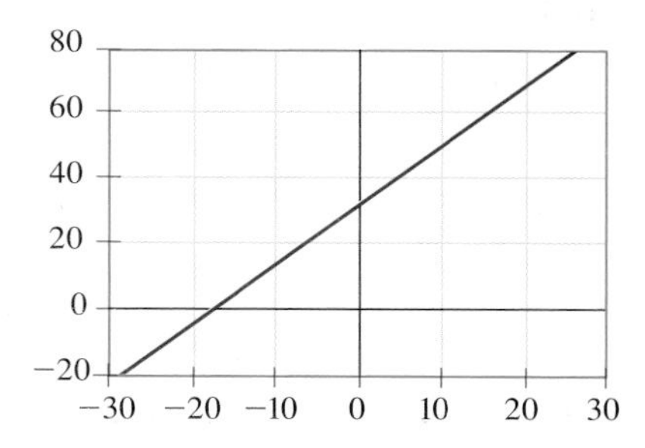

(c) An appropriate viewing rectangle. The essential features of the graph are clear.

Figure 4

Finding an appropriate viewing rectangle for $y = \frac{9}{5}x + 32$ (or $F = \frac{9}{5}C + 32$)

We can use Figure 4 to introduce some common notation that is often used in describing viewing rectangles. (Even if you are not working with a graphing utility, you are apt to see this notation in other math books or on a friend's math paper.) In Figure 4(a), x runs from -10 to 10 in increments of 5. A notation for this is

`Xmin = -10     Xmax = 10     Xscl = 5`

In the rightmost equation, "Xscl" refers to the *scale* on the x-axis, and the equation Xscl = 5 tells you that it is five units between adjacent tick marks or grid lines. In Figure 4(a), y also runs from -10 to 10 in increments of 5, so we write `Ymin = -10`, `Ymax = 10`, and `Yscl = 5`. As another example of this notation, for Figure 4(c) we have (as you should check)

`Xmin = -30`		`Ymin = -20`
`Xmax = 30`	and	`Ymax = 80`
`Xscl = 10`		`Yscl = 20`

Another type of shorthand that you'll see is based on the closed interval notation $[a, b]$ described in Section 1.1. With this notation we specify the size of the viewing rectangle in Figure 4(c) as

$$[-30, 30] \text{ by } [-20, 80]$$

In this notation the x-interval precedes the y-interval.

The only drawback to this notation is that the scale or number of units between tick marks is not specified. We remedy this by inserting a third number, denoting scale, in each bracket. The viewing rectangle in Figure 4(c) is then completely specified by writing

$$[-30, 30, 10] \text{ by } [-20, 80, 20]$$

EXAMPLE 3 Specifying viewing rectangles on a graphing utility

Figure 5 shows two views of the graph of the equation $y = 2x^3 - x^2$. Describe each viewing rectangle using the two types of notation discussed above (that is, the min-max-scl equations and the bracket notation).

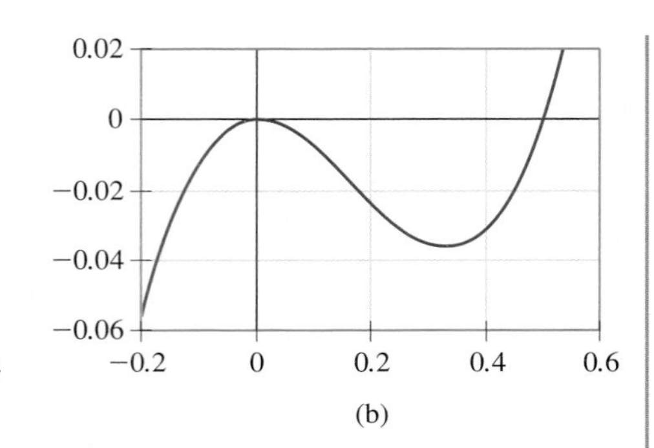

Figure 5
Two views of $y = 2x^3 - x^2$

SOLUTION
For Figure 5(a) we have

```
Xmin = -4              Ymin = -40
Xmax = 4      and      Ymax = 60
Xscl = 2               Yscl = 20
```

The bracket notation describing the viewing rectangle in Figure 5(a) is

$$[-4, 4, 2] \text{ by } [-40, 60, 20]$$

For Figure 5(b) we have

```
Xmin = -0.2            Ymin = -0.06
Xmax = 0.6    and      Ymax = 0.02
Xscl = 0.2             Yscl = 0.02
```

The bracket notation describing the viewing rectangle in Figure 5(b) is

$$[-0.2, 0.6, 0.2] \text{ by } [-0.06, 0.02, 0.02]$$

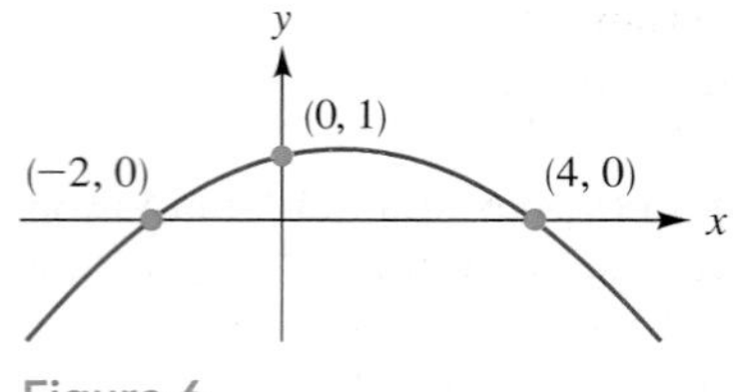

Figure 6
x-intercepts: −2 and 4
y-intercept: 1

When we graph an equation, it's helpful to know where the curve intersects the x- or y-axis. By an ***x*-intercept** of a graph we mean the x-coordinate of a point where the graph intersects the x-axis. For instance, in Figure 6 there are two x-intercepts: −2 and 4. We define y-intercepts in a similar manner: A ***y*-intercept** of a graph is the y-coordinate of a point where the graph intersects the y-axis. Thus, the y-intercept of the graph in Figure 6 is 1.

EXAMPLE 4 Computing x- and y-intercepts

Figure 7 shows the graph of the equation $3x + 5y = 15$. Find the x- and y-intercepts.

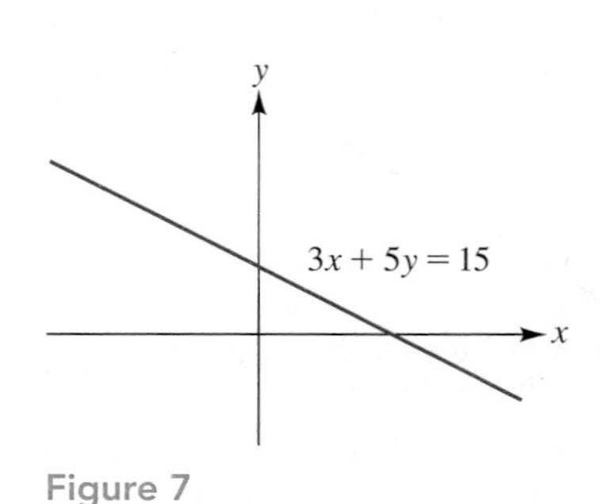

Figure 7

SOLUTION
At the point where the graph crosses the x-axis, the y-coordinate is zero. So, to find the x-intercept, we set $y = 0$ in the given equation $3x + 5y = 15$ and solve for x:

$$\begin{aligned} 3x + 5(0) &= 15 \\ 3x &= 15 \\ x &= 5 \end{aligned}$$

Thus, the x-intercept is 5. Similarly, to find the y-intercept, we set $x = 0$ in the given equation $3x + 5y = 15$ and solve for y:

$$\begin{aligned} 3(0) + 5y &= 15 \\ 5y &= 15 \\ y &= 3 \end{aligned}$$

The y-intercept is 3.

In the next example we show two different ways to determine x-intercepts. First we use a graphing utility to obtain approximate values for the intercepts; then we use algebra to find exact values.

EXAMPLE 5 Finding approximate and exact values for x-intercepts

GRAPHICAL PERSPECTIVE

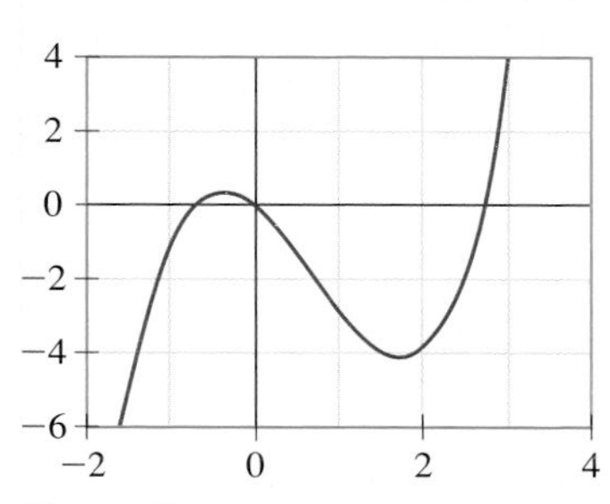

Figure 8
$y = x^3 - 2x^2 - 2x$
viewing rectangle: $[-4, 4, 2]$
by $[-6, 4, 2]$

Figure 8 shows a graph of the equation $y = x^3 - 2x^2 - 2x$ obtained with a graphing utility.

(a) Use a graphing utility to estimate (to one decimal place) the x-intercepts of the graph.

(b) Use algebra to determine the exact values for the x-intercepts. Then use a calculator to check that the answers are consistent with the estimates obtained in part (a).

SOLUTION

(a) According to Figure 8, there appear to be three x-intercepts: one between -1 and 0, one that is either zero or very close to zero, and one between 2 and 3. To estimate the intercept that is between -1 and 0 as well as the intercept near (or possibly at) zero, we'll change the viewing rectangle shown in Figure 8 so that x runs from -1 to 1.1 in increments of 0.1. After some experimenting with the y-values, we obtain the graph in Figure 9(a), which shows that the leftmost intercept is between -0.8 and -0.7. Is it closer to -0.8 or -0.7? On a graphing calculator screen (as opposed to a larger picture on a computer monitor) it may not be clear that the intercept is closer to -0.7 than to -0.8. In Figure 9(b) we've adjusted the viewing rectangle for a closer view, which shows that the intercept is indeed closer to -0.7 than to -0.8. So to one decimal place, we can say that the leftmost intercept is

GRAPHICAL PERSPECTIVE

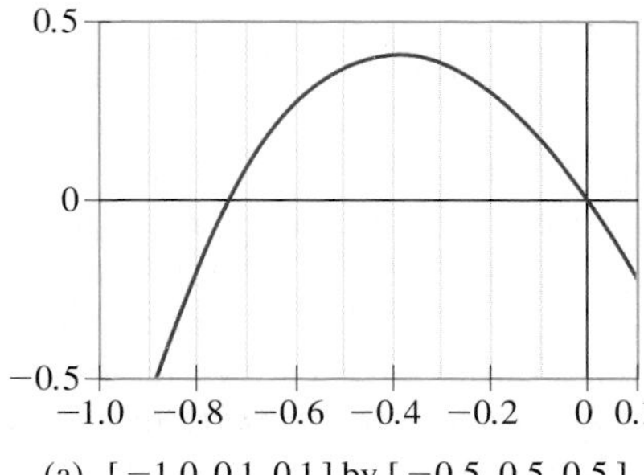

(a) $[-1.0, 0.1, 0.1]$ by $[-0.5, 0.5, 0.5]$

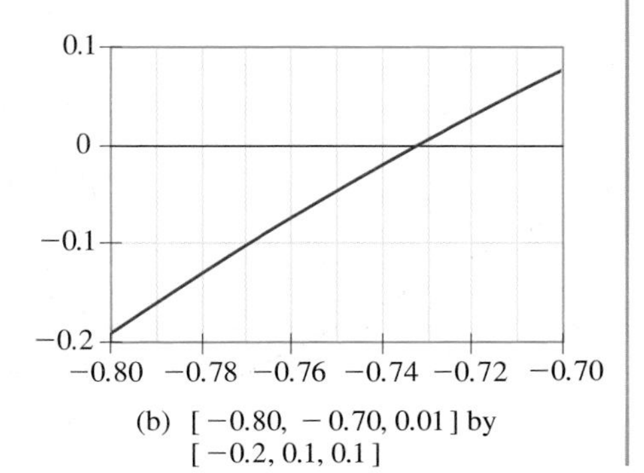

(b) $[-0.80, -0.70, 0.01]$ by $[-0.2, 0.1, 0.1]$

Figure 9
Adjusting the viewing rectangle to estimate an x-intercept for $y = x^3 - 2x^2 - 2x$

-0.7. [Actually, Figure 9(b) tells us more than this. The intercept is between -0.74 and -0.73.]

Regarding the intercept that appears in Figure 8 to be at or near zero, Figure 9(a) tells us that to one decimal place, the intercept is 0.0. In part (b) we'll show that this intercept is indeed exactly zero. (You cannot conclude this fact from Figure 9(a); see Exercise 35 for an example along these lines.)

There is one more x-intercept to estimate, the one in Figure 8 that is between 2 and 3. Exercise 36 asks you to use the methods we've just demonstrated to show that this rightmost x-intercept is approximately 2.7.

(b) To compute the exact values for the x-intercepts of the graph, we set y equal to zero in the given equation $y = x^3 - 2x^2 - 2x$ to obtain $0 = x^3 - 2x^2 - 2x$. We then have

$$x^3 - 2x^2 - 2x = 0 \qquad (3)$$

$$x(x^2 - 2x - 2) = 0 \qquad \text{factoring out the common factor } x$$

The zero product property from Section 1.3 now tells us that $x = 0$ or $x^2 - 2x - 2 = 0$. We can use the quadratic formula (from Section 1.3) to solve the equation $x^2 - 2x - 2 = 0$:

$$\begin{aligned} x &= \frac{-b \pm \sqrt{b^2 - 4ac}}{2a} \\ &= \frac{-(-2) \pm \sqrt{(-2)^2 - 4(1)(-2)}}{2(1)} && \text{using } a = 1, b = -2, c = -2 \\ &= \frac{2 \pm \sqrt{12}}{2} \\ &= \frac{2 \pm 2\sqrt{3}}{2} = \frac{2(1 \pm \sqrt{3})}{2} = 1 \pm \sqrt{3} && \text{See Appendix B.2 if you need to review working with radicals.} \end{aligned}$$

Putting things together now, we have found three roots for equation (3): 0, $1 - \sqrt{3}$, and $1 + \sqrt{3}$. The largest (and the only positive root) of these three is $1 + \sqrt{3}$; that is the rightmost x-intercept for the graph in Figure 8. The smallest (and the only negative root) is $1 - \sqrt{3}$; that is the leftmost x-intercept for the graph in Figure 8. Finally, the root $x = 0$ tells us that the middle x-intercept is indeed exactly zero. As you should check for yourself using a calculator, we have $1 - \sqrt{3} \approx -0.73$ and $1 + \sqrt{3} \approx 2.73$. Note that these values are indeed consistent with the estimates obtained in part (a).

In the graphing utility portion of the example just concluded, we started with the viewing rectangle in Figure 8 and then adjusted it to obtain a closer look at the x-intercepts. Most graphing utilities also allow you to obtain a closer look in another way, by using a ZOOM feature and a TRACE feature. For details, see the user's manual for your graphing utility. Another technology note: In part (b) of Example 5 we used algebra to determine the exact roots of the equation $x^3 - 2x^2 - 2x = 0$. Most graphing utilities have a SOLVE feature that can be used to obtain roots that, while only approximate, are accurate to several decimal places. Again, for details consult your user's manual.

As was mentioned previously in this section, when we first graph a given equation (either by hand or with a graphing utility), we try to present a view that shows all the essential features of the graph. The next example deals with one aspect of this issue.

EXAMPLE 6 Obtaining information about a graph through its equation

Figure 10 shows a graph of the equation $x^2y - x^3 + 3 = 0$.

(a) From the portion of the graph shown in Figure 10, it appears that there is no y-intercept. Use the given equation to show that this is indeed the case.

(b) As is indicated in Figure 10, there is an x-intercept between 1 and 2. Find this intercept. Give two forms for the answer: an exact expression and a calculator approximation rounded to two decimal places.

Figure 10
Does the graph ever cross the y-axis?

SOLUTION

(a) If there were a y-intercept (perhaps somewhere below the portion of the graph displayed in Figure 10), we could find it by setting x equal to zero in the given equation. But with $x = 0$, the given equation becomes $3 = 0$, which is clearly impossible. We thus conclude that there is no y-intercept; the graph never crosses the y-axis.

(b) The x-intercept of the graph is determined by setting y equal to zero in the given equation. That gives us $x^2(0) - x^3 + 3 = 0$, and therefore $x^3 = 3$. To solve this equation for x, just take the cube root of both sides to obtain

$$x = \sqrt[3]{3}$$
$$\approx 1.44 \quad \text{using a calculator}$$

As you can see, the value $x = 1.44$ is consistent with the graph shown in Figure 10. Also, see the Graphical Perspective in Figure 11. Note that in the caption for Figure 11 we've solved the given equation for y in terms of x. Most graphing utilities require that you enter the equation in this form.

GRAPHICAL PERSPECTIVE

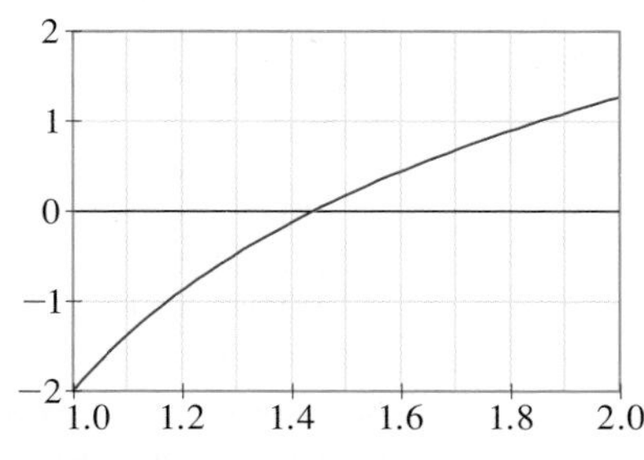

(a) Zoom-in view of $y = (x^3 - 3)/x^2$ shows that the graph has an x-intercept between 1.4 and 1.5.

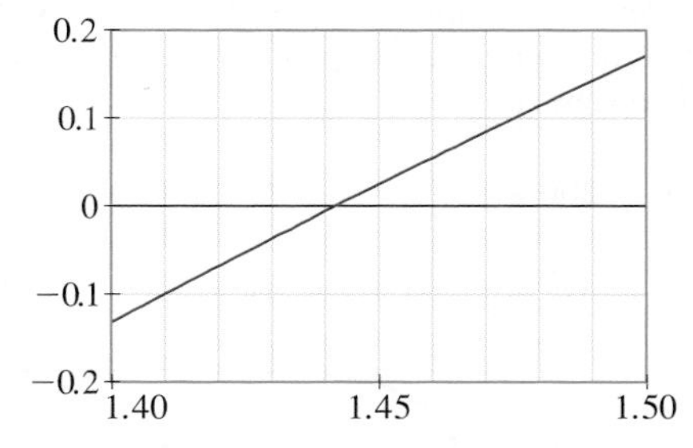

(b) Further magnification shows that the x-intercept is between 1.44 and 1.45 (closer to 1.44 than 1.45).

Figure 11
To graph $x^2y - x^3 + 3 = 0$, first solve for y:

$$x^2y = x^3 - 3$$
$$y = \frac{x^3 - 3}{x^2}$$

EXERCISE SET 1.5

A

In Exercises 1–6, determine whether the given point lies on the graph of the equation, as in Example 1. Note: You are not asked to draw the graph.

1. $(8, 6); y = \frac{1}{2}x + 3$

2. $(\frac{3}{5}, -\frac{17}{5}); y = -\frac{2}{3}x - 3$

3. $(4, 3); 3x^2 + y^2 = 52$

4. $(4, -2); 3x^2 + y^2 = 52$

5. $(a, 4a); y = 4x$

6. $(a - 1, a + 1); y = x + 2$

7. (a) Solve the equation $2x - 3y = -3$ for y and then complete the following table.

x	−6	−3	0	3	6
y					

(b) Use your table from part (a) to graph the equation $2x - 3y = -3$.

8. (a) Solve the equation $3x + 2y = 6$ for y and then complete the following table.

x	−4	−2	0	2	4
y					

(b) Use your table from part (a) to graph the equation $3x + 2y = 6$.

In Exercises 9–14, the graph of each equation is a straight line. Graph the equation after finding the x- and the y-intercepts. (Since you are given that the graph is a line, you need only plot two points before drawing the line.)

9. $3x + 4y = 12$

10. $3x - 4y = 12$

11. $y = 2x - 4$

12. $x = 2y - 4$

13. $x + y = 1$

14. $2x - 3y = 6$

For Exercises 15 and 16: As in Example 5, describe each viewing rectangle using the two types of notation discussed in the text (i.e., the min-max-scl equations and the bracket notation).

15. Two views of $y = x^5 - x^4$

(a)

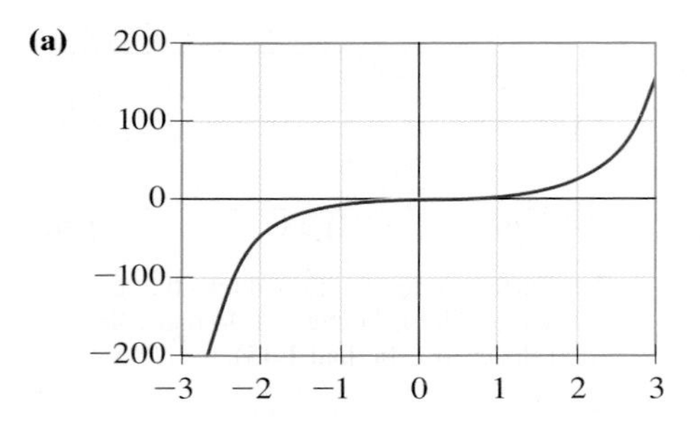

(b)

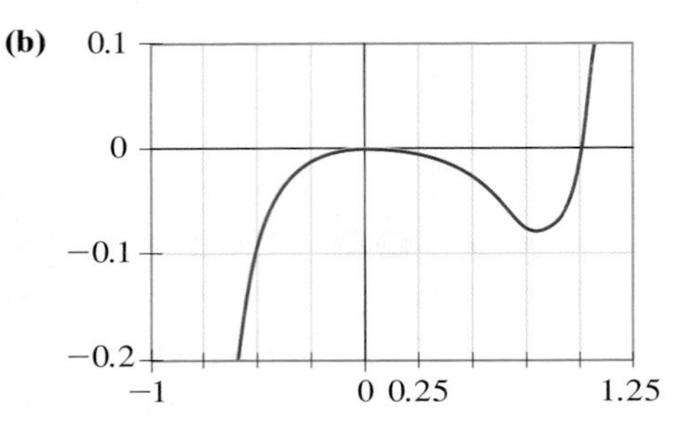

16. Two views of $y = \sqrt{1 - x^2}$

(a)

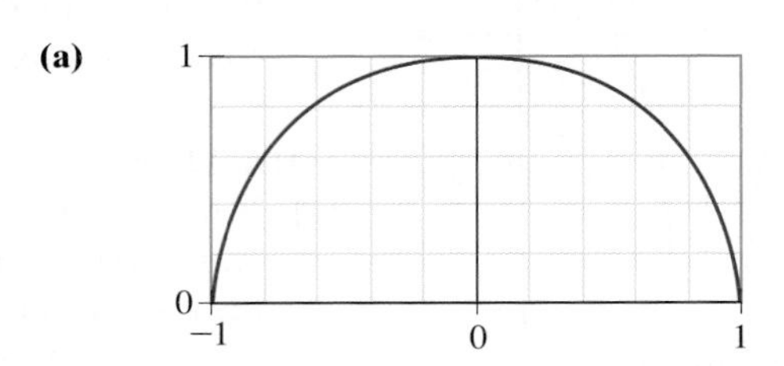

(b)

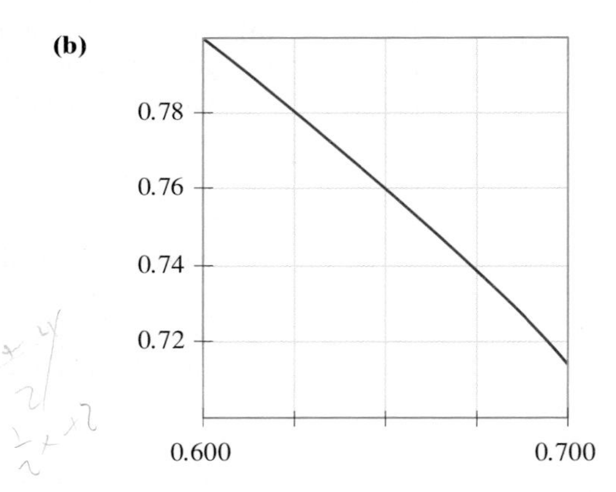

In Exercises 17–20, determine any x- or y-intercepts for the graph of the equation. Note: You're not asked to draw the graph.

17. (a) $y = x^2 + 3x + 2$
(b) $y = x^2 + 2x + 3$

18. (a) $y = x^2 - 4x - 12$
(b) $y = x^2 - 4x + 12$

19. (a) $y = x^2 + x - 1$
(b) $y = x^2 + x + 1$

20. (a) $y = 6x^3 + 9x^2 + x$
(b) $y = 9x^3 + 6x^2 + x$

In Exercises 21–24, each figure shows the graph of an equation. Find the x- and y-intercepts of the graph. If an intercept involves a radical, give both the radical form of the answer and a calculator approximation rounded to two decimal places. (Check to see that your answer is consistent with the given figure.)

21.

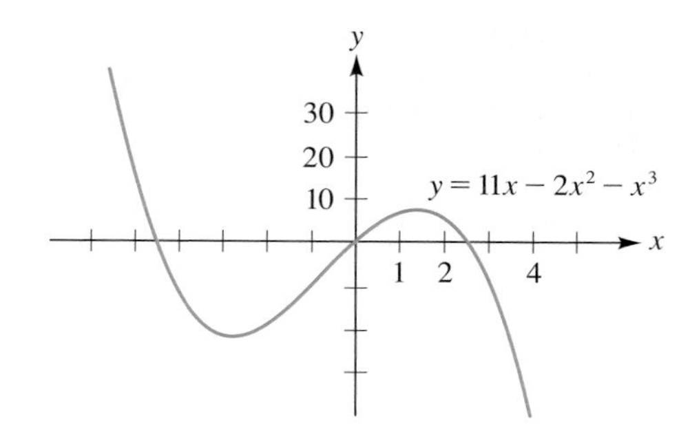

22.

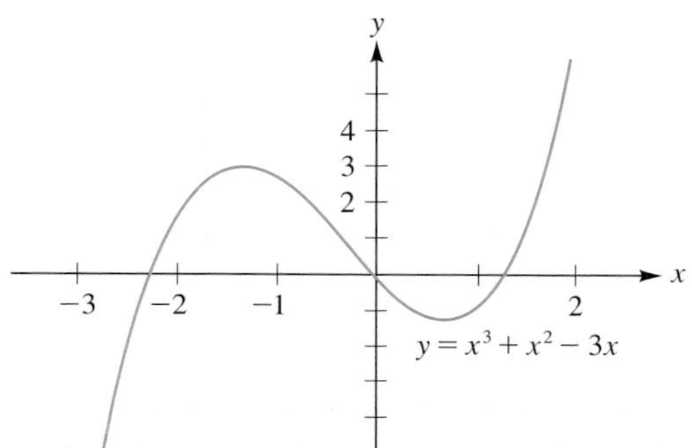

23.

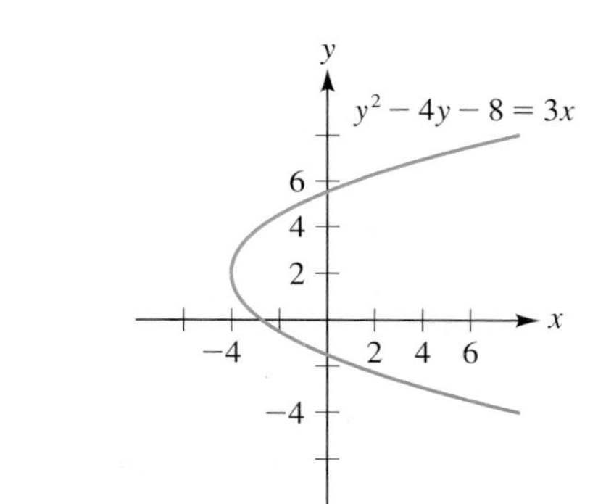

24.

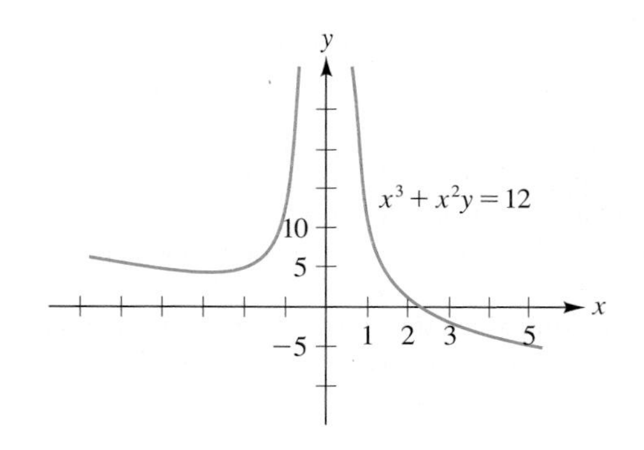

For Exercises 25–30:

Ⓖ **(a)** *Use a graphing utility to graph the equation.*

Ⓖ **(b)** *Use a graphing utility, as in Example 5, to estimate to one decimal place the x-intercepts.*

(c) *Use algebra to determine the exact values for the x-intercepts. Then use a calculator to check that the answers are consistent with the estimates obtained in part (b).*

25. $y = x^2 - 2x - 2$

26. $y = 2x^2 + x - 5$

27. $y = 2x^3 - 5x$

28. $y = 3x^3 + 5x^2 + x$

29. $2xy - x^3 - 5 = 0$

Hint: See Example 6(b) and the caption for Figure 11.

30. $xy - 12y = x^2 - x - 1$

Hint for part (a): To solve the equation for y, first factor out the common term y on the left-hand side. Regarding the essential features of the graph, you should find that there are are two distinct pieces. You may have to adjust the viewing rectangle.

Ⓖ *In Exercises 31–34, use a graphing utility to graph the equations and to approximate the x-intercepts. In approximating the x-intercepts, use a "solve" key or a sufficiently magnified view to ensure that the values you give are correct in the first three decimal places.* Remark: *None of the x-intercepts for these four equations can be obtained using factoring techniques.)*

31. $y = x^3 - 3x + 1$

32. $y = 8x^3 - 6x - 1$

33. $y = x^5 - 6x^4 + 3$

34. $y = 2x^5 - 5x^4 + 5$

Ⓖ **35. (a)** Graph the equation $y = x^3 + 10x + 2$ in the standard viewing rectangle. On the basis of this view, what can you say about the location of any x-intercepts?

(b) The following viewing rectangle shows that $y = x^3 + 10x + 2$ has an x-intercept either at -0.2 or very close to -0.2.

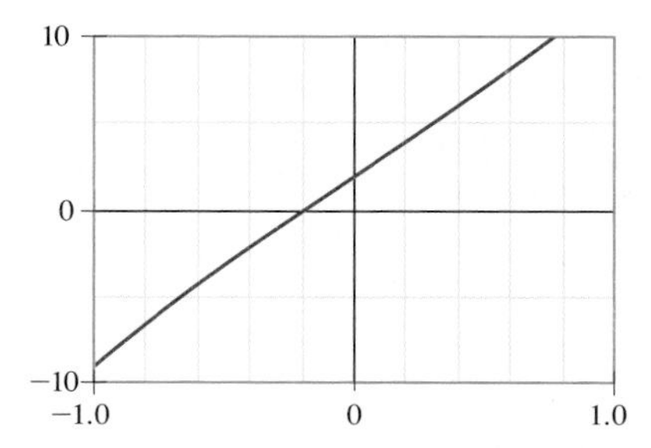

(c) Find a viewing rectangle to demonstrate that -0.2 is *not* an x-intercept.

(d) Looking at the picture in part (b) and your result in part (c), what would you say is the moral of the story?

Ⓖ **36.** This exercise refers to Example 5 and Figure 8.

(a) According to Figure 8, the graph of $y = x^3 - 2x^2 - 2x$ has an x-intercept between 2 and 3, slightly closer to 3 than to 2. Graph the equation in the viewing rectangle [2.5, 3, 0.1] by [−2, 3, 1] to see that the intercept is between 2.7 and 2.8, closer to 2.7 than to 2.8. Thus, to one decimal place, the intercept is 2.7, as stated in Example 5.

(b) Estimate the x-intercept to three decimal places by finding a suitable viewing rectangle.

B

For Exercises 37 and 38, refer to Figure 4(c) on page 39, which shows a graph of the equation $F = \frac{9}{5}C + 32$. The equation relates temperature C on the Celsius scale to temperature F on the Fahrenheit scale. [In Figure 4(c), C is the variable for the horizontal axis and F is the variable for the vertical axis.]

37. (a) Use Figure 4(c) to estimate (to the nearest 5 degrees) the Celsius temperature corresponding to 0 degrees on the Fahrenheit scale. When you do this, which intercept are you looking at, the C-intercept or the F-intercept?

(b) Use the equation $F = \frac{9}{5}C + 32$ to give the exact temperature on the Celsius scale corresponding to 0 degrees on the Fahrenheit scale.

38. (a) Use Figure 4(c) to estimate (to the nearest 10 degrees) the temperature on the Fahrenheit scale corresponding to 0 degrees on the Celsius scale. When you do this, which intercept are you looking at, the C-intercept or the F-intercept?

(b) Use the equation $F = \frac{9}{5}C + 32$ to give the exact temperature on the Fahrenheit scale corresponding to 0 degrees on the Celsius scale.

39. The following figure shows the graph of $y = \sqrt{x}$. Use this graph to estimate the following quantities (to one decimal place).

(a) $\sqrt{2}$
(b) $\sqrt{3}$
(c) $\sqrt{6}$

Hint: $\sqrt{ab} = \sqrt{a}\sqrt{b}$

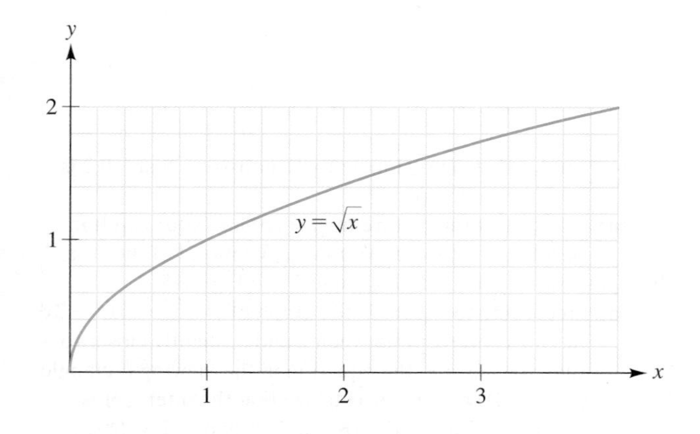

40. The following figure shows the graph of $y = \sqrt[3]{x}$; use it to estimate the following quantities. **(a)** $\sqrt[3]{2}$; **(b)** $\sqrt[3]{5}$; **(c)** $\sqrt[3]{6.5}$. In each case, compare your estimate with a value obtained from a calculator.

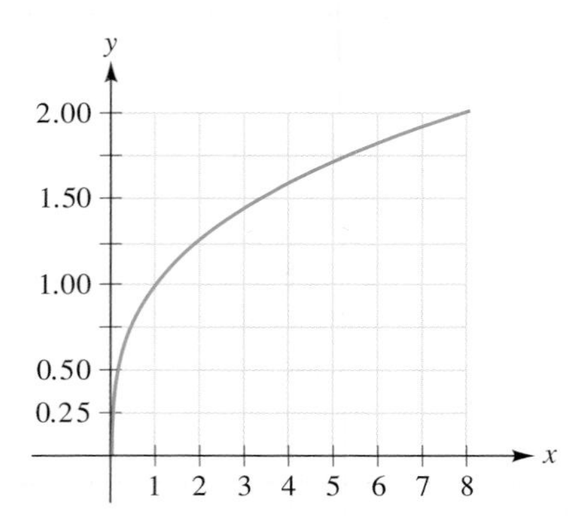

41. In a certain biology experiment, the number N of bacteria increases with time t in hours as indicated in the following figure.

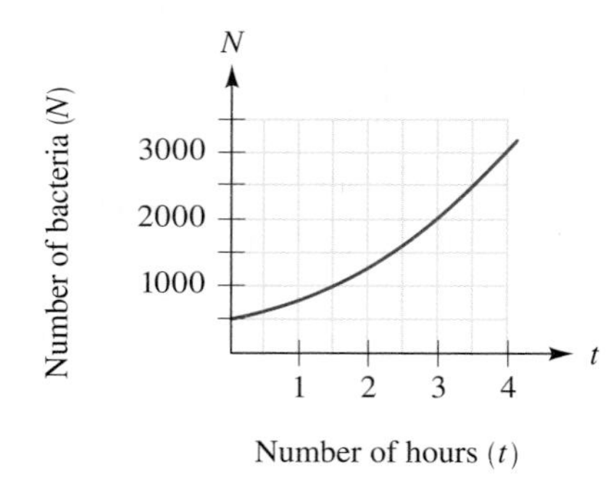

(a) How many bacteria are initially present when $t = 0$?

(b) Approximately how long does it take for the original colony to double in size?

(c) For which value of t is the population approximately 2500?

(d) During which time interval does the population increase more rapidly, between $t = 0$ and $t = 1$ or between $t = 3$ and $t = 4$?

In Exercises 42 and 43, determine the x-coordinates of the points A and B. (Each dashed line is parallel to a coordinate axis.)

42.

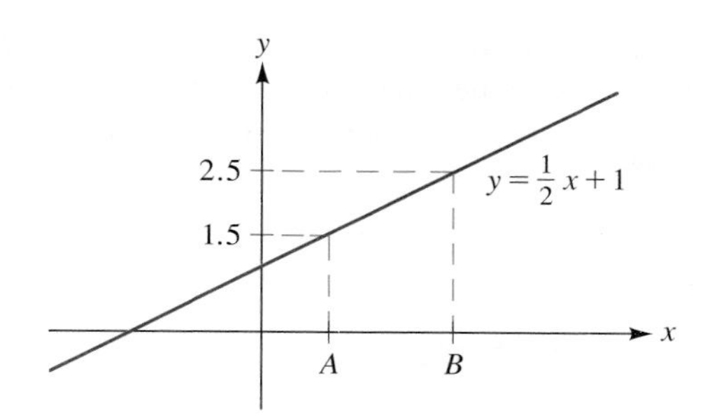

43.

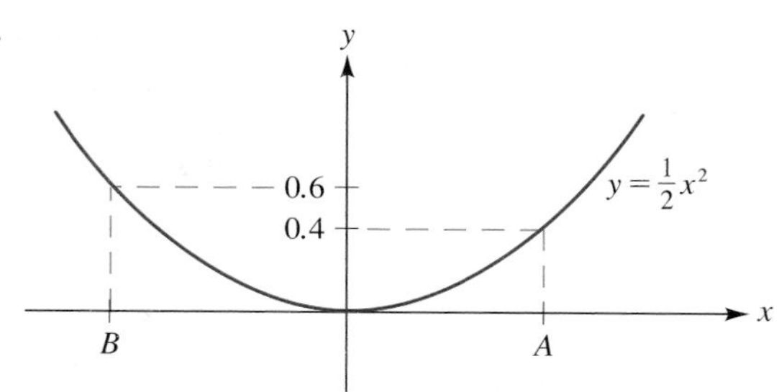

MINI PROJECT

DRAWING CONCLUSIONS FROM VISUAL EVIDENCE

This discussion concerns x-intercepts and requires that at least one person in the group have a graphing utility. How much graphical evidence do you feel is necessary before you are "reasonably certain" about the exact value of an x-intercept? Or would graphical evidence alone never leave you reasonably certain? To focus the discussion, begin by considering the two views of the graph of the equation $y = x^3 + 550x + 11$ in Figures A and B.

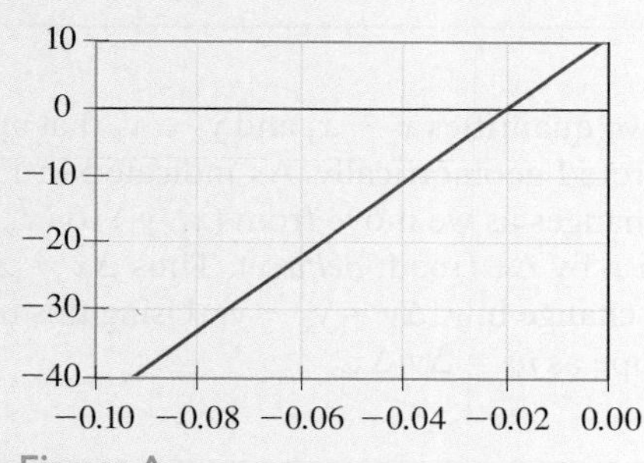

Figure A
Close-up view of $y = x^3 + 550x + 11$ near an x-intercept. The intercept is either -0.02 or something very close to that.

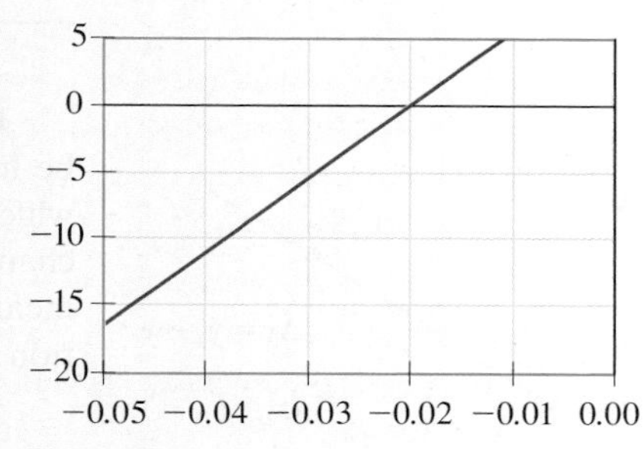

Figure B
Still closer view of the x-intercept in Figure A (zooming in by a factor of 2). Would you say now that it is quite likely that the intercept is exactly -0.02?

1.6 EQUATIONS OF LINES

He [Pierre de Fermat (1601–1665)] *introduced perpendicular axes and found the general equations of straight lines and circles and the simplest equations of parabolas, ellipses, and hyperbolas; and he further showed in a fairly complete and systematic way that every first- or second-degree equation can be reduced to one of these types.*—George F. Simmons in *Calculus Gems, Brief Lives and Memorable Moments* (New York: McGraw-Hill Book Company, 1992)

For sufficiently short periods of time, the graphs of many real-world quantities can be closely approximated by a straight-line graph. In this section we take a systematic look at equations of lines and their graphs. (In Section 4.1 we'll see applications that use the material developed here.) We begin by recalling the concept of *slope,* which you've seen in previous courses. As defined in the box that follows, the slope of a nonvertical line is a number that measures the slant or direction of the line.

DEFINITION **Slope**

The slope of a nonvertical line passing through the two points (x_1, y_1) and (x_2, y_2) is the number m defined by

$$m = \frac{y_2 - y_1}{x_2 - x_1}$$

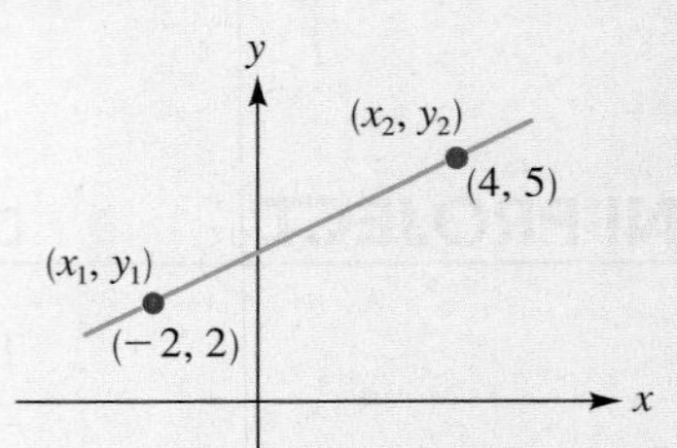

Calculating m using $(-2, 2)$ and $(4, 5)$:

$$m = \frac{5 - 2}{4 - (-2)} = \frac{3}{4 + 2} = \frac{3}{6} = \frac{1}{2}$$

The two quantities $x_2 - x_1$ and $y_2 - y_1$ that appear in the definition of slope can be interpreted geometrically. As indicated in Figure 1, $x_2 - x_1$ is the amount by which x changes as we move from (x_1, y_1) to (x_2, y_2) along the line. We denote this change in x by Δx (read: *delta x*). Thus $\Delta x = x_2 - x_1$. Similarly, Δy is defined to mean the change in y: $\Delta y = y_2 - y_1$. Using this notation, we can rewrite our definition of slope as $m = \Delta y/\Delta x$.

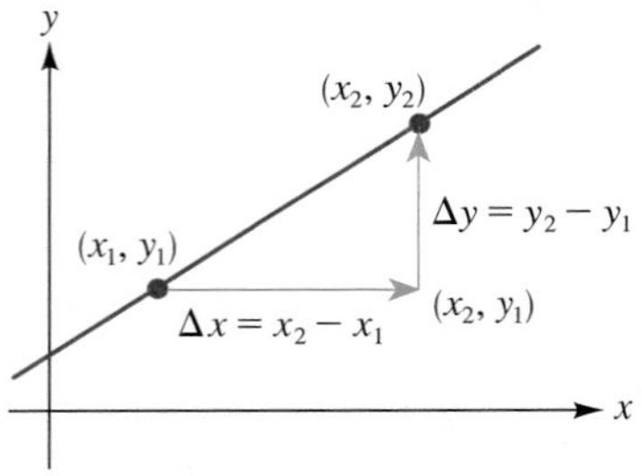

Figure 1
The delta notation

EXAMPLE 1 **Using the delta notation in computing slope**

Suppose a cardiologist asks a patient to walk on a treadmill at a slow steady pace and Table 1 lists some of the data collected. In Table 1, x represents time, in seconds, with $x = 0$ corresponding to the instant that the patient starts walking. The variable y stands for the total distance covered by the patient after x seconds. Figure 2 shows a graph obtained from these data. Compute Δx, Δy, and the slope m for each of the following intervals:
(a) from $x = 5$ seconds to $x = 10$ seconds;
(b) from $x = 10$ seconds to $x = 30$ seconds.

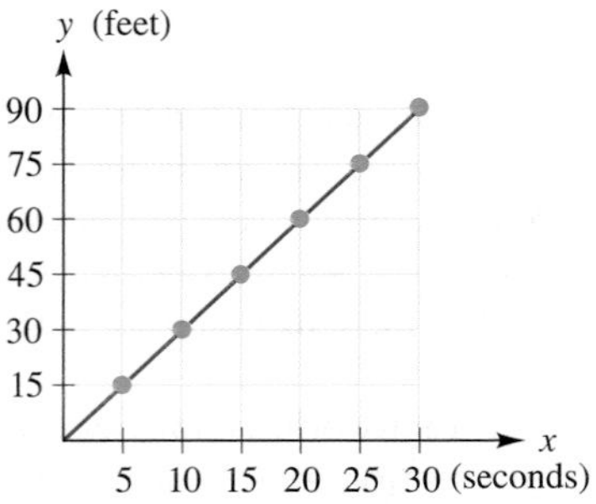

TABLE 1

x (time in seconds)	0	5	10	15	20	25	30
y (distance in feet)	0	15	30	45	60	75	90

Figure 2

SOLUTION

(a) From $x = 5$ seconds to $x = 10$ seconds:

$$\Delta x = x_2 - x_1 = 10 \text{ sec} - 5 \text{ sec} = 5 \text{ sec}$$
$$\Delta y = y_2 - y_1 = 30 \text{ ft} - 15 \text{ ft} = 15 \text{ ft}$$
$$\frac{\Delta y}{\Delta x} = \frac{15 \text{ ft}}{5 \text{ sec}} = 3 \text{ ft/sec}$$

(b) From $x = 10$ seconds to $x = 30$ seconds:

$$\Delta x = x_2 - x_1 = 30 \text{ sec} - 10 \text{ sec} = 20 \text{ sec}$$
$$\Delta y = y_2 - y_1 = 90 \text{ ft} - 30 \text{ ft} = 60 \text{ ft}$$
$$\frac{\Delta y}{\Delta x} = \frac{60 \text{ ft}}{20 \text{ sec}} = 3 \text{ ft/sec}$$

Note that both parts of this exercise lead to the same value for slope. Exercises 5 and 6 ask you to check other instances of this, and in the next paragraph we prove that for a given line, the same value is obtained for slope no matter which pair of points on the line is used. Another point to observe: Slope is a rate of change. In this example, it is the rate of change of distance with respect to time, or *velocity*. We'll talk more about slope as a rate of change in Section 4.1.

To see why slope of a line does not depend on which two particular points on the line are used in the calculation, consider Figure 3. The two right triangles are similar (because the corresponding angles are equal). This implies that the corresponding sides of the two triangles are proportional, and so we have

$$\frac{a}{d} = \frac{b}{c}$$

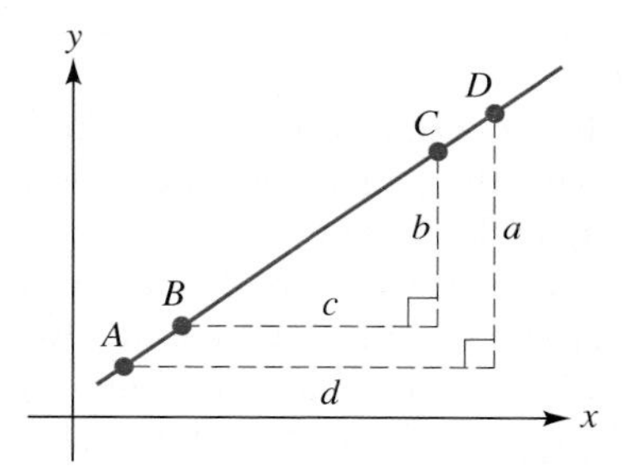

Figure 3

Now notice that the left-hand side of this equation represents the slope $\Delta y/\Delta x$ calculated using the points A and D, and the right-hand side represents the slope calculated using the points B and C. Thus the values we obtain for the slope are indeed equal.

EXAMPLE 2 Comparing slopes

Compute and compare the slopes of the three lines shown in Figure 4.

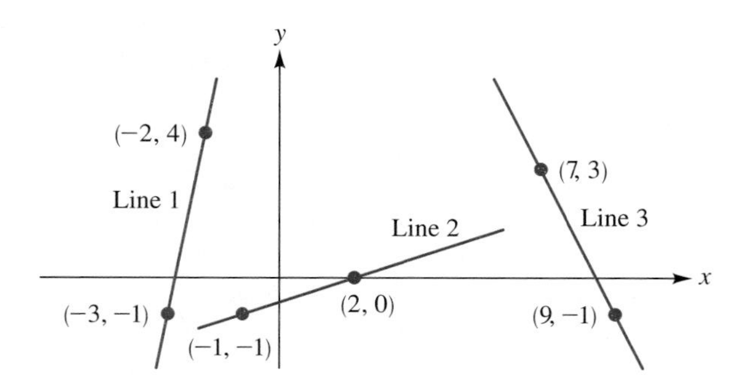

Figure 4

SOLUTION
First, we will calculate the slope of line 1, using the formula $m = \dfrac{y_2 - y_1}{x_2 - x_1}$. Which point should we use as (x_1, y_1) and which as (x_2, y_2)? In fact, it doesn't matter how we label our points. Using $(-3, -1)$ for (x_1, y_1) and $(-2, 4)$ for (x_2, y_2), we have

$$m = \frac{y_2 - y_1}{x_2 - x_1} = \frac{4 - (-1)}{-2 - (-3)} = 5$$

So the slope of line 1 is 5. If, instead, we had used $(-2, 4)$ for (x_1, y_1) and $(-3, -1)$ for (x_2, y_2), then we'd have $m = (-1 - 4)/(-3 + 2) = 5$, the same result. This is not accidental because, in general,

$$\frac{y_2 - y_1}{x_2 - x_1} = \frac{y_1 - y_2}{x_1 - x_2}$$

(Exercise 55 asks you to verify this identity.) Next, we calculate the slopes of lines 2 and 3.

Slope of line 2: $\dfrac{0 - (-1)}{2 - (-1)} = \dfrac{1}{3}$ Slope of line 3: $\dfrac{3 - (-1)}{7 - 9} = -2$

Lines 1 and 2 both have positive slopes and slant upward to the right. Note that line 1 is steeper than line 2 and correspondingly has the larger slope, 5. Line 3 has a negative slope, -2, and slants downward to the right.

The observations made in Example 2 are true in general. Lines with a positive slope slant upward to the right, the steeper line having the larger slope. Likewise, lines with a negative slope slant downward to the right. See Figure 5.

We have yet to mention slopes for horizontal or vertical lines. In Figure 6 line 1 is horizontal; it passes through the points (a, b) and (c, b). Note that the two y-coordinates must be the same for the line to be horizontal. Line 2 in Figure 6 is vertical; it passes through (d, e) and (d, f). Note that the two x-coordinates must be the same for the line to be vertical. For the slope of line 1 we have $m = \dfrac{b - b}{c - a} = \dfrac{0}{c - a} = 0$ (provided that $a \neq c$). Thus the slope of line 1, a horizontal line, is zero. For the vertical line in Figure 6 the calculation of slope begins with

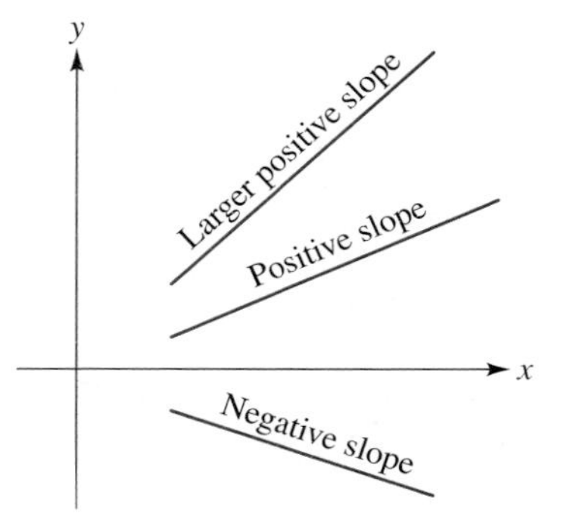

Figure 5

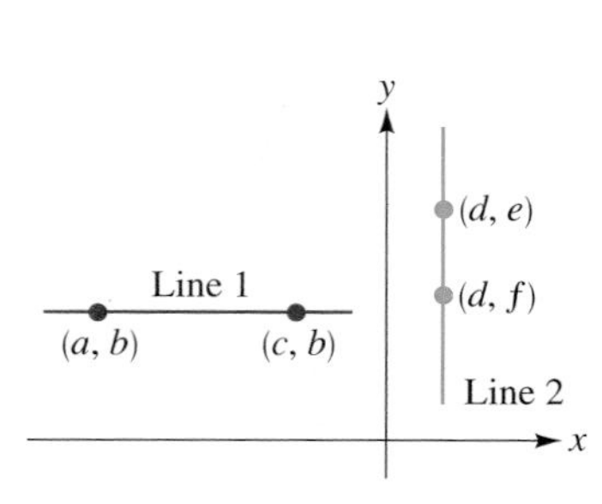

Figure 6

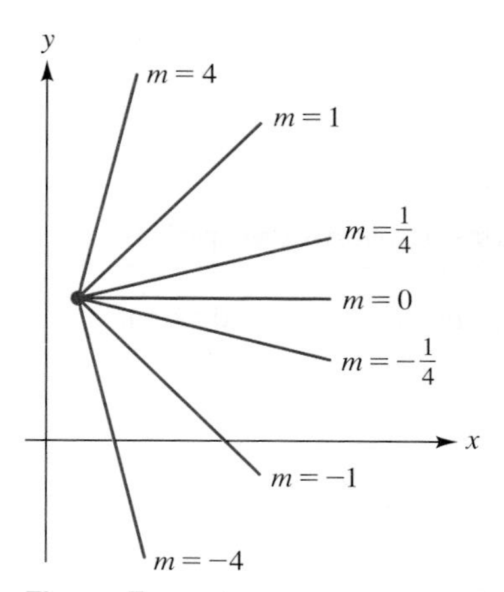

Figure 7

writing $\frac{e - f}{d - d}$, but the denominator is zero, and since division by zero is undefined, we conclude that slope is undefined for vertical lines. We summarize these results in the box that follows. Figure 7 shows for comparison some values of m for various lines.

Horizontal and Vertical Lines

1. The slope of a horizontal line is zero.
2. Slope is not defined for vertical lines.

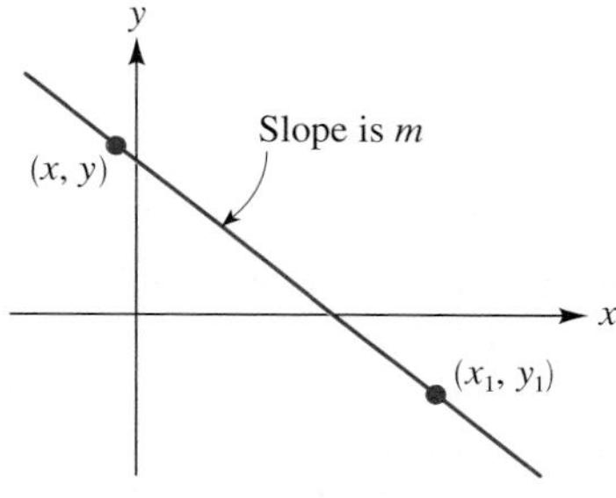

Figure 8

We can use the concept of slope to find the equation of a line. Suppose we have a line with slope m and passing through the point (x_1, y_1), as shown in Figure 8. Let (x, y) be any other point on the line, as in Figure 8. Then the slope of the line is given by $m = (y - y_1)/(x - x_1)$, and therefore, $y - y_1 = m(x - x_1)$. Note that the given point (x_1, y_1) also satisfies this last equation, because in that case we have $y_1 - y_1 = m(x_1 - x_1)$, or $0 = 0$. The equation $y - y_1 = m(x - x_1)$ is called the **point–slope formula.** We have shown that any point on the line satisfies this equation. (Conversely, it can be shown that if a point satisfies this equation, then the point does lie on the given line.)

The Point–Slope Formula

An equation of the line with slope m passing through the point (x_1, y_1) is

$$y - y_1 = m(x - x_1)$$

EXAMPLE 3 Using the point–slope formula

Write an equation of the line passing through $(-3, 1)$ with a slope of -2. Sketch a graph of the line.

SOLUTION

Since the slope and a point are given, we use the point–slope formula:

$$\begin{aligned} y - y_1 &= m(x - x_1) \\ y - 1 &= -2[x - (-3)] \\ y - 1 &= -2x - 6 \\ y &= -2x - 5 \end{aligned}$$

This is the desired equation. Since two points determine a line, we need to find one more point on the line. For example, for $x = 0$, $y = -5$. So the y-intercept is -5. A table and graph are displayed in Figure 9.

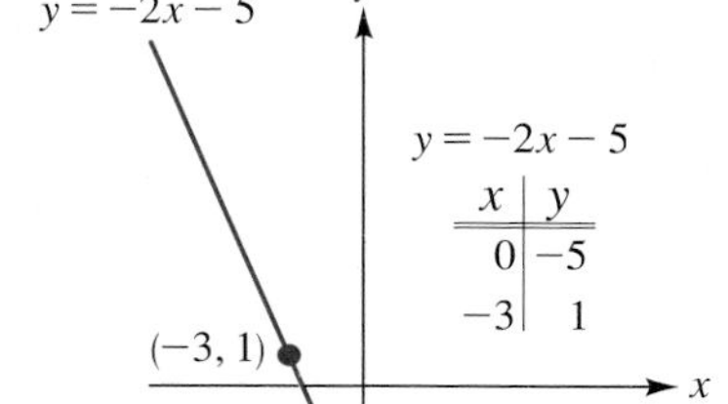

Figure 9
$y = -2x - 5$

There is another way to go about graphing the line $y = -2x - 5$ in Example 3. Since slope is (change in y)/(change in x), a slope of -2 (or $-2/1$) can be interpreted as telling us that if we start at $(-3, 1)$ and let x *increase* by one unit, then y must *decrease* by two units to bring us back to the line. Following this path in

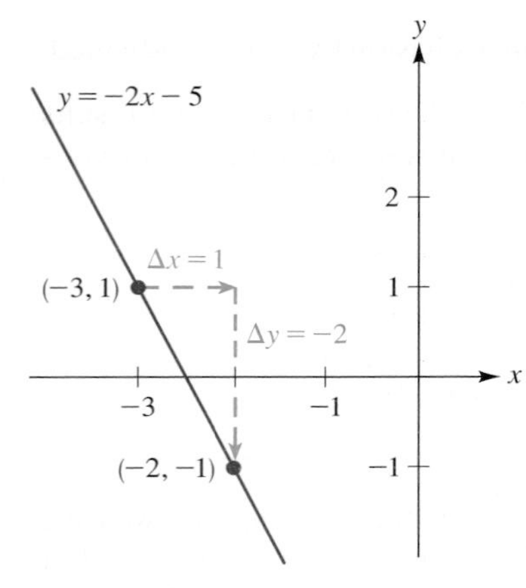

Figure 10

Figure 10 takes us from $(-3, 1)$ to $(-2, -1)$. We now draw the line through these two points, as shown in Figure 10.

EXAMPLE 4 Finding the equation of a line through two points

Find an equation of the line passing through the points $(-2, -3)$ and $(2, 5)$.

SOLUTION

The slope of the line is

$$m = \frac{y_2 - y_1}{x_2 - x_1} = \frac{5 - (-3)}{2 - (-2)} = \frac{8}{4} = 2$$

Knowing the slope, we can apply the point–slope formula, making use of either $(-2, -3)$ or $(2, 5)$. Using the point $(2, 5)$ as (x_1, y_1), we have

$$\begin{aligned} y - y_1 &= m(x - x_1) \\ y - 5 &= 2(x - 2) \\ y - 5 &= 2x - 4 \\ y &= 2x + 1 \end{aligned}$$

Thus the required equation is $y = 2x + 1$. You should check for yourself that the same answer is obtained using the point $(-2, -3)$ instead of $(2, 5)$ in the last set of calculations.

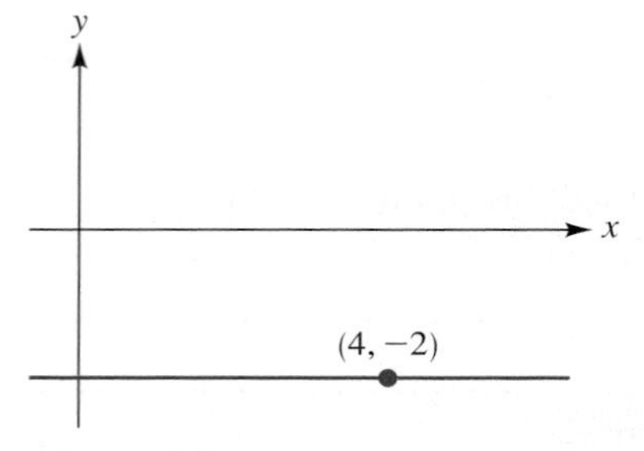

Figure 11

EXAMPLE 5 Finding the equation of a horizontal line

Find an equation of the horizontal line passing through the point $(4, -2)$. See Figure 11.

SOLUTION

Since the slope of a horizontal line is zero, we have

$$\begin{aligned} y - y_1 &= m(x - x_1) \\ y - (-2) &= 0(x - 4) \\ y &= -2 \end{aligned}$$

Thus the equation of the horizontal line passing through $(4, -2)$ is $y = -2$.

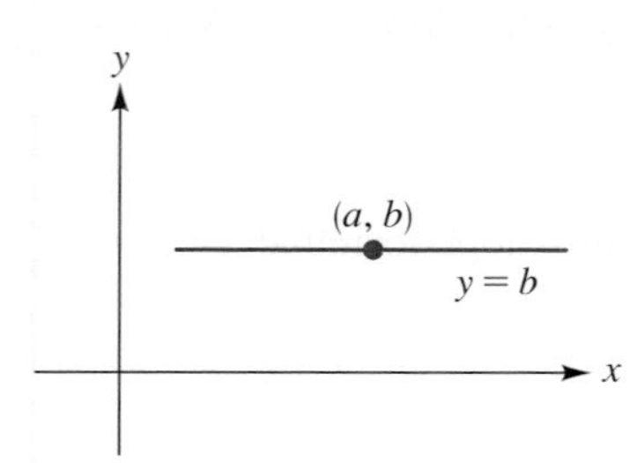

Figure 12

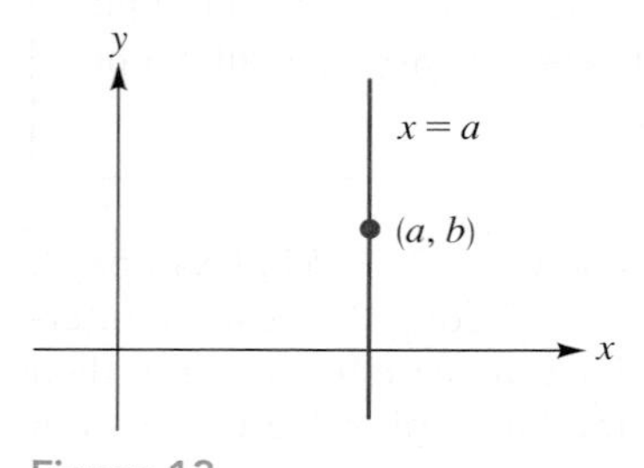

Figure 13

By using the point–slope formula exactly as we did in Example 5, we can show more generally that the equation of the horizontal line in Figure 12 is $y = b$. What about the equation of the vertical line in Figure 13 passing through the point (a, b)? Because slope is not defined for vertical lines, the point–slope formula does not apply. However, note that as we move along the vertical line, the x-coordinate is always a, since only the y-coordinate varies. The equation $x = a$ expresses exactly these two facts; it says that x must always be a, and it places no restrictions on y.

In the box that follows, we summarize our results concerning horizontal and vertical lines.

Equations of Horizontal and Vertical Lines

1. The equation of a horizontal line through the point (a, b) is $y = b$. (See Fig. 12.)
2. The equation of a vertical line through the point (a, b) is $x = a$. (See Fig. 13.)

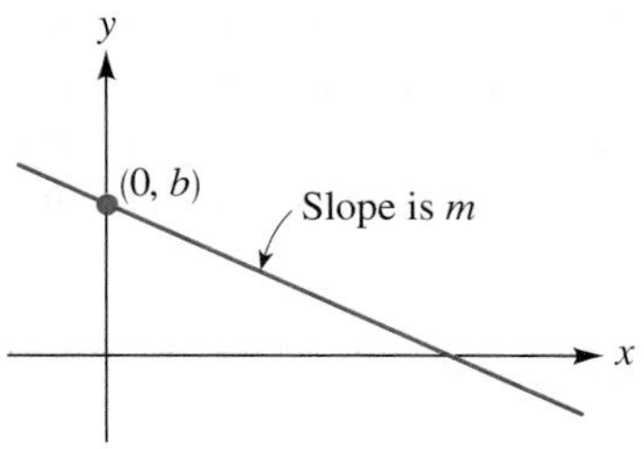

Another basic form for the equation of a line is the *slope–intercept form.* We are given the slope m and the y-intercept b, as shown in Figure 14, and we want to find an equation of the line. To say that the line has a y-intercept of b is the same as saying that the line passes through $(0, b)$. The point–slope formula is applicable now, using the slope m and the point $(0, b)$. We have

$$\begin{aligned} y - y_1 &= m(x - x_1) \\ y - b &= m(x - 0) \\ y &= mx + b \end{aligned}$$

Figure 14

This last equation is called the **slope–intercept formula.**

The Slope–Intercept Formula

An equation of the line with slope m and y-intercept b is

$$y = mx + b$$

EXAMPLE 6 Using the slope–intercept formula

Write an equation of the line with slope 4 and y-intercept 1. Graph the line.

SOLUTION
Substituting $m = 4$ and $b = 1$ in the equation $y = mx + b$ yields

$$y = 4x + 1$$

This is the required equation. We could draw the graph by first setting up a simple table, but for purposes of emphasis and review we proceed as we did just after Example 3. Starting from the point $(0, 1)$, we interpret the slope of 4 as saying that if x increases by 1, then y increases by 4. This takes us from $(0, 1)$ to the point $(1, 5)$, and the line can now be sketched as in Figure 15.

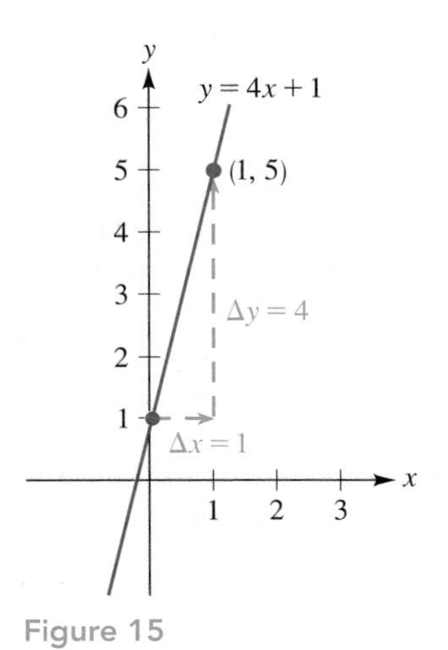

Figure 15

EXAMPLE 7 Determining the slope and y-intercept

Find the slope and y-intercept of the line $3x - 5y = 15$.

SOLUTION
First we solve for y to write the equation in the form $y = mx + b$:

$$\begin{aligned} 3x - 5y &= 15 \\ -5y &= -3x + 15 \end{aligned}$$

and therefore

$$y = \frac{3}{5}x - 3$$

The slope m and the y-intercept b can now be read directly from the equation: $m = 3/5$ and $b = -3$.

As a consequence of our work up to this point, we can say that the graph of any **linear equation** $Ax + By + C = 0$, where A and B are not both zero, is a line, since an equation of this type can always be rewritten in one of the following three forms: $y = mx + b$, $x = a$, or $y = b$. In the box that follows, we summarize our basic results on equations of lines.

PROPERTY SUMMARY Equations of Lines

Equation	Comment
$y - y_1 = m(x - x_1)$ (the point–slope formula)	This is an equation of the line with slope m, passing through the point (x_1, y_1).
$y = mx + b$ (the slope–intercept formula)	This is an equation of the line with slope m and y-intercept b.
$Ax + By + C = 0$	The graph of any equation of this form (where A and B are not both zero) is a line. Special cases: If $A = 0$, then the equation can be written $y = -C/B$, which is the equation of the horizontal line with y-intercept $-C/B$. If $B = 0$, then the equation can be written $x = -C/A$, which represents the vertical line with x-intercept $-C/A$.

GRAPHICAL PERSPECTIVE

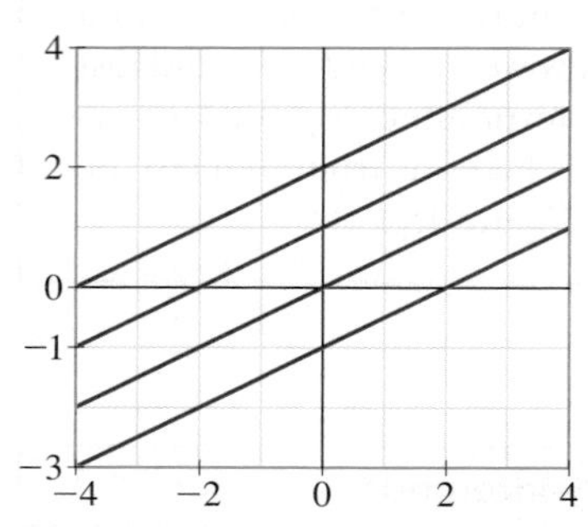

Figure 16
Parallel lines have the same slope. Each line here has a slope of 0.5. The equations of the lines, from top to bottom, are $y = 0.5x + 2$, $y = 0.5x + 1$, $y = 0.5x$, and $y = 0.5x - 1$.

We conclude this section by discussing two useful relationships regarding the slopes of parallel lines and the slopes of perpendicular lines. First, nonvertical parallel lines have the same slope. (See Figure 16.) This should seem reasonable if you recall that slope is a number indicating the direction or slant of a line. The relationship concerning slopes of perpendicular lines is not so obvious. The slopes of two nonvertical perpendicular lines are negative reciprocals of each other. That is, if m_1 and m_2 denote the slopes of the two perpendicular lines, then $m_1 = -1/m_2$ or, equivalently, $m_1m_2 = -1$. For example, if a line has a slope of $2/3$, then any line perpendicular to it must have a slope of $-3/2$. For reference we summarize these facts in the box that follows. (Proofs of these facts are outlined in detail in Exercises 53 and 54.)

Parallel and Perpendicular Lines

Let m_1 and m_2 denote the slopes of two distinct nonvertical lines. Then:

1. The lines are parallel if and only if $m_1 = m_2$;
2. The lines are perpendicular if and only if $m_1 = -1/m_2$.

EXAMPLE 8 Determining whether two lines are parallel

Determine whether the two lines $3x - 6y - 8 = 0$ and $2y = x + 1$ are parallel.

SOLUTION

By solving each equation for y, we can see what the slopes are:

$$\begin{aligned} 3x - 6y - 8 &= 0 \\ -6y &= -3x + 8 \\ y &= \frac{-3}{-6}x + \frac{8}{-6} \\ y &= \frac{1}{2}x - \frac{4}{3} \end{aligned} \qquad \begin{aligned} 2y &= x + 1 \\ y &= \frac{1}{2}x + \frac{1}{2} \end{aligned}$$

From this we see that both lines have the same slope ($m = 1/2$) and different y-intercepts. It follows therefore that the lines are parallel.

NOTE In the solution to part (b) of Example 9 we discuss how to get your graphing utility to show true proportions so that when you display the graphs of two perpendicular lines they actually look perpendicular on the display screen.

EXAMPLE 9 Finding parallel lines and perpendicular lines

In each case, find an equation for the line satisfying the given conditions, and then use a graphing utility to check that the result appears reasonable.

(a) The line passes through the point (2, 0) and is parallel to the line $5x + 6y = 30$.

(b) The line passes through the point (2, 0) and is perpendicular to $5x + 6y = 30$.

SOLUTION

(a) The line we are looking for is parallel to $5x + 6y = 30$ and therefore has the same slope. To find that slope, we solve the equation $5x + 6y = 30$ for y as follows:

$$\begin{aligned} 6y &= -5x + 30 \\ y &= -\frac{5}{6}x + 5 \end{aligned}$$

The slope is the x-coefficient in this last equation, so $m = -5/6$. Knowing now that the slope of the required line is $-5/6$ and that it passes through the given point (2, 0), we can use the point–slope formula $y - y_1 = m(x - x_1)$ to obtain $y - 0 = -\frac{5}{6}(x - 2)$, and consequently,

$$y = -\frac{5}{6}x + \frac{5}{3}$$

This is the required equation of the line passing through the point (2, 0) and parallel to $5x + 6y = 30$. See the graphical perspective in Figure 17.

GRAPHICAL PERSPECTIVE

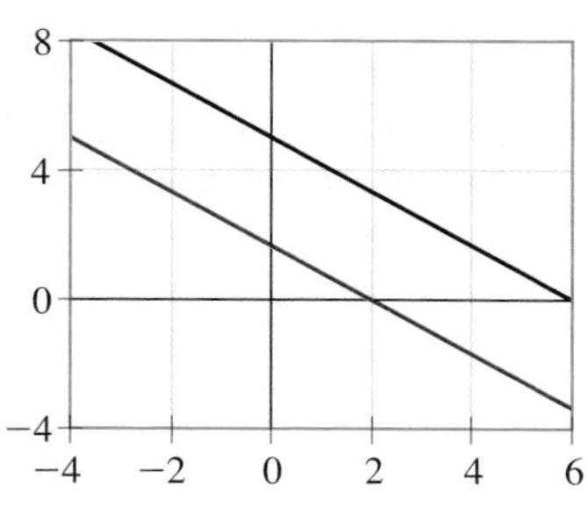

Figure 17
The line $y = -\frac{5}{6}x + \frac{5}{3}$ (in red) passes through the point (2, 0) and is parallel to $5x + 6y = 30$ (in black).

(b) In part (a) we found that the slope of the line $5x + 6y = 30$ is $-5/6$. The slope of a perpendicular line is the opposite reciprocal of this, which is $6/5$. Knowing now that the slope of the required line is $6/5$ and that it passes through the point $(2, 0)$, we use the point–slope formula to obtain

$$y - 0 = \frac{6}{5}(x - 2)$$
$$y = \frac{6}{5}x - \frac{12}{5}$$

This is an equation of the line that passes through the point $(2, 0)$ and that is perpendicular to $5x + 6y = 30$. For a quick visual check to see whether our result is reasonable, we've used a graphing utility in Figure 18(a) to plot the given line $5x + 6y = 30$ (entering it as $y = -\frac{5}{6}x + 5$) and the perpendicular through $(2, 0)$ that we found to be $y = \frac{6}{5}x - \frac{12}{5}$. The latter line, shown in red in Figure 18(a), does appear to pass through $(2, 0)$. It does not, however, appear to be perpendicular to $5x + 6y = 30$. Assuming that our graphing utility work is correct, does this mean we've made an error in our algebra? The answer is "No, not necessarily." The problem with the graph in Figure 18(a) is that it does not show true proportions. The actual length (or number of pixels) for one unit on the x-axis is not the same as that on the y-axis. In Figure 18(b) we show a view using true proportions, and the two lines do indeed appear to be perpendicular. (If you need help obtaining a view showing true proportions, see the user's manual that came with your graphing utility. Example: On the Texas Instruments graphing calculators, the zoom choice "ZSquare" yields a viewing screen showing true proportions, albeit often with awkward numbers for Xmin and Xmax.)

GRAPHICAL PERSPECTIVE

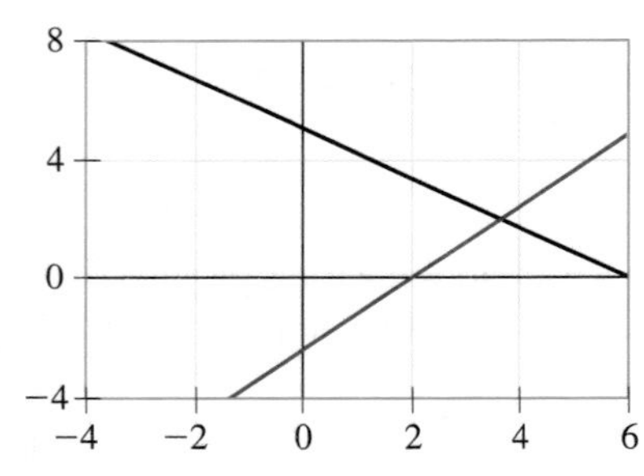

(a) This view does not show true proportions.

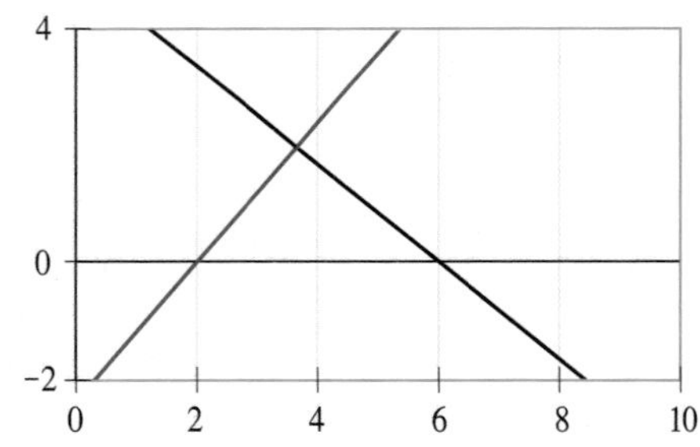

(b) A *true proportions* command was used to generate this view. The two lines do appear to be perpendicular.

Figure 18
Two views of the lines $y = \frac{6}{5}x - \frac{12}{5}$ (in red) and $y = -\frac{5}{6}x + 5$ (in black)

EXERCISE SET 1.6

A

In Exercises 1–3, compute the slope of the line passing through the two given points. In Exercise 3, include a sketch with your answers.

1. (a) $(-3, 2), (1, -6)$
(b) $(2, -5), (4, 1)$
(c) $(-2, 7), (1, 0)$
(d) $(4, 5), (5, 8)$

2. (a) $(-3, 0), (4, 9)$
(b) $(-1, 2), (3, 0)$
(c) $(\frac{1}{2}, -\frac{3}{5}), (\frac{3}{2}, \frac{3}{4})$
(d) $(\frac{17}{3}, -\frac{1}{2}), (-\frac{1}{2}, \frac{17}{3})$

3. **(a)** $(1, 1), (-1, -1)$
   (b) $(0, 5), (-8, 5)$
4. Compute the slope of the line in the following figure using each pair of points indicated.
   (a) A and B **(b)** B and C **(c)** A and C
   The principle involved here is that no matter which pair of points you choose, the slope is the same.

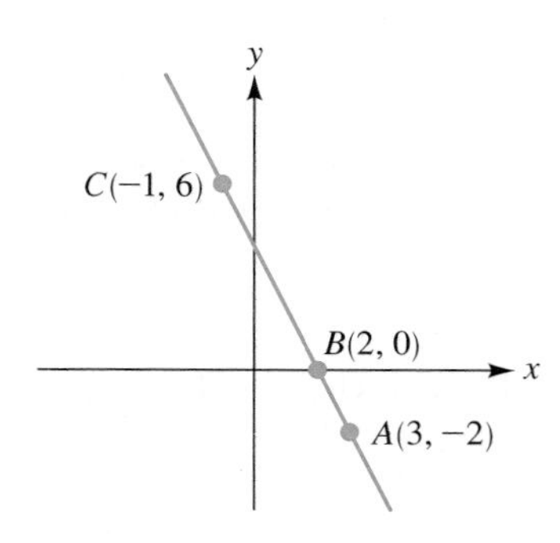

For Exercises 5 and 6, refer to Figure 2 on page 49 and find Δx, Δy, and $m = \Delta y/\Delta x$ for each of the indicated intervals. (Be sure to include units with your answers.)

5. **(a)** $x = 10$ sec to $x = 15$ sec
   (b) $x = 10$ sec to $x = 25$ sec
   (c) $x = 5$ sec to $x = 30$ sec
6. **(a)** $x = 0$ sec to $x = 5$ sec
   (b) $x = 0$ sec to $x = 30$ sec

For Exercises 7 and 8, refer to Figure A, which follows. At the beginning of this section it was mentioned that for sufficiently short periods of time, the graphs of many real-world quantities can be approximated quite accurately by a straight line. For instance, the following graph provides a close approximation to U.S. population data for the years 1970–1990.

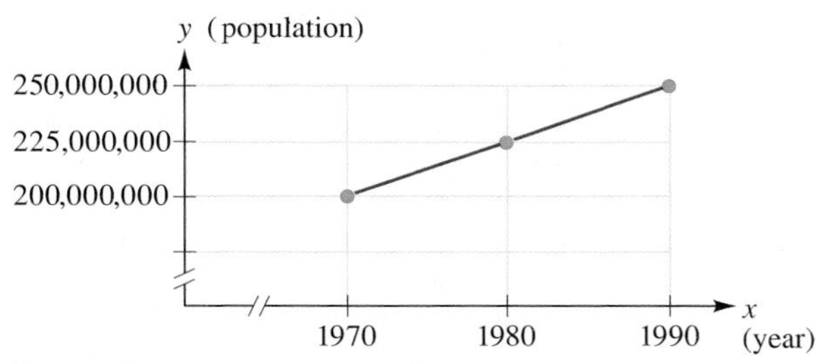

Figure A
U.S. population (as of July 1) for years 1970–1990
Adapted from U.S. Bureau of the Census data

7. Specify Δx and Δy for each of the following intervals.
   (a) $x = 1970$ to $x = 1980$ **(c)** $x = 1980$ to $x = 1990$
   (b) $x = 1970$ to $x = 1990$
8. What is the slope of the population graph in Figure A? (Be sure to include units with your answer.)
9. The following straight-line graph summarizes the data for world cigarette production over the years 1990–1993. Specify Δt and ΔN for the period 1990–1993. (Include units with each answer.)

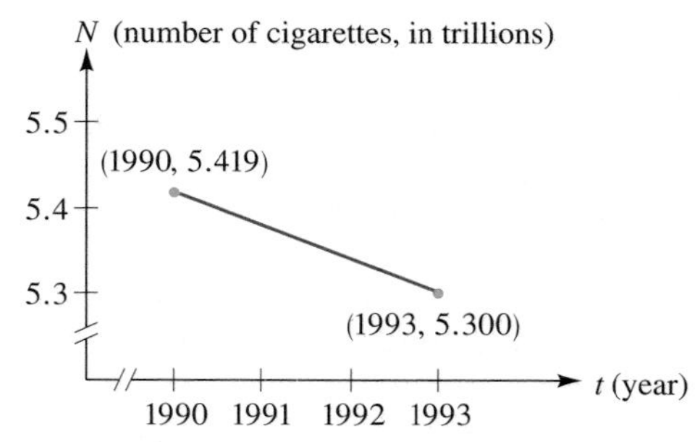

World cigarette production, 1990–1993.

Adapted from data compiled in *Vital Signs, 1998,* Lester R. Brown et al. (New York: W. W. Norton & Company, 1998)

10. Compute the slope $\dfrac{\Delta N}{\Delta t}$ for the straight-line graph shown in Exercise 9. (Be sure to include units as part of your answer.)
11. The slopes of four lines are indicated in the figure. List the slopes m_1, m_2, m_3, and m_4 in order of increasing value.

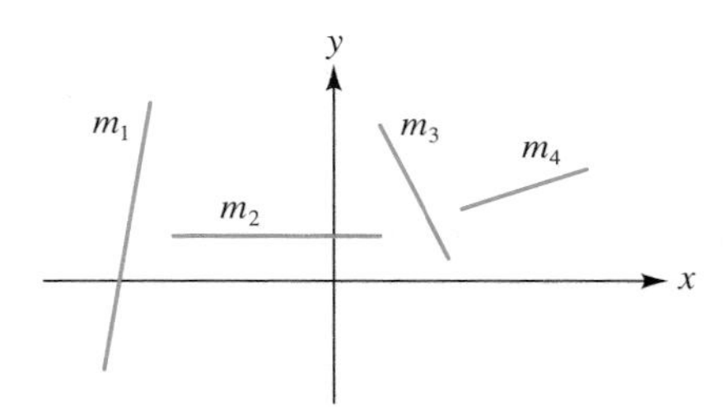

12. Refer to the accompanying figure.
    (a) List the slopes m_1, m_2, and m_3 in order of increasing size.
    (b) List the numbers b_1, b_2, and b_3 in order of increasing size.

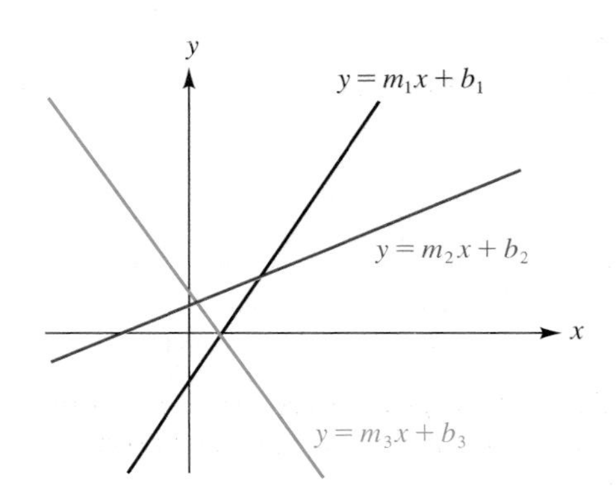

In Exercises 13 and 14(a), three points A, B, and C are specified. Determine whether A, B, and C are **collinear** *(lie on the same line) by checking to see whether the slope of $\overline{AB}$ equals the slope of $\overline{BC}$.*

13. $A(-8, -2)$; $B(2, \frac{1}{2})$; $C(11, -1)$

14. (a) $A(0, -5)$; $B(3, 4)$; $C(-1, -8)$

(b) If the area of the "triangle" formed by three points is zero, then the points must in fact be collinear. Use this observation, along with the formula in Exercise 14(b) of Exercise Set 1.4, to rework part (a).

In Exercises 15 and 16, find an equation for the line having the given slope and passing through the given point. Write your answers in the form $y = mx + b$.

15. (a) $m = -5$; through $(-2, 1)$

(b) $m = \frac{1}{3}$; through $(-6, -\frac{2}{3})$

16. (a) $m = 22$; through $(0, 0)$

(b) $m = -222$; through $(0, 0)$

In Exercises 17 and 18, find an equation for the line passing through the two given points. Write your answer in the form $y = mx + b$.

17. (a) $(4, 8)$ and $(-3, -6)$

(b) $(-2, 0)$ and $(3, -10)$

(c) $(-3, -2)$ and $(4, -1)$

18. (a) $(7, 9)$ and $(-11, 9)$

(b) $(5/4, 2)$ and $(3/4, 3)$

(c) $(12, 13)$ and $(13, 12)$

In Exercises 19 and 20, write an equation of:

(a) *a vertical line passing through the given point;*

(b) *a horizontal line passing through the given point.*

19. $(-3, 4)$

20. $(5, 8)$

21. Is the graph of the line $x = 0$ the x-axis or the y-axis?

22. Is the graph of the line $y = 0$ the x-axis or the y-axis?

In Exercises 23 and 24, find an equation of the line with the given slope and y-intercept.

23. (a) slope -4; y-intercept 7

(b) slope 2; y-intercept 3/2

24. (a) slope 0; y-intercept 14

(b) slope 14; y-intercept 0

In Exercises 25–28, find an equation for the line that is described, and sketch the graph. For Exercises 25 and 26, write the final answer in the form $y = mx + b$; for Exercises 27 and 28, write the answer in the form $Ax + By + C = 0$.

25. (a) Passes through $(-3, -1)$ and has slope 4

(b) Passes through $(5/2, 0)$ and has slope 1/2

(c) Has x-intercept 6 and y-intercept 5

(d) Has x-intercept -2 and slope 3/4

(e) Passes through $(1, 2)$ and $(2, 6)$

26. (a) Passes through $(-7, -2)$ and $(0, 0)$

(b) Passes through $(6, -3)$ and has y-intercept 8

(c) Passes through $(0, -1)$ and has the same slope as the line $3x + 4y = 12$

(d) Passes through $(6, 2)$ and has the same x-intercept as the line $-2x + y = 1$

(e) Has x-intercept -6 and y-intercept $\sqrt{2}$

27. Passes through $(-3, 4)$ and is parallel to the x-axis

28. Passes through $(-3, 4)$ and is parallel to the y-axis

In Exercises 29 and 30, find the x- and y-intercepts of the line, and find the area and the perimeter of the triangle formed by the line and the axes.

29. (a) $3x + 5y = 15$

(b) $3x - 5y = 15$

30. (a) $5x + 4y = 40$

(b) $2x + 4y = \sqrt{2}$

31. Determine whether each pair of lines is parallel, perpendicular, or neither.

(a) $3x - 4y = 12$; $4x - 3y = 12$

(b) $y = 5x - 16$; $y = 5x + 2$

(c) $5x - 6y = 25$; $6x + 5y = 0$

(d) $y = -\frac{2}{3}x - 1$; $y = \frac{3}{2}x - 1$

32. Are the lines $y = x + 1$ and $y = 1 - x$ parallel, perpendicular, or neither?

In Exercises 33–36, find an equation for the line that is described. Write the answer in the two forms $y = mx + b$ and $Ax + By + C = 0$.

33. Is parallel to $2x - 5y = 10$ and passes through $(-1, 2)$

34. Is parallel to $4x + 5y = 20$ and passes through $(0, 0)$

35. Is perpendicular to $4y - 3x = 1$ and passes through $(4, 0)$

36. Is perpendicular to $x - y + 2 = 0$ and passes through $(3, 1)$

Ⓖ **37. (a)** Use a graphing utility to graph the following three parallel lines in the standard viewing rectangle: $y + 4 = -0.5(x - 2)$; $y - 3 = -0.5(x + 2)$; $y = -0.5x$.

(b) Experiment with different settings for `Xmin`, `Xmax`, `Ymin`, and `Ymax`. In each case, do the three lines still appear to be parallel?

Ⓖ **38. (a)** The lines $y = 4x$ and $y = -0.25x$ are perpendicular because their slopes are negative reciprocals. Use a graphing utility to graph these two lines in the standard viewing rectangle. Unless your graphing utility automatically shows true proportions, the lines will not appear to be perpendicular.

(b) If necessary, modify the viewing rectangle in part (a) so that true proportions are used and the two lines indeed appear perpendicular.

39. (a) Find an equation of the line that passes through the origin and is perpendicular to the line $3x + 4y = 12$.

Ⓖ **(b)** Use a graphing utility to check that your answer in part (a) is reasonable. (That is, graph the two lines using true proportions; the line you found should appear to pass through the origin and be perpendicular to $3x + 4y = 12$.)

G **40.** In each of parts (a) through (d), first solve the equation for y so that you can enter it in your graphing utility. Then use the graphing utility to graph the equation in an appropriate viewing rectangle. In each case, the graph is a line. Given that the x- and y-intercepts are (in every case here) integers, read their values off the screen and write them down for easy reference when you get to part (e).

(a) $\frac{x}{2} + \frac{y}{3} = 1$ **(c)** $\frac{x}{6} + \frac{y}{5} = 1$

(b) $\frac{x}{-2} + \frac{y}{-3} = 1$ **(d)** $\frac{x}{-6} + \frac{y}{-5} = 1$

(e) On the basis of your results in parts (a) through (d), describe, in general, the graph of the equation

$\frac{x}{a} + \frac{y}{b} = 1$, where a and b are nonzero constants.

B

41. After analyzing sales figures for a particular model of CD player, the accountant for College Sound Company has produced the following graph relating the selling price P (in dollars) and the number y of units that can be sold each month at that price. For instance, as the graph shows, setting the selling price at \$225 yields sales of 260 units per month.

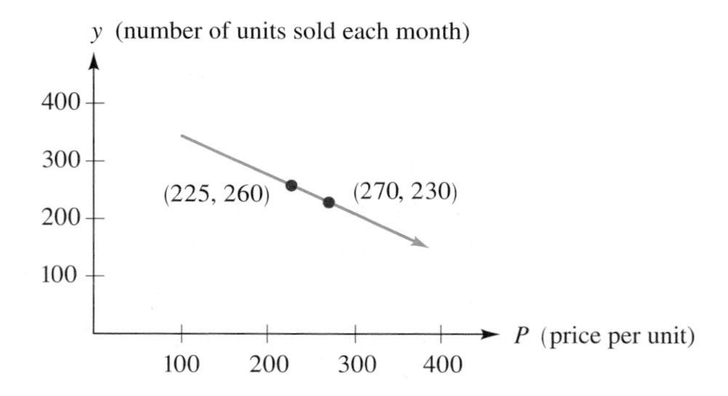

(a) Find an equation of the line. (Remember to use the letter P instead of the usual x.)

(b) Use the equation that you found in part (a) to determine how many units can be sold in a month when the price is \$303 per unit.

(c) What should the price be to sell 288 units per month?

42. Imagine that you own a grove of orange trees, and suppose that from past experience you know that when 100 trees are planted, each tree will yield approximately 240 oranges per year. Furthermore, you've noticed that when additional trees are planted in the grove, the yield per tree decreases. Specifically, you have noted that the yield per tree decreases by about 20 oranges for each additional tree planted.

(a) Let y denote the yield per tree when x trees are planted. Find a linear equation relating x and y. *Hint:* You are given that the point (100, 240) is on the line. What is given about $\Delta y/\Delta x$?

(b) Use the equation in part (a) to determine how many trees should be planted to obtain a yield of 400 oranges per tree.

(c) If the grove contains 95 trees, what yield can you expect from each tree?

43. Show that the slope of the line passing through the two points (a, a^2) and (x, x^2) is $x + a$. *Hint:* You'll need to use difference of squares factoring from intermediate algebra. If you need a review, see Appendix B.4.

44. Show that the slope of the line passing through the two points (3, 9) and $(3 + h, (3 + h)^2)$ is $6 + h$.

45. Show that the slope of the line passing through the two points (a, a^3) and (x, x^3) is $x^2 + ax + a^2$. *Hint:* You'll need to use difference of cubes factoring from intermediate algebra. If you need a review, see Appendix B.4.

46. Write down, and then simplify as much as possible, an expression for the slope of the line passing through the two points $(a, 1/a)$ and $(x, 1/x)$.

47. A line with a slope of -5 passes through the point (3, 6). Find the area of the triangle in the first quadrant formed by this line and the coordinate axes.

48. The y-intercept of the line in the figure is 6. Find the slope of the line if the area of the shaded triangle is 72 square units.

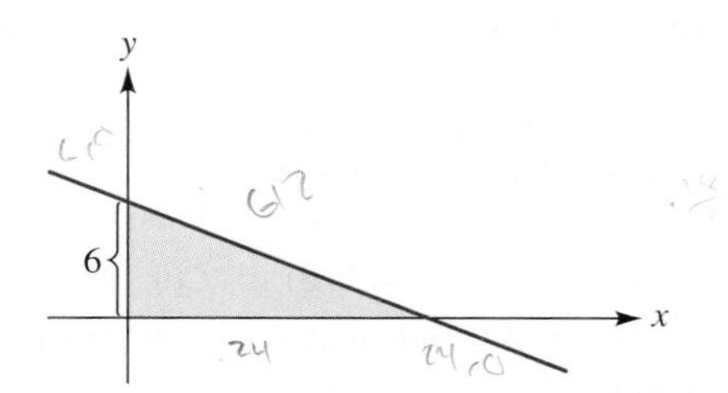

49. **(a)** Sketch the line $y = \frac{1}{2}x - 5$ and the point $P(1, 3)$. Follow parts (b)–(d) to calculate the perpendicular distance from point $P(1, 3)$ to the line.

(b) Find an equation of the line that passes through $P(1, 3)$ and is perpendicular to the line $y = \frac{1}{2}x - 5$.

(c) Find the coordinates of the point where these two lines intersect. *Hint:* From intermediate algebra, to find where two lines $y = mx + b$ and $y = Mx + B$ intersect, set the expressions $mx + b$ and $Mx + B$ equal to each other, and solve for x.

(d) Use the distance formula to find the perpendicular distance from $P(1, 3)$ to the line $y = \frac{1}{2}x - 5$.

50. The following figure shows a circle centered at the origin and a line that is tangent to the circle at the point $(3, -4)$.

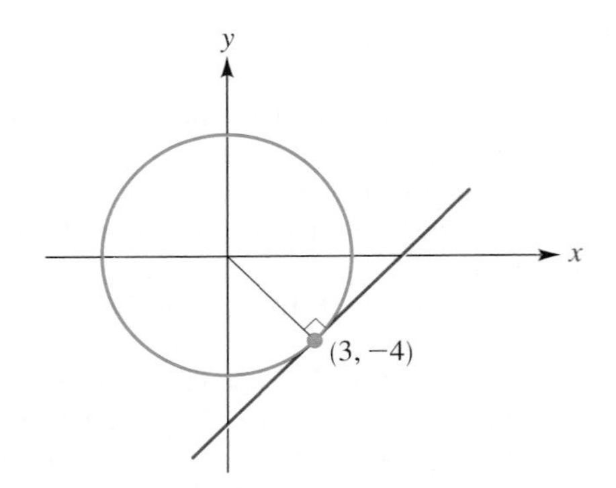

(a) Find an equation of the tangent line. *Hint*: Make use of the theorem from elementary geometry stating that the tangent line is perpendicular to the radius drawn to the point of contact.
(b) Find the intercepts of the tangent line.
(c) Find the length of the portion of the tangent line in quadrant IV.

For Exercises 51 and 52, you'll need to recall the following definitions and results from elementary geometry. In a triangle, a line segment drawn from a vertex to the midpoint of the opposite side is called a ***median.*** *The three medians of a triangle are* ***concurrent;*** *that is, they intersect in a single point. This point of intersection is called the* ***centroid*** *of the triangle. A line segment drawn from a vertex perpendicular to the opposite side is an* ***altitude.*** *The three altitudes of a triangle are concurrent; the point where the altitudes intersect is the* ***orthocenter*** *of the triangle.*

Ⓖ **51.** This exercise provides an example of the fact that the medians of a triangle are concurrent.
(a) The vertices of $\triangle ABC$ are as follows:

$$A(-4, 0) \qquad B(2, 0) \qquad C(0, 6)$$

Use a graphing utility to draw $\triangle ABC$. (Since $\overline{AB}$ coincides with the x-axis, you won't need to draw a line segment for this side.) *Note*: If the graphing utility you use does not have a provision for drawing line segments, you will need to determine an equation for the line in each case and then graph the line.
(b) Find the coordinates of the midpoint of each side of the triangle, then include the three medians in your picture from part (a). Note that the three medians do appear to intersect in a single point. Use the graphing utility to estimate the coordinates of the centroid.
(c) Using paper and pencil, find the equation of the medians from A to $\overline{BC}$ and from B to $\overline{AC}$. Then (using simultaneous equations from intermediate algebra), determine the exact coordinates of the centroid. How do these numbers compare with your estimates in part (b)?

Ⓖ **52.** This exercise illustrates the fact that the altitudes of a triangle are concurrent. Again, we'll be using $\triangle ABC$ with vertices $A(-4, 0)$, $B(2, 0)$, and $C(0, 6)$. Note that one of the altitudes of this triangle is just the portion of the y-axis extending from $y = 0$ to $y = 6$; thus, you won't need to graph this altitude; it will already be in the picture.
(a) Using paper and pencil, find the equations for the three altitudes. (Actually, you are finding equations for the lines that coincide with the altitude segments.)
(b) Use a graphing utility to draw $\triangle ABC$ along with the three altitude lines that you determined in part (a). Note that the altitudes appear to intersect in a single point. Use the graphing utility to estimate the coordinates of this point.
(c) Using simultaneous equations (from intermediate algebra), find the exact coordinates of the orthocenter. Are your estimates in part (b) close to these values?

53. This exercise outlines a proof of the fact that two nonvertical lines are parallel if and only if their slopes are equal. The proof relies on the following observation for the given figure: The lines $y = m_1x + b_1$ and $y = m_2x + b_2$ will be parallel if and only if the two vertical distances AB and CD are equal. (In the figure, the points C and D both have x-coordinate 1.)

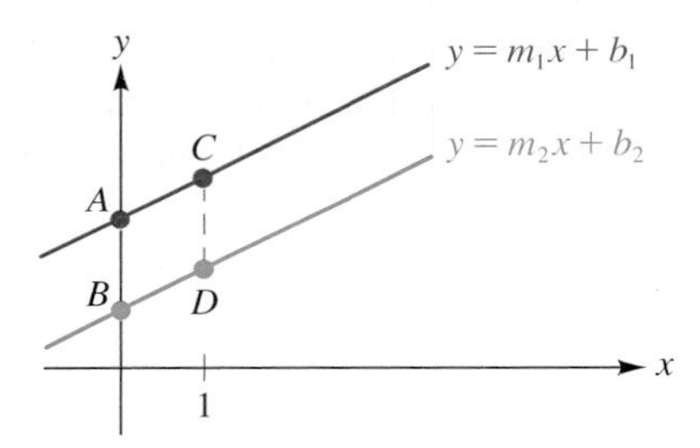

(a) Verify that the coordinates of A, B, C, and D are

$$A(0, b_1) \quad B(0, b_2) \quad C(1, m_1 + b_1) \quad D(1, m_2 + b_2)$$

(b) Using the coordinates in part (a), check that

$$AB = b_1 - b_2 \quad \text{and} \quad CD = (m_1 + b_1) - (m_2 + b_2)$$

(c) Use part (b) to show that the equation $AB = CD$ is equivalent to $m_1 = m_2$.

54. This exercise outlines a proof of the fact that two nonvertical lines with slopes m_1 and m_2 are perpendicular if and only if $m_1m_2 = -1$. In the following figure, we've assumed that our two nonvertical lines $y = m_1x$ and $y = m_2x$ intersect at the origin. [If they did not intersect there, we could just as well work with lines parallel to these that do intersect at (0, 0), recalling that parallel lines have the same slope.] The proof relies on the following geometric fact:

$$\overline{OA} \perp \overline{OB} \quad \text{if and only if} \quad (OA)^2 + (OB)^2 = (AB)^2$$

(a) Verify that the coordinates of A and B are $A(1, m_1)$ and $B(1, m_2)$.
(b) Show that

$$OA^2 = 1 + m_1^2$$
$$OB^2 = 1 + m_2^2$$
$$AB^2 = m_1^2 - 2m_1m_2 + m_2^2$$

(c) Use part (b) to show that the equation

$$OA^2 + OB^2 = AB^2$$

is equivalent to $m_1 m_2 = -1$.

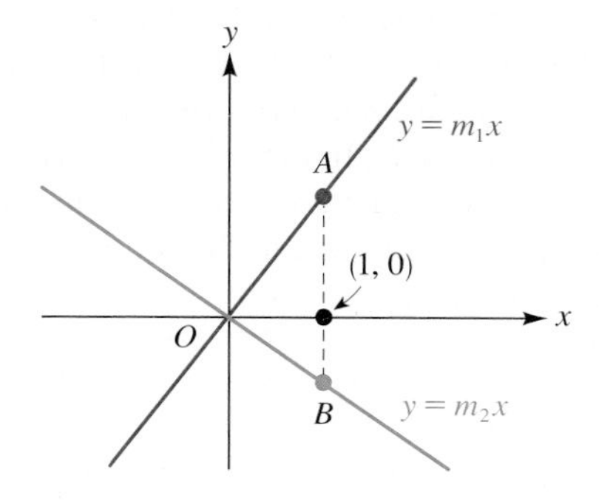

55. Verify the identity

$$(y_2 - y_1)/(x_2 - x_1) = (y_1 - y_2)/(x_1 - x_2)$$

What does this identity tell you about calculating slope?

C

56. Find the slope m of the line in the following figure.

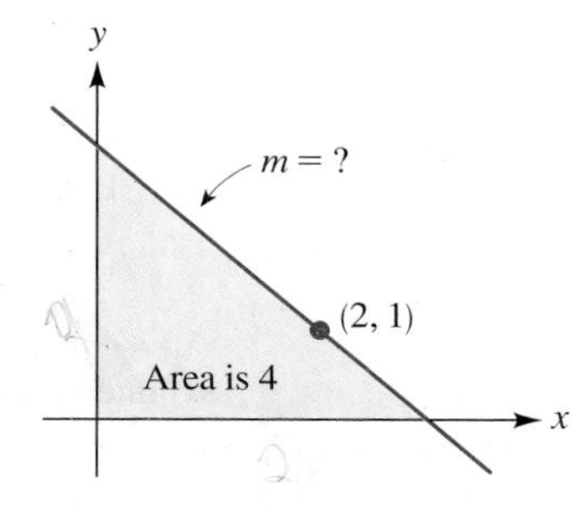

MINI PROJECT

THINKING ABOUT SLOPE

Working within a group or with the class at large, carry out the two activities indicated below. Then, on your own, write a summary of what you have learned.

1. What if the definition of slope that we gave in this section had instead been $m = \dfrac{y_2 + y_1}{x_2 + x_1}$? Try reworking Example 1 using this version of slope and describe what happens. Why is this unsatisfactory?
2. Why, where, and when did the use of the letter m for slope originate? See what you can find out by reading comments from mathematicians and math historians at the following Internet website:

   http://mathforum.org/epigone/math-history-list

   On the right-hand side of that web page, click on "search this discussion," and type in the word *slope.* As with many historical issues in many fields, don't expect one simple definitive answer. Summarize what you find out.

1.7 SYMMETRY AND GRAPHS. CIRCLES

Symmetry is a working concept. If all the object is symmetrical, then the parts must be halves (or some other rational fraction) and the amount of information necessary to describe the object is halved (etc.).—Alan L. Macay, Department of Crystallography, University of London

As you will see throughout this text, there are some basic techniques that help us to understand the essential features of a graph. In the box that follows, we introduce three types of *symmetry* that are useful in analyzing graphs.

THREE TYPES OF SYMMETRY

Type of Symmetry	Example	Definition
1. Symmetry about the x-axis	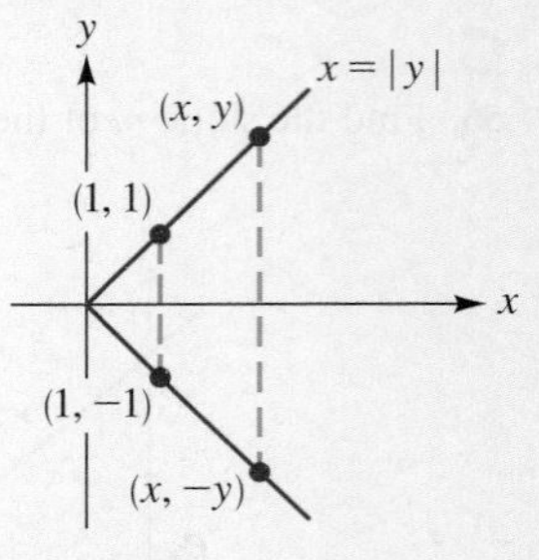 Figure 1 Symmetry about the x-axis	*For each point (x, y) on the graph, the point $(x, -y)$ is also on the graph.* We say that the points (x, y) and $(x, -y)$ are *reflections of one another about* (or *in*) *the x-axis.* In Figure 1 the portions of the graph above and below the x-axis also are said to be reflections of one another about the x-axis.
2. Symmetry about the y-axis	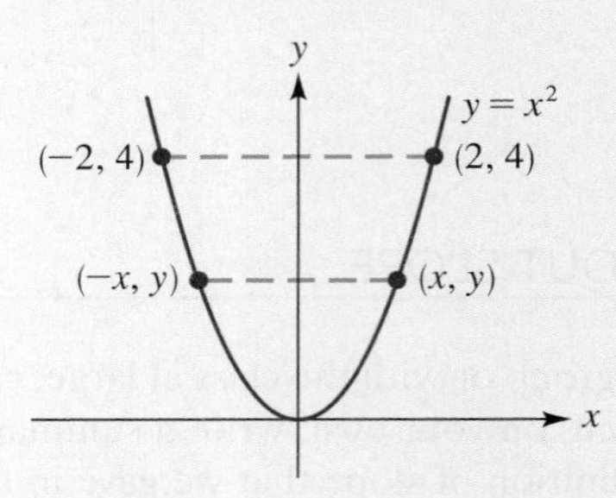 Figure 2 Symmetry about the y-axis	*For each point (x, y) on the graph, the point $(-x, y)$ is also on the graph.* We say that the points (x, y) and $(-x, y)$ are *reflections of one another about* (or *in*) *the y-axis.* In Figure 2 the portions of the graph to the right and left of the y-axis also are said to be reflections of one another about the y-axis.
3. Symmetry about the origin	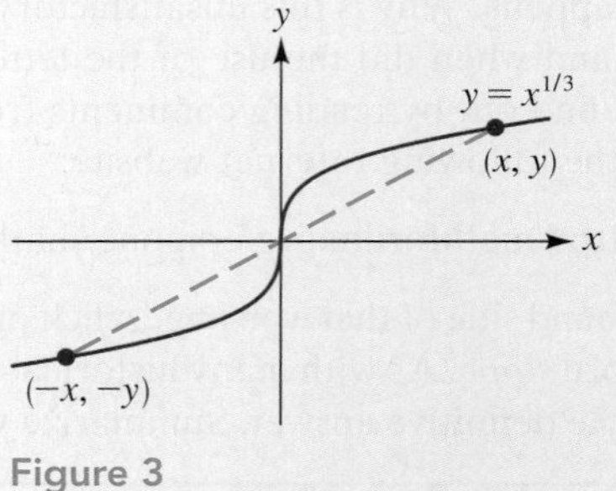 Figure 3 Symmetry about the origin	*For each point (x, y) on the graph, the point $(-x, -y)$ is also on the graph.* We say that the points (x, y) and $(-x, -y)$ are *reflections of one another about the origin.* In Figure 3 the first- and third-quadrant portions of the curve also are said to be reflections of one another about the origin. In terms of the two previous symmetries, the point $(-x, -y)$ can be obtained from (x, y) as follows: First reflect (x, y) about the y-axis to obtain $(-x, y)$, then reflect $(-x, y)$ about the x-axis to obtain $(-x, -y)$.

EXAMPLE 1 Sketching reflections

A line segment $\mathcal{L}$ has endpoints (1, 2) and (5, 3). Sketch the reflection of $\mathcal{L}$ about: **(a)** the x-axis; **(b)** the y-axis; and **(c)** the origin.

SOLUTION

(a) First reflect the endpoints (1, 2) and (5, 3) about the x-axis to obtain the new endpoints $(1, -2)$ and $(5, -3)$, respectively. Then join these two points as indicated in Figure 4(a), in quadrant IV.

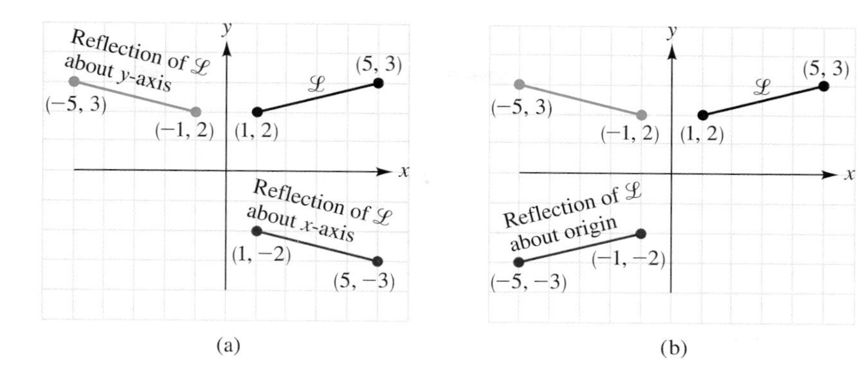

Figure 4

(b) Reflect the given endpoints about the y-axis to obtain the new endpoints $(-1, 2)$ and $(-5, 3)$; then join these two points as shown in Figure 4(a), in quadrant II.

(c) As described in the box preceding this example, reflection about the origin can be carried out in two steps: First reflect about the y-axis, then reflect about the x-axis. In part (b) we obtained the reflection of $\mathcal{L}$ in the y-axis, so now we need only reflect the line segment obtained in part (b) about the x-axis. See Figure 4(b).

In the box that follows, we list three rules for testing whether the graph of an equation possesses any of the types of symmetry we've been discussing. (The validity of each rule follows directly from the definitions of symmetry.)

Three Tests for Symmetry

1. The graph of an equation is symmetric about the y-axis if replacing x with $-x$ yields an equivalent equation.
2. The graph of an equation is symmetric about the x-axis if replacing y with $-y$ yields an equivalent equation.
3. The graph of an equation is symmetric about the origin if replacing x and y with $-x$ and $-y$, respectively, yields an equivalent equation.

EXAMPLE 2 Testing for symmetry about the x-axis and the y-axis

(a) In Figure 5(a), obtained with a graphing utility, the graph of $y = x^4 - 3x^2 + 1$ appears to be symmetric about the y-axis. Use an appropriate symmetry test to find out whether the graph indeed possesses this type of symmetry.

(b) In Figure 5(b), obtained with a graphing utility, the graph of $x = y^2 - \frac{1}{10}y - 15$ appears to be symmetric about the x-axis. Use an appropriate symmetry test to find out whether the graph indeed possesses this type of symmetry.
Remark: Most graphing utilities require that you solve the equation for y before you enter it. See Exercise 61 for details.

GRAPHICAL PERSPECTIVE

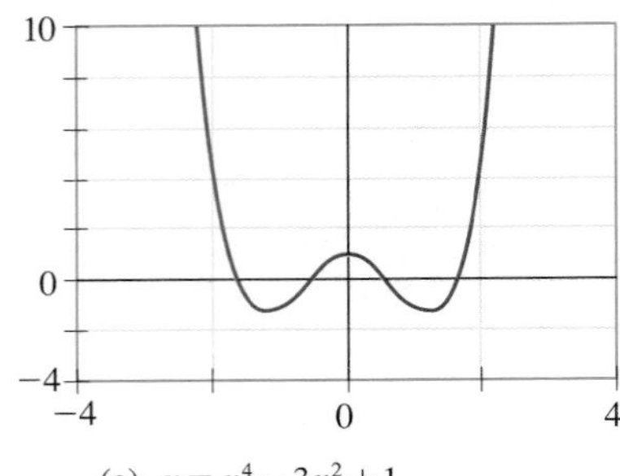

(a) $y = x^4 - 3x^2 + 1$
$[-4, 4, 2]$ by $[-4, 10, 2]$

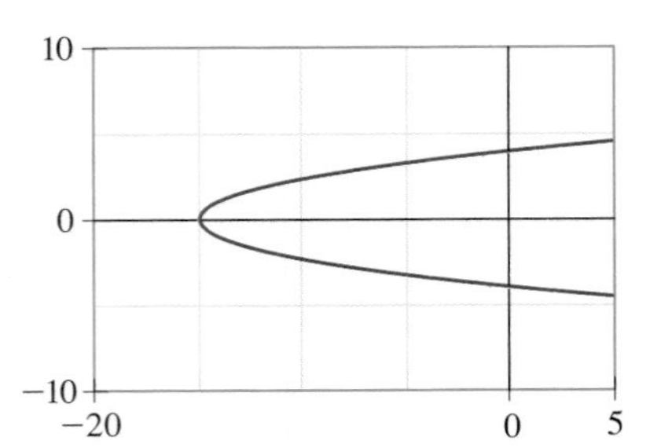

(b) $x = y^2 - \frac{1}{10}y - 15$
$[-20, 5, 5]$ by $[-10, 10, 5]$

Figure 5

SOLUTION

(a) To test for symmetry about the y-axis, we replace x with $-x$ in the given equation, to obtain

$$y = (-x)^4 - 3(-x)^2 + 1 \qquad \text{or} \qquad y = x^4 - 3x^2 + 1$$

Since this last equation is the same as the original equation, we conclude that the graph is indeed symmetric about the y-axis.

(b) To test for symmetry about the x-axis, we replace y with $-y$ in the given equation, to obtain

$$x = (-y)^2 - \frac{1}{10}(-y) - 15 \qquad \text{or} \qquad x = y^2 + \frac{1}{10}y - 15$$

This last equation is not equivalent to the given equation. This tells us that the graph is not symmetric about the x-axis. (Exercise 61(c), a graphing utility exercise, asks for a viewing rectangle that clearly shows this lack of symmetry.)

EXAMPLE 3 Testing for symmetry about the origin

In Figure 6 it appears that the graph of $y = 2x^3 - 3x$ is symmetric about the origin. Use a symmetry test to find out whether this is indeed the case.

GRAPHICAL PERSPECTIVE

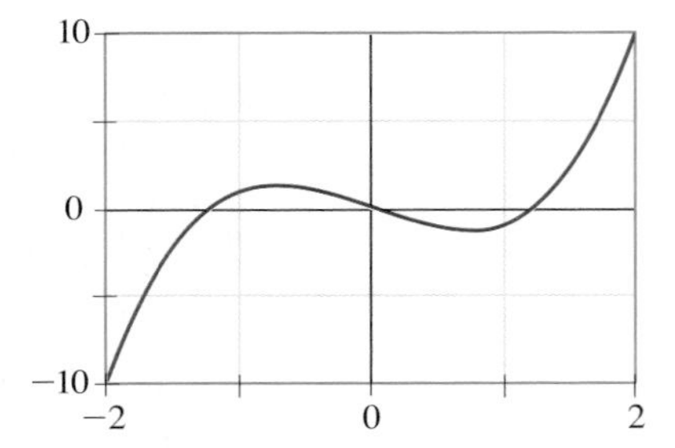

Figure 6
$y = 2x^3 - 3x$,
$[-2, 2, 1]$ by $[-10, 10, 5]$
The graph appears to be symmetric about the origin. Is this really the case?

SOLUTION

Replacing x and y with $-x$ and $-y$, respectively, gives us

$$-y = 2(-x)^3 - 3(-x) = -2x^3 + 3x$$
$$y = 2x^3 - 3x \quad \text{multiplying through by } -1$$

This last equation is identical to the given equation, and we conclude that in this case appearances are not deceiving; the graph is indeed symmetric about the origin.

In the next two examples we use the notions of symmetry and intercepts as guides for drawing graphs. True, the graphs could be obtained more quickly using a graphing utility. The idea here, however, is to *understand* why the graphs look as they do.

EXAMPLE 4 Using symmetry and intercepts as aids in graphing

Graph the equation $y = -x^2 + 5$.

SOLUTION

The domain of the variable x in the expression $-x^2 + 5$ is the set of all real numbers. However, since the graph must be symmetric about the y-axis (why?), we need only sketch the graph to the right of the y-axis; the portion to the left will then be the mirror image of this (with the y-axis as the mirror). The x- and y-intercepts (if any) are computed as follows:

x-INTERCEPTS

$$-x^2 + 5 = 0$$
$$x^2 = 5$$
$$x = \pm\sqrt{5} \ (\approx \pm 2.2)$$

y-INTERCEPTS

$$y = -(0)^2 + 5$$
$$y = 5$$

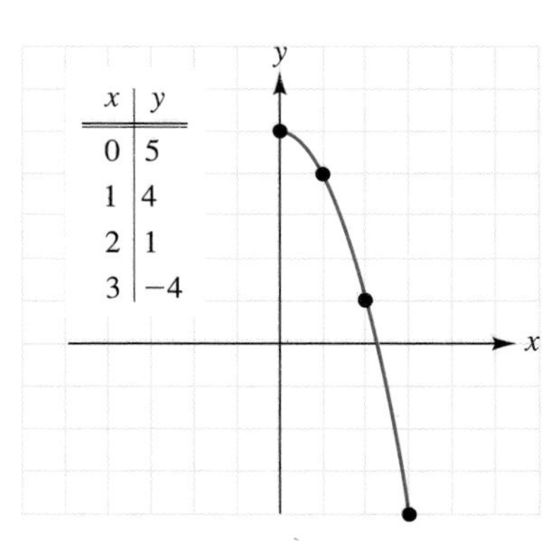

(a) $y = -x^2 + 5, x \geq 0$

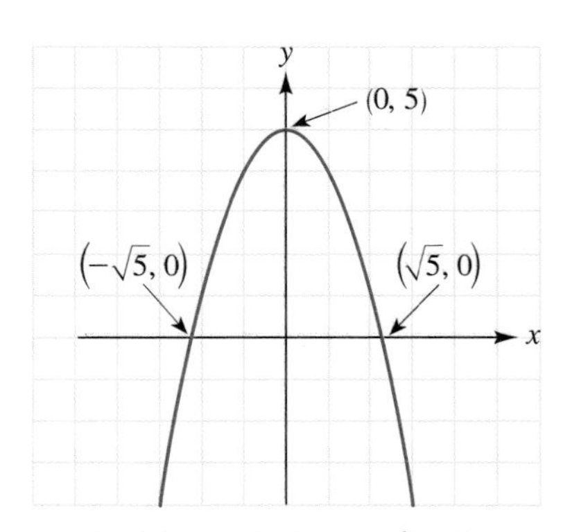

(b) The graph of $y = -x^2 + 5$ is symmetric about the y-axis.

Figure 7

In Figure 7(a) we've set up a short table of values and sketched the graph for $x \geq 0$. The complete graph is then obtained by reflection about the y-axis, as shown in Figure 7(b). [The curve in Figure 7(b) is a *parabola*. This type of curve will be considered in detail in later chapters.]

EXAMPLE 5 Using symmetry and domain as aids in graphing

Graph: $y = -4/x$.

SOLUTION

Before doing any calculations, note that the domain of the variable x consists of all real numbers other than zero. So, however the resulting graph might look, it cannot contain a point with an x-coordinate of zero. In other words, the graph cannot cross the y-axis. (Check for yourself that the graph cannot cross the x-axis either.) Now, since the graph is symmetric about the origin (why?), we need only sketch the graph for $x > 0$; the portion corresponding to $x < 0$ can then be obtained by reflection about the origin. In Figure 8(a) we've set up a table of values and sketched the graph for $x > 0$. (Note the fractional x-values at the end of the table: since y is undefined when x is zero, it's informative to pick x-values near zero and look at the corresponding y-values.) In Figure 8(b) we have reflected the fourth-quadrant portion of the graph first about the y-axis and then about the x-axis to obtain the required reflection about the origin. Figure 9 shows our final graph of $y = -4/x$.

x	1	2	3	4	5	6	7	8	$\frac{1}{2}$	$\frac{1}{3}$
y	-4	-2	$-\frac{4}{3}$	-1	$-\frac{4}{5}$	$-\frac{2}{3}$	$-\frac{4}{7}$	$-\frac{1}{2}$	-8	-12

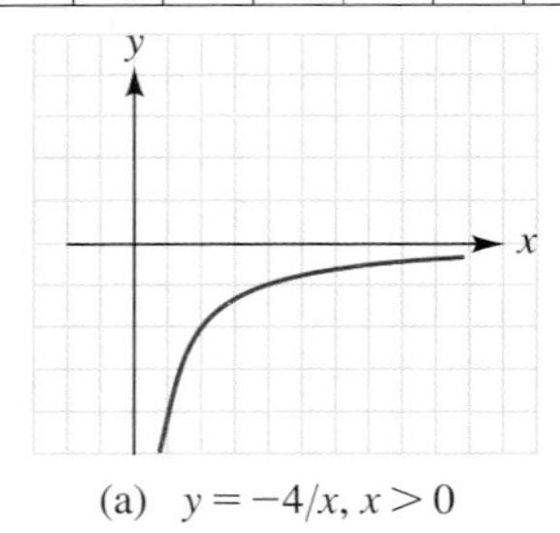

(a) $y = -4/x, x > 0$

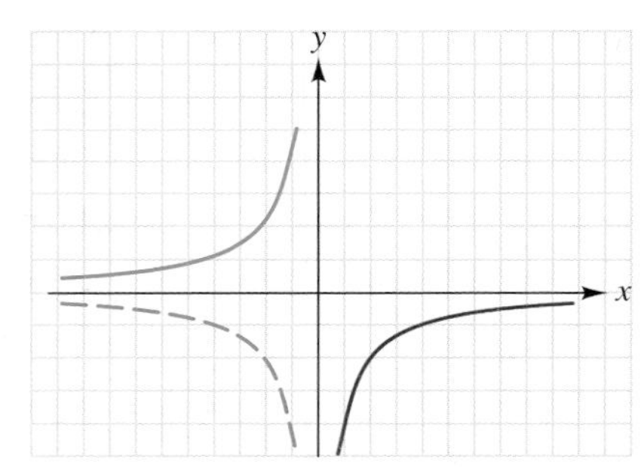

(b) Reflections about the y-axis and then the x-axis yield a reflection about the origin.

Figure 8

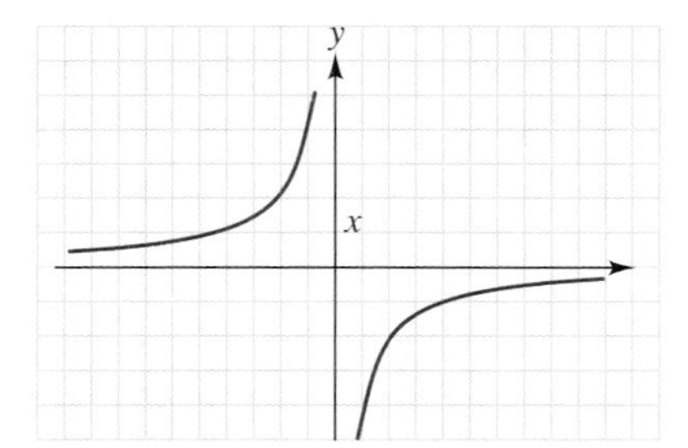

Figure 9
The graph of $y = -4/x$ is symmetric about the origin.

We can use the graph in Figure 8(a) to introduce the idea of an *asymptote.* A line is an **asymptote** for a curve if the distance between the line and the curve approaches zero as we move farther and farther out along the line. So for the curve in Figure 8(a) both the x-axis and the y-axis are asymptotes. (We'll return to this idea several times later in the text. For other pictures of asymptotes, see Figure 1 in Section 4.7.)

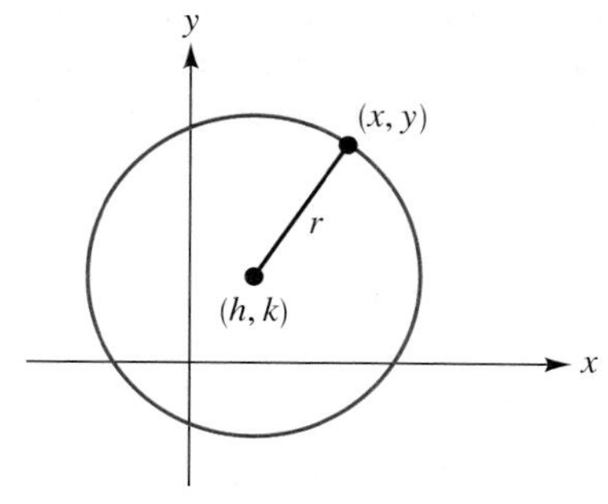

Figure 10

Since we've been discussing symmetry in this section, it seems appropriate to conclude with a discussion of the circle, because in some sense, this is the most symmetric curve. We can use the distance formula to obtain *the equation of a circle.* Figure 10 shows a circle with center (h, k) and radius r. By definition, a point (x, y) is on this circle if and only if the distance from (x, y) to (h, k) is r. Thus we have

$$\sqrt{(x - h)^2 + (y - k)^2} = r$$

or, equivalently,

$$(x - h)^2 + (y - k)^2 = r^2 \tag{1}$$

(These last two equations are equivalent because two nonnegative quantities are equal if and only if their squares are equal.)

The work in the previous paragraph tells us two things. First, if a point (x, y) lies on the circle in Figure 10, then x and y together satisfy equation (1). Second, if a pair of numbers x and y satisfies equation (1), then the point (x, y) lies on the circle in Figure 10. (Does it sound to you as if the previous two sentences say the same thing? They don't! Think about it.) Equation (1) is called the **standard form for the equation of a circle** or the **standard equation of a circle.** For reference, we record the result in the box that follows.

The Equation of a Circle in Standard Form

The standard equation of the circle with center (h, k) and radius r is

$$(x - h)^2 + (y - k)^2 = r^2$$

EXAMPLE 6 Finding the equation of a circle

(a) Write the equation of the circle with center $(-2, 1)$ and radius 3.
(b) Does the point $(-4, 3)$ lie on this circle?

SOLUTION

(a) In the equation $(x - h)^2 + (y - k)^2 = r^2$, we substitute the given values $h = -2$, $k = 1$, and $r = 3$. This yields

$$[x - (-2)]^2 + (y - 1)^2 = 3^2$$

or

$$(x + 2)^2 + (y - 1)^2 = 9 \tag{2}$$

This is the standard form for the equation of the given circle.

(b) To find out whether the point $(-4, 3)$ lies on the circle, we check to see if the coordinates $x = -4$ and $y = 3$ satisfy equation (2). We have

$$(-4 + 2)^2 + (3 - 1)^2 \stackrel{?}{=} 9$$
$$(-2)^2 + 2^2 \stackrel{?}{=} 9$$
$$8 \stackrel{?}{=} 9 \qquad \text{False}$$

This shows that the values $x = -4$ and $y = 3$ do not satisfy equation (2). Consequently, the point $(-4, 3)$ does not lie on the circle.

In the example just concluded we found that an equation for the circle with center $(-2, 1)$ and radius 3 is $(x + 2)^2 + (y - 1)^2 = 9$. There is another way to write this equation that is useful for work with graphing utilities. It involves solving the equation for y in terms of x, as follows:

$$(y - 1)^2 = 9 - (x + 2)^2$$
$$y - 1 = \pm\sqrt{9 - (x + 2)^2}$$
$$y = 1 \pm \sqrt{9 - (x + 2)^2}$$

Thus the equation of the circle is equivalent to the *pair* of equations $y = 1 + \sqrt{9 - (x + 2)^2}$ and $y = 1 - \sqrt{9 - (x + 2)^2}$. In working with a graphics calculator or a computer, both of these equations are graphed to yield the required circle. See Figure 11 and the GRAPHICAL PERSPECTIVE in Figure 12. In Figure 11, note that the equation with the positive square root represents the upper semicircle, while the equation with the negative square root represents the lower semicircle.

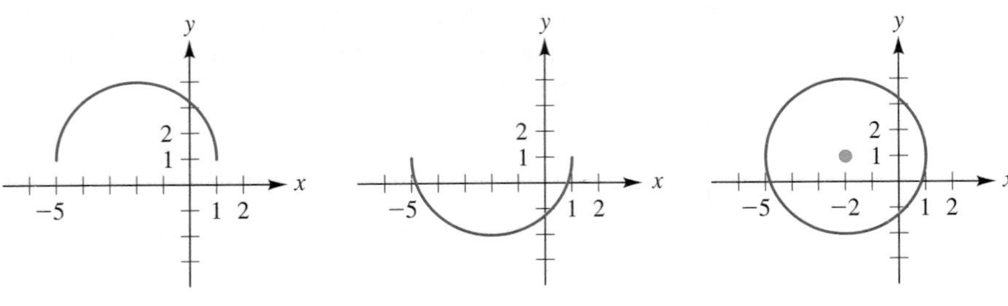

Figure 11 (a) $y = 1 + \sqrt{9 - (x + 2)^2}$ (b) $y = 1 - \sqrt{9 - (x + 2)^2}$ (c) $(x + 2)^2 + (y - 1)^2 = 9$

GRAPHICAL PERSPECTIVE

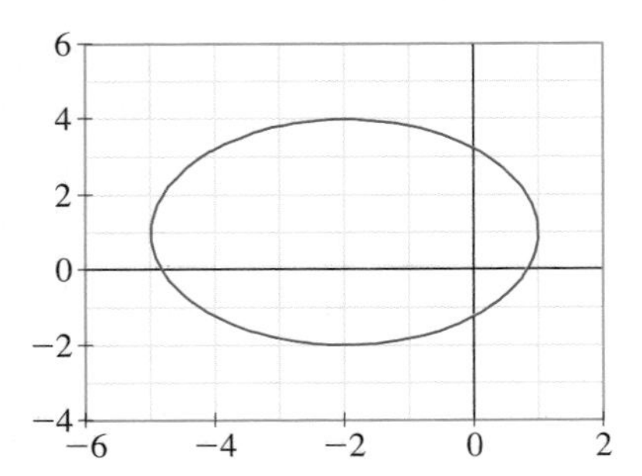

Figure 12
A distorted view of the circle $(x + 2)^2 + (y - 1)^2 = 9$ results when a *true proportions* option is not selected on the graphing utility.

Again, let us go back to the equation obtained in Example 6 for the circle with center $(-2, 1)$ and radius 3: $(x + 2)^2 + (y - 1)^2 = 9$. An alternative form for this equation is found by carrying out the indicated algebra:

$$(x + 2)^2 + (y - 1)^2 = 9$$
$$x^2 + 4x + 4 + y^2 - 2y + 1 = 9$$
$$x^2 + 4x + y^2 - 2y - 4 = 0 \qquad (3)$$

The disadvantage of this last equation is that the center and radius are no longer readily apparent. For this reason it is useful to have a systematic procedure for converting equations such as equation (3) back into standard form. The algebraic technique from intermediate algebra known as *completing the square* allows us to accomplish this. We'll review (or reintroduce) this technique for you with two examples. (We'll have several other uses for completing the square later in this text. It's also useful at times in calculus.)

As a first example in completing the square, we convert the equation $x^2 + y^2 + 10x - 6y - 4 = 0$ into the standard form for a circle. We begin by grouping the x-terms, grouping the y-terms, and moving the constant to the other side (that is, adding 4 to both sides):

$$(x^2 + 10x \qquad) + (y^2 - 6y \qquad) = 4 \tag{4}$$

We've left the extra space within each set of parentheses because we are going to add something there. (You'll see in a moment why we want to do this.) To determine the number that we want to add in the first set of parentheses, we follow the completing-the-square procedure:

Take half of the coefficient of x and square it.

From equation (4), the coefficient of x is 10. Taking half of 10 and then squaring it gives us 5^2, or 25. That's the number we want to add in the first set of parentheses. Of course, to keep the equation in balance, we have to add 25 on the other side too. So equation (4) becomes

$$(x^2 + 10x + 25) + (y^2 - 6y \qquad) = 4 + 25 \tag{5}$$

In the same way now, to determine the number that we want to add in the second set of parentheses, we

Take half of the coefficient of y and square it.

From equation (5), the coefficient of y is -6. Taking half of -6 and squaring gives us $(-3)^2$, or 9. That's the number we want to add in the second set of parentheses. To keep the equation in balance, we also add 1 to the right side to obtain

$$(x^2 + 10x + 25) + (y^2 - 6y + 9) = 4 + 25 + 9 = 38 \tag{6}$$

The whole point in completing the square is that the expressions within the parentheses are now perfect squares, and we can rewrite equation (6) as

$$(x + 5)^2 + (y - 3)^2 = 38 \qquad \text{Verify this factoring for yourself.}$$

This last equation is the standard form for the equation of a circle. From the equation we see that the center of the circle is $(-5, 3)$. What about the radius? In standard form, the right side of the equation is r^2. So we have $r^2 = 38$, and consequently $r = \sqrt{38}$.

The process that we've just described for completing the square requires that the coefficients of x^2 and y^2 both be 1. If this is not the case at the start, then you need to divide both sides of the equation by an appropriate constant, as indicated in the next example. We will modify this technique in Chapter 11 when we discuss equations in which the x^2 and y^2 terms have different coefficients.

EXAMPLE 7 Completing the square and determining x-intercepts

Refer to Figure 13, which shows a graph of the circle

$$4x^2 + 4y^2 - 24x - 12y + 29 = 0$$

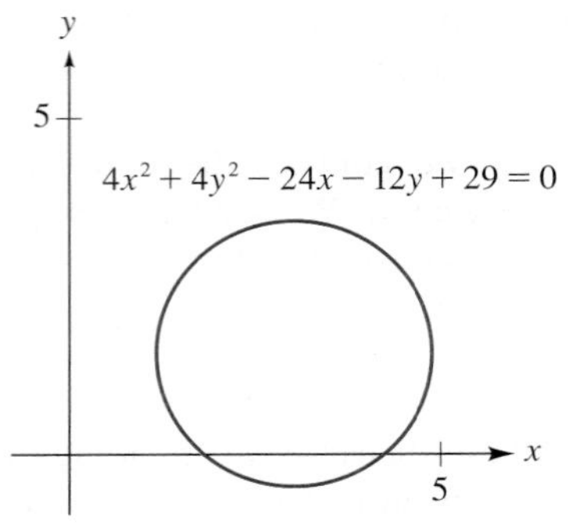

Figure 13

(a) Find the center and radius of the circle.

(b) Use a graphing utility to estimate (to two decimal places) the x-intercepts of the circle.

(c) Use algebra to find exact values for the x-intercepts.

SOLUTION

(a) To obtain coefficients of 1 in front of both x^2 and y^2, we divide both sides of the equation by 4:

$$x^2 + y^2 - 6x - 3y + \frac{29}{4} = 0$$

Next, we group the x-terms, group the y-terms, and subtract the constant 29/4 from both sides. This gives us

$$(x^2 - 6x \qquad) + (y^2 - 3y \qquad) = -\frac{29}{4}$$

To complete the square within the first set of parentheses we need to add $(\frac{-6}{2})^2$, which is 9. Likewise in the second set of parentheses, we want to add $(\frac{-3}{2})^2$ or $\frac{9}{4}$. So we obtain

$$(x^2 - 6x + 9) + \left(y^2 - 3y + \frac{9}{4}\right) = -\frac{29}{4} + 9 + \frac{9}{4}$$

or, equivalently,

$$(x - 3)^2 + \left(y - \frac{3}{2}\right)^2 = 4 \qquad \text{Check the arithmetic.} \qquad (7)$$

Looking at this last equation, we see that the center of the circle is $(3, \frac{3}{2})$ and the radius is 2.

(b) To use a graphing utility to graph the circle, we solve equation (7) for y in terms of x as follows.

$$\left(y - \frac{3}{2}\right)^2 = 4 - (x - 3)^2$$

$$y - \frac{3}{2} = \pm\sqrt{4 - (x - 3)^2}$$

$$y = \frac{3}{2} + \sqrt{4 - (x - 3)^2} \quad \text{or} \quad y = \frac{3}{2} - \sqrt{4 - (x - 3)^2}$$

Figure 14(a) shows a graph of this pair of equations (using true proportions). The view in Figure 14(a) indicates that the smaller x-intercept is

GRAPHICAL PERSPECTIVE

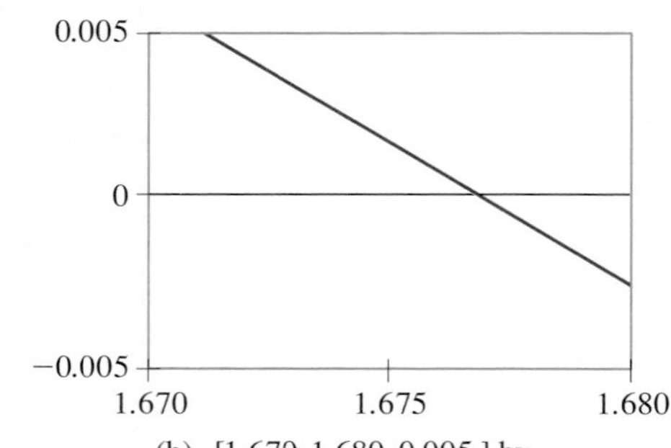

(a) [0, 6, 1] by [−1, 4, 1] True proportions

(b) [1.670, 1.680, 0.005] by [−0.005, 0.005, 0.005]

Figure 14
A graph of the circle $(x - 3)^2 + (y - \frac{3}{2})^2 = 4$ along with a magnified view near the smaller of the two x-intercepts

between 1 and 2, closer to 2 than to 1; the larger x-intercept is between 4 and 5, closer to 4 than to 5. Figure 14(b) shows a close-up view of the smaller x-intercept. Evidently, it is closer to 1.680 than to 1.670, and so our estimate to two decimal places is 1.68. Exercise 62 asks you to use a graphing utility in a similar manner to show that the larger x-intercept is approximately 4.32.

(c) The easiest way to find exact expressions for the x-intercepts is to use equation (7). Setting $y = 0$ in equation (7) yields

$$(x-3)^2 + \frac{9}{4} = 4$$

$$(x-3)^2 = \frac{7}{4} \qquad \text{Check the arithmetic.}$$

$$x - 3 = \pm\sqrt{\frac{7}{4}} = \pm\frac{\sqrt{7}}{2}$$

$$x = 3 \pm \frac{\sqrt{7}}{2}$$

In summary, the two x-intercepts are $3 - \sqrt{7}/2$ and $3 + \sqrt{7}/2$. As you can check for yourself with a calculator, these last two expressions are approximately 1.68 and 4.32, respectively. These values are consistent with the numbers that we obtained graphically in part (b).

EXERCISE SET 1.7 

A

In Exercises 1–6, the endpoints of a line segment $\overline{AB}$ are given. Sketch the reflection of $\overline{AB}$ about **(a)** *the x-axis;* **(b)** *the y-axis; and* **(c)** *the origin.*

1. $A(1, 4)$ and $B(3, 1)$
2. $A(-1, -2)$ and $B(-5, -2)$
3. $A(-2, -3)$ and $B(2, -1)$
4. $A(-3, -3)$ and $B(-3, -1)$
5. $A(0, 1)$ and $B(3, 1)$
6. $A(-2, -2)$ and $B(0, 0)$

In Exercises 7–24, graph the equation after determining the x- and y-intercepts and whether the graph possesses any of the three types of symmetry described on page 58.

7. $y = 4 - x^2$
8. $y = -x^3$
9. $y = -1/x$
10. $x = y^2 - 1$
11. $y = -x^2$
12. $y = 1/x^2$
13. $y = -1/x^3$
14. $y = |x| - 2$
15. $y = \sqrt{x^2}$
16. $y = x + 1$
17. $y = x^2 - 2x + 1$
18. $x = y^3 - 1$
19. $y^2 = 2x - 4$
20. $|y| = 2x - 4$
21. $y = 2x^2 + x - 4$
22. $y = 2x^2$
23. **(a)** $x + y = 2$ **(b)** $|x| + y = 2$
24. **(a)** $x + |y| = 2$ **(b)** $|x + y| = 2$

Ⓖ *In Exercises 25–38:*

(a) *Use a graphing utility to graph each equation in the standard viewing rectangle.*
(b) *Does the graph in part (a) appear to possess any of the three types of symmetry defined on page 58?*
(c) *In cases in which your answer to part (b) is Yes, adjust your viewing rectangle for a second, more careful inspection. (In Exercises 25–31, suggestions for the second look are provided.)*
(d) *In cases in which, after the second look, it still appears that the graph possesses symmetry, use an appropriate symmetry test from page 59 to settle the matter.*

25. $y = x^2 - 3x$ (second view: x from -2 to 5; y from -4 to 10)
26. $y = x^3 - 3x$ (second view: x from -3 to 3; y unchanged)
27. $y = 2^x$ (second view: x from -3 to 3; y from 0 to 8)
28. $y = 2^{|x|}$ (second view: x from -3 to 3; y from 0 to 8)
29. $y = 1/(x^2 - x)$ (second view: x from -2 to 2; y unchanged)
30. $y = 1/(x^3 - x)$ (second view: x from -2 to 2; y unchanged)
31. $y = x^2 - 0.2x - 15$ *Hint*: Look closely at the x-intercepts.
32. $y = x^2 - 2x^3$
33. $y = \sqrt{|x|}$
34. $y = \sqrt{|x|^3}$
35. $y = 2x - x^3 - x^5 + x^7$
36. $y = |2x - x^3 - x^5 + x^7|$
37. $y = x^4 - 10x^2 + \frac{1}{4}x$ *Suggestion*: For the second view, try zooming out in the y-direction.
38. $y = x^4 - 10x^2 + \frac{1}{4}$

In Exercises 39–42, specify the center and radius of each circle. Also, determine whether the given point lies on the circle.

39. $(x - 1)^2 + (y - 5)^2 = 169$; $(6, -7)$
40. $(x + 4)^2 + (y + 2)^2 = 20$; $(0, 1)$
41. $(x + 8)^2 + (y - 5)^2 = 13$; $(-5, 2)$
42. $x^2 + y^2 = 1$; $(1/2, \sqrt{3}/2)$

In Exercises 43–48, determine the center and the radius for the circle. Also, find the y-coordinates of the points (if any) where the circle intersects the y-axis.

43. $x^2 + y^2 = \sqrt{2}$
44. $x^2 + y^2 - 10x + 2y + 17 = 0$
45. $x^2 + y^2 + 8x - 6y = -24$
46. $4x^2 - 4x + 4y^2 - 63 = 0$
47. $9x^2 + 54x + 9y^2 - 6y + 64 = 0$
48. $3x^2 + 3y^2 + 5x - 4y = 1$

In Exercises 49 and 50, use the techniques shown in Example 7 to carry out the following procedures.

(a) *Find the center and radius of the circle.*
Ⓖ **(b)** *Use a graphing utility to graph the circle and to estimate (to two decimal places) the x-intercepts.*
(c) *Use algebra to find exact values for the x-intercepts, and then use a calculator to check that the results are consistent with the estimates in part (b).*

49. $16x^2 - 64x + 16y^2 + 48y - 69 = 0$
50. $3x^2 + x + 3y^2 + 3y - 1 = 0$

Ⓖ **51.** In the text we said that a line is an *asymptote* for a curve if the distance between the line and the curve approaches zero as we move farther and farther out along the line. In terms of graphing, this means that as we zoom out, the curve and the line eventually appear indistinguishable. In this exercise, we'll demonstrate this using the curve $y = -4/x$ (which we graphed in Figure 9). As indicated in the text, both the x- and y-axes are asymptotes for this curve. First, graph $y = -4/x$ using a viewing rectangle that extends from -5 to 5 in both the x- and the y-directions. Then take a second look using a viewing rectangle that extends from -30 to 30 in both the x- and y-directions. At this scale, you'll see that the curve is virtually indistinguishable from an asymptote when either $|x| > 8$ or $|y| > 8$.

Ⓖ **52. (a)** Graph the equation $y = 20/x$ using a standard viewing rectangle.
(b) Although both the x- and the y-axes are asymptotes for this curve, the graph in part (a) does not show this clearly. Take a second look, using a viewing rectangle that extends from -100 to 100 in both the x- and the y-directions. Note that the curve indeed appears indistinguishable from an asymptote when either $|x|$ or $|y|$ is sufficiently large.

B

53. The center of a circle is the point (3, 2). If the point $(-2, -10)$ lies on this circle, find the standard equation for the circle.

54. Find the standard equation of the circle tangent to the x-axis and with center (3, 5). *Hint:* First draw a sketch.

55. Find the standard equation of the circle tangent to the y-axis and with center (3, 5).

56. Find the standard equation of the circle passing through the origin and with center (3, 5).

57. The points $A(-1, 6)$ and $B(3, -2)$ are the endpoints of a diameter of a circle, as indicated in the accompanying figure. Find the y-intercepts of the circle. *Hint:* Could you do the problem if you had the equation of the circle?

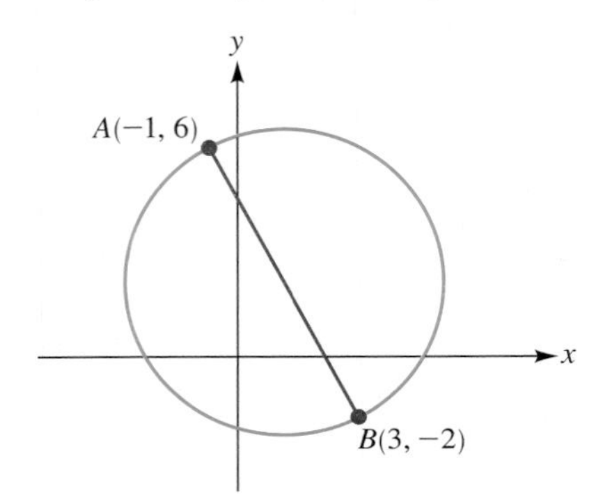

58. (a) Verify that the point (3, 7) is on the circle

$$x^2 + y^2 - 2x - 6y - 10 = 0$$

(b) Find the equation of the line tangent to this circle at the point (3, 7). *Hint:* A result from elementary geometry says that the tangent to a circle is perpendicular to the radius drawn to the point of contact.

59. The accompanying figure shows the graphs of $y = \frac{3}{4}x - 2$ and $y = |\frac{3}{4}x - 2|$.

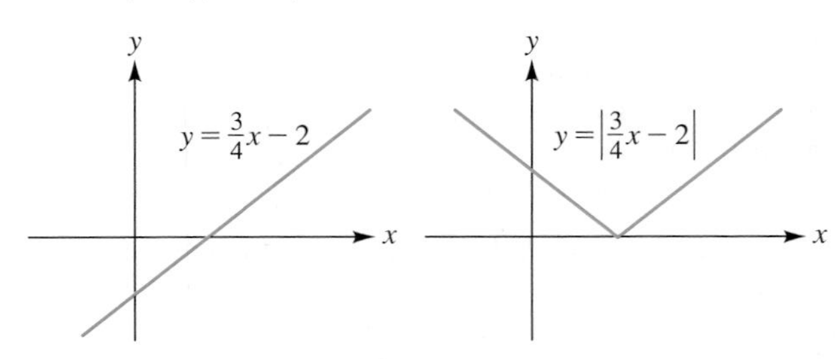

(a) Determine the x- and y-intercepts for each graph.
(b) Which portions of the two graphs are identical? (Give your answer in terms of an interval along the x-axis.)
(c) Explain how the graph of $y = |\frac{3}{4}x - 2|$ can be obtained from that of $y = \frac{3}{4}x - 2$ by means of reflection.

60. The accompanying figure shows the graphs of the two equations $y = x^2 - 6x + 8$ and $y = |x^2 - 6x + 8|$.

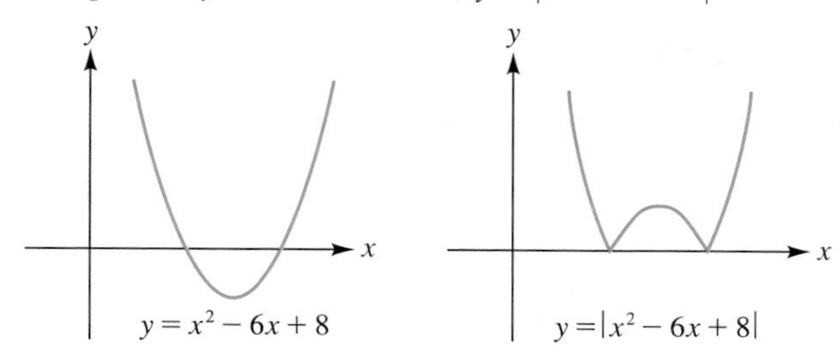

(a) Determine the x- and y-intercepts for each graph.
(b) Which portions of the two graphs are identical? (Give your answer in terms of intervals along the x-axis.)
(c) Explain how the graph of $y = |x^2 - 6x + 8|$ can be obtained from that of $y = x^2 - 6x + 8$ by means of reflection.

G **61.** This exercise relates to Figure 5(b) on page 60. That figure shows a graph of the equation $x = y^2 - 0.1y - 15$. Note that this equation is solved for x rather than y. In the standard Cartesian graphing mode, however, most graphing utilities require that the equation be solved for y. This exercise shows one method for obtaining the graph. (A remark at the end of this exercise mentions a second method.)

(a) Rewrite the equation as

$$y^2 - 0.1y - (15 + x) = 0$$

Then solve for y in terms of x by using the quadratic formula with $a = 1$, $b = -0.1$, and $c = -(15 + x)$. After simplifying, you should obtain

$$y = \frac{0.1 \pm \sqrt{60.01 + 4x}}{2}$$

(b) Enter and graph the pair of equations

$$y = \frac{0.1 - \sqrt{60.01 + 4x}}{2} \quad \text{and} \quad y = \frac{0.1 + \sqrt{60.01 + 4x}}{2}$$

using the viewing rectangle $[-20, 5, 5]$ by $[-10, 10, 5]$. Check that your graph agrees with that shown in Figure 5(b).

(c) Experiment to find a viewing rectangle in which it is clear that the graph of the equation $x = y^2 - 0.1y - 15$ is *not* symmetric about the x-axis.

Remark: A second method for graphing $x = y^2 - 0.1y - 15$ requires the *parametric mode* on your graphing utility. Consult the user's manual for your graphing utility to see how to access and operate this mode. Entering the following pair of equations in the parametric mode will yield the graph (or a portion of the graph): $x = t^2 - 0.1t - 15$; $y = t$.

G **62.** Use a graphing utility and the method shown in the solution of Example 7(b) to show that the larger of the two x-intercepts of the circle $4x^2 + 4y^2 - 24x - 12y + 29 = 0$ is (to two decimal places) 4.32.

63. Suppose that the circle $x^2 + 2Ax + y^2 + 2By = C$ has two x-intercepts, x_1 and x_2, and two y-intercepts, y_1 and y_2. Prove each statement.

(a) $\dfrac{x_1 + x_2}{y_1 + y_2} = \dfrac{A}{B}$
(b) $x_1x_2 - y_1y_2 = 0$
(c) $x_1x_2 + y_1y_2 = -2C$

MINI PROJECT

THINKING ABOUT SYMMETRY

This mini project works well in the context of a small group discussion but, of course, it can also be carried out by individuals. Begin by studying and/or discussing the following example involving a regular hexagon. Then work on the exercise that follows. For each part of the exercise, label each figure that you create (e.g., "symmetric about the y- but not the x-axis").

Example: The following figure shows a regular hexagon.

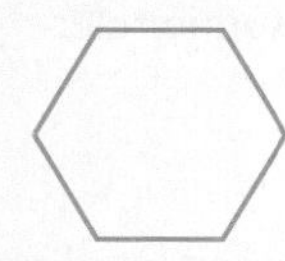

The next four figures show some of the (infinitely many) ways to locate the regular hexagon with respect to an x-y coordinate system.

(a) Symmetric about y-axis, but not about x-axis

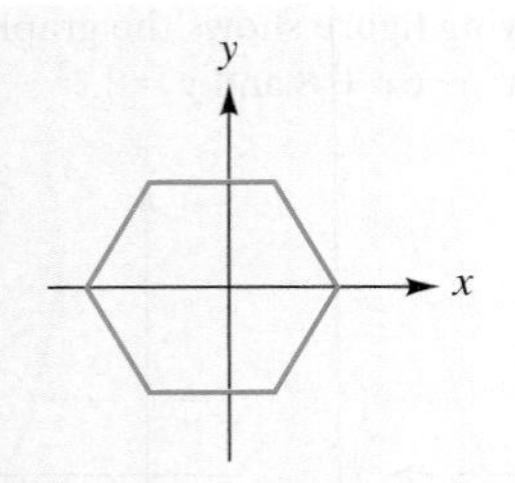

(b) Symmetric about x-axis, but not about y-axis

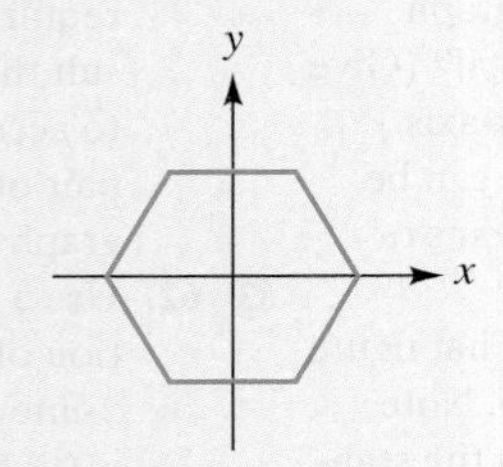

(c) Symmetric about origin and about both axes

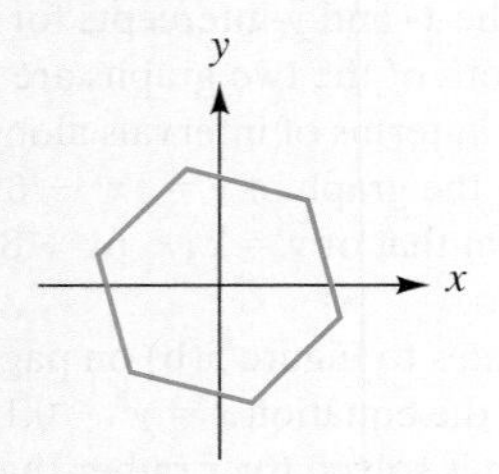

(d) Symmetric about origin and not about either axis

In each case below, you are given a geometric figure. Try to find a way to draw the figure with an x-y coordinate system so that the figure is
(a) symmetric about the y-axis, but not about the x-axis;
(b) symmetric about the x-axis, but not about the y-axis;
(c) symmetric about the origin and both axes;
(d) symmetric about the origin but not about either axis.
(In some cases there are many ways, in others none.)

1. equilateral triangle:

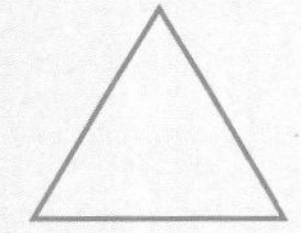

2. rectangle:

3. regular pentagon:

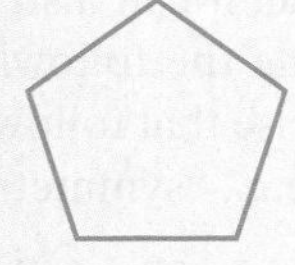

4. semicircle:

5. connected semicircles:

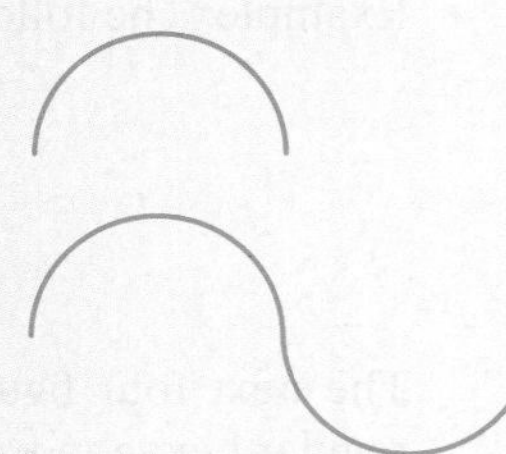

Chapter 1 Summary of Principal Terms and Formulas

Terms or notation	Page reference	Comments
1. Natural numbers, integers, rational numbers, and irrational numbers	2	The box on page 2 provides both definitions and examples. Also note the theorem in the box, which explains how to distinguish between rationals and irrationals in terms of their decimal representations.
2. $a < b$ $b > a$	3	a is less than b. b is greater than a.
3. (a, b) $[a, b]$	2	The open interval (a, b) consists of all real numbers between a and b, excluding a and b. The closed interval $[a, b]$ consists of all real numbers between a and b, including a and b.
4. (a, ∞)	4	The unbounded interval (a, ∞) consists of all real numbers x such that $x > a$. The infinity symbol, ∞, does not denote a real number. It is used in the context (a, ∞) to indicate that the interval has no right-hand boundary. For the definitions of the other types of unbounded intervals, see the box on page 4.
5. $\lvert x \rvert$	6	The absolute value of a real number x is the distance from x to the origin. This is equivalent to the following algebraic definition: $$\lvert x \rvert = \begin{cases} x & \text{if } x \geq 0 \\ -x & \text{if } x < 0 \end{cases}$$
6. Constant, variable, and the domain convention	10	By a constant we mean either a particular number (such as -8 or π) or a letter that remains fixed (although perhaps unspecified) throughout a given discussion. In contrast, a variable is a letter for which we can substitute any number from a given set of numbers. The given set is called the domain of the variable. According to the domain convention, the domain of a variable in a given expression consists of all real numbers for which the expression is defined.
7. Linear (or first-degree) equation in one variable	11	These are equations that can be written in the form $ax + b = 0$, where a and b are constants with $a \neq 0$.
8. Solution (or root) of an equation	11	A solution or root is a number that, when substituted for the variable in an equation, yields a true statement.
9. Equivalent equations	12	Two equations are equivalent if they have the same set of solutions.
10. Extraneous solution (or extraneous root)	14	In solving equations, certain processes can lead to answers that do not check in the original equation. These numbers are called extraneous solutions (or extraneous roots).

Terms or notation	Page reference	Comments
11. Quadratic (or second-degree) equation	14	A quadratic equation is one that can be written in the form $ax^2 + bx + c = 0$, where a, b, and c are constants with $a \neq 0$.
12. Zero-product property of real numbers	15	If $pq = 0$ then $p = 0$ or $q = 0$; conversely, if $p = 0$ or $q = 0$, then $pq = 0$. (We used this property in solving quadratic equations by factoring.)
13. Quadratic formula $x = \dfrac{-b \pm \sqrt{b^2 - 4ac}}{2a}$	16	This is the quadratic formula; it provides the solutions of the quadratic equation $ax^2 + bx + c = 0$, where $a \neq 0$. The formula is derived in Section 2.1.
14. Polynomial equation in one variable	16	A polynomial equation is an equation of the form $$a_nx^n + a_{n-1}x^{n-1} + \cdots + a_1x + a_0 = 0$$ where the subscripted letter a's represent constants and the exponents on the variable are nonnegative integers. The degree of the polynomial equation is the largest exponent of the variable that appears in the equation.
15. Pythagorean theorem $a^2 + b^2 = c^2$	20	In a right triangle, the lengths of the sides are related by this equation, where c is the length of the hypotenuse. Conversely, if the lengths of the sides of a triangle are related by an equation of the form $a^2 + b^2 = c^2$, then the triangle is a right triangle, and c is the length of the hypotenuse.
16. Distance formula $d = \sqrt{(x_2 - x_1)^2 + (y_2 - y_1)^2}$	21	d is the distance between the points (x_1, y_1) and (x_2, y_2).
17. Midpoint formula $\left(\dfrac{x_1 + x_2}{2}, \dfrac{y_1 + y_2}{2}\right)$	24	This is the midpoint of the line segment joining (x_1, y_1) and (x_2, y_2).
18. Graph of an equation	32	The graph of an equation in the variables x and y is the set of all points (x, y) with coordinates that satisfy the equation.
19. Graphing utility and standard viewing rectangle	33	By a graphing utility we mean either a graphing calculator or a computer with software for graphing and analyzing equations or data. The standard viewing rectangle on a graphing utility is the default view screen and coordinate system for graphing. For many graphing utilities, the standard viewing rectangle extends from -10 to 10 in both the x- and y-directions.
20. x-intercept and y-intercept	36	An x-intercept of a graph is the x-coordinate of a point where the graph intersects the x-axis. Similarly, a y-intercept is the y-coordinate of a point where the graph intersects the y-axis.

Terms or notation	Page reference	Comments
21. Slope $m = \dfrac{y_2 - y_1}{x_2 - x_1}$	44	m is the slope of a nonvertical line passing through the points (x_1, y_1) and (x_2, y_2).
22. Δx, Δy	44	If the value of a variable x changes from x_1 to x_2, then Δx denotes the amount by which x changes: $\Delta x = x_2 - x_1$. Similarly, if y changes from y_1 to y_2 then $\Delta y = y_2 - y_1$. With this notation, the slope formula becomes $m = \Delta y/\Delta x$.
23. Point–slope formula $y - y_1 = m(x - x_1)$	47	This is the point–slope form for the equation of the line passing through the point (x_1, y_1) with slope m.
24. Slope–intercept formula $y = mx + b$	49	This is the slope–intercept form for the equation of the line with slope m and y-intercept b.
25. Parallel lines $m_1 = m_2$	50	Nonvertical parallel lines have the same slope.
26. Perpendicular lines $m_1 = -1/m_2$	50	Two nonvertical lines are perpendicular if and only if the slopes are negative reciprocals of one another.
27. Symmetry about the x-axis	58	A graph is symmetric about the x-axis if, for each point (x, y) on the graph, the point $(x, -y)$ is also on the graph. The points (x, y) and $(x, -y)$ are *reflections* of each other about (or in) the x-axis.
28. Symmetry about the y-axis	58	A graph is symmetric about the y-axis if, for each point (x, y) on the graph, the point $(-x, y)$ is also on the graph. The points (x, y) and $(-x, y)$ are *reflections* of each other about (or in) the y-axis.
29. Symmetry about the origin	58	A graph is symmetric about the origin if, for each point (x, y) on the graph, the point $(-x, -y)$ is also on the graph. The points (x, y) and $(-x, -y)$ are *reflections* of each other about the origin.
30. Symmetry tests	59	(i) The graph of an equation is symmetric about the y-axis if replacing x with $-x$ yields an equivalent equation. (ii) The graph of an equation is symmetric about the x-axis if replacing y with $-y$ yields an equivalent equation. (iii) The graph of an equation is symmetric about the origin if replacing x and y with $-x$ and $-y$, respectively, yields an equivalent equation.
31. Equation of a circle $(x - h)^2 + (y - k)^2 = r^2$	63	This is the standard form for the equation of the circle with center (h, k) and radius r.

Writing Mathematics

1. The following geometric method for solving quadratic equations of the form $x^2 - ax + b = 0$ is due to the British writer Thomas Carlyle (1795–1881):

   To solve the equation $x^2 - ax + b = 0$, plot the points $A(0, 1)$ and $B(a, b)$. Then draw the circle with $\overline{AB}$ as diameter. The x-intercepts of this circle are the roots of the equation.

   Use the techniques of this chapter to verify for yourself that this method indeed yields the roots for the equation $x^2 - 6x + 5 = 0$. (You need to find the equation of the circle, determine the x-intercepts, and then check that the numbers you obtain are the roots of the given equation.) After you have done this, carefully write out your verification in detail (using complete sentences), as if you were explaining the method to a classmate. Be sure to make explicit reference to any formulas or equations that you use from the chapter.
2. In Section 1.6 we developed the point–slope formula, $y - y_1 = m(x - x_1)$, and the slope–intercept formula, $y = mx + b$. Although these two formulas look quite different from one another, both are merely restatements of the definition of slope. Write a paragraph (or two, at the most) to explain this last sentence. Make use of the following two figures.

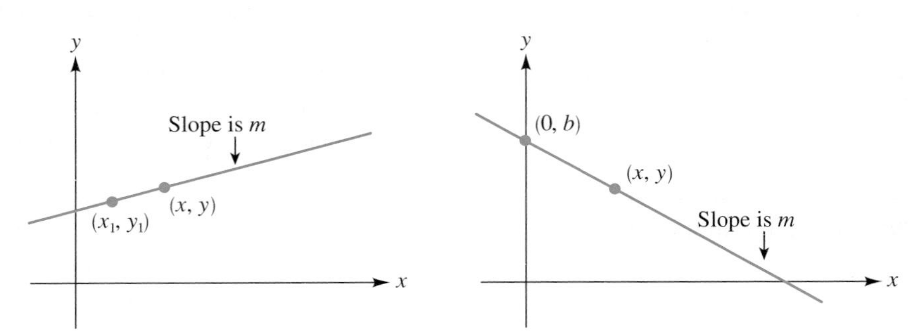

Chapter 1 Review Exercises 

For Exercises 1–6 rewrite the statements using absolute values and inequalities or equalities.

1. The distance between x and 6 is 2.
2. The distance between x and a is less than $1/2$.
3. The distance between a and b is 3.
4. The distance between x and -1 is 5.
5. The distance between x and 0 exceeds 10.
6. What can you say about x if $|x - 5| = 0$?

Rewrite each of the expressions in Exercises 7–12 in a form that does not contain absolute values.

7. $|\sqrt{6} - 2|$
8. $|2 - \sqrt{6}|$
9. $|x^4 + x^2 + 1|$
10. **(a)** $|x - 3|$, if $x < 3$ **(b)** $|x - 3|$, if $x > 3$
11. $|x - 2| + |x - 3|$, if: **(a)** $x < 2$ **(b)** $2 < x < 3$ **(c)** $x > 3$
12. $|x + 2| + |x - 1|$, if: **(a)** $x < -2$ **(b)** $-2 < x < 1$ **(c)** $x > 1$

In Exercises 13–18, express each interval in inequality notation and sketch the interval on a number line.

13. $(3, 5)$
14. $(3, 5]$

15. The set of all negative real numbers that are in the interval $[-5, 2]$

16. $(-\infty, 4)$

17. $[-1, \infty)$

18. The set of real numbers that belongs to either of the intervals $(0, \sqrt{2}]$ or $(\sqrt{2}, 2]$

In Exercises 19–23, sketch the intervals described by the given inequalities.

19. $|x - 6| < 3$

20. $|x - \frac{1}{2}| < 1$

21. $|x + 1| \geq 1$

22. $|x| \geq 5$

23. **(a)** $0 < |x - 4| < 5$ **(b)** $|x - 4| < 5$

24. Determine whether each of the following is true or false. (You should be able to do these without a calculator.)
(a) $\pi \leq \sqrt{3}$ **(b)** $\frac{1}{5} > -\sqrt{2}$

In Exercises 25–40, find all the real solutions of each equation.

25. $5 - 9x = 2$

26. $\dfrac{x}{3} - \dfrac{3x}{5} = \dfrac{x}{6} - 13$

27. $(t - 4)(t + 3) = (t + 5)^2$

28. $\dfrac{1 - x}{1 + x} = 1$

29. $\dfrac{2t - 1}{t + 2} = 5$

30. $\dfrac{1}{1 + \dfrac{1}{x + 1}} = \dfrac{5}{6}$

31. $\dfrac{2y - 5}{4y + 1} = \dfrac{y - 1}{2y + 5}$

32. $\dfrac{2x - 3}{x - 2} = \dfrac{1}{x - 2}$

33. $12x^2 + 2x - 2 = 0$

34. $4y^2 - 21y = 18$

35. $\frac{1}{2}x^2 + x - 12 = 0$

36. $x^2 + \frac{13}{2}x + 10 = 0$

37. $\dfrac{x}{5 - x} = \dfrac{-2}{11 - x}$

38. $4x^2 + x - 2 = 0$

39. $t^2 + t - \frac{1}{2} = 0$

40. $x^3 - 8x^2 - 9x = 0$

In Exercises 41–57, find an equation for the line satisfying the given conditions. For Exercises 41–49, write the answer in the form $y = mx + b$; in Exercises 50–57, write the answer in the form $Ax + By + C = 0$.

41. Passes through $(-4, 2)$ and $(-6, 6)$

42. $m = -2$; y-intercept 5

43. $m = 1/4$; passes through $(-2, -3)$

44. $m = 1/3$; x-intercept -1

45. x-intercept -4; y-intercept 8

46. $m = -10$; y-intercept 0

47. y-intercept -2; parallel to the x-axis

48. Passes through $(0, 0)$ and is parallel to $6x - 3y = 5$

49. Passes through $(1, 2)$ and is perpendicular to the line $x + y + 1 = 0$

50. Passes through $(1, 1)$ and through the center of the circle $(x - 2)^2 + (y - 4)^2 = 20$

51. Passes through the centers of the circles $(x + 2)^2 + (y + 1)^2 = 5$ and $(x - 2)^2 + (y - 8)^2 = 68$

52. $m = 3$ and has the same x-intercept as the line $3x - 8y = 12$

53. Passes through the origin and the midpoint of the line segment joining the points $(-2, -3)$ and $(6, -5)$

54. Is tangent to the circle $x^2 + y^2 = 20$ at the point $(-2, 4)$

55. Is tangent to the circle $(x - 3)^2 + (y + 4)^2 = 25$ at the point $(0, 0)$

Hint: In Exercises 56 and 57 you might try the intercept form of the equation for a line: $\dfrac{x}{a} + \dfrac{y}{b} = 1$, where a and where a and b are the x and y intercepts. See Exercise 40 in Section 1.6.

56. Passes through $(2, 4)$; the y-intercept is twice the x-intercept

57. Passes through $(2, -1)$; the sum of the x- and y-intercepts is 2. (There are two answers.)

58. **(a)** Find the perimeter of the triangle with vertices $A(3, 1)$, $B(7, 4)$, and $C(-2, 13)$.
(b) Find the perimeter of the triangle formed by joining the midpoints of the sides of the triangle in part (a).

In Exercises 59–62, test each equation for symmetry about the x-axis, the y-axis, and the origin.

59. $y = x^4 - 2x^2$

60. $y = x^3 + 5x + 1$

61. $y = 2^x + 2^{-x}$

62. $y = 2^x - 2^{-x}$

For Exercises 63–68, tell whether each graph appears to be symmetric about the x-axis, the y-axis, or the origin.

63.

64.

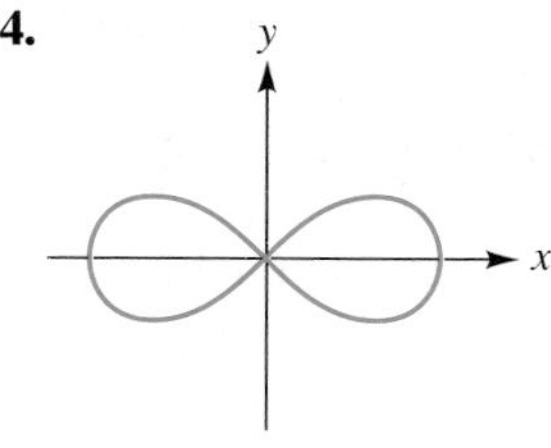

65.

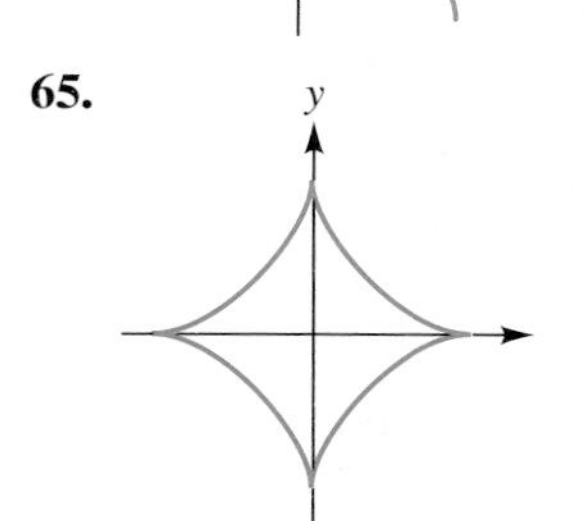

66.

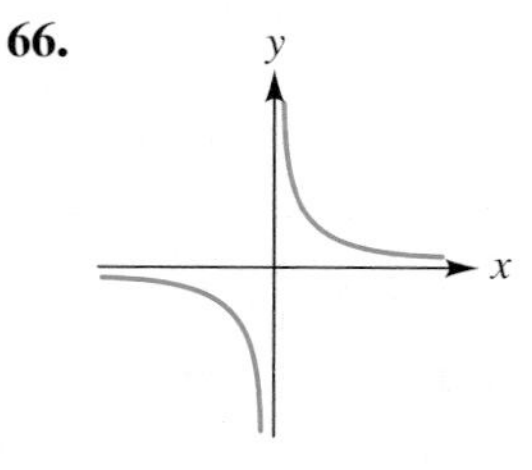

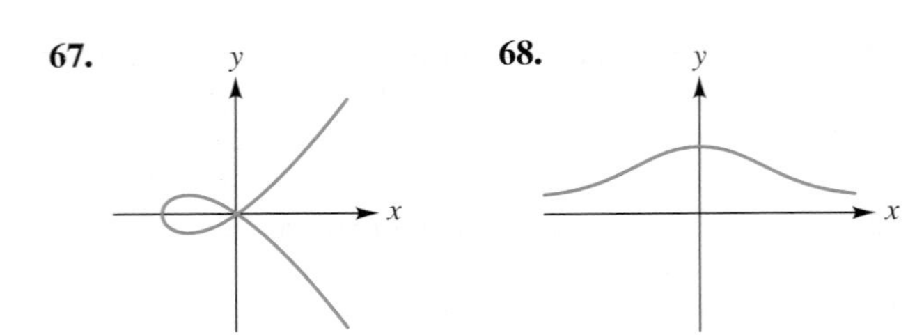

Graph the equations in Exercises 69–74.

69. $x = 9 - y^2$

70. $(x + 2)^2 + y^2 = 1$

71. $y = 1 - |x|$

72. $y = |x - 2| + 2$

73. $(4x - y + 4)(4x + y - 4) = 0$
Hint: Use the zero-product property.

74. $(y - x)(y - x + 2)(y - x - 2) = 0$

You know from the text that the graph of the equation $x^2 + y^2 = 9$ is a circle with center (0, 0) and radius 3. In Exercises 75–81, use a graphing utility to investigate what happens when the equation is changed slightly. (As in graphing circles, you'll first need to solve the given equations for y.)

75. (a) $x^2 + 2y^2 = 9$ (The resulting curve is an *ellipse*. We'll study this curve in a later chapter.)
(b) On the same set of axes, graph the ellipse $x^2 + 2y^2 = 9$ and the circle $x^2 + y^2 = 9$. At which points do the two curves appear to intersect?
(c) Carry out the calculations to verify that the points indicated in part (b) are indeed the exact intersection points.

76. $2x^2 + y^2 = 9$

77. (a) $x^2 - y^2 = 9$ (The resulting curve is a *hyperbola*. We'll study this curve in a later chapter.)
(b) $x^2 - y^2 = 0$
(c) On the same set of axes, graph the equations given in parts (a) and (b). Are there any intersection points?

78. $x^2 - y^3 = 9$

79. $x^3 - y^2 = 9$

80. (a) $x^3 + y^3 = 9$
(b) $x^5 + y^5 = 9$
(c) On the same set of axes, graph the equations given in parts (a) and (b). On the basis of the pattern you see, what do you think the graph of the equation $x^7 + y^7 = 9$ will look like? Draw a sketch (by hand) to show your prediction. Then use a graphing utility to see how accurate your prediction is.

81. (a) $x^4 + y^4 = 9$
(b) $x^6 + y^6 = 9$
(c) On the same set of axes, graph the equations given in parts (a) and (b). Based on the pattern you see, what do you think the graph of the equation $x^8 + y^8 = 9$ will look like? Draw a sketch (by hand) to show your prediction. Then use a graphing utility to see how accurate your prediction is.

82. Find a value for t such that the slope of the line passing through (2, 1) and (5, t) is 6.

83. The vertices of a right triangle are $A(0, 0)$, $B(0, 2b)$, and $C(2c, 0)$. Let M be the midpoint of the hypotenuse. Compute the three distances MA, MB, and MC. What do you observe?

84. The vertices of parallelogram $ABCD$ are $A(-4, -1)$, $B(2, 1)$, $C(3, 3)$, and $D(-3, 1)$.
(a) Compute the sum of the squares of the two diagonals.
(b) Compute the sum of the squares of the four sides. What do you observe?

In Exercises 85 and 86, two points are given. In each case compute **(a)** *the distance between the two points,* **(b)** *the slope of the line segment joining the two points, and* **(c)** *the midpoint of the line segment joining the two points.*

85. (2, 5) and (5, −6)

86. $(\sqrt{3}/2, 1/2)$ and $(-\sqrt{3}/2, -1/2)$

87. A line passes through the points (1, 2) and (4, 1). Find the area of the triangle bounded by this line and the coordinate axes.

88. Let $P_1(x_1, y_1)$ and $P_2(x_2, y_2)$ be two given points. Let Q be the point $(\frac{1}{3}x_1 + \frac{2}{3}x_2, \frac{1}{3}y_1 + \frac{2}{3}y_2)$.
(a) Show that the points P_1, Q, and P_2 are collinear. *Hint*: Compute the slope of $\overline{P_1Q}$ and the slope of $\overline{QP_2}$.
(b) Show that $P_1Q = \frac{2}{3}P_1P_2$. (In other words, Q is on the line segment $\overline{P_1P_2}$ and two-thirds of the way from P_1 to P_2.)

89. (a) Let the vertices of $\triangle ABC$ be $A(-5, 3)$, $B(7, 7)$, and $C(3, 1)$. Find the point on each median that is two-thirds of the way from the vertex to the midpoint of the opposite side. (Recall that a median of a triangle is a line segment drawn from a vertex to the midpoint of the opposite side.) What do you observe? *Hint*: Use the result in Exercise 88.
(b) Follow part (a) but take the vertices to be $A(0, 0)$, $B(2a, 0)$, and $C(2b, 2c)$. What do you observe? What does this prove?

90. In the following figure, points P and Q trisect the hypotenuse in $\triangle ABC$. Prove that the square of the hypotenuse is equal to 9/5 the sum of the squares of the distances from the trisection points to the vertex A of the right angle.

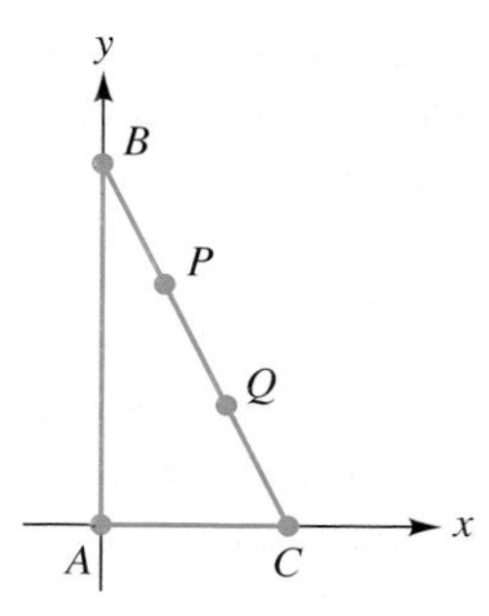

Hint: Let the coordinates of B and C be $(0, 3b)$ and $(3c, 0)$, respectively. Then the coordinates of P and Q are $(c, 2b)$ and $(2c, b)$, respectively.

91. Figure A shows a triangle with sides of lengths s, t, and u and a median of length m. Prove that

$$m^2 = \frac{1}{2}(s^2 + t^2) - \frac{1}{4}u^2$$

Hint: Set up a coordinate system as indicated in Figure B. Then each of the quantities m^2, s^2, t^2, and u^2 can be computed in terms of a, b, and c.

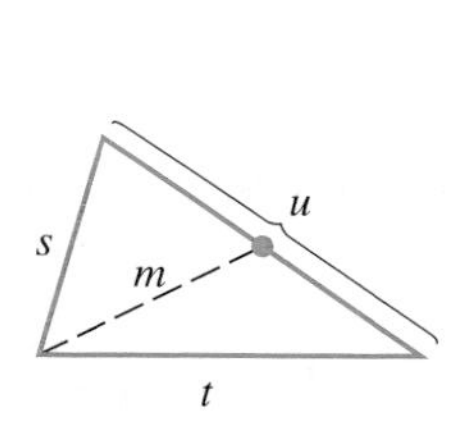

Figure A　　Figure B

In Exercises 92 and 93, the endpoints of a line segment $\overline{AB}$ are given. Sketch the reflection of $\overline{AB}$ about **(a)** *the origin;* **(b)** *the x-axis; and* **(c)** *the y-axis.*

92. $A(-2, -3)$ and $B(0, 2)$

93. $A(3, 1)$ and $B(3, -2)$

94. In this exercise you'll derive a useful formula for the (perpendicular) distance d from the point (x_0, y_0) to the line $y = mx + b$. The formula is

$$d = \frac{|y_0 - mx_0 - b|}{\sqrt{1 + m^2}}$$

(a) Refer to the figure. Use similar triangles to show that

$$\frac{d}{AB} = \frac{1}{\sqrt{1 + m^2}}$$

Therefore, $d = AB/\sqrt{1 + m^2}$.

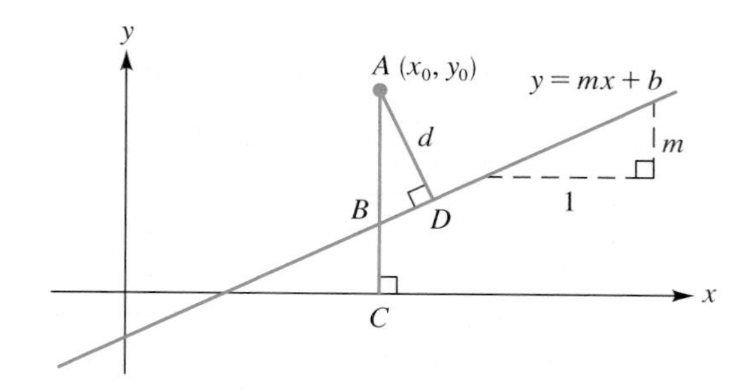

(b) Check that $AB = AC - BC = y_0 - mx_0 - b$.

(c) Conclude from parts (a) and (b) that

$$d = \frac{y_0 - mx_0 - b}{\sqrt{1 + m^2}}$$

For the general case (in which the point and line may not be situated as in our figure), we need to use the absolute value of the quantity in the numerator to ensure that AB and d are nonnegative.

95. Use the formula given in Exercise 94 to find the distance from the point $(1, 2)$ to the line $y = \frac{1}{2}x - 5$.

96. Use the formula given in Exercise 94 to demonstrate that the distance from the point (x_0, y_0) to the line $Ax + By + C = 0$ is

$$d = \frac{|Ax_0 + By_0 + C|}{\sqrt{A^2 + B^2}}$$

97. Use the formula given in Exercise 96 to find the distance from the point $(-1, -3)$ to the line $2x + 3y - 6 = 0$.

98. Find the equation of the circle that has center $(2, 3)$ and is tangent to the line $x + y - 1 = 0$.
Hint: Use the formula given in Exercise 96 to find the radius.

99. In the figure, the circle is tangent to the x-axis, to the y-axis, and to the line $3x + 4y = 12$.

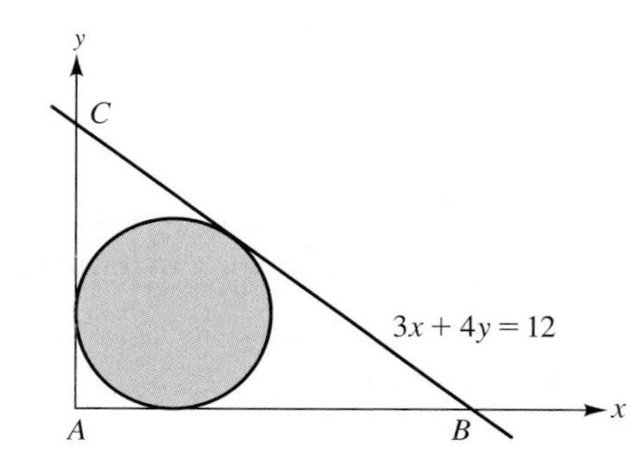

(a) Find the equation of the circle.
Suggestion: First decide what the relationships must be between h, k, and r in the equation

$(x - h)^2 + (y - k)^2 = r^2$. Then find a way to apply the formula given in Exercise 94.

(b) Let S, T, and U denote the points where the circle touches the x-axis, the line $3x + 4y = 12$, and the y-axis, respectively. Find the equation of the line through A and T; through B and U; through S and C.

(c) Where do the line segments $\overline{AT}$ and $\overline{CS}$ intersect? Where do $\overline{AT}$ and $\overline{BU}$ intersect? What do you observe? *Remark*: The point determined here is called the *Gergonne point* of triangle ABC, so named in honor of its discoverer, French mathematician Joseph Diaz Gergonne (1771–1859).

100. This exercise outlines a proof of the Pythagorean theorem that was discovered by James A. Garfield, the twentieth President of the United States. Garfield published the proof in 1876, when he was the Republican leader in the House of Representatives. We start with a right triangle with legs of length a and b and hypotenuse of length c. We want to prove that $a^2 + b^2 = c^2$.

(a) Take two copies of the given triangle and arrange them as shown in Figure A. Explain why the angle marked θ is a right angle. (θ is the Greek letter "theta.")

(b) Draw the line segment indicated in Figure B. Notice that the outer quadrilateral in Figure B is a trapezoid. (Two sides are parallel.) The area of this trapezoid can be computed in two distinct ways: using the formula for the area of a trapezoid (given on the inside front cover of this book) or adding the areas of the three right triangles in Figure B. Compute the area of the trapezoid in each of these two ways. When you equate the two answers and simplify, you should obtain $a^2 + b^2 = c^2$.

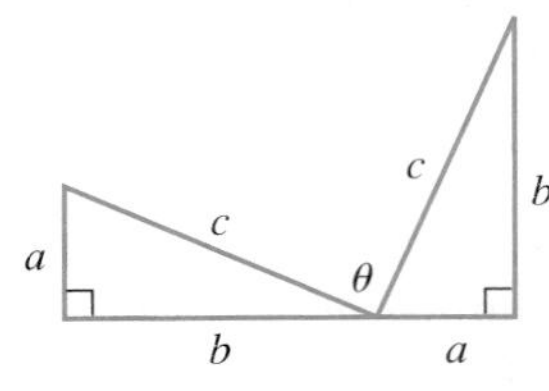

Figure A

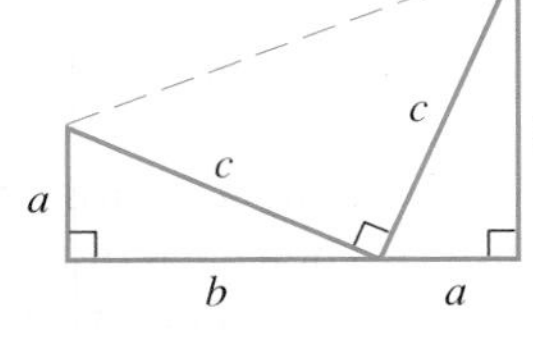

Figure B

Chapter 1 Test

1. Simplify $|x - 6| + |x - 7|$, where $6 < x < 7$.

2. Specify the intervals that are described by the inequalities.

(a) $|x - 4| < \frac{1}{10}$ **(b)** $x \geq 2$

In Problems 3–5, find all the real solutions of each equation.

3. $\dfrac{2}{x + 4} - \dfrac{1}{x - 4} = \dfrac{-7}{x^2 - 16}$

4. $\dfrac{1}{1 - x} + \dfrac{4}{2 - x} = \dfrac{11}{6}$

5. **(a)** $x^2 + 4x = 5$ **(b)** $x^2 + 4x = 1$

6. Solve the following equation for x in terms of the other letters:

$$\frac{ax + b}{cx + d} = e \qquad \text{where } a \neq ce$$

7. Refer to the following graph. As indicated in the graph, the number N of U.S. radio stations on the air increased in (approximately) a straight-line fashion from 1990 to 1994. Source of data: *The Universal Almanac 1997,* John W. Wright (ed.) (Kansas City: Andrews and McMeel, 1997)

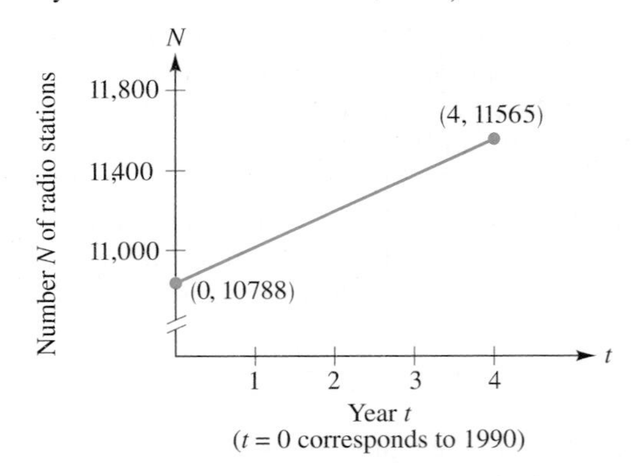

(a) As t increases from 0 to 4, we have $\Delta t = 4$ years. What is the corresponding value of ΔN?

(b) Find the slope $\Delta N/\Delta t$ for the line. Round the answer to the nearest integer. (Include units with your answer.)

(c) Use your result in part (b) to answer the following question. If the number of radio stations had continued to grow in a straight-line fashion, how many stations would there have been in 1995?

8. The following data are from a classic experiment in biology that measured the average height of a group of sunflower plants at seven-day intervals.

Growth in (Mean) Height of Sunflower Plants

Day t	Mean height h (cm)
7	17.93
21	67.76

Source: H. S. Reed and R. H. Holland in *Proceedings of the National Academy of Sciences,* vol. 5 (1919), p. 140.

(a) Plot the two data points in the following coordinate system.

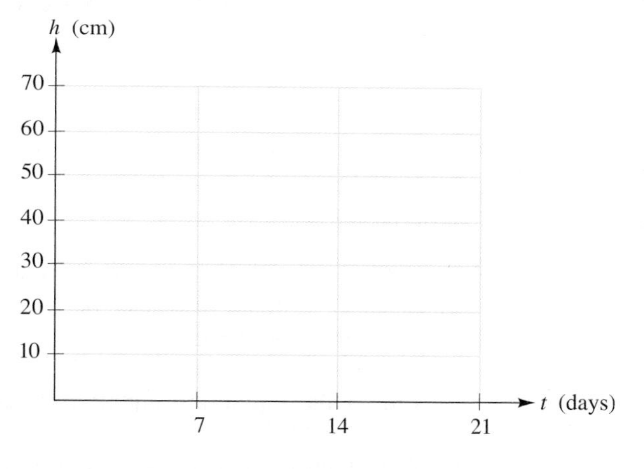

(b) Use the midpoint formula and the data for $t = 7$ days and $t = 21$ days to estimate what the average height of the sunflowers might have been at $t = 14$ days. (Round the answer to one decimal place.)

(c) Compute the percentage error in the estimation in part (b), given that the actual average height at $t = 14$ days was 36.36 cm. Round the answer to the nearest whole percent. *Check*: You should obtain a substantial percentage error. In Chapter 5 we'll study some types of equations that do a much better job than the midpoint formula in modeling biological growth.)

In Problems 9 and 10, find an equation for the line satisfying the given conditions. Write the answer in the form $y = mx + b$.

9. Passes through $(1, -2)$ and the y-intercept is 6

10. Passes through $(2, -1)$ and is perpendicular to the line $5x + 6y = 30$

11. Which point is farther from the origin, $(3, 9)$ or $(5, 8)$?

12. Test each equation for symmetry about the x-axis, the y-axis, and the origin:
(a) $y = x^3 + 5x$
(b) $y = 3^x + 3^{-x}$
(c) $y^2 = 5x^2 + x$

In Problems 13 and 14, graph the equations and specify all x- and y-intercepts.

13. $3x - 5y = 15$

14. $(x - 1)^2 + (y + 2)^2 = 9$

Ⓖ **15. (a)** Use a graphing utility to graph the equation $y = x^3 - 2x^2 - 9x$ in an appropriate viewing rectangle.
Ⓖ **(b)** Use a graphing utility to estimate to one decimal place the x-intercepts of the graph.
(c) Use algebra to find exact values for the x-intercepts.

16. Does the point $(-1/2, 5)$ lie on the graph of the equation $y = 4x^2 - 8x$?

17. The endpoints of line segment $\overline{AB}$ are $A(3, -1)$ and $B(-1, 2)$. Sketch the reflection of $\overline{AB}$ about **(a)** the x-axis; **(b)** the y-axis; and **(c)** the origin.

18. A line passes through the points $(6, 2)$ and $(1, 4)$. Find the area of the triangle bounded by this line and the x- and y-axes.

2 CHAPTER Equations and Inequalities

We continue the work begun in Section 1.3 on solving equations, and then we adapt those skills in solving inequalities. Some of the applications that you'll see in this chapter are:

- Tracking population growth in the United States (Example 1 in Section 2.1)
- Predicting world records for the men's and women's 10,000-meter race (Exercises 21 and 22 in Section 2.1)
- Predicting sulfur dioxide emissions for the United States and Asia (Exercises 42 and 43 in Section 2.3) (Sulfur dioxide emissions are the main contributor to acid rain.)

2.1 QUADRATIC EQUATIONS: THEORY AND EXAMPLES

A wealth of quadratic equations applied to genetic problems can be found in . . . [the textbook by Albert Jacquard, *The Genetic Structure of Populations*].—Edward Batschelet, *Introduction to Mathematics for Life Scientists,* 3rd ed. (New York: Springer-Verlag, 1979)

By 1900 B.C. the Babylonians had a well-established algebra. They could solve quadratic equations (positive coefficients and positive solutions only) and some types of higher degree equations as well.—Stuart Hollingdale in *Makers of Mathematics* (London: Penguin Books, 1989)

['Abd-al-Hamid ibn-Turk, a ninth-century Persian mathematician] *. . . gives geometric figures to prove if the discriminant is negative, a quadratic equation has no* [real] *solution.*—Carl B. Boyer in *A History of Mathematics,* 2nd ed. (Uta C. Merzbach, revision editor) (New York: John Wiley & Sons, 1991)

Recall (from Section 1.3, or from a previous course) that a **quadratic equation** is an equation that can be written in the form $ax^2 + bx + c = 0$, where a, b, and c are real numbers and a is not zero. In Section 1.3 we solved quadratic equations in two ways: by factoring and by the quadratic formula. The factoring method is the simpler of the two, but it can't be used in every instance. (Example: Try solving $x^2 - 2x - 4 = 0$ by factoring.) The quadratic formula, on the other hand, can be used to solve *any* quadratic equation. In this section we take a more careful look at the quadratic formula: We derive the formula and we use it, not only to solve equations, but also to *analyze* them. The technique that we'll use to derive the quadratic formula is completing the square (introduced in Section 1.7).

We'll review the technique of completing the square by solving the equation $x^2 - 2x - 4 = 0$. First, we rewrite the equation in the form

$$x^2 - 2x = 4 \tag{1}$$

with the x-terms isolated on the left-hand side of the equation. To complete the square, we follow these two steps:

Step 1 Take half of the coefficient of x and square it.
Step 2 Add the number obtained in Step 1 to both sides of the equation.

For equation (1), the coefficient of x is -2. Taking half of -2 and then squaring it gives us $(-1)^2$, or 1. Now, as directed in Step 2, we add 1 to both sides of equation (1). This yields

$$\begin{aligned} x^2 - 2x + 1 &= 4 + 1 \\ (x - 1)^2 &= 5 \\ x - 1 &= \pm\sqrt{5} \\ x &= 1 \pm \sqrt{5} \end{aligned}$$

We have now obtained the two solutions, $1 + \sqrt{5}$ and $1 - \sqrt{5}$. Let's check that these numbers are indeed solutions of $x^2 - 2x - 4 = 0$. (We'll show the work for $x = 1 + \sqrt{5}$; you should carry out the calculations for $1 - \sqrt{5}$ on your own.) We have

$$\begin{aligned} (1 + \sqrt{5})^2 - 2(1 + \sqrt{5}) - 4 &= (1 + 2\sqrt{5} + 5) - 2 - 2\sqrt{5} - 4 \\ &= 6 - 2 - 4 \\ &= 0 \qquad \text{as required} \end{aligned}$$

In the box that follows, we summarize the process of completing the square and we indicate how it got its name.

Algebraic Procedure for Completing the Square in the Expression $x^2 + bx$

Add the square of half of the x-coefficient:

$$(x^2 + bx) + \left(\frac{b}{2}\right)^2 = \left(x + \frac{b}{2}\right)^2$$

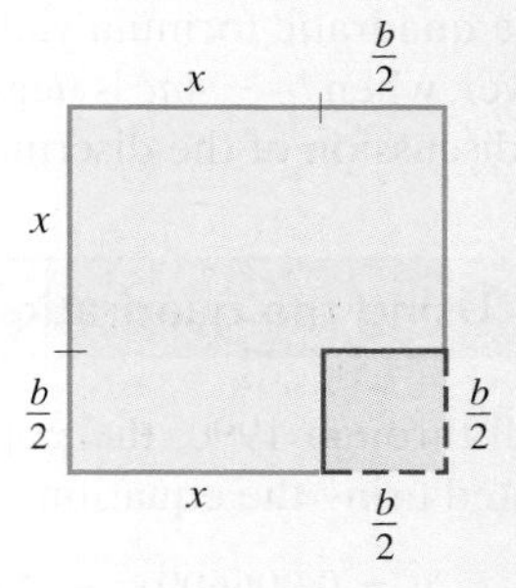

Geometric Interpretation of Completing the Square for $x^2 + bx$

The blue region in the figure represents the quantity $x^2 + bx$, since the area is

$$x^2 + \left(\frac{b}{2}\right)x + \left(\frac{b}{2}\right)x$$

By adding the red square to the blue region, we fill out or "complete" the larger square. The area of the red region that completes the square is

$$\frac{b}{2} \cdot \frac{b}{2} = \frac{b^2}{4}$$

The technique of completing the square can be used to derive the quadratic formula. We start with the general quadratic equation $ax^2 + bx + c = 0$ $(a \neq 0)$ and divide both sides by a (so that the coefficient of x^2 will be 1, as in the box above). This yields

$$x^2 + \frac{b}{a}x + \frac{c}{a} = 0$$

Subtracting c/a from both sides yields

$$x^2 + \frac{b}{a}x = -\frac{c}{a}$$

Now, to complete the square, we add $[\frac{1}{2}(b/a)]^2$, or $b^2/4a^2$, to both sides. That gives us

$$x^2 + \frac{b}{a}x + \frac{b^2}{4a^2} = \frac{b^2}{4a^2} - \frac{c}{a}$$

$$\left(x + \frac{b}{2a}\right)^2 = \frac{b^2 - 4ac}{4a^2}$$

$$x + \frac{b}{2a} = \pm\sqrt{\frac{b^2 - 4ac}{4a^2}} = \pm\frac{\sqrt{b^2 - 4ac}}{2|a|}$$

$$= \pm\frac{\sqrt{b^2 - 4ac}}{2a}$$

This last equality follows from the fact that for any real number $a \neq 0$, the expressions $\pm 2|a|$ and $\pm 2a$ both represent the same two numbers. We now conclude that the solutions are

$$x = -\frac{b}{2a} + \frac{\sqrt{b^2 - 4ac}}{2a} \quad \text{and} \quad x = -\frac{b}{2a} - \frac{\sqrt{b^2 - 4ac}}{2a}$$

For reference, we summarize our work in the box that follows.

The Quadratic Formula

The solutions of the equation $ax^2 + bx + c = 0$ $(a \neq 0)$ are given by

$$x = \frac{-b \pm \sqrt{b^2 - 4ac}}{2a}$$

COMMENT The quadratic formula yields the solutions of a quadratic equation in *all* cases. However, when $b^2 - 4ac$ is negative, the solutions are *not* real numbers. See page 87 for a discussion of the discriminant.

EXAMPLE 1 Using the quadratic formula

For the years 1850 through 1990, the population of the United States can be closely approximated using the equation

$$y = 0.006609x^2 - 23.771x + 21{,}382 \tag{2}$$

where y is the population in millions, and x is the year.

(a) To get a feeling for working with equation (2), use the equation to estimate the U.S. population in 1900. (Round the answer to one decimal place.)

(b) Use equation (2) and the quadratic formula to estimate the year in which the U.S. population reached 200 million.

SOLUTION

(a) Substituting $x = 1900$ in equation (2) yields

$$y = 0.006609(1900)^2 - 23.771(1900) + 21{,}382$$
$$= 75.6 \text{ million} \quad \text{using a calculator and rounding to one decimal place}$$

Thus our estimate for the U.S. population in 1900 is 76.6 million. *Remark*: This estimate is very good, for according to the U.S. Bureau of the Census, the population in 1900 was approximately 76.2 million.

(b) Using equation (2), we want to find the year x in which the population y reached 200 million. Substituting $y = 200$ in equation (2) yields

$$0.006609x^2 - 23.771x + 21{,}382 = 200$$

This is a quadratic equation, and to put it in the form $ax^2 + bx + c = 0$ (so that the quadratic formula can be used), we subtract 200 from both sides to obtain

$$0.006609x^2 - 23.771x + 21{,}182 = 0$$

From this last equation we see that

$$a = 0.006609 \qquad b = -23.771 \qquad c = 21{,}182$$

and, consequently,

$$\begin{aligned} x &= \frac{-b \pm \sqrt{b^2 - 4ac}}{2a} \\ &= \frac{-(-23.771) \pm \sqrt{(-23.771)^2 - 4(0.006609)(21{,}182)}}{2(0.006609)} \end{aligned}$$

Using a calculator now and rounding to the nearest integer, we obtain the two values $x \approx 1969$ and $x \approx 1628$. We choose the value in the required range between 1850 and 1990. Thus our estimate is that 1969 was the year in which the U.S. population reached 200 million. *Remark*: This is a reasonable estimate, for according to Census Bureau data, the 1960 population was approximately 179 million, whereas the 1970 population was approximately 203 million.

COMMENT The range of validity of the approximation formula (2) covers the years from 1850 through 1990. The year 1628 is so far outside of this range that we do not expect to get a reasonable approximation from equation (2), and we don't.

EXAMPLE 2 Using the quadratic formula

Consider the quadratic equation $x^2 = 5 - 3x$.

(a) Use the quadratic formula to find the roots of the equation.

(b) Compute the product of the roots.

(c) Compute the sum of the roots.

SOLUTION

(a) Just as we did in the previous example, we first write the equation in the form $ax^2 + bx + c = 0$:

$$x^2 + 3x - 5 = 0$$

From this last equation we see that $a = 1$, $b = 3$, and $c = -5$. The quadratic formula then yields

$$x = \frac{-b \pm \sqrt{b^2 - 4ac}}{2a} = \frac{-3 \pm \sqrt{3^2 - 4(1)(-5)}}{2(1)}$$

$$= \frac{-3 \pm \sqrt{9 + 20}}{2} = \frac{-3 \pm \sqrt{29}}{2}$$

(b) Computing the product of the roots, we have

$$\left(\frac{-3 + \sqrt{29}}{2}\right)\left(\frac{-3 - \sqrt{29}}{2}\right) = \frac{9 - 29}{4} = \frac{-20}{4} = -5$$

So the product of the two roots is -5.

(c) For the sum of the roots we have

$$\frac{-3 + \sqrt{29}}{2} + \frac{-3 - \sqrt{29}}{2} = \frac{-3 + \sqrt{29} + (-3) - \sqrt{29}}{2}$$

$$= \frac{-6}{2} = -3$$

Therefore the sum of the roots is -3.

If you look over the results in Example 2, you'll see that there appears to be something of a coincidence. We found that the product of the roots of the equation $x^2 + 3x - 5 = 0$ is -5, which happens to be the "c" term in the equation. Furthermore, the sum of the roots is -3, which happens to be the negative of the "b" term in the equation. In fact, however, these are not coincidences. Rather, they are examples of the following general result.

THEOREM The Product and the Sum of the Roots of $x^2 + bx + c = 0$

Let r_1 and r_2 be the roots of the quadratic equation $x^2 + bx + c = 0$. Then

$$r_1 r_2 = c \quad \text{and} \quad r_1 + r_2 = -b$$

In words: The product of the roots of $x^2 + bx + c = 0$ is the constant term in the equation, and the sum of the roots is the negative of the coefficient of the x-term.

The proof of this theorem follows exactly the same steps that we used in Example 2. For the equation $x^2 + bx + c = 0$, we have $a = 1$, and therefore the quadratic formula gives us

$$x = \frac{-b \pm \sqrt{b^2 - 4(1)c}}{2(1)} = \frac{-b \pm \sqrt{b^2 - 4c}}{2}$$

So the two roots are $r_1 = \dfrac{-b + \sqrt{b^2 - 4c}}{2}$ and $r_2 = \dfrac{-b - \sqrt{b^2 - 4c}}{2}$. Then, for the product $r_1 r_2$ we have

$$r_1 r_2 = \frac{-b + \sqrt{b^2 - 4c}}{2} \cdot \frac{-b - \sqrt{b^2 - 4c}}{2}$$

$$= \frac{b^2 - (b^2 - 4c)}{4} = \frac{4c}{4} = c \qquad \text{as required}$$

For the sum of the roots we have

$$r_1 + r_2 = \frac{-b + \sqrt{b^2 - 4c}}{2} + \frac{-b - \sqrt{b^2 - 4c}}{2}$$

$$= \frac{-b + \sqrt{b^2 - 4c} + (-b) - \sqrt{b^2 - 4c}}{2} = \frac{-2b}{2} = -b$$

as we wished to show. The next two examples show how these results are applied.

EXAMPLE 3 **Finding the product and sum of the roots of a quadratic equation**

Find the product and the sum of the roots of the quadratic equation $2x^2 + 6x - 7 = 0$.

SOLUTION

To apply the theorem on the product and sum of roots, the coefficient of x^2 in the equation must be 1. To arrange this, we divide both sides of the given equation by 2. This yields the equivalent equation

$$x^2 + 3x - \frac{7}{2} = 0$$

In this equation we have $b = 3$ and $c = -7/2$. Therefore

$$r_1 r_2 = c = -\frac{7}{2}$$

and

$$r_1 + r_2 = -b = -3$$

That is, the product of the roots is $-7/3$ and the sum of the roots is -3.

EXAMPLE 4 **Finding a quadratic equation with given roots**

Find a quadratic equation with roots

$$r_1 = \frac{1 + \sqrt{5}}{4} \quad \text{and} \quad r_2 = \frac{1 - \sqrt{5}}{4}$$

Write the equation in a form involving integer coefficients only.

SOLUTION

If r_1 and r_2 are the roots of the equation $x^2 + bx + c = 0$, we have

$$c = r_1 r_2 = \frac{1 + \sqrt{5}}{4} \cdot \frac{1 - \sqrt{5}}{4} = \frac{1 - 5}{16} = -\frac{1}{4}$$

and

$$b = -(r_1 + r_2) = -\left(\frac{1 + \sqrt{5}}{4} + \frac{1 - \sqrt{5}}{4}\right) = -\frac{2}{4} = -\frac{1}{2}$$

With these values for b and c the quadratic equation is $x^2 - \frac{1}{2}x - \frac{1}{4} = 0$. To obtain an equation with integer coefficients (but still the same roots), we multiply both sides of this last equation by 4 to obtain

$$4x^2 - 2x - 1 = 0$$

This is the equation with the required roots and with integer coefficients.

Suggestion: For practice in using the quadratic formula, solve this last equation to verify that the roots are indeed $(1 \pm \sqrt{5})/4$.

In Examples 1 and 2 we found that each quadratic equation had two real roots (although in Example 1 we were interested only in one root). As the next example indicates, however, it is also possible for quadratic equations to have only one real root, or even no real roots.

EXAMPLE 5 Using the quadratic formula

Use the quadratic formula to solve each equation:
(a) $4x^2 - 12x + 9 = 0$; **(b)** $x^2 + x + 2 = 0$.

SOLUTION

(a) Here $a = 4$, $b = -12$, and $c = 9$, so

$$x = \frac{12 \pm \sqrt{(-12)^2 - 4(4)(9)}}{2(4)} = \frac{12 \pm \sqrt{144 - 144}}{8}$$
$$= \frac{12 \pm 0}{8} = \frac{12}{8} = \frac{3}{2}$$

In this case our only solution is 3/2. We refer to 3/2 as a **double root.** *Note*: We've used the quadratic formula here only for an illustration; in this case we can solve the original equation more efficiently by factoring.

(b) Since $a = 1$, $b = 1$, and $c = 2$, we have

$$x = \frac{-1 \pm \sqrt{(1)^2 - 4(1)(2)}}{2(1)} = \frac{-1 \pm \sqrt{-7}}{2}$$

From this we conclude that the given equation has no real-number solutions because the expression $\sqrt{-7}$ is undefined within the real-number system. [However, if we take a broader point of view and work within the *complex-number system,* then there are indeed two solutions, albeit not real-number solutions. As you'll see in Section 12.1, or as you perhaps already know from a previous course, these two complex-number solutions can be written $(-1 \pm i\sqrt{7})/2$.]

The results in Example 5 can be interpreted in terms of graphs. For as we know from Section 1.5, the x-intercepts of the graph of $y = ax^2 + bx + c$ are found by solving the equation $ax^2 + bx + c = 0$. As indicated in Figure 1(a), the graph of

GRAPHICAL PERSPECTIVE

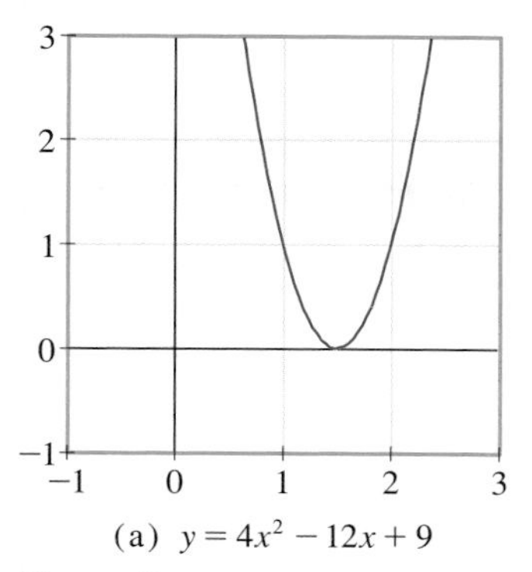

(a) $y = 4x^2 - 12x + 9$

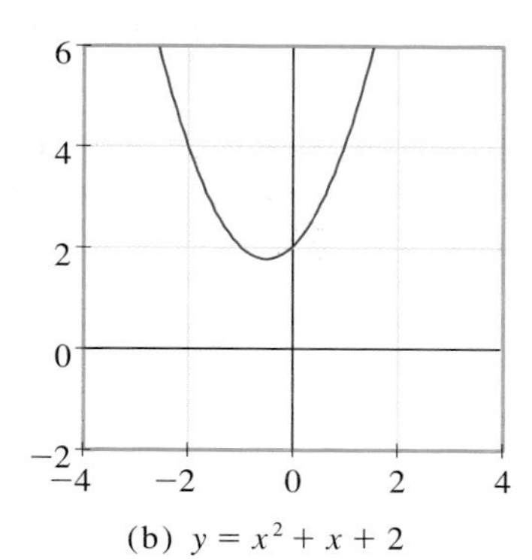

(b) $y = x^2 + x + 2$

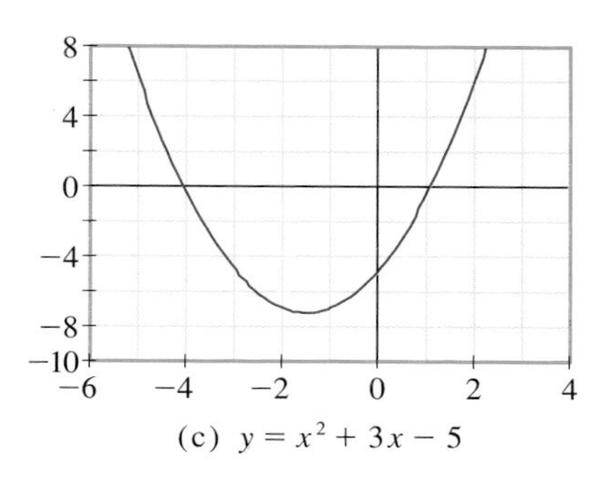

(c) $y = x^2 + 3x - 5$

Figure 1

In Figure 1(a) the graph of $y = 4x^2 - 12x + 9$ has only one x-intercept; the equation $4x^2 - 12x + 9 = 0$ has only one root. In Figure 1(b) the graph of $y = x^2 + x + 2$ has no x-intercepts; the equation $x^2 + x + 2 = 0$ has no real roots. In Figure 1(c) the graph of $y = x^2 + 3x - 5$ has two x-intercepts; the equation $x^2 + 3x - 5 = 0$ has two real roots.

$y = 4x^2 - 12x + 9$ has only one x-intercept; this intercept is the root of the equation $4x^2 - 12x + 9 = 0$ that we solved in Example 5(a). Figure 1(b) shows that the graph of $y = x^2 + x + 2$ has no x-intercepts; this reflects the fact that the equation $x^2 + x + 2 = 0$ has no real solutions. For completeness, we've also shown a graph [Figure 1(c)] that has two x-intercepts. We computed these two x-intercepts in Example 2(a) when we solved the equation $x^2 + 3x - 5 = 0$.

The quantity $b^2 - 4ac$ that appears under the radical sign in the quadratic formula is called the *discriminant.* (The term "discriminant" was coined by the nineteenth-century British mathematician James Joseph Sylvester.) If you look back at Example 2(a), you'll see that the discriminant is positive and that this leads to two real solutions. In Example 5(a), the discriminant is zero, which leads to only one real solution. Finally, in Example 5(b), the discriminant is negative and, consequently, there are no real solutions. These observations are generalized in the box that follows.

The Discriminant $b^2 - 4ac$

Consider the quadratic equation $ax^2 + bx + c = 0$, where a, b, and c are real numbers and $a \neq 0$. The expression $b^2 - 4ac$ is called the **discriminant.**

1. If $b^2 - 4ac > 0$, then the equation has two distinct real roots.
2. If $b^2 - 4ac = 0$, then the equation has exactly one real root, referred to as a **double root** or a root of **multiplicity two.**
3. If $b^2 - 4ac < 0$, then the equation has no real root. [There are two (nonreal) complex-number roots. This is discussed in Chapter 12.]

EXAMPLE 6 Using the discriminant

(a) Compute the discriminant to determine how many real solutions there are for the equation $x^2 + x - 1 = 0$.

(b) Find a value for k such that the quadratic equation $x^2 + \sqrt{2}x + k = 0$ has exactly one real solution.

SOLUTION

(a) Here $a = 1$, $b = 1$, and $c = -1$. Therefore,

$$b^2 - 4ac = 1^2 - 4(1)(-1) = 5$$

Since the discriminant is positive, the equation has two distinct real solutions.

(b) For the equation to have exactly one real solution, the discriminant must be zero; that is,

$$\begin{aligned} b^2 - 4ac &= 0 \\ (\sqrt{2})^2 - 4(1)(k) &= 0 \\ 2 - 4k &= 0 \\ k &= \frac{1}{2} \end{aligned}$$

The required value for k is 1/2.

EXERCISE SET 2.1

A

In Exercises 1–18, solve the quadratic equations. If an equation has no real roots, state this. In cases where the solutions involve radicals, give both the radical form of the answer and a calculator approximation rounded to two decimal places.

1. $x^2 + 8x - 2 = 0$
2. $x^2 - 6x - 2 = 0$
3. $x^2 + 4x + 1 = 0$
4. $x^2 + 12x + 18 = 0$
5. $2y^2 - 5y - 2 = 0$
6. $3y^2 - 3y - 4 = 0$
7. $4y^2 + 8y + 5 = 0$
8. $y^2 + y + 1 = 0$
9. $4s^2 - 20s + 25 = 0$
10. $16s^2 + 8s + 1 = 0$
11. $x^2 = 8x - 6$
12. $x^2 = 8x + 9$
13. $-3x^2 + x = -3$
14. $(x - 5)(x + 3) = 1$
15. $-y^2 - 8y = 1$
16. $5y(y - 2) = 2$
17. $t^2 = -3t - 4$
18. $t^2 = -3t + 4$

19. (a) Use the formula given in Example 1 to estimate the year that the U.S. population reached 100 million.
(b) Use the following U.S. Census Bureau data to say whether your estimate in part (a) is reasonable. In 1910 the population was approximately 92.2 million; in 1920 it was approximately 106.0 million.

20. (a) Use the formula given in Example 1 to estimate the year that the U.S. population reached 225 million.
(b) According to the U.S. Census Bureau, the U.S. population in 1980 was 226.5 million. Is your estimate in part (a) reasonable?

21. The chart that follows shows the world records for the men's 10,000 meter run in the years 1972 and 1998.

Year	Time (minutes and seconds)	Runner
1972	27:38.4	Lasse Viren (Finland)
1998	26:22.75	Haile Gebrselassie (Ethiopia)

For the years between 1972 and 1998, the world record can be approximated by the equation

$$y = -0.09781t^2 + 385.8336t - 378{,}850.4046 \quad (1972 \le t \le 1998)$$

where y is the world-record time (in seconds) in the year t.
(a) Use the given equation, the quadratic formula, and your calculator to estimate the year in which the record might have been 27 minutes (= 1620 seconds). (You'll get two solutions for the quadratic; be sure to pick the appropriate one.) Then say by how many years your prediction is off, given the following information: On July 5, 1993, Richard Chelimo of Kenya ran a record time of 27:07.91; five days after that, Yobes Ondieki, also of Kenya, ran 26:58.38.
(b) Estimate the year in which the record might have been 26:30 (= 1590 seconds). Then say by how many years your prediction is off, given the following information: On July 4, 1997, Haile Gebrselassie ran a record time of 27:07.91; eighteen days later, Paul Tergat of Kenya ran 26:27.85.

22. The chart below shows the world records for the women's 10,000 meter run in the years 1970 and 1993.

Year	Time (minutes and seconds)	Runner
1970	35:30.5	Paola Pigni (Italy)
1993	29:31.78	Wang Junxia (China)

For the years between 1970 and 1993, the world record can be approximated by the equation

$$y = 0.37553t^2 - 1503.7154t + 1{,}507{,}042.699 \quad (1970 \le t \le 1993)$$

where y is the world-record time (in seconds) in the year t. Use the given equation, the quadratic formula, and your calculator to estimate the year in which the record might have been 32 minutes (= 1920 seconds).

(As in Exercise 21, you'll get two solutions for the quadratic; pick the appropriate one.) Then say by how many years your prediction is off, given the following information: On September 19, 1981, Yekena Sipatova of the former Soviet Union ran a record time of 32:17.2; in the following year on July 16, Mary Decker of the United States ran 31:35.3.

Ⓖ *In each of Exercises 23–28, you are given an equation of the form $y = ax^2 + bx + c$.* **(a)** *Use a graphing utility to graph the equation and to estimate the x-intercepts. (Use a zoom-in process to obtain the estimates; keep zooming in until the first three decimal places of the estimate remain the same as you progress to the next step.)* **(b)** *Determine the exact values of the intercepts by using the quadratic formula. Then use a calculator to evaluate the expressions that you obtain. Round off the results to four decimal places. [Check to see that your results are consistent with the estimates in part (a).]*

23. $y = x^2 - 4x + 1$
24. $y = x^2 - 10x + 15$
25. $y = 0.5x^2 + 8x - 3$
26. $y = 2x^2 + 2x - 1$
27. $y = 2x^2 + 2\sqrt{26}x + 13$
28. $y = 3x^2 - 12\sqrt{3}x + 36$

In Exercises 29–32, find the sum and the product of the roots of each quadratic equation.

29. $x^2 + 8x - 20 = 0$
30. $x^2 - 3x + 12 = 0$
31. $4y^2 - 28y + 9 = 0$
32. $\frac{1}{2}y^2 = 4y - 5$

In Exercises 33–38, find a quadratic equation with the given roots r_1 and r_2. Write each answer in the form $ax^2 + bx + c = 0$, where a, b, and c are integers and $a > 0$.

33. $r_1 = 3$ and $r_2 = 11$
34. $r_1 = -4$ and $r_2 = -9$
35. $r_1 = 1 - \sqrt{2}$ and $r_2 = 1 + \sqrt{2}$
36. $r_1 = 2 - \sqrt{5}$ and $r_2 = 2 + \sqrt{5}$
37. $r_1 = \frac{1}{2}(2 + \sqrt{5})$ and $r_2 = \frac{1}{2}(2 - \sqrt{5})$
38. $r_1 = \frac{1}{3}(4 - \sqrt{5})$ and $r_2 = \frac{1}{3}(4 + \sqrt{5})$

In Exercises 39 and 40, solve the equations. Hint: Look before you leap.

39. $x^2 - (\sqrt{2} + \sqrt{5})x + \sqrt{10} = 0$
40. $x^2 + (\sqrt{2} - 1)x = \sqrt{2}$

41. A ball is thrown straight upward. Suppose that the height of the ball at time t is given by the formula $h = -16t^2 + 96t$, where h is in feet and t is in seconds, with $t = 0$ corresponding to the instant that the ball is first tossed.
(a) How long does it take before the ball lands?
(b) At what time is the height 80 ft? Why does this question have two answers?

42. During a flu epidemic in a small town, a public health official finds that the total number of people P who have caught the flu after t days is closely approximated by the formula $P = -t^2 + 26t + 106$, where $1 \le t \le 13$.
(a) How many have caught the flu after 10 days?
(b) After approximately how many days will 250 people have caught the flu?

In Exercises 43–50, use the discriminant to determine how many real roots each equation has.

43. $x^2 - 12x + 16 = 0$
44. $2x^2 - 6x + 5 = 0$
45. $4x^2 - 5x - \frac{1}{2} = 0$
46. $4x^2 - 28x + 49 = 0$
47. $x^2 + \sqrt{3}x + \frac{3}{4} = 0$
48. $\sqrt{2}x^2 + \sqrt{3}x + 1 = 0$
49. $y^2 - \sqrt{5}y = -1$
50. $\frac{m^2}{4} - \frac{4m}{3} + \frac{16}{9} = 0$

In each of Exercises 51–54, find the value(s) of k such that the equation has exactly one real root.

51. $x^2 + 12x + k = 0$
52. $3x^2 + (\sqrt{2k})x + 6 = 0$
53. $x^2 + kx + 5 = 0$
54. $kx^2 + kx + 1 = 0$

B

In Exercises 55–58, solve for the indicated letter.

55. $2\pi r^2 + 2\pi rh = 20\pi$; for r
Hint: Rewrite the equation as $(2\pi)r^2 + (2\pi h)r - 20\pi = 0$ and use the quadratic formula with $a = 2\pi$, $b = 2\pi h$, and $c = -20\pi$.

56. $2\pi y^2 + \pi yx = 12$; for y

57. $-16t^2 + v_0t = 0$; for t

58. $-\frac{1}{2}gt^2 + v_0t + h_0 = 0$; for t

Ⓖ **59. (a)** On the same set of axes, graph the equations $y = x^2 + 8x + 16$ and $y = x^2 - 8x + 16$.
(b) Use the graphs to estimate the roots of the two equations $x^2 + 8x + 16 = 0$ and $x^2 - 8x + 16 = 0$. How do the roots appear to be related?
(c) Solve the two equations in part (b) to determine the exact values of the roots. Do your results support the response you gave to the question at the end of part (b)?

Ⓖ **60. (a)** Figure 1(c) in the text shows a graph of the equation $y = x^2 + 3x - 5$. Use a graphing utility to reproduce the graph. [Use the same viewing rectangle that is used in Figure 1(c).]
(b) Add the graph of the equation $y = x^2 - 3x - 5$ to the picture that you obtained in part (a). (This new equation is the same as the one in part (a) except that the sign of the coefficient of x has been reversed.) Note that the x-intercepts of the two graphs appear to be negatives of one another.
(c) Use the quadratic formula to determine exact expressions for the roots of the two equations $x^2 + 3x - 5 = 0$ and $x^2 - 3x - 5 = 0$. You'll find that the roots of one equation are the opposites of the roots of the other equation. [In general, the graphs of the two equations $y = ax^2 + bx + c$ and $y = ax^2 - bx + c$ are symmetric about the y-axis. Thus, the x-intercepts (when they exist) will always be opposites of one another.]

61. (a) Use the quadratic formula to show that the roots of the equation $x^2 + 3x + 1 = 0$ are $\frac{1}{2}(-3 \pm \sqrt{5})$.
(b) Show that $\frac{1}{2}(-3 + \sqrt{5}) = 1/[\frac{1}{2}(-3 - \sqrt{5})]$.
Hint: Rationalize the denominator on the right-hand side of the equation.

(c) The result in part (b) shows that the roots of the equation $x^2 + 3x + 1 = 0$ are reciprocals. Can you find another, much simpler way to establish this fact?

62. If r_1 and r_2 are the roots of the quadratic equation $ax^2 + bx + c = 0$, show that $r_1 + r_2 = -b/a$ and $r_1r_2 = c/a$.

63. Show that the quadratic equation

$$ax^2 + bx - a = 0 \qquad (a \neq 0)$$

has two distinct real roots.

64. Show that the quadratic equation

$$(x - p)(x - q) = r^2 \qquad (p \neq q)$$

has two distinct real roots.

In Exercises 65 and 66, determine the value(s) of the constant k for which the equation has equal roots (that is, only one distinct root).

65. $x^2 = 2x(3k + 1) - 7(2k + 3)$

66. $x^2 + 2(k + 1)x + k^2 = 0$

67. Here is an outline for a slightly different derivation of the quadratic formula. The advantage of this method is that fractions are avoided until the very last step. Fill in the details.

(a) Beginning with $ax^2 + bx = -c$, multiply both sides by $4a$. Then add b^2 to both sides.

(b) Now factor the resulting left-hand side and take square roots.

68. In this section and in Section 1.3, we solved quadratic equations by factoring and by using the quadratic formula. This exercise shows how to solve a quadratic equation by the **method of substitution.** As an example, we use the equation

$$x^2 + x - 1 = 0 \qquad (1)$$

(a) In equation (1), make the substitution $x = y + k$. Show that the resulting equation can be written

$$y^2 + (2k + 1)y = 1 - k - k^2 \qquad (2)$$

(b) Find a value for k so that the coefficient of y in equation (2) is 0. Then, using this value of k, show that equation (2) becomes $y^2 = 5/4$.

(c) Solve the equation $y^2 = 5/4$. Then use the equation $x = y + k$ to obtain the solutions of equation (1).

69. Use the substitution method (explained in Exercise 68) to solve the quadratic equation $2x^2 - 3x + 1 = 0$.

70. Assume that a and b are the roots of the equation $x^2 + px + q = 0$.

(a) Find the value of $a^2b + ab^2$ in terms of p and q.
Hint: Factor the expression $a^2b + ab^2$.

(b) Find the value of $a^3b + ab^3$ in terms of p and q.
Hint: Factor. Then use the fact that $a^2 + b^2 = (a + b)^2 - 2ab$.

Ⓖ **71.** In this exercise we investigate the effect of the constant c upon the roots of the quadratic equation $x^2 - 6x + c = 0$. We do this by looking at the x-intercepts of the graphs of the corresponding equations $y = x^2 - 6x + c$.

(a) Set a viewing rectangle that extends from 0 to 5 in the x-direction and from -2 to 3 in the y-direction. Then (on the same set of axes) graph the equations $y = x^2 - 6x + c$ with c running from 8 to 10 at increments of 0.25. In other words, graph the equations $y = x^2 - 6x + 8$, $y = x^2 - 6x + 8.25$, $y = x^2 - 6x + 8.50$, and so on, up through $y = x^2 - 6x + 10$.

(b) Note from the graphs in part (a) that, initially, as c increases, the x-intercepts draw closer and closer together. For which value of c do the two x-intercepts seem to merge into one?

(c) Use algebra as follows to check your observation in part (b). Using that value of c for which there appears to be only one intercept, solve the quadratic equation $x^2 - 6x + c = 0$. How many roots do you obtain?

(d) Some of the graphs in part (a) have no x-intercepts. What are the corresponding values of c in these cases? Pick any one of these values of c and use the quadratic formula to solve the equation $x^2 - 6x + c = 0$. What happens?

C

72. Find nonzero real numbers A and B so that the roots of the equation $x^2 + Ax + B = 0$ are A and B.

PROJECT — PUT THE QUADRATIC EQUATION IN ITS PLACE!

This is easier as a group project (divide and conquer) but, of course, is open to individuals too.

Despite the critical influence of science and mathematics in today's culture, many high school history books tell very little about the history of these fields. In mathematics, for example, it's easy to develop the false impression that mathematics is strictly a "Western" achievement, developed rather exclusively in

Greece (Pythagoras and Euclid), France (René Descartes), and England (Isaac Newton). A mere glance, however, at the following list of a few mathematicians (from the period 0 to A.D. 1400) will tell you otherwise. (All right, one Greek name does appear in the list!) The common element linking the names is that, among other mathematical achievements, each person solved quadratic equations. By checking math history books or encyclopedias in your college library, locate these names on the time line that follows. The World Wide Web is also a useful source here. *Some hints*: The listing here is in no particular order. In some cases, you'll find that only very approximate dates are known. Also, be prepared to see numerous alternate spellings.

Omar Khayyam	Abraham bar Hiyya (or Chiya)
Mahāvira	Jia Xian
Bhāskara	Muhammad al-Khwārizmi
Diophantus	Āryabhata
Brahmagupta	Joranus de Nemore

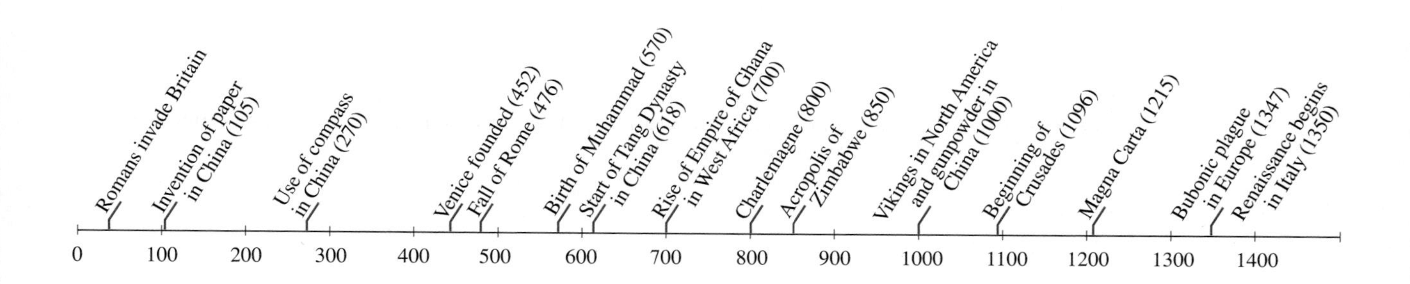

2.2 OTHER TYPES OF EQUATIONS

In this chapter we propose to consider some miscellaneous equations; it will be seen that many of these can be solved by the ordinary rules for quadratic equations. . . .—H. S. Hall and S. R. Knight, *Higher Algebra* [London: Macmillan and Co., 1946 (first edition published in 1887)]

In Sections 1.3 and 2.1 we developed techniques for solving linear and quadratic equations. In this section we consider some other types of equations that can be solved by applying those techniques. As in the previous section, we will consider only solutions that are real numbers. (Solutions involving the complex-number system are discussed in Chapter 12.) We begin with two examples involving absolute values.

EXAMPLE 1 **Solving an equation where the variable is inside absolute value signs**

Solve: $|2x - 1| = 7$.

SOLUTION

There are two cases to consider.

If $2x - 1 \geq 0$, the equation becomes	If $2x - 1 < 0$, the equation becomes
$2x - 1 = 7$	$-(2x - 1) = 7$
$2x = 8$	$-2x + 1 = 7$
$x = 4$	$-2x = 6$
	$x = -3$

We've now obtained the values $x = 4$ and $x = -3$. You should check for yourself that both of these numbers indeed satisfy the original equation.

The solution in Example 1 proceeds by considering two separate cases. There is another way to solve the equation in Example 1 that avoids this two-case technique. This alternate method relies on the following basic property of the absolute value (from Section 1.2):

For any real number a, we have $|a|^2 = a^2$.

To apply this property in solving the equation $|2x - 1| = 7$, we square both sides of the equation to obtain

$$|2x - 1|^2 = (2x - 1)^2 = 49$$
$$4x^2 - 4x - 48 = 0$$
$$x^2 - x - 12 = 0$$

This last equation is a quadratic equation that can be solved by factoring (as in Section 1.3). We have

$$(x - 4)(x + 3) = 0$$

$x - 4 = 0$	$x + 3 = 0$
$x = 4$	$x = -3$

This yields the two roots $x = 4$ and $x = -3$, in agreement with the result in Example 1.

CAUTION We obtained these two roots by squaring both sides of the original equation. However, it is not always the case that squaring both sides yields roots that check in the original equation. Consider, for example, the equation $|x| = -1$. This equation has no solution (why?) yet squaring both sides yields $x^2 = 1$, with roots $x = \pm 1$. However, neither 1 nor -1 checks in the original equation $|x| = -1$. The roots 1 and -1 are extraneous roots. We'll return to this point again, later in this section.

The solutions in Example 1 can be checked graphically. One way to do this is first to rewrite the given equation $|2x - 1| = 7$ in the equivalent form $|2x - 1| - 7 = 0$. Then, as we saw in previous sections, the roots of this last equation are the x-intercepts of the graph of

$$y = |2x - 1| - 7$$

As indicated in Figure 1(a), the x-intercepts of the graph appear to be -3 and 4, which are indeed the two roots obtained in Example 1. A different way to check the solutions of the equation $|2x - 1| = 7$ involves graphing the *two* equations $y = |2x - 1|$ and $y = 7$ on the same set of axes. As indicated in Figure 1(b), the graphs intersect in two points. The x-coordinates of those two points appear to be -3 and 4, which again are the two roots determined in Example 1.

GRAPHICAL PERSPECTIVE

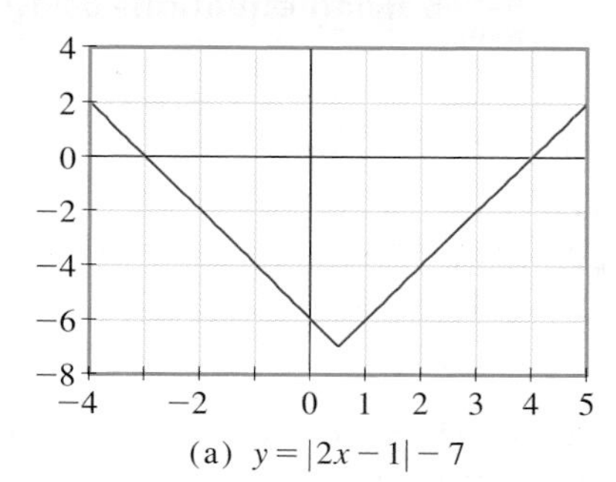

(a) $y = |2x - 1| - 7$

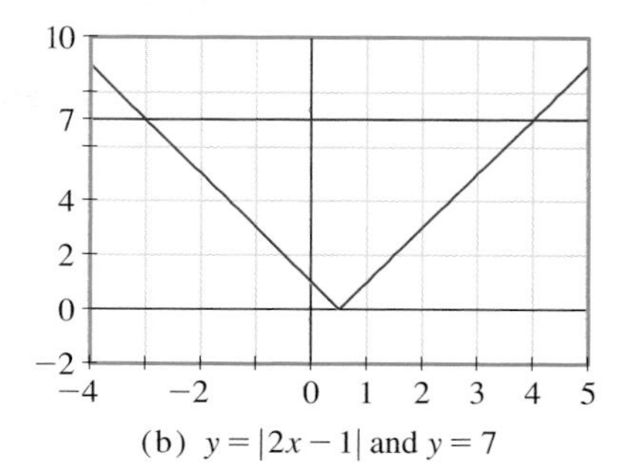

(b) $y = |2x - 1|$ and $y = 7$

Figure 1
Two ways to check the solutions in Example 1. In Figure 1(a) the x-intercepts of the graph of $y = |2x - 1| - 7$ are the solutions of the equation $|2x - 1| = 7$. In Figure 1(b) the x-coordinates of the intersection points of the two graphs $y = |2x - 1|$ and $y = 7$ are the solutions of the equation $|2x - 1| = 7$.

CAUTION Some care needs to be taken in drawing conclusions from a graph, even a graph obtained with a computer or graphing calculator. For instance, taken in isolation, the graphs in Figure 1 do not *prove* that the roots we are looking for are -3 and 4. Indeed, the only honest conclusion from the graphs themselves is that the two roots we are looking for must be very close to -3 and 4. The only way to tell whether the roots are exactly -3 and 4 is to use an algebraic method, as in Example 1, or to substitute each purported root in the given equation to see whether it checks. Figure 2 serves to emphasize this point. In Figure 2(a) we are looking at a portion of the curve $y = x^2 - 2.2x$ and the line $y = 0.8$ in order to find a root of the equation $x^2 - 2.2x = 0.8$. On the basis of the evidence in Figure 2(a), it appears that $x = 2.5$ may be a root. But, if you replace x with 2.5 in the equation $x^2 - 2.2x = 0.8$, you'll see that it does *not* check; 2.5 is not a root. Indeed, as is indicated in the magnified view in Figure 2(b), the graphs of $y = x^2 - 2.2x$ and $y = 0.8$ do not intersect at $x = 2.5$. That is, 2.5 is not a root. *Note*: Figure 2(b) does more than just tell us that 2.5 is not a root. It tells us that the root lies in the open interval (2.5, 2.55), so certainly the digit in the first decimal place of the root is a 5. More zooming in with a graphing calculator or a computer would yield additional decimal places.

When n is a natural number, equations of the form $x^n = a$ are solved simply by rewriting the equation in terms of the appropriate nth root or roots. For example, if $x^3 = 34$, then $x = \sqrt[3]{34}$. If the exponent n is even, however, we have to remember that there are *two* real nth roots for each positive number. For instance, if $x^4 = 81$, then $x = \pm\sqrt[4]{81} = \pm 3$. *Note*: If you feel that you need to review the subject of nth roots from intermediate algebra, see Appendix B.2 at the back of this book.

GRAPHICAL PERSPECTIVE

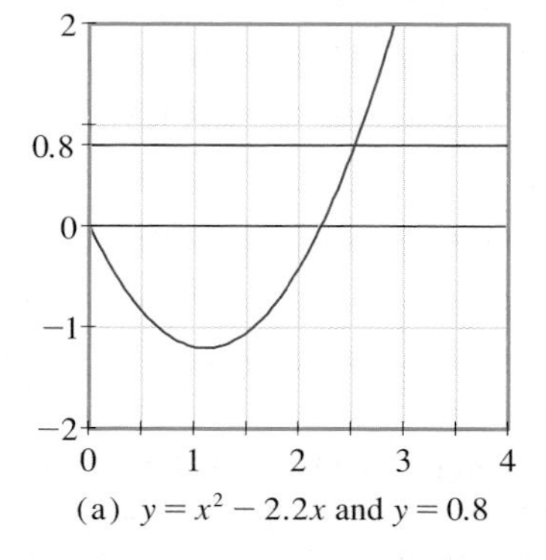

(a) $y = x^2 - 2.2x$ and $y = 0.8$

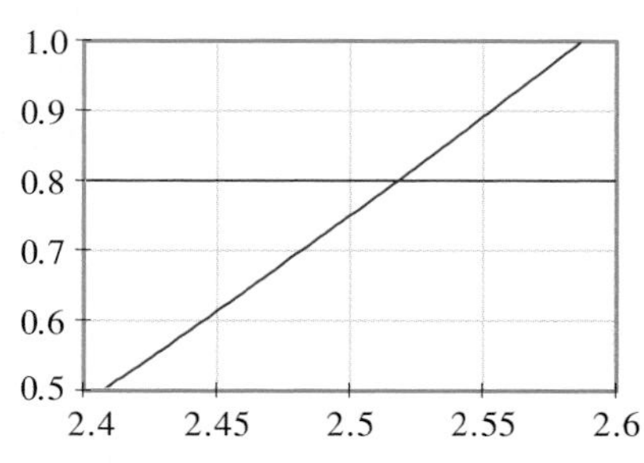

(b) Zoom-in view of $y = x^2 - 2.2x$ and $y = 0.8$

Figure 2
Looking for a root of the equation $x^2 - 2.2x = 0.8$

EXAMPLE 2 Solving equations using *n*th roots

Solve:
(a) $(x - 1)^4 = 15$; **(b)** $3x^4 = -48$; **(c)** $3x^5 = -48$.

SOLUTION

(a)
$$(x - 1)^4 = 15$$
$$x - 1 = \pm\sqrt[4]{15}$$
$$x = 1 \pm \sqrt[4]{15}$$

As you can check for yourself now, both of the values, $1 + \sqrt[4]{15}$ and $1 - \sqrt[4]{15}$, satisfy the given equation.

Remark: We can estimate these values without a calculator. For instance, for $x = 1 + \sqrt[4]{15}$ we have

$$1 + \sqrt[4]{15} \approx 1 + \sqrt[4]{16} = 1 + 2 = 3$$

Therefore, $1 + \sqrt[4]{15}$ is a little less than 3. (In fact, a calculator shows that $1 + \sqrt[4]{15} \approx 2.97$.)

(b) For any real number x, the quantity x^4 is nonnegative. Therefore, the left-hand side of the equation is nonnegative, whereas the right-hand side is negative. Consequently, there are no real numbers satisfying the given equation.

(c) Dividing both sides of the given equation by 3 yields $x^5 = -16$, and therefore

$$x = \sqrt[5]{-16} \qquad \text{or, equivalently,} \qquad x = -\sqrt[5]{16}$$

Calculator note: To approximate $-\sqrt[5]{16}$ using a calculator, first recall that $\sqrt[5]{16}$ can be rewritten in the equivalent form $16^{1/5}$. (See Appendix B.3 if you need a review of rational exponents.) We then have

$$x = -\sqrt[5]{16} = -16^{1/5} \approx -1.74 \qquad \text{using a calculator and rounding to two decimal places}$$

In Example 3 we solve two equations by factoring. As with quadratic equations, the factoring method is justified by the zero-product property of real numbers (from Section 1.3).

EXAMPLE 3 Using factoring to solve equations

Find the real-number solutions for each equation:
(a) $3x^3 - 12x^2 - 15x = 0$; **(b)** $x^4 + x^2 - 6 = 0$.

SOLUTION

(a) First we factor the left-hand side

$$3x^3 - 12x^2 - 15x = 0$$
$$3x(x^2 - 4x - 5) = 0$$
$$3x(x - 5)(x + 1) = 0$$

Now, by setting each factor equal to zero, we obtain the three values $x = 0$, $x = 5$, and $x = -1$. As you can check, each of these numbers indeed satisfies the original equation. So we have three solutions: 0, 5, and -1.

Note: If initially we had divided both sides of the given equation by $3x$ to obtain $x^2 - 4x - 5 = 0$, then the solution $x = 0$ would have been overlooked.

(b) Again we factor the left-hand side

$$x^4 + x^2 - 6 = 0$$
$$(x^2 - 2)(x^2 + 3) = 0$$

$$\begin{array}{r|r} x^2 - 2 = 0 & x^2 + 3 = 0 \\ x^2 = 2 & x^2 = -3 \\ x = \pm\sqrt{2} & \end{array}$$

By setting the first factor equal to zero, we've obtained the two values $x = \pm\sqrt{2}$. As you can check, these values do satisfy the given equation. When we set the second factor equal to zero, however, we obtained $x^2 = -3$, which has no real solutions (because the square of a real number must be nonnegative). We conclude therefore that the given equation in this case has only two real solutions, $\sqrt{2}$ and $-\sqrt{2}$.

The equation $x^4 + x^2 - 6 = 0$ that we considered in Example 3(b) is called **an equation of quadratic type.** This means that with a suitable substitution, the resulting equation becomes quadratic. In particular here, suppose that we let

$$x^2 = t \qquad \text{and therefore} \qquad x^4 = t^2$$

Then the equation $x^4 + x^2 - 6 = 0$ becomes $t^2 + t - 6 = 0$, which is "quadratic in t." Solving for t, we obtain $t = -3$ or 2. Then substituting x^2 for t, we obtain $x^2 = -3$ or $x^2 = 2$ as before. This idea is exploited in the next three examples.

EXAMPLE 4 Solving an equation of quadratic type

As shown in Figure 3, the graph of $y = x^4 - 6x^2 + 4$ has four x-intercepts. Determine these x-intercepts. Give two forms for each answer: an exact expression involving radicals and a calculator approximation rounded to two decimal places.

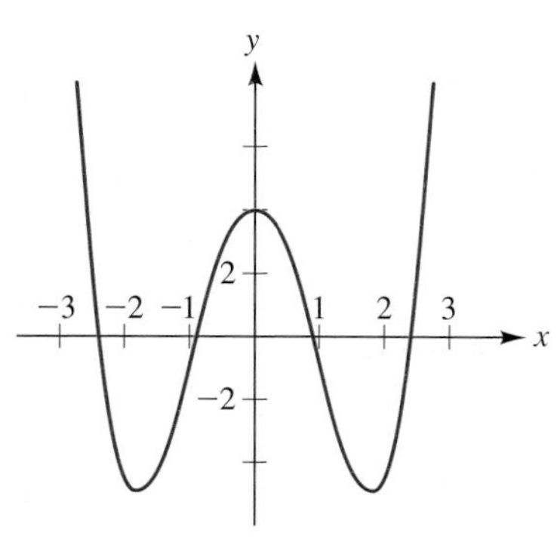

Figure 3
$y = x^4 - 6x^2 + 4$

SOLUTION

As explained in Section 1.5, the x-intercepts are obtained by setting y equal to zero in the given equation and then solving for x. So we need to solve the equation

$$x^4 - 6x^2 + 4 = 0 \tag{1}$$

This equation cannot readily be solved by factoring; however, it is of quadratic type. We let $x^2 = t$; then $x^4 = t^2$, and the equation becomes

$$t^2 - 6t + 4 = 0$$

The quadratic formula then yields

$$t = \frac{6 \pm \sqrt{36 - 4(4)}}{2} = \frac{6 \pm \sqrt{20}}{2}$$
$$= \frac{6 \pm 2\sqrt{5}}{2} = 3 \pm \sqrt{5}$$

Consequently, we have (in view of the definition of t)

$$x^2 = 3 \pm \sqrt{5}$$

and therefore

$$x = \pm\sqrt{3 \pm \sqrt{5}}$$

We've now found four values for x:

$$\sqrt{3+\sqrt{5}} \qquad -\sqrt{3+\sqrt{5}} \qquad \sqrt{3-\sqrt{5}} \qquad -\sqrt{3-\sqrt{5}}$$

These are the required x-intercepts for the graph in Figure 3. [Exercise 87 asks you to check that these values indeed satisfy equation (1).] Which is which, though? Without using a calculator, but with a little thought, you should be able to list these four quantities in order of increasing size and thereby assign them to the appropriate locations in Figure 3. (Try it before reading on!) Using a calculator now, and referring to Figure 3, we obtain the following results.

the x-intercept between -3 and -2: $-\sqrt{3+\sqrt{5}} \approx -2.29$
the x-intercept between -1 and 0: $-\sqrt{3-\sqrt{5}} \approx -0.87$
the x-intercept between 0 and 1: $\sqrt{3-\sqrt{5}} \approx 0.87$
the x-intercept between 2 and 3: $\sqrt{3+\sqrt{5}} \approx 2.29$.

EXAMPLE 5 Solving an equation of quadratic type

Solve: $6x^{-2} - x^{-1} - 2 = 0$.

SOLUTION

We'll show two methods. The first method depends on recognizing the equation as one of quadratic type.

FIRST METHOD

Let $x^{-1} = t$. Then $x^{-2} = t^2$ and the equation becomes

$$6t^2 - t - 2 = 0$$
$$(3t-2)(2t+1) = 0$$
$$3t - 2 = 0 \quad \Big| \quad 2t + 1 = 0$$
$$t = \frac{2}{3} \quad \Big| \quad t = -\frac{1}{2}$$

If $t = 2/3$, then $x^{-1} = 2/3$. Therefore $\frac{1}{x} = \frac{2}{3}$, and consequently, $x = 3/2$. On the other hand, if $t = -1/2$, then $\frac{1}{x} = -\frac{1}{2}$, and consequently, $x = -2$. In summary, the two solutions are $3/2$ and -2.

ALTERNATIVE METHOD

Multiplying both sides of the given equation by x^2 (*Note*: $x \neq 0$) yields

$$6 - x - 2x^2 = 0$$
$$2x^2 + x - 6 = 0$$
$$(2x-3)(x+2) = 0$$
$$2x - 3 = 0 \quad \Big| \quad x + 2 = 0$$
$$x = \frac{3}{2} \quad \Big| \quad x = -2$$

Thus we have $x = 3/2$ or $x = -2$, as we obtained using the first method.

In the next example we first use a graphing utility to approximate the solutions of an equation, then we use algebra to determine the solutions exactly. Note, however, that even in the graphing utility portion of the problem, algebra is useful. It turns out that the graph is symmetric about the y-axis, and that cuts the work in half.

EXAMPLE 6 Solving an equation with fractional exponents

Consider the equation $2x^{4/3} - x^{2/3} - 6 = 0$.

(a) Use a graphing utility to approximate the roots to the nearest hundredth.

(b) Use algebra to determine the exact solutions. Then check that the answers are consistent with the approximations obtained in part (a).

SOLUTION

(a) The roots of the given equation are the x-intercepts of the graph of $y = 2x^{4/3} - x^{2/3} - 6$. After experimenting with different viewing rectangles, we find that the standard viewing rectangle in Figure 4(a) does show the essential features of the graph.

GRAPHICAL PERSPECTIVE

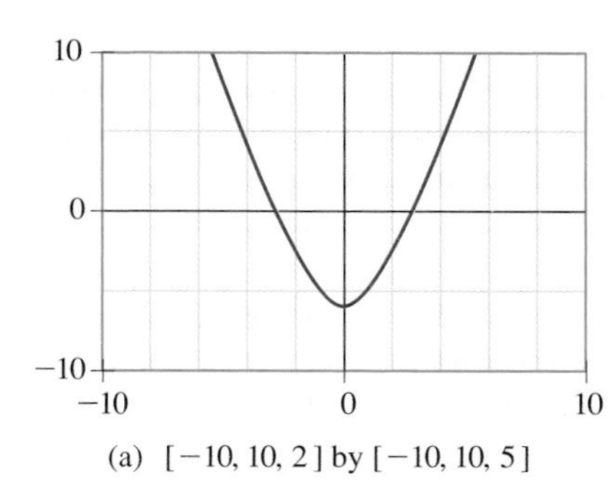

(a) $[-10, 10, 2]$ by $[-10, 10, 5]$

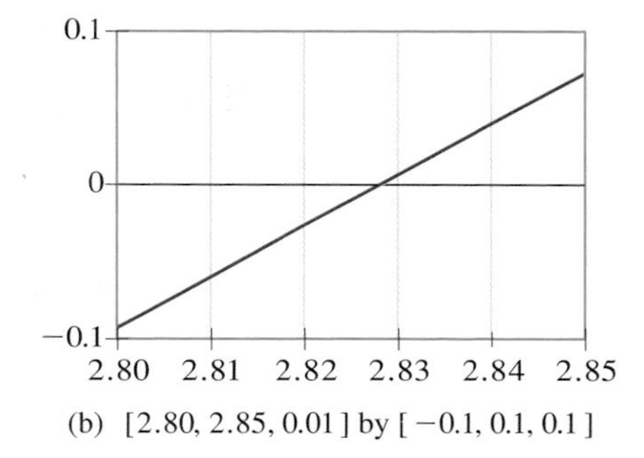

(b) $[2.80, 2.85, 0.01]$ by $[-0.1, 0.1, 0.1]$

Figure 4
Two views of $y = 2x^{4/3} - x^{2/3} - 6$

In looking at Figure 4(a), it appears that there are two roots: one between 2 and 4, the other between -2 and -4. It also appears that the graph may be symmetric about the y-axis. If so, once we find the first root between 2 and 4, then we'll automatically know the second root: It will just be the opposite of the first. To show that the graph is indeed symmetric about the y-axis, we can rewrite the equation $y = 2x^{4/3} - x^{2/3} - 6$ as $y = 2(x^4)^{1/3} - (x^2)^{1/3} - 6$. In this form, because of the even integer exponents on x, replacing x with $-x$ will make no difference. So, according to the first symmetry test on page 59, the graph is symmetric about the y-axis. In Figure 4(b) we've zoomed in on the positive root. Since we are supposed to approximate the root to the nearest hundredth, we've zoomed in far enough to allow us to mark off the x-scale in units of 0.01. Figure 4(b) shows that the positive root is 2.83, to the nearest hundredth. By symmetry, the other root then is -2.83, also to the nearest hundredth. (We note in passing that many graphing utilities also have "solve" keys for approximating roots of equations. For details, consult your user's manual.)

(b) In the given equation $2x^{4/3} - x^{2/3} - 6 = 0$, we let $x^{2/3} = t$. Then $x^{4/3} = t^2$, and the equation becomes

$$2t^2 - t - 6 = 0$$
$$(2t + 3)(t - 2) = 0$$
$$2t + 3 = 0 \quad\Big|\quad t - 2 = 0$$
$$t = -\frac{3}{2} \quad\Big|\quad t = 2$$

We now have two cases to consider: $t = -3/2$ and $t = 2$. With $t = -3/2$, we have (in view of the definition of t)

$$x^{2/3} = -\frac{3}{2} \qquad \text{or, equivalently,} \qquad (x^{1/3})^2 = -\frac{3}{2}$$

This last equation has no real-number solutions, because the square of a real number is never negative. Turning our attention then to the other case where $t = 2$, we have

$$x^{2/3} = 2$$
$$(x^{2/3})^3 = 2^3 \qquad \text{Two real numbers are equal if and only if their cubes are equal.}$$
$$x^2 = 8$$
$$x = \pm\sqrt{8} = \pm 2\sqrt{2}$$

As Exercise 88 asks you to verify, each of the values $2\sqrt{2}$ and $-2\sqrt{2}$ satisfies the original equation. These are the required roots. As you can check now by picking up a calculator, we have $2\sqrt{2} \approx 2.828$, which is indeed consistent with the approximations in part (a).

Some equations can be solved by raising both sides to the same power. In fact, if you review the solution given for Example 6, you'll see that we've already made use of this idea in cubing both sides of one of the equations there. As another example, consider the equation

$$\sqrt{x - 3} = 5$$

By squaring both sides of this equation we obtain $x - 3 = 25$ and, consequently, $x = 28$. As you can easily check, the value $x = 28$ does satisfy the original equation.

There is a complication, however, that can arise in raising both sides of an equation to the same power. Consider, for example, the equation

$$x - 2 = \sqrt{x}$$

By squaring both sides, we obtain

$$x^2 - 4x + 4 = x$$
$$x^2 - 5x + 4 = 0$$
$$(x - 4)(x - 1) = 0$$

Therefore

$$x = 4 \qquad \text{or} \qquad x = 1$$

The value $x = 4$ checks in the original equation, but the value $x = 1$ does not. (Verify this.) Therefore $x = 1$ is an extraneous solution, and the only solution of the given equation is $x = 4$.

The extraneous solution in this case arises from the fact that if $a^2 = b^2$, then it need not be true that $a = b$. For instance, $(-3)^2 = 3^2$, but certainly $-3 \neq 3$. In the box that follows, we summarize this remark about extraneous solutions. For reference, we also include the results from Section 1.3.

PROPERTY SUMMARY Extraneous Solutions

1. Squaring both sides of an equation (or raising both sides to an even integral power) may introduce extraneous solutions that do not check in the original equation.
2. Multiplying sides of an equation by an expression involving the variable may introduce extraneous solutions.

Therefore it is always necessary to check any candidates for solutions that you obtain in either of these ways.

In the next example we solve equations involving radicals. In each case, notice that *before* squaring, we first isolate one of the radicals on one side of the equation. Otherwise things become more complicated, rather than less. For instance, in Example 7(a) we're given the equation $2 + \sqrt{10 - x} = -x$. If we square both sides without first isolating the radical, we obtain

$$(2 + \sqrt{10 - x})^2 = (-x)^2$$

and, consequently,

$$4 + 4\sqrt{10 - x} + 10 - x = x^2 \qquad \text{or} \qquad 14 + 4\sqrt{10 - x} - x = x^2$$

As you can see, this last equation is indeed more complicated than the original.

EXAMPLE 7 Solving equations with radicals

Find all real-number solutions of each equation.

(a) $2 + \sqrt{10 - x} = -x$ **(b)** $\sqrt{2x + 3} - 2\sqrt{x - 2} = 1$

SOLUTION

(a) First isolate the radical by rewriting the given equation as

$$\sqrt{10 - x} = -x - 2$$

Squaring both sides now yields

$$\begin{aligned} (\sqrt{10 - x})^2 &= (-x - 2)^2 \\ 10 - x &= x^2 + 4x + 4 \\ 0 &= x^2 + 5x - 6 \\ 0 &= (x + 6)(x - 1) \end{aligned}$$

$$x + 6 = 0 \quad \bigg| \quad x - 1 = 0$$
$$x = -6 \quad \bigg| \quad x = 1$$

Check for extraneous roots: With $x = -6$ the *original* equation becomes

$$2 + \sqrt{10 - (-6)} \stackrel{?}{=} -(-6)$$
$$2 + 4 \stackrel{?}{=} 6 \qquad \text{True}$$

With $x = 1$ the original equation becomes

$$\begin{aligned} 2 + \sqrt{10 - 1} &\stackrel{?}{=} -1 \\ 2 + 3 &\stackrel{?}{=} -1 \qquad \text{False} \end{aligned}$$

In summary, the given equation has but one root: $x = -6$.

(b) The given equation contains two radicals, so initially the best that we can do is to isolate one of them before squaring. (It does not matter which.) Rewriting the given equation as $\sqrt{2x + 3} = 2\sqrt{x - 2} + 1$ and then squaring yields

$$\begin{aligned} (\sqrt{2x + 3})^2 &= (2\sqrt{x - 2} + 1)^2 \\ 2x + 3 &= 4(x - 2) + 4\sqrt{x - 2} + 1 \end{aligned}$$

You should check now for yourself that if we combine like terms in this last equation, then it can be rewritten as

$$-2x + 10 = 4\sqrt{x - 2}$$

Next, don't square both sides quite yet! It won't be wrong, but you can work with smaller numbers if you first divide both sides by 2 to obtain

$$-x + 5 = 2\sqrt{x - 2}$$

Now square both sides:

$$\begin{aligned} (-x + 5)^2 &= (2\sqrt{x - 2})^2 \\ x^2 - 10x + 25 &= 4x - 8 \\ x^2 - 14x + 33 &= 0 \\ (x - 3)(x - 11) &= 0 \end{aligned}$$

From this last equation we conclude that $x = 3$ or $x = 11$.
Check for extraneous roots: With $x = 3$ the *original* equation becomes

$$\begin{aligned} \sqrt{2(3) + 3} - 2\sqrt{3 - 2} &\stackrel{?}{=} 1 \\ 3 - 2 &\stackrel{?}{=} 1 \qquad \text{True} \end{aligned}$$

Next, with $x = 11$ the original equation becomes

$$\begin{aligned} \sqrt{2(11) + 3} - 2\sqrt{11 - 2} &\stackrel{?}{=} 1 \\ 5 - 2(3) &\stackrel{?}{=} 1 \qquad \text{False} \end{aligned}$$

So the only solution of the given equation is $x = 3$. For a graphical perspective, see Figure 5.

GRAPHICAL PERSPECTIVE

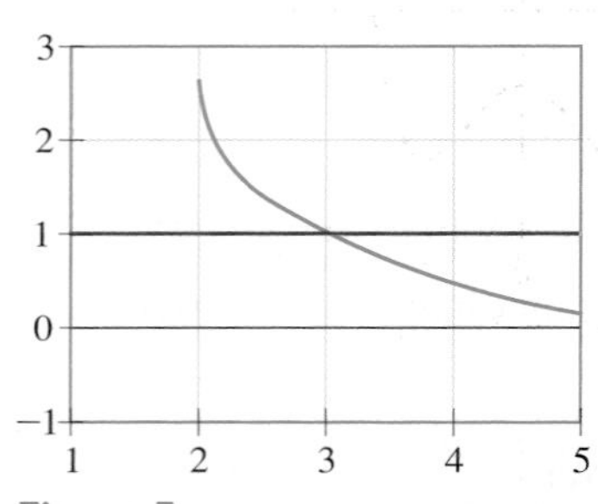

Figure 5
The curve $y = \sqrt{2x + 3} - 2\sqrt{x - 3}$ meets the horizontal line $y = 1$ at $x = 3$. That is, $x = 3$ is a solution of the equation $\sqrt{2x + 3} - 2\sqrt{x - 3} = 1$.

EXERCISE SET 2.2

A

In Exercises 1–6, solve each equation.

1. $|x - 5| = 1$
2. $|x - 4| - 5 = 2$
3. $|x + 6| = 1/2$
4. $|x + 6| + 1/2 = 0$
5. $|6x - 5| = 25$
6. $|5 - 6x| = 0$

In Exercises 7–10, solve each equation by using the squaring technique that is discussed following Example 1 in the text.

7. $|x + 3| = 2x - 2$
8. $4|x - 2| = 3x - 4$
9. $|2x - 1| = 1 - \frac{1}{2}x$
10. $|x + 1| - |3x - 2| = 0$

Ⓖ **11. (a)** On the same set of axes, graph the equations $y = x^2 - 2.2x$ and $y = 0.8$. Use the same viewing rectangle that is shown in Figure 2(a) in the text. Check that your graphs are consistent with those drawn in Figure 2(a). The graphs show that the line $y = 0.8$ intersects the curve $y = x^2 - 2.2x$ at a point with an x-coordinate very close to 2.5. In other words, one of the roots of the equation $x^2 - 2.2x = 0.8$ is approximately 2.5.

(b) Take a closer look at the intersection point using the same viewing rectangle that is shown in Figure 2(b) in the text. As is pointed out in the text, this view shows that the root we are looking for is actually slightly larger than 2.5; it lies in the open interval (2.5, 2.55). Thus, the first decimal place of the root must indeed be 5, and the second decimal place must be a digit between 0 and 4, inclusive. Continue to zoom in on this intersection point until you are sure of the first three decimal places of the root.

(c) Use the quadratic formula to find an expression for the exact value of the root in part (b). Then evaluate the expression and round to four decimal places. Check that your answer is consistent with the result obtained graphically in part (b).

Ⓖ **12. (a)** Graph the equation $y = 3x^3 - 12x^2 - 15x$.

(b) On the basis of the graph, how many roots are there for the equation $3x^3 - 12x^2 - 15x = 0$? Use the graph to estimate these roots. Then compare your estimates to the actual values obtained in Example 3(a).

Ⓖ **13. (a)** Graph the equation $y = x^4 + x^2 - 6$.

(b) On the basis of the graph, how many roots are there for the equation $x^4 + x^2 - 6 = 0$? Use the graph to estimate these roots. Zoom in until you are sure about the first three decimal places of each root. Then compare your estimates to the actual values obtained in Example 3(b).

Ⓖ **14. (a)** Using a viewing rectangle that extends from −5 to 5 in both the x- and y-directions, graph the equation $y = 6x^{-2} - x^{-1} - 2$. Use the graph to estimate the roots of the equation $6x^{-2} - x^{-1} - 2 = 0$. How do your estimates compare to the actual values determined in Example 5?

(b) Using the viewing rectangle specified in part (a), graph the equation $y = 6x^{-2} - x^{-1} + 2$. On the basis of the graph, make a statement about the roots of the equation $6x^{-2} - x^{-1} + 2 = 0$. Then solve the equation algebraically to confirm your statement.

In Exercises 15–58, find all real solutions of each equation. For Exercises 31–36, give two forms for each answer: an exact answer (involving a radical) and a calculator approximation rounded to two decimal places.

15. $3x^2 - 48x = 0$
16. $3x^3 - 48x = 0$
17. $t^3 - 125 = 0$
18. $t - t^3 = 0$
19. $7x^4 - 28x^2 = 0$
20. $y^4 - 81 = 0$
21. $t^4 + 2t^3 - 3t^2 = 0$
22. $2t^5 + 5t^4 - 12t^3 = 0$
23. $6x = 23x^2 + 4x^3$
24. $x^5 = 36x$
25. $x^4 - x^2 = 6$
26. $x^4 - 5x^2 = -6$
27. $4y^2 = 5 - y^4$
28. $6y^2 = -5 - y^4$
29. $3t^2 + 2 = 9t^4$
30. $5t^2 - 1 = 4t^4$
31. $(x - 2)^3 - 5 = 0$
32. $(x + 4)^3 + 2 = 0$
33. $(x + 4)^5 + 16 = 0$
34. $(1 - x)^5 - 40 = 0$
35. (a) $(x - 3)^4 - 30 = 0$ **(b)** $(x - 3)^4 + 30 = 0$
36. (a) $(x^2 - 1)^4 - 81 = 0$ **(b)** $(x^2 - 1)^4 + 81 = 0$
37. $x^6 - 10x^4 + 24x^2 = 0$
38. $2x^5 - 15x^3 - 27x = 0$
39. $x^4 + x^2 - 1 = 0$
40. $x^4 - x^2 + 1 = 0$
41. $x^4 + 3x^2 - 2 = 0$
42. $2x^4 + x^2 + 2 = 0$
43. $x^6 + 7x^3 = 8$
44. $x^6 + 27 = -28x^3$
45. $t^{-2} - 7t^{-1} + 12 = 0$
46. $8t^{-2} - 17t^{-1} + 2 = 0$
47. $12y^{-2} - 23y^{-1} = -5$
48. $y^{-2} - y^{-1} = 0$
49. $4x^{-4} - 33x^{-2} - 27 = 0$
50. $x^{-4} + x^{-2} + 1 = 0$
51. $t^{2/3} = 9$
52. $t^{3/2} = 8$
53. $(y - 1)^3 = 7$
54. $(2y + 3)^4 = 5$
55. $(t + 1)^5 = -243$
56. $(t + 3)^4 = 625$
57. $9x^{4/3} - 10x^{2/3} + 1 = 0$
58. $x^{4/3} + 3x^{2/3} - 28 = 0$

In Exercises 59 and 60, find the x-intercepts of the graphs. Round each answer to two decimal places. Check to see that your answers are consistent with the graphs.

59.

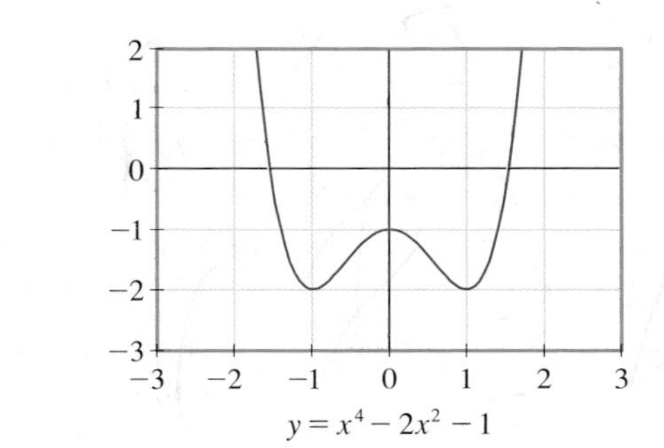

$y = x^4 - 2x^2 - 1$

60.

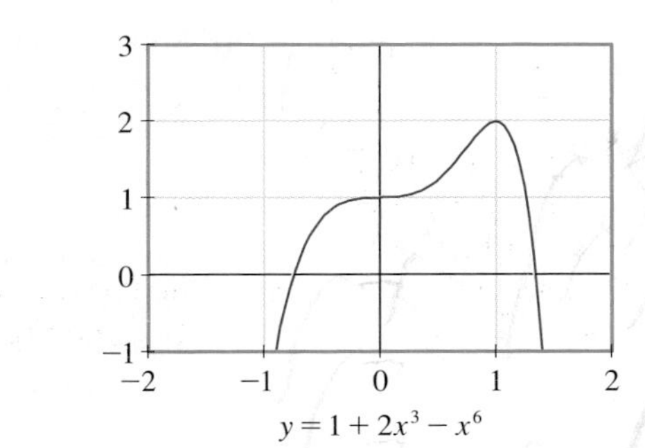

$y = 1 + 2x^3 - x^6$

In Exercises 61–76, determine all of the real-number solutions for each equation. (Remember to check for extraneous solutions.)

61. $\sqrt{1 - 3x} = 2$
62. $\sqrt{x^2 + 5x - 2} = 2$
63. $\sqrt{x + 6} = x$
64. $x - \sqrt{x} = 20$
65. $x - \sqrt{3 - x} = -3$
66. $\sqrt{2 - x} - 10 = x$
67. $4x + \sqrt{2x + 5} = 0$
68. $\sqrt{7 - 3x} - 6(x + 1) = 1$
69. $\sqrt{x^4 - 13x^2 + 37} = 1$
70. $\sqrt{y + 2} = y - 4$
71. $\sqrt{1 - 2x} + \sqrt{x + 5} = 4$
72. $\sqrt{x - 5} - \sqrt{x + 4} + 1 = 0$

73. $\sqrt{3 + 2t} + \sqrt{-1 + 4t} = 1$

74. $\sqrt{2t + 5} - \sqrt{8t + 25} + \sqrt{2t + 8} = 0$

75. $\sqrt{2y - 3} - \sqrt{3y + 3} + \sqrt{3y - 2} = 0$

76. $\sqrt{a - x} + \sqrt{b - x} = \sqrt{a + b - 2x}$, where $b > a > 0$

Ⓖ *In Exercises 77–86,* **(a)** *use zoom-in techniques to estimate the roots of each equation to the nearest hundredth, as in Example 6; and* **(b)** *use algebraic techniques to determine an exact expression for each root, then evaluate the expression and round to four decimal places. Check to see that your answers are consistent with the graphical results obtained in part (a).*

77. (a) $x^4 - 5x^2 = -6.2$
(b) $x^4 - 5x^2 = -6.3$

78. (a) $x^4 - 5x^2 = -25/4$
(b) $x^4 - 5x^2 = -13/2$

79. $x - 5\sqrt{x} = -3$

80. $x^{-2} - 5x^{-1} = -3$

81. $\sqrt{2x - 1} - \sqrt{x - 2} = 2$

82. $\sqrt{2x - 1} - \sqrt{x - 2} = 1$

83. $\dfrac{\sqrt{x} - 4}{\sqrt{x} + 3} + \dfrac{1}{2} = 0$

84. $\dfrac{\sqrt[4]{x} - 4}{\sqrt[4]{x} + 3} + \dfrac{1}{2} = 0$

85. $x^{2/3} - x^{1/3} = 1$

86. $x^{-2/3} - x^{-1/3} = 1$

87. Verify that the four solutions obtained in Example 4 indeed satisfy the original equation.

88. Verify that the values $x = \pm 2\sqrt{2}$ obtained in Example 6(b) satisfy the equation $2x^{4/3} - x^{2/3} - 6 = 0$.

B

In Exercises 89–92, solve each equation.

89. $\sqrt{\sqrt{x} + \sqrt{a}} + \sqrt{\sqrt{x} - \sqrt{a}} = \sqrt{2\sqrt{x} + 2\sqrt{b}}$

90. $\sqrt{x^2 - x - 1} - \dfrac{2}{\sqrt{x^2 - x - 1}} = 1$

Hint: Let $t = x^2 - x - 1$.

91. $\sqrt{\dfrac{x - a}{x}} + 4\sqrt{\dfrac{x}{x - a}} = 5$, where $a \neq 0$

Hint: Let $t = \dfrac{x - a}{x}$; then $\dfrac{1}{t} = \dfrac{x}{x - a}$.

92. $\sqrt{x^2 + 3x - 4} - \sqrt{x^2 - 5x + 4} = x - 1$, where $x > 4$

Hint: Factor the expressions beneath the radicals. Then note that $\sqrt{x - 1}$ is a factor of both sides of the equation.

MINI PROJECT — FLYING THE FLAG

According to Clint Brookhart in his book *Go Figure!* (Lincolnwood, Ill.: Contemporary Books, 1998),

> *A flag the area of A in square feet, furled in a wind with a velocity of v in miles* per hour, *exerts a force of F in pounds on the flagpole equal to:* $F = 0.0003Av^{1.9}$.

(a) Solve the given equation for the velocity v in terms of the force F.

(b) According to the *Encyclopedia Americana,* "A 5-foot by 6-foot 9-inch U.S. flag flown in bad weather [at a U.S. military post] is known as a storm flag." Suppose that the flagpole for a storm flag can withstand a force of 40 pounds before it breaks away from the wall to which it is attached. Use the result in part (a) to determine the wind speed that produces a 40-pound force on the flagpole. Round your answer to the nearest integer. (Actually, in a more realistic model, the length of the flagpole, not just the force on it, would have to be taken into account.)

PROJECT — SPECIFIC OR GENERAL? WHATEVER WORKS!

Sometimes in math and science, when we're stuck on a problem, we can make progress by turning from the specific to the general. This approach sometimes surprises students because of a mistaken belief that the general or more abstract case is always tougher than the specific. The specific equation that you'll work to solve in this project is

$$\sqrt{4 - \sqrt{4 + x}} - x = 0 \qquad (1)$$

(a) Use a graphing utility to graph the equation $y = \sqrt{4 - \sqrt{4 + x}} - x$. Then, by zooming in on the x-intercept, approximate the root of equation (1) to the nearest hundredth.

(b) Now let's try to find the exact value for the root. By using the process of isolating the radical and then squaring, show that equation (1) eventually becomes

$$x^4 - 8x^2 - x + 12 = 0$$

(c) At this point we are stuck; the equation obtained in part (b) can't be solved by using any of the elementary factoring techniques from intermediate algebra. Here is where we shift our focus to a slightly more general problem. Instead of equation (1), we consider

$$\sqrt{k - \sqrt{k + x}} - x = 0 \qquad (2)$$

where k is a constant. Use the process of isolating the radical and then squaring to show that equation (2) eventually becomes

$$x^4 - 2kx^2 - x + k^2 - k = 0 \qquad (3)$$

(d) Although equation (3) is a fourth-degree equation in x, it is only a quadratic equation in k. That is, equation (3) can be written

$$k^2 - (2x^2 + 1)k + (x^4 - x) = 0 \qquad \text{Verify that this is equivalent to equation (3).}$$

Use the quadratic formula to solve this last equation for k in terms of x. You should obtain (after simplifying) two solutions:

$$k = x^2 + x + 1 \qquad \text{or} \qquad k = x^2 - 1$$

(e) Finally now, in both of the equations obtained in part (d), replace k with 4 [that's what it is in equation (1)], and solve each equation for x.

(f) Three of the four values that you obtain in part (e) are extraneous solutions for equation (1). Use your approximation in part (a) and a calculator to decide which value is the actual root of equation (1).

2.3 INEQUALITIES

The fundamental results of mathematics are often inequalities *rather than* equalities.—E. Beckenbach and R. Bellman in *An Introduction to Inequalities* (New York: Random House, 1961)

If we replace the equal sign in an equation with any one of the four symbols $<$, $\leq$, $>$, or $\geq$, we obtain an **inequality.** As with equations in one variable, a real number is a **solution** of an inequality if we obtain a true statement when the variable is replaced by the real number. For example, the value $x = 5$ is a solution of the inequality $2x - 3 < 8$, because when $x = 5$ we have

$$2(5) - 3 < 8$$
$$7 < 8 \qquad \text{which is true}$$

We also say in this case that the value $x = 5$ **satisfies** the inequality. To **solve** an inequality means to find all of the solutions. The set of all solutions of an inequality is called (naturally enough) the **solution set.**

Recall that two equations are said to be equivalent if they have exactly the same solutions. Similarly, two inequalities are **equivalent** if they have the same solution set. Most of the procedures used for solving inequalities are similar to those for equalities. For example, adding or subtracting the same number on both sides of an inequality produces an equivalent inequality. We need to be careful, however, in multiplying or dividing both sides of an inequality by the same nonzero number. For instance, suppose that we start with the inequality $2 < 3$ and multiply both sides by 5. That yields $10 < 15$, which is certainly true. But if we multiply both sides of the inequality $2 < 3$ by -5, we obtain $-10 < -15$, which is false. Multiplying both sides of an inequality by the same *positive* number preserves the inequality, whereas multiplying by a *negative* number reverses the inequality. In the following box, we list some of the principal properties of inequalities. In general, whenever we use Property 1 or 2 in solving an inequality, we obtain an equivalent inequality. Also, note that each property can be rewritten to reflect the fact that $a < b$ is equivalent to $b > a$. For example, Property 3 can just as well be written this way: If $b > a$ and $c > b$, then $c > a$.

PROPERTY SUMMARY Properties of Inequalities

For real numbers a, b, and c:

Property	Example
1. If $a < b$, then $a + c < b + c$ and $a - c < b - c$.	If $x - 3 < 0$, then $(x - 3) + 3 < 0 + 3$ and, consequently, $x < 3$.
2. (a) If $a < b$ and c is positive, then $ac < bc$ and $\frac{a}{c} < \frac{b}{c}$.	If $\frac{1}{2}x < 4$, then $2\left(\frac{1}{2}x\right) < 2(4)$ and, consequently, $x < 8$.
(b) If $a < b$ and c is negative, then $ac > bc$ and $\frac{a}{c} > \frac{b}{c}$.	If $-\frac{x}{5} < 6$, then $(-5)\left(-\frac{x}{5}\right) > (-5)(6)$ and, consequently, $x > -30$.
3. The transitive property: If $a < b$ and $b < c$, then $a < c$.	If $a < x$ and $x < 2$, then $a < 2$.

EXAMPLE 1 Solving two simple inequalities

Solve each of the following inequalities, state the solution in interval notation, and graph the solution set:
(a) $2x - 3 < 5$; **(b)** $5t + 8 \le 7(1 + t)$.

SOLUTION

(a) Our work follows the same pattern that we would use to solve the equation $2x - 3 = 5$. We have

$$2x - 3 < 5$$
$$2x < 8 \quad \text{adding 3 to both sides}$$
$$x < 4 \quad \text{dividing both sides by 2}$$

The solution set is therefore $(-\infty, 4)$; see Figure 1.

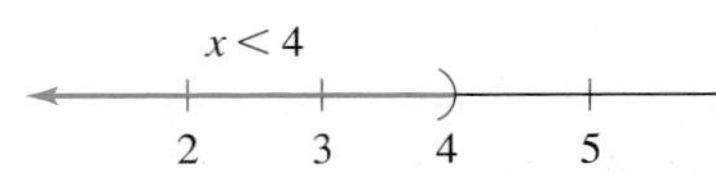

Figure 1

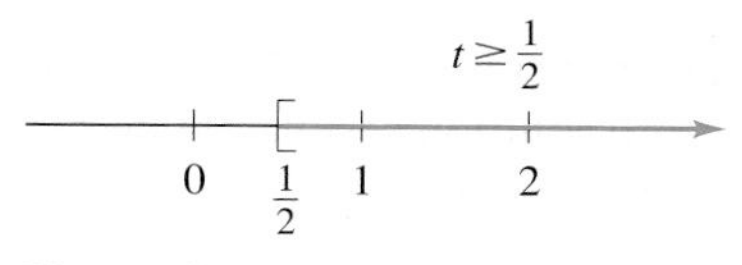

Figure 2

(b) Again, we follow the pattern that we would use to solve the corresponding equation.

$$5t + 8 \leq 7 + 7t$$

$$-2t \leq -1 \qquad \text{subtracting } 7t \text{ and } 8 \text{ from both sides}$$

$$t \geq \frac{1}{2} \qquad \text{dividing both sides by } -2\text{, which reverses the inequality}$$

The solution set is therefore $[1/2, \infty)$; see Figure 2.

In Section 1.5 we saw that there is a close connection between solving equations and graphing. (Example: The roots of the equation $x^2 - 4 = 0$ are the x-intercepts of the graph of $y = x^2 - 4$.) It's also true that solving inequalities and graphing are closely related. Consider, for instance, the inequality $2x - 3 < 5$ that we solved in Example 1(a). To interpret this inequality geometrically, we graph the two equations $y = 2x - 3$ and $y = 5$. Then, as indicated in Figure 3(a), the graph of $y = 2x - 3$ is below the line $y = 5$ as long as x is to the left of 4. The previous sentence translates algebraically into

$$2x - 3 < 5 \qquad \text{provided that} \qquad x < 4$$

This confirms the result in Example 1(a).

Similarly, for a geometric interpretation of the inequality in Example 1(b), we graph the two lines $y = 5t + 8$ and $y = 7(1 + t)$ in a t-y coordinate system. As indicated in Figure 3(b), the line $y = 5t + 8$ is below the line $y = 7(1 + t)$ provided that t is to the right of $1/2$. Furthermore, at $t = 1/2$, the lines intersect. That is, for $t = 1/2$, the quantities $5t + 8$ and $7(1 + t)$ are equal. Algebraically, we can restate this information by saying

$$5t + 8 \leq 7(1 + t) \qquad \text{provided that} \qquad t \geq \frac{1}{2}$$

This confirms the result in Example 1(b).

In the next example, we solve the inequality

$$-\frac{1}{2} < \frac{3 - x}{-4} < \frac{1}{2}$$

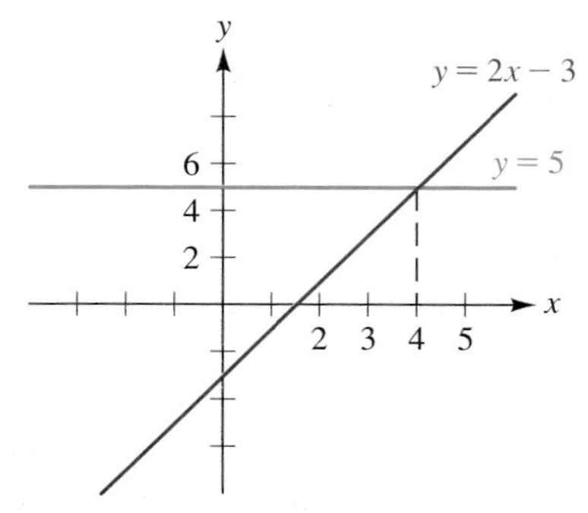

(a) To the left of $x = 4$, the line $y = 2x - 3$ is below the line $y = 5$.

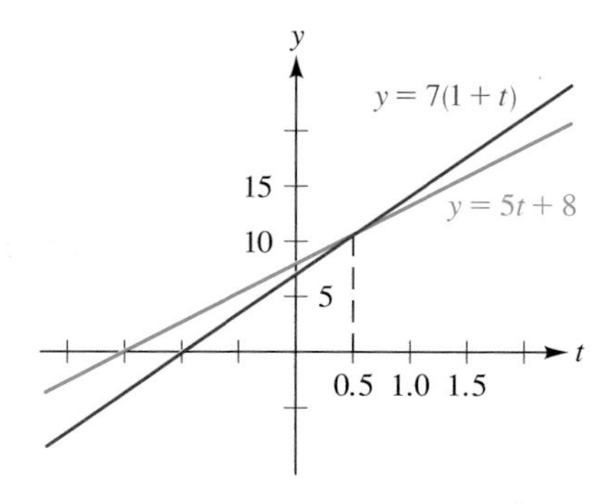

(b) The graph indicates that if $t \geq 1/2$ then $5t + 8 \leq 7(1 + t)$.

Figure 3

By definition, this is equivalent to the pair of inequalities

$$-\frac{1}{2} < \frac{3-x}{-4} \quad \text{and} \quad \frac{3-x}{-4} < \frac{1}{2}$$

One way to proceed here would be first to determine the solution set for each inequality. Then the set of real numbers common to both solution sets would be the solution set for the original inequality. However, the method shown in Example 2 is more efficient.

EXAMPLE 2 Solving a compound inequality

Solve: $-\frac{1}{2} < \frac{3-x}{-4} < \frac{1}{2}$.

SOLUTION

We begin by multiplying through by -4. Remember that this will reverse the inequalities:

$$2 > 3 - x > -2$$

Next, with a view toward isolating x, we first subtract 3 to obtain

$$-1 > -x > -5$$

Finally, multiplying through by -1, we have

$$1 < x < 5$$

The solution set is therefore the open interval $(1, 5)$.

The next example refers to the Celsius (C) and Fahrenheit (F) scales for measuring temperature.* The formula relating the temperature readings on the two scales is

$$F = \frac{9}{5}C + 32$$

EXAMPLE 3 Converting a Fahrenheit temperature range into Celsius

Over the temperature range $32° \le F \le 39.2°$ on the Fahrenheit scale, water contracts (rather than expands) with increasing temperature. What is the corresponding temperature range on the Celsius scale?

SOLUTION

$$\begin{aligned}
32 &\le F \le 39.2 && \text{given}\\
32 &\le \frac{9}{5}C + 32 \le 39.2 && \text{substituting } \tfrac{9}{5}C + 32 \text{ for } F\\
0 &\le \frac{9}{5}C \le 7.2 && \text{subtracting 32}\\
0 &\le C \le \frac{5}{9}(7.2) && \text{multiplying by } \tfrac{5}{9}\\
0 &\le C \le 4
\end{aligned}$$

*The Celsius scale was devised in 1742 by the Swedish astronomer Anders Celsius. The Fahrenheit scale was first used by the German physicist Gabriel Fahrenheit in 1724.

Thus a range of 32°F to 39.2°F on the Fahrenheit scale corresponds to 0°C to 4°C on the Celsius scale.

In the next three examples we solve inequalities that involve absolute values. The following theorem is very useful in this context.

THEOREM Absolute Value and Inequalities

If $a > 0$, then

$$|u| < a \quad \text{if and only if} \quad -a < u < a$$

and

$$|u| > a \quad \text{if and only if} \quad u < -a \quad \text{or} \quad u > a$$

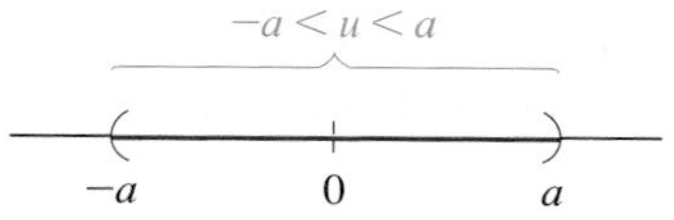

You can see why this theorem is valid if you think in terms of distance and position on a number line. The condition $-a < u < a$ means that u lies between $-a$ and a, as indicated in Figure 4. But this is the same as saying that the distance from u to zero is less than a, which in turn can be written $|u| < a$. (The second part of the theorem can be justified in a similar manner.)

$-a < u < a$

$-a$ 0 a

Figure 4

EXAMPLE 4 Solving inequalities containing absolute values

Solve:
(a) $|x| < 1$; **(b)** $|x| \geq 1$.

SOLUTION

(a) By the theorem we just discussed, the condition $|x| < 1$ is equivalent to

$$-1 < x < 1$$

The solution set is therefore the open interval $(-1, 1)$; see Figure 5(a).

(b) In view of the second part of the theorem, the inequality $|x| \geq 1$ is satisfied when x satisfies either of the inequalities

$$x \leq -1 \quad \text{or} \quad x \geq 1$$

So the inequality is satisfied if x is in either of the intervals $(-\infty, -1]$ or $[1, \infty)$; see Figure 5(b).

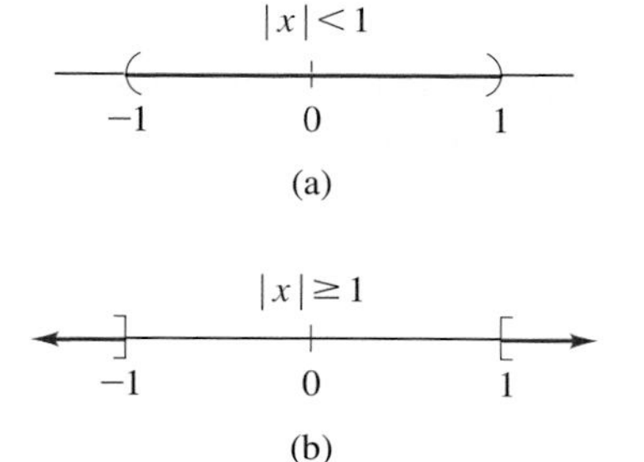

Figure 5

In Example 4(b) we found that the solution set consisted of two intervals on the number line. We have a convenient notation for describing such sets. Given any two sets A and B, we define the set $A \cup B$ (read **A union B**) to be the set of all elements that are in A or in B (or in both). For example, if $A = \{1, 2, 3\}$ and $B = \{4, 5\}$, then $A \cup B = \{1, 2, 3, 4, 5\}$. As another example, the union of the two closed intervals $[3, 5]$ and $[4, 7]$ is given by

$$[3, 5] \cup [4, 7] = [3, 7]$$

because the numbers in the interval [3, 7] are precisely those numbers that are in [3, 5] or [4, 7] (or in both). Using this notation, we can write the solution set for Example 4(b) as

$$(-\infty, -1] \cup [1, \infty)$$

EXAMPLE 5 Two different methods for solving an inequality containing absolute values

Solve: $|x - 3| < 1$.

SOLUTION
We'll show two methods.

FIRST METHOD
We use the theorem preceding Example 4. With $u = x - 3$ and $a = 1$, the theorem tells us that the given inequality is equivalent to

$$-1 < x - 3 < 1$$

or (by adding 3)

$$2 < x < 4$$

The solution set is therefore the open interval (2, 4).

ALTERNATIVE METHOD
The given inequality tells us that x must be less than one unit away from 3 on the number line. Looking one unit to either side of 3, then, we see that x must lie strictly between 2 and 4. The solution set is therefore (2, 4), as obtained using the first method.

EXAMPLE 6 Another inequality with absolute values

Solve: $\left|1 - \frac{t}{2}\right| > 5$.

SOLUTION
Referring again to the theorem, we use the fact that $|u| > a$ means that either $u < -a$ or $u > a$. So, in the present example, the given inequality means that either

$$1 - \frac{t}{2} < -5 \qquad \text{or} \qquad 1 - \frac{t}{2} > 5$$

$$-\frac{t}{2} < -6 \qquad\qquad -\frac{t}{2} > 4$$

$$t > 12 \qquad\qquad t < -8$$

This tells us that the given inequality is satisfied when t is in either of the intervals $(12, \infty)$ or $(-\infty, -8)$. Thus the solution set is $(-\infty, -8) \cup (12, \infty)$.

Earlier in this section we saw that there is a close connection between inequalities and graphs. The next example emphasizes this point again.

EXAMPLE 7 A graphical approach to solving an inequality

GRAPHICAL PERSPECTIVE

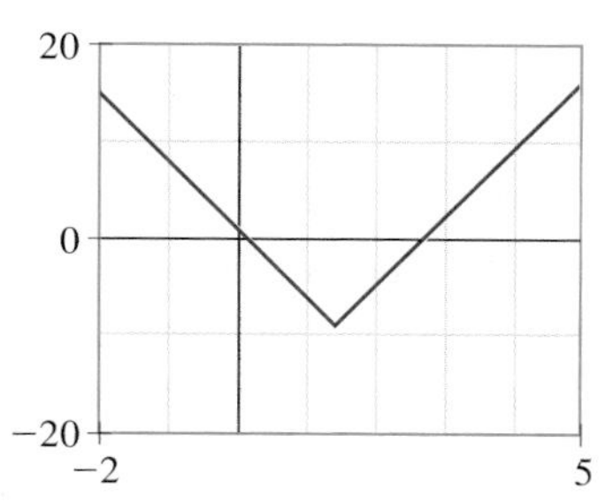

Figure 6
$y = |7x - 10| - 9$
$[-2, 5, 1]$ by $[-20, 20, 10]$

As is indicated in Figure 6, the graph of the equation $y = |7x - 10| - 9$ has two x-intercepts. Let a denote the x-intercept between 0 and 1, and let b denote the x-intercept between 2 and 3.

(a) Use the graph in Figure 6 to specify the general form of the solution set for the inequality

$$|7x - 10| - 9 \leq 0$$

Use interval notation. (You can assume that Figure 6 shows all of the essential features of the graph.)

(b) Use a graphing utility and the ZOOM feature to estimate to the nearest hundredth the x-intercepts a and b. Then state the corresponding estimate for the solution set of the inequality $|7x - 10| - 9 \leq 0$.

(c) Use algebra to solve the inequality $|7x - 10| - 9 \leq 0$. Check that the solution is consistent with the graphical results in parts (a) and (b).

SOLUTION

(a) The graph has two x-intercepts, which we are calling a and b (with $a < b$). At each of these x-intercepts, the quantity $|7x - 10| - 9$ is zero, while for x between a and b, the quantity $|7x - 10| - 9$ is negative (because the graph is *below* the x-axis). In other words, the inequality $|7x - 10| - 9 \leq 0$ is satisfied by all values of x in the closed interval $[a, b]$. Furthermore, the graph shows that these are the only values of x satisfying the required inequality (because the graph is *above* the x-axis when x is outside of this closed interval). In summary then, the solution set for the inequality $|7x - 10| - 9 \leq 0$ has the form

$[a, b]$ where a is between 0 and 1 and b is between 2 and 3

(b) After some experimenting with viewing rectangles of various sizes, we obtain the views displayed in Figure 7. Figure 7(a) shows that the smaller of the two intercepts is between 0.14 and 0.15, closer to 0.14 than to 0.15, so $a \approx 0.14$ to the nearest hundredth. Similarly, for the larger intercept b, Figure 7(b) indicates that $b \approx 2.71$, to the nearest hundredth. Our corresponding estimate for the solution set of the given inequality is then the closed interval $[0.14, 2.71]$.

GRAPHICAL PERSPECTIVE

0.05
0
−0.05
0.140 0.145 0.150

(a) $[0.140, 0.150, 0.005]$ by $[-0.05, 0.05, 0.05]$

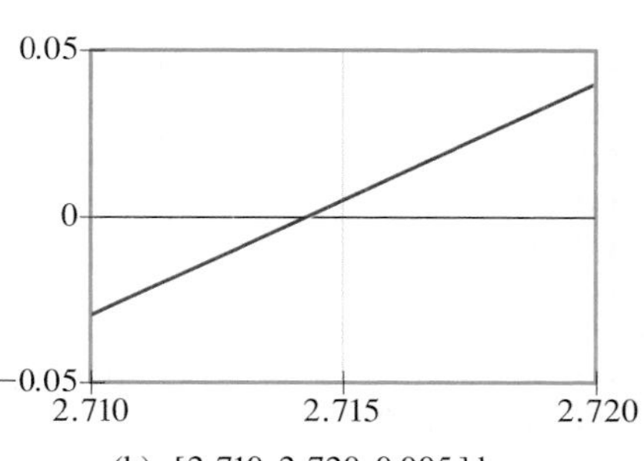

(b) $[2.710, 2.720, 0.005]$ by $[-0.05, 0.05, 0.05]$

Figure 7
Zoom-in views of the x-intercepts for $y = |7x - 10| - 9$

(c) So that we can apply the theorem on page 133, we first rewrite the given inequality in the equivalent form $|7x - 10| \le 9$. Then we have

$$-9 \le 7x - 10 \le 9 \quad \text{using the first part of the theorem on page 107}$$
$$1 \le 7x \le 19 \quad \text{adding 10}$$
$$\frac{1}{7} \le x \le \frac{19}{7} \quad \text{dividing by 7}$$

The solution set is therefore the closed interval $[\frac{1}{7}, \frac{19}{7}.]$ The fact that it's a closed interval agrees with our graphical work in part (a). Furthermore, if you now use a calculator, you'll find that $\frac{1}{7} \approx 0.143$ and $\frac{19}{7} \approx 2.714$, values that are indeed consistent with the estimates obtained in part (b).

EXERCISE SET 2.3

A

In Exercises 1–30, solve the inequality and specify the answer using interval notation.

1. $2x - 7 < 11$
2. $6 - 4x \le 22$
3. $4x + 6 < 3(x - 1) - x$
4. $1 - 2(t + 3) - t \le 1 - 2t$
5. $\frac{3x}{5} - \frac{x - 1}{3} < 1$
6. $\frac{x - 1}{4} - \frac{2x + 3}{5} \le x$
7. $\frac{2x + 1}{2} - \frac{x - 1}{3} < x + \frac{1}{2}$
8. $-3 \le 2x + 1 \le 5$
9. $-1 \le \frac{1 - 4t}{3} \le 1$
10. $\frac{2}{3} \le \frac{5 - 3t}{-2} \le \frac{3}{4}$
11. $0.99 < \frac{x}{2} - 1 < 0.999$
12. $\frac{9}{10} < \frac{3x - 1}{-2} < \frac{91}{100}$

Hint: In Exercises 13 and 14 treat the compound inequality as two separate inequalities.

13. $x - 5 \le 2x + 3 < 10 - 3x$
14. $x - 3 < 3x + 1 < 17 - x$
15. **(a)** $|x| \le \frac{1}{2}$ **(b)** $|x| > \frac{1}{2}$
16. **(a)** $|x| > 2$ **(b)** $|x| \le 2$
17. **(a)** $|x| > 0$ **(b)** $|x| < 0$
18. **(a)** $|t| \ge 0$ **(b)** $|t| \le 0$
19. **(a)** $x - 2 < 1$ **(b)** $|x - 2| < 1$ **(c)** $|x - 2| > 1$
20. **(a)** $x - 4 \ge 4$ **(b)** $|x - 4| \ge 4$ **(c)** $|x - 4| \le 4$
21. **(a)** $1 - x \le 5$ **(b)** $|1 - x| \le 5$ **(c)** $|1 - x| > 5$
22. **(a)** $3x + 5 < 17$ **(b)** $|3x + 5| < 17$ **(c)** $|3x + 5| > 17$
23. **(a)** $a - x < c$ **(b)** $|a - x| < c$ **(c)** $|a - x| \ge c$
24. (Assume $b < c$ throughout this exercise.)
(a) $|x - a| + b < c$
(b) $|x + a| + b > c$
(c) $|x + a| + b < c$
25. $\left|\frac{x - 2}{3}\right| < 4$
26. $\left|\frac{4 - 5x}{2}\right| > 1$
27. $\left|\frac{x + 1}{2} - \frac{x - 1}{3}\right| < 1$ *Hint*: Combine the fractions.
28. $\left|\frac{3(x - 2)}{4} + \frac{4(x - 1)}{3}\right| \le 2$
29. **(a)** $|(x + h)^2 - x^2| < 3h^2$, where $h > 0$
(b) $|(x + h)^2 - x^2| < 3h^2$, where $h < 0$
30. **(a)** $|3(x + 2)^2 - 3x^2| < \frac{1}{10}$
(b) $|3(x + 2)^2 - 3x^2| < \varepsilon$, where $\varepsilon > 0$

For Exercises 31–34:

Ⓖ **(a)** *Use a graph [as in Example 7(a)] to determine which of the following general forms describes the solution set of the given inequality:*

$$[a, b] \qquad (-\infty, a) \cup (b, \infty) \qquad [a, \infty) \qquad (a, \infty)$$

Ⓖ **(b)** *Use a graphing utility [as in Example 7(b)] to estimate to the nearest hundredth the value of a, and where appropriate, b.*

(c) *Solve the inequality algebraically and write the solution set using interval notation. Check that your answers are consistent with the graphical results in parts (a) and (b).*

31. $7x - 2 \ge 0$
32. $6 - 13x < 0$
33. $|8x - 3| - 2 \le 0$
34. $1 - |15x - 3| < 0$
35. Data from the *Apollo 11* moon mission in July 1969 showed temperature readings on the lunar surface varying over the interval $-183° \le C \le 112°$ on the Celsius scale. What is the corresponding interval on the Fahrenheit scale? (Round the numbers you obtain to the nearest integers.)
36. The temperature of the variable star Delta Cephei varies over the interval $5100° \le C \le 6500°$ on the Celsius scale. What is the corresponding range on the Fahrenheit scale? (Round the numbers in your answer to the nearest 100°F.)
37. Data from the *Mariner 10* spacecraft (launched November 3, 1973) indicate that the surface temperature on the planet Mercury varies over the interval $-170° \le C \le 430°$

on the Celsius scale. What is the corresponding interval on the Fahrenheit scale? (Round the values that you obtain to the nearest 10°F.)

B

38. Solve the inequality $|x - 1| + |x - 2| < 3$.
Hint: Begin by considering three cases: $x < 1$; $1 \le x < 2$; $x \ge 2$.

39. Solve the inequality $6 - |x + 3| - |x - 2| < 0$.
Hint: Adapt the hint in Exercise 38.

40. Given two positive numbers a and b, we define the **geometric mean** (G.M.) and the **arithmetic mean** (A.M.) as follows:

$$\text{G.M.} = \sqrt{ab} \qquad \text{A.M.} = \frac{a+b}{2}$$

(a) Complete the table, using a calculator as necessary so that the entries in the third and fourth columns are in decimal form.

a	b	$\sqrt{ab}$ (G.M.)	$(a + b)/2$ (A.M.)	Which is larger, G.M. or A.M.?
1	2			
1	3			
1	4			
2	3			
3	4			
5	10			
9	10			
99	100			
999	1000			

(b) Prove that for all nonnegative numbers a and b we have

$$\sqrt{ab} \le \frac{a+b}{2} \qquad (1)$$

Ⓖ *Hint*: Use the following property of inequalities: If x and y are nonnegative, then the inequality $x \le y$ is equivalent to $x^2 \le y^2$.

(c) Assuming that $a = b$ (and that a and b are nonnegative), show that inequality (1) becomes an equality.

(d) Assuming that a and b are nonnegative and that $\sqrt{ab} = \frac{a+b}{2}$, show that $a = b$.

Remark: Parts (b) through (d) can be summarized as follows. For all nonnegative numbers a and b, we have $\sqrt{ab} \le \frac{a+b}{2}$, with equality holding if and only if $a = b$. This result is known as the *arithmetic-geometric mean inequality* for two numbers. The mini project at the end of this section shows an application of this result.

41. Use the result in Exercise 40(b) to prove each of the following inequalities. Assume that p, q, r, and s all are nonnegative.
(a) $\sqrt{pqr} \le (pq + r)/2$ **(b)** $\sqrt{pqrs} \le (pq + rs)/2$

Exercises 42 and 43 concern the environmental problem known as ***acid rain.*** *The principal source contributing to acid rain is sulfur dioxide, emitted into the atmosphere through the burning of fossil fuels. Once in the clouds, the sulfur dioxide combines with water vapor to form one of the nasty ingredients of acid rain—sulfuric acid.*

42. In the United States over the years 1980–2000, sulfur dioxide emissions due to the burning of fossil fuels can be approximated by the equation

$$y = -0.4743t + 24.086$$

where y represents the sulfur dioxide emissions (in millions of tons) for the year t, with $t = 0$ corresponding to 1980. *Source*: This equation (and the equation in Exercise 48) were computed using data from the book *Vital Signs 1999*, Lester Brown et al. (New York: W. W. Norton & Co., 1999).

Ⓖ **(a)** Use a graphing utility to graph the equation $y = -0.4743t + 24.086$ in the viewing rectangle [0, 25, 5] by [0, 30, 5]. According to the graph, sulfur dioxide emissions are decreasing. What piece of information in the equation $y = -0.4743t + 24.086$ tells you this even before looking at the graph?

(b) Assuming this equation remains valid, estimate the year in which sulfur dioxide emissions in the United States might fall below 10 million tons per year. (You need to solve the inequality $-0.4743t + 24.086 \le 10$.)

43. In Asia over the years 1980–2000, sulfur dioxide emissions due to the burning of fossil fuels can be approximated by the equation

$$y = 1.84t + 14.8$$

where y represents the sulfur dioxide emissions (in millions of tons) for the year t, with $t = 0$ corresponding to 1980.

(a) Use a graphing utility to graph the equation $y = 1.84t + 14.8$ in the viewing rectangle [0, 25, 5] by [0, 60, 20]. According to the graph, sulfur dioxide emissions are increasing. What piece of information in the equation $y = 1.84t + 14.8$ tells you this even before looking at the graph?

(b) Assuming that this equation remains valid, estimate the year in which sulfur dioxide emissions in Asia might exceed 65 million tons per year.

C

44. Let a, b, and c be nonnegative numbers. Follow steps (a) through (e) to show that

$$\sqrt[3]{abc} \le \frac{a+b+c}{3}$$

with equality holding if and only if $a = b = c$.

This result is known as the *arithmetic-geometric mean inequality* for three numbers. (Applications are developed in the projects at the ends of Sections 4.6 and 4.7.)

(a) By multiplying out the right-hand side, show that the following equation holds for all real numbers A, B, and C.

$$3ABC = A^3 + B^3 + C^3 - \frac{1}{2}(A + B + C) \times [(A - B)^2 + (B - C)^2 + (C - A)^2] \tag{1}$$

(b) Now assume for the remainder of this exercise that A, B, and C are nonnegative numbers. Use equation (1) to explain why

$$3ABC \leq A^3 + B^3 + C^3 \tag{2}$$

(c) Make the following substitutions in inequality (2): $A^3 = a$, $B^3 = b$, and $C^3 = c$. Show that the result can be written

$$\sqrt[3]{abc} \leq \frac{a + b + c}{3} \tag{3}$$

(d) Assuming that $a = b = c$, show that inequality (3) becomes an equality.

(e) Assuming $\sqrt[3]{abc} = \dfrac{a + b + c}{3}$, show that $a = b = c$.

Hint: In terms of A, B, and C, the assumption becomes $ABC = \dfrac{A^3 + B^3 + C^3}{3}$. Use this to substitute for ABC on the left-hand side of equation (1). Then use the resulting equation to deduce that $A = B = C$, and consequently $a = b = c$.

MINI PROJECT AN INEQUALITY FOR THE GARDEN

Part of a landscape architect's plans call for a rectangular garden with a perimeter of 150 ft. The architect wants to find the width w and length l that maximize the area A of the garden. Two preliminary sketches are shown in the figure that follows. (In both cases, note that the perimeter is 150 ft, as required.) Of the two sketches, the one on the right yields the larger area. In this project you'll use the inequality given in Exercise 40(b) to find exactly which combination of w and l yields the largest possible area.

(a) As background, first review or rework Exercise 40(b). If you are working in a group, assign one person to do this and then present the result to the others.

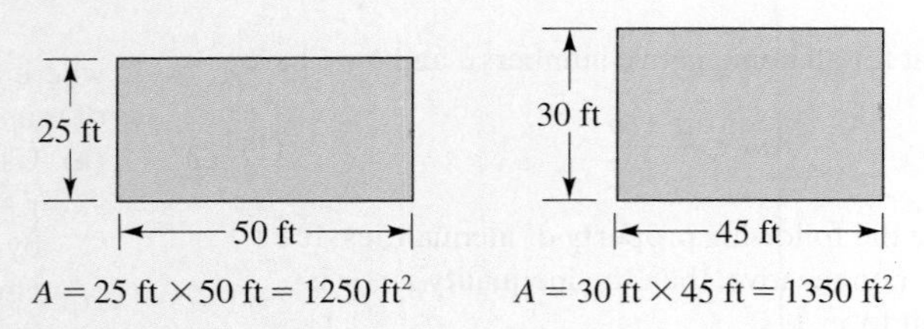

(b) Letting w denote the width of the rectangular garden, and assuming that the perimeter is 150 ft, show or explain why the area A is given by $A = w(75 - w)$.

(c) Use the inequality in Exercise 40(b) to show that $A \leq 5625/4 = 1406.25$. In words: Given a perimeter of 150 ft, no matter how we pick w and l, the area can never exceed 1406.25 ft^2. *Hint*: Apply the inequality to the product $w(75 - w)$.

Ⓖ **(d)** From part (c) we know that the area can't exceed 1406.25 ft^2, so now the job is to find a combination of w and l yielding that area. To accomplish this, take the equation obtained in part (b), replace A with 1406.25, and then use a graphing utility to solve for w.

(e) As a check on the work with the graphing utility (and to stay in shape with the algebra), use the quadratic formula to solve the same equation.

(f) Now that you know w, find l and describe the proportions of the required rectangle.

2.4 MORE ON INEQUALITIES

The symbols of inequality $>$ *and* $<$ *were introduced by* [Thomas] *Harriot* [1560–1621]. *The signs* $\geq$ *and* $\leq$ *were first used about a century later by the Parisian hydrographer Pierre Bouguer.* —Florian Cajori in *A History of Mathematics,* 4th ed. (New York: Chelsea Publishing Co., 1985)

It is rather surprising to think that the man who surveyed and mapped Virginia was one of the founders of algebra as we know the subject today. Such however is the case, for Thomas Harriot was sent by Sir Walter Raleigh to accompany Sir Richard Grenville (1585) to the New World, where he made the survey of that portion of American territory. —David Eugene Smith in *History of Mathematics,* vol. I (New York: Ginn and Co., 1923)

The last example in Section 2.3 emphasized the connection between inequalities and graphs. This is our starting point for the present section. We begin with an example in which polynomial inequalities are solved by relying on a graph. Then we'll use the ideas there to explain a method for solving inequalities even when a graph is not given.

EXAMPLE 1 A graphical approach to solving inequalities

Figure 1 shows the graph of the equation $y = x^2 - 4x + 3$. Use the graph to solve each of the following inequalities:

(a) $x^2 - 4x + 3 < 0$; **(b)** $x^2 - 4x + 3 \geq 0$.

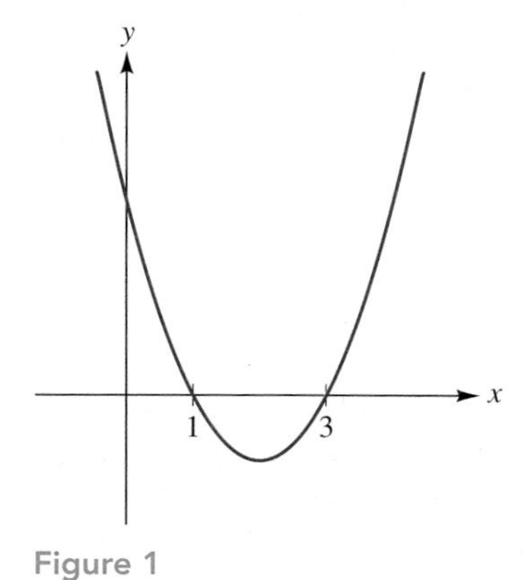

Figure 1
$y = x^2 - 4x + 3$

SOLUTION

(a) As indicated in Figure 1, the graph of $y = x^2 - 4x + 3$ is below the x-axis when (and only when) the x-values are in the open interval $(1, 3)$. This is the same as saying that the quantity $x^2 - 4x + 3$ is negative when (and only when) the x-values are in the open interval $(1, 3)$. Or, rewording once more, the solution set for the inequality $x^2 - 4x + 3 < 0$ is the open interval $(1, 3)$.

(b) The graph in Figure 1 is above the x-axis when (and only when) the x-values are in either of the two intervals $(-\infty, 1)$ or $(3, \infty)$. This is the same as saying that the quantity $x^2 - 4x + 3$ is positive when (and only when) the x-values are in the set $(-\infty, 1) \cup (3, \infty)$. Furthermore, the graph shows that the quantity $x^2 - 4x + 3$ is equal to 0 when $x = 1$ and when $x = 3$. In summary, the solution set for the inequality $x^2 - 4x + 3 \geq 0$ is the set $(-\infty, 1] \cup [3, \infty)$.

We can summarize the essential features of our work in Example 1 as follows. The x-intercepts of the graph in Figure 1 divide the number line into three intervals: $(-\infty, 1)$, $(1, 3)$, and $(3, \infty)$. Within each of these intervals, the algebraic sign of $x^2 - 4x + 3$ stays the same. (For example, to the left of $x = 1$, that is, on the interval $(-\infty, 1)$, the quantity $x^2 - 4x + 3$ is always positive.) More generally, it can be shown (using calculus) that this same type of behavior, regarding *persistence of sign,* occurs with all polynomials and, indeed, with quotients of polynomials as well. This important fact, along with the definition of a *key number,* is presented in the box that follows.

PROPERTY SUMMARY Key Numbers and Persistence of Sign

Let P and Q be polynomials with no common factors (other than, possibly, constants), and consider the following four inequalities:

$$\frac{P}{Q} < 0 \qquad \frac{P}{Q} \le 0 \qquad \frac{P}{Q} > 0 \qquad \frac{P}{Q} \ge 0$$

The **key numbers** for each of these inequalities are the real numbers for which $P = 0$ or $Q = 0$. (Geometrically speaking, the key numbers are the x-intercepts for the graphs of the equations $y = P$ and $y = Q$.) It can be proved (using calculus) that the algebraic sign of P/Q does not change within each of the intervals determined by the key numbers.

EXAMPLES

1. The key numbers for the inequality $(x - 7)/(x - 8) \ge 0$ are 7 and 8.
2. The key numbers for the inequality $(x - 3)(x + 3) < 0$ are ± 3. (In this case, the polynomial Q is the constant 1.)

We'll show how this result is applied by solving the polynomial inequality

$$x^3 - 2x^2 - 3x > 0$$

First, using factoring techniques from basic algebra, we rewrite the inequality in the equivalent form

$$x(x + 1)(x - 3) > 0$$

(If you need a review of factoring, see Appendix B.4.) The key numbers, then, are the solutions of the equation $x(x + 1)(x - 3) = 0$. That is, the key numbers are $x = 0$, $x = -1$, and $x = 3$. Next, we locate these numbers on a number line. As indicated in Figure 2, this divides the number line into four distinct intervals.

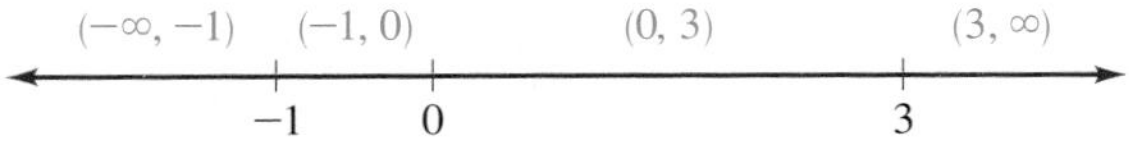

Figure 2

Now, according to the result stated in the box just before this example, no matter what x-value we choose in the interval $(-\infty, -1)$, the resulting sign of $x^3 - 2x^2 - 3x$ $[= x(x + 1)(x - 3)]$ will always be the same. Thus to see what that sign is, we first choose any convenient *test number* in the interval $(-\infty, -1)$, for example, $x = -2$. Then, using $x = -2$, we determine the sign of $x(x + 1)(x - 3)$ simply by considering the sign of each factor, as indicated in Table 1.

Looking at Table 1, we conclude that the values of $x^3 - 2x^2 - 3x$ are negative *throughout* the interval $(-\infty, -1)$ and, consequently, no number in this interval satisfies the given inequality. Next, we carry out similar analyses for the remaining

TABLE 1

Interval	Test number	x	$x + 1$	$x - 3$	$x(x + 1)(x - 3)$
$(-\infty, -1)$	-2	neg.	neg.	neg.	neg.

On the interval $(-\infty, -1)$, the sign of $x^3 - 2x^2 - 3x$ [$= x(x + 1)(x - 3)$] is negative because it is the product of three negative factors.

TABLE 2

Interval	Test number	x	$x + 1$	$x - 3$	$x(x + 1)(x - 3)$
$(-1, 0)$	$-\frac{1}{2}$	neg.	pos.	neg.	pos.
$(0, 3)$	1	pos.	pos.	neg.	neg.
$(3, \infty)$	4	pos.	pos.	pos.	pos.

On the interval $(-1, 0)$, the product $x(x + 1)(x - 3)$ is positive because it has two negative factors and one positive factor. On $(0, 3)$, the product is negative because it has two positive factors and one negative factor. And on $(3, \infty)$, the product is positive because all the factors are positive.

three intervals, as shown in Table 2. (You should verify for yourself that the entries in the table are correct.)

Looking at the two tables now, we conclude that the sign of $x(x + 1)(x - 3)$ is positive throughout both of the intervals $(-1, 0)$ and $(3, \infty)$ and, consequently, all the numbers in these intervals satisfy the given inequality. Moreover, our work also shows that the other two intervals that we considered are not part of the solution set. As you can readily check, the key numbers themselves do not satisfy the given inequality for this example. In summary, then, the solution set for the inequality $x^3 - 2x^2 - 3x > 0$ is $(-1, 0) \cup (3, \infty)$. Furthermore, it is important to note that the work we just carried out also provides us with three additional pieces of information.

The solution set for $x^3 - 2x^2 - 3x \geq 0$ is $[-1, 0] \cup [3, \infty)$.
The solution set for $x^3 - 2x^2 - 3x < 0$ is $(-\infty, -1) \cup (0, 3)$.
The solution set for $x^3 - 2x^2 - 3x \leq 0$ is $(-\infty, -1] \cup [0, 3]$.

In the box that follows, we summarize the steps for solving polynomial inequalities.

Steps for Solving Polynomial Inequalities

1. If necessary, rewrite the inequality so that the polynomial is on the left-hand side and zero is on the right-hand side.
2. Find the key numbers for the inequality and locate them on a number line.
3. List the intervals determined by the key numbers.
4. From each interval, choose a convenient test number. Then use the test number to determine the sign of the polynomial throughout the interval.
5. Use the information obtained in the previous step to specify the required solution set. [Don't forget to take into account whether the original inequality is strict ($<$ or $>$) or nonstrict ($\leq$ or $\geq$).]

EXAMPLE 2 Using factoring and key numbers to solve an inequality

Solve: $x^4 \leq 14x^3 - 48x^2$.

SOLUTION

First we rewrite the inequality so that zero is on the right-hand side. Then we factor the left-hand side as follows:

$$x^4 - 14x^3 + 48x^2 \leq 0$$
$$x^2(x^2 - 14x + 48) \leq 0$$
$$x^2(x - 6)(x - 8) \leq 0$$

From this last line, we see that the key numbers are $x = 0$, 6, and 8. As indicated in Figure 3, these numbers divide the number line into four distinct intervals.

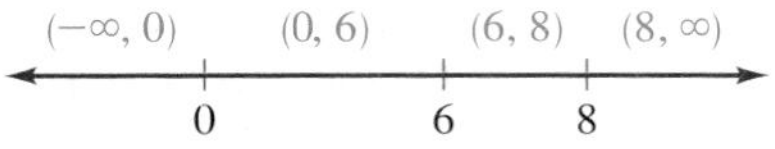

$(-\infty, 0)$ $(0, 6)$ $(6, 8)$ $(8, \infty)$

0 6 8

Figure 3

We need to choose a test number from each interval and see whether the polynomial is positive or negative on the interval. This work is carried out in the following table.

Interval	Test number	x^2	$x - 6$	$x - 8$	$x^2(x - 6)(x - 8)$
$(-\infty, 0)$	-1	pos.	neg.	neg.	pos.
$(0, 6)$	1	pos.	neg.	neg.	pos.
$(6, 8)$	7	pos.	pos.	neg.	neg.
$(8, \infty)$	9	pos.	pos.	pos.	pos.

The table shows that the quantity $x^2(x - 6)(x - 8)$ is negative only for x-values in the interval $(6, 8)$. Also, as noted at the start, the quantity is equal to zero when $x = 0$, 6, or 8. Thus the solution set consists of the numbers in the closed interval $[6, 8]$, along with the number 0. We can write this set

$$[6, 8] \cup \{0\}$$

where $\{0\}$ denotes the set that has zero as its only member.

EXAMPLE 3 An inequality with no key numbers

Solve:

(a) $x^2 - 4x + 5 > 0$; **(b)** $x^2 - 4x + 5 < 0$.

SOLUTION

(a) The equation $x^2 - 4x + 5 = 0$ has no real solution (because the discriminant of the quadratic is -4, which is negative), so there is no key number. Consequently, the polynomial $x^2 - 4x + 5$ never changes sign. To see what that sign is, choose a convenient test number, for example, $x = 0$, and evaluate the polynomial: $0^2 - 4(0) + 5 > 0$. So the polynomial is positive for every value of x, and the solution set is $(-\infty, \infty)$, the set of all real numbers.

GRAPHICAL PERSPECTIVE

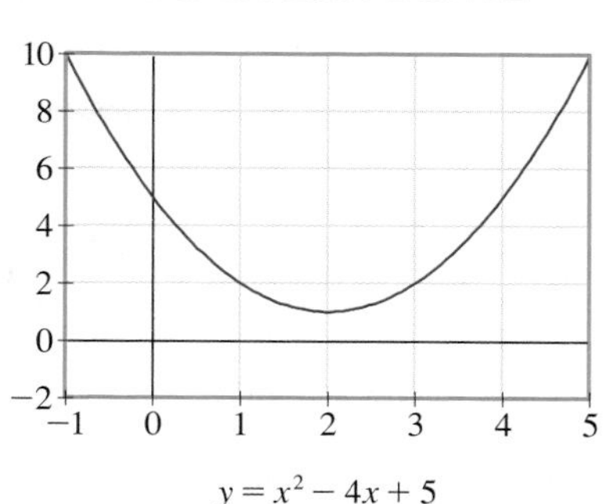

Figure 4
The graph of $y = x^2 - 4x + 5$ lies entirely above the x-axis. Algebraically this means that the quantity $x^2 - 4x + 5$ is positive for all values of x. Consequently, the solution set for $x^2 - 4x + 5 > 0$ is the set of all real numbers. Equivalently, there are no real numbers satisfying $x^2 - 4x + 5 < 0$.

(b) Our work in part (a) shows that no real number satisfies the inequality $x^2 - 4x + 5 < 0$. See Figure 4 for a graphical perspective on this and on the result in part (a).

The technique used in the previous examples can also be used to solve inequalities involving quotients of polynomials. For these cases recall that the definition of a key number includes the x-values for which the denominator is zero. For example, the key numbers for the inequality $\dfrac{x+3}{x-4} \geq 0$ are -3 and 4.

EXAMPLE 4 An inequality involving a quotient of polynomials

Solve: $\dfrac{x+3}{x-4} \geq 0$.

SOLUTION

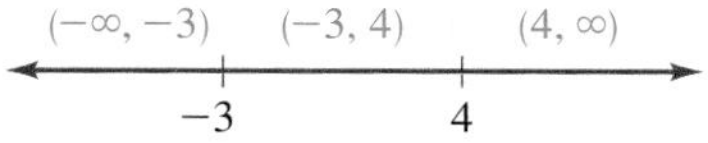

$(-\infty, -3)$ $(-3, 4)$ $(4, \infty)$
-3 4

Figure 5

The key numbers are -3 and 4. As indicated in Figure 5, these numbers divide the number line into three intervals. In the table that follows, we've chosen a test number from each interval and determined the sign of the quotient $(x + 3)/(x - 4)$ for each interval.

Interval	Test number	$x + 3$	$x - 4$	$\dfrac{x+3}{x-4}$
$(-\infty, -3)$	-4	neg.	neg.	pos.
$(-3, 4)$	0	pos.	neg.	neg.
$(4, \infty)$	5	pos.	pos.	pos.

From these results, we conclude that the solution set for $\dfrac{x+3}{x-4} \geq 0$ contains the two intervals $(-\infty, -3)$ and $(4, \infty)$. However, we still need to consider the two endpoints, -3 and 4. As you can easily check, the value $x = -3$ does satisfy the given inequality, but $x = 4$ does not. In summary, then, the solution set is $(-\infty, -3] \cup (4, \infty)$. See Figure 6 for a graphical interpretation of this result. [In

GRAPHICAL PERSPECTIVE

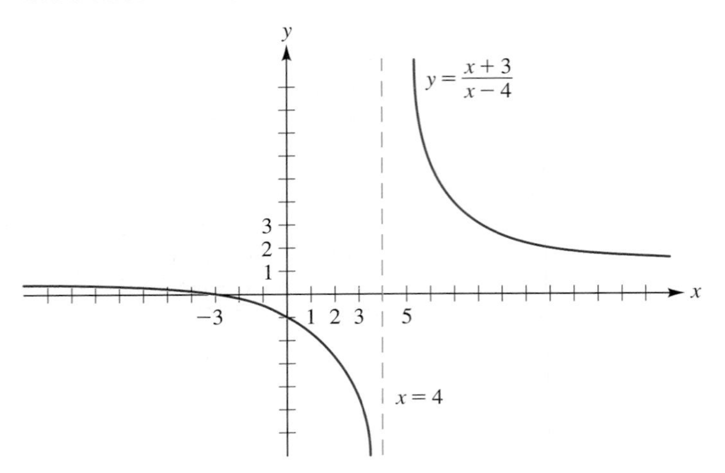

Figure 6
The graph of $y = (x + 3)/(x - 4)$ is above the x-axis when x is to the left of -3 and also when x is to the right of 4. Furthermore, the graph intersects the x-axis (where $y = 0$) when $x = -3$. Consequently, the solution set for the inequality $(x + 3)/(x - 4) \geq 0$ is the set $(-\infty, -3] \cup (4, \infty)$.

Figure 6 the graph of the equation $y = (x + 3)/(x - 4)$ has two distinct pieces, or *branches,* separated by the vertical line $x = 4$. The points on the vertical line itself are not a part of the graph of $y = (x + 3)/(x - 4)$. We'll study graphs like this in Section 4.7, *Rational Functions.*]

EXAMPLE 5 Another inequality involving quotients of polynomials

Solve: $\dfrac{2x+1}{x-1} - \dfrac{2}{x-3} < 1.$

SOLUTION

Our first inclination here might be to multiply through by $(x - 1)(x - 3)$ to eliminate fractions. This strategy is faulty, however, since we don't know whether the quantity $(x - 1)(x - 3)$ is positive or negative. Thus, we begin by rewriting the inequality in an equivalent form, with zero on the right-hand side and a single fraction on the left-hand side.

$$\frac{2x+1}{x-1} - \frac{2}{x-3} - 1 < 0$$

$$\frac{(2x+1)(x-3) - 2(x-1) - 1(x-1)(x-3)}{(x-1)(x-3)} < 0$$

$$\frac{x^2 - 3x - 4}{(x-1)(x-3)} < 0 \qquad \text{Check the algebra!} \tag{1}$$

$$\frac{(x+1)(x-4)}{(x-1)(x-3)} < 0$$

The key numbers are those x-values for which the denominator or the numerator is zero. By inspection, then, we see that these numbers are $-1, 4, 1,$ and 3. As Figure 7 indicates, these numbers divide the number line into five distinct intervals.

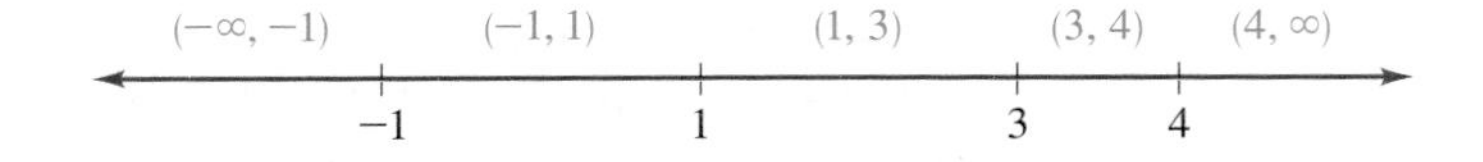

Figure 7

Now, just as in the previous examples, we choose a test number from each interval and determine the sign of the quotient for that interval. (You should check each entry in the following table for yourself.)

Interval	Test number	$(x+1)(x-4)$	$(x-1)(x-3)$	$\dfrac{(x+1)(x-4)}{(x-1)(x-3)}$
$(-\infty, -1)$	-2	pos.	pos.	pos.
$(-1, 1)$	0	neg.	pos.	neg.
$(1, 3)$	2	neg.	neg.	pos.
$(3, 4)$	$7/2$	neg.	pos.	neg.
$(4, \infty)$	5	pos.	pos.	pos.

From these results we can see that the quotient on the left-hand side of inequality (1) is negative (as required) on the two intervals $(-1, 1)$ and $(3, 4)$. Now we need to check the endpoints of these intervals. When $x = -1$ or $x = 4$, the quotient is zero, and so, in view of the original inequality, we exclude these two x-values from the solution set. Furthermore, the quotient is undefined when $x = 1$ or $x = 3$, so we must also exclude those two values from the solution set. In summary, then, the solution set is $(-1, 1) \cup (3, 4)$.

EXERCISE SET 2.4

A

In Exercises 1–6, use the graph to solve each inequality. (Assume that each figure shows all of the essential features of the graph of the equation; that is, that there are no surprises "out of camera range.")

1. (a) $x^2 - 3x - 4 \geq 0$
(b) $x^2 - 3x - 4 \leq 0$

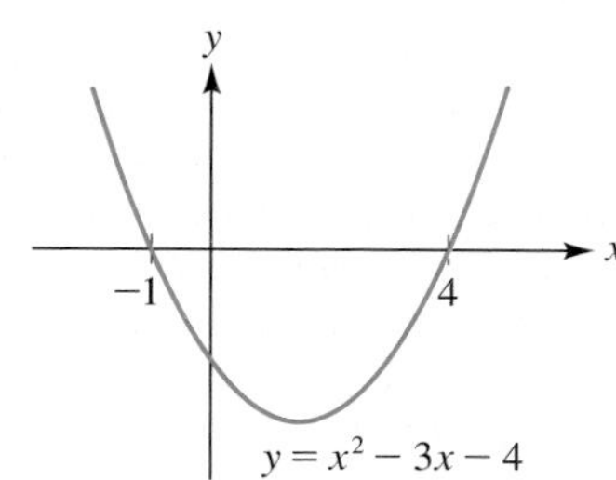

2. (a) $-\frac{1}{2}x^2 - \frac{7}{2}x - 5 > 0$
(b) $-\frac{1}{2}x^2 - \frac{7}{2}x - 5 < 0$

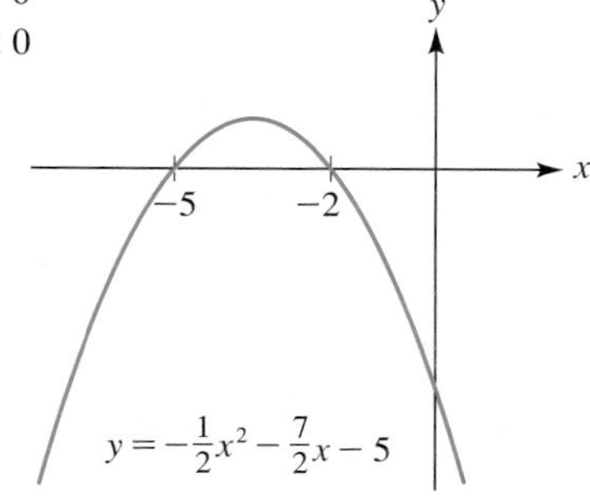

3. (a) $x^4 - 4x^3 + 6x^2 - 4x + 2 < 0$
(b) $x^4 - 4x^3 + 6x^2 - 4x + 2 > 0$

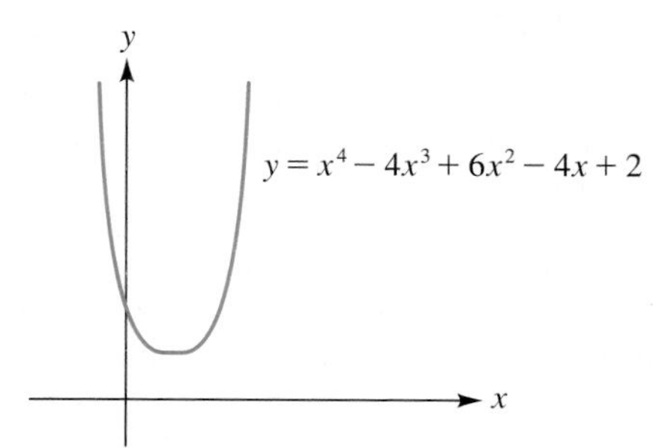

4. (a) $-3(x + 3)^2 \leq 0$
(b) $-3(x + 3)^2 < 0$

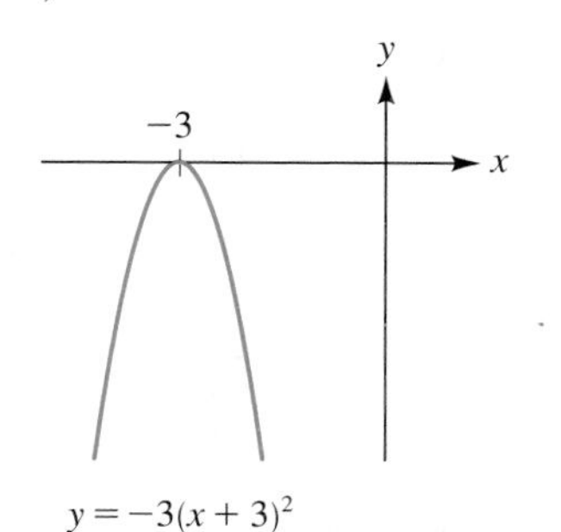

5. **(a)** $x^3 - 3x^2 - x + 3 \geq 0$
   (b) $x^3 - 3x^2 - x + 3 < 0$

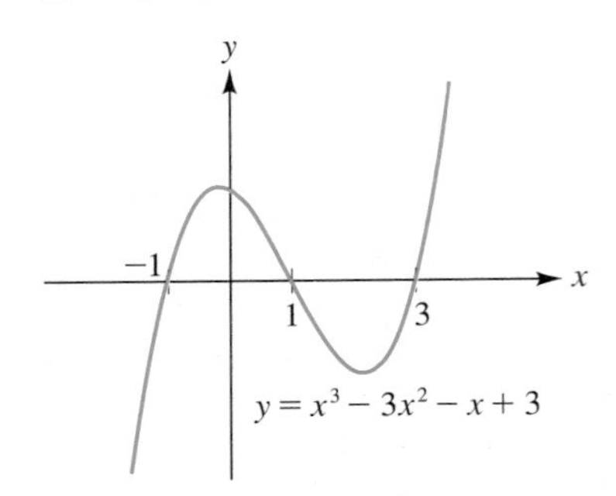

6. **(a)** $-x^3 + 6x^2 - 12x + 9 < 0$
   (b) $-x^3 + 6x^2 - 12x + 9 \geq 0$

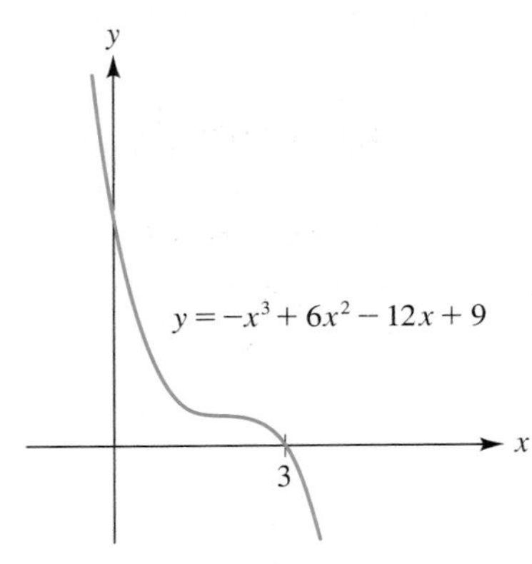

Ⓖ **7.** In the text (pages 114–115) we solved the inequality $x^3 - 2x^2 - 3x > 0$. Graph the equation $y = x^3 - 2x^2 - 3x$ and explain or describe (in complete sentences) the relationship between the graph and the solution set of the given inequality.

Ⓖ **8.** In Example 2 in the text, we solved the inequality $x^4 \leq 14x^3 - 48x^2$. Graph the equation $y = x^4 - 14x^3 + 48x^2$ and explain or describe (in complete sentences) the relationship between the graph and the solution set of the given inequality.

Solve the inequalities in Exercises 9–60. Suggestion: A calculator may be useful for approximating key numbers.

9. $x^2 + x - 6 < 0$
10. $x^2 + 4x - 32 < 0$
11. $x^2 - 11x + 18 > 0$
12. $2x^2 + 7x + 5 > 0$
13. $9x - x^2 \leq 20$
14. $3x^2 + x \leq 4$
15. $x^2 - 16 \geq 0$
16. $24 - x^2 \geq 0$
17. $16x^2 + 24x < -9$
18. $x^4 - 16 < 0$
19. $x^3 + 13x^2 + 42x > 0$
20. $2x^3 - 9x^2 + 4x \geq 0$
21. $2x^2 + 1 \geq 0$
22. $1 + x^2 < 0$
23. $12x^3 + 17x^2 + 6x < 0$
24. $8x^4 < x^2 - 2x^3$
25. $x^2 + x - 1 > 0$
26. $2x^2 + 9x - 1 > 0$
27. $x^2 - 8x + 2 \leq 0$
28. $3x^2 - x + 5 \leq 0$
29. $(x - 1)(x + 3)(x + 4) \geq 0$
30. $x^4(x - 2)(x - 16) \geq 0$
31. $(x + 4)(x + 5)(x + 6) < 0$
32. $(x - \frac{1}{2})(x + \frac{1}{2})(x + \frac{3}{2}) < 0$
33. $(x - 2)^2(3x + 1)^3(3x - 1) > 0$
34. $(2x - 1)^3(2x - 3)^5(2x - 5) > 0$
35. $(x - 3)^2(x + 1)^4(2x + 1)^4(3x + 2) \leq 0$
36. $x^4 - 25x^2 + 144 \leq 0$
37. $20 \geq x^2(9 - x^2)$
38. $x^2(3x^2 + 11) \geq 4$
39. $9(x - 4) - x^2(x - 4) < 0$
40. $(x + 1)^2 - 5(x + 1) > 14$
41. $4(x^2 - 9) - (x^2 - 9)^2 > -5$
42. $x(1 - x^2)^4 + (x + 3)(1 - x^2)^4 \geq 0$
43. $\dfrac{x - 1}{x + 1} \leq 0$
44. $\dfrac{x + 4}{2x - 5} \leq 0$
45. $\dfrac{2 - x}{3 - 2x} \geq 0$
46. $\dfrac{x^2 - 1}{x^2 + 8x + 15} \geq 0$
47. $\dfrac{x^2 - 8x - 9}{x} < 0$
48. $\dfrac{x^2 - 3x + 1}{1 - x} < 0$
49. $\dfrac{2x^3 + 5x^2 - 7x}{3x^2 + 7x + 4} > 0$
50. $\dfrac{x^2 - x - 1}{x^2 + x - 1} > 0$
51. $\dfrac{x}{x + 1} > 1$
52. $\dfrac{2x}{x - 2} < 3$
53. $\dfrac{1}{x} \leq \dfrac{1}{x + 1}$
54. $\dfrac{2}{x} < \dfrac{x}{2}$
55. $\dfrac{1}{x - 2} - \dfrac{1}{x - 1} \geq \dfrac{1}{6}$
56. $\dfrac{2x}{x + 5} + \dfrac{x - 1}{x - 5} < \dfrac{1}{5}$
57. $\dfrac{1 + x}{1 - x} - \dfrac{1 - x}{1 + x} < -1$
58. $\dfrac{x + 1}{x + 2} > \dfrac{x - 3}{x + 4}$
59. $\dfrac{x^2 - x - 2}{x^2 - 3x + 2} > 0$
60. $\dfrac{x^2 + 3x}{x^2 + 8x + 15} < 0$

For Exercises 61–76:

Ⓖ **(a)** *Use a graph to estimate the solution set for each inequality. Zoom in far enough so that you can estimate the relevant endpoints to the nearest thousandth.*

(b) *Exercises 61–70 can be solved algebraically using the techniques presented in this section. Carry out the algebra to obtain exact expressions for the endpoints that you estimated in part (a). Then use a calculator to check that your results are consistent with the previous estimates.*

61. $x^2 - 5x + 3 \leq 0$
62. $x^2 + x - 4 \leq 0$
63. $0.25x^2 - 6x - 2 < 0$
64. $0.25x^3 - 6x^2 - 2x < 0$
65. $x^4 - 2x^2 - 1 > 0$
66. $x^4 - 2x^2 + 1 > 0$
67. $(x^2 - 5)/(x^2 + 1) \leq 0$
68. $(x^3 - 5)/(x^2 + 1) \leq 0$
69. $(x^2 + 1)/(x^2 - 5) \leq 0$
70. $(x^2 + 1)/(x^3 - 5) \leq 0$
71. $x^3 + 2x \geq -1$
72. $x^3 - 2x \geq -1$
73. $x^4 - 2x > -1$
74. $x^5 - 2x > -1$
75. $x - \dfrac{x^3}{3!} + \dfrac{x^5}{5!} - \dfrac{x^7}{7!} < 0$ *Note*: For natural numbers n, the symbol $n!$ (read *n factorial*) denotes the product of the first n natural numbers. For instance, $3! = 1 \times 2 \times 3 = 6$.
76. $1 - \dfrac{x^2}{2!} + \dfrac{x^4}{4!} - \dfrac{x^6}{6!} + \dfrac{x^8}{8!} < 0$

B

Ⓖ **77.** Suppose that after studying a corporation's records, a business analyst predicts that the corporation's monthly revenues R for the near future can be closely approximated by the equation

$$R = -0.0217x^5 + 0.626x^4 - 6.071x^3 + 25.216x^2 - 57.703x + 159.955 \qquad (1 \le x \le 12)$$

where R is the revenue (in thousands of dollars) for the month x, with $x = 1$ denoting January, $x = 2$ denoting February, and so on.

(a) According to this model, for which months will the monthly revenue be no more than \$80,000?
Hint: You need to solve the inequality $R \le 80$. Round each key number to the nearest integer. (Why?)

(b) For which months, if any, will the monthly revenue be at least \$120,000?

Ⓖ **78.** (Continuation of Exercise 77)

(a) Solve the inequality $R > 165$ to determine the months, if any, that the revenue will exceed \$165,000.

(b) Are there any months when the revenue will fall below \$45,000?

79. For which values of b will the equation $x^2 + bx + 1 = 0$ have real solutions?

80. The sum of the first n natural numbers is given by

$$1 + 2 + 3 + \cdots + n = \frac{n(n+1)}{2}$$

For which values of n will the sum be less than 1225?

81. For which values of a is $x = 1$ a solution of the following inequality?

$$\frac{2a + x}{x - 2a} < 1$$

82. Solve $\dfrac{ax + b}{\sqrt{x}} > 2\sqrt{ab}$, where a and b are positive constants.

83. The two shorter sides in a right triangle have lengths x and $1 - x$, where $x > 0$. For which values of x will the hypotenuse be less than $\sqrt{17}/5$?

84. A piece of wire 12 cm long is cut into two pieces. Denote the lengths of the two pieces by x and $12 - x$. Both pieces are then bent into squares. For which values of x will the combined areas of the squares exceed 5 cm^2?

C

85. Find a nonzero value for c so that the solution set for the inequality

$$x^2 + 2cx - 6c < 0$$

is the open interval $(-3c, c)$.

86. Solve $(x - a)^2 - (x - b)^2 > (a - b)^2/4$, where a and b are constants and $a > b$.

PROJECT — WIND POWER

The relatively recent [in terms of human history] *transition to coal that began in Europe in the seventeenth century marked a major shift to dependence on a finite stock of fossilized fuels whose remaining energy is now equivalent to less than 11 days of sunshine. From a millennial perspective, today's hydrocarbon-based civilization is but a brief interlude in human history.*—Christopher Flavin and Seth Dunn, "Reinventing the Energy System" in *Vital Signs, 1999,* Lester R. Brown et al. (New York: W. W. Norton & Co., 1999)

As the world's nonrenewable energy resources such as oil and coal start to become more scarce, alternative sources such wind and solar become increasingly important. In this project you'll calculate some estimates concerning wind power, an industry and technology that has grown rapidly over the past two decades. (Later, in Chapter 5, after studying exponential and logarithmic functions, we'll be able to make some rough estimates for possible depletion dates regarding resources such as oil and coal.) In working this project, you'll be reviewing math from the following sections of this book:

- 1.3 Solving Equations
- 1.6 Equations of Lines
- 2.3 Inequalities
- 2.4 More on Inequalities

(*continues*)

Wind turbines in Altamont, California. (Photo by Glen Allison, PhotoDisc.)

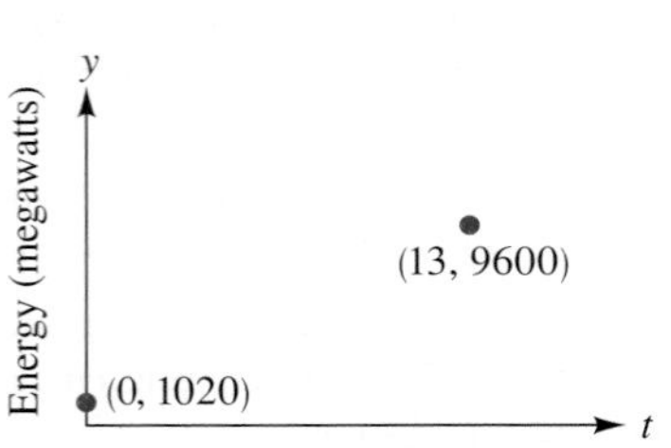

World wind-energy generating capacity, 1985 ($t = 0$) and 1998 ($t = 13$).

Source: Data from *BTM Consult,* published in *Vital Signs, 1999,* p. 53, Lester R. Brown et al. (New York: W. W. Norton & Co., 1999)

(a) The two data points in the above figure show world wind-energy generating capacity for the years 1985 and 1998. What is the slope of the line joining the two points? What are the units for the slope?

(b) Find the equation of the line joining the two given points. Write the answer in the form $y = mt + b$.

(c) Use the result in part (b) to estimate the years in which world wind-energy generating capacity was between 3000 megawatts and 7500 megawatts (inclusive). *Hint*: You need to solve the inequality $3000 \leq y \leq 7500$; round the answers to the nearest integers.

Ⓖ **(d)** Your estimates in part (c) are based on the assumption that world wind-energy generating capacity increased in a straight-line fashion from 1985 to 1998. By using techniques from statistics (and more detailed data), it can be shown that a more accurate summary of world wind-energy generating capacity over these years is given by the equation

$$y = 6.977t^3 - 64.786t^2 + 325.4t + 1012$$

where again, $t = 0$ corresponds to 1985. Using this expression for y, estimate the years in which world wind-energy generating capacity was between 3000 megawatts and 7500 megawatts (inclusive). (Round the key numbers to the nearest integers.) Compare your results to those in part (c).

Ⓖ **(e)** Although increasing rapidly, wind-energy generation is currently but a small portion of the world energy picture. For instance, whereas world wind-energy generating capacity in 1998 was 9600 megawatts, the corresponding figure for nuclear energy in 1998 was 343,000 megawatts. When might wind-energy reach this level? Use the equation in part (d) to obtain one possible answer to this question. Round your answer to the nearest five years. *Hint*: In an appropriate viewing rectangle, you want to look for the point where the curve $y = 6.977t^3 - 64.786t^2 + 325.4t + 1012$ meets the horizontal line $y = 343{,}000$.

(f) Rework part (e), but instead of using the cubic equation from part (d), use the straight-line equation obtained in part (b). Compare the results; which scenario do you think is more realistic?

Chapter 2 Summary of Principal Terms and Formulas

Terms or notation	Page reference	Comments
1. Quadratic equation	80	A quadratic equation is an equation that can be written in the form $ax^2 + bx + c = 0$, where a, b, and c are constants with $a \neq 0$.
2. Quadratic formula $x = \dfrac{-b \pm \sqrt{b^2 - 4ac}}{2a}$	82	This is the quadratic formula; it provides the solutions of the quadratic equation $ax^2 + bx + c = 0$. The formula is derived on pages 81–82 by means of the useful technique of completing the square.
3. Product-and-sum-of-roots theorem	84	The product and the sum of the roots of the quadratic equation $x^2 + bx + c = 0$ are c and $-b$, respectively.
4. Discriminant	87	The discriminant of the quadratic equation $ax^2 + bx + c = 0$ is the number $b^2 - 4ac$. As indicated in the box on page 87, the discriminant provides information about the roots of the equation.
5. Extraneous solution (or extraneous root)	99	In solving equations, certain processes (such as squaring both sides) can lead to answers that do not check in the original equation. These numbers are called extraneous solutions (or extraneous roots).
6. Solution set of an inequality	103	This is the set of numbers that satisfy the inequality.
7. $A \cup B$	107	The set $A \cup B$ consists of all elements that are in the set A or the set B (or both).
8. Key numbers of an inequality	114	Let P and Q denote polynomials, with no common factors (except possibly constants), and consider the following four inequalities: $\dfrac{P}{Q} < 0 \quad \dfrac{P}{Q} \leq 0 \quad \dfrac{P}{Q} > 0 \quad \dfrac{P}{Q} \geq 0$ The key numbers for each of these inequalities are those real numbers for which $P = 0$ or $Q = 0$. It can be proved that the algebraic sign of P/Q is constant within each of the intervals determined by these key numbers.

Writing Mathematics

1. More than 3000 years ago, the ancient Babylonian mathematicians solved quadratic equations. A method they used is demonstrated (using modern notation) in the following example.

> To find the positive root of $x^2 + 8x = 84$:
>
> Rewrite the equation as $x(x + 8) = 84$, and let $y = x + 8$. Then the equation to be solved becomes $xy = 84$. Now take half of the coefficient of x in the original equation, which is 4, and define another variable t by $t = x + 4$. Then we have $x = t - 4$ and $y = t + 4$. Therefore
>
> $$\begin{aligned}(t - 4)(t + 4) &= 84\\ t^2 &= 100\\ t &= 10\end{aligned}$$
>
> With $t = 10$ we get $x = 10 - 4 = 6$, the required positive root.

On your own or with a group of classmates, work through the above example, filling in any missing details if necessary. Then (strictly on your own), use the Babylonian method to find the positive root of each of the following equations. Write out your solutions in detail, as if you were explaining the method to another student who had not seen it before. This will involve a combination of English composition and algebra. Also, in part (b), check your answer by using the quadratic formula.

(a) $x^2 + 14x = 72$ **(b)** $x^2 + 2Ax = B$, where $B > 0$

2. The ancient Greek mathematicians (2500 years ago) used geometric methods to solve quadratic equations or, rather, to construct line segments whose lengths were the roots of the equations. According to historian Howard Eves in *An Introduction to the History of Mathematics*, 6th ed. (Philadelphia: Saunders College Publishing, 1990):

> *Imbued with the representation of a number by a length and completely lacking any adequate algebraic notation, the early Greeks devised ingenious geometrical processes for carrying out algebraic operations.*

One of the methods used by the ancient Greeks to solve a quadratic equation is described in the following example (using modern algebraic notation).

To construct a line segment whose length is equal to the (positive) root of the equation $x^2 + 8x = 84$:

> Begin with a line segment $\overline{AB}$ of length 8. At B, construct a line segment $\overline{BC} \perp \overline{AB}$ such that the length of $\overline{BC}$ is $\sqrt{84}$. Next, let M be the midpoint of $\overline{AB}$. With M as center, draw a circular arc of radius $\overline{MC}$ intersecting $\overline{AB}$ (extended) at P. Then the length of $\overline{AP}$ is the required root.

On your own or with a group of classmates, work through the preceding construction. That is, sketch the appropriate figure and verify that the construction indeed yields the positive root of the equation. Then, on your own, use the Greek method to determine the positive root of each of the following equations. Write out your work in detail, as if you were explaining it to a student who had not seen it before. Be sure to include the appropriate geometric figures and an explanation of why the method works.

(a) $x^2 + 14x = 72$ **(b)** $x^2 + 2cx = d \quad (c, d > 0)$

3. The example in Exercise 1 demonstrates a Babylonian method for solving a quadratic equation. According to Professor Victor J. Katz in *A History of Mathematics* (New York: HarperCollins College Publishers, 1993), "... whatever the ultimate origin of this method, a close reading of the wording of the [ancient clay] tablets seems to indicate that the scribe had in mind a geometric procedure [completing-the-square]. ..."

One of the earliest *explicit* uses of the completing-the-square technique to solve a quadratic is due to the ninth-century Persian mathematician and astronomer Muhammed al-Khwārizmī. The following example demonstrates al-Khwārizmī's use of completing the square to solve the equation $x^2 + 8x = 84$:

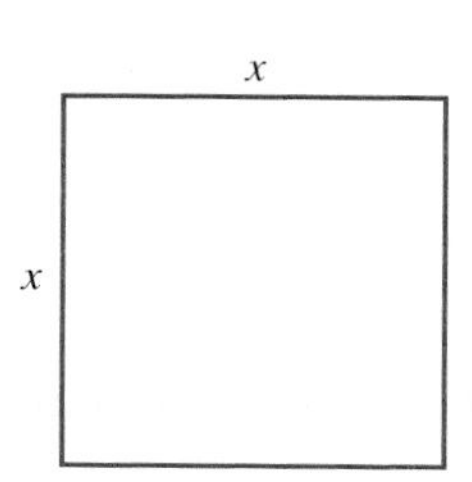

Figure A

To find the positive root of $x^2 + 8x = 84$:

Begin with a square of side x, as in Figure A. Take half of the coefficient of x: this is one-half of 8, or 4. Now form two rectangles, each with dimensions 4 by x, and adjoin them to the square, as indicated in Figure B. Then, as in Figure C, draw the dashed lines to complete the (outer) square.

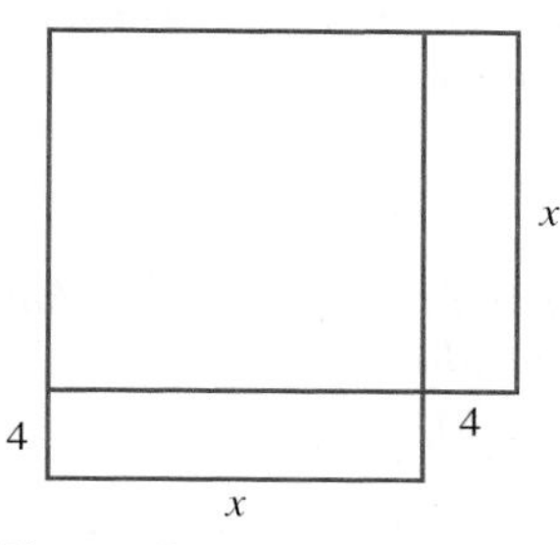

Figure B

In Figure B, the combined area of the square and the two rectangles is $x^2 + 8x$. But from the equation we wish to solve, $x^2 + 8x$ is equal to 84. In Figure C the area of the small square in the lower right-hand corner is 16. Thus the area of the entire outer square is $16 + 84$, or 100. But the area of the outer square is also $(x + 4)^2$. Therefore

$$(x + 4)^2 = 100 \qquad \text{and, consequently,} \qquad x + 4 = 10$$

The positive root we are looking for is therefore $x = 6$.

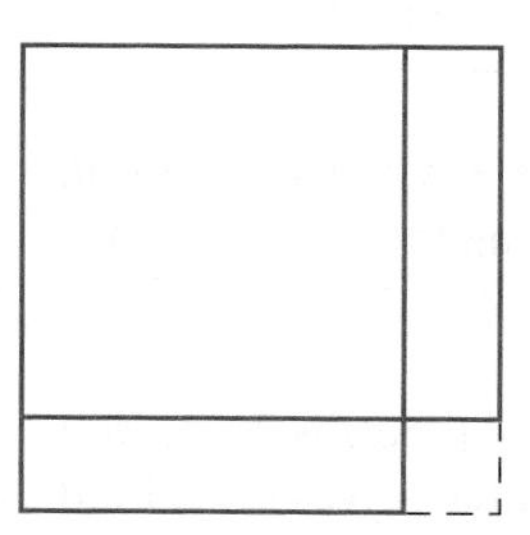
Figure C

On your own or with a group of classmates, work through the preceding example, filling in the missing details as necessary. Then, on your own, use this completing-the-square process to find the positive root of each of the following equations. Write out your solutions in detail, as if you were explaining the method to another student who had not seen it before. Be sure to include the appropriate geometric figures.

(a) $x^2 + 14x = 72$ **(b)** $x^2 + 2Ax = B$, where $B > 0$

Chapter 2 Review Exercises

In Exercises 1–12, answer TRUE *if the statement is true without exception. Otherwise, answer* FALSE.

1. If $x < 3$, then $x + 7 < 10$.

2. If $x < y$, then $x^2 < y^2$.

3. If $x \geq -4$, then $x \leq 4$.

4. If $-x > y$, then $x < -y$.

5. If $x < 2$, then $1/x < 1/2$.

6. $x \leq x^2$

7. If $\sqrt{x + 1} = 3$, then $x + 1 = 9$.

8. If $0 < x < 1$, then $1/x > x$.

9. If $x < 2$ and $y < 3$, then $x < y$.

10. If $x < 2$ and $y > 3$, then $x < y$.
11. If $a - b \le a^2 - b^2$, then $1 \le a + b$.
12. If $0 < a - b \le a^2 - b^2$, then $1 \le a + b$.

In Exercises 13–32, find all of the real solutions of each equation.

13. $12x^2 + 2x - 2 = 0$
14. $4y^2 - 21y = 18$
15. $\frac{1}{2}x^2 + x - 12 = 0$
16. $x^2 + \frac{13}{2}x + 10 = 0$
17. $\dfrac{x}{5 - x} = \dfrac{-2}{11 - x}$
18. $\dfrac{x^2}{(x - 1)(x + 1)} = \dfrac{4}{x + 1} + \dfrac{4}{(x - 1)(x + 1)}$
19. $\dfrac{1}{3x - 7} - \dfrac{2}{5x - 5} - \dfrac{3}{3x + 1} = 0$
20. $4x^2 + x - 2 = 0$
21. $t^2 + t - \frac{1}{2} = 0$
22. $x^3 - 6x^2 + 7x = 0$
23. $1 + 14x^{-1} + 48x^{-2} = 0$
24. $x^{-2} - x^{-1} - 1 = 0$
25. $x^{1/2} - 13x^{1/4} + 36 = 0$
26. $y^{2/3} - 21y^{1/3} + 80 = 0$
27. $\sqrt{4x + 3} = \sqrt{11 - 8x} - 1$
28. $2 - \sqrt{3\sqrt{2x - 1} + x} = 0$
29. $\dfrac{2}{\sqrt{x^2 - 36}} + \dfrac{1}{\sqrt{x + 6}} - \dfrac{1}{\sqrt{x - 6}} = 0$
30. $\sqrt{x - \sqrt{1 - x}} + \sqrt{x} - 1 = 0$
31. $\sqrt{x + 7} - \sqrt{x + 2} = \sqrt{x - 1} - \sqrt{x - 2}$
32. $\dfrac{\sqrt{x} - 4}{\sqrt{x} - 18} = 3$

In Exercises 33–38, solve each equation for x in terms of the other letters.

33. $4x^2y^2 - 4xy = -1$, where $y \ne 0$
34. $4x^4y^4 + 2x^2y^2 + \frac{1}{4} = 0$, where $y \ne 0$
35. $x + \dfrac{1}{a} - \dfrac{1}{b} = \dfrac{2}{a^2x} + \dfrac{2}{abx}$
36. $\dfrac{a}{x - b} + \dfrac{b}{x - a} = \dfrac{x - a}{b} + \dfrac{x - b}{a}$, where $a \ne -b$
Suggestion: Before clearing fractions, carry out the indicated additions on each side of the equation.
37. $\dfrac{1}{x + a + b} = \dfrac{1}{x} + \dfrac{1}{a} + \dfrac{1}{b}$, where $a + b \ne 0$
38. $\dfrac{1}{x} + \dfrac{1}{a - x} + \dfrac{1}{x + 3a} = 0$, where $a \ne 0$

Solve the inequalities in Exercises 39–55.

39. $-1 < \dfrac{1 - 2(1 + x)}{3} < 1$
40. $3 < \dfrac{x - 1}{-2} < 4$
41. $|x| \le 1/2$
42. $|3 - 5x| < 2$
43. $|2x - 1| \ge 5$
44. $x^2 - 21x + 108 \le 0$
45. $x^2 + 3x - 40 < 0$
46. $x^2 \ge 15x$
47. $x^2 - 6x - 1 < 0$
48. $(x + 12)(x - 1)(x - 8) < 0$
49. $x^4 - 34x^2 + 225 < 0$
50. $\dfrac{x + 12}{x - 5} > 0$
51. $\dfrac{(x - 7)^2}{(x + 2)^3} \ge 0$
52. $\dfrac{(x - 6)^2(x - 8)(x + 3)}{(x - 3)^2} \le 0$
53. $\dfrac{3x + 1}{x - 4} < 1$
54. $\dfrac{1 - 2x}{1 + 2x} \le \dfrac{1}{2}$
55. $x^2 + \dfrac{1}{x^2} > 3$ *Suggestion*: Use a calculator to evaluate the key numbers.

For Exercises 56 and 57, find the values of k for which the roots of the equations are real numbers.

56. $kx^2 - 6x + 5 = 0$
57. $x^2 + (k + 1)x + 2k = 0$

In Exercises 58 and 59, find a quadratic equation that has integer coefficients and the given values as roots.

58. $r_1 = (1 + \sqrt{5})/3, r_2 = (1 - \sqrt{5})/3$
59. $r_1 = -3, r_2 = -4$
60. Solve for x:

$$5\sqrt[3]{b^2 - x^2} = \sqrt[3]{(b + x)^2} + 4\sqrt[3]{(b - x)^2},$$

where $b \ne 0$.
Hint: Divide through by $\sqrt[3]{(b - x)^2}$. The resulting equation can be solved by using the substitution $t = \sqrt[3]{\dfrac{b + x}{b - x}}$.

61. The sum of the cubes of two numbers is 2071, while the sum of the two numbers themselves is 19. Find the two numbers.
62. The sum of the digits in a certain two-digit number is 11. If the order of the digits is reversed, the number is increased by 27. Find the original number.
63. The four corners of a square $ABCD$ have been cut off to form a regular octagon, as shown in the figure. If each side of the square is 1 cm long, how long is each side of the octagon?

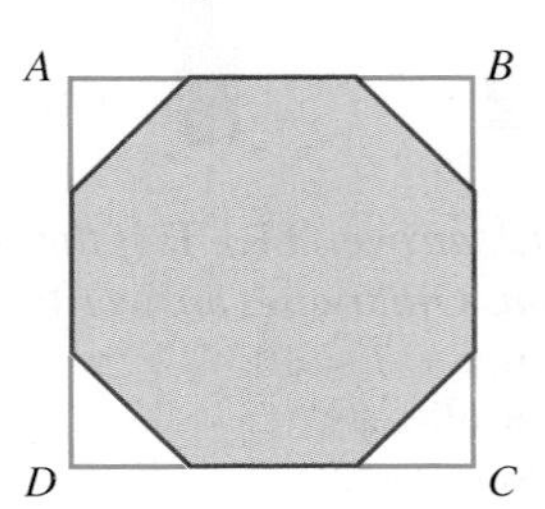

64. Determine p if the larger root of the equation $x^2 + px - 2 = 0$ is p.

65. A piece of wire x cm long is bent into a square. For which values of x will the area be (numerically) greater than the perimeter?

66. A piece of wire $6x$ cm long is bent into an equilateral triangle. For which values of x will the area be (numerically) less than the perimeter?

67. Find three consecutive positive integers such that the sum of their squares is 1454.

68. The length of a rectangular piece of tin exceeds the width by 8 cm. A 1 cm square is cut from each corner of the piece of tin, and then the resulting flaps are turned up to form a box with no top. What are the dimensions of the box if its volume is 48 cm^3?

69. The length and width of a rectangular flower garden are a and b, respectively. The garden is bordered on all four sides by a gravel path of uniform width. Find the width of the path, given that the area of the garden equals the area of the path.

If an object is thrown vertically upward from a height of h_0 ft with an initial speed of v_0 ft/sec, then its height h (in feet) after t seconds is given by

$$h = -16t^2 + v_0t + h_0$$

Make use of this formula in working Exercises 70–72.

70. A ball is thrown vertically upward from ground level with an initial speed of 64 ft/sec.

(a) At what time will the height of the ball be 15 ft? (Two answers.)

(b) For how long an interval of time will the height exceed 63 ft?

71. One ball is thrown vertically upward from a height of 50 ft with an initial speed of 40 ft/sec. At the same instant, another ball is thrown vertically upward from a height of 100 ft with an initial speed of 5 ft/sec. Which ball hits the ground first?

72. An object is projected vertically upward. Suppose that its height is H ft at t_1 sec and again at t_2 sec. Express the initial speed in terms of t_1 and t_2.

Answer: $16(t_1 + t_2)$

73. A rectangle is inscribed in a semicircle of radius 1 cm, as shown. For which value of x is the area of the rectangle 1 cm^2? *Note:* x is defined in the figure.

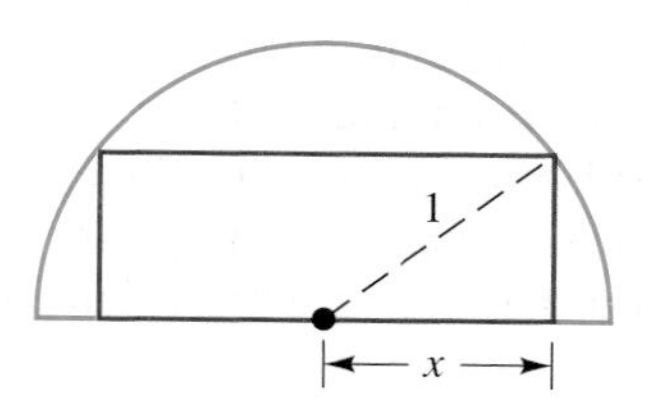

74. The height of an isosceles triangle is $a + b$ units, where $a > b > 0$, and the area does not exceed $a^2 - b^2$ square units. What is the range of possible values for the base of the triangle?

75. A circle is inscribed in a quadrant of a larger circle of radius r (as shown in the figure). Find the radius of the inscribed circle.

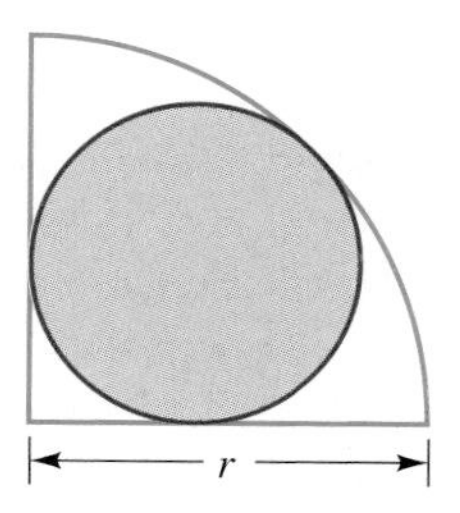

Chapter 2  Test 

In Problems 1–4, find all of the real solutions of each equation.

1. (a) $x^2 + 4x = 5$

(b) $x^2 + 4x = 1$

2. $x^2(x^2 - 7) + 12 = 0$

3. $1 + \sqrt{2 - x} - \sqrt{5 - 2x} = 0$

4. $|3x - 1| = 2$

5. Find the sum of the roots of the equation $2x^2 + 8x - 9 = 0$.

6. Find a quadratic equation with integer coefficients and with roots $r_1 = 2 + 3\sqrt{7}$ and $r_2 = 2 - 3\sqrt{7}$.

7. As indicated in the following figure, the graph of $y = x^4 - 3x^2 - 1$ has two x-intercepts, one between 1.5 and 2 and one between -1.5 and -2. Find exact expressions (containing radicals) for these intercepts and also calculator approximations rounded to three decimal places. Check that your calculator values are consistent with the graph.

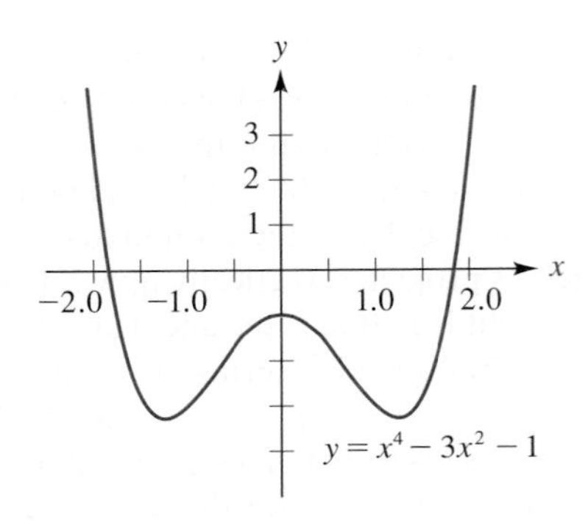

8. Consider the equation $2x^{4/3} - x^{2/3} - 6 = 0$.

Ⓖ **(a)** Use a graphing utility to approximate the roots to the nearest hundredth.

(b) Use algebra to determine the exact solutions. Then check that the answers are consistent with the approximations obtained in part (a).

In Problems 9–13, solve the inequalities. Write the answers using interval notation.

9. $4(1 + x) - 3(2x - 1) \geq 1$

10. $\dfrac{3}{5} < \dfrac{3 - 2x}{-4} < \dfrac{4}{5}$

11. $|3x - 8| \leq 1$

12. $(x - 4)^2(x + 8)^3 \geq 0$

13. $\dfrac{1}{x} + \dfrac{1}{x + 1} + \dfrac{1}{x + 2} \geq 0$

14. Find the values of k for which the roots of the equation $x^2 + 3x + k^2 = 0$ are real numbers.

15. The point (a, b) lies in the third quadrant on the graph of the line $y = 2x + 1$. Find a and b given that the distance from (a, b) to the origin is $\sqrt{65}$.

16. After solving the equation $1 + \sqrt{3x + 7} = x$, give an example of what is meant by an extraneous root, and explain (in complete sentences) how, in this case, the extraneous root was generated.

3 CHAPTER Functions

To make a start out of particulars and make them general . . . —William Carlos Williams (1883–1963) in his poem *Paterson*

The concept of a function is one of the most useful and broad-ranging ideas in all of mathematics. After introducing the definition of a function and function notation in Section 3.1, we use graphs in the next three sections to understand and to analyze functions. In Sections 3.5 and 3.6 we find that two functions can be combined to produce new functions. One of these ways of combining two functions is known as *composition of functions.* As you'll see, this is the unifying theme between Sections 3.5 and 3.6. In this chapter we use functions to

- Estimate how many units of a particular item a store can expect to sell at a given price (Example 7 in Section 3.1)
- Compute the average rate of increase or decrease of a patient's temperature over given time periods (Example 3 in Section 3.3)
- Analyze and compare changing patterns of consumption of red meat in China and in the United States (Exercises 19 and 20 in Section 3.3)
- Determine the rate of spread of an oil spill (Example 5 in Section 3.5)

3.1 THE DEFINITION OF A FUNCTION

The word "function" was introduced into mathematics by Leibniz, who used the term primarily to refer to certain kinds of mathematical formulas. It was later realized that Leibniz's idea of function was much too limited in scope, and the meaning of the word has since undergone many steps of generalization. —Tom M. Apostol in *Calculus,* 2nd ed. (New York: John Wiley & Sons, 1967)

There are numerous instances in mathematics, business, and the sciences in which one quantity corresponds to or depends on another according to some definite *rule.* Consider, for example, the equation $y = 2x + 1$. Each time we select an x-value, a corresponding unique y-value is determined, in this case according to the rule *multiply by 2, then add 1.* In this sense, the equation $y = 2x + 1$ represents a *function.* It is a rule specifying a y-value corresponding to each x-value. It is useful to think of the x-values as inputs and the corresponding y-values as outputs. The function or rule then tells us what output results from a given input, as indicated schematically in Figure 1.

In general, a function can be represented in one or more of the following forms:

- Algebraic, using an equation or formula (as in the previous paragraph)
- Verbal (also as in the previous paragraph)
- Tabular, using a two-column table
- Graphical, using a coordinate system and graph

After looking at some examples of these ways of representing functions, we'll give an "official" definition of the term *function* on page 132.

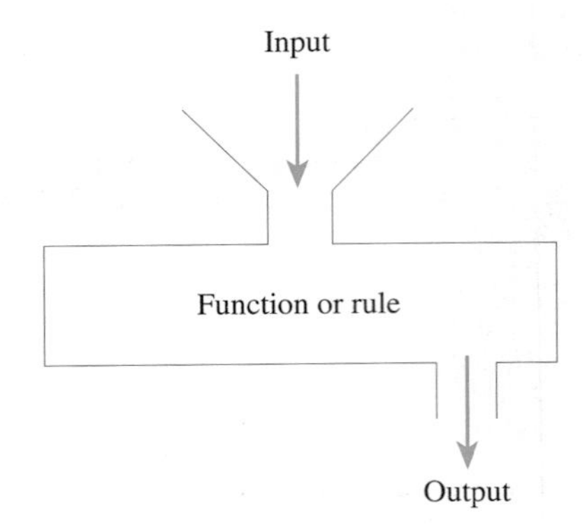

Figure 1

Algebraic. Many of the familiar formulas from geometry represent functions. The area A of a circle depends on the radius r according to the formula $A = \pi r^2$. For each value of r, a corresponding value for A is determined by this formula or rule. The values for r are the inputs, and the corresponding values for A are the outputs. The field of statistics provides another rich source of examples of functions. For instance, over the years 1850–1900, the population of the United States can be closely approximated by using the equation

$$y = 0.006609x^2 - 23.771x + 21{,}382$$

where y is the population in millions and x is the year. (We worked with this equation in Example 1 in Section 2.1.) For this rule or function, the inputs are the numbers 1850, 1851, . . . 1900, and the corresponding outputs are the population estimates (in millions).

Verbal. If you count the number of times that a cricket chirps in one minute, you can estimate the air temperature (Fahrenheit) using the following rule or function: *Divide the number of chirps by 4, then add 40*; the answer will be a good estimate for the air temperature. (We're assuming here that at a given temperature, all crickets chirp at the same rate.) For instance, suppose that you count 80 chirps in one minute. This is our input. Applying the rule, we divide 80 by 4, which is 20, and then add 40 to obtain 60. Our estimate for the air temperature is 60°F; that's the output corresponding to an input of 80 chirps/min.

The terminology of proportion and variation introduced in intermediate algebra is sometimes used in the sciences to represent functions verbally. Here's an example from physics. If you drop a ball off the roof of a building, then before it hits the ground (and neglecting air resistance), *the distance the ball falls is directly proportional to the square of the elapsed time.* The equivalent algebraic representation for this verbal rule or function is

$$d = kt^2$$

where d is the distance that the ball falls in time t, and k is a constant. (If the distance d is measured in feet and the time t in seconds, then the numerical value for k turns out to be 16, and the algebraic form of the rule or function becomes $d = 16t^2$.)

Tabular. In many applications a function is specified by means of a two-column table. Unless stated otherwise, we usually think of the entries in the left-hand column as the inputs and the entries in the right-hand column as the outputs. In Table 1, for example, the input 1975 yields the output \$2.00. The inputs and outputs for a function needn't always be numbers. Notice that in Table 2, the inputs aren't

TABLE 1 Federal Minimum Wage

Year	Minimum wage (dollars)
1975	2.00
1980	3.10
1985	3.35
1990	3.80
1995	4.25
2000	5.15

Source: U.S. Bureau of Labor Statistics

TABLE 2 Number of Moons Each Planet in the Solar System Is Known to Have as of 2000

Planet	Number of moons
Mercury	0
Venus	0
Earth	1
Mars	2
Jupiter	16
Saturn	18
Uranus	18
Neptune	8
Pluto	1

TABLE 3 Original Names of Some Movie Stars

Movie star	Original name
Kate Capshaw	Kathleen Sue Nail
Tom Cruise	Thomas Cruise Mapother, IV
Demi Moore	Demetria Guynes
Kevin Spacey	Kevin Fowler
Arnold Schwarzenegger	Arnold Schwarzenegger

numbers, but rather names of the planets in our solar system. In Table 3, neither the inputs nor the outputs are numbers. Another item to note: As indicated in the last row of Table 3, it is possible for an input and output to be the same item. For an algebraic example of this, consider the rule defined by $y = x^2$. For the input $x = 1$, the corresponding output is $y = 1^2 = 1$. That is, the input and output are both 1.

Graphical. In newspapers, business, and the sciences, functions are often presented graphically. Sometimes, but certainly not always, this is in conjunction with an explicit table of values or an equation. For example, consider the minimum wage information given in Table 1. Each row of the table gives us an ordered pair of numbers that we can plot in a coordinate system. Looking at Table 1, we see that the first row gives us the point with coordinates (1975, 2.00), the second row gives us the point (1980, 3.10), and so on. The resulting graph is shown in Figure 2.

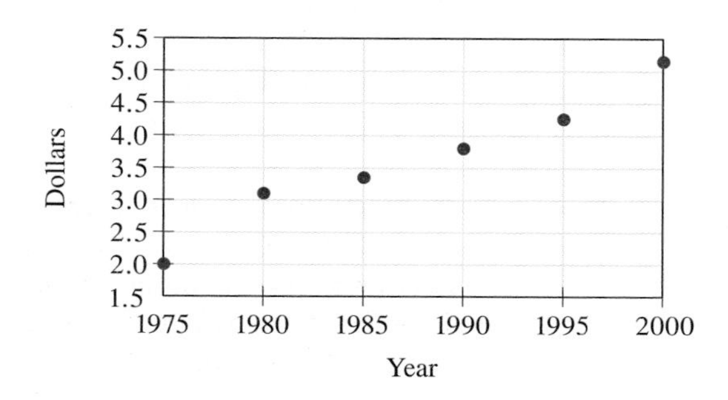

Figure 2
Federal minimum wage. A graphical presentation of the input/output pairs in Table 1.

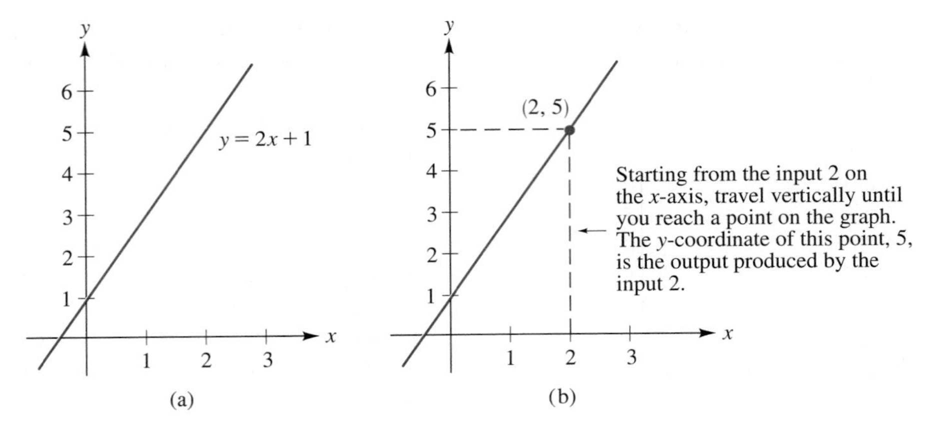

Figure 3
The graph of $y = 2x + 1$ represents a function.

In the first paragraph of this section we used the equation $y = 2x + 1$ as an example of a function represented algebraically. A graphical representation of this function is shown in Figure 3(a). (We've used the techniques of Section 1.6 to graph the equation $y = 2x + 1$.) Given this graphical representation of the function (and ignoring, for the moment, the fact that we have the equation $y = 2x + 1$), what is the underlying rule here for obtaining the output y from a given input x? One way to state the rule is this:

> Given an input on the x-axis, for example $x = 2$ as in Figure 3(b), travel vertically until you reach a point on the graph. The y-coordinate of this point is the required output. According to Figure 3(b), the input $x = 2$ produces the output $y = 5$.

A less dynamic but more succinct (and more general) way to describe the relationship between the graph of a function and the "input-output" point of view is this:

> Suppose we have a graph of a function in an x-y coordinate system. Then a point (x, y) is on the graph if and only if the function assigns the output y to the input x.

(Actually, when we take a more careful look at graphs of functions in Section 3.2, we'll adapt this last version for the very definition of the graph of a function.)

Now that we've looked at a few introductory examples, it's time to state the definition of the term *function* as it is used in mathematics. We need this definition before we can answer the following two questions, which may have already occurred to you:

> Does every rule or table represent a function?
> Does every graph represent a function?

As you'll see later in this section, the answer to the first question is *no*. For the second question, the answer is also negative, but we postpone that discussion until the next section. The following definition of the term *function* is broad enough to encompass all of the examples that we've looked at.

DEFINITION **Function**

Let A and B be two nonempty sets. A function from A to B is a rule of correspondence that assigns to each element in set A exactly one element in B.

The set A in the definition just given is called the **domain** of the function. Think of the domain as the set of all inputs. For each input, the function gives us an output (from the set B mentioned in the definition). The set of all outputs is called the **range** of the function. For example, for the minimum wage function represented in Table 1 (on page 131), the domain is the set {1975, 1980, 1985, 1990, 1995, 2000}, and the range is {$2.00, $3.10, $3.35, $3.80, $4.25, $5.15}.

In a moment we'll talk more about domain and range, but first reread the definition of function in the box above and note the use of the word "exactly." The inclusion of that word implies that some rules will *not* qualify as functions. For example, the rule represented in Table 4(a) doesn't qualify as a function; for the input year 1981 there are two outputs (namely, 18¢ and 20¢), whereas the definition of a function requires exactly one output. Table 4(b) remedies this problem; the rule in Table 4(b) does represent a function.

TABLE 4(a) Price of U.S. First-Class Stamp and the Year the Price Went into Effect

Year (inputs)	1975	1978	1981	1981	1985	1988	1991	1995	1999
Price in cents (outputs)	13	15	18	20	22	25	29	32	33

TABLE 4(b) Price of U.S. First-Class Stamp and the Date the Price Went into Effect

Date (inputs)	12/31/75	5/29/78	3/22/81	11/1/81	2/17/85	4/3/88	2/3/91	1/1/95	1/10/99
Price in cents (outputs)	13	15	18	20	22	25	29	32	33

For an algebraic example of this same issue, consider square roots and the symbol $\sqrt{\ }$. (*Note*: The following discussion is intended to be self-contained. If after you read it, however, things don't seem clear, you should refer to the detailed review of roots in Appendix B.2 at the back of this book.) Recall from basic algebra that each positive number has two square roots. For instance, the square roots of 16 are 4 and -4. Thus the rule in Table 5(a) does not represent a function because for each input there are two outputs. To remedy this, the symbol $\sqrt{\ }$ is defined in algebra to mean the *positive* square root only. Thus $\sqrt{16} = 4$, and if you are thinking of the other root, you write $-\sqrt{16} = -4$. Consequently, the correspondences in Table 5(b) do represent a function; for each input there is exactly one output.

TABLE 5(a) Not a Function

x	Square roots of x
4	2, -2
9	3, -3
16	4, -4

TABLE 5(b) A Function

x	$\sqrt{x}$
4	2
9	3
16	4

EXAMPLE 1 Using the definition of a function

Let $A = \{b, g\}$ and $B = \{s, t, u, z\}$. Which of the four correspondences in Figure 4 represent functions from A to B? For those correspondences that do represent functions, specify the range in each case.

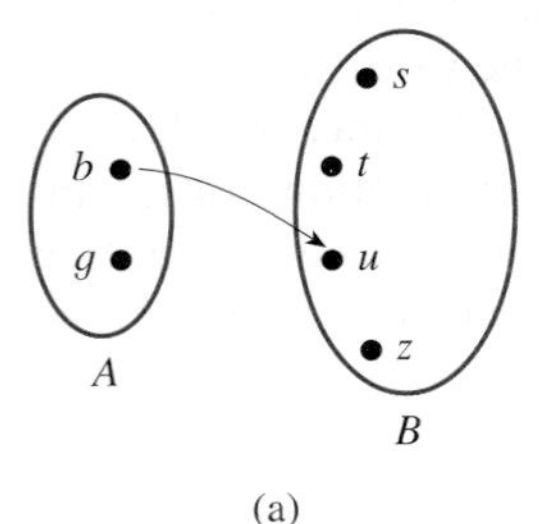

(a)

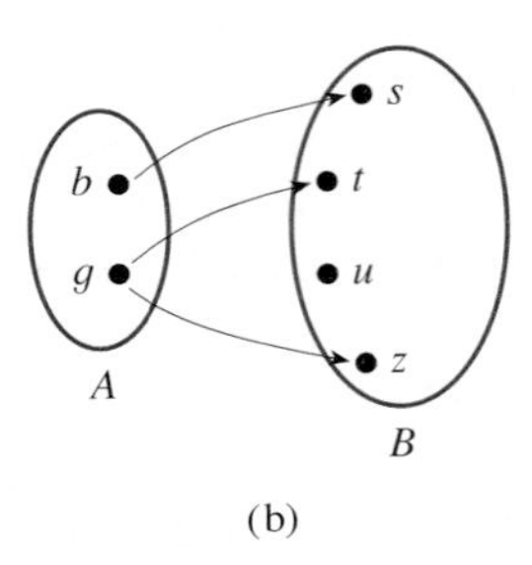

(b)

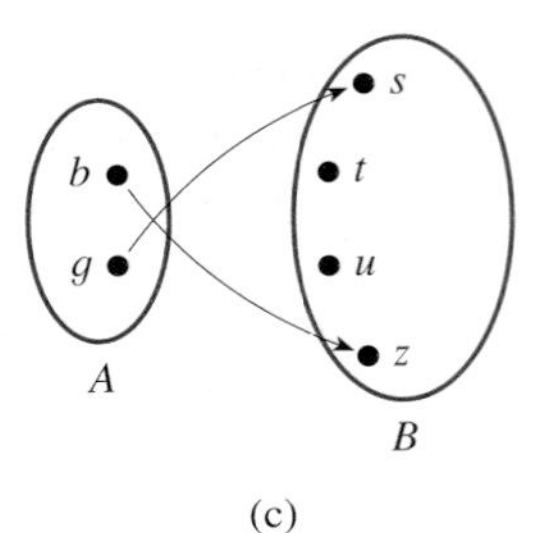

(c)

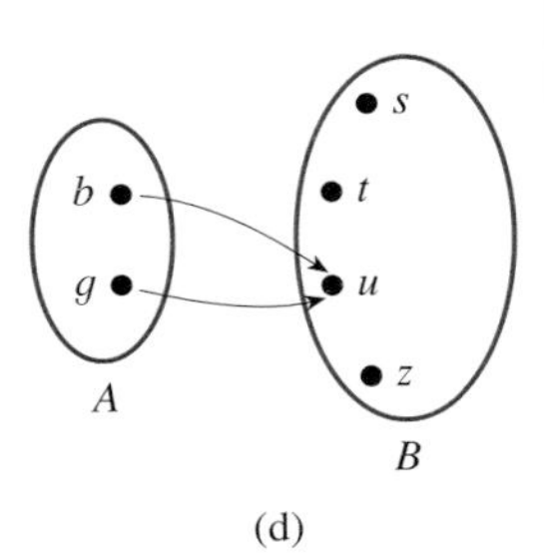

(d)

Figure 4

SOLUTION

(a) This is not a function. The definition requires that *each* element in A be assigned an element in B. The element g in this case has no assignment.

(b) This is not a function. The definition requires *exactly one* output for a given input. In this case there are two outputs for the input g.

(c) and **(d)** Both of these rules qualify as functions from A to B. For each input there is exactly one output. (Regarding the function in (d) in particular, notice that nothing in the definition of a function prohibits two different inputs from producing the same output.) For the function in (c), the outputs are s and z, and so the range is the set $\{s, z\}$. For the function in (d), the only output is u; consequently, the range is the set $\{u\}$.

For functions defined by equations, we'll agree on the following convention regarding the domain: Unless otherwise indicated, the domain is assumed to be the set of all real numbers that lead to unique real-number outputs. (This is essentially the domain convention described in Section 1.3.) Thus, the domain of the function defined by $y = 3x - 2$ is the set of all real numbers, whereas the domain of the function defined by $y = 1/(x - 5)$ is the set of all real numbers except 5 (since, $1/(x - 5)$ is undefined when the denominator is zero, namely, when $x = 5$). In general, the letter representing elements from the domain (that is, the inputs) is called the **independent variable.** For example, in the equation $y = 3x - 2$, the independent variable is x. The letter representing elements from the range (that is, the outputs) is called the **dependent variable.** In the equation $y = 3x - 2$, the dependent variable is y; its value *depends* on x. This is also expressed by saying that *y is a function of x*.

EXAMPLE 2 Determining the domain of a function

Find the domain of the function defined by each equation:

(a) $y = \sqrt{2x + 6}$; **(b)** $s = 1/(t^2 - 6t - 7)$.

SOLUTION

(a) The quantity under the radical sign must be nonnegative, so we have

$$\begin{aligned} 2x + 6 &\geq 0 \\ 2x &\geq -6 \\ x &\geq -3 \end{aligned}$$

The domain is therefore the interval $[-3, \infty)$.

(b) Since division by zero is undefined, the domain of this function consists of all real numbers t except those for which the denominator is zero. Thus to find out which values of t to exclude, we solve the equation $t^2 - 6t - 7 = 0$. We have

$$t^2 - 6t - 7 = 0$$
$$(t - 7)(t + 1) = 0$$
$$t - 7 = 0 \quad \Big| \quad t + 1 = 0$$
$$t = 7 \quad \Big| \quad t = -1$$

It follows now that the domain of the function defined by $s = 1/(t^2 - 6t - 7)$ is the set of all real numbers except $t = 7$ and $t = -1$.

EXAMPLE 3 Determining the domain of a function

Find the domain of the function defined by each equation:

(a) $y = \sqrt{\dfrac{x+3}{x-4}}$; **(b)** $y = \sqrt[3]{\dfrac{x+3}{x-4}}$.

SOLUTION

(a) The quantity underneath the square root sign must be nonnegative, so we require

$$\frac{x+3}{x-4} \geq 0$$

This inequality can be solved by using the techniques of Section 2.4. In fact, we did solve exactly this inequality in Example 4 of Section 2.4. As you can check (preferably on your own, without looking back), the solution set is $(-\infty, -3] \cup (4, \infty)$. This is the domain of the given function.

(b) In contrast to the situation with square roots, cube roots are defined for all real numbers. Thus the only trouble spot for $y = \sqrt[3]{(x+3)/(x-4)}$ occurs when the denominator is zero, that is, when x is 4. Consequently, the domain of the given function consists of all real numbers except $x = 4$. Using interval notation, this set can be written $(-\infty, 4) \cup (4, \infty)$.

EXAMPLE 4 Determining the range of a function

Find the range of the function defined by $y = \dfrac{x+2}{x-3}$.

SOLUTION

Since the range of a function depends on the domain, we first find the domain. The domain of the given function is the set of all real numbers except 3. The range of this function is the set of all outputs y. One way to see what restrictions the given equation imposes on y is to solve the equation for x as follows: y is in the range of this function if there is an x in the domain such that

$$\begin{aligned} y &= \frac{x+2}{x-3} \\ y(x-3) &= x+2 \qquad \text{multiplying by } x-3 \text{ (for } x \neq 3) \\ xy - 3y &= x+2 \\ xy - x &= 3y+2 \\ x(y-1) &= 3y+2 \\ x &= \frac{3y+2}{y-1} \end{aligned}$$

From this last equation we see that given a value for y, there will be an x in the domain provided that y is not equal to 1. (The denominator is zero when $y = 1$.) The range therefore consists of all real numbers except $y = 1$. Using interval notation, this is the set $(-\infty, 1) \cup (1, \infty)$.

We often use single letters to name functions. If f is a function and x is an input for the function, then the resulting output is denoted by $f(x)$. This is read *f of x* or *the value of f at x*. As an example of this notation, suppose that f is the function defined by

$$f(x) = x^2 - 3x + 1 \tag{1}$$

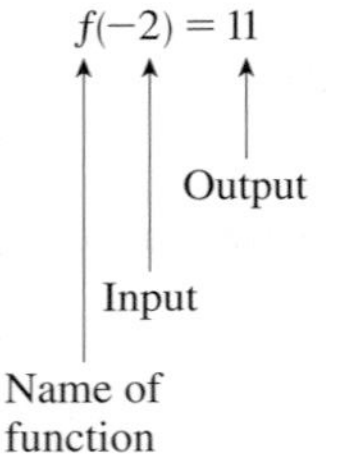

Figure 5

Then $f(-2)$ denotes the output that results when the input is -2. To calculate this output, replace x with -2 throughout equation (1). This yields

$$\begin{aligned} f(-2) &= (-2)^2 - 3(-2) + 1 \\ &= 4 + 6 + 1 = 11 \end{aligned}$$

That is, $f(-2) = 11$. Figure 5 summarizes this result and the notation.

EXAMPLE 5 Using function notation

Let g denote the minimum wage function defined by Table 1 (on page 131).

(a) Find $g(1975)$.

(b) Find $g(1990) - g(1980)$ and interpret the result.

SOLUTION

(a) The notation $g(1975)$ stands for the output corresponding to the input 1975. According to Table 1, that output is \$2.00, and so $g(1975) = \$2.00$.

(b) We have

$$\begin{aligned} g(1990) - g(1980) &= \$3.80 - \$3.10 \\ &= \$0.70 \end{aligned}$$

This represents the increase in the minimum wage from 1980 to 1990.

EXAMPLE 6 Using function notation

Let $f(x) = \dfrac{x}{x-4}$.

(a) Compute each of the following: $f(2)$, $f(t)$, and $f(a^2)$.
(b) Find an input x such that $f(x) = 3$.

SOLUTION

(a) For $f(2)$, replace each occurrence of x in the given equation by 2 to obtain

$$f(2) = \frac{2}{2-4} = -1$$

Similarly, for $f(t)$ and for $f(a^2)$ we have

$$f(t) = \frac{t}{t-4} \quad \text{and} \quad f(a^2) = \frac{a^2}{a^2-4}$$

(b) To solve the equation $f(x) = 3$, replace $f(x)$ by the given expression $x/(x-4)$. This yields

$$\frac{x}{x-4} = 3$$

$$x = 3(x-4) \quad \text{multiplying both sides by the quantity } x-4 \text{ (for } x \neq 4)$$

$$x = 3x - 12$$

$$-2x = -12$$

$$x = 6$$

CHECK Using the input $x = 6$, we have

$$f(6) = \frac{6}{6-4} = 3 \quad \text{as required}$$

So $x = 6$ is indeed the input that we were looking for.

As background for the next example (and for work in later sections), we introduce the concept of a *demand function* from the field of economics. As you well know, the price of an item that you're interested in has a strong influence on whether or not you actually buy that item. A function relating the selling price of an item to the number of units sold at that price is called a **demand function.** Such a function might be obtained after analyzing either a company's sales records or the results of a marketing survey. As a simple example, suppose that an auto manufacturer invites 12 people to participate in a focus group. At some point in the session, the manufacturer shows them pictures and specifications for a proposed new model. Then each of the 12 people fills out a questionnaire asking whether he or she would purchase the car at various price levels. The results, shown in Table 6, represent a demand function. Figure 6 provides a graphical representation of this same function.

TABLE 6 Tabular Representation of a Demand Function

Price (1000s of dollars)	Number of cars that can be sold
12	12
14	10
16	6
18	5
20	2
22	1

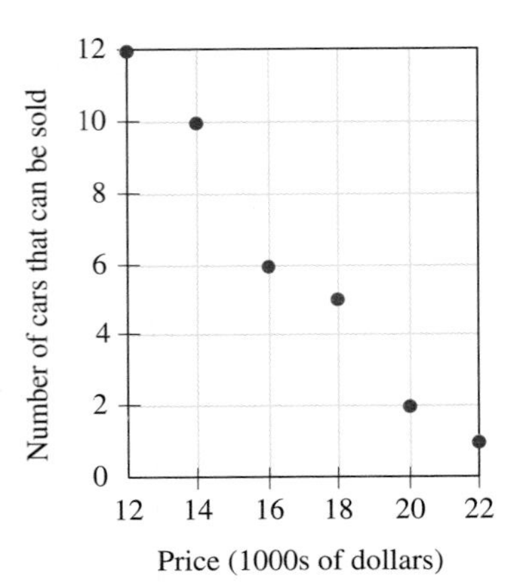

Figure 6
Graphical representation of the demand function

The demand function represented in Table 6 and Figure 6 involves only six data points, whereas an actual marketing survey might well produce 600 or even 6000 points. To detect patterns in the data or to make predictions, it's useful to have an equation with a relatively simple graph that somehow "fits" or summarizes the data. In Figure 7(a), for instance, you can see that the blue line does seem to fit the data points from Figure 6 quite well.

Assuming that we are using x on the horizontal axis for price and y on the vertical axis for the number of cars, the equation for the line in Figure 7(a) is $y = -1.143x + 25.43$. The equation was obtained by using statistical techniques that are built into many types of graphing utilities and spreadsheets. (More about that in Section 4.1.) The line is known variously as a *trend line, regression line,* or *least squares line.* We say in this case that the function defined by $y = -1.143x + 25.43$ is a *mathematical model* for the demand data displayed in Table 6 and in Figure 6.

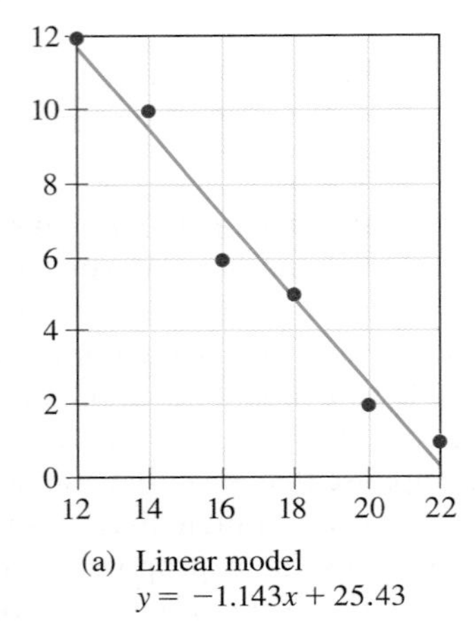

(a) Linear model
$y = -1.143x + 25.43$

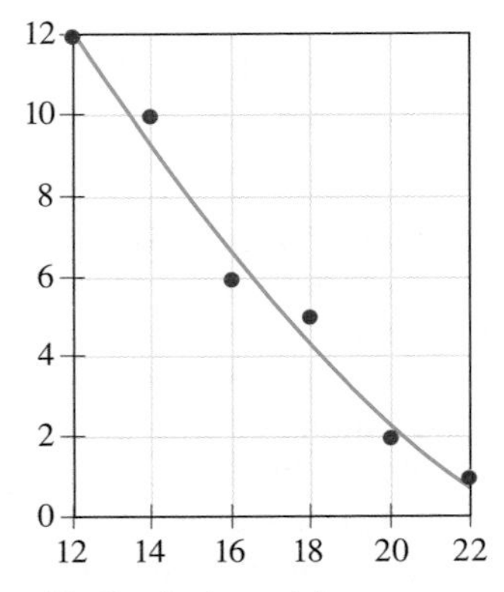

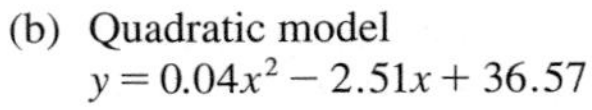

(b) Quadratic model
$y = 0.04x^2 - 2.51x + 36.57$

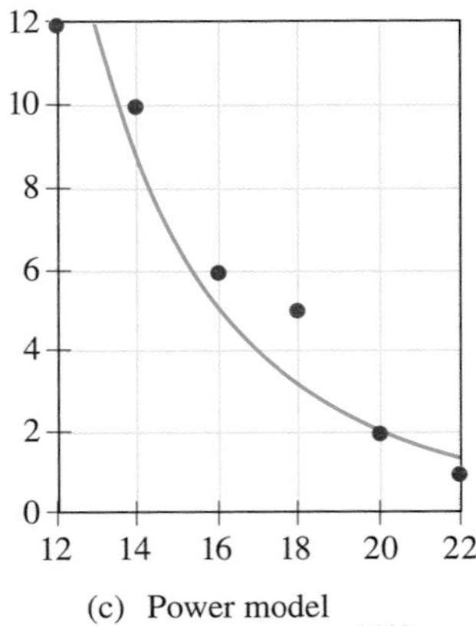

(c) Power model
$y = 359684x^{-4.023}$

Figure 7
Three models for the demand data in Table 6

More generally, any function used to describe or summarize a real-world situation is called a **mathematical model,** and the process of creating or developing that function is called **mathematical modeling.** In Figures 7(b) and 7(c) we show two other functions that can be used to model the demand data. We'll study these and other types of functions used in mathematical modeling in subsequent chapters.

For the demand function represented in Figure 6, prices are represented on the horizontal axis, and the numbers of items sold are on the vertical axis. However, this is not a standard convention. Economists sometimes put prices on the vertical axis and numbers of items on the horizontal axis. Example 7 shows one instance of this.

EXAMPLE 7 Using a demand function and function notation

Suppose that an economist for a factory outlet store determines that the demand function p for a certain type of tee shirt is given by

$$p(n) = -0.012n + 20.49 \qquad (100 \leq n \leq 1500)$$

where n is the number of tee shirts that can be sold per month at a price of $p(n)$ dollars per shirt. Solve the equation $p(n) = 16$ and interpret the result.

SOLUTION

Substituting the given expression for $p(n)$ into the equation $p(n) = 16$ yields

$$\begin{aligned} -0.012n + 20.49 &= 16 \\ -0.012n &= 16 - 20.49 \\ n &= \frac{16 - 20.49}{-0.012} = 374.166\ldots \qquad \text{using a calculator} \end{aligned}$$

Now since n represents the number of tee shirts, we round to the nearest integer to obtain $n = 374$. Interpretation: When the price is set at \$16 per shirt, the store can expect to sell 374 tee shirts per month.

Before going on to the next example, we offer a word of caution to help you avoid a common error in using function notation.

NOTE OF CAUTION REGARDING FUNCTION NOTATION It is not in general true that $f(a + b) = f(a) + f(b)$.

As an example, suppose we take f to be the function defined by $f(x) = x^2$ and use the two inputs $a = 1$ and $b = 2$. Then we have

$$f(1 + 2) = (1 + 2)^2 = 3^2 = 9$$

whereas

$$f(1) = 1^2 = 1, f(2) = 2^2 = 4 \qquad \text{and, consequently,} \qquad f(1) + f(2) = 1 + 4 = 5$$

In summary, we have $f(1 + 2) = 9$ and $f(1) + f(2) = 5$, so certainly,

$$f(1 + 2) \neq f(1) + f(2)$$

EXAMPLE 8 Using function notation

Compute $g(x + 1)$ for the function g defined by $g(x) = 1 - x - x^2$.

SOLUTION

We're given $g(x)$. To find $g(x + 1)$, we go to the given equation $g(x) = 1 - x - x^2$ and replace each occurrence of x with the quantity $x + 1$. This yields

$$\begin{aligned} g(x + 1) &= 1 - (x + 1) - (x + 1)^2 \\ &= 1 - x - 1 - (x^2 + 2x + 1) \\ &= -x^2 - 3x - 1 \qquad \text{(Check the algebra.)} \end{aligned}$$

Here's a slightly different perspective that we can apply in Example 8 and in similar situations. Instead of writing $g(x) = 1 - x - x^2$, we can write

$$g(\quad) = 1 - (\quad) - (\quad)^2 \tag{1}$$

with the understanding that whatever quantity goes into the parentheses on the left-hand side of equation (1) must also be placed into each set of parentheses on the right-hand side. In particular then, placing the quantity $x + 1$ into all three sets of parentheses in equation (1) yields $g(x + 1) = 1 - (x + 1) - (x + 1)^2$. From here on, the algebra is the same as in Example 8.

EXERCISE SET 3.1

A

1. As in Example 5, let g denote the minimum wage function represented in Table 1.
   (a) Find $g(1975)$.
   (b) Find $g(1995) - g(1975)$ and interpret the result.
2. Let f denote the function represented in Table 3.
   (a) Find f(Kate Capshaw).
   (b) For which input x are x and $f(x)$ identical?
3. Let h denote the function represented in Table 2.
   (a) Is it the domain or the range of h that consists of real numbers?
   (b) Find h(Mars).
   (c) Which is larger: h(Neptune) or h(Pluto)? Interpret the result (using a complete sentence).
4. Again let h denote the function represented in Table 2.
   (a) List those inputs x for which $h(x) = 1$.
   (b) For which real number t will it be true that $h(\text{Jupiter}) + t = h(\text{Saturn})$?

For Exercises 5 and 6, two sets A and B are defined as follows: $A = \{x, y, z\}$; $B = \{1, 2, 3\}$.

(a) *Which of the rules displayed in the figures represent functions from A to B?*

(b) *For each rule that does represent a function, specify the range.*

5.

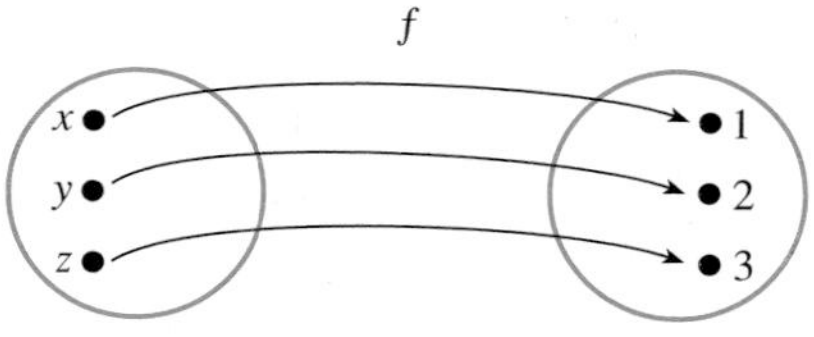

f

x y z 1 2 3

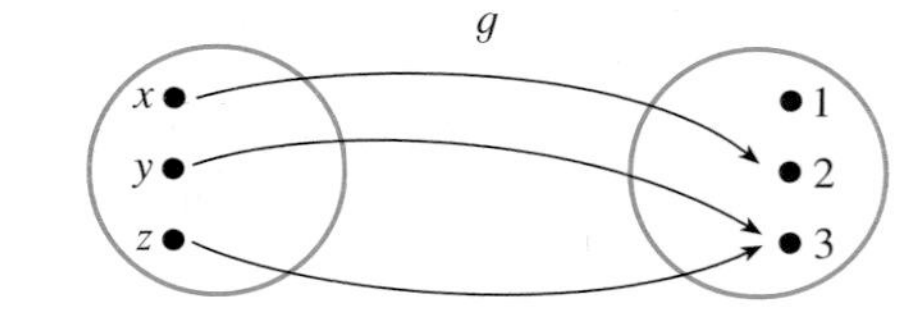

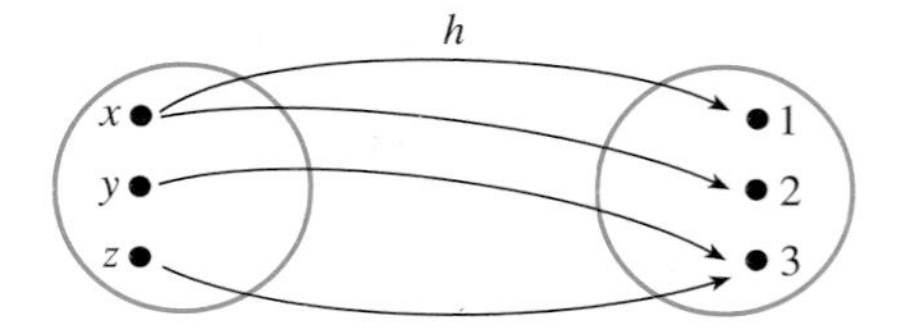

6.

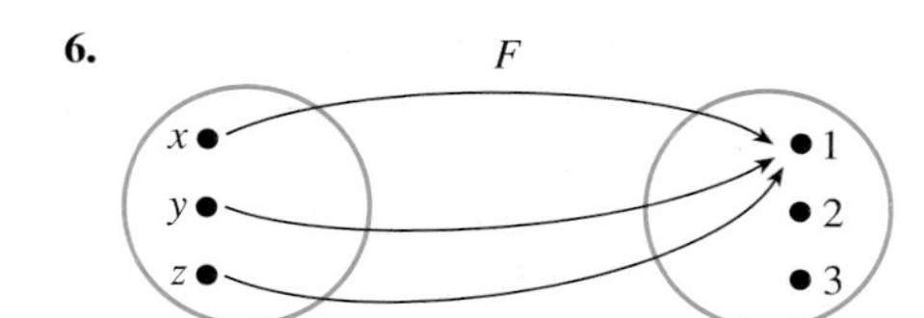

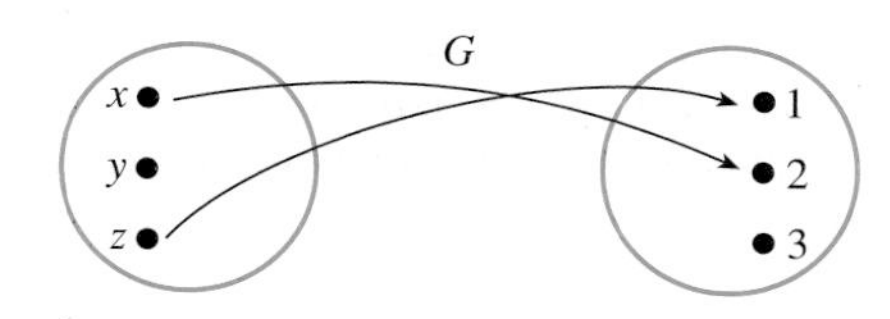

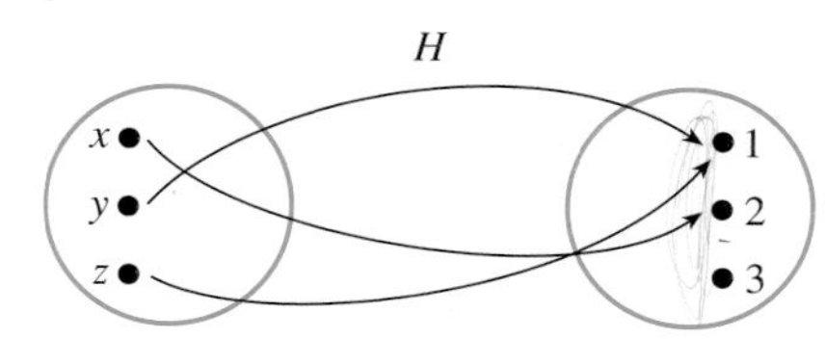

For Exercises 7 and 8, two sets D and C are defined as follows: $D = \{a, b\}$; $C = \{i, j, k\}$.

(a) *Which of the rules displayed in the figures represent functions from D to C?*

(b) *For each rule that represents a function, specify the range.*

7.

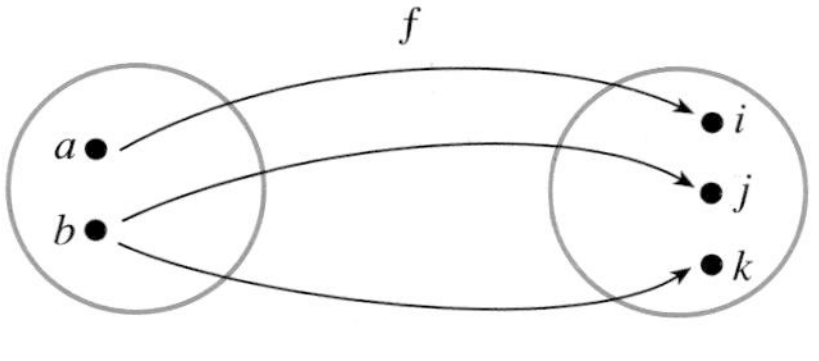

f

a • → • j, • k; b • → • i ...

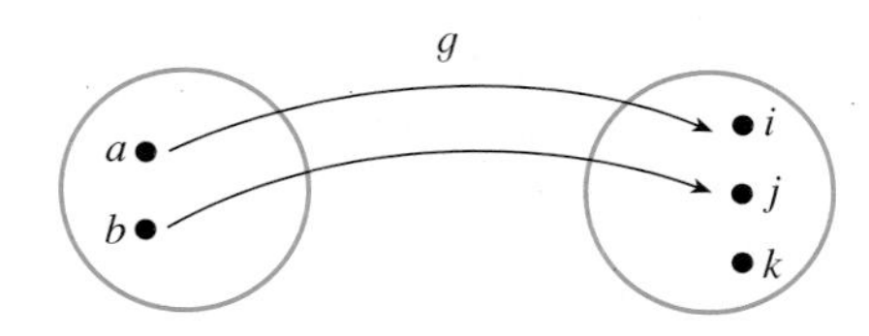

8.

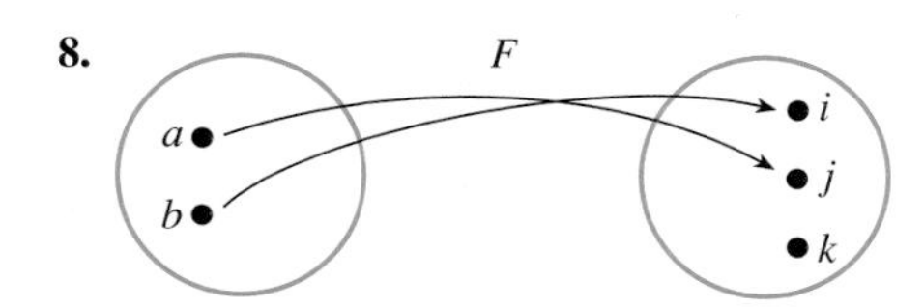

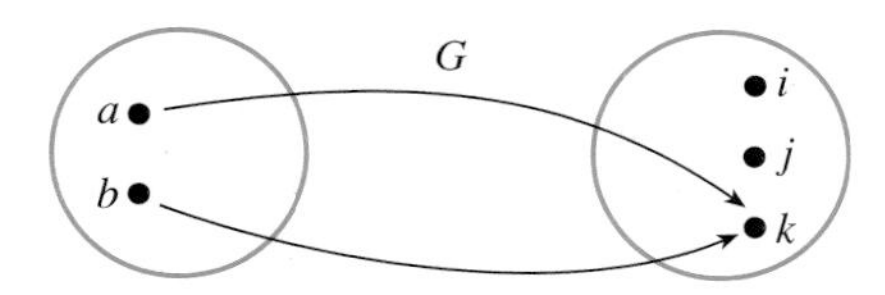

In Exercises 9–16, determine the domain of each function.

9. **(a)** $y = -5x + 1$
(b) $y = 1/(-5x + 1)$
(c) $y = \sqrt{-5x + 1}$
(d) $y = \sqrt[3]{-5x + 1}$

10. **(a)** $s = 3t + 12$
(b) $s = 1/(3t + 12)$
(c) $s = \sqrt{3t + 12}$
(d) $s = \sqrt[3]{3t + 12}$

11. **(a)** $f(x) = x^2 - 9$
(b) $g(x) = 1/(x^2 - 9)$
(c) $h(x) = \sqrt{x^2 - 9}$
(d) $k(x) = \sqrt[3]{x^2 - 9}$

12. **(a)** $F(t) = t^2 + 4t$
(b) $G(t) = 1/(t^2 + 4t)$
(c) $H(t) = \sqrt{t^2 + 4t}$
(d) $K(t) = \sqrt[3]{t^2 + 4t}$

13. **(a)** $f(t) = t^2 - 8t + 15$
(b) $g(t) = 1/(t^2 - 8t + 15)$
(c) $h(t) = \sqrt{t^2 - 8t + 15}$
(d) $k(t) = \sqrt[3]{t^2 - 8t + 15}$

14. **(a)** $F(x) = 2x^2 + x - 6$
(b) $G(x) = 1/(2x^2 + x - 6)$
(c) $H(x) = \sqrt{2x^2 + x - 6}$
(d) $K(x) = \sqrt[3]{2x^2 + x - 6}$

15. **(a)** $f(x) = (x - 2)/(2x + 6)$
(b) $g(x) = \sqrt{(x - 2)/(2x + 6)}$
(c) $h(x) = \sqrt[3]{(x - 2)/(2x + 6)}$

16. **(a)** $F(t) = (3t - 4)/(7 - 2t)$
(b) $G(t) = \sqrt{(3t - 4)/(7 - 2t)}$
(c) $H(t) = \sqrt[3]{(3t - 4)/(7 - 2t)}$

In Exercises 17–26, determine the domain and the range of each function.

17. $y = 4x - 5$

18. $y = 125 - 12x$

19. $y = 4x^3 - 5$

20. $y = 125 - 12x^3$

21. $g(x) = \dfrac{4x - 20}{3x - 18}$

22. $f(x) = \dfrac{1 - x}{x}$

23. **(a)** $f(x) = \dfrac{x + 3}{x - 5}$
(b) $F(x) = \dfrac{x^3 + 3}{x^3 - 5}$

24. **(a)** $g(x) = \dfrac{2x - 7}{3x + 24}$
(b) $G(x) = \dfrac{2x^3 - 7}{3x^3 + 24}$

25. $s = t^2 + 4$

26. $s = 2t^2 - 10$

27. Each of the following rules defines a function with domain the set of all real numbers. Express each rule in the form of an equation.

EXAMPLE

The rule *For each real number, compute its square* can be written $y = x^2$.

(a) For each real number, subtract 3 and then square the result.
(b) For each real number, compute its square and then subtract 3 from the result.
(c) For each real number, multiply it by 3 and then square the result.
(d) For each real number, compute its square and then multiply the result by 3.

28. Each of the following rules defines a function with domain equal to the set of all real numbers. Express each rule in words.

(a) $y = 2x^3 + 1$
(b) $y = 2(x + 1)^3$
(c) $y = (2x + 1)^3$
(d) $y = (2x)^3 + 1$

29. Let $f(x) = x^2 - 3x + 1$. Compute the following.
(a) $f(1)$ **(b)** $f(0)$ **(c)** $f(-1)$ **(d)** $f(3/2)$ **(e)** $f(z)$ **(f)** $f(x + 1)$ **(g)** $f(a + 1)$ **(h)** $f(-x)$ **(i)** $|f(1)|$ **(j)** $f(\sqrt{3})$ **(k)** $f(1 + \sqrt{2})$ **(l)** $|1 - f(2)|$

30. Let $H(x) = 1 - x + x^2 - x^3$.
(a) Which number is larger, $H(0)$ or $H(1)$?
(b) Find $H(\frac{1}{2})$. Does $H(\frac{1}{2}) + H(\frac{1}{2}) = H(1)$?

31. Let $f(x) = 3x^2$. Find the following.
(a) $f(2x)$ **(b)** $2f(x)$ **(c)** $f(x^2)$ **(d)** $[f(x)]^2$ **(e)** $f(x/2)$ **(f)** $f(x)/2$
For checking: No two answers are the same.

32. Let $f(x) = 4 - 3x$. Find the following.
(a) $f(2)$ **(b)** $f(-3)$ **(c)** $f(2) + f(-3)$ **(d)** $f(2 + 3)$ **(e)** $f(2x)$ **(f)** $2f(x)$ **(g)** $f(x^2)$ **(h)** $f(1/x)$ **(i)** $f[f(x)]$ **(j)** $x^2 f(x)$ **(k)** $1/f(x)$ **(l)** $f(-x)$ **(m)** $-f(x)$ **(n)** $-f(-x)$

33. Let $H(x) = 1 - 2x^2$. Find the following.
(a) $H(\sqrt{2})$ **(b)** $H(5/6)$ **(c)** $H(x + 1)$ **(d)** $H(x + h)$

34. **(a)** If $f(x) = 2x + 1$, does $f(3 + 1) = f(3) + f(1)$?
(b) If $f(x) = 2x$, does $f(3 + 1) = f(3) + f(1)$?
(c) If $f(x) = \sqrt{x}$, does $f(3 + 1) = f(3) + f(1)$?

35. Let $g(x) = 2$, for all x. Find each output.
(a) $g(0)$ **(b)** $g(5)$ **(c)** $g(x + h)$

36. Let $g(t) = |t - 4|$. Find $g(3)$. Find $g(x + 4)$.

37. Let $f(x) = x^2 - 6x$. In each case, find all real numbers x (if any) that satisfy the given equation.
(a) $f(x) = 16$ **(b)** $f(x) = -10$ **(c)** $f(x) = -9$

38. Let $g(t) = (4t - 6)/(t - 4)$. In each case, find all the real number solutions (if any) for the given equation.
(a) $g(t) = 14$ **(b)** $g(t) = 4$ **(c)** $g(t) = 0$

For Exercises 39 and 40, refer to the demand function given in Example 7. In each case, solve the indicated equation for n. Round each answer to the nearest integer and interpret the result.

39. $p(n) = 8$ **40.** $p(n) = 18$

In Exercises 41 and 42, p refers to the demand function given in Example 7. Now assume that a second economist proposes an alternative model f for the demand function:

$$f(n) = 4 + \frac{3000}{n + 100} \qquad (100 \le n \le 1500)$$

where n is the number of tee shirts that can be sold per month at a price of f(n) dollars per shirt.

41. At a price level of \$19 per shirt, compare the predictions for monthly sales obtained using each model. *Hint*: You need to solve each of the equations $p(n) = 19$ and $f(n) = 19$ and interpret the results.

42. Show that at a price level of \$10 per shirt, the model p predicts more than twice as many sales per month as does the model f. *Hint*: Solve each of the equations $p(n) = 10$ and $f(n) = 10$, and interpret the results.

In Exercises 43 and 44, let $T(x) = 2x^2 - 3x$. Find (and simplify) each expression.

43. **(a)** $T(x + 2)$ **(b)** $T(x - 2)$ **(c)** $T(x + 2) - T(x - 2)$

44. **(a)** $T(x + h)$ **(b)** $T(x - h)$ **(c)** $T(x + h) - T(x - h)$

In Exercises 45–48, refer to the following table. The left-hand column of the table lists four errors to avoid in working with function notation. In each case, use the function $f(x) = x^2 - 1$ and give a numerical example showing that the expressions on each side of the equation are not equal.

Errors to avoid	Numerical example showing that the equation is not, in general, valid
45. $f(a + b) = f(a) + f(b)$	
46. $f(ab) = f(a) \cdot f(b)$	
47. $f\left(\frac{1}{a}\right) = \frac{1}{f(a)}$	
48. $\frac{f(a)}{f(b)} = \frac{a}{b}$	

B

49. Let $f(x) = (x - a)/(x + a)$.
(a) Find $f(a)$, $f(2a)$, and $f(3a)$. Is it true that $f(3a) = f(a) + f(2a)$?
(b) Show that $f(5a) = 2f(2a)$.

50. Let $k(x) = 5x^3 + \dfrac{5}{x^3} - x - \dfrac{1}{x}$. Show that $k(x) = k(1/x)$.

51. Let $f(x) = 2x + 3$. Find values for a and b such that the equation $f(ax + b) = x$ is true for all values of x. *Hint*: Use the fact that if two polynomials (in the variable x) are equal for all values of x, then the corresponding coefficients are equal.

52. Let $f(t) = (t - x)/(t + y)$. Show that

$$f(x + y) + f(x - y) = \frac{-2y^2}{x^2 + 2xy}$$

53. Let $f(z) = \dfrac{3z - 4}{5z - 3}$. Find $f\left(\dfrac{3z - 4}{5z - 3}\right)$.

54. Let $F(x) = \dfrac{ax + b}{cx - a}$. Show that $F\left(\dfrac{ax + b}{cx - a}\right) = x$, where $a^2 + bc \neq 0$.

55. If $f(x) = -2x^2 + 6x + k$ and $f(0) = -1$, find k.

56. If $g(x) = x^2 - 3xk - 4$ and $g(1) = -2$, find k.

57. Let $h(x) = x^2 - 4x - c$. Find a nonzero value for c such that $h(c) = c$.

58. Let the function L be defined by the following rule: $L(x)$ *is the exponent to which 2 must be raised to yield x.* (For the moment, we won't concern ourselves with the domain and range.) Then $L(8) = 3$, for example, since the exponent to which 2 must be raised to yield 8 is 3 (that is, $8 = 2^3$). Find the following outputs.

(a) $L(1)$ **(e)** $L(1/2)$
(b) $L(2)$ **(f)** $L(1/4)$
(c) $L(4)$ **(g)** $L(1/64)$
(d) $L(64)$ **(h)** $L(\sqrt{2})$

The function L is called a *logarithmic function.* The usual notation for $L(x)$ in this example is $\log_2 x$. Logarithmic functions will be studied in Chapter 5.

59. Let $q(x) = ax^2 + bx + c$. Evaluate

$$q\left(\frac{-b + \sqrt{b^2 - 4ac}}{2a}\right)$$

60. By definition, a **fixed point** for the function f is a number x_0 such that $f(x_0) = x_0$. For instance, to find any fixed points for the function $f(x) = 3x - 2$, we write $3x_0 - 2 = x_0$. On solving this last equation, we find that $x_0 = 1$. Thus, 1 is a fixed point for f. Calculate the fixed points (if any) for each function.

(a) $f(x) = 6x + 10$ **(c)** $S(t) = t^2$
(b) $g(x) = x^2 - 2x - 4$ **(d)** $R(z) = (z + 1)/(z - 1)$

61. Consider the following two rules, F and G, where F is the rule that assigns to each person his or her birth-mother and G is the rule that assigns to each person his or her aunt. Explain why F is a function but G is not.

In Exercises 62–64, use this definition: ***A prime number*** *is a positive whole number with no factors other than itself and* 1. *For example,* 2, 13, *and* 37 *are primes, but* 24 *and* 39 *are not. By convention* 1 *is not considered prime, so the list of the first few primes is as follows:*

$$2, 3, 5, 7, 11, 13, 17, 19, 23, 29, \ldots$$

62. Let G be the rule that assigns to each positive integer the nearest prime. For example, $G(8) = 7$, since 7 is the prime nearest 8. Explain why G is not a function. How could you alter the definition of G to make it a function? *Note*: There is more than one way to do this.

63. Let f be the function that assigns to each natural number x the number of primes that are less than or equal to x. For example, $f(12) = 5$ because, as you can easily check, five primes are less than or equal to 12. Similarly, $f(3) = 2$, because two primes are less than or equal to 3. Find $f(8)$, $f(10)$, and $f(50)$.

64. (a) If $P(x) = x^2 - x + 17$, find $P(1)$, $P(2)$, $P(3)$, and $P(4)$. Can you find a natural number x for which $P(x)$ is not prime?

(b) If $Q(x) = x^2 - x + 41$, find $Q(1)$, $Q(2)$, $Q(3)$, and $Q(4)$. Can you find a natural number x for which $Q(x)$ is not prime?

65. $\pi = 3.141592653589793\ldots$ *and so on!*
For each natural number n, let $G(n)$ be the digit in the nth decimal place of π. For instance, according to the expression for π given above, we have $G(1) = 1$, $G(2) = 4$, and $G(5) = 9$.

(a) Use the expression for π given above to evaluate $G(10)$ and $G(14)$.

(b) Use the Internet to help you evaluate $G(100)$, $G(750)$, and $G(1000)$. *Suggestions*: Using any of the common search engines on the World Wide Web, under the categories of mathematics or science, search for "pi." Here, for example, are two sites that contain the information you need. (They were accessible at the time of this writing, March 2004.)

(University of Exeter)
http://www.ex.ac.uk/cimt/general/pi10000.htm

(The Pi-Search Page)
http://www.angio.net/pi/piquery

C

66. If $f(x) = mx + b$, show that $\dfrac{f(x_1) + f(x_2)}{2} = f\left(\dfrac{x_1 + x_2}{2}\right)$.

67. Let $f(x) = ax^2 + bx + c$, where $a < 0$. Show that $\dfrac{f(x_1) + f(x_2)}{2} \le f\left(\dfrac{x_1 + x_2}{2}\right)$.

PROJECT A PRIME FUNCTION

Primes fuel cryptography, the technological field behind the secret coding of computer communication as well as communication between governments.—The Oregonian, June 26, 1999

. . . the security of large international data networks nowadays relies on the inability of mathematicians to find an efficient method of factoring large numbers (while at the same time being able to produce large primes with ease).—Keith Devlin, *Mathematics: The New Golden Age* (New York: Columbia University Press, 1999)

. . . why does a large supercomputer manufacturer like Cray Research invest so much money . . . [in searching for huge prime numbers]*? The answer is that the computation required . . . is a heavy one, stretching over days or weeks, and so it provides an excellent way to test the efficiency and accuracy of a new computer system.*—Keith Devlin, *op. cit.*

As indicated by the quotes above, prime numbers do have real-world applications. (For the definition of a prime number, see the instructions preceding Exercise 62 on page 143. For a readable explanation of how prime numbers are used in secret codes, refer to Chapter 2 in the book by Keith Devlin cited above.)

In this project you'll work with a function F involving prime numbers: *For each natural number n, let F(n) be the nth prime number.* For instance, since the first, second, third, and fourth primes are 2, 3, 5, and 7, respectively, we have

$$F(1) = 2 \qquad F(2) = 3 \qquad F(3) = 5 \qquad F(4) = 7$$

At the outset, we remark that there is no simple algebraic formula for F.

(a) After creating a list of the first 40 prime numbers, complete the following table. (Though certainly not conceptually difficult, creating the list takes some time, and each error may throw off your ordering. Therefore, this is a good small-group project; some people should work as checkers.)

n	5	10	20	30	40
$F(n)$					

(b) Make use of the Internet or a library to complete the following table.

n	100	250	500	750	1000
$F(n)$					

Suggestion: Using any of the common search engines on the World Wide Web and typing *prime numbers* can lead to websites that have the information you need here. Here are two that were accessible at the time of this writing, March 2004:

http://www.utm.edu/research/primes
(University of Tennessee at Martin website)

and

http://www.svobodat.com/primes

An advantage to the latter site is that a list of primes is available in spreadsheet format, in which it's immediately clear which is the nth prime.

(c) Although there is no simple algebraic formula for F, there are simple functions that do a good job of approximating F along relatively small portions of its domain. For example, look at the graphs in Figures A and B. Figure A shows the graph of F plotted from actual prime number data. Figure B shows the graph of a function g used to model these data:

$$g(n) = 0.0014n^2 + 6.712n - 141.85$$

As you can see, at this scale at least, it's hard to discern any difference between the two functions.

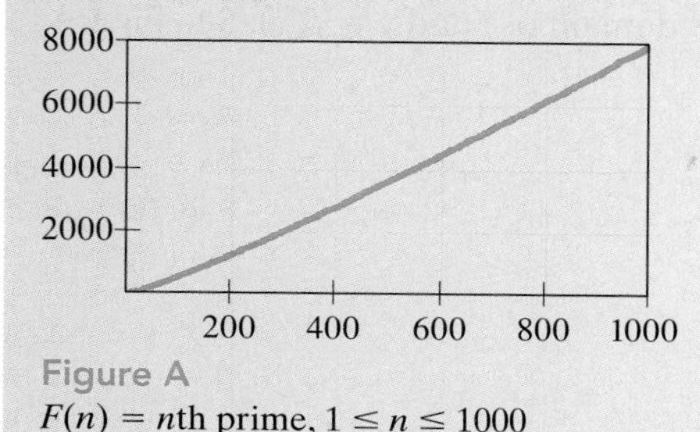

Figure A
$F(n) = n$th prime, $1 \le n \le 1000$

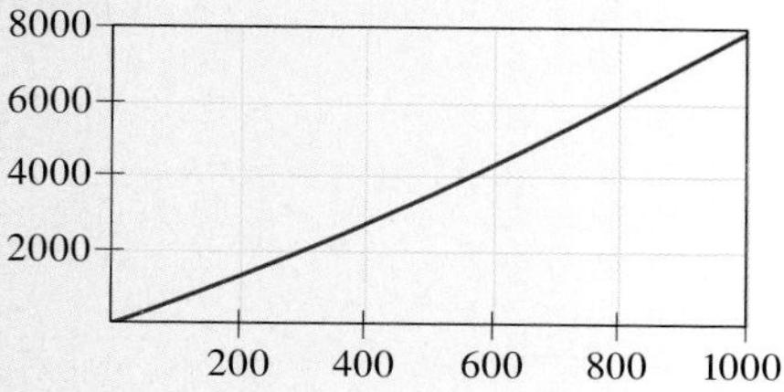

Figure B
$g(n) = 0.0014n^2 + 6.7121n - 141.85$,
$1 \le n \le 1000$

Complete the following table to see some examples of how well the function g approximates the nth prime. Round the values for $g(n)$ to the nearest integer. Round the percentage error to one decimal place. Recall that percentage error is defined by

$$\left| \frac{(\text{actual value}) - (\text{approximate value})}{\text{actual value}} \right| \times 100$$

n	100	250	500	750	1000
$g(n)$					
Percentage error in approximation $g(n) \approx F(n)$					

3.2 THE GRAPH OF A FUNCTION

In my own case, I got along fine without knowing the name of the distributive law until my sophomore year in college; meanwhile I had drawn lots of graphs.—Professor Donald E. Knuth (one of the world's preeminent computer scientists) in *Mathematical People* (Boston: Birkhäuser, 1985)

When the domain and range of a function are sets of real numbers, we can graph the function in the same way that we graphed equations in Chapters 1 and 2. In graphing functions, the usual practice is to reserve the horizontal axis for the independent variable (the inputs) and the vertical axis for the dependent variable (the outputs). In terms of the familiar x-y coordinate system, for each x-coordinate or input in the domain of the function, the function or rule tells you the corresponding y-coordinate or output in the range of the function. These ideas are summarized in the following definition.

DEFINITION Graph of a Function

The graph of a function f in the x-y plane consists of those points (x, y) such that x is in the domain of f and $y = f(x)$. See Figure 1.

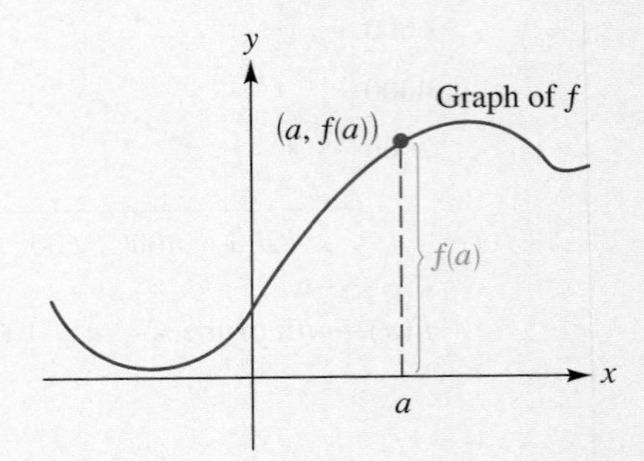

Figure 1

EXAMPLE 1 Using the definition of the graph of a function

In Figures 2(a) and 2(b), specify the y-coordinates of the points P and Q, respectively. In each case, give an exact expression and also a calculator approximation rounded to two decimal places. (We'll look at the graphs of these two functions in more detail later in this section.)

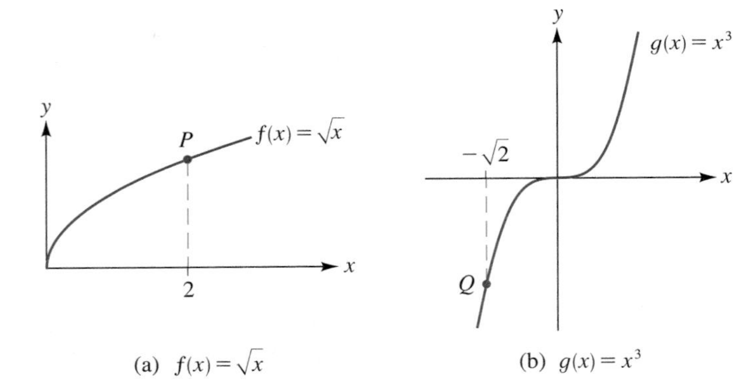

Figure 2
What are the y-coordinates of the points P and Q?

SOLUTION

In Figure 2(a), which shows a graph of the function f defined by $f(x) = \sqrt{x}$, the x-coordinate of P is 2. The y-coordinate is therefore $f(2) = \sqrt{2}$. Using a calculator, we find $\sqrt{2} \approx 1.41$. In Figure 2(b), which shows a graph of the function g defined by $g(x) = x^3$, the x-coordinate of Q is $-\sqrt{2}$. The y-coordinate is therefore

$$\begin{aligned} g(-\sqrt{2}) &= (-\sqrt{2})^3 = (-\sqrt{2})^2(-\sqrt{2})^1 \\ &= 2(-\sqrt{2}) = -2\sqrt{2} \\ &\approx -2.28 \qquad \text{using a calculator} \end{aligned}$$

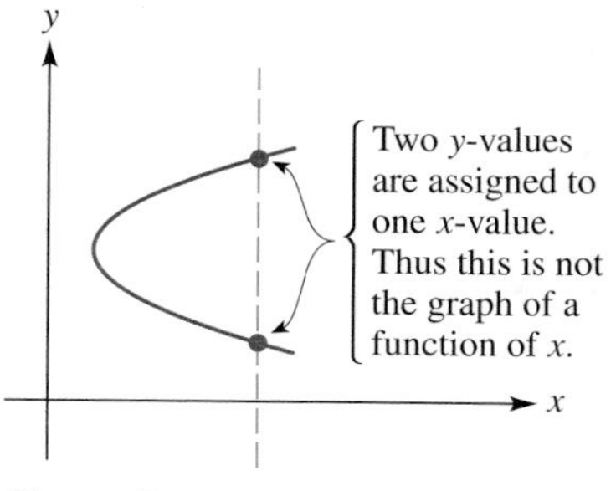
y
Two y-values are assigned to one x-value. Thus this is not the graph of a function of x.
x

Figure 3

All of the graphs that we looked at in Section 3.1 are graphs of functions. So too are most of the graphs in Chapter 1. However, it's important to understand that not every graph represents a function. Consider, for example, the graph in Figure 3 (the red curve). Figure 3 shows a vertical line intersecting the graph in two distinct points. The specific coordinates of the two points are unimportant. What the vertical line helps us to see is that two different y-values (outputs) have been assigned to the same x-value (an input), and therefore the graph does not represent a function of x. These remarks are generalized in the box that follows.

Vertical Line Test

A graph in the x-y plane represents a function of x provided that any vertical line intersects the graph in at most one point.

EXAMPLE 2 Applying the vertical line test

The vertical line test implies that the graph in Figure 4(a) represents a function and that the graph in Figure 4(b) does not.

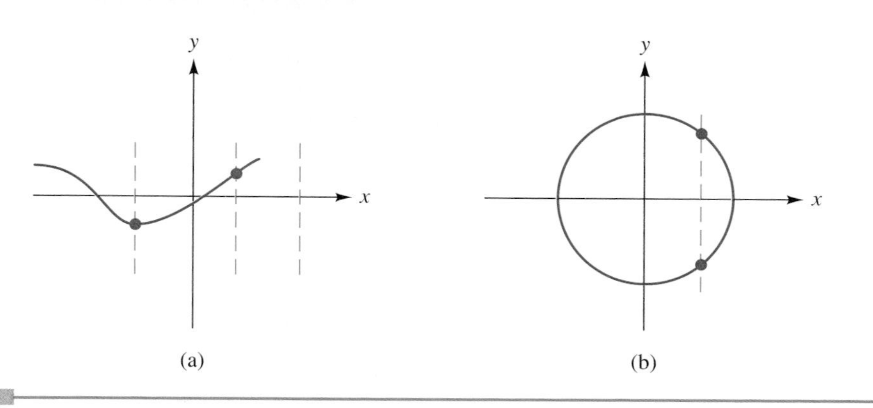
y
x
(a)
y
x
(b)

Figure 4

In Examples 2 through 4 in the previous section we used algebra to determine the domain and range of functions. If the graph of a function is given, however, we can read off that information directly. The next two examples show instances of this.

EXAMPLE 3 Determining the domain and range of a function from its graph

Specify the domain and the range of the function g graphed in Figure 5(a).

SOLUTION

The domain of g is just that portion of the x-axis (the inputs) utilized in graphing g. As Figure 5(b) indicates, this amounts to all real numbers x from 1 to 5, inclusive: $1 \leq x \leq 5$. In interval notation, this set of numbers is denoted by $[1, 5]$. To find the range of g, we need to check which part of the y-axis is utilized in graphing g. As Figure 5(b) indicates, this is the set of all real numbers y

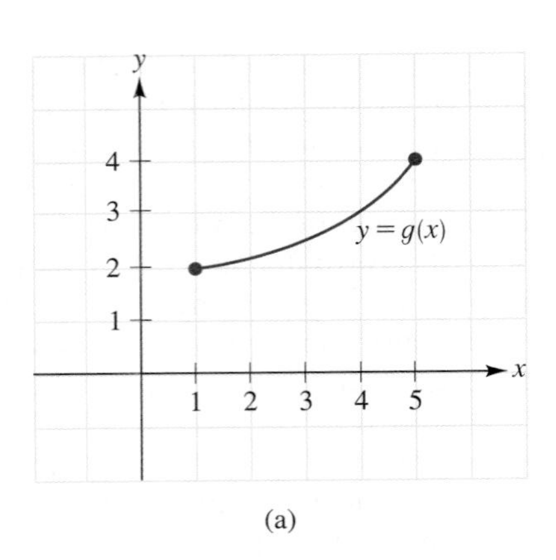

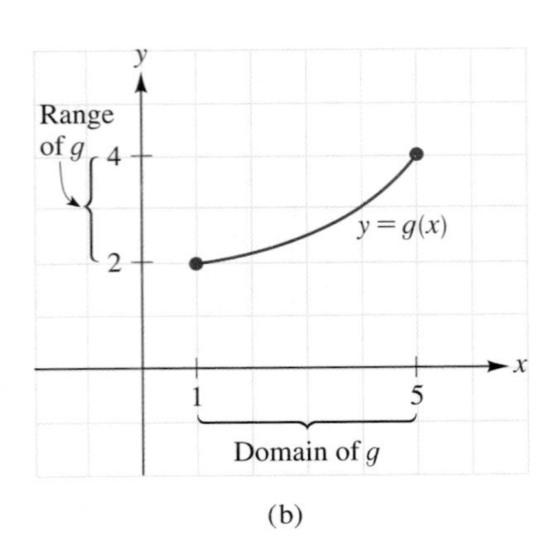

Figure 5 (a) (b)

between 2 and 4, inclusive: $2 \le y \le 4$. Interval notation for this set of numbers is $[2, 4]$.

EXAMPLE 4 Determining domain, range, and function values from a graph

The graph of a function h is shown in Figure 6. The open circle in the figure is used to indicate that the point (3, 3) does not belong to the graph of h.

(a) Specify the domain and the range of the function h.

(b) Determine each value: **(i)** $h(-2)$; **(ii)** $h(3)$; **(iii)** $|h(-4)|$.

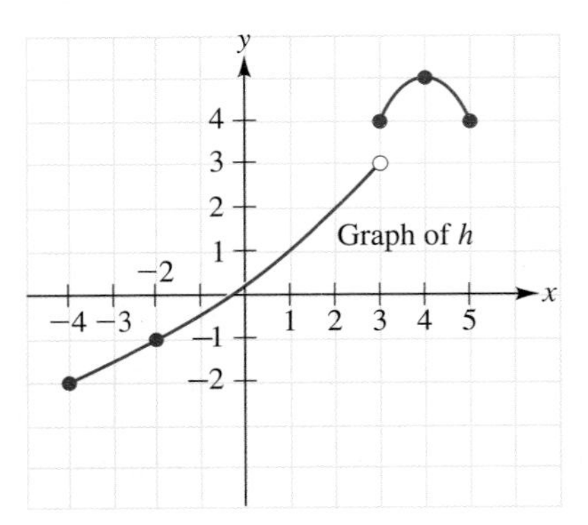

Figure 6

SOLUTION

(a) The domain of h is the interval $[-4, 5]$. The range is the set $[-2, 3) \cup [4, 5]$.

(b) **(i)** The function notation $h(-2)$ stands for the y-coordinate of that point on the graph of h whose x-coordinate is -2. Since the point $(-2, -1)$ is on the graph of h, we conclude that $h(-2) = -1$.

(ii) We have $h(3) = 4$ because the point (3, 4) lies on the graph of h. Note that h would not be considered a function if the point (3, 3) were also part of the graph. (Why?)

(iii) Since the point $(-4, -2)$ lies on the graph of h, we write $h(-4) = -2$. Thus $|h(-4)| = |-2| = 2$.

Given a function represented by an equation $y = f(x)$, how do we obtain the graph? From Chapter 1 (or from previous courses) we have these three options:

1. Set up a table, plot points, and then draw the line or curve that the points seem to describe.
2. If the equation has the form $y = mx + b$, the graph is a line (as explained in Section 1.6), and we can draw the graph using slope-intercept information.
3. Use a graphing utility.

As a practical matter, option 1 above has a drawback. How many points must we plot before we're sure about the essential features of the graph? In later sections of this text you'll see that there are a number of techniques and concepts that make it unnecessary to plot a large number of points. For now, however, we point out that some functions arise frequently enough to make it worth memorizing the basic shapes and features of their graphs. True, you can always reach for a graphing calculator, but sometimes it helps to be able to think on your feet, so to speak. Also, if you've memorized the graph of a function, then you automatically know the domain and the range. Figure 7 displays the graphs of six basic functions. Exercises 21

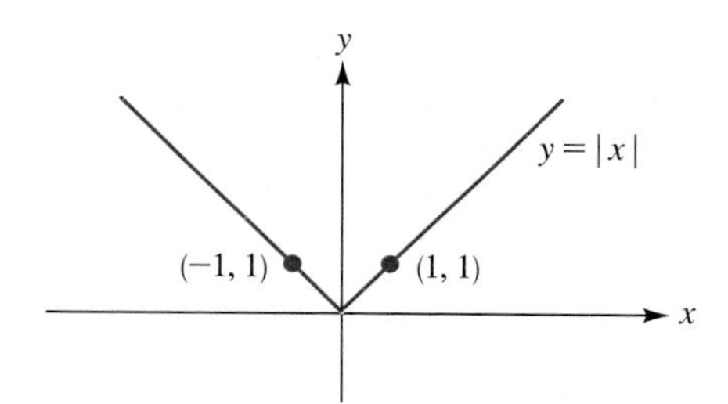

(a) The absolute value function
$y = |x|$
Domain: $(-\infty, \infty)$
Range: $[0, \infty)$

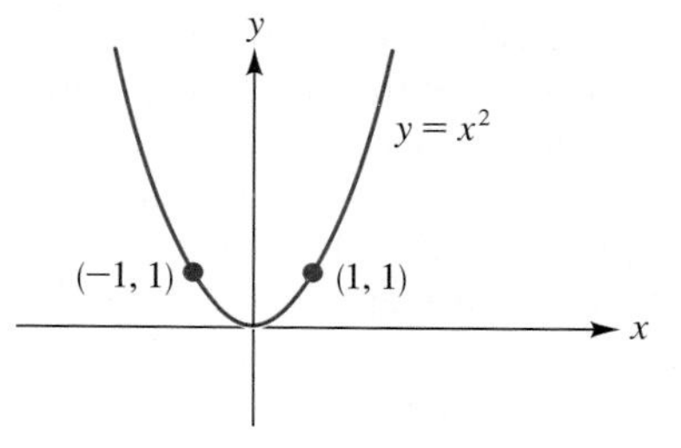

(b) The squaring function
$y = x^2$
Domain: $(-\infty, \infty)$
Range: $[0, \infty)$

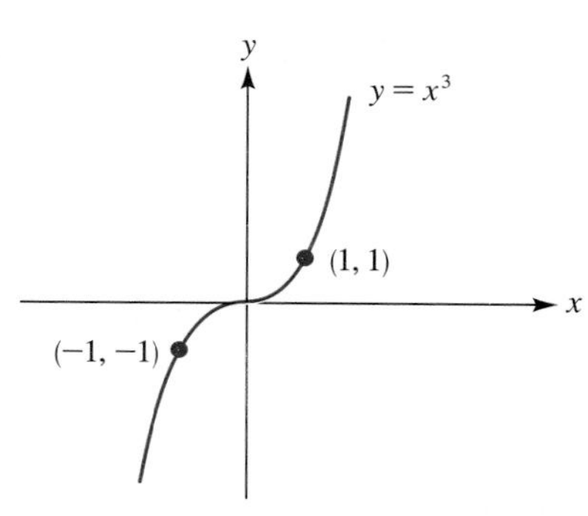

(c) The cubing function
$y = x^3$
Domain: $(-\infty, \infty)$
Range: $(-\infty, \infty)$

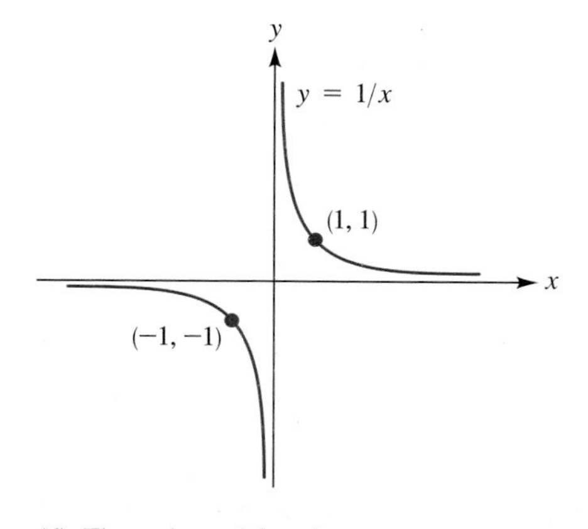

(d) The reciprocal function
$y = 1/x$
Domain: $(-\infty, 0) \cup (0, \infty)$
Range: $(-\infty, 0) \cup (0, \infty)$

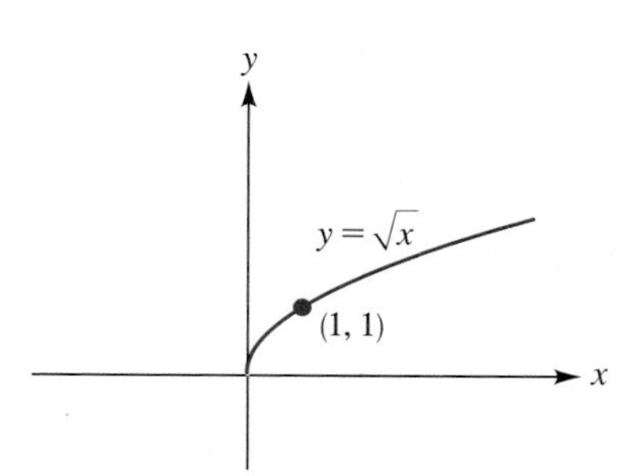

(e) The square root function
$y = \sqrt{x}$
Domain: $[0, \infty)$
Range: $[0, \infty)$

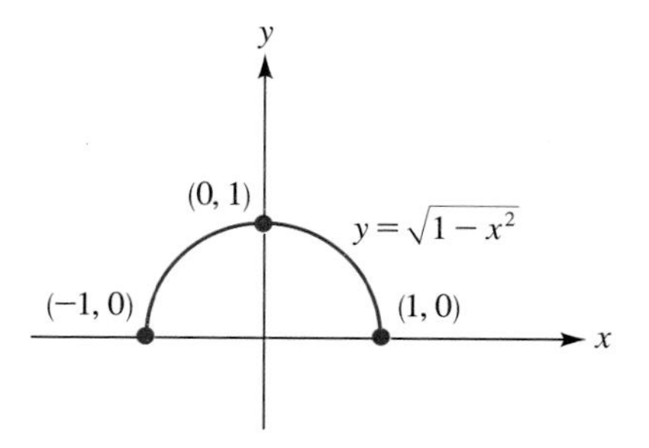

(f) The semicircle function
$y = \sqrt{1 - x^2}$
Domain: $[-1, 1]$
Range: $[0, 1]$

Figure 7
The graphs of six basic functions

and 22 at the end of this section ask you to set up tables and verify for yourself that the graphs in Figure 7 are indeed correct. From now on (with the exception of Exercises 21 and 22), if you need to sketch by hand a graph of one of these basic six functions, you should do so from memory.

All the graphs in Figure 7 represent functions defined by rather simple equations. In some instances, however, functions may be defined by different equations on different parts of the domain. This is illustrated in the next two examples, in which we graph *piecewise-defined functions.*

EXAMPLE 5 Graphing a piecewise-defined function

A function g is defined by $g(x) = \begin{cases} x^2 & \text{if } x < 2 \\ 1/x & \text{if } x \geq 2 \end{cases}$.

(a) Find $g(3/2)$, $g(2)$, and $g(3)$.
(b) Sketch the graph of the function g.

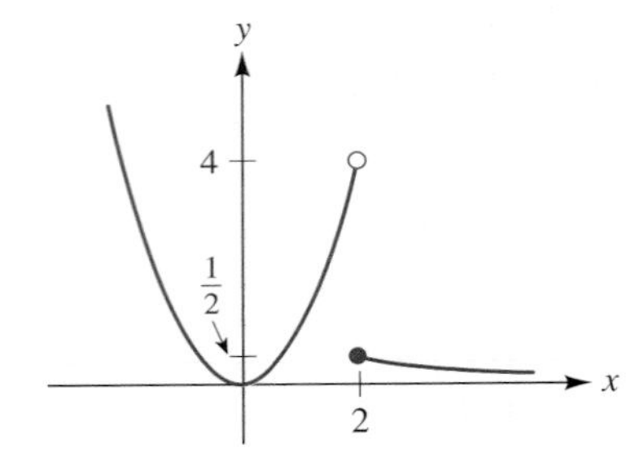

Figure 8

A graph of $g(x) = \begin{cases} x^2, & x < 2 \\ \dfrac{1}{x}, & x \geq 2 \end{cases}$

SOLUTION

(a) To find $g(3/2)$, which of the two equation do we use: $g(x) = x^2$ or $g(x) = 1/x$? According to the given inequalities, we should use $g(x) = x^2$ whenever the x-values are less than 2. So for $x = 3/2$ (which is surely less than 2), we have $g(3/2) = (3/2)^2 = 9/4$. On the other hand, the inequalities tell us to use the equation $g(x) = 1/x$ whenever the x-values are greater than or equal to 2. Thus we have $g(2) = 1/2$ and $g(3) = 1/3$.

(b) For the graph, we look back at Figures 7(b) and (d) and choose the appropriate portion of each. The result is displayed in Figure 8. The open circle in the figure is used to indicate that the point (2, 4) does not belong to the graph. The filled-in circle, on the other hand, is used to indicate that the point (2, 1/2) does belong to the graph.

EXAMPLE 6

Sketch a graph of each function:

(a) $y = \begin{cases} |x| & \text{if } -1 \leq x \leq 1 \\ \sqrt{x} & \text{if } 1 \leq x \leq 4 \end{cases}$

(b) $y = \begin{cases} |x| & \text{if } -1 \leq x < 1 \\ \sqrt{x} & \text{if } 1 < x \leq 4 \end{cases}$

SOLUTION

(a) We want to "paste together" portions of the graphs of $y = |x|$ and $y = \sqrt{x}$. The appropriate portions of each graph are indicated in color in Figures 9(a) and 9(b). Using Figures 9(a) and 9(b) as a guide, we sketch the required graph in Figure 9(c).

(b) Comparing the given equations and inequalities with those in part (a), we see that there is only one difference: in part (b), the number 1 is excluded from the domain of the variable x. To graph the equations in part (b), we

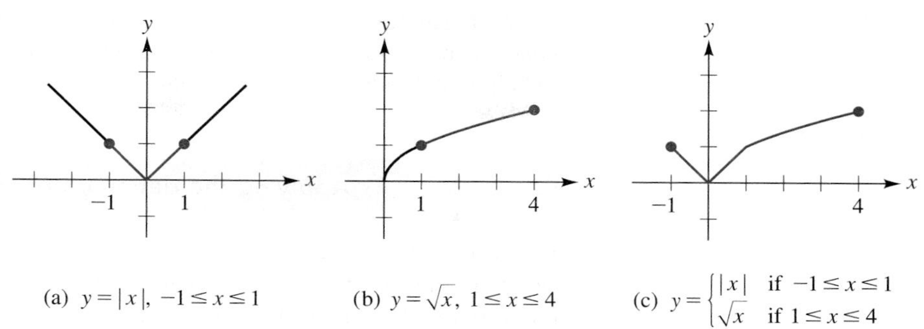

(a) $y = |x|,\ -1 \le x \le 1$ (b) $y = \sqrt{x},\ 1 \le x \le 4$ (c) $y = \begin{cases} |x| & \text{if } -1 \le x \le 1 \\ \sqrt{x} & \text{if } 1 \le x \le 4 \end{cases}$

Figure 9

need only take the graph obtained in Figure 9(c) and delete the point that corresponds to $x = 1$. That is, we delete the point (1, 1); see Figure 10.

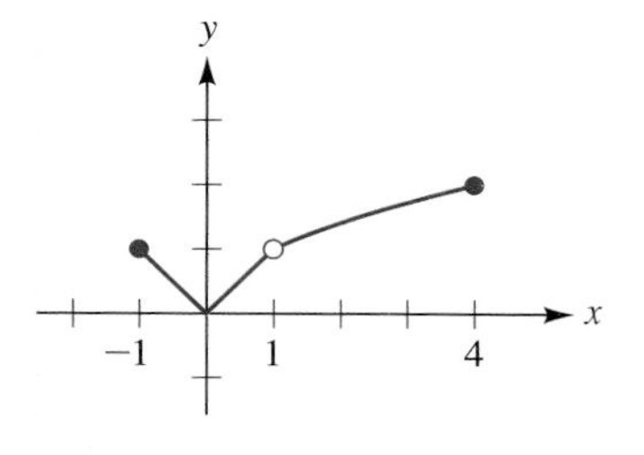

$y = \begin{cases} |x|, & \text{if } -1 \le x < 1 \\ \sqrt{x}, & \text{if } 1 < x \le 4 \end{cases}$

Figure 10

It is useful to compare the graphs in Figures 8, 9(c), and 10. Each graph was obtained by combining portions of the basic graphs in Figure 7. In Figures 8 and 10 the graphs have a break or gap in them. We say that the graph (or more precisely, the *function*) in Figure 8 is *discontinuous* when $x = 2$. (The two pieces do not meet at $x = 2$.) The function in Figure 10 is discontinuous when $x = 1$. (There is a gap or missing point when $x = 1$.) In Figure 9(c), however, the two portions of the graph form a graph with no break or gap. We say that the function in Figure 9(c) is *continuous* when $x = 1$. A rigorous definition of continuity is properly a subject for calculus. However, even at the intuitive level at which we've presented the idea here, you'll find that the concept is useful in helping you to organize your thoughts about graphs.

The graph of a piecewise-defined function can also be obtained using a graphing utility. For details, consult the owner's manual that accompanies your calculator or software. We will, however, point out one problem to watch for. In Figure 11(a), we've used a graphing utility to graph the function g from Example 5. Note the extraneous vertical line segment that was generated, connecting what should be two disjoint curves. Not all graphing utilities make this error; but if yours does, it can be corrected by switching the curve-drawing mode from *connected* to *dotted.* See Figure 11(b).

GRAPHICAL PERSPECTIVE

Figure 11

$g(x) = \begin{cases} x^2 & \text{if } x < 2 \\ 1/x & \text{if } x \ge 2 \end{cases}$

On a TI-83 graphing calculator, for example, this function g can be entered with the keystrokes $(x < 2)(x^2) + (x \ge 2)(x^{-1})$.

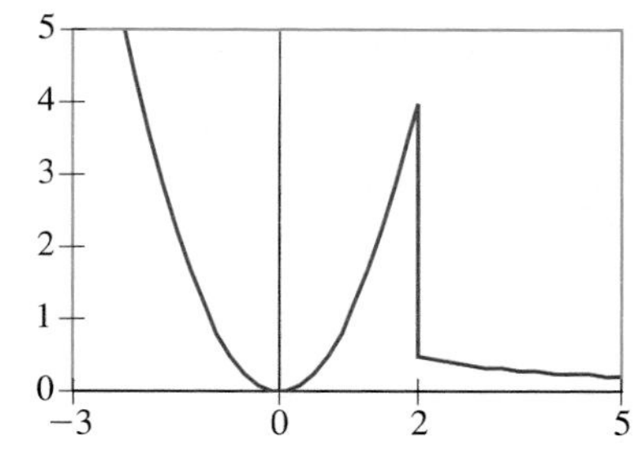

(a) Graphing utility produces an extraneous vertical segment in the graph of g.

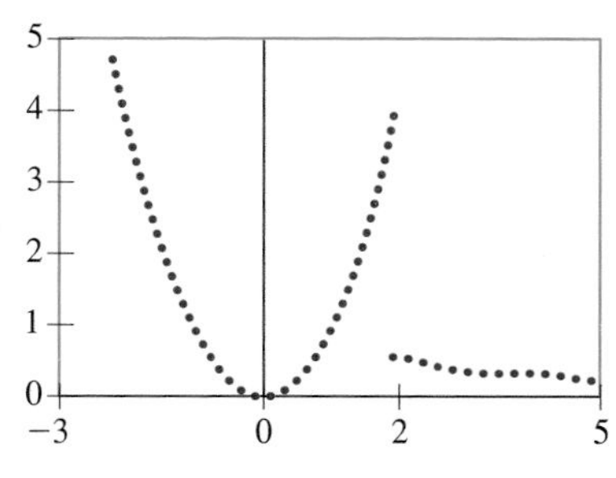

(b) Switching to dot mode eliminates the extraneous segment.

We began this section with a definition of what is meant by the *graph of a function*. The next example uses one of the six basic functions along with the line $y = x$ to help you develop some additional insight into that definition.

EXAMPLE 7 Applying the definition of the graph of a function

Figure 12 shows portions of the graphs of the functions $f(x) = \sqrt{x}$ and $y = x$. What are the coordinates of the points P, Q, and R in Figure 12? (Each dashed line is parallel to one of the coordinate axes.) For those coordinates involving radicals, also supply a calculator approximation rounded to two decimal places.

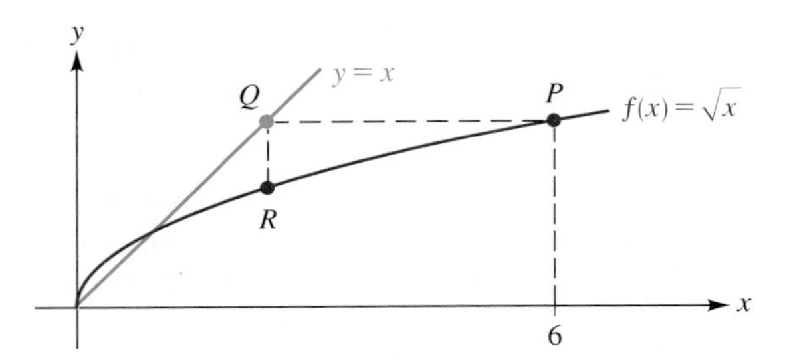

Figure 12
What are the coordinates of P, Q, and R?

SOLUTION

In Figure 12 the point P lies on the graph of $f(x) = \sqrt{x}$. Since the x-coordinate of P is 6, the y-coordinate is $f(6)$, which, according to the rule for f, is $\sqrt{6}$. Thus the coordinates of P are $(6, \sqrt{6})$. Next, we want the coordinates of Q. Figure 12 shows that the points P and Q have the same y-coordinate, so the y-coordinate of Q must also be $\sqrt{6}$. What about the x-coordinate of Q? Since Q lies on the graph of $y = x$, the x-coordinate of Q must be the same as its y-coordinate, namely, $\sqrt{6}$. Thus, the coordinates of Q are $(\sqrt{6}, \sqrt{6})$.

Finally, let us determine the coordinates of the point R. According to Figure 12, the x-coordinate of R is the same as that of Q, which we know to be $\sqrt{6}$. The y-coordinate of R is then found by taking the square root of this x-coordinate, to obtain $\sqrt{\sqrt{6}}$. (We've used the fact that R lies on the graph of $f(x) = \sqrt{x}$.)

Summarizing now, the required coordinates are as follows. (You should check for yourself that the calculator values given are correct.)

P: $(6, \sqrt{6}) \approx (6, 2.45)$

Q: $(\sqrt{6}, \sqrt{6}) \approx (2.45, 2.45)$

R: $(\sqrt{6}, \sqrt{\sqrt{6}}) \approx (2.45, 1.57)$

EXERCISE SET 3.2 

A

In Exercises 1–4, specify the y-coordinate of the point P on the graph of the given function. In each case, give an exact expression and a calculator approximation rounded to three decimal places.

1.

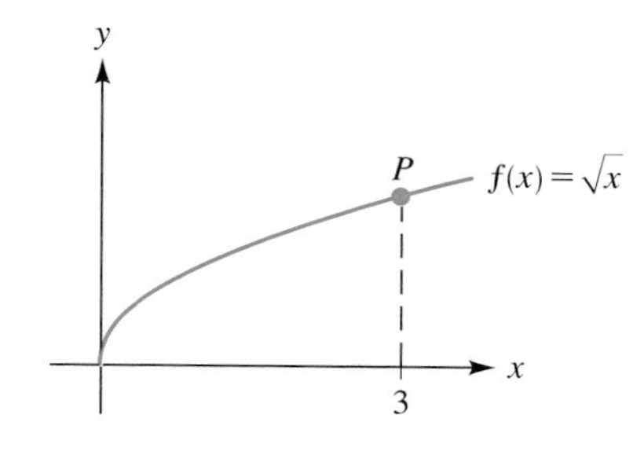

2.

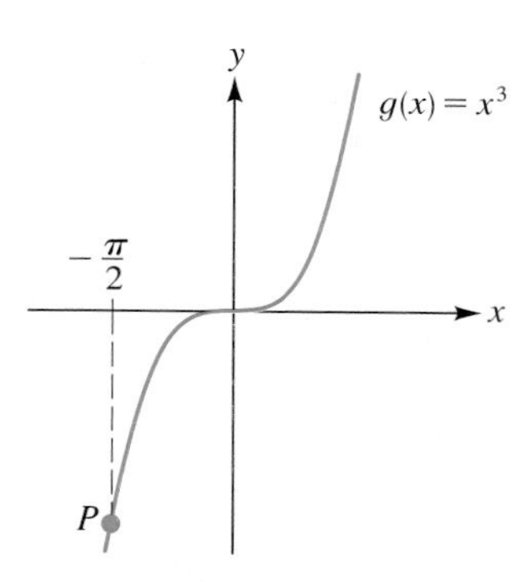

3.

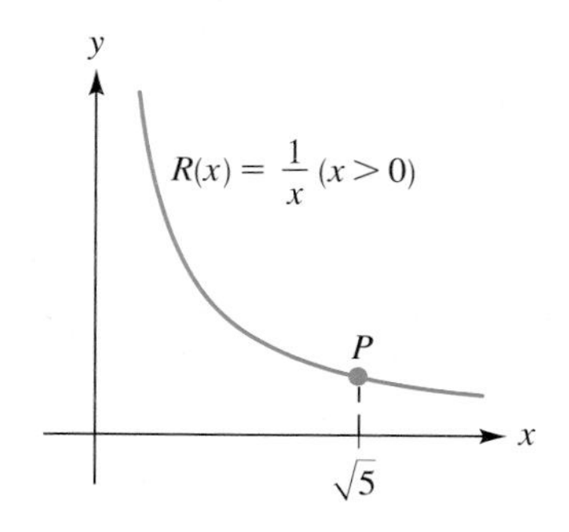

4.

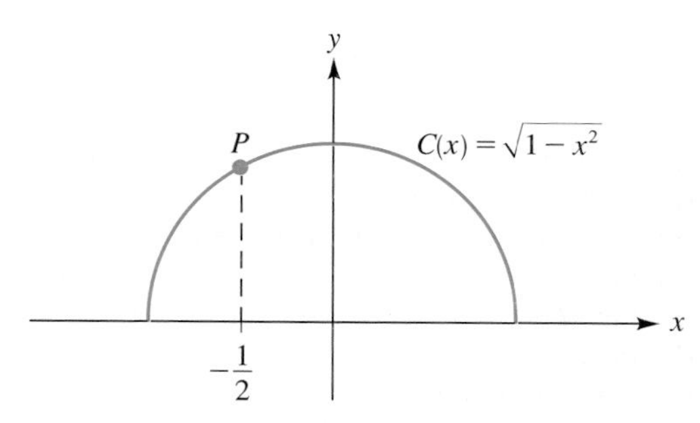

In Exercises 5 and 6, use the vertical line test to determine whether each graph represents a function of x.

5.

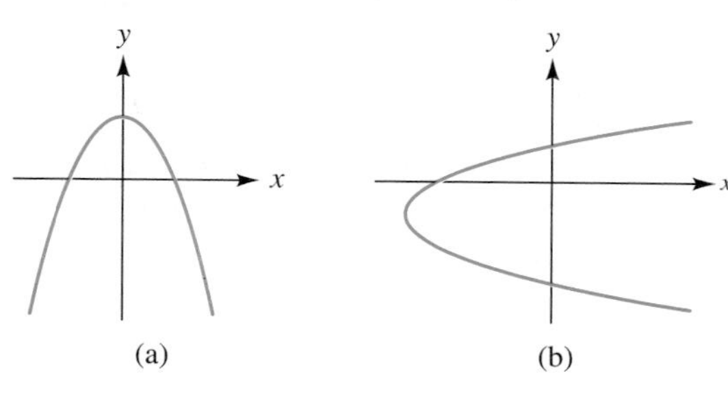

(c)

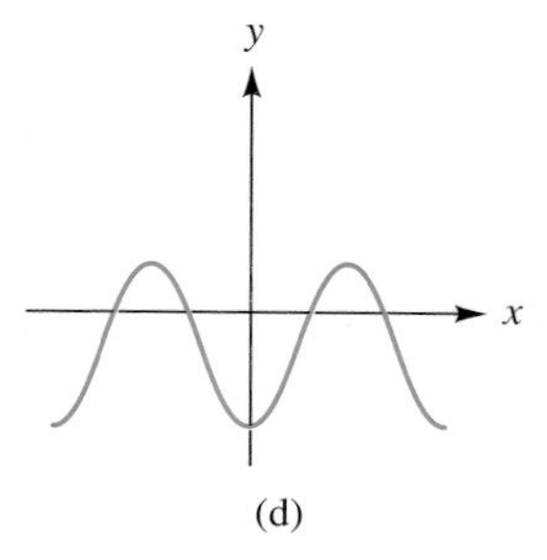

(d)

6.

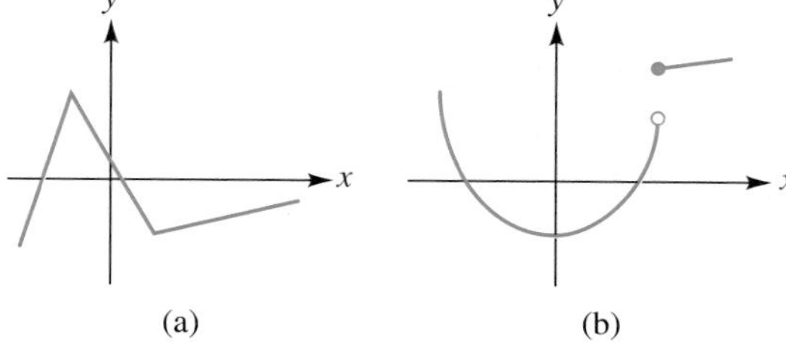

(c)

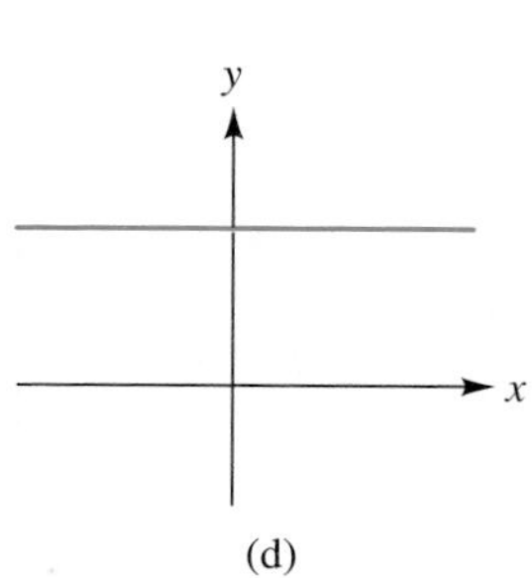

(d)

In Exercises 7–14, the graph of a function is given. In each case, specify the domain and the range of the function. (The axes are marked off in one-unit intervals.)

7.

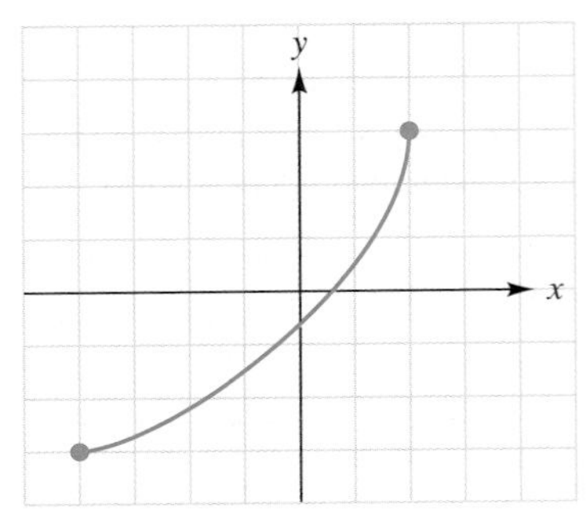

8.

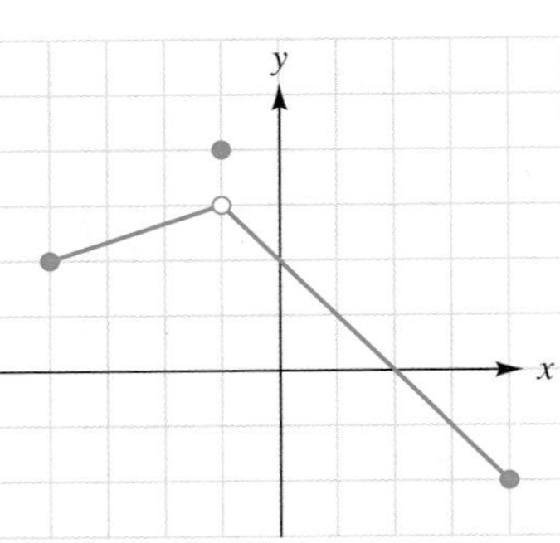

9.

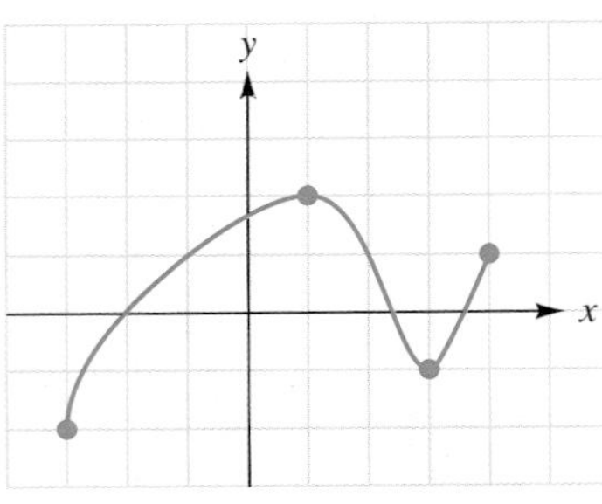

10.

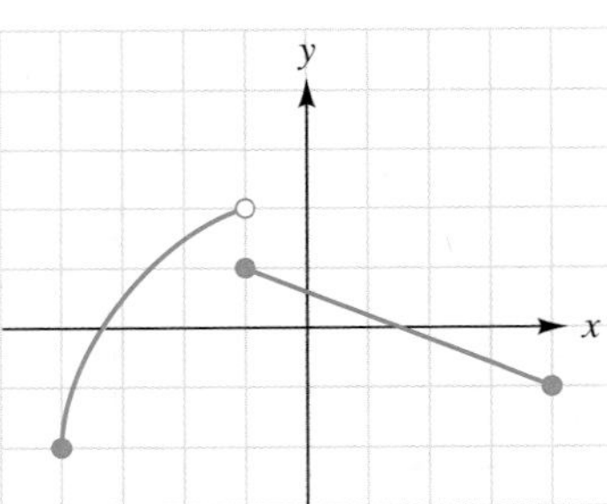

11.

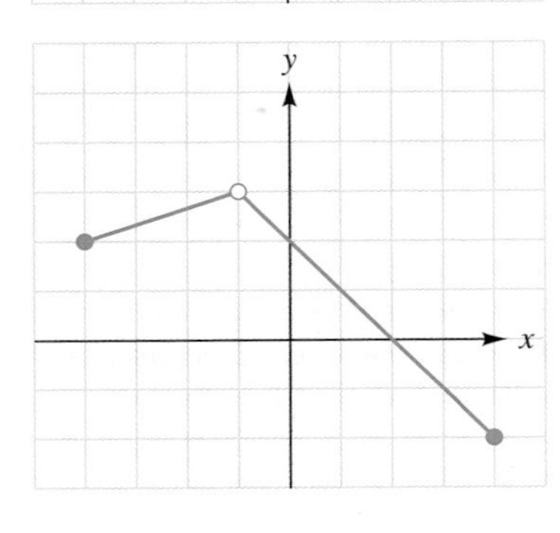

12.

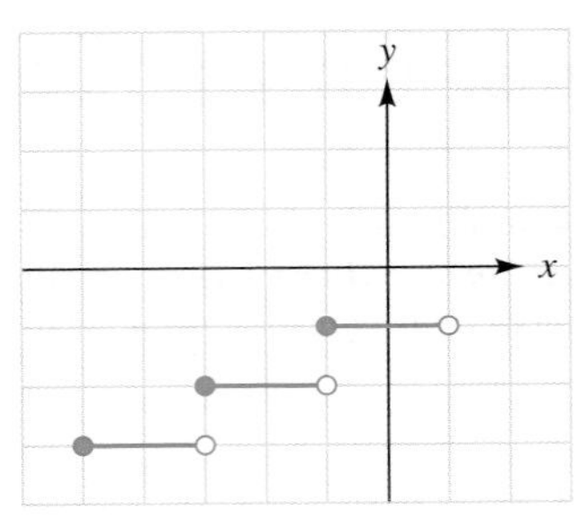

13.

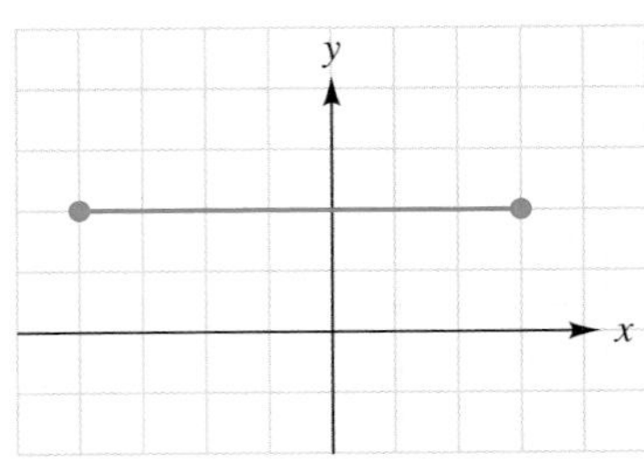

14.

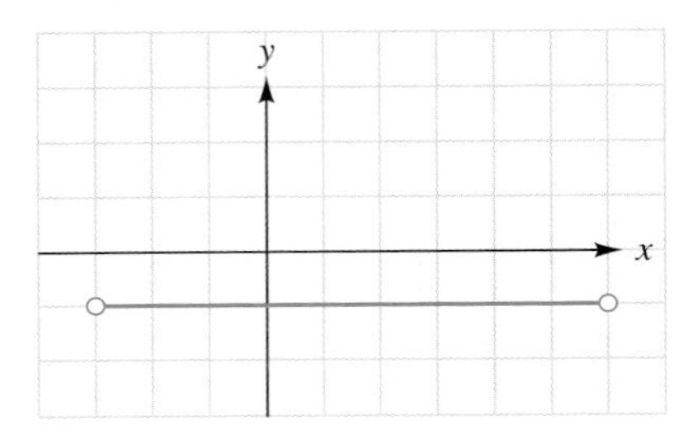

In Exercises 15 and 16, refer to the graph of the function F in the figure. (Assume that the axes are marked off in one-unit intervals.)

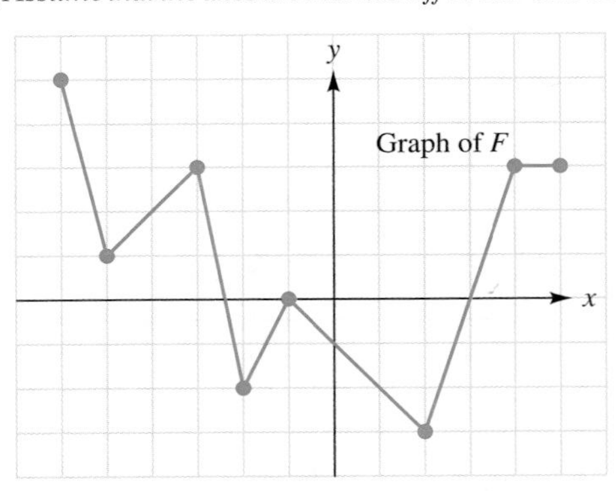

15. (a) Find $F(-5)$
(b) Find $F(2)$.
(c) Is $F(1)$ positive?
(d) For which value of x is $F(x) = -3$?
(e) Find $F(2) - F(-2)$.

16. (a) Find $F(4)$.
(b) Find $F(-1)$.
(c) Is $F(-4)$ positive?
(d) For which value of x is $F(x) = 5$?
(e) Find $F(5) - F(-3)$.

17. The following figure displays the graph of a function f.

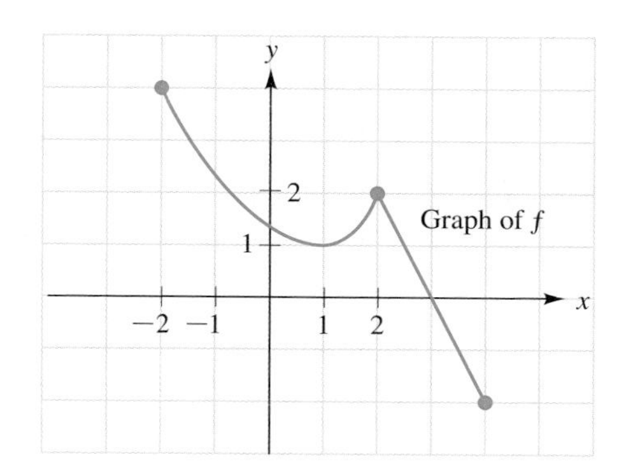

(a) Is $f(0)$ positive or negative?
(b) Find $f(-2)$, $f(1)$, $f(2)$, and $f(3)$.
(c) Which is larger, $f(2)$ or $f(4)$?
(d) Find $f(4) - f(1)$.
(e) Find $|f(4) - f(1)|$.
(f) Write the domain and range of f using the interval notation $[a, b]$.

18. The following figure shows the graph of a function h.
(a) Find $h(a)$, $h(b)$, $h(c)$, and $h(d)$.
(b) Is $h(0)$ positive or negative?
(c) For which values of x does $h(x) = 0$?
(d) Which is larger, $h(b)$ or $h(0)$?
(e) As x increases from c to d, do the corresponding values of $h(x)$ increase or decrease?
(f) As x increases from a to b, do the corresponding values of $h(x)$ increase or decrease?

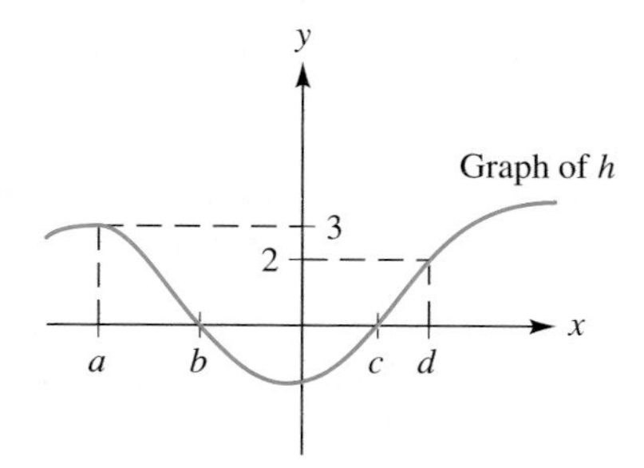

In Exercises 19 and 20, refer to the graphs of the functions f and g in the figure. Assume that the domain of each function is $[-3, 3]$ *and that the axes are marked off in one-unit intervals.*

19. **(a)** Which is larger, $f(-2)$ or $g(-2)$?
(b) Compute $f(0) - g(0)$.
(c) Which among the following three quantities is the smallest?

$$f(1) - g(1) \qquad f(2) - g(2) \qquad f(3) - g(3)$$

(d) For which value(s) of x does $g(x) = f(1)$?
(e) Is the number 4 in the range of f or in the range of g?

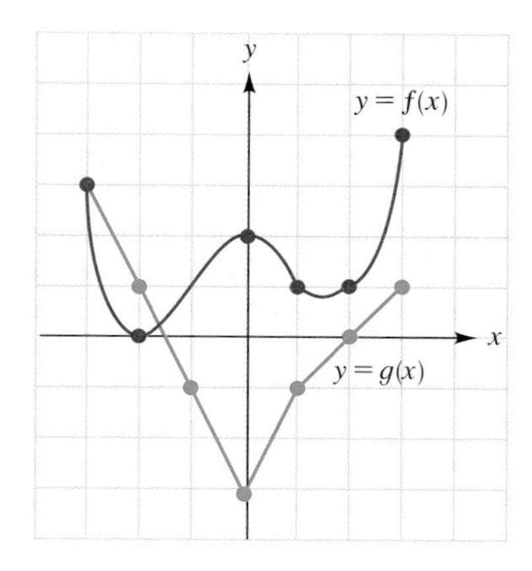

20. **(a)** For the interval $[0, 3]$, is the quantity $g(x) - f(x)$ positive or negative?
(b) For the interval $(-3, -2)$, is the quantity $g(x) - f(x)$ positive or negative?

In Exercises 21 and 22, set up a table and graph each function. [In Exercises 22(b) and 22(c), use a calculator to compute the square roots.]

21. **(a)** $y = |x|$
(b) $y = x^2$
(c) $y = x^3$

22. **(a)** $y = 1/x$
(b) $y = \sqrt{x}$
(c) $y = \sqrt{1 - x^2}$

In Exercises 23–30, sketch the graphs, as in Examples 5 and 6.

23. $A(x) = \begin{cases} x^3 & \text{if } -2 \le x \le -1 \\ x^2 & \text{if } x > -1 \end{cases}$

24. $B(x) = \begin{cases} \sqrt{1 - x^2} & \text{if } -1 \le x < 1 \\ 1/x & \text{if } x \ge 1 \end{cases}$

25. $C(x) = \begin{cases} x^3 & \text{if } x < 1 \\ \sqrt{x} & \text{if } x > 1 \end{cases}$

26. $y = \begin{cases} |x| & \text{if } x \le 0 \\ x^2 & \text{if } x > 0 \end{cases}$

27. **(a)** $y = \begin{cases} \sqrt{x} & \text{if } 0 \le x \le 1 \\ 1/x & \text{if } 1 < x < 2 \end{cases}$

(b) $y = \begin{cases} \sqrt{x} & \text{if } 0 \le x < 1 \\ 1/x & \text{if } 1 < x < 2 \end{cases}$

28. **(a)** $y = \begin{cases} x^3 & \text{if } x \le 0 \\ |x| & \text{if } x > 0 \end{cases}$

(b) $y = \begin{cases} |x| & \text{if } x \le 0 \\ x^3 & \text{if } x > 0 \end{cases}$

29. **(a)** $y = \begin{cases} \sqrt{1 - x^2} & \text{if } -1 \le x \le 0 \\ x^2 & \text{if } 0 < x \le 2 \end{cases}$

(b) $y = \begin{cases} \sqrt{1 - x^2} & \text{if } -1 \le x < 0 \\ x^2 & \text{if } 0 < x \le 2 \end{cases}$

30. $y = \begin{cases} x^3 & \text{if } -1 \le x < 0 \\ \sqrt{x} & \text{if } 0 \le x < 1 \\ 1/x & \text{if } 1 \le x \le 3 \end{cases}$

Ⓖ *In Exercises 31 and 32, use a graphing utility to obtain the graphs of the piecewise-defined functions. If there are cases in which the graphing utility generates extraneous segments, use the dot mode to correct the situation.*

31. (a) $f(x) = \begin{cases} x^3 - x & \text{if } -1 \le x < 0 \\ x^3 + x & \text{if } 0 \le x \le 1 \end{cases}$

(b) $g(x) = \begin{cases} x^3 - x & \text{if } -1 \le x < 0 \\ x^3 + x + 1 & \text{if } 0 \le x \le 1 \end{cases}$

32. (a) $h(x) = \begin{cases} \sqrt{1 - x^6} & \text{if } -1 \le x < 1 \\ \sqrt{1 - (x-2)^6} & \text{if } 1 \le x \le 3 \end{cases}$

(b) $k(x) = \begin{cases} \sqrt{1 - x^6} & \text{if } -1 \le x < 1 \\ \sqrt{1 - (x-2)^6} + 3 & \text{if } 1 \le x \le 3 \end{cases}$

In Exercises 33–36, determine the coordinates of the points P, Q, and R in each figure; give an exact expression and also a calculator approximation rounded to three decimal places. Assume that each dashed line is parallel to one of the coordinate axes. (In Exercise 36, note that the line is $y = x/2$ rather than $y = x$.)

33.

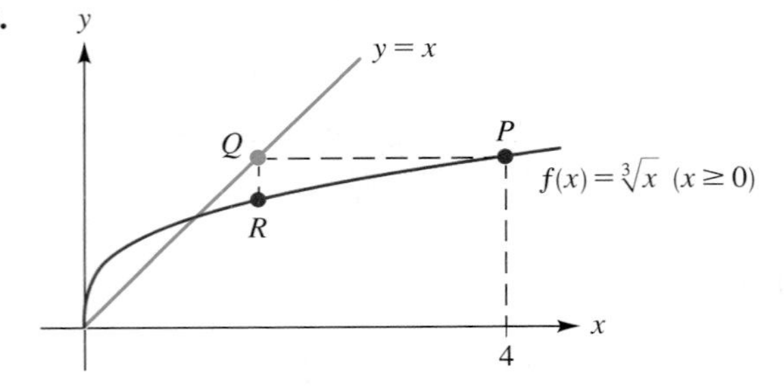

34.

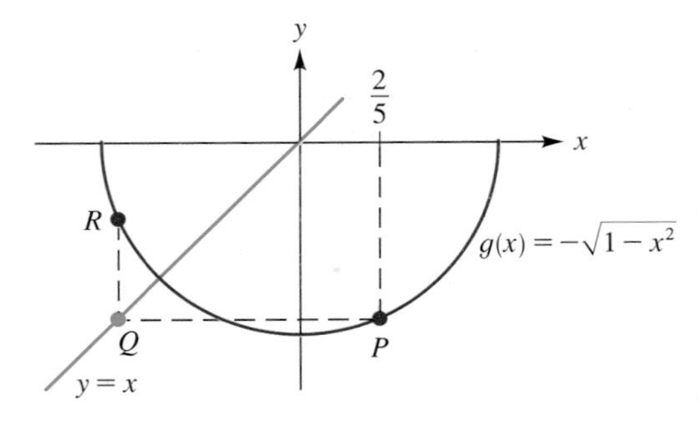

35.

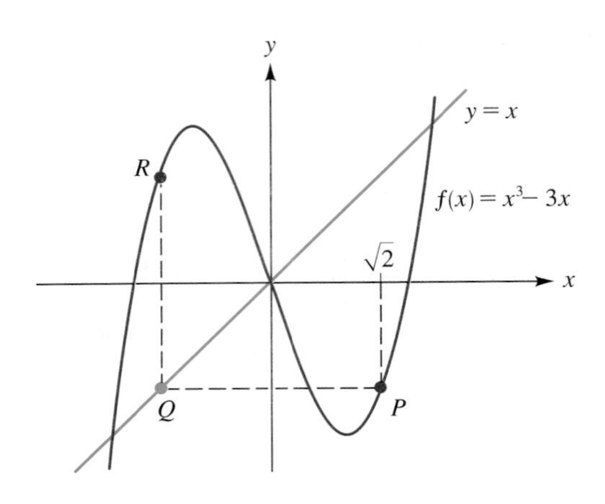

36.

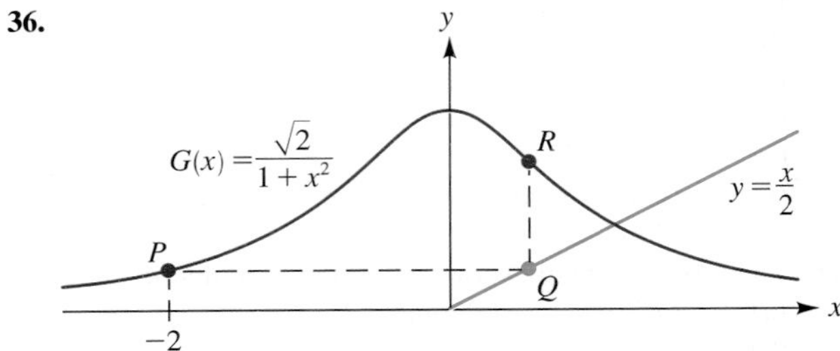

B

Ⓖ **37. (a)** Graph each of the following functions on the same screen, using the viewing rectangle $[0, 1, 0.2] \times [0, 1, 0.2]$.

$y = x \qquad y = x^2 \qquad y = x^3 \qquad y = x^4 \qquad y = x^5$

(b) Describe, in complete sentences, the pattern you see in your results in part (a). On the basis of your results in part (a), draw (by hand) a sketch of what you think the graph of $y = x^{100}$ must look like in this interval. Then use a graphing utility to see how accurate your sketch was.

38. (Although this is not a graphing utility exercise, the previous exercise, which does use a graphing utility, provides some useful background here.)

(a) Suppose that the thickness of the line drawn by your pencil or pen is 0.05 cm, and suppose that the common unit of length marked off along both axes is 1 cm. Under these drawing conditions, where is the graph of $y = x^2$ indistinguishable from the x-axis? That is, for which x-values will the corresponding y-values be less than 0.05?

(b) Follow part (a), using $y = x^4$ instead of $y = x^2$.

PROJECT

IMPLICIT FUNCTIONS: BATTERIES REQUIRED?

From Chapter 1 we know that the graph of the equation $x^2 + y^2 = 1$ is a circle, as shown in Figure A. From our work in this section, we know the graph does not represent a function. However, certain pieces of the graph do represent functions. In particular, solving the equation for y yields $y = \pm\sqrt{1 - x^2}$. Figures B and C show the graphs of $y = \sqrt{1 - x^2}$ and $y = -\sqrt{1 - x^2}$, respectively. Note that the graphs in Figures B and C indeed represent functions. We say that the initial equation $x^2 + y^2 = 1$ defines these functions *implicitly,* whereas each of the equations $y = \sqrt{1 - x^2}$ and $y = -\sqrt{1 - x^2}$ defines a function *explicitly.*

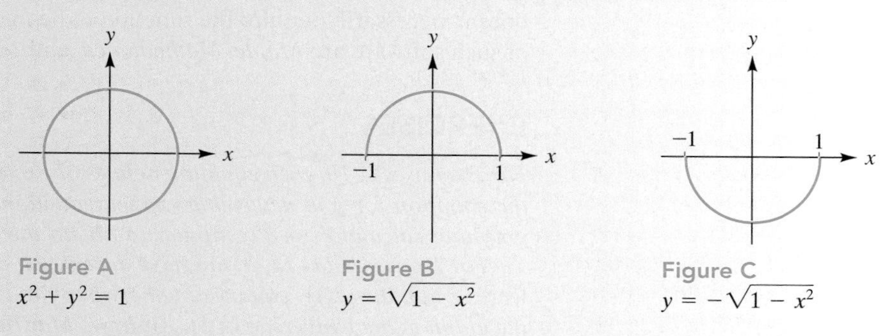

Figure A
$x^2 + y^2 = 1$

Figure B
$y = \sqrt{1 - x^2}$

Figure C
$y = -\sqrt{1 - x^2}$

Here's another example of this terminology. One of the basic functions in Figure 7 on page 149 is defined by the equation $y = 1/x$. In this form, the function is defined explicitly. If the equation is given in the equivalent form $xy = 1$, then the function is defined implicitly.

The reason we ask "Batteries Required?" in the title of this project is that a graphing calculator or a computer is often the option of choice in graphing implicitly defined functions. In some cases, it's the only option (assuming the tools of calculus are not at our disposal). As examples, consider the two equations graphed in Figures D and E. We'll talk about how these graphs were obtained.

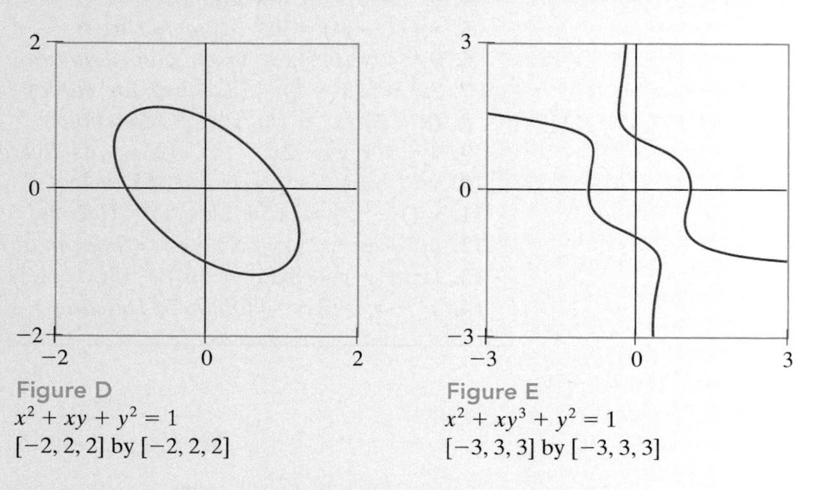

Figure D
$x^2 + xy + y^2 = 1$
$[-2, 2, 2]$ by $[-2, 2, 2]$

Figure E
$x^2 + xy^3 + y^2 = 1$
$[-3, 3, 3]$ by $[-3, 3, 3]$

For Figure D we can rewrite the equation in the form $y^2 + xy + (x^2 - 1) = 0$. In this form, the quadratic formula can be used to solve for y in terms of x. As you should check for yourself (using $a = 1$, $b = x$, and $c = x^2 - 1$), the result is $y = \frac{1}{2}(-x \pm \sqrt{4 - 3x^2})$. The graph of $x^2 + xy + y^2 = 1$ is now obtained by using a graphing utility to graph the *two* explicit functions,

$$y_1 = \frac{-x + \sqrt{4 - 3x^2}}{2} \quad \text{and} \quad y_1 = \frac{-x - \sqrt{4 - 3x^2}}{2}$$

In Figure E the equation cannot be solved for y in terms of x by elementary means. (Try it!) Furthermore, as of this writing, most graphing calculators require that functions be entered in explicit rather than implicit form. The graph in Figure E was obtained by using a computer with commercial software that doesn't necessarily require the function to be in explicit form. (Three examples of such software are *Maple, Mathematica,* and *LiveMath Maker.*)

EXERCISES

For Exercises 1–10, each equation defines one or more functions implicitly. Solve the equation for y to determine explicit equations for the functions. Then use a graphing calculator or a computer to obtain the graph of the original equation.

For Exercises 11–14, if you have access to a computer with appropriate software, graph the given equation. Otherwise, you can look up the general form of the graph at the University of St. Andrews MacTutor website:

http://www-groups.dcs.st-and.ac.uk/~history/Curves/Curves.html

(Actually, in Exercises 11–13, the equations can be solved for y explicitly, using elementary means, and then a graphing calculator could be used. But we are not asking for that here.)

1. $xy = 0.1x^3 - 2x^2 - 3x - 4$ (*trident of Newton*)
2. $x^4 - 10(x^2 - y^2) = 0$ (*figure eight curve*)
3. $x^2y + 4y - 36x = 0$ (*serpentine*)
4. $9x^2 + 16y^2 = 144$ (*ellipse*)
5. $y^2(x^2 + y^2) = 9x^2$ (*kappa curve*)
6. $9y^2 = x^3(6 - x)$ (*pear-shaped quartic*)
7. $3y^2 = x(x - 1)^2$ (*Tschirnhaus' cubic*)
8. $(x - 3)^2(x^2 + y^2) = 8x^2$ (*conchoid*)
9. $4y^2 = x(x^2 - 2x - 10)$ (*Newton's diverging parabolas*)
10. $y^2(7 - x) = x^3$ (*cissoid of Diocles*)
11. $y^2(1 - x^2) = (x^2 + 2y - 1)^2$ (*bicorn*)
12. $(x^2 + y^2 - 2x)^2 = x^2 + y^2$ (*limaçon of Pascal*)
13. $(x^2 + y^2)^2 - 8x(x^2 - 3y^2) + 18(x^2 + y^2) = 27$ (*deltoid*)
14. $x^3 + y^3 = 3xy$ (*folium of Descartes*)

3.3 SHAPES OF GRAPHS. AVERAGE RATE OF CHANGE

In many applications, one of the most basic questions regarding a function is this: What are the highest and lowest points (if any) on the graph? We'll use the graph of the function G in Figure 1 to introduce some terminology that is useful here.

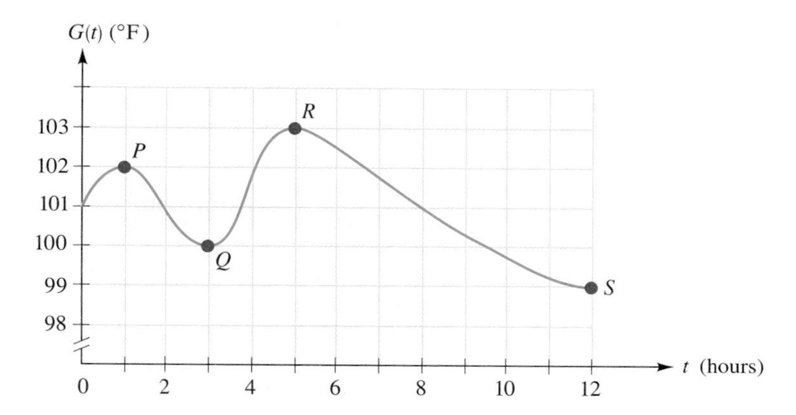

Figure 1
The graph of the function G is a fever graph: $G(t)$ is a patient's temperature t hours after midnight, where $0 \leq t \leq 12$.

In the figure, each of the points, P, Q, and R is called a *turning point.* At a **turning point,** a graph changes from rising to falling, or vice versa. In Figure 1, the point R, with coordinates (5, 103), is the highest point on the graph of the function G. We say that the *maximum value* of the function G is 103, and this maximum occurs at $t = 5$. In terms of the temperature interpretation (described in the caption for Figure 1), the patient's temperature reaches a maximum of 103°F, and this maximum occurs 5 hours after midnight. More generally, we say that a number $f(x_0)$ is the **maximum value** of a function f if the inequality $f(x_0) \geq f(x)$ holds for every x in the domain of f. Minimum values are defined similarly: $f(x_0)$ is the **minimum value** for a function f if the inequality $f(x_0) \leq f(x)$ holds for every x in the domain of f. Assuming that the domain of the function in Figure 1 is [0, 12], the minimum value of the function G is 99 and it occurs at $t = 12$. (How would you express this in terms of the temperature interpretation?) *Caution:* Not every function has a maximum or minimum value. See Figure 2 for an example.

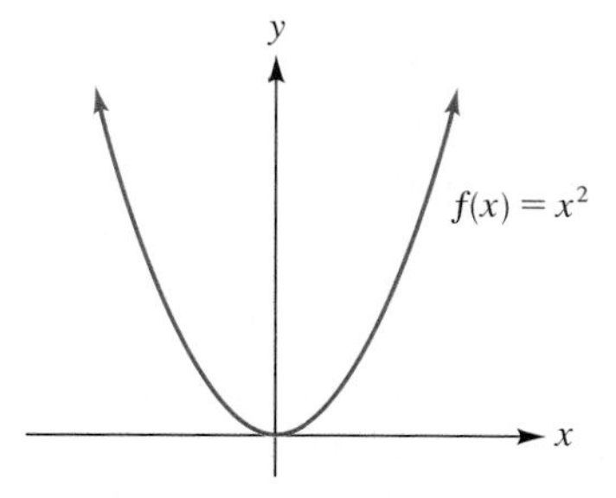

Figure 2
There is no maximum value for the function f defined by $f(x) = x^2$, assuming that the domain is $(-\infty, \infty)$. The minimum value of the function is 0, and it occurs at $x = 0$.

Returning to Figure 1, there are two time intervals when the patient's temperature is rising: from midnight to 1 A.M. and from 3 A.M. to 5 A.M. We say that the function G is *increasing* on each of the intervals [0, 1] and [3, 5]. Similarly (again looking at the graph), the patient's temperature is falling from 1 A.M. to 3 A.M. and again from 5 A.M. to 12 noon. We say that the function G is *decreasing* on each of the intervals [1, 3] and [5, 12]. For another example of this terminology, look at the graph of $f(x) = x^2$ in Figure 2. The function is decreasing on the interval $(-\infty, 0]$ and increasing on the interval $[0, \infty)$. For analytical work involving functions, the terms increasing and decreasing are defined in terms of inequalities. A function f is **increasing** on an interval provided the following condition holds: For all pairs of numbers a and b in the interval, if $a < b$, then $f(a) < f(b)$. Similarly, a function f is **decreasing** on an interval if the following condition holds: For all pairs of numbers a and b in the interval, if $a < b$, then $f(a) > f(b)$.

In general, for functions defined by equations, the techniques of calculus are required to compute exact coordinates for turning points, maximum or minimum values, and intervals where the function is increasing or decreasing. However, these quantities can be easily approximated by using a graphing utility to zoom in on the turning points. We do this in Example 1. *Technology note*: On a graphing calculator, the TRACE key can be used in conjunction with a ZOOM key to facilitate this process. Alternatively, graphing calculators have MIN and MAX operations that produce very accurate approximations for the minimum or maximum value of a function within any interval that the user requests. For details refer to the owner's manual that came with your calculator. *Algebra note*: In Sections 4.2 and 4.5 you'll see that in certain cases we can use the methods of algebra to compute *exact* maximum and minimum values.

EXAMPLE 1 Approximating the coordinates of a turning point

In Figure 3 a graphing utility was used to generate a graph of the function f defined by $f(x) = x^4 - 6x$. Assume that the chosen viewing rectangle shows all the essential features of the graph. Use a graphing utility to estimate to the nearest hundredth the coordinates of the turning point. What are the corresponding estimates for the minimum value of the function and the intervals where the function is increasing or decreasing?

GRAPHICAL PERSPECTIVE

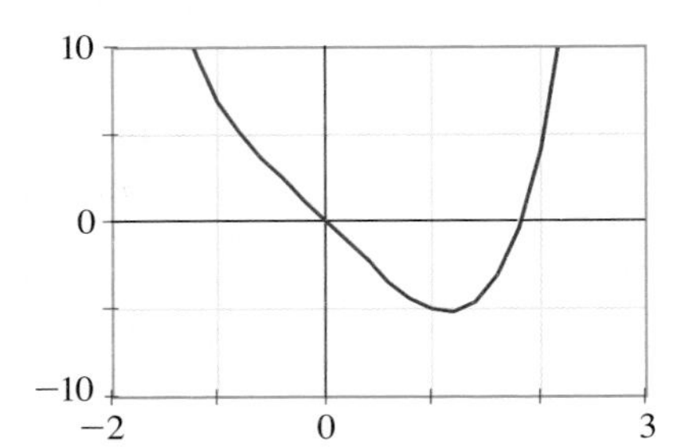

Figure 3
$f(x) = x^4 - 6x$
$[-2, 3, 1]$ by $[-10, 10, 5]$

SOLUTION

According to Figure 3, the turning point is located in quadrant IV, quite close to the point $(1, -5)$. Our goal is to obtain views that allow us to estimate the coordinates of the turning point to the nearest hundredth. We'll do this by repeatedly zooming in and, when appropriate, fine-tuning the viewing rectangle. At the outset, we remark that some trial and error is usually required here and that many different approaches are possible.

The view obtained in Figure 4(a) (after some experimenting) shows that the y-coordinate of the turning point is -5.15, to the nearest hundredth. The x-coordinate appears to be something between 1.14 and 1.15, but evidently we need a closer look to say whether it's nearer to 1.14 or 1.15. With that in mind, we try adjusting the x-specifications of the viewing rectangle so that x runs from 1.140 to 1.150.

As indicated in Figure 4(b), however, adjusting only the x-specifications isn't very helpful; much more magnification in the y-direction is required. [Are you

GRAPHICAL PERSPECTIVE

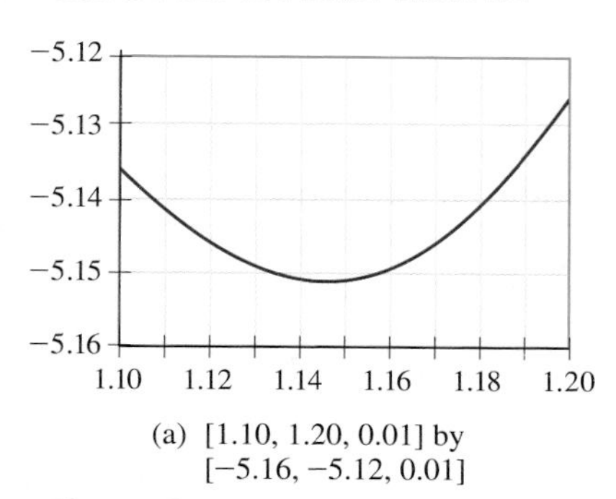

(a) [1.10, 1.20, 0.01] by [−5.16, −5.12, 0.01]

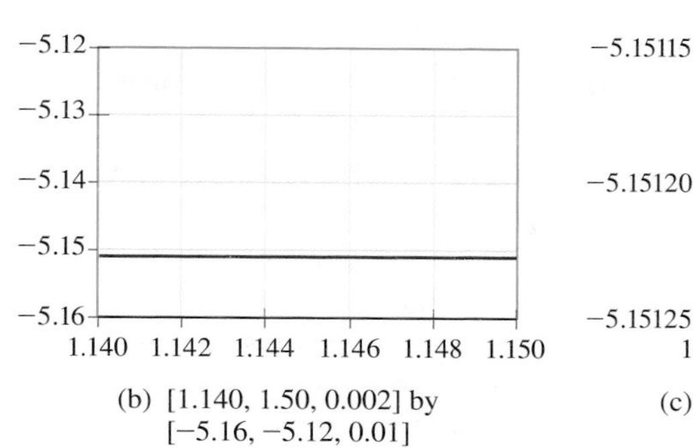

(b) [1.140, 1.50, 0.002] by [−5.16, −5.12, 0.01]

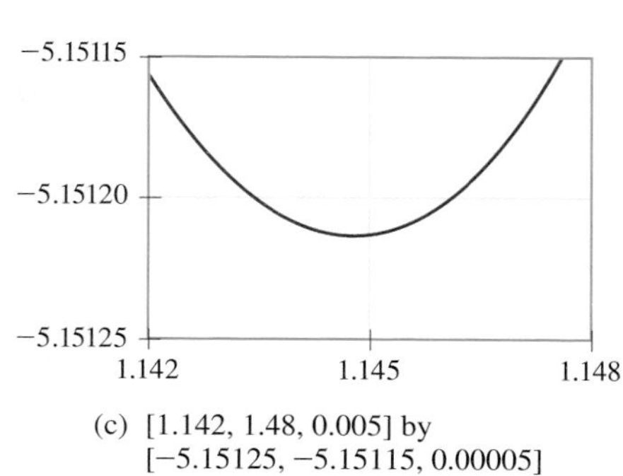

(c) [1.142, 1.48, 0.005] by [−5.15125, −5.15115, 0.00005]

Figure 4
Three views of $f(x) = x^4 - 6x$

surprised by the seemingly horizontal line obtained in Figure 4(b)? We'll say more about it after this example.] In Figure 4(c), after more zooming and then fine-tuning of the viewing rectangle, we finally are able to see that the x-coordinate of the turning point is closer to 1.14 than to 1.15 (because it is to the *left* of 1.145).

In summary now, to the nearest hundredth, the coordinates of the turning point are $(1.14, -5.15)$. Thus our *estimates* for the minimum value and intervals of increase or decrease are as follows:

$$\begin{aligned} \text{minimum value of } f&: \quad -5.15 \\ f \text{ is decreasing on}&: \quad (-\infty, 1.14] \\ f \text{ is increasing on}&: \quad [1.14, \infty) \end{aligned}$$

AFTERWORD ON ZOOMING FOR TURNING POINTS In the process of trying to estimate the coordinates of the turning point in Example 1, we produced the view in Figure 4(b), which seems to show a horizontal line segment rather than a curve. This is actually a common occurrence when zooming in on turning points. There are two factors at work here, one having to do with the actual shape of the graph, the other with the limitations of the display screen and technology. For many types of functions, including those defined by polynomial expressions (as in Example 1), the graph does very closely resemble a horizontal line in the immediate vicinity of a turning point. We say "resemble" here because the "true" graph, as opposed to the approximation displayed on your calculator or computer screen, is slightly curved. Remember that the calculator display or the computer screen has only a finite number of dots or *pixels* with which to display a graph. Thus, in general, the display on the screen must necessarily be an approximation, because the true graph contains infinitely many points (one point for each x-value in the domain). In Figure 4(b), the calculator display shows a horizontal segment because, evidently, the actual curve differs from the horizontal segment by less than one pixel.

We can use the temperature graph in Figure 1 (on page 159) to introduce another important concept relating to functions. According to Figure 1, the patient's temperature is increasing from midnight to 1 A.M. and again from 3 A.M. to 5 A.M. During which of these two time intervals is the temperature increasing faster? That is, how can we measure the *rate* at which the temperature is rising during a given time interval? Similarly, how can we measure and thereby compare the rates at which the temperature is falling during the time intervals from 1 A.M. to 3 A.M. and from 5 A.M. to 12 noon? One way to measure this is to use the *average rate of change,* which we define in the box that follows. (In Example 3 we'll answer the questions raised in this paragraph regarding the temperature graph.)

DEFINITION The Average Rate of Change of a Function

Refer to Figure 5. The **average rate of change** of a function f on the interval $[a, b]$ is the slope of the line joining the two points $(a, f(a))$ and $(b, f(b))$. If $y = f(x)$, the average rate of change is denoted by $\Delta y/\Delta x$ or $\Delta f/\Delta x$ and we have

$$\frac{\Delta y}{\Delta x} = \frac{f(b) - f(a)}{b - a}$$

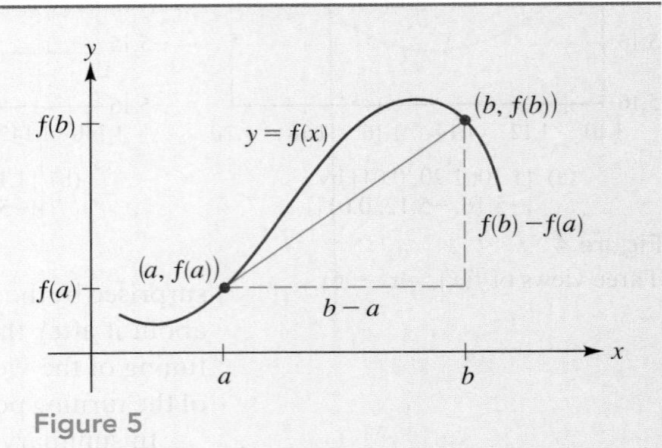

Figure 5

EXAMPLE 2 Computing and comparing average rates of change

Compute and compare the average rates of change for the function $f(x) = x^2$ on the following intervals:
(a) $[-2, 0]$; **(b)** $[0, 1]$; **(c)** $[1, 2]$.

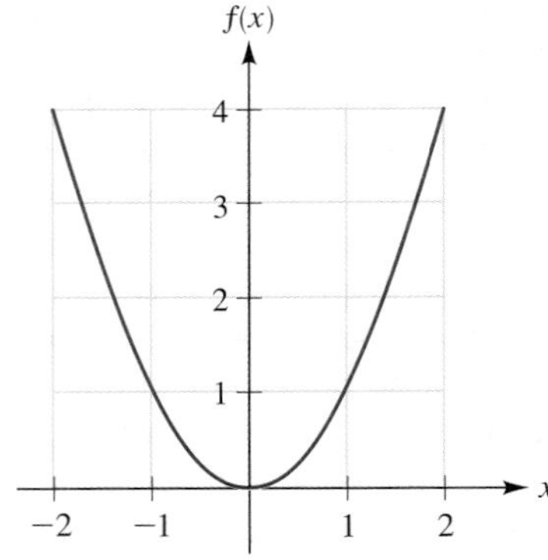

Figure 6
$f(x) = x^2$

SOLUTION

(a) On $[-2, 0]$: $\frac{\Delta f}{\Delta x} = \frac{f(0) - f(-2)}{0 - (-2)} = \frac{0^2 - (-2)^2}{2} = -2$

(b) On $[0, 1]$: $\frac{\Delta f}{\Delta x} = \frac{f(1) - f(0)}{1 - 0} = \frac{1^2 - 0^2}{1} = 1$

(c) On $[1, 2]$: $\frac{\Delta f}{\Delta x} = \frac{f(2) - f(1)}{2 - 1} = \frac{2^2 - 1^2}{1} = 3$

(Refer to Figure 6.) On the interval $[-2, 0]$ the function is decreasing, and the average rate of change is negative. On both of the intervals $[0, 1]$ and $[1, 2]$ the function is increasing, and the average rates of change are positive. The graph rises more steeply on the interval $[1, 2]$ than on $[0, 1]$, and correspondingly, the average rate of change is greater for the interval $[1, 2]$ than for $[0, 1]$.

A WORD OF CAUTION Like other types of averages occurring in arithmetic and statistics, the average rate of change can mask or hide some details. For instance, the fact that the average rate of change is positive on an interval does not guarantee that the function is increasing throughout that interval. For example, look back at Figure 5; for the function shown there, the average rate of change on $[a, b]$ is positive (because the slope of the solid blue line is positive). The function, however, is increasing only on a portion of $[a, b]$, not throughout the entire interval.

EXAMPLE 3 Computing and interpreting average rates of change

Compute and compare the average rates of change for the patient's temperature function G in Figure 1 (page 159) on each of the following pairs of intervals. Keep track of the units in the computations.

(a) $t = 0$ to $t = 1$ hr and $t = 3$ to $t = 5$ hr
(b) $t = 1$ to $t = 3$ hr and $t = 5$ to $t = 12$ hr

SOLUTION

(a) On the interval $t = 0$ to $t = 1$ hr:

$$\begin{aligned}\frac{\Delta G}{\Delta t} &= \frac{G(1) - G(0)}{1 - 0} \\ &= \frac{102° - 101°}{1 \text{ hr}} \qquad \text{reading the coordinates from the graph in Figure 1} \\ &= \frac{1°}{1 \text{ hr}} = 1 \text{ degree/hr}\end{aligned}$$

Similarly, on the interval $t = 3$ to $t = 5$ hr, we have

$$\begin{aligned}\frac{\Delta G}{\Delta t} &= \frac{G(5) - G(3)}{5 - 3} \\ &= \frac{103° - 100°}{2 \text{ hr}} = \frac{3°}{2 \text{ hr}} = 1.5 \text{ degrees/hr}\end{aligned}$$

The calculations we've just made show that, on average, the temperature is increasing slightly faster from 3 A.M. to 5 A.M. than it is from midnight to 1 A.M. This is consistent with the graph in Figure 1: The curve looks slightly steeper from 3 A.M. to 5 A.M. than from midnight to 1 A.M.

(b) Before computing, let us note that both answers will be negative here because the graph shows the function is decreasing throughout each of the given intervals.

On the interval $t = 1$ to $t = 3$ hr:

$$\begin{aligned}\frac{\Delta G}{\Delta t} &= \frac{G(3) - G(1)}{3 - 1} \\ &= \frac{100° - 102°}{2 \text{ hr}} = \frac{-2°}{2 \text{ hr}} = -1 \text{ degree/hr}\end{aligned}$$

On the interval $t = 5$ to $t = 12$ hr:

$$\begin{aligned}\frac{\Delta G}{\Delta t} &= \frac{G(12) - G(5)}{12 - 5} \\ &= \frac{99° - 103°}{7 \text{ hr}} = \frac{-4°}{7 \text{ hr}} = -\frac{4}{7} \text{ degree/hr} \approx -0.6 \text{ degree/hr}\end{aligned}$$

These answers tell us that, on average, the patient's temperature is dropping faster between 1 A.M. and 3 A.M. than it is between 5 A.M. and noon. Again, this is consistent with what we see in Figure 1: The portion of the curve between points P and Q appears to be steeper, in general, than that between R and S.

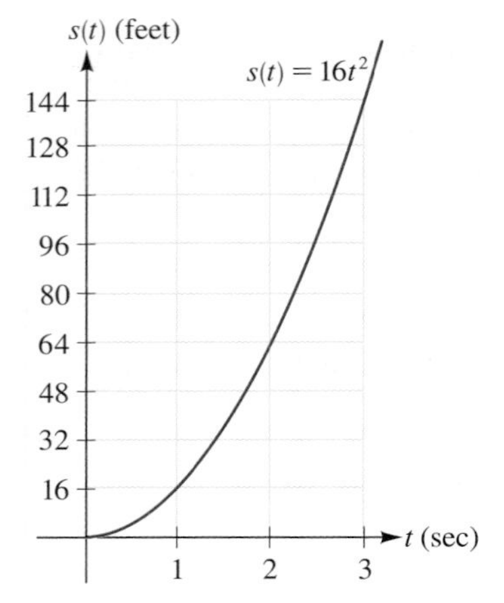

Figure 7

For an example using the average rate of change in a different context, consider the function s represented in Figure 7, with defining equation $s(t) = 16t^2$, and $t \geq 0$. This function relates the distance $s(t)$ and the time t for an object falling in a vacuum. Here t is measured in seconds, $s(t)$ is in feet, and $t = 0$ corresponds to the instant that the object begins to fall. For instance, after 1 sec, the object falls a distance of $s(1) = 16(1)^2 = 16$ ft. After 2 sec, the total distance will be $s(2) = 16(2)^2 = 64$ ft. Let's calculate the average rate of change of this function from, say, $t = 1$ to $t = 3$ sec. As in the temperature example, we'll keep track of the units. We have

$$\begin{aligned}\frac{\Delta s}{\Delta t} &= \frac{s(3) - s(1)}{3 - 1}\\ &= \frac{16(3)^2\text{ ft} - 16(1)^2\text{ ft}}{2\text{ sec}} = \frac{128\text{ ft}}{2\text{ sec}} = 64\text{ ft/sec}\end{aligned}$$

Notice that the units here have the form *distance per unit time.* So in this case $\Delta s/\Delta t$ gives us the *average velocity* of the object. More generally, whenever we have a function expressing distance $s(t)$ in terms of time t, the **average velocity** over an interval is defined to be the average rate of change $\Delta s/\Delta t$ over that interval. We'll return to this idea in Example 6 at the end of this section.

Let's talk for a moment about notation. We defined the average rate of change of a function f on the interval $[a, b]$ to be the quantity

$$\frac{f(b) - f(a)}{b - a} \tag{1}$$

If, instead of $[a, b]$, we have the interval $[a, x]$, then, by definition, the average rate of change is

$$\frac{f(x) - f(a)}{x - a} \tag{2}$$

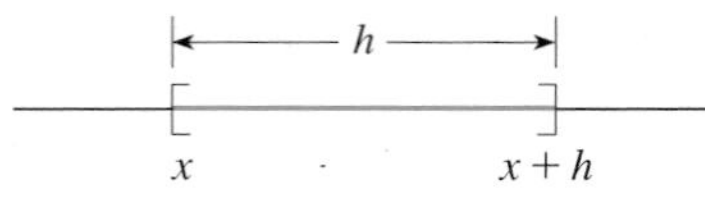

Figure 8
The closed interval $[x, x + h]$

Finally, if the interval is $[x, x + h]$ as displayed in Figure 8, then the average rate of change is $\dfrac{f(x + h) - f(x)}{(x + h) - x}$, which simplifies to

$$\frac{f(x + h) - f(x)}{h} \tag{3}$$

Each of the algebraic expressions given in (1), (2), or (3) is known as a **difference quotient.** (But whatever the name, remember, it represents an average rate of change.) Our last three examples for this section show how difference quotients can be simplified once a specific function is given. (This skill is required in calculus when studying rates of change.)

EXAMPLE 4 Simplifying difference quotients of the form $\dfrac{f(x) - f(a)}{x - a}$

(a) If $f(x) = x^2$, find $\dfrac{f(x) - f(5)}{x - 5}$.

(b) If $g(x) = x^2 + 3x$, find $\dfrac{g(x) - g(a)}{x - a}$.

SOLUTION

PRELIMINARY NOTE The solutions that follow make use of factoring techniques from intermediate algebra. If you find that you need a review of this topic, either now or in working the exercises, refer to Appendix B.4 at the back of this book.

(a)
$$\begin{aligned}\frac{f(x)-f(5)}{x-5} &= \frac{x^2-5^2}{x-5} \\ &= \frac{(x-5)(x+5)}{x-5} && \text{using difference of squares factoring} \\ &= x+5 && \text{simplifying}\end{aligned}$$

(b) Since $g(x) = x^2 + 3x$ and $g(a) = a^2 + 3a$, we have

$$\begin{aligned}\frac{g(x)-g(a)}{x-a} &= \frac{(x^2+3x)-(a^2+3a)}{x-a} = \frac{x^2+3x-a^2-3a}{x-a} \\ &= \frac{x^2-a^2+3x-3a}{x-a} && \text{rearranging} \\ &= \frac{(x-a)(x+a)+3(x-a)}{x-a} \\ &= \frac{(x-a)[(x+a)+3]}{x-a} && \text{factoring by grouping} \\ &= x+a+3 && \text{simplifying}\end{aligned}$$

EXAMPLE 5 Simplifying a difference quotient of the form $\dfrac{f(x+h)-f(x)}{h}$

Let $f(x) = 2/x$. Find $\dfrac{f(x+h)-f(x)}{h}$.

SOLUTION

We have $\dfrac{f(x+h)-f(x)}{h} = \dfrac{\frac{2}{x+h}-\frac{2}{x}}{h}$. This last expression needs to be simplified because the numerator itself contains fractions. An easy way to simplify here is to multiply by $\dfrac{x(x+h)}{x(x+h)}$, which equals 1. This yields

$$\begin{aligned}\frac{f(x+h)-f(x)}{h} &= \frac{x(x+h)}{x(x+h)} \cdot \frac{\frac{2}{x+h}-\frac{2}{x}}{h} \\ &= \frac{2x-2(x+h)}{xh(x+h)} \\ &= \frac{2x-2x-2h}{xh(x+h)} = \frac{-2h}{xh(x+h)} \\ &= \frac{-2}{x(x+h)}\end{aligned}$$

EXAMPLE 6 Computing average velocity

Refer to the distance function s in Figure 7, $s = 16t^2$.

(a) Find a general expression for the average velocity $\Delta s/\Delta t$ over the interval $[1, 1 + h]$.

(b) Use the result in part (a) to complete the following table.

h (seconds)	Average velocity $\Delta s/\Delta t$ on the interval $[1, 1 + h]$ (feet/second)
1.0	
0.5	
0.1	

SOLUTION

(a) On the interval $[1, 1 + h]$ we have

$$\begin{aligned}\frac{\Delta s}{\Delta t} &= \frac{16(1+h)^2 - 16(1)^2}{(1+h) - 1}\\ &= \frac{16 + 32h + 16h^2 - 16}{h} = \frac{32h + 16h^2}{h}\\ &= \frac{16h(2+h)}{h} = 16(2+h) = 32 + 16h\end{aligned}$$

(b) From part (a) we have $\Delta s/\Delta t = 32 + 16h$. Using this result, we complete the required table as follows.

h (seconds)	Average velocity $\Delta s/\Delta t$ on the interval $[1, 1 + h]$ (feet/second)
1.0	$32 + 16(1) = 48$
0.5	$32 + 16(0.5) = 40$
0.1	$32 + 16(0.1) = 33.6$

NOTE Exercise 36 asks you to carry out these calculations several steps further, using shorter and shorter time intervals. In calculus, this procedure leads to the concept of *instantaneous* (as opposed to *average*) *velocity.*

EXERCISE SET 3.3

A

1. Using the functions in Figure 7 on page 149 for example, explain what is meant by each of the following. Your answers should be in complete sentences.
   (a) turning point
   (b) maximum value
   (c) minimum value
   (d) f is increasing on an interval.
   (e) f is decreasing on an interval.
2. What is the definition of the average rate of change of a function g on an interval $[c, d]$? What does this have to do with slope?

In Exercises 3–6, you are given functions with domain [0, 4]. *Specify:*

(a) *the range of each function;*
(b) *the maximum value of the function;*
(c) *the minimum value of the function;*
(d) *interval(s) where the function is increasing; and*
(e) *interval(s) where the function is decreasing.*

3.

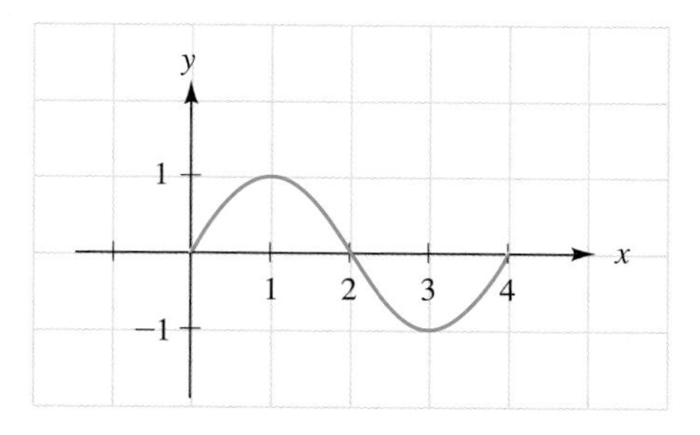

4.

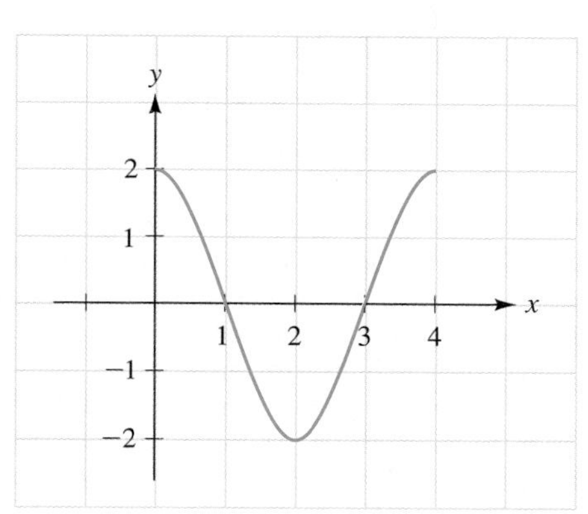

5.

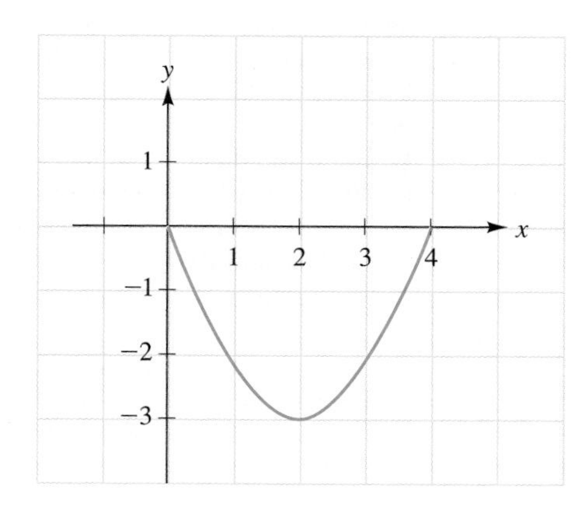

6.

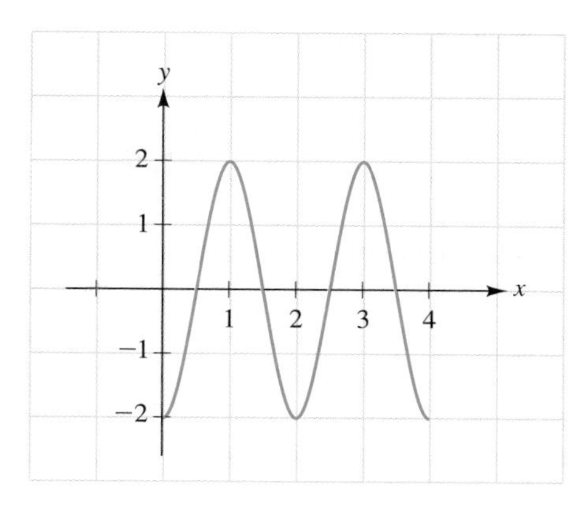

Ⓖ **7.** Assume that the accompanying viewing rectangle shows the essential features of the graph of $f(x) = x^3 - 4x - 2$. Use a graphing utility to estimate to the nearest hundredth the coordinates of the turning points. What are the corresponding estimates for the intervals where the function is increasing or decreasing?

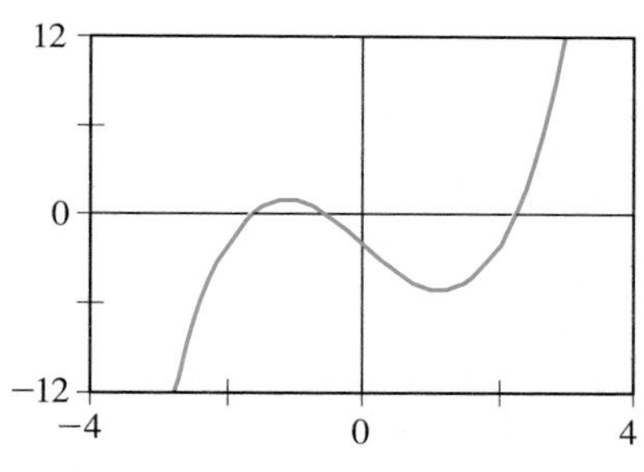

$f(x) = x^3 - 4x - 2$
[−4, 4, 2] by [−12, 12, 6]

8. Assume that the accompanying viewing rectangle shows the essential features of the graph of $f(x) = -x^4 + 3x + 5$. Use a graphing utility to estimate to the nearest hundredth the coordinates of the turning point. What is the corresponding estimate for the maximum value of the function? Estimate the intervals where the function is increasing or decreasing.

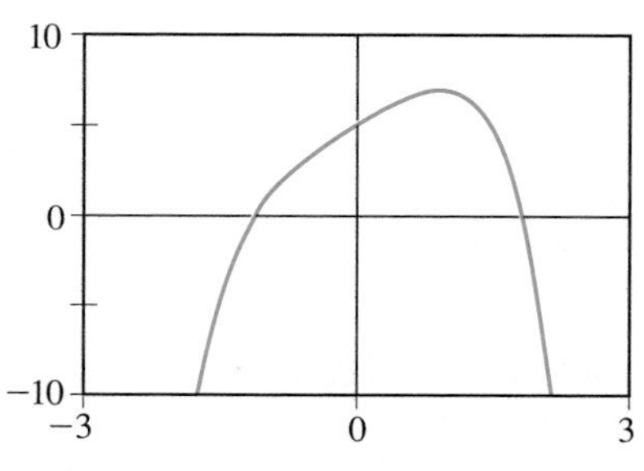

$f(x) = -x^4 + 3x + 5$
[−3, 3, 3] by [−10, 10, 5]

In Exercises 9–14, compute the average rate of change of the function on the given interval.

9. $f(x) = x^2 + 2x$ on $[3, 5]$

10. $f(x) = \sqrt{x}$ on $[4, 9]$

11. $g(x) = 2x^2 - 4x$ on $[-1, 3]$

12. $g(x) = x^3 - x$ on $[1, 2]$

13. $h(t) = 2t - 6$ on $[5, 12]$

14. $h(t) = 16 - 7t$ on $[-\sqrt{2}, 2\sqrt{2}]$

15. The following graph shows the temperature $G(t)$ of a solution during the first 8 minutes of a chemistry experiment.

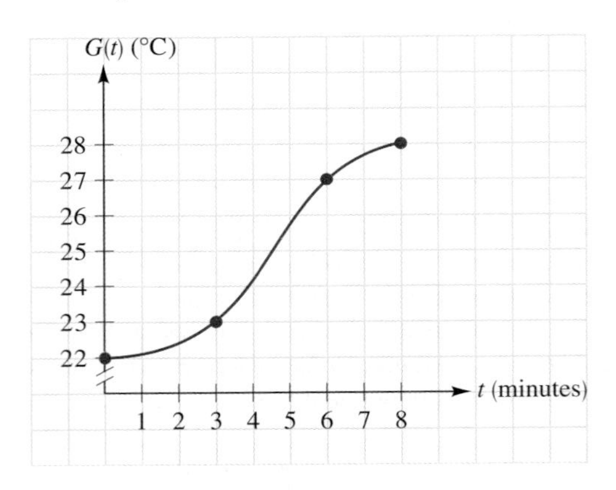

Compute the average rate of change of temperature, $\Delta G/\Delta t$, over the following intervals. (Be sure to specify the units as part of each answer.)

(a) $t = 0$ min to $t = 3$ min

(b) $t = 3$ min to $t = 6$ min

(c) $t = 6$ min to $t = 8$ min

16. Iodine-131 is a radioactive substance. The accompanying graph shows how an initial one-gram sample decays over a 32-day period; $f(t)$ represents the number of grams present after t days.

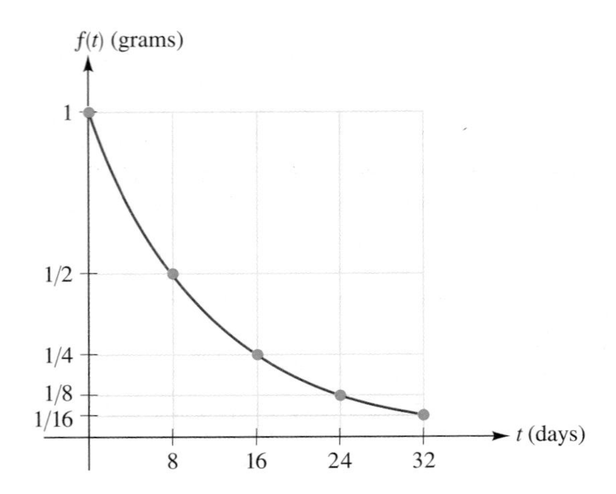

(a) Compute $\Delta f/\Delta t$ over each of the following intervals: $t = 0$ days to $t = 8$ days; $t = 8$ days to $t = 16$ days; $t = 16$ days to $t = 24$ days; $t = 24$ days to $t = 32$ days. (Be sure to specify the units with each answer.)

(b) Find the average of the four answers in part (a).

(c) Compute $\Delta f/\Delta t$ over the interval from $t = 0$ to $t = 32$. Is the answer the same as that obtained in part (b)?

For Exercises 17 and 18, let $y = P(t)$ denote the percentage of U.S. households in the year t with at least one VCR (video cassette recorder). The following figure shows a graph of this function P over the period $1978 \le t \le 1997$.

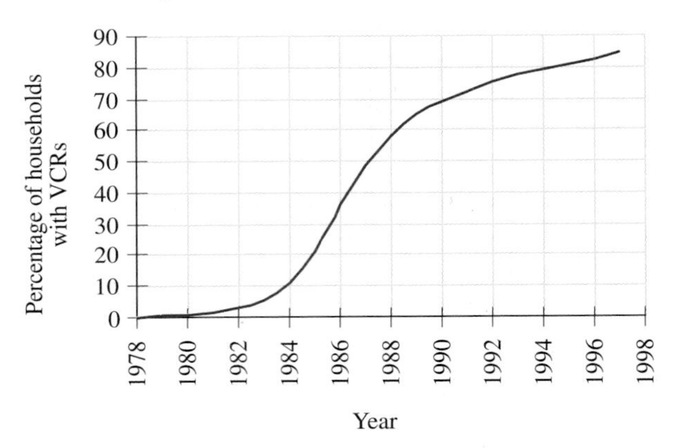

Source of data used to create graph: A. C. Nielsen and Nielsen Media Research

17. (a) According to the graph, is $\Delta P/\Delta t$ greater over the period 1978–1984 or over the period 1984–1990? After answering, use rough numerical values obtained from the graph to calculate estimates for these two rates of change. Include units with answers.

(b) Using the following data, compute $\Delta P/\Delta t$ for the periods mentioned in part (a). Round the answers to one decimal place. $P(1978) = 0.3\%$; $P(1984) = 10.6\%$; $P(1990) = 68.6\%$.

18. (a) In the given graph (or in a photocopy), use a ruler to draw two line segments, one connecting the two points (1978, $P(1978)$) and (1988, $P(1988)$), the other connecting the two points (1984, $P(1984)$) and (1997, $P(1997)$). You'll find that the two line segments appear to be parallel (or nearly so). Now complete the following sentence (concerning $\Delta P/\Delta t$ on the two intervals $1978 \le t \le 1988$ and $1984 \le t \le 1997$). Because parallel lines have equal slopes, we can estimate that ______________________.

(b) Using the following data, compute $\Delta P/\Delta t$ for each of the two periods mentioned in part (a). Round the answers to one decimal place. Do your answers support the estimate you made in part (a)?

$P(1978) = 0.3\%$; $P(1988) = 58.0\%$;
$P(1984) = 10.6\%$; $P(1997) = 84.2\%$

For Exercises 19 and 20, refer to the following figure, which shows consumption of red meat in the United States and in China over the period 1980–1996. Let $y = f(t)$ denote the consumption function for the United States, and let $y = g(t)$ denote the consumption function for China.

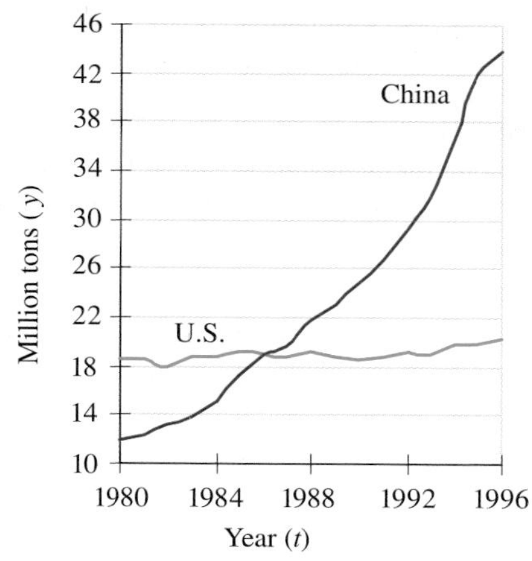

Consumption of red meat in United States and in China, 1980–1996

Source of data used to create graph: Worldwatch Institute

19. **(a)** Using values that you estimate from the graph, calculate $\Delta f/\Delta t$, the average rate of change of red meat consumption in the United States over the period 1980–1996. Express the answer as a fraction and as a decimal rounded to the nearest tenth.

(b) Use the data in the following table to obtain a more accurate value for $\Delta f/\Delta t$ over the period 1980–1996. Round the answer to two decimal places.

Year (t)	1980	1996
Consumption (y) in U.S. (million tons)	18.68	20.32

20. **(a)** Using values that you estimate from the graph, calculate $\Delta g/\Delta t$, the average rate of change of red meat consumption in China over the period 1980–1996.

(b) Use the data in the following table to obtain a more accurate value for $\Delta g/\Delta t$ over the period 1980–1996. Round the answer to two decimal places.

Year (t)	1980	1996
Consumption (y) in China (million tons)	11.90	43.83

21. Let $f(x) = 2x^2$.

(a) Find $\dfrac{f(x) - f(3)}{x - 3}$.

(b) Find $\dfrac{f(x) - f(a)}{x - a}$.

22. Let $f(x) = 4x^2$.

(a) Find $\dfrac{f(x) - f(-2)}{x + 2}$.

(b) Find $\dfrac{f(x) - f(a)}{x - a}$.

For Exercises 23 and 24, let $f(x) = x^2$. Find and simplify the indicated difference quotient.

23. **(a)** $\dfrac{f(2 + h) - f(2)}{h}$ **(b)** $\dfrac{f(x + h) - f(x)}{h}$

24. **(a)** $\dfrac{f(1 + h) - f(1)}{h}$ **(b)** $\dfrac{f(t + h) - f(t)}{h}$

In Exercises 25–32:

(a) *Find the difference quotient* $\dfrac{f(x) - f(a)}{x - a}$ *for each function, as in Example 4.*

(b) *Find the difference quotient* $\dfrac{f(x + h) - f(x)}{h}$ *for each function, as in Example 5.*

25. $f(x) = 8x - 3$

26. $f(x) = -2x + 5$

27. $f(x) = x^2 - 2x + 4$

28. $f(x) = 2x^2 - x + 1$

29. $f(x) = 1/x$

30. $f(x) = -3/x^2$

31. $f(x) = 2x^3$

32. $f(x) = 1 - x^3$

Hint: For Exercises 31 and 32, you'll need to use difference-of-cubes factoring from intermediate algebra. See the inside back cover for the relevant formula. For a more detailed review of the topic, refer to Appendix B.4.

For Exercises 33–36, use the distance function $s(t) = 16t^2$ discussed on page 164 and in Example 6. Recall that this function relates the distance $s(t)$ and the time t for a freely falling object (neglecting air resistance). The time t is measured in seconds, with $t = 0$ corresponding to the instant that the object begins to fall; the distance $s(t)$ is in feet.

33. Find the average velocity $\Delta s/\Delta t$ over the time interval $1 \leq t \leq 2$.

34. **(a)** Find the average velocity over each of the following time intervals: [2, 3], [3, 4], and [2, 4].

(b) Let a, b, and c denote the three average velocities that you computed in part (a), in the order given. Is it true that the arithmetical average of a and b is c?

35. **(a)** Follow the method of Example 6(a) to find a general expression for the average velocity $\Delta s/\Delta t$ over the interval $[2, 2 + h]$.

(b) Complete a table similar to the one shown in the solution of Example 6(b); for the h-values in the left-hand column use 0.1, 0.01, 0.001, 0.0001, and 0.00001.

(c) Looking at your results in part (b), answer the following question. As h approaches zero, what value does the average velocity in the right-hand column seem to be approaching? This target value or limit is the *instantaneous* (as opposed to average) *velocity* of the object when $t = 2$ sec.

36. **(a)** After rereading Example 6, extend the results in part (b) of the example by completing the following table. Don't round your answers.

h (seconds)	0.01	0.001	0.0001	0.00001
Average velocity $\Delta s/\Delta t$ on interval $[1, 1 + h]$ (ft/second)				

(b) As h approaches zero, what value does the average velocity $\Delta s/\Delta t$ seem to be approaching? This target value or limit is called the *instantaneous velocity* of the object when $t = 1$ sec.

37. Suppose that the demand function for a certain item is given by $p(x) = \dfrac{24}{2^{x/100}}$, where x is the number of items that can be sold when the price of each item is $p(x)$ dollars. Compute $\Delta p/\Delta x$ over each of the intervals $0 \le x \le 100$ and $300 \le x \le 400$. Include units in your answers. Why does it make sense (economically) that the answers are negative?

38. Suppose that during the first few hours of a laboratory experiment, the temperature of a certain substance is closely approximated by the function $f(t) = t^3 - 6t^2 + 9t$, where t is measured in hours, with $t = 0$ corresponding to the instant the experiment begins, and $f(t)$ is the temperature (°F) of the substance after t hours.

(a) Find an expression for $\Delta f/\Delta t$. What are the units for $\Delta f/\Delta t$?

(b) Use the result in part (a) to complete the following table. Round the answers to four decimal places.

Interval	[0, 0.1]	[0, 0.01]	[0, 0.001]	[0, 0.0001]
$\Delta f/\Delta t$				

(c) In part (b), as the right-hand endpoint gets closer and closer to 0, what value does $\Delta f/\Delta t$ seem to be approaching? This target value tells us the rate at which the temperature is changing at the *instant* the experiment begins.

39. Complete the following table.

Function	$\lvert x\rvert$	x^2	x^3
Domain			
Range			
Turning point			
Maximum value			
Minimum value			
Interval(s) where increasing			
Interval(s) where decreasing			

40. Set up and complete a table like the one in Exercise 39 for the three functions $1/x$, $\sqrt{x}$, and $\sqrt{1 - x^2}$.

B

41. Let $f(x) = 1/x$. Find a number b so that the average rate of change of f on the interval $[1, b]$ is $-1/5$.

42. Let $f(x) = \sqrt{x}$. Find a number b so that the average rate of change of f on the interval $[1, b]$ is $1/7$.

43. Let $f(x) = ax^2 + bx + c$. Show that

$$\frac{f(x + h) - f(x)}{h} = 2ax + ah + b.$$

44. Find a number a between 0 and 1 so that the average rate of change of $f(x) = x^2$ on the interval $[a, 1/a]$ is $10a$.

Ⓖ *In Exercises 45 and 46 (as opposed to Exercises 7 and 8), the viewing rectangles do not show all of the essential features of the graphs. In each case, use a graphing utility and experiment with different viewing rectangles to determine approximate x- and y-coordinates for all turning points. What are the corresponding intervals of increase and decrease for the function?*

45.

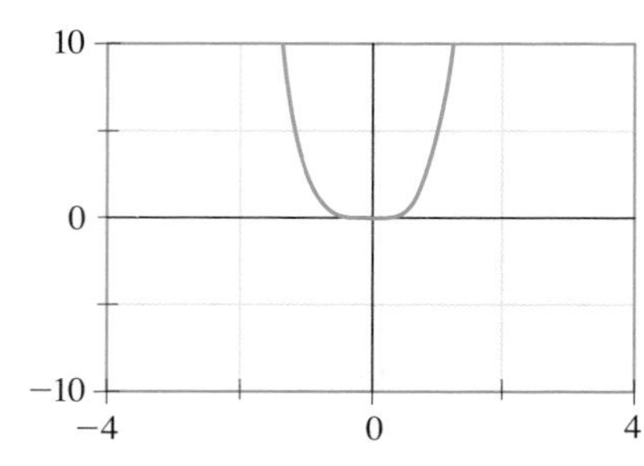

$f(x) = 4x^4 + x^3$
$[-4, 4, 2]$ by $[-10, 10, 5]$

46.

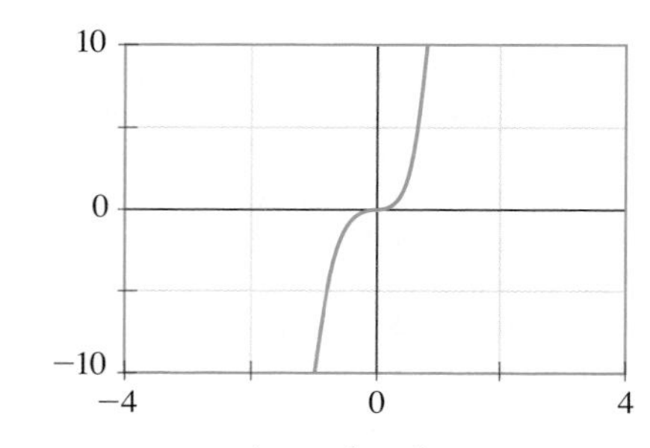

$g(x) = x^4 + 12x^3 + x^2$
$[-4, 4, 2]$ by $[-10, 10, 5]$

47. Let $h(x) = \sqrt{x}$. Two functions f and g are defined in terms of h as follows:

$$f(x) = \frac{h(x) - h(2)}{x - 2} \qquad g(x) = \frac{1}{h(x) + h(2)}$$

Ⓖ **(a)** Using a graphing utility, graph the two functions f and g in the same viewing screen. What do you observe?

(b) Use algebra to explain the result in part (a). That is, either derive or verify the identity $f(x) = g(x)$. Are there any positive values of x for which this equation does not hold?

48. Consider the function f defined by

$$f(x) = x^2 + \frac{2}{x^2} \qquad (x > 0)$$

(a) Complete the table. (Round the results to four decimal places.)

x	1	1.05	1.10	1.15	1.20	1.25
f(x)						

(b) Which $f(x)$-value in the table is the smallest? What is the corresponding input?

Ⓖ **(c)** Use a graphing utility to graph the function f. Then use the ZOOM and/or TRACE features to find an output $f(x)$ that is less than the output you indicated in part (b).

(d) It can be shown using calculus that the input x yielding the minimum value for the function f is $\sqrt[4]{2}$. Compute $f(\sqrt[4]{2})$. Which value of $f(x)$ in the table is closest to $f(\sqrt[4]{2})$? To how many decimal places does your estimate in part (c) agree with $f(\sqrt[4]{2})$?

3.4 TECHNIQUES IN GRAPHING

. . . geometrical figures are graphic formulas. —David Hilbert (1862–1943)

The simple geometric concepts of reflection and translation can be used to great advantage in graphing. We discussed the idea of reflection in the x-axis and in the y-axis in Section 1.7. By a **translation** of a graph, we mean a shift in its location such that every point of the graph is moved the same distance in the same direction. The size and the shape of a graph are unchanged by a translation.

EXAMPLE 1 Sketching combinations of translations and reflections

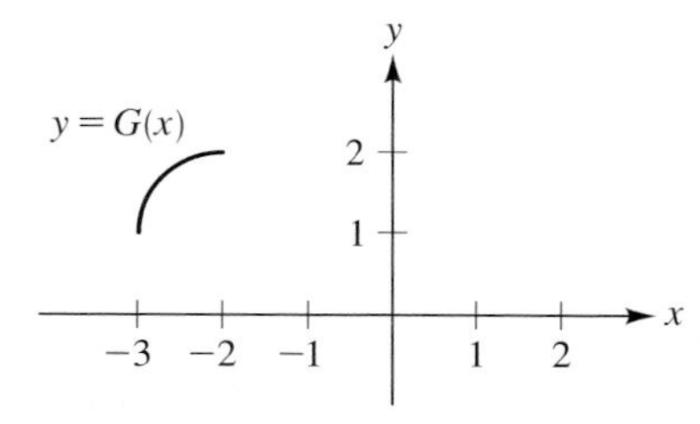

Figure 1

The graph of a function G is the quarter-circle shown in Figure 1. In each case, sketch the resulting graph after the following operations are carried out on the graph of G:

(a) a translation of four units to the right followed by a translation of one unit vertically downward;

(b) a translation of three units to the right followed by a reflection in the y-axis;

(c) a reflection in the y-axis followed by a translation of three units to the right. [Note that these are the same two operations used in part (b), but here the order is reversed.]

SOLUTION

(a) Figure 2 shows the result of translating the graph of G four units to the right and then one unit down. (As you can check by drawing a sketch, the same end result is obtained if we first translate one unit down and then four units to the right.) *Fact*: If several translations are to be carried out in succession, the end result will be the same, no matter in what order the individual translations are carried out.

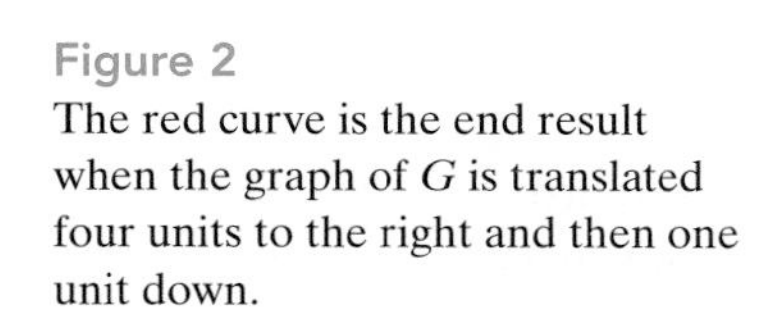

Figure 2
The red curve is the end result when the graph of G is translated four units to the right and then one unit down.

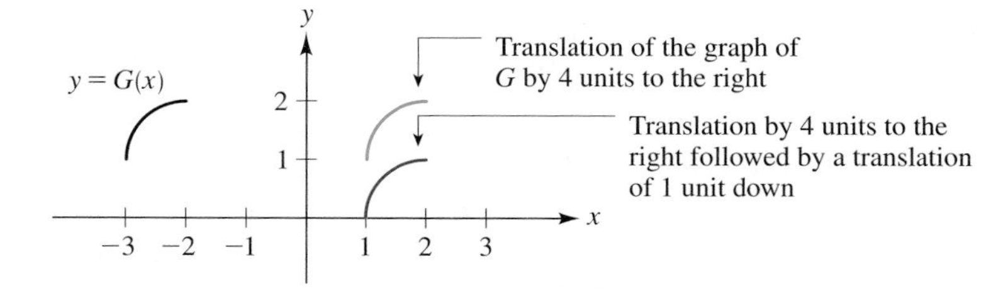

(b) Figure 3(a) shows the graph of G after a translation of three units to the right. When this is followed by a reflection in the y-axis, we obtain the result shown in Figure 3(b).

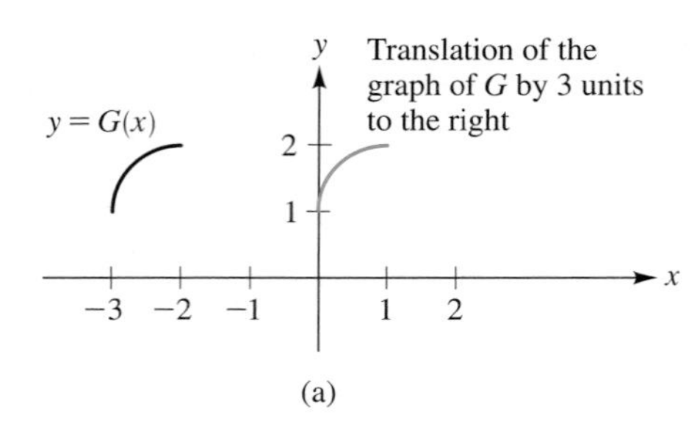

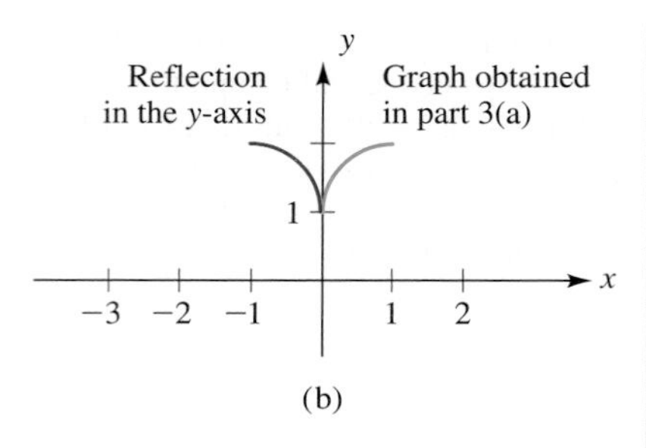

Figure 3
Figure (a) shows the result of translating the graph of G three units to the right. When this is followed by a reflection in the y-axis, we obtain the red curve in Figure (b).

(c) Figure 4 shows what happens to the graph of G if we use the same two operations that were used in part (b) but reverse the order in which they are carried out. Note that the end result is quite different from that obtained in part (b).

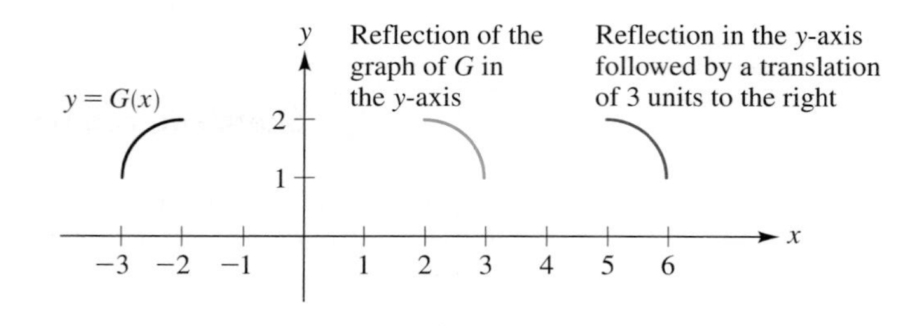

Figure 4
The reflection of the graph of G in the y-axis followed by a translation of three units to the right.

Given the graph of a function f, it is useful to know how to sketch efficiently the graphs of the following closely related functions. (In the following equations, c denotes a positive constant.)

$$y = f(x) + c \qquad y = f(x + c) \qquad y = -f(x)$$
$$y = f(x) - c \qquad y = f(x - c) \qquad y = f(-x)$$

Each of these can be graphed by translating or reflecting the graph of $y = f(x)$. In the box that follows, we summarize the techniques that are involved. (Exercises 47–50 will help you to see why these techniques are valid, and Section 11.2 will explain these techniques from a more general point of view.)

PROPERTY SUMMARY Translations and Reflections

(In Items 1–4, the letter c denotes a positive constant.)

Equation	How to Obtain the Graph from That of $y = f(x)$
1. $y = f(x) + c$	Translate c units vertically upward
2. $y = f(x) - c$	Translate c units vertically downward
3. $y = f(x + c)$	Translate c units to the left
4. $y = f(x - c)$	Translate c units to the right
5. $y = -f(x)$	Reflect in the x-axis
6. $y = f(-x)$	Reflect in the y-axis

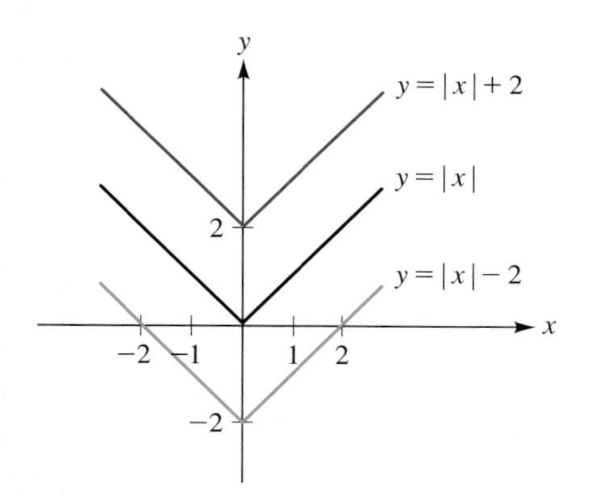

(a) To graph $y = |x| + 2$, translate $y = |x|$ up 2 units. To graph $y = |x| - 2$, translate $y = |x|$ down 2 units.

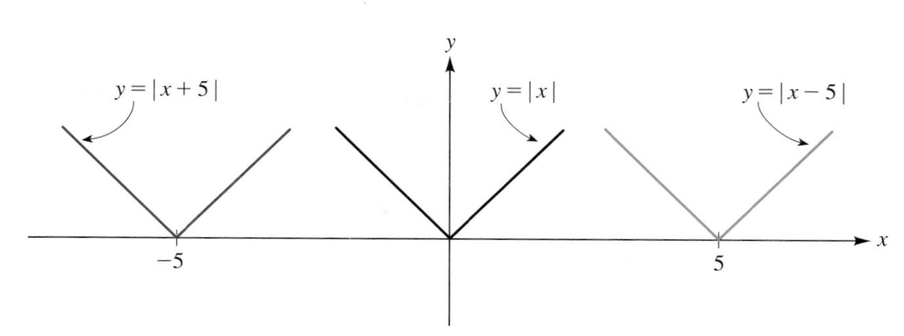

(b) To graph $y = |x + 5|$, translate $y = |x|$ to the left 5 units. To graph $y = |x - 5|$, translate $y = |x|$ to the right 5 units.

Figure 5

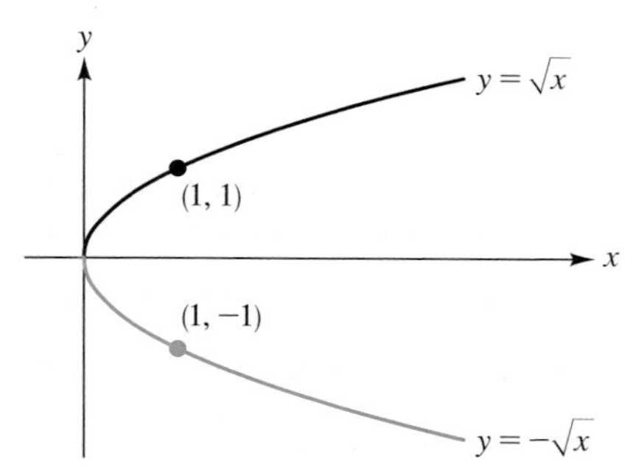

(a) To graph $y = -\sqrt{x}$, reflect $y = \sqrt{x}$ in the x-axis. More generally, to graph $y = -f(x)$, reflect the graph of $y = f(x)$ in the x-axis.

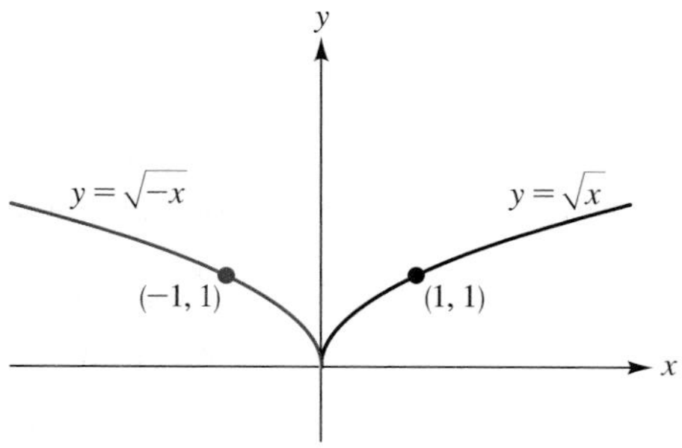

(b) To graph $y = \sqrt{-x}$, reflect $y = \sqrt{x}$ in the y-axis. More generally, to graph $y = f(-x)$, reflect the graph of $y = f(x)$ in the y-axis.

Figure 6

In Figure 5 we show examples of the first four techniques in the box, the techniques involving translation. Figure 6 displays examples of the remaining two techniques, the ones involving reflection. You should look over these figures carefully before going on to Examples 2 through 4, in which the various graphing techniques are combined.

EXAMPLE 2 Using translations of $y = x^2$

Graph: $y = (x - 2)^2 + 1$.

SOLUTION

Begin with the graph of $y = x^2$ in Figure 7(a). Move the curve two units to the right to obtain the graph of $y = (x - 2)^2$; see Figure 7(b). Then move the curve in Figure 7(b) up one unit to obtain the graph of $y = (x - 2)^2 + 1$; see Figure 7(c).

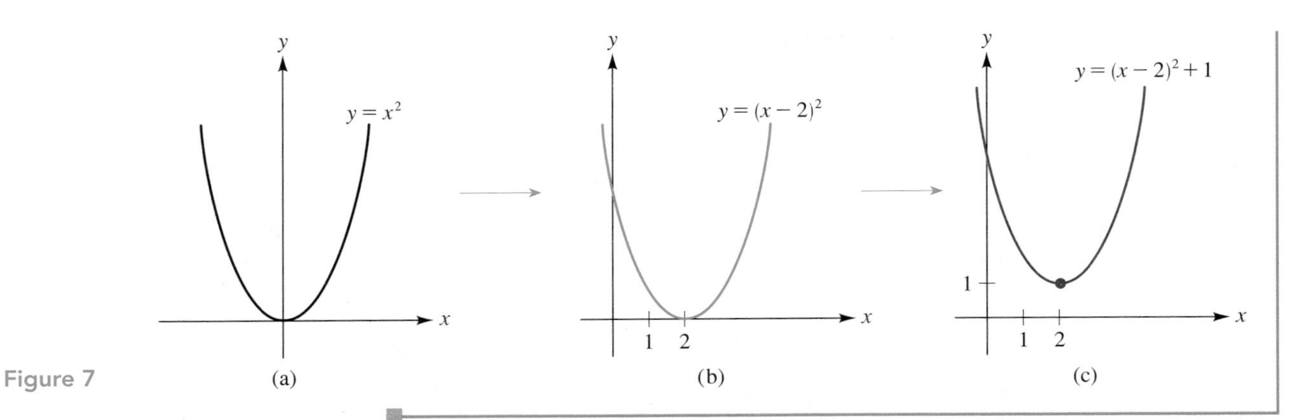

Figure 7

EXAMPLE 3 Using translations of $y = 1/x$

Graph the functions $y = \dfrac{1}{x-1}$ and $y = \dfrac{1}{x-1} + 1$.

SOLUTION

Begin with the graph of $y = 1/x$ in Figure 8(a). The x- and y-axes are asymptotes for this graph. Moving this graph to the right one unit yields the graph of $y = \dfrac{1}{x-1}$, shown in Figure 8(b). Note that the vertical asymptote moves one unit to the right also, but the horizontal asymptote is unchanged. Next, we move the graph in Figure 8(b) up one unit (why?) to obtain the graph of $y = \dfrac{1}{x-1} + 1$, as shown in Figure 8(c). (You should verify for yourself that the y-intercepts in Figures 8(b) and 8(c) are -1 and 0, respectively.)

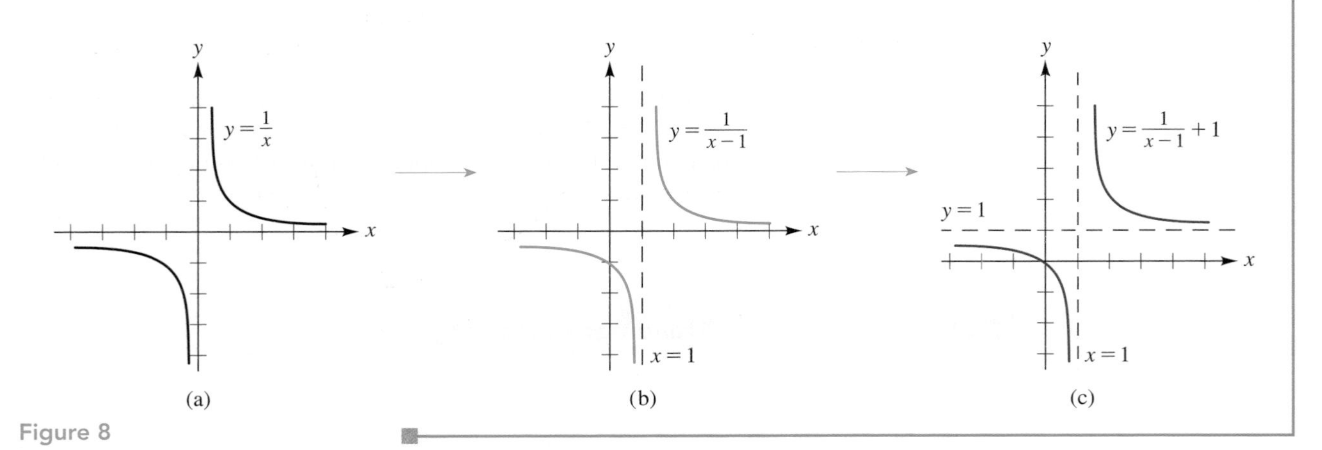

Figure 8

EXAMPLE 4 Applying reflection and translations to $y = |x|$

Graph each equation:

(a) $y = -|x|$; **(b)** $y = -|x-2| + 3$.

SOLUTION
See Figure 9.

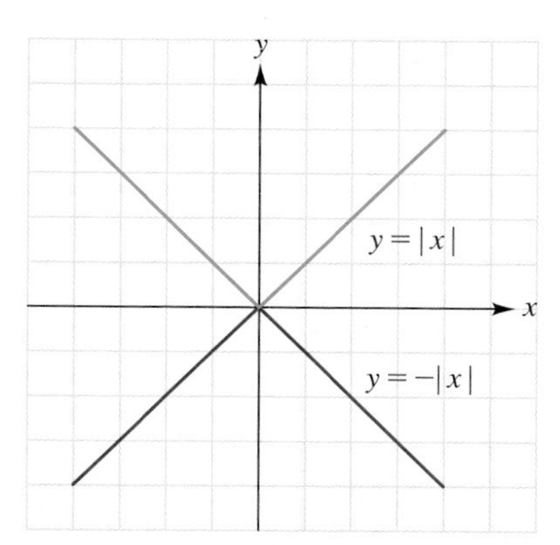

(a) Reflecting the graph of $y = |x|$ about the x-axis yields the graph of $y = -|x|$. Note that both the x- and y-intercepts are zero.

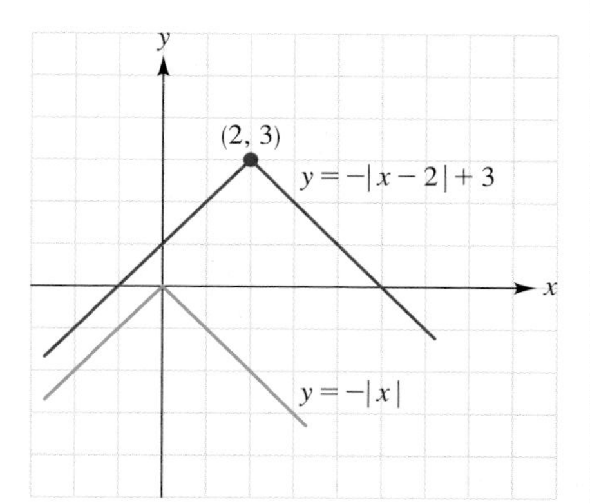

(b) Translating the graph of $y = -|x|$ two units to the right and three units up yields the graph of $y = -|x - 2| + 3$.
To get the y-intercept: $x = 0$ implies that $y = -|0 - 2| + 3 = 1$.
To get the x-intercept: $y = 0$ implies that

$$-|x-2| + 3 = 0 \text{ so } |x-2| = 3$$
$$x - 2 = 3 \text{ or } -(x-2) = 3$$
$$x = 5 \text{ or } -1$$

Figure 9

EXAMPLE 5 Using algebra to determine how two graphs are related

Figure 10 shows a graph of the function k defined by $k(x) = x^4 - 3x + 1$. In each case, say how the graph of the given function can be obtained from the graph of k. Then use a graphing utility to verify your statement:
(a) $f(x) = -x^4 + 3x - 1$; **(b)** $g(x) = x^4 + 3x + 1$; **(c)** $h(x) = x^4 - 3x + 3$.

GRAPHICAL PERSPECTIVE

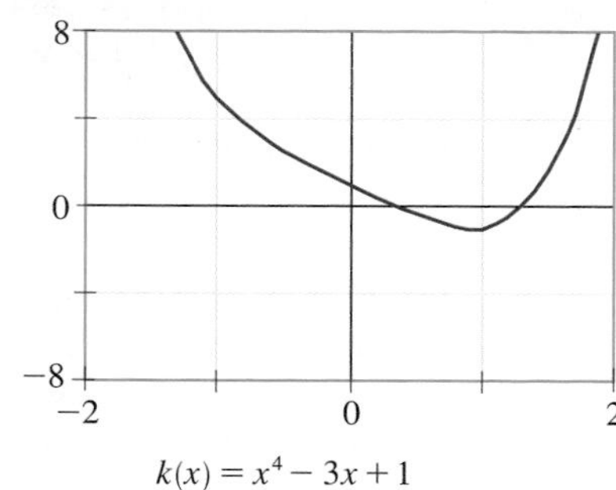

$k(x) = x^4 - 3x + 1$
$[-2, 2, 1]$ by $[-8, 8, 4]$

Figure 10

SOLUTION
(a) Note that the given expression for $f(x)$ is just the negative of $k(x)$. That is,

$$f(x) = -x^4 + 3x - 1 = -(x^4 - 3x + 1) = -k(x)$$

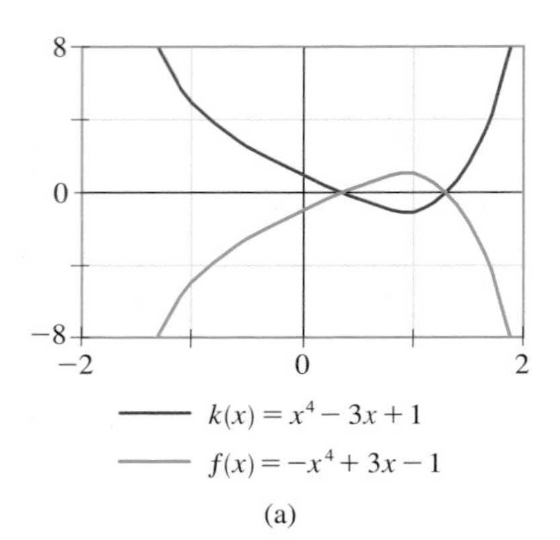

(a)

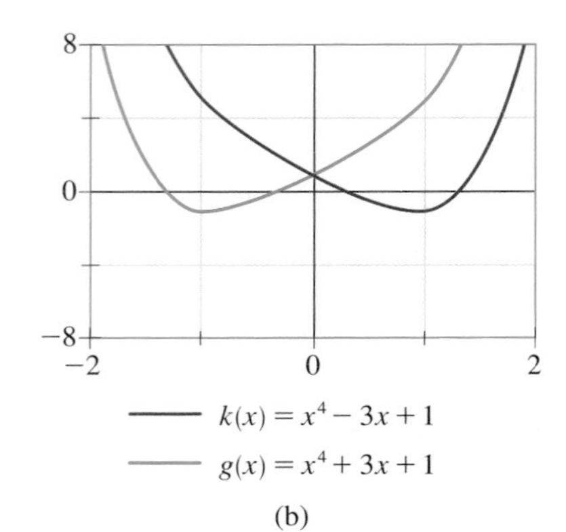

(b)

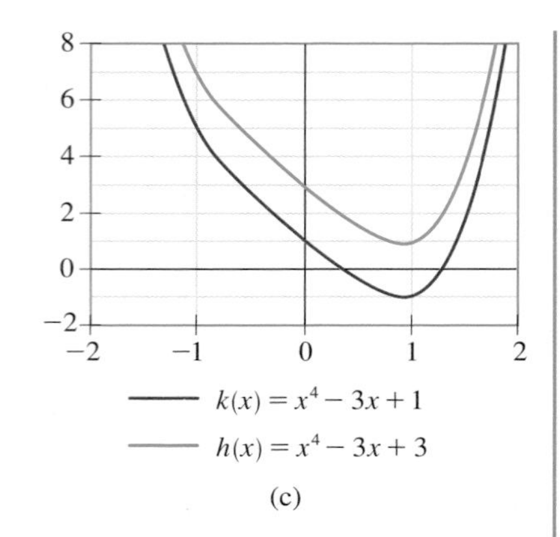

(c)

Figure 11
Some graphs obtained by reflecting or translating the graph of $k(x) = x^4 - 3x + 1$

Therefore the graph of f can be obtained by reflecting the graph of k in the x-axis. See Figure 11(a).

(b) The expressions for $g(x)$ and $k(x)$ are the same except for the sign in front of the $3x$-term. After some thought (or perhaps some trial and error) we find that

$$\begin{aligned} k(-x) &= (-x)^4 - 3(-x) + 1 \\ &= x^4 + 3x + 1 \\ &= g(x) \end{aligned}$$

That is, we have $k(-x) = g(x)$, and therefore the graph of g can be obtained by reflecting the graph of k in the y-axis. See Figure 11(b).

(c) Comparing the two equations $k(x) = x^4 - 3x + 1$ and $h(x) = x^4 - 3x + 3$, we see that

$$h(x) = k(x) + 2$$

Thus the graph of h is obtained by translating the graph of k up two units in the y-direction. See Figure 11(c).

EXAMPLE 6 A case where the translation rule for $y = f(x + c)$ is not directly applicable

(a) Graph: $y = \sqrt{-x + 2}$. **(b)** Graph: $y = |1 - x|$.

SOLUTION

(a) Our translation rule for $y = f(x + c)$ is not directly applicable here because in the generic equation $y = f(x + c)$, the sign of the x-term is positive; the rule provides no information about $y = f(-x + c)$. Rather than introducing yet another rule, we'll explain how to proceed using what you've already learned. To graph $y = \sqrt{-x + 2}$, first ignore the negative sign and graph the simpler equation $y = \sqrt{x + 2}$, as shown in Figure 12(a). Then for the graph of $y = \sqrt{-x + 2}$, just reflect the graph of $y = \sqrt{x + 2}$ in the y-axis, as shown in Figure 12(b). (We are using the fact that replacing x with $-x$ throughout an equation reflects the graph in the y-axis.)

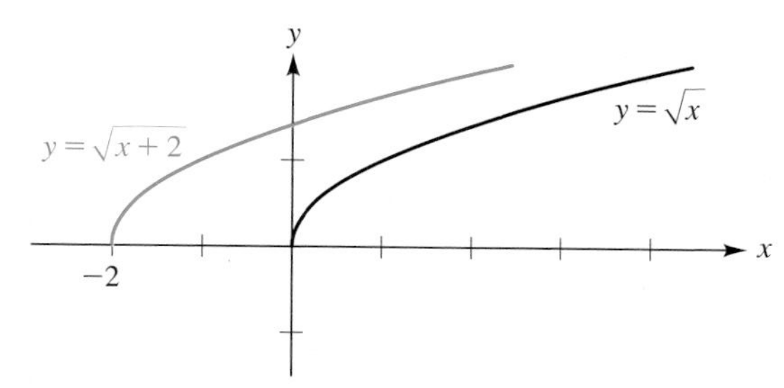

(a) Translate the graph of $y = \sqrt{x}$ to the left 2 units to obtain the graph of $y = \sqrt{x+2}$.

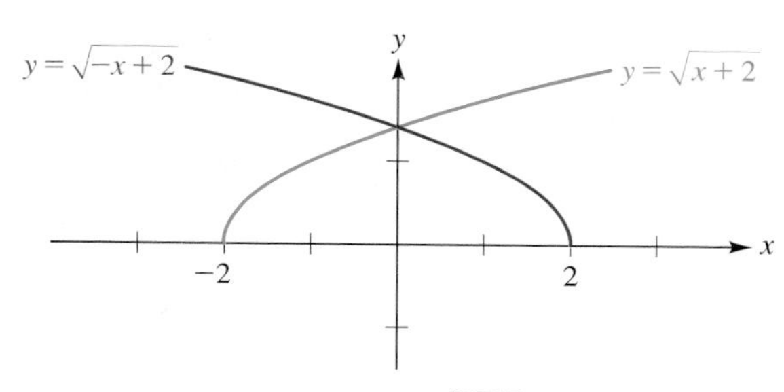

(b) Reflect the graph of $y = \sqrt{x+2}$ in the y-axis to obtain the graph of $y = \sqrt{-x+2}$.

Figure 12
Steps in graphing $y = \sqrt{-x+2}$: First graph $y = \sqrt{x+2}$; then reflect in the y-axis to obtain the graph of $y = \sqrt{-x+2}$.

(b) Write the given equation in the equivalent but somewhat more familiar form $y = |-x + 1|$. Now apply the strategy explained in part (a). That is, first graph $y = |x + 1|$, and then reflect in the y-axis to obtain the graph of $y = |-x + 1|$. See Figure 13.

NOTE We can use one of the properties of absolute value to obtain the graph in a different way and thus check the result in Figure 13. We have

$$\begin{aligned} y = |-x + 1| = |-1(x - 1)| &= |-1||x - 1| \\ &= |x - 1| \end{aligned}$$

This shows that the required graph can be obtained by graphing $y = |x - 1|$ —that is, by translating $y = |x|$ one unit to the right. As you can see, this agrees with the end result in Figure 13.

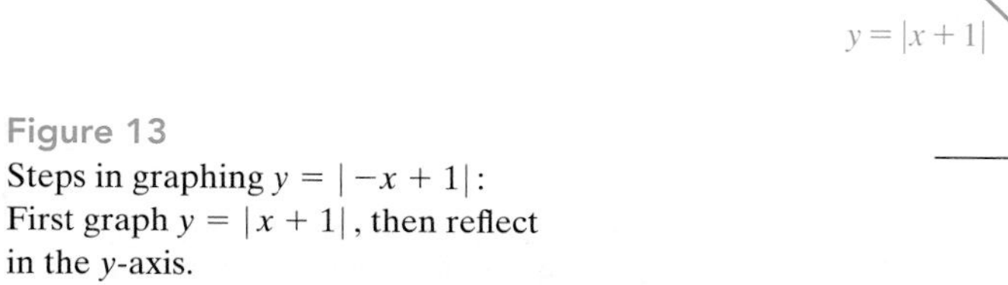

Figure 13
Steps in graphing $y = |-x + 1|$: First graph $y = |x + 1|$, then reflect in the y-axis.

The following example makes use of all the graphing techniques presented in this section. When you reach the point where you can work each part of this example on your own, then you know the material in this section.

EXAMPLE 7 A comprehensive example

The graph of a function f is a line segment joining the points $(-3, 1)$ and $(2, 4)$. Graph each of the following functions:

(a) $y = f(-x)$; **(b)** $y = -f(x)$; **(c)** $y = -f(-x)$;
(d) $y = f(x + 1)$; **(e)** $y = -f(x + 1)$; **(f)** $y = f(1 - x)$.

SOLUTION
See Figure 14.

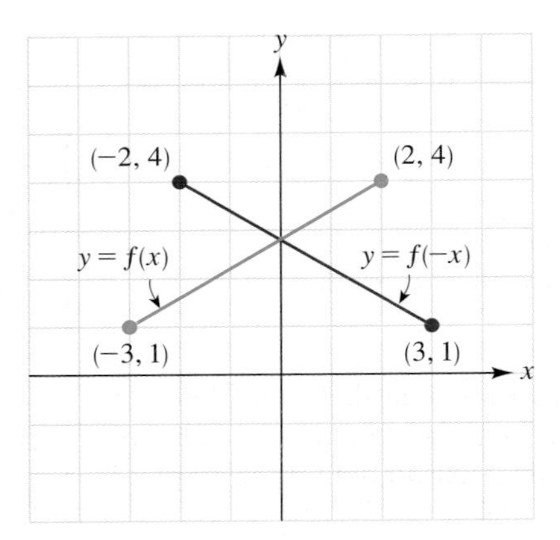

(a) Reflect the graph of $y = f(x)$ in the y-axis to obtain the graph of $y = f(-x)$.

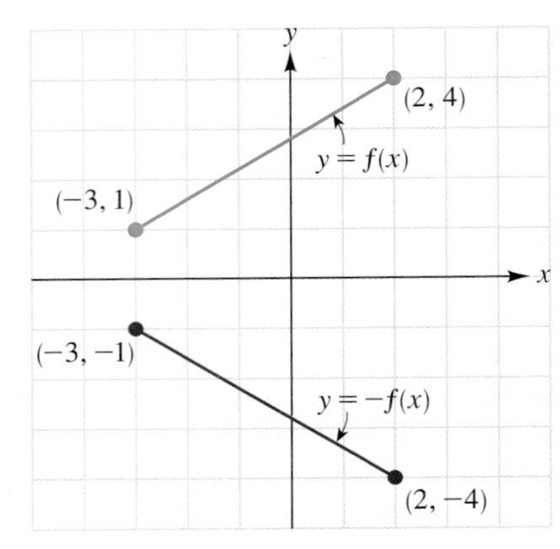

(b) Reflect the graph of $y = f(x)$ in the x-axis to obtain the graph of $y = -f(x)$.

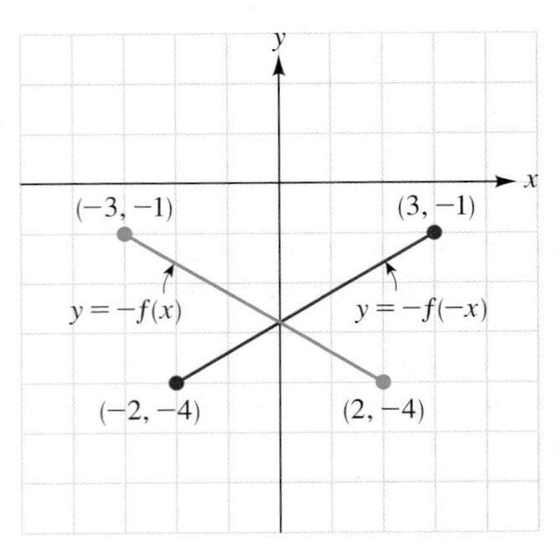

(c) Reflect the graph of $y = -f(x)$ in the y-axis to obtain the graph of $y = -f(-x)$. [Or reflect the graph of $y = f(-x)$ in the x-axis.]

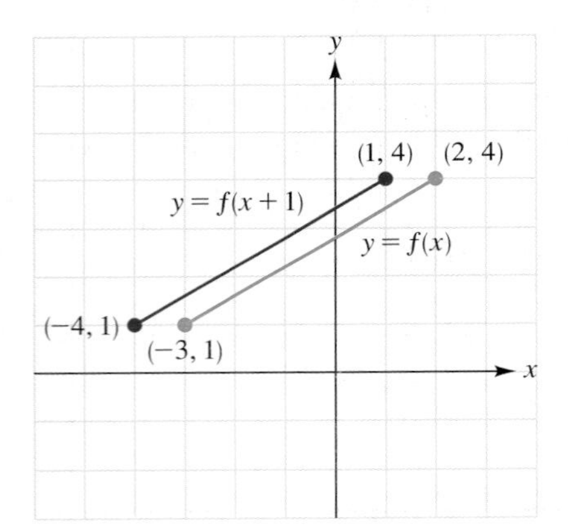

(d) Translate the graph of $y = f(x)$ one unit to the left to obtain the graph of $y = f(x + 1)$.

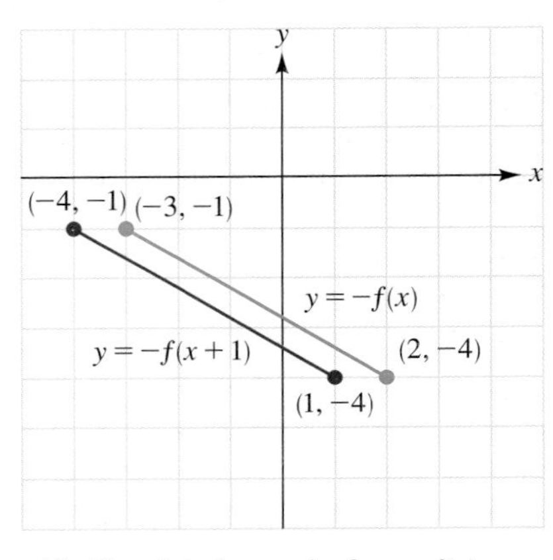

(e) Translate the graph of $y = -f(x)$ one unit to the left to obtain the graph of $y = -f(x + 1)$. [Or reflect the graph of $y = f(x + 1)$ in the x-axis.]

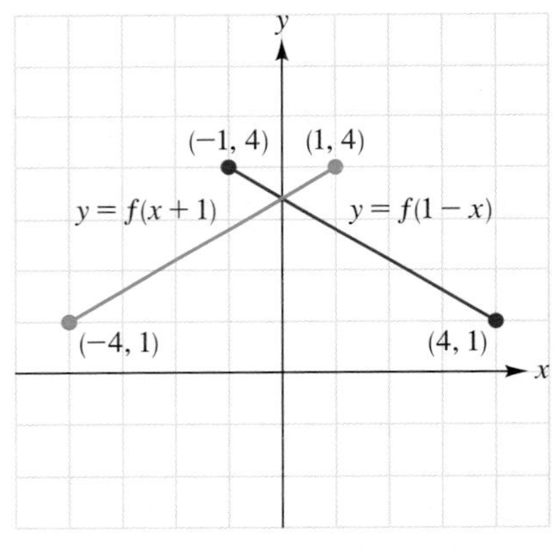

(f) Reflect the graph of $y = f(x + 1)$ in Figure 14(d) in the y-axis to obtain the graph of $y = f(1 - x)$. [Reason: Replacing x with $-x$ in $y = f(x + 1)$ yields $y = f(1 - x)$.]

Figure 14

EXERCISE SET 3.4

A

In Exercises 1 and 2, the right-hand column contains instructions for translating and/or reflecting the graph of $y = f(x)$. Match each equation in the left-hand column with an appropriate set of instructions in the right-hand column.

1. **(a)** $y = f(x - 1)$
(b) $y = f(x) - 1$
(c) $y = f(x) + 1$
(d) $y = f(x + 1)$

(A) Translate left 1 unit.
(B) Reflect in the x-axis, then translate left 1 unit.
(C) Translate right 1 unit.
(D) Reflect in the x-axis, then translate up 1 unit.

(e) $y = f(-x) + 1$	**(E)** Reflect in the x-axis, then translate down 1 unit.
(f) $y = f(-x) - 1$	**(F)** Translate down 1 unit.
(g) $y = -f(x) + 1$	**(G)** Reflect in the x-axis, reflect in the y-axis, then translate up 1 unit.
(h) $y = -f(x + 1)$	**(H)** Translate left 1 unit, then reflect in the y-axis, then translate up 1 unit.
(i) $y = -f(x) - 1$	**(I)** Translate up 1 unit.
(j) $y = f(1 - x) + 1$ *Hint*: See Example 6.	**(J)** Reflect in the y-axis, then translate up 1 unit.
(k) $y = -f(-x) + 1$	**(K)** Reflect in the y-axis, then translate down 1 unit.

2.

(a) $y = f(x + 2) + 3$	**(A)** Translate left 2 units, then translate down 3 units.
(b) $y = f(x + 3) + 2$	**(B)** Translate left 3 units, then translate up 2 units.
(c) $y = f(x - 2) + 3$	**(C)** Translate right 3 units, then translate up 2 units.
(d) $y = f(x - 2) - 3$	**(D)** Translate left 3 units, then translate down 2 units.
(e) $y = f(x + 2) - 3$	**(E)** Translate right 3 units and down 2 units.
(f) $y = f(x - 3) + 2$	**(F)** Reflect in the y-axis, then translate up 2 units.
(g) $y = f(x - 3) - 2$	**(G)** Reflect in the x-axis, then translate right 2 units.
(h) $y = f(x + 3) - 2$	**(H)** Reflect in the x-axis, then translate left 2 units.
(i) $y = -f(x + 2)$	**(I)** Translate left 2 units, then reflect in the y-axis.
(j) $y = -f(x - 2)$	**(J)** Translate right 2 units, then translate up 3 units.
(k) $y = f(2 - x)$	**(K)** Translate left 2 units, then translate up 3 units.
(l) $y = f(-x) + 2$	**(L)** Translate right 2 units and down 3 units.

In Exercises 3–24, sketch the graph of the function. Hint: Start with the basic graphs in Figure 7 on page 149.

3. $y = x^3 - 3$
4. $y = x^2 + 3$
5. $y = (x + 4)^2$
6. $y = (x + 4)^2 - 3$
7. $y = (x - 4)^2$
8. $y = (x - 4)^2 + 1$
9. $y = -x^2$
10. $y = -x^2 - 3$
11. $y = -(x - 3)^2$
12. $y = -(x - 3)^2 - 3$
13. $y = \sqrt{x - 3}$
14. $y = \sqrt{x - 3} + 1$
15. $y = \sqrt{-x + 3}$
16. $y = \sqrt{-x + 3} + 2$
17. $y = \dfrac{1}{x + 2} + 2$
18. $y = \dfrac{1}{x - 3} - 1$
19. $y = (x - 2)^3$
20. $y = (x - 2)^3 + 1$
21. $y = -x^3 + 4$
22. $y = -(x - 1)^3 + 4$
23. **(a)** $y = |x + 4|$ **(b)** $y = |4 - x|$ **(c)** $y = -|4 - x| + 1$
24. **(a)** $y = \sqrt{x + 2}$ **(b)** $y = \sqrt{2 - x}$ **(c)** $y = -\sqrt{2 - x}$

In Exercises 25–40, sketch the graph of the function, given that f, F, and g are defined as follows. (Hint: Start with the basic graphs in Figure 7 on page 149.)

$$f(x) = |x| \qquad F(x) = 1/x \qquad g(x) = \sqrt{1 - x^2}$$

25. $y = f(x - 5)$
26. $y = -f(x - 5)$
27. $y = f(5 - x)$
28. $y = -f(5 - x)$
29. $y = 1 - f(x - 5)$
30. $y = f(-x)$
31. $y = F(x + 3)$
32. $y = F(x) + 3$
33. $y = -F(x + 3)$
34. $y = F(-x) + 3$
35. $y = g(x - 2)$
36. $y = -g(x - 2)$
37. $y = 1 - g(x - 2)$
38. $y = g(-x)$
39. $y = g(2 - x)$
40. $y = -g(2 - x)$

Ⓖ *As background for Exercises 41–46, you should review Example 5. In Exercises 41–46, a function f is given. Say how the graph of each of the related functions can be obtained from the graph of f, and then use a graphing utility to verify your statement (as in Figure 11).*

41. $f(x) = 3x - 2$
(a) $y = -3x - 2$
(b) $y = -3x + 2$

42. $f(x) = 4 - x^2$
(a) $y = x^2 - 4$
(b) $y = 1 - x^2$

43. $f(x) = x^2 + 4x + 2$
(a) $y = x^2 - 4x + 2$
(b) $y = x^2 + 4x$

44. $f(x) = x^3 + 3x^2 - 4$
(a) $y = -x^3 + 3x^2 - 4$
(b) $y = -x^3 + 3x^2 + 1$

45. $f(x) = x^4 - 3x + 3$
(a) $y = x^4 + 3x + 3$
(b) $y = -x^4 + 3x - 3$
(c) $y = -x^4 + 3x$

46. $f(x) = -x^3 + 3x^2 - 3x + 1$
(a) $y = -x^3 + 3x^2 - 3x - 1$
(b) $y = x^3 + 3x^2 + 3x + 1$
(c) $y = x^3 - 3x^2 + 3x - 1$

47. **(a)** Complete the following table.

x	x^2	$x^2 - 1$	$x^2 + 1$
0			
±1			
±2			
±3			

(b) Using the results in the table, graph the functions $y = x^2$, $y = x^2 - 1$, and $y = x^2 + 1$ on the same set of axes. How are the graphs related?

48. **(a)** Complete the given table.

x	x^2	$(x - 1)^2$	$(x + 1)^2$
0			
1			
2			
3			
−1			
−2			
−3			

(b) Using the results in the table, graph the functions $y = x^2$, $y = (x - 1)^2$, and $y = (x + 1)^2$ on the same set of axes. How are the graphs related?

49. (a) Complete the following table. (Use a calculator where necessary.)

x	$\sqrt{x}$	$-\sqrt{x}$
0		
1		
2		
3		
4		
5		

(b) Using the results in the table, graph the functions $y = \sqrt{x}$ and $y = -\sqrt{x}$ on the same set of axes. How are the two graphs related?

50. (a) Complete the given tables. (Use a calculator where necessary.)

x	0	1	2	3	4	5
$\sqrt{x}$						

x	0	−1	−2	−3	−4	−5
$\sqrt{-x}$						

(b) Using the tables, graph the functions $y = \sqrt{x}$ and $y = \sqrt{-x}$ on the same set of axes. How are the graphs related?

In Exercises 51–53 you'll use a graphing utility to provide examples of the six graphing techniques listed in the box on page 172.

Ⓖ **51. (a)** Graph the two functions $f(x) = \sqrt{1 + x^2}$ and $g(x) = \sqrt{1 + x^2} + 3$. Observe that the graph of g is obtained by translating the graph of f *up* three units. This illustrates Property 1 on page 172.

(b) Graph the two functions $f(x) = \sqrt{1 + x^2}$ and $h(x) = \sqrt{1 + x^2} - 3$. Observe that the graph of h is obtained by translating the graph of f *down* three units. This illustrates Property 2 on page 172.

Ⓖ **52.** Using the viewing rectangle $[-5, 3]$ by $[0, 5]$, graph the three functions

$$F(x) = \sqrt{1 - x^3}$$
$$L(x) = \sqrt{1 - (x + 2)^3}$$
$$R(x) = \sqrt{1 - (x - 2)^3}$$

Observe that the graph of L is obtained by translating the graph of F to the *left* 2 units. Also, the graph of R is obtained by translating the graph of F to the *right* 2 units. This illustrates Properties 3 and 4 on page 172.

Ⓖ **53.** For this exercise, use the viewing rectangle $[-4, 4]$ by $[-4, 4]$.

(a) On the same set of axes, graph the two functions $f(x) = 2^x$ and $g(x) = -2^x$. [*Algebra reminder*: -2^x means $-(2^x)$, not $(-2)^x$]. Observe that the graph of g is obtained by reflecting the graph of f in the x-axis. This illustrates Property 5 on page 172.

(b) On the same set of axes graph the two functions $f(x) = 2^x$ and $h(x) = 2^{-x}$. Observe that the graph of h is obtained by reflecting the graph of f in the y-axis. This illustrates Property 6 on page 172.

B

54. Reflect the graph of $y = \sqrt{x}$ in the y-axis and then translate that two units to the left. What is the equation of the resulting graph? *Hint*: The answer is not $y = \sqrt{-x + 2}$.

55. Let P be a point with coordinates (a, b), and assume that c and d are positive numbers. (The condition that c and d are positive isn't really necessary in this problem, but it will help you to visualize things.)

(a) Translate the point P by c units in the x-direction to obtain a point Q, then translate Q by d units in the y-direction to obtain a point R. What are the coordinates of the point R?

(b) Translate the point P by d units in the y-direction to obtain a point S, then translate S by c units in the x-direction to obtain a point T. What are the coordinates of the point T?

(c) Compare your answers for parts (a) and (b). What have you demonstrated? (Answer in complete sentences.)

56. Let P be a point with coordinates (a, b).

(a) Reflect P in the x-axis to obtain a point Q, then reflect Q in the y-axis to obtain a point R. What are the coordinates of the point R?

(b) Reflect P in the y-axis to obtain a point S, then reflect S in the x-axis to obtain a point T. What are the coordinates of the point T?

(c) Compare your answers for parts (a) and (b). What have you demonstrated? (Answer in complete sentences.)

Ⓖ **57. (a)** Use a graphing utility to graph the function $y = x/(x - 1)$.

(b) From the screen display, it appears that the graph may be a translation of the basic $y = 1/x$ graph. Prove that this is indeed the case by using algebra to show that

$$\frac{x}{x - 1} = \frac{1}{x - 1} + 1.$$

Hint: Combine the quantities on the right side of the equation.

(c) Beginning with the graph of $y = 1/x$, what are the translations used to obtain the graph of $y = x/(x - 1)$?

For Exercises 58 and 59, assume that (a, b) is a point on the graph of $y = f(x)$, and specify the corresponding point on the graph of each equation. [For example, the point that corresponds to (a, b) on the graph of $y = f(x - 1)$ is $(a + 1, b)$.]

58. (a) $y = f(x - 3)$
(b) $y = f(x) - 3$
(c) $y = f(x - 3) - 3$
(d) $y = -f(x)$

(e) $y = f(-x)$
(f) $y = -f(-x)$
(g) $y = f(3 - x)$
(h) $y = -f(3 - x) + 1$

59. **(a)** $y = f(-x) + 2$
(b) $y = -f(-x) + 2$
(c) $y = -f(x - 3)$
(d) $y = 1 - f(x + 1)$
(e) $y = f(1 - x)$
(f) $y = -f(1 - x) + 1$

60. **(a)** A function f is said to be **even** if the equation $f(-x) = f(x)$ is satisfied by all values of x in the domain of f. Explain why the graph of an even function must be symmetric about the y-axis.
(b) Show that each function is even by computing $f(-x)$ and then noting that $f(x)$ and $f(-x)$ are equal.
(i) $f(x) = x^2$
(ii) $f(x) = 2x^4 - 6$
(iii) $f(x) = 3x^6 - \dfrac{4}{x^2} + 1$

61. **(a)** A function f is said to be **odd** if the equation $f(-x) = -f(x)$ is satisfied by all values of x in the domain of f. Show that if (x, y) is a point on the graph of an odd function f, then the point $(-x, -y)$ is also on the graph. (This implies that the graph of an odd function must be symmetric about the origin.)
(b) Show that each function is odd by computing $f(-x)$ as well as $-f(x)$ and then noting that the two expressions obtained are equal.
(i) $f(x) = x^3$
(ii) $f(x) = -2x^5 + 4x^3 - x$
(iii) $f(x) = |x|/(x + x^7)$

62. Is each function odd, even, or neither? (See Exercises 60 and 61 for definitions.)
(a) $f(x) = \dfrac{1 - x^2}{2 + x^2}$
(b) $g(x) = \dfrac{x - x^3}{2x + x^3}$
(c) $h(x) = x^2 + x$
(d) $F(x) = (x^2 + x)^2$
(e) $G(x) = \begin{cases} 1 & \text{if } x > 0 \\ 0 & \text{if } x = 0 \\ -1 & \text{if } x < 0 \end{cases}$

Suggestion for part (e): Look at the graph.

MINI PROJECT

CORRECTING A GRAPHING UTILITY DISPLAY

As background for this discussion, you need to recall the definitions from intermediate algebra for expressions of the form $\sqrt[n]{x}$ and $x^{1/n}$, where n is a positive integer. See Appendices B.2 and B.3 if you need a reminder, or have someone in the group consult the appendix and present the definitions and relevant examples to the group at large.

Some graphing utilities produce incomplete graphs for root functions $y = \sqrt[n]{x}$ when n is an odd integer. To find out how your graphing utility operates in this respect, try the function $y = \sqrt[3]{x}$ on your graphing utility, and compare your result to Figures A and B. (You might need to enter the function in the form $y = x^{1/3}$.)

The reason for the incomplete graph in Figure A is that the graphing utility used requires nonnegative inputs in computing roots, even cube roots. Suppose for this discussion that you have this type of graphing utility. Find a way to use the techniques in this section to produce a correct graph, one similar to that shown in Figure B.

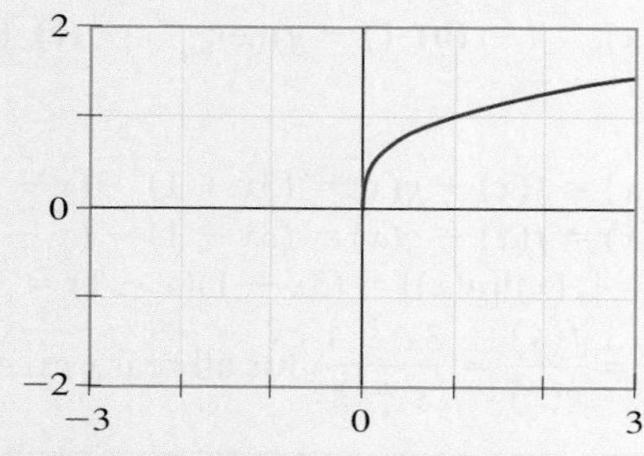

Figure A
An incomplete graph of $y = \sqrt[3]{x}$

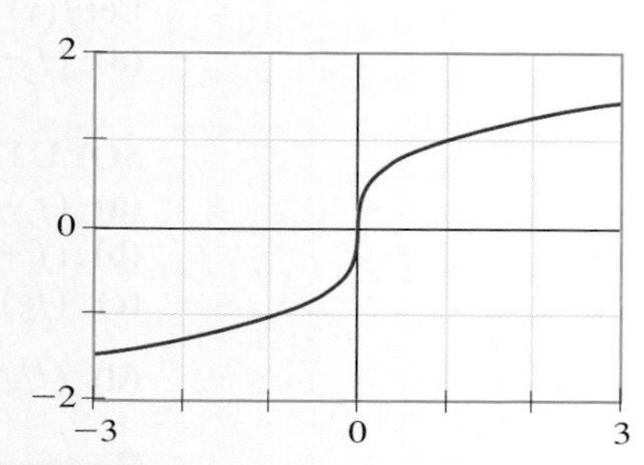

Figure B
A complete graph of $y = \sqrt[3]{x}$

3.5 METHODS OF COMBINING FUNCTIONS. ITERATION

The notation ϕx to indicate a function of x was introduced by . . . [John Bernoulli] in 1718, . . . but the general adoption of symbols like f, F, ϕ, and ψ . . . to represent functions, seems to be mainly due to Euler and Lagrange. —W. W. Rouse Ball in *A Short Account of the History of Mathematics* (New York: Dover Publications, 1960)

The composition of functions is basic to the study of iteration and chaos. —Hartmut Jürgens et al., *Fractals for the Classroom: Strategic Activities,* vol. 2 (New York: Springer-Verlag, 1992)

Two given numbers a and b can be combined in various ways to produce a third number. For instance, we can form the sum $a + b$ or the difference $a - b$ or the product ab. Also, if $b \neq 0$, we can form the quotient a/b. Similarly, two functions can be combined in various ways to produce a third function. Suppose, for example, that we start with the two functions $y = x^2$ and $y = x^3$. It seems natural to define their sum, difference, product, and quotient as follows:

$$\text{sum:} \quad y = x^2 + x^3$$
$$\text{difference:} \quad y = x^2 - x^3$$
$$\text{product:} \quad y = x^2x^3 = x^5$$
$$\text{quotient:} \quad y = \frac{x^2}{x^3} = \frac{1}{x} \qquad (\text{if } x \neq 0)$$

Indeed, this is the idea behind the formal definitions we now give.

DEFINITION Arithmetical Operations with Functions

Let f and g be two functions. Then the **sum** $f + g$, the **difference** $f - g$, the **product** fg, and the **quotient** f/g are functions defined by the following equations:

$$(f + g)(x) = f(x) + g(x) \tag{1}$$
$$(f - g)(x) = f(x) - g(x) \tag{2}$$
$$(fg)(x) = f(x) \cdot g(x) \tag{3}$$
$$(f/g)(x) = f(x)/g(x) \qquad \text{provided that } g(x) \neq 0 \tag{4}$$

For the functions defined by equations (1), (2), and (3), the domain is the set of all inputs x belonging to both the domain of f and the domain of g. For the quotient function in equation (4), we impose the additional restriction that the domain exclude all inputs x for which $g(x) = 0$.

EXAMPLE 1 Combining functions arithmetically

Let $f(x) = 3x + 1$ and $g(x) = x - 1$. Compute each of the following:
(a) $(f + g)(x)$; **(b)** $(f - g)(x)$; **(c)** $(fg)(x)$; **(d)** $(f/g)(x)$.

SOLUTION

(a) $(f + g)(x) = f(x) + g(x) = (3x + 1) + (x - 1) = 4x$, for all real x
(b) $(f - g)(x) = f(x) - g(x) = (3x + 1) - (x - 1) = 2x + 2$, for all real x
(c) $(fg)(x) = [f(x)][g(x)] = (3x + 1)(x - 1) = 3x^2 - 2x - 1$, for all real x
(d) $(f/g)(x) = \dfrac{f(x)}{g(x)} = \dfrac{3x + 1}{x - 1}$, for all real x *except* $x = 1$

For the remainder of this section, we are going to discuss a method of combining functions known as **composition of functions.** As you will see, this method is based on the familiar algebraic process of substitution. Suppose, for example, that f and g are two functions defined by

$$f(x) = x^2 \qquad g(x) = 3x + 1$$

Choose any number in the domain of g, for example, $x = -2$. We can compute $g(-2)$:

$$g(-2) = 3(-2) + 1 = -5$$

Now let's use the output -5 that g has produced as an *input* for f. We obtain

$$f(-5) = (-5)^2 = 25$$

Consequently,

$$f[g(-2)] = 25$$

So, beginning with the input -2, we've successively applied g and then f to obtain the output 25. Similarly, we could carry out this same procedure for any other number in the domain of g, provided that its output is in the domain of f. Here is a summary of the procedure:

1. Start with an input x and calculate $g(x)$.
2. Use $g(x)$ as an input for f; that is, calculate $f[g(x)]$.

We use the notation $f \circ g$ to denote the function, or rule, that tells us to assign the output $f[g(x)]$ to the initial input x. In other words, $f \circ g$ denotes the rule consisting of two steps: *First apply g; then apply f.* We read the notation $f \circ g$ as *f circle g* or *f composed with g*. In Figure 1, we summarize these ideas.

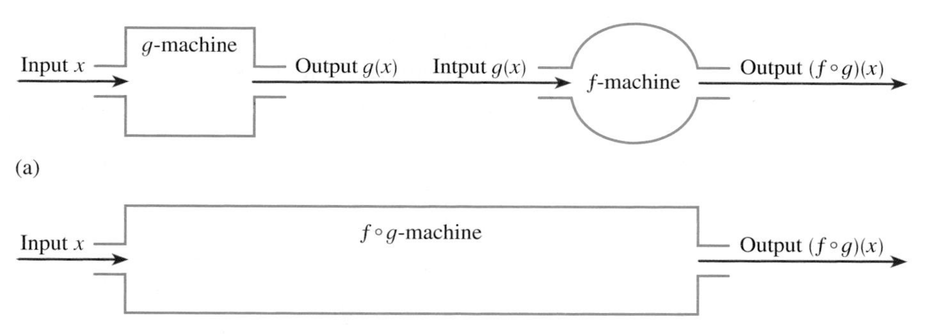

(b) Figure 1(b) indicates that, by incorporating the two steps in part (a), we can think of the composite function as one machine.

Figure 1
Diagram for the function $f \circ g$

When we write $g(x)$, we assume that x is in the domain of the function g. Likewise, for the notation $f[g(x)]$ to make sense, the outputs $g(x)$ must themselves be acceptable inputs for the function f. Our formal definition, then, for the composite function $f \circ g$ is as follows.

DEFINITION **Composition of Functions: $f \circ g$**

Given two functions f and g, the function $f \circ g$ is defined by

$$(f \circ g)(x) = f[g(x)]$$

The domain of $f \circ g$ consists of those inputs x (in the domain of g) for which $g(x)$ is in the domain of f.

EXAMPLE 2 Comparing $f \circ g$ to $g \circ f$

Let $f(x) = x^2$ and $g(x) = 3x + 1$. Compute $(f \circ g)(x)$ and $(g \circ f)(x)$.

SOLUTION

$$\begin{aligned}
(f \circ g)(x) &= f[g(x)] && \text{definition of } f \circ g \\
&= f(3x + 1) && \text{definition of } g \\
&= (3x + 1)^2 && \text{definition of } f \\
&= 9x^2 + 6x + 1
\end{aligned}$$

$$\begin{aligned}
(g \circ f)(x) &= g[f(x)] && \text{definition of } g \circ f \\
&= g(x^2) && \text{definition of } f \\
&= 3(x^2) + 1 && \text{definition of } g \\
&= 3x^2 + 1
\end{aligned}$$

Notice that the two results obtained in Example 2 are not the same. This shows that, in general, $f \circ g$ and $g \circ f$ represent different functions.

EXAMPLE 3 Two ways to compute a composite value

Let f and g be defined as in Example 2: $f(x) = x^2$ and $g(x) = 3x + 1$. Compute $(f \circ g)(-2)$.

SOLUTION
We will show two methods.

FIRST METHOD Using the formula for $(f \circ g)(x)$ developed in Example 2, we have

$$(f \circ g)(x) = 9x^2 + 6x + 1$$

and, therefore,

$$\begin{aligned}
(f \circ g)(-2) &= 9(-2)^2 + 6(-2) + 1 \\
&= 36 - 12 + 1 \\
&= 25
\end{aligned}$$

ALTERNATIVE METHOD Working directly from the definition of $f \circ g$, we have

$$\begin{aligned}
(f \circ g)(-2) &= f[g(-2)] \\
&= f[3(-2) + 1] \\
&= f(-5) \\
&= (-5)^2 \\
&= 25
\end{aligned}$$

In Examples 2 and 3 the domain of both f and g is the set of all real numbers. And as you can easily check, the domain of both $f \circ g$ and $g \circ f$ is also the set of all real numbers. In Example 4, however, some care needs to be taken in describing the domain of the composite function.

EXAMPLE 4 Finding the domain of a composite function

Let f and g be defined as follows:

$$f(x) = x^2 + 1 \qquad g(x) = \sqrt{x}$$

Compute $(f \circ g)(x)$. Find the domain of $f \circ g$ and sketch its graph.

SOLUTION

$$(f \circ g)(x) = f[g(x)]$$
$$= f(\sqrt{x}) = (\sqrt{x})^2 + 1 = x + 1$$

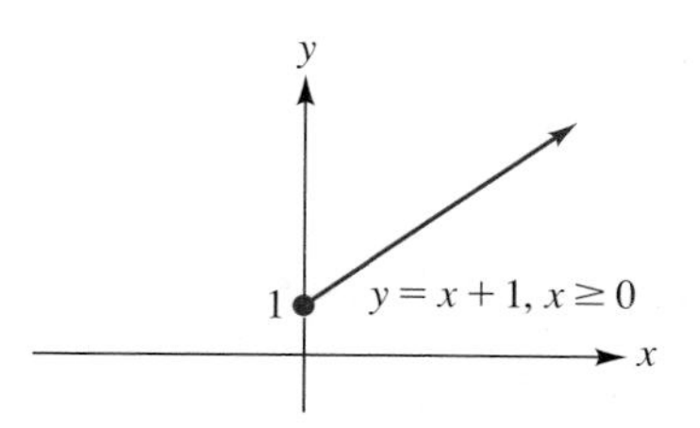

Figure 2
The graph of the function $f \circ g$ in Example 4

So we have $(f \circ g)(x) = x + 1$. Now, what about the domain of $f \circ g$? Our first inclination might be to say (incorrectly!) that the domain is the set of all real numbers, since any real number can be used as an input in the expression $x + 1$. However, the definition of $f \circ g$ on page 183 tells us that the inputs for $f \circ g$ must first of all be acceptable inputs for g. Given the definition of g, then, we must require that x be nonnegative. On the other hand, for any nonnegative input x, the number $g(x)$ will be an acceptable input for f. (Why?) In summary, then, the domain of $f \circ g$ is the interval $[0, \infty)$. The graph of $f \circ g$ is shown in Figure 2.

EXAMPLE 5 An application of composition of functions

Suppose that an offshore oil rig is leaking and that the oil forms a circular region whose radius r increases over the first 12 hours according to the function

$$r = f(t) = \frac{t}{2t + 4} \qquad 0 \le t \le 12$$

where r is measured in miles, t is in hours, and $t = 0$ corresponds to the instant that the leak begins.

(a) Using function notation, the area of a circle of radius r is given by $A(r) = \pi r^2$. Find $(A \circ f)(t)$ and interpret the result.

(b) Compute the rate of change of the function $A \circ f$ over the interval $0 \le t \le 6$.

SOLUTION

(a) $$(A \circ f)(t) = A[f(t)] = A\left(\frac{t}{2t + 4}\right)$$
$$= \pi\left(\frac{t}{2t + 4}\right)^2$$

After squaring the fraction in this last expression, we obtain

$$(A \circ f)(t) = \frac{\pi t^2}{4t^2 + 16t + 16} \tag{1}$$

Since $f(t)$ expresses the radius in terms of time, our result in equation (1) gives the area of the oil spill in terms of time.

(b) By definition (from Section 3.3) the average rate of change of the function $A \circ f$ on the interval $[0, 6]$ is given by

$$\frac{\Delta(A \circ f)}{\Delta t} = \frac{(A \circ f)(6) - (A \circ f)(0)}{6 - 0} \tag{2}$$

To evaluate the difference quotient on the right side of equation (2), we need to find $(A \circ f)(6)$ and $(A \circ f)(0)$. Exercise 33 asks you to use equation (1) to check that $(A \circ f)(6) = 9\pi/64$ and $(A \circ f)(0) = 0$. Assuming these results, equation (2) becomes

$$\frac{\Delta(A \circ f)}{\Delta t} = \frac{9\pi/64}{6}$$
$$= 3\pi/128 \qquad \text{Check the arithmetic.}$$
$$\approx 0.07 \text{ (sq. miles)/hour}$$

That is, on average over the first six hours, the area of the oil spill is increasing at approximately 0.07 (sq. miles)/hour. (Exercise 34 asks you to compute the rate of change of the area over the next six-hour period and to compare it to this result for the first six hours.)

AFTERWORD ON THE SOLUTION TO EXAMPLE 5(b) A graphing utility can also be used for the computations. For instance, with a TI-83 graphing calculator, in the [Y=] screen, one enters the three functions

$$Y_1 = \frac{x}{2x + 4} \qquad Y_2 = \pi x^2 \qquad Y_3 = Y_1(Y_2)$$

Then, in the home screen, the required rate of change is obtained by evaluating the expression $\frac{Y_3(6) - Y_3(0)}{6 - 0}$ or, more simply, $\frac{Y_3(6)}{6}$, since in this example $Y_3(0) = 0$. For other types of graphing utilities, consult your user's manual.

One reason for studying the composition of functions is that it lets us express a given function in terms of simpler functions. This procedure is often useful in calculus. Suppose, for example, that we wish to express the function C defined by

$$C(x) = (2x^3 - 5)^2$$

as a composition of simpler functions. That is, we want to come up with two functions f and g so that the equation

$$C(x) = (f \circ g)(x)$$

holds for every x in the domain of C.

We begin by thinking what we would do to compute $(2x^3 - 5)^2$ for a given value of x. First, we would compute the quantity $2x^3 - 5$, then we would square the result. Therefore, recalling that the rule $f \circ g$ tells us to do g *first,* we let $g(x) = 2x^3 - 5$. Then, since the next step is squaring, we let $f(x) = x^2$. Now let's see whether these choices for f and g are correct; that is, let us calculate $(f \circ g)(x)$ and see whether it really is the same as $C(x)$.

Using $f(x) = x^2$ and $g(x) = 2x^3 - 5$, we have

$$(f \circ g)(x) = f[g(x)] = f(2x^3 - 5) = (2x^3 - 5)^2 = C(x)$$

This equation shows that our choices for f and g were indeed correct, and we have expressed C as a composition of two simpler functions.*

Note that in expressing C as $f \circ g$, we chose g to be the "inner" function, that is, the quantity inside the parentheses: $g(x) = 2x^3 - 5$. This observation is used in Example 6.

*Other answers are possible, too. For instance, if $F(x) = (x - 5)^2$ and $G(x) = 2x^3$, then (as you should verify for yourself) $C(x) = F[G(x)]$.

EXAMPLE 6 Expressing a function in terms of two simpler functions

Let $s(x) = \sqrt{1 + x^4}$. Express the function s as a composition of two simpler functions f and g.

SOLUTION

Let g be the "inner" function; that is, let $g(x)$ be the quantity inside the radical:

$$g(x) = 1 + x^4$$

and let's take f to be the square root function:

$$f(x) = \sqrt{x}$$

Now we need to verify that these are the appropriate choices for f and g; that is, we need to check that the equation $(f \circ g)(x) = s(x)$ is true for every x in the domain of s. We have

$$(f \circ g)(x) = f[g(x)] = f(1 + x^4) = \sqrt{1 + x^4} = s(x)$$

Thus $(f \circ g)(x) = s(x)$, as required. (What is the domain of $s(x)$?)

We conclude this section by describing the process of *iteration* for a function. This simple process is of fundamental importance in modern mathematics and science in the study of fractals and chaos.* The process of iteration is sequential; it proceeds step by step. We begin with a function f and an initial input x_0. In the first step, we compute the output $f(x_0)$. For the second step, we use the number $f(x_0)$ as an *input* for f and compute $f(f(x_0))$. The process then continues in this way: At each step, we use the output from the previous step as the new input.

For example, suppose that we start with the function $f(x) = x/2$ and the input $x_0 = 6$. Then, for the first three steps in the iteration process, we have

$$f(6) = \frac{6}{2} = 3$$

$$f(3) = \frac{3}{2} = 1.5$$

$$f(1.5) = \frac{1.5}{2} = 0.75$$

After this a calculator becomes convenient. Check for yourself that the iterations run as follows:

$$6 \to 3 \to 1.5 \to 0.75 \to 0.375 \to 0.1875 \to 0.09375 \to 0.046875 \to \cdots$$

This list of numbers is called the *orbit* of 6 under the function f. In the list or orbit, the number 3 is the *first iterate* of 6, the number 1.5 is the *second iterate* of 6, and so on. In the box that follows, we summarize these ideas and introduce a useful notation for the iterates.

*See, for example, pp. 314–363 in *The Nature and Power of Mathematics* by Donald M. Davis (Princeton, N.J.: Princeton University Press, 1993). For a wider view, see the book by James Gleick: *Chaos: Making a New Science* (New York: Penguin Books, 1988).

DEFINITION **Iterates**

Given a function f and an input x_0, the **iterates** of x_0 are the numbers $f(x_0), f(f(x_0)), f(f(f(x_0)))$, and so on. The number $f(x_0)$ is called the **first iterate,** the number $f(f(x_0))$ is called the **second iterate,** and so on. Subscript notation is used to denote the iterates as follows.

$$x_1 = f(x_0) \qquad \text{the first iterate}$$
$$x_2 = f(f(x_0)) \qquad \text{the second iterate}$$
$$x_3 = f(f(f(x_0))) \qquad \text{the third iterate}$$
$$\vdots$$

The **orbit** of x_0 under the function f is the list of numbers $x_0, x_1, x_2, x_3, \ldots$.

Notice that the subscripts in this notation tell you how many times to apply the function f. For example, x_1 indicates that f is applied once to obtain $f(x_0)$, and x_2 indicates that the function is applied twice to obtain $f(f(x_0))$.

EXAMPLE 7 **Computing iterates**

Compute the first four iterates in each case. Use a calculator as necessary. In part (b), round the final answers to three decimal places.

(a) $f(x) = x^2, x_0 = -2$ **(b)** $g(x) = \sqrt{x}, x_0 = 0.1$

SOLUTION

(a)
$$x_1 = (-2)^2 = 4$$
$$x_2 = (4)^2 = 16$$
$$x_3 = (16)^2 = 256$$
$$x_4 = (256)^2 = 65{,}536$$

(b) We use the square root key (or keys) on the calculator repeatedly. As you should check for yourself, this yields

$$x_1 = \sqrt{0.1} = 0.316227\ldots$$
$$x_2 = \sqrt{0.316227\ldots} = 0.562341\ldots$$
$$x_3 = \sqrt{0.562341\ldots} = 0.749894\ldots$$
$$x_4 = \sqrt{0.749989\ldots} = 0.865964\ldots \approx 0.866 \qquad \text{rounding to three decimal places}$$

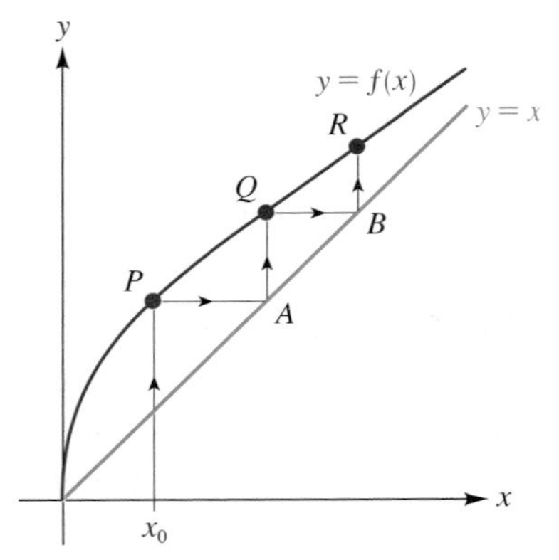

Figure 3
The y-coordinates of the points P, Q, and R are, respectively, the first three iterates of x_0 under the function f.

The process of iterating a function can be visualized in an x-y coordinate system. We sketch the function $y = f(x)$ along with the line $y = x$, as in Figure 3. Then, starting with the initial input x_0 on the x-axis, we follow the horizontal and vertical segments to determine points P, Q, and R on the graph of f. As the next paragraph explains, the y-coordinates of the points P, Q, and R are the first, second, and third iterates, respectively, of x_0. Additional iterates are obtained by continuing this pattern. In Figure 4 we show how this looks in a specific case. The y-coordinates of the five points P, Q, R, S, and T are the first five iterates of $x_0 = 0.1$ under the function $g(x) = -4x^2 + 4x$. Although Figure 4 at first appears more complicated than Fig-

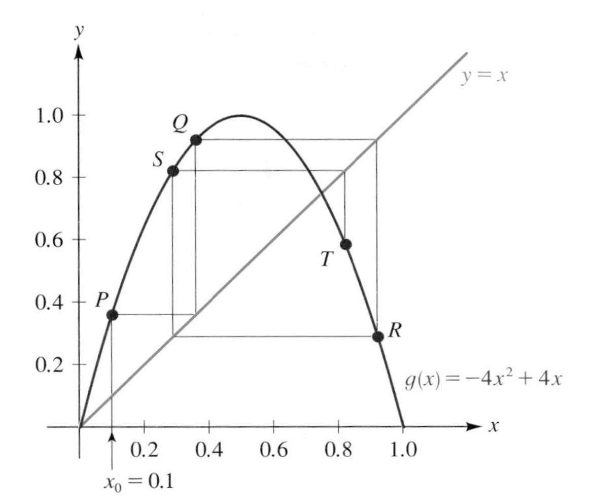

Figure 4
Graphical iteration of $g(x) = -4x^2 + 4x$ with initial input $x_0 = 0.1$. The y-coordinates of the five points P, Q, R, S, and T are, respectively, the first five iterates of x_0.

ure 3, it is important to understand that the pattern is formed in exactly the same way. We start with the initial input on the x-axis and travel vertically to determine the point P on the curve. After that, each successive point on the curve is generated in the same way: Go horizontally to the line $y = x$ and then vertically to the curve.

As we've said, this paragraph explains why the y-coordinates of the three points P, Q, and R in Figure 3 are, respectively, the first three iterates of x_0 under the function f. In Figure 3 the x-coordinate of P is given to be x_0. So (according to the definition of the graph of a function), the y-coordinate of P is $f(x_0)$. In other words, the y-coordinate of P is the first iterate of x_0. Next in Figure 3, we move horizontally from the point P to the point A. Because the movement is horizontal, the y-coordinate doesn't change; it's the y-coordinate of P, or $f(x_0)$. Now, what about the x-coordinate of A? Since the point A lies on the line $y = x$, the x- and y-coordinates of A must be identical. Consequently, the x-coordinate of A must be the number $f(x_0)$. Finally, we move vertically in Figure 3 from A to Q. The point Q has the same x-coordinate as the point A, namely, $f(x_0)$. Now, since Q lies on the graph of the function f, we conclude that the y-coordinate of Q is $f(f(x_0))$. That is, the y-coordinate of Q is indeed the second iterate of x_0. The reasoning to show that the y-coordinate of R is the third iterate of x_0 is entirely similar, so we omit giving the details here.

EXERCISE SET 3.5 

A

In Exercises 1–8, compute each expression, given that the functions f, g, h, k, and m are defined as follows:

$$f(x) = 2x - 1 \qquad k(x) = 2, \quad \text{for all } x$$
$$g(x) = x^2 - 3x - 6 \qquad m(x) = x^2 - 9$$
$$h(x) = x^3$$

1. (a) $(f + g)(x)$
(b) $(f - g)(x)$
(c) $(f - g)(0)$

2. (a) $(fh)(x)$
(b) $(h/f)(x)$
(c) $(f/h)(1)$

3. (a) $(m - f)(x)$
(b) $(f - m)(x)$

4. (a) $(fg)(x)$
(b) $(fg)(1/2)$

5. (a) $(fk)(x)$
(b) $(kf)(x)$
(c) $(fk)(1) - (kf)(2)$

6. (a) $(g + m)(x)$
(b) $(g + m)(x) - (g - m)(x)$

7. (a) $(f/m)(x) - (m/f)(x)$
(b) $(f/m)(0) - (m/f)(0)$

8. (a) $[h \cdot (f + m)](x)$ *Note:* h and $(f + m)$ are two functions; the notation $h \cdot (f + m)$ denotes the product function.
(b) $(hf)(x) + (hm)(x)$

9. Let $f(x) = 3x + 1$ and $g(x) = -2x - 5$. Compute the following.
(a) $(f \circ g)(x)$
(b) $(f \circ g)(10)$
(c) $(g \circ f)(x)$
(d) $(g \circ f)(10)$

10. Let $f(x) = 1 - 2x^2$ and $g(x) = x + 1$. Compute the following.
(a) $(f \circ g)(x)$
(b) $(f \circ g)(-1)$
(c) $(g \circ f)(x)$
(d) $(g \circ f)(-1)$
(e) $(f \circ f)(x)$
(f) $(g \circ g)(-1)$

11. Compute $(f \circ g)(x)$, $(f \circ g)(-2)$, $(g \circ f)(x)$, and $(g \circ f)(-2)$ for each pair of functions.
(a) $f(x) = x^2 - 3x - 4$; $g(x) = 2 - 3x$
(b) $f(x) = 2^x$; $g(x) = x^2 + 1$
(c) $f(x) = x$; $g(x) = 3x^5 - 4x^2$
(d) $f(x) = 3x - 4$; $g(x) = (x + 4)/3$

12. Let $h(x) = 4x^2 - 5x + 1$, $k(x) = x$, and $m(x) = 7$ for all x. Compute the following.
(a) $h[k(x)]$
(b) $k[h(x)]$
(c) $h[m(x)]$
(d) $m[h(x)]$
(e) $k[m(x)]$
(f) $m[k(x)]$

13. Let $F(x) = \dfrac{3x - 4}{3x + 3}$ and $G(x) = \dfrac{x + 1}{x - 1}$. Compute the following.
(a) $(F \circ G)(x)$
(b) $F[G(t)]$
(c) $(F \circ G)(2)$
(d) $(G \circ F)(x)$
(e) $G[F(y)]$
(f) $(G \circ F)(2)$

14. Let $f(x) = (1/x^2) + 1$ and $g(x) = 1/(x - 1)$
(a) Compute $(f \circ g)(x)$.
(b) What is the domain of $f \circ g$?
(c) Graph the function $f \circ g$.

15. Let $M(x) = (2x - 1)/(x - 2)$.
(a) Compute $M(7)$ and then $M[M(7)]$.
(b) Compute $(M \circ M)(x)$.
(c) Compute $(M \circ M)(7)$, using the formula you obtained in part (b). Check that your answer agrees with that obtained in part (a).

16. Let $F(x) = (x + 1)^5$, $f(x) = x^5$, and $g(x) = x + 1$. Which of the following is true for all x?

$$(f \circ g)(x) = F(x) \quad \text{or} \quad (g \circ f)(x) = F(x)$$

17. Refer to the graphs of the functions f, g, and h to compute the required quantities. Assume that all the axes are marked off in one-unit intervals

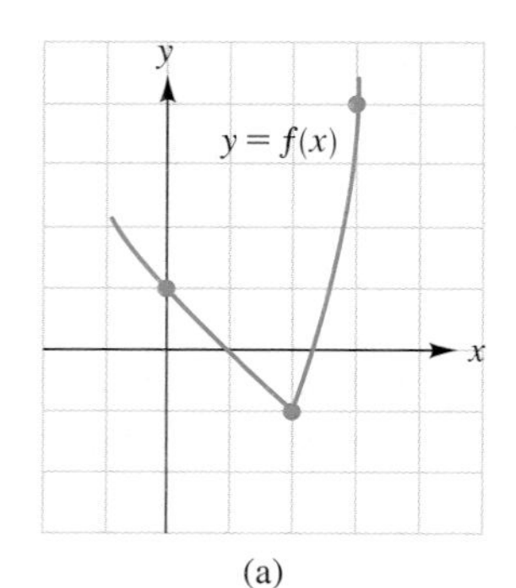

(a)

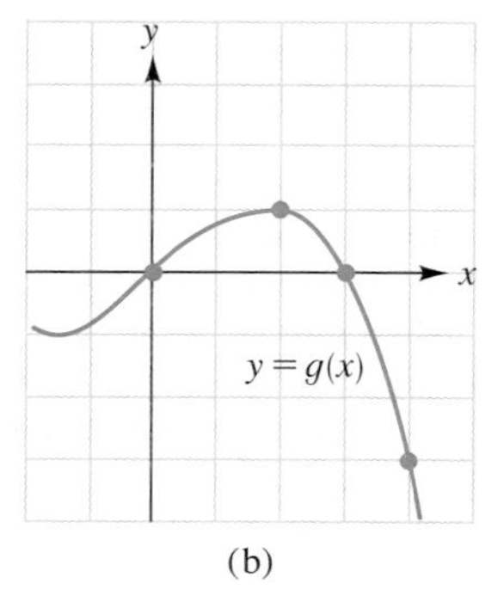

(b)

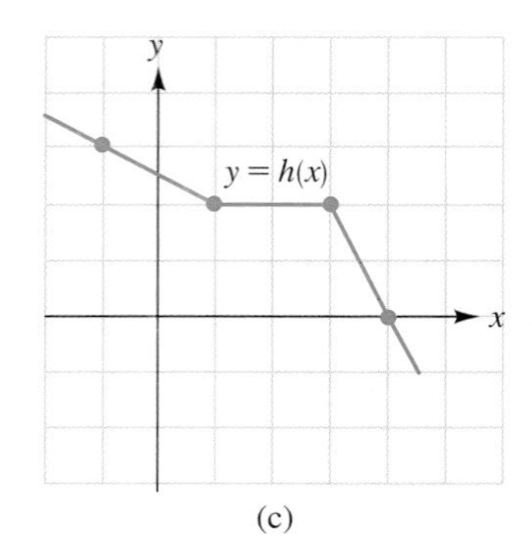

(c)

(a) $f[g(3)]$
(b) $g[f(3)]$
(c) $f[h(3)]$
(d) $(h \circ g)(2)$
(e) $h\{f[g(3)]\}$
(f) $(g \circ f \circ h \circ f)(2)$

Note: The notation in part (f) means first do f, then h, then f, then g.

18. **(a)** Let $T(x) = 4x^3 - 3x^2 + 6x - 1$ and $I(x) = x$. Find $(T \circ I)(x)$ and $(I \circ T)(x)$.
(b) Let $G(x) = ax^2 + bx + c$ and $I(x) = x$. Find $(G \circ I)(x)$ and $(I \circ G)(x)$.
(c) What general conclusion do you arrive at from the results of parts (a) and (b)?

19. The domain of a function f consists of the numbers -1, 0, 1, 2, and 3. The following table shows the output that f assigns to each input.

x	1	0	1	2	3
f(x)	2	2	0	3	1

The domain of a function g consists of the numbers 0, 1, 2, 3, and 4. The following table shows the output that g assigns to each input.

x	0	1	2	3	4
g(x)	3	2	0	4	−1

Use this information to complete the following tables for $f \circ g$ and $g \circ f$. *Note*: Two of the entries will be undefined.

x	0	1	2	3	4
(f ∘ g)(x)					

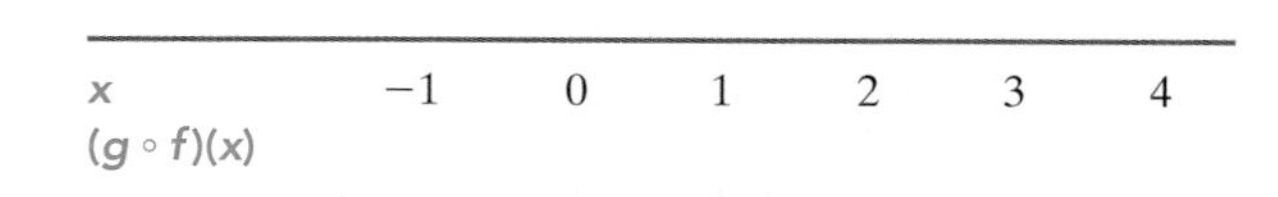

x	−1	0	1	2	3	4
(g ∘ f)(x)						

20. The following two tables show certain pairs of inputs and outputs for functions f and g.

x	0	$\pi/6$	$\pi/4$	$\pi/3$	$\pi/2$
f(x)	0	1/2	$\sqrt{2}/2$	$\sqrt{3}/2$	1

y	0	1/4	$\sqrt{2}/4$	1/2	$\sqrt{2}/2$	3/4	$\sqrt{3}/2$	1
g(y)	$\pi/2$	π	0	$\pi/3$	$\pi/4$	0	$\pi/6$	0

Use this information to complete the following table of values for $(g \circ f)(x)$.

x	0	$\pi/6$	$\pi/4$	$\pi/3$	$\pi/2$
$(g \circ f)(x)$					

21. Let $f(x) = x^3 - 2x$ and $g(x) = x + 4$.
(a) What is the relationship between the graphs of the two functions f and $f \circ g$? (The idea here is to answer without looking at the graphs; use a concept from Section 3.4.)
Ⓖ **(b)** Use a graphing utility to check your answer in part (a).

22. Let $f(x) = \sqrt{x^3 + 2x + 17}$ and $g(x) = x + 6$.
(a) What is the relationship between the graphs of the two functions f and $f \circ g$? (As in Exercise 21, the idea here is to answer without looking at the graphs.)
Ⓖ **(b)** Use a graphing utility to check your answer in part (a).

23. Let $g(x) = \sqrt{x} - 3$ and $f(x) = x - 1$.
(a) Sketch a graph of g. Specify the domain and range.
(b) Sketch a graph of f. Specify the domain and range.
(c) Compute $(f \circ g)(x)$. Graph the function $f \circ g$ and specify its domain and range.
(d) Find a formula for $g[f(x)]$. Which values of x are acceptable inputs here? That is, what is the domain of $g \circ f$?
(e) Use the results of part (d) to sketch a graph of the function $g \circ f$.

24. Let $F(x) = -x^2$ and $G(x) = \sqrt{x}$. Determine the domains of $F \circ G$ and $G \circ F$.

25. Suppose that an oil spill in a lake covers a circular area and that the radius of the circle is increasing according to the formula $r = f(t) = 15 + t^{1.65}$, where t represents the number of hours since the spill was first observed and the radius r is measured in meters. (Thus when the spill was first discovered, $t = 0$ hr, and the initial radius was $r = f(0) = 15 + 0^{1.65} = 15$ m.)
(a) Let $A(r) = \pi r^2$, as in Example 5. Compute a table of values for the composite function $A \circ f$ with t running from 0 to 5 in increments of 0.5. (Round each output to the nearest integer.) Then use the table to answer the questions that follow in parts (b) through (d).
(b) After one hour, what is the area of the spill (rounded to the nearest 10 m²)?
(c) Initially, what was the area of the spill (when $t = 0$)? Approximately how many hours does it take for this area to double?
(d) Compute the average rate of change of the area of the spill from $t = 0$ to $t = 2.5$ and from $t = 2.5$ to $t = 5$. Over which of the two intervals is the area increasing faster?

26. A spherical weather balloon is being inflated in such a way that the radius is given by

$$r = g(t) = \frac{1}{2}t + 2$$

Assume that r is in meters and t is in seconds, with $t = 0$ corresponding to the time that inflation begins. If the volume of a sphere of radius r is given by

$$V(r) = \frac{4}{3}\pi r^3$$

compute $V[g(t)]$ and use this to find the time at which the volume of the balloon is 36π m³.

27. Suppose that a manufacturer knows that the daily production cost to build x bicycles is given by the function C, where

$$C(x) = 100 + 90x - x^2 \qquad (0 \le x \le 40)$$

That is, $C(x)$ represents the cost in dollars of building x bicycles. Furthermore, suppose that the number of bicycles that can be built in t hr is given by the function f, where

$$x = f(t) = 5t \qquad (0 \le t \le 8)$$

(a) Compute $(C \circ f)(t)$.
(b) Compute the production cost on a day that the factory operates for $t = 3$ hr.
(c) If the factory runs for 6 hr instead of 3 hr, is the cost twice as much?

28. Suppose that in a certain biology lab experiment, the number of bacteria is related to the temperature T of the environment by the function

$$N(T) = -2T^2 + 240T - 5400 \qquad (40 \le T \le 90)$$

Here, $N(T)$ represents the number of bacteria present when the temperature is T degrees Fahrenheit. Also, suppose that t hr after the experiment begins, the temperature is given by

$$T(t) = 10t + 40 \qquad (0 \le t \le 5)$$

(a) Compute $N[T(t)]$.
(b) How many bacteria are present when $t = 0$ hr? When $t = 2$ hr? When $t = 5$ hr?

29. Express each function as a composition of two functions.
(a) $F(x) = \sqrt[3]{3x + 4}$
(b) $G(x) = |2x - 3|$
(c) $H(x) = (ax + b)^5$
(d) $T(x) = 1/\sqrt{x}$

30. Let $a(x) = x^2$, $b(x) = |x|$, and $c(x) = 3x - 1$. Express each of the following functions as a composition of two of the given functions.
(a) $f(x) = (3x - 1)^2$
(b) $g(x) = |3x - 1|$
(c) $h(x) = 3x^2 - 1$

31. Let $a(x) = 1/x$, $b(x) = \sqrt[3]{x}$, $c(x) = 2x + 1$, and $d(x) = x^2$. Express each of the following functions as a composition of two of the given functions.

(a) $f(x) = \sqrt[3]{2x + 1}$
(b) $g(x) = 1/x^2$
(c) $h(x) = 2x^2 + 1$
(d) $K(x) = 2\sqrt[3]{x} + 1$
(e) $l(x) = \dfrac{2}{x} + 1$
(f) $m(x) = \dfrac{1}{2x + 1}$

32. Express $n(x) = x^{2/3}$ as a composition of two of the functions given at the begining of Exercise 31. (If you don't recall the definition for fractional exponents, see Appendix B.3 at the back of this book.)

33. For this exercise, refer to equation (1) in Example 5.
(a) Show that $(A \circ f)(0) = 0$ and that $(A \circ f)(6) = 9\pi/64$.
Ⓖ **(b)** Use a graphing utility to graph the average-rate-of-change function $y = (A \circ f)(t)$. Note that the graph appears to pass through the point (0, 0), which supports the first result in part (a).
Ⓖ **(c)** By zooming in, estimate the y-coordinate of the point on the graph corresponding to $t = 6$. What does this have to do with part (a) of this exercise?

34. Refer to Example 5(b).
(a) Compute the average rate of change of $A \circ f$ over the time interval from $t = 6$ to $t = 12$ hours. On average, was the area of the oil leak increasing faster over the first six hours or the second six hours?
Ⓖ **(b)** Graph the average-rate-of-change function $y = (A \circ f)(t)$ in the viewing rectangle [0, 12] by [0, 0.8]. Then draw two line segments, one connecting the two points on the graph corresponding to $t = 0$ and $t = 6$, the other connecting the two points corresponding to $t = 6$ and $t = 12$. (If your graphing utility does not have a draw feature, use hard copy and draw using a ruler.) Explain in complete sentences how the picture supports your answer to the question in part (a).

In Exercises 35–40, use the given function and compute the first six iterates of each initial input x_0. In cases in which a calculator answer contains four or more decimal places, round the final answer to three decimal places. (However, during the calculations, work with all of the decimal places that your calculator affords.)

35. $f(x) = 2x$
(a) $x_0 = 1$
(b) $x_0 = 0$
(c) $x_0 = -1$

36. $f(x) = \frac{1}{4}x$
(a) $x_0 = 16$
(b) $x_0 = 0$
(c) $x_0 = -16$

37. $g(x) = 2x + 1$
(a) $x_0 = -2$
(b) $x_0 = -1$
(c) $x_0 = 1$

38. $g(x) = \frac{1}{4}x + 3$
(a) $x_0 = 3$
(b) $x_0 = 4$
(c) $x_0 = 5$

39. $F(x) = x^2$
(a) $x_0 = 0.9$
(b) $x_0 = 1$
(c) $x_0 = 1.1$

40. $G(x) = x^2 + 0.25$
(a) $x_0 = 0.4$
(b) $x_0 = 0.5$
(c) $x_0 = 0.6$

41. The accompanying figure shows a portion of the iteration process for $f(x) = \sqrt{x}$ with initial input $x_0 = 0.1$. Use the figure to estimate (to within 0.05) the first four iterates of $x_0 = 0.1$. Then, use a calculator to compute the four iterates. (Round to three decimal places, where appropriate.) Check that your calculator results are consistent with the estimates obtained from the graph.

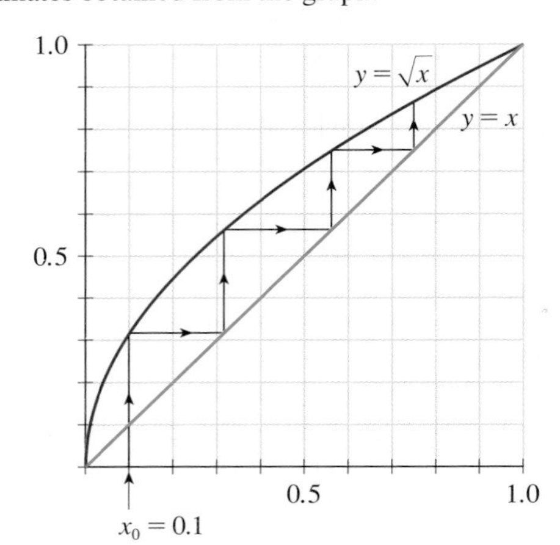

42. The following figure shows a portion of the iteration process for $f(x) = 3.6(x - x^2)$ with initial input $x_0 = 0.3$. Use the figure to estimate (to within 0.1, or closer if it seems appropriate) the first seven iterates of $x_0 = 0.3$. Then, use a calculator to compute the seven iterates. (Round the final answers to three decimal places.) Check that your calculator results are consistent with the estimates obtained from the graph.

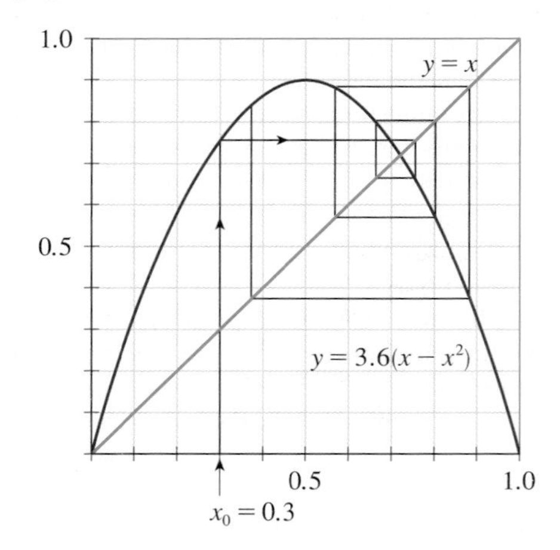

B

43. *The* $3x + 1$ *conjecture* Define a function f, with domain the positive integers, as follows:

$$f(x) = \begin{cases} 3x + 1 & \text{if } x \text{ is odd} \\ x/2 & \text{if } x \text{ is even} \end{cases}$$

(a) Compute $f(1), f(2), f(3), f(4), f(5)$, and $f(6)$.
(b) Compute the first three iterates of $x_0 = 1$.
(c) Compute the iterates of $x_0 = 3$ until you obtain the value 1. [After this, the iterates will recycle through the simple pattern obtained in part (b).]
(d) *The* $3x + 1$ *conjecture* asserts that for any positive integer x_0, the iterates eventually return to the value 1. Verify that this conjecture is valid for each of the following values of x_0: 2, 4, 5, 6, and 7.

Remark: At present, the $3x + 1$ conjecture is indeed a conjecture, not a theorem. No one yet has found a proof that the assertion is valid for *every* positive integer. Computer checks, however, have verified the conjecture on a case-by-case basis for very large values of x_0. As of April 2000, the conjecture had been verified for all values of x_0 up to approximately 1.8×10^{16}.

44. *The* $3x + 1$ *conjecture* (continued from Exercise 43) If you have access to the Internet, use Alfred Wassermann's $3x + 1$ on-line calculator located at

http://did.mat.uni-bayreuth.de/personen/wassermann/fun/3np1.html

to answer the following questions. For which n does x_n first reach 1 if $x_0 = 100$? If $x_0 = 1000$? If $x_0 = 10^4$? (The Web address above was accessible at the time of this writing, March 2004.)

45. Let $g(x) = 4x - 1$. Find $f(x)$, given that the equation $(g \circ f)(x) = x + 5$ is true for all values of x.

46. Let $g(x) = 2x + 1$. Find $f(x)$, given that $(g \circ f)(x) = 10x - 7$.

47. Let $f(x) = -2x + 1$ and $g(x) = ax + b$. Find a and b so the equation $f[g(x)] = x$ holds for all values of x.

48. Let $f(x) = (3x - 4)/(x - 3)$.
(a) Compute $(f \circ f)(x)$.
(b) Find $f[f(113/355)]$. (Try not to do it the hard way.)

49. Let $f(x) = x^2$ and $g(x) = 2x - 1$.
(a) Compute $\dfrac{f[g(x)] - f[g(a)]}{g(x) - g(a)}$.
(b) Compute $\dfrac{f[g(x)] - f[g(a)]}{x - a}$.

Exercises 50 and 51 indicate how iteration is used in finding roots of numbers and roots of equations. (The functions that are given in each exercise were determined using Newton's method, a process studied in calculus.)

50. Let $f(x) = 0.5\left(x + \dfrac{3}{x}\right)$.
(a) Compute the first ten iterates of $x_0 = 1$ under the function f. What do you observe?
(b) Use your calculator to evaluate $\sqrt{3}$ and compare the answer to your results in part (a). What do you observe?
(c) It can be shown that for any positive number x_0, the iterates of x_0 under the function $f(x) = 0.5(x + 3/x)$ always approach the number $\sqrt{3}$. (You'll see the reasons for this in Section 4.3.) Looking at your results in parts (a) and (b), which is the first iterate that agrees with $\sqrt{3}$ through the first three decimal places? Through the first eight decimal places?
(d) Compute the first ten iterates of $x_0 = 50$ under the function f, then answer the questions presented in part (c).

51. Let $f(x) = \dfrac{2x^3 + 7}{3x^2}$.
(a) Compute the first ten iterates of $x_0 = 1$ under the function f. What do you observe?
(b) Evaluate the expression $\sqrt[3]{7}$ and compare the answer to your results in part (a). What do you observe?
(c) It can be shown that for any positive number x_0, the iterates of x_0 under the function $f(x) = \dfrac{2x^3 + 7}{3x^2}$ always approach the number $\sqrt[3]{7}$. Looking at your results in parts (a) and (b), which is the first iterate that agrees with $\sqrt[3]{7}$ through the first three decimal places? Through the first eight decimal places?

C

52. Let $f(x) = \dfrac{x - 3}{x + c}$, where c denotes a constant.
(a) If $c = -1$, show that $f(f(x)) = x$.
Ⓖ **(b)** Use a graphing utility to support the result in part (a). That is, enter the function $f(x) = \dfrac{x - 3}{x - 1}$, and then have the machine graph the two functions $y = x$ and $y = f(f(x))$ in the same picture.
(c) What does the result in part (a) tell you about the iteration process for the function $y = (x - 3)/(x - 1)$? That is, what pattern emerges in the iterates? Answer in complete sentences.
(d) Now assume $c = 1$, instead of -1. Show that $f(f(f(x))) = x$.
Ⓖ **(e)** Use a graphing utility to support the result in part (d).
(f) What does the result in part (d) tell you about the iteration process for the function $y = (x - 3)/(x + 1)$? That is, what pattern emerges in the iterates? Answer in complete sentences.

MINI PROJECT A GRAPHICAL APPROACH TO COMPOSITION OF FUNCTIONS

As we saw in this section, the iteration process can be carried out either algebraically or graphically. Composition of functions can also be represented graphically. The accompanying figure shows the graphs of two functions f and g, the line $y = x$, and an input a. Use the figure and the ideas of this section to discover and then explain how to find the point $(a, g[f(a)])$ on the graph of the composite function $g \circ f$.

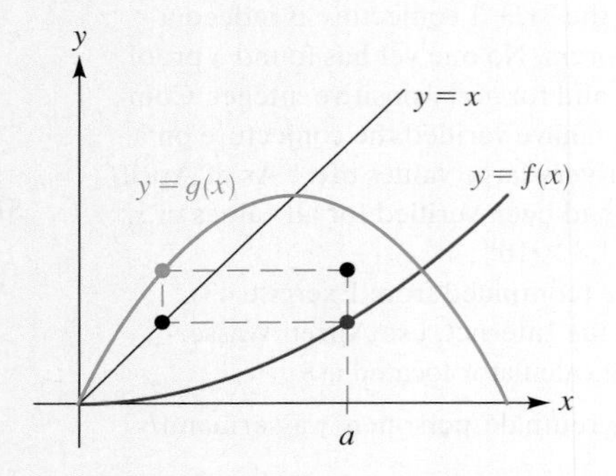

3.6 INVERSE FUNCTIONS

If anybody ever told me why the graph of $y = x^{1/2}$ is the reflection of the graph of $y = x^2$ in a 45° line, it didn't sink in. To this day, there are textbooks that expect students to think that it is so obvious as to need no explanation. This is a pity, if only because [in calculus] it is such a common practice to define the natural logarithm first and then define the exponential function as its inverse.—Professor Ralph P. Boas, recalling his student days in the article "Inverse Functions" in *The College Mathematics Journal,* vol. 16 (1985), p. 42.

PROBLEM SOLUTION

© 1981 Scott Kim. From *Inversions* (W. H. Freeman, 1989).

We shall introduce the idea of inverse functions through an easy example. Consider the function f defined by $f(x) = 2x$. A short table of values for f and the graph of f are displayed in Figure 1.

Now we ask this question: What happens if we take the table for f (in Figure 1) and interchange the entries in the x and y columns to obtain a new table that looks like the following?

x	y
-4	-2
-2	-1
0	0
2	1
4	2

← New table formed by interchanging the inputs and outputs for f

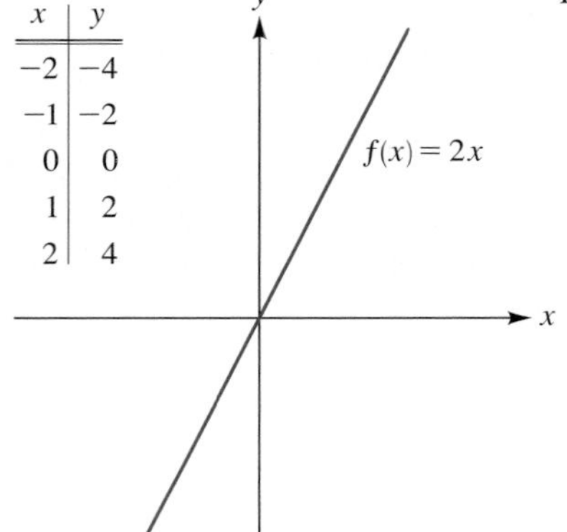

x	y
-2	-4
-1	-2
0	0
1	2
2	4

Figure 1

Several observations are in order. Our new table does itself represent a function (in this case): for each input x there is exactly one output y. Let's call this new function g. Note that each output is one-half of the corresponding input. So the function g can be described by the formula $g(x) = x/2$. In summary, we began with the "doubling function," $f(x) = 2x$; then, by interchanging the inputs with the outputs, we obtained the "halving function," $g(x) = x/2$.

The doubling function f and the halving function g are in a certain sense opposites; each reverses, or undoes, the effect of the other. That is, if you begin with x and then calculate $g[f(x)]$, you get x again; similarly, $f[g(x)]$ is also equal to x. (Verify these last two statements for yourself.)

The two functions f and g that we've been discussing are an example of a pair of *inverse functions.* The general definition for inverse functions is given in the box that follows.

DEFINITION Inverse Functions

Two functions f and g are **inverses** of one another provided that

$$f[g(x)] = x \qquad \text{for each } x \text{ in the domain of } g$$

and

$$g[f(x)] = x \qquad \text{for each } x \text{ in the domain of } f$$

EXAMPLE 1 Applying the definition of inverse functions

Suppose that f and g are a pair of inverse functions.
(a) If 13 is in the domain of g, find $f[g(13)]$.
(b) If $\sqrt{5}$ is in the domain of f, find $g[f(\sqrt{5})]$.

SOLUTION
(a) By the definition of inverse functions, the equation $f[g(x)] = x$ is true for each x in the domain of g. In particular then, using $x = 13$, we have $f[g(13)] = 13$.
(b) Since $\sqrt{5}$ is in the domain of f, the definition of inverse functions gives us $g[f(\sqrt{5})] = \sqrt{5}$.

EXAMPLE 2 Applying the definition of inverse functions

Verify that the functions f and g are inverses, where

$$f(x) = \frac{1}{3}x + 2 \qquad \text{and} \qquad g(x) = 3x - 6$$

SOLUTION
In view of the definition, since both f and g have domain all real numbers, we must check that $f[g(x)] = x$ and $g[f(x)] = x$, for all real numbers x. We have

$$\begin{aligned} f[g(x)] &= f(3x - 6) \\ &= \frac{1}{3}(3x - 6) + 2 = x \qquad \text{(Check the algebra.)} \end{aligned}$$

Thus, $f[g(x)] = x$. Now we still need to check that $g[f(x)] = x$. We have

$$\begin{aligned} g[f(x)] &= g\left(\frac{1}{3}x + 2\right) \\ &= 3\left(\frac{1}{3}x + 2\right) - 6 = x \qquad \text{Again, check the algebra.} \end{aligned}$$

Having shown that $f[g(x)] = x$ and $g[f(x)] = x$, we conclude that f and g are indeed inverse functions.

EXAMPLE 3 Two perspectives on inverse functions

Suppose that f and g are a pair of inverse functions. If $f(2) = 3$, what is $g(3)$?

SOLUTION

We can use either of two methods.

FIRST METHOD The quantities 3 and $f(2)$ are declared to be equal. Thus whether we use 3 or $f(2)$ as an input for g, the result must be the same. We then have

$$\begin{aligned} 3 &= f(2) \\ g(3) &= g[f(2)] \\ g(3) &= 2 \qquad \text{using the fact that } g[f(x)] = x \text{ for any } x \text{ in the domain of } f \end{aligned}$$

Thus $g(3) = 2$.

ALTERNATIVE METHOD Think of a table of values for the function f. Since $f(2) = 3$, one entry in the table must look like this:

x			2		
y			3		

Now, g reverses the roles of x and y, so in the table for g, one entry must look like this:

x			3		
y			2		

Thus $g(3) = 2$, as obtained previously.

It is customary to use the notation f^{-1} (read *f inverse*) for the function that is the inverse of f. So in Example 2, for instance, we have

$$f(x) = \frac{1}{3}x + 2 \qquad \text{and} \qquad f^{-1}(x) = 3x - 6 \tag{1}$$

CAUTION In the context of functions, $f^{-1}(x)$ does not in general mean $\dfrac{1}{f(x)}$. For instance, regarding the pair of inverse functions in equations (1), it's certainly not true in general that $3x - 6 = \dfrac{1}{\frac{1}{3}x + 2}$. (For other examples, see Exercises 11 and 12 at the end of this section.) For reference now, in the box that follows, we rewrite the defining equations for inverse functions using the f^{-1} notation. Figure 2 provides a graphic summary of these equations.

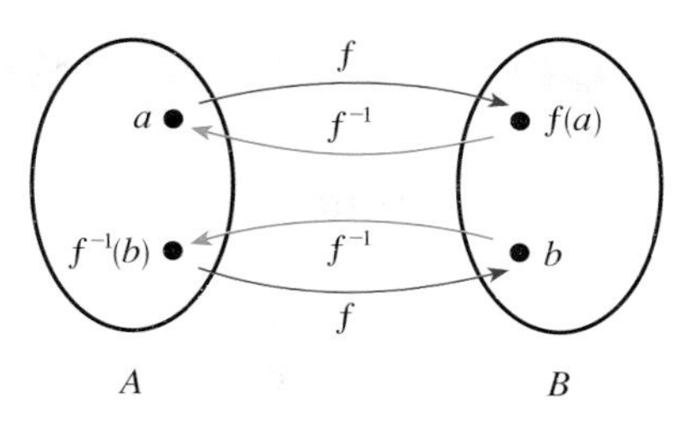

Figure 2
The action of inverse functions: The set A in the figure is both the domain of f and the range of f^{-1}. The set B is both the domain of f^{-1} and the range of f.

$$f[f^{-1}(x)] = x \qquad \text{for each } x \text{ in the domain of } f^{-1}$$

and

$$f^{-1}[f(x)] = x \qquad \text{for each } x \text{ in the domain of } f$$

EXAMPLE 4 Using the definition of inverse functions

Suppose $f(x) = 4x^3 + 7$. Find $f[f^{-1}(5)]$. (Assume that f^{-1} exists and that 5 is in its domain.)

SOLUTION

As in Example 1, this is just an application of the definition of inverse functions. We have, by definition, $f[f^{-1}(x)] = x$ for every x in the domain of f^{-1}. Replacing x with 5 throughout this last equation, we immediately obtain

$$f[f^{-1}(5)] = 5$$

Note that we did not need to use the equation $f(x) = 4x^3 + 7$, nor did we need to find a formula for $f^{-1}(x)$.

EXAMPLE 5 An equation to solve that involves inverse functions

Solve the following equation for x, given that the domain of both f and f^{-1} is $(-\infty, \infty)$ and that $f(1) = -2$:

$$3 + f^{-1}(x - 1) = 4$$

SOLUTION

$$
\begin{aligned}
3 + f^{-1}(x - 1) &= 4 \\
f^{-1}(x - 1) &= 1 \\
f[f^{-1}(x - 1)] &= f(1) && \text{applying } f \text{ to both sides} \\
x - 1 &= f(1) && \text{definition of inverse function} \\
x &= 1 + f(1) = 1 + (-2) = -1
\end{aligned}
$$

We'll postpone a discussion until the end of this section as to which functions have inverses and which do not. For functions that do have inverses, however, there's a simple method that can often be used to determine those inverses. This method is illustrated in Example 6.

EXAMPLE 6 Determining the inverse function

Let $f(x) = \dfrac{x - 1}{3x + 5}$. Find $f^{-1}(x)$.

SOLUTION

Step 1 We begin by writing the given equation as

$$y = \frac{x - 1}{3x + 5} \tag{2}$$

We know that f^{-1} interchanges the inputs and outputs of f, so to determine $y = f^{-1}(x)$, we first interchange the x's and y's in equation (2) and then solve for y. Interchanging gives us

$$x = \frac{y - 1}{3y + 5} \tag{3}$$

Step 2 Now we solve equation (3) for y.

$$\begin{aligned} 3xy + 5x &= y - 1 && \text{multiplying both sides of equation (3) by } 3y + 5 \\ 3xy - y &= -5x - 1 && \text{rearranging} \\ y(3x - 1) &= -5x - 1 \\ y &= \frac{-5x - 1}{3x - 1} \end{aligned}$$

Thus the inverse function is $y = (-5x - 1)/(3x - 1)$. Using function notation, we can write this result as

$$f^{-1}(x) = \frac{-5x - 1}{3x - 1}$$

REMARK Actually, we should call this result only a candidate for f^{-1}, because certain technical matters remain to be discussed. However, if you compute $f[f^{-1}(x)]$ and $f^{-1}[f(x)]$ and find they both do equal x, then the matter is settled, by definition. Exercise 21 at the end of this section asks you to carry out those calculations.

For reference, in the box that follows we summarize our procedure for calculating $f^{-1}(x)$.

To Find $f^{-1}(x)$ for the Function $y = f(x)$

Step 1 Interchange x and y in the equation $y = f(x)$.

Step 2 Solve the resulting equation for y.

CAUTION This method does not work for every function with an inverse. See the discussion of exponential and logarithmic functions in Chapter 5 and the discussions of trigonometric and inverse trigonometric functions in Chapters 6–8.

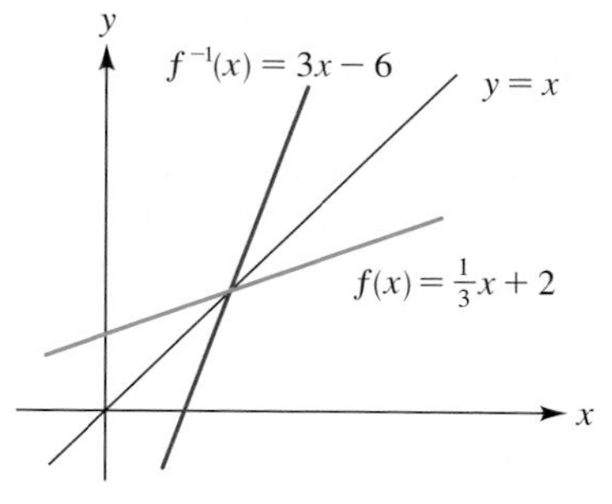

Figure 3

There is a certain type of symmetry that always occurs when we graph a function and its inverse. Consider, for example, the pair of inverse functions $f(x) = \frac{1}{3}x + 2$ and $f^{-1}(x) = 3x - 6$ that are graphed in Figure 3. (In Example 2 we verified that these were indeed inverse functions.) Note that the graphs of f and f^{-1} appear to be *symmetric about the line* $y = x$. That is, if the page were folded along the dashed line $y = x$ and held up to the light, then the graphs of f and f^{-1} would coincide. To say it differently, the graphs of f and f^{-1} are mirror images of one another about the line $y = x$. As background for the remainder of our work on inverse functions, we give a definition of symmetry about the line $y = x$ in the box that follows. (After you read the definition, note that the ideas involved are the same as those for symmetry about the x- or y-axis, as presented in Section 1.7.)

DEFINITION **Symmetry about the Line y = x**

Refer to Figure 4. Two points P and Q are symmetric about the line $y = x$ provided that

1. $\overline{PQ}$ is perpendicular to the line $y = x$; and
2. the points P and Q are equidistant from the line $y = x$.

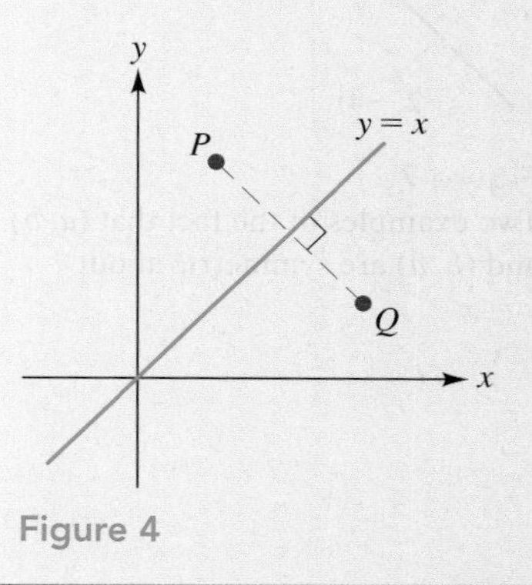

Figure 4

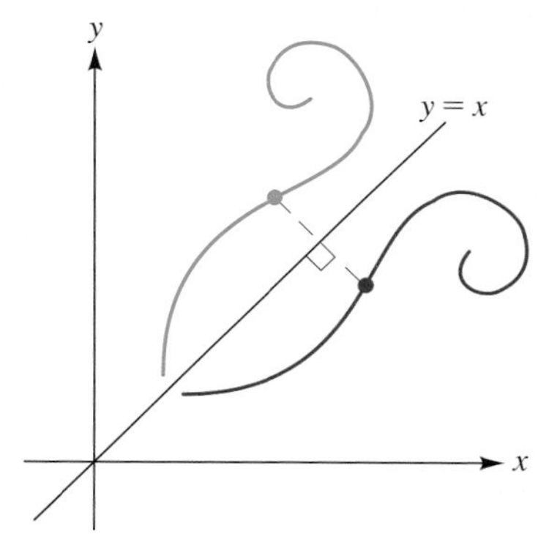

Figure 5
The two curves are symmetric about the line $y = x$.

This definition says that P and Q are symmetric about the line $y = x$ if $y = x$ is the perpendicular bisector of line segment $\overline{PQ}$. In Figure 4, we say that the two points P and Q are **reflections** of each other about the line $y = x$ and that $y = x$ is the **axis of symmetry.** In addition, we say that two curves are symmetric about $y = x$ if each point on one curve is the reflection of a corresponding point on the other curve, and vice versa; see Figure 5.

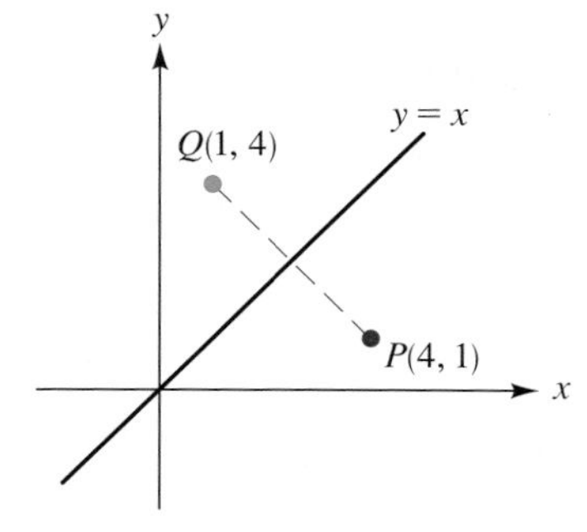

Figure 6

EXAMPLE 7 **Showing two given points are symmetric about y = x**

Verify that the points $P(4, 1)$ and $Q(1, 4)$ are symmetric about the line $y = x$; see Figure 6.

SOLUTION

We have to show that the line $y = x$ is the perpendicular bisector of the line segment $\overline{PQ}$. Now, the slope of $\overline{PQ}$ is

$$m = \frac{4 - 1}{1 - 4} = -1$$

We know that the slope of the line $y = x$ is 1. Since these two slopes are negative reciprocals, we conclude that the line $y = x$ is perpendicular to $\overline{PQ}$. Next, we must show that the line $y = x$ passes through the midpoint of $\overline{PQ}$. However (as you should check for yourself), the midpoint of $\overline{PQ}$ is $(\frac{5}{2}, \frac{5}{2})$, which does lie on the line $y = x$. (Why?) In summary, then, we've shown that the line $y = x$ passes through the midpoint of $\overline{PQ}$ and is perpendicular to $\overline{PQ}$. Thus P and Q are symmetric about the line $y = x$, as we wished to show.

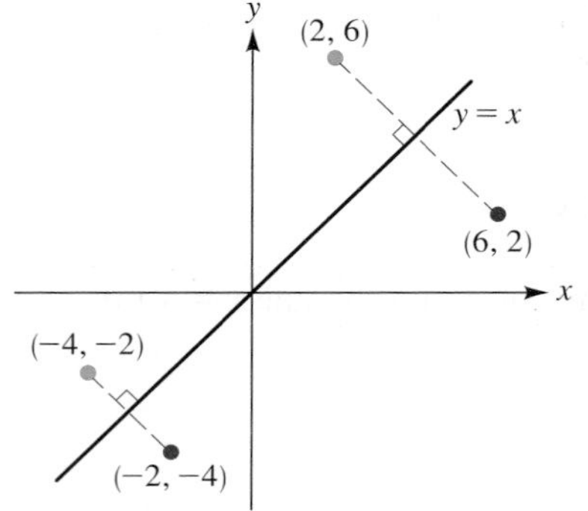

Figure 7
Two examples of the fact that (a, b) and (b, a) are symmetric about $y = x$

Just as we've shown that (4, 1) and (1, 4) are symmetric about the line $y = x$, we can also show that in general, (a, b) and (b, a) are always symmetric about the line $y = x$. Figure 7 displays two examples of this, and Exercise 51 at the end of this section asks you to supply a proof for the general situation.

Now let's see why the graphs of f and f^{-1} are always mirror images of each other about the line $y = x$. First, recall that the function f^{-1} interchanges the inputs and outputs of f. Thus, (a, b) is on the graph of f if and only if (b, a) is on the graph of f^{-1}. But as we have just stated, the points (a, b) and (b, a) are symmetric about the line $y = x$. It follows therefore that the graphs of f and f^{-1} are symmetric about the line $y = x$. For reference, we restate this useful fact in the box that follows.

PROPERTY SUMMARY The Graphs of f and f^{-1}

The graphs of $y = f(x)$ and $y = f^{-1}(x)$ are symmetric about the line $y = x$.

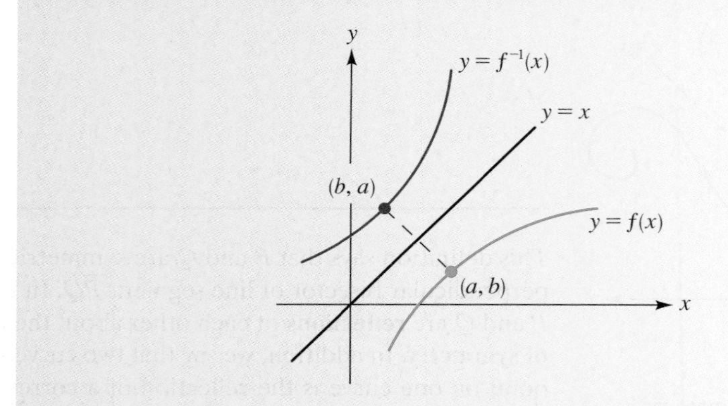

EXAMPLE 8 Using symmetry to sketch the graph of an inverse function

The graph of a function f consists of the line segment joining the points $(-2, -3)$ and $(-1, 4)$, as shown in Figure 8. Sketch a graph of f^{-1}.

SOLUTION

The graph of f^{-1} is obtained by reflecting the graph of f about the line $y = x$. The reflection of $(-2, -3)$ is $(-3, -2)$, and the reflection of $(-1, 4)$ is $(4, -1)$. To graph f^{-1}, then, we plot the reflected points $(-3, -2)$ and $(4, -1)$ and connect them with a line segment, as shown in Figure 9.

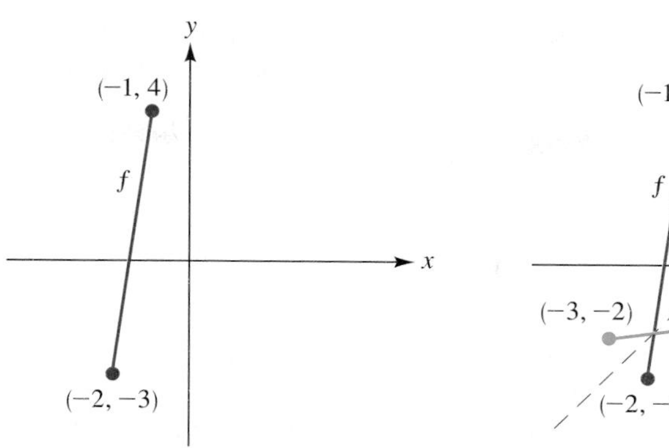

Figure 8

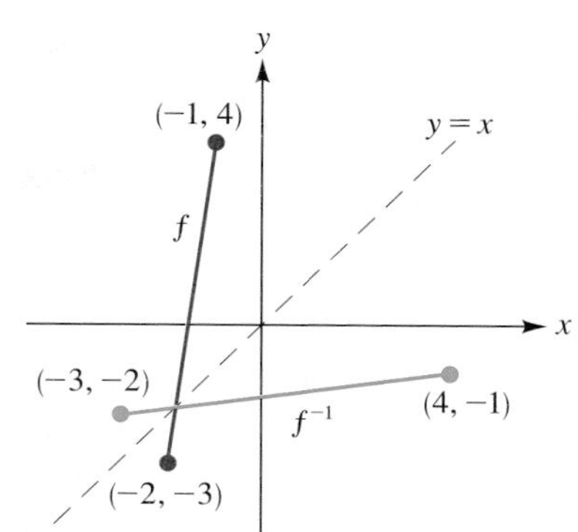

Figure 9

NOTE When a linear function has an inverse the inverse function is also linear. See Exercise 52 at the end of this section. For reference, Figure 9 also shows the graphs of f and $y = x$.

EXAMPLE 9 Finding an inverse function and its graph

Let $g(x) = x^3$. Find $g^{-1}(x)$, and then, on the same set of axes, sketch the graphs of g, g^{-1}, and $y = x$.

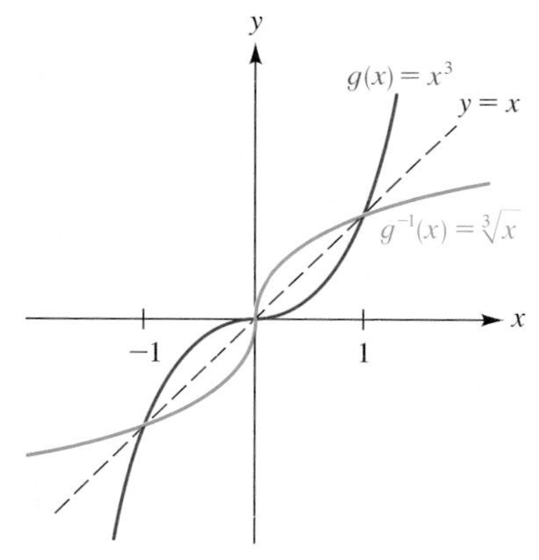

Figure 10

SOLUTION

We begin by writing $y = x^3$, then switching x and y and solving for y. We first have

$$x = y^3$$

To solve this equation for y, we take the cube root of both sides to obtain

$$\sqrt[3]{x} = y$$

So the inverse function is $g^{-1}(x) = \sqrt[3]{x}$. We could graph the function g^{-1} by plotting points, but the easier way is to reflect the graph of $g(x) = x^3$ about the line $y = x$; see Figure 10.

Earlier in this section, we mentioned that not every function has an inverse function. For instance, Tables 1(a) and 1(b) indicate what happens if we begin with the function defined by $F(x) = x^2$ and interchange the inputs and outputs. The resulting table, Table 1(b), does not define a function. For instance, the input 4 in Table 1(b) produces two distinct outputs, 2 and -2, but the definition of a function requires exactly one output for a given input.

Why didn't this difficulty arise when we looked for the inverse of $f(x) = 2x$ at the beginning of this section? The answer is that there is an essential difference between the functions defined by $f(x) = 2x$ and $F(x) = x^2$. With $f(x) = 2x$, different inputs never yield the same output. To put it another way, for the function $f(x) = 2x$, the only time it can happen that $f(a) = f(b)$ is when $a = b$. This condition guarantees

TABLE 1(a)
$F(x) = x^2$

x	y
1	1
2	4
3	9
−1	1
−2	4
−3	9

TABLE 1(b)
Reversing the Inputs and Outputs

x	y
1	1
4	2
9	3
1	−1
4	−2
9	−3

that interchanging the inputs and outputs of f will yield a function. With $F(x) = x^2$, however, it *can* happen that $F(a) = F(b)$ and yet $a \neq b$. For instance,

$$F(2) = F(-2) \qquad \text{because } 2^2 = 4 = (-2)^2$$

but, of course, $2 \neq -2$.

It's useful to have a name for functions such as $f(x) = 2x$, in which distinct inputs always yield distinct outputs. We call these types of functions *one-to-one* functions.

DEFINITION **One-to-One**

A function f is **one-to-one** provided that the following condition holds for all a and b in the domain of f:

$$\text{If } f(a) = f(b) \quad \text{then } a = b$$

EXAMPLES

$f(x) = 2x$ is one-to-one.

$F(x) = x^2$ is not one-to-one because $F(2) = F(-2)$, yet $2 \neq -2$.

Using graphs, there is an easy way to tell which functions are one-to-one:

Horizontal Line Test

A function f is one-to-one if and only if each horizontal line intersects the graph of $y = f(x)$ in at most one point.

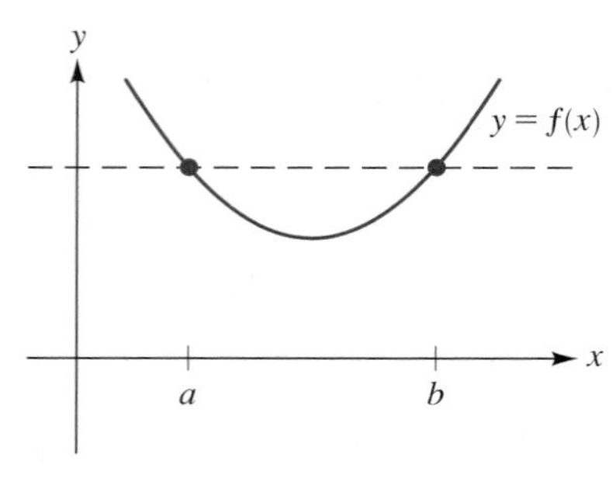

Figure 11

Figure 11 will help you to see why the horizontal line test is valid. The figure shows a horizontal line intersecting the graph of a function f at two distinct points with x-coordinates a and b. From the graph, we see that $f(a) = f(b)$, even though $a \neq b$. Thus the function represented in Figure 11 is not one-to-one.

EXAMPLE 10 Using the horizontal line test

Use the horizontal line test to determine which functions in Figure 12 are one-to-one.

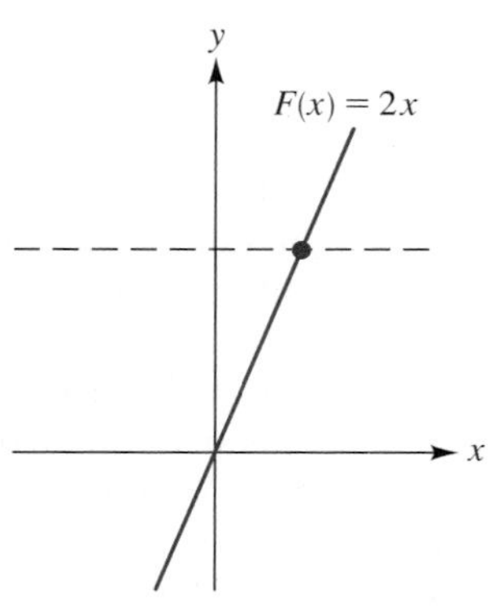

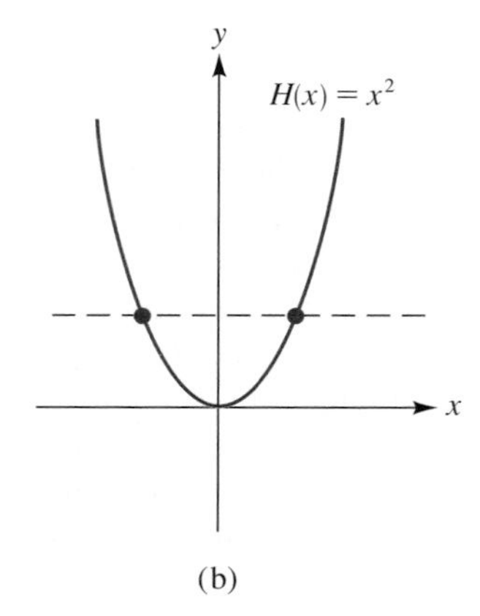

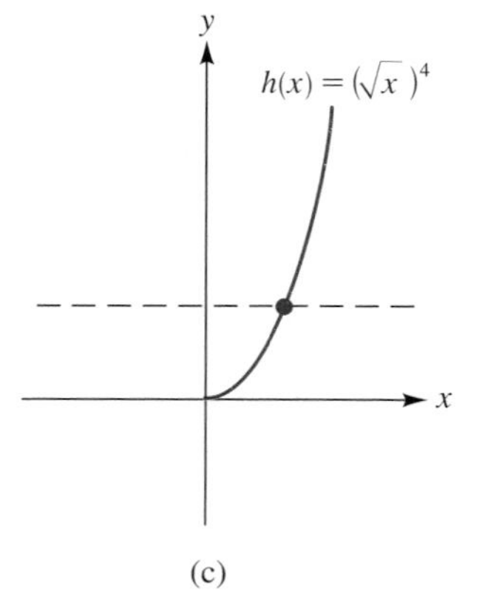

Figure 12 (a) (b) (c)

SOLUTION

The horizontal line test tells us that F and h are one-to-one, but H is not.

The following theorem tells us that the functions with inverses are precisely the one-to-one functions. (The proof of the theorem isn't difficult, but we'll omit it here.)

THEOREM

A function f has an inverse function f^{-1} if and only if f is one-to-one.

EXAMPLE 11 Using one-to-one to determine whether a function has an inverse

Which of the three functions in Figure 12 have inverse functions?

SOLUTION

According to our theorem, the functions with inverses are those that are one-to-one. In Example 10, we saw that F and h were one-to-one but H was not. So F and h each have inverses, but H does not.

EXERCISE SET 3.6

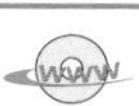

A

1. If h and k are a pair of inverse functions, then (fill in the blanks).
  (a) $h[k(x)] =$ _____ for every x in the domain of ______.
  (b) $k[h(x)] =$ _____ for every x in the domain of ______.

2. Let $f(x) = x^2$ and $g(x) = 1/x$. Compute $f[g(x)]$ and $g[f(x)]$, and note that the results are identical. Then say why f and g do *not* qualify as a pair of inverse functions.

3. Verify that the given pairs of functions are inverse functions, as in Example 2.
  (a) $f(x) = 3x$; $g(x) = x/3$
  (b) $f(x) = 4x - 1$; $g(x) = \frac{1}{4}x + \frac{1}{4}$
  (c) $g(x) = \sqrt{x}$; $h(x) = x^2$ [Assume that the domain of both g and h is $[0, \infty)$.]

4. Which pairs of functions are inverses?
  (a) $f(x) = -3x + 2$; $g(x) = \frac{2}{3} - \frac{1}{3}x$
  (b) $F(x) = 2x + 1$; $G(x) = \frac{1}{2}x - 1$
  (c) $G(x) = x^3$; $H(x) = 1 - x^3$
  (d) $f(t) = t^3$; $g(t) = \sqrt[3]{t}$

Ⓖ **5. (a)** Use a graphing utility to graph the following three functions on the same set of axes.

$$f(x) = \sqrt[3]{x^3 + 3} - 1$$
$$g(x) = \sqrt[3]{x^3 + 1} - 3$$
$$h(x) = \sqrt[3]{(x + 1)^3 - 3}$$

If necessary, adjust the picture so that it shows true proportions.
  (b) Which two graphs appear to be symmetric about the line $y = x$? Use algebra (as in Example 2) to verify that the two functions are indeed inverses.

Ⓖ **6. (a)** Use a graphing utility to graph both $g(x) = \sqrt[3]{2 - x^3}$ and $y = x$ on the same set of axes.
  (b) What kind of symmetry do you observe? What does this tell you about the inverse for g? [If you are uncertain how to answer this last question, try coming back to it after working part (c).]
  (c) Starting with the equation $g(x) = \sqrt[3]{2 - x^3}$, follow the two-step procedure on page 198 to find $g^{-1}(x)$. What do you observe?

In Exercises 7 and 8, suppose that f and g are a pair of inverse functions.

7. If $f(7) = 12$, what is $g(12)$? (If you need a hint, reread Example 3.)

8. If $g(-2) = 0$, what is $f(0)$?

In Exercises 9 and 10, let $f(x) = x^3 + 2x + 1$, and assume that f^{-1} exists and has domain $(-\infty, \infty)$. Simplify each expression (as in Example 4).

9. (a) $f[f^{-1}(4)]$
  (b) $f^{-1}[f(-1)]$
  (c) $(f \circ f^{-1})(\sqrt{2})$
  (d) $f[f^{-1}(t + 1)]$

10. **(a)** $f(0)$
(b) $f^{-1}(1)$ *Hint*: Use the result in part (a).
(c) $f(-1)$ **(d)** $f^{-1}(-2)$

11. Let $f(x) = 2x + 1$.
(a) Find $f^{-1}(x)$.
(b) Calculate $f^{-1}(5)$ and $1/f(5)$. Are your answers the same?

12. Let $g(t) = \dfrac{1}{t} + 1$.
(a) Find $g^{-1}(t)$.
(b) Calculate $g^{-1}(2)$ and $1/g(2)$, and note that the answers are not the same.

13. Let $f(x) = 3x - 1$.
(a) Compute $f^{-1}(x)$.
(b) Verify that $f[f^{-1}(x)] = x$ and that $f^{-1}[f(x)] = x$.
(c) On the same set of axes, sketch the graphs of f, f^{-1}, and the line $y = x$. Note that the graphs of f and f^{-1} are symmetric about the line $y = x$.

14. Follow Exercise 13, but use $f(x) = \frac{1}{3}x - 2$.

15. Follow Exercise 13, but use $f(x) = \sqrt{x - 1}$. [The domain of f^{-1} will be $[0, \infty)$.]

16. Follow Exercise 13, but use $f(x) = 1/x$.

17. Let $f(x) = (x + 2)/(x - 3)$.
(a) Find the domain and range of the function f.
(b) Find $f^{-1}(x)$.
(c) Find the domain and range of the function f^{-1}. What do you observe?

18. Let $f(x) = (2x - 3)/(x + 4)$. Find $f^{-1}(x)$. Find the domain and range for f and f^{-1}. What do you observe?

19. Let $f(x) = 2x^3 + 1$. Find $f^{-1}(x)$.

20. In our discussion at the beginning of this section, we considered the functions $f(x) = 2x$ and $g(x) = x/2$. Verify that $f[g(x)] = x$ and that $g[f(x)] = x$. On the same set of axes, sketch the graphs of f, g, and $y = x$.

21. This exercise refers to the comments made at the end of Example 6. Compute $f[f^{-1}(x)]$ and $f^{-1}[f(x)]$ using the functions $f(x) = (x - 1)/(3x + 5)$ and $f^{-1}(x) = (-5x - 1)/(3x - 1)$. By actually carrying out the calculations, you'll find that the result in each case is x, which *proves* these are indeed inverse functions.

22. Let $f(x) = (x - 3)^3 - 1$.
(a) Compute $f^{-1}(x)$.
(b) On the same set of axes, sketch the graphs of f, f^{-1}, and the line $y = x$. *Hint*: Sketch f using the ideas of Section 3.4. Then reflect the graph of $y = f(x)$ in the line $y = x$ to obtain the graph of $y = f^{-1}(x)$.

For Exercises 23 and 24, refer to the following figure, which shows the graph of a function f. (The axes are marked off in one-unit intervals.) In each case, graph the indicated function.

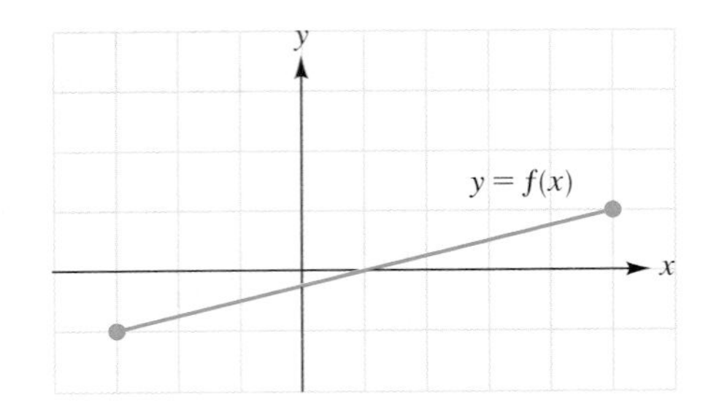

23. **(a)** $y = f^{-1}(x)$ **(b)** $y = f^{-1}(x - 2)$ **(c)** $y = f^{-1}(x) - 2$ **(d)** $y = f^{-1}(x - 2) - 2$

24. **(a)** $y = f^{-1}(-x)$ **(b)** $y = -f^{-1}(x)$ **(c)** $y = -f^{-1}(-x)$ **(d)** $y = -f^{-1}(-x) - 1$

In Exercises 25–28, assume that the domain of f and f^{-1} is $(-\infty, \infty)$. Solve the equation for x or for t (whichever is appropriate) using the given information.

25. **(a)** $7 + f^{-1}(x - 1) = 9$; $f(2) = 6$
(b) $4 + f(x + 3) = -3$; $f^{-1}(-7) = 0$

26. **(a)** $f^{-1}(2x + 3) = 5$; $f(5) = 13$
(b) $f(1 - 2x) = -4$; $f^{-1}(-4) = -5$

27. $f^{-1}\left(\dfrac{t + 1}{t - 2}\right) = 12$; $f(12) = 13$

28. $f\left(\dfrac{1 - 2t}{1 + 2t}\right) = 7$; $f^{-1}(7) = -3$

For Exercises 29–33, use the horizontal line test to determine whether the function is one-to-one (and therefore has an inverse). (You should be able to sketch the graph of each function on your own, without using a graphing utility.)

29. $f(x) = -x^2 + 1$ **30.** $g(x) = x^3 - 1$

31. $g(x) = 5$ (for all x)

32. $f(x) = \begin{cases} x^2 & \text{if } -1 \le x \le 0 \\ x^2 + 1 & \text{if } x > 0 \end{cases}$

33. $g(x) = \begin{cases} x^2 & \text{if } -1 \le x < 0 \\ x^2 + 1 & \text{if } x \ge 0 \end{cases}$

34. Which of the six basic functions graphed in Figure 7 in Section 3.2 are one-to-one?

Ⓖ *In Exercises 35–40, use a graphing utility to graph each function and then apply the horizontal line test to see whether the function is one-to-one.*

35. $y = x^2 + 2x$ **36.** $y = x^3 + 2x$

37. $y = 2x^3 + x^2$ *Suggestion*: Begin with the standard viewing rectangle and then use a zoom-in view. (First looks can be deceiving.)

38. $y = 0.01x^4 - 1$ **39.** $y = 2x^5 + x - 1$

40. **(a)** $f(x) = x^3 + x^2 + x$
(b) $g(x) = x^3 - x^2 + x$
(c) $h(x) = x^3 - x^2 - x$

B

Ⓖ **41. (a)** In the standard viewing rectangle, graph two functions

$$f(x) = x^3 + 1.73x^2 + x + 2$$

and

$$g(x) = x^3 + 1.74x^2 + x + 2$$

(b) Use the zoom feature of your graphing utility to find out whether either function is one-to-one.

42. Let $f(x) = \sqrt{x}$.

(a) Find $f^{-1}(x)$. What is the domain of f^{-1}? [The domain is not $(-\infty, \infty)$.]

(b) In each case, determine whether the given point lies on the graph of f or f^{-1}.

(i) $(4, 2)$
(ii) $(2, 4)$
(iii) $(5, \sqrt{5})$
(iv) $(\sqrt{5}, 5)$
(v) $(a, f(a))$, where $a \geq 0$
(vi) $(f(a), a)$
(vii) $(b, f^{-1}(b))$, where $b \geq 0$
(viii) $(f^{-1}(b), b)$

43. In the following figure, determine the coordinates of the points A, B, C, and D. Express your answers in terms of the function f and the number a. (Each dashed line through A is parallel to an axis.)

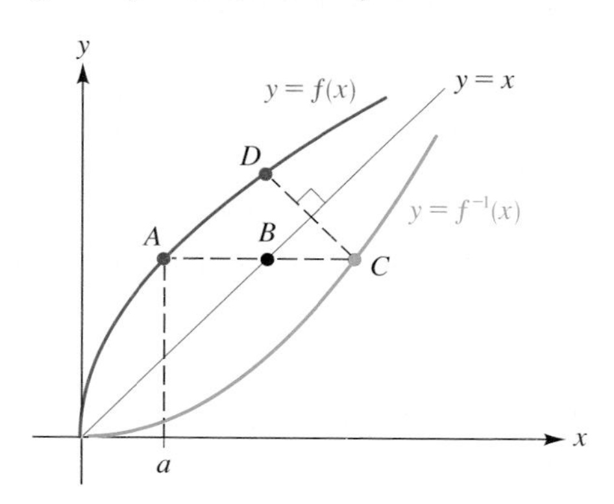

44. In the following figure, determine the coordinates of the points A, B, C, D, E, and F. Specify the answers in terms of the functions f and f^{-1} and the number b.

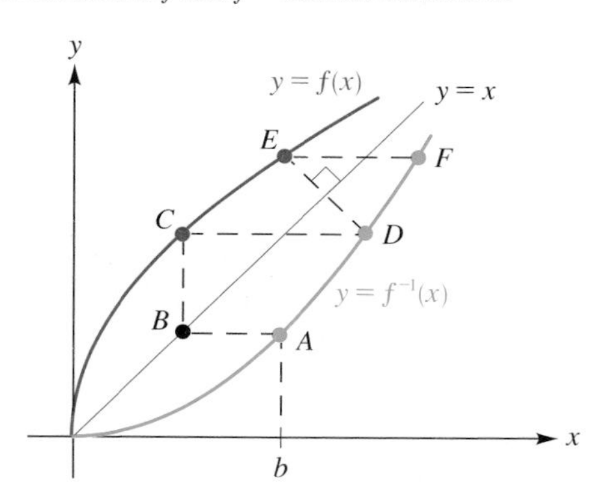

45. Let $f(x) = 3/(x - 1)$.

(a) Find the average rate of change of f on the interval $[4, 9]$.

(b) Find $f^{-1}(x)$ and then compute the average rate of change of f^{-1} on the interval $[f(4), f(9)]$. What do you observe?

46. The accompanying figure shows the graph of a function f on the interval $[a, b]$. If the average rate of change of f on the interval $[a, b]$ is k, show that the average rate of change of the function f^{-1} on the interval $[f(a), f(b)]$ is $1/k$, where $k \neq 0$.

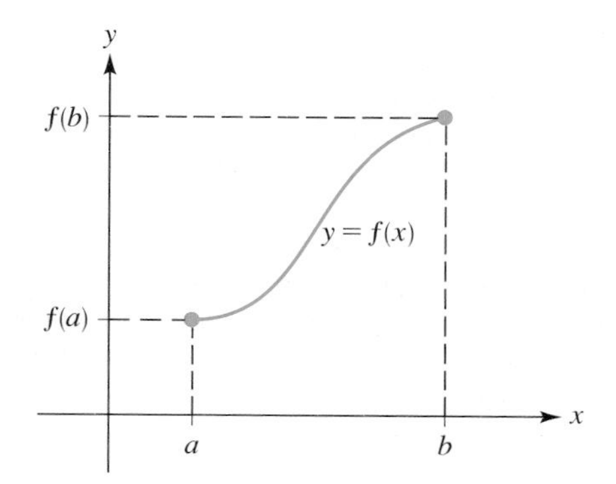

47. Suppose that (a, b) is a point on the graph of $y = f(x)$. Match the functions defined in the left-hand column with the points in the right-hand column. For example, the appropriate match for (a) in the left-hand column is determined as follows. The graph of $y = f(x) + 1$ is obtained by translating the graph of $y = f(x)$ up one unit. Thus the point (a, b) moves up to $(a, b + 1)$, and consequently, (E) is the appropriate match for (a).

(a) $y = f(x) + 1$	**(A)** $(-a, b)$
(b) $y = f(x + 1)$	**(B)** (b, a)
(c) $y = f(x - 1) + 1$	**(C)** $(a - 1, b)$
(d) $y = f(-x)$	**(D)** $(-b, a + 1)$
(e) $y = -f(x)$	**(E)** $(a, b + 1)$
(f) $y = -f(-x)$	**(F)** $(1 - a, b)$
(g) $y = f^{-1}(x)$	**(G)** $(-a, -b)$
(h) $y = f^{-1}(x) + 1$	**(H)** $(-b, 1-a)$
(i) $y = f^{-1}(x - 1)$	**(I)** $(b, -a)$
(j) $y = f^{-1}(-x) + 1$	**(J)** $(a, -b)$
(k) $y = -f^{-1}(x)$	**(K)** $(b + 1, a)$
(l) $y = -f^{-1}(-x) + 1$	**(L)** $(a + 1, b + 1)$
(m) $y = 1 - f^{-1}(x)$	**(M)** $(b, a + 1)$
(n) $y = f(1 - x)$	**(N)** $(b, 1 - a)$

48. Let $f(x) = \dfrac{1}{1 - x}$. Show that the function $f \circ f$ is the inverse of f.

49. Let $f(x) = -\dfrac{2x + 2}{x}$.

(a) Find $f[f(x)]$.

Ⓖ **(b)** Use a graphing utility to graph $y = f[f(x)]$. Display the graph using true proportions. What type of symmetry does the graph appear to have?

(c) The result in part (b) suggests that the inverse of the function $f \circ f$ is again $f \circ f$. Use algebra to show that this is indeed correct.

50. In this exercise you'll investigate the inverse of a composite function. In parts (b) and (c), which involve graphing, be sure to use the same size unit and scale on both axes so that symmetry about the line $y = x$ can be checked visually.

(a) Let $f(x) = 2x + 1$ and $g(x) = \frac{1}{4}x - 3$. Compute each of the following:

(i) $f(g(x))$
(ii) $g(f(x))$
(iii) $f^{-1}(x)$
(iv) $g^{-1}(x)$
(v) $f^{-1}(g^{-1}(x))$
(vi) $g^{-1}(f^{-1}(x))$

Ⓖ **(b)** On the same set of axes, graph the two answers that you obtained in (i) and (v) of part (a). Note that the graphs are *not* symmetric about $y = x$. The conclusion here is that the inverse function for $f(g(x))$ is not $f^{-1}(g^{-1}(x))$.

Ⓖ **(c)** On the same set of axes, graph the two answers that you obtained in (i) and (vi) of part (a); also put the line $y = x$ into the picture. Note that the two graphs *are* symmetric about the line $y = x$. The conclusion here is that the inverse function for $f(g(x))$ is $g^{-1}(f^{-1}(x))$. In fact, it can be shown that this result is true in general. For reference, then, we summarize this fact about the inverse of a composite function in the box that follows.

> **Inverse of a Composition**
>
> $$(f \circ g)^{-1} = g^{-1} \circ f^{-1}$$

51. Use the method of Example 7 to show that the points (a, b) and (b, a) are symmetric about the line $y = x$.

52. In this exercise you will show that if a linear function has an inverse, then the inverse function is also linear. Let $f(x) = mx + b$, where m and b are constants, with $m \neq 0$.

(a) Show f^{-1} exists. *Hint:* Show f is one-to-one.
(b) Find a formula for $f^{-1}(x)$.
(c) Explain why f^{-1} is linear. In particular, what are the slope and y-intercept of the graph of $y = f^{-1}(x)$?
(d) What happens when $m = 0$?

MINI PROJECT

A FREQUENTLY ASKED QUESTION ABOUT INVERSES

Student: If $f[g(x)] = x$, isn't it automatically true then that $g[f(x)] = x$?
Professor: No, that's only when you have a pair of inverse functions.
Student: But in Example 2 and in similar homework problems, whenever I find that $f[g(x)] = x$, it *always* turns out that $g[f(x)] = x$.

This project provides an example of two functions f and g satisfying the following two conditions:

1. The equation $g[f(x)] = x$ holds for every real number x.
2. There are some real numbers x for which $f[g(x)] \neq x$.

Figures A and B show these two functions f and g; but you are not told which is which. Determine which function is f and which is g. In investigating $f[g(x)]$ and $g[f(x)]$, be sure you cover all cases.

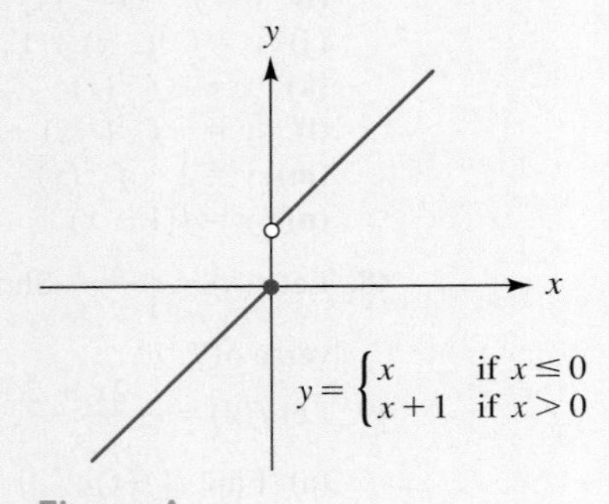

Figure A

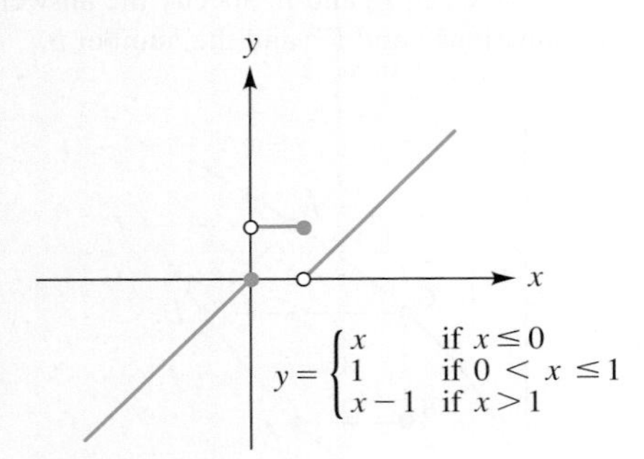

Figure B

Chapter 3 Summary of Principal Terms and Formulas

Terms or notation	Page reference	Comments
1. Function	132	Given two nonempty sets A and B, a function from A to B is a rule of correspondence that assigns to each element of A exactly one element in B.
2. Domain	133	The domain of a function is the set of all inputs for that function. If f is a function from A to B, then the domain is the set A.
3. Range	133	The range of a function from A to B is the set of all elements in B that are actually used as outputs.
4. $f(x)$	136	Given a function f, the notation $f(x)$ denotes the output that results from the input x.
5. Graph of a function	146	The graph of a function f consists of those points (x, y) such that x is in the domain of f and $y = f(x)$.
6. Vertical line test	147	A graph in the x-y plane represents a function $y = f(x)$ provided that any vertical line intersects the graph in at most one point.
7. Turning point	159	A turning point on a graph is a point where the graph changes from rising to falling, or vice versa. See, for example, Figure 1 on page 159.
8. Maximum value and minimum value	159	An output $f(a)$ is a maximum value of the function f if $f(a) \geq f(x)$ for every x in the domain of f. An output $f(b)$ is a minimum value if $f(b) \leq f(x)$ for every x in the domain of f.
9. Increasing function	159	A function f is increasing on an interval if the following condition holds: If a and b are in the interval and $a < b$, then $f(a) < f(b)$. Geometrically, this means that the graph is rising as we move in the positive x-direction.
10. Decreasing function	159	A function f is decreasing on an interval if the following condition holds: If a and b are in the interval and $a < b$, then $f(a) > f(b)$. Geometrically, this means that the graph is falling as we move in the positive x-direction.
11. Average rate of change	162	The average rate of change, $\Delta f/\Delta x$, of a function f on an interval $[a, b]$ is the slope of the line segment through the two points $(a, f(a))$ and $(b, f(b))$: $$\frac{\Delta f}{\Delta x} = \frac{f(b) - f(a)}{b - a}$$

Terms or notation	Page reference	Comments
12. Techniques for graphing functions	172	The box on page 172 lists six techniques involving translation and reflection. For a comprehensive example, see Example 7 on pages 177–178.
13. $f \circ g$	183	The composition of two functions f and g: $(f \circ g)(x) = f[g(x)]$
14. Iterates	188	Given a function f and an input x_0, the iterates of x_0 are the numbers $f(x_0), f(f(x_0)), f(f(f(x_0))), \ldots$. The number $f(x_0)$ is called the first iterate, the number $f(f(x_0))$ is called the second iterate, and so on. The orbit of x_0 under the function f is the list of numbers $$x_0, f(x_0), f(f(x_0)), \ldots$$
15. Inverse functions	195	Two functions f and g are inverses of one another provided that the following two conditions are met. First, $f[g(x)] = x$ for each x in the domain of g. Second, $g[f(x)] = x$ for each x in the domain of f.
16. f^{-1}	196	f^{-1} denotes the inverse function for f. *Note*: In this context, f^{-1} does not mean $1/f$.
17. One-to-one function	202	A function f is said to be one-to-one provided that distinct inputs always yield distinct outputs. Geometrically, this means that every horizontal line intersects the graph of $y = f(x)$ in at most one point. Algebraically, this means that f satisfies the following condition: If a and b are in the domain of f, and $f(a) = f(b)$, then $a = b$. The relationship between inverse functions and one-to-one functions is this: A function f has an inverse if and only if f is one-to-one.

Writing Mathematics

1. Section 3.6 on inverse functions begins with the following quotation from Professor R. P. Boas:

> If anybody ever told me why the graph of $y = x^{1/2}$ is the reflection of the graph of $y = x^2$ $[x \geq 0]$ in a 45° line, it didn't sink in. To this day, there are textbooks that expect students to think that it is so obvious as to need no explanation.

Show your instructor that you are not in the dark about inverse functions. Write out an explanation of why the graphs of $y = x^{1/2}$ and $y = x^2$ $(x \geq 0)$ are indeed reflections of one another in a 45° line. (As preparation for your writing, you'll probably need to look over Section 3.6 and make a few notes.)

2. Decide which of the following rules are functions. Write out your reasons in complete sentences. (In each case, assume that the domain is the set of students in your school.)
   (a) F is the rule that assigns to each person his or her brother.
   (b) G is the rule that assigns to each person his or her aunt.
   (c) H is the rule that assigns to each person his or her birth-mother.
   (d) K is the rule that assigns to each person his or her mother or father.

Chapter 3 Review Exercises

1. **(a)** Find the domain of the function defined by $f(x) = \sqrt{15 - 5x}$.
   (b) Find the range of the function defined by $g(x) = (3 + x)/(2x - 5)$.
2. Let $f(x) = 3x^2 - 4x$ and $g(x) = 2x + 1$. Compute each of the following:
   (a) $(f - g)(x)$
   (b) $(f \circ g)(x)$
   (c) $f[g(-1)]$
3. A **linear function** is a function defined by an equation of the form $F(x) = ax + b$, where a and b are real numbers. Suppose f and g are linear functions. Is the function $f \circ g$ a linear function?
4. A **quadratic function** is a function defined by an equation of the form $F(x) = ax^2 + bx + c$, where a, b, and c are real numbers and $a \neq 0$.
   (a) Give an example of quadratic functions f and g for which $f + g$ is also a quadratic function.
   (b) Give an example of quadratic functions f and g for which $f + g$ is not a quadratic function.
   (c) Suppose f and g are quadratic functions. Is the function $f \circ g$ a quadratic function?
5. **(a)** Compute $\dfrac{F(x) - F(a)}{x - a}$ given that $F(x) = 1/x$.
   (b) Compute $\dfrac{g(x + h) - g(x)}{h}$ for the function defined by $g(x) = x - 2x^2$.
6. The y-intercept for the graph of $y = f(x)$ is 4. What is the y-intercept for the graph of $y = -f(x) + 1$?
7. **(a)** Find $g^{-1}(x)$ given that $g(x) = \dfrac{1 - 5x}{3x}$.
   (b) The figure displays the graph of a function f. Sketch the graph of f^{-1}.

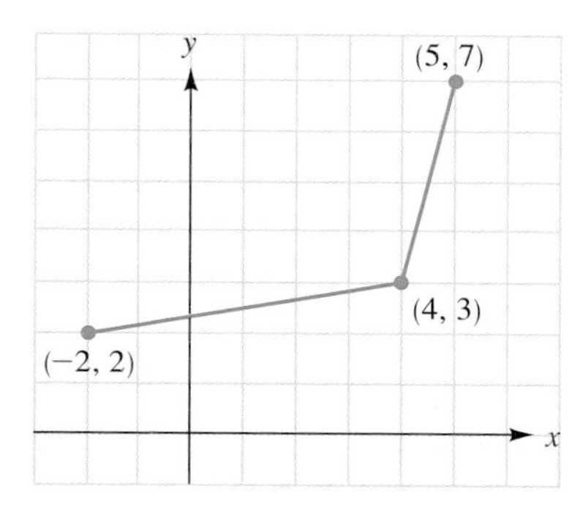

8. Refer to the function f in Exercise 7(b).
   (a) Compute $\Delta f/\Delta x$ on $[-2, 5]$.
   (b) Find the average rate of change of f^{-1} on $[2, 7]$.
9. Graph each function and specify the intercepts.
   (a) $y = |x + 2| - 3$
   (b) $y = \dfrac{1}{x + 2} - 1$
10. Refer to the following graph of $y = g(x)$. The domain of g is $[-5, 2]$.
    (a) What are the coordinates of the turning point(s)?
    (b) What is the maximum value of g?
    (c) Which input yields a minimum value for g?
    (d) On which interval(s) is g increasing?
    (e) Compute $\Delta g/\Delta x$ for each of the following intervals: $[-5, -3]$; $[-3, -2]$; $[-5, 2]$.

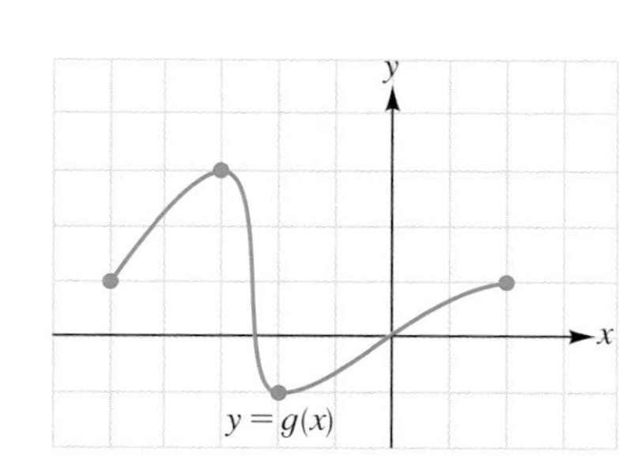

11. Let $f(x) = 3x^2 - 2x$.
(a) Find $f(-1)$. **(b)** Find $f(1 - \sqrt{2})$.

12. Graph the function G defined by

$$G(x) = \begin{cases} \sqrt{1 - x^2} & \text{if } -1 \leq x < 0 \\ \sqrt{x} & \text{if } x \geq 0 \end{cases}$$

13. Express the slope of a line passing through the points $(5, 25)$ and $(5 + h, (5 + h)^2)$ as a function of h.

14. Graph the function $y = |9 - x^2|$. Specify symmetry and intercepts.

15. The following figure shows the graph of a function $y = f(x)$. Sketch the graph of $y = f(-x)$.

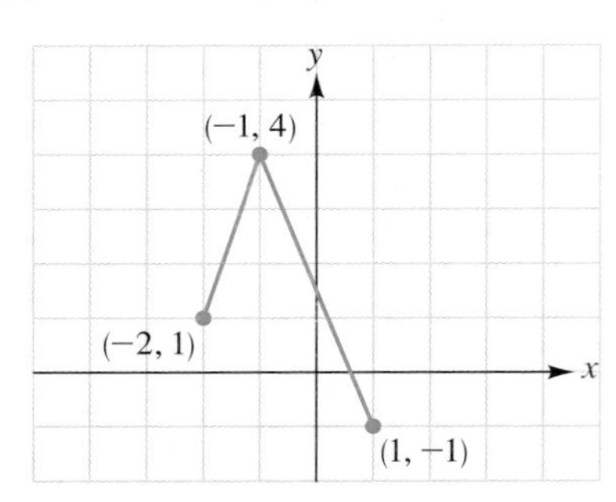

16. Given that the domains of f and f^{-1} are both $(-\infty, \infty)$ and that $f^{-1}(1) = -4$, solve the following equation for t: $2 + f(3t + 5) = 3$.

17. Let $f(x) = \frac{1}{2}x + 2$.
(a) Compute the first three iterates of $x_0 = 1$.
(b) Refer to the following figure. Explain why the y-coordinates of the points A, B, and C are the iterates that you computed in part (a).
(c) Determine the y-coordinate of the point where the lines $y = x$ and $y = \frac{1}{2}x + 2$ intersect.
(d) If we continue the pattern of dashed lines in the figure, it appears that the iterates of $x_0 = 1$ will approach 4. Verify this empirically by using your calculator to compute the fourth through tenth iterates. (Round your final answers to three decimal places.)

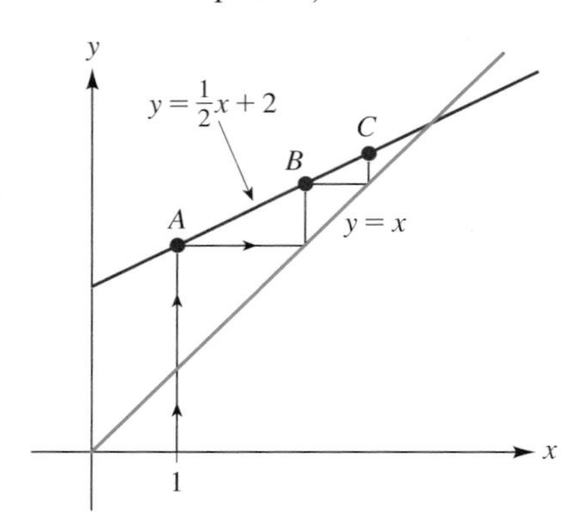

In Exercises 18–34, sketch the graph and specify any x- or y-intercepts.

18. $y = \dfrac{1}{x} + 1$

19. $f(x) = \dfrac{1}{x + 1}$

20. $y = \dfrac{1}{x + 1} + 1$

21. $y = |x + 3|$

22. $g(x) = -\sqrt{x - 4}$

23. $h(x) = \sqrt{1 - x^2}$

24. $y = 4 - \sqrt{-x}$

25. $y = (\sqrt{x})^2$

26. $f \circ g$, where $g(x) = x + 3$ and $f(x) = x^2$

27. $f \circ g$, where $g(x) = \sqrt{x - 1}$ and $f(x) = -x^2$

28. $f(x) = \begin{cases} \sqrt{1 - x^2} & \text{if } -1 \leq x \leq 0 \\ \sqrt{x} + 1 & \text{if } x > 0 \end{cases}$

29. $y = \begin{cases} 1/x & \text{if } 0 < x \leq 1 \\ 1/(x - 1) & \text{if } 1 < x \leq 2 \end{cases}$

30. $(y + |x| - 1)(y - |x| + 1) = 0$

31. f^{-1}, where $f(x) = \frac{1}{2}(x + 1)$

32. g^{-1}, where $g(x) = \sqrt[3]{x + 2}$

33. $f \circ f^{-1}$, where $f(x) = \sqrt{x - 2}$

34. $f^{-1} \circ f$, where $f(x) = \sqrt{x - 2}$

In Exercises 35–40, find the domain of the function.

35. $y = 1/(x^2 - 9)$

36. $y = x^3 - x^2$

37. $y = \sqrt{8 - 2x}$

38. $y = \dfrac{x}{6x^2 + 7x - 3}$

39. $y = \sqrt{x^2 - 2x - 3}$

40. $y = \sqrt{5 - x^2}$

In Exercises 41–44, determine the range of the function.

41. $y = \dfrac{x + 4}{3x - 1}$

42. $f \circ g$, where $f(x) = 1/x$ and $g(x) = 3x + 4$

43. $g \circ f$, where $f(x) = \dfrac{x + 2}{x - 1}$ and $g(x) = \dfrac{x + 1}{x + 4}$

44. f^{-1}, where $f(x) = \dfrac{x}{3x - 6}$

In Exercises 45–52, express the function as a composition of two or more of the following functions:

$$f(x) = \frac{1}{x} \quad g(x) = x - 1 \quad F(x) = |x| \quad G(x) = \sqrt{x}$$

45. $a(x) = \dfrac{1}{x - 1}$

46. $b(x) = \dfrac{1}{x} - 1$

47. $c(x) = \sqrt{x - 1}$

48. $d(x) = \sqrt{x} - 1$

49. $A(x) = (1/\sqrt{x}) - 1$

50. $B(x) = |x - 2|$

51. $C(x) = \sqrt[4]{x} - 1$

52. $D(x) = 1/\sqrt{x - 3}$

For Exercises 53–80, compute the indicated quantity using the functions f, g, and F defined as follows:

$$f(x) = x^2 - x \qquad g(x) = 1 - 2x \qquad F(x) = \frac{x-3}{x+4}$$

53. $f(-3)$
54. $f(1+\sqrt{2})$
55. $F(3/4)$
56. $f(t)$
57. $f(-t)$
58. $g(2x)$
59. $f(x-2)$
60. $g(x+h)$
61. $f(x^2)$
62. $f(x)/x \quad (x \neq 0)$
63. $[f(x)][g(x)]$
64. $f[f(x)]$
65. $f[g(x)]$
66. $g[f(3)]$
67. $(g \circ f)(x)$
68. $(g \circ f)(x) - (f \circ g)(x)$
69. $(F \circ g)(x)$
70. $\dfrac{g(x+h) - g(x)}{h}$
71. $\dfrac{f(x+h) - f(x)}{h}$
72. $\dfrac{F(x) - F(a)}{x - a}$
73. $F^{-1}(x)$
74. $F[F^{-1}(x)]$
75. $(g \circ g^{-1})(x)$
76. $g^{-1}(x)$
77. $g^{-1}(-x)$
78. $\dfrac{g^{-1}(x+h) - g^{-1}(x)}{h}$
79. $F^{-1}[F(22/7)]$
80. $T^{-1}(x)$, where $T(x) = f(x)/x \quad (x \neq 0)$

In Exercises 81–94, refer to the graph of the function f in the figure. (The axes are marked off in one-unit intervals.)

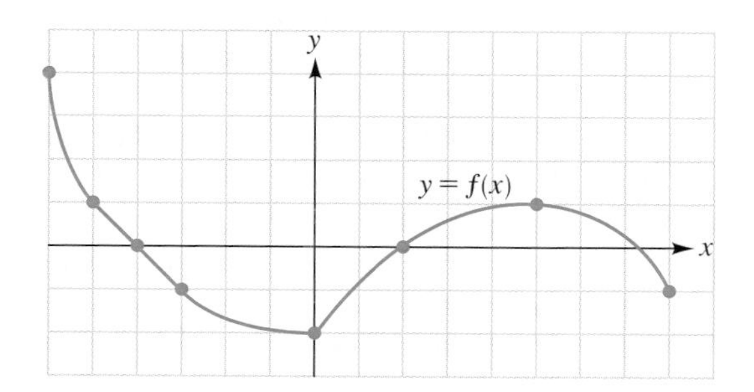

81. Is $f(0)$ positive or negative?
82. Specify the domain and range of f.
83. Find $f(-3)$.
84. Which is larger, $f(-5/2)$ or $f(-1/2)$?
85. Compute $f(0) - f(8)$.
86. Compute $|f(0) - f(8)|$.
87. Specify the coordinates of the turning points.
88. What are the minimum and the maximum values of f?
89. On which interval(s) is f decreasing?
90. For which x-values is it true that $1 \leq f(x) \leq 4$?
91. What is the largest value of $f(x)$ when $|x| \leq 2$?
92. Is f a one-to-one function?
93. Does f possess an inverse function?
94. Compute $f[f(-4)]$.

For Exercises 95–110, refer to the graphs of the functions f and g in the following figure. Assume that the domain of each function is [0, 10].

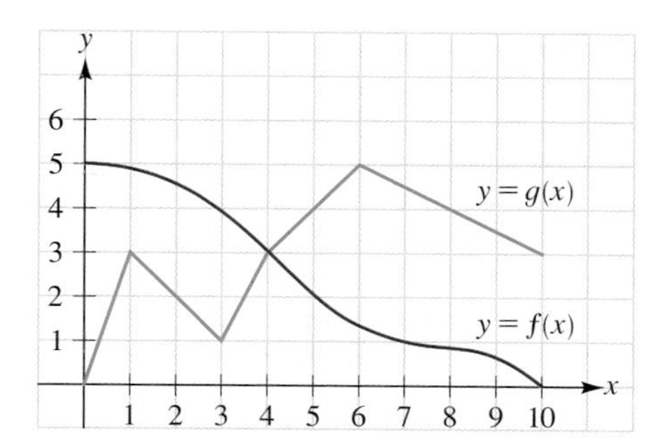

95. For which x-value is $f(x) = g(x)$?
96. For which x-values is it true that $g(x) \leq f(x)$?
97. **(a)** For which x-value is $f(x) = 0$?
(b) For which x-value is $g(x) = 0$?
98. Compute $f(0) + g(0)$.
99. Compute each of the following.
(a) $(f+g)(8)$ **(b)** $(f-g)(8)$ **(c)** $(fg)(8)$ **(d)** $(f/g)(8)$
100. Compute each of the following.
(a) $g[f(5)]$ **(b)** $f[g(5)]$ **(c)** $(g \circ f)(5)$ **(d)** $(f \circ g)(5)$
101. Which is larger, $(f \circ f)(10)$ or $(g \circ g)(10)$?
102. Compute $g[f(10)] - f[g(10)]$.
103. For which x-values is it true that $f(x) \geq 3$?
104. For which x-values is it true that $|f(x) - 3| \leq 1$?
105. What is the largest number in the range of g?
106. Specify the coordinates of the highest point on the graph of each of the following equations.
(a) $y = g(-x)$ **(b)** $y = -g(x)$ **(c)** $y = g(x-1)$ **(d)** $y = f(-x)$ **(e)** $y = -f(x)$ **(f)** $y = -f(-x)$
107. On which intervals is the function g decreasing?
108. What are the coordinates of the turning points of g?
109. For which values of x in the interval (4, 7) is the quantity $\dfrac{f(x) - f(5)}{x - 5}$ negative?
110. For which values of x in the interval (0, 5) is the quantity $\dfrac{f(x) - f(2)}{x - 2}$ positive?

Chapter 3 Test

1. Find the domain of the function defined by $f(x) = \sqrt{x^2 - 5x - 6}$.
2. Find the range of the function defined by $g(x) = \dfrac{2x - 8}{3x + 5}$.
3. Let $f(x) = 2x^2 - 3x$ and $g(x) = 2 - x$. Compute each of the following:
   (a) $(f - g)(x)$ **(b)** $(f \circ g)(x)$ **(c)** $f[g(-4)]$
4. Let $f(t) = 2/t$. Compute $\dfrac{f(t) - f(a)}{t - a}$.
5. Let $g(x) = 2x^2 - 5x$. Compute $\dfrac{g(x + h) - g(x)}{h}$.
6. Find $g^{-1}(x)$ given that $g(x) = \dfrac{-4x}{6x + 1}$.
7. The graph of a function f is a line segment joining the two points $(-3, 1)$ and $(5, 6)$. Determine the slope of the line segment that results from graphing the equation $y = -f^{-1}(x)$.
8. Graph each function and specify the intercepts:
   (a) $y = -|x - 3| + 1$ **(b)** $y = \dfrac{1}{x + 3} - 2$
9. The figure shows the graph of a function g with domain $[-4, 2]$.

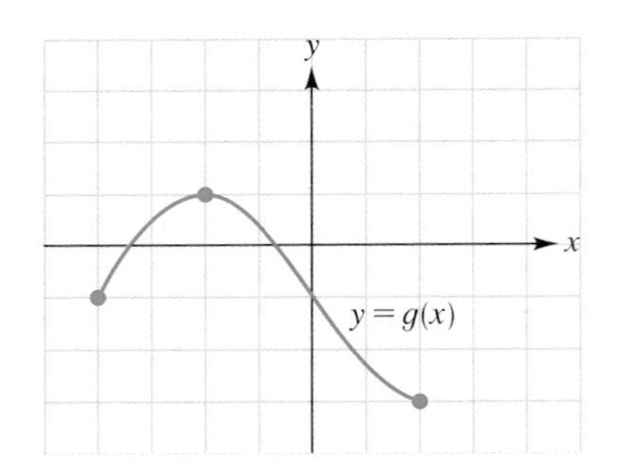

   (a) Specify the range of g.
   (b) What are the coordinates of the turning point?
   (c) What is the minimum value of g?
   (d) Which input yields a maximum value for g? What is that maximum value?
   (e) On which interval is g decreasing?
   (f) Use the horizontal line test to say whether or not g is one-to-one.
10. Let $f(x) = x^2 - 3x - 1$. Compute each of the following:
   (a) $f(-3/2)$ **(b)** $f(\sqrt{3} - 2)$
11. Graph the function F defined by

$$F(x) = \begin{cases} |x + 1| & \text{if } x < -1 \\ -x^2 + 1 & \text{if } x > -1 \end{cases}$$

   What is the domain of this function?
12. In the figure below, the graph of $f(x) = 1 - |2x - 1|$ is shown in red ($y = x$ is included also).

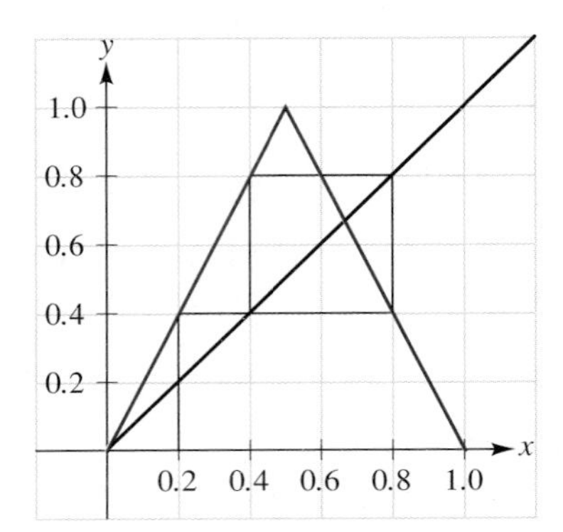

   (a) Use the figure to specify the first six iterates of $x_0 = 0.2$.
   (b) Use a calculator to compute the first six iterates of $x_0 = 0.2$. Are your results consistent with those in part (a)?
13. The graph of f is a line segment joining the points $(1, 3)$ and $(5, -2)$. Sketch the graph of $y = f(-x)$.
14. Given that the domains of f and f^{-1} are both $(-\infty, \infty)$ and that $f^{-1}(-3) = 1$, solve the equation $5 + f(4t - 3) = 2$ for t.
15. Let $f(x) = 1/x$. Find a value for b so that on the interval $[1, b]$, we have $\Delta f/\Delta x = -1/10$.
16. During the first five hours of a lab experiment, the temperature of a solution is given by

$$F(t) = 0.16t^2 - 1.6t + 35 \qquad (0 \le t \le 5)$$

   where t is measured in hours, $t = 0$ corresponds to the time that the experiment begins, and $F(t)$ is the temperature (in degrees Celsius) at time t.
   (a) Find (and simplify) a formula for $\Delta F/\Delta t$ over the interval $[a, 5]$.
   (b) Use the result in part (a) to compute the average rate of change of temperature during the period from $t = 4$ hr to $t = 5$ hr. (Be sure to include the units as part of the answer.)

CHAPTER 4 Polynomial and Rational Functions. Applications to Optimization

POLYNOMIAL was used by François Viéta (1540–1603). The term RATIONAL FUNCTION was used by Joseph Louis Lagrange (1736–1813) in "Réflexions sur la résolution algébrique des équations," . . .—From Jeff Miller's website *Earliest Known Uses of Some of the Words of Mathematics* (http://members.aol.com/jeff570/mathword.html)

In the previous chapter we studied some rather general rules for working with functions and graphs. With that as background we are ready to focus our attention on a few specific types of functions and their applications. Some contexts in which we'll use the functions and ideas introduced in this chapter are:

- Modeling data relating smoking and lung cancer (Example 5 in Section 4.1)
- Summarizing global trends in airline passenger miles using a quadratic model (Exercise 1 in Exercise Set 4.2)
- Modeling the spread of AIDS using linear and quadratic functions (Example 1 in Section 4.2)
- Finding the minimum value of a quadratic function to gain insight into how least-squares lines are defined and determined (Project at end of Section 4.5)
- Determining long-term behavior in a population model by computing the horizontal asymptote of a rational function (Example 7 in Section 4.7)

4.1 LINEAR FUNCTIONS

All decent functions are practically linear.—Professor Andrew Gleason

I had a moment of mixed joy and anguish, when my mind took over. It raced well ahead of my body and drew my body compellingly forward. I felt that the moment of a lifetime had come.—Dr. Roger Bannister, first person to run the mile in under four minutes, recalling the last lap of his record-breaking run at the Ilffley Road Track, Oxford, England, May 6, 1954. The quotation is from Roger Bannister, *The Four Minute Mile* (New York: Dodd, Mead & Co., 1955).

By a **linear function** we mean a function defined by an equation of the form

$$f(x) = Ax + B$$

where A and B are constants. In this chapter the constants A and B will always be real numbers. From our work in Chapter 1 we know that the graph of $y = Ax + B$ is a straight line.

EXAMPLE 1 Finding an equation defining a linear function

Suppose that f is a linear function. If $f(1) = 0$ and $f(2) = 3$, find an equation defining f.

SOLUTION

From the statement of the problem we know that the graph of f is a straight line passing through the points (1, 0) and (2, 3). Thus the slope of the line is

$$m = \frac{y_2 - y_1}{x_2 - x_1} = \frac{3 - 0}{2 - 1} = 3$$

Now we can use the point–slope formula using the point (1, 0) to find the required equation. We have

$$\begin{aligned} y - y_1 &= m(x - x_1) \\ y - 0 &= 3(x - 1) \\ y &= 3x - 3 \end{aligned}$$

This is the equation defining f. If we wish, we can rewrite it using function notation: $f(x) = 3x - 3$.

One basic application of linear functions that occurs in business and economics is *linear* or *straight-line depreciation.* In this situation we assume that the value $V(t)$ of an asset (such as a machine or an apartment building) decreases linearly over time t:

$$V(t) = mt + b$$

where $t = 0$ corresponds to the time when the asset is new (or when its value is first assessed), and the slope m is negative.

EXAMPLE 2 Using a linear model for depreciation

A factory owner buys a new machine for \$8000. After ten years, the machine has a salvage value of \$500.

(a) Assuming linear depreciation (as indicated in Figure 1), find a formula for the value $V(t)$ of the machine after t yr, where $0 \le t \le 10$.

(b) Use the depreciation function determined in part (a) to find the value of the machine after seven years.

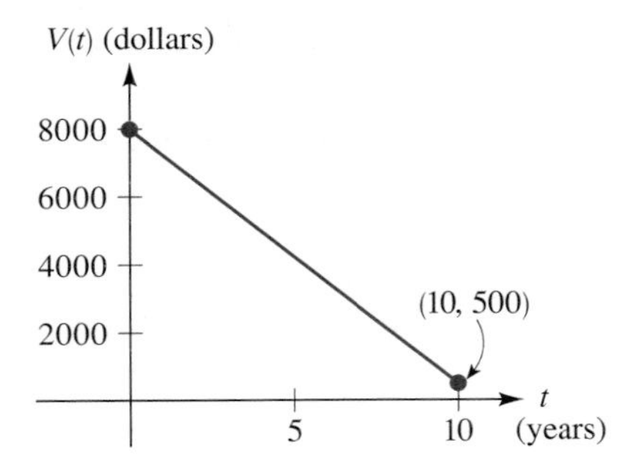

Figure 1

SOLUTION

(a) We need to determine m and b in the equation $V(t) = mt + b$. From Figure 1 we see that the V-intercept of the line segment is $b = 8000$. Furthermore, since the line segment passes through the two points (10, 500) and (0, 8000), the slope m is

$$\frac{8000 - 500}{0 - 10} = -750$$

So we have $m = -750$ and $b = 8000$, and the required equation is

$$V(t) = -750t + 8000 \qquad (0 \leq t \leq 10)$$

(b) Substituting $t = 7$ in the equation $V(t) = -750t + 8000$ yields

$$V(7) = -750(7) + 8000 = -5250 + 8000 = 2750$$

Thus the value of the machine after seven years is \$2750.

It is instructive to keep track of the units associated with the slope in Example 2. Repeating the slope calculation and keeping track of the units, we have

$$m = \frac{\$8000 - \$500}{0\text{ yr} - 10\text{ yr}} = -\$750/\text{yr} \tag{1}$$

Thus slope is a *rate of change.* (We met this idea earlier, back in Section 1.6.) In this example the slope represents the rate of change of the value of the machine. Notice that the calculations in (1) are exactly those that we would have used in Chapter 3 to compute the *average* rate of change $\Delta V/\Delta t$. In the context of linear functions we often suppress the word "average" and refer simply to the *rate of change,* because no matter what two points we choose on the graph of a linear function, the slope will always be the same.

As a second example of slope as a rate of change, let us suppose that a small manufacturer of handmade running shoes knows that her total cost in dollars, $C(x)$, for producing x pairs of shoes each business day is given by the linear function

$$C(x) = 10x + 50$$

Figure 2(a) displays a table and the graph for this function. Actually, since x represents the daily number of pairs of shoes, x can assume only whole-number values. So technically, the graph that should be given is the one in Figure 2(b). However,

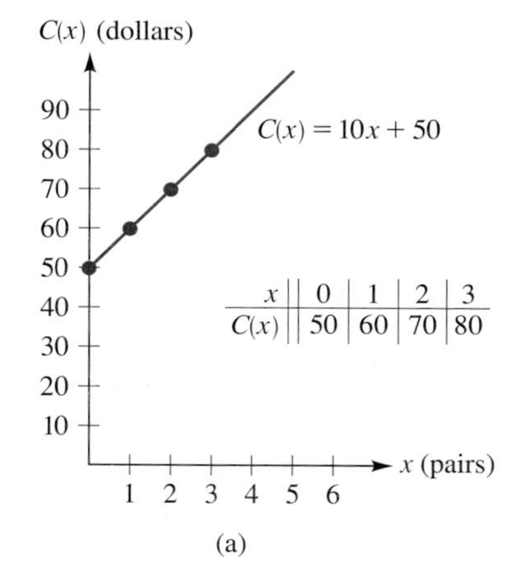

x	0	1	2	3
$C(x)$	50	60	70	80

(a)

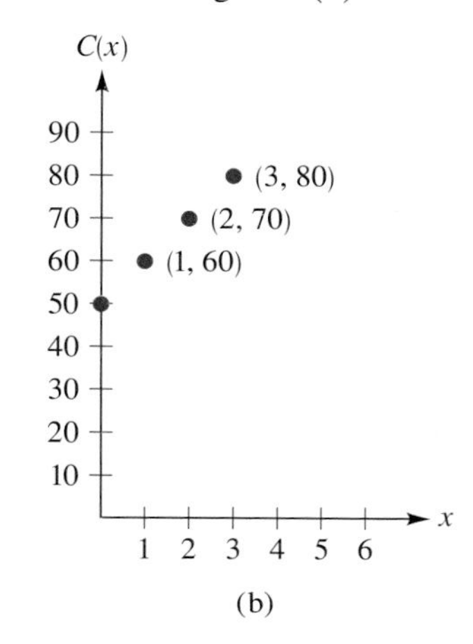

(b)

Figure 2

the graph in Figure 2(a) turns out to be useful in practice, and we shall follow this convention.

The slope of the line $C(x) = 10x + 50$ in Figure 2(a) is 10, the coefficient of x. To understand the units involved, let's calculate the slope using two of the points in Figure 2(a). Using the points (0, 50) and (1, 60) and keeping track of the units, we have

$$m = \frac{\$60 - \$50}{1 \text{ pair} - 0 \text{ pairs}} = \frac{\$10}{1 \text{ pair}} = \$10/\text{pair}$$

Again the slope is a rate. In this case the slope represents the rate of increase of cost; each additional pair of shoes produced costs the manufacturer $10.

In the study of economics a function that gives the cost $C(x)$ for producing x units of a commodity is called a **cost function.** We define the **marginal cost** as the additional cost to produce one more unit. When the cost function is linear, the marginal cost is the slope of the corresponding line. In the preceding example the marginal cost is $10 per pair, and we see that the slope of the line in Figure 2(a) is 10 (dollars per pair).

EXAMPLE 3 A cost function that is linear

Suppose that the cost $C(x)$ in dollars of producing x bicycles is given by the linear function

$$C(x) = 625 + 45x$$

(a) Find the cost of producing 10 bicycles.
(b) What is the marginal cost?
(c) Use the answers in parts (a) and (b) to find the cost of producing 11 bicycles. Then check the answer by evaluating $C(11)$.

SOLUTION

(a) Using $x = 10$ in the cost equation, we have

$$C(10) = 625 + 45(10) = 1075$$

Thus the cost of producing ten bicycles is $1075.

(b) Since C is a linear function, the marginal cost is the slope, and we have

$$\text{marginal cost} = \$45 \text{ per bicycle}$$

(c) According to the result in part (b), each additional bicycle costs $45. Therefore we can compute the cost of 11 bicycles by adding $45 to the cost for 10 bicycles:

$$\begin{aligned} \text{cost of 11 bicycles} &= \text{cost of 10 bicycles} + \text{marginal cost} \\ &= \$1075 + \$45 = \$1120 \end{aligned}$$

So the cost of producing 11 bicycles is $1120. We can check this result by using the cost function $C(x) = 625 + 45x$ to compute $C(11)$ directly:

$$\begin{aligned} C(11) &= 625 + 45(11) \\ &= 625 + 495 \\ &= \$1120 \qquad \text{as obtained previously} \end{aligned}$$

Interpreting slope as a rate of change is not restricted to applications in business or economics. Suppose, for example, that you are driving a car at a steady rate of 50 mph. Using the distance formula from elementary mathematics,

$$\text{distance} = \text{rate} \times \text{time}$$

or

$$d = rt$$

We have in this case $d = 50t$, where d represents the distance traveled (in miles) in t hours. In Figure 3 we show a graph of the linear function $d = 50t$. The slope of this line is 50, the coefficient of t. But 50 is also the given rate of speed, in miles per hour. So again, slope is a rate of change. In this case, slope is the **velocity,** or *rate of change of distance with respect to time.*

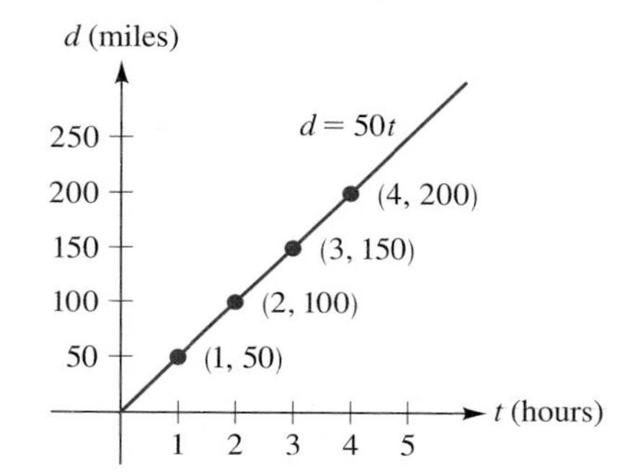

Figure 3
The slope of the line is the rate of change of distance with respect to time.

We conclude this section by indicating one way that linear functions are used in statistical applications. Tables 1 and 2 show the evolution of the world record for the one-mile run during the years 1911–1993. In case you're wondering why Table 1

TABLE 1 Evolution of the Record for the Mile Run, 1911–1954

x (year)	y (time)	
1911	4:15.4	(John Paul Jones, United States)
1913	4:14.6	(John Paul Jones, United States)
1915	4:12.6	(Norman Taber, United States)
1923	4:10.4	(Paavo Nurmi, Finland)
1931	4:09.2	(Jules Ladoumegue, France)
1933	4:07.6	(Jack Lovelock, New Zealand)
1934	4:06.8	(Glen Cunningham, United States)
1937	4:06.4	(Sidney Wooderson, Great Britain)
1942	4:06.2	(Gunder Haegg, Sweden)
1942	4:06.2	(Arne Andersson, Sweden)
1942	4:04.6	(Gunder Haegg, Sweden)
1943	4:02.6	(Arne Andersson, Sweden)
1944	4:01.6	(Arne Andersson, Sweden)
1945	4:01.4	(Gunder Haegg, Sweden)
1954	3:59.4	(Roger Bannister, Great Britain)
1954	3:58.0	(John Landy, Australia)

TABLE 2 Evolution of the Record for the Mile Run, 1957–1993

x (year)	y (time)	
1957	3:57.2	(Derek Ibbotson, Great Britain)
1958	3:54.5	(Herb Elliott, Australia)
1962	3:54.4	(Peter Snell, New Zealand)
1964	3:54.1	(Peter Snell, New Zealand)
1965	3:53.6	(Michel Jazy, France)
1966	3:51.3	(Jim Ryun, United States)
1967	3:51.1	(Jim Ryun, United States)
1975	3:51.0	(Filbert Bayi, Tanzania)
1975	3:49.4	(John Walker, New Zealand)
1979	3:49.0	(Sebastian Coe, Great Britain)
1980	3:48.8	(Steve Ovett, Great Britain)
1981	3:48.53	(Sebastian Coe, Great Britain)
1981	3:48.40	(Steve Ovett, Great Britain)
1981	3:47.33	(Sebastian Coe, Great Britain)
1985	3:46.32	(Steve Cram, Great Britain)
1993	3:44.39	(Noureddine Morceli, Algeria)

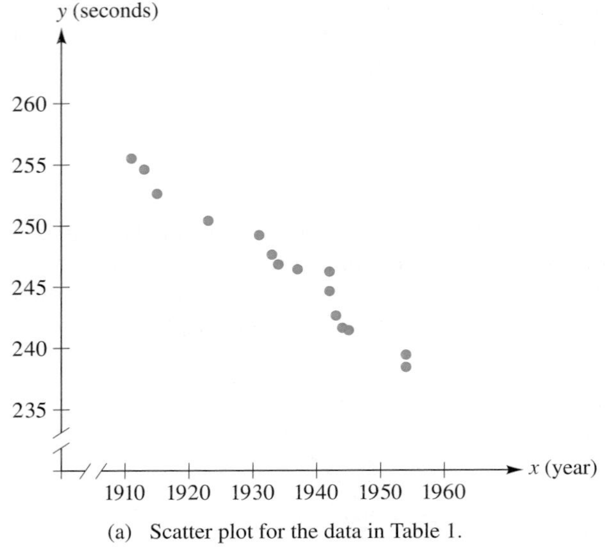

(a) Scatter plot for the data in Table 1.

(b) Scatter plot and regression line for the data in Table 1.

Figure 4

begins with 1911, John Paul Jones was the first twentieth-century runner to break the previous century's record of 4:15.6, set in 1895. Table 1 ends with the year 1954, the first year in which the "four-minute barrier" was broken. (See the quote by Roger Bannister at the beginning of this section.)

Let's begin by looking at the data in Table 1 graphically. In Figure 4(a) we've plotted the (x, y) pairs given in the table. The resulting plot is called a **scatter diagram** or **scatter plot.**

A striking feature of the scatter diagram in Figure 4(a) is that the records do not appear to be leveling off; rather, they seem to be decreasing in an approximately linear fashion. Using the *least-squares technique* from the field of statistics, it can be shown that the linear function that best fits the data in Table 1 is

$$f(x) = -0.370x + 962.041 \tag{2}$$

The graph of this line is shown in Figure 4(b). The line itself is referred to as the **regression line,** or the **least-squares line.**

General formulas for determining the regression line are given in Exercise 45. (The derivation, however, is not given, because that requires calculus.) Many graphing utilities and spreadsheet applications have features for computing regression lines when two columns or lists of numbers are entered. We'll cite but one example here using the popular spreadsheet program Microsoft *Excel*™. For more details or for other types of graphing utilities, see the user's manual that came with your graphing utility.

Suppose, for example, that we want to obtain a scatter plot and regression line for the four data points

$$(0, 1) \qquad (1, 2) \qquad (3, 6) \qquad (6, 7)$$

As indicated in Figure 5(a), we enter the x-y pairs using adjacent columns of the *Excel* spreadsheet. (In the figure, the x-values are in column A and the y-values in column B.) Next, of the many chart types that *Excel* provides for displaying data vi-

	A	B
1	0	1
2	1	2
3	3	6
4	6	7

(a) Data

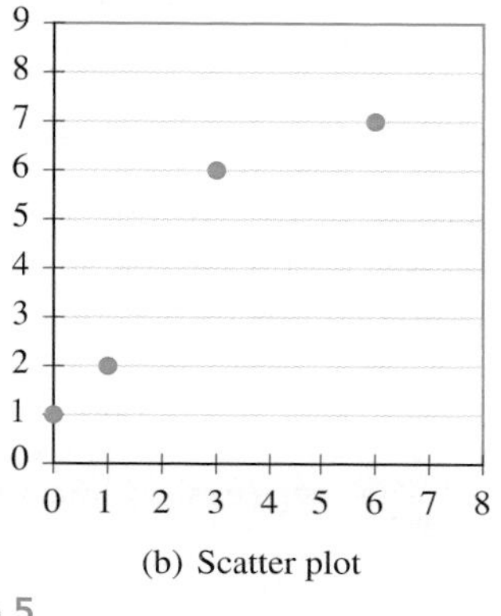

(b) Scatter plot

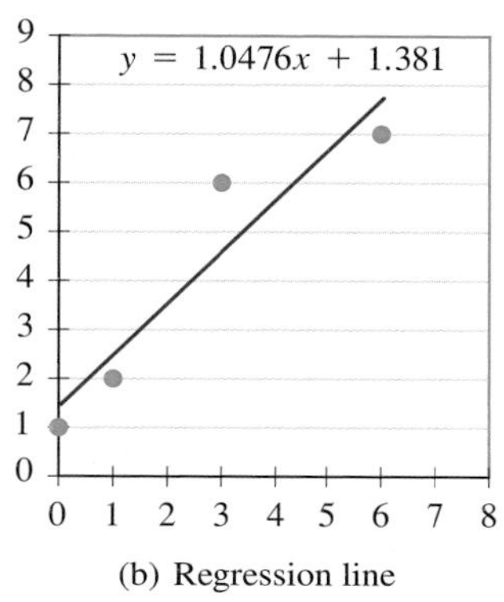

(b) Regression line

Figure 5
Using a spreadsheet application to obtain a regression line

sually, we choose the "Scatter" option and obtain the result shown in Figure 5(b). Finally, for the required regression line, we go to the "Add Trendline" menu and choose "Linear" along with "Display equation on chart." The final result is shown in Figure 5(c).

EXAMPLE 4 Using a regression line

(a) The regression line given by equation (2) is based on records from 1911 through 1954. Use this regression line to estimate what the world record for the mile run might have been in 1999.
(b) Check the prediction against the actual 1999 record 3:43.13, set by Hicham El Guerrouj of Morocco. Compute the percentage error.

SOLUTION
(a) Substituting the value $x = 1999$ in equation (2) yields

$$f(1999) = -0.370(1999) + 962.041 \approx 222.41 \text{ sec}$$
$$= 3 \text{ min and } 42.41 \text{ sec}$$

(b) The prediction obtained in part (a) is 3:42.41, and the actual record is 3:43.13. Thus the actual record is slightly higher than the prediction, but the difference is less than one second. For computing the percentage error, let's first express the actual record of 3:43.13 in seconds. [From part (a) we already know the predicted record expressed in seconds.] We have

$$\begin{aligned}\text{3 min and 43.13 sec} &= 3(60)\text{ sec} + 43.13\text{ sec}\\ &= 223.13\text{ sec}\end{aligned}$$

Now we're ready to compute the percentage error in the prediction:

$$\begin{aligned}\text{percentage error} &= \frac{|(\text{actual value}) - (\text{predicted value})|}{\text{actual value}} \times 100\\ &= \frac{|223.13 - 222.41|}{223.13} \times 100 \approx 0.32\%\end{aligned}$$

The percentage error in the prediction is 0.32%.

REMARK The prediction in this example has turned out to be extremely good. However, see part (c) in the next example for an important word of caution regarding predictions such as this.

EXAMPLE 5 A regression line in a medical application

The data in Table 3 are extracted from a 1954 paper on smoking and lung cancer that appeared in the journal *Danish Medical Bulletin.* In Table 3 the left-hand column of numbers shows annual cigarette consumption in 1930 for four countries. The right-hand column gives the mortality rates from lung cancer in those same countries 20 years later.

TABLE 3 Cigarette Consumption and Lung Cancer

	Cigarette consumption for 1930 (per person)	Lung cancer deaths in 1950 (per 100,000 males)
Sweden	320	11.1
Netherlands	444	28.3
Finland	1106	35.3
England-Wales	1200	53.0

Data extracted from "Bronchial Carcinoma—A Pandemic," by A. Nielsen and J. Clemmennsen, *Danish Medical Bulletin,* vol. 1 (1954), pp. 194–199. In some instances the indicated years 1930 and 1950 are only approximate. For instance, for Sweden, the mortality rate pertains to 1951, rather than 1950.

(a) Use a graphing utility to create a scatter plot and to find and graph the regression line. (Use the consumption data for the x-values and the mortality rates for the corresponding y-values.)

(b) The regression line gives a functional relationship between per capita cigarette consumption and lung cancer death rates. Use this relationship to obtain estimates for the mortality rates in 1950 for Denmark and Norway, given that the 1930 cigarette consumption figures for Denmark and Norway were 373 and 257, respectively.

(c) Compute the percentage error in each estimate, given that the actual death rates (per 100,000) in Denmark and Norway were 18.4 and 9.2, respectively.

Smokescreen? The figure is from a 1948 magazine ad for Camel cigarettes. A 1929 ad for Lucky Strike cigarettes announced "Many prominent athletes smoke Luckies all day long with no harmful effects to wind or physical condition." A 1953 ad by Liggett & Meyers said "It's so satisfying to know that a doctor reports no adverse effects to the nose, throat and sinuses from smoking Chesterfield."

Source: The University of Alabama Center for the Study of Tobacco and Society (Alan Blum, M.D.)

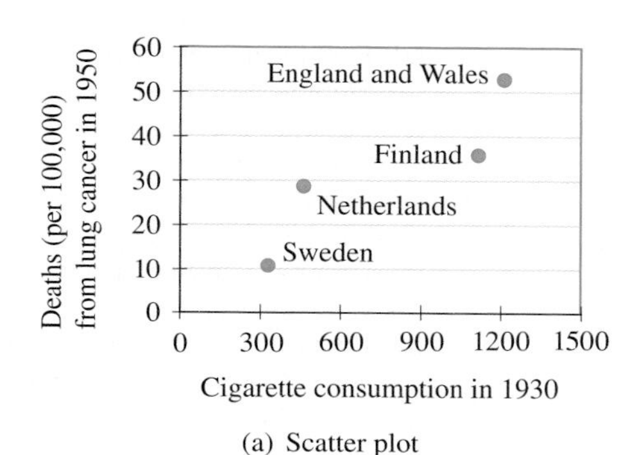

(a) Scatter plot

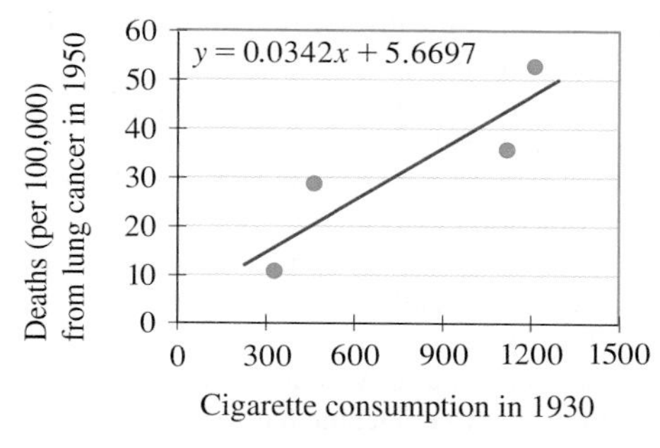

(b) Scatter plot with regression line

Figure 6

SOLUTION

(a) See Figure 6 for the scatter plot and regression line.

(b) As indicated in Figure 6(b), the regression line is $y = 0.0342x + 5.697$.

For Denmark with $x = 373$:	For Norway with $x = 257$:
$y = 0.0342(373) + 5.6697$	$y = 0.0342(257) + 5.6697$
≈ 18.4	≈ 14.5

According to the calculations, the projected mortality rate for Denmark in 1950 is 18.4 deaths per 100,000; for Norway the projection is 14.5 per 100,000.

(c) For Denmark the projection is 18.4 and so is the actual death rate. So in this case we happen to have 0% error. For Norway we have

$$\text{percentage error} = \frac{|(\text{actual value}) - (\text{projected value})|}{\text{actual value}} \times 100$$
$$= \frac{|9.2 - 14.5|}{9.2} \times 100 \approx 58\%$$

In summary, for Denmark the projected death rate from lung cancer agrees with the actual rate. For Norway, however, there is a relatively large discrepancy between the projection and actual value; the percentage error there is 58%. This highlights one of the shortcomings in using a regression line. Without further analysis and the use of more advanced statistical concepts (and perhaps more data) it's difficult or impossible to know ahead of time how much one should rely on these projections.

EXERCISE SET 4.1 

A

In Exercises 1–8, find the linear functions satisfying the given conditions.

1. $f(-1) = 0$ and $f(5) = 4$

2. $f(3) = 2$ and $f(-3) = -4$

3. $g(0) = 0$ and $g(1) = \sqrt{2}$

4. The graph passes through the points (2, 4) and (3, 9).

5. $f(\frac{1}{2}) = -3$ and the graph of f is a line parallel to the line $x - y = 1$.

6. $g(2) = 1$ and the graph of g is perpendicular to the line $6x - 3y = 2$.

7. The graph of the inverse function passes through the points $(-1, 2)$ and $(0, 4)$.

8. The x- and y-intercepts of the inverse function are 5 and -1, respectively.

9. Let $f(x) = 3x - 4$ and $g(x) = 1 - 2x$. Determine whether the function $f \circ g$ is linear.

10. Explain why there is no linear function with a graph that passes through all three of the points $(-3, 2)$, $(1, 1)$, and $(5, 2)$.

11. A factory owner buys a new machine for \$20,000. After eight years, the machine has a salvage value of \$1000. Find a formula for the value of the machine after t years, where $0 \le t \le 8$.

12. A manufacturer buys a new machine costing \$120,000. It is estimated that the machine has a useful lifetime of ten years and a salvage value of \$4000 at that time.

(a) Find a formula for the value of the machine after t years, where $0 \le t \le 10$.

(b) Find the value of the machine after eight years.

13. A factory owner installs a new machine costing \$60,000. Its expected lifetime is five years, and at the end of that time the machine has no salvage value.

(a) Find a formula for the value of the machine after t years, where $0 \le t \le 5$.

(b) Complete the following depreciation schedule.

End of year	Yearly depreciation	Accumulated depreciation	Value V
0	0	0	60,000
1			
2			
3			
4			
5		60,000	0

14. Let x denote a temperature on the Celsius scale, and let y denote the corresponding temperature on the Fahrenheit scale.

(a) Find a linear function relating x and y; use the facts that 32°F corresponds to 0°C and 212°F corresponds to 100°C. Write the function in the form $y = Ax + B$.

(b) What Celsius temperature corresponds to 98.6°F?

(c) Find a number z for which z°F = z°C.

15. Suppose that the cost $C(x)$, in dollars, of producing x electric fans is given by $C(x) = 450 + 8x$.

(a) Find the cost to produce 10 fans.

(b) Find the cost to produce 11 fans.

(c) Use your answers in parts (a) and (b) to find the marginal cost. (Then check that your answer is the slope of the line.)

16. Suppose that the cost to a manufacturer of producing x units of a certain motorcycle is given by $C(x) = 220x + 4000$, where $C(x)$ is in dollars.

(a) Find the marginal cost.

(b) Find the cost of producing 500 motorcycles.

(c) Use your answers in parts (a) and (b) to find the cost of producing 501 motorcycles.

17. Suppose that the cost $C(x)$, in dollars, of producing x compact discs is given by $C(x) = 0.5x + 500$.

(a) Graph the given equation.

(b) Compute $C(150)$.

(c) If you add the marginal cost to the answer in part (b), you obtain a certain dollar amount. What does this amount represent?

18. The following graphs each relate distance and time for a moving object. Determine the velocity in each case.

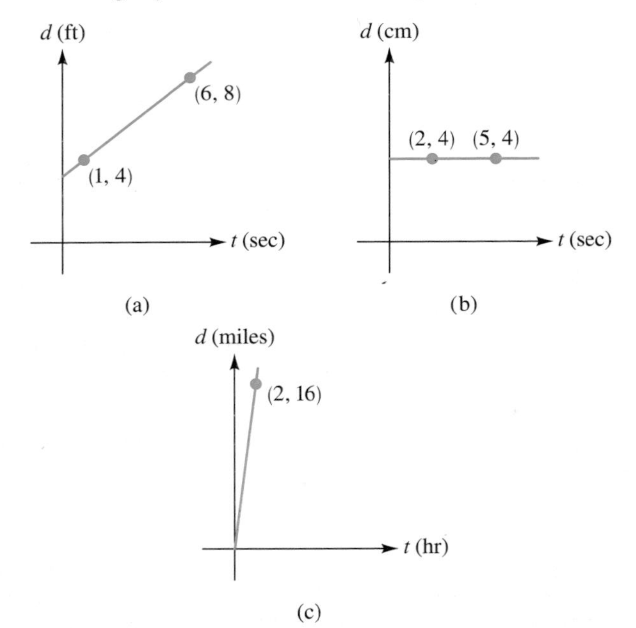

19. Two points A and B move along the x-axis. After t sec, their positions are given by the equations

$$A:\quad x = 3t + 100$$
$$B:\quad x = 20t - 36$$

(a) Which point is traveling faster, A or B?

(b) Which point is farther to the right when $t = 0$?
(c) At what time t do A and B have the same x-coordinate?

20. A point moves along the x-axis, and its x-coordinate after t sec is $x = 4t + 10$. (Assume that x is in centimeters.)
(a) What is the velocity?
(b) What is the x-coordinate when $t = 2$ sec?
(c) Use your answers in parts (a) and (b) to find the x-coordinate when $t = 3$ sec. *Hint*: What are the units of the velocity in part (a)? Check your answer by letting $t = 3$ in the given equation.

21. The following table gives the population of California in 1995 and in 1997.

x (year)	y (population)
1995	31,493,525
1997	32,217,708

(a) Find the equation of the linear function whose graph passes through the two (x, y) points given in the table.
(b) Use the linear function determined in part (a) to make a projection for the population of California in 2000. (Round your answer to the nearest 1000.)
(c) The actual California population for 2000 was 33,871,648. Does the linear function yield a projection that is too high or too low? Compute the percentage error.

22. The following table gives the population of Florida in 1985 and 1990.

x (year)	y (population)
1985	11,351,118
1990	13,018,365

Follow the steps in Exercise 21 to make a projection for the population of Florida in 2000. For part (c) the actual population of Florida was 15,982,378.

23. The following table indicates total motion picture receipts (including video tape rentals) in the United States for the years 1994 and 1995.

Motion Picture Receipts

x (year)	y (receipts) (in millions of dollars)
1994	53,504
1995	57,184

Source: U.S. Census Bureau, *Statistical Abstract of the United States: 1999*

(a) Find the equation of the linear function whose graph passes through the two (x, y) points given in the table.
Ⓖ **(b)** Use a graphing utility to graph the line in part (a). Then use a TRACE or ZOOM feature to estimate what motion picture receipts might have been for the year 1997.
(c) Compute the percentage error in the estimate in part (b), given that the actual figure for 1997 was $63,010 millions.

24. During the 1990s the percentage of TV households viewing cable and satellite TV programs increased while the percentage viewing network affiliate shows (ABC, CBS, NBC, and FOX) generally decreased. The following table shows the primetime ratings for the network affiliates in the years 1993 and 1995. (In the table the *rating percentage* is defined as the percentage of TV households viewing a TV program in an average minute.)

Primetime Ratings for Network Affiliates

x (year)	y (rating percentage)
1993	40.9
1995	37.3

Source: Nielsen Media Research, *1998 Report on Television*

(a) Find the equation of the linear function whose graph passes through the two (x, y) points given in the table.
Ⓖ **(b)** Using a graphing utility to graph the line in part (a). Then use a TRACE or ZOOM feature to estimate what the rating percentage might have been for the year 1997.
(c) Compute the percentage error in the estimate in part (b), given that the actual figure for 1997, according to Nielsen Media Research, was 33.1.

In general, the growth of plants or animals does not follow a linear pattern. For relatively short intervals of time, however, a linear function may provide a reasonable description of the growth. Exercises 25 and 26 provide examples of this.

25. (a) In an experiment with sunflower plants, H. S. Reed and R. H. Holland measured the height of the plants every seven days for several months. [The experiment is reported in *Proceedings of the National Academy of Sciences,* vol. 5 (1919), p. 140.] The following data are from this experiment.

x (number of days)	21	49
y [average height (cm) of plants after x days]	67.76	205.50

Find the linear function whose graph passes through the two points given in the table. (Round each number in the answer to two decimal places.)
(b) Use the linear function determined in part (a) to estimate the average height of the plants after 28 days. (Round your answer to two decimal places.)

(continues)

(c) In the experiment, Reed and Holland found that the average height after 28 days was 98.10 cm. Is your estimate in part (b) too high or too low? Compute the percentage error in your estimate.

(d) Follow parts (b) and (c) using $x = 14$ days. For the computation of percentage error, you need to know that Reed and Holland found that the average height after 14 days was 36.36 cm.

(e) As you've seen in parts (c) and (d), your estimates are quite close to the actual values obtained in the experiment. This indicates that for a relatively short interval, the growth function is nearly linear. Now repeat parts (b) and (c) using $x = 84$ days, which is a longer interval of time. You'll find that the linear function does a poor job in describing the growth of the sunflower plants. To compute the percentage error, you need to know that Reed and Holland determined the average height after 84 days to be 254.50 cm.

26. (a) The biologist R. Pearl measured the population of a colony of fruit flies (*Drosophila melanogaster*) over a period of 39 days. [The experiment is discussed in Pearl's book, *The Biology of Population Growth* (New York: Alfred Knopf, 1925).] Two of the measurements made by Pearl are given in the following table.

x (number of days)	12	18
y (population)	105	225

Find the linear function whose graph passes through the two points given in the table.

(b) Use the linear function determined in part (a) to estimate the population after 15 days. Then compute the percentage error in your estimate, given that the actual population after 15 days, as found by Pearl, was 152.

(c) Use the linear function determined in part (a) to estimate the population after 9 days. Then compute the percentage error in your estimate, given that after 9 days, Pearl found the actual population to be 39.

(d) As you've seen in parts (b) and (c), your estimates are quite close to the actual values obtained in the experiment. This indicates that for relatively short intervals of time, the growth is nearly linear. Now repeat part (b) using $x = 39$ days, which covers a longer interval of time. You'll find that the linear function does a poor job in describing the population growth. For the computation of percentage error, you need to know that Pearl determined the population after 39 days to be 938.

27. (a) On graph paper, plot the following points: (1, 0), (2, 3), (3, 6), (4, 7).

(b) In your scatter diagram from part (a), sketch a line that best seems to fit the data. Estimate the slope and the y-intercept of the line.

(c) The actual regression line in this case is $y = 2.4x - 2$. Add the graph of this line to your sketch from parts (a) and (b).

28. (a) On graph paper, plot the following points: (1, 2), (3, 2), (5, 4), (8, 5), (9, 6).

(b) In your scatter diagram from part (a), sketch a line that best seems to fit the data.

(c) Using your sketch from part (b), estimate the y-intercept and the slope of the regression line.

(d) The actual regression line is $y = 0.518x + 1.107$. Check to see whether your estimates in part (c) are consistent with the actual y-intercept and slope. Graph this line along with the points given in part (a). (In sketching the line, use the approximation $y = 0.5x + 1.1$.)

29. The table in the following figure shows the population of Los Angeles over the years 1930–1990. The accompanying graph shows the corresponding scatter plot and regression line. The equation of the regression line is

$$f(x) = 37{,}546.068x - 71{,}238{,}863.429$$

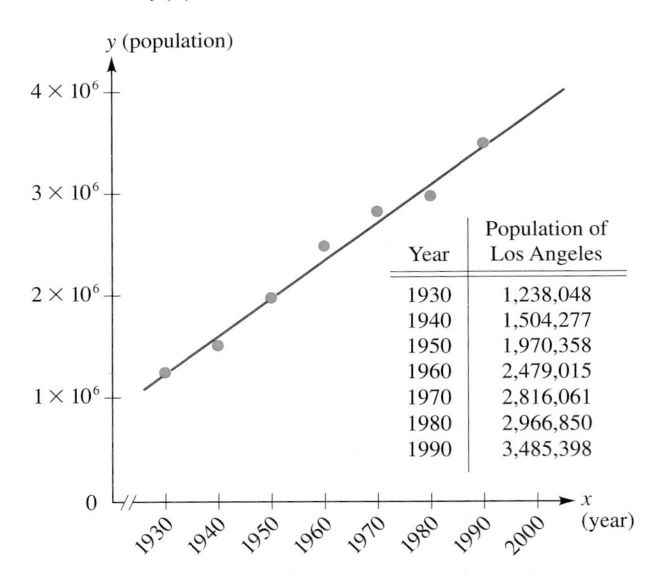

Year	Population of Los Angeles
1930	1,238,048
1940	1,504,277
1950	1,970,358
1960	2,479,015
1970	2,816,061
1980	2,966,850
1990	3,485,398

(a) Use the regression line to compute an estimate for what the population of Los Angeles might have been in 2000. (Round the answer to the nearest thousand.) Then compute the percentage error in the estimate, given that the actual figure for 2000 is 3.823 million.

(b) Find $f^{-1}(x)$.

(c) Use your answer in part (b) to estimate the year in which the population of Los Angeles might reach 4 million. *Hint*: For the function f, the inputs are years and the outputs are populations; for f^{-1}, the inputs are populations and the outputs are years.

30. Perhaps foreshadowing the end of the Cold War in the early 1990s, the number of nuclear warheads worldwide began to decrease after the year 1986 (when the number was at an all-time high). The following table indicates the global number of nuclear warheads over the years 1986–1992. The graph shows the corresponding scatter plot and regression line. The equation for the regression line is

$$f(x) = -2810.96x + 5{,}653{,}063.25$$

Year	Number of nuclear warheads
1986	69,075
1987	67,302
1988	65,932
1989	63,645
1990	60,642
1991	57,017
1992	53,136

Source of data: R. S. Norris and W. M. Arkin, *Bulletin of Atomic Scientists,* Nov./Dec. 1997

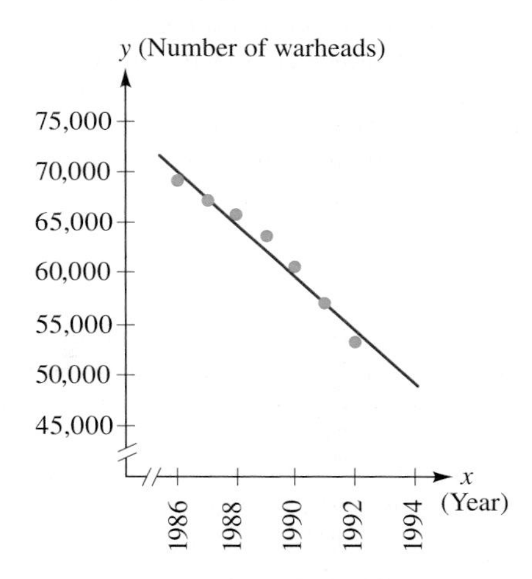

(a) Use the regression line to make a projection for the number of nuclear warheads in 1997. Then compute the percentage error, given that the actual number was 36,110.

(b) When might there be only (!) 10,000 nuclear warheads worldwide? Round the answer to the nearest integer. Also, if the trend were to continue, when would there be no warheads remaining?

31. The following table shows world grain production for selected years over the period 1950–1990.

Year	World grain production (million tons)
1950	631
1960	824
1970	1079
1980	1430
1990	1769

Source: Lester R. Brown et al. in *Vital Signs 1999* (New York: W. W. Norton & Co, 1999)

Ⓖ **(a)** Use a graphing utility or spreadsheet to determine the equation of the regression line. For the x-y data pairs, use x for the year and y for the grain production. Create a graph showing both the scatter plot and regression line.

(b) Use the equation of the regression line to make a projection for the world grain production in 1993. Then compute the percentage error in your projection, given that the actual grain production in 1993 was 1714 million tons. *Remark*: Your projection will turn out to be higher than the actual figure. One (among many) reasons for this: In 1993 there was a drop in world grain production due largely to the effects of poor weather on the U.S. corn crop.

(c) Use the equation of the regression line to make a projection for the world grain production in 1998. Then compute the percentage error in your projection, given that the actual world grain production in 1998 was 1844 million tons. *Remark*: Again, your projection will turn out to be too optimistic. According to Lester Brown's *Vital Signs: 1999,* world grain production dropped in 1998 "due largely to severe drought and heat in Russia on top of an overall deterioration of that country's economy."

32. The following table shows global natural gas production for the years 1990–1996. The abbreviation "tcf" stands for *trillion cubic feet.*

Year	Natural gas production (tcf)
1990	71.905
1991	73.037
1992	73.219
1993	74.570
1994	75.190
1995	76.614
1996	80.045

Source: *BP Amoco Statistical Review of World Energy* (49th ed.) http://www.bp.com/worldenergy

Ⓖ **(a)** Use a graphing utility or spreadsheet to determine the equation of the regression line. For the x-y data pairs, use x for the year and y for the natural gas production. Create a graph showing both the scatter plot and regression line.

(b) What are the units associated with the slope of the regression line in part (a)?

Ⓖ **(c)** In your graph of the regression line, use a TRACE or ZOOM feature to make an estimate for natural gas production in the year 1999.

(d) The actual production figure for 1999 is 83.549 tcf. Is your estimate high or low? Compute the percentage error.

(e) The table above shows that global production of natural gas increased over the period 1990–1996, and in fact it still continues to increase each year. However, for purposes of making a very conservative estimate, let's assume for the moment that natural gas production levels off at its 1999 value of 83.549 tcf per year. According to the *BP Amoco Statistical Review of World Energy,* as of the end of 1999, proved world reserves of natural gas were 5171.8 tcf. Carry out the following calculation and interpret your answer. *Hint*: Keep track of the units.

$$\frac{5171.8 \text{ tcf}}{83.549 \text{ tcf/yr}} = \cdots$$

33. This exercise illustrates one of the pitfalls that can arise in using a regression line: large discrepancies can occur when the regression line is used to make long-term projections. The table in the following figure shows the population of California during the years 1860–1900. The graph displays the scatter plot of the data; the equation of the regression line is

$$f(x) = 28{,}632.69x - 52{,}928{,}780$$

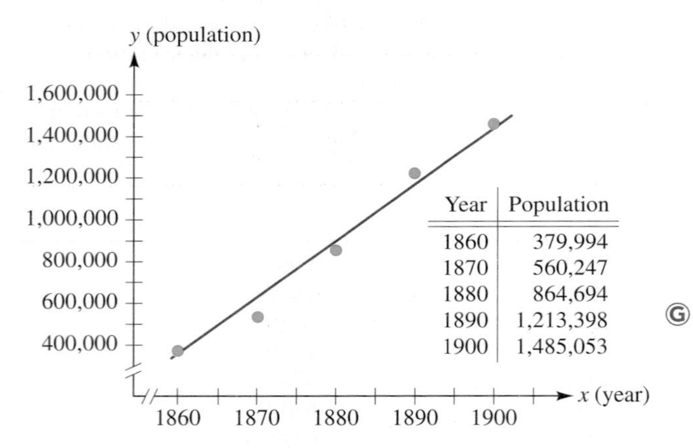

Year	Population
1860	379,994
1870	560,247
1880	864,694
1890	1,213,398
1900	1,485,053

As you can see from the scatter plot, the population growth over this period is very nearly linear.

(a) Use the regression line to estimate the population of California in 1990.

(b) According to the U.S. Bureau of the Census, the population of California in 1990 was 29,839,250. Is your estimate in part (a) close to this figure?

34. (a) The June 1976 issue of *Scientific American* magazine contained an article entitled "Future Performance in Footracing" by H. W. Ryder, H. J. Carr, and P. Herget. According to the authors of this article, "It appears likely that within 50 years the record [for the mile] will be down to 3:30." Use the regression line defined in equation (2) on page 218 to make a projection for the mile time in the year 2026 (which will be 50 years after the article appeared). Is your projection close to the one given in the *Scientific American* article?

(b) The regression line defined in equation (2) is based only on the data from Table 1. If we use all the data from both Table 1 and Table 2, then the least-squares technique can be used to derive the following regression line:

$$f(x) = -0.400x + 1019.472$$

Use this equation to make a projection for the mile time in the year 2026. Which projection is closer to the value predicted in the *Scientific American* article, this one or the one calculated in part (a)?

35. The data in Table 2 on page 217 covers the years 1957–1993. The equation of the regression line for the data is

$$y = -0.318x + 858.955$$

(a) Use this equation to estimate what the mile record might have been for the year 1954. Then check Table 1 to see the actual record for 1954, which was set by John Landy. Is your estimate too high or too low? Compute the percentage error in your estimate.

(b) Use the given equation to estimate what the mile record might have been back in 1911. Check your estimate against the actual record as shown in Table 1, and compute the percentage error. How does the percentage error here compare to that in part (a)? Why is this to be expected?

B

36. Linear functions can be used to approximate more complicated functions. This is one of the meanings or implications of the quotation by Professor Gleason on page 213. This exercise illustrates that idea.

Ⓖ **(a)** Using calculus, it can be shown that the equation of the line that is tangent to the curve $y = x^2$ at the point $(1, 1)$ is $y = 2x - 1$. Verify this visually by graphing the two functions $y = x^2$ and $y = 2x - 1$ on the same set of axes. (*Suggestion*: Use a viewing rectangle that extends from -2 to 3 in the x-direction and from -3 to 4 in the y-direction.) Note that the tangent line is virtually indistinguishable from the curve in the immediate vicinity of the point $(1, 1)$.

(b) For numerical rather than visual evidence of how well the linear function $y = 2x - 1$ approximates the function $y = x^2$ in the immediate vicinity of $(1, 1)$, complete the following tables.

x	0.9	0.99	0.999
x^2			
$2x - 1$			

x	1.1	1.01	1.001
x^2			
$2x - 1$			

In Exercises 37–40, let f and g be the linear functions defined by

$f(x) = Ax + B \quad (A \neq 0) \quad$ *and* $\quad g(x) = Cx + D \quad (C \neq 0)$

In each case, compute the average rate of change of the given function on the interval $[a, b]$.

37. **(a)** f
(b) $f \circ f$
(c) $g \circ f$
(d) $f \circ g$

38. **(a)** f^{-1}
(b) g^{-1}

39. **(a)** $(f \circ g)^{-1}$
(b) $(g \circ f)^{-1}$

40. **(a)** $f^{-1} \circ g^{-1}$ **(b)** $g^{-1} \circ f^{-1}$

41. Show that the linear function $f(x) = mx$ satisfies the following identities:
(a) $f(a + b) = f(a) + f(b)$ **(b)** $f(ax) = af(x)$

42. Let f be a linear function such that

$$f(a + b) = f(a) + f(b)$$

for all real numbers a and b. Show that the graph of f passes through the origin. *Hint*: Let $f(x) = Ax + B$ and show that $B = 0$.

43. Find a linear function $f(x) = mx + b$ such that m is positive and $(f \circ f)(x) = 9x + 4$.

44. **(a)** Let f be a linear function. Show that

$$f\left(\frac{x_1 + x_2}{2}\right) = \frac{f(x_1) + f(x_2)}{2}$$

(In words: The output of the average is the average of the outputs.)

(b) Show (by using an example) that the equation in part (a) does not hold for the function $f(x) = x^2$.

45. This exercise shows how to compute the slope and the y-intercept of the regression line. As an example, we'll work with the simple data set given in Exercise 27.

x	1	2	3	4
y	0	3	6	7

(a) Let Σx denote the sum of the x-coordinates in the data set, and let Σy denote the sum of the y-coordinates. Check that $\Sigma x = 10$ and $\Sigma y = 16$.

(b) Let Σx^2 denote the sum of the squares of the x-coordinates, and let Σxy denote the sum of the products of the corresponding x- and y-coordinates. Check that $\Sigma x^2 = 30$ and $\Sigma xy = 52$.

(c) The slope m and the y-intercept b of the regression line satisfy the following pair of simultaneous equations [in the first equation, n denotes the number of points (x, y) in the data set]:

$$\begin{cases} nb + (\Sigma x)m = \Sigma y \\ (\Sigma x)b + (\Sigma x^2)m = \Sigma xy \end{cases}$$

In the present example these equations become

$$\begin{cases} 4b + 10m = 16 \\ 10b + 30m = 52 \end{cases}$$

Solve this pair of equations for m and b, and check that your answers agree with the values in Exercise 27(c).

In Exercises 46–49:

(a) *Use the method described in Exercise 45 to find the equation of the regression line for the given data set.*

(b) *For Exercises 48 and 49, if your graphing utility has a feature for computing regression lines, use it to check your answer in part (a).*

46.

x	2	4	8	10
y	−7	−5	−2	−1

47.

x	1	2	3	4	5
y	2	3	9	9	11

48.

x	1	2	3	4	5
y	16	13.1	10.5	7.5	2

49.

x	520	740	560	610	650
y	81	98	83	88	95

50. Suppose that f is a linear function satisfying the condition $f(kx) = kf(x)$ for all real numbers k. Prove that the graph of f passes through the origin.

51. **(a)** Find all linear functions f satisfying the identity $f(f(x)) = 2x + 1$. (For your answer, rationalize any denominators containing radicals.)

(b) Find all linear functions f satisfying the identity

$$f(f(f(x))) = 2x + 1$$

(Rationalize any denominators containing radicals.) *Hint*: Try the factoring formula, $x^3 - y^3 = (x - y)(x^2 + xy + y^2)$ from Appendix B.4.

52. Let a and b be real numbers, and suppose that the inequality $ax + b \leq 0$ has no solutions. What can you say about the linear function $f(x) = ax + b$? Answer in complete sentences and justify what you say.

53. Are there any linear functions f satisfying the identity $f^{-1}(x) = f(f(x))$? If so, list them; if not, explain why not.

54. Let $f(x) = ax + b$, where a and b are positive numbers, and assume that the following equation holds for all values of x: $f(f(x)) = bx + a$. Show that $a + b = 1$.

MINI PROJECT

WHO ARE BETTER RUNNERS, MEN OR WOMEN?

An oversimplified answer here would be to say that men are better because in any given event, the men's world record time is less than the women's. For instance, as of the year 2000, the men's and women's world records for the 100-m dash were 9.79 sec and 10.49 sec, respectively. This reasoning, however, ignores the fact that, historically, women have been training and competing on the world stage for far less time than men. Likewise, it's only relatively recently that women have been supported or sponsored to a degree approaching that for men. Indeed, currently, women's records are falling at a faster rate than men's. Complete the following research and calculations, then discuss this issue.

The following table gives the names of athletes who set world records in the 100-m dash over the years 1968–1999. (The table is restricted to times that were recorded electronically, rather than manually; 1968 was the year in which electric timing made its major debut (in the 1968 Olympics in Mexico City).

Selected Record-Setting Performances in 100-m Dash, 1968–1999*

Year	Men	Women
1968	James Hines, USA (10/14/68)	Wyomia Tyus, USA (10/15/68)
1972		Renate Stecher, DDR (East Germany) (9/2/72)
1976		Annegret Richter, FRG (West Germany) (7/25/76)
1977		Marlies Oelsner (Göhr) DDR (7/1/77)
1983	Calvin Smith, USA (7/3/83)	Evelyn Ashford, USA (7/3/83)
1984		Evelyn Ashford, USA (8/22/84)
1988	Carl Lewis, USA (9/24/88)	Florence Griffith-Joyner, USA (7/17/88)
1991	Carl Lewis, USA (8/25/91)	
1994	Leroy Burrell, USA (7/6/94)	
1996	Donovan Bailey, CAN (7/29/96)	
1999	Maurice Greene, USA (6/16/99)	

*In cases in which the record was broken more than once in a given year, only the fastest performance is indicated.

(a) Using the library or the Internet, look up the 100-m times for each athlete in the table.

(b) For the men's records: Let x represent the year, with $x = 0$ corresponding to 1960, and let y represent the record time in that year. Find the regression line $y = f(x)$.

(c) For the women's records: Again, let x represent the year, with $x = 0$ corresponding to 1960, and let y represent the record time in that year. Find the regression line $y = g(x)$.

(d) Solve the equation $f(x) = g(x)$ to compute a projection for the year in which the men's and women's records might be equal. What would that common record be?

4.2 QUADRATIC FUNCTIONS

In the text and exercises for the previous section we saw examples in which data sets and their scatter plots were modeled using a linear function. Clearly, however, linear functions with their straight line graphs cannot be appropriate models for every data set. Consider, for instance, the scatter plot in Figure 1, which shows the daily trading volumes for Amazon.com stock in a particular month. It appears that no line could adequately summarize the situation. Next, in Figure 2, look at the data and scatter plot relating to the spread of AIDS. Here it appears that some sort of curve, rather than straight line, would be better for summarizing the trend in the data. As you'll see in a moment, one type of curve that does fit the data in Figure 2 quite well is obtained by graphing a *quadratic function.*

x years after 1980	4	5	6	7	8	9	10
y AIDS cases (millions)	0.2	0.4	0.7	1.1	1.6	2.3	3.2

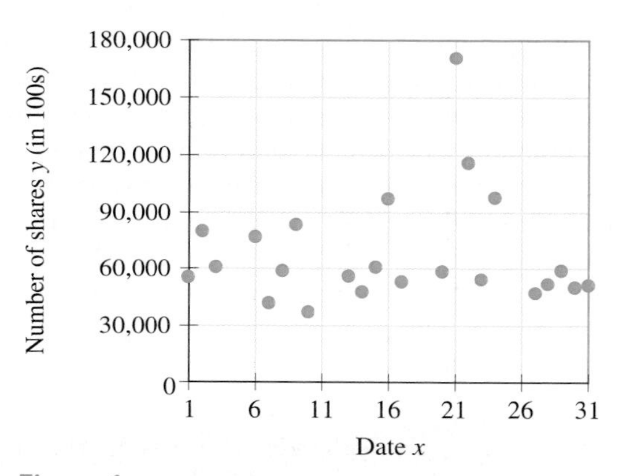

Figure 1
Daily number of shares of Amazon.com stock traded (i.e., sold) in March 2000. $x = 1$ corresponds to March 1.

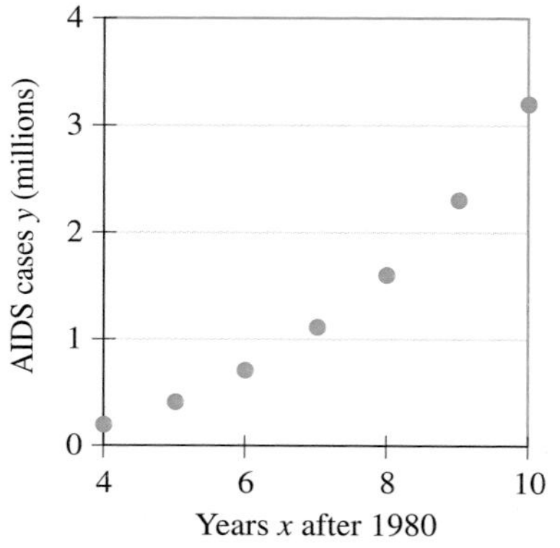

Figure 2
Estimated cumulative number of AIDS cases worldwide, 1984–1990

Source: Joint United Nations Programme on HIV/AIDS

After the linear functions, the next simplest functions are the **quadratic functions,** which are defined by equations of the form

$$f(x) = ax^2 + bx + c \qquad (a \neq 0)$$

where a, b, and c are constants and a is not zero. (The word "quadratic" is derived from the Latin word for a square, *quadratus.*) We will see that the graph of any quadratic function is a curve called a **parabola** that is similar in shape to the basic $y = x^2$ graph. Figure 3 displays the graphs of two typical quadratic functions. Subsequent examples will demonstrate that the parabola opens upward when $a > 0$ and downward when $a < 0$. As Figure 3 indicates, the turning point on the parabola is called the **vertex.** The **axis of symmetry** of the parabola $y = ax^2 + bx + c$ is the vertical line passing through the vertex. (Mini Project 1 at the end of this section indicates how to *prove* that the graph does indeed have this symmetry.)

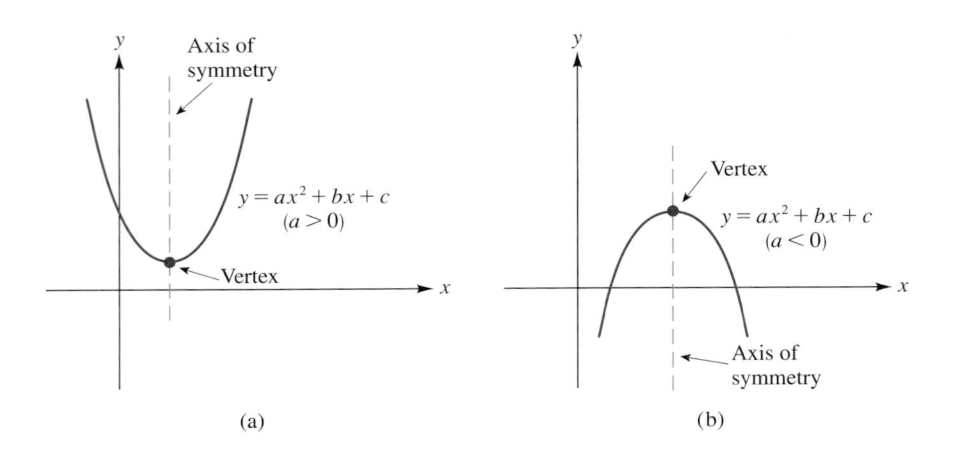

Figure 3

We mentioned in Section 4.1 that many graphing utilities have features for determining the regression line for a given set of data points. This is the line that best fits the data points. (We are using the phrase "best fits" in an intuitive sense here; the technical meaning is explained in a project at the end of Section 4.5.) In addition to fitting a linear function to a data set, many of these graphing utilities can fit other types of functions as well, including quadratics. In Example 1 we find and compare linear and quadratic models for the AIDS data in Figure 2. For details on using a graphing calculator to determine a quadratic function that best fits a data set, see the owner's manual that came with your graphing utility. [On a TI graphing calculator, go to the "STAT menu," choose "CALC," and then select "QuadReg" (which stands for *quadratic regression*). In Microsoft *Excel*™, go to the "Chart" menu, select "Add Trendline," then choose "Polynomial," and type in "2" for the order.]

EXAMPLE 1 **Using linear and quadratic functions to model AIDS data**

(a) Use a graphing utility to find and graph the linear and quadratic models for the AIDS data in Figure 2 on page 229.

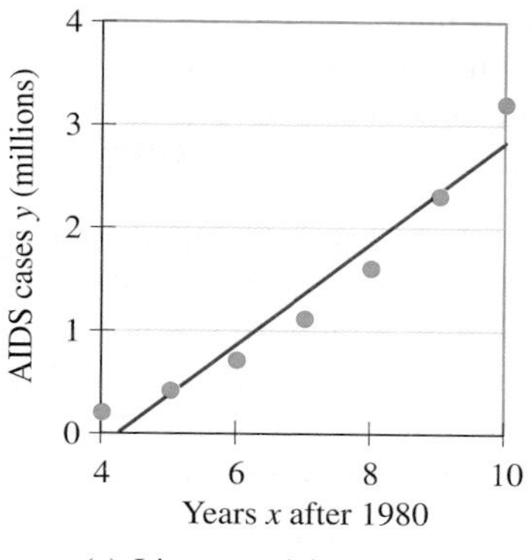

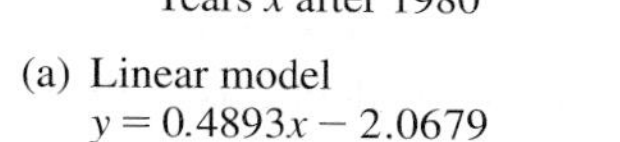
(a) Linear model
$y = 0.4893x - 2.0679$

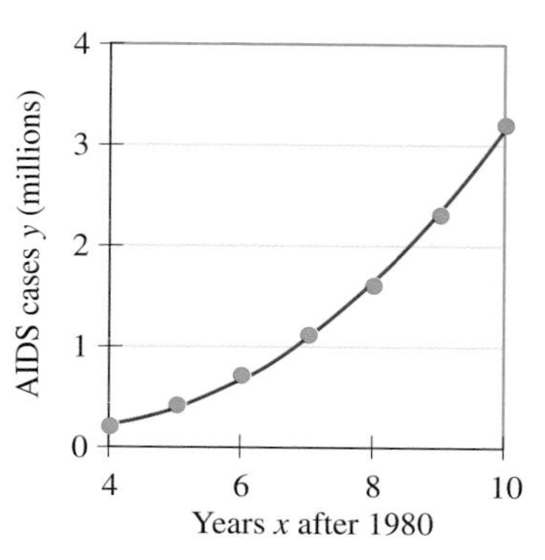

(b) Quadratic model
$y = 0.0679x^2 - 0.4607x + 0.9857$

Figure 4
Linear and quadratic models for AIDS data from Figure 2

(b) Use each function to make projections for the years 1992 and 1997. Then use the following information to see which model is more accurate in each case. The cumulative numbers of AIDS cases in 1992 and 1997 were 5.5 million and 15.1 million, respectively.

SOLUTION

(a) See Figure 4.

(b) The results of the calculations (rounded to one decimal place) are shown in Table 1. As perspective in evaluating these models, keep in mind that they are obtained using the data in Figure 2, which covers the years 1984–1990. So we are using the models to project two years and seven years beyond the data, that is, beyond 1990. For 1992 both projections fall short of the actual figure, but the quadratic model is much closer. As you should compute for yourself, the percentage errors for the linear and quadratic projections for 1992 are approximately 31% and 5%, respectively. For 1997 both projections again fall short of the actual figure. The linear model is even farther off this time; it predicts less than half of the actual figure. The quadratic model is again closer to reality, but as you can check, the percentage error in this case is a relatively large 15%. In summary: The quadratic model works quite well in the short run (two years), but it misses the mark considerably for the longer term (seven years). The linear model misses badly in both cases. For a health professional or a United Nations administrator writing a budget proposal to deal with the AIDS epidemic in 1990, evidently even the quadratic model doesn't grow fast enough and so might mislead the

TABLE 1 Comparison of Linear and Quadratic Models for Cumulative Number of AIDS Cases Worldwide in 1992 and 1997

Year	Linear model $y = 0.4893x - 2.0679$ (millions)	Quadratic model $y = 0.0679x^2 - 0.4607x + 0.9857$ (millions)	Actual figures (millions)
1992 ($x = 12$)	3.8	5.2	5.5
1997 ($x = 17$)	6.3	12.8	15.1

person into underfunding the project. (In later work we look at types of functions that grow faster than any quadratic. See, for example, Exercise 64 in Section 4.6, or see Section 5.1.)

In the next set of examples the main focus will be on graphing quadratic functions by hand, rather than with a graphing utility. The goal is to *understand* why the graphs look as they do and to see how they relate to the basic $y = x^2$. The techniques that we'll use in analyzing quadratic functions have already been developed in previous chapters. In particular, the following two topics are prerequisites for understanding the examples in this section:

- Completing the square—for a review see either Section 1.7 or 2.1
- Translations and reflections—for a review see Section 3.4

EXAMPLE 2 Graphing a quadratic function using completing the square and translation

Graph the function $y = x^2 - 2x + 3$.

SOLUTION

The idea here is to use the technique of completing the square; this will enable us to obtain the required graph simply by shifting the basic $y = x^2$ graph. We begin by writing the given equation

$$y = x^2 - 2x \qquad + 3$$

To complete the square for the x-terms we want to add 1. (Check this.) Of course, to keep the equation in balance, we have to account for this by writing

$$y = (x^2 - 2x + 1) + 3 - 1 \qquad \text{adding zero to the right side}$$

or

$$y = (x - 1)^2 + 2$$

Now, as we know from Section 3.4, the graph of this last equation is obtained by moving the parabola $y = x^2$ one unit in the positive x-direction and two units in the positive y-direction. This shifts the vertex from the origin to the point (1, 2). See Figure 5.

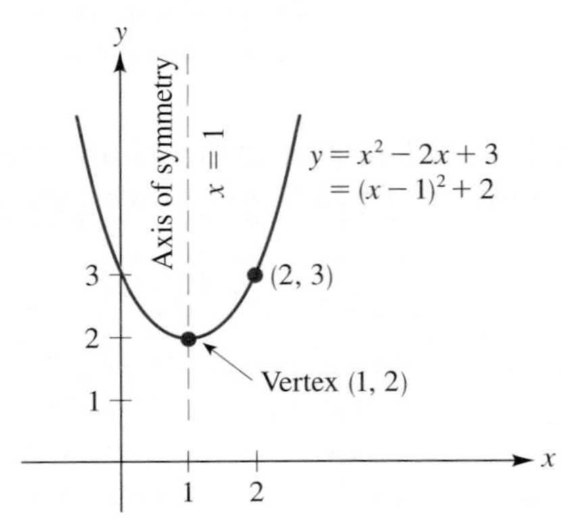

Figure 5

NOTE As a guide to sketching the graph, you'll want to know the y-intercept. To find the y-intercept, substitute $x = 0$ in the given equation to obtain $y = 3$. Then, given the vertex (1, 2) and the point (0, 3), a reasonably accurate graph can be quickly sketched. [Actually, once you find that (0, 3) is on the graph, you also know that the reflection of this point about the axis of symmetry is on the graph. This is why the point (2, 3) is shown in Figure 5.]

Now we want to compare the graphs of $y = x^2$, $y = 2x^2$, and $y = \frac{1}{2}x^2$. The last two functions were not specifically discussed in the previous chapters, so for now you can graph them by first setting up tables. If you do this, rather than using a graphing utility, you'll gain some insight into why the graphs in Figure 6 look as they do. All three graphs are parabolas that open upward, but $y = 2x^2$ appears to be narrower than $y = x^2$, while $y = \frac{1}{2}x^2$ appears wider than $y = x^2$. Another way (besides setting up tables) to see why the graphs appear this way is to start from the inequalities $\frac{1}{2} < 1 < 2$. Then, assuming for the moment that $x \neq 0$, we can multiply through by the *positive* quantity x^2 to obtain

$$\frac{1}{2}x^2 < x^2 < 2x^2 \qquad \text{(provided that } x \neq 0\text{)}$$

This tells us that with the exception of the origin, the graph of $y = \frac{1}{2}x^2$ is *below* that of $y = x^2$, which in turn is *below* $y = 2x^2$. This is why, on any given interval (for example $-2 \le x \le 2$, as in Figure 6), the graph of $y = \frac{1}{2}x^2$ appears wider than $y = x^2$, which in turn appears wider than $y = 2x^2$.

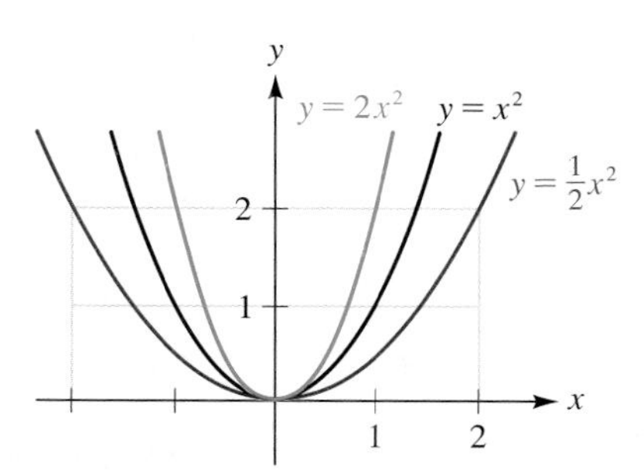

Figure 6

The observations that we've just made about the relative shapes of three parabolas also apply to $y = -x^2$, $y = -2x^2$, and $y = -\frac{1}{2}x^2$, except in these cases the parabolas open downward rather than upward. In the box that follows we summarize and generalize our observations up to this point.

PROPERTY SUMMARY The Graph of $y = ax^2$

1. The graph of $y = ax^2$ is a parabola with vertex at the origin. It is similar in shape to $y = x^2$.
2. The parabola $y = ax^2$ opens upward if $a > 0$, downward if $a < 0$.
3. The parabola $y = ax^2$ is narrower than $y = x^2$ if $|a| > 1$, wider than $y = x^2$ if $|a| < 1$.

EXAMPLE 3 Graphing parabolas that appear narrower than the basic $y = x^2$

Sketch the graphs of the following quadratic functions:
(a) $y = 3(x - 1)^2$; **(b)** $y = -3(x - 1)^2$.

SOLUTION

(a) Because of the $x - 1$, we shift the basic parabola $y = x^2$ one unit to the right. The factor of 3 in the given equation tells us that we want to draw a parabola that is narrower than $y = x^2$ but that has the same vertex, (1, 0). To see exactly how narrow to draw $y = 3(x - 1)^2$, we need to know another point on the graph other than the vertex, (1, 0). An easy point to obtain is the y-intercept. Setting $x = 0$ in the equation yields $y = 3(0 - 1)^2 = 3$. Now that we know the vertex, (1, 0), and the y-intercept, 3, we can sketch a reasonably accurate graph; see Figure 7(a).

(b) In part (a) we sketched the graph of $y = 3(x - 1)^2$. By reflecting that graph in the x-axis, we obtain the graph of $y = -3(x - 1)^2$, as shown in Figure 7(b).

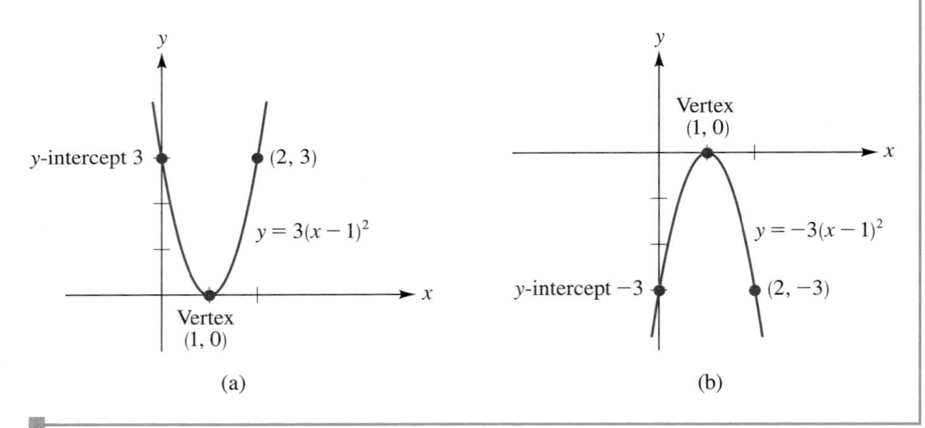

Figure 7

EXAMPLE 4 Analyzing a quadratic function

Graph the function $f(x) = -2x^2 + 4x + 6$ and specify the vertex, axis of symmetry, maximum or minimum value of f, and x- and y-intercepts.

SOLUTION

The idea is to complete the square, as in Example 2. We have

$$\begin{aligned} y &= -2x^2 + 4x \qquad + 6 \\ &= -2(x^2 - 2x \qquad) + 6 \\ &= -2(x^2 - 2x + 1) + 6 + 2 && \text{adding } 0 = (-2)(1) + 2 \text{ to the right-hand side} \\ &= -2(x - 1)^2 + 8 && (1) \end{aligned}$$

From equation (1) we see that the required graph is obtained simply by shifting the graph of $y = -2x^2$ "right 1, up 8," so that the vertex is (1, 8). As a guide to sketching the graph, we want to compute the intercepts. The y-intercept is 6. (Why?) For the x-intercepts we replace y with 0 in equation (1) to obtain

$$\begin{aligned} -2(x-1)^2 &= -8 \\ (x-1)^2 &= 4 \\ x-1 &= \pm 2 \\ x = 1 \pm 2 &= -1 \quad \text{or} \quad 3 \end{aligned}$$

Thus the x-intercepts are $x = -1$ and $x = 3$. Knowing these intercepts and the vertex, we can sketch the graph as in Figure 8. You should check for yourself that the information accompanying Figure 8 is correct.

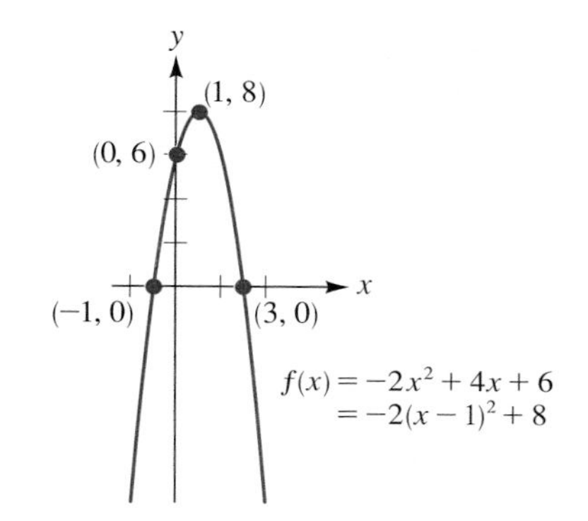

$f(x) = -2x^2 + 4x + 6$	
Vertex:	(1, 8)
Axis of symmetry:	$x = 1$
Maximum value of f:	8
y-intercept:	6
x-intercepts:	−1 and 3

Figure 8

The next three examples involve maximum and minimum values of functions. As background for this, we first summarize our basic technique for graphing parabolas.

PROPERTY SUMMARY
The Graph of the Parabola $y = ax^2 + bx + c$

By completing the square, the equation of the parabola $y = ax^2 + bx + c$ can always be rewritten in the form

$$y = a(x - h)^2 + k$$

In this form, the vertex of the parabola is (h, k) and the axis of symmetry is the line $x = h$. The parabola opens upward if $a > 0$ and downward if $a < 0$.

EXAMPLE 5 Determining the input at which a quadratic function has a minimum value

The graph of the quadratic function $g(x) = x^2 - \frac{6}{7}x + 2$ is a parabola that opens upward because the coefficient of x^2 is positive. Thus it makes sense to talk about a minimum rather than maximum value for this function. Let x_0 denote the input for which $g(x_0)$ is minimum.

(a) Use a TRACE and/or ZOOM feature on a graphing utility to estimate x_0 to the nearest one-tenth. What is the corresponding estimate for $g(x_0)$?

(b) Use algebra to determine the exact value of x_0 and the minimum value of the function. Check that the results are consistent with the visual evidence in part (a).

SOLUTION

(a) From the preliminary view obtained in Figure 9(a), we can see that the input x_0 at which the function has a minimum is roughly 0.5, a bit less actually. The zoom-in view in Figure 9(b) reveals that x_0 is closer to 0.4 than to 0.5. So we have $x_0 \approx 0.4$, to the nearest tenth. Figure 9(b) shows that the output $g(x_0)$ is between 1.81 and 1.82, so certainly to the nearest tenth we can say that $g(x_0) \approx 1.8$. *Suggestion*: Before working this example with your own graphing utility, reread the comments on page 161 regarding zooming in on turning points.

GRAPHICAL PERSPECTIVE

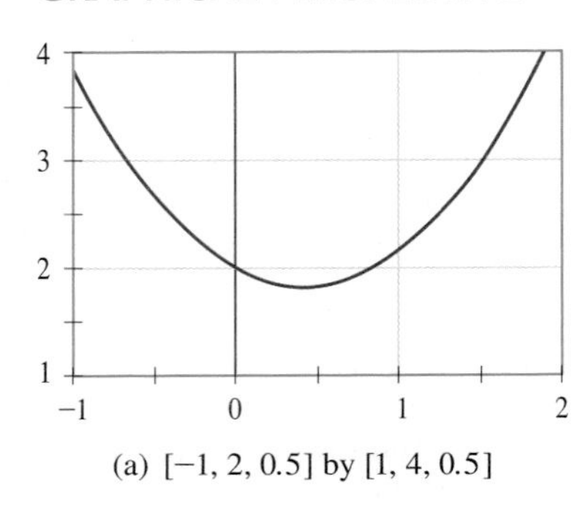

(a) $[-1, 2, 0.5]$ by $[1, 4, 0.5]$

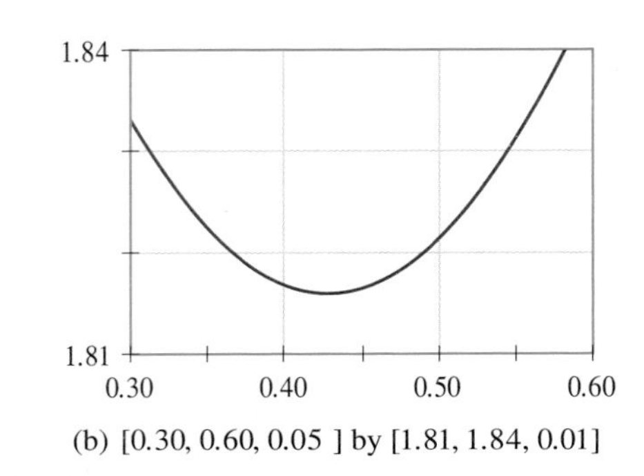

(b) $[0.30, 0.60, 0.05]$ by $[1.81, 1.84, 0.01]$

Figure 9
Zooming in on the turning point of $g(x) = x^2 - \frac{6}{7}x + 2$

(b) By completing the square as in Example 4, we find that the given equation can be rewritten as

$$g(x) = \left(x - \frac{3}{7}\right)^2 + \frac{89}{49}$$

(You should verify this for yourself, using Example 4 as a model.) This last equation tells us that the vertex of the parabola is the point $(\frac{3}{7}, \frac{89}{49})$. Thus the minimum value of the function is $\frac{89}{49}$ and this minimum occurs at $x_0 = \frac{3}{7}$. Using a calculator now, you'll find that $\frac{3}{7} \approx 0.429$ and $\frac{89}{49} \approx 1.816$. Both of these values are consistent with the graphical estimates in part (a).

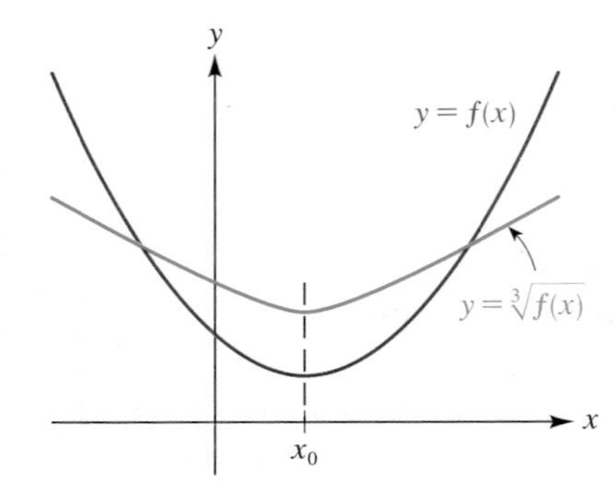

Figure 10
The minimum value occurs at x_0 for both of the functions $y = \sqrt[3]{f(x)}$ and $y = f(x)$.

For the previous two examples, keep in mind that we were able to find the maximum or minimum easily only because the functions were quadratics. In contrast, you cannot expect to find the minimum of $y = x^4 - 8x$ using the method of Examples 4 and 5 because it is not a quadratic function. In general, the techniques of calculus are required to find maxima and minima for functions other than quadratics. There are some cases, however, in which our present method can be adapted to functions that are closely related to quadratics. For instance, in the next example, we look for an input that minimizes a function of the form $y = \sqrt[3]{f(x)}$, where f is a quadratic function. As indicated in Figure 10, we need only find the input that minimizes the quadratic function $y = f(x)$, because this same input also minimizes $y = \sqrt[3]{f(x)}$.

(The outputs of the two functions are, of course, different, but the point here is that the same *input* does the job for both functions.)

EXAMPLE 6 **Finding the minimum of a function that's related to a quadratic**

Let $f(x) = \sqrt[3]{x^2 - x + 1}$. Which x-value yields the minimum value for the function f? What is this minimum value?

SOLUTION

According to the remarks just prior to this example, the x-value that minimizes the function $f(x) = \sqrt[3]{x^2 - x + 1}$ will be the same x-value that minimizes the function $y = x^2 - x + 1$. To find this x-value, we complete the square, just as we've done previously in this section. We have

$$\begin{aligned} y &= x^2 - x \qquad + 1 \\ &= x^2 - x + \frac{1}{4} + 1 - \frac{1}{4} \\ &= \left(x - \frac{1}{2}\right)^2 + \frac{3}{4} \end{aligned}$$

This shows that the vertex of the parabola $y = x^2 - x + 1$ is $(1/2, 3/4)$. Since this parabola opens upward, we conclude that the input $x = 1/2$ will produce the minimum value for $y = x^2 - x + 1$ and also for $f(x) = \sqrt[3]{x^2 - x + 1}$. The minimum value of the function f is

$$\begin{aligned} f(1/2) &= \sqrt[3]{(1/2)^2 - (1/2) + 1} \\ &= \sqrt[3]{3/4} \approx 0.91 \end{aligned}$$

In summary: The minimum value of the function f occurs when $x = 1/2$. This minimum value is $\sqrt[3]{3/4}$, which is approximately 0.91. See Figure 11 for a graphical perspective.

GRAPHICAL PERSPECTIVE

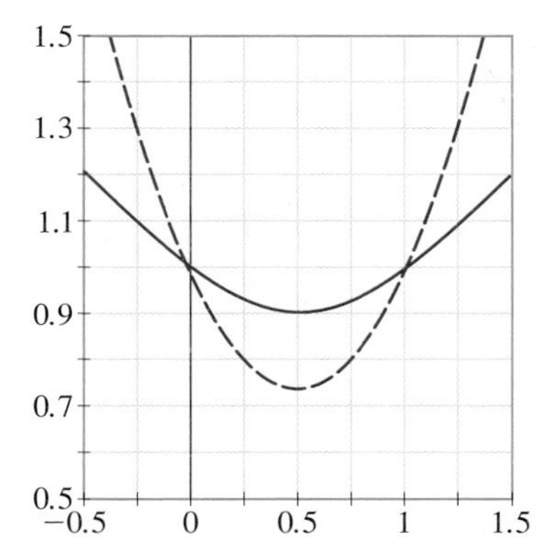

Figure 11
The dashed graph is $y = x^2 - x + 1$; the other is $f(x) = \sqrt[3]{x^2 - x + 1}$. The minimum value for both functions occurs when $x = 0.5$. The minimum value of f is $\sqrt[3]{3/4}$, which is approximately 0.9.

EXAMPLE 7 Finding inputs that maximize a function related to a quadratic

As indicated in Figure 12, the maximum value of the function $y = -t^4 + 3t^2$ occurs when t is between 1.0 and 1.5 and also when t is between -1.0 and -1.5. Determine these t-values exactly, then obtain calculator approximations rounded to three decimal places.

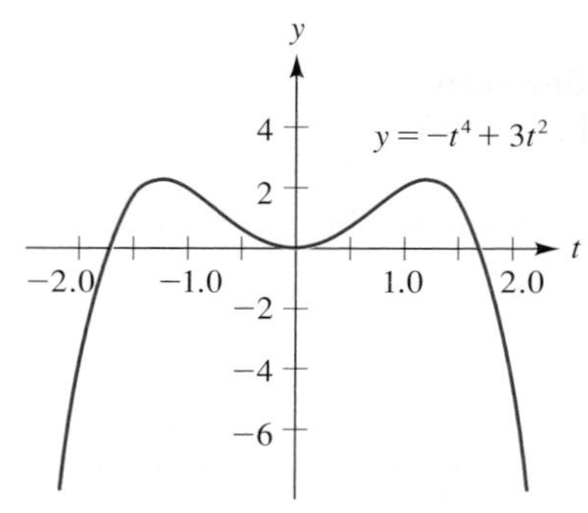

Figure 12

SOLUTION

The substitution $t^2 = x$ will reduce this question to one about a quadratic function. If $t^2 = x$, then $t^4 = x^2$ and we have

$$\begin{aligned} y &= -t^4 + 3t^2 \\ &= -x^2 + 3x \\ &= -\left(x - \frac{3}{2}\right)^2 + \frac{9}{4} \end{aligned}$$

completing the square, as we've done throughout this section (Check the algebra!)

The graph of this last equation is a parabola that opens downward, and the maximum occurs when x is $3/2$. So we have

$$t^2 = x = \frac{3}{2}$$

and therefore

$$t = \pm\sqrt{\frac{3}{2}} = \pm\frac{\sqrt{3}}{\sqrt{2}} = \pm\frac{\sqrt{6}}{2} \approx \pm 1.225$$

Note that these calculator values are consistent with Figure 12.

In many real-life applications in science and business, data are collected that can be arranged in an x-y table of values. (We've seen examples in this and the previous section.) We conclude this section with a useful theorem that tells how to determine whether a given x-y table can be generated by a linear function or a quadratic function (or neither). As background for the theorem, we explain what is meant by *first differences* and *second differences.* Suppose we have a list of numbers, say, 2, 7, 8, 4, 15. Then the list of **first differences** is a new list formed by subtracting adjacent members of the given list as follows:

$$7 - 2(=5), \quad 8 - 7(= 1), \quad 4 - 8(= -4), \quad 15 - 4 = 11$$

So, for this example, the list of first differences is 5, 1, -4, 11. The process can be summarized conveniently as shown in Figure 13.

Once we have the list of first differences, we can follow the same procedure to form the list of **second differences** as shown in Figure 14. For the second differences we are subtracting adjacent numbers in the list of first differences.

With the terminology we've now introduced, we are ready to state the theorem for determining whether or not a given table of x-y values can be generated by a lin-

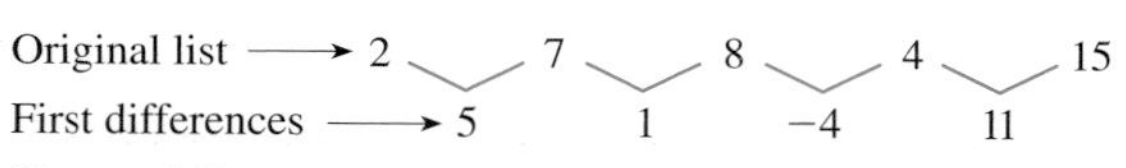

Figure 13

Original list → 2 7 8 4 15
First differences → 5 1 −4 11
Second differences → −4 −5 15

Figure 14

ear or quadratic function. In the statement of the theorem, reference is made to x values that are *equally spaced.* That means, for example, we could have $x = 1, 2, 3,$ and so on, or $x = 5, 10, 15,$ and so on, but not $x = 1, 2, 4, \ldots$.

THEOREM How to Determine Whether a Data Set Can Be Generated by a Linear or Quadratic Function

Suppose we have a table of x-y values and the x-values are equally spaced. Then:

(a) The data can be generated by a linear function if and only if the first differences of the y-values are constant.

(b) The data can be generated by a quadratic function if and only if the second differences of the y-values are constant.

The proof for part (a) of this theorem is outlined in Exercises 63 and 64. We omit the proof of part (b) because of the lengthy algebra that is involved.

EXAMPLE 8 Using first and second differences

The x-y values in Figure 15 are generated by a linear function because the first differences of the y's are constant. The x-y values in Figure 16 are generated by neither a linear function nor a quadratic function because neither the first nor second differences are constant.

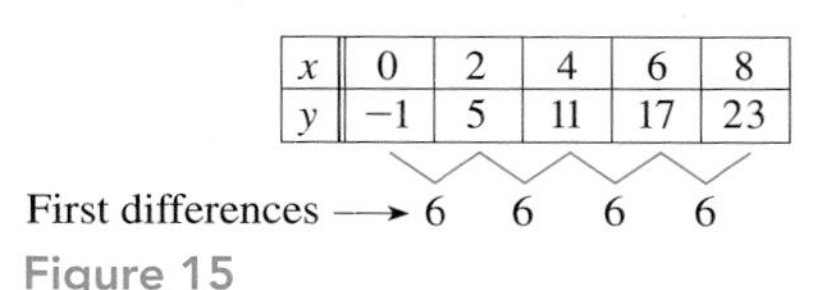

Figure 15

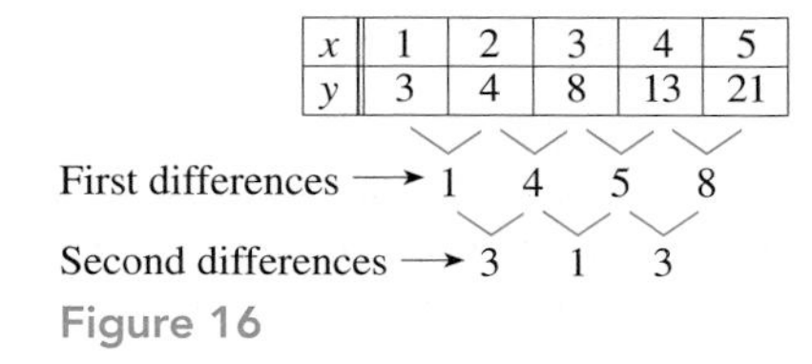

Figure 16

EXERCISE SET 4.2 

A

Exercises 1 and 2 require graphing utilities that can create scatter plots and compute linear and quadratic regression models.

Ⓖ **1.** The following table shows the number of air passenger miles flown worldwide for selected years from 1950 to 1985.

(a) Use a graphing utility to create a scatter plot for the data. Then determine the linear model and the quadratic model that fit the data as closely as possible. Add the graph of each model to your scatter plot.

(b) Use each model to make projections for the number of passenger miles for 1990 and for 1998.

(continues)

Year x ($x = 0$ is 1950)	0	5	10	15	20	25	30	35
Passenger miles y (billions)	16.94	36.90	65.94	119.78	278.57	421.83	658.88	827.19

Data computed from *Vital Signs 1999,* Lester R. Brown et al. (New York: W. W. Norton & Co., 1999)

(c) Use the following information to compute the percentage errors in the projections in part (b). For each case, which model produces the smaller percentage error? The number of passenger miles for 1990 and 1998 were 1145.94 billion and 1585.74 billion, respectively.

2. The following table shows the number of cellular phone subscribers worldwide over the years 1990–1995.

Year x (x = 0 is 1990)	0	1	2	3	4	5
Cell phone subscribers (millions)	11	16	23	34	55	91

Source: *Vital Signs, 2000,* Lester R. Brown et al. (New York: W. W. Norton & Co., 2000)

Ⓖ **(a)** Use a graphing utility to create a scatter plot for the data. Then determine the quadratic model that fit the data as closely as possible. Add the graph of the quadratic function to your scatter plot.

(b) Use the model to make projections for the number of cell phone subscribers for 1996 and for 1998. Which projection do you think might be more accurate? Now compute the percentage error in each projection using the following information. The number of cell phone subscribers in 1996 and 1998 were 142 million and 319 million, respectively.

For Exercises 3 and 4, refer to the following figures. Figure A shows a scatter plot and quadratic model for the U.S. population over the period 1790–1950. Figure B shows similar information for the world population over the period 1750–1950. For both figures, x represents the number of years after 1750. (Thus x = 0 corresponds to the year 1750, and x = 200 corresponds to 1950.)

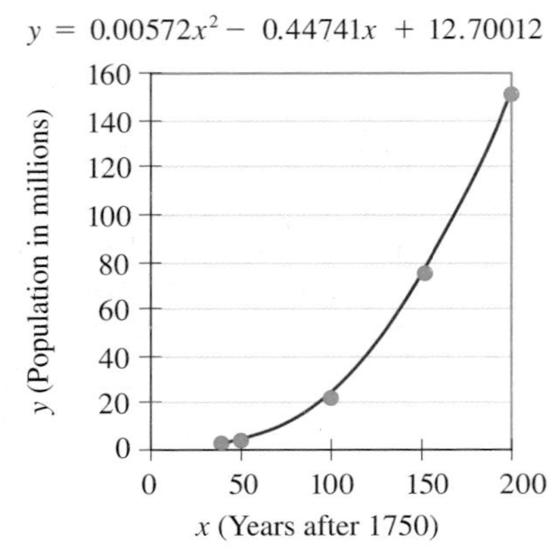

Figure A
Scatter plot and quadratic model for U.S. population, 1790–1950

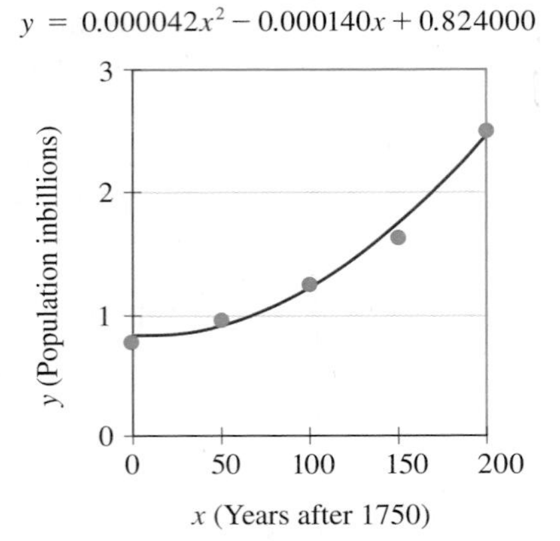

Figure B
Scatter plot and quadratic model for world population, 1750–1950

3. Use the quadratic model in Figure A to complete the following table. Round each projection to two decimal places; round the percentage error to the nearest integer.

	Projected U.S. population (millions)	Actual U.S. population (millions)	Projection too high or too low?	% error
1970		203.30		
2000		275.60		

4. Use the quadratic model in Figure B to complete the following table. Round each projection to two decimal places; round the percentage error to the nearest integer.

	Projected world population (billions)	Actual world population (billions)	Projection too high or too low?	% error
1970		3.708		
2000		6.080		

Afterword on Exercises 3 and 4: For the time spans shown in Figures A and B, quadratic models describe both the U.S. and world populations quite well. That is, the graph of each quadratic function comes very close to the data points in the scatter plot. Exercise 3 shows that a quadratic model for the U.S. population does a reasonably good job of projecting the population figures 20 and even 50 years beyond the base data. Exercise 4, on the other hand, shows that the quadratic model is inappropriate for similar projections of the world population, because the quadratic model grows much too slowly. In Chapter 5 we study *exponential functions,* which can be much more appropriate for modeling very rapid growth.

In Exercises 5–20, graph the quadratic function. Specify the vertex, axis of symmetry, maximum or minimum value, and intercepts.

5. $y = (x + 2)^2$
6. $y = -(x + 2)^2$
7. $y = 2(x + 2)^2$
8. $y = 2(x + 2)^2 + 4$
9. $y = -2(x + 2)^2 + 4$
10. $y = x^2 + 6x - 1$
11. $f(x) = x^2 - 4x$
12. $F(x) = x^2 - 3x + 4$
13. $g(x) = 1 - x^2$
14. $y = 2x^2 + \sqrt{2}x$
15. $y = x^2 - 2x - 3$
16. $y = 2x^2 + 3x - 2$
17. $y = -x^2 + 6x + 2$
18. $y = -3x^2 + 12x$
19. $s = 2 + 3t - 9t^2$
20. $s = -\frac{1}{4}t^2 + t - 1$

For Exercises 21–26, determine the input that produces the largest or smallest output (whichever is appropriate). State whether the output is largest or smallest.

21. $y = 2x^2 - 4x + 11$
22. $f(x) = 8x^2 + x - 5$
23. $g(x) = -6x^2 + 18x$
24. $s = -16t^2 + 196t + 80$
25. $f(x) = x^2 - 10$
26. $h(x) = x^2 - 10x$

In Exercises 27–30, find the maximum or minimum value for each function (whichever is appropriate). State whether the value is a maximum or minimum.

27. $y = x^2 - 8x + 3$
28. $y = \frac{1}{2}x^2 + x + 1$
29. $y = -2x^2 - 3x + 2$
30. $y = -\frac{1}{3}x^2 - 2x$

In Exercises 31–34, you are given a quadratic function.

(a) *By looking at the coefficient of the square term, state whether the function has a maximum or a minimum value.*

Ⓖ **(b)** *Use a* TRACE *and/or* ZOOM *feature on a graphing utility to estimate the input x_0 for which the function obtains its maximum or minimum value. (Estimate to the nearest one-tenth, as in Example 5.) What is the corresponding estimate for the maximum or minimum value?*

(c) *Use algebra to determine the exact value of x_0 and the corresponding maximum or minimum value of the function. Check to see that the results are consistent with the graphical estimates obtained in part (b).*

31. $f(x) = 6x^2 - x - 4$
32. $g(t) = 40t - 7t^2$
33. $y = -9t^2 + 40t + 1$
34. $y = \sqrt{2}x^2 + 4x + 3$

35. How far from the origin is the vertex of the parabola $y = x^2 - 6x + 13$?

36. Find the distance between the vertices of the parabolas $y = -\frac{1}{2}x^2 + 4x$ and $y = 2x^2 - 8x - 1$.

For Exercises 37–42, the functions f, g, and h are defined as follows:

$$f(x) = 2x - 3 \qquad g(x) = x^2 + 4x + 1 \qquad h(x) = 1 - 2x^2$$

In each exercise, classify the function as linear, quadratic, or neither.

37. $f \circ g$
38. $g \circ f$
39. $g \circ h$
40. $h \circ g$
41. $f \circ f$
42. $h \circ h$

In Exercises 43 and 44, determine the inputs that yield the minimum values for each function. Compute the minimum value in each case.

43. (a) $f(x) = \sqrt{x^2 - 6x + 73}$
(b) $g(x) = \sqrt[3]{x^2 - 6x + 73}$
(c) $h(x) = x^4 - 6x^2 + 73$

44. (a) $F(x) = (4x^2 - 4x + 109)^{1/2}$
(b) $G(x) = (4x^2 - 4x + 109)^{1/3}$
(c) $H(x) = 4x^4 - 4x^2 + 109$

Ⓖ *For Exercises 45 and 46, first, use a graphing utility to estimate to the nearest one-tenth the maximum value of the function. Then use algebra to determine the exact value, and check that your answer is consistent with the graphical estimate.*

45. (a) $f(x) = \sqrt{-x^2 + 4x + 12}$
(b) $g(x) = \sqrt[3]{-x^2 + 4x + 12}$
(c) $h(x) = -x^4 + 4x^2 + 12$

46. (a) $F(x) = (27x - x^2)^{1/2}$
(b) $G(x) = (27x - x^2)^{1/3}$
(c) $H(x) = 27x^2 - x^4$

In Exercises 47–50, determine whether the x-y values are generated by a linear function, a quadratic function, or neither.

47.

x	−4	−2	0	2	4
y	25	3	−4	7	33

48.

x	1	2	3	4	5
y	−12	−8.5	−5	−1.5	2

49.

x	0	1	2	3	4
y	−21	−3	7	9	3

50.

x	0.25	0.50	0.75	1.00	1.25
y	−0.40	−0.16	0.08	0.32	0.62

Ⓖ **51.** On the same set of axes, graph the four parabolas $y = x^2$, $2x^2$, $3x^2$, and $8x^2$. Relative to your graphs, where do you think the graph of $y = 50x^2$ would fit in? After answering, check by adding the graph of $y = 50x^2$ to the picture.

Ⓖ **52.** Graph the four parabolas $y = x^2$, $0.5x^2$, $0.25x^2$, and $0.125x^2$. Relative to your graphs, where do you think the graph of $y = 0.02x^2$ would fit in? After answering, check by adding the graph of $y = 0.02x^2$ to the picture.

B

53. Let $f(x) = x^2$. Find the average rate of change $\Delta f/\Delta x$ on the interval $[a, x]$.

54. If $f(x) = ax^2 + bx + c$, show that
$$\frac{f(x+h) - f(x)}{h} = 2ax + ah + b.$$

55. Find the x-coordinate of the vertex of the parabola $y = (x - a)(x - b)$. (Your answer will be in terms of the constants a and b.) *Hint*: It's easier here to rely on symmetry than on completing the square.

Ⓖ **56. (a)** Graph the two functions $y = x^2 + 4x$ and $y = x^2 - 4x$. How are the two graphs related (in terms of symmetry)?
(b) Follow part (a) using the two functions $y = -2x^2 + 3x + 4$ and $y = -2x^2 - 3x + 4$.
(c) Which one of the graphing techniques from Section 3.4 relates to what you have observed in parts (a) and (b)?

In Exercises 57–60, find quadratic functions satisfying the given conditions.

57. The graph passes through the origin, and the vertex is the point (2, 2). *Hint*: What do h and k stand for in the general equation $y = a(x - h)^2 + k$?

58. The graph is obtained by translating $y = x^2$ four units in the negative x-direction and three units in the positive y-direction.

59. The vertex is $(3, -1)$, and one x-intercept is 1.

60. The axis of symmetry is the line $x = 1$. The y-intercept is 1. There is only one x-intercept.

61. Let $g(x) = x^2 + bx$. Are there any values for b for which the minimum value of this function is -1? If so, what are they? If not, explain why.

62. For which value of c will the minimum value of the function $f(x) = x^2 + 2x + c$ be $\sqrt{2}$?

63. This exercise shows that if we have a table generated by a linear function and the x-values are equally spaced, then the first differences of the y-values are constant.
(a) In the following data table, the three x-entries are equally spaced. Compute the three entries in the $f(x)$ row assuming that f is the linear function given by $f(x) = mx + b$. (Don't worry about the fact that your answers contain all four of the letters m, b, a, and h.)

x	a	$a + h$	$a + 2h$
f(x)			

(b) Compute the first differences for the three quantities that you listed in the $f(x)$ row in part (a). (The two first differences that you obtain should turn out to be equal, as required.)

64. Suppose we have a table of x-y values with the x-values equally spaced. This exercise shows that if the first differences of the y-values are constant, then the table can be generated by a linear function. *Note*: To show that the data points are generated by a linear function, it is enough to show that the points all lie on one nonvertical line.

Consider the following table with the x-values equally spaced by an amount $h \neq 0$. (We are assuming that h is nonzero to guarantee that the three x-values are distinct.)

x	a	$a + h$	$a + 2h$
y	y_1	y_2	y_3

Assuming that the first differences of the y-values are constant, we have $y_2 - y_1 = y_3 - y_2 = k$, where k is a constant.
(a) Check that the slope of the line joining the two data points (a, y_1) and $(a + h, y_2)$ is k/h.
(b) Likewise, check that the slope of the line joining the two data points $(a + h, y_2)$ and $(a + 2h, y_3)$ is k/h.
From parts (a) and (b), we conclude that the three data points lie on a nonvertical line, as required. (The slope k/h is well defined because we assumed $h \neq 0$.)

65. The following table and scatter plot show global coal consumption for the years 1990–1995.

Global Coal Consumption 1990–1995

Year x x = 0 ↔ 1990	Coal consumption y (billion tons)
0	3.368
1	3.285
2	3.258
3	3.243
4	3.261
5	3.311

Source: World Resources Institute

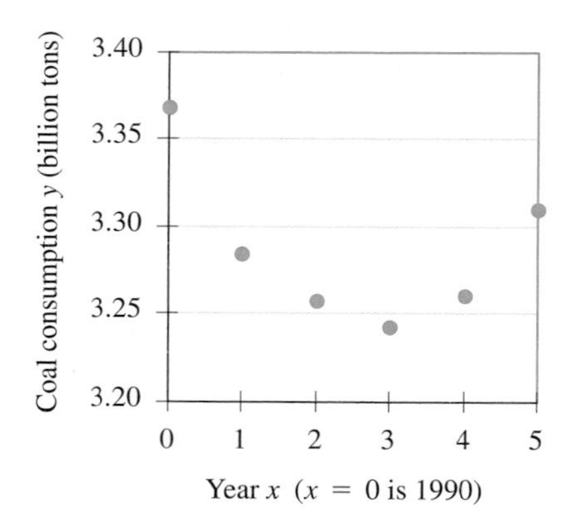

(a) Use a graphing utility to find a quadratic model for the data. Then use the model to make estimates for global coal consumption in 1989 and 1996.
(b) Use the following information to show that, in terms of percentage error, the 1996 estimate is better than the 1989 estimate, but in both cases the percentage error is less than 2%. The actual figures for coal consumption in 1989 and 1996 are 3.408 and 3.428 billion tons, respectively.

(c) Use the model to project worldwide coal consumption in 1998. Then show that the percentage error is more than 9%, given that the actual 1998 consumption was 3.329 billion tons.

66. *Driven by rising consumer demand and growing dissatisfaction with conventional farming practices, the organic agriculture industry is soaring.—Vital Signs 2000,* Lester R. Brown et al. (New York: W. W. Norton & Co., 2000)

The following table and scatter plot show the area devoted to organic farming in the European Union over the period 1988–1998. (One hectare equals 2.471 acres or 10,000 square meters.)

Year x	Area y (million hectares)
1988	0.17
1990	0.34
1992	0.69
1994	1.15
1996	1.90
1998	3.17

Data from *Vital Signs 2000*

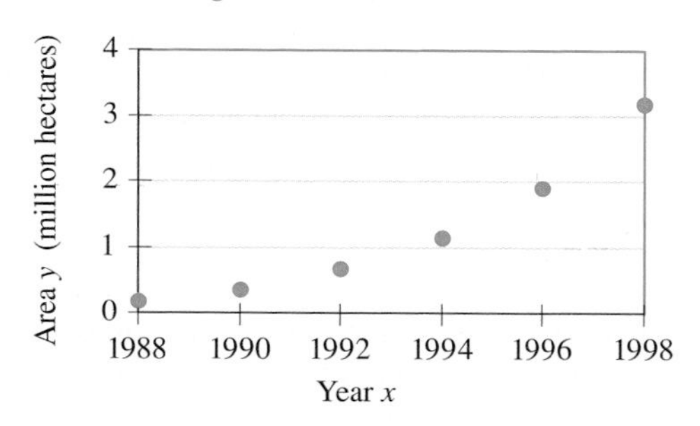

(a) In view of the scatter plot and the opening quotation, a quadratic model might be useful here, at least in making short-term predictions. Indeed, if you look at the scatter plot without knowing that it was generated from real-life data, in the context of this section you might think that the points all lie exactly on the graph of a parabola. Show that this is not the case: Compute the second differences of the y-values in the given table, and note that they are not equal. (In fact, no two are equal.)

(b) In the following table, fill in the missing entries so that all six data points are generated by one quadratic function. *Hint:* Work back from the second differences; you want them all to be equal.

1988	0.17
1990	0.34
1992	0.69
1994	?
1996	?
1998	?

C

67. By completing the square, show that the coordinates of the vertex of the parabola $y = ax^2 + bx + c$ are $(-b/2a, -D/4a)$, where $D = b^2 - 4ac$.

68. Compute the average of the two x-intercepts of the graph of $y = ax^2 + bx + c$. (Assume $b^2 - 4ac > 0$.) How does your answer relate to the result in Exercise 67?

69. Consider the quadratic function $y = px^2 + px + r$, where $p \neq 0$.

(a) Show that if the vertex lies on the x-axis, then $p = 4r$.

(b) Show that if $p = 4r$, then the vertex lies on the x-axis.

70. Let $f(x) = ax^2 + bx$, where $a \neq 0$ and $b \neq 0$. Find a value for b such that the equation $f(f(x)) = 0$ has exactly three real roots.

MINI PROJECT 1 — HOW DO YOU KNOW THAT THE GRAPH OF A QUADRATIC FUNCTION IS ALWAYS SYMMETRIC ABOUT A VERTICAL LINE?

(a) A short answer that sweeps everything under the rug: The graph of the basic parabola $y = x^2$ is symmetric about a vertical axis, and the graph of $f(x) = ax^2 + bx + c$ inherits this type of symmetry. In a group, discuss the reasons supporting this short answer. (The discussion won't be complete without taking into account the following three concepts: completing the square, translations, and reflection in the y-axis.) After the discussion, write a paragraph or two answering, as carefully as you can, the question raised in the title of this mini project.

(b) Here's a completely different way to go about answering the question in the title of this mini project. An advantage to this alternative approach is that, as a by-product, it gives you the equation of the axis of symmetry without the need for completing the square. We start with the definition of what it means for the graph of $y = f(x)$ to be symmetric about a vertical line, say, $x = x_0$.

DEFINITION **Symmetry about the Line $x = x_0$**

The graph of a function f is symmetric about the line $x = x_0$ provided the following condition holds. Whenever $x_0 + h$ is in the domain of f, then so is $x_0 - h$ and

$$f(x_0 + h) = f(x_0 - h)$$

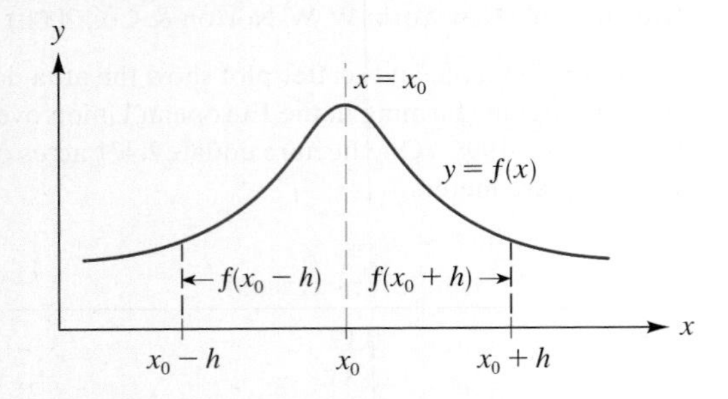

Now suppose we have a quadratic function

$$f(x) = ax^2 + bx + c \qquad (a \neq 0) \tag{1}$$

Because the domain is the set of all real numbers, we don't have to worry about the part of the definition in the box above regarding domain. Thus, to show that the graph of the quadratic function is symmetric about a line $x = x_0$ we need to find a value for x_0 so that the following equation holds for all values of h:

$$f(x_0 + h) = f(x_0 - h) \tag{2}$$

Now your job: Evaluate each side of equation (2) using the definition of f in equation (1). Then solve the resulting equation for x_0. You should obtain $x_0 = -b/2a$. This shows that the graph of the quadratic function is indeed symmetric about a vertical line, namely, $x = -b/2a$.

MINI PROJECT 2 WHAT'S LEFT IN THE TANK?

(a) The following table shows global oil consumption for the years 1990–1995, 1998, and 1999. Using only the data for the years 1990–1995, use a graphing utility to find a linear model and a quadratic model for global oil consumption.

(b) Provide the detailed calculations to back up the following statements regarding the use of these models in projecting global oil consumption for the years 1998 and 1999.

For both 1998 and 1999 the linear model underestimates oil consumption, while the quadratic model overestimates. For both models, however, the percentage error is always less than 5%. For both years, the quadratic model gives the better projection (as measured by percentage error).

Global Oil Consumption, 1990–1995, 1998, 1999

Year x ($x = 0$ corresponds to 1990)	0	1	2	3	4	5	8	9
Global oil consumption y (billion barrels)	23.886	23.913	24.105	23.997	24.506	24.893	26.251	26.723

Source: U.S. Energy Information Administration

(c) Oil, as you know, is a nonrenewable energy resource; at some point in the future, it will be used up. When this will occur will depend on, among other things, consumption patterns and the amount of oil that exists on the planet. In this portion of the mini project you'll compute estimates for the *depletion date* for oil using three different consumption scenarios. On the basis of figures from the U.S. Geological Survey and the Energy Information Administration, we'll assume that in 1990 the total amount of oil remaining was 2886 billion barrels. [This estimate takes into account reserves and resources as of 1990, plus resources discovered in the 1990s, plus current estimates (at the end of the 1990s) for as yet undiscovered resources.]

A consequence of the linear yearly consumption model: Suppose that the consumption y (in billions of barrels) in year x is given by a linear function, $y = mx + b$. Then, using a result from Chapter 13 (Exercise 1 in Section 13.1), it can be shown that the total amount A of oil consumed over the years $x = 0$ through $x = T$ is given by the quadratic model

$$A = \frac{m}{2}T^2 + \left(\frac{m}{2} + b\right)T + b$$

On the left side of this equation, replace A with 2886; on the right side, replace m and b with the values that you obtained for the linear model in part (a). Now find the positive root of the resulting quadratic equation. (Use either the quadratic formula or a graphing utility.) The root represents the "life expectancy" of oil under this scenario. Round the answer to the nearest ten years and add to 1990 to get a depletion date for oil.

A consequence of the quadratic yearly consumption model: Suppose that the consumption y (in billions of barrels) in year x is given by a quadratic function, $y = ax^2 + bx + c$. Then, using a result from Chapter 13 (Exercise 5 in Section 13.1), it can be shown that the total amount A of oil consumed over the years $x = 0$ through $x = T$ is given by the cubic model

$$A = \frac{a}{3}T^3 + \frac{a + b}{2}T^2 + \left(\frac{a + 3b + 6c}{6}\right)T + c$$

On the left side of this equation, replace A with 2886; on the right side, replace a, b, and c with the values that you obtained for the quadratic model in part (a). Now use a graphing utility to find the positive root of the resulting cubic equation. The root gives you the "life expectancy" of oil under this scenario. Round the answer to the nearest ten years and add to 1990 to get a depletion date for oil.

Constant consumption model: If you've done the preceding calculations correctly, you will have found that the depletion dates under both models occur within the present century. Now try the following "what if." Go back to the linear consumption model and redo the calculations using for m the value 0 rather than the value from part (a). In terms of consumption, what's the interpretation of $m = 0$? What depletion date do you obtain?

(d) Write a paragraph or two summarizing the results from part (c). (Don't forget to include what the assumptions are in each case.)

4.3 USING ITERATION TO MODEL POPULATION GROWTH (Optional Section)

Although I shall henceforth adopt the habit of referring to the variable X as "the population," there are countless situations outside population biology where . . . [iteration of functions] *applies. . . . Examples in economics include models for the relationship between commodity quantity and price, for the theory of business cycles, and for the temporal sequences generated by various other economic quantities.*

. . . I would therefore urge that people be introduced to, say, [the iteration process for $f(x) = kx(1 - x)$] *early in their mathematical education. This equation can be studied phenomenologically by iterating it on a calculator, or even by hand. Its study does not involve as much conceptual sophistication as does elementary calculus. Such study would greatly enrich the student's intuition about nonlinear systems.*—Biologist Robert M. May, "Simple mathematical models with very complicated dynamics," *Nature,* vol. 261 (1976), pp. 459–467.

The size of a population or its genetic makeup may change from one generation to the next. In the study of *discrete dynamics* we use functions and the iteration process (from Section 3.5) to investigate and analyze changes such as these that occur over discrete intervals of time. We begin by introducing the notion of a *fixed point* of a function.

If we start with a function f and an input x, it's usually not the case that $f(x)$ turns out to be the same as x itself. That is, usually, the output is not the same as the input. But sometimes this does happen. Take, for example, the function $f(x) = 3x - 2$ and the input $x = 1$. Then we have

$$f(1) = 3(1) - 2 = 1$$

So for this particular function the input $x = 1$ is an instance where "input = output." The input $x = 1$ in this case is called a *fixed point* of the function $f(x) = 3x - 2$. In the box that follows, we give the general definition of a fixed point.

DEFINITION **Fixed Point of a Function**

A fixed point of a function f is an input x in the domain of f such that

$$f(x) = x$$

EXAMPLE

Both 0 and 1 are fixed points for $f(x) = x^2$ because $f(0) = 0^2 = 0$ and $f(1) = 1^2 = 1$.

EXAMPLE 1 Finding fixed points

Find the fixed points (if any) for each function:
(a) $f(x) = 1 - x$; **(b)** $g(x) = 1 + x$; **(c)** $h(x) = x^2 - x - 3$.

SOLUTION

(a) We're looking for a number x such that $f(x) = x$. In view of the definition of f, this last equation becomes

$$1 - x = x$$

and therefore

$$x = \frac{1}{2}$$

This result shows that the function f has one fixed point; it is $x = 1/2$.

(b) If x is a fixed point of $g(x) = 1 + x$, we have

$$1 + x = x$$

But then subtracting x from both sides of this last equation yields $1 = 0$, which is impossible. We conclude from this that there is no fixed point for the function $g(x) = 1 + x$.

(c) The fixed points (if any) are the solutions of the quadratic equation $x^2 - x - 3 = x$. Subtracting x from both sides of this equation, we have

$$x^2 - 2x - 3 = 0$$
$$(x - 3)(x + 1) = 0$$

Looking at this last equation, we can see that there are two roots: 3 and -1. Each of these numbers is a fixed point for the given function $h(x) = x^2 - x - 3$. That is, $h(3) = 3$ and $h(-1) = -1$. [You should verify each of these last two statements for yourself by actually computing $h(3)$ and $h(-1)$.]

A fixed point of a function can be interpreted geometrically: It is the x-coordinate of a point where the graph of the given function intersects the line $y = x$. Figure 1 shows the fixed points for the functions in the example that we've just completed.

Figure 1
Fixed points for the functions in Example 1

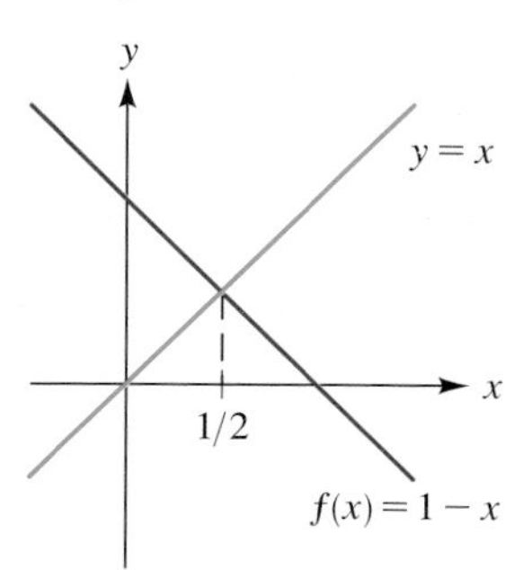

(a) The fixed point of f is $1/2$.

(b) The function g has no fixed points.

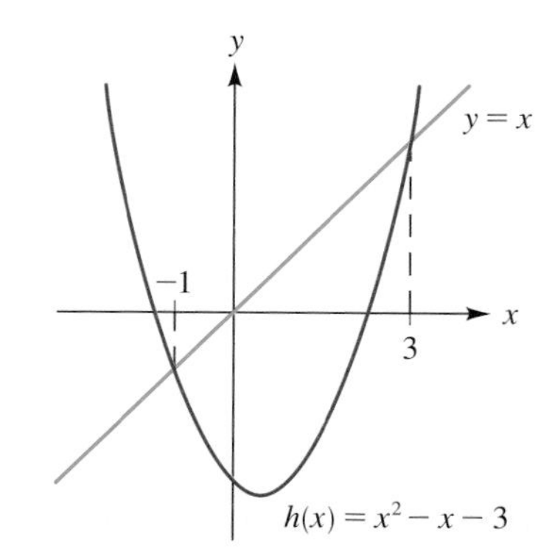

(c) The fixed points of h are -1 and 3.

Fixed points are related to the iteration process in several ways. Suppose that a number a is a fixed point of the function f. Then by definition we have $f(a) = a$, which says that the first iterate of a is equal to a itself. Similarly, all of the subsequent iterates of the fixed point a will be equal to a. For instance, for the second iterate we have

$$f(f(a)) = f(a) \qquad \text{substituting } a \text{ for } f(a) \text{ on the left-hand side}$$
$$= a \qquad \text{again because } f(a) = a$$

This shows that the second iterate is equal to a. The same type of calculation will show that any subsequent iterate of the fixed point a is equal to a.

Another connection between fixed points and iteration is this: for some functions, a fixed point can be a "target value" for other iterates. We'll explain this using Figure 2.

Figure 2 shows the first four steps in the iteration process for $f(x) = \frac{1}{2}x + 2$ with $x_0 = 1$. (See Section 3.5 if you need to review graphical iteration.) As indicated in Figure 2, the input 4 is a fixed point for the function f, and the iteration process follows a staircase pattern that approaches the point (4, 4). We say in this case that the iterates of $x_0 = 1$ **approach** the fixed point 4 and that this target value 4 is an **attracting fixed point** of the function f. Table 1 gives you a more numerical look at what is meant by saying that the iterates approach the target value 4.

In the table, notice, for example, that

x_5 differs from 4 by less than 0.1

x_{10} differs from 4 by less than 0.01

and x_{15} differs from 4 by less than 0.0001

What's important here is that the differences between the iterates and 4 can be made as small as we please by carrying out the iteration process sufficiently far. The idea of a target value or *limit* is made more precise in calculus. But for our purposes, Figure 2 and Table 1 will certainly give you an intuitive understanding of the idea and what we mean by saying that the iterates approach 4.

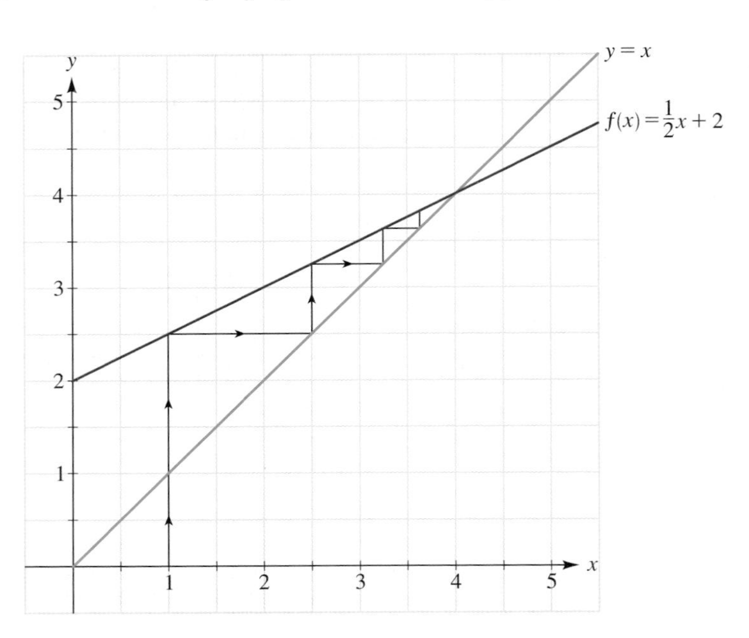

Figure 2
The first four steps in the iteration process for $f(x) = \frac{1}{2}x + 2$ with $x_0 = 1$

TABLE 1 The Iterates Approach 4

x_1	2.5		
x_2	3.25		
x_3	3.625		
x_4	3.8125	x_{11}	3.9985 . . .
x_5	3.906 . . .	x_{12}	3.99926 . . .
x_6	3.953 . . .	x_{13}	3.99963 . . .
x_7	3.976 . . .	x_{14}	3.99981 . . .
x_8	3.988 . . .	x_{15}	3.999908 . . .
x_9	3.9941 . . .	x_{20}	3.9999971 . . .
x_{10}	3.9970 . . .	x_{25}	3.999999910 . . .

Figures 3 and 4 show two more ways in which the iteration process may relate to fixed points. In Figure 3 there is an attracting fixed point for the iterates of $x_0 = -0.1$, but this time the iteration process approaches the fixed point through a spiral pattern rather than a staircase pattern. To find the fixed point (and thereby determine the number that the iterates are approaching), we need to solve the quadratic equation $x^2 - 0.5 = x$. As you should check for yourself by means of the quadratic formula and then a calculator, the relevant root here is $(1 - \sqrt{3})/2 \approx -0.366$. Notice that this value is consistent with Figure 3. In Figure 4 the fixed point 1 is a **repelling fixed point** for the iterates of $x_0 = 1.25$. As Figure 4 indicates, the iterates of $x_0 = 1.25$ move farther and farther away from the value 1. Indeed, as you can check with a calculator, the first five iterates of 1.25 are as follows (we're rounding to two decimal places):

$$x_1 \approx 1.56 \qquad x_2 \approx 2.44 \qquad x_3 \approx 5.96 \qquad x_4 \approx 35.53 \qquad x_5 \approx 1262.18$$

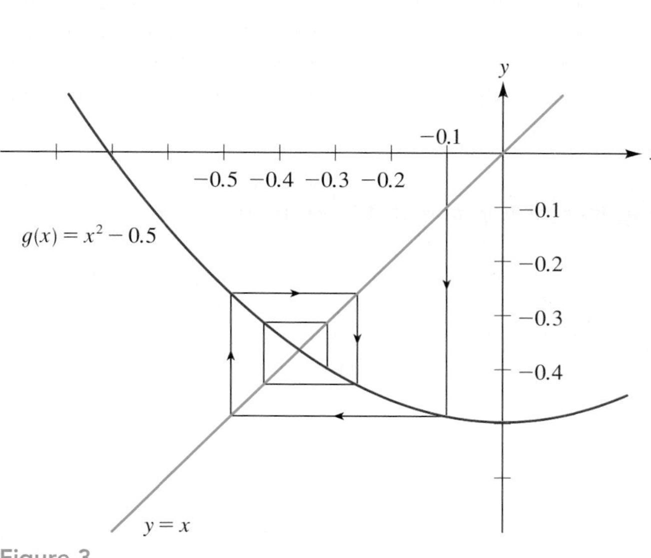

Figure 3
The iterates of $x_0 = -0.1$ under the function $g(x) = x^2 - 0.5$ approach the attracting fixed point $\frac{1}{2}(1 - \sqrt{3}) \approx -0.366$.

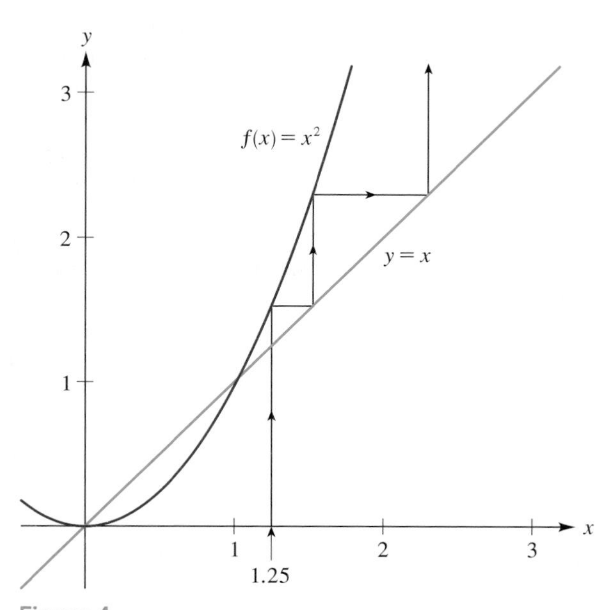

Figure 4
The iterates of $x_0 = 1.25$ under the function $f(x) = x^2$ move away from the repelling fixed point 1.

The iteration process for functions is often applied in the study of population growth. The word "population" here is used in a general sense. It needn't refer only to human populations. For instance, biological or ecological studies may involve animal, insect, or bacterial populations. (Also, see the quotation at the beginning of this section.) The following equation defines one type of quadratic function that has been studied extensively in this context:

$$f(x) = kx(1 - x) \tag{1}$$

In using this idealized model, we assume that the population size is measured by a number between 0 and 1, where 1 corresponds to the maximum possible population size in the given environment and 0 corresponds to the case in which the population has become extinct. We start with a given input x_0 $(0 \leq x_0 \leq 1)$ that represents the fraction of the maximum population size that is initially present. For instance, if the maximum possible population of catfish in a pond is 100 and initially there were 70 catfish, then we would have $x_0 = 70/100 = 0.7$.

The next basic assumption in using equation (1) to model population size is that the iterates of x_0 represent the fraction of the maximum possible population present after each successive time interval. That is,

$f(x_0) = x_1$ is the fraction of the maximum population after the first time interval

$f(x_1) = x_2$ is the fraction of the maximum population after the second time interval

and, in general,

$f(x_{n-1}) = x_n$ is the fraction of the maximum population after the nth time interval

It is important to note that the function $f(x)$ does *not* represent the size of the population. The population size after n time intervals is given by

$$x_n \cdot (\text{the maximum population}) = f(x_{n-1}) \cdot (\text{maximum population})$$

In a given study, the time intervals might be measured, for example, in years, in months, or in breeding seasons. The constant k in equation (1) is the *growth parameter*; it is related to the rate of growth of the particular population being studied. Science writer James Gleick has described k this way: "In a pond, it might correspond to the fecundity of the fish, the propensity of the population not just to boom but also to bust. . . ." [*Chaos: Making a New Science* (New York: Viking Penguin, Inc., 1988)]

EXAMPLE 2 Using iteration in analyzing population size

In the Mississippi Delta region, many farmers have replaced unproductive cotton fields with catfish ponds. Suppose that a farmer has a catfish pond with a maximum population size of 500 and that initially the pond is stocked with 50 catfish. Also, assume that the growth parameter for this population is $k = 2.9$, so that equation (1) becomes

$$f(x) = 2.9x(1 - x) \tag{2}$$

Finally, assume that the time intervals here are breeding seasons.

(a) What is the value for x_0?

(b) Use Figure 5 to estimate the iterates x_1 through x_5. Then use a calculator to compute these values. Round the final answers to three decimal places. Interpret the results.

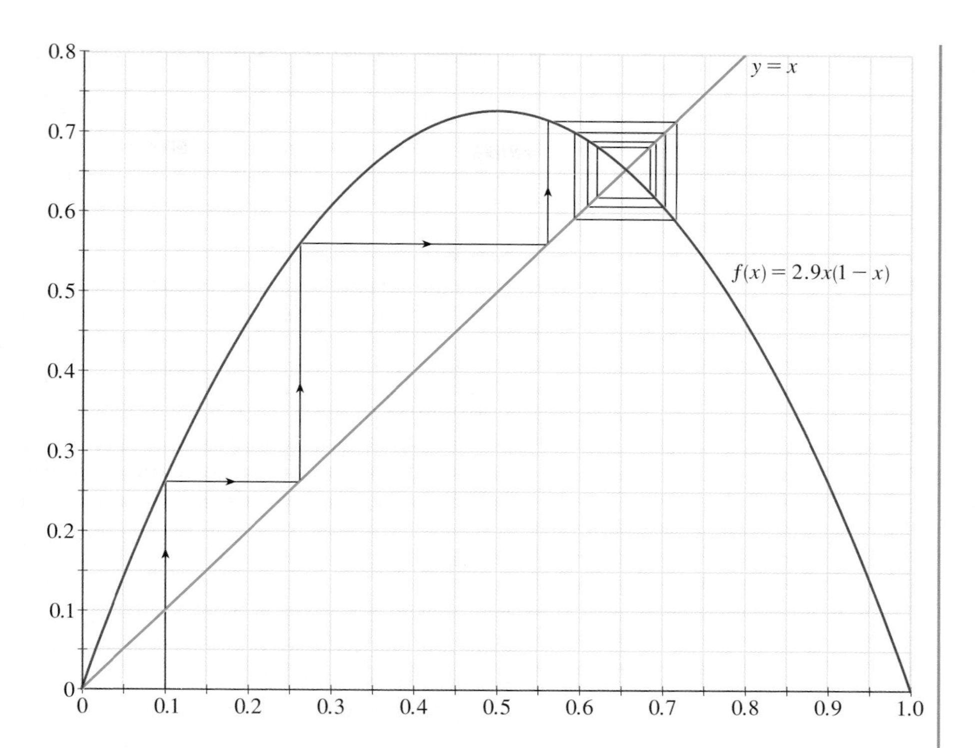

Figure 5
The first ten iterations of $x_0 = 0.1$ under $f(x) = 2.9x(1 - x)$

(c) As indicated in Figure 5, the iteration process is spiraling in on a fixed point of the function. (Figure 6 later in this section will demonstrate this in greater detail.) Find this fixed point and interpret the result.

SOLUTION

(a) $x_0 = \dfrac{\text{initial population}}{\text{maximum population}} = \dfrac{50}{500} = 0.1$

(b) In looking at Figure 5, it appears that x_1, the first iterate of x_0, is between 0.25 and 0.30, much closer to the former number than the latter. As an estimate, let's say that x_1 is about 0.26. This and the other estimates are given in Table 2. *Suggestion*: Make the estimates for yourself before looking at the estimates we give. Some slight discrepancies are okay. In the bottom row of Table 2 are the values of the iterates obtained using a calculator. You should verify these for yourself. (There should be no discrepancies here.)

The results in Table 2 tell us what is happening to the population through the first five breeding seasons. From an initial population of 50 catfish, the population size steadily increases through the first three breeding seasons

TABLE 2
Iterates of $x_0 = 0.1$ under the Function $f(x) = 2.9x(1 - x)$ (calculator values rounded to three decimal places)

	x_1	x_2	x_3	x_4	x_5
From graph	0.26	0.56	0.71	0.58	0.70
From calculator	0.261	0.559	0.715	0.591	0.701

TABLE 3 Catfish Population

n	0	1	2	3	4	5
Number of fish after n breeding seasons	50	131	280	358	296	351

(the numbers in the table are getting bigger). The population size then drops after the fourth breeding season and goes back up after the fifth season. (These facts can be deduced from Figure 5, as well as from Table 2.) To compute the actual numbers of fish, we need to multiply each number in the bottom row of Table 2 by 500. (Remember that the iterates in Table 2 represent fractions of the maximum possible population size 500.) For example, to compute the actual number of fish at the end of the third breeding season, we multiply the initial population size of 500 by x_3:

$$\begin{aligned} x_3 \times 500 &\approx 0.715 \times 500 \\ &\approx 358 \text{ catfish} \end{aligned}$$

The number of catfish at the end of each of the other breeding seasons is obtained in the same manner. See Table 3; use your calculator to check each of the entries in the table.

(c) The fixed point we are looking for occurs when the parabola in Figure 5 intersects the line $y = x$. So, following the method in Example 1, we need to solve the equation $2.9x(1 - x) = x$. We have

$$\begin{aligned} 2.9x(1 - x) &= x \\ -2.9x^2 + 1.9x &= 0 \\ x(-2.9x + 1.9) &= 0 \end{aligned}$$

Therefore, $x = 0$ or $-2.9x + 1.9 = 0$, that is,

$$x = \frac{-1.9}{-2.9} = \frac{19}{29} \approx 0.655 \qquad \text{using a calculator and rounding to three decimal places}$$

This shows that there are two fixed points for the function f, namely, 0 and 19/29. Looking at Figure 5, we know that 19/29 is the fixed point that we are interested in here, not 0. So the iterates of $x_0 = 0.1$ are approaching the value 19/29, which is approximately 0.655.

We now summarize and interpret the results. There were initially 50 catfish in a pond that could hold at most 500. As we saw in part (b), the population size increases over the first three breeding seasons. After this, as Figure 5 shows, the population oscillates up and down but draws closer and closer to an equilibrium population corresponding to the fixed point $x = 19/29$. This equilibrium population is

$$\frac{19}{29} \times 500 \approx 328 \text{ catfish}$$

We conclude this section with some pictures indicating only three of the many possibilities that can arise in the iteration of quadratic functions of the form

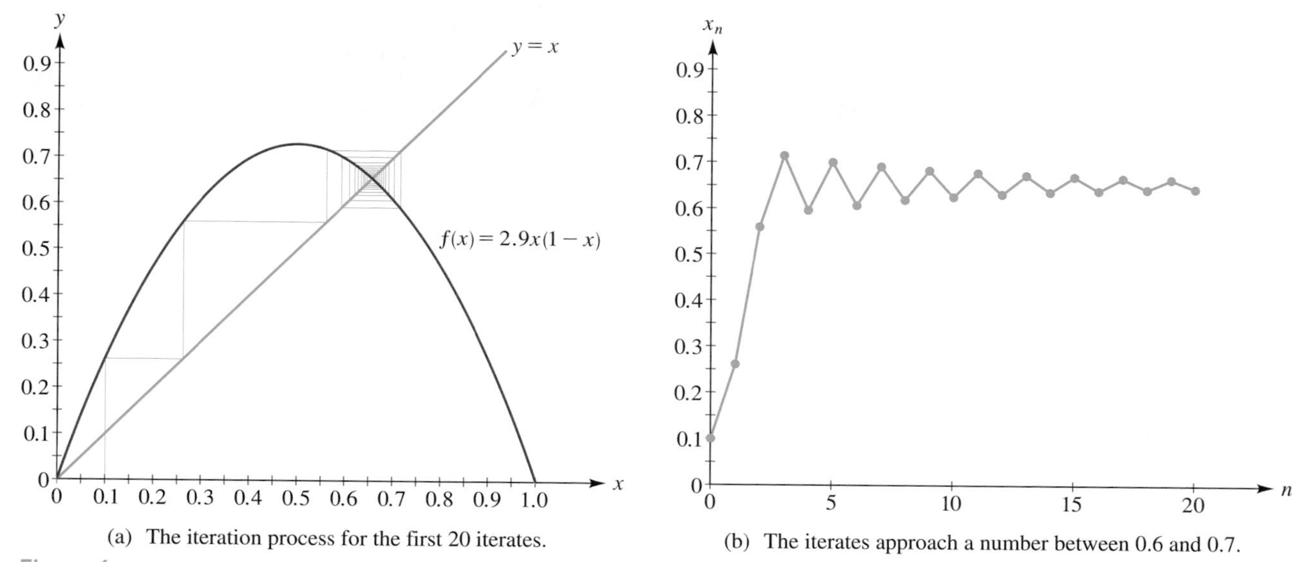

(a) The iteration process for the first 20 iterates.

(b) The iterates approach a number between 0.6 and 0.7.

Figure 6
The iteration of $f(x) = kx(1 - x)$ with growth parameter $k = 2.9$ and $x_0 = 0.1$

$f(x) = kx(1 - x)$. Figure 6 concerns the function with growth parameter $k = 2.9$ that we used in the catfish example: $f(x) = 2.9x(1 - x)$. In Figure 6(a) we've carried out the iteration of $x_0 = 0.1$ through the twentieth iterate. (In the catfish example, Figure 5 goes only as far as the tenth iterate.) Figure 6(a) indicates quite clearly that the iterates are indeed approaching a fixed point of the function. Figure 6(b) presents another way to visualize the long-term behavior of the iterates. Values of n are marked on the horizontal axis, values of the iterates x_n are marked on the vertical axis, and the points with coordinates (n, x_n) are then plotted. For example, since $x_0 = 0.1$, we plot the point $(0, 0.1)$; and since $x_1 = 0.261$, we plot the point $(1, 0.261)$. The line segments in Figure 6(b) are drawn in only to help the eye see the pattern that is emerging. Three facts that can be inferred from Figure 6(b) are as follows: After the first few iterates, the iterates oscillate up and down; the magnitude of the oscillations is decreasing; and, in the long run, the iterates are approaching a number between 0.6 and 0.7. (In Example 2 we found this value to be approximately 0.655.)

Unlike the iteration pictured in Figure 6(a), Figure 7(a) shows a case in which the iteration process is spiraling away from, rather than toward, a fixed point. Again, we've used the initial input $x_0 = 0.1$, but this time the growth parameter is $k = 3.2$. As indicated in Figure 7(b), the long-term behavior of the iterates becomes quite predictable: they alternate between two values. Figure 7(b) shows that the smaller of these two values is between 0.5 and 0.6, while the larger is approximately 0.8. Exercise 36 gives you formulas for computing these two limiting values for the iterates. (They turn out to be, approximately, 0.513 and 0.799.)

In Figure 8, once again we take the initial input to be $x_0 = 0.1$, but this time the growth parameter is $k = 3.9$. Now the iterates seem to fluctuate widely with no apparent pattern, in sharp contrast to the previous two figures, in which there were clear patterns. Phenomena such as this are the subject of **chaos theory,** a new branch of twentieth-century mathematics with wide application. [For a readable and nontechnical introduction to this relatively new subject, see James Gleick's *Chaos: Making a New Science* (New York: Viking Penguin, Inc., 1988). For a little more detail on the mathematics, see the paperback by Donald M. Davis, *The Nature and Power of Mathematics* (Princeton, N.J.: Princeton University Press, 1993), pp. 314–363.]

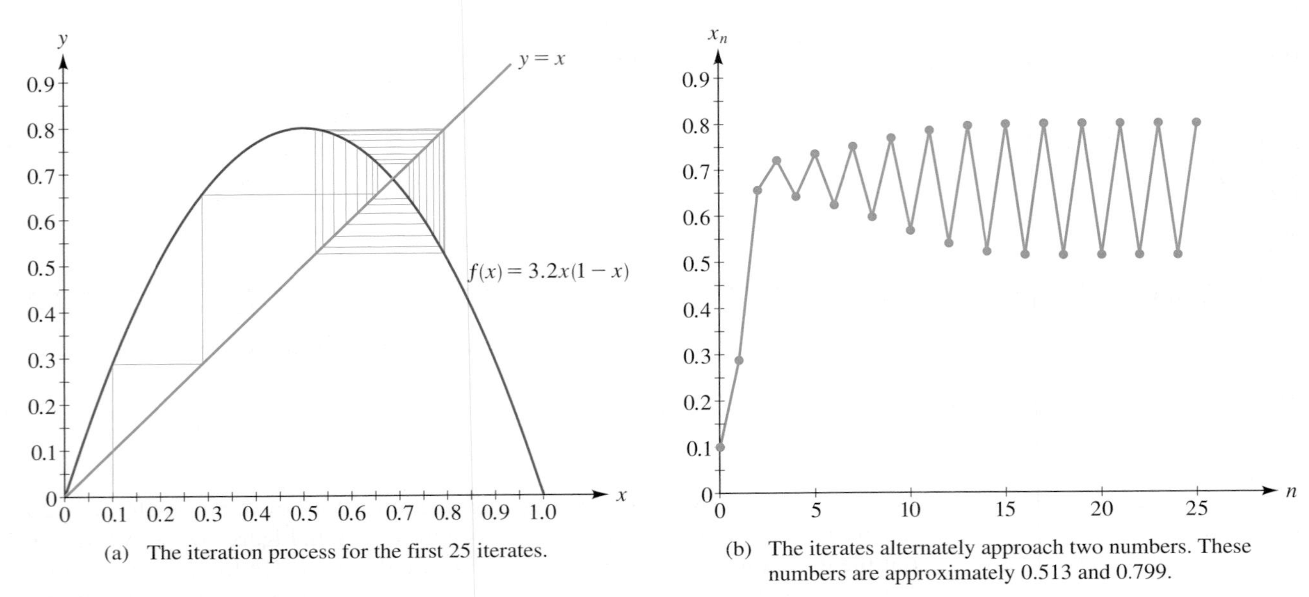

(a) The iteration process for the first 25 iterates.

(b) The iterates alternately approach two numbers. These numbers are approximately 0.513 and 0.799.

Figure 7
The iteration of $f(x) = kx(1 - x)$ with growth parameter $k = 3.2$ and $x_0 = 0.1$

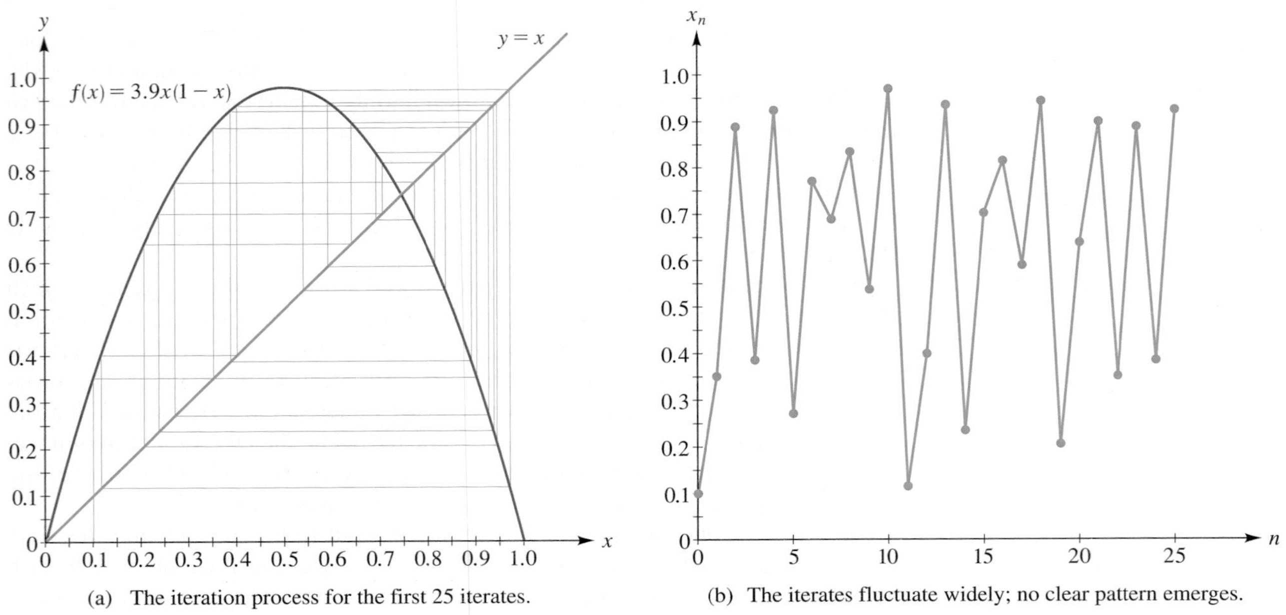

(a) The iteration process for the first 25 iterates.

(b) The iterates fluctuate widely; no clear pattern emerges.

Figure 8
The iteration of $kx(1 - x)$ with growth parameter $k = 3.9$ and $x_0 = 0.1$

EXERCISE SET 4.3 

A

In Exercises 1–16, find all real numbers (if any) that are fixed points for the given functions.

1. $f(x) = -4x + 5$
2. $g(x) = 3x - 14$
3. $G(x) = \frac{1}{2} + x$
4. $F(x) = (7 - 2x)/8$
5. $h(x) = x^2 - 3x - 5$
6. $H(t) = 3t^2 + 18t - 6$
7. $f(t) = t^2 - t + 1$
8. $F(t) = t^2 - t - 1$
9. $k(t) = t^2 - 12$
10. $K(t) = t^2 + 12$
11. $T(x) = 1.8x(1 - x)$
12. $T(y) = 3.4y(1 - y)$
13. $g(u) = 2u^2 + 3u - 4$
14. $G(u) = 3u^2 - 4u - 2$
15. $f(x) = 7 + \sqrt{x - 1}$
16. $f(x) = \sqrt{10 + 3x} - 4$

Ⓖ *In Exercises 17–22:*

(a) *Graph each function along with the line $y = x$. Use the graph to determine how many (if any) fixed points there are for the given function.*

(b) *For those cases in which there are fixed points, use the zoom-in capability of the graphing utility to estimate the fixed point. (In each case, continue the zoom-in process until you are sure about the first three decimal places.)*

17. $f(x) = x^3 + 3x + 2$
18. $g(x) = x^3 - 3x + 2$
19. $h(x) = x^3 - 3x - 3.07$
20. $k(x) = x^3 - 3x - 3.08$
21. $s(t) = t^4 + 3t - 2$
22. $u(t) = t^4 + 3t + 2$

23. This exercise refers to the function $g(x) = x^2 - 0.5$ in Figure 3 in the text.

(a) Use the quadratic formula to verify that one of the fixed points of this function is $(1 - \sqrt{3})/2$, then use your calculator to check that this is approximately -0.366.

(b) According to the text, the iterates of -0.1 approach the value determined in part (a). Use your calculator: which is the first iterate to have the digit 3 in the first decimal place?

(c) Use your calculator: Which is the first iterate to have the digit 6 in the second decimal place?

24. This exercise refers to the function $f(x) = x^2$ in Figure 4 in the text. According to the text, the iterates of 1.25 move farther and farther away from the fixed point 1. In this exercise you'll see that iterates of other points even closer to the fixed point 1 nevertheless are still "repelled" by 1.

(a) Let $x_0 = 1.1$. Use your calculator: Which is the first iterate to exceed 10? Which is the first iterate to exceed one million?

(b) Let $x_0 = 1.001$. Use your calculator: Which is the first iterate to exceed 2? Which is the first iterate to exceed one million?

(c) Let $x_0 = 0.99$. Compute x_1 through x_{10} to see that the iterates are indeed moving farther and farther away from the fixed point 1. What value are the iterates approaching? Is this value a fixed point of the function?

25. The accompanying figure shows the first eight steps in the iteration process for $f(x) = -0.7x + 2$, with $x_0 = 0.4$.

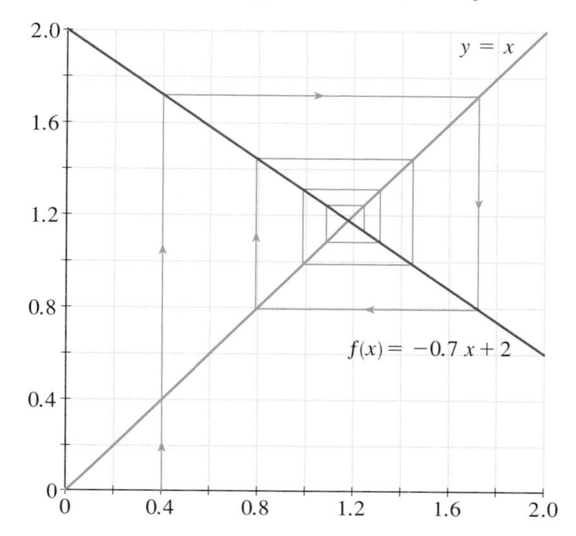

(a) Complete the following table. For the values obtained from the graph, estimate to the nearest one-tenth; for the calculator values, round the final answers to three decimal places.

	x_1	x_2	x_3	x_4	x_5	x_6	x_7	x_8
From graph								
From calculator								

(b) The figure shows that the iterates are approaching a fixed point of the function. Determine the exact value of this fixed point and then give a calculator approximation rounded to three decimal places.

(c) In the table for part (a), you used a calculator to compute the first eight iterates. Which of these iterates is the first to have the same digit in the first decimal place as the fixed point?

26. The figure on the next page shows the first six steps in the iteration process for $f(x) = 2.9x(1 - x)$, with $x_0 = 0.2$.

(a) Complete the following table. For the values obtained from the graph, estimate to the nearest 0.05, or closer if you can. For instance, to the nearest 0.05, the first iterate is 0.45. But, since the graph shows the iterate a bit above 0.45, the estimate 0.46 would be better. For the calculator work, round the final answers to three decimal places.

	x_1	x_2	x_3	x_4	x_5	x_6
From graph						
From calculator						

(continues)

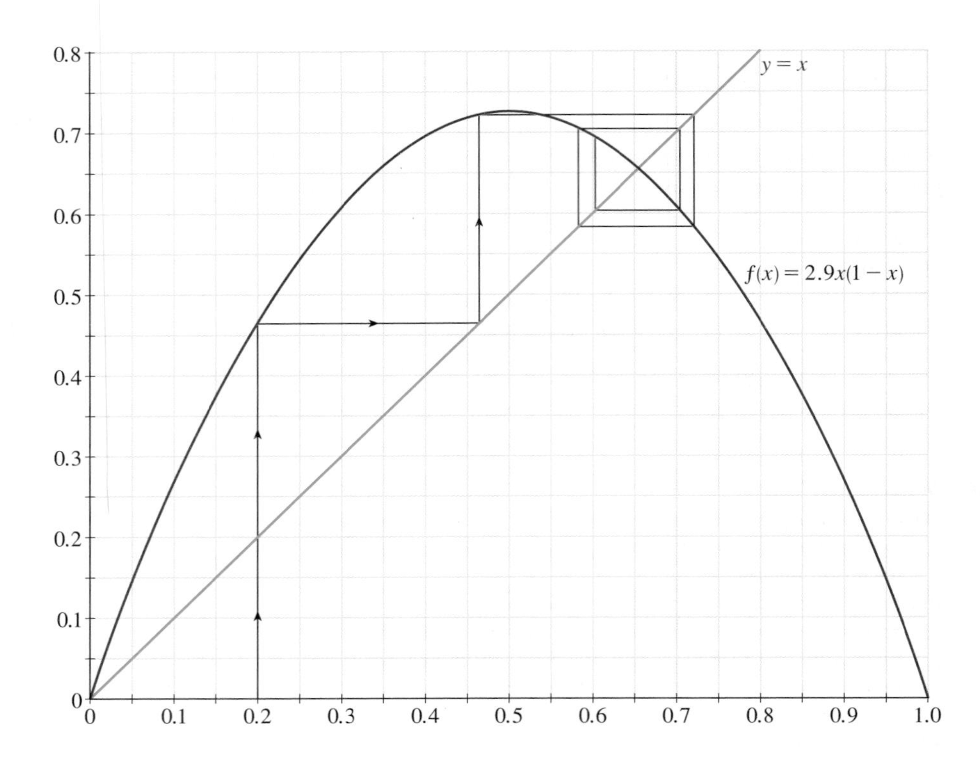

Figure for Exercise 26

(b) The figure shows that the iterates are approaching a fixed point of the function. Use the figure to estimate, to the nearest 0.05, a value for this fixed point. Then determine the exact value of this fixed point, and also give a calculator approximation rounded to three decimal places.

(c) In the table for part (a), you used a calculator to compute the first six iterates. Which of these iterates is the first to have the same digit in the first decimal place as the fixed point? *Remark*: Agreement in the second decimal place doesn't occur until the 26th iterate.

For Exercises 27 and 28, refer to the figure on page 257, which shows the first nine steps in the iteration process for $f(x) = 4x(1 - x)$, with $x_0 = 0.9$. (Qualitatively, note that this figure is quite different from those in the previous two exercises or in Figures 2 through 7 in the text. Here, no clear pattern in the iterates seems to emerge.)

27. Complete the following table. For the values obtained from the graph, estimate to the nearest 0.05, or closer if you can. For instance, to the nearest 0.05, the first iterate is 0.35. But since the graph shows the iterate a bit above 0.35, the estimate 0.36 would be better. For the calculator work, round the final answers to three decimal places.

	x_1	x_2	x_3	x_4	x_5	x_6	x_7	x_8	x_9
From graph									
From calculator									

28. As in Exercise 27, we work with the function $f(x) = 4x(1 - x)$. Furthermore, in part (a) we use an input that is very close to the one in Exercise 27. As you will see, however, the results will be remarkably different. This phenomenon, whereby a very small change in the initial input results in a completely different pattern in the iterates, is called **sensitivity to initial conditions.** Sensitivity to initial conditions is one of the characteristic behaviors studied in chaos theory.

(a) For the initial input, use $x_0 = \frac{1}{8}(5 + \sqrt{5}) \approx 0.9045$. Note that this differs from the input in Exercise 27 by less than 0.01. (Exercise 36 shows how this seemingly off-the-wall input was obtained.) Use algebra (not a calculator!) to compute exact expressions for x_1 and x_2. What do you observe? What are x_3 and x_4? What's the general pattern here? *Note*: If you were to use a calculator rather than algebra for all of this, due to rounding errors you could miss seeing the patterns.

(b) Take $x_0 = 0.905$. Use a calculator to compute x_1 through x_{10}. Round the final results to three decimal places. Is the behavior of the iterates more like that in Exercise 27 or in part (a) of this exercise?

As background for Exercises 29–32, you need to have read Example 2 in this section. As in Example 2, assume that there is a catfish pond with a maximum population size of 500 catfish. Also assume, unless stated otherwise, that the initial population size is 50 catfish (so that $x_0 = 0.1$).

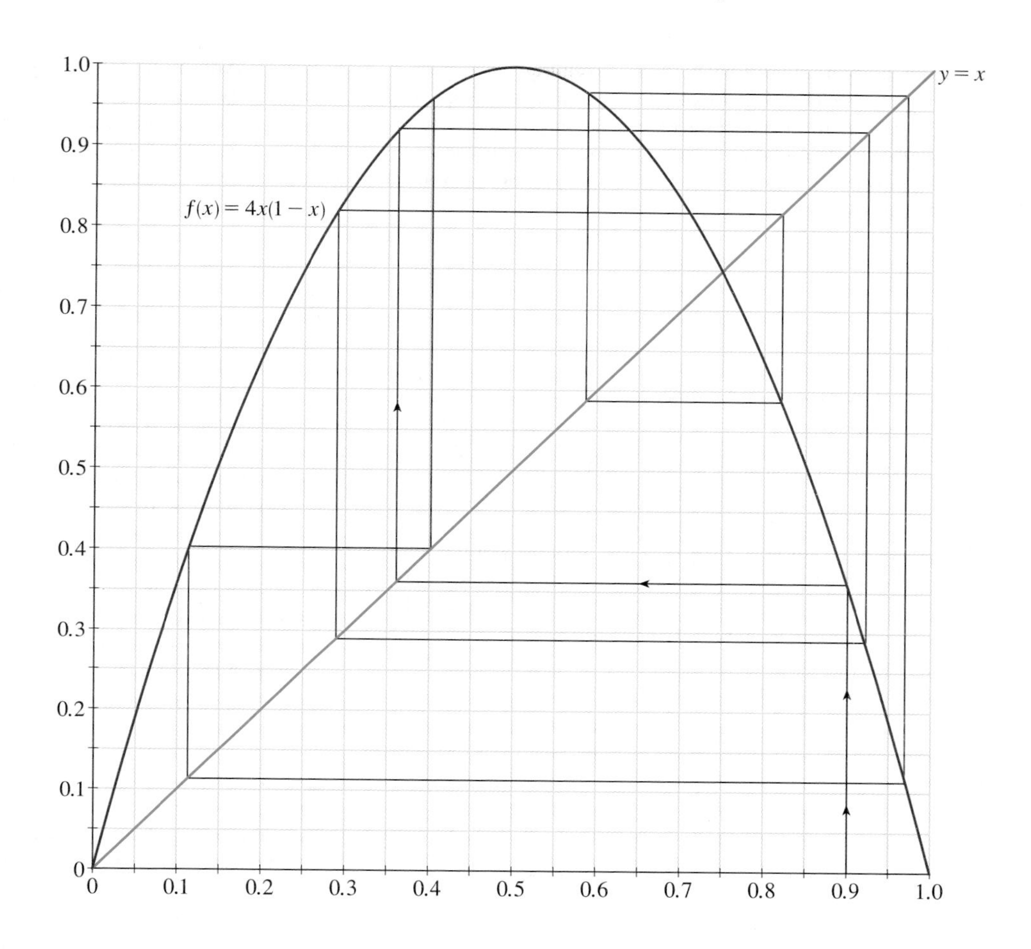

Figure for Exercises 27 and 28

29. (a) In Example 2 we used a growth parameter of $k = 2.9$ and we computed the first five iterates of $x_0 = 0.1$. Now (under these same assumptions) compute x_{21} through x_{25}, given that $x_{20} \approx 0.64594182$. Round your final answers to four decimal places.

(b) Use the results in part (a) to complete the following table. [Compare your table to Table 3 in the text; note that the population size continues to oscillate up and down, but now the sizes of the oscillations are much smaller. This provides additional evidence that the population size is approaching an equilibrium value. (In Example 2 we determined this equilibrium size to be about 328 catfish.)]

n	20	21	22	23	24	25
Number of fish after n breeding seasons						

(c) In parts (a) and (b) and in Example 2 we worked with a growth parameter of 2.9, and we found that the iterates of $x_0 = 0.1$ were approaching a fixed point of the function. Now assume instead that the growth parameter is $k = 0.75$ (but, still, that $x_0 = 0.1$). Determine the iterates x_1 through x_5. (Round to five decimal places.) Are the iterates approaching a fixed point of the function $f(x) = 0.75x(1 - x)$? Interpret your results.

30. As in Example 2, take $x_0 = 0.1$, but now assume that the growth parameter is $k = 3$, so that equation (1) in the text becomes $f(x) = 3x(1 - x)$.

(a) Complete the following tables. (Round the final answers for the iterates to three decimal places.) Notice that after the third breeding season, the population oscillates up and down, as in Example 2.

n	0	1	2	3	4	5
x_n	0.1					
Number of fish after n breeding seasons	50					

n	6	7	8	9	10
x_n					
Number of fish after n breeding seasons					

(b) Complete the following table, given that $x_{100} \approx 0.643772529$. Round your final answers for the iterates to four decimal places. In your results, note that the population continues to oscillate up and down, but that the sizes of the oscillations are less than those observed in part (a).

n	101	102	103	104	105	106
x_n						
Number of fish after n breeding seasons						

(c) The iterates that you computed in parts (a) and (b) are approaching a fixed point of the function $f(x) = 3x(1 - x)$. Find this fixed point and the corresponding equilibrium population size.

31. (a) As in Example 2, take $x_0 = 0.1$, but now assume that the growth parameter is $k = 3.1$, so that equation (1) in the text becomes $f(x) = 3.1x(1 - x)$. Complete the following three tables. For the third table, use the fact that $x_{20} \approx 0.56140323$. Round your final answers for the iterates to four decimal places.

n	0	1	2	3	4	5
x_n	0.1					
Number of fish after n breeding seasons	50					

n	6	7	8	9	10
x_n					
Number of fish after n breeding seasons					

n	21	22	23	24	25	26
x_n						
Number of fish after n breeding seasons						

(b) Your results in part (a) will show that the population size is oscillating up and down, but that the iterates don't seem to be approaching a fixed point. Indeed, determine the (nonzero) fixed point of the function $f(x) = 3.1x(1 - x)$. Then note that the successive iterates in part (a) actually move farther and farther away from this fixed point.

(c) It can be shown that the long-term behavior of the iterates in this case resembles the pattern in Figure 7(b) on page 254; that is, the iterates alternately approach two values. Use the following formulas, with $k = 3.1$, to determine these two values a and b that the iterates are alternately approaching. (The formulas are developed in Exercise 36.) Round the answers to four decimal places. Check to see that your answers are consistent with the results in part (a).

$$a = \frac{1 + k + \sqrt{(k-3)(k+1)}}{2k}$$

$$b = \frac{1 + k - \sqrt{(k-3)(k+1)}}{2k}$$

(d) What are the two populations corresponding to these two numbers a and b?

32. As in Example 2, assume that the maximum population size of the pond is 500 catfish, but now suppose that the growth parameter is $k = 0.6$.

(a) Suppose that the initial population is again 50 catfish, so that $x_0 = 0.1$. Compute the first ten iterates. Do they appear to be approaching a fixed point? Interpret the results.

(b) Follow part (a), but assume that the initial population is 450, so that $x_0 = 450/500 = 0.9$.

(c) What relationship do you see between the iterates in part (a) and in part (b)?

B

33. Suppose that c and d are inputs for a function g and that $g(c) = d$ and $g(d) = c$. Then we say the set $\{c, d\}$ is a **2-cycle** for the function g.

(a) Assuming that $\{c, d\}$ is a 2-cycle for the function g, list the first six iterates of c and the first six iterates of d. Describe in a complete sentence or two the pattern you see.

(b) The following figure shows the iteration process for the function $T(x) = 1 - |2x - 1|$, with initial input $x_0 = 0.4$. Use the figure to list the first six iterates of 0.4 under the function T.

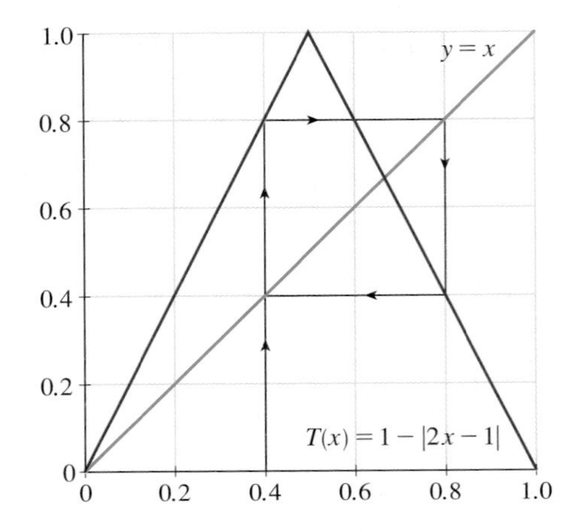

(c) In the following sentence, fill in the two blank spaces with numbers. The work in part (b) shows that {__, __} is a 2-cycle for the function T.

(d) In part (b) you used a graph to list the first six iterates of $x_0 = 0.4$ under the function $T(x) = 1 - |2x - 1|$. Now, using a calculator (or just simple arithmetic), actually compute x_1 and x_2 and thereby check your results.

34. As background for this exercise, you need to have worked part (a) in the previous exercise so that you know the definition of a 2-cycle.

(a) The curve in the following figure is the graph of the function $Q(x) = x^2 - 7$. Use the figure to list the first six iterates of -3 under the function Q. Also, list the first six iterates of 2.

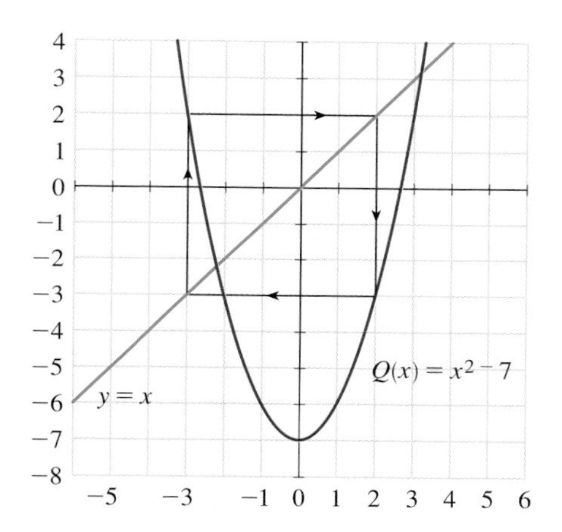

(b) In the following sentence, fill in the two blank spaces with numbers. The work in part (a) shows that {__, __} is a 2-cycle for the function Q.

(c) In part (a) you used a graph to list the first six iterates of -3 under the function Q. Now, using the formula $Q(x) = x^2 - 7$, actually compute x_1 and x_2 and thereby check your results.

(d) Use your calculator to compute the first six iterates of $x_0 = -2.99$ under the function $Q(x) = x^2 - 7$. Note that the behavior of the iterates is vastly different than that observed in part (a), even though the initial inputs differ by only 0.01. (As pointed out in Exercise 28, this type of behavior is referred to as *sensitivity to initial conditions.*)

35. Let $f(x) = 4x(1 - x)$. In this exercise we find distinct inputs a and b such that $f(a) = b$ and $f(b) = a$. As indicated in Exercise 33, the set $\{a, b\}$ is called a *2-cycle for the function f.*

(a) From the equation $f(a) = b$ and the definition of f, we have

$$4a(1 - a) = b \qquad (1)$$

Likewise, from the equation $f(b) = a$ and the definition of f, we have

$$4b(1 - b) = a \qquad (2)$$

Subtract equation (2) from equation (1) and show that the resulting equation can be written

$$4(b - a)(b + a - 1) = b - a \qquad (3)$$

(b) Divide both sides of equation (3) by the quantity $b - a$. (The quantity $b - a$ is nonzero because we are assuming that a and b are distinct.) Then solve the resulting equation for b in terms of a. You should obtain

$$b = \frac{5}{4} - a \qquad (4)$$

(c) Use equation (4) to substitute for b in equation (1). After simplifying, you should obtain

$$16a^2 - 20a + 5 = 0 \qquad (5)$$

(d) Use the quadratic formula to solve equation (5) for a. You should obtain

$$a = (5 \pm \sqrt{5})/8$$

(e) Using the positive root for the moment, suppose $a = (5 + \sqrt{5})/8$. Use this expression to substitute for a in equation (4). Show that the result is $b = (5 - \sqrt{5})/8$. Now check that these values of a and b satisfy the conditions of the problem. That is, given that $f(x) = 4x(1 - x)$, show that

$$f\left[\frac{1}{8}(5 + \sqrt{5})\right] = \frac{1}{8}(5 - \sqrt{5})$$

and

$$f\left[\frac{1}{8}(5 - \sqrt{5})\right] = \frac{1}{8}(5 + \sqrt{5})$$

Note: If we'd begun part (e) by using the other root of the quadratic, namely, $a = (5 - \sqrt{5})/8$, then we would have found $b = (5 + \sqrt{5})/8$, so no new information would have been obtained. In summary, the 2-cycle for the function f consists of the two numbers that are the roots of equation (5).

36. Let $f(x) = kx(1 - x)$ and assume that $k > 0$. Follow the method of Exercise 35 to show that the values of a and b for which $f(a) = b$ and $f(b) = a$ are given by the formulas

$$a = \frac{1 + k + \sqrt{(k - 3)(k + 1)}}{2k}$$

and

$$b = \frac{1 + k - \sqrt{(k - 3)(k + 1)}}{2k}$$

Note that for $k > 3$, both of these expressions represent real numbers (because the quantity beneath the radical sign is then positive). In summary: For all values of k greater than 3, the function $f(x) = kx(1 - x)$ has a unique 2-cycle $\{a, b\}$, where a and b are given by the preceding formulas.

37. **(a)** Let $f(x) = 3.5x(1 - x)$. Use the formulas in Exercise 36 to find values for a and b such that $\{a, b\}$ is a 2-cycle for this function [that is, so that $f(a) = b$ and $f(b) = a$]. Exact answers are required, not calculator approximations.

(b) The figure on the right shows the iteration process for the 2-cycle determined in part (a). Use the answers in part (a) to specify the coordinates of the four points P, Q, R, and S.

38. Let $f(x) = 3.2x(1 - x)$. Use the formulas in Exercise 36 to determine the values of a and b such that $\{a, b\}$ is a 2-cycle for this function. Use a calculator to evaluate the answers and round to three decimal places. [You'll find that these are the two numbers referred to in the caption for Figure 7(b) in this section.]

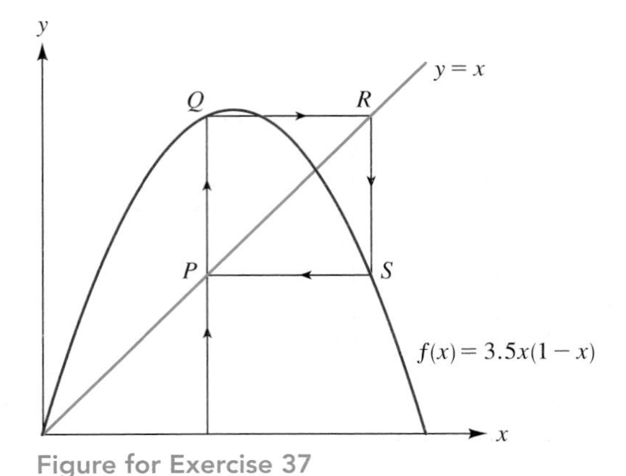

Figure for Exercise 37

4.4 SETTING UP EQUATIONS THAT DEFINE FUNCTIONS

I hope that I shall shock a few people in asserting that the most important single task of mathematical instruction in the secondary schools is to teach the setting up of equations to solve word problems. —George Polya (1887–1985)

Each problem that I solved became a rule which afterwards served to solve other problems. —René Descartes (1596–1650)

One of the first steps in problem solving often involves defining a function. The function then serves to describe or summarize a given situation in a way that is both concise and (one hopes) revealing. In this section we practice setting up equations that define such functions. In Section 4.5 we will use this skill in solving an important class of applied problems involving quadratic functions.

For many of the examples in this section we'll rely on the following four-step procedure to set up the required equation. You may eventually want to modify this procedure to fit your own style. The important point, however, is that it *is* possible to approach these problems in a systematic manner. A word of advice: You're accustomed to working mathematics problems in which the answers are numbers. In this section, the answers are functions (or, more precisely, equations defining functions); you may need to get used to this.

Steps for Setting Up Equations

Step 1 After reading the problem carefully, draw a picture that conveys the given information.

Step 2 State in your own words, as specifically as you can, what the problem is asking for. (This usually requires rereading the problem.) Now, assuming that the problem asks you to find a particular quantity (or a formula for a particular quantity), assign a variable to denote that key quantity.

Step 3 Label any other quantities in your figure that appear relevant. Are there equations relating these quantities?

Step 4 Find an equation involving the key variable that you identified in Step 2. (Some people prefer to do this right after Step 2.) Now, as necessary, substitute in this equation using the auxiliary equations from Step 3 to obtain an equation involving only the required variables.

EXAMPLE 1 The area of a rectangle as a function of its width

The perimeter of a rectangle is 100 cm. Express the area of the rectangle in terms of the width.

SOLUTION
Let's follow our four-step procedure.

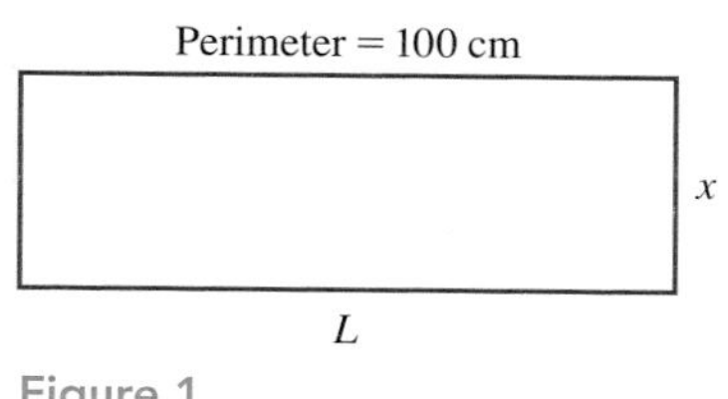

Figure 1

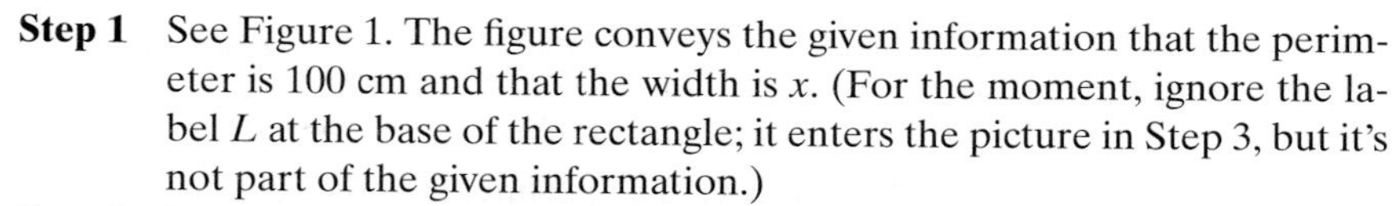

Step 1 See Figure 1. The figure conveys the given information that the perimeter is 100 cm and that the width is x. (For the moment, ignore the label L at the base of the rectangle; it enters the picture in Step 3, but it's not part of the given information.)

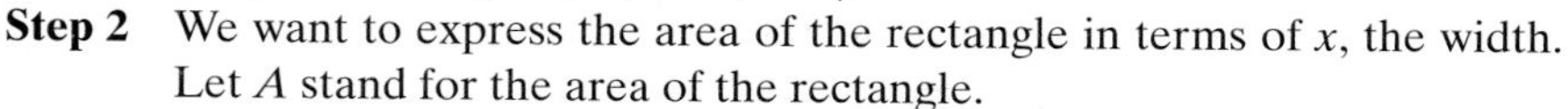

Step 2 We want to express the area of the rectangle in terms of x, the width. Let A stand for the area of the rectangle.

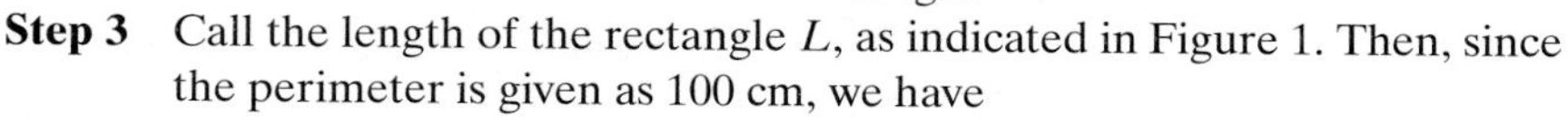

Step 3 Call the length of the rectangle L, as indicated in Figure 1. Then, since the perimeter is given as 100 cm, we have

$$\begin{aligned} 2x + 2L &= 100 \\ x + L &= 50 \\ L &= 50 - x \qquad (1) \end{aligned}$$

Step 4 The area of a rectangle equals width times length:

$$\begin{aligned} A &= x \cdot L \\ &= x(50 - x) \qquad \text{substituting for } L \text{ using equation (1)} \\ &= 50x - x^2 \qquad (2) \end{aligned}$$

This is the required equation expressing the area of the rectangle in terms of the width x.

To emphasize this dependence of A on x, we can use function notation to rewrite equation (2):

$$A(x) = 50x - x^2$$

The domain of this area function is the open interval (0, 50). To see why this is so, first note that $x > 0$ because x denotes a width. Furthermore, in view of equation (1), we must have $x < 50$; otherwise, L, the length, would be zero or negative.

EXAMPLE 2 The perimeter of a rectangle as a function of its width

A rectangle is inscribed in a circle of diameter 8 cm. Express the perimeter of the rectangle as a function of its width.

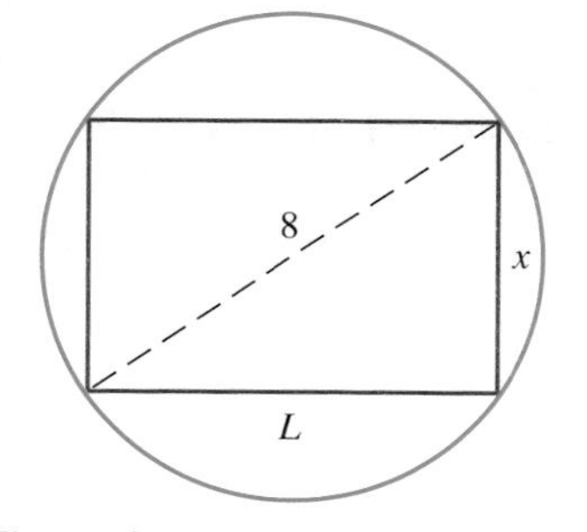

Figure 2

SOLUTION
We follow our four-step procedure.

Step 1 See Figure 2. Notice that the diagonal of the rectangle is a diameter of the circle.

Step 2 The problem asks us to come up with a formula or a function that gives us the perimeter of the rectangle in terms of x, the width. Let P denote the perimeter.

Step 3 Let L denote the length of the rectangle, as shown in Figure 2. Then, by the Pythagorean theorem, we have

$$L^2 + x^2 = 8^2 = 64$$

This equation relates the length L and the width x. Rather than leaving the equation in this form, however, we'll solve for L in terms of x (because the instructions for the problem ask for the perimeter in terms of the width x):

$$L^2 = 64 - x^2$$
$$L = \sqrt{64 - x^2} \qquad (3)$$

Step 4 The perimeter of the rectangle is

$$P = 2x + 2L \qquad (4)$$

This equation expresses P in terms of x and L. However, the problem asks for P in terms of just x. Using equation (3) to substitute for L in equation (4), we have

$$P = 2x + 2\sqrt{64 - x^2} \qquad (5)$$

This is the required equation. It expresses the perimeter P as a function of the width x, so if you know x, you can calculate P.

To emphasize this dependence of P on x, we can employ function notation to rewrite equation (5):

$$P(x) = 2x + 2\sqrt{64 - x^2} \qquad (6)$$

Before leaving this example, we need to specify the domain of the perimeter function in equation (6). An easy way to do this is to look again at Figure 2. Since x represents a width, we certainly want $x > 0$. Furthermore, Figure 2 tells us that $x < 8$, because in any right triangle, a leg is always shorter than the hypotenuse. Putting these observations together, we conclude that the domain of the perimeter function in equation (6) is the open interval (0, 8).

Note: Although it would make sense *algebraically* to use an input such as $x = -1$ in equation (6), it does not make sense in our *geometric* context, where x denotes the width.

NOTE In the following example we introduce the notation $P(x, y)$, which stands for the phrase "a point P with coordinates (x, y)."

EXAMPLE 3 Expressing a distance as a function of one variable

Let $P(x, y)$ be a point on the curve $y = \sqrt{x}$. Express the distance from P to the point (1, 0) as a function of one variable.

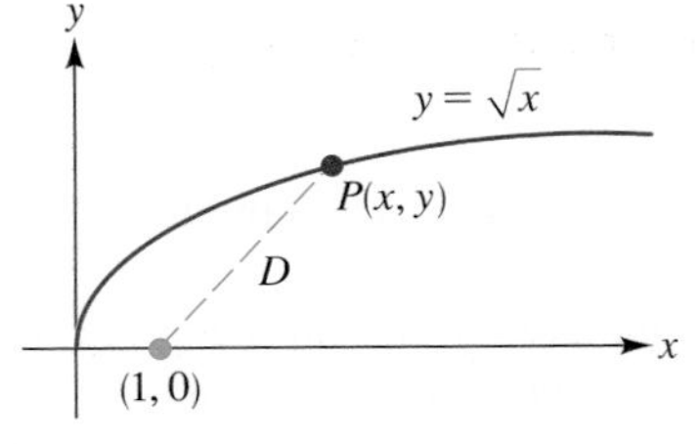

Figure 3

SOLUTION

Step 1 See Figure 3.

Step 2 We want to express the length of the dashed line in Figure 3 in terms of x. Call this length D.

Step 3 There are no other quantities in Figure 3 that need labeling. But don't forget that we are given

$$y = \sqrt{x} \tag{7}$$

Step 4 By the distance formula we have

$$\begin{aligned} D &= \sqrt{(x-1)^2 + (y-0)^2} \\ &= \sqrt{x^2 - 2x + 1 + y^2} \end{aligned} \tag{8}$$

Now we can use equation (7) to eliminate y in equation (8):

$$\begin{aligned} D(x) &= \sqrt{x^2 - 2x + 1 + (\sqrt{x})^2} \\ &= \sqrt{x^2 - 2x + 1 + x} \\ &= \sqrt{x^2 - x + 1} \end{aligned} \tag{9}$$

Equation (9) expresses the distance as a function of x, as required. What about the domain of this distance function? Since the x-coordinate of a point on the curve $y = \sqrt{x}$ can be any nonnegative number, the domain of the distance function is $[0, \infty)$.

EXAMPLE 4 **Expressing the area of a triangle as a function of one variable**

A point $P(x, y)$ lies in the first quadrant on the parabola $y = 16 - x^2$, as indicated in Figure 4. Express the area of the triangular region in Figure 4 as a function of one variable.

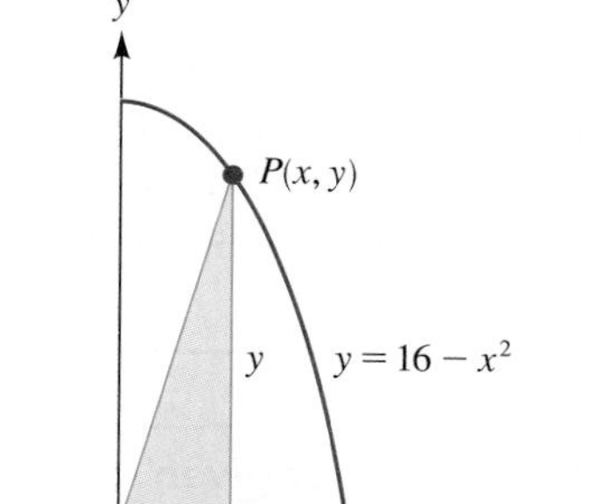

Figure 4

SOLUTION

Step 1 See Figure 4.

Step 2 We want to express the area of the shaded triangle in terms of x. Let A denote the area of this triangle.

Step 3 Since the coordinates of P are (x, y), the base of our triangle is x and the height is y. Also, x and y are related by the given equation

$$y = 16 - x^2 \tag{10}$$

Step 4 The area of a triangle equals $\frac{1}{2}$(base)(height):

$$A = \frac{1}{2}(x)(y)$$

so

$$\begin{aligned} A(x) &= \frac{1}{2}(x)(16 - x^2) \quad \text{substituting for } y \text{ using equation (10)} \\ &= 8x - \frac{1}{2}x^3 \end{aligned}$$

This last equation expresses the area of the triangle as a function of x, as required. (Exercise 50 at the end of this section will ask you to specify the domain of this area function.)

EXAMPLE 5 Expressing the area of a circle as a function of the circumference

A piece of wire x in. long is bent into the shape of a circle. Express the area of the circle in terms of x.

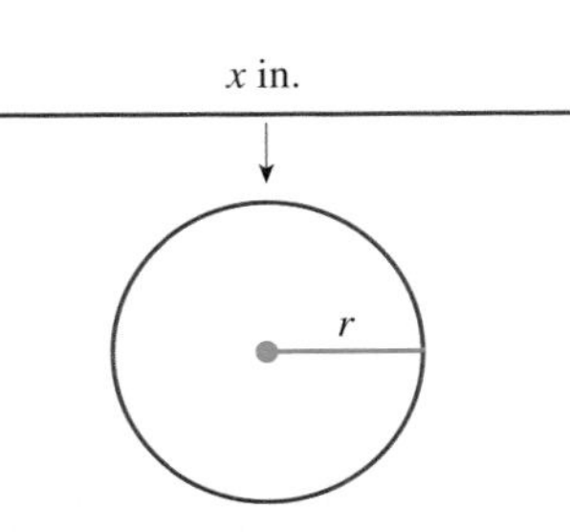

Circumference is x in.

Figure 5

SOLUTION

Step 1 See Figure 5.

Step 2 We are supposed to express the area of the circle in terms of x, the circumference. Let A denote the area.

Step 3 The general formula for circumference in terms of radius is $C = 2\pi r$. Since in our case the circumference is given as x, our equation becomes

$$x = 2\pi r \qquad \text{or} \qquad r = \frac{x}{2\pi} \tag{11}$$

Step 4 The general formula for the area of a circle in terms of the radius is

$$A = \pi r^2 \tag{12}$$

This expresses A in terms of r, but we want A in terms of x. So we replace r in equation (12) by the quantity given in equation (11). This yields

$$A(x) = \pi\left(\frac{x}{2\pi}\right)^2 = \frac{\pi x^2}{4\pi^2} = \frac{x^2}{4\pi}$$

Thus we have

$$A(x) = \frac{x^2}{4\pi}$$

This is the required equation. It expresses the area of the circle in terms of the circumference x. In other words, if we know the length of the piece of wire that is to be bent into a circle, we can use that length to calculate what the area of the circle will be. *Note*: In this example a reasonable domain for the function $A(x)$ is $(0, \infty)$.

EXAMPLE 6 Expressing the surface area of a cylinder of given volume as a function of the radius of the base

Figure 6 displays a right circular cylinder, along with the formulas for its volume V and total surface area S. Given that the volume is 10 cm^3, express the surface area S as a function of r, the radius of the base.

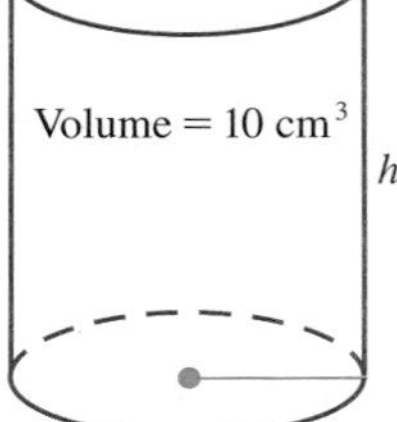

$V = \pi r^2 h$
$S = 2\pi r^2 + 2\pi r h$

Figure 6

SOLUTION

Step 1 See Figure 6.

Step 2 We are given a formula that expresses the surface area S in terms of both r and h. We want to express S in terms of just r.

Step 3 We are given that $V = 10$ and also that $V = \pi r^2 h$. Thus

$$\pi r^2 h = 10$$

and consequently, expressing h in terms of r,

$$h = \frac{10}{\pi r^2} \tag{13}$$

Step 4 We take the given formula for S, namely,

$$S = 2\pi r^2 + 2\pi rh$$

and replace h with the quantity given in equation (13). We obtain

$$\begin{aligned} S(r) &= 2\pi r^2 + 2\pi r\left(\frac{10}{\pi r^2}\right) \\ &= 2\pi r^2 + \frac{20}{r} \end{aligned}$$

This is the required equation. It expresses the total surface area in terms of the radius r. Since the only restriction on r is that it be positive, the domain of the area function is $(0, \infty)$.

We have followed the same four-step procedure in Examples 1 through 6. Of course, no single method can cover all possible cases. As usual, common sense and experience are often necessary. Also, you should not feel compelled to follow this procedure at any cost. Keep this in mind as you study the last two examples in this section.

EXAMPLE 7 Expressing a certain product as a function of a single variable

Two numbers add up to 8. Express the product P of these two numbers in terms of a single variable.

SOLUTION

If we call the two numbers x and $8 - x$, then their product P is given by

$$P(x) = x(8 - x) = 8x - x^2$$

That's it. This last equation expresses the product as a function of the variable x. Since there are no restrictions on x (other than its being a real number), the domain of this function is $(-\infty, \infty)$.

EXAMPLE 8 Expressing revenue as a function of one variable

In economics, the revenue R generated by selling x units at a price of p dollars per unit is given by

$$R = x \cdot p$$

(x: number of units; p: price per unit)

In Figure 7, we are given a hypothetical function relating the selling price of a certain item to the number of units sold. Such a function is called a **demand function.** Express the revenue as a function of x.

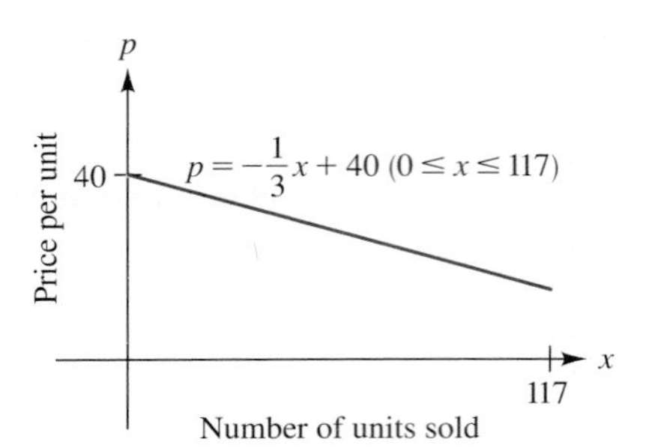

Figure 7

SOLUTION

More than anything else, this problem is an exercise in reading. After reading the problem several times, we find that it comes down to this:

given: $R = x \cdot p$ and $p = -\frac{1}{3}x + 40 \quad (0 \le x \le 117)$

find: an equation expressing R in terms of x

In view of this, we write

$$R = x \cdot p = x\left(-\frac{1}{3}x + 40\right)$$

Thus we have

$$R(x) = -\frac{1}{3}x^2 + 40x \qquad (0 \le x \le 117)$$

This is the required function. It allows us to calculate the revenue when we know the number of units sold. Note that this revenue function is a quadratic function.

QUESTION FOR REVIEW AND ALSO PREVIEW How would you find the maximum revenue in this case?

EXERCISE SET 4.4

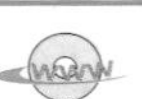

A

In each exercise in Exercise Set 4.4 you are asked to express one variable as a function of another. Be sure to state a domain for the function that reflects the constraints of the problem.

1. (a) The perimeter of a rectangle is 16 cm. Express the area of the rectangle in terms of the width x. *Suggestion*: First reread Example 1.

(b) The area of a rectangle is 85 cm^2. Express the perimeter as a function of the width x.

2. A rectangle is inscribed in a circle of diameter 12 in.

(a) Express the perimeter of the rectangle as a function of its width x. *Suggestion*: First reread Example 2.

(b) Express the area of the rectangle as a function of its width x.

3. A point $P(x, y)$ is on the curve $y = x^2 + 1$.

(a) Express the distance from P to the origin as a function of x. *Suggestion*: First reread Example 3.

(b) In part (a), you expressed the length of a certain line segment as a function of x. Now express the slope of that line segment in terms of x.

4. A point $P(x, y)$ lies on the curve $y = \sqrt{x}$, as shown in the figure.

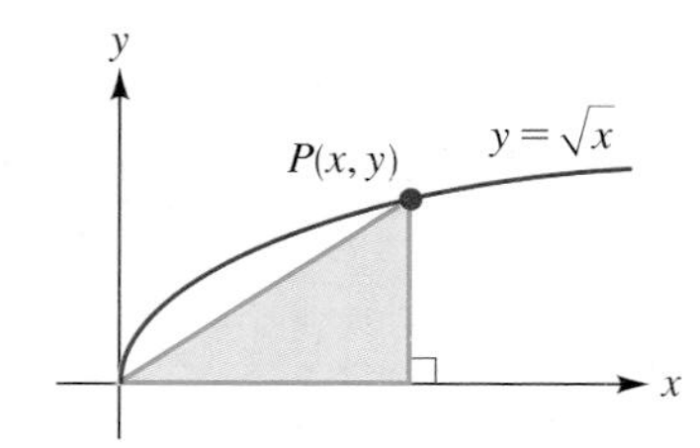

(a) Express the area of the shaded triangle as a function of x. *Suggestion*: First reread Example 4.

(b) Express the perimeter of the shaded triangle in terms of x.

5. A piece of wire πy inches long is bent into a circle.

(a) Express the area of the circle as a function of y. *Suggestion*: First reread Example 5.

(b) If the original piece of wire were bent into a square instead of a circle, how would you express the area in terms of y?

6. The volume of a right circular cylinder is 12π in.^{3}.

(a) Express the height as a function of the radius.

(b) Express the total surface area as a function of the radius.

7. Two numbers add to 16.

(a) Express the product of the two numbers in terms of a single variable. *Suggestion*: First reread Example 7.

(b) Express the sum of the squares of the two numbers in terms of a single variable.

(c) Express the difference of the cubes of the two numbers in terms of a single variable. (There are two answers.)

(d) What happens when you try to express the average of the two numbers in terms of one variable?

8. The product of two numbers is 16. Express the sum of the squares of the two numbers as a function of a single variable.

9. Suppose we are given the following demand function relating the price p in dollars to the number of units x sold of a certain commodity:

$$p = 5 - \frac{x}{4} \quad (p \text{ in dollars})$$

(a) Graph this demand function.

(b) How many units can be sold when the unit price is \$3? Locate the point on the graph of the demand function that conveys this information.

(c) To sell 12 items, how should the unit price be set? Locate the point on the graph of the demand function that conveys this information.

(d) Find the revenue function corresponding to the given demand function. (Use the formula $R = x \cdot p$.) Graph the revenue function.

(e) Find the revenue when $x = 2$, when $x = 8$, and when $x = 14$.

(f) According to your graph in part (d), which x-value yields the greatest revenue? What is that revenue? What is the corresponding unit price?

10. In Example 1 we considered a rectangle with perimeter 100 cm. We found that the area of such a rectangle is given by $A(x) = 50x - x^2$, where x is the width of the rectangle. Compute the numbers $A(1)$, $A(10)$, $A(20)$, $A(25)$, and $A(35)$. Which width x seems to yield the largest area $A(x)$?

11. In Example 2 we considered a rectangle inscribed in a circle of diameter 8 cm. We found that the perimeter of such a rectangle is given by

$$P(x) = 2x + 2\sqrt{64 - x^2}$$

where x is the width of the rectangle.

(a) Use a calculator to complete the following table, rounding each answer to two decimal places.

x	1	2	3	4	5	6	7
$P(x)$							

(b) In your table, what is the largest value for $P(x)$? What is the width x in this case?

(c) Using calculus, it can be shown that among all possible widths x, the width $x = 4\sqrt{2}$ cm yields the largest possible perimeter. Use a calculator to compute the perimeter in this case, and check to see that the value you obtain is indeed larger than all of the values obtained in part (a).

12. Let $2s$ denote the length of the side of an equilateral triangle.

(a) Express the height of the triangle as a function of s.

(b) Express the area of the triangle as a function of s.

(c) Use the function you found in part (a) to determine the height of an equilateral triangle, each side of which is 8 cm long.

(d) Use the function you found in part (b) to determine the area of an equilateral triangle, each side of which is 5 in. long.

13. If x denotes the length of a side of an equilateral triangle, express the area of the triangle as a function of x.

14. The height of a right circular cylinder is twice the radius. Express the volume as a function of the radius.

15. Using the information given in Exercise 14, express the radius as a function of the volume.

16. The total surface area of a right circular cylinder is 14 in.^{2}. Express the volume as a function of the radius.

17. The volume V and the surface area S of a sphere of radius r are given by the formulas $V = \frac{4}{3}\pi r^3$ and $S = 4\pi r^2$. Express V as a function of S.

18. The base of a rectangle lies on the x-axis, while the upper two vertices lie on the parabola $y = 10 - x^2$. Suppose that the coordinates of the upper right vertex of the rectangle are (x, y). Express the area of the rectangle as a function of x.

19. The hypotenuse of a right triangle is 20 cm. Express the area of the triangle as a function of the length x of one of the legs.

20. (a) Express the area of the shaded triangle in the figure on the next page as a function of x.

(b) Express the perimeter of the shaded triangle as a function of x.

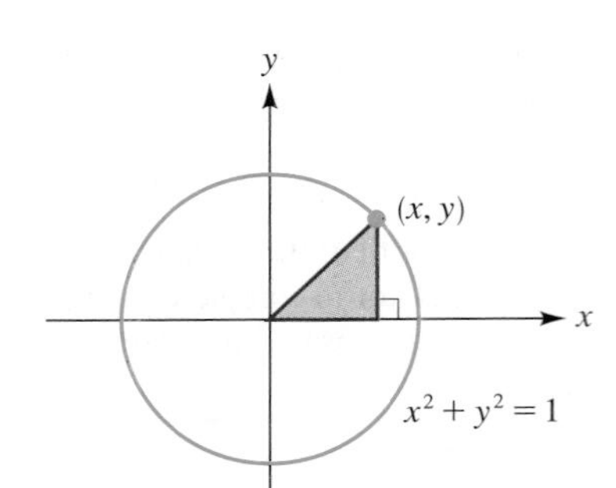

Figure for Exercise 20

21. For the following figure, express the length AB as a function of x. *Hint:* Note the similar triangles.

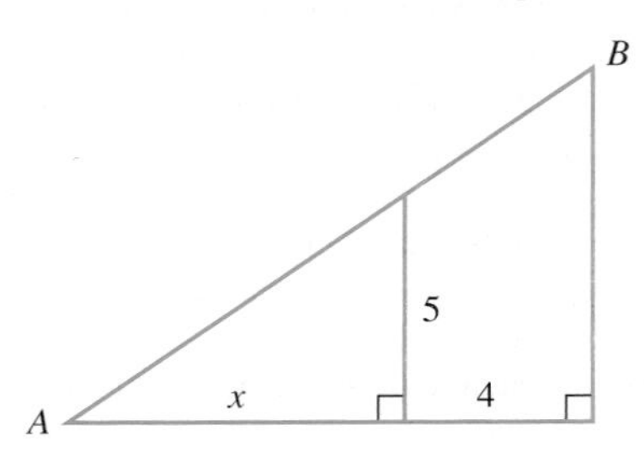

22. Five hundred feet of fencing is available to enclose a rectangular pasture alongside a river, which serves as one side of the rectangle (so only three sides require fencing—see the figure). Express the area of the rectangular pasture as a function of x.

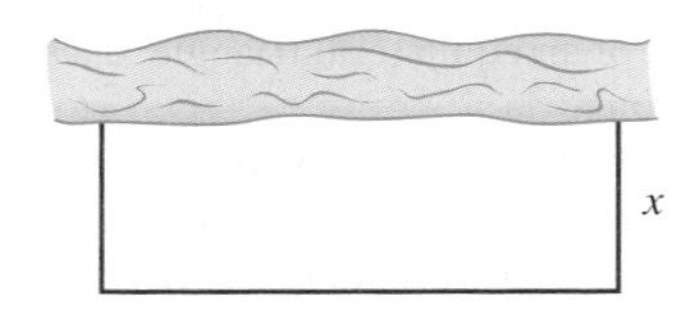

Ⓖ *For Exercises 23–30:*

(a) *Is this a quadratic function? Use a graphing utility to draw the graph.*

(b) *How many turning points are there within the given interval?*

(c) *On the given interval, does the function have a maximum value? A minimum value?*

23. $A(x) = 50x - x^2$, $0 < x < 50$ (This is the function that we obtained in Example 1.)

24. $P(x) = 2x + 2\sqrt{64 - x^2}$, $0 < x < 8$ (from Example 2)

25. $D(x) = \sqrt{x^2 - x + 1}$, $x \geq 0$ (from Example 3)

26. $A(x) = 8x - \frac{1}{2}x^3$, $0 < x < 4$ (from Example 4)

27. $S(r) = 2\pi r^2 + 20/r$, $r > 0$ (from Example 6)

28. $R(x) = x(-\frac{1}{3}x + 40)$, $x \geq 0$ (from Example 8)

29. **(a)** $f(x) = x + \dfrac{1}{x}$, $x > 0$ **(b)** $g(x) = x - \dfrac{1}{x}$, $x > 0$

30. **(a)** $F(x) = \sqrt{x} - \dfrac{1}{x}$ **(b)** $G(x) = \sqrt{x} + \dfrac{1}{x}$

Ⓖ **31.** **(a)** Suppose that the product of two positive numbers is $\sqrt{11}$. Express the sum of the two numbers as a function of a single variable, and then use a graphing utility to draw the graph. Based on the graph, does the sum have a minimum value or maximum value?

(b) Suppose that the sum of two positive numbers is $\sqrt{11}$. Express the product of the two numbers as a function of a single variable. Without drawing a graph, explain why the product has a maximum value.

B

32. A piece of wire 4 m long is cut into two pieces, then each piece is bent into a square. Express the combined area of the two squares in terms of one variable.

33. A piece of wire 3 m long is cut into two pieces. Let x denote the length of the first piece and $3 - x$ the length of the second. The first piece is bent into a square and the second into a rectangle in which the width is half the length. Express the combined area of the square and the rectangle as a function of x. Is the resulting function a quadratic function?

In Exercises 34–37, refer to the following figure, which displays a right circular cone along with the formulas for the volume V and the lateral surface area S.

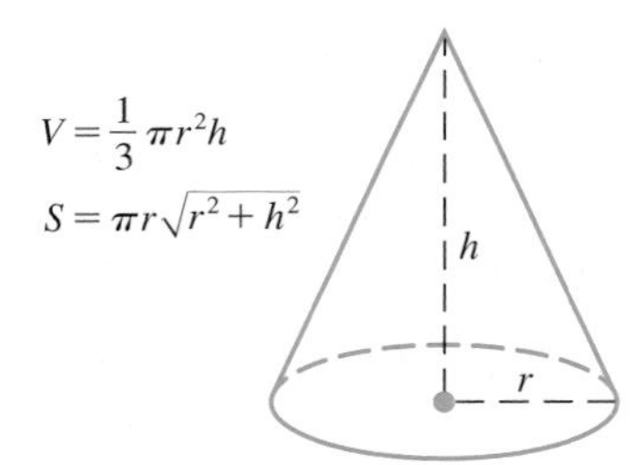

34. The volume of a right circular cone is 12π cm^3.

(a) Express the height as a function of the radius.

(b) Express the radius as a function of the height.

35. Suppose that the height and radius of a right circular cone are related by the equation $h = \sqrt{3}r$.

(a) Express the volume as a function of r.

(b) Express the lateral surface area as a function of r.

36. The volume of a right circular cone is 2 ft.^{3} Show that the lateral surface area as a function of r is given by

$$S = \frac{\sqrt{\pi^2 r^6 + 36}}{r}$$

37. In a certain right circular cone the volume is numerically equal to the lateral surface area.

(a) Express the radius as a function of the height.

(b) Express the height as a function of the radius.

38. A line is drawn from the origin O to a point $P(x, y)$ in the first quadrant on the graph of $y = 1/x$. From point P, a line is drawn perpendicular to the x-axis, meeting the x-axis at B.

(a) Draw a figure of the situation described.

(b) Express the perimeter of $\triangle OPB$ as a function of x.

(c) Try to express the area of $\triangle OPB$ as a function of x. What happens?

39. A piece of wire 14 in. long is cut into two pieces. The first piece is bent into a circle, the second into a square. Express the combined total area of the circle and the square as a function of x, where x denotes the length of the wire that is used for the circle.

40. A wire of length L is cut into two pieces. The first piece is bent into a square, the second into an equilateral triangle. Express the combined total area of the square and the triangle as a function of x, where x denotes the length of wire used for the triangle. (Here, L is a constant, not another variable.)

41. An athletic field with a perimeter of $\frac{1}{4}$ mile consists of a rectangle with a semicircle at each end, as shown in the figure. Express the area of the field as a function of r, the radius of the semicircle.

42. A square of side x is inscribed in a circle. Express the area of the circle as a function of x.

43. An equilateral triangle of side x is inscribed in a circle. Express the area of the circle as a function of x.

44. (a) Refer to the accompanying figure. An offshore oil rig is located at point A, which is 10 miles out to sea. An oil pipeline is to be constructed from A to a point C on the shore and then to an oil refinery at point D, farther up the coast. If it costs \$8000 per mile to lay the pipeline in the sea and \$2000 per mile on land, express the cost of laying the pipeline in terms of x, where x is the distance from B to C.

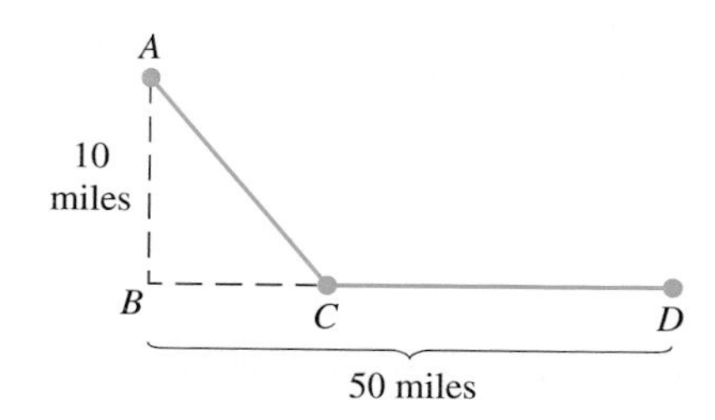

(b) Use a calculator and your result in part (a) to complete the following table, rounding the answers to the nearest 100 dollars.

x (miles)	0	10	20	30	40	50
Cost (dollars)						

(c) On the basis of the results in part (b), does it appear that the cost function is increasing, decreasing, or neither on the interval $0 \le x \le 50$? Which of the x-values in the table yields the lowest cost?

(d) Use a calculator and your result in part (a) to complete the following table, rounding the answers to the nearest 100 dollars.

x (miles)	0	4	8	12	16	20
Cost (dollars)						

(e) Again, answer the questions in part (c), but now take into account the results in part (d) also.

45. (a) An open-top box is constructed from a 6-by-8-inch rectangular sheet of tin by cutting out equal squares at each corner and then folding up the flaps, as shown in the figure. Express the volume of the box as a function of x, the length of the side of each cutout square.

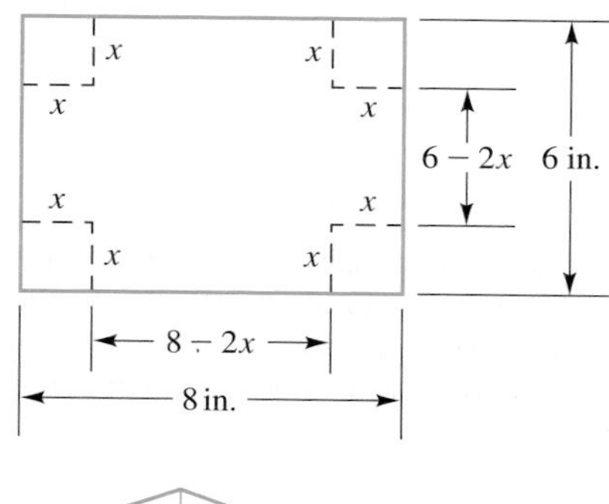

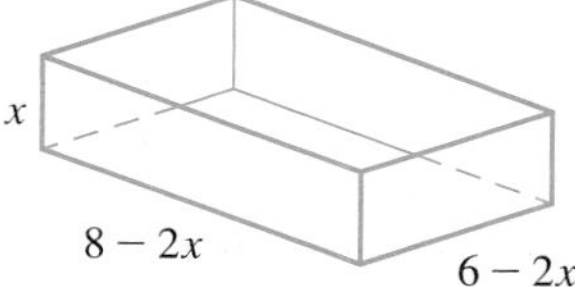

(b) Use a calculator and your result in part (a) to complete the following table. Round the answers to the nearest 0.5 in.^{3}.

x (in.)	0	0.5	1.0	1.5	2.0	2.5	3.0
Volume (in.^{3})							

(c) On the basis of the results in part (b), which x-value in the interval $0 \le x \le 3$ appears to yield the largest volume for the box?

(d) Use a calculator and your result in part (a) to complete the following table. Round off the answers to the nearest 0.1 in.^{3}.

x (in.)	0.8	0.9	1.0	1.1	1.2	1.3	1.4
Volume (in.^{3})							

(e) Again, answer the questions in part (c), but now take into account the results in part (d), too.

46. Follow Exercise 45(a), but assume that the original piece of tin is a square, 12 in. on each side.

47. A Norman window is in the shape of a rectangle surmounted by a semicircle, as shown in the figure. Assume that the perimeter of the window is 32 ft.

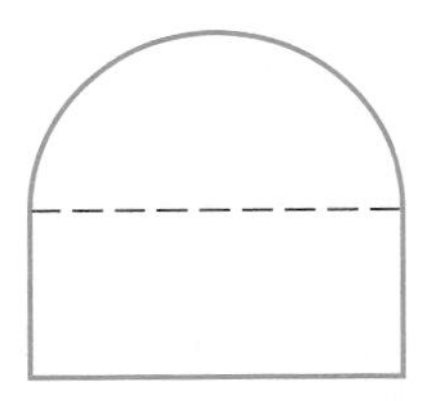

(a) Express the area of the window as a function of r, the radius of the semicircle.

(b) The function you were asked to find in part (a) is a quadratic function, so its graph is a parabola. Does the parabola open upward or downward? Does it pass through the origin? Show that the vertex of the parabola is

$$\left(\frac{32}{\pi + 4}, \frac{512}{\pi + 4}\right)$$

48. Refer to the following figure. Express the lengths CB, CD, BD, and AB in terms of x. *Hint:* Recall the theorem from geometry stating that in a 30°-60°-90° right triangle, the side opposite the 30° angle is half the hypotenuse.

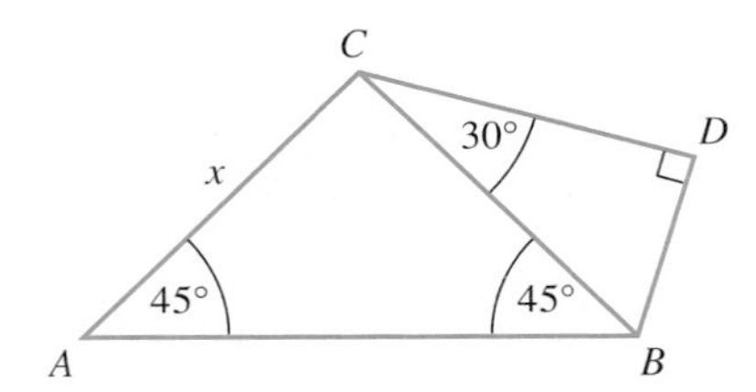

49. Refer to the following figure; let s denote the ratio of y to z.

(a) Express y as a function of s.

(b) Express s as a function of y.

(c) Express z as a function of s.

(d) Express s as a function of z.

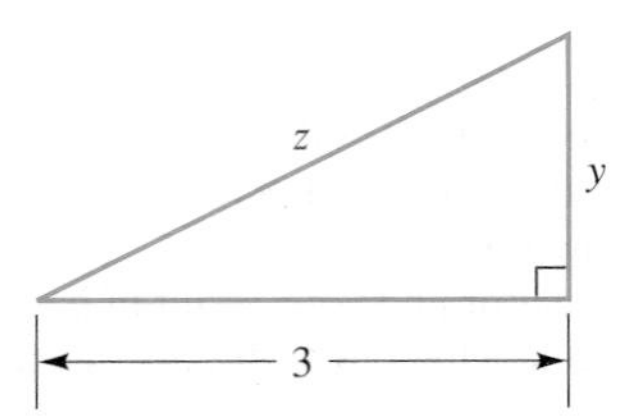

50. Refer to Example 4.

(a) What is the x-intercept of the curve $y = 16 - x^2$ in Figure 4?

(b) What is the domain of the area function in Example 4, assuming that the point P does not lie on the x- or y-axis?

51. The following figure shows the parabola $y = x^2$ and a line segment $\overline{AP}$ drawn from the point $A(0, -1)$ to the point $P(a, a^2)$ on the parabola.

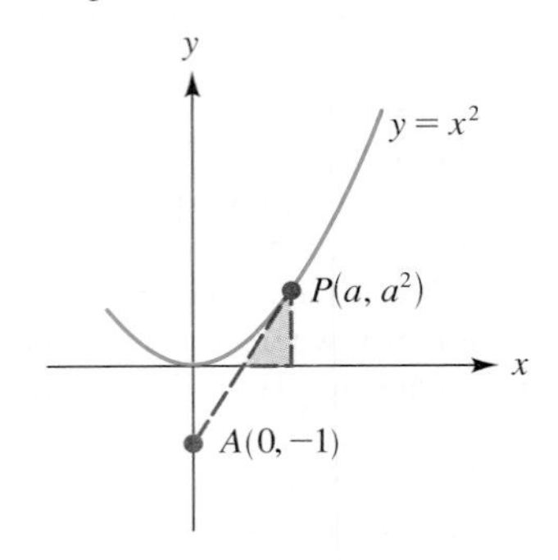

(a) Express the slope of $\overline{AP}$ in terms of a.

(b) Show that the area of the shaded triangle in the figure is given by

$$\text{area} = \frac{a^5}{2(a^2 + 1)}$$

52. A rancher who wishes to fence off a rectangular area finds that the fencing in the east-west direction will require extra reinforcement due to strong prevailing winds. Fencing in the east-west direction will therefore cost \$12 per (linear) yard, as opposed to a cost of \$8 per yard for fencing in the north-south direction. Given that the rancher wants to spend \$4800 on fencing, express the area of the rectangle as a function of its width x. (*Note:* In this problem by *width* we mean *the measure in yards of a side running in the east-west direction.*) The required function is in fact a quadratic, so its graph is a parabola. Does the parabola open upward or downward? By considering this graph, find which width x yields the rectangle of largest area. What is this maximum area?

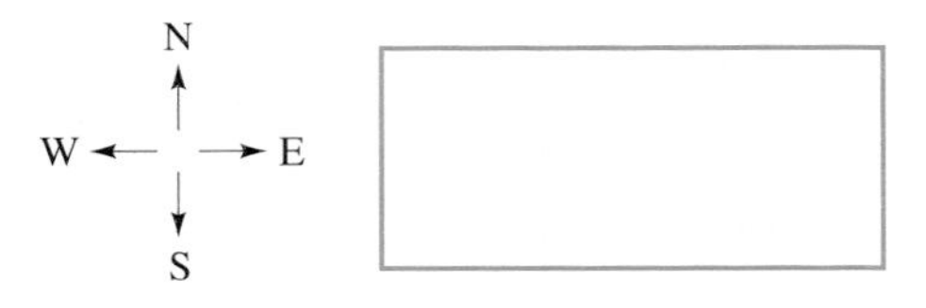

C

53. The following figure shows two concentric squares. Express the area of the shaded triangle as a function of x.

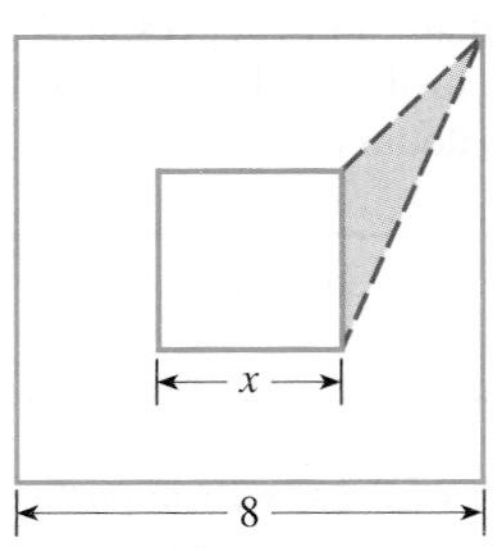

54. The following figure shows two concentric circles of radii r and R. Let A denote the area within the larger circle but outside the smaller one. Express A as a function of x (where x is defined in the figure).

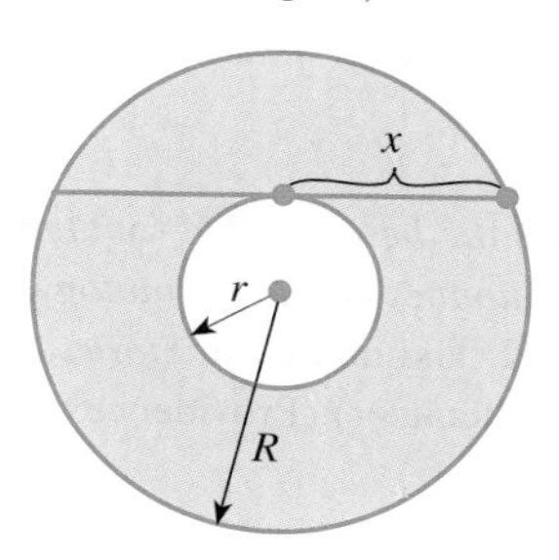

55. A straight line with slope m ($m < 0$) passes through the point $(1, 2)$ and intersects the line $y = 4x$ at a point in the first quadrant. Let A denote the area of the triangle bounded by $y = 4x$, the x-axis, and the given line of slope m. Express A as a function of m.

56. A line with slope m ($m < 0$) passes through the point (a, b) in the first quadrant and intersects the line $y = Mx$ ($M > 0$) at another point in the first quadrant. Let A denote the area of the triangle bounded by $y = Mx$, the x-axis, and the given line with slope m. Show that A can be written as

$$A = \frac{M(am - b)^2}{2m(m - M)}$$

57. A line with slope m ($m < 0$) passes through the point (a, b) in the first quadrant. Express the area of the triangle bounded by this line and the axes in terms of m.

58. One corner of a sheet of paper of width a is folded over and just reaches the opposite side, as indicated in the figure. Express L, the length of the crease, in terms of x and a.

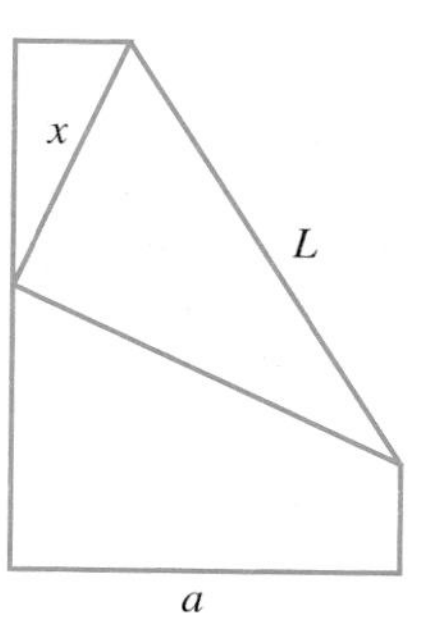

MINI PROJECT

GROUP WORK ON FUNCTIONS OF TIME

In each of the examples and exercises of this section, the goal was to express a certain quantity as a function of a single variable. In the three problems that follow, that single variable will be time t. For each problem, after the group finds a solution, each individual should write up the solution on his or her own. (The problems are in order of increasing difficulty.)

1. (a) This problem is based on Example 1. So first, have someone in your group present the example.

(b) Under the conditions of Example 1, assume the following additional information. The width x is increasing at the rate of 3 cm/min, and at time $t = 0$ min, the width x is 2 cm. Express the area of the rectangle as a function of time t.

(c) During the time interval from $t = 0$ min to $t = 16$ min, when is the area of the rectangle increasing? When is it decreasing? Why are we using the value $t = 16$ min?

2. Starting at time $t = 0$ hr, two cars move away from the same intersection. Car A goes north at 40 miles/hr, and car B goes east at 30 miles/hr. Express the distance between the cars as a function of the time t. Is your result a linear function, a quadratic function, or neither?

3. Car C passes through an intersection, heading north at 40 miles/hr. One quarter hour later, car D goes through the intersection, heading east at 30 miles/hr. Express the distance between the two cars as a function of time. Is your result a linear function, a quadratic function, or neither?

4.5 MAXIMUM AND MINIMUM PROBLEMS

... problems on maxima and minima, although new features in an English textbook, stand so little in need of apology with the scientific public that I offer none.—G. Chrystal in his preface to *Textbook of Algebra* (1886)

In daily life it is constantly necessary to choose the best possible (optimal) solution. A tremendous number of such problems arise in economics and in technology. In such cases it is frequently useful to resort to mathematics.—V. K. Tikhomirov in *Stories about Maxima and Minima* (translated from the Russian by Abe Shenitzer) (Providence, R.I.: The American Mathematical Society, 1990)

TABLE 1

x	y	$x + y$	xy
−2	11	9	−22
−1	10	9	−10
0	9	9	0
1	8	9	8
2	7	9	14
3	6	9	18
4	5	9	(20)

TABLE 2

x	y	$x + y$	xy
1	8	9	8
1.5	7.5	9	11.25
2	7	9	14
2.5	6.5	9	16.25
3	6	9	18
3.5	5.5	9	19.25
4	5	9	20
4.5	4.5	9	(20.25)

You have already seen several examples of maximum and minimum problems in Section 4.2. Before reading further in the present section, you should first review Examples 5 through 7 on pages 235–238.

Actually, we begin this section's discussion of maximum and minimum problems at a more intuitive level. Consider, for example, the following question: If two numbers add to 9, what is the largest possible value of their product? To gain some insight here, we carry out a few preliminary calculations. See Table 1.

We have circled the 20 in the right-hand column of Table 1 because it *appears* to be the largest product. We say "appears" because our table is incomplete. For instance, what if we allowed x- and y-values that are not whole numbers? Might we get a product exceeding 20? Table 2 shows the results of some additional calculations along these lines.

As you can see from Table 2, there is a product exceeding 20, namely, 20.25. Now the question is, if we further expand our tables, can we find an even larger product, one exceeding 20.25? And here we have come about as far as we want to go using this approach involving arithmetic and tables. For no matter what candidate we come up with for the largest product, there will always be the question of whether we might do still better using a larger table.

Nevertheless, this approach was useful, for it showed us what is really at the heart of a typical maximum or minimum problem. Essentially, we are trying to sort through an infinite number of possible cases and pick out the required extreme case. In the example at hand, there are infinitely many pairs of numbers x and y adding to 9. We are asked to look at the products of all of these pairs and see which (if any!) is the largest. Example 1 shows how to apply our knowledge of quadratic functions to solve this problem in a definitive manner.

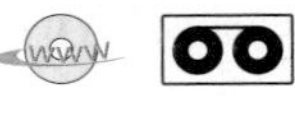

EXAMPLE 1 Maximizing a product

Two numbers add to 9. What is the largest possible value for their product?

SOLUTION

Call the two numbers x and $9 - x$. Then their product P is given by

$$P = x(9 - x) = 9x - x^2$$

The graph of this quadratic function is the parabola in Figure 1. Note the accompanying calculations for the vertex. As Figure 1 and the accompanying

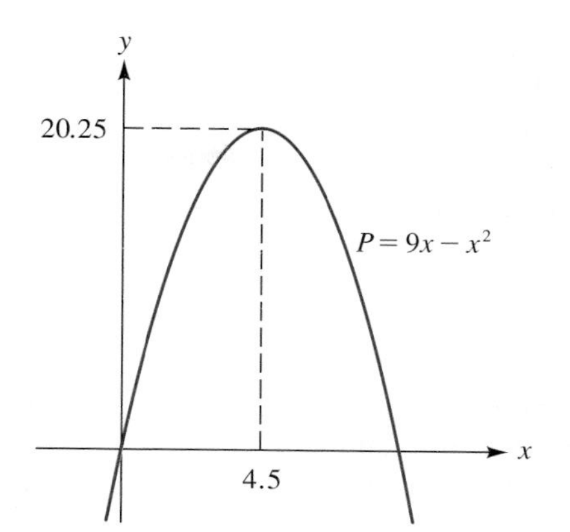

Figure 1

Vertex calculation (by completing the square):

$$P = -(x^2 - 9x)$$
$$= -\left(x^2 - 9x + \frac{81}{4}\right) + \frac{81}{4}$$
$$= -\left(x - \frac{9}{2}\right)^2 + \frac{81}{4}$$

The vertex is $\left(\frac{9}{2}, \frac{81}{4}\right)$

calculations show, the largest value of the product P is 20.25. This is the required solution. (Note from the graph that there is no smallest value of P.)

In the example just completed, we used the technique of completing the square to determine the vertex of a parabola. Another way to determine the vertex, approximately at least, is to use a graphing utility, as in Example 5(a) in Section 4.2. There is a third method and we shall make use of it for the remaining examples in this section. This third method uses the *vertex formula* given in the box that follows. You should check with your instructor to see which of the three methods he or she wants you to use in a given context. We're going to use the vertex formula in the exposition here so that you can better focus on the problem-solving aspects of the examples.

THEOREM The Vertex Formula

The x-coordinate of the vertex of the parabola $y = ax^2 + bx + c$ $(a \neq 0)$ is given by

$$x = \frac{-b}{2a}$$

EXAMPLE

For the parabola $y = 2x^2 - 12x + 5$ the x-coordinate of the vertex is

$$x = \frac{-b}{2a} = \frac{-(-12)}{2(2)} = 3$$

The vertex formula can be derived by completing the square (see Exercise 67 on p. 243) or by using the method outlined in the first mini project at the end of Section 4.2. Here's another way, perhaps the simplest. The parabola $y = ax^2 + bx + c$ will have the same x-coordinate for its vertex as does $y = ax^2 + bx$ because the two graphs are just vertical translates of one another by a distance $|c|$. But due to symmetry, it's easy to find the x-coordinate of the vertex of $y = ax^2 + bx$. As Figure 2 indicates, the x-coordinate of the vertex is midway between the two x-intercepts. To find these x-intercepts, we have

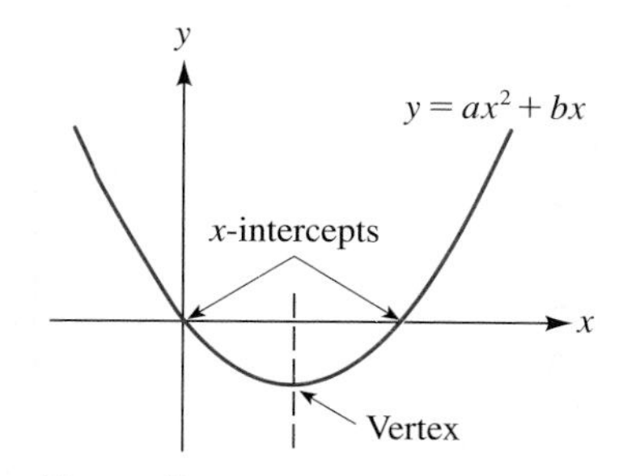

Figure 2

$$ax^2 + bx = 0$$
$$x(ax + b) = 0$$
$$x = 0 \quad \Big| \quad ax + b = 0$$
$$x = -\frac{b}{a}$$

Now we want the x-value that is halfway between $x = 0$ and $x = -b/a$, namely, $\frac{1}{2}(-b/a)$. Thus the x-coordinate of the vertex is $-b/2a$, as we wished to show.

Except for the fact that we completed the square rather than using the vertex formula, Example 1 indicates the general strategy that we're going to follow for solving the maximum and minimum problems in this section.

Strategy for Solving the Maximum and Minimum Problems in This Section

1. Express the quantity to be maximized or minimized in terms of a single variable. For instance, in Example 1 we found $P = 9x - x^2$. In setting up such functions, you'll want to keep in mind the four-step procedure used in Section 4.4.
2. Assuming that the function you have determined is a quadratic, note whether its graph, a parabola, opens upward or downward. Check whether this is consistent with the requirements of the problem. For instance, the parabola in Figure 1 opens downward; so it makes sense to look for a *largest,* not a smallest, value of P. Now use the vertex formula $x = -b/2a$ to locate the x-coordinate of the vertex. (If the function is not a quadratic but is closely related to a quadratic, these ideas may still apply. See, for instance, Examples 6 and 7 in Section 4.2.)
3. After you have determined the x-coordinate of the vertex, you must relate that information to the original question. In Example 1, for instance, we were asked for the product P, not for x.

EXAMPLE 2 Finding the dimensions of the rectangle with maximum area

Among all rectangles having a perimeter of 10 ft, find the dimensions (length and width) of the one with the greatest area.

SOLUTION

First we want to set up a function that expresses the area of the rectangle in terms of a single variable. In doing this, we'll be guided by the four-step procedure that we used in Section 4.4. Figure 3(a) displays the given information. Our problem is to determine the dimensions of the rectangle that has the greatest area. As Figure 3(b) indicates, we can label the dimensions x and y. Since the perimeter is given as 10 ft, we have

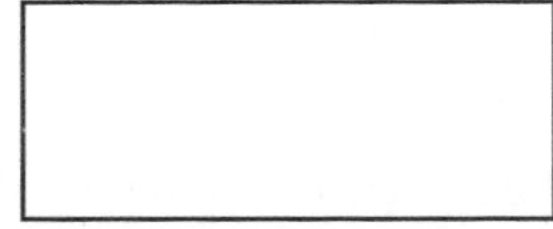

Perimeter is 10 ft.

(a)

$$\begin{aligned} 2x + 2y &= 10 \\ x + y &= 5 \\ y &= 5 - x \end{aligned} \qquad (1)$$

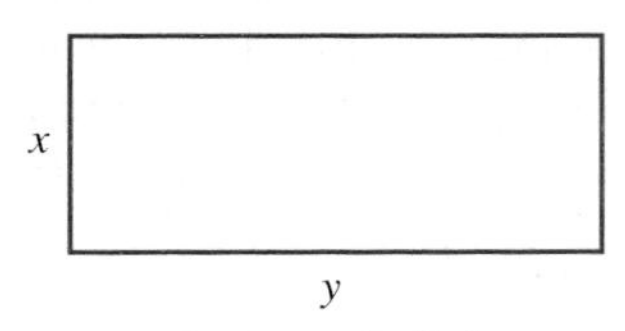

Perimeter is 10 ft.

(b)

Figure 3

Letting A denote the area of the rectangle, we can write

$$\begin{aligned} A &= xy \\ A(x) &= x(5 - x) \qquad \text{substituting for } y \text{ using equation (1)} \\ &= 5x - x^2 \end{aligned}$$

GRAPHICAL PERSPECTIVE

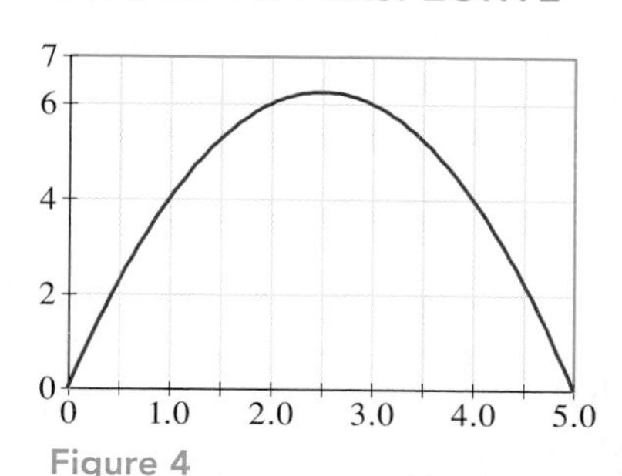

Figure 4
The graph of the area function, $A(x) = 5x - x^2$, is a parabola that opens downward. The maximum area occurs when $x = 2.5$ ft.

This expresses the area of the rectangle in terms of the width x. Since the graph of this quadratic function is a parabola that opens downward, it does make sense to talk about the maximum. We calculate the x-coordinate of the vertex:

$$\begin{aligned} x &= \frac{-b}{2a} \\ &= \frac{-5}{2(-1)} = 2.5 \text{ ft} \end{aligned}$$

This is the width of the rectangle with the greatest area. Now, the problem asks for the length and width of this rectangle, not its area. So we don't want to calculate $A(x)$; rather, we want y. Using equation (1), we have

$$y = 5 - x = 5 - 2.5 = 2.5 \text{ ft}$$

Thus among all rectangles having a perimeter of 10 ft, the one with the greatest area is actually the square with dimensions of 2.5 ft by 2.5 ft. See Figure 4 for a graphical view of this result.

EXAMPLE 3 Finding the maximum height of a baseball

Suppose that a baseball is tossed straight up and that its height as a function of time is given by the function

$$h(t) = -16t^2 + 64t + 6$$

where $h(t)$ is measured in feet, t is in seconds, and $t = 0$ corresponds to the instant that the ball is released. What is the maximum height of the ball?

SOLUTION

The given function tells us how the height $h(t)$ of the ball depends upon the time t. We want to know the largest possible value for $h(t)$. Since the graph of the given height function is a parabola opening downward, we can determine the largest value of $h(t)$ just by finding the vertex of the parabola. We use the vertex formula $x = -b/2a$, letting t play the role of x. This yields

$$t = \frac{-b}{2a} = \frac{-64}{2(-16)} = 2$$

Thus when $t = 2$ sec, the ball reaches its maximum height. To find that maximum height (and the problem does ask for height, not time), we substitute the value $t = 2$ in the equation $h(t) = -16t^2 + 64t + 6$ to obtain

$$\begin{aligned} h(2) &= -16(2)^2 + 64(2) + 6 \\ &= 70 \text{ ft} \qquad \text{(Check the arithmetic.)} \end{aligned}$$

The maximum height of the ball is 70 ft. See Figure 5 for a graphical perspective. *Caution*: The path of the baseball is not the parabola. Why?

GRAPHICAL PERSPECTIVE

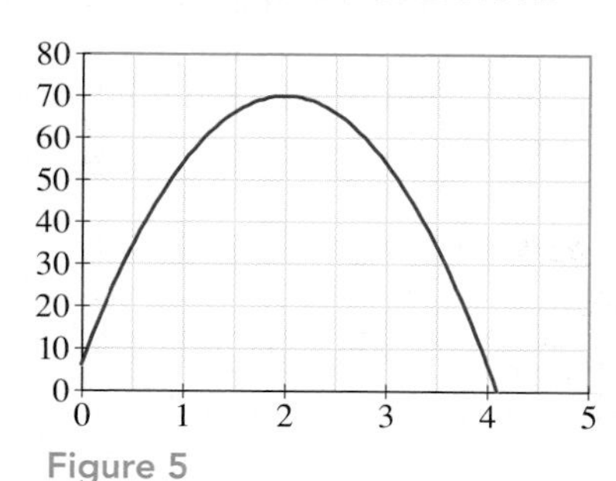

Figure 5
The graph of the height function $h(t) = -16t^2 + 64t + 6$ is a parabola that opens downward. The maximum height is 70 ft, and this occurs when $t = 2$ sec.

In the next example we look for an input to minimize a function of the form $y = \sqrt{f(x)}$, where f is a quadratic function with values that are always nonnegative.

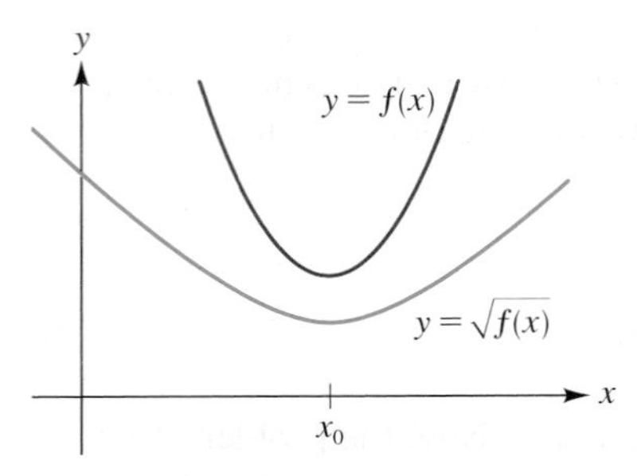

Figure 6
The same input x_0 minimizes both functions $y = f(x)$ and $y = \sqrt{f(x)}$ [provided $f(x)$ is nonnegative for all x].

(There was an example similar to this in Section 4.2.) As indicated in Figure 6, we need only find the input that minimizes the quadratic function $y = f(x)$, because this same input also minimizes $y = \sqrt{f(x)}$. (The *outputs* of the two functions are different; but the same *input* serves to minimize both functions.)

EXAMPLE 4 Working with a function that is closely related to a quadratic

Which point on the curve $y = \sqrt{x}$ is closest to the point (1, 0)?

SOLUTION

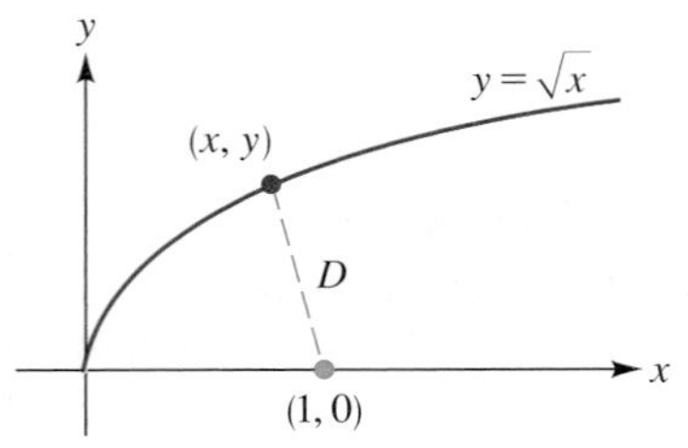

Figure 7

In Figure 7 we let D denote the distance from a point (x, y) on the curve to the point $(1, 0)$. We are asked to find out exactly which point (x, y) will make the distance D as small as possible. Using the distance formula, we have

$$\begin{aligned} D &= \sqrt{(x-1)^2 + (y-0)^2} \\ &= \sqrt{x^2 - 2x + 1 + y^2} \end{aligned} \tag{2}$$

This expresses D in terms of both x and y. To express D in terms of x alone, we use the given equation $y = \sqrt{x}$ to substitute for y in equation (3). This yields

$$\begin{aligned} D(x) &= \sqrt{x^2 - 2x + 1 + (\sqrt{x})^2} \\ &= \sqrt{x^2 - x + 1} \end{aligned}$$

Now, according to the remarks just prior to this example, the x-value that minimizes the function $D(x) = \sqrt{x^2 - x + 1}$ is the same x-value that minimizes $y = x^2 - x + 1$. (Actually, to apply those remarks, we need to know that the expression $x^2 - x + 1$ is nonnegative for all inputs x. See Exercise 43 for this.) Since the graph of $y = x^2 - x + 1$ is a parabola that opens upward, the required x-value is the x-coordinate of the vertex, and we have

$$x = \frac{-b}{2a} = \frac{-(-1)}{2(1)} = \frac{1}{2}$$

Now that we have the x-coordinate of the vertex, we want to calculate the y-coordinate. [We don't want to calculate $D(x)$ because this particular problem asks us to specify a *point,* not a distance. Go back and reread the wording of the question if you are not clear on this.] Substituting $x = 1/2$ in the equation $y = \sqrt{x}$ yields

$$y = \sqrt{\frac{1}{2}} = \sqrt{\frac{1}{2} \cdot \frac{2}{2}} = \frac{\sqrt{2}}{\sqrt{2^2}} = \frac{\sqrt{2}}{2}$$

Thus, among all points on the curve $y = \sqrt{x}$, the point closest to (1, 0) is $(1/2, \sqrt{2}/2)$. *Question*: Why would it not have made sense to ask instead for the point farthest from (1, 0)?

EXAMPLE 5 Finding the dimensions that maximize an area

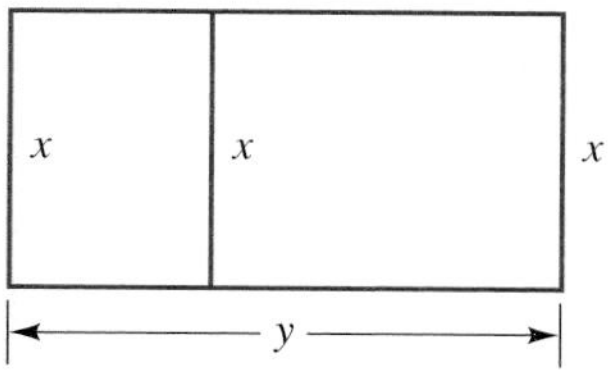

Figure 8

Suppose that you have 600 m of fencing with which to build two adjacent rectangular corrals. The two corrals are to share a common fence on one side, as shown in Figure 8. Find the dimensions x and y so that the total enclosed area is as large as possible.

SOLUTION

You have 600 m of fencing to be set up as shown in Figure 8. The question is how to choose x and y so that the total area is maximum. Since the total length of fencing is 600 m, we can relate x and y by writing

$$\begin{aligned} x + x + x + y + y &= 600 \\ 3x + 2y &= 600 \\ 2y &= 600 - 3x \\ y &= 300 - \frac{3}{2}x \qquad (4) \end{aligned}$$

Letting A denote the total area, we have

$$\begin{aligned} A &= xy \\ A(x) &= x\left(300 - \frac{3}{2}x\right) \qquad \text{using equation (4) to substitute for } y \\ &= 300x - \frac{3}{2}x^2 \end{aligned}$$

This last equation expresses the area as a function of x. Note that the graph of this function is a parabola opening downward, so it does make sense to talk about a maximum. The x-coordinate of the vertex is

$$x = \frac{-b}{2a} = \frac{-300}{2(-3/2)} = \frac{300}{3} = 100$$

Now that we have the x-value that maximizes the area $A(x)$, the remaining dimension y can be calculated by using equation (4):

$$y = 300 - \frac{3}{2}x = 300 - \frac{3}{2}(100) = 150$$

Thus by choosing x to be 100 m and y to be 150 m, the total area in Figure 8 will be as large as possible. Incidentally, note that the exact location of the fence dividing the two corrals does not influence our work or the final answer.

In the next example a demand function is given, and we want to maximize the revenue. As background, you should review Example 8 in Section 4.4.

EXAMPLE 6 A maximization problem involving revenue

Suppose that the following demand function relates the selling price, p, of an item to the quantity sold, x:

$$p = -\frac{1}{3}x + 40 \qquad (p \text{ in dollars})$$

For which value of x will the revenue be a maximum? Compute the corresponding unit price and maximum revenue.

SOLUTION
First, recall the formula for revenue given on page 265.

$$R = \text{numbers of units} \times \text{price per unit}$$

Using this, we have

$$R = x \cdot p$$
$$R(x) = x\left(-\frac{1}{3}x + 40\right) = -\frac{1}{3}x^2 + 40x \tag{5}$$

We want to know which value of x yields the largest revenue $R(x)$. Since the graph of the revenue function in equation (5) is a parabola opening downward, the required x-value is the x-coordinate of the vertex of the parabola. We have

$$x = \frac{-b}{2a} = \frac{-40}{2(-1/3)} = \frac{20}{1/3} = 60$$

This tells us that to earn the maximum revenue, 60 items should be sold. We calculate the corresponding unit price by substituting $x = 60$ in the given equation $p = -\frac{1}{3}x + 40$. This yields

$$p = -\frac{1}{3}(60) + 40$$
$$= -20 + 40 = 20$$

Thus setting the price at \$20 per item will maximize the revenue. That maximum revenue can be calculated by using $x = 60$ in equation (5). More simply, since we now have both values $x = 60$ and $p = 20$, we can compute

$$R = x \cdot p = 60 \times 20 = 1200$$

The maximum revenue is \$1200.

GRAPHICAL PERSPECTIVE

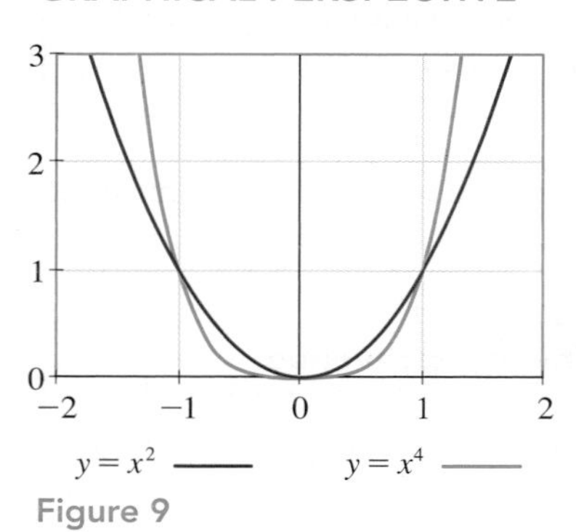

Figure 9

The next example, the last for this section, is similar to Example 3 in that we are looking for the minimum value of a function that, although not a quadratic function, is closely related to one. Also, in this example you'll see a translation of the function $y = x^4$. Since we have not yet studied this function, for background we show its graph along with that of the familiar $y = x^2$ in Figure 9. In the next section we do study functions of the form $y = x^n$, where n is a positive integer.

EXAMPLE 7 Minimizing the length of a connecting road

Figure 10 shows portions of two train routes. One route follows the curve $y = x^4 + 10$, and the other follows $y = 4x^2$. Assume that distance along both axes is measured in miles. The railroad wants to construct a north-south maintenance road $\overline{PQ}$ between the two routes, as indicated in the figure. Where should

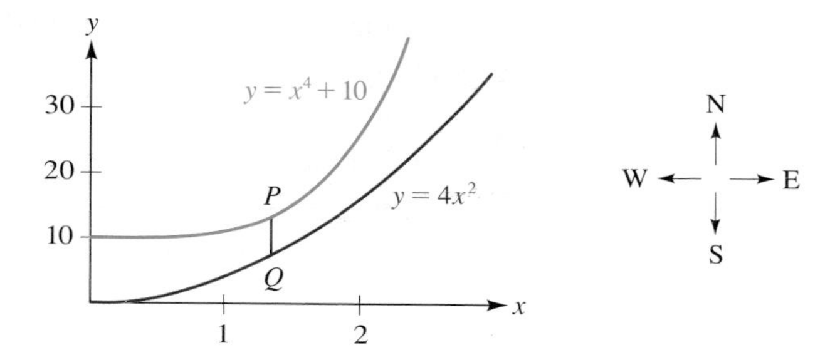

Figure 10
Portions of two train routes and a north-south maintenance road $\overline{PQ}$

the road be located so that it is as short as possible? What is the minimum length?

SOLUTION

We want to determine points P and Q on the curves $y = x^4 + 10$ and $y = 4x^2$, respectively, so that the distance from P to Q is as small as possible. Since P and Q lie on a north-south line, they have the *same* x-coordinate, and we can write the coordinates of P and Q as

$$P(x, x^4 + 10) \qquad \text{and} \qquad Q(x, 4x^2)$$

The vertical distance PQ between these two points is just the difference in the y-coordinates:

$$PQ = (x^4 + 10) - 4x^2 = x^4 - 4x^2 + 10$$

So the function that we want to minimize is

$$y = x^4 - 4x^2 + 10 \tag{6}$$

The substitution $x^2 = t$ will transform this into a quadratic function. (You saw this substitution technique before, in Example 7 of Section 4.2.) Note that if $x^2 = t$, then $x^4 = t^2$, and equation (6) becomes

$$y = t^2 - 4t + 10 \tag{7}$$

Since the graph of equation (7) will be a parabola opening upward, the input t that yields a minimum value for this function is

$$t = \frac{-b}{2a} = \frac{-(-4)}{2(1)} = 2$$

That gives us t, but what about x? Substituting the value $t = 2$ in the equation $x^2 = t$ gives us $x^2 = 2$ and consequently $x = \sqrt{2}$. (We're using the positive square root because x is in the first quadrant.) With this value for x, we can calculate the y-coordinates of P and Q as follows:

y-coordinate of P:	y-coordinate of Q:
$(\sqrt{2})^4 + 10 = 4 + 10 = 14$	$4(\sqrt{2})^2 = 8$

Putting things together now, the road should connect the points P and Q given by

$$P: (\sqrt{2}, 14) \approx (1.41, 14) \qquad \text{and} \qquad Q: (\sqrt{2}, 8) \approx (1.41, 8)$$

The length of the road is $14 - 8 = 6$ mi.

EXERCISE SET 4.5

A

1. Two numbers add to 5. What is the largest possible value of their product?
2. Find two numbers adding to 20 such that the sum of their squares is as small as possible.
3. The difference of two numbers is 1. What is the smallest possible value for the sum of their squares?
4. For each quadratic function, state whether it would make sense to look for a highest or a lowest point on the graph. Then determine the coordinates of that point.
   (a) $y = 2x^2 - 8x + 1$
   (b) $y = -3x^2 - 4x - 9$
   (c) $h = -16t^2 + 256t$
   (d) $f(x) = 1 - (x + 1)^2$
   (e) $g(t) = t^2 + 1$
   (f) $f(x) = 1000x^2 - x + 100$
5. Among all rectangles having a perimeter of 25 m, find the dimensions of the one with the largest area.
6. What is the largest possible area for a rectangle with a perimeter of 80 cm?
7. What is the largest possible area for a right triangle in which the sum of the lengths of the two shorter sides is 100 in.?
8. The perimeter of a rectangle is 12 m. Find the dimensions for which the diagonal is as short as possible.
9. Two numbers add to 6.
   (a) Let T denote the sum of the squares of the two numbers. What is the smallest possible value for T?
   (b) Let S denote the sum of the first number and the square of the second. What is the smallest possible value for S?
   (c) Let U denote the sum of the first number and twice the square of the second number. What is the smallest possible value for U?
   (d) Let V denote the sum of the first number and the square of twice the second number. What is the smallest possible value for V?
10. Suppose that the height of an object shot straight up is given by $h = 512t - 16t^2$. (Here h is in feet and t is in seconds.) Find the maximum height and the time at which the object hits the ground.
11. A baseball is thrown straight up, and its height as a function of time is given by the formula $h = -16t^2 + 32t$ (where h is in feet and t is in seconds).
   (a) Find the height of the ball when $t = 1$ sec and when $t = 3/2$ sec.
   (b) Find the maximum height of the ball and the time at which that height is attained.
   (c) At what times is the height 7 ft?
12. Find the point on the curve $y = \sqrt{x}$ that is nearest to the point $(3, 0)$.
13. Which point on the curve $y = \sqrt{x - 2} + 1$ is closest to the point $(4, 1)$? What is this minimum distance?
14. Find the coordinates of the point on the line $y = 3x + 1$ closest to $(4, 0)$.
15. **(a)** What number exceeds its square by the greatest amount?
   (b) What number exceeds twice its square by the greatest amount?
16. Suppose that you have 1800 m of fencing with which to build three adjacent rectangular corrals, as shown in the figure. Find the dimensions so that the total enclosed area is as large as possible.

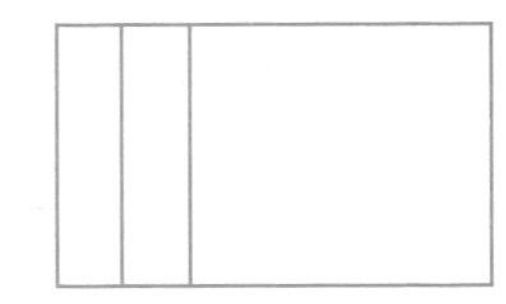

17. Five hundred feet of fencing is available for a rectangular pasture alongside a river, the river serving as one side of the rectangle (so only three sides require fencing). Find the dimensions yielding the greatest area.
18. Let $A = 3x^2 + 4x - 5$ and $B = x^2 - 4x - 1$. Find the minimum value of $A - B$.
19. Let $R = 0.4x^2 + 10x + 5$ and $C = 0.5x^2 + 2x + 101$. For which value of x is $R - C$ a maximum?
20. Suppose that the revenue generated by selling x units of a certain commodity is given by $R = -\frac{1}{5}x^2 + 200x$. Assume that R is in dollars. What is the maximum revenue possible in this situation?

21. Suppose that the function $p = -\frac{1}{4}x + 30$ relates the selling price p of an item to the number of units x that are sold. Assume that p is in dollars. For which value of x will the corresponding revenue be a maximum? What is this maximum revenue and what is the unit price?

22. The action of sunlight on automobile exhaust produces air pollutants known as *photochemical oxidants*. In a study of cross-country runners in Los Angeles, it was shown that running performances can be adversely affected when the oxidant level reaches 0.03 part per million. Suppose that on a given day, the oxidant level L is approximated by the formula

$$L = 0.059t^2 - 0.354t + 0.557 \qquad (0 \le t \le 7)$$

where t is measured in hours, with $t = 0$ corresponding to 12 noon, and L is in parts per million. At what time is the oxidant level L a minimum? At this time, is the oxidant level high enough to affect a runner's performance?

23. **(a)** Find the smallest possible value of the quantity $x^2 + y^2$ under the restriction that $2x + 3y = 6$.
(b) Find the radius of the circle whose center is at the origin and that is tangent to the line $2x + 3y = 6$. How does this answer relate to your answer in part (a)?

24. **(a)** Find the coordinates of the vertex of the parabola $y = 2x^2 - 4x + 7$. Then, on the same set of axes, graph this parabola along with $y = x^2$.
(b) Rework Example 7 using the two curves in part (a) rather than the two used in Example 7.

Ⓖ 25. **(a)** Using a graphing utility, graph the two functions $y = x^4 + 2$ and $y = x^2$ in the viewing rectangle $[0, 4, 1] \times [0, 16, 2]$. If your graphing utility allows, include the y-axis gridlines in the picture; they'll be helpful in part (b).
Ⓖ **(b)** Suppose that we use the two functions in part (a) as the given functions in Example 7. On the basis of the graph in part (a), make a rough estimate for the minimum length of the road.
(c) Use the technique of Example 7 to find the exact value for the minimum length.

B

26. Through a type of chemical reaction known as *autocatalysis*, the human body produces the enzyme trypsin from the enzyme trypsinogen. (Trypsin then breaks down proteins into amino acids, which the body needs for growth.) Let r denote the rate of the chemical reaction in which trypsin is formed from trypsinogen. It has been shown experimentally that $r = kx(a - x)$, where k is a positive constant, a is the initial amount of trypsinogen, and x is the amount of trypsin produced (so x increases as the reaction proceeds). Show that the reaction rate r is a maximum when $x = a/2$. In other words, the speed of the reaction is greatest when the amount of trypsin formed is half the original amount of trypsinogen.

27. **(a)** Let $x + y = 15$. Find the minimum value of the quantity $x^2 + y^2$.
(b) Let C be a constant and $x + y = C$. Show that the minimum value of $x^2 + y^2$ is $C^2/2$. Then use this result to check your answer in part (a).

28. Suppose that A, B, and C are positive constants and that $x + y = C$. Show that the minimum value of $Ax^2 + By^2$ occurs when $x = BC/(A + B)$ and $y = AC/(A + B)$.

29. The following figure shows a square inscribed within a unit square. For which value of x is the area of the inner square a minimum? What is the minimum area? *Hint*: Denote the lengths of the two segments that make up the base of the unit square by t and $1 - t$. Now use the Pythagorean theorem and congruent triangles to express x in terms of t.

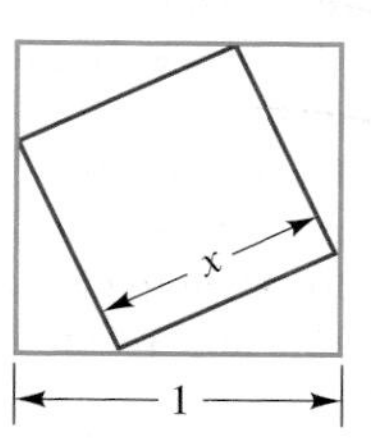

30. **(a)** Show that the coordinates of the point on the line $y = mx + b$ that is closest to the origin are given by

$$\left(\frac{-mb}{1 + m^2}, \frac{b}{1 + m^2}\right)$$

(b) Show that the perpendicular distance from the origin to the line $y = mx + b$ is $|b|/\sqrt{1 + m^2}$.
Suggestion: Use the result in part (a).
(c) Use part (b) to show that the perpendicular distance from the origin to the line $Ax + By + C = 0$ is $|C|/\sqrt{A^2 + B^2}$.

31. The point P lies in the first quadrant on the graph of the line $y = 7 - 3x$. From the point P, perpendiculars are drawn to both the x-axis and the y-axis. What is the largest possible area for the rectangle thus formed?

32. Show that the largest possible area for the shaded rectangle shown in the figure is $-b^2/4m$. Then use this to check your answer to Exercise 31.

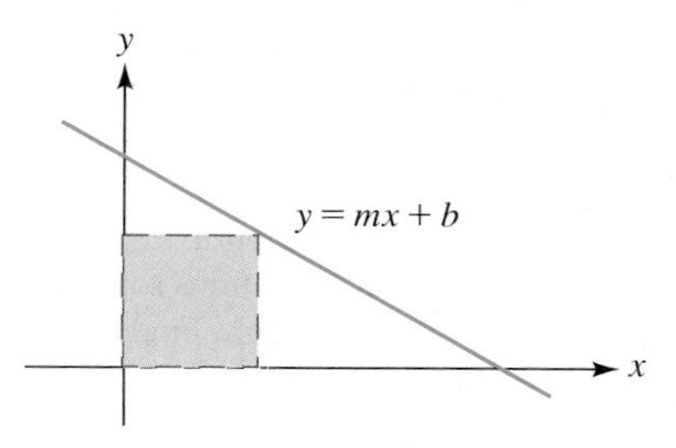

33. Show that the maximum possible area for a rectangle inscribed in a circle of radius R is $2R^2$. *Hint*: Maximize the square of the area.

34. An athletic field with a perimeter of $\frac{1}{4}$ mile consists of a rectangle with a semicircle at each end, as shown in the figure. Find the dimensions x and r that yield the greatest possible area for the rectangular region.

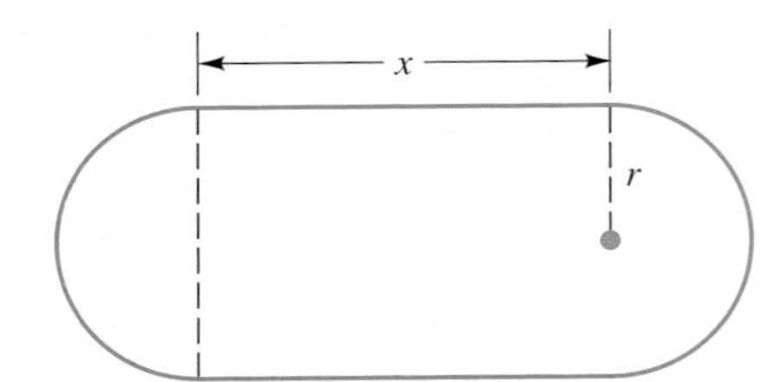

35. A rancher who wishes to fence off a rectangular area finds that the fencing in the east-west direction will require extra reinforcement due to strong prevailing winds. Because of this, the cost of fencing in the east-west direction will be \$12 per (linear) yard, as opposed to a cost of \$8 per yard for fencing in the north-south direction. Find the dimensions of the largest possible rectangular area that can be fenced for \$4800.

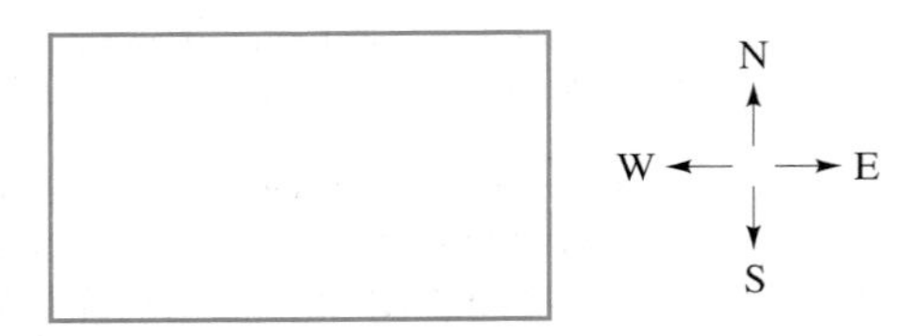

36. Let $f(x) = (x - a)^2 + (x - b)^2 + (x - c)^2$, where a, b, and c are constants. Show that $f(x)$ will be a minimum when x is the average of a, b, and c.

37. Let $y = a_1(x - x_1)^2 + a_2(x - x_2)^2$, where a_1, a_2, x_1, and x_2 are all constants. In addition, suppose that a_1 and a_2 are both positive. Show that the minimum of this function occurs when

$$x = \frac{a_1x_1 + a_2x_2}{a_1 + a_2}$$

38. Among all rectangles with a given perimeter P, find the dimensions of the one with the shortest diagonal.

39. By analyzing sales figures, the economist for a stereo manufacturer knows that 150 units of a compact disc player can be sold each month when the price is set at $p = \$200$ per unit. The figures also show that for each \$10 hike in price, five fewer units are sold each month.

(a) Let x denote the number of units sold per month and let p denote the price per unit. Find a linear function relating p and x. *Hint:* $\Delta p/\Delta x = 10/(-5) = -2$

(b) The revenue R is given by $R = xp$. What is the maximum revenue? At what level should the price be set to achieve this maximum revenue?

40. Let $f(x) = x^2 + px + q$, and suppose that the minimum value of this function is 0. Show that $q = p^2/4$.

41. Among all possible inputs for the function $f(t) = -t^4 + 6t^2 - 6$, which ones yield the largest output?

42. Let $f(x) = x - 3$ and $g(x) = x^2 - 4x + 1$.

(a) Find the minimum value of $g \circ f$.

(b) Find the minimum value of $f \circ g$.

(c) Are the results in parts (a) and (b) the same?

43. This exercise completes a detail mentioned in Example 4. In that example we used the result that the input minimizing the function $D(x) = \sqrt{x^2 - x + 1}$ is the same input that minimizes the function $y = x^2 - x + 1$. For this result to be applicable, we need to know that the quantity $x^2 - x + 1$ is nonnegative for all x. Parts (a) and (b) each suggest a way to verify that this is indeed the case. [Actually, parts (a) and (b) both show that the quantity $x^2 - x + 1$ is positive for all x.]

(a) Use the fact that $x^2 - x + 1$ can be written as $(x - \frac{1}{2})^2 + \frac{3}{4}$ to explain (in complete sentences) why the quantity $x^2 - x + 1$ is positive for all inputs x.

(b) Compute the discriminant of the quadratic $x^2 - x + 1$ and explain why the graph of $y = x^2 - x + 1$ has no x-intercepts. Then use this fact to explain why the quantity $x^2 - x + 1$ is always positive. *Hint:* Is the graph of the parabola $y = x^2 - x + 1$ U-shaped up or U-shaped down?

44. Let $f(x) = (x - 1)^2 - 4$.

(a) Sketch the graph of the function f and note that the minimum value occurs when x is 1.

(b) Although the minimum value of the function f occurs when $x = 1$, the minimum value of the function $y = \sqrt{f(x)}$ does not occur when $x = 1$. Explain why this does not contradict the statement in the caption to Figure 6 on page 276.

(c) Find the value(s) of x that minimize the function $y = \sqrt{f(x)}$. *Hint:* What is the domain of $y = \sqrt{f(x)}$?

45. A piece of wire 16 in. long is to be cut into two pieces. Let x denote the length of the first piece and $16 - x$ the length of the second. The first piece is to be bent into a circle and the second piece into a square.

(a) Express the total combined area A of the circle and the square as a function of x.

(b) For which value of x is the area A a minimum?

(c) Using the x-value that you found in part (b), find the ratio of the lengths of the shorter to the longer piece of wire. *Answer:* $\pi/4$

46. A 30-in. piece of string is to be cut into two pieces. The first piece will be formed into the shape of an equilateral triangle and the second piece into a square. Find the length of the first piece if the combined area of the triangle and the square is to be as small as possible.

Exercises 47–52 are maximum-minimum problems in which the function that you set up will not turn out to be a quadratic, nor will it be closely related to one. Thus the vertex formula will not be applicable.

(a) *Set up the appropriate function to be maximized or minimized just as you have practiced in this section and Section 4.4.*

Ⓖ **(b)** *Use a graphing utility to draw the graph of the function and determine, as least approximately, the number or numbers that the problem asks for. (General techniques for determining the exact values are developed in calculus.)*

Exercises 47–52 are quoted from the following calculus textbooks with the permission of the publishers.

EXERCISE	TEXT
47.	*Calculus: A New Horizon,* 6th ed., Howard Anton (New York: John Wiley & Sons, 1998)
48.	*Calculus,*, 6th ed., Roland E. Larson et al. (Boston: Houghton Mifflin, 1997)
49.	*Calculus and Analytic Geometry,* 5th ed., Sherman K. Stein and Anthony Barcellos (New York: McGraw-Hill, 1992)
50.	*Calculus with Analytic Geometry,* 5th ed., C. H. Edwards, Jr. and David E. Penney (Upper Saddle River, N.J.: Prentice-Hall, 1997)
51.	*Calculus: One and Several Variables,* 8th ed., Saturnino L. Salas and Garret J. Etgen (New York: John Wiley & Sons, 1998)
52.	*Calculus,* 5th ed., James Stewart (Pacific Grove, Calif.: Brooks/Cole, 2003)

47. A cylindrical can, open at the top, is to hold 500 cm^3 of liquid. Find the height and radius that minimize the amount of material needed to manufacture the can.

48. A power station is on one side of a river that is 1/2 mi. wide, and a factory is 6 mi. downstream on the other side. It costs $6 per foot to run power lines overland and $8 per foot to run them underwater. Find the most economical path for the transmission line from the power station to the factory.

49. What point on the parabola $y = x^2$ is closest to the point $(3, 0)$?

50. A rectangle of fixed perimeter 36 is rotated about one of its sides, thus sweeping out a figure in the shape of a right circular cylinder. [Refer to the accompanying figure.] What is the maximum possible volume of that cylinder?

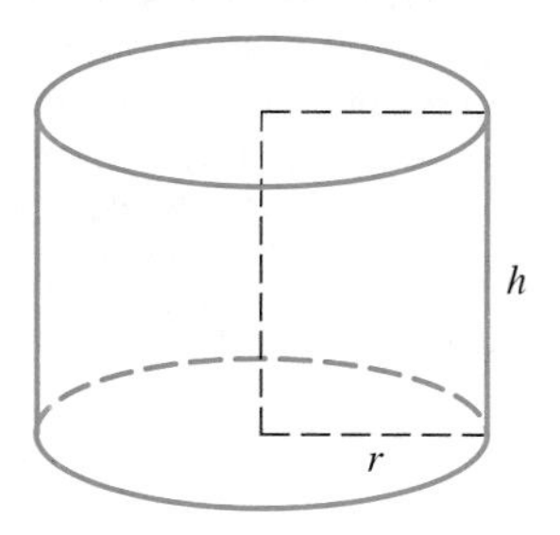

51. Let ABC be a triangle with vertices $A(-3, 0)$, $B(0, 6)$, $C(3, 0)$. Let P be a point on the line segment that joins B to the origin. Find the position of P that minimizes the sum of the distances between P and the vertices.

52. A fence 8 ft tall runs parallel to a tall building at a distance of 4 ft from the building. What is the length of the shortest ladder that will reach from the ground over the fence to the wall of the building?

C

Ⓖ **53.** **(a)** Use a graphing utility to graph the parabolas $y = x^2 - 2x$, $y = x^2 + 2x$, $y = x^2 - 3x$, and $y = x^2 + 3x$. Check visually that, in each case, the vertex of the parabola appears to lie on the curve $y = -x^2$.

(b) Prove that for all real numbers k, the vertex of the parabola $y = x^2 + kx$ lies on the curve $y = -x^2$.

Ⓖ **54.** **(a)** Graph the four parabolas $y = x^2 + 2kx + 1$ corresponding to $k = 2, 3, 0.75, 1.5$. Among the four parabolas, which one appears to have the vertex closest to the origin?

(b) Let $f(x) = x^2 + 2kx + 1$. Find a positive value for k so that the distance from the origin to the vertex of the parabola is as small as possible. Check that your answer is consistent with your observations in part (a).

55. The figure shows a rectangle inscribed in a given triangle of base b and height h. Find the ratio of the area of the triangle to the area of the rectangle when the area of the rectangle is maximum.

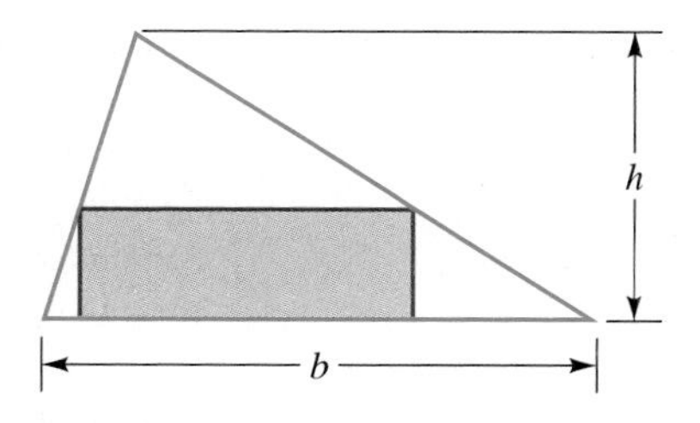

56. A Norman window is in the shape of a rectangle surmounted by a semicircle, as shown in the figure. Assume that the perimeter of the window is P, a constant. Show that the area of the window is a maximum when both x and r are equal to $P/(\pi + 4)$. Show that this maximum area is $\frac{1}{2}P^2/(\pi + 4)$.

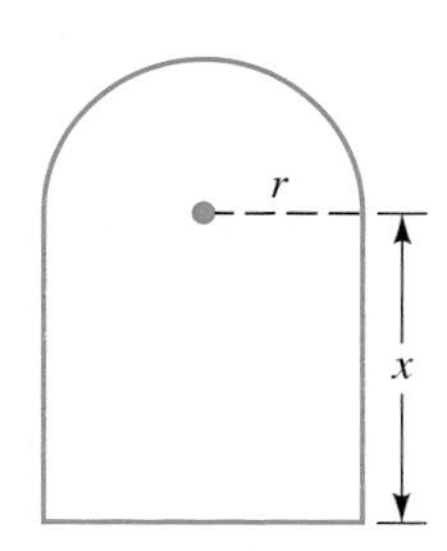

57. A triangle is inscribed in a semicircle of diameter $2R$, as shown in the figure. Show that the smallest possible value for the area of the shaded region is $(\pi - 2)R^2/2$.

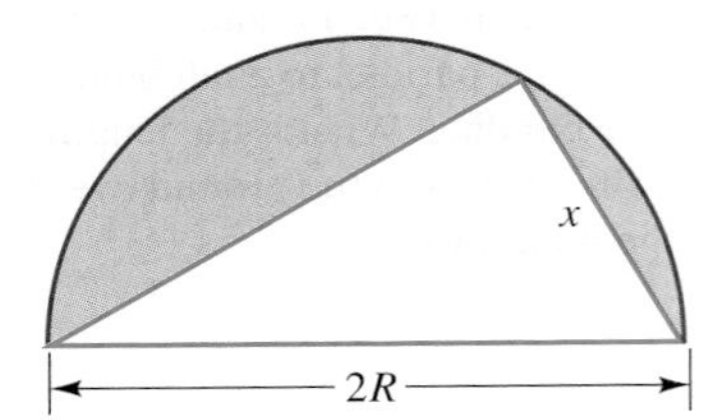

Hint: The area of the shaded region is a minimum when the area of the triangle is a maximum. Find the value of x that maximizes the *square* of the area of the triangle. This will be the same x that maximizes the area of the triangle.

58. (a) Complete the following table. Which x-y pair in the table yields the smallest sum $x + y$?

x	0.5	1	1.5	2	2.5	3	3.5
y							
xy	12	12	12	12	12	12	12
$x + y$							

(b) Find two positive numbers with a product of 12 and as small a sum as possible. *Hint*: The quantity that you need to minimize is $x + (12/x)$, where $x > 0$. But

$$x + \frac{12}{x} = \left(\sqrt{x} - \sqrt{\frac{12}{x}}\right)^2 + 2\sqrt{12}$$

This last expression is minimized when the quantity within parentheses is zero. Why?

(c) Use a calculator to verify that the two numbers obtained in part (b) produce a sum that is smaller than any of the sums obtained in part (a).

59. What is the smallest possible value for the sum of a positive number and its reciprocal? *Hint*: After setting up the appropriate function, adapt the hint given in Exercise 58(b).

60. Suppose that a and b are positive numbers whose sum is 1.

(a) Find the maximum possible value of the product ab.

(b) Prove that $\left(1 + \frac{1}{a}\right)\left(1 + \frac{1}{b}\right) \geq 9$.

PROJECT

THE LEAST-SQUARES LINE

The great advances in mathematical astronomy made during the early years of the nineteenth century were due in no small part to the development of the method of least squares. The same method is the foundation for the calculus of errors of observation now occupying a place of great importance in the scientific study of social, economic, biological, and psychological problems.—*A Source Book in Mathematics,* by David Eugene Smith (New York: Dover Publications, 1959)

Of all the principles which can be proposed . . . [for finding the line or curve that best fits a data set], *I think there is none more general, more exact, and more easy of application than . . . rendering the sum of the squares of the errors a minimum.*—Adrien Marie Legendre (1752–1833) [Legendre and Carl Friedrich Gauss (1777–1855) developed the least-squares method independently.]

When we worked with the least-squares (or regression) line in Section 4.1, we said that this is the line that "best fits" the given set of data points. In this project we explain the meaning of "best fits," and we use quadratic functions to gain some insight into how the least-squares line is calculated. *Suggestion*: Two or three people could volunteer to study the text here and then present the material to another group or to the class at large. The presentation will be deemed successful if the audience can then work on their own the two exercises at the end of this project.

Suppose that we have a data point (x_i, y_i) and a line $y = f(x) = mx + b$ (not necessarily the least-squares line). As indicated in Figure A, we define the *deviation* e_i of this line from the data point as

$$e_i = y_i - f(x_i)$$

Figure A

In Figure A, where the data point is above the line, the deviation e_i represents the vertical distance between the data point and the line. If the data point were below the line, then the deviation $y_i - f(x_1)$ would be the negative of that distance. Think of e_i as the *error* made if $f(x_i)$ were used as a prediction for the value of y_i.

Figure B and the accompanying calculations show two examples using this definition. For the data point $(x_1, y_1) = (4, 3)$, the result $e_1 = 2$ does represent the vertical distance between the data point and the line. For the data point $(8, 2)$, the deviation is $e_2 = -1$, and the absolute value of this gives the vertical distance between the data point and the line.

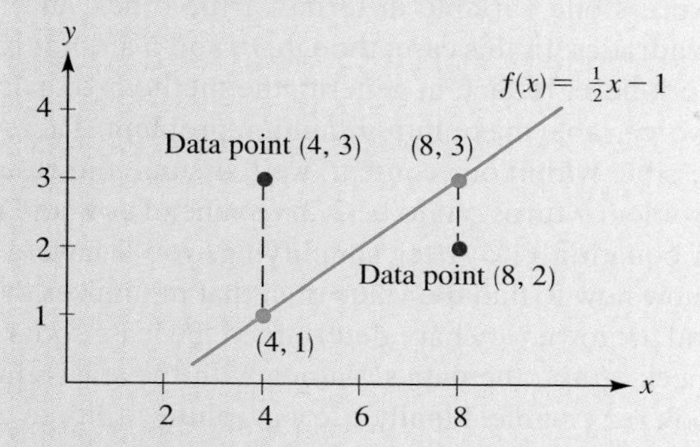

For the data point $(x_1, y_1) = (4, 3)$:

$$e_1 = y_1 - f(x_1)$$
$$= 3 - \left(\tfrac{1}{2}(4) - 1\right) = 2$$

For the data point $(x_2, y_2) = (8, 2)$:

$$e_2 = y_2 - f(x_2)$$
$$= 2 - \left(\tfrac{1}{2}(8) - 1\right) = -1$$

Figure B

Now suppose we have a given data set and a line. One way to measure how closely this line fits the data would be to simply add up all the deviations. The problem with this, however, is that positive and negative deviations would tend to cancel one another out. For this reason, we use the sum of the squares of the deviations as the measure of how closely a line fits a data set. (The sum of the absolute values of the deviations would be another alternative, but in calculus, squares are easier to work with than absolute values.)

As an example, suppose we have the following data set:

x	1	2	3
y	2	4	5

or equivalently

$$(x_1, y_1) = (1, 2)$$
$$(x_2, y_2) = (2, 4)$$
$$(x_3, y_3) = (3, 5)$$

DEFINITION **Line of Best Fit**

Suppose we have a set of n data points $(x_1, y_1), (x_2, y_2), (x_3, y_3), \ldots, (x_n, y_n)$. For any line, let E denote the sum of the squares of the deviations:

$$E = e_1^2 + e_2^2 + e_3^2 + \cdots + e_n^2$$

Then the **line of best fit** is that line for which E is as small as possible. (It can be shown that there is only one such line.) The line of best fit is also known as the **least-squares line** or the **regression line.**

Then, for any line $y = mx + b$ we have

$$e_1 = 2 - [m(1) + b] = 2 - m - b \qquad e_2 = 4 - [m(2) + b] = 4 - 2m - b \qquad e_3 = 5 - [m(3) + b] = 5 - 3m - b$$

Now, by definition, for the least-squares line we want to find the values for m and b so that the quantity $E = e_1^2 + e_2^2 + e_3^2$ is as small as possible. Using the expressions we've just calculated for e_1, e_2, and e_3, we have

$$\begin{aligned} E &= e_1^2 + e_2^2 + e_3^2 \\ &= (2 - m - b)^2 + (4 - 2m - b)^2 + (5 - 3m - b)^2 \end{aligned} \qquad (1)$$

Equation (1) expresses E as a function of *two* variables, m and b. In all the work we've done previously on maximum-minimum problems, we were always able to express one variable in terms of the other, and then apply our knowledge of quadratics. In this case, though, m and b are not related, so that strategy doesn't work here. In fact, in general, the methods of calculus are required to handle a two-variable maximum-minimum problem. Therefore to make this problem accessible within our context, we'll assume the result (from calculus) that in this problem b turns out to be $2/3$. Go ahead now and make the substitution $b = 2/3$ in equation (1). After simplifying, you'll have a quadratic function, and you know how to find the value of m that minimizes the function. Then, with both m and b known, you have determined the least-squares line $y = mx + b$. As a visual check, graph the data set along with the least-squares line. See whether things look reasonable. Finally, use a graphing utility to compute the least-squares line and check your answer.

EXERCISES

In each case you are given a data set and the value for b in the least-squares line $y = mx + b$. Find the equation of the least-squares line using the method indicated above. Check your answer using an appropriate graphing utility. Then graph the data set along with the least-squares line. In Exercise 2, in reporting the final answer, round both b and m to two decimal places.

1. (*Easy numbers*)

x	1	2	3
y	2	5	11

$b = -3$

2. (*"Actual" numbers*) The data in the following table are from a study relating air pollution and incidence of respiratory disease. The data were collected in five cities over the years 1955, 1957, and 1958. The subjects were women working in RCA factories. In the table, x denotes the average concentration of particulate sulfates ("sulfur dust") in the air of the given city, measured in micrograms per cubic meter ($\mu g/m^3$). The variable y denotes the number of absences due to respiratory disease per 1000 employees per year. (Only absences lasting more than seven days were counted.)

$b = -20.8117$

City	x Concentration of suspended particulate sulfates ($\mu g/m^3$)	y Number of respiratory disease related absences per 1000 employees per year
Cincinnati, Oh.	7.4	18.5
Indianapolis, Ind.	13.2	44.2
Woodbridge, N.J.	13.6	50.3
Camden, N.J.	17.1	58.3
Harrison, N.J.	19.8	85.7

Source: F. Curtis Dohan, "Air Pollutants and Incidence of Respiratory Disease," *Archives of Environmental Health,* 3(1961), pp. 387–395

4.6 POLYNOMIAL FUNCTIONS

After a few hitches or twitches, the polynomial functions settle down to behavior in which things simply get bigger or smaller. An example is $f(x) = x^2$. Just once, as the values of x approach 0, the graph of this function dips downward to alter its shape; thereafter it ascends solemnly like a helium-filled balloon.—David Berlinski, *A Tour of the Calculus* (New York: Vintage Books, 1995)

We can rephrase the definitions of linear and quadratic functions using the terminology for polynomials that was reviewed in Section 1.3. A function f is *linear* if $f(x)$ is a polynomial of degree 1:

$$f(x) = a_1x + a_0 \qquad \text{where } a_0 \text{ and } a_1 \text{ are constants} \tag{1}$$

A function f is *quadratic* if $f(x)$ is a polynomial of degree 2:

$$f(x) = a_2x^2 + a_1x + a_0 \qquad \text{where } a_0, a_1 \text{ and } a_2 \text{ are constants, with } a_2 \text{ not zero} \tag{2}$$

In the box that follows, we give the general definition for a polynomial function of degree n.

DEFINITION **Polynomial Function**

A **polynomial function** is a function defined by an equation of the form

$$f(x) = a_n x^n + a_{n-1}x^{n-1} + \cdots + a_1 x + a_0$$

where n is a nonnegative integer and the a_i's are constants. If $a_n \neq 0$, the **degree** of the polynomial function is n, the largest exponent on the input variable.

EXAMPLES

$f(x) = 2x^3 - 3x^2 + x - 4$ (degree 3)

$g(u) = 3u^7 - \sqrt{3}u^2 - \pi$ (degree 7)

$L(x) = 5x + 1$ (degree 1)

A nonzero constant function, for example, $y = 7$ for all x, is a polynomial function of *degree zero.* (Think of the defining equation $y = 7$ as $y = 7x^0$.) The one polynomial function for which degree is not defined is the *zero function:* $f(x) = 0$ for all x. (What does the graph of that function look like?)

A function of the form $y = x^n$, where n is any real-number constant, is called a **power function.** If we only consider exponents n that are nonnegative integers, then these power functions are the simplest polynomial functions. In a sense, they are the building blocks for all other polynomial functions. Figure 1 shows examples of power functions with which we are already familiar. Our focus here is on the power functions in Figures 1(a) through 1(d), where n is a nonnegative integer.

When n is an integer greater than 3, the graph of $y = x^n$ resembles the graph of $y = x^2$ or $y = x^3$, depending on whether n is even or odd. Consider, for instance, the graph of $y = x^4$, shown in Figure 2 along with the graph of $y = x^2$. Just as with

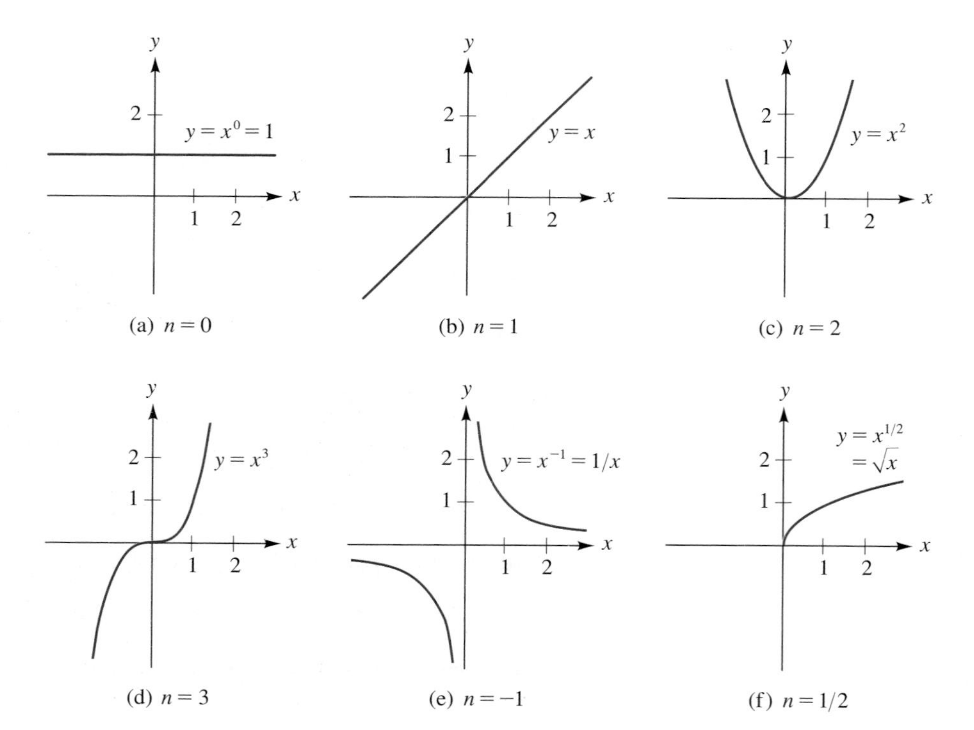

Figure 1
Some familiar power functions $y = x^n$, where n is a real number

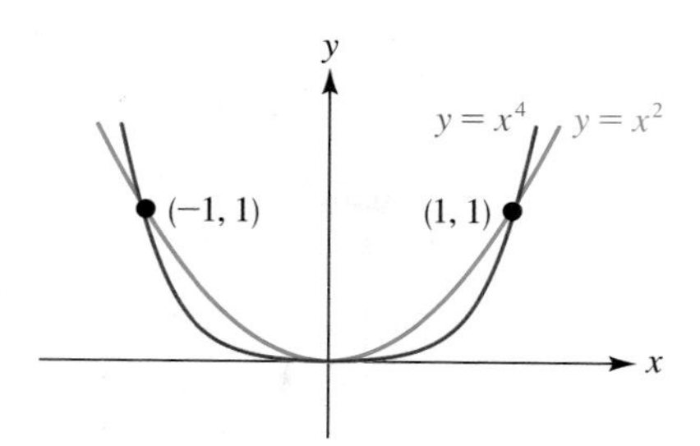

Figure 2

$y = x^2$, the graph of $y = x^4$ is a symmetric, U-shaped curve passing through the three points $(-1, 1)$, $(1, 1)$, and $(0, 0)$. However, in the interval $-1 < x < 1$, the graph of $y = x^4$ is flatter than that of $y = x^2$. Similarly, the graph of $y = x^6$ in this interval would be flatter still. (*Note*: Exercise 65 at the end of this section asks you to show why this is true.) The data in Table 1 illustrate this behavior. Figure 3 displays the graphs of $y = x^2$, $y = x^4$, and $y = x^6$ for the interval $0 \le x \le 1$.

TABLE 1

x	0.2	0.4	0.6	0.8	1.0
x^2	0.04	0.16	0.36	0.64	1.0
x^4	0.0016	0.0256	0.1296	0.4096	1.0
x^6	0.000064	0.004096	0.046656	0.262144	1.0

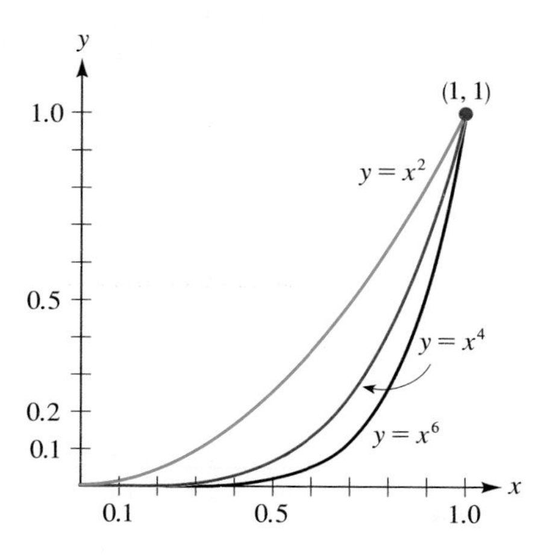

Figure 3

Incidentally, Figure 3 indicates one of the practical difficulties you may encounter in trying to draw an accurate graph of $y = x^n$. Suppose, for instance, that you want to graph $y = x^6$ and the lines you draw are 0.01 cm thick. Also suppose that you use the same scale on both axes, taking the common unit to be 1 cm. Then, in the first quadrant, your graph of $y = x^6$ will be indistinguishable from the x-axis when $x^6 < 0.01$, or $x < \sqrt[6]{0.01} \approx 0.46$ cm (using a calculator). This explains why sections of the graphs in Figure 3 appear horizontal.

For $|x| > 1$, the graph of $y = x^4$ rises more rapidly than that of $y = x^2$. Similarly, the graph of $y = x^6$ rises still more rapidly. This is shown in Figures 4(a) and (b). (Note the different scales used on the y-axes in the two figures.)

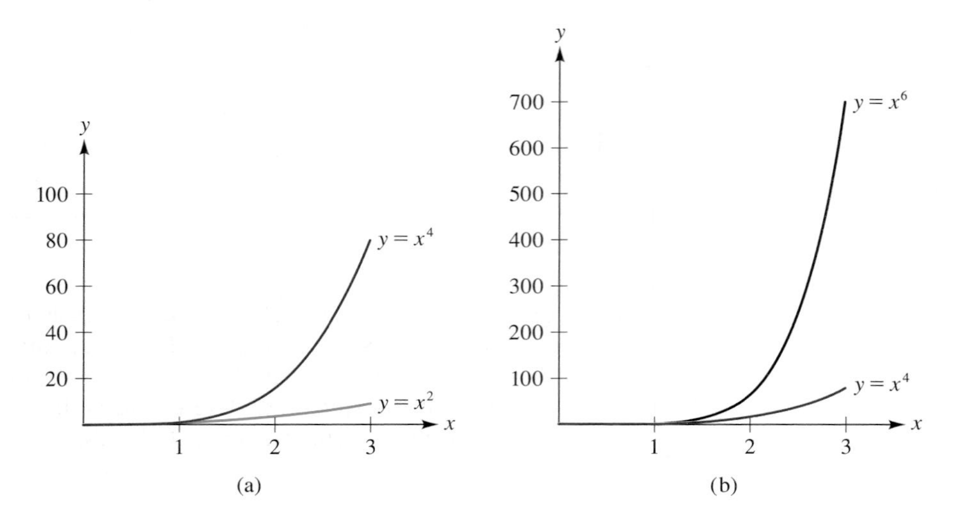

Figure 4

EXAMPLE 1 Comparing average rates of change of two power functions

Let $f(x) = x^4$ and $g(x) = x^8$.

(a) Use Figure 5 to estimate and compare the averages rates of change of f and g on the interval $[0, 0.8]$ and on the interval $[0, 1.25]$.

(b) Use the given equations for f and g to compute the rates of change mentioned in part (a). Round the answers to three decimal places.

GRAPHICAL PERSPECTIVE

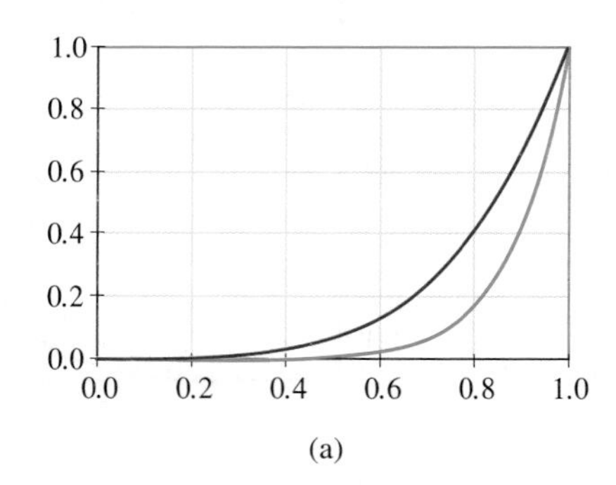

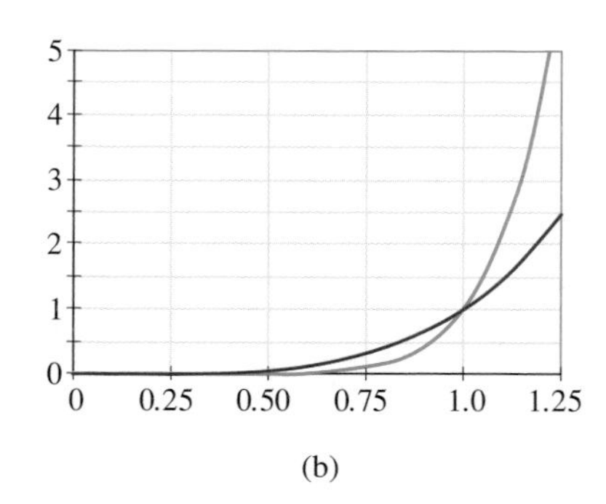

Figure 5
Two views of $f(x) = x^4$ (red) and $g(x) = x^8$ (blue)

SOLUTION

(a) Refer to Figure 5(a). On the interval $[0, 0.8]$ we have $g(0) = 0$, and from the graph, $g(0.8)$ is roughly 0.2. So on this interval we have

$$\frac{\Delta g}{\Delta x} = \frac{g(0.8) - g(0)}{0.8 - 0} \approx \frac{0.2 - 0}{0.8} = 0.25$$

Also on the interval $[0, 0.8]$, $f(0) = 0$ and $f(0.8) \approx 0.4$. Thus we have

$$\frac{\Delta f}{\Delta x} = \frac{f(0.8) - f(0)}{0.8 - 0} \approx \frac{0.4 - 0}{0.8} = 0.5$$

In summary, on the interval [0, 0.8], the average rate of change of g is approximately 0.25, which is half of the value estimated for the average rate of change of f.

Exercise 45 asks you to follow this same procedure for the interval [0, 1.25] shown in Figure 5(b). The exercise asks you to show that the estimates for this interval are $\Delta g/\Delta x \approx 4$ and $\Delta f/\Delta x \approx 2$. So, on this interval, it's g rather than f that has the larger average rate of change. The estimate for the average rate of change of g is twice the value estimated for the average rate of change of f.

(b) For the interval [0, 0.8],

$$\frac{\Delta g}{\Delta x} = \frac{g(0.8) - g(0)}{0.8 - 0} = \frac{(0.8)^8 - 0^8}{0.8}$$
$$= (0.8)^7 \approx 0.210 \quad \text{rounding to three decimal places}$$

$$\frac{\Delta f}{\Delta x} = \frac{f(0.8) - f(0)}{0.8 - 0} = \frac{(0.8)^4 - 0^4}{0.8}$$
$$= (0.8)^3 = 0.512 \quad \text{(exactly)}$$

Exercise 46(a) asks you to carry out similar calculations for the interval [0, 1.25] to obtain the results $\Delta g/\Delta x \approx 4.768$ and $\Delta f/\Delta x \approx 1.953$.

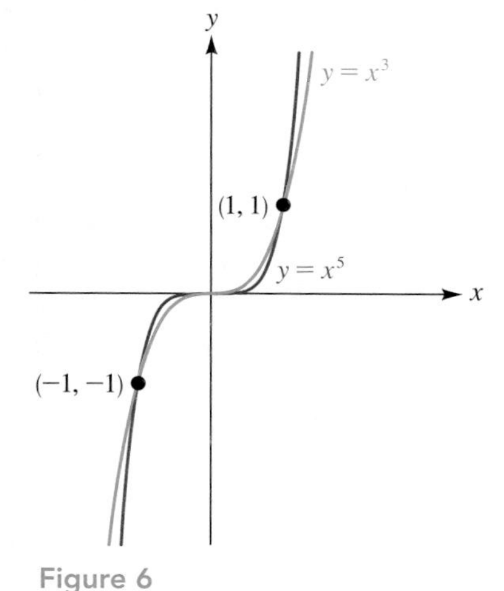

Figure 6

As was mentioned earlier, when n is odd, the graph of $y = x^n$ resembles that of $y = x^3$. In Figure 6 we compare the graphs of $y = x^3$ and $y = x^5$. Notice that both curves pass through (0, 0), (1, 1), and $(-1, -1)$. For reasons similar to those explained for even n, the graph of $y = x^5$ is flatter than that of $y = x^3$ in the interval $-1 < x < 1$, and the graph of $y = x^7$ is flatter still. For $|x| > 1$, the graph of $y = x^5$ is steeper than that of $y = x^3$, and that of $y = x^7$ is steeper still. (See Exercise 65 at the end of this section.)

EXAMPLE 2 Translating the graph of a power function

Sketch the graph of $y = (x + 2)^5$ and specify the y-intercept.

SOLUTION

The graph of $y = (x + 2)^5$ is obtained by moving the graph of $y = x^5$ two units to the left. As a guide to drawing the curve, we recall that $y = x^5$ passes through the points (0, 0), (1, 1), and $(-1, -1)$. Thus $y = (x + 2)^5$ must pass through $(-2, 0)$, $(-1, 1)$, and $(-3, -1)$, as shown in Figure 7.

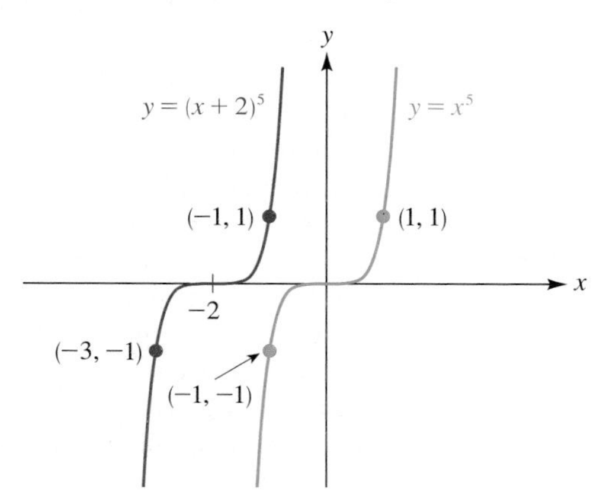

Figure 7

Although the curve rises and falls very sharply, it is important to realize that it is never really vertical. For instance, the curve eventually crosses the y-axis. To find the y-intercept, we set $x = 0$ to obtain $y = 2^5 = 32$. Thus the y-intercept is 32.

In Section 4.2 we observed the effect of the constant a on the graph of $y = ax^2$. Those same comments apply to the graph of $y = ax^n$. For instance, the graph of $y = \frac{1}{2}x^4$ is wider than that of $y = x^4$, whereas the graph of $y = -\frac{1}{2}x^4$ is obtained by reflecting the graph of $y = \frac{1}{2}x^4$ in the x-axis.

EXAMPLE 3 Translating a reflection of a power function

Graph the function $y = -2(x - 3)^4$.

SOLUTION

We begin with the graph of $y = -2x^4$, in Figure 8(a). The points $(1, -2)$ and $(-1, -2)$ are obtained by substituting $x = 1$ and $x = -1$, respectively, in the equation $y = -2x^4$. Now if we replace x with $x - 3$ in the equation $y = -2x^4$, we have $y = -2(x - 3)^4$, which we can graph by translating the graph in Figure 8(a) to the right 3 units; see Figure 8(b).

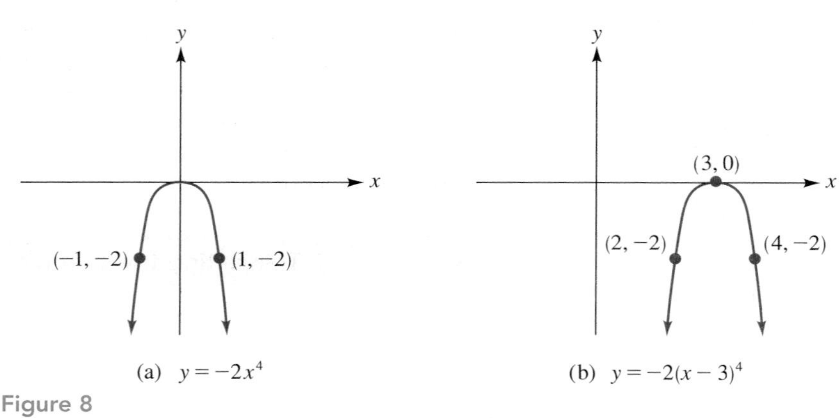

(a) $y = -2x^4$ (b) $y = -2(x - 3)^4$

Figure 8

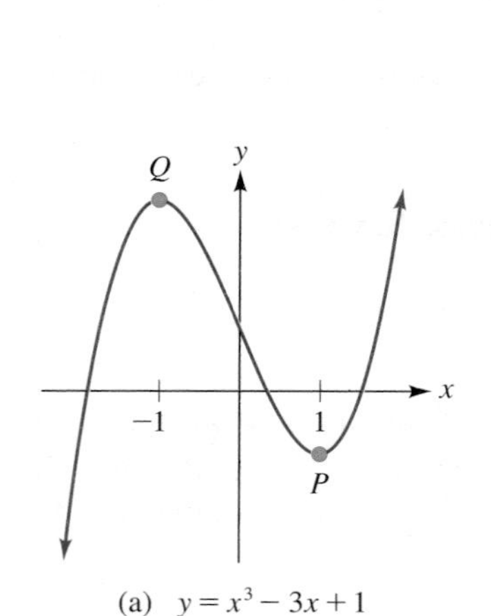

(a) $y = x^3 - 3x + 1$

(b) $y = -(x + 1)^3 + 1$

Figure 9

In principle, we can obtain the graph of any polynomial function by setting up a table and plotting a sufficient number of points. Indeed, this is just the way a graphing calculator or computer operates. However, to *understand* why the graphs look as they do, we want to discuss some additional methods for graphing polynomial functions.

There are three facts that we shall need. By way of example, look at the graphs of the polynomial functions in Figure 9. First, notice that both graphs are unbroken, smooth curves with no "corners." As in shown in calculus, this is true for the graph of every polynomial function. In contrast, the graphs in Figures 10 and 11 cannot represent polynomial functions. The graph in Figure 10 has a break in it, and the graph in Figure 11 has what's called a **cusp.**

Now look back at the graph in Figure 9(a). Recall (from Section 3.2) that points such as P and Q are called **turning points.** These are points where the graph changes from rising to falling or vice versa. It is a fact (proved in calculus) that the graph of a polynomial function of degree n has *at most* $n - 1$ turning points. For instance, in

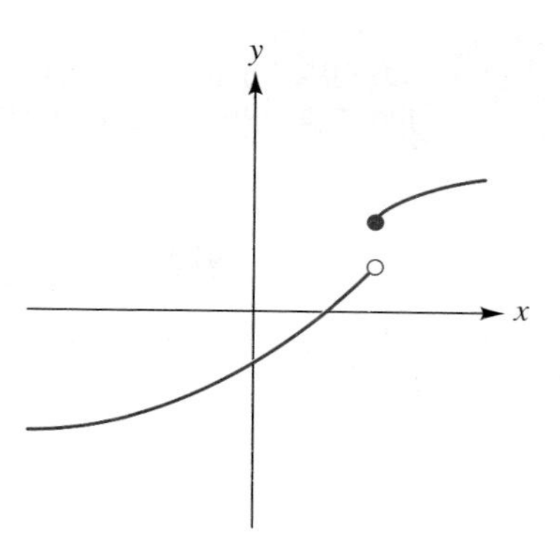

Figure 10
Since the graph has a break, it cannot represent a polynomial function.

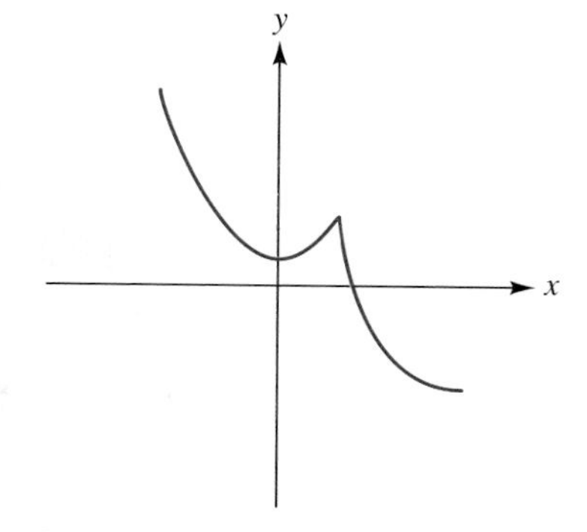

Figure 11
Since the graph has a cusp, it cannot represent a polynomial function.

Figure 9(a) there are two turning points, and the degree of the polynomial is 3. However, as Figure 9(b) indicates, we needn't have any turning points at all.

A third property of polynomial functions concerns their behavior when $|x|$ is very large. We'll illustrate this property using the function $y = x^3 - 3x + 1$, graphed in Figure 9(a). Now, in Figure 9(a), the x-values are relatively small; for instance, the x-coordinates of P and Q are 1 and -1, respectively. In Figure 12, however, we show the graph of this same function using units of 100 on the x-axis. On this scale, the graph appears indistinguishable from that of $y = x^3$. In particular, note that as $|x|$ gets very large, $|y|$ grows very large.

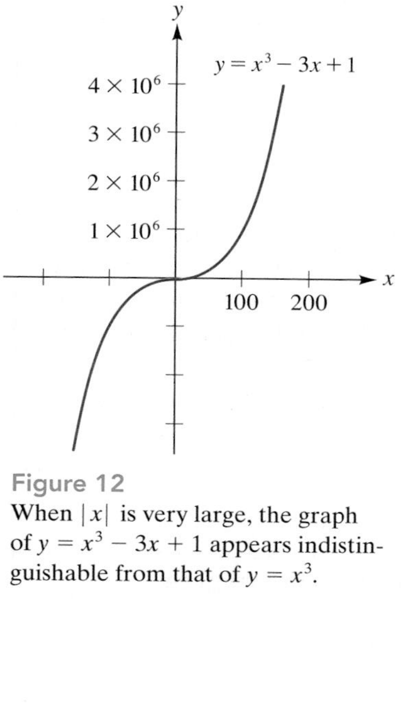

Figure 12
When $|x|$ is very large, the graph of $y = x^3 - 3x + 1$ appears indistinguishable from that of $y = x^3$.

It's easy to see why the function $y = x^3 - 3x + 1$ resembles $y = x^3$ when $|x|$ is very large. First, let's rewrite the equation $y = x^3 - 3x + 1$ as

$$y = x^3\left(1 - \frac{3}{x^2} + \frac{1}{x^3}\right)$$

Now, when $|x|$ is very large, both $3/x^2$ and $1/x^3$ are close to zero. So we have

$$\begin{aligned} y &\approx x^3(1 - 0 + 0) \\ &\approx x^3 \qquad \text{when } |x| \text{ is very large} \end{aligned}$$

The same technique that we've just used in analyzing $y = x^3 - 3x + 1$ can be applied to any (nonconstant) polynomial function. The result is summarized in item 3 in the following box.

PROPERTY SUMMARY Graphs of Polynomial Functions

1. The graph of a polynomial function of degree 2 or greater is an unbroken smooth curve. (For degrees 1 and 0, the graph is a line.)
2. The graph of a polynomial function of degree n has at most $n - 1$ turning points.
3. For the graph of any polynomial function (other than a constant function), as $|x|$ gets very large, $|y|$ grows very large. If

$$f(x) = a_n x^n + a_{n-1}x^{n-1} + \cdots + a_1 x + a_n \qquad (a_n \neq 0)$$

then

$$f(x) \approx a_n x^n \qquad \text{when } |x| \text{ is very large}$$

EXAMPLE 4 Choosing a plausible graph for a given polynomial function

A function f is defined by

$$f(x) = -x^3 + x^2 + 9x + 9$$

Which of the graphs in Figure 13 might represent this function?

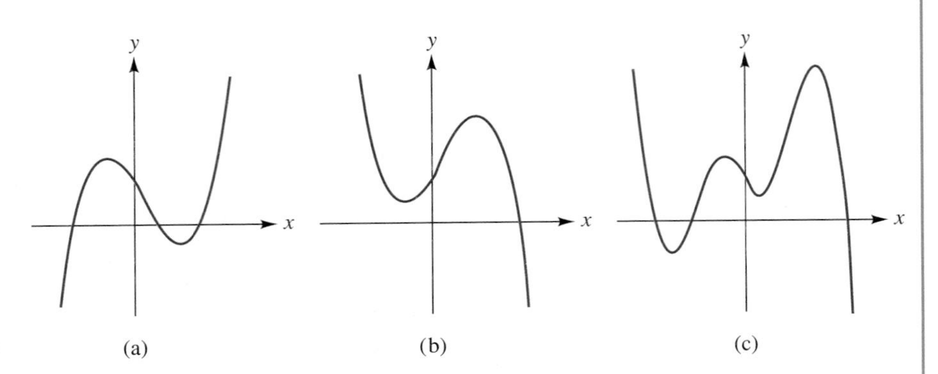

Figure 13

SOLUTION

When $|x|$ is very large, $f(x) \approx -x^3$. This rules out the graph in Figure 13(a). The graph in Figure 13(c) can also be ruled out, but for a different reason. That graph has four turning points, whereas the graph of the cubic function f can have at most two turning points. The graph in Figure 13(b), on the other hand, does have two turning points; furthermore, that graph does behave like $y = -x^3$ when $|x|$ is very large. The graph in Figure 13(b) might be (in fact, it is) the graph of the given function f.

EXAMPLE 5 Comparing polynomials of the same degree

Consider the two polynomial functions f and g defined by

$$f(x) = x^4 \qquad g(x) = x^4 - 3x^3 + 5x + 4$$

(a) With a graphing utility, find a viewing rectangle that highlights the differences between these two functions.

(b) Find a sequence of viewing rectangles demonstrating that as x gets larger and larger, the graph of g looks more and more like the graph of f.

GRAPHICAL PERSPECTIVE

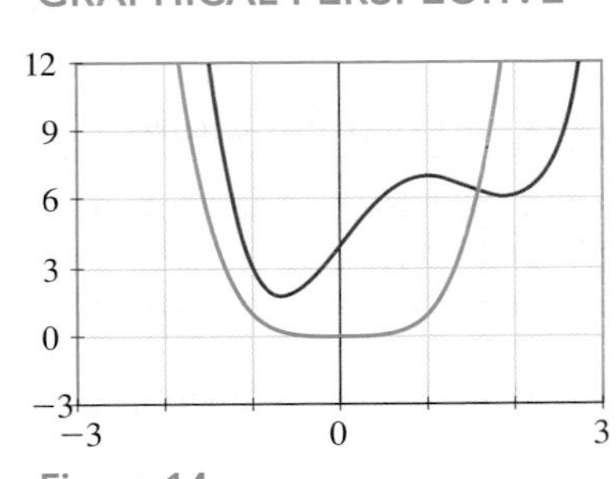

Figure 14
$f(x) = x^4$ (blue) and $g(x) = x^4 - 3x^3 + 5x + 4$ (red) in the viewing rectangle $[-3, 3, 1]$ by $[-3, 12, 3]$

SOLUTION

(a) The view in Figure 14 indicates that the graph of g has three turning points, while the graph of f has but one. The graph of f is symmetric about the y-axis; the graph of g is not.

(b) See the three views in Figure 15. For the scale and size used in Figure 15(c), the two graphs are virtually indistinguishable from one another.

GRAPHICAL PERSPECTIVE

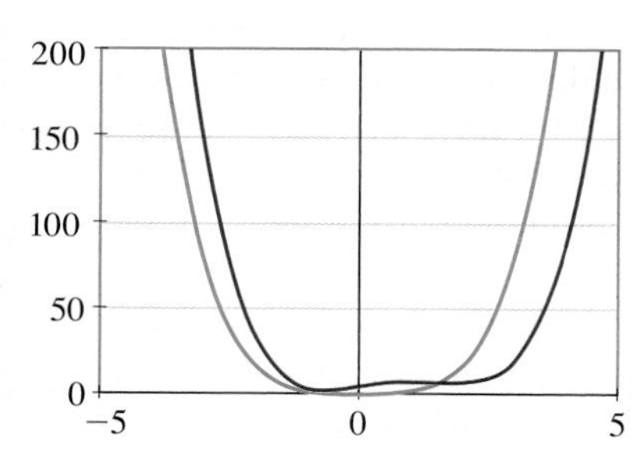

(a) [−5, 5, 5] by [0, 200, 50]

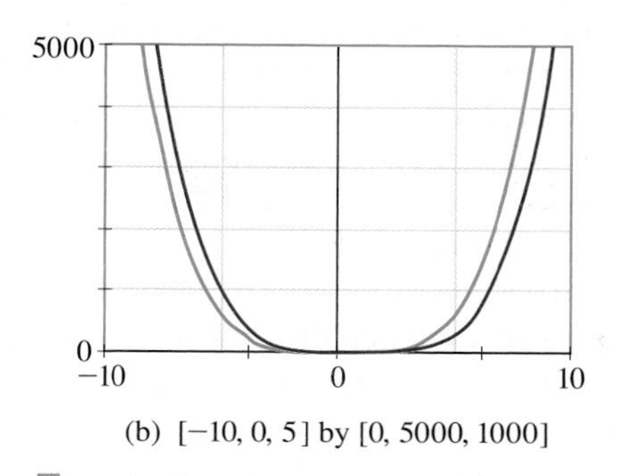

(b) [−10, 0, 5] by [0, 5000, 1000]

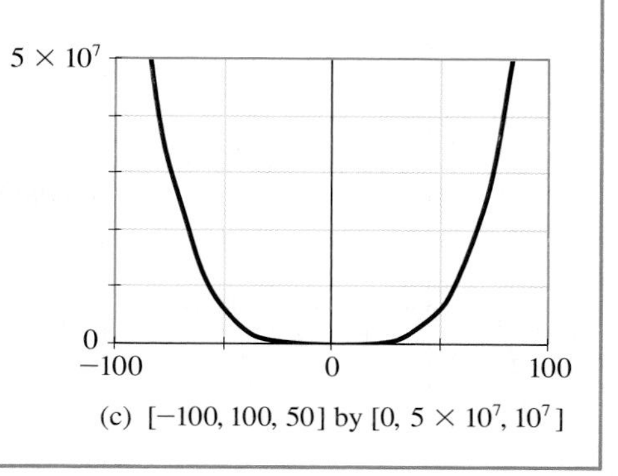

(c) [−100, 100, 50] by [0, 5 × 10⁷, 10⁷]

Figure 15
In views (a) and (b), the blue graph is $f(x) = x^4$ and the red is $g(x) = x^4 - 3x^2 + 5x + 4$. In view (c), both graphs are black; they are indistinguishable from one another at this scale and size.

We can use our work on solving inequalities (in Section 2.6) to graph polynomial functions that are in factored form. Consider, for example, the function

$$f(x) = x(x + 1)(x - 3)$$

First of all, by inspection we see that $f(x) = 0$ when $x = 0$, $x = -1$, or $x = 3$. These are the x-intercepts for the graph. Also, note that the y-intercept is 0. (Why?) Next, we want to know what the graph looks like in the intervals between the x-intercepts. To do this, we solve the two inequalities $f(x) > 0$ and $f(x) < 0$ using the technique in Section 2.6. Table 2 shows the results. (You should check these results for yourself; if you need a review, the details of this example are worked out on pages 114–115.)

TABLE 2

Interval	$x(x + 1)(x - 3)$
$(-\infty, -1)$	negative
$(-1, 0)$	positive
$(0, 3)$	negative
$(3, \infty)$	positive

Now we interpret the results in Table 2 graphically. When x is in either of the intervals $(-\infty, -1)$ or $(0, 3)$, the graph lies below the x-axis (because $f(x) < 0$); and when x is in either of the intervals $(-1, 0)$ or $(3, \infty)$, the graph lies above the x-axis (because $f(x) > 0$). This information is summarized in Figure 16(a). The three dots in the figure indicate the x-intercepts of the graph. The shaded regions are the **excluded regions** through which the graph cannot pass. The graph must pass only through the unshaded regions (and through the three x-intercepts). In Figure 16(b) we have drawn a rough sketch of a curve satisfying these conditions.

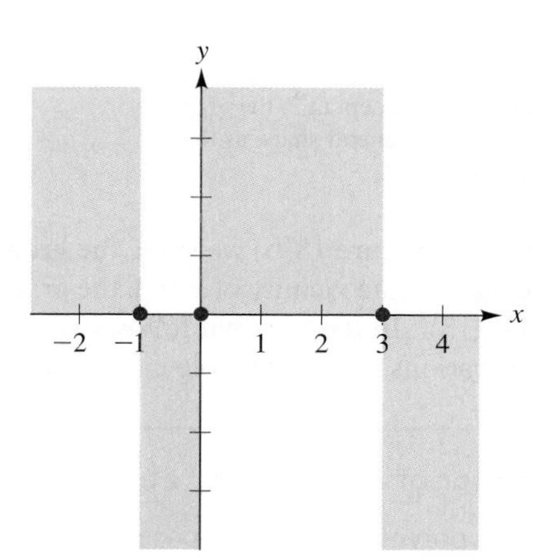

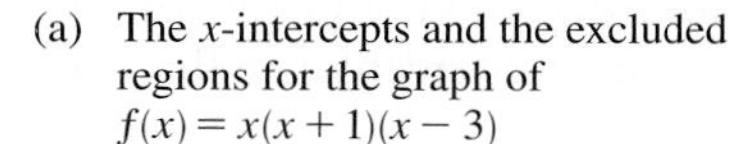
(a) The x-intercepts and the excluded regions for the graph of $f(x) = x(x + 1)(x - 3)$

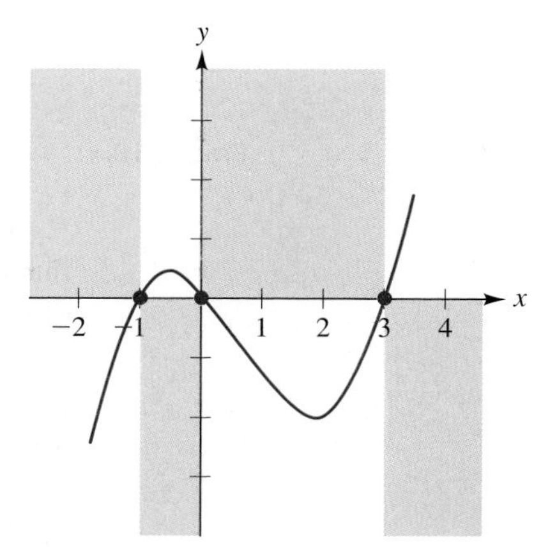

(b) A rough graph of $f(x) = x(x + 1)(x - 3)$

Figure 16

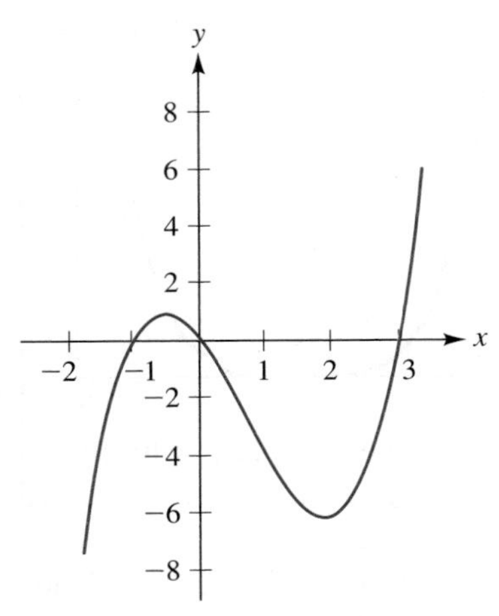

Figure 17
$f(x) = x(x + 1)(x - 3)$

Notice that to draw a smooth curve satisfying the conditions of Figure 16(a), we need at least two turning points: one between $x = -1$ and $x = 0$ and another between $x = 0$ and $x = 3$. On the other hand, since the degree of $f(x)$ is 3, there can be no more than two turning points. Thus Figure 16(b) has exactly two turning points. As another check on our rough sketch, we note that for large values of $|x|$, the graph indeed resembles that of $y = x^3$. (We are using the third property in the summary box on page 293.)

Although the precise location of the turning points is a matter for calculus, we can nevertheless improve upon the sketch in Figure 16(b) by computing $f(x)$ for some specific values of x. Some reasonable choices in this case are the inputs $-2, -\frac{1}{2}, 1, 2,$ and 4. As you can check, the resulting points on the graph are $(-2, -10)$, $(-\frac{1}{2}, \frac{7}{8})$, $(1, -4)$, $(2, -6)$, and $(4, 20)$. We can now sketch the graph, as shown in Figure 17.

In the example just concluded, the polynomial $f(x) = x(x + 1)(x - 3)$ has no *repeated factors.* That is, none of the factors is squared, cubed, or raised to a higher power. Now let us look at two examples in which there are repeated factors. The observations we make will help us in determining the general shape of a graph without the need for plotting a large number of individual points. In Figure 18(a), we show the graph of $g(x) = x(x + 1)(x - 3)^2$. Notice that in the immediate vicinity of the intercept at $x = 3$, the graph has the same general shape as that of $y = A(x - 3)^2$.

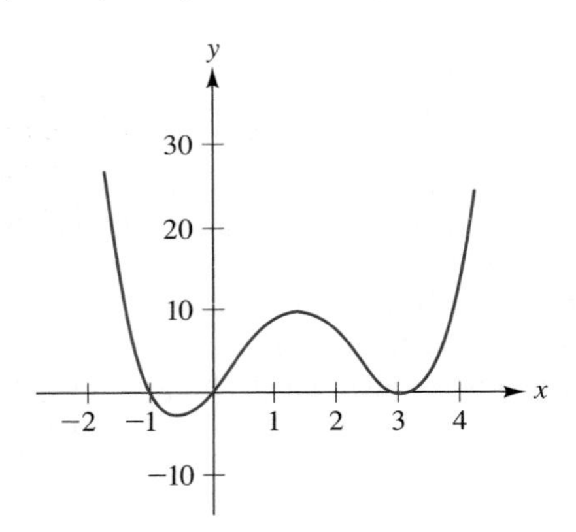

(a) $g(x) = x(x + 1)(x - 3)^2$
Near the x-intercept at 3, the graph has the same general shape as that of $y = A(x - 3)^2$.

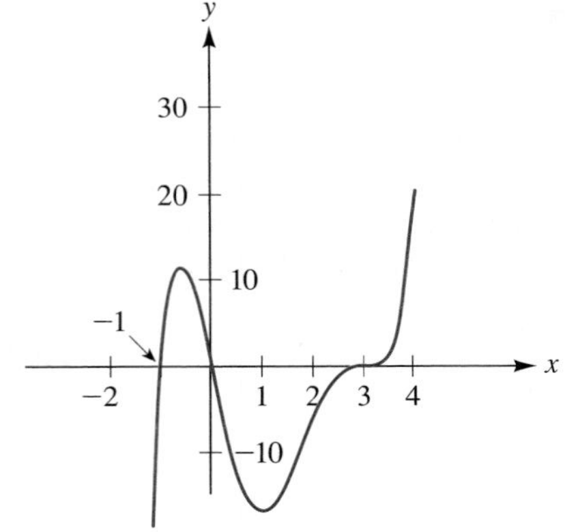

(b) $h(x) = x(x + 1)(x - 3)^3$
Near the x-intercept at 3, the graph has the same general shape as that of $y = A(x - 3)^3$.

Figure 18

Similarly, in Figure 18(b) we show the graph of $h(x) = x(x + 1)(x - 3)^3$; notice that in the immediate vicinity of $x = 3$, the graph has the same general shape as that of $y = A(x - 3)^3$. In the box that follows, we state the general principle underlying these observations. (The principle can be justified using calculus.)

The Behavior of a Polynomial Function Near an x-Intercept

Let $f(x)$ be a polynomial and suppose that $(x - a)^n$ is a factor of $f(x)$. [Furthermore, assume that none of the other factors of $f(x)$ contains $(x - a)$.] Then, in the immediate vicinity of the x-intercept at a, the graph of $y = f(x)$ closely resembles that of $y = A(x - a)^n$.

The principle that we have just stated is easy to apply because we already know how to graph functions of the form $y = A(x - a)^n$. The next example shows how this works.

EXAMPLE 6 **Determining the shape of the graph of a polynomial function near an x-intercept**

Describe the behavior of each function in the immediate vicinity of the indicated x-intercept.

(a) $f(x) = \frac{1}{2}(x - 3)(x - 1)^3$; intercept: $x = 1$

(b) $g(x) = (x + 1)(x + 4)(x + 3)^2$; intercept: $x = -3$

SOLUTION

(a) We make the following observation:

$$f(x) = \frac{1}{2}\underbrace{(x - 3)}(x - 1)^3$$

When x is close to 1, this factor is close to $1 - 3$, or -2.

So if x is very close to 1, we have the approximation

$$f(x) \approx \frac{1}{2}(1 - 3)(x - 1)^3 = \frac{1}{2}(-2)(x - 1)^3 = -(x - 1)^3$$

Thus in the immediate vicinity of $x = 1$ the graph of f closely resembles $y = -(x - 1)^3$. Notice the technique we used to obtain this result. We retained the factor corresponding to the intercept $x = 1$, and we approximated the remaining factor using the value $x = 1$. See Figure 19.

(b) We use the approximation technique shown in part (a).

$$g(x) = \underbrace{(x + 1)}\underbrace{(x + 4)}(x + 3)^2$$

When x is close to -3, this factor is close to $-3 + 1$, or -2.

When x is close to -3, this factor is close to $-3 + 4$, or 1.

Thus when x is close to -3, we have the approximation

$$g(x) \approx (-3 + 1)(-3 + 4)(x + 3)^2 = -2(x + 3)^2$$

This tells us that in the immediate vicinity of $x = -3$, the graph of g resembles $y = -2(x + 3)^2$. See Figure 20.

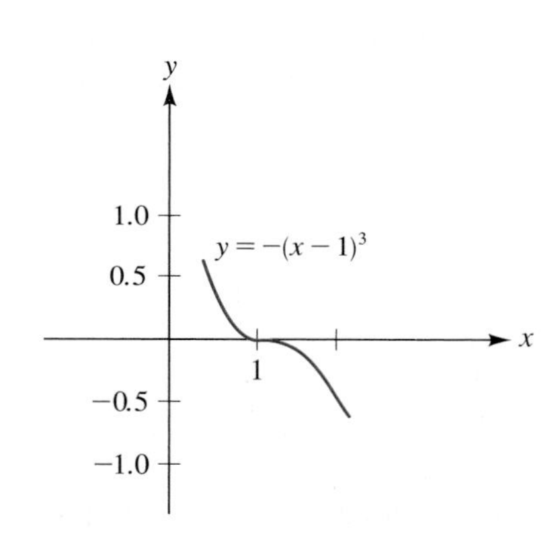

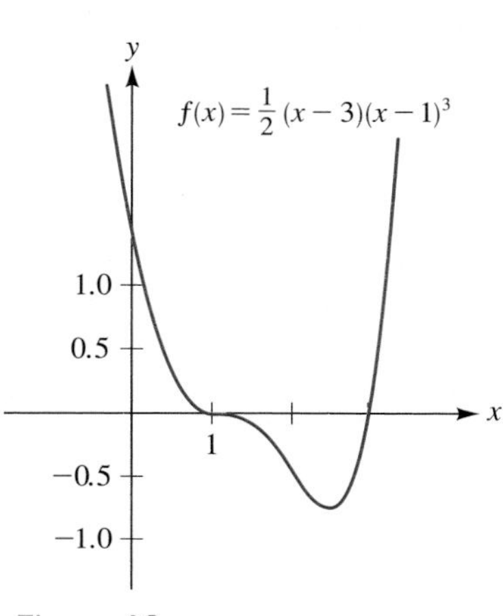

Figure 19
When x is close to 1, the graph of $f(x) = \frac{1}{2}(x - 3)(x - 1)^3$ closely resembles $y = -(x - 1)^3$.

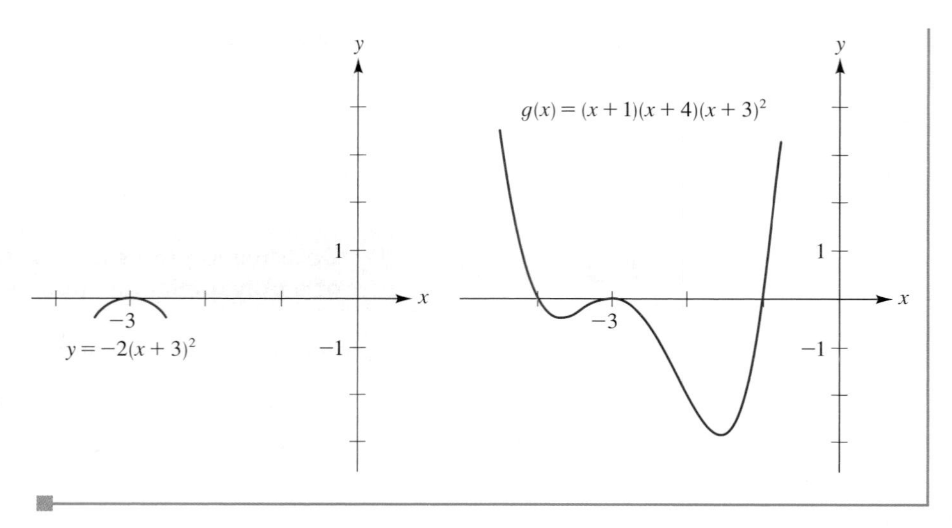

Figure 20
When x is close to -3, the graph of $g(x) = (x + 1)(x + 4)(x + 3)^2$ closely resembles $y = -2(x + 3)^2$.

The technique introduced in Example 6, whereby we approximate the behavior of the factored polynomial near each x-intercept, allows us to draw by hand a reasonably accurate graph. (Of course, some judicious point plotting is necessary too, as mentioned previously.) It's interesting though, to use a graphing utility to confirm visually just how well that simple approximation technique does work. As an example, consider the polynomial function f defined by

$$f(x) = 2(x + 1)(x - 2)^3$$

Near the x-intercept $x = 2$, our approximation technique yields

$$f(x) \approx 2(2 + 1)(x - 2)^3$$
$$f(x) \approx 6(x - 2)^3 \qquad \text{near } x = 2$$

In Figure 21 we've used a graphing utility to plot both the graph of f and the graph of the approximating polynomial $y = 6(x - 2)^2$. As you can see, in the immediate vicinity of $x = 2$, the two graphs are remarkably close.

GRAPHICAL PERSPECTIVE

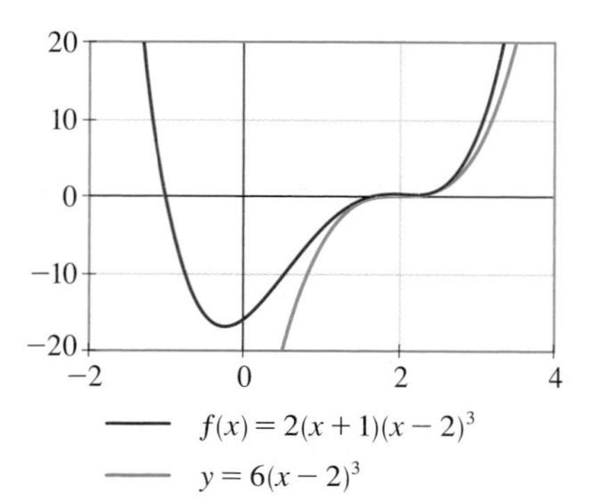

Figure 21
$f(x) = 2(x + 1)(x - 2)^3$ and $y = 6(x - 2)^3$

EXERCISE SET 4.6 

A

1. Which of the functions in Figure 1 on page 288 are polynomial functions?
2. **(a)** Give an example of a power function that is not a polynomial function.
   (b) Give an example of a polynomial function that is not a power function.
3. Let $f(x) = x^2$ and $g(x) = x^3$.
   (a) Either by hand or with a graphing utility, on the same set of axes draw the graphs of f and g on the interval $[0, 1]$.
   (b) Compute and compare the average rates of change of f and g on the interval $[0, 1]$ and on the interval $[0, 1/2]$.
4. (G) Let $f(t) = t^4$ and $g(t) = t^5$.
   (a) Using a graphing utility, draw the graphs of f and g on the same set of axes.
   (b) Which of the three types of symmetry discussed on page 58 does the graph of each function possess?
   (c) Compute the average rate of change of each function on the interval $[-2, 2]$.
   (d) Explain why each answer in part (c) is predictable from the symmetry of the graph.

In Exercises 5–16, sketch the graph of each function and specify all x- and y-intercepts.

5. $y = (x - 2)^2 + 1$
6. $y = -3x^4$
7. $y = -(x - 1)^4$
8. $y = -(x + 2)^3$
9. $y = (x - 4)^3 - 2$
10. $y = -(x - 4)^3 - 2$
11. $y = -2(x + 5)^4$
12. $y = -2x^4 + 5$
13. $y = (x + 1)^5/2$
14. $y = \frac{1}{2}x^5 + 1$
15. $y = -(x - 1)^3 - 1$
16. $y = x^8$

In Exercises 17–20, give a reason (as in Example 4) why each graph cannot represent a polynomial function of degree 3.

17.

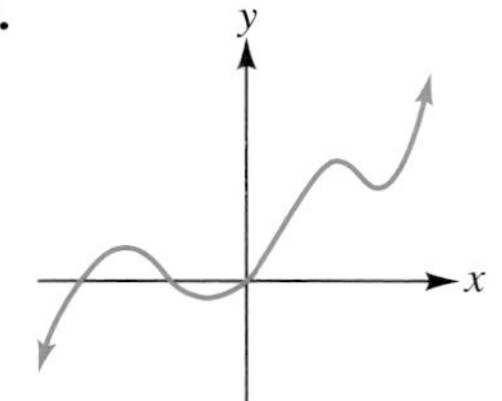

18.

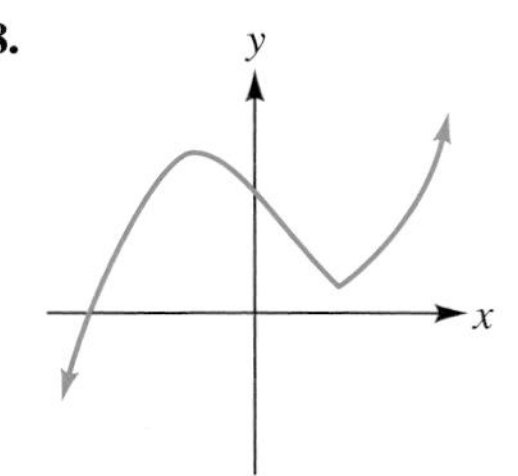

19.

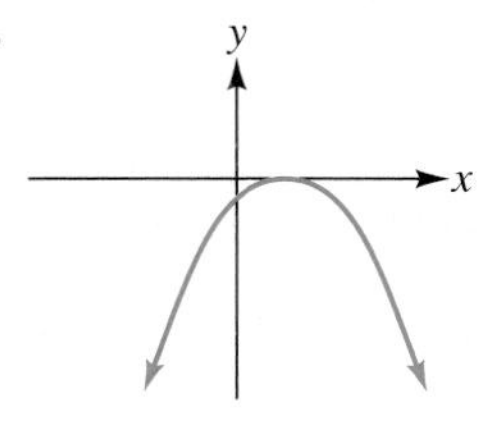

20.

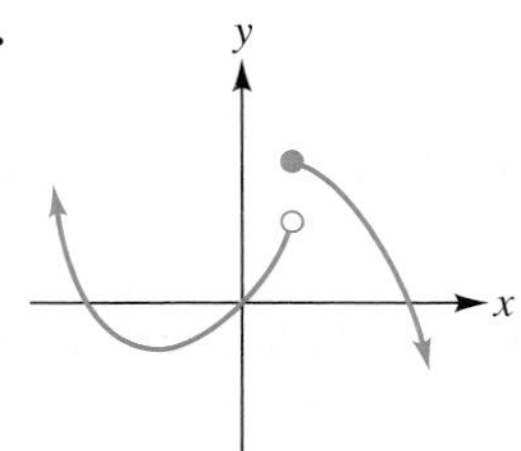

In Exercises 21–24, give a reason why each graph cannot represent a polynomial function that has the highest-degree term $2x^5$.

21.

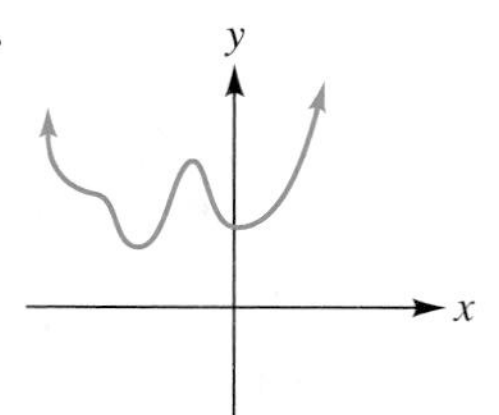

22.

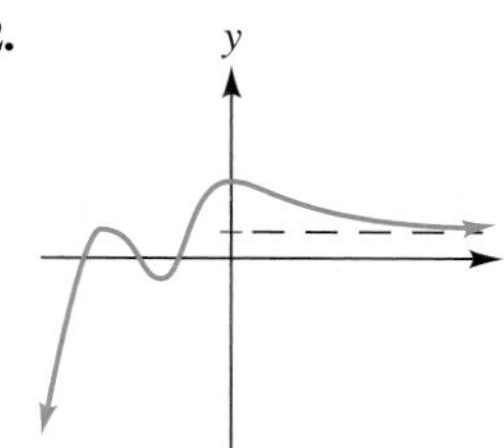

23.

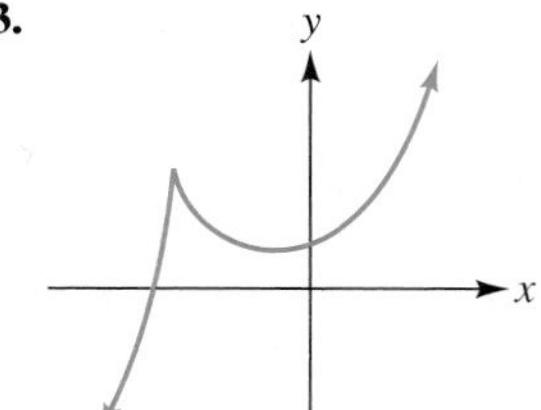

24.

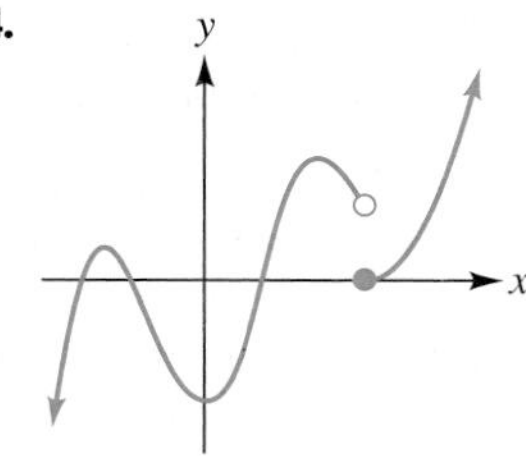

(G) *For Exercises 25 and 26:*
(a) *With a graphing utility, find a viewing rectangle that highlights the differences between the two functions, as in Example 5.*
(b) *Find a sequence of viewing rectangles demonstrating that as x gets larger and larger, the graph of g looks more and more like the graph of f.*

25. $f(x) = 2x^2, g(x) = 2x^2 - 12x - 5$
26. $f(x) = 4x^3, g(x) = 4x^3 + 72x^2 + 420x + 805$

In Exercises 27–34:
(a) *Determine the x- and y-intercepts and the excluded regions for the graph of the given function. Specify your results using a sketch similar to Figure 16(a). In Exercises 31–34, you will first need to factor the polynomial.*
(b) *Graph each function.*

27. $y = (x - 2)(x - 1)(x + 1)$
28. $y = (x - 3)(x + 2)(x + 1)$
29. $y = 2x(x - 2)(x - 1)$
30. $y = (x - 3)(x - 2)(x + 2)$
31. $y = x^3 - 4x^2 - 5x$
32. $y = x^3 - 9x$
33. $y = x^3 + 3x^2 - 4x - 12$
34. $y = x^3 - 5x^2 - x + 5$

In Exercises 35–44:
(a) *Determine the x- and y-intercepts and the excluded regions for the graph of the given function. Specify your results using a sketch similar to Figure 16(a).*
(b) *Describe the behavior of the function at each x-intercept that corresponds to a repeated factor. Specify your results using a sketch similar to the left-hand portion of Figure 20.*
(c) *Graph each function.*

35. $y = x^3(x + 2)$
36. $y = (x - 1)(x - 4)^2$
37. $y = 2(x - 1)(x - 4)^3$
38. $y = (x - 1)^2(x - 4)^2$
39. $y = (x + 1)^2(x - 1)(x - 3)$
40. $y = x^2(x - 4)(x + 2)$

41. $y = -x^3(x - 4)(x + 2)$ **42.** $y = 4(x - 2)^2(x + 2)^3$

43. $y = -4x(x - 2)^2(x + 2)^3$ **44.** $y = -3x^3(x + 1)^4$

45. This exercise refers to Example 1(a) in the text. Use Figure 5(b) to obtain the following estimates: on the interval $[0, 1.25]$, $\Delta g/\Delta x \approx 4$, and $\Delta f/\Delta x \approx 2$.

B

46. **(a)** This exercise refers to Example 1(b), in which $f(x) = x^4$ and $g(x) = x^8$. Show that on the interval $[0, 1.25]$, we have $\Delta g/\Delta x \approx 4.768$ and $\Delta f/\Delta x \approx 1.953$.

(b) Let $f(x) = x^4$ and $g(x) = x^8$, as in part (a). Find a positive constant b so that on the interval $[0, b]$, the average rate of change of g is 100 times the average rate of change of f. Give two forms for the answer: an exact expression and a decimal approximation rounded to three decimal places.

In Exercises 47–50, first use the graph to estimate the x-intercepts. Then use algebra to determine each x-intercept. If an intercept involves a radical, give that answer as well as a calculator approximation rounded to three decimal places. Be sure to check that your results are consistent with the initial graphical estimates.

47.

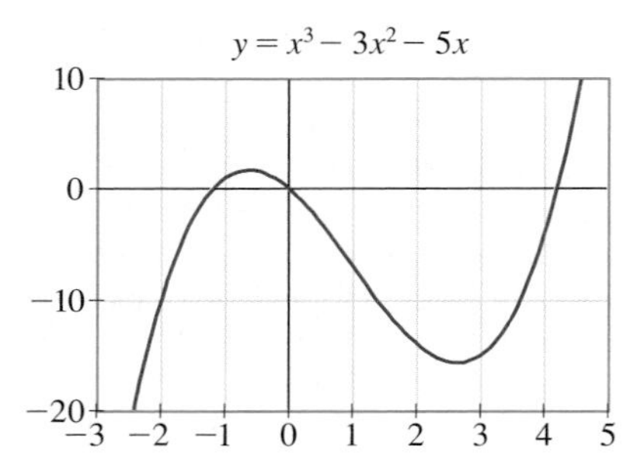

48.

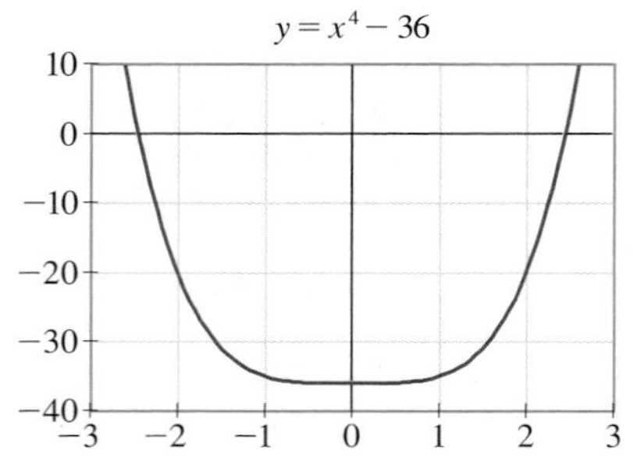

49.

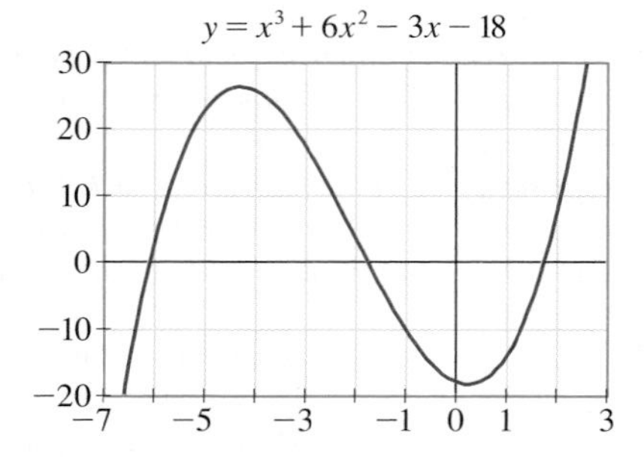

Hint: Factor $x^3 + 6x^2 - 3x - 18$ by grouping.

50.

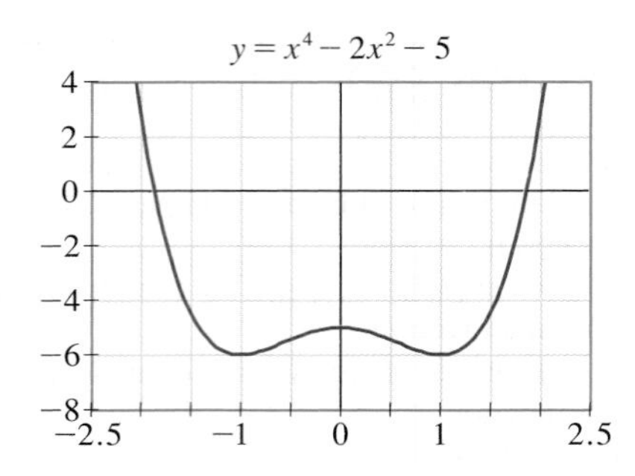

For Exercises 51 and 52:

Ⓖ **(a)** *Use a graphing utility to draw a graph of each function.*

Ⓖ **(b)** *For each x-intercept, zoom in until you can estimate it accurately to the nearest one-tenth.*

(c) *Use algebra to determine each x-intercept. If an intercept involves a radical, give that answer as well as a calculator approximation rounded to three decimal places. Check to see that your results are consistent with the graphical estimates obtained in part (b).*

51. $N(t) = t^7 + 8t^4 + 16t$ **52.** $W(u) = 2u^4 - 17u^2 + 35$

In Exercises 53–58, six functions are defined as follows:

$$f(x) = x \qquad g(x) = x^2 \qquad h(x) = x^3$$
$$F(x) = x^4 \qquad G(x) = x^5 \qquad H(x) = x^6$$

Refer also to the following figure.

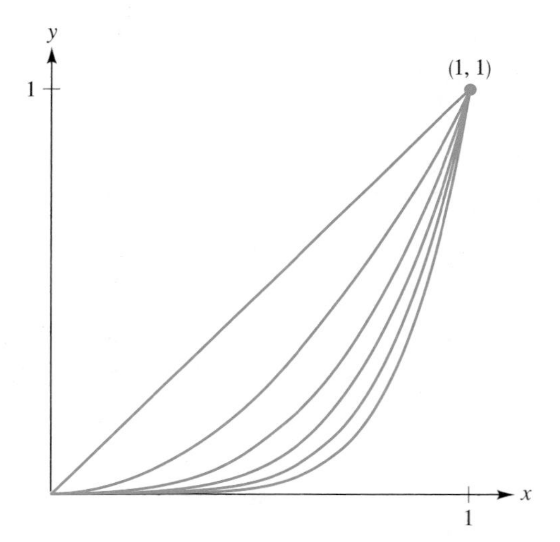

53. The six graphs in the figure are the graphs of the six given functions for the interval $[0, 1]$, but the graphs are not labeled. Which is which?

54. For which x-values in $[0, 1]$ will the graph of g lie strictly below the horizontal line $y = 0.1$? Use a calculator to evaluate your answer. Round off the result to two significant figures.

55. Follow Exercise 54, using the function H instead of g.

56. Find a number t in $[0, 1]$ such that the vertical distance between $f(t)$ and $g(t)$ is $1/4$.

57. Is there a number t in $[0, 1]$ such that the vertical distance between $g(t)$ and $F(t)$ is 0.26?

58. Find all numbers t in $[0, 1]$ such that $F(t) = G(t) + H(t)$.

59. The figure shows the graphs of $y = 4x$ and $y = x^2/100$ in a standard viewing rectangle.

(a) Do the graphs intersect anywhere other than at the origin?

Ⓖ **(b)** Use a graphing utility to support your answer in part (a).

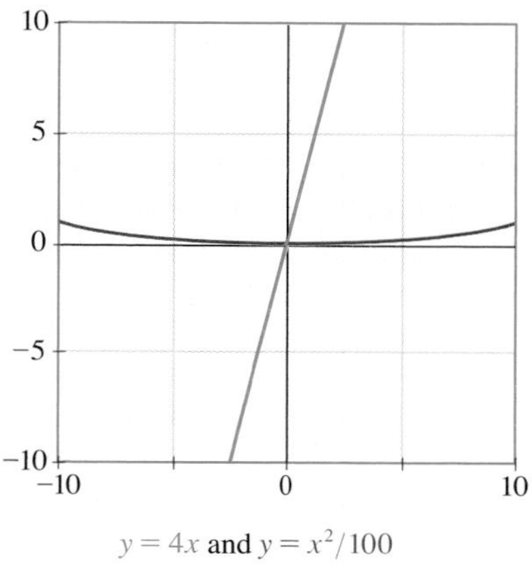

$y = 4x$ and $y = x^2/100$

$[-10, 10, 5]$ by $[-10, 10, 5]$

60. (a) Factor the expression $4x^2 - x^4$. Then use the techniques explained in this section to graph the function defined by $y = 4x^2 - x^4$.

(b) Find the coordinates of the turning points. *Hint*: As in previous sections, use the substitution $x^2 = t$.

61. (a) Graph the function $D(x) = x^2 - x^4$.

(b) Find the turning points of the graph. (See the hint in Exercise 60.)

(c) On the same set of axes, sketch the graphs of $y = x^2$ and $y = x^4$ for $0 \le x \le 1$. What is the maximum vertical distance between the graphs?

62. (a) An open-top box is to be constructed from a 6-by-8-in. rectangular sheet of tin by cutting out equal squares at each corner and then folding up the resulting flaps. Let x denote the length of the side of each cutout square. Show that the volume $V(x)$ is

$$V(x) = x(6 - 2x)(8 - 2x)$$

(b) What is the domain of the volume function in part (a)? [The answer is *not* $(-\infty, \infty)$.]

Ⓖ **(c)** Use a graphing utility to graph the volume function, taking into account your answer in part (b).

Ⓖ **(d)** By zooming in on the turning point, estimate to the nearest one-hundredth the maximum volume.

63. The point P is in the first quadrant on the graph of $y = 1 - x^4$. From P, perpendiculars are drawn to the x- and y-axes, thus forming a rectangle.

(a) Express the area of the rectangle as a function of a single variable.

Ⓖ **(b)** Use a graphing utility to graph the area function. Then, using the ZOOM feature, estimate to the nearest hundredth the maximum possible area for the rectangle.

64. In Example 1 in Section 4.2 we used a graphing utility to determine linear and quadratic functions to model the AIDS data in Figure 2 on page 229. We then used these models to make projections for the years 1992 and 1997. For both models and both years we found that the projections were too low. We also saw that the quadratic projections were more accurate than the linear ones, as measured by percentage error.

Ⓖ **(a)** Use a graphing utility to find a polynomial of degree 3 (a *cubic model*) that best fits the data in Figure 2 on page 229.

(b) Use the cubic model to make projections for the cumulative number of AIDS cases in 1992 and 1997.

(c) Are the projections higher or lower than the actual figures given in Example 1 in Section 4.2? Compute the percentage error for each projection. In each case, which is the better predictor, the quadratic or the cubic model?

C

65. (a) Show that for m and n positive integers with $m < n$ the following inequalities hold:

$$\text{if } 0 \le x \le 1 \quad \text{then } 0 \le x^n \le x^m$$

and

$$\text{if } x > 1 \quad \text{then } x^m < x^n$$

(b) Using part (a), explain why for m and n positive *even* integers, with $m < n$, the graph of $y = x^n$ lies below the graph of $y = x^m$ for $-1 < x < 1$ while the graph of $y = x^n$ lies above the graph of $y = x^m$ for $|x| > 1$.

(c) Using part (a), explain why for m and n positive *odd* integers, with $m < n$, the graph of $y = x^n$ lies above the graph of $y = x^m$ for $x > 1$ and $-1 < x < 0$ and lies below the graph of $y = x^m$ for $x < -1$ and $0 < x < 1$.

PROJECT FINDING SOME MAXIMUM VALUES WITHOUT USING CALCULUS

In Section 4.2 it was stated that, *in general*, the techniques of calculus are required to find maximum or minimum values for functions that are not quadratics. Nevertheless, there are a number of cases involving polynomials or quotients of polynomials where algebra, rather than calculus, will suffice. The following theorem can be used to find maximum values in certain types of problems. (One way to establish this theorem is indicated at the end of this project.)

THEOREM

If the sum of three positive quantities s, t, and u is constant, then their product stu is maximum if and only if $s = t = u$.

We'll give three examples that show how to use this theorem; then you can try your hand at the three problems that follow. First sample problem: *The sum of three positive numbers is* 1. *What is the maximum possible value for their product?* Letting s, t, and u denote the three numbers, we have $s + t + u = 1$. According to the theorem, the product is maximum when the three numbers are equal. So in that case, the equation $s + t + u = 1$ is equivalent to $s + s + s = 1$, which implies $s = 1/3$. Therefore t and u are also equal to $1/3$, and the maximum possible value for the product is $stu = (\frac{1}{3})(\frac{1}{3})(\frac{1}{3}) = \frac{1}{27}$. As a quick empirical check, pick any three positive fractions or decimals that add to 1. Compute the product, and check that it's less than $\frac{1}{27}$ (assuming, of course, that you didn't pick all three numbers to be $1/3$).

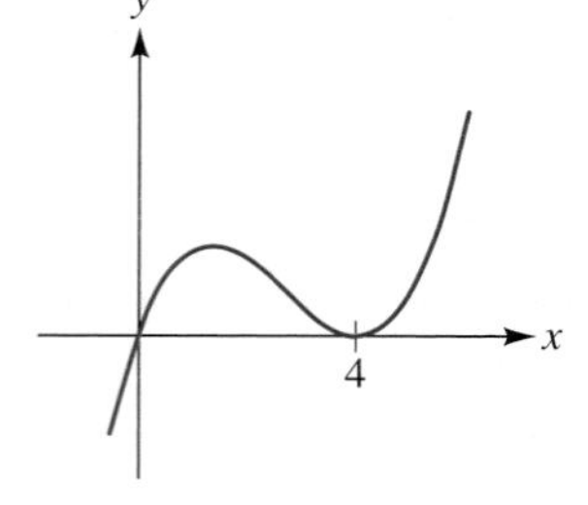

Figure A
$f(x) = 2x(4 - x)^2$

Second sample problem: *Let* $f(x) = 2x(4 - x)^2$. *By using the standard graphing techniques discussed in Section 4.6, we find that the general shape for the graph of f is as shown in Figure A. One turning point,* (4, 0), *is on the x-axis; the second turning point is in the first quadrant, with an x-coordinate between 0 and 4. Find the exact coordinates of this second turning point.* First, note that the three factors comprising $f(x)$, namely, $2x$, $(4 - x)$, and $(4 - x)$ again, do have a constant sum: $2x + (4 - x) + (4 - x) = 8$. Furthermore, on the open interval $0 < x < 4$, each of these factors is positive. Thus our theorem in the box is applicable. The maximum value of the product $f(x) = 2x(4 - x)^2$ on the open interval $0 < x < 4$ occurs when the three factors are equal. Setting $2x = 4 - x$, we obtain $3x = 4$, and consequently, $x = 4/3$. This is the required x-coordinate of the turning point. For the y-coordinate we have $f(\frac{4}{3}) = 2(\frac{4}{3})(4 - \frac{4}{3})^2$. As you can verify, that works out to $\frac{512}{27}$. In summary, the coordinates of the turning point are $(\frac{4}{3}, \frac{512}{27})$.

Third sample problem: *According to Figure B, the graph of the polynomial function defined by* $g(x) = x^2(1 - x)$ *has a turning point in the first quadrant. Find the x-coordinate of that turning point.* First, note that on the open interval $0 < x < 1$, each of the three factors x, x again, and $1 - x$ is positive. However, as you can check, the sum of the three factors is not constant. We can work around this by writing

$$(x)(x)(1 - x) = 4\left[\left(\frac{x}{2}\right)\left(\frac{x}{2}\right)(1 - x)\right]$$

GRAPHICAL PERSPECTIVE

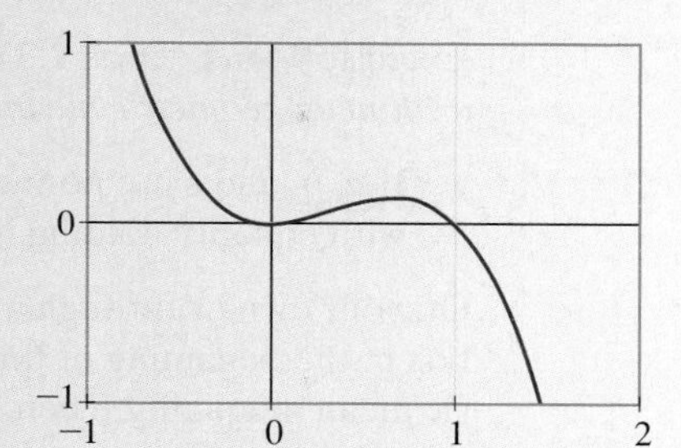

Figure B
$g(x) = x^2(1 - x)$

Now notice that the three factors in brackets do have a constant sum. For a maximum then, we require that $\frac{x}{2} = 1 - x$. Solving this last equation yields $x = 2/3$, the required x-coordinate.

EXERCISES

1. **(a)** First, sketch a graph of the function $f(x) = x^2(6 - 2x)$ using the standard techniques from Section 4.6. Then, use the ideas presented above to find the exact coordinates of the turning point in the first quadrant.
(b) Follow part (a), but use $g(x) = x^2(6 - 3x)$.

2. An open-top box is to be constructed from a 16-inch square sheet of cardboard by cutting out congruent squares at each corner and then folding up the flaps. See the following figure. Express the volume of the box as a function of a single variable. Then find the maximum possible volume for the box.

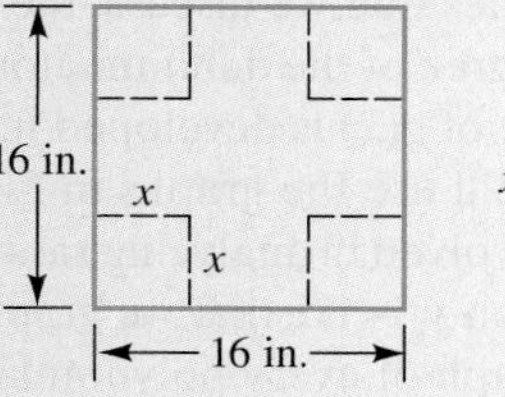

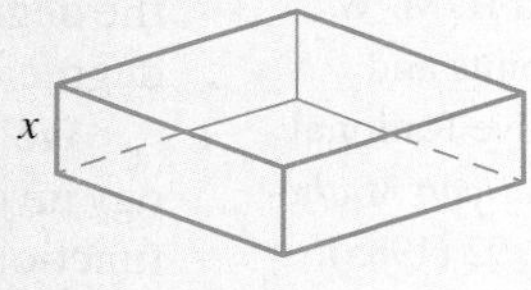

3. When a person coughs, the radius r of the trachea (windpipe) decreases. In a paper, "The Human Cough" (Lexington, Mass.: COMAP, Inc., 1979), Philip Tuchinsky developed the following model for the average velocity of air through the trachea during a cough:

$$v(r) = c(r_0 - r)r^2 \qquad (\tfrac{1}{2}r_0 \leq r \leq r_0)$$

In this formula the variable r represents the radius of the trachea; the constant r_0 is the normal radius when the person is not coughing; c is a positive constant (which, among other things, depends upon the length of the person's trachea); and $v(r)$ is the average velocity of the air in the trachea. Show that $v(r)$ is a maximum when $r = \frac{2}{3}r_0$. In other words, according to this model, in coughing to clear the air passages, the most effective cough occurs when the radius contracts to two-thirds its normal size.

ESTABLISHING THE THEOREM

Exercise 44 in Section 2.3 outlines a proof of the following result, known as the *arithmetic-geometric mean inequality* for three numbers:

> If a, b, and c are nonnegative real numbers, then $\sqrt[3]{abc} \leq (a + b + c)/3$, with equality holding if and only if $a = b = c$.

On your own or through a group discussion, determine why our theorem in the box at the beginning of this project is a consequence of the arithmetic-geometric mean inequality. Then, on your own, write a paragraph carefully explaining this. (You are not being asked here to prove the arithmetic-geometric mean inequality.)

4.7 RATIONAL FUNCTIONS

We shall not attempt to explain the numerous situations in life sciences where the study of such graphs is important nor the chemical reactions which give rise to the rational functions $r(x)$. . . . Let it suffice to mention that, to the experimental biochemist, theoretical results concerning the shapes of rational functions are of considerable interest.
—W. G. Bardsley and R. M. W. Wood in "Critical Points and Sigmoidicity of Positive Rational Functions," *The American Mathematical Monthly,* vol. 92 (1985), pp. 37–42.

After the polynomial functions, the next simplest functions are the **rational functions.** These are functions defined by equations of the form

$$y = \frac{f(x)}{g(x)}$$

where $f(x)$ and $g(x)$ are polynomials. In general, throughout this section, when we write a function such as $y = f(x)/g(x)$, we assume that $f(x)$ and $g(x)$ contain no common factors (other than constants). (Exercises 44 and 45 ask you to consider several cases in which $f(x)$ and $g(x)$ do contain common factors.) Also, for each of the examples that we discuss, the degree of the numerator $f(x)$ is less than or equal to the degree of the denominator $g(x)$. (A case in which the degree of $f(x)$ exceeds the degree of $g(x)$ is developed in the exercises.)

We'll use the graphs in Figure 1 to introduce some of the ideas and terminology involved in analyzing rational functions. Figure 1(a) shows the familiar rational function $y = 1/x$ that we graphed in Section 3.2. The graph of $y = 1/x$ differs from the graph of every polynomial function in two important aspects. First, the graph has a break in it; it is composed of two distinct pieces or **branches.** (Recall from Sec-

GRAPHICAL PERSPECTIVE

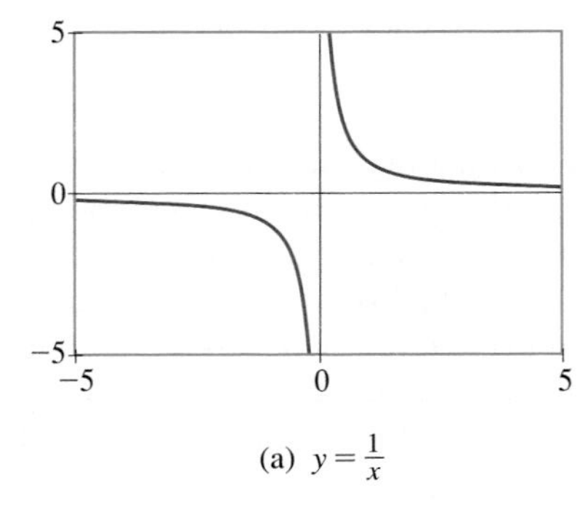

(a) $y = \frac{1}{x}$

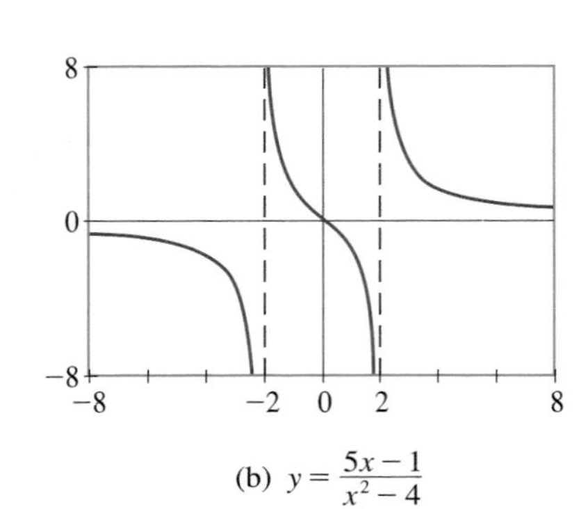

(b) $y = \frac{5x - 1}{x^2 - 4}$

Figure 1
Graphs of two rational functions

tion 4.6 that the graph of a polynomial function never has a break or gap in it.) In general, the graph of a rational function has one more branch than the number of distinct real values for which the denominator is zero. In Figure 1(b), for example, there are two distinct numbers for which the denominator is zero (what are they?), and the graph has three branches.

The second way that the graph of $y = 1/x$ differs from that of a polynomial function is related to the *asymptotes.* A line is an **asymptote** for a curve if the distance between the line and the curve approaches zero as we move out farther and farther along the line. In Figure 1(a) the x-axis is a horizontal asymptote for the curve; the y-axis is a vertical asymptote. In Figure 1(b) the x-axis is a horizontal asymptote, and each of the lines $x = 2$ and $x = -2$ is a vertical asymptote. Note that 2 and -2 are also the x-values for which the denominator $x^2 - 4$ is zero. In general, this type of correspondence will always occur for a rational function $y = f(x)/g(x)$ as long as the two polynomials have no common factors (other than constants). Thus to find the vertical asymptotes for the graph of $y = f(x)/g(x)$, just find the x-values for which the denominator is zero. What if there are no real-number values for x that make the denominator zero? Then the graph of the rational function has no vertical asymptotes. Figure 2 shows such an example.

GRAPHICAL PERSPECTIVE

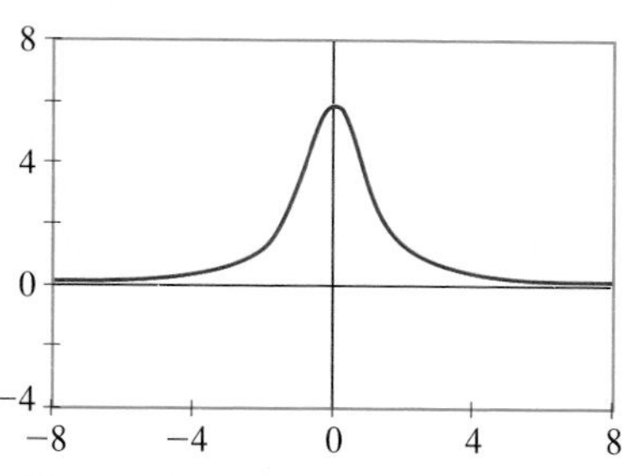

Figure 2
$y = 6/(1 + x^2)$
There are no real numbers x for which the denominator is zero. The graph has no vertical asymptotes, and there is no break in the graph.

We can use the functions in Figures 1(b) and 2 as examples in determining domain and intercepts. For the domain of the function $y = (5x - 1)/(x^2 - 4)$ in Figure 1(b), rewrite the equation with the denominator in factored form:

$$y = \frac{5x - 1}{(x - 2)(x + 2)}$$

In this form, we can see than any x-value other than -2 or 2 will yield an unambiguous output value for y. Thus the domain consists of all real numbers except for -2 and 2. [Note that the graph in Figure 1(b) supports this conclusion.] Using interval notation, the domain is

$$(-\infty, -2) \cup (-2, 2) \cup (2, \infty)$$

For the function in Figure 2, $y = 6/(x^2 + 1)$, the denominator is never zero. (Why?) Thus any real number x is an appropriate input, and the domain is the set of all real numbers.

Now look back at Figure 1(b). Here we can see that the x- and y-intercepts are each close to zero, if not zero. To calculate the y-intercept, set $x = 0$ in the given equation. That yields

$$y = \frac{5(0) - 1}{0^2 - 4} = \frac{1}{4}$$

Thus, the y-intercept is 1/4. For the x-intercept in Figure 1(b), set $y = 0$ in the given equation:

$$0 = \frac{5x - 1}{x^2 - 4}$$

The fraction in this last equation, like any fraction, can only be zero if the numerator is zero (and the denominator isn't zero). Setting $5x - 1$ equal to zero gives us $x = 1/5$, which is the required x-intercept. By the way, this simple idea that a fraction can only be zero when the numerator is zero can tell you right away, even before seeing a graph, that some rational functions have no x-intercepts. For instance, the numerator of $y = 6/(x^2 + 1)$ is certainly never zero, and therefore the graph of $y = 6/(x^2 + 1)$ has no x-intercepts. (See Figure 2.)

If k is a positive constant, the graph of $y = k/x$ resembles that of $y = 1/x$. Figure 3 shows the graphs of $y = 1/x$ and $y = 4/x$.

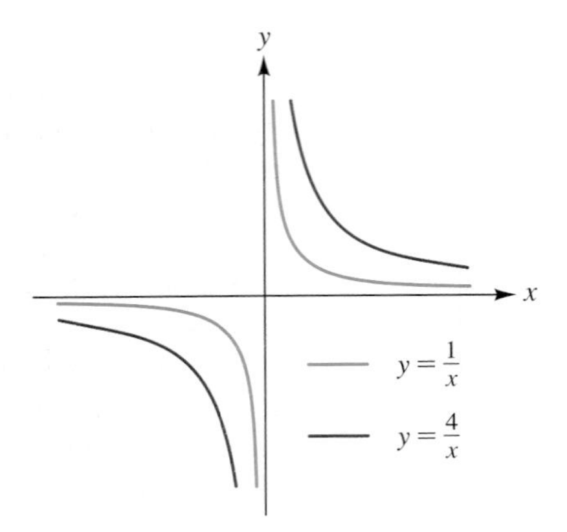

Figure 3

Once we know about the graph of $y = k/x$, we can graph any rational function of the form

$$y = \frac{ax + b}{cx + d}$$

The next three examples show how this is done using the translation and reflection techniques explained in Section 3.4.

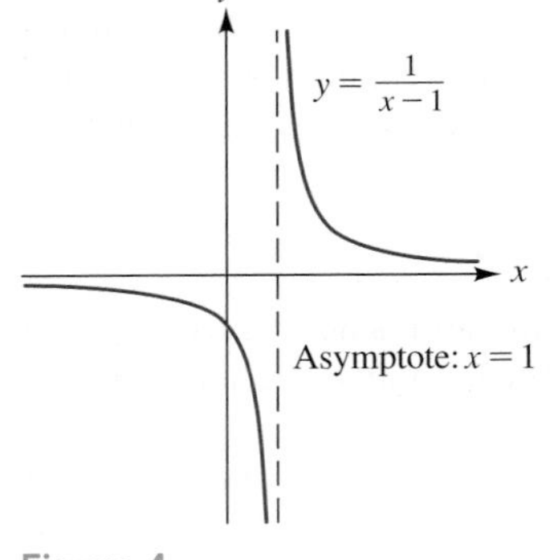

Figure 4
The graph of $y = 1/(x - 1)$

EXAMPLE 1 Graphing a translate of $y = 1/x$

Graph: $y = 1/(x - 1)$.

SOLUTION

By translating the graph of $y = 1/x$ one unit to the right, we obtain the graph of $y = 1/(x - 1)$, shown in Figure 4. Notice that the horizontal asymptote is unchanged by the translation. However, the vertical asymptote moves one unit to the right. (The y-intercept is -1; why?)

EXAMPLE 2 Two graphs obtained from that of $y = 2/x$

Graph: $y = 2/(x - 1)$ and $y = -2/(x - 1)$.

SOLUTION

The graph of $y = 2/(x - 1)$ has the same basic shape and location as the graph of $y = 1/(x - 1)$ shown in Figure 4. As a further guide to sketching $y = 2/(x - 1)$, we can pick several convenient x-values on either side of the asymptote $x = 1$ and then compute the corresponding y-values. After doing this, we obtain the graph shown in Figure 5(a). By reflecting this graph about the x-axis, we obtain the graph of $y = -2/(x - 1)$, which is shown in Figure 5(b).

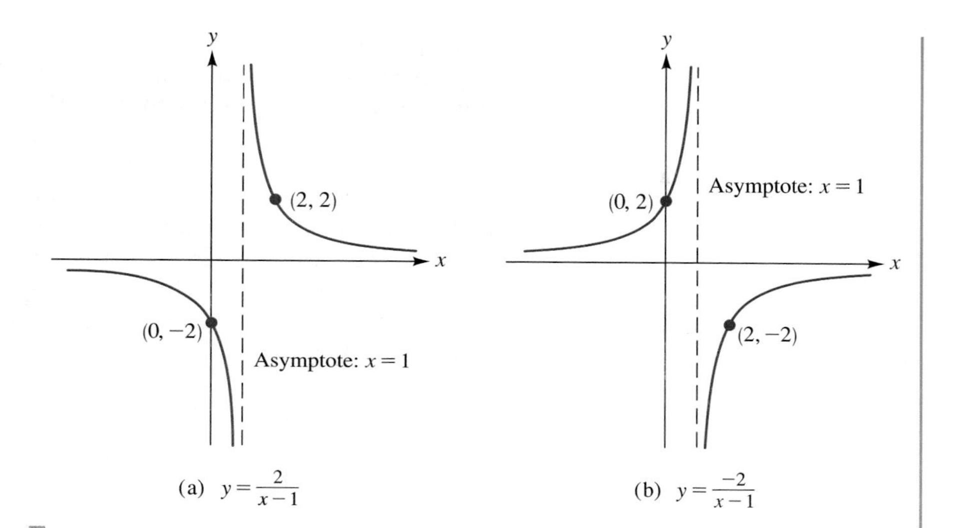

Figure 5

EXAMPLE 3 Long division as an aid in graphing

Graph: $y = \dfrac{4x - 2}{x - 1}$.

SOLUTION

First, as you can readily check, the x- and y-intercepts are $1/2$ and 2, respectively. Next, using long division, we find that

$$\frac{4x - 2}{x - 1} = 4 + \frac{2}{x - 1}$$

[If you need a review of (or reintroduction to) the long division process from basic or intermediate algebra, see Section 12.2.] We conclude that the required graph can be obtained by moving the graph of $y = 2/(x - 1)$ up four units in the y-direction; see Figure 6. Notice that the vertical asymptote is still $x = 1$, but the horizontal asymptote is now $y = 4$ instead of the x-axis.

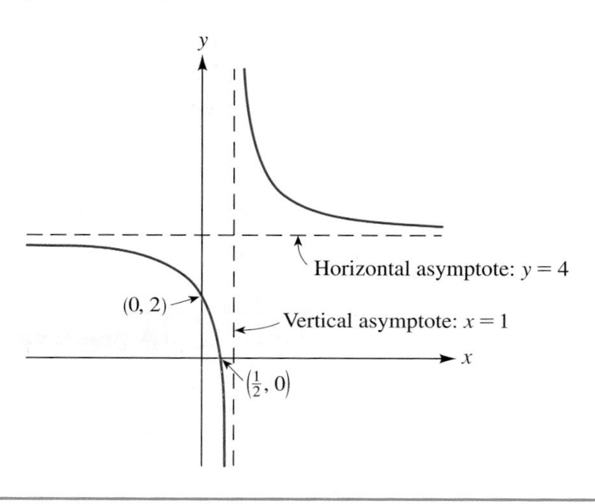

Figure 6

The graph of $y = \dfrac{4x - 2}{x - 1}$

Now let's look at rational functions of the form $y = 1/x^n$. First we'll consider $y = 1/x^2$. As with $y = 1/x$, the domain consists of all real numbers except $x = 0$. For $x \neq 0$ the quantity $1/x^2$ is always positive. This means that the graph will always lie above the x-axis. Furthermore, the graph will be symmetric about the y-axis. (Why?) As $|x|$ becomes very large, the quantity $1/x^2$ approaches zero; this is true whether x itself is positive or negative. So when $|x|$ is very large, we expect the graph of $y = 1/x^2$ to look as shown in Figure 7(a). On the other hand, when x is a very small fraction, close to zero, either negative or positive, the quantity $1/x^2$ is very large. For instance, if $x = 1/10$, we find that $y = 100$, and if $x = 1/100$, we find that $y = 10{,}000$. Thus as x approaches zero, from the right or from the left, the graph of $y = 1/x^2$ must look as shown in Figure 7(b).

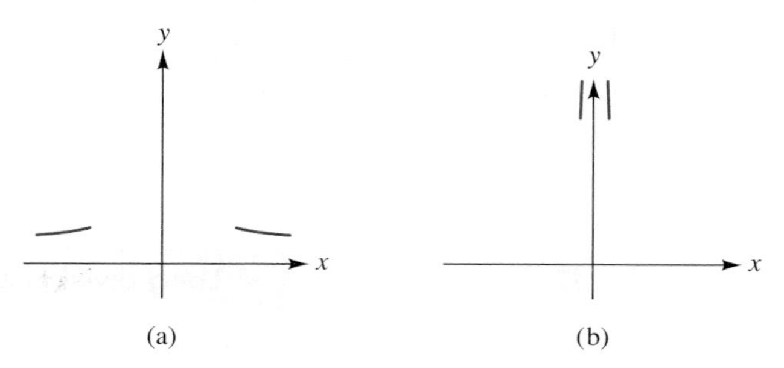

Figure 7 (a) (b)

Now, by plotting several points and taking Figures 7(a) and 7(b) into account, we obtain the graph of $y = 1/x^2$, shown in Figure 8. Also, by following a similar line of reasoning, we find that the graph of $y = 1/x^3$ looks as shown in Figure 9. Note that the graph of $y = 1/x^3$ is symmetric about the origin, and the graph of $y = 1/x^2$ is symmetric about the y-axis.

In general, when n is an even integer greater than 2, the graph of $y = 1/x^n$ resembles that of $y = 1/x^2$. When n is an odd integer greater than 3, the graph of $y = 1/x^n$ resembles that of $y = 1/x^3$.

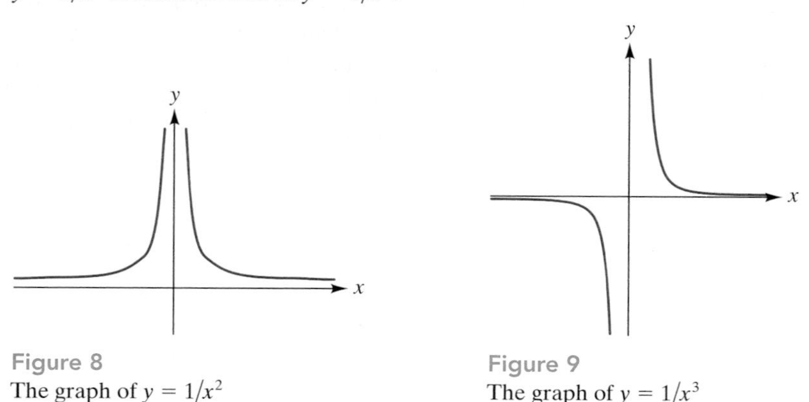

Figure 8
The graph of $y = 1/x^2$

Figure 9
The graph of $y = 1/x^3$

EXAMPLE 4 A graph based on that of $y = 1/x^2$

Graph: $y = -1/(x + 3)^2$.

SOLUTION

Refer to Figure 10. Begin with the graph of $y = 1/x^2$. By moving the graph three units to the left, we obtain the graph of $y = 1/(x + 3)^2$. Then by re-

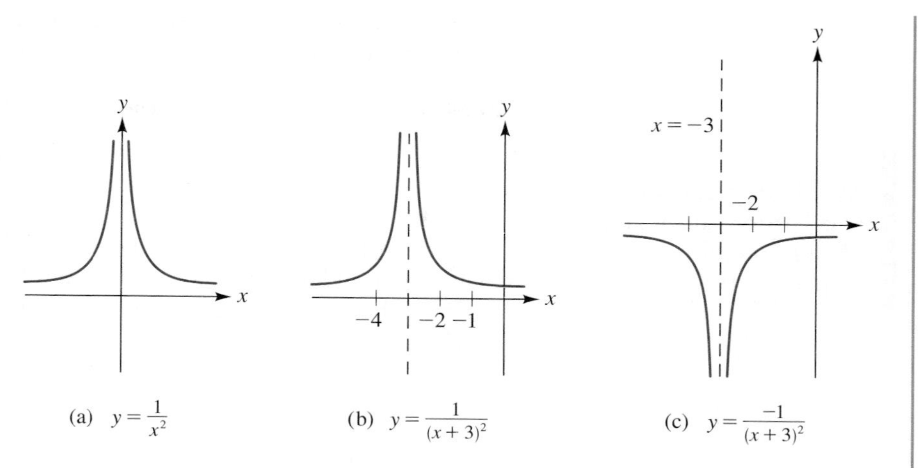

Figure 10

flecting the graph of $y = 1/(x + 3)^2$ about the x-axis, we get the graph of $y = -1/(x + 3)^2$.

Question: What are the y-intercepts in Figures 10(b) and 10(c)?

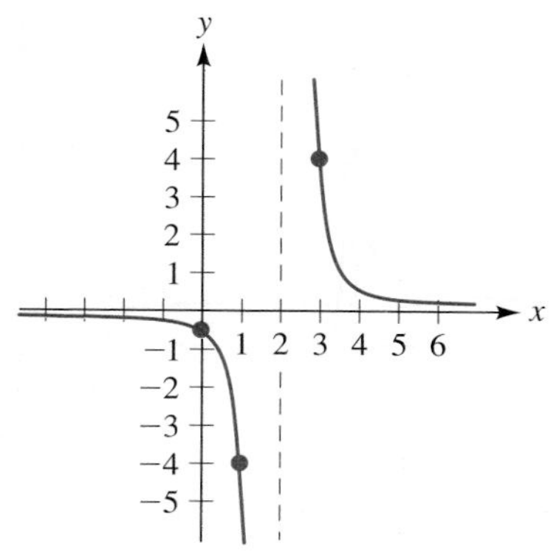

Figure 11
The graph of $y = 4/(x - 2)^3$

EXAMPLE 5 A graph based on that of $y = 1/x^3$

Graph: $y = 4/(x - 2)^3$.

SOLUTION
Moving the graph of $y = 1/x^3$ two units to the right gives us the graph of the function $y = 1/(x - 2)^3$. The graph of $y = 4/(x - 2)^3$ will have the same basic shape and location. As a further guide to sketching the required graph, we can pick several convenient x-values near the asymptote $x = 2$ and compute the corresponding y-values. Using $x = 0$, $x = 1$, and $x = 3$, we find that the points $(0, -\frac{1}{2})$, $(1, -4)$, and $(3, 4)$ are on the graph. With this information, the graph can be sketched as in Figure 11.

As we have seen in Examples 1 through 5, the vertical asymptotes of $y = f(x)/g(x)$ are found by solving the equation $g(x) = 0$, provided that f and g are polynomials with no common factors other than constants. In Example 6 we determine the horizontal asymptote by comparing the highest powers of x in the numerator and denominator.

EXAMPLE 6 Finding a horizontal asymptote

Determine the horizontal asymptote for the graph of $y = \dfrac{x^2 - 5x + 6}{2x^2 - 4x + 3}$.

SOLUTION
The strategy is to factor out the highest power of x that occurs in the numerator and do the same for the denominator. Then reduce the ratio of powers of x fac-

tored out of the numerator and denominator. Finally, determine the behavior of the remaining complex fraction as $|x|$ gets larger and larger. We have

$$y = \frac{x^2 - 5x + 6}{2x^2 - 4x + 3} = \frac{x^2\left(1 - \frac{5}{x} + \frac{6}{x^2}\right)}{x^2\left(2 - \frac{4}{x} + \frac{3}{x^2}\right)} = \frac{1 - \frac{5}{x} + \frac{6}{x^2}}{2 - \frac{4}{x} + \frac{3}{x^2}}$$

TABLE 1
As x Grows Larger and Larger, the Values of the Function Approach 0.5

x	$y = \frac{x^2 - 5x + 6}{2x^2 - 4x + 3}$
10	0.343558 . . .
10^2	0.484925 . . .
10^3	0.498499 . . .
10^4	0.499849 . . .
10^5	0.499984 . . .
10^6	0.499998 . . .

Now, in that last expression, look at the four resulting fractions appearing within the numerator and denominator of the main fraction: $5/x$, $6/x^2$, $4/x$, and $3/x^2$. The point is that as $|x|$ gets larger and larger, all four of those fractions approach zero. We therefore have

$$y = \frac{1 - \frac{5}{x} + \frac{6}{x^2}}{2 - \frac{4}{x} + \frac{3}{x^2}} \approx \frac{1 - 0 + 0}{2 - 0 + 0} = \frac{1}{2} \qquad \text{as } |x| \text{ gets larger and larger}$$

This tells us that the line $y = 1/2$ is a horizontal asymptote for the graph of the given function. See Table 1 and Figure 12 for numerical and graphical perspectives.

GRAPHICAL PERSPECTIVE

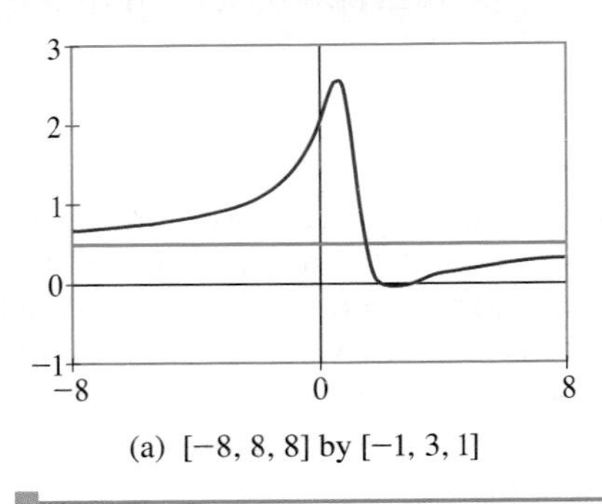

(a) [−8, 8, 8] by [−1, 3, 1]

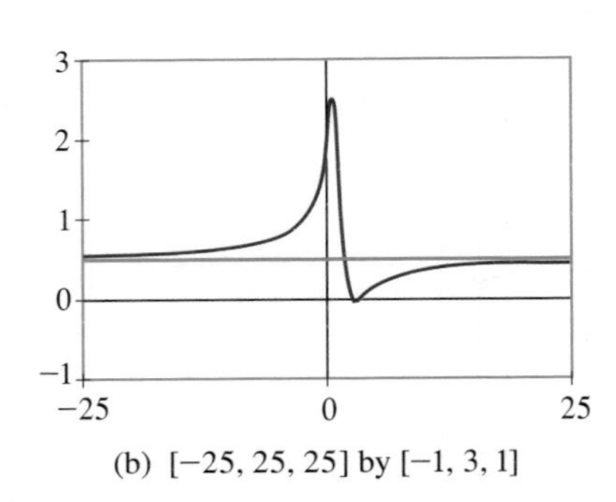

(b) [−25, 25, 25] by [−1, 3, 1]

Figure 12
Two views of the rational function $y = \frac{x^2 - 5x + 6}{2x^2 - 4x + 3}$ (red) with horizontal asymptote $y = 1/2$ (blue)

EXAMPLE 7 Using rational functions to model bacterial growth

A group of agricultural scientists has been studying how the growth of a particular type of bacteria is affected by the acidity level of the soil. One colony of the bacteria is placed in a soil that is slightly acidic. A second colony of the same size is placed in a neutral soil. Suppose that after analyzing the data, the scientists determine that the size of each population over time can be modeled by the following functions.

colony in neutral soil: $y = (2t + 1)/(t + 1) \quad t \geq 0$
colony in acidic soil: $y = (4t + 3)/(t^2 + 3) \quad t \geq 0$

In both cases, y represents the population, in thousands, after t hours. Figure 13 shows the graphs of these two population functions over the first hour of the

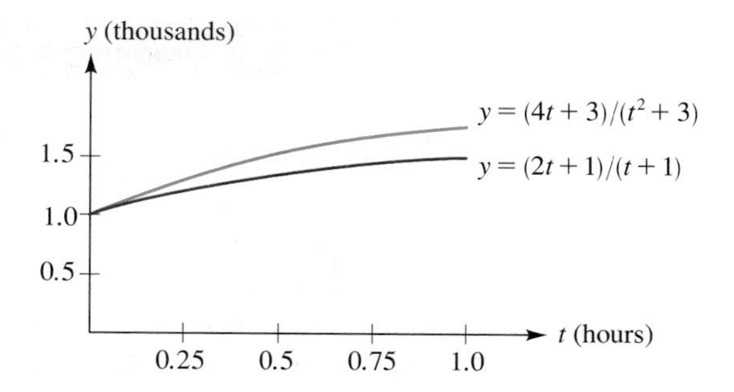

Figure 13
Population models for two colonies of bacteria

experiment. (Note from both the graphs and the equations that each colony begins with 1000 bacteria.) Find the horizontal asymptotes for each graph and thereby determine the long-term behavior of each colony.

SOLUTION

Colony in neutral soil:

$$y = \frac{2t+1}{t+1}$$

$$= \frac{t(2+1/t)}{t(1+1/t)} \quad \text{factoring out } t \text{ from the top and bottom}$$

$$= \frac{2+1/t}{1+1/t}$$

$$\approx \frac{2+0}{1+0} = 2 \quad \text{as } t \text{ gets larger and larger}$$

Colony in acidic soil:

$$y = \frac{4t+3}{t^2+3}$$

$$= \frac{t^2(4/t+3/t^2)}{t^2(1+3/t^2)} \quad \text{factoring out } t^2 \text{ from the top and bottom}$$

$$= \frac{4/t+3/t^2}{1+3/t^2}$$

$$\approx \frac{0+0}{1+0} = 0 \quad \text{as } t \text{ gets larger and larger}$$

For the colony in neutral soil, the horizontal asymptote for the growth function is $y = 2$. So in the long run, this colony approaches a population of 2000. For the colony in acidic soil, the horizontal asymptote is $y = 0$ and the population approaches extinction (or becomes extinct, depending on the interpretation). See Figure 14.

GRAPHICAL PERSPECTIVE

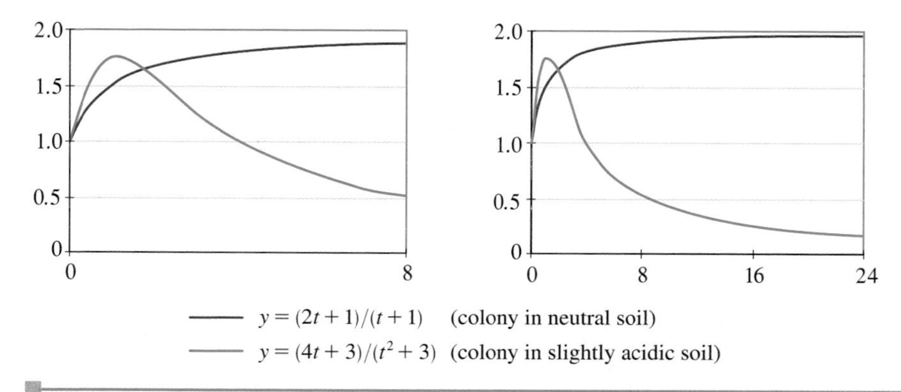

Figure 14
Snapshots of the population functions after $t = 8$ hr and $t = 24$ hr. The vertical axis is the population y, in thousands.

EXAMPLE 8 Graphing a rational function

Graph the function $y = \frac{x^2 - x - 6}{x^2 - x - 2} = \frac{(x-3)(x+2)}{(x+1)(x-2)}$.

SOLUTION

By looking at the factored form of the fraction, we see that the x-intercepts are 3 and -2. Also, the vertical asymptotes are the lines $x = -1$ and $x = 2$. For the y-intercept it's a bit simpler to set $x = 0$ in the unfactored, rather than factored, expression. As you can check, the result in either case is that the y-intercept is 3. The horizontal asymptote works out to be the line $y = 1$; you should verify this for yourself using the method demonstrated in the previous two examples.

Now we want to see how the graph looks in the immediate vicinity of the x-intercepts and the vertical asymptotes. To do this, we use the approximation technique explained in the previous section for polynomial functions. Let's start with the x-intercept at $x = 3$. As in Section 4.6, we'll retain the factor $x - 3$ and approximate the remaining factors using $x = 3$. So for x near 3 we have

$$y = \frac{(x-3)(x+2)}{(x+1)(x-2)} \approx \frac{(x-3)(3+2)}{(3+1)(3-2)} = \frac{5}{4}(x-3)$$

Thus in the immediate vicinity of the x-intercept $x = 3$, the required graph will closely resemble the line $y = \frac{5}{4}(x-3)$. The remaining calculations for approximating the graph near the other x-intercept and near the two vertical asymptotes are carried out in exactly the same manner. As Exercise 43 will ask you to verify, the results are

x near 3: $\quad y \approx \frac{5}{4}(x-3)$	x near -1: $\quad y \approx \frac{4/3}{x+1}$
x near -2: $\quad y \approx -\frac{5}{4}(x+2)$	x near 2: $\quad y \approx \frac{-4/3}{x-2}$

We summarize these results as follows. As the graph passes through the points $(3, 0)$ and $(-2, 0)$, it resembles the lines $y = \frac{5}{4}(x-3)$ and $y = -\frac{5}{4}(x+2)$, respectively. Near the vertical asymptote $x = -1$, the graph has the same basic shape as $y = 1/(x+1)$, and near the vertical asymptote $x = 2$, the graph resembles $y = -1/(x-2)$; see Figure 15.

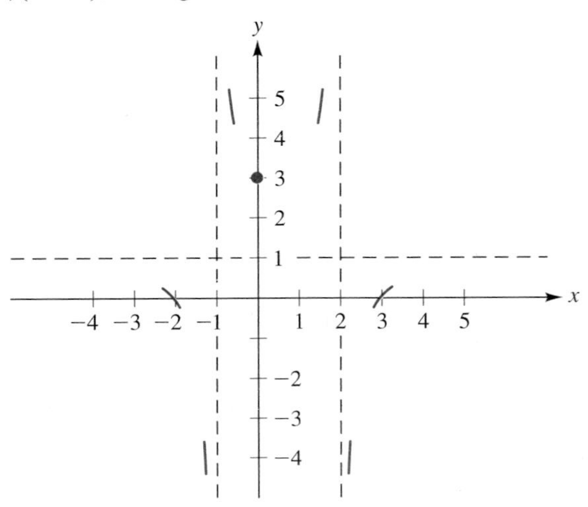

Figure 15

The graph of $y = \frac{(x-3)(x+2)}{(x+1)(x-2)}$ near its x-intercepts and vertical asymptotes

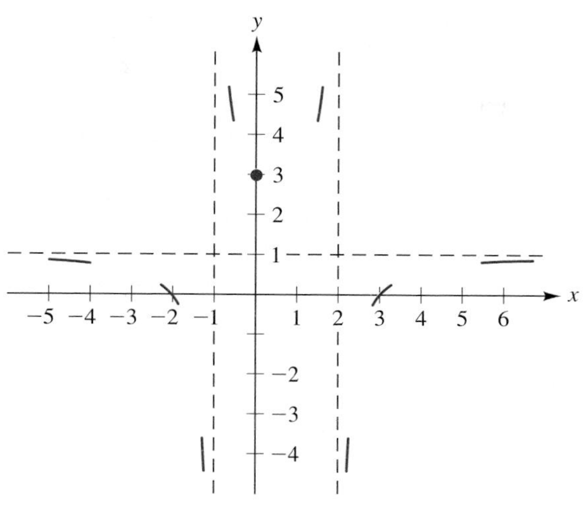

Figure 16

The graph of $y = \dfrac{(x-3)(x+2)}{(x+1)(x-2)}$ near x-intercepts and its horizontal and vertical asymptotes

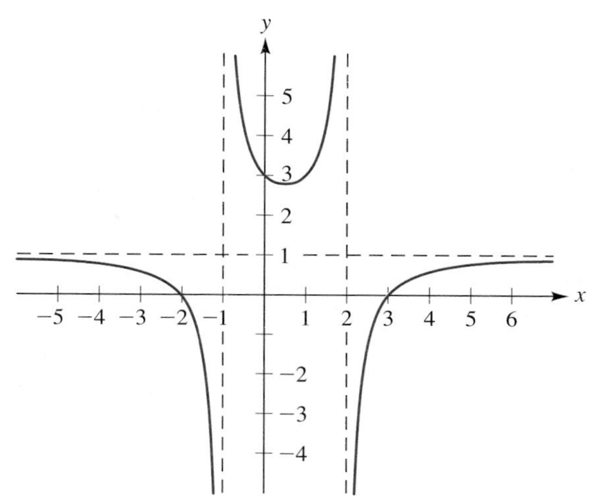

Figure 17

The graph of $y = \dfrac{(x-3)(x+2)}{(x+1)(x-2)}$

TABLE 2
When $|x|$ Is Large, y Is Less Than 1

x	y
10	0.95
100	0.99
−10	0.96
−100	0.9996

Finally, we want to find out how the graph approaches the horizontal asymptote $y = 1$. Perhaps the simplest way is to do some calculations. In Table 2 we have computed the outputs for some relatively large values of $|x|$. From the table we see that when $|x|$ is large, y is less than 1. Graphically, this means that the curve approaches the asymptote $y = 1$ from below. In Figure 16 we summarize what we have discovered up to this point. Then, using Figure 16 as a guide, we can sketch the required graph, as shown in Figure 17.

NOTE In this example the coordinates of the lowest point on the graph in the interval $-1 < x < 2$ can be found by applying algebraic techniques. See Exercise 46 for the details.

EXERCISE SET 4.7

A

In Exercises 1–8:

(a) *Find the domain, x- and y-intercepts, vertical asymptotes, and horizontal asymptotes for each rational function.*

Ⓖ **(b)** *Use a graphing utility to graph the function. Check to see that the graph is consistent with your results in part (a).*

1. $y = (3x + 15)/(4x - 12)$
2. $y = (x + 6)(x + 4)/(x - 1)^2$
3. $y = (6x^2 - 5x + 1)/(2x^2)$
4. $y = x/(x^2 + x + 1)$
5. $y = (x^2 - 9)/(4x^2 - 1)$
6. $y = (x^2 - 9)/(4x^2 + 1)$
7. $y = (3x^2 - 2x - 8)/(2x^3 + x^2 - 3x)$
8. $y = (x^3 - 27)/(x^4 - 2x^3 + 9x^2 - 18x)$

In Exercises 9–34, sketch the graph of each rational function. Specify the intercepts and the asymptotes.

9. $y = 1/(x + 4)$
10. $y = -1/(x + 4)$
11. $y = 3/(x + 2)$
12. $y = -3/(x + 2)$
13. $y = -2/(x - 3)$
14. $y = (x - 1)/(x + 1)$
15. $y = (x - 3)/(x - 1)$
16. $y = 2x/(x + 3)$
17. $y = (4x - 2)/(2x + 1)$
18. $y = (3x + 2)/(x - 3)$
19. $y = 1/(x - 2)^2$
20. $y = -1/(x - 2)^2$
21. $y = 3/(x + 1)^2$
22. $y = -3/(x + 1)^2$
23. $y = 1/(x + 2)^3$
24. $y = -1/(x + 2)^3$
25. $y = -4/(x + 5)^3$
26. $y = x/[(x + 1)(x - 3)]$
27. $y = -x/[(x + 2)(x - 2)]$
28. $y = 2x/(x + 1)^2$
29. **(a)** $y = 3x/[(x - 1)(x + 3)]$
(b) $y = 3x^2/[(x - 1)(x + 3)]$

30. **(a)** $y = (4x^2 + 1)/(x^2 - 1)$
(b) $y = (4x^2 + 1)/(x^2 + 1)$

31. $y = (4x^2 + x - 5)/(2x^2 - 3x - 5)$

32. $y = (4x^2 - x - 3)/(2x^2 - 3x - 5)$

Ⓖ **33.** **(a)** $f(x) = (x - 2)(x - 4)/[x(x - 1)]$
(b) $g(x) = (x - 2)(x - 4)/[x(x - 3)]$
[Compare the graphs you obtain in parts (a) and (b). Notice how a change in only one constant can radically alter the nature of the graph.]

Ⓖ **34.** **(a)** $f(x) = (x - 1)(x + 2.75)/[(x + 1)(x + 3)]$
(b) $g(x) = (x - 1)(x + 3.25)/[(x + 1)(x + 3)]$
[Compare the graphs you obtain in parts (a) and (b). Notice how a relatively small change in one of the constants can radically alter the graph.]

35. The population y of a colony of bacteria after t hr is given by

$$y = (t + 12)/(0.0004t + 0.024) \qquad \text{where } t \geq 0$$

(a) Find the initial population (that is, the population when $t = 0$ hr).
(b) Determine the long-term behavior of the population (as in Example 7).

36. The population y (in thousands) of a colony of bacteria after t hr is given by

$$y = (6t + 12)/(2t + 1) \qquad \text{where } t \geq 0$$

(a) Find the initial population and the long-term population. Which is larger?
Ⓖ **(b)** Use a graphing utility to graph the population function. Is the function increasing or decreasing? Check that your response here is consistent with your answers in part (a).

B

37. A desktop publisher designing a small rectangular poster decides to make the area 500 in.² The margins on the top and bottom of the poster are to be 3 in. and 4 in., respectively. The left and right margins are each to be 1.5 in.

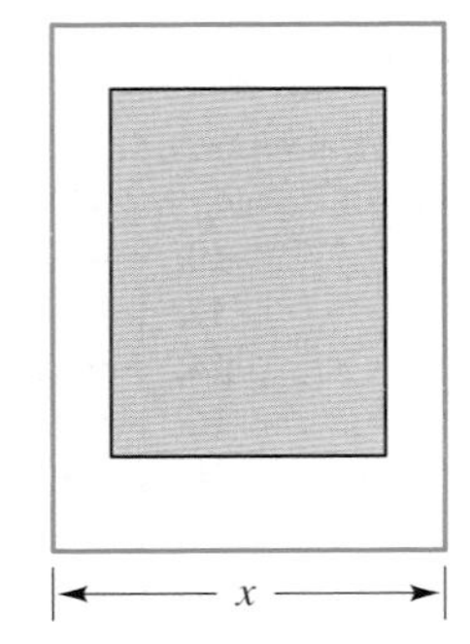

(a) Express the area of the printed portion of the poster as a function of x, the width of the entire poster. (The gray region in the figure represents the printed area.)
Ⓖ **(b)** Use a graphing utility to graph the area function that you found in part (a).
Ⓖ **(c)** Use a ZOOM feature to estimate (to the nearest one-tenth) the width x for which the printed area is a maximum. What is the corresponding length of the poster in this case?

38. The accompanying figure gives formulas for the volume V and total surface area S of a circular cylinder with radius r and height h. For a cylinder of given volume we are interested in finding the dimensions that minimize the surface area.

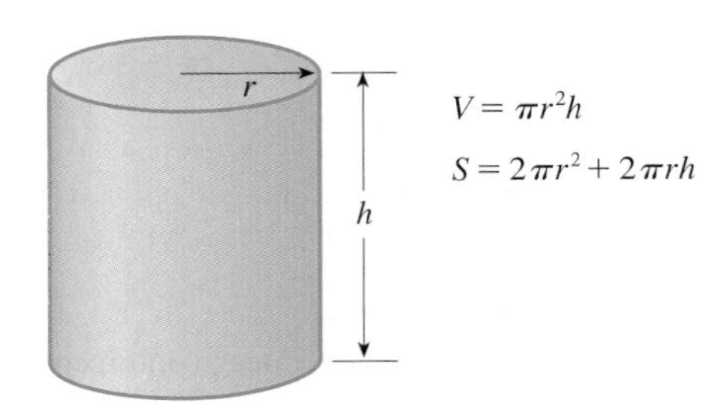

(a) Assume that the volume of the cylinder is 1000 cm³. Express the surface area as a function of the radius. After combining terms in your answer, show that the resulting function can be written

$$S(r) = \frac{2\pi r^3 + 2000}{r} \qquad (r > 0)$$

Ⓖ **(b)** Use a graphing utility to graph the surface area function obtained in part (a).
Ⓖ **(c)** Estimate, to the nearest one-hundredth, the radius r that minimizes the surface area. What is the corresponding value for h?

In Exercises 39–42, graph the functions. Note: In each case, the graph crosses its horizontal asymptote once. To find the point where the rational function $y = f(x)$ crosses its horizontal asymptote $y = k$, you'll need to solve the equation $f(x) = k$.

39. $y = \dfrac{(x - 4)(x + 2)}{(x - 1)(x - 3)}$ **40.** $y = \dfrac{(x - 1)(x - 3)}{(x + 1)^2}$

41. $y = \dfrac{(x + 1)^2}{(x - 1)(x - 3)}$ **42.** $y = \dfrac{2x^2 - 3x - 2}{x^2 - 3x - 4}$

43. (This exercise refers to Example 8.) Let

$$y = \frac{(x - 3)(x + 2)}{(x + 1)(x - 2)}$$

Verify each of the following approximations.
(a) When x is close to -2, then $y \approx -\frac{5}{4}(x + 2)$.
(b) When x is close to -1, then $y \approx \dfrac{4/3}{x + 1}$.
(c) When x is close to 2, then $y \approx \dfrac{-4/3}{x - 2}$.

In Exercises 44 and 45, graph the functions. Notice in each case that the numerator and denominator contain at least one common factor. Thus you can simplify each quotient; but don't lose track of the domain of the function as it was initially defined.

44. **(a)** $y = \dfrac{x+2}{x+2}$ **(c)** $y = \dfrac{x-1}{(x-1)(x-2)}$

(b) $y = \dfrac{x^2-4}{x-2}$

45. **(a)** $y = \dfrac{x^2-9}{x+3}$

(b) $y = \dfrac{x^2-5x+6}{x^2-2x-3}$

(c) $y = \dfrac{(x-1)(x-2)(x-3)}{(x-1)(x-2)(x-3)(x-4)}$

46. This exercise shows you how to determine the coordinates of the lowest point on the middle branch of the curve in Figure 17. The basic idea is as follows. Suppose that the required coordinates are (h, k). Then the horizontal line $y = k$ is the unique horizontal line intersecting the curve in one and only one point. (Any other horizontal line intersects the curve either in two points or not at all.) In steps (a) through (c) that follow, we use these observations to determine the point (h, k).

(a) Given any horizontal line $y = k$, its intersection with the curve in Figure 17 is determined by solving the following pair of simultaneous equations:

$$\begin{cases} y = k \\ y = \dfrac{x^2-x-6}{x^2-x-2} \end{cases}$$

In the second equation of the system, replace y with k and show that the resulting equation can be written

$$(k-1)x^2 - (k-1)x + (6-2k) = 0 \quad (1)$$

(b) If k is indeed the required y-coordinate, then equation (1) must have exactly one real solution. Set the discriminant of the quadratic equation equal to zero to obtain

$$(k-1)^2 - 4(k-1)(6-2k) = 0$$

and deduce from this that $k = 1$ or $k = 25/9$. The solution $k = 1$ can be discarded. (To see why, look at Figure 17.)

(c) Using the value $y = 25/9$, show that the corresponding x-coordinate is $1/2$. Thus, the required point is $(\frac{1}{2}, \frac{25}{9})$.

47. Graph the function $y = x/(x-3)^2$. Use the technique explained in Exercise 46 to find the coordinates of any turning points on the graph.

48. Graph the function $y = 2/(x - x^2)$. Use the technique explained in Exercise 46 to find the coordinates of any turning points on the graph.

An asymptote that is neither horizontal nor vertical is called a ***slant*** *or* ***oblique asymptote.*** *For example, as indicated in the following figure, the line $y = x$ is a slant asymptote for the graph of $y = (x^2+1)/x$. To understand why the line $y = x$ is an asymptote, we carry out the indicated division and write the function in the form*

$$y = x + \frac{1}{x} \quad (2)$$

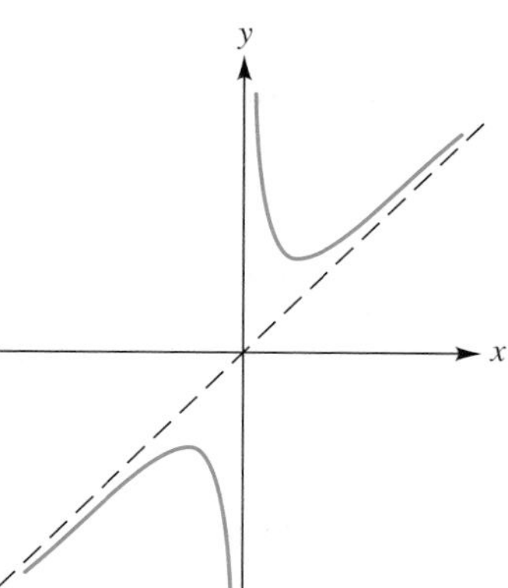

From equation (2) we see that if $|x|$ is very large then $y \approx x + 0$; that is, $y \approx x$, as we wished to show. Equation (2) actually tells us more than this. When $|x|$ is very close to zero, equation (2) yields $y \approx 0 + (1/x)$. In other words, as we approach the y-axis, the curve looks more and more like the graph of $y = 1/x$. In general, if we have a rational function $f(x)/g(x)$ in which the degree of $f(x)$ is 1 greater than the degree of $g(x)$, then the graph has a slant asymptote that is obtained as follows. Divide $f(x)$ by $g(x)$ to obtain an equation of the form

$$\frac{f(x)}{g(x)} = (mx + b) + \frac{h(x)}{g(x)}$$

where the degree of $h(x)$ is less than the degree of $g(x)$. Then the equation of the slant asymptote is $y = mx + b$. (For instance, using the previous example, we have $y = mx + b = x$ and $h(x)/g(x) = 1/x$.) In Exercises 49–51, you are asked to graph functions that have slant asymptotes.

49. Let $y = F(x) = \dfrac{x^2+x-6}{x-3}$.

(a) Use long division to show that

$$\frac{x^2+x-6}{x-3} = (x+4) + \frac{6}{x-3}$$

(b) The result in part (a) shows that the line $y = x + 4$ is a slant asymptote for the graph of the function F. Verify this fact empirically by completing the following two tables.

x	$x + 4$	$\dfrac{x^2+x-6}{x-3}$
10		
100		
1000		

(continues)

x	$x + 4$	$\dfrac{x^2 + x - 6}{x - 3}$
-10		
-100		
-1000		

(c) Determine the vertical asymptote and the x- and y-intercepts of the graph of F.

(d) Graph the function F. (Use the techniques in this section along with the fact that $y = x + 4$ is a slant asymptote.)

(e) Use the technique explained in Exercise 46 to find the coordinates of the two turning points on the graph of F.

50. (a) Show that the line $y = x - 2$ is a slant asymptote for the graph of $F(x) = x^2/(x + 2)$.

(b) Sketch the graph of F.

51. Show that the line $y = -x$ is a slant asymptote for the graph of $y = (1 - x^2)/x$. Then sketch the graph of this function.

52. Let $f(x) = (x^3 + 2x^2 + 1)/(x^2 + 2x)$.

Ⓖ **(a)** Graph the function f using a viewing rectangle extending from -5 to 5 in both the x- and y-directions.

Ⓖ **(b)** Add the graph of the line $y = x$ to your picture in part (a). Note that to the right of the origin, as x increases, the graph of f begins to look more and more like the line $y = x$. This also occurs to the left of the origin as x decreases.

(c) The results in part (b) suggest that the line $y = x$ may be an asymptote for the graph of f. Verify this visually by changing the viewing rectangle so that it extends from -20 to 20 in both the x- and the y-directions. What do you observe?

(d) Using algebra, verify the identity

$$\frac{x^3 + 2x^2 + 1}{x^2 + 2x} = \frac{1}{x^2 + 2x} + x$$

Then explain why, for large values of $|x|$, the graph of f looks more and more like the line $y = x$. *Hint*: Substitute some large numbers (such as 100 or 1000) into the expression $1/(x^2 + 2x)$. What happens?

53. Let $f(x) = (x^5 + 1)/x^2$.

Ⓖ **(a)** Graph the function f using a viewing rectangle that extends from -4 to 4 in the x-direction and from -8 to 8 in the y-direction.

Ⓖ **(b)** Add the graph of the curve $y = x^3$ to your picture in part (a). Note that as $|x|$ increases (that is, as x moves away from the origin), the graph of f looks more and more like the curve $y = x^3$. For additional perspective, first change the viewing rectangle so that y extends from -20 to 20. (Retain the x-settings for the moment.) Describe what you see. Next, adjust the viewing rectangle so that x extends from -10 to 10 and y extends from -100 to 100. Summarize your observations.

(c) In the text we said that a line is an asymptote for a curve if the distance between the line and the curve approaches zero as we move further and further out along the curve. The work in part (b) illustrates that a *curve* can behave like an asymptote for another curve. In particular, part (b) illustrates that the distance between the curve $y = x^3$ and the graph of the given function f approaches zero as we move further and further out along the graph of f. That is, the curve $y = x^3$ is an "asymptote" for the graph of the given function f. Complete the following two tables for a numerical perspective on this. In the tables, d denotes the vertical distance between the curve $y = x^3$ and the graph of f:

$$d = \left|\frac{x^5 + 1}{x^2} - x^3\right|$$

x	5	10	50	100	500
d					

x	-5	-10	-50	-100	-500
d					

(d) Parts (b) and (c) have provided both a graphical and a numerical perspective. For an algebraic perspective that ties together the previous results, verify the following identity, and then use it to explain why the results in parts (b) and (c) were inevitable:

$$\frac{x^5 + 1}{x^2} = x^3 + \frac{1}{x^2}$$

PROJECT

FINDING SOME MINIMUM VALUES WITHOUT USING CALCULUS

In this project you'll solve some minimum problems that involve rational functions. The theorem that we'll rely on is given in the box that follows. Methods for establishing this theorem are indicated in Question 3 at the end of this project. (This theorem can be stated in a more general form for n positive quantities, rather than two or three, but we won't need that for the applications in this project.)

THEOREM On Minimum Sums

(a) If the product of two positive quantities s and t is constant, then their sum $s + t$ is a minimum if and only if $s = t$.

(b) If the product of three positive quantities s, t, and u is constant, then their sum $s + t + u$ is a minimum if and only if $s = t = u$.

First, just to get a feeling for what part (a) of the theorem is saying, complete the following two tables, and in each case note which pair of numbers a and b yields the smallest sum.

a	b	ab	$a + b$
1	36		
2	18		
3	12		
4	9		
6	6		

a	b	ab	$a + b$
1	12		
2	6		
3	4		
$3\frac{1}{2}$	$3\frac{3}{7}$		
$2\sqrt{3}$	$2\sqrt{3}$		

1. Use part (a) of the theorem to solve the following problems.

(a) Find the smallest possible value for the sum of a positive number and its reciprocal.

(b) The accompanying figure shows a line with slope m $(m < 0)$ passing through the point (2, 1). Find the smallest possible area for the triangle that is formed. *Hint*: First use the point–slope formula to write the equation of a line with slope m that passes through the point (2, 1). Next, find the x- and y-intercepts of the line in terms of m. Then express the area of the triangle in terms of m and find a way to apply part (a) of the theorem.

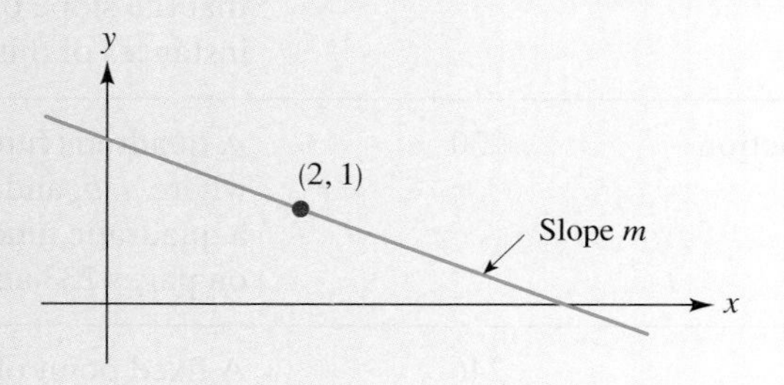

2. Use part (b) of the theorem to solve the following problems.

(a) After using a graphing utility to estimate the minimum value of the function $f(x) = x^2 + \frac{1}{x}$ for $x > 0$, find the exact value for this minimum. *Hint*: Write $x^2 + \frac{1}{x}$ as $x^2 + \frac{1}{2x} + \frac{1}{2x}$.

(b) Use a graphing utility to estimate the minimum value of the function $g(x) = (2x^3 + 3)/x$ for $x > 0$. Then find the exact minimum. (Adapt the hint in the previous exercise.)

(c) A box with a square base and no top is to be constructed. The volume is to be 27 ft^3. Find the dimensions of the box so that the surface area is a minimum.

(d) In Exercise 38 on page 314, a graphing utility is used to estimate the minimum possible surface area for a circular cylinder with a volume 1000 cm^3. Find the exact value for this minimum. Also confirm the following fact: For the values of r and h that yield the minimum surface area, we have $2r = h$.

3. (a) *Proof for part (a) of the theorem:* Verify that the following simple identity holds for all real numbers a and b.

$$(a + b)^2 = 4ab + (a - b)^2$$

Then use this identity to explain why part(a) of the theorem is valid. *Suggestion*: First discuss the reasoning within a group; then, on your own, write a paragraph carefully explaining the reasoning in your own words.

(b) *Proof for part (b) of the theorem:* Use the result in Exercise 44 of Section 2.3 to explain why part (b) of the theorem is valid.

Chapter 4 Summary of Principal Terms

Terms	Page reference	Comments
1. Linear function	213	A linear function is a function of the form $f(x) = Ax + B$, where A and B are constants. The graph of a linear function is a straight line. An important idea that arose in several of the examples is that the slope of a line can be interpreted as a rate of change. Two instances of this are marginal cost and velocity.
2. Quadratic function	230	A quadratic function is a function of the form $f(x) = ax^2 + bx + c$, where a, b, and c are constants and a is not zero. The graph of a quadratic function is a parabola. See the Property Summaries on pages 233 and 235.
3. Fixed point	246	A fixed point of a function f is an input x in the domain of f such that $f(x) = x$. For example, $x = 1$ is a fixed point for $f(x) = \sqrt{x}$ because $f(1) = \sqrt{1} = 1$.

Terms	Page reference	Comments
4. Vertex formula	273	The x-coordinate of the vertex of the parabola $y = ax^2 + bx + c$ is given by $x = -b/2a$.
5. Polynomial function	288	A polynomial function is a function of the form $$f(x) = a_n x^n + a_{n-1}x^{n-1} + \cdots + a_1 x + a_0$$ where n is a nonnegative integer and $a_0, a_1, \ldots,$ and a_n are constants. Three basic properties of polynomial functions are summarized in the box on page 293. If $a_n \neq 0$, the degree of the polynomial is n, the largest exponent on the input variable.
6. Rational function	304	A rational function is a function of the form $y = f(x)/g(x)$, where $f(x)$ and $g(x)$ are polynomials.
7. Asymptote	305	A line is said to be an asymptote for a curve if the distance between the line and the curve approaches zero as we move out farther and farther along the line.

Writing Mathematics

1. At the start of Section 4.6 is a quotation by David Berlinski describing polynomial functions. Reread the quotation, and then write a paragraph or two explaining its contents using the ideas and terminology from Section 4.6. With the aid of a graphing utility, use examples as appropriate to illustrate your points.

2. Consider the following two problems.

PROBLEM 1

The perimeter of a rectangle is 2 m.

(i) Express the area as a function of the width w.

(ii) Find the maximum possible area.

PROBLEM 2

The area of a rectangle is 2 m^2.

(i) Express the perimeter as a function of the width w.

(ii) Find the minimum possible perimeter.

(a) After working out Problem 1 for yourself, write out the solution in complete sentences, as if you were explaining it to a classmate or to your instructor. Be sure to let the reader know where you are headed and why each of the main steps is necessary.

(b) After working out part (i) of Problem 2 for yourself, write out the solution to part (i) in complete sentences.

(c) Explain why the methods of Section 4.5 are not applicable for solving part (ii) of Problem 2.

(d) The following result is known as the *arithmetic-geometric mean inequality:*

For all nonnegative real numbers a and b we have $a + b \geq 2\sqrt{ab}$, with equality holding if and only if $a = b$.

By working with classmates or your instructor, find a way to apply this result in part (ii) of Problem 2. Then write out your solution in complete sentences. How does your final answer here compare with that in Problem 1?

Chapter 4 Review Exercises

1. Find $G(0)$ if G is a linear function such that $G(1) = -2$ and $G(-2) = -11$.
2. **(a)** Let $f(x) = 3x^2 + 6x - 10$. For which input x is the value of the function a minimum? What is that minimum value?
   (b) Let $g(t) = 6t^2 - t^4$. For which input t is the value of the function a maximum?
3. Suppose the function $p = -\frac{1}{8}x + 100 \quad (0 \leq x \leq 12)$ relates the selling price p of an item to the quantity x that is sold. Assume that p is in dollars. What is the maximum revenue possible in this situation?
4. Graph the function $y = -1(x + 1)^3$.
5. Graph the function $y = (x - 4)(x - 1)(x + 1)$.
6. Graph the function $f(x) = x^2 + 4x - 5$. Specify the vertex, the x- and y-intercepts, and the axis of symmetry.
7. A factory owner buys a new machine for \$1000. After five years, the machine has a salvage value of \$100. Assuming linear depreciation, find a formula for the value V of the machine after t yr, where $0 \leq t \leq 5$.
8. Graph the function $y = 2(x - 3)^4$. Does the graph cross the y-axis? If so, where?
9. Graph the function $y = (3x + 5)/(x + 2)$. Specify all intercepts and asymptotes.
10. What is the largest area possible for a right triangle in which the sum of the lengths of the two shorter sides is 12 cm?
11. Graph the function $y = x/[(x + 2)(x - 4)]$.
12. Let $P(x, y)$ be a point [other than $(-1, -1)$] on the graph of $f(x) = x^3$. Express the slope of the line passing through the points P and $(-1, -1)$ as a function of x. Simplify your answer as much as possible.
13. A rectangle is inscribed in a circle. The circumference of the circle is 12 cm. Express the perimeter of the rectangle as a function of its width w.
14. Give a reason why each of the following two graphs cannot represent a polynomial function with highest-degree term $-\frac{1}{3}x^3$.

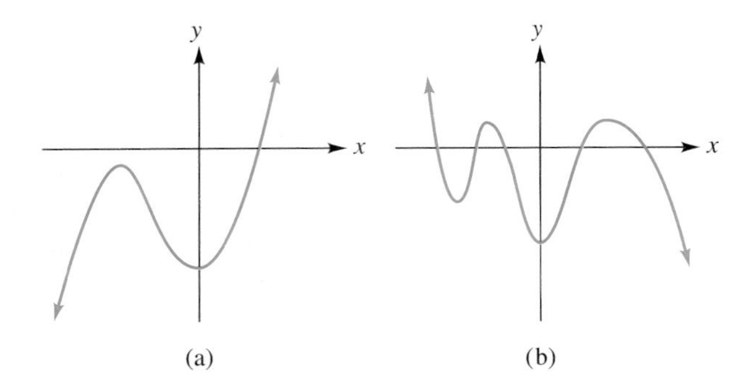

In Exercises 15–18, find equations for the linear functions satisfying the given conditions. Write each answer in the form $f(x) = mx + b$.

15. $f(4) = -1$ and the graph of f is parallel to the line $3x - 8y = 16$.
16. The graph passes through $(6, 1)$ and the x-intercept is twice the y-intercept.
17. $f(-3) = 5$, and the graph of the inverse function passes through $(2, 1)$.
18. The graph of f passes through the vertices of the two parabolas $y = x^2 + 4x + 1$ and $y = \frac{1}{2}x^2 + 9x + \frac{81}{2}$.

In Exercises 19–22, graph the quadratic functions. In each case, specify the vertex and the x- and y-intercepts.

19. $y = x^2 + 2x - 3$ **20.** $f(x) = x^2 - 2x - 15$

21. $f(x) = 2x^2 - 2x + 1$ **22.** $y = -3x^2 + 12x$

23. Find the distance between the vertices of the two parabolas $y = x^2 - 4x + 6$ and $y = -x^2 - 4x - 5$.

24. Find the value of a, given that the maximum value of the function $f(x) = ax^2 + 3x - 4$ is 5.

25. Suppose that an object is thrown vertically upward (from ground level) with an initial velocity of v_0 ft/sec. It can be shown that the height h (in feet) after t sec is given by the formula $h = v_0 t - 16t^2$.

(a) At what time does the object reach its maximum height? What is that maximum height?

(b) At what time does the object strike the ground?

26. Let $f(x) = 4x^2 - x + 1$ and $g(x) = (x - 3)/2$.

(a) For which input will the value of the function $f \circ g$ be a minimum?

(b) For which input will the value of $g \circ f$ be a minimum?

27. Find all values of b such that the minimum distance from the point $(2, 0)$ to the line $y = \frac{4}{3}x + b$ is 5. *Hint*: Use the formula from Exercise 94 on page 77.

28. What number exceeds one-half its square by the greatest amount?

29. Suppose that $x + y = \sqrt{2}$. Find the minimum value of the quantity $x^2 + y^2$.

30. For which numbers t will the value of $9t^2 - t^4$ be as large as possible?

31. Find the maximum area possible for a right triangle with a hypotenuse of 15 cm. *Hint*: Let x denote the length of one leg. Show that the area is $A = x\sqrt{225 - x^2}/2$. Now work with A^2.

32. For which point (x, y) on the curve $y = 1 - x^2$ is the sum $x + y$ a maximum?

33. Let $f(x) = x^2 - (a^2 + 2a)x + 2a^3$, where $0 < a < 2$. For which value of a will the distance between the x-intercepts of the graph of $y = f(x)$ be a maximum?

34. Suppose that the function $p = 160 - \frac{1}{5}x$ relates the selling price p of an item to the quantity x that is sold. Assume that p is in dollars. For which value of x will the revenue R be a maximum? What is the selling price p in this case?

35. A piece of wire 16 cm long is cut into two pieces. Let x denote the length of the first piece and $16 - x$ the length of the second. The first piece is formed into a rectangle in which the length is twice the width. The second piece of wire is also formed into a rectangle, but with the length three times the width. For which value of x is the total area of the two rectangles a minimum?

36. Find all real numbers that are fixed points of the given functions.

(a) $f(x) = x^2 - 8$ **(c)** $y = 4x - x^3$

(b) $g(x) = x^2 + 8$ **(d)** $y = 8x^2 - x - 15$

37. In the following figure, $PQRS$ is a square with sides parallel to the coordinate axes. The coordinates of points A and B are $A(a, 0)$ and $B(b, 0)$. Show that $f(a) = b$ and $f(b) = a$.

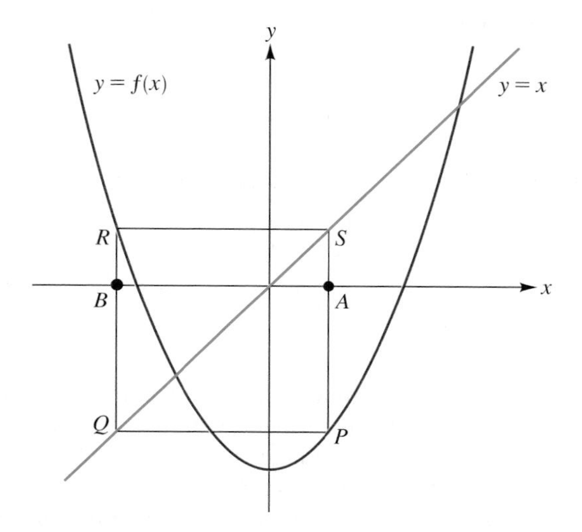

38. Let $f(x) = x^2 - 2$. Find two numbers a and b $(a \neq b)$ such that $f(a) = b$ and $f(b) = a$. *Hint*: To simplify the algebra, see part (a) of Exercise 35 on page 259.

In Exercises 39–50, graph each function and specify the x- and y-intercepts and asymptotes, if any.

39. $y = (x + 4)(x - 2)$ **40.** $y = (x + 4)(x - 2)^2$

41. $y = -x^2(x + 1)$ **42.** $y = -x^3(x + 1)$

43. $y = x(x - 2)(x + 2)$ **44.** $y = \dfrac{1 - 2x}{x}$

45. $y = \dfrac{-1}{(x - 1)^2}$ **46.** $y = \dfrac{x + 1}{x + 2}$

47. $y = \dfrac{x - 2}{x - 3}$ **48.** $y = \dfrac{x}{(x - 2)(x + 4)}$

49. $y = \dfrac{x^2 - 2x + 1}{x^2 - 4x + 4}$ **50.** $y = \dfrac{x(x - 2)}{(x - 4)(x + 4)}$

51. The range of the function $y = x^2 - 2x + k$ is the interval $[5, \infty)$. Find the value of k.

52. Find a value for b such that the range of the function $f(x) = x^2 + bx + b$ is the interval $[-15, \infty)$.

53. Find the range of the function $y = \dfrac{(x-1)(x-3)}{x-4}$.

Hint: Solve the equation for x in terms of y using the quadratic formula. If you're careful with the algebra, you will find that the expression under the resulting radical sign is $y^2 - 8y + 4$. The range of the given function can then be found by solving the inequality $y^2 - 8y + 4 \geq 0$.

54. In the following figure, triangle OAB is equilateral and $\overline{AB}$ is parallel to the x-axis. Find the length of a side and the area of the triangle OAB.

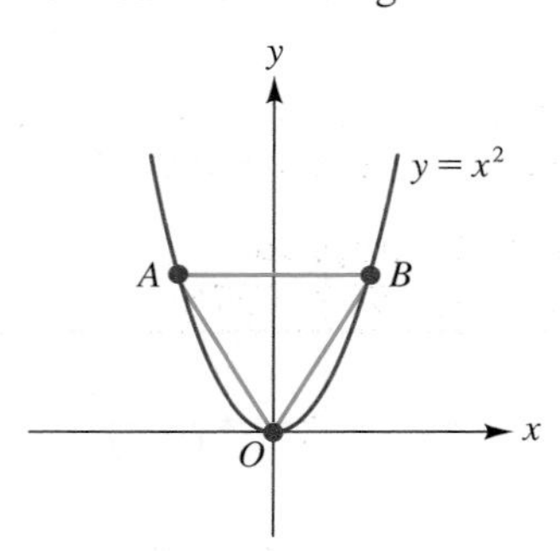

55. Let A denote the area of the right triangle in the first quadrant that is formed by the y-axis and the lines $y = mx$ and $y = m$. (Assume $m > 0$.) Express the area of the triangle as a function of m.

56. **(a)** The figure at the top of the next column shows the parabola $y = x^2 - 6x$ and a circle that passes through the vertex and x-intercepts of the parabola. Find the center and radius of the circle.

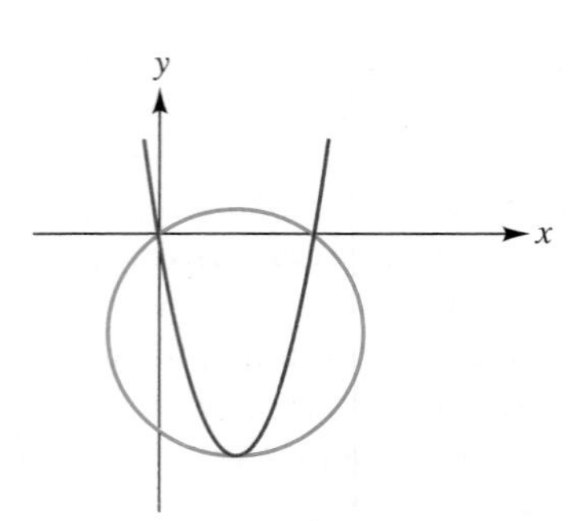

(b) Suppose that a circle passes through the vertex and x-intercepts of a parabola $y = x^2 - 2bx$, where $b > 0$. Show that the circumference of the circle as a function of b is $c = \pi b^2 + \pi$.

57. In the accompanying figure, the radius of the circle is $OC = 1$. Express the area of ΔABC as a function of x.

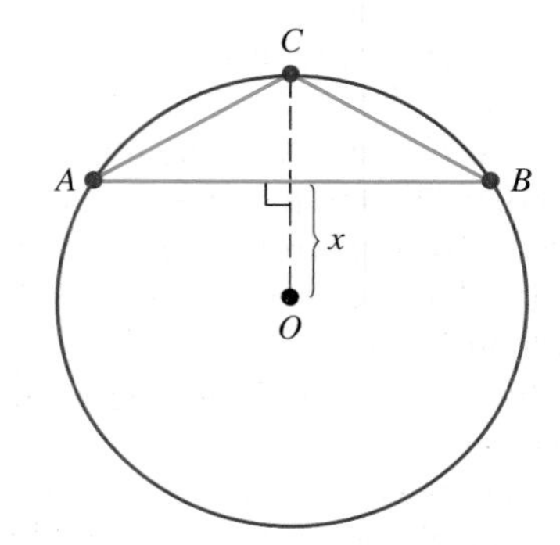

58. **(a)** Factor the expression $x^3 - 3x^2 + 4$.
Hint: Subtract and add 1, then factor by grouping.

(b) Use the factorization from part (a) to graph the function $y = x^3 - 3x^2 + 4$. Check your result using a graphing utility.

Chapter 4 Test

1. A linear function L satisfies the conditions $L(-2) = -4$ and $L(5) = 1$. Find $L(0)$.

2. **(a)** Let $F(x) = 4x - 2x^2$. What is the maximum value for this function? On which interval is this function increasing?

(b) Let $G(t) = 9t^4 + 6t^2 + 2$. For which value of t is this function a minimum?

3. Let $f(x) = (x - 3)(x + 4)^2$.

(a) Determine the intercepts and the excluded regions for the graph of f.

(b) Determine the behavior of f when x is very close to -4.

(c) Sketch the graph of f.

4. Graph each function:
(a) $y = -1/(x-3)^3$
(b) $y = -1/(x-3)^2$

5. Graph the function $y = -x^2 + 7x + 6$. Specify the intercepts, the axis of symmetry, and the coordinates of the turning point.

6. Suppose that the function $p = -\frac{1}{6}x + 80$ $(0 \le x \le 400)$ relates the selling price of an item to the quantity x that is sold. Assume that p is in dollars. What is the maximum revenue possible in this situation? Which price p generates this maximum revenue?

7. The price of a new machine is $14,000. After ten years, the machine has a salvage value of $750. Assuming linear depreciation, find a formula for the value of the machine after t yr, where $0 \le t \le 10$.

8. Graph the function $y = -\frac{1}{2}(3-x)^3$. Specify the intercepts. *Hint*: First graph $y = -\frac{1}{2}(x-3)^3$.

9. Graph the function $y = (2x-3)/(x+1)$. Specify the intercepts and asymptotes.

10. **(a)** Suppose that $P(x, y)$ is a point on the line $y = 3x - 1$ and Q is the point $(-1, 3)$. Express the length PQ as a function of x.
(b) For which value of x is the length PQ a minimum?

11. Let $f(x) = x(x-2)/(x^2-9)$.
(a) Find the vertical and the horizontal asymptotes.
(b) Use a sketch to show the behavior of f near the x-intercepts.
(c) Use a sketch to show the behavior of f near the asymptotes.
(d) Sketch the graph of f.

12. A rectangle is inscribed in a semicircle of diameter 8 cm. (See the accompanying figure.) Express the area of the rectangle as a function of the width w of the rectangle.

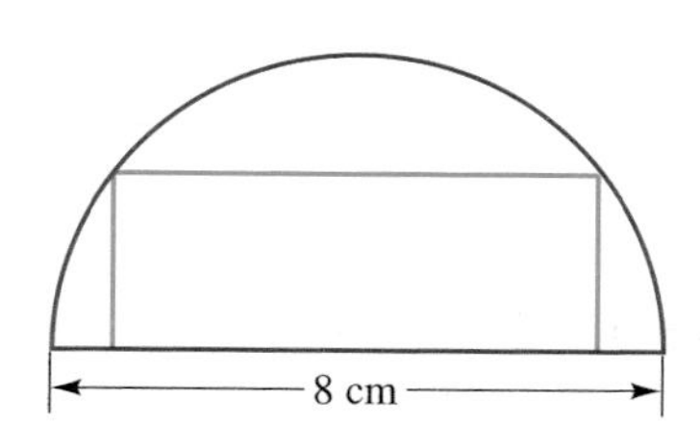

13. Explain why each of the following graphs cannot represent a polynomial function with highest-degree term $-x^3$.

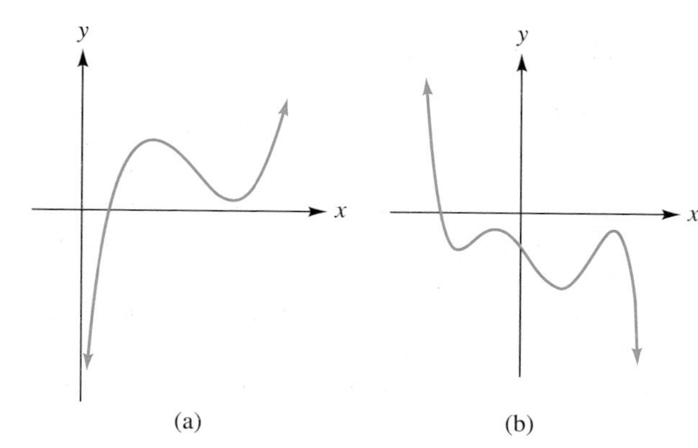

14. The following table shows U.S. expenditures for national defense for the years 1992–1995.

Federal Budget Outlays for National Defense, 1992–1995

Year t ($t = 0$ is 1990)	2	3	4	5
Expenditures y (billions of dollars)	298.4	291.1	281.6	272.1

Source: U.S. Office of Management and Budget

(a) Draw (by hand) a scatter plot of the data. (Show the portion of the y-axis from 270 to 300, in increments of 5.) You'll see that the data points appear to fall roughly on a straight line. Sketch in a line that seems to fit the data. Estimate the slope of the line. (Include the units for the slope.)

Ⓖ **(b)** Use a graphing utility to determine the equation of the line that best fits the data. Is your slope estimate in part (a) close to the slope obtained here?

(c) Use the least-squares equation obtained in part (b) to make estimates for what the expenditures might have been for 1996 and for 1999. Then compare the percentage errors in these two estimates, given that the actual expenditures (in billions of dollars) for 1996 and 1999 were 265.8 and 276.7, respectively.

Problems 15 and 16, which follow, are based on the work in Optional Section 4.3.

15. Find the real numbers (if any) that are fixed points of the given functions.
(a) $f(x) = 2x(1-x)$
(b) $g(x) = (4x-1)/(3x+6)$

16. The following figure shows the first six steps in the iteration process for the function $f(x) = 0.6 - x^2$ with initial input $x_0 = 0.2$.

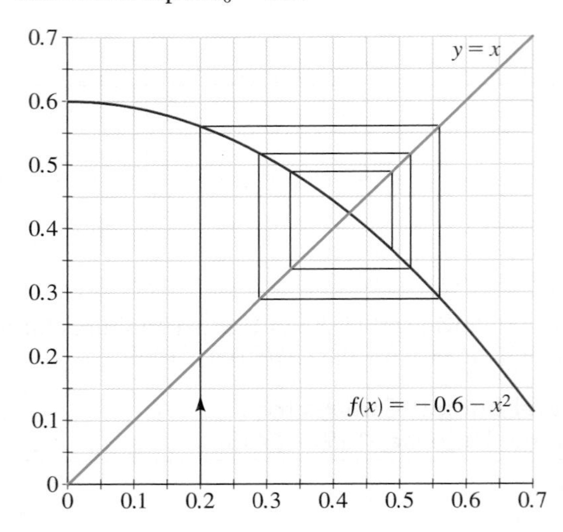

(a) Use the figure and your calculator to complete the following table. For your answers from the graph, try to estimate to the nearest 0.02 (or 0.01). For the calculator answers, round to three decimal places.

	x_1	x_2	x_3	x_4	x_5	x_6
From graph						
From calculator						

(b) Compute the two fixed points of the function $f(x) = 0.6 - x^2$. Give two forms for each answer: an exact expression involving a radical and a calculator approximation rounded to four decimal places. Which fixed point do the iterates in part (a) approach?

(c) Use your calculator to compute the first six iterates of $x_0 = 1$. Do the iterates seem to approach either of the fixed points determined in part (b)?

5 CHAPTER Exponential and Logarithmic Functions

. . . in 1958 . . . C. D. Keeling of The Scripps Institution of Oceanography . . . began a series of painstaking measurements of CO_2 concentration on a remote site at what is now the Mauna Loa Observatory in Hawaii. His observations, continued to the present, show an exponential growth of atmospheric carbon dioxide. . . . Keeling's observations have been duplicated at other stations in various parts of the world over shorter periods of time. In all sets of observations, the exponential increase is clear. . . . —Gordon MacDonald in "Scientific Basis for the Greenhouse Effect," from *The Challenge of Global Warming,* Dean Edwin Abrahamson, ed. (Washington, D.C.: Island Press, 1989)

In this chapter we study two major types of functions with widespread applications: the *exponential functions* and their inverses, the *logarithmic functions.* Exponential functions are functions of the form $y = ab^{cx}$ where a, b, and c are constants (with b positive and not equal to 1). As an introductory example, suppose that when you were born your parents opened a savings account in your name. Assume that the initial deposit was $100, with an annual interest rate of 8%, compounded twice a year. (If you are unfamiliar with compound interest, the details are in Section 5.6; but you won't need that to understand the point of the present discussion.) Under these conditions the dollar amount y in the account after x years is given by the exponential function $y = 100 \times 1.04^{2x}$. In Table 1 we show the amount in the account at five-year intervals for the first 30 years. Figure 1(a) shows a graph of the function $y = 100 \times 1.04^{2x}$ and the data points from Table 1. For comparison, Figure 1(b) shows the same scatter plot along with a quadratic function fitted to the data (using the quadratic regression option on a graphing utility). Over this relatively short interval of time, it appears that there is not much of a difference between the two models. For longer time intervals, however, the situation changes drastically.

Table 2 shows additional values for the exponential function $y = 100 \times 1.04^{2x}$ at ten-year intervals. In Figure 2 we've plotted these data points along with the original data set, and we show the exponential and quadratic models. As you can see, over the longer time interval, the quadratic function is no longer a good model for the amount of money in your account; the quadratic grows too slowly to keep pace with the exponential. Indeed, this is one of the key features of exponential growth: Eventually, it will outpace not only a quadratic, but any polynomial, no matter how high the degree.

Some of the applications you will see in this chapter for logarithmic or exponential functions involve:

- Modeling the temperature of a cup of coffee (Project in Section 5.2)
- Using the Richter scale to quantify the size of an earthquake (Example 8 in Section 5.3)

TABLE 1 $100 Invested at 8% per Year, Compounded Semiannually

x (years)	0	5	10	15	20	25	30
y (dollars)	100	148.02	219.11	324.34	480.10	710.67	1051.96

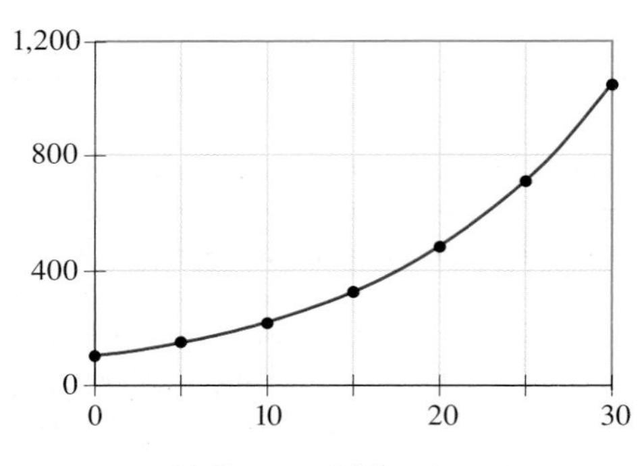

(a) Exponential function $y = 100 \times 1.04^{2x}$

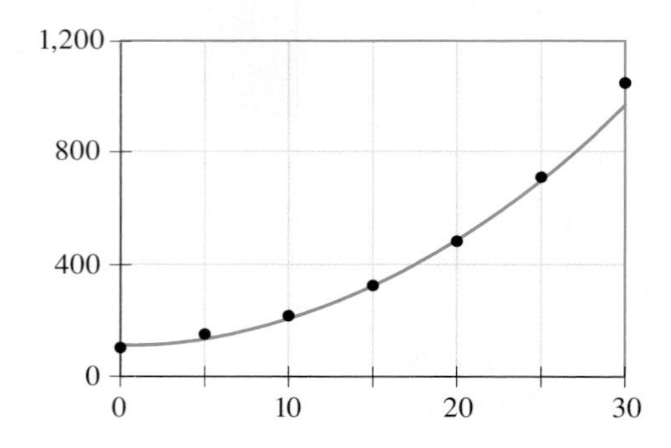

(b) Quadratic function fitted to the data in Table 1. $y = 0.894x^2 + 1.395x + 108.09$

Figure 1
Initially, the exponential and quadratic models appear quite similar (but see Figure 2).

TABLE 2 Amount in the Account After 40, 50, and 60 Years

x (years)	40	50	60
y (dollars)	2304.98	5050.50	11066.26

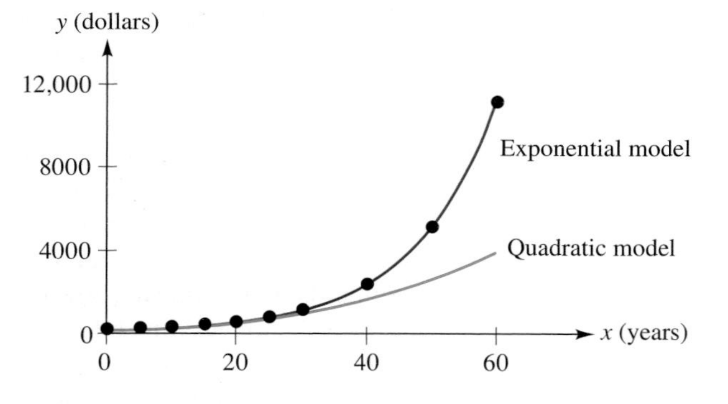

Figure 2
The exponential function eventually grows much faster than the quadratic.

- Computing the annual yield (= effective rate) for money earning compound interest (Example 4 in Section 5.6)
- Analyzing world population growth (Examples 2–4 in Section 5.7)
- Describing the increasing levels of carbon dioxide in the atmosphere (Example 5 and Exercises 15 and 16 in Section 5.7)

5.1 EXPONENTIAL FUNCTIONS

TABLE 1

x (day number)	y (amount earned that day)
1	2¢ $(= 2^1)$
2	4¢ $(= 2^2)$
3	8¢ $(= 2^3)$
4	16¢ $(= 2^4)$
5	32¢ $(= 2^5)$
⋮	⋮
x	2^x

We begin with an example. Suppose that your mathematics instructor, in an effort to improve classroom attendance, offers to pay you each day for attending class! Suppose you are to receive 2¢ on the first day you attend class, 4¢ the second day, 8¢ the third day, and so on, as shown in Table 1. How much money will you receive for attending class on the 30th day?

As you can see by looking at Table 1, the amount y earned on day x is given by the rule, or *exponential function,*

$$y = 2^x$$

Thus on the 30th day (when $x = 30$) you will receive

$$y = 2^{30} \text{ cents}$$

If you use a calculator, you will find this amount to be well over 10 million dollars! The point here is simply this: Although we begin with a small amount, $y = 2$¢, repeated doubling quickly leads to a very large amount. In other words, the exponential function grows very rapidly.

Before leaving this example, we mention a simple method for quickly estimating numbers such as 2^{30} (or any power of two) in terms of the more familiar powers of ten. Begin by observing that

$$2^{10} \approx 10^3 \qquad \text{(a useful coincidence, worth remembering)}$$

Now just cube both sides to obtain

$$(2^{10})^3 \approx (10^3)^3 \qquad \text{or} \qquad 2^{30} \approx 10^9$$

Thus 2^{30} is about one billion. To convert this number of cents to dollars, we divide by 100 or 10^2 to obtain

$$\frac{10^9}{10^2} = 10^7 \text{ dollars}$$

which is 10 million dollars, as mentioned before.

EXAMPLE 1 Estimating a power of 2

Estimate 2^{40} in terms of a power of 10.

SOLUTION

Take the basic approximation $2^{10} \approx 10^3$ and raise both sides to the fourth power. This yields

$$(2^{10})^4 \approx (10^3)^4$$

or

$$2^{40} \approx 10^{12} \qquad \text{as required}$$

EXAMPLE 2 Estimating an exponent

Estimate the power to which 10 must be raised to yield 2.

SOLUTION

We begin with our approximation

$$10^3 \approx 2^{10}$$

Raising both sides to the power 1/10 yields

$$(10^3)^{1/10} \approx (2^{10})^{1/10}$$

and, consequently, $$10^{3/10} \approx 2$$

Thus the power to which 10 must be raised to yield 2 is approximately 3/10.

In Appendix B.3 we define the expression b^x, where x is a rational number. We also state that if x is irrational, then b^x can be defined so that the usual properties of exponents continue to hold. Although a rigorous definition of irrational exponents requires concepts from calculus, we can nevertheless convey the basic idea by means of an example. (We need to do this before we give the general definition for exponential functions.)

How shall we assign a meaning to $2^{\sqrt{2}}$, for example? The basic idea is to evaluate the expression 2^x successively by using rational numbers x that are closer and closer to $\sqrt{2}$. Table 2 displays the results of some calculations along these lines.

TABLE 2 Values of 2^x for Rational Numbers x Approaching $\sqrt{2}$ (= 1.41421356 . . .)

x	1.4	1.41	1.414	1.4142	1.41421	1.414213
2^x	2.6 . . .	2.65 . . .	2.664 . . .	2.6651 . . .	2.66514 . . .	2.665143 . . .

The data in the table suggest that as x approaches $\sqrt{2}$, the corresponding values of 2^x approach a unique real number, call it t, with a decimal expansion that begins as 2.665. Furthermore, by continuing this process we can obtain (in theory, at least) as many places in the decimal expansion of t as we wish. The value of the expression $2^{\sqrt{2}}$ is then defined to be this number t. The following results (stated here without proof) summarize this discussion and also pave the way for the definition we will give for exponential functions.

PROPERTY SUMMARY Real Number Exponents

Let b denote an arbitrary positive real number. Then:

1. For each real number x, the quantity b^x is a unique real number.
2. When x is irrational, we can approximate b^x as closely as we wish by evaluating b^r, where r is a rational number sufficiently close to the number x.
3. The properties of rational exponents continue to hold for irrational exponents.
4. If $b^x = b^y$ and $b \neq 1$, then $x = y$.

EXAMPLE 3 Simplifying expressions containing irrational exponents

Use the properties of exponents to simplify each expression:
(a) $(3^{\sqrt{2}})^{\sqrt{2}}$; **(b)** $(3^{\sqrt{2}})^2$.

SOLUTION

(a) $(3^{\sqrt{2}})^{\sqrt{2}} = 3^{\sqrt{2}\times\sqrt{2}} = 3^2 = 9$

(b) $(3^{\sqrt{2}})^2 = (3^2)^{\sqrt{2}}$ (Why?)
$= 9^{\sqrt{2}}$

EXAMPLE 4 An equation with the unknown in the exponent

Solve the equation $4^x = 8$.

SOLUTION

First let's estimate x, just to get a feeling for what kind of answer to expect. Since $4^1 = 4$, which is less than 8, and $4^2 = 16$, which is more than 8, we know that our final answer should be a number between 1 and 2. To obtain this answer, we take advantage of the fact that both 4 and 8 are powers of 2. Using this fact, we can write the given equation as

$$\begin{aligned} (2^2)^x &= 2^3 \\ 2^{2x} &= 2^3 \\ 2x &= 3 \qquad \text{using Property 4 on page 328} \\ x &= \frac{3}{2} \end{aligned}$$

Note that the answer $x = 3/2$ is indeed between 1 and 2, as we estimated.

For the remainder of this section, b denotes an arbitrary positive constant other than 1. In the box that follows, we define the exponential function with base b.

DEFINITION The Exponential Function with Base b ($b > 0$, $b \neq 1$)

Let b denote an arbitrary positive constant other than 1. The **exponential function with base b** is defined by the equation

$$y = b^x$$

Note: In many applications, functions of the form $y = ab^{kx}$, where a, b, and k are constants, with $b \neq 1$, are also called exponential functions.

EXAMPLES

1. The equations $y = 2^x$ and $y = 3^x$ define the exponential functions with bases 2 and 3, respectively.
2. The equation $y = (1/2)^x$ defines the exponential functions with base $1/2$.
3. The equations $y = x^2$ and $y = x^3$ do not define exponential functions.

To help with our analysis of exponential functions, let's set up a table and use it to graph the exponential function $y = 2^x$. This is done in Figure 1. In drawing a

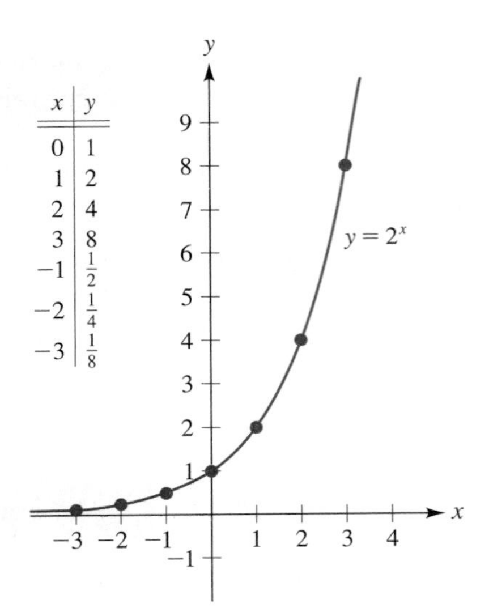

x | y
0 | 1
1 | 2
2 | 4
3 | 8
−1 | 1/2
−2 | 1/4
−3 | 1/8
y = 2^x
x
y

Figure 1

smooth and unbroken curve, we are actually relying on the results in the Property Summary box on the previous page. The key features of the exponential function $y = 2^x$ and its graph are:

1. The domain of $y = 2^x$ is the set of all real numbers. The range is the set of all *positive* real numbers.
2. The y-intercept of the graph is 1. The graph has no x-intercept.
3. For $x > 0$, the function increases or grows very rapidly. For $x < 0$, the graph rapidly approaches the x-axis; the x-axis is a horizontal asymptote for the graph. (Recall from Section 4.7 that a line is an **asymptote** for a curve if the separation distance between the curve and the line approaches zero as we move farther and farther out along the line.)

You should memorize the basic shape and features of the graph of $y = 2^x$ so that you can sketch it as needed without first setting up a table. The next example shows why this is useful.

EXAMPLE 5 Sketching graphs related to that of $y = 2^x$

Graph each of the following functions. In each case specify the domain, the range, the intercept(s), and the asymptote:
(a) $y = -2^x$; **(b)** $y = 2^{-x}$; **(c)** $y = (1/2)^x$; **(d)** $y = 2^{-x} - 2$.

SOLUTION

(a) Recall that -2^x means $-(2^x)$, not $(-2)^x$. The graph of $y = -2^x$ is obtained by reflecting the graph of $y = 2^x$ in the x-axis. See Figure 2(a).

(b) Similarly, the graph of $y = 2^{-x}$ is obtained from the graph of $y = 2^x$ by reflection in the y-axis. See Figure 2(b).

(c) Next, regarding $y = (1/2)^x$, observe that

$$\left(\frac{1}{2}\right)^x = \frac{1^x}{2^x} = \frac{1}{2^x} = 2^{-x}$$

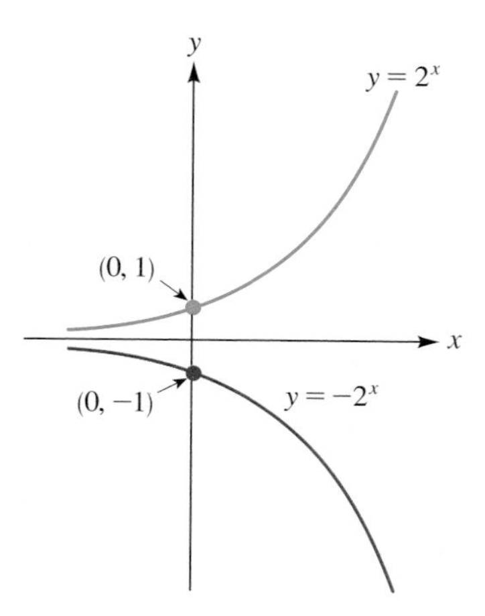

$y = -2^x$	
Domain:	$(-\infty, \infty)$
Range:	$(-\infty, 0)$
y-intercept:	-1
x-intercept:	none
Asymptote:	x-axis

(a)

$y = 2^{-x} = \left(\frac{1}{2}\right)^x$

$y = 2^x$

$y = 2^{-x}$ (also $y = (\frac{1}{2})^x$)	
Domain:	$(-\infty, \infty)$
Range:	$(0, \infty)$
y-intercept:	1
x-intercept:	none
Asymptote:	x-axis

(b)

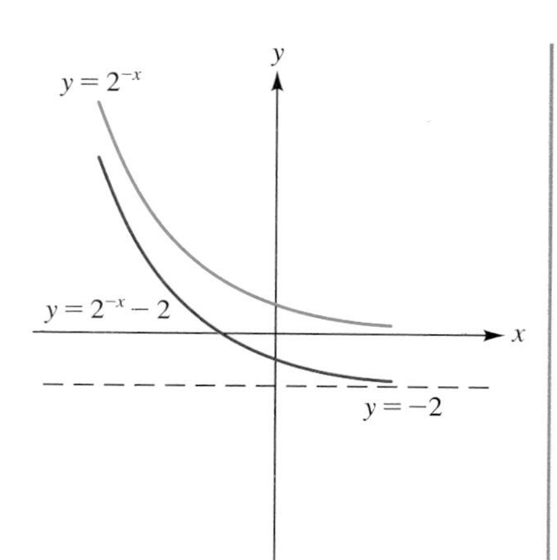

$y = 2^{-x} - 2$	
Domain:	$(-\infty, \infty)$
Range:	$(-2, \infty)$
y-intercept:	-1
x-intercept:	-1
Asymptote:	$y = -2$

(c)

Figure 2

In other words, $y = (\frac{1}{2})^x$ is really the same function as $y = 2^{-x}$, which we already graphed in Figure 2(b).

(d) Finally, to graph $y = 2^{-x} - 2$, take the graph of $y = 2^{-x}$ in Figure 2(b) and move it two units in the negative y-direction, as shown in Figure 2(c). Note that the asymptote and y-intercept will also move down two units. To find the x-intercept, we set $y = 0$ in the given equation to obtain

$$\begin{aligned} 2^{-x} - 2 &= 0 \\ 2^{-x} &= 2^1 \\ -x &= 1 \qquad \text{using Property 4 on page 328} \\ x &= -1 \end{aligned}$$

Thus the x-intercept is -1.

In the next example we apply our knowledge about the graph of $y = 2^x$ to solve an equation. In particular, we use the fact that the graph of $y = 2^x$ always lies above the x-axis; for no value of x is 2^x ever zero.

EXAMPLE 6 **Using the fact that 2^x is always positive to solve an equation**

Consider the equation

$$x^3 2^x - 3(2^x) = 0 \tag{1}$$

The graph in Figure 3 indicates that equation (1) has a positive root between 0 and 2. However, the viewing rectangle does not tell us whether equation (1) has

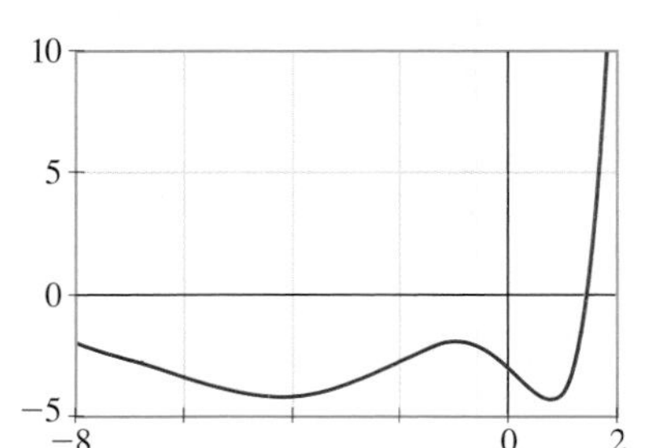

Figure 3
$y = x^3 2^x - 3(2^x)$
$[-8, 2, 2]$ by $[-5, 10, 5]$

other roots. (For instance, perhaps there is a negative root somewhere to the left of -8). By solving the equation, find out whether there are any roots other than the one between 0 and 2. For each root, give both an exact expression and a calculator approximation rouded to two decimal places.

SOLUTION

(a) On the left-hand side of equation (1) the common term 2^x can be factored out. This yields

$$2^x(x^3 - 3) = 0 \tag{2}$$

Now applying the zero-product property in equation (2), we have

$$2^x = 0 \quad \text{or} \quad x^3 - 3 = 0$$

The equation $2^x = 0$ has no solutions because 2^x is positive for all values of x. From the equation $x^3 - 3 = 0$ we conclude that $x^3 = 3$, and therefore $x = \sqrt[3]{3}$. With a calculator we find that $\sqrt[3]{3} \approx 1.44$. This is the root between 0 and 2 in Figure 3. The algebra that we've carried out indicates that this is the only root of the given equation.

GRAPHICAL PERSPECTIVE

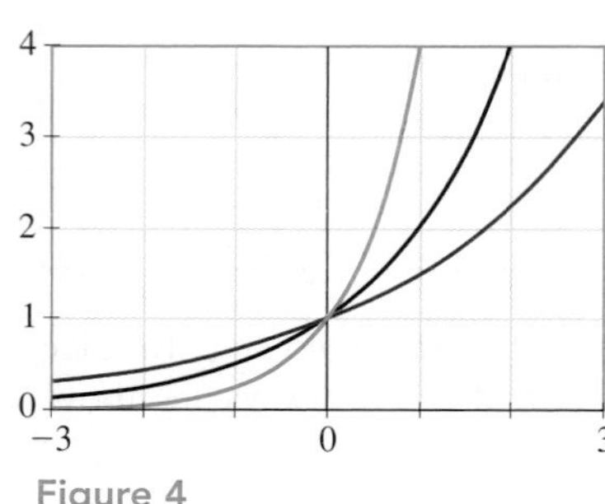

Figure 4
$y = 1.5^x$ $y = 2^x$ $y = 4^x$
$[-3, 3, 1]$ by $[0, 4, 1]$

Now what about exponential functions with bases other than 2? As Figure 4 indicates, the graphs are similar to $y = 2^x$. The graph of $y = 4^x$ rises more rapidly than $y = 2^x$ when x is positive. For negative x-values, the graph of $y = 4^x$ is below that of $y = 2^x$. You can see why this happens by taking $x = -1$, for example, and comparing the values of 4^x and 2^x. If $x = -1$, then

$$2^x = 2^{-1} = \frac{1}{2} \quad \text{but} \quad 4^x = 4^{-1} = \frac{1}{4}$$

Therefore $4^x < 2^x$ when $x = -1$. In general $0 < 2^x < 4^x$ for $x > 0$, so $0 < 1/4^x < 1/2^x$ or $0 < 4^{-x} < 2^{-x}$ for $x > 0$. Reflecting in the y-axis by replacing x by its opposite we obtain $0 < 4^x < 2^x$ when $x < 0$. Notice also in Figure 4 all three graphs have the same y-intercept of 1. This follows from the fact that $b^0 = 1$ for any positive number b.

The exponential functions in Figure 4 all have bases larger than 1. To see examples in which the bases are in the interval $0 < b < 1$, we need only reflect the graphs in Figure 4 in the y-axis. For instance, the reflection of $y = 4^x$ in the y-axis gives us the graph of $y = 4^{-x} = 1/4^x$. (We discussed the idea behind this in Example 5; see Figure 2(b), for instance.)

In the box that follows, we summarize what we've learned up to this point regarding the exponential function $y = b^x$.

PROPERTY SUMMARY The Exponential Function $y = b^x$

Domain: $(-\infty, \infty)$
Range: $(0, \infty)$
y-intercept: 1
x-intercept: none
Asymptote: x-axis

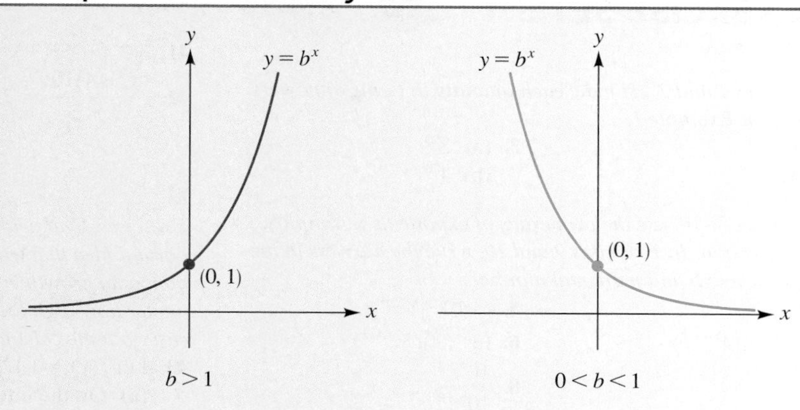

Note: The function defined by the equation $y = b^x$ for positive constant $b \neq 1$ is called the *exponential function with base b.*

EXAMPLE 7 Reflecting and translating the graph of an exponential function

Graph the function $y = -3^{-x} + 1$.

SOLUTION

The required graph is obtained by reflecting and translating the graph of $y = 3^x$, as shown in Figure 5.

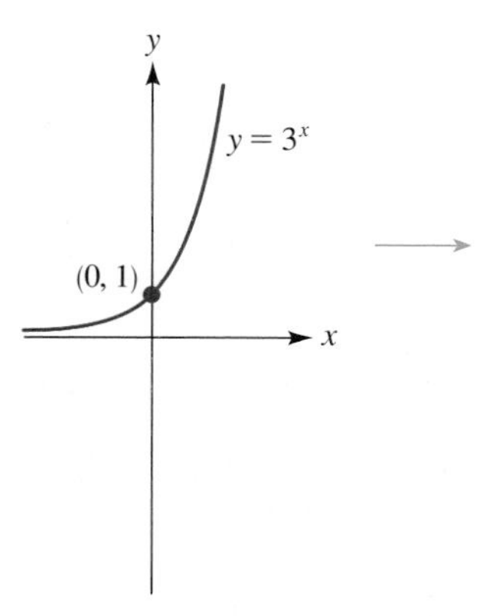

Begin with $y = 3^x$.

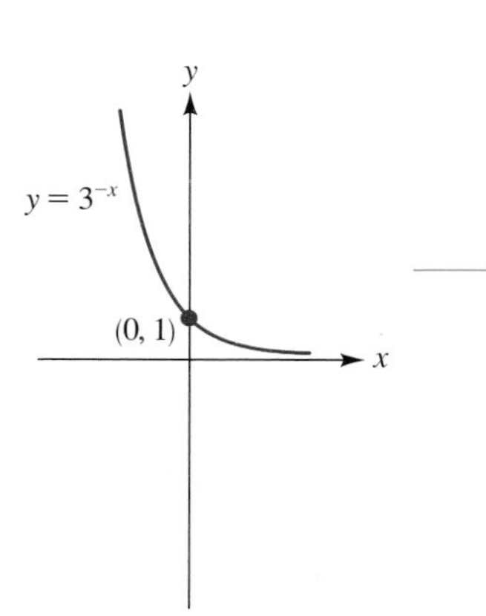

Reflect $y = 3^x$ in the y-axis to obtain $y = 3^{-x}$.

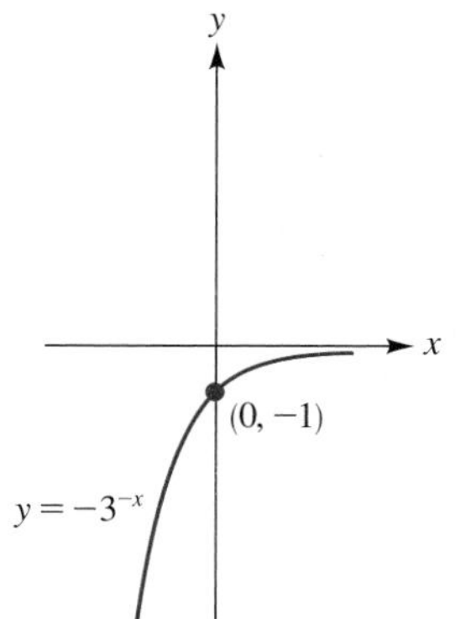

Reflect $y = 3^{-x}$ in the x-axis to obtain $y = -3^{-x}$.

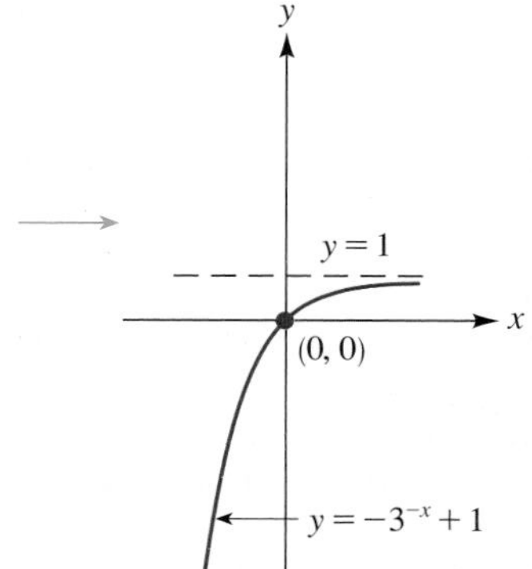

Translate $y = -3^{-x}$ up one unit to obtain $y = -3^{-x} + 1$. Note that the y-intercept is 0 and the asymptote is the line $y = 1$.

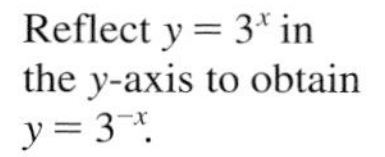

Figure 5

EXERCISE SET 5.1 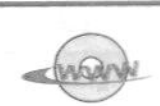

A

In Exercises 1 and 2, estimate each quantity in terms of powers of ten, as in Example 1.

1. (a) 2^{30}
(b) 2^{50}

2. (a) 2^{90}
(b) 4^{50}

In Exercises 3–10, use the properties of exponents to simplify each expression. In Exercises 9 and 10, write the answers in the form b^n, where b and n are real numbers.

3. $(5^{\sqrt{3}})^{\sqrt{3}}$

4. $(\sqrt{2}^{\sqrt{2}})^{\sqrt{2}}$

5. $(4^{1+\sqrt{2}})(4^{1-\sqrt{2}})$

6. $(3^{2+\sqrt{5}})(3^{2-\sqrt{5}})$

7. $\dfrac{2^{4+\pi}}{2^{1+\pi}}$

8. $\dfrac{10^{\pi+2}}{10^{\pi-2}}$

9. $(\sqrt{5}^{\sqrt{2}})^2$

10. $[(\sqrt{3})^{\pi}]^4$

In Exercises 11 and 12, solve each equation, as in Example 4.

11. (a) $3^x = 27$
(b) $9^t = 27$
(c) $3^{1-2y} = \sqrt{3}$
(d) $3^z = 9\sqrt{3}$

12. (a) $2^x = 32$
(b) $2^t = 1/4$
(c) $2^{3y+1} = \sqrt{2}$
(d) $8^{z+1} = 32\sqrt{2}$

In Exercises 13–16, specify the domain of the function.

13. $y = 2^x$

14. $y = 1/2^x$

15. $y = 1/(2^{x-1})$

16. $y = 1/(2^x - 1)$

In Exercises 17–24, graph the pair of functions on the same set of axes.

17. $y = 2^x; y = 2^{-x}$

18. $y = 3^x; y = 3^{-x}$

19. $y = 3^x; y = -3^x$

20. $y = 4^x; y = -4^x$

21. $y = 2^x; y = 3^x$

22. $y = (1/3)^x; y = 3^x$

23. $y = (1/2)^x; y = (1/3)^x$

24. $y = (1/2)^{-x}; y = (1/3)^{-x}$

For Exercises 25–32, graph the function and specify the domain, range, intercept(s), and asymptote.

25. $y = -2^x + 1$

26. $y = -3^x + 3$

27. $y = 3^{-x} + 1$

28. $y = 3^{-x} - 3$

29. $y = 2^{x-1}$

30. $y = 2^{x-1} - 1$

31. $y = 3^{x+1} + 1$

32. $y = 1 - 3^{x-1}$

For Exercises 33–38:

(a) *Use paper and pencil to determine the intercepts and asymptotes for the graph of each function.*

Ⓖ **(b)** *Use a graphing utility to graph each function. Your results in part (a) will be helpful in choosing an appropriate viewing rectangle that shows the essential features of the graph.*

33. $y = -3^{x-2}$

34. $y = -3^{x-2} + 1$

35. $y = 4^{-x} - 4$

36. $y = 4 + (1/4)^x$

37. $y = 10^{x-1}$

38. $y = -10^{x-1} + 0.5$

For Exercises 39–42, solve the equation, as in Example 6.

39. $3x(10^x) + 10^x = 0$

40. $4x^2(2^x) - 9(2^x) = 0$

41. $3(3^x) - 5x(3^x) + 2x^2(3^x) = 0$

42. $\dfrac{(x+4)10^x}{x-3} = 2x(10^x)$

B

Exercises 43 and 44 provide some simple examples of an important idea that was presented in the chapter introduction: Exponential growth eventually outpaces polynomial growth (no matter how high the degree of the polynomial). In each case, carry out the calculations needed to verify the given statement.

43. Let $f(x) = 1.125x^2 - 0.75x + 1$ and $g(x) = 2^x$.

(a) On the interval [0, 2], both functions have the same average rate of change. This is also true for the interval [2, 4].

(b) On the interval [4, 6], however, $\Delta g/\Delta x$ is more than twice $\Delta f/\Delta x$. And on the interval [6, 8], $\Delta g/\Delta x$ is more than six times $\Delta f/\Delta x$.

(c) On the interval [10, 12], the average rate of change of the exponential function is 64 times that of the quadratic function.

44. Let $g(x) = x^4$ and $h(x) = 3^x$.

(a) On the interval [0, 5], the average rate of change of g is more than two and one-half times the average rate of change of h. On the interval [5, 10], however, the average rate of change of h is more than six times that for g.

(b) On the interval [10, 15] the average rate of change of the exponential function is more than 350 times that of the polynomial function.

45. Let $f(x) = 2^x$. Show that

$$\frac{f(x+h) - f(x)}{h} = 2^x\left(\frac{2^h - 1}{h}\right)$$

46. Let $\phi(t) = 1 + a^t$. Show that $\dfrac{1}{\phi(t)} + \dfrac{1}{\phi(-t)} = 1$.

47. Let $f(x) = 2^x$ and let g denote the function that is the inverse of f.

(a) On the same set of axes, sketch the graphs of f, g, and the line $y = x$.

(b) Using the graph you obtained in part (a), specify the domain, range, intercept, and asymptote for the function g.

48. Let $S(x) = (2^x - 2^{-x})/2$ and $C(x) = (2^x + 2^{-x})/2$. Compute $[C(x)]^2 - [S(x)]^2$.

For Exercises 49–56, refer to the following figure, which shows portions of the graphs of $y = 2^x$, $y = 3^x$, and $y = 5^x$. In each case: **(a)** *use the figure to estimate the indicated quantity; and* **(b)** *use a calculator to compute the indicated quantity, rounding the result to two decimal places.*

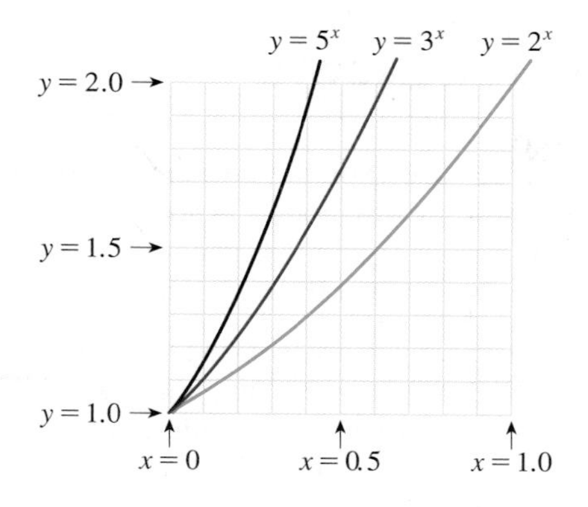

49. $\sqrt{2}$ *Hint:* $\sqrt{2} = 2^{1/2}$
50. $\sqrt[5]{2}$
51. $2^{3/5}$
52. $\sqrt[5]{8}$ *Hint:* $8^{1/5} = 2^?$
53. $\sqrt{3}$
54. $\sqrt[3]{3}$
55. $5^{3/10}$
56. $\sqrt[4]{5}$

In Exercises 57–60, refer to the following graph of the exponential function $y = 10^x$. Use the graph to estimate (to the nearest tenth) the solution of each equation. (After we've studied logarithmic functions later in the chapter, we will be able to obtain precise solutions.)

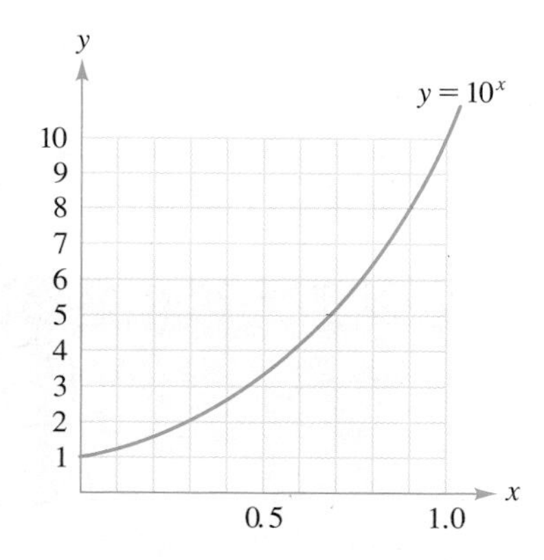

57. $10^x = 2$ **58.** $10^x = 4$ **59.** $10^x = 5$ **60.** $10^x = 8$

Ⓖ **61.** If you add two quadratic functions, the result is again a quadratic function (assuming that the x^2-terms don't add to zero). For example, if $f(x) = x^2 + 2x$ and $g(x) = 2x^2 - 1$, then the sum is $(f + g)(x) = 3x^2 + 2x - 1$, another quadratic. But this is not the case for exponential functions. To explore a visual example, consider the function $y = 2^x + 2^{-x}$. This is the sum of the two exponentials $y = 2^x$ and $y = (1/2)^x$.

(a) Graph the function $y = 2^x + 2^{-x}$ in the standard viewing rectangle. As you can see, the resulting graph is U-shaped.

(b) For comparison, add the graph of $y = x^2$ to your picture. Note that (at this scale) the graphs are very similar for $|x| > 2$.

(c) Actually, the graph of $y = 2^x + 2^{-x}$ rises much more steeply than does $y = x^2$. To demonstrate this, change the viewing rectangle so that x extends from -100 to 100 and y extends from 0 to 100,000.

Ⓖ **62.** Follow the general procedure in Exercise 61, but use the two functions $y = 2^x - 2^{-x}$ and $y = x^3$.

Ⓖ **63.** As background for this exercise, reread Example 2 in the text. (The example shows how to estimate the value of x for which $10^x = 2$.)

(a) In the same picture, graph the two functions $y = 10^x$ and $y = 2$. Use a viewing rectangle in which x runs from 0 to 1 and y runs from 0 to 10.

(b) Explain how your picture in part (a) supports the answer 3/10 that was obtained in Example 2.

C

64. This exercise serves as a preview for the work on logarithms in Section 5.3. Follow Steps (a)–(f) to complete the table. (Notice that one entry in the table is already filled in. Reread Example 2 in the text to see how that entry was obtained.)

(a) Fill in the entries in the right-hand column corresponding to $x = 1$ and $x = 10$.

(b) Note that 4 and 8 are powers of 2. Use this information along with the approximation $10^{0.3} \approx 2$ to find the entries in the table corresponding to $x = 4$ and $x = 8$.

(c) Find the entry corresponding to $x = 5$.
Hint: $5 = 10/2 \approx 10/10^{0.3}$

(d) Find the entry corresponding to $x = 7$.
Hint: $7^2 \approx 50 = 5 \times 10$; now make use of your answer in part (c).

(e) Find the entry corresponding to $x = 3$.
Hint: $3^4 \approx 80 = 8 \times 10$

(f) Find the entries corresponding to $x = 6$ and $x = 9$.
Hint: $6 = 3 \times 2$ and $9 = 3^2$

Remark: This table is called a *table of logarithms to the base* 10. We say, for example, that the logarithm of 2 to the base 10 is (about) 0.3. We write this symbolically as $\log_{10} 2 \approx 0.3$.

x	Exponent to which 10 must be raised to yield x
1	
2	≈ 0.3
3	
4	
5	
6	
7	
8	
9	
10	

MINI PROJECT

USING DIFFERENCES TO COMPARE EXPONENTIAL AND POLYNOMIAL GROWTH

In the chapter introduction (on pages 325–326), you saw an example comparing and contrasting exponential and quadratic growth models. Another way to compare exponential growth and quadratic growth (or, more generally, polynomial growth) uses the idea of *differences,* as discussed at the end of Section 4.2.

(a) Appoint a member of the group in which you are working to summarize for the group the definitions and results on differences on page 239.

(b) Just as second differences are obtained by subtracting adjacent first differences, *third differences* are obtained by subtracting adjacent second differences. For the cubic polynomial function $y = x^3$, set up a table of x-y values using $x = 0, 1, 2, \ldots, 6$. Then compute the first, second, and third differences. What do you observe?

(c) For the exponential function $y = 2^x$, set up a table of x-y values using $x = 0, 1, 2, \ldots, 6$. Then compute the first, second, and third differences. What do you observe? Write a paragraph summarizing your observations and contrasting the results with what occurs in the case of linear and quadratic functions and the cubic function in part (b).

5.2 THE EXPONENTIAL FUNCTION $y = e^x$

Why did he [the Swiss mathematician Leonhard Euler (1707–1783)] *choose the letter* e*? There is no general consensus. According to one view, Euler chose it because it is the first letter of the word* exponential. *More likely, the choice came to him naturally as the first "unused" letter of the alphabet, since the letters* a, b, c, *and* d *frequently appear elsewhere in mathematics. It seems unlikely that Euler chose the letter because it is the initial of his own name, as occasionally has been suggested: he was an extremely modest man . . .*

—Eli Maor, *e: The Story of a Number* (Princeton, N.J.: Princeton University Press, 1994)

The shape of the Gateway Arch in St. Louis is based on the *catenary curve,* which is defined in terms of the exponential function $y = e^x$. For details see Exercises 55 and 56. (Photo from Scenics of America/Photo Link)

From the standpoint of calculus and scientific applications, one particular base for exponential functions is by far the most useful. This base is a certain irrational

TABLE 1

x	$f(x) = (1 + \frac{1}{x})^x$
10^1	2.59374246
10^2	2.70481382
10^3	2.71692393
10^4	2.71814592
10^5	2.71826823
10^6	2.71828046
10^7	2.71828169

As x grows larger and larger, the values of $[1 + (1/x)]^x$ approach the number e. [Compare the digits in the right column with those in the expression for e given in equation (1).]

number that lies between 2 and 3 and is denoted by the letter e. To ten decimal places, without rounding off, the value of e is

$$e = 2.7182818284\ldots \tag{1}$$

For most purposes of approximation and simple estimation, all you need to remember is that

$$e \approx 2.7$$

At the precalculus level, it's hard to escape the feeling that $y = 2^x$ or $y = 10^x$ is by far more simple and more natural than $y = e^x$. At the end of this section we'll explain why the function $y = e^x$ makes life simpler, not more complex.

There are several different (but equivalent) ways that the number e can be defined. One way involves investigating the values of the function $f(x) = [1 + (1/x)]^x$ as x becomes larger and larger. As indicated in Table 1, the values of this function get closer and closer to e as x gets larger and larger. Indeed, in some calculus books, the number e is *defined* as the target value or *limit* of the function $f(x) = [1 + (1/x)]^x$ as x grows ever larger. Now admittedly, we have not defined here the meaning of the terms *target value* or *limit;* that's a topic for calculus. Nevertheless, Table 1 will give you a reasonable, if intuitive, appreciation of the idea. In Figure 1 we give a graphical interpretation.

One more comment about Table 1 and the number e: In Section 5.6 you'll see that the data in Table 1 can be interpreted in terms of banking and compound interest. For this reason the constant e is sometimes called "the banker's constant."

In calculus the number e is sometimes introduced in a way that involves slopes of lines. You know that the graph of each exponential function $y = b^x$ passes through the point (0, 1). Figure 2(a) shows the exponential function $y = 2^x$ along with a line that is tangent to the curve at the point (0, 1). By carefully measuring rise and run, it can be shown that the slope of this tangent line is about 0.7. Figure 2(b) shows a similar situation with the curve $y = 3^x$. Here the slope of the tangent line through (0, 1) is approximately 1.1. Now, since the slope of the tangent to $y = 2^x$ is a bit less than 1 while that for $y = 3^x$ is a bit more than 1, it seems reasonable to suppose that there is a number between 2 and 3, call it e, with the property that the

GRAPHICAL PERSPECTIVE

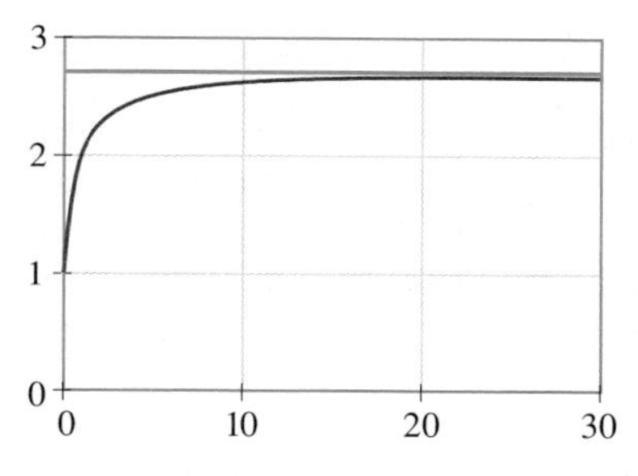

$f(x) = [1 + (1/x)]^x$ for $x > 0$

$y = e$

Figure 1
The line $y = e$ is an asymptote for the graph of $f(x) = [1 + (1/x)]^x$. As x gets larger and larger, the values of $[1 + (1/x)]^x$ approach the number e.

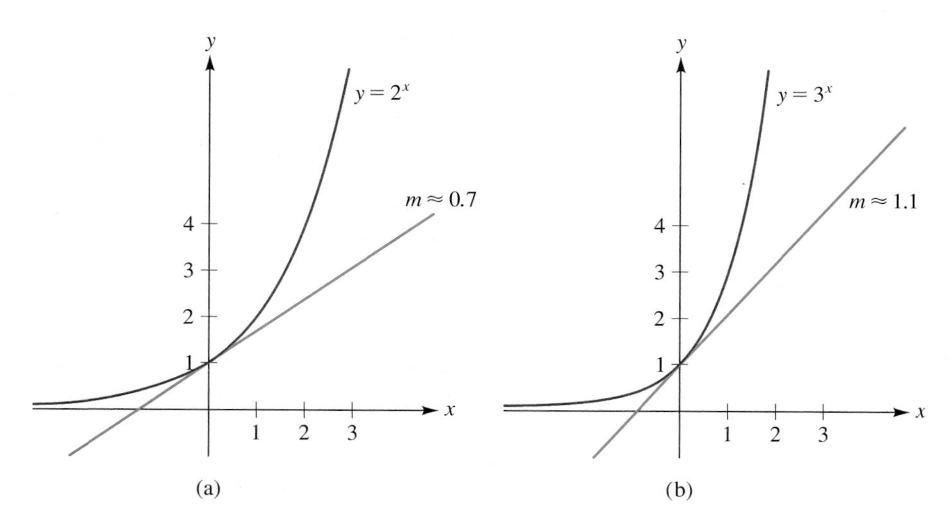

Figure 2

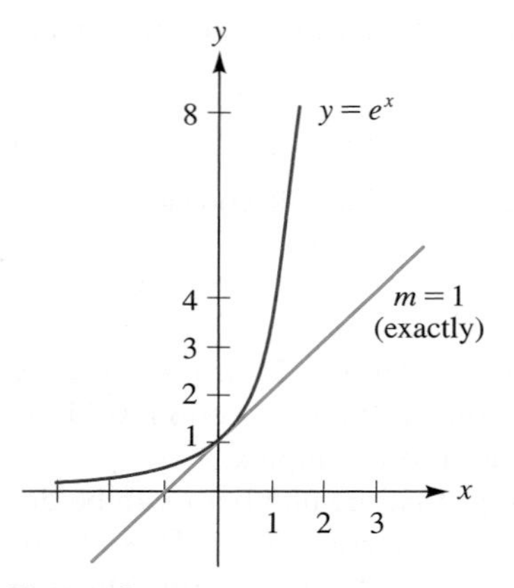

Figure 3
The slope of the tangent to the curve $y = e^x$ at the point (0, 1) is $m = 1$.

slope of the tangent through (0, 1) is exactly 1. See Figure 3 at left and the Property Summary box that follows (see Figure 4).

PROPERTY SUMMARY The Exponential Function $y = e^x$

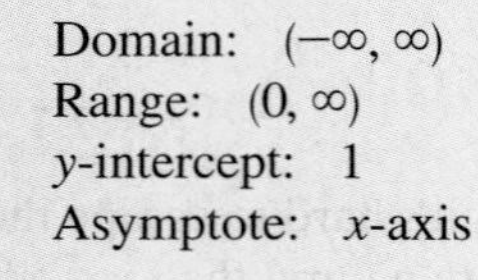

Domain: $(-\infty, \infty)$
Range: $(0, \infty)$
y-intercept: 1
Asymptote: x-axis

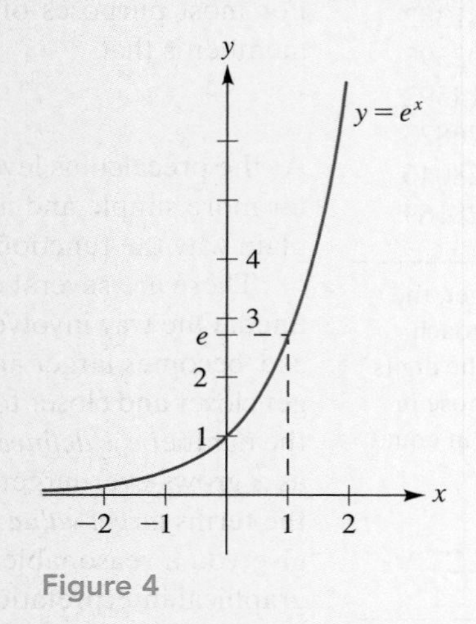

Figure 4

REMARK It's certainly not obvious that defining e to be the unique number with the property that the slope of the tangent line to the graph of $y = e^x$ at the point (0, 1) is exactly 1 is equivalent to the definition of e as the limit of the function $f(x) = (1 + 1/x)^x$ as x grows ever larger. The methods of calculus are required to demonstrate that the definitions are indeed equivalent. (For an informal justification, however, see Exercise 62.)

EXAMPLE 1 Sketching graphs related to that of $y = e^x$

Graph each of the following functions, specifying the domain, range, intercept, and asymptote:
(a) $y = e^{x-1}$; **(b)** $y = -e^{x-1}$; **(c)** $y = -e^{x-1} + 1$.

SOLUTION

(a) We begin with the graph of $y = e^x$ (as in Figure 4). Moving the graph to the right one unit yields the graph of $y = e^{x-1}$, shown in Figure 5(a). The x-axis is still an asymptote for this translated graph, but the y-intercept will no longer be 1. To find the y-intercept, we replace x with 0 in the given equation to obtain

$$y = e^{-1} = \frac{1}{e} \approx 0.37 \qquad \text{using a calculator}$$

(b) Reflecting the graph from part (a) in the x-axis yields the graph of $y = -e^{x-1}$; see Figure 5(b). Note that under this reflection, the y-intercept moves from $1/e$ to $-1/e$.

(c) Translating the graph in Figure 5(b) up one unit produces the graph of $y = -e^{x-1} + 1$, shown in Figure 5(c). Under this translation, the asymptote moves from $y = 0$ (the x-axis) to $y = 1$. Also, the y-intercept moves from

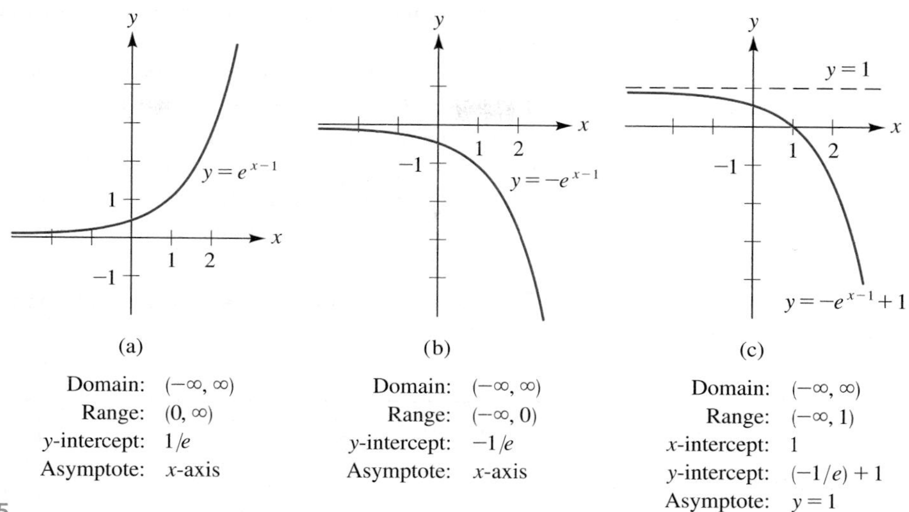

Figure 5

$-1/e$ to $(-1/e) + 1 \approx 0.63$. The x-intercept in Figure 5(c) is obtained by setting $y = 0$ in the equation $y = -e^{x-1} + 1$. This yields

$$\begin{aligned} 0 &= -e^{x-1} + 1 \\ e^{x-1} &= 1 \\ e^{x-1} &= e^0 \\ x - 1 &= 0 \quad \text{(Why?)} \\ x &= 1 \end{aligned}$$

Near the beginning of this section we said that using e as the base for an exponential function makes things simpler. To explain the context in which this is true, we introduce the concept of the *instantaneous rate of change* of a function. We are going to rely on examples and numerical calculations. A more rigorous development properly belongs to calculus. Our first two examples will involve simple polynomial functions; then we'll move on to the exponential function $f(x) = e^x$. In Table 2(a) we've calculated the average rates of change of the function $f(x) = x^2$ on a sequence of increasingly short intervals, each interval having 1 as the left endpoint. Table 2(b) shows similar calculations with intervals that have 1 as the right endpoint.

TABLE 2 The Average Rates of Change Approach 2

Interval	$\Delta f/\Delta x$ for $f(x) = x^2$	Interval	$\Delta f/\Delta x$ for $f(x) = x^2$
[1, 1.1]	2.1	[0.9, 1]	1.9
[1, 1.01]	2.01	[0.99, 1]	1.99
[1, 1.001]	2.001	[0.999, 1]	1.999
[1, 1.0001]	2.0001	[0.9999, 1]	1.9999
[1, 1.00001]	2.00001	[0.99999, 1]	1.99999
(a)		(b)	

From the tables we make the following observation: As the lengths of the intervals get smaller and smaller, the average rates of change get closer and closer to 2. We say in this case that the **instantaneous rate of change** of the function $f(x) = x^2$ at $x = 1$ is 2.

In Tables 3(a) and (b), we are looking for the instantaneous rate of change of the function $g(x) = x^3$ at $x = 2$. From the tables, it appears that as the lengths of the intervals get smaller and smaller, the average rates of change get closer and closer to 12. We say that the instantaneous rate of change of $g(x) = x^3$ at $x = 2$ is 12.

TABLE 3 The Average Rates of Change Approach 12

Interval	$\Delta g/\Delta x$ for $g(x) = x^3$
[2, 2.1]	12.61
[2, 2.01]	12.0601
[2, 2.001]	12.006001
[2, 2.0001]	12.00060001
[2, 2.00001]	12.0000600001

(a)

Interval	$\Delta g/\Delta x$ for $g(x) = x^3$
[1.9, 2]	11.41
[1.99, 2]	11.9401
[1.999, 2]	11.994001
[1.9999, 2]	11.99940001
[1.99999, 2]	11.9999400001

(b)

For a more intuitive example of instantaneous rate of change, something closer to your experience, suppose you are driving a car, and at a particular instant the speedometer reads 45 mi/hr. At that instant the instantaneous rate of change of distance with respect to time is 45 mi/hr. We say that the *instantaneous velocity* at that instant is 45 mi/hr. (For a specific example relating average velocity and instantaneous velocity, go back to Section 3.3 and see Example 6 and the related Exercise 36.)

Now we are prepared to say what is so special about the exponential function $f(x) = e^x$. For reference, we state this result in the box that follows. (Remember, we are leaving the precise definitions and the proofs to calculus; we content ourselves here with examples.)

THEOREM The Instantaneous Rate of Change of the Function $f(x) = e^x$

Let a be any real number. Then the instantaneous rate of change of the function $f(x) = e^x$ at $x = a$ is e^a. In other words, at each point on the graph of $f(x) = e^x$, the instantaneous rate of change is just the y-coordinate of the point. Furthermore, the *only* functions with this property are the exponential function $f(x) = e^x$ and constant multiples of this function (that is, functions of the form $y = ce^x$, where c is a constant).

EXAMPLES

1. At $x = 1$ the instantaneous rate of change of $f(x) = e^x$ is $e^1 = e$
2. At $x = 2$ the instantaneous rate of change of $f(x) = e^x$ is e^2

Table 4 provides a numerical check on Example 2 in the box above. (Exercise 29 asks you to complete a similar table for intervals in which 2 is the right, rather than left, endpoint.)

TABLE 4
The Average Rates of Change Approach $e^2 = 7.3890560\ldots$

Interval	$\Delta f/\Delta x$ for $f(x) = e^x$ (7 decimal places, no rounding)
[2, 2.1]	7.7711381
[2, 2.01]	7.4261248
[2, 2.001]	7.3927518
[2, 2.0001]	7.3894255
[2, 2.00001]	7.3890930
[2, 2.000001]	7.3890597

EXAMPLE 2 Average and instantaneous rates of change of a population

In a biology experiment, let $N(t)$ be the number of bacteria in a colony after t hours, where $t = 0$ corresponds to the time the experiment begins. Suppose that during the period from $t = 4$ hours to $t = 8$ hours the number of bacteria is modeled by the exponential function $N(t) = e^t$. (See Figure 6.)

GRAPHICAL PERSPECTIVE

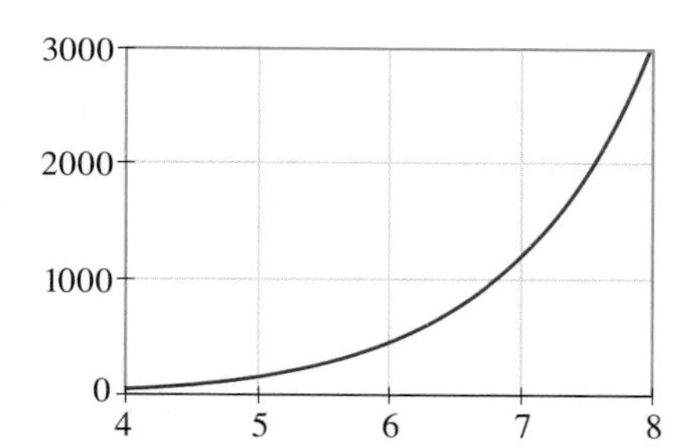

Figure 6
$N(t) = e^t, 4 \leq t \leq 8$

(a) Find the average rate of change of the population over the time period $5 \leq t \leq 7$.

(b) What is the instantaneous rate of change of the population at the instant $t = 7$ hours?

SOLUTION

(a) On the interval from $t = 5$ to $t = 7$ hours, we have

$$\frac{\Delta N}{\Delta t} = \frac{e^7 - e^5}{7 - 5} \quad \text{bacteria/hour}$$
$$\approx 474 \quad \text{bacteria/hour} \qquad \text{using a calculator and rounding}$$

Thus on average over this 2-hour time period, the population is increasing at a rate of 474 bacteria/hour.

(b) Applying the theorem in the box on page 340, the instantaneous rate of change of the function $N(t) = e^t$ at $t = 7$ is e^7. Thus at the instant when $t = 7$ hours, the population is changing at a rate of

$$e^7 \quad \text{bacteria/hour}$$
$$\approx 1097 \quad \text{bacteria/hour} \qquad \text{using a calculator and rounding}$$

EXERCISE SET 5.2

A

Ⓖ *In Exercises 1–8, answer* True *or* False. *You do not need a calculator for these exercises. Rather, use the fact that e is approximately 2.7.*

1. $e < 1/2$ **2.** $e < 5/2$ **3.** $\sqrt{e} < 1$ **4.** $e^2 < 4$
5. $e^2 < 9$ **6.** $e^3 < 27$ **7.** $e^{-1} < 0$ **8.** $e^0 = 1$

For Exercises 9 and 10, you are given some simple rational approximations for the irrational numbers π and e. In each case, use a calculator to find out how many decimal places of the approximation agree with the irrational number. Remark: *Each of the approximations in Exercise 10 is the best rational approximation using numerators and denominators less than 1000.*

9. (a) $\pi \approx 22/7$
(b) $e \approx 19/7$

10. (a) $\pi \approx 355/113$
(b) $e \approx 878/323$

In Exercises 11–20, graph the function and specify the domain, range, intercept(s), and asymptote.

11. $y = e^x$ **12.** $y = e^{-x}$
13. $y = -e^x$ **14.** $y = -e^{-x}$
15. $y = e^x + 1$ **16.** $y = e^{x+1}$
17. $y = e^{x+1} + 1$ **18.** $y = e^{x-1} - 1$
19. $y = e - e^x$ **20.** $y = e^{-x} - e$

Ⓖ *In Exercises 21 and 22, first tell what translations or reflections are required to obtain the graph of the given function from that of $y = e^x$. Then graph the given function along with $y = e^x$ and check that the picture is consistent with what you've said.*

21. (a) $y = e^{x+2}$
(b) $y = e^{-x+2}$

22. (a) $y = 1 - e^{-x}$
(b) $y = 1 - e^{-x+1}$

In Exercises 23 and 24, let $f(x) = x^2$, $g(x) = 2^x$, and $h(x) = e^x$. In each case, carry out the calculations needed to verify the given statement.

23. On the interval [0, 1]: The average rate of change of h is about 1.7 times that of g. The functions f and g have the same average rate of change on the interval [0, 1].

24. On the interval [9, 10]: The average rate of change of g is more than 25 times greater than that of f. The average rate of change of h is approximately 733 times that of f.

Ⓖ **25.** In this exercise we compare three functions:

$$y = 2^x \qquad y = e^x \qquad y = 3^x$$

(a) Begin by graphing all three functions in the standard viewing rectangle. The picture confirms three facts that you know from the text: on the positive x-axis, the functions increase very rapidly; on the negative x-axis, the graphs approach the asymptote (which is the x-axis) as you move to the left; the y-intercept in each case is 1.

(b) To compare the functions for positive values of x, use a viewing rectangle in which x extends from 0 to 3 and y extends from 0 to 10. Note that the graph of e^x is bounded between the graphs of 2^x and 3^x, just as the number e is between 2 and 3. In particular, the picture that you obtain demonstrates the following fact: For positive values of x,

$$2^x < e^x < 3^x$$

(c) To see the graphs more clearly when x is negative, change the viewing rectangle so that x extends from -3 to 0 and y extends from 0 to 1. Again, note that the graph of e^x is bounded between the graphs of 2^x and 3^x, but now the graph of 3^x is the bottom (rather than the top) curve in the picture. This demonstrates the following fact: For negative values of x,

$$3^x < e^x < 2^x$$

(d) Explain how the result in part (c) follows from the result in part (b).

26. This exercise introduces the approximation

$$e^x \approx x + 1 \qquad \text{provided that } x \text{ is close to zero}$$

This approximation has an important consequence when we study population growth in Section 5.7. (The details are in Exercise 61 in Section 5.7.)

(a) Use the information in Figure 3 to explain why the equation of the line tangent to the curve $y = e^x$ at the point (0, 1) is $y = x + 1$.

Ⓖ **(b)** Verify visually that the line $y = x + 1$ is tangent to the curve $y = e^x$ at the point (0, 1) by using a graphing utility to display the graphs of both functions in the same standard viewing rectangle. Note that the curve and the line are virtually indistinguishable in the immediate vicinity of the point (0, 1).

Ⓖ **(c)** Zoom in on the point (0, 1); use a viewing rectangle in which x extends from -0.05 to 0.05 and y extends from 0.95 to 1.05. Again, note that the line and the curve are virtually indistinguishable in the immediate vicinity of the point (0, 1). For a numerical look at this, complete the following table. In the columns for e^x and $e^x - (x + 1)$, round each entry to four decimal places. When you are finished, observe that the closer x is to 0, the closer the agreement between the two quantities $x + 1$ and e^x.

x	$x + 1$	e^x	$e^x - (x + 1)$
−0.04			
−0.03			
−0.02			
−0.01			
0.00			
0.01			
0.02			
0.03			
0.04			

27. Complete the following two tables. On the basis of the results you obtain, what would you say is the instantaneous rate of change of the function $f(x) = x^2$ at $x = 3$?

Interval	$\Delta f/\Delta x$ for $f(x) = x^2$
[3, 3.1]	
[3, 3.01]	
[3, 3.001]	
[3, 3.0001]	
[3, 3.00001]	

Interval	$\Delta f/\Delta x$ for $f(x) = x^2$
[2.9, 3]	
[2.99, 3]	
[2.999, 3]	
[2.9999, 3]	
[2.99999, 3]	

28. Tables 2(a) and 2(b) on page 339 were used to determine the instantaneous rate of change of $f(x) = x^2$ at $x = 1$. Set up and complete similar tables to find the instantaneous rate of change of $g(x) = x^3$ at $x = 1$.

29. Let $f(x) = e^x$. Complete the following table. In the right column, give the average rates of change to six decimal places without rounding. Note that the average rates of change seem to approach $e^2 = 7.3890560\ldots$. This result along with Table 4 in the text supports a statement in the box on page 340: The instantaneous rate of change of $f(x) = e^x$ at $x = 2$ is e^2.

Interval	$\Delta f/\Delta x$ for $f(x) = e^x$
[1.9, 2]	
[1.99, 2]	
[1.999, 2]	
[1.9999, 2]	
[1.99999, 2]	
[1.999999, 2]	
[1.9999999, 2]	

30. Complete a table similar to the one shown in Exercise 29, but use the function $h(x) = 3^x$, rather than $f(x) = e^x$. Note that the average rates of change do *not* approach $3^2 = 9$. What does this have to do with the statements in the box on page 340? (Explain using complete sentences.)

Exercises 31 and 32 refer to Example 2 in the text. Round all answers to the nearest integer. (Don't forget to include units with your answers.)

31. (a) Find the average rate of change of the population over the time period $5 \le t \le 6$.

(b) What is the instantaneous rate of change of the population at $t = 5$ hr? At $t = 5.5$ hr?

32. (a) Find the average rate of change of the population over the time period $4 \le t \le 6$.

(b) What is the instantaneous rate of change of the population at $t = 6$ hr?

33. As is indicated in the accompanying figure, the three points P, Q, and R lie on the graph of $y = e^x$, and the x-coordinates of these points are -1, 0, and 0.5, respectively. Use the theorem in the box on page 340 to specify the instantaneous rate of change of the function $y = e^x$ at each of these points. For one of the points, a numerical answer can be obtained without a calculator. For the other points, give two forms of each answer: an exact expression involving e and a calculator approximation rounded to two decimal places.

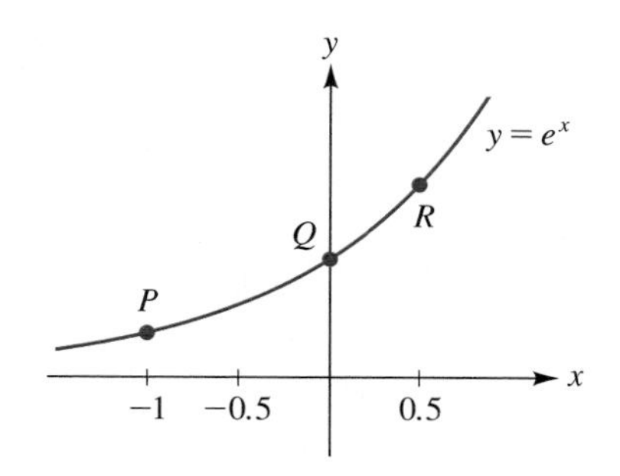

34. Suppose that during the first hour and 15 minutes of a physics experiment, the surface temperature of a small iron block is modeled by the exponential function $f(t) = 15e^t$, where $f(t)$ is the Celsius temperature t hours after the experiment begins.

Ⓖ **(a)** Use a graphing utility to graph the function $f(t) = 15e^t$ over the interval $0 \le t \le 1.25$.

(b) Compute the average rate of change of temperature over the second half hour of the experiment (i.e., over the interval $0.5 \le t \le 1$). Round the answer to one decimal place.

(c) Use one of the facts in the box on page 340 to determine the instantaneous rate of change of temperature: 30 minutes after the start of the experiment; 1 hour after the start of the experiment. Round each answer to one decimal place.

B

In Exercises 35–42, decide which of the following properties apply to each function. (More than one property may apply to a function.)

A. *The function is increasing for* $-\infty < x < \infty$.
B. *The function is decreasing for* $-\infty < x < \infty$.
C. *The function has a turning point.*
D. *The function is one-to-one.*
E. *The graph has an asymptote.*
F. *The function is a polynomial function.*
G. *The domain of the function is* $(-\infty, \infty)$.
H. *The range of the function is* $(-\infty, \infty)$.

35. $y = e^x$ **36.** $y = e^{-x}$
37. $y = -e^x$ **38.** $y = e^{|x|}$
39. $y = x + e$ **40.** $y = x^2 + e$
41. $y = x/e$ **42.** $y = e/x$

In Exercises 43–50, refer to the following graph of $y = e^x$. In each case, use the graph to estimate the indicated quantity to the nearest tenth (or closer, if it seems appropriate). Also, use a calculator to obtain a second estimate. Round the calculator values to three decimal places.

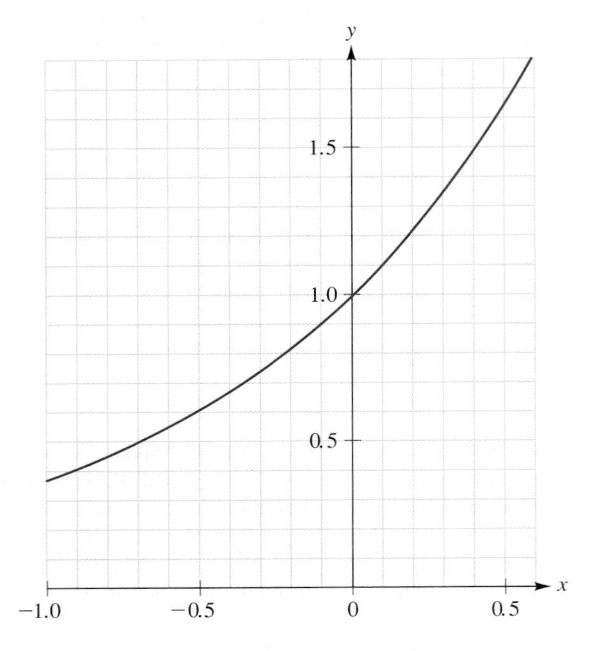

43. $e^{0.1}$ **44.** $e^{-0.1}$ **45.** $e^{-0.3}$
46. $e^{0.4}$ **47.** e^{-1} **48.** $\sqrt{e}$

49. $1/\sqrt{e}$ *Hint:* Rewrite this expression using a rational exponent, that is, $1/\sqrt{e} = e^{?}$.

50. $\sqrt[5]{e}$

51. (a) Use the graph preceding Exercise 43 to estimate the value of x for which $e^x = 1.5$.

(b) One of the keys on your calculator will allow you to solve the equation in part (a) for x; it is the "ln" key. Use your calculator to compute ln(1.5). Round your answer to three decimal places and check to see that it is consistent with the value obtained graphically in part (a). *Remark:* The "ln" function, which is called the *natural logarithm* function, is discussed in detail in the next section.

52. Follow Exercise 51 using the equation $e^x = 0.6$ (rather than $e^x = 1.5$).

53. Follow Exercise 51 using the equation $e^x = 1.8$.

54. Follow Exercise 51 using the equation $e^x = 0.4$.

55. The **hyperbolic cosine function,** denoted by **cosh,** is defined by the equation

$$\cosh(x) = \frac{1}{2}(e^x + e^{-x})$$

(a) Without a calculator, find cosh(0). Using a calculator, find cosh(1) and cosh(−1), rounding the answers to two decimal places.

(b) What is the domain of the function cosh? *Hint:* Visualize the graphs of $y = e^x$ and $y = e^{-x}$.

(c) Show that $\cosh(-x) = \cosh(x)$. What does this say about the graph of $y = \cosh(x)$?

Ⓖ **(d)** Use a graphing utility to graph $y = \cosh(x)$. Check that the picture is consistent with your answer in part (c).

Ⓖ **56.** (Continuation of Exercise 55.) Suppose that a flexible cable of uniform density is suspended between two supports, as indicated in Figure A. (Example: a telephone line between two phone poles.) Using calculus and physics, it can be shown that the curve formed is a portion of the graph of $y = a\cosh(x/a)$, where a is a constant. See Figure B. (The actual value for a depends on the tension in the cable and the density of the cable.) The graph of an equation of the form $y = a\cosh(x/a)$ is called a **catenary** (from the Latin *catena,* meaning *chain*).

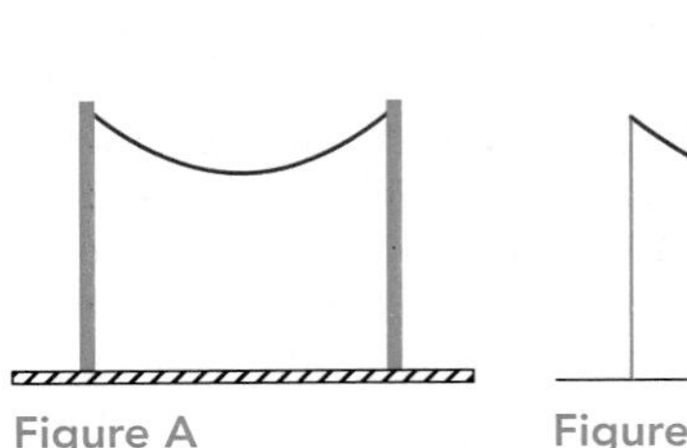
Figure A

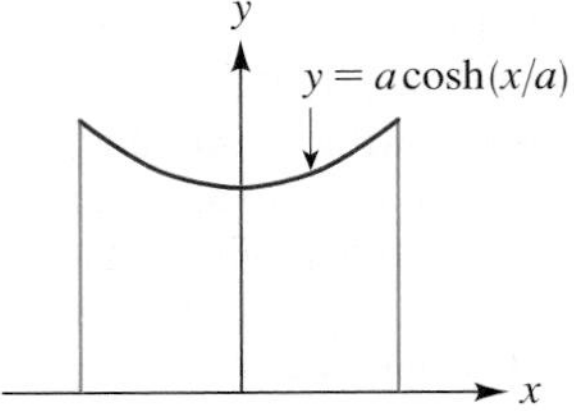

Figure B

(a) Use a graphing utility to graph the four catenaries $y = a\cosh(x/a)$ corresponding to $a = 0.2$, 1, 3, and 5. Describe (in complete sentences) how the shape and location of the curve changes as the value of a increases.

(b) The Gateway Arch in St. Louis (shown on page 336) is built in the form of an inverted catenary. The equation for the curve is given in Figure C. In this equation both x and y are measured in feet. Find the height of the Gateway Arch, then determine the width by using a graphing utility to graph the equation given in Figure C. *Remark:* The Gateway Arch is the tallest monument in the U.S. National Park System. Designed by architect Eero Saarinen and completed in 1965, the Gateway Arch commemorates the Louisiana Purchase, and more generally, the westward expansion of the United States over the period 1803–1890.

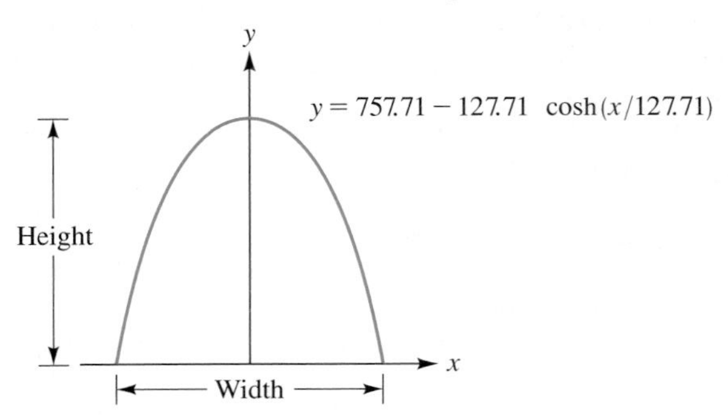

Figure C
Gateway Arch design. Distance on axes is measured in feet.

57. The **hyperbolic sine function,** denoted by **sinh,** is defined by the equation

$$\sinh(x) = \frac{1}{2}(e^x - e^{-x})$$

Note: For speaking and reading purposes, *sinh* is pronounced as "cinch."

(a) Without a calculator, find sinh(0). Using a calculator, find sinh(1) and sinh(−1), rounding the answers to two decimal places.

(b) What is the domain of the function sinh?

(c) Show that $\sinh(-x) = -\sinh(x)$. What does this say about the graph of $y = \sinh(x)$?

Ⓖ **(d)** Use a graphing utility to graph $y = \sinh(x)$. Check that the picture is consistent with your answer in part (c).

Ⓖ **(e)** How many points are there where the two curves $y = \sinh(x)$ and $y = x^3$ intersect?

58. For this exercise you need to know the definitions of the functions cosh and sinh, which were given in Exercises 55 and 57, respectively.

Ⓖ **(a)** Use a graphing utility to graph the equation $y = \cosh(x) + \sinh(x)$. What do you observe?

(b) Use algebra to show why the graph in part (a) looks as it does. [That is, use the definitions of cosh(x) and sinh(x) to simplify the quantity cosh(x) + sinh(x).]

59. The following figure shows portions of the graphs of the exponential functions $y = e^x$ and $y = \pi^x$.

(a) Use the figure (not your calculator) to determine which number is larger, e^π or π^e.

(b) Use your calculator to check your answer in part (a).

Remark: The number e^π is known to be irrational. This was proven by the Russian mathematician A. O. Gelfond in 1929. It is still not known, however, whether π^e is irrational.

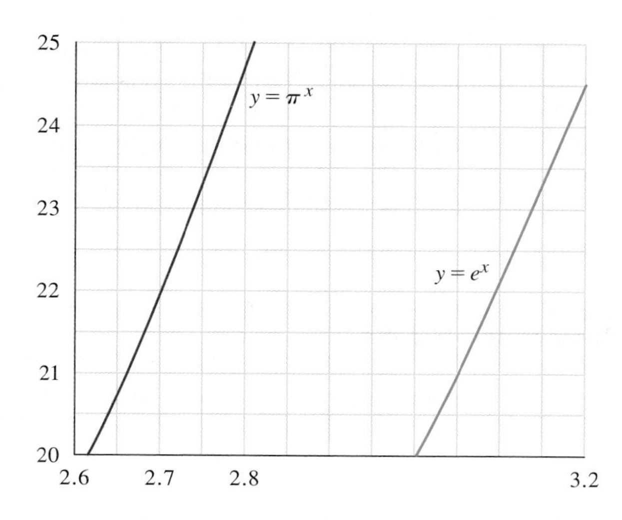

60. (a) Use your calculator to approximate the numbers $e\pi$ and $e + \pi$. *Remark*: It is not known whether these numbers are rational or irrational.

(b) Use your calculator to approximate $(e^{\pi\sqrt{163}})^{1/3}$.

Remark: Contrary to the evidence on your calculator, it is known that this number is irrational.

61. Let $f(x) = e^x$. Let L denote the function that is the inverse of f.

(a) On the same set of axes, sketch the graphs of f and L.

Hint: You do not need the equation for $L(x)$.

(b) Specify the domain, range, intercept, and asymptote for the function L and its graph.

(c) Graph each of the following functions. Specify the intercept and asymptote in each case.

(i) $y = -L(x)$ **(ii)** $y = L(-x)$ **(iii)** $y = L(x-1)$

62. The text mentioned two ways to define the number e. One involved the expression $(1 + 1/x)^x$ and the other involved a tangent line. In this exercise you'll see that these two approaches are, in fact, related. For convenience, we'll write the expression $(1 + 1/x)^x$ using the letter n rather than x (so that we can use x for something else in a moment.) Thus, we have

$$(1 + 1/n)^n \approx e \quad \text{as } n \text{ becomes larger and larger without bound} \tag{1}$$

Working from this approximation, we'll obtain evidence that the slope of the tangent to the curve $y = e^x$ at $x = 0$ is 1. (A rigorous proof requires calculus.)

(a) Define x by the equation $x = 1/n$. As n becomes larger and larger without bound, what happens to the corresponding values of x? Complete the following table.

n	10^2	10^3	10^6	10^9
x				

(b) Substitute $x = 1/n$ in approximation (1) to obtain

$$(1 + x)^{1/x} \approx e \quad \text{as } x \text{ approaches } 0$$

Next, raise both sides to the power x to obtain

$$(1 + x) \approx e^x \quad \text{as } x \text{ approaches } 0 \tag{2}$$

Approximation (2) says that the curve $y = e^x$ looks more and more like the line $y = 1 + x$ as x approaches 0. For numerical perspective on this, complete the following tables. These results suggest (but do not *prove*) that the line $y = 1 + x$ is tangent to the curve $y = e^x$ at $x = 0$. Thus, the results suggest that the slope of the tangent to $y = e^x$ at $x = 0$ is 1.

x	0.3	0.2	10^{-1}	10^{-2}	10^{-3}	10^{-4}
x + 1						
e^x						

x	−0.3	−0.2	-10^{-1}	-10^{-2}	-10^{-3}	-10^{-4}
x + 1						
e^x						

C

For Exercises 63 and 64, you need to know the definitions of the functions cosh and sinh given in Exercises 55 and 57, respectively.

63. Consider the following two equations:

$$[\cosh(x)]^2 - 2[\sinh(x)]^2 = 1 \qquad [\cosh(x)]^2 - [\sinh(x)]^2 = 1$$

G **(a)** One of the given equations is an identity, and the other is not. Use a graphing utility to determine which equation is *not* an identity. For that equation, make a conjecture about the root(s). Then use the definitions of $\cosh(x)$ and $\sinh(x)$ to verify algebraically that your conjecture is valid.

(b) For the other equation, use the definitions of $\cosh(x)$ and $\sinh(x)$ to prove that the equation is indeed an identity.

64. Consider the following three equations:

$$[\cosh(x)]^2 = \frac{1}{2}[\cosh(2x)+1] \qquad [\sinh(x)]^2 = \frac{1}{2}[\cosh(2x)-1]$$

$$[\sinh(x)]^2 = \frac{1}{2}[\sinh(2x)+1]$$

G **(a)** Only one of the three equations is not an identity. Use a graphing utility to find out which. For that equation, are there any values of x for which the equation becomes a true statement? If so, use the graphing utility to estimate to the nearest hundredth each such value.

(b) For each of the other two equations, use the definitions of $\cosh(x)$ and $\sinh(x)$ to prove that the equation is indeed an identity.

PROJECT COFFEE TEMPERATURE

Suppose that a cup of hot coffee at a temperature of T_0 is set down to cool in a room where the temperature is kept at T_1. Then the temperature of the coffee as it cools can be modeled by the function

$$f(t) = (T_0 - T_1)e^{kt} + T_1 \qquad (1)$$

Here, $f(t)$ is the temperature of the coffee after t minutes; $t = 0$ corresponds to the initial instant when the temperature of the hot coffee is T_0; and k is a (negative) constant that depends, among other factors, on the dimensions of the cup and the material from which it is constructed. [This model is derived in calculus. It is based on *Newton's law of cooling*: The rate of change of temperature of a cooling object is *proportional* to the difference between the temperature of the object and the surrounding temperature. (*Note*: *y is proportional to x* means $y = kx$ for some constant k.)]

After studying the following example, solve Problem A below by applying equation (1) and using the technique shown in the example. Then try your hand at Problem B. You won't be able to complete Problem B using the techniques developed in this text up to now. In complete sentences, explain exactly at which point you get stuck. What would you have to know how to do to complete the solution? (In the next section we discuss *logarithms,* which will allow us to complete Problem B.) For now, use a graphing utility to obtain an approximate answer for Problem B.

EXAMPLE Coffee Temperature Project

Given that the graph of the function $y = ae^{kx}$ passes through the points $(0, 2)$ and $(3, 5)$, find y when $x = 7$.

SOLUTION

Substituting $x = 0$ and $y = 2$ in the given equation yields $2 = ae^0$, and therefore $a = 2$. Next, substituting $x = 3$ and $y = 5$ in the equation $y = 2e^{kx}$ yields $5 = 2e^{k(3)}$, or $e^{3k} = 2.5$. This last equation can be rewritten $(e^k)^3 = 2.5$. Taking the cube root of both sides gives us

$$e^k = 2.5^{1/3} \tag{2}$$

The original function can now be written as follows

$$\begin{aligned} y = 2e^{kx} &= 2(e^k)^x \\ &= 2(2.5^{1/3})^x \qquad \text{using equation (2)} \end{aligned}$$

That is, $y = 2(2.5^{x/3})$. In this form, we can calculate y when x is given. In particular then, when $x = 7$, we obtain

$$\begin{aligned} y &= 2(2.5^{7/3}) \\ &\approx 16.97 \qquad \text{using a calculator} \end{aligned}$$

PROBLEM A

Suppose that a cup of hot coffee at a temperature of 185°F is set down to cool in a room where the temperature is kept at 70°F. What is the temperature of the coffee ten minutes later?

PROBLEM B

Suppose that a cup of hot coffee at a temperature of 185°F is set down to cool in a room where the temperature is kept at 70°F. How long will it take for the coffee to cool to 140°F?

5.3 LOGARITHMIC FUNCTIONS

But if logarithms have lost their role as the centerpiece of computational mathematics, the logarithmic function *remains central to almost every branch of mathematics, pure or applied. It shows up in a host of applications, ranging from physics and chemistry to biology, psychology, art, and music.* —Eli Maor in *e: The Story of a Number* (Princeton, N.J.: Princeton University Press, 1994)

[John Napier] *hath set my head and hands a work with his new and admirable logarithms. I hope to see him this summer, if it please God, for I never saw book that pleased me better, or made me more wonder.* —Henry Briggs, March 10, 1615

It has been thought that the earliest reference to the logarithmic curve was made by the Italian Evangelista Torricelli in a letter on the year 1644, but Paul Tannery made it practically certain that Descartes knew the curve in 1639. —Florian Cajori in *A History of Mathematics,* 4th ed. (New York: Chelsea Publishing Co., 1985)

For the number whose logarithm is unity, let e *be written, . . .* —Leonhard Euler in a letter written in 1727 or 1728

In the previous two sections we studied exponential functions. Now we consider functions that are inverses of exponential functions. These inverse functions are called *logarithmic functions.*

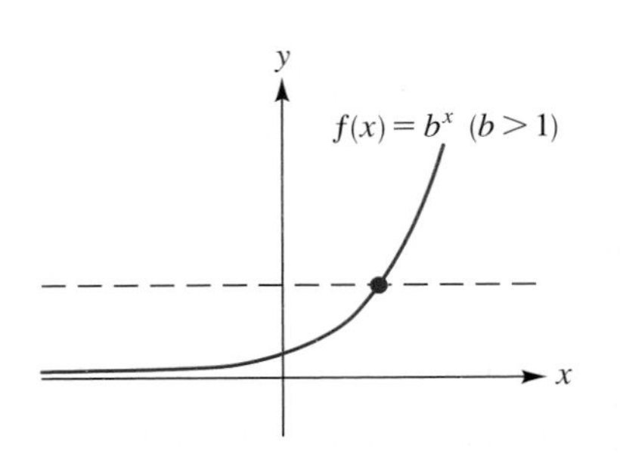

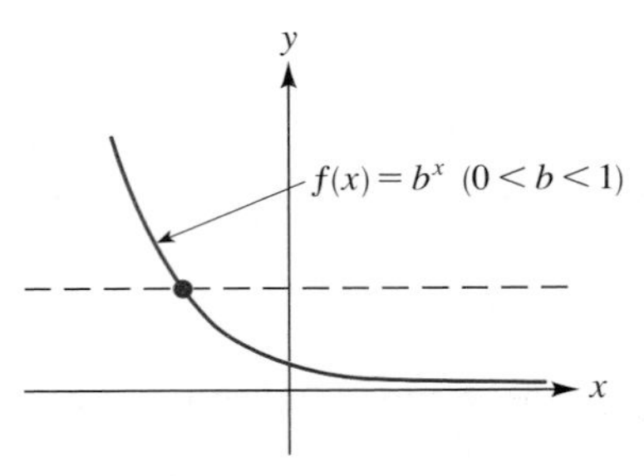

Figure 1
The exponential function $y = b^x$ is one-to-one.

Having said this, let's back up for a moment to review briefly some of the basic ideas behind inverse functions (as discussed in Section 3.6). We start with a given function F, for example, $F(x) = 3x$, that is one-to-one. (That is, for each output there is only one input.) Then, by interchanging the inputs and outputs, we obtain a new function, the so-called inverse function. In the case of $F(x) = 3x$, it's easy to find an equation defining the inverse function. We just interchange x and y in the equation $y = 3x$ to obtain $x = 3y$. Solving for y in this last equation then gives us $y = \frac{1}{3}x$, which defines the inverse function. Using function notation, we can summarize the situation by writing $F(x) = 3x$ and $F^{-1}(x) = \frac{1}{3}x$.

In the preceding paragraph we saw that a particular linear function had an inverse, and we found a formula for that inverse. Now let's repeat that same reasoning, beginning with an exponential function. First, we must make sure the exponential function f defined by $f(x) = b^x$ is one-to-one. We can see this by applying the horizontal line test, as indicated in Figure 1. Next, since $f(x) = b^x$ is one-to-one, it has an inverse function. Let's study this inverse.

We begin by writing the exponential function $f(x) = b^x$ in the form

$$y = b^x \tag{1}$$

Then, to obtain an equation for f^{-1}, we interchange x and y in equation (1). This gives us

$$x = b^y \tag{2}$$

The crucial step now is to express equation (2) in words:

$$y \text{ is the exponent to which } b \text{ must be raised to yield } x. \tag{3}$$

Statement (3) defines the function that is the inverse of $y = b^x$. Now we introduce a notation that will allow us to write this statement in a more compact form.

DEFINITION $\log_b x$

We define the expression $\mathbf{log}_b\, \boldsymbol{x}$ to mean "the exponent to which b must be raised to yield x." ($\log_b x$ is read *log base b of x* or *the logarithm of x to the base b.*)

EXAMPLES

(a) $\log_2 8 = 3$, since 3 is the exponent to which 2 must be raised to yield 8.

(b) $\log_{10}(1/10) = -1$, since -1 is the exponent to which 10 must be raised to yield 1/10.

(c) $\log_5 1 = 0$, since 0 is the exponent to which 5 must be raised to yield 1.

Using this notation, statement (3) becomes

$$y = \log_b x$$

Since (2) and (3) are equivalent, we have the following important relationship.

$$y = \log_b x \quad \text{is equivalent to} \quad x = b^y$$

We say that the equation $y = \log_b x$ is in **logarithmic form** and that the equivalent equation $x = b^y$ is in **exponential form.** Table 1 displays some examples.

TABLE 1

Exponential form of equation	Logarithmic form of equation
$8 = 2^3$	$\log_2 8 = 3$
$\frac{1}{9} = 3^{-2}$	$\log_3 \frac{1}{9} = -2$
$1 = e^0$	$\log_e 1 = 0$
$a = b^c$	$\log_b a = c$

Let us now summarize our discussion up to this point.

1. According to the horizontal line test, the function $f(x) = b^x$ is one-to-one and therefore possesses an inverse. This inverse function is written
$$f^{-1}(x) = \log_b x$$
2. $\log_b a = c$ means that $a = b^c$.

To graph the function $y = \log_b x$, we recall from Section 3.5 that the graph of a function and its inverse are reflections of one another about the line $y = x$. Thus to graph $y = \log_b x$, we need only reflect the curve $y = b^x$ about the line $y = x$. This is shown in Figure 2 for the case $b > 1$.

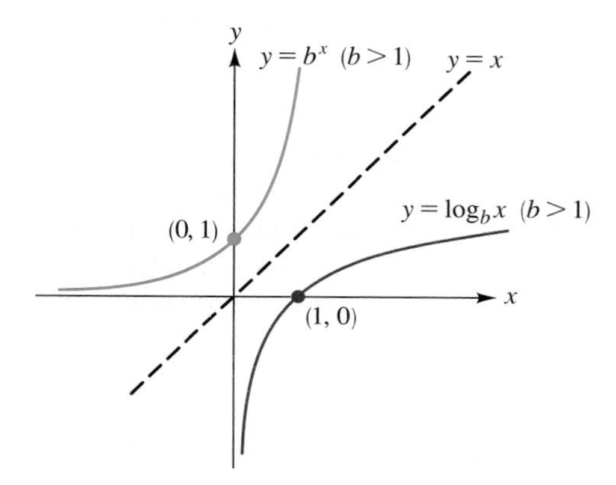

Figure 2

NOTE For the rest of this chapter we assume that the base b is greater than 1 when we use the expression $\log_b x$.

With the aid of Figure 2 we can make the following observations about the function $y = \log_b x$.

PROPERTY SUMMARY
The Logarithmic Function $y = \log_b x$ $(b > 1)$

Domain: $(0, \infty)$
Range: $(-\infty, \infty)$
y-intercept: none
x-intercept: 1
Asymptote: y-axis

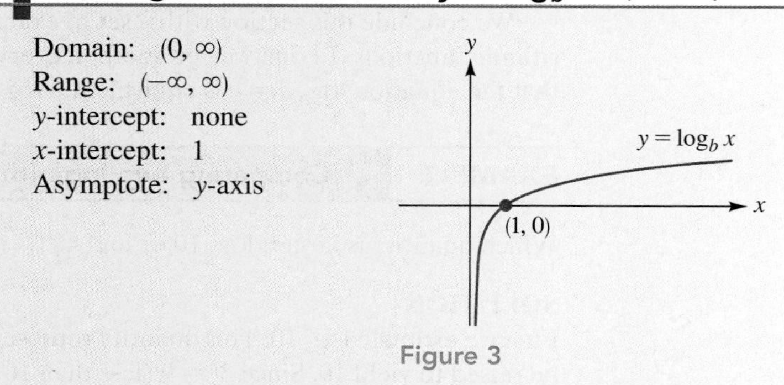

Figure 3

One aspect of the function $y = \log_b x$ may not be immediately apparent to you from Figures 2 and 3: The function grows or increases *very* slowly. Consider $y = \log_2 x$, for example. Let's ask how large x must be before the curve reaches the height $y = 10$ (see Figure 4).

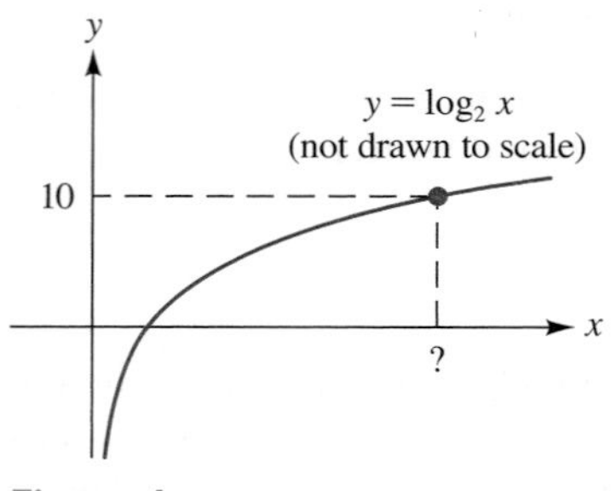

Figure 4

To answer this question, we substitute $y = 10$ in the equation $y = \log_2 x$:

$$10 = \log_2 x$$

Writing this equation in exponential form yields

$$x = 2^{10} = 1024$$

In other words, we must go out beyond 1000 on the x-axis before the curve $y = \log_2 x$ reaches a height of 10 units. Exercise 63 at the end of this section asks you to show that the graph of $y = \log_2 x$ doesn't reach a height of 100 until x is greater than 10^{30}. (Numbers as large as 10^{30} rarely occur in any of the sciences. For instance, the distance in inches to the Andromeda galaxy is less than 10^{24}.) The point we are emphasizing is this: The graph of $y = \log_2 x$ (or $\log_b x$) is always rising, but very slowly.

Before going on to consider some numerical examples, let's pause for a moment to think about the notation we've been using. In the expression

$$\log_b x$$

the name of the function is $\log_b$ and the input is x. To emphasize this, we might be better off writing $\log_b x$ as $\log_b(x)$, so that the similarity to the familiar $f(x)$ notation is clear. However, for historical reasons* the convention is to suppress the parentheses, and we follow that convention here. In the box that follows, we indicate some errors that can occur upon forgetting that $\log_b$ is the name of a function, not a number.

ERRORS TO AVOID

Error	Correction	Comment
$\dfrac{\log_2 8}{\log_2 4} = \dfrac{8}{4}$	$\dfrac{\log_2 8}{\log_2 4} = \dfrac{3}{2}$	$\log_2$ is the name of a function. It is not a factor that can be "cancelled" from the numerator and denominator.
$\dfrac{\log_2 16}{16} = \log_2$	$\dfrac{\log_2 16}{16} = \dfrac{4}{16} = \dfrac{1}{4}$	This "equation" is nonsense. On the left-hand side there is a quotient of two real numbers, which is a real number. On the right-hand side is the name of a function.

We conclude this section with a set of examples involving logarithms and logarithmic functions. In one way or another, every example makes use of the key fact that the equation $\log_b a = c$ is equivalent to $b^c = a$.

EXAMPLE 1 Comparing two logarithms by using the definition

Which quantity is larger: $\log_3 10$ or $\log_7 40$?

SOLUTION

First we estimate $\log_3 10$. This quantity represents the exponent to which 3 must be raised to yield 10. Since $3^2 = 9$ (less than 10), but $3^3 = 27$ (more than 10), we conclude that the quantity $\log_3 10$ lies between 2 and 3. In a similar way we can

*The notation *log* was introduced in 1624 by the astronomer Johannes Kepler (1571–1630). In Leonhard Euler's text *Introduction to Analysis of the Infinite,* first published in 1748, appears the statement, "It has been customary to designate the logarithm of y by the symbol $\log y$."

estimate $\log_7 40$; this quantity represents the exponent to which 7 must be raised to yield 40. Since $7^1 = 7$ (less than 40) while $7^2 = 49$ (more than 40), we conclude that the quantity $\log_7 40$ lies between 1 and 2. It now follows from these two estimates that $\log_3 10$ is larger than $\log_7 40$.

EXAMPLE 2 Evaluating a logarithmic expression

Evaluate $\log_4 32$.

SOLUTION

Let $y = \log_4 32$. The exponential form of this equation is $4^y = 32$. Now, since both 4 and 32 are powers of 2, we can rewrite the equation $4^y = 32$ using the same base on both sides:

$$\begin{aligned} (2^2)^y &= 2^5 \\ 2^{2y} &= 2^5 \\ 2y &= 5 && \text{using Property 4 on page 328} \\ y &= \frac{5}{2} && \text{as required} \end{aligned}$$

EXAMPLE 3 Sketching the graph of a logarithmic function and its reflection

Graph the following equations:

(a) $y = \log_{10} x$; **(b)** $y = -\log_{10} x$.

SOLUTION

(a) The function $y = \log_{10} x$ is the inverse function for the exponential function $y = 10^x$. Thus we obtain the graph of $y = \log_{10} x$ by reflecting the graph of $y = 10^x$ in the line $y = x$; see Figure 5(a).

(b) To graph $y = -\log_{10} x$, we reflect the graph of $y = \log_{10} x$ in the x-axis; see Figure 5(b).

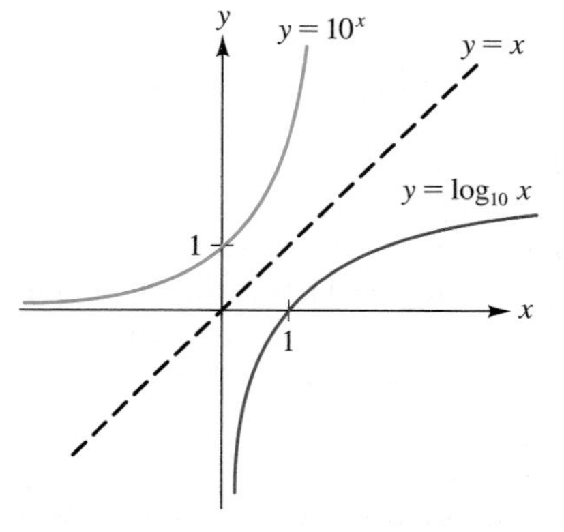

(a)

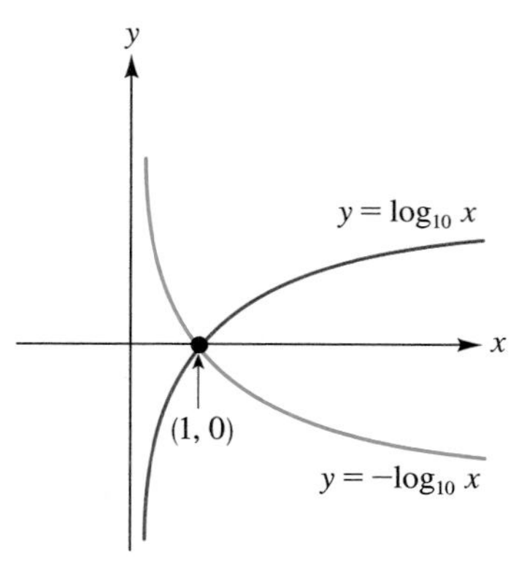

(b)

Figure 5

EXAMPLE 4 **Finding the domain of a function defined by a logarithm**

Find the domain of the function $f(x) = \log_2(12 - 4x)$.

SOLUTION

As you can see by looking back at Figure 3 on page 349, the inputs for the logarithmic function must be positive. So, in the case at hand, we require that the quantity $12 - 4x$ be positive. Consequently, we have

$$\begin{aligned} 12 - 4x &> 0 \\ -4x &> -12 \\ x &< 3 \end{aligned}$$

Therefore the domain of the function $f(x) = \log_2(12 - 4x)$ is the interval $(-\infty, 3)$.

The next example concerns the exponential function $y = e^x$ and its inverse function, $y = \log_e x$. Many books, as well as calculators, abbreviate the expression $\log_e x$ by $\ln x$, read *natural log of x*.* For reference and emphasis we repeat this fact in the following box. (Incidentally, on most calculators, "log" is an abbreviation for $\log_{10}$.)

DEFINITION **The "ln" Notation for Base *e* Logarithms**

$\ln x$ means $\log_e x$

EXAMPLES

1. $\ln e = 1$ because $\ln e$ stands for $\log_e e$, which equals 1.
2. $\ln(e^2) = 2$ because $\ln(e^2)$ stands for $\log_e(e^2)$, which equals 2.
3. $\ln 1 = 0$ because $\ln 1$ stands for $\log_e 1$, which equals 0. (The exponential form of the equation $\ln 1 = 0$ is $e^0 = 1$.)

EXAMPLE 5 **Sketching the graph of ln x and a translation**

Graph the following functions:

(a) $y = \ln x$; **(b)** $y = \ln(x - 1)$.

SOLUTION

(a) The function $y = \ln x$ $(=\log_e x)$ is the inverse of $y = e^x$. Thus its graph is obtained by reflecting the graph of $y = e^x$ in the line $y = x$, as in Figure 6(a).

(b) To graph $y = \ln(x - 1)$, we take the graph of $y = \ln x$ and move it one unit in the positive x-direction; see Figure 6(b).

*According to the historian Florian Cajori, the notation $\ln x$ was used by (and perhaps first introduced by) Irving Stringham in his text *Uniplanar Algebra* (San Francisco: University Press, 1893).

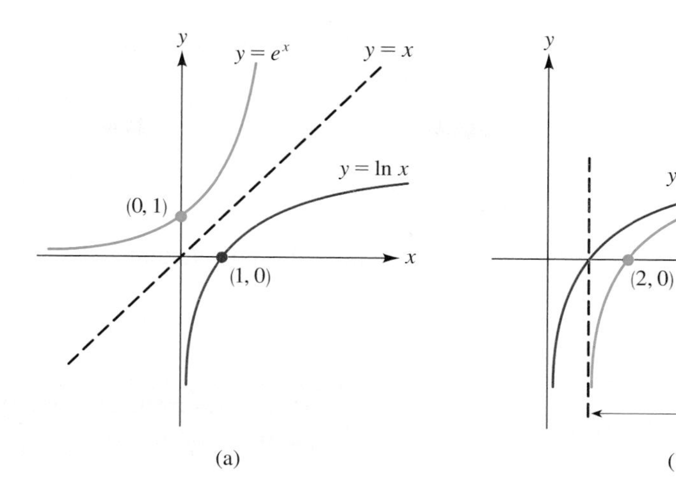

Figure 6

EXAMPLE 6 Solving an equation involving an exponential function

Consider the equation $10^{x/2} = 16$.

(a) Use a graphing utility to show that the equation has exactly one root. Then estimate the root to the nearest one-tenth (by finding an appropriate viewing rectangle).

(b) Solve the given equation algebraically by rewriting it in logarithmic form. Give two forms for the answer: an exact expression and a calculator approximation rounded to three decimal places. Check to see that the result is consistent with the graphical estimate obtained in part (a).

SOLUTION

(a) Figure 7(a) shows a viewing rectangle in which the graphs of $y = 10^{x/2}$ and $y = 16$ intersect. The x-coordinate of this intersection point is a root of the given equation. As you can check by zooming out from this view, there are no other intersection points, and thus no other roots of the given equation.

From Figure 7(a) we know that the required root is between 2 and 3, a bit closer to 2 than to 3. This suggests that we try a viewing rectangle with x running from 2 to 2.5, as indicated in Figure 7(b). This view shows that the

GRAPHICAL PERSPECTIVE

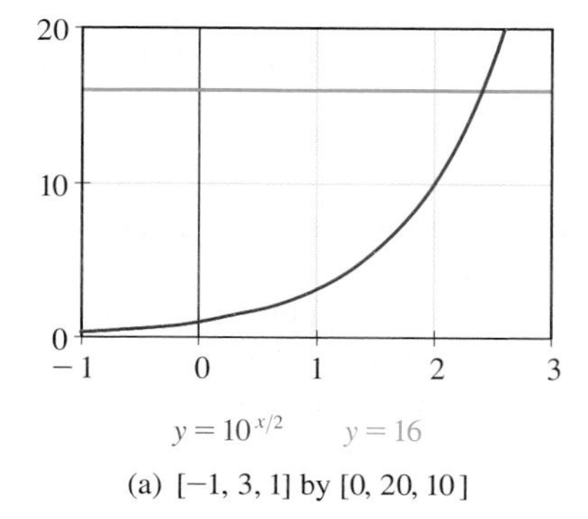

(a) [−1, 3, 1] by [0, 20, 10]

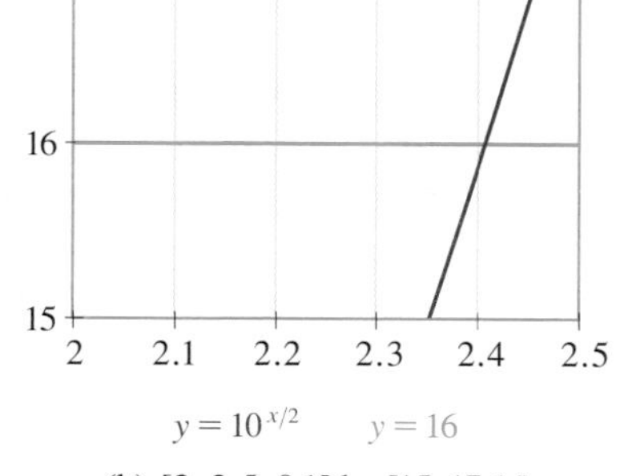

(b) [2, 2.5, 0.1] by [15, 17, 1]

Figure 7

root is closer to 2.4 than to either 2.3 or 2.5. Thus to the nearest one-tenth the required root is $x \approx 2.4$.

(b) Rewriting the equation $10^{x/2} = 16$ in the equivalent logarithmic form gives us $x/2 = \log_{10} 16$, and consequently,

$$\begin{aligned} x &= 2 \log_{10} 16 && \text{multiplying both sides by 2} \\ &\approx 2.408 && \text{using a calculator} \end{aligned}$$

Note that this value is consistent with the graphical estimate 2.4 that was obtained in part (a).

EXAMPLE 7 Using the natural logarithm function to solve an equation

Solve the equation $e^{2t-5} = 5000$. Give two forms for the answer: an exact expression and a calculator approximation rounded to three decimal places.

SOLUTION

Rewriting the equation in logarithmic form yields $2t - 5 = \ln 5000$. Consequently, we have

$$\begin{aligned} 2t &= 5 + \ln 5000 && \\ t &= (5 + \ln 5000)/2 && \text{dividing by 2} \\ &\approx 6.759 && \text{using a calculator} \end{aligned}$$

Remark: This result says that, for the function $y = e^{2t-5}$, a relatively small input of about 6.8 produces a very large output of 5000. This is a reminder of the fact that exponential functions grow very rapidly (when the exponents are positive and the base is greater than 1).

The next example indicates one of the many ways that logarithmic functions occur in applications. Geologists record the vibrations from an earthquake on a device known as a *seismograph*. Figure 8 shows a (simplified) output from a seismograph during an earthquake.

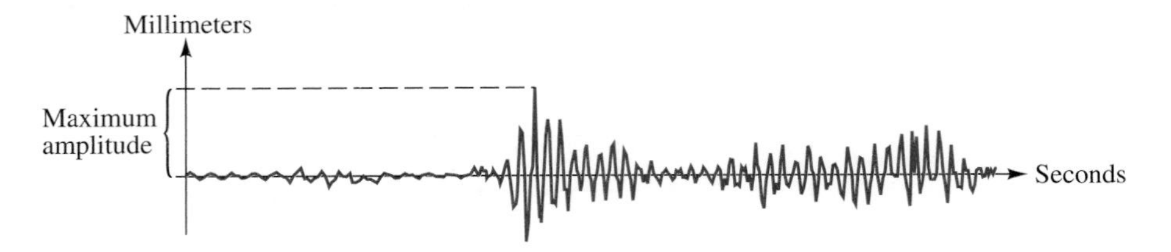

Figure 8

The maximum amplitude of the shock wave provides a measure of the intensity or strength of the particular earthquake. One of the problems faced by the American geologist Charles F. Richter (1900–1985) in devising a meaningful scale for comparing the sizes of earthquakes had to do with the fact that the maximum amplitudes for different earthquakes can vary greatly, by factors of thousands or more. For instance, an earthquake so small that most people would not feel it might register a maximum amplitude of 0.2 mm, whereas a destructive (yet not major) quake might register a maximum amplitude of 1000 mm. But a comparison scale for rating earthquakes that ranged from 0.2 to 1000, say, would not be especially useful; the wide range of numbers would make it nonintuitive. For instance, you might ask a friend to rate on a scale of 1 to 10 a movie he or she had seen, but you wouldn't want to use a scale of 1 to 1000. It would be nonintuitive and not especially informative in most cases. On the other hand, if we introduce a logarithm function in this example, we have $\log_{10} 0.2 \approx -0.7$ and $\log_{10} 1000 = 3$. So while the original amplitudes range numerically from (approximately) 0 to 1000, the logs range from (approximately) -1 to 3, a much smaller, more manageable scale in this case. This is the observation that Dr. Richter used in defining his scale for comparing earthquakes. In his words:

> *. . . the range between the largest and smallest magnitudes seemed unmanageably large. Dr. Beno Gutenberg* [1889–1960] *then made the natural suggestion to plot the amplitudes logarithmically.*
>
> Charles F. Richter in an interview by Henry Spall of the U.S. Geological Survey, http://wwwneic.cr.usgs.gov/neis/seismology/people/int_richter.html

In its simplest form, the Richter magnitude M of an earthquake is defined by the equation

$$M = \log_{10}\left(\frac{A}{A_0}\right)$$

where A is the maximum amplitude recorded on a standard seismograph located 100 km from the epicenter of the earthquake, and the constant A_0 is the maximum amplitude of a certain standard intensity earthquake. In Example 8 we use this definition to compare the intensities of two earthquakes. *Remark*: Richter introduced this formula for the magnitude of an earthquake in 1935. Beginning in 1936, he worked with Gutenberg to modify the formula to account for different types of seismographs located at arbitrary distances from the epicenter of the quake. For more information about earthquakes and hands-on experience in computing the Richter magnitude of an earthquake, see the following *VirtualEarthquake* web pages by Professor Gary Novak:

http://vcourseware.calstatela.edu/VirtualEarthquake/VQuakeIntro.html

EXAMPLE 8 Using the Richter magnitude scale

The Richter magnitudes for two large earthquakes are given in Table 2. Use the formula $M = \log_{10}(A/A_0)$ to determine how many times stronger was the earthquake in India than the earthquake in California. Assume that A (the maximum amplitude on the seismograph) is a measure of the strength of an earthquake.

TABLE 2 Data for Two Earthquakes

Location of earthquake	Date	Richter magnitude M
Northridge, California	January 17, 1994	6.8
Gujarat Province, India	January 26, 2001	7.9

SOLUTION

We begin by solving the Richter magnitude equation $M = \log_{10}(A/A_0)$ for A. We want to do this so that we can compare the values of A for the two quakes. Rewriting the equation in its equivalent exponential form yields $A/A_0 = 10^M$, and therefore

$$A = A_0 10^M$$

Now, using this last equation and the values of M given in Table 2, we have

For Northridge quake:	For Gujarat quake:
$A = A_0 10^{6.8}$	$A = A_0 10^{7.9}$

Computing the ratio of these two A-values gives us

$$\frac{A\text{-value for Gujarat}}{A\text{-value for Northridge}} = \frac{A_0 10^{7.9}}{A_0 10^{6.8}} = 10^{7.9-6.8} = 10^{1.1}$$
$$\approx 12.6 \qquad \text{using a calculator}$$

In summary, the quake in India was more than 12 times stronger than the one in California, even though the Richter magnitudes differ by only 1.1. *Remark*: The Richter scale is an example of what we call a *logarithmic scale*. Exercise 45 asks you to show that if the Richter magnitudes of two quakes differ by an amount d, then the larger quake is 10^d (not just d) times stronger than the smaller quake. Other applications of logarithmic scales are given in Exercises 46–52.

EXERCISE SET 5.3

A

In Exercises 1–4, fill in the blanks.

1. If f and g are inverse functions, then:
(a) $f[g(x)] =$ _____ for each x in the domain of g.
(b) $g[f(x)] =$ _____ for each x in the domain of _____.

2. Suppose that f and g are inverse functions, 5 is in the domain of f, and 3 is in the domain of g. Then $f[g(3)] =$ _____, and $g[f(5)] =$ _____.

3. **(a)** The functions $f(x) = 2^x$ and $g(x) =$ _____ are inverse functions.
(b) In view of part (a) we have $2^{\log_2 x} =$ _____ for each positive number x, and $\log_2(2^x) =$ _____ for all real numbers x.
(c) So, for example, without calculating, I know that $2^{\log_2 99} =$ _____ and also that $\log_2(2^{-\pi}) =$ _____.

4. **(a)** The functions $f(x) = e^x$ and $g(x) =$ _____ are inverse functions.
(b) In view of part (a) we have $e^{\ln x} =$ _____ for each positive number x, and $\ln(e^x) =$ _____ for all real numbers x.
(c) So, for example, without calculating, I know that $e^{\ln 99} =$ _____ and also that $\ln(e^{-\pi}) =$ _____.

Exercises 5–8 provide additional review on the inverse function concept. The exercises do not explicitly involve logarithmic functions (but may be helpful to you because we've defined logarithmic functions in terms of inverse functions).

5. Which of the following functions are one-to-one and therefore possess an inverse?
(a) $y = x + 1$
(b) $y = (x + 1)^2$
(c) $y = (x + 1)^3$

6. Let $f(x) = 4x - 5$.
(a) Find $f^{-1}(x)$.
(b) Find $f[f^{-1}(x)]$.
(c) Find $[f(x)][f^{-1}(x)]$.
(d) Note that your answers for parts (b) and (c) are not the same (or at least they shouldn't be). Of the two expressions given in parts (b) and (c), which one is involved in the definition of an inverse function?

7. Let $f(x) = x^3 + 5x + 1$. Given that f^{-1} exists and 6 is in the domain of f^{-1}, evaluate $f[f^{-1}(6)]$.

8. The graph of $y = f(x)$ is a line segment joining the two points $(3, -2)$ and $(-1, 5)$. Specify the corresponding endpoints for the graph of:
(a) $y = f^{-1}(x)$ **(b)** $y = f^{-1}(x - 1)$

In Exercises 9 and 10, write each equation in logarithmic form.

9. **(a)** $9 = 3^2$
(b) $1000 = 10^3$
(c) $7^3 = 343$
(d) $\sqrt{2} = 2^{1/2}$

10. **(a)** $1/125 = 5^{-3}$
(b) $e^0 = 1$
(c) $5^x = 6$
(d) $e^{3t} = 8$

In Exercises 11 and 12, write each equation in exponential form.

11. **(a)** $\log_2 32 = 5$
(b) $\log_{10} 1 = 0$
(c) $\log_e \sqrt{e} = 1/2$
(d) $\ln(1/e) = -1$

12. **(a)** $\ln u = s$
(b) $\log_a b = c$

In Exercises 13 and 14, complete the tables.

13.

x	1	10	10^2	10^3	10^{-1}	10^{-2}	10^{-3}
$\log_{10} x$							

14.

x	1	e	e^2	e^3	e^{-1}	e^{-2}	e^{-3}
$\ln x$							

In Exercises 15 and 16, rely on the definition of $\log_b x$ *(in the box on page 348) to decide which of the two quantities is larger.*

15. **(a)** $\log_5 30$ or $\log_8 60$
(b) $\ln 17$ or $\log_{10} 17$

16. **(a)** $\log_9 80$ or $\ln(e^2)$
(b) $\log_2 3$ or $\log_3 2$

In Exercises 17 and 18, evaluate each expression.

17. **(a)** $\log_9 27$
(b) $\log_4(1/32)$
(c) $\log_5 5\sqrt{5}$

18. **(a)** $\log_{25}(1/625)$
(b) $\log_{16}(1/64)$
(c) $\log_{10} 10$
(d) $\log_2 8\sqrt{2}$

In Exercises 19 and 20, solve each equation for x by converting to exponential form. In Exercises 19(b) and 20, give two forms for each answer: one involving e and the other a calculator approximation rounded to two decimal places.

19. **(a)** $\log_4 x = -2$
(b) $\ln x = -2$

20. **(a)** $\log_5 x = e$
(b) $\ln x = -e$

In Exercises 21 and 22, find the domain of each function.

21. **(a)** $y = \log_4 5x$
(b) $y = \log_{10}(3 - 4x)$
(c) $y = \ln(x^2)$
(d) $y = (\ln x)^2$
(e) $y = \ln(x^2 - 25)$

22. **(a)** $y = \ln(2 - x - x^2)$
(b) $y = \log_{10} \dfrac{2x + 3}{x - 5}$

23. In the accompanying figure, what are the coordinates of the four points A, B, C, and D?

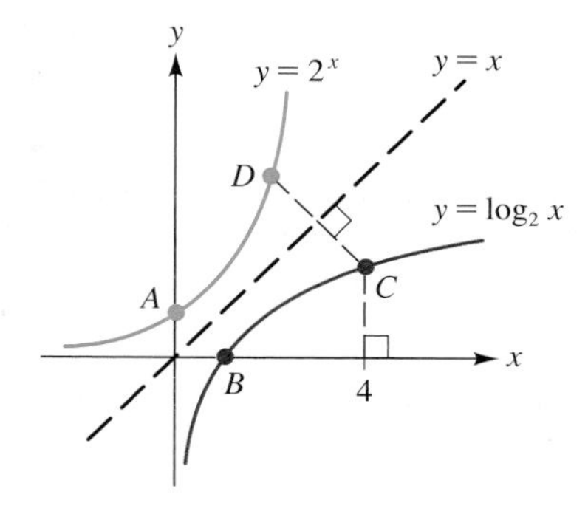

24. In the accompanying figure, what are the coordinates of the points A, B, C, and D?

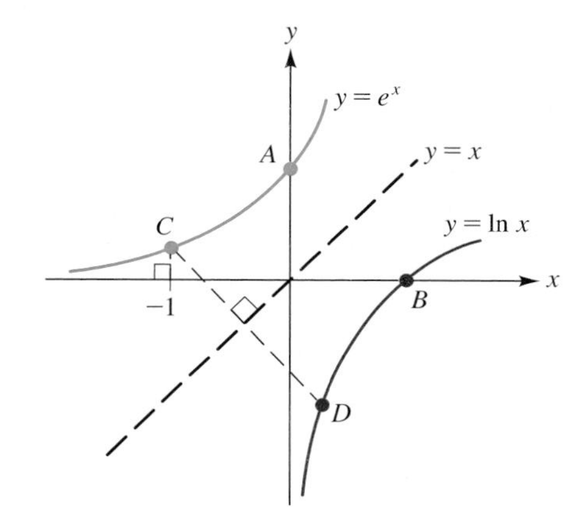

In Exercises 25–30, graph each function and specify the domain, range, intercept(s), and asymptote.

25. **(a)** $y = \log_2 x$
(b) $y = -\log_2 x$
(c) $y = \log_2(-x)$
(d) $y = -\log_2(-x)$

26. **(a)** $y = \ln x$
(b) $y = -\ln x$
(c) $y = \ln(-x)$
(d) $y = -\ln(-x)$

27. $y = -\log_3(x - 2) + 1$

28. $y = -\log_{10}(x + 1)$

29. $y = \ln(x + e)$

30. $y = \ln(-x) + e$

In Exercises 31 and 32, simplify each expression.

31. **(a)** $\ln e^4$
(b) $\ln(1/e)$
(c) $\ln\sqrt{e}$

32. **(a)** $\ln e$
(b) $\ln e^{-2}$
(c) $(\ln e)^{-2}$

For Exercises 33–40:

Ⓖ **(a)** *Use a graphing utility to estimate the root(s) of the equation to the nearest one-tenth (as in Example 6).*
(b) *Solve the given equation algebraically by first rewriting it in logarithmic form. Give two forms for each answer: an exact expression and a calculator approximation rounded to three decimal places. Check to see that each result is consistent with the graphical estimate obtained in part (a).*

33. $10^x = 25$

34. $10^{2x-1} = 145$

35. $10^{x^2} = 40$

36. $(10^x)^2 = 40$

37. $e^{2t+3} = 10$
38. $e^{t-1} = 16$
39. $e^{1-4t} = 12.405$
40. $e^{3x^2} = 112$

Ⓖ **41.** Use graphs to help answer the following questions.
(a) For which x-values is $\ln x < \log_{10} x$?
(b) For which x-values is $\ln x > \log_{10} x$?
(c) For which x-values is $e^x < 10^x$?
(d) For which x-values is $e^x > 10^x$?

Ⓖ **42.** This exercise demonstrates the very slow growth of the natural logarithm function $y = \ln x$. We consider the following question: How large must x be before the graph of $y = \ln x$ reaches a height of 10?
(a) Graph the function $y = \ln x$ using a viewing rectangle that extends from 0 to 10 in the x-direction and 0 to 12 in the y-direction. Note how slowly the graph rises. Use the graphing utility to estimate the height of the curve (the y-coordinate) when $x = 10$.
(b) Since we are trying to see when the graph of $y = \ln x$ reaches a height of 10, add the horizontal line $y = 10$ to your picture. Next, adjust the viewing rectangle so that x extends from 0 to 100. Now use the graphing utility to estimate the height of the curve when $x = 100$. [As both the picture and the y-coordinate indicate, we're still not even halfway to 10. Go on to part (c).]
(c) Change the viewing rectangle so that x extends to 1000, then estimate the y-coordinate corresponding to $x = 1000$. (You'll find that the height of the curve is almost 7. We're getting closer.)
(d) Repeat part (c) with x extending to 10,000. (You'll find that the height of the curve is over 9. We're almost there.)
(e) The last step: Change the viewing rectangle so that x extends to 100,000, then use the graphing utility to estimate the x-value for which $\ln x = 10$. As a check on your estimate, rewrite the equation $\ln x = 10$ in exponential form, and evaluate the expression that you obtain for x.

In Exercises 43 and 44, Richter magnitudes of earthquakes are given. Follow the method of Example 8 to determine how many times stronger one quake was than the other.

43.

Location of earthquake	Date	Richter magnitude M
Bombay, India	September 29, 1993	6.4
(15 mi. east of) San Salvador, El Salvador	February 13, 2001	6.6

44.

Location of earthquake	Date	Richter magnitude M
Napa, California	September 3, 2000	5.2
Sumatra, Indonesia	June 4, 2000	8.0

B

As background for Exercises 45 and 46, you need to have read Example 8.

45. Suppose that an earthquake has a Richter magnitude of M_0, and a second earthquake has a magnitude of $M_0 + d$ (where $d > 0$). Show that the second earthquake is 10^d times stronger than the first earthquake.

46. In 1956 the geologists Gutenberg and Richter developed the following formula for estimating the amount of energy E released in an earthquake: $\log_{10} E = 4.4 + (1.5)M$, where E is the energy in joules and M is the Richter magnitude. Now refer to the following table, and let E_1 denote the energy of the quake in the Philippines and E_2 the energy of the quake in Washington. Compute the ratio E_1/E_2 to compare the energies of the two quakes.

Location of earthquake	Date	Richter magnitude M
Mindanao, Philippines	January 1, 2001	7.5
(northeast of) Olympia, Washington	February 28, 2001	6.8

47. The intensity of the sounds that the human ear can detect varies over a very wide range of values. For instance, a whisper from 1 meter away has an intensity of approximately 10^{-10} watts per square meter (W/m^2), whereas, from a distance of 50 meters, the intensity of a launch of the Space Shuttle is approximately 10^8 W/m^2. For a sound with intensity I, the sound level β is defined by

$$\beta = 10 \log_{10}(I/I_0)$$

where the constant I_0 is the sound intensity of a barely audible sound at the threshold of hearing. The units for the sound level β are *decibels,* abbreviated dB.
(a) Solve the equation $\beta = 10 \log_{10}(I/I_0)$ for I by first dividing by 10 and then converting to exponential form.
(b) The sound level for a power lawnmower is $\beta = 100$ db, and that for a cat purring is $\beta = 10$ db. Use your result in part (a) to determine how many times more intense is the power mower sound than the cat's purring.

48. A sound level of $\beta = 120$ db is at the threshold of pain. (Some loud rock concerts reach this level.) The sound intensity that corresponds to $\beta = 120$ db is 1 W/m^2. Use this information and the equation $\beta = 10 \log_{10}(I/I_0)$ to determine I_0, the intensity of a barely audible sound at the threshold of hearing. What is the decibel level, β, of a barely audible sound?

In Exercises 49–52, use the following information on pH. Chemists define pH by the formula $pH = -\log_{10}[H^+]$, where $[H^+]$ is the hydrogen ion concentration measured in moles per liter. For example, if $[H^+] = 10^{-5}$, then $pH = 5$. Solutions with a pH of 7 are said to be neutral; a pH below 7 indicates an acid; and a pH above 7 indicates a base. (A calculator is helpful for Exercises 49 and 50.)

49. **(a)** For some fruit juices, $[H^+] = 3 \times 10^{-4}$. Determine the pH and classify these juices as acid or base.
(b) For sulfuric acid, $[H^+] = 1$. Find the pH.

50. An unknown substance has a hydrogen ion concentration of 3.5×10^{-9}. Classify the substance as acid or base.

51. What is the hydrogen ion concentration for black coffee if the pH is 5.9?

52. A chemist adds some acid to a solution changing the pH from 6 to 4. By what factor does the hydrogen ion concentration change? *Note*: Lower pH corresponds to higher hydrogen ion concentration.

In Exercises 53–60, decide which of the following properties apply to each function. (More than one property may apply to a function.)

A. *The function is increasing for $-\infty < x < \infty$.*
B. *The function is decreasing for $-\infty < x < \infty$.*
C. *The function has a turning point.*
D. *The function is one-to-one.*
E. *The graph has an asymptote.*
F. *The function is a polynomial function.*
G. *The domain of the function is $(-\infty, \infty)$.*
H. *The range of the function is $(-\infty, \infty)$.*

53. $y = \ln x$ **54.** $y = \ln(-x)$
55. $y = -\ln x$ **56.** $y = \ln|x|$
57. $y = \ln x + e$ **58.** $y = x + \ln e$
59. $y = -\ln(-x)$ **60.** $y = (\ln x)^2$

61. Let $f(x) = e^{x+1}$. Find $f^{-1}(x)$ and sketch its graph. Specify any intercept or asymptote.

62. Let $g(t) = \ln(t - 1)$. Find $g^{-1}(t)$ and draw its graph. Specify any intercept or asymptote.

63. Estimate a value for x such that $\log_2 x = 100$. Use the approximation $10^3 \approx 2^{10}$ to express your answer as a power of 10. *Answer*: 10^{30}

64. **(a)** How large must x be before the graph of $y = \ln x$ reaches a height of $y = 100$?
(b) How large must x be before the graph of $y = e^x$ reaches a height of **(i)** $y = 100$? **(ii)** $y = 10^6$?

Ⓖ **65.** This exercise uses the natural logarithm and a regression line (i.e., a linear function) to find a simple formula relating the following two quantities:

x: the average distance of a planet from the sun
y: the period of a planet (i.e., the time required for a planet to make one complete revolution around the sun)

(a) Complete the following table. (In the table the abbreviation AU stands for *astronomical unit*, a unit of distance. By definition, one AU is the average distance from the earth to the sun.)

Planet	x Average distance from the sun (AU)	y Period (years)	ln x	ln y
Mercury	0.387	0.241		
Venus	0.723	0.615		
Earth	1.000	1.000		
Mars	1.523	1.881		
Jupiter	5.202	11.820		

(b) Use a graphing utility to obtain a scatter plot and a regression equation $Y = AX + B$ for the pairs of numbers $(\ln x, \ln y)$. After some rounding, you should obtain $A = 1.5$ and $B = 0$. Thus the quantities $\ln x$ and $\ln y$ are related as follows:

$$\ln y = 1.5 \ln x$$

(c) Show that the equation obtained in part (b) can be simplified to

$$y = x^{1.5} \tag{1}$$

(d) Equation (1) was obtained using data for the first five planets from the sun. Complete the following table to see how well equation (1) fits the data for the remaining planets.

Planet	x (AU)	y (years) [calculated from eqn. (1)]	y (years) (from astronomical observations)
Saturn	9.555		29.46
Uranus	19.22		84.01
Neptune	30.11		164.79
Pluto	39.44		248.50

Remark: Equation (1) is approximately *Kepler's third law of planetary motion.* According to George F. Simmons in *Calculus Gems* (New York: McGraw-Hill, Inc., 1992), "Kepler had struggled for more than twenty years to find this connection between a planet's *distance* from the sun and the *time* required to complete its orbit. He published his discovery in 1619 in a work entitled *Harmonices Mundi* (The Harmonies of the World)."

Ⓖ **66.** This exercise uses the natural logarithm and a regression line to find a simple formula relating the following two quantities.

x: the average distance of a planet from the sun
y: the average orbital velocity of the planet

(a) Complete the following table.

Planet	x Average distance from the sun (millions of miles)	y Average orbital velocity (miles/sec)	ln x	ln y
Mercury	35.98	29.75		
Venus	67.08	21.76		
Earth	92.96	18.51		
Mars	141.64	14.99		

(b) Use a graphing utility to obtain a scatter plot and a regression equation $Y = AX + B$ for the pairs of numbers ($\ln x$, $\ln y$). You will find that $A = -0.500$ and $B = 5.184$. Thus the quantities $\ln x$ and $\ln y$ are related as follows:

$$\ln y = -0.500 \ln x + 5.184$$

(c) Show that the equation obtained in part (b) can be rewritten in the form

$$y = \frac{e^{5.184}}{\sqrt{x}} \qquad (2)$$

Hint: Convert from logarithmic to exponential form.

(d) Equation (2) was obtained using data for the first four planets from the sun. Complete the following table to see how well equation (2) fits the data for the remaining planets. (Round the answers to two decimal places.)

Planet	x Average distance from the sun (millions of miles)	y Average orbital velocity (miles/sec) [calculated from eqn. (2)]	y Average orbital velocity (miles/sec) (from astronomical observations)
Jupiter	483.63		8.12
Saturn	888.22		5.99
Uranus	1786.55		4.23
Neptune	2799.06		3.38
Pluto	3700.75		2.95

Ⓖ **67.** *Logarithmic regression:* In Sections 4.1 and 4.2 we looked at applications in which linear functions or quadratic functions were used to model data sets. In each case we used a graphing utility to obtain the equation of the linear function or quadratic function that "best fitted" the data. In this exercise we use a function of the form $y = a + b \ln x$ to model a given data set, and we compare a projection made with this model to one based on a linear model.

(a) The following table gives the population of Los Angeles County at ten-year intervals over the period 1950–1990. Use a graphing utility to obtain a scatter plot for the data, and then use the *logarithmic regression* option on the graphing utility to find a function of the form $y = a + b \ln x$ that best fits the data. (Let $x = 50$ correspond to 1950. The reason we don't use the seemingly simpler choice $x = 0$ for 1950 is that $x = 0$ is not in the domain of the function $y = a + b \ln x$.) Display the graph of the function and the scatter plot on the same screen.

Population of Los Angeles County

Year	1950	1960	1970	1980	1990
Population (millions)	4.152	6.040	7.032	7.478	8.863

Source: U.S. Bureau of the Census

(b) Follow part (a), but use a linear model. Display the scatter plot and the graphs of the linear and logarithmic models on the same screen.

(c) Use each model to make a projection for what the population might have been in the year 2000.

(d) According to the U.S. Bureau of the Census, the population of Los Angeles County in 2000 was 9.519 million. Which answer in part (c) is closer to this? Compute the percentage error for each projection.

68. This exercise indicates one of the ways the natural logarithm function is used in the study of prime numbers. Recall that a prime number is a natural number greater than 1 with no factors other than itself and 1. For example, the first ten prime numbers are 2, 3, 5, 7, 11, 13, 17, 19, 23, and 29.

(a) Let $P(x)$ denote the number of prime numbers that do not exceed x. For instance, $P(6) = 3$, since there are three prime numbers (2, 3, and 5) that do not exceed 6. Compute $P(10)$, $P(18)$, and $P(19)$.

(b) According to the **prime number theorem,** $P(x)$ can be approximated by $x/(\ln x)$ when x is large and, in fact, the ratio $\dfrac{P(x)}{x/(\ln x)}$ approaches 1 as x grows larger and larger. Verify this empirically by completing the following table. Round your results to three decimal places. (These facts were discovered by Carl Friedrich Gauss in 1792, when he was 15 years old. It was not until 1896, more than 100 years later, that the prime number theorem was formally proved by the French mathematician J. Hadamard and also by the Belgian mathematician C. J. de la Vallée-Poussin.)

x	$P(x)$	$\dfrac{x}{\ln x}$	$\dfrac{P(x)}{x/\ln x}$
10^2	25		
10^4	1229		
10^6	78498		
10^8	5761455		
10^9	50847534		
10^{10}	455052512		

(c) In 1808 A. M. Legendre found that he could improve upon Gauss's approximation for $P(x)$ by using the expression $x/(\ln x - 1.08366)$ rather than $x/(\ln x)$. Complete the following table to see how well Legendre's expression approximates $P(x)$. Round your results to four decimal places.

x	$P(x)$	$\frac{x}{\ln x - 1.08366}$	$\frac{P(x)}{x/(\ln x - 1.08366)}$
10^2	25		
10^4	1229		
10^6	78498		
10^8	5761455		
10^9	50847534		
10^{10}	455052512		

Ⓖ

C

69. **(a)** Find the domain of the function f defined by $f(x) = \ln(\ln x)$.
(b) Find $f^{-1}(x)$ for the function f in part (a).

70. Find the domain of the function g defined by $g(x) = \ln(\ln(\ln x))$.

71. Let $g(x) = \ln(\ln(\ln x))$.
(a) Using a graphing utility, display the graph of g, first in the standard viewing rectangle, then in the viewing rectangle $[0, 100, 10]$ by $[-5, 5, 1]$.
(b) Find the range of the function g. (See Example 4 in Section 3.1 if you need a hint on how to begin.)
(c) Could you have guessed the answer in part (b) from the graphs in part (a)?

MINI PROJECT

MORE COFFEE

As a prerequisite for this mini project, someone in your group or in the class at large needs to have completed the Section 5.2 project that dealt with the temperature of a cooling cup of coffee. This is a continuation (a completion, actually) of Problem B in that project. As a start here, have that person present a detailed summary of the questions, methods, and results from the Section 5.2 project. Then, for this mini project: Use the techniques from both Sections 5.2 and 5.3 to work out a complete solution to Problem B. Round the answer to the nearest one second. Check that your result is consistent with the approximation from the graphing utility obtained in the Section 5.2 project.

5.4 PROPERTIES OF LOGARITHMS

... he [John Napier] *invented the word "logarithms," using two Greek words, ...* arithmos, *"number"* and logos, *"ratio." It is impossible to say exactly what he had in mind when making up this word.* —Alfred Hooper in *Makers of Mathematics* (New York: Random House, Inc., 1948)

A few basic properties of logarithms are used repeatedly. In this section we will state these properties, discuss their proofs, and then look at examples.

Properties of Logarithms

1. (a) $\log_b b = 1$ (b) $\log_b 1 = 0$
2. $\log_b PQ = \log_b P + \log_b Q$
   The log of a product is the sum of the logs of the factors.
3. $\log_b(P/Q) = \log_b P - \log_b Q$
   The log of a quotient is the log of the numerator minus the log of the denominator.
   As a useful particular case, we have $\log_b(1/Q) = -\log_b Q$.
4. $\log_b P^n = n \log_b P$
5. $b^{\log_b P} = P$
6. $\log_b b^x = x$, for all real numbers x.

Note: P and Q are assumed to be positive in Properties 2–6.

In essence, each of these properties follows from the equivalence of the two equations $y = \log_b x$ and $b^y = x$. For instance, the equivalent exponential forms for Properties 1(a) and 1(b) are $b^1 = b$ and $b^0 = 1$, respectively, both of which are certainly valid.

To prove Property 2, we begin by letting $x = \log_b P$. The equivalent exponential form of this equation is

$$P = b^x \tag{1}$$

Similarly, we let $y = \log_b Q$. The exponential form of this equation is

$$Q = b^y \tag{2}$$

If we multiply equation (1) by equation (2), we get

$$PQ = b^x b^y$$

and therefore

$$PQ = b^{x+y} \tag{3}$$

Next we write equation (3) in its equivalent logarithmic form. This yields

$$\log_b PQ = x + y$$

But using the definitions of x and y, this last equation is equivalent to

$$\log_b PQ = \log_b P + \log_b Q$$

That completes the proof of Property 2.

The proof of Property 3 is quite similar to the proof given for Property 2. Exercise 72(a) asks you to carry out the proof of Property 3.

We turn now to the proof of Property 4. We begin by letting $x = \log_b P$. In exponential form, this last equation becomes

$$b^x = P \tag{4}$$

Now we raise both sides of equation (4) to the power n. This yields

$$(b^x)^n = b^{nx} = P^n$$

The logarithmic form of this last equation is

$$\log_b P^n = nx$$

or (from the definition of x)

$$\log_b P^n = n \log_b P$$

The proof of Property 4 is now complete.

Property 5 is again just a restatement of the meaning of $\log_b P$. To derive this property, let $x = \log_b P$. Therefore

$$b^x = P$$

Now, in this last equation, we simply replace x with $\log_b P$. The result is

$$b^{\log_b P} = P \qquad \text{as required}$$

Finally, Property 6 follows from the meaning of the logarithm $\log_b b^x = x$ since x is the power to which b must be raised to get b^x.

Now let's see how these properties are used. To begin with, we display some simple numerical examples in the box that follows.

PROPERTY SUMMARY Properties of Logarithms

Property	Example
$\log_b P + \log_b Q = \log_b PQ$	Simplify: $\log_{10} 50 + \log_{10} 2$ Solution: $\log_{10} 50 + \log_{10} 2 = \log_{10}(50 \cdot 2) = \log_{10} 100 = 2$
$\log_b P - \log_b Q = \log_b \frac{P}{Q}$	Simplify: $\log_8 56 - \log_8 7$ Solution: $\log_8 56 - \log_8 7 = \log_8 \frac{56}{7} = \log_8 8 = 1$
$\log_b P^n = n \log_b P$	Simplify: $\log_2 \sqrt[5]{16}$ Solution: $\log_2 \sqrt[5]{16} = \log_2(16^{1/5}) = \frac{1}{5} \log_2 16 = \frac{4}{5}$
$b^{\log_b P} = P$	Simplify: $3^{\log_3 7}$ Solution: $3^{\log_3 7} = 7$

The examples in the box showed how we can simplify or shorten certain expressions involving logarithms. The next example is also of this type.

EXAMPLE 1 Using log properties to shorten expressions

Express as a single logarithm with a coefficient of 1:

$$\frac{1}{2}\log_b x - \log_b(1 + x^2)$$

SOLUTION

$$\begin{aligned} \frac{1}{2}\log_b x - \log_b(1 + x^2) &= \log_b x^{1/2} - \log_b(1 + x^2) && \text{using Property 4 on page 361} \\ &= \log_b \frac{x^{1/2}}{1 + x^2} && \text{using Property 3} \end{aligned}$$

This last expression is the required answer.

Property 2 says that the logarithm of a product of two factors is equal to the sum of the logarithms of the two factors. This can be generalized to any number of factors. For instance, with three factors we have

$$\begin{aligned} \log_b(ABC) &= \log_b[A(BC)] \\ &= \log_b A + \log_b BC && \text{using Property 2} \\ &= \log_b A + \log_b B + \log_b C && \text{using Property 2 again} \end{aligned}$$

The next example makes use of this idea.

EXAMPLE 2 Using log properties to shorten expressions

Express as a single logarithm with a coefficient of 1:

$$\ln(x^2 - 9) + 2\ln\frac{1}{x+3} + 4\ln x \qquad (x > 3)$$

SOLUTION

$$\begin{aligned}\ln(x^2 - 9) + 2\ln\frac{1}{x+3} + 4\ln x &= \ln(x^2 - 9) + \ln\left[\left(\frac{1}{x+3}\right)^2\right] + \ln x^4\\ &= \ln(x^2 - 9) + \ln\frac{1}{(x+3)^2} + \ln x^4\\ &= \ln\left[(x^2 - 9)\cdot\frac{1}{(x+3)^2}\cdot x^4\right]\\ &= \ln\frac{(x^2 - 9)x^4}{(x+3)^2}\end{aligned}$$

This last expression can be simplified still further by factoring the quantity $x^2 - 9$ as $(x - 3)(x + 3)$. Then a factor of $x + 3$ can be divided out of the numerator and denominator of the fraction. The result is

$$\begin{aligned}\ln(x^2 - 9) + 2\ln\frac{1}{x+3} + 4\ln x &= \ln\frac{(x-3)x^4}{x+3}\\ &= \ln\frac{x^5 - 3x^4}{x+3}\end{aligned}$$

In the examples we have considered so far, we've used the properties of logarithms to shorten given expressions. We can also use these properties to expand an expression. (This is useful in calculus.)

EXAMPLE 3 Using log properties to expand expressions

Write each of the following quantities as sums and differences of simpler logarithmic expressions. Express each answer in such a way that no logarithm of products, quotients, or powers appears.

(a) $\log_{10}\sqrt{3x}$ **(b)** $\log_{10}\sqrt[3]{\dfrac{2x}{3x^2+1}}$ **(c)** $\ln\dfrac{x^2\sqrt{2x-1}}{(2x+1)^{3/2}}$

SOLUTION

(a)
$$\begin{aligned}\log_{10}\sqrt{3x} &= \log_{10}(3x)^{1/2} = \tfrac{1}{2}\log_{10}(3x)\\ &= \tfrac{1}{2}(\log_{10}3 + \log_{10}x)\end{aligned}$$

(b)
$$\begin{aligned}\log_{10}\sqrt[3]{\frac{2x}{3x^2+1}} &= \log_{10}\left[\left(\frac{2x}{3x^2+1}\right)^{1/3}\right]\\ &= \tfrac{1}{3}\log_{10}\frac{2x}{3x^2+1}\\ &= \tfrac{1}{3}[\log_{10}2x - \log_{10}(3x^2+1)]\\ &= \tfrac{1}{3}[\log_{10}2 + \log_{10}x - \log_{10}(3x^2+1)]\end{aligned}$$

(c) $\ln \dfrac{x^2\sqrt{2x-1}}{(2x+1)^{3/2}} = \ln \dfrac{x^2(2x-1)^{1/2}}{(2x+1)^{3/2}}$

$= \ln x^2 + \ln(2x-1)^{1/2} - \ln(2x+1)^{3/2}$

$= 2\ln x + \frac{1}{2}\ln(2x-1) - \frac{3}{2}\ln(2x+1)$

EXAMPLE 4 Practice applying the log properties

Given that $\log_b 2 = A$ and $\log_b 6 = B$, express each of the following in terms of A and/or B.

(a) $\log_b 8$ **(c)** $\log_b 12$ **(e)** $\log_{\sqrt{b}} 2$
(b) $\log_b \sqrt{6}$ **(d)** $\log_b 3$ **(f)** $\log_b(b/36)$

SOLUTION

(a) $\log_b 8 = \log_b 2^3 = 3\log_b 2 = 3A$
(b) $\log_b \sqrt{6} = \log_b 6^{1/2} = \frac{1}{2}\log_b 6 = \frac{1}{2}B$
(c) $\log_b 12 = \log_b(2 \times 6) = \log_b 2 + \log_b 6 = A + B$
(d) $\log_b 3 = \log_b(6/2) = \log_b 6 - \log_b 2 = B - A$
(e) Let $\log_{\sqrt{b}} 2 = x$. (So we want to express x in terms of A and/or B.) Writing this equation in exponential form, we have $(\sqrt{b})^x = 2$, and therefore

$$(b^{1/2})^x = 2$$
$$b^{x/2} = 2$$

Now we rewrite this last equation in logarithmic form to obtain $\log_b 2 = x/2$, and therefore

$$x = 2\log_b 2 = 2A$$

(f) $\log_b(b/36) = \log_b b - \log_b 36$
$= 1 - \log_b 6^2 = 1 - 2\log_b 6 = 1 - 2B$

EXAMPLE 5 Using the property $\log_b P^n = n\log_b P$ to find an x-intercept

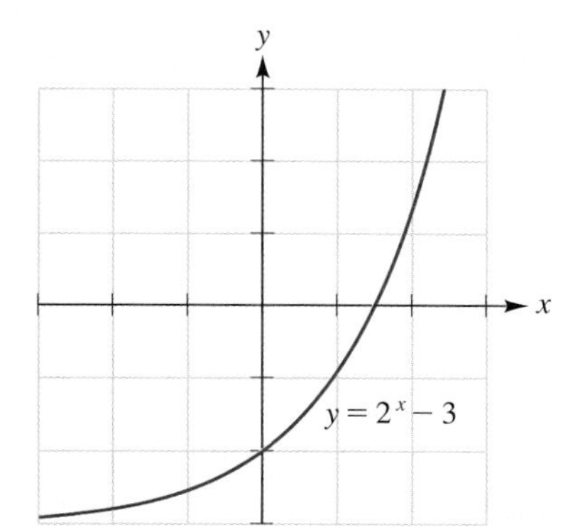

Figure 1

Figure 1 shows the graph of $y = 2^x - 3$. (The axes are marked off in one-unit intervals.) As indicated in the figure, the x-intercept of the curve is between 1 and 2. Determine both an exact expression for this x-intercept and a calculator approximation rounded to two decimal places.

SOLUTION

Setting y equal to zero in the given equation yields $2^x - 3 = 0$, or $2^x = 3$. Notice that the unknown, x, appears in the exponent. To solve for x, we take the logarithm of both sides of the equation. (Base e logarithms are used in the following computations; Exercise 71 asks you to check that the use of base 10 logarithms produces an equivalent answer.) We have

$$\ln 2^x = \ln 3$$
$$x \ln 2 = \ln 3 \quad \text{using Property 4 on page 361}$$
$$x = \frac{\ln 3}{\ln 2} \approx 1.58 \quad \text{CAUTION: } \frac{\ln 3}{\ln 2} \neq \ln 3 - \ln 2$$

Alternatively, we could express $2^x = 3$ in logarithmic form to obtain the solution $x = \log_2 3$ and then use the change of base formula on page 367 to obtain a decimal approximation.

EXAMPLE 6 **Using the property $\log_b P^n = n \log_b P$ to solve an equation**

Solve for x: $4^{5x+2} = 70$. Use a calculator to evaluate the final answer; round to three decimal places.

SOLUTION

The fact that the unknown appears in the exponent suggests that Property 4 (the property that "brings down" the exponent) may be useful. To put Property 4 into play, we'll take the logarithm of both sides of the given equation. Either base 10 or base e logarithms can be used here (because they're both on the calculator). Since we used base e logarithms in the previous example, let's use base 10 here just for the sake of variety. (You can check for yourself at the end that base e logarithms will yield the same numerical answer.)

Taking the base 10 logarithm of both sides of the given equation, we have

$$\log_{10} 4^{5x+2} = \log_{10} 70$$
$$(5x + 2)\log_{10} 4 = \log_{10} 70 \quad \text{using Property 4}$$
$$5x + 2 = (\log_{10} 70)/(\log_{10} 4)$$
$$x = \frac{[(\log_{10} 70)/(\log_{10} 4)] - 2}{5}$$
$$x \approx 0.213$$

The required solution is approximately 0.213. Figure 2 shows a graphical interpretation of this result.

GRAPHICAL PERSPECTIVE

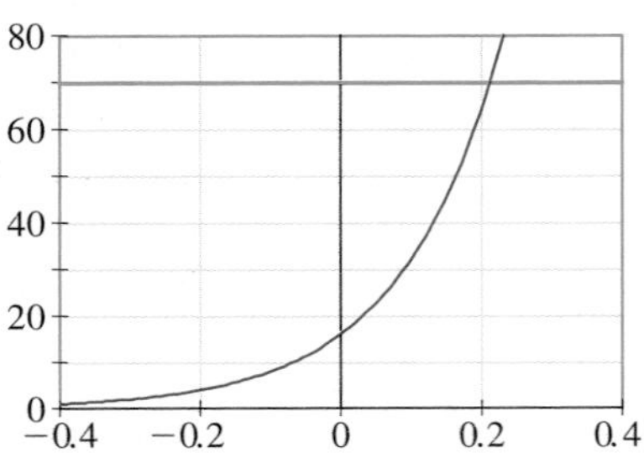

Figure 2
The curve $y = 4^{5x+2}$ intersects the line $y = 70$ at a point whose x-coordinate is slightly greater than 0.2. Example 6 shows that this x-coordinate is 0.213 (rounded to three decimal places).

It is sometimes necessary to convert logarithms in one base to logarithms in another base. After the next example, we will state a formula for this. However, as the next example indicates, it is easy to work this type of problem from the basics, without relying on a formula.

EXAMPLE 7 Changing bases

Express $\log_2 5$ in terms of base 10 logarithms.

SOLUTION

Let $z = \log_2 5$. The exponential form of this equation is

$$2^z = 5$$

We now take the base 10 logarithm of each side of this equation to obtain

$$\log_{10} 2^z = \log_{10} 5$$

$$z\log_{10} 2 = \log_{10} 5 \qquad \text{using Property 4 on page 361}$$

$$z = \frac{\log_{10} 5}{\log_{10} 2}$$

Given our definition of z, this last equation can be written

$$\log_2 5 = \frac{\log_{10} 5}{\log_{10} 2}$$

This is the required answer.

The method shown in Example 7 can be used to convert between any two bases. Exercise 72(b) at the end of this section asks you to follow this method, using letters rather than numbers, to arrive at the following general formula.

PROPERTY SUMMARY Change of Base for Logarithms

Formula	Examples	
$\log_a x = \dfrac{\log_b x}{\log_b a}$	$\log_2 3 = \dfrac{\log_{10} 3}{\log_{10} 2} = \dfrac{\ln 3}{\ln 2}$	$\log_{10} e = \dfrac{\ln e}{\ln 10} = \dfrac{1}{\ln 10}$

EXAMPLE 8 Changing bases to simplify an expression

Simplify the product $(\log_2 10)(\log_{10} 2)$.

SOLUTION

We'll use the change-of-base formula so that both factors in the given product involve logarithms with the same base. Expressing the first factor, $\log_2 10$, in terms of base 10 logarithms, we have

$$\log_2 10 = \frac{\log_{10} 10}{\log_{10} 2} = \frac{1}{\log_{10} 2}$$

and therefore $$(\log_2 10)(\log_{10} 2) = \frac{1}{\log_{10} 2}(\log_{10} 2) = 1$$

So the given product is equal to 1. [Using the same method, one can show that, in general, $(\log_a b)(\log_b a) = 1$.]

Examples 1–8 have dealt with applications of the properties of logarithms. However, you also need to understand what the properties *don't* say. For instance, Property 3 does not apply to an expression such as $\dfrac{\log_{10} 5}{\log_{10} 2}$. [Property 3 *would* apply if the expression were $\log_{10}(5/2)$.] In the box that follows, we list some errors to avoid in working with logarithms.

ERRORS TO AVOID

Error	Correction	Comment
$\log_b(x + y) = \log_b x + \log_b y$	$\log_b(xy) = \log_b x + \log_b y$	In general, there is no simple identity involving $\log_b(x + y)$. This is similar to the situation for $\sqrt{x + y}$ (which is not equal to $\sqrt{x} + \sqrt{y}$).
$\dfrac{\log_b x}{\log_b y} = \log_b x - \log_b y$	$\log_b\left(\dfrac{x}{y}\right) = \log_b x - \log_b y$	In general, there is no simple identity involving the quotient $(\log_b x)/(\log_b y)$.
$(\ln x)^3 = 3 \ln x$	$(\ln x)^3 = (\ln x)(\ln x)(\ln x)$	$(\ln x)^3$ is not the same as $\ln(x^3)$. Regarding the latter, we do have $\ln(x^3) = 3 \ln x$, where $x > 0$.
$\ln \dfrac{x}{2} = \dfrac{\ln x}{2}$	$\ln \dfrac{x}{2} = \ln x - \ln 2 \quad \text{for } x > 0$ and $\dfrac{\ln x}{2} = \dfrac{1}{2} \ln x = \ln \sqrt{x} \quad \text{for } x > 0$	The confusion between $\ln \dfrac{x}{2}$ and $\dfrac{\ln x}{2}$ is sometimes due simply to careless handwriting.

EXERCISE SET 5.4

A

In Exercises 1–10, simplify the expression by using the definition and properties of logarithms.

1. $\log_{10} 70 - \log_{10} 7$

2. $\log_{10} 40 + \log_{10}(5/2)$

3. $\log_7 \sqrt{7}$

4. $\log_9 25 - \log_9 75$

5. $\log_3 108 + \log_3(3/4)$

6. $\ln e^3 - \ln e$

7. $-\frac{1}{2} + \ln \sqrt{e}$

8. $e^{\ln 3} + e^{\ln 2} - e^{\ln e}$

9. $2^{\log_2 5} - 3 \log_5 \sqrt[3]{5}$

10. $\log_b b^b$

In Exercises 11–19, write the expression as a single logarithm with a coefficient of 1.

11. $\log_{10} 30 + \log_{10} 2$

12. $2 \log_{10} x - 3 \log_{10} y$

13. $\log_5 6 + \log_5(1/3) + \log_5 10$

14. $p \log_b A - q \log_b B + r \log_b C$

15. (a) $\ln 3 - 2 \ln 4 + \ln 32$
(b) $\ln 3 - 2(\ln 4 + \ln 32)$

16. (a) $\log_{10}(x^2 - 16) - 3 \log_{10}(x + 4) + 2 \log_{10} x$
(b) $\log_{10}(x^2 - 16) - 3[\log_{10}(x + 4) + 2 \log_{10} x]$

17. $\log_b 4 + 3[\log_b(1 + x) - \frac{1}{2} \log_b(1 - x)]$

18. $\ln(x^3 - 1) - \ln(x^2 + x + 1)$

19. $4 \log_{10} 3 - 6 \log_{10}(x^2 + 1) + \frac{1}{2}[\log_{10}(x + 1) - 2 \log_{10} 3]$

In Exercises 20–26, write the quantity using sums and differences of simpler logarithmic expressions. Express the answer so that logarithms of products, quotients, and powers do not appear.

20. (a) $\log_{10}\sqrt{(x + 1)(x + 2)}$
(b) $\ln \sqrt{\dfrac{(x + 1)(x + 2)}{(x - 1)(x - 2)}}$

21. (a) $\log_{10}\dfrac{x^2}{1 + x^2}$
(b) $\ln \dfrac{x^2}{\sqrt{1 + x^2}}$

22. (a) $\log_b \dfrac{\sqrt{1 - x^2}}{x}$
(b) $\ln \dfrac{x\sqrt[3]{4x + 1}}{\sqrt{2x - 1}}$

23. (a) $\log_{10}\sqrt{9 - x^2}$
(b) $\ln \dfrac{\sqrt{4 - x^2}}{(x - 1)(x + 1)^{3/2}}$

24. (a) $\log_b \sqrt[3]{\dfrac{x + 3}{x}}$
(b) $\ln \dfrac{1}{\sqrt{x^2 + x + 1}}$

25. **(a)** $\log_b \sqrt{x/b}$
(b) $2 \ln \sqrt{(1 + x^2)(1 + x^4)(1 + x^6)}$

26. **(a)** $\log_b \sqrt[3]{\dfrac{(x-1)^2(x-2)}{(x+2)^2(x+1)}}$ **(b)** $\ln\left(\dfrac{e-1}{e+1}\right)^{3/2}$

In Exercises 27–36, suppose b is a positive constant greater than 1, and let A, B, and C be defined as follows:

$$\log_b 2 = A \qquad \log_b 3 = B \qquad \log_b 5 = C$$

In each case, use the properties of logarithms to evaluate the given expression in terms of A, B, and/or C. (In Exercises 31–36, use the change-of-base formula.)

27. **(a)** $\log_b 6$
(b) $\log_b(1/6)$
(c) $\log_b 27$
(d) $\log_b(1/27)$

28. **(a)** $\log_b 10$
(b) $\log_b 100$
(c) $\log_b 0.01$
(d) $\log_b 0.3$

29. **(a)** $\log_b(5/3)$
(b) $\log_b 0.6$
(c) $\log_b(5/9)$
(d) $\log_b(5/16)$

30. **(a)** $\log_b \sqrt{5}$
(b) $\log_b \sqrt{15}$
(c) $\log_b \sqrt[3]{0.4}$
(d) $\log_b \sqrt[4]{60}$

31. **(a)** $\log_3 b$
(b) $\log_3(10b)$

32. **(a)** $\log_{b^2} 5$
(b) $\log_{\sqrt{b}} 2$

33. **(a)** $\log_{3b} 2$
(b) $\log_{3b} 15$

34. **(a)** $\log_{5b} 1.2$
(b) $\log_{5b} 2.5$

35. **(a)** $(\log_b 5)(\log_5 b)$
(b) $(\log_b 6)(\log_6 b)$

36. **(a)** $\log_{2b} 6 + \log_{2b}(1/6)$
(b) $\log_{18}(1/b)$

In Exercises 37 and 38, suppose that $\log_{10} A = a$, $\log_{10} B = b$, and $\log_{10} C = c$. *Express the following logarithms in terms of a, b, and c.*

37. **(a)** $\log_{10} AB^2C^3$
(b) $\log_{10} 10\sqrt{A}$
(c) $\log_{10}\sqrt{10ABC}$
(d) $\log_{10}(10A/\sqrt{BC})$

38. **(a)** $\log_{10} A + 2\log_{10}(1/A)$
(b) $\log_{10}(A/10)$
(c) $\log_{10}\dfrac{100\,A^2}{B^4\sqrt[3]{C}}$
(d) $\log_{10}\dfrac{(4B)^5}{C}$

In Exercises 39 and 40, suppose that $\ln x = t$ *and* $\ln y = u$. *Write each expression in terms of t and u.*

39. **(a)** $\ln(ex)$
(b) $\ln xy - \ln(x^2)$
(c) $\ln\sqrt{xy} + \ln(x/e)$
(d) $\ln(e^2x\sqrt{y})$

40. **(a)** $\ln(e^{\ln x})$
(b) $e^{\ln(\ln xy)}$
(c) $\ln\left(\dfrac{ex}{y}\right) - \ln\left(\dfrac{y}{ex}\right)$
(d) $\dfrac{(\ln x)^3 - \ln(x^4)}{\left(\ln \dfrac{x}{e^2}\right)\ln(xe^2)}$

In Exercises 41 and 42, graph the equations and determine the x-intercepts (as in Example 5).

41. **(a)** $y = 2^x - 5$
(b) $y = 2^{x/2} - 5$

42. **(a)** $y = 3^{x-1} - 2$
(b) $y = 3^{1-x} - 4$

In Exercises 43–48, solve the equations. Express the answers in terms of natural logarithms.

43. $5 = 2e^{2x-1}$ *Suggestion*: First divide by 2.
44. $3e^{1+t} = 2$ **45.** $2^x = 13$ **46.** $5^{3x-1} = 27$
47. $10^x = e$ **48.** $10^{2x+3} = 280$

In Exercises 49 and 50, solve the equations. Give two forms for each answer: one involving base 10 logarithms and the other a calculator approximation rounded to three decimal places.

49. $3^{x^2-1} = 12$ **50.** $2^{9-x^2} = 430.5389$

In Exercises 51–56, express the quantity in terms of base 10 *logarithms.*

51. $\log_2 5$ **52.** $\log_5 10$ **53.** $\ln 3$
54. $\ln 10$ **55.** $\log_b 2$ **56.** $\log_2 b$

In Exercises 57–61, express the quantity in terms of natural logarithms.

57. $\log_{10} 6$ **58.** $\log_2 10$ **59.** $\log_{10} e$
60. $\log_b 2$, where $b = e^2$ **61.** $\log_{10}(\log_{10} x)$

62. Give specific examples showing that each statement is *false.*
(a) $\log(x + y) = \log x + \log y$
(b) $(\log x)/(\log y) = \log x - \log y$
(c) $(\log x)(\log y) = \log x + \log y$
(d) $(\log x)^k = k \log x$

63. True or false?
(a) $\log_{10} A + \log_{10} B - \frac{1}{2}\log_{10} C = \log_{10}(AB/\sqrt{C})$
(b) $\log_e \sqrt{e} = 1/2$
(c) $\ln \sqrt{e} = 1/2$
(d) $\ln x^3 = \ln 3x$
(e) $\ln x^3 = 3 \ln x$
(f) $\ln 2x^3 = 3 \ln 2x$
(g) $\log_a c = b$ means $a^b = c$.
(h) $\log_5 24$ is between 5^1 and 5^2.
(i) $\log_5 24$ is between 1 and 2.
(j) $\log_5 24$ is closer to 1 than to 2.
(k) The domain of $g(x) = \ln x$ is the set of all real numbers.
(l) The range of $g(x) = \ln x$ is the set of all real numbers.
(m) The function $g(x) = \ln x$ is one-to-one.

Use a calculator for Exercises 64–66.

64. **(a)** Check Property 2 using the values $b = 10$, $P = \pi$, and $Q = \sqrt{2}$.
(b) Let $P = 3$ and $Q = 4$. Show that $\ln(P + Q) \neq \ln P + \ln Q$.
(c) Check Property 3 using the values $b = 10$, $P = 2$, and $Q = 3$.
(d) If $P = 10$ and $Q = 20$, show that $\ln(PQ) \neq (\ln P)(\ln Q)$.
(e) Check Property 3 using natural logarithms and the values $P = 17$ and $Q = 76$.
(f) Show that $(\log_{10} 17)/(\log_{10} 76) \neq \log_{10} 17 - \log_{10} 76$.

65. **(a)** Check Property 4 using the values $b = 10$, $P = \pi$, and $n = 7$.
(b) Using the values given for b, P, and n in part (a), show that $\log_b P^n \neq (\log_b P)^n$.

(continues)

(c) Verify Property 5 using the values $b = 10$ and $P = 1776$.
(d) Verify that $\ln 2 + \ln 3 + \ln 4 = \ln 24$.
(e) Verify that

$$\log_{10} A + \log_{10} B + \log_{10} C = \log_{10}(ABC)$$

using the values $A = 11$, $B = 12$, and $C = 13$.
(f) Let $f(x) = e^x$ and $g(x) = \ln x$. Compute $f[g(2345.6)]$.
(g) Let $f(x) = 10^x$ and $g(x) = \log_{10} x$. Compute $g[f(0.123456)]$.

66. Evaluate $[\frac{1}{\pi}\ln(640320^3 + 744)]^2$. *Remark*: Contrary to the empirical evidence from your calculator, it is known that this number is irrational.

B

67. *The approximation* $\ln(1 + x) \approx x$: As indicated in the accompanying graph, the values of $\ln(1 + x)$ and x are very close to one another for small positive values of x.

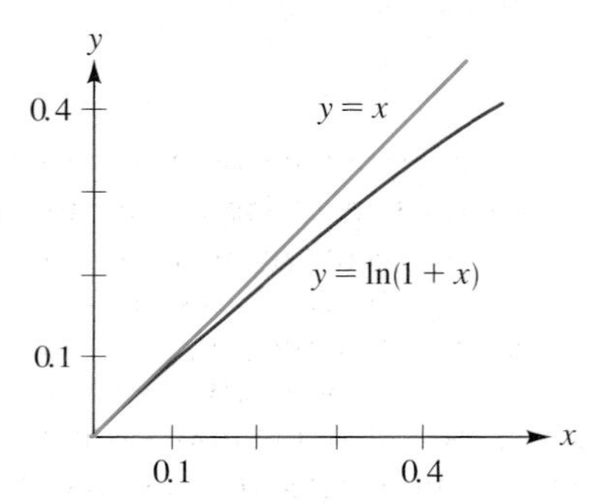

Using your calculator, complete the table to obtain numerical evidence of this. For your answers, report the first six decimal places of the calculator display (don't round off).

x	0.1	0.05	0.005	0.0005
$\ln(1 + x)$				

68. As indicated in the figure, the graph of $y = ae^{bx}$ passes through the two points (2, 10) and (8, 80). Follow steps (a) through (c) to determine the values of the constants a and b.

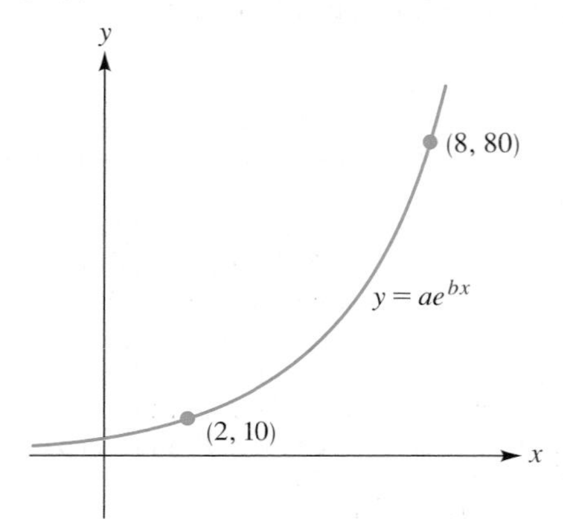

(a) Since the point (8, 80) lies on the graph, the pair of values $x = 8$ and $y = 80$ satisfies the equation $y = ae^{bx}$; that is,

$$80 = ae^{8b} \tag{1}$$

Similarly, since the point (2, 10) lies on the graph, we have

$$10 = ae^{2b} \tag{2}$$

Now use equations (1) and (2) to show that $b = (\ln 8)/6$.
(b) In equation (2), substitute for b using the expression obtained in part (a). Show that the resulting equation can be written $10 = a(e^{\ln 8})^{1/3}$.
(c) Use one of the properties of logarithms to simplify the right-hand side of the equation in part (b), then solve for a. (You should obtain $a = 5$.)

In Exercises 69 and 70, you are given the coordinates of two points on the graph of the curve $y = ae^{bx}$. In each case, determine the values of a and b. Hint: Use the method explained in Exercise 68.

69. $(-2, 324)$ and $(1/2, 4/3)$ **70.** $(1, 2)$ and $(4, 8)$

71. (a) Use base 10 logarithms to solve the equation $2^x - 3 = 0$.
(b) Use your calculator to evaluate the answer in part (a). Also use the calculator to evaluate $(\ln 3)/(\ln 2)$.
(c) Prove that $(\log_{10} 3)/(\log_{10} 2) = (\ln 3)/(\ln 2)$.

72. (a) Prove that $\log_b(P/Q) = \log_b P - \log_b Q$.
Hint: Study the proof of Property 2 in the text.
(b) Prove the **change-of-base formula:**

$$\log_a x = \frac{\log_b x}{\log_b a}$$

Hint: Use the method of Example 7 in the text.

73. Show that $\log_b \dfrac{\sqrt{3} + \sqrt{2}}{\sqrt{3} - \sqrt{2}} = 2\log_b(\sqrt{3} + \sqrt{2})$.

74. (a) Show that $\log_b(P/Q) + \log_b(Q/P) = 0$.
(b) Simplify: $\log_a x + \log_{1/a} x$.

75. Simplify: $b^{3\log_b x}$

76. Is there a constant k such that the equation $e^x = 2^{kx}$ holds for all values of x?

77. Prove that $\log_b a = 1/(\log_a b)$.

78. Simplify: $(\log_2 3)(\log_3 4)(\log_4 5)$.

79. (a) Without using your calculator, show that

$$\frac{1}{\log_2 \pi} + \frac{1}{\log_5 \pi} > 2.$$

Hint: Convert to base π logarithms.
(b) Without using your calculator, show that

$$\log_\pi 2 + \frac{1}{\log_\pi 2} > 2$$

Hint: First explain, in complete sentences, how you know that the quantity $\log_\pi 2$ is positive. Then apply the inequality given in Exercise 40(b) in Section 2.3.
(c) Use your calculator (and the change-of-base formula) to find out which of the two quantities is larger:

$$\frac{1}{\log_2 \pi} + \frac{1}{\log_5 \pi} \quad \text{or} \quad \log_\pi 2 + \frac{1}{\log_\pi 2}$$

80. A function f with domain $(1, \infty)$ is defined by the equation $f(x) = \log_x 2$.
(a) Find a value for x such that $f(x) = 2$.
(b) Is the number that you found in part (a) a fixed point of the function f?

C

81. Prove that $(\log_a x)/(\log_{ab} x) = 1 + \log_a b$.

82. Simplify a^x when $x = \log_b(\log_b a)/\log_b a$.

83. If $a^2 + b^2 = 7ab$, where a and b are positive, show that

$$\log\left[\frac{1}{3}(a + b)\right] = \frac{1}{2}(\log a + \log b)$$

no matter which base is used for the logarithms (but it is understood that the same base is used throughout).

84. Let $f(x) = \ln\left(1 - \frac{1}{x^2}\right)$.
(a) Use the properties of logarithms (and some algebra) to show that

$$f(2) + f(3) + f(4) = \ln\frac{5}{8}$$

(b) Use a calculator to check the result in part (a).

5.5 EQUATIONS AND INEQUALITIES WITH LOGS AND EXPONENTS

Initially, [John] *Napier* [1550–1617] *called logarithms "artificial numbers" but later coined the term logarithm, meaning "number of the ratio."*—Ronald Calinger, ed., *Classics of Mathematics* (Englewood Cliffs, N.J.: Prentice Hall, 1995)

. . . we would like to give a value for z, such that $a^z = y$. This value of z, insofar as it is viewed as a function of y, it is called the LOGARITHM of y.—Leonhard Euler (1707–1783), *Introductio in analysin infinitorum* (Lausanne: 1748) [This classic text has been translated by John D. Blanton (Berlin: Springer-Verlag, 1988).]

In Examples 2 through 9 in this section we illustrate some of the more common approaches for solving equations and inequalities involving exponential and logarithmic expressions. Although there is no single technique that can be used to solve every equation or inequality of this type, the methods in this section do have much in common: they all rely on the definition and the basic properties of logarithms.

Actually, in the previous two sections we've already solved some equations involving exponential expressions. So as background for the work to follow, you might want to review the following four examples. *Study suggestion*: Try solving for yourself the following four equations before you turn back to the text's solution.

EXAMPLE	EQUATION
Example 6 in Section 5.3	$10^{x/2} = 16$
Example 7 in Section 5.3	$e^{2t-5} = 5000$
Example 5 in Section 5.4	$2^x - 3 = 0$
Example 6 in Section 5.4	$4^{5x+2} = 70$

EXAMPLE 1 Reviewing some possibilities for roots

Consider the following two equations:

$$(\ln x)^2 = 2 \ln x \tag{1}$$
$$\ln(x^2) = 2 \ln x \tag{2}$$

(a) Is either one of the values $x = 1$ or $x = 2$ a root of equation (1)?
(b) Is either one of the values $x = 1$ or $x = 2$ a root of equation (2)?

SOLUTION

(a) To see whether the value $x = 1$ satisfies equation (1), we have

$$(\ln 1)^2 = 2 \ln 1$$
$$0^2 = 2(0) \qquad \text{True}$$

Thus $x = 1$ is a root of equation (1). To see whether $x = 2$ is a root, we write

$$(\ln 2)^2 = 2 \ln 2$$
$$\ln 2 = 2 \qquad \text{dividing through by } \ln 2\ (\neq 0)$$

This last equation is not valid. You can see this by using a calculator or, more directly, by rewriting the equation in exponential form to obtain $e^2 = 2$, which is clearly false. (Why?) Consequently, the value $x = 2$ is not a solution of equation (1).

(b) You've seen equation (2) before, or at least one very much like it. It's an example of one of the basic properties of logarithms that we studied in the previous section. That is, we know from the previous section that the equation $\ln(x^2) = 2 \ln x$ holds for every value of x in the domain of the function $y = \ln x$. In particular, then, since both $x = 1$ and $x = 2$ are in the domain of the function $y = \ln x$, we can conclude immediately, without any work required, that both of the values 1 and 2 satisfy equation (2).

The example that we have just completed serves to remind us of the difference between a *conditional equation* and an *identity.* An identity is true for all values of the variable in its domain. For example, the equation $\ln(x^2) = 2 \ln x$ is an identity; it is true for every positive real number x. In contrast to this, a conditional equation is true only for some (or perhaps even none) of the values of the variable. Equation (1) in Example 1 is a conditional equation; we saw that it is true when $x = 1$ and false when $x = 2$. The equation $2^x = -1$ is an example of a conditional equation that has no real-number root. (Why?) Most of the equations that we solve in this section are conditional equations, but watch for a few identities to pop up in the examples and exercises. It can make your work much easier.

EXAMPLE 2 An equation where unknown appears in exponent

Find all real-number roots of the equation $4^x = 3^{2x+1}$.

SOLUTION

Taking the natural logarithm of both sides of the given equation, we have

$$\ln(4^x) = \ln(3^{2x+1})$$

and therefore

$$\begin{aligned} x \ln 4 &= (2x + 1)\ln 3 \\ x \ln 4 &= 2x \ln 3 + \ln 3 \\ x(\ln 4 - 2 \ln 3) &= \ln 3 \\ x &= \frac{\ln 3}{\ln 4 - 2 \ln 3} \end{aligned}$$

Using a calculator now, we find that x is approximately -1.355. In Figure 1 we show a graphical check of this solution. Exercise 88 asks you to check the solution algebraically.

GRAPHICAL PERSPECTIVE

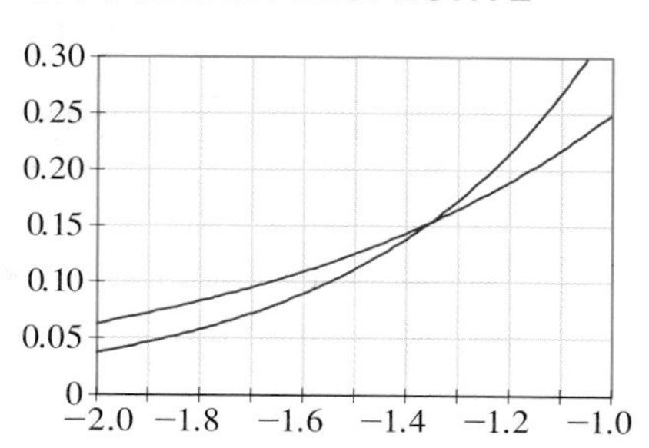

Figure 1
The curves $y = 4^x$ and $y = 3^{2x+1}$ intersect at a point whose x-coordinate is between -1.4 and -1.3. In Example 2 we found this x-value to be -1.355 (rounded to three decimal places). (Question: Which of the two graphs in the figure represents $y = 4^x$?)

EXAMPLE 3 Some equations involving the natural logarithm

Find all real-number roots of the following equations:
(a) $\ln(\ln x) = 2$; **(b)** $e^{\ln x} = -2$; **(c)** $e^{\ln x} = 2$.

SOLUTION

(a) For convenience we let t stand for $\ln x$. Then the given equation becomes

$$\ln t = 2$$

Rewriting this equation in exponential form, we have

$$t = e^2$$

or, in view of the definition of t,

$$\ln x = e^2$$

Now we rewrite this last equation in exponential form to obtain

$$\begin{aligned} x &= e^{e^2} \\ &\approx 1618.2 \qquad \text{using a calculator} \end{aligned}$$

As Exercise 85 asks you to verify, the value $x = e^{e^2}$ indeed checks in the original equation.

(b) Think for a moment about the graph of the basic exponential function $y = e^x$. The graph is always above the x-axis. This tells us that e raised to any exponent is never negative. Consequently, the equation $e^{\ln x} = -2$ has no real-number solution.

(c) From Section 5.2 (as well as Section 5.3) we know that $e^{\ln x} = x$ for all positive numbers x, so the given equation $e^{\ln x} = 2$ becomes simply $x = 2$, and we are done.

In the next example we solve equations involving logarithms by converting them to exponential form. For part (b) of the example you need to recall the quadratic formula.

EXAMPLE 4 Equations involving base 10 logarithms

Find all real-number roots of the following equations:
(a) $\log_{10}(x^2 - 2) = 3$; **(b)** $\log_{10}(x^2 - 2x) = 3$.

SOLUTION

(a) Converting to exponential form, we have

$$x^2 - 2 = 10^3$$

and therefore

$$x^2 = 1002$$

or

$$x = \pm\sqrt{1002} \approx \pm 31.65$$

As Exercise 86(a) asks you to check, both of the values $\pm\sqrt{1002}$ satisfy the given equation.

(b) Again we begin by converting to exponential form. This yields

$$x^2 - 2x = 1000$$

and therefore

$$x^2 - 2x - 1000 = 0$$

Now we apply the quadratic formula:

$$x = \frac{-b \pm \sqrt{b^2 - 4ac}}{2a} = \frac{2 \pm \sqrt{(-2)^2 - 4(1)(-1000)}}{2}$$
$$= \frac{2 \pm \sqrt{4004}}{2} = \frac{2 \pm \sqrt{4(1001)}}{2}$$
$$= \frac{2 \pm 2\sqrt{1001}}{2} = 1 \pm \sqrt{1001}$$

As Exercise 86(b) asks you to verify, both of the numbers $1 + \sqrt{1001}$ and $1 - \sqrt{1001}$ satisfy the given equation. Using a calculator, we find that these two values are approximately 32.6 and -30.6, respectively. In Figure 2 we show a graphical check of the solutions that we've determined in this example.

GRAPHICAL PERSPECTIVE

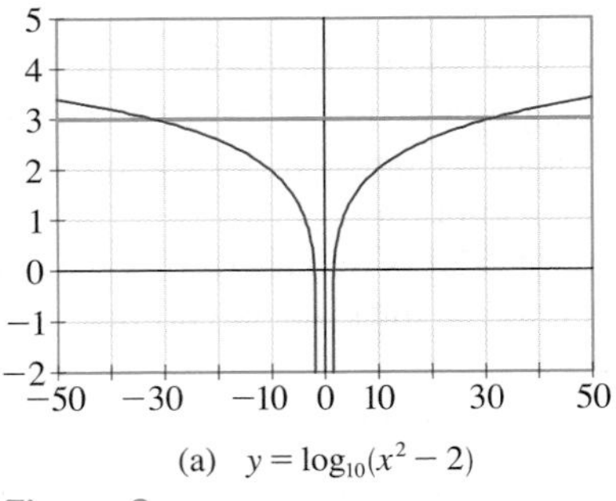

(a) $y = \log_{10}(x^2 - 2)$

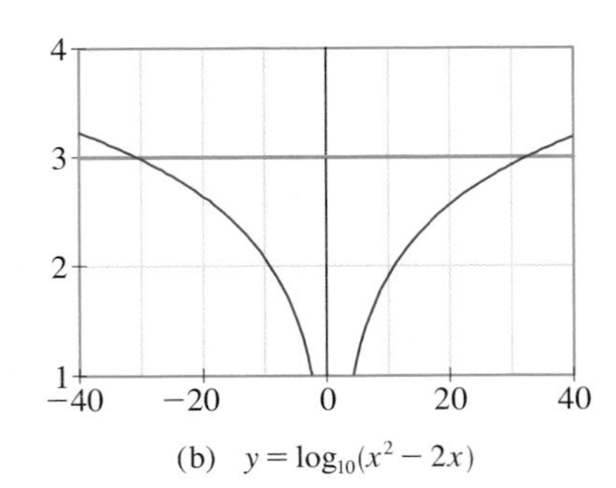

(b) $y = \log_{10}(x^2 - 2x)$

Figure 2

In (a) the figure shows that the curve $y = \log_{10}(x^2 - 2)$ intersects the horizontal line $y = 3$ at two points whose x-coordinates are approximately 30 and -30. In part (a) of Example 4 we found these two numbers to be 31.65 and -31.65, respectively (rounded to two decimal places). Similarly, in (b) the figure shows that the curve $y = \log_{10}(x^2 - 2x)$ intersects the horizontal line $y = 3$ at two points, one with an x-coordinate slightly greater than 30, the other with an x-coordinate very close to -30. In part (b) of the example we found these two x-coordinates to be 32.6 and -30.6, respectively (rounded to one decimal place).

If you look over both parts of the example just completed, you'll see that we used the same general procedure in both cases: Convert from logarithmic to exponential form, then solve the resulting quadratic equation. The next example works in the other direction: First we solve a quadratic equation, then we isolate the variable by converting from exponential to logarithmic form.

EXAMPLE 5 An equation in quadratic form that involves exponentials

As indicated in Figure 3, the graph of the function $y = 2(3^{2x}) - 3^x - 3$ has an x-intercept between 0.25 and 0.50. Find the exact value of this intercept and also a calculator approximation rounded to three decimal places.

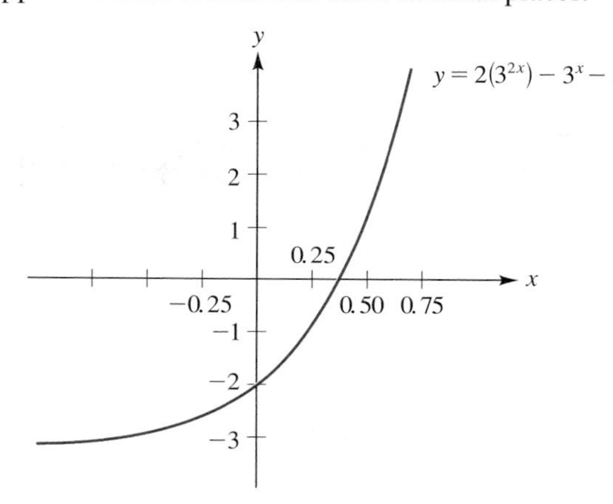

Figure 3

SOLUTION

We need to solve the equation $2(3^{2x}) - 3^x - 3 = 0$. The observation that helps here is that 3^{2x} can be written as $(3^x)^2$. This lets us rewrite the equation as

$$2(3^x)^2 - 3^x - 3 = 0$$

Now, for convenience, we let $t = 3^x$ so that the equation becomes

$$2t^2 - t - 3 = 0$$

or

$$(2t - 3)(t + 1) = 0 \qquad \text{Check the factoring.}$$

$$2t - 3 = 0 \quad \Big| \quad t + 1 = 0$$

$$t = \frac{3}{2} \quad \Big| \quad t = -1$$

Using the value $t = 3/2$ in the equation $t = 3^x$ gives us

$$3^x = \frac{3}{2}$$

We can solve this last equation by taking the logarithm of both sides. Using base e logarithms, we have

$$\ln(3^x) = \ln 1.5$$

and therefore

$$x \ln 3 = \ln 1.5$$

Dividing both sides of this last equation by the quantity ln 3, we obtain $x = (\ln 1.5)/(\ln 3)$. As Exercise 87 asks you to verify, this value for x indeed checks in the original equation.

Now, what about the other value for t that we found? With $t = -1$ we have $3^x = -1$, but that is impossible. (Why?) Consequently, there is only one root of the given equation, namely, $x = (\ln 1.5)/(\ln 3)$. This is the required x-intercept for the graph shown in Figure 3. Using a calculator, we find that $\ln(1.5)/(\ln 3) \approx 0.369$. Note that this value is consistent with Figure 3, in which the x-intercept appears to be roughly halfway between 0.25 and 0.50.

When we solved equations in earlier chapters of this text, we learned that some techniques, such as squaring both sides of an equation, may introduce extraneous roots. The next example shows another type of situation in which an extraneous root may be introduced. We'll return to this point after the example.

EXAMPLE 6 Solving and then checking for extraneous roots

Solve for x: $\log_3 x + \log_3(x + 2) = 1$.

SOLUTION

Using Property 2 (on page 361), we can write the given equation as

$$\log_3[x(x + 2)] = 1 \qquad \text{or} \qquad \log_3(x^2 + 2x) = 1$$

Writing this last equation in exponential form yields

$$\begin{aligned} x^2 + 2x &= 3^1 \\ x^2 + 2x - 3 &= 0 \\ (x + 3)(x - 1) &= 0 \end{aligned}$$

Thus we have

$$x + 3 = 0 \qquad \text{or} \qquad x - 1 = 0$$

and, consequently,

$$x = -3 \qquad \text{or} \qquad x = 1$$

Now let's check these values in the original equation to see whether they are indeed solutions.

If $x = -3$, the equation becomes

$$\underbrace{\log_3(-3)} + \underbrace{\log_3(-1)} \stackrel{?}{=} 1$$

Neither expression is defined, since the domain of the logarithm function does not contain negative numbers.

If $x = 1$, the equation becomes

$$\begin{aligned} \log_3 1 + \log_3 3 &\stackrel{?}{=} 1 \\ 0 + 1 &\stackrel{?}{=} 1 \qquad \text{True} \end{aligned}$$

Thus the value $x = 1$ is a solution of the original equation, but $x = -3$ is not.

In the example we just concluded, an extraneous solution ($x = -3$) was generated along with the correct solution, $x = 1$. How did this happen? It happened because in the second line of the solution, we used the property $\log_b PQ = \log_b P + \log_b Q$, which is valid only when both P and Q are positive. In the box that follows, we generalize this remark about extraneous solutions.

PROPERTY SUMMARY
Extraneous Solutions for Logarithmic Equations

Using the properties of logarithms (on page 361) to solve logarithmic equations may introduce extraneous solutions that do not check in the original equation. (This occurs because the logarithm function requires positive inputs, but in solving an equation, we may not know ahead of time the sign of an input involving a variable.) Therefore it is always necessary to check any candidates for solutions that are obtained in this manner.

Earlier in this book, in Sections 2.3 and 2.4, we learned how to solve inequalities involving polynomials. Those same techniques, along with the properties of logarithms, are often useful in solving inequalities involving logarithmic and exponential expressions. In the box that follows, we list some additional facts that are helpful in solving these types of inequalities.

PROPERTY SUMMARY Inequalities Involving Exponential and Logarithmic Functions

For each of the following, assume that b is a positive constant greater than 1.

1. (Refer to Figure 4.)
   (a) If $p < q$ then $b^p < b^q$.
   (b) Conversely, if $b^p < b^q$ then $p < q$.

2. (Refer to Figure 5.) Assume that p and q are positive.
   (a) If $p < q$ then $\log_b p < \log_b q$.
   (b) Conversely, if $\log_b p < \log_b q$ then $p < q$.

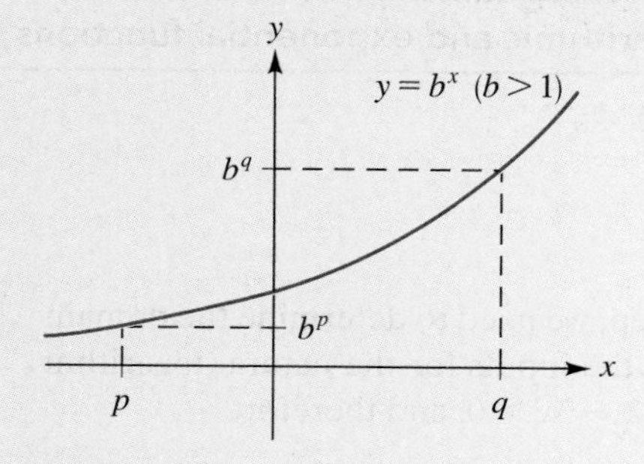

Figure 4

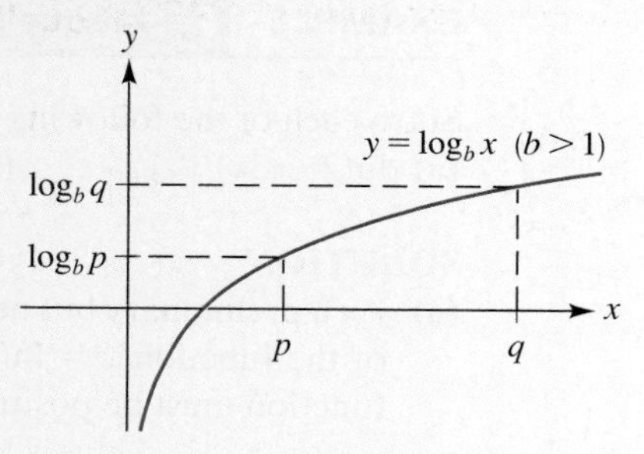

Figure 5

EXAMPLE 7 An inequality involving an exponential function

Solve the inequality $2(1 + 0.4^x) < 5$.

SOLUTION

Dividing both sides by 2, we have

$$1 + 0.4^x < 5/2$$
$$0.4^x < 3/2$$
$$\ln(0.4^x) < \ln(3/2) \quad \text{using Property 2(a) above}$$
$$x \ln(0.4) < \ln(3/2)$$

Now, to isolate x, we want to divide both sides of this last inequality by the quantity $\ln(0.4)$, which is negative. (Without relying on a calculator, how do you know that this quantity is negative?) Since dividing both sides by a negative quantity reverses the inequality, we obtain

$$x > \frac{\ln(3/2)}{\ln(0.4)}$$

So the solution set consists of all real numbers greater than $\dfrac{\ln(3/2)}{\ln(0.4)}$.

Using a calculator to evaluate this last expression, we find that it is approximately -0.443. With this approximation, we can use interval notation to write the solution set as $(-0.443, \infty)$. See Figure 6 for a graphical view of this solution.

GRAPHICAL PERSPECTIVE

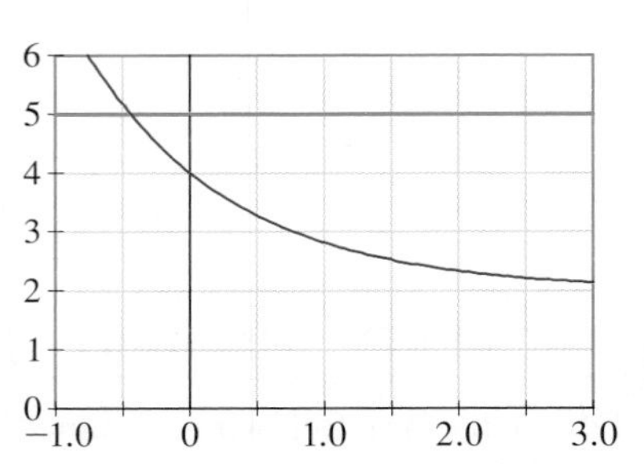

Figure 6
The graph of the equation $y = 2(1 + 0.4^x)$ indicates that the solution set of the inequality $2(1 + 0.4^x) < 5$ is an interval of the form (a, ∞), where a is approximately -0.5. In Example 7 we found that this value of a is $(\ln 1.5)/(\ln 0.4) \approx -0.443$ (rounded to three decimal places).

EXAMPLE 8 Inequalities with logarithmic and exponential functions

Solve each of the following inequalities:
(a) $\ln(2 - 3x) \leq 1$; **(b)** $e^{2-3x} \leq 1$.

SOLUTION

(a) As a preliminary but necessary first step, we need to determine the domain of the function $y = \ln(2 - 3x)$. Since the inputs for the natural logarithm function must be positive, we require $2 - 3x > 0$, and therefore

$$-3x > -2 \quad \text{or} \quad x < \frac{2}{3}$$

Now we turn to the given inequality $\ln(2 - 3x) \leq 1$. Applying Property 1(a) on the previous page to this inequality, we can write

$$e^{\ln(2-3x)} \leq e^1$$

and, consequently,

$$\begin{aligned} 2 - 3x &\leq e \\ -3x &\leq e - 2 \\ x &\geq \frac{e-2}{-3} = \frac{2-e}{3} \end{aligned}$$

Putting things together now, we want x to be greater than or equal to $(2 - e)/3$ but less than $2/3$. Thus the solution set is the interval $\left[\dfrac{2-e}{3}, \dfrac{2}{3}\right)$.

NOTE For this interval notation to make sense, the fraction $(2 - e)/3$ must be smaller than $2/3$. Although you can verify this using a calculator, you can also explain it without relying on a calculator. Exercise 89 asks you to do this.

(b) Since the domain of the exponential function is the set of all real numbers, there is no preliminary restriction on x, as there was in part (a). Taking the natural logarithm of both sides of the given inequality [as in Property 2(a) on page 377], we obtain

$$\ln(e^{2-3x}) \le \ln(1)$$

and therefore

$$\begin{aligned} 2 - 3x &\le 0 \\ -3x &\le -2 \\ x &\ge \frac{2}{3} \end{aligned}$$

The solution set of the given inequality is the interval $[2/3, \infty)$.

EXAMPLE 9 Taking domain into account in determining the solution set

(a) Determine the domain of the function $f(x) = \log_{10} x + \log_{10}(x - 2)$.
(b) Solve the inequality $\log_{10} x + \log_{10}(x - 2) \le \log_{10} 24$.

SOLUTION

(a) As you know, the logarithm function requires positive inputs. So for the expression $\log_{10} x$ we require $x > 0$, whereas for the expression $\log_{10}(x - 2)$ we require

$$x - 2 > 0 \qquad \text{and therefore} \qquad x > 2$$

In summary, then, we want x to be greater than zero, and at the same time, we need to have x greater than 2. Notice now that if x is greater than 2, it's automatically true that x is greater than zero. Thus the domain of the given function f consists of all real numbers greater than 2.

(b) On the left-hand side of the given inequality, we can use one of the basic properties of logarithms to write

$$\log_{10} x + \log_{10}(x - 2) = \log_{10}[x(x - 2)] = \log_{10}(x^2 - 2x)$$

With this result the given inequality becomes

$$\log_{10}(x^2 - 2x) \le \log_{10} 24$$

and therefore

$$x^2 - 2x \le 24 \qquad \text{using Property 2(b) on page 377}$$

or

$$x^2 - 2x - 24 \le 0 \tag{1}$$

Inequality (1) is one of the types of inequalities that you learned to solve earlier in this text (in Section 2.4). As Exercise 90 at the end of this section asks you to show, the solution set for inequality (1) is the closed interval

$[-4, 6]$. This closed interval, however, is not the solution set for the original inequality; we need to take into account the result in part (a). Putting things together, then, we want only those numbers in the interval $[-4, 6]$ that are greater than 2. Consequently, the solution set for the given inequality is the interval $(2, 6]$. Figure 7 shows a graphical view of this result.

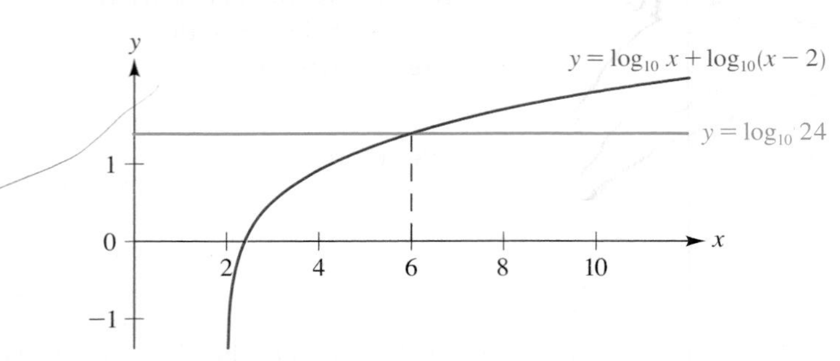

Figure 7
The curve $y = \log_{10} x + \log_{10}(x-2)$ is below the horizontal line $y = \log_{10} 24$ when $2 < x < 6$. At $x = 6$ the curve and the line intersect. Thus the solution set of the inequality $\log_{10} x + \log_{10}(x-2) \le \log_{10} 24$ is the interval $(2, 6]$.

EXERCISE SET 5.5

A

To help you get started, Exercises 1–10 correlate directly with Examples 2–6 as shown in the chart. (Thus, if you need help in any of Exercises 1–10, first consult the indicated example.)

Exercise	1, 2	3, 4	5, 6	7, 8	9, 10
Based on Example No.	2	3(a)	4(a)	4(b)	5

In Exercises 1–41, find all the real-number roots of each equation. In each case, give an exact expression for the root and also (where appropriate) a calculator approximation rounded to three decimal places.

1. $5^x = 3^{2x-1}$
2. $7^{-4x} = 2^{1+3x}$
3. $\ln(\ln x) = 1.5$
4. $\log_3[\log_3(2x)] = -2$
5. $\log_{10}(x^2 + 36) = 2$
6. $\log_2(2x^2 - 4) = 5$
7. $\log_{10}(2x^2 - 3x) = 2$
8. $\log_9(x^2 + x) = 0.5$
9. $10^{2x} + 3(10^x) - 10 = 0$
10. $3(2^{2x}) - 11(2^x) - 4 = 0$
11. **(a)** $\ln(x^3) = 3 \ln x$
(b) $(\ln x)^3 = 3 \ln x$
12. **(a)** $(\log_{10} x)^2 = 2 \log_{10} x$
(b) $\log_{10}(x^2) = 2 \log_{10} x$
13. **(a)** $\log_3 6x = \log_3 6 + \log_3 x$
(b) $\log_3 6x = 6 \log_3 x$
14. **(a)** $\ln x = (\log_{10} x)/(\log_{10} e)$
(b) $\ln x = (\log_{10} e)/(\log_{10} x)$
15. $7^{\log_7 2x} = 2x$
16. $7^{\log_7 2x} = 7$
17. $\log_2(\log_3 x) = -1$
18. $\ln[\ln(\ln x)] = 1$
19. $\ln 4 - \ln x = (\ln 4)/(\ln x)$
20. $\log_5 x - \log_5 10 = \log_5(x/10)$
21. $\ln(3x^2) = 2 \ln(3x)$
22. $\ln(3x^2) = 2 \ln(\sqrt{3}x)$
23. $\log_{16}\dfrac{x+3}{x-1} = \dfrac{1}{2}$
24. $\log_{\sqrt{2}}\dfrac{1-4x}{1+4x} = 4$
25. **(a)** $e^{2x} + 2e^x + 1 = 0$
(b) $e^{2x} - 2e^x + 1 = 0$
(c) $e^{2x} - 2e^x - 3 = 0$
(d) $e^{2x} - 2e^x - 4 = 0$
26. **(a)** $4e^{6x} - 12e^{3x} + 9 = 0$
(b) $4e^{6x} + 12e^{3x} + 9 = 0$
(c) $4e^{6x} - 16e^{3x} - 9 = 0$
(d) $e^{6x} - 12e^{3x} - 9 = 0$
27. $e^x - e^{-x} = 1$ *Hint*: Multiply both side by e^x.
28. $e^x + e^{-x} = 2$
29. $2^{5x} = 3^x(5^{x+3})$
30. $e^{3x} = 10^{2x}(2^{1-x})$
31. $\log_6 x + \log_6(x+1) = 1$
32. $\log_6 x + \log_6(x+1) = 0$
33. $\log_9(x+1) = \frac{1}{2} + \log_9 x$
34. $\log_2(x+4) = 2 - \log_2(x+1)$
35. $\log_{10}(2x+4) + \log_{10}(x-2) = 1$
36. $\ln x + \ln(x+1) = \ln 12$
37. $\log_{10}(x+3) - \log_{10}(x-2) = 2$
38. $\ln(x+1) = 2 + \ln(x-1)$
39. $\log_{10}(x+1) = 2 \log_{10}(x-1)$
40. $\log_2(2x^2 + 4) = 5$
41. $\log_{10}(x-6) + \log_{10}(x+3) = 1$
42. Solve for x in terms of a:
$\log_2(x+a) - \log_2(x-a) = 1$.
43. Solve for x in terms of y:
(a) $\log_{10} x - y = \log_{10}(3x-1)$;
(b) $\log_{10}(x-y) = \log_{10}(3x-1)$.
44. Solve for x in terms of b: $\log_b(1-3x) = 3 + \log_b x$.

In Exercises 45–50, you are given an equation of the form $\ln x = f(x)$. *In each case the equation cannot be solved using the algebraic techniques of this section.*

(a) *Without using a graphing utility, determine how many roots the equation has by sketching the graphs of* $y = \ln x$ *and* $y = f(x)$ *in the same coordinate system.*

Ⓖ **(b)** *Use a graphing utility to draw the graphs on the same screen. Then, by zooming in on each point where the graphs intersect, estimate to the nearest hundredth each root of the equation* $\ln x = f(x)$.

45. $\ln x = -x$
46. $\ln x = (x-2)^2$
47. $\ln x = 1/(x+1)$
48. $\ln x = e^{-x}$
49. $\ln x = |x-2|$
50. $\ln x = -|x-2|$

In Exercises 51–66, solve the inequalities. Where appropriate, give an exact answer as well as a decimal approximation.

51. $3(2 - 0.6^x) \le 1$
52. $6(5 - 1.6^x) \ge 13$
53. $4(10 - e^x) \le -3$
54. $\frac{2}{3}(1 - e^{-x}) \le -3$
55. $\ln(2 - 5x) > 2$
56. $3 \log_{10}(4x + 3) < 1$
57. $e^{2+x} \ge 100$
58. $4^{5-x} > 15$
59. $2^x > 0$
60. $\log_2 x \ge 0$
61. $\log_2 \dfrac{2x-1}{x-2} < 0$
62. $\ln \dfrac{3x-2}{4x+1} > \ln 4$
63. $e^{x^2-4x} \ge e^5$
64. $10^{-x^2} \le 10^{-12}$
65. $e^{(1/x)-1} > 1$
66. $e^{1/(x-1)} > 1$

67. **(a)** Specify the domain of the function $y = \ln x + \ln(x-4)$.
(b) Solve the inequality $\ln x + \ln(x-4) \le \ln 21$.

68. **(a)** Specify the domain of the function $y = \ln x + \ln(x+2)$.
(b) Solve the inequality $\ln x + \ln(x+2) \le \ln 35$.

69. Solve the inequality
$\log_2 x + \log_2(x+1) - \log_2(2x+6) < 0$.

70. Solve the inequality $\log_{10}(x^2 - 6x - 6) > 0$.

B

71. Find all roots of the equation $\log_2 x = \log_x 2$, or explain why there are none.

72. Solve the equation $\log_2 x = \log_x 3$. For each root, give an exact expression and a calculator approximation rounded to two decimal places.

Ⓖ **73.** Use a graphing utility to graph the two functions $y = \ln(x^2)$ and $y = 2 \ln x$, first separately, then in the same viewing rectangle. What do you observe? What does this have to do with the statements made in the text in the solution of Example 1(b)?

Ⓖ **74.** **(a)** Graph the two functions $f(x) = (\ln x)/(\ln 3)$ and $g(x) = \ln x - \ln 3$. (Use a viewing rectangle in which x extends from 0 to 10 and y extends from -5 to 5.) Why aren't the two graphs identical? That is, doesn't one of the basic log identities say that $(\ln a)/(\ln b) = \ln a - \ln b$?

(b) Your picture in part (a) indicates that

$$\frac{\ln x}{\ln 3} > \ln x - \ln 3 \quad (0 < x \le 10)$$

Find a viewing rectangle in which $(\ln x)/(\ln 3) \le \ln x - \ln 3$.

(c) Use the picture that you obtain in part (b) to estimate the value of x for which $(\ln x)/(\ln 3) = \ln x - \ln 3$.

(d) Solve the equation $(\ln x)/(\ln 3) = \ln x - \ln 3$ algebraically and use the result to check your estimate in part (c).

In Exercises 75–84, solve each equation. In Exercises 79–84, solve for x in terms of the other letters.

75. $3(\ln x)^2 - \ln(x^2) - 8 = 0$
76. $x^{1+\log_x 16} = 4x^2$
77. $\log_6 x = \dfrac{1}{\dfrac{1}{\log_2 x} + \dfrac{1}{\log_3 x}}$
78. $\dfrac{\ln(\sqrt{x+4}+2)}{\ln \sqrt{x}} = 2$
79. $\alpha \ln x + \ln \beta = 0$
80. $3 \ln x = \alpha + 3 \ln \beta$
81. $y = Ae^{kx}$
82. $\beta = 10 \log_{10}(x/x_0)$
83. $y = a/(1 + be^{-kx})$
84. $T = T_1 + (T_0 - T_1)e^{-kx}$

In Exercises 85–88, you are given an equation and a root that was obtained in an example in the text. In each case: **(a)** *verify (algebraically) that the root indeed satisfies the equation; and* **(b)** *use a calculator to check that the root satisfies the equation.*

85. [*From Example 3(a)*] $\ln(\ln x) = 2$; $x = e^{e^2}$

86. **(a)** [*From Example 4(a)*] $\log_{10}(x^2 - 2) = 3$; $x = \pm\sqrt{1002}$
(b) [*From Example 4(b)*] $\log_{10}(x^2 - 2x) = 3$; $x = 1 \pm \sqrt{1001}$

87. (*From Example 5*) $2(3^{2x}) - 3^x - 3 = 0$; $x = (\ln 1.5)/\ln 3$
Hint: Use the change of base formula on page 367 to express $(\ln 1.5)/\ln 3$ as a logarithm base 3.

88. (*From Example 2*) $4^x = 3^{2x+1}$; $x = (\ln 3)/(\ln 4 - 2 \ln 3)$
Hint: First show that $4^{\ln 3} = 3^{\ln 4}$ and then use it to write the left-hand side as a power of 3.

89. [*From Example 8(a)*] Explain, in one or two complete sentences, how you know (without using a calculator) that the fraction $(2 - e)/3$ is less than $2/3$.

90. [*From Example 9(b)*] Solve the inequality $x^2 - 2x - 24 \le 0$. You should find that the solution set is the closed interval $[-4, 6]$.

C

In Exercises 91 and 92, solve the inequalities.

91. $\log_\pi[\log_4(x^2 - 5)] < 0$ *Hint*: In the expression $\log_b y$, y must be positive.

92. $\dfrac{1}{\log_2 x} + \dfrac{1}{\log_3 x} + \dfrac{1}{\log_4 x} > 2 \quad (\text{for } x > 1)$

In Exercises 93 and 94, solve the equations.

93. $x^{(x^x)} = (x^x)^x$ *Hint*: There are two solutions.

94. $(\pi x)^{\log_{10} \pi} = (ex)^{\log_{10} e}$

95. Solve for x (assuming that $a > b > 0$):

$$(a^4 - 2a^2b^2 + b^4)^{x-1} = (a-b)^{2x}(a+b)^{-2}$$

Answer: $x = \dfrac{\ln(a-b)}{\ln(a+b)}$

96. Let $f(x) = \ln(x + \sqrt{x^2 + 1})$. Find $f^{-1}(x)$.

97. Suppose that $\log_{10} 2 = a$ and $\log_{10} 3 = b$. Solve for x in terms of a and b:

$$6^x = \frac{10}{3} - 6^{-x}$$

5.6 COMPOUND INTEREST

$S = Pe^{rt}$. This result is remarkable both because of its simplicity and the occurrence of e. (Who would expect that number to pop up in finance theory?) —Philip Gillett in *Calculus and Analytic Geometry,* 3rd ed. (Lexington, Mass.: D.C. Heath, 1988)

We begin this section by considering how money accumulates in a savings account. Eventually, this will lead us back to the number e and to functions of the form $y = ae^{bx}$.

The following idea from arithmetic is a prerequisite for our discussion. To increase a given quantity by, say, 15%, we multiply the quantity by 1.15. For instance, suppose that we want to increase \$100 by 15%. The calculations can be written as

$$100 + 0.15(100) = 100(1 + 0.15) = 100(1.15)$$

Similarly, to increase a quantity by 30%, we would multiply by 1.30, and so on. The next example displays some calculations involving percentage increase. The results might surprise you unless you're already familiar with this topic.

EXAMPLE 1 Computing percentage increase

An amount of \$100 is increased by 15%, and then the new amount is increased by 15%. Is this the same as an overall increase of 30%?

SOLUTION

To increase \$100 by 15%, we multiply by 1.15 to obtain \$100(1.15). Now to increase this new amount by 15%, we multiply it by 1.15 to obtain

$$[(\$100)(1.15)](1.15) = \$100(1.15)^2 = \$132.25$$

Alternatively, if we increase the original \$100 by 30%, we obtain

$$\$100(1.30) = \$130$$

Comparing, we see that the result of two successive 15% increases is greater than the result of a single 30% increase.

Now let's look at another example and use it to introduce some terminology. Suppose that you place \$1000 in a savings account at 10% interest *compounded annually.* This means that at the end of each year, the bank contributes to your account 10% of the amount that is in the account at that time. Interest compounded in this manner is called **compound interest.** The original deposit of \$1000 is called the **principal,** denoted P. The interest rate, expressed as a decimal, is denoted by r. Thus, $r = 0.10$ in this example. The variable A is used to denote the **amount** in the account at any given time. The calculations displayed in Table 1 show how the account grows.

We can learn several things from Table 1. First, consider how much interest is paid each year.

Interest paid for first year: $\$1100 - \$1000 = \$100$

Interest paid for second year: $\$1210 - \$1100 = \$110$

Interest paid for third year: $\$1331 - \$1210 = \$121$

Thus the interest earned each year is not a constant; it increases each year.

TABLE 1

Time period	Algebra	Arithmetic
After 1 year	$A = P(1 + r)$	$A = 1000(1.10)$ $= \$1100$
After 2 years	$A = [P(1 + r)](1 + r)$ $= P(1 + r)^2$	$A = 1100(1.10)$ $= \$1210$
After 3 years	$A = [P(1 + r)^2](1 + r)$ $= P(1 + r)^3$	$A = 1210(1.10)$ $= \$1331$

If you look at the algebra in Table 1, you can see what the general formula should be for the amount after t years.

Compound Interest Formula (interest compounded annually)

Suppose that a principal of P dollars is invested at an annual rate r that is compounded annually. Then the amount A after t years is given by

$$A = P(1 + r)^t$$

EXAMPLE 2 Interest compounded annually

Suppose that \$2000 is invested at $7\frac{1}{2}\%$ interest compounded annually. How many years will it take for the money to double?

SOLUTION

In the formula $A = P(1 + r)^t$, we use the given values $P = \$2000$ and $r = 0.075$. We want to find how long it will take for the money to double; that is, we want to find t when $A = \$4000$. Making these substitutions in the formula, we obtain

$$4000 = 2000(1 + 0.075)^t$$

and therefore

$$2 = 1.075^t \qquad \text{dividing by 2000}$$

We can solve this exponential equation by taking the logarithm of both sides. We use base e logarithms. (Base 10 would also be convenient here.) This yields

$$\ln 2 = \ln 1.075^t$$

and, consequently,

$$\ln 2 = t \ln 1.075 \quad \text{(Why?)}$$

To isolate t, we divide both sides of this last equation by ln 1.075. This yields

$$t = \frac{\ln 2}{\ln 1.075} \approx 9.6 \text{ years}$$

Now, assuming that the bank computes the compound interest only at the end of the year, we must round the preliminary answer of 9.6 years and say that when

TABLE 2
$A = 2000(1.075)^t$

t (years)	A (dollars)
9	3834.48
10	4122.06

$t = 10$ years, the initial \$2000 will have *more than* doubled. Table 2 adds some perspective to this. The table shows that after 9 years, something less than \$4000 is in the account; whereas after 10 years, the amount exceeds \$4000.

In Example 2 the interest was compounded annually. In practice, though, the interest is usually computed more often. For instance, a bank may advertise a rate of 8% per year *compounded semiannually*. This means that after half a year, the interest is compounded at 4%, and then after another half year, the interest is again compounded at 4%. (If you review Example 1, you'll see that two compoundings, each at a rate of $r/2$, is not the same as one compounding at the rate r. The former scheme yields more money overall than the latter.) In the case of a rate of 8% per year compounded semiannually, we say that the *periodic interest rate* is 4%. Similarly, if the interest rate of 8% per year were compounded four times per year, then the periodic interest rate would be 2 (= 8/4) percent. In Table 3 we show examples of this terminology, assuming an annual rate of r%. Note that, in general, the **periodic interest rate** is equal to r/n, where r is the annual interest rate and n is the number of times per year that the interest is compounded.

TABLE 3 Periodic Interest Rates, Assuming an Annual Rate of r%*

	n (number of times per year that interest is compounded)	Periodic interest rate (%)
Annually	1	r
Semiannually	2	$r/2$
Quarterly	4	$r/4$
Monthly	12	$r/12$
Daily (assuming 360-day year)	360	$r/360$
Daily (assuming 365-day year)	365	$r/365$

*Regarding the last two rows of the table: Some banks base their calculations on a 360-day rather than 365-day year.

The compound interest formula

$$A = P(1 + r)^t \tag{1}$$

that we worked with in Example 2 can be generalized to cover the cases that we've just been discussing, in which the interest is compounded more than once per year. The general formula is given in the box that follows.

Compound Interest Formula (interest compounded n times per year)

Suppose that a principal of P dollars is invested at an annual rate r that is compounded n times per year. Then the amount A after t years is given by

$$A = P\left(1 + \frac{r}{n}\right)^{nt} \tag{2}$$

Note that the quantity r/n in parentheses is the periodic interest rate, and the exponent nt represents the total number of compoundings.

Although we won't derive equation (2) from scratch, note that it is obtained from equation (1) as follows: In equation (1), replace the annual interest rate r by the periodic interest rate r/n; in equation (1), replace t (which not only represents time, but also the number of compoundings) with the quantity nt, which also represents the total number of compoundings.

EXAMPLE 3 Comparing annual to quarterly compounding

Suppose that \$1000 is placed in a savings account at 10% per annum. How much is in the account at the end of 1 year if the interest is: **(a)** compounded once each year ($n = 1$); and **(b)** compounded quarterly ($n = 4$)?

SOLUTION

We use the formula $A = P[1 + (r/n)]^{nt}$.

(a) For $n = 1$ we obtain

$$A = 1000\left(1 + \frac{0.10}{1}\right)^{(1)(1)} = 1000(1.1) = \$1100$$

(b) For $n = 4$ we obtain

$$A = 1000\left(1 + \frac{0.10}{4}\right)^{4(1)} = 1000(1.025)^4 = \$1103.81 \quad \text{using a calculator}$$

Notice that compounding the interest quarterly rather than annually yields the greater amount. This is in agreement with our observations in Example 1.

The results in Example 3 will serve to illustrate some additional terminology used by financial institutions. In that example, the interest for the year under quarterly compounding was

$$\$1103.81 - \$1000 = \$103.81$$

Now, \$103.81 is 10.381% of \$1000. We say in this case that the **effective rate** of interest is 10.381%. The given rate of 10% per annum compounded once a year is called the **nominal rate.** (The nominal rate and the effective rate are also referred to as the *annual rate* and the *annual yield,* respectively.) The next example further illustrates these ideas.

EXAMPLE 4 Computing the effective rate

A bank offers a nominal interest rate of 6% per annum for certain accounts. Compute the effective rate if interest is compounded monthly.

SOLUTION

Let P denote the principal, which earns 6% ($r = 0.06$) compounded monthly ($n = 12$). Then with $t = 1$, our formula yields

$$A = P\left(1 + \frac{0.06}{12}\right)^{12(1)} \approx P(1.0617) \quad \text{using a calculator}$$

This shows that the effective interest rate is about 6.17%.

There are two rather natural questions to ask on first encountering compound interest calculations:

Question 1 For a fixed period of time (say, one year), does more and more frequent compounding of interest continue to yield greater and greater amounts?

Question 2 Is there a limit on how much money can accumulate in a year when interest is compounded more and more frequently?

The answer to both of these questions is yes. If you look back over Example 1, you'll see evidence for the affirmative answer to Question 1. For additional evidence, and for the answer to Question 2, let's do some calculations. To keep things as simple as possible, suppose a principal of \$1 is invested for 1 year at the nominal rate of 100% per annum. (More realistic figures could be used here, but the algebra becomes more cluttered.) With these data our formula becomes

$$A = 1\left(1 + \frac{1}{n}\right)^{n(1)} \quad \text{or} \quad A = \left(1 + \frac{1}{n}\right)^{n}$$

Now, as we know from Section 5.2, the expression $\left(1 + \frac{1}{n}\right)^{n}$ approaches the value e as n grows larger and larger; see Table 4.

TABLE 4 Results of Compounding Interest More and More Frequently

Number of compoundings, n	Amount, $\left(1 + \frac{1}{n}\right)^{n}$
$n = 1$ (annually)	$[1 + (1/1)]^{1} = 2$
$n = 2$ (semiannually)	$[1 + (1/2)]^{2} = 2.25$
$n = 4$ (quarterly)	$[1 + (1/4)]^{4} \approx 2.44$
$n = 12$ (monthly)	$[1 + (1/12)]^{12} \approx 2.61$
$n = 365$ (daily)	$[1 + (1/365)]^{365} \approx 2.7146$
$n = 8760$ (hourly)	$[1 + (1/8760)]^{8760} \approx 2.7181$
$n = 525{,}600$ (each minute)	$[1 + (1/525{,}600)]^{525{,}600} \approx 2.71827$
$n = 31{,}536{,}000$ (each second)	$[1 + (1/31{,}536{,}000)]^{31{,}536{,}000} \approx 2.71828$

Table 4 shows that the amount does increase with the number of compoundings. But assuming that the bank rounds to the nearest penny, Table 4 also shows that there is no difference between compounding hourly, compounding each minute, and compounding each second. In each case, the rounded amount is \$2.72.

Some banks advertise interest compounded not monthly, daily, or even hourly, but *continuously,* that is, at each instant. The formula (derived in calculus) for the amount earned under continuous compounding of interest is as follows.

Compound Interest Formula (interest compounded continuously)

Suppose that a principal of P dollars is invested at an annual rate r that is compounded continuously. Then the amount A after t years is given by

$$A = Pe^{rt}$$

EXAMPLE 5 Continuous compounding of interest

A principal of \$1600 is placed in a savings account at 5% per annum compounded continuously. Assuming no subsequent withdrawals or deposits, when will the balance reach \$2400?

SOLUTION

Substitute the values $A = 2400$, $P = 1600$, and $r = 0.05$ in the formula $A = Pe^{rt}$ to obtain $2400 = 1600e^{0.05t}$. To isolate the time t in this last equation, first divide both sides by 1600. This yields

$$1.5 = e^{0.05t}$$

and therefore

$$0.05t = \ln 1.5 \quad \text{converting from exponential to logarithmic form}$$

$$t = \frac{\ln 1.5}{0.05}$$

Using a calculator now to evaluate this last expression, we find $t \approx 8.1$ years. In summary, it will take slightly longer that 8 years 1 month for the balance to reach \$2400. *Suggestion*: You can use a graphing utility to check this result. Graph the function $A = 1600e^{0.05t}$ along with the horizontal line $A = 2400$ in an appropriate viewing rectangle. Is it then the first or second coordinate of the intersection point that you want to estimate?

In the next example we compare the results of continuous compounding and annual compounding of interest.

EXAMPLE 6 Comparing continuous to annual compounding

A principal of \$100 is deposited in Account #1 at 6% per annum, compounded continuously. At the same time, another principal of \$100 is deposited in Account #2 at 6% per annum, compounded only once a year. Use a graphing utility to determine how long it will take until the amount in Account #1 exceeds that in Account #2 by (at least) ten dollars.

GRAPHICAL PERSPECTIVE

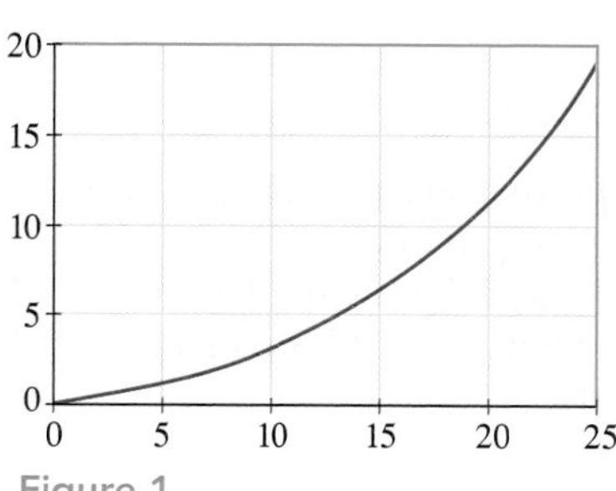

Figure 1
$f(t) = 100e^{0.06t} - 100(1.06^t)$
[0, 25, 5] by [0, 20, 5]

SOLUTION

The amount in Account #1 after t years is given by $A = 100e^{0.06t}$; for Account #2 the amount is $A = 100(1 + 0.06)^t = 100(1.06)^t$. Define $f(t)$ to be the *difference* between the amounts in each account at time t:

$$f(t) = 100e^{0.06t} - 100(1.06)^t$$

We want to know when the difference $f(t)$ exceeds ten dollars. Using a graphing utility as in Figure 1, we find that when $t = 15$, the difference is less than 10, whereas when $t = 20$, the difference is more than 10.

At this point, we can either zoom in on the portion of the graph corresponding to $15 \leq t \leq 20$, or we can use the TABLE feature of the graphing utility. For this example we'll use the TABLE feature to evaluate the amount in each account

TABLE 5

t	Account #1 $A = 100e^{0.06t}$	Account #2 $A = 100(1.06)^t$	Difference $100e^{0.06t} - 100(1.06^t)$
15	245.96	239.66	6.30
16	261.17	254.04	7.13
17	277.32	269.28	8.04
18	294.47	285.43	9.03
19	312.68	302.56	10.12
20	332.01	320.71	11.30

and the difference $f(t)$ corresponding to the years $t = 15$ to $t = 20$. The results are shown in Table 5. You should verify these results for yourself. (For help on the TABLE feature, consult the owner's manual for your graphing utility.)

As Table 5 indicates, after 18 years, the accounts differ by less than ten dollars, but after 19 years the difference is more than ten dollars. In summary, it takes 19 years (to the nearest year) before the accounts differ by at least ten dollars. *Remark*: An algebraic solution for this example would involve solving the inequality $100e^{0.06t} - 100(1.06^t) \geq 10$. This can't be done, however, using any of the algebraic techniques of this chapter. (Try it!)

In the next example we compare the nominal rate with the effective rate under continuous compounding of interest.

EXAMPLE 7 Nominal and effective rates under continuous compounding

(a) Given a nominal rate of 8% per annum compounded continuously, compute the effective interest rate.

(b) Given an effective rate of 8% per annum, compute the nominal rate compounded continuously.

SOLUTION

(a) With the values $r = 0.08$ and $t = 1$, the formula $A = Pe^{rt}$ yields

$$A = Pe^{0.08(1)}$$
$$A \approx P(1.08329) \qquad \text{using a calculator}$$

This shows that the effective interest rate is approximately 8.33% per year.

(b) We now wish to compute the nominal rate r, given an effective rate of 8% per year. An effective rate of 8% means that the initial principal P grows to $P(1.08)$ by the end of the year. Thus in the formula $A = Pe^{rt}$ we make the substitutions $A = P(1.08)$ and $t = 1$. This yields

$$P(1.08) = Pe^{r(1)}$$

Dividing both sides of this last equation by P, we have

$$1.08 = e^r$$

TABLE 6
Comparison of Nominal and Effective Rates in Example 7

Nominal rate (% per annum)	Effective rate (% per annum)
8	8.33
7.70	8

To solve this equation for r, we rewrite it in its equivalent logarithmic form:

$$r = \ln(1.08)$$
$$r \approx 0.07696 \qquad \text{using a calculator}$$

Thus, a nominal rate of about 7.70% per annum compounded continuously yields an effective rate of 8%. Table 6 summarizes these results.

Now we come to one of the remarkable and characteristic features of compound interest and of growth governed by the formula $A = Pe^{rt}$. By the **doubling time** we mean, as the name implies, the amount of time required for a given principal to double. The surprising fact here is that the doubling time does not depend on the principal P. To see why this is so in the continuous compounding case, we begin with the formula

$$A = Pe^{rt}$$

We are interested in the time t at which $A = 2P$. Replacing A by $2P$ in the formula yields

$$2P = Pe^{rt}$$
$$2 = e^{rt}$$
$$rt = \ln 2$$
$$t = \frac{\ln 2}{r}$$

Denoting the doubling time by T_2, we have the following formula.

Doubling Time under Continuous Compounding

$$\text{Doubling time} = T_2 = \frac{\ln 2}{r}$$

As you can see, the formula for the doubling time T_2 does not involve P, but only r. Thus at a given rate under continuous compounding, $2 and $2000 would both take the same amount of time to double. (This idea takes some getting used to.) *Note*: For compound interest other than continuous compounding, the doubling time is also independent of the principal, but the formula is a little more complicated.

EXAMPLE 8 Computing the doubling time

Compute the doubling time T_2 when a sum is invested at an interest rate of 4% per annum compounded continuously.

SOLUTION

$$T_2 = \frac{\ln 2}{r} = \frac{\ln 2}{0.04} \approx 17.3 \text{ years} \qquad \text{using a calculator}$$

There is a convenient approximation that allows us to estimate doubling times easily. As you can check with a calculator, we have $\ln 2 \approx 0.7$. Using this approximation in the doubling time formula, we have

$$T_2 \approx \frac{0.7}{r}$$

where r (as usual) is the annual interest rate expressed as a decimal. Now let R be the annual interest rate expressed as a percentage, so that $R = 100r$. Then we can write

$$T_2 \approx \frac{0.7}{r} = \frac{0.7}{r} \times \frac{100}{100} = \frac{70}{100r} = \frac{70}{R}$$

That is,

$$T_2 \approx \frac{70}{R} \qquad \text{where } R \text{ is the annual rate expressed as a percentage}$$

This is the so-called *rule of 70* for estimating doubling times. For instance, for an annual percentage rate of 10%, we estimate the doubling time as $T_2 \approx 70/10 = 7$ years. As another example, let's use this rule of 70 to rework Example 8, in which the annual interest rate is 4%. We have

$$T_2 \approx \frac{70}{4} = \frac{35}{2} = 17.5 \text{ years}$$

This is quite close to the actual doubling time of 17.3 years computed in Example 8. *Remark*: A slightly less accurate approximation that's often used in business is the *rule of 72*, $T_2 \approx 72/R$. The number 72 is used instead of 70 simply because there are more numbers that divide into 72 evenly than into 70.

The rule of 70 for estimating doubling times is useful when you want to sketch a graph of the function $A = Pe^{rt}$ without relying on a graphing utility. As an example, suppose that a principal of $1000 is invested at 10% per annum compounded continuously. Then the function that we wish to graph is

$$A = 1000e^{0.1t}$$

Approximating the doubling time using the rule of 70 gives us

$$T_2 \approx \frac{70}{R} = \frac{70}{10} = 7 \text{ years}$$

Now we just set up a table showing the results of doubling $1000 every 7 years, plot the points, and then join them with a smooth curve. See Figure 2.

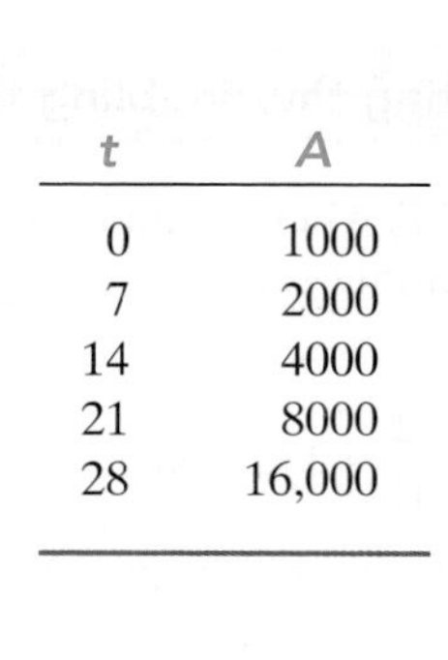

t	A
0	1000
7	2000
14	4000
21	8000
28	16,000

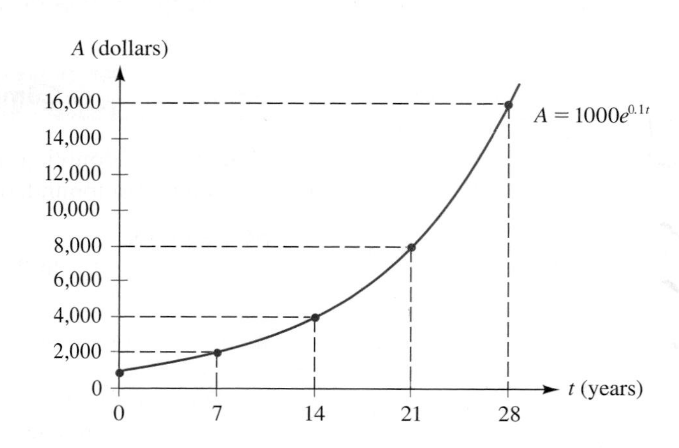

Figure 2
Applying the rule of 70 to sketch a graph

EXERCISE SET 5.6 

A

1. You invest $800 at 6% interest compounded annually. How much is in the account after 4 years, assuming that you make no subsequent withdrawal or deposit?
2. A sum of $1000 is invested at an interest rate of $5\frac{1}{2}$% compounded annually. How many years will it take until the sum exceeds $2500? (First find out when the amount equals $2500; then round off as in Example 2.)
3. At what interest rate (compounded annually) will a sum of $4000 grow to $6000 in 5 years?
4. A bank pays 7% interest compounded annually. What principal will grow to $10,000 in 10 years?
5. You place $500 in a savings account at 5% compounded annually. After 4 years you withdraw all your money and take it to a different bank, which advertises a rate of 6% compounded annually. What is the balance in this new account after 4 more years? (As usual, assume that no subsequent withdrawal or deposit is made.)
6. A sum of $3000 is placed in a savings account at 6% per annum. How much is in the account after 1 year if the interest is compounded **(a)** annually? **(b)** semiannually? **(c)** daily?
7. A sum of $1000 is placed in a savings account at 7% per annum. How much is in the account after 20 years if the interest is compounded **(a)** annually? **(b)** quarterly?
8. Your friend invests $2000 at $5\frac{1}{4}$% per annum compounded semiannually. You invest an equal amount at the same yearly rate, but compounded daily. How much larger is your account than your friend's after 8 years?
9. You invest $100 at 6% per annum compounded quarterly. How long will it take for your balance to exceed $120? (Round your answer up to the next quarter.)
10. A bank offers an interest rate of 7% per annum compounded daily. What is the effective rate?
11. What principal should you deposit at $5\frac{1}{2}$% per annum compounded semiannually so as to have $6000 after 10 years?
12. You place a sum of $800 in a savings account at 6% per annum compounded continuously. Assuming that you make no subsequent withdrawal or deposit, how much is in the account after 1 year? When will the balance reach $1000?
13. A principal of $600 dollars is placed in a savings account at 6.5% per annum compounded continuously.
    - Ⓖ **(a)** Use a graphing utility to estimate how long it will take for the balance to reach $800. (Adapt the suggestion at the end of Example 5.)
    - **(b)** Use algebra (as in Example 5) to determine how long it will take for the balance to reach $800.
14. A principal of $4000 dollars is invested at 8% per annum compounded continuously.
    - Ⓖ **(a)** Use a graphing utility to estimate how long it will take for the balance to increase by 25%. (That is, you want a balance of $4000 + $1000 = $5000.) (Adapt the suggestion at the end of Example 5.)
    - **(b)** Use algebra, rather than a graphing utility, to solve the problem in part (a).
15. Ⓖ A principal of $3500 is deposited in Account #1 at 7% per annum, compounded continuously. At the same time, another principal of $3500 is deposited in Account #2 at 7% per annum, compounded semiannually. Use a graphing utility to determine how long it will take until the amount in Account #1 exceeds that in Account #2 by (at least) $200. (Follow the method of Example 6).
16. Ⓖ A principal of $1000 is deposited in Account #1 at 8% per annum, compounded continuously. At the same time, a principal of $1200 is deposited in Account #2 at 6% per annum, compounded once per year. Use a graphing utility to determine how long it will take until the amount in Account #1 exceeds that in Account #2. Give your answer to the nearest whole number of years, rounded upward.
17. A bank offers an interest rate of $6\frac{1}{2}$% per annum compounded continuously. What principal will grow to $5000 in 10 years under these conditions?
18. Given a nominal rate of 6% per annum, compute the effective rate under continuous compounding of interest.
19. Suppose that under continuous compounding of interest, the effective rate is 6% per annum. Compute the nominal rate.
20. You have two savings accounts, each with an initial principal of $1000. The nominal rate on both accounts is $5\frac{1}{4}$% per annum. In the first account, interest is compounded semiannually. In the second account, interest is compounded continuously. How much more is in the second account after 12 years?
21. You want to invest $10,000 for 5 years, and you have a choice between two accounts. The first pays 6% per annum compounded annually. The second pays 5% per annum compounded continuously. Which is the better investment?
22. Suppose that a certain principal is invested at 6% per annum compounded continuously.
    - **(a)** Use the rule of 70, $T_2 \approx 70/R$, to estimate the doubling time.
    - **(b)** Compute the doubling time using the formula $T_2 = (\ln 2)/r$.
    - **(c)** Do your answers in (a) and (b) differ by more than 2 months?
23. A sum of $1500 is invested at 5% per annum compounded continuously.
    - **(a)** Estimate the doubling time.
    - **(b)** Compute the actual doubling time.
    - **(c)** Let d_1 and d_2 denote the actual and estimated doubling times, respectively. Define d by $d = |d_1 - d_2|$. What percentage is d of the actual doubling time?
24. A sum of $5000 is invested at 10% per annum compounded continuously.
    - **(a)** Estimate the doubling time.

(continues)

(b) Estimate the time required for the $5000 to grow to $40,000.

25. After carrying out the calculations in this problem, you'll see one of the reasons why some governments impose inheritance taxes and why laws are passed to prohibit savings accounts from being passed from generation to generation without restriction. Suppose that a family invests $1000 at 8% per annum compounded continuously. If this account were to remain intact, being passed from generation to generation, for 300 years, how much would be in the account at the end of those 300 years?

26. A principal of $500 is invested at 7% per annum compounded continuously.
(a) Estimate the doubling time.
(b) Sketch a graph similar to the one in Figure 2, showing how the amount increases with time.

27. A principal of $7000 is invested at 5% per annum compounded continuously.
(a) Estimate the doubling time.
(b) Sketch a graph showing how the amount increases with time.

28. In one savings account, a principal of $1000 is deposited at 5% per annum. In a second account, a principal of $500 is deposited at 10% per annum. Both accounts compound interest continuously.
(a) Estimate the doubling time for each account.
(b) On the same set of axes, sketch graphs showing the amount of money in each account over time. Give the (approximate) coordinates of the point where the two curves meet. In financial terms, what is the significance of this point? (In working this problem, assume that the initial deposits in each account were made at the same time.)
(c) During what period of time does the first account have the larger balance?

PROJECT LOAN PAYMENTS

Amortize: Gradual paying off of a debt by making regular equal payments to cover interest and principal.—http://www.moneywords.com

Amortize: (obsolete) To deaden, render as if dead, destroy.—Oxford English Dictionary

The goode werkes that men don whil thay ben in good lif ben al amortised by synne folwyng.
—Chaucer (ca. 1386), as quoted in the Oxford English Dictionary

(a) In this first part of the project, we'll imagine a loan scenario and explain how the calculations work from month to month. Suppose that when you graduate college, you are shopping for a new car. You decide that you'll need a loan of $15,000 toward the purchase of the particular model that you are interested in. The car dealer tells you that he can arrange this loan for you. It's a four-year loan at 6% per annum, compounded monthly. The dealer says that the monthly payments will each be $283.40.

In the following table we will explain how the entries in the row for monthly payment #1 are obtained. Then, as a check on your understanding of the process, you should fill in the next two rows yourself. (Tables of this form are called *amortization schedules.*)

Monthly payment #	Monthly payment	Portion of payment toward interest	Portion of payment toward principal	Loan balance
0				$15,000.00
1	$283.40	$75.00	$208.40	$14,791.60
2	$283.40			
3	$283.40			

Portion of monthly payment #1 toward interest: Each month, a portion of your \$283.40 payment is used to pay off the previous month's interest on the unpaid loan balance. Then, whatever is left from the \$283.40 is used as a payment on the actual loan balance. We are assuming a rate of 6% per annum compounded monthly. So the interest on an unpaid balance of P dollars for one month is

$$\begin{aligned} \text{interest} &= P(1 + r/n)^{nt} - P \\ &= P[(1 + 0.06/12)^{12(1/12)} - 1] = P[(1 + 0.06/12) - 1] \\ &= 0.005P \end{aligned}$$

For payment #1 then, we have

$$\text{interest} = 0.005P = 0.005(\$15{,}000) = \$75.00$$

Portion of monthly payment #1 toward principal:
\$283.40 − \$75.00 = \$208.40

Loan balance: \$15,000 − \$208.40 = \$14,791.60

The calculations corresponding to monthly payments #2 and #3 are now carried out in the same manner. In starting your work for monthly payment #2, remember that your unpaid balance now is $P = \$14{,}791.60$, rather than $P = \$15{,}000$. As a check on your work, after monthly payment #2 the loan balance should turn out to be \$14,582.16.

(b) There are formulas (that we won't derive) from the mathematics of finance relating the loan amount, the monthly payments, and the interest rate. One such formula, which applies to our example (with monthly compounding and monthly payments) is

$$L = M\frac{1 - (1 + p)^{-n}}{p} \tag{1}$$

where L is the loan amount, p is the periodic interest rate, and n is the total number of compoundings in the life of the loan (or, equivalently in our case, the length of the loan, in months). Use equation (1) to calculate what the monthly payments M should be for the example in part (a). Did the car dealer give you the correct figure?

(c) Solve equation (1) for n in terms of the other variables. Then use the result to compute an answer for the following problem: You are thinking of taking out a bank loan of \$315,848 toward the purchase of a beachfront condominium. In your notes, you've written down that the interest rate is $7\frac{1}{2}$% per annum, compounded monthly, and that the monthly payments would be \$953.90. You forgot, however, to write down the amount of time required to pay off the loan. Compute the required time.

Ⓖ **(d)** You graduate, and the government reminds you that the time has come to begin repaying your student loan of $L = \$5800$. You have 4 years to repay this loan; the interest is compounded monthly; and the monthly payments are $M = \$113.88$. Find the periodic interest rate p and then the annual rate $r = 12p$. *Hint*: Substitute the given numerical values for L and M into equation (1) and then convince yourself that you cannot solve for p using the algebraic techniques you learned in this chapter. Now use a graphing utility and the ZOOM feature to find p. It may take some work to find an appropriate viewing rectangle.

Drawing by Professor Ann Jones, University of Colorado, Boulder. From the cover of *The Physics Teacher,* vol. 14, no. 7 (October 1976).

5.7 EXPONENTIAL GROWTH AND DECAY

In newspapers and in everyday speech, the term "exponential growth" is used rather loosely to describe any situation involving rapid growth. In the sciences, however, **exponential growth** refers specifically to growth governed by functions of the form $y = ae^{bx}$, where a and b are positive constants. For example, since the function $A = Pe^{rt}$ (discussed in the previous section) has this general form, we say that money grows exponentially under continuous compounding of interest. Similarly, in the sciences, **exponential decay** refers specifically to decrease or decay governed by functions of the form $y = ae^{bx}$, where a is positive and b is negative. As examples in this section, we shall consider population growth, global warming, and radioactive decay.

Under ideal conditions involving unlimited food and space, the size of a population of bacteria is modeled by a function of the form

$$\mathcal{N}(t) = \mathcal{N}_0 e^{kt}$$

In this *growth law,* $\mathcal{N}(t)$ is the population at time t, and k is a positive constant related to (but not equal to) the growth rate of the population. The constant k is referred to as the **growth constant.** The number $\mathcal{N}_0$ is also a constant; it represents the size of the population at time $t = 0$. You can see that this is the case by substituting $t = 0$ in the equation $\mathcal{N}(t) = \mathcal{N}_0 e^{kt}$. This yields

$$\mathcal{N}(0) = \mathcal{N}_0 e^0 = \mathcal{N}_0 \cdot 1 = \mathcal{N}_0$$

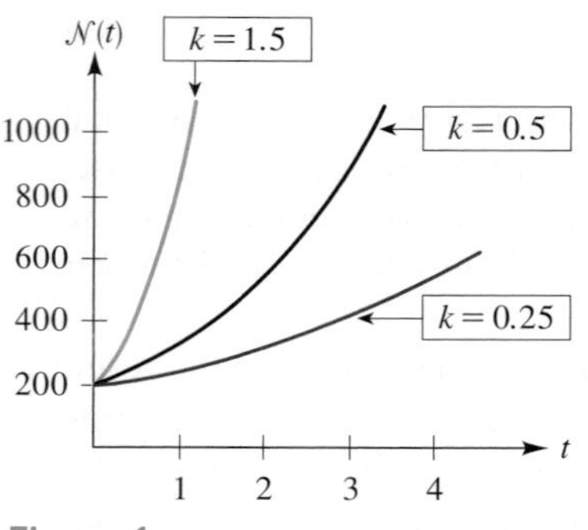

Figure 1
The graph of $\mathcal{N}(t) = 200e^{kt}$ for $k = 0.25$, $k = 0.5$, and $k = 1.5$

That is, $\mathcal{N}(0) = \mathcal{N}_0$, which says that the population at time $t = 0$ is indeed $\mathcal{N}_0$. Figure 1 shows three examples. In each case the initial population size is $\mathcal{N}_0 = 200$, and we've graphed the growth laws using three different values for the growth constant k.

A remark about notation: In science texts the growth law is often written as

$$\mathcal{N} = \mathcal{N}_0 e^{kt} \qquad \text{rather than} \qquad \mathcal{N}(t) = \mathcal{N}_0 e^{kt}$$

In both cases the quantity on the left-hand side of the equation represents the population at time t. For ease in writing, we follow this convention too, in Examples 5 and 7. (For the record: The difference between the use of $\mathcal{N}$ in the two equations is this. On the left-hand side of the equation $\mathcal{N}(t) = \mathcal{N}_0 e^{kt}$, $\mathcal{N}$ is the name of the function. On the left-hand side of the equation $\mathcal{N} = \mathcal{N}_0 e^{kt}$, $\mathcal{N}$ serves as a dependent variable, just like the "y" in $y = x^2$.)

EXAMPLE 1 Modeling bacterial growth

Suppose that at the start of an experiment in a biology lab, 1200 bacteria are present in a colony. Two hours later, the size of the population is found to be 1440. Assume that the population size grows exponentially.

(a) Determine the growth constant k and the growth law for this population.

(b) How many bacteria were there 1.5 hours after the experiment began? Round the answer to the nearest 5 (that is, nearest multiple of 5).
(c) When will the population size reach 4800?

SOLUTION

(a) The initial population size $\mathcal{N}_0$ is 1200, so the equation $\mathcal{N}(t) = \mathcal{N}_0 e^{kt}$ becomes

$$\mathcal{N}(t) = 1200e^{kt}$$

We are also told that after 2 hours, the size of the population is 1440. Using this information in equation (1) gives us

$$\begin{aligned} \mathcal{N}(2) &= 1200e^{k(2)} \\ 1440 &= 1200e^{2k} \end{aligned}$$

and therefore

$$e^{2k} = \frac{1400}{1200} = \frac{6}{5}$$

To isolate k, we rewrite this last equation in logarithmic form (or take the natural logarithm of both sides) to obtain $2k = \ln(1440/1200)$, and consequently,

$$\begin{aligned} k &= \frac{\ln(6/5)}{2} && (2) \\ &\approx 0.09116 && \text{using a calculator and rounding to five decimal places} \end{aligned}$$

The growth law for the population is therefore

$$\mathcal{N}(t) = 1200e^{kt} \qquad \text{where } k \text{ is given by equation (2)} \qquad (3)$$

Notice that in our statement of the growth law we referenced the exact expression for k rather than the calculator approximation. In parts (b) and (c) of this problem we don't round during the intermediate steps. This is standard procedure in scientific calculations because rounding errors can, in general, build up and adversely effect the accuracy of the final result.

(b) Substituting $t = 1.5$ in the growth law [that's equation (3)] yields

$$\mathcal{N}(1.5) = 1200e^{k(1.5)} \qquad \text{where } k = \frac{\ln(6/5)}{2}$$

With a calculator now, you should verify that $\mathcal{N}(1.5) \approx 1375.8$. In our context, $\mathcal{N}(1.5)$ must be a whole number because it represents the number of bacteria. Furthermore, the problem asks us to round to the nearest 5. In summary then, after 1.5 hours the population is 1375.

(c) We want to find out when the population reaches 4800. Replacing $\mathcal{N}(t)$ with 4800 in equation (3), we have $4800 = 1200e^{kt}$, and therefore $4 = e^{kt}$. Rewriting this last equation in logarithmic form gives us

$$\begin{aligned} kt &= \ln 4 \\ t &= \frac{\ln 4}{k} \approx 15.207 && \text{using the expression for } k \text{ in equation (2) and a calculator} \end{aligned}$$

Thus it takes approximately 15.2 hours for the size of the population to reach 4800. As you can check with arithmetic, this is 15 hours 12 minutes. See Figure 2 for a graphical solution to this part of the example.

GRAPHICAL PERSPECTIVE

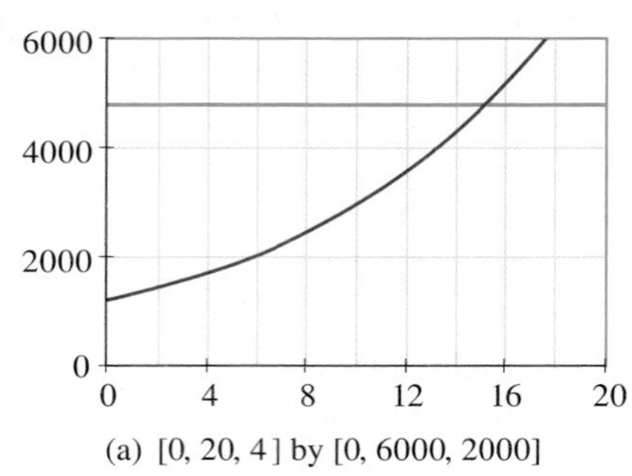

(a) [0, 20, 4] by [0, 6000, 2000]

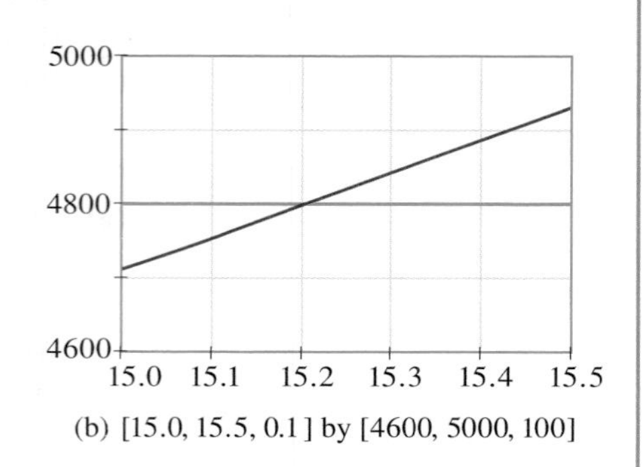

(b) [15.0, 15.5, 0.1] by [4600, 5000, 100]

Figure 2
In both viewing rectangles the red graph is the population growth function determined in part (a) of Example 1. The blue graph is the horizontal line $\mathcal{N}(t) = 4800$. The viewing rectangle in Figure 2(a) shows that the population reaches 4800 at a time that is roughly halfway between $t = 14$ and $t = 16$ hours. The zoom-in view in Figure 2(b) shows the time to be 15.2 hours, to the nearest tenth of an hour.

In Example 2 and in exercises we are going to use the function $\mathcal{N}(t) = \mathcal{N}_0 e^{kt}$ as a model for the size of human populations. But in order to do that in a reasonably honest way (short of bringing in calculus), first we need to introduce the concept of a *relative growth rate.* Suppose that in a biology lab you obtain the following population figures in successive hours for a colony of bacteria:

3 P.M.: 12,000 bacteria 4 P.M.: 15,000 bacteria

Let us compute the percentage increase in the population over this 1-hour interval. We have

$$\text{percentage increase} = \frac{\text{change in population}}{\text{initial population}} \times 100$$

$$= \frac{15{,}000 - 12{,}000}{12{,}000} \times 100 = \frac{1}{4} \times 100 = 25\%$$

We say in this case that the **average relative growth rate** of the population is 25%/hour. In this section and in the exercises we will shorten the phrase *average relative growth rate* to simply **relative growth rate.** Just as the interest rate in banking is sometimes expressed as a percentage and sometimes as a decimal, so too with the relative growth rate. In the example we are using, then, the decimal form of the relative growth rate is 0.25/hour. Whether expressed as a percentage or a decimal, the relative growth rate over a unit of time is a measure of how the population increases (or decreases) compared to its starting value.

There is a relationship between the growth constant k in the equation $\mathcal{N}(t) = \mathcal{N}_0 e^{kt}$ and the relative growth rate. To demonstrate this, we'll use the growth function obtained in Example 1. In Table 1, we've computed the relative growth rates (in decimal form) for this function over successive 1-hour time intervals.

We make two observations about the results in Table 1.

- The relative growth rates all are equal.
- The relative growth rate (0.095) is approximately equal to the growth constant k (0.091).

TABLE 1 Relative Growth Rates for the Function in Example 1 over Successive 1-Hour Intervals*

t	Relative growth rate on interval [t, t + 1]	Calculator value (rounded to 3 places)
0	$\dfrac{\mathcal{N}(1) - \mathcal{N}(0)}{\mathcal{N}(0)}$	0.095
1	$\dfrac{\mathcal{N}(2) - \mathcal{N}(1)}{\mathcal{N}(1)}$	0.095
2	$\dfrac{\mathcal{N}(3) - \mathcal{N}(2)}{\mathcal{N}(2)}$	0.095
3	$\dfrac{\mathcal{N}(4) - \mathcal{N}(3)}{\mathcal{N}(3)}$	0.095

*The function is given by $\mathcal{N}(t) = 1200e^{kt}$, where $k = \dfrac{\ln(1440/1200)}{2} = 0.09116\ldots$

The observation that the relative growth rates all turn out to be the same seems surprising at first. This is, however, one of the characteristic features of exponential growth and not only a coincidence. To see why, we compute $[\mathcal{N}(t+1) - \mathcal{N}(t)]/\mathcal{N}(t)$ for the function $\mathcal{N}(t) = \mathcal{N}_0 e^{kt}$. We have

$$\begin{aligned}\frac{\mathcal{N}(t+1) - \mathcal{N}(t)}{\mathcal{N}(t)} &= \frac{\mathcal{N}_0 e^{k(t+1)} - \mathcal{N}_0 e^{kt}}{\mathcal{N}_0 e^{kt}} \\ &= \frac{\mathcal{N}_0 e^{kt}(e^k - 1)}{\mathcal{N}_0 e^{kt}} \qquad \text{since } e^{k(t+1)} = e^{kt+k} = e^{kt}e^k \qquad (4) \\ &= e^k - 1\end{aligned}$$

This shows that the relative growth rate over every interval $[t, t+1]$ is the constant $e^k - 1$; the value does not depend upon t.

The second of the two observations we made was that the relative growth rate was nearly the same as the growth constant k. It turns out that this is true for every exponential growth function provided k is close to zero. (Exercise 61 will help you to see why this is so.) For this reason, in using the exponential growth law $\mathcal{N}(t) = \mathcal{N}_0 e^{kt}$ to model human populations, we'll simply use the relative growth rate for the value of k. This is convenient to do, since most of the available data report growth in terms of a relative growth rate. Furthermore, since the census figures themselves are, in general, not precise, it wouldn't be particularly meaningful to quibble over a fraction of a percent in this context. See Table 2.

EXAMPLE 2 Using an exponential growth model for world population growth

The Population Reference Bureau, www.prb.org, estimates that in the year 1900 the size of the world's population was 1.628 billion, with a relative growth rate of about 0.7%/year.

TABLE 2 Comparison of Relative Growth Rate $e^k - 1$ [from Equation (4)] to Growth Constant k (computed assuming exponential growth)*

	Relative growth rate for year 2000		
Country	as percentage	as decimal	k
Liberia	3.2	0.032	0.03149...
United States	0.6	0.006	0.00598...
Sweden	−0.1	−0.001	−0.00099...

* When the numbers are close to zero, the relative growth rate (in decimal form) is a good approximation for k.
Source for relative growth rates: Population Reference Bureau, www.prb.org

TABLE 3

Year	World population
1925	1.963
2000	6.067

Source: Population Reference Bureau

(a) Use the exponential growth model to make population projections for the years 1925 and 2000.
(b) Use the data in Table 3 to compute the percentage errors in the projections, and summarize the results.

SOLUTION
(a) Let $t = 0$ correspond to 1900. The initial population is $\mathcal{N}_0 = 1.628$, in units of one billion. For the value of k, we convert the given relative growth rate 0.7%/year into its decimal form, which is 0.007/year. With these values, the growth law becomes $\mathcal{N}(t) = 1.628e^{0.007t}$. Now we're ready to compute.

For 1925, $t = 25$

$$\mathcal{N}(25) = 1.628e^{0.007(25)} \approx 1.939 \quad \text{using a calculator}$$

For 2000, $t = 100$

$$\mathcal{N}(100) = 1.628e^{0.007(100)} \approx 3.278 \quad \text{using a calculator}$$

The exponential model projects a world population of 1.939 billion in 1925 and 3.278 billion in 2000.

(b) For 1925 the projection is 1.939 billion, whereas the figure from Table 3 is 1.963 billion. The percentage error in the projection is then

$$\begin{aligned}\text{percentage error} &= \frac{|\text{actual value} - \text{estimate}|}{\text{actual value}} \times 100\\ &= \frac{|1.963 \text{ billion} - 1.939 \text{ billion}|}{1.963 \text{ billion}} \times 100\\ &\approx 1.2\% \quad \text{using a calculator}\end{aligned}$$

As you should check now for yourself using the year 2000 figures, the percentage error in the projection for 2000 is 46.0%. *Remark*: In both cases the exponential projection is too low. However, the projection for 1925 *is* quite close, with a percentage error of only about 1%. In the case of the year 2000 projection, the model is way off; the percentage error is 46%.

EXAMPLE 3 More on world population growth

In the previous example we found that from 1900 to 2000, the world population grew faster than what was predicted by an exponential growth model with a relative growth rate of 0.7%/year. Actually, as you might well expect, the relative growth rate was not constant over the course of the entire century. But, assuming that it had been, find the value of the growth constant k that would yield an accurate projection for the year 2000. (The year 2000 population is given in Table 3.)

SOLUTION

Again we let $t = 0$ correspond to 1900, and we have $\mathcal{N}_0 = 1.628$, in units of one billion. Our exponential growth law is $\mathcal{N}(t) = 1.628e^{kt}$. We want to find k so that the population in the year 2000 is 6.067 billion (according to Table 3). The year 2000 corresponds to $t = 100$, so we have

$$\begin{aligned} \mathcal{N}(100) &= 1.628e^{k(100)} \\ 6.067 &= 1.628e^{100k} \\ \frac{6.067}{1.628} &= e^{100k} \end{aligned}$$

Writing this last equation in logarithmic form gives us

$$100\,k = \ln \frac{6.067}{1.628}$$

and therefore

$$k = \frac{\ln \dfrac{6.067}{1.628}}{100} \approx 0.0131\ldots \qquad \text{using a calculator}$$

Thus rounding to three decimal places, the required value of k is approximately 0.013. The corresponding relative growth rate in percent form is then 1.3%/year. *Note*: If you try checking this result using the approximation $k = 0.013$, you'll find that the population projection falls short of the required 6.067 billion. Don't round until the end. A good way to do this with your calculator is to first evaluate $k = (\ln \frac{6.067}{1.628})/100$, then *store* the result, and use the stored value in computing $1.628e^{k(100)}$. (If necessary, see your calculator owner's manual on how to store and then recall a constant.)

In the next example we compute the *doubling time* for a population that is growing exponentially. As the name indicates, the **doubling time** for a population is the amount of time required for the size of the population to double. Now, in the previous section we developed a formula for the doubling time T_2 of an amount of money under continuous compounding of interest. That formula is $T_2 = (\ln 2)/r$, where r is annual interest rate. The same formula is applicable now, because in both cases the underlying assumption is that of exponential growth. In the current context, the growth constant k plays the role of the annual interest rate r. Thus for a population that grows exponentially, the doubling time is $T_2 = (\ln 2)/k$. Note that the doubling time depends only upon the growth constant k and not upon the initial population size $\mathcal{N}_0$. For reference we state the formula for doubling time in the box that follows.

Doubling Time Formula

Suppose that a population grows exponentially. Then the doubling time T_2 is given by

$$T_2 = \frac{\ln 2}{k}$$

where k is the growth constant.

EXAMPLE 4 Computing a doubling time

According to the *Global Data Files* of the U.S. Bureau of the Census, the size of the world population in 1950 was approximately 2.555 billion, with an annual growth rate of 1.47%/year. Estimate the doubling time, assuming exponential growth at 1.47%/year.

SOLUTION

Converting 1.47% to a decimal, we have $k = 0.0147$. The doubling time is therefore

$$T_2 = \frac{\ln 2}{k} = \frac{\ln 2}{0.0147}$$
$$\approx 47 \text{ years} \quad \text{using a calculator and rounding to the nearest year}$$

The exponential model predicts a doubling time of approximately 47 years. That would be the year 1950 + 47 = 1997. Note the fact that the 1950 population size was not needed in this computation. *Remark*: It's interesting to check this model against the actual data. The model predicts that the size of the population in 1997 is 2×2.555 billion = 5.110 billion. The actual figure from the U.S. census is 5.847 billion. So the model projects too low a population for 1997. To put it another way, it actually took less than 47 years for the 1950 population to double in size. Indeed, a further check of the census data shows that the size of the population reached (our projected) 5.110 billion back in 1988.

In the next example we use the exponential growth model to describe the concentration of the so-called greenhouse gas carbon dioxide in the earth's atmosphere. As background, we quote from an article on the front page of the *New York Times,* June 7, 2001. The headline under which the article appears is "Panel Tells Bush Global Warming Getting Worse."

> A panel of top American scientists declared today that global warming was a real problem, and was getting worse . . . . In a much anticipated report from the National Academy of Sciences, 11 leading atmospheric scientists, including previous skeptics about global warming, reaffirmed the mainstream scientific view that the earth's atmosphere was getting warmer and that human activity was largely responsible. "Greenhouse gases are accumulating in the earth's atmosphere as a result of human activities, causing surface air temperatures and subsurface ocean temperature to rise," the report said.
>
> *New York Times* article by Katharine Q. Seelye with Andrew C. Revkin, June 7, 2001

EXAMPLE 5 **Using an exponential model for carbon dioxide levels in the atmosphere**

The following statements about the increasing levels of carbon dioxide in the atmosphere appear in the book *The Challenge of Global Warming,* edited by Dean Edwin Abrahamson (Washington, D.C.: Island Press, 1989).

> Carbon dioxide, the single most important greenhouse gas, accounts for about half of the warming that has been experienced as a result of past emissions* and also for half of the projected future warming. The present [1988] concentration is now about 350 parts per million (ppm) and is increasing about 0.4% . . . per year. . . .

Assuming that the concentration of carbon dioxide in the atmosphere continues to increase exponentially at 0.4% per year, estimate when the concentration might reach 600 ppm. (This would be roughly twice the level estimated to exist prior to the Industrial Revolution.)

SOLUTION

Let $t = 0$ correspond to the year 1988. In the formula $\mathcal{N} = \mathcal{N}_0 e^{kt}$, we want to determine the time t when $\mathcal{N} = 600$. Using the values $\mathcal{N}_0 = 350$, $\mathcal{N} = 600$, and $k = 0.004$, we have

$$600 = 350\, e^{0.004t} \qquad \text{or} \qquad \frac{600}{350} = \frac{12}{7} = e^{0.004t}$$

and, consequently,

$$\ln(12/7) = 0.004t$$

$$t = \frac{\ln(12/7)}{0.004} \approx 135 \qquad \text{using a calculator and rounding to the nearest whole number}$$

Now, 135 years beyond the base year of 1988 is 2123. Rounding once more, we summarize our result this way: If the carbon dioxide levels continue to increase exponentially at 0.4% per year, then the concentration will reach 600 ppm by approximately 2125. (*Note*: We round our result to the nearest five years because of the uncertainty in both the initial concentration of 350 ppm and the 0.4% rate of increase.)

In Examples 1 through 4 we used the function $\mathcal{N}(t) = \mathcal{N}_0 e^{kt}$ to describe population growth. In Example 5 we used the function to model the increasing levels of carbon dioxide in the atmosphere. And in the previous section a function of this form was used to describe how a sum of money grows under continuous compounding of interest. It is a remarkable fact that the same basic function, but now with $k < 0$, describes radioactive decay. How do scientists know that this is the appropriate model for radioactive decay? We mention two reasons: one empirical, one theoretical. As early as 1900–1903, the physicist Ernest Rutherford (Nobel prize, 1908) and the chemist Frederick Soddy (Nobel prize, 1921) carried out experiments measuring radioactive decay. They found that an equation of the form

*The two principal sources of these emissions are the burning of fossil fuels and the burning of vegetation in the tropics.

$\mathcal{N} = \mathcal{N}_0 e^{kt}$, with $k < 0$, indeed aptly described their data. Alternatively, on theoretical grounds, it can be argued that the decay rate at time t must be proportional to the amount $\mathcal{N}(t)$ of the radioactive substance present. Under this condition, calculus can be used to show that a law of exponential decay applies. (Exercise 63 demonstrates the converse: Assuming that the law of exponential decay applies, then the rate of decay must be proportional to the amount of the substance present.)

In discussing radioactive decay, it is convenient to introduce the term *half-life.* As you'll see, this is analogous to the concept of *doubling time* for exponential growth.

DEFINITION **Half-life**

The **half-life** of a radioactive substance is the time required for half of a given sample to disintegrate. The half-life is an intrinsic property of the substance; it does not depend on the given sample size.

EXAMPLE

Iodine-131 is a radioactive substance with a half-life of 8 days. Suppose that 2 g are present initially. Then:

at $t = 0$, 2 g are present;

at $t = 8$ days, 1 g is left;

at $t = 16$ days, $\frac{1}{2}$ g is left;

at $t = 24$ days, $\frac{1}{4}$ g is left;

at $t = 32$ days, $\frac{1}{8}$ g is left;

Just as we used the idea of doubling time to graph an exponential growth function in the previous section, we can use the half-life to graph an exponential decay function. Consider, for example, the radioactive substance iodine-131, which has a half-life of 8 days. Table 4 shows what fraction of an initial amount remains at 8-day intervals. Using the data in this table, we can draw the graph of the decay function $\mathcal{N}(t) = \mathcal{N}_0 e^{kt}$ for iodine-131 (see Figure 3). Notice that we are able to construct this graph without specifically evaluating the decay constant k. (The next example shows how to determine k.)

TABLE 4

t (days)	$\mathcal{N}(t)$ (amount)
0	$\mathcal{N}_0$
8	$\mathcal{N}_0/2$
16	$\mathcal{N}_0/4$
24	$\mathcal{N}_0/8$
32	$\mathcal{N}_0/16$

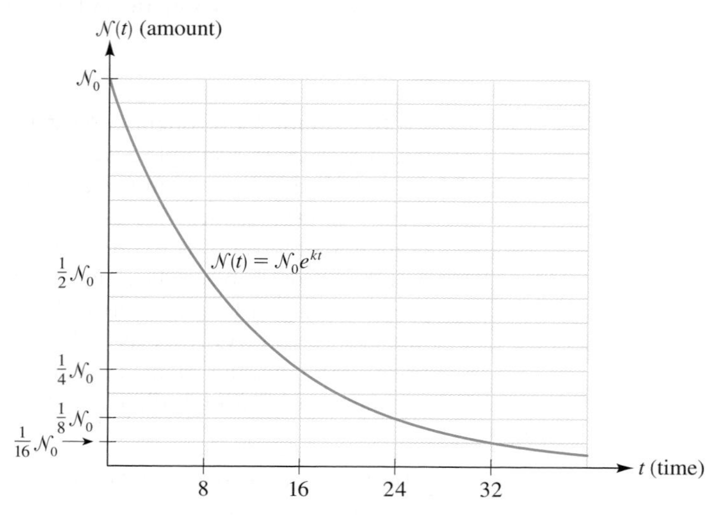

Figure 3
The exponential decay function for iodine-131

EXAMPLE 6 Using the exponential decay formula

Hospitals utilize the radioactive substance iodine-131 in the diagnosis of the thyroid gland. The half-life of iodine-131 is 8 days.

(a) Determine the *decay constant k* for iodine-131.

(b) If a hospital acquires 2 g of iodine-131, how much of this sample will remain after 20 days?

(c) How long will it be until only 0.01 g remains?

SOLUTION

(a) We use the half-life information to find the value of the decay constant k. Substituting $t = 8$ in the decay law $\mathcal{N}(t) = \mathcal{N}_0 e^{kt}$ yields

$$\mathcal{N}(8) = \mathcal{N}_0 e^{k(8)}$$

$$\frac{1}{2}\mathcal{N}_0 = \mathcal{N}_0 e^{8k} \qquad \text{A half-life of 8 days means that } \mathcal{N}(8) = \tfrac{1}{2}\mathcal{N}_0.$$

$$\frac{1}{2} = e^{8k} \qquad \text{dividing both sides by } \mathcal{N}_0$$

In logarithmic form, this last equation becomes

$$8k = \ln\frac{1}{2} \qquad \text{and therefore} \qquad k = \frac{\ln\frac{1}{2}}{8} = \frac{-\ln 2}{8}$$

$$\approx -0.08664 \qquad \text{rounding to five places}$$

(b) We are given $\mathcal{N}_0 = 2$ g and we want to find $\mathcal{N}(20)$, the amount remaining after 20 days. First, before using algebra and a calculator, let's estimate the answer to get a feeling for the situation. The half-life is 8 days. This means that after 8 days, 1 g remains; after 16 days, 0.5 g remains; and after 24 days, 0.25 g remains. Since 20 is between 16 and 24, it follows that after 20 days, the amount remaining will be between 0.5 g and 0.25 g. Now for the actual calculations: Substitute the values $\mathcal{N}_0 = 2$ and $t = 20$ in the decay law $\mathcal{N}(t) = \mathcal{N}_0 e^{kt}$ to obtain

$$\mathcal{N}(20) = 2e^{k(20)} \qquad \text{where } k = \frac{-\ln 2}{8}$$

$$\approx 0.354 \text{ g} \qquad \text{using a calculator}$$

Thus after 20 days, approximately 0.35 g of the iodine-131 remains. Note that this amount is indeed between 0.5 g and 0.25 g as we first estimated.

(c) We want to find the time t for which $\mathcal{N}(t) = 0.01$ g. Substituting 0.01 for $\mathcal{N}(t)$ in the decay law gives us

$$0.01 = 2e^{kt} \qquad \text{and therefore} \qquad 0.005 = e^{kt}$$

The logarithmic form of this last equation is $kt = \ln(0.005)$, from which we conclude that

$$t = \frac{\ln(0.005)}{k} \qquad \text{where } k = \frac{-\ln 2}{8}$$

$$\approx 61.2 \text{ days} \qquad \text{using a calculator}$$

In Example 6(a) we found that the decay constant for iodine-131 is given by $k = -(\ln 2)/8$. Notice that the denominator in this expression is the half-life of iodine-131. By following the same reasoning used in Example 6(a), we find that the decay constant for any radioactive substance is always $-\ln 2$ divided by the half-life. For reference, this useful fact is restated in the box that follows.

PROPERTY SUMMARY A Formula for the Decay Constant k

Formula	Example
$k = \dfrac{-\ln 2}{\text{half-life}}$	The half-life of strontium-90 is 28 years, and therefore $k = -(\ln 2)/28$.

EXAMPLE 7 Calculations and estimates for radioactivity

An article on nuclear energy appeared in the January 1976 issue of *Scientific American*. The author of the article was Hans Bethe (1906–1992), a Nobel prize winner in physics. At one point in the article, Professor Bethe discussed the disposal (through burial) of radioactive waste material from a nuclear reactor. The particular waste product under discussion was plutonium-239.

> . . . Plutonium-239 has a half-life of nearly 25,000 years, and 10 half-lives are required to cut the radioactivity by a factor of 1000. Thus, the buried wastes must be kept out of the biosphere for 250,000 years.

(a) Supply the detailed calculations to support the statement that 10 half-lives are required before the radioactivity is reduced by a factor of 1000.

(b) Show how the figure of 10 half-lives can be obtained by estimation, as opposed to detailed calculation.

SOLUTION

(a) Let $\mathcal{N}_0$ denote the initial amount of plutonium-239 at time $t = 0$. Then the amount $\mathcal{N}$ present at time t is given by $\mathcal{N} = \mathcal{N}_0 e^{kt}$. We wish to determine t when $\mathcal{N} = \frac{1}{1000}\mathcal{N}_0$. First, since the half-life is 25,000 years, we have $k = -(\ln 2)/25{,}000$. Now, substituting $\frac{1}{1000}\mathcal{N}_0$ for $\mathcal{N}$ in the decay law gives us

$$\frac{1}{1000}\mathcal{N}_0 = \mathcal{N}_0 e^{kt} \qquad \text{where } k \text{ is } -(\ln 2)/25{,}000$$

$$\frac{1}{1000} = e^{kt}$$

$$kt = \ln\frac{1}{1000} \qquad \text{converting from exponential to logarithmic form}$$

$$t = \frac{-\ln 1000}{k} = \frac{-\ln 1000}{\dfrac{-\ln 2}{25{,}000}} = \frac{25{,}000 \ln 1000}{\ln 2}$$

Using a calculator, we obtain the value 249,114.6 years; however, given the time scale involved, it would be ludicrous to announce the answer in this form. Instead, we round off the answer to the nearest thousand years and say that after 249,000 years, the radioactivity will have decreased by a factor of 1000. Notice that this result is consistent with Professor Bethe's ballpark estimate of 10 half-lives, or 250,000 years

(b) After 1 half-life: $\mathcal{N} = \dfrac{\mathcal{N}_0}{2}$

After 2 half-lives: $\mathcal{N} = \dfrac{1}{2}\left(\dfrac{\mathcal{N}_0}{2}\right) = \dfrac{\mathcal{N}_0}{2^2}$

After 3 half-lives: $\mathcal{N} = \dfrac{1}{2}\left(\dfrac{\mathcal{N}_0}{2^2}\right) = \dfrac{\mathcal{N}_0}{2^3}$

Following this pattern, we see that after 10 half-lives we should have

$$\mathcal{N} = \frac{\mathcal{N}_0}{2^{10}}$$

However, as we noted in the first section of this chapter, 2^{10} is approximately 1000. Therefore we have

$$\mathcal{N} \approx \frac{\mathcal{N}_0}{1000} \qquad \text{after 10 half-lives}$$

This is in agreement with Professor Bethe's statement.

EXERCISE SET 5.7 

A

In Exercises 1 and 2, assume that the populations grow exponentially, that is, according to the law $\mathcal{N}(t) = \mathcal{N}_0 e^{kt}$.

1. At the start of an experiment, 2000 bacteria are present in a colony. Two hours later, the population is 3800.
   (a) Determine the growth constant k.
   (b) Determine the population five hours after the start of the experiment.
   (c) When will the population reach 10,000?

2. At the start of an experiment, 2×10^4 bacteria are present in a colony. Eight hours later, the population is 3×10^4.
   (a) Determine the growth constant k.
   (b) What was the population two hours after the start of the experiment?
   (c) How long will it take for the population to triple?

3. The figure in the next column shows the graph of an exponential growth function $\mathcal{N}(t) = \mathcal{N}_0 e^{kt}$. Determine the values of $\mathcal{N}_0$ and k.

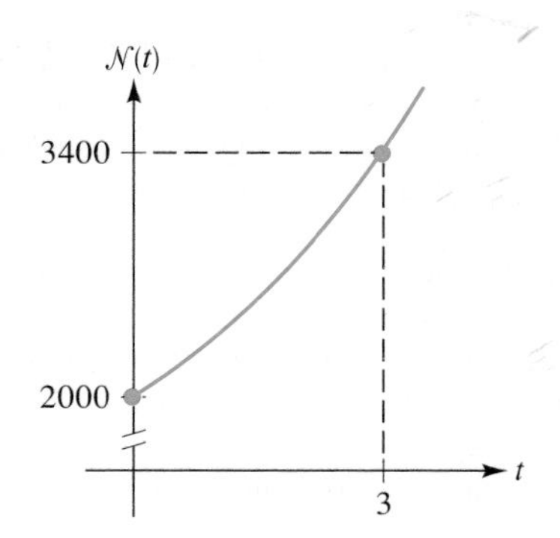

4. Suppose that you are helping a friend with his homework on the growth law, $\mathcal{N}(t) = \mathcal{N}_0 e^{kt}$, and on his paper you see the equation $e^k = -0.75$. How do you know at this point that your friend must have made an error?

TABLE A

Region	1995 Population (billions)	Percentage of population in 1995	Relative growth rate (%/year)	Year 2000 population (billions)	Percentage of world population in 2000
World	5.702	100	1.5	?	?
More developed regions	1.169	?	0.2	?	?
Less developed regions	4.533	?	1.9	?	?

TABLE B

Country	1995 Population (millions)	Relative growth rate (%/year)	Year 2000 population (millions)	Percentage increase in population
United States	263.2	0.7	?	?
People's Republic of China	1218.8	1.1	?	?
Mexico	93.7	2.2	?	?

In Exercises 5 and 6: **(a)** *Complete the table; and* **(b)** *use the information that is given to compute the percentage errors in the population projections in part (a). Round each answer to one decimal place. [The populations projections are in the second column from the right in Tables A and B.] The data in Exercises 5 and 6 are from The Population Reference Bureau,* www.prb.org.

5. **(a)** See Table A.
(b) Year 2000 population in billions: world, 6.067; more developed, 1.184; less developed 4.883

6. **(a)** See Table B.
(b) Year 2000 population in millions: United States, 275.6; People's Republic of China, 1264.5; Mexico, 99.6

7. As of the year 2000, the African nation of Chad had one of the highest population growth rates in the world, 3.3%/year. At the other extreme in 2000, the United Kingdom had one of the lowest (positive) growth rates, 0.1%/year. (There are countries with zero or with negative growth rates.)
(a) In 2000 the sizes of the populations of Chad and the United Kingdom were 8.0 million and 59.8 million, respectively. Write the exponential growth law for each country, letting $t = 0$ correspond to 2000.
(b) Assuming exponential growth, when would the two countries have populations of the same size? Round the answer to the nearest five years. *Hint*: Equate the two expressions for $\mathcal{N}(t)$ obtained in part (a).

8. In 2000 the nations of Mali and Cuba had similar size populations: Mali 11.2 million, Cuba 11.1 million. However, the relative growth rate for Mali was 3.1%/year, whereas that for Cuba was 0.7%/year.
(a) Assuming exponential growth at the given rates, make projections for each population in the year 2015.
(b) When might the population of Mali reach 20 million? What would the population of Cuba be at that time?

9. In 2000 the nations of Niger and Portugal had similar size populations: Niger 10.1 million, Portugal 10.0 million. However, the relative growth rate for Niger was 3.0%/year, whereas that for Portugal was only 0.1%/year.
(a) Assuming exponential growth at the given rates, make projections for each population in the year 2015.
(b) When might the population of Niger reach 15 million? What would the population of Portugal be at that time?

10. The population of Guatemala in 2000 was 12.7 million.
(a) Assuming exponential growth, what value for k would lead to a population of 20 million one quarter of a century later (that is, in 2025)? *Remark*: The answer you obtain is, in fact, less than the actual year 2000 growth rate, which was about 2.9%/year.
(b) Again, assuming a population of 12.7 million in 2000, what value for k would lead to a population of 20 million, one century later (that is, in 2100)?

As background for Exercises 11 and 12, here's a reminder of what we found in Example 2. Starting with data for the year 1900, we saw two cases where the world population had grown faster than what was predicted by an exponential model. In Exercises 11 and 12, you'll work with more recent data for the base year and find that now the population is, in fact, growing slower than the exponential model predicts.

11. According to the U.S. Bureau of the Census, the world population in 1975 was 4.088 billion, with a relative growth rate of 1.75%/year.
(a) Use an exponential growth model to predict the year 2000 population.
(b) According to Table 3 in the text, the world population in 2000 was 6.067 billion. Is your projection in part (a) higher or lower than this? Compute the percentage

error in the projection and round the answer to the nearest one percent.

12. Follow Exercise 11, but start from the following data. In 1990, the world population was 5.284 billion, with a relative growth rate of 1.56%/year.

13. In each case, you are given the relative growth rate of a country or region in the year 2000. Compute the doubling time for the population (assuming an exponential growth model). Round the answer to the nearest whole number of years.

(a) United States: 0.6%/year
(b) Tajikistan: 1.6%/year
(c) Cambodia: 2.6%/year
(d) Palestinian Territory: 3.7%/year

14. Refer to the following table, which gives global populations and relative growth rates for the year 2000. Assume an exponential growth model.

	2000 Population (billions)	Relative growth rate (%/year)
World	6.067	1.4
More developed regions	1.184	0.1
Less developed regions	4.883	1.7

Source: Population Reference Bureau

(a) Compute the doubling time for the world population, and give the year in which that population would be reached. (Round each answer to the nearest year.)
(b) Follow part (a) for the more developed regions.
(c) Follow part (a) for the less developed regions.

15. In Example 5 in the text we modeled the concentration of carbon dioxide in the atmosphere using an exponential growth model with $k = 0.4\%$/year and $\mathcal{N}_0 = 350$ ppm, where $t = 0$ corresponds to 1988.

(a) A more precise value for $\mathcal{N}_0$ is 351.31 ppm. Using this value in the exponential growth model, complete the following table of projected values for the atmospheric carbon dioxide concentrations. Round the answers to one decimal place.

Year	1998 ($t = 10$)	2000 ($t = 12$)
Atmospheric concentration of carbon dioxide (ppm)		

Source for data: C. D. Keeling and T. P. Whorf, Scripps Institution of Oceanography

(b) The actual values for 1998 and 2000 were 366.7 ppm and 368.4 ppm, respectively. For each of these two years, say whether the exponential model projects too high or two low a value, and compute the percentage error.

16. For this exercise refer to the following table, which gives the atmospheric concentrations of carbon dioxide (CO_2) for selected years over the latter half of the 20th century.

Year	1960	1985	1990	1995	1999
Atmospheric concentration of carbon dioxide (ppm)	316.75	345.73	354.04	360.91	368.37

Source: C. D. Keeling and T. P. Whorf, Scripps Institution of Oceanography

(a) Let $t = 0$ correspond to 1960, and assume an exponential growth model $\mathcal{N} = 316.75e^{kt}$. Use the information for 1999 in the table to find the growth constant k.
(b) Use the growth law determined in part (a) to make a projection for the CO_2 concentration in 1985. Then, referring to the table, say whether the exponential model predicts too high or low a figure, and compute the percentage error.
(c) Follow part (b) for the year 1990 and 1995.
(d) Why do you think that the year 1999 was not included in part (c)?

17. According to figures from the U.S. Bureau of the Census, in 2000, the size of the population of the state of New York was more than three times larger than that of Arizona. However, New York had one of the lower growth rates in the nation, and Arizona had the second highest.

(a) Use the data in the following table to specify an exponential growth model for each state. (Let $t = 0$ correspond to the year 2000.)

State	Population in 2000 (millions)	Relative growth rate (%/year)
New York	18.976	0.6
Arizona	5.131	4.0

(b) Assuming continued exponential growth, when would the two states have populations of the same size? Round the answer to the nearest five years. *Hint*: Equate the two expressions for $\mathcal{N}(t)$ obtained in part (a).

18. The following table gives the population size and relative growth rates in 2000 for three states. Use exponential models to project the population of each state in the year 2010, listing the results from largest to smallest. Does the order remain the same as it was in 2000?

State	Population in 2000 (millions)	Relative growth rate (%/year)
Iowa	2.926	0.54
Arkansas	2.673	1.37
Nevada	1.998	6.63

19. Lester R. Brown, in his book *State of the World 1985* (New York: W. W. Norton and Co., 1985), makes the following statement:

> The projected [population] growth for North America, all of Europe, and the Soviet Union is less than the additions expected in either Bangladesh or Nigeria.

In this exercise, you are asked to carry out the type of calculations that could be used to support Lester Brown's projection for the period 1990–2025. The source of the data in the following table is the Population Division of the United Nations.

Region	1990 Population (millions)	Projected relative growth rate (%/year)	2025 Projected population (millions)
North America	275.2	0.7	?
(Former) Soviet	291.3	0.7	?
Union	291.3	0.7	?
Europe	499.5	0.2	?
Nigeria	113.3	3.1	?

(a) Complete the table, assuming that the populations grow exponentially and that the indicated growth rates are valid over the period 1990–2025.
(b) According to the projections in part (a), what will be the net increase in Nigeria's population over the period 1990–2025?
(c) According to the projections in part (a), what will be the net increase in the combined populations of North America, the (former) Soviet Union, and Europe over the period 1990–2025?
(d) Compare your answers in parts (b) and (c). Do your results support or contradict Lester Brown's projection?

20. The *2000 World Population Data Sheet* (published by the Population Reference Bureau, Washington, D.C.) lists ten nations with populations exceeding 100 million. (See Exercise 21 regarding a possible eleventh nation.) The populations, in order of decreasing size, along with relative growth rates, are listed in the accompanying table. Over a decade ago, in 1987, the U.S. Bureau of the Census made the following statement in its analysis of world population data: "The latest projections suggest that India's population may surpass China's in less than 60 years, or before today's youngsters in both countries reach old age." Using the data in the table, and assuming exponential growth, make projections for the populations of India and China in the year 2050 (which would be roughly 60 years after that statement by the Bureau of the Census). Do your results for China and India support the projection by the Bureau of the Census?

Country	2000 Population (millions)	Relative growth rate (%/year)
1. China	1264.5	0.9
2. India	1002.1	1.8
3. United States	275.6	0.6
4. Indonesia	212.2	1.6
5. Brazil	170.1	1.5
6. Pakistan	150.6	2.8
7. Russia	145.2	−0.6
8. Bangladesh	128.1	1.8
9. Japan	126.9	0.2
10. Nigeria	123.3	2.8

Source: Population Reference Bureau, www.prb.org

21. The population figures in Exercise 20 are actually for mid-2000. As of that time, the population of Mexico was 99.6 million, with a relative growth rate of 2.0%/year. Let $t = 0$ correspond to June 2000 and use the exponential growth model $\mathcal{N}(t) = 99.6e^{0.02t}$ to verify that $\mathcal{N}(0.5) > 100$. Interpret the result.

22. In 2000 the Philippines and Germany had similar size populations, but very different growth rates. The population of the Philippines was 80.3 million, with a relative growth rate of 2.0%/year. The population of Germany was 82.1 million, with a relative "growth" rate of −0.1%/year. Using exponential models, make a projection for the population of Germany in the year when the Philippine population has doubled.

In Exercises 23 and 24, use the half-life information to complete each table. (The formula $\mathcal{N} = \mathcal{N}_0e^{kt}$ is not required.)

23. **(a)** Uranium-228: half-life = 550 seconds

t (seconds)	0	550	1100	1650	2200
$\mathcal{N}$ (grams)	8				

(b) Uranium-238: half-life = 4.9×10^9 years

t (years)	0				
$\mathcal{N}$ (grams)	10	5	2.5	1.25	0.625

24. **(a)** Polonium-210: half-life = 138.4 days

t (years)	0	138.4	276.8	415.2	
$\mathcal{N}$ (grams)	0.4				0.025

(b) Polonium-214: half-life = 1.63×10^{-4} second

t (seconds)	0				6.52×10^{-4}
$\mathcal{N}$ (grams)	0.1	0.05	0.025	0.0125	

25. The half-life of iodine-131 is 8 days. How much of a one-gram sample will remain after 7 days?

26. The half-life of strontium-90 is 28 years. How much of a 10-g sample will remain after **(a)** 1 year? **(b)** 10 years?

27. The radioactive isotope sodium-24 is used as a tracer to measure the rate of flow in an artery or vein. The half-life of sodium-24 is 14.9 hours. Suppose that a hospital buys a 40-g sample of sodium-24.

(a) How much of the sample will remain after 48 hours?

(b) How long will it be until only 1 gram remains?

28. The radioactive isotope carbon-14 is used as a tracer in medical and biological research. Compute the half-life of carbon-14 given that the decay constant k is -1.2097×10^{-4}. (The units for k here are such that your half-life answer will be in years.)

29. (a) The half-life of radium-226 is 1620 years. Sketch a graph of the decay function for radium-226, similar to that shown in Figure 3.

(b) The half-life of radium-A is 3 min. Sketch a graph of the decay function for radium-A.

30. (a) The half-life of thorium-232 is 1.4×10^{10} years. Sketch a graph of the decay function.

(b) The half-life of thorium-A is 0.16 sec. Sketch a graph of the decay function.

31. The half-life of plutonium-241 is 13 years.

(a) How much of an initial 2-g sample remains after 5 years?

(b) Find the time required for 90% of the 2-g sample to decay. *Hint*: If 90% has decayed, then 10% remains.

32. The half-life of radium-226 is 1620 years.

(a) How much of a 2-g sample remains after 100 years?

(b) Find the time required for 80% of the 2-g sample to decay.

33. The half-life of thorium-229 is 7340 years.

(a) Compute the time required for a given sample to be reduced by a factor of 1000. Show detailed calculations, as in Example 7(a).

(b) Express your answer in part (a) in terms of half-lives.

(c) As in Example 7(b), estimate the time required for a given sample for thorium-229 to be reduced by a factor of 1000. Compare your answer with that obtained in part (b).

34. The Chernobyl nuclear explosion (in the former Soviet Union, on April 26, 1986) released large amounts of radioactive substances into the atmosphere. These substances included cesium-137, iodine-131, and strontium-90. Although the radioactive material covered many countries, the actual amount and intensity of the fallout varied greatly from country to country, due to vagaries of the weather and the winds. One area that was particularly hard hit was Lapland, where heavy rainfall occurred just when the Chernobyl cloud was overhead.

(a) Many of the pastures in Lapland were contaminated with cesium-137, a radioactive substance with a half-life of 33 years. If the amount of cesium-137 was found to be ten times the normal level, how long would it take until the level returned to normal? *Hint*: Let $\mathcal{N}_0$ be the amount that is ten times the normal level. Then you want to find the time when $\mathcal{N}(t) = \mathcal{N}_0/10$.

(b) Follow part (a), but assume that the amount of cesium-137 was 100 times the normal level.

Remark: Several days after the explosion, it was reported that the level of cesium-137 in the air over Sweden was 10,000 times the normal level. Fortunately there was little or no rainfall.

35. Strontium-90, with a half-life of 28 years, is a radioactive waste product from nuclear fission reactors. One of the reasons great care is taken in the storage and disposal of this substance stems from the fact that strontium-90 is, in some chemical respects, similar to ordinary calcium. Thus strontium-90 in the biosphere, entering the food chain via plants or animals, would eventually be absorbed into our bones.

(a) Compute the decay constant k for strontium-90.

(b) Compute the time required if a given quantity of strontium-90 is to be stored until the radioactivity is reduced by a factor of 1000.

(c) Using half-lives, estimate the time required for a given sample to be reduced by a factor of 1000. Compare your answer with that obtained in (b).

36. (a) Suppose that a certain country violates the ban against above-ground nuclear testing and, as a result, an island is contaminated with debris containing the radioactive substance iodine-131. A team of scientists from the United Nations wants to visit the island to look for clues in determining which country was involved. However, the level of radioactivity from the iodine-131 is estimated to be 30,000 times the safe level. Approximately how long must the team wait before it is safe to visit the island? The half-life of iodine-131 is 8 days.

(b) Rework part (a), assuming instead that the radioactive substance is strontium-90 rather than iodine-131. The half-life of strontium-90 is 28 years. Assume, as before, that the initial level of radioactivity is 30,000 times the safe level. (This exercise underscores the difference between a half-life of 8 days and one of 28 years.)

37. An article that appeared in the August 13, 1994, *New York Times* reported

> German authorities have discovered . . . a tiny sample of weapons-grade nuclear material believed to have been smuggled out of Russia to interest foreign governments or terrorist groups that might want to build atomic bombs. . . . [the police] said they had seized the material, .028 ounces of highly enriched uranium-235, in June in . . . Bavaria . . . and have since arrested . . . [six] suspects. . . .

Suppose that the suspects, in an attempt to avoid arrest, had thrown the 0.028 ounces of uranium-235 into the Danube River, where it would sink to the bottom. How many ounces of the uranium-235 would still be in the river after 1000 years? The half-life of uranium-235 is 7.1×10^8 years.

38. In 1969 the United States National Academy of Sciences issued a report entitled *Resources and Man.* One conclusion in the report is that a world population of 10 billion "is close to (if not above) the maximum that an intensively managed world might hope to support with some degree of comfort and individual choice." (The figure "10 billion" is sometimes referred to as the *carrying capacity* of the Earth.)

(a) When the report was issued in 1969, the world population was about 3.6 billion, with a relative growth rate of 2% per year. Assuming continued exponential growth at this rate, estimate the year in which the Earth's carrying capacity of 10 billion might be reached.

(b) Repeat the calculations in part (a) using the following more recent data: In 2000 the world population was about 6.0 billion, with a relative growth rate of 1.4% per year. How does your answer compare with that in part (a)?

39. The following extract is from an article by Kim Murphy that appeared in the *Los Angeles Times* on September 14, 1994.

> CAIRO—Over a chorus of reservations from Latin America and Islamic countries still troubled about abortion and family issues, nearly 180 nations adopted a wide-ranging plan Tuesday on global population, the first in history to obtain partial endorsement from the Vatican.
>
> The plan, approved on the final day of the U.N. population conference here, for the first time tries to limit the growth of the world's population by preventing it from exceeding 7.2 billion people over the next two decades.

(a) In 1995 the world population was 5.7 billion, with a relative growth rate of 1.6%/year. Assuming continued exponential growth at this rate, make a projection for the world population in the year 2020. Round off the answer to one decimal place. How does your answer compare to the target value of 7.2 billion mentioned in the article?

(b) As in part (a), assume that in 1995 the world population was 5.7 billion. Determine a value for the growth constant k so that exponential growth throughout the years 1995–2020 leads to a world population of 7.2 billion in the year 2020.

Ⓖ **40.** Economists define the **gross domestic product (GDP)** as the total market value of a nation's goods and services produced (within the borders of the nation) over a specified period of time. The GDP is one of the key measures of a nation's economic health. Table A gives the annual GDP for the United States for the period 1950–1990, at 10-year intervals. In Figures A and B we show scatter plots for the data along with regression functions that model the data. Figure A uses a linear model, Figure B an exponential model. In each case, the years are on the horizontal axis, with $t = 0$ corresponding to 1900. In this exercise you'll use a graphing utility to obtain the specific equations for these models and for two other models as well. You'll also compare projections using the various models.

TABLE A Gross Domestic Product (GDP) in Constant 1996 Dollars

Year t ($t = 0 \leftrightarrow 1900$)	50	60	70	80	90
GDP (billions of dollars)	1686.6	2376.7	3578.0	4900.9	6707.9

Source: U.S. Bureau of Economic Analysis, http://www.bea.gov

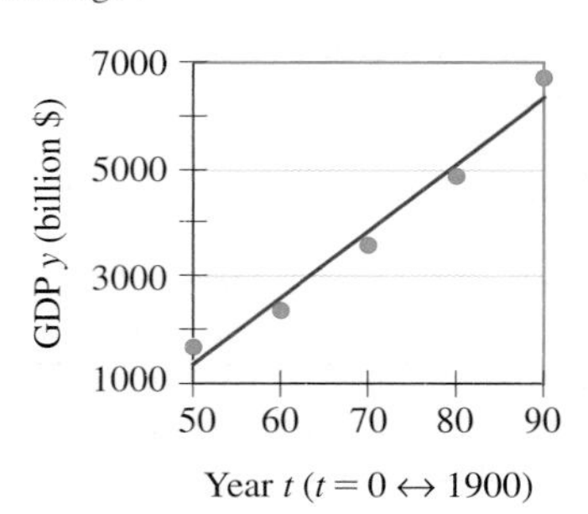

Figure A
Linear model $y = at + b$

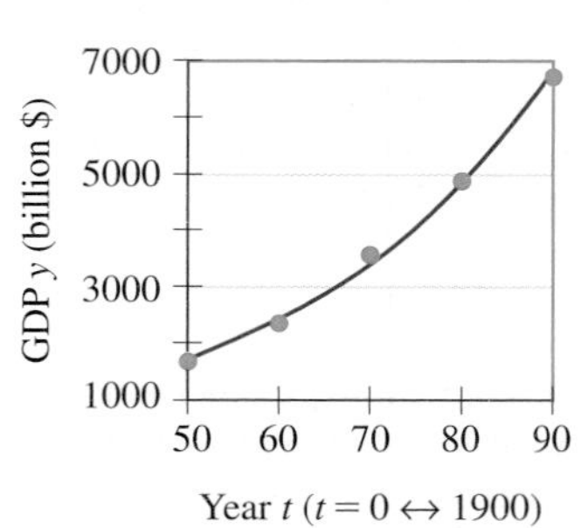

Figure B
Exponential model $y = ab^t$ or $y = Ae^{Bt}$

(a) Use a graphing utility to create your own scatter plot of the data in Table A. Then, use the regression or trend line options of the graphing utility to obtain the specific equations for the linear model in Figure A and the exponential model in Figure B. (For the exponential model, you can report your answer in either one of the two forms indicated in the caption for Figure B, depending on what your graphing utility provides.) For comparison, graph both models in the same viewing rectangle.

(b) Use the regression or trend line options of the graphing utility to obtain the equation for a quadratic model $y = at^2 + bt + c$. Show the graph of the quadratic model with the scatter plot. Now repeat this process for a power model $y = at^b$.

(c) Use the functions that you determined in parts (a) and (b) to complete Table B, showing what each model projects for the gross domestic product in the indicated year. (Round each projection to one decimal place.) Then, in each row, circle the projection that is closest to the actual value given at the end of that row. Finally, compute the percentage error for each projection that you circled.

TABLE B Gross Domestic Product (GDP) Projections Using Various Models

Year	GDP (billion \$) linear model	GDP (billion \$) exponential model	GDP (billion \$) quadratic model	GDP (billion \$) power model	GDP (billion \$) actual
1995					7543.8
1999					8875.8
2001					9333.8

B

41. *Depletion of Nonrenewable Resources*: Suppose that the world population grows exponentially. Then, as a first approximation, it is reasonable to assume that the use of nonrenewable resources, such as petroleum and coal, also grows exponentially. Under these conditions, the following formula can be derived (using calculus):

$$A = \frac{A_0}{k}(e^{kT} - 1)$$

where A is the amount of the resource consumed from time $t = 0$ to $t = T$, the quantity A_0 is the amount of the resource consumed during the year $t = 0$, and k is the relative growth rate of annual consumption.

(a) Show that solving the formula for T yields

$$T = \frac{\ln[(Ak/A_0) + 1]}{k}$$

This formula gives the "life expectancy" T for a given resource. In the formula, A_0 and k are as previously defined, and A represents the total amount of the resource available.

(b) Over the years 1965–1972, world oil consumption grew exponentially. In this part of the exercise, you'll compute the life expectancy of oil, assuming continued exponential growth. Let $t = 0$ correspond to 1972. In that year, worldwide consumption of oil was approximately 18.7 billion barrels, with a relative growth rate of 7%/year. The global reserves of oil at that time were estimated to be 700 billion barrels. Compute the life expectancy T of oil under this exponential growth scenario and specify the *depletion year* ($= 1972 + T$). If the relative growth had been only 1%/year, what would the depletion year be?

(c) Beginning with the first Arab oil embargo of 1973–1974, worldwide oil consumption ceased to grow exponentially. In fact, oil consumption actually decreased over the years 1973–1975, and it decreased again for 1979–1983, following the second Arab oil embargo. More recently, as indicated by the following data, scatter plot, and regression line, worldwide oil consumption has been increasing in a linear fashion. Use a graphing utility and the data in the table to determine the equation $y = mt + b$ for the regression line.

Year t ($t = 0 \leftrightarrow 1993$)	0	1	2	3
Global oil consumption y (billion barrels)	23.997	24.506	24.893	25.450

Year t ($t = 0 \leftrightarrow 1993$)	4	5	6
Global oil consumption y (billion barrels)	26.123	26.251	26.723

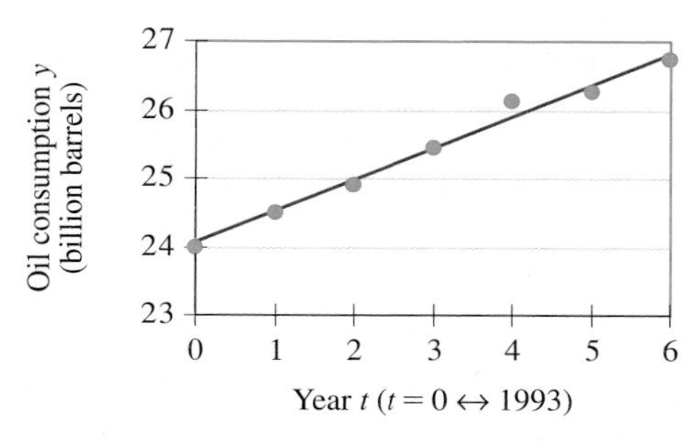

(d) If yearly consumption of a nonrenewable resource is given by a linear function $y = mt + b$, then it can be shown that the life expectancy T of the resource is given by the following formula,

$$T = \frac{-b + \sqrt{b^2 + 2mA}}{m}$$

where A represents the total amount of the resource available (that is, the world reserves) in the year corresponding to $t = 0$. (The derivation does not require calculus, but we shall omit giving it here.) In 1993, world reserves of oil were estimated to be 1007 billion barrels. Using this value for A and the values of m and b determined in part (c), estimate the life expectancy for oil and the depletion date ($=1993 + T$). (Aside: The figure we've given here for the estimated world reserves in 1993 is greater than the one we gave for 1972. This reflects improved technology in locating oil fields and in extracting the oil.)

For Exercises 42 and 43, refer to the following table. (Data compiled from United States Geological Survey, World Resources Institute, and the World Almanac, 2000.)

Resource	1999 world consumption (million metric tons)	1999 world reserves (million metric tons)	Relative growth rate (%/year)
Aluminum (in bauxite)	23.1	34,000	3.0
Copper	12.6	650	2.4

42. (a) Suppose that the consumption of aluminum were growing exponentially. Use the formula for T in Exercise 41 to compute the life expectancy of aluminum and the depletion year.

(b) What if the world reserves for aluminum were actually twice that listed in the table, and the relative growth rate were half that in the table? Compute the depletion year under this scenario.

(c) Suppose that from 1999 onward, consumption of aluminum stopped increasing exponentially and stabilized at the 1999 level of 23.1 million metric tons per year. How many years would it take to deplete the world reserves of 34,000 million metric tons?

43. Follow Exercise 42, using copper instead of aluminum.

Ⓖ **44.** The following table and scatter plots display data for the number of fixed-line phone connections worldwide and the number of cell phone subscribers over the years 1985–1998. As indicated by the scatter plots, the number of cell phone subscribers is far less than the number of phone lines, but it is increasing at a faster rate.

(a) Use the regression option on a graphing utility or spreadsheet to find a quadratic model $y = at^2 + bt + c$ describing the number of phone lines over the period 1985–1998 ($t = 0$ to $t = 13$). Similarly, find an exponential model of the form $y = ab^t$ or $y = Ae^{Bt}$ describing the number of cell phone subscribers. Then use the graphing utility to display the scatter plots and the two models in the same viewing rectangle.

(b) By finding an appropriate viewing rectangle, make a projection for the year in which there would be as many cell phone subscribers as phone lines.

Year t t (t = 0 ↔ 1985)	Phone lines (millions)	Cell phone subscribers (millions)
0	407	1
1	426	1
2	446	2
3	469	4
4	493	7
5	520	11
6	546	16
7	574	23
8	606	34
9	645	55
10	691	91
11	738	142
12	788	215
13	844	319

Source: International Telecommunications Union data as reported in *Vital Signs 2000,* Lester Brown et al. (New York: W. W. Norton & Co., 2000)

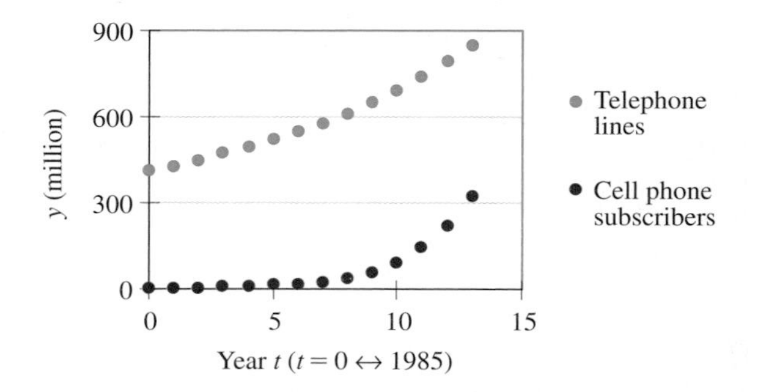

45. The age of some rocks can be estimated by measuring the ratio of the amounts of certain chemical elements within the rock. The method known as the *rubidium–strontium method* will be discussed here. This method has been used in dating the moon rocks brought back on the Apollo missions.

Rubidium-87 is a radioactive substance with a half-life of 4.7×10^{10} years. Rubidium-87 decays into the substance strontium-87, which is stable (nonradioactive). We are going to derive the following formula for the age of a rock:

$$T = \frac{\ln[(\mathcal{N}_s/\mathcal{N}_r) + 1]}{-k}$$

where T is the age of the rock, k is the decay constant for rubidium-87, $\mathcal{N}_s$ is the number of atoms of strontium-87 now present in the rock, and $\mathcal{N}_r$ is the number of atoms of rubidium-87 now present in the rock.

(a) Assume that initially, when the rock was formed, there were $\mathcal{N}_0$ atoms of rubidium-87 and none of strontium-87. Then, as time goes by, some of the rubidium atoms decay into strontium atoms, but the total number of atoms must still be $\mathcal{N}_0$. Thus, after T years, we have $\mathcal{N}_0 = \mathcal{N}_r + \mathcal{N}_s$ or, equivalently,

$$\mathcal{N}_s = \mathcal{N}_0 - \mathcal{N}_r \qquad (1)$$

However, according to the law of exponential decay for the rubidium-87, we must have $\mathcal{N}_r = \mathcal{N}_0 e^{kT}$. Solve this equation for $\mathcal{N}_0$ and then use the result to eliminate $\mathcal{N}_0$ from equation (1). Show that the result can be written

$$\mathcal{N}_s = \mathcal{N}_r e^{-kT} - \mathcal{N}_r \qquad (2)$$

(b) Solve equation (2) for T to obtain the formula given at the beginning of this exercise.

46. (Continuation of Exercise 45)

(a) The half-life of rubidium-87 is 4.7×10^{10} years. Compute the decay constant k.

(b) Analysis of lunar rock samples taken on the Apollo 11 mission showed the strontium–rubidium ratio to be

$$\frac{\mathcal{N}_s}{\mathcal{N}_r} = 0.0588$$

Estimate the age of these lunar rocks.

47. (Continuation of Exercise 45) Analysis of the so-called genesis rock sample taken on the Apollo 15 mission revealed a strontium–rubidium ratio of 0.0636. Estimate the age of this rock.

48. *Radiocarbon Dating*: Because rubidium-87 decays so slowly, the technique of rubidium–strontium dating is generally considered effective only for objects older than 10 million years. In contrast, archeologists and geologists rely on the *radiocarbon dating* method in assigning ages ranging from 500 to 50,000 years.

Two types of carbon occur naturally in our environment: carbon-12, which is nonradioactive, and carbon-14, which has a half-life of 5730 years. All living plant and animal tissue contains both types of carbon, always in the same ratio. (The ratio is one part carbon-14 to 10^{12} parts carbon-12.) As long as the plant or animal is living, this ratio is maintained. When the organism dies, however, no new carbon-14 is absorbed, and the amount of carbon-14 begins to decrease exponentially. Since the amount of carbon-14 decreases exponentially, it follows that the level of radioactivity also must decrease exponentially. The formula describing this situation is

$$\mathcal{N} = \mathcal{N}_0 e^{kT}$$

where T is the age of the sample, $\mathcal{N}$ is the present level of radioactivity (in units of disintegrations per hour per gram of carbon), and $\mathcal{N}_0$ is the level of radioactivity T years ago, when the organism was alive. Given that the half-life of carbon-14 is 5730 years and that $\mathcal{N}_0 = 920$ disintegrations per hour per gram, show that the age T of a sample is given by

$$T = \frac{5730 \ln(\mathcal{N}/920)}{\ln(1/2)}$$

In Exercises 49–54, use the formula derived in Exercise 48 to estimate the age of each sample. Note: Some technical complications arise in interpreting such results. Studies have shown that the ratio of carbon-12 to carbon-14 in the air (and therefore in living matter) has not in fact been constant over time. For instance, air pollution from factory smokestacks tends to increase the level of carbon-12. In the other direction, nuclear bomb testing increases the level of carbon-14.

49. Prehistoric cave paintings were discovered in the Lascaux cave in France. Charcoal from the site was analyzed and the level of radioactivity was found to be $\mathcal{N} = 141$ disintegrations per hour per gram. Estimate the age of the paintings.

50. Before radiocarbon dating was used, historians estimated that the tomb of Vizier Hemaka, in Egypt, was constructed about 4900 years ago. After radiocarbon dating became available, wood samples from the tomb were analyzed, and it was determined that the radioactivity level was 510 disintegrations per hour per gram. Estimate the age of the tomb on the basis of this reading and compare your answer to the figure already mentioned.

51. Before radiocarbon dating, scholars believed that agriculture (farming, as opposed to the hunter-gatherer existence) in the Middle East began about 6500 years ago. However, when radiocarbon dating was used to study an ancient farming settlement at Jericho, the radioactivity level was found to be in the range $\mathcal{N} = 348$ disintegrations per hour per gram of carbon. On the basis of this evidence, estimate the age of the site, and compare your answer to the figure mentioned at the beginning of this exercise. Round your answer to the nearest 100 years. (Similar analyses in Iraq, Turkey, and other countries have since shown that agriculture was firmly established at least 9000 years ago.)

52. (a) Analyses of some ancient campsites in the Western Hemisphere reveal a carbon-14 radioactivity level of $\mathcal{N} = 226$ disintegrations per hour per gram of carbon. Show that this implies an age of 11,500 years, to the nearest 500 years. Remark: An age of 11,500 years corresponds to the last Ice Age, when the sea level was significantly lower than it is today. According to the *Clovis hypothesis,* this was the time humans first entered the Western Hemisphere, across what would have been a land bridge extending over what is now the Bering Strait. (The name "Clovis" refers to Clovis, New Mexico. In 1933 a spearpoint and bones were discovered there that were subsequently found to be about 11,500 years old.)

(b) In the article "Coming to America" (*Time Magazine,* May 3, 1993), Michael D. Lemonick describes some of the evidence indicating that the Clovis hypothesis is not valid:

> A team led by University of Kentucky archeologist Tom Dillehay discovered indisputable traces . . . [in Monte Verde, in southern Chile] of a human settlement that was inhabited between 12,800 and

12,300 years ago. Usually all scientists can find from that far back are stones and bones. In this case, thanks to a peat layer that formed during the late Pleistocene era, organic matter [to which radiocarbon dating can be applied] was mummified and preserved as well.

What range for the carbon-14 radioactivity level $\mathcal{N}$ corresponds to the ages mentioned in this article? (That is, compute the values of $\mathcal{N}$ corresponding to $T = 12{,}800$ and $T = 12{,}300$.)

53. The Dead Sea Scrolls are a collection of ancient manuscripts discovered in caves along the west bank of the Dead Sea. (The discovery occurred by accident when an Arab herdsman of the Taamireh tribe was searching for a stray goat.) When the linen wrappings on the scrolls were analyzed, the carbon-14 radioactivity level was found to be 723 disintegrations per hour per gram. Estimate the age of the scrolls using this information. Historical evidence suggests that some of the scrolls date back somewhere between 150 B.C. and A.D. 40. How do these dates compare with the estimate derived using radiocarbon dating?

54. (a) According to Exercise 48, the formula for the age T of a sample in terms of the radioactivity level $\mathcal{N}$ is $T = \dfrac{5730 \ln(\mathcal{N}/920)}{\ln(1/2)}$. Solve this equation for $\mathcal{N}$ and show that the result can be written $\mathcal{N} = 920(2^{-T/5730})$.

(b) The famous prehistoric stone monument Stonehenge is located 8 miles north of Salisbury, England. Excavations and radiocarbon dating have led anthropologists and archeologists to distinguish three periods in the building of Stonehenge: period I, 2800 years ago; period II, 2100 years ago; and period III, 2000 years ago. Use the formula in part (a) to compute the carbon-14 radioactivity levels that would have led the scientists to assign these ages.

Exercises 55–60 introduce a model for population growth that takes into account limitations on food and the environment. This is the **logistic growth model,** *named and studied by the nineteenth century Belgian mathematician and sociologist Pierre Verhulst. (The word "logistic" has Latin and Greek origins meaning "calculation" and "skilled in calculation," respectively. However, that is not why Verhulst named the curve as he did. See Exercise 56 for more about this.)*

In the logistic model that we'll study, the initial population growth resembles exponential growth. But then, at some point, owing perhaps to food or space limitations, the growth slows down and eventually levels off, and the population approaches an equilibrium level. The basic equation that we'll use for logistic growth is

$$\mathcal{N} = \frac{P}{1 + ae^{-bt}}$$

where $\mathcal{N}$ is the population at time t, P is the equilibrium population (or the upper limit for population), and a and b are positive constants.

55. The following figure shows the graph of the logistic function $\mathcal{N}(t) = 4/(1 + 8e^{-t})$. Note that in this equation the equilibrium population P is 4 and that this corresponds to the asymptote $\mathcal{N} = 4$ in the graph.

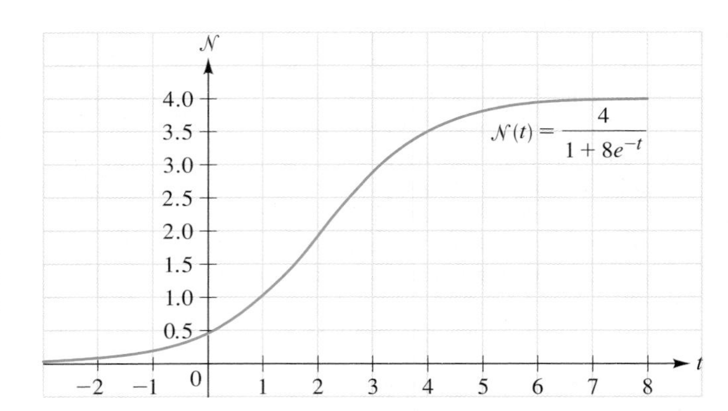

(a) Use the graph and your calculator to complete the following table. For the values that you read from the graph, estimate to the nearest 0.25. For the calculator values, round to three decimal places.

	$\mathcal{N}(-1)$	$\mathcal{N}(0)$	$\mathcal{N}(1)$	$\mathcal{N}(4)$	$\mathcal{N}(5)$
From graph					
From calculator					

(b) As indicated in the graph, the line $\mathcal{N} = 4$ appears to be an asymptote for the curve. Confirm this empirically by computing $\mathcal{N}(10)$, $\mathcal{N}(15)$, and $\mathcal{N}(20)$. Round each answer to eight decimal places.

(c) Use the graph to estimate, to the nearest integer, the value of t for which $\mathcal{N}(t) = 3$.

(d) Find the exact value of t for which $\mathcal{N}(t) = 3$. Evaluate the answer using a calculator, and check that it is consistent with the result in part (c).

Ⓖ **56.** (Continuation of Exercise 55) The author's ideas for this exercise are based on Professor Bonnie Shulman's article "Math-Alive! Using Original Sources to Teach Mathematics in Social Context," *Primus,* vol. VIII (March 1998).

(a) The function $\mathcal{N}$ in Exercise 55 expresses population as a function of time. But as pointed out by Professor Shulman, in Verhulst's original work it was the other way around; he expressed time as a function of population. In terms of our notation, we would say that he was studying the function $\mathcal{N}^{-1}$. Given $\mathcal{N}(t) = 4/(1 + 8e^{-t})$, find $\mathcal{N}^{-1}(t)$.

(b) Use a graphing utility to draw the graphs of $\mathcal{N}$, $\mathcal{N}^{-1}$, and the line $y = x$ in the viewing rectangle $[-3, 8, 2]$ by $[-3, 8, 2]$. Use true portions. (Why?)

(c) In the viewing rectangle $[0, 5, 1]$ by $[-3, 2, 1]$, draw the graphs of $y = \mathcal{N}^{-1}(t)$ and $y = \ln t$. Note that the two graphs have the same general shape and characteristics. In other words, Verhulst's logistic function (our $\mathcal{N}^{-1}$) appears log-like, or *logistique,* as Verhulst actually named it in French. (For details, both historical and mathematical, see the paper by Professor Shulman cited above.)

57. The following figure shows the graph of a logistic function $\mathcal{N}(t) = P/(1 + ae^{-bt})$. As indicated in the figure, the graph passes through the two points (0, 1) and (1, 2), and the asymptote is $\mathcal{N} = 5$. In this exercise we determine the values of the constants P, a, and b.

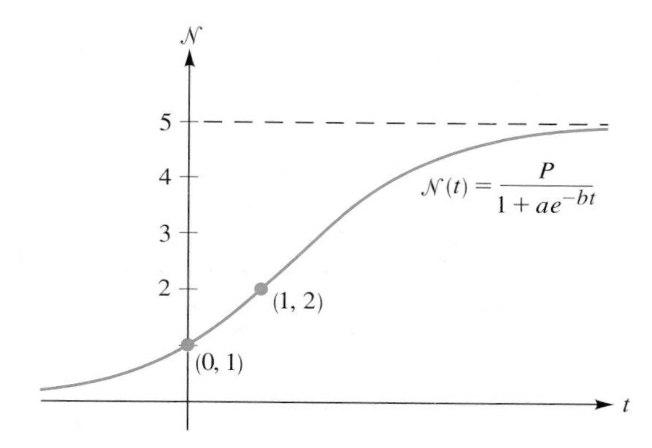

(a) As indicated in the figure, the logistic curve approaches an equilibrium population of 5. By definition, this is the value of the constant P, so the equation becomes

$$\mathcal{N}(t) = 5/(1 + ae^{-bt})$$

Now use the fact that the graph passes through the point (0, 1) to obtain an equation that you can solve for a. Solve the equation; you should obtain $a = 4$.

(b) With $P = 5$ and $a = 4$, the logistic equation becomes $\mathcal{N}(t) = 5/(1 + 4e^{-bt})$. Use the fact that the graph of this equation passes through the point (1, 2) to show that $b = \ln(8/3) \approx 0.9808$.

58. Biologist H. G. Thornton carried out an experiment in the 1920s to measure the growth of a colony of bacteria in a closed environment. As is common in the biology lab, Thornton measured the area of the colony, rather than count the number of individuals. (The reasoning is that the actual population size is directly proportional to the area, and the area is much easier to measure.) The following table and scatter plot summarize Thornton's measurements. Notice that the data points in the scatter plot suggest a logistic growth curve.

Days	0	1	2	3	4	5
Area of colony (cm^2)	0.24	2.78	13.53	36.30	47.50	49.40

Source: H. G. Thornton, *Annals of Applied Biology,* 1922, p. 265

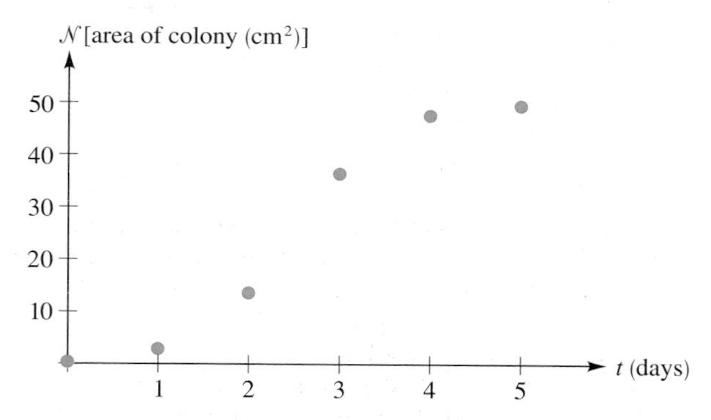

(a) Use the technique shown in Exercise 57 to determine a logistic function $\mathcal{N}(t) = P/(1 + ae^{-bt})$ with a graph that passes through Thornton's two points (0, 0.24) and (2, 13.53). Assume that the equilibrium population is $P = 50$. Round the final values of a and b to two decimal places.

(b) Use the logistic function that you determined in part (a) to estimate the area of the colony after 1, 3, 4, and 5 days. Round your estimates to one decimal place. If you compare your estimates to the results that Thornton obtained in the laboratory, you'll see there is good agreement.

(c) Use the logistic function that you determined in part (a) to estimate the time at which the area of the colony was 10 cm^2. Express your answer in terms of days and hours, and round to the nearest half-hour.

59. The following data on the growth of *Lupinus albus,* a plant in the pea family, are taken from an experiment that was summarized in the classic text by Sir D'Arcy Wentworth Thompson, *On Growth and Form* (New York: Dover Publications, Inc., 1992).

Day	4	6	8	10	12
length (mm)	10.5	23.3	42.2	77.9	107.4
Day	14	16	18	20	21
length (mm)	132.3	149.7	158.1	161.4	161.6

(a) Follow Exercise 57 to determine an equation for a logistic function that models these data. Assume $P = 162$, $\mathcal{N}(0) = 1$, and $\mathcal{N}(10) = 77.9$.

(b) Using the model that you determined in part (a), compute $\mathcal{N}(4)$, $\mathcal{N}(8)$, $\mathcal{N}(12)$, and $\mathcal{N}(16)$. Round the answers to one decimal place. In each case, say whether your answer is higher or lower than the actual value.

(c) For the logistic model in general, it can be shown (using calculus) that the population is growing fastest at the instant when it reaches half the equilibrium level P. Use this fact to determine when the population in this experiment is growing fastest. Give your answer in the form d days, h hours, with the number of hours rounded to the nearest half-hour.

Ⓖ **60.** This exercise requires a graphing utility with a logistic regression feature. (This feature is included on the Texas Instruments *TI-83 Plus,* for example.)

(a) Use a graphing utility to create a scatter plot for the data in Exercise 59. Next, use the logistic regression option to find the equation of the logistic model that best fits the data.

(b) On the basis of the model that you determined in part (a), what is the equilibrium population P? (Remember that P has to be an integer.) Add the graph of the horizontal line $\mathcal{N} = P$ to the picture obtained in part (a).

(c) In the initial stages of logistic growth (when the population is small compared to the equilibrium level P), the growth resembles exponential growth. Demonstrate this as follows. Use a graphing utility to find an exponential growth model fitting only the first three data pairs in the table in Exercise 59. Add the graph of this exponential model to the picture that you obtained in part (b). Also include the graph of the horizontal line $\mathcal{N} = P/2$. Now (in complete sentences), summarize what you see.

61. (a) Let $\mathcal{N}(t) = \mathcal{N}_0e^{kt}$. Show that $[\mathcal{N}(t+1) - \mathcal{N}(t)]/\mathcal{N}(t) = e^k - 1$. (This is actually done in detail in the text. So, ideally, you should look back only if you get stuck or want to check your answer.)

(b) Assume as given the following approximation, which was introduced in Exercise 26 of Section 5.2.

$$e^x \approx x + 1 \qquad \text{provided } x \text{ is close to zero}$$

Use this approximation to explain why $e^k - 1 \approx k$, provided that k is close to zero. *Remark*: Combining this result with that in part (a), we conclude that the relative growth rate for the function $\mathcal{N}(t) = \mathcal{N}_0e^{kt}$ is approximately equal to the growth constant k. As explained in the text, this is one of the reasons why in applications we've not distinguished between the relative growth rate and the decay constant k.

62. In the text we showed that the relative growth rate for the function $\mathcal{N}(t) = \mathcal{N}_0e^{kt}$ is constant for all time intervals of unit length, $[t, t+1]$. Recall that we did this by computing the relative change $[\mathcal{N}(t+1) - \mathcal{N}(t)]/\mathcal{N}(t)$ and noting that the result was a constant, independent of t. (If you've completed the previous exercise, you've done this calculation for yourself.) Now consider a time interval of arbitrary length, $[t, t+d]$. The relative change in the function $\mathcal{N}(t) = \mathcal{N}_0e^{kt}$ over this time interval is $[\mathcal{N}(t+d) - \mathcal{N}(t)]/\mathcal{N}(t)$. Show that this quantity is a constant, independent of t. (The expression that you obtain for the constant will contain e and d, but not t. As a check on your work, replace d by 1 in the expression you obtain and make sure the result is the same as that in the text where we worked with intervals of length $d = 1$.)

63. Let $\mathcal{N} = \mathcal{N}_0e^{kt}$. In this exercise we show that if Δt is very small, then $\Delta\mathcal{N}/\Delta t \approx k\mathcal{N}$. In other words, over very small intervals of time, the average rate of change of $\mathcal{N}$ is proportional to $\mathcal{N}$ itself.

(a) Show that the average rate of change of the function $\mathcal{N} = \mathcal{N}_0e^{kt}$ on the interval $[t, t+\Delta t]$ is given by

$$\frac{\Delta\mathcal{N}}{\Delta t} = \frac{\mathcal{N}_0e^{kt}(e^{k\Delta t} - 1)}{\Delta t} = \frac{\mathcal{N}(e^{k\Delta t} - 1)}{\Delta t}$$

(b) In Exercise 26 of Section 5.2 we saw that $e^x \approx x + 1$ when x is close to zero. Thus, if Δt is sufficiently small, we have $e^{k\Delta t} \approx k\,\Delta t + 1$. Use this approximation and the result in part (a) to show that $\Delta\mathcal{N}/\Delta t \approx kN$ when Δt is sufficiently close to zero.

PROJECT

A VARIABLE GROWTH CONSTANT?

Ⓖ As background for this project you need to have worked Exercise 39 in this section regarding the United Nation's proposed population limits. Or, if you are working in a group, have a group member present a summary of the problem and the results.

Over the past few decades, world population growth rates have, in general, been decreasing. (Over the period 1970–2000, for example, the rate fell from 2.0%/year to 1.4%/year.) In this project we consider a modification of the exponential model that takes this declining growth rate into account. In particular, we consider a growth equation of the form

$$\mathcal{N}(t) = \mathcal{N}_0 e^{[f(t)]t} \qquad (1)$$

where the function f in the exponent describes the changing relative growth rate.

(a) The following table lists some relative growth rates for the world's population over the period 1985–2000. Set up a scatter plot for the information in the table, subject to the following specifications. For the horizontal time-axis, let $t = 0$ correspond to 1980. For the relative growth rates, use the decimal form, rather than the given percentage. Then use a graphing utility to determine the equation of the regression line $f(t) = mt + b$.

Year	1980	1985	1990	1995	2000
Relative growth rate (%/year)	1.8	1.8	1.7	1.6	1.4

Source: Population Reference Bureau

(b) In 1980 the world population was 4.448 billion. Use this value for $\mathcal{N}_0$ in equation (1). Also, in equation (1), replace $f(t)$ by the quantity $mt + b$ determined in part (a). Now, use a graphing utility to graph the resulting population growth model for the period 1980 to 2020 (that is, $t = 0$ to $t = 40$). Use the graph to decide whether or not (according to this model) the projected population in 2020 exceeds United Nation's proposed limit of 7.2 billion.

(c) For purposes of comparison, add the graph of an exponential growth model to the viewing rectangle. For the model, use the value of $\mathcal{N}_0$ given in part (b) and choose a year from the table to obtain k. (The value of k obtained will differ slightly depending on the year you choose.) Does this model project a population excess of 7.2 billion in 2020?

(d) Switch to the viewing rectangle [0, 50, 200] by [0, 8, 4]. Give a description, in complete sentences, of the general behavior of the population after 2020 as projected by the model $\mathcal{N}(t) = \mathcal{N}_0 e^{[f(t)]t}$ in part (b). What do you think about this scenario?

Chapter 5 Summary of Principal Terms and Formulas

Term, notation, or formula	Page reference	Comments
1. Exponential function with base b	329	The exponential function with base b is defined by the equation $y = b^x$. It is understood here that the base b is a positive number other than 1. More generally, functions of the form $y = ab^{cx}$ are also referred to as exponential functions.
2. The number e	337	The irrational number e is one of the basic constants in mathematics, as is the irrational number π. To five decimal places, the value of e is 2.71828. In calculus e is the base most commonly used for exponential functions. The graph of the exponential function $y = e^x$ is shown in the box on page 338.
3. Instantaneous rate of change of the function $y = e^x$	340	The concept of the instantaneous rate of change of a function is introduced on pages 339–340. One of the distinguishing characteristics of the function $y = e^x$ is that the instantaneous rate of change at any point on the graph is just the y-coordinate of that point. See the examples in the box on page 340.
4. $\log_b x$	348	The expression $\log_b x$ denotes the exponent to which b must be raised to yield x. The equation $\log_b x = y$ is equivalent to $b^y = x$.
5. $\ln x$	352	The expression $\ln x$ means $\log_e x$. Logarithms to the base e are known as *natural logarithms*. For the graph of $y = \ln x$, see Figure 6(a) in Section 5.3.
6. $\log_a x = \dfrac{\log_b x}{\log_b a}$	367	This is the change-of-base formula for converting logarithms from one base to another.
7. $A = P\left(1 + \dfrac{r}{n}\right)^{nt}$	384	This formula gives the amount A that accumulates after t years when a principal of P dollars is invested at an annual rate r that is compounded n times per year.
8. $A = Pe^{rt}$	386	This formula gives the amount A that accumulates after t years when a principal of P dollars is invested at an annual rate r that is compounded continuously.
9. Doubling time; rule of 70	389	For money earning compound interest, the doubling time T_2 is the length of time required for an amount to double. As can be seen from the formula $T_2 = (\ln 2)/r$, the doubling time does not depend on the initial amount, but only on the interest rate r. (In the formula, the interest rate is expressed as a decimal.) The *rule of 70* gives an approximate value for the doubling time: $T_2 \approx 70/R$, where R is the interest rate expressed as a percentage.

Term, notation, or formula	Page reference	Comments
10. $\mathcal{N} = \mathcal{N}_0 e^{kt}$ $(k > 0)$	394	This is the exponential growth model, where $\mathcal{N}_0$ is the size of the population at time $t = 0$, and k is a constant called the *growth constant.* In the applications we studied, we used the value of the relative growth rate for k. (As is explained in the text, the two quantities are approximately equal when their values are near zero.)
11. $\mathcal{N} = \mathcal{N}_0 e^{kt}$ $(k < 0)$	401	In this exponential decay model for radioactive substances, $\mathcal{N}_0$ is the amount of the substance present at $t = 0$, and k is the decay constant.
12. Half-life	402	The half-life of a radioactive substance is the time required until only half of the initial amount remains. For a given radioactive substance the half-life is independent of the amount initially present, as evidenced by the formula *half-life* $= -(\ln 2)/k$.

Writing Mathematics

1. In Section 5.3 we defined logarithmic functions in terms of inverse functions.
   (a) Explain in general what is meant by a pair of inverse functions.
   (b) Assuming a knowledge of the function $f(x) = 2^x$, explain how to define and then graph the function $g(x) = \log_2 x$.

2. A student who was trying to simplify the expression $\dfrac{\ln(e^3)}{\ln(e^2)}$ wrote

$$\frac{\ln(e^3)}{\ln(e^2)} = \ln(e^3) - \ln(e^2) = 3 - 2 = 1$$

   (a) What is the error here? What do you think is the most probable reason for making this error?
   (b) Give the correct solution.

3. A student who wanted to simplify the expression $\dfrac{100}{\log_{10} 100}$ wrote

$$\frac{100}{\log_{10} 100} = \frac{\overset{1}{\cancel{100}}}{\log_{10} \underset{1}{\cancel{100}}} = \frac{1}{\log_{10}}$$

Explain why this is nonsense, and then indicate the correct solution.

4. Look over the following method for solving the equation $\ln x^2 = 6$:

$$2 \ln x = 6$$
$$\ln x = 3$$
$$x = e^3$$

As you can check, e^3 is a root of the given equation. The value $-e^3$, however, is also a root. Why is it that the demonstrated method fails to produce this root? Find a method that does produce both roots.

Chapter 5 Review Exercises

1. Which is larger, $\log_5 126$ or $\log_{10} 999$?
2. Graph the function $y = 3^{-x} - 3$. Specify the domain, range, intercept(s), and asymptote.
3. Suppose that the population of a colony of bacteria increases exponentially. If the population at the start of an experiment is 8000, and 4 hours later it is 10,000, how long (from the start of the experiment) will it take for the population to reach 12,000? (Express the answer in terms of base e logarithms.)
4. Express $\log_{10} 2$ in terms of base e logarithms.
5. Let f be the function defined by

$$f(x) = \begin{cases} 2^{-x} & \text{if } x < 0 \\ x^2 & \text{if } x \geq 0 \end{cases}$$

Sketch the graph of f and then use the horizontal line test to determine whether f is one-to-one.
6. Estimate 2^{60} in terms of an integral power of 10.
7. Solve for x: $\ln(x + 1) - 1 = \ln(x - 1)$.
8. Suppose that \$5000 is invested at 8% interest compounded annually. How many years will it take for the money to double? Make use of the approximations $\ln 2 \approx 0.7$ and $\ln 1.08 \approx 0.08$ to obtain a numerical answer.
9. On the same set of axes, sketch the graphs of $y = e^x$ and $y = \ln x$. Specify the domain and range for each function.
10. Solve for x: $xe^x - 2e^x = 0$.
11. Simplify: $\log_9(1/27)$.
12. Given that $\ln A = a$, $\ln B = b$, and $\ln C = c$, express $\ln[(A^2\sqrt{B})/C^3]$ in terms of a, b, and c.
13. The half-life of plutonium-241 is 13 years. What is the decay constant? Use the approximation $\ln(1/2) \approx -0.7$ to obtain a numerical answer.
14. Express as a single logarithm with a coefficient of 1: $3 \log_{10} x - \log_{10}(1 - x)$.
15. Solve for x, leaving your answer in terms of base e logarithms: $5e^{2-x} = 12$.
16. Let $f(x) = e^{x+1}$. Find a formula for $f^{-1}(x)$ and specify the domain of f^{-1}.
17. Suppose that in 1995 the population of a certain country was 2 million and increasing with a relative growth rate of 2%/year. Estimate the year in which the population will reach 3 million.
18. Simplify: $\ln e + \ln\sqrt{e} + \ln 1 + \ln e^{\ln 10}$.
19. A principal of \$1000 is deposited at 10% per annum, compounded continuously. Estimate the doubling time and then sketch a graph that shows how the amount increases with time.
20. Simplify: $\ln(\log_8 56 - \log_8 7)$.

In Exercises 21–32, graph the function and specify the asymptote(s) and intercepts(s).

21. $y = e^x$
22. $y = -e^{-x}$
23. $y = \ln x$
24. $y = \ln(x + 2)$
25. $y = 2^{x+1} + 1$
26. $y = \log_{10}(-x)$
27. $y = (1/e)^x$
28. $y = (1/2)^{-x}$
29. $y = e^{x+1} + 1$
30. $y = -\log_2(x + 1)$
31. $y = \ln(e^x)$
32. $y = e^{\ln x}$

In Exercises 33–49, solve the equation for x. (When logarithms appear in your answer, leave the answer in that form, rather than using a calculator.)

33. $\log_4 x + \log_4(x - 3) = 1$
34. $\log_3 x + \log_3(2x + 5) = 1$
35. $\ln x + \ln(x + 2) = \ln 15$
36. $\log_6 \dfrac{x + 4}{x - 1} = 1$
37. $\log_2 x + \log_2(3x + 10) - 3 = 0$
38. $2 \ln x - 1 = 0$
39. $3 \log_9 x = 1/2$
40. $e^{2x} = 6$
41. $e^{1-5x} = 3\sqrt{e}$
42. $2^x = 100$
43. $\log_{10} x - 2 = \log_{10}(x - 2)$
44. $\log_{10}(x^2 - x - 10) = 1$
45. $\ln(x + 2) = \ln x + \ln 2$
46. $\ln(2x) = \ln 2 + \ln x$
47. $\ln x^4 = 4 \ln x$
48. $(\ln x)^4 = 4 \ln x$
49. $\log_{10} x = \ln x$
50. Solve for x: $(\ln x)/(\ln 3) = \ln x - \ln 3$. Use a calculator to evaluate your result and round to the nearest integer. *Answer:* 206,765

In Exercises 51–66, simplify the expression without using a calculator.

51. $\log_{10}\sqrt{10}$
52. $\log_7 1$
53. $\ln \sqrt[5]{e}$
54. $\log_3 54 - \log_3 2$
55. $\log_{10} \pi - \log_{10} 10\pi$
56. $\log_2 2$
57. 10^t, where $t = \log_{10} 16$
58. $e^{\ln 5}$
59. $\ln(e^4)$
60. $\log_{10}(10^{\sqrt{2}})$
61. $\log_{12} 2 + \log_{12} 18 + \log_{12} 4$

62. $(\log_{10} 8)/(\log_{10} 2)$

63. $(\ln 100)/(\ln 10)$

64. $\log_5 2 + \log_{1/5} 2$

65. $\log_2 \sqrt[7]{16\sqrt[3]{2\sqrt{2}}}$

66. $\log_{1/8} (1/16) + \log_5 0.02$

In Exercises 67–70, express the quantity in terms of a, b, and c, where $a = \log_{10} A$, $b = \log_{10} B$, and $c = \log_{10} C$.

67. $\log_{10} A^2B^3\sqrt{C}$

68. $\log_{10}\sqrt[3]{AC/B}$

69. $16 \log_{10}\sqrt{A}\sqrt[4]{B}$

70. $6 \log_{10}[B^{1/3}/(A\sqrt{C})]$

In Exercises 71–76, find consecutive integers n and n + 1 such that the given expression lies between n and n + 1. Do not use a calculator.

71. $\log_{10} 209$

72. $\ln 2$

73. $\log_6 100$

74. $\log_{10} (1/12)$

75. $\log_{10} 0.003$

76. $\log_3 244$

77. (a) On the same set of axes, graph the curves $y = \ln(x + 2)$ and $y = \ln(-x) - 1$. According to your graph, in which quadrant do the two curves intersect?
(b) Find the x-coordinate of the intersection point.

78. The curve $y = ae^{bt}$ passes through the point (3, 4) and has a y-intercept of 2. Find a and b.

79. A certain radioactive substance has a half-life of T years. Find the decay constant (in terms of T).

80. Find the half-life of a certain radioactive substance if it takes T years for one-third of a given sample to disintegrate. (Your answer will be in terms of T.)

81. A radioactive substance has a half-life of M minutes. What percentage of a given sample will remain after $4M$ minutes?

82. At the start of an experiment, a colony of bacteria has initial population a. After b hours, there are c bacteria present. Determine the population d hours after the start of the experiment.

83. The half-life of a radioactive substance is d days. If you begin with a sample weighing b grams, how long will it be until c grams remain?

84. Find the half-life of a radioactive substance if it takes D days for P percent of a given sample to disintegrate.

In Exercises 85–90, write the expression as a single logarithm with a coefficient of 1.

85. $\log_{10} 8 + \log_{10} 3 - \log_{10} 12$

86. $4 \log_{10} x - 2 \log_{10} y$

87. $\ln 5 - 3 \ln 2 + \ln 16$

88. $\ln(x^4 - 1) - \ln(x^2 - 1)$

89. $a \ln x + b \ln y$

90. $\ln(x^3 + 8) - \ln(x + 2) - 2 \ln(x^2 - 2x + 4)$

In Exercises 91–98, write the quantity using sums and differences of simpler logarithmic expressions. Express the answer so that logarithms of products, quotients, and powers do not appear.

91. $\ln\sqrt{(x - 3)(x + 4)}$

92. $\log_{10}\dfrac{x^2 - 4}{x + 3}$

93. $\log_{10}\dfrac{3}{\sqrt{1 + x}}$

94. $\ln\dfrac{x^2\sqrt{2x + 1}}{2x - 1}$

95. $\ln\left(\dfrac{1 + 2e}{1 - 2e}\right)^3$

96. $\log_{10}\sqrt[3]{x\sqrt{1 + y^2}}$

97. Suppose that A dollars are invested at R% compounded annually. How many years will it take for the money to double?

98. A sum of \$2800 is placed in a savings account at 9% per annum. How much is in the account after 2 years if the interest is compounded quarterly?

99. A bank offers an interest rate of 9.5% per annum, compounded monthly. Compute the effective interest rate.

100. A sum of D dollars is placed in an account at R% per annum, compounded continuously. When will the balance reach E dollars?

101. Compute the doubling time for a sum of D dollars invested at R% per annum, compounded continuously.

102. Your friend invests D dollars at R% per annum, compounded semiannually. You invest an equal amount at the same yearly rate, but compounded daily. How much larger is your account than your friend's after T years?

103. (a) You invest \$660 at 5.5% per annum, compounded quarterly. How long will it take for your balance to reach \$1000? Round off your answer to the next quarter of a year.
(b) You invest D dollars at R% per annum, compounded quarterly. How long will it take for your balance to reach nD dollars? (Assume that $n > 1$.)

In Exercises 104–107, find the domain of each function.

104. (a) $y = \ln x$
(b) $y = e^x$

105. (a) $y = \log_{10}\sqrt{x}$
(b) $y = \sqrt{\log_{10} x}$

106. (a) $f \circ g$, where $f(x) = \ln x$ and $g(x) = e^x$
(b) $g \circ f$, where f and g are as in part (a)

107. $y = \dfrac{2 + \ln x}{2 - \ln x}$

108. Find the range of the function defined by $y = \dfrac{e^x + 1}{e^x - 1}$.

Chapter 5  Test 

1. Graph the function $y = 2^{-x} - 3$. Specify the domain, range, intercepts, and asymptote.
2. Suppose that the population of a colony of bacteria increases exponentially. At the start of an experiment there are 6000 bacteria, and 1 hour later the population is 6200. How long (from the start of the experiment) will it take for the population to reach 10,000? Give two forms for the answer: one in terms of base e logarithms and the other a calculator approximation rounded to the nearest hour.
3. Which is larger, $\log_2 17$ or $\log_3 80$? Explain the reasons for your answer.
4. Express $\log_2 15$ in terms of base e logarithms.
5. Estimate 2^{40} in terms of an integral power of 10.
6. Suppose that \$9500 is invested at 6% interest, compounded annually. How long will it take for the amount in the account to reach \$12,000?
7. **(a)** For which values of x is the identity $e^{\ln x} = x$ valid?
   (b) On the same set of axes, graph $y = e^x$ and $y = \ln x$.
8. Given that $\log_{10} A = a$ and $\log_{10} B = b$, express $\log_{10}(A^3/\sqrt{B})$ in terms of a and b.
9. Simplify each expression:
   (a) $\log_5(1/\sqrt{5})$
   (b) $\ln e^2 + \ln 1 - e^{\ln 3}$
10. Solve for x: $x^3e^x - 4xe^x = 0$.
11. The half-life of a radioactive substance is 4 days.
    (a) Find the decay constant.
    (b) How much of an initial 2-g sample will remain after 10 days?
12. Express as a single logarithm with a coefficient of 1: $2\ln x - \ln\sqrt[3]{x^2 + 1}$.
13. Solve for x, leaving your answer in terms of base e logarithms: $-2e^{3x-1} = 9$.
14. Let $g(x) = \log_{10}(x - 1)$. Find a formula for $g^{-1}(x)$ and specify the range of g^{-1}.
15. A principal of \$12,000 is deposited at 6% per annum, compounded continuously.
    (a) Use the rule of 70 to estimate the doubling time.
    (b) Sketch a graph showing how the amount increases with time.
16. **(a)** For which values of x is it true that $\ln(x^2) = 2\ln x$?
    (b) For which values of x is it true that $(\ln x)^2 = 2\ln x$?
17. Let $f(x) = e^x$.
    (a) Compute the average rate of change of the function f over the interval $[\ln 3, \ln 4]$. Give two forms for the answer: an expression simplified as much as possible and a calculator approximation rounded to two decimal places.
    (b) Specify the instantaneous rate of change of the function f at the point on the graph where $x = \ln 4$.
18. Solve for x: $\frac{1}{2} - \log_{16}(x - 3) = \log_{16} x$.
19. Solve for x: $6e^{2x} - 5e^x = 6$.
20. Solve each inequality:
    (a) $5(4 - 0.3^x) > 12$
    (b) $\ln x + \ln(x - 3) \le \ln 4$

CHAPTER 6 The Trigonometric Functions

Trigonometric functions were originally introduced and tabulated in calculations in astronomy and navigation and were used extensively in surveying. What is more important [now], *they describe a periodic process, like the vibration of a string, the tides, planetary motion, alternating currents, or the emission of light by atoms.*—From *The Mathematical Sciences,* edited by the Committee on Support of Research in the Mathematical Sciences with the collaboration of George Boehm (Cambridge, Mass.: The M.I.T. Press, 1969)

Chapters 6 through 9 form a unit on trigonometry. In general, there are two approaches to trigonometry at the precalculus level. In a sense, these two approaches correspond to the two historical roots of the subject mentioned in Alfred Hooper's opening quotation. One approach centers on the study of triangles. Indeed, the word "trigonometry" is derived from two Greek words, *trigonon,* meaning triangle, and *metria,* meaning measurement. Although you may have already been introduced to this triangle approach in high school geometry or algebra, we'll postpone its discussion until Section 6.5, where, as you'll see, it follows in a natural way from the more modern *unit-circle* approach that we introduce in Section 6.2.

Trigonometry. as we know the subject today, is a branch of mathematics that is linked to algebra. As such it dates back only to the eighteenth century.

When treated purely as a development of geometry, however, it goes back to the time of the great Greek mathematician-astronomers. . . .—Alfred Hooper in *Makers of Modern Mathematics* (New York: Random House, Inc., 1948)

6.1 RADIAN MEASURE

Degree measure, traditionally used to measure the angles of a geometric figure, has one serious drawback. It is artificial. There is no intrinsic connection between a degree and the geometry of a circle. Why 360 degrees for one revolution? Why not 400? or 100?—*Calculus, One and Several Variables,* 7th ed., by S. L. Salas and Einar Hille, revised by Garret J. Etgen (New York: John Wiley and Sons, Inc., 1995)

We use radian measure routinely in calculus for the sake of simplicity, in order to avoid the repeated occurence of the nuisance factor $\pi/180$.—George F. Simmons in his text *Calculus with Analytic Geometry,* 2nd ed. (New York: McGraw-Hill Book Co., 1996)

In elementary geometry an **angle** is a figure formed by two rays with a common endpoint. As is indicated in Figure 1, the common endpoint is called the **vertex** of the angle. There are several conventions used in naming angles; Figure 2 indicates some of these. In Figure 2 the symbol θ is the lowercase Greek letter *theta.* Greek

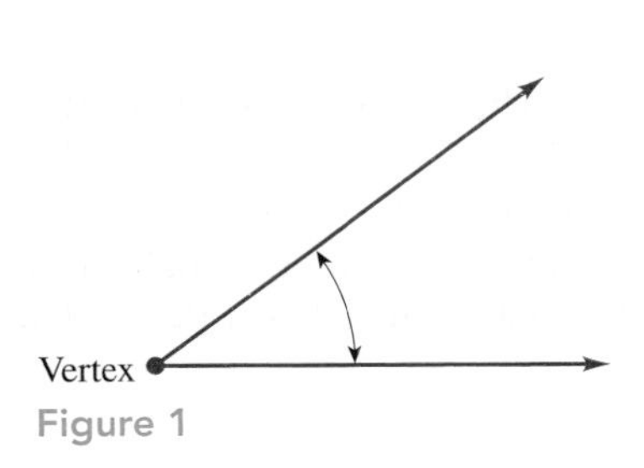

Figure 1

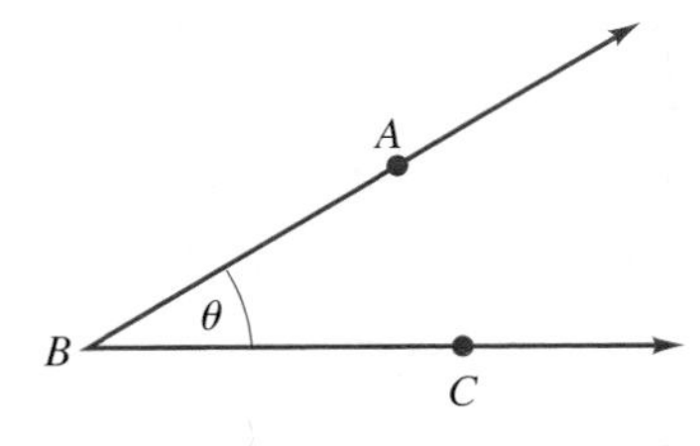

Figure 2

letters are often used to name angles. For reference, the Greek alphabet is given in the endpapers at the back of this book. The $\angle$ symbol that you see in Figure 2 stands for the word "angle." When three letters are used in naming an angle, as with $\angle ABC$ in Figure 2, the middle letter always indicates the vertex of the angle.

There are two principal systems used to indicate the size of (that is, the amount of rotation in) an angle: *degree measure,* which you used in previous math courses, and *radian measure,* which we'll introduce in a moment. Recall that 360 degrees (abbreviated 360°) is the measure of an angle obtained by rotating a ray through one complete circle. For comparison, Figure 3 shows angles of various degrees. In this section, we will not make any distinction whether the rotation generating an angle is clockwise or conterclockwise; however, in the next section this distinction will be important. The angles in Figures 3(a) and 3(b) are **acute** angles; these are angles with degree measure θ satisfying $0° < \theta < 90°$. The angle in Figure 3(d) is an **obtuse** angle, that is, an angle with degree measure θ satisfying $90° < \theta < 180°$.

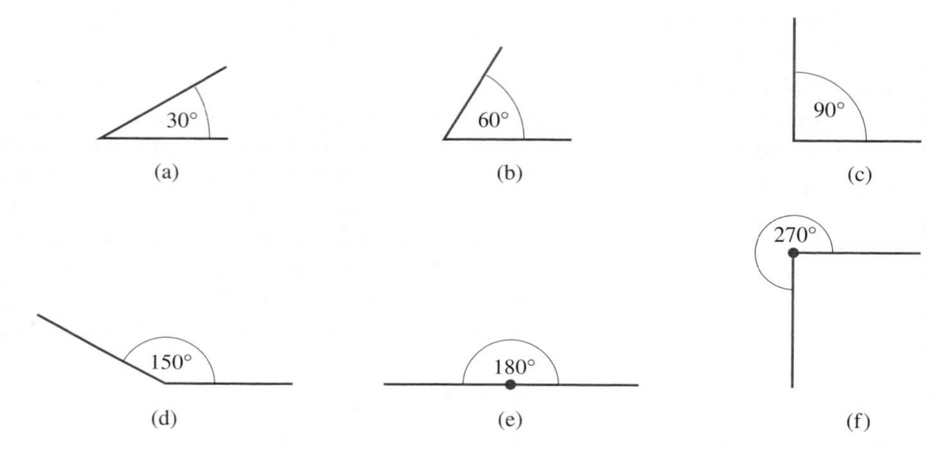

Figure 3

One more convention: Suppose, for example, that the measure of angle θ is 70°. We can write this as

$$\text{measure } \angle\theta = 70° \qquad \text{or just} \qquad m\,\angle\theta = 70°$$

For ease of notation and speech, however, we will usually write

$$\angle\theta = 70° \qquad \text{or simply} \qquad \theta = 70°$$

as shorthand for "θ is an angle whose measure is 70°."

For the portion of trigonometry dealing with angles and geometric figures, the units of degrees are quite suitable for measuring angles. However, for the more analytical portions of trigonometry and for calculus, *radian* measure is most often used. The radian measure of an angle is defined as follows.

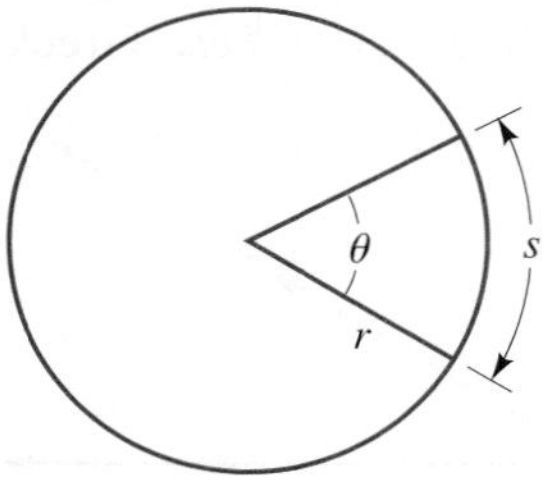

Figure 4
The radian measure θ is defined by the equation $\theta = s/r$.

DEFINITION **The Radian Measure of an Angle**

Place the vertex of the angle at the center of a circle of radius r. Let s denote the length of the arc intercepted by the angle, as indicated in Figure 4. The **radian measure** θ of the angle is the ratio of the arc length s to the radius r.

In symbols,

$$\theta = \frac{s}{r} \tag{1}$$

In this definition it is assumed that s and r have the same linear units.

EXAMPLE 1 Calculating the radian measure of an angle

Determine the radian measure for each angle in Figure 5.

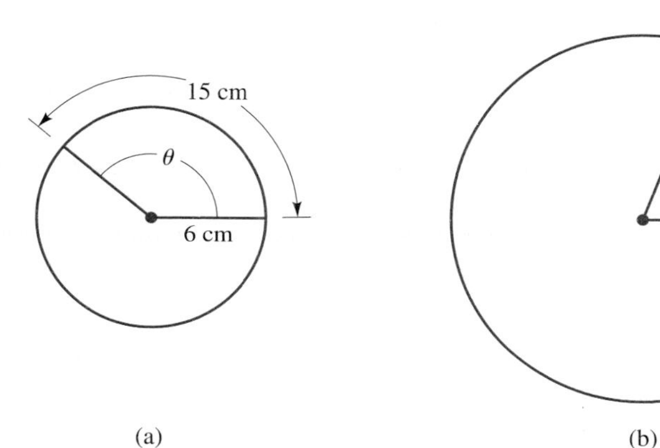

Figure 5 (a) (b)

SOLUTION

In Figure 5(a) we have $r = 6$ cm, $s = 15$ cm, and therefore

$$\theta = \frac{s}{r} = \frac{15 \text{ cm}}{6 \text{ cm}} = \frac{5}{2}$$

So for the angle in Figure 5(a) we obtain $\theta = 5/2$ radians. In the calculations just completed, notice that although both s and r have the dimensions of length, the resulting radian measure s/r is simply a real number with no dimensions. (The dimensions "cancel out." *Note*: When we write $\theta = 5/2$ radians, we do not think of radians as a dimension, but we are indicating that we are interpreting the *ratio* 5/2 as the measure of an angle.)

Before carrying out the calculations for the radian measure in Figure 5(b), we have to arrange matters so that the same unit of length is used for the radius and the arc length. Converting the arc length s to centimeters, we have $s = 12$ cm, and, consequently,

$$\theta = \frac{s}{r} = \frac{12 \text{ cm}}{10 \text{ cm}} = \frac{6}{5}$$

Thus in Figure 5(b) we have $\theta = 6/5$ radians. Again, note that the radian measure s/r is dimensionless. (Check for yourself that the same answer is obtained if we use meters rather than centimeters for the common unit of measurement.)

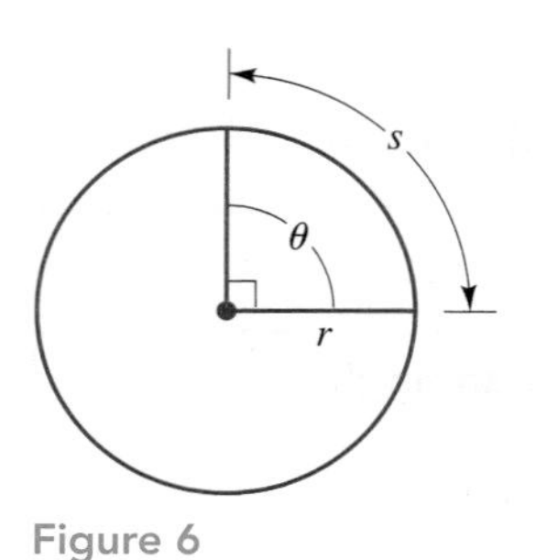

Figure 6

At first it might appear to you that the radian measure depends on the radius of the particular circle that we use. But as you will see, this is not the case. To gain some experience in working with the definition of radian measure, let's calculate the radian measure of the right angle in Figure 6. We begin with the formula $\theta = s/r$. Now, since θ is a right angle, the arc length s is one-quarter of the entire circumference. Thus

$$s = \frac{1}{4}(2\pi r) = \frac{\pi r}{2} \tag{2}$$

Using equation (2) to substitute for s in equation (1), we get

$$\theta = \frac{\pi r/2}{r} = \frac{\pi r}{2} \times \frac{1}{r} = \frac{\pi}{2}$$

We commonly write

$$90° = \frac{\pi}{2} \text{ radians} \tag{3}$$

So the right angle θ in Figure 6 has degree measure 90° and radian measure $\pi/2$ radians. (Notice that the radius r does not appear in our answer.)

For practical reasons we would like to be able to convert rapidly between degree and radian measure. Multiplying both sides of equation (3) by 2 yields

$$180° = \pi \text{ radians} \tag{4}$$

Equation (4) is useful and should be memorized. For instance, dividing both sides of equation (4) by 6 yields

$$\frac{180°}{6} = \frac{\pi}{6} \text{ radians}$$

or

$$30° = \frac{\pi}{6} \text{ radians}$$

Similarly, dividing both sides of equation (4) by 4 and 3, respectively, yields

$$45° = \frac{\pi}{4} \text{ radians} \qquad \text{and} \qquad 60° = \frac{\pi}{3} \text{ radians}$$

And multiplying both sides of equation (4) by 2 gives us

$$360° = 2\pi \text{ radians}$$

Figure 7 summarizes some of these results.

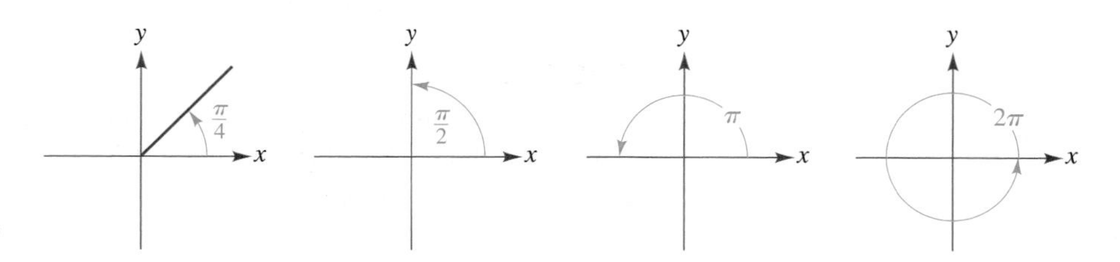

Figure 7

EXAMPLE 2 An important observation

(a) Express 1° in radian measure.
(b) Express 1 radian in terms of degrees.

SOLUTION
(a) We can solve equation (4) for 1 degree by dividing both sides by 180 to yield

$$1^\circ = \frac{\pi}{180} \text{ radian} \qquad (\approx 0.017 \text{ radian})$$

(b) Similarly, we can solve equation (4) for 1 radian by dividing both sides by π to yield

$$1 \text{ radian} = \frac{180^\circ}{\pi}$$

In other words, 1 radian is approximately 180°/3.14, or 57.3°.

From the results in Example 2 we have the following rules for converting between radians and degrees.

To convert from degrees to radians, multiply by $\pi/180^\circ$. To convert from radians to degrees, multiply by $180^\circ/\pi$.

EXAMPLE 3 Converting from degrees to radians

Convert 150° to radians.

SOLUTION

$$150^\circ\left(\frac{\pi}{180^\circ}\right) = \frac{5\pi}{6} \qquad \text{reducing the fraction}$$

Thus

$$150^\circ = \frac{5\pi}{6} \text{ radians}$$

EXAMPLE 4 Converting from radians to degrees

Convert $11\pi/6$ radians to degrees.

SOLUTION

$$\frac{11\pi}{6}\left(\frac{180^\circ}{\pi}\right) = \frac{(11)(180^\circ)}{6} = 11(30^\circ) = 330^\circ$$

So $11\pi/6$ radians $= 330^\circ$.

We saw in Example 2(b) that 1 radian is approximately 57°. It is also important to be able to visualize an angle of 1 radian without thinking in terms of degree measure. This is done as follows. In the equation $\theta = s/r$, we let $\theta = 1$. This yields $1 = s/r$, and, consequently, $r = s$. In other words, in any circle, 1 radian is the measure of the central angle that intercepts an arc equal in length to the radius of the circle. In the box that follows we summarize this result. Figure 8 displays angles of 1, 2, and 3 radians.

PROPERTY SUMMARY Radian Measure

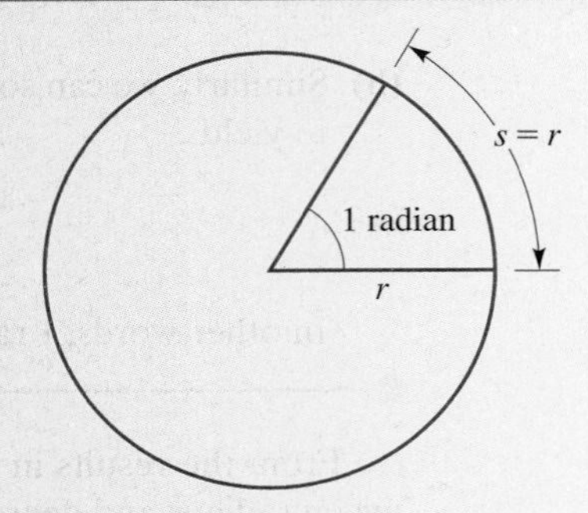

In a circle, 1 radian is the measure of the central angle that intercepts an arc equal in length to the radius of the circle.

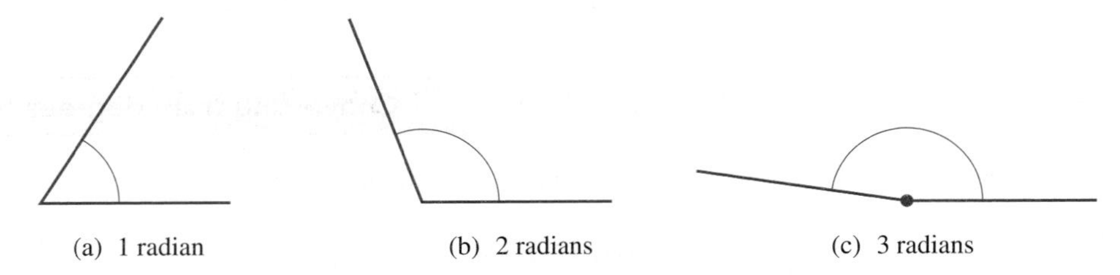

Figure 8

From now on, when we specify the measure of an angle, we will assume that the units are radians unless the degree symbol is explicitly used. (This convention is also used in calculus.) For instance, the equation $\theta = 2$ means that θ is 2 radians.

One of the advantages in using radian measure in precalculus and calculus is that many formulas then take on especially simple forms. The two basic formulas in the box that follows are examples of this.

PROPERTY SUMMARY Formulas for Arc Length and Sector Area

1. *Arc Length Formula*
   (Refer to Figure 9.) In a circle of radius r, the arc length s determined by a central angle of radian measure θ is given by

$$s = r\theta$$

In this formula it is assumed that s and r have the same linear units.

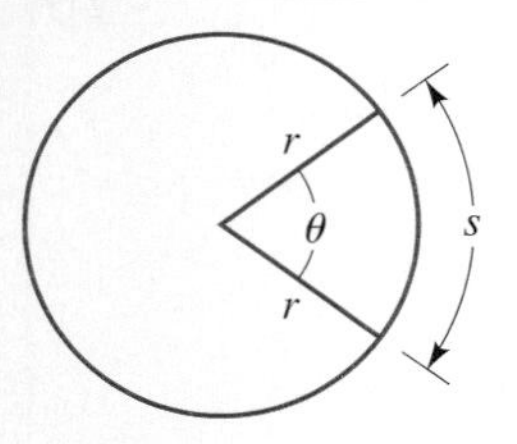

Figure 9

2. *Sector Area Formula*
(Refer to Figure 10, in which the shaded region is a *sector.*) In a circle of radius r, the area A of a sector with central angle of radian measure θ is given by

$$A = \frac{1}{2}r^2\theta$$

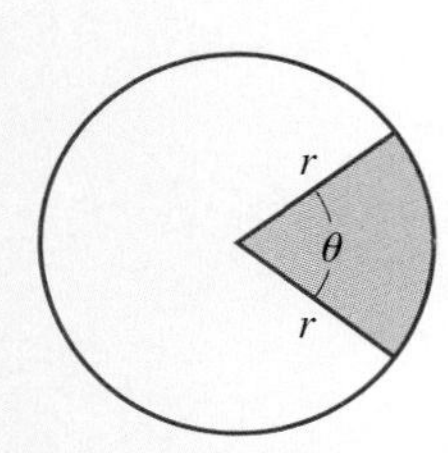

Figure 10

The arc length formula is a direct consequence of the definition of radian measure. Recall from the previous section that the defining equation for radian measure is $\theta = s/r$. Multiplying both sides of this equation by r yields $\theta r = s$; that is, $s = r\theta$, which is the arc length formula.

To derive the sector area formula, we'll rely on the following two facts from elementary geometry. First, the area of a circle of radius r is πr^2. Second, the area A of a sector is directly proportional to the measure θ of its central angle. (For example, if you double the angle, the area of the sector is doubled.) The statement that the area A is directly proportional to the angle θ is expressed algebraically as

$$A = k\theta \tag{5}$$

To determine the value of the constant k, we use the fact that when $\theta = 2\pi$, the sector is actually a full circle with area πr^2. Substituting the values $\theta = 2\pi$ and $A = \pi r^2$ in equation (5) yields

$$\pi r^2 = k(2\pi)$$

and therefore

$$k = \frac{1}{2}r^2$$

Using this last value for k in equation (5), we obtain $A = \frac{1}{2}r^2\theta$, as required. The next three examples indicate how the arc length formula and the sector area formula are used in calculations.

EXAMPLE 5 Calculating arc length

Compute the indicated arc lengths in Figures 11(a) and 11(b). Express the answers both in terms of π and as decimal approximations rounded to two decimal places.

SOLUTION

(a) We use the formula $s = r\theta$ with $r = 3$ m and $\theta = 4\pi/5$. This yields

$$s = (3\text{ m})\left(\frac{4\pi}{5}\right) = \frac{12\pi}{5}\text{ m}$$

Using a calculator, we find that this is approximately 7.54 m.

(b) To apply the formula $s = r\theta$, the angle measure θ must first be expressed in radians (because the formula is just a restatement of the definition of radian measure). We've seen previously that $30° = \pi/6$ radians. So we have

$$s = r\theta = (10\text{ cm})\left(\frac{\pi}{6}\right) = \frac{5\pi}{3}\text{ cm}$$

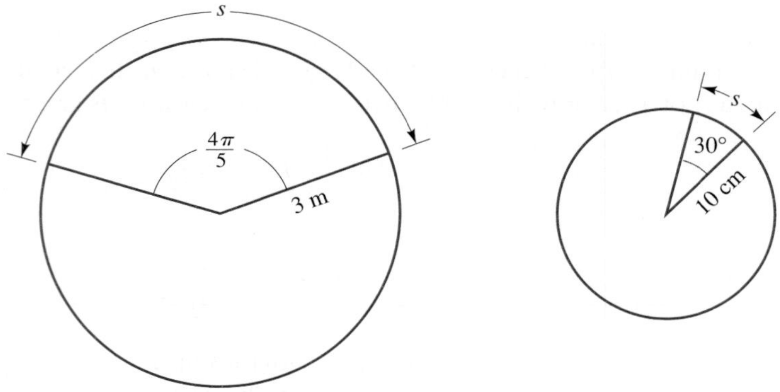

Figure 11 (a) (b)

Before picking up the calculator, note that we can obtain a quick approximation here: Since $\pi \approx 3$, we have $s \approx 5(3/3) = 5$ cm. With a calculator now, we obtain $s = \frac{5\pi}{3}$ cm ≈ 5.24 cm.

EXAMPLE 6 Calculating the area of a sector

Compute the area of the sector in Figure 12.

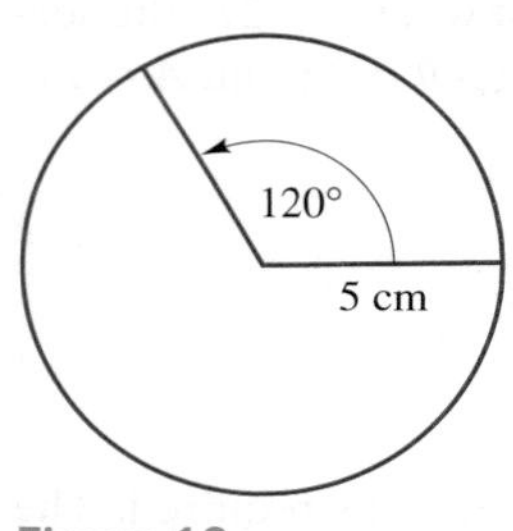

Figure 12

SOLUTION

We first convert 120° to radians:

$$120°\left(\frac{\pi}{180°}\right) = \frac{2\pi}{3} \text{ radians}$$

We then have

$$\begin{aligned} A &= \frac{1}{2}r^2\theta \\ &= \frac{1}{2}(5 \text{ cm})^2\left(\frac{2\pi}{3}\right) = \frac{25\pi}{3} \text{ cm}^2 \qquad (\approx 26.18 \text{ cm}^2) \end{aligned}$$

The next example indicates how radian measure is used in the study of rotating objects. As a prerequisite for this example, we first define the terms *angular speed* and *linear speed.*

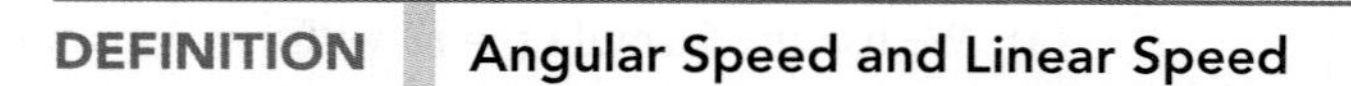

DEFINITION Angular Speed and Linear Speed

Suppose that a wheel rotates about its axis at a constant rate.

1. Refer to Figure 13(a) on the next page. If a radial line turns through an angle θ in time t, then the **angular speed** of the wheel [denoted by the Greek letter ω (*omega*)] is defined to be

$$\text{angular speed} = \omega = \frac{\theta}{t}$$

2. Refer to Figure 13(b). If a point P on the rotating wheel travels a distance d in time t, then the **linear speed** of P, denoted by v, is defined to be

$$\text{linear speed} = v = \frac{d}{t}$$

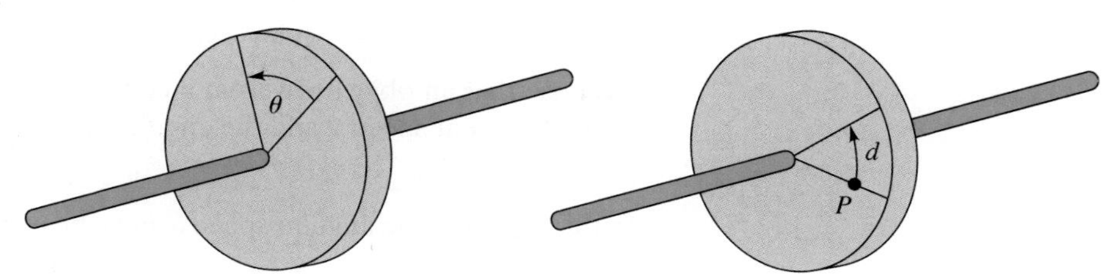

Figure 13

(a) A radial line turns through an angle θ in time t.

(b) The point P travels a distance d in time t.

EXAMPLE 7 Calculating angular and linear speed

A circular gear in a motor rotates at the rate of 100 rpm (revolutions per minute).

(a) What is the angular speed of the gear in radians per minute?

(b) Find the linear speed of a point on the gear 4 cm from the center.

SOLUTION

(a) Each revolution of the gear is 2π radians. So in 100 revolutions there are

$$\theta = 100(2\pi) = 200\pi \text{ radians}$$

Consequently, we have

$$\omega = \frac{\theta}{t} = \frac{200\pi \text{ radians}}{1 \text{ min}} = 200\pi \text{ radians/min}$$

(b) We can use the formula $s = r\theta$ to find the distance traveled by the point in 1 minute. Using $r = 4$ cm and $\theta = 200\pi$, we obtain

$$s = r\theta = 4(200\pi) = 800\pi \text{ cm}$$

The linear speed is therefore

$$v = \frac{d}{t} = \frac{800\pi \text{ cm}}{1 \text{ min}}$$
$$= 800\pi \text{ cm/min} \qquad (\approx 2513 \text{ cm/min})$$

There is a simple equation relating linear speed and angular speed. [Actually, as you'll see, the explanation we are about to give is just a repetition of the reasoning used in Example 7(b).] We refer back to Figure 13, in which the radial line turns through an angle θ in time t. Now we make two additional assumptions. First, we assume that the angle θ is measured in radians (so that we can apply the formula $s = r\theta$). Second, we assume that the point P is a distance r from the center of the circle. That is, the radius of the circular path is r. Then, according to our arc length formula, the distance d in Figure 13(b) is given by $d = r\theta$. So we have

$$v = \frac{d}{t} = \frac{r\theta}{t} = r \times \frac{\theta}{t}$$
$$= r\omega \qquad \text{using the definition of } \omega$$

We have now shown that $v = r\omega$. For reference we record this result in the box that follows.

Linear Speed and Angular Speed

Suppose that an object travels at a constant rate along a circular path of radius r. Then the linear speed v and the angular speed ω are related by the equation

$$v = r\omega$$

In this equation the angular speed ω must be expressed in radians per unit time.

We can use the formula $v = r\omega$ to rework Example 7(b) in a more concise fashion. Recall that we wanted to find the linear speed v of a point 4 cm from the center. To do this, we need only substitute the values $r = 4$ cm and $\omega = 200\pi$ radians/min [from part (a) of the example] in the formula $v = r\omega$. This yields

$$v = r\omega$$
$$= (4 \text{ cm})(200\pi \text{ radians/min}) \tag{6}$$
$$= 800\pi \text{ cm/min} \qquad \text{(as obtained previously)} \tag{7}$$

Notice how the units work in equations (6) and (7). Although radians appear in equation (6), they do not appear in the final answer in equation (7); the units for the linear speed are simply cm/min. We drop the radians from the final result because, as was pointed out earlier, radian measure is unitless.*

EXERCISE SET 6.1

A

In Exercises 1–4, use the definition $\theta = s/r$ to determine the radian measure of each angle.

1.

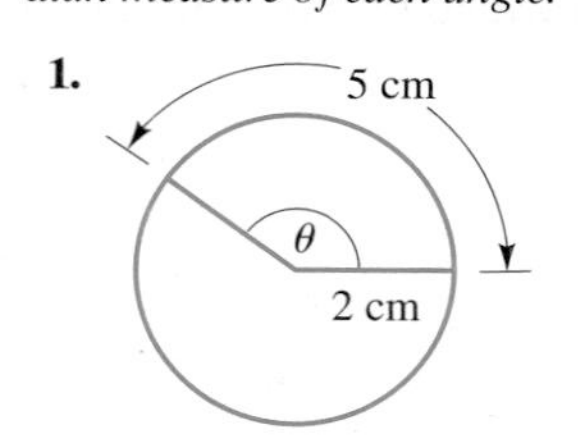

2.

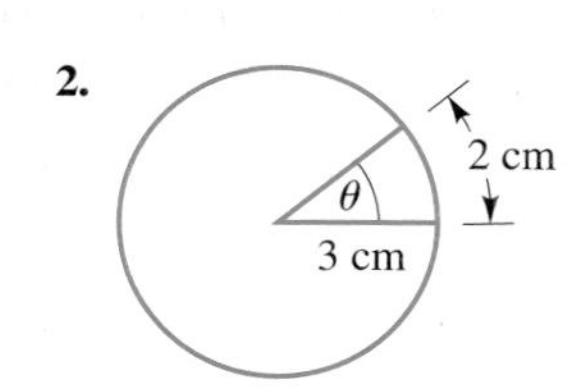

3.

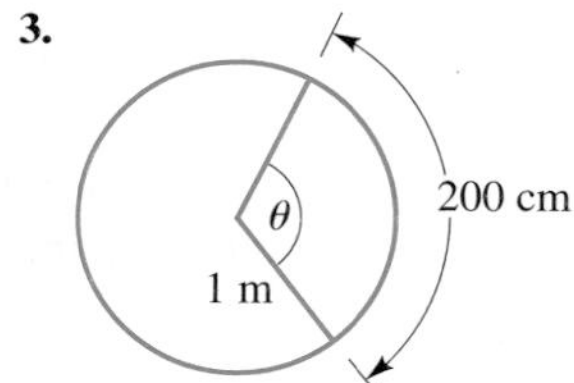

4.

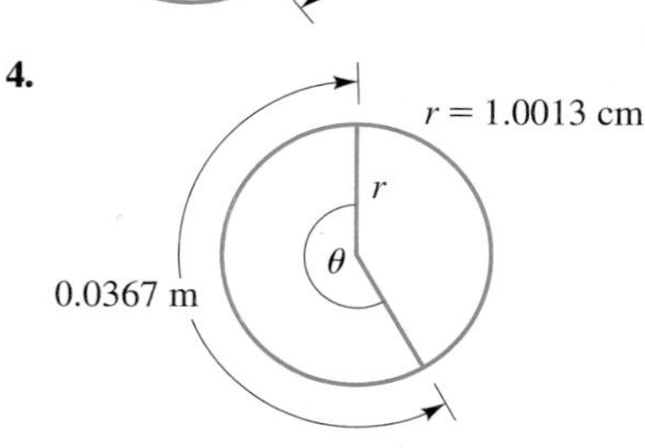

---

*Many physics texts gloss over this point in the calculations. One text that does provide sufficient emphasis here is *Physics,* 7th ed., by John D. Cutnell and Kenneth W. Johnson (New York: John Wiley & Sons, Inc., 1995). In that text, the authors state, "In calculations, therefore, the radian is treated as a unitless number and has no effect on other units that it multiplies or divides."

In Exercises 5–8, convert to radian measure. Express your answers both in terms of π and as decimal approximations rounded to two decimal places.

5. (a) 45° **(b)** 90° **(c)** 135°
6. (a) 30° **(b)** 150° **(c)** 300°
7. (a) 0° **(b)** 360° **(c)** 450°
8. (a) 36° **(b)** 35° **(c)** 720°

In Exercises 9–12, convert the radian measures to degrees.

9. (a) $\pi/12$ **(b)** $\pi/6$ **(c)** $\pi/4$
10. (a) 3π **(b)** $3\pi/2$ **(c)** 2π
11. (a) $\pi/3$ **(b)** $5\pi/3$ **(c)** 4π
12. (a) $5\pi/6$ **(b)** $11\pi/6$ **(c)** 0

In Exercises 13 and 14, convert the radian measures to degrees. Round the answers to two decimal places.

13. (a) 2 **(b)** 3 **(c)** π^2
14. (a) 1.32 **(b)** 0.96 **(c)** $1/\pi$

15. Suppose that the radian measure of an angle is 3/2. Without using a calculator or tables, determine if this angle is larger or smaller than a right angle. *Hint*: What is the radian measure of a right angle?

16. Two angles in a triangle have radian measure $\pi/5$ and $\pi/6$. What is the radian measure of the third angle?

For Exercises 17 and 18, refer to the following figure, which shows all of the angles from 0° to 360° that are multiples of 30° or 45°.

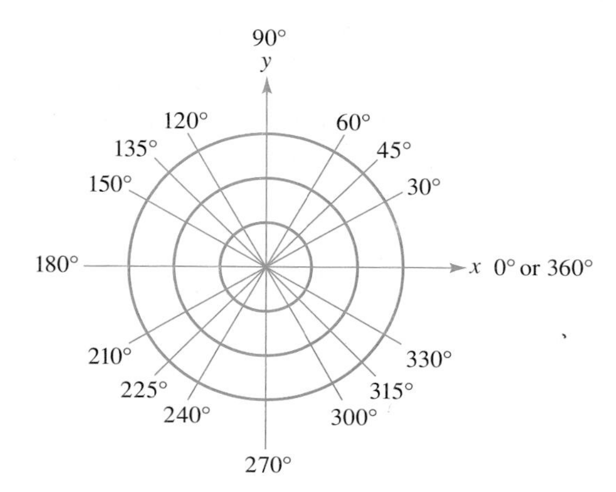

17. In the figure, relabel the angles in Quadrants I and II using radian measure.

18. In the figure, relabel the angles in Quadrants III and IV using radian measure.

In Exercises 19–22, find the arc length s in each case.

19.

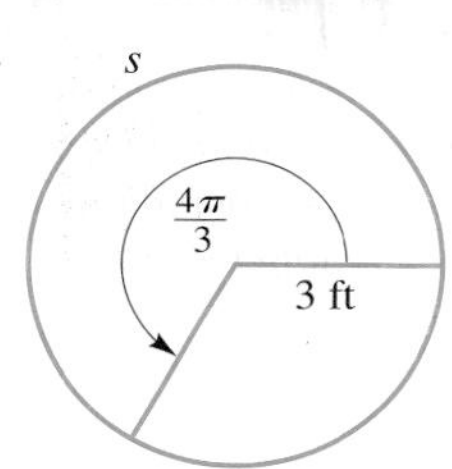

20.

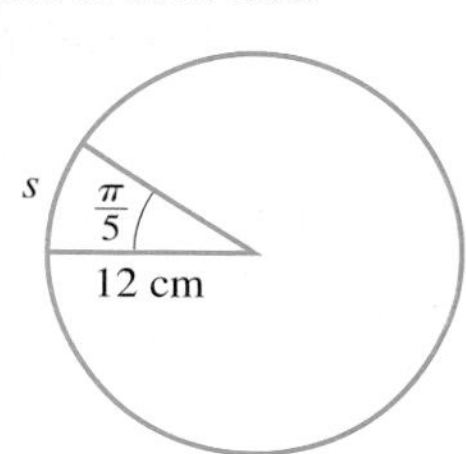

21.

45°
2 cm
s

22.

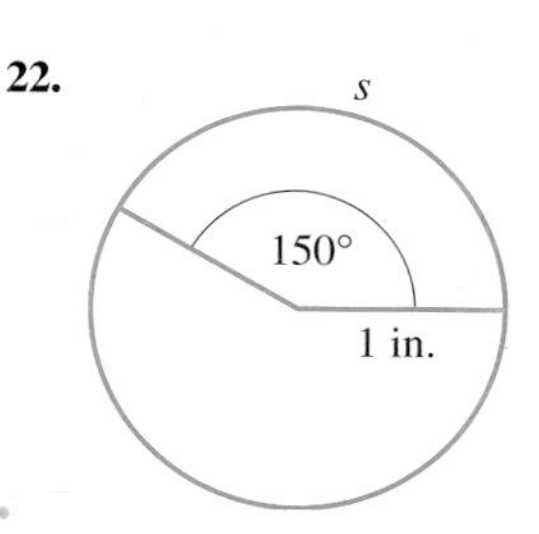

In Exercises 23 and 24, find the area of the sector determined by the given radius r and central angle θ. Express the answer both in terms of π and as a decimal approximation rounded to two decimal places.

23. (a) $r = 6$ cm; $\theta = 2\pi/3$ **(c)** $r = 24$ m; $\theta = \pi/20$
(b) $r = 5$ m; $\theta = 80°$ **(d)** $r = 1.8$ cm; $\theta = 144°$

24. (a) $r = 4$ cm; $\theta = \pi/10$ **(c)** $r = 21$ ft; $\theta = 11\pi/6$
(b) $r = 16$ m; $\theta = 5°$ **(d)** $r = 4.2$ in.; $\theta = 170°$

25. In a circle of radius 1 cm, the area of a certain sector is $\pi/5$ cm^2. Find the radian measure of the central angle. Express the answer in terms of π rather than as a decimal approximation.

26. In a circle of radius 3 m, the area of a certain sector is 20 m^2. Find the degree measure of the central angle. Round the answer to two decimal places.

In Exercises 27 and 28, find **(a)** *the perimeter of the sector; and* **(b)** *the area of the sector. In each case, use a calculator to evaluate the answer and round to two decimal places.*

27.

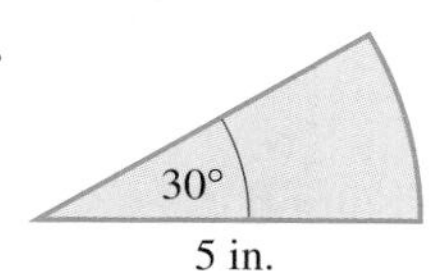

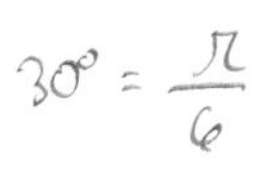

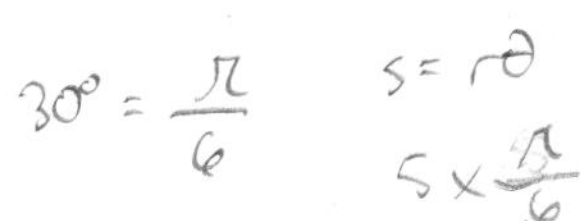

28.

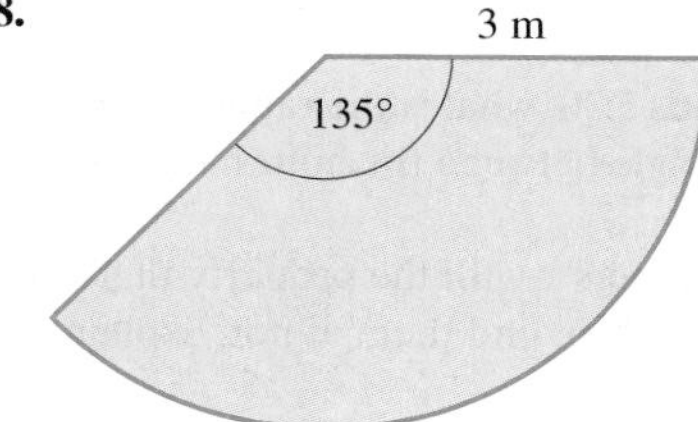

In Exercises 29–34, you are given the rate of rotation of a wheel as well as its radius. In each case, determine the following: **(a)** *the angular speed, in units of radians/sec;* **(b)** *the linear speed, in units of cm/sec, of a point on the circumference of the wheel; and* **(c)** *the linear speed, in cm/sec, of a point halfway between the center of the wheel and the circumference.*

29. 6 revolutions/sec; $r = 12$ cm
30. 15 revolutions/sec; $r = 20$ cm
31. 1080°/sec; $r = 25$ cm
32. 2160°/sec; $r = 60$ cm
33. 500 rpm; $r = 45$ cm
34. 1250 rpm; $r = 10$ cm
35. For this problem, assume that the earth is a sphere with a radius of 3960 miles and a rotation rate of 1 revolution per 24 hours.
(a) Find the angular speed. Express your answer in units of radians/sec, and round to two significant digits.
(b) Find the linear speed of a point on the equator. Express the answer in units of miles per hour, and round to the nearest 10 mph.
36. A wheel 3 ft in diameter makes x revolutions. Find x, given that the distance traveled by a point on the circumference of the wheel is 22619 ft. (Round your answer to the nearest whole number.)

B

37. Suppose that you have two sticks and a piece of wire, each of length 1 ft, fastened at the ends to form an equilateral triangle; see Figure A. If side $\overline{BC}$ is bent out to form an arc of a circle with center A, then the angle at A will decrease from 60° to something less. See Figure B. What is the measure of this new angle at A in both radians and degrees?

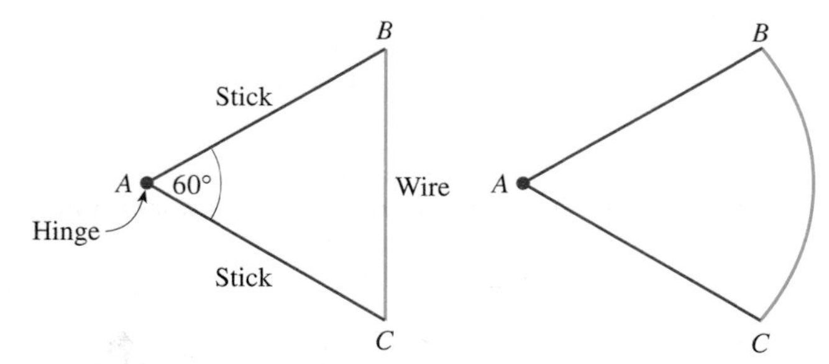

Figure A Figure B

38. (a) When a clock reads 4:00, what is the radian measure of the (smaller) angle between the hour hand and the minute hand?
(b) When a clock reads 5:30, what is the radian measure of the (smaller) angle between the hour hand and the minute hand?
39. Are there any real numbers x with the property that x degrees equals x radians? If so, find them; if not, explain why not.
40. Are there any real numbers x with the property that x degrees equals $2x$ radians? If so, find them; if not, explain why not.

In Exercises 41 and 42, suppose that a belt drives two wheels of radii r and R, as indicated in the figure.

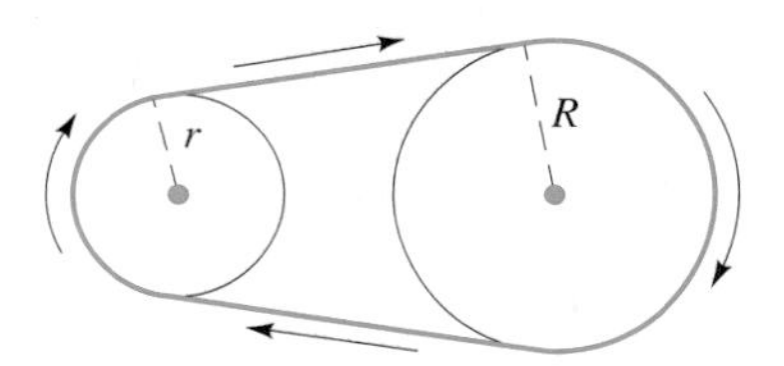

41. If $r = 6$ cm, $R = 10$ cm, and the angular speed of the larger wheel is 100 rpm, determine each of the following:
(a) the angular speed of the larger wheel in radians per minute;
(b) the linear speed of a point on the circumference of the larger wheel;
(c) the angular speed of the smaller wheel in radians per minute. *Hint*: Because of the belt, the linear speed of a point on the circumference of the larger wheel is equal to the linear speed of a point on the circumference of the smaller wheel.
(d) The angular speed of the smaller wheel in rpm.
42. Follow Exercise 41, assuming that $r = 5$ cm, $R = 15$ cm, and the angular speed of the larger wheel is 1800 rpm.

The latitude of a point P on the surface of the Earth is specified by means of the angle θ in the figure. For instance, the latitude of Paris, France, is 48°52′ N. The letter N is used here to indicate that the location is north of, rather than south of, the equator. (Recall that the notation 52′ indicates 52/60 of one degree.) In Exercises 43–48, use the arc length formula (and your calculator) to determine the distance $\widehat{PE}$ from the given location P to the equator. Assume that the Earth is a sphere with radius OP = OE = 3960 miles. Round each answer to the nearest 10 miles.

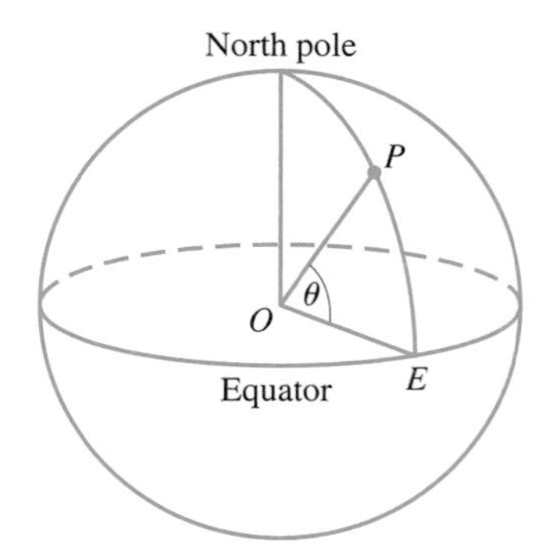

43. Point Barrow, Alaska: 71°23′ N
44. Singapore: 1°17′ N

45. Honolulu: 21°19′ N
46. Lagos: 6°27′ N
47. Washington, D.C.: 38°54′ N
48. Fairbanks, Alaska: 64°51′ N

Exercises 49 and 50 provide geometric results that you will need in working Exercises 51–54. For Exercise 50, you need to know that a ***segment*** *of a circle is the region bounded by an arc of the circle and its chord. In the accompanying figure, the red region is a segment. (The white region also is a segment.)*

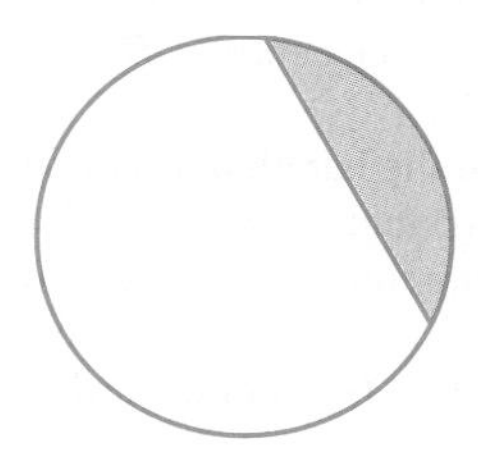

49. Show that the area of an equilateral triangle of side s is given by

$$\frac{\sqrt{3}}{4}s^2$$

Hint: Draw an altitude and use the Pythagorean theorem. (This exercise does not require any knowledge of radian measure.)

50. In the accompanying figure, $\triangle ABC$ is equilateral and s denotes the length of each side. The arc in the figure is a portion of a circle with center A and radius $AB = s$. Use the result in Exercise 49 and a formula from this section to show that the area of the shaded segment in the figure is given by

$$s^2\left(\frac{2\pi - 3\sqrt{3}}{12}\right)$$

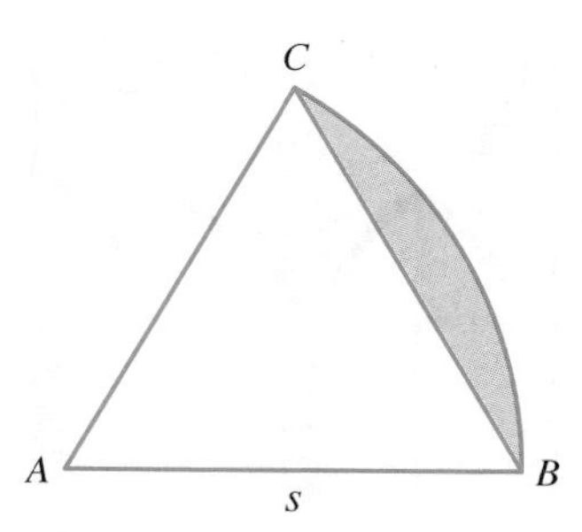

51. Many of the window designs used in gothic architecture involve circles, sectors, and segments of circles. The **equilateral arch** in Figure A is an example of this. Figure B shows how the arch is designed. Starting with the equilateral triangle ABC, circular arc $\widehat{AC}$ is drawn with center B and radius AB. Similarly, circular arc $\widehat{BC}$ is drawn with center A and radius AB.

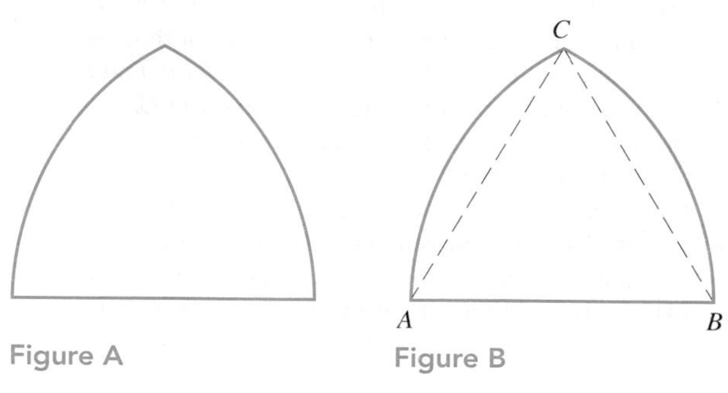

Figure A Figure B

(a) Let s denote the length of a side of the equilateral triangle in Figure B. Express the perimeter of the equilateral arch in terms of s.

(b) Express the area of the equilateral arch in terms of s.

52. Closely related to the equilateral arch in Exercise 51 is the **equilateral curved triangle** shown in Figure C. Just as with the equilateral arch in Exercise 51, the design begins with the equilateral triangle ABC. Circular arcs are then constructed on each side of the triangle, following the method explained in Exercise 51.

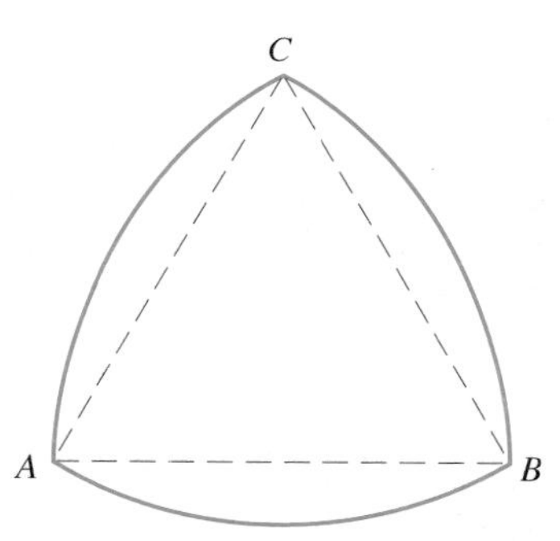

Figure C

(a) In Figure C, let s denote the length of a side of the equilateral triangle ABC. Express the perimeter and the area of the equilateral curved triangle in terms of s.

(b) Show that the area of the equilateral triangle ABC is approximately 61% of the area of the equilateral curved triangle ABC.

53. Figure D shows one of the gothic window designs used in Wells Cathedral in England. (The cathedral was constructed in the mid-thirteenth century.) Figure E indicates how the design is formed. We start with the equilateral $\triangle ABC$ and construct the equilateral arch ABC. Next, the midpoints of the three sides of $\triangle ABC$ are joined to create four smaller equilateral triangles. The two equilateral triangles ADF and FEC are then used to construct the two smaller equilateral arches shown in Figure E. And finally, the equilateral triangle DBE is used to construct the equilateral curved triangle within the top half of the figure.

(a) Let s denote the length of a side of the equilateral triangle ABC. Express the area of each of the equilateral arches ABC and ADF in terms of s. Also, find the ratio of the area of arch ADF to arch ABC.

(b) Express the area of the equilateral curved triangle DBE in terms of s.

(c) Express the area of the curved figure DEF in terms of s. (By "the curved figure DEF" we mean the region bounded by the circular arcs $\widehat{DE}$, $\widehat{EF}$, and $\widehat{FD}$.)

(d) Express the area of the curved figure BEC in terms of s.

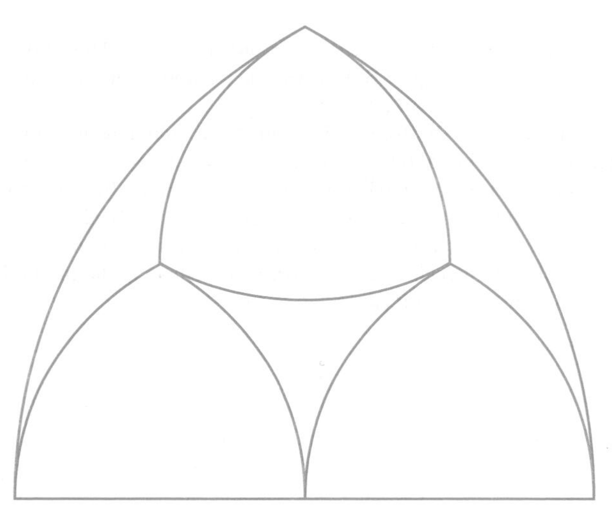

Figure D

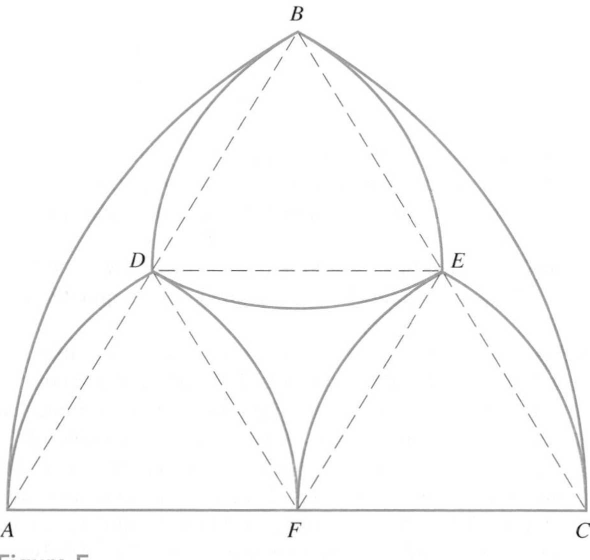

Figure E

54. Figure F shows a gothic window design from the cathedral at Reims, France. (The cathedral was constructed during the years 1211–1311.) Figure G indicates how the design is formed. Triangle ABC is equilateral and arch ABC is the corresponding equilateral arch. Equilateral triangle GID, congruent to triangle ABC, is constructed such that D is the midpoint of $\overline{AB}$, and $\overline{GI}$ is parallel to $\overline{AB}$. The points E and F are the midpoints of segments $\overline{DG}$ and $\overline{DI}$, respectively. The two small arches at the bottom of the figure are equilateral arches corresponding to the equilateral triangles GHE and HIF. For the circle in the figure, the center is D and the radius is $AD(= ED = FD)$.

(a) Let s denote the length of a side in each of the two equilateral triangles ABC and GID. Find the area and the perimeter (in terms of s) of the curved figure $AJBCA$. *Hint*: For the area, subtract the area of the semicircle AJB from the area of the equilateral arch ABC.

(b) Express (in terms of s) the area and the perimeter of equilateral arch GHE.

(c) Show that the area of the curved figure EHF is $s^2(2\sqrt{3} - \pi)/8$.

(d) Express (in terms of s) the perimeter and the area of the curved figure FBI.

Figure F
A window design in the cathedral at Reims, France. From Hans H. Hofstätter, *Living Architecture: Gothic* (Office du Livre, S. A., 1970)

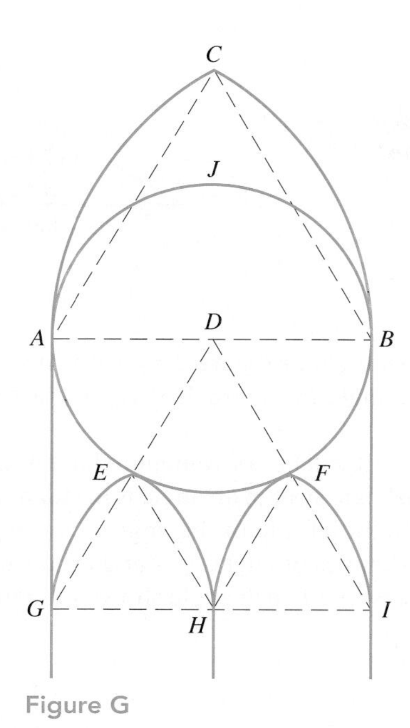

Figure G

55. The accompanying figure shows a circular sector with radius r cm and central angle θ (radian measure). The perimeter of the sector is 12 cm.

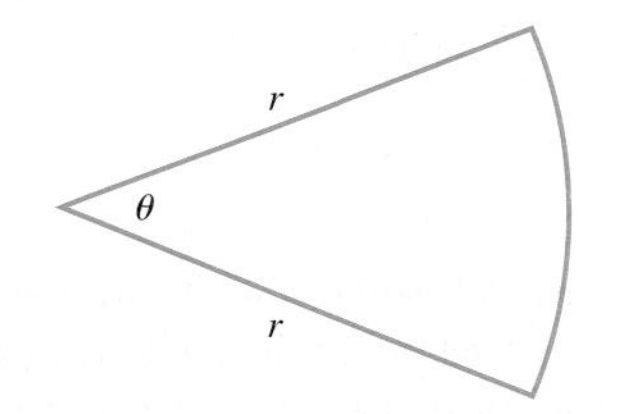

(a) Express r as a function of θ.
(b) Express the area A of the sector as a function of θ. Is this a quadratic function?
(c) Express θ as a function of r.
(d) Express the area A of the sector as a function of r. Is this a quadratic function?
(e) For which value of r is the area A a maximum? What is the corresponding value of θ in this case?

56. The following figure shows a semicircle of radius 1 unit and two adjacent sectors, AOC and COB.
(a) Show that the product P of the areas of the two sectors is given by

$$P = \frac{\pi\theta}{4} - \frac{\theta^2}{4}$$

Is this a quadratic function?
(b) For what value of θ is P a maximum?

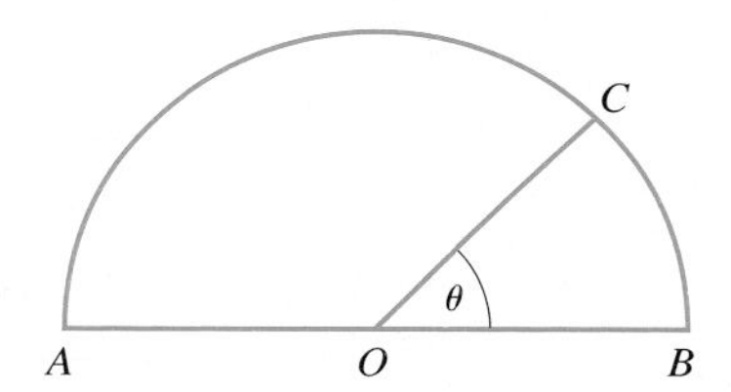

6.2 TRIGONOMETRIC FUNCTIONS OF ANGLES

. . . *many natural phenomena are repetitive or cyclical—for example, the motion of the planets in our solar system, earthquake vibrations, and the natural rhythm of the heart. Thus, the* [sine and cosine] *functions introduced in this chapter add considerably to our capacity to describe physical processes.*—Professor Larry J. Goldstein in his text *Calculus and Its Applications,* 7th ed. (Upper Saddle River, New Jersey: Prentice Hall, 1996)

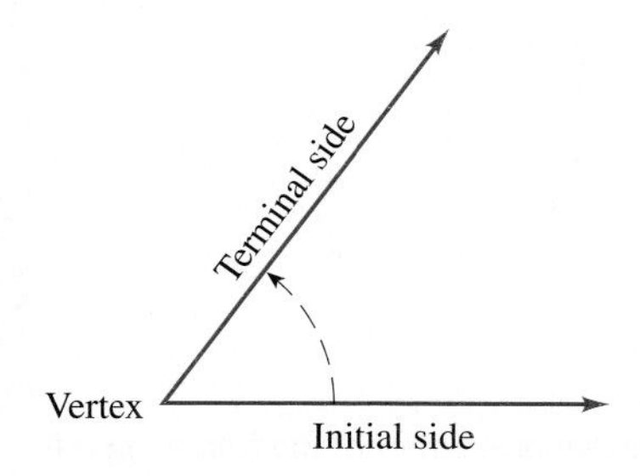

Figure 1

We began the previous section by saying that an angle is a figure formed by two rays with a common endpoint, the vertex. Now, for analytical purposes, it is useful to think of the two rays that form the angle as having been originally coincident; then, while one ray is held fixed, the other is rotated to create the given angle. As Figure 1 indicates, the fixed ray is called the **initial side** of the angle, and the rotated ray is the **terminal side.** By convention, we take the measure of an angle to be **positive** if the rotation is counterclockwise (as in Figure 1) and **negative** if the rotation is clock-

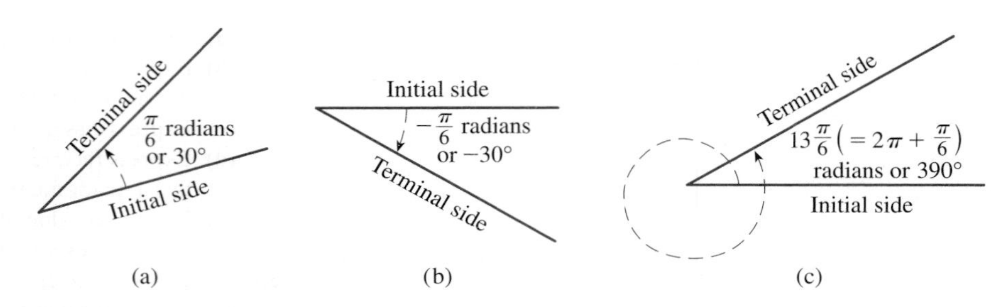

Figure 2
Angles generated by a counterclockwise rotation have positive measure. Angles generated by a clockwise rotation have negative measure.

wise. For example, the measure of the angle in Figure 2(a) is positive $\pi/6$ radian (or 30°), whereas the measure of the angle in Figure 2(b) is negative $\pi/6$ radian (or −30°).

In our development of trigonometry it will be convenient to have a *standard position* for angles. In a rectangular coordinate system an angle is in **standard position** if the vertex is located at (0, 0) and the initial side of the angle lies along the positive horizontal axis. Figure 3 shows examples of angles in standard position. *Question for review*: What is the degree measure of each angle shown in Figure 3?

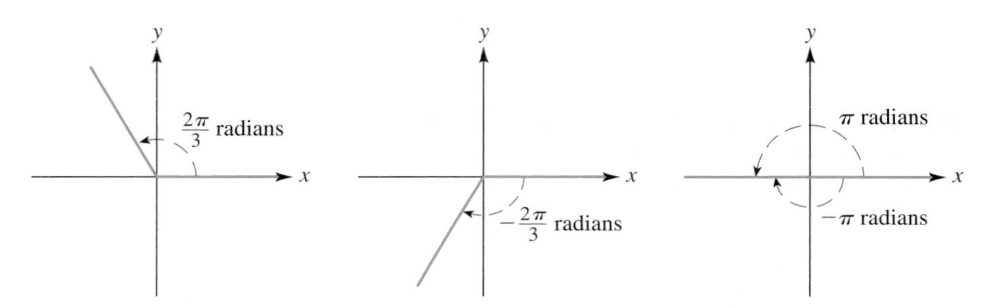

Figure 3
Examples of angles in standard position

We are going to define the six *trigonometric functions.* As indicated by the title of this section, the inputs for these functions will be angles. The outputs will be real numbers. In Chapter 7 we'll make an important transition to real-number inputs. But for now, as we've said, the inputs will be angles.

For reasons that are more historical* than mathematical, the trigonometric functions have names that are words rather than single letters such as *f*. Using words rather than single letters for naming functions is by no means peculiar to trigonometry. For example, $\log_b x$ denotes the logarithmic function with base b. The names of the six trigonometric functions, along with their abbreviations, are as follows.

NAME OF FUNCTION	ABBREVIATION
cosine	cos
sine	sin
tangent	tan
secant	sec
cosecant	csc
cotangent	cot

*For a discussion of the names of the trigonometric functions, see either of the following references: Howard Eves, *An Introduction to the History of Mathematics,* 6th ed., pp. 236–237 (Philadelphia: Saunders College Publishing, 1990); D. E. Smith, *History of Mathematics,* Vol. II pp. 614–622 (New York: Dover Publications, Inc., 1953).

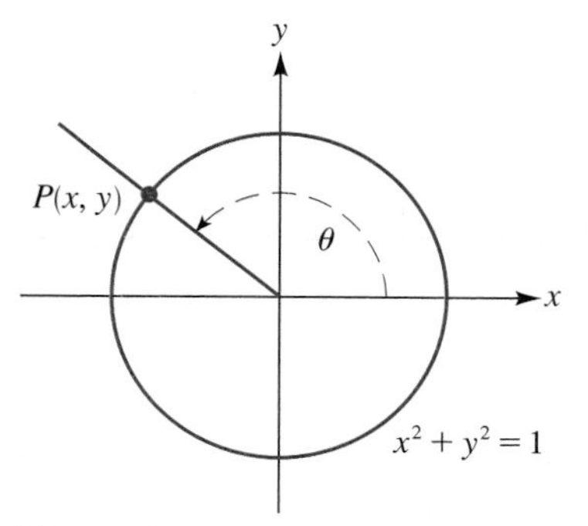

Figure 4
$P(x, y)$ denotes the point where the terminal side of angle θ intersects the unit circle.

To define the trigonometric functions, we begin by placing the angle θ in standard position and drawing in the **unit circle** $x^2 + y^2 = 1$, as shown in Figure 4. (Recall from Chapter 1 that the equation $x^2 + y^2 = 1$ represents the circle of radius 1, with center at the origin.) Notice the notation $P(x, y)$ in Figure 4; this stands for the point P, with coordinates (x, y), where the terminal side of angle θ intersects the unit circle. With this notation, we define the six trigonometric functions of θ as follows.

DEFINITION **Trigonometric Functions of Angles**

$$\cos\theta = x \qquad \sec\theta = \frac{1}{x} \quad (x \neq 0)$$

$$\sin\theta = y \qquad \csc\theta = \frac{1}{y} \quad (y \neq 0)$$

$$\tan\theta = \frac{y}{x} \quad (x \neq 0) \qquad \cot\theta = \frac{x}{y} \quad (y \neq 0)$$

Unit circle $x^2 + y^2 = 1$

Much of our subsequent work in trigonometry will be devoted to exploring the consequences of these definitions. Two initial observations that will help you in memorizing the definitions are these:

1. $\cos\theta$ is the first coordinate of the point where the terminal side of angle θ intersects the unit circle; $\sin\theta$ is the second coordinate. (You can remember this by noting that, alphabetically, cosine comes before sine.)
2. There are three pairs of reciprocals in the definitions: $\cos\theta$ and $\sec\theta$; $\sin\theta$ and $\csc\theta$; $\tan\theta$ and $\cot\theta$.

EXAMPLE 1 **Using the unit circle to calculate trigonometric functions of $\pi/2$ radians**

(a) Evaluate the trigonometric functions of $\pi/2$ radians. That is, determine $\cos(\pi/2)$, $\sin(\pi/2)$, $\tan(\pi/2)$, $\sec(\pi/2)$, $\csc(\pi/2)$, and $\cot(\pi/2)$.
(b) Evaluate $\cos 90^\circ$ and $\sin 90^\circ$.

SOLUTION

(a) We place the angle $\theta = \pi/2$ in standard position. Then, as indicated in Figure 5, the terminal side of the angle meets the unit circle at the point $(0, 1)$. Now we apply the definitions.

$(0, 1)$

By definition, $\cos(\pi/2)$ is this number. — By definition, $\sin(\pi/2)$ is this number.

Thus,

$$\cos\frac{\pi}{2} = 0 \qquad \text{and} \qquad \sin\frac{\pi}{2} = 1$$

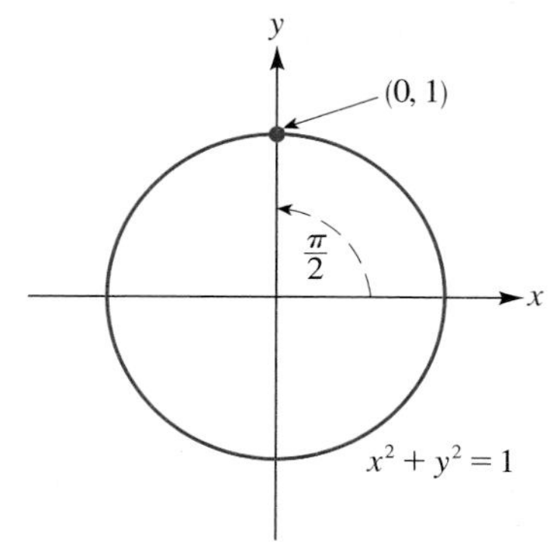

Figure 5

For the remaining trigonometric functions of $\pi/2$, we have

$$\tan\frac{\pi}{2} = \frac{y}{x} = \frac{1}{0} \quad \text{undefined}$$
$$\sec\frac{\pi}{2} = \frac{1}{x} = \frac{1}{0} \quad \text{undefined}$$
$$\csc\frac{\pi}{2} = \frac{1}{y} = \frac{1}{1} = 1$$
$$\cot\frac{\pi}{2} = \frac{x}{y} = \frac{0}{1} = 0$$

(b) Because $90° = \pi/2$ radians, we've already found the required values in part (a). That is,

$$\cos 90° = \cos(\pi/2) = 0 \qquad \text{and} \qquad \sin 90° = \sin(\pi/2) = 1$$

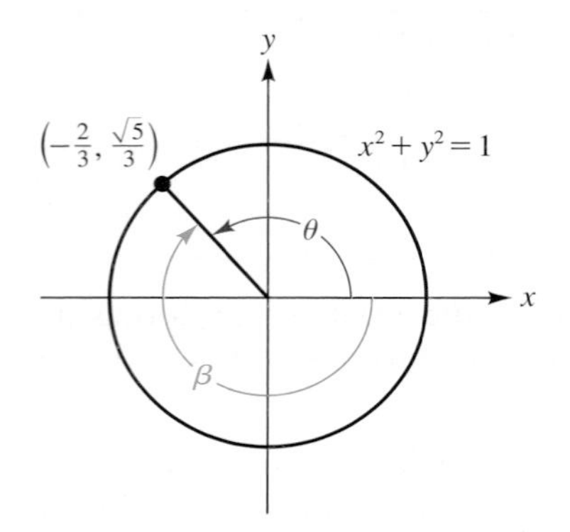

Figure 6

As preparation for the next example take a look at Figure 6. The figure shows the point $(-\frac{2}{3}, \frac{\sqrt{5}}{3})$ on the unit circle. How do we know for certain that this point really lies on the unit circle (aside from the fact that the picture is drawn that way)? Although this may seem to you a frivolous question, it focuses on the fundamental connection between graphs and equations in the *xy*-plane: A point is on the graph of an equation if and only if the coordinates of the point satisfy that equation. Thus, to verify that the point $(-\frac{2}{3}, \frac{\sqrt{5}}{3})$ in Figure 6 actually lies on the unit circle, we need to check that the pair of coordinates $x = -2/3$ and $y = \sqrt{5}/3$ satisfies the equation $x^2 + y^2 = 1$. Substituting these coordinates for x and y in the equation yields

$$\left(-\frac{2}{3}\right)^2 + \left(\frac{\sqrt{5}}{3}\right)^2 = \frac{4}{9} + \frac{5}{9} = 1$$

So the given point $(-\frac{2}{3}, \frac{\sqrt{5}}{3})$ indeed lies on the unit circle.

EXAMPLE 2 Using the unit circle to calculate trigonometric functions of an angle

Refer to Figure 6.
(a) Specify $\cos\theta$, $\sin\theta$, and $\tan\theta$.
(b) Specify $\cos\beta$, $\sin\beta$, and $\tan\beta$.

SOLUTION

(a) This is another exercise in using the definitions of the trigonometric functions. Since the point $(-\frac{2}{3}, \frac{\sqrt{5}}{3})$ lies on the unit circle, we have

$$\cos\theta = x = -\frac{2}{3} \qquad \sin\theta = y = \frac{\sqrt{5}}{3}$$

and

$$\tan\theta = \frac{y}{x} = \frac{\sqrt{5}/3}{-2/3} = -\frac{\sqrt{5}}{2}$$

(b) Again, using the definitions, we have

$$\cos\beta = -\frac{2}{3} \qquad \sin\beta = \frac{\sqrt{5}}{3}$$

and

$$\tan \beta = \frac{\sqrt{5}/3}{-2/3} = -\frac{\sqrt{5}}{2}$$

Note that these values are identical to the corresponding values obtained in part (a). This is because the angles θ and β have the same terminal side, and the values of the trigonometric functions depend only upon where the terminal side of the angle intersects the unit circle.

Together, the two angles θ and β in Figure 6 are an example of a pair of *coterminal angles,* that is, angles with a common terminal side. As indicated in Example 2, the corresponding trigonometric values are always identical for a pair of coterminal angles. The next example uses this observation.

EXAMPLE 3 Calculating trigonometric functions of coterminal angles

Evaluate the trigonometric functions of:
(a) $-\pi$; **(b)** π.

SOLUTION

(a) As indicated in Figure 7, the terminal side of $-\pi$ radians meets the unit circle at the point $(-1, 0)$. Applying the definitions, then, we have

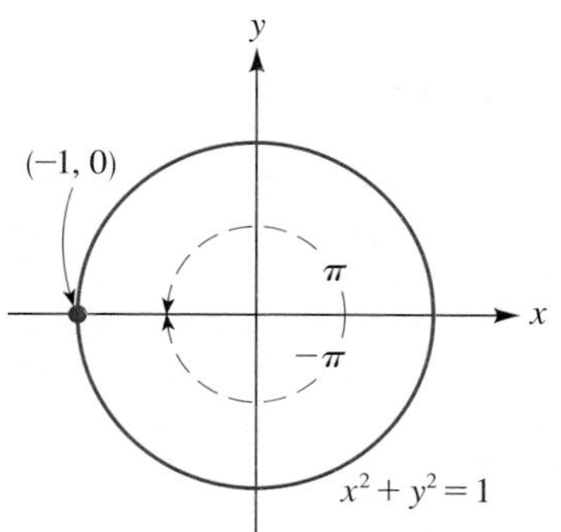

Figure 7

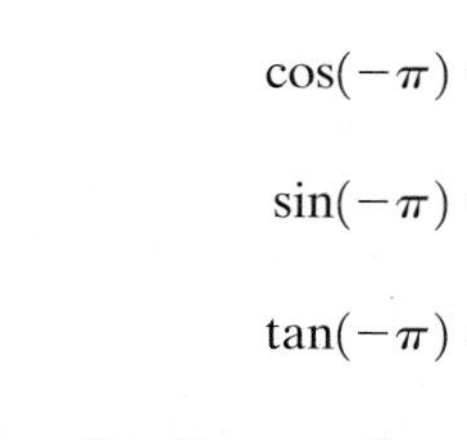

$$\cos(-\pi) = x = -1 \qquad \sec(-\pi) = \frac{1}{x} = \frac{1}{-1} = -1$$

$$\sin(-\pi) = y = 0 \qquad \csc(-\pi) = \frac{1}{y} = \frac{1}{0} \quad \text{undefined}$$

$$\tan(-\pi) = \frac{y}{x} = \frac{0}{-1} = 0 \qquad \cot(-\pi) = \frac{x}{y} = \frac{-1}{0} \quad \text{undefined}$$

(b) Because the angles π and $-\pi$ are coterminal, their corresponding trigonometric values are identical and we have [using the results in part (a)]

$$\begin{array}{ll} \cos \pi = \cos(-\pi) = -1 & \sec \pi = \sec(-\pi) = -1 \\ \sin \pi = \sin(-\pi) = 0 & \csc \pi \quad \text{undefined} \\ \tan \pi = \tan(-\pi) = 0 & \cot \pi \quad \text{undefined} \end{array}$$

Note: Instead of relying on part (a), we could have obtained the answers for part (b) directly by referring to Figure 7 and applying the unit-circle definitions to the angle of π radians. However, the point here was simply to make use of the idea of coterminal angles.

Using the method shown in Examples 1 and 3, we can easily evaluate the trigonometric functions for any angle that is an integral multiple of $\pi/2$ (90°). Table 1 shows the results of such calculations. Exercise 25 asks you to make these calculations for yourself. *Question for review*: What is the corresponding degree measure for each entry in the first column of Table 1?

You can use a calculator to verify the results in Examples 1 and 3. Better yet, you can use those results to make sure you know how to operate your calculator with respect to the trigonometric functions. For the moment, set your calculator to

TABLE 1

θ	$\cos\theta$	$\sin\theta$	$\tan\theta$	$\sec\theta$	$\csc\theta$	$\cot\theta$
0	1	0	0	1	undefined	undefined
$\pi/2$	0	1	undefined	undefined	1	0
π	-1	0	0	-1	undefined	undefined
$3\pi/2$	0	-1	undefined	undefined	-1	0
2π	1	0	0	1	undefined	undefined

radian mode. (If necessary, consult the user's manual on this point.) Now suppose, for example, that you want to use the calculator to evaluate $\cos\pi$. For most calculators you enter the name of the function before the input. On these calculators, the usual keystrokes for evaluating $\cos\pi$ are as follows:

[cos] π [ENTER]   name of function before input

Use your calculator now to compute $\cos\pi$. As you know from Example 3(b), the answer should be -1. (If you don't get this result, it may be that your calculator was not set to the radian mode.)

No matter which type of calculator you have, you'll find that the keystrokes for parentheses can be crucial. Suppose, for instance that you want to use the calculator to compute $\sin(\pi/2)$. The usual keystrokes are

[sin] [(] π [÷] 2 [)] [ENTER]

Use your own calculator now to evaluate $\sin(\pi/2)$. As you know from Example 1, the answer should be 1. If you do not use the parentheses in this case, the calculator will misinterpret your intentions and evaluate something other than $\sin(\pi/2)$. (Exercise 91 provides details.) As a rule of thumb, when in doubt use parentheses. *Note*: If your calculator does not give the correct answers with the keystrokes described in the previous discussion, you should consult the section in the user's manual on evaluating trigonometric functions.

EXAMPLE 4 Determining the sign of a value of a trigonometric function

First, without using a calculator, determine whether the given value is positive or negative. Then use a calculator to verify the answer.

(a) $\cos 3$ **(b)** $\sin 1$ **(c)** $\tan 6$

SOLUTION

(a) Since π radians is 180°, we estimate that 3 radians is slightly less than 180° (because 3 is slightly less than π). Thus, in standard position, the terminal side of an angle of 3 radians lies in the second quadrant. Therefore, $\cos 3$ is negative. Indeed, a calculator check shows that $\cos 3 \approx -0.99$.

(b) One radian is approximately 57° (as we saw in Example 2 in the previous section). So, in standard position, the terminal side of an angle lies in Quadrant I. Thus, $\sin 1$ is positive. A calculator check shows that $\sin 1 \approx 0.84$.

(c) 6 radians is slightly less than 2π radians. [Reason: $2\pi \approx 2(3.14) = 6.28$.] Now, 2π radians is one complete revolution, or 360°, and therefore 6 radians is slightly less than 360°. Consequently, in standard position, the terminal side of an angle of 6 radians lies in Quadrant IV. But in the fourth quadrant, x is positive, y is negative, and therefore $\tan 6 = y/x$ must be negative. A check of the calculator shows that $\tan 6 \approx -0.29$.

In the next example we evaluate the sine and cosine functions using both the unit-circle definitions and a calculator. In the example, degree measure is used for the angles. In verifying for yourself the results obtained there, be sure that your calculator is set to degree mode.

EXAMPLE 5 Using the unit circle to visually estimate values of cosine and sine

Use Figure 8 to approximate the following trigonometric values to within successive tenths. Then use a calculator to check your answers. Round the calculator values to two decimal places.

(a) $\cos 160°$ and $\sin 160°$ **(b)** $\cos(-40°)$ and $\sin(-40°)$

SOLUTION

(a) Figure 8 shows that (the terminal side of) an angle of 160° in standard position meets the unit circle at a point in Quadrant II. Letting (x, y) denote the coordinates of that point, we have (from Figure 8)

$$-1.0 < x < -0.9 \quad \text{and} \quad 0.3 < y < 0.4$$

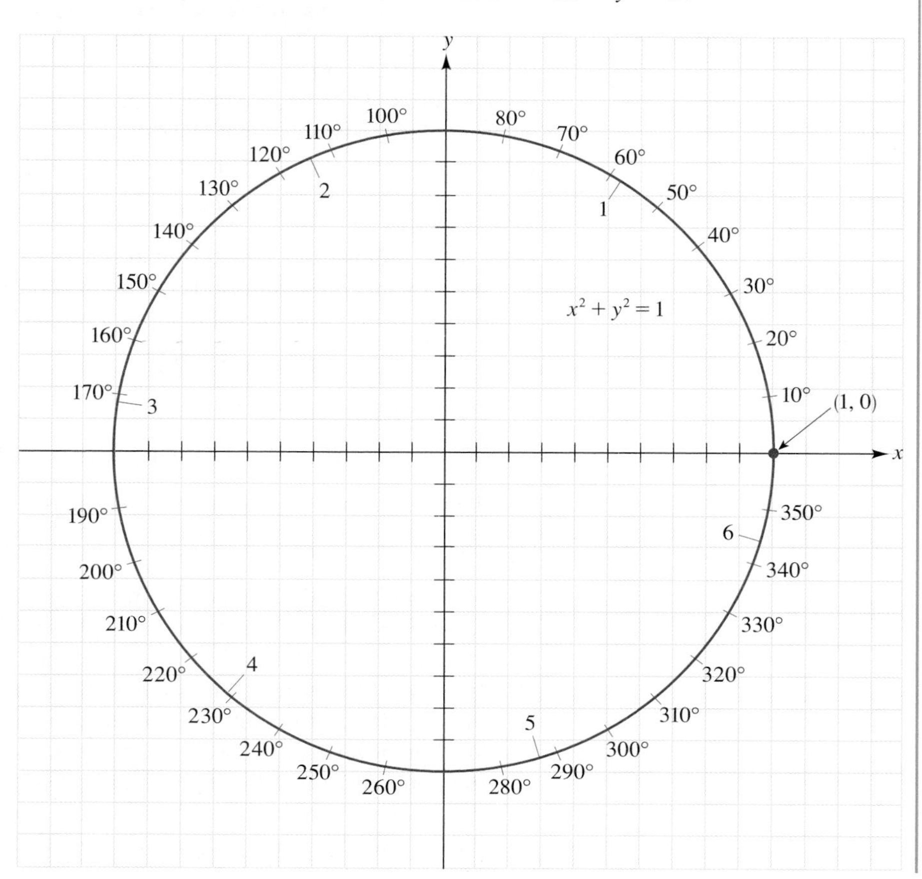

Figure 8
Radian measure and degree measure on the unit circle for angles in standard position

But by definition, $\cos 160° = x$ and $\sin 160° = y$, and, consequently,

$$-1.0 < \cos 160° < -0.9 \qquad \text{and} \qquad 0.3 < \sin 160° < 0.4$$

The corresponding calculator results are $\cos 160° \approx -0.94$ and $\sin 160° \approx 0.34$. Note that the estimations we obtained from Figure 8 are consistent with these calculator values.

(b) As you can verify using Figure 8, the terminal side of an angle of $-40°$ intersects the unit circle at the same point as does the terminal side of an angle of 320°. Letting (x, y) denote the coordinates of that point, we have (from Figure 8)

$$0.7 < x < 0.8 \qquad \text{and} \qquad -0.7 < y < -0.6$$

and, consequently,

$$0.7 < \cos(-40°) < 0.8 \qquad \text{and} \qquad -0.7 < \sin(-40°) < -0.6$$

The corresponding calculator values here are $\cos(-40°) \approx 0.77$ and $\sin(-40°) \approx -0.64$. Again, note that the estimations we obtain from Figure 8 are consistent with these calculator values.

EXAMPLE 6 Using the unit circle to visually estimate values of cosine and sine

Use Figure 8 to approximate to within successive tenths the following trigonometric values. Then use a calculator to check your answers. Round off the calculator values to two decimal places:

(a) $\cos 3$ and $\sin 3$; **(b)** $\cos(-3)$ and $\sin(-3)$.

SOLUTION

(a) Figure 8 shows that (the terminal side of) an angle of 3 radians in standard position meets the unit circle at a point in the second quadrant. Letting (x, y) denote the coordinates of that point, we have (using the grid in Figure 8)

$$-1.0 < x < -0.9 \qquad \text{and} \qquad 0.1 < y < 0.2$$

But by definition, $\cos 3 = x$ and $\sin 3 = y$, so, consequently,

$$-1.0 < \cos 3 < -0.9 \qquad \text{and} \qquad 0.1 < \sin 3 < 0.2$$

The corresponding calculator values here are

$$\cos 3 \approx -0.99 \qquad \text{and} \qquad \sin 3 \approx 0.14$$

Note that the estimates we obtained from Figure 8 are consistent with these calculator values. (When you verify these calculator values for yourself, be sure that the calculator is set in radian mode.)

(b) In Figure 8 we want to know where the terminal side of an angle of -3 radians intersects the unit circle. There is no marker for this point in Figure 8. We can, nevertheless, locate the point by using symmetry. Let (x, y) denote the point corresponding to 3 radians, determined in part (a). By reflecting the point in the x-axis, we obtain the point corresponding to -3 radians. This means that the required point will have the same x-coordinate as the point in part (a), while the y-coordinate will be the negative of that in part (a). So, given that

$$-1.0 < \cos 3 < -0.9 \qquad \text{and} \qquad 0.1 < \sin 3 < 0.2$$

we conclude that

$$-1.0 < \cos(-3) < -0.9 \qquad \text{and} \qquad -0.2 < \sin(-3) < -0.1$$

For the corresponding calculator values here, we obtain

$$\cos(-3) \approx -0.99 \qquad \text{and} \qquad \sin(-3) \approx -0.14$$

Again, note that the estimates from Figure 8 are consistent with these calculator values. When you verify these calculator values for yourself, remember that the calculator should be in radian mode.

EXAMPLE 7 Evaluating trigonometric functions

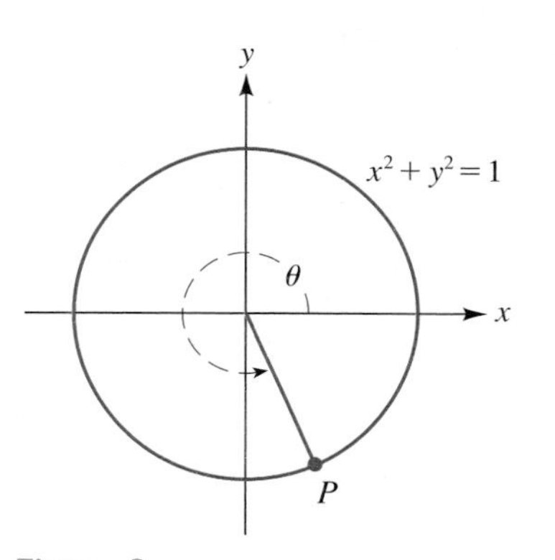

Figure 9

In Figure 9 the x-coordinate of the point P is $2/5$. Evaluate the trigonometric functions of θ.

SOLUTION

In order to evaluate the trigonometric functions of θ, we need to know the coordinates of the point P in Figure 9. The x-coordinate of P is given to be $2/5$. Because P lies on the unit circle, the y-coordinate of P satisfies the equation $(2/5)^2 + y^2 = 1$. Consequently,

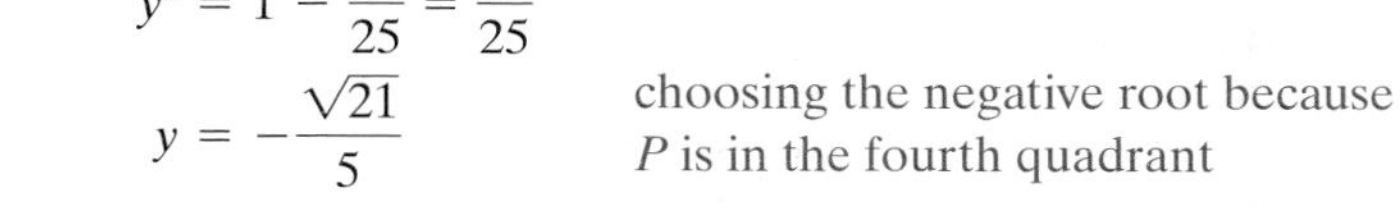

$$y^2 = 1 - \frac{4}{25} = \frac{21}{25}$$

$$y = -\frac{\sqrt{21}}{5} \qquad \text{choosing the negative root because } P \text{ is in the fourth quadrant}$$

Now that we know the coordinates of P, we can evaluate the trigonometric functions as follows.

$$\cos\theta = x = \frac{2}{5} \qquad \sec\theta = \frac{1}{x} = \frac{5}{2}$$

$$\sin\theta = y = -\frac{\sqrt{21}}{5} \qquad \csc\theta = \frac{1}{y} = \frac{1}{-\sqrt{21}/5} = -\frac{5}{\sqrt{21}}$$

$$\tan\theta = \frac{y}{x} = -\frac{\sqrt{21}/5}{2/5} = -\frac{\sqrt{21}}{2} \qquad \cot\theta = \frac{x}{y} = \frac{2/5}{-\sqrt{21}/5} = -\frac{2}{\sqrt{21}}$$

In our final example for this section, we fill in some calculator-related details for the three functions secant, cosecant, and cotangent. These three functions are referred to as the **reciprocal functions.** This is because, from the definitions on page 439, it follows that

$$\sec\theta = \frac{1}{\cos\theta} \qquad \csc\theta = \frac{1}{\sin\theta} \qquad \cot\theta = \frac{1}{\tan\theta}$$

So, for example, although most calculators do not have a key labeled "sec," we can nevertheless evaluate the secant function by using the cosine key [cos] and then the reciprocal key [1/x] or [x^{-1}].

EXAMPLE 8 Using a calculator to evaluate reciprocal functions

Use a calculator to evaluate sec 10°, csc 10°, and cot 10°. Round the results to three decimal places.

SOLUTION

As in the previous example, we first check that the calculator is in the degree mode. Then the sequence of keystrokes and the results are as follows (again, you should verify each of these results for yourself):

EXPRESSION	KEYSTROKES	OUTPUT
sec 10°	[(] [cos] 10 [)] [x^{-1}] [ENTER]	1.015
csc 10°	[(] [sin] 10 [)] [x^{-1}] [ENTER]	5.759
cot 10°	[(] [tan] 10 [)] [x^{-1}] [ENTER]	5.671

EXERCISE SET 6.2

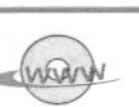

A

In Exercises 1–8, sketch each angle in standard position.

1. **(a)** $\pi/4$ **(b)** $-\pi/4$ **(c)** $3\pi/4$
2. **(a)** $\pi/6$ **(b)** $-\pi/6$ **(c)** $-5\pi/6$
3. **(a)** $\pi/3$ **(b)** $-5\pi/3$ **(c)** $-7\pi/3$
4. **(a)** $3\pi/2$ **(b)** $-3\pi/2$ **(c)** $-5\pi/2$
5. **(a)** 210° **(b)** −210° **(c)** −570°
6. **(a)** 45° **(b)** −225° **(c)** 315°
7. **(a)** 120° **(b)** −120° **(c)** 300°
8. **(a)** 450° **(b)** −180° **(c)** 270°

In Exercises 9–22, use the definitions (not a calculator) to evaluate the six trigonometric functions of each angle. If a value is undefined, state this.

9. π **10.** $-\pi/2$ **11.** -2π
12. $3\pi/2$ **13.** $-3\pi/2$ **14.** -3π
15. 0 **16.** 4π **17.** 90°
18. 450° **19.** −270° **20.** −630°
21. 180° **22.** −540°

For Exercises 23 and 24: **(a)** *Verify algebraically that the given point in the figure indeed lies on the unit circle (Hint: See the discussion preceding Example 2);* **(b)** *Evaluate the six trigonometric functions for the indicated angle.*

23.

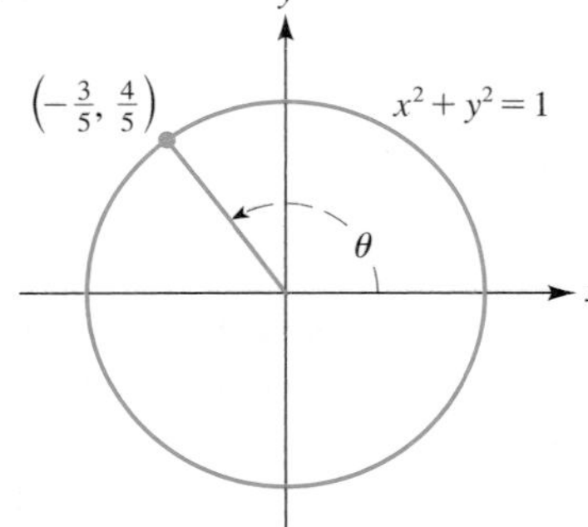

24.

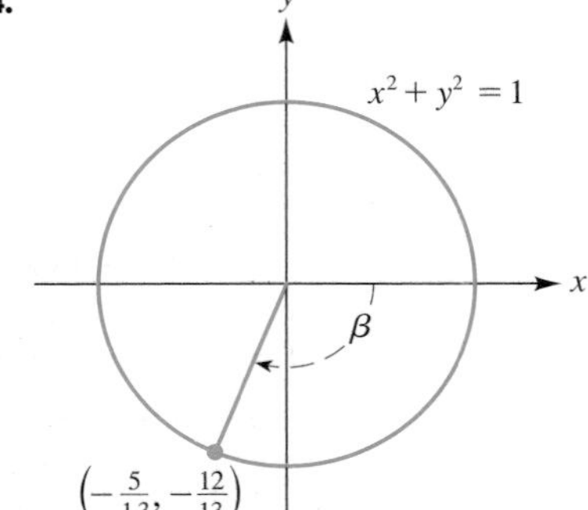

In Exercises 25 and 26, use the definitions of the trigonometric functions (as in Example 1) to complete the tables. (For Exercise 25, when you are finished check your answer against the values shown in Table 1 on page 422.)

25.

θ	$\cos\theta$	$\sin\theta$	$\tan\theta$	$\sec\theta$	$\csc\theta$	$\cot\theta$
0						
$\pi/2$						
π						
$3\pi/2$						
2π						

26.

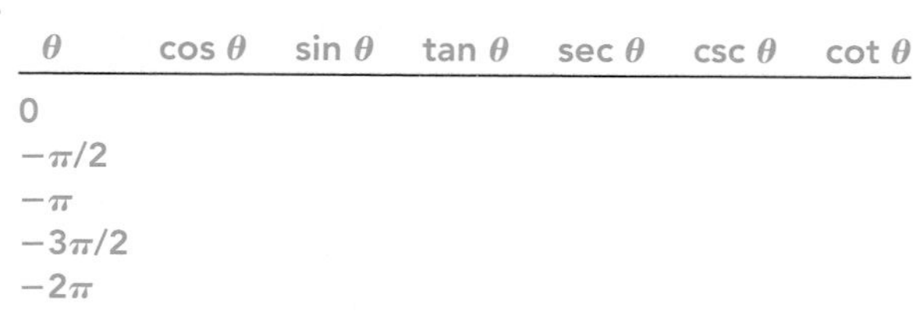

θ	$\cos\theta$	$\sin\theta$	$\tan\theta$	$\sec\theta$	$\csc\theta$	$\cot\theta$
0						
$-\pi/2$						
$-\pi$						
$-3\pi/2$						
-2π						

In Exercises 27–36, refer to the following figure, which indicates radian measure on the unit circle for angles in standard position. For Exercises 27–30, use the figure (and the unit circle definitions) to determine whether the given quantity is positive or negative. For Exercises 31–36, use the figure to determine which of the given quantities is the larger.

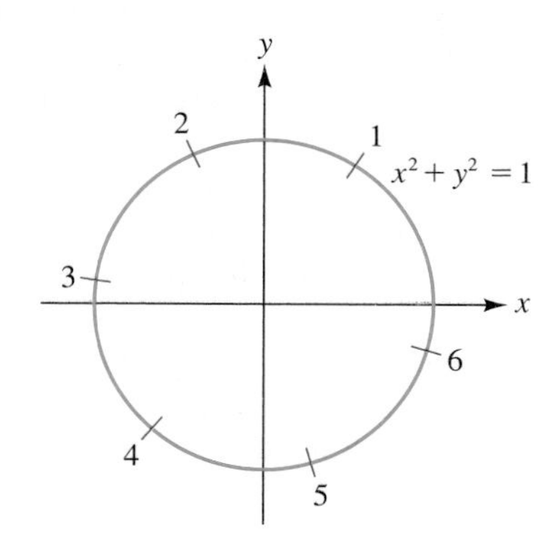

27. **(a)** $\sin 2$ **(b)** $\cos 2$ **(c)** $\tan 2$
28. **(a)** $\sin 4$ **(b)** $\cos 4$ **(c)** $\tan 4$
29. **(a)** $\cos 1 + \cos 6$ **(b)** $\cos 1 - \cos 6$
30. **(a)** $\sin 1 + \sin 6$ **(b)** $\sin 1 - \sin 6$
31. $\sin 2$ or $\sin 3$
32. $\sin 5$ or $\sin 6$
33. $\cos 2$ or $\cos 3$
34. $\cos 3$ or $\cos 4$
35. $\tan 2$ or $\tan 4$
36. $(\tan 4)(\tan 5)$ or $(\tan 5)(\tan 6)$

In Exercises 37–58, use Figure 8 on page 443 to approximate the given trigonometric values to within successive tenths (as in Example 5). Then use a calculator to compute the values to the nearest hundredth.

37. $\cos 1$ and $\sin 1$
38. $\cos 2$ and $\sin 2$
39. $\cos(-1)$ and $\sin(-1)$
40. $\cos(-2)$ and $\sin(-2)$
41. $\cos 4$ and $\sin 4$
42. $\cos 5$ and $\sin 5$
43. $\cos(-4)$ and $\sin(-4)$
44. $\cos(-5)$ and $\sin(-5)$
45. $\sin 10^\circ$ and $\sin(-10^\circ)$
46. $\cos 10^\circ$ and $\cos(-10^\circ)$
47. $\cos 80^\circ$ and $\cos(-80^\circ)$
48. $\sin 80^\circ$ and $\sin(-80^\circ)$
49. $\sin 120^\circ$ and $\sin(-120^\circ)$
50. $\cos 120^\circ$ and $\cos(-120^\circ)$
51. $\sin 150^\circ$ and $\sin(-150^\circ)$
52. $\cos 150^\circ$ and $\cos(-150^\circ)$
53. $\cos 220^\circ$ and $\cos(-220^\circ)$
54. $\sin 220^\circ$ and $\sin(-220^\circ)$
55. $\cos 310^\circ$ and $\cos(-310^\circ)$
56. $\sin 310^\circ$ and $\sin(-310^\circ)$
57. $\sin(1 + 2\pi)$
58. $\sin(2 + 2\pi)$

In Exercises 59–74, let $P(x, y)$ denote the point where the terminal side of angle θ (in standard position) meets the unit circle (as in Figure 4). Use the given information to evaluate the six trigonometric functions of θ.

59. P is in Quadrant I and $x = 1/3$.
60. P is in Quadrant IV and $x = 1/3$.
61. P is in Quadrant III and $x = -3/5$.
62. P is in Quadrant I and $x = 3/5$.
63. P is in Quadrant II and $y = 5/13$.
64. P is in Quadrant III and $y = -5/13$.
65. P is in Quadrant III and $y = -3/4$.
66. P is in Quadrant II and $y = 3/4$.
67. $x = -8/15$ and $\pi/2 < \theta < \pi$
68. $x = -8/15$ and $\pi < \theta < 3\pi/2$
69. $y = -2/9$ and $3\pi/2 < \theta < 2\pi$
70. $y = 2/9$ and $0 < \theta < \pi/2$
71. $x = 7/25$ and $270^\circ < \theta < 360^\circ$
72. $x = -7/25$ and $180^\circ < \theta < 270^\circ$
73. $x = 1/2$ and $0^\circ < \theta < 90^\circ$
74. $x = -1/2$ and $90^\circ < \theta < 180^\circ$

In Exercises 75–89, use a calculator to evaluate $\sec\theta$, $\csc\theta$, and $\cot\theta$ for the given value of θ. Round the answers to two decimal places.

75. 2.06
76. 5.23
77. 9
78. -9
79. -0.55
80. 0.55
81. $\pi/6$
82. $-6\pi/5$
83. 1400
84. $1400 + 2\pi$
85. 33°
86. 393°
87. -125°
88. -179°
89. 225°

90. **(a)** Complete the following table, using the words "positive" or "negative" as appropriate.

	Terminal side of angle θ lies in			
	Quadrant I	Quadrant II	Quadrant III	Quadrant IV
$\cos\theta$ and $\sec\theta$	positive	negative		
$\sin\theta$ and $\csc\theta$				
$\tan\theta$ and $\cot\theta$				

(b) The mnemonic (memory device) ASTC (*a*ll *s*tudents *t*ake *c*alculus) is sometimes used to recall the signs of the trigonometric values in each quadrant:

A All are positive in Quadrant I.
S Sine is positive in Quadrant II.
T Tangent is positive in Quadrant III.
C Cosine is positive in Quadrant IV.

Check the validity of this mnemonic against your chart in part (a).

91. This exercise completes the discussion in the text concerning the use of parentheses in calculator work.

(a) Use the unit circle definitions to briefly explain (in complete sentences) why $\sin(\pi/2) = 1$ and $\sin \pi = 0$.

(b) Set the calculator to the radian mode and enter the following sequence of keystrokes.

sin π ÷ 2 ENTER

Your calculator will show an output of 0, which, as you know, is not the value of $\sin(\pi/2)$. This is because, in the absence of parentheses, the calculator interprets the sequence of keystrokes sin π ÷ 2 as follows: First compute $\sin \pi$, then divide the result by 2. That is, the calculator computes $0 \div 2$, which, of course, results in the 0 output. *Conclusion*: If you want the calculator to compute $\sin(\pi/2)$, you must use parentheses and enter the sequence of keystrokes

sin (π ÷ 2) ENTER

6.3 EVALUATING THE TRIGONOMETRIC FUNCTIONS

. . . when we know the sines and cosines of angles less than half a right angle, then we also have sines and cosines of greater angles.—Leonhard Euler *in Introductio in Analysin Infinitorum* (Lausanne: 1748)

Be clear about the signs of the functions. Your calculator will show them to you correctly, but you still need to be able to figure them out for yourself, quadrant by quadrant. —David Halliday and Robert Resnick, in *Fundamentals of Physics,* 3rd ed. (New York: John Wiley and Sons, Inc., 1988)

In the previous section we pointed out that for angles that are multiples of $\pi/2$ radians or 90°, the trigonometric functions can be evaluated without a calculator. In this section we explain how to obtain the trigonometric values without a calculator for angles that are multiples of either $\pi/4$ radians (=45°) or $\pi/6$ radians (=30°). We focus on these particular angles for now, not because they are somehow more fundamental than others, but because a ready knowledge of their trigonometric values will provide a useful source for examples in trigonometry and calculus. Also, in the course of obtaining these trigonometric values, we will introduce the concept of a *reference angle* or *reference number,* which is useful in other portions of trigonometry as well.

We can use Figure 1 to obtain the trigonometric values for $\theta = \pi/4$. Note that the terminal side of the 45° angle in Figure 1 coincides with the line $y = x$; so the coordinates of the point P can be labeled (x, x). Substituting these coordinates in the equation of the unit circle, we have

$$x^2 + x^2 = 1$$
$$2x^2 = 1$$
$$x = \frac{1}{\sqrt{2}} \quad \text{choosing the positive root because } P \text{ is in Quadrant I}$$

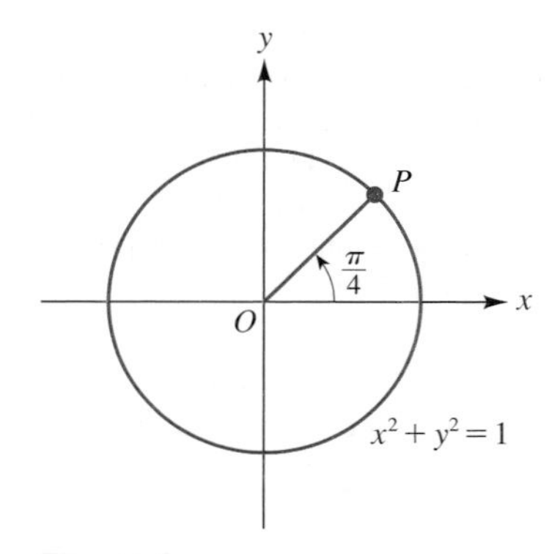

Figure 1
An angle of $\pi/4$ radians (=45°) in standard position

The coordinates of P are therefore $x = y = 1/\sqrt{2}$, and consequently we have

$$\cos\frac{\pi}{4} = x = \frac{1}{\sqrt{2}} = \frac{\sqrt{2}}{2} \qquad \sec\frac{\pi}{4} = \frac{1}{x} = \sqrt{2}$$
$$\sin\frac{\pi}{4} = y = \frac{1}{\sqrt{2}} = \frac{\sqrt{2}}{2} \qquad \csc\frac{\pi}{4} = \frac{1}{y} = \sqrt{2}$$
$$\tan\frac{\pi}{4} = \frac{y}{x} = \frac{1/\sqrt{2}}{1/\sqrt{2}} = 1 \qquad \cot\frac{\pi}{4} = \frac{x}{y} = \frac{1/\sqrt{2}}{1/\sqrt{2}} = 1$$

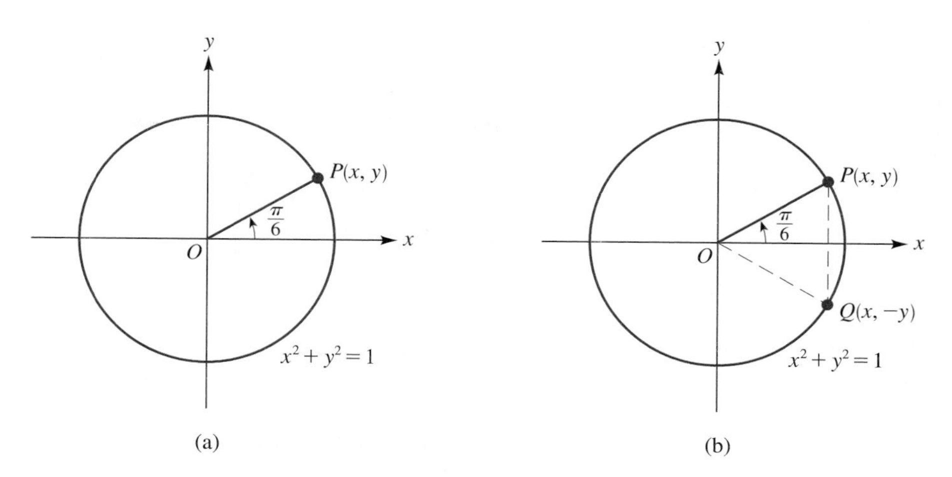

Figure 2

Next, let us consider an angle of $\pi/6$ radians in standard position, as shown in Figure 2(a). A simple way to compute the coordinates of the point P is first to reflect P in the x-axis, as indicated in Figure 2(b). Then $\triangle OPQ$ is equilateral (why?) and therefore

$$PQ = OP$$

Now, $\overline{OP}$ is a radius, so $OP = 1$. Also, for the vertical distance PQ, we have $PQ = y - (-y) = 2y$. Using these values in equation (1), we have

$$2y = 1 \qquad \text{or} \qquad y = \frac{1}{2}$$

Thus, the y-coordinate of P is $1/2$. To find the corresponding x-coordinate, we substitute the value $y = 1/2$ in the equation of the unit circle to obtain

$$x^2 + \left(\frac{1}{2}\right)^2 = 1$$

$$x^2 = \frac{3}{4}$$

$$x = \frac{\sqrt{3}}{2} \qquad \text{choosing the positive root because } P \text{ is in the first quadrant}$$

We have now found that the coordinates of P are $\left(\frac{\sqrt{3}}{2}, \frac{1}{2}\right)$, and so the trigonometric functions of $\pi/6$ are evaluated as follows.

$$\cos\frac{\pi}{6} = x = \frac{\sqrt{3}}{2}$$

$$\sin\frac{\pi}{6} = y = \frac{1}{2}$$

$$\tan\frac{\pi}{6} = \frac{y}{x} = \frac{1/2}{\sqrt{3}/2} = \frac{1}{\sqrt{3}} = \frac{\sqrt{3}}{3}$$

$$\sec\frac{\pi}{6} = \frac{1}{x} = \frac{1}{\sqrt{3}/2} = \frac{2}{\sqrt{3}} = \frac{2\sqrt{3}}{3}$$

$$\csc\frac{\pi}{6} = \frac{1}{y} = \frac{1}{1/2} = 2$$

$$\cot\frac{\pi}{6} = \frac{x}{y} = \frac{\sqrt{3}/2}{1/2} = \sqrt{3}$$

We can use the concept of symmetry to evaluate the trigonometric functions of $\theta = \pi/3$. Figure 3(a) displays angles of $\pi/3$ and $\pi/6$ radians in standard position.

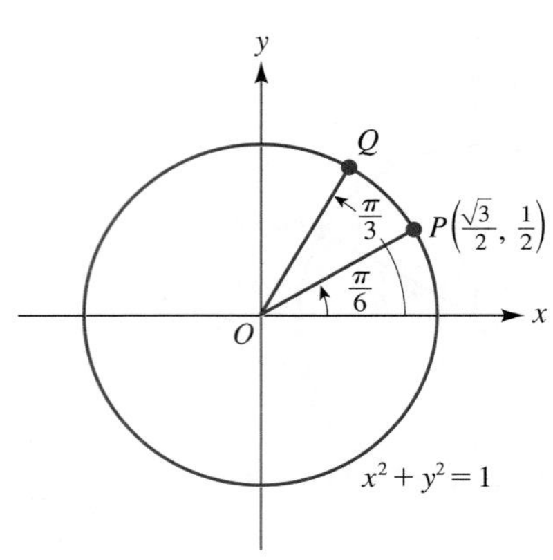

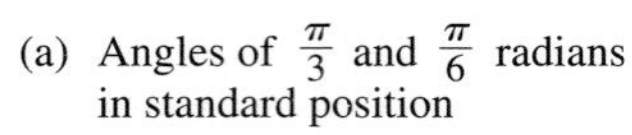
(a) Angles of $\frac{\pi}{3}$ and $\frac{\pi}{6}$ radians in standard position

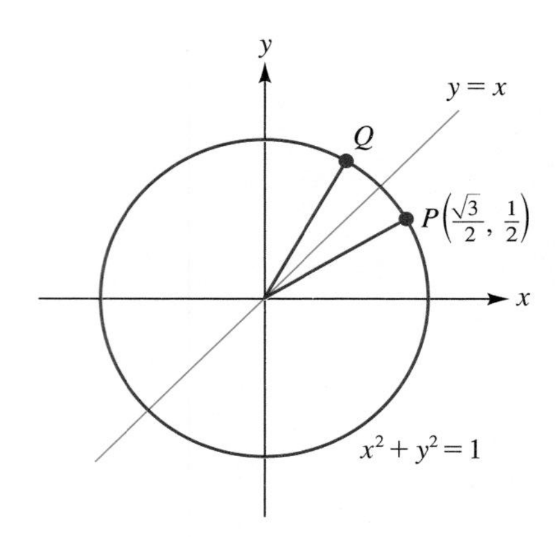

(b) The points P and Q are symmetric about the line $y = x$.

Figure 3

In the figure, note that the coordinates of P are $(\frac{\sqrt{3}}{2}, \frac{1}{2})$, as obtained in the previous paragraph.

Now, as Figure 3(b) indicates (and Exercise 32 shows you how to prove), the points P and Q are symmetric about the line $y = x$. Therefore (according to our discussion of symmetry in Section 3.6), the coordinates of Q are $(\frac{1}{2}, \frac{\sqrt{3}}{2})$, and the trigonometric functions of $\pi/3$ can be evaluated as follows.

$$\cos\frac{\pi}{3} = x = \frac{1}{2}$$
$$\sin\frac{\pi}{3} = y = \frac{\sqrt{3}}{2}$$
$$\tan\frac{\pi}{3} = \frac{y}{x} = \frac{\sqrt{3}/2}{1/2} = \sqrt{3}$$

$$\sec\frac{\pi}{3} = \frac{1}{x} = \frac{1}{1/2} = 2$$
$$\csc\frac{\pi}{3} = \frac{1}{y} = \frac{1}{\sqrt{3}/2} = \frac{2}{\sqrt{3}} = \frac{2\sqrt{3}}{3}$$
$$\cot\frac{\pi}{3} = \frac{x}{y} = \frac{1/2}{\sqrt{3}/2} = \frac{1}{\sqrt{3}} = \frac{\sqrt{3}}{3}$$

TABLE 1

θ	$\cos\theta$	$\sin\theta$	$\tan\theta$
$\pi/6$ radians or 30°	$\frac{\sqrt{3}}{2}$	$\frac{1}{2}$	$\frac{\sqrt{3}}{3}$
$\pi/4$ radians or 45°	$\frac{\sqrt{2}}{2}$	$\frac{\sqrt{2}}{2}$	1
$\pi/3$ radians or 60°	$\frac{1}{2}$	$\frac{\sqrt{3}}{2}$	$\sqrt{3}$

In Table 1 we summarize the results we have now obtained for angles of $\pi/6$, $\pi/4$, and $\pi/3$ radians. In the table we list only the values for cosine, sine, and tangent, because the remaining three values are just reciprocals of these. Check with your instructor whether she or he wants you to memorize the results in Table 1.

To evaluate the trigonometric functions for angles that are not multiples of 90° or $\pi/2$ radians, we introduce the concept of a *reference angle* or *reference number.*

DEFINITION **The Reference Angle or Reference Number**

Let θ be an angle in standard position, and suppose that θ is not a multiple of 90° or $\pi/2$ radians. The **reference angle** associated with θ is the acute angle (with positive measure) formed by the x-axis and the terminal side of the angle θ. When radian measure is used, the reference angle is sometimes referred to as the **reference number** [because a radian angle measure (such as $\pi/4$ or 3) is a real number].

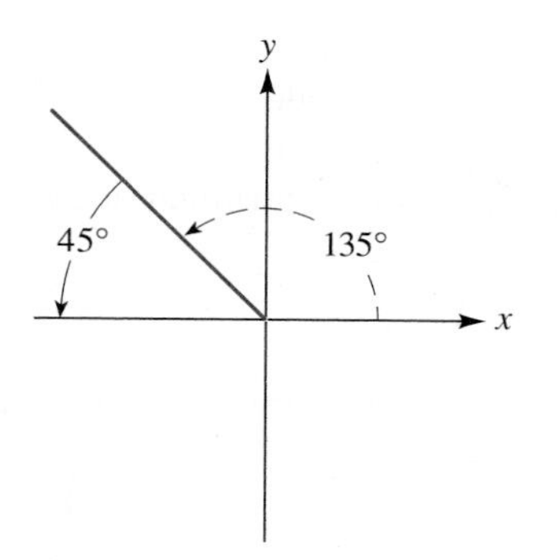

The reference angle for 135° is 45° [180° − 135° = 45°].

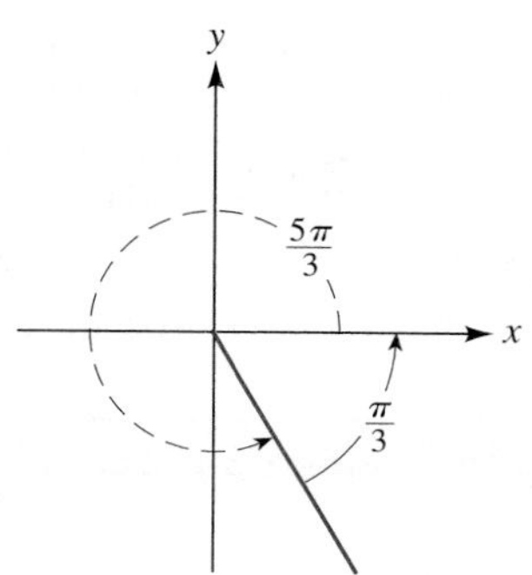

The reference number for $\frac{5\pi}{3}$ is $\frac{\pi}{3}$ $[2\pi - \frac{5\pi}{3} = \frac{\pi}{3}]$.

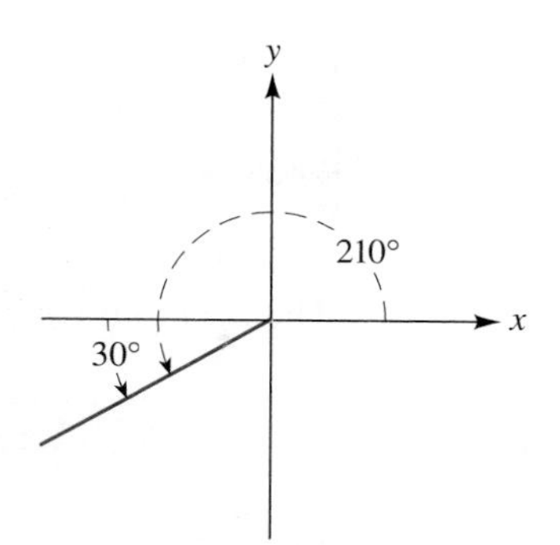

The reference angle for 210° is 30° [210° − 180° = 30°].

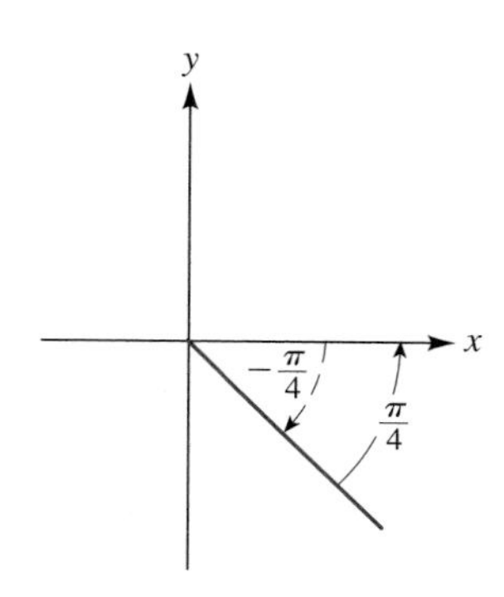

The reference number for $-\frac{\pi}{4}$ is $\frac{\pi}{4}$.

Figure 4

In Figure 4 we show four examples of angles and their respective reference angles. The first part of Figure 4 shows how to find the reference angle for $\theta = 135°$. First we place the angle $\theta = 135°$ in standard position. Then we find the acute angle between the x-axis and the terminal side of θ. As you can see in this case, the acute angle is 45°, the reference angle associated with $\theta = 135°$. In the same way, you should work through the remaining three parts of Figure 4.

Now let's look at an example to see how reference angles are used in evaluating the trigonometric functions. Suppose that we want to evaluate cos 150°. In Figure 5(a) we've placed the angle $\theta = 150°$ in standard position. As you can see, the reference angle for 150° is 30°.

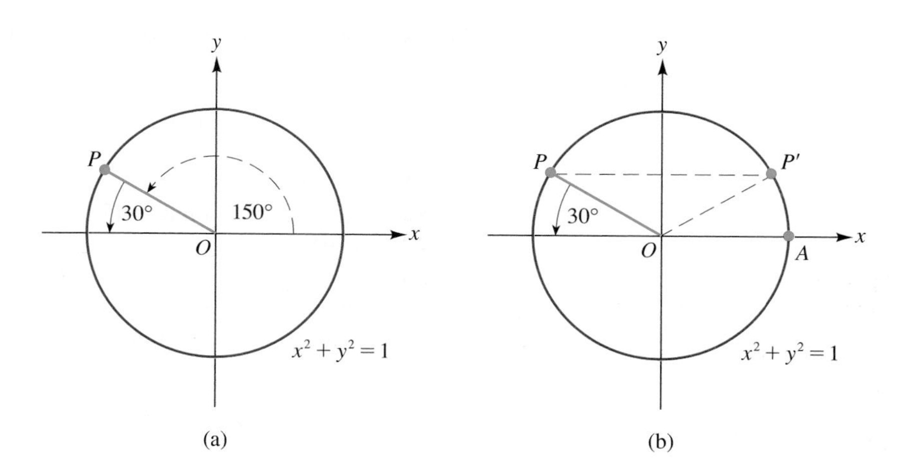

Figure 5

By definition the value of cos 150° is the x-coordinate of the point P in Figure 5(a). To find this x-coordinate, we reflect the line segment $\overline{OP}$ in the y-axis; the reflected line segment is the segment $\overline{OP'}$ in Figure 5(b). Since $\angle P'OA = 30°$, the x-coordinate of the point P' is by definition cos 30°, or $\sqrt{3}/2$. The x-coordinate of P is then the negative of this, that is, the x-coordinate of P is $-\sqrt{3}/2$. It follows now, again by definition, that the value of cos 150° is $-\sqrt{3}/2$.

The same method that we have just used to evaluate cos 150° can be used to evaluate any of the trigonometric functions when the angles are not multiples of 90°. The following four steps summarize this method.

Step 1 Draw (or refer to) a sketch of the angle and reference angle.
Step 2 Determine the reference angle associated with the given angle.
Step 3 Evaluate the given trigonometric function of the reference angle.
Step 4 Affix the appropriate sign determined by the quadrant of the terminal side of the angle in standard position.

The next four examples illustrate this procedure.

EXAMPLE 1 Using the four-step procedure

Evaluate the following quantities:
(a) $\sin 135^\circ$; **(b)** $\cos 135^\circ$; **(c)** $\tan 135^\circ$.

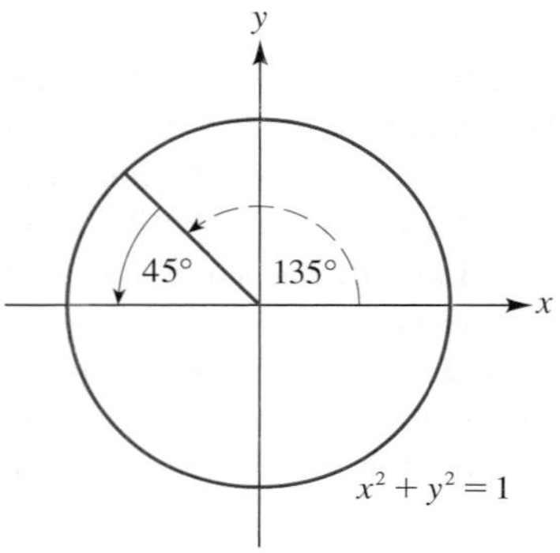

SOLUTION
As Figure 6 indicates, the reference angle associated with 135° is 45°.

(a) Step 1 See Figure 6.
Step 2 The reference angle is 45°.
Step 3 $\sin 45^\circ = \sqrt{2}/2$
Step 4 The terminal side of $\theta = 135^\circ$ lies in Quadrant II, where the y-coordinate is positive, so $\sin 135^\circ$ is positive. We therefore have

$$\sin 135^\circ = \sin 45^\circ = \frac{\sqrt{2}}{2}$$

(b) Step 1 See Figure 6.
Step 2 The reference angle is 45°.
Step 3 $\cos 45^\circ = \sqrt{2}/2$
Step 4 The terminal side of $\theta = 135^\circ$ lies in Quadrant II, where the x-coordinate is negative, so $\cos 135^\circ$ is negative. We therefore have

$$\cos 135^\circ = -\cos 45^\circ = -\frac{\sqrt{2}}{2}$$

(c) Step 1 See Figure 6.
Step 2 The reference angle is 45°.
Step 3 $\tan 45^\circ = 1$
Step 4 By definition, $\tan \theta = y/x$. The terminal side of $\theta = 135^\circ$ lies in Quadrant II, where the y-coordinate is positive and the x-coordinate is negative, so $\tan 135^\circ$ is negative. We therefore have

$$\tan 135^\circ = -\tan 45^\circ = -1$$

Figure 6

EXAMPLE 2 Finding the values of trigonometric functions of an angle with negative measure

Evaluate the following quantities:
(a) $\cos(-120^\circ)$; **(b)** $\cot(-120^\circ)$; **(c)** $\sec(-120^\circ)$.

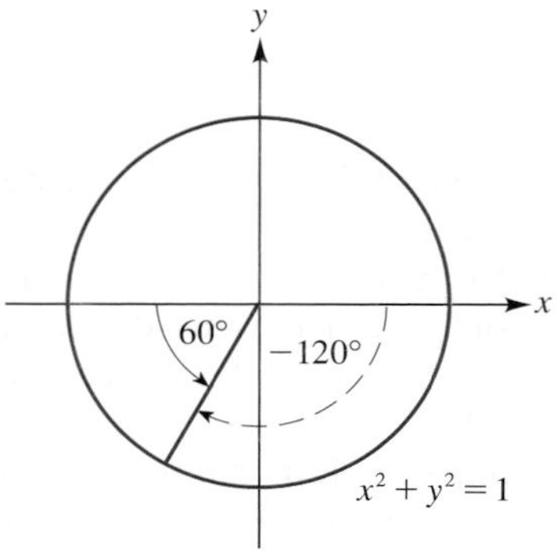

SOLUTION
As Figure 7 shows, the reference angle for -120° is 60°.

(a) Step 1 See Figure 7.
Step 2 The reference angle for -120° is 60°.

Figure 7

Step 3 $\cos 60° = 1/2$

Step 4 The terminal side of $\theta = -120°$ lies in Quadrant III, where the x-coordinate is negative, so $\cos(-120°)$ is negative. We therefore have

$$\cos(-120°) = -\cos(60°) = -\frac{1}{2}$$

(b) Step 1 See Figure 7.

Step 2 The reference angle for $-120°$ is $60°$.

Step 3 $\cot 60° = \sqrt{3}/3$

Step 4 By definition, $\cot\theta = x/y$. Now, the terminal side of $\theta = -120°$ lies in Quadrant III, where the x-coordinate is negative and the y-coordinate is negative. Thus $\cot(-120°)$ is positive. We therefore have

$$\cot(-120°) = \cot(60°) = \frac{\sqrt{3}}{3}$$

(c) We could follow our four-step procedure here, but in this case there is a faster method. In part (a) of this example we found that $\cos(-120°) = -1/2$. Therefore, since $\sec\theta$ is the reciprocal of $\cos\theta$, we have

$$\sec(-120°) = \frac{1}{\cos(-120°)} = \frac{1}{(-1/2)} = -2$$

EXAMPLE 3 Using the reference angle to evaluate trigonometric functions

Evaluate the following expressions:
(a) $\sin(11\pi/6)$; **(b)** $\cos(11\pi/6)$; **(c)** $\tan(11\pi/6)$.

SOLUTION

As Figure 8 shows, the reference angle for $11\pi/6$ is $\pi/6$.

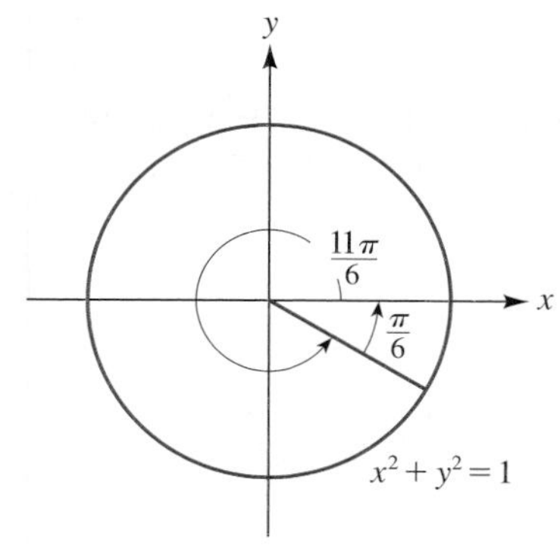

Figure 8

(a) Step 1 See Figure 8.

Step 2 The reference angle for $11\pi/6$ is $\pi/6$.

Step 3 $\sin(\pi/6) = 1/2$

Step 4 By definition, $\sin\theta$ is the y-coordinate. The terminal side of $11\pi/6$ lies in Quadrant IV, where the y-coordinate is negative, so $\sin(11\pi/6)$ is negative. Hence we have

$$\sin\frac{11\pi}{6} = -\sin\frac{\pi}{6} = -\frac{1}{2}$$

(b) Step 1 See Figure 8.

Step 2 The reference angle for $11\pi/6$ is $\pi/6$.

Step 3 $\cos(\pi/6) = \sqrt{3}/2$

Step 4 By definition, $\cos\theta$ is the x-coordinate. The terminal side of $11\pi/6$ lies in Quadrant IV, where the x-coordinate is positive, so $\cos(11\pi/6)$ is positive. We therefore have

$$\cos\frac{11\pi}{6} = \cos\frac{\pi}{6} = \frac{\sqrt{3}}{2}$$

(c) Step 1 See Figure 8.
Step 2 The reference angle for $11\pi/6$ is $\pi/6$.
Step 3 $\tan(\pi/6) = \sqrt{3}/3$
Step 4 By definition, $\tan\theta = y/x$. The terminal side of $11\pi/6$ lies in Quadrant IV, where the x-coordinate is positive and the y-coordinate is negative, so $\tan(11\pi/6)$ is negative. Therefore

$$\tan\frac{11\pi}{6} = -\tan\frac{\pi}{6} = -\frac{\sqrt{3}}{3}$$

EXAMPLE 4 Using the reference angle to calculate values of trigonometric functions

Evaluate the following expressions:
(a) $\sec(-7\pi/4)$; **(b)** $\csc(-7\pi/4)$; **(c)** $\cot(-7\pi/4)$.

SOLUTION
As Figure 9 indicates, the reference angle for $-7\pi/4$ is $\pi/4$.

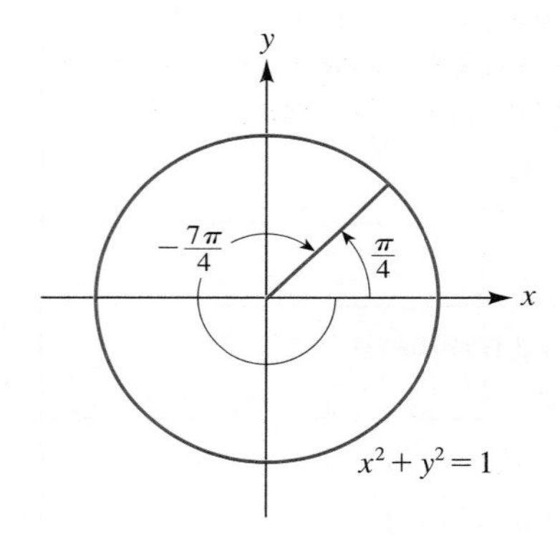

Figure 9

(a) Step 1 See Figure 9.
Step 2 The reference angle for $-7\pi/4$ is $\pi/4$.
Step 3 $\sec\left(\frac{\pi}{4}\right) = \frac{1}{\cos(\pi/4)} = \frac{1}{1/\sqrt{2}} = \sqrt{2}$
Step 4 By definition, $\sec\theta = 1/x$. The terminal side of $-7\pi/4$ lies in Quadrant I, where the x-coordinate is positive, so $1/x$ is positive. We therefore have

$$\sec\left(-\frac{7\pi}{4}\right) = \sec\frac{\pi}{4} = \sqrt{2}$$

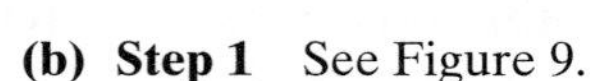

(b) Step 1 See Figure 9.
Step 2 The reference angle for $-7\pi/4$ is $\pi/4$.
Step 3 $\csc\left(\frac{\pi}{4}\right) = \frac{1}{\sin(\pi/4)} = \frac{1}{1/\sqrt{2}} = \sqrt{2}$
Step 4 By definition, $\csc\theta = 1/y$. The terminal side of $-7\pi/4$ lies in Quadrant I, where the y-coordinate is positive, so $1/y$ is positive, and we have

$$\csc\left(-\frac{7\pi}{4}\right) = \csc\frac{\pi}{4} = \sqrt{2}$$

(c) Step 1 See Figure 9.
Step 2 The reference angle for $-7\pi/4$ is $\pi/4$.
Step 3 $\cot\left(\frac{\pi}{4}\right) = \frac{1}{\tan(\pi/4)} = \frac{1}{1} = 1$
Step 4 By definition, $\cot\theta = x/y$. The terminal side of $-7\pi/4$ lies in Quadrant I, where both the x- and y-coordinates are positive, so x/y is positive. Thus we have

$$\cot\left(-\frac{7\pi}{4}\right) = \cot\frac{\pi}{4} = 1$$

EXERCISE SET 6.3

A

In Exercises 1–4, sketch each angle in standard position and specify the reference angle or reference number.

1. **(a)** $110°$ **(b)** $240°$ **(c)** $60°$ **(d)** $-60°$
2. **(a)** $300°$ **(b)** $1000°$ **(c)** $-15°$ **(d)** $15°$
3. **(a)** $3\pi/4$ **(b)** $-5\pi/6$ **(c)** $5\pi/3$ **(d)** $7\pi/6$
4. **(a)** $5\pi/6$ **(b)** $-\pi/3$ **(c)** $2\pi/3$ **(d)** $5\pi/4$

In Exercises 5 and 6, match an appropriate value from the right-hand column with each expression in the left-hand column.

5. **(a)** $\cos(\pi/3)$ **(A)** $\sqrt{3}/2$
   (b) $\sin(\pi/3)$ **(B)** $1/2$
   (c) $\tan(\pi/3)$ **(C)** $\sqrt{3}/3$
   (d) $\cos(\pi/6)$ **(D)** $\sqrt{3}$
   (e) $\sin(\pi/6)$ **(E)** $\sqrt{2}$
   (f) $\tan(\pi/6)$ **(F)** $\sqrt{2}/2$
6. **(a)** $\sec 45°$ **(A)** $\sqrt{3}$
   (b) $\csc 45°$ **(B)** $\sqrt{2}$
   (c) $\csc 30°$ **(C)** $2/\sqrt{3}$
   (d) $\sec 30°$ **(D)** $\sqrt{2}/2$
   (E) 2
   (F) $1/2$

In Exercises 7–22, evaluate the expression using the method shown in Examples 1–4.

7. **(a)** $\cos 300°$ **(c)** $\sin 300°$
   (b) $\cos(-300°)$ **(d)** $\sin(-300°)$
8. **(a)** $\cos 150°$ **(c)** $\sin 150°$
   (b) $\cos(-150°)$ **(d)** $\sin(-150°)$
9. **(a)** $\cos 210°$ **(c)** $\sin 210°$
   (b) $\cos(-210°)$ **(d)** $\sin(-210°)$
10. **(a)** $\cos 585°$ **(c)** $\sin 585°$
    (b) $\cos(-585°)$ **(d)** $\sin(-585°)$
11. **(a)** $\cos 390°$ **(c)** $\sin 390°$
    (b) $\cos(-390°)$ **(d)** $\sin(-390°)$
12. **(a)** $\cos 405°$ **(c)** $\sin 405°$
    (b) $\cos(-405°)$ **(d)** $\sin(-405°)$
13. **(a)** $\sec 600°$ **(c)** $\tan 600°$
    (b) $\csc(-600°)$ **(d)** $\cot(-600°)$
14. **(a)** $\sec 330°$ **(c)** $\tan 330°$
    (b) $\csc(-330°)$ **(d)** $\cot(-330°)$
15. **(a)** $\cos(4\pi/3)$ **(c)** $\sin(4\pi/3)$
    (b) $\cos(-4\pi/3)$ **(d)** $\sin(-4\pi/3)$
16. **(a)** $\cos(2\pi/3)$ **(c)** $\sin(2\pi/3)$
    (b) $\cos(-2\pi/3)$ **(d)** $\sin(-2\pi/3)$
17. **(a)** $\cos(5\pi/4)$ **(c)** $\sin(5\pi/4)$
    (b) $\cos(-5\pi/4)$ **(d)** $\sin(-5\pi/4)$
18. **(a)** $\cos(17\pi/6)$ **(c)** $\sin(17\pi/6)$
    (b) $\cos(-17\pi/6)$ **(d)** $\sin(-17\pi/6)$
19. **(a)** $\sec(4\pi/3)$ **(c)** $\tan(4\pi/3)$
    (b) $\csc(-4\pi/3)$ **(d)** $\cot(-4\pi/3)$
20. **(a)** $\sec(7\pi/4)$ **(c)** $\tan(7\pi/4)$
    (b) $\csc(-7\pi/4)$ **(d)** $\cot(-7\pi/4)$
21. **(a)** $\sec(17\pi/6)$ **(c)** $\tan(17\pi/6)$
    (b) $\csc(-17\pi/6)$ **(d)** $\cot(-17\pi/6)$
22. **(a)** $\sec(3\pi/4)$ **(c)** $\tan(3\pi/4)$
    (b) $\csc(-3\pi/4)$ **(d)** $\cot(-3\pi/4)$
23. List three angles (in radian measure) that have a cosine of $-1/2$.
24. List three angles (in radian measure) that have a sine of $-1/2$.
25. List three angles (in degree measure) that have a cosine of $\sqrt{3}/2$.
26. List three angles (in degree measure) that have a cosine of $-\sqrt{2}/2$.

B

In Exercises 27 and 28 refer to the following figure, in which the x-coordinate of P is 3/4 and the y-coordinate of Q is 1/5.

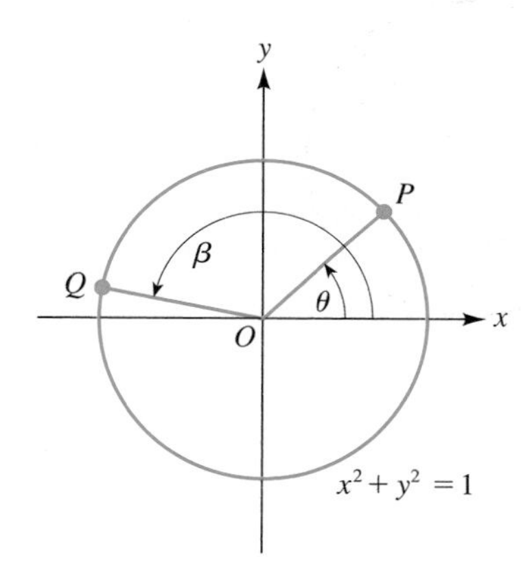

27. **(a)** Evaluate $\cos\theta$.
    (b) Evaluate $\sin\theta$ and $\tan\theta$.
    (c) Which is larger: $\cos(\beta - \frac{\pi}{2})$ or $\cos(\theta + \frac{\pi}{2})$?
    Hint: Sketch both angles in standard position.
28. **(a)** Evaluate $\sin\beta$.
    (b) Evaluate $\cos\beta$ and $\cot\beta$.
    (c) Which is larger: $\cos(\beta + \pi)$ or $\sin(\theta + \pi)$?
    Hint: Sketch both angles in standard position.

In Exercises 29 and 30, evaluate each expression given that.

$$f(x) = \sin x \qquad g(x) = \cos x$$
$$h(x) = \tan x \qquad k(x) = 2x$$

In Exercise 29, exact values or expressions are required, not calculator approximations. In Exercise 30, use a calculator and round the final answers to one decimal place.

29. **(a)** $f(\frac{5\pi}{6} + \frac{\pi}{6})$ **(e)** $\dfrac{\Delta f}{\Delta x}$ on $[\frac{\pi}{4}, \frac{\pi}{2}]$

(b) $f(\frac{5\pi}{6}) + f(\frac{\pi}{6})$ **(f)** $\dfrac{\Delta g}{\Delta x}$ on $[\frac{\pi}{4}, \frac{\pi}{2}]$

(c) $g[k(\frac{3\pi}{4})]$ **(g)** $\dfrac{\Delta f}{\Delta x}$ on $[\frac{5\pi}{6}, \frac{7\pi}{6}]$

(d) $k[g(\frac{3\pi}{4})]$ **(h)** $\dfrac{\Delta g}{\Delta x}$ on $[\frac{5\pi}{6}, \frac{7\pi}{6}]$

30. **(a)** $h(1) + h(2)$ **(e)** $\dfrac{\Delta h}{\Delta x}$ on $[\frac{\pi}{4}, 1]$

(b) $h(1 + 2)$ **(f)** $\dfrac{\Delta h}{\Delta x}$ on $[\frac{\pi}{4}, 1.5]$

(c) $(h \circ k)(\frac{\pi}{5})$ **(g)** $\dfrac{\Delta h}{\Delta x}$ on $[\frac{\pi}{4}, 1.57]$

(d) $(k \circ h)(\frac{\pi}{5})$ **(h)** $\dfrac{\Delta h}{\Delta x}$ on $[\frac{\pi}{4}, 1.58]$

31. **(a)** Use a calculator to complete the following table. (Set your calculator in radian mode.)

θ	$\sin\theta$	Which is larger, θ or $\sin\theta$?
0.1		
0.2		
0.3		
0.4		
0.5		

(b) From the following figure, explain why $PQ < PR < \theta$. *Hint*: What is the length of the arc $\widehat{PR}$?

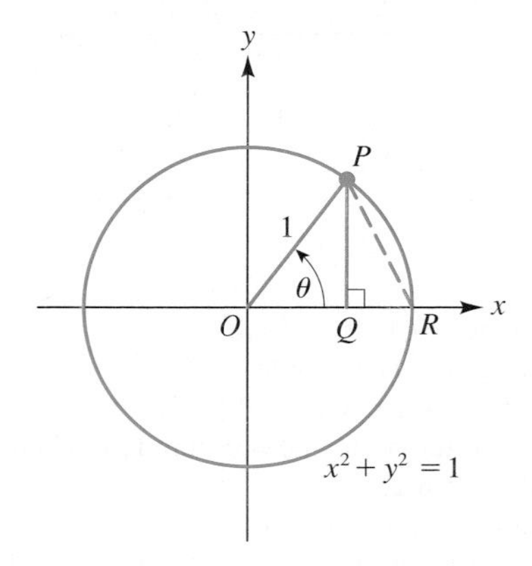

(c) Use the result in part (b) to show that if $0 < \theta < \pi/2$, then

$$\sin\theta < \theta$$

32. For this exercise, refer to Figure 3(b) on page 450. We want to prove that the points P and Q are symmetric about the line $y = x$.

(a) In Figure 3(b), label the origin O, and draw the line segment $\overline{PQ}$. Also, let R denote the point where $\overline{PQ}$ meets the line $y = x$. Explain why $\angle QOR = \angle POR$.

(b) Show that $\triangle QOR$ is congruent to $\triangle POR$.

(c) Use the result in part (b) to show that

(i) angles ORQ and ORP are right angles;

(ii) $QR = PR$.

This shows that the line $y = x$ is the perpendicular bisector of $\overline{PQ}$ and, consequently, that P and Q are symmetric about the line $y = x$.

33. This exercise requires no trigonometry. (The result will be needed in Exercise 34.) We are going to show that the coordinates of the point Q in the following figure are $(-b, a)$.

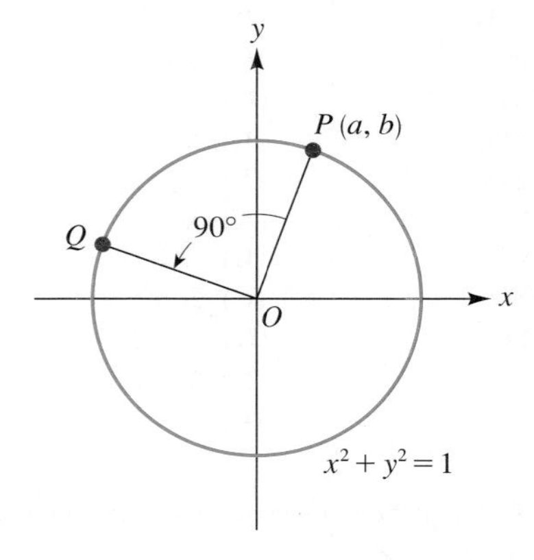

(a) Show that the equation of the line through the points O and Q in the given figure is $y = (-a/b)x$. *Hint*: What is the slope of line segment OP?

(b) As indicated in the figure, Q is the point in the second quadrant where the line segment OQ intersects the unit circle. Determine the coordinates of Q by solving the simultaneous equations

$$\begin{cases} y = (-a/b)x \\ x^2 + y^2 = 1 \end{cases}$$

Answer: $\left(\dfrac{-b}{\sqrt{a^2 + b^2}}, \dfrac{a}{\sqrt{a^2 + b^2}}\right)$

(c) First [without reference to parts (a) or (b)], explain why $a^2 + b^2 = 1$. Then use the result in part (b) to conclude that the coordinates of Q are $(-b, a)$, as required.

34. Go back to page 448 and read the Euler quotation. This exercise explains why knowing the sines and cosines of *acute* angles is sufficient to find the sines and cosines of non-acute angles.

(a) Refer to the following figure. According to the unit circle definitions, what are the coordinates (in terms of θ) of the point P?

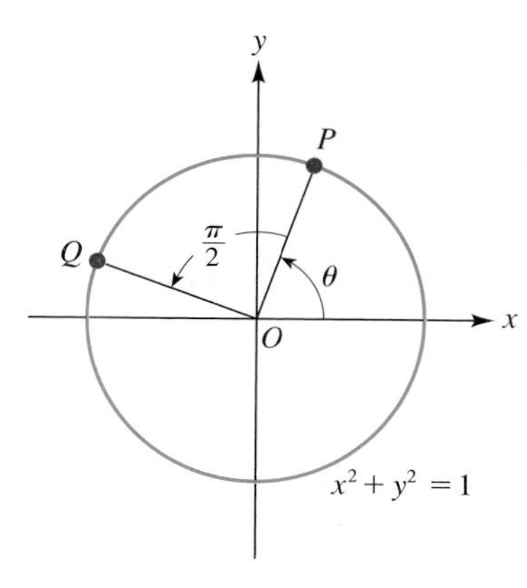

(b) Use your answer in part (a) and the result in Exercise 33 to show (or explain) why

$$\cos(\theta + \tfrac{\pi}{2}) = -\sin\theta \quad \text{and} \quad \sin(\theta + \tfrac{\pi}{2}) = \cos\theta$$

Remark: These two formulas show that if you know the cosine and the sine for a first-quadrant angle θ, then you know the cosine and the sine for the second-quadrant angle $\theta + \pi/2$. (In fact, these formulas are valid when θ is in any quadrant. Although the figures we used in Exercises 33 and 34 show a first-quadrant angle, the method of proof used applies equally well to the other quadrants.)

(c) As examples for the formulas in part (b), use your calculator to verify each of the following equations.

(i) $\cos(1 + \frac{\pi}{2}) = -\sin 1$

(ii) $\sin(1 + \frac{\pi}{2}) = \cos 1$

(iii) $\cos(17^\circ + 90^\circ) = -\sin 17^\circ$

(iv) $\sin(17^\circ + 90^\circ) = \cos 17^\circ$

PROJECT

A LINEAR APPROXIMATION FOR THE SINE FUNCTION

In this project we derive the following very important property of the sine function:

For real numbers θ, the ratio $\dfrac{\sin\theta}{\theta}$ approaches 1 as θ approaches zero.

In the phrase "as θ approaches zero" (in the domain of $\sin\theta$) it is important to understand that θ cannot be zero because we can't substitute zero for θ in our trigonometric ratio.

To begin we need a preliminary result:

For real numbers θ, $\cos\theta$ approaches 1 as θ approaches zero.

EXERCISE 1 Convince yourself that this preliminary result is true in each of the following ways.

Ⓖ **(a)** By looking at the graph of $y = \cos\theta$, for $-\pi/4 < \theta < \pi/4$, in a θ-y coordinate system.

(b) By looking at a point $(\cos\theta, \sin\theta)$ on the unit circle.

In both parts (a) and (b), be sure to consider θ approaching zero through both positive and negative values.

Now consider real numbers θ approaching zero through positive values. Eventually $\theta < \pi/2$ so we may assume $0 < \theta < \pi/2$. In Figure A on the next page we have the unit circle with various labeled points:

O is the origin.

A is the point $(1, 0)$.

P is the point on the unit circle corresponding to the number θ, or, equivalently, the angle with radian measure θ.

B is the point at the foot of the perpendicular from P to the x-axis.

Q is the vertex of $\triangle OAQ$ lying on the terminal side of angle θ outside the unit circle; the line through Q and A is tangent to the circle.

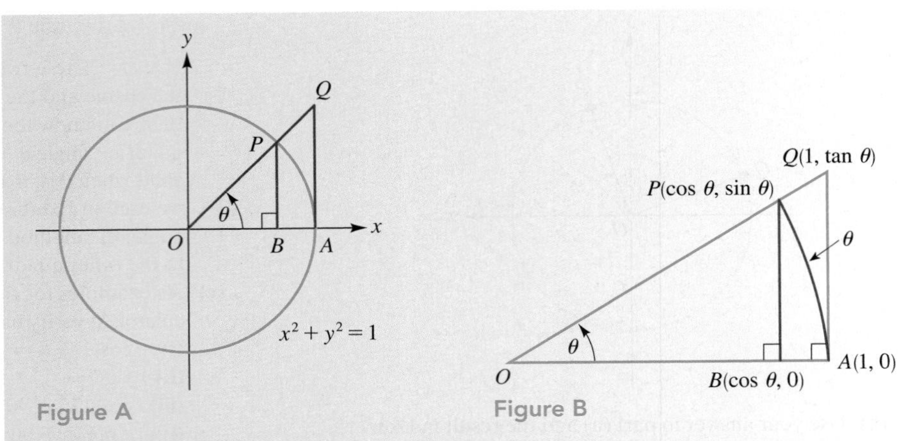

Figure A

Figure B

EXERCISE 2 In Figure A explain why $\angle OAQ$ is a right angle. Then explain why

$$P = (\cos\theta, \sin\theta), \quad B = (\cos\theta, 0), \quad \text{and} \quad Q = (1, \tan\theta)$$

The information from Exercise 2 is shown in Figure B.

EXERCISE 3 In Figure B (where $0 < \theta < \pi/2$) explain why $\sin\theta \leq \theta$. (*Hint*: Draw the line segment from P to A and argue that its length is less than θ, the length of the circular arc from P to A. Use $\triangle PBA$ to complete your explanation.) Then show that

$$\frac{\sin\theta}{\theta} \leq 1 \tag{i}$$

Inequality (i) provides a nice upper bound for $(\sin\theta)/\theta$. Considering the arc length interpretation of radian measure our derivation of (i) seems quite natural. However, we do not know of a similar elementary method for obtaining a lower bound. The standard approach uses an area inequality.

EXERCISE 4 In Figure B explain why

$$\text{area of circular sector } OAP \leq \text{area of triangle } OAQ \tag{ii}$$

EXERCISE 5 Show the following:

(a) area of circular sector $OAP = \frac{1}{2}\theta$

(b) area of triangle $OAQ = \frac{1}{2}\tan\theta$

Substituting the values from Exercise 5 into inequality (ii) we have

$$\frac{1}{2}\theta \leq \frac{1}{2}\tan\theta \tag{iii}$$

EXERCISE 6 Use inequality (iii) to show

$$\cos\theta \leq \frac{\sin\theta}{\theta} \tag{iv}$$

Be careful to explain why the sense of inequality (iii) is preserved.

Combining inequalities (i) and (iv) we have

$$\cos\theta \le \frac{\sin\theta}{\theta} \le 1 \tag{v}$$

As θ approaches zero through positive values, the quantity $(\sin\theta)/\theta$ is "sandwiched" between the two quantities $\cos\theta$ and 1, which both approach 1. This leads us to conclude that

$$\frac{\sin\theta}{\theta} \text{ approaches 1 as } \theta \text{ approaches zero through positive values} \tag{vi}$$

EXERCISE 7 As θ approaches zero through negative values, eventually $-\pi/2 < \theta < 0$. Let $\theta = -\alpha$.
(a) Explain why $0 < \alpha < \pi/2$.
(b) Show that $(\sin\theta)/\theta = (\sin\alpha)/\alpha$.

Continuing with the notation of Exercise 7, as θ approaches zero through negative values, α approaches zero through positive values. Then $(\sin\theta)/\theta$ approaches the same quantity that $(\sin\alpha)/\alpha$ approaches. We know by the argument preceding Exercise 7 that $(\sin\alpha)/\alpha$ approaches 1 as α approaches zero through positive values. Thus

$$\frac{\sin\theta}{\theta} \text{ approaches 1 as } \theta \text{ approaches zero through negative values} \tag{vii}$$

Combining (vi) and (vii), we conclude that

$$\text{For real numbers } \theta, \frac{\sin\theta}{\theta} \text{ approaches 1 as } \theta \text{ approaches zero.} \tag{viii}$$

An important interpretation of (viii) is that for a nonzero real number θ close to zero, $(\sin\theta)/\theta$ is close to 1, or equivalently, $\sin\theta$ is close to θ. We say

$$\sin\theta \approx \theta \qquad \text{for } |\theta| \approx 0 \tag{ix}$$

We interpret (ix) by saying $\sin\theta$ can be approximated by the linear function θ for θ near zero.

EXERCISE 8
(a) For what real numbers θ "near" zero is the percentage error introduced by using θ as an approximation for $\sin\theta$ at most 1%? At most 5%? What are the corresponding angles in degrees?
(b) In Figure B, if the degree measure of the angle θ is 30°, what are the lengths of the circular arc from P to A and the line segment from P to B? What is the percentage error introduced by using the length of the arc as an approximation for the length of the vertical line segment?

Ⓖ **EXERCISE 9** For $-\pi/4 < \theta < \pi/4$, graph $y = \sin\theta$ and $y = \theta$ in a θ-y coordinate system. Note that the graph of $y = \sin\theta$ is very close to the graph of $y = \theta$ for θ near 0. This is the graphical interpretation of the approximation in (ix).

EXERCISE 10

(a) Show that

$$\frac{\tan\theta}{\theta} = \frac{\sin\theta}{\theta}\cdot\frac{1}{\cos\theta} \qquad \text{for } \theta \neq 0$$

(b) Use the identity in part (a) to show that $(\tan\theta)/\theta$ approaches 1 as θ approaches zero.

(c) Explain why

$$\tan\theta \approx \theta \qquad \text{for} \qquad |\theta| \approx 0 \tag{x}$$

What is the functional interpretation of (x)?

Ⓖ **(d)** For $-\pi/4 < \theta < \pi/4$, graph $y = \tan\theta$ and $y = \theta$. Use these graphs to explain the graphical interpretation of the approximation in (x).

6.4 ALGEBRA AND THE TRIGONOMETRIC FUNCTIONS

Albert Girard (1595–1632), who seems to have lived chiefly in Holland, . . . interested himself in spherical trigonometry and trigonometry. In 1626, he published a treatise on trigonometry that contains the earliest use of our abbreviations sin, tan, *and* sec *for* sine, tangent, *and* secant.—Howard Eves in *An Introduction to the History of Mathematics,* 6th ed. (Philadelphia: Saunders College Publishing, 1990)

It was Robert of Chester's translation from the Arabic that resulted in our word "sine." The Hindus had given the name jiva *to the half chord in trigonometry, and the Arabs had taken this over as* jiba. *In the Arabic language there is also a word* jaib *meaning "bay" or "inlet." When Robert of Chester came to translate the technical word* jiba, *he seems to have confused this with the word* jaib *(perhaps because vowels are omitted); hence he used the word* sinus, *the Latin word for "bay" or "inlet."*—Carl B. Boyer in *A History of Mathematics,* 2nd ed., revised by Uta C. Merzback (New York: John Wiley and Sons, 1991)

Sin²φ is odious to me, even though Laplace made use of it . . .—Carl Freidrich Gauss (1777–1855) in a letter to astronomer Heinrich Christian Schumacher (1780–1850)

In this section we are going to practice the algebra involved in working with the trigonometric functions. This will help to pave the way for the more analytical parts of trigonometry in the next chapter. We begin by listing some common notational conventions.

Notational Conventions

1. An expression such as $\sin\theta$ really means $\sin(\theta)$, where sin or sine is the name of the function and θ is an input. It is for historical rather than mathematical reasons that the parentheses are suppressed. An exception to this, however, occurs in expressions such as $\sin(A + B)$, where the parentheses are necessary.
2. Parentheses are often omitted in multiplication. For example, the product $(\sin\theta)(\cos\theta)$ is usually written $\sin\theta\cos\theta$. Similarly, $2(\sin\theta)$ is written $2\sin\theta$.
3. The quantity $(\sin\theta)^n$ is usually written $\sin^n\theta$. For example, $(\sin\theta)^2$ is written $\sin^2\theta$. The same convention also applies to the other five trigonometric functions.*

* A single exception to this convention occurs when $n = -1$. The meaning of the expression $\sin^{-1} x$ will be explained in Section 4 in Chapter 8.

EXAMPLE 1 Algebraic simplification of trigonometric expressions

(a) Combine like terms: $3 \sin^2 \theta \cos \theta - 5 \sin^2 \theta \cos \theta$.
(b) Carry out the multiplication: $(2 \sin \theta - 3 \cos \theta)^2$.

SOLUTION

(a) We do this in the same way we would simplify the algebraic expression $3S^2C - 5S^2C$. Since

$$3S^2C - 5S^2C = -2S^2C$$

we have

$$3 \sin^2 \theta \cos \theta - 5 \sin^2 \theta \cos \theta = -2 \sin^2 \theta \cos \theta$$

(b) We do this in the same way that we would expand $(2S - 3C)^2$. Since

$$(2S - 3C)^2 = 4S^2 - 12SC + 9C^2$$

we have

$$(2 \sin \theta - 3 \cos \theta)^2 = 4 \sin^2 \theta - 12 \sin \theta \cos \theta + 9 \cos^2 \theta$$

EXAMPLE 2 Factoring trigonometric expressions

Factor: $\tan^2 A + 5 \tan A + 6$.

SOLUTION

PRELIMINARY SOLUTION To help us focus on the algebra that is actually involved, let's replace each occurrence of the quantity $\tan A$ by the letter T. Then

$$T^2 + 5T + 6 = (T + 3)(T + 2)$$

ACTUAL SOLUTION

$$\tan^2 A + 5 \tan A + 6 = (\tan A + 3)(\tan A + 2)$$

Note: After you are accustomed to working with trigonometric expressions, you should be able to eliminate the preliminary step.

EXAMPLE 3 Simplifying trigonometric expressions

Add: $\sin \theta + \dfrac{1}{\cos \theta}$.

SOLUTION

PRELIMINARY SOLUTION

$$S + \frac{1}{C} = \frac{S}{1} + \frac{1}{C} = \frac{S}{1} \cdot \frac{C}{C} + \frac{1}{C}$$

$$= \frac{SC}{C} + \frac{1}{C} = \frac{SC + 1}{C}$$

ACTUAL SOLUTION

$$\begin{aligned}\sin\theta + \frac{1}{\cos\theta} &= \frac{\sin\theta}{1} + \frac{1}{\cos\theta} = \frac{\sin\theta}{1}\cdot\frac{\cos\theta}{\cos\theta} + \frac{1}{\cos\theta}\\ &= \frac{\sin\theta\cos\theta}{\cos\theta} + \frac{1}{\cos\theta} = \frac{\sin\theta\cos\theta + 1}{\cos\theta}\end{aligned}$$

One of the most useful techniques for simplifying a trigonometric expression is first to rewrite it in terms of sines and cosines and then to carry out the usual algebraic simplifications. The next example shows how this works.

EXAMPLE 4 Simplifying trigonometric expressions

Simplify the expression $\dfrac{\sec A + 1}{\sin A + \tan A}$.

SOLUTION

$$\begin{aligned}\frac{\sec A}{\sin A + \tan A} &= \frac{\dfrac{1}{\cos A} + 1}{\sin A + \dfrac{\sin A}{\cos A}}\\ &= \frac{\cos A\left(\dfrac{1}{\cos A} + 1\right)}{\cos A\left(\sin A + \dfrac{\sin A}{\cos A}\right)} \qquad \text{multiplying numerator and denominator by } \cos A\\ &= \frac{1 + \cos A}{\cos A \sin A + \sin A} = \frac{\overset{1}{\cancel{1 + \cos A}}}{\sin A(\underset{1}{\cancel{\cos A + 1}})} = \frac{1}{\sin A}\end{aligned}$$

This last expression can also be written as $\csc A$.

EXAMPLE 5 Determining when an equation is not an identity

Show that the statement $\cos A + \cos B = \cos(A + B)$ is not true in general.

SOLUTION

Consider a related question. How would you convince a beginning algebra student that the statement $(x + y)^2 = x^2 + y^2$ is not true in general? One way is simply to pick specific values for x and y and then show that the equation fails for these values. For instance, using $x = 1$ and $y = 1$ gives

$$\begin{aligned}(1 + 1)^2 &\overset{?}{=} 1^2 + 1^2\\ 4 &\overset{?}{=} 2 \qquad \text{No}\end{aligned}$$

We can do the same thing in this example. Let $A = 30°$ and $B = 60°$. Then we have

$$\cos A + \cos B \stackrel{?}{=} \cos(A + B)$$
$$\cos 30^\circ + \cos 60^\circ \stackrel{?}{=} \cos(30^\circ + 60^\circ)$$
$$\frac{\sqrt{3}}{2} + \frac{1}{2} \stackrel{?}{=} \cos 90^\circ$$
$$\frac{\sqrt{3} + 1}{2} \stackrel{?}{=} 0 \qquad \text{No}$$

We conclude that the statement $\cos A + \cos B = \cos(A + B)$ is *not* true in general. It follows from the discussion below that the equation is not an identity.

Although the equation in the last example is not an identity, there are numerous identities involving the trigonometric functions. Recall that an **identity** is an equation that is satisfied by all relevant values of the variables concerned. Two examples of identities are $(x - y)(x + y) = x^2 - y^2$ and $x^3 = x^4/x$. The first of these is true no matter what real numbers are used for x and y; the second is true for all real numbers except $x = 0$. For now, we'll consider only a few of the most basic identities.

PROPERTY SUMMARY Basic Trigonometric Identities

Identity	Examples
1. $\sin^2\theta + \cos^2\theta = 1$	1. $\sin^2 30^\circ + \cos^2 30^\circ = 1$ $\sin^2 19^\circ + \cos^2 19^\circ = 1$
2. $\dfrac{\sin\theta}{\cos\theta} = \tan\theta$	2. $\dfrac{\sin 60^\circ}{\cos 60^\circ} = \tan 60^\circ$
3. $\sec\theta = \dfrac{1}{\cos\theta}$; $\csc\theta = \dfrac{1}{\sin\theta}$; $\cot\theta = \dfrac{1}{\tan\theta}$	3. $\sec 2 = \dfrac{1}{\cos 2}$; $\csc 45^\circ = \dfrac{1}{\sin 45^\circ}$; $\cot(-3) = \dfrac{1}{\tan(-3)}$

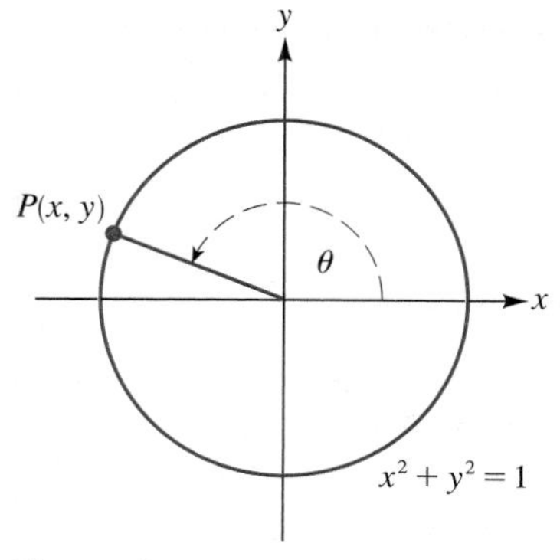

Figure 1

To see why the first identity in the box is valid, consider Figure 1. Since the angle θ in standard position determines the point $P(x, y)$ on the unit circle, we have

$$x^2 + y^2 = 1$$

But by definition, $x = \cos\theta$ and $y = \sin\theta$. Therefore we have

$$(\cos\theta)^2 + (\sin\theta)^2 = 1 \qquad \text{or, equivalently,} \qquad \sin^2\theta + \cos^2\theta = 1$$

which is essentially what we wished to show. Incidentally, you should also become familiar with the equivalent forms of this identity:

$$\cos^2\theta = 1 - \sin^2\theta \qquad \text{and} \qquad \sin^2\theta = 1 - \cos^2\theta$$

The second and the third identities in the box are immediate consequences of the unit-circle definitions of the trigonometric functions. For example,

$$\tan\theta = \frac{y}{x} = \frac{\sin\theta}{\cos\theta}, \quad \text{assuming } x \neq 0 \qquad \text{or, equivalently,} \qquad \cos\theta \neq 0$$

EXAMPLE 6 Using identities to calculate trigonometric functions

Given $\sin\theta = 2/3$ and $\pi/2 < \theta < \pi$, find $\cos\theta$ and $\tan\theta$.

SOLUTION

Substituting $\sin\theta = 2/3$ into the identity $\sin^2\theta + \cos^2\theta = 1$ yields

$$\left(\frac{2}{3}\right)^2 + \cos^2\theta = 1$$

$$\cos^2\theta = 1 - \left(\frac{2}{3}\right)^2 = \frac{5}{9}$$

$$\cos\theta = \pm\sqrt{\frac{5}{9}} = \frac{\pm\sqrt{5}}{3}$$

(Note that we could have started with $\cos^2\theta = 1 - \sin^2\theta$.)

To decide whether to choose the positive or the negative value, note that the given inequality $\pi/2 < \theta < \pi$ tells us that the terminal side of θ lies in the second quadrant. Since x-coordinates are negative in Quadrant II, we choose the negative value here. Thus

$$\cos\theta = \frac{-\sqrt{5}}{3}$$

For $\tan\theta$ we have

$$\tan\theta = \frac{y}{x} = \frac{\sin\theta}{\cos\theta} = \frac{2/3}{-\sqrt{5}/3}$$

$$= -\frac{2}{\sqrt{5}} = -\frac{2\sqrt{5}}{5}$$

The next example on simplifying a trigonometric expression is similar to Examples 1 through 4. This time, as we work the problem, we come across the expression $\sin^2\theta + \cos^2\theta$, which is then replaced by 1.

EXAMPLE 7 Using identities to simplify trigonometric expressions

Combine and simplify: $\dfrac{\sin A}{\cos A} + \dfrac{\cos A}{\sin A}$.

SOLUTION

The common denominator is $\cos A \sin A$. Therefore we have

$$\frac{\sin A}{\cos A} + \frac{\cos A}{\sin A} = \frac{\sin A}{\cos A}\cdot\frac{\sin A}{\sin A} + \frac{\cos A}{\sin A}\cdot\frac{\cos A}{\cos A}$$

$$= \frac{\sin^2 A}{\cos A \sin A} + \frac{\cos^2 A}{\sin A \cos A}$$

$$= \frac{\sin^2 A + \cos^2 A}{\cos A \sin A}$$

$$= \frac{1}{\cos A \sin A}$$

This is the required result. If we wish, we can write this answer in an alternative form that doesn't involve fractions:

$$\frac{1}{\cos A \sin A} = \frac{1}{\cos A} \cdot \frac{1}{\sin A} = \sec A \csc A$$

In the next three examples we show that certain trigonometric equations are, in fact, identities. This type of problem is similar to Examples 1–4 and 7, except now we are given an answer toward which to work. The identities in these examples should not be memorized; they are too specialized. Instead, concentrate on the proofs themselves, noting where the fundamental identities (such as $\sin^2 \theta + \cos^2 \theta = 1$) come into play. Remember that the identities are true only for values of the variable for which all expressions are defined.

EXAMPLE 8 Proving an identity by expressing all trigonometric functions in terms of sine and cosine

Prove that the following equation is an identity:

$$\csc A \tan A \cos A = 1$$

SOLUTION

We begin with the left-hand side and express each factor in terms of sines or cosines:

$$\begin{aligned} \csc A \tan A \cos A &= \frac{1}{\cancel{\sin A}_{1}} \cdot \frac{\cancel{\sin A}^{1}}{\cancel{\cos A}} \cdot \cancel{\cos A}^{1} \\ &= 1 \quad \text{as required} \end{aligned}$$

EXAMPLE 9 Proving an identity by expressing all trigonometric functions in terms of sine and cosine

Prove that $\cos^2 B - \sin^2 B = \dfrac{1 - \tan^2 B}{1 + \tan^2 B}$.

SOLUTION

We begin with the right-hand side this time; it is the more complicated expression, and it is easier to express $\tan^2 B$ in terms of $\sin B$ and $\cos B$ than it is to express $\sin^2 B$ and $\cos^2 B$ in terms of $\tan B$. As in previous examples we write everything in terms of sines and cosines.

$$\begin{aligned} \frac{1 - \tan^2 B}{1 + \tan^2 B} &= \frac{1 - \dfrac{\sin^2 B}{\cos^2 B}}{1 + \dfrac{\sin^2 B}{\cos^2 B}} = \frac{\cos^2 B\left(1 - \dfrac{\sin^2 B}{\cos^2 B}\right)}{\cos^2 B\left(1 + \dfrac{\sin^2 B}{\cos^2 B}\right)} \\ &= \frac{\cos^2 B - \sin^2 B}{\cos^2 B + \sin^2 B} = \frac{\cos^2 B - \sin^2 B}{1} = \cos^2 B - \sin^2 B \end{aligned}$$

EXAMPLE 10 An algebraic technique for creating a difference of squares

Prove that $\dfrac{\cos\theta}{1-\sin\theta} = \dfrac{1+\sin\theta}{\cos\theta}$.

SOLUTION

The suggestions given in the previous examples are not applicable here. Everything is already in terms of sines and cosines. Furthermore, neither side appears more complicated than the other. A technique that does work here is to begin with the left-hand side and "rationalize" the denominator to obtain a perfect square. We do this by multiplying numerator and denominator by the same quantity, namely, $1 + \sin\theta$. This process is analogous to that of rationalizing a denominator using the factoring formula for the difference of squares. Doing so gives us

$$\frac{\cos\theta}{1-\sin\theta} = \frac{\cos\theta}{1-\sin\theta}\cdot\frac{1+\sin\theta}{1+\sin\theta} = \frac{(\cos\theta)(1+\sin\theta)}{1-\sin^2\theta}$$
$$= \frac{(\cos\theta)(1+\sin\theta)}{\cos^2\theta} = \frac{1+\sin\theta}{\cos\theta}$$

This identity could also be proven by starting on the right-hand side and rationalizing the numerator.

The general strategy for each of the proofs in Examples 8 through 10 was the same. In each case we worked with one side of the given equation, and we transformed it into equivalent expressions until it was identical to the other side of the equation. This is not the only strategy that can be used.

Before we describe an alternative strategy, a warning is in order. Establishing an identity is *not* like solving an equation. When solving an equation, we begin with an assumption that there is a value for the variable that makes the left-hand and right-hand sides equal. Then, using operations such as adding the same quantity to both sides or multiplying both sides by the same nonzero quantity, we derive a series of equivalent equations until the original equation is solved.

On the other hand, an identity is an assertion that two functions are equal, that is, their function values are equal for *all* values of the variable in their common domain. To prove an identity, we *cannot assume* that the left-hand and right-hand sides are equal, since that is precisely what we are trying to prove; so we cannot use the same techniques that we use to solve an equation.

In proving an identity, it is important to *separately* transform the left-hand side into the right-hand side or the right-hand side into the left-hand side, as in Examples 8 through 10. Another strategy that can be used in establishing identities is to separately transform each side to a common expression. We use this strategy in the next example.

EXAMPLE 11 Proving an identity by showing that both sides equal the same expression

Prove that $\dfrac{1}{1-\cos\beta} + \dfrac{1}{1+\cos\beta} = 2 + 2\cot^2\beta$.

SOLUTION

$$\text{Left-hand side} = \frac{1(1+\cos\beta)+1(1-\cos\beta)}{(1-\cos\beta)(1+\cos\beta)} = \frac{2}{1-\cos^2\beta}$$

$$\text{Right-hand side} = 2 + \frac{2\cos^2\beta}{\sin^2\beta} = \frac{2\sin^2\beta + 2\cos^2\beta}{\sin^2\beta} = \frac{2(\sin^2\beta+\cos^2\beta)}{1-\cos^2\beta} = \frac{2}{1-\cos^2\beta}$$

We've now established the required identity by showing that both sides are equal to the same expression.

EXERCISE SET 6.4

A

In Exercises 1–12, carry out the indicated operations.

1. (a) $-SC + 12SC$
(b) $-\sin\theta\cos\theta + 12\sin\theta\cos\theta$

2. (a) $10SC + 4SC - 16SC$
(b) $10\sin\theta\cos\theta + 4\sin\theta\cos\theta - 16\sin\theta\cos\theta$

3. (a) $4C^3S - 12C^3S$
(b) $4\cos^3\theta\sin\theta - 12\cos^3\theta\sin\theta$

4. (a) $-C^2S^2 + (2SC)^2$
(b) $-\cos^2\theta\sin^2\theta + (2\sin\theta\cos\theta)^2$

5. (a) $(1+T)^2$
(b) $(1+\tan\theta)^2$

6. (a) $(3-2T)^2$
(b) $(3-2\tan\theta)^2$

7. (a) $(T+3)(T-2)$
(b) $(\tan\theta+3)(\tan\theta-2)$

8. (a) $(S^2-3)(S^2+3)$
(b) $(\sec^2\theta-3)(\sec^2\theta+3)$

9. (a) $\dfrac{S-C}{C-S}$
(b) $\dfrac{\sin\theta-\cos\theta}{\cos\theta-\sin\theta}$

10. (a) $\dfrac{5-2T}{2T-5}$
(b) $\dfrac{5-2\tan\theta}{2\tan\theta-5}$

11. (a) $C + \dfrac{2}{S}$
(b) $\cos A + \dfrac{2}{\sin A}$

12. (a) $\dfrac{1}{S} - \dfrac{3}{C}$
(b) $\dfrac{1}{\sin A} - \dfrac{3}{\cos A}$

In Exercises 13–18, factor each expression.

13. (a) $T^2 + 8T - 9$
(b) $\tan^2\beta + 8\tan\beta - 9$

14. (a) $3S^2 + 2S - 8$
(b) $3\sec^2\beta + 2\sec\beta - 8$

15. (a) $4C^2 - 1$
(b) $4\cos^2 B - 1$

16. (a) $16S^3 - 9S^2$
(b) $16\sin^3 B - 9\sin^2 B$

17. (a) $9S^2T^3 + 6ST^2$
(b) $9\sec^2 B\tan^3 B + 6\sec B\tan^2 B$

18. (a) $5C^2c^2 - 15Cc.$
(b) $5\csc^2 B\cot^2 B - 15\csc B\cot B$

In Exercises 19–32, write in terms of sine and cosine and simplify each expression.

19. $\dfrac{\sin^2 A - \cos^2 A}{\sin A - \cos A}$ *Hint*: Factor the numerator.

20. $\dfrac{\sin^4 A - \cos^4 A}{\cos A - \sin A}$

21. $\sin^2\theta\cos\theta\csc^3\theta\sec\theta$

22. $\sin\theta\csc\theta\tan\theta$

23. $\cot B\sin^2 B\cot B$

24. $\dfrac{3\sin\theta + 6}{\sin^2\theta - 4}$

25. $\dfrac{\cos^2 A + \cos A - 12}{\cos A - 3}$

26. $\dfrac{\cos A - 2\sin A\cos A}{\cos^2 A - \sin^2 A + \sin A - 1}$

27. $\dfrac{\tan\theta}{\sec\theta - 1} + \dfrac{\tan\theta}{\sec\theta + 1}$

28. $\cot\theta + \dfrac{1 - 2\cos^2\theta}{\sin\theta\cos\theta}$

29. $\sec A\csc A - \tan A - \cot A$

30. $(\sec A + \tan A)(\sec A - \tan A)$

31. $\dfrac{\cot^2\theta}{\csc^2\theta} + \dfrac{\tan^2\theta}{\sec^2\theta}$

32. $\dfrac{\tan\theta + \tan\theta\sin\theta - \cos\theta\sin\theta}{\sin\theta\tan\theta}$

33. If $\sin\theta = -3/5$ and $\pi < \theta < 3\pi/2$, compute $\cos\theta$ and $\tan\theta$.

34. If $\cos\theta = 5/13$ and $3\pi/2 < \theta < 2\pi$, compute $\sin\theta$ and $\cot\theta$.

35. If $\sin t = \sqrt{3}/4$ and $\pi/2 < t < \pi$, compute $\tan t$.

36. If $\sec\theta = -\sqrt{13}/2$ and $\sin\theta > 0$, compute $\tan\theta$.

37. If $\sec\beta = -17/15$ and $\pi/2 < \beta < \pi$, compute $\csc\beta$ and $\cot\beta$.

38. If $\csc\alpha = \sqrt{3}$ and $2\pi < \alpha < 5\pi/2$, compute $\cos\alpha$ and $\sin\alpha$.

In Exercises 39–44, use the given information to determine the remaining five trigonometric values.

39. $\sin\theta = 1/5$, $90° < \theta < 180°$

40. $\sin\theta = -24/25$, $180° < \theta < 270°$

41. $\cos\theta = -3/5$, $180° < \theta < 270°$

42. $\cos\theta = 1/4$, $270° < \theta < 360°$

43. $\sec \beta = 3, \quad 0^\circ < \beta < 90^\circ$

44. $\csc \beta = -\sqrt{5}, \quad 270^\circ < \beta < 360^\circ$

In Exercises 45–66, prove that the equations are identities.

45. $\sin \theta \cos \theta \sec \theta \csc \theta = 1$

46. $\tan^2 A + 1 = \sec^2 A$

47. $(\sin \theta \sec \theta)/(\tan \theta) = 1$

48. $\tan \beta \sin \beta = \sec \beta - \cos \beta$

49. $(1 - 5 \sin x)/\cos x = \sec x - 5 \tan x$

50. $\dfrac{1}{\sin \theta} - \sin \theta = \cot \theta \cos \theta$

51. $(\cos A)(\sec A - \cos A) = \sin^2 A$

52. $\dfrac{\sin \theta}{\csc \theta} + \dfrac{\cos \theta}{\sec \theta} = 1$

53. $(1 - \sin \theta)(\sec \theta + \tan \theta) = \cos \theta$

54. $(\cos \theta - \sin \theta)^2 + 2 \sin \theta \cos \theta = 1$

55. $(\sec \alpha - \tan \alpha)^2 = \dfrac{1 - \sin \alpha}{1 + \sin \alpha}$

56. $\dfrac{\sin B}{1 + \cos B} + \dfrac{1 + \cos B}{\sin B} = 2 \csc B$

57. $\sin A + \cos A = \dfrac{\sin A}{1 - \cot A} - \dfrac{\cos A}{\tan A - 1}$

58. $(1 - \cos C)(1 + \sec C) = \tan C \sin C$

59. $\csc^2 \theta + \sec^2 \theta = \csc^2 \theta \sec^2 \theta$

60. $\cos^2 \theta - \sin^2 \theta = 1 - 2 \sin^2 \theta$

61. $\sin A \tan A = \dfrac{1 - \cos^2 A}{\cos A}$

62. $\dfrac{\cot A - 1}{\cot A + 1} = \dfrac{1 - \tan A}{1 + \tan A}$

63. $\cot^2 A + \csc^2 A = -\cot^4 A + \csc^4 A$

64. $\dfrac{\cot^2 A - \tan^2 A}{(\cot A + \tan A)^2} = 2 \cos^2 A - 1$

65. $\dfrac{\sin A - \cos A}{\sin A} + \dfrac{\cos A - \sin A}{\cos A} = 2 - \sec A \csc A$

66. $\tan A \tan B = \dfrac{\tan A + \tan B}{\cot A + \cot B}$

In Exercises 67 and 68, refer to the following figure.

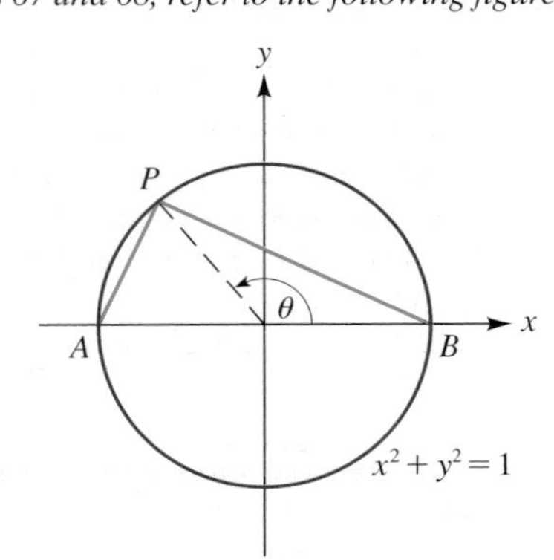

67. Express the area of $\triangle ABP$ as a function of θ.

68. **(a)** Express the slope of $\overline{PB}$ as a function of θ.

(b) Express the slope of $\overline{PA}$ as a function of θ.

(c) Using the expressions obtained in parts (a) and (b), compute the product of the two slopes.

(d) What can you conclude from your answer in part (c)?

69. **(a)** Refer to the figure. Express the slope m of the line as a function of θ. *Hint*: What are the coordinates of the point P?

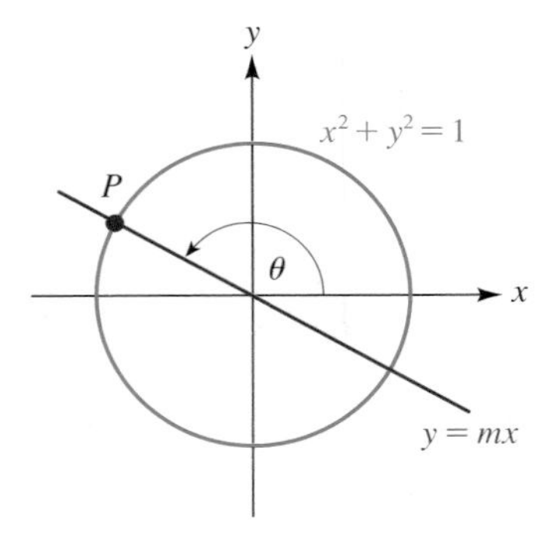

(b) Use your result from part (a) to specify the slope of each of the following lines. For line (ii), use a calculator and round the answer to one decimal place.

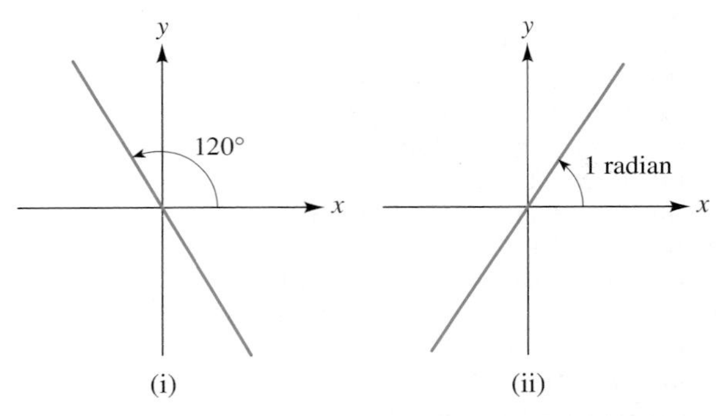

(i) (ii)

70. **(a)** Let P be a point on the unit circle. As indicated in the following figure, rotating the point P about the origin through an angle of π radians yields a point P' (on the unit circle) that is the reflection of P through the origin. Use this observation and the definition of symmetry about the origin (on page 58) to explain in complete sentences why

$$\sin(\theta + \pi) = -\sin \theta \qquad \text{and} \qquad \cos(\theta + \pi) = -\cos \theta$$

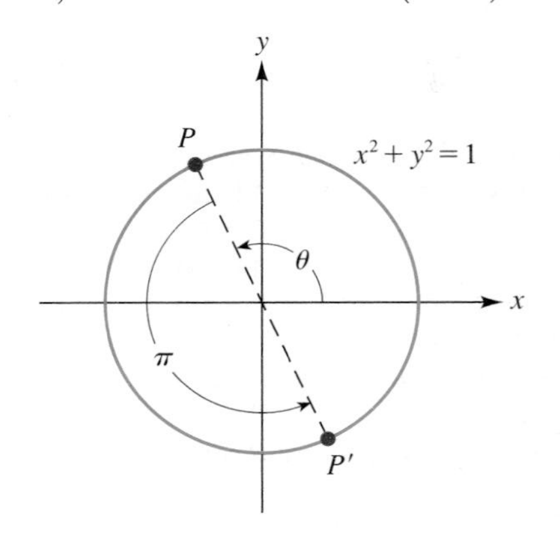

(b) Use the results in part (a) to show that $\tan(\theta + \pi) = \tan\theta$.

(c) As examples of the results in parts (a) and (b), use a calculator to verify each of the following statements:

$$\sin(2+\pi) = -\sin 2$$
$$\cos(2+\pi) = -\cos 2$$
$$\tan(2+\pi) = \tan 2$$

71. Only one of the following two equations is an identity. Decide which equation this is, and give a proof to show that it is, indeed, an identity. For the other equation, give an example showing that it is not an identity. (For example, to show that the equation $\sin\theta + \cos\theta = 1$ is not an identity, let $\theta = 30°$. Then the equation becomes $1/2 + \sqrt{3}/2 = 1$, which is false.)

(a) $\dfrac{\csc^2\alpha - 1}{\csc^2\alpha} = \cos\alpha$

(b) $(\sec^2\alpha - 1)(\csc^2\alpha - 1) = 1$

72. Follow the directions given in Exercise 71.

(a) $(\csc\beta - \cot\beta)^2 = \dfrac{1+\cos\beta}{1-\cos\beta}$

(b) $\cot\beta + \dfrac{\sin\beta}{1+\cos\beta} = \csc\beta$

73. Prove the identity $\dfrac{\sin\theta}{1-\cos\theta} = \dfrac{1+\cos\theta}{\sin\theta}$ in two ways.

(a) Adapt the method of Example 10.

(b) Begin with the left-hand side and multiply numerator and denominator by $\sin\theta$.

In Exercises 74–78, prove that the equations are identities.

74. $\dfrac{2\sin^3\beta}{1-\cos\beta} = 2\sin\beta + 2\sin\beta\cos\beta$

Hint: Write $\sin^3\beta$ as $(\sin\beta)(\sin^2\beta)$.

75. $\dfrac{\sec\theta - \csc\theta}{\sec\theta + \csc\theta} = \dfrac{\tan\theta - 1}{\tan\theta + 1}$

76. $1 - \dfrac{\sin^2\theta}{1+\cot\theta} - \dfrac{\cos^2\theta}{1+\tan\theta} = \sin\theta\cos\theta$

77. $(\sin^2\theta)(1 + n\cot^2\theta) = (\cos^2\theta)(n + \tan^2\theta)$

78. $(r\sin\theta\cos\phi)^2 + (r\sin\theta\sin\phi)^2 + (r\cos\theta)^2 = r^2$

79. (a) Factor the expression $\cos^3\theta - \sin^3\theta$.

(b) Prove the identity

$$\frac{\cos\phi\cot\phi - \sin\phi\tan\phi}{\csc\phi - \sec\phi} = 1 + \sin\phi\cos\phi$$

80. Prove the following identities. (These two identities, along with $\sin^2\theta + \cos^2\theta = 1$, are known as the *Pythagorean identities.* They will be discussed in Chapter 8.)

(a) $\tan^2\theta = \sec^2\theta - 1$ **(b)** $\cot^2\theta = \csc^2\theta - 1$

81. If $\tan\alpha\tan\beta = 1$ and α and β are acute angles, show that $\sec\alpha = \csc\beta$. *Hint:* Make use of the identities in Exercise 80.

82. Suppose that

$$A\sin\theta + \cos\theta = 1 \quad \text{and} \quad B\sin\theta - \cos\theta = 1$$

Show that $AB = 1$. *Hint:* Solve the first equation for A, the second for B, and then compute AB.

83. If $\sin\alpha + \cos\alpha = a$ and $\sin\alpha - \cos\alpha = b$, show that

$$\tan\alpha = \frac{a+b}{a-b}$$

84. If $a\sin^2\theta + b\cos^2\theta = 1$, show that

$$\sin^2\theta = \frac{1-b}{a-b} \quad \text{and} \quad \tan^2\theta = \frac{b-1}{1-a}$$

As background for Exercises 85–90, you need to have studied logarithms in Sections 5.3 and 5.4.

85. (a) Choose (at random) an angle θ such that $0° < \theta < 90°$. Then with this value of θ, use your calculator to verify that $\log_{10}(\sin^2\theta) = 2\log_{10}(\sin\theta)$.

(b) For which values of θ in the interval $0° \le \theta \le 180°$ is the equation in part (a) valid?

86. (a) Choose (at random) an angle θ such that $0° < \theta < 90°$. Then with this value of θ, use your calculator to verify that $\log_{10}(\cos^2\theta) = 2\log_{10}(\cos\theta)$.

(b) For which values of θ in the interval $0° \le \theta \le 180°$ is the equation in part (a) valid?

For Exercises 87 and 88, let

$$\left.\begin{array}{l} S(\theta) = \sin\theta \\ C(\theta) = \cos\theta \end{array}\right\} \quad 0° \le \theta \le 360°$$
$$L(x) = \ln x$$

87. What is the domain of the function $(L \circ S)(\theta)$?

88. What is the domain of the function $(L \circ C)(\theta)$?

89. (a) Choose (at random) an angle θ such that $0° < \theta < 90°$. Then with this value of θ, use your calculator to verify that

$$\ln\sqrt{1-\cos\theta} + \ln\sqrt{1+\cos\theta} = \ln(\sin\theta)$$

(b) Use the properties of logarithms to prove that if $0° < \theta < 90°$, then

$$\ln\sqrt{1-\cos\theta} + \ln\sqrt{1+\cos\theta} = \ln(\sin\theta)$$

(c) For which values of θ in the interval $0° \le \theta \le 360°$ is the equation in part (b) valid?

90. (a) Choose (at random) an angle θ such that $0° < \theta < 90°$. Then with this value of θ, use your calculator to verify that

$$\ln\sqrt{1-\sin\theta} + \ln\sqrt{1+\sin\theta} = \ln(\cos\theta)$$

(b) Use the properties of logarithms to prove that the equation in part (a) holds for all acute angles $(0° < \theta < 90°)$.

(c) Does the equation in part (a) hold if $\theta = 90°$? If $\theta = 0°$?

(d) For which values of θ in the interval $0° \le \theta \le 360°$ is the equation in part (a) valid?

C

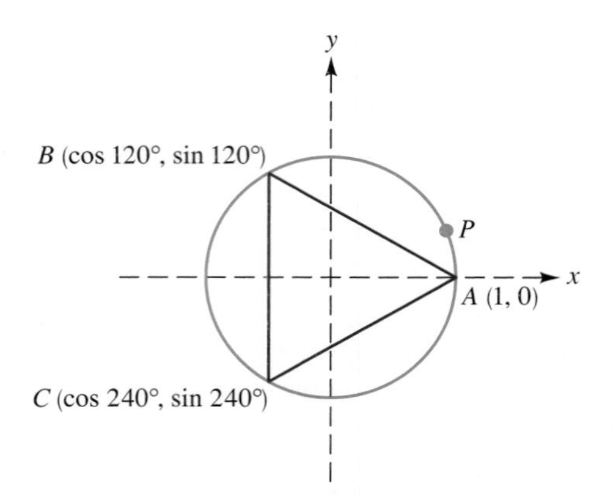

91. In this exercise, you'll use the unit circle definitions of sine and cosine, along with the identity $\sin^2\theta + \cos^2\theta = 1$ to prove a surprising geometric result. In the figure at the right, we show an equilateral triangle inscribed in the unit circle $x^2 + y^2 = 1$. Prove that for any point P on the unit circle, the sum of the squares of the distances from P to the three vertices is 6. *Hint*: Let the coordinates of P be $(\cos\theta, \sin\theta)$.

92. Prove the identity $\dfrac{\tan\theta + \sec\theta - 1}{\tan\theta - \sec\theta + 1} = \dfrac{1 + \sin\theta}{\cos\theta}$.

MINI PROJECT

IDENTITIES AND GRAPHS

An identity is an assertion that two functions are equal for every value in their common domain. In this mini project you will use your graphing utility to visualize identities.

Ⓖ EXERCISE

Graph each pair of equations. If a pair of graphs suggests an identity, then state the identity clearly with a carefully stated domain. If a pair does not suggest an identity then exhibit a counterexample, that is, a value of the domain variable at which the function values are not the same. For trigonometric functions the variables represent angles in *degrees.* So, in parts (e) through (j), make sure you are working in degree mode and try starting with a window of [−360, 360, 30] by [−1, 1, 0.25] and then vary the size.

(a) $y = x^2 + 2x + 1$ and $y = (x + 1)^2$

(b) $y = \dfrac{t^2 - 9}{t + 3}$ and $y = t - 3$

(c) $y = \sqrt{x^2 + 9}$ and $y = x + 3$

(d) $y = \dfrac{|p - 1|}{p - 1}$ and $y = \begin{cases} -1, & \text{if } p < 1 \\ 1, & \text{if } p > 1 \end{cases}$

(e) $y = \cos^2\theta + \sin^2\theta$ and $y = 1$

(f) $y = \sec^2\alpha - 1$ and $y = \tan^2\alpha$

(g) $y = \sin 2\theta$ and $y = 2\sin\theta$

(h) $y = \cos 2\beta$ and $y = \cos^2\beta - \sin^2\beta$

(i) $y = \cos(\alpha - 60°)$ and $y = \cos\alpha - \cos 60°$

(j) $y = \cos 3\theta + \sqrt{3}\sin 3\theta$ and $y = 2\cos(3\theta - 60°)$

6.5 RIGHT-TRIANGLE TRIGONOMETRY

A special name for the function which we call the sine is first found in the works of Āryabhata (c. 510). . . . It is further probable from the efforts made to develop simple tables that the Hindus were acquainted with the principles which we represent by the . . . [formula]

$\sin^2 \phi + \cos^2 \phi = 1. \ldots$—David Eugene Smith in *History of Mathematics,* vol. II (New York: Ginn and Company, 1925)

Trigonometry was developed into an independent branch of mathematics by Islamic writers, notably by Nasir ed-dīn at-Tūsī (or Nasir Eddin, 1201–1274). The first publication in Latin Europe to achieve the same goal was Regiomontanus' De triangulis omnimodis (On triangles of all kinds); *Nuremburg, 1533.—A Source Book in Mathematics,* 1200–1800, edited by D. J. Struik (Princeton: Princeton University Press, 1986)

Now an acute angle can always be placed in a right triangle as shown in Figure 1. The six trigonometric functions are defined in the box that follows.

DEFINITIONS **Trigonometric Functions of an Acute Angle**

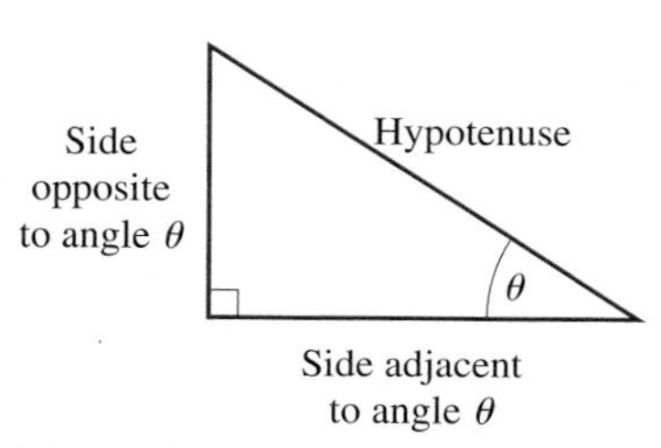

Figure 1

Let θ be an acute angle placed in a right triangle (as shown in Figure 1); then

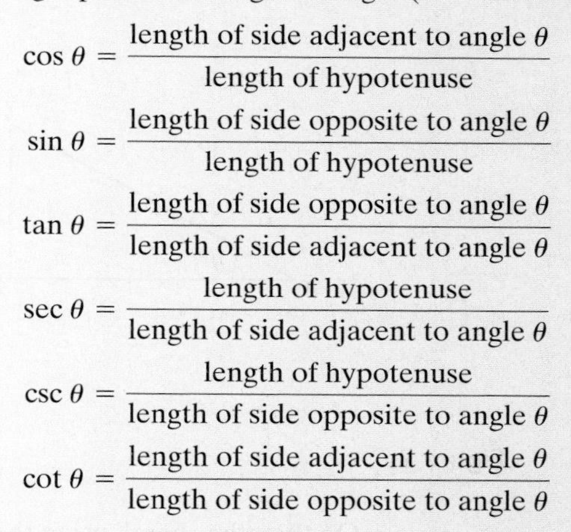

$$\cos \theta = \frac{\text{length of side adjacent to angle } \theta}{\text{length of hypotenuse}}$$

$$\sin \theta = \frac{\text{length of side opposite to angle } \theta}{\text{length of hypotenuse}}$$

$$\tan \theta = \frac{\text{length of side opposite to angle } \theta}{\text{length of side adjacent to angle } \theta}$$

$$\sec \theta = \frac{\text{length of hypotenuse}}{\text{length of side adjacent to angle } \theta}$$

$$\csc \theta = \frac{\text{length of hypotenuse}}{\text{length of side opposite to angle } \theta}$$

$$\cot \theta = \frac{\text{length of side adjacent to angle } \theta}{\text{length of side opposite to angle } \theta}$$

The definitions that we have just given are often called the right triangle definitions of the trigonometric functions. Our work for the next several sections will be devoted to exploring these definitions and their consequences. Here are four preliminary observations that will help you to understand and memorize the definitions.

1. An expression such as $\cos \theta$ really means $\cos(\theta)$, where cos or cosine is the name of the function and θ is an input. It is for historical rather than mathematical reasons that the parentheses are suppressed.
2. For convenience and ease of memorization the phrases in the definition that describe the sides of the triangle are often shortened to one word. For example, the expression for the cosine would be written

$$\cos \theta = \frac{\text{adjacent}}{\text{hypotenuse}}$$

3. There are three pairs of reciprocals in the definitions: cos and sec, sin and csc, and tan and cot. To be more explicit, we have

$$\sec \theta = \frac{\text{hypotenuse}}{\text{adjacent}} = \frac{1}{\dfrac{\text{adjacent}}{\text{hypotenuse}}} = \frac{1}{\cos \theta}$$

Similarly,

$$\csc\theta = \frac{1}{\sin\theta} \qquad \text{and} \qquad \cot\theta = \frac{1}{\tan\theta}$$

4. For acute angles the values of the trigonometric functions are always positive since they are ratios of lengths.

Even before looking at some simple numerical examples, we need to check that these definitions are consistent with the unit-circle definitions in Section 6.2. In Figure 2(a) we show an acute angle θ in right triangle ABC. Figure 2(b) displays the same right triangle, but with an x-y coordinate system and the unit circle superimposed. On the one hand, applying the right triangle definition of $\sin\theta$ in Figure 2(a), we have

$$\sin\theta = \frac{\text{opposite}}{\text{hypotenuse}} = \frac{BC}{AC}$$

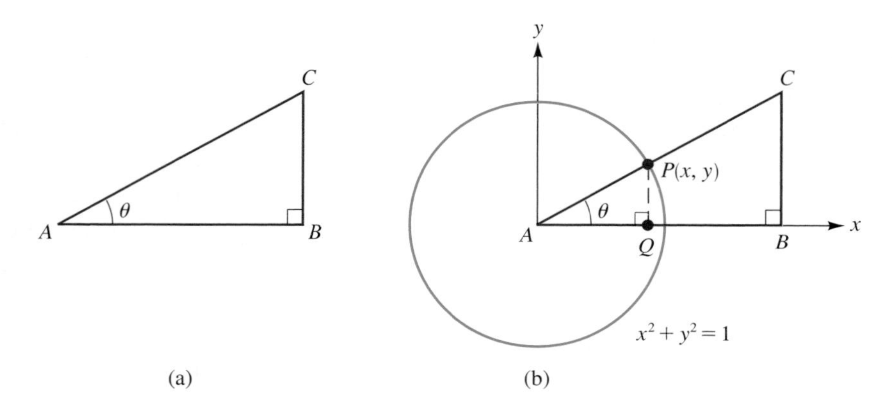

Figure 2

On the other hand, using the unit circle definition in Figure 2(b), we have

$$\sin\theta = y = QP$$

So, to show that the two definitions agree (for $\sin\theta$, at least), we must show that $BC/AC = QP$. Now, in Figure 2(b), the right triangles ABC and AQP are similar. (Why?) So we have

$$\frac{BC}{AC} = \frac{QP}{AP} = \frac{QP}{1} \qquad \text{In Figure 2(b), } AP = 1.$$

That is,

$$\frac{BC}{AC} = QP \qquad \text{as we wished to show}$$

In the same manner, it can be shown that the definitions are consistent for the remaining five trigonometric functions.

EXAMPLE 1 Comparing trigonometric ratios for similar triangles

Figure 3 shows two right triangles. The first right triangle has sides 5, 12, and 13. The second right triangle is similar to the first (the angles are the same), but

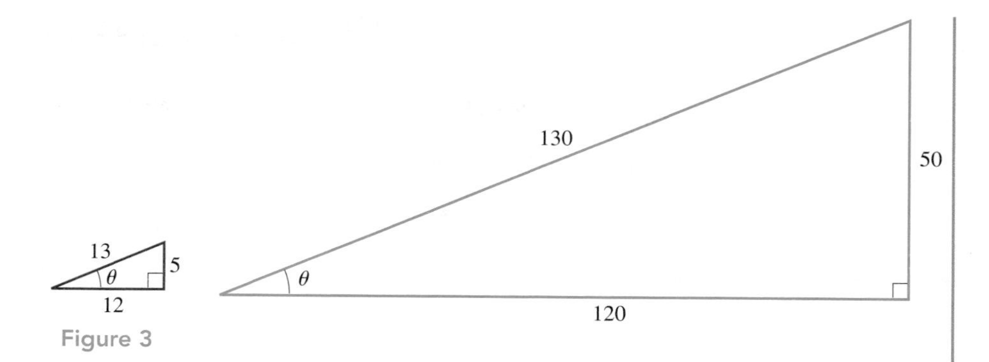

Figure 3

each side is 10 times longer than the corresponding side in the first triangle. Calculate and compare the values of $\sin\theta$, $\cos\theta$, and $\tan\theta$ for both triangles.

SOLUTION

SMALL TRIANGLE

$$\sin\theta = \frac{\text{opposite}}{\text{hypotenuse}} = \frac{5}{13}$$

$$\cos\theta = \frac{\text{adjacent}}{\text{hypotenuse}} = \frac{12}{13}$$

$$\tan\theta = \frac{\text{opposite}}{\text{adjacent}} = \frac{5}{12}$$

LARGE TRIANGLE

$$\sin\theta = \frac{50}{130} = \frac{5}{13}$$

$$\cos\theta = \frac{120}{130} = \frac{12}{13}$$

$$\tan\theta = \frac{50}{120} = \frac{5}{12}$$

OBSERVATION The corresponding values for $\sin\theta$, $\cos\theta$, and $\tan\theta$ are the same for both triangles.

EXAMPLE 2 Calculating trigonometric functions of an acute angle

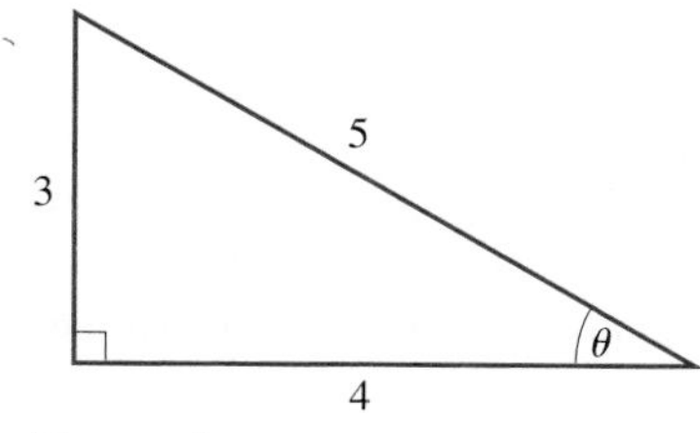

Figure 4

Let θ be the acute angle indicated in Figure 4. Determine the six quantities $\cos\theta$, $\sin\theta$, $\tan\theta$, $\sec\theta$, $\csc\theta$, and $\cot\theta$.

SOLUTION

We use the definitions:

$$\cos\theta = \frac{\text{adjacent}}{\text{hypotenuse}} = \frac{4}{5} \qquad \sec\theta = \frac{\text{hypotenuse}}{\text{adjacent}} = \frac{5}{4}$$

$$\sin\theta = \frac{\text{opposite}}{\text{hypotenuse}} = \frac{3}{5} \qquad \csc\theta = \frac{\text{hypotenuse}}{\text{opposite}} = \frac{5}{3}$$

$$\tan\theta = \frac{\text{opposite}}{\text{adjacent}} = \frac{3}{4} \qquad \cot\theta = \frac{\text{adjacent}}{\text{opposite}} = \frac{4}{3}$$

Note the pairs of answers that are reciprocals. (As was mentioned before, this helps in memorizing the definitions.)

Figure 5

EXAMPLE 3 Calculating sine and cosine

Let β be the acute angle indicated in Figure 5. Find $\sin \beta$ and $\cos \beta$.

SOLUTION

In view of the definitions we need to know the length of the hypotenuse in Figure 5. If we call this length h, then by the Pythagorean theorem we have

$$h^2 = 3^2 + 1^2 = 10$$
$$h = \sqrt{10}$$

Therefore

$$\sin \beta = \frac{\text{opposite}}{\text{hypotenuse}} = \frac{3}{\sqrt{10}} = \frac{3\sqrt{10}}{10}$$

and

$$\cos \beta = \frac{\text{adjacent}}{\text{hypotenuse}} = \frac{1}{\sqrt{10}} = \frac{\sqrt{10}}{10}$$

As we mentioned in the previous section, there are numerous identities involving the trigonometric functions. In the box that follows, we list some of these that are the most useful for right-triangle trigonometry.

PROPERTY SUMMARY Basic Right-Triangle Identities for Sine and Cosine

Identity			Examples
1. $\sin^2 \theta + \cos^2 \theta = 1$			1. $\sin^2(10^\circ) + \cos^2(10^\circ) = 1$ $\sin^2(\pi/5) + \cos^2(\pi/5) = 1$
2. $\dfrac{\sin \theta}{\cos \theta} = \tan \theta$			2. $\dfrac{\sin 45^\circ}{\cos 45^\circ} = \tan 45^\circ$
3. $\sin(90^\circ - \theta) = \cos \theta$	or	$\sin(\frac{\pi}{2} - \theta) = \cos \theta$	3. $\sin 70^\circ = \cos 20^\circ$
$\cos(90^\circ - \theta) = \sin \theta$	or	$\cos(\frac{\pi}{2} - \theta) = \sin \theta$	$\cos(3\pi/10) = \sin(\pi/5)$

Although we've already seen identities 1 and 2 in the previous section, we are going to prove them again now. This will serve to review and reinforce the right triangle definitions. All three of these identities can be proved by referring to Figure 6.

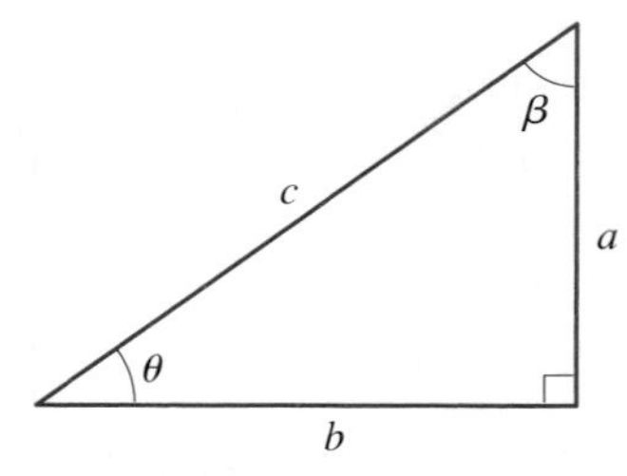

Figure 6

Proof That $\sin^2 \theta + \cos^2 \theta = 1$. Looking at Figure 6, we have $\sin \theta = a/c$ and $\cos \theta = b/c$. Thus

$$\sin^2 \theta + \cos^2 \theta = \left(\frac{a}{c}\right)^2 + \left(\frac{b}{c}\right)^2 = \frac{a^2}{c^2} + \frac{b^2}{c^2} = \frac{a^2 + b^2}{c^2}$$

$$= \frac{c^2}{c^2} \quad \text{using } a^2 + b^2 = c^2 \text{, the Pythagorean theorem}$$

$$= 1 \quad \text{as required}$$

Proof That $(\sin\theta)/(\cos\theta) = \tan\theta$. Again with reference to Figure 6, we have, by definition, $\sin\theta = a/c$, $\cos\theta = b/c$ and $\tan\theta = a/b$. Therefore

$$\frac{\sin\theta}{\cos\theta} = \frac{a/c}{b/c} = \frac{a}{c}\cdot\frac{c}{b}$$
$$= \frac{a}{b} = \tan\theta \qquad \text{as required}$$

Proof That $\sin(90^\circ - \theta) = \cos\theta$. First of all, since the sum of the angles in any triangle is 180°, we have

$$\theta + \beta + 90^\circ = 180^\circ$$
$$\beta = 90^\circ - \theta$$

Then $\sin(90^\circ - \theta) = \sin\beta = b/c$. But also (by definition) $\cos\theta = b/c$. Thus

$$\sin(90^\circ - \theta) = \cos\theta$$

since both expressions equal b/c. This is what we wanted to prove.

The proof that $\cos(90^\circ - \theta) = \sin\theta$ is entirely similar, so we omit it. We can conveniently summarize these last two results by recalling the notion of complementary angles. Two acute angles are said to be **complementary** provided that their sum is 90°. Thus the two angles θ and $90^\circ - \theta$ are complementary. In view of this, we can restate the last two results as follows:

If two angles are complementary, then the sine of (either) one equals the cosine of the other.

Incidentally, this result gives us an insight into the origin of the term "cosine": it is a shortened form of the phrase "complement's sine."

EXAMPLE 4 Using the value of one trigonometric function to find the value of other trigonometric functions

Suppose that B is an acute angle and $\cos B = \frac{2}{5}$. Find $\sin B$ and $\tan B$.

SOLUTION

We'll show two methods. The first uses the identities $\sin^2 B + \cos^2 B = 1$ and $(\sin B)/(\cos B) = \tan B$. The second makes direct use of the Pythagorean theorem.

First Method (using identities). Replace $\cos B$ with $\frac{2}{5}$ in the identity $\sin^2 B + \cos^2 B = 1$. This yields

$$\sin^2 B + \left(\frac{2}{5}\right)^2 = 1$$
$$\sin^2 B = 1 - \frac{4}{25} = \frac{21}{25}$$
$$\sin B = \sqrt{\frac{21}{25}} = \frac{\sqrt{21}}{5}$$

Notice that we've chosen the positive square root here. This is because the values of the trigonometric functions of an acute angle are by definition positive.

For tan B we have

$$\tan B = \frac{\sin B}{\cos B} = \frac{\sqrt{21}/5}{2/5} = \frac{\sqrt{21}}{2}$$

Second Method (using the Pythagorean Theorem). Since $\cos B = \frac{2}{5} = \frac{\text{adjacent}}{\text{hypotenuse}}$, we can work with a right triangle labeled as in Figure 7. Using the Pythagorean theorem, we have $2^2 + x^2 = 5^2$, from which it follows that $x = \sqrt{21}$. Consequently,

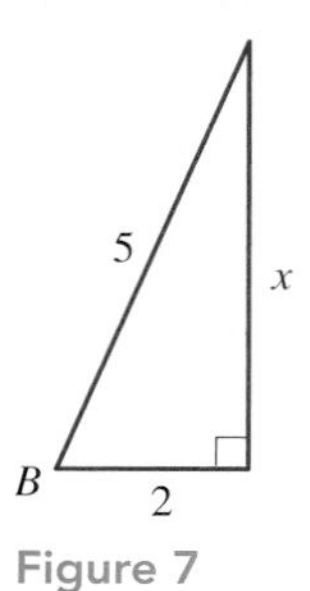

Figure 7

$$\sin B = \frac{\text{opposite}}{\text{hypotenuse}} = \frac{x}{5} = \frac{\sqrt{21}}{5}$$

and

$$\tan B = \frac{\text{opposite}}{\text{adjacent}} = \frac{x}{2} = \frac{\sqrt{21}}{2} \qquad \text{as obtained previously}$$

EXAMPLE 5 Using the value of one trigonometric function to find the value of other trigonometric functions

If θ is an acute angle and $\sin\theta = t$, express the other five trigonometric values as functions of t. (Note that in the context of right-triangle trigonometry, t, which is $\sin\theta$, and the other five trigonometric functions must all be positive.)

SOLUTION

We can use either of the methods shown in the previous example. We'll demonstrate the second method here. The right triangle in Figure 8 conveys the given information, $\sin\theta = t/1 = t$. Using the Pythagorean theorem, the length of the side adjacent to θ is $\sqrt{1-t^2}$. So we have

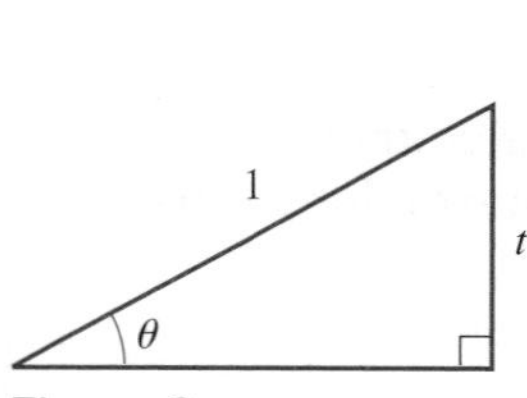

Figure 8

$$\cos\theta = \frac{\text{adjacent}}{\text{hypotenuse}} = \frac{\sqrt{1-t^2}}{1} = \sqrt{1-t^2}$$

and

$$\tan\theta = \frac{\text{opposite}}{\text{adjacent}} = \frac{t}{\sqrt{1-t^2}}$$

The values of $\csc\theta$, $\sec\theta$, and $\cot\theta$ are, by definition, the reciprocals of $\sin\theta$, $\cos\theta$, and $\tan\theta$, respectively. Thus we have

$$\csc\theta = \frac{1}{t} \qquad \sec\theta = \frac{1}{\sqrt{1-t^2}} \qquad \cot\theta = \frac{\sqrt{1-t^2}}{t}$$

Question: Why are the positive roots appropriate here?

EXAMPLE 6 **Using the value of one trigonometric function to find the value of other trigonometric functions**

Suppose that

$$\cos\theta = \frac{t}{2}$$

where $270° < \theta < 360°$. Express the other five trigonometric values as functions of t. Note that for $270° < \theta < 360°$, t must be positive. Why? In fact, $0 < t < 2$. Why?

SOLUTION

Replacing $\cos\theta$ with the quantity $t/2$ in the identity $\sin^2\theta = 1 - \cos^2\theta$ yields

$$\sin^2\theta = 1 - \left(\frac{t}{2}\right)^2 = 1 - \frac{t^2}{4} = \frac{4 - t^2}{4}$$

and, consequently,

$$\sin\theta = \pm\frac{\sqrt{4 - t^2}}{2}$$

To decide whether to choose the positive or the negative value here, note that the given inequality $270° < \theta < 360°$ tells us that the terminal side of θ lies in the fourth quadrant. Since y-coordinates are negative in Quadrant IV, we choose the negative value here. Thus

$$\sin\theta = -\frac{\sqrt{4 - t^2}}{2}$$

To obtain $\tan\theta$, we use the identity $\tan\theta = (\sin\theta)/(\cos\theta)$. This yields

$$\tan\theta = \frac{-\sqrt{4 - t^2}/2}{t/2} = -\frac{\sqrt{4 - t^2}}{t}$$

We can now find the remaining three values simply by taking reciprocals:

$$\sec\theta = \frac{1}{\cos\theta} = \frac{2}{t}$$

$$\csc\theta = \frac{1}{\sin\theta} = -\frac{2}{\sqrt{4 - t^2}} = -\frac{2\sqrt{4 - t^2}}{4 - t^2} \qquad \text{rationalizing the denominator}$$

$$\cot\theta = \frac{1}{\tan\theta} = -\frac{t}{\sqrt{4 - t^2}} = -\frac{t\sqrt{4 - t^2}}{4 - t^2} \qquad \text{rationalizing the denominator}$$

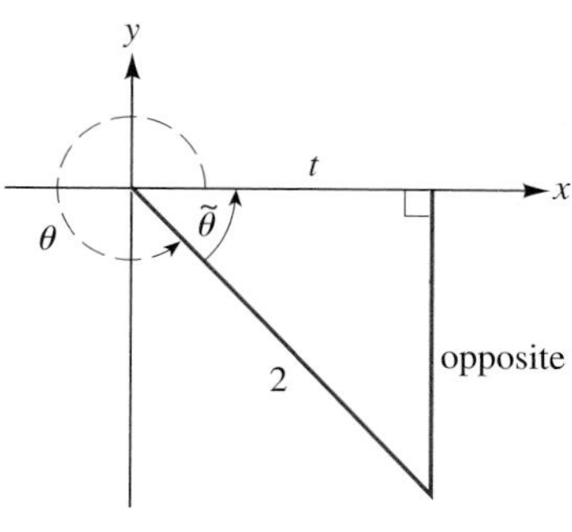

Figure 9

We give an alternative solution for Example 2 based on a right-triangle picture rather than identities. We draw an angle θ between 270° and 360° in standard position and complete a right triangle as shown in Figure 9. Let $\tilde{\theta}$ denote the reference angle. For convenience, since the trigonometric functions are ratios, we can use a length of 2 units for the hypotenuse. Since the terminal side of θ is in Quadrant IV, $\cos\theta = t/2$ is positive, so t is positive; also $\cos\theta = \cos\tilde{\theta}$. So we label the length of the adjacent side t, and we have

$$\cos\theta = \cos\tilde{\theta} = \frac{t}{2}$$

From the right triangle,

$$\text{opposite} = \sqrt{2^2 - t^2} = \sqrt{4 - t^2}$$

So

$$\sin\theta = -\sin\tilde{\theta} = -\frac{\text{opposite}}{\text{hypotenuse}} = -\frac{\sqrt{4-t^2}}{2}$$

$$\tan\theta = -\tan\tilde{\theta} = -\frac{\text{opposite}}{\text{adjacent}} = -\frac{\sqrt{4-t^2}}{t}$$

The rest follows as in the previous solution.

EXERCISE SET 6.5 

A

In Exercises 1–4, use the definitions (as in Example 2) to evaluate the six trigonometric functions of **(a)** θ *and* **(b)** β. *In cases in which a radical occurs in a denominator, rationalize the denominator.*

1.

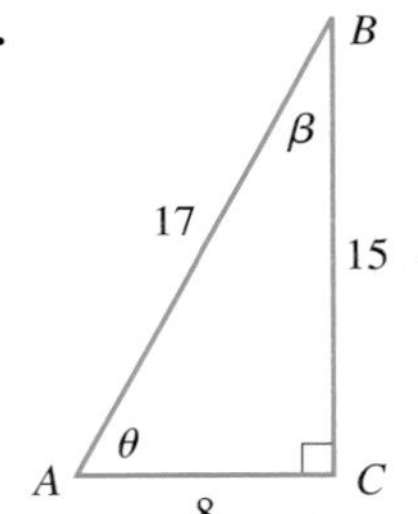

2.

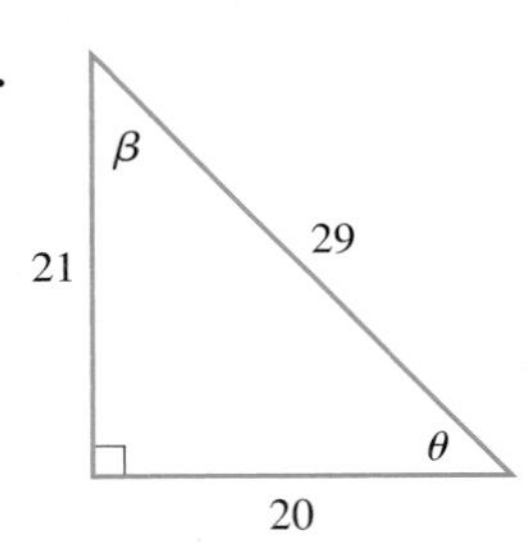

3.

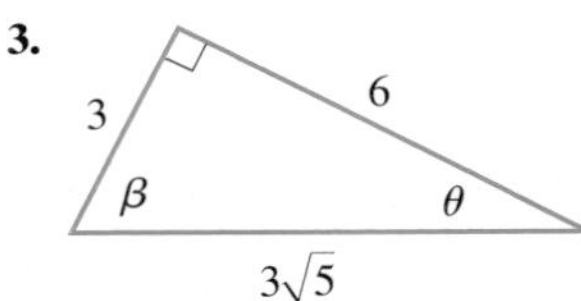

4.

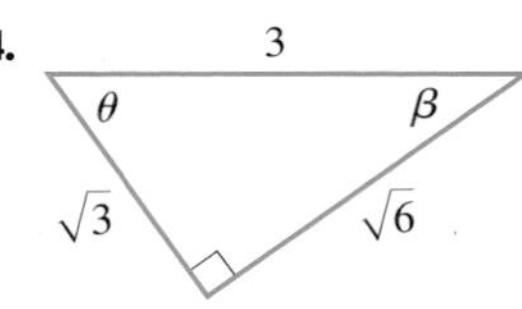

In Exercises 5–12, suppose that $\triangle ABC$ is a right triangle with $\angle C = 90°$.

5. If $AC = 3$ and $BC = 2$, find the following quantities.
 (a) $\cos A$, $\sin A$, $\tan A$
 (b) $\sec B$, $\csc B$, $\cot B$

6. If $AC = 6$ and $BC = 2$, find the following quantities.
 (a) $\cos A$, $\sin A$, $\tan A$
 (b) $\sec B$, $\csc B$, $\cot B$

7. If $AB = 13$ and $BC = 5$, compute the values of the six trigonometric functions of angle B.

8. If $AB = 3$ and $AC = 1$, compute the values of the six trigonometric functions of angle A.

9. If $AC = 1$ and $BC = 3/4$, compute each quantity.
 (a) $\sin B$, $\cos A$
 (b) $\sin A$, $\cos B$
 (c) $(\tan A)(\tan B)$

10. If $AC = BC = 4$, compute the following.
 (a) $\sec A$, $\csc A$, $\cot A$
 (b) $\sec B$, $\csc B$, $\cot B$
 (c) $(\cot A)(\cot B)$

11. If $AB = 25$ and $AC = 24$, compute each of the required quantities.
 (a) $\cos A$, $\sin A$, $\tan A$
 (b) $\cos B$, $\sin B$, $\tan B$
 (c) $(\tan A)(\tan B)$

12. If $AB = 1$ and $BC = \sqrt{3}/2$, compute the following.
 (a) $\cos A$, $\sin B$
 (b) $\tan A$, $\cot B$
 (c) $\sec A$, $\csc B$

In Exercises 13–24, verify that each equation is correct by evaluating each side. Do not use a calculator. The purpose of Exercises 13–24 is twofold. First, doing the problems will help you to review the values in Table 1 of Section 6.3. Second, the exercises serve as an algebra review.

13. $\cos 60° = \cos^2 30° - \sin^2 30°$

14. $\cos 60° = 1 - 2\sin^2 30°$

15. $\sin^2 30° + \sin^2 45° + \sin^2 60° = 3/2$

16. $\sin 30° \cos 60° + \cos 30° \sin 60° = 1$

17. $2 \sin 30° \cos 30° = \sin 60°$

18. $2 \sin 45° \cos 45° = 1$

19. $\sin 30° = \sqrt{(1 - \cos 60°)/2}$

20. $\cos 30° = \sqrt{(1 + \cos 60°)/2}$

21. $\tan 30° = \dfrac{\sin 60°}{1 + \cos 60°}$

22. $\tan 30° = \dfrac{1 - \cos 60°}{\sin 60°}$

23. $1 + \tan^2 45° = \sec^2 45°$

24. $1 + \cot^2 60° = \csc^2 60°$

For Exercises 25–28, refer to the following figures. In each case, express the indicated trigonometric values as functions of x. Rationalize any denominators containing radicals.

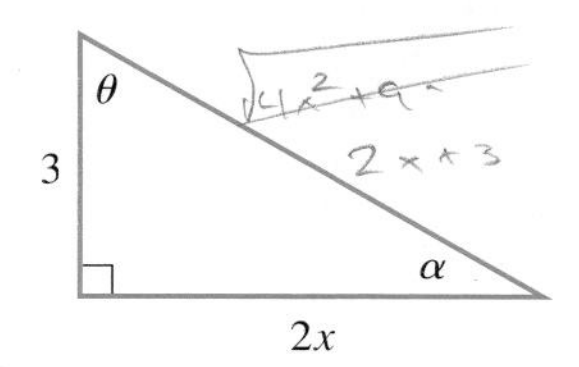

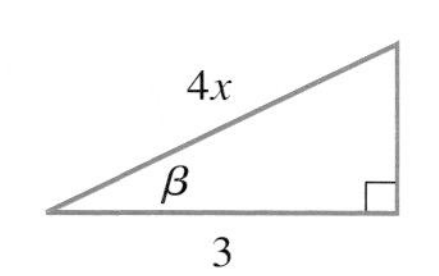

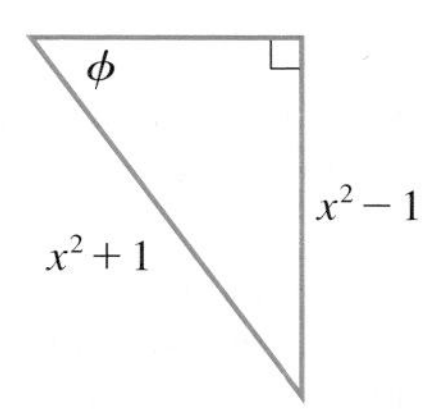

25. **(a)** $\sin\theta$, $\cos\theta$, $\tan\theta$
(b) $\sin^2\theta$, $\cos^2\theta$, $\tan^2\theta$
(c) $\sin(90° - \theta)$, $\cos(90° - \theta)$, $\tan(90° - \theta)$

26. **(a)** $\csc\alpha$, $\sec\alpha$, $\cot\alpha$
(b) $\sin^2\alpha + \cos^2\alpha + \tan^2\alpha$

27. **(a)** $\sin\beta$, $\cos\beta$, $\tan\beta$
(b) $\csc\beta$, $\sec\beta$, $\cot\beta$
(c) $\sin(90° - \beta)$, $\cos(90° - \beta)$, $\tan(90° - \beta)$

28. **(a)** $\sin\phi$, $\cos\phi$, $\tan\phi$
(b) $(\csc\phi)(\sec\phi)(\cot\phi)$
(c) $\sin(90° - \phi)$, $\cos(90° - \phi)$, $\tan(90° - \phi)$

In Exercises 29–34, use the given information to determine the values of the remaining five trigonometric functions. (The angles are assumed to be acute angles.)

29. $\cos B = 4/7$

30. $\cos B = 3/8$

31. $\sin\theta = 2\sqrt{3}/5$

32. $\sin\theta = 7/25$

33. $\tan A = \dfrac{\sqrt{2} - 1}{\sqrt{2} + 1}$

34. $\tan A = \dfrac{2 - \sqrt{3}}{2 + \sqrt{3}}$

In Exercises 35–38, use the given information to express the remaining five trigonometric values of the angle θ in terms of x. (Rationalize any denominators containing radicals.)

35. $\sin\theta = x/2$
(a) if $0° < \theta < 90°$
(b) if $180° < \theta < 270°$

36. $\sin\theta = 3x/5$
(a) if $0° < \theta < 90°$
(b) if $180° < \theta < 270°$

37. $\cos\theta = x^2$
(a) if $0° < \theta < 90°$
(b) if $270° < \theta < 360°$

38. $\cos\theta = (x - 1)/(x + 1)$
(a) if $0° < \theta < 90°$
(b) if $270° < \theta < 360°$

B

For Exercises 39–46, four functions S, C, T, and D are defined as follows:

$$\left.\begin{aligned} S(\theta) &= \sin\theta \\ C(\theta) &= \cos\theta \\ T(\theta) &= \tan\theta \\ D(\theta) &= 2\theta \end{aligned}\right\} \quad 0° < \theta < 90°$$

In each case, use the values in Table 1 (in Section 6.3) to decide if the statement is true or false. A calculator is not required.

39. $2[S(30°)] = s(60°)$

40. $T(45°) - (C \circ D)(30°) > 0$

41. $(T \circ D)(30°) > 1$

42. $T(60°) = 2[T(30°)]$

43. $S(45°) - C(45°) = 0$

44. $S(45°) - C(45°) = 0°$

45. $(C \circ D)(30°) = S(30°)$

46. $(T \circ D)(15°) - C(30°) > 0$

In Exercises 47 and 48, refer to the following figure. In the figure, the arc is a portion of a circle with center O and radius r.

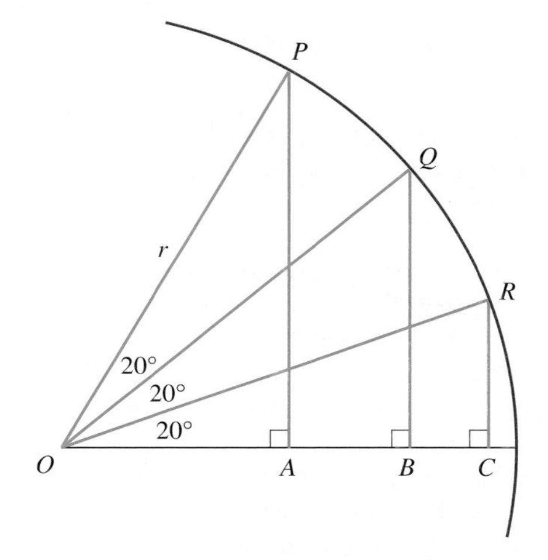

47. **(a)** Use the figure and the right-triangle definition of sine to explain (in complete sentences) why $\sin 20° < \sin 40° < \sin 60°$.
(b) Use a calculator to verify that $\sin 20° < \sin 40° < \sin 60°$.

48. **(a)** Use the figure and the right-triangle definition of cosine to explain (in complete sentences) why $\cos 20° > \cos 40° > \cos 60°$.
(b) Use a calculator to verify that $\cos 20° > \cos 40° > \cos 60°$.

49. In the accompanying figure, each of the line segments $\overline{OE}$, $\overline{AC}$, and $\overline{BF}$ is perpendicular to $\overline{OB}$; $\overline{DE}$ is perpendicular to $\overline{OE}$; $OB = OE = 1$; and the arc is a portion of a circle with center O. Use the right triangle definitions of the trigonometric functions to show that each of the equations accompanying the figure is correct.

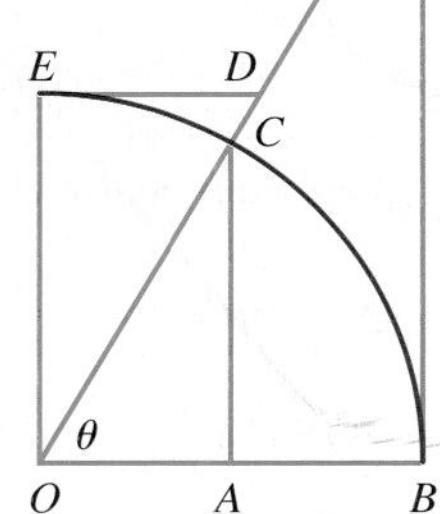

(a) $OA = \cos\theta$
(b) $AC = \sin\theta$
(c) $BF = \tan\theta$
(d) $ED = \cot\theta$
(e) $OF = \sec\theta$
(f) $OD = \csc\theta$

50. Refer to the following figure. Show that

$$\frac{\sin^2 A}{\sin^2 B} - \frac{\cos^2 A}{\cos^2 B} = \frac{a^4 - b^4}{a^2b^2}$$

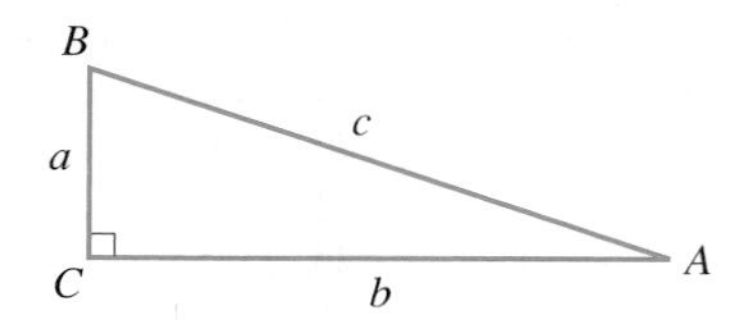

51. Suppose that $\tan\theta = p/q$, where p and q are positive and $0° < \theta < 90°$. Show that

$$\frac{p\sin\theta - q\cos\theta}{p\sin\theta + q\cos\theta} = \frac{p^2 - q^2}{p^2 + q^2}$$

52. This exercise shows how to obtain radical expressions for sin 15° and cos 15°. In the figure, assume that $AB = BD = 2$.

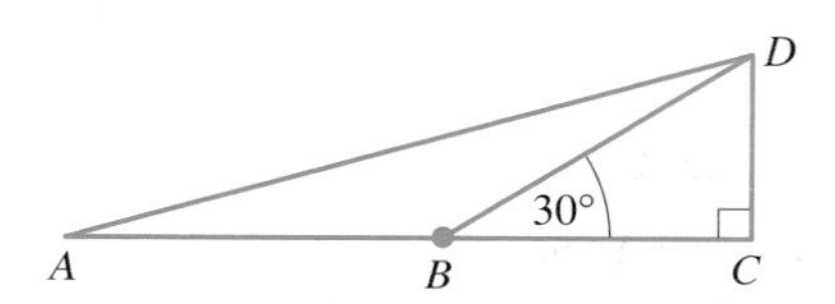

(a) In the right triangle BCD, note that $DC = 1$ because $\overline{DC}$ is opposite the 30° angle and $BD = 2$. Use the Pythagorean theorem to show that $BC = \sqrt{3}$.

(b) Use the Pythagorean theorem to show that $AD = 2\sqrt{2 + \sqrt{3}}$.

(c) Show that the expression for AD in part (b) is equal to $\sqrt{6} + \sqrt{2}$. *Hint:* Two nonnegative quantities are equal if and only if their squares are equal.

(d) Explain why $\angle BAD = \angle BDA$.

(e) According to a theorem from geometry, an exterior angle of a triangle is equal to the sum of the two nonadjacent interior angles. Apply this to $\triangle ABD$ with exterior angle $DBC = 30°$, and show that $\angle BAD = 15°$.

(f) Using the figure and the values that you have obtained for the lengths, conclude that

$$\sin 15° = \frac{1}{\sqrt{6} + \sqrt{2}} \qquad \cos 15° = \frac{2 + \sqrt{3}}{\sqrt{6} + \sqrt{2}}$$

(g) Rationalize the denominators in part (f) to obtain

$$\sin 15° = \frac{\sqrt{6} - \sqrt{2}}{4} \qquad \cos 15° = \frac{\sqrt{6} + \sqrt{2}}{4}$$

(h) Use your calculator to check the results in part (g).

53. The following figure shows a regular eleven-sided polygon inscribed in a circle of radius r.

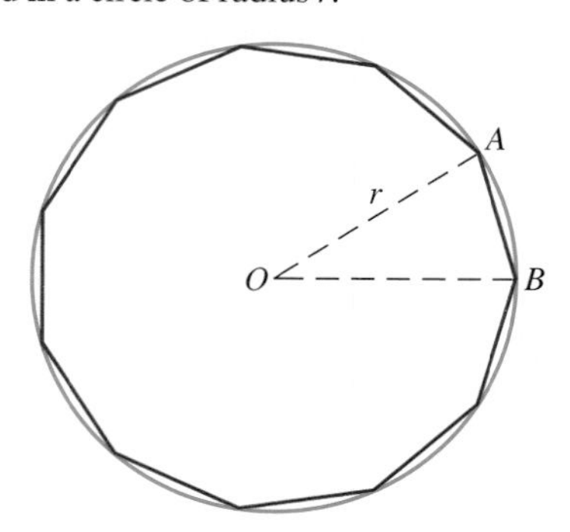

(a) Show that the length of a side of the polygon is $2r\sin(360°/22) \approx 0.5635r$. *Hint:* Draw a perpendicular from O to $\overline{AB}$.

(b) The Renaissance artist Albrecht Dürer (1471–1528) gave the following geometric construction for approximating the length of a side of a regular eleven-sided polygon inscribed in a circle: "To construct an eleven-sided figure by means of a compass, I take a quarter of a circle's diameter, extend it by one eighth of its length, and use this for construction [that is, for the side] of the eleven-sided figure." Show that this recipe yields a side of length $0.5625r$.

(c) Show, by computing the percentage error, that Dürer's approximation is very good. Percentage error is defined as

$$\left|\frac{\text{actual value} - \text{approximate value}}{\text{actual value}}\right| \times 100$$

54. This exercise shows how to obtain radical expressions for sin 18° and cos 18°, using the following figure.

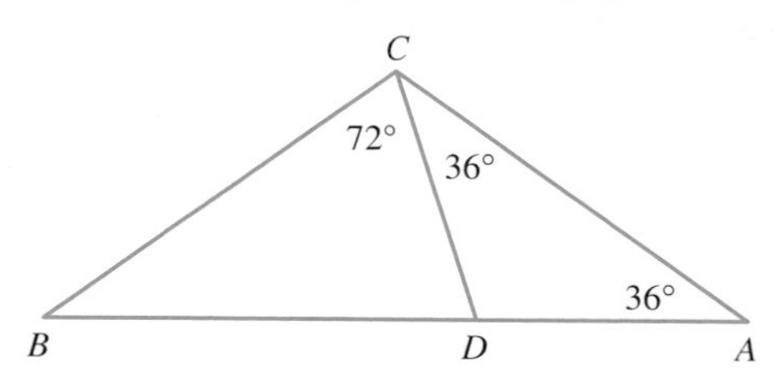

(a) Find $\angle B$, $\angle BDC$, and $\angle ADC$.

(b) Why does $AC = BC = BD$?

For the rest of this problem, assume that $AD = 1$.

(c) Why does $CD = 1$?

(d) Let x denote the common lengths AC, BC, and BD. Use similar triangles to deduce that $x/(1 + x) = 1/x$. Then show that $x = (1 + \sqrt{5})/2$.

(e) In $\triangle BDC$, draw an altitude from B to $\overline{DC}$, meeting $\overline{DC}$ at F. Use right triangle BFC to conclude that $\sin 18° = 1/(1 + \sqrt{5})$.

(f) Rationalize the denominator in part (e) to obtain $\sin 18° = \frac{1}{4}(\sqrt{5} - 1)$.

(g) Use the identity $\sin^2 \theta + \cos^2 \theta = 1$, along with part (f), to show that

$$\cos 18° = \frac{1}{4}\sqrt{10 + 2\sqrt{5}}$$

(h) Use your calculator to check the results in parts (f) and (g).

Note: In the project at the end of this section the result in part (f) is used as the basis for a compass and straightedge construction of a regular pentagon.

55. (a) Use the expression for sin 18° given in Exercise 54(f) to show that the number sin 18° is a root of the quadratic equation $4x^2 + 2x - 1 = 0$.

(b) Use the expression for sin 15° given in Exercise 52(g) to show that the number sin 15° is a root of the equation $16x^4 - 16x^2 + 1 = 0$.

56. *Formula for* $\sin(\alpha + \beta)$ In the following figure, $\overline{AD} \perp \overline{BC}$ and $AD = 1$.

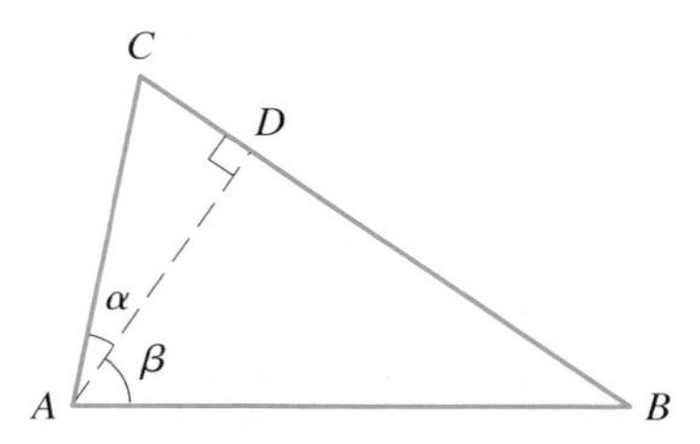

(a) Show that $AC = \sec \alpha$ and $AB = \sec \beta$.

(b) Show that

$$\text{area } \triangle ADC = \frac{1}{2}\sec \alpha \sin \alpha$$

$$\text{area } \triangle ADB = \frac{1}{2}\sec \beta \sin \beta$$

$$\text{area } \triangle ABC = \frac{1}{2}\sec \alpha \sec \beta \sin(\alpha + \beta)$$

(c) The sum of the areas of the two smaller triangles in part (b) equals the area of $\triangle ABC$. Use this fact and the expressions given in part (b) to show that

$$\sin(\alpha + \beta) = \sin \alpha \cos \beta + \cos \alpha \sin \beta$$

(d) Use the formula in part (c) to compute sin 75°. *Hint*: 75° = 30° + 45°

(e) Show that sin 75° ≠ sin 30° + sin 45°.

(f) Compute sin 105° and then check that sin 105° ≠ sin 45° + sin 60°.

(g) Use the formula in part (c) and the values for sin 18° and cos 18° in Exercise 54 to show that $\sin 36° = \sqrt{10 - 2\sqrt{5}}/4$. Then use a calculator to verify this last equation. *Hint*: The resulting algebra turns out to be much easier if you first find $\sin^2 36°$, simplify, and then take the square root.

C

57. Given that $\dfrac{\sin \alpha}{\sin \beta} = p$ and $\dfrac{\cos \alpha}{\cos \beta} = q$, express $\tan \alpha$ and $\tan \beta$ in terms of p and q. (Assume that α and β are acute angles.)

58. This exercise is adapted from a problem that appears in the classic text *A Treatise on Plane and Advanced Trigonometry,* 7th ed., by E. W. Hobson (New York: Dover Publications, 1928). (The first edition of the book was published by Cambridge University Press in 1891.)

Given: A, B, and C are acute angles such that

$$\cos A = \tan B \qquad \cos B = \tan C \qquad \cos C = \tan A$$

Prove: $\sin A = \sin B = \sin C = 2 \sin 18°$

Follow steps (a) through (e) to obtain this result.

(a) In each of the three given equations, use the identity $\tan \theta = (\sin \theta)/(\cos \theta)$ so that the equations contain only sines and cosines.

(b) In each of the three equations obtained in part (a), square both sides. Then use the identity $\sin^2 \theta = 1 - \cos^2 \theta$ so that each equation contains only the cosine function.

(c) For ease in writing, replace $\cos^2 A$, $\cos^2 B$, and $\cos^2 C$ by a, b, and c, respectively. Now you have a system of three equations in the three unknowns a, b, and c. Solve for a, b, and c.

(d) Using the results in part (c), show that

$$\sin A = \sin B = \sin C = \sqrt{\frac{3 - \sqrt{5}}{2}}$$

(e) From Exercise 54(f) we know that $\sin 18° = (\sqrt{5} - 1)/4$. Show that the expression obtained in part (d) is equal to twice this expression for sin 18°. This completes the proof. (Use the fact that two nonnegative quantities are equal if and only if their squares are equal.)

PROJECT

TRANSITS OF VENUS AND THE SCALE OF THE SOLAR SYSTEM

Almost every High School child knows that the Sun is 93 million miles (or 150 million Kilometres) away from the Earth. Despite the incredible immensity of this figure in comparison with everyday scales—or perhaps even because *it is so hard to grasp—astronomical data of this kind is accepted on trust by most educated people. Very few pause to consider how it could be possible to measure such a distance . . . and few are aware of the heroic efforts which attended early attempts at measuring it.*—David Sellers, *The Transit of Venus & the Quest for the Solar Parallax* (Maga Velda Press, 2001)

A transit of Venus occurs when the planet Venus crosses a line of sight from Earth to the Sun. For over four hundred years, transits of Venus have caused excitement among astronomers, explorers, and the interested public. From the seventeenth through the twenty-first century transits occurred in 1631, 1639, 1761, 1769, 1874, 1882, and 2004. The next transit, in 2012, will be the last of the current century and will be visible from the western United States. Inspired by a paper presented by Edmond Halley (1656–1742) to the Royal Society in 1716, expeditions set out to distant locations all over the Earth to observe the transits of 1761 and 1769. Among famous explorers and mapmakers of the eighteenth century, Charles Mason and Jeremiah Dixon observed the 1761 transit from South Africa, and Captain James Cook observed the 1769 transit from Tahiti. When the data from the observations was processed, the value of the solar parallax was determined to be between 8.5 and 8.9 seconds of arc. The modern value is 8.794148 arc seconds.

Halley's method used complicated tools from spherical trigonometry. In this project, we will see how observations of a transit of Venus and some basic plane trigonometry can be used to estimate the solar parallax and the distance from Earth to the Sun. This project can be done as an individual or group activity but may be more fun with a group.

Figure A shows the geometric relationship between the **solar parallax,** angle α, the distance r_e, from Earth to the Sun, and the radius R of the Earth. The notation r_e comes from thinking of the distance from Earth to the Sun as the radius of the Earth's orbit. Figure A is not drawn to scale. It greatly exaggerates angle α, and r_e is really about 25,000 times larger than R.

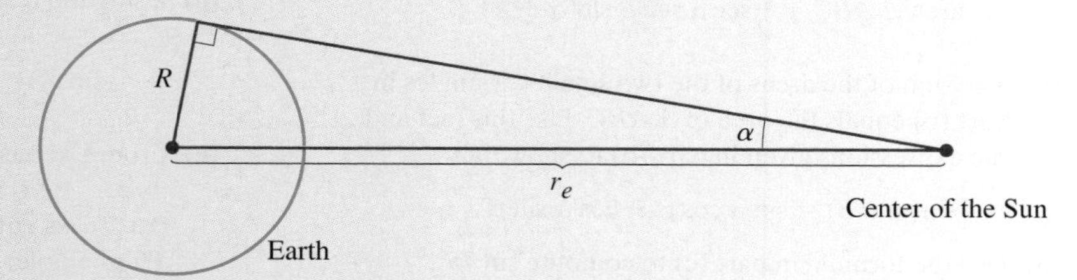

Figure A

EXERCISE 1 Use the right triangle definition of the sine function to derive the formula

$$r_e = \frac{R}{\sin \alpha} \tag{i}$$

for the distance from Earth to the Sun in terms of the solar parallax and the radius of Earth.

What modern school-children learn about him is that he invented "the sieve of Eratosthenes"—a method for sifting through all the numbers to find which are prime numbers—and that he discovered a way to measure the circumference of the earth with astounding accuracy.
—Kitty Ferguson, *Measuring the Universe: Our Historic Quest to Chart the Horizons of Space and Time* (New York: Walker and Company, 1999)

More than 2000 years ago Eratosthenes calculated that the circumference of Earth was about 250,000 stades. While the precise length of a stade is not known, an estimate of about 160 meters gives a circumference of 40,000 kilometers. (For a short, very readable account of Eratosthenes' life and work, see pp. 13–21 of Kitty Ferguson's book quoted in the margin.)

EXERCISE 2 Assuming a circumference of 40,000 kilometers, show that the radius R of the Earth is about 6400 kilometers. (The modern value is 6371 kilometers.)

EXERCISE 3 After the data from the expeditions of 1761 and 1769 were analyzed, the best estimate for the solar parallax was about 8.7 arc seconds. Assuming α is 8.7 arc seconds, use formula (i) to find the distance from Earth to the Sun. (Don't forget to convert 8.7 arc seconds to degrees.) How does it compare with a modern value of 149,600,000 km?

A transit of Venus occurs when Earth, Venus, and the Sun lie along a straight line. This happens because the orbits of both planets lie in planes containing the Sun, but it happens only rarely because the orbital plane of Venus is tilted about 3.4 degrees to the plane of Earth's orbit. See Figure B. A transit occurs when the three bodies line up along the **line of nodes** as illustrated in Figure B and Figure C. Transits occur in pairs, about eight years apart, about once per century. Although the orbits of Earth and Venus are not quite circular, in our discussion we'll assume that they are circles centered at the center of the Sun; a top view would resemble Figure C.

Figure D, like Figure A, is greatly exaggerated. It shows the geometric situation during a transit of Venus as seen by two symmetrically placed observers on Earth. The observer at point A sees Venus as a dark spot on the Sun at point A', and the observer at B sees Venus at the point B'.

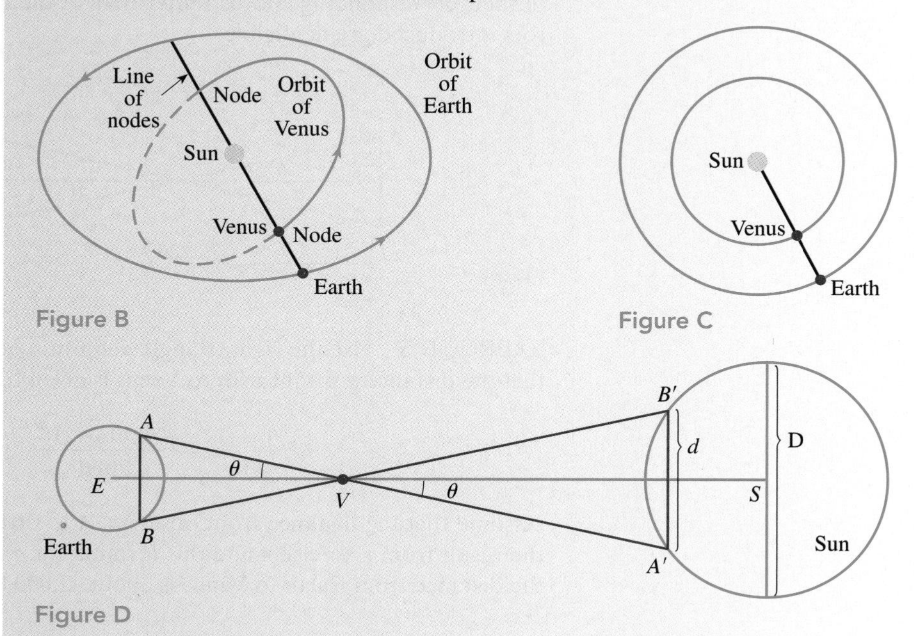

Figure B

Figure C

Figure D

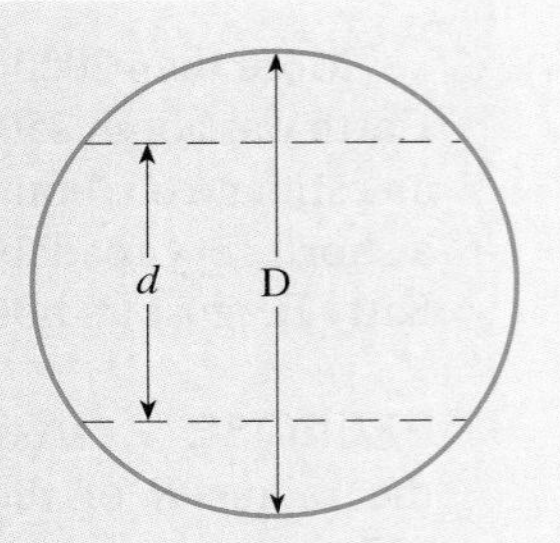

Figure E

Observer B would see Venus transit the Sun, that is, move across the face of the Sun, along the path indicated by the upper dashed line in Figure E. Similarly, observer A would see Venus transit the Sun along the lower dashed line. In principle, the ratio d/D can be measured from a careful drawing. The eighteenth-century observers recorded the time of the first contact of Venus with the solar disk, the time that Venus separated from the solar disk, and their geographical position (latitude and longitude). This data was used to calculate d/D.

EXERCISE 4 The value of the ratio d/D calculated as a result of the eighteenth-century measurements was about 0.0260. To an observer on Earth the full diameter of the Sun subtends an angle of about 31.5 minutes of arc. Use this information to show that angle θ is about 24.6 arc seconds (0.00683 degrees). *Hint*: Use a proportion.

Since the radii of both the Earth and the Sun are very small compared to the distance between their centers we can capture the geometric information needed to estimate the radius of Earth's orbit in the simplified triangle diagram in Figure F, where the points have the same meaning as in Figure D. The simplification involves positioning the centers of the Earth and Sun at the midpoints of their corresponding chords rather than at the centers of the bodies. The errors introduced are negligible.

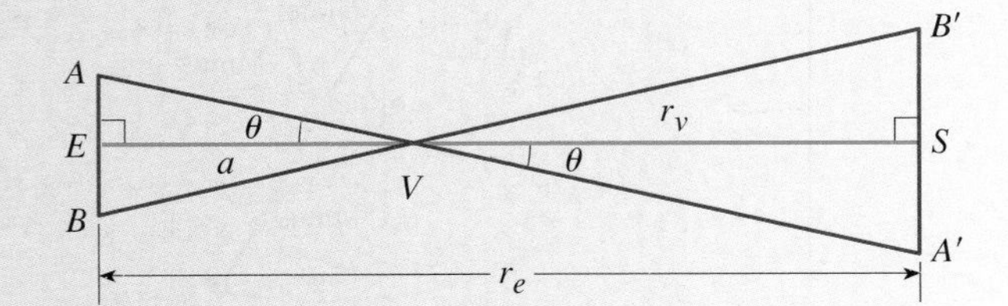

Figure F

EXERCISE 5 Use the right triangle definition of the tangent function to show that the distance a from Earth to Venus is given by

$$a = \frac{\text{length of } AE}{\tan \theta}$$

Assume that the distance from observer A to observer B is 10,000 km and use the result from Exercise 4 and this formula for a to show that, during a transit, the distance from Earth to Venus is about 42,000,000 km.

It is clear from Figure F and the result of Exercise 5 that with our simplifications

$$r_e - r_v = a = 42{,}000{,}000 \qquad \text{(ii)}$$

where r_v = the distance from Venus to the Sun.

To get another equation we move from geometry to physics. We use Kepler's third law, first published in 1616 as an empirical result, and later derived from Newton's laws of motion and gravity. In the case of circular orbits, for two planets in orbit around the Sun, Kepler's third law states that the cube of the ratio of the orbital radii equals the square of the ratio of the orbital periods. For Earth and Venus this implies

$$\left(\frac{r_e}{r_v}\right)^3 = \left(\frac{T_e}{T_v}\right)^2 \qquad \text{(iii)}$$

where T_e and T_v are the orbital periods of Earth and Venus, respectively.

EXERCISE 6 Use $T_e = 365$ (Earth) days for the orbital period of Earth and $T_v = 225$ (Earth) days for the orbital period of Venus in equation (iii) to show

$$r_e = 1.381 r_v \qquad \text{(iv)}$$

EXERCISE 7 Solve the system of equations (ii) and (iv) and compare your result for the radius of Earth's orbit with your result from Exercise 3.

This completes our work on using the observations of a transit of Venus to determine the radii of the orbits of Earth and Venus. You might want to review your work on this project shortly before the next transit of Venus on June 6, 2012.

We conclude with a historical note. From centuries of observations the orbital periods of the six innermost planets were known by Kepler's time. It follows from the derivation of equation (ii) and Kepler's third law that if we can find a value for the distance between any two planets in the solar system, then we can find values for *all* of the orbital radii. The first measurement of an interplanetary distance, from Earth to Mars (not to Venus), was made almost 100 years before the transits of Venus discussed in this project. Based on a 1672 measurement of the "parallax of Mars" (the angle with vertex at the center of Mars and rays to two widely separated observers on Earth) by Cassini in Paris and Rocher in what is now French Guiana, Cassini determined a value of 140,000,000 km for the distance from Earth to the Sun. One hundred years later the expeditions for the transits of Venus significantly improved this value.

PROJECT

CONSTRUCTING A REGULAR PENTAGON

This is intended as a group project to create a poster explaining a compass and straightedge construction of a regular pentagon. (Recall that *regular* means that all of the sides are equal and all of the angles are equal.) In the first part you are going to construct a regular pentagon by first constructing a regular decagon and then connecting every other vertex. The second part asks you to write a short paper explaining why the construction is valid. Although the construction given here is somewhat awkward, its justification is fairly straightforward. In the Writing Mathematics section at the end of Chapter 8 a shorter and more elegant construction is described. The third part provides some history of the construction of regular n-gons.

(a) As background you should review the ruler and compass construction for the perpendicular bisector of a line segment. The construction proceeds as follows.

> Draw a line segment, and label the endpoints A and D. Construct the perpendicular bisector, and denote the midpoint of AD by C. Set your compass to the length of AC, which is 1 unit for this construction, and mark a point two units from C along the perpendicular bisector. Label this point B, and draw the line segment AB. With your compass still set to the length of AC, draw a unit circle somewhere away from triangle ABC. Set the needle tip of your compass at B, and let E denote the point on AB that is 1 unit from B. Bisect the line segment AE, and label the midpoint F. The length of AF will be one side of a regular decagon inscribed in the unit circle. Mark ten consecutive points around the unit circle spaced this distance apart. These are the vertices of a regular decagon. Connecting every other vertex completes this construction of a regular pentagon.

(b) As background for this part you should review parts (a) though (f) of Exercise 60 in Exercise Set 6.5. To complete this part of the project, write a careful justification for the construction described in part (a). You should use a mixture of English sentences, equations, and figures similar to the exposition in this textbook. A brief sketch of the beginning of a justification follows:

> It is straightforward to show that the line segment AF described in part (a) has length $(\sqrt{5} - 1)/2$. From Exercise 60(f) in Section 6.5, $\sin 18° = (\sqrt{5} - 1)/4$. This implies, in a few steps, that an isosceles triangle with two sides of length 1 and the third side of length $(\sqrt{5} - 1)/2$ has a vertex angle of 36°.

(c) Carl Friedrich Gauss (1777–1855) was one of the greatest mathematicians in history. In the last section of his great work on number theory, *Disquisitiones Arithmeticae,* published in 1801, he discusses the problem of constructing regular n-gons. He first deals with regular polygons with a prime

WE CAN SHOW WITH ALL RIGOR THAT THESE HIGHER-DEGREE EQUATIONS CANNOT BE AVOIDED IN ANY WAY NOR CAN THEY BE REDUCED TO LOWER-DEGREE EQUATIONS. The limits of the present work exclude this demonstration here, but we issue this warning lest anyone attempt to achieve geometric constructions for sections other than the one suggested by our theory (e.g. sections into 7, 11, 13, 19, etc. parts) and so spend his time uselessly.
—Carl Friedrich Gauss in *Disquisitions Arithmeticae,* 1801. [Translation by Arthur A. Clarke, S. J. (New Haven and London: Yale University Press, 1966)]

number, p, of sides and proves that if p is of the form $p = 2^{2^k} + 1$, where k is a nonnegative integer, then a regular p-gon is constructible, that is, can be constructed using a compass and straightedge. Then he gives an explicit construction for a regular 17-gon. In the quotation to the left Gauss warns the reader that a regular p-gon is *not* constructible if p is not a prime of the form $p = 2^{2^k} + 1$ but that the limits of the book do not allow him to present the proof. The first published proof was by Pierre L. Wantzel in 1837. The numbers $F_k = 2^{2^k} + 1$ are called Fermat numbers after Pierre de Fermat (1601–1665). F_0 through F_4 are prime, but it remains an open problem as to whether or not there are any other *prime* Fermat numbers. Gauss also notes that constructions for regular polygons with 3, 4, 5, and 15 sides and those that are easily constructed from these, with $2^m \cdot 3$, $2^m \cdot 4$, $2^m \cdot 5$, or $2^m \cdot 15$ sides, for m a positive integer, were known since Euclid's time but that no new constructible polygons had been found for 2000 years (until his discoveries of 1796).

Assuming a regular n-gon is constructible, explain why a regular polygon of $2^m \cdot n$ sides, where m is a positive integer, is constructible.

Gauss's *Disquisitiones Arithmeticae* consists of 366 numbered articles separated into seven sections. In the last article he makes a statement equivalent to the assertion that a regular polygon of n sides is constructible if and only if

$$n = 2^k p_1 p_2 \cdots p_l$$

where $p_1, p_2, \ldots, p_l$ are distinct (no two are equal) prime Fermat numbers and k is a nonnegative integer. He finishes the article with a list of the 37 constructible regular polygons with 300 or fewer sides.

List the 12 constructible regular polygons with 25 or fewer sides. When you finish, you might try to find the 25 remaining constructible regular polygons with 300 or fewer sides, thus completing Gauss's list.

Chapter 6 Summary of Principal Formulas and Terms

Terms or formulas	Page reference	Comments
1. $\theta = \frac{s}{r}$	424–425	This equation defines the *radian measure* θ of an angle. We assume here that the vertex of the angle is placed at the center of a circle of radius r and that s is the length of the intercepted arc. See Figure 4 in Section 6.1.

Terms or formulas	Page reference	Comments
2. $s = r\theta$	428	This formula expresses the arc length s on a circle in terms of the radius r and the radian measure θ of the central angle subtended by the arc.
3. $A = \frac{1}{2}r^2\theta$	429	This formula expresses the area of a sector of a circle in terms of the radius r and the radian measure θ of the central angle.
4. $\omega = \frac{\theta}{t}$ $v = \frac{d}{t}$ $v = r\omega$	430–432	Suppose that a wheel rotates about its axis at a constant rate. If a radial line on the wheel turns through an angle of measure θ in time t, then the *angular speed* ω of the wheel is defined by $\omega = \theta/t$. If a point on the rotating wheel travels a distance d in time t, then the linear speed v of the point is defined by $v = d/t$. If the angular speed ω is expressed in radians per unit time, then the linear and angular speeds are related by $v = r\omega$.
5. Initial side of an angle Terminal side of an angle	437	For analytical purposes, we think of the two rays that form an angle as originally coincident. Then, while one ray is held fixed, the other is rotated to create the given angle. As is indicated in Figure 1 in Section 6.2, the fixed ray is called the *initial side* of the angle and the rotated ray is called the *terminal side.* The measure of an angle is positive if the rotation is counterclockwise and negative if the rotation is clockwise.
6. Standard position	438	In a rectangular coordinate system, an angle is in *standard position* if the vertex is located at (0, 0) and the initial side of the angle lies along the positive horizontal axis. For examples, see Figure 3 in Section 6.2.
7. $\cos\theta$ $\sin\theta$ $\tan\theta$ $\sec\theta$ $\csc\theta$ $\cot\theta$	439	If θ is an angle in standard position and $P(x, y)$ is the point where the terminal side of the angle meets the unit circle, then the six trigonometric functions of θ are defined as follows. $\cos\theta = x \qquad \sec\theta = \frac{1}{x} \quad (x \neq 0)$ $\sin\theta = y \qquad \csc\theta = \frac{1}{y} \quad (y \neq 0)$ $\tan\theta = \frac{y}{x} \quad (x \neq 0) \qquad \cot\theta = \frac{x}{y} \quad (y \neq 0)$
8. Reference angle or reference number	450	Let θ be an angle in standard position in an x-y coordinate system, and suppose that θ is not a multiple of 90° or $\pi/2$ radians. The reference angle associated with θ is the acute angle (with positive measure) formed by the x-axis and the terminal side of the angle θ. In work with radians, the reference angle is sometimes referred to as the reference number [because a radian angle measure (such as $\pi/4$ or 2) is a real number].

Terms or formulas	Page reference	Comments
9. The four-step procedure for evaluating the trigonometric functions	451–452	The following four-step procedure can be used to evaluate the trigonometric functions of angles that are not multiples of 90° or $\pi/2$ radians. **Step 1** Draw or refer to a sketch of the angle and reference angle. **Step 2** Determine the reference angle associated with the given angle. **Step 3** Evaluate the given trigonometric function using the reference angle for the input. **Step 4** Affix the appropriate sign to the number found in Step 2. (See Examples 1 through 4 on pages 452–454.) For the right-triangle version of the four-step procedure, remember that each side has *positive* length and each *non*-right angle is *acute*.
10. $\sin^2\theta + \cos^2\theta = 1$ $\frac{\sin\theta}{\cos\theta} = \tan\theta$ $\sec\theta = \frac{1}{\cos\theta}$; $\csc\theta = \frac{1}{\sin\theta}$ $\cot\theta = \frac{1}{\tan\theta}$	463	These are some of the most basic trigonometric identities. The proofs of the first two identities are given on page 463 (The other identities in this list are immediate consequences of the definitions of the trigonometric functions.)
11. Right-triangle definitions of the trigonometric functions	471	When θ is an acute angle in a right triangle, the six trigonometric functions of θ can be defined as follows: $\cos\theta = \frac{\text{adjacent}}{\text{hypotenuse}}$ $\sec\theta = \frac{\text{hypotenuse}}{\text{adjacent}}$ $\sin\theta = \frac{\text{opposite}}{\text{hypotenuse}}$ $\csc\theta = \frac{\text{hypotenuse}}{\text{opposite}}$ $\tan\theta = \frac{\text{opposite}}{\text{adjacent}}$ $\cot\theta = \frac{\text{adjacent}}{\text{opposite}}$
12. $\sin(90° - \theta) = \cos\theta$ $\cos(90° - \theta) = \sin\theta$	474–475	If two angles are complementary, then the sine of (either) one equals the cosine of the other.

Writing Mathematics

1. A student who wanted to simplify the expression $\frac{\pi/6}{\sin(\pi/6)}$ wrote

$$\frac{\pi/6}{\sin(\pi/6)} = \frac{\cancel{\pi/6}}{\sin(\cancel{\pi/6})} = \frac{1}{\sin}$$

Explain why this is nonsense and then indicate the correct solution.

2. A student who was asked to simplify the expression $\sin^2(\pi/5) + \sin^2(3\pi/10)$ wrote

$$\sin^2(\pi/5) + \sin^2(3\pi/10) = \sin^2\left(\frac{\pi}{5} + \frac{3\pi}{10}\right) = \sin^2(\pi/2) = 1^2 = 1$$

(a) Where is the error?
(b) What is the correct answer?

3. Determine if each statement is TRUE or FALSE. In each case, write out your reason or reasons in complete sentences. If you draw a diagram to accompany your writing, be sure that you clearly label any parts of the diagram to which you refer in the writing.

(a) If θ is an angle in standard position and $180° < \theta < 270°$, then $\tan\theta < \sin\theta$.
(b) If $\theta < \pi$, then $\sin\theta$ is positive.
(c) For all angles θ, we have $\sin\theta = \sqrt{1 - \cos^2\theta}$.
(d) If θ is an acute angle in a right triangle, then $\sin\theta < 1$.
(e) If θ is an acute angle in a right triangle, then $\tan\theta < 1$.

4. (a) Use a calculator to evaluate the quantity $4 \sin 18° \cos 36°$.
(b) The following *very* short article, with the accompanying figure, appeared in *The Mathematical Gazette,* Vol. XXIII (1939), p. 211. On your own or with a classmate, study the article and fill in the missing details. Then, strictly on your own, rewrite the article in a paragraph (or two at the most). Write as if you were explaining to a friend or classmate why $4 \sin 18° \cos 36° = 1$.

To prove that $4 \sin 18° \cos 36° = 1$.
In the figure,
$\sin 18° = \frac{1}{2}a/b$, $\cos 36° = \frac{1}{2}b/a$.
Multiply.

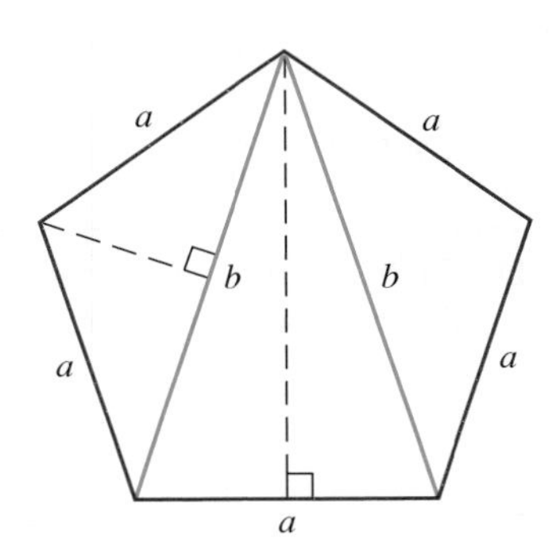

Chapter 6 Review Exercises 

In Exercises 1 and 2, complete the tables.

1.

θ	cos θ	sin θ	tan θ	sec θ	csc θ	cot θ
0						
$\pi/6$						
$\pi/4$						
$\pi/3$						
$\pi/2$						
$2\pi/3$						
$3\pi/4$						
$5\pi/6$						
π						

2.

θ	cos θ	sin θ	tan θ	sec θ	csc θ	cot θ
π						
$7\pi/6$						
$5\pi/4$						
$4\pi/3$						
$3\pi/2$						
$5\pi/3$						
$7\pi/4$						
$11\pi/6$						

In Exercises 3–8, the lengths of the three sides of a triangle are denoted a, b, and c; the angles opposite these sides are A, B, and C, respectively. In each exercise, use the given information to find the required quantities.

	GIVEN	FIND
3.	$B = 90°, A = 30°, b = 1$	a and c
4.	$B = 90°, A = 60°, a = 1$	c and b
5.	$B = 90°, \sin A = 2/5, a = 7$	b
6.	$B = 90°, \sec C = 4, c = \sqrt{2}$	b
7.	$B = 90°, \cos A = 3/8$	$\sin A$ and $\cot A$
8.	$B = 90°, b = 1, \tan C = \sqrt{5}$	a

9. For the following figure, show that

$$y = x[\tan(\alpha + \beta) - \tan \beta]$$

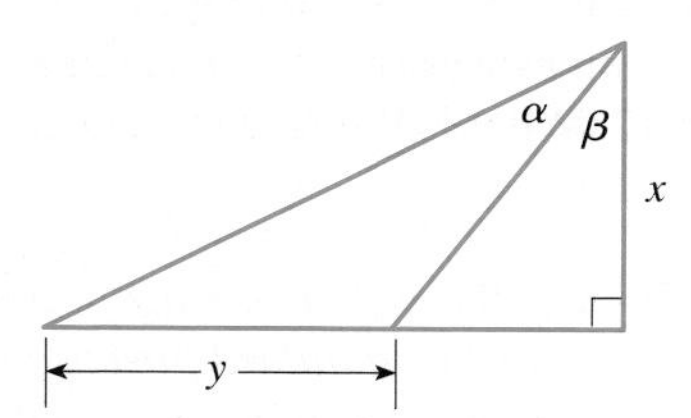

10. For the following figure, show that

$$y = \frac{x}{\cot \alpha - \cot \beta}$$

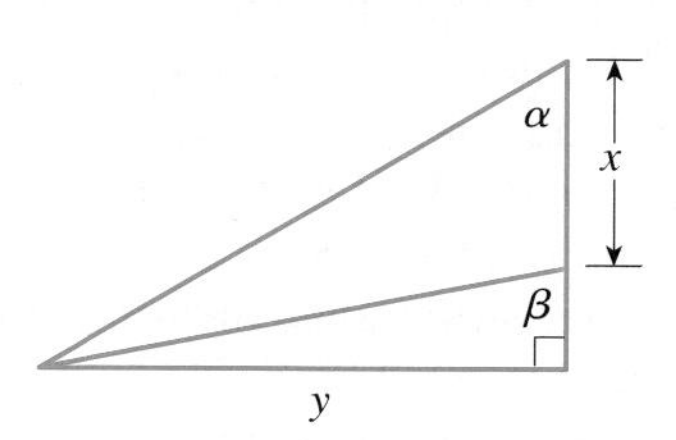

11. For the following figure, show that

$$\cot \theta = \frac{a}{b} + \cot \alpha$$

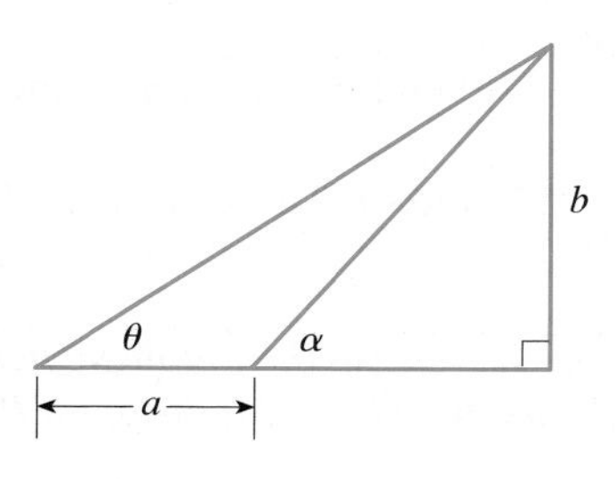

12. Suppose that θ is an acute angle in a right triangle and $\sin \theta = \dfrac{2p^2q^2}{p^4 + q^4}$. Find $\cos \theta$ and $\tan \theta$.

13. This problem is adapted from the text *An Elementary Treatise on Plane Trigonometry,* by R. D. Beasley, first published in 1884.

(a) Prove the following identity, which will be used in part (b): $1 + \tan^2 \alpha = \sec^2 \alpha$.

(b) Suppose that α and θ are acute angles and $\tan \theta = \dfrac{1 + \tan \alpha}{1 - \tan \alpha}$. Express $\sin \theta$ and $\cos \theta$ in terms of α. *Hint*: Draw a right triangle, label one of the angles θ, and let the lengths of the sides opposite and adjacent to θ be $1 + \tan \alpha$ and $1 - \tan \alpha$, respectively.

Answer: $\sin \theta = (\cos \alpha + \sin \alpha)/\sqrt{2}$
$\cos \theta = (\cos \alpha - \sin \alpha)/\sqrt{2}$

14. If $\theta = \pi/4$, evaluate each of the following. Exact answers are required, not calculator approximations.

(a) $\cos \theta$ **(b)** $\cos^3 \theta$ **(c)** $\cos 2\theta$ **(d)** $\cos 3\theta$ **(e)** $\cos(2\theta/3)$ **(f)** $(\cos 3\theta)/3$ **(g)** $\cos(-\theta)$ **(h)** $\cos^3(5\theta)$

In Exercises 15–22, convert each expression into one involving only sines and cosines and then simplify. (Leave your answers in terms of sines and/or cosines.)

15. $\dfrac{\sin A + \cos A}{\sec A + \csc A}$

16. $\dfrac{\csc A \sec A}{\sec^2 A + \csc^2 A}$

17. $\dfrac{\sin A \sec A}{\tan A + \cot A}$

18. $\cos A + \tan A \sin A$

19. $\dfrac{\cos A}{1 - \tan A} + \dfrac{\sin A}{1 - \cot A}$

20. $\dfrac{1}{\sec A - 1} \div \dfrac{1}{\sec A + 1}$

21. $(\sec A + \csc A)^{-1}[(\sec A)^{-1} + (\csc A)^{-1}]$

22. $\dfrac{\dfrac{\tan^2 A - 1}{\tan^3 A + \tan A}}{\dfrac{\tan A + 1}{\tan^2 A + 1}}$

In Exercises 23–34, use Figure 8 in Section 6.2 to approximate to within successive tenths the given trigonometric values. Then use a calculator to compute the values to the nearest hundredth.

23. (a) $\cos 6$ **(b)** $\cos(-6)$

24. (a) $\sin 6$ **(b)** $\sin(-6)$

25. (a) $\cos 140°$ **(b)** $\cos(-140°)$

26. (a) $\sin 140°$ **(b)** $\sin(-140°)$

27. (a) $\sin(\pi/4)$ **(b)** $\sin(-\pi/4)$

28. **(a)** $\cos(\pi/4)$ **(b)** $\cos(-\pi/4)$
29. **(a)** $\sin 250°$ **(b)** $\sin(-250°)$
30. **(a)** $\cos 250°$ **(b)** $\cos(-250°)$
31. **(a)** $\cos 4$ **(b)** $\cos(-4)$
32. **(a)** $\sin 4$ **(b)** $\sin(-4)$
33. **(a)** $\cos(4 + 2\pi)$ **(b)** $\cos(-4 + 2\pi)$
34. **(a)** $\sin(4 + 2\pi)$ **(b)** $\sin(-4 + 2\pi)$

In Exercises 35–44, evaluate each expression in terms of a, where $a = \cos 20°$.

35. $\sin 20°$
36. $\tan 20°$
37. $\cos 70°$
38. $\sin 70°$
39. $\cos 160°$
40. $\cos 340°$
41. $\cos(-160°)$
42. $\cos 200°$
43. $\sin 200°$
44. $\cot 200°$

45. Simplify: $\dfrac{\sin^4\theta - \cos^4\theta}{\sin^2\theta - \cos^2\theta} \div \dfrac{1 + \sin\theta\cos\theta}{\sin^3\theta - \cos^3\theta}$.

46. Simplify: $\dfrac{1 + \sin\theta + \sin^2\theta}{1 - \sin^3\theta}$.

In Exercises 47–64, show that each equation is an identity.

47. $\dfrac{1 - \sin\theta\cos\theta}{(\cos\theta)(\sec\theta - \csc\theta)} \cdot \dfrac{\sin^2\theta - \cos^2\theta}{\sin^3\theta + \cos^3\theta} = \sin\theta$

48. $\sec A - 1 = (\sec A)(1 - \cos A)$

49. $\dfrac{\cot A - 1}{\cot A + 1} = \dfrac{\cos A - \sin A}{\cos A + \sin A}$

50. $\dfrac{\sin A}{\csc A - \cot A} = 1 + \cos A$

51. $\cos^2\theta - \sin^2\theta = 2\cos^2\theta - 1$

52. $\cos^2\theta - \sin^2\theta = 1 - 2\sin^2\theta$

53. $\sin A \tan A = \dfrac{1 - \cos^2 A}{\cos A}$

54. $\dfrac{\cot A - 1}{\cot A + 1} = \dfrac{1 - \tan A}{1 + \tan A}$

55. $\tan A \tan B = \dfrac{\tan A + \tan B}{\cot A + \cot B}$

56. $\dfrac{\sin A}{1 + \cos A} + \dfrac{1 + \cos A}{\sin A} = 2\csc A$

57. $\tan A - \dfrac{\sec A \sin^3 A}{1 + \cos A} = \sin A$

58. $\dfrac{1}{1 - \cos A} + \dfrac{1}{1 + \cos A} = 2 + 2\cot^2 A$

59. $\dfrac{1}{\csc A - \cot A} - \dfrac{1}{\csc A + \cot A} = 2\cot A$

60. $\dfrac{\sin^3 A}{\cos A - \cos^3 A} = \tan A$

61. $\dfrac{\sec A - \csc A}{\sec A + \csc A} = \dfrac{\sin A - \cos A}{\sin A + \cos A}$

62. $\sin^6 A + \cos^6 A = 1 - 3\sin^2 A\cos^2 A$

63. $\sin A \cos A = \dfrac{\tan A}{1 + \tan^2 A}$

64. $\dfrac{2\tan A}{1 - \tan^2 A} + \dfrac{1}{\cos^2 A - \sin^2 A} = \dfrac{\cos A + \sin A}{\cos A - \sin A}$

In Exercises 65–72, use the arc length formula $s = r\theta$ and the sector area formula $A = \frac{1}{2}r^2\theta$ to find the required quantities.

	GIVEN	FIND
65.	$\theta = \pi/8$; $r = 16$ cm,	s and A
66.	$\theta = 120°$; $r = 12$ cm	s and A
67.	$r = s = 1$ cm	A
68.	$r = s = \sqrt{3}$ cm	θ
69.	$s = 4$ cm; $\theta = 36°$	r and A
70.	$\theta = \pi/10$; $A = 200\pi$ cm^2	r
71.	$s = 12$ cm; $\theta = r + 1$	r and θ
72.	$\theta = 50°$; $A = 20\pi$ cm^2	s

73. Two angles in a triangle are 40° and 70°. What is the radian measure of the third angle?

74. The radian measures of two angles in a triangle are 1/6 and 5/12. What is the radian measure of the third angle?

In Exercises 75–82, $P(x, y)$ denotes the point where the terminal side of angle θ in standard position meets the unit circle. Use the given information to evaluate the six trigonometric functions of θ.

75. P is in the second quadrant and $y = 2/5$
76. P is in the third quadrant and $x = -7/9$
77. P is in the third quadrant and $x = -5/7$
78. P is in the second quadrant and $y = \sqrt{3}/4$
79. $x = -15/17$ and $\pi/2 < \theta < \pi$
80. $x = -8/9$ and $\pi < \theta < 3\pi/2$
81. $y = -7/25$ and $3\pi/2 < \theta < 2\pi$
82. $y = 0.6$ and $0 < \theta < \pi/2$
83. List three angles (in radian measure) for which the cosine of each is $-\sqrt{3}/2$.
84. List three angles (in radian measure) for which the cosine of each is $\sqrt{3}/2$.
85. List four angles (in degree measure) for which the sine of each is -1.
86. List four angles (in degree measure) for which the sine of each is 1.
87. If $\cos\theta = -5/13$ and $\pi < \theta < 3\pi/2$, find $\sin\theta$ and $\tan\theta$.
88. If $\sec\theta = 25/7$ and $\sin\theta$ is negative, find $\sin\theta$ and $\tan\theta$.

In Exercises 89–92, refer to the following figure, which shows a highly magnified view of the point P, where the terminal side of an angle of 10° (in standard position) meets the unit circle.

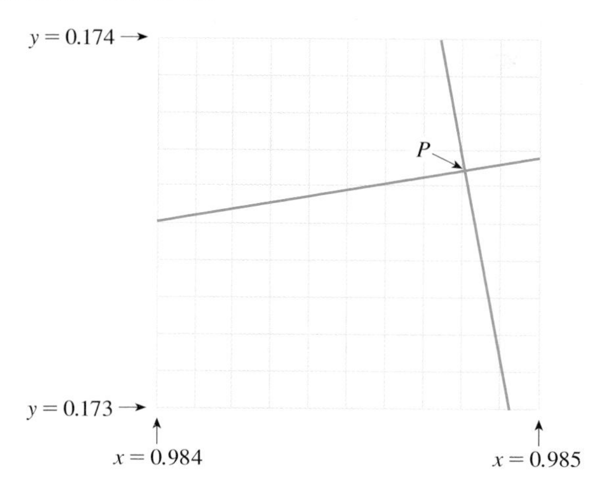

89. Using the figure, estimate the value of cos 10° to four decimal places. (Then use a calculator to check your estimate.)

90. Use your estimate in Exercise 89 to evaluate each of the following.
(a) cos 170°
(b) cos 190°
(c) cos 350°

91. Using the figure, estimate the value of sin 10° to four decimal places. (Then use a calculator to check your estimate.)

92. Use your estimate in Exercise 91 to evaluate each of the following.
(a) $\sin(-10°)$
(b) $\sin(-190°)$
(c) $\sin(-370°)$

Chapter 6 Test

1. Evaluate each expression.
(a) $\sin(-\pi/2)$
(b) cos 540°
(c) cot 450°

2. Specify the value for each expression.
(a) $\cos(\pi/6)$
(b) sin 45°
(c) $\sin^2(7°) + \cos^2(7°)$

In Exercises 3 and 4, use the given information to determine the other five trigonometric values.

3. $\sin\theta = -1/5$ and $3\pi/2 < \theta < 2\pi$

4. $\cos\theta = -\sqrt{5}/6$ and $90° < \theta < 180°$

In Exercises 5–8, evaluate each expression.

5. $\sin(5\pi/3)$ **6.** $\cot(-5\pi/4)$

7. cos 300° **8.** $\csc(-135°)$

9. Without using a calculator, determine which of the expressions, sin 2 or cos 2, is larger. Explain your reasoning (using complete sentences).

10. **(a)** Convert 165° to radian measure.
(b) Convert 3 radians to degree measure.

In Exercises 11 and 12, the two points B and C are on a circle of radius 5 cm. The center of the circle is A and angle BAC is 75°.

11. Find the length of the (shorter) arc of the circle from B to C.

12. Find the area of the (smaller) sector determined by angle BAC.

13. Simplify: $\dfrac{1 + \dfrac{\tan\theta + 1}{\tan\theta}}{-1 + \dfrac{\tan\theta - 1}{\tan\theta}}$.

14. Prove that the following equation is an identity:

$$\frac{\cos\theta + 1}{\csc\theta + \cot\theta} = \sin\theta$$

15. Suppose that a belt drives two wheels of radii 15 cm and 25 cm, as shown in the figure.

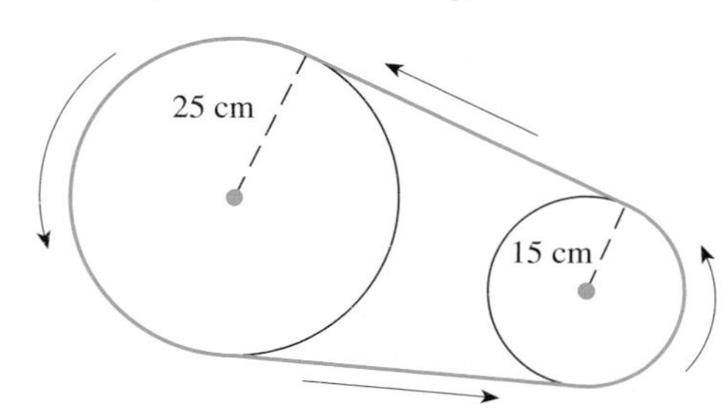

Determine each of the following quantities, given that the angular speed of the smaller wheel is 600 rpm.

(a) The angular speed of the smaller wheel in radians per minute

(b) The linear speed of a point on the circumference of the smaller wheel

(c) The angular speed of the larger wheel in radians per minute
Hint: Because of the belt, the linear speed of a point on the circumference of the smaller wheel is equal to the linear speed of a point on the circumference of the larger wheel.

(d) The angular speed of the larger wheel in rpm

16. If θ is an acute angle in a right triangle and $\tan \theta = t$, express the other five trigonometric values in terms of t.

17. In $\triangle ABC$, $AB = \sqrt{17} - 1$, $BC = \sqrt{17} + 1$, and $\angle B = 90°$. Find $\cos A$.

18. Refer to the following figure. In each case, decide which of the two given quantities is larger. Explain your reasons in complete sentences.

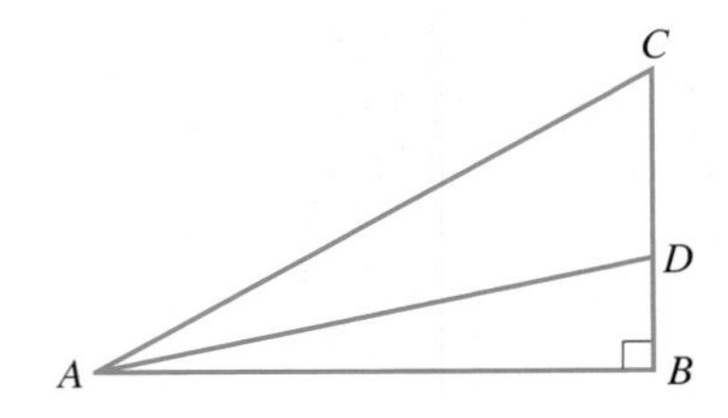

(a) $\tan(\angle DAB)$; $\tan(\angle CAB)$

(b) $\cos(\angle DAB)$; $\cos(\angle CAB)$

7 CHAPTER Graphs of the Trigonometric Functions

Yet another spiral [i.e., curve] *can be made from a proper circle. It is used by stonemasons and is commonly referred to as a "screw-line." But whatever it is called, it is a useful line, and I want to explain how to construct it. Whoever explores its possibilities will find that it has many applications. First, draw a circle with center point* a. *Then divide it with a vertical line through center* a *into two equal parts, and again* [i.e., now] *mark the point at the upper periphery 12, and the bottom 6. Then extend this line upward as far as you wish and mark the topmost point* a. *Draw a horizontal line near the periphery of the circle and mark it* cd. *The point at which it crosses the vertical line is point* b. *Then divide the circular ground plan into twelve equal parts, numbering each point, next to point 12, 1, 2, 3, etc., until you again reach 12. But thereafter the numbers must continue, as far as needed in the projection, i.e., 13 above 1, 14 above 2, etc. Accordingly, one can use the ground plan numbers three, four, or five times, or as many times as one desires to extend the projection. Now that the ground plan has been prepared, mark off the vertical line* ab *into as many points as you wish, beginning to number them from point* b *upward, 1, 2, 3, 4, etc. Then draw a vertical line upward from point 1 on the circle and a horizontal line from point 1 on the vertical line* ab. *Where these two meet, mark the corner of the rectangle 1. Then proceed likewise with all the points on the circle and all points of the vertical line* ab, *even where the numbers on the ground plan are used repeatedly. Once the points of the spiral have been established throughout, draw in the line by hand, as I have done in the diagram below.*—Albrecht Dürer (1471–1528)

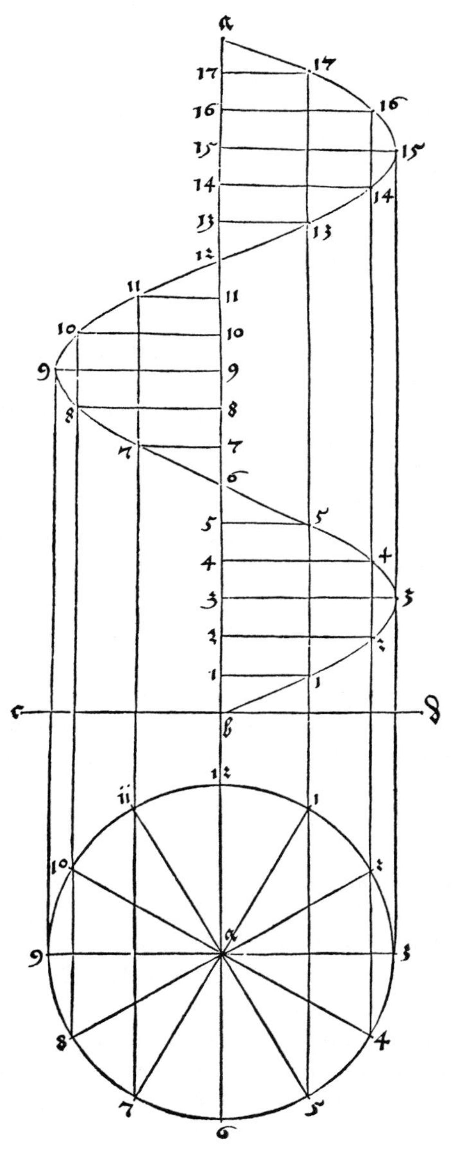

One hundred years before the development of coordinate geometry by Fermat and Descartes, and well over two hundred years before Euler standardized the unit circle definitions of the sine and cosine, the Renaissance artist Albrect Dürer constructed what we would now call a sine curve by projecting points from a circle. The figure and text explaining the construction are from Dürer's book, *Underweysung Der Messung* (Nuremburg: 1525). The translation here is from *The Painter's Manual,* a translation with commentary by Walter L. Strauss (New York: Abaris Books, Inc., 1977).

The trigonometric functions play an important role in many of the modern applications of mathematics. In these applications, the inputs for these functions are real numbers, rather than angles as in the previous chapter. In Section 7.1 of this chapter, we use radian measure to restate the definitions of the trigonometric functions in such a way that the domains are indeed sets of real numbers. Among other advantages, this allows us to analyze the trigonometric functions using the graphing techniques that we developed in Chapter 3.

7.1 TRIGONOMETRIC FUNCTIONS OF REAL NUMBERS

The ***sine*** *and* ***cosine*** *functions may now be observed bursting buoyantly from their chrysalis to take up a new identity as circular trigonometric functions. The angles of old are represented by arcs or arc lengths. Those right triangles in which the trigonometric functions were imprisoned may like dry-husks be allowed to disappear along with their degrees. . . .*—David Berlinski, *A Tour of the Calculus* (New York: Pantheon Books, 1995)

We always assume that the radius of the circle is 1 and let z be an arc of this circle . We are especially interested in the sine and cosine of this arc z.—Leonhard Euler (1707–1783) in *Introductio in Analysin Infinitorum* (Lausanne, 1748)

In calculus and in the sciences many of the applications of the trigonometric functions require that the inputs be real numbers, rather than angles. By making a small but crucial change in our viewpoint, we can define the trigonometric functions in such a way that the inputs are real numbers. As background for our discussion, recall that radian measure is defined by the equation $\theta = s/r$. In this definition, s and r are assumed to have the same linear units (for example, centimeters) and therefore the ratio s/r is a real number with no units. For example, in Figure 1 the radius is 3 cm, the arc length is 6 cm, and so the radian measure θ is given by

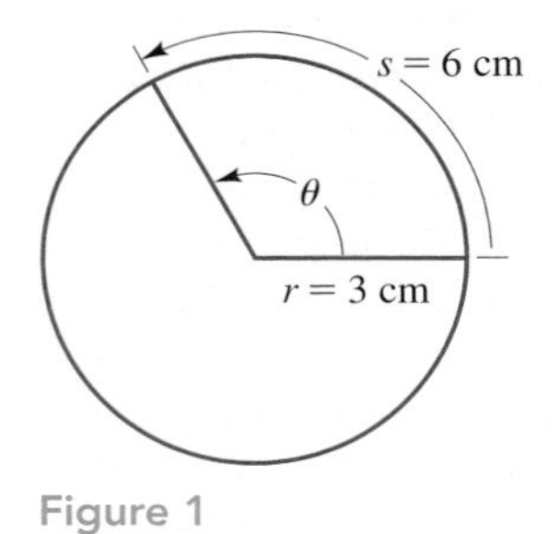

Figure 1

$$\theta = \frac{6\text{ cm}}{3\text{ cm}} = 2 \qquad \text{(a real number)}$$

In the previous chapter (Section 6.2) the definitions of the trigonometric functions were based on the unit circle, so let's look at radian measure in that context. In the equation $\theta = s/r$, we set $r = 1$. This yields

$$\theta = \frac{s}{1} = s \qquad \text{(just as before, a real number)}$$

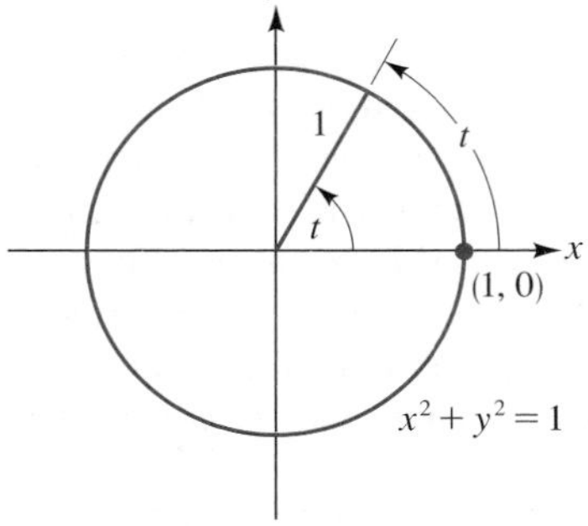

Figure 2
In the unit circle, the radian measure of an angle equals the measure of the intercepted arc.

Thus, in the unit circle, the radian measure of the angle is equal to the measure of the intercepted arc. For reference we summarize this important observation in Figure 2. (We've used the letter t, rather than our usual θ, to emphasize the fact that both the radian measure of the angle and the measure of the intercepted arc are given by the same *real number.*)

The conventions regarding the measurement of arc length on the unit circle are the same as those introduced in the previous chapter for angles. As Figure 3 (on the next page) indicates, we measure from the point (1, 0), and we assume that the positive direction is counterclockwise.

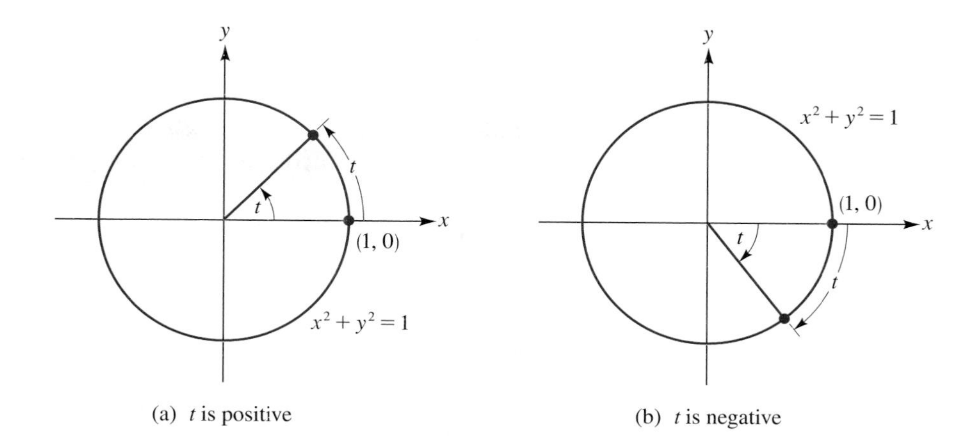

Figure 3
The positive direction is counter-clockwise.

In the definitions that follow, you may think of t as either the measure of an arc or the radian measure of an angle. But in both cases, and this is the point, t denotes a real number.

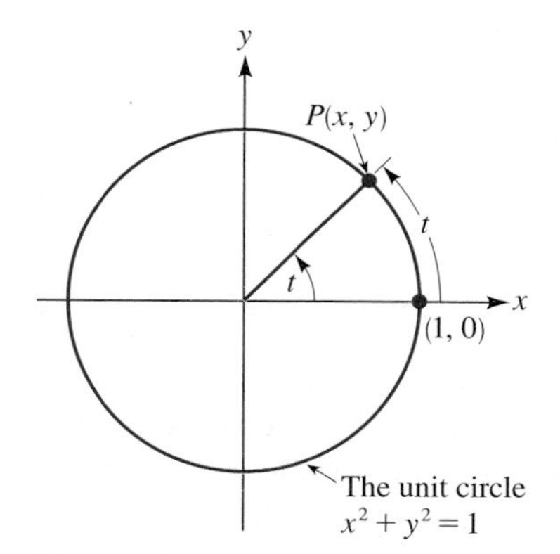

Figure 4

DEFINITION Trigonometric Functions of Real Numbers

(Refer to Figure 4.) Let $P(x, y)$ denote the point on the unit circle that has arc length measure t from $(1, 0)$. (Note that in Figure 4, $0 < t < \pi/2$.) Let $P(x, y)$ denote the point where the terminal side of the angle with radian measure t intersects the unit circle. Then the six trigonometric functions of the real number t are defined as follows.

$$\cos t = x \qquad \sec t = \frac{1}{x} \quad (x \neq 0)$$

$$\sin t = y \qquad \csc t = \frac{1}{y} \quad (y \neq 0)$$

$$\tan t = \frac{y}{x} \quad (x \neq 0) \qquad \cot t = \frac{x}{y} \quad (y \neq 0)$$

Note that there is nothing essentially new here as far as evaluating the trigonometric functions. For instance, $\sin(\pi/2)$ is still equal to 1. What is different now is that the inputs are real numbers:

$$\sin \frac{\pi}{2} = 1$$

name of function ↑ (sin); input (a real number) ↑ ($\frac{\pi}{2}$); output (a real number) ↑ (1)

As was explained in Section 6.3, the four-step procedure for evaluating the trigonometric functions is valid whether we are working in degrees or radians. When radian measure is used, the reference angle is sometimes referred to as the **reference number** to emphasize the fact that a radian angle measure (such as $\pi/4$ or 3) is indeed a real number. In the next example we evaluate trigonometric functions with real number inputs. As you'll see, however, there are no new techniques for you to learn in this context. As we've said, the new definition amounts to a change in viewpoint more than anything else.

EXAMPLE 1 Calculating trigonometric functions of real numbers

Evaluate each of the following expressions:

(a) $\cos \frac{2\pi}{3}$; **(b)** $\sec \frac{2\pi}{3}$.

SOLUTION

(a) The cosine of the real number $2\pi/3$ is, by definition, the cosine of $2\pi/3$ radians, which we can evaluate by means of our four-step procedure.

Step 1 See Figure 5.

Step 2 The reference angle for $2\pi/3$ is $\pi/3$.

Step 3 $\cos(\pi/3) = 1/2$

Step 4 $\cos t$ is the x-coordinate, and in Quadrant II, x-coordinates are negative. Thus since the terminal side of $t = 2\pi/3$ is in Quadrant II, $\cos(2\pi/3)$ is negative. Consequently, we have

$$\cos \frac{2\pi}{3} = -\cos \frac{\pi}{3} = -\frac{1}{2}$$

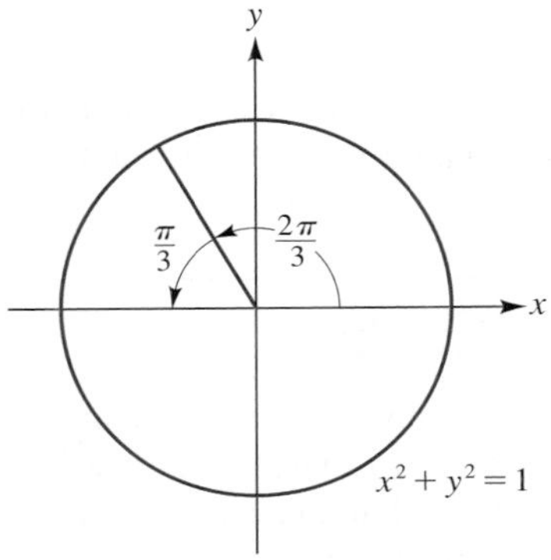

(b) We could use the four-step procedure to evaluate $\sec(2\pi/3)$ but in this case it is much more direct simply to note that $\sec(2\pi/3)$ is the reciprocal of $\cos(2\pi/3)$, which we evaluated in part (a). Thus we have

$$\sec \frac{2\pi}{3} = -2$$

Figure 5

For the remainder of this section we discuss some of the more fundamental identities for the trigonometric functions of real numbers. As we said in Section 6.4, an **identity** is an equation that is satisfied by all values of the variable in its domain. As a demonstration of what this definition really means, consider the equation

$$\sin^2 t + \cos^2 t = 1$$

As we'll prove in a moment (and as you should already expect from the previous chapter), this equation is an identity. The domain of the variable t in this case is the set of all real numbers. So if we pick a real number at random, say, $t = 17$, then it follows that the equation

$$\sin^2 17 + \cos^2 17 = 1$$

is true. You should take a minute now to confirm this empirically with your calculator. That is, actually compute the left-hand side of the preceding equation and check that the result is indeed 1. The point is, no matter what real number we choose for t, the resulting equation will be true.

As an example of an equation that is not an identity, consider the *conditional equation*

$$2 \sin t = 1 \tag{1}$$

As you can check (without a calculator), this equation is true when $t = \pi/6$ but false when $t = \pi/3$. Thus equation (1) is *not* satisfied by *all* values of the variable in its domain. In other words, equation (1) is not an identity. In a later section, we'll work with conditional equations such as equation (1). When we do that, we will be trying to solve these equations; that is, we'll want to find just which values of the variable (if any) do satisfy the equation.

QUESTION FOR REVIEW AND PREVIEW Can you find some real numbers other than $\pi/6$ that also satisfy equation (1)?

There are five identities that are immediate consequences of the unit-circle definitions of the trigonometric functions. For example, using the unit-circle definitions, we have

$$\begin{aligned} \tan t &= \frac{y}{x} && \text{definition of } \tan t \\ &= \frac{\sin t}{\cos t} && \text{definitions of } \sin t \text{ and } \cos t \end{aligned}$$

Thus we have $\tan t = (\sin t)/(\cos t)$. This identity and four others are listed in the box that follows.

Consequences of the Definitions

$$\sec t = \frac{1}{\cos t} \qquad \csc t = \frac{1}{\sin t} \qquad \cot t = \frac{1}{\tan t}$$

$$\tan t = \frac{\sin t}{\cos t} \qquad \cot t = \frac{\cos t}{\sin t}$$

The next identities that we consider are the three *Pythagorean identities.* The term "Pythagorean" is used here because, as you'll see, the proofs of the identities rely on the unit-circle definitions of the trigonometric functions; and the equation of the unit circle essentially is derived from the Pythagorean theorem.

The Pythagorean Identities

$$\sin^2 t + \cos^2 t = 1$$

$$\tan^2 t + 1 = \sec^2 t$$

$$\cot^2 t + 1 = \csc^2 t$$

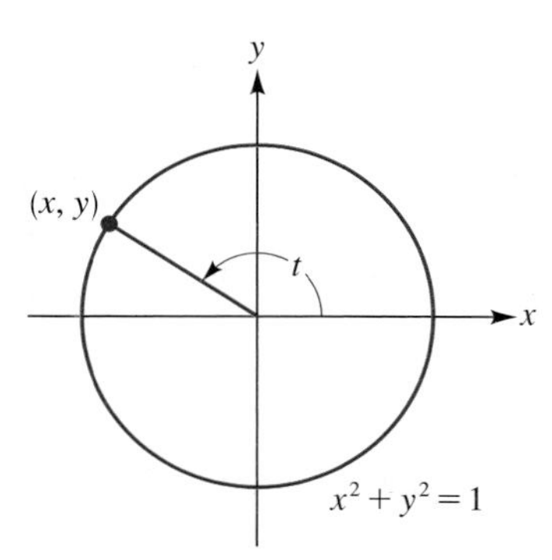

Figure 6
A real number t interpreted as the radian measure of an angle in standard position

To establish the identity $\sin^2 t + \cos^2 t = 1$, we think of the real number t, for the moment, as the radian measure of an angle in standard position, as indicated in Figure 6. Now we proceed exactly as in the previous chapter, when we proved this identity in the context of angles. Since the point (x, y) in Figure 6 lies on the unit circle, we have

$$x^2 + y^2 = 1$$

Now, replacing x by $\cos t$ and y by $\sin t$ gives us

$$\cos^2 t + \sin^2 t = 1$$

which is what we wished to show. You should also be familiar with two other ways of writing this identity:

$$\cos^2 t = 1 - \sin^2 t \qquad \text{and} \qquad \sin^2 t = 1 - \cos^2 t$$

To prove the second of the Pythagorean identities, we begin with the identity $\sin^2 t + \cos^2 t = 1$ and divide both sides by the quantity $\cos^2 t$ to obtain

$$\frac{\sin^2 t}{\cos^2 t} + \frac{\cos^2 t}{\cos^2 t} = \frac{1}{\cos^2 t} \qquad \text{assuming that } \cos t \neq 0$$

and, consequently,

$$\tan^2 t + 1 = \sec^2 t \qquad \text{as required}$$

Since the proof of the third Pythagorean identity is similar, we omit it here.

EXAMPLE 2 Finding values of trigonometric functions

If $\sin t = 2/3$ and $\pi/2 < t < \pi$, compute $\cos t$ and $\tan t$.

SOLUTION

$$\begin{aligned} \cos^2 t &= 1 - \sin^2 t && \text{using the first Pythagorean identity} \\ &= 1 - \left(\frac{2}{3}\right)^2 && \text{substituting} \\ &= 1 - \frac{4}{9} = \frac{5}{9} \end{aligned}$$

Consequently,

$$\cos t = \frac{\sqrt{5}}{3} \qquad \text{or} \qquad \cos t = -\frac{\sqrt{5}}{3}$$

Now, since $\pi/2 < t < \pi$, it follows that $\cos t$ is negative. (Why?) Thus

$$\cos t = -\frac{\sqrt{5}}{3} \qquad \text{as required}$$

To compute $\tan t$, we use the identity $\tan t = (\sin t)/(\cos t)$ to obtain

$$\tan t = \frac{2/3}{-\sqrt{5}/3} = \frac{2}{3} \times \frac{3}{-\sqrt{5}} = -\frac{2}{\sqrt{5}}$$

If required, we can rationalize the denominator (by multiplying by $\sqrt{5}/\sqrt{5}$) to obtain

$$\tan t = -\frac{2\sqrt{5}}{\sqrt{5}}$$

EXAMPLE 3 Finding values of trigonometric functions

If $\sec t = -5/3$ and $\pi < t < 3\pi/2$, compute $\cos t$ and $\tan t$.

SOLUTION

Since $\cos t$ is the reciprocal of $\sec t$, we have $\cos t = -3/5$. We can compute $\tan t$ using the second Pythagorean identity as follows:

$$\begin{aligned}\tan^2 t &= \sec^2 t - 1\\ &= \left(-\frac{5}{3}\right)^2 - 1 = \frac{16}{9}\end{aligned}$$

Therefore

$$\tan t = \frac{4}{3} \qquad \text{or} \qquad \tan t = -\frac{4}{3}$$

Since t is between π and $3\pi/2$, $\tan t$ is positive. (Why?) Thus

$$\tan t = \frac{4}{3}$$

Example 4 shows how certain radical expressions can be simplified with an appropriate trigonometric substitution. (This technique is often useful in calculus.)

EXAMPLE 4 Using a trigonometric substitution to rationalize a square root

In the expression $u/\sqrt{u^2 - 1}$, make the substitution $u = \sec\theta$ and show that the resulting expression is equal to $\csc\theta$. (Assume that $0 < \theta < \pi/2$.)

SOLUTION

Replacing u by $\sec\theta$ in the given expression yields

$$\begin{aligned}\frac{u}{\sqrt{u^2 - 1}} &= \frac{\sec\theta}{\sqrt{\sec^2\theta - 1}}\\ &= \frac{\sec\theta}{\sqrt{\tan^2\theta}} \qquad \text{using the second Pythagorean identity}\\ &= \frac{\sec\theta}{\tan\theta} = \frac{1/\cos\theta}{\sin\theta/\cos\theta}\\ &= \frac{1}{\sin\theta} = \csc\theta\end{aligned}$$

Question: Where did we use the condition $0 < \theta < \pi/2$?

As we saw in Chapter 6, one technique for simplifying trigonometric expressions involves first writing everything in terms of sines and cosines. However, this is not necessarily the most efficient method. As the next example indicates, the second and third Pythagorean identities can be quite useful in this context.

EXAMPLE 5 Using trigonometric identities to simplify trigonometric expressions

Simplify the expression $\dfrac{\csc t + \csc t \cot^2 t}{\sec^2 t - \tan^2 t}$

SOLUTION

$$\frac{\csc t + \csc t \cot^2 t}{\sec^2 t - \tan^2 t} = \frac{(\csc t)(1 + \cot^2 t)}{1}$$ factoring out the common factor in the numerator and applying the second Pythagorean identity in the denominator

$$= (\csc t)(\csc^2 t)$$ applying the third Pythagorean identity

$$= \csc^3 t$$

As background for the next set of identities, we define the terms "even function" and "odd function." A function f is said to be an **even function** provided

$$f(-t) = f(t) \qquad \text{for every value of } t \text{ in the domain of } f$$

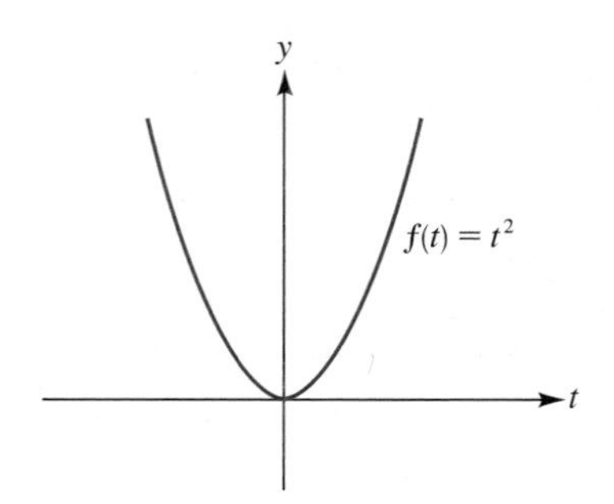

Figure 7
The graph of an even function is symmetric about the y-axis.

The function defined by $f(t) = t^2$ serves as a simple example of an even function. Whether we use t or $-t$ for the input, the output is the same. For instance,

$$f(5) = 5^2 = 25 \qquad \text{and also} \qquad f(-5) = (-5)^2 = 25$$

The graph of an even function is always symmetric about the y-axis. (This follows directly from the symmetry tests in Chapter 1 on page 59.) For example, in Figure 7 we show the graph of the even function $f(t) = t^2$.

A function f is said to be an **odd function** provided

$$f(-t) = -f(t) \qquad \text{for every value of } t \text{ in the domain of } f$$

The function defined by $g(t) = t^3$ is an odd function. For instance, if you compute $g(5)$ and $g(-5)$, you'll find that the outputs are the negatives of one another. (Check this.) The graph of an odd function is always symmetric about the origin. (Again, this follows from the symmetry tests in Chapter 1.) In Figure 8 we display the graph of the odd function $g(t) = t^3$.

Note: The domains of both even and odd functions must have the property that if t is in the domain, then $-t$ is also in the domain.

Now let's return to our development of trigonometric identities. The three identities in the box that follows can be interpreted using the terminology we've just been discussing.

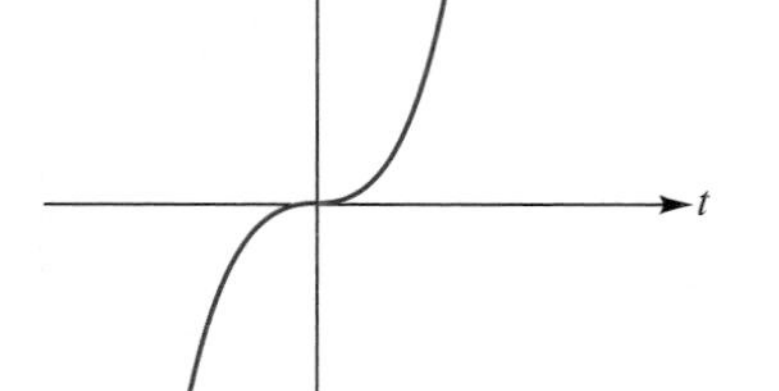

Figure 8
The graph of an odd function is symmetric about the origin.

The Opposite-Angle Identities

$$\cos(-t) = \cos t$$

$$\sin(-t) = -\sin t$$

$$\tan(-t) = -\tan t$$

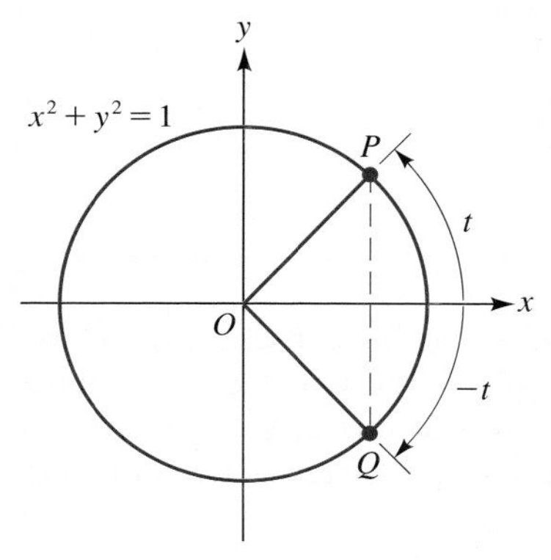

Figure 9

To see why the first two of these identities are true, consider Figure 9. We'll assume that $0 < t < \pi/2$, so Figure 9 shows an arc of length t terminating in Quadrant I. By definition the coordinates of the points P and Q in Figure 9 are as follows:

$$P: \quad (\cos t, \sin t)$$
$$Q: \quad (\cos(-t), \sin(-t))$$

However, as you can see by looking at Figure 9, the x-coordinates of P and Q are the same, while the y-coordinates are negatives of each other. Thus,

$$\cos(-t) = \cos t$$

and

$$\sin(-t) = -\sin t \qquad \text{as we wished to show}$$

Now we can establish the third identity, involving $\tan(-t)$, as follows:

$$\tan(-t) = \frac{\sin(-t)}{\cos(-t)}$$
$$= \frac{-\sin t}{\cos t} = -\frac{\sin t}{\cos t} = -\tan t$$

Although we assumed that the arc terminated in Quadrant I the argument we used will work no matter where the arc terminates. To check this, you should try the last argument with a t such that P is not in Quadrant I.

The opposite-angle identities tell us that the cosine function is an even function and that the sine and tangent functions are odd functions.

EXAMPLE 6 Using the opposite-angle identities

(a) If $\sin t = -0.76$, find $\sin(-t)$.
(b) If $\cos s = -0.29$, find $\cos(-s)$.
(c) If $\sin u = 0.54$, find $\sin^2(-u) + \cos^2(-u)$.

SOLUTION

(a) $\sin(-t) = -\sin t$
$= -(-0.76) = 0.76$

(b) $\cos(-s) = \cos s = -0.29$

(c) The opposite-angle identities are unnecessary here, and so too is the given value of $\sin u$, for, according to the first Pythagorean identity, we have

$$\sin^2(-u) + \cos^2(-u) = 1$$

The final identities that we are going to discuss in this section are simply consequences of the fact that the circumference C of the unit circle is 2π. (Substituting $r = 1$ in the formula $C = 2\pi r$ gives us $C = 2\pi$.) Thus if we begin at any point P on the unit circle and travel a distance of 2π units along the perimeter, we return to the same point P. In other words, arc lengths of t and $t + 2\pi$ [measured from $(1, 0)$, as usual] yield the same terminal point on the unit circle. Since the trigonometric functions are defined in terms of the coordinates of that point P, we obtain the following identities.

Periodicity of Sine and Cosine

$$\sin(t + 2\pi) = \sin t$$
$$\cos(t + 2\pi) = \cos t$$

These two identities are true for all real numbers t. As you'll see in the next section, they provide important information about the graphs of the sine and cosine functions; the graphs of both functions repeat themselves at intervals of 2π. Similar identities hold for the other trigonometric functions in their respective domains:

$$\tan(t + 2\pi) = \tan t \qquad \csc(t + 2\pi) = \csc t$$
$$\cot(t + 2\pi) = \cot t \qquad \sec(t + 2\pi) = \sec t$$

EXAMPLE 7 Using periodicity to find a value of sine

Evaluate: $\sin \dfrac{5\pi}{2}$.

SOLUTION

First, simply as a matter of arithmetic, we observe that $5\pi/2 = \pi/2 + 2\pi$. Thus in view of our earlier remarks we have

$$\sin \frac{5\pi}{2} = \sin \frac{\pi}{2} = 1$$

The preceding set of identities can be generalized as follows. If we start at a point P on the unit circle and make two complete counterclockwise revolutions, the arc length we travel is $2\pi + 2\pi = 4\pi$. Similarly, for three complete revolutions the arc length traversed is $3(2\pi) = 6\pi$. And, in general, if k is any integer, the arc length for k complete revolutions is $2|k|\pi$. (When k is positive, the revolutions are counterclockwise; when k is negative, the revolutions are clockwise.) Consequently, we have the following identities.

For any real number t and any integer k, the following identities hold:

$$\sin(t + 2k\pi) = \sin t \quad \text{and} \quad \cos(t + 2k\pi) = \cos t$$

EXAMPLE 8 Using periodicity to find a value of cosine

Evaluate: $\cos(-17\pi)$.

SOLUTION

$$\cos(-17\pi) = \cos(\pi - 18\pi) = \cos \pi = -1$$

EXERCISE SET 7.1

A

In Exercises 1–8, evaluate each expression (as in Example 1).

1. **(a)** $\cos(11\pi/6)$ **(b)** $\cos(-11\pi/6)$ **(c)** $\sin(11\pi/6)$ **(d)** $\sin(-11\pi/6)$
2. **(a)** $\cos(2\pi/3)$ **(b)** $\cos(-2\pi/3)$ **(c)** $\sin(2\pi/3)$ **(d)** $\sin(-2\pi/3)$
3. **(a)** $\cos(\pi/6)$ **(b)** $\cos(-\pi/6)$ **(c)** $\sin(\pi/6)$ **(d)** $\sin(-\pi/6)$
4. **(a)** $\cos(13\pi/4)$ **(b)** $\cos(-13\pi/4)$ **(c)** $\sin(13\pi/4)$ **(d)** $\sin(-13\pi/4)$
5. **(a)** $\cos(5\pi/4)$ **(b)** $\cos(-5\pi/4)$ **(c)** $\sin(5\pi/4)$ **(d)** $\sin(-5\pi/4)$
6. **(a)** $\cos(9\pi/4)$ **(b)** $\cos(-9\pi/4)$ **(c)** $\sin(9\pi/4)$ **(d)** $\sin(-9\pi/4)$
7. **(a)** $\sec(5\pi/3)$ **(b)** $\csc(-5\pi/3)$ **(c)** $\tan(5\pi/3)$ **(d)** $\cot(-5\pi/3)$
8. **(a)** $\sec(7\pi/4)$ **(b)** $\csc(-7\pi/4)$ **(c)** $\tan(7\pi/4)$ **(d)** $\cot(-7\pi/4)$
9. **(a)** List four positive real-number values of t for which $\cos t = 0$.
   (b) List four negative real-number values of t for which $\cos t = 0$.
10. **(a)** List four positive real numbers t such that $\sin t = 1/2$.
    (b) List four positive real numbers t such that $\sin t = -1/2$.
    (c) List four negative real number t such that $\sin t = 1/2$.
    (d) List four negative real numbers t such that $\sin t = -1/2$.

In Exercises 11–14, use a calculator to evaluate the six trigonometric functions using the given real-number input. (Round the results to two decimal places.)

11. **(a)** 2.06 **(b)** -2.06
12. **(a)** 0.55 **(b)** -0.55
13. **(a)** $\pi/6$ **(b)** $\pi/6 + 2\pi$
14. **(a)** 1000 **(b)** $1000 - 2\pi$

In Exercises 15–22, check that both sides of the identity are indeed equal for the given values of the variable t. For part (c) of each problem, use your calculator.

15. $\sin^2 t + \cos^2 t = 1$
    (a) $t = \pi/3$
    (b) $t = 5\pi/4$
    (c) $t = -53$
16. $\tan^2 t + 1 = \sec^2 t$
    (a) $t = 3\pi/4$
    (b) $t = -2\pi/3$
    (c) $t = \sqrt{5}$
17. $\cot^2 t + 1 = \csc^2 t$
    (a) $t = -\pi/6$
    (b) $t = 7\pi/4$
    (c) $t = 0.12$
18. $\cos(-t) = \cos t$
    (a) $t = \pi/6$
    (b) $t = -5\pi/3$
    (c) $t = -4$
19. $\sin(-t) = -\sin t$
    (a) $t = 3\pi/2$
    (b) $t = -5\pi/6$
    (c) $t = 13.24$
20. $\tan(-t) = -\tan t$
    (a) $t = -4\pi/3$
    (b) $t = \pi/4$
    (c) $t = 1000$
21. $\sin(t + 2\pi) = \sin t$
    (a) $t = 5\pi/3$
    (b) $t = -3\pi/2$
    (c) $t = \sqrt{19}$
22. $\cos(t + 2\pi) = \cos t$
    (a) $t = -5\pi/3$
    (b) $t = \pi$
    (c) $t = -\sqrt{3}$

In Exercises 23 and 24, show that the equation is not an identity by evaluating both sides using the given value of t and noting that the results are unequal.

23. $\cos 2t = 2 \cos t$; $t = \pi/6$
24. $\sin 2t = 2 \sin t$; $t = \pi/2$
25. If $\sin t = -3/5$ and $\pi < t < 3\pi/2$, compute $\cos t$ and $\tan t$.
26. If $\cos t = 5/13$ and $3\pi/2 < t < 2\pi$, compute $\sin t$ and $\cot t$.
27. If $\sin t = \sqrt{3}/4$ and $\pi/2 < t < \pi$, compute $\tan t$.
28. If $\sec s = -\sqrt{13}/2$ and $\sin s > 0$, compute $\tan s$.
29. If $\tan \alpha = 12/5$ and $\cos \alpha > 0$, compute $\sec \alpha$, $\cos \alpha$, and $\sin \alpha$.
30. If $\cot \theta = -1/\sqrt{3}$ and $\cos \theta < 0$, compute $\csc \theta$ and $\sin \theta$.
31. In the expression $\sqrt{9 - x^2}$, make the substitution $x = 3 \sin \theta$ $(0 < \theta < \frac{\pi}{2})$, and show that the result is $3 \cos \theta$.
32. Make the substitution $u = 2 \cos \theta$ in the expression $1/\sqrt{4 - u^2}$, and simplify the result. (Assume that $0 < \theta < \pi$.)
33. In the expression $1/(u^2 - 25)^{3/2}$, make the substitution $u = 5 \sec \theta$ $(0 < \theta < \frac{\pi}{2})$, and show that the result is $(\cot^3 \theta)/125$.
34. In the expression $1/(x^2 + 5)^2$, replace x by $\sqrt{5} \tan \theta$ and show that the result is $(\cos^4 \theta)/25$.
35. In the expression $1/\sqrt{u^2 + 7}$, let $u = \sqrt{7} \tan \theta$, where $0 < \theta < \pi/2$, and simplify the result.
36. In the expression $\sqrt{x^2 - a^2}/x$ $(a > 0)$, let $x = a \sec \theta$ $(0 < \theta < \frac{\pi}{2})$, and simplify the result.
37. **(a)** If $\sin t = 2/3$, find $\sin(-t)$.
    (b) If $\sin \phi = -1/4$, find $\sin(-\phi)$.
    (c) If $\cos \alpha = 1/5$, find $\cos(-\alpha)$.
    (d) If $\cos s = -1/5$, find $\cos(-s)$.
38. **(a)** If $\sin t = 0.35$, find $\sin(-t)$.
    (b) If $\sin \phi = -0.47$, find $\sin(-\phi)$.
    (c) If $\cos \alpha = 0.21$, find $\cos(-\alpha)$.
    (d) If $\cos s = -0.56$, find $\cos(-s)$.
39. If $\cos t = -1/3$ $(\frac{\pi}{2} < t < \pi)$, compute the following:
    (a) $\sin(-t) + \cos(-t)$ **(b)** $\sin^2(-t) + \cos^2(-t)$
40. If $\sin(-s) = 3/5$ $(\pi < s < \frac{3\pi}{2})$, compute:
    (a) $\sin s$ **(b)** $\cos(-s)$ **(c)** $\cos s$ **(d)** $\tan s + \tan(-s)$

In Exercises 41 and 42, use one of the identities $\cos(t + 2\pi k) = \cos t$ or $\sin(t + 2\pi k) = \sin t$ to evaluate each expression.

41. **(a)** $\cos(\frac{\pi}{4} + 2\pi)$ **(b)** $\sin(\frac{\pi}{3} + 2\pi)$ **(c)** $\sin(\frac{\pi}{2} - 6\pi)$

42. **(a)** $\sin(17\pi/4)$
(b) $\sin(-17\pi/4)$
(c) $\cos 11\pi$
(d) $\cos(53\pi/4)$
(e) $\tan(-7\pi/4)$
(f) $\cos(7\pi/4)$
(g) $\sec(\frac{11\pi}{6} + 2\pi)$
(h) $\csc(2\pi - \frac{\pi}{3})$

In Exercises 43–46, use the Pythagorean identities to simplify the given expressions.

43. $\dfrac{\sin^2 t + \cos^2 t}{\tan^2 t + 1}$

44. $\dfrac{\sec^2 t - 1}{\tan^2 t}$

45. $\dfrac{\sec^2 \theta - \tan^2 \theta}{1 + \cot^2 \theta}$

46. $\dfrac{\csc^4 \theta - \cot^4 \theta}{\csc^2 \theta + \cot^2 \theta}$

In Exercises 47–54, prove that the equations are identities.

47. $\csc t = \sin t + \cot t \cos t$

48. $\sin^2 t - \cos^2 t = \dfrac{1 - \cot^2 t}{1 + \cot^2 t}$

49. $\dfrac{1}{1 + \sec s} + \dfrac{1}{1 - \sec s} = -2\cot^2 s$

50. $\dfrac{1 + \tan s}{1 - \tan s} = \dfrac{\sec^2 s + 2\tan s}{2 - \sec^2 s}$

51. $\dfrac{\sec s + \cot s \csc s}{\cos s} = \csc^2 s \sec^2 s$

52. $(\tan \theta)(1 - \cot^2 \theta) + (\cot \theta)(1 - \tan^2 \theta) = 0$

53. $(\cos \alpha \cos \beta - \sin \alpha \sin \beta)(\cos \alpha \cos \beta + \sin \alpha \sin \beta)$
$= \cos^2 \alpha - \sin^2 \beta$

54. $\cot \theta + \tan \theta + 1 = \dfrac{\cot \theta}{1 - \tan \theta} + \dfrac{\tan \theta}{1 - \cot \theta}$

55. If $\sec t = 13/5$ and $3\pi/2 < t < 2\pi$, evaluate

$$\frac{2\sin t - 3\cos t}{4\sin t - 9\cos t}$$

56. If $\sec t = (b^2 + 1)/2b$ and $\pi < t < 3\pi/2$, find $\tan t$ and $\sin t$. (*Note*: b is negative. Why?) You should assume that $b < -1$.

B

57. Use the accompanying figure to explain why the following four identities are valid. (The identities can be used to provide an algebraic foundation for the reference-angle technique that we've used to evaluate the trigonometric functions.)

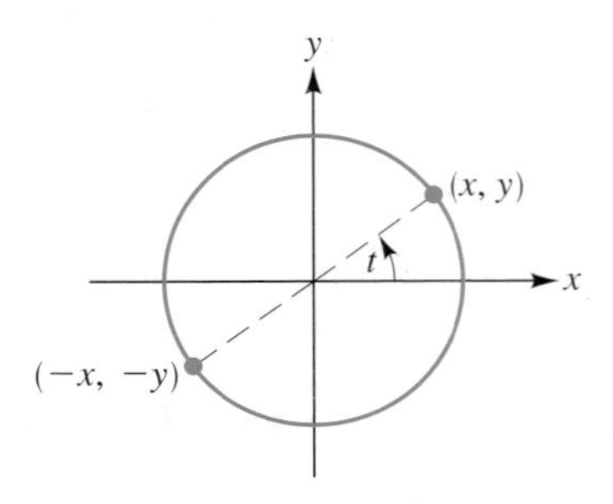

(i) $\sin(t + \pi) = -\sin t$
(ii) $\sin(t - \pi) = -\sin t$
(iii) $\cos(t + \pi) = -\cos t$
(iv) $\cos(t - \pi) = -\cos t$

58. Use two of the results in Exercise 57 to verify the identity $\tan(t + \pi) = \tan t$. (You'll see the graphical aspect of this identity in Section 7.5.)

59. In the equation $x^4 + 6x^2y^2 + y^4 = 32$, make the substitutions

$$x = X\cos\frac{\pi}{4} - Y\sin\frac{\pi}{4} \quad \text{and} \quad y = X\sin\frac{\pi}{4} + Y\cos\frac{\pi}{4}$$

and show that the result simplifies to $X^4 + Y^4 = 16$. (*Hint*: Evaluate the trigonometric functions, simplify the expressions for x and y, take out the common factor, and then substitute.)

60. Suppose that $\tan \theta = 2$ and $0 < \theta < \pi/2$.
(a) Compute $\sin \theta$ and $\cos \theta$.
(b) Using the values obtained in part (a), make the substitutions

$$x = X\cos\theta - Y\sin\theta \quad \text{and} \quad y = X\sin\theta + Y\cos\theta$$

in the expression $7x^2 - 8xy + y^2$, and simplify the result.

61. In this exercise, we are going to find the minimum value of the function

$$f(t) = \tan^2 t + 9\cot^2 t \qquad 0 < t < \frac{\pi}{2}$$

(a) Set your calculator in the radian mode and complete the table. Round the values you obtain to two decimal places.

t	0.2	0.4	0.6	0.8	1.0	1.2	1.4
f(t)							

(b) Of the seven outputs you calculated in part (a), which is the smallest? What is the corresponding input?
(c) Prove that $\tan^2 t + 9\cot^2 t = (\tan t - 3\cot t)^2 + 6$.
(d) Use the identity in part (c) to explain why $\tan^2 t + 9\cot^2 t \geq 6$.
(e) The inequality in part (d) tells us that $f(t)$ is never less than 6. Furthermore, in view of part (c), $f(t)$ will equal 6 when $\tan t - 3\cot t = 0$. From this last equation, show that $\tan^2 t = 3$, and conclude that $t = \pi/3$. In summary, the minimum value of f is 6, and this occurs when $t = \pi/3$. How do these values compare with your answers in part (b)?

62. Let $f(\theta) = \sin\theta\cos\theta\ (0 \leq \theta \leq \frac{\pi}{2})$.
(a) Set your calculator in the radian mode and complete the table. Round the results to two decimal places.

θ	0	$\frac{\pi}{10}$	$\frac{\pi}{5}$	$\frac{\pi}{4}$	$\frac{3\pi}{10}$	$\frac{2\pi}{5}$	$\frac{\pi}{2}$
$f(\theta)$							

(b) What is the largest value of $f(\theta)$ in your table in part (a)?

(c) Show that $\sin\theta\cos\theta \le 1/2$ for all real numbers θ in the interval $0 \le \theta \le \pi/2$. *Hint*: Use the inequality $\sqrt{ab} \le (a+b)/2$ [given in Exercise 40(b) on page 111], with $a = \sin\theta$ and $b = \cos\theta$.

(d) Does the inequality $\sin\theta\cos\theta \le \frac{1}{2}$ hold for all real numbers θ?

63. Consider the equation

$$2\sin^2 t - \sin t = 2\sin t\cos t - \cos t$$

(a) Evaluate each side of the equation when $t = \pi/6$.

(b) Evaluate each side of the equation when $t = \pi/4$.

(c) Is the given equation an identity?

64. Suppose that

$$f(t) = (\sin t\cos t)(2\sin t - 1)(2\cos t - 1)(\tan t - 1)$$

(a) Compute each of the following: $f(0), f(\pi/6), f(\pi/4), f(\pi/3)$, and $f(\pi/2)$.

(b) Is the equation $f(t) = 0$ an identity?

In Section 6.1 we pointed out that one of the advantages in using radian measure is that many formulas then take on particularly simple forms. Another reason for using radian measure is that the trigonometric functions can be closely approximated by very simple polynomial functions. To see examples of this, complete the tables in Exercises 65–68. Round (or, for exact values, simply report) the answers to six decimal places. In Exercises 67 and 68, note that the higher-degree polynomial provides the better approximation. Note: The approximating polynomials in Exercises 65–68 are known as ***Taylor polynomials,*** *after the English mathematician Brook Taylor (1685–1731). The theory of Taylor polynomials is developed in calculus.*

65. (a)

t	$1 - \frac{1}{2}t^2$	$\cos t$
0.02		
0.05		
0.1		
0.2		
0.3		

Ⓖ **(b)** On the same set of axes, graph the functions $1 - \frac{1}{2}t^2$ and $\cos t$. Use a window extending from $-\pi$ to π in the x-direction and -1 to 1 in the y-direction. What do you observe?

66. (a)

t	$t - \frac{1}{6}t^3$	$\sin t$
0.02		
0.05		
0.1		
0.2		
0.3		

Ⓖ **(b)** On the same set of axes, graph the functions $t - \frac{1}{6}t^3$ and $\sin t$. Use a window extending from $-\pi$ to π in the x-direction and -1 to 1 in the y-direction. What do you observe?

67. (a)

x	$\frac{1}{3}x^3 + x$	$\frac{2}{15}x^5 + \frac{1}{3}x^3 + x$	$\tan x$
0.1			
0.2			
0.3			
0.4			
0.5			

Ⓖ **(b)** On the same set of axes, graph the functions $\frac{1}{3}x^3 + x$, $\frac{5}{12}x^5 + \frac{1}{3}x^3 + x$, and $\tan x$. Use a window extending from $-\pi$ to π in the x-direction and -10 to 10 in the y-direction. What do you observe?

68. (a)

x	$x^2 + x$	$\frac{1}{3}x^3 + x^2 + x$	$e^x \sin x$
0.1			
0.2			
0.3			
0.4			
0.5			

Ⓖ **(b)** On the same set of axes, graph the functions $x^2 + x$, $\frac{1}{3}x^3 + x^2 + x$, and $e^x \sin x$. Use a window extending from $-\pi$ to π in the x-direction and -10 to 10 in the y-direction. What do you observe?

69. The figure on the following page shows two x-y coordinate systems. (The same unit of length is used on all four axes.) In the coordinate system on the left, the curve is a portion of the unit circle

$$x^2 + y^2 = 1$$

and A is the point (1, 0). The points B, C, D, E, and F are located on the circle according to the information in the following table.

arc	$\widehat{AB}$	$\widehat{AC}$	$\widehat{AD}$	$\widehat{AE}$	$\widehat{AF}$
length	$\frac{\pi}{12}$	$\frac{\pi}{6}$	$\frac{\pi}{4}$	$\frac{\pi}{3}$	$\frac{5\pi}{12}$

Determine the y-coordinates of the points P, Q, R, S, and T. Give an exact expression for each answer and, where appropriate, a calculator approximation rounded to three decimal places.

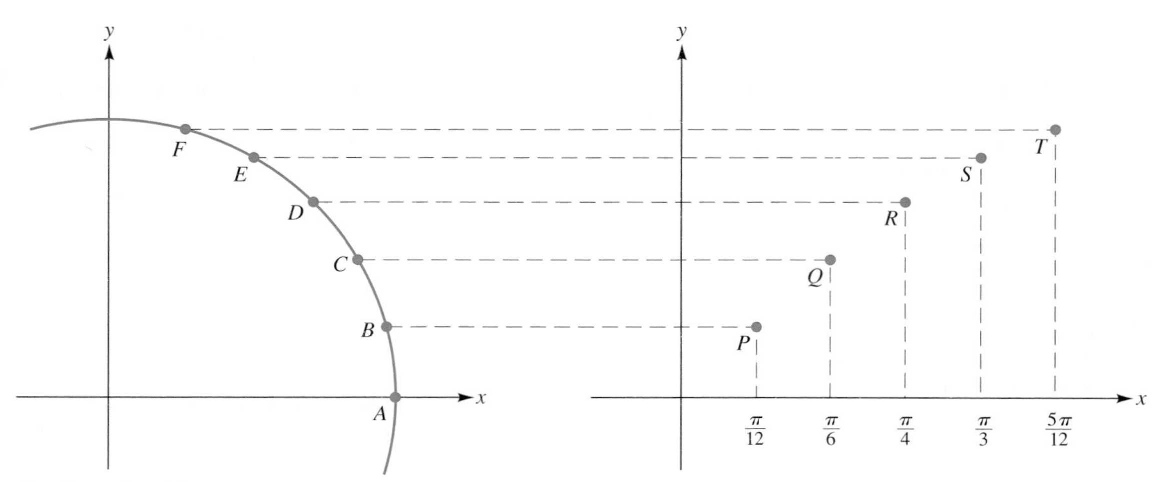

Figure for Exercise 69

7.2 GRAPHS OF THE SINE AND COSINE FUNCTIONS

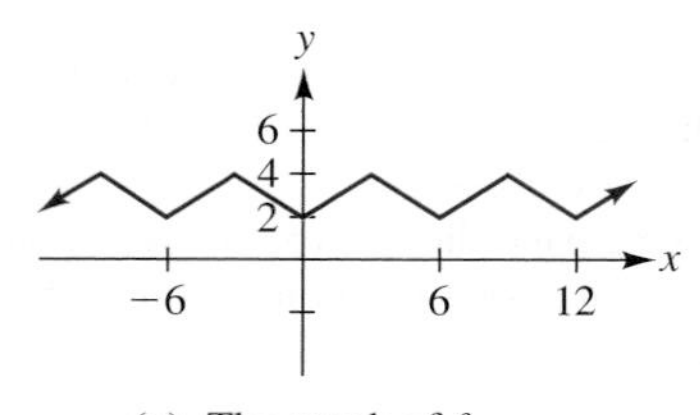

(a) The graph of f

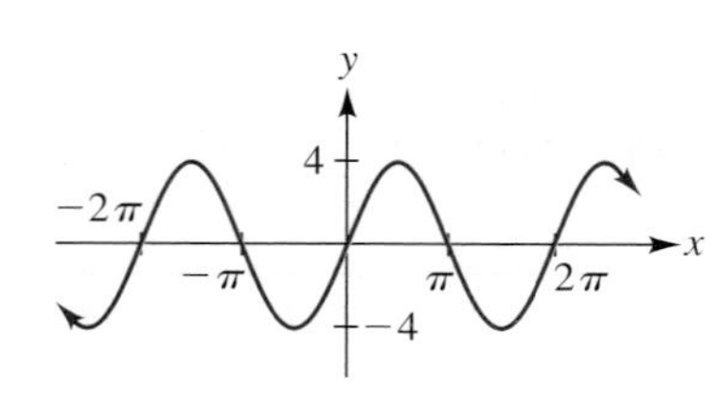

(b) The graph of g

Figure 1

Our focus in this section is on the sine and cosine functions. As preparation for the discussion, we want to understand what is meant by the term "periodic function." By way of example both of the functions in Figure 1 are **periodic.** That is, their graphs display patterns that repeat themselves at regular intervals.

In Figure 1(a) the graph of the function f repeats itself every six units. We say that *the period of f is 6.* Similarly, the period of g in Figure 1(b) is 2π. In both cases, notice that the period represents the minimum number of units that we must travel along the horizontal axis before the graph begins to repeat itself. Because of this, we can state the definition of a periodic function as follows.

DEFINITION A Periodic Function and Its Period

A nonconstant function f is said to be **periodic** if there is a number $p > 0$ such that

$$f(x + p) = f(x)$$

for all x in the domain of f. The smallest such number p is called the **period** of f.

We also want to define the term "amplitude" as it applies to periodic functions. For a function such as g in Figure 1(b), in which the graph is centered about the horizontal axis, the amplitude is simply the maximum height of the graph above the horizontal axis. Thus the amplitude of g is 4. More generally, we define the amplitude of any periodic function as follows.

DEFINITION **Amplitude**

Let f be a periodic function and let m and M denote, respectively, the minimum and maximum values of the function. Then the **amplitude** of f is the number

$$\frac{M - m}{2}$$

For the function g in Figure 1(b) this definition tells us that the amplitude is $\frac{4 - (-4)}{2} = 4$, which agrees with our previous value. Check for yourself now that the amplitude of the function in Figure 1(a) is 1.

Periodic functions are used throughout the sciences to analyze or describe a variety of phenomena ranging from the vibrations of an electron to the variations in the size of an animal population as it interacts with its environment. Figures 2 through 4 display some examples of periodic functions.

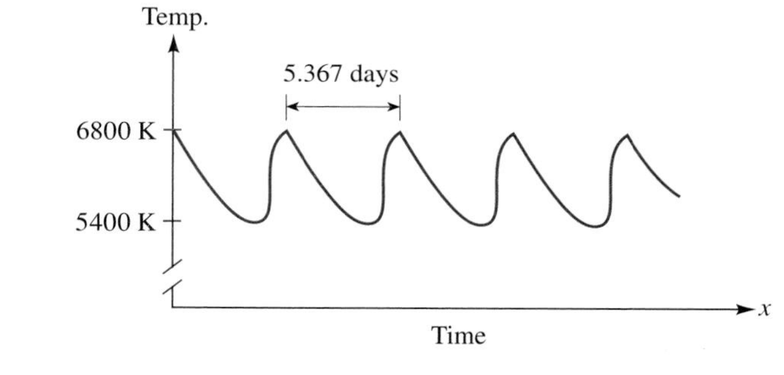

Figure 2
The surface temperature of the star Delta Cephei is a periodic function of time. The period is 5.367 days. The amplitude is $(6800 - 5400)/2 = 700$ degrees Kelvin.

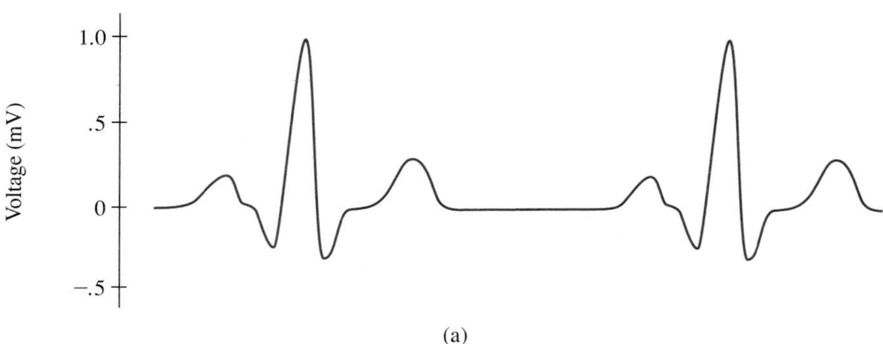

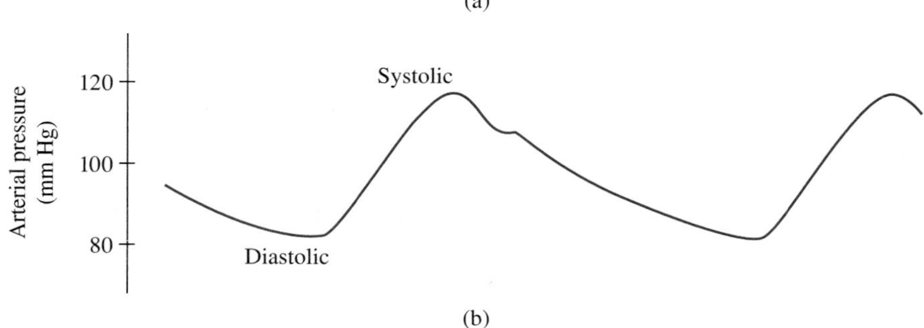

Figure 3
Electrical activity of the heart and blood pressure as periodic functions of time. The figure shows (a) a typical ECG (electrocardiogram) and (b) the corresponding graph of arterial blood pressure. [Adapted from *Physics for the Health Sciences* by C. R. Nave and B. C. Nave (Philadelphia: W. B. Saunders Co., 1975)]

The sound wave generated by a note played on the bamboo flute is a periodic function. The sound wave is recorded on an oscilloscope. (Photograph by Professor Vern Ostdiek)

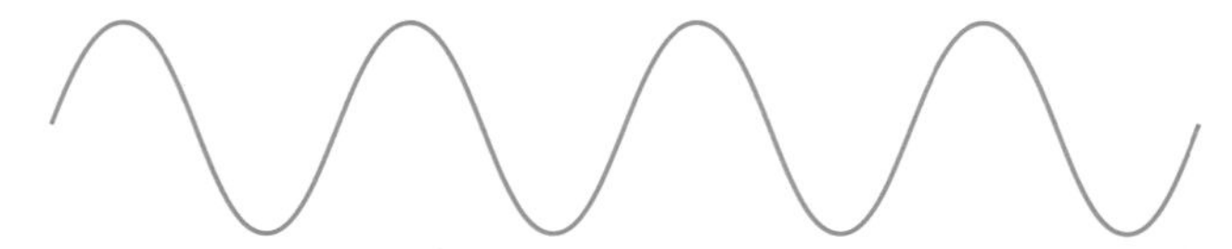

(a) The sound wave generated by a tuning fork vibrating at 440 vibrations per second (On the piano, this note corresponds to the first A above middle C.)

(b) The sound wave generated by a tuning fork vibrating at 880 vibrations per second [This note is an A one octave above that in part (a).]

(c) The sound wave generated by simultaneously striking the tuning forks in parts (a) and (b)

Figure 4
Sound waves

Let us now graph the sine function $f(\theta) = \sin\theta$. We are assuming that θ is a real number, so the domain of the sine function is the set of all real numbers. Even before making any calculations, we can gain strong intuitive insight into how the graph must look by carrying out the following experiment. After drawing the unit circle, $x^2 + y^2 = 1$, place your fingertip at the point $(1, 0)$ and then move your finger

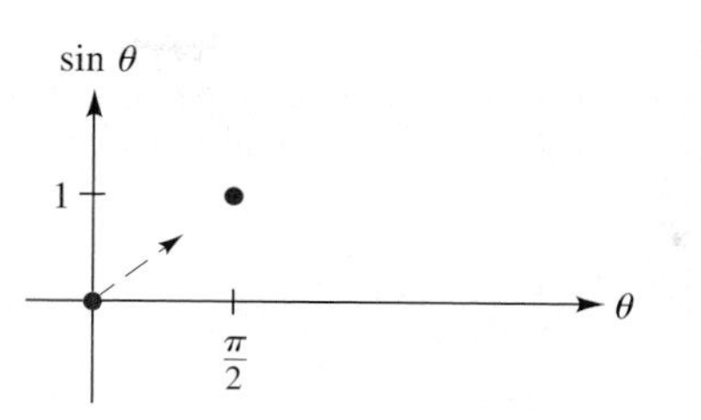

(a) As θ increases from 0 to $\pi/2$, the y-coordinate (sin θ) increases from 0 to 1.

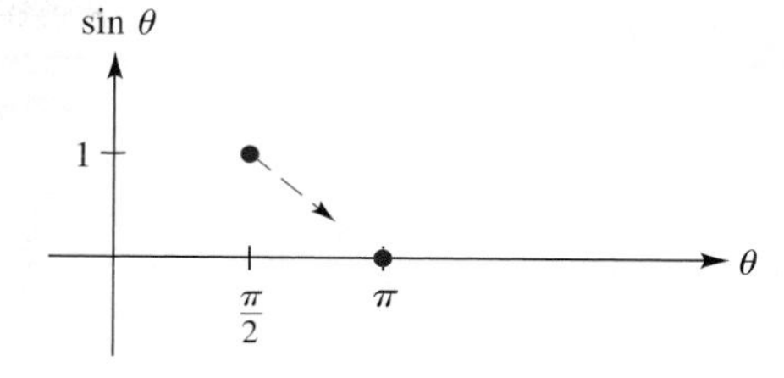

(b) As θ increases from $\pi/2$ to π, the y-coordinate decreases from 1 back down to 0.

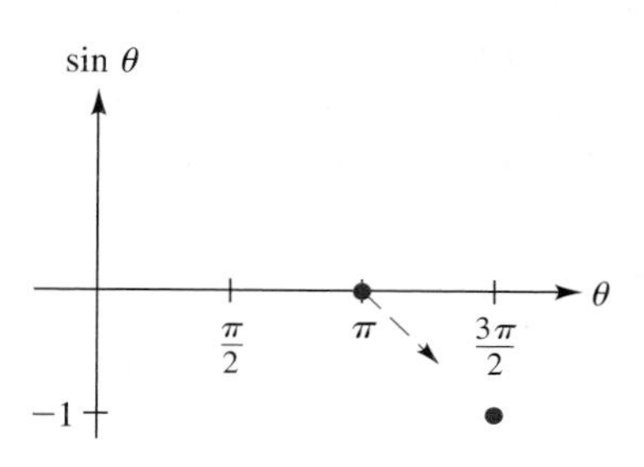

(c) As θ increases from π to $3\pi/2$, the y-coordinate decreases from 0 to -1.

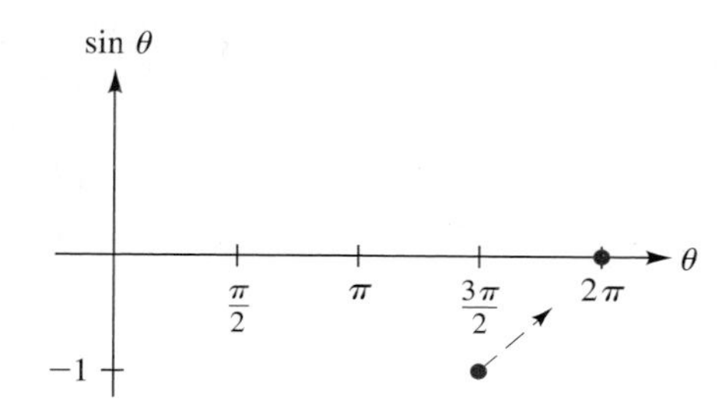

(d) As θ increases from $3\pi/2$ to 2π, the y-coordinate increases from -1 back up to 0.

Figure 5

counterclockwise around the circle. As you do this, keep track of what happens to the y-coordinate of your fingertip. (The y-coordinate is $\sin\theta$.) If we think of θ as the radian measure of an angle, the y-coordinate of your fingertip is $\sin\theta$.

Figure 5 tells us a great deal about the sine function: where the function is increasing and decreasing, where the graph crosses the x-axis, and where the high and low points of the graph occur. Furthermore, since at $\theta = 2\pi$ we've returned to our starting point (1, 0), additional counterclockwise trips around the unit circle will just result in repetitions of the pattern established in Figure 5. This insight can be used to prove that the period of the function $y = \sin\theta$ is indeed 2π.

As informative as Figure 5 is, however, there is not enough information there to tell us the precise shape of the required graph. For instance, all three of the curves in Figure 6 fit the specifications described in Figure 5.

Now let's set up a table so that we can accurately sketch the graph of $y = \sin\theta$. Since the sine function is periodic (as we've just observed) with period 2π, our

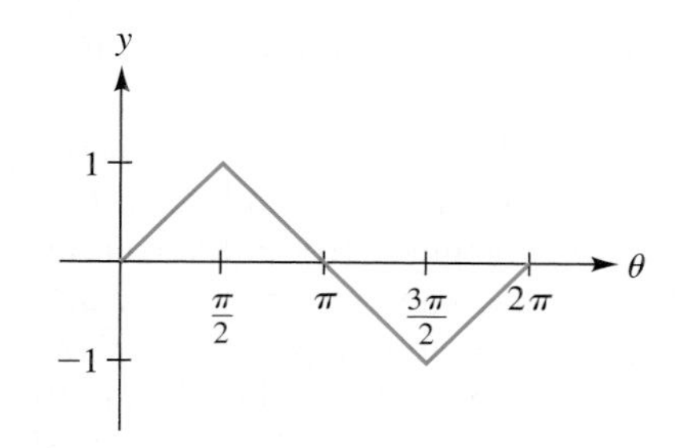

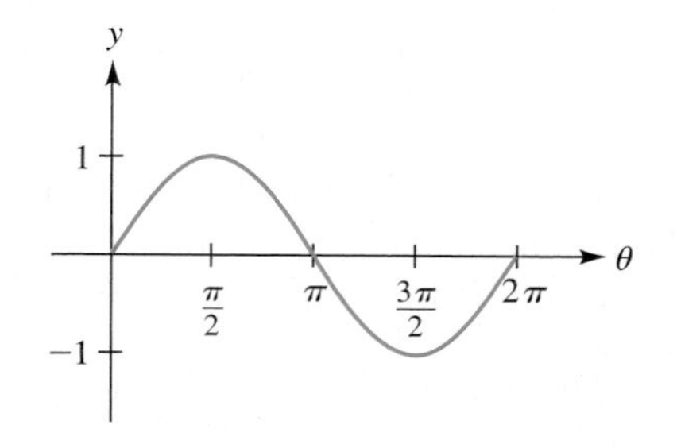

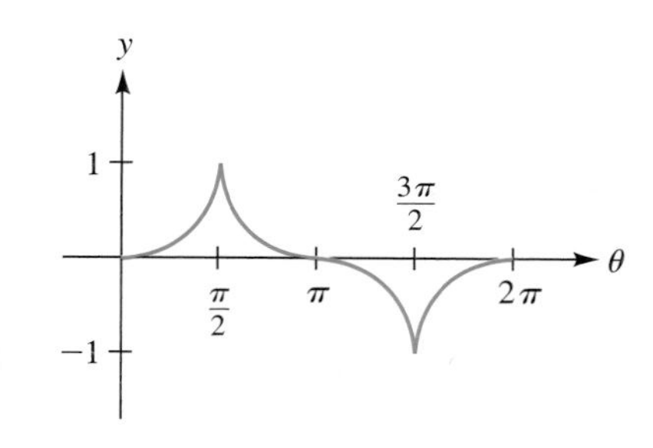

Figure 6

TABLE 1

θ	0	$\frac{\pi}{6}$	$\frac{\pi}{3}$	$\frac{\pi}{2}$	$\frac{2\pi}{3}$	$\frac{5\pi}{6}$	π	$\frac{7\pi}{6}$	$\frac{4\pi}{3}$	$\frac{3\pi}{2}$	$\frac{5\pi}{3}$	$\frac{11\pi}{6}$	2π
$\sin\theta$	0	$\frac{1}{2}$	$\frac{\sqrt{3}}{2}$	1	$\frac{\sqrt{3}}{2}$	$\frac{1}{2}$	0	$-\frac{1}{2}$	$-\frac{\sqrt{3}}{2}$	-1	$-\frac{\sqrt{3}}{2}$	$-\frac{1}{2}$	0

table need contain only values of θ between 0 and 2π, as shown in Table 1. This will establish the basic pattern for the graph.

In plotting the points obtained from Table 1, we use the approximation $\sqrt{3}/2 \approx 0.87$. Rather than approximating π, however, we mark off units on the horizontal axis in terms of π. The resulting graph is shown in Figure 7. By continuing this same pattern to the left and right, we obtain the complete graph of $f(\theta) = \sin\theta$, as indicated in Figure 8.

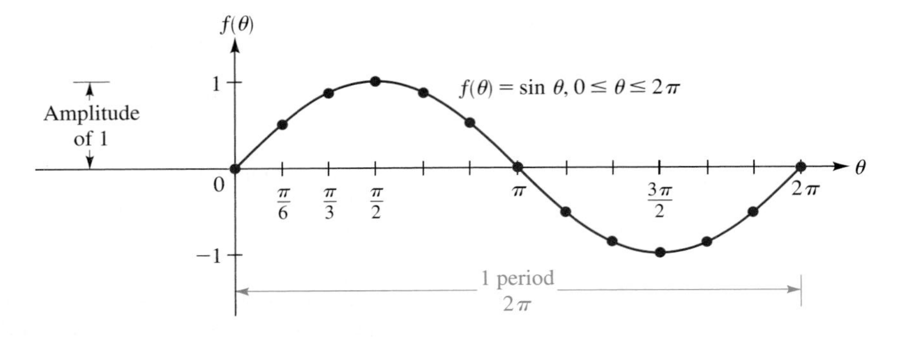

Figure 7

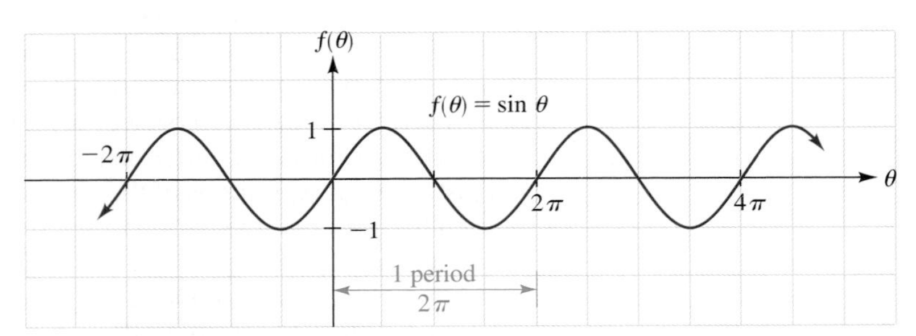

Figure 8

Before going on to analyze the sine function or to study the graphs of the other trigonometric functions, we're going to make a slight change in the notation we've been using. To conform with common usage, we will use x instead of θ on the horizontal axis, and we will use y for the vertical axis. The sine function is then written simply as $y = \sin x$, where the real number x denotes the radian measure of an angle or, equivalently, the length of the corresponding arc on the unit circle. For reference we redraw Figures 7 and 8 using this familiar x-y notation, as shown in Figure 9.

You should memorize the graph of the sine curve in Figure 9(b) so that you can sketch it without first setting up a table. Of course, once you know the shape and the location of the basic cycle shown in Figure 9(a), you automatically know the graph of the full sine curve in Figure 9(b). *Note*: The graph of $y = \sin x$ on an interval of length 2π, the period of the sine function, is called a *cycle* of the graph of $y = \sin x$.

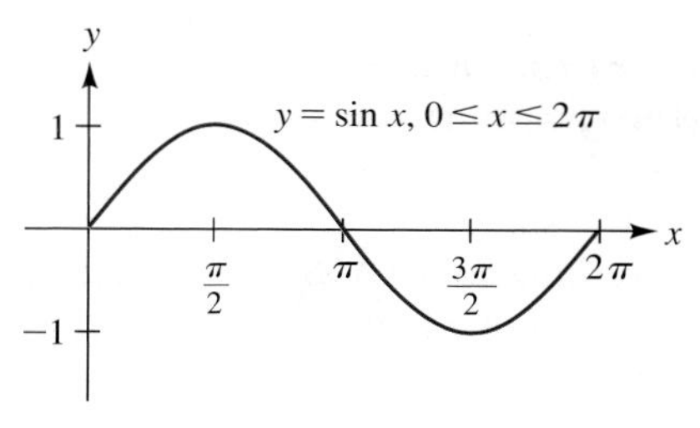

(a) The basic cycle of $y = \sin x$

(b) The period of $y = \sin x$ is 2π

Figure 9

We can use the graphs in Figure 9 to help us list some of the key properties of the sine function.

PROPERTY SUMMARY The Sine Function: $y = \sin x$

1. The domain of the sine function is the set of all real numbers. The range of the sine function is the closed interval $[-1, 1]$, and we have
$$-1 \leq \sin x \leq 1 \qquad \text{for all } x$$
2. The sine function is an odd periodic function with period 2π. The amplitude is 1.
3. The graph of $y = \sin x$ consists of repetitions, over the entire domain, of the **basic sine wave** shown in Figure 10. The basic sine wave crosses the x-axis at the beginning, middle, and end of the cycle. The curve reaches its highest point one-quarter of the way through the cycle and its lowest point three-quarters of the way through the cycle.

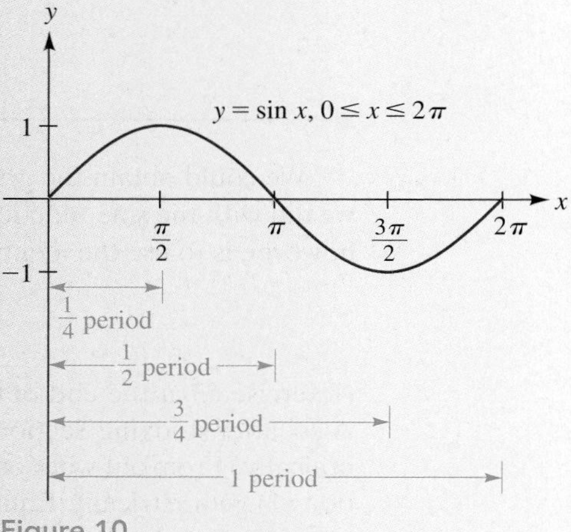

Figure 10
The basic sine wave

Note: The graph of $y = \sin x$ is symmetric about the origin.

EXAMPLE 1 Finding coordinates of turning points and intercepts of the graph of $y = \sin x$

Figure 11 shows a portion of the graph of $y = \sin x$.

(a) What are the coordinates of the three turning points A, B, and C? Here, and in part (b), report the answers both in terms of π and as calculator approximations rounded to three decimal places.

(b) What are the x-intercepts at D and E?

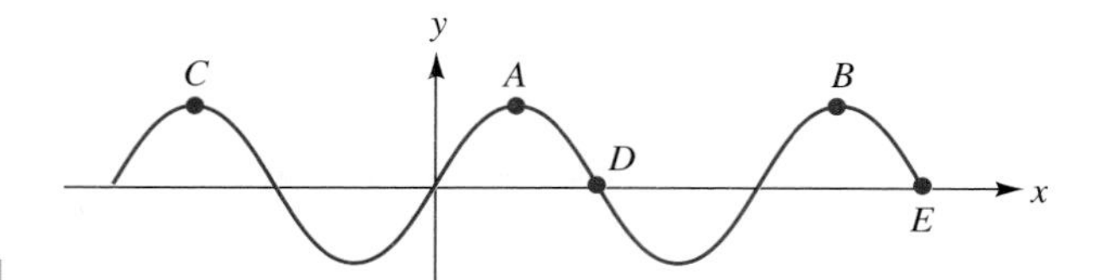

Figure 11

SOLUTION

(a) From Figure 10 (which you should also memorize) we know that the coordinates of the point A are $(\pi/2, 1)$. Therefore since the period of the sine function is 2π, the coordinates of B and C are as follows:

$$\text{Coordinates of } B: \quad \left(\frac{\pi}{2} + 2\pi, 1\right) \quad \text{or} \quad \left(\frac{5\pi}{2}, 1\right)$$

$$\text{Coordinates of } C: \quad \left(\frac{\pi}{2} - 2\pi, 1\right) \quad \text{or} \quad \left(-\frac{3\pi}{2}, 1\right)$$

As you should now check for yourself, the calculator approximations here are

$$A(1.571, 1) \qquad B(7.854, 1) \qquad C(-4.712, 1)$$

(b) From Figure 10 we also know the x-coordinate at D; it is π. Since the point E is 2π units to the right of D, the x-coordinate of E is $\pi + 2\pi = 3\pi$. Using a calculator, we have

$$x\text{-intercept at } D: \quad 3.142$$
$$x\text{-intercept at } E: \quad 9.425$$

We could obtain the graph of the cosine function by setting up a table, just as we did with the sine function. A more interesting and informative way to proceed, however, is to use the identity

$$\cos x = \sin\left(x + \frac{\pi}{2}\right) \tag{1}$$

(Exercise 63 at the end of this section outlines a geometric proof of this identity. Also, after studying Section 8.1, you'll see a simple way to prove this identity algebraically.) From our work on graphing functions in Chapter 3, we can interpret equation (1) geometrically. Equation (1) tells us that the graph of $y = \cos x$ is obtained by translating the sine curve $\pi/2$ units to the left. The result is shown in Figure 12(a). Figure 12(b) displays one complete cycle of the cosine curve, from $x = 0$ to $x = 2\pi$.

As with the sine function, you should memorize the graph and the basic features of the cosine function, which are summarized in the box that follows.

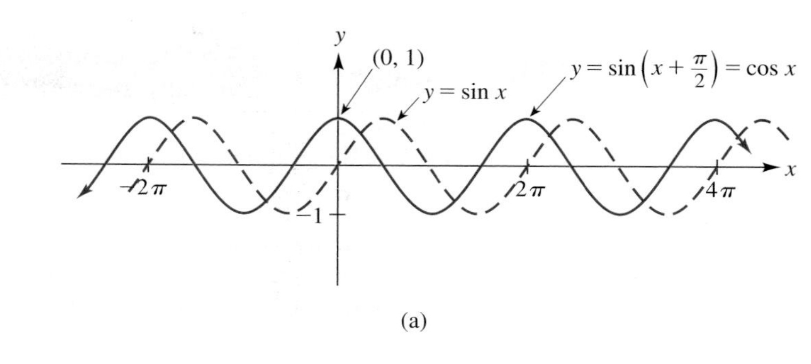

(a)

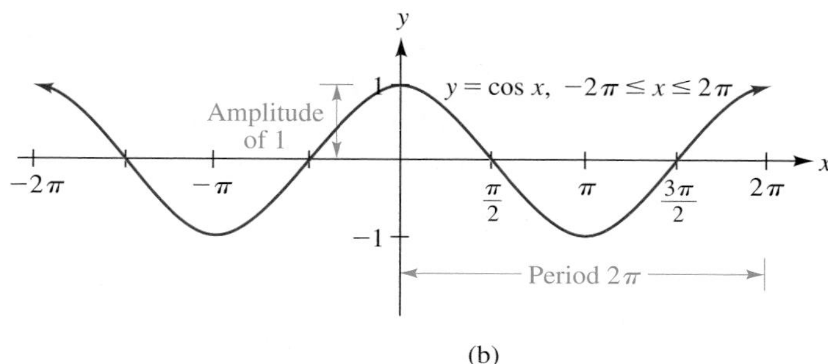

Figure 12

(b)

PROPERTY SUMMARY The Cosine Function: $y = \cos x$

1. The domain of the cosine function is the set of all real numbers. The range of the cosine function is the closed interval $[-1, 1]$, and we have

$$-1 \le \cos x \le 1 \qquad \text{for all } x$$

2. The cosine function is an even periodic function with period 2π. The amplitude is 1.
3. The graph of $y = \cos x$ consists of repetitions, over the entire domain, of the **basic cosine wave** shown in Figure 13. The basic cosine wave crosses the x-axis one-quarter of the way and again three-quarters of the way through the basic cycle. The curve reaches its highest point at the beginning and end of the basic cycle and reaches its lowest point half-way through the basic cycle.

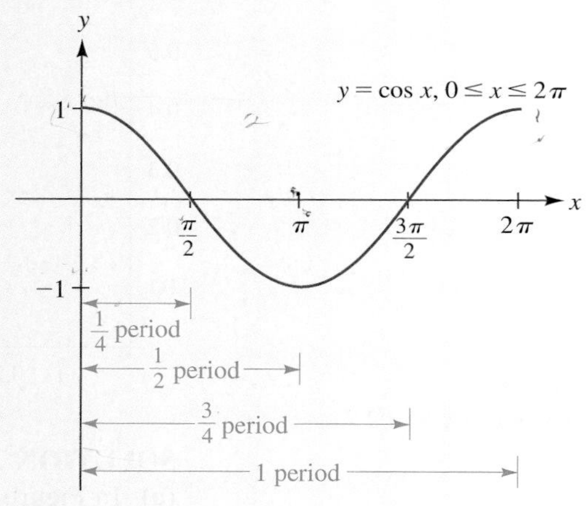

Figure 13
The basic cosine wave

Note: The graph of $y = \cos x$ is symmetric about the y-axis.

In the next example we use a graph to estimate roots of equations involving the cosine function. We also use a calculator to obtain more accurate results. As you'll see, the calculator portion of the work involves more than just button pushing. You'll need to use the reference-angle concept, and you'll need to apply the following property of inverse functions (from Section 3.6).

$$f^{-1}(f(x)) = x \qquad \text{for every } x \text{ in the domain of } f$$

EXAMPLE 2 Finding approximate solutions of cos x = ±0.8 visually and with a calculator

(a) Use the graph in Figure 14 to estimate a root of the equation

$$\cos x = 0.8$$

in the interval $0 \leq x \leq \pi/2$.

(b) Use a calculator to obtain a more accurate value for the root in part (a). Round the answer to three decimal places.

(c) Use the reference-angle concept to find all solutions to the equation $\cos x = 0.8$ in the interval $0 \leq x \leq 2\pi$. Round the answer to three decimal places.

(d) Use the reference-angle concept and a calculator to find all solutions of the equation $\cos x = -0.8$ within the interval $0 \leq x \leq 2\pi$. Again, round the answer to three decimal places.

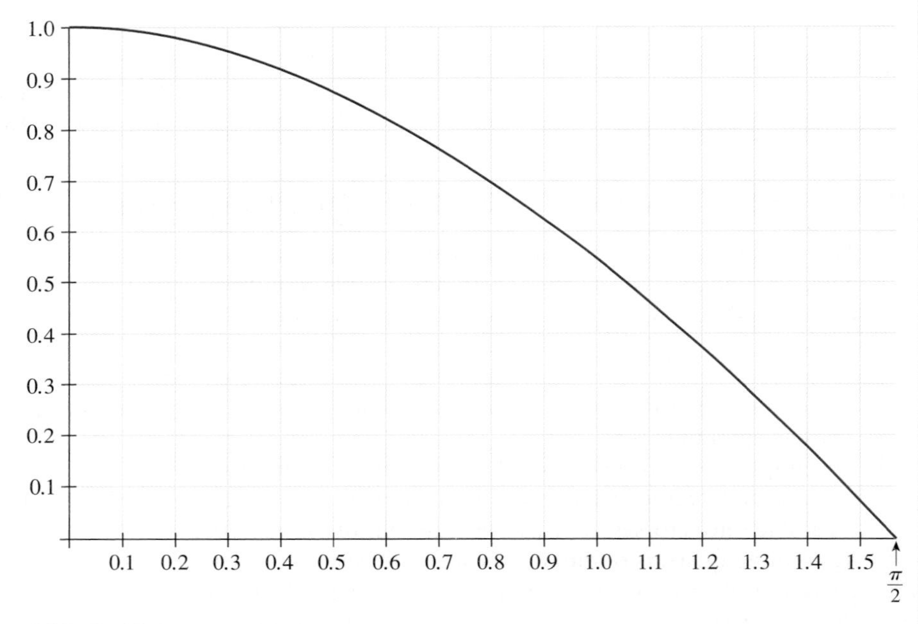

Figure 14
$y = \cos x, 0 \leq x \leq \pi/2$

SOLUTION

(a) In Figure 14, look at the point where the horizontal line $y = 0.8$ intersects the curve $y = \cos x$. The x-coordinate of this point, call it x_1, is the root we are looking for. (Why?) Using Figure 14, we estimate that x_1 is approximately halfway between $x = 0.6$ and $x = 0.7$. So our estimate for the root is $x_1 \approx 0.65$.

(b) We can determine the root x_1 by using the *inverse cosine function.* We will discuss the inverse cosine function in detail in Chapter 8. Just as the notation f^{-1} denotes the inverse of a function f, so the notation $\cos^{-1}$ is often used to denote the inverse cosine function. Starting with the equation $\cos x_1 = 0.8$, we apply the inverse cosine to both sides to obtain

$$\cos^{-1}(\cos x_1) = \cos^{-1}(0.8)$$

and therefore

$$x_1 = \cos^{-1}(0.8)$$

We use a calculator, set in the radian mode, to evaluate the expression $\cos^{-1}(0.8)$. The keystrokes are as follows. (If these keystrokes don't work on your calculator, be sure to check the user's manual.)

EXPRESSION	KEYSTROKES	OUTPUT
$\cos^{-1}(0.8)$	2nd cos 0.8 ENTER	0.644

Note that the value 0.644 is consistent with the estimate $x_1 \approx 0.65$ that was obtained graphically in part (a).

(c) To find another root of the equation, we use the fact that the cosine is positive in Quadrant IV as well as in Quadrant 1. The root $x_1 \approx 0.644$ is the radian measure of a first-quadrant angle (because $0.644 < \pi/2$). As is indicated in Figure 15, we let x_2 denote the radian measure of the fourth-quadrant angle that has x_1 for its reference angle. Then we have $\cos x_2 = \cos x_1 = 0.8$. That is, x_2 is also a solution of the equation $\cos x = 0.8$. The calculations for x_2 now run as follows:

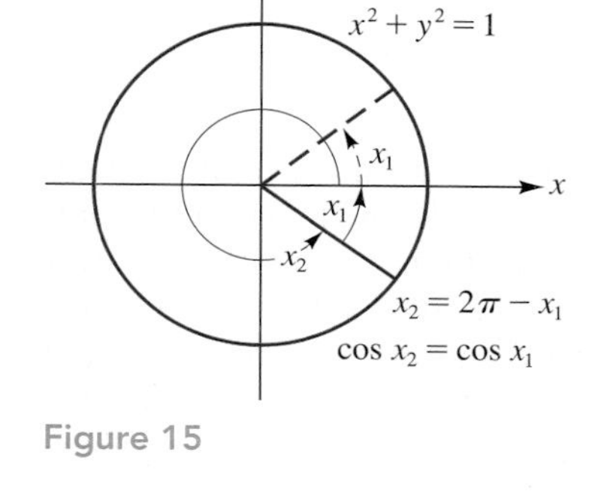

Figure 15

$$\begin{aligned} x_2 &= 2\pi - x_1 \\ &= 2\pi - \cos^{-1}(0.8) \\ &\approx 5.640 \quad \text{using a calculator (radian mode) and rounding to three decimal places} \end{aligned}$$

We have now determined two roots of the equation $\cos x = 0.8$ in the interval $0 \le x \le 2\pi$. There are no other roots in this interval because the cosine is negative in Quadrants II and III.

(d) In part (b) we determined a first-quadrant angle x_1 satisfying the equation $\cos x = 0.8$. Now we want solutions for the equation $\cos x = -0.8$. Since the cosine is negative in Quadrants II and III, we therefore want angles in Quadrants II and III that have x_1 for the reference angle. As is indicated in Figure 16, these angles are $\pi - x_1$ and $\pi + x_1$. Computing, we have

$$\pi - x_1 = \pi - \cos^{-1}(0.8) \approx 2.498$$

and

$$\pi + x_1 = \pi + \cos^{-1}(0.8) \approx 3.785$$

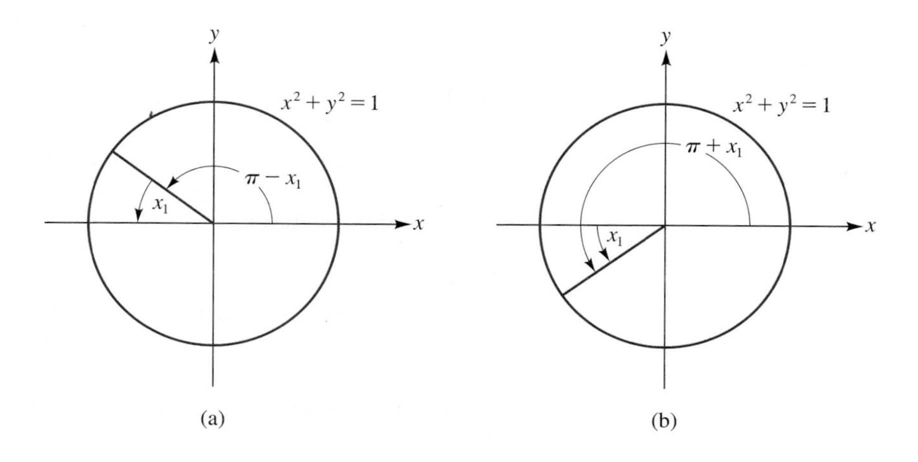

Figure 16
The reference angle for both $\pi - x_1$ and $\pi + x_1$ is x_1.

In summary, the equation $\cos x = -0.8$ has two roots in the closed interval $[0, 2\pi]$. These roots are approximately 2.498 and 3.785. In Figure 17,

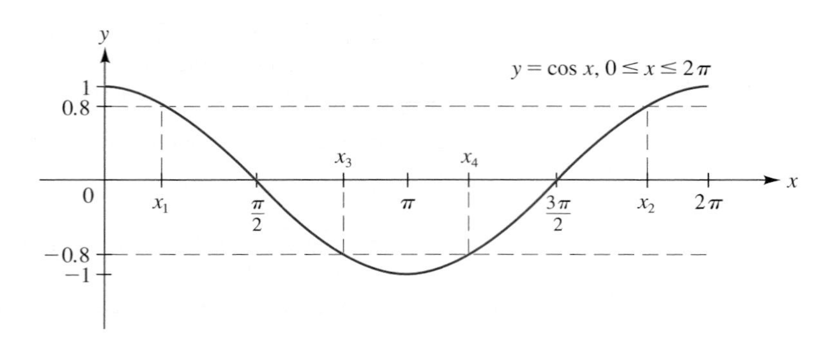

Figure 17
Summary of Example 2. The roots of the equation $\cos x = 0.8$ in the interval $[0, 2\pi]$ are $x_1 \approx 0.644$ and $x_2 \approx 5.640$. The roots of the equation $\cos x \approx -0.8$ in the interval $[0, 2\pi]$ are $x_3 \approx 2.498$ and $x_4 \approx 3.785$.

which summarizes this example, these two roots are denoted by x_3 and x_4, respectively.

In part (b) of the example just concluded, we used a calculator to find a number x such that $\cos x = 0.8$. As was indicated in the example, the function that outputs such a number is called the inverse cosine function. There are two standard abbreviations for the name of this function, and a third abbreviation that is used on some types of calculators. The two standard abbreviations are $\cos^{-1}$ and arccos (read *arc cos*). In the next chapter we will analyze the inverse cosine function in some detail. But for our present purposes you need only know the following definition:

$\cos^{-1}(x)$ denotes the unique number in the interval $[0, \pi]$ whose cosine is x.

As examples of this notation, we have

$$\cos^{-1}\left(\frac{1}{2}\right) = \frac{\pi}{3} \qquad \text{because } \cos\tfrac{\pi}{3} = \tfrac{1}{2} \text{ and } 0 < \tfrac{\pi}{3} < \pi$$

$$\cos^{-1}\left(\frac{1}{2}\right) \neq \frac{5\pi}{3} \qquad \text{because although } \cos\tfrac{5\pi}{3} = \tfrac{1}{2}, \text{ the number } \tfrac{5\pi}{3} \text{ is not in the required interval } [0, \pi]$$

$$\cos^{-1}\left(-\frac{1}{2}\right) = \frac{2\pi}{3} \qquad \text{because } \cos\tfrac{2\pi}{3} = -\tfrac{1}{2} \text{ and } 0 < \tfrac{2\pi}{3} < \pi,$$

$$\cos^{-1}(0.8) \approx 0.644 \qquad \text{as we saw in Example 2}$$

In Example 2, Figure 17 tells us that the number x_1 is in the interval $[0, \pi]$ but x_2 is not. That is why the keystrokes on page 517 gave us the value for x_1 rather than x_2. (We then used the reference-angle or reference-number concept to obtain x_2.)

For some of the exercises in this section you will need to use the *inverse sine function,* rather than the inverse cosine function that we have just been discussing. For reference, we define both functions in the box that follows. (These functions are discussed at greater length in the next chapter.)

DEFINITION	Inverse Cosine Function and Inverse Sine Function	
NAME OF FUNCTION	ABBREVIATION	DEFINITION
Inverse cosine	$\cos^{-1}$	$\cos^{-1}(x)$ is the (unique) number in the interval $[0, \pi]$ whose cosine is x.
Inverse sine	$\sin^{-1}$	$\sin^{-1}(x)$ is the (unique) number in the interval $[-\frac{\pi}{2}, \frac{\pi}{2}]$ whose sine is x.

EXERCISE SET 7.2

A

In Exercises 1–8, specify the period and amplitude for each function.

1.

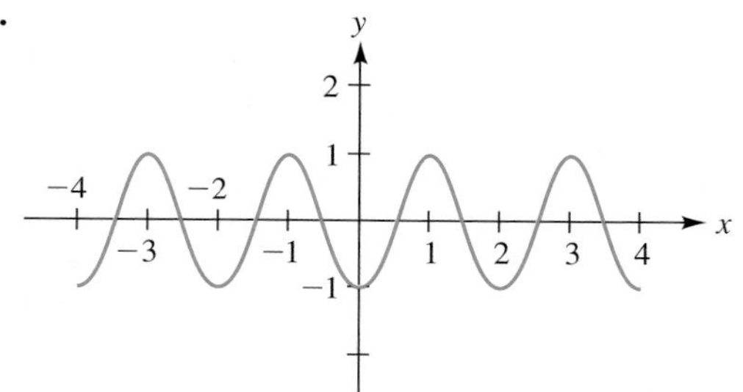

2.

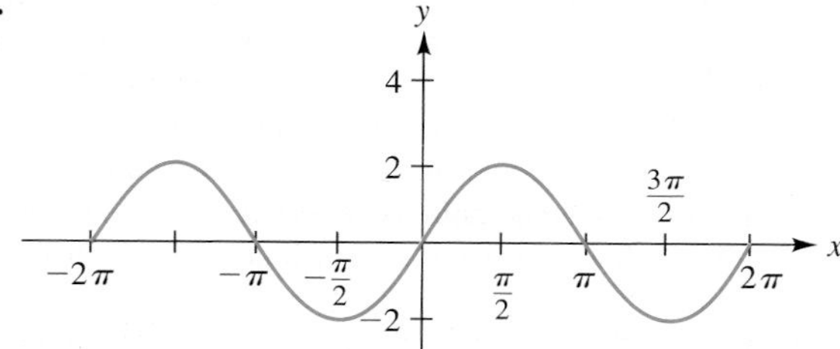

3.

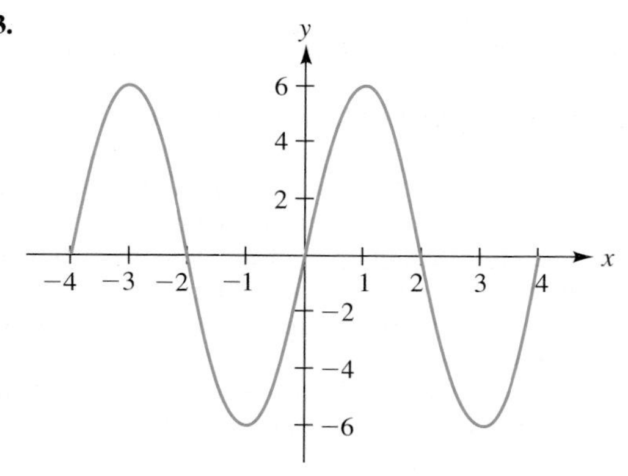

4.

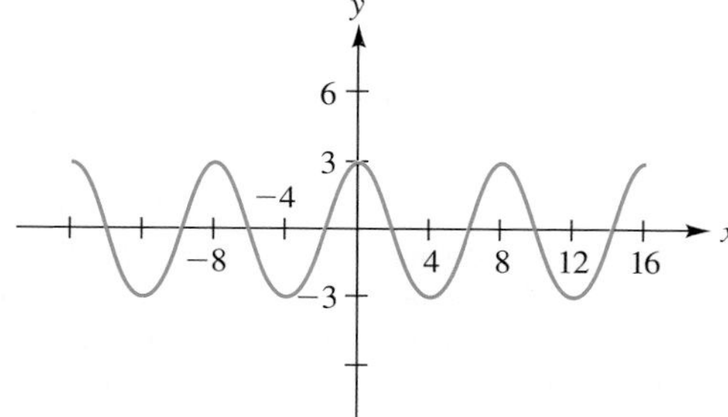

5.

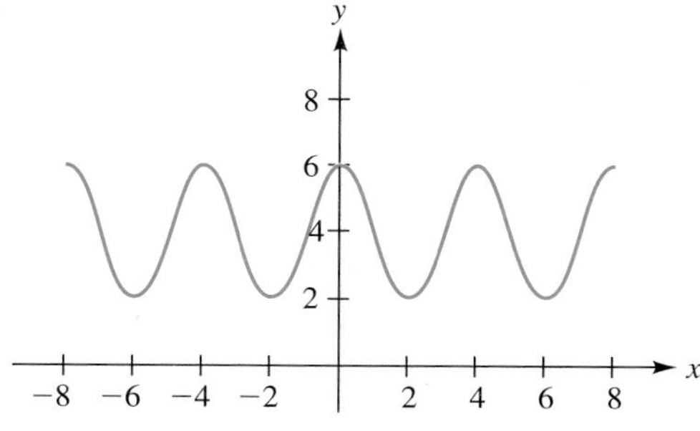

6.

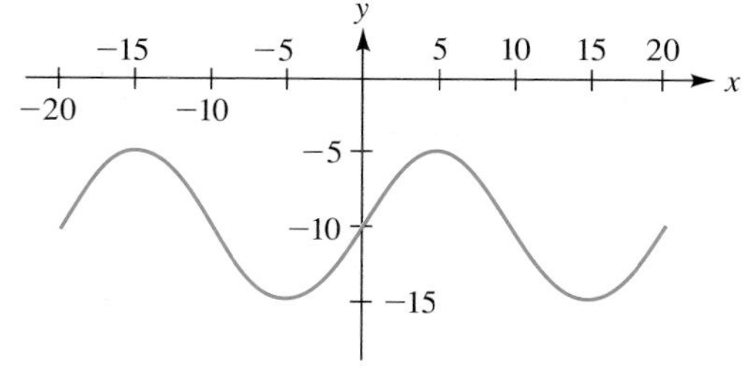

7. 

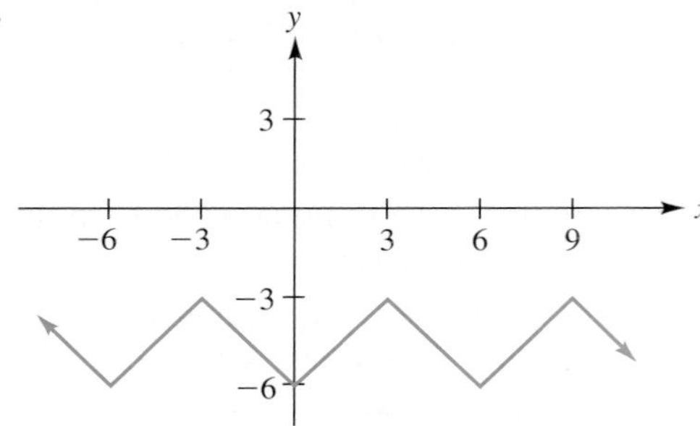

8.

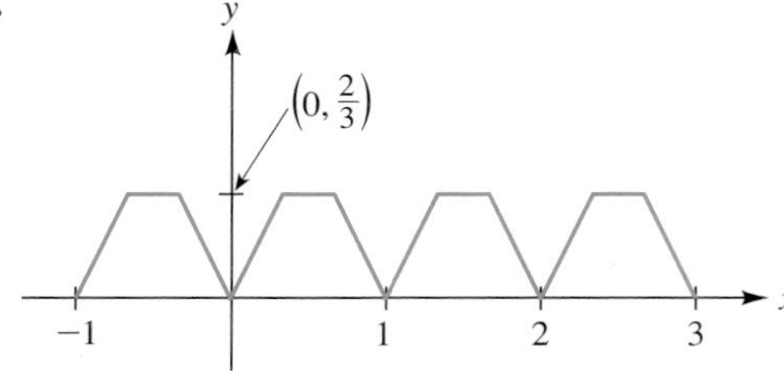

In Exercises 9–18, refer to the graph of $y = \sin x$ *in the following figure. Specify the coordinates of the indicated points. Give the x-coordinates both in terms of* π *and as calculator approximations rounded to three decimal places.*

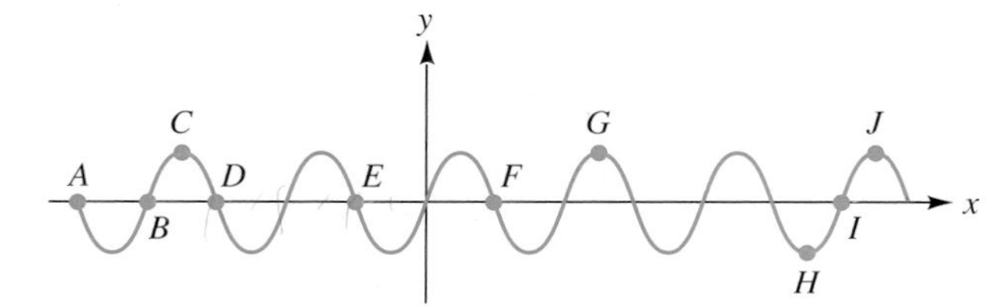

9. *C*
10. *F*
11. *G*
12. *A*
13. *B*
14. *J*
15. *D*
16. *H*
17. *E*
18. *I*

In Exercises 19–22, state whether the function $y = \sin x$ *is increasing or decreasing on the given interval. (The terms* increasing *and* decreasing *are explained on page 159.)*

19. $3\pi/2 < x < 2\pi$
20. $-\pi/2 < x < \pi/2$
21. $5\pi/2 < x < 7\pi/2$
22. $-5\pi/2 < x < -2\pi$

In Exercises 23–32, refer to the graph of $y = \cos x$ *in the following figure. Specify the coordinates of the indicated points. Give the x-coordinates both in terms of* π *and as calculator approximations rounded to three decimal places.*

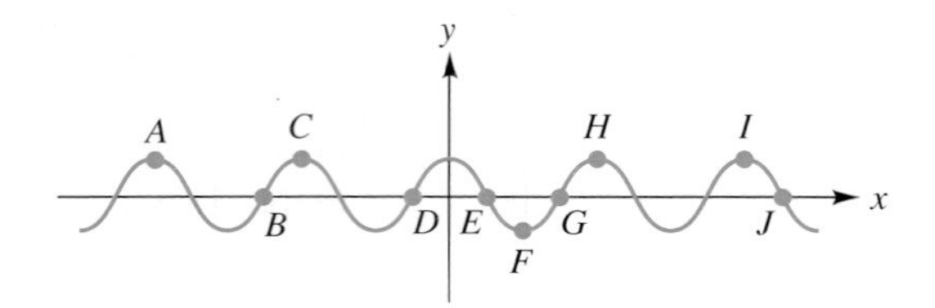

23. *J*
24. *H*
25. *A*
26. *G*
27. *E*
28. *D*
29. *I*
30. *F*
31. *B*
32. *C*

In Exercises 33–36, state whether the function $y = \cos x$ *is increasing or decreasing on the given interval.*

33. $0 < x < \pi$
34. $6\pi < x < 7\pi$
35. $-\pi/2 < x < 0$
36. $-5\pi/2 < x < -2\pi$

In Exercises 37–39, use a graphing utility to obtain several different views of $y = \sin x$ *and* $y = \cos x$*. If you are using a graphics calculator, make sure it is set for the radian mode (rather than the degree mode).*

Ⓖ **37. (a)** Graph $y = \sin x$ using a viewing rectangle that extends from -7 to 7 in the x-direction and from -3 to 3 in the y-direction. Note that there is an x-intercept between 3 and 4. Using your knowledge of the sine function (and not the graphing utility), what is the exact value for this x-intercept?

(b) Refer to the graph that you obtained in part (a). How many turning points do you see? Note that one of the turning points occurs when x is between 1 and 2. What is the exact x-coordinate for this turning point?

(c) Add the graph of $y = \cos x$ to your picture. How many turning points do you see for $y = \cos x$? Note that one of the turning points occurs between $x = 6$ and $x = 7$. What is the exact x-coordinate for this turning point?

(d) Your picture in part (c) indicates that the graphs of $y = \sin x$ and $y = \cos x$ are just translates of one another. By what distance would we have to shift the graph of $y = \sin x$ to the left for it to coincide with the graph of $y = \cos x$?

(e) Most graphing utilities have an option that will let you mark off the units on the x-axis in terms of π. Check your instruction manual if necessary, and then, for the picture that you obtained in part (c), change the x-axis units to multiples of $\pi/2$. Use the resulting picture to confirm your answers to the questions in parts (a) through (c) regarding x-intercepts and turning points.

Ⓖ **38.** This exercise presents an interesting fact about the graph of $y = \sin x$ that is useful in numerical work.

(a) In the standard viewing rectangle, graph the function $y = \sin x$ along with the line $y = x$. Notice that the two graphs appear to be virtually identical in the vicinity of the origin. Actually, the only point that the two graphs have in common is $(0, 0)$, but very near the origin, the distance between the two graphs is less than the thickness of the lines or dots that your graphing utility draws. To underscore this fact, zoom in on the origin several times. What do you observe?

(b) The work in part (a) can be summarized as follows. (Recall that the symbol $\approx$ means "is approximately equal to.")

$$\sin x \approx x \qquad \text{when } |x| \text{ is close to } 0$$

To see numerical evidence that supports this result, complete the following tables.

x	0.253	0.0253	0.00253	0.000253
$\sin x$				
$x - \sin x$				

x	-0.253	-0.0253	-0.00253	-0.000253
$\sin x$				
$x - \sin x$				

(c) The numerical evidence in part (b) suggests that for x positive and close to 0, $\sin x < x$. State the corresponding result for x negative and close to 0. Now on the same set of axes draw, without a calculator, the graphs of $y = x$ and $y = \sin x$.

Ⓖ **39. (a)** Graph the function $y = \cos x$ in the standard viewing rectangle, and look at the arch-shaped portion of the curve between $-\pi/2$ and $\pi/2$. This portion of the cosine curve has the general shape of a parabola. Could it actually be a portion of a parabola? Go on to part (b).

(b) Calculus shows that the answer to the question raised in part (a) is "no." But calculus also shows that there is a parabola, $y = 1 - 0.5x^2$, that closely resembles the cosine curve in the vicinity of $x = 0$. To see this, graph the parabola $y = 1 - 0.5x^2$ and the curve $y = \cos x$ in the standard viewing rectangle. Describe, in a complete sentence or two, what you observe.

(c) Complete the following table to see numerical evidence that strongly supports your observations in part (b).

x	1	0.5	0.1	0.01	0.001
$\cos x$					
$1 - 0.5x^2$					

In Exercises 40–45 you are given an equation of the form

$$\cos x = k \qquad \text{where } k \geq 0$$

(a) *Use the graph in Figure 14 (on page 516) to estimate (to the nearest 0.05) a root of the equation within the interval $0 \leq x \leq \pi/2$.*

(b) *Use a calculator to obtain a more accurate value for the root in part (a). Round the answer to four decimal places.*

(c) *Use the reference-angle concept and a calculator to find another root of the equation (within the interval $0 \leq x \leq 2\pi$).*

(d) *Use the reference-angle concept and a calculator to find all solutions of the equation $\cos x = -k$ within the interval $0 \leq x \leq 2\pi$. Again, round the answers to four decimal places.*

40. $\cos x = 0.7$ **41.** $\cos x = 0.9$

42. $\cos x = 0.4$ **43.** $\cos x = 0.3$

44. $\cos x = 0.6$ **45.** $\cos x = 0.55$

In Exercises 46–51 you are given an equation of the form

$$\sin x = k \qquad \text{where } k > 0$$

(a) *Use the graph in Figure A on the next page to estimate (to the nearest 0.05) a root of the equation within the interval $0 \leq x \leq \pi/2$.*

(b) *Use a calculator (with the inverse sine function) to obtain a more accurate value for the root in part (a). Round the answer to four decimal places.*

(c) *Use the reference-angle concept and a calculator to find another root of the equation (within the interval $0 \leq x \leq 2\pi$). Round the answer to four decimal places.*

(d) *Use the reference-angle concept and a calculator to find all solutions of the equation $\sin x = -k$ within the interval $0 \leq x \leq 2\pi$. Round the answer to four decimal places.*

46. $\sin x = 0.1$ **47.** $\sin x = 0.6$

48. $\sin x = 0.4$ **49.** $\sin x = 0.2$

50. $\sin x = 0.85$ **51.** $\sin x = 0.7$

Ⓖ **52.** In Example 2 of this section we determined two roots x_1 and x_2 of the equation $\cos x = 0.8$ in the interval $0 \leq x \leq 2\pi$. The values were $x_1 \approx 0.644$ and $x_2 \approx 5.640$. Check these results graphically as follows. Using a viewing rectangle extending from $x = 0$ to $x = 2\pi$, graph the curve $y = \cos x$ along with the horizontal line $y = 0.8$. Then use the graphing utility (with repeated zooms) to estimate the x-coordinates of the two points where the curve $y = \cos x$ intersects the line $y = 0.8$. What do you observe?

Ⓖ **53.** In Example 2(d) we determined two roots x_3 and x_4 of the equation $\cos x = -0.8$ in the interval $0 \leq x \leq 2\pi$. The values were $x_3 \approx 2.498$ and $x_4 \approx 3.785$. Use the procedure indicated in Exercise 52 to check these results graphically.

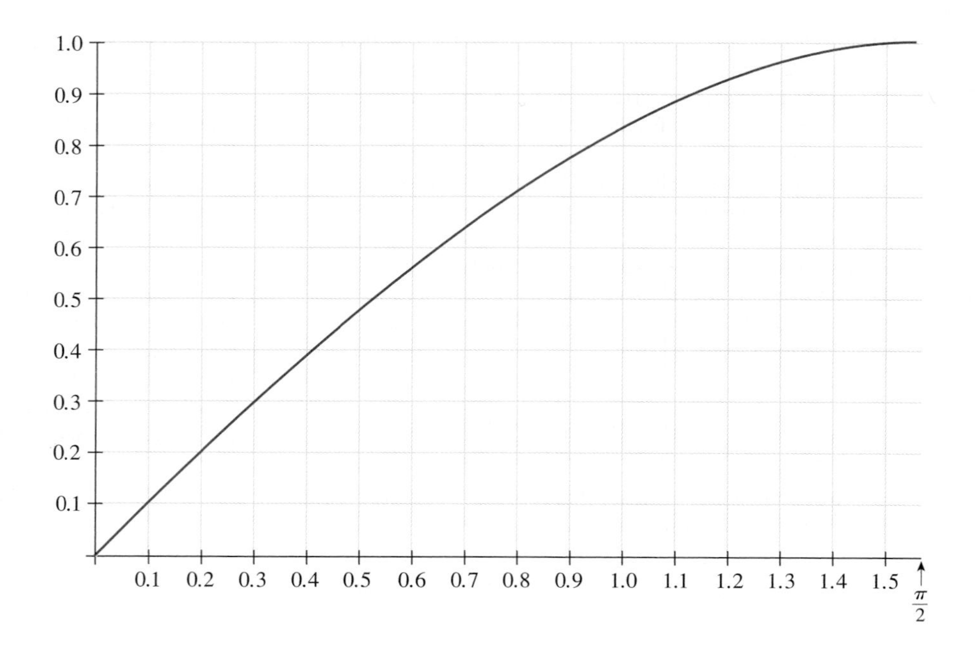

Figure A
$y = \sin x, 0 \le x \le \pi/2$

In Exercises 54 and 55, use graphs (as in Example 2) to estimate the roots of each equation for $0 \le x \le 2\pi$. Zoom in close enough on the intersection points until you are sure about the first two decimal places in each root. Then use a calculator, as in Example 2, to determine more accurate values for the roots. Round the calculator values to four decimal places.

Ⓖ **54. (a)** $\cos x = 0.351$
(b) $\cos x = -0.351$

Ⓖ **55. (a)** $\sin x = 0.687$
(b) $\sin x = -0.687$

B

In Exercises 56–60 refer to the following figure, which shows the graphs of $y = \sin x$ and $y = \cos x$ on the closed interval $[0, 2\pi]$.

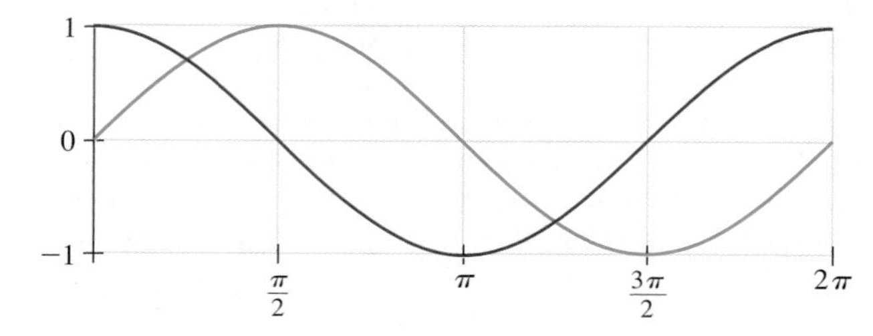

56. (a) When the value of $\sin x$ is a maximum, what is the corresponding value of $\cos x$?
(b) When the value of $\cos x$ is a maximum, what is the corresponding value of $\sin x$?

57. Follow Exercise 56, but replace the word "maximum" with "minimum."

58. (a) For which x-values in the interval $0 \le x \le 2\pi$ is $\sin x = \cos x$? *Hint*: Refer to Table 1 on page 512.
(b) For which x-values in the interval $0 \le x \le 2\pi$ is $\sin x < \cos x$?

59. Specify an open interval in which both the sine and cosine functions are decreasing.

60. Specify an open interval in which the sine function is decreasing but the cosine function is increasing.

As a prerequisite for Exercises 61 and 62, you need to have studied Section 4.3 on iteration and population growth. In particular, you should be familiar with Example 2 on pages 250–252.

61. As in Example 2 in Section 4.3, suppose a farmer has a fishpond with a maximum population size of 500, and suppose that initially the pond is stocked with 50 fish. Unlike Example 2, however, assume that the growth equation is

$$f(x) = \cos x \qquad (0 \le x \le 1)$$

Finally, as in Example 2, assume that the time intervals are breeding seasons.

(a) Use the iteration diagram on the next page and your calculator to complete the following table. For the values obtained from the graph, estimate to the nearest 0.05 (or closer, if it seems appropriate). For the calculator values, round the final answers to five decimal places. As usual, check to see that the calculator results are consistent with the estimates obtained from the graph.

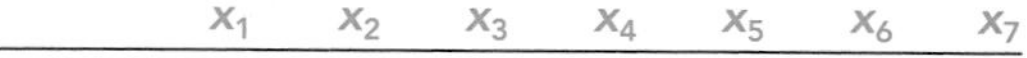

	x_1	x_2	x_3	x_4	x_5	x_6	x_7
From graph							
From calculator							

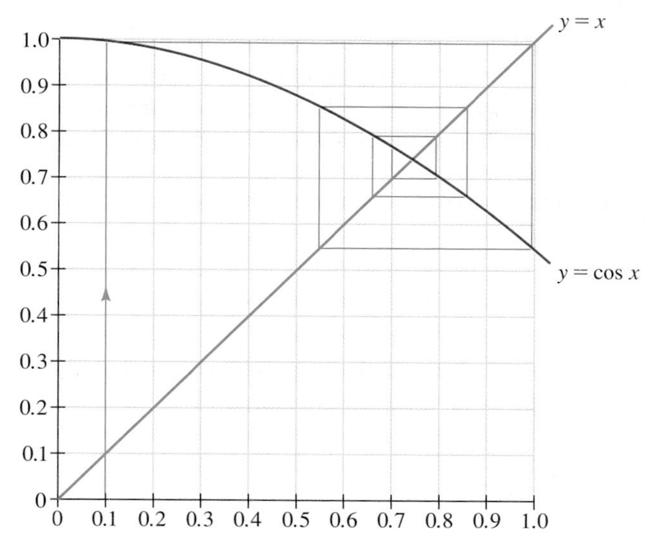

(b) Use the calculator results in part (a) to complete the following table.

n	0	1	2	3	4	5	6	7
Number of fish after n breeding seasons	50							

(c) As indicated in the figure accompanying part (a), the iteration process is spiraling in on a fixed point between 0.7 and 0.8. Using calculus (or simply a calculator), it can be shown that this fixed point is 0.7391 (rounded to four decimal places). What is the corresponding equilibrium population?

62. Suppose that in Exercise 61, instead of $f(x) = \cos x$, we use $g(x) = \sin x$ $(0 \le x \le 1)$ for the growth function.

(a) Complete a table similar to the one in Exercise 61(b) (assuming $x_0 = 50$).

(b) What do you think would be the long-term behavior of this population? *Hint*: Think graphically. Is there a fixed point for the function $g(x) = \sin x$?

63. In the text we used the identity $\cos\theta = \sin(\theta + \pi/2)$ in obtaining the graph of $y = \cos x$ from that of $y = \sin x$. In this exercise you'll derive this identity. Refer to the figure below. (Although the figure shows the angle of radian measure θ in the first quadrant, the proof can be easily carried over for the other quadrants as well.)

(a) What are the coordinates of C?

(b) Show that $\triangle AOB$ and $\triangle COD$ are congruent.

(c) Use the results in parts (a) and (b) to show that the coordinates of A are $(-\sin\theta, \cos\theta)$.

(d) Since the radian measure of $\angle DOA$ is $\theta + \pi/2$, the coordinates of A (by definition) are $(\cos(\theta + \frac{\pi}{2}), \sin(\theta + \frac{\pi}{2}))$. Now explain why $\cos\theta = \sin(\theta + \frac{\pi}{2})$ and $-\sin\theta = \cos(\theta + \frac{\pi}{2})$.

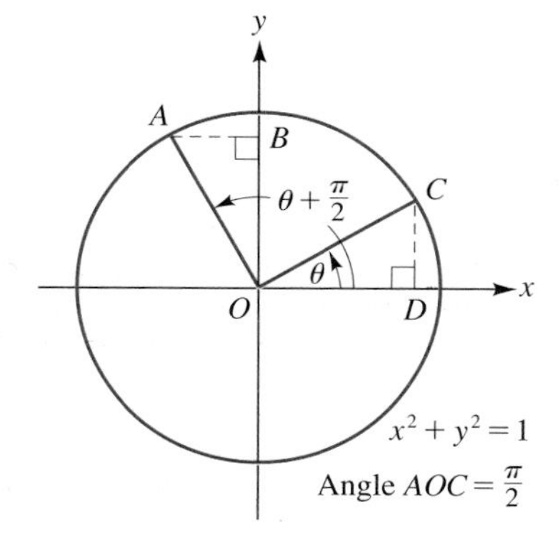

PROJECT

MAKING WAVES

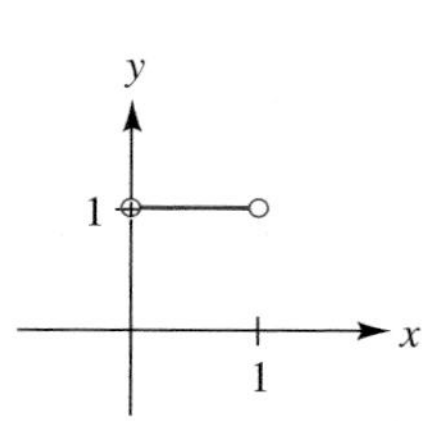

Figure A
A graph of $f(x) = 1$ for $0 < x < 1$

The cosine function is the prototypical example of a function that is both even and periodic with domain all real numbers. Similarly, the sine function is the prototypical example of a function that is both odd and periodic with domain all real numbers. In many applications, especially in physics and engineering, even and odd periodic phenomena are often represented by more basic waves. In this project we examine square waves, sawtooth waves, and triangular waves.

We begin with a square wave. Consider the function f defined by $f(x) = 1$, for $0 < x < 1$, graphed in Figure A. We want to extend this function to be an odd and periodic function with domain all real numbers. First we extend this function to be an odd function. We get

$$f(x) = \begin{cases} 1, & 0 < x < 1 \\ -1, & -1 < x < 0 \end{cases}$$

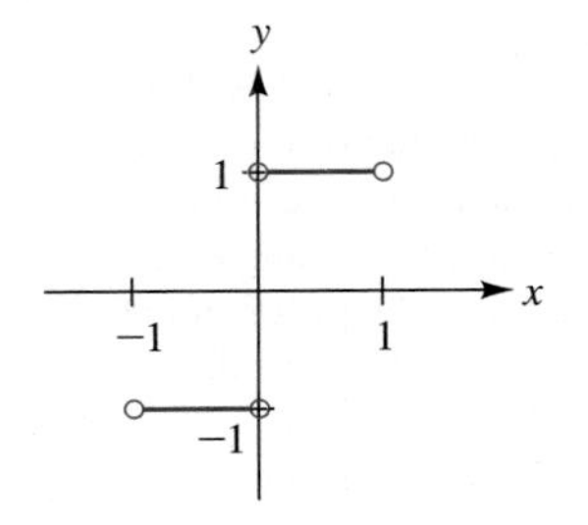

Figure B
A graph of
$f(x) = \begin{cases} 1, & 0 < x < 1 \\ -1, & -1 < x < 0 \end{cases}$

This extended version of f is an odd function, since its domain, $(-1, 0) \cup (0, 1)$, is symmetric about zero and for each x in its domain $f(-x) = -f(x)$. For example, if $x = \frac{1}{2}$, then $f(\frac{1}{2}) = 1$ and $f(-\frac{1}{2}) = -1 = -f(\frac{1}{2})$. Notice the graph of this extended version of f, shown in Figure B, is symmetric about the origin as is true for any odd function.

Next, we extend again to obtain a periodic version of this odd function. We have

$$f(x) = \begin{cases} 1, & 0 < x < 1 \\ -1, & -1 < x < 0 \end{cases} \quad \text{and} \quad f(x + 2) = f(x)$$

for all noninteger real numbers x. Note that f is still an odd function and is also a periodic function with period 2. Its graph is shown in Figure C.

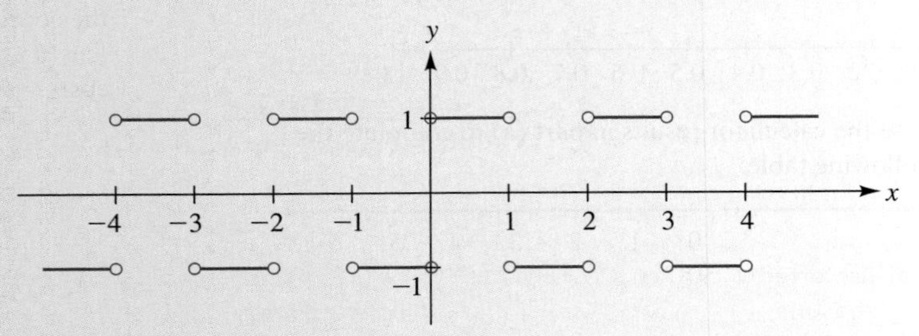

Figure C
A graph of
$f(x) = \begin{cases} 1, & 0 < x < 1 \\ -1, & -1 < x < 0 \end{cases}$ and $f(x + 2) = f(x)$ for noninteger real numbers x

To complete our task, we define $f(x) = 0$ for all integers x. So we obtain $f(x) = \begin{cases} 1, & 0 < x < 1 \\ -1, & -1 < x < 0 \end{cases}$, $f(x + 2) = f(x)$, for noninteger real number x, and $f(x) = 0$ for integer x.

This final version of f is an odd periodic function of period 2 with domain all real numbers. Its graph, shown in Figure D, is called a **square wave.**

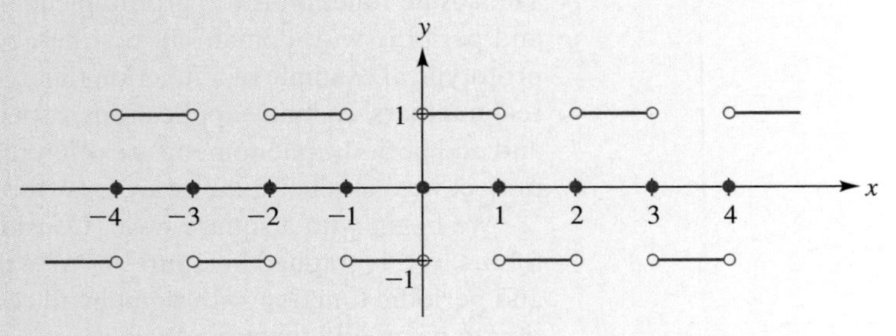

Figure D
A graph of
$f(x) = \begin{cases} 1, & 0 < x < 1 \\ -1, & -1 < x < 0 \end{cases}$, $f(x) = 0$ for integer x, and $f(x + 2) = f(x)$

The next two problems guide you through the construction of a sawtooth wave and a triangular wave.

1. *Sawtooth wave:* Consider the function g defined by $g(x) = x$ for $0 \le x < 1$.
   (a) Graph $y = g(x)$.
   (b) Extend g to be an odd function for $-1 < x < 1$. Graph this odd version of g.
   (c) Extend the odd version of g to be a periodic function with period 2. Graph this odd and periodic version of g.
   (d) Finally extend the odd and periodic version of g to have domain all real numbers. Graph this odd and periodic version of g with domain all real numbers. This graph is called a **sawtooth wave.**

2. *Triangular wave:* Consider the function h defined by $h(x) = x$ for $0 \le x \le 1$.
   (a) Graph $y = h(x)$.
   (b) Extend h to be an even function for $-1 \le x \le 1$. Graph this even version of h.
   (c) Extend the even version of h to be a periodic function with period 2. Notice the domain is all real numbers. Graph this even and periodic version of h with domain all real numbers. This graph is called a **triangular wave.**

7.3 GRAPHS OF $y = A \sin(Bx - C)$ AND $y = A \cos(Bx - C)$

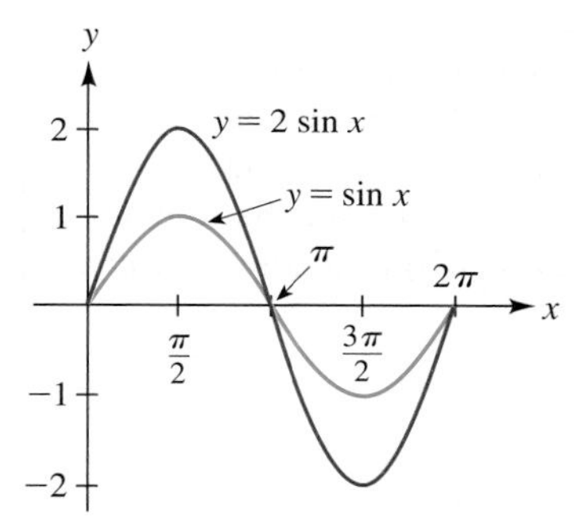

Figure 1
The amplitude of $y = 2 \sin x$ is 2. Both $y = \sin x$ and $y = 2 \sin x$ have a period of 2π.

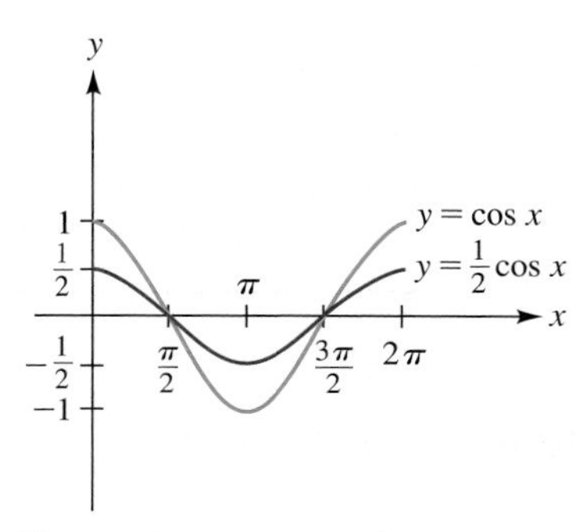

Figure 2
The amplitude of $y = \frac{1}{2} \cos x$ is $\frac{1}{2}$. Both $y = \cos x$ and $y = \frac{1}{2} \cos x$ have a period of 2π.

The graphs of $y = \sin x$ and $y = \cos x$ are the building blocks we need for graphing functions of the form

$$y = A \sin(Bx - C) \qquad \text{and} \qquad y = A \cos(Bx - C)$$

As a first example, consider $y = 2 \sin x$. To obtain the graph of $y = 2 \sin x$ from that of $y = \sin x$, we multiply each y-coordinate on the graph of $y = \sin x$ by 2. As is indicated in Figure 1, this changes the amplitude from 1 to 2, but it does not affect the period, which remains 2π. As a second example, Figure 2 shows the graphs of $y = \cos x$ and $y = \frac{1}{2} \cos x$. Note that the amplitude of $y = \frac{1}{2} \cos x$ is $\frac{1}{2}$ and the period is, again, 2π. More generally, graphs of functions of the form $y = A \sin x$ and $y = A \cos x$ always have an amplitude of $|A|$ and a period of 2π.

Note: The complete graph of $y = 2 \sin x$ is obtained by repeating, over the entire domain, the basic cycle shown in Figure 1. The complete graph for $y = \frac{1}{2} \cos x$ is obtained similarly.

EXAMPLE 1 Graphing a function of the form $y = A \sin x$

Graph the function $y = -2 \sin x$ over one period. On which interval(s) is the function decreasing?

SOLUTION

In Section 3.4 we saw that the graph of $y = -f(x)$ is obtained from that of $y = f(x)$ by reflection about the x-axis. Thus we need only take the graph of $y = 2 \sin x$ from Figure 1 and reflect it about the x-axis; see Figure 3. Note that

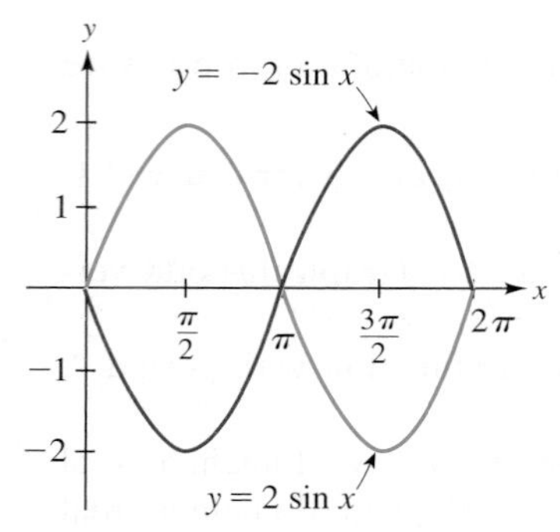

Figure 3

both functions have an amplitude of 2 and a period of 2π. From the graph in Figure 3, we can see that the function $y = -2 \sin x$ is decreasing on the intervals $(0, \frac{\pi}{2})$ and $(\frac{3\pi}{2}, 2\pi)$.

We have just seen that functions of the form $y = A \sin x$ and $y = A \cos x$ have an amplitude of $|A|$ and a period of 2π. The next two examples show how to analyze functions of the form $y = A \sin Bx$ and $y = A \cos Bx$ $(B > 0)$. As you'll see, these functions have an amplitude of $|A|$ and a period of $2\pi/B$.

EXAMPLE 2 Graphing a function of the form $y = \cos Bx$

Graph the function $y = \cos 3x$ over one period.

SOLUTION

We know that the cosine curve $y = \cos x$ begins its basic pattern when $x = 0$ and completes that pattern when $x = 2\pi$. Thus $y = \cos 3x$ will begin its basic pattern when $3x = 0$, and it will complete that pattern when $3x = 2\pi$. From the equation $3x = 0$ we conclude that $x = 0$, and from the equation $3x = 2\pi$ we conclude that $x = 2\pi/3$. Thus the graph of $y = \cos 3x$ begins its basic pattern at $x = 0$ and completes the pattern at $x = 2\pi/3$. This tells us that the period is $2\pi/3$. Next, in preparation for drawing the graph, we divide the period into quarters, as shown in Figure 4(a). In Figure 4(b) we've plotted the points with x-coordinates shown in Figure 4(a). (We've also plotted the point on the curve corresponding to $x = 0$, where the basic pattern is to begin.) From Figure 4(b) we can see that the amplitude is going to be 1. Now, with the points in Figure 4(b) as a guide, we can sketch one cycle of $y = \cos 3x$, as shown in Figure 4(c).

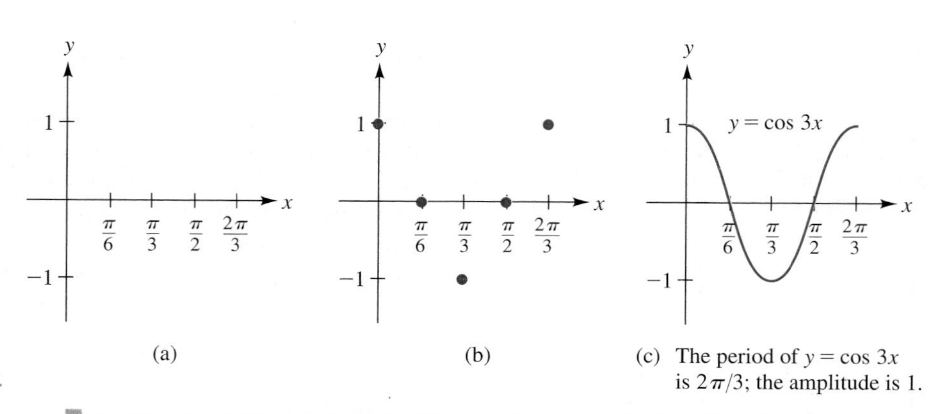

Figure 4

(a) (b) (c) The period of $y = \cos 3x$ is $2\pi/3$; the amplitude is 1.

EXAMPLE 3 Graphing functions of the form $y = A \cos Bx$

Graph each function over one period:

(a) $y = \dfrac{1}{2} \cos 3x$; **(b)** $y = -\dfrac{1}{2} \cos 3x$.

SOLUTION

(a) In Example 2 we graphed $y = \cos 3x$. To obtain the graph of $y = \frac{1}{2}\cos 3x$ from that of $y = \cos 3x$, we multiply each y-coordinate on the graph of $y = \cos 3x$ by $1/2$. As is indicated in Figure 5(a), this changes the amplitude from 1 to $1/2$, but it does not affect the period, which remains $2\pi/3$.

(b) The graph of $y = -\frac{1}{2}\cos 3x$ is obtained by reflecting the graph of $y = \frac{1}{2}\cos 3x$ about the x-axis, as is indicated in Figure 5(b). Both functions have a period of $2\pi/3$ and an amplitude of $1/2$.

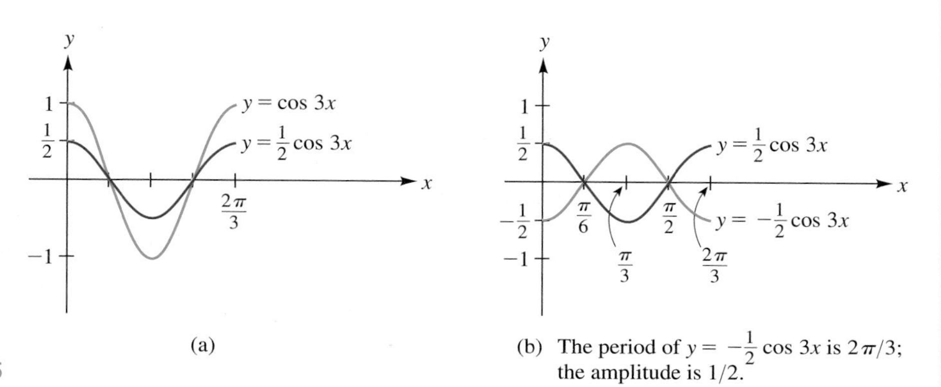

(a)

(b) The period of $y = -\frac{1}{2}\cos 3x$ is $2\pi/3$; the amplitude is $1/2$.

Figure 5

Before looking at more examples, let's take a moment to summarize where we are. Our work in Examples 2 and 3(a) shows how to graph $y = \frac{1}{2}\cos 3x$. The same technique that we used in those examples can be applied to any function of the form $y = A \cos Bx$ or $y = A \sin Bx$. As indicated in the box that follows, for both functions the amplitude is $|A|$, and the period is $2\pi/B$. (Exercise 56 asks you to use the method of Example 2 to show that the period is indeed $2\pi/B$.)

PROPERTY SUMMARY The Graphs of (a) $y = A \sin Bx$ and (b) $y = A \cos Bx$ $(B > 0)$

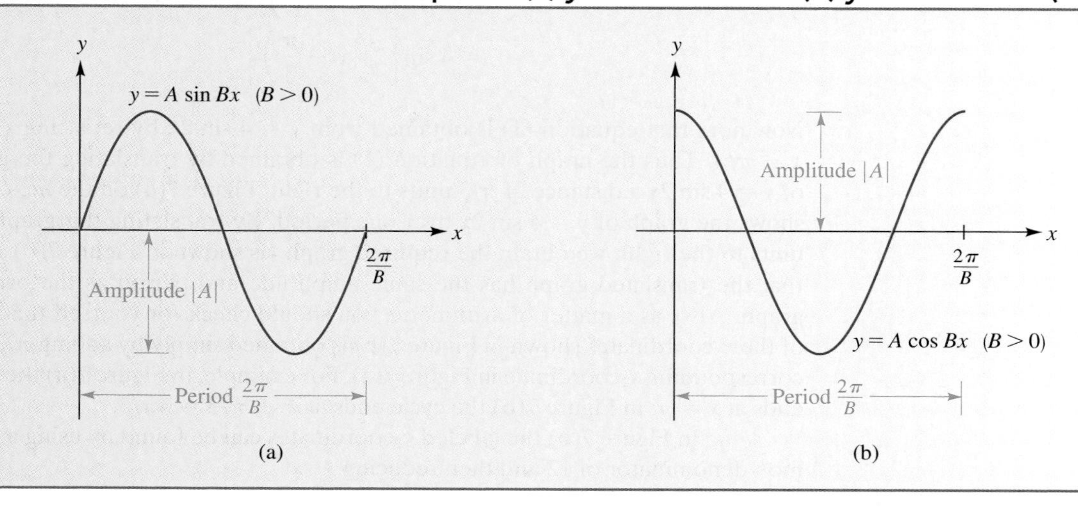

(a)

(b)

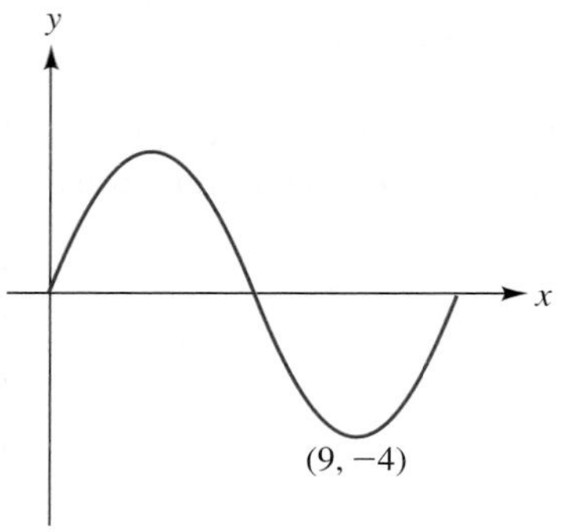

EXAMPLE 4 Using a graph of $y = A \sin Bx$ to determine A and B

y

x

(9, −4)

Figure 6

In Figure 6 a function of the form $y = A \sin Bx$ $(B > 0)$ is graphed for one period. Determine the values of A and B.

SOLUTION

From the figure we see that the amplitude is 4. Also from the figure we know that three-fourths of the period is 9, so

$$\frac{3}{4}\left(\frac{2\pi}{B}\right) = 9$$

$$\frac{\pi}{2B} = 3$$

$$B = \frac{\pi}{6} \qquad \text{(Check the algebra in the last two lines.)}$$

In summary, we have $A = 4$ and $B = \pi/6$; the equation of the curve is $y = 4 \sin(\pi x/6)$.

EXAMPLE 5 Graphing a function of the form $y = A \sin(Bx - C)$

Graph the function $y = 4 \sin\left(2x - \dfrac{2\pi}{3}\right)$ over one period.

SOLUTION

The technique here is to factor the quantity within parentheses so that the coefficient of x is 1. We'll then be able to graph the function using a simple translation, as in Chapter 3. We have

$$\begin{aligned} y &= 4 \sin\left(2x - \frac{2\pi}{3}\right) \\ &= 4 \sin\left[2\left(x - \frac{\pi}{3}\right)\right] \end{aligned} \tag{1}$$

Now note that equation (1) is obtained from $y = 4 \sin 2x$ by replacing x with $x - \pi/3$. Thus the graph of equation (1) is obtained by translating the graph of $y = 4 \sin 2x$ a distance of $\pi/3$ units to the right. Figure 7(a) on the next page shows the graph of $y = 4 \sin 2x$ over one period. By translating this graph $\pi/3$ units to the right, we obtain the required graph, as shown in Figure 7(b). Note that the translated graph has the same amplitude and period as the original graph. Also, as a matter of arithmetic, you should check for yourself that each of the x-coordinates shown in Figure 7(b) is obtained simply by adding $\pi/3$ to a corresponding x-coordinate in Figure 7(a). For example, in Figure 7(a) the cycle ends at $x = \pi$; in Figure 7(b) the cycle ends at $\pi + \pi/3 = 4\pi/3$.

Note: In Figure 7(b) the labeled x-coordinates can be found by using a common denominator of 12 and then reducing.

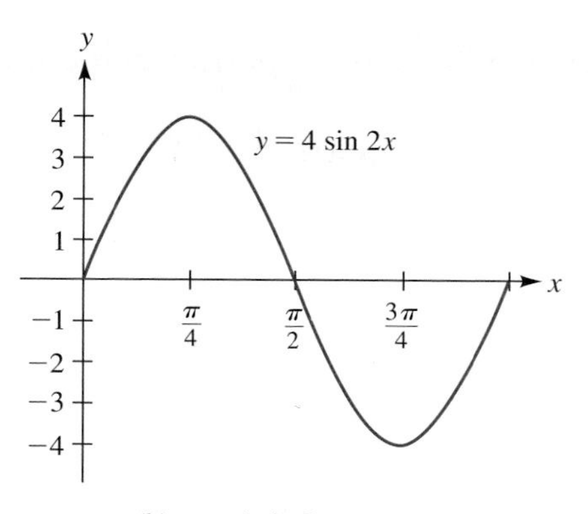

(b) $y = 4\sin 2x$
Period: $2\pi/B = \pi$
Amplitude: $|A| = 4$

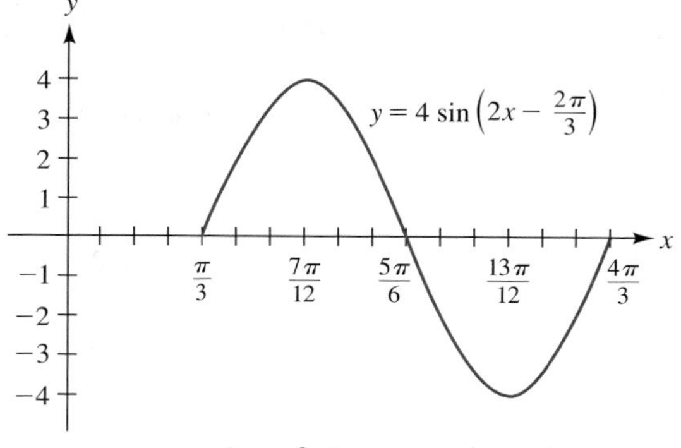

(b) $y = 4\sin\left(2x - \frac{2\pi}{3}\right) = 4\sin\left[2\left(x - \frac{\pi}{3}\right)\right]$
The graph is obtained by translating the graph of $y = 4\sin 2x$ a distance of $\pi/3$ units to the right. The period and amplitude are still π and 4, respectively.

Figure 7

In Example 5 we used translation to graph the function $y = 4\sin(2x - \frac{2\pi}{3})$. In particular, we translated the graph of $y = 4\sin 2x$ so that the starting point of the basic cycle was shifted from $x = 0$ to $x = \pi/3$. The number $\pi/3$ in this case is called the *phase shift* of the function. In the box that follows, we define phase shift, and we generalize the results of the graphing technique used in Example 5.

PROPERTY SUMMARY $y = A\sin(Bx - C)$ and $y = A\cos(Bx - C)$ $(B > 0, C \neq 0)$

The graphs of (a) $y = A\sin(Bx - C)$ and (b) $y = A\cos(Bx - C)$ are obtained by horizontally translating the graphs of $y = A\sin Bx$ and $y = A\cos Bx$, respectively, so that the starting point of the basic cycle is shifted from $x = 0$ to $x = C/B$. The number C/B is called the **phase shift** for each of the functions $y = A\sin(Bx - C)$ and $y = A\cos(Bx - C)$. The amplitude and the period for these functions are $|A|$ and $2\pi/B$, respectively.

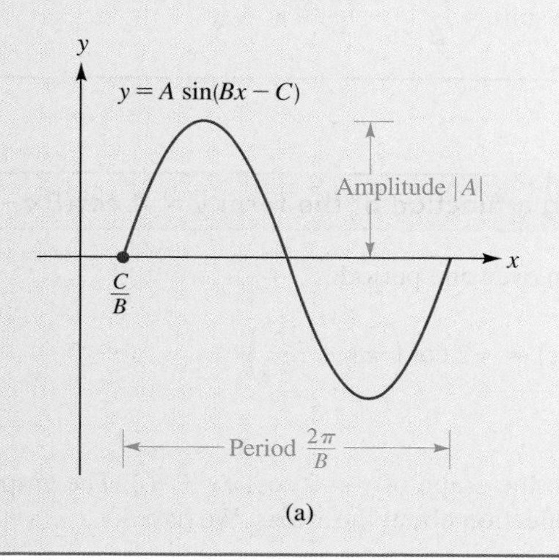

(a)

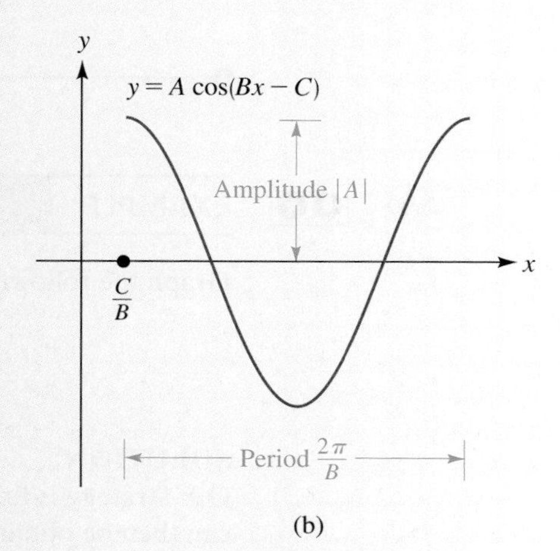

(b)

EXAMPLE 6 Finding amplitude, period, and phase shift

Specify the amplitude, period, and phase shift for each function:

(a) $f(x) = 3\cos(4x - 5)$; **(b)** $g(x) = -2\cos\left(\pi x + \dfrac{2\pi}{3}\right)$.

SOLUTION

(a) By comparing the given equation with $y = A\cos(Bx - C)$, we see that $A = 3$, $B = 4$, and $C = 5$. Consequently, we have

$$\text{amplitude} = |A| = 3$$
$$\text{period} = \frac{2\pi}{B} = \frac{2\pi}{4} = \frac{\pi}{2}$$
$$\text{phase shift} = \frac{C}{B} = \frac{5}{4}$$

For purposes of review, let's also calculate the phase shift without explicitly relying on the expression C/B. In the equation $f(x) = 3\cos(4x - 5)$, we can factor a 4 out of the parentheses to obtain

$$f(x) = 3\cos\left[4\left(x - \frac{5}{4}\right)\right]$$

This last equation tells us that we can obtain the graph of f by translating the graph of $y = 3\cos 4x$. In particular, the translation shifts the starting point of the basic cycle from $x = 0$ to $x = 5/4$. The number $5/4$ is the phase shift, as obtained previously.

(b) We have $A = -2$, $B = \pi$, and $C = -2\pi/3$, and therefore

$$\text{amplitude} = |A| = 2$$
$$\text{period} = \frac{2\pi}{B} = \frac{2\pi}{\pi} = 2$$
$$\text{phase shift} = \frac{C}{B} = \frac{-2\pi/3}{\pi} = -\frac{2}{3}$$

EXAMPLE 7 Graphing a function of the form $y = A\cos(Bx - C)$

Graph the following function over one period:

$$g(x) = -2\cos\left(\pi x + \frac{2\pi}{3}\right)$$

SOLUTION

Our strategy is first to obtain the graph of $y = 2\cos(\pi x + \frac{2\pi}{3})$. The graph of g can then be obtained by a reflection about the x-axis. We have

$$y = 2\cos\left(\pi x + \frac{2\pi}{3}\right) = 2\cos\left[\pi\left(x + \frac{2}{3}\right)\right]$$

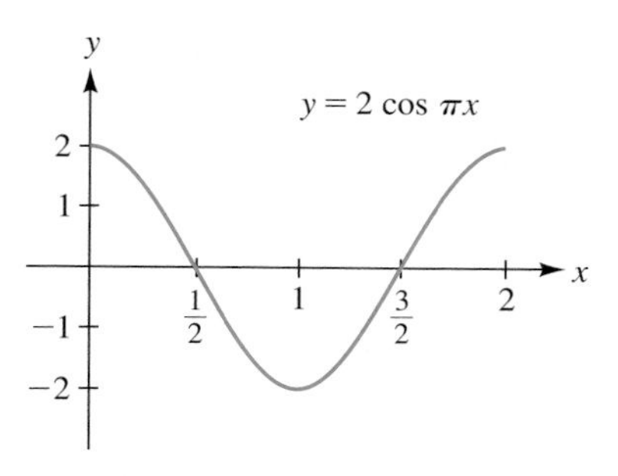

(a) $y = 2\cos\pi x$
Amplitude: $|A| = 2$
Period: $2\pi/B = 2$

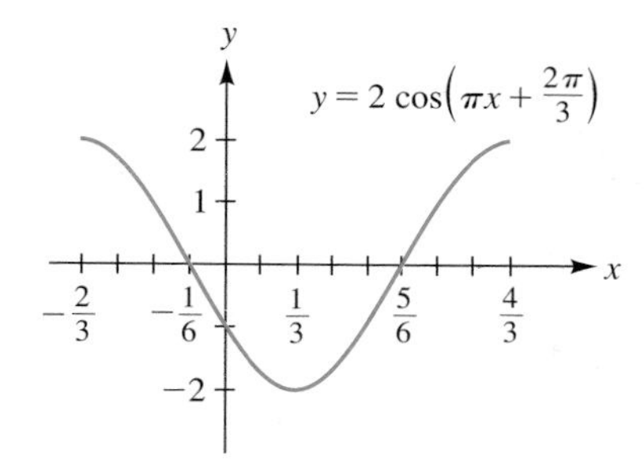

(b) $y = 2\cos\left(\pi x + \frac{2\pi}{3}\right)$
Amplitude: 2
Period: 2
Phase shift: $-2/3$

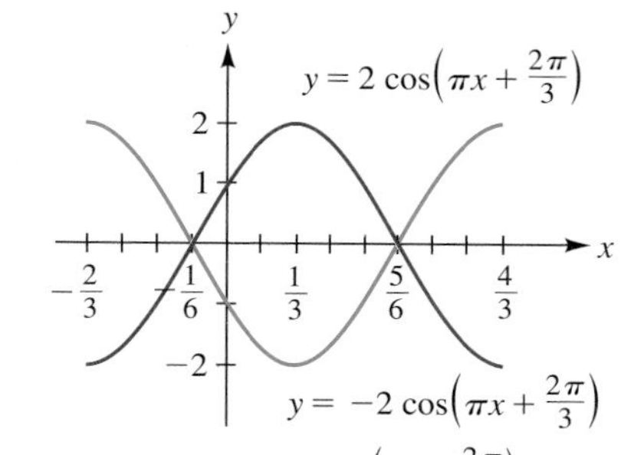

(c) $y = -2\cos\left(\pi x + \frac{2\pi}{3}\right)$
Amplitude: 2
Period: 2
Phase shift: $-2/3$

Figure 8

Now the graph of this last equation is obtained by translating the graph of $y = 2\cos\pi x$ a distance of $2/3$ unit to the left. Figures 8(a) and 8(b) show the graphs of $y = 2\cos\pi x$ and $y = 2\cos[\pi(x + \frac{2}{3})]$. By reflecting the graph of this last equation about the x-axis, we obtain the graph of g. See Figure 8(c).

In previous chapters we've used several types of functions to model real-life data sets: Linear and quadratic functions were used in Chapter 4; exponential and log functions were used in Chapter 5. The functions that we've graphed in this section are often used in modeling data where a variable repeats itself at regular or near-regular intervals. In the next example, we use a trigonometric function of the form $y = A\sin(Bt - C) + D$ to approximate average monthly temperatures.

EXAMPLE 8 Modeling data using a trigonometric function

Table 1 shows the average monthly temperatures for Minneapolis–St. Paul, and Figure 9 shows a scatter plot based on this data. The numbers on the horizontal

TABLE 1 Average Monthly Temperatures for Minneapolis–St. Paul (The monthly averages were computed using daily maximum temperatures.)

Month	Jan.	Feb.	Mar.	Apr.	May	June	July	Aug.	Sept.	Oct.	Nov.	Dec.
Temperature (°F)	20.7	26.6	39.2	56.5	69.4	78.8	84.0	80.7	70.7	58.8	41.0	25.5

Source: Robert B. Thomas, *The Old Farmers' Almanac, 1996* (Dublin, New Hampshire: Yankee Publishing, Inc., 1995)

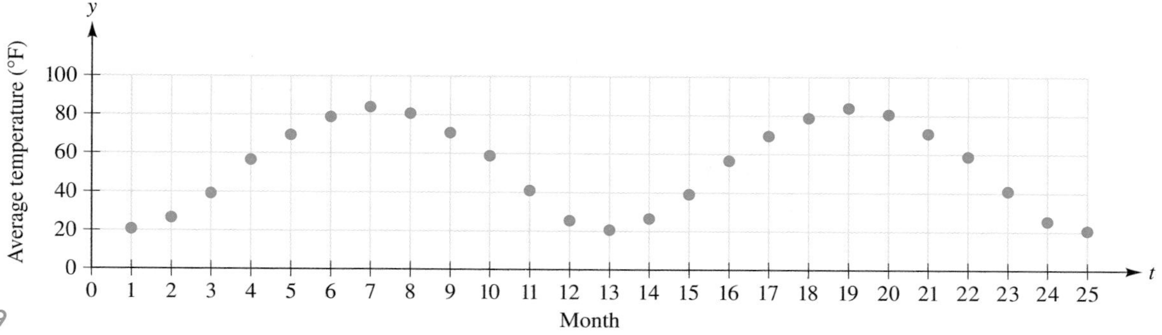

Figure 9

t-axis indicate months: January is $t = 1$, February is $t = 2$, and so on; the next January is $t = 13$. Find a periodic function of the form $y = A\sin(Bt - C) + D$ whose values approximate the monthly temperatures in Table 1.

SOLUTION

We need to compute the four constants A, B, C, and D in the equation $y = A\sin(Bt - C) + D$. We'll do this in four steps, one step for each required constant.

Step 1 *Computing A* For the amplitude A, we'll use the equation $A = (M - m)/2$ (from Section 7.2). According to Table 1, the maximum average temperature is $M = 84.0°$ and the minimum is $m = 20.7°$. Therefore

$$A = \frac{M - m}{2} = \frac{84.0 - 20.7}{2} = 31.65$$

Step 2 *Computing B* We know from our work in this section that the period is given by $2\pi/B$. So, assuming that the period of the average temperature function is 12 months, we have $2\pi/B = 12$, and therefore

$$12B = 2\pi \qquad \text{or} \qquad B = \frac{\pi}{6}$$

Step 3 *Computing C* Now that we know B, we can compute C by considering the phase shift C/B. If there were no phase shift, a maximum for the sine wave would occur one-quarter of the way through the period at $t = \frac{1}{4}(12)$, which is 3. As you can see in Figure 9, however, a maximum occurs when $t = 7$, four units to the right of $t = 3$. Thus, the phase shift here is 4 and we have

$$\frac{C}{B} = 4 \qquad \text{or} \qquad C = 4B$$

Substituting $B = \pi/6$ in this last equation then yields

$$C = 4\left(\frac{\pi}{6}\right) = \frac{2\pi}{3}$$

Step 4 *Computing D* Substituting the values we've found for A, B, and C in the equation $y = A\sin(Bt - C) + D$ yields

$$y = 31.65\sin\left(\frac{\pi}{6}t - \frac{2\pi}{3}\right) + D \tag{2}$$

One way to determine a value for D is to use the data pair $t = 1, y = 20.7$ from Table 1 and substitute in equation (2). This yields

$$\begin{aligned} 20.7 &= 31.65\sin\left(\frac{\pi}{6} - \frac{2\pi}{3}\right) + D \\ &= 31.65\sin\left(-\frac{\pi}{2}\right) + D \\ &= -31.65 + D \qquad \text{using } \sin(-\pi/2) = -1 \end{aligned}$$

and consequently, $D = 20.7 + 31.65 = 52.35$.

In summary now, our function that approximates the average monthly temperatures is

$$y = 31.65\sin\left(\frac{\pi}{6}t - \frac{2\pi}{3}\right) + 52.35$$

Figure 10(a) shows the graph of this equation along with the scatter plot from Figure 9. As you can see, the curve does a fairly good job in approximating the values from the scatter plot. *Remark*: More advanced techniques from the field of statistics yield a slightly different sine wave that does an even better job in modeling the temperature data. See Figure 10(b).

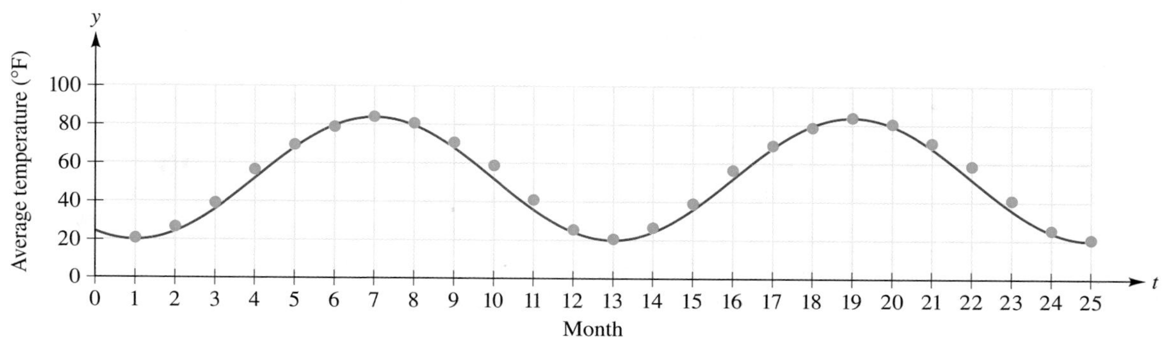

Figure 10(a)
The scatter plot for average monthly temperatures along with the graph of the function obtained in Example 8, $y = 31.65\sin(\frac{\pi}{6}t - \frac{2\pi}{3}) + 52.35$

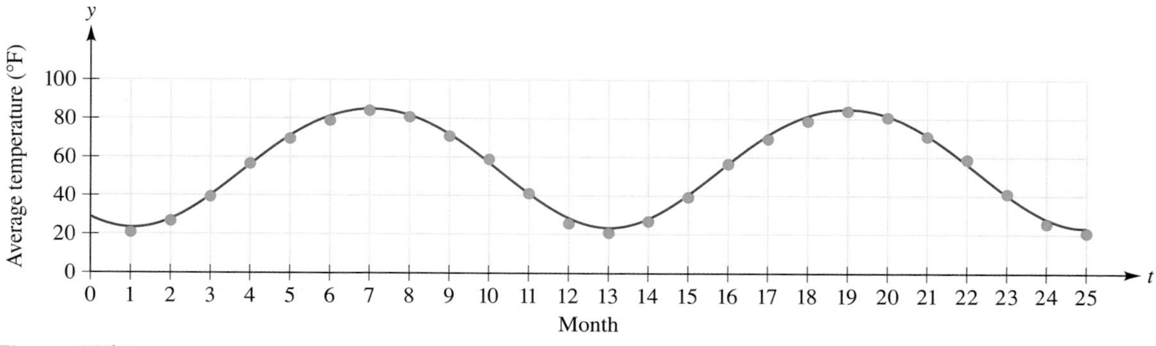

Figure 10(b)
The scatter plot for average monthly temperatures along with the graph of the function $y = 31.139\sin(0.521t - 2.086) + 54.157$. This function was obtained using the SinReg (sine regression) algorithm on the Texas Instruments TI-83 Graphing Calculator.

EXERCISE SET 7.3

A

In Exercises 1–8, graph the functions for one period. In each case, specify the amplitude, period, x-intercepts, and interval(s) on which the function is increasing.

1. (a) $y = 2\sin x$
(b) $y = -\sin 2x$

2. (a) $y = 3\sin x$
(b) $y = \sin 3x$

3. (a) $y = \cos 2x$
(b) $y = 2\cos 2x$

4. (a) $y = \cos(x/2)$
(b) $y = -\frac{1}{2}\cos(x/2)$

5. (a) $y = 3\sin(\pi x/2)$
(b) $y = -3\sin(\pi x/2)$

6. (a) $y = 2\sin \pi x$
(b) $y = -2\sin \pi x$

7. (a) $y = \cos 2\pi x$
(b) $y = -4\cos 2\pi x$

8. (a) $y = -2\cos(x/4)$
(b) $y = -2\cos(\pi x/4)$

Ⓖ **9.** Set the viewing rectangle so that it extends from 0 to 2π in the x-direction and from -4 to 4 in the y-direction. On the same set of axes, graph the four functions $y = \sin x$, $y = 2\sin x$, $y = 3\sin x$, and $y = 4\sin x$. What is the amplitude in each case? What is the period?

Ⓖ **10. (a)** Without using a graphing utility, specify the amplitude and the period for each of the following four functions: $y = \cos x$, $y = 2\cos x$, $y = 3\cos x$, and $y = 4\cos x$.

(b) Check your answers in part (a) by graphing the four functions. (Use the viewing rectangle specified in Exercise 1.)

Ⓖ **11. (a)** Without using a graphing utility, specify the amplitude and the period for $y = 2\sin \pi x$ and for $y = \sin 2\pi x$.

(b) Check your answers in part (a) by graphing the two functions. (Use a viewing rectangle that extends from 0 to 2 in the x-direction and from -2 to 2 in the y-direction.)

In Exercises 12–15, graph the function for one period. Specify the amplitude, period, x-intercepts, and interval(s) on which the function is increasing.

12. $y = 1 + \sin 2x$ **13.** $y = \sin(x/2) - 2$

14. $y = 1 - \cos(\pi x/3)$ **15.** $y = -2 - 2\cos 3\pi x$

In Exercises 16–31, determine the amplitude, period, and phase shift for the given function. Graph the function over one period. Indicate the x-intercepts and the coordinates of the highest and lowest points on the graph.

16. $f(x) = \sin(x - \frac{\pi}{6})$ **17.** $g(x) = \cos(x + \frac{\pi}{3})$

18. $F(x) = -\cos(x + \frac{\pi}{4})$ **19.** $G(x) = -\sin(x + 2)$

20. $y = \sin(2x - \frac{\pi}{2})$ **21.** $y = \sin(3x + \frac{\pi}{2})$

22. $y = \cos(2x - \pi)$ **23.** $y = \cos(x - \frac{\pi}{2})$

24. $y = 3\sin(\frac{1}{2}x + \frac{\pi}{6})$ **25.** $y = -2\sin(\pi x + \pi)$

26. $y = 4\cos(3x - \frac{\pi}{4})$ **27.** $y = \cos(x + 1)$

28. $y = \frac{1}{2}\sin(\frac{\pi x}{2} - \pi^2)$ **29.** $y = \cos(2x - \frac{\pi}{3}) + 1$

30. $y = 1 - \cos(2x - \frac{\pi}{3})$ **31.** $y = 3\cos(\frac{2x}{3} + \frac{\pi}{6})$

For Exercises 32–39:

(a) *Using pencil and paper, not a graphing utility, determine the amplitude, period, and (where appropriate) phase shift for each function.*

(b) *Use a graphing utility to graph each function for two complete cycles. [In choosing an appropriate viewing rectangle, you will need to use the information obtained in part (a).]*

(c) *Use the graphing utility to estimate the coordinates of the highest and the lowest points on the graph.*

(d) *Use the information obtained in part (a) to specify the exact values for the coordinates that you estimated in part (c).*

Ⓖ **32.** $y = -2.5\cos(3x + 4)$ Ⓖ **33.** $y = -2.5\cos(3\pi x + 4)$

Ⓖ **34.** $y = -2.5\cos(\frac{1}{3}x + 4)$ Ⓖ **35.** $y = -2.5\cos(\frac{1}{3}\pi x + 4)$

Ⓖ **36.** $y = \sin(0.5x - 0.75)$ Ⓖ **37.** $y = \sin(0.5x + 0.75)$

Ⓖ **38.** $y = 0.02\cos(100\pi x - 4\pi)$

Ⓖ **39.** $y = 0.02\cos(0.01\pi x - 4\pi)$

In Exercises 40–45, determine whether the equation for the graph has the form $y = A\sin Bx$ or $y = A\cos Bx$ (with $B > 0$) and then find the values of A and B.

40.

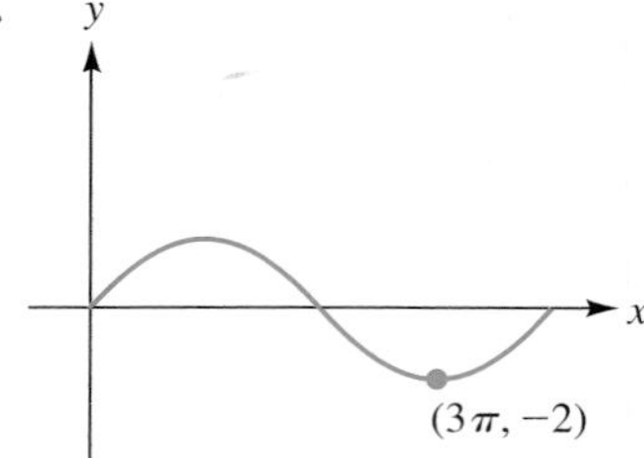

41.

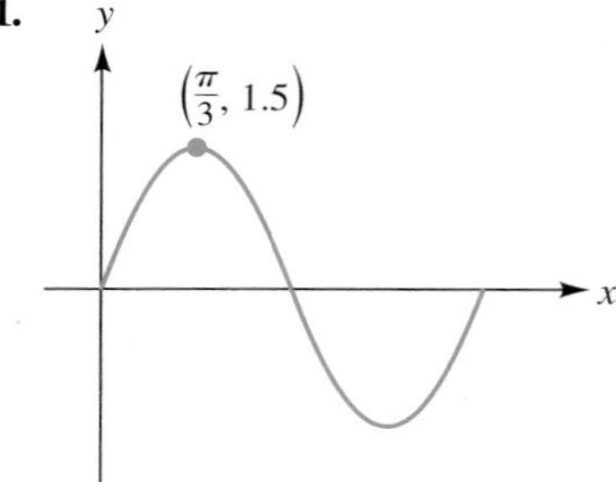

42.

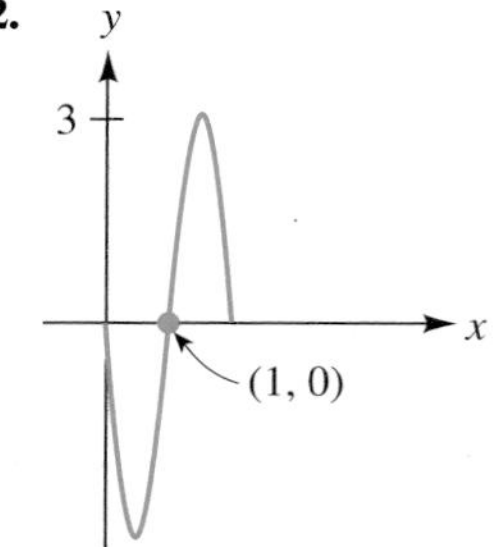

43.

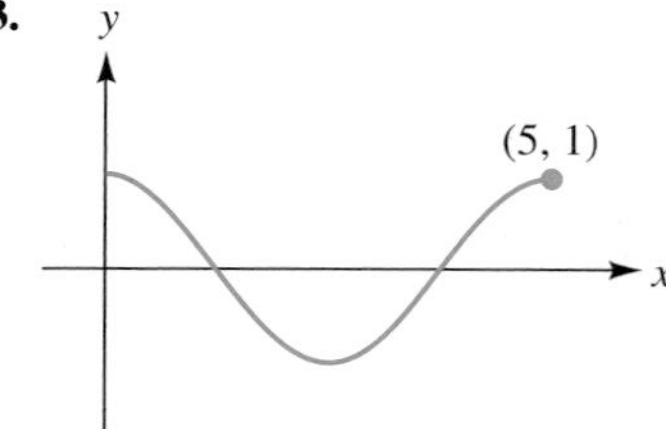

44.

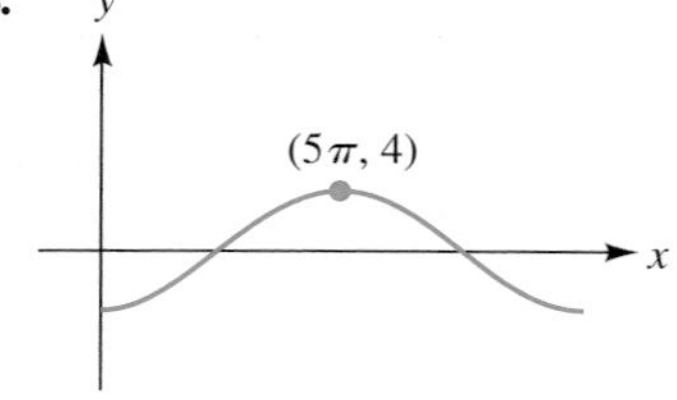

45.

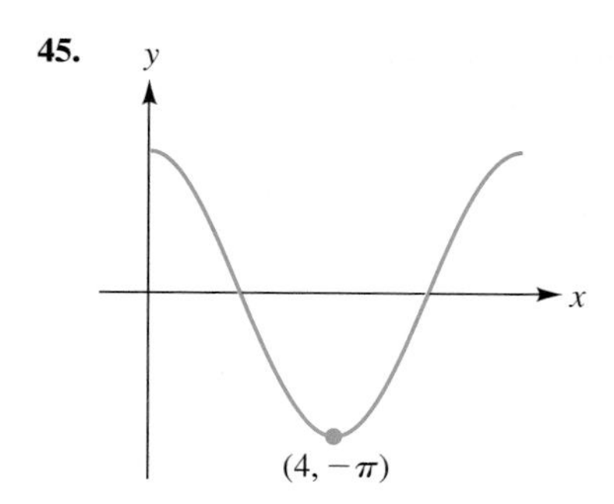

In each of Exercises 46–50, you are given a table of average monthly temperatures and a scatter plot based on the data. Use the methods of Example 8 to find a periodic function of the form $y = A \sin(Bt - C) + D$ whose values approximate the monthly temperatures. The data in these exercises, as well as in Exercise 51, are from the Global Historical Climatology Network, and can be accessed on the internet through the following website created by Robert Hoare: http://www.worldclimate.com/.

46.

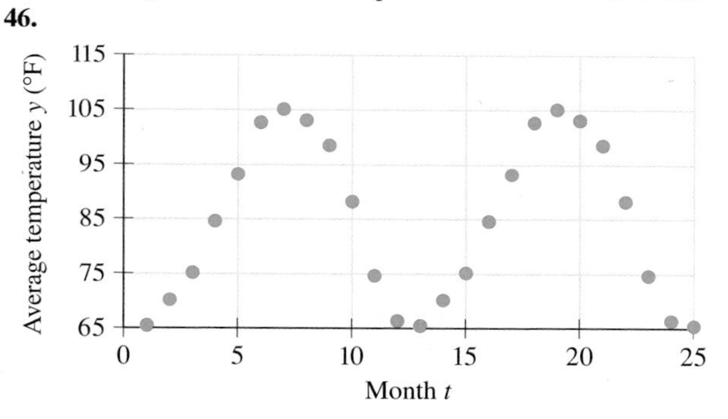

Average Monthly Temperatures for Phoenix, Arizona (based on daily maximums)

Month	Temperature (°F)
Jan.	65.5
Feb.	70.2
Mar.	75.2
Apr.	84.6
May	93.2
June	102.7
July	105.1
Aug.	103.1
Sept.	98.6
Oct.	88.2
Nov.	74.7
Dec.	66.4

47.

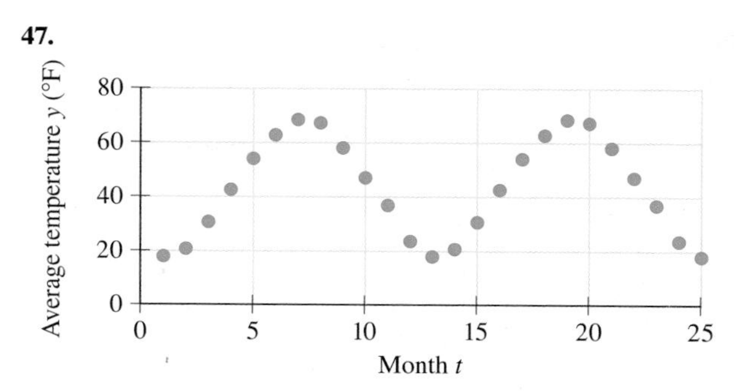

Average Monthly Temperatures for Bangor, Maine

Month	Temperature (°F)
Jan.	18.0
Feb.	20.8
Mar.	30.7
Apr.	42.6
May	54.1
June	62.8
July	68.5
Aug.	67.3
Sept.	58.1
Oct.	47.1
Nov.	36.9
Dec.	23.7

48.

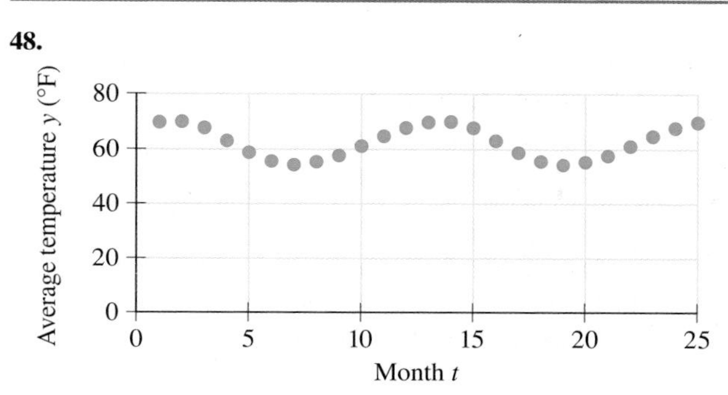

Average Monthly Temperatures for Cape Town, South Africa

Month	Temperature (°F)
Jan.	69.8
Feb.	70.0
Mar.	67.8
Apr.	63.1
May	58.8
June	55.6
July	54.3
Aug.	55.4
Sept.	57.7
Oct.	61.2
Nov.	64.8
Dec.	67.8

49.

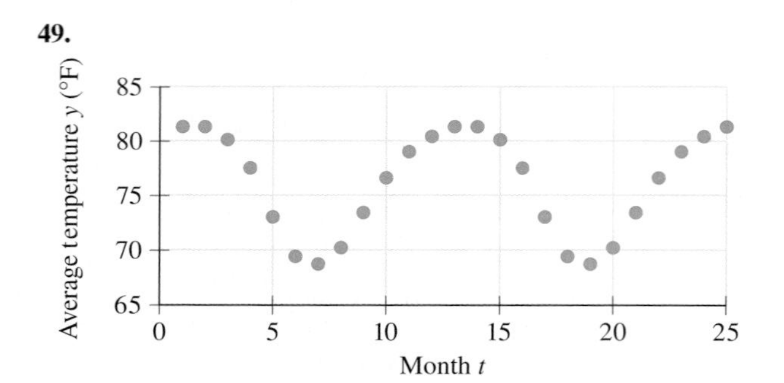

Average Monthly Temperatures for Beira, Mozambique

Month	Temperature (°F)
Jan.	81.3
Feb.	81.3
Mar.	80.1
Apr.	77.5
May	73.0
June	69.4
July	68.7
Aug.	70.2
Sept.	73.4
Oct.	76.6
Nov.	79.0
Dec.	80.4

Hint: In Step 3 of the solution, since both Jan. ($t = 1$) and Feb. ($t = 2$) yield the same maximum temperature, use $t = 1.5$ for the corresponding input.

50. The following two scatter plots and table display average temperature data for two locations with very different climates: Dar es Salaam is south of the Equator, in Tanzania, on the Indian Ocean; Tiksi is in Russia, far north of the Arctic Circle, on the Arctic Ocean.

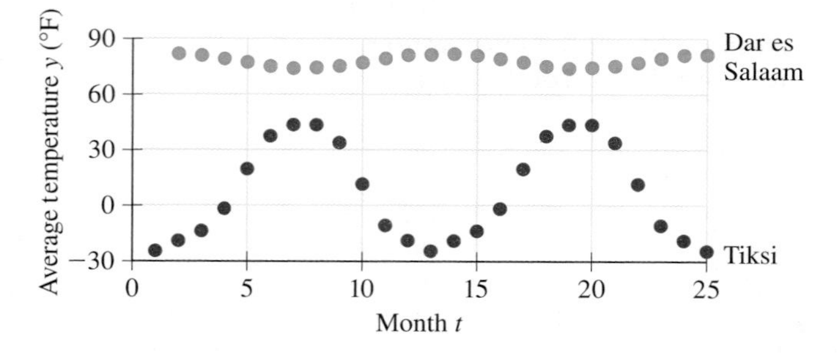

Average Monthly Temperatures for Dar es Salaam, Tanzania, and Tiksi, Russia

Month	Temperature (°F) Dar es Salaam	Temperature (°F) Tiksi
Jan.	81.3	−24.2
Feb.	81.7	−18.7
Mar.	80.8	−13.4
Apr.	79.0	−1.4
May	77.2	19.9
June	75.0	37.8
July	73.9	43.9
Aug.	74.3	43.9
Sept.	75.2	34.2
Oct.	77.0	11.8
Nov.	79.2	−10.4
Dec.	81.1	−18.5

(a) Use the methods of Example 8 to find a periodic function of the form $y = A\sin(Bt - C) + D$ whose values approximate the monthly temperatures for Dar es Salaam.

(b) Follow part (a) for Tiksi.

(c) By looking at the scatter plot for temperatures in Dar es Salaam, say if the average rate of change of temperature over the interval from January through July is positive or negative.

(d) Use the table of values to compute the average rate of change of temperature in Dar es Salaam over the interval from January through July. Be sure to include units as part of your answer, and check that the sign is consistent with your response in part (c).

(e) By looking at the scatter plot for temperatures in Tiksi, say if the average rate of change of temperature over the interval from January through July is positive or negative.

(f) Use the table of values to compute the average rate of change of temperature in Tiksi over the interval from January through July. As before, be sure to include units as part of your answer, and check that the sign is consistent with your response in part (e).

51. The scatter plot on the next page displays average monthly temperatures for Death Valley, California. (The averages were obtained using daily maximums.) Use the methods of Example 8 to find a periodic function of the form $y = A\sin(Bt - C) + D$ whose values approximate the monthly temperatures. (No table is given; you will need to rely on the scatter plot and make estimates.)

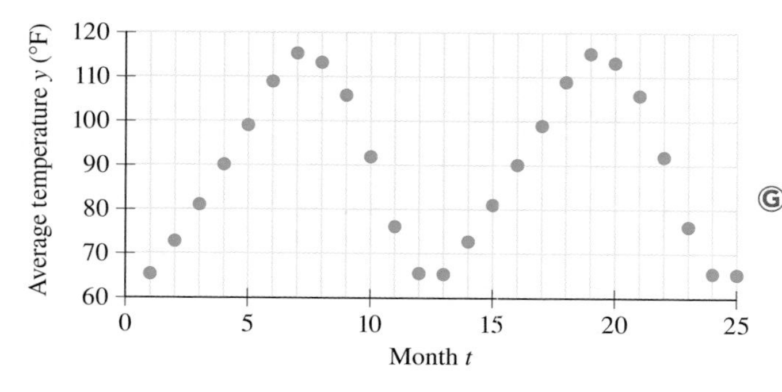

B

52. In Section 8.2 you'll see the identity $\sin^2 x = \frac{1}{2} - \frac{1}{2}\cos 2x$. Use this identity to graph the function $y = \sin^2 x$ for one period.

53. In Section 8.2 you'll see the identity $\cos^2 x = \frac{1}{2} + \frac{1}{2}\cos 2x$. Use this identity to graph the function $y = \cos^2 x$ for one period.

54. In Section 8.2 we derive the identity $\sin 2x = 2 \sin x \cos x$. Use this to graph $y = \sin x \cos x$ for one period.

55. In Section 8.2 we derive the identity $\cos 2x = \cos^2 x - \sin^2 x$. Use this to graph $y = \cos^2 x - \sin^2 x$ for one period.

56. In Example 2 we showed that the period of $y = \cos 3x$ is $2\pi/3$. Use the same method to show that the period of $y = A \cos Bx$ is $2\pi/B$.

Ⓖ **57.** Let $F(x) = \sin x$, $G(x) = x^2$, and $H(x) = x^3$. Which, if any, of the following four composite functions have graphs that do not go below the x-axis? First, try to answer without using a graphing utility, then use the graphing utility to check yourself. (You will learn more this way than if you were to draw the graphs immediately.)

$$y = G(F(x)) \qquad y = F(G(x))$$
$$y = F(H(x)) \qquad y = H(F(x))$$

Ⓖ **58. (a)** Graph the two functions $y = \sin x$ and $y = \sin(\sin x)$ in the standard viewing rectangle. Then for a closer look, switch to a viewing rectangle extending from 0 to 2π in the x-direction and from -1 to 1 in the y-direction. Compare the two graphs; write out your observations in complete sentences.

(b) Use the graphing utility to estimate the amplitude of the function $y = \sin(\sin x)$.

(c) Using your knowledge of the sine function, explain why the amplitude of the function $y = \sin(\sin x)$ is the number sin 1. Then evaluate sin 1 and use the result to check your approximation in part (b).

Ⓖ **59.** Let $f(x) = e^{x/20}(\sin x)$.

(a) Graph the function f using a viewing rectangle that extends from -5 to 5 in both the x- and the y-directions. Note that the resulting graph resembles a sine curve.

(b) Change the viewing rectangle so that x extends from 0 to 50 and y extends from -10 to 10. Describe what you see. Is the function periodic?

(c) Add the graphs of the two functions $y = e^{x/20}$ and $y = -e^{x/20}$ to your picture in part (b). Describe what you see.

Ⓖ **60.** For this exercise, use the standard viewing rectangle.

(a) Graph the function $y = \ln(\sin^2 x)$.

(b) Graph the function $y = \ln(1 - \cos x) + \ln(1 + \cos x)$.

(c) Explain why the two graphs are identical.

PROJECT FOURIER SERIES

In the project at the end of Section 7.2 we discussed square waves, sawtooth waves, and triangular waves. In this project we use the idea of superposition of waves to find a "trigonometric" way to describe a square wave.

We start with the square wave given by $y = f(x)$ where

$$f(x) = \begin{cases} 1, & 0 < x < 1 \\ -1, & -1 < x < 0 \end{cases}$$

$$f(x) = 0 \quad \text{for all integers } x \qquad \text{and} \qquad f(x + 2) = f(x) \qquad \text{(i)}$$

This square wave is shown in Figure A.

Now, f is an odd function, periodic with period 2, and has domain all real numbers. That f is periodic suggests it might be related to more familiar periodic waves, for example, sine waves and cosine waves. Since f is an odd function we might be able to describe f by using only sine waves. And since the domain of f is all real numbers, our description of f using sine waves should coincide with f for most, if not all, real numbers.

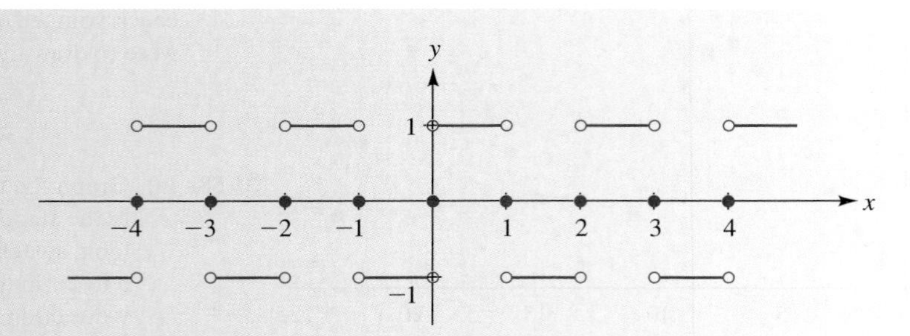

Figure A

A graph of $f(x) = \begin{cases} 1, & 0 < x < 1 \\ -1, & -1 < x < 0 \end{cases}$, $f(x) = 0$ for integer x, and $f(x + 2) = f(x)$

It turns out that f can be expressed by adding together *infinitely* many sine waves. In fact, using methods from calculus, it can be shown that

$$f(x) = \frac{4}{\pi} \sin \pi x + \frac{4}{3\pi} \sin 3\pi x + \frac{4}{5\pi} \sin 5\pi x + \frac{4}{7\pi} \sin 7\pi x + \cdots \qquad \text{(ii)}$$

for all real numbers x. The symbol "$\cdots$" means there are infinitely many more terms that in this case follow the given pattern. So we are expressing $f(x)$ as an *infinite series,* a sum of infinitely many terms. (While we will study some infinite series of numbers in Chapter 13, it should be noted that the right-hand side of equation (ii) is an infinite series of *functions.*) The particular infinite series on the right-hand side of equation (ii) is called the **Fourier series** for the function f. Notice that the Fourier series for f consists of sine waves of different periods, but that 2, the period of the given square wave, is an integer multiple of each period. When we superimpose or add up the waves, we obtain the function f of period 2.

Figure B shows the graphs of the first three partial sums of the Fourier series for f on the closed interval $[-3, 3]$ along with the graph of the square wave.

Notice that as we add more and more terms of the Fourier series, we appear to obtain better approximations to f.

It turns out that equation (ii) is valid for all real numbers x, that is, for each real number x, the Fourier series for f yields the function value $f(x)$. This fact leads us to a famous formula for π discovered by Gottfried Wilhelm Leibniz (ca. 1700), who along with Isaac Newton is credited with the invention of calculus. Using equation (ii), let $x = 1/2$. Then for the square wave $f(1/2) = 1$. And

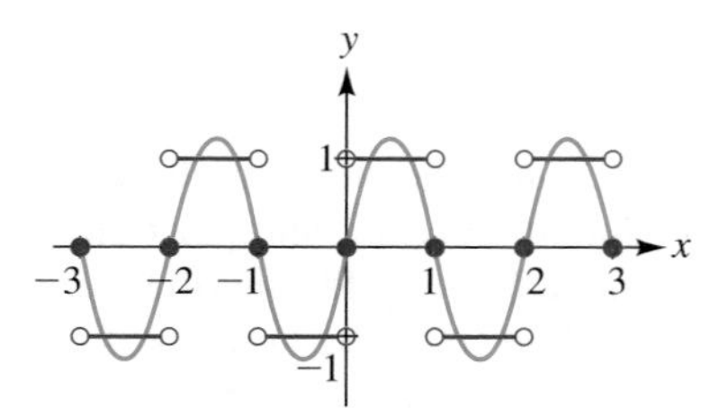

Graph of $y = \frac{4}{\pi} \sin \pi x$

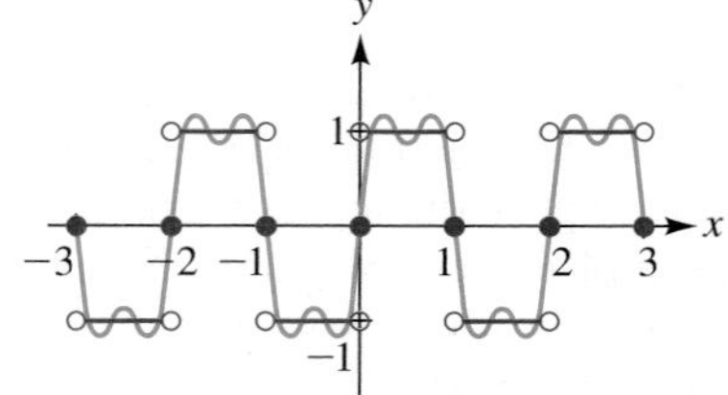

Graph of $y = \frac{4}{\pi} \sin \pi x + \frac{4}{3\pi} \sin 3\pi x$

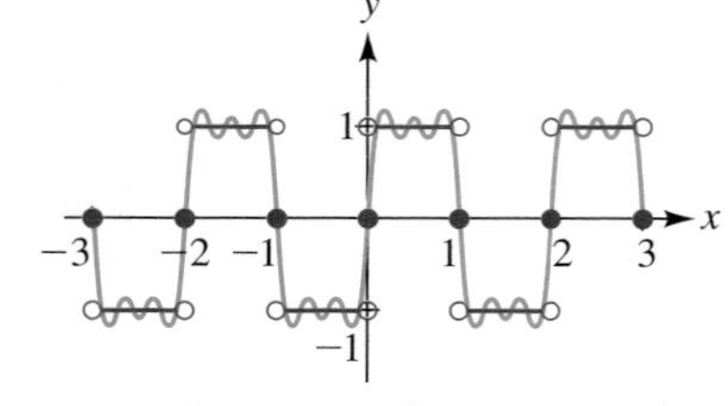

Graph of $y = \frac{4}{\pi} \sin \pi x + \frac{4}{3\pi} \sin 3\pi x + \frac{4}{5\pi} \sin 5\pi x$

Figure B

the Fourier series with $x = 1/2$ becomes

$$\frac{4}{\pi}\sin\frac{\pi}{2} + \frac{4}{3\pi}\sin\frac{3\pi}{2} + \frac{4}{5\pi}\sin\frac{5\pi}{2} + \frac{4}{7\pi}\sin\frac{7\pi}{2} + \cdots$$
$$= \frac{4}{\pi} - \frac{4}{3\pi} + \frac{4}{5\pi} - \frac{4}{7\pi} + \cdots = \frac{4}{\pi}\left(1 - \frac{1}{3} + \frac{1}{5} - \frac{1}{7} + \cdots\right)$$

So

$$f\left(\frac{1}{2}\right) = \frac{4}{\pi}\left(1 - \frac{1}{3} + \frac{1}{5} - \frac{1}{7} + \cdots\right) \quad \text{or} \quad 1 = \frac{4}{\pi}\left(1 - \frac{1}{3} + \frac{1}{5} - \frac{1}{7} + \cdots\right)$$

Multiplying each side of this last equation by $\pi/4$, we obtain Leibniz's formula for π,

$$\frac{\pi}{4} = 1 - \frac{1}{3} + \frac{1}{5} - \frac{1}{7} + \cdots$$

(It should be noted that Leibniz's formula is an inefficient way to calculate π. See also Exercises 86 and 87 in Section 8.5. Also Leibniz derived his formula by a much different method than the one just presented.)

For many years Fourier series were commonly used by mathematicians, chiefly Daniel Bernoulli and Leonhard Euler, to analyze wave phenomena, among other things. But, it was not until the publication of Joseph Fourier's classic treatise, *Théorie Analytique de la Chlaleur* ("The Analytic Theory of Heat") in 1822 that the series that now bear his name were used in a systematic way to solve problems in the theory and application of heat conduction. Much of the development of mathematics in the nineteenth and well into the twentieth century was driven by the desire to place Fourier's methods on a rigorous foundation.

Fourier's idea that almost all important functions that arise in mathematical models of real-world phenomena can be represented as a superposition of sine and cosine functions has been of central importance in many physical theories and engineering applications. Modern optics, information theory, communication technology, celestial mechanics (and yes, rocket science), geophysics, meteorology, analysis of structures, and sound systems all rely on Fourier's idea. Terms such as "frequency response" and "bandwidth" from the mathematical theory of Fourier analysis have entered into common usage. The "bass" and "treble" controls on your sound system directly adjust the amplitudes of terms in a Fourier series.

EXERCISES

1. (a) Look at equation (ii) on page 538. What are the next three terms of the Fourier series for f?

(b) On the same set of axes, for $0 \le x \le 1$, graph carefully *by hand* the equations

$$y = \frac{4}{\pi}\sin \pi x, \qquad y = \frac{4}{3\pi}\sin 3\pi x, \qquad y = \frac{4}{\pi}\sin \pi x + \frac{4}{3\pi}\sin 3\pi x$$

and compare with Figure B.

Ⓖ **(c)** Using a graphing utility, graph each of the first seven partial sums. (Figure B shows graphs of the first three.) For each graph, use the zoom facility to estimate the slope near the origin and the deviation of the bump at $x = 1/2$ from the square wave value, one. As the number of terms of the partial sums increase, what happens to the slopes near the origin and to the deviation of the bump?

Ⓖ **(d)** Compare your graph of the seventh partial sum with the graphs in Figure B. What do you notice?

2. Consider the triangular wave defined by

$$g(x) = \begin{cases} x, & 0 \le x \le 1 \\ -x, & -1 \le x < 0 \end{cases} \quad \text{and} \quad g(x+2) = g(x) \quad \text{for all real numbers } x$$

(a) Graph $y = g(x)$.

(b) Is g an even function, an odd function, or neither even nor odd? Is g a periodic function? If yes what is the period of g? What is the domain of g?

(c) The Fourier series of g should consist only of cosine waves. Why? What is true about the periods of all of the cosine waves?

Ⓖ **(d)** It can be shown that

$$g(x) = \frac{1}{2} - \frac{4}{\pi^2}\cos \pi x - \frac{4}{3^2\pi^2}\cos 3\pi x - \frac{4}{5^2\pi^2}\cos 5\pi x - \frac{4}{7^2\pi^2}\cos 7\pi x - \cdots$$

for all real numbers x. Graph the fourth partial sum of the Fourier series for g, and compare your graph with the triangular wave. The fit of the fourth partial sum of this series to the triangular wave should be comparable to the fit of the third partial sum in Figure B to the square wave. Which do you think gives a better fit? Why?

(e) In the series in part (d), let $x = 1$ and obtain a well-known formula due to Daniel Bernoulli (among others).

$$\frac{\pi^2}{8} = 1 + \frac{1}{9} + \frac{1}{25} + \frac{1}{49} + \frac{1}{81} + \frac{1}{121} + \cdots$$

3. In this exercise we sketch a (nonrigorous) derivation of one of the most celebrated formulas of eighteenth-century mathematics, discovered by Euler in 1736.

(a) Start with the series

$$S = 1 + \frac{1}{2^2} + \frac{1}{3^2} + \frac{1}{4^2} + \frac{1}{5^2} + \frac{1}{6^2} + \cdots$$

and regroup it into a sum of two series, one, S_1, with odd denominators and the other, S_2, with even denominators. So $S = S_1 + S_2$.

(b) Note that $S_1 = \pi^2/8$ by Exercise 2(e). Also note that $S_2 = (1/4)S$. So $S = (\pi^2/8) + (1/4)S$.

(c) Solve the last equation for S to obtain $S = \pi^2/6$, that is,

$$\frac{\pi^2}{6} = 1 + \frac{1}{2^2} + \frac{1}{3^2} + \frac{1}{4^2} + \frac{1}{5^2} + \frac{1}{6^2} + \cdots$$

which is Euler's formula for the sum of the squares of the reciprocals of the positive integers.

7.4 SIMPLE HARMONIC MOTION

Amongst the most important classes of motions which we have to consider in Natural Philosophy, there is one, namely, Harmonic Motion, *which is of such immense use, not only in ordinary kinetics, but in the theories of sound, light, heat, etc., that we make no apology for entering here into considerable detail regarding it.*—Sir William Thomson (Lord Kelvin) and Peter Tait in *Treatise on Natural Philosophy* (Cambridge University Press, 1879)

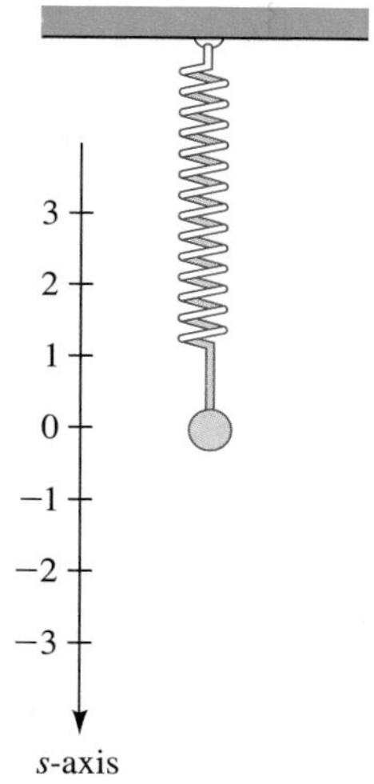

s-axis

Figure 1

Figure 1 shows a mass attached to a spring hung from the ceiling. If we pull the mass down a bit and then release it, the resulting up-and-down oscillations are referred to as **simple harmonic motion.** (We are neglecting the effects of friction and air resistance here.) Other examples of simple harmonic motion (or combinations of simple harmonic motions) include the vibration of a string or of a column of air in a musical instrument and the vibration of an atom in a solid. Furthermore, the mathematics used in describing or analyzing these mechanical oscillations is the same as that used in studying electromagnetic oscillations (such as radio waves, microwaves, or the alternating electrical current in your house).

To analyze the motion of a mass attached to a spring, we set up a coordinate system as shown in Figure 1, in which the equilibrium position of the mass (the position before we pull it down) corresponds to $s = 0$. Now suppose that we pull the mass down to $s = -2$ and release it. If we take a sequence of "snapshots" at one-second time intervals, we will obtain the type of result shown in Figure 2.

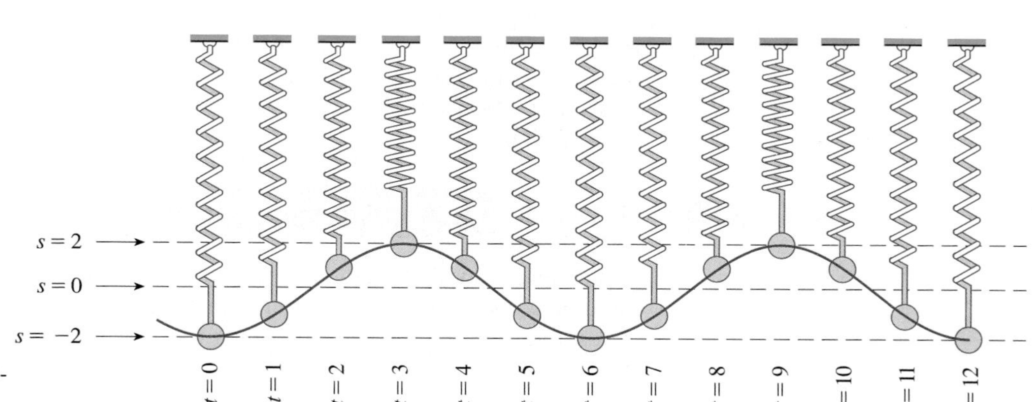

Figure 2
Snapshots of a spring–mass system taken at one-second intervals

The curve in Figure 2 shows how the coordinate s of the mass changes over time. As you can see, the curve resembles the graphs of $y = A\sin(Bx - C)$ and $y = A\cos(Bx - C)$ that we considered in the previous section. Indeed, using calculus, it can be shown that simple harmonic motion is characterized by either one of these types of equations. For example, for the simple harmonic motion depicted in Figure 2, it can be shown that an appropriate equation relating the position s and the time t is

$$s = -2\cos\frac{\pi t}{3} \qquad (1)$$

Using equation (1) and Figure 2, we can give a physical interpretation to the term *period.* From Figure 2 we see that it takes 6 seconds for the mass to return to its starting position at $s = -2$. Then in the next 6 seconds, the same motion is repeated, and so on. We say that the **period** of the motion is 6 seconds. That is, it takes 6 seconds for the motion to go through one complete cycle. Notice that this agrees numerically with the period we calculate using equation (1) and the expression $2\pi/B$ from Section 7.3:

$$\text{period} = \frac{2\pi}{B} = \frac{2\pi}{\pi/3} = 6$$

In simple harmonic motion the **frequency** f is the number of complete cycles per unit time, and it is given by

$$\text{frequency} = f = \frac{1}{\text{period}}$$

For instance, for the motion in Figure 2 we have

$$f = \frac{1}{\text{period}} = \frac{1}{6}\text{ cycles per second}$$

We mention in passing that 1 cycle per second (cps) is known as a *hertz,* abbreviated Hz. This unit is named after the German physicist Heinrich Hertz (1857–1894), who was the first person to produce and study radio waves.* Although you might not have realized it, you have probably heard this unit mentioned (implicitly, at least) many times on the radio. For example, when a radio station advertises itself as "98.1 on the FM dial," this refers to radio waves with a frequency of 98.1 million hertz.

From Figure 2 we can see that the mass moves back and forth between $s = -2$ and $s = 2$. In other words, the maximum displacement of the mass from its equilibrium position (at $s = 0$) is two units. We say that the **amplitude** of the motion is two units. Notice that this agrees with the amplitude we would calculate using equation (1).

EXAMPLE 1 A spring–mass system

A mass on a smooth tabletop is attached to a spring, as shown in Figure 3. The coordinate system has been chosen so that the equilibrium position of the mass corresponds to $s = 0$. Assume that the mass moves in simple harmonic motion described by

$$s = 5\cos\frac{\pi t}{4}$$

where s is in centimeters and t is in seconds.

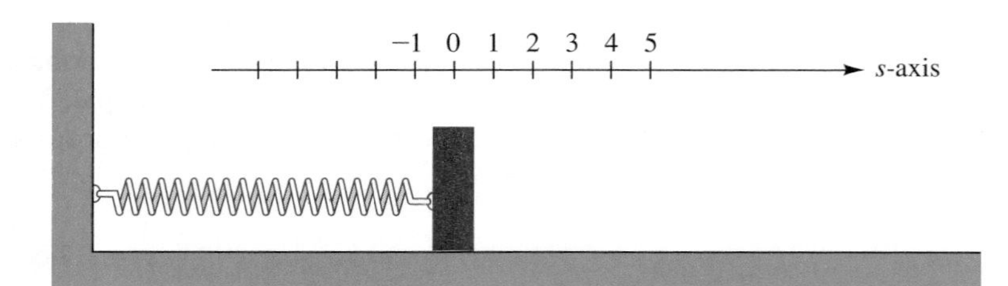

Figure 3

*For background and details, see the interesting article "Heinrich Hertz" by Philip and Emily Morrison in the December 1957 issue of *Scientific American.*

(a) Graph the function $s = 5\cos(\pi t/4)$ over the interval $0 \le t \le 16$. Specify the amplitude, the period, and the frequency of the motion.
(b) Use the graph to determine the times in this interval at which the mass is farthest from the origin.
(c) When during this interval of time is the mass passing through the equilibrium position?

SOLUTION

(a) Using the techniques developed in Section 7.3, we obtain the graph shown in Figure 4. From the graph (or from the given equation) we have

$$\text{amplitude} = 5 \text{ cm}$$
$$\text{period} = 8 \text{ sec}$$
$$\text{frequency} = \frac{1}{8} \text{ cycles/sec}$$

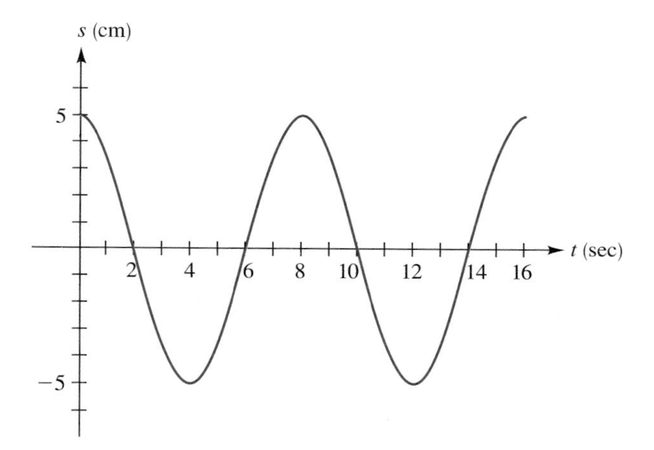

Figure 4

(b) The distance of the mass from the origin is given by $|s|$. According to Figure 4, the maximum value of $|s|$ is 5, and this occurs when $t = 0$ sec, 4 sec, 8 sec, 12 sec, and 16 sec.
(c) We are given that the equilibrium position of the mass is $s = 0$. From Figure 4 we see that s is zero when $t = 2$ sec, 6 sec, 10 sec, and 14 sec.

EXERCISE SET 7.4

A

For Exercises 1 and 2, suppose that we have a spring–mass system as shown in Figure 3 on page 542.

1. Assume that the simple harmonic motion is described by the equation $s = 4\cos(\pi t/2)$, where s is in centimeters and t is in seconds.

(a) Specify the s-coordinate of the mass at each of the following times: $t = 0$ sec, 0.5 sec, 1 sec, and 2 sec. (One of these coordinates will involve a radical sign; for this case, use a calculator and round the final answer to two decimal places.)

(b) Find the amplitude, the period, and the frequency of this motion. Sketch the graph of $s = 4\cos(\pi t/2)$ over the interval $0 \le t \le 8$.

(c) Use your graph to determine the times in this interval at which the mass is farthest from the origin.

(d) When during the interval of time $0 \le t \le 8$ is the mass passing through the origin?

(e) When during the interval of time $0 \le t \le 8$ is the mass moving to the right? *Hint:* The mass is moving to the right when the s-coordinate is increasing. Use the graph to see when s is increasing.

2. Assume that the simple harmonic motion is described by the equation $s = -6\cos(\pi t/6)$, where s is in centimeters and t is in seconds.

(a) Complete the table. (For coordinates that involve radical signs, use a calculator and round the result to two decimal places.)

t (sec)	1	2	3	4	5	6	7	8	9	10	11	12
s (cm)												

(b) Find the amplitude, period, and frequency of this motion. Sketch the graph of $s = -6\cos(\pi t/6)$ over the interval $0 \le t \le 24$.

(c) Use your graph to determine the times in this interval at which the mass is farthest from the equilibrium position.

(d) When during the interval of time $0 \le t \le 24$ is the mass passing through the origin?

(e) When during the interval of time $0 \le t \le 24$ is the mass moving to the left? *Hint*: The mass is moving to the left when the s-coordinate is decreasing. Use the graph to see when s is decreasing.

In Exercises 3 and 4, suppose that we have a spring–mass system, as shown in Figure 1 on page 541.

3. Assume that the simple harmonic motion is described by the equation $s = -3\cos(\pi t/3)$, where s is in feet, t is in seconds, and the equilibrium position of the mass is $s = 0$.

(a) Specify the amplitude, period, and frequency for this simple harmonic motion, and sketch the graph of the function $s = -3\cos(\pi t/3)$ over the interval $0 \le t \le 12$.

(b) When during the interval of time $0 \le t \le 12$ is the mass moving upward? *Hint*: The mass is moving upward when the s-coordinate is increasing. Use the graph to see when s is increasing.

(c) When during the interval of time $0 \le t \le 12$ is the mass moving downward? *Hint*: The mass is moving downward when the s-coordinate is decreasing. Use the graph to see when s is decreasing.

(d) For this harmonic motion it can be shown (using calculus) that the velocity v of the mass is given by

$$v = \pi \sin \frac{\pi t}{3}$$

where t is in seconds and v is in feet per second. Graph this velocity function over the interval $0 \le t \le 12$.

(e) Use your graph of the velocity function from part (d) to find the times during this interval when the velocity is zero. At these times, where is the mass? (That is, what are the s-coordinates?)

(f) Use your graph of the velocity function to find the times when the velocity is maximum. Where is the mass at these times?

(g) Use your graph of the velocity function to find the times when the velocity is minimum. Where is the mass at these times?

(h) On the same set of axes, graph the velocity function $v = \pi \sin(\pi t/3)$ and the position function $s = -3\cos(\pi t/3)$ for $0 \le t \le 12$.

4. Assume that the simple harmonic motion is described by the equation $s = 4\cos(2t/3)$, where s is in feet, t is in seconds, and the equilibrium position of the mass is $s = 0$.

(a) Specify the amplitude, period, and frequency for this simple harmonic motion, and sketch the graph of the function $s = 4\cos(2t/3)$ over the interval $0 \le t \le 6\pi$.

(b) When during the interval of time $0 \le t \le 6\pi$ is the mass moving upward? *Hint*: The mass is moving upward when the s-coordinate is increasing. Use the graph to see when s is increasing.

(c) When during the interval of time $0 \le t \le 6\pi$ is the mass moving downward? *Hint*: The mass is moving downward when the s-coordinate is decreasing. Use the graph to see when s is decreasing.

(d) For this harmonic motion, it can be shown (using calculus) that the velocity v of the mass is given by $v = -\frac{8}{3}\sin(2t/3)$, where t is in seconds and v is in ft/sec. Graph this velocity function over the interval $0 \le t \le 6\pi$.

(e) Use your graph of the velocity function from part (d) to find the times during this interval when the velocity is zero. At these times, where is the mass? (That is, what are the s-coordinates?)

(f) Use your graph of the velocity function to find the times when the velocity is maximum. Where is the mass at these times?

(g) Use your graph of the velocity function to find the times when the velocity is minimum. Where is the mass at these times?

5. The voltage in a household electrical outlet is given by

$$V = 170\cos(120\pi t)$$

where V is measured in volts and t in seconds.

(a) Specify the amplitude and the frequency for this oscillation.

(b) Graph the function $V = 170\cos(120\pi t)$ for two complete cycles beginning at $t = 0$.

(c) For which values of t in part (b) is the voltage maximum?

B

6. The following figure (on the next page) shows a simple pendulum consisting of a string with a weight attached at one end and the other end suspended from a fixed point. As indicated in the figure, the angle between the vertical and the pendulum is denoted by θ (where θ is in radians). Suppose that we pull the pendulum out from the equilibrium position (where $\theta = 0$) to a position $\theta = \theta_0$. Now we release the pendulum so that it swings back and forth. Then (neglecting friction and assuming that θ_0 is a small angle), it can be shown that the angle θ at time t is very closely approximated by

$$\theta = \theta_0 \cos(t\sqrt{g/L})$$

where L is the length of the pendulum, g is a constant (the acceleration due to gravity), and t is the time in seconds.

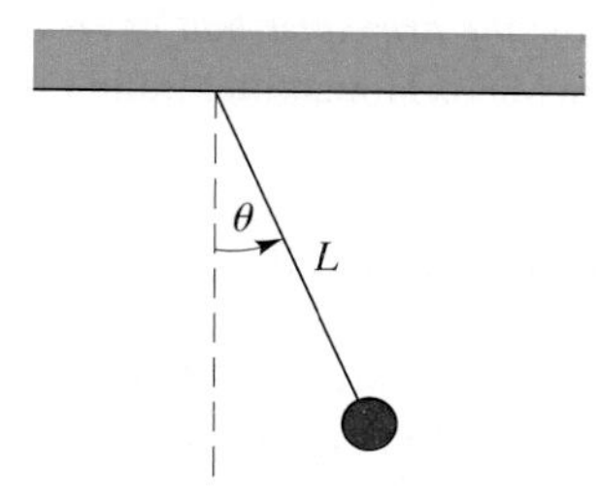

(a) What are the amplitude, the period, and the frequency for the motion defined by the equation $\theta = \theta_0 \cos(t\sqrt{g/L})$? Assume that t is in seconds, L is in meters, and g is in m/sec^2.

(b) Use your results in part (a) to answer these two questions. Does the period of the pendulum depend upon the amplitude? Does the period depend upon the length L?

(c) Graph the function $\theta = \theta_0 \cos(t\sqrt{g/L})$ for two complete cycles beginning at $t = 0$ and using the following values for the constants:

$$\theta_0 = 0.1 \text{ radian} \qquad L = 1 \text{ m} \qquad g = 9.8 \text{ m/sec}^2$$

(d) For which values of t during these two cycles is the weight moving to the right? *Hint*: The weight is moving to the right when θ is increasing; use your graph in part (c) to see when this occurs.

(e) The velocity V of the weight as it oscillates back and forth is given by

$$V = -\theta_0\sqrt{g/L}\sin(t\sqrt{g/L})$$

where V is in m/sec. Graph this function for two complete cycles using the values of the constants given in part (c).

(f) At which times during these two cycles is the velocity maximum? What is the corresponding value of θ in each case?

(g) At which times during these two cycles is the velocity minimum? What is the corresponding value of θ in each case?

(h) At which times during these two cycles is the velocity zero? What is the corresponding value of θ in each case?

7. Refer to Figure A. Suppose that the point P travels counterclockwise around the unit circle at a constant angular speed of $\pi/3$ radians/sec. Assume that at time $t = 0$ sec, the location of P is $(1, 0)$.

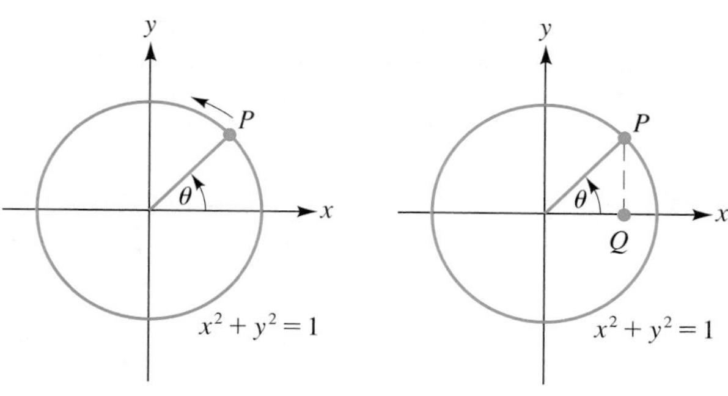

Figure A Figure B

(a) Complete the table.

t (sec)	0	1	2	3	4	5	6	7
θ (radians)								

(b) Now (for each position of the point P) suppose that we draw a perpendicular from P to the x-axis, meeting the x-axis at Q, as indicated in Figure B. The point Q is called the *projection* of the point P on the x-axis. As the point P moves around the circle, the point Q will move back and forth along the diameter of the circle. What is the x-coordinate of the point Q at each of the times listed in the table for part (a)?

(c) Draw a sketch showing the location of the points P and Q when $t = 1$ sec.

(d) Draw sketches as in part (c) for $t = 2$, 3, and 4 sec.

(e) The x-coordinate of Q is always equal to $\cos\theta$. Why?

(f) If you look back at your table in part (a), you'll see that the relation between θ and t is $\theta = \pi t/3$. Thus the x-coordinate of Q at time t is given by

$$x = \cos\frac{\pi t}{3}$$

This tells us that as the point P moves around the circle at a constant angular speed, the point Q oscillates in simple harmonic motion along the x-axis. Graph this function for two complete cycles, beginning at $t = 0$. Specify the amplitude, the period, and the frequency for the motion.

(g) Using calculus, it can be shown that the velocity of the point Q at time t is given by

$$V = -\frac{\pi}{3}\sin\frac{\pi t}{3}$$

Graph this function for two complete cycles, beginning at $t = 0$.

(h) Use the graph in part (g) to determine the times (during the first two cycles) when the velocity of the point Q is zero. What are the corresponding x-coordinates of Q in each case?

(i) Use the graph in part (g) to determine the times (during the first two cycles) when the velocity of the point Q is maximum. Where is Q located at these times?

(j) Use the graph in part (g) to determine the times (during the first two cycles) when the velocity of the point Q is minimum. Where is Q located at these times?

PROJECT THE MOTION OF A PISTON

To see the way a machine works, you can take the the covers off and look inside. But to understand what goes on, you need to get to know the principles that govern its actions.— David Macaulay, *The Way Things Work* (Boston Mass.: Houghton Mifflin Company, 1988)

In this project we examine a simple linkage that exhibits an oscillatory motion that is somewhat more complicated than simple harmonic motion. The tools that we've developed in the last two chapters will enable us to derive a function describing this motion. This function provides a very good mathematical model for the motion of a piston in a conventional internal combustion car engine. Although you won't need to be familiar with car engines to follow the mathematics, your appreciation of its applicability would be greatly enhanced if someone in your group could explain how a crankshaft piston and cylinder move in a car engine. The internet or a reference such as the one quoted above might be useful.

Figure A shows a line segment OC of length R rotating counterclockwise about the fixed origin O and another line segment CP of length L, with L greater than R. The line segments are thought of as being linked together at point C as if they were flat rods with a pin through them so that they can rotate freely about C, their common endpoint. The point P is constrained to move along the x-axis. As the line segment OC rotates about the origin, the point P moves back and forth along the x-axis. The two figures illustrate the configuration at two different rotation positions. In the language of car engines the origin would be the center of a cross-section of the crankshaft, the segment OC would be a crank arm, the segment CP would be a piston rod, and point P would be the center of a cross-section of a wrist pin.

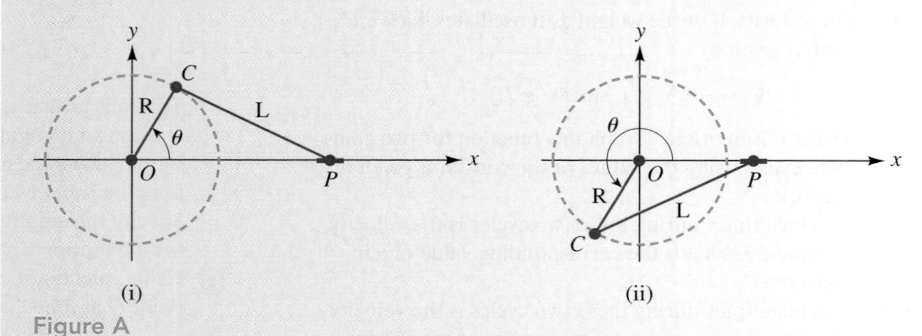

Figure A

EXERCISE 1 Let $f(\theta)$ equal the x-coordinate of the point P when the line segment OC is at an angle θ (measured in radians) from the positive x-axis. What are the maximum and minimum values of f and at what values of θ do they occur?

EXERCISE 2 Figure B can be used to derive a formula for $f(\theta)$. Express the lengths of OA and CA in terms of R and θ. Now use the Pythagorean theorem

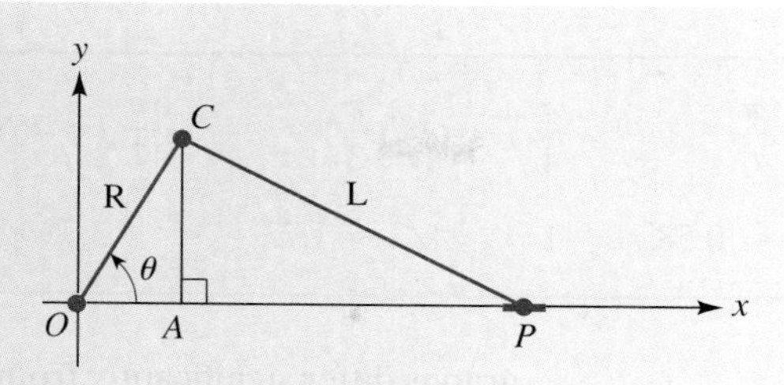

Figure B

to express the length of AP in terms of R, L, and θ. Since, in Figure B, $f(\theta)$ equals the length of line segment OP we have

$$f(\theta) = \mathrm{R}\cos\theta + \sqrt{\mathrm{L}^2 - \mathrm{R}^2\sin^2\theta}$$

How do the maximum and minimum value for f compare with your answers from Exercise 1? Explain why this derivation would work when point C is in the second quadrant.

Ⓖ **EXERCISE 3** For given values of L and R, graph the indicated functions, for $0 \le \theta \le 4\pi$, on the same set of axes and zoom in near the maximum, minimum, and any other interesting points. Write short notes of your observations. $f(\theta)$ is the function derived in the previous exercise.

(a) Let L = 20 cm and R = 4 cm. Graph $y = \mathrm{L}$, $y = \mathrm{L} + \mathrm{R}\cos\theta$, and $y = f(\theta)$.
(b) Let L = 20 cm and R = 10 cm. Graph $y = \mathrm{L}$, $y = \mathrm{L} + \mathrm{R}\cos\theta$, and $y = f(\theta)$.
(c) Let L = 20 cm and R = 10 cm. Graph $y = 1.868\,\mathrm{R} + \mathrm{R}\cos\theta$, and $y = f(\theta)$.
(d) Let L = 20 cm and R = 10 cm. Graph $y = 1.868\,\mathrm{R} + \mathrm{R}\cos\theta + 0.1339\,\mathrm{R}\cos 2\theta$, and $y = f(\theta)$.

In parts (c) and (d) the functions graphed with f are the second and third partial sums of the Fourier series for f. If you did the project on Fourier series at the end of Section 7.3, note how much better the third partial sum of the Fourier series is as an approximation to f here than in Figure B of that project.

Finally, let's apply the result of Exercise 2 to a typical automotive situation.

EXERCISE 4 Given that segment OC is rotating at 3000 revolutions per minute (rpm) let $g(t)$ be the x-coordinate of the point P at time t seconds and find a formula for $g(t)$. [*Hint*: Find θ in terms of t and substitute into the formula for $f(\theta)$.] What would the formula be for k revolutions per minute?

7.5 GRAPHS OF THE TANGENT AND THE RECIPROCAL FUNCTIONS

A third . . . function, the tangent of θ, or tan *θ, is of secondary importance, in that it is not associated with wave phenomena. Nevertheless, it enters into the body of analysis so prominently that we cannot ignore it.*—Samuel E. Urner and William B. Orange in *Elements of Mathematical Analysis* (Boston: Ginn and Co., 1950)

We have seen in the previous sections that the sine and cosine functions are periodic. The remaining four trigonometric functions are also periodic, but their

TABLE 1

x	0	$\frac{\pi}{6}$	$\frac{\pi}{4}$	$\frac{\pi}{3}$	$\frac{5\pi}{12}\ (= 75°)$	$\frac{17\pi}{36}\ (= 85°)$	$\frac{89\pi}{180}\ (= 89°)$	$\frac{\pi}{2}$
$\tan x$	0	$\frac{\sqrt{3}}{3} \approx 0.58$	1	$\sqrt{3} \approx 1.73$	3.73	11.43	57.29	undefined

graphs differ significantly from those of sine and cosine. In particular, the graphs of $y = \tan x$, $\cot x$, $\csc x$, and $\sec x$ all possess vertical asymptotes.

We'll obtain the graph of the tangent function by a combination of both point-plotting and symmetry considerations. Table 1 displays a list of values for $y = \tan x$ using x-values in the interval $[0, \pi/2)$.

Note: As x increases from 0 to $\pi/2$, the values of $\tan x$ increase slowly at first then more and more rapidly.

As is indicated in Table 1, $\tan x$ is undefined when $x = \pi/2$. This follows from the identity

$$\tan x = \frac{\sin x}{\cos x} \tag{1}$$

When $x = \pi/2$, the denominator in this identity is zero. Indeed, when x is equal to any odd integral multiple of $\pi/2$ (for example, $\pm 3\pi/2$, $\pm 5\pi/2$), the denominator in equation (1) will be zero, and, consequently, $\tan x$ will be undefined.

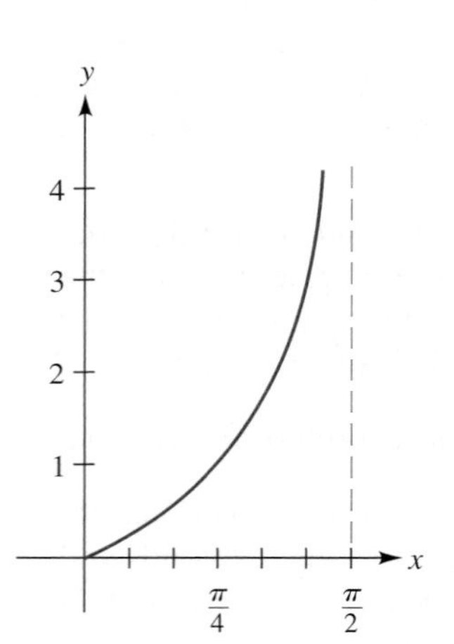

Figure 1
$y = \tan x,\ 0 \le x < \pi/2$

Because $\tan x$ is undefined when $x = \pi/2$, we want to see how the graph behaves as x gets closer and closer to $\pi/2$. This is why the x-values $5\pi/12$, $17\pi/36$, and $89\pi/180$ are used in Table 1. In Figure 1 we've used the data in Table 1 to draw the graph of $y = \tan x$ for $0 \le x < \pi/2$. As the figure indicates, the vertical line $x = \pi/2$ is an asymptote for the graph.

The graph of $y = \tan x$ can now be completed without further need for tables or a calculator. First, the identity $\tan(-x) = -\tan x$ (from Section 7.1) tells us that the tangent is an odd function so the graph of $y = \tan x$ is symmetric about the origin. After reflecting the graph in Figure 1 about the origin, we can draw the graph of $y = \tan x$ on the interval $(-\frac{\pi}{2}, \frac{\pi}{2})$, as shown in Figure 2.

Now, to complete the graph of $y = \tan x$, we use the identity

$$\tan(s + \pi) = \tan s \tag{2}$$

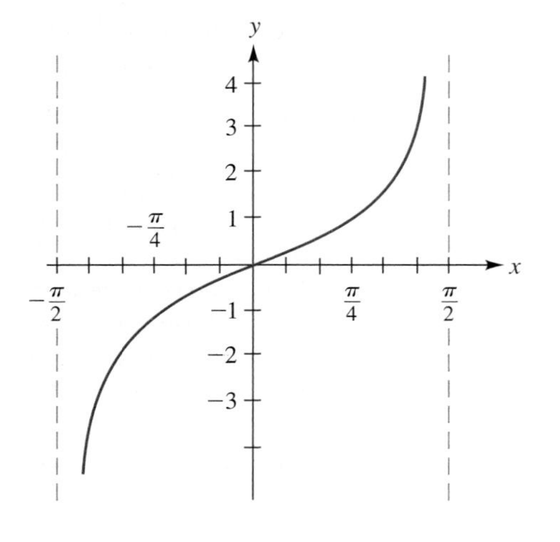

Figure 2
$y = \tan x,\ -\pi/2 < x < \pi/2$
The graph is symmetric about the origin. The lines $x = \pm\pi/2$ are asymptotes.

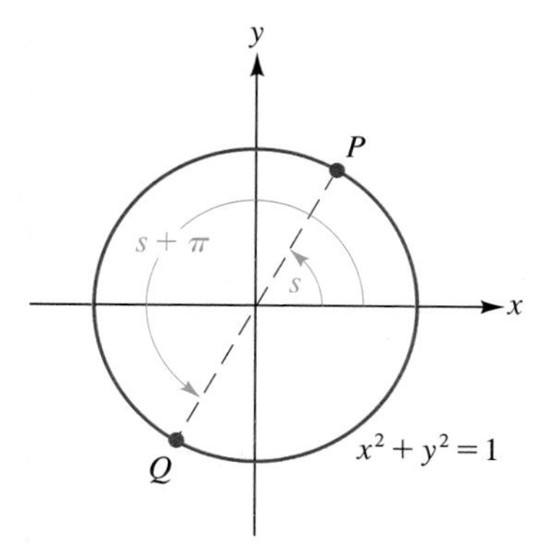

Figure 3

Looking at Figure 3, we can see why this identity is valid. By definition, the coordinates of P and Q are

$$P(\cos s, \sin s) \qquad \text{and} \qquad Q(\cos(s+\pi), \sin(s+\pi))$$

On the other hand, the points P and Q are symmetric about the origin, so the coordinates of Q are just the negatives of the coordinates of P. That is,

$$\cos(s+\pi) = -\cos s \qquad \text{and} \qquad \sin(s+\pi) = -\sin s$$

Consequently, we have

$$\tan(s+\pi) = \frac{\sin(s+\pi)}{\cos(s+\pi)} = \frac{-\sin s}{-\cos s}$$

$$= \frac{\sin s}{\cos s} = \tan s \qquad \text{as required}$$

[Although Figure 3 shows the angle with radian measure s terminating in Quadrant I, our proof is valid for the other quadrants as well. (Draw a figure for yourself and verify this.)]

Identity (2) tells us that the graph of $y = \tan x$ must repeat itself at intervals of length π. Taking this fact into account, along with Figure 2, we conclude that the period of $y = \tan x$ is exactly π. Our final graph of $y = \tan x$ is shown in the Property Summary Box that follows.

PROPERTY SUMMARY The Tangent Function: $y = \tan x$

Domain: The set of all real numbers other than
$\pm\frac{\pi}{2}, \pm\frac{3\pi}{2}, \pm\frac{5\pi}{2}, \ldots$

Range: $(-\infty, \infty)$

Period: π

Asymptotes: $x = \pm\frac{\pi}{2}, x = \pm\frac{3\pi}{2}, x = \pm\frac{5\pi}{2}, \ldots$

x-intercepts: $0, \pm\pi, \pm 2\pi, \pm 3\pi, \ldots$

Note: The x-intercepts occur midway between consecutive vertical asymptotes

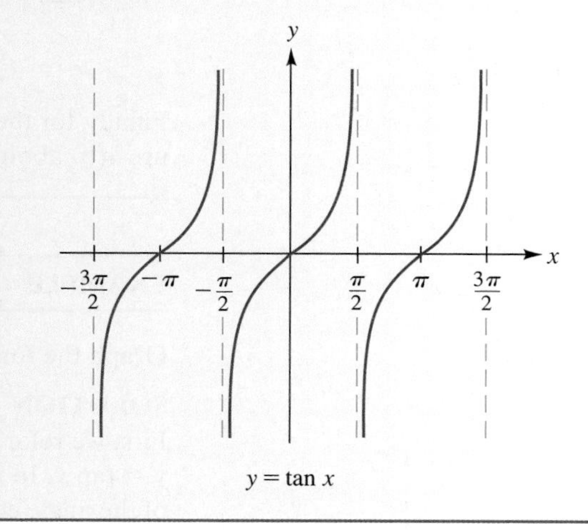

Note: The tangent is an odd function so the graph of $y = \tan x$ is symmetric about the origin.

EXAMPLE 1 Finding periods, asymptotes, and intercepts

Graph the following functions for one period. In each case, specify the period, the asymptotes, and the intercepts:

(a) $y = \tan\left(x - \frac{\pi}{4}\right)$; **(b)** $y = -\tan\left(x - \frac{\pi}{4}\right)$.

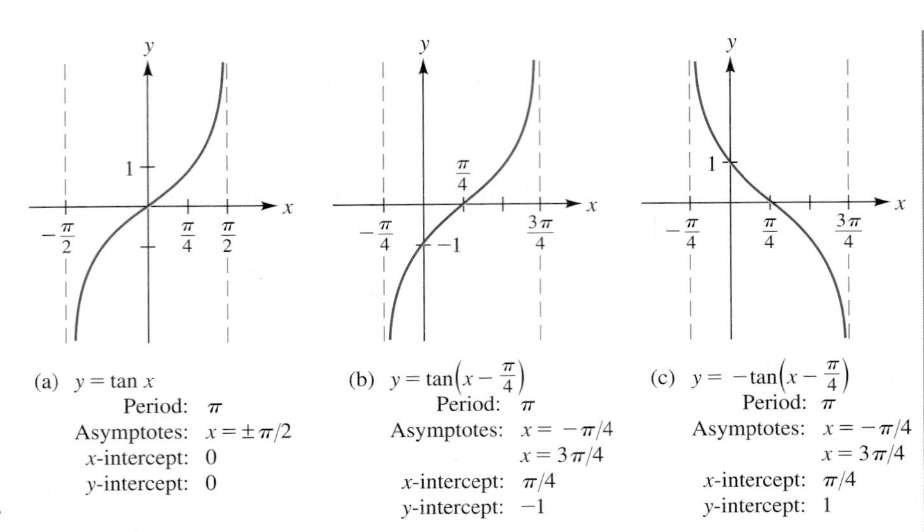

(a) $y = \tan x$
Period: π
Asymptotes: $x = \pm\pi/2$
x-intercept: 0
y-intercept: 0

(b) $y = \tan\left(x - \frac{\pi}{4}\right)$
Period: π
Asymptotes: $x = -\pi/4$
$x = 3\pi/4$
x-intercept: $\pi/4$
y-intercept: -1

(c) $y = -\tan\left(x - \frac{\pi}{4}\right)$
Period: π
Asymptotes: $x = -\pi/4$
$x = 3\pi/4$
x-intercept: $\pi/4$
y-intercept: 1

Figure 4

SOLUTION

We begin with the graph of one period of $y = \tan x$, as shown in Figure 4(a). By translating this graph $\pi/4$ units to the right, we obtain the graph of $y = \tan(x - \frac{\pi}{4})$, in Figure 4(b). With this translation, note that the left asymptote shifts from $x = -\pi/2$ to $x = -\pi/2 + \pi/4 = -\pi/4$; the right asymptote shifts from $x = \pi/2$ to $x = \pi/2 + \pi/4 = 3\pi/4$; and the x-intercept shifts from 0 to $\pi/4$.

For the y-intercept of $y = \tan(x - \frac{\pi}{4})$ we replace x with 0 in the equation. This yields

$$y = \tan\left(0 - \frac{\pi}{4}\right) = \tan\left(-\frac{\pi}{4}\right) = -\tan\frac{\pi}{4} = -1$$

Finally, for the graph of $y = -\tan(x - \frac{\pi}{4})$ we need only reflect the graph in Figure 4(b) about the x-axis; see Figure 4(c).

EXAMPLE 2 Graphing a function of the form $y = \tan Bx$

Graph the function $y = \tan(x/2)$ for one period.

SOLUTION

First we refer back to Figure 2, which shows the basic pattern for one period of $y = \tan x$. In this basic pattern the asymptotes occur when the radian measure of the angle equals $-\pi/2$ or $\pi/2$. Consequently, for $y = \tan(x/2)$, the asymptotes occur when $x/2 = -\pi/2$ and when $x/2 = \pi/2$. From the equation $x/2 = -\pi/2$ we conclude that $x = -\pi$, and from the equation $x/2 = \pi/2$ we conclude that $x = \pi$. Thus the asymptotes for $y = \tan(x/2)$ are $x = -\pi$ and $x = \pi$. The distance between these asymptotes, namely, 2π, is the period of $y = \tan(x/2)$. This is twice the period of $y = \tan x$, so basically, we want to draw a curve with the same general shape as $y = \tan x$ but twice as wide; see Figure 5. Note that the graph in Figure 5 passes through the origin, since when $x = 0$, we have

$$y = \tan\frac{0}{2} = \tan 0 = 0$$

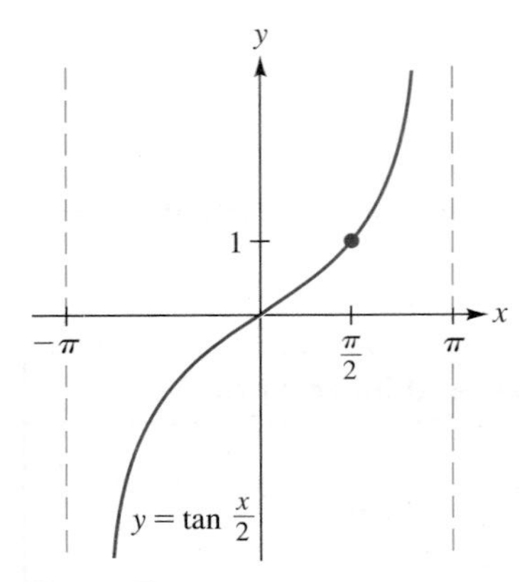

Figure 5

The graph of the cotangent function can be obtained from that of the tangent function by means of the identity

$$\cot x = -\tan\left(x - \frac{\pi}{2}\right) \tag{3}$$

(Exercise 57 shows how to derive this identity.) According to identity (3), the graph of $y = \cot x$ can be obtained by first translating the graph of $y = \tan x$ to the right $\pi/2$ units and then reflecting the translated graph about the x-axis. When this is done, we obtain the graph shown in the box that follows.

PROPERTY SUMMARY The Cotangent Function: $y = \cot x$

Domain: The set of all real numbers other than $0, \pm\pi, \pm 2\pi, \ldots$
Range: $(-\infty, \infty)$
Period: π
Asymptotes: $x = 0, x = \pm\pi, x = \pm 2\pi, \ldots$
x-intercepts: $\pm\frac{\pi}{2}, \pm\frac{3\pi}{2}, \pm\frac{5\pi}{2}, \ldots$

Note: The x-intercepts occur midway between consecutive vertical asymptotes.

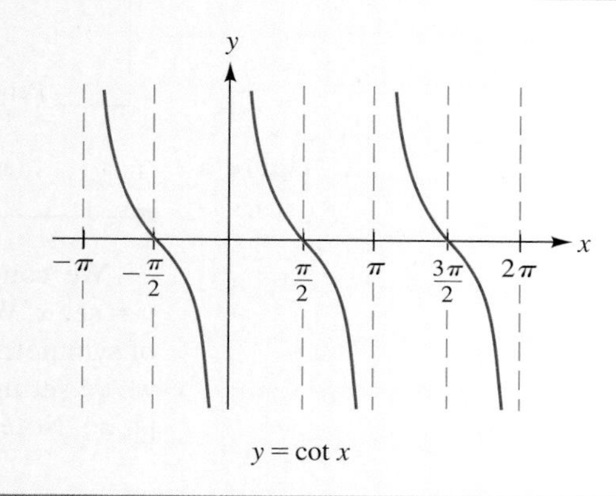

Note: The cotangent is an odd function so the graph of $y = \cot x$ is symmetric about the origin.

EXAMPLE 3 Graphing functions of the form $y = A \cot Bx$

Graph each of the following functions for one period:

(a) $y = \cot \pi x$; **(b)** $y = \frac{1}{2}\cot \pi x$; **(c)** $y = -\frac{1}{2}\cot \pi x$.

SOLUTION

(a) Looking at the graph of $y = \cot x$ in the Property Summary above, we see that one complete pattern or cycle of the graph occurs between the asymptotes $x = 0$ and $x = \pi$. Now, for the function we are given, x has been replaced by πx. Thus the corresponding asymptotes occur when $\pi x = 0$ and when $\pi x = \pi$, in other words, when $x = 0$ and when $x = 1$; see Figure 6(a).

(b) The graph of $y = \frac{1}{2}\cot \pi x$ will have the same general shape as that of $y = \cot \pi x$, but each y-coordinate on $y = \frac{1}{2}\cot \pi x$ will be one-half of the corresponding coordinate on $y = \cot \pi x$; see Figure 6(b).

(c) The graph of $y = -\frac{1}{2}\cot \pi x$ is obtained by reflecting the graph of $y = \frac{1}{2}\cot \pi x$ about the x-axis; see Figure 6(c).

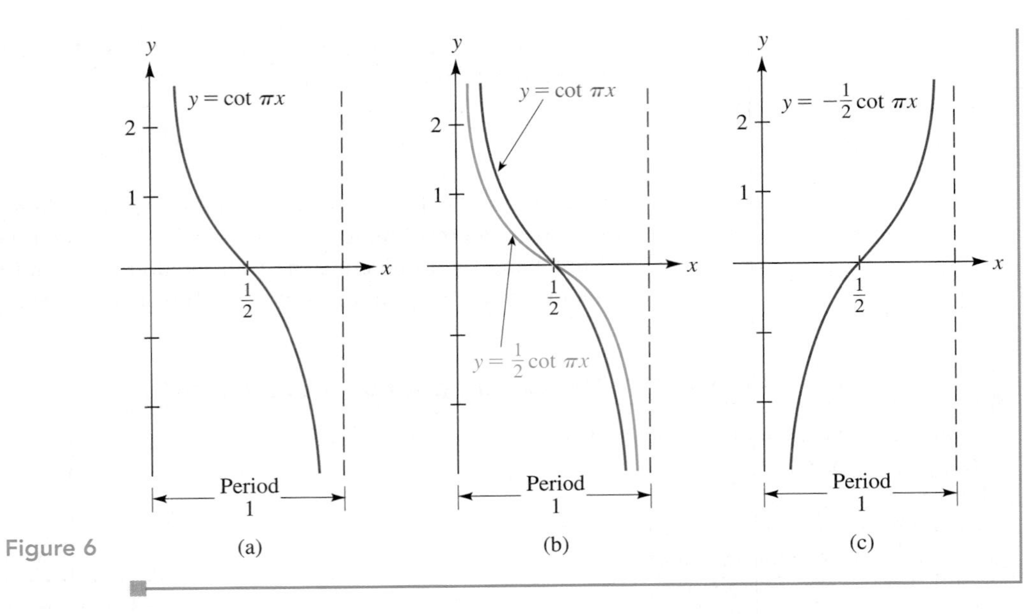

Figure 6

We conclude this section with a discussion of the graphs of $y = \csc x$ and $y = \sec x$. We will obtain these graphs in a series of easy steps, relying on the ideas of symmetry and translation. First consider the function $y = \csc x$. In Figure 7(a) we've set up a table and used it to sketch the graph of $y = \csc x$ on the interval $[\frac{\pi}{2}, \pi)$. Note that $\csc x$ is undefined when $x = \pi$. (Why?)

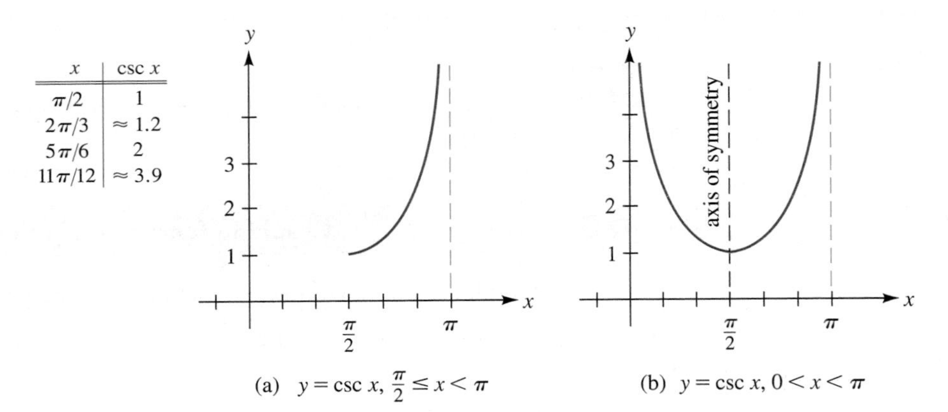

x	$\csc x$
$\pi/2$	1
$2\pi/3$	≈ 1.2
$5\pi/6$	2
$11\pi/12$	≈ 3.9

Figure 7

(a) $y = \csc x,\ \frac{\pi}{2} \le x < \pi$ (b) $y = \csc x,\ 0 < x < \pi$

As Figure 7(a) indicates, the vertical line $x = \pi$ is an asymptote for the graph. The graph in Figure 7(a) can be extended to the interval $(0, \pi)$ by means of the identity

$$\csc\left(\frac{\pi}{2} + s\right) = \csc\left(\frac{\pi}{2} - s\right)$$

(Exercise 59 shows you how to verify this identity.) This identity tells us that, starting at $x = \pi/2$, whether we travel a distance s to the right or a distance s to the left, the value of $y = \csc x$ is the same. In other words, the graph of $y = \csc x$ is symmetric about the line $x = \pi/2$. In view of this symmetry we can sketch the graph of $y = \csc x$ on the interval $(0, \pi)$, as shown in Figure 7(b).

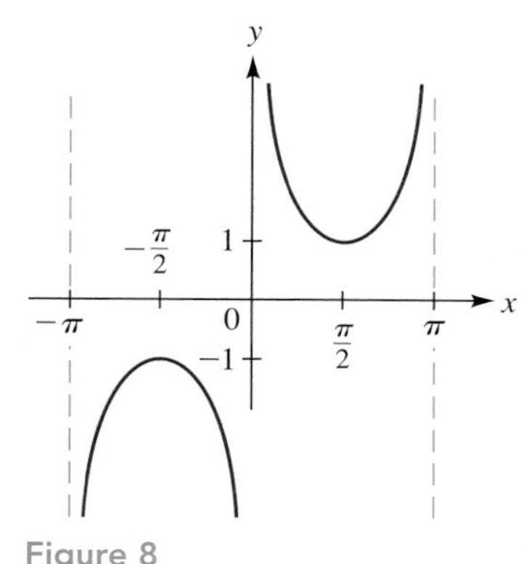

Figure 8
$y = \csc x,\ -\pi < x < \pi$
The graph is symmetric about the origin.

The next step in obtaining the graph of $y = \csc x$ is to use the fact that the graph is symmetric about the origin. To verify this, we need to check that $\csc(-x) = -\csc x$. We have

$$\csc(-x) = \frac{1}{\sin(-x)} = \frac{1}{-\sin x} = -\csc x \qquad \text{as required}$$

Now, taking into account this symmetry about the origin and the portion of the graph that we've already obtained in Figure 7(b), we can sketch the graph of $y = \csc x$ over the interval $(-\pi, \pi)$, as shown in Figure 8.

To complete the graph of $y = \csc x$, we observe that the values of $\csc x$ must repeat themselves at intervals of 2π. This is because $\csc x = 1/(\sin x)$, and the sine function has a period of 2π. In view of this we can draw the graph of $y = \csc x$ as shown in the box that follows.

PROPERTY SUMMARY The Cosecant Function: $y = \csc x$

Domain: All real numbers other than $0, \pm\pi, \pm 2\pi, \ldots$
Range: $(-\infty, -1] \cup [1, \infty)$
Period: 2π
Asymptotes: $x = 0, x = \pm\pi, x = \pm 2\pi, \ldots$
Intercepts: None

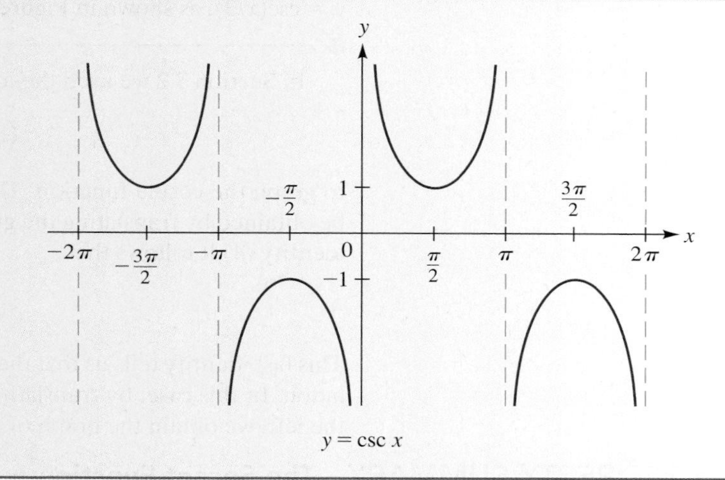

Note: The cosecant is an odd function so the graph of $y = \csc x$ is symmetric about the origin.

EXAMPLE 4 Graphing a function of the form $y = \csc Bx$

Graph the function $y = \csc(x/3)$ for one period.

SOLUTION

Since $\csc(x/3) = 1/\sin(x/3)$, it will be helpful first to graph one period of $y = \sin(x/3)$. This is done in Figure 9(a) using the techniques of Section 7.3. Note that the period of $y = \sin(x/3)$ is 6π. This is also the period of $y = \csc(x/3)$, because $\csc(x/3)$ and $\sin(x/3)$ are just reciprocals. The asymptotes for $y = \csc(x/3)$ occur when $\sin(x/3) = 0$. From Figure 9(a) we see that $\sin(x/3) = 0$ when $x = 0$, when $x = 3\pi$, and when $x = 6\pi$. These asymptotes are sketched in Figure 9(b). The colored points in Figure 9(b) indicate where the value of $\sin(x/3)$ is 1 or -1;

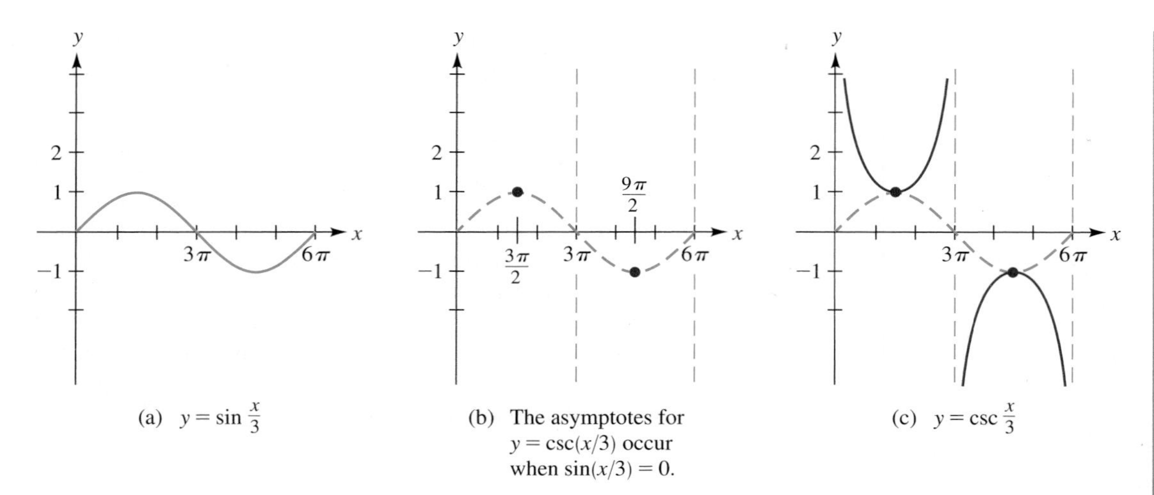

Figure 9

(a) $y = \sin \frac{x}{3}$

(b) The asymptotes for $y = \csc(x/3)$ occur when $\sin(x/3) = 0$.

(c) $y = \csc \frac{x}{3}$

the graph of $y = \csc(x/3)$ must pass through these points. (Why?) Finally, using the points and the asymptotes in Figure 9(b), we can sketch the graph of $y = \csc(x/3)$, as shown in Figure 9(c).

In Section 7.2 we used the identity

$$\cos x = \sin\left(x + \frac{\pi}{2}\right) \tag{4}$$

to graph the cosine function. This identity tells us that the graph of $y = \cos x$ can be obtained by translating the graph of $y = \sin x$ to the left by $\pi/2$ units. Now, from identity (4) it follows that

$$\sec x = \csc\left(x + \frac{\pi}{2}\right)$$

This last identity tells us that the graph of $y = \sec x$ can also be obtained by a translation. In this case, by translating the graph of $y = \csc x$ a distance of $\pi/2$ units to the left, we obtain the graph of $y = \sec x$, as shown in the box that follows.

PROPERTY SUMMARY The Secant Function: $y = \sec x$

Domain: All real numbers other than

$\pm\frac{\pi}{2}, \pm\frac{3\pi}{2}, \pm\frac{5\pi}{2}, \ldots$

Range: $(-\infty, -1] \cup [1, \infty)$

Period: 2π

Asymptotes: $x = \pm\frac{\pi}{2}, x = \pm\frac{3\pi}{2},$

$x = \pm\frac{5\pi}{2}, \ldots$

y-intercept: 1

x-intercept: None

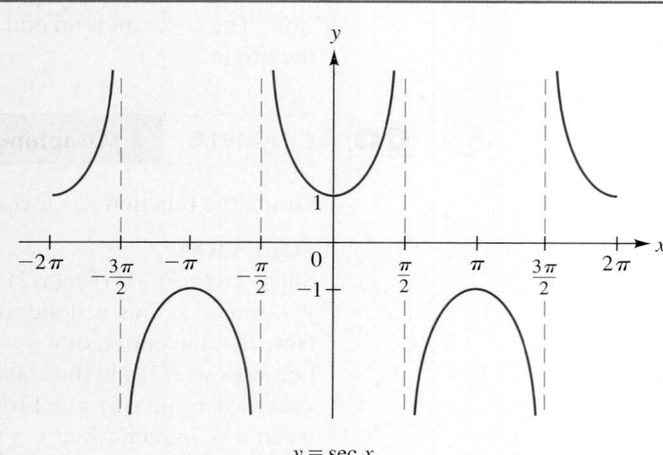

$y = \sec x$

Note: The secant is an even function so the graph of $y = \sec x$ is symmetric about the y-axis.

EXERCISE SET 7.5

A

In Exercises 1–12, graph each function for one period, and show (or specify) the intercepts and asymptotes.

1. (a) $y = \tan(x + \frac{\pi}{4})$
(b) $y = -\tan(x + \frac{\pi}{4})$

2. (a) $y = \tan(x - \frac{\pi}{3})$
(b) $y = -\tan(x - \frac{\pi}{3})$

3. (a) $y = \tan(x/3)$
(b) $y = -\tan(x/3)$

4. (a) $y = 2\tan \pi x$
(b) $y = -2\tan \pi x$

5. $y = \frac{1}{2}\tan(\pi x/2)$

6. $y = -\frac{1}{2}\tan 2\pi x$

7. $y = \cot(\pi x/2)$

8. $y = \cot 2\pi x$

9. $y = -\cot(x - \frac{\pi}{4})$

10. $y = \cot(x + \frac{\pi}{6})$

11. $y = \frac{1}{2}\cot 2x$

12. $y = -\frac{1}{2}\cot(x/2)$

Ⓖ **13.** Graph the function $y = \tan x$. Use a viewing rectangle that extends from -5 to 5 in both the x- and the y-directions. What are the exact values for the x-intercepts shown in your graph?

Ⓖ **14.** In graphing the tangent function in the text, we used the identity $\tan(x + \pi) = \tan x$. Check this identity by graphing the two equations $y = \tan x$ and $y = \tan(x + \pi)$ and noting that the graphs indeed appear to be identical.

In Exercises 15–20, graph each function. Adjust the viewing rectangle as necessary so that the graph is shown for at least two periods.

Ⓖ **15. (a)** $y = \tan(x/4)$
(b) $y = \tan(4x)$

Ⓖ **16. (a)** $y = \cot(2x)$
(b) $y = \cot(x/2)$

Ⓖ **17. (a)** $y = 0.5\tan \pi x$
(b) $y = 0.5\tan(\pi x + \frac{\pi}{3})$
(c) $y = 0.5\tan(\pi x + 1)$

Ⓖ **18. (a)** $y = 0.25\cot x$
(b) $y = 0.25\cot(x + \frac{\pi}{4})$
(c) $y = -0.25\cot(x + \frac{\pi}{4})$

Ⓖ **19. (a)** $y = 0.4\tan(x/2)$
(b) $y = 0.4\tan(x/3)$
(c) $y = 0.4\tan(x/5)$

Ⓖ **20. (a)** $y = 0.2\cot(3x)$
(b) $y = -0.2\cot(4x)$
(c) $y = -0.2\cot(12x)$

In Exercises 21–32, graph each function for one period, and show (or specify) the intercepts and asymptotes.

21. $y = \csc(x - \frac{\pi}{4})$

22. $y = \csc(x - \frac{\pi}{6})$

23. $y = -\csc(x/2)$

24. $y = 2\csc x$

25. $y = \frac{1}{3}\csc \pi x$

26. $y = -\frac{1}{2}\csc 2\pi x$

27. $y = -\sec x$

28. $y = -2\sec x$

29. $y = \sec(x - \pi)$

30. $y = \sec(x + 1)$

31. $y = 3\sec(\pi x/2)$

32. $y = -2\sec(\pi x/3)$

Ⓖ **33.** Graph the functions $y = \sin x$ and $y = \csc x$ in the standard viewing rectangle. [For $\csc x$, use $1 \div (\sin x)$.] Observe that $|\sin x| \le 1$, while $|\csc x| \ge 1$. At which points in the picture do we have $\sin x = \csc x$? Why? (*Hint*: Which two numbers are their own reciprocals?) There are no points where $\sin x = -\csc x$. Why?

Ⓖ **34.** Graph the functions $y = \cos x$ and $y = \sec x$ in the standard viewing rectangle. [For $\sec x$, use $1 \div (\cos x)$.] Observe that $|\cos x| \le 1$, while $|\sec x| \ge 1$. At which points in the picture do we have $\cos x = \sec x$? Why? (*Hint*: Which two numbers are their own reciprocals?) There are no points where $\cos x = -\sec x$. Why?

In Exercises 35–38, graph each function for two periods. Specify the intercepts and the asymptotes.

35. (a) $y = 2\sin(3\pi x - \frac{\pi}{6})$
(b) $y = 2\csc(3\pi x - \frac{\pi}{6})$

36. (a) $y = -\frac{1}{2}\sin(\pi x + \frac{\pi}{3})$
(b) $y = -\frac{1}{2}\csc(\pi x + \frac{\pi}{3})$

37. (a) $y = -3\cos(2\pi x - \frac{\pi}{4})$
(b) $y = -3\sec(2\pi x - \frac{\pi}{4})$

38. (a) $y = \cos(3x + \frac{\pi}{3})$
(b) $y = \sec(3x + \frac{\pi}{3})$

In Exercises 39–42, graph each pair of functions on the same set of axes. Adjust the viewing rectangle as necessary so that the graphs are shown for at least two periods.

Ⓖ **39.** $Y_1 = 0.6\sin(x/2)$
$Y_2 = 0.6\csc(x/2)$

Ⓖ **40.** $Y_1 = -1.2\sin(\pi x/3)$
$Y_2 = -1.2\csc(\pi x/3)$

Ⓖ **41.** $Y_1 = 3\cos(2x - \frac{\pi}{6})$
$Y_2 = 3\sec(2x - \frac{\pi}{6})$

Ⓖ **42.** $Y_1 = -1.5\cos(\pi x - \frac{\pi}{6})$
$Y_2 = -1.5\sec(\pi x - \frac{\pi}{6})$

B

For Exercises 43–46, six functions are defined as follows:

$$f(x) = \sin x \qquad g(x) = \csc x \qquad h(x) = \pi x - \frac{\pi}{6}$$

$$F(x) = \cos x \qquad G(x) = \sec x \qquad H(x) = \pi x + \frac{\pi}{4}$$

In each case, graph the indicated function for one period.

43. (a) $f \circ h$
(b) $g \circ h$

44. (a) $F \circ H$
(b) $G \circ H$

45. (a) $f \circ H$
(b) $g \circ H$

46. (a) $F \circ h$
(b) $G \circ h$

For Exercises 47–50, four functions are defined as follows:

$$f(x) = \csc x \qquad T(x) = \tan x$$

$$g(x) = \sec x \qquad A(x) = |x|$$

In each case, graph the indicated function over the interval $[-2\pi, 2\pi]$.

47. $A \circ T$

48. $A \circ g$

49. $A \circ f$

50. $f \circ A$

Ⓖ **51.** In this exercise we compare the graphs of $f(x) = \tan x$ and $g(x) = x^3$ on the open interval $-\pi/2 < x < \pi/2$. On the same set of axes, graph these two functions for $-\pi/2 < x < \pi/2$. Then answer the following questions.

(a) Which of the two graphs appears to be horizontal as it passes through the origin?

(b) When x is in the open interval $(0, \frac{\pi}{2})$, which quantity is larger, $\tan x$ or x^3? Why is your choice reasonable? (*Hint*: x^3 is finite at $x = \pi/2$.)

(c) When x is in the open interval $(-\frac{\pi}{2}, 0)$, which quantity is larger, $\tan x$ or x^3? Why is your choice reasonable?

Ⓖ **52.** Graph the two functions $y = \tan^2 x$ and $y = \sec^2 x - 1$. What do you observe? What does this demonstrate?

Ⓖ **53.** Graph the two functions $y = \cot^2 x$ and $y = \csc^2 x - 1$. What do you observe? What does this demonstrate?

Ⓖ **54. (a)** Graph the two functions $y = \tan x$ and $y = x$ in the standard viewing rectangle. Observe that $\tan x$ and x are very close to one another when x is close to zero. In fact, the approximation $\tan x \approx x$ is often used in applications when it is known that x is close to zero.

(b) To obtain a better view of $y = \tan x$ and $y = x$ near the origin, adjust the viewing rectangle so that x extends from $-\pi/2$ to $\pi/2$ and y extends from -2 to 2. Again, note that when x is close to zero, the values of $\tan x$ are indeed close to x. To see numerical evidence of this, complete the following table.

x	0.000123	0.01	0.05	0.1	0.2	0.3	0.4	0.5
$\tan x$								

x	−0.000123	−0.01	−0.05	−0.1	−0.2	−0.3	−0.4	−0.5
$\tan x$								

(c) The numerical evidence in part (b) suggests that for x positive and close to 0, $\tan x > x$. State the corresponding result for x negative and close to 0. Then on the same set of axes, draw, without a calculator, the graphs of $y = x$ and $y = \tan x$.

(d) If you study *Taylor polynomials* in calculus, you'll see that an even better approximation to $\tan x$ is

$$\tan x \approx x + \frac{1}{3}x^3 \qquad \text{when } x \text{ is close to } 0$$

Add the graph of $y = x + \frac{1}{3}x^3$ to the picture that you obtained in part (b). Describe what you see.

(e) To see numerical evidence of how well $\tan x$ is approximated by $x + \frac{1}{3}x^3$, add a third row to the table you worked out in part (b); in this third row, show the values for $x + \frac{1}{3}x^3$. When you've completed the table, note that the new values are much closer to $\tan x$ than were the values of x.

Ⓖ **55. (a)** Graph the equations $y = \tan x$ and $y = 2$ in the standard viewing rectangle.

(b) Use the graph to give a rough estimate for the smallest positive root of the equation $\tan x = 2$.

Answer: $x \approx 1$

(c) Use the graphing utility to determine the root more accurately, say, through the first four decimal places.

(d) Let r denote the root that you determined in part (c). Is the number $r + \pi$ also a root of the equation $\tan x = 2$?

Ⓖ **56. (a)** Graph the equations $y = \tan x$ and $y = x$ in the standard viewing rectangle. Use the graph to give a rough estimate for the smallest positive root of the equation $\tan x = x$.

Answer: Something between 4 and 5, call it 4.5

(b) Use the graphing utility to determine the root more accurately, say, through the first four decimal places.

(c) Let r denote the root that you determined in part (b). Is the number $r + \pi$ also a root of the equation $\tan x = x$?

57. In the text we used the identity

$$\cot s = -\tan\left(s - \frac{\pi}{2}\right)$$

to obtain the graph of $y = \cot x$ from that of $y = \tan x$. The following steps show one way to derive this identity. (Although the accompanying figure shows the angle with radian measure s terminating in the first quadrant, the proof is valid no matter where the angle terminates.) In the figure, $\overline{PO} \perp \overline{QO}$.

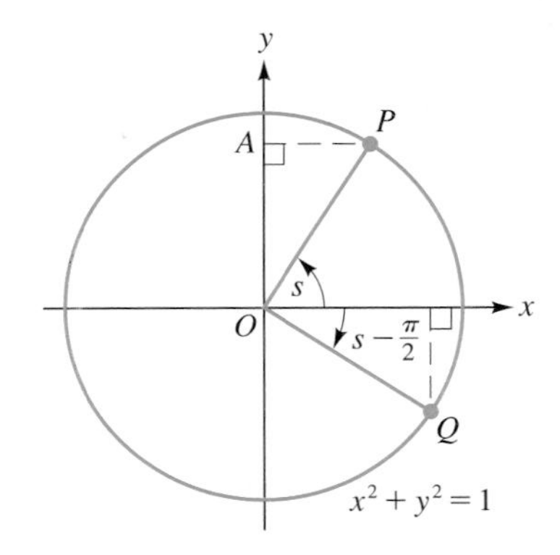

(a) Why are the coordinates of P and Q as follows?

$$P(\cos s, \sin s) \quad \text{and} \quad Q\left(\cos\left(s - \frac{\pi}{2}\right), \sin\left(s - \frac{\pi}{2}\right)\right)$$

(b) Using congruent triangles [and without reference to part (a)], explain why the y-coordinate of Q is the negative of the x-coordinate of P, and the x-coordinate of Q equals the y-coordinate of P.

(c) Use the results in parts (a) and (b) to conclude that

$$\sin\left(s - \frac{\pi}{2}\right) = -\cos s \quad \text{and} \quad \cos\left(s - \frac{\pi}{2}\right) = \sin s$$

(d) Use the result in part (c) to show that

$$\cot s = -\tan\left(s - \frac{\pi}{2}\right)$$

58. In the text we obtained the graph of the cotangent function from that of the tangent by means of the identity $\cot x = -\tan(x - \frac{\pi}{2})$. Verify this identity graphically by graphing the two functions $y = \cot x$ and $y = -\tan(x - \frac{\pi}{2})$ and noting that the graphs indeed appear to be identical.

59. In this exercise we verify the identity $\csc(\frac{\pi}{2} + s) = \csc(\frac{\pi}{2} - s)$. Refer to the figure at the right, in which $\angle AOB = \angle BOC = s$ radians. [Although the figure shows s in the interval $0 < s < \pi/2$, a similar proof will work for other intervals as well. (A proof that does not depend upon a picture can be given using the formula for $\sin(s + t)$, which is developed in Section 8.1.)]

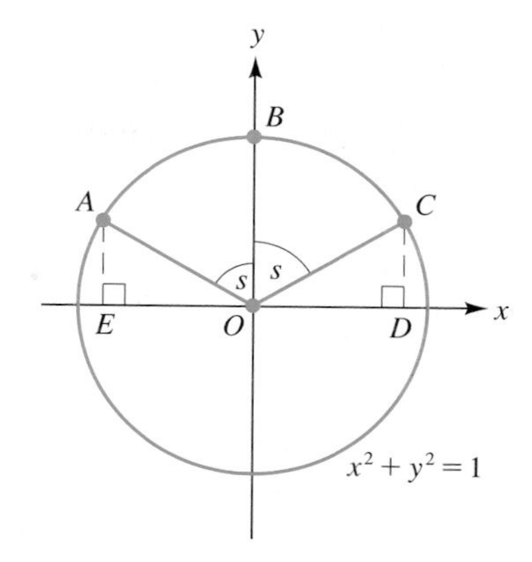

(a) Show that the triangles AOE and COD are congruent and, consequently, that $AE = CD$.

(b) Explain why the y-coordinates of the points C and A are $\sin(\frac{\pi}{2} - s)$ and $\sin(\frac{\pi}{2} + s)$, respectively.

(c) Use parts (a) and (b) to conclude that $\sin(\frac{\pi}{2} + s) = \sin(\frac{\pi}{2} - s)$. It follows from this that $\csc(\frac{\pi}{2} + s) = \csc(\frac{\pi}{2} - s)$, as required.

Chapter 7 Summary of Principal Formulas and Terms

Terms or formulas	Page reference	Comments
1. $\cos t$ $\sin t$ $\tan t$ $\sec t$ $\csc t$ $\cot t$	497	Let $P(x, y)$ denote the point on the unit circle such that the arc length from $(1, 0)$ is t. Or equivalently, let $P(x, y)$ denote the point where the terminal side of the angle with radian measure t intersects the unit circle. Then the six trigonometric functions of the real number t are defined as follows: $\cos t = x \qquad \sec t = \frac{1}{x} \quad (x \neq 0)$ $\sin t = y \qquad \csc t = \frac{1}{y} \quad (y \neq 0)$ $\tan t = \frac{y}{x} \quad (x \neq 0) \qquad \cot t = \frac{x}{y} \quad (y \neq 0)$
2. $\sec t = \frac{1}{\cos t} \quad \csc t = \frac{1}{\sin t}$ $\cot t = \frac{1}{\tan t} \quad \tan t = \frac{\sin t}{\cos t}$ $\cot t = \frac{\cos t}{\sin t}$	499	These five identities are direct consequences of the definitions of the trigonometric functions.

Terms or formulas	Page reference	Comments
3. $\sin^2 t + \cos^2 t = 1$ $\tan^2 t + 1 = \sec^2 t$ $\cot^2 t + 1 = \csc^2 t$	499	These are the three *Pythagorean identities.*
4. Even function, odd function	502	A function is an *even function* if $f(-t) = f(t)$ for every value of t in the domain of f. A function is an *odd function* if $f(-t) = -f(t)$ for every value of t in the domain of f.
5. $\cos(-t) = \cos t$ $\sin(-t) = -\sin t$ $\tan(-t) = -\tan t$	502	These three identities are sometimes referred to as the *opposite-angle identities.* In each case, a trigonometric function of $-t$ is expressed in terms of that same function of t. Cosine is an *even* function; sine and tangent are *odd* functions.
6. Periodic function	508	A function f is periodic if there is a positive number p such that the equation $f(x + p) = f(x)$ holds for all x in the domain of f. The smallest such number p is called the *period* of f. Important examples: The period of $y = \sin x$ is 2π; the period of $y = \cos x$ is 2π, the period of $y = \tan x$ is π.
7. Amplitude	509	Let m and M denote the smallest and the largest values, respectively, of the periodic function f. Then the *amplitude* of f is defined to be the number $\frac{1}{2}(M - m)$. Important examples: The amplitude for both $y = \sin x$ and $y = \cos x$ is 1; amplitude is not defined for the function $y = \tan x$.
8. Period $= \dfrac{2\pi}{B}$	526–527	This formula gives the period for the functions $y = A\sin(Bx - C)$ and $y = A\cos(Bx - C)$.
9. Phase shift $= \dfrac{C}{B}$	529	The phase shift serves as a guide in graphing functions of the form $y = A\sin(Bx - C)$ or $y = A\cos(Bx - C)$. For instance, to graph one complete cycle of $y = A\sin(Bx - C)$, first sketch one complete cycle of $y = A\sin Bx$, beginning at $x = 0$. Then draw a curve with exactly the same shape, but beginning at $x = C/B$ rather than $x = 0$. This will represent one cycle of $y = A\sin(Bx - C)$.
10. Simple harmonic motion	541	Oscillation described by an equation of the form $y = A\sin(Bx - C)$ or $y = A\cos(Bx - C)$ is referred to as *simple harmonic motion.* The standard example for simple harmonic motion is the motion of a mass on the end of a spring (neglecting friction), as depicted in Figures 1 and 3 in Section 7.4.
11. Period, frequency, and amplitude for simple harmonic motion	541–542	The amplitude for simple harmonic motion is the maximum displacement from the equilibrium position. The period is the time required for one complete cycle of the motion. The frequency f, measured in cycles per second, is computed from the equation $f = 1/\text{period}$.

Writing Mathematics

1. Say whether the statement is TRUE or FALSE. Write out your reason or reasons in complete sentences. If you draw a diagram to accompany your writing, be sure that you clearly label any parts of the diagram to which you refer.
   (a) If t is a real number less than π, then $\cos t$ is positive.
   (b) The equation $1 + \tan^2 t = \sec^2 t$ is true for every real number t.
   (c) If t is a positive real number, then $\cos^2 t$ is a positive real number.
   (d) Between any two consecutive turning points, the graph of $y = \sin x$ is decreasing.
   (e) If a and b are real numbers and $\ln(\sin a) = \ln(\sin b)$, then $\sin a = \sin b$.
   (f) If a and b are real numbers and $\ln(\sin a) = \ln(\sin b)$, then $a = b$.
   (g) If a and b are real numbers and $\sin(\ln a) = \sin(\ln b)$, then $\ln a = \ln b$.
2. Let $f(x) = x + \sin x$. This function has an interesting property: The iterates can be used to obtain better and better approximations to the number π. The following table, for instance, shows the first five iterates of $x_0 = 0.8$. As you can see, the fifth iterate is indeed an excellent approximation to π. In fact, the fifth iterate, x_5, agrees with π through the first 15 decimal places. [As you'll see in part (a), this behavior is not peculiar to $x_0 = 0.8$. Any value for x_0 between 0 and π will produce similar results.]

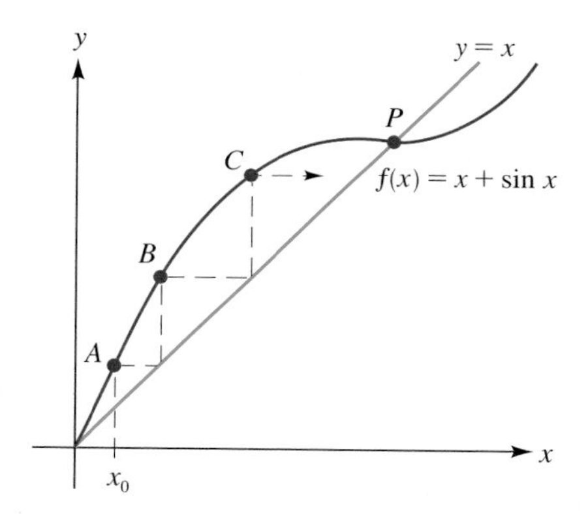

x_1	1.5173560908 . . .
x_2	2.5159285012 . . .
x_3	3.1015642481 . . .
x_4	3.1415819650 . . .
x_5	3.1415926535 . . .

Iterates of $x_0 = 0.8$ for the function $f(x) = x + \sin x$

   (a) Use your calculator to set up three tables of iterates, similar to the given table, using $x_0 = 0.9$, $x_0 = 1.5$, and $x_0 = 2$. Describe the results.
   (b) Use the figure in the margin to explain why the successive iterates approach the number π.

Chapter 7 Review Exercises

1. Evaluate the following.
   (a) $\sin(5\pi/3)$ **(b)** $\cot(11\pi/6)$
2. Graph the function $y = 3\cos 3\pi x$ over the interval $-\frac{1}{3} \le x \le \frac{1}{3}$. Specify the x-intercepts and the coordinates of the highest point on the graph.
3. Simplify the following expression:
$$\cos t - \cos(-t) + \sin t - \sin(-t)$$
4. In the expression $1/\sqrt{4 - t^2}$, make the substitution $t = 2\cos x$ and simplify the result. Assume that $0 < x < \pi$.
5. Graph the equation $y = \sec(2\pi x - 3)$ for one complete cycle.
6. If $\sec t = -5/3$ and $\pi < t < 3\pi/2$, compute $\cot t$.
7. Graph the function $y = -\sin(2x - \pi)$ for one complete period. Specify the amplitude, period, and phase shift.
8. Graph $y = -\frac{1}{2}\tan(\pi x/3)$ over one period.

In Exercises 9–20, evaluate each expression without using a calculator or tables.

9. $\cos \pi$ **10.** $\sin(-3\pi/2)$ **11.** $\csc(2\pi/3)$
12. $\tan(\pi/3)$ **13.** $\tan(11\pi/6)$ **14.** $\cos 0$
15. $\sin(\pi/6)$ **16.** $\sec(3\pi/4)$
17. $\cot(5\pi/4)$ **18.** $\tan(-7\pi/4)$
19. $\csc(-5\pi/6)$ **20.** $\sin^2(\pi/7) + \cos^2(\pi/7)$

Exercises 21–30 are calculator exercises. (Set your calculator to the radian mode.) In Exercises 21–26, where numerical answers are required, round your results to three decimal places.

21. Evaluate $\sin 1$. **22.** Evaluate $\cos 2$.
23. Evaluate $\sin(3\pi/2)$. **24.** Evaluate $\sin(0.78)$.
25. Evaluate $\sin(\sin 0.0123)$.
26. Evaluate $\sin[\sin(\sin 0.0123)]$.
27. Verify that $\sin^2 1776 + \cos^2 1776 = 1$.
28. Verify that $\sin 14 = 2 \sin 7 \cos 7$.
29. Verify that $\cos(0.5) = \cos^2(0.25) - \sin^2(0.25)$.
30. Verify that $\cos(0.3) = [\frac{1}{2}(1 + \cos 0.6)]^{1/2}$.
31. In the expression $\sqrt{25 - x^2}$, make the substitution $x = 5 \sin \theta$ $(0 < \theta < \frac{\pi}{2})$ and simplify the result.
32. In the expression $(49 + x^2)^{1/2}$, make the substitution $x = 7 \tan \theta$ $(0 < \theta < \frac{\pi}{2})$ and simplify the result.
33. In the expression $(x^2 - 100)^{1/2}$, make the substitution $x = 10 \sec \theta$ $(0 < \theta < \frac{\pi}{2})$ and simplify the result.
34. In the expression $(x^2 - 4)^{-3/2}$, make the substitution $x = 2 \sec \theta$ $(0 < \theta < \frac{\pi}{2})$ and simplify the result.
35. In the expression $(x^2 + 5)^{-1/2}$, make the substitution $x = \sqrt{5} \tan \theta$ $(0 < \theta < \frac{\pi}{2})$ and simplify the result.
36. If $\sin \theta = -5/13$ and $\pi < \theta < 3\pi/2$, compute $\cos \theta$.
37. If $\cos \theta = 8/17$ and $\sin \theta$ is negative, compute $\tan \theta$.
38. If $\sec \theta = -25/7$ and $\pi < \theta < 3\pi/2$, compute $\cot \theta$.

39. In the accompanying figure, P is the center of the circle, which has radius $\sqrt{2}$ units. If the radian measure of angle BPA is θ, express the area of the shaded region in terms of θ. Simplify your answer as much as possible.

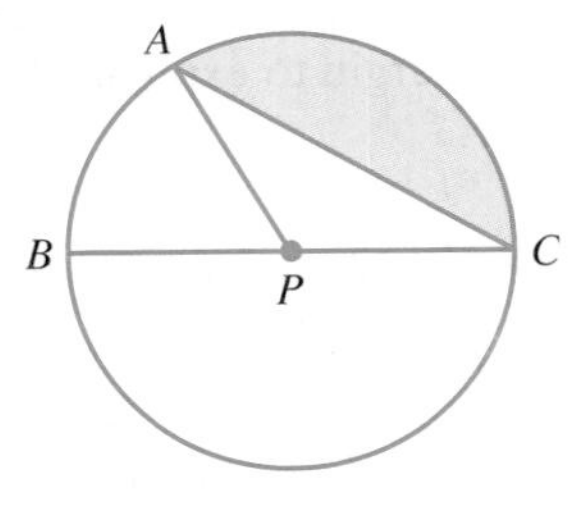

40. Express the area of the shaded region in the accompanying figure in terms of r and θ. (Assume that θ is in radians.)

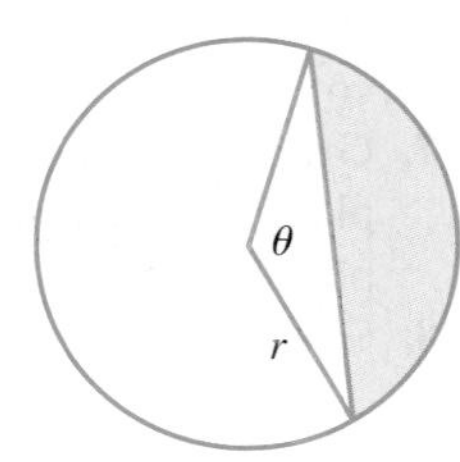

41. In the following figure, $ABCD$ is a square, each side of which is 1 cm. The two arcs are portions of circles with radii of 1 cm and with centers A and C. Find the area of the shaded region. *Hint:* Draw $\overline{BD}$ and use your result from Exercise 40.

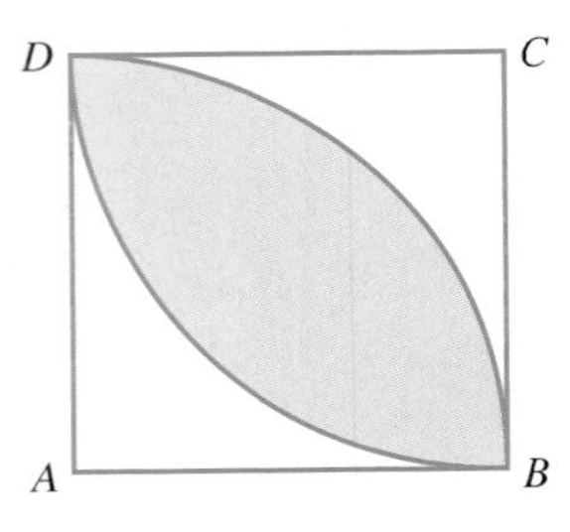

In Exercises 42–46, a function of the form $y = A \sin Bx$ or $y = A \cos Bx$ is graphed for one period. Determine the equation in each case. (Assume that $B > 0$.)

42.

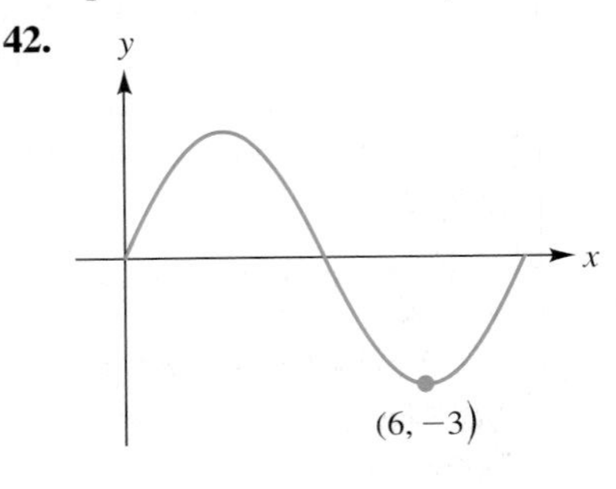

43.

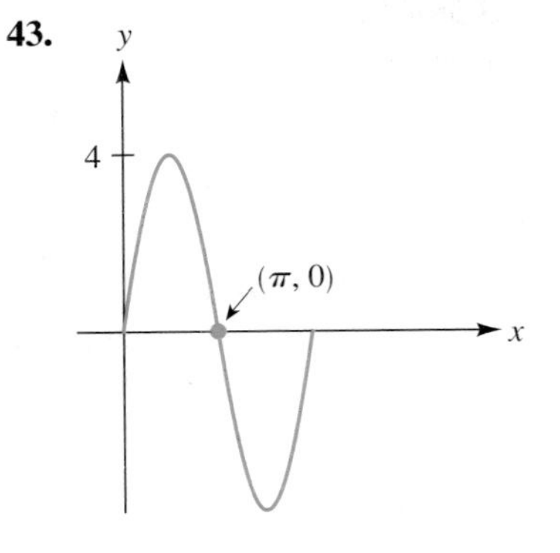

44.

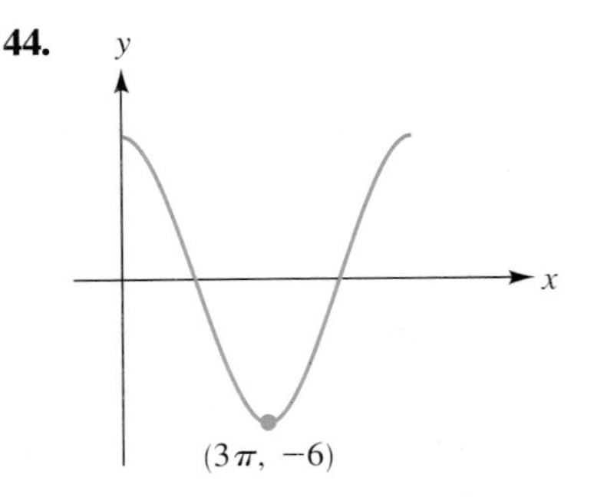

45.

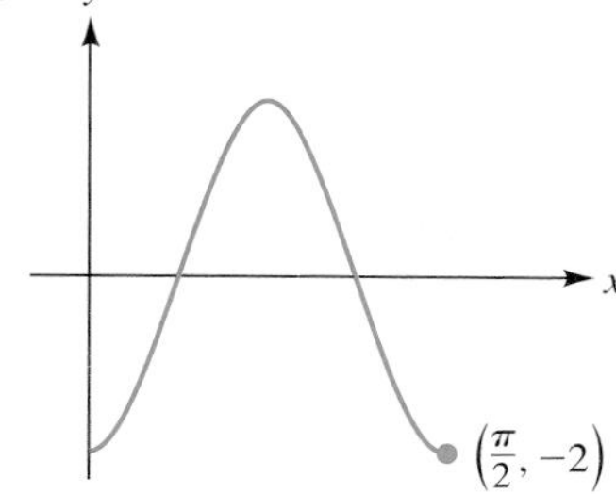

46.

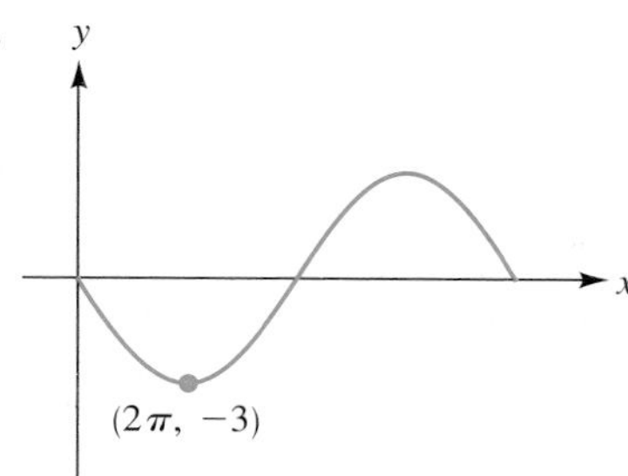

In Exercises 47–52, sketch the graph of each function for one complete cycle. In each case, specify the x-intercepts and the coordinates of the highest and lowest points on the graph.

47. $y = -3\cos 4x$

48. $y = 2\sin(3\pi x/4)$

49. $y = 2\sin\left(\frac{\pi x}{2} - \frac{\pi}{4}\right)$

50. $y = -\sin(x - 1)$

51. $y = 3\cos\left(\frac{\pi x}{3} - \frac{\pi}{3}\right)$

52. $y = -2\cos(x + \pi)$

In Exercises 53–56, sketch the graph of each function for one period.

53. **(a)** $y = \tan(\pi x/4)$
(b) $y = \cot(\pi x/4)$

54. **(a)** $y = 2\cot 2x$
(b) $y = 2\tan 2x$

55. **(a)** $y = 3\sec(x/4)$
(b) $y = 3\csc(x/4)$

56. **(a)** $y = \sec \pi x$
(b) $y = \csc \pi x$

57. A mass on a tabletop is attached to a spring, as shown in the figure. The coordinate system has been chosen so that the equilibrium position of the mass corresponds to $s = 0$. Assume that the mass moves in simple harmonic motion described by the equation $s = -2.5\cos(\pi t/8)$, where s is in centimeters and t is in seconds.

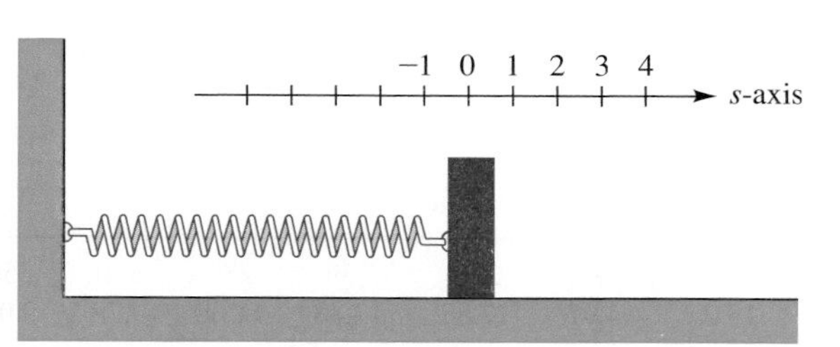

(a) Graph two complete cycles of the function $s = -2.5\cos(\pi t/8)$ beginning at $t = 0$. Specify the amplitude, period, and frequency of the motion.

(b) Use the graph to determine the times in this interval at which the mass is farthest from the equilibrium position.

(c) When during this interval of time is the mass passing through the equilibrium position?

58. In the following figure, the arc is a portion of a circle with center O and radius 1. Use the figure to prove the following result: If α and β are acute angles (radian measure) with $\beta > \alpha$, then $\sin\beta - \sin\alpha < \beta - \alpha$. *Hint*: The area of the segment of the circle determined by chord $\overline{AB}$ is less than the area of the segment determined by chord $\overline{AC}$.

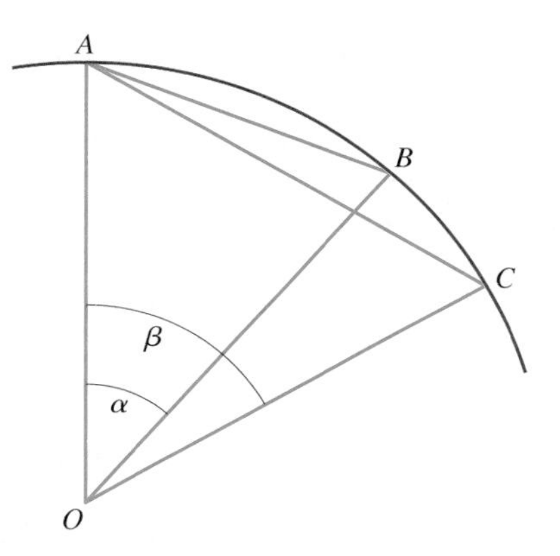

59. In the following figure, the arc is a portion of a circle with center O and radius 1, and $\angle OAE = \pi/2$. Use the figure to prove the following result: If α and β are acute angles (radian measure) with $\beta > \alpha$, then $\tan\beta - \tan\alpha > \beta - \alpha$. *Hint*: The area of region $ABDA$ is less than the area of region $ABCEDA$.

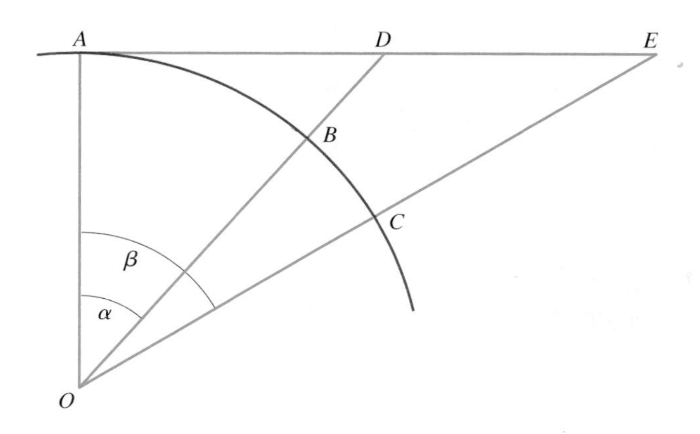

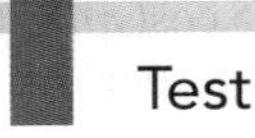

Chapter 7 Test

1. Evaluate each expression without using a calculator or tables.
   (a) $\cos(4\pi/3)$
   (b) $\csc(-5\pi/6)$
   (c) $\sin^2(3\pi/4) + \cos^2(3\pi/4)$
2. In the expression $1/\sqrt{16 - t^2}$, make the substitution $t = 4 \sin u$ and simplify the result. Assume that $0 < u < \pi/2$.
3. Graph the function $y = 0.5 \sec(4\pi x - 1)$ for one complete cycle.
4. Graph the function $y = -\sin(3x - \frac{\pi}{4})$ for one complete cycle. Specify the amplitude, period, and phase shift.
5. Graph the function $y = 3 \tan(\pi x/4)$ on the interval $0 \le x \le 4$.
6. A wheel rotates about its axis with an angular speed of 25 revolutions/sec.
   (a) Find the angular speed of the wheel in radians/sec.
   (b) Find the linear speed of a point on the wheel that is 5 cm from the center.
7. Prove that the following equation is an identity:
$$\frac{\cot\theta}{1 + \tan(-\theta)} + \frac{\tan\theta}{1 + \cot(-\theta)} = \cot\theta + \tan\theta + 1$$
8. A point moves in simple harmonic motion along the x-axis. The x-coordinate of the point at time t is given by $x = 10 \cos(\pi t/3)$, where t is in seconds and x is in centimeters.
   (a) Graph this function for two complete cycles, beginning at $t = 0$.
   (b) At what times during these two cycles is the point passing through the origin? At what times is the point farthest from the origin?
9. Evaluate each expression. (Show your work or supply reasons; *don't* use a calculator.)
   (a) $\sin^2 13 + \cos^2 13$
   (b) $\sin 5 + \sin(-5)$
   (c) $\tan 1 + \tan(-1 - 2\pi)$
10. The figure at the top of the next column shows the graph of $y = \sin x$ for $0 \le x \le 1.6$.

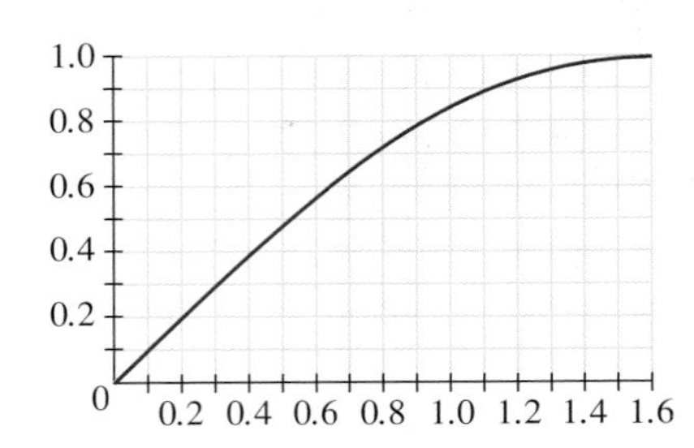

   (a) Use the graph to estimate, to the nearest tenth, a root of the equation $\sin x = 0.9$.
   (b) Use a calculator to obtain a more accurate value for the root in part (a). Round the answer to four decimal places.
   (c) Use the reference-angle concept and a calculator to find another root of the equation $\sin x = 0.9$ in the interval $0 \le x \le 2\pi$. Round the answer to four decimal places.
11. Table 1 shows length-of-day statistics for Boston over the year 1996. ("Length of day" refers to the number of hours that the sun is above the horizon.) Figure A shows a scatter plot based on this data. The numbers on the horizontal t-axis indicate days of the year: January 1 is $t = 1$, January 2 is $t = 2$, January 31 is $t = 31$, February 1 is $t = 32$, and so on. (The numbers in the top row of Table 1 are not as arbitrary as they might first appear; they correspond to the fifteenth day of the month.) Find a periodic function of the form $y = A \sin(Bt - C) + D$ whose values approximate the length-of-day values in Table 1. (Assume that the period of the function is 366 days; 1996 was a leap year.)

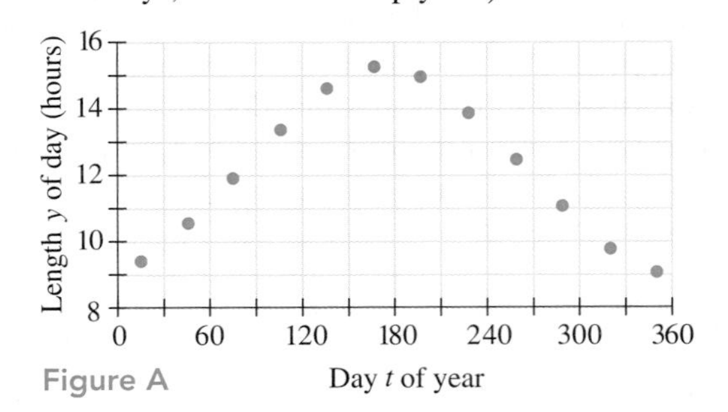

Figure A

TABLE 1
Length of day for Boston in 1996

Day t	15	46	75	106	136	167	197	228	259	289	320	350
Length y of Day (hours)	9.42	10.57	11.92	13.38	14.62	15.27	14.97	13.88	12.48	11.08	9.78	9.08

Source: Robert B. Thomas, *The Old Farmers' Almanac, 1996* (Dublin, New Hampshire: Yankee Publishing, Inc., 1995).

8 CHAPTER Analytical Trigonometry

. . . through the improvements in algebraic symbolism . . . trigonometry became, in the 17th century, largely an analytic science, and as such it entered the field of higher mathematics. —David Eugene Smith in *History of Mathematics* (New York: Ginn and Company, 1925)

This chapter is devoted to some of the more algebraic (as opposed to geometric) portions of trigonometry. In Section 8.1 we develop six basic identities known as the *addition formulas.* Then, in the next two sections, we consider a number of identities that follow directly from these addition formulas. In Section 8.4 we return to a topic that was introduced briefly in the previous chapter: solving trigonometric equations. In solving many of these equations, we'll use the identities developed in the previous sections. We also make use of the inverse trigonometric functions that were introduced in the previous chapter (in Section 7.2). In the last section of this chapter, Section 8.5, we take a more careful look at the inverse trigonometric functions and their properties.

8.1 THE ADDITION FORMULAS

It has long been recognized that the addition formulas are the heart of trigonometry. Indeed, Professor Rademacher and others have shown that the entire body of trigonometry can be derived from the assumption that there exist functions S and C such that

1. $S(x - y) = S(x)C(y) - C(x)S(y)$
2. $C(x - y) = C(x)C(y) + S(x)S(y)$
3. $\lim_{x \to 0^+} \frac{S(x)}{x} = 1$

—Professor Frederick H. Young in his article, "The Addition Formulas" from *The Mathematics Teacher,* vol. L (1957), pp. 45–48.

For any real numbers r, s, and t it is always true that $r(s + t) = rs + rt$. This is the so-called **distributive law** for real numbers. If f is a function, however, it is not true in general that $f(s + t) = f(s) + f(t)$. For example, consider the cosine function. It is not true in general that $\cos(s + t) = \cos s + \cos t$. For instance, with $s = \pi/6$ and $t = \pi/3$ we have

$$\cos\left(\frac{\pi}{6} + \frac{\pi}{3}\right) \stackrel{?}{=} \cos\frac{\pi}{6} + \cos\frac{\pi}{3}$$
$$\cos\frac{\pi}{2} \stackrel{?}{=} \cos\frac{\pi}{6} + \cos\frac{\pi}{3}$$
$$0 \stackrel{?}{=} \frac{\sqrt{3}}{2} + \frac{1}{2} \quad \text{(No!)}$$

In this section we will see just what $\cos(s + t)$ does equal. The correct formula for $\cos(s + t)$ is one of a group of important trigonometric identities called the **addition formulas.** We begin with the four addition formulas for sine and cosine.

The Addition Formulas for Sine and Cosine

$$\sin(s + t) = \sin s \cos t + \cos s \sin t$$
$$\sin(s - t) = \sin s \cos t - \cos s \sin t$$
$$\cos(s + t) = \cos s \cos t - \sin s \sin t$$
$$\cos(s - t) = \cos s \cos t + \sin s \sin t$$

Our strategy for deriving these formulas will be as follows. First we'll prove the fourth formula, which takes some effort. The other three formulas are then relatively easy to derive from the fourth one. As background for our discussion, you will need to recall that the distance d between two points (x_1, y_1) and (x_2, y_2) is given by $d = \sqrt{(x_2 - x_1)^2 + (y_2 - y_1)^2}$.

To prove the fourth formula in the box, we use Figure 1. The idea behind the proof is as follows.* We begin in Figure 1(a) with the unit circle and the angles s, t, and $s - t$. Then we rotate $\triangle OPQ$ about the origin until the point P coincides with the point $(1, 0)$, as indicated in Figure 1(b). Although this rotation changes the coordinates for the points P and Q, it certainly has no effect upon the length of the line segment $\overline{PQ}$. Thus whether we calculate PQ using the coordinates in Figure 1(a) or those in Figure 1(b), the results must be the same. As you'll see, by equating the two expressions for PQ, we will obtain the required formula.

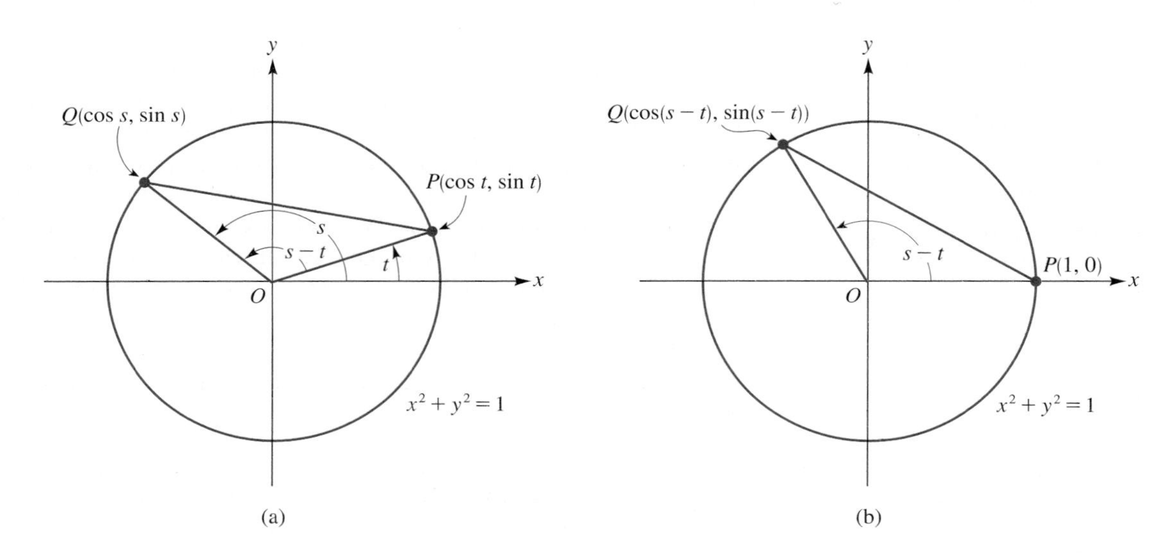

Figure 1

Applying the distance formula in Figure 1(a), we have

$$\begin{aligned} PQ &= \sqrt{(\cos s - \cos t)^2 + (\sin s - \sin t)^2} \\ &= \sqrt{\cos^2 s - 2\cos s\cos t + \cos^2 t + \sin^2 s - 2\sin s\sin t + \sin^2 t} \\ &= \sqrt{2 - 2\cos s\cos t - 2\sin s\sin t} \quad \text{(Why?)} \end{aligned} \tag{1}$$

Next, applying the distance formula in Figure 1(b), we have

$$\begin{aligned} PQ &= \sqrt{[\cos(s - t) - 1]^2 + [\sin(s - t) - 0]^2} \\ &= \sqrt{\cos^2(s - t) - 2\cos(s - t) + 1 + \sin^2(s - t)} \end{aligned}$$

The right-hand side of this last equation can be simplified by using the fact that $\cos^2(s - t) + \sin^2(s - t) = 1$. We then have

$$PQ = \sqrt{2 - 2\cos(s - t)} \tag{2}$$

*This idea for the proof can be traced back to the great French mathematician Augustin-Louis Cauchy (1789–1857).

From equations (1) and (2) we conclude that

$$2 - 2\cos(s - t) = 2 - 2\cos s \cos t - 2\sin s \sin t$$
$$-2\cos(s - t) = -2\cos s \cos t - 2\sin s \sin t$$
$$\cos(s - t) = \cos s \cos t + \sin s \sin t$$

This completes the proof of the fourth addition formula. Before deriving the other three addition formulas, let's look at some applications of this result.

EXAMPLE 1 Using an addition formula: the cosine of a difference

Simplify the expression $\cos 2\theta \cos \theta + \sin 2\theta \sin \theta$.

SOLUTION

According to the identity that we just proved, we have

$$\cos 2\theta \cos \theta + \sin 2\theta \sin \theta = \cos(2\theta - \theta) = \cos \theta$$

Thus the given expression is equal to $\cos \theta$.

EXAMPLE 2 The cosine of a difference

Simplify $\cos(\theta - \pi)$.

SOLUTION

We use the formula for $\cos(s - t)$ with s and t replaced by θ and π, respectively. This yields

$$\begin{aligned} \cos(s - t) &= \cos s \cos t + \sin s \sin t \\ &\updownarrow \quad \updownarrow \\ \cos(\theta - \pi) &= \cos \theta \cos \pi + \sin \theta \sin \pi \\ &= (\cos \theta)(-1) + (\sin \theta)(0) = -\cos \theta \end{aligned}$$

Thus the required simplification is $\cos(\theta - \pi) = -\cos \theta$.

In Example 2 we found that $\cos(\theta - \pi) = -\cos \theta$. This type of identity is often referred to as a **reduction formula.** The next example develops two basic reduction formulas that we will need to use later in this section.

EXAMPLE 3 Cofunction identities

Prove the following identities:

(a) $\cos\left(\frac{\pi}{2} - \alpha\right) = \sin \alpha$; **(b)** $\sin\left(\frac{\pi}{2} - \beta\right) = \cos \beta$.

SOLUTION

(a) $$\begin{aligned} \cos\left(\frac{\pi}{2} - \alpha\right) &= \cos \frac{\pi}{2} \cos \alpha + \sin \frac{\pi}{2} \sin \alpha \\ &= (0)\cos \alpha + (1)\sin \alpha = \sin \alpha \end{aligned}$$

This proves the identity. Incidentally, if we use degree measure instead of radian measure, this identity states that $\cos(90^\circ - \alpha) = \sin \alpha$ as we saw in Section 6.5.

(b) Since the identity $\cos(\frac{\pi}{2} - \alpha) = \sin \alpha$ holds for all values of α, we can simply replace α by the quantity $\frac{\pi}{2} - \beta$ to obtain

$$\cos\left[\frac{\pi}{2} - \left(\frac{\pi}{2} - \beta\right)\right] = \sin\left(\frac{\pi}{2} - \beta\right)$$

$$\cos\left(\frac{\pi}{2} - \frac{\pi}{2} + \beta\right) = \sin\left(\frac{\pi}{2} - \beta\right)$$

$$\cos \beta = \sin\left(\frac{\pi}{2} - \beta\right)$$

This proves the identity.

The two identities in Example 3 are worth memorizing. For simplicity we replace both α and β by the single θ to get the following formulas.

$$\cos\left(\frac{\pi}{2} - \theta\right) = \sin \theta \quad \text{and} \quad \sin\left(\frac{\pi}{2} - \theta\right) = \cos \theta$$

In the formulas in the preceding box, the arguments on the left-hand side and right-hand side, $\pi/2 - \theta$, and θ, respectively, add up to $\pi/2$, so the angles or numbers are complementary. Then the boxed formulas express the idea that cosine and sine are *cofunctions:* The function value of one function at a number is equal to the cofunction's value at the complementary number. Similar formulas hold for tangent and cotangent as well as for secant and cosecant.

In Examples 2 and 3, radian measure was used. However, if you look back at Figure 1 and the derivation of the formula for $\cos(s - t)$, you will see that the derivation makes no reference, implicit or explicit, to a specific system of angle measurement. (For instance, the derivation does not involve the formula $s = r\theta$, which does require radian measure.) Thus the formula for $\cos(s - t)$ is also valid when angles are measured in degrees. In the next example we apply the formula in just such a case.

EXAMPLE 4 Using the cosine of a difference

Use the formula for $\cos(s - t)$ to determine the exact value of $\cos 15^\circ$.

SOLUTION

First observe that $15^\circ = 45^\circ - 30^\circ$. Then we have

$$\begin{aligned} \cos 15^\circ &= \cos(45^\circ - 30^\circ) \\ &= \cos 45^\circ \cos 30^\circ + \sin 45^\circ \sin 30^\circ \quad && \text{using the formula for } \cos(s - t) \text{ with } s = 45^\circ \text{ and } t = 30^\circ \\ &= \left(\frac{\sqrt{2}}{2}\right)\left(\frac{\sqrt{3}}{2}\right) + \left(\frac{\sqrt{2}}{2}\right)\left(\frac{1}{2}\right) \\ &= \frac{\sqrt{6}}{4} + \frac{\sqrt{2}}{4} = \frac{\sqrt{6} + \sqrt{2}}{4} \end{aligned}$$

Thus the exact value of $\cos 15^\circ$ is $(\sqrt{6} + \sqrt{2})/4$.

Now let's return to our derivations of the addition formulas. Using the fourth addition formula, we can easily derive the third formula as follows. In the formula

$$\cos(s - t) = \cos s \cos t + \sin s \sin t$$

we replace t by the quantity $-t$. This is permissible because the formula holds for all real numbers. We obtain

$$\cos[s - (-t)] = \cos s \cos(-t) + \sin s \sin(-t)$$

On the right-hand side of this equation we can use the identities developed for $\cos(-t)$ and $\sin(-t)$ in Section 7.1. Doing this yields

$$\cos(s + t) = (\cos s)(\cos t) + (\sin s)(-\sin t)$$

which is equivalent to

$$\cos(s + t) = \cos s \cos t - \sin s \sin t$$

This is the third addition formula, as we wished to prove.

Next we derive the formula for $\sin(s + t)$. We have

$$\sin(s + t) = \cos\left[\frac{\pi}{2} - (s + t)\right] \qquad \text{replacing } \theta \text{ by } s + t \text{ in the identity } \sin\theta = \cos(\tfrac{\pi}{2} - \theta)$$

$$= \cos\left[\left(\frac{\pi}{2} - s\right) - t\right]$$

$$= \cos\left(\frac{\pi}{2} - s\right)\cos t + \sin\left(\frac{\pi}{2} - s\right)\sin t \quad \text{(Why?)}$$

$$= \sin s \cos t + \cos s \sin t \quad \text{(Why?)}$$

This proves the first addition formula.

Finally, we can use the first addition formula to prove the second one as follows:

$$\begin{aligned} \sin(s - t) &= \sin[s + (-t)] \\ &= \sin s \cos(-t) + \cos s \sin(-t) \\ &= (\sin s)(\cos t) + (\cos s)(-\sin t) \\ &= \sin s \cos t - \cos s \sin t \end{aligned}$$

This completes the proofs of the four addition formulas for sine and cosine.

EXAMPLE 5 Using an addition formula to calculate a value of a trigonometric function

If $\sin s = 3/5$ $(0 < s < \frac{\pi}{2})$ and $\sin t = -\sqrt{3}/4$ $(\pi < t < \frac{3\pi}{2})$, compute $\sin(s - t)$.

SOLUTION

$$\sin(s - t) = \underbrace{\sin s}_{\text{given as } 3/5} \cos t - \cos s \underbrace{\sin t}_{\text{given as } -\sqrt{3}/4} \qquad (3)$$

In view of equation (3) we need to find only $\cos t$ and $\cos s$. These can be determined by using the Pythagorean identity $\cos^2\theta = 1 - \sin^2\theta$. We have

$$\cos^2 t = 1 - \sin^2 t$$
$$= 1 - \left(\frac{-\sqrt{3}}{4}\right)^2 = 1 - \frac{3}{16} = \frac{13}{16}$$

Therefore

$$\cos t = \frac{\sqrt{13}}{4} \quad \text{or} \quad \cos t = -\frac{\sqrt{13}}{4}$$

We choose the negative value here for cosine, since it is given that $\pi < t < 3\pi/2$. Thus

$$\cos t = -\frac{\sqrt{13}}{4}$$

Similarly, to find $\cos s$, we have

$$\cos^2 s = 1 - \sin^2 s = 1 - \left(\frac{3}{5}\right)^2 = \frac{16}{25}$$

Therefore

$$\cos s = \frac{4}{5} \qquad \cos s \text{ is positive, since } 0 < s < \pi/2$$

Finally, we substitute the values we've obtained for $\cos t$ and $\cos s$, along with the given data, back into equation (3). This yields

$$\sin(s - t) = \left(\frac{3}{5}\right)\left(\frac{-\sqrt{13}}{4}\right) - \left(\frac{4}{5}\right)\left(\frac{-\sqrt{3}}{4}\right)$$
$$= \frac{-3\sqrt{13}}{20} + \frac{4\sqrt{3}}{20} = \frac{-3\sqrt{13} + 4\sqrt{3}}{20}$$

In the box that follows, we list the two addition formulas for the tangent function. As you'll see, these two formulas follow directly from the addition formulas for sine and cosine.

Addition Formulas for Tangent

$$\tan(s + t) = \frac{\tan s + \tan t}{1 - \tan s \tan t}$$
$$\tan(s - t) = \frac{\tan s - \tan t}{1 + \tan s \tan t}$$

Note: These formulas are valid only for values of s and t that are in the domain of the tangent function and for which the denominators are nonzero.

To prove the formula for $\tan(s + t)$, we begin with

$$\tan(s + t) = \frac{\sin(s + t)}{\cos(s + t)}$$
$$= \frac{\sin s \cos t + \cos s \sin t}{\cos s \cos t - \sin s \sin t} \qquad (4)$$

Now we divide both numerator and denominator on the right-hand side of equation (4) by the quantity $\cos s \cos t$. This yields

$$\tan(s+t) = \frac{\dfrac{\sin s \cancel{\cos t}}{\cos s \cancel{\cos t}} + \dfrac{\cancel{\cos s} \sin t}{\cancel{\cos s} \cos t}}{\dfrac{\cancel{\cos s \cos t}}{\cancel{\cos s \cos t}} - \dfrac{\sin s \sin t}{\cos s \cos t}} = \frac{\tan s + \tan t}{1 - \tan s \tan t}$$

This proves the formula for $\tan(s+t)$. The formula for $\tan(s-t)$ can be deduced from this with the aid of the identity $\tan(-t) = -\tan t$, which was derived in Section 7.1. We have

$$\begin{aligned}
\tan(s-t) &= \tan[s + (-t)] \\
&= \frac{\tan s + \tan(-t)}{1 - \tan s \tan(-t)} = \frac{\tan s + (-\tan t)}{1 - (\tan s)(-\tan t)} \\
&= \frac{\tan s - \tan t}{1 + \tan s \tan t} \qquad \text{as required}
\end{aligned}$$

EXAMPLE 6 Using an addition formula for tangent

Simplify the expression $\dfrac{\tan \frac{\pi}{9} + \tan \frac{2\pi}{9}}{1 - \tan \frac{\pi}{9} \tan \frac{2\pi}{9}}$.

SOLUTION

$$\begin{aligned}
\frac{\tan \frac{\pi}{9} + \tan \frac{2\pi}{9}}{1 - \tan \frac{\pi}{9} \tan \frac{2\pi}{9}} &= \tan\left(\frac{\pi}{9} + \frac{2\pi}{9}\right) \qquad \text{using the formula for } \tan(s+t) \\
&= \tan \frac{3\pi}{9} = \tan \frac{\pi}{3} = \sqrt{3}
\end{aligned}$$

EXAMPLE 7 Using an addition formula for tangent

Compute $\tan \frac{\pi}{12}$, using the fact that $\frac{\pi}{12} = \frac{\pi}{3} - \frac{\pi}{4}$.

SOLUTION

$$\begin{aligned}
\tan \frac{\pi}{12} &= \tan\left(\frac{\pi}{3} - \frac{\pi}{4}\right) \\
&= \frac{\tan \frac{\pi}{3} - \tan \frac{\pi}{4}}{1 + \tan \frac{\pi}{3} \tan \frac{\pi}{4}} \qquad \text{using the formula for } \tan(s-t) \text{ with } s = \pi/3 \text{ and } t = \pi/4 \\
&= \frac{\sqrt{3} - 1}{1 + \sqrt{3}(1)}
\end{aligned}$$

So $\tan \dfrac{\pi}{12} = \dfrac{\sqrt{3} - 1}{\sqrt{3} + 1}$. We can write this answer in a more compact form by rationalizing the denominator. As you can check, the result is $\tan \dfrac{\pi}{12} = 2 - \sqrt{3}$.

EXERCISE SET 8.1

A

In Exercises 1–10, use the addition formulas for sine and cosine to simplify the expression.

1. $\sin\theta\cos 2\theta + \cos\theta\sin 2\theta$
2. $\sin\frac{\pi}{6}\cos\frac{\pi}{3} + \cos\frac{\pi}{6}\sin\frac{\pi}{3}$
3. $\sin 3\theta\cos\theta - \cos 3\theta\sin\theta$
4. $\sin 110°\cos 20° - \cos 110°\sin 20°$
5. $\cos 2u\cos 3u - \sin 2u\sin 3u$
6. $\cos 2u\cos 3u + \sin 2u\sin 3u$
7. $\cos\frac{2\pi}{9}\cos\frac{\pi}{18} + \sin\frac{2\pi}{9}\sin\frac{\pi}{18}$
8. $\cos\frac{3\pi}{10}\cos\frac{\pi}{5} - \sin\frac{3\pi}{10}\sin\frac{\pi}{5}$
9. $\sin(A+B)\cos A - \cos(A+B)\sin A$
10. $\cos(s-t)\cos t - \sin(s-t)\sin t$

In Exercises 11–14, simplify each expression (as in Example 2).

11. $\sin(\theta - \frac{3\pi}{2})$
12. $\cos(\frac{3\pi}{2} + \theta)$
13. $\cos(\theta + \pi)$
14. $\sin(\theta - \pi)$
15. Expand $\sin(t + 2\pi)$ using the appropriate addition formula, and check to see that your answer agrees with the formula in the first box on page 504.
16. Follow the directions in Exercise 15, but use $\cos(t + 2\pi)$.
17. Use the formula for $\cos(s + t)$ to compute the exact value of $\cos 75°$.
18. Use the formula for $\sin(s - t)$ to compute the exact value of $\sin\frac{\pi}{12}$.
19. Use the formula for $\sin(s + t)$ to find $\sin\frac{7\pi}{12}$.
20. Determine the exact value of **(a)** $\sin 105°$ and **(b)** $\cos 105°$.

In Exercises 21–24, use the addition formulas for sine and cosine to simplify each expression.

21. $\sin(\frac{\pi}{4} + s) - \sin(\frac{\pi}{4} - s)$
22. $\sin(t + \frac{\pi}{6}) - \sin(t - \frac{\pi}{6})$
23. $\cos(\frac{\pi}{3} - \theta) - \cos(\frac{\pi}{3} + \theta)$
24. $\cos(\theta - \frac{\pi}{4}) + \cos(\theta + \frac{\pi}{4})$

In Exercises 25–28, compute the indicated quantity using the following data.

$$\sin\alpha = \frac{12}{13} \quad \text{where } \frac{\pi}{2} < \alpha < \pi$$

$$\cos\beta = -\frac{3}{5} \quad \text{where } \pi < \beta < \frac{3\pi}{2}$$

$$\cos\theta = \frac{7}{25} \quad \text{where } -2\pi < \theta < -\frac{3\pi}{2}$$

25. **(a)** $\sin(\alpha + \beta)$ **(b)** $\cos(\alpha + \beta)$
26. **(a)** $\sin(\alpha - \beta)$ **(b)** $\cos(\alpha - \beta)$
27. **(a)** $\sin(\theta - \beta)$ **(b)** $\sin(\theta + \beta)$
28. **(a)** $\cos(\alpha + \theta)$ **(b)** $\cos(\alpha - \theta)$
29. Suppose that $\sin\theta = 1/5$ and $0 < \theta < \pi/2$.
    (a) Compute $\cos\theta$.
    (b) Compute $\sin 2\theta$. *Hint:* $\sin 2\theta = \sin(\theta + \theta)$
30. Suppose that $\cos\theta = 12/13$ and $3\pi/2 < \theta < 2\pi$.
    (a) Compute $\sin\theta$.
    (b) Compute $\cos 2\theta$. *Hint:* $\cos 2\theta = \cos(\theta + \theta)$
31. Given $\tan\theta = -2/3$, where $\pi/2 < \theta < \pi$, and $\csc\beta = 2$, where $0 < \beta < \pi/2$, find $\sin(\theta + \beta)$ and $\cos(\beta - \theta)$.
32. Given $\sec s = 5/4$, where $\sin s < 0$, and $\cot t = -1$, where $\pi/2 < t < \pi$, find $\sin(s - t)$ and $\cos(s + t)$.

In Exercises 33–36, prove that each equation is an identity.

33. $\sin(t + \frac{\pi}{4}) = (\sin t + \cos t)/\sqrt{2}$
34. $\cos(t + \frac{\pi}{4}) = (\cos t - \sin t)/\sqrt{2}$
35. $\sin(t + \frac{\pi}{4}) + \cos(t + \frac{\pi}{4}) = \sqrt{2}\cos t$
36. $\sec(\alpha + \beta) = \dfrac{\sec\alpha\sec\beta}{1 - \tan\alpha\tan\beta}$

In Exercises 37–40, use the given information to compute $\tan(s + t)$ and $\tan(s - t)$.

37. $\tan s = 2$ and $\tan t = 3$
38. $\tan s = 1/2$ and $\tan t = 1/3$
39. $s = 3\pi/4$ and $\tan t = -4$
40. $s = 7\pi/4$ and $\tan t = -2$

In Exercises 41–46, use the addition formulas for tangent to simplify each expression.

41. $\dfrac{\tan t + \tan 2t}{1 - \tan t\tan 2t}$
42. $\dfrac{\tan\frac{\pi}{5} - \tan\frac{\pi}{30}}{1 + \tan\frac{\pi}{5}\tan\frac{\pi}{30}}$
43. $\dfrac{\tan 70° - \tan 10°}{1 + \tan 70°\tan 10°}$
44. $\dfrac{2\tan\frac{\pi}{12}}{1 - \tan^2\frac{\pi}{12}}$
45. $\dfrac{\tan(x - y) + \tan y}{1 - \tan(x - y)\tan y}$
46. $[\tan(\theta + \pi)][\tan(\theta - \pi)] + 1$
47. Compute $\tan\frac{7\pi}{12}$ and rationalize the answer.
    Hint: $\frac{7\pi}{12} = \frac{\pi}{3} + \frac{\pi}{4}$
48. Compute $\tan 15°$ using the fact that $15° = 45° - 30°$. Then check that your answer is consistent with the result in Example 7.

B

In Exercises 49–58, prove that each equation is an identity.

49. $\dfrac{\sin(s + t)}{\cos s\cos t} = \tan s + \tan t$
50. $\dfrac{\cos(s - t)}{\cos s\sin t} = \cot t + \tan s$
51. $\cos(A - B) - \cos(A + B) = 2\sin A\sin B$
52. $\sin(A - B) + \sin(A + B) = 2\sin A\cos B$
53. $\cos(A + B)\cos(A - B) = \cos^2 A - \sin^2 B$
54. $\sin(A + B)\sin(A - B) = \cos^2 B - \cos^2 A$
55. $\cos(\alpha + \beta)\cos\beta + \sin(\alpha + \beta)\sin\beta = \cos\alpha$
56. $\cos(\theta + \frac{\pi}{4}) + \sin(\theta - \frac{\pi}{4}) = 0$
57. $\tan 2\theta = \dfrac{2\tan\theta}{1 - \tan^2\theta}$
58. $\tan(\frac{\pi}{4} + \theta) - \tan(\frac{\pi}{4} - \theta) = 2\tan 2\theta$
    Hint: Use the addition formulas for tangent and the result in Exercise 57.

In Exercises 59–61, you are asked to derive expressions for the average rates of change of the functions $\sin x$, $\cos x$, *and* $\tan x$. *In each case, assume that the interval is* $[x, x + h]$. *(The results are used in calculus in the study of derivatives.)*

59. Let $f(x) = \sin x$. Show that

$$\frac{\Delta f}{\Delta x} = (\sin x)\left(\frac{\cos h - 1}{h}\right) + (\cos x)\left(\frac{\sin h}{h}\right)$$

60. Let $g(x) = \cos x$. Show that

$$\frac{\Delta g}{\Delta x} = (\cos x)\left(\frac{\cos h - 1}{h}\right) - (\sin x)\left(\frac{\sin h}{h}\right)$$

61. Let $T(x) = \tan x$. Show that

$$\frac{\Delta T}{\Delta x} = \frac{\tan h}{h} \cdot \frac{\sec^2 x}{1 - \tan h \tan x}$$

62. Let θ be the acute angle defined by the following figure.

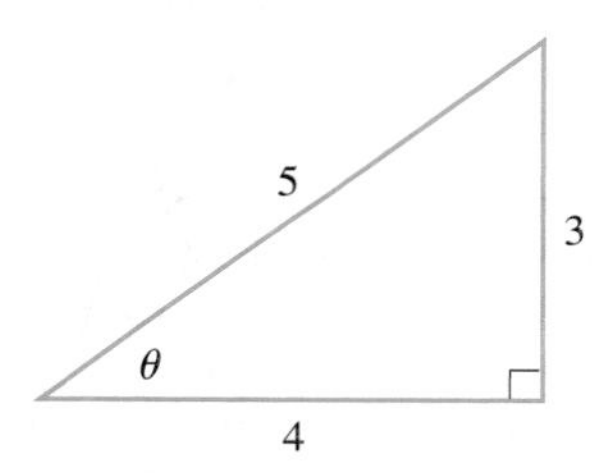

Use an addition formula and the figure to show that $5\sin(x + \theta) = 4\sin x + 3\cos x$.

63. Let a and b be positive constants, and let θ be the acute angle (in radian measure) defined by the following figure.

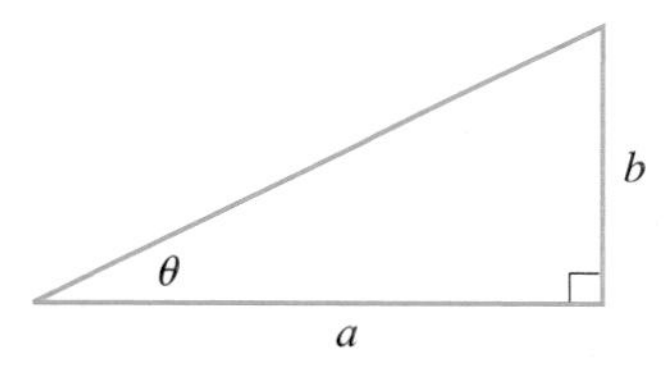

(a) Use an addition formula and the figure to show that $\sqrt{a^2 + b^2}\sin(x + \theta) = a\sin x + b\cos x$.

(b) Use the result in part (a) to specify the maximum value of the function $f(x) = a\sin x + b\cos x$.

64. (a) Use an addition formula to show that $2\sin(x + \frac{\pi}{6}) = \cos x + \sqrt{3}\sin x$.

(b) Use the result in part (a) to graph the function $f(x) = \cos x + \sqrt{3}\sin x$ for one period.

65. (a) Use an addition formula to show that $\sqrt{2}\cos(x - \frac{\pi}{4}) = \cos x + \sin x$.

(b) Use the result in part (a) to graph the function $f(x) = \cos x + \sin x$ for one period.

66. Let A, B, and C be the angles of a triangle, so that $A + B + C = \pi$.

(a) Show that $\sin(A + B) = \sin C$.

(b) Show that $\cos(A + B) = -\cos C$.

(c) Show that $\tan(A + B) = -\tan C$.

67. Suppose that A, B, and C are the angles of a triangle, so that $A + B + C = \pi$. Show that

$$\cos^2 A + \cos^2 B + \cos^2 C + 2\cos A\cos B\cos C = 1$$

68. Prove that

$$\frac{\sin(\alpha - \beta)}{\cos\alpha\cos\beta} + \frac{\sin(\beta - \gamma)}{\cos\beta\cos\gamma} + \frac{\sin(\gamma - \alpha)}{\cos\gamma\cos\alpha} = 0$$

69. Suppose that $a^2 + b^2 = 1$ and $c^2 + d^2 = 1$. Prove that $|ac + bd| \le 1$. *Hint:* Let $a = \cos\theta$, $b = \sin\theta$, $c = \cos\phi$, and $d = \sin\phi$.

In Exercises 70–72, simplify the expression.

70. $\cos(\frac{\pi}{6} + t)\cos(\frac{\pi}{6} - t) - \sin(\frac{\pi}{6} + t)\sin(\frac{\pi}{6} - t)$
Hint: If your solution relies on four separate addition formulas, then you are doing this the hard way.

71. $\sin(\frac{\pi}{3} - t)\cos(\frac{\pi}{3} + t) + \cos(\frac{\pi}{3} - t)\sin(\frac{\pi}{3} + t)$

72. $\dfrac{\tan(A + 2B) - \tan(A - 2B)}{1 + \tan(A + 2B)\tan(A - 2B)}$

73. If $\alpha + \beta = \pi/4$, show that $(1 + \tan\alpha)(1 + \tan\beta) = 2$.

Exercises 74 and 75 outline simple geometric derivations of the formulas for $\sin(\alpha + \beta)$ *and* $\cos(\alpha + \beta)$ *in the case in which* α *and* β *are acute angles, with* $\alpha + \beta < 90°$. *The exercises rely on the accompanying figures, which are constructed as follows. Begin, in Figure A, with* $\alpha = \angle GAD$, $\beta = \angle HAG$, *and* $AH = 1$. *Then, from H, draw perpendiculars to* $\overline{AD}$ *and to* $\overline{AG}$, *as shown in Figure B. Finally, draw* $\overline{FE} \perp \overline{BH}$ *and* $\overline{FC} \perp \overline{AD}$.

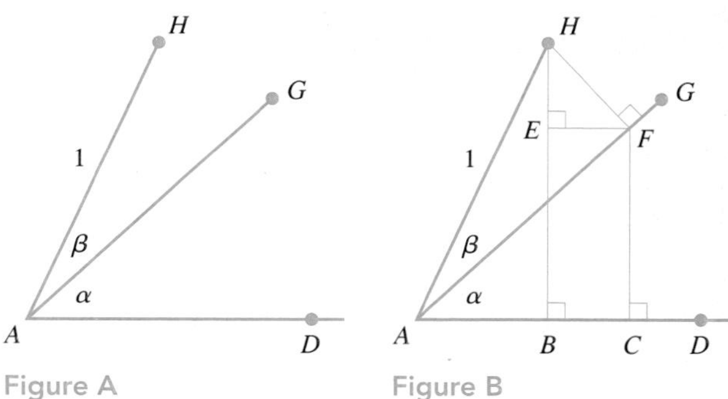

Figure A Figure B

74. *Formula for* $\sin(\alpha + \beta)$. Supply the reasons or steps behind each statement.

(a) $BH = \sin(\alpha + \beta)$

(b) $FH = \sin\beta$

(c) $\angle BHF = \alpha$

(d) $EH = \cos\alpha\sin\beta$ *Hint:* Use $\triangle EFH$ and the result in part (b).

(e) $AF = \cos\beta$

(f) $CF = \sin\alpha\cos\beta$

(g) $\sin(\alpha + \beta) = \sin\alpha\cos\beta + \cos\alpha\sin\beta$
Hint: $\sin(\alpha + \beta) = BH = EH + CF$

75. *Formula for* $\cos(\alpha + \beta)$. Supply the reasons or steps behind each statement.

(a) $\cos(\alpha + \beta) = AB$

(b) $AC = \cos\alpha\cos\beta$
Hint: Use $\triangle ACF$ and the result in Exercise 74(e).

(continues)

(c) $EF = \sin \alpha \sin \beta$
(d) $\cos(\alpha + \beta) = \cos \alpha \cos \beta - \sin \alpha \sin \beta$
Hint: $AB = AC - BC$

76. Let S and C be two functions. Assume that the domain for both S and C is the set of all real numbers and that S and C satisfy the following two identities.

$$S(x - y) = S(x)C(y) - C(x)S(y) \quad (1)$$
$$C(x - y) = C(x)C(y) + S(x)S(y) \quad (2)$$

Also, suppose that the function S is not identically zero. That is,

$$S(x) \neq 0 \quad \text{for at least one real number } x \quad (3)$$

(a) Show that $S(0) = 0$. *Hint*: In identity (1), let $x = y$.
(b) Show that $C(0) = 1$. *Hint*: In identity (1), let $y = 0$.
(c) Explain (in complete sentences) why it was necessary to use condition (3) in the work for part (b).
(d) Prove the identity $[C(x)]^2 + [S(x)]^2 = 1$.
Hint: In identity (2), let $y = x$.
(e) Show that C is an even function and S is an odd function. That is, prove the identities $C(-x) = C(x)$ and $S(-x) = -S(x)$.
Hint: Write $-x$ as $0 - x$.

In Exercises 77–80, prove the identities.

77. $\dfrac{\sin(A + B)}{\sin(A - B)} = \dfrac{\tan A + \tan B}{\tan A - \tan B}$

78. $\dfrac{\cos(A + B)}{\cos(A - B)} = \dfrac{1 - \tan A \tan B}{1 + \tan A \tan B}$

79. $\cot(A + B) = \dfrac{\cot A \cot B - 1}{\cot A + \cot B}$

80. $\cot(A - B) = \dfrac{\cot A \cot B + 1}{\cot B - \cot A}$

81. Let $f(t) = \cos^2 t + \cos^2(t + \frac{2\pi}{3}) + \cos^2(t - \frac{2\pi}{3})$.
(a) Complete the table. (Use a calculator.)

t	1	2	3	4
$f(t)$				

(b) On the basis of your results in part (a), make a conjecture about the function f. Prove that your conjecture is correct.

C

82. (a) Use your calculator to check that $\tan 50° - \tan 40° = 2 \tan 10°$.
(b) Part (a) is a specific example of a more general identity. State the identity and prove it.

83. If $\tan B = \dfrac{n \sin A \cos A}{1 - n \sin^2 A}$, show that $\tan(A - B) = (1 - n)\tan A$.

84. If triangle ABC is not a right triangle, and $\cos A = \cos B \cos C$, show that $\tan B \tan C = 2$.

85. (a) The angles of a triangle are $A = 20°$, $B = 50°$, and $C = 110°$. Use your calculator to compute the sum $\tan A + \tan B + \tan C$ and then the product $\tan A \tan B \tan C$. What do you observe?
(b) The angles of a triangle are $\alpha = \pi/10$, $\beta = 3\pi/10$, and $\gamma = 3\pi/5$. Use your calculator to compute $\tan \alpha + \tan \beta + \tan \gamma$ and $\tan \alpha \tan \beta \tan \gamma$.
(c) If triangle ABC is not a right triangle, prove that $\tan A + \tan B + \tan C = \tan A \tan B \tan C$.

PROJECT THE DESIGN OF A FRESNEL LENS

Before the invention of the Fresnel lens lighthouse beacons were visible from a distance of only four or five miles out at sea. To form a beam of light visible from significantly further than five miles required construction of a lens system, based on the then conventional design, that would be unfeasibly large and heavy—not to mention expensive—and was beyond the manufacturing capabilities of the time. In 1822 Augustin Fresnel revolutionized the design and construction of lighthouse beacons by inventing a lens that was both thin and lightweight, which could form a beam of light that could be seen from twenty or more miles away. Today Fresnel lenses are used in traffic lights, overhead projectors, and on the rear windows of buses and motor homes.

In this group project we use trigonometric identities and Snell's law to design a Fresnel lens. As background, you will need to read the discussion of geometric optics and Snell's law in the project at the end of Section 9.1.

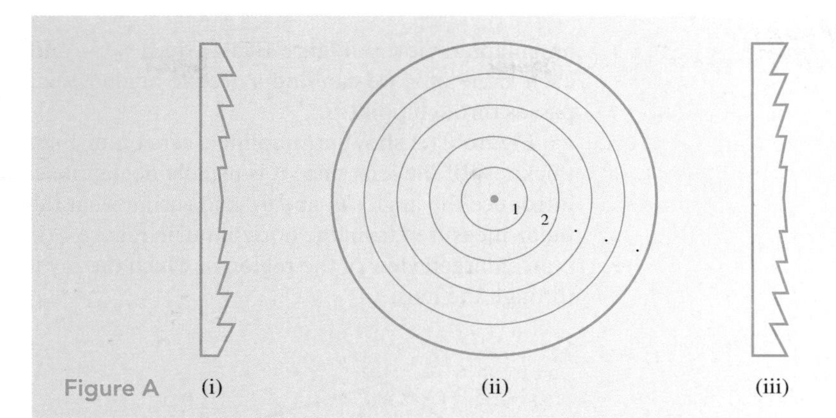

Figure A (i) (ii) (iii)

Figure A shows cross-sectional views of a *positive* or *converging* Fresnel lens on the left and a *negative* or *diverging* lens on the right. The cross-sections are taken along a diameter of the lens. The front view (ii) would be the same for both. A **groove** is the annular region between two consecutive concentric circles seen when looking at the front of a Fresnel lens. In Figure A(ii) groove number 1 lies between the center dot and the smallest circle, groove number 2 lies between the smallest circle and the next one out, and so on.

The lenses commonly seen on the the rear windows of vehicles are diverging lenses, but most uses of Fresnel lenses, as in traffic lights and overhead projectors, require a converging lens. The optical situation is a little simpler for converging lenses, so that's what we'll study in this project. However, the groove angle formula we derive applies to both converging and diverging Fresnel lenses.

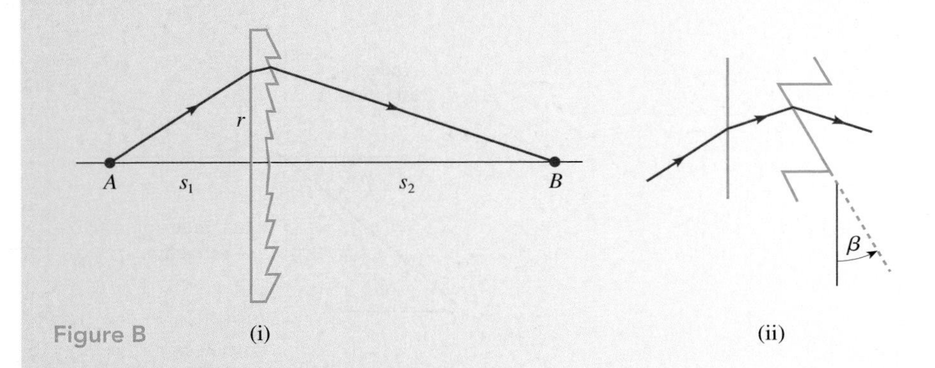

Figure B (i) (ii)

Figure B shows the basic design problem. We are given two points lying along an axis through the center of the lens and perpendicular to the flat surface of the lens. Point A is at a distance s_1 to the left of the lens and point B is at a distance s_2 to the right of the lens. The lens is made of a material of index of refraction n. The design goal is to ensure that light rays emerging from point A go through point B after refraction by the lens. Figure B(i) shows a ray from A striking the flat surface of the lens at a distance r from the center of the lens and, after refraction at that surface, striking the slanted groove surface where it is again refracted. The **groove angle** β is measured from the vertical as is shown in an enlarged view in Figure B(ii). Given s_1, s_2, and n, we need to show how, for each value of r, we can find a groove angle β such that after refraction the ray passes through point B.

Figure C(i) shows a simplified version of Figure B, in which we ignore the thickness of the lens since it is usually negligible compared to s_1 and s_2. We also introduce the angles θ_1 and θ_2, and (using what theorem from geometry?) show them, measured from the horizontal, in Figure C(ii). As in Figure B, Figure C(ii) is an enlarged view of the region in which the ray indicated in Figure C(i) passes through the lens.

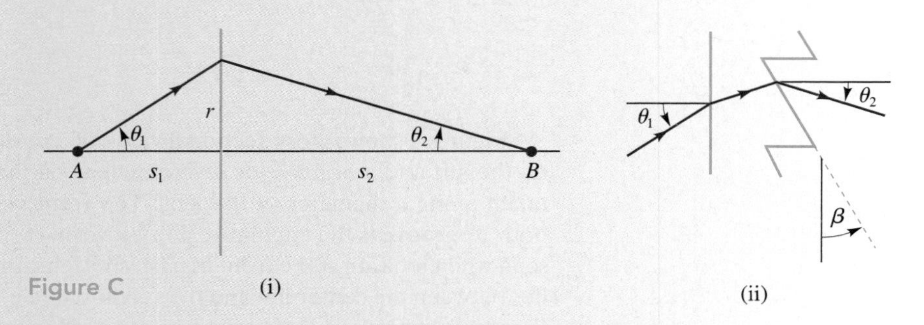

Figure C

We are now ready to derive a formula for β. Figure D shows the path of the ray through the lens with all the relevant angles labeled. Angles φ_1 and φ_1' denote the angles of incidence and refraction at the first surface, and angles φ_2' and φ_2 denote the angles of incidence and refraction at the second surface.

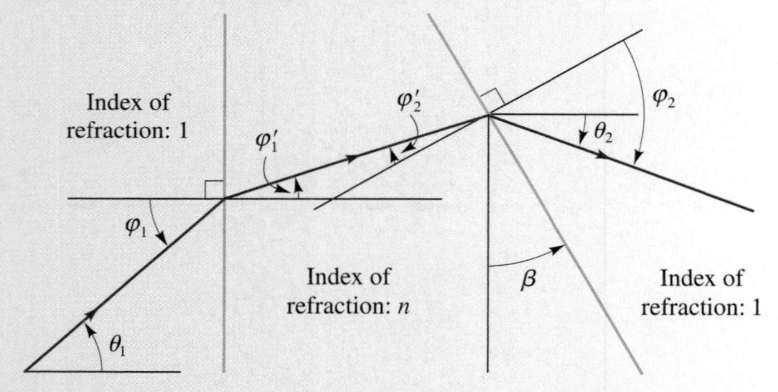

Figure D

EXERCISE 1

(a) Given r, s_1, and s_2, explain how to use a calculator to find θ_1 and θ_2.

(b) Explain why $\varphi_1 = \theta_1$, $\varphi_2 = \beta + \theta_2$, and $\varphi_1' + \varphi_2' = \beta$.

(c) Using Snell's law and part (b) show that

$$\sin\theta_1 = n\sin\varphi_1' \qquad \text{(i)}$$
$$n\sin\varphi_2' = \sin(\beta + \theta_2) \qquad \text{(ii)}$$
$$n\sin\varphi_2' = n\sin(\beta - \varphi_1') \qquad \text{(iii)}$$

and conclude that

$$n\sin(\beta - \varphi_1') = \sin(\beta + \theta_2) \qquad \text{(iv)}$$

(d) Use an addition formula for sine and a Pythagorean identity to show that

$$n \sin(\beta - \varphi_1') = (\sin \beta)(n\sqrt{1 - \sin^2 \varphi_1'}) - (\cos \beta)(n \sin \varphi_1') \qquad \text{(v)}$$

Use equation (i) to obtain

$$n \sin(\beta - \varphi_1') = (\sin \beta)\sqrt{n^2 - \sin^2 \theta_1} - \cos \beta \sin \theta_1 \qquad \text{(vi)}$$

Then use equations (iv) and (vi) and an addition formula for sine to obtain

$$(\sin \beta)\sqrt{n^2 - \sin^2 \theta_1} - \cos \beta \sin \theta_1 = \sin \beta \cos \theta_2 + \cos \beta \sin \theta_2 \qquad \text{(vii)}$$

(e) It is surprising that equation (vii) can be be solved for β in a few steps. Divide both sides by $\cos \beta$ and solve for $\tan \beta$ to derive the Fresnel lens groove angle formula

$$\tan \beta = \frac{\sin \theta_1 + \sin \theta_2}{\sqrt{n^2 - \sin^2 \theta_1} - \cos \theta_2} \qquad \text{(viii)}$$

EXERCISE 2 An optical designer must design a Fresnel lens 80 millimeters in diameter to gather light from a lamp 100 mm from the lens and cause it to pass through a small opening 400 mm from the lens on the side opposite the lamp. The index of refraction of the lens material is 1.50. She lets $s_1 = 100$ mm and $s_2 = 400$ mm, decides to make the width of each groove 0.4 mm, and calculates that there will be 100 grooves. (How?)

She writes a computer program to generate a file containing the 100 groove angles, in degrees, which will be the input to another computer program that will generate commands for a computer-controlled machine (a lathe in this case) that will produce the lens. Before trying to make the lens she decides to calculate the groove angle for grooves numbered 1, 25, 50, and 100, and use the calculated values to check the output of her program. She makes a table with column headings r in mm and θ_1, θ_2, and β in degrees. Her first r-value is 0.200 mm and her second is 9.800 mm and she calculates θ_1, θ_2, and β to four decimal places. Your job as her assistant is to produce your own version of the table as another check of her work. Use the results of Exercise 1 parts (a) and (e) to produce the table.

8.2 THE DOUBLE-ANGLE FORMULAS

Ptolemy (ca. 150) knew substantially the sine of half an angle . . . and it is probable that Hipparchus (ca. 140 B.C.) and certain that Varahamihira (ca. 505) knew the relation that we express as $\sin \frac{\phi}{2} = \sqrt{(1 - \cos \phi)/2}$.—David Eugene Smith in *History of Mathematics* (New York: Ginn and Company, 1925)

There are a number of basic identities that follow from the addition formulas. In the following two boxes we summarize some of the most useful of these.

The Double-Angle Formulas

1. $\sin 2\theta = 2 \sin \theta \cos \theta$
2. $\cos 2\theta = \cos^2 \theta - \sin^2 \theta$
3. $\tan 2\theta = \dfrac{2 \tan \theta}{1 - \tan^2 \theta}$

The Half-Angle Formulas

1. $\sin\frac{s}{2} = \pm\sqrt{\frac{1-\cos s}{2}}$
2. $\cos\frac{s}{2} = \pm\sqrt{\frac{1+\cos s}{2}}$
3. $\tan\frac{s}{2} = \frac{\sin s}{1+\cos s}$

Note: In the half-angle formulas the $\pm$ symbol is intended to mean either positive or negative but not both, and the sign before the radical is determined by the quadrant in which the angle (or arc) $s/2$ terminates.

The identities for $\sin 2\theta$, $\cos 2\theta$, and $\tan 2\theta$ are all derived in the same way: We replace 2θ by $(\theta + \theta)$ and use the appropriate addition formula. For instance, for $\sin 2\theta$ we have

$$\begin{aligned}\sin 2\theta &= \sin(\theta+\theta) = \sin\theta\cos\theta + \cos\theta\sin\theta\\ &= 2\sin\theta\cos\theta\end{aligned}$$

This establishes the formula for $\sin 2\theta$. (Exercise 33 asks you to carry out the corresponding derivations for $\cos 2\theta$ and $\tan 2\theta$.)

EXAMPLE 1 Using double-angle formulas

If $\sin\theta = 4/5$ and $\pi/2 < \theta < \pi$, find the quantities $\cos\theta$, $\sin 2\theta$, and $\cos 2\theta$.

SOLUTION
We have

$$\cos^2\theta = 1 - \sin^2\theta = 1 - \left(\frac{4}{5}\right)^2 = \frac{9}{25}$$

Consequently, $\cos\theta = 3/5$ or $\cos\theta = -3/5$. We want the negative value for the cosine here, since $\pi/2 < \theta < \pi$. Thus

$$\cos\theta = -\frac{3}{5}$$

Now that we know the values of $\cos\theta$ and $\sin\theta$, the double-angle formulas can be used to determine $\sin 2\theta$ and $\cos 2\theta$. We have

$$\sin 2\theta = 2\sin\theta\cos\theta = 2\left(\frac{4}{5}\right)\left(-\frac{3}{5}\right) = -\frac{24}{25}$$

and

$$\begin{aligned}\cos 2\theta &= \cos^2\theta - \sin^2\theta\\ &= \left(-\frac{3}{5}\right)^2 - \left(\frac{4}{5}\right)^2 = \frac{9}{25} - \frac{16}{25} = -\frac{7}{25}\end{aligned}$$

The required quantities are therefore $\cos\theta = -3/5$, $\sin 2\theta = -24/25$, and $\cos 2\theta = -7/25$.

EXAMPLE 2 Using a double-angle formula

If $x = 4 \sin \theta, 0 < \theta < \pi/2$, express $\sin 2\theta$ in terms of x.

SOLUTION

The given equation is equivalent to $\sin \theta = x/4$, so we have

$$\begin{aligned}\sin 2\theta &= 2 \sin \theta \cos \theta = 2\left(\frac{x}{4}\right)\cos \theta\\ &= \frac{x}{2}\sqrt{1 - \sin^2 \theta} \qquad \text{(Why is the positive root appropriate?)}\\ &= \frac{x}{2}\sqrt{1 - \frac{x^2}{16}} = \frac{x}{2}\sqrt{\frac{16 - x^2}{16}} = \frac{x\sqrt{16 - x^2}}{8}\end{aligned}$$

At the start of this section we listed formulas for $\cos 2\theta$ and for $\sin 2\theta$. There are also formulas for $\cos 3\theta$ and for $\sin 3\theta$:

$$\cos 3\theta = 4 \cos^3 \theta - 3 \cos \theta \qquad \sin 3\theta = 3 \sin \theta - 4 \sin^3 \theta$$

In the next example we derive the formula for $\cos 3\theta$, and Exercise 40 asks you to derive the formula for $\sin 3\theta$. Although these formulas needn't be memorized, they are useful. For instance, Exercise 102 in Section 8.4 shows how the formula for $\cos 3\theta$ can be used to solve certain types of cubic equations. *Historical note*: The identities for $\cos 3\theta$ and $\sin 3\theta$ are usually attributed to the French mathematician François Viète (1540–1603). Recent research, however, has shown that a geometric version of the formula for $\sin 3\theta$ was developed much earlier by the Persian mathematican and astronomer Jashmid al-Kāshi (d. 1429). (For background and details, see the article by Professor Farhad Riahi, "An Early Iterative Method for the Determination of sin 1°," in *The College Mathematics Journal,* vol. 26, January 1995, pp. 16–21.)

EXAMPLE 3 Using double-angle formulas to prove an identity

Prove the following identity:

$$\cos 3\theta = 4 \cos^3 \theta - 3 \cos \theta$$

SOLUTION

$$\begin{aligned}\cos 3\theta &= \cos(2\theta + \theta)\\ &= \underbrace{\cos 2\theta}_{(\cos^2 \theta - \sin^2 \theta)} \cos \theta - \underbrace{\sin 2\theta}_{(2 \sin \theta \cos \theta)} \sin \theta\\ &= (\cos^2 \theta - \sin^2 \theta)\cos \theta - (2 \sin \theta \cos \theta)\sin \theta\\ &= \cos^3 \theta - \sin^2 \theta \cos \theta - 2 \sin^2 \theta \cos \theta\end{aligned}$$

Collecting like terms now gives us

$$\cos 3\theta = \cos^3 \theta - 3 \sin^2 \theta \cos \theta$$

Finally, we replace $\sin^2\theta$ by the quantity $1 - \cos^2\theta$. This yields

$$\begin{aligned}\cos 3\theta &= \cos^3\theta - 3(1 - \cos^2\theta)\cos\theta\\ &= \cos^3\theta - 3\cos\theta + 3\cos^3\theta\\ &= 4\cos^3\theta - 3\cos\theta \qquad \text{as required}\end{aligned}$$

In the box that follows, we list several alternate ways of writing the formula for $\cos 2\theta$. Formulas (3) and (4) are quite useful in calculus.

Equivalent Forms of the Formula $\cos 2\theta = \cos^2\theta - \sin^2\theta$

1. $\cos 2\theta = 2\cos^2\theta - 1$
2. $\cos 2\theta = 1 - 2\sin^2\theta$
3. $\cos^2\theta = \dfrac{1 + \cos 2\theta}{2}$
4. $\sin^2\theta = \dfrac{1 - \cos 2\theta}{2}$

One way to prove identity (1) is as follows:

$$\begin{aligned}\cos 2\theta &= \cos^2\theta - \sin^2\theta\\ &= \cos^2\theta - (1 - \cos^2\theta)\\ &= \cos^2\theta - 1 + \cos^2\theta\\ &= 2\cos^2\theta - 1 \qquad \text{as required}\end{aligned}$$

In this last equation, if we add 1 to both sides and then divide by 2, the result is identity (3). (Verify this.) The proofs for (2) and (4) are similar; see Exercise 34.

EXAMPLE 4 Expressing a power of cosine in terms of cosines of multiple angles

Express $\cos^4 t$ in a form that does not involve powers of the trigonometric functions.

SOLUTION

$$\begin{aligned}\cos^4 t = (\cos^2 t)^2 &= \left(\frac{1 + \cos 2t}{2}\right)^2 && \text{using the formula for } \cos^2\theta\\ &= \frac{1 + 2\cos 2t + \cos^2 2t}{4}\\ &= \frac{1 + 2\cos 2t + \frac{1}{2}(1 + \cos 4t)}{4} && \text{using the formula for } \cos^2\theta \text{ with } \theta = 2t\end{aligned}$$

An easy way to simplify this last expression is to multiply both the numerator and the denominator by 2. As you should check for yourself, the final result is

$$\cos^4 t = \frac{3 + 4\cos 2t + \cos 4t}{8}$$

The last three formulas we are going to prove in this section are the **half-angle formulas** which follow.

$$\cos\frac{s}{2} = \pm\sqrt{\frac{1+\cos s}{2}}$$

$$\sin\frac{s}{2} = \pm\sqrt{\frac{1-\cos s}{2}}$$

$$\tan\frac{s}{2} = \frac{\sin s}{1+\cos s}$$

To derive the formula for $\cos(s/2)$, we begin with one of the alternative forms of the cosine double-angle formula:

$$\cos^2\theta = \frac{1+\cos 2\theta}{2} \qquad \text{or} \qquad \cos\theta = \pm\sqrt{\frac{1+\cos 2\theta}{2}}$$

Since this identity holds for all values of θ, we may replace θ by $s/2$ to obtain

$$\cos\frac{s}{2} = \pm\sqrt{\frac{1+\cos 2(s/2)}{2}} = \pm\sqrt{\frac{1+\cos s}{2}}$$

This is the required formula for $\cos(s/2)$. To derive the formula for $\sin(s/2)$, we follow exactly the same procedure, except that we begin with the identity $\sin^2\theta = \frac{1}{2}(1-\cos 2\theta)$. [Exercise 34(c) asks you to complete the proof.] In both formulas the sign before the radical is determined by the quadrant in which the angle or arc $s/2$ terminates.

EXAMPLE 5 Using a half-angle formula

Evaluate cos 105° using a half-angle formula.

SOLUTION

$$\cos 105^\circ = \cos\frac{210^\circ}{2} = \pm\sqrt{\frac{1+\cos 210^\circ}{2}} \qquad \text{using the formula for } \cos(s/2) \text{ with } s = 210^\circ$$

$$= \pm\sqrt{\frac{1+(-\sqrt{3}/2)}{2}}$$

$$= \pm\sqrt{\frac{1-(\sqrt{3}/2)}{2}\cdot\frac{2}{2}} = \pm\sqrt{\frac{2-\sqrt{3}}{4}}$$

$$= \frac{\pm\sqrt{2-\sqrt{3}}}{2}$$

So

$$\cos\theta = \frac{\sqrt{2-\sqrt{3}}}{2} \qquad \text{or} \qquad \frac{-\sqrt{2-\sqrt{3}}}{2} \qquad \text{(Which one?)}$$

We choose the negative value here, since the terminal side of 105° lies in Quadrant II. Thus we finally obtain

$$\cos 105^\circ = \frac{-\sqrt{2-\sqrt{3}}}{2}$$

Our last task is to establish the formula for $\tan(s/2)$. To do this, we first prove the equivalent identity, $\tan\theta = \dfrac{\sin 2\theta}{1+\cos 2\theta}$.

Proof That $\tan\theta = \dfrac{\sin 2\theta}{1+\cos 2\theta}$.

$$\frac{\sin 2\theta}{1+\cos 2\theta} = \frac{2\sin\theta\cos\theta}{2\cos^2\theta} \qquad \text{using the identity } \cos 2\theta = 2\cos^2\theta - 1 \text{ in the denominator}$$

$$= \frac{\sin\theta}{\cos\theta} = \tan\theta$$

If we now replace θ by $s/2$ in the identity $\tan\theta = \dfrac{\sin 2\theta}{1+\cos 2\theta}$, the result is

$$\tan\frac{s}{2} = \frac{\sin s}{1+\cos s}$$

This is the half-angle formula for the tangent.

We conclude this section with a summary of the principal trigonometric identities developed in this section and in Chapter 7. For completeness the list also includes two sets of trigonometric identities that we did not discuss in this section. These are the so-called **product-to-sum formulas** and **sum-to-product formulas.** Proofs and applications of these formulas are discussed in the next section.

PROPERTY SUMMARY Principal Trigonometric Identities

1. Consequences of the definitions
   (a) $\csc\theta = \dfrac{1}{\sin\theta}$ (b) $\sec\theta = \dfrac{1}{\cos\theta}$ (c) $\cot\theta = \dfrac{1}{\tan\theta}$
   (d) $\tan\theta = \dfrac{\sin\theta}{\cos\theta}$ (e) $\cot\theta = \dfrac{\cos\theta}{\sin\theta}$
2. The Pythagorean identities
   (a) $\sin^2\theta + \cos^2\theta = 1$ (b) $\tan^2\theta + 1 = \sec^2\theta$ (c) $\cot^2\theta + 1 = \csc^2\theta$
3. The opposite-angle formulas
   (a) $\sin(-\theta) = -\sin\theta$ (b) $\cos(-\theta) = \cos\theta$ (c) $\tan(-\theta) = -\tan\theta$
4. The reduction formulas
   (a) $\sin(\theta + 2\pi k) = \sin\theta$ (b) $\cos(\theta + 2\pi k) = \cos\theta$ (c) $\sin(\frac{\pi}{2} - \theta) = \cos\theta$
   (d) $\cos(\frac{\pi}{2} - \theta) = \sin\theta$
5. The addition formulas
   (a) $\sin(s+t) = \sin s\cos t + \cos s\sin t$ (b) $\sin(s-t) = \sin s\cos t - \cos s\sin t$
   (c) $\cos(s+t) = \cos s\cos t - \sin s\sin t$ (d) $\cos(s-t) = \cos s\cos t + \sin s\sin t$
   (e) $\tan(s+t) = \dfrac{\tan s + \tan t}{1 - \tan s\tan t}$ (f) $\tan(s-t) = \dfrac{\tan s - \tan t}{1 + \tan s\tan t}$
6. The double-angle formulas
   (a) $\sin 2\theta = 2\sin\theta\cos\theta$ (b) $\cos 2\theta = \cos^2\theta - \sin^2\theta$ (c) $\tan 2\theta = \dfrac{2\tan\theta}{1-\tan^2\theta}$

7. The half-angle formulas

(a) $\sin\frac{\theta}{2} = \pm\sqrt{\frac{1-\cos\theta}{2}}$ (b) $\cos\frac{\theta}{2} = \pm\sqrt{\frac{1+\cos\theta}{2}}$ (c) $\tan\frac{\theta}{2} = \frac{\sin\theta}{1+\cos\theta}$

8. The product-to-sum formulas

(a) $\sin A\sin B = \frac{1}{2}[\cos(A-B) - \cos(A+B)]$ (b) $\sin A\cos B = \frac{1}{2}[\sin(A+B) + \sin(A-B)]$

(c) $\cos A\cos B = \frac{1}{2}[\cos(A+B) + \cos(A-B)]$

9. The sum-to-product formulas

(a) $\sin\alpha + \sin\beta = 2\sin\frac{\alpha+\beta}{2}\cos\frac{\alpha-\beta}{2}$ (b) $\sin\alpha - \sin\beta = 2\cos\frac{\alpha+\beta}{2}\sin\frac{\alpha-\beta}{2}$

(c) $\cos\alpha + \cos\beta = 2\cos\frac{\alpha+\beta}{2}\cos\frac{\alpha-\beta}{2}$ (d) $\cos\alpha - \cos\beta = -2\sin\frac{\alpha+\beta}{2}\sin\frac{\alpha-\beta}{2}$

EXERCISE SET 8.2

A

In Exercises 1–8, use the given information to evaluate each expression.

1. $\cos\varphi = 7/25$ $(0° < \varphi < 90°)$
(a) $\sin 2\varphi$ **(b)** $\cos 2\varphi$ **(c)** $\tan 2\varphi$

2. $\cos\varphi = 3/5$ $(0° < \varphi < 90°)$
(a) $\sin 2\varphi$ **(b)** $\cos 2\varphi$ **(c)** $\tan 2\varphi$

3. $\tan u = -4$ $(\frac{3\pi}{2} < u < 2\pi)$
(a) $\sin 2u$ **(b)** $\cos 2u$ **(c)** $\tan 2u$

4. $\cot s = 2$ $(\pi < s < \frac{3\pi}{2})$
(a) $\sin 2s$ **(b)** $\cos 2s$ **(c)** $\tan 2s$

5. $\sin\alpha = \sqrt{3}/2$ $(0° < \alpha < 90°)$
(a) $\sin(\alpha/2)$ **(b)** $\cos(\alpha/2)$ **(c)** $\tan(\alpha/2)$

6. $\cos\beta = -1/8$ $(180° < \beta < 270°)$
(a) $\sin(\beta/2)$ **(b)** $\cos(\beta/2)$ **(c)** $\tan(\beta/2)$

7. $\cos\theta = -7/9$ $(\frac{\pi}{2} < \theta < \pi)$
(a) $\sin(\theta/2)$ **(b)** $\cos(\theta/2)$ **(c)** $\tan(\theta/2)$

8. $\cos\theta = 12/13$ $(\frac{3\pi}{2} < \theta < 2\pi)$
(a) $\sin(\theta/2)$ **(b)** $\cos(\theta/2)$ **(c)** $\tan(\theta/2)$

In Exercises 9–12, use the given information to compute each of the following:

(a) $\sin 2\theta$ **(c)** $\sin(\theta/2)$
(b) $\cos 2\theta$ **(d)** $\cos(\theta/2)$

9. $\sin\theta = 3/4$ and $\pi/2 < \theta < \pi$

10. $\cos\theta = 2/5$ and $3\pi/2 < \theta < 2\pi$

11. $\cos\theta = -1/3$ and $180° < \theta < 270°$

12. $\sin\theta = -1/10$ and $270° < \theta < 360°$

In Exercises 13–16, use an appropriate half-angle formula to evaluate each quantity.

13. (a) $\sin(\pi/12)$ **(b)** $\cos(\pi/12)$ **(c)** $\tan(\pi/12)$

14. (a) $\sin(\pi/8)$ **(b)** $\cos(\pi/8)$ **(c)** $\tan(\pi/8)$

15. (a) $\sin 105°$ **(b)** $\cos 105°$ **(c)** $\tan 105°$

16. (a) $\sin 165°$ **(b)** $\cos 165°$ **(c)** $\tan 165°$

In Exercises 17–24, refer to the two triangles and compute the quantities indicated.

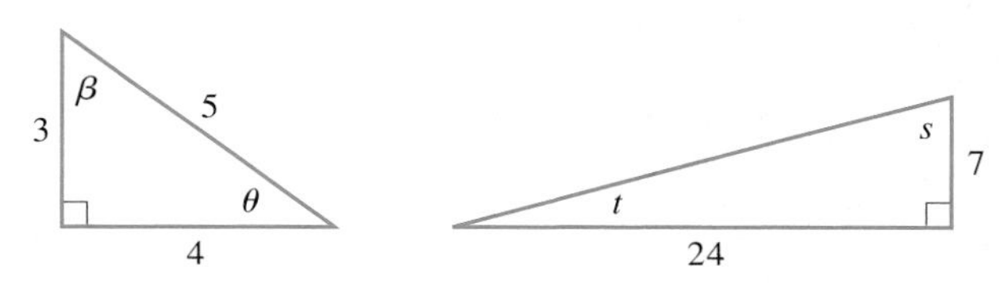

17. (a) $\sin 2\theta$ **(b)** $\cos 2\theta$ **(c)** $\tan 2\theta$

18. (a) $\sin 2t$ **(b)** $\cos 2t$ **(c)** $\tan 2t$

19. (a) $\sin 2\beta$ **(b)** $\cos 2\beta$ **(c)** $\tan 2\beta$

20. (a) $\sin 2s$ **(b)** $\cos 2s$ **(c)** $\tan 2s$

21. (a) $\sin(\theta/2)$ **(b)** $\cos(\theta/2)$ **(c)** $\tan(\theta/2)$

22. (a) $\sin(s/2)$ **(b)** $\cos(s/2)$ **(c)** $\tan(s/2)$

23. (a) $\sin(\beta/2)$ **(b)** $\cos(\beta/2)$ **(c)** $\tan(\beta/2)$

24. (a) $\sin(t/2)$ **(b)** $\cos(t/2)$ **(c)** $\tan(t/2)$

In Exercises 25–28, use the given information to express $\sin 2\theta$ *and* $\cos 2\theta$ *in terms of x.*

25. $x = 5\sin\theta$ $(0 < \theta < \frac{\pi}{2})$

26. $x = \sqrt{2}\cos\theta$ $(0 < \theta < \frac{\pi}{2})$

27. $x - 1 = 2\sin\theta$ $(0 < \theta < \frac{\pi}{2})$

28. $x + 1 = 3\sin\theta$ $(\frac{\pi}{2} < \theta < \pi)$

In Exercises 29–32, express each quantity in a form that does not involve powers of the trigonometric functions (as in Example 4).

29. $\sin^4\theta$ **30.** $\cos^6\theta$

31. $\sin^4(\theta/2)$ **32.** $\sin^6(\theta/4)$

33. Prove each of the following double-angle formulas. *Hint:* As in the text, replace 2θ with $\theta + \theta$, and use an appropriate addition formula.

(a) $\cos 2\theta = \cos^2\theta - \sin^2\theta$

(b) $\tan 2\theta = \dfrac{2\tan\theta}{1 - \tan^2\theta}$

34. (a) Beginning with the identity $\cos 2\theta = \cos^2\theta - \sin^2\theta$, prove that $\cos 2\theta = 1 - 2\sin^2\theta$.

(b) Using the result in part (a), prove that $\sin^2\theta = (1 - \cos 2\theta)/2$.

(c) Derive the formula for $\sin(s/2)$ as follows: using the identity in part (b), replace θ with $s/2$, and then take square roots.

B

In Exercises 35–50, prove that the given equations are identities.

35. $\cos 2s = \dfrac{1 - \tan^2 s}{1 + \tan^2 s}$

36. $1 + \cos 2t = \cot t \sin 2t$

37. $\cos\theta = 2\cos^2(\theta/2) - 1$

38. $\dfrac{\sin 2\theta}{\sin\theta} - \dfrac{\cos 2\theta}{\cos\theta} = \sec\theta$

39. $\sin^4\theta = \dfrac{3 - 4\cos 2\theta + \cos 4\theta}{8}$

40. $\sin 3\theta = 3\sin\theta - 4\sin^3\theta$

41. $\sin 2\theta = \dfrac{2\tan\theta}{1 + \tan^2\theta}$

42. $2\csc 2\theta = \dfrac{\csc^2\theta}{\cot\theta}$

43. $\sin 2\theta = 2\sin^3\theta\cos\theta + 2\sin\theta\cos^3\theta$

44. $\cot\theta = \dfrac{1 + \cos 2\theta}{\sin 2\theta}$

45. $\dfrac{1 + \tan(\theta/2)}{1 - \tan(\theta/2)} = \tan\theta + \sec\theta$

46. $\tan\theta + \cot\theta = 2\csc 2\theta$

47. $2\sin^2(45^\circ - \theta) = 1 - \sin 2\theta$

48. $(\sin\theta - \cos\theta)^2 = 1 - \sin 2\theta$

49. $1 + \tan\theta\tan 2\theta = \tan 2\theta\cot\theta - 1$

50. $\tan(\frac{\pi}{4} + \theta) - \tan(\frac{\pi}{4} - \theta) = 2\tan 2\theta$

51. If $\tan\alpha = 1/11$ and $\tan\beta = 5/6$, find $\alpha + \beta$, given that $0 < \alpha < \pi/2$ and $0 < \beta < \pi/2$. *Hint:* Compute $\tan(\alpha + \beta)$.

52. Let $z = \tan\theta$ for $-\pi/2 < \theta < \pi/2$. Show that

(a) $\cos 2\theta = \dfrac{1 - z^2}{1 + z^2}$ and $\sin 2\theta = \dfrac{2z}{1 + z^2}$

(b) Explain why these formulas give the correct signs for $\cos 2\theta$ and $\sin 2\theta$.

53. (a) Use a calculator to verify that the value $x = \cos 20^\circ$ is a root of the cubic equation $8x^3 - 6x - 1 = 0$.

(b) Use the identity $\cos 3\theta = 4\cos^3\theta - 3\cos\theta$ (from Example 3 on page 577) to prove that $\cos 20^\circ$ is a root of the cubic equation $8x^3 - 6x - 1 = 0$. *Hint:* In the given identity, substitute $\theta = 20^\circ$.

54. (a) Use a calculator to verify that the value $x = \sin\frac{5\pi}{18}$ is a root of the cubic equation $8x^3 - 6x + 1 = 0$.

(b) Use the identity $\sin 3\theta = 3\sin\theta - 4\sin^3\theta$ (from Exercise 40) to prove that $\sin\frac{5\pi}{18}$ is a root of the equation $8x^3 - 6x + 1 = 0$. *Hint:* In the identity, substitute $\theta = 5\pi/18$.

55. The following figure shows a semicircle with radius $AO = 1$.

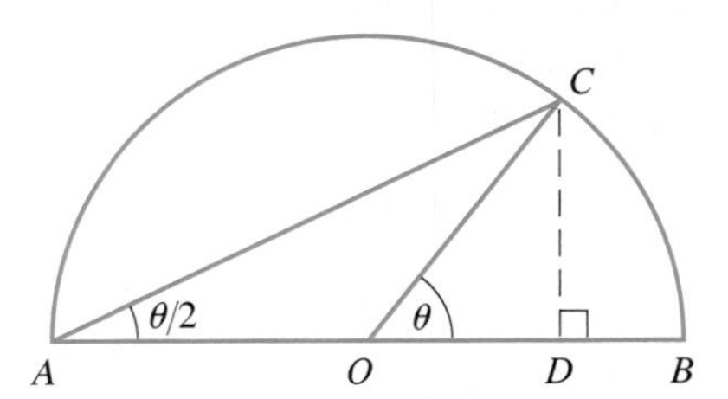

(a) Use the figure to derive the formula

$$\tan\frac{\theta}{2} = \frac{\sin\theta}{1 + \cos\theta} \qquad \left(0 < \theta < \frac{\pi}{2}\right)$$

Hint: Show that $CD = \sin\theta$ and $OD = \cos\theta$. Then look at right triangle ADC to find $\tan(\theta/2)$.

(b) Use the formula developed in part (a) to show that

(i) $\tan 15^\circ = \dfrac{1}{2 + \sqrt{3}} = 2 - \sqrt{3}$

(ii) $\tan(\pi/8) = \sqrt{2} - 1$

56. In this exercise we'll use the accompanying figure to prove the following identities:

$$\cos 2\theta = 2\cos^2\theta - 1$$
$$\sin 2\theta = 2\sin\theta\cos\theta$$
$$\cos 3\theta = 4\cos^3\theta - 3\cos\theta$$
$$\sin 3\theta = 3\sin\theta - 4\sin^3\theta$$

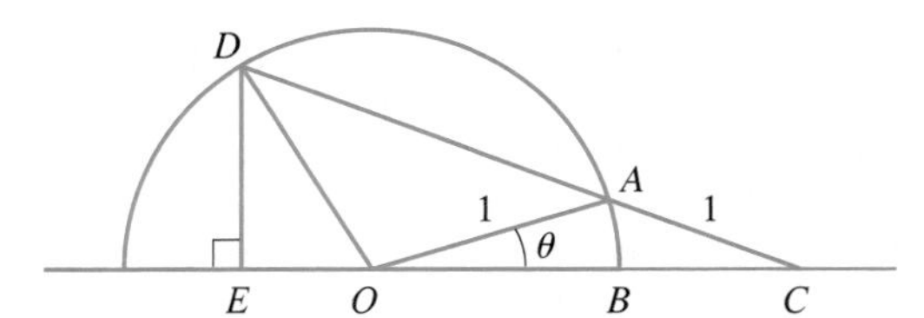

[The figure and the technique in this exercise are adapted from the article by Wayne Dancer, "Geometric Proofs of Multiple Angle Formulas," in *American Mathematical Monthly,* vol. 44 (1937), pp. 366–367.] The figure is constructed as follows. We start with $\angle AOB = \theta$ in standard position in the unit circle, as shown. The point C is chosen on the extended diameter such that $CA = 1$. Then $\overline{CA}$ is extended to meet the circle at D, and radius $\overline{DO}$ is drawn. Finally, from D a perpendicular is drawn to the diameter, as shown.

Supply the reason or reasons that justify each of the following statements.

(a) $\angle ACO = \theta$

(b) $\angle DAO = 2\theta$ *Hint:* $\angle DAO$ is an exterior angle to $\triangle AOC$.

(c) $\angle ODA = 2\theta$ **(d)** $\angle DOE = 3\theta$

(e) From O, draw a perpendicular to $\overline{CD}$, meeting $\overline{CD}$ at F. From A, draw a perpendicular to $\overline{OC}$, meeting $\overline{OC}$ at G. Then $GC = OG = \cos\theta$ and $FA = DF = \cos 2\theta$.

(f) $\cos\theta = \dfrac{1 + \cos 2\theta}{2\cos\theta}$ *Hint*: Use $\triangle CFO$.

(g) $\cos 2\theta = 2\cos^2\theta - 1$ *Hint*: In the equation in part (f), solve for $\cos 2\theta$.

(h) $OF = \sin 2\theta$

(i) $\sin 2\theta = 2\sin\theta\cos\theta$ *Hint*: Find $\sin\theta$ in $\triangle CFO$, and then solve the resulting equation for $\sin 2\theta$.

(j) $DC = 1 + 2\cos 2\theta$ and $EO = \cos 3\theta$

(k) $\cos\theta = \dfrac{2\cos\theta + \cos 3\theta}{1 + 2\cos 2\theta}$ *Hint*: Compute $\cos\theta$ in $\triangle CDE$, and then use part (j).

(l) $\cos 3\theta = 4\cos^3\theta - 3\cos\theta$ *Hint*: Use the results in parts (k) and (g).

(m) $DE = \sin 3\theta$

(n) $\sin 3\theta = 3\sin\theta - 4\sin^3\theta$ *Hint*: Compute $\sin\theta$ in $\triangle CDE$.

57. Prove the following identities involving products of cosines. *Suggestion*: In each case, begin with the right-hand side and use the double-angle formula for the sine.

(a) $\cos\theta\cos 2\theta = \dfrac{\sin 4\theta}{4\sin\theta}$

(b) $\cos\theta\cos 2\theta\cos 4\theta = \dfrac{\sin 8\theta}{8\sin\theta}$

(c) $\cos\theta\cos 2\theta\cos 4\theta\cos 8\theta = \dfrac{\sin 16\theta}{16\sin\theta}$

58. (a) Use your calculator to evaluate the expression $\cos 72^\circ \cos 144^\circ$. Then follow steps (b) through (d) to *prove* that $\cos 72^\circ \cos 144^\circ = -1/4$.

(b) Multiply the expression $\cos 72^\circ \cos 144^\circ$ by the quantity $(\sin 72^\circ)/(\sin 72^\circ)$, which equals 1. Show that the result can be written

$$\frac{\frac{1}{2}\sin 144^\circ \cos 144^\circ}{\sin 72^\circ}$$

(c) Explain why the expression obtained in part (b) is equal to

$$\frac{\frac{1}{4}\sin 288^\circ}{\sin 72^\circ}$$

(d) Use the reference-angle concept to explain why the expression in part (c) is equal to $-1/4$, as required.

59. (a) Use your calculator to evaluate the expression $\cos 72^\circ + \cos 144^\circ$. Then follow steps (b) through (d) to *prove* that $\cos 72^\circ + \cos 144^\circ = -1/2$.

(b) Use the observation

$$\cos 72^\circ + \cos 144^\circ = \cos(108^\circ - 36^\circ) + \cos(108^\circ + 36^\circ)$$

and the addition formulas for cosine to show that

$$\cos 72^\circ + \cos 144^\circ = 2\cos 108^\circ \cos 36^\circ$$

(c) Use the reference-angle concept to explain why $\cos 108^\circ \cos 36^\circ = \cos 72^\circ \cos 144^\circ$.

(d) Use parts (b) and (c) and the identity in Exercise 58(a) to conclude that $\cos 72^\circ + \cos 144^\circ = -1/2$, as required.

60. In the figure below, the points A_1, A_2, A_3, A_4, and A_5 are the vertices of a regular pentagon. Follow steps (a) through (c) to show that

$$(A_1A_2)(A_1A_3)(A_1A_4)(A_1A_5) = 5$$

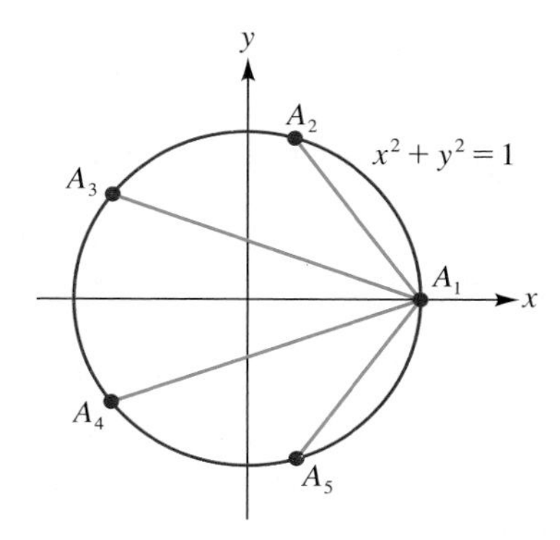

[This is a particular case of a general result due to Roger Cotes (1682–1716): Suppose that a regular n-gon is inscribed in the unit circle. Let the vertices of the n-gon be $A_1, A_2, A_3, \ldots, A_n$. Then the product $(A_1A_2)(A_1A_3)\cdots(A_1A_n)$ is equal to n, the number of sides of the polygon.]

(a) What are the coordinates of the points A_2, A_3, A_4, and A_5? (Give your answers in terms of sines and cosines.)

(b) Show that

$$(A_1A_2)(A_1A_3)(A_1A_4)(A_1A_5) = (2 - 2\cos 72^\circ)(2 - 2\cos 144^\circ)$$

(c) Show that the expression on the right-hand side of the equation in part (b) is equal to 5. *Hint*: Use the equations given in Exercises 58(a) and 59(a).

For Exercises 61 and 62, refer to the following figures. Figure A shows an equilateral triangle inscribed in the unit circle. Figure B shows a regular pentagon inscribed in the unit circle. In both figures, the coordinates of the point P are (x, 0).

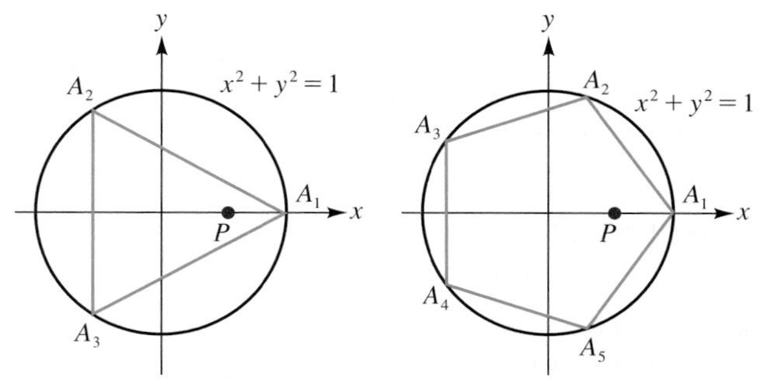

Figure A Figure B

61. For Figure A, show that $(PA_1)(PA_2)(PA_3) = 1 - x^3$.

62. For Figure B, show that

$$(PA_1)(PA_2)(PA_3)(PA_4)(PA_5) = 1 - x^5$$

Hints: First find the coordinates of A_2 and A_3 in terms of sines and cosines. Next, show that

$$\begin{aligned}&(PA_1)(PA_2)(PA_3)(PA_4)(PA_5)\\ &\quad = (1 - x)[(1 - 2x \cos 72^\circ + x^2)(1 - 2x \cos 144^\circ + x^2)]\end{aligned}$$

Then expand the expression within the brackets and simplify using the equations in Exercises 58(a) and 59(a). *Remark*: The results in Exercises 61 and 62 are particular cases of the following theorem of Roger Cotes: Suppose that a regular n-gon $A_1A_2A_3 \cdots A_n$ is inscribed in the unit circle. Suppose that A_1 is the point $(1, 0)$ and that P is a point with coordinates $(x, 0)$, where $0 \le x \le 1$. Then the product $(PA_1)(PA_2)(PA_3) \cdots (PA_n)$ is equal to $1 - x^n$.

In Exercises 63 and 64, the results and the techniques are taken from the article by Zalman Usiskin, "Products of Sines," which appeared in The Two-Year College Mathematics Journal, *vol. 10 (1979), pp. 334–340.*

63. (a) Use your calculator to check that

$$\sin 18^\circ \sin 54^\circ = 1/4$$

(b) Supply reasons for each of the following steps to prove that the equation in part (a) is indeed correct.

(i) $\sin 72^\circ = 2 \sin 36^\circ \cos 36^\circ = 2 \sin 36^\circ \sin 54^\circ$

(ii) $\sin 72^\circ = 4 \sin 18^\circ \cos 18^\circ \sin 54^\circ$
$= 4 \sin 18^\circ \sin 72^\circ \sin 54^\circ$

(iii) $1/4 = \sin 18^\circ \sin 54^\circ$

64. (a) Use your calculator to check that

$$\sin 10^\circ \sin 50^\circ \sin 70^\circ = 1/8$$

(b) Prove that the equation in part (a) is indeed correct. *Hint*: Use the technique in Exercise 63; begin with the equation $\sin 80^\circ = 2 \sin 40^\circ \cos 40^\circ$.

65. (a) Use two of the addition formulas from the previous section to show that

$$\cos(60^\circ - \theta) \cos(60^\circ + \theta) = (4 \cos^2 \theta - 3)/4$$

(b) Show that

$$\cos \theta \cos(60^\circ - \theta) \cos(60^\circ + \theta) = (\cos 3\theta)/4$$

Hint: Use the result in part (a) and the identity for $\cos 3\theta$ that was derived in Example 3 on page 577.

(c) Use the result in part (b) to show that

$$\cos 20^\circ \cos 40^\circ \cos 80^\circ = 1/8$$

(d) Use a calculator to check that

$$\cos 20^\circ \cos 40^\circ \cos 80^\circ = 1/8$$

66. (a) Use two of the addition formulas from the previous section to show that

$$\sin(60^\circ - \theta) \sin(60^\circ + \theta) = \frac{3 - 4 \sin^2 \theta}{4}$$

(b) Show that

$$\sin \theta \sin(60^\circ - \theta) \sin(60^\circ + \theta) = \frac{\sin 3\theta}{4}$$

Hint: Use the result in part (a) and the identity for $\sin 3\theta$ in Exercise 40.

(c) Use the result in part (b) to show that $\sin 20^\circ \sin 40^\circ \sin 80^\circ = \sqrt{3}/8$.

(d) Use a calculator to check that $\sin 20^\circ \sin 40^\circ \sin 80^\circ = \sqrt{3}/8$.

C

67. In this exercise we show that the irrational number $\cos \frac{2\pi}{7}$ is a root of the cubic equation

$$8x^3 + 4x^2 - 4x - 1 = 0$$

(a) Prove the following two identities:

$$\cos 3\theta = 4 \cos^3 \theta - 3 \cos \theta$$
$$\cos 4\theta = 8 \cos^4 \theta - 8 \cos^2 \theta + 1$$

(b) Let $\theta = 2\pi/7$. Use the reference-angle concept [not the formulas in part (a)] to explain why $\cos 3\theta = \cos 4\theta$.

(c) Now use the formulas in part (a) to show that if $\theta = 2\pi/7$, then

$$8 \cos^4 \theta - 4 \cos^3 \theta - 8 \cos^2 \theta + 3 \cos \theta + 1 = 0$$

(d) Show that the equation in part (c) can be written

$$(\cos \theta - 1)(8 \cos^3 \theta + 4 \cos^2 \theta - 4 \cos \theta - 1) = 0$$

Conclude that the value $x = \cos \frac{2\pi}{7}$ satisfies the equation $8x^3 + 4x^2 - 4x - 1 = 0$.

(e) Use your calculator (in the radian mode) to check that $x = \cos \frac{2\pi}{7}$ satisfies the cubic equation $8x^3 + 4x^2 - 4x - 1 = 0$. *Remark*: An interesting fact about the real number $\cos \frac{2\pi}{7}$ is that it cannot be expressed in terms of radicals within the real-number system.

68. *Calculation of* $\sin 18^\circ$, $\cos 18^\circ$, *and* $\sin 3^\circ$.

(a) Prove that $\cos 3\theta = 4 \cos^3 \theta - 3 \cos \theta$.

(b) Supply a reason for each statement.

(i) $\sin 36^\circ = \cos 54^\circ$

(ii) $2 \sin 18^\circ \cos 18^\circ = 4 \cos^3 18^\circ - 3 \cos 18^\circ$

(iii) $2 \sin 18^\circ = 4 \cos^2 18^\circ - 3$

(c) In equation (iii), replace $\cos^2 18^\circ$ by $1 - \sin^2 18^\circ$ and then solve the resulting equation for $\sin 18^\circ$. Thus show that $\sin 18^\circ = \frac{1}{4}(\sqrt{5} - 1)$.

(d) Show that $\cos 18^\circ = \frac{1}{4}\sqrt{10 + 2\sqrt{5}}$.

(e) Show that $\sin 3^\circ$ is equal to

$$\frac{1}{16}[(\sqrt{5} - 1)(\sqrt{6} + \sqrt{2}) - 2(\sqrt{3} - 1)\sqrt{5 + \sqrt{5}}]$$

Hint: $3^\circ = 18^\circ - 15^\circ$

(f) Use your calculator to check the results in parts (c), (d), and (e).

8.3 THE PRODUCT-TO-SUM AND SUM-TO-PRODUCT FORMULAS

In the summer of 1580, . . . [the Polish mathematician Paul Wittich] *went for a short time to Uraniborg to work with Tycho Brache. He soon showed himself to be a skillful mathematician, for with Tycho he discovered—or, more precisely, rediscovered—the method of prosthaphaeresis, by which the products and quotients of trigonometric functions . . . can be replaced by simpler sums and differences.*

The method of prosthaphaeresis originated with Johann Werner [a German astronomer, mathematician, and geographer (1468–1522)], *who developed it in conjunction with the law of cosines for sides of a spherical triangle.*—Charles C. Gillipsie (ed.), *Dictionary of Scientific Biography,* vol. XIV (New York: Charles Scribner's Sons, 1976)

The addition formulas for sine and cosine (from Section 8.1) can be used to establish identities concerning sums and products of sines and cosines. These identities are useful at times for simplifying expressions in trigonometry and in calculus. We begin with the three *product-to-sum formulas.*

The Product-to-Sum Formulas

$$\sin A \sin B = \frac{1}{2}[\cos(A - B) - \cos(A + B)]$$

$$\sin A \cos B = \frac{1}{2}[\sin(A - B) + \sin(A + B)]$$

$$\cos A \cos B = \frac{1}{2}[\cos(A - B) + \cos(A + B)]$$

To derive the first identity in the box, we write down the addition formulas for $\cos(A - B)$ and for $\cos(A + B)$:

$$\cos A \cos B + \sin A \sin B = \cos(A - B) \quad (1)$$

$$\cos A \cos B - \sin A \sin B = \cos(A + B) \quad (2)$$

If we subtract equation (2) from equation (1), we have

$$2 \sin A \sin B = \cos(A - B) - \cos(A + B)$$

Now, dividing both sides of this last equation by 2, we obtain the required identity:

$$\sin A \sin B = \frac{1}{2}[\cos(A - B) - \cos(A + B)]$$

The derivations of the remaining two product-to-sum formulas are entirely similar. (See Exercise 42.)

EXAMPLE 1 Changing a product of sines to a difference of cosines

Use the formula for $\sin A \sin B$ to simplify the expression $\sin 15° \sin 75°$.

SOLUTION

In the identity for $\sin A \sin B$ we substitute the values $A = 15°$ and $B = 75°$. This yields

$$\begin{aligned} \sin 15° \sin 75° &= \frac{1}{2}[\cos(15° - 75°) - \cos(15° + 75°)] \\ &= \frac{1}{2}[\cos(-60°) - \cos(90°)] \\ &= \frac{1}{2}\left(\frac{1}{2} - 0\right) = \frac{1}{4} \end{aligned}$$

EXAMPLE 2 **Changing a product of a sine and a cosine to a sum of sines**

Convert the product $\sin 4x \cos 3x$ to a sum.

SOLUTION
Using the formula for $\sin A \cos B$ with $A = 4x$ and $B = 3x$, we have

$$\begin{aligned}\sin 4x \cos 3x &= \frac{1}{2}[\sin(4x - 3x) + \sin(4x + 3x)]\\ &= \frac{1}{2}[\sin x + \sin 7x]\\ &= \frac{1}{2}\sin x + \frac{1}{2}\sin 7x\end{aligned}$$

Note: This result is *not* equal to $\sin(x/2) + \sin(7x/2)$.

EXAMPLE 3 **Changing a product of cosines to a sum of cosines**

Simplify the following expression.

$$\cos\left(s + \frac{\pi}{4}\right)\cos\left(s - \frac{\pi}{4}\right)$$

SOLUTION
In the formula

$$\cos A \cos B = \frac{1}{2}[\cos(A - B) + \cos(A + B)]$$

we will let $A = s + \pi/4$ and $B = s - \pi/4$. Notice that with these values for A and B, we have

$$A - B = \frac{\pi}{2} \qquad \text{and} \qquad A + B = 2s$$

Consequently,

$$\begin{aligned}\cos\left(s + \frac{\pi}{4}\right)\cos\left(s - \frac{\pi}{4}\right) &= \frac{1}{2}\left(\cos\frac{\pi}{2} + \cos 2s\right)\\ &= \frac{1}{2}(0 + \cos 2s) = \frac{1}{2}\cos 2s\end{aligned}$$

As we saw in Examples 1 and 3, converting a product into a sum can produce in some instances a much simpler form of a given expression. Of course, there are times when it may be more useful to proceed in the other direction. That is, there are times when we want to convert a sum (or a difference) into a product. The following *sum-to-product formulas* are useful here.

The Sum-to-Product Formulas

$$\sin \alpha + \sin \beta = 2 \sin \frac{\alpha + \beta}{2} \cos \frac{\alpha - \beta}{2}$$

$$\sin \alpha - \sin \beta = 2 \cos \frac{\alpha + \beta}{2} \sin \frac{\alpha - \beta}{2}$$

$$\cos \alpha + \cos \beta = 2 \cos \frac{\alpha + \beta}{2} \cos \frac{\alpha - \beta}{2}$$

$$\cos \alpha - \cos \beta = -2 \sin \frac{\alpha + \beta}{2} \sin \frac{\alpha - \beta}{2}$$

Each of these formulas can be derived from one of the product-to-sum formulas that are listed on page 585. For instance, to derive the formula for $\cos \alpha - \cos \beta$, we begin with

$$\sin A \sin B = \frac{1}{2}[\cos(A - B) - \cos(A + B)] \tag{3}$$

If we let

$$A + B = \alpha \tag{4}$$

and

$$A - B = \beta \tag{5}$$

then we have

$$A = \frac{\alpha + \beta}{2} \qquad \text{adding equations (4) and (5) and then dividing by 2}$$

and

$$B = \frac{\alpha - \beta}{2} \qquad \text{subtracting equation (5) from equation (4) and then dividing by 2}$$

Now we use these last two equations and also equations (4) and (5) to substitute in equation (3). The result is

$$\sin \frac{\alpha + \beta}{2} \sin \frac{\alpha - \beta}{2} = \frac{1}{2}(\cos \beta - \cos \alpha)$$

or, after multiplying by 2,

$$2 \sin \frac{\alpha + \beta}{2} \sin \frac{\alpha - \beta}{2} = \cos \beta - \cos \alpha$$

This last equation is equivalent to

$$\cos \alpha - \cos \beta = -2 \sin \frac{\alpha + \beta}{2} \sin \frac{\alpha - \beta}{2}$$

This is the fourth sum-to-product formula, as required. (For the derivation of the remaining three sum-to-product formulas, see Exercises 43–45.)

EXAMPLE 4 Using the sum-to-product formulas

Convert each expression to a product and simplify where possible:
(a) $\sin 50^\circ + \sin 70^\circ$; **(b)** $\cos\theta - \cos 5\theta$.

SOLUTION

(a) In the formula

$$\sin\alpha + \sin\beta = 2\sin\frac{\alpha+\beta}{2}\cos\frac{\alpha-\beta}{2}$$

we set $\alpha = 50^\circ$ and $\beta = 70^\circ$. Then we have

$$\begin{aligned}\sin 50^\circ + \sin 70^\circ &= 2\sin\frac{50^\circ + 70^\circ}{2}\cos\frac{50^\circ - 70^\circ}{2}\\ &= 2\sin 60^\circ\cos(-10^\circ)\\ &= \sqrt{3}\cos 10^\circ \qquad \text{using the identity } \cos(-\theta) = \cos\theta\end{aligned}$$

(b) Using the formula

$$\cos\alpha - \cos\beta = -2\sin\frac{\alpha+\beta}{2}\sin\frac{\alpha-\beta}{2}$$

we have

$$\begin{aligned}\cos\theta - \cos 5\theta &= -2\sin\frac{\theta + 5\theta}{2}\sin\frac{\theta - 5\theta}{2}\\ &= -2\sin 3\theta\sin(-2\theta)\\ &= 2\sin 3\theta\sin 2\theta \qquad \text{using the identity } \sin(-t) = -\sin t\end{aligned}$$

EXAMPLE 5 Expressing a sum of a sine and a cosine as a scaled and shifted cosine function

(a) Use a sum-to-product formula to show that

$$\sin x + \cos x = \sqrt{2}\cos\left(x - \frac{\pi}{4}\right)$$

(b) Use the identity in part (a) to graph the function $y = \sin x + \cos x$.

SOLUTION

(a) Using the identity $\cos x = \sin(\frac{\pi}{2} - x)$, we have

$$\begin{aligned}\sin x + \cos x &= \sin x + \sin\left(\frac{\pi}{2} - x\right)\\ &= 2\sin\frac{x + (\frac{\pi}{2} - x)}{2}\cos\frac{x - (\frac{\pi}{2} - x)}{2}\\ &= 2\sin\frac{\pi}{4}\cos\left(x - \frac{\pi}{4}\right) \qquad \text{Check the algebra.}\\ &= \sqrt{2}\cos\left(x - \frac{\pi}{4}\right)\end{aligned}$$

Actually, there is an easier way to obtain this identity. Expand $\cos(x - \pi/4)$ using an addition formula, then multiply the result by $\sqrt{2}$; the result will be $\cos x + \sin x$. We use the sum-to-product formula in this example because we want to begin with $\cos x + \sin x$ and work toward a form that can be graphed using the techniques of the previous chapter.

(b) In view of the identity established in part (a), we need to graph only the function $y = \sqrt{2}\cos(x - \frac{\pi}{4})$. This is a function of the form $y = A\cos(Bx - C)$, which we studied in Section 7.3. In Figure 1 we show the graph of $y = \sqrt{2}\cos(x - \frac{\pi}{4})$. You should verify for yourself (using the techniques of Section 7.3) that the amplitude, period, and phase shift indicated in Figure 1 are indeed correct.

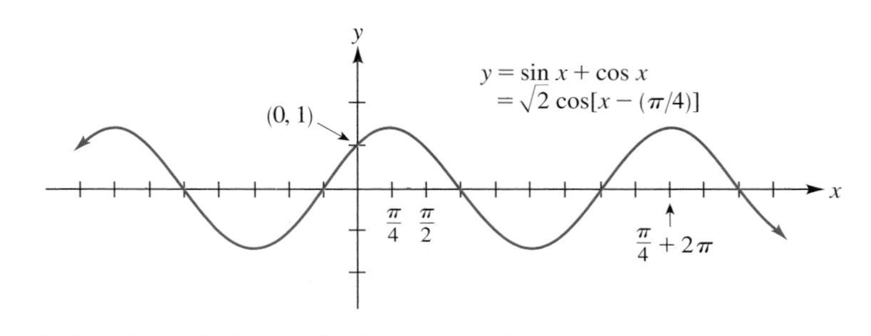

Figure 1
The graph of $y = \sin x + \cos x$ or $y = \sqrt{2}\cos(x - \pi/4)$
Amplitude: $\sqrt{2}$
Period: 2π
Phase shift: $\pi/4$

EXERCISE SET 8.3 

A

In Exercises 1–22, use a product-to-sum formula to convert each expression to a sum or difference. Simplify where possible.

1. $\cos 70° \cos 20°$
2. $\cos 50° \cos 40°$
3. $\sin 5° \sin 85°$
4. $\sin 130° \sin 10°$
5. $\sin 20° \cos 10°$
6. $\cos 18° \sin 72°$
7. $\cos \frac{\pi}{5} \cos \frac{4\pi}{5}$
8. $\cos \frac{5\pi}{12} \cos \frac{\pi}{12}$
9. $\sin \frac{2\pi}{7} \sin \frac{5\pi}{7}$
10. $\sin \frac{3\pi}{8} \sin \frac{\pi}{8}$
11. $\sin \frac{7\pi}{12} \cos \frac{\pi}{12}$
12. $\cos \frac{7\pi}{8} \sin \frac{\pi}{8}$
13. $\sin 3x \sin 4x$
14. $\cos 5x \cos 2x$
15. $\sin 6\theta \cos 5\theta$
16. $\sin \frac{2\theta}{3} \sin \frac{5\theta}{3}$
17. $\cos \frac{\theta}{2} \sin \frac{3\theta}{2}$
18. $\cos \frac{t}{4} \cos \frac{3t}{4}$
19. $\sin(2x + y)\sin(2x - y)$
20. $\cos(\theta + \frac{\pi}{6})\cos(\theta - \frac{\pi}{6})$
21. $\sin 2t \cos(s - t)$
22. $\cos(\alpha + 2\beta)\sin(2\alpha - \beta)$

In Exercises 23–34, convert each expression into a product and simplify where possible.

23. $\cos 35° + \cos 55°$
24. $\cos 50° - \cos 10°$
25. $\sin \frac{\pi}{5} - \sin \frac{3\pi}{10}$
26. $\sin \frac{\pi}{12} + \sin \frac{11\pi}{12}$
27. $\cos 5\theta - \cos 3\theta$
28. $\sin \frac{5\theta}{2} - \sin \frac{\theta}{2}$
29. $\sin 35° + \cos 65°$ *Hint*: Use the identity $\cos\theta = \sin(90° - \theta)$.
30. $\cos \frac{3\pi}{8} - \sin \frac{\pi}{8}$ *Hint*: Use the identity $\cos\theta = \sin(\frac{\pi}{2} - \theta)$.
31. $\sin(\frac{\pi}{3} + 2\theta) - \sin(\frac{\pi}{3} - 2\theta)$
32. $\cos(\frac{5\pi}{12} + \theta) + \cos(\frac{\pi}{12} - \theta)$
33. $\dfrac{\cos \frac{5\pi}{12} + \sin \frac{5\pi}{12}}{\cos \frac{\pi}{12} - \sin \frac{\pi}{12}}$
34. $\dfrac{\sin 47° + \cos 17°}{\cos 47° + \sin 17°}$

In Exercises 35–38, prove that the equations are identities.

35. $\dfrac{\sin s + \sin t}{\cos s + \cos t} = \tan\left(\dfrac{s + t}{2}\right)$
36. $\dfrac{\cos[(n - 2)\theta] - \cos n\theta}{\sin[(n - 2)\theta] + \sin n\theta} = \tan\theta$
37. $\dfrac{\sin 2x + \sin 2y}{\cos 2x + \cos 2y} = \tan(x + y)$
38. $\sin(\theta + \phi)\sin(\theta - \phi) = \sin^2\theta - \sin^2\phi$

In Exercises 39 and 40, convert each sum to a product.

39. $\cos 7\theta + \cos 5\theta + \cos 3\theta + \cos\theta$
Hint: The given expression can be written $(\cos 7\theta + \cos 5\theta) + (\cos 3\theta + \cos\theta)$. After converting the quantities in parentheses to products, look for a common term to factor out.

40. $\sin 2\theta + \sin 4\theta + \sin 6\theta + \sin 8\theta$

41. (a) Express the quantity $\sqrt{2}\,[\sin(x/2) + \cos(x/2)]$ in the form $A\cos(Bx - C)$.
(b) Use your result in part (a) to graph the function $f(x) = \sqrt{2}[\sin(x/2) + \cos(x/2)]$ for two complete cycles. Specify the amplitude, period, and phase shift.

B

42. **(a)** Derive the product-to-sum formula for $\cos A \cos B$.
Hint: Add equations (1) and (2) in the text on page 585.
(b) Derive the product-to-sum formula for $\sin A \cos B$.
Hint: Start by writing down the two addition formulas for $\sin(A + B)$ and $\sin(A - B)$, then add the two equations.

43. Derive the sum-to-product formula for $\cos\alpha + \cos\beta$.
Hint: Follow the method used in the text to derive the formula for $\cos\alpha - \cos\beta$, but rather than beginning with equation (3), begin instead with $\cos A \cos B = \frac{1}{2}[\cos(A - B) + \cos(A + B)]$.

44. Derive the sum-to-product formula for $\sin\alpha + \sin\beta$.
Hint: Follow the method used in the text to derive the formula for $\cos\alpha - \cos\beta$, but rather than beginning with equation (3), begin instead with $\sin A \cos B = \frac{1}{2}[\sin(A - B) + \sin(A + B)]$.

45. Derive the sum-to-product formula for $\sin\alpha - \sin\beta$.
Hint: In the formula for $\sin\alpha + \sin\beta$ (which we derived in Exercise 44), replace each occurrence of β with $-\beta$.

46. For this exercise, follow steps (a) through (c) to show that

$$\cos\frac{\pi}{7}\cos\frac{2\pi}{7}\cos\frac{3\pi}{7} = \frac{1}{8}$$

(a) Start with the expression on the left side of the given equation and multiply both the numerator and the denominator by the quantity $\sin\frac{\pi}{7}$. Show that the resulting expression can be written as

$$\frac{\frac{1}{2}\sin\frac{2\pi}{7}\cos\frac{2\pi}{7}\cos\frac{3\pi}{7}}{\sin\frac{\pi}{7}}$$

(b) Explain why the expression obtained in part (a) is equal to

$$\frac{\frac{1}{4}\sin\frac{4\pi}{7}\cos\frac{3\pi}{7}}{\sin\frac{\pi}{7}}$$

(c) Now use a product-to-sum formula to simplify the numerator of the expression in part (b). Then conclude that the original expression is indeed equal to 1/8, as we wished to show.

47. The following problem appears in *Problems in Elementary Mathematics* by V. Lidsky et al. (Moscow: MIR Publishers, 1973).

Simplify the following expression:

$$\frac{1}{2\sin 10^\circ} - 2\sin 70^\circ$$

C

In Exercises 48–50, assume that $A + B + C = 180^\circ$. Prove that each equation is an identity.

48. $\sin 2A + \sin 2B + \sin 2C = 4\sin A \sin B \sin C$

49. $\cos A + \cos B + \cos C = 1 + 4\sin(A/2)\sin(B/2)\sin(C/2)$

50. $\sin A + \sin B + \sin C = 4\cos(A/2)\cos(B/2)\cos(C/2)$

51. **(a)** The accompanying figure shows three triangles. For each triangle, use a calculator to verify that the sum of the cosines of the angles of the triangle is less than 3/2.

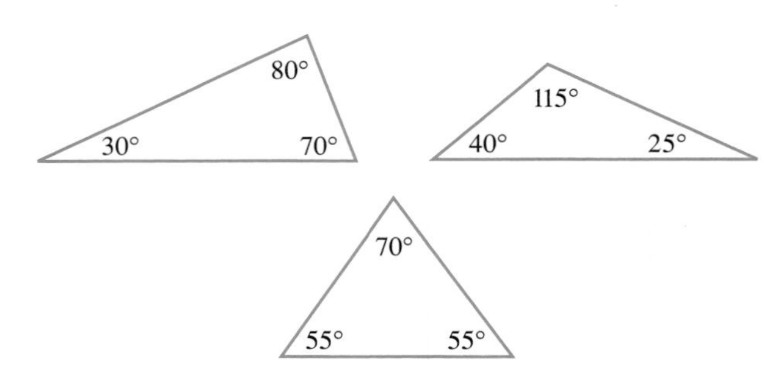

(b) What is the sum of the cosines of the three angles in an equilateral triangle?

(c) Let A, B, and C denote the three angles of a triangle, so that $A + B + C = 180^\circ$. The following sequence of steps proves the inequality $\cos A + \cos B + \cos C \le 3/2$. Supply the reasons or calculations that support each step. [The proof, by W. O. J. Moser, appeared in *The American Mathematical Monthly*, vol. 67 (1960), p. 695.]

(i) $\cos A + \cos B + \cos C$
$= 2\cos[(A + B)/2]\cos[(A - B)/2] + \cos C$

(ii) $\le 2\cos[(A + B)/2] + \cos C$

(iii) $= 2\cos[(180^\circ - C)/2] + \cos C$

(iv) $= 2\cos[90^\circ - (C/2)] + \cos C$

(v) $= 2\sin(C/2) + [1 - 2\sin^2(C/2)]$

(vi) $= (3/2) - 2[\sin(C/2) - (1/2)]^2$

(vii) $\le 3/2$

PROJECT

SUPERPOSITION

In Example 5 of Section 8.3 we saw how to use a sum-to-product formula to combine a cosine wave and a sine wave with the same amplitude, the same period, and no phase shift into a single cosine wave with the same period but a bigger amplitude and a phase shift. By using exactly the same argument, you can show that

$$a \cos Bx + a \sin Bx = \sqrt{2}a \cos\left(Bx - \frac{\pi}{4}\right)$$

What happens if the original waves have different amplitudes? The previous approach does not apply because to use a sum-to-product formula requires that the original waves have the same amplitude. It is not hard to imagine that we should obtain a single wave with a bigger amplitude and the same period. It turns out that the new cosine wave will exhibit a phase shift.

Let's assume that there is a way to combine the waves as desired and see what that requires. So suppose that

$$a \cos Bx + b \sin Bx = A \cos(Bx + C) \qquad \text{(i)}$$

Here we assume that a, b, and B are constants with B positive and a and b nonzero. We want to express A and C in terms of a and b. Using the cosine addition formula to expand $\cos(Bx + C)$, we have

$$\cos(Bx + C) = \cos Bx \cos C - \sin Bx \sin C$$

Substituting this into the right-hand side of equation (i) and rearranging, we obtain

$$a \cos Bx + b \sin Bx = (A \cos C) \cos Bx + (-A \sin C) \sin Bx \qquad \text{(ii)}$$

We want to choose A and C so that equation (ii) is true for all real numbers, that is, equation (ii) is an identity. Comparing the numerical coefficients of $\cos Bx$ and $\sin Bx$ on both sides of equation (ii), we choose A and C such that

$$A \cos C = a \qquad \text{and} \qquad -A \sin C = b \qquad \text{(iii)}$$

For equation (ii) to be an identity, it can be shown that these are in fact the *only* possible choices for A and C. (We will use a similar argument in our discussion of partial fractions in Chapter 12.) Equations (iii) imply that

$$\begin{aligned} a^2 + b^2 &= (A \cos C)^2 + (-A \sin C)^2 \\ &= A^2 \cos^2 C + A^2 \sin^2 C \\ &= A^2 (\cos^2 C + \sin^2 C) \\ &= A^2 \end{aligned}$$

Let's take $A = \sqrt{a^2 + b^2}$. Equations (iii) also imply that

$$\frac{b}{a} = \frac{-A \sin C}{A \cos C} = -\tan C$$

Let's take C in the open interval between $-\pi/2$ and $\pi/2$ so that $\tan C = -b/a$. These conditions allow us to find a unique number for C. (See Section 8.5 on inverse trigonometric functions for a discussion of the inverse tangent function.) The notation $C = \tan^{-1}(-b/a)$ means "C is the unique number in the interval $(-\frac{\pi}{2}, \frac{\pi}{2})$ such that $\tan C = -b/a$." C can be numerically approximated using the inverse tangent function on your calculator. We summarize

$$a \cos Bx + b \sin Bx = A \cos(Bx + C)$$

where

$$A = \sqrt{a^2 + b^2} \qquad \text{and} \qquad C = \tan^{-1}\left(-\frac{b}{a}\right) \qquad \text{(iv)}$$

The process of adding wave functions is called **superposition** and (iv) is an example of a **superposition formula.** Superposition is an essential feature of many physical phenomena involving waves, including acoustics (sound waves), optics (electromagnetic waves), radio (also electromagnetic waves), and surface waves (water waves). The exercises below use and extend the concepts developed on superposition.

EXERCISES

1. Use the method of Example 5 in Section 8.3 to show

$$a \cos Bx + a \sin Bx = \sqrt{2}a \cos\left(Bx - \frac{\pi}{4}\right)$$

2. Use formula (iv) to redo problem 1.

3. (a) Use formula (iv) to show that $3 \cos 4x + \sqrt{3} \sin 4x = 2\sqrt{3} \cos(4x - \frac{\pi}{6})$.

Ⓖ **(b)** Use a graphing utility to graph the left-hand side of the identity in part (a), and estimate the amplitude and phase shift. How do your estimates compare with the exact values?

4. (a) Use formula (iv) to show $2 \cos 3x + 5 \sin 3x = \sqrt{29} \cos(3x + C)$, where C is in the interval $(-\frac{\pi}{2}, \frac{\pi}{2})$ and $\tan C = -5/2$.

(b) Use a calculator to find C correct to two decimal places.

Ⓖ **(c)** Use a graphing utility to graph the left-hand side of the identity in part (a) and estimate the amplitude and phase shift. How do your estimates compare with the values from (a) and (b)?

5. What happens if the waves have different phase shifts? Use the method of derivation of formula (iv) to show that

$$a \cos(Bx + D) + b \sin(Bx + E) = A \cos(Bx + C)$$

where

$$A = \sqrt{a^2 + b^2 + 2ab \sin(E - D)}$$

and

$$\tan C = -\frac{b \cos E - a \sin D}{a \cos D + b \sin E} \qquad \text{with } C \text{ in } (-\tfrac{\pi}{2}, \tfrac{\pi}{2})$$

8.4 TRIGONOMETRIC EQUATIONS

In this section we consider some techniques for solving equations involving the trigonometric functions. As usual, by a **solution** or **root** of an equation we mean a value of the variable for which the equation becomes a true statement.

EXAMPLE 1 Determining whether a given number is a solution of a trigonometric equation

Consider the trigonometric equation $\sin x + \cos x = 1$. Is $x = \pi/4$ a solution? Is $x = \pi/2$ a solution?

SOLUTION

To see whether the value $x = \pi/4$ satisfies the given equation, we write

$$\sin\frac{\pi}{4} + \cos\frac{\pi}{4} \stackrel{?}{=} 1$$

$$\frac{\sqrt{2}}{2} + \frac{\sqrt{2}}{2} \stackrel{?}{=} 1$$

$$\sqrt{2} \stackrel{?}{=} 1 \quad \text{(No)}$$

Thus $x = \pi/4$ is not a solution. In a similar fashion we can check to see whether $x = \pi/2$ is a solution:

$$\sin\frac{\pi}{2} + \cos\frac{\pi}{2} \stackrel{?}{=} 1$$

$$1 + 0 \stackrel{?}{=} 1 \quad \text{(Yes)}$$

Thus $x = \pi/2$ is a solution.

The example that we've just concluded serves to remind us of the difference between a *conditional equation* and an *identity*. An identity is true for all values of the variable in its domain. For example, the equation $\sin^2 t + \cos^2 t = 1$ is an identity: It is true for every real number t. In contrast to this, a conditional equation is true only for some (or perhaps even none) of the values of the variable. The equation in Example 1 is a conditional equation; we saw that it is false when $x = \pi/4$ and true when $x = \pi/2$. The equation $\sin t = 2$ is an example of a conditional equation that has no solution. (Why?) The equations that we are going to solve in this section are conditional equations that involve the trigonometric functions. In general, there is no single technique that can be used to solve every trigonometric equation. In the examples that follow, we illustrate some of the more common approaches to solving trigonometric equations. As background for Example 2, you should review Example 2 in Section 7.2. (In that example we obtained solutions for the trigonometric equation $\cos x = 0.8$.)

EXAMPLE 2 Solving a trigonometric equation

Consider the equation

$$\sin x = \frac{1}{2} \tag{1}$$

(a) Use your knowledge of the sine function (and not a calculator) to find all solutions of equation (1) in the open interval $(0, 2\pi)$.

(b) Use a calculator to find all solutions of equation (1) in the open interval $(0, 2\pi)$. Check that the answers are consistent with those in part (a).

(c) Find all real-number solutions of equation (1).

SOLUTION

(a) First of all, we know from earlier work that $\sin(\pi/6) = 1/2$. So one solution of equation (1) is certainly $x_1 = \pi/6$. To find another solution, we note that in Quadrant II, as well as in Quadrant I, $\sin x$ is positive. In Quadrant II the angle with a reference angle of $\pi/6$ is $5\pi/6$; see Figure 1. Thus $\sin(5\pi/6) = 1/2$, and we conclude that $x_2 = 5\pi/6$ is also a solution of equation (1). Since $\sin x$ is negative in Quadrants III and IV, we needn't look there for solutions. In summary, then, the solutions of the equation $\sin x = 1/2$ in the open interval $(0, 2\pi)$ are $x_1 = \pi/6$ and $x_2 = 5\pi/6$.

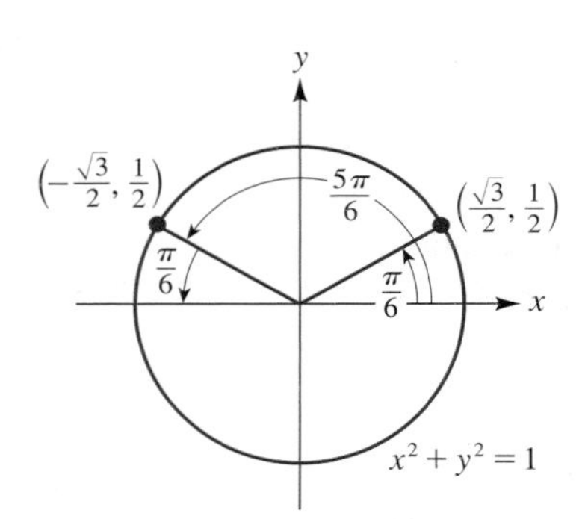

Figure 1
$\sin\frac{\pi}{6} = \sin\frac{5\pi}{6} = \frac{1}{2}$

(b) We can determine a root of equation (1) by using a calculator and the *inverse sine function,* denoted by $\sin^{-1}$. Applying the inverse sine function to both sides of equation (1) yields

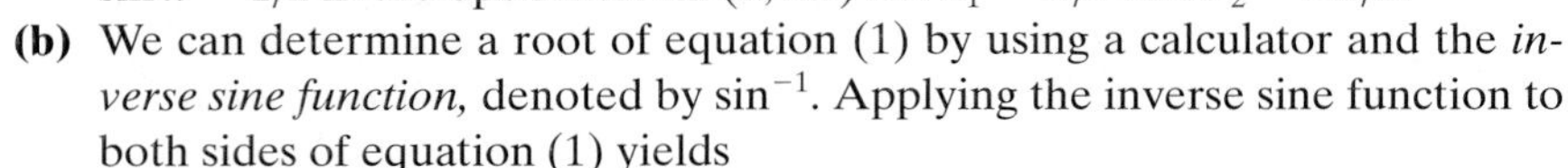

$$\sin^{-1}(\sin x) = \sin^{-1}\left(\frac{1}{2}\right)$$

and therefore

$$x = \sin^{-1}\left(\frac{1}{2}\right)$$

Now we use a calculator, set in the radian mode, to evaluate the expression $\sin^{-1}(1/2)$. (If these keystrokes don't work on your calculator, be sure to check the instruction manual.)

EXPRESSION	KEYSTROKES	OUTPUT
$\sin^{-1}(0.5)$	2nd sin 0.5 ENTER	0.52359 . . .

This gives us a root 0.52359 . . . for equation (1). Indeed, this is the decimal approximation for the root $x_1 = \pi/6$ that we determined in part (a). Take a moment to confirm this fact; that is, use your calculator to verify that $\pi/6 = 0.52359\ldots$.

As explained in part (a), there is another root, one in Quadrant II. Taking the calculator value 0.52359 . . . for our reference angle, the second-quadrant root then is

$$\pi - 0.52359\ldots = 2.61799\ldots$$

This is the decimal approximation for the root $x_2 = 5\pi/6$ that we determined in part (a). (Use your calculator to verify that $5\pi/6 = 2.61799\ldots$.)

(c) We can use the results in part (a), along with the fact that the sine function is periodic, to specify all real number solutions for the given equation. Since the period of the sine function is 2π, we have

$$\sin(x + 2k\pi) = \sin x \qquad \text{for every integer } k$$

Consequently, since $\sin\frac{\pi}{6} = \frac{1}{2}$, we know that $\sin(\frac{\pi}{6} + 2k\pi)$ is also equal to $1/2$. In other words, for every integer k the quantity $\pi/6 + 2k\pi$ is a solution of the given equation. Following the same reasoning, the quantity $5\pi/6 + 2k\pi$ is also a solution of the given equation for every integer k. In summary, there are infinitely many real-number solutions of the equation $\sin x = 1/2$. These solutions are given by the expressions

$$\frac{\pi}{6} + 2k\pi \quad \text{and} \quad \frac{5\pi}{6} + 2k\pi \quad \text{where } k \text{ is an integer}$$

Figure 2 shows some of these solutions; they are the x-coordinates of the points where the sine curve intersects the line $y = 1/2$.

GRAPHICAL PERSPECTIVE

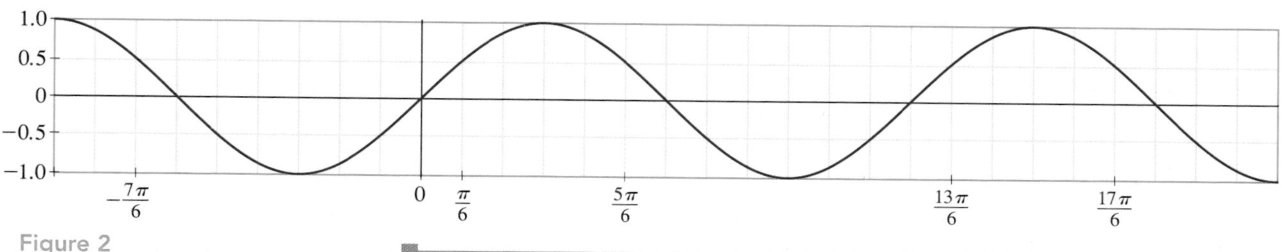

Figure 2
A view of the curve $y = \sin x$ for $-1.5\pi \le x \le 3.5\pi$
The x-coordinates of the points where the curve $y = \sin x$ intersects the line $y = 1/2$ are solutions of the equation $\sin x = 1/2$. This view shows five of the solutions: $-7\pi/6$, $\pi/6$, $5\pi/6$, $13\pi/6$, and $17\pi/6$.

In part (b) of the example just concluded, we indicated how the inverse sine function can be used to solve a simple trigonometric equation. In the next section the inverse sine function is analyzed in detail. But for our present work with this function, you need to know only the following definition:

$\sin^{-1}(x)$ denotes the unique number in the interval $[-\frac{\pi}{2}, \frac{\pi}{2}]$ whose sine is x.

As examples of this definition, we have

$$\sin^{-1}\left(\frac{\sqrt{3}}{2}\right) = \frac{\pi}{3}$$ since $\pi/3$ is the unique number in the interval $[-\frac{\pi}{2}, \frac{\pi}{2}]$ whose sine is $\sqrt{3}/2$

$$\sin^{-1}\left(\frac{\sqrt{3}}{2}\right) \neq \frac{2\pi}{3}$$ Although $\sin(2\pi/3) = \sqrt{3}/2$, the number $2\pi/3$ is not in the required interval $[-\frac{\pi}{2}, \frac{\pi}{2}]$.

For some of the exercises in this section you'll need to use the inverse cosine function (which we worked with in Section 7.2) or the *inverse tangent function,* rather than the inverse sine function that we've just been discussing. For reference we define the three functions in the box that follows. (All three functions are discussed at greater length in the next section.)

DEFINITION **Inverse Trigonometric Functions**

NAME OF FUNCTION	ABBREVIATION	DEFINITION
Inverse cosine	$\cos^{-1}$	$\cos^{-1}(x)$ is the unique number in the interval $[0, \pi]$ whose cosine is x.
Inverse sine	$\sin^{-1}$	$\sin^{-1}(x)$ is the unique number in the interval $[-\frac{\pi}{2}, \frac{\pi}{2}]$ whose sine is x.
Inverse tangent	$\tan^{-1}$	$\tan^{-1}(x)$ is the unique number in the interval $(-\frac{\pi}{2}, \frac{\pi}{2})$ whose tangent is x.

The next two examples show how factoring can be used to solve a trigonometric equation.

EXAMPLE 3 Solving a trigonometric equation by factoring

Find all real-number solutions of the equation

$$\cos^2 x + \cos x - 2 = 0$$

SOLUTION
By factoring the expression on the left-hand side, we obtain

$$(\cos x + 2)(\cos x - 1) = 0$$

Therefore

$$\cos x = -2 \qquad \text{or} \qquad \cos x = 1$$

We discard the result $\cos x = -2$ because the value of $\cos x$ is never less than -1. From the equation $\cos x = 1$ we conclude that $x = 0$ is the only solution in the interval $[0, 2\pi)$. Since the cosine function has period 2π, all solutions are given by $x = 0 + 2\pi k = 2\pi k$, where k is an integer.

EXAMPLE 4 Solving a trigonometric equation by factoring

Find all solutions of the equation $\tan^2 \theta + \tan \theta - 6 = 0$ in the interval $[0, 2\pi]$. Round the final answers to three decimal places.

SOLUTION
By factoring, we have

$$(\tan \theta - 2)(\tan \theta + 3) = 0$$

and therefore

$$\tan \theta = 2 \qquad \text{or} \qquad \tan \theta = -3$$

Applying the inverse tangent function to both sides of the equation $\tan \theta = 2$ gives us

$$\tan^{-1}(\tan \theta) = \tan^{-1}(2)$$

and, consequently,

$$\begin{aligned} \theta &= \tan^{-1}(2) \\ &\approx 1.107 \qquad \text{using a calculator set in the radian mode} \end{aligned}$$

Now, since the period of the tangent function is π, we can find another root of the equation $\tan \theta = 2$ just by adding π to the root $\tan^{-1}(2)$. This gives us

$$\tan^{-1}(2) + \pi \approx 4.249 \qquad \text{using a calculator set in the radian mode}$$

Note that this root is also in the required interval $[0, 2\pi]$. However, adding π again to this value will take us out of the interval $[0, 2\pi]$. So let's turn to the other equation that we obtained through factoring: $\tan\theta = -3$. Applying the inverse tangent function to both sides of the equation $\tan\theta = -3$ gives us

$$\tan^{-1}(\tan\theta) = \tan^{-1}(-3)$$

and therefore

$$\begin{aligned}\theta &= \tan^{-1}(-3)\\ &= -1.2490457\ldots \qquad \text{using a calculator}\end{aligned}$$

Although this last value is a root of the equation $\tan\theta = -3$, it is not in the required interval $[0, 2\pi]$. To generate a root that does belong to this interval, we add π (the period of the tangent function) to the quantity $\tan^{-1}(-3)$ to obtain

$$\tan^{-1}(-3) + \pi \approx 1.893 \qquad \text{using a calculator}$$

Still another root can be obtained by again adding π.

$$\tan^{-1}(-3) + 2\pi \approx 5.034$$

Adding π again would take us out of the interval $[0, 2\pi]$, so we stop here. Let's summarize our results. In the interval $[0, 2\pi]$ there are four roots of the equation $\tan^2\theta + \tan\theta - 6 = 0$. They are

$$\begin{aligned}\tan^{-1}(2) &\approx 1.107\\ \tan^{-1}(2) + \pi &\approx 4.249\\ \tan^{-1}(-3) + \pi &\approx 1.893\\ \tan^{-1}(-3) + 2\pi &\approx 5.034\end{aligned}$$

For a graphical check of these results, see Figure 3, which shows the graph of the equation $y = \tan^2\theta + \tan\theta - 6$. The θ-intercepts of the graph are the roots of the equation $\tan^2\theta + \tan\theta - 6 = 0$. In the figure, note that each intercept is consistent with one of the roots that we have determined. For instance, the largest intercept in Figure 3 is approximately 5.0. This corresponds to the root $\tan^{-1}(-3) + 2\pi \approx 5.034$.

GRAPHICAL PERSPECTIVE

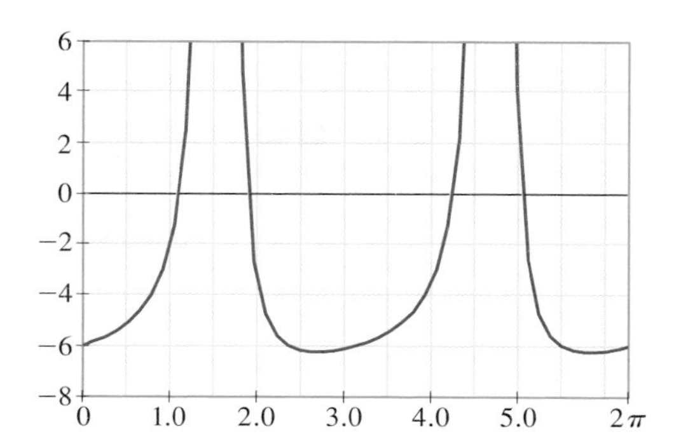

Figure 3
A view of $y = \tan^2\theta + \tan\theta - 6$ for $0 \le \theta \le 2\pi$

In some equations more than one trigonometric function is present. A common approach here is to express the various functions in terms of a single function. The next example demonstrates this technique.

EXAMPLE 5 Using a trigonometric identity to solve an equation

Find all real-number solutions of the equation

$$3 \tan^2 x - \sec^2 x - 5 = 0$$

SOLUTION

We use the Pythagorean identity $\tan^2 x + 1 = \sec^2 x$ to substitute for $\sec^2 x$ in the given equation. This gives us

$$\begin{aligned} 3 \tan^2 x - (\tan^2 x + 1) - 5 &= 0 \\ 2 \tan^2 x - 6 &= 0 \\ 2 \tan^2 x &= 6 \\ \tan^2 x &= 3 \\ \tan x &= \pm\sqrt{3} \end{aligned}$$

Since the period of the tangent function is π, we need to find only those values of x between 0 and π that satisfy $\tan x = \pm\sqrt{3}$. The other solutions will then be obtained by adding multiples of π to these solutions. Now, in Quadrant I the tangent is positive, and we know that $\tan x = \sqrt{3}$ when $x = \pi/3$. In Quadrant II $\tan x$ is negative, and we know that $\tan 2\pi/3 = -\sqrt{3}$, since the reference angle for $2\pi/3$ is $\pi/3$. Thus the solutions between 0 and π are $x = \pi/3$ and $x = 2\pi/3$. It follows that all of the solutions to the equation $3 \tan^2 x - \sec^2 x - 5 = 0$ are given by

$$x = \frac{\pi}{3} + \pi k \qquad \text{and} \qquad x = \frac{2\pi}{3} + \pi k \qquad \text{where } k \text{ is an integer}$$

The technique used in Example 5, that is, expressing the various functions in terms of a single function, is most useful when it does not involve introducing a radical expression. For instance, consider the equation $\sin s + \cos s = 1$. Although we could begin by replacing $\cos s$ by the expression $\pm\sqrt{1 - \sin^2 s}$, it turns out to be easier in this situation to begin by squaring both sides of the given equation. This is done in the next example.

EXAMPLE 6 Squaring both sides to solve a trigonometric equation

Find all solutions of the equation $\sin s + \cos s = 1$ satisfying $0° \leq s < 360°$.

SOLUTION

Squaring both sides of the equation yields

$$\begin{aligned} (\sin s + \cos s)^2 &= 1^2 \\ \sin^2 s + 2 \sin s \cos s + \cos^2 s &= 1 \end{aligned}$$

(The $\sin^2 s$ and $\cos^2 s$ terms: These add to 1.)

Consequently, we have

$$2 \sin s \cos s = 0$$

From this last equation we conclude that $\sin s = 0$ or $\cos s = 0$. When $\sin s = 0$, we know that $s = 0°$ or $s = 180°$. And when $\cos s = 0$, we know that $s = 90°$ or $s = 270°$. Now we must go back and check which (if any) of these values is a solution to the *original* equation. This must be done because squaring both sides in the process of solving an equation may introduce extraneous roots.

$s = 0°$	$\sin 0° + \cos 0° \stackrel{?}{=} 1$	
	$0 + 1 \stackrel{?}{=} 1$	True
$s = 90°$	$\sin 90° + \cos 90° \stackrel{?}{=} 1$	
	$1 + 0 \stackrel{?}{=} 1$	True
$s = 180°$	$\sin 180° + \cos 180° \stackrel{?}{=} 1$	
	$0 + (-1) \stackrel{?}{=} 1$	False
$s = 270°$	$\sin 270° + \cos 270° \stackrel{?}{=} 1$	
	$-1 + 0 \stackrel{?}{=} 1$	False

We conclude that the only solutions of the equation $\sin s + \cos s = 1$ on the interval $0° \le s < 360°$ are $s = 0°$ and $s = 90°$.

Note: The equation $2 \sin s \cos s = 0$ is equivalent to $\sin 2s = 0$. (Why?) Then $2s = 0°$ or $180°$. So $s = 0°$ or $90°$. As before, both are solutions of the equation.

In the example that follows, we consider an equation that involves a multiple of the unknown angle.

EXAMPLE 7 Solving a trigonometric equation that involves a multiple of an angle

Solve the equation $\sin 3x = 1$ on the interval $0 \le x \le 2\pi$.

SOLUTION

We know that $\sin(\pi/2) = 1$. Thus one solution can be found by writing $3x = \pi/2$, from which we conclude that $x = \pi/6$. We can look for other solutions in the required interval by writing, more generally,

$$3x = \frac{\pi}{2} + 2\pi k = \frac{\pi + 4\pi k}{2}$$

and therefore

$$x = \frac{\pi + 4\pi k}{6}$$

Thus when $k = 1$, we obtain

$$x = \frac{\pi + 4\pi(1)}{6} = \frac{5\pi}{6}$$

With $k = 2$ we have

$$x = \frac{\pi + 4\pi(2)}{6} = \frac{9\pi}{6} = \frac{3\pi}{2}$$

And with $k = 3$ we have

$$x = \frac{\pi + 4\pi(3)}{6} = \frac{13\pi}{6} \qquad \text{which is greater than } 2\pi$$

We conclude that the solutions of $\sin 3x = 1$ on the interval $0 \le x \le 2\pi$ are $\pi/6$, $5\pi/6$, and $3\pi/2$.

In Example 7, notice that we did not need to make use of a formula for $\sin 3x$, even though the expression $\sin 3x$ did appear in the given equation. In the next example, however, we do make use of the identity $\sin 2x = 2 \sin x \cos x$.

EXAMPLE 8 Transforming an equation in order to use a trigonometric identity

Solve the equation $\sin x \cos x = 1$.

SOLUTION

We could begin by squaring both sides. (See Exercise 83.) However, with the double-angle formula for sine in mind, we can proceed as follows. We multiply both sides of the given equation by 2. This yields

$$2 \sin x \cos x = 2$$

and, consequently,

$$\sin 2x = 2 \qquad \text{using the double-angle formula}$$

This last equation has no solution, since the value of the sine function never exceeds 1. Thus the equation $\sin x \cos x = 1$ has no solution.

For the last example in this section we look at another case in which a calculator is required.

EXAMPLE 9 Transforming an equation in both sine and cosine to an equation in only the tangent function

Find all angles θ in the interval $0° \le \theta \le 360°$ that satisfy the following equation: $2 \sin \theta = \cos \theta$.

SOLUTION

We first want to rewrite the given equation using a single function, rather than both sine and cosine. The easiest way to do this is to divide both sides by $\cos \theta$. Nothing is lost here in assuming $\cos \theta \ne 0$. (If $\cos \theta$ were zero, then θ would be $90°$ or $270°$; but neither of those angles is a solution of the given equation.) Dividing both sides of the given equation by $\cos \theta$, we have

$$\frac{2 \sin \theta}{\cos \theta} = \frac{\cos \theta}{\cos \theta}$$
$$2 \tan \theta = 1$$
$$\tan \theta = 0.5$$

From experience we know that none of the angles with which we are familiar (the multiples of 30° and 45°) has a value of 0.5 for the tangent. Therefore, as in Example 4, we use the inverse tangent function and a calculator. First, applying the inverse tangent function to both sides of the equation $\tan\theta = 0.5$, we have

$$\tan^{-1}(\tan\theta) = \tan^{-1}(0.5)$$

and therefore

$$\theta = \tan^{-1}(0.5)$$

Now, according to the statement of the problem, our answers are to be in degrees, not radians. So before using the calculator to evaluate the expression $\tan^{-1}(0.5)$, we first set the calculator in the degree mode. Then, as you should check for yourself, we obtain

$$\theta = \tan^{-1}(0.5) \approx 26.6^\circ \qquad \text{using a calculator set in the degree mode}$$

To find another angle θ satisfying the equation $\tan\theta = 0.5$, we use the fact that the period of the tangent is π radians, which is 180°. Thus a second solution (rounded to one decimal place) is

$$26.6^\circ + 180^\circ = 206.6^\circ$$

Adding another 180° to this last value will take us out of the specified interval $0^\circ \le \theta \le 360^\circ$, so we stop here. In summary, there are two angles between 0° and 360° satisfying the given equation $2\sin\theta = \cos\theta$. They are approximately, to one decimal place, 26.6° and 206.6°. For a graphical check of these results, see Figure 4, which shows the graphs of the equations $y = 2\sin\theta$ and $y = \cos\theta$ for $0^\circ \le \theta \le 360^\circ$. (In Figure 4 we are viewing the horizontal axis as the θ-axis, where θ is measured in degrees.) The graphs intersect at two points. The θ-coordinates of these points are solutions of the equation $2\sin\theta = \cos\theta$. The figure shows that at one of the intersection points, θ is slightly less than 30°, while at the other intersection point, θ is slightly less than 210°. These observations are consistent with the numerical values that we have obtained for the roots.

GRAPHICAL PERSPECTIVE

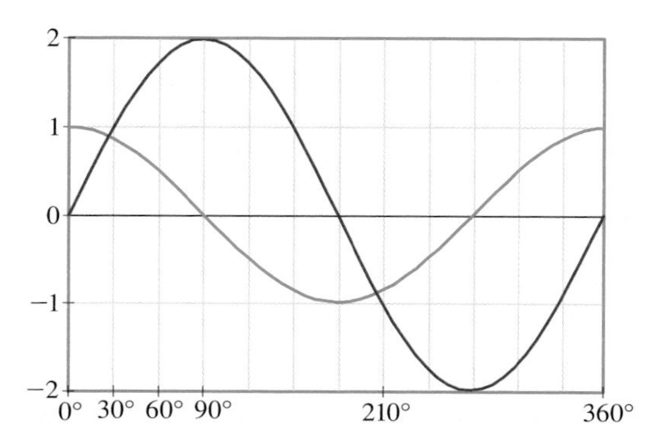

Figure 4
Graphs of $y = 2\sin\theta$ and $y = \cos\theta$ for $0^\circ \le \theta \le 360^\circ$

EXERCISE SET 8.4

A

1. Is $\theta = \pi/2$ a solution of $2\cos^2\theta - 3\cos\theta = 0$?
2. Is $x = 15°$ a solution of $(\sqrt{3}/3)\cos 2x + \sin 2x = 1$?
3. Is $x = 3\pi/4$ a solution of $\tan^2 x - 3\tan x + 2 = 0$?
4. Is $t = 2\pi/3$ a solution of $2\sin t + 2\cos t = \sqrt{3} - 1$?

In Exercises 5–22, determine all solutions of the given equations. Express your answers using radian measure.

5. $\sin\theta = \sqrt{3}/2$
6. $\sin\theta = \sqrt{2}/2$
7. $\sin\theta = -1/2$
8. $\sin\theta + (\sqrt{2}/2) = 0$
9. $\cos\theta = -1$
10. $\cos\theta = 1/2$
11. $\tan\theta = \sqrt{3}$
12. $\tan\theta + (\sqrt{3}/3) = 0$
13. $\tan x = 0$
14. $2\sin^2 x - 3\sin x + 1 = 0$
15. $2\cos^2\theta + \cos\theta = 0$
16. $\sin^2 x - \sin x - 6 = 0$
17. $\cos^2 t \sin t - \sin t = 0$
18. $\cos\theta + 2\sec\theta = -3$
19. $2\cos^2 x - \sin x - 1 = 0$
20. $2\cot^2 x + \csc^2 x - 2 = 0$
21. $\sqrt{3}\sin t - \sqrt{1 + \sin^2 t} = 0$
22. $\sec\alpha + \tan\alpha = \sqrt{3}$

Ⓖ 23. **(a)** Graph the equation $y = \cos^2 x + \cos x - 2$. Use a viewing rectangle extending from 0 to 10 in the x-direction and from -3 to 3 in the y-direction.
(b) The picture obtained in part (a) indicates that the graph has an x-intercept between 6 and 7. By zooming in on this x-intercept, or by using a solve key, obtain the first two decimal places of this intercept.
(c) According to Example 3, what is the exact value of the x-intercept that you approximated in part (b)?

Ⓖ 24. **(a)** Graph the equation $y = \tan^2 x + \tan x - 6$ for $0 \le x \le 2\pi$.
(b) By zooming in on each x-intercept of the graph in part (a), estimate the roots of the equation $\tan^2 x + \tan x - 6 = 0$. Then check that your estimates are consistent with the values obtained in Example 4.

In Exercises 25–38, use a calculator to find all solutions in the interval $(0, 2\pi)$. Round the answers to two decimal places.

25. $\cos x = 0.184$
26. $\cos t = -0.567$
27. $\sin x = 1/\sqrt{5}$
28. $\sin t = -0.301$
29. $\tan x = 6$
30. $\tan t = -5.25$
31. $\sin t = 5\cos t$
32. $\sin x \cos x = 0.035$
33. $\sec t = 2.24$ *Hint*: The equation is equivalent to $\cos t = 1/(2.24)$.
34. $\cot x = -3.27$
35. $\tan^2 x + \tan x - 12 = 0$
36. $15\sin^2 x - 26\sin x + 8 = 0$
37. $16\sin^3 x - 12\sin^2 x + 36\sin x - 27 = 0$ *Hint*: Factor by grouping.
38. $3\cos^3 x - 9\cos^2 x + \cos x - 3 = 0$ *Hint*: Factor by grouping.

In Exercises 39–44, find all solutions in the interval $0° \le \theta \le 360°$. Where necessary, use a calculator and round to one decimal place.

39. $\sin\theta = 1/4$
40. $\cos\theta = -4/5$
41. $9\tan^2\theta - 16 = 0$
42. $5\sin^2\theta + 13\sin\theta - 6 = 0$
43. $\cos^2\theta - \cos\theta - 1 = 0$ *Hint*: You'll need to use the quadratic formula.
44. $\tan^2\theta - \tan\theta - 1 = 0$

In Exercises 45–52, determine all of the solutions in the interval $0° \le \theta < 360°$.

45. $\cos 3\theta = 1$
46. $\tan 2\theta = -1$
47. $\sin 3\theta = -\sqrt{2}/2$
48. $\sin(\theta/2) = 1/2$
49. $\sin 2\theta = -2\cos\theta$
50. $2\sin^2\theta - \cos 2\theta = 0$
51. $\sin 2\theta = \sqrt{3}\cos 2\theta$ *Hint*: Divide by $\cos 2\theta$.
52. $\sin\theta = \cos(\theta/2)$. *Hint*: $\sin\theta = 2\sin(\theta/2)\cos(\theta/2)$

Ⓖ *In Exercises 53–74, solve the equations on the interval $[0, 2\pi]$ as follows. Graph the expression on each side of the equation and then zoom in on the intersection points until you are certain of the first three decimal places in each answer. For instance, for Exercise 53, when you graph the two equations $y = \cos x$ and $y = 0.623$ on the interval $[0, 2\pi]$, you'll see that there are two intersection points. The x-coordinates of these points are roots of the equation $\cos x = 0.623$.*

53. $\cos x = 0.623$
54. $\sin x = -0.438$
55. $\cos x = \tan x$
56. $\cos x = \tan 2x$
57. $\cos^2 x = 2\sin x$
58. $\cos^3 x = 2\sin x$
59. $\cos 2x + 1 = \cos(2x + 1)$
60. $\cos 2x + 0.9 = \cos(2x + 1)$
61. $\cos(x/2) = \cos(x^2/2)$
62. $\cos(x/2) = \cos(x^2/12)$
63. $2\sin x - 3\cos x = \tan(x/4)$
64. $2\sin x + 3\cos x = \tan(x/4)$
65. $\sqrt{x} = \tan x$
66. $x^2 = \tan x$
67. $\sin(\cos x) = \sin x$
68. $\cos(\sin x) = \sin x$
69. $\tan x = x$
70. $\cos x = x$
71. $\sin^3 x + \cos^3 x = 0.5$
72. $\sin^3 x - \cos^3 x = 0.5$
73. $1 - \tan^2 x = 2\sin(x/5)$
74. $1 - \tan^2 x = 2\sin(x/6)$

B

75. Find all solutions of the equation $\tan 3x - \tan x = 0$ in the interval $0 \le x < 2\pi$. *Hint*: Write $\tan 3x = \tan(2x + x)$ and use the addition formula for tangent.
76. Find all solutions of the equation $2\sin x = 1 - \cos x$ in the interval $0° \le x < 360°$. Use a calculator and round the answer(s) to one decimal place.
77. Find all solutions of the equation $\cos(x/2) = 1 + \cos x$ in the interval $0 \le x < 2\pi$.
78. Find all solutions of the equation
$$\sin 3x \cos x + \cos 3x \sin x = \frac{\sqrt{3}}{2}$$
in the interval $0 < x < 2\pi$.

79. Find all real numbers θ for which

$$\sec 4\theta + 2 \sin 4\theta = 0$$

80. Consider the equation $\sin^2 x - \cos^2 x = 7/25$.

(a) Solve the equation for $\cos x$.

(b) Find all of the solutions of the equation satisfying $0 < x < \pi$.

(c) Solve the original equation by means of a double-angle formula and use the result to check your answers in part (b).

81. Find a solution of the equation $4 \sin \theta - 3 \cos \theta = 2$ in the interval $0° < \theta < 90°$. *Hint:* Add $3 \cos \theta$ to both sides, then square.

82. Find all solutions of the equation

$$\sin^3 \theta \cos \theta - \sin \theta \cos^3 \theta = -\frac{1}{4}$$

in the interval $0 < \theta < \pi$. *Hint:* Factor the left-hand side, then use the double-angle formulas.

83. Consider the equation $\sin x \cos x = 1$.

(a) Square both sides and then replace $\cos^2 x$ by $1 - \sin^2 x$. Show that the resulting equation can be written $\sin^4 x - \sin^2 x + 1 = 0$.

(b) Show that the equation $\sin^4 x - \sin^2 x + 1 = 0$ has no solutions. Conclude from this that the original equation has no real-number solutions.

84. The accompanying figure shows a portion of the graph of the periodic function

$$f(x) = \sin(1 + \sin x)$$

Find the x-coordinates of the turning points P, Q, and R. Round the answers to three decimal places. *Hint:* You may use the fact that the x-coordinate of Q is halfway between the x-coordinates of P and R.

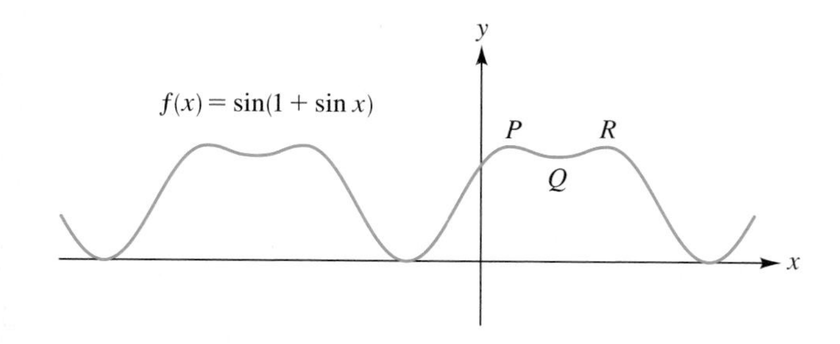

85. (a) Find the smallest solution of $\cos x = 0.412$ in the interval $(1000, \infty)$. (Round the answer to three decimal places.) *Hint:* Use a calculator to approximate a solution of $\cos x = 0.412$. Use this solution to approximate the other solution in the interval $[0, 2\pi]$. Use these solutions to finish the problem. (Don't round off your intermediate results.)

(b) Find the smallest solution of $\cos x = -0.412$ in the interval $(1000, \infty)$.

Ⓖ **86. (a)** Graph the function $y = \sin(\pi x/180)$ in the standard viewing rectangle. Describe what you see.

(b) Change the viewing rectangle so that x extends from 0 to 360. The graph that you obtain is the graph of $y = \sin x$, where x is now measured in degrees.

(c) According to Example 6, there are two solutions of the equation $\sin s + \cos s = 1$ in the interval $0° \le s < 360°$. Confirm this by graphing the two equations

$$y = \sin \frac{\pi x}{180} + \cos \frac{\pi x}{180}$$

and

$$y = 1 \qquad \text{for } 0 \le x \le 360$$

and noting that there are two intersection points. (In Example 6 the intersection point at $x = 360$ is excluded.)

(d) In Example 6 we saw that 90° is a solution of the equation $\sin s + \cos s = 1$ in the interval $0° \le s < 360°$. Confirm this by zooming in on the appropriate intersection point in part (c) and checking the x-coordinate.

87. The accompanying figure shows a graph of the function $f(x) = x + 0.4 \sin(2\pi x)$ on the interval $0 \le x \le 1.5$. (Functions of this form occur in mathematical biology in the study of rhythmic behaviors, such as heartbeat.) Using calculus, it can be shown that the x-coordinates of the turning points of this function are found by solving the equation

$$1 + 0.8\pi \cos(2\pi x) = 0$$

Use this fact to find the x-coordinates of the turning points P, Q, and R in the figure. Round the answers to three decimal places.

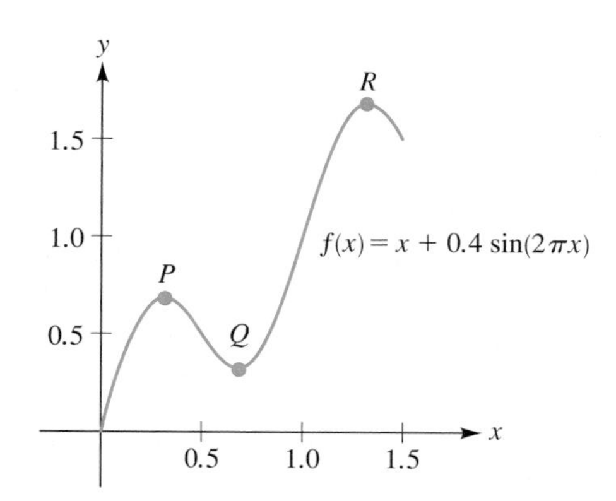

88. (As background for this exercise, you need to know the definitions of *fixed point* and *iterate* from Section 4.3.) The figure on the next page shows a view of the function $f(x) = x + 0.4 \sin(2\pi x)$ along with the line $y = x$.

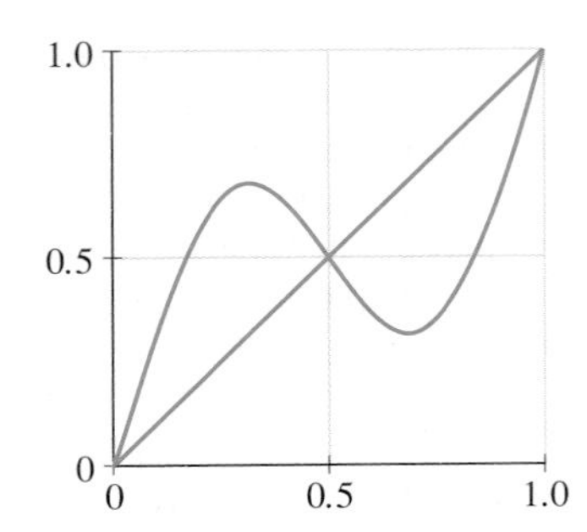

(a) From the figure it appears that the value $x = 0.5$ is a fixed point for the function f. Confirm this algebraically by solving the equation $f(x) = x$ on the interval $0 \leq x \leq 1$.

(b) Compute the first ten iterates of $x_0 = 0.25$ under the function f. Round the answers to four decimal places. Describe any patterns that you detect. Are the iterates approaching the fixed point $x = 0.5$?

(c) Follow part (b) using $x_0 = 0.45$.

For Exercises 89–92, refer to the following figure and formulas. The figure shows the trajectory of an object projected upward from ground level at an angle α. Neglecting air resistance, the trajectory is approximately a parabola. The maximum height h_{max} and the range r are given by the formulas adjacent to the figure. In the formulas, v_0 is the initial velocity; the constant g is the acceleration due to gravity at the earth's surface. If distance is measured in feet, the value for g is 32 ft/sec²; if distance is measured in meters, the value for g is 9.8 m/sec².

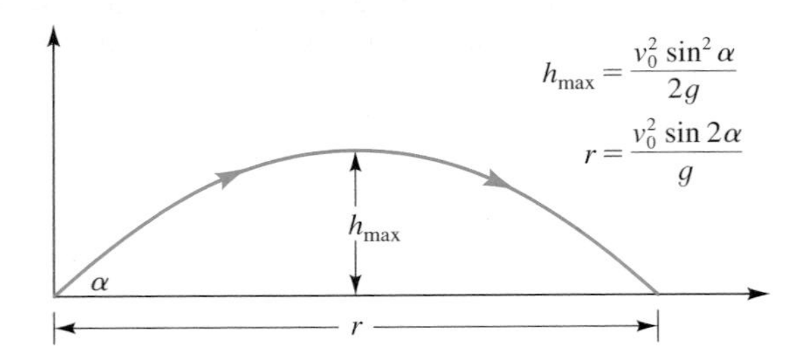

$h_{max} = \frac{v_0^2 \sin^2 \alpha}{2g}$
$r = \frac{v_0^2 \sin 2\alpha}{g}$
h_{max}
α
r

89. **(a)** Solve the range formula $r = (v_0^2 \sin 2\alpha)/g$ for α in terms of the other letters.

(b) Assuming an initial velocity of 80 ft/sec, use the result in part (a) to find an angle that yields a range of 100 ft. Express the answer both in radians, rounded to two decimal places, and in degrees, rounded to one decimal place.

(c) Prove the following trigonometric identity [which will be needed in part (d)]:

$$\sin 2\alpha = \sin[2(\tfrac{\pi}{2} - \alpha)]$$

(d) In part (b) you found an angle that yields a range of 100 ft, given an initial velocity of 80 ft/sec. Now use the result in part (c) to specify another angle that yields the same range (still assuming an initial velocity of 80 ft/sec).

(e) Again, assume an initial velocity of 80 ft/sec. Find all values for α that yield a range of 200 ft. Express the answer(s) in degrees, rounded to one decimal place.

90. As background for this exercise, you need to have worked Exercise 89. Assume that the initial velocity is 25 m/sec. Find the two angles that yield a range of 60 m. Express the answers in degrees, rounded to one decimal place.

91. **(a)** Suppose that the initial velocity is 40 m/sec. Find an angle α for which the maximum height is the same as the range. Express the answer in degrees rounded to one decimal place.

(b) Follow part (a), but assume that the initial velocity is 88 ft/sec.

(c) After looking over your answers and your work in parts (a) and (b), what conclusion can you draw?

92. **(a)** Solve the equation $h_{max} = (v_0^2 \sin^2 \alpha)/(2g)$ for α in terms of the other letters.

(b) Use the result in part (a) to find an angle that yields a maximum height of 55 ft if the initial velocity is 70 ft/sec. Express the answer in radians rounded to two decimal places.

In Exercises 93–98, each exercise results in an equation that you will need to solve using a graphing utility, as in Exercises 53–74.

Ⓖ **93.** As indicated in the accompanying figure, $P(x, y)$ is a point on the graph of $y = \sin x$ for $0 < x < \pi$ and the distance from P to the origin is 2 units. Find the coordinates of P. Round your answers to two decimal places.

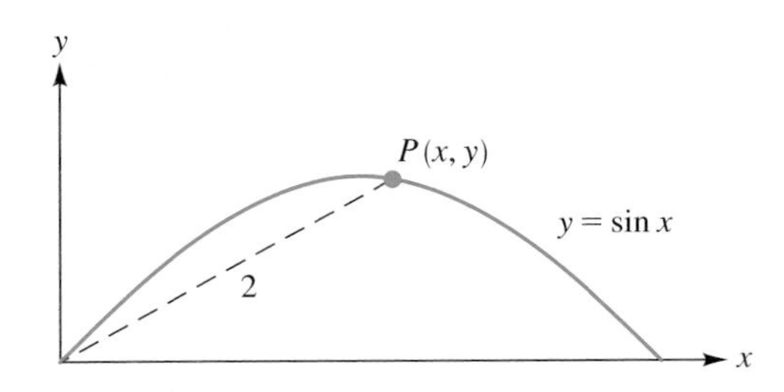

Ⓖ **94.** The point $P(x, y)$ is on the graph of $y = \cos x$ for $0 < x < \pi$. From P, perpendiculars are drawn to the x-axis and the y-axis to form a rectangle, as indicated in the figure.

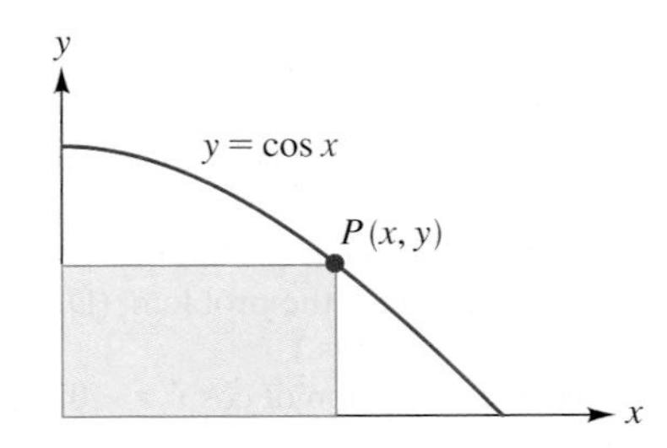

(a) Express the area of the rectangle as a function of x.
(b) Find a value for x so that the area of the rectangle is 0.5 square units.
(c) Can you find a value for x so that the area of the rectangle is 0.6 square units?

Ⓖ **95.** The accompanying figure shows an isosceles triangle ABC inscribed in the unit circle. Legs $\overline{AB}$ and $\overline{AC}$ are the congruent legs.

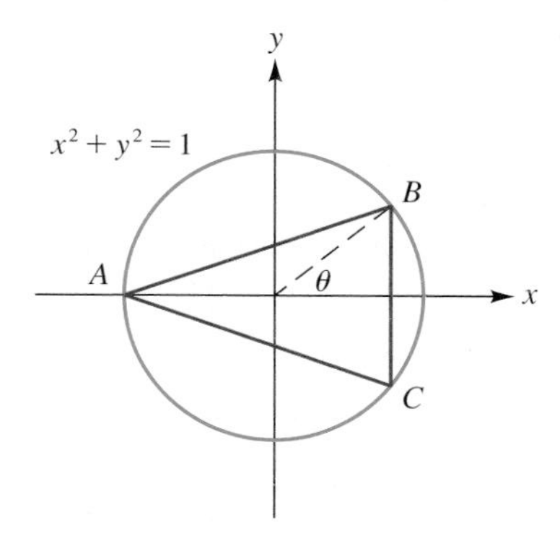

(a) Express the area of $\triangle ABC$ as a function of θ.
(b) Find a value for θ so that the area of $\triangle ABC$ is 40% of the area of the unit circle. Give the answer in degrees, rounded to two decimal places.
(c) Can you find a value for θ so that the area of $\triangle ABC$ is 42% of the area of the unit circle? If so, round the result to two decimal places. If not, explain why.

Ⓖ **96.** Refer to the accompanying figure. An offshore oil rig is located at point A, which is 1.5 miles out to sea. An oil pipeline is to be constructed from A to a point C on the shore and then to an oil refinery at point D, farther up the coast. It costs \$1000 /mile to lay the pipeline in the sea and \$500/mile on land.

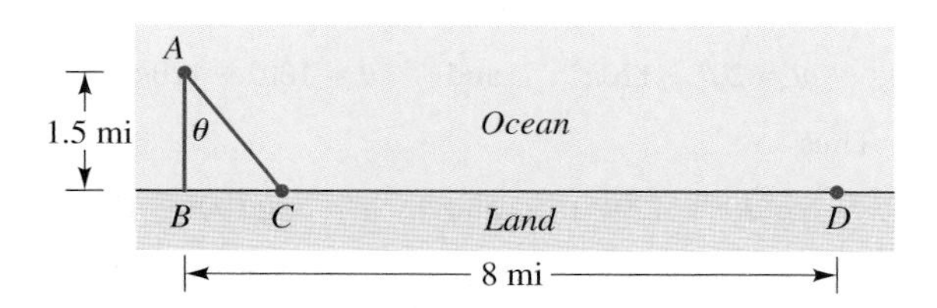

(a) Compare the cost C of laying the pipeline in terms of the angle θ. (The angle θ is defined in the figure.)
(b) Find the values of θ for which $C = \$5400$. Express the answers in radians, rounded to three decimal places. (If you are using a zooming technique, you will need to zoom in until you are certain of the first four decimal places; then you can round to three places.)
(c) Express the answers in part (b) using degree measure; round to one place.
(d) Can the job be done for $C = \$5300$? If so, specify the value(s) of θ; if not, explain why.

Ⓖ **97.** The figure shows a right circular cone with a slant height of 1 meter.

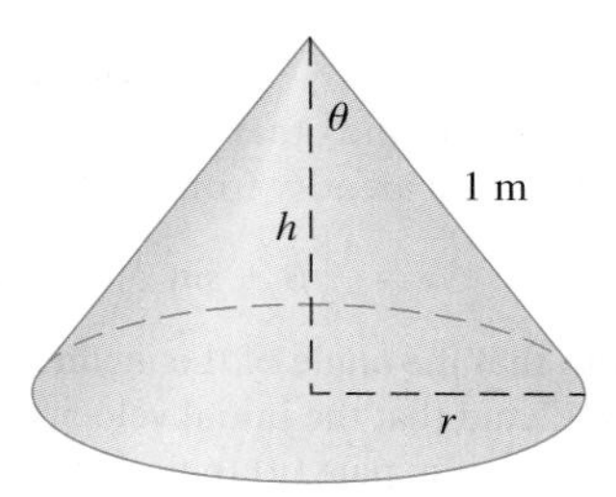

(a) Express the volume of the cone as a function of θ.
(b) Graph the volume function obtained in part (a).
(c) Find θ so that the volume of the cone is 0.4 m^3. (There are two answers.)
(d) Demonstrate by means of a graph that the volume of the cone cannot be 0.41 m^3.

Ⓖ **98.** Let $f(x) = \dfrac{4}{\sin^3 x} - \dfrac{1}{\sin^6 x}$ for $0 < x \le \pi/2$.
(a) Find all values of x in the given interval such that $f(x) = 3.5$. Round the answer(s) to two decimal places.
(b) Find all values of x in the given interval such that $f(x) = 4$. Round the answer(s) to two decimal places.
(c) If you graph the given function f on the interval $0 < x \le \pi/2$, you can see that the maximum value of the function appears to be 4. Use the methods of Section 4.5 (rather than a graphing utility) to *prove* that the maximum value is indeed 4. What is the corresponding x-value here? Give both an exact expression and a calculator approximation rounded to two decimal places. Check that your answer is consistent with the result in part (b).

Hint: In the expression $\dfrac{4}{\sin^3 x} - \dfrac{1}{\sin^6 x}$, make the substitution $t = (1/\sin^3 x)$.

For Exercises 99–101, refer to the following figure and formula. The figure shows the trajectory of an object projected upward from a point P at the base of an inclined plane. The inclined plane makes an angle β with the horizontal; the initial direction of the object is at an angle α with the horizontal; and the initial velocity is v_0. The object lands at point Q, and r (=PQ) denotes the range of the object along the inclined plane.

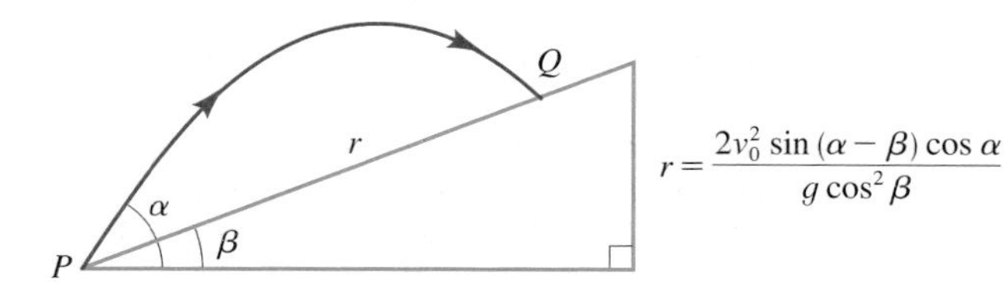

99. (a) Solve the range equation.

$$r = \frac{2v_0^2 \sin(\alpha - \beta)\cos\alpha}{g\cos^2\beta}$$

for α in terms of the other letters. *Hint*: Use a product-to-sum identity from Section 8.3.

Answer: $\alpha = \frac{1}{2}\left[\beta + \sin^{-1}\left(\frac{gr\cos^2\beta}{v_0^2} + \sin\beta\right)\right]$

(b) Suppose that the angle of the inclined plane is $\beta = \pi/12$ and that the initial velocity is 20 ft/sec. Use the result in part (a) to find an angle that yields a range of $r = 6$ ft. Express the answer both in radians, rounded to two decimal places, and in degrees, rounded to one decimal place.

(c) Let $f(\alpha) = \sin(\alpha - \beta)\cos\alpha$. Prove the following trignometric identity [which will be needed in part (d)]: $f(\alpha) = f[\beta + (\frac{\pi}{2} - \alpha)]$.

(d) In part (b) you found an angle that yields a range of 6 ft, assuming an initial velocity of 20 ft/sec. Now use the result in part (c) to specify another angle that yields the same range (still assuming the same initial velocity and the same angle for the inclined plane). Express the answer in degrees, rounded to one decimal place.

100. As in Exercise 99, assume that the angle of the inclined plane is $\beta = \pi/12$ and that the initial velocity is $v_0 = 20$ ft/sec. Show that there is no value for α that produces a range of 10 ft. *Hint*: Substitute the given values in the formula obtained in Exercise 99(a); what happens?

101. As in the previous two exercises, assume that $\beta = \pi/12$ and $v_0 = 20$ ft/sec. In this exercise you'll determine the maximum range under these conditions, as well as the angle α that produces that range.

(a) Substitute the given values $\beta = \pi/12$ and $v_0 = 20$ ft/sec in the range formula $r = [2v_0^2\sin(\alpha - \beta)\cos\alpha]/(g\cos^2\beta)$. Then use a product-to-sum identity and show that the formula becomes

$$r = \frac{400[\sin(2\alpha - \frac{\pi}{12}) - \sin\frac{\pi}{12}]}{g\cos^2(\frac{\pi}{12})} \tag{1}$$

(b) As you know, the maximum value for the sine function is 1, and this value is obtained when the input is $\pi/2$. Use this observation to conclude that the maximum value for r in equation (1) is given by

$$r_{max} = \frac{400[1 - \sin(\pi/12)]}{g\cos^2(\pi/12)}$$

and that the corresponding input is $\alpha = 7\pi/24$, which is about 52.5°.

(c) *Loose ends.* Evaluate the expression for r_{max} obtained in part (b). *Check*: Your answer should be less than 10 ft (because in Exercise 100 we found that a range of 10 ft is unobtainable under the given conditions). Also check that the input $\alpha = 7\pi/24$ is halfway between $\beta = \pi/12$ and $\pi/2$. This is an instance of the following general fact: For maximum range up an inclined plane of angle β, the launch angle α should be halfway between β and the vertical.

102. In this exercise you will see how certain cubic equations can be solved by using the following identity (which we proved in Example 3 in Section 8.2):

$$4\cos^3\theta - 3\cos\theta = \cos 3\theta \tag{1}$$

For example, suppose that we wish to solve the equation

$$8x^3 - 6x - 1 = 0 \tag{2}$$

To transform this equation into a form in which the stated identity is useful, we make the substitution $x = a\cos\theta$, where a is a constant to be determined. With this substitution, equation (2) can be written

$$8a^3\cos^3\theta - 6a\cos\theta = 1 \tag{3}$$

In equation (3) the coefficient of $\cos^3\theta$ is $8a^3$. Since we want this coefficient to be 4 [as it is in equation (1)], we divide both sides of equation (3) by $2a^3$ to obtain

$$4\cos^3\theta - \frac{3}{a^2}\cos\theta = \frac{1}{2a^3} \tag{4}$$

Next, a comparison of equations (4) and (1) leads us to require that $3/a^2 = 3$. Thus $a = \pm 1$. For convenience we choose $a = 1$; equation (4) then becomes

$$4\cos^3\theta - 3\cos\theta = \frac{1}{2} \tag{5}$$

Comparing equation (5) with the identity in (1) leads us to the equation

$$\cos 3\theta = \frac{1}{2}$$

As you can check, the solutions here are of the form

$$\theta = 20° + 120k° \qquad \text{and} \qquad \theta = 100° + 120k°$$

Thus

$$x = \cos(20° + 120k°) \qquad \text{and} \qquad x = \cos(100° + 120k°)$$

Now, however, as you can again check, only three of the angles yield distinct values for $\cos\theta$, namely, $\theta = 20°$, $\theta = 140°$, and $\theta = 260°$. Thus the solutions of the equation $8x^3 - 6x - 1 = 0$ are given by $x = \cos 20°$, $x = \cos 140°$, and $x = \cos 260°$. *Note*: If you choose $a = -1$, your solutions will be equivalent to those we found with $a = 1$.

Use the method just described to solve the following equations.

(a) $x^3 - 3x + 1 = 0$

Answers: $2\cos 40°, -2\cos 20°, 2\cos 80°$

(b) $x^3 - 36x - 72 = 0$

(c) $x^3 - 6x + 4 = 0$ *Answers*: $2, -1 \pm \sqrt{3}$

(d) $x^3 - 7x - 7 = 0$ (Round your answers to three decimal places.)

MINI PROJECT

ASTIGMATISM AND EYEGLASS LENSES

In the manufacture of eyeglass lenses it is important to control a geometric property called *surface astigmatism.* We will not need to know what surface astigmatism is in order to do this project, but we do need to know that there is a direction on the lens associated with surface astigmatism called the *axis of astigmatism.* The angle that this line makes with a horizontal reference line is the **angle of the astigmatism axis** and is shown as angle θ in Figure A. A measurement of θ is called an axis measurement; it is given in degrees and always lies in the closed interval from 0 to 180 degrees.

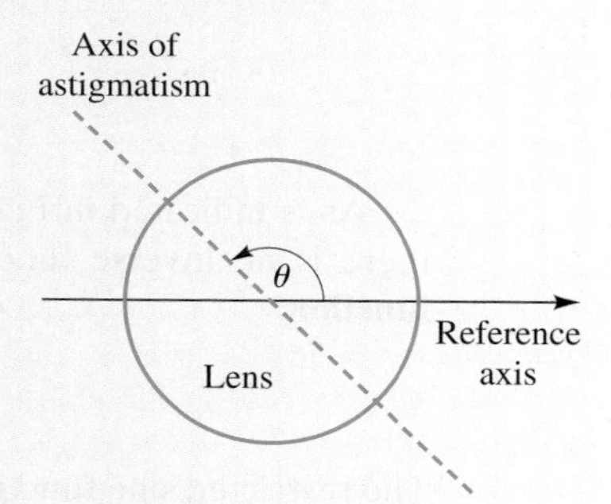

Figure A

A certain manufacturer uses an accurate measuring instrument to measure surface astigmatism, but instead of measuring the axis angle θ, the device produces data containing two numbers a and b that are equal to $\sin 2\theta$ and $\cos 2\theta$, respectively. An engineer has decided that a computer program is needed to take the output data from the measuring instrument as its input and produce the axis angle θ as its output. Your group has been asked to come up with a step-by-step procedure for finding the angle θ given the data a and b.

One approach is to find all of the solutions in an appropriate interval to the pair of equations $\sin 2\theta = a$ and $\cos 2\theta = b$. Then, among the values of θ that satisfy both equations choose one that is between 0 and 180 degrees. Another approach uses an identity that arose in the proof of the half-angle formula for the tangent in Section 8.2. Choose one of these approaches and write a step-by-step procedure to find, to the nearest degree, the correct value of θ between 0 and 180 degrees. Use your procedure on the following four test cases: $a = 0.5000$ and $b = 0.8660$, $a = 0.5000$ and $b = -0.8660$, $a = -0.5000$ and $b = 0.8660$, and $a = -0.5000$ and $b = -0.8660$. Show for each case what values result at each step of your step-by-step procedure.

8.5 THE INVERSE TRIGONOMETRIC FUNCTIONS

The notation $\cos^{-1}\theta$ *must not be understood to signify* $1/\cos\theta$.—John Herschel in *Philosophical Transactions of London,* 1813

In previous sections we introduced the three inverse trigonometric functions $y = \sin^{-1} x$, $y = \cos^{-1} x$, and $y = \tan^{-1} x$. Because we were using these functions as tools to solve equations, our emphasis there was more computational than theoretical. Now we want to take a second look at these functions, this time from a more conceptual point of view. As background for this work, you should be familiar with the material on inverse functions in Section 3.6.

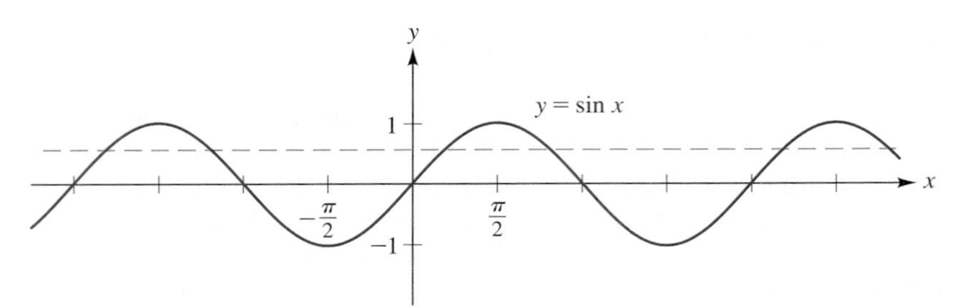

(a) The horizontal line test shows that the sine function is not one-to-one.

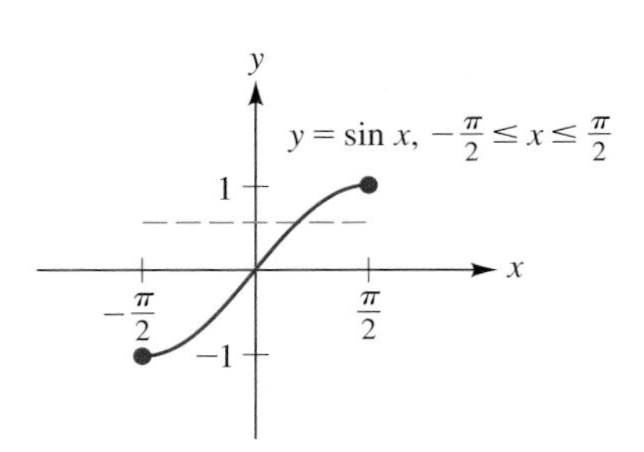

(b) By restricting the domain of the sine function to the closed interval $\left[-\frac{\pi}{2}, \frac{\pi}{2}\right]$, we obtain the restricted sine function. The horizontal line test shows that the restricted sine function is one-to-one.

Figure 1

As is indicated in Figure 1(a), the sine function in not one-to-one. Therefore there is no inverse function. However, let us now consider the **restricted sine function:**

$$y = \sin x \qquad \left(-\frac{\pi}{2} \leq x \leq \frac{\pi}{2}\right)$$

The restricted sine function has domain $[-\frac{\pi}{2}, \frac{\pi}{2}]$ and range $[-1, 1]$. And, as is indicated in Figure 1(b), the restricted sine function *is* one-to-one, and therefore the inverse function does exist in this case. We refer to the inverse of the restricted sine function as the **inverse sine function.**

Two notations are commonly used to denote the inverse sine function:

$$y = \sin^{-1} x \qquad \text{and} \qquad y = \arcsin x$$

Initially, at least, we will use the notation $y = \sin^{-1} x$. Basic facts about inverse functions tell us that:

the domain of $\sin^{-1} x$ = the range of the restricted sine function = $[-1, 1]$, and

the range of $\sin^{-1} x$ = the domain of the restricted sine function = $[-\frac{\pi}{2}, \frac{\pi}{2}]$

The graph of $y = \sin^{-1} x$ is easily obtained by using the fact that the graphs of a function and its inverse are reflections of one another about the line $y = x$. Figure 2(a) shows the graph of the restricted sine function and its inverse, $y = \sin^{-1} x$. Figure 2(b) shows the graph of $y = \sin^{-1} x$ alone. From Figure 2(b) we see that the domain of $y = \sin^{-1} x$ is the interval $[-1, 1]$, while the range is $[-\frac{\pi}{2}, \frac{\pi}{2}]$.

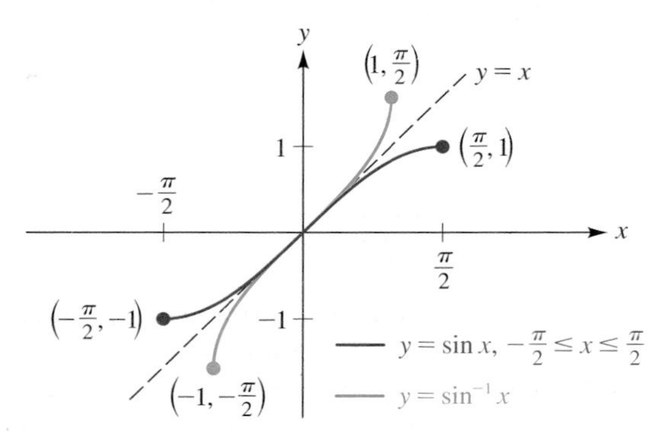

(a) The graphs of a function and its inverse are mirror images of one another about the line $y = x$.

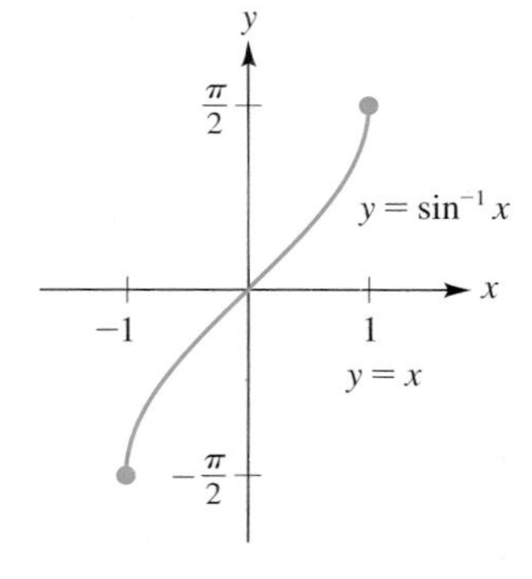

(b) The inverse sine function.

Figure 2

What exactly is $\sin^{-1} x$? To see, we begin with the restricted sine function:

$$y = \sin x \qquad \left(-\frac{\pi}{2} \le x \le \frac{\pi}{2}\right)$$

As was explained in Section 3.6, the inverse function is obtained by interchanging x and y (the inputs and the outputs). So for the inverse function $y = \sin^{-1} x$ we have

$$x = \sin y \qquad \left(-\frac{\pi}{2} \le y \le \frac{\pi}{2}\right) \tag{1}$$

Equation (1) tells us that $y = \sin^{-1} x$, is the number in the interval $[-\frac{\pi}{2}, \frac{\pi}{2}]$ whose sine is x. So values of $\sin^{-1} x$ are computed according to the following rule.

$\sin^{-1} x$ is that number in the interval $[-\frac{\pi}{2}, \frac{\pi}{2}]$ whose sine is x.

EXAMPLE 1 Finding some exact values of the inverse sine

Evaluate:

(a) $\sin^{-1}\left(\frac{1}{2}\right)$; **(b)** $\sin^{-1}\left(-\frac{1}{2}\right)$.

SOLUTION

(a) $\sin^{-1}(1/2)$ is that number in the interval $[-\frac{\pi}{2}, \frac{\pi}{2}]$ whose sine is $1/2$. Since $\pi/6$ is in $[-\frac{\pi}{2}, \frac{\pi}{2}]$ and $\sin(\pi/6) = 1/2$, we conclude that $\sin^{-1}(1/2) = \pi/6$.

(b) Since $\sin(-\pi/6) = -1/2$ and $-\pi/6$ is in the interval $[-\frac{\pi}{2}, \frac{\pi}{2}]$, we conclude that $\sin^{-1}(-1/2) = -\pi/6$.

Note: $\sin(5\pi/6) = 1/2$, but $5\pi/6$ is not in the interval $[-\frac{\pi}{2}, \frac{\pi}{2}]$, so $\sin^{-1}(1/2) \ne 5\pi/6$. Similarly, $\sin(7\pi/6) = -1/2$, but $\sin^{-1}(-1/2) \ne 7\pi/6$. (Why?)

We can use the result in Example 1(a) to see why the notation *arcsin* is used for the inverse sine function. With the arcsin notation the result in Example 1(a) can be written

$$\arcsin \frac{1}{2} = \frac{\pi}{6}$$

Now, as Figure 3 indicates, the length of the arc whose sine is $1/2$ is $\pi/6$. This is the idea behind the arcsin notation.

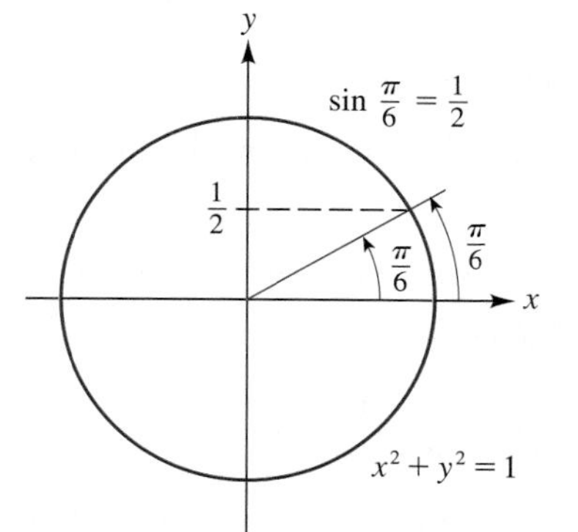

Figure 3

EXAMPLE 2 Finding approximate values of the inverse sine

Evaluate $\arcsin(3/4)$. Round the answer to three decimal places.

SOLUTION

The quantity $\arcsin(3/4)$ is that number (or arc length, or angle in radians) in the interval $[-\frac{\pi}{2}, \frac{\pi}{2}]$ whose sine is $3/4$. Since we're not familiar with an angle with a sine of $3/4$, we use a calculator. As in previous sections, we show keystrokes for the most common type of calculator. Make sure your calculator is in radian mode. (If these keystrokes don't work on your calculator, be sure to check the user's manual.)

EXPRESSION	KEYSTROKES	OUTPUT
arcsin(0.75)	2nd sin 0.75 ENTER	0.84806 . . .

Now, rounding to three decimal places, our result is

$$\arcsin\frac{3}{4} \approx 0.848$$

EXAMPLE 3 Explaining a common notational confusion

Show that the following expressions are not equal:

$$\sin^{-1} 0 \qquad \text{and} \qquad \frac{1}{\sin 0}$$

SOLUTION

The quantity $\sin^{-1} 0$ is that number in the interval $[-\frac{\pi}{2}, \frac{\pi}{2}]$ whose sine is 0. Since 0 is in the interval $[-\frac{\pi}{2}, \frac{\pi}{2}]$ and $\sin 0 = 0$, we conclude that

$$\sin^{-1} 0 = 0$$

On the other hand, since $\sin 0 = 0$, the expression $1/(\sin 0)$ is not even defined. Thus the two given expressions certainly are not equal.

If f and f^{-1} are any pair of inverse functions, then by definition,

$$f[f^{-1}(x)] = x \qquad \text{for every } x \text{ in the domain of } f^{-1}$$

and

$$f^{-1}[f(x)] = x \qquad \text{for every } x \text{ in the domain of } f$$

Applying these facts to the restricted sine function, $y = \sin x$, and its inverse, $y = \sin^{-1} x$, we obtain the following two basic identities.

$$\sin(\sin^{-1} x) = x \qquad \text{for every } x \text{ in the interval } [-1, 1]$$

$$\sin^{-1}(\sin x) = x \qquad \text{for every } x \text{ in the interval } [-\tfrac{\pi}{2}, \tfrac{\pi}{2}]$$

You should convince yourself in words that these two identities are true. For example, $\sin(\sin^{-1} x)$ = the sine of (the number between $-\pi/2$ and $\pi/2$, inclusive, whose sine is x) = x, provided that x is in the domain of $\sin^{-1} x$ = the range of $\sin x$ = the interval $[-1, 1]$. Now try to verify in words the second identity.

The following example indicates that the domain restrictions accompanying these two identities cannot be ignored.

EXAMPLE 4 Calculations involving the inverse sine

Compute each quantity that is defined.

(a) $\sin^{-1}\left(\sin\dfrac{\pi}{4}\right)$ **(b)** $\sin^{-1}(\sin 2)$

(c) $\sin(\sin^{-1} 2)$ **(d)** $\sin[\sin^{-1}(-1/\sqrt{5})]$

SOLUTION

(a) Since $\pi/4$ lies in the domain of the restricted sine function, the identity $\sin^{-1}(\sin x) = x$ is applicable here. Thus

$$\sin^{-1}\left(\sin\frac{\pi}{4}\right) = \frac{\pi}{4}$$

CHECK $\sin\frac{\pi}{4} = \sqrt{2}/2$.

Therefore

$$\sin^{-1}\left(\sin\frac{\pi}{4}\right) = \sin^{-1}\left(\frac{\sqrt{2}}{2}\right) = \frac{\pi}{4}$$

(b) The number 2 is not in the domain of the restricted sine function, so the identity $\sin^{-1}(\sin x) = x$ does not apply in this case. However, $\pi - 2$ is in the first quadrant (see Figure 4), so it is in the domain of the restricted sine function and $\sin 2 = \sin(\pi - 2)$, so

$$\sin^{-1}(\sin 2) = \sin^{-1}[\sin(\pi - 2)] = \pi - 2$$

Thus $\sin^{-1}(\sin 2)$ is equal to $\pi - 2$, not 2.

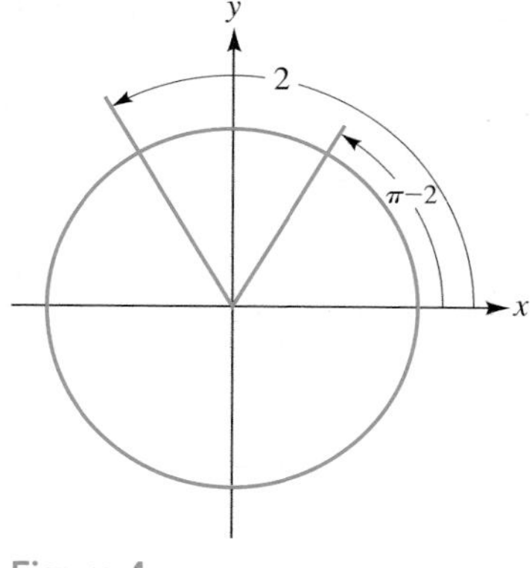

Figure 4

(c) The number 2 is not in the domain of the inverse sine function. Thus the expression $\sin(\sin^{-1} 2)$ is undefined.

(d) The identity $\sin(\sin^{-1} x) = x$ is applicable here. (Why?) Thus

$$\sin\left[\sin^{-1}\left(-\frac{1}{\sqrt{5}}\right)\right] = -\frac{1}{\sqrt{5}}$$

EXAMPLE 5 From a trigonometric to an algebraic expression

If $\sin\theta = x/3$ and $0 < \theta < \pi/2$, express the quantity $\theta - \sin 2\theta$ as a function of x.

SOLUTION

The given conditions tell us that $\theta = \sin^{-1}(x/3)$. That expresses θ in terms of x. To express $\sin 2\theta$ in terms of x, we have

$$\sin 2\theta = 2\sin\theta\cos\theta = 2(\sin\theta)\sqrt{1 - \sin^2\theta}$$ since $\cos\theta = +\sqrt{1 - \sin^2\theta}$ (Why?)

$$= 2\left(\frac{x}{3}\right)\sqrt{1 - \frac{x^2}{9}} = \frac{2x}{3}\sqrt{\frac{9 - x^2}{9}}$$

$$= \frac{2x\sqrt{9 - x^2}}{9}$$

Now, combining the results, we have

$$\theta - \sin 2\theta = \sin^{-1}\left(\frac{x}{3}\right) - \frac{2x\sqrt{9 - x^2}}{9}$$

We turn now to the inverse cosine function. We'll define it by following the same general procedure that we used to define the inverse sine. We begin by defining the **restricted cosine function:**

$$y = \cos x \qquad (0 \le x \le \pi) \qquad \text{with domain } [0, \pi] \text{ and range } [-1, 1]$$

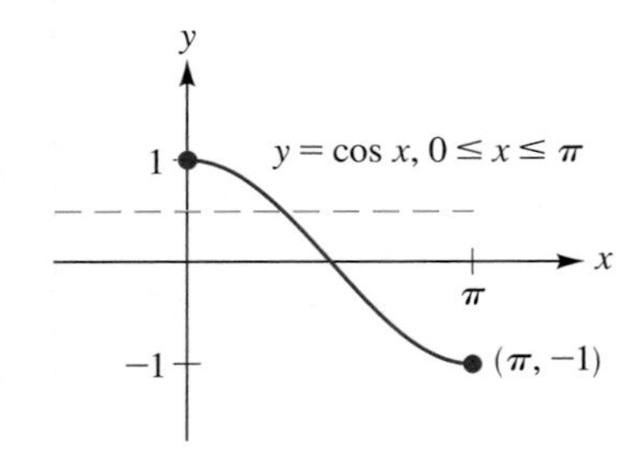

Figure 5
The restricted cosine function is one-to-one.

As is indicated by the horizontal line test in Figure 5, the restricted cosine function is one-to-one, and therefore the inverse function exists. We denote the **inverse cosine function** by

$$y = \cos^{-1} x \qquad \text{or} \qquad y = \arccos x$$

Then the domain of $\cos^{-1} x$ = the range of the restricted cosine function = $[-1, 1]$, and the range of $\cos^{-1} x$ = the domain of the restricted cosine function = $[0, \pi]$.

The graph of $y = \cos^{-1} x$ is obtained by reflecting the graph of the restricted cosine function about the line $y = x$. Figure 6(a) displays the graph of the restricted cosine function along with $y = \cos^{-1} x$. Figure 6(b) shows the graph of $y = \cos^{-1} x$ alone. From Figure 6(b) we see that the domain of $y = \cos^{-1} x$ is the interval $[-1, 1]$, while the range is $[0, \pi]$.

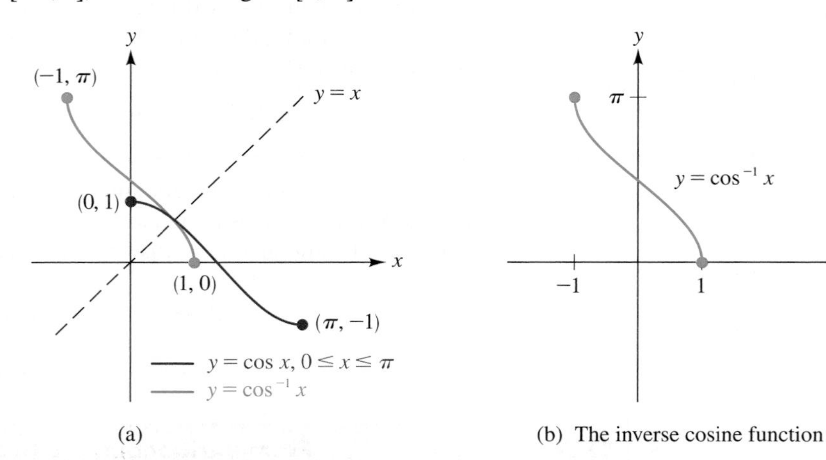

Figure 6 (a) (b) The inverse cosine function

As was stated in the previous section, the values of $\cos^{-1} x$ are computed according to the following rule.

$\cos^{-1} x$ is that number in the interval $[0, \pi]$ whose cosine is x.

To see why this is so, we begin with the restricted cosine function:

$$y = \cos x \qquad (0 \le x \le \pi)$$

Then for the inverse function, $y = \cos^{-1} x$, we interchange x and y (the inputs and the outputs) to obtain

$$x = \cos y \qquad (0 \le y \le \pi)$$

This last equation tells us that $y = \cos^{-1} x$ is that number in the interval whose cosine is x. This is what we wished to show.

As we mentioned previously, the defining equations for inverse functions are

$$f[f^{-1}(x)] = x \qquad \text{for every } x \text{ in the domain of } f^{-1}$$

and

$$f^{-1}[f(x)] = x \qquad \text{for every } x \text{ in the domain of } f$$

In the particular case of the restricted cosine function and the inverse cosine function, these two identities read as follows.

$$\cos(\cos^{-1} x) = x \qquad \text{for every } x \text{ in the interval } [-1, 1]$$

$$\cos^{-1}(\cos x) = x \qquad \text{for every } x \text{ in the interval } [0, \pi]$$

Again you should convince yourself in words that these two identities are true.

EXAMPLE 6 Calculations involving the inverse cosine

Compute each of the following:

(a) $\cos^{-1}(0)$; **(b)** $\cos\left(\arccos\frac{2}{3}\right)$; **(c)** $\arccos(\cos 4)$.

SOLUTION

(a) $\cos^{-1}(0)$ is that number in the interval $[0, \pi]$ whose cosine is 0. Since $\pi/2$ is in $[0, \pi]$ and $\cos(\pi/2) = 0$, we have

$$\cos^{-1}(0) = \frac{\pi}{2}$$

(b) Since the number $2/3$ is in the domain of the inverse cosine function, the identity $\cos(\arccos x) = x$ is applicable. Thus,

$$\cos\left(\arccos\frac{2}{3}\right) = \frac{2}{3}$$

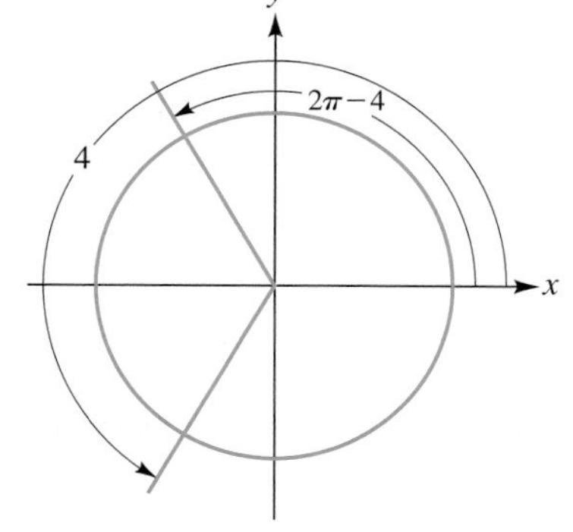

Figure 7

(c) The number 4 is not in the domain of the restricted cosine function, so the identity $\cos^{-1}(\cos x) = x$ does not apply in this case. However, since $2\pi - 4$ is in the second quadrant (see Figure 7), it is in the domain of the restricted cosine function and $\cos 4 = \cos(2\pi - 4)$, so

$$\cos^{-1}(\cos 4) = \cos^{-1}[\cos(2\pi - 4)] = 2\pi - 4$$

Thus $\cos^{-1}(\cos 4)$ is equal to $2\pi - 4$, not 4.

EXAMPLE 7 An identity involving the inverse cosine

Show that $\sin(\cos^{-1} x) = \sqrt{1 - x^2}$ for $-1 \le x \le 1$.

SOLUTION

We use the identity $\sin y = \sqrt{1 - \cos^2 y}$, which is valid for $0 \le y \le \pi$. Substituting $\cos^{-1} x$ for y in this identity, we obtain

$$\sin(\cos^{-1}x) = \sqrt{1 - [\cos(\cos^{-1} x)]^2}$$

$$= \sqrt{1 - x^2} \qquad \text{as required}$$

Before leaving this example, we point out an alternate method of solution that is useful when the restriction on x is $0 < x < 1$. In this case we let $\theta = \cos^{-1} x$. Then θ is the radian measure of the acute angle whose cosine is x, and we can sketch θ as shown in Figure 8. The sides of the triangle in Figure 8 have been labeled in such a way that

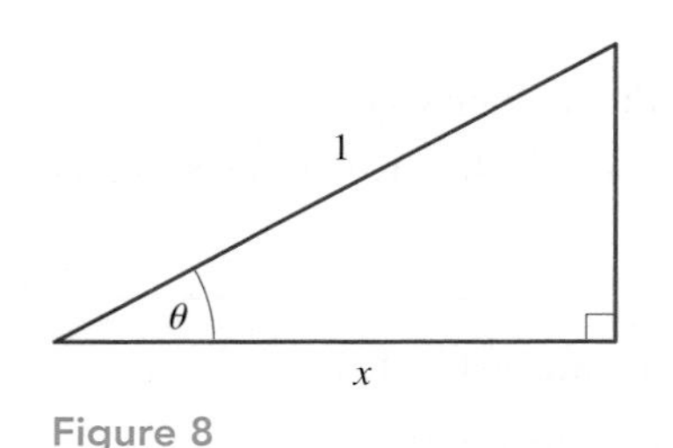

Figure 8

$$\cos\theta = \frac{\text{adjacent}}{\text{hypotenuse}} = \frac{x}{1} = x$$

Then, by the Pythagorean theorem we find that the third side of the triangle is $\sqrt{1 - x^2}$. We therefore have

$$\sin\theta = \frac{\text{opposite}}{\text{hypotenuse}} = \frac{\sqrt{1 - x^2}}{1} = \sqrt{1 - x^2}$$

Just as there is a basic identity connecting $\sin x$ and $\cos x$, $\sin^2 x + \cos^2 x = 1$, there is also an identity connecting $\sin^{-1} x$ and $\cos^{-1} x$.

$$\sin^{-1} x + \cos^{-1} x = \frac{\pi}{2} \qquad \text{for every } x \text{ in the closed interval } [-1, 1]$$

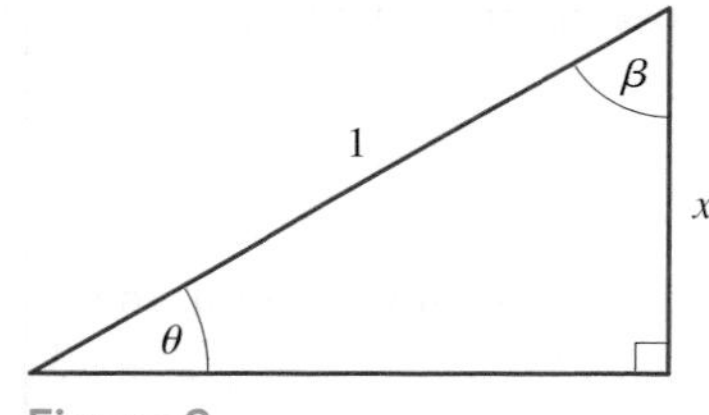

Figure 9

We can use Figure 9 to see why this identity is valid when $0 < x < 1$. (Exercise 61 shows you how to establish this identity for all values of x from -1 to 1.) In Figure 9, assume that θ and β are the radian measures of the indicated acute angles. Then we have

$$\sin\theta = \frac{x}{1} = x \qquad \text{and} \qquad \cos\beta = \frac{x}{1} = x$$

and therefore

$$\theta = \sin^{-1} x \qquad \text{and} \qquad \beta = \cos^{-1} x$$

But from Figure 9 we know that $\theta + \beta = \pi/2$. So, we have $\sin^{-1} x + \cos^{-1} x = \pi/2$, as required.

Now let us turn to the definition of the inverse tangent function. We begin by defining the **restricted tangent function:**

$$y = \tan x \qquad \left(-\frac{\pi}{2} < x < \frac{\pi}{2}\right) \qquad \text{with domain } (-\tfrac{\pi}{2}, \tfrac{\pi}{2}) \text{ and range } R$$

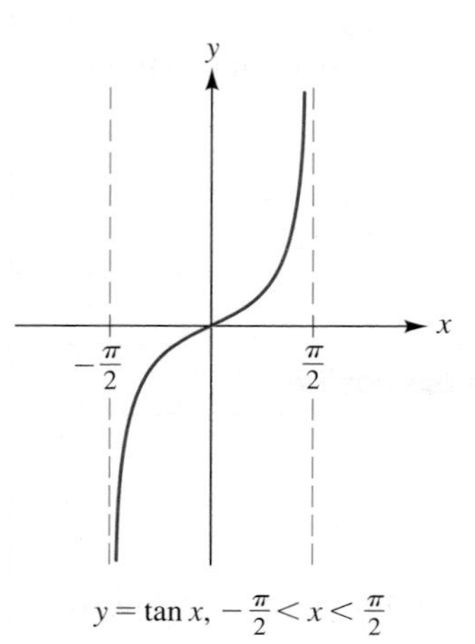

Figure 10
The restricted tangent function

Figure 10 shows the graph of the restricted tangent function. As you can check by applying the horizontal line test, the restricted tangent function is one-to-one. This tells us that the inverse function exists. The two common notations for this **inverse tangent function** are

$$y = \tan^{-1} x \qquad \text{or} \qquad y = \arctan x$$

The domain of $\tan^{-1} x = R$, and the range of $\tan^{-1} x =$ the interval $(-\frac{\pi}{2}, \frac{\pi}{2})$. Why? Furthermore, if we follow the same line of reasoning used previously for the inverse sine and the inverse cosine, we obtain the following properties of the inverse tangent function.

$\tan^{-1} x$ is that number in the interval $(-\frac{\pi}{2}, \frac{\pi}{2})$ whose tangent is x.

$$\tan(\tan^{-1} x) = x \qquad \text{for every real number } x$$

$$\tan^{-1}(\tan x) = x \qquad \text{for every } x \text{ in the open interval } (-\tfrac{\pi}{2}, \tfrac{\pi}{2})$$

The graph of $y = \tan^{-1} x$ is obtained in the same way that we obtained the graphs of the inverse sine and the inverse cosine. That is, we use the fact that the graphs of a function and its inverse are mirror images of one another about the line $y = x$. In Figure 11(a) we've reflected the graph of the restricted tangent function about the line $y = x$ to obtain the graph of $y = \tan^{-1} x$. In Figure 11(b) we show the graph of $y = \tan^{-1} x$ alone.

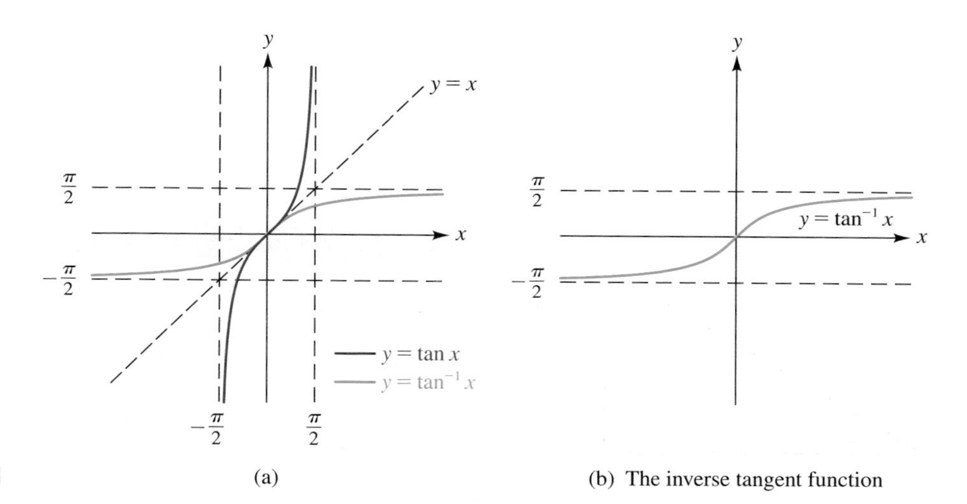

Figure 11 (a) (b) The inverse tangent function

From the graph in Figure 11(b) we can see that the domain of the inverse tangent function is the set of all real numbers, and the range is the open interval $(-\frac{\pi}{2}, \frac{\pi}{2})$.

EXAMPLE 8 Calculations involving the inverse tangent

Evaluate:

(a) $\tan^{-1}(-1)$; **(b)** $\tan(\tan^{-1}\sqrt{5})$.

SOLUTION

(a) The quantity $\tan^{-1}(-1)$ is that number in the interval $(-\frac{\pi}{2}, \frac{\pi}{2})$ whose tangent is -1. Since $-\pi/4$ is in $(-\frac{\pi}{2}, \frac{\pi}{2})$ and $\tan(-\pi/4) = -1$, we have $\tan^{-1}(-1) = -\pi/4$.

(b) The identity $\tan(\tan^{-1} x) = x$ holds for all real numbers x. We therefore have $\tan(\tan^{-1} \sqrt{5}) = \sqrt{5}$.

EXAMPLE 9 Transforming a trigonometric expression to an algebraic expression

If $\tan\theta = x/3$ and $0 < \theta < \pi/2$, express the quantity $\theta - \tan 2\theta$ as a function of x.

SOLUTION

The given equation tells us that $\theta = \tan^{-1}(x/3)$. That expresses θ in terms of x. To express $\tan 2\theta$ in terms of x, we have

$$\tan 2\theta = \frac{2\tan\theta}{1-\tan^2\theta} = \frac{2(x/3)}{1-(x/3)^2}$$

$$= \frac{2x/3}{1-(x^2/9)} = \frac{6x}{9-x^2} \qquad \text{multiplying numerator and denominator by 9}$$

Now, combining the results, we have

$$\theta - \tan 2\theta = \tan^{-1}\frac{x}{3} - \frac{6x}{9-x^2}$$

EXAMPLE 10 An identity involving the inverse tangent

Simplify the quantity $\sec(\tan^{-1} x)$, where $x > 0$.

SOLUTION

We let $\theta = \tan^{-1} x$ or, equivalently, $\tan\theta = x$. Then $0 < \theta < \pi/2$. Why? Now, as is shown in Figure 12, we sketch a right triangle with an acute angle θ (in radians) whose tangent is x. The Pythagorean theorem tells us that the length of the hypotenuse in this triangle is $\sqrt{1+x^2}$. Consequently, we have

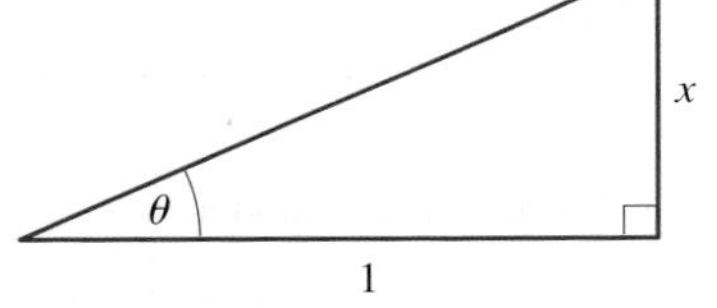

Figure 12
$\theta = \tan^{-1} x\ (x > 0)$ or $\tan\theta = x$

$$\sec(\tan^{-1} x) = \sec\theta = \frac{\text{hypotenuse}}{\text{adjacent}} = \frac{\sqrt{1+x^2}}{1} = \sqrt{1+x^2}$$

Note: In both Examples 7 and 10 we found that a trigonometric function with an inverse trigonometric function for its input could be expressed in terms of quadratic and square root functions. This is explored further in Exercises 38 through 41 at the end of this section.

When suitable restrictions are placed on the domains of the secant, cosecant, and cotangent, corresponding inverse functions can be defined. However, with the exception of the inverse secant, these are rarely, if ever, encountered in calculus. Therefore we omit a discussion of these functions here. (For the inverse secant function, see the project at the end of this section.)

Let's take a moment now to review our work in this section. The sine, cosine, and tangent functions are not one-to-one functions. However, with suitable restrictions on the domains, we do obtain one-to-one functions. The graphs of these restricted versions of the sine, cosine, and tangent functions are shown in Figures 1(b), 5, and 10, respectively. It is the inverses of these restricted functions that are the focus in this section. In the three boxes that follow, we summarize the key properties of the inverse sine, the inverse cosine, and the inverse tangent.

PROPERTY SUMMARY Inverse Sine Function: $y = \sin^{-1} x$ or $y = \arcsin x$

1. *Domain, range, and graph:* The domain of the inverse sine function is the closed interval $[-1, 1]$. The range is the closed interval $[-\frac{\pi}{2}, \frac{\pi}{2}]$.

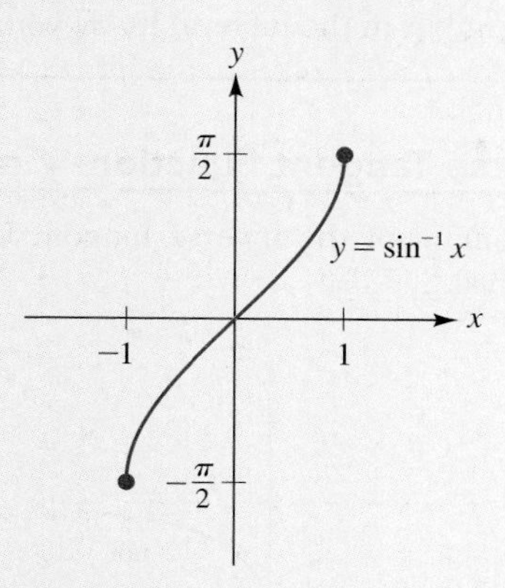

2. *Defining equations:* The function $y = \sin^{-1} x$ is the inverse of the restricted sine function. (The restricted sine function is defined by the equation $y = \sin x$ along with the restriction $-\pi/2 \le x \le \pi/2$.) Since the restricted sine function and the function $y = \sin^{-1} x$ are inverses of each other, we have

$$\sin(\sin^{-1}x) = x \qquad \text{for every } x \text{ in the interval } [-1, 1]$$

and

$$\sin^{-1}(\sin x) = x \qquad \text{for every } x \text{ in the interval } [-\tfrac{\pi}{2}, \tfrac{\pi}{2}]$$

3. *Computing* $\sin^{-1} x$*:* $\sin^{-1} x$ is that number in the interval $[-\frac{\pi}{2}, \frac{\pi}{2}]$ whose sine is x.

PROPERTY SUMMARY Inverse Cosine Function $y = \cos^{-1} x$ or $y = \arccos x$

1. *Domain, range, and graph:* The domain of the inverse cosine function is the closed interval $[-1, 1]$. The range is the closed interval $[0, \pi]$.

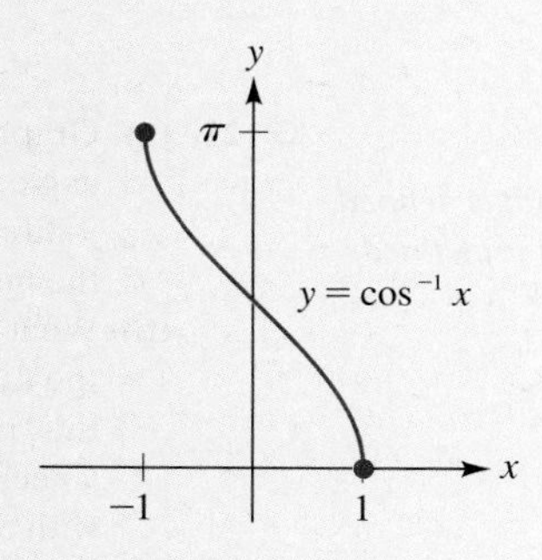

2. *Defining equations:* The function $y = \cos^{-1} x$ is the inverse of the restricted cosine function. (The restricted cosine function is defined by the equation $y = \cos x$ along with the restriction $0 \le x \le \pi$.) Since the restricted cosine function and the function $y = \cos^{-1} x$ are inverses of each other, we have

$$\cos(\cos^{-1} x) = x \qquad \text{for every } x \text{ in the interval } [-1, 1]$$

(continues)

and

$$\cos^{-1}(\cos x) = x \qquad \text{for every } x \text{ in the interval } [0, \pi]$$

3. *Computing* $\cos^{-1} x$: $\cos^{-1} x$ is that number in the interval $[0, \pi]$ whose cosine is x.

PROPERTY SUMMARY Inverse Tangent Function: $y = \tan^{-1} x$ or $y = \arctan x$

1. *Domain, range, and graph:* The domain of the inverse tangent function is the set of all real numbers, $(-\infty, \infty)$. The range is the open interval $(-\frac{\pi}{2}, \frac{\pi}{2})$.

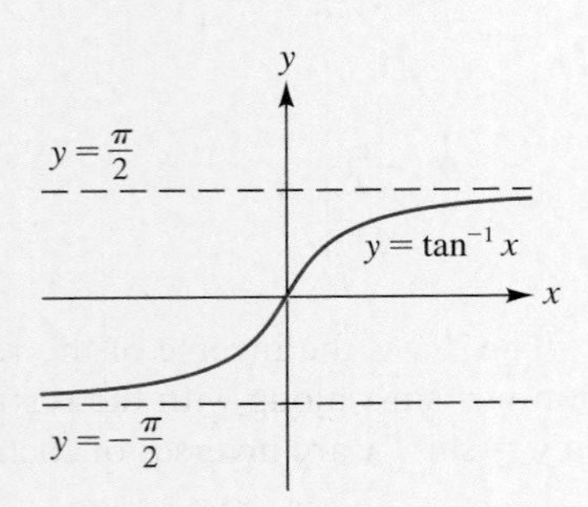

2. *Defining equations:* The function $y = \tan^{-1} x$ is the inverse of the restricted tangent function. (The restricted tangent function is defined by the equation $y = \tan x$ along with the restriction $-\pi/2 < x < \pi/2$.) Since the restricted tangent function and the function $y = \tan^{-1} x$ are inverses of each other, we have

$$\tan(\tan^{-1} x) = x \qquad \text{for every real number } x$$

and

$$\tan^{-1}(\tan x) = x \qquad \text{for every } x \text{ in the interval } (-\tfrac{\pi}{2}, \tfrac{\pi}{2})$$

3. *Computing* $\tan^{-1} x$: $\tan^{-1} x$ is that number in the interval $(-\frac{\pi}{2}, \frac{\pi}{2})$ whose tangent is x.

EXERCISE SET 8.5

A

In Exercises 1–20, evaluate each of the quantities that is defined, but do not use a calculator or tables. If a quantity is undefined, say so.

1. $\sin^{-1}(\sqrt{3}/2)$
2. $\cos^{-1}(-1)$
3. $\tan^{-1}\sqrt{3}$
4. $\arccos(-\sqrt{2}/2)$
5. $\arctan(-1/\sqrt{3})$
6. $\arcsin(-1)$
7. $\tan^{-1} 1$
8. $\sin^{-1} 0$
9. $\cos^{-1} 2\pi$
10. $\arctan 0$
11. $\sin[\sin^{-1}(\frac{1}{4})]$
12. $\cos[\cos^{-1}(\frac{4}{3})]$
13. $\cos[\cos^{-1}(\frac{3}{4})]$
14. $\tan(\arctan 3\pi)$
15. $\arctan[\tan(-\frac{\pi}{7})]$
16. $\sin(\arcsin 2)$
17. $\arcsin(\sin\frac{\pi}{2})$
18. $\arccos(\cos\frac{\pi}{8})$
19. $\arccos(\cos 2\pi)$
20. $\sin^{-1}(\sin\frac{3\pi}{2})$

Ⓖ **21. (a)** Graph the function $y = \arcsin x$ using a viewing rectangle that extends from -2 to 2 in both the x- and the y-directions. Then use the graphing utility to estimate the maximum and the minimum values of the function.
(b) What are the exact maximum and minimum values of the function $y = \arcsin x$? (If you need help answering this, refer to the graph in the first box on page 617.) Check that your estimates in part (a) are consistent with these results.

Ⓖ **22.** Using the viewing rectangle specified in Exercise 21, graph the function $f(x) = \sin(\arcsin x)$. What do you observe? What identity does this demonstrate?

Ⓖ **23. (a)** In the standard viewing rectangle, graph the function $y = \arctan x$.

(b) According to the text, the graph in part (a) has two horizontal asymptotes. What are the equations for these two asymptotes? Add the graphs of the two asymptotes to the picture obtained in part (a). Finally, to emphasize the fact that the two lines are indeed asymptotes, change the viewing rectangle so that x extends from -50 to 50. What do you observe?

In Exercises 24–33, evaluate the given quantities without using a calculator or tables.

24. $\tan[\sin^{-1}(\frac{4}{5})]$
25. $\cos(\arcsin\frac{2}{7})$
26. $\sin(\tan^{-1} 1)$
27. $\sin[\tan^{-1}(-1)]$
28. $\tan(\arccos\frac{5}{13})$
29. $\cos[\sin^{-1}(\frac{2}{3})]$
30. $\cos(\arctan\sqrt{3})$
31. $\sin[\cos^{-1}(\frac{1}{3})]$
32. $\sin[\arccos(-\frac{1}{3})]$
33. $\tan(\arcsin\frac{20}{29})$

34. Use a calculator to evaluate each of the following quantities. Express your answers to two decimal places without rounding them.
(a) $\sin^{-1}(\frac{3}{4})$
(b) $\cos^{-1}(\frac{2}{3})$
(c) $\tan^{-1}\pi$
(d) $\tan^{-1}(\tan^{-1}\pi)$

In Exercises 35–37, evaluate the given expressions without using a calculator or tables.

35. $\csc[\sin^{-1}(\frac{1}{2}) - \cos^{-1}(\frac{1}{2})]$
36. $\sec[\cos^{-1}(\sqrt{2}/2) + \sin^{-1}(-1)]$
37. $\cot[\cos^{-1}(-1/2) + \cos^{-1}(0) + \tan^{-1}(1/\sqrt{3})]$

Ⓖ **38.** In Example 7 we proved the identity $\sin(\cos^{-1} x) = \sqrt{1 - x^2}$ for $-1 \le x \le 1$. Demonstrate this identity visually by graphing the two functions $y = \sin(\cos^{-1} x)$ and $y = \sqrt{1 - x^2}$ and noting that the graphs appear to be identical.

Ⓖ **39.** In Example 10 we found that $\sec(\tan^{-1} x) = \sqrt{1 + x^2}$ for $x > 0$. Actually, this identity is valid for all real numbers. Demonstrate this visually by graphing the two functions $y = \sec(\tan^{-1} x)$ and $y = \sqrt{1 + x^2}$.

40. Show that $\cos(\sin^{-1} x) = \sqrt{1 - x^2}$ for $-1 \le x \le 1$. *Suggestion*: Use the method of Example 7 in the text.

Ⓖ **41.** Demonstrate the identity $\cos(\sin^{-1} x) = \sqrt{1 - x^2}$ for $-1 \le x \le 1$ by graphing the two functions $y = \cos(\sin^{-1} x)$ and $y = \sqrt{1 - x^2}$ and noting that the graphs appear to be identical.

42. If $\sin\theta = 2x$ and $0 < \theta < \pi/2$, express $\theta + \cos 2\theta$ as a function of x.

43. If $\sin\theta = 3x/2$ and $0 < \theta < \pi/2$, express $\frac{1}{4}\theta - \sin 2\theta$ as a function of x.

44. If $\cos\theta = x - 1$ and $0 < \theta < \pi/2$, express $2\theta - \cos 2\theta$ as a function of x.

45. If $\tan\theta = \frac{1}{2}(x - 1)$ and $0 < \theta < \pi/2$, express $\theta - \cos\theta$ as a function of x.

46. If $\tan\theta = \frac{1}{3}(x + 1)$ and $0 < \theta < \pi/2$, express $2\theta + \tan 2\theta$ as a function of x.

B

In Exercises 47–56, use the graphing techniques from Section 3.4 to sketch the graph of each function.

47. (a) $y = -\sin^{-1} x$
(b) $y = \sin^{-1}(-x)$
(c) $y = -\sin^{-1}(-x)$

48. (a) $y = -\cos^{-1} x$
(b) $y = \cos^{-1}(-x)$
(c) $y = -\cos^{-1}(-x)$

49. (a) $f(x) = \arccos(x + 1)$
(b) $g(x) = \arccos x + \frac{\pi}{2}$

50. (a) $F(x) = \arcsin(x - 1)$
(b) $G(x) = \arcsin x - \frac{\pi}{4}$

51. (a) $y = \arcsin(2 - x) + \frac{\pi}{2}$
(b) $y = -\arcsin(2 - x) + \frac{\pi}{2}$

52. (a) $y = \cos^{-1}(1 - x)$
(b) $y = -\cos^{-1}(1 - x) - \pi$

53. (a) $y = -\tan^{-1} x$
(b) $y = \tan^{-1}(-x)$
(c) $y = -\tan^{-1}(-x)$

54. (a) $y = \arctan(x + 3)$
(b) $y = \arctan x + \pi/2$
(c) $y = -\arctan(-x) + \pi/2$

55. $f(x) = -\arctan(1 - x) - \pi/2$

56. $y = \tan^{-1}(2 - x) + \pi/4$

57. Evaluate $\sin(2\tan^{-1} 4)$. *Hint*: $\sin 2\theta = 2\sin\theta\cos\theta$

58. Evaluate $\cos[2\sin^{-1}(\frac{5}{13})]$.

59. Evaluate $\sin(\arccos\frac{3}{5} - \arctan\frac{7}{13})$. *Suggestion*: Use the formula for $\sin(x - y)$.

60. Show that $\sin[\sin^{-1}(\frac{1}{3}) + \sin^{-1}(\frac{1}{4})] = (\sqrt{15} + 2\sqrt{2})/12$.

61. In the text we showed that $\sin^{-1} x + \cos^{-1} x = \pi/2$ for x in the open interval $(0, 1)$. Follow steps (a) and (b) to show that this identity actually holds for every x in the closed interval $[-1, 1]$.
(a) Let $\alpha = \sin^{-1} x$ and $\beta = \cos^{-1} x$. Explain why $-\pi/2 \le \alpha + \beta \le 3\pi/2$. *Hint*: What are the ranges of the inverse sine and the inverse cosine functions?
(b) Use the addition formula for sine to show that $\sin(\alpha + \beta) = 1$. Conclude [with the help of part (a)] that $\alpha + \beta = \pi/2$, as required.

Ⓖ **62.** Using the viewing rectangle specified in Exercise 21, graph the function $g(x) = \sin^{-1} x + \cos^{-1} x$. What do you observe? What is the exact value for the y-intercept of the graph? What identity does this demonstrate?

In Exercises 63 and 64, solve the given equations.

63. $\cos^{-1} t = \sin^{-1} t$ *Hint*: Compute the cosine of both sides.

64. $\sin^{-1}(3t - 2) = \sin^{-1} t - \cos^{-1} t$

65. Show that $\arctan x + \arctan y = \arctan\dfrac{x + y}{1 - xy}$ when x and y are positive and $xy < 1$.

66. Use the identity in Exercise 65 to show that the following equations are correct.
(a) $\arctan\frac{1}{2} + \arctan\frac{1}{3} = \frac{\pi}{4}$
(b) $\arctan\frac{1}{4} + \arctan\frac{3}{5} = \frac{\pi}{4}$
(c) $2\arctan\frac{1}{3} + \arctan\frac{1}{7} = \frac{\pi}{4}$
Hint: $2\arctan\frac{1}{3} = \arctan\frac{1}{3} + \arctan\frac{1}{3}$
(d) $\arctan\frac{1}{2} + \arctan\frac{1}{5} + \arctan\frac{1}{8} = \frac{\pi}{4}$

In Exercises 67 and 68, solve the given equations.

67. $2\tan^{-1} x = \tan^{-1}\dfrac{1}{4x}$ *Hint*: Compute the tangent of both sides.

68. $2\tan^{-1}\sqrt{t - t^2} = \tan^{-1} t + \tan^{-1}(1 - t)$

69. Consider the equation $\cos^{-1} x = \tan^{-1} x$.

(a) Explain why x cannot be negative or zero.

(b) As you can see in the accompanying figure, the graphs of $y = \cos^{-1} x$ and $y = \tan^{-1} x$ intersect at a point in Quadrant I. By solving the equation $\cos^{-1} x = \tan^{-1} x$, show that the x-coordinate of this intersection point is given by

$$x = \sqrt{\frac{\sqrt{5} - 1}{2}}$$

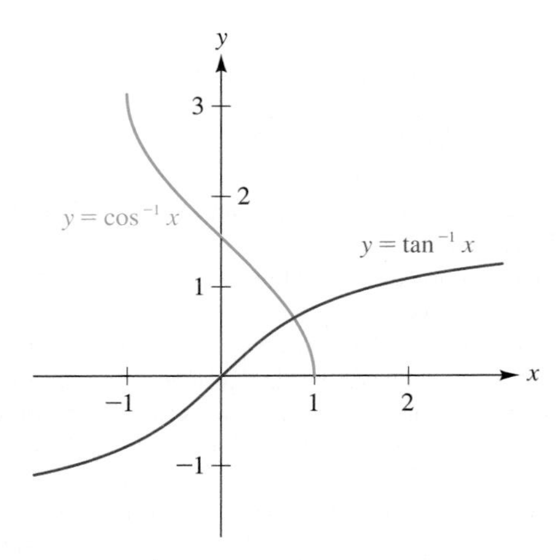

(c) Use the result in part (b) along with your calculator to specify the coordinates of the intersection point.

Ⓖ *In Exercises 70–85, use graphs to determine whether there are solutions for each equation in the interval* $[0, 1]$. *If there are solutions, use the graphing utility to find them accurately to two decimal places.*

70. $\cos^{-1} x = \tan^{-1} x$

71. $x = \arccos x$

72. $\cos^{-1} x = x^2$

73. (a) $1.3(x - \frac{1}{2})^2 = \cos^{-1} x$
(b) $1.4(x - \frac{1}{2})^2 = \cos^{-1} x$

74. $\tan^{-1} x = \sin 3x$

75. (a) $\arccos x = 2\sin 3x$
(b) $\arccos x = 2\sin 4x$

76. (a) $\sin(2.3x) = \arctan x$
(b) $\sin(2.2x) = \arctan x$

77. (a) $1/(\tan^{-1} x + \sin^{-1} x) = \sin 2x$
(b) $1/(\tan^{-1} x + \sin^{-1} x) = \sin 3x$

78. (a) $1/(\sin^{-1} x + \cos^{-1} x) = 4x^3$
(b) $1/(\sin^{-1} x + \cos^{-1} x) = 5x^3$

79. $\sin^{-1} x = \cos^{-1} x$

80. $\sin^{-1} x = \sin^{-1}(\sin^{-1} x)$

81. $\cos^{-1} x = \cos^{-1}(\cos^{-1} x)$

82. $\cos^{-1}(\sin^{-1} x) = \sin^{-1}(\cos^{-1} x)$

83. $1/(\sin x) = \sin^{-1} x$

84. $1/(\cos x) = \cos^{-1} x$

85. $1/(\tan x) = \tan^{-1} x$

Ⓖ **86.** Let $P(x, y)$ denote a point on the curve $y = \sqrt{x}$. As indicated in the accompanying figure, a line is drawn passing through P and the point $(-1, 0)$, and θ is the acute angle between this line and the x-axis.

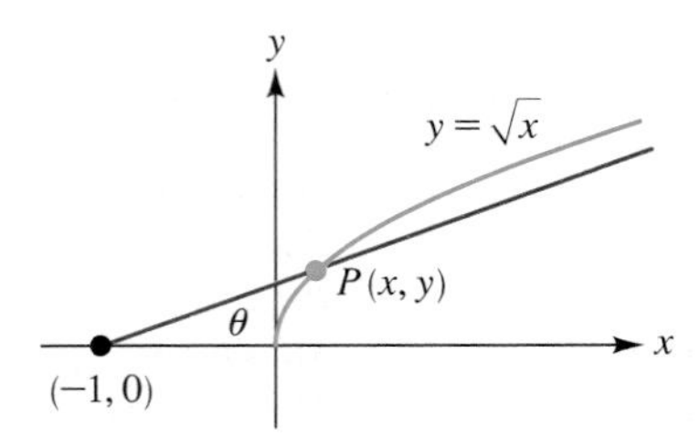

(a) Express θ as a function of x. (Use the inverse tangent function.)

(b) Use a graphing utility to obtain a graph of the function in part (a).

(c) Zoom in on the turning point of the graph in part (b) until you know the first three decimal places in the maximum possible value for θ. What is the corresponding x input?

(d) Using the x-value determined in part (c), find the equation of the line passing through $(-1, 0)$ and $P(x, y)$. Then graph this line along with the curve $y = \sqrt{x}$. What do you observe?

Ⓖ **87.** Refer to the accompanying diagrams. Figure A shows an observer looking at a picture on the wall. The base of the picture is 3 ft above the level of the person's eyes, the picture itself is 2 feet high, and the person is x ft from the wall. Figure B isolates the geometry of the situation.

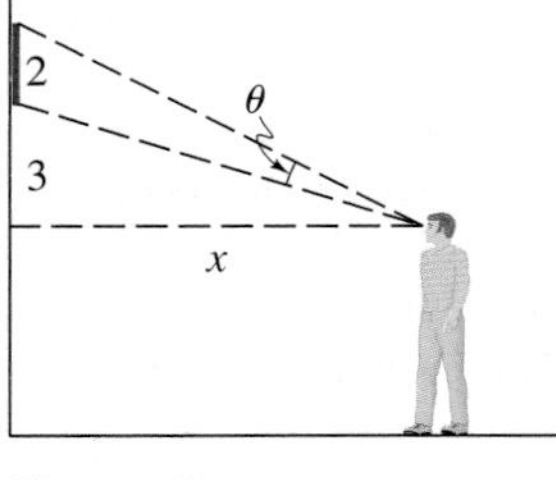

Figure A

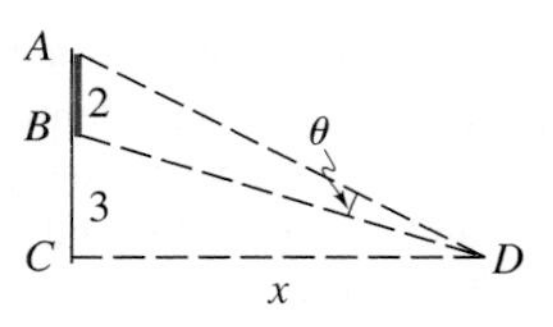

Figure B

(a) Show that $\theta = \tan^{-1}(5/x) - \tan^{-1}(3/x)$.
Hint: In Figure B, the angle $\theta = \angle ADC - \angle BDC$.

(b) Graph the equation in part (a). Then use the graph to explain why, in viewing the picture, the observer should stand about 4 ft away from the wall.

In Exercises 88 and 89 we make use of some simple geometric figures to evaluate sums involving the inverse tangent function. The idea here is taken from the note by Professor Edward M.

Harris entitled Behold! Sums of Arctan (*The College Mathematics Journal,* vol. 18 no. 2, p. 141).

88. In this exercise we use the following figure to show that $\tan^{-1}(\frac{2}{3}) + \tan^{-1}(\frac{1}{5}) = \frac{\pi}{4}$.

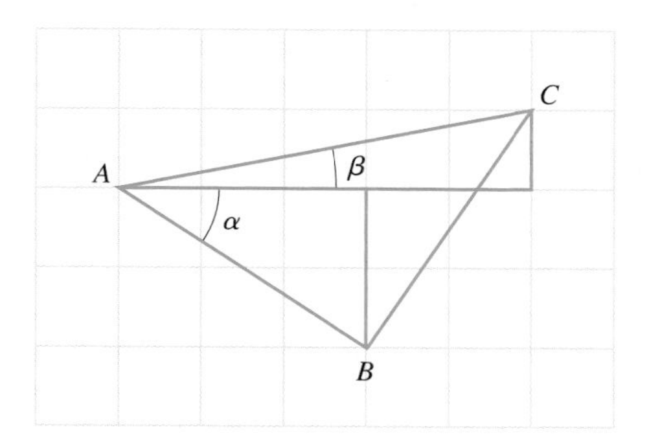

(a) Show that $\angle ABC = \pi/2$. *Hint*: Compute the slopes of $\overline{AB}$ and $\overline{BC}$. (Assume that the grid lines are marked off at one-unit intervals.)

(b) Show that $AB = BC$ and conclude that $\angle BAC = \pi/4$.

(c) Now explain why $\alpha = \tan^{-1}(\frac{2}{3})$, $\beta = \tan^{-1}(\frac{1}{5})$, and $\tan^{-1}(\frac{2}{3}) + \tan^{-1}(\frac{1}{5}) = \frac{\pi}{4}$.

89. In this exercise we use the accompanying figure to show that $\tan^{-1} 1 + \tan^{-1} 2 + \tan^{-1} 3 = \pi$.

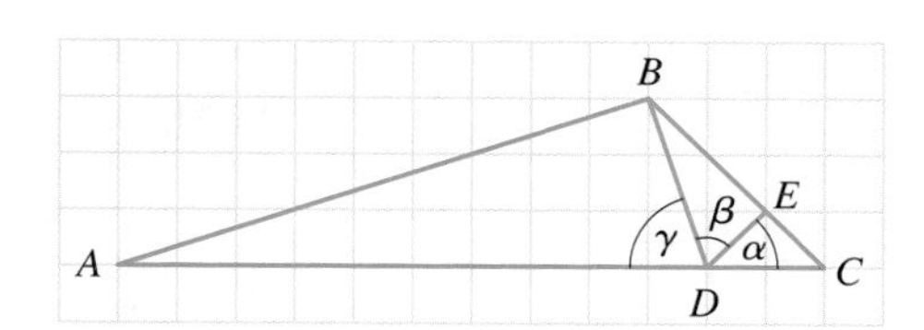

(a) By computing slopes, show that $\overline{DE} \perp \overline{BC}$ and $\overline{AB} \perp \overline{BD}$. (Assume that the grid lines are at one-unit intervals.)

(b) Determine the following lengths: DE, CE, BE, AB, and BD.

(c) Show that $\tan \alpha = 1$, $\tan \beta = 2$, and $\tan \gamma = 3$. Then explain why $\tan^{-1} 1 + \tan^{-1} 2 + \tan^{-1} 3 = \pi$.

Note: It's possible to do part (c) without first doing part (b). Try it.

90. In the following figure, $\triangle ABC$ is a right triangle and the four lengths AB, BP, PQ, and QC are all equal. Use the result in Exercise 89 to show that $\angle BAP + \angle BAQ + \angle BAC = 180°$.

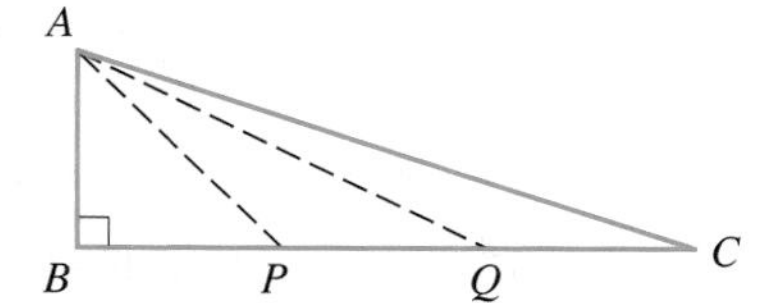

PROJECT — INVERSE SECANT FUNCTIONS

It is interesting to note that the property of periodicity, fundamental to the importance and usefulness of the sine, cosine, and tangent functions, is precisely what prevents these functions from being one-to-one and therefore having inverse functions. So in Section 8.5 we restricted the domains of these functions to obtain one-to-one versions and proceeded to construct the inverse sine, inverse cosine, and inverse tangent functions, systematically following our development of inverse functions from Chapter 3. The domains we chose are the ones that are traditionally used for this purpose.

In the case of the secant function there are two common choices of domain. These result in two one-to-one versions of the secant function, each with its own distinct inverse secant function. Before we start, keep in mind that our choice of domain is made to satisfy two conditions: We need a version of the function that is both one-to-one and has the same range as the unrestricted secant function.

The secant function is not one-to-one. Why? Looking at the graph of $y = \sec x$ where the secant's range is obtained, we can restrict the domain to the union of intervals $[0, \frac{\pi}{2}) \cup [\pi, \frac{3\pi}{2})$ to obtain a version of secant with the correct range. We have our first restricted secant function:

$$y = \sec x$$

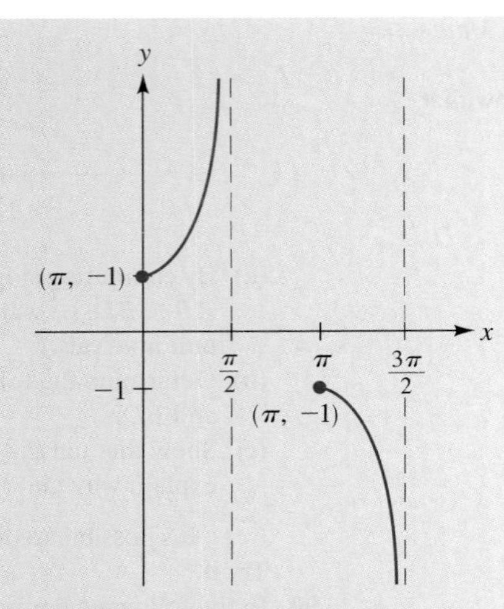

(i) The graph of $y = \sec x$ for x in $\left[0, \frac{\pi}{2}\right) \cup \left[\pi, \frac{3\pi}{2}\right)$

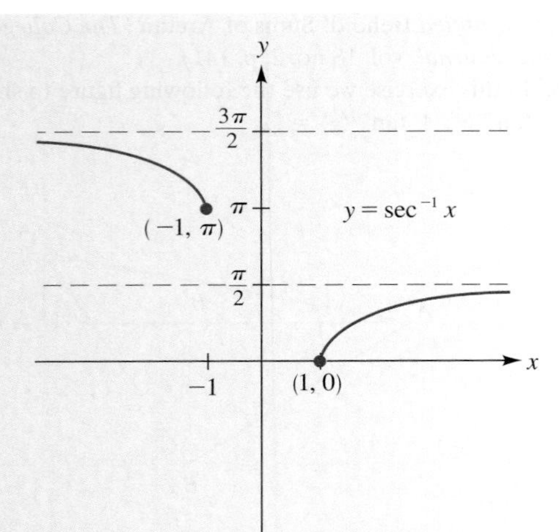

(ii) The graph of $y = \sec^{-1} x$

Figure A

with

$$\text{domain} = [0, \tfrac{\pi}{2}) \cup [\pi, \tfrac{3\pi}{2}) \qquad \text{and} \qquad \text{range} = (-\infty, -1] \cup [1, \infty)$$

which is one-to-one. Its graph is shown in Figure A(i).

This secant function has an inverse, denoted $\sec^{-1} x$ or arcsec, with

$$\text{domain of } \sec^{-1} = \text{range of the restricted secant} = (-\infty, -1] \cup [1, \infty)$$

and

$$\text{range of } \sec^{-1} = \text{domain of the restricted secant} = [0, \tfrac{\pi}{2}) \cup [\pi, \tfrac{3\pi}{2})$$

The graph of $y = \sec^{-1} x$ is shown in Figure A(ii) and can be obtained by reflecting the graph in Figure A(i) about the line with equation $y = x$. (Try it.)

What is $\sec^{-1} x$? Start with the equation $y = \sec x$. Interchange x and y to get $x = \sec y$. What does the y in this last equation represent? The last equation tells us that $y = \sec^{-1} x$ is the unique number in the domain of the restricted secant, $[0, \frac{\pi}{2}) \cup [\pi, 3\frac{\pi}{2})$, whose secant is x.

EXAMPLE 1 Values of some expressions involving an inverse secant

(a) $\sec^{-1} 2$ = the unique number in $[0, \frac{\pi}{2}) \cup [\pi, \frac{3\pi}{2})$ whose secant is $2 = \frac{\pi}{3}$

(b) $\text{arcsec}(-\sqrt{2})$ = the unique number in $[0, \frac{\pi}{2}) \cup [\pi, \frac{3\pi}{2})$ whose secant is $-\sqrt{2} = \frac{5\pi}{4}$

(c) $\sec[\sec^{-1}(5)]$ = the secant of (the unique number in $[0, \frac{\pi}{2}) \cup [\pi, \frac{3\pi}{2})$ whose secant is 5) = 5

(d) $\sec^{-1}(\sec \frac{2}{3})$ = the unique number in $[0, \frac{\pi}{2}) \cup [\pi, \frac{3\pi}{2})$ whose secant is $\sec \frac{2}{3} = \frac{2}{3}$

■

EXAMPLE 11 Evaluating expressions involving an inverse secant

(a) $\sin[\sec^{-1}(\frac{5}{2})]$

Let $\theta = \sec^{-1}(\frac{5}{2})$, then θ is in $[0, \frac{\pi}{2})$ and $\sec\theta = \frac{5}{2}$. Using the right triangle labeled in Figure B, we have

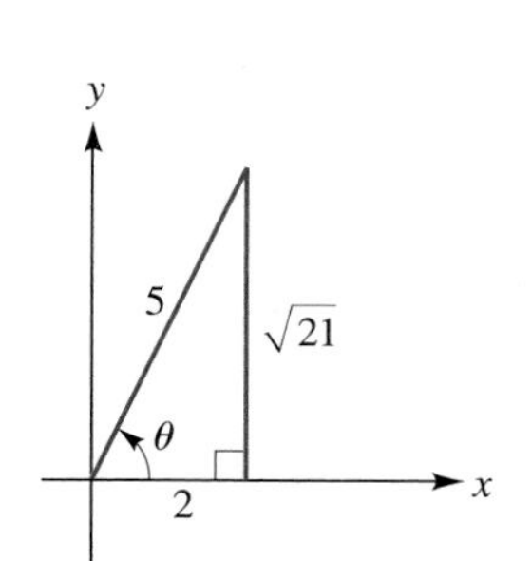

Figure B

$$\sin\left[\sec^{-1}\left(\frac{5}{2}\right)\right] = \sin\theta = \frac{\text{opposite}}{\text{hypotenuse}} = \frac{\sqrt{21}}{5}$$

Alternatively, using identities,

$$\cos\theta = \frac{1}{\sec\theta} = \frac{1}{5/2} = \frac{2}{5}$$

So

$$\sin\theta = \sqrt{1 - \cos^2\theta} = \sqrt{1 - (2/5)^2} = \frac{\sqrt{21}}{5}$$

Why the positive square root?

(b) $\tan[\sec^{-1}(-\frac{8}{3})]$

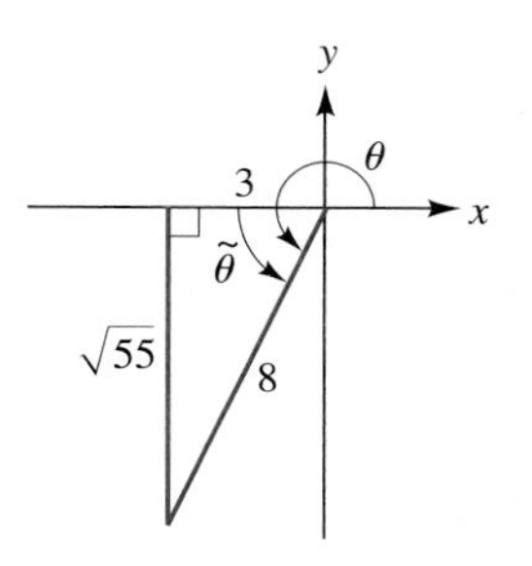

Figure C

Let $\theta = \sec^{-1}(-\frac{8}{3})$ and let $\tilde{\theta}$ be its reference angle. Then θ is in Quadrant III and $\sec\theta = -\frac{8}{3}$. Using the right triangle labeled in Figure C, we have

$$\tan\left[\sec^{-1}\left(-\frac{8}{3}\right)\right] = \tan\theta = \tan\tilde{\theta} = \frac{\text{opposite}}{\text{adjacent}} = \frac{\sqrt{55}}{3}$$

Alternatively, using identities,

$$\sec\theta = -\frac{8}{3}$$

So

$$\tan\theta = \sqrt{\sec^2\theta - 1} = \sqrt{(-8/3)^2 - 1} = \frac{\sqrt{55}}{3}$$

Why the positive square root?

Now we construct a second version of the inverse secant by restricting the secant's domain in a different way. We'll use a capital S to distinguish this version from our previous version of a restricted secant. Let

$$y = \operatorname{Sec} x$$

with

$$\text{domain} = [0, \tfrac{\pi}{2}) \cup (\tfrac{\pi}{2}, \pi] \quad \text{and} \quad \text{range} = (-\infty, -1] \cup [1, \infty)$$

Then $\operatorname{Sec} x$ is one-to-one as can be seen from its graph in Figure D(i). So this Secant function has an inverse denoted Sec^{-1} or Arcsec, with

$$\text{domain of } \operatorname{Sec}^{-1} = \text{range of the restricted Secant} = (-\infty, -1] \cup [1, \infty)$$

and

$$\text{range of } \operatorname{Sec}^{-1} = \text{domain of the restricted Secant} = [0, \tfrac{\pi}{2}) \cup (\tfrac{\pi}{2}, \pi]$$

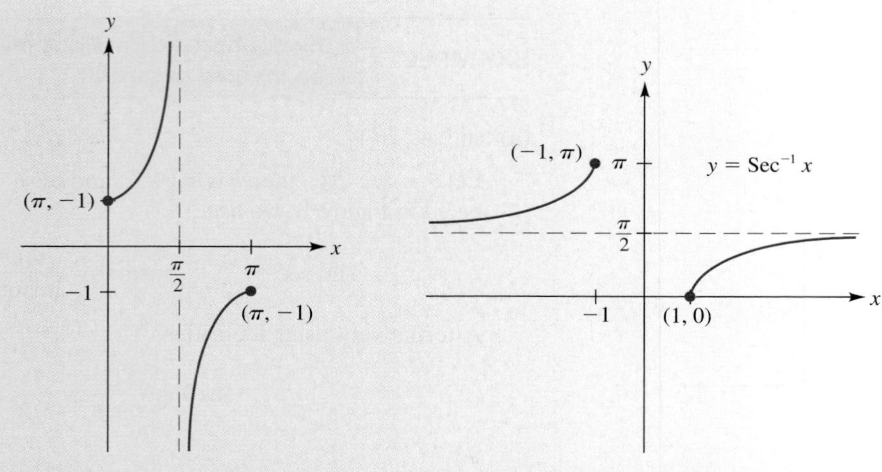

(i) The graph of $y = \text{Sec } x$ for x in $\left[0, \frac{\pi}{2}\right) \cup \left(\frac{\pi}{2}, \pi\right]$

(ii) The graph of $y = \text{Sec}^{-1} x$

Figure D

The graph of $y = \text{Sec}^{-1} x$ is shown in Figure D(ii) and can be obtained by reflecting the graph in Figure D(i) about the line with equation $y = x$. (Try it.)

What is $\text{Sec}^{-1} x$? This time, $y = \text{Sec}^{-1} x =$ the unique number in the restricted domain of Secant, $[0, \frac{\pi}{2}) \cup (\frac{\pi}{2}, \pi]$, whose *Secant* is x.

EXAMPLE III Values of some expressions involving the second version of the inverse secant

You can show that $\text{Sec}^{-1} 2 = \frac{\pi}{3}$, $\text{Sec}[\text{Sec}^{-1}(5)] = 5$, and $\text{Sec}^{-1}(\text{Sec } \frac{2}{3}) = \frac{2}{3}$ as in Example I. However, $\text{Arcsec}(-\sqrt{2}) =$ the unique number in $[0, \frac{\pi}{2}) \cup (\frac{\pi}{2}, \pi]$ whose secant is $-\sqrt{2} = \frac{3\pi}{4}$, *not* $\frac{5\pi}{4}$ as in Example I.

NOTE $\text{Sec}^{-1} x = \sec^{-1} x$ for $x \geq 1$, but $\text{Sec}^{-1} x = 2\pi - \sec^{-1} x$ for $x \leq -1$.

EXAMPLE IV Evaluating expressions involving the second version of the inverse secant

$\sin[\text{Sec}^{-1}(\frac{5}{2})] = \sqrt{21}/5$ as in Example II(a). Or, by the previous note,

$$\sin\left[\text{Sec}^{-1}\left(\frac{5}{2}\right)\right] = \sin\left[\sec^{-1}\left(\frac{5}{2}\right)\right] = \frac{\sqrt{21}}{5}$$

But, $\text{Sec}^{-1}(-\frac{8}{3})$ is in $(\frac{\pi}{2}, \pi]$, $\sec^{-1}(-\frac{8}{3})$ is in $[\pi, \frac{3\pi}{2})$, and neither equals π. (Why?) So $\text{Sec}^{-1}(-\frac{8}{3}) \neq \sec^{-1}(-\frac{8}{3})$, and in fact $\tan[\text{Sec}^{-1}(-\frac{8}{3})] \neq \tan[\sec^{-1}(-\frac{8}{3})]$. To evaluate $\tan[\text{Sec}^{-1}(-\frac{8}{3})]$, let $\theta = \text{Sec}^{-1}(-\frac{8}{3})$ and let $\tilde{\theta}$ be its reference angle. Then θ is in $(\frac{\pi}{2}, \pi]$ and $\text{Sec } \theta = -\frac{8}{3}$. Using the right triangle labeled in Figure E, we have

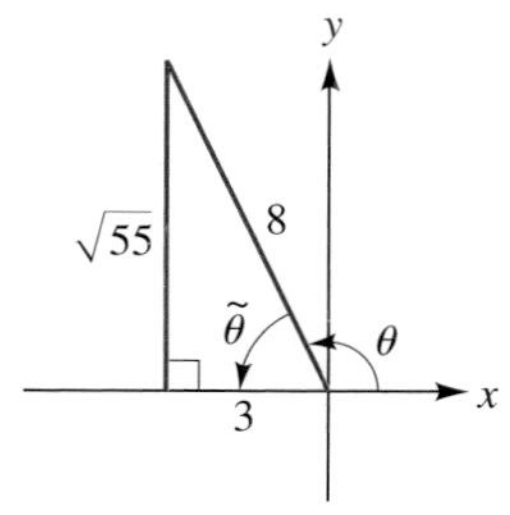

Figure E

$$\tan[\text{Sec}^{-1}(-\tfrac{8}{3})] = \tan\theta = -\tan\tilde{\theta} = -\frac{\text{opposite}}{\text{adjacent}} = -\frac{\sqrt{55}}{3}$$

Why is $\tan\theta = -\tan\tilde{\theta}$?

Alternatively, using identities, $\sec\theta = -\frac{8}{3}$, so

$$\tan\theta = -\sqrt{\sec^2\theta - 1} = -\sqrt{(-8/3)^2 - 1} = -\frac{\sqrt{55}}{3}$$

Why the negative square root?

EXERCISES

In each of these exercises the secant and the Secant are the restricted versions defined in this project, so the inverse secant and the inverse Secant are the corresponding inverse functions.

1. Evaluate the following quantities:
(a) $\sec^{-1}(0)$ **(b)** $\sec^{-1}(-2/\sqrt{3})$ **(c)** $\operatorname{arcsec}(-1)$ **(d)** $\sec[\sec^{-1}(10)]$ **(e)** $\operatorname{arcsec}\sqrt{2}$ **(f)** $\sec^{-1}(\sec 4)$

2. Evaluate the following quantities:
(a) $\cos[\sec^{-1}(\frac{6}{5})]$ **(b)** $\sin[\operatorname{arcsec}(-5)]$ **(c)** $\tan[\sec^{-1}(-\frac{7}{3})]$

3. Evaluate the following quantities:
(a) $\mathrm{Sec}^{-1}(1)$ **(b)** $\mathrm{Sec}^{-1}(\mathrm{Sec}\,2)$ **(c)** $\mathrm{Arcsec}(-2)$ **(d)** $\sin[\mathrm{Sec}^{-1}(\frac{13}{5})]$ **(e)** $\mathrm{Sec}[\mathrm{Sec}^{-1}(15)]$ **(f)** $\tan[\mathrm{Sec}^{-1}(-\frac{15}{7})]$

4. Prove

$$\mathrm{Sec}^{-1} x = \sec^{-1} x \qquad \text{for } x \geq 1$$

and

$$\mathrm{Sec}^{-1} x = 2\pi - \sec^{-1} x \qquad \text{for } x \leq -1$$

5. Use the result in Exercise 4 to show that

$$\tan[\mathrm{Sec}^{-1}(-\tfrac{8}{3})] = \tan[\sec^{-1}(-\tfrac{8}{3})] = -\sqrt{55}/3$$

6. Explain why each of the following statements is true:
(a) $\sec[\mathrm{Sec}^{-1}(3)] = 3$
(b) $\sec[\mathrm{Sec}^{-1}(-3)]$ is undefined
(c) $\mathrm{Sec}[\sec^{-1}(8)] = 8$
(d) $\mathrm{Sec}[\sec^{-1}(-8)]$ is undefined
(e) $\mathrm{Sec}^{-1}[\sec(\frac{4}{3})] = \frac{4}{3}$
(f) $\sec^{-1}[\mathrm{Sec}\,\frac{3\pi}{4}] = \frac{5\pi}{4}$
(g) $\sec^{-1}(\mathrm{Sec}\,2) = 2\pi - 2$

Chapter 8 Summary of Principal Terms and Formulas

Terms or formulas	Page reference	Comments
1. $\sin(s+t) = \sin s\cos t + \cos s\sin t$ $\sin(s-t) = \sin s\cos t - \cos s\sin t$ $\cos(s+t) = \cos s\cos t - \sin s\sin t$ $\cos(s-t) = \cos s\cos t + \sin s\sin t$	563	These identities are referred to as the *addition formulas* for sine and cosine.

Terms or formulas	Page reference	Comments
2. $\cos(\frac{\pi}{2} - \theta) = \sin\theta$ $\sin(\frac{\pi}{2} - \theta) = \cos\theta$	565–566	These two identities, called *reduction formulas,* hold for all real numbers θ. In terms of angles, the identities state that the sine of an angle is equal to the cosine of its complement.
3. $\tan(s + t) = \dfrac{\tan s + \tan t}{1 - \tan s \tan t}$ $\tan(s - t) = \dfrac{\tan s - \tan t}{1 + \tan s \tan t}$	568	These are the *addition formulas* for tangent.
4. $\sin 2\theta = 2\sin\theta\cos\theta$ $\cos 2\theta = \cos^2\theta - \sin^2\theta$ $\tan 2\theta = \dfrac{2\tan\theta}{1 - \tan^2\theta}$	575	These are the *double-angle formulas.* There are four other forms of the double-angle formula for cosine. They appear in the box on page 578.
5. $\sin\dfrac{s}{2} = \pm\sqrt{\dfrac{1 - \cos s}{2}}$ $\cos\dfrac{s}{2} = \pm\sqrt{\dfrac{1 + \cos s}{2}}$ $\tan\dfrac{s}{2} = \dfrac{\sin s}{1 + \cos s}$	576	These three identities are referred to as the *half-angle formulas.* In the case of the half-angle formulas for sine and cosine the sign before the radical is determined by the quadrant in which the angle or arc $s/2$ terminates.
6. $\sin A \sin B$ $= \frac{1}{2}[\cos(A - B) - \cos(A + B)]$ $\sin A \cos B$ $= \frac{1}{2}[\sin(A - B) + \sin(A + B)]$ $\cos A \cos B$ $= \frac{1}{2}[\cos(A - B) + \cos(A + B)]$	585	These *product-to-sum* identities are derived from the addition formulas for sine and cosine.
7. $\sin\alpha + \sin\beta$ $= 2\sin\dfrac{\alpha + \beta}{2}\cos\dfrac{\alpha - \beta}{2}$ $\sin\alpha - \sin\beta$ $= 2\cos\dfrac{\alpha + \beta}{2}\sin\dfrac{\alpha - \beta}{2}$ $\cos\alpha + \cos\beta$ $= 2\cos\dfrac{\alpha + \beta}{2}\cos\dfrac{\alpha - \beta}{2}$ $\cos\alpha - \cos\beta$ $= -2\sin\dfrac{\alpha + \beta}{2}\sin\dfrac{\alpha - \beta}{2}$	587	These are the *sum-to-product* identities.

Terms or formulas	Page reference	Comments
8. The restricted sine function	608	The domain of the function $y = \sin x$ is the set of all real numbers. By allowing inputs only from the closed interval $[-\frac{\pi}{2}, \frac{\pi}{2}]$. we obtain the restricted sine function. (See Figure 1 on page 608.) The motivation for restricting the domain is that the restricted sine function is one-to-one, and therefore the inverse function exists.
9. $\sin^{-1} x$ or $\arcsin x$	608	$\sin^{-1} x$ is that number in the interval $[-\frac{\pi}{2}, \frac{\pi}{2}]$ whose sine is x. The basic properties of the inverse sine function are summarized in the first box on page 612.
10. The restricted cosine function	612	The domain of the function $y = \cos x$ is the set of all real numbers. By allowing inputs only from the closed interval $[0, \pi]$, we obtain the restricted cosine function. (See Figure 5 on page 612.) The motivation for restricting the domain is that the restricted cosine function is one-to-one, and therefore the inverse function exists.
11. $\cos^{-1} x$ or $\arccos x$	612	$\cos^{-1} x$ is that number in the interval $[0, \pi]$ whose cosine is x. The basic properties of the inverse cosine function are summarized in a box on pages 617–618.
12. The restricted tangent function	614	The tangent function, $y = \tan x$, is not one-to-one. However, by allowing inputs only from the open interval $(-\frac{\pi}{2}, \frac{\pi}{2})$ we obtain the restricted tangent function, which is one-to-one. (See Figure 10 on page 614.) Since the restricted tangent function is one-to-one, the inverse function exists.
13. $\tan^{-1} x$ or $\arctan x$	615	$\tan^{-1} x$ is that number in the interval $(-\frac{\pi}{2}, \frac{\pi}{2})$ whose tangent is x. The basic properties of the inverse tangent function are summarized in a box on page 618.

Writing Mathematics

1. Say whether the statement is TRUE or FALSE. Write out your reason or reasons in complete sentences. If you draw a diagram to accompany your writing, be sure that you clearly label any parts of the diagram to which you refer.

(a) The equation $\tan^2 t + 1 = \sec^2 t$ is true for every real number t.

(b) There is no real number x satisfying the equation $\cos(\frac{\pi}{4} + x) = 2$.

(c) There is no real number x satisfying the equation $\cos(\frac{\pi}{4} + x) = \cos x$.

(d) For every number x in the closed interval $[-1, 1]$, we have $\sin^{-1} x = 1/\sin x$.

(e) There is no real number x for which $\sin^{-1} x = 1/\sin x$. *Hint*: Draw a careful sketch of the graphs of the inverse sine function and the cosecant function on the interval $0 < x \leq 1$.

(f) The equation $\sin(x + y) = \sin x \cos y + \cos x \sin y$ holds for all real numbers x and y.

(g) The equation $\tan(x + y) = \dfrac{\tan x + \tan y}{1 - \tan x \tan y}$ holds for all real numbers x and y.

2. There is a formula for calculating the angle between two given lines in the x–y plane. The derivation of this formula relies on the identity for $\tan(s - t)$. Find a book on analytic geometry in the library. Look up this formula; find out how it is derived and how to use it. You can work with a classmate or your instructor. Then on your own, write a summary of what you have learned. Include an example (like the ones in this precalculus text) explaining how the formula is applied in a specific case.

3. This exercise consists of two parts. In the first part you are going to follow some simple instructions to construct a regular pentagon. In the second part, you'll write a paper explaining why the construction is valid.

(a) Use the following instructions to make a poster showing a regular pentagon inscribed in the unit circle. The poster should also show the steps used in the construction.

> Draw a unit circle. Label the origin O; let B and C' denote the points $(1, 0)$ and $(0, 1)$, respectively; let C be the midpoint of OC'; and let A denote the point where the bisector of $\angle OCB$ meets the x-axis. From A, draw a line segment straight up to the unit circle, meeting the unit circle at a point P. Then the line segment joining B and P will be one side of a regular pentagon inscribed in the unit circle.

Remark: This is perhaps the simplest geometric construction known for the regular pentagon. The ancient Greek mathematicians had a more complicated method. The construction given here was discovered (only) a century ago by H. W. Richmond; it appeared in the *Quarterly Journal of Mathematics,* vol. 26 (1893), pp. 296–297.

(b) With a group of classmates or your instructor, work out the details in the following terse justification for the construction in part (a). Then, on your own, carefully write out the justification in full detail. This will involve a mixture of English sentences and equations, much like the exposition in this textbook. At each of the main steps, be sure to tell the reader where you are going and what that step will accomplish.

> We want to show that $\angle POB = 72°$. Let $\angle OCB = \theta$. From right triangle OCB we get $\sin\theta = 2\sqrt{5}/5$ and $\cos\theta = \sqrt{5}/5$. Using these values in the half-angle formula for tangent then yields (after simplifying) $\tan(\theta/2) = (\sqrt{5} - 1)/2$. Next, from right triangle OCA we have $OA = \frac{1}{2}\tan(\theta/2)$. From these last two equations we conclude that $OA = (\sqrt{5} - 1)/4$. However, according to Exercise 68(c) in Section 8.2, $\sin 18° = (\sqrt{5} - 1)/4$. It now follows that $OA = \cos 72°$, and therefore $\angle POB = 72°$.

(In working out the details, you can assume the result from Exercise 68(c) in Section 8.2.)

Chapter 8 Review Exercises

In Exercises 1–32, prove that the equations are identities.

1. $\cot(x+y) = \dfrac{\cot x \cot y - 1}{\cot x + \cot y}$

2. $\cos 2x = \dfrac{1 - \tan^2 x}{1 + \tan^2 x}$

3. $\sin 2x = \dfrac{2 \tan x}{1 + \tan^2 x}$

4. $\sin^2 x - \sin^2 y = \sin(x+y)\sin(x-y)$

5. $\tan^2 x - \tan^2 y = \dfrac{\sin(x+y)\sin(x-y)}{\cos^2 x \cos^2 y}$

6. $2 \csc 2x = \sec x \csc x$

7. $(\sin x)[\tan(x/2) + \cot(x/2)] = 2$

8. $\tan(x + \frac{\pi}{4}) = (1 + \tan x)/(1 - \tan x)$

9. $\tan(\frac{\pi}{4} + x) - \tan(\frac{\pi}{4} - x) = 2 \tan 2x$

10. $\dfrac{\cot x - 1}{\cot x + 1} = \dfrac{1 - \sin 2x}{\cos 2x}$

11. $2 \sin\left(\dfrac{\pi}{4} - \dfrac{x}{2}\right) \cos\left(\dfrac{\pi}{4} - \dfrac{x}{2}\right) = \cos x$

12. $\dfrac{\tan(x+y) - \tan y}{1 + \tan(x+y)\tan y} = \tan x$

13. $\dfrac{1 - \tan(\frac{1}{4}\pi - t)}{1 + \tan(\frac{1}{4}\pi - t)} = \tan t$

14. $\tan(\frac{1}{4}\pi + \frac{1}{2}t) = \tan t + \sec t$

15. $\dfrac{\tan(\alpha - \beta) + \tan \beta}{1 - \tan(\alpha - \beta)\tan \beta} = \tan \alpha$

16. $(1 + \tan \theta)[1 + \tan(\frac{1}{4}\pi - \theta)] = 2$

17. $\tan 3\theta = \dfrac{3t - t^3}{1 - 3t^2}$, where $t = \tan \theta$

18. $\sin 2t = \sin(t + \frac{2}{3}\pi)\cos(t - \frac{2}{3}\pi) + \cos(t + \frac{2}{3}\pi)\sin(t - \frac{2}{3}\pi)$

19. $\tan 2x + \sec 2x = \dfrac{\cos x + \sin x}{\cos x - \sin x}$

20. $\cos^4 x - \sin^4 x = \cos 2x$

21. $2 \sin x + \sin 2x = \dfrac{2 \sin^3 x}{1 - \cos x}$

22. $1 + \tan x \tan(x/2) = \sec x$

23. $\tan \dfrac{x}{2} = \dfrac{1 - \cos x + \sin x}{1 + \cos x + \sin x}$

24. $\dfrac{\sin 3x}{\sin x} - \dfrac{\cos 3x}{\cos x} = 2$

25. $\sin(x+y)\cos y - \cos(x+y)\sin y = \sin x$

26. $\dfrac{\sin x + \sin 2x}{\cos x - \cos 2x} = \cot \dfrac{x}{2}$

27. $\dfrac{1 - \tan^2(x/2)}{1 + \tan^2(x/2)} = \cos x$

28. $4 \sin(x/4)\cos(x/4)\cos(x/2) = \sin x$

29. $\sin 4x = 4 \sin x \cos x - 8 \sin^3 x \cos x$

30. $\cos 4x = 8 \cos^4 x - 8 \cos^2 x + 1$

31. $\sin 5x = 16 \sin^5 x - 20 \sin^3 x + 5 \sin x$

32. $\cos 5x = 16 \cos^5 x - 20 \cos^3 x + 5 \cos x$

In Exercises 33–41, establish the identities by applying the sum-to-product formulas.

33. $\sin 80° - \sin 20° = \cos 50°$

34. $\sin 65° + \sin 25° = \sqrt{2} \cos 20°$

35. $\dfrac{\cos x - \cos 3x}{\sin x + \sin 3x} = \tan x$

36. $\dfrac{\sin 3° + \sin 33°}{\cos 3° + \cos 33°} = \tan 18°$

37. $\sin(5\pi/12) + \sin(\pi/12) = \sqrt{6}/2$

38. $\cos 10° - \sin 10° = \sqrt{2} \sin 35°$

39. $\dfrac{\cos 3y + \cos(2x - 3y)}{\sin 3y + \sin(2x - 3y)} = \cot x$

40. $\dfrac{\sin 10° - \sin 50°}{\cos 50° - \cos 10°} = \sqrt{3}$

41. $\dfrac{\sin 40° - \sin 20°}{\cos 20° - \cos 40°} = \dfrac{\sin 10° - \sin 50°}{\cos 50° - \cos 10°}$

42. Suppose that a and b are in the open interval $(0, \pi/2)$ and $a \neq b$. If $\sin a + \sin b = \cos a + \cos b$, show that $a + b = \pi/2$. *Hint*: Begin with the sum-to-product formulas.

43. Refer to the figure. Using calculus, it can be shown that the area of the first-quadrant region under the curve $y = 1/(1 + x^2)$ from $x = 0$ to $x = a$ is given by the expression $\tan^{-1} a$. Use this fact to carry out the calculations on the following page.

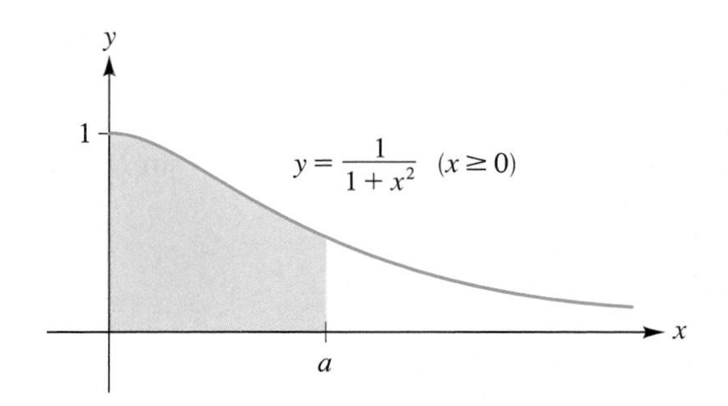

(a) Find the area of the first-quadrant region under the curve $y = 1/(1 + x^2)$ from $x = 0$ to $x = 1$.

(b) Find a value of a so that the area of the first-quadrant region under the curve from $x = 0$ to $x = a$ is: (i) 1.5; (ii) 1.56; (iii) 1.57. In each case, round your answer to the nearest integer.

44. The figure below shows the graph of the curve $y = 1/\sqrt{1 - x^2}$ for $0 \le x < 1$.

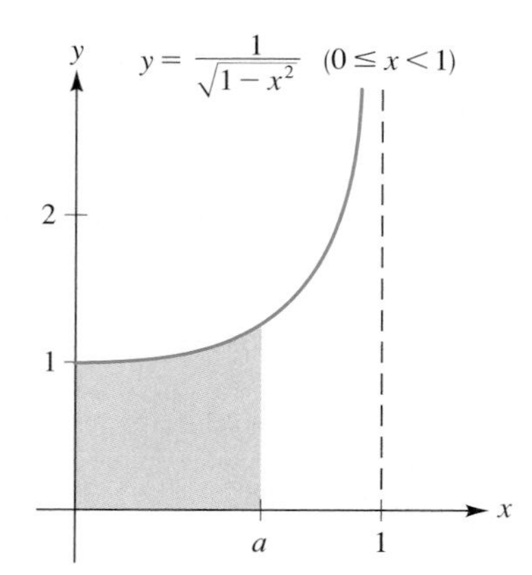

Using calculus it can be shown that the area of the first-quadrant region bounded by this curve, the coordinate axes, and the line $x = a$ is given by $\sin^{-1} a$. Use this fact to carry out the following calculations.

(a) Find the area of the first-quadrant region under this curve from $x = 0$ to $x = \sqrt{2}/2$. Give both the exact form of the answer and a calculator approximation rounded to three decimal places.

(b) Find a value of a so that the area of the first-quadrant region under the curve from $x = 0$ to $x = a$ is 1.5. Round your answer to three decimal places.

In Exercises 45–61, find all solutions of each equation in the interval $[0, 2\pi)$. In cases in which a calculator is necessary, round the answers to two decimal places.

45. $\tan x = 4.26$

46. $\tan x = -4.26$

47. $\csc x = 2.24$

48. $\sin(\sin x) = \pi/6$

49. $\tan^2 x - 3 = 0$

50. $\cot^2 x - \cot x = 0$

51. $1 + \sin x = \cos x$

52. $2 \sin 3x - \sqrt{3} = 0$

53. $\sin x - \cos 2x + 1 = 0$

54. $\sin x + \sin 2x = 0$

55. $3 \csc x - 4 \sin x = 0$

56. $2 \sin^2 x + \sin x - 1 = 0$

57. $2 \sin^4 x - 3 \sin^2 x + 1 = 0$

58. $\sec^2 x - \sec x - 2 = 0$

59. $\sin^4 x + \cos^4 x = 5/8$

60. $4 \sin^2 2x + \cos 2x - 2 \cos^2 x - 2 = 0$

61. $\cot x + \csc x + \sec x = \tan x$ *Suggestion*: If you use sines and cosines, the given equation becomes $\cos^2 x - \sin^2 x + \cos x + \sin x = 0$, which can be factored.

62. If A and B both are solutions of the equation $a \cos x + b \sin x = c$, show that

$$\tan\left[\frac{1}{2}(A + B)\right] = \frac{b}{a}$$

Hint: The given information yields two equations. After subtracting one of those equations from the other and rearranging, you will have $\frac{\cos A - \cos B}{\sin A - \sin B} = -\frac{b}{a}$. Now use the sum-to-product formulas.

63. Evaluate $\cos \tan^{-1} \sin \tan^{-1}(\sqrt{2}/2)$.

In Exercises 64–87, evaluate each expression (without using a calculator or tables).

64. $\cos^{-1}(-\sqrt{2}/2)$

65. $\arctan(\sqrt{3}/3)$

66. $\sin^{-1} 0$

67. $\arcsin \frac{1}{2}$

68. $\arctan \sqrt{3}$

69. $\cos^{-1}(\frac{1}{2})$

70. $\tan^{-1}(-1)$

71. $\cos^{-1}(-\frac{1}{2})$

72. $\sin(\sin^{-1} 1)$

73. $\cos[\cos^{-1}(\frac{2}{7})]$

74. $\sin^{-1}(\sin \frac{\pi}{7})$

75. $\cos^{-1}(\cos 5)$

76. $\sin^{-1}(\sin 2)$

77. $\sin[\tan^{-1}(-1)]$

78. $\sin[\arccos(-\frac{1}{2})]$

79. $\sec[\cos^{-1}(\sqrt{2}/3)]$

80. $\cot[\cos^{-1}(\frac{1}{2})]$

81. $\tan[\frac{\pi}{4} + \sin^{-1}(\frac{5}{13})]$

82. $\sin(\frac{3\pi}{2} + \arccos \frac{3}{5})$

83. $\tan(2 \tan^{-1} 2)$

84. $\sin[2 \sin^{-1}(\frac{4}{5})]$

85. $\cos[\frac{1}{2} \cos^{-1}(\frac{4}{5})]$

86. In this exercise we investigate the relationship between the variables x and θ in the accompanying figure. (Assume that θ is in radians.)

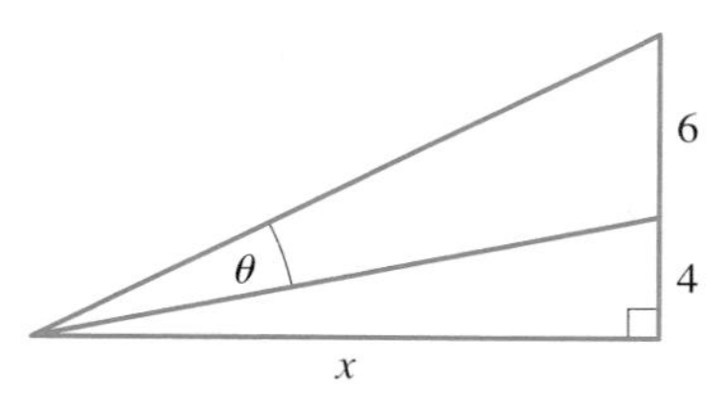

(a) Refer to the figure. Show that

$$\theta = \tan^{-1}\left(\frac{10}{x}\right) - \tan^{-1}\left(\frac{4}{x}\right)$$

(b) Use your calculator to complete the following table. (Round the results to two decimal places.) Which x-value in the table yields the largest value for θ?

x	0.1	1	2	3	10	100
θ						

(c) As indicated in the following graph, the value of x that makes θ as large as possible is a number between 5 and 10, closer to 5 than to 10.

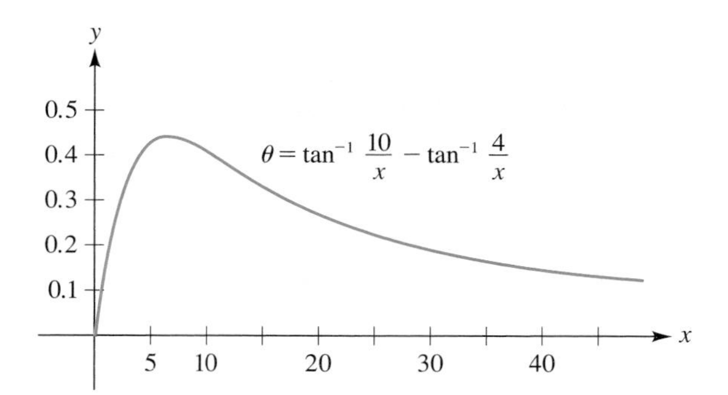

Using calculus, it can be shown that this value of x is, in fact, $2\sqrt{10}$. Use your calculator to evaluate $2\sqrt{10}$; check that the result is consistent with the given graph. What is the corresponding value of θ in this case? Also, give the coordinates of the highest point on the accompanying graph. Round both coordinates to two decimal places.

(d) What is the degree measure for the angle θ obtained in part (c)? Round the answer to one decimal place.

In Exercises 87–91, show that each equation is an identity.

87. $\tan(\tan^{-1} x + \tan^{-1} y) = (x + y)/(1 - xy)$

88. $\tan^{-1}(x/\sqrt{1 - x^2}) = \sin^{-1} x$

89. $\sin(2 \arctan x) = 2x/(1 + x^2)$

90. $\cos(2 \cos^{-1} x) = 2x^2 - 1$

91. $\sin[\frac{1}{2} \sin^{-1}(x^2)] = \sqrt{\frac{1}{2} - \frac{1}{2}\sqrt{1 - x^4}}$

In Exercises 92–94, without using a calculator, show that each statement is true.

92. $\tan^{-1}(\frac{1}{3}) + \tan^{-1}(\frac{1}{5}) = \tan^{-1}(\frac{4}{7})$

93. $\arcsin(4\sqrt{41}/41) + \arcsin(\sqrt{82}/82) = \pi/4$

94. $\tan[\sin^{-1}(\frac{1}{3}) + \cos^{-1}(\frac{1}{2})] = \frac{1}{5}(8\sqrt{2} + 9\sqrt{3})$

95. (a) Use a calculator to compute the quantity $\cos 20° \cos 40° \cos 60° \cos 80°$. Give your answer to as many decimal places as is shown on your calculator.

(b) Now use a product-to-sum formula to *prove* that the display on your calculator is the exact value of the given expression, not an approximation.

96. In this exercise we will use the accompanying figure to derive the half-angle formula for sine:

$$\sin\frac{\theta}{2} = \pm\sqrt{\tfrac{1}{2}(1 - \cos\theta)}$$

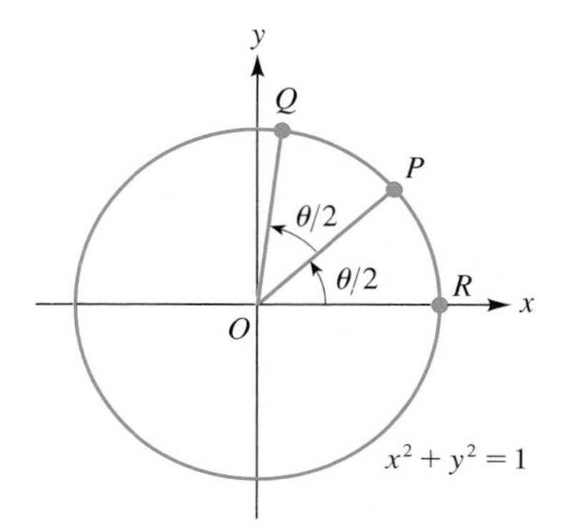

Our derivation will make use of the formula for the distance between two points and the identity $\sin^2 t + \cos^2 t = 1$. However, we will not rely on an addition formula for sine, as we did in Section 8.2.

(a) Explain why the coordinates of P and Q are $P(\cos\frac{\theta}{2}, \sin\frac{\theta}{2})$ and $Q(\cos\theta, \sin\theta)$.

(b) Use the distance formula to show that

$$(PQ)^2 = 2 - 2\cos\frac{\theta}{2}\cos\theta - 2\sin\frac{\theta}{2}\sin\theta$$

and

$$(PR)^2 = 2 - 2\cos\frac{\theta}{2}$$

(c) Explain why $\triangle POR$ is congruent to $\triangle QOP$.

(d) From part (c), it follows that $(PQ)^2 = (PR)^2$. By equating the expressions for $(PQ)^2$ and $(PR)^2$ [obtained in part (b)], show that

$$\sin\frac{\theta}{2}\sin\theta = \left(\cos\frac{\theta}{2}\right)(1 - \cos\theta)$$

(e) Square both sides of the equation obtained in part (d); then replace $\cos^2(\theta/2)$ by $1 - \sin^2(\theta/2)$ and show that the resulting equation can be written

$$[\sin^2(\theta/2)](2 - 2\cos\theta) = (1 - \cos\theta)^2$$

(continues)

(f) Solve the equation in part (e) for the quantity $\sin\frac{\theta}{2}$. You should obtain

$$\sin\frac{\theta}{2} = \pm\sqrt{\tfrac{1}{2}(1-\cos\theta)} \qquad \text{as required}$$

In Exercises 97–99, prove that the equations are identities. (These identities appear in Trigonometry and Double Algebra, *by August DeMorgan, published in 1849.)*

97. $\sin 2\theta = 2/(\cot\theta + \tan\theta)$

98. $\sin 2\theta = \dfrac{\tan(45^\circ+\theta) - \tan(45^\circ-\theta)}{\tan(45^\circ+\theta) + \tan(45^\circ-\theta)}$

99. $\cos 2\theta = 1/(1 + \tan 2\theta \tan\theta)$

100. Prove the following two identities. These identities were given by the Swiss mathematician Leonhard Euler in 1748.

(a) $\tan 2\theta = \dfrac{2\tan\theta}{1-\tan^2\theta}$

(b) $\cot 2\theta = \dfrac{\cot\theta - \tan\theta}{2}$

101. Prove the following two identities. These identities were given by the Swiss mathematician Johann Heinrich Lambert in 1765.

(a) $\sin 2\theta = \dfrac{2\tan\theta}{1+\tan^2\theta}$

(b) $\cos 2\theta = \dfrac{1-\tan^2\theta}{1+\tan^2\theta}$

102. Prove the following identity, which was essentially given by the German mathematician Johann Müller (known as "Regiomontanus") around 1464.

$$\frac{\sin A + \sin B}{\sin A - \sin B} = \frac{\tan[\frac{1}{2}(A+B)]}{\tan[\frac{1}{2}(A-B)]}$$

103. Prove the following identity, given by the Austrian mathematician George Joachim Rhaeticus in 1569.

$$\cos n\theta = \cos[(n-2)\theta] - 2\sin\theta\sin[(n-1)\theta]$$

104. Suppose that $x + \dfrac{1}{x} = 2\cos\theta$.

(a) Show that $x^2 + \dfrac{1}{x^2} = 2\cos 2\theta$.

(b) Show that $x^3 + \dfrac{1}{x^3} = 2\cos 3\theta$.

Chapter 8  Test 

1. Use an appropriate addition formula to simplify the expression $\sin(\theta + \frac{3\pi}{2})$.

2. Compute $\cos 2t$ given that $\sin t = -2\sqrt{5}/5$ and $3\pi/2 < t < 2\pi$.

3. Compute $\tan(\theta/2)$ given that $\cos\theta = -5/13$ and $\pi < \theta < 3\pi/2$.

4. Use a calculator to find all solutions of the equation $\sin x = 3\cos x$ in the interval $(0, 2\pi)$.

5. Find all solutions of the equation

$$2\sin^2 x + 7\sin x + 3 = 0$$

on the interval $0 \le x \le 2\pi$.

6. If $\cos\alpha = 2/\sqrt{5}$ $(\frac{3\pi}{2} < \alpha < 2\pi)$ and $\sin\beta = 4/5$ $(\frac{\pi}{2} < \beta < \pi)$, compute $\sin(\beta - \alpha)$.

7. Find all solutions of the equation $\sin(x + 30^\circ) = \sqrt{3}\sin x$ on the interval $0^\circ < x < 90^\circ$.

8. If $\csc\theta = -3$ and $\pi < \theta < 3\pi/2$, compute $\sin(\theta/2)$.

9. On the same set of axes, sketch the graphs of the restricted sine function and the function $y = \sin^{-1} x$. Specify the domain and the range for each function.

10. Compute each of the following quantities:
(a) $\sin^{-1}[\sin(\pi/10)]$ **(b)** $\sin^{-1}(\sin 2\pi)$

11. Compute $\cos(\arcsin\frac{3}{4})$.

12. Prove that the following equation is an identity:

$$\tan\left(\frac{\pi}{4} + \frac{\theta}{2}\right) = \frac{1 + \cos\theta + \sin\theta}{1 + \cos\theta - \sin\theta}$$

13. Use a product-to-sum formula to simplify the expression $\sin(7\pi/24)\cos(\pi/24)$.

14. Use the sum-to-product formulas to simplify the expression

$$\frac{\sin 3\theta + \sin 5\theta}{\cos 3\theta + \cos 5\theta}$$

15. Simplify the following expression: $\sec(\arctan\sqrt{x^2-1})$. (Assume that $x > 1$.)

16. Sketch a graph of the function $y = \tan^{-1} x$, and specify the domain and the range.

9 CHAPTER Additional Topics In Trigonometry

The subject of trigonometry is an excellent example of a branch of mathematics . . . which was motivated by both practical and intellectual interests—surveying, map-making, and navigation on the one hand, and curiosity about the size of the universe on the other. With it the Alexandrian mathematicians triangulated the universe and rendered precise their knowledge about the Earth and the heavens. —Morris Klein in *Mathematics in Western Culture* (New York: Oxford University Press, 1953)

This is our fourth and final chapter on trigonometry. Some of the subject matter here takes us back to the historical roots of the subject: the study of the relationships between the sides and the angles in a triangle. In Section 9.1 we look at some applications of right-triangle trigonometry (which was introduced in Section 6.5). Section 9.2 presents the law of sines and the law of cosines. These two laws relate the lengths of the sides and the angles for any triangle. In Sections 9.3 and 9.4 we introduce the important topic of vectors, first from a geometric standpoint, then from an algebraic standpoint. In Sections 9.5 through 9.7 we expand upon some of the ideas in Chapters 1 and 3 on graphs and equations. The topics presented here are parametric equations and polar coordinates.

9.1 RIGHT-TRIANGLE APPLICATIONS

We continue the work we began in Section 6.5 on right-triangle trigonometry. As prerequisites for this section, you'll need to have memorized the definitions of the trigonometric functions, given in the box on page 471, and the table of trigonometric values for 30°, 45°, and 60° on page 450.

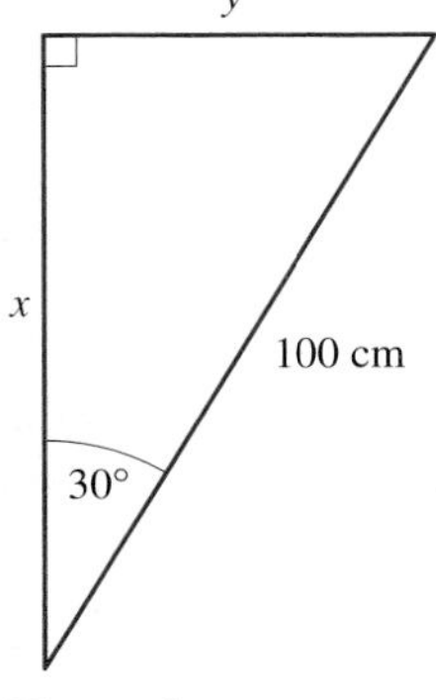

Figure 1

EXAMPLE 1 Using trigonometric functions to find a side in a right triangle

Use one of the trigonometric functions to find x in Figure 1.

SOLUTION

Relative to the given 30° angle, x is the adjacent side. The length of the hypotenuse is 100 cm. Since the adjacent side and the hypotenuse are involved, we use the cosine function here:

$$\cos 30^\circ = \frac{\text{adjacent}}{\text{hypotenuse}} = \frac{x}{100}$$

Consequently,

$$x = 100 \cos 30° = (100) \cdot (\sqrt{3}/2) = 50\sqrt{3} \text{ cm}$$

This is the result we are looking for.

We used the cosine function in Example 1 because the adjacent side and the hypotenuse were involved. We could instead use the secant. In that case, again with reference to Figure 1, the calculations look like this:

$$\sec 30° = \frac{\text{hypotenuse}}{\text{adjacent}} = \frac{100}{x}$$

$$\frac{2}{\sqrt{3}} = \frac{100}{x}$$

$$2x = 100\sqrt{3}$$

$$x = \frac{100\sqrt{3}}{2} = 50\sqrt{3} \text{ cm} \qquad \text{as was obtained previously}$$

EXAMPLE 2 Using trigonometric functions to find a side in a right triangle

Find y in Figure 1.

SOLUTION

As you can see from Figure 1, the side of length y is opposite the 30° angle. Furthermore, we are given the length of the hypotenuse. Since the opposite side and the hypotenuse are involved, we use the sine function. This yields

$$\sin 30° = \frac{\text{opposite}}{\text{hypotenuse}} = \frac{y}{100}$$

and therefore

$$y = 100 \sin 30° = (100) \cdot (1/2) = 50 \text{ cm}$$

This is the required answer. Actually, we could have obtained this particular result much faster by recalling that in the 30°–60° right triangle, the side opposite the 30° angle, namely, y, is half of the hypotenuse. That is $y = 100/2 = 50$ cm, as was obtained previously.

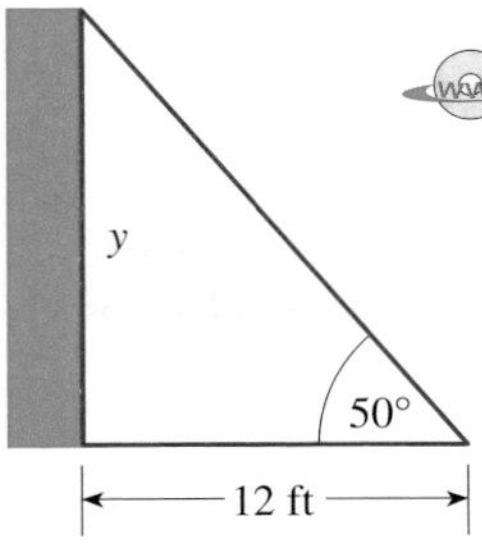

EXAMPLE 3 A right-triangle application

A ladder that is leaning against the side of a building forms an angle of 50° with the ground. If the foot of the ladder is 12 ft from the base of the building, how far up the side of the building does the ladder reach? See Figure 2.

y
50°
12 ft

Figure 2

SOLUTION

In Figure 2 we have used y to denote the required distance. Notice that y is opposite the 50° angle, while the given side is adjacent to that angle. Since the opposite and adjacent sides are involved, we'll use the tangent function. (The cotangent function could also be used.) We have

$$\tan 50^\circ = \frac{y}{12} \qquad \text{and therefore} \qquad y = 12 \tan 50^\circ$$

Without the use of a calculator or tables, this is our final answer. On the other hand, using a calculator, we find that $y = 14$ ft, to the nearest foot.

EXAMPLE 4 Finding an angle given the sides of a right triangle

Figure 3 shows a 3-4-5 right triangle. Compute $\sin\theta$ and θ. For θ, express the answer both in radians, rounded to two decimal places, and in degrees, rounded to one decimal place.

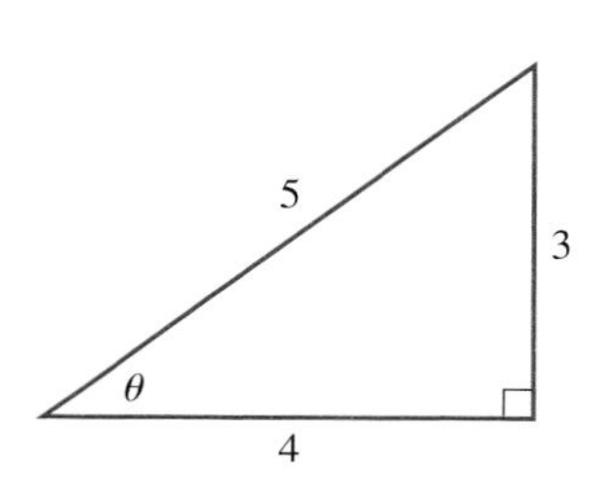

Figure 3

SOLUTION

Applying the definition $\sin\theta = \text{opposite}/\text{hypotenuse}$ in Figure 3, we have

$$\sin\theta = \frac{3}{5} \qquad (1)$$

Next, to isolate θ, we take the inverse sine of both sides of equation (1) to obtain

$$\sin^{-1}(\sin\theta) = \sin^{-1}\left(\frac{3}{5}\right)$$
$$\theta = \sin^{-1}\left(\frac{3}{5}\right)$$

Now, as you should verify for yourself with a calculator set in the radian mode, we find $\theta = \sin^{-1}(3/5) \approx 0.64$ radians; and in the degree mode, $\theta = \sin^{-1}(3/5) \approx 36.9^\circ$.

In the example just completed, we used the inverse sine function to determine the angle θ in Figure 3. It's worth noting, however, that the other inverse trig functions could be used equally well. For instance, from Figure 3 we have $\tan\theta = 3/4$ and, consequently,

$$\tan^{-1}(\tan\theta) = \tan^{-1}\left(\frac{3}{4}\right)$$

or

$$\theta = \tan^{-1}\left(\frac{3}{4}\right)$$

If you use a calculator now to evalute this last expression, you'll find that the answer indeed agrees with the results in Example 4.

EXAMPLE 5 A trigonometric formula for the area of a triangle

Show that the area of the triangle in Figure 4 is given by

$$A = \frac{1}{2}ab\sin\theta$$

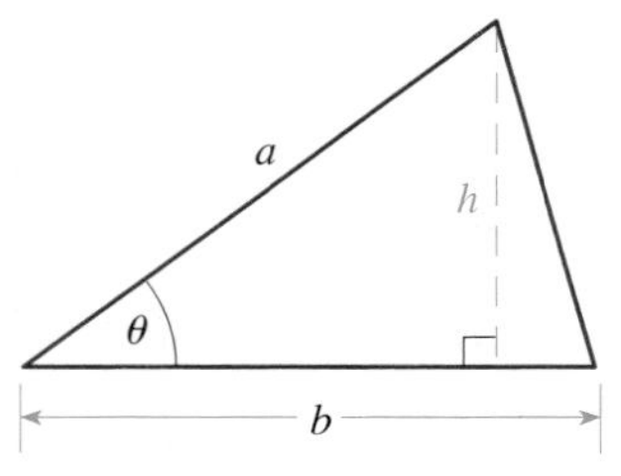

Figure 4

SOLUTION
In Figure 4, if h denotes the length of the altitude, then we have

$$\sin\theta = \frac{h}{a} \qquad \text{and therefore} \qquad h = a\sin\theta$$

This value for h can now be used in the usual formula for the area of a triangle:

$$A = \frac{1}{2}bh = \frac{1}{2}b(a\sin\theta) = \frac{1}{2}ab\sin\theta$$

Our derivation in the example just completed relies on Figure 4, in which θ is an acute angle. The formula that we obtained, however, is valid even if θ is not an acute angle. (See Exercise 36 for the proof.) In the box that follows we summarize this useful result.

Formula for the Area of a Triangle

If a and b are lengths of two sides of a triangle and θ is the angle included between those two sides, then the area of the triangle is given by

$$\text{area} = \frac{1}{2}ab\sin\theta$$

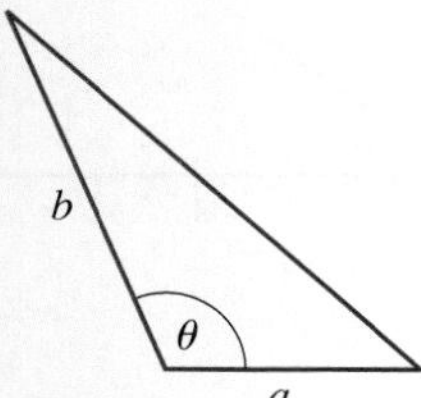

In Words The area of a triangle equals one-half the product of the lengths of two sides times the sine of the included angle.

EXAMPLE 6 Using the trigonometric formula to find the area of a triangle

Find the area of the triangle in Figure 5.

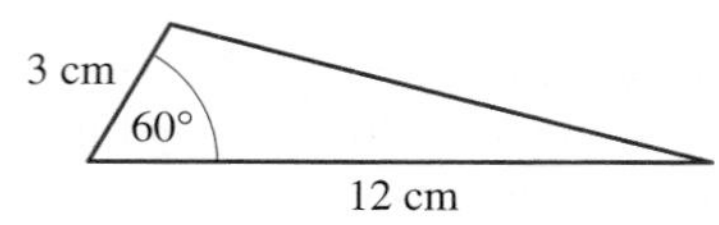

Figure 5

SOLUTION
From Example 5 the area is given by the formula $A = \frac{1}{2}ab\sin\theta$. Here we let $a = 3$ cm, $b = 12$ cm, and $\theta = 60°$. Then

$$\begin{aligned} A &= \frac{1}{2}(3)(12)\sin 60° = (18)\frac{\sqrt{3}}{2} \\ &= 9\sqrt{3}\ \text{cm}^2 \end{aligned}$$

This is the required area.

EXAMPLE 7 Finding the area of a segment of a circle

A **segment** of a circle is a region bounded by an arc of the circle and its chord. Compute the area of the segment (the shaded region) in Figure 6. Give two forms for the answer: an exact expression involving π and a calculator approximation rounded to two decimal places.

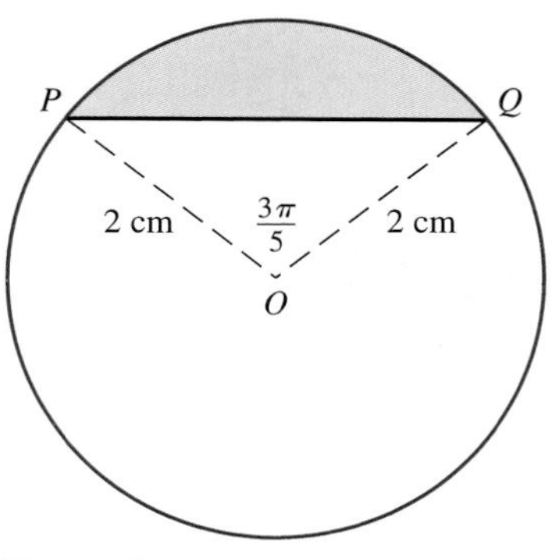

Figure 6

SOLUTION

Although we have not developed an explicit formula for the area of a segment, notice in Figure 6 that $\triangle OPQ$ and the shaded segment, taken together, form a sector of the circle. So we have

$$\text{area of segment} = (\text{area of sector } OPQ) - (\text{area of } \Delta OPQ)$$

For the area of the sector, we compute

$$\begin{aligned}\text{area of sector } OPQ &= \frac{1}{2}r^2\theta = \frac{1}{2}(2^2)\left(\frac{3\pi}{5}\right)\\ &= \frac{6\pi}{5}\text{ cm}^2\end{aligned}$$

For the area of the triangle, we use the formula $A = \frac{1}{2}ab\sin\theta$:

$$\text{area of } \triangle OPQ = \frac{1}{2}(2)(2)\sin\frac{3\pi}{5} = 2\sin\frac{3\pi}{5}\text{ cm}^2$$

Putting things together now, we obtain

$$\begin{aligned}\text{area of segment} &= \left(\frac{6\pi}{5} - 2\sin\frac{3\pi}{5}\right)\text{cm}^2\\ &\approx 1.87\text{ cm}^2\end{aligned}$$

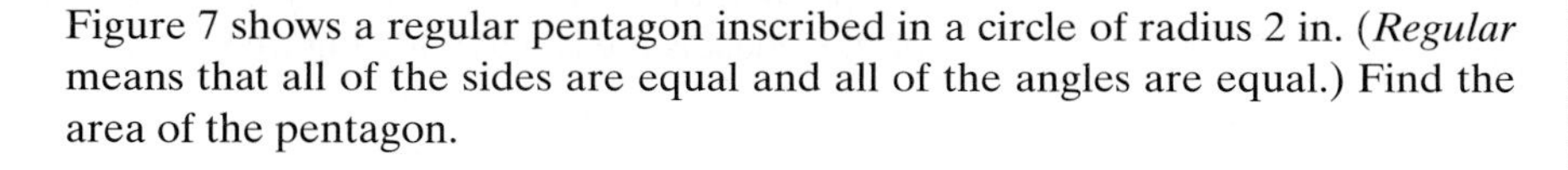

EXAMPLE 8 The area of a regular pentagon

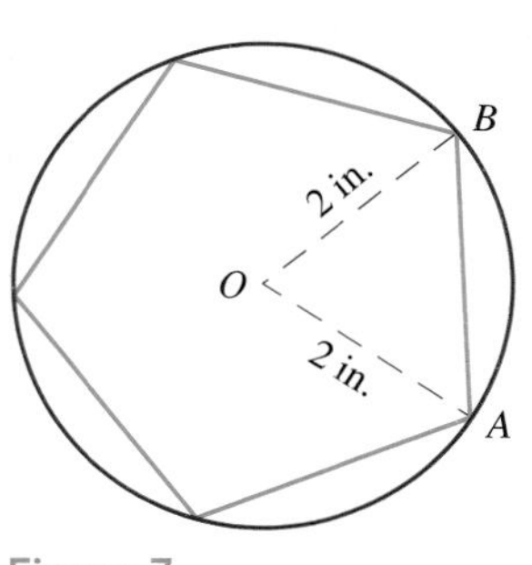

Figure 7

Figure 7 shows a regular pentagon inscribed in a circle of radius 2 in. (*Regular* means that all of the sides are equal and all of the angles are equal.) Find the area of the pentagon.

SOLUTION

The idea here is first to find the area of triangle BOA by using the area formula from Example 5. Then, since the pentagon is composed of five such identical triangles, the area of the pentagon will be five times the area of triangle BOA. We will make use of the result from geometry that, in a regular n-sided polygon, the central angle is $360°/n$. In our case we therefore have

$$\angle BOA = \frac{360°}{5} = 72°$$

We can now find the area of triangle BOA:

$$\begin{aligned}\text{area} &= \frac{1}{2}ab\sin\theta\\ &= \frac{1}{2}(2)(2)\sin 72° = 2\sin 72°\text{ in.}^2\end{aligned}$$

The area of the pentagon is five times this, or $10\sin 72°$ in.^{2}. (Using a calculator, this is about 9.51 in.^{2}.)

Now we introduce some terminology that will be used in the next two examples. Suppose that a surveyor sights an object at a point above the horizontal, as indicated in Figure 8(a). Then the angle between the line of sight and the horizontal is

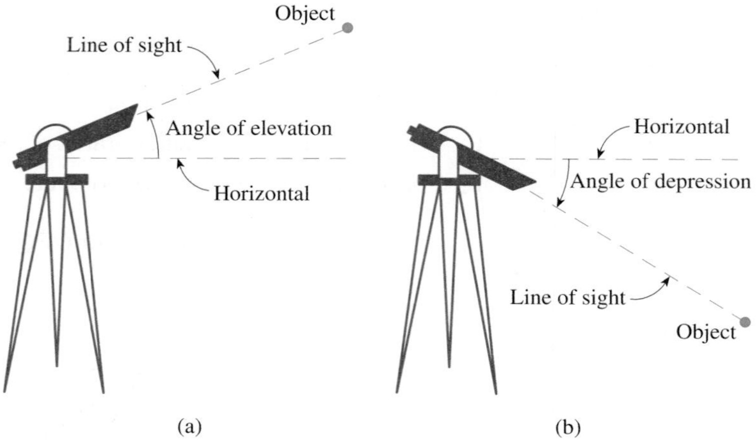

Figure 8 (a) (b)

called the **angle of elevation.** The **angle of depression** is similarly defined for an object below the horizontal, as shown in Figure 8(b).

EXAMPLE 9 Using the angle of depression to find a distance

A helicopter hovers 800 ft directly above a small island that is off the California coast. From the helicopter the pilot takes a sighting to a point P directly ashore on the mainland, at the water's edge. If the angle of depression is 35°, how far off the coast is the island? See Figure 9.

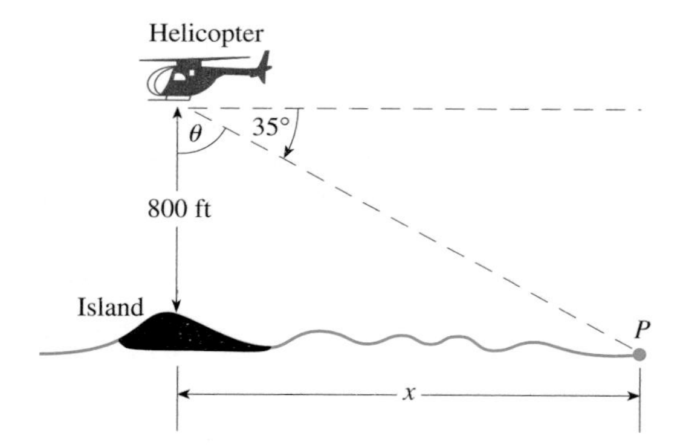

Figure 9

SOLUTION

Let x denote the distance from the island to the mainland. Then, as you can see from Figure 9, we have $\theta + 35° = 90°$, from which it follows that $\theta = 55°$. Now we can write

$$\tan 55° = \frac{x}{800}$$

or

$$x = 800 \tan 55° \approx 1150 \text{ ft}$$ using a calculator and rounding to the nearest 50 feet

EXAMPLE 10 Finding the altitude of a satellite

Two satellite-tracking stations, located at points A and B in California's Mojave Desert, are 200 miles apart. At a prearranged time, both stations measure the angle of elevation of a satellite as it crosses the vertical plane containing A and B. This means that A, B, and S lie in a plane perpendicular to the ground. (See Figure 10.) If the angles of elevation from A and from B are α and β, respectively, express the altitude h of the satellite in terms of α and β.

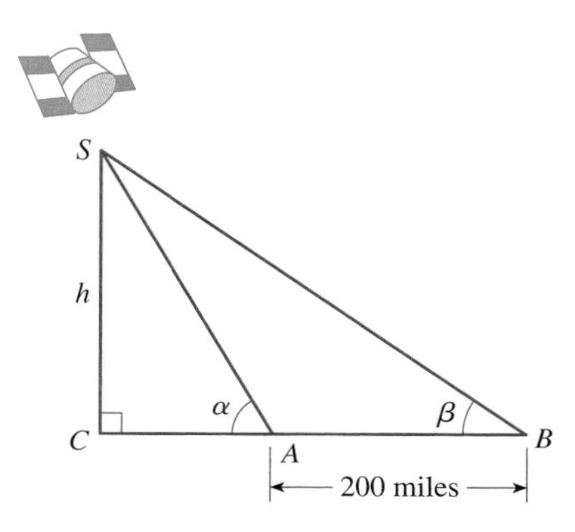

Figure 10

SOLUTION

We want to express the length $h = SC$ in Figure 10 in terms of the angles α and β. Note that $\overline{SC}$ is a side of both of the right triangles SCA and SCB. Working first in the right triangle SCB, we have

$$\cot \beta = \frac{CA + 200}{h}$$

or

$$h \cot \beta = CA + 200 \tag{2}$$

We can eliminate CA from equation (2) as follows. Looking at right triangle SCA, we have

$$\cot \alpha = \frac{CA}{h} \qquad \text{and thus} \qquad CA = h \cot \alpha$$

Using this last equation to substitute for CA in equation (2), we obtain

$$\begin{aligned} h \cot \beta &= h \cot \alpha + 200 \\ h \cot \beta - h \cot \alpha &= 200 \\ h(\cot \beta - \cot \alpha) &= 200 \\ h &= \frac{200}{\cot \beta - \cot \alpha} \text{ miles} \qquad \text{as required} \end{aligned}$$

EXAMPLE 11 Expressing lengths and area using trigonometric functions

The arc shown in Figure 11 is a portion of the unit circle, $x^2 + y^2 = 1$. Express the following quantities in terms of θ:

(a) OA; **(b)** AB; **(c)** OC; **(d)** The area of $\triangle OAC$.

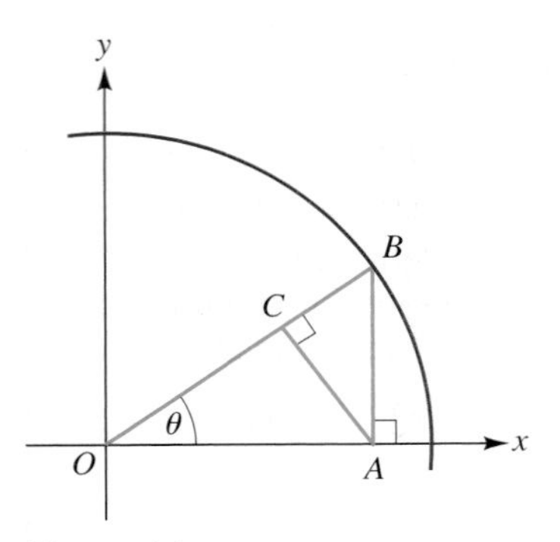

Figure 11

SOLUTION

(a) In right triangle OAB,

$$\cos\theta = \frac{OA}{OB} = \frac{OA}{1}$$
$$OA = \cos\theta$$

(b) In right triangle OAB,

$$\sin\theta = \frac{AB}{OB} = \frac{AB}{1}$$
$$AB = \sin\theta$$

(c) In right triangle OAC, we have $\cos\theta = OC/OA$, and therefore

$$OC = OA\cos\theta = (\cos\theta)(\cos\theta) = \cos^2\theta$$

(d)

$$\text{area}\ \triangle OAC = \tfrac{1}{2}(OA)(OC)(\sin\theta) \qquad \text{using the formula area} = \tfrac{1}{2}ab\sin\theta$$
$$= \tfrac{1}{2}(\cos\theta)(\cos^2\theta)(\sin\theta) \qquad \text{using the results from parts (a) and (c)}$$
$$= \tfrac{1}{2}\cos^3\theta\sin\theta$$

EXERCISE SET 9.1 

A

For Exercises 1–6, refer to the following figure. (However, each problem is independent of the others.)

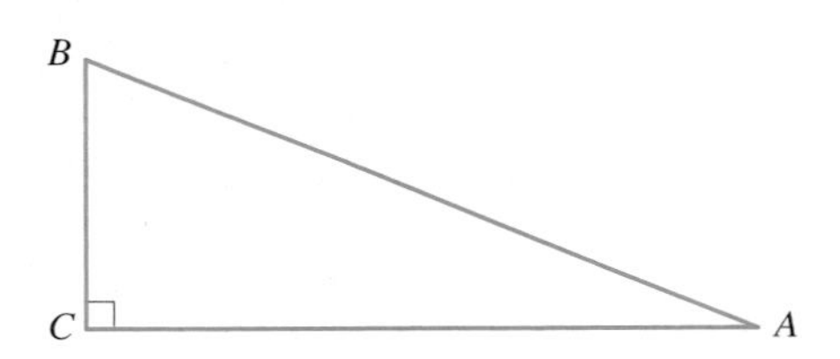

1. If $\angle A = 30^\circ$ and $AB = 60$ cm, find AC and BC.
2. If $\angle A = 60^\circ$ and $AB = 12$ cm, find AC and BC.
3. If $\angle B = 60^\circ$ and $AC = 16$ cm, find BC and AB.
4. If $\angle B = 45^\circ$ and $AC = 9$ cm, find BC and AB.
5. If $\angle B = 50^\circ$ and $AB = 15$ cm, find BC and AC. (Round your answers to one decimal place.)
6. If $\angle A = 25^\circ$ and $AC = 100$ cm, find BC and AB. (Round your answers to one decimal place.)
7. A ladder 18 ft long leans against a building. The ladder forms an angle of 60° with the ground.
   (a) How high up the side of the building does the ladder reach? [Here and in part (b), give two forms for your answers: one with radicals and one (using a calculator) with decimals, rounded to two places.]
   (b) Find the horizontal distance from the foot of the ladder to the base of the building.
8. From a point level with and 1000 ft away from the base of the Washington Monument, the angle of elevation to the top of the monument is 29.05°. Determine the height of the monument to the nearest half foot.
9. Refer to the following figure.

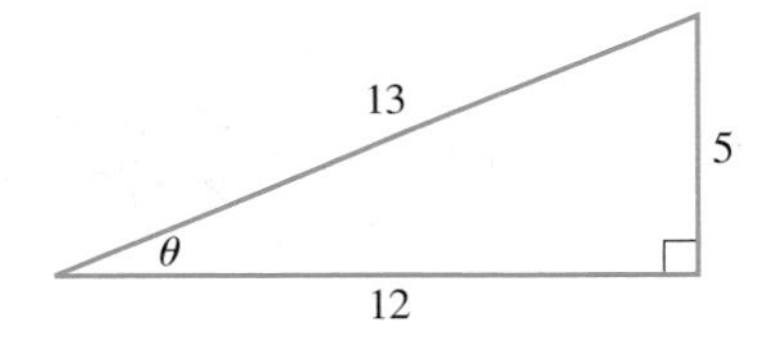

   (a) Use the inverse sine function, as in Example 4, to find θ. Express the answer in degrees, rounded to one decimal place.
   (b) Follow part (a), but use the inverse cosine function. Check that your answer agrees with the result in part (a).
   (c) Follow part (a), but use the inverse tangent function. Again, check that your answer agrees with the result in part (a).
10. In isosceles triangle ABC, the sides are of length $AC = BC = 8$ and $AB = 4$. Find the angles of the triangle. Express the answers both in radians, rounded to two decimal places, and in degrees, rounded to one decimal place. *Hints:* To find $\angle A$, start by drawing an altitude from C to side $\overline{AB}$. Then for $\angle C$, use the fact that the sum of the angles in a triangle is π radians or 180°.

11. Refer to the following figure. At certain times, the planets Earth and Mercury line up in such a way that $\angle EMS$ is a right angle. At such times, $\angle SEM$ is found to be 21.16°. Use this information to estimate the distance MS of Mercury from the Sun. Assume that the distance from the Earth to the Sun is 93 million miles. (Round your answer to the nearest million miles. Because Mercury's orbit is not really circular, the actual distance of Mercury from the Sun varies from about 28 million miles to 43 million miles.)

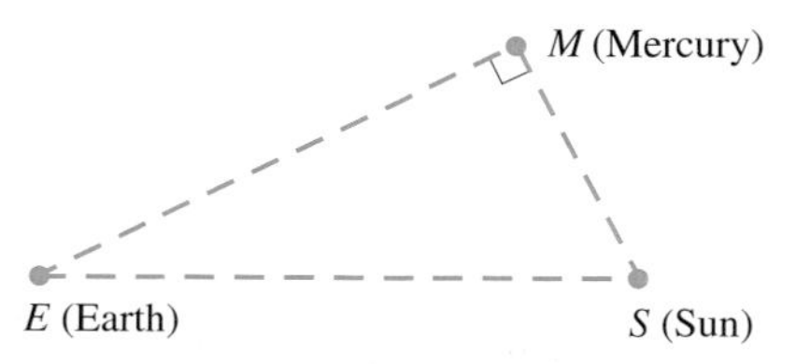

12. Determine the distance AB across the lake shown in the figure, using the following data: $AC = 400$ m, $\angle C = 90°$, and $\angle CAB = 40°$. Round the answer to the nearest meter.

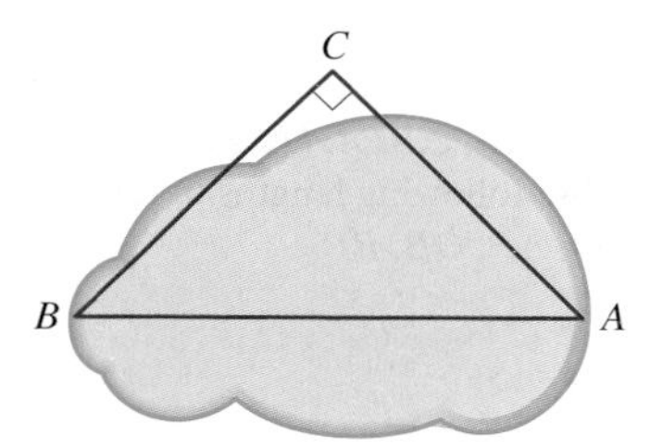

13. A building contractor wants to put a fence around the perimeter of a flat lot that has the shape of a right triangle. One angle of the triangle is 41.4°, and the length of the hypotenuse is 58.5 m. Find the length of fencing required. Round the answer to one decimal place.

14. Suppose that the contractor in Exercise 13 reviews his notes and finds that it is not the hypotenuse that is 58.5 m but rather the side opposite the 41.4° angle. Find the length of fencing required in this case. Again, round the answer to one decimal place.

For Exercises 15 and 16, refer to the following diagram for the roof of a house. In the figure, x is the length of a rafter measured from the top of a wall to the top of the roof; θ is the acute angle between a rafter and the horizontal; and h is the vertical distance from the top of the wall to the top of the roof.

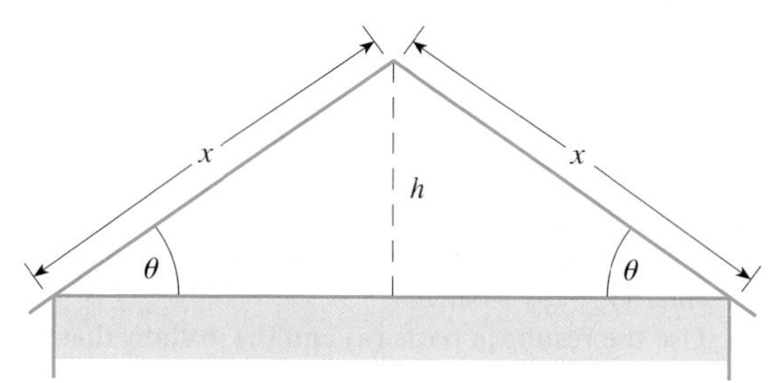

15. Suppose that $\theta = 39.4°$ and $x = 43.0$ ft.
(a) Determine h. Round the answer to one decimal place.
(b) The *gable* is the triangular region bounded by the rafters and the attic floor. Find the area of the gable. Round the final answer to one decimal place.

16. Suppose that $\theta = 34°$ and $h = 36.5$ ft.
(a) Determine x. Round the answer to one decimal place.
(b) Find the area of the gable. Round the final answer to one decimal place. [See Exercise 15(b) for the definition of gable.]

In Exercises 17 and 18, find the area of the triangle. In Exercise 18, use a calculator and round the final answer to two decimal places.

17. **18.**

In Exercises 19–24, determine the area of the shaded region, given that the radius of the circle is 1 unit and the inscribed polygon is a regular polygon. Give two forms for each answer: an expression involving radicals or the trigonometric functions; a calculator approximation rounded to three decimal places.

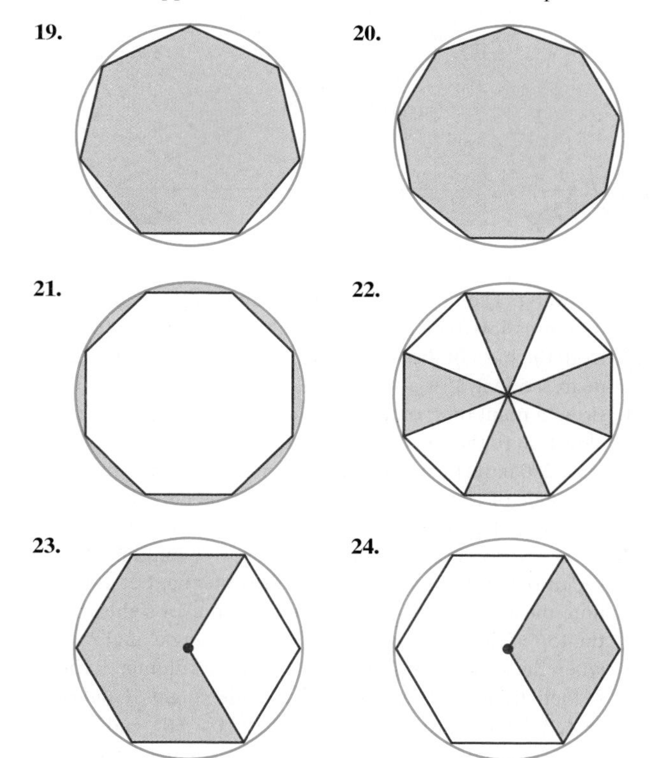

In Exercises 25 and 26, compute the area of the shaded segment of the circle, as in Example 7. Give two forms for each answer: an exact expression and a calculator approximation rounded to two decimal places.

25.

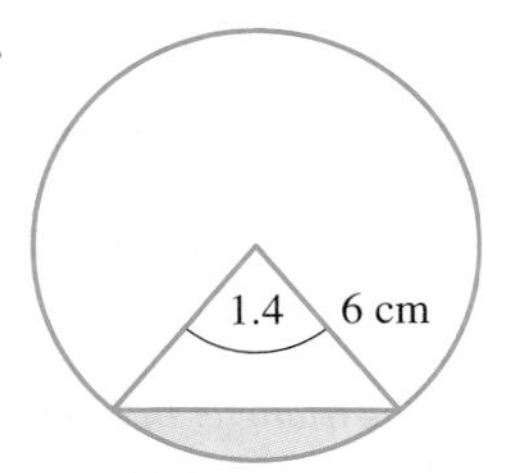

26.

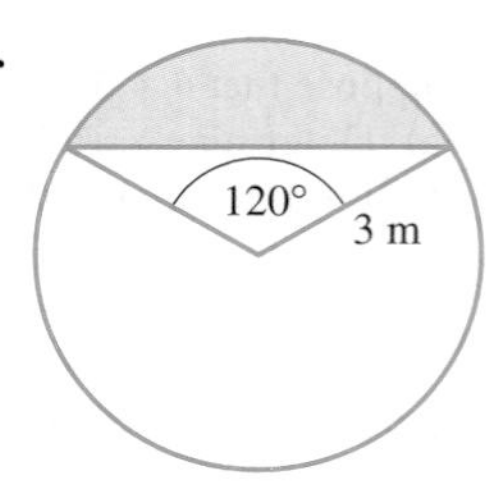

27. Show that the perimeter of the pentagon in Example 8 is 20 sin 36°. *Hint:* In Figure 7, draw a perpendicular from O to $\overline{AB}$.

28. In triangle OAB, lengths $OA = OB = 6$ in. and $\angle AOB = 72°$. Find AB. *Hint:* Draw a perpendicular from O to AB. Round the answer to one decimal place.

29. The accompanying figure shows two ships at points P and Q, which are in the same vertical plane as an airplane at point R. When the height of the airplane is 3500 ft, the angle of depression to P is 48°, and that to Q is 25°. Find the distance between the two ships. Round the answer to the nearest 10 feet.

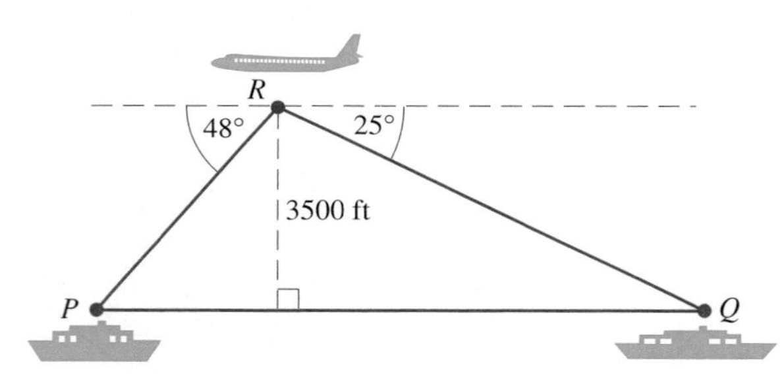

30. An observer in a lighthouse is 66 ft above the surface of the water. The observer sees a ship and finds the angle of depression to be 0.7°. Estimate the distance of the ship from the base of the lighthouse. Round the answer to the nearest 5 feet.

31. From a point on ground level, you measure the angle of elevation to the top of a mountain to be 38°. Then you walk 200 m farther away from the mountain and find that the angle of elevation is now 20°. Find the height of the mountain. Round the answer to the nearest meter.

32. A surveyor stands 30 yd from the base of a building. On top of the building is a vertical radio antenna. Let α denote the angle of elevation when the surveyor sights to the top of the building. Let β denote the angle of elevation when the surveyor sights to the top of the antenna. Express the length of the antenna in terms of the angles α and β.

33. In $\triangle ACD$, you are given $\angle C = 90°$, $\angle A = 60°$, and $AC = 18$ cm. If B is a point on $\overline{CD}$ and $\angle BAC = 45°$, find BD. Express the answer in terms of a radical (rather than using a calculator).

34. The radius of the circle in the following figure is 1 unit. Express the lengths OA, AB, and DC in terms of α.

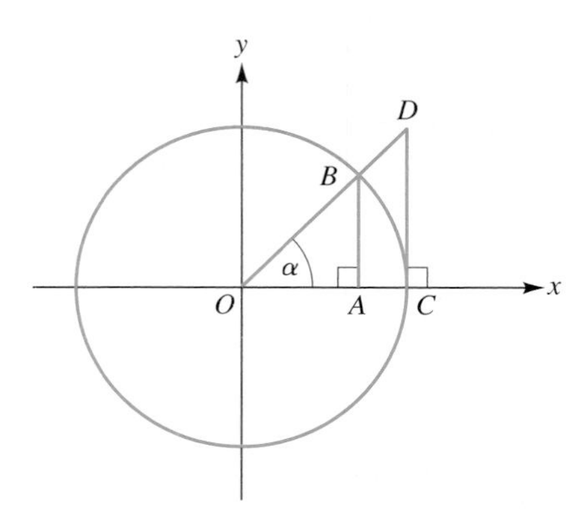

35. The arc in the next figure is a portion of the unit circle, $x^2 + y^2 = 1$.

(a) Express the following angles in terms of θ: $\angle BOA$, $\angle OAB$, $\angle BAP$, $\angle BPA$. (Assume that θ is in degrees.)

(b) Express the following lengths in terms of $\sin\theta$ and $\cos\theta$: AO, AP, OB, BP.

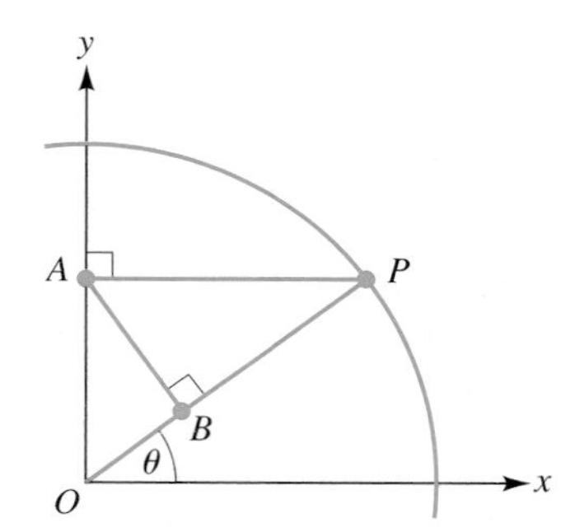

36. The following figure shows $\triangle ABC$, in which $\angle BCA = \theta$ is an obtuse angle. Complete Steps (a)–(c) to show that the area of the triangle is $\frac{1}{2}ab\sin\theta$.

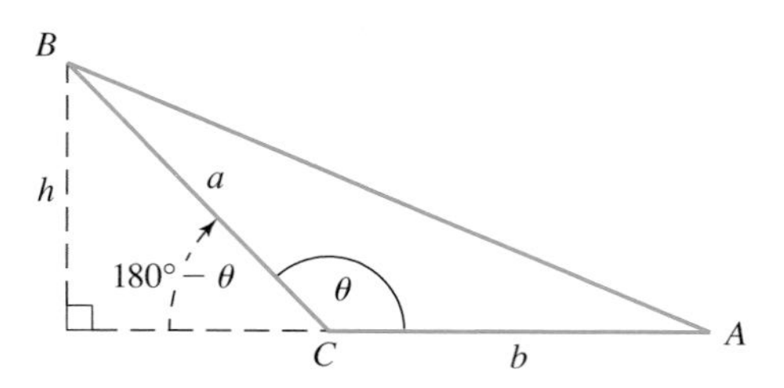

(a) Show that $h = a\sin(180° - \theta)$.

(b) Use one of the addition formulas to verify that $\sin(180° - \theta) = \sin\theta$.

(c) Use the results in parts (a) and (b) to show that the area of $\triangle ABC$ is given by $\frac{1}{2}ab\sin\theta$.

B

37. Refer to the figure. Express each of the following lengths as a function of θ.

(a) BC **(b)** AB **(c)** AC

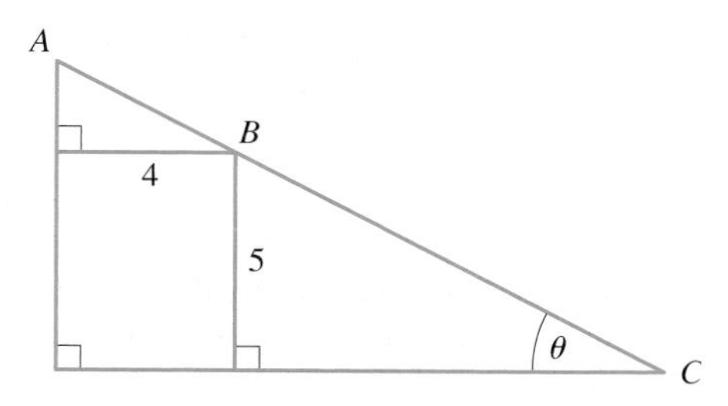

38. In the following figure, $AB = 8$ in. Express x as a function of θ. *Hint:* First work Exercise 37.

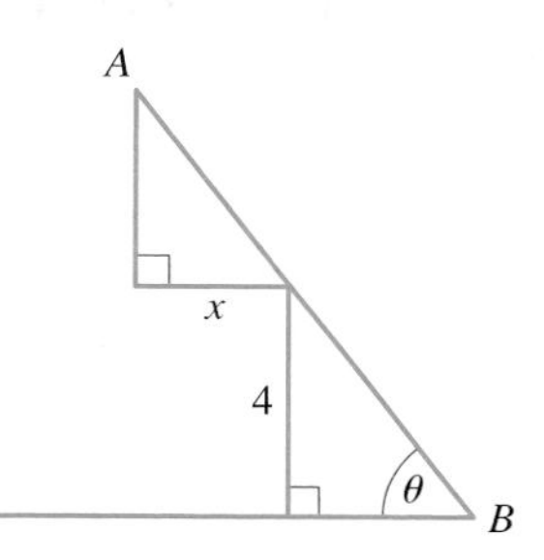

39. In the figure, line segment $\overline{BA}$ is tangent to the unit circle at A. Also, $\overline{CF}$ is tangent to the circle at F. Express the following lengths in terms of θ.

(a) DE **(c)** CF **(e)** AB
(b) OE **(d)** OC **(f)** OB

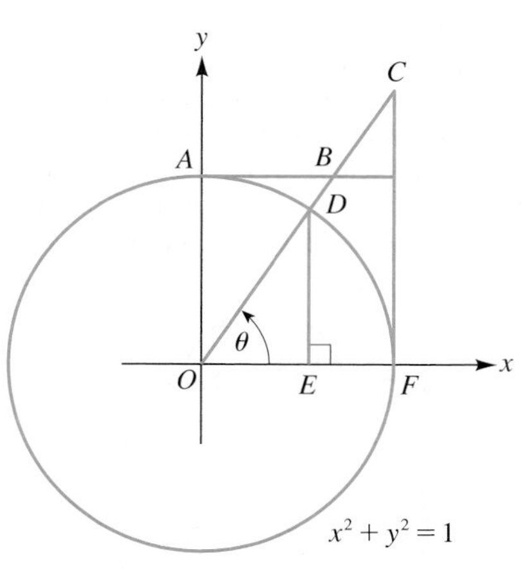

40. At point P on Earth's surface, the moon is observed to be directly overhead, while at the same time at point T, the moon is just visible. See the following figure.

(a) Show that $MP = \dfrac{OT}{\cos\theta} - OP$.

(b) Use a calculator and the following data to estimate the distance MP from the earth to the moon: $\theta = 89.05°$ and $OT = OP = 4000$ miles. Round your answer to the nearest thousand miles. (Because the moon's orbit is not really circular, the actual distance varies from about 216,400 miles to 247,000 miles.)

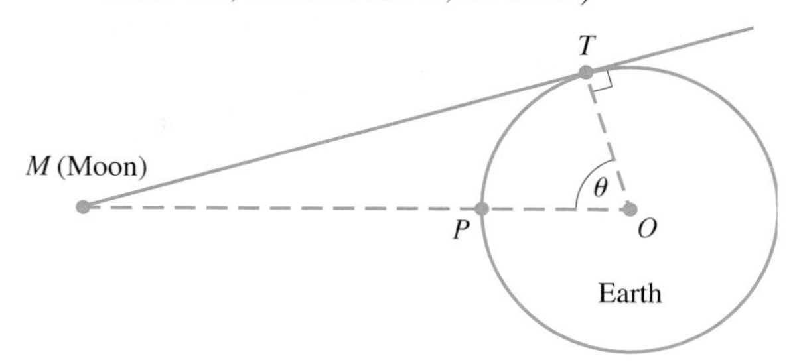

41. Refer to the figure below. Let r denote the radius of the moon.

(a) Show that $r = \left(\dfrac{\sin\theta}{1 - \sin\theta}\right)PS$.

(b) Use a calculator and the following data to estimate the radius r of the moon: $PS = 238{,}857$ miles and $\theta = 0.257°$. Round your answer to the nearest 10 miles.

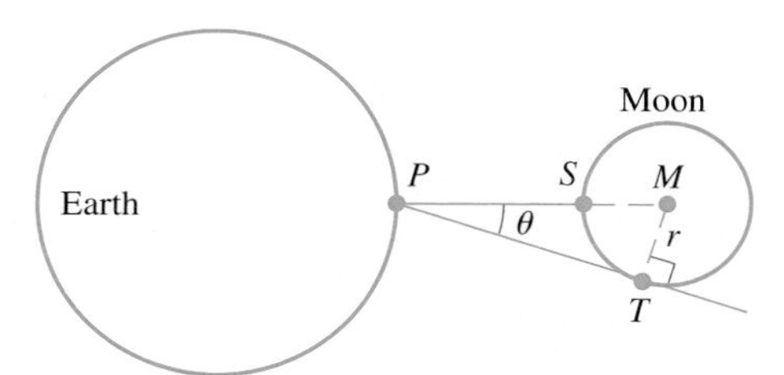

42. Figure A shows a regular hexagon inscribed in a circle of radius 1. Figure B shows a regular heptagon (seven-sided polygon) inscribed in a circle of radius 1. In Figure A, a line segment drawn from the center of the circle perpendicular to one of the sides is called an **apothem** of the polygon.

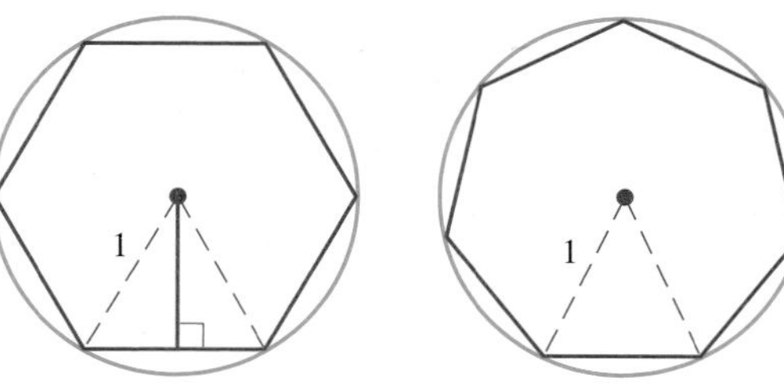

Figure A Figure B

(a) Show that the length of the apothem in Figure A is $\sqrt{3}/2$.

(b) Show that the length of one side of the heptagon in Figure B is $2\sin(180°/7)$.

(c) Use a calculator to evaluate the expressions in parts (a) and (b). Round each answer to four decimal places, and note how close the two values are. Approximately two thousand years ago, Heron of Alexandria made use of this coincidence when he used the length of the apothem of the hexagon to approximate the length of the side of the heptagon. (The apothem of the hexagon can be constructed with ruler and compass; the side of the regular heptagon cannot.)

43. The following figure shows a regular seven-sided polygon inscribed in a circle of radius 1.

(a) Explain why the area of $\triangle AOB$ is $\frac{1}{2}\sin(2\pi/7)$. Hint: Use the area formula from Example 5.

(b) Explain why the area of the entire polygon is $\frac{7}{2}\sin(2\pi/7)$.

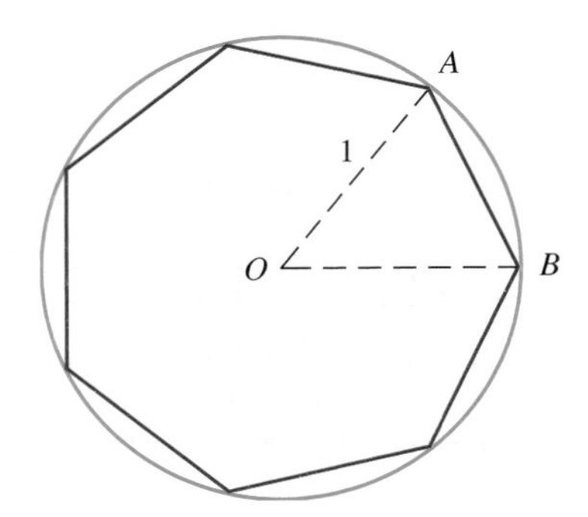

(c) Let a_n denote the area of a regular n-sided polygon inscribed in a circle of radius 1. Use the ideas from parts (a) and (b) to show that $a_n = \frac{1}{2}n\sin(2\pi/n)$.

(d) Use the formula from part (c) and a calculator to complete the following table. Round each result to eight decimal places.

n	5	10	50	100	10^3	10^4	10^5
a_n							

(e) Explain (in complete sentences) why the values of a_n in your table get closer and closer to π. (The value of π, correct to ten decimal places, is 3.1415926535.)

44. (a) The following figure shows a segment with central angle α in a circle of radius r. (Assume α is in radians.) Show that the area A of the segment is given by

$$A = \frac{1}{2}r^2(\alpha - \sin\alpha).$$

(b) In the following figure, the arc is a semicircle with diameter AOB, where radius $OB = 1$. Use the formula in part (a) to show that the sum of the areas of the two shaded segments is $\pi/2 - \sin\theta$, while the difference (larger minus smaller) is $\pi/2 - \theta$.

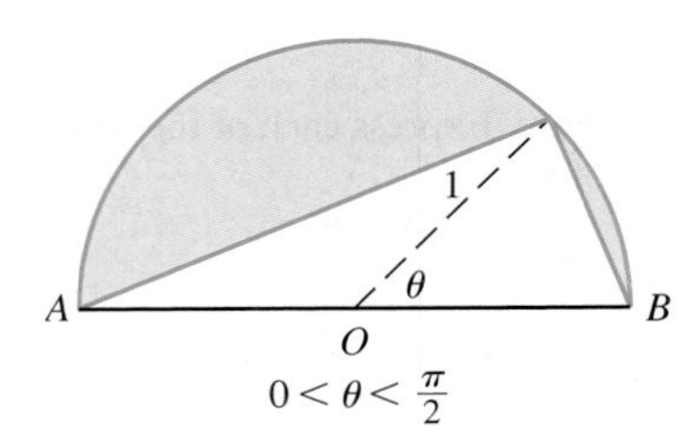

45. In the following figure, arc ABC is a portion of a circle with center $D(0, -1)$ and radius $\overline{DC}$. The shaded crescent-shaped region is called a **lune.** Verify the following result, which was discovered (and proved) by the Greek mathematician Hippocrates of Chios approximately 2500 years ago: The area of the lune is equal to the area of the square $OCED$. *Hint*: In computing the area of the lune, make use of the formula given in Exercise 44(a) for the area of a segment of a circle.

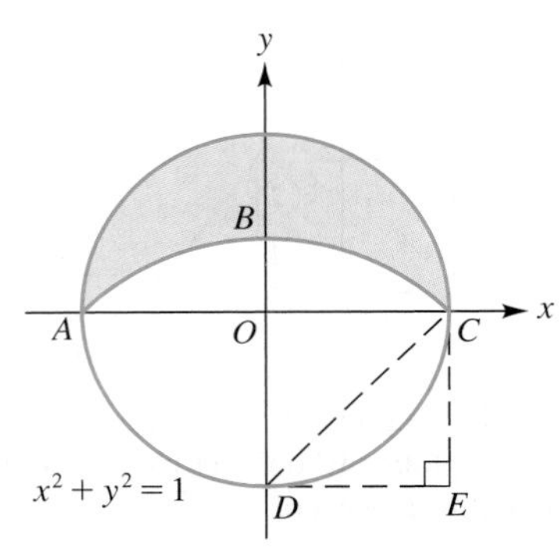

46. In this exercise, you'll verify another result about lunes that was discovered by Hippocrates of Chios. The figure shows a regular hexagon inscribed in a circle of radius 1. Outward from each side of the hexagon, (congruent) semicircles are constructed with the sides of the hexagon as diameters. Follow steps (a) through (f) to show that the area of the hexagon is equal to the sum of the areas of the six (congruent) lunes plus twice the area of one of the semicircles.

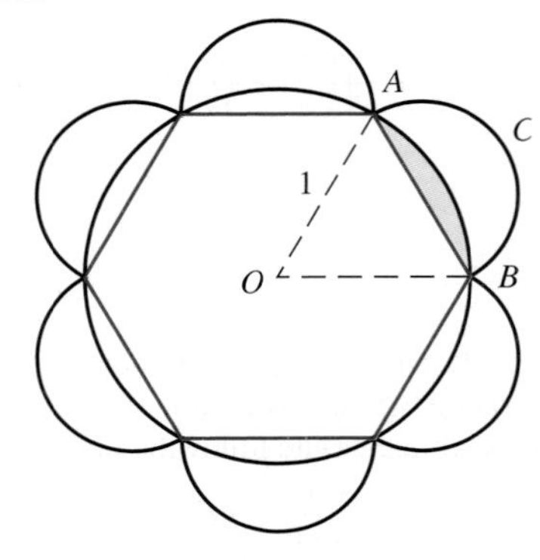

(a) What is the radian measure of $\angle AOB$?

(b) Show that the area of the hexagon is $3\sqrt{3}/2$.

(c) Show that the area of the shaded segment is $(2\pi - 3\sqrt{3})/12$.
(d) Show that the area of semicircle ACB is $\pi/8$.
(e) Use the results in parts (c) and (d) to show that the area of lune ACB is $\dfrac{6\sqrt{3} - \pi}{24}$.
(f) Use the results in parts (d) and (e) to verify Hippocrates' result:

$$\textit{area of hexagon} = 6 \times (\textit{area of lune } ACB) + 2 \times (\textit{area of semicircle } ACB)$$

47. In the following figure, $\overline{AB}$ is a chord in a circle of radius 1. The length of $\overline{AB}$ is d, and $\overline{AB}$ subtends an angle θ at the center of the circle, as shown.

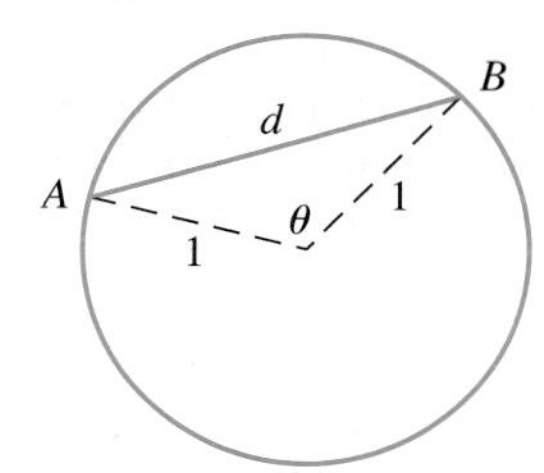

In this exercise we derive the following formula for the length d of the chord in terms of the angle θ:

$$d = \sqrt{2 - 2\cos\theta}$$

(The derivation of this formula does not require any new material from this section. It is developed here for use in subsequent exercises.)

(a) We place the figure in an x-y coordinate system and orient it so that the angle θ is in standard position and the point B is located at $(1, 0)$. (See the following figure.) What are the coordinates of the point A (in terms of θ)?

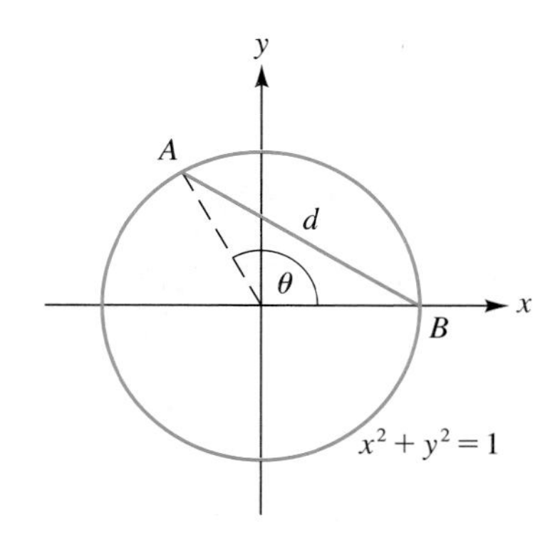

(b) Use the formula for the distance between two points to show that $d = \sqrt{2 - 2\cos\theta}$.

48. This exercise provides practice in using the chord-length formula developed in Exercise 47.

(a) Use the formula from Exercise 47 to compute the length d in the following figure. Round the answer to one decimal place.

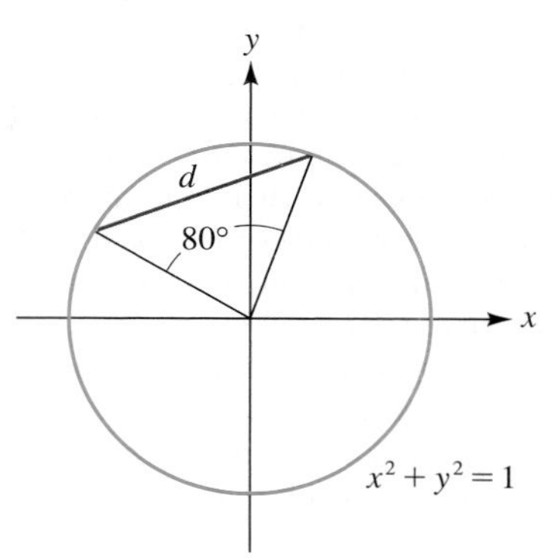

(b) In the following figure, $AB = 1.2$. Use the formula developed in Exercise 47 to compute the angle θ. Express the answer in radians, rounded to one decimal place.

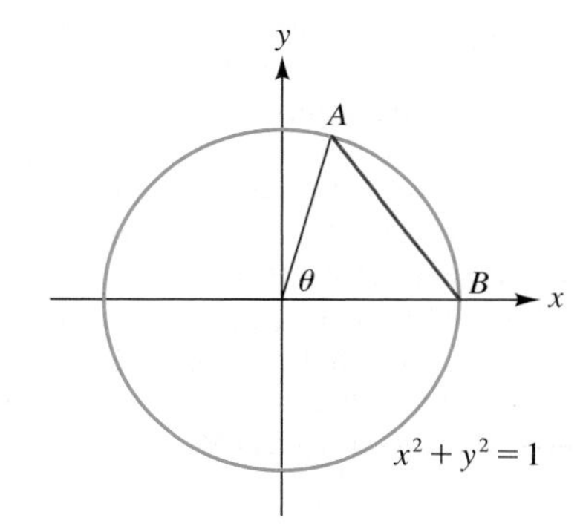

49. In the following figure, arc ABC is a semicircle with diameter $\overline{AC}$, and arc CDE is a semicircle with diameter $\overline{CE}$.

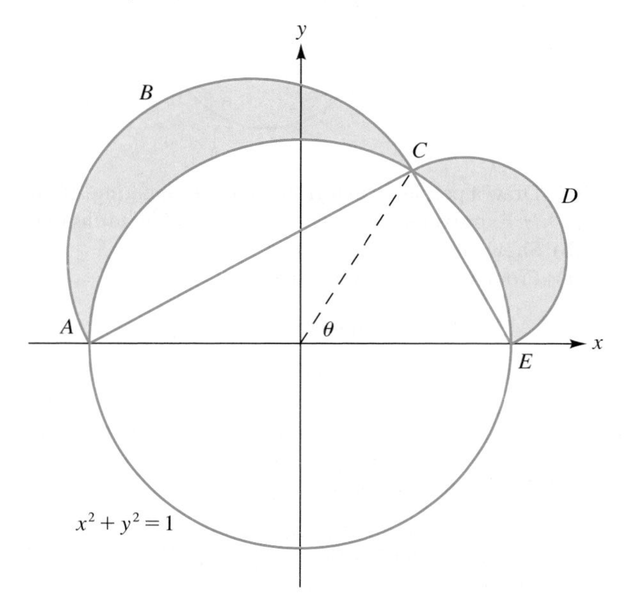

(continues)

(a) Show that the area of semicircle CDE is $\pi(1 - \cos\theta)/4$.
Hint: Use the chord-length formula in Exercise 47.

(b) Express the area of lune CDE in terms of θ.
Hint: Use the result in part (a) along with the formula in Exercise 44(a) for the area of a segment.

(c) Express the area of lune ABC in terms of θ.

(d) Express the area of $\triangle ACE$ in terms of θ.

(e) Use the results in parts (b), (c), and (d) to verify that the area of $\triangle ACE$ is equal to the sum of the areas of the two lunes CDE and ABC.

Remark: As with the results in Exercises 45 and 46, this result about lunes was discovered and proved by the ancient Greek mathematician Hippocrates of Chios. According to Professor George F. Simmons in his book *Calculus Gems* (New York: McGraw-Hill Book Co., 1992), these results appear "to be the earliest precise determination of the area of a region bounded by curves."

50. Using the ruler-and-compass constructions of elementary geometry, there is a well known method for bisecting any angle. (Do you remember this from a geometry class?) However, there is no similar method for *trisecting* an angle. This exercise demonstrates a geometric method for the approximate trisection of small acute angles. [The origins of the method can be traced back to the German cleric Nicolaus Cusanus (1401–1464) and the Dutch physicist Willebrord Snell (1580–1626).]

In the following figure, O is the center of unit circle and $\angle DOC = \theta$ is the angle to be trisected. Radius $\overline{OB}$ is extended to a point A so that $AB = OB = 1$. Then line segment $\overline{AD}$ is drawn, creating $\angle DAC = \beta$.

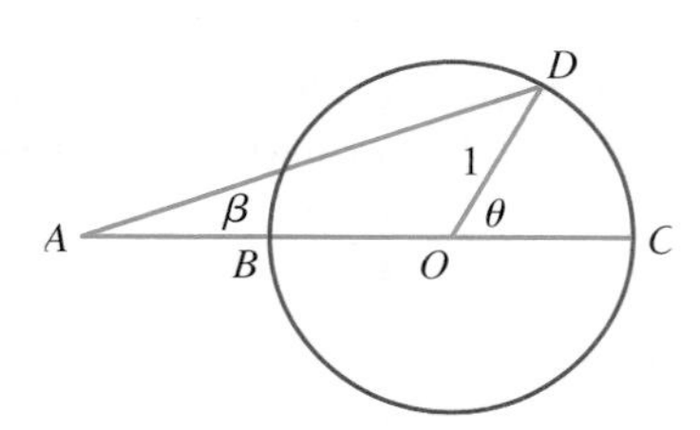

(a) Draw a perpendicular from D to $\overline{AC}$, meeting $\overline{AC}$ at E. Express the two lengths DE and OE in terms of θ.

(b) Show that $\tan\beta = (\sin\theta)/(2 + \cos\theta)$.

(c) From part (b) it follows that

$$\beta = \tan^{-1}[(\sin\theta)/(2 + \cos\theta)]$$

Use this formula to complete the following table. As you will see by completing the table, $\beta \approx \theta/3$. In the table, express β in degrees, rounded to four decimal places. For the percentage error in the approximation $\beta \approx \theta/3$, use the formula

$$\text{percentage error} = \frac{\theta/3 - \beta}{\theta/3} \times 100$$

Round the percentage error to two decimal places.

θ	$\theta/3$	β	Percentage error in approximation $\beta \approx \theta/3$
30°			
15°			
9°			
6°			
3°			

51. In this exercise we prove the following trigonometric identity:

$$\sin 3\theta = \sin\theta + 2\sin\theta\cos 2\theta$$

[This identity is valid for all angles θ, but in this exercise we use right-triangle trigonometry and the resulting proof is valid only when $0° < 3\theta < 90°$. The idea for the proof is due to Professors J. Chris Fisher and E. L. Koh (*Mathematics Magazine,* vol. 65 no. 2 (April 1992).]

(a) In the following figure, O is the center of the circle and the radius is 1. Show that $AB = 2\sin\theta$. *Hint*: Draw a perpendicular from O to $\overline{AB}$.

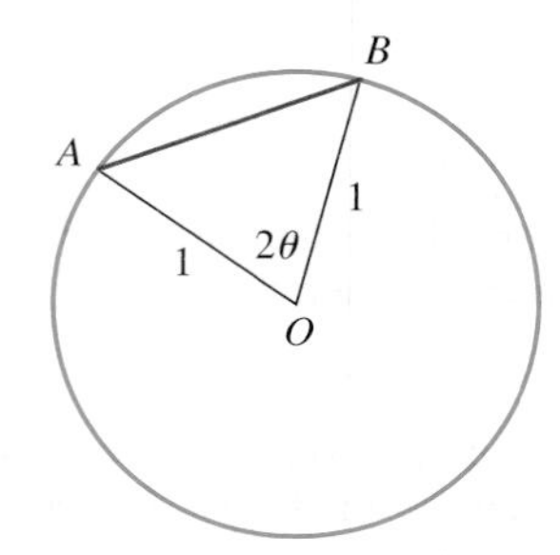

(b) For the remainder of this exercise refer to the following figure in which the arc is a portion of the unit circle, lines $\overline{AB}$ and $\overline{DC}$ are parallel to the y-axis, and $\overline{DB}$ is parallel to the x-axis. Why does $AB = \sin\theta$? Why does $CB = 2\sin\theta$?

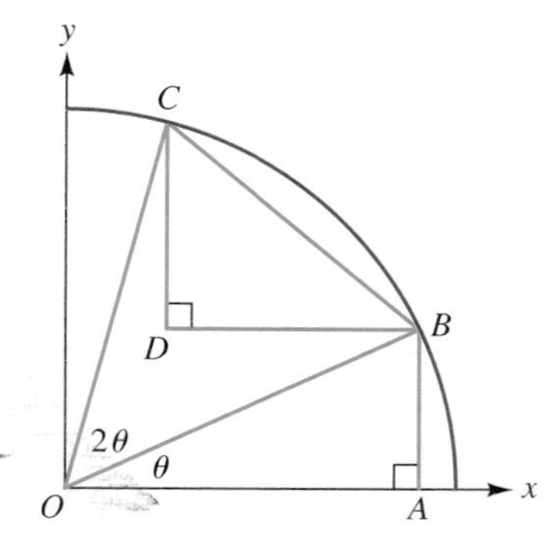

(c) Use the fact that $\triangle OBC$ is isosceles to show that $\angle OBC = 90° - \theta$.

(d) From elementary geometry we know that alternate interior angles are equal. Consequently $\angle DBO = \angle BOA = \theta$. Use this observation and the result in part (c) to show that $\angle DBC = 90° - 2\theta$.

(e) By referring to $\triangle CDB$ and using two of the previous results, show that $CD = 2\sin\theta\cos 2\theta$.

(f) From the figure, you can see that the y-coordinate of point C is equal to $AB + CD$. But independent of that fact, why is the y-coordinate of point C also equal to $\sin 3\theta$? After you've answered this, use these observations to conclude that $\sin 3\theta = \sin \theta + 2 \sin \theta \cos 2\theta$, as required.

(g) Substituting $\theta = 10°$ in the identity $\sin 3\theta = \sin \theta + 2 \sin \theta \cos 2\theta$ yields the statement $\frac{1}{2} = \sin 10° + 2 \sin 10° \cos 20°$. Use a calculator to check this last equation.

(h) Substituting $\theta = \pi/9$ in the identity $\sin 3\theta = \sin \theta + 2 \sin \theta \cos 2\theta$ yields the statement

$$\frac{\sqrt{3}}{2} = \sin \frac{\pi}{9} + 2 \sin \frac{\pi}{9} \cos \frac{2\pi}{9}$$

Use a calculator to check this last equation.

C

52. In the accompanying figure, the smaller circle is tangent to the larger circle. Ray PQ is a common tangent and ray PR passes through the centers of both circles. If the radius of the smaller circle is a and the radius of the larger circle is b, show that $\sin \theta = (b - a)/(a + b)$. Then, using the identity $\sin^2 \theta + \cos^2 \theta = 1$, show that $\cos \theta = 2\sqrt{ab}/(a + b)$.

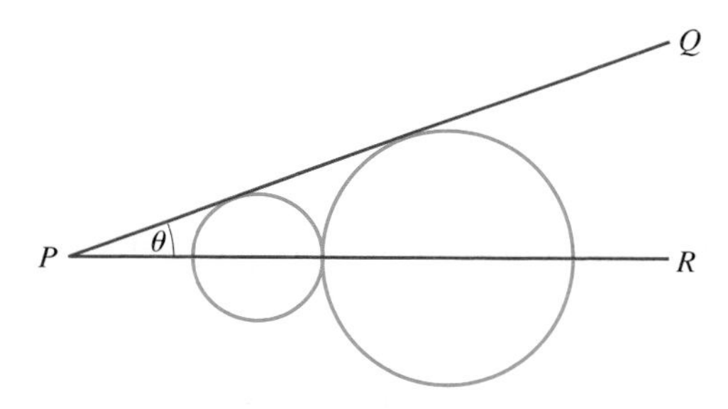

53. A vertical tower of height h stands on level ground. From a point P at ground level and due south of the tower, the angle of elevation to the top of the tower is θ. From a point Q at ground level and due west of the tower, the angle of elevation to the top of the tower is β. If d is the distance between P and Q, show that

$$h = \frac{d}{\sqrt{\cot^2 \theta + \cot^2 \beta}}$$

54. The following problem is taken from *An Elementary Treatise on Plane Trigonometry,* 8th ed., by R. D. Beasley (London: Macmillan and Co.; first published in 1884):

The [angle of] elevation of a tower standing on a horizontal plane is observed; a feet nearer it is found to be 45°; b feet nearer still it is the complement of what it was at the first station; shew that the height of the tower is $ab/(a - b)$ feet.

55. (a) The following problem is taken from *Plane Trigonometry,* 5th ed., by Isaac Todhunter (London: Macmillan and Co., 1874).

$\overline{AB}$ is the diameter of a circle, C its centre; a straight line $\overline{AP}$ is drawn dividing the [area of the] semicircle into two equal parts; θ is the circular [radian] measure of the complement of $\angle PCB$: shew that $\cos \theta = \theta$.

Hint: Use the figure below. Let r denote the radius of the circle and note that $\angle PCB = \frac{\pi}{2} - \theta$.

(b) Use Figure 14 in Section 7.2 to estimate, to the nearest 0.05, the value for θ for which $\cos \theta = \theta$.

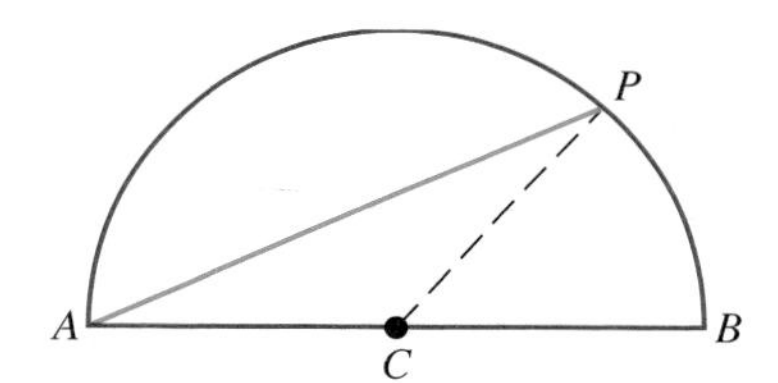

(c) Use a graphing calculator to show that the actual value for the root in part (b), rounded to three decimal places, is $\theta = 0.739$. Use this result to compute the percentage error for the estimate in part (b). Also, use the value $\theta = 0.739$ to compute $\angle PCB$. Express that answer in degrees, rounded to the nearest one degree.

PROJECT

SNELL'S LAW AND AN ANCIENT EXPERIMENT

"The 'lifting' effect produced by refraction is the basis for one of the earliest recorded experiments in optics, one which was known to the ancient Greeks. A coin is placed in the bottom of an empty vessel and the eye of an observer is placed in such a position that the coin is hidden below the edge of the vessel. If water is poured into the vessel the coin appears to rise and come into view."—Francis Weston Sears in *Optics* (Reading, Mass.: Addison-Wesley Publishing Company, Inc., 1958)

In this group project you will learn some of the basic ideas of geometrical optics, a subject first studied by the ancient Greeks, advanced by artists and mathematicians of the Renaissance, and further developed by mathematicians and physicists from the seventeenth through the twentieth and into the twenty-first century. Major contributors include Archimedes, Descartes, Fermat, Huygens, Newton, Gauss, Hamilton, Fresnel, Einstein, and Born.

We now discuss some of the basic principles. The first one is the notion of a light ray. When traveling through a uniform transparent medium (such as a vacuum, water, plastic, or glass) light travels along straight line paths called **light rays.**

The second principle is that light rays bend when passing from one medium to another, for example, when passing from water to air. The amount of bending is determined by the Law of Refraction, usually referred to as Snell's Law after its discoverer, Willebrod Snell (1580–1626). When a light ray traveling in a transparent medium intersects the boundary with another transparent medium, its direction changes as indicated in Figure A.

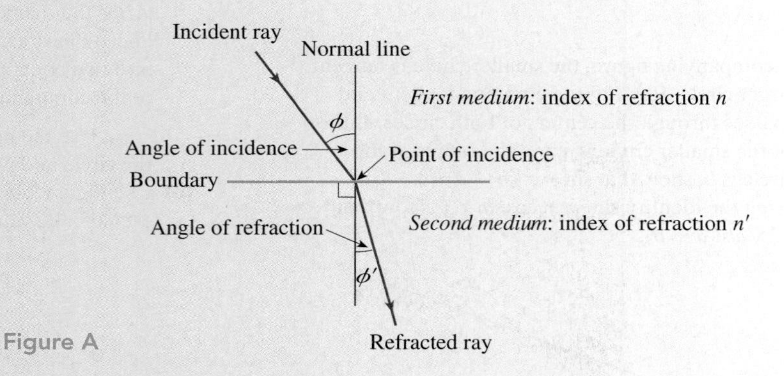

Figure A

Using the terminology introduced in Figure A we can state **Snell's Law** in two parts:

I. The incident ray, the refracted ray and the line normal (or perpendicular) to the boundary at the point of incidence all lie in the same plane.

II. The angle of incidence and the angle of refraction are related by the equation

$$n \sin \phi = n' \sin \phi'$$

The **index of refraction** of a vacuum is defined to be 1. For other materials it may be determined empirically by precise measurement of angles. Air has an index of refraction very close to 1, for water it's about 4/3, and for many glasses and plastics it's about 3/2.

Note: Although it is often left out of statements of Snell's Law, part I is essential because light rays travel in three-dimensional space where three lines can intersect at a point, but not lie in the same plane. (You should think of an example.)

The third principle concerns vision. The eye can see a light source if and only if light rays pass from the source into the pupil of the eye. In Figure B, the eye labeled E can see the light source S because rays pass from the source to the eye through the gap in the blackened surrounding sphere. However, the eye labeled

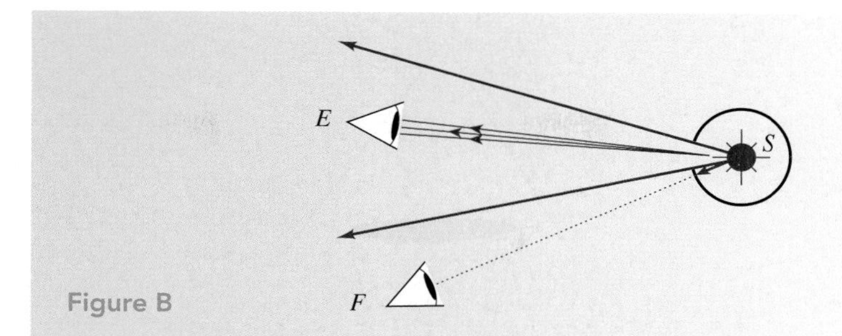

Figure B

F can't see S because light traveling along the line of sight SF is blocked before reaching the eye.

Our fourth principle is that an object appears to be located in the direction determined by the rays entering the eye from the object. This is illustrated in Figure C(i), where one ray from the fish's mouth at A appears to come from the direction of A'.

While the single ray in Figure C(i) suffices to give the apparent direction to the fish's mouth, it is not sufficient to determine the distance to A'. To determine this distance we must consider several rays coming from point A. Figure C(ii) shows several such rays which, after bending at the water's surface, all appear to come from point A'. The fact that all of the rays entering the eye from point A appear to come from a single point after refraction is quite surprising. We will examine this in more detail below.

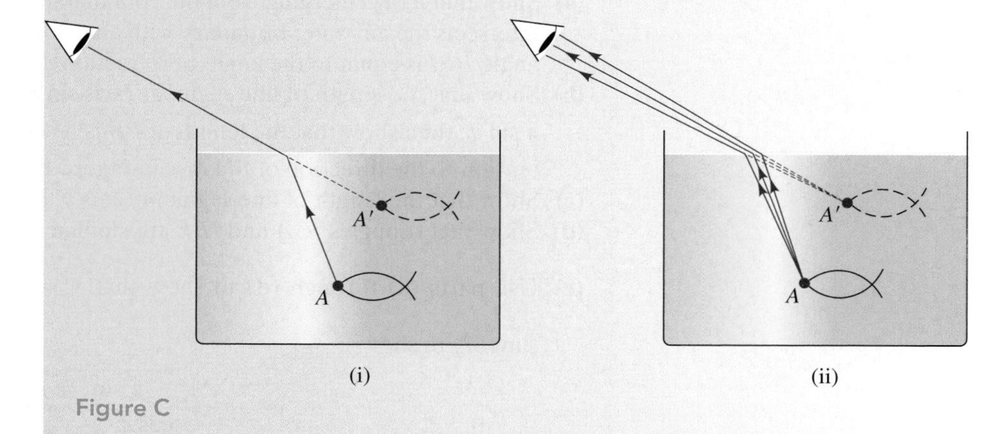

Figure C

Although the principles we've stated do not completely define geometric optics, they are sufficient to allow us to analyze the experiment discussed in the opening quotation. Figure D shows a greatly exaggerated eye looking vertically downward to the point at the center of a coin lying at the bottom of a cup of water. The solid red rays emerge from this point and enter the eye after bending at the water's surface. The ray emerging from the center of the coin to the center of the eye is called the **chief ray.** On the right side of the figure, one ray emerging from the center of the coin is shown refracted at the surface of the water and entering the pupil of the eye. The dashed line extends the part of this ray in air *back* through the water intersecting the chief ray. We could perform this same construction for any other non-chief ray in Figure D. Our goal is to show that all

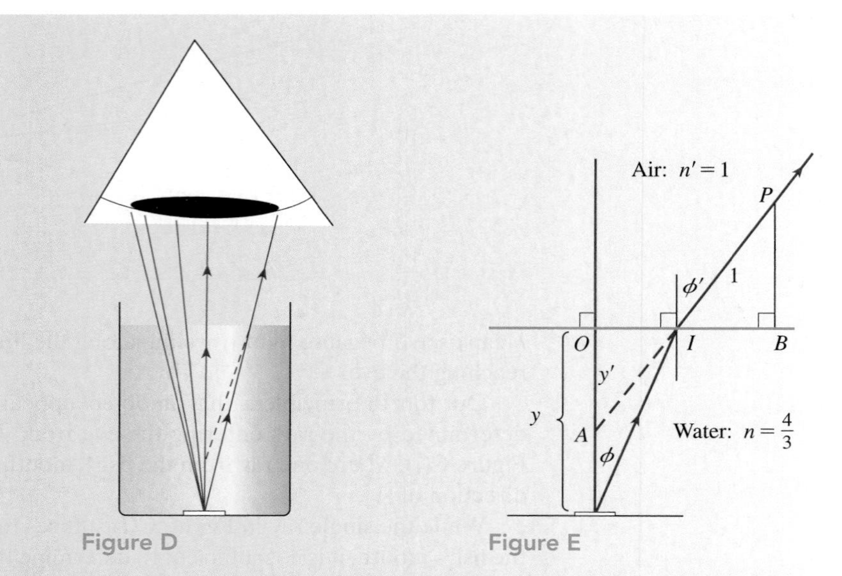

Figure D

Figure E

the dashed extensions would intersect the chief ray at approximately the same point.

In Figure E we lay out the geometry for a typical ray in great detail.

EXERCISE 1

(a) Show that a ray emerging from the coin making an angle ϕ with the vertical intersects the air-water boundary with angle of incidence ϕ. Then show that angle BPI is equal to the angle of refraction ϕ'.

(b) Show that the length of line segment IB is $\sin \phi'$, which by Snell's law is $n \sin \phi$, then show that the length of PB is $\sqrt{1 - n^2 \sin^2 \phi}$. *Caution*: Pay attention to the direction of the rays in Figure E.

(c) Show that the length of line segment OI is $y \tan \phi$.

(d) Show that triangles IOA and IBP are similar.

(e) Use parts (a) through (d) to show that $y' = \dfrac{y \tan \phi \sqrt{1 - n^2 \sin^2 \phi}}{n \sin \phi}$ and simplify to show

$$y' = \frac{\sqrt{1 - n^2 \sin^2 \phi}}{\cos \phi} \frac{1}{n} y$$

The last formula allows us to compute the point at which the center of the coin would appear to be located if we just consider the chief ray and a particular ray emerging from the center point at an angle ϕ to the vertical. To complete your investigation, you need to see if other rays entering the eye from the center of the coin appear to be coming from the same (or nearly the same) point.

EXERCISE 2 Letting n equal $4/3$ compute, to four decimal places, the value of $\dfrac{\sqrt{1 - n^2 \sin^2 \phi}}{\cos \phi} \dfrac{1}{n}$ for $\phi = 0.25, 0.5, 1$, and 2 degrees. Based on these values, is it reasonable to conclude that the center of a coin in a cup filled with water to a depth of y centimeters would appear to be at a depth of $y' = (3/4)y$ centimeters? If the water is 10 centimeters deep how far above the table will the coin appear to rise? What do you think would happen if instead of water we use an oil with an index of refraction of 3/2?

The situation of the ancient Greek experiment is illustrated in Figure F. An observer looking into the empty cup cannot see the coin, but looking into the full cup will see the coin. Although the observer is not looking straight down, it turns out that the refraction at the water's surface will still cause the coin to appear to rise from the bottom of the cup to a position about three-fourths of the actual depth below the surface.

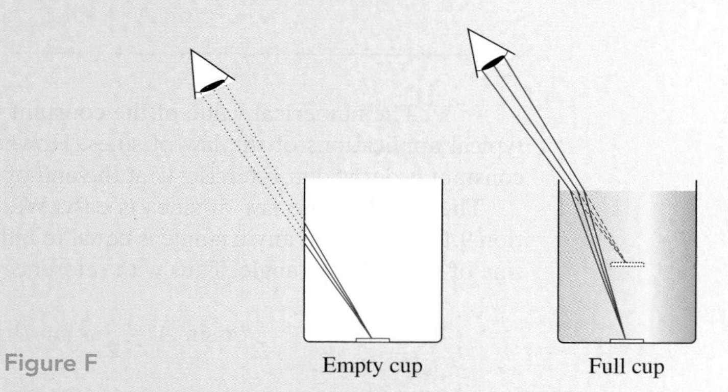

Figure F

EXERCISE 3 Do the experiment. Get an empty cup. Place a coin in the bottom of the cup (a little gum or honey to keep it from moving might be helpful). Have an observer stand where he or she can see the coin and then move back until the coin is just barely hidden by the rim of the cup. Have another member of your group pour water into the cup, stopping frequently to allow the observer to describe what he or she sees. Have yet another member write down these observations.

9.2 THE LAW OF SINES AND THE LAW OF COSINES

The ratio of the sides of a triangle to each other is the same as the ratio of the sines of the opposite angles. —Bartholomaus Pitiscus (1561–1613) in his text, *Trigonometriae sive de dimensione triangulorum libri quinque.* This text was first published in Frankfort, Germany, in 1595, and according to several historians of mathematics, it was the first satisfactory textbook on trigonometry.

In this section we discuss two formulas relating the sides and the angles in any triangle: the *law of sines* and the *law of cosines.* These formulas can be used to determine an unknown side or angle using given information about the triangle. As you will see, which formula to apply in a particular case depends on what data are initially given.

We will usually follow the convention of denoting the angles of a triangle by A, B, and C and the lengths of the corresponding opposite sides by a, b, and c; see Figure 1 on the next page. With this notation we are ready to state the law of sines.

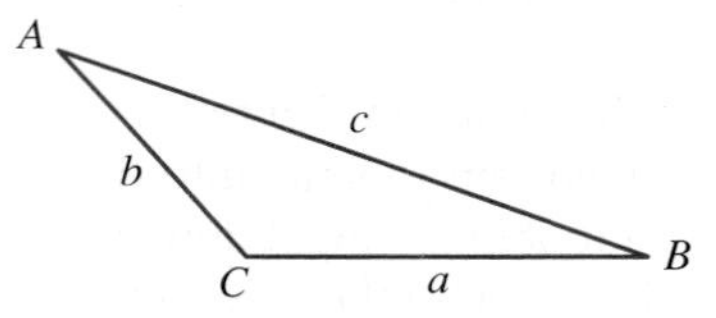

Figure 1

The Law of Sines

In any triangle the ratio of the sine of an angle to the length of its opposite side is constant:

$$\frac{\sin A}{a} = \frac{\sin B}{b} = \frac{\sin C}{c}$$

Equivalently,

$$\frac{a}{\sin A} = \frac{b}{\sin B} = \frac{c}{\sin C}$$

Note: The numerical value of the constant ratio $a/\sin A$ is not used directly in typical applications of the law of sines. However, the geometric meaning of this constant is derived in Exercise 47 at the end of this section.

The proof of the law of sines is easy. We use the following result from Section 9.1: The area of any triangle is equal to half the product of two sides times the sine of the included angle. Thus with reference to Figure 1 we have

$$\frac{1}{2}bc \sin A = \frac{1}{2}ac \sin B = \frac{1}{2}ab \sin C$$

since each of these three expressions equals the area of triangle ABC. Now we just multiply through by the quantity $2/abc$ to obtain

$$\frac{\sin A}{a} = \frac{\sin B}{b} = \frac{\sin C}{c}$$

which completes the proof.

Note: To use the law of sines effectively, we must know one angle and the length of its opposite side plus one additional angle or side. Why?

EXAMPLE 1 Using the law of sines to find a length

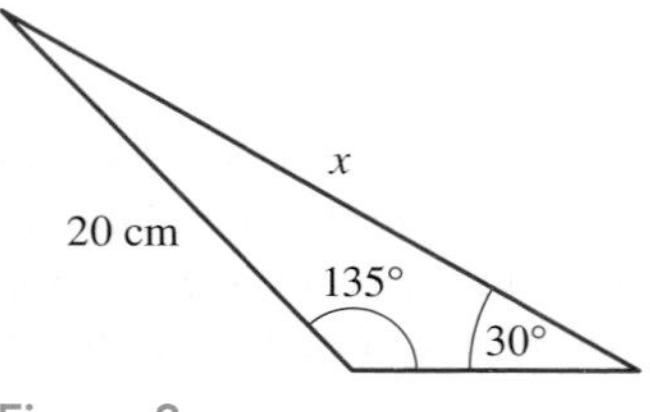

Figure 2

Find the length x in Figure 2.

SOLUTION

We have an angle, 30°; the length of its opposite side, 20 cm; and one more angle, 135°. So the law of sines applies. We have

$$\frac{\sin 30^\circ}{20} = \frac{\sin 135^\circ}{x}$$

(20: length of side opposite the 30° angle; x: length of side opposite the 135° angle)

$$x \sin 30^\circ = 20 \sin 135^\circ$$

$$x = \frac{20 \sin 135^\circ}{\sin 30^\circ} = \frac{20(\sqrt{2}/2)}{1/2} = 20\sqrt{2} \text{ cm}$$

EXAMPLE 2 Using the law of sines to find a length

Find the length y in Figure 3.

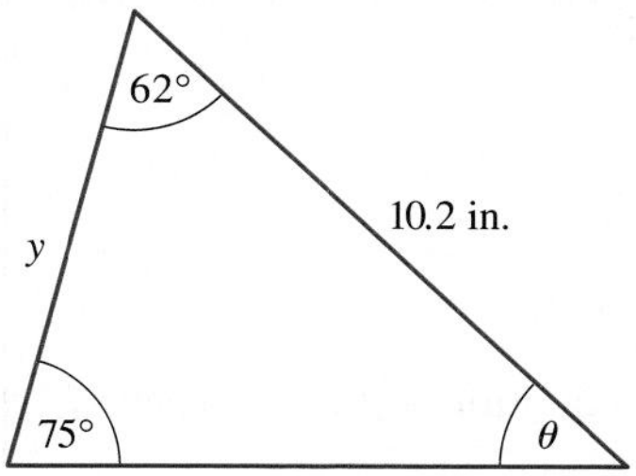

Figure 3

SOLUTION
We have an angle, 75°; the length of its opposite side, 10.2 in.; and one more angle, 62°. So the law of sines applies. To find y, we need to determine the angle θ in Figure 3. Since the sum of the angles in any triangle is 180°, we have

$$\theta = 180^\circ - (75^\circ + 62^\circ) = 43^\circ$$

Now, using the law of sines, we obtain

$$\frac{y}{\sin 43^\circ} = \frac{10.2}{\sin 75^\circ}$$

$$y = \frac{10.2 \sin 43^\circ}{\sin 75^\circ}$$

This is the exact answer. For an approximation a calculator yields

$$y \approx 7.2 \text{ in.}$$

Using given measurements in a triangle to find the remaining measurements is called solving the triangle. This is what we are asked to do in the next example.

EXAMPLE 3 Using the law of sines to solve a triangle

In $\triangle ABC$ we are given $\angle C = 45^\circ$, $b = 4\sqrt{2}$ ft, and $c = 8$ ft. Solve the triangle; that is, determine the remaining side and angles.

SOLUTION
First let's draw a preliminary sketch conveying the given data; see Figure 4. (The sketch must be considered tentative. At the outset we don't know whether the other angles are acute or even whether the given data are compatible.) We have an angle, 45°, and the length of its opposite side, 8 ft. Since we have $b = 4\sqrt{2}$ ft, we can use the law of sines to find B. Thus

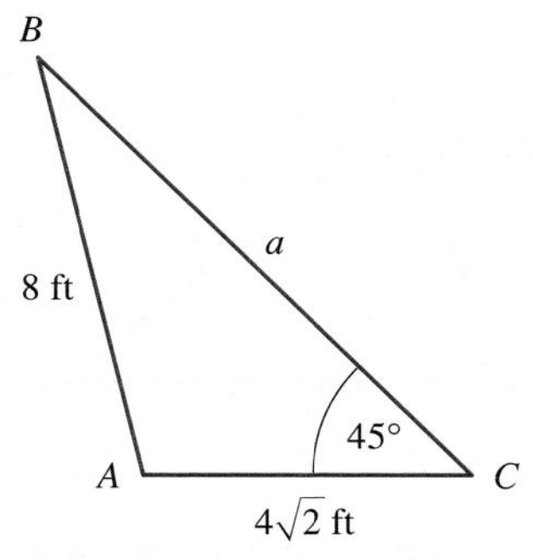

Figure 4

$$\frac{\sin B}{4\sqrt{2}} = \frac{\sin 45^\circ}{8}$$

Then

$$\sin B = \frac{4\sqrt{2} \sin 45^\circ}{8} = \frac{4\sqrt{2}(\sqrt{2}/2)}{8} = \frac{1}{2}$$

So

$$\sin B = \frac{1}{2}$$

From our previous work we know that one possibility for B is 30°, since $\sin 30^\circ = 1/2$. However, there is another possibility. Since the reference angle for

150° is 30°, we know that sin 150° is also equal to 1/2. Which angle do we want? For the problem at hand this is easy to answer. Since angle C is given as 45°, angle B cannot equal 150°, for the sum of 45° and 150° exceeds 180°. We conclude that

$$\angle B = 30^\circ$$

Next, since $\angle B = 30^\circ$ and $\angle C = 45^\circ$, we have

$$\angle A = 180^\circ - (30^\circ + 45^\circ) = 105^\circ$$

Finally, we use the law of sines to find a. From the equation $a/\sin A = c/\sin C$ we have

$$\begin{aligned} a &= \frac{c \sin A}{\sin C} = \frac{8 \sin 105^\circ}{\sin 45^\circ} = \frac{8 \sin 105^\circ}{1/\sqrt{2}} \\ &= 8\sqrt{2} \sin 105^\circ \text{ ft} \approx 10.9 \text{ ft} \end{aligned}$$

In the preceding example, two possibilities arose for the angle B: both 30° and 150°. However, it turned out that the value 150° was incompatible with the given information in the problem. In Exercise 13 at the end of this section you will see a case in which both of two possibilities are compatible with the given data. This results in two distinct solutions to the problem. In contrast to this, Exercise 11(a) shows a case in which there is no triangle fulfilling the given conditions. For these reasons the case in which we are given two sides and an angle opposite one of them is sometimes referred to as the **ambiguous case.**

Now we turn to the law of cosines.

The Law of Cosines

In any triangle the square of the length of any side equals the sum of the squares of the lengths of the other two sides minus twice the product of the lengths of those other two sides times the cosine of their included angle.

$$a^2 = b^2 + c^2 - 2bc \cos A$$

$$b^2 = c^2 + a^2 - 2ca \cos B$$

$$c^2 = a^2 + b^2 - 2ab \cos C$$

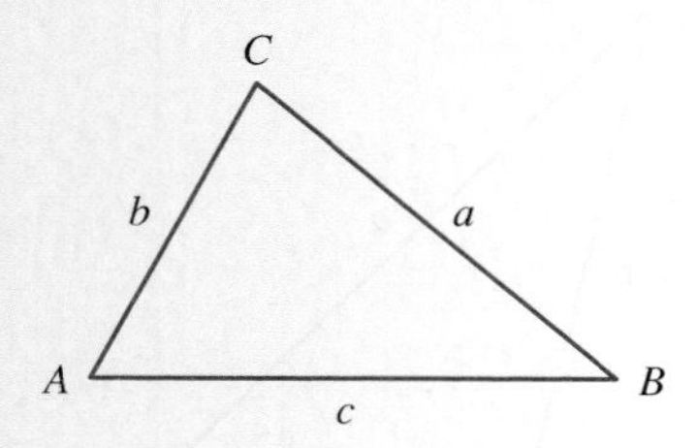

Before looking at a proof of this law, we make two preliminary comments. First, it is important to understand that the three equations in the box all follow the same pattern. For example, look at the first equation:

side and opposite angle

$$a^2 = b^2 + c^2 - 2bc \cos A$$

sides that include this angle

Now check for yourself that the other two equations also follow this pattern. It is the pattern that is important here; after all, not every triangle is labeled ABC.

The second observation is that the law of cosines is a generalization of the Pythagorean theorem. In fact, look what happens to the equation

$$a^2 = b^2 + c^2 - 2bc \cos A$$

when angle A is a right angle:

$$a^2 = b^2 + c^2 - 2bc \underbrace{\cos 90^\circ}_{0}$$

$$a^2 = b^2 + c^2 \qquad \text{which is the Pythagorean theorem}$$

Now let us prove the law of cosines:

$$a^2 = b^2 + c^2 - 2bc \cos A$$

(The other two equations can be proved in the same way. Indeed, just relabeling the figure would suffice.) The proof that we give uses coordinate geometry in a very nice way to complement the trigonometry. We begin by placing angle A in standard position, as indicated in Figure 5. (So in the figure, angle A is then identified with angle CAB.) Then if u and v denote the lengths indicated in Figure 5, the coordinates of C are (u, v), and we have

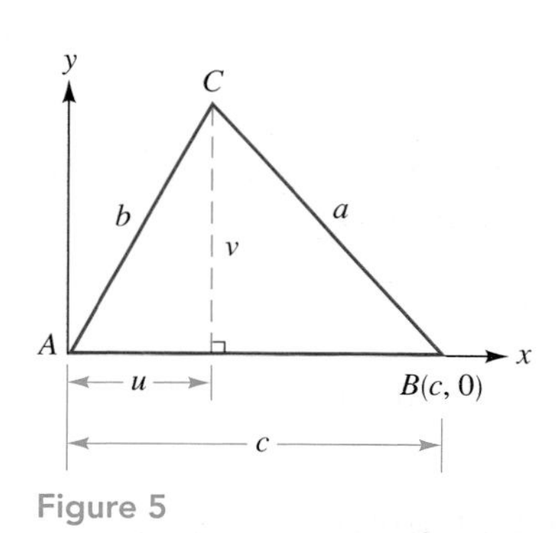

Figure 5

$$\cos A = \frac{\text{adjacent}}{\text{hypotenuse}} = \frac{u}{b} \qquad \text{and therefore} \qquad u = b \cos A$$

Similarly, we have

$$\sin A = \frac{\text{opposite}}{\text{hypotenuse}} = \frac{v}{b} \qquad \text{and therefore} \qquad v = b \sin A$$

Thus the coordinates of C are

$$(b \cos A, b \sin A)$$

(Exercise 44 at the end of this section asks you to check that these represent the coordinates of C even when angle A is not acute.)

Now we use the distance formula,

$$d = \sqrt{(x_2 - x_1)^2 + (y_2 - y_1)^2}$$

or, equivalently,

$$d^2 = (x_2 - x_1)^2 + (y_2 - y_1)^2$$

to compute the square of the distance a between the points $C\,(b \cos A, b \sin A)$ and $B\,(c, 0)$. We have

$$\begin{aligned} a^2 &= (b \cos A - c)^2 + (b \sin A - 0)^2 \\ &= b^2 \cos^2 A - 2bc \cos A + c^2 + b^2 \sin^2 A \\ &= b^2 \underbrace{(\cos^2 A + \sin^2 A)}_{1} - 2bc \cos A + c^2 \\ &= b^2 + c^2 - 2bc \cos A \end{aligned}$$

completing our proof of the law of cosines.

Figure 6

EXAMPLE 4 Using the law of cosines to find a length

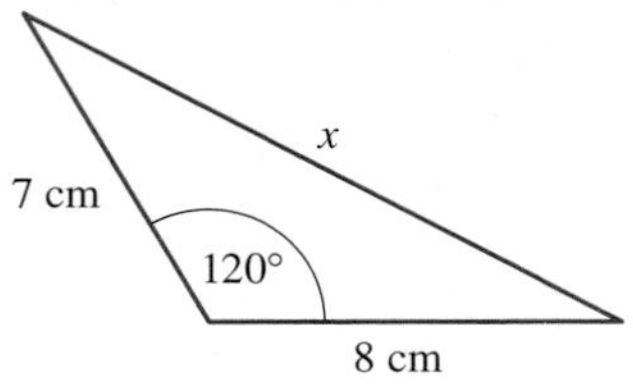

Compute the length x in Figure 6.

SOLUTION

Note the law of sines does not apply. Why? The law of cosines is directly applicable. We have

$$x^2 = 7^2 + 8^2 - 2(7)(8)\cos 120^\circ$$

$$= 49 + 64 - 112\left(-\frac{1}{2}\right) = 169$$

$$x = \sqrt{169} = 13 \text{ cm}$$

The law of cosines can be used effectively if we know the lengths of the sides of a triangle. For example, if the equation $a^2 = b^2 + c^2 - 2bc \cos A$ is solved for $\cos A$, the result is

$$\cos A = \frac{b^2 + c^2 - a^2}{2bc}$$

This expresses the cosine of an angle in a triangle in terms of the lengths of the sides. In a similar fashion we obtain the corresponding formulas

$$\cos B = \frac{c^2 + a^2 - b^2}{2ca} \quad \text{and} \quad \cos C = \frac{a^2 + b^2 - c^2}{2ab}$$

These alternative forms for the law of cosines are used in the next example.

EXAMPLE 5 Using the law of cosines to find angles

In triangle ABC the sides are $a = 3$ units, $b = 5$ units, and $c = 7$ units. Find the angles. (Use degree measure.)

SOLUTION

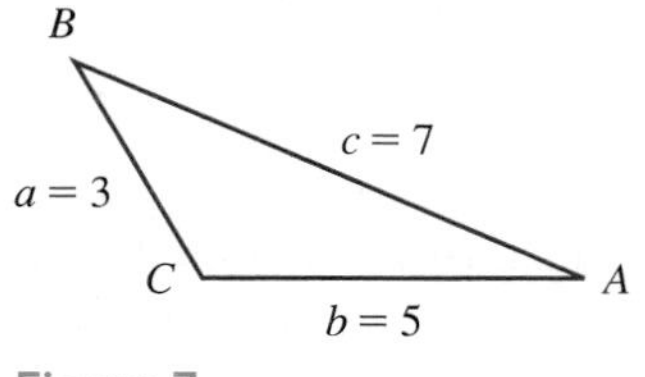

Figure 7

Figure 7 summarizes the given data. We have

$$\cos A = \frac{b^2 + c^2 - a^2}{2bc} = \frac{5^2 + 7^2 - 3^2}{2(5)(7)} = \frac{65}{70} = \frac{13}{14}$$

Now that we know $\cos A = 13/14$, we can find $\angle A$ by using a calculator to compute $\cos^{-1}(13/14)$. Since we are required to give the answer in degree measure, we first set the calculator to the degree mode. Then we obtain

$$\cos^{-1}(13/14) \approx 21.7867893 \qquad \text{using a calculator set in the degree mode}$$

and therefore

$$\angle A \approx 21.8^\circ \qquad \text{rounding to one decimal place}$$

Why is there no second possible angle $\angle A$? In a similar manner we have

$$\cos B = \frac{c^2 + a^2 - b^2}{2ca} = \frac{7^2 + 3^2 - 5^2}{2(7)(3)} = \frac{33}{42} = \frac{11}{14}$$

So $\cos B = 11/14$, and a calculator then yields $\angle B \approx 38.2132107° \approx 38.2°$.

At this point we can find $\angle C$ in either of two ways. Each has its advantage. The first way is to begin by computing $\cos C$ in the same way that we found $\cos A$ and $\cos B$. As you can check, the result is $\cos C = -1/2$. A calculator is not needed in this case. We know from previous work that $\angle C$ must be 120°. The second method that can be used relies on the fact that the sum of the angles in a triangle is 180°. Thus we have

$$\begin{aligned}\angle C &= 180° - \angle A - \angle B \\ &\approx 180° - 21.7867893° - 38.2132107° \\ &\approx 120.0°\end{aligned}$$

This way is quicker than the first method. The disadvantage, however, is that we know that $\angle C$ is only approximately 120°. The first method (using the cosine law) is longer, but it tells us that $\angle C$ is exactly 120°. In summary, then, the three required angles are

$$\angle A \approx 21.8° \qquad \angle B \approx 38.2° \qquad \angle C = 120°$$

Alternatively, after finding angle $\angle A$, we could have used the law of sines to find a second angle, say, $\angle B$. In this example, given the lengths of all three sides, there will be no ambiguity in $\angle B$. Why? Finally, we could find $\angle C$ using the law of sines, the law of cosines, or the fact that the sum of the angles in a triangle is 180°.

We conclude this section with an example indicating how the law of sines and the law of cosines are used in navigation. In this example you'll see the term *bearing* used in specifying the location of one point relative to another. To explain this term, we refer to Figure 8.

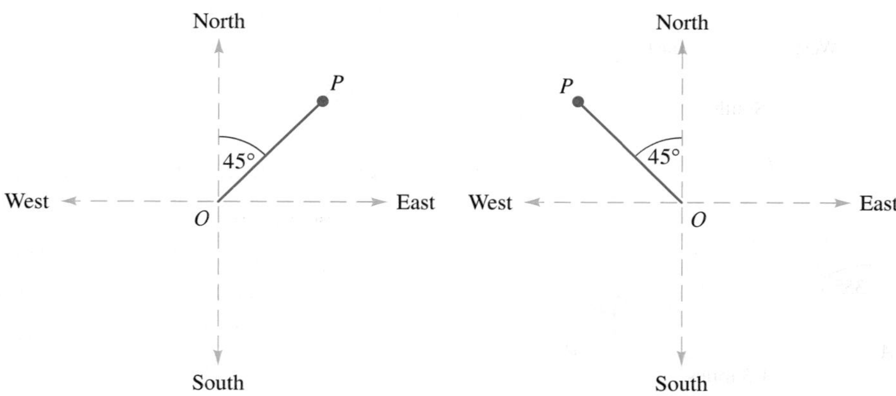

Figure 8 (a) The bearing of P from O is N45°E. (b) The bearing of P from O is N45°W.

In Figure 8(a) the bearing of P from O is N45°E (read "north, 45° east"). This bearing tells us the *acute* angle between line segment $\overline{OP}$ and the north–south line through O. In Figure 8(b) the bearing of P from O is N45°W (read "north, 45° west"). Again, note that the bearing gives us the acute angle between line segment $\overline{OP}$ and the north–south line through O. Figure 9 provides additional examples of this terminology.

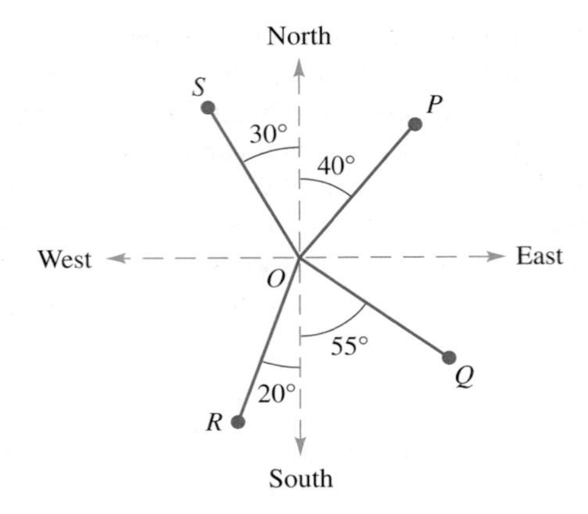
North
S
30°
P
40°
West
East
O
55°
Q
20°
R
South
The bearing of P from O is N40°E.
The bearing of Q from O is S55°E.
The bearing of R from O is S20°W.
The bearing of S from O is N30°W.

Figure 9
The bearing is specified by means of the acute angle measured from the north–south line.

EXAMPLE 6 Using the law of sines and the law of cosines in navigation

A small fire is sighted from ranger stations A and B. The bearing of the fire from station A is N35°E, and the bearing of the fire from station B is N49°W. Station A is 1.3 miles due west of station B.

(a) How far is the fire from each ranger station?

(b) At fire station C, which is 1.5 miles from A, there is a helicopter that can be used to drop water on the fire. If the bearing of C from A is S42°E, find the distance from C to the fire, and find the bearing of the fire from C.

SOLUTION

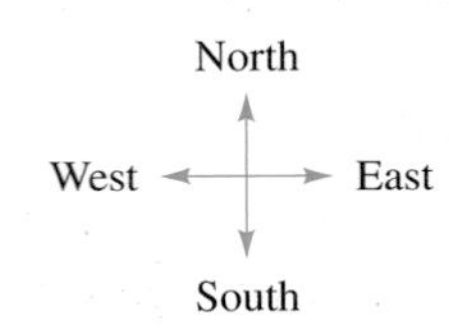
North
West
East
South

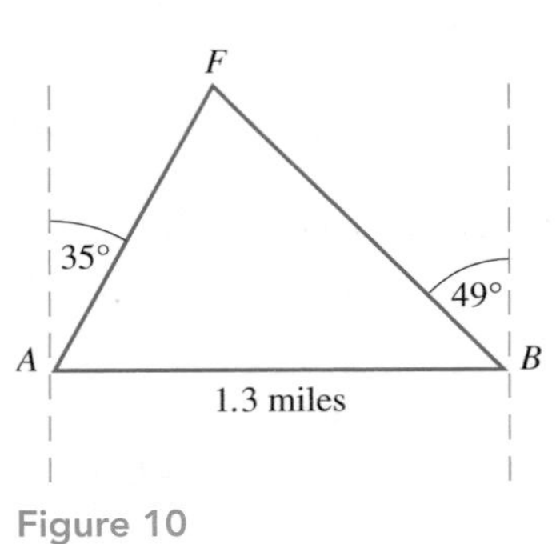
F
35°
49°
A
B
1.3 miles

Figure 10

(a) In Figure 10 we have sketched the situation involving ranger stations A and B and the fire (denoted by F).

We compute the angles of $\triangle ABF$ as follows.

$$\angle FAB = 90° - 35° = 55° \qquad \angle FBA = 90° - 49° = 41°$$
$$\angle F = 180° - (55° + 41°) = 84°$$

We can now use the law of sines in $\triangle ABF$ to compute AF and BF.

$$\frac{AF}{\sin \angle FBA} = \frac{AB}{\sin \angle F} \qquad\qquad \frac{BF}{\sin \angle FAB} = \frac{AB}{\sin \angle F}$$

and therefore | and therefore

$$AF = \frac{AB \sin \angle FBA}{\sin \angle F} \qquad\qquad BF = \frac{AB \sin \angle FAB}{\sin \angle F}$$

or | or

$$AF = \frac{(1.3)\sin 41°}{\sin 84°} \qquad\qquad BF = \frac{(1.3)\sin 55°}{\sin 84°}$$

We use a calculator to evaluate these expressions for AF and BF. As you should check for yourself, the results (rounded to the nearest tenth of a mile) are

$$AF \approx 0.9 \text{ miles} \qquad \text{and} \qquad BF \approx 1.1 \text{ miles}$$

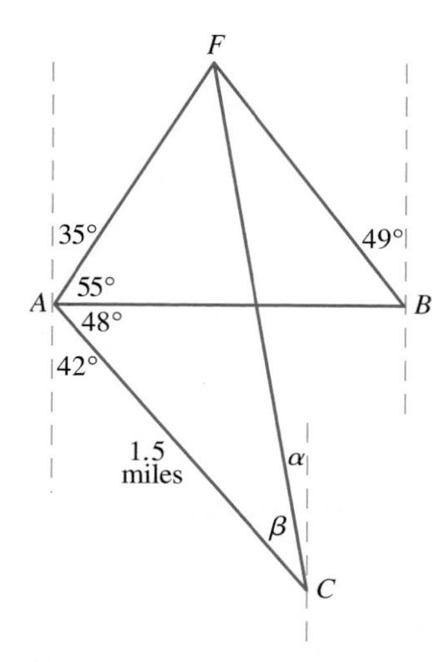

Figure 11

(b) We draw a sketch of the situation, as shown in Figure 11. In Figure 11 we can compute CF, the distance from the helicopter to the fire, using the law of cosines in $\triangle CAF$. First, note that $\angle CAF = 48° + 55° = 103°$. So we have

$$CF = \sqrt{AC^2 + AF^2 - 2 \cdot AC \cdot AF \cdot \cos 103°}$$
$$= \sqrt{1.5^2 + \left(\frac{1.3 \sin 41°}{\sin 84°}\right)^2 - 2(1.5)\left(\frac{1.3 \sin 41°}{\sin 84°}\right) \cos 103°}$$
$$\approx 1.9 \text{ miles} \qquad \text{using a calculator and rounding to one decimal place}$$

To find the bearing of the fire at F from the fire station at C, we need to determine the angle α in Figure 11. First we find the angle β. Using the law of sines in $\triangle CAF$, we have

$$\frac{\sin \beta}{AF} = \frac{\sin 103°}{CF}$$

and therefore

$$\sin \beta = \frac{AF \cdot \sin 103°}{CF}$$

Before using the inverse sine function and a calculator to compute β, we note from Figure 11 that β is an acute angle (because β is an angle in $\triangle CAF$, and in that triangle, $\angle CAF$ is greater than 90°). So in this particular application of the law of sines there is no ambiguity. We have then

$$\beta = \sin^{-1}\left(\frac{AF \cdot \sin 103°}{CF}\right)$$

Now, on the right-hand side of this last equation we substitute the expressions we obtained previously for AF and CF; then, using a calculator set in the degree mode, we obtain

$$\beta \approx 26° \qquad \text{rounding to the nearest degree}$$

(You should verify this calculator value for yourself.) Now that we know β, we can determine the bearing of the fire from station C. From Figure 11 we have

$$\alpha + \beta = 42° \quad \text{(Why?)}$$

and therefore

$$\alpha = 42° - \beta = 42° - 26° = 16° \qquad \text{to the nearest degree}$$

In summary now, fire station C is approximately 1.9 miles from the fire, and the bearing of the fire from station C is N16°W.

EXERCISE SET 9.2

A

In Exercises 1–8, assume that the vertices and the lengths of the sides of a triangle are labeled as in Figure 1 on page 652. For Exercises 1–4, leave your answers in terms of radicals or the trigonometric functions; that is, don't use a calculator. In Exercises 5–8, use a calculator and round your final answers to one decimal place.

1. If $\angle A = 60°$, $\angle B = 45°$, and $BC = 12$ cm, find AC.
2. If $\angle A = 30°$, $\angle B = 135°$, and $BC = 4$ cm, find AC.
3. If $\angle B = 100°$, $\angle C = 30°$, and $AB = 10$ cm, find BC.

4. If $\angle A = \angle B = 35°$, and $AB = 16$ cm, find AC and BC.

5. If $\angle A = 36°$, $\angle B = 50°$, and $b = 12.61$ cm, find a and c.

6. If $\angle B = 81°$, $\angle C = 55°$, and $b = 6.24$ cm, find c and a.

7. If $a = 29.45$ cm, $b = 30.12$ cm, and $\angle B = 66°$, find the remaining side and angles of the triangle.

8. If $a = 52.15$ cm, $c = 42.90$ cm, and $\angle A = 125°$, find the remaining side and angles of the triangle.

In Exercises 9 and 10, use degree measure for your answers. In parts (c) and (d), use a calculator and round the results to one decimal place.

9. (a) In $\triangle ABC$, $\sin B = \sqrt{2}/2$. What are the possible values for $\angle B$?

(b) In $\triangle DEF$, $\cos E = \sqrt{2}/2$. What are the possible values for $\angle E$?

(c) In $\triangle GHI$, $\sin H = 1/4$. What are the possible values for $\angle H$?

(d) In $\triangle JKL$, $\cos K = -2/3$. What are the possible values for $\angle K$?

10. (a) In $\triangle ABC$, $\sin B = \sqrt{3}/2$. What are the possible values for $\angle B$?

(b) In $\triangle DEF$, $\cos E = -\sqrt{3}/2$. What are the possible values for $\angle E$?

(c) In $\triangle GHI$, $\sin H = 2/9$. What are the possible values for $\angle H$?

(d) In $\triangle JKL$, $\cos K = 2/3$. What are the possible values for $\angle K$?

11. (a) Show that there is no triangle satisfying the conditions $a = 2.0$ ft, $b = 6.0$ ft, and $\angle A = 23.1°$
Hint: Try computing $\sin B$ using the law of sines.

(b) If $a = 2.0$ ft, $b = 3.0$ ft, $\angle A = 23.1°$, and $\angle B$ is obtuse, show that $c = 1.1$ ft.

12. (a) Show that there is no triangle with $a = 2$, $b = 3$, and $\angle A = 42°$.

(b) Is there any triangle in which $a = 2$, $b = 3$, and $\angle A = 41°$?

13. Let $b = 1$, $a = \sqrt{2}$, and $\angle B = 30°$.

(a) Use the law of sines to show that $\sin A = \sqrt{2}/2$. Conclude that $\angle A = 45°$ or $\angle A = 135°$.

(b) Assuming that $\angle A = 45°$, determine the remaining parts of $\triangle ABC$.

(c) Assuming that $\angle A = 135°$, determine the remaining parts of $\triangle ABC$.

(d) Find the areas of the two triangles.

14. Let $a = 30$, $b = 36$, and $\angle A = 20°$.

(a) Show that $\angle B = 24.23°$ or $\angle B = 155.77°$ (rounding to two decimal places).

(b) Determine the remaining parts for each of the two possible triangles. Round your final results to two decimal places. [However, in your calculations, do *not* work with rounded values. (Why?)]

(c) Find the areas of the two triangles.

15. Find the lengths a, b, c, and d in the following figure.

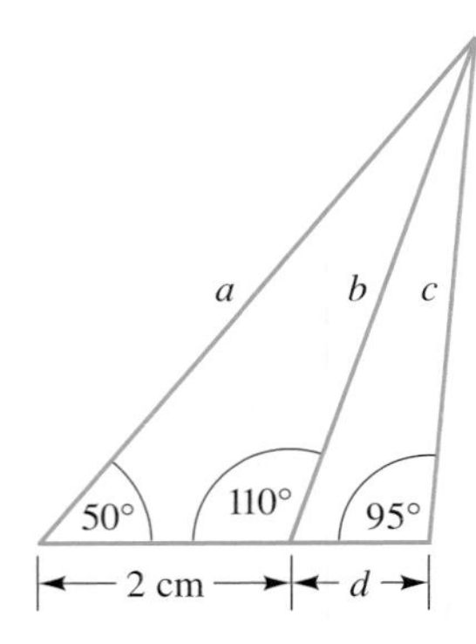

Leave your answers in terms of trigonometric functions (rather than using a calculator).

16. In the following figure, $\overline{PQR}$ is a straight line segment. Find the distance PR. Round your final answer to two decimal places.

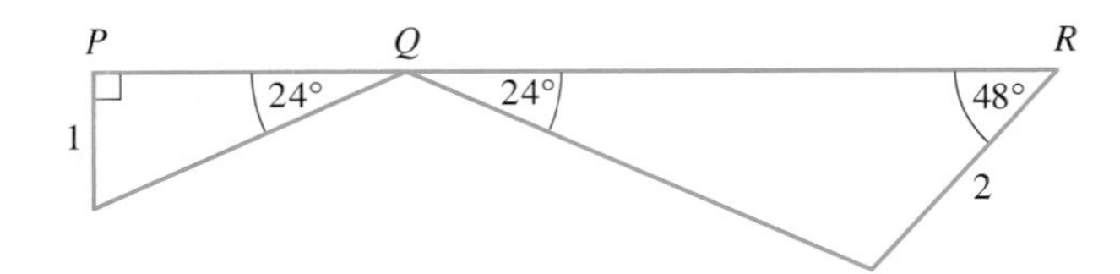

17. Two points P and Q are on opposite sides of a river (see the sketch). From P to another point R on the same side is 300 ft. Angles PRQ and RPQ are found to be 20° and 120°, respectively. Compute the distance from P to Q, across the river. (Round your answer to the nearest foot.)

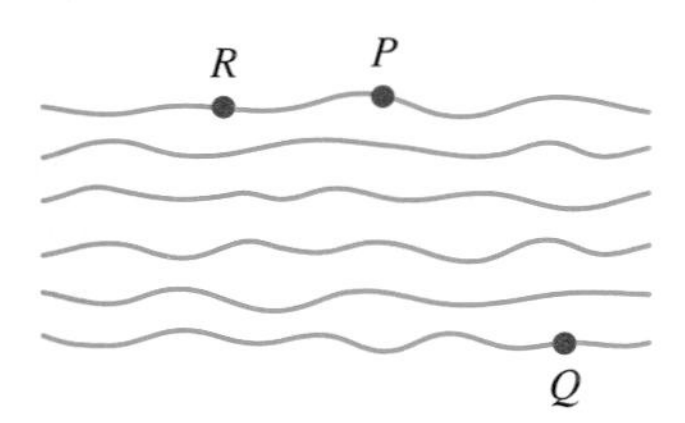

18. Determine the angle θ in the accompanying figure. Round your answer to two decimal places.

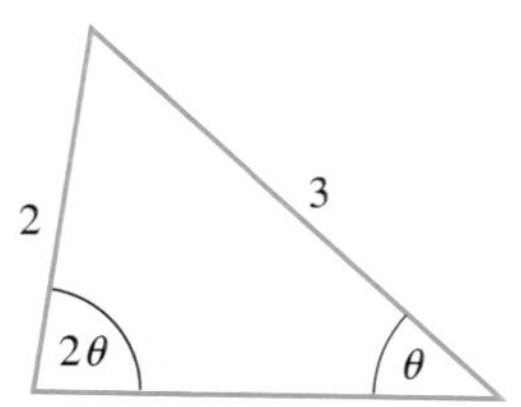

In Exercises 19–22, use the law of cosines to determine the length x in each figure. For Exercises 19 and 20, leave your answers in terms of radicals. In Exercises 21 and 22, use a calculator and round the answers to one decimal place.

19. (a)

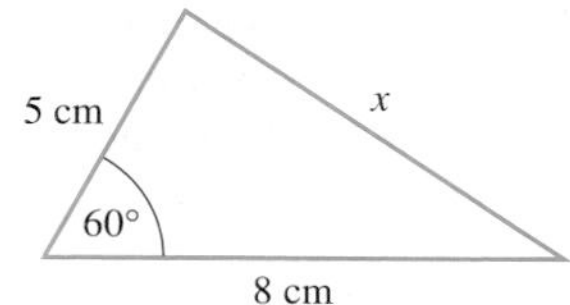

(b)

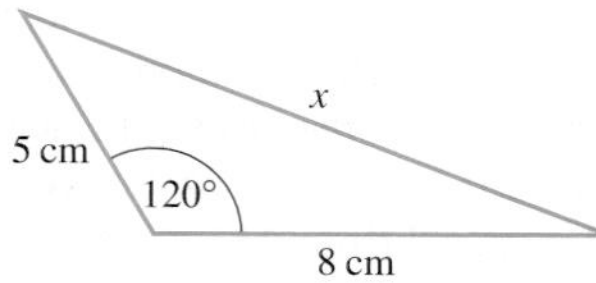

20. (a)

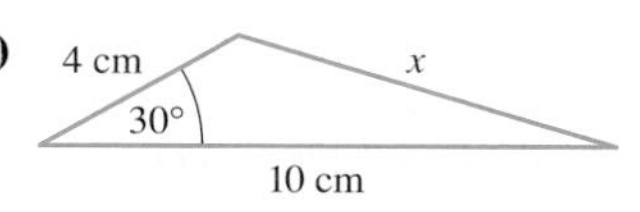

(b)

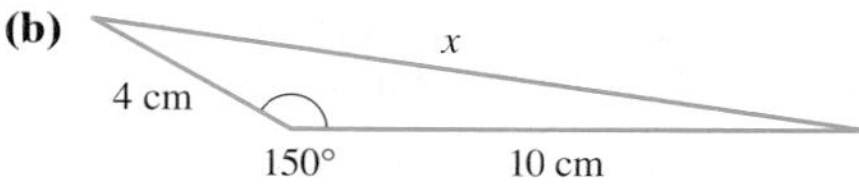

21. (a)

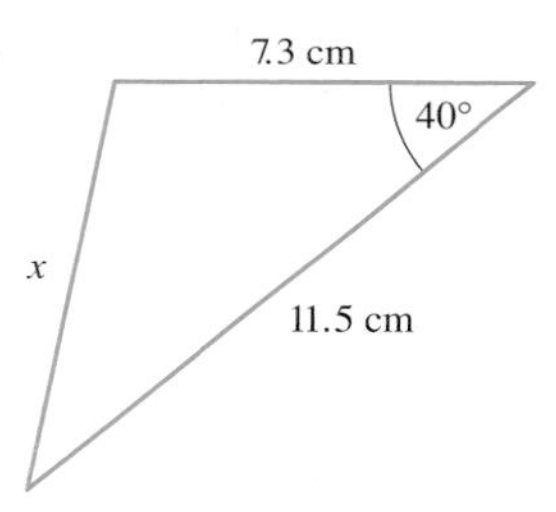

(b)

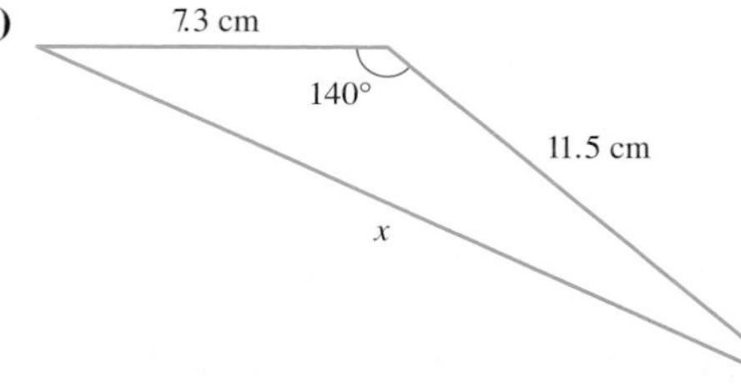

22. (a)

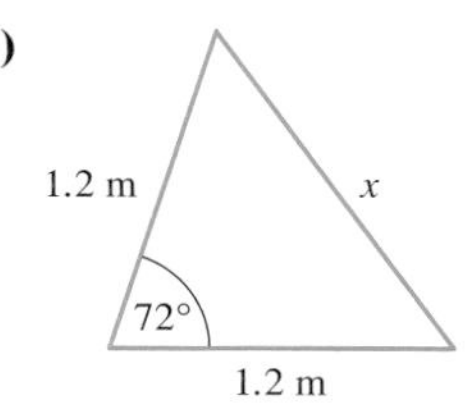

(b)

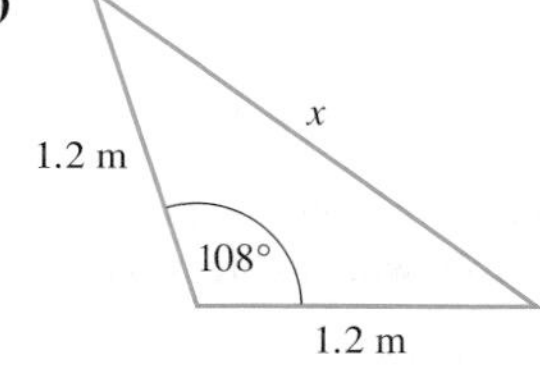

In Exercises 23 and 24, refer to the following figure.

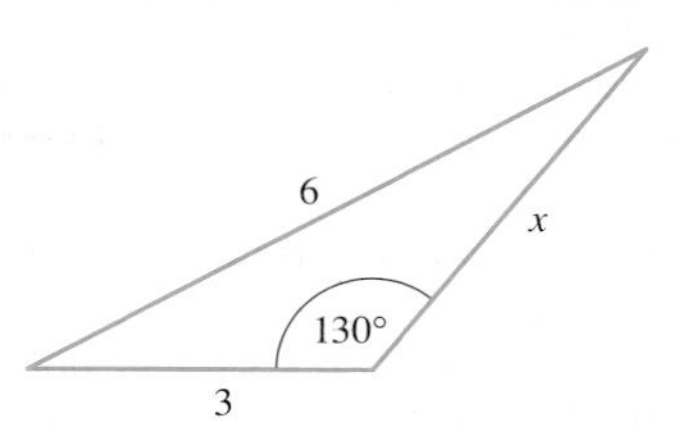

23. In applying the law of cosines to the figure, a student incorrectly writes $x^2 = 3^2 + 6^2 - 2(3)(6)\cos 130°$. Why is this incorrect? What is the correct equation?

24. In applying the law of cosines to the figure, a student writes $6^2 = 3^2 + x^2 + 6x \cos 50°$. Why is this correct?

In Exercises 25 and 26, use the given information to find the cosine of each angle in $\triangle ABC$.

25. $a = 6$ cm, $b = 7$ cm, $c = 10$ cm

26. $a = 17$ cm, $b = 8$ cm, $c = 15$ cm (For this particular triangle, you can check your answers, because there is an alternative method of solution that does not require the law of cosines.)

In Exercises 27–30, compute each angle of the given triangle. Where necessary, use a calculator and round to one decimal place.

27. $a = 7, b = 8, c = 13$

28. $a = 33, b = 7, c = 37$

29. $a = b = 2/\sqrt{3}, c = 2$

30. $a = 36, b = 77, c = 85$

In Exercises 31–34, round each answer to one decimal place.

31. A regular pentagon is inscribed in a circle of radius 1 unit. Find the perimeter of the pentagon. *Hint:* First find the length of a side using the law of cosines.

32. Find the perimeter of a regular nine-sided polygon inscribed in a circle of radius 4 cm. (See the hint for Exercise 31.)

33. In $\triangle ABC$, $\angle A = 40°$, $b = 6.1$ cm, and $c = 3.2$ cm.

(a) Find a using the law of cosines.

(b) Find $\angle C$ using the law of sines.

(c) Find $\angle B$.

34. In parallelogram $ABCD$ you are given $AB = 6$ in., $AD = 4$ in., and $\angle A = 40°$. Find the length of each diagonal.

35. Town B is 26 miles from town A at a bearing of S15°W. Town C is 54 miles from town A at a bearing of S7°E. Compute the distance from town B to town C. Round your final answer to the nearest mile.

36. Town C is 5 miles due east of town D. Town E is 12 miles from town C at a bearing (from C) of N52°E.

(a) How far apart are towns D and E? (Round to the nearest one-half mile.)

(b) Find the bearing of town E from town D. (Round the angle to the nearest degree.)

37. An airplane crashes in a lake and is spotted by observers at lighthouses A and B along the coast. Lighthouse B is 1.50 miles due east of lighthouse A. The bearing of the airplane from lighthouse A is S20°E; the bearing of the plane from lighthouse B is S42°W. Find the distance from each lighthouse to the crash site. (Round your final answers to two decimal places.)

38. (Continuation of Exercise 37) A rescue boat is in the lake, three-fourths of a mile from lighthouse B and at a bearing of S35°E from lighthouse B.

(a) Find the distance from the rescue boat to the airplane. Express your answer using miles and feet, with the portion in feet rounded to the nearest 10 feet.

(b) Find the bearing of the plane from the rescue boat. (Your answer should have the form of Sθ°W. Round θ to two decimal places.)

39. (Refer to the following figure.) When the Sun is viewed from Earth, it subtends an angle of $\theta = 32'$ ($= 32/60$ degree). Assuming that the distance d from Earth to the Sun is 92,690,000 miles, use the law of cosines to compute the diameter D of the Sun. Round the answer to the nearest ten thousand miles.

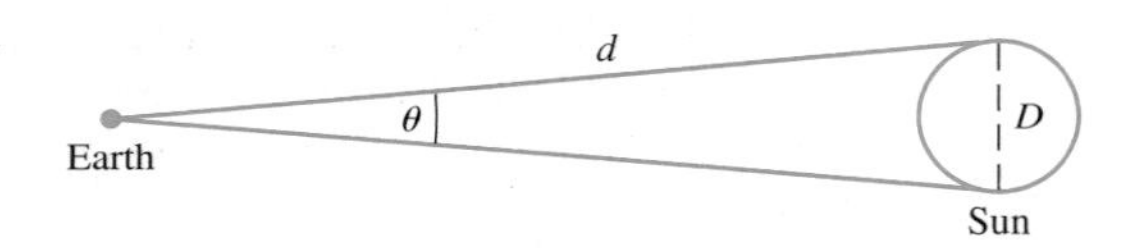

40. Compute the lengths CD and CE in the accompanying figure. Round the final answers to two decimal places.

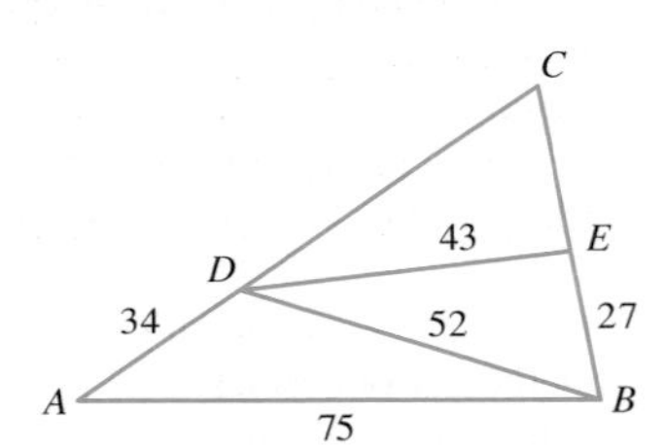

B

41. (a) Let m and n be positive numbers, with $m > n$. Furthermore, suppose that in triangle ABC the lengths a, b, and c are given by

$$a = 2mn + n^2 \qquad b = m^2 - n^2$$
$$c = m^2 + n^2 + mn$$

Show that $\cos C = -1/2$, and conclude that $\angle C = 120°$.

(b) Give an example of a triangle in which the lengths of the sides are whole numbers and one of the angles is 120°. (Specify the three sides; you needn't find the other angles.)

42. If the lengths of two adjacent sides of a parallelogram are a and b, and if the acute angle formed by these two sides is θ, show that the product of the lengths of the two diagonals is given by the expression

$$\sqrt{(a^2 + b^2)^2 - 4a^2b^2 \cos^2 \theta}$$

43. Two trains leave the railroad station at noon. The first train travels along a straight track at 90 mph. The second train travels at 75 mph along another straight track that makes an angle of 130° with the first track. At what time are the trains 400 miles apart? Round your answer to the nearest minute.

44. In this exercise you will complete a detail mentioned in the text in the proof of the law of cosines. Let the positive numbers u and v denote the lengths indicated in the figure at the top of the next column, so that the coordinates of C are $(-u, v)$. Show that $u = -b \cos A$ and $v = b \sin A$. Conclude from this that the coordinates of C are

$$(b \cos A, b \sin A)$$

Hint: Use the right-triangle definitions for cosine and sine along with the addition formulas for $\cos(180° - \theta)$ and $\sin(180° - \theta)$.

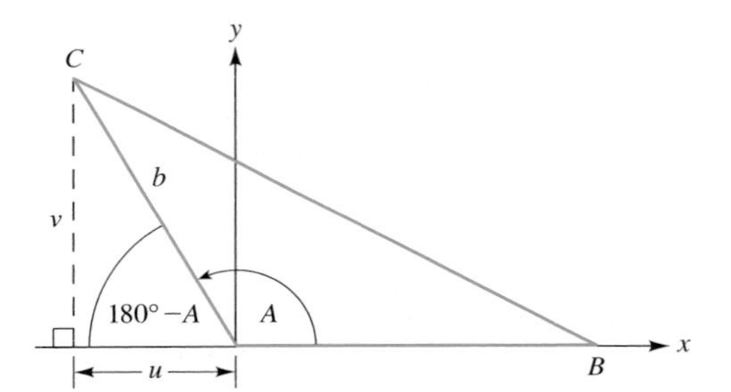

45. In the following figure, $ABCD$ is a square, $AB = 1$, and $\angle EAB = \angle EBA = 15°$. Show that $\triangle CDE$ is equilateral. *Hint:* First use the law of sines to find AE. Then use the law of cosines to find DE.

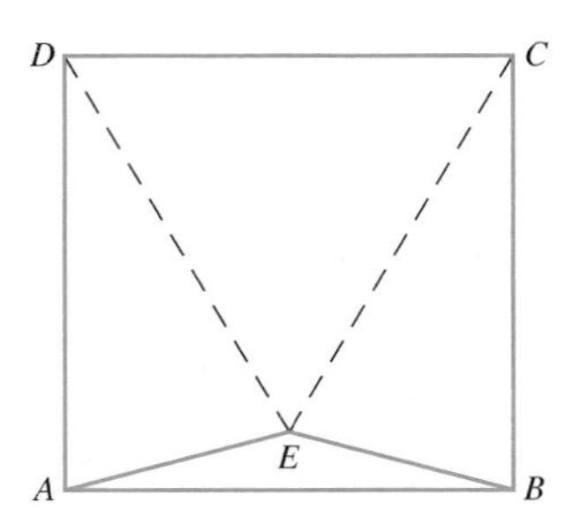

46. Use steps (a) through (c) to show that the area of any triangle ABC is given by the following expression:

$$\frac{a^2 - b^2}{2} \cdot \frac{\sin A \sin B}{\sin(A - B)}$$

(a) Use the law of sines to show that $(a^2 - b^2)\sin A \sin B = ab(\sin^2 A - \sin^2 B)$.

(b) Prove the trigonometric identity $\sin(A - B)\sin(A + B) = \sin^2 A - \sin^2 B$.

(c) Use the results in parts (a) and (b) to show that $\frac{a^2 - b^2}{2} \cdot \frac{\sin A \sin B}{\sin(A - B)} = \frac{1}{2} ab \sin C$, which is the area of $\triangle ABC$, as required.

As background for Exercises 47 and 48, refer to the figure below. The smaller circle in the figure is the ***inscribed circle*** *for $\triangle ABC$. Each side of $\triangle ABC$ is tangent to the inscribed circle. The larger circle is the* ***circumscribed circle*** *for $\triangle ABC$. The circumscribed circle is the circle passing through the three vertices of the triangle.*

In Exercises 47 and 48 you will derive expressions for the radii of the circumscribed circle and the inscribed circle for $\triangle ABC$. In these exercises, assume as given the following two results from geometry:

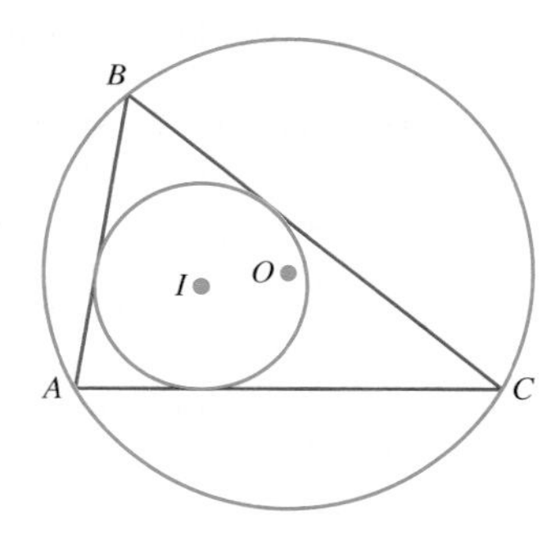

i. *The three angle bisectors of the angles of a triangle meet in a point. This point (labeled I in the figure) is the center of the inscribed circle.*

ii. *The perpendicular bisectors of the sides of a triangle meet in a point. This point (labeled O in the figure) is the center of the circumscribed circle.*

47. The following figure shows the circumscribed circle for $\triangle ABC$. The point O is the center of the circle, and $\mathcal{R}$ is the radius.

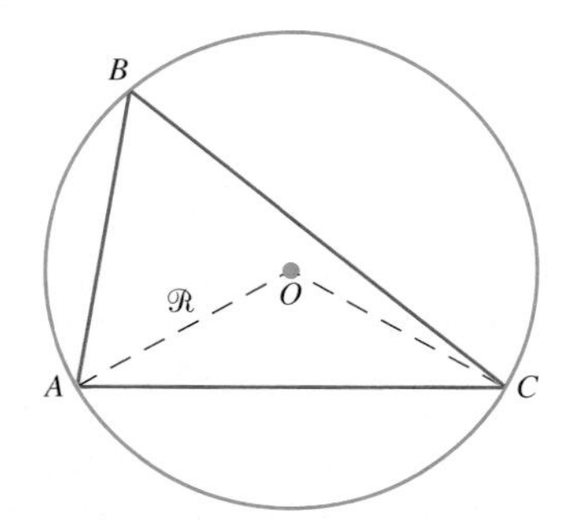

In this exercise you will derive the formulas for the radius $\mathcal{R}$ of the circumscribed circle for $\triangle ABC$:

$$\mathcal{R} = \frac{a}{2 \sin A} = \frac{b}{2 \sin B} = \frac{c}{2 \sin C}$$

(a) According to a theorem from geometry, the measure of $\angle AOC$ is twice the measure of $\angle B$. What theorem is this? (State the theorem using complete sentences.)

(b) Draw a perpendicular from O to $\overline{AC}$, meeting $\overline{AC}$ at T. Explain why $AT = TC = b/2$. (As usual, b denotes the length of the side opposite angle B.)
Hint: Result (i) or (ii) above may be useful.

(c) Explain why $\triangle ATO$ is congruent to $\triangle CTO$.

(d) Use the results in parts (a) and (c) to show that $\angle COT = \angle B$.

(e) Use the result in part (d) to show that $\mathcal{R} = b/(2 \sin B)$. From this we can conclude (by the law of sines) that

$$\mathcal{R} = \frac{a}{2 \sin A} = \frac{b}{2 \sin B} = \frac{c}{2 \sin C} \quad \text{as required}$$

(f) In the figure that we used for parts (a) through (e), the center of the circle falls within $\triangle ABC$. Draw a figure in which the center lies outside of $\triangle ABC$ and prove that the equation $\mathcal{R} = b/(2 \sin B)$ is true in this case, too.

(g) Two triangles have the same numerical value for the ratio that appears in the law of sines (the length of a side to the sine of its opposite angle). What geometric property do the two triangles have in common?

48. Let $\mathcal{A}$ denote the area of $\triangle ABC$. Show that the radius $\mathcal{R}$ of the circumscribed circle is given by

$$\mathcal{R} = \frac{abc}{4\mathcal{A}}$$

Hint: From Exercise 47 we have $\mathcal{R} = a/(2 \sin A)$. Multiply the right-hand side of this equation by bc/bc.

49. Let r denote the radius of the inscribed circle for $\triangle ABC$ and (as in the previous exercise) let $\mathcal{A}$ denote the area of $\triangle ABC$. Follow parts (a) through (c) to show that

$$r = \frac{2\mathcal{A}}{a + b + c}$$

(a) In the following figure, the inscribed circle (with center I) is tangent to side $\overline{AC}$ at the point D. According to a theorem from geometry, $\overline{ID}$ is perpendicular to $\overline{AC}$. What theorem is this? (State the theorem in complete sentences.) Then explain why the area of $\triangle AIC$ is $\frac{1}{2}rb$.

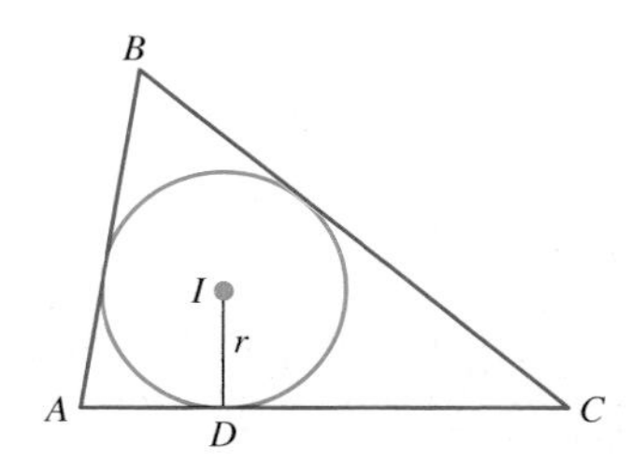

(*continues*)

(b) In the figure accompanying part (a), draw line segments $\overline{IA}$, $\overline{IB}$, and $\overline{IC}$, and then explain why $\mathcal{A} = \frac{1}{2}rb + \frac{1}{2}rc + \frac{1}{2}ra$.

(c) Solve the equation in part (b) for r. You should obtain $r = 2\mathcal{A}/(a + b + c)$, as required.

50. The following figure shows the inscribed and the circumscribed circles for $\triangle ABC$ with $A = 90°$, $C = 30°$, and $a = 2$.

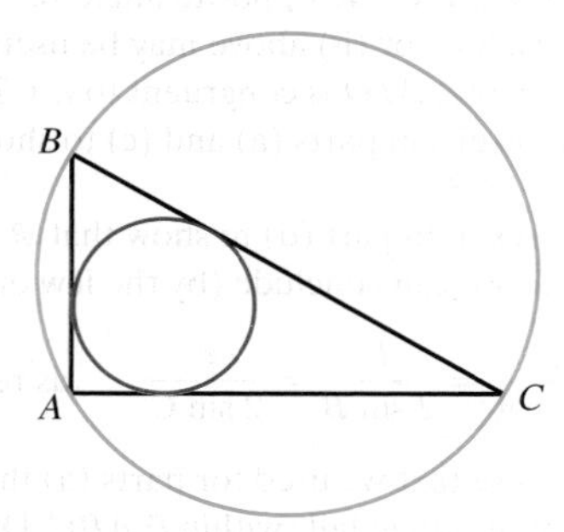

(a) Find b, c, and the area $\mathcal{A}$ of the triangle.

(b) Use the formula for $\mathcal{R}$ in Exercise 48 to find the radius of the circumscribed circle. Then check your answer by recomputing $\mathcal{R}$ using a formula from Exercise 47.

(c) Use the formula for r in Exercise 49 to find the radius of the inscribed circle.

(d) Use the values obtained in this exercise to check that $r\mathcal{R} = abc/[2(a + b + c)]$. (In the next exercise, you'll prove that this result holds in general.)

51. Use the formulas for $\mathcal{R}$ and r from Exercises 48 and 49, respectively, to show that $r\mathcal{R} = abc/[2(a + b + c)]$.

52. In this exercise you will show that the radius r of the inscribed circle for $\triangle ABC$ is given by

$$r = 4\mathcal{R} \sin\frac{A}{2} \sin\frac{B}{2} \sin\frac{C}{2}$$

In the figure below, the inscribed circle (with center I) is tangent to side $\overline{AC}$ at the point D.

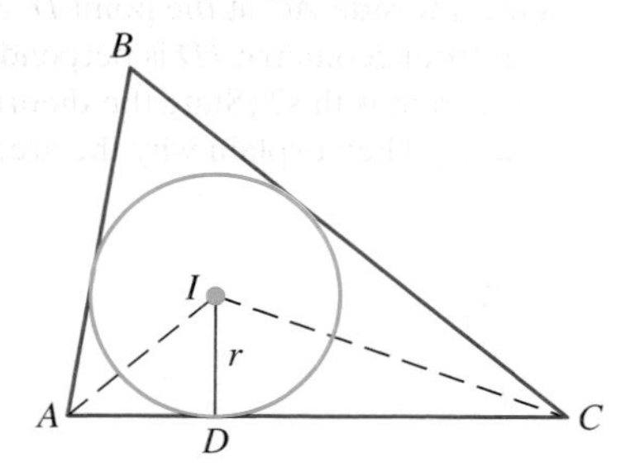

(a) According to a theorem from geometry, $\overline{ID}$ is perpendicular to $\overline{AC}$. What theorem is this? (State the theorem using complete sentences.)

(b) Show that $AD = r\cot(A/2)$ and $DC = r\cot(C/2)$. *Hint:* Result (i) or (ii) on page 663 may be useful.

(c) Use the results in part (b) to show that

$$r = \frac{b}{\cot(A/2) + \cot(C/2)}$$

Hint: From the figure we have $b = AD + DC$. (As usual, the letter b denotes the length AC in $\triangle ABC$.)

(d) Show that

$$r = \frac{b\sin(A/2)\sin(C/2)}{\sin[(A + C)/2]}$$

Hint: Begin with the expression for r in part (c), and convert to sines and cosines. Then simplify and use an addition formula.

(e) Show that

$$r = \frac{b\sin(A/2)\sin(C/2)}{\cos(B/2)}$$

Hint: Begin with the expression for r in part (d), and use the fact that $A + B + C = 180°$.

(f) From Exercise 47 we have $b = 2\mathcal{R}\sin B$. Use this to substitute for b in part (e). Show that the resulting equation can be written

$$r = 4\mathcal{R} \sin\frac{A}{2} \sin\frac{B}{2} \sin\frac{C}{2} \quad \text{as required}$$

53. The following figure shows a quadrilateral with sides a, b, c, and d inscribed in a circle.

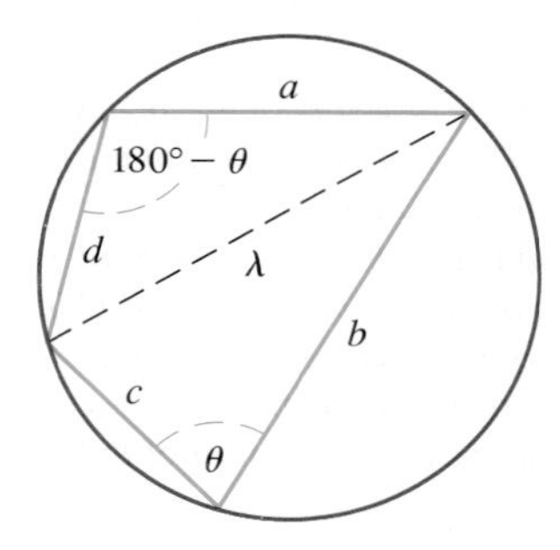

If λ denotes the length of the diagonal indicated in the figure, prove that

$$\lambda^2 = \frac{(ab + cd)(ac + bd)}{bc + ad}$$

This result is known as **Brahmagupta's theorem.** It is named after its discoverer, a seventh-century Hindu mathematician. *Hint:* Assume as given the theorem from geometry stating that when a quadrilateral is inscribed in a circle, the opposite angles are supplementary. Apply the law of cosines in both of the triangles in the figure to obtain expressions for λ^2. Then eliminate $\cos\theta$ from one equation.

54. Prove the following identity for $\triangle ABC$:

$$\frac{\cos A}{a} + \frac{\cos B}{b} + \frac{\cos C}{c} = \frac{a^2 + b^2 + c^2}{2abc}$$

Suggestion: Use the law of cosines to substitute for a^2, for b^2, and for c^2 in the numerator of the expression on the right-hand side.

55. In $\triangle ABC$, suppose that $a^4 + b^4 + c^4 = 2(a^2 + b^2)c^2$. Find $\angle C$. (There are two answers.) *Hint*: Solve the given equation for c^2.

56. In this section we have seen that the cosines of the angles in a triangle can be expressed in terms of the lengths of the sides. For instance, for cos A in $\triangle ABC$, we obtained $\cos A = (b^2 + c^2 - a^2)/2bc$. This exercise shows how to derive corresponding expressions for the sines of the angles. For ease of notation in this exercise, let us agree to use the letter T to denote the following quantity:

$$T = 2(a^2b^2 + b^2c^2 + c^2a^2) - (a^4 + b^4 + c^4)$$

Then the sines of the angles in $\triangle ABC$ are given by

$$\sin A = \frac{\sqrt{T}}{2bc} \qquad \sin B = \frac{\sqrt{T}}{2ac} \qquad \sin C = \frac{\sqrt{T}}{2ab}$$

In the steps that follow, we'll derive the first of these three formulas, the derivations for the other two being entirely similar.

(a) In $\triangle ABC$, why is the positive root always appropriate in the formula $\sin A = \sqrt{1 - \cos^2 A}$?

(b) In the formula in part (a), replace cos A by $(b^2 + c^2 - a^2)/2bc$ and show that the result can be written

$$\sin A = \frac{\sqrt{4b^2c^2 - (b^2 + c^2 - a^2)^2}}{2bc}$$

(c) On the right-hand side of the equation in part (b), carry out the indicated multiplication. After combining like terms, you should obtain $\sin A = \sqrt{T}/2bc$, as required.

57. In the two easy steps that follow, we derive the law of sines by using the formulas obtained in Exercise 56. (Since the formulas in Exercise 56 were obtained using the law of cosines, we are, in essence, showing how to derive the law of sines from the law of cosines.)

(a) Use the formulas in Exercise 56 to check that each of the three fractions $(\sin A)/a$, $(\sin B)/b$, and $(\sin C)/c$ is equal to $\sqrt{T}/2abc$.

(b) Conclude from part (a) that

$$(\sin A)/a = (\sin B)/b = (\sin C)/c$$

58. In this exercise you are going to use the law of cosines and the law of sines to determine the area of the shaded equilateral triangle in Figure A. Begin by labeling points, as shown in Figure B.

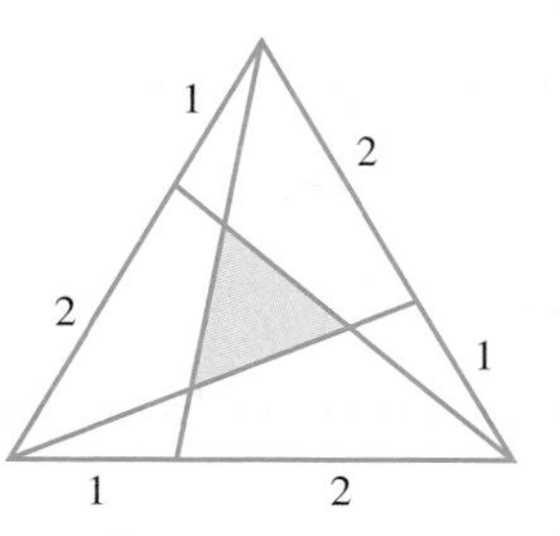

Figure A

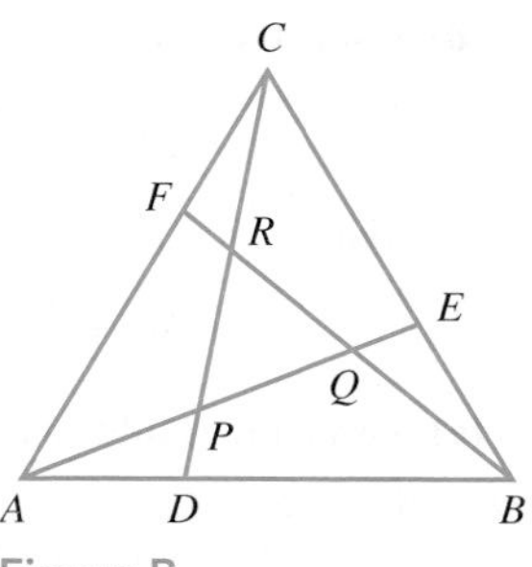

Figure B

(a) Apply the law of cosines in $\triangle ABE$ to show that $AE = \sqrt{7}$.

(b) Apply the law of sines in $\triangle ABE$ to show that $\sin(\angle AEB) = (3\sqrt{3})/(2\sqrt{7})$.

(c) Apply the law of sines in $\triangle CFB$ to show that $\sin(\angle FBC) = (\sqrt{3})/(2\sqrt{7})$. *Hint*: $\triangle AEB$ is congruent to $\triangle BFC$, so $BF = AE$.

(d) Apply the law of sines in $\triangle QEB$ to show that $QE = 1/\sqrt{7}$.

(e) Apply the law of sines in $\triangle QEB$ to show that $QB = 3/\sqrt{7}$.

(f) Show that $PQ = 3\sqrt{7}/7$.
Hint: $PQ = AE - (QE + AP)$, and by symmetry, $AP = QB$.

(g) Use the result in part (f) to find the area of equilateral triangle PQR.

59. This exercise is adapted from a problem proposed by Professor Norman Schaumberger in the May 1990 issue of *The College Mathematics Journal*. In the accompanying figure, radius $OA = 1$ and $\angle AFC = 60°$. Follow steps (a) through (f) to prove that

$$AD \cdot AB - AE \cdot AC = BC$$

(a) Let $\angle BAF = \alpha$; show that $AD = 2\cos\alpha$.
Hint: In isosceles triangle AOD, drop a perpendicular from O to side $\overline{AD}$.

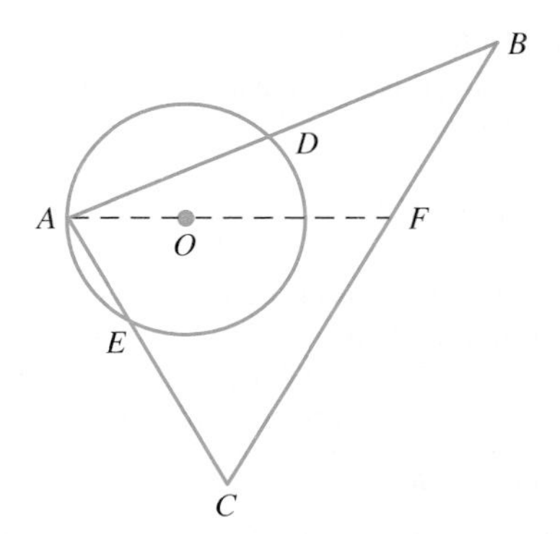

(b) Let $\angle CAF = \beta$; show that $AE = 2\cos\beta$.

(c) Explain why $\angle B = 60° - \alpha$ and $\angle C = 120° - \beta$.

(continues)

(d) Using the law of sines, show that

$$AB = \frac{BC \sin(120° - \beta)}{\sin(\alpha + \beta)}$$

and

$$AC = \frac{BC \sin(60° - \alpha)}{\sin(\alpha + \beta)}$$

(e) Using the results in parts (a), (b), and (d), verify that

$$AD \cdot AB - AE \cdot AC = BC \cdot \left[\frac{2 \cos \alpha \sin(120° - \beta) - 2 \cos \beta \sin(60° - \alpha)}{\sin(\alpha + \beta)}\right]$$

(f) To complete the proof, you need to show that the quantity in brackets in part (e) is equal to 1. In other words, you want to show that

$$2 \cos \alpha \sin(120° - \beta) - 2 \cos \beta \sin(60° - \alpha) = \sin(\alpha + \beta)$$

Use the addition formulas for sine to prove that this last equation is indeed an identity.

C

60. *Heron's formula:* Approximately 2000 years ago, Heron of Alexandria derived a formula for the area of a triangle in terms of the lengths of the sides. A more modern derivation of Heron's formula is indicated in the steps that follow.

(a) Use the expression for $\sin A$ in Exercise 56(b) to show that

$$\sin^2 A = \frac{(a - b + c)(a + b - c)(b + c - a)(b + c + a)}{4b^2c^2}$$

Hint: Use difference-of-squares factoring repeatedly.

(b) Let s denote one-half of the perimeter of $\triangle ABC$. That is, let $s = \frac{1}{2}(a + b + c)$. Using this notation (which is due to Euler), verify that

(i) $a + b + c = 2s$
(ii) $-a + b + c = 2(s - a)$
(iii) $a - b + c = 2(s - b)$
(iv) $a + b - c = 2(s - c)$

Then, using this notation and the result in part (a), show that

$$\sin A = \frac{2\sqrt{s(s - a)(s - b)(s - c)}}{bc}$$

Note: Since $\sin A$ is positive (Why?), the positive root is appropriate here.

(c) Use the result in part (b) and the formula area $\triangle ABC = \frac{1}{2}bc \sin A$ to conclude that

$$\text{area } \triangle ABC = \sqrt{s(s - a)(s - b)(s - c)}$$

This is Heron's formula. For historical background and a purely geometric proof, see *An Introduction to the History of Mathematics,* 6th ed., by Howard Eves (Philadelphia: Saunders College Publishing, 1990), pp. 178 and 194.

61. In this exercise you will derive a formula for the length of an angle bisector in a triangle. (The formula will be needed in Exercise 62.) Let f denote the length of the bisector of angle C in $\triangle ABC$, as shown in the following figure.

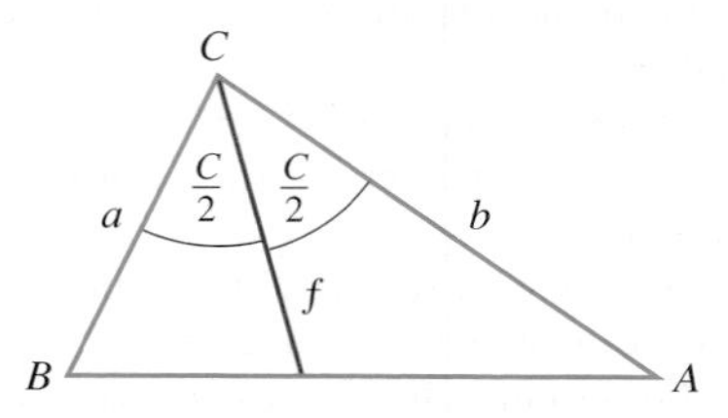

(a) Explain why

$$\frac{1}{2}af \sin \frac{C}{2} + \frac{1}{2}bf \sin \frac{C}{2} = \frac{1}{2}ab \sin C$$

Hint: Use areas.

(b) Show that

$$f = \frac{2ab \cos(C/2)}{a + b}$$

(c) By the law of cosines, $\cos C = (a^2 + b^2 - c^2)/2ab$. Use this to show that

$$\cos \frac{C}{2} = \frac{1}{2}\sqrt{\frac{(a + b - c)(a + b + c)}{ab}}$$

(d) Show that the length of the angle bisector in terms of the sides is given by

$$f = \frac{\sqrt{ab}}{a + b}\sqrt{(a + b - c)(a + b + c)}$$

62. In triangle XYZ (in the accompanying figure), $\overline{SX}$ bisects angle ZXY and $\overline{TY}$ bisects angle ZYX. In this exercise you are going to prove the following theorem, known as the **Steiner–Lehmus theorem:** If the lengths of the angle bisectors $\overline{SX}$ and $\overline{TY}$ are equal, then $\triangle XYZ$ is isosceles (with $XZ = YZ$).

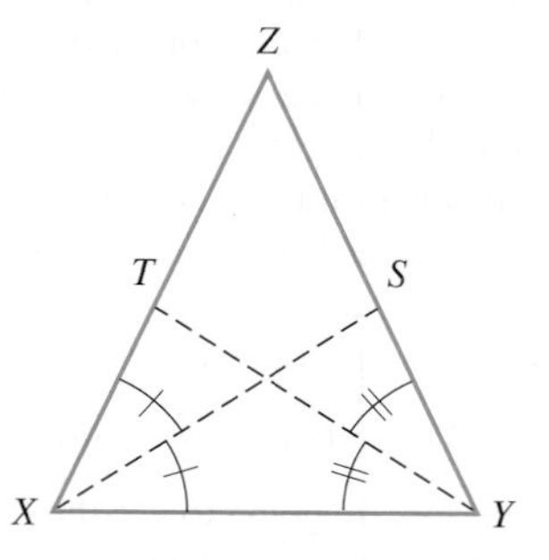

(a) Let x, y, and z denote the lengths of the sides $\overline{YZ}$, $\overline{XZ}$, and $\overline{XY}$, respectively. Use the formula in Exercise 61(d) to show that the equation $TY = SX$ is equivalent to

$$\frac{\sqrt{xz}}{x+z}\sqrt{(x+z-y)(x+z+y)}$$
$$= \frac{\sqrt{yz}}{y+z}\sqrt{(y+z-x)(y+z+x)} \quad (1)$$

(b) What common factors do you see on both sides of equation (1)? Divide both sides of equation (1) by those common factors. You should obtain

$$\frac{\sqrt{x}}{x+z}\sqrt{x+z-y} = \frac{\sqrt{y}}{y+z}\sqrt{y+z-x} \quad (2)$$

(c) Clear equation (2) of fractions, and then square both sides. After combining like terms and then grouping, the equation can be written

$$(3x^2yz - 3xy^2z) + (x^3y - xy^3)$$
$$+ (x^2z^2 - y^2z^2) + (xz^3 - yz^3) = 0 \quad (3)$$

(d) Show that equation (3) can be written

$$(x-y)[3xyz + xy(x+y) + x^2(x+y) + z^3] = 0$$

Now notice that the quantity in brackets in this last equation must be positive. (Why?) Consequently, $x - y = 0$, and so $x = y$, as required.

Remark: This theorem has a fascinating history, beginning in 1840 when C. L. Lehmus first proposed the theorem to the great Swiss geometer Jacob Steiner (1796–1863). For background (and much shorter proofs!), see either of the following references: *Scientific American,* vol. 204 (1961), pp. 166–168; *American Mathematical Monthly,* vol. 70 (1963), pp. 79–80.

63. Show that

$$\text{area}\,\triangle ABC = \frac{a^2 \sin 2B + b^2 \sin 2A}{4}$$

64. For any triangle ABC, show that

$$\frac{\sin(A-B)}{\sin(A+B)} = \frac{a^2 - b^2}{c^2}$$

9.3 VECTORS IN THE PLANE: A GEOMETRIC APPROACH

The idea of a parallelogram of velocities may be found in various ancient Greek authors, and the concept of a parallelogram of forces was not uncommon in the sixteenth and seventeenth centuries.—Michael J. Crowe in *A History of Vector Analysis* (Notre Dame, Ind.: University of Notre Dame Press, 1967)

Vector notation is compact. If we can express a law of physics in vector form we usually find it easier to understand and to manipulate mathematically.—David Halliday and Robert Resnick in *Fundamentals of Physics,* 3rd ed. (New York: John Wiley and Sons, Inc., 1988)

Certain quantities, such as temperature, length, and mass, can be specified by means of a single number (assuming that a system of units has been agreed on). We call these quantities **scalars.** On the other hand, quantities such as force and velocity are characterized by both a *magnitude* (a positive number) and a *direction.* We call these quantities **vectors.**

Geometrically, a vector is a directed line segment or arrow. The vector in Figure 1, for instance, represents a wind velocity of 5 mph from the west. The length of this vector represents the magnitude of the wind velocity, while the direction of the vector indicates the direction of the wind velocity. As another example, the vector in Figure 2 represents a force acting on an object: The magnitude of the force is 3 pounds, and the force acts at an angle of 135° with the horizontal.

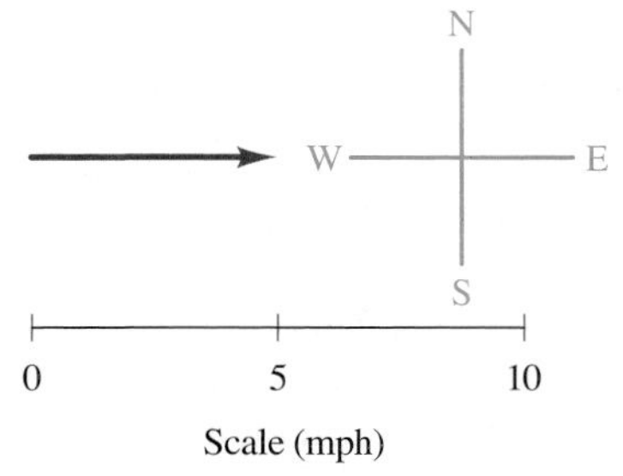

Figure 1
A vector representing a wind velocity of 5 mph from the west

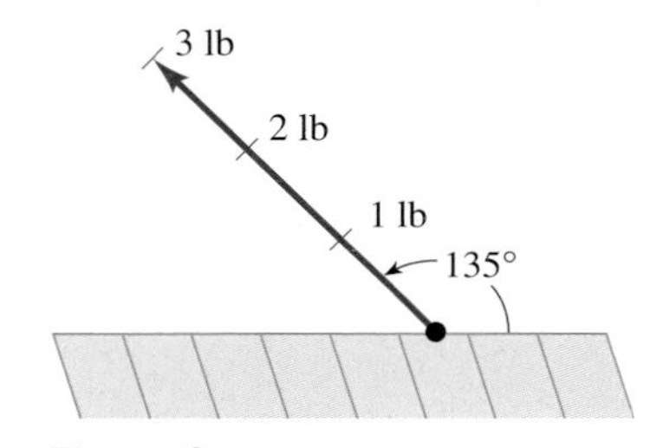

Figure 2
A vector representing a force of 3 lb acting at an angle of 135° with the horizontal

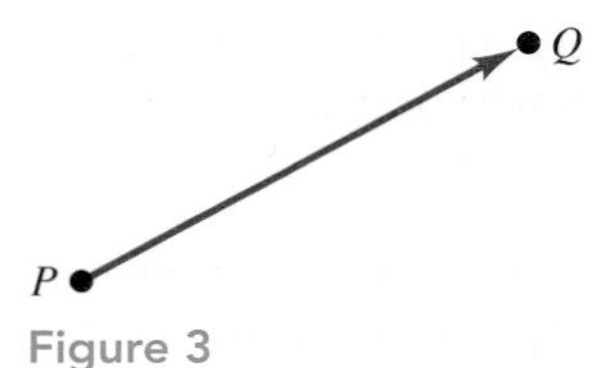

Figure 3

In a moment we are going to discuss the important concept of vector addition, but first let us agree on some matters of notation. Suppose that we have a vector drawn from a point P to a point Q, as shown in Figure 3. The point P in Figure 3 is called the **initial point** of the vector, and Q is the **terminal point.** We can denote this vector by the notation

$$\overrightarrow{PQ}$$

We sometimes think of the vector $\overrightarrow{PQ}$ as the *directed* line segment from P to Q.

The length or **magnitude** of the vector $\overrightarrow{PQ}$ is denoted by $|\overrightarrow{PQ}|$. On the printed page, vectors are often indicated by boldface letters, such as **a**, **A**, and **v**.

A word about notation. As has perhaps already occurred to you, the same vertical bars that we are using to denote the length of a vector are also used to denote the absolute value of a real number. It will be clear from the context which meaning is intended. Some books avoid this situation by using double bars to indicate the length of a vector: $\|\mathbf{v}\|$. In this text we use the notation $|\mathbf{v}|$ simply because that is the one found in most calculus books.

If two vectors **a** and **b** have the same length and the same direction, we say that they are *equal*, and we write $\mathbf{a} = \mathbf{b}$; see Figure 4. Notice that our definition for vector equality involves magnitude and direction but not location. Thus when it is convenient to do so, we are free to move a given vector to another location, provided that we do not alter the magnitude or the direction.

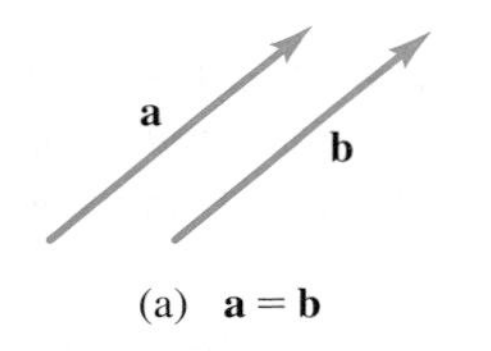

(a) $\mathbf{a} = \mathbf{b}$

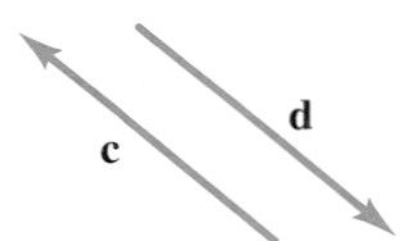

(b) $\mathbf{c} \neq \mathbf{d}$
The magnitudes are the same but the directions are not.

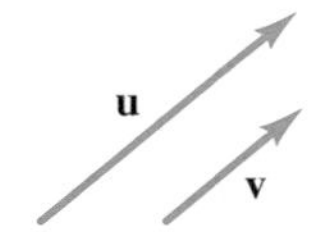

(c) $\mathbf{u} \neq \mathbf{v}$
The directions are the same but the magnitudes are not.

(d) $\mathbf{p} \neq \mathbf{q}$
Neither the magnitudes nor the directions are the same.

Figure 4

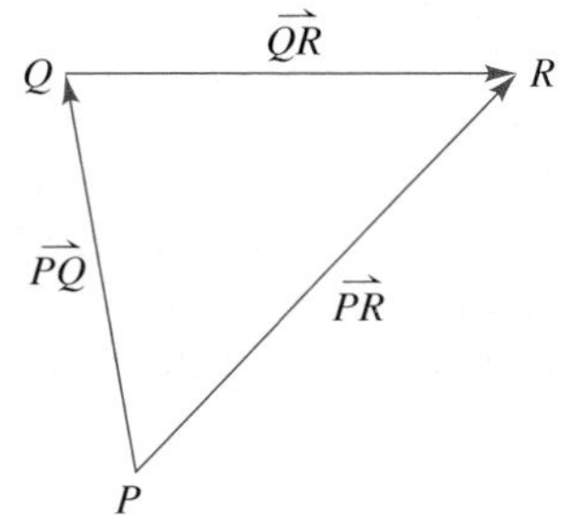

Figure 5

As motivation for the definition of vector addition, let's suppose that an object moves from a point P to a point Q. Then we can represent this *displacement* by the vector $\overrightarrow{PQ}$. (Indeed, the word "vector" is derived from the Latin *vectus,* meaning "carried.") Now suppose that after moving from P to Q, the object moves from Q to R. We represent this displacement by the vector $\overrightarrow{QR}$. Then, as you can see in Figure 5, the net effect is a displacement from P to R. We say in this case that the vector $\overrightarrow{PR}$ is the **sum** or **resultant** of the vectors $\overrightarrow{PQ}$ and $\overrightarrow{QR}$, and we write

$$\overrightarrow{PQ} + \overrightarrow{QR} = \overrightarrow{PR}$$

These ideas are formalized in the definition that follows.

DEFINITION **Vector Addition**

Let **u** and **v** be two vectors. Position **v** (without changing its magnitude or direction) so that its initial point coincides with the terminal point of **u**, as in Figure 6(a). Then, as is indicated in Figure 6(b), the vector $\mathbf{u} + \mathbf{v}$ is the directed line segment from the initial point of **u** to the terminal point of **v**. The vector $\mathbf{u} + \mathbf{v}$ is called the **sum** or **resultant** of **u** and **v**.

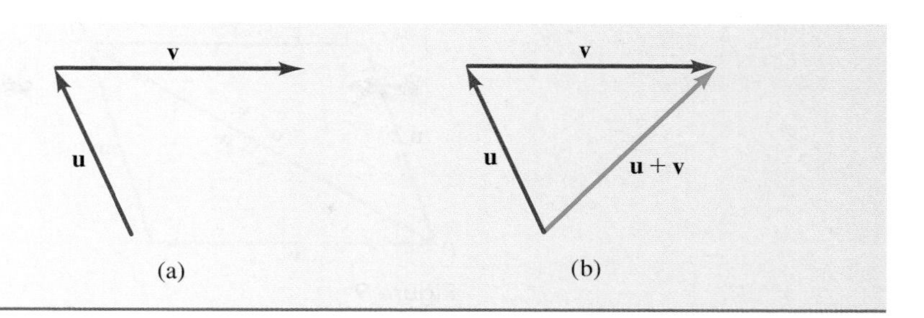

Figure 6

EXAMPLE 1 Adding two vectors geometrically

Refer to Figure 7.

(a) Determine the initial and terminal points of $\mathbf{u} + \mathbf{v}$. **(b)** Compute $|\mathbf{u} + \mathbf{v}|$.

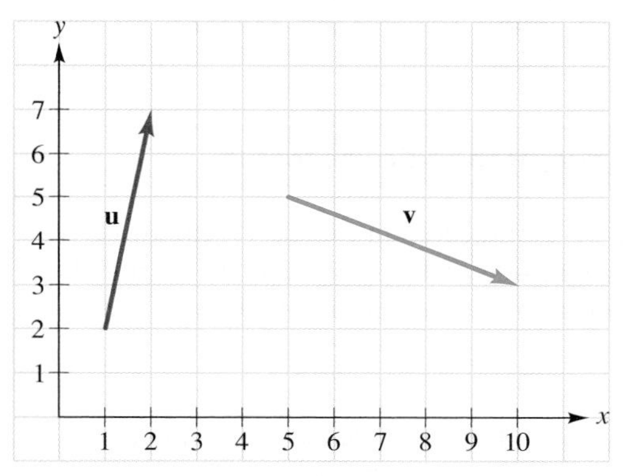

Figure 7

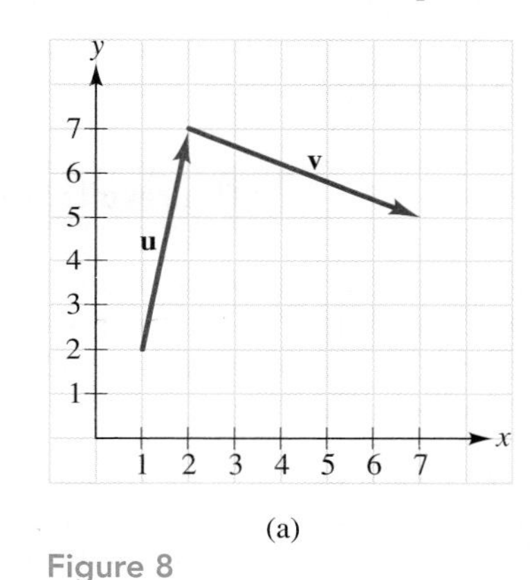

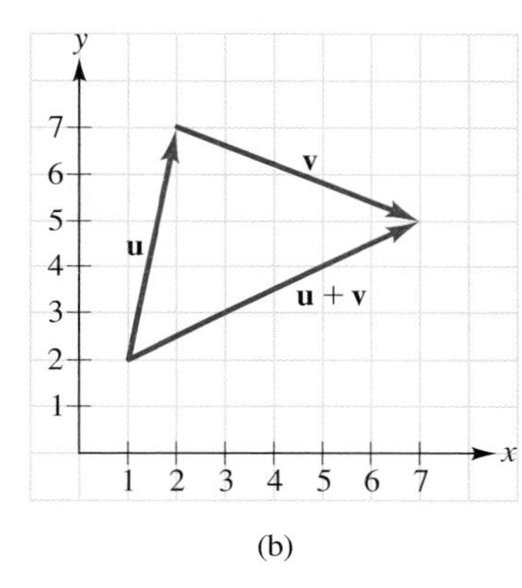

Figure 8

SOLUTION

(a) According to the definition, we first need to move $\mathbf{v}$ (without changing its length or direction) so that its initial point coincides with the terminal point of $\mathbf{u}$. From Figure 7 we see that this can be accomplished by moving each point of $\mathbf{v}$ three units in the negative x-direction and two units in the positive y-direction. Figure 8(a) shows the new location of $\mathbf{v}$, and Figure 8(b) indicates the sum $\mathbf{u} + \mathbf{v}$. From Figure 8(b) we see that the initial and terminal points of $\mathbf{u} + \mathbf{v}$ are $(1, 2)$ and $(7, 5)$, respectively.

(b) We can use the distance formula to determine $|\mathbf{u} + \mathbf{v}|$. Using the points $(1, 2)$ and $(7, 5)$ that were obtained in part (a), we have

$$|\mathbf{u} + \mathbf{v}| = \sqrt{(7-1)^2 + (5-2)^2} = \sqrt{45} = \sqrt{9 \cdot 5} = 3\sqrt{5}$$

One important consequence of our definition for vector addition is that this operation is *commutative.* That is, for any two vectors $\mathbf{u}$ and $\mathbf{v}$ we have

$$\mathbf{u} + \mathbf{v} = \mathbf{v} + \mathbf{u}$$

Figure 9 indicates why this is so. Vector addition can also be carried out by using the **parallelogram law.** In Figure 9, to determine $\mathbf{u} + \mathbf{v}$, position $\mathbf{u}$ and $\mathbf{v}$ so that their initial points coincide. Then, as is indicated in Figure 10, the vector $\mathbf{u} + \mathbf{v}$ is the directed diagonal of the parallelogram determined by $\mathbf{u}$ and $\mathbf{v}$.

It is a fact, verified experimentally, that if two forces $\mathbf{F}$ and $\mathbf{G}$ act on an object, the net effect is the same as if just the resultant $\mathbf{F} + \mathbf{G}$ acted on the object.

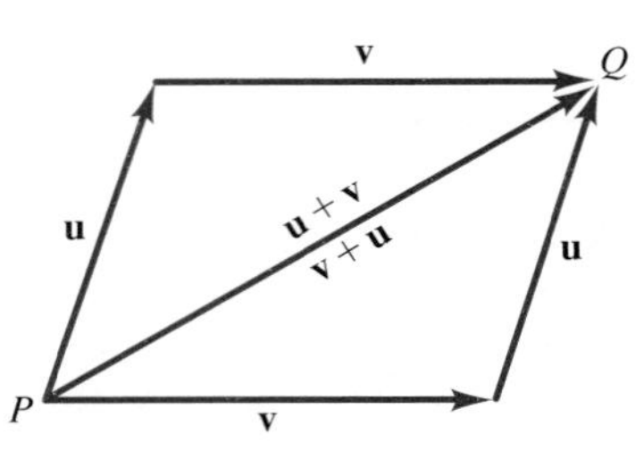

Figure 9
The upper triangle shows the sum **u** + **v**, while the lower triangle shows the sum **v** + **u**. Since in both cases the resultant is $\overrightarrow{PQ}$, it follows that **u** + **v** = **v** + **u**.

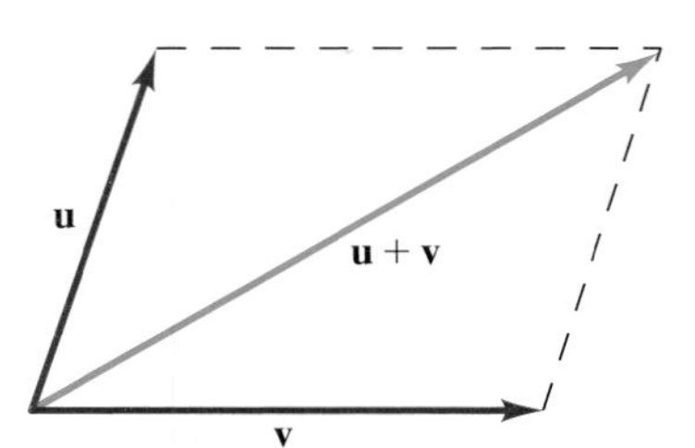

Figure 10
The parallelogram law for vector addition

In Example 2 we use the parallelogram law to compute the resultant of two forces. Note that the units of force used in this example are **newtons** (N), where 1 N ≈ 0.2248 lb.

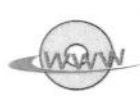

EXAMPLE 2 The resultant of two perpendicular forces

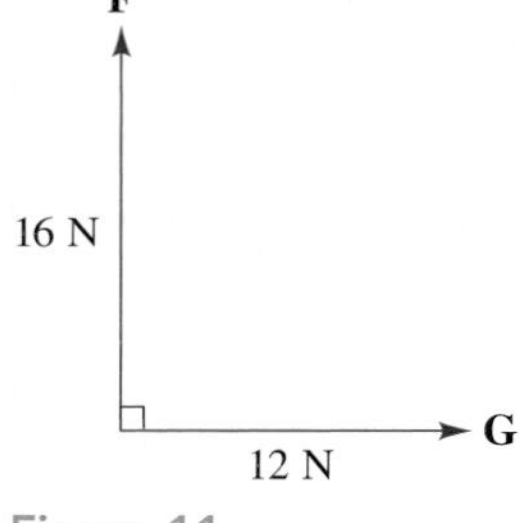

Figure 11

Two forces **F** and **G** act on an object. As is indicated in Figure 11, the force **G** acts horizontally to the right with a magnitude of 12 N, while **F** acts vertically upward with a magnitude of 16 N. Determine the magnitude and the direction of the resultant force.

SOLUTION
We complete the parallelogram, as shown in Figure 12. Now we need to calculate the length of **F** + **G** and the angle θ. Applying the Pythagorean theorem in Figure 12, we have

$$|\mathbf{F} + \mathbf{G}| = \sqrt{12^2 + 16^{16}} = \sqrt{144 + 256} = \sqrt{400} = 20$$

Also from Figure 12 we have

$$\tan\theta = \frac{16}{12} = \frac{4}{3}$$

Consequently,

$$\theta = \tan^{-1}\frac{4}{3} \approx 53.1° \qquad \text{using a calculator set in degree mode}$$

Summarizing our results, the magnitude of **F** + **G** is 20 N, and the angle θ between **F** + **G** and the horizontal is (approximately) 53.1°.

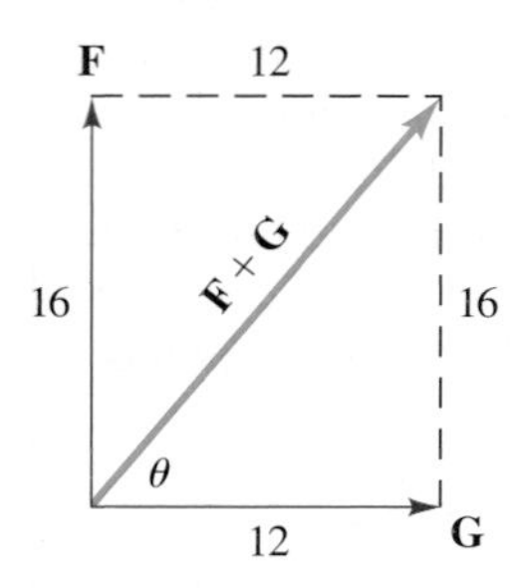

Figure 12

In Example 2 we determined the resultant for two perpendicular forces. The next example shows how to compute the resultant when the forces are not perpendicular. Our calculations will make use of both the law of sines and the law of cosines.

EXAMPLE 3 The resultant of two forces

Determine the resultant of the two forces in Figure 13. (Round the answers to one decimal place.)

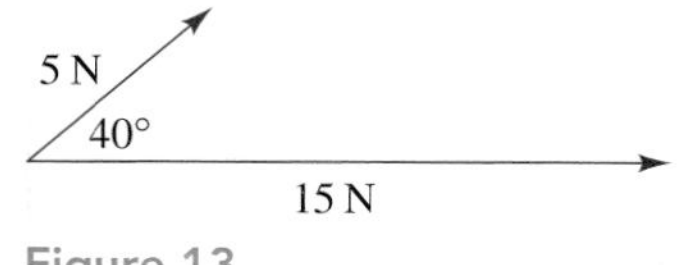

Figure 13

SOLUTION

As in the previous example, we complete the parallelogram. In Figure 14 the angle in the lower right-hand corner of the parallelogram is 140°. This is because the sum of two adjacent angles in any parallelogram is always 180°. Letting d denote the length of the diagonal in Figure 14, we can use the law of cosines to write

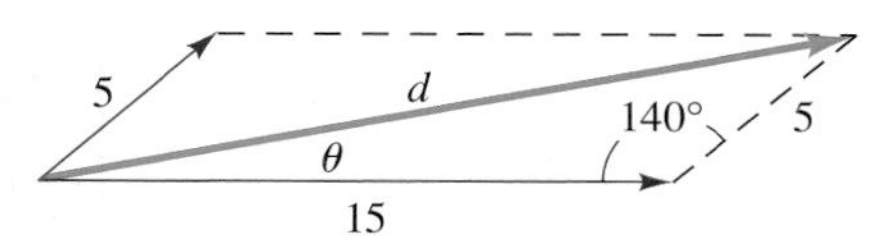

Figure 14

$$\begin{aligned} d^2 &= 15^2 + 5^2 - 2(15)(5)\cos 140^\circ \\ d &= \sqrt{250 - 150\cos 140^\circ} = \sqrt{250 + 150\cos 40^\circ} \quad \text{(Why?)} \\ &= \sqrt{25(10 + 6\cos 40^\circ)} = 5\sqrt{10 + 6\cos 40^\circ} \\ &\approx 19.1 \qquad \text{using a calculator} \end{aligned}$$

So the magnitude of the resultant is 19.1 N (to one decimal place). To specify the direction of the resultant, we need to determine the angle θ in Figure 14. Using the law of sines, we have

$$\frac{\sin\theta}{5} = \frac{\sin 140^\circ}{d}$$

and, consequently,

$$\begin{aligned} \sin\theta &= \frac{5\sin 40^\circ}{d} \quad \text{(Why?)} \\ &= \frac{5\sin 40^\circ}{5\sqrt{10 + 6\cos 40^\circ}} = \frac{\sin 40^\circ}{\sqrt{10 + 6\cos 40^\circ}} \end{aligned}$$

Using a calculator now, we obtain

$$\theta = \sin^{-1}\left(\frac{\sin 40^\circ}{\sqrt{10 + 6\cos 40^\circ}}\right) \approx 9.7^\circ$$

In summary, the magnitude of the resultant force is about 19.1 N, and the angle θ between the resultant and the 15 N force is approximately 9.7°.

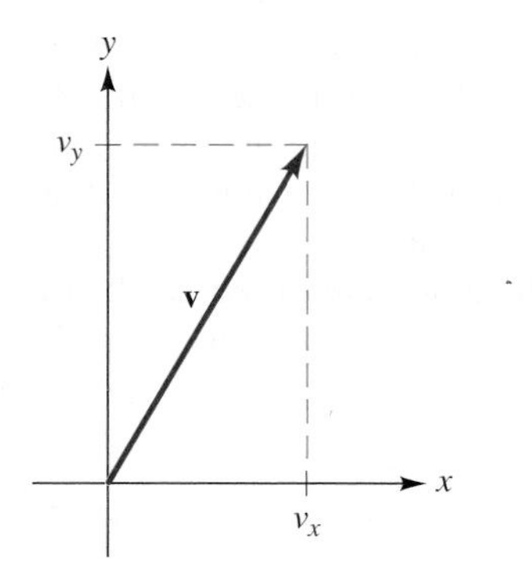

Figure 15

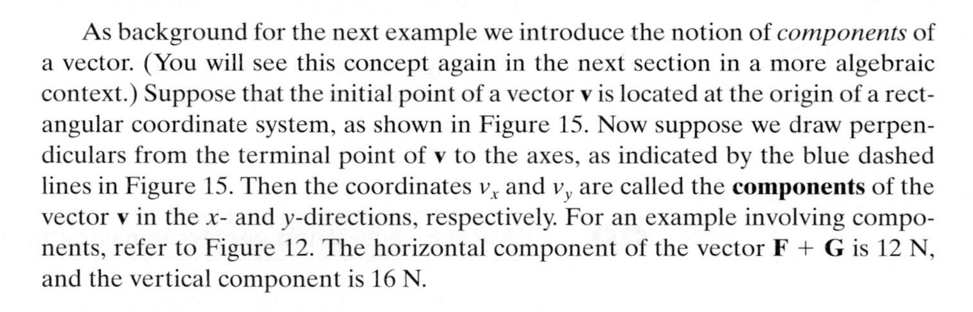

As background for the next example we introduce the notion of *components* of a vector. (You will see this concept again in the next section in a more algebraic context.) Suppose that the initial point of a vector **v** is located at the origin of a rectangular coordinate system, as shown in Figure 15. Now suppose we draw perpendiculars from the terminal point of **v** to the axes, as indicated by the blue dashed lines in Figure 15. Then the coordinates v_x and v_y are called the **components** of the vector **v** in the x- and y-directions, respectively. For an example involving components, refer to Figure 12. The horizontal component of the vector **F** + **G** is 12 N, and the vertical component is 16 N.

EXAMPLE 4 Finding vertical and horizontal components

Determine the horizontal and vertical components of the velocity vector **v** in Figure 16.

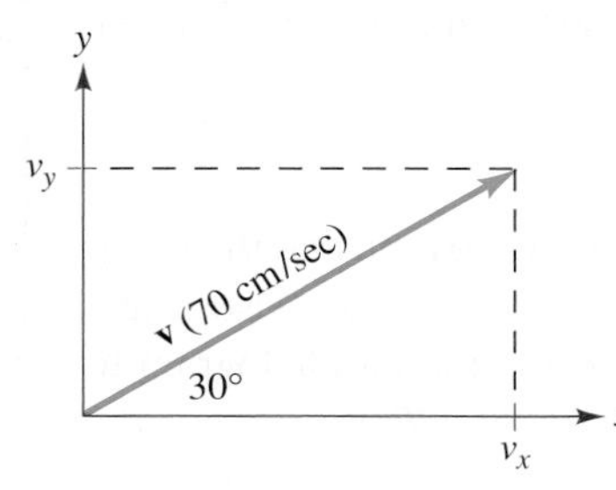

Figure 16

SOLUT ION

From Figure 16 we can write

$$\cos 30^\circ = \frac{\text{adjacent}}{\text{hypotenuse}} = \frac{v_x}{70}$$

and, consequently,

$$v_x = (\cos 30°)(70) = \frac{\sqrt{3}}{2}(70)$$
$$= 35\sqrt{3} \approx 61 \text{ cm/sec} \qquad \text{to two significant digits}$$

Similarly, we have

$$\sin 30° = \frac{v_y}{70}$$

and therefore

$$v_y = (\sin 30°)(70) = 35 \text{ cm/sec}$$

So the x-component of the velocity is about 61 cm/sec, and the y-component is 35 cm/sec.

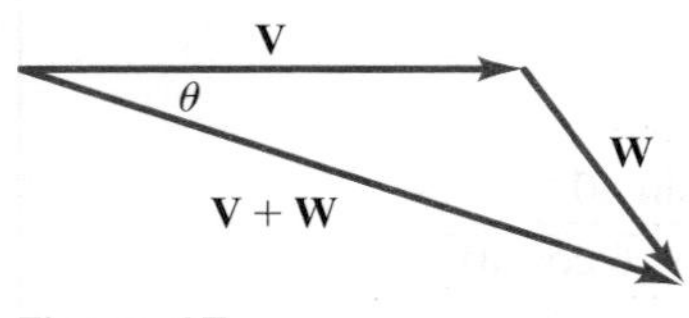

Figure 17

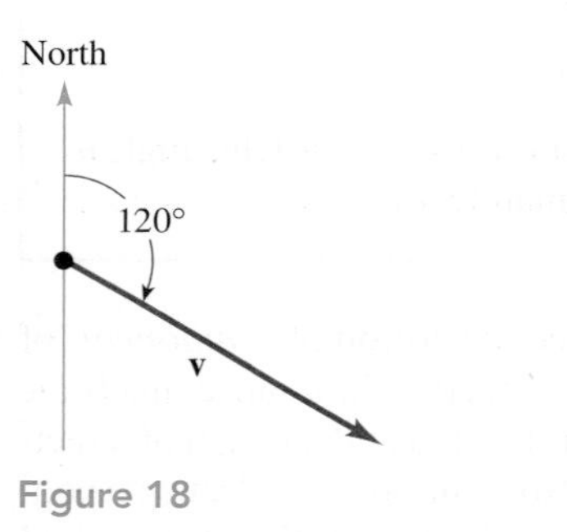

Figure 18

Our last example in this section will indicate how vectors are used in navigation. First, however, let's introduce some terminology. Suppose that an airplane has a **heading** of *due east.* This means that the airplane is pointed directly east, and if there were no wind effects, the plane would indeed travel due east with respect to the ground. The **air speed** is the speed of the airplane relative to the air, whereas the **ground speed** is the plane's speed relative to the ground. Again, if there were no wind effects, then the air speed and the ground speed would be equal. Now suppose that the heading and air speed of an airplane are represented by the velocity vector **V** in Figure 17. (The direction of **V** is the heading; the magnitude of **V** is the air speed.) Also suppose that the wind velocity is represented by the vector **W** in Figure 17. Then the vector sum **V** + **W** represents the actual velocity of the plane with respect to the ground. The direction of **V** + **W** is called the **course;** it is the direction in which the airplane is moving with respect to the ground. The magnitude of **V** + **W** is the ground speed (which was defined previously). The angle θ in Figure 17, from the heading to the course, is called the **drift angle.**

In navigation, directions are given in terms of the angle measured clockwise from true north. For example, the direction of the velocity vector **v** in Figure 18 is 120°.

EXAMPLE 5 Using vectors in navigation

[Refer to Figure 19(a).] The heading and air speed of an airplane are 60° and 250 mph, respectively. If the wind is 40 mph from 150°, find the ground speed, the drift angle, and the course.

SOLUTION

In Figure 19(a) the vector $\overrightarrow{OA}$ represents the air speed of 250 mph and the heading of 60°. The vector $\overrightarrow{OW}$ represents a wind of 40 mph from 150°. Also, angle $WOA = 90°$. (Why?) In Figure 19(b) we have completed the parallelogram to obtain the vector sum $\overrightarrow{OA} + \overrightarrow{OW} = \overrightarrow{OB}$. The length of $\overrightarrow{OB}$ represents the

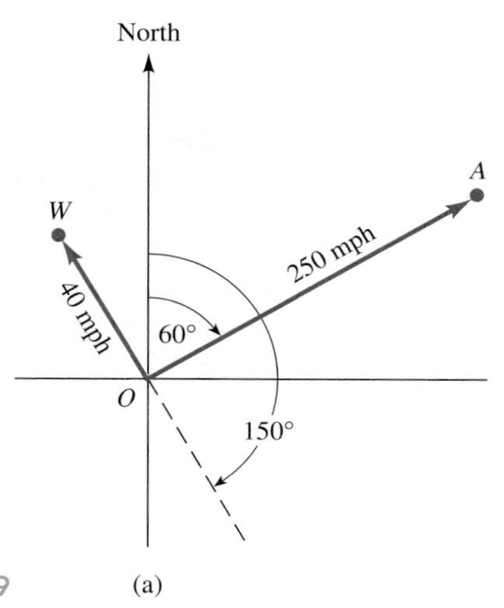

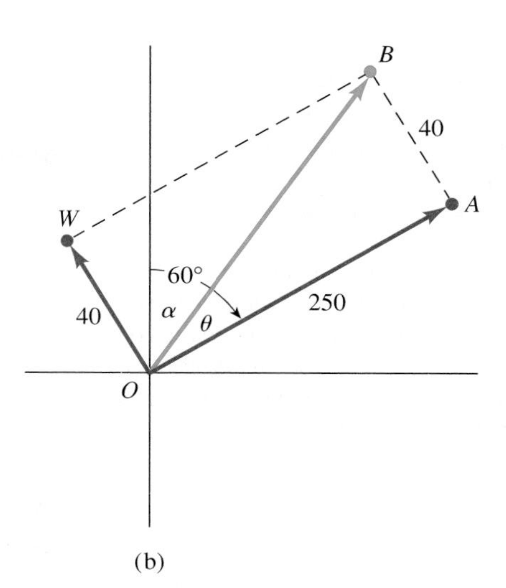

Figure 19 (a) (b)

ground speed, and θ is the drift angle. Because triangle BOA is a right triangle, we have

$$\tan\theta = \frac{40}{250} = \frac{4}{25} \qquad |\overrightarrow{OB}| = \sqrt{250^2 + 40^2}$$
$$\theta = 9.1^\circ \qquad \approx 253.2$$

From these calculations we conclude that the ground speed is approximately 253.2 mph and the drift angle is about 9.1°. We still need to compute the course (that is, the direction of vector $\overrightarrow{OB}$). From Figure 19(b) we have

$$\alpha = 60^\circ - \theta \approx 60^\circ - 9.1^\circ = 50.9^\circ$$

Thus the course is 50.9°, to one decimal place. (We are using the convention, mentioned previously, that directions are given in terms of the angle measured clockwise from true north.)

EXERCISE SET 9.3

A

In Exercises 1–26, assume that the coordinates of the points P, Q, R, S, and O are as follows:

$P(-1, 3)$ $\quad Q(4, 6)$ $\quad R(4, 3)$ $\quad S(5, 9)$ $\quad O(0, 0)$

For each exercise, draw the indicated vector (using graph paper) and compute its magnitude. In Exercises 7–20, compute the sums using the definition given on page 668. In Exercises 21–26, use the parallelogram law to compute the sums.

1. $\overrightarrow{PQ}$
2. $\overrightarrow{QP}$
3. $\overrightarrow{SQ}$
4. $\overrightarrow{QS}$
5. $\overrightarrow{OP}$
6. $\overrightarrow{PO}$
7. $\overrightarrow{PQ} + \overrightarrow{QS}$
8. $\overrightarrow{SQ} + \overrightarrow{QP}$
9. $\overrightarrow{OP} + \overrightarrow{PQ}$
10. $\overrightarrow{OS} + \overrightarrow{SQ}$
11. $(\overrightarrow{OS} + \overrightarrow{SQ}) + \overrightarrow{QP}$
12. $(\overrightarrow{OS} + \overrightarrow{SP}) + \overrightarrow{PR}$
13. $\overrightarrow{OP} + \overrightarrow{QS}$
14. $\overrightarrow{QS} + \overrightarrow{PO}$
15. $\overrightarrow{SR} + \overrightarrow{PO}$
16. $\overrightarrow{OS} + \overrightarrow{QO}$
17. $\overrightarrow{OP} + \overrightarrow{RQ}$
18. $\overrightarrow{OP} + \overrightarrow{QR}$
19. $\overrightarrow{SQ} + \overrightarrow{RO}$
20. $\overrightarrow{SQ} + \overrightarrow{OR}$
21. $\overrightarrow{OP} + \overrightarrow{OR}$
22. $\overrightarrow{OP} + \overrightarrow{OQ}$
23. $\overrightarrow{RP} + \overrightarrow{RS}$
24. $\overrightarrow{QP} + \overrightarrow{QR}$
25. $\overrightarrow{SO} + \overrightarrow{SQ}$
26. $\overrightarrow{SQ} + \overrightarrow{SR}$

In Exercises 27–32 the vectors **F** *and* **G** *denote two forces that act on an object:* **G** *acts horizontally to the right, and* **F** *acts vertically upward. In each case, use the information that is given to compute* $|\mathbf{F} + \mathbf{G}|$ *and* θ*, where* θ *is the angle between* **G** *and the resultant.*

27. $|\mathbf{F}| = 4$ N, $|\mathbf{G}| = 5$ N
28. $|\mathbf{F}| = 15$ N, $|\mathbf{G}| = 6$ N
29. $|\mathbf{F}| = |\mathbf{G}| = 9$ N
30. $|\mathbf{F}| = 28$ N, $|\mathbf{G}| = 1$ N
31. $|\mathbf{F}| = 3.22$ N, $|\mathbf{G}| = 7.21$ N
32. $|\mathbf{F}| = 4.06$ N, $|\mathbf{G}| = 26.83$ N

In Exercises 33–38 the vectors **F** *and* **G** *represent two forces acting on an object, as indicated in the following figure. In each case, use the given information to compute (to two decimal places) the magnitude and direction of the resultant. (Give the direction of the resultant by specifying the angle θ between* **F** *and the resultant.)*

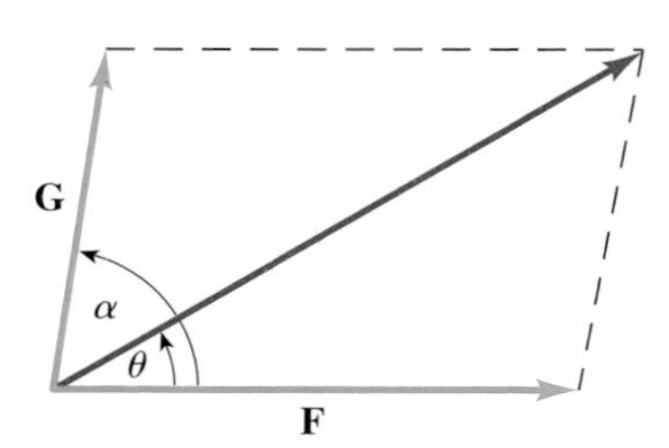

33. $|\mathbf{F}| = 5$ N, $|\mathbf{G}| = 4$ N, $\alpha = 80°$
34. $|\mathbf{F}| = 8$ N, $|\mathbf{G}| = 10$ N, $\alpha = 60°$
35. $|\mathbf{F}| = 16$ N, $|\mathbf{G}| = 25$ N, $\alpha = 35°$
36. $|\mathbf{F}| = 4.24$ N, $|\mathbf{G}| = 9.01$ N, $\alpha = 45°$
37. $|\mathbf{F}| = 50$ N, $|\mathbf{G}| = 25$ N, $\alpha = 130°$
38. $|\mathbf{F}| = 1.26$ N, $|\mathbf{G}| = 2.31$ N, $\alpha = 160°$

In Exercises 39–46 the initial point for each vector is the origin, and θ denotes the angle (measured counterclockwise) from the x-axis to the vector. In each case, compute the horizontal and vertical components of the given vector. (Round your answers to two decimal places.)

39. The magnitude of **V** is 16 cm/sec, and $\theta = 30°$.
40. The magnitude of **V** is 40 cm/sec, and $\theta = 60°$.
41. The magnitude of **F** is 14 N, and $\theta = 75°$.
42. The magnitude of **F** is 23.12 N, and $\theta = 52°$.
43. The magnitude of **V** is 1 cm/sec, and $\theta = 135°$.
44. The magnitude of **V** is 12 cm/sec, and $\theta = 120°$.
45. The magnitude of **F** is 1.25 N, and $\theta = 145°$.
46. The magnitude of **F** is 6.34 N, and $\theta = 175°$.

In Exercises 47–50, use the given flight data to compute the ground speed, the drift angle, and the course. (Round your answers to two decimal places.)

47. The heading and air speed are 30° and 300 mph, respectively; the wind is 25 mph from 120°.
48. The heading and air speed are 45° and 275 mph, respectively; the wind is 50 mph from 135°.
49. The heading and air speed are 100° and 290 mph, respectively; the wind is 45 mph from 190°.
50. The heading and air speed are 90° and 220 mph, respectively; the wind is 80 mph from **(a)** 180°; **(b)** 90°.

B

51. A block weighing 12 lb rests on an inclined plane, as indicated in Figure A. Determine the components of the weight perpendicular to and parallel to the plane. Round your answers to two decimal places.
Hint: In Figure B the component of the weight perpendicular to the plane is $|\overrightarrow{OR}|$; the component parallel to the plane is $|\overrightarrow{OP}|$. Why does angle QOR equal 35°? (Observing that angle POR is a right angle, that line segment PO is parallel to AC, and that OQ is perpendicular to AB may be helpful.)

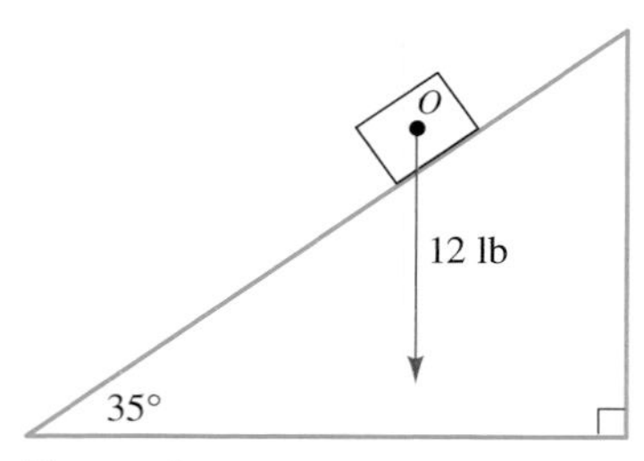

Figure A

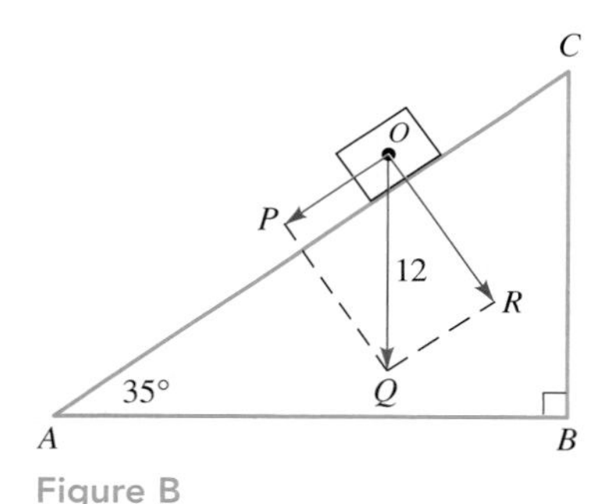

Figure B

In Exercises 52 and 53 you are given the weight of a block on an inclined plane, along with the angle θ that the inclined plane makes with the horizontal. In each case, determine the components of the weight perpendicular to and parallel to the plane. (Round your answers to two decimal places where necessary.)

52. 15 lb; $\theta = 30°$
53. 12 lb; $\theta = 10°$
54. A block rests on an inclined plane that makes an angle of 20° with the horizontal. The component of the weight parallel to the plane is 34.2 lb.
(a) Determine the weight of the block. (Round your answer to one decimal place.)
(b) Determine the component of the weight perpendicular to the plane. (Round your answer to one decimal place.)
55. In Section 9.4 we will see that vector addition is associative. That is, for any three vectors **A**, **B**, and **C**, we have $(\mathbf{A} + \mathbf{B}) + \mathbf{C} = \mathbf{A} + (\mathbf{B} + \mathbf{C})$. In this exercise you are going to check that this property holds in a particular case.

Let **A**, **B**, and **C** be the vectors with initial and terminal points as follows:

Vector	Initial point	Terminal point
A	$(-1, 2)$	$(2, 4)$
B	$(1, 2)$	$(3, 0)$
C	$(6, 2)$	$(4, -3)$

(a) Use the definition of vector addition on page 668 to determine the initial and terminal points of $(\mathbf{A} + \mathbf{B}) + \mathbf{C}$. *Suggestion:* Use graph paper.

(b) Use the definition of vector addition to determine the initial and terminal points of $\mathbf{A} + (\mathbf{B} + \mathbf{C})$. [Your answers should agree with those in part (a).]

PROJECT

VECTOR ALGEBRA USING VECTOR GEOMETRY

In the previous section we defined vectors and vector addition geometrically and showed that vector addition is commutative, that is, if $\mathbf{u}$ and $\mathbf{v}$ are vectors then $\mathbf{u} + \mathbf{v} = \mathbf{v} + \mathbf{u}$. What happens if we want to add three vectors? Let $\mathbf{u}$, $\mathbf{v}$, and $\mathbf{w}$ be vectors. How should we define the sum $\mathbf{u} + \mathbf{v} + \mathbf{w}$? Addition was defined for two vectors at a time. That leads to two possible "groupings" of the three terms. We might add $\mathbf{u}$ and $\mathbf{v}$ first, then add $\mathbf{w}$ to their sum. In symbols we write $(\mathbf{u} + \mathbf{v}) + \mathbf{w}$, where the parentheses are grouping symbols, as in algebra, indicating what operation is done first. Alternatively we might add $\mathbf{v}$ and $\mathbf{w}$ first, then add their sum to $\mathbf{u}$. In symbols we write $\mathbf{u} + (\mathbf{v} + \mathbf{w})$. Of course, we would like $(\mathbf{u} + \mathbf{v}) + \mathbf{w}$ to equal $\mathbf{u} + (\mathbf{v} + \mathbf{w})$. If this is true, then we can write the sum $\mathbf{u} + \mathbf{v} + \mathbf{w}$ without ambiguity.

EXERCISE 1 Draw three distinct (unequal) vectors $\mathbf{u}$, $\mathbf{v}$, and $\mathbf{w}$. Use the definition of vector addition to compute $(\mathbf{u} + \mathbf{v}) + \mathbf{w}$ and $\mathbf{u} + (\mathbf{v} + \mathbf{w})$. Convince yourself that the resulting sums are equal. So we have

$$(\mathbf{u} + \mathbf{v}) + \mathbf{w} = \mathbf{u} + (\mathbf{v} + \mathbf{w}) \qquad \text{(i)}$$

the *associative property* of vector addition.

Now we want to define a vector with the property that when we add it to any vector $\mathbf{v}$, it leaves $\mathbf{v}$ unchanged. This new vector is called the **zero vector**, denoted $\mathbf{O}$. So

$$\mathbf{v} + \mathbf{O} = \mathbf{v} \qquad \text{(ii)}$$

According to equation (ii), the zero vector is a vector with no direction and whose magnitude is zero. It can be shown that the zero vector is unique, that is, $\mathbf{O}$ is the *only* vector with the property that $\mathbf{v} + \mathbf{O} = \mathbf{v}$ for *all* vectors $\mathbf{v}$.

EXERCISE 2 Explain why the zero vector must have no direction and have magnitude zero.

The next exercise provides a motivation for a definition of the multiplication of a vector by a positive real number.

EXERCISE 3 Draw a nonzero vector $\mathbf{v}$. Then draw the vector $\mathbf{v} + \mathbf{v}$. What are the direction and magnitude of $\mathbf{v} + \mathbf{v}$?

On the basis of Exercise 3 it is natural to call the vector $\mathbf{v} + \mathbf{v}$, $2\mathbf{v}$, that is, $\mathbf{v} + \mathbf{v} = 2\mathbf{v}$. Similarly, for any *positive* real number k the vector $k\mathbf{v}$ is *defined* to be the vector with the same direction as $\mathbf{v}$ and whose magnitude is k times the magnitude of $\mathbf{v}$. In symbols, $|k\mathbf{v}| = k|\mathbf{v}|$.

Now start with a nonzero vector $\mathbf{v}$. Consider a new vector with the "opposite" direction of $\mathbf{v}$ and the same magnitude as $\mathbf{v}$. By opposite direction we mean that this new vector is represented by an arrow whose initial point is the terminal point of $\mathbf{v}$ and whose terminal point is the initial point of $\mathbf{v}$. This vector is called the **opposite** of $\mathbf{v}$, denoted $-\mathbf{v}$. Clearly, given $\mathbf{v}$, the vector $-\mathbf{v}$ is unique.

EXERCISE 4 Draw a nonzero vector $\mathbf{v}$ and its opposite, $-\mathbf{v}$. Then draw the vector $\mathbf{v} + (-\mathbf{v})$ to show

$$\mathbf{v} + (-\mathbf{v}) = \mathbf{O} \qquad \text{(iii)}$$

Now we want to define $k\mathbf{v}$, where k is a *negative* real number.

EXERCISE 5 Draw a nonzero vector $\mathbf{v}$. Then draw a vector with the opposite direction of $\mathbf{v}$ and with magnitude twice that of $\mathbf{v}$. On the basis of our definition of $-\mathbf{v}$, show that we can construct the desired vector by adding two copies of $-\mathbf{v}$, that is, $(-\mathbf{v}) + (-\mathbf{v})$.

Exercises 5 and 3 suggest it is natural to denote this new vector by the symbol $-2\mathbf{v}$, that is, $(-\mathbf{v}) + (-\mathbf{v}) = -2\mathbf{v}$. The -2 indicates the vector $-2\mathbf{v}$ has direction opposite that of $\mathbf{v}$ and magnitude $|-2| = 2$ times that of $\mathbf{v}$. This leads us to define, for a negative real number k, the vector $k\mathbf{v}$ as the vector with direction opposite that of $\mathbf{v}$ and magnitude $|k|$ times that of $\mathbf{v}$. In symbols, $|k\mathbf{v}| = |k||\mathbf{v}|$. Note that $|k|$ is the absolute value of the number k and $|\mathbf{v}|$ is the magnitude of the vector $\mathbf{v}$. Note also that for k positive we have $|k\mathbf{v}| = k|\mathbf{v}| = |k||\mathbf{v}|$, too.

This leads us to a definition.

DEFINITION Scalar Multiplication

Given a nonzero vector $\mathbf{v}$ and a nonzero real number k, the vector $k\mathbf{v}$ is the vector of magnitude $|k\mathbf{v}| = |k||\mathbf{v}|$ whose direction is the same as that of $\mathbf{v}$ if k is positive and opposite that of $\mathbf{v}$ if k is negative. If either $\mathbf{v} = \mathbf{O}$ or $k = 0$, then we define $k\mathbf{v}$ to be the zero vector, that is, we define $k\mathbf{O} = \mathbf{O}$ and $0\mathbf{v} = \mathbf{O}$. "Multiplying" a vector by a real number is called **scalar multiplication.**

EXERCISE 6 For any vector $\mathbf{v}$, explain why $1\mathbf{v} = \mathbf{v}$ and $-1\mathbf{v} = -\mathbf{v}$.

Now we want to define *subtraction* of vectors. In arithmetic we define the difference of two numbers to be the number you add to the number subtracted to get the number you started with. For example, $8 - 5 = 3$, since $5 + 3 = 8$, that is, 3 [or $(8 - 5)$] is what you add to 5 to get 8. Now let $\mathbf{u}$ and $\mathbf{v}$ be vectors and consider a new vector called the **difference** of $\mathbf{u}$ and $\mathbf{v}$, denoted by $\mathbf{u} - \mathbf{v}$. By analogy to subtraction of numbers, we define the difference $\mathbf{u} - \mathbf{v}$ to be the vector that you add to $\mathbf{v}$ to get $\mathbf{u}$. See Figure A.

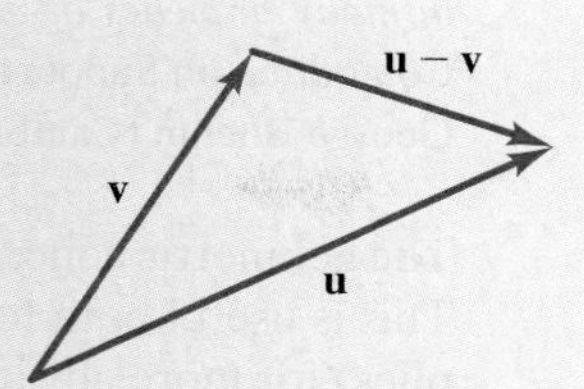

Figure A
$\mathbf{u} - \mathbf{v}$ is the vector that you add to $\mathbf{v}$ to get $\mathbf{u}$.

EXERCISE 7 Draw two vectors $\mathbf{u}$ and $\mathbf{v}$ with different directions and with $\mathbf{v}$ not equal to the opposite of $\mathbf{u}$. Show

$$\mathbf{u} - \mathbf{v} = \mathbf{u} + (-\mathbf{v}) \qquad \text{(iv)}$$

So we can think of subtracting as adding the opposite.

EXERCISE 8 We outline an algebraic derivation of the formula $\mathbf{u} - \mathbf{v} = \mathbf{u} + (-\mathbf{v}$.) According to the definition of $\mathbf{u} - \mathbf{v}$, we need to verify that adding $\mathbf{u} + (-\mathbf{v})$ to $\mathbf{v}$ results in $\mathbf{u}$. Give reasons for each step.

$$\begin{aligned} \mathbf{v} + [\mathbf{u} + (-\mathbf{v})] &= \mathbf{v} + [(-\mathbf{v}) + \mathbf{u}] \\ &= [\mathbf{v} + (-\mathbf{v})] + \mathbf{u} \\ &= \mathbf{O} + \mathbf{u} \\ &= \mathbf{u} \end{aligned}$$

EXERCISE 9 Here are *examples* of more vector algebra properties. Draw two nonzero vectors $\mathbf{u}$ and $\mathbf{v}$ with different directions and $\mathbf{v}$ not equal to the opposite of $\mathbf{u}$.

(a) Show $2(\mathbf{u} + \mathbf{v}) = 2\mathbf{u} + 2\mathbf{v}$.
(b) Show $(2 + 3)\mathbf{u} = 2\mathbf{u} + 3\mathbf{u}$.
(c) Show $2(3\mathbf{u}) = (2 \cdot 3)\mathbf{u}$.

Parts (a), (b), and (c) are specific examples of the following vector algebra properties. For any two vectors $\mathbf{u}$ and $\mathbf{v}$ and any two scalars k and l,

$$k(\mathbf{u} + \mathbf{v}) = k\mathbf{u} + k\mathbf{v} \qquad \text{(v)}$$
$$(k + l)\mathbf{u} = k\mathbf{u} + l\mathbf{u} \qquad \text{(vi)}$$
$$k(l\mathbf{u}) = (kl)\mathbf{u} \qquad \text{(vii)}$$

EXERCISE 10 For any vector $\mathbf{v}$, show that $-3\mathbf{v} = (-3)\mathbf{v}$. *Hint*: $-3\mathbf{v}$ is the opposite of $3\mathbf{v}$, that is, $-3\mathbf{v}$ is the unique vector to add to $3\mathbf{v}$ to get the zero vector. Use the result from Exercise 9(b) to show that $3\mathbf{v} + [(-3)\mathbf{v}] = \mathbf{O}$. Then $(-3)\mathbf{v}$ must be $-3\mathbf{v}$.

9.4 VECTORS IN THE PLANE: AN ALGEBRAIC APPROACH

A great many of the mathematical ideas that apply to physics and engineering are collected in the concept of vector spaces. This branch of mathematics has applications in such practical problems as calculating the vibrations of bridges and airplane wings. Logical extensions to spaces of infinitely many dimensions are widely used in modern theoretical physics as well as in many branches of mathematics itself.—From *The Mathematical Sciences,* edited by the Committee on Support of Research in the Mathematical Sciences with the collaboration of George Boehm (Cambridge, Mass.: The M.I.T. Press, 1969)

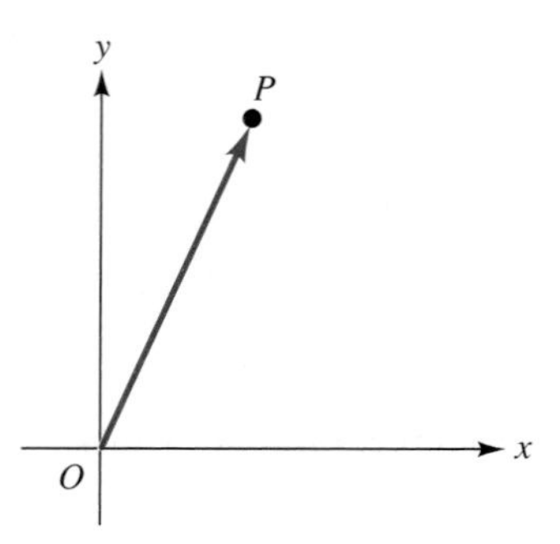

Figure 1

The geometric concept of a vector in the plane can be recast in an algebraic setting. This is useful both for computational purposes and (as our opening quotation implies) for more advanced work.

Consider an x-y coordinate system and a vector $\overrightarrow{OP}$ with initial point the origin, as shown in Figure 1. We call $\overrightarrow{OP}$ the **position vector** (or **radius vector**) of the point P. Most of our work in this section will involve such position vectors. There is no loss of generality in focusing on these types of vectors, for, as is indicated in Figure 2, each vector $\mathbf{v}$ in the plane is equal to a unique position vector $\overrightarrow{OP}$.

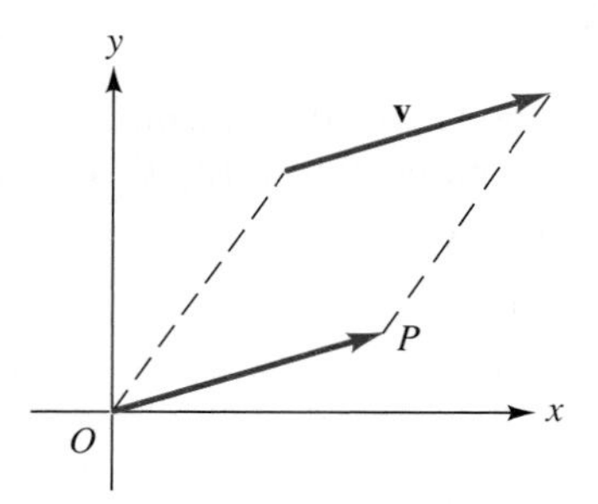

Figure 2

If the coordinates of the point P are (a, b), we call a and b the **components** of the vector $\overrightarrow{OP}$, and we use the notation

$$\langle a, b \rangle$$

to denote this vector (see Figure 3). The number a is the **horizontal component** or ***x*-component** of the vector; b is the **vertical component** or ***y*-component.**

In the previous section we said that two vectors are equal provided that they have the same length and the same direction. For vectors $\langle a, b \rangle$ and $\langle c, d \rangle$ this implies that

$$\langle a, b \rangle = \langle c, d \rangle \qquad \text{if any only if} \qquad a = c \quad \text{and} \quad b = d$$

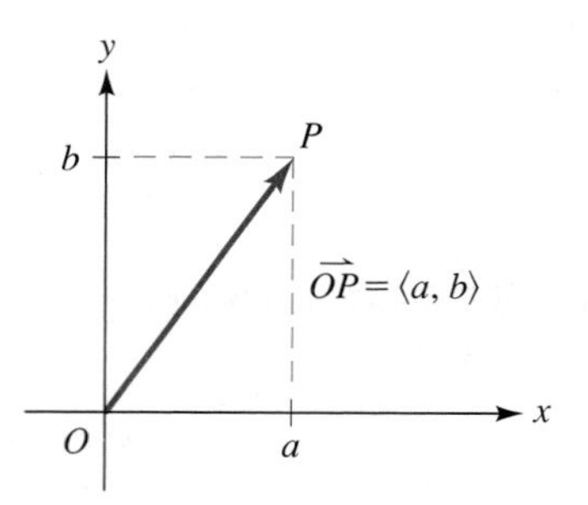

Figure 3
The vector $\overrightarrow{OP}$ is denoted by $\langle a, b \rangle$. The coordinates a and b are the components of the vector.

(If you've studied complex numbers in a previous course, notice the similarity here. Two complex numbers are equal provided that their corresponding real and imaginary parts are equal; two vectors are equal provided that their corresponding components are equal.)

It's easy to calculate the length of a vector $\mathbf{v}$ when its components are given. Suppose that $\mathbf{v} = \langle v_1, v_2 \rangle$. Applying the Pythagorean theorem in Figure 4, we have

$$|\mathbf{v}|^2 = v_1^2 + v_2^2$$

and consequently,

$$|\mathbf{v}| = \sqrt{v_1^2 + v_2^2}$$

Although Figure 4 shows the point (v_1, v_2) in Quadrant I, you can check for yourself that the same formula results when (v_1, v_2) is located in any of the other three quadrants. We therefore have the following general formula.

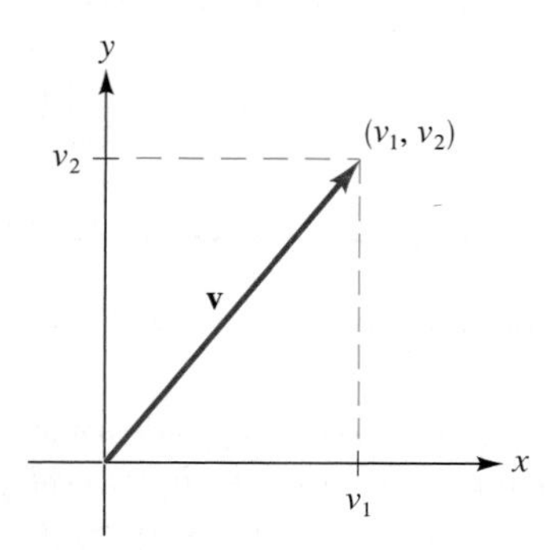

Figure 4

The Length of a Vector

If $\mathbf{v} = \langle v_1, v_2 \rangle$, then

$$|\mathbf{v}| = \sqrt{v_1^2 + v_2^2}$$

EXAMPLE 1 Finding the length of a vector

Compute the length of the vector $\mathbf{v} = \langle 2, -4 \rangle$.

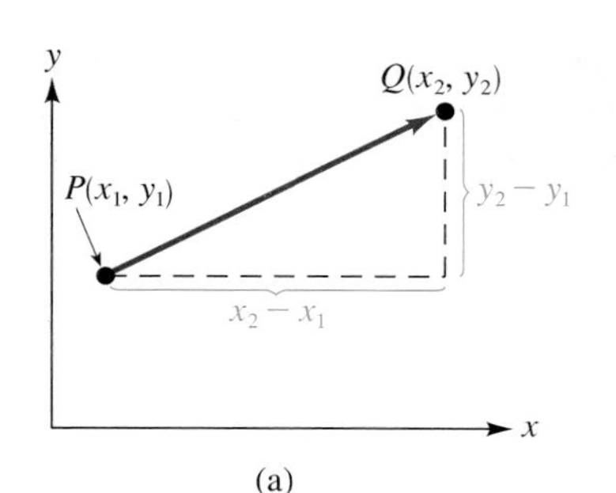

(a)

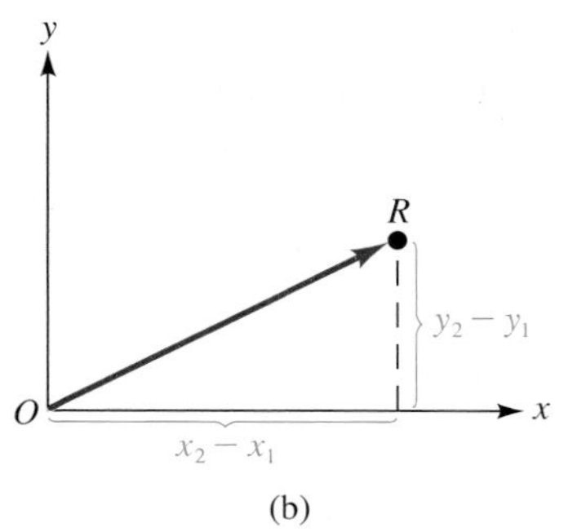

(b)

Figure 5

SOLUTION

$$|\mathbf{v}| = \sqrt{v_1^2 + v_2^2} = \sqrt{2^2 + (-4)^2} = \sqrt{20} = 2\sqrt{5}$$

If the coordinates of the points P and Q are $P(x_1, y_1)$ and $Q(x_2, y_2)$, then

$$\overrightarrow{PQ} = \langle x_2 - x_1, y_2 - y_1 \rangle$$

To derive this formula, we first construct the right triangle shown in Figure 5(a). Now we let R denote the point $(x_2 - x_1, y_2 - y_1)$ and draw the position vector $\overrightarrow{OR}$ shown in Figure 5(b). Since the right triangles in Figures 5(a) and 5(b) are congruent (Why?) and have corresponding legs that are parallel, we have

$$\overrightarrow{PQ} = \overrightarrow{OR} = \langle x_2 - x_1, y_2 - y_1 \rangle$$

as required. We can summarize this result as follows. For any vector $\mathbf{v}$ we have

x-component of $\mathbf{v}$:
(x-coordinate of terminal point) − (x-coordinate of initial point)
y-component of $\mathbf{v}$:
(y-coordinate of terminal point) − (y-coordinate of initial point)

EXAMPLE 2 Finding the components of a vector

Let P and Q be the points $P(3, 1)$ and $Q(7, 4)$. Sketch $\overrightarrow{PQ}$, find the components of $\overrightarrow{PQ}$, and sketch the position vector for $\overrightarrow{PQ}$.

SOLUTION

Plot the points P and Q and draw the vector $\overrightarrow{PQ}$ as shown in Figure 6.

$$\begin{aligned} x\text{-component of } \overrightarrow{PQ} &= x\text{-coordinate of } Q - x\text{-coordinate of } P \\ &= 7 - 3 = 4 \\ y\text{-component of } \overrightarrow{PQ} &= y\text{-coordinate of } Q - y\text{-coordinate of } P \\ &= 4 - 1 = 3 \end{aligned}$$

Consequently, $\overrightarrow{PQ} = \langle 4, 3 \rangle$. The position vector for $\overrightarrow{PQ}$ is shown in Figure 6.

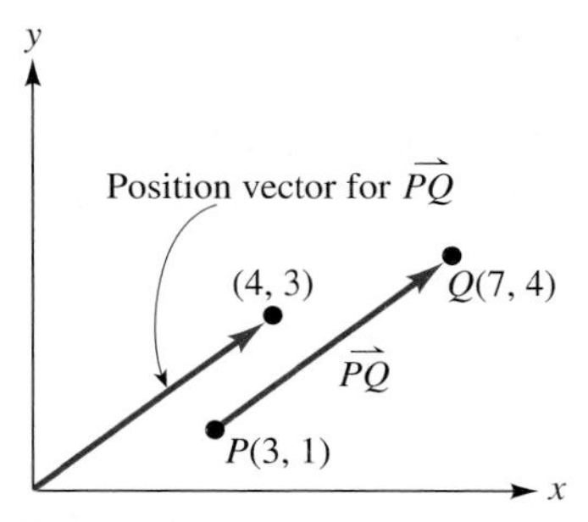

Figure 6

Vector addition is particularly simple to carry out when the vectors are in component form. Indeed, we can use the parallelogram law to verify the following result.

THEOREM Vector Addition

If $\mathbf{u} = \langle u_1, u_2 \rangle$ and $\mathbf{v} = \langle v_1, v_2 \rangle$, then $\mathbf{u} + \mathbf{v} = \langle u_1 + v_1, u_2 + v_2 \rangle$.

This theorem tells us that vector addition can be carried out *componentwise;* in other words, to add two vectors, just add the corresponding components. For example,

$$\langle 1, 2 \rangle + \langle 3, 7 \rangle = \langle 4, 9 \rangle$$

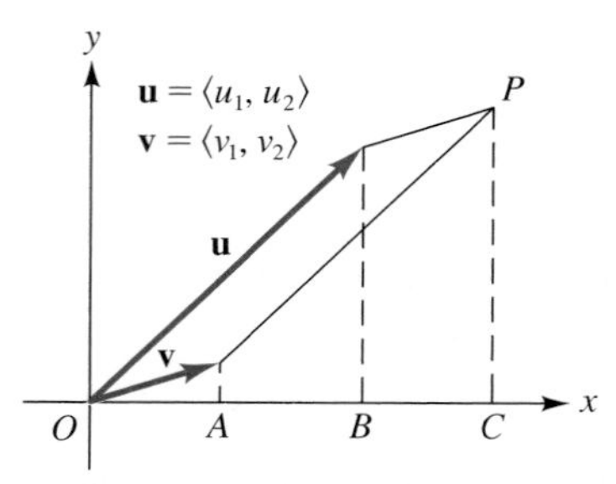

Figure 7

To see why this theorem is valid, consider Figure 7, where we have completed the parallelogram. (*Caution*: In reading the derivation that follows, don't confuse notation such as OA with $\overrightarrow{OA}$; recall that OA denotes the length of the line segment $\overline{OA}$.) Since the x-component of $\mathbf{u}$ is u_1, we have

$$OB = u_1$$

Also,

$$BC = OA = v_1$$

Therefore,

$$OC = OB + BC = u_1 + v_1$$

But OC is the x-component of the vector $\overrightarrow{OP}$, and by the parallelogram law, $\overrightarrow{OP} = \mathbf{u} + \mathbf{v}$. In other words, the x-component of $\mathbf{u} + \mathbf{v}$ is $u_1 + v_1$, as we wished to show. The fact that the y-component of $\mathbf{u} + \mathbf{v}$ is $u_2 + v_2$ is proved in a similar fashion.

The vector $\langle 0, 0\rangle$ is called the **zero vector,** and it is denoted by $\mathbf{0}$. Notice that for any vector $\mathbf{v} = \langle v_1, v_2\rangle$ we have

$$\mathbf{v} + \mathbf{0} = \mathbf{v} \qquad \text{because} \qquad \langle v_1, v_2\rangle + \langle 0, 0\rangle = \langle v_1, v_2\rangle$$

and

$$\mathbf{0} + \mathbf{v} = \mathbf{v} \qquad \text{because} \qquad \langle 0, 0\rangle + \langle v_1, v_2\rangle = \langle v_1, v_2\rangle$$

Thus the sum of a given vector and the zero vector is the given vector. In other words, adding the zero vector to a given vector leaves the given vector unchanged. So for the operation of vector addition the zero vector plays the same role as does the real number zero in addition of real numbers. There are, in fact, several other ways in which vector addition resembles ordinary addition of real numbers. We'll return to this point again near the end of this section.

In the box that follows, we define an operation called **scalar multiplication,** in which a vector is "multiplied" by a real number (a **scalar**) to obtain another vector.

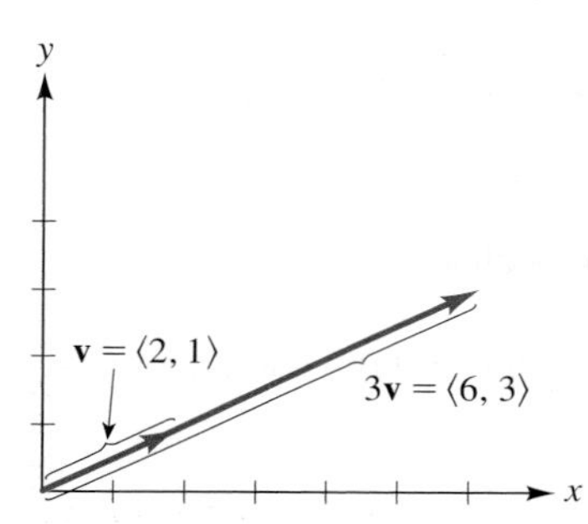

(a) The vectors $\mathbf{v}$ and $3\mathbf{v}$ have the same direction. The length of $3\mathbf{v}$ is three times that of $\mathbf{v}$.

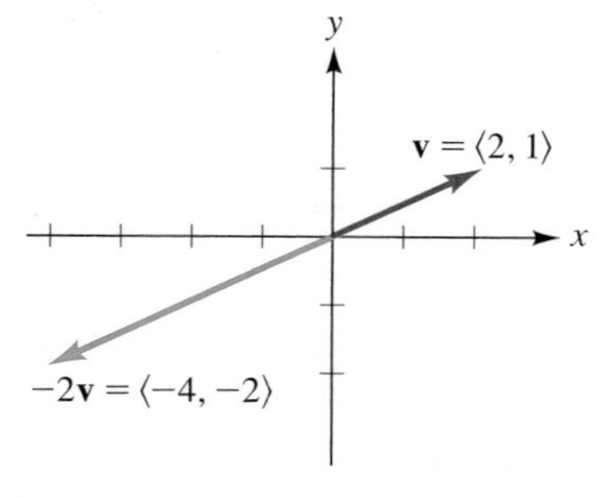

(b) The vectors $\mathbf{v}$ and $-2\mathbf{v}$ have opposite directions. The length of $-2\mathbf{v}$ is twice that of $\mathbf{v}$.

Figure 8

DEFINITION Scalar Multiplication

For each real number k and each vector $\mathbf{v} = \langle x, y\rangle$, we define a vector $k\mathbf{v}$ by the equation

$$k\mathbf{v} = k\langle x, y\rangle = \langle kx, ky\rangle$$

EXAMPLES

If $\mathbf{v} = \langle 2, 1\rangle$, then

$2\mathbf{v} = \langle 4, 2\rangle$ $\qquad 3\mathbf{v} = \langle 6, 3\rangle$

$0\mathbf{v} = \langle 0, 0\rangle = \mathbf{0}$

$-1\mathbf{v} = \langle -2, -1\rangle$

In geometric terms, the length of $k\mathbf{v}$ is $|k|$ times the length of $\mathbf{v}$. (See Exercise 60 at the end of this section.) The vectors $\mathbf{v}$ and $k\mathbf{v}$ have the same direction if $k > 0$ and opposite directions if $k < 0$. For example, let $\mathbf{v} = \langle 2, 1\rangle$. In Figure 8(a) we show the vectors $\mathbf{v}$ and $3\mathbf{v}$, while in Figure 8(b) we show $\mathbf{v}$ and $-2\mathbf{v}$.

EXAMPLE 3 Vector algebra

Let $\mathbf{v} = \langle 3, 4\rangle$ and $\mathbf{w} = \langle -1, 2\rangle$. Compute each of the following:

(a) $\mathbf{v} + \mathbf{w}$; **(b)** $-2\mathbf{v} + 3\mathbf{w}$; **(c)** $|-2\mathbf{v} + 3\mathbf{w}|$.

SOLUTION

(a) $\mathbf{v} + \mathbf{w} = \langle 3, 4\rangle + \langle -1, 2\rangle = \langle 3 - 1, 4 + 2\rangle = \langle 2, 6\rangle$

(b) $-2\mathbf{v} + 3\mathbf{w} = -2\langle 3, 4\rangle + 3\langle -1, 2\rangle = \langle -6, -8\rangle + \langle -3, 6\rangle$
$= \langle -9, -2\rangle$

(c) $|-2\mathbf{v} + 3\mathbf{w}| = |\langle -9, -2\rangle| = \sqrt{(-9)^2 + (-2)^2} = \sqrt{85}$

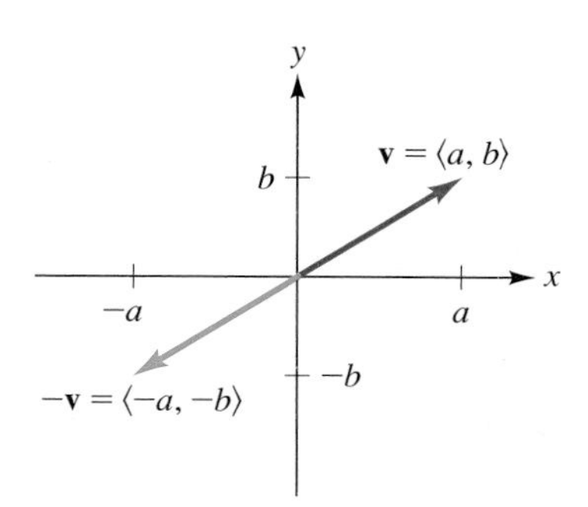

Figure 9

For each vector $\mathbf{v}$ we define a vector $-\mathbf{v}$, called the **negative** or **opposite** of $\mathbf{v}$, by the equation

$$-\mathbf{v} = -1\mathbf{v}$$

Thus if $\mathbf{v} = \langle a, b\rangle$, then $-\mathbf{v} = \langle -a, -b\rangle$. As is indicated in Figure 9, the vectors $\mathbf{v}$ and $-\mathbf{v}$ have the same length but opposite directions.

We can use the ideas in the preceding paragraph to define **vector subtraction.** Given two vectors $\mathbf{u}$ and $\mathbf{v}$, we define a vector $\mathbf{u} - \mathbf{v}$ by the equation

$$\mathbf{u} - \mathbf{v} = \mathbf{u} + (-\mathbf{v}) \tag{1}$$

First let's see what equation (1) is saying in terms of components; then we will indicate a simple geometric interpretation of vector subtraction.

If $\mathbf{u} = \langle u_1, u_2\rangle$ and $\mathbf{v} = \langle v_1, v_2\rangle$, then equation (1) tells us that

$$\begin{aligned}\mathbf{u} - \mathbf{v} &= \langle u_1, u_2\rangle + \langle -v_1, -v_2\rangle \\ &= \langle u_1 + (-v_1), u_2 + (-v_2)\rangle\end{aligned}$$

That is,

$$\mathbf{u} - \mathbf{v} = \langle u_1 - v_1, u_2 - v_2\rangle \tag{2}$$

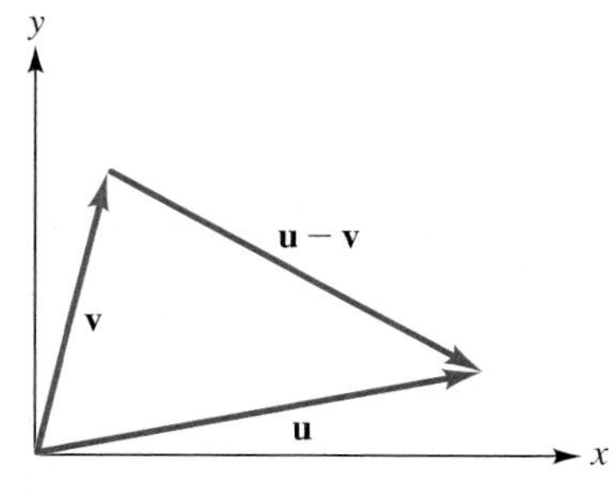

Figure 10

In other words, to subtract one vector from another, just subtract the corresponding components.

Vector subtraction can be interpreted geometrically. According to the formula in the first box on page 679, the right-hand side of equation (2) represents a vector drawn from the terminal point of $\mathbf{v}$ to the terminal point of $\mathbf{u}$. Figure 10 summarizes this fact, and Figure 11 provides a geometric comparison of vector addition and vector subtraction.

Figure 10 also indicates the algebraic significance of the difference $\mathbf{u} - \mathbf{v}$ as the vector we need to *add* to what is subtracted, $\mathbf{v}$, to get back what we started with, $\mathbf{u}$.

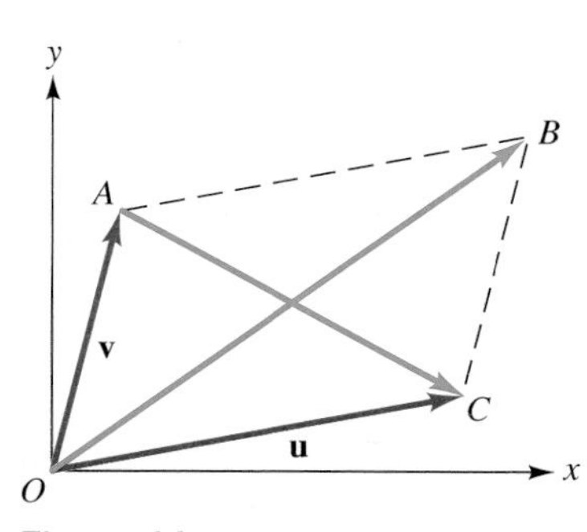

Figure 11
$\mathbf{u} + \mathbf{v}$ and $\mathbf{u} - \mathbf{v}$ are the directed diagonals of parallelogram $OABC$: $\mathbf{u} + \mathbf{v} = \overrightarrow{OB}$ and $\mathbf{u} - \mathbf{v} = \overrightarrow{AC}$.

EXAMPLE 4 Vector subtraction

Let $\mathbf{u} = \langle 5, 3\rangle$ and $\mathbf{v} = \langle -1, 2\rangle$. Compute $3\mathbf{u} - \mathbf{v}$.

SOLUTION

$$\begin{aligned}3\mathbf{u} - \mathbf{v} &= 3\langle 5, 3\rangle - \langle -1, 2\rangle \\ &= \langle 15, 9\rangle - \langle -1, 2\rangle = \langle 15 - (-1), 9 - 2\rangle = \langle 16, 7\rangle\end{aligned}$$

The next three examples deal with unit vectors. By definition any vector of length 1 is called a **unit vector.** Two particularly useful unit vectors are as follows.

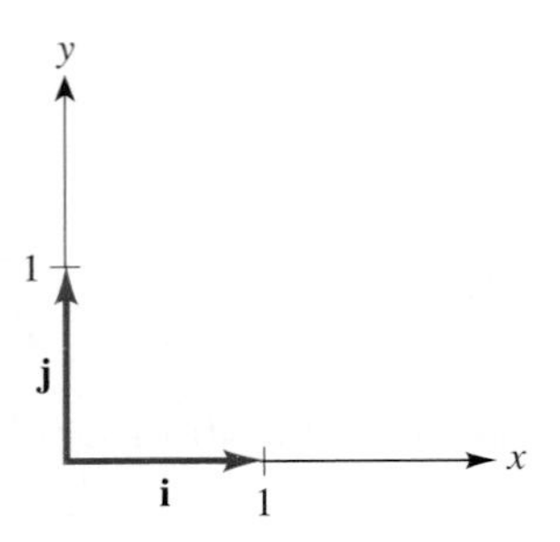

Figure 12

$$\mathbf{i} = \langle 1, 0 \rangle \qquad \text{and} \qquad \mathbf{j} = \langle 0, 1 \rangle$$

These are shown in Figure 12.

Any vector $\mathbf{v} = \langle x, y \rangle$ can be uniquely expressed in terms of the unit vectors **i** and **j** as follows:

$$\langle x, y \rangle = x\mathbf{i} + y\mathbf{j} \tag{3}$$

To verify equation (3), we have

$$\begin{aligned} x\mathbf{i} + y\mathbf{j} &= x\langle 1, 0 \rangle + y\langle 0, 1 \rangle \\ &= \langle x, 0 \rangle + \langle 0, y \rangle = \langle x, y \rangle \end{aligned}$$

EXAMPLE 5 Unit vectors and components

(a) Express the vector $\langle 3, -7 \rangle$ in terms of the unit vectors **i** and **j**.
(b) Express the vector $\mathbf{v} = -4\mathbf{i} + 5\mathbf{j}$ in component form.

SOLUTION
(a) Using equation (3), we can write

$$\langle 3, -7 \rangle = 3\mathbf{i} + (-7)\mathbf{j} = 3\mathbf{i} - 7\mathbf{j}$$

Note: In the last step we used the fact that $(-7)\mathbf{j} = -7\mathbf{j}$ and the definition of vector subtraction.
(b) $\mathbf{v} = -4\mathbf{i} + 5\mathbf{j} = -4\langle 1, 0 \rangle + 5\langle 0, 1 \rangle$
$= \langle -4, 0 \rangle + \langle 0, 5 \rangle = \langle -4, 5 \rangle$

Thus the component form of **v** is $\langle -4, 5 \rangle$.

EXAMPLE 6 Finding a unit vector

Find a unit vector **u** that has the same direction as the vector $\mathbf{v} = \langle 3, 4 \rangle$.

SOLUTION
First, let's determine the length of **v**:

$$|\mathbf{v}| = \sqrt{3^2 + 4^2} = \sqrt{25} = 5$$

So we want a vector whose length is one-fifth that of **v** and whose direction is the same as that of **v**. Such a vector is

$$\mathbf{u} = \frac{1}{5}\mathbf{v} = \frac{1}{5}\langle 3, 4 \rangle = \left\langle \frac{3}{5}, \frac{4}{5} \right\rangle$$

(You should check for yourself now that the length of this vector is 1.) *Note*: $-\mathbf{u} = \langle -\frac{3}{5}, -\frac{4}{5} \rangle$ is a unit vector that has the opposite direction **v**. (You should check this for yourself as well.)

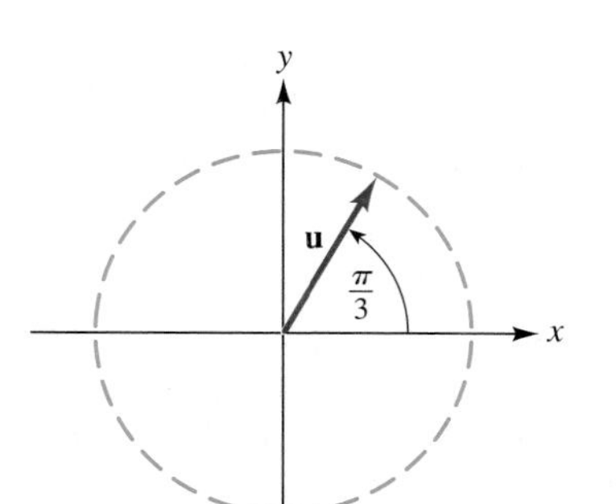

Figure 13

EXAMPLE 7 Finding a unit vector given a direction

The angle from the positive x-axis to the unit vector $\mathbf{u}$ is $\pi/3$, as indicated in Figure 13. Determine the components of $\mathbf{u}$.

SOLUTION

Let P denote the terminal point of $\mathbf{u}$. Since P lies on the unit circle, the coordinates of P are by definition $(\cos \frac{\pi}{3}, \sin \frac{\pi}{3})$. We therefore have

$$\mathbf{u} = \left\langle \frac{1}{2}, \frac{\sqrt{3}}{2} \right\rangle$$

It was mentioned earlier that there are several ways in which vector addition resembles ordinary addition of real numbers. In the previous section, for example, we saw that vector addition is commutative. That is, for any two vectors $\mathbf{u}$ and $\mathbf{v}$,

$$\mathbf{u} + \mathbf{v} = \mathbf{v} + \mathbf{u}$$

By using components, we can easily verify that this property holds. (In the previous section we used a geometric argument to establish this property.) We begin by letting $\mathbf{u} = \langle u_1, u_2 \rangle$ and $\mathbf{v} = \langle v_1, v_2 \rangle$. Then we have

$$\begin{aligned}
\mathbf{u} + \mathbf{v} &= \langle u_1, u_2 \rangle + \langle v_1, v_2 \rangle \\
&= \langle u_1 + v_1, u_2 + v_2 \rangle \\
&= \langle v_1 + u_1, v_2 + u_2 \rangle \quad \text{Addition of real numbers is commutative.} \\
&= \langle v_1, v_2 \rangle + \langle u_1, u_2 \rangle \\
&= \mathbf{v} + \mathbf{u}
\end{aligned}$$

which is what we wanted to show.

There are a number of other properties of vector addition and scalar multiplication that can be proved in a similar fashion. In the following box we list a particular collection of these properties, known as the **vector space properties.** (Exercises 55–58 ask that you verify these properties by using components, just as we did for the commutative property.)

Properties of Vector Addition and Scalar Multiplication

For all vectors $\mathbf{u}$, $\mathbf{v}$, and $\mathbf{w}$, and for all scalars (real numbers) a and b, the following properties hold.

1. $\mathbf{u} + (\mathbf{v} + \mathbf{w}) = (\mathbf{u} + \mathbf{v}) + \mathbf{w}$
2. $\mathbf{0} + \mathbf{v} = \mathbf{v} + \mathbf{0} = \mathbf{v}$
3. $\mathbf{v} + (-\mathbf{v}) = \mathbf{0}$
4. $\mathbf{u} + \mathbf{v} = \mathbf{v} + \mathbf{u}$
5. $a(\mathbf{u} + \mathbf{v}) = a\mathbf{u} + a\mathbf{v}$
6. $(a + b)\mathbf{v} = a\mathbf{v} + b\mathbf{v}$
7. $(ab)\mathbf{v} = a(b\mathbf{v})$
8. $1\mathbf{v} = \mathbf{v}$

EXERCISE SET 9.4

A

In Exercises 1–6, sketch each vector in an x-y coordinate system, and compute the length of the vector.

1. $\langle 4, 3 \rangle$ **2.** $\langle 5, 12 \rangle$ **3.** $\langle -4, 2 \rangle$

4. $\langle -6, -6 \rangle$ **5.** $\langle \frac{3}{4}, -\frac{1}{2} \rangle$ **6.** $\langle -3, 0 \rangle$

In Exercises 7–12 the coordinates of two points P and Q are given. In each case, determine the components of the vector $\overrightarrow{PQ}$. Write your answers in the form $\langle a, b \rangle$.

7. $P(2, 3)$ and $Q(3, 7)$

8. $P(5, 1)$ and $Q(4, 9)$

9. $P(-2, -3)$ and $Q(-3, -2)$
10. $P(0, -4)$ and $Q(0, -8)$
11. $P(-5, 1)$ and $Q(3, -4)$
12. $P(1, 0)$ and $Q(0, 1)$

In Exercises 13–32, assume that the vectors **a**, **b**, **c**, *and* **d** *are defined as follows:*

$$\mathbf{a} = \langle 2, 3 \rangle \qquad \mathbf{b} = \langle 5, 4 \rangle \qquad \mathbf{c} = \langle 6, -1 \rangle \qquad \mathbf{d} = \langle -2, 0 \rangle$$

Compute each of the indicated quantities.

13. $\mathbf{a} + \mathbf{b}$
14. $\mathbf{c} + \mathbf{d}$
15. $2\mathbf{a} + 4\mathbf{b}$
16. $-2\mathbf{c} + 2\mathbf{d}$
17. $|\mathbf{b} + \mathbf{c}|$
18. $|5\mathbf{b} + 5\mathbf{c}|$
19. $|\mathbf{a} + \mathbf{c}| - |\mathbf{a}| - |\mathbf{c}|$
20. $1/|\mathbf{d}|$
21. $\mathbf{a} + (\mathbf{b} + \mathbf{c})$
22. $(\mathbf{a} + \mathbf{b}) + \mathbf{c}$
23. $3\mathbf{a} + 4\mathbf{a}$
24. $|4\mathbf{b} + 5\mathbf{b}|$
25. $\mathbf{a} - \mathbf{b}$
26. $\mathbf{b} - \mathbf{c}$
27. $3\mathbf{b} - 4\mathbf{d}$
28. $\dfrac{1}{|3\mathbf{b} - 4\mathbf{d}|}(3\mathbf{b} - 4\mathbf{a})$
29. $\mathbf{a} - (\mathbf{b} + \mathbf{c})$
30. $(\mathbf{a} - \mathbf{b}) - \mathbf{c}$
31. $|\mathbf{c} + \mathbf{d}|^2 - |\mathbf{c} - \mathbf{d}|^2$
32. $|\mathbf{a} + \mathbf{b}|^2 + |\mathbf{a} - \mathbf{b}|^2 - 2|\mathbf{a}|^2 - 2|\mathbf{b}|^2$

In Exercises 33–38, express each vector in terms of the unit vectors **i** *and* **j**.

33. $\langle 3, 8 \rangle$
34. $\langle 4, -2 \rangle$
35. $\langle -8, -6 \rangle$
36. $\langle -9, 0 \rangle$
37. $3\langle 5, 3 \rangle + 2\langle 2, 7 \rangle$
38. $|\langle 12, 5 \rangle|\langle 3, 4 \rangle + |\langle 3, 4 \rangle|\langle 12, 5 \rangle$

In Exercises 39–42, express each vector in the form $\langle a, b \rangle$.

39. $\mathbf{i} + \mathbf{j}$
40. $\mathbf{i} - 2\mathbf{j}$
41. $5\mathbf{i} - 4\mathbf{j}$
42. $\dfrac{1}{|\mathbf{i} + \mathbf{j}|}(\mathbf{i} + \mathbf{j})$

In Exercises 43–48, find a unit vector having the same direction as the given vector.

43. $\langle 4, 8 \rangle$
44. $\langle -3, 3 \rangle$
45. $\langle 6, -3 \rangle$
46. $\langle -12, 5 \rangle$
47. $8\mathbf{i} - 9\mathbf{j}$
48. $\langle 7, 3 \rangle - \mathbf{i} + \mathbf{j}$

In Exercises 49–54, you are given an angle θ *measured counterclockwise from the positive x-axis to a unit vector* $\mathbf{u} = \langle u_1, u_2 \rangle$. *In each case, determine the components* u_1 *and* u_2.

49. $\theta = \pi/6$
50. $\theta = \pi/4$
51. $\theta = 2\pi/3$
52. $\theta = 3\pi/4$
53. $\theta = 5\pi/6$
54. $\theta = 3\pi/2$

B

In Exercises 55–58, let $\mathbf{u} = \langle u_1, u_2 \rangle$, $\mathbf{v} = \langle v_1, v_2 \rangle$, *and* $\mathbf{w} = \langle w_1, w_2 \rangle$. *Refer to the box on page 683.*

55. Verify Properties 1 and 2.
56. Verify Properties 3 and 4.
57. Verify Properties 5 and 6.
58. Verify Properties 7 and 8.
59. Verify that $\mathbf{v} + (\mathbf{u} - \mathbf{v}) = \mathbf{u}$.
60. Show that $|k\mathbf{v}| = |k||\mathbf{v}|$ for any real number k.

In Exercises 61–77 we study the dot product of two vectors. Given two vectors $\mathbf{A} = \langle x_1, y_1 \rangle$ *and* $\mathbf{B} = \langle x_2, y_2 \rangle$, *we define the* ***dot product*** $\mathbf{A} \cdot \mathbf{B}$ *as follows:*

$$\mathbf{A} \cdot \mathbf{B} = x_1x_2 + y_1y_2$$

For example, if $\mathbf{A} = \langle 3, 4 \rangle$ *and* $\mathbf{B} = \langle -2, 5 \rangle$, *then* $\mathbf{A} \cdot \mathbf{B} = (3)(-2) + (4)(5) = 14$. *Notice that the dot product of two vectors is a real number. For this reason, the dot product is also known as the* ***scalar product.*** *For Exercises 61–63 the vectors* **u**, **v**, *and* **w** *are defined as follows:*

$$\mathbf{u} = \langle -4, 5 \rangle \qquad \mathbf{v} = \langle 3, 4 \rangle \qquad \mathbf{w} = \langle 2, -5 \rangle$$

61. **(a)** Compute $\mathbf{u} \cdot \mathbf{v}$ and $\mathbf{v} \cdot \mathbf{u}$.
(b) Compute $\mathbf{v} \cdot \mathbf{w}$ and $\mathbf{w} \cdot \mathbf{v}$.
(c) Show that for any two vectors **A** and **B**, we have $\mathbf{A} \cdot \mathbf{B} = \mathbf{B} \cdot \mathbf{A}$. That is, show that the dot product is commutative. *Hint:* Let $\mathbf{A} = \langle x_1, y_1 \rangle$, and let $\mathbf{B} = \langle x_2, y_2 \rangle$.

62. **(a)** Compute $\mathbf{v} + \mathbf{w}$.
(b) Compute $\mathbf{u} \cdot (\mathbf{v} + \mathbf{w})$.
(c) Compute $\mathbf{u} \cdot \mathbf{v} + \mathbf{u} \cdot \mathbf{w}$.
(d) Show that for any three vectors **A**, **B**, and **C** we have $\mathbf{A} \cdot (\mathbf{B} + \mathbf{C}) = \mathbf{A} \cdot \mathbf{B} + \mathbf{A} \cdot \mathbf{C}$.

63. **(a)** Compute $\mathbf{v} \cdot \mathbf{v}$ and $|\mathbf{v}|^2$.
(b) Compute $\mathbf{w} \cdot \mathbf{w}$ and $|\mathbf{w}|^2$.

64. Show that for any vector **A** we always have $|\mathbf{A}|^2 = \mathbf{A} \cdot \mathbf{A}$. That is, the square of the length of a vector is equal to the dot product of the vector with itself. *Hint:* Let $\mathbf{A} = \langle x, y \rangle$.

Let θ *(where* $0 \le \theta \le \pi$*) denote the angle between the two nonzero vectors* **A** *and* **B**. *Then it can be shown that the cosine of* θ *is given by the formula*

$$\cos\theta = \frac{\mathbf{A} \cdot \mathbf{B}}{|\mathbf{A}||\mathbf{B}|}$$

(See Exercise 77 for the derivation of this result.) In Exercises 65–70, sketch each pair of vectors as position vectors, then use this formula to find the cosine of the angle between the given pair of vectors. Also, in each case, use a calculator to compute the angle. Express the angle using degrees and using radians. Round the values to two decimal places.

65. $\mathbf{A} = \langle 4, 1 \rangle$ and $\mathbf{B} = \langle 2, 6 \rangle$
66. $\mathbf{A} = \langle 3, -1 \rangle$ and $\mathbf{B} = \langle -2, 5 \rangle$
67. $\mathbf{A} = \langle 5, 6 \rangle$ and $\mathbf{B} = \langle -3, -7 \rangle$
68. $\mathbf{A} = \langle 3, 0 \rangle$ and $\mathbf{B} = \langle 1, 4 \rangle$
69. **(a)** $\mathbf{A} = \langle -8, 2 \rangle$ and $\mathbf{B} = \langle 1, -3 \rangle$
(b) $\mathbf{A} = \langle -8, 2 \rangle$ and $\mathbf{B} = \langle -1, 3 \rangle$
70. **(a)** $\mathbf{A} = \langle 7, 12 \rangle$ and $\mathbf{B} = \langle 1, 2 \rangle$
(b) $\mathbf{A} = \langle 7, 12 \rangle$ and $\mathbf{B} = \langle -1, -2 \rangle$

71. **(a)** Compute the cosine of the angle between the vectors $\langle 2, 5\rangle$ and $\langle -5, 2\rangle$.
(b) What can you conclude from your answer in part (a)?
(c) Draw a sketch to check your conclusion in part (b).

72. Follow Exercise 71, but use the vectors $\langle 6, -8\rangle$ and $\langle -4, -3\rangle$.

73. Let **A** and **B** be nonzero vectors.
(a) If $\mathbf{A}\cdot\mathbf{B} = 0$, explain why **A** and **B** are perpendicular.
(b) If **A** and **B** are perpendicular, explain why $\mathbf{A}\cdot\mathbf{B} = 0$.
(a) and (b) together show that two nonzero vectors are perpendicular if and only if their dot product is zero.

74. Find a value for t such that the vectors $\langle 15, -3\rangle$ and $\langle -4, t\rangle$ are perpendicular.

75. Find a unit vector that is perpendicular to the vector $\langle -12, 5\rangle$. (There are two answers.)

C

76. Suppose that **A** and **B** be nonzero vectors.
(a) Show that $|\mathbf{A} + \mathbf{B}|^2 = |\mathbf{A}|^2 + 2\mathbf{A}\cdot\mathbf{B} + |\mathbf{B}|^2$
Hint: Start with $|\mathbf{A} + \mathbf{B}|^2 = (\mathbf{A} + \mathbf{B})\cdot(\mathbf{A} + \mathbf{B})$ and use Exercises 62(d), 64, and 61(c). *Alternatively,* let $\mathbf{A} = \langle x_1, y_1\rangle$ and $\mathbf{B} = \langle x_2, y_2\rangle$. Then separately compute each side of the equation in part (a).
(b) Show that $|\mathbf{A} + \mathbf{B}|^2 = |\mathbf{A}|^2 + |\mathbf{B}|^2$ if and only if **A** and **B** are perpendicular. Draw a sketch with **A** and **B** joined "head to tail." This is a vector statement of the **Pythagorean theorem.**

77. Refer to the figure at the top of the next column. In this exercise we are going to derive the following formula for the cosine of the angle θ between two nonzero vectors **A** and **B**:

$$\cos\theta = \frac{\mathbf{A}\cdot\mathbf{B}}{|\mathbf{A}||\mathbf{B}|}$$

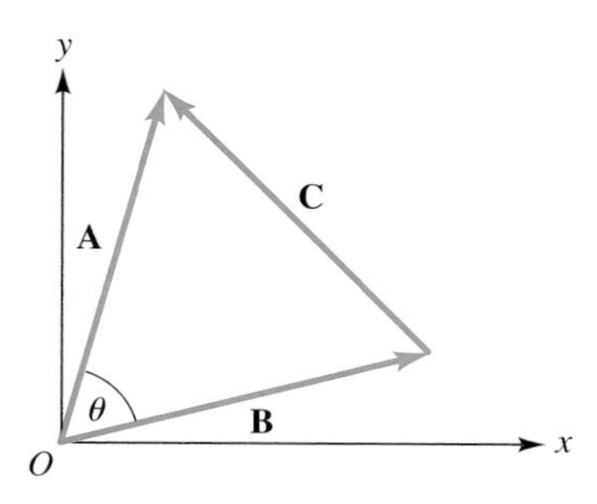

(a) Let **A** and **B** be nonzero vectors, and let $\mathbf{C} = \mathbf{A} - \mathbf{B}$. Show that

$$|\mathbf{C}|^2 = |\mathbf{A}|^2 + |\mathbf{B}|^2 - 2\mathbf{A}\cdot\mathbf{B}$$

Hint: Use the result or method of Exercise 76(a).
(b) According to the law of cosines, we have

$$|\mathbf{C}|^2 = |\mathbf{A}|^2 + |\mathbf{B}|^2 - 2|\mathbf{A}|\,|\mathbf{B}|\cos\theta$$

Set this expression for $|\mathbf{C}|^2$ equal to the expression obtained in part (a), and then solve for $\cos\theta$ to obtain the required formula.

Note: The formula that we just derived can be written in the form

$$\mathbf{A}\cdot\mathbf{B} = |\mathbf{A}|\,|\mathbf{B}|\cos\theta$$

This expresses the dot product in purely geometric terms: The dot product of two vectors is the product of the lengths of the two vectors and the cosine of their included angle. So the easy-to-compute algebraic expression in the components of **A** and **B** that is used to define the dot product on page 684 has important geometric applications. In particular, dot products can be used to compute lengths and angles and to determine whether two vectors are mutually perpendicular.

PROJECT

LINES, CIRCLES, AND RAY TRACING WITH VECTORS

In the x-y plane, a line and its equations are uniquely determined by the line's slope (or undefined slope if the line is vertical) and the coordinates of one point on the line. We know an equation of the line with slope m and y-intercept b is $y = mx + b$.

Our first goal in this project is to find a vector equation for a line. Let P_0 be a point in the x-y plane with **A** a nonzero vector. In Figure A we draw the position vector $\mathbf{P}_0$ from the origin to the point P_0 and the vector **A** with initial point P_0, and then draw the vectors $\mathbf{P}_0 + \mathbf{A}$, $\mathbf{P}_0 + 2\mathbf{A}$, $\mathbf{P}_0 + 3\mathbf{A}$, $\mathbf{P}_0 + (-1)\mathbf{A}$, and $\mathbf{P}_0 + (-2)\mathbf{A}$.

Figure A shows that each point P on the line has a description of the form $\mathbf{P}_0 + t\mathbf{A}$ for *some* real number t. As t varies over all real numbers we sweep out all points on the line. These observations give us a vector equation for the line

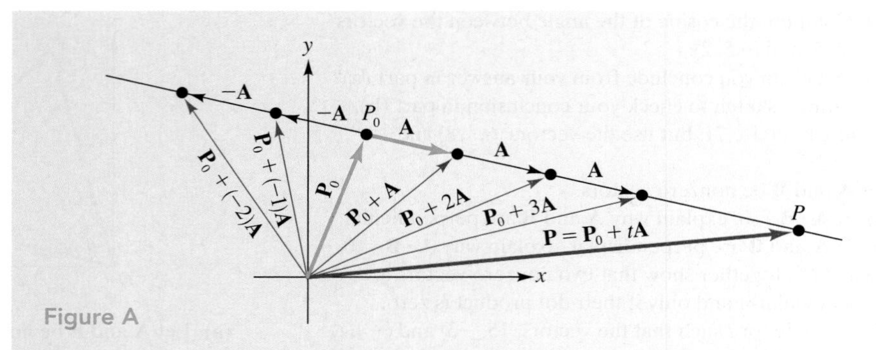

Figure A

through the point P_0 with **direction vector A**. (*Note*: A direction vector for a line is a nonzero vector that is parallel to the line.) Letting **P** denote the position vector from the origin to the arbitrary point P on the line we have

$$\mathbf{P} = \mathbf{P}_0 + t\mathbf{A} \qquad \text{for } -\infty < t < \infty \tag{i}$$

EXAMPLE **A vector equation for a line**

Consider the line in the x-y plane passing through the point $(-2, 1)$ with slope 2.

(a) Sketch the line and find its slope-intercept equation.

(b) Find a vector equation for the line.

(c) Find the points on the line corresponding to $t = -1, 0,$ and 2.

SOLUTION

(a) The line is sketched in Figure B. It has an equation of the form

$$y = mx + b$$

Here $m = 2$, so $y = 2x + b$. $(-2, 1)$ lies on the line, so

$$1 = 2(-2) + b$$
$$b = 5$$

Therefore $y = 2x + 5$.

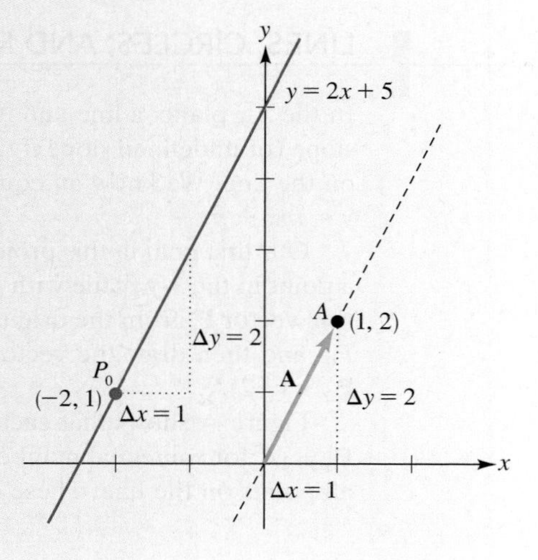

Figure B

(b) Let $P = (x, y)$ be any point on the line and let $P_0 = (-2, 1)$. So the position vectors from the origin to these points are $\mathbf{P} = \langle x, y\rangle$ and $\mathbf{P}_0 = \langle -2, 1\rangle$. Given the slope 2, we can find the line's *direction* by drawing the line of slope 2 passing through the origin and picking a convenient point (different from the origin) on this line. See Figure B. Using the point $A = (1, 2)$, we see that the vector $\mathbf{A} = \langle 1, 2\rangle$ is parallel to the line. So a vector equation for the line is

$$\mathbf{P} = \mathbf{P}_0 + t\mathbf{A} \qquad \text{for } -\infty < t < \infty$$

which in component form becomes

$$\langle x, y\rangle = \langle -2, 1\rangle + t\langle 1, 2\rangle \qquad \text{for } -\infty < t < \infty$$

(c) For $t = -1$, $\langle x, y\rangle = \langle -2, 1\rangle + (-1)\langle 1, 2\rangle = \langle -3, -1\rangle$. So the vector $\langle -3, -1\rangle$ goes from the origin to the point P on the line, with $P = (-3, -1)$. Similarly, for $t = 0$ the point on the line is $P = (-2, 1)$ and for $t = 2$ the point is $P = (0, 5)$.

EXERCISE 1 Consider the line in the x-y plane passing through the point $(1, -2)$ with slope 3.

(a) Sketch the line and find its slope-intercept equation.

(b) Find a vector equation for the line. Label three vectors on the line in part (a).

(c) Find the points on this line corresponding to $t = -2, -1, 0, 1$, and 2.

EXERCISE 2 Consider the line in the x-y plane passing through the points $A = (-3, 1)$ and $B = (2, 4)$.

(a) Sketch the line and then find a vector equation for it.

(b) Find a vector equation for the line *segment* from A to B. *Hint*: You need to restrict the t values to a finite interval $k \le t \le l$ where $t = k$ corresponds to point A and $t = l$ to point B.

We next look at an optical application of our vector equation for a line. Ray tracing techniques have been used in the design of camera lenses, microscopes, telescopes, and other optical devices for hundreds of years. More recently ray tracing has been used in computer graphics applications, particularly in the creation of realistic scenes for games and special effects for movies. In the last 40 years, computationally efficient vector ray tracing techniques have replaced older trigonometric methods. A basic problem in optical ray tracing is to find the point of intersection of a ray, determined by a point and a direction vector, with a spherical lens surface of known center and radius (see Figure C). (Of course it is possible that a given ray does not intersect a particular lens surface.)

In two dimensions, in the x-y plane, the problem becomes as follows: Given two points P_0 and C, a nonzero vector $\mathbf{A}$, and a positive real number r, find the point P of intersection of the line through P_0 in the direction of $\mathbf{A}$ with the circle of radius r and center C. See Figure D, where we let $\mathbf{P}_0$, $\mathbf{P}$, and $\mathbf{C}$ denote the vectors from the origin to the points P_0, P, and C respectively, and $\mathbf{P} - \mathbf{C}$ denotes the vector of length r from the center of the circle to the point of intersection.

The point of intersection P lies on the line through P_0 in the direction of $\mathbf{A}$ so

$$\mathbf{P} = \mathbf{P}_0 + t\mathbf{A} \qquad \text{for } \textit{some} \text{ real number } t \tag{ii}$$

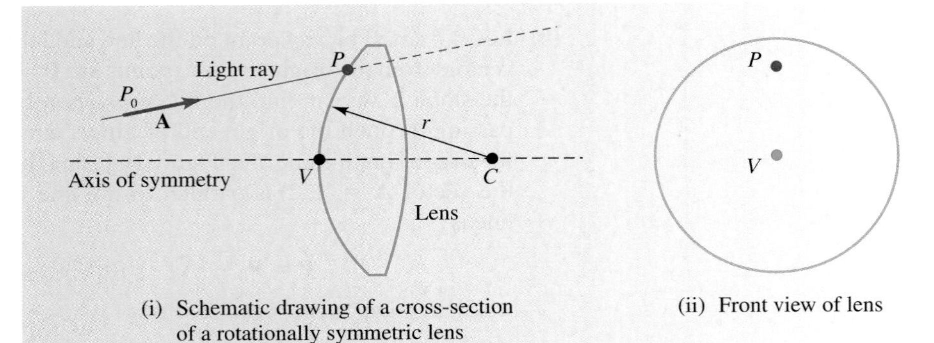

(i) Schematic drawing of a cross-section of a rotationally symmetric lens

(ii) Front view of lens

Figure C

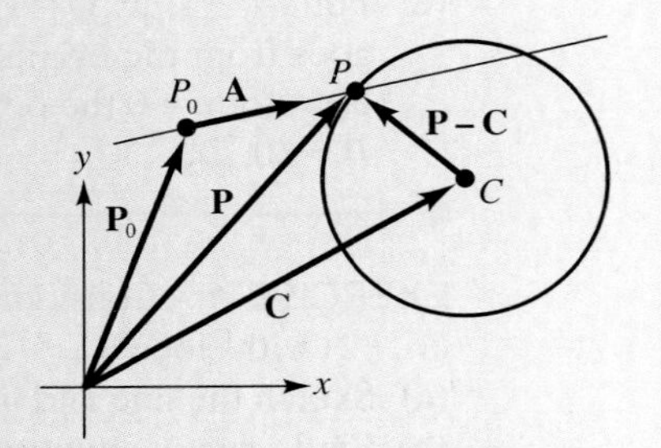

Figure D

The point P also lies on the circle with center C and radius r, so the distance from C to P is r. Using vectors, the length of the vector $\mathbf{P} - \mathbf{C}$ is r, that is

$$|\mathbf{P} - \mathbf{C}| = r \qquad \text{(iii)}$$

We can solve the system of equations (ii) and (iii) for t by substituting the expression for $\mathbf{P}$ from equation (ii) into equation (iii) and rearranging to get

$$|t\mathbf{A} + (\mathbf{P}_0 - \mathbf{C})| = r \qquad \text{(iv)}$$

Squaring both sides of equation (iv) we get

$$|t\mathbf{A} + (\mathbf{P}_0 - \mathbf{C})|^2 = r^2 \qquad \text{(v)}$$

Now (from Exercise 64 in Section 9.4) the squared magnitude of a vector is the dot product of the vector with itself. So

$$[t\mathbf{A} + (\mathbf{P}_0 - \mathbf{C})] \cdot [t\mathbf{A} + (\mathbf{P}_0 - \mathbf{C})] = r^2 \qquad \text{(vi)}$$

Then [using Exercises 61(c) and 62(d) in Section 9.4] we have

$$t\mathbf{A} \cdot [t\mathbf{A} + (\mathbf{P}_0 - \mathbf{C})] + (\mathbf{P}_0 - \mathbf{C}) \cdot [t\mathbf{A} + (\mathbf{P}_0 - \mathbf{C})] = r^2 \qquad \text{(vii)}$$

EXERCISE 3 Expand the left-hand side of equation (vii) and regroup the terms to obtain

$$(t\mathbf{A}) \cdot (t\mathbf{A}) + (t\mathbf{A}) \cdot (\mathbf{P}_0 - \mathbf{C}) + (\mathbf{P}_0 - \mathbf{C}) \cdot (t\mathbf{A}) + (\mathbf{P}_0 - \mathbf{C}) \cdot (\mathbf{P}_0 - \mathbf{C}) = r^2 \qquad \text{(viii)}$$

EXERCISE 4

(a) Show that if $\mathbf{A}$ and $\mathbf{B}$ are any two vectors and s is a real number, then $(s\mathbf{A}) \cdot \mathbf{B} = s(\mathbf{A} \cdot \mathbf{B}) = \mathbf{A} \cdot (s\mathbf{B})$. *Hint:* Let $\mathbf{A} = \langle x_1, y_1 \rangle$ and $\mathbf{B} = \langle x_2, y_2 \rangle$.

(b) Use part (a) to put equation (viii) in the form of a quadratic equation in t to obtain

$$(\mathbf{A}\cdot\mathbf{A})t^2 + [2\mathbf{A}\cdot(\mathbf{P}_0 - \mathbf{C})]t + [(\mathbf{P}_0 - \mathbf{C})\cdot(\mathbf{P}_0 - \mathbf{C}) - r^2] = 0 \qquad \text{(ix)}$$

Then show that t satisfies the quadratic equation

$$|\mathbf{A}|^2t^2 + [2\mathbf{A}\cdot(\mathbf{P}_0 - \mathbf{C})]t + (|\mathbf{P}_0 - \mathbf{C}|^2 - r^2) = 0 \qquad \text{(x)}$$

Note that the coefficients in equation (ix) are written completely in terms of dot products, which is most convenient for computer computations. In equation (x) two of the coefficients look a little simpler in terms of lengths of vectors.

EXERCISE 5 Use the quadratic formula to solve equation (x) for t and simplify to obtain

$$t = \frac{-\mathbf{A}\cdot(\mathbf{P}_0 - \mathbf{C}) \pm \sqrt{[\mathbf{A}\cdot(\mathbf{P}_0 - \mathbf{C})]^2 - |\mathbf{A}|^2(|\mathbf{P}_0 - \mathbf{C}|^2 - r^2)}}{|\mathbf{A}|^2} \qquad \text{(xi)}$$

Then explain the geometric significance of the numerical value of the expression $[\mathbf{A}\cdot(\mathbf{P}_0 - \mathbf{C})]^2 - |\mathbf{A}|^2(|\mathbf{P}_0 - \mathbf{C}|^2 - r^2)$ being positive, negative, or zero.

EXERCISE 6 In this exercise we will use formula (xi) to find the intersection of a light ray and a lens surface. In your own sketch of Figure C introduce a coordinate system such that the x-axis coincides with the axis of symmetry and the y-axis goes through the point P_0. Assume that P_0 is 1 cm above the x-axis, so its coordinates are (0, 1), that the point V is on the x-axis 13 cm to the right of the origin, that the radius of the first (spherical) surface of the lens is 5 cm, and that the center of the sphere is on the axis of symmetry. If $\mathbf{A} = \langle 7, 1\rangle$ is a vector in the direction of the ray through P_0 and the top edge of the lens is 4 cm above the x-axis, find the point P where the ray intersects the first surface of the lens.

Hint: From the given information and your figure conclude that $\mathbf{A} = \langle 7, 1\rangle$, $\mathbf{P}_0 = \langle 0, 1\rangle$, $r = 5$ and $\mathbf{C} = \langle 18, 0\rangle$. Then find $\mathbf{P}_0 - \mathbf{C}$, and use formula (xi) to show that the two values for t are $t = 2$ or 3. Then use equation (i) to find the points corresponding to the two values for t. Which, if any, is the point of intersection of the ray with the first surface? What is the geometric significance of the other point?

Real optical systems live in three-dimensional space, not in a two-dimensional plane, and not all rays in a rotationally symmetric system lie in a plane containing the axis of symmetry. It is surprising that our vector solution to the problem of finding the intersection of a line and a circle in the plane is, through the step-by-step derivation of equation (xi), exactly the same as the vector solution for the problem of the intersection of a line with a spherical surface in three-dimensional space. In three dimensions points have three coordinates and vectors have three components, but a line is still determined by a point it passes through and a vector in the direction of the line. So equation (i) is also the equation for a line in three dimensions. Every point on a spherical surface is the same distance from the center of the sphere so equation (iii) holds. The rest is vector algebra, which works the same in *any* dimension: This is a major advantage of a vector approach.

9.5 PARAMETRIC EQUATIONS

We will introduce the idea of parametric equations through a simple example. Suppose that we have a point $P(x, y)$ that moves in the x-y plane and that the x- and y-coordinates of P at time t (in seconds) are given by the following pair of equations:

$$x = 2t \qquad \text{and} \qquad y = \frac{1}{2}t^2 \qquad (t \geq 0)$$

This pair of *parametric equations* tells us the location of the point P at any time t. For instance, when $t = 1$, we have

$$x = 2t = 2(1) = 2 \qquad \text{and} \qquad y = \frac{1}{2}t^2 = \frac{1}{2}(1)^2 = \frac{1}{2}$$

TABLE 1

t	x	y
0	0	0
1	2	$\frac{1}{2}$
2	4	2
3	6	$\frac{9}{2}$
4	8	8
5	10	$\frac{25}{2}$

In other words, after 1 second, the coordinates of P are $(2, \frac{1}{2})$. Let's see where P is after 2 seconds. We have

$$x = 2t = 2(2) = 4 \qquad \text{and} \qquad y = \frac{1}{2}t^2 = \frac{1}{2}(2)^2 = 2$$

So, after 2 seconds, the location of the point P is $(4, 2)$. In Table 1 we have computed the values of x and y (and hence the location of P) for integral values of t running from 0 to 5. Figure 1(a) shows the points that are determined. (To avoid cluttering the figure, we have labeled only the points for $t = 2, \ldots, 5$.) In Figure 1(b) we have joined the points with a smooth curve. [If we were not certain of the pattern emerging in Figure 1(a), we could have computed additional points using fractional values for t.]

The curve in Figure 1(b) appears to be a portion of a parabola. We can confirm this as follows. From the equation $x = 2t$ we obtain $t = x/2$. Now we use this result to substitute for t in the equation $y = \frac{1}{2}t^2$. This yields

$$y = \frac{1}{2}\left(\frac{x}{2}\right)^2 = \frac{1}{8}x^2$$

We conclude from this that the curve in Figure 1(b) is a portion of the parabola $y = \frac{1}{8}x^2$. The equations $x = 2t$ and $y = \frac{1}{2}t^2$ (with the restriction $t \geq 0$) are **parametric equations** for the curve. The variable t is called the **parameter.** In our initial example the parameter t represented time. In other cases the parameter might have other interpretations, for example, as the slope of a certain line, as an angle, or as the length of a curve. See the project at the end of this section. In still other cases there may be no immediate interpretation for the parameter.

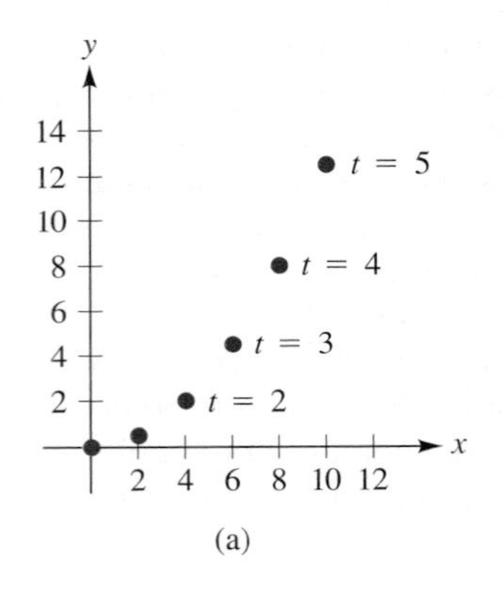

(a)

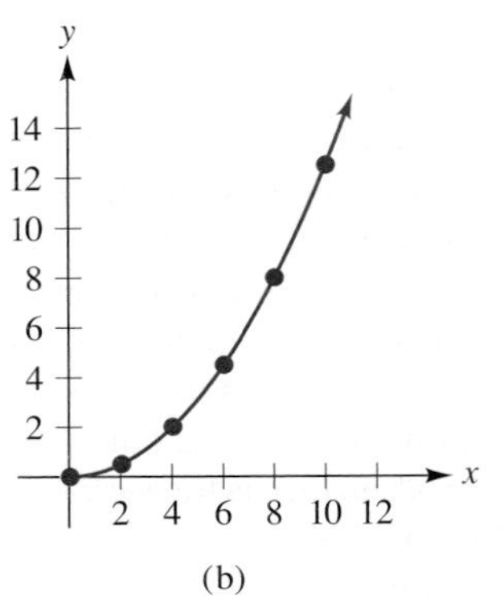

(b)

Figure 1
$x = 2t,\ y = \frac{1}{2}t^2 \quad (t \geq 0)$.

Before looking at additional examples, we point out two advantages in using parametric equations to describe curves. First, by restricting the values of the parameter (as we did in our initial example), we can focus on specific portions of a curve. Second, parametric equations let us think of a curve as a path traced out by a moving point; as the parameter t increases, a definite direction of motion, called the **orientation** of the curve, is established.

In physics parametric equations are often used to describe the motion of an object. Suppose, for example, that a ball is thrown from a height of 6 feet, with an initial speed of 88 ft/sec and at an angle of 35° with the horizontal, as shown in Fig-

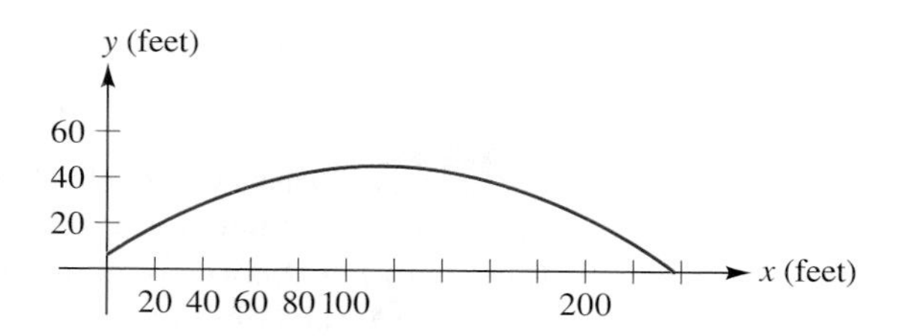

Figure 2
The path of a ball thrown from a height of 6 ft, with an initial speed of 88 ft/s, at an angle of 35° with the horizontal

ure 2. Then (neglecting air resistance and spin), it can be shown that the parametric equations for the path of the ball are

$$x = (88 \cos 35°)t \tag{1}$$

$$y = 6 + (88 \sin 35°)t - 16t^2 \tag{2}$$

In these equations x and y are measured in feet and t is in seconds, with $t = 0$ corresponding to the instant the ball is thrown.

With equations (1) and (2) we can calculate the location of the ball at any time t. For instance, to determine the location when $t = 1$ second, we substitute $t = 1$ in the equations to obtain

$$x = (88 \cos 35°)(1) \approx 72.1 \text{ ft}$$

and

$$y = 6 + (88 \sin 35°)(1) - 16(1)^2 \approx 40.5 \text{ ft}$$

So after one second the ball has traveled a horizontal distance of approximately 72.1 ft, and the height of the ball is approximately 40.5 ft. In Figure 3 we display this information along with the results for similar calculations corresponding to $t = 2$ and $t = 3$ seconds. [You should use equations (1) and (2) and your calculator to check the results in Figure 3 for yourself.]

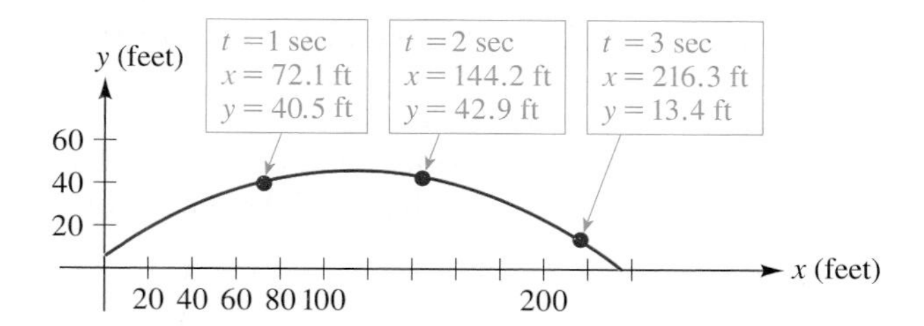

Figure 3
The location of the ball after 1, 2, and 3 seconds

From Figure 3 we can see that the total horizontal distance traveled by the ball in flight is something between 220 ft and 240 ft. Using the parametric equations for the path of the ball, we can determine this distance exactly. We can also find out how long the ball is in the air. When the ball does hit the ground, its y-coordinate is zero. Replacing y by zero in equation (2), we have

$$0 = 6 + (88 \sin 35°)t - 16\,t^2$$

This is a quadratic equation in the variable t. To solve for t, we use the quadratic formula

$$t = \frac{-b \pm \sqrt{b^2 - 4ac}}{2a}$$

with the following values for a, b, and c:

$$a = -16 \qquad b = 88 \sin 35° \qquad c = 6$$

As Exercise 19 asks you to check, the results are $t \approx 3.27$ and $t \approx -0.11$. We discard the negative root here because in equations (1) and (2) and in Figure 3, t represents time, with $t = 0$ corresponding to the instant that the ball is thrown. Now, using $t = 3.27$, we can compute x:

$$x = (88 \cos 35°)t \approx (88 \cos 35°)(3.27) \approx 235.7 \text{ ft}$$

In summary, the ball is in the air for about 3.3 seconds, and the total horizontal distance is approximately 236 feet.

Just as in Figure 1, the curve in Figures 2 and 3 is a parabola. Exercise 22 asks you to verify this by solving equation (1) for t in terms of x and then substituting the result in equation (2). This process of obtaining an explicit equation relating x and y from the parametric equations is referred to as *eliminating the parameter.* The next example shows a technique that is often useful in this context. As background for this example, we point out that the graph of an equation of the form $\frac{x^2}{a^2} + \frac{y^2}{b^2} = 1$ is an *ellipse,* as is indicated in Figure 4. (We will study this curve in detail in Chapter 11.)

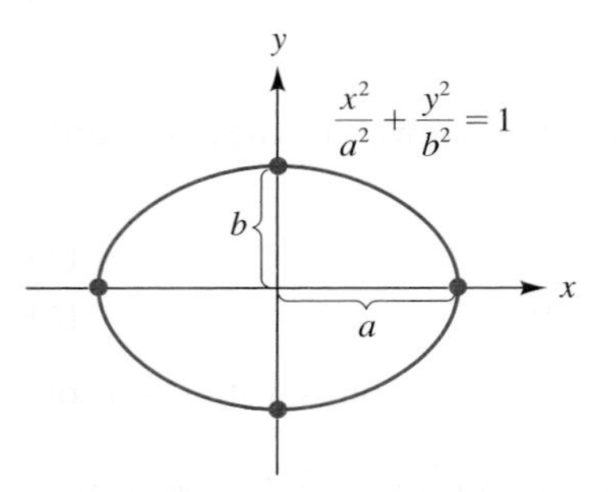

Figure 4
The ellipse $(x^2/a^2) + (y^2/b^2) = 1$ is symmetric about both coordinate axes. The intercepts of the ellipse are $x = \pm a$ and $y = \pm b$.

EXAMPLE 1 Parametric equations for an ellipse

The parametric equations of an ellipse are

$$x = 6 \cos t \qquad \text{and} \qquad y = 3 \sin t \quad (0 \leq t \leq 2\pi)$$

(a) Eliminate the parameter t to obtain an x-y equation for the curve.
(b) Graph the ellipse and indicate the points corresponding to $t = 0, \pi/2, \pi, 3\pi/2$, and 2π. As t increases, what is the direction of travel along the curve?

SOLUTION

(a) So that we can apply the identity $\cos^2 t + \sin^2 t = 1$, we divide the first equation by 6 and the second by 3. This gives us

$$\frac{x}{6} = \cos t \qquad \text{and} \qquad \frac{y}{3} = \sin t$$

Squaring and then adding these two equations yields

$$\left(\frac{x}{6}\right)^2 + \left(\frac{y}{3}\right)^2 = \cos^2 t + \sin^2 t = 1$$

or

$$\frac{x^2}{6^2} + \frac{y^2}{3^2} = 1$$

This is the x-y equation for the ellipse.

(b) Figure 5 shows the graph of the ellipse $(x^2/6^2) + (y^2/3^2) = 1$. When $t = 0$, the parametric equations yield

$$x = 6 \cos 0 = 6 \qquad \text{and} \qquad y = 3 \sin 0 = 0$$

Thus with $t = 0$ we obtain the point (6, 0) on the ellipse, as is indicated in Figure 5. The points corresponding to $t = \pi/2, \pi, 3\pi/2$, and 2π are ob-

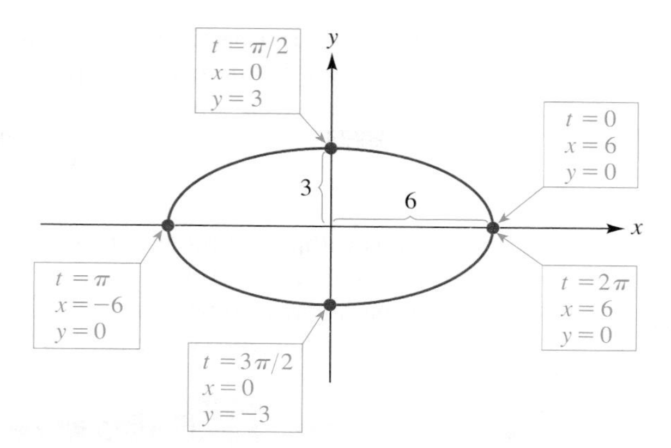

Figure 5
Parametric equations for this ellipse are $x = 6 \cos t$ and $y = 3 \sin t$. The x-y equation is $(x^2/6^2) + (y^2/3^2) = 1$.

tained similarly. Note that as t increases, the direction of travel around the ellipse is counterclockwise and that when $t = 2\pi$, we are back to the starting point, (6, 0).

The parametric equations $x = 6 \cos t$ and $y = 3 \sin t$ that we graphed in Figure 5 are by no means the only parametric equations yielding that ellipse. Indeed, there are numerous parametric equations that give rise to the same ellipse. One such pair is

$$x = 6 \sin t \qquad \text{and} \qquad y = 3 \cos t \quad (0 \le t \le 2\pi)$$

As you can check for yourself, these equations lead to the same x-y equation as before, namely, $(x^2/6^2) + (y^2/3^2) = 1$. The difference now is that in tracing out the curve from the parametric equations, we start (when $t = 0$) and end (when $t = 2\pi$) with the point (0, 3) on the ellipse and we travel clockwise rather than counterclockwise. Similarly, the equations

$$x = 6 \sin 2t \qquad \text{and} \qquad y = 3 \cos 2t \quad (0 \le t \le 2\pi)$$

also produce the ellipse. In this case, as you can check for yourself, as t runs from 0 to 2π, we make two complete trips clockwise around the ellipse. If we think of t as time, then the parametric equations $x = 6 \sin 2t$ and $y = 3 \cos 2t$ describe a point traveling around the ellipse $(x^2/6^2) + (y^2/3^2) = 1$ twice as fast as would be the case with the equations $x = 6 \sin t$ and $y = 3 \cos t$.

In Example 1 we saw that the parametric equations $x = 6 \cos t$ and $y = 3 \sin t$ represent an ellipse. The same method that we used in Example 1 can be used to establish the following general results.

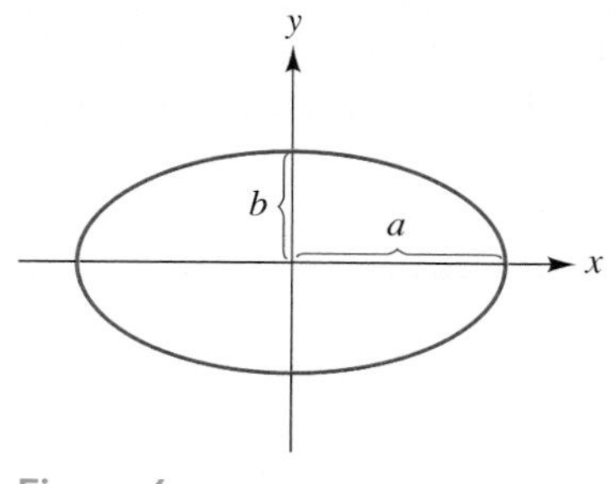

Figure 6
$x = a \cos t$ and $y = b \sin t$
or
$x = a \sin t$ and $y = b \cos t$,
both for $0 \le t \le 2\pi$

Parametric Equations for the Ellipse

Let a and b be positive constants. Then the parametric equations

$$x = a \cos t \qquad \text{and} \qquad y = b \sin t \quad (0 \le t \le 2\pi)$$

represent the ellipse shown in Figure 6. As t increases, we move around the ellipse in a counterclockwise direction. The parametric equations

$$x = a \sin t \qquad \text{and} \qquad y = b \cos t \quad (0 \le t \le 2\pi)$$

also represent the ellipse in Figure 6. Here, the direction of motion is clockwise.

There is a particular case involving the parametric equations $x = a\cos t$ and $y = b\sin t$ that deserves mention. When a and b are equal, we have $x = a\cos t$ and $y = a\sin t$, from which we deduce that

$$\begin{aligned} x^2 + y^2 &= a^2\cos^2 t + a^2\sin^2 t \\ &= a^2(\cos^2 t + \sin^2 t) = a^2 \end{aligned}$$

Thus with $a = b$ the parametric equations describe a circle of radius a. And specializing further still, if $a = b = 1$, we obtain $x = \cos t$ and $y = \sin t$, for $0 \le t \le 2\pi$, as parametric equations for the unit circle. This agrees with our unit circle definitions for sine and cosine back in Section 7.1.

EXAMPLE 2 Finding an x-y equation given parametric equations

The position of a point $P(x, y)$ at time t is given by the parametric equations

$$x = 1 + 2t \qquad \text{and} \qquad y = 2 + 4t \quad (t \ge 0)$$

Find an x-y equation for the path traced out by the point P.

SOLUTION

From the parametric equation for x we obtain $t = (x - 1)/2$. Using this to substitute for t in the second parametric equation, we have

$$y = 2 + 4\left(\frac{x-1}{2}\right) = 2 + 2(x - 1) = 2x$$

This tells us that the point P moves along the line $y = 2x$. However, the entire line is not traced out, but only a portion of it. This is because of the original restriction $t \ge 0$. To see which portion of the line is described by the equations, we can successively let $t = 0, 1$, and 2 in the parametric equations. As you can check, the points obtained are (1, 2), (3, 6), and (5, 10). So as t increases, we move to the right along the line $y = 2x$, starting from the point (1, 2); see Figure 7.

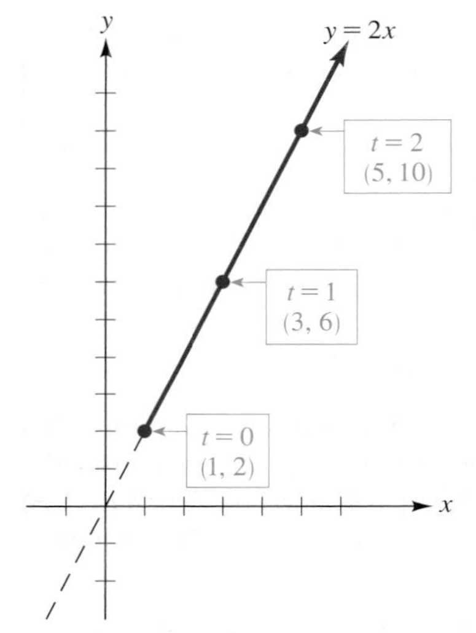

Figure 7

The parametric equations $x = 1 + 2t$ and $y = 2 + 4t$, with the restriction $t \ge 0$, describe the portion of the line $y = 2x$ to the right of and including the point (1, 2).

One of the recurrent techniques that we have used for analyzing parametric equations in this section has been that of eliminating the parameter. It is important to note, however, that it is not always a simple matter to eliminate the parameter. Indeed, in most cases it is not even possible. When this occurs, we have several techniques available. We can set up a table with t, x, and y and plot points; we can use the techniques of calculus; or we can use a graphing utility to obtain the graph. Even when we use a graphing utility, however, the techniques of calculus are helpful in understanding why the graph looks as it does.

EXERCISE SET 9.5

A

In Exercises 1–6 you are given the parametric equations of a curve and a value for the parameter t. Find the coordinates of the point on the curve corresponding to the given value of t.

1. $x = 2 - 4t,\ y = 3 - 5t;\ t = 0$

2. $x = 3 - t^2,\ y = 4 + t^3;\ t = -1$

3. $x = 5\cos t,\ y = 2\sin t;\ t = \pi/6$

4. $x = 4\cos 2t,\ y = 6\sin 2t;\ t = \pi/3$

5. $x = 3\sin^3 t,\ y = 3\cos^3 t;\ t = \pi/4$

6. $x = \sin t - \sin 2t,\ y = \cos t + \cos 2t;\ t = 2\pi/3$

Ⓖ **7.** Use the standard viewing rectangle for this exercise.

(a) Graph the parametric equations $x = 2t$, $y = 0.5t^2$ with t-values running from 0 to 1; from 0 to 3; from 0 to 4. Compare the results (in a complete sentence or two).

(b) Graph the parametric equations given in part (a) using t-values running from -5 to 5. Compare your picture to the graph shown in Figure 1(b) on page 690. What are the restrictions on t in Figure 1(b)?

In Exercises 8–17, graph the parametric equations after eliminating the parameter t. Specify the direction on the curve corresponding to increasing values of t. For Exercises 8–11, t can be any real number; for Exercises 12–17, $0 \le t \le 2\pi$.

8. $x = t + 1,\ y = t^2$

9. $x = 2t - 1,\ y = t^2 - 1$

10. $x = t^2 - 1,\ y = t + 1$

11. $x = t - 4,\ y = |t|$

12. $x = 5 \cos t,\ y = 2 \sin t$

13. $x = 2 \sin t,\ y = 3 \cos t$

14. $x = 4 \cos 2t,\ y = 6 \sin 2t$

15. $x = 2 \cos(t/2),\ y = \sin(t/2)$

16. **(a)** $x = 2 \cos t,\ y = 2 \sin t$
(b) $x = 4 \cos t,\ y = 2 \sin t$

17. **(a)** $x = 3 \sin t,\ y = 3 \cos t$
(b) $x = 5 \sin t,\ y = 3 \cos t$

Ⓖ **18.** **(a)** Graph the parametric equations $x = (88 \cos 35°)t$, $y = 6 + (88 \sin 35°)t - 16t^2$. Use settings that will give you a picture similar to Figure 3 on page 691.
(b) Use the graphing utility to estimate the x-intercept of the graph. Check that your answer is consistent with the value determined in the text.

19. In the text we said that the solutions of the quadratic equation $0 = 6 + (88 \sin 35°)t - 16t^2$ are $t \approx 3.27$ and $t \approx -0.11$. Use the quadratic formula and your calculator to verify these results.

20. The following figure shows the parametric equations and the path for a ball thrown from a height of 5 ft, with an initial speed of 100 ft/sec and at an angle of 70° with the horizontal.

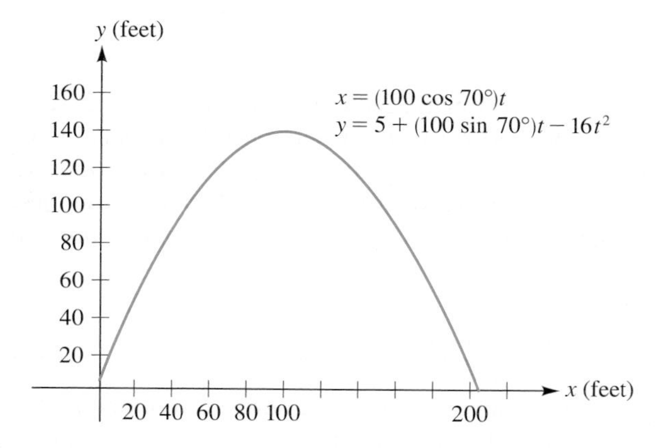

(a) Compute the x- and y-coordinates of the ball when $t = 1$, 2, and 3 seconds. (Round the answers to one decimal place.)
(b) How long is the ball in flight? (Round the answer to two decimal places.) What is the total horizontal distance traveled by the ball before it lands? (Round to the nearest foot.) Check that your answer is consistent with the figure.

21. The figure shows the parametric equations and the path for a ball thrown from a height of 5 ft, with an initial speed of 100 ft/s and at an angle of 45° with the horizontal. (So except for the initial angle, the data are the same as in Exercise 20.)

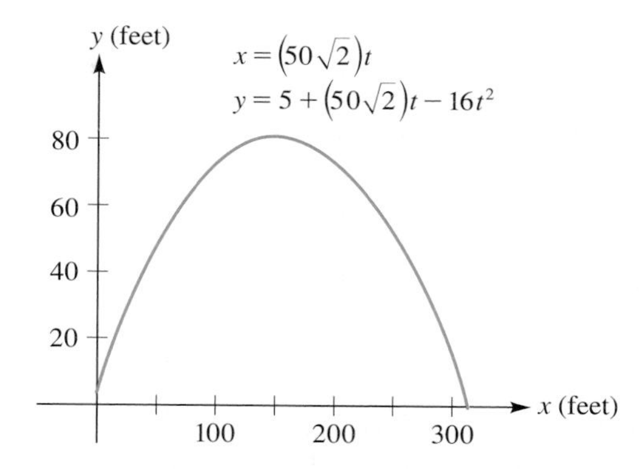

(a) Compute the x- and y-coordinates of the ball when $t = 1$, 2, and 3 seconds. (Round the answers to one decimal place.)
(b) How long is the ball in flight? (Round the answer to two decimal places.) What is the total horizontal distance traveled by the ball before it lands? (Round to the nearest foot.)

22. Refer to parametric equations (1) and (2) in the text. By eliminating the parameter t, show that the equations describe a parabola. (That is, show that the resulting x-y equation has the form $y = ax^2 + bx + c$.)

In Exercises 23–39, graph the parametric equations using the given range for the parameter t. In each case, begin with the standard viewing rectangle and then make adjustments, as necessary, so that the graph utilizes as much of the viewing screen as possible. For example, in graphing the circle given by $x = \cos t$ and $y = \sin t$, it would be natural to choose a viewing rectangle extending from -1 to 1 in both the x- and y-directions.

Ⓖ **23.** **(a)** $x = 3t + 2,\ y = 3t - 2,\ (-2 \le t \le 2)$
(b) $x = 3t + 2,\ y = 3t - 2,\ (-3 \le t \le 3)$

Ⓖ **24.** $x = \ln(3t + 2),\ y = \ln(3t - 2)$ using:
(a) $2/3 < t \le 1$
(b) $2/3 < t \le 10$
(c) $2/3 < t \le 100$
(d) $2/3 < t \le 1000$

Ⓖ **25.** $x = 4 \cos t,\ y = 3 \sin t,\ 0 \le t \le 2\pi$ *(ellipse)*

Ⓖ **26.** $x = 4 \cos t,\ y = -3 \sin t,\ 0 \le t \le 2\pi$ (the same as the ellipse in Exercise 25 but traced out in the opposite direction)

Ⓖ **27.** $x = 4 \cos t,\ y = 3 \sin t,\ 0 \le t \le \pi/2$ (one-quarter of an ellipse)

B

28. **(a)** The curve in the accompanying figure is called an *astroid* (or a *hypocycloid of four cusps*). A pair of parametric equations for the curve is $x = \cos^3 t$, $y = \sin^3 t$. By eliminating the parameter t, find the x-y equation for the curve. *Hint*: In each equation, raise both sides to the two-thirds power.

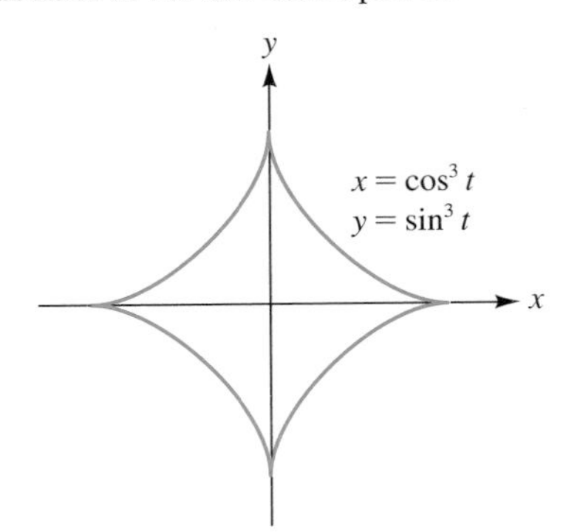

Ⓖ **(b)** Graph the curve in part (a) for $0 \le t \le 2\pi$ to reproduce the figure above. [The hypocycloid of four cusps was first studied by the Danish astronomer Olaf Roemer (1644–1710) and by the Swiss mathematician Jacob Bernoulli (1654–1705).]

Ⓖ **29.** $x = 2\cos t + \cos 2t$, $y = 2\sin t - \sin 2t$, $0 \le t \le 2\pi$ [This curve is the *deltoid*. It was first studied by the Swiss mathematician Leonhard Euler (1707–1783).]

Ⓖ **30.** $x = \dfrac{\cos t}{1 + \sin^2 t}$, $y = \dfrac{\sin t \cos t}{1 + \sin^2 t}$, $0 \le t \le 2\pi$
[This curve is the *lemniscate of Bernoulli*. The Swiss mathematician Jacob Bernoulli (1654–1705) studied the curve and took the name *lemniscate* from the Greek *lemniskos*, meaning "ribbon."]

Ⓖ **31.** $x = 2\tan t$, $y = 2\cos^2 t$, $0 \le t \le 2\pi$ *Remark*: If you eliminate the parameter t, you'll find that the Cartesian form of the curve is $y = 8/(x^2 + 4)$. (Verify this last statement, first algebraically, then graphically.) The curve is known as the *witch of Agnesi*, named after the Italian mathematician and scientist Maria Gaetana Agnesi (1718–1799). The word "witch" in the name of the curve is the result of a mistranslation from Italian to English. In Agnesi's time, the curve was known as *la versiera*, an Italian name with a Latin root meaning "to turn." In translation, the word *versiera* was confused with another Italian word *avversiera*, which means "wife of the devil" or "witch."

Ⓖ **32.** $x = \dfrac{3t}{1 + t^3}$, $y = \dfrac{3t^2}{1 + t^3}$, $-\infty < t < \infty$
Hint: Use $-10 \le t \le 10$. [This curve is the *folium of Descartes*. The word *folium* means "leaf." When Descartes drew the curve in 1638, he did not use negative values for coordinates. Thus he obtained only the first-quadrant portion of the curve, which resembles a leaf or loop. The complete graph of the curve was first given by the Dutch mathematician and scientist Christian Huygens in 1692. Huygens's graph is reproduced on page 108 in John Stillwell's *Mathematics and Its History*, Second Edition (New York: Springer Verlag, 2002).]

Ⓖ **33.** $x = 3t^2$, $y = 2t^3$, $-2 \le t \le 2$ (*semicubical parabola*)

Ⓖ **34.** $x = \sec t$, $y = 2\tan t$, $0 \le t \le 2\pi$ (*hyperbola*)

Ⓖ **35.** $x = 3(t^2 - 3)$, $y = t(t^2 - 3)$, $-3 \le t \le 3$ (*Tschirnhausen's cubic*)

Ⓖ **36.** $x = 8\cos t + 2\cos 4t$, $y = 8\sin t - 2\sin 4t$, $0 \le t \le 2\pi$ (*hypocycloid with five cusps*)

Ⓖ **37.** $x = 8\cos t + \cos 8t$, $y = 8\sin t - \sin 8t$, $0 \le t \le 2\pi$ (*hypocycloid with nine cusps*)

Ⓖ **38.** $x = \cos t + t\sin t$, $y = \sin t - t\cos t$ (*involute of a circle*)
(a) $-\pi \le t \le \pi$
(b) $-2\pi \le t \le 2\pi$
(c) $-4\pi \le t \le 4\pi$

Ⓖ **39.** $x = \sin(0.8t + \pi)$, $y = \sin t$, $0 \le t \le 10\pi$ (*Bowditch curve*)

NOTE Historical background and information about many of the curves mentioned in Exercises 28 through 39 (and throughout Section 9.7) is available on the Internet through the World Wide Web. The address is http://www-groups.dcs.st-and.ac.uk/~history/Curves/. Another excellent source for information on these curves, and many others, is J. Dennis Lawrence's book *A Catalogue of Special Plane Curves* (Dover Publications, 1972).

PROJECT PARAMETERIZATIONS FOR LINES AND CIRCLES

In the previous project on lines, circles, and ray tracing we derived a vector equation for a line in the x-y plane. You and your group might want to review that derivation. In particular, let P_0 be a point in the x-y plane, $\mathbf{P}_0$ be the position vector with endpoint P_0, and $\mathbf{A}$ be a nonzero vector. We showed that a vector equation for the line in the x-y plane passing through the point P_0 with direction vector $\mathbf{A}$ is

$$\mathbf{P} = \mathbf{P}_0 + t\mathbf{A} \quad \text{for } -\infty < t < \infty \tag{i}$$

where $\mathbf{P}$ is the position vector of a point P on the line (see Figure A).

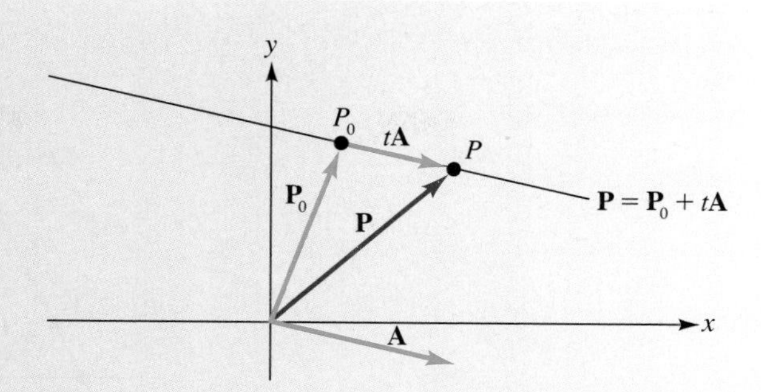

Figure A

Using equation (i) and vector components we can derive two useful scalar equations for the same line. Let $P_0 = (x_0, y_0)$ be the given point on the line, $\mathbf{A} = \langle a, b\rangle \neq \langle 0, 0\rangle$ be the nonzero direction vector of the line, and $P = (x, y)$ be an arbitrary point on the line. The corresponding position vectors for the points P_0 and P are $\mathbf{P}_0 = \langle x_0, y_0\rangle$ and $\mathbf{P} = \langle x, y\rangle$, respectively. Substituting into equation (i) we get

$$\begin{aligned} \mathbf{P} &= \mathbf{P}_0 + t\mathbf{A} \qquad \text{for } -\infty < t < \infty \\ \langle x, y\rangle &= \langle x_0, y_0\rangle + t\langle a, b\rangle \\ \langle x, y\rangle &= \langle x_0, y_0\rangle + \langle ta, tb\rangle \\ \langle x, y\rangle &= \langle x_0 + ta, y_0 + tb\rangle \end{aligned}$$

So

$$\begin{cases} x = x_0 + ta \\ y = y_0 + tb \end{cases} \qquad \text{for } -\infty < t < \infty \tag{ii}$$

Equations (ii) are called **parametric equations** for this line. If both a and b are nonzero then we can solve each equation in (ii) for t to get

$$\frac{x - x_0}{a} = t = \frac{y - y_0}{b}$$

or more simply,

$$\frac{x - x_0}{a} = \frac{y - y_0}{b} \tag{iii}$$

Equation (iii) is called the **symmetric equation** for this line. If either a or b is zero then the line has no symmetric equation.

Let's apply equations (i), (ii), and (iii) to find vector, parametric, and symmetric equations for the line in the x-y plane passing through the point $P_0 = (-2, 1)$ with direction $\mathbf{A} = \langle 1, 2\rangle$. See Figure B. (*Note*: This is a continuation of Example 1 from the previous project on lines, circles, and ray tracing.) A vector equation for this line is

$$\begin{aligned} \mathbf{P} &= \mathbf{P}_0 + t\mathbf{A} \qquad \text{for } -\infty < t < \infty \\ \langle x, y\rangle &= \langle -2, 1\rangle + t\langle 1, 2\rangle \qquad \text{for } -\infty < t < \infty \end{aligned}$$

Then

$$\langle x, y\rangle = \langle -2 + t, 1 + 2t\rangle \qquad \text{for } -\infty < t < \infty$$

And we obtain parametric equations for the line

$$\begin{cases} x = -2 + t \\ y = 1 + 2t \end{cases} \qquad \text{for } -\infty < t < \infty$$

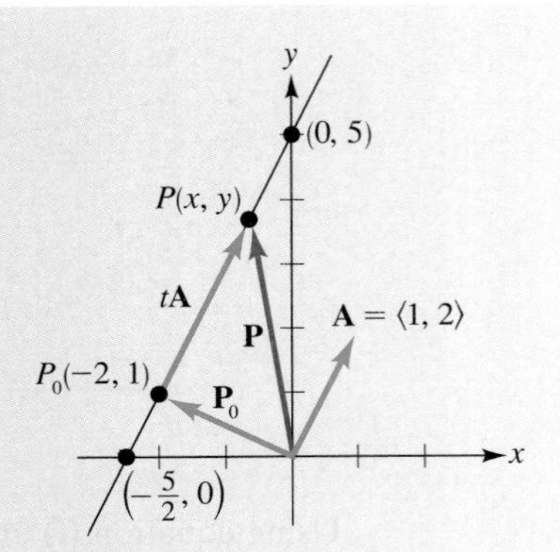

Figure B

Let's use the parametric equations to find the x and y-intercepts of the line. For the x-intercept, $y = 0$ implies $1 + 2t = 0$ or $t = -1/2$. Then the x-intercept is $x = -2 - 1/2 = -5/2$. For the y-intercept, $x = 0$ implies $-2 + t = 0$ or $t = 2$. Then the y-intercept is $y = 1 + 2(2) = 5$. See Figure B. Finally, eliminating the parameter t in the parametric equations gives us the symmetric equation for this line,

$$\frac{x + 2}{1} = \frac{y - 1}{2}$$

From the symmetric equation we have

$$\begin{aligned} x + 2 &= \frac{y - 1}{2} \\ 2(x + 2) &= y - 1 \\ y &= 2x + 5 \end{aligned}$$

the slope-intercept equation for this line.

It is worth noting that all of our work with vector, parametric, and symmetric equations for lines generalizes in a natural way to three-dimensional and higher-dimensional Cartesian "spaces." We also note that there are other parametric equations for lines, but they do not concern us here.

Now we want to derive parametric equations for a circle in the x-y plane. Let's start with the fundamental case of the circle of radius a centered at the origin. Stand at the center of this circle and look along the positive x-axis to see the point $(a, 0)$ on the circle. Rotate your head (and body!) counterclockwise to view each point on the circle until your view returns to the point $(a, 0)$. Each point you see on the circle is at a distance a from you. A convenient way to keep track of which point on the circle you are viewing is to use the angle θ of counterclockwise rotation from the positive x-axis to the point (see Figure C). As we know from basic trigonometry, each point P on the circle has coordinates $(a \cos \theta, a \sin \theta)$ for some angle θ, $0 \leq \theta \leq 2\pi$. So the x and y-coordinates of P are $a \cos \theta$ and $a \sin \theta$, respectively, and we have

$$\begin{cases} x = a \cos \theta \\ y = a \cos \theta \end{cases} \quad \text{for } 0 \leq \theta \leq 2\pi \tag{iv}$$

Equations (iv) are called parametric equations for the circle in the x-y plane centered at the origin with radius a.

As with lines, these are not the only parametric equations for this circle. We can use equations (iv) and a little geometry to derive more "intrinsic" paramet-

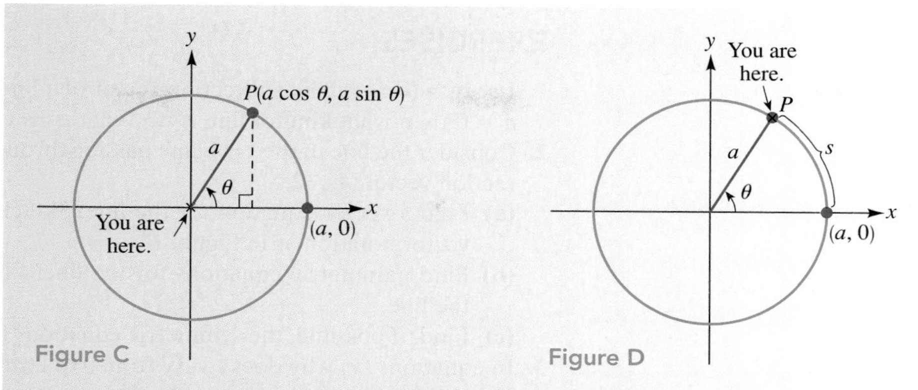

Figure C

Figure D

ric equations for this circle. Suppose, instead of being lucky enough to be standing at the center of the circle, we are standing at a point P *on* the circle. Consider the arc-length s from the point $(a, 0)$ to P. (See Figure D.) We know from geometry that $s = a\theta$, so $\theta = s/a$. Substituting into equations (iv) we obtain an **arc-length parameterization** for this circle.

$$\begin{cases} x = a\cos\left(\dfrac{s}{a}\right) \\ y = a\cos\left(\dfrac{s}{a}\right) \end{cases} \quad \text{for } 0 \leq s \leq 2\pi a \tag{v}$$

In Exercise 3 we ask you to explain why s in equations (v) varies from 0 to $2\pi a$.

Finally we mention another pair of parametric equations for the *unit* circle, that is, the circle in the x-y plane centered at the origin with radius 1. Consider the line passing through the point $(-1, 0)$ on the unit circle with slope m. In Exercise 6 you will show that this line intersects the unit circle at the point $P\left(\dfrac{1 - m^2}{1 + m^2}, \dfrac{2m}{1 + m^2}\right)$. See Figure E. This construction is equivalent to the one discussed in the Writing Mathematics section at the end of Chapter 10, and can also be used to generate all Pythagorean triples. From this construction we obtain our third pair of parametric equations for the unit circle:

$$\begin{cases} x = \left(\dfrac{1 - m^2}{1 + m^2}\right) \\ y = \left(\dfrac{2m}{1 + m^2}\right) \end{cases} \quad \text{for } -\infty < m < \infty \tag{vi}$$

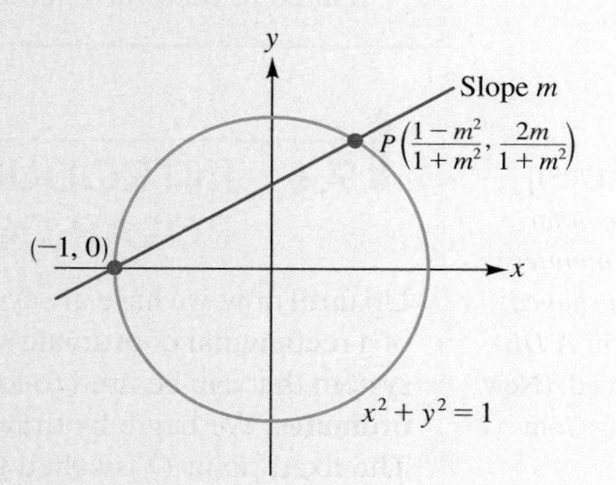

Figure E

EXERCISES

1. Let $\mathbf{A} = \langle a, b \rangle \neq \mathbf{0}$ be a direction vector of a line through the point (x_0, y_0). If $a = 0$ then what kind of line is it? What is its equation? What about $b = 0$?
2. Consider the line in the x-y plane passing through the point $(2, -3)$ with a direction vector $\langle 4, -2 \rangle$.
   (a) Find a vector equation for the line. Sketch the line with Cartesian and vector notation as in Figure B.
   (b) Find parametric equations for the line. Find the x- and y-intercepts of the line.
   (c) Find, if possible, the symmetric equations for the line.
3. In equations (v) why does s vary from 0 to $2\pi a$?
4. Consider the circle in the x-y plane centered at the origin with radius 12.
   (a) Use equations (iv) to find parametric equations for this circle.
   (b) Find an arc-length parameterization for this circle.
5. Consider the circle in the x-y plane centered at the point (h, k) with radius a.
   (a) Use a fixed distance and a variable angle to find parametric equations for this circle. *Hint*: Modify Figure C and equations (iv) by "translation." Sketch this circle and label it as in Figure C.
   (b) Find an arc-length parameterization for this circle.
6. Show that the line with slope m passing through the point $(-1, 0)$ intersects the unit circle at the point $P\left(\dfrac{1 - m^2}{1 + m^2}, \dfrac{2m}{1 + m^2}\right)$.
7. In Figure E let A be the point $(1, 0)$, P be the point $\left(\dfrac{1 - m^2}{1 + m^2}, \dfrac{2m}{1 + m^2}\right)$, and O be the origin, and let angle POA be θ.
   (a) Explain why $\cos\theta = \dfrac{1 - m^2}{1 + m^2}$ and $\sin\theta = \dfrac{2m}{1 + m^2}$.
   (b) Prove: $\tan(\theta/2) = m$.
   (c) Let Q be the point $(-1, 0)$ and explain why angle PQA equals $\theta/2$.
   (d) Prove: $\cos\theta = \dfrac{1 - \tan^2(\theta/2)}{1 + \tan^2(\theta/2)}$ and $\sin\theta = \dfrac{2\tan(\theta/2)}{1 + \tan^2(\theta/2)}$.

   Comment: In the study of *techniques of integration* in calculus the results of this exercise are the basis for a substitution that transforms a rational expression in sines and cosines to a rational expression in a single variable m. Exercises 55 and 52 on page 582 and Exercise 101 on page 632 cover results related to those in Exercises 6 and 7 above.

9.6 INTRODUCTION TO POLAR COORDINATES

[Jacob Bernoulli (1654–1705)] *was one of the first to use polar coordinates in a general manner, and not simply for spiral shaped curves.* —Florian Cajori in *A History of Mathematics,* 4th ed. (New York: Chelsea Publishing Company, 1985)

Up until now, we have always specified the location of a point in the plane by means of a rectangular coordinate system. In this section we introduce another coordinate system that can be used to locate points in the plane. This is the system of **polar coordinates.** We begin by drawing a half-line or ray emanating from a fixed point O. The fixed point O is called the **pole** or **origin,** and the half-line is called the **polar**

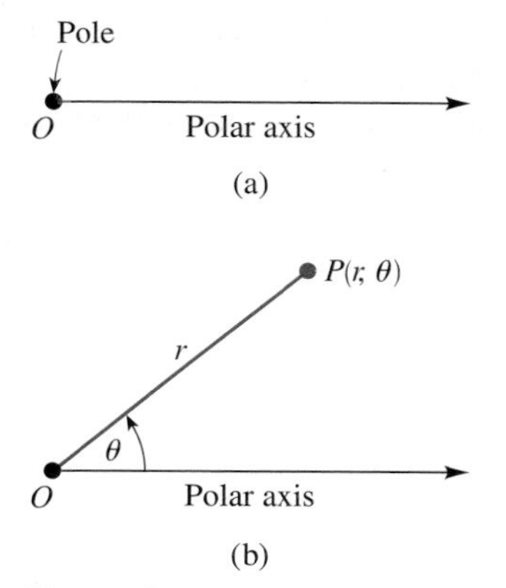

Figure 1

axis. The polar axis is usually depicted as being horizontal and extending to the right, as indicated in Figure 1(a). Now let P be any point in the plane. As is indicated in Figure 1(b), we initially let r denote the distance from O to P, and we let θ denote the angle measured from the polar axis to $\overline{OP}$. (Just as in our earlier work with angles in standard position, we take the measure of θ to be positive if the rotation is counterclockwise and negative if it is clockwise.) Then the ordered pair (r, θ) serves to locate the point P with respect to the pole and the polar axis. We refer to r and θ as polar coordinates of P, and we write $P(r, \theta)$ to indicate that P is the point with polar coordinates (r, θ).

Plotting points in polar coordinates is facilitated by the use of polar coordinate graph paper, such as that shown in Figure 2(a). Figure 2(b) shows the points with polar coordinates $A(2, \frac{\pi}{6})$, $B(3, \frac{2\pi}{3})$, and $C(4, 0)$.

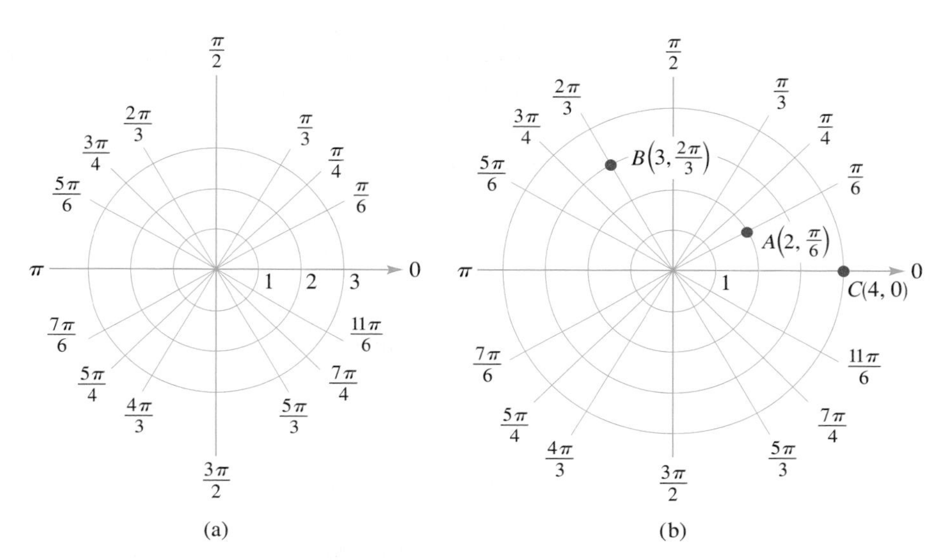

Figure 2

There is a minor complication that arises in using polar rather than rectangular coordinates. Consider, for example, the point $C(4, 0)$ in Figure 2(b). This point could just as well have been labeled with the coordinates $(4, 2\pi)$ or $(4, 2k\pi)$ for any integral value of k. Similarly, the coordinates (r, θ) and $(r, \theta + 2k\pi)$ represent the same point for all integral values of k. This is in marked contrast to the situation with rectangular coordinates, where the coordinate representation of each point is unique. Also, what polar coordinates should we assign to the origin? For if $r = 0$ in Figure 1(b), we cannot really define an angle θ. To cover this case we agree on the convention that the coordinates $(0, \theta)$ denote the origin for all values of θ. Finally, we point out that in working with polar coordinates, it is sometimes useful to let r take on negative values. For example, consider the point $P(2, \frac{5\pi}{4})$ in Figure 3. The coordinates $(2, \frac{5\pi}{4})$ indicate that to reach P from the origin, we go two units in the direction $5\pi/4$. Alternatively, we can describe this as -2 units in the $\pi/4$ direction. (Refer again to Figure 3.) For reasons such as this we will adhere to the convention that the polar coordinates (r, θ) and $(-r, \theta + \pi)$ represent the same point. (Since r can now, in fact, be positive, negative, or zero, r is referred to as a *directed distance.*)

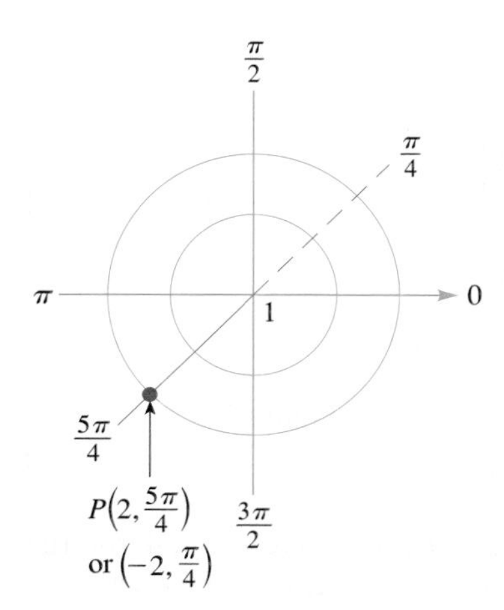

Figure 3

It is often useful to consider both rectangular and polar coordinates simultaneously. To do this, we draw the two coordinate systems so that the origins coincide

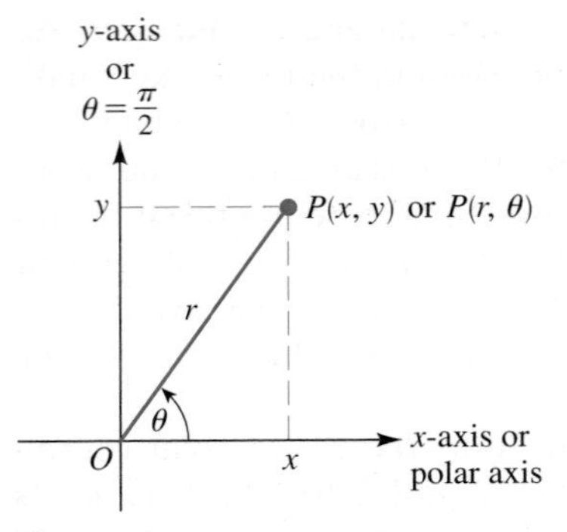

Figure 4

and the positive x-axis coincides with the polar axis; see Figure 4. Suppose now that a point P, other than the origin, has rectangular coordinates (x, y) and polar coordinates (r, θ), as indicated in Figure 4. We wish to find equations relating the two sets of coordinates. From Figure 4 we see that

$$\cos\theta = \frac{x}{r} \quad \text{so} \quad x = r\cos\theta$$

$$\sin\theta = \frac{y}{r} \quad \text{so} \quad y = r\sin\theta$$

$$x^2 + y^2 = r^2$$

$$\tan\theta = \frac{y}{x}$$

Although Figure 4 displays the point P in the first quadrant, it can be shown that these same equations hold when P is in any quadrant. For reference we summarize these equations as follows.

Relations Between Polar and Rectangular Coordinates

$$x = r\cos\theta$$
$$y = r\sin\theta$$
$$x^2 + y^2 = r^2$$
$$\tan\theta = \frac{y}{x}$$

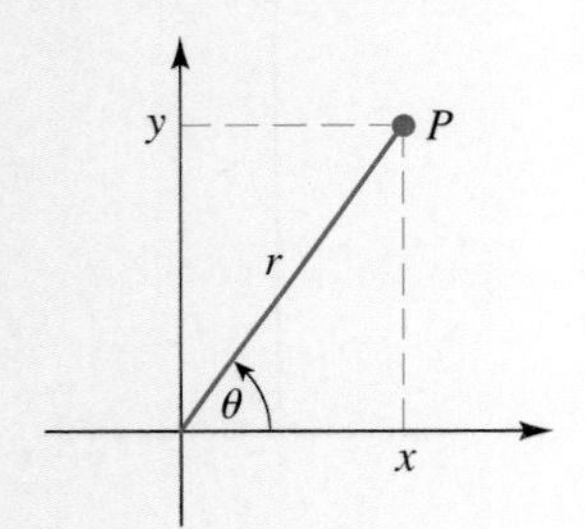

EXAMPLE 1 Changing from polar to rectangular coordinates

The polar coordinates of a point are $(5, \frac{\pi}{6})$. Find the rectangular coordinates.

SOLUTION

We are given that $r = 5$ and $\theta = \pi/6$. Thus

$$x = r\cos\theta = 5\cos\frac{\pi}{6} = 5\left(\frac{\sqrt{3}}{2}\right)$$

and

$$y = r\sin\theta = 5\sin\frac{\pi}{6} = 5\left(\frac{1}{2}\right)$$

The rectangular coordinates are therefore $(\frac{5}{2}\sqrt{3}, \frac{5}{2})$. See Figure 5.

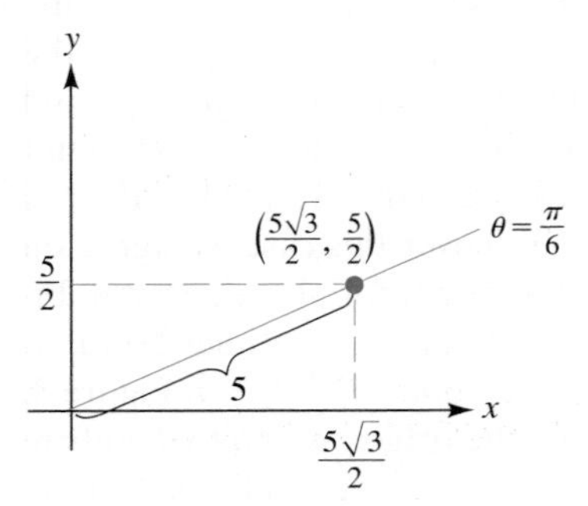

Figure 5

The definition of the graph of an equation in polar coordinates is similar to the corresponding definition for rectangular coordinates. The **graph** of an equation in polar coordinates is the set of all points (r, θ) with coordinates that satisfy the given equation. It is often the case that the equation of a curve is simpler in one coordinate system than in another. The next two examples show instances of this.

EXAMPLE 2 Converting a polar equation to rectangular form

Convert each polar equation to rectangular form:
(a) $r = \cos\theta + 2\sin\theta$; **(b)** $r^2 = \sin 2\theta$.

SOLUTION

(a) In view of the transformation equations $x = r\cos\theta$ and $y = r\sin\theta$, we multiply both sides of the given equation by r, for $r \neq 0$, to obtain

$$r^2 = r\cos\theta + 2r\sin\theta$$

and therefore

$$x^2 + y^2 = x + 2y \qquad \text{or} \qquad x^2 - x + y^2 - 2y = 0$$

Before concluding that we have a rectangular form of the original equation, we need to check what happens when $r = 0$. This means checking that the origin is on (or not on) the graph of both the original and the transformed equation. Since $r = 0$ for $\theta = \arctan(-1/2)$, the origin is on the graph of the original equation. The origin is also on the graph of the equation $x^2 - x + y^2 - 2y = 0$, so it is a rectangular form of the given equation.
Question for review: What is the graph of this last equation?

(b) Using the double-angle formula for $\sin 2\theta$, we have

$$r^2 = 2\sin\theta\cos\theta$$

Now, to obtain the expressions $r\sin\theta$ and $r\cos\theta$ on the right-hand side of the equation, we multiply both sides by r^2. This yields

$$r^4 = 2(r\sin\theta)(r\cos\theta)$$

and, consequently,

$$(x^2 + y^2)^2 = 2yx$$

or

$$x^4 + 2x^2y^2 + y^4 - 2xy = 0$$

Since the origin lies on the graph of both the given equation and the last equation, this is a rectangular form of the given equation. Notice how much simpler the equation is in its polar coordinate form.

EXAMPLE 3 Converting a rectangular equation to polar form

Convert the rectangular equation $x^2 + y^2 + ax = a\sqrt{x^2 + y^2}$ to polar form, expressing r as a function of θ. Assume that a is a constant.

SOLUTION

Using the relations $x^2 + y^2 = r^2$ and $x = r\cos\theta$, we obtain

$$r^2 + ar\cos\theta = ar$$

Notice that this equation is satisfied by $r = 0$. In other words, the graph of this equation will pass through the origin. This is consistent with the fact that the original equation is satisfied when x and y are both zero. Now assume for the moment that $r \neq 0$. Then we can divide both sides of the last equation by r to obtain

$$r + a\cos\theta = a \qquad \text{or} \qquad r = a - a\cos\theta$$

This expresses r as a function of θ, as required. Note that when $\theta = 0$, we obtain

$$r = a - a(1) = 0$$

That is, nothing has been lost in dividing through by r; the graph will still pass through the origin. Again, notice how much simpler the equation is in polar form.

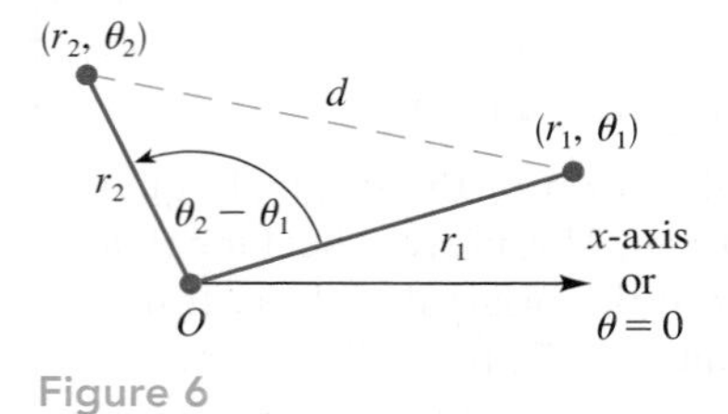

Figure 6

When we work in an x-y coordinate system, one of the basic tools is the formula for the distance between two points: $d = \sqrt{(x_2 - x_1)^2 + (y_2 - y_1)^2}$. There is a corresponding distance formula for use with polar coordinates. In Figure 6 we let d denote the distance between the points with polar coordinates (r_1, θ_1) and (r_2, θ_2). Then, applying the law of cosines, we have $d^2 = r_1^2 + r_2^2 - 2r_1r_2\cos(\theta_2 - \theta_1)$. This is the distance formula in polar coordinates. For ease of reference we repeat this result in the box that follows.

PROPERTY SUMMARY
Distance Formula for Polar Coordinates

Let d denote the distance between two points with polar coordinates (r_1, θ_1) and (r_2, θ_2). Then

$$d^2 = r_1^2 + r_2^2 - 2r_1r_2\cos(\theta_2 - \theta_1)$$

EXAMPLE 4 Finding the distance between points in polar coordinates

Compute the distance between the points with polar coordinates $(2, \frac{5\pi}{6})$ and $(4, \frac{\pi}{6})$.

SOLUTION

We use the distance formula, taking the points (r_1, θ_1) and (r_2, θ_2) to be $(2, \frac{5\pi}{6})$ and $(4, \frac{\pi}{6})$, respectively (see Figure 7). This yields

$$\begin{aligned} d^2 &= r_1^2 + r_2^2 - 2r_1r_2\cos(\theta_2 - \theta_1) \\ &= 2^2 + 4^2 - 2(2)(4)\cos\left(\frac{\pi}{6} - \frac{5\pi}{6}\right) \\ &= 20 - 16\cos\left(-\frac{2\pi}{3}\right) = 20 - 16\left(-\frac{1}{2}\right) = 28 \\ d &= \sqrt{28} = 2\sqrt{7} \end{aligned}$$

You should check for yourself that the same result is obtained if we choose (r_1, θ_1) and (r_2, θ_2) to be $(4, \frac{\pi}{6})$ and $(2, \frac{5\pi}{6})$, respectively. This is because, in general, $\cos(\theta_2 - \theta_1) = \cos(\theta_1 - \theta_2)$. (Why?)

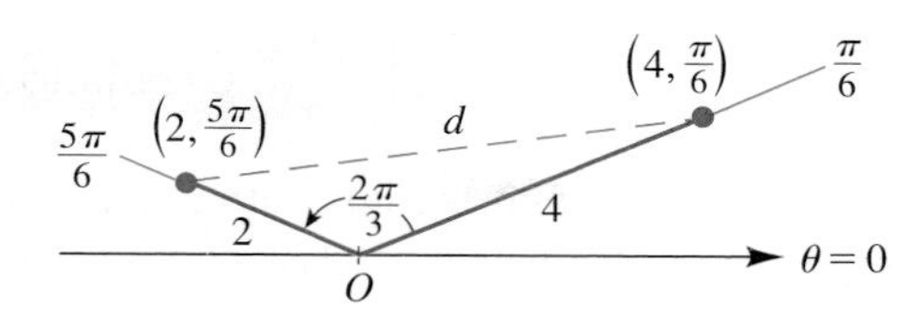

Figure 7

The distance formula can be used to find an equation for a circle. Suppose that the radius of the circle is a and that the polar coordinates of the center are (r_0, θ_0). As is indicated in Figure 8, we let (r, θ) denote an arbitrary point on the circle. Then, since the distance between the points (r, θ) and (r_0, θ_0) is a, the radius of the circle, we have

$$r^2 + r_0^2 - 2rr_0\cos(\theta - \theta_0) = a^2 \tag{1}$$

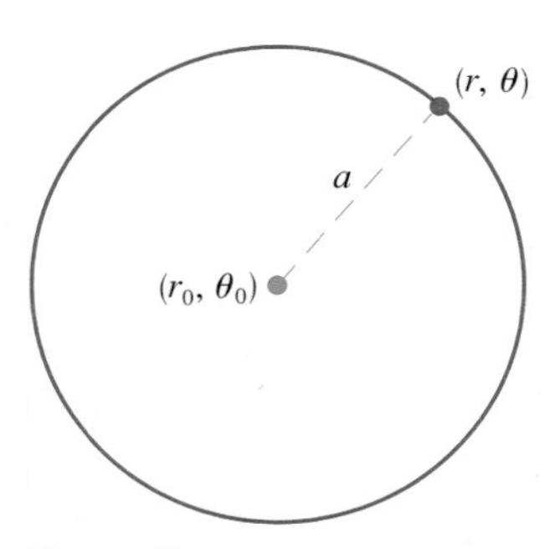

Figure 8

This is the required equation. There is a special case of this equation that is worth noting. If the center of the circle is located at the origin, then $r_0 = 0$, and equation (1) becomes simply $r^2 = a^2$. Therefore

$$r = a \qquad \text{or} \qquad r = -a$$

The graph of each of these equations is a circle of radius a, with center at the origin.

EXAMPLE 5 Finding a polar equation for a circle

Determine a polar equation for the circle satisfying the given conditions.

(a) The radius is 2, and the polar coordinates of the center are $(4, \frac{\pi}{5})$.

(b) The radius is 5, and the center is the origin.

SOLUTION

(a) Using equation (1), we have

$$r^2 + 4^2 - 2r(4)\cos\left(\theta - \frac{\pi}{5}\right) = 2^2$$

and therefore

$$r^2 - 8r\cos\left(\theta - \frac{\pi}{5}\right) = -12$$

(b) A polar equation for a circle of radius a and with center at the origin is $r = a$. So if the radius is 5, the required equation is $r = 5$. (The equation $r = -5$ would yield the same graph.)

We conclude this section by discussing equation of lines in polar coordinates. There are two cases to consider: lines that pass through the origin and lines that do not pass through the origin. First, consider a line through the origin, and suppose that the line makes an angle θ_0 with the positive x-axis; see Figure 9. The polar equation of this line is simply

$$\theta = \theta_0$$

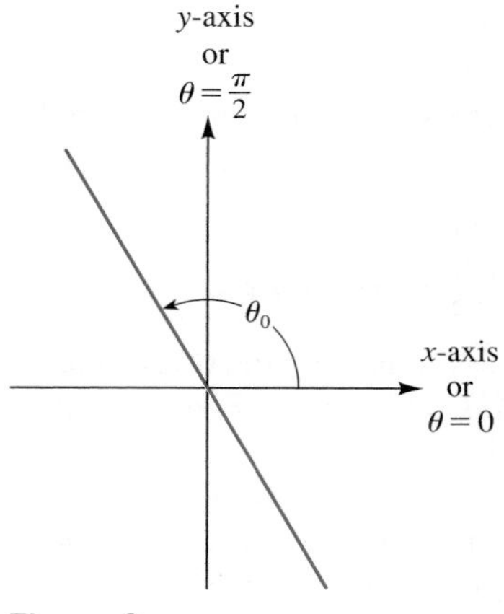

Figure 9

Notice that this equation poses no restrictions on r. In other words, for every real number r, the point with polar coordinates (r, θ_0) lies on this line.

EXAMPLE 6 Graphing a line through the pole

Graph the line with polar equation $\theta = \pi/3$, and locate the points on this line for which $r = 1, 2,$ and -1.

SOLUTION

See Figure 10.

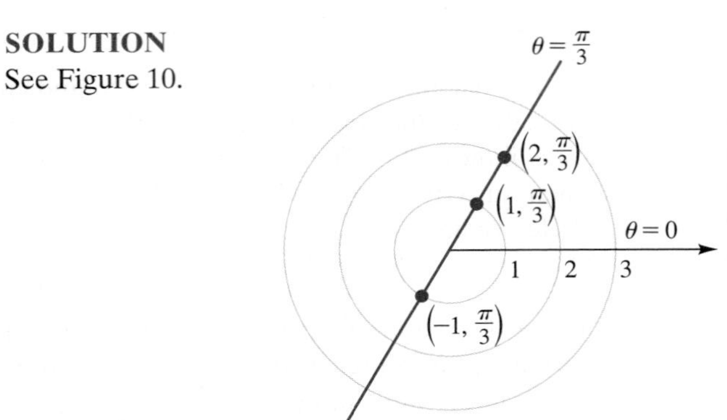

Figure 10

Now let us determine the polar equation for a line $\mathscr{L}$ that does not pass through the origin. As is indicated in Figure 11, we assume that the perpendicular distance from the origin to $\mathscr{L}$ is d and that the polar coordinates of the point N (in Figure 11) are (d, α). The point $P(r, \theta)$ in Figure 11 denotes an arbitrary point on line $\mathscr{L}$. We want to find an equation relating r and θ.

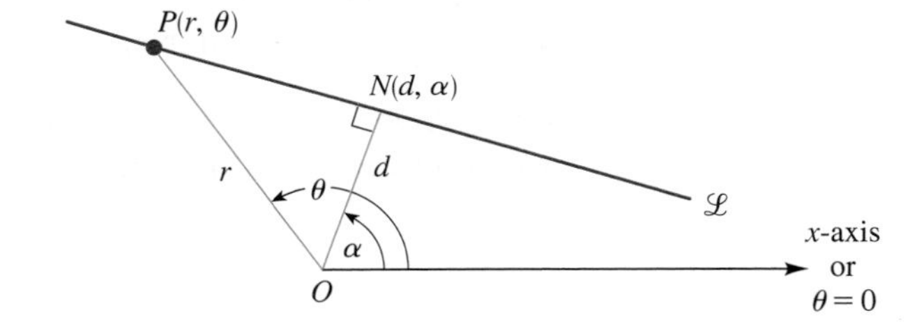

Figure 11

In right triangle ONP, note that $\angle PON = \theta - \alpha$, and therefore

$$\cos(\theta - \alpha) = \frac{d}{r} \quad \text{or} \quad r\cos(\theta - \alpha) = d$$

This is the polar equation of the line $\mathscr{L}$. (*Note*: Since cosine is an even function the equation is correct when the point P is to "the right" of the point N.) For reference, we summarize our results about lines in the box that follows.

PROPERTY SUMMARY Polar Equations for Lines

1. *Line through the origin:* Suppose that a line passes through the origin and makes an angle θ_0 with the positive x-axis (as indicated in Figure 9 on page 705). Then an equation for the line is $\theta = \theta_0$.
2. *Line not passing through the origin:* Suppose that the perpendicular distance from the origin to the line is d and that the point (d, α) is the foot of the perpendicular from the origin to the line (as indicated in Figure 11). Then a polar equation for the line is $r\cos(\theta - \alpha) = d$.

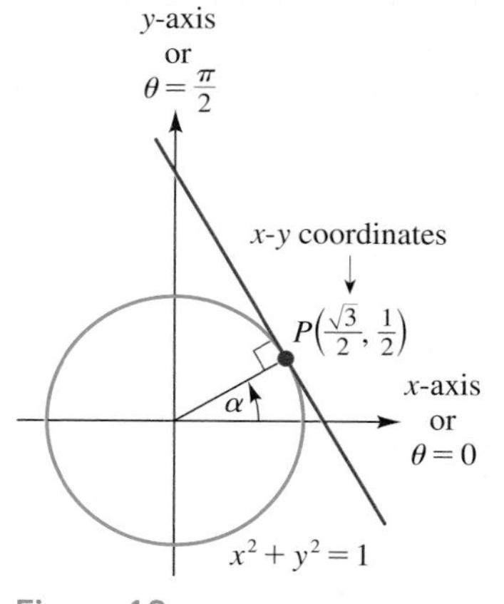

Figure 12

EXAMPLE 7 Finding a polar equation for a tangent line to a circle

Figure 12 shows a line tangent to the unit circle at the point P. The x-y coordinates of P are $\left(\frac{\sqrt{3}}{2}, \frac{1}{2}\right)$.

(a) What are the polar coordinates of the point P?
(b) Find the polar equation for the tangent line.
(c) Use the polar equation to find the x- and y-intercepts of the tangent line.
(d) Find a rectangular equation for the line.

SOLUTION

(a) Using the unit-circle definition for cosine, we have $\cos\alpha = \sqrt{3}/2$ and, consequently, $\alpha = \pi/6$. The polar coordinates of P are therefore $(1, \frac{\pi}{6})$.

(b) In the general equation $r\cos(\theta - \alpha) = d$ we use the values $\alpha = \pi/6$ and $d = 1$ to obtain

$$r\cos\left(\theta - \frac{\pi}{6}\right) = 1 \tag{2}$$

This is the polar equation for the tangent line.

(c) The x-axis corresponds to $\theta = 0$. Substituting $\theta = 0$ in equation (2) gives us $r\cos\left(0 - \frac{\pi}{6}\right) = 1$, and therefore

$$r\cos\frac{\pi}{6} = 1 \quad \text{(Why?)}$$
$$r = \frac{1}{\sqrt{3}/2} = \frac{2}{\sqrt{3}} = \frac{2\sqrt{3}}{3}$$

This tells us that the tangent line meets the x-axis at the point with polar coordinates $\left(\frac{2\sqrt{3}}{3}, 0\right)$. So the x-intercept of the line is $2\sqrt{3}/3$. For the y-intercept we set $\theta = \pi/2$ in equation (2). As you should check for yourself, this yields $r = 2$. Thus the y-intercept of the tangent line is 2.

(d) From part (b),

$$r\cos\left(\theta - \frac{\pi}{6}\right) = 1$$
$$r\left(\cos\theta\cos\frac{\pi}{6} + \sin\theta\sin\frac{\pi}{6}\right) = 1$$
$$\frac{\sqrt{3}}{2}r\cos\theta + \frac{1}{2}r\sin\theta = 1$$
$$\frac{\sqrt{3}}{2}x + \frac{1}{2}y = 1 \quad \text{or} \quad \sqrt{3}x + y = 2$$

It is also instructive to find a rectangular equation for the line by using the point-slope formula for a line. We have a point $\left(\frac{\sqrt{3}}{2}, \frac{1}{2}\right)$ on the line. The slope of the line is $\tan\left(\frac{\pi}{6} + \frac{\pi}{2}\right)$. (Why?) Then the slope is

$$m = \tan\left(\frac{2\pi}{3}\right) = -\sqrt{3}$$

So the line has equation

$$y - \frac{1}{2} = -\sqrt{3}\left(x - \frac{\sqrt{3}}{2}\right)$$

which is equivalent to

$$\sqrt{3}x + y = 2$$

Finally, you can find the same equation by using the x- and y-intercepts from part (c).

EXERCISE SET 9.6 

A

In Exercises 1–3, graph each point in a polar coordinate system then convert the given polar coordinates to rectangular coordinates.

1. (a) $(3, \frac{2\pi}{3})$ **(b)** $(4, \frac{11\pi}{6})$ **(c)** $(4, -\frac{\pi}{6})$

2. (a) $(5, \frac{\pi}{4})$ **(b)** $(-5, \frac{\pi}{4})$ **(c)** $(-5, -\frac{\pi}{4})$

3. (a) $(1, \frac{\pi}{2})$ **(b)** $(1, \frac{5\pi}{2})$ **(c)** $(-1, \frac{\pi}{8})$

In Exercises 4–6, convert the given rectangular coordinates to polar coordinates. Express your answers in such a way that r is nonnegative and $0 \le \theta < 2\pi$.

4. $(3, \sqrt{3})$ **5.** $(-1, -1)$ **6.** $(0, -2)$

In Exercises 7–16, convert to rectangular form.

7. $r = 2\cos\theta$

8. $2\sin\theta - 3\cos\theta = r$

9. $r = \tan\theta$

10. $r = 4$

11. $r = 3\cos 2\theta$

12. $r = 4\sin 2\theta$

13. $r^2 = 8/(2 - \sin^2\theta)$

14. $r^2 = 1/(3 + \cos^2\theta)$

15. $r\cos\left(\theta - \frac{\pi}{6}\right) = 2$

16. $r\sin\left(\theta + \frac{\pi}{4}\right) = 6$

In Exercises 17–24, convert to polar form.

17. $3x - 4y = 2$

18. $x^2 + y^2 = 25$

19. $y^2 = x^3$

20. $y = x^2$

21. $2xy = 1$

22. $x^2 + 4x + y^2 + 4y = 0$

23. $9x^2 + y^2 = 9$

24. $x^2(x^2 + y^2) = y^2$

In Exercises 25–30 you are given a polar equation and its graph. Use the equation to determine polar coordinates of the points labeled with capital letters. (For the r-values, where necessary, use a calculator and round to two decimal places.)

25. $r = \dfrac{4}{1 + \sin\theta}$

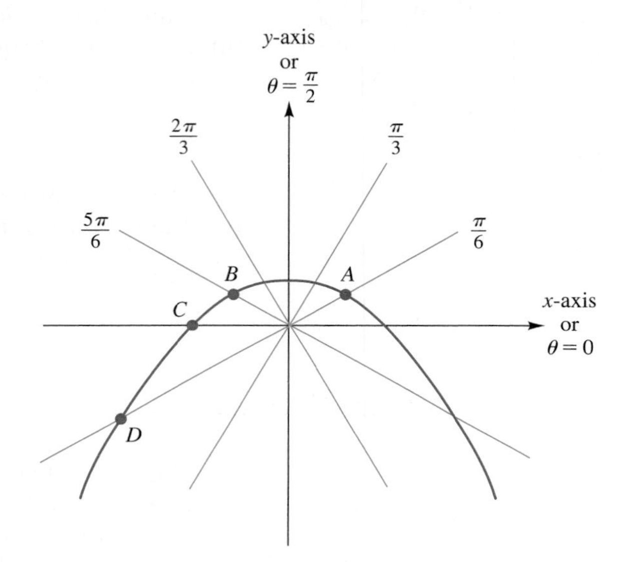

26. 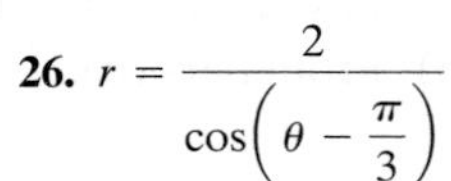 $r = \dfrac{2}{\cos\left(\theta - \frac{\pi}{3}\right)}$

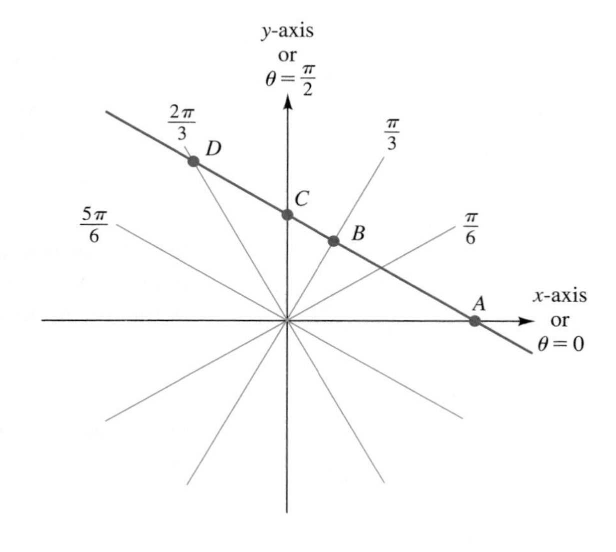

27. $r = 2\cos 2\theta$

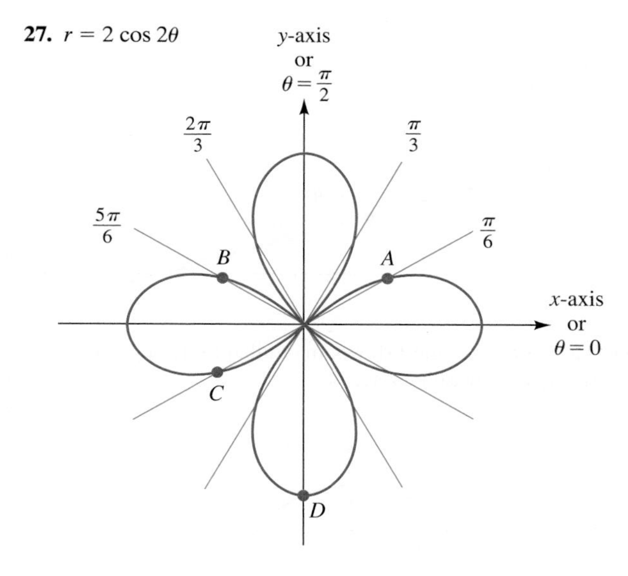

28. $r = -1 - \sin\theta$

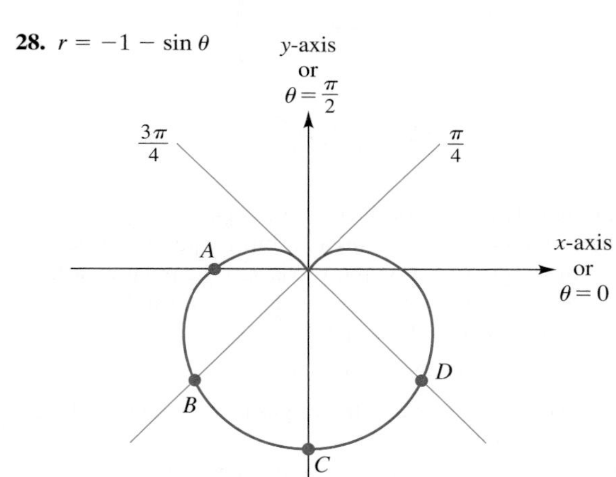

29. $r = e^{\theta/6}$

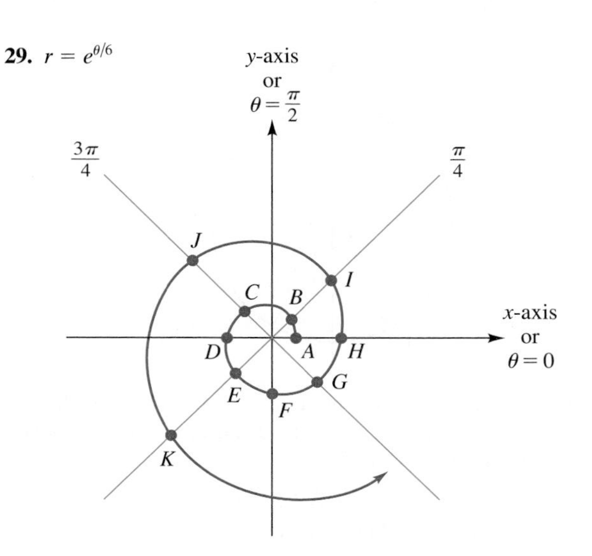

30. $r^2 = \cos 2\theta$

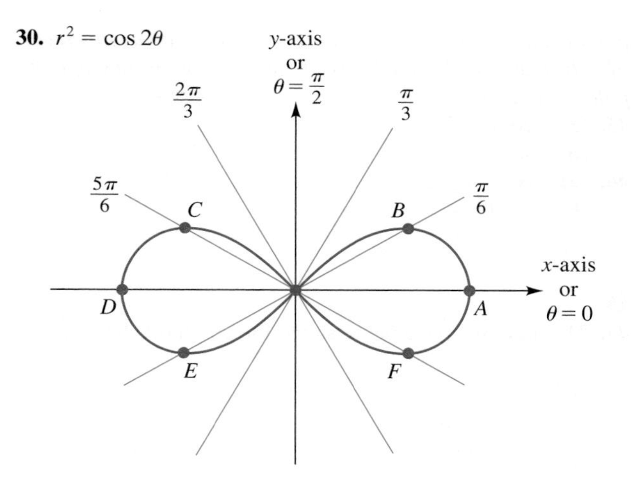

In Exercises 31–34, compute the distance between the given points. (The coordinates are polar coordinates.)

31. $(2, \frac{2\pi}{3})$ and $(4, \frac{\pi}{6})$

32. $(4, \pi)$ and $(3, \frac{7\pi}{4})$

33. $(4, \frac{4\pi}{3})$ and $(1, 0)$

34. $(3, \frac{5\pi}{6})$ and $(5, \frac{5\pi}{3})$

In Exercises 35–38, determine a polar equation for the circle satisfying the given conditions.

35. The radius is 2, and the polar coordinates for the center are: **(a)** $(4, 0)$ **(b)** $(4, \frac{2\pi}{3})$ **(c)** $(0, 0)$

36. The radius is $\sqrt{6}$, and the polar coordinates of the center are: **(a)** $(2, \pi)$ **(b)** $(2, \frac{3\pi}{4})$ **(c)** $(0, 0)$

37. The radius is 1, and the polar coordinates of the center are: **(a)** $(1, \frac{3\pi}{2})$ **(b)** $(1, \frac{\pi}{4})$

38. The radius is 6, and the polar coordinates of the center are: **(a)** $(-3, \frac{5\pi}{4})$ **(b)** $(-2, \frac{\pi}{4})$

In Exercises 39–42 the polar equation of a line is given. In each case: **(a)** *specify the perpendicular distance from the origin to the line;* **(b)** *determine the polar coordinates of the points on the line corresponding to $\theta = 0$ and $\theta = \pi/2$;* **(c)** *specify the polar coordinates of the foot of the perpendicular from the origin to the line;* **(d)** *use the results in parts (a), (b), and (c) to sketch the line; and* **(e)** *find a rectangular form for the equation of the line.*

39. $r\cos(\theta - \frac{\pi}{6}) = 2$

40. $r\cos(\theta - \frac{\pi}{4}) = 1$

41. $r\cos(\theta + \frac{2\pi}{3}) = 4$

42. $r\cos(\theta - \frac{\pi}{2}) = \sqrt{2}$

43. A line is tangent to the circle $x^2 + y^2 = 4$ at a point P with x-y coordinates $(-\sqrt{3}, 1)$.

(a) What are the polar coordinates of P?

(b) Find a polar equation for the tangent line.

(c) Use the polar equation to determine the x- and y-intercepts of the line.

(d) Find a rectangular equation for the line.

44. Follow Exercise 43 using the circle $x^2 + y^2 = 36$ and the point P with x-y coordinates $(-3, -3\sqrt{3})$.

Notice that the rectangular form of the polar equation $r\cos\theta = a$ is $x = a$. Thus the graph of the polar equation $r\cos\theta = a$ is a vertical line with an x-intercept of a. Similarly, the graph of the *polar equation* $r \sin \theta = b$ *is a horizontal line with a y-intercept of b. In Exercises 45 and 46, use these observations to graph the polar equations.*

45. **(a)** $r \cos \theta = 3$
(b) $r \sin \theta = 3$

46. **(a)** $r \cos \theta = -2$
(b) $r \sin \theta = 4$

47. The accompanying figure shows the graph of a line $\mathscr{L}$.

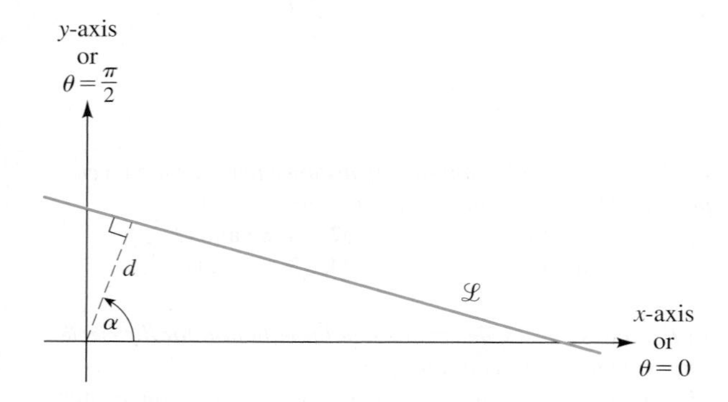

(a) According to the text, what is the polar equation for $\mathscr{L}$?

(b) By converting the equation in part (a) to rectangular form, show that the *x-y* equation for $\mathscr{L}$ is $x \cos \alpha + y \sin \alpha = d$. (This equation is called the **normal form** for a line.) *Hint*: First use one of the addition formulas for cosine.

48. By converting the polar equation

$$r = a \cos \theta + b \sin \theta$$

to rectangular form, show that the graph is a circle, and find the center and the radius.

49. Show that the rectangular form of the equation $r = a \sin 3\theta$ is $(x^2 + y^2)^2 = ay\,(3x^2 - y^2)$.

50. Show that the rectangular form of the equation

$$r = \frac{ab}{(1 - a \cos \theta)} \qquad (a < 1)$$

is $(1 - a^2)x^2 + y^2 - 2a^2bx - a^2b^2 = 0$.

51. Show that the polar form of the equation $(x^2/a^2) - (y^2/b^2) = 1$ is

$$r^2 = \frac{a^2b^2}{b^2 \cos^2 \theta - a^2 \sin^2 \theta}$$

52. Show that a polar equation for the line passing through the two points $A(r_1, \theta_1)$ and $B(r_2, \theta_2)$ is $rr_1 \sin(\theta - \theta_1) - rr_2 \sin(\theta - \theta_2) + r_1r_2 \sin(\theta_1 - \theta_2) = 0$. *Hint*: Use the following figure and the observation that

$$\text{area } \triangle OAB = (\text{area } \triangle OBP) + (\text{area } \triangle OPA)$$

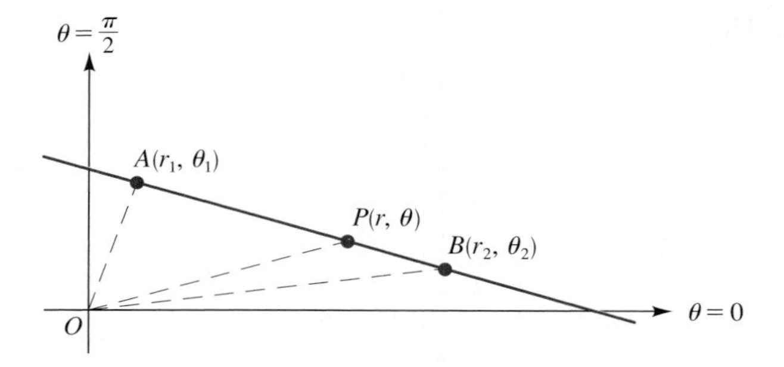

53. The following figure shows an equilateral triangle inscribed in the unit circle.

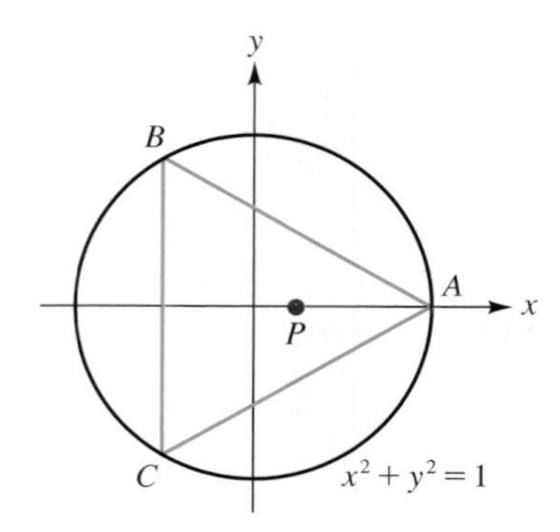

(a) Specify polar coordinates for the points A, B, and C.

(b) In the figure, P denotes an arbitrary point between the origin and A on the x-axis. Let the polar coordinates of P be $(r, 0)$. Use the distance formula for polar coordinates to show that

$$(PA)(PB)(PC) = 1 - r^3$$

[This result is due to the English mathematician Roger Cotes (1682–1716).]

54. The following figure shows an equilateral triangle inscribed in the unit circle.

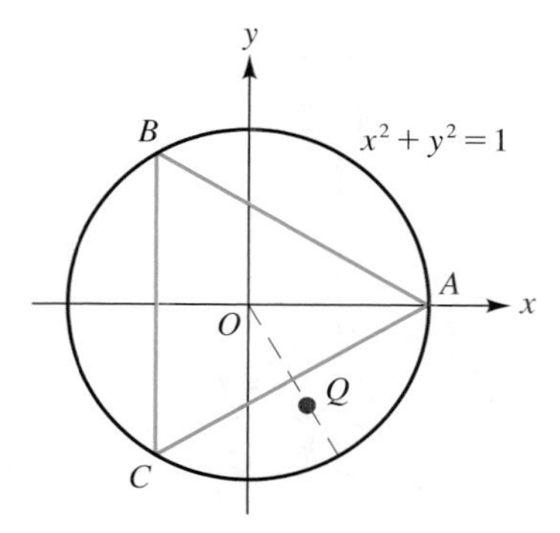

The point Q lies on the bisector of $\angle AOC$, and the distance from O to Q is r. Use the distance formula for polar coordinates to show that

$$(QA)(QB)(QC) = 1 + r^3$$

(This result also is due to Roger Cotes.)

9.7 CURVES IN POLAR COORDINATES

In principle, the graph of any polar equation $r = f(\theta)$ can be obtained by setting up a table and plotting a sufficient number of points. Indeed, this is just the way a graphing calculator or a computer operates. Figure 1, for example, shows computer-generated graphs of the polar equations $r = (\sin 4\theta)^4 + \cos 3\theta$ and $r = \cos^2 5\theta + \sin 3\theta + 0.3$. In this section we'll concentrate on a few basic types of polar equations and their graphs. With the exception of Example 1 the polar equations that we consider involve trigonometric functions. To *understand* why the graphs look as they do, we will often use symmetry and the basic properties of the sine and cosine functions.

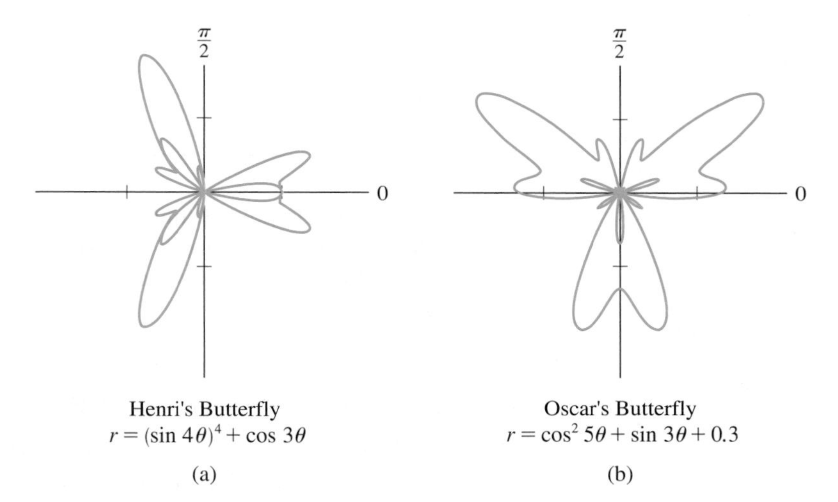

Figure 1
(a) A polar coordinate graph discovered by Henri Berger, a student in one of the author's mathematics classes at UCLA in Spring of 1988; (b) a polar coordinate graph discovered by Oscar Ramirez, a precalculus student at UCLA in Fall of 1991

EXAMPLE 1 Graphing the spiral of Archimedes

Graph the polar equation $r = \theta/\pi$ for $\theta \geq 0$.

SOLUTION

The equation shows that as θ increases, so does r. Geometrically, this means that as we plot points, moving counterclockwise with increasing θ, the points will be farther and farther from the origin. In Table 1 we have computed values for r corresponding to some convenient values of θ. In Figure 2 we have plotted the points in the table and connected them with a smooth curve. The curve is known as the **spiral of Archimedes.**

TABLE 1

θ	0	$\frac{\pi}{4}$	$\frac{\pi}{2}$	$\frac{3\pi}{4}$	π	$\frac{5\pi}{4}$	$\frac{3\pi}{2}$	$\frac{7\pi}{4}$	2π	$\frac{5\pi}{2}$	3π	4π
r	0	$\frac{1}{4}$	$\frac{1}{2}$	$\frac{3}{4}$	1	$\frac{5}{4}$	$\frac{3}{2}$	$\frac{7}{4}$	2	$\frac{5}{2}$	3	4

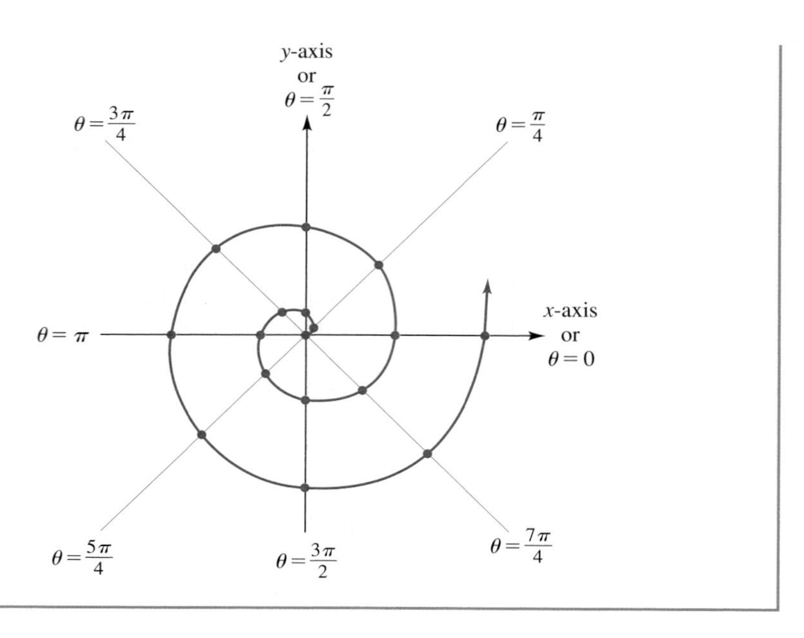

Figure 2
$r = \frac{\theta}{\pi} \quad (\theta \geq 0)$

We saw in Chapters 1 and 3 that symmetry considerations can often be used to reduce the amount of work involved in graphing equations. This is also true for polar equations. In the box that follows, we list four tests for symmetry in polar coordinates.

Symmetry Tests in Polar Coordinates

1. If substituting $-\theta$ for θ yields an equivalent equation, then the graph is symmetric about the polar axis $\theta = 0$ (the x-axis).
2. If substituting $-\theta$ for θ and $-r$ for r yields an equivalent equation, then the graph is symmetric about the line $\theta = \pi/2$ (the y-axis).
3. If substituting $\pi - \theta$ for θ yields an equivalent equation, then the graph is symmetric about the line $\theta = \pi/2$ (the y-axis).
4. If substituting $-r$ for r yields an equivalent equation, then the graph is symmetric about the pole $r = 0$ (the origin).

CAUTION As opposed to the case with the x-y symmetry tests (in Chapter 1), if a polar equation fails a symmetry test, then it is still possible for the graph to possess the indicated symmetry. This is a consequence of the fact that the polar coordinates of a point are not unique. (For an example, see Example 3.)

The validity of the first test follows from the fact that the points (r, θ) and $(r, -\theta)$ are reflections of each other about the polar axis (the x-axis); see Figure 3. The validity of test 2 follows from the fact that the points (r, θ) and $(-r, -\theta)$ are reflections of each other about the line $\theta = \pi/2$ (the y-axis), as indicated in Figure 3. The other tests can be justified in a similar manner.

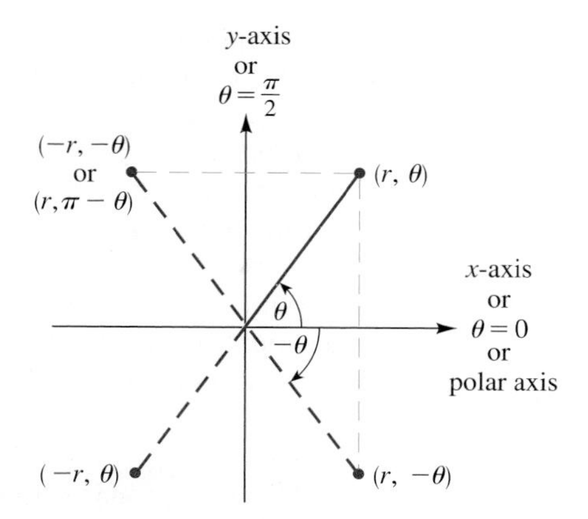

Figure 3

EXAMPLE 2 Graphing a cardioid

Graph the polar equation $r = 1 - \cos\theta$.

SOLUTION

We know (from Section 7.1) that the cosine function satisfies the identity $\cos(-\theta) = \cos\theta$. So according to symmetry test 1, the graph of $r = 1 - \cos\theta$ will be symmetric about the line $\theta = 0$. In Figure 4(a) we have set up a table (using a calculator as necessary) with some convenient values of θ running from 0 to π. Plotting the points in the table leads to the curve shown in Figure 4(a). Reflecting this curve in the line $\theta = 0$, we obtain the graph in Figure 4(b). The fact that the period of the cosine function is 2π implies that we need not consider values of θ beyond 2π. (Why?) Thus the curve in Figure 4(b) is the required graph of $r = 1 - \cos\theta$. The curve is a **cardioid.**

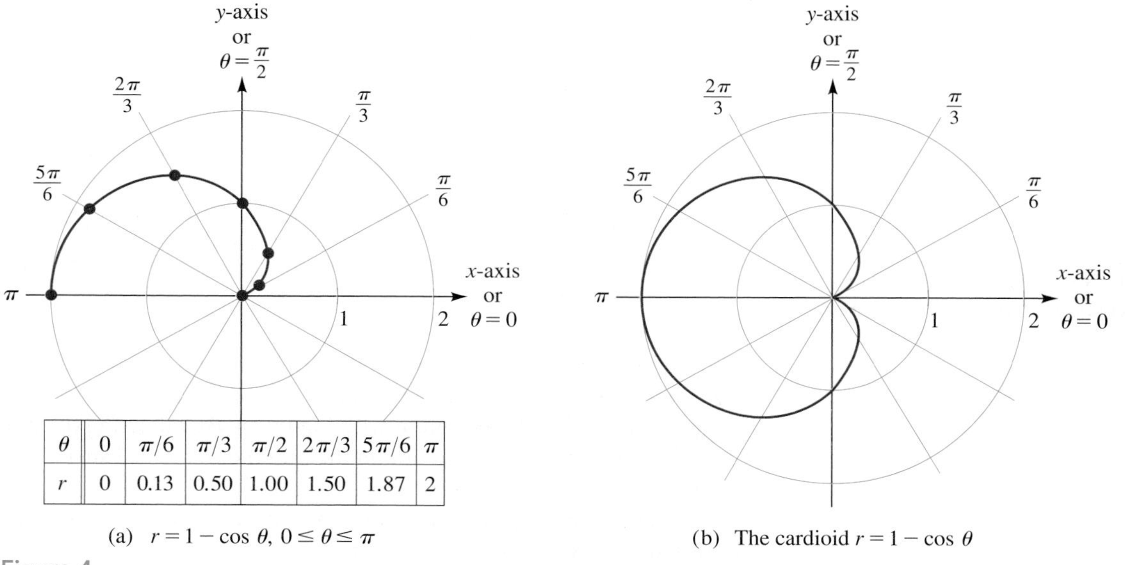

θ	0	$\pi/6$	$\pi/3$	$\pi/2$	$2\pi/3$	$5\pi/6$	π
r	0	0.13	0.50	1.00	1.50	1.87	2

(a) $r = 1 - \cos\theta,\ 0 \leq \theta \leq \pi$

(b) The cardioid $r = 1 - \cos\theta$

Figure 4

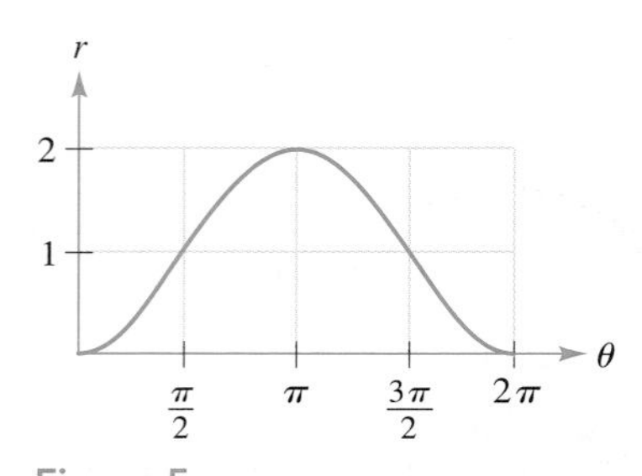

Figure 5
The graph of $r = 1 - \cos\theta$ in a rectangular θ-r coordinate system

In graphing polar equations, just as in graphing Cartesian x-y equations, there is always the question of whether we have plotted a sufficient number of points to reveal the essential features of the graph. As a check on our work in Figure 4(a) we can graph the equation $r = 1 - \cos\theta$ in a *rectangular* θ-r coordinate system, as shown in Figure 5. (This is easy to do using reflection and translation.) From the graph in Figure 5 we see that as θ increases from 0 to π, the values of r increase from 0 to 2. Now, recall that in polar coordinates, r represents the directed distance from the origin. So on the polar graph, as we go from $\theta = 0$ to π, the points that we obtain should be farther and farther from the origin, starting with $r = 0$ and ending with $r = 2$. The polar graph in Figure 4(a) is consistent with these observations.

EXAMPLE 3 Graphing a limaçon

Graph the polar curve $r = 1 + 2\sin\theta$.

SOLUTION

Using symmetry test 3, we replace θ by $\pi - \theta$ in the given equation:

$$\begin{aligned} r = 1 + 2\sin(\pi - \theta) &= 1 + 2(\sin\pi\cos\theta - \cos\pi\sin\theta) \\ &= 1 + 2\sin\theta \qquad \text{because } \sin\pi = 0 \text{ and } \cos\pi = -1 \end{aligned}$$

This shows that the graph is symmetric about the line $\theta = \pi/2$. (Note that symmetry test 2 fails for this equation, yet the graph is symmetric about the y-axis.)

Next, to see what to expect in polar coordinates, we first sketch the graph of the given equation in a rectangular θ-r coordinate system; see Figure 6. In Figure 6, notice that the maximum value of $|r|$ is 3. This implies that the required polar graph is contained within a circle of radius 3 about the origin. (We'll refer to this sketch several more times in the course of the solution.)

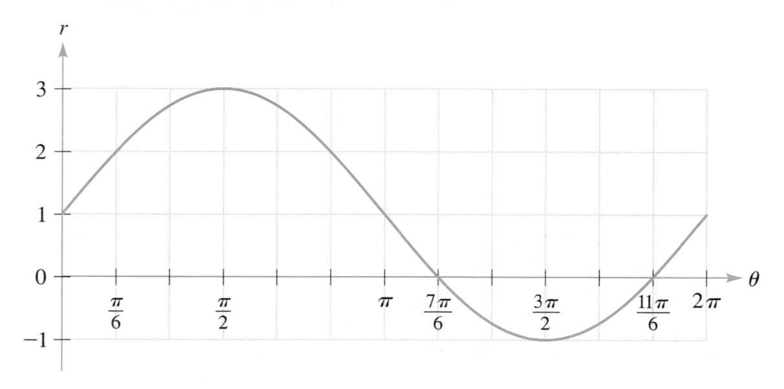

Figure 6
The graph of $r = 1 + 2\sin\theta$ in a rectangular θ-r coordinate system

In Figure 7(a) we have set up a table and graphed the given polar equation for θ running from 0 to $\pi/2$. Note that as θ increases from 0 to $\pi/2$, the r-values increase from 1 to 3 (in agreement with Figure 6). By reflecting the curve in Figure 7(a) in the line $\theta = \pi/2$, we obtain the graph of $r = 1 + 2\sin\theta$ for $0 \le \theta \le \pi$; see Figure 7(b).

Next, let's consider the values of r as θ increases from π to $3\pi/2$. Figure 6 tells us what to expect on this interval:

As θ increases from π to $7\pi/6$, the r-values decrease from 1 to 0; as θ increases from $7\pi/6$ to $3\pi/2$, the r-values decrease from 0 to -1. However

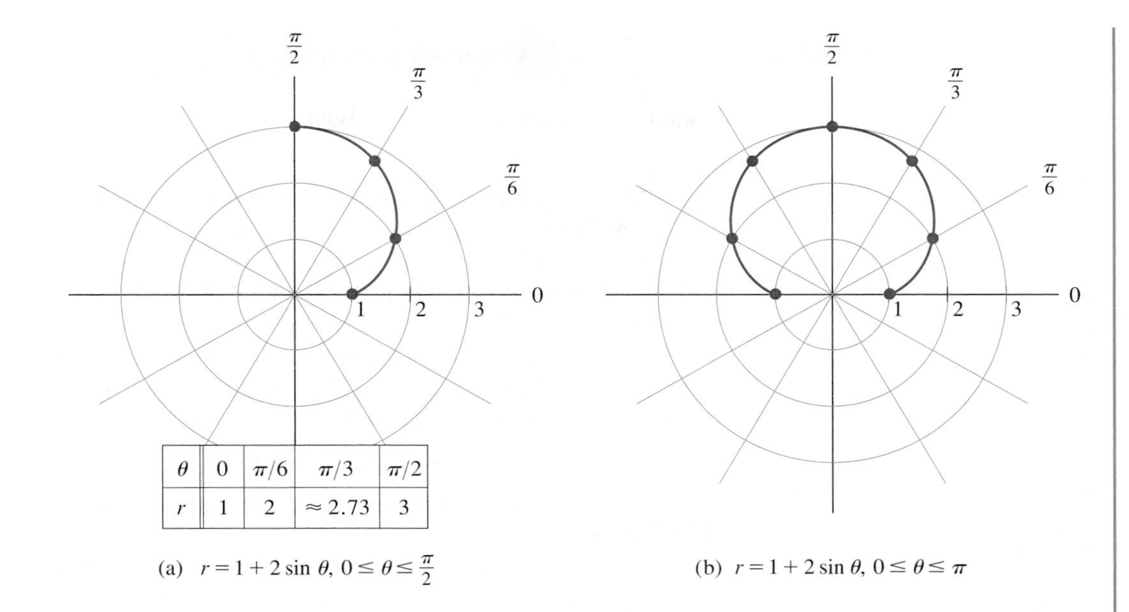

θ	0	$\pi/6$	$\pi/3$	$\pi/2$
r	1	2	≈ 2.73	3

Figure 7 (a) $r = 1 + 2\sin\theta,\ 0 \leq \theta \leq \frac{\pi}{2}$ (b) $r = 1 + 2\sin\theta,\ 0 \leq \theta \leq \pi$

(in view of the convention regarding negative r-values), this will mean that the points on the polar graph are, in fact, moving farther and farther from the origin as θ increases from $7\pi/6$ to $3\pi/2$.

In Figure 8(a) we have set up a table for $\pi \leq \theta \leq 3\pi/2$ and added the corresponding points from Figure 7(b). Finally, again taking into account the symmetry to the line $\theta = \pi/2$, we sketch the graph for $0 \leq \theta \leq 2\pi$, as indicated in Figure 8(b). The curve in Figure 8(b) is a **limaçon.** (Not all limaçons have inner loops. See, for example, Exercises 27 and 28.)

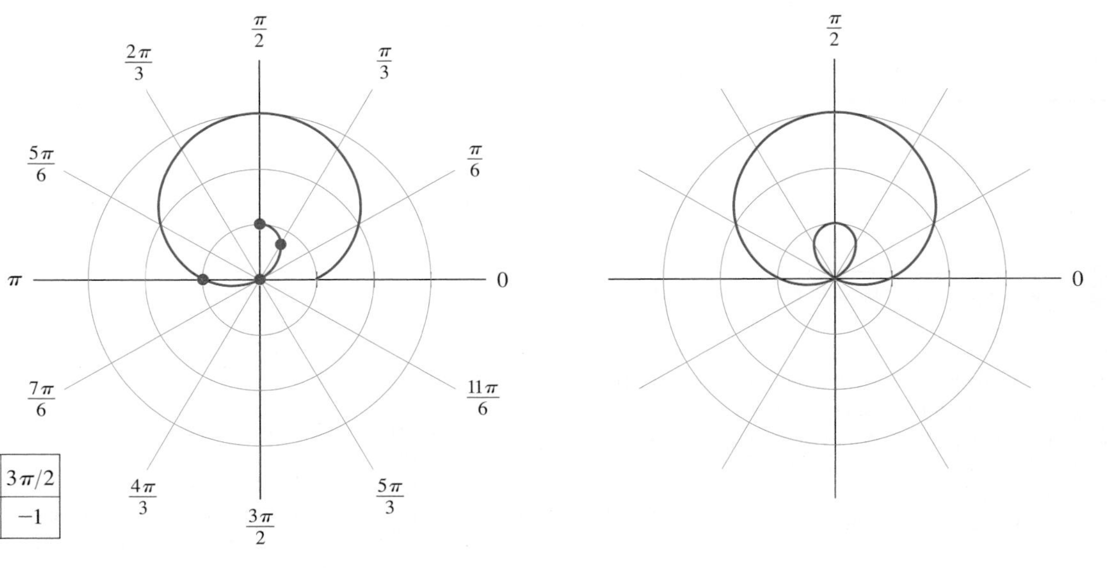

θ	π	$7\pi/6$	$4\pi/3$	$3\pi/2$
r	1	0	≈ -0.73	-1

Figure 8 (a) $r = 1 + 2\sin\theta,\ 0 \leq \theta \leq \frac{3\pi}{2}$ (b) The limaçon $r = 1 + 2\sin\theta$

EXAMPLE 4 Graphing a lemniscate

Graph the polar equation $r^2 = 4\cos 2\theta$.

SOLUTION

As a guide for graphing the given polar equation, we first sketch the curve $y = 4\cos 2\theta$ in a rectangular θ-y coordinate system. As is indicated in Figure 9, the period of $4\cos 2\theta$ is π. So in graphing the given polar equation, it suffices to consider only the values of θ from 0 to π. Figure 9 also shows that $4\cos 2\theta$ is negative on the interval $(\frac{\pi}{4} < \theta < \frac{3\pi}{4})$. The value of r^2, however, cannot be negative. Consequently, we do not need to consider values of θ in the open interval from $\pi/4$ to $3\pi/4$. Finally, Figure 9 shows that the absolute value of $4\cos 2\theta$ is always less than or equal to 4. Consequently, for the polar graph we have

$$r^2 \le 4 \qquad \text{and therefore} \qquad |r| \le 2$$

This tells us that the polar graph will be contained within a circle of radius 2 about the origin.

For the polar graph we begin by computing a table of values with θ running from 0 to $\pi/4$; see Table 2. Plotting the points in Table 2 and joining them with a smooth curve leads to the graph shown in Figure 10(a). Then, rather than setting up another table with θ running from $3\pi/4$ to π, we rely on symmetry to complete the graph. According to the first two symmetry tests, the curve is symmetric about both the line $\theta = 0$ and the line $\theta = \pi/2$. Thus we obtain the graph shown in Figure 10(b). The curve is called a **lemniscate.**

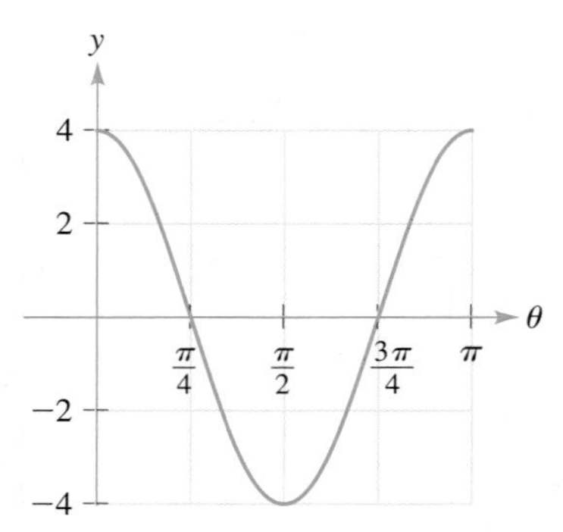

Figure 9
$y = 4\cos 2\theta$

TABLE 2

θ	$r = \pm 2\sqrt{\cos 2\theta}$
0	± 2
$\frac{\pi}{12}$	± 1.86
$\frac{\pi}{8}$	± 1.68
$\frac{\pi}{6}$	± 1.41
$\frac{\pi}{4}$	0

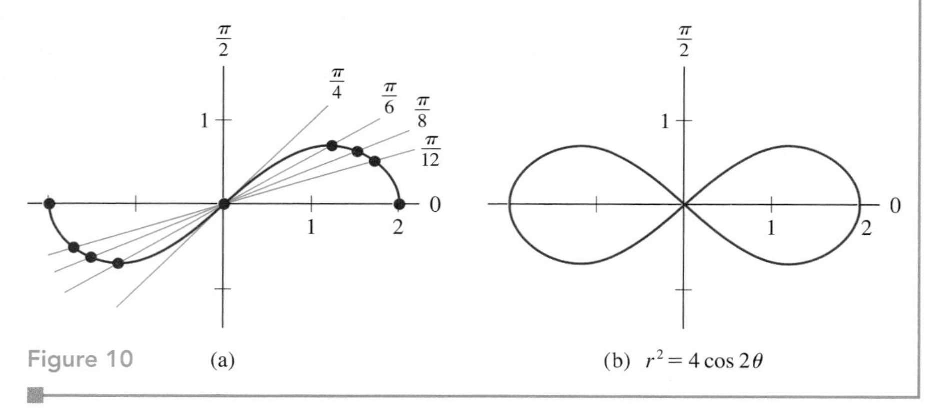

Figure 10 (a) (b) $r^2 = 4\cos 2\theta$

EXERCISE SET 9.7

A

1. In the same picture, graph the four polar equations $r = 2$, $r = 4$, $r = 6$, and $r = 8$. Describe the graphs.

Ⓖ 2. In the same picture, graph the polar equations $r = 1$ and $r\cos[\theta - (\pi/6)] = 1$. What do you observe? Check that your result is consistent with Figure 12 on page 707. *Hint*: Write the equation

$$r\cos[\theta - (\pi/6)] = 1 \qquad \text{as} \qquad r = \frac{1}{\cos[\theta - (\pi/6)]}$$

In Exercises 3–30, graph the polar equations.

3. $r = \theta/(2\pi)$, for $\theta \ge 0$
4. $r = \theta/\pi$, for $-4\pi \le \theta \le 0$
5. $r = \ln\theta$, for $1 \le \theta \le 3\pi$
6. **(a)** $r = e^{\theta/2\pi}$, for $0 \le \theta \le 2\pi$
   (b) $r = e^{-\theta/2\pi}$, for $0 \le \theta \le 2\pi$

Ⓖ 7. $r = \theta$ (*spiral of Archimedes*) *Suggestion*: Use a viewing rectangle extending from -30 to 30 in both the x- and y-directions. Let θ run from 0 to 2π, then from 0 to 4π, and finally from 0 to 8π.

Ⓖ 8. $r = 1/\theta$ (*hyperbolic spiral*) *Suggestion*: Use a viewing rectangle extending from -1 to 1 in both the x- and

y-directions. Let θ run from 0 to 2π, then from 0 to 4π, and finally from 0 to 8π.

Ⓖ **9.** $r = 1/\sqrt{\theta}$ (*spiral of Cotes,* or *lituus*) *Suggestion*: Use the guidelines in Exercise 8.

Ⓖ **10.** $r = \sqrt{\theta}$ (*spiral of Fermat*) *Suggestion*: Use a viewing rectangle extending from -5 to 5 in both the x- and y-directions. Let θ run from θ to 2π, then from 0 to 4π, and finally from 0 to 8π.

11. $r = 1 + \cos\theta$

12. $r = 1 - \sin\theta$

13. $r = 2 - 2\sin\theta$

14. $r = 3 + 3\cos\theta$

15. $r = 1 - 2\sin\theta$

16. $r = 1 + 2\cos\theta$

17. $r = 2 + 4\cos\theta$

18. $r = 2 - 4\sin\theta$

19. $r^2 = 4\sin 2\theta$

20. $r^2 = 9\cos 2\theta$

21. $r^2 = \cos 4\theta$

22. $r^2 = 3\sin 4\theta$

23. $r = \cos 2\theta$ (*four-leafed rose*)

24. $r = 2\sin 2\theta$ (*four-leafed rose*)

25. $r = \sin 3\theta$ (*three-leafed rose*)

26. $r = 2\cos 5\theta$ (*five-leafed rose*)

27. $r = 4 + 2\sin\theta$ (*limaçon with no inner loop*)

28. $r = 1.5 - \cos\theta$ (*limaçon with no inner loop*)

29. $r = 8\tan\theta$ (*kappa curve*)

30. $r = \csc\theta + 2$ (*conchoid of Nicomedes*)

Ⓖ **31. (a)** Use one of the polar symmetry tests to show that the graph of $r = \cos^2\theta - 2\cos\theta$ is symmetric about the x-axis.

(b) Graph the equation given in part (a) and note that the curve is indeed symmetric about the x-axis.

Ⓖ **32.** As background for this exercise, read the caution note at the bottom of the box on page 712.

(a) Graph the polar equation $r = \sin^2\theta - 2\sin\theta$. What type of symmetry do you observe?

(b) Check (algebraically) that the equation $r = \sin^2\theta - 2\sin\theta$ satisfies symmetry test 3 on page 712, but not symmetry test 2.

In Exercises 33–35, graph the polar equations.

Ⓖ **33.** $r = \cos 3\theta$ (*three-leafed rose*)

Ⓖ **34.** $r = \sin 5\theta$

Ⓖ **35. (a)** $r = \cos 4\theta$ (*eight-leafed rose*)

(b) $r = \sin 4\theta$

In Exercises 36 and 37, match one of the graphs with each equation. Hint: Compute the values of r when $\theta = 0$, $\pi/2$, π, and $3\pi/2$.

36. (a) $r = 3 + 3\sin\theta$

(b) $r = 3 - 3\sin\theta$

(c) $r = 3 - 3\cos\theta$

(d) $r = 3 + 3\cos\theta$

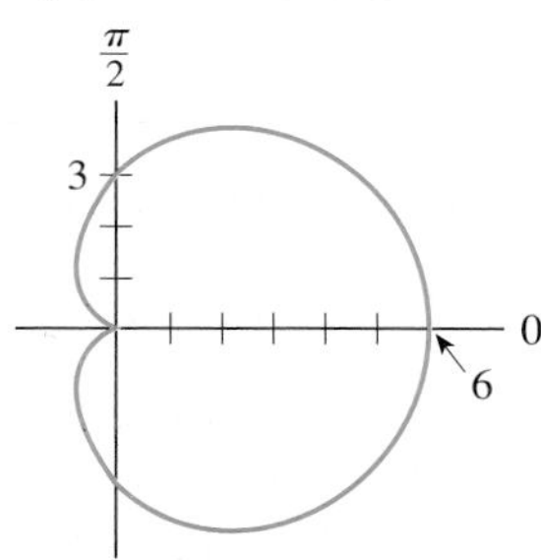

Graph A

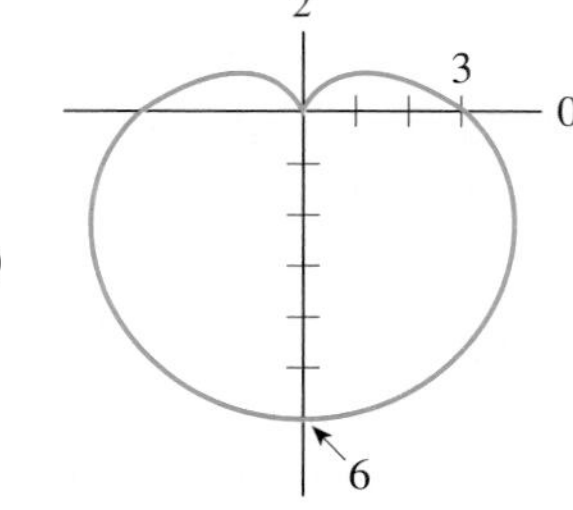

Graph B

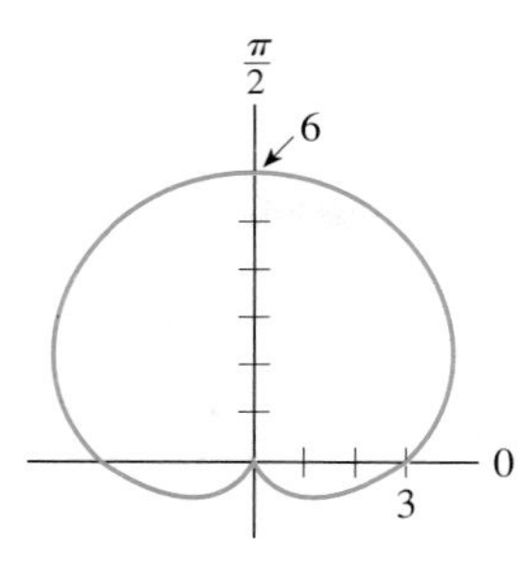

Graph C

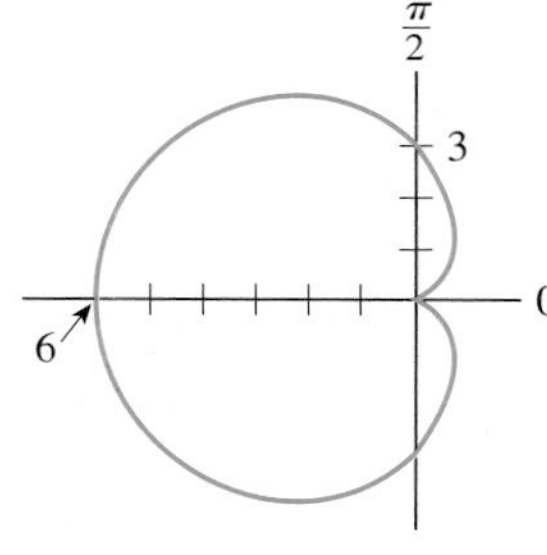

Graph D

37. (a) $r = 3 + 2\cos\theta$

(b) $r = 2 + 3\cos\theta$

(c) $r = 2 - 3\cos\theta$

(d) $r = 3 - 2\cos\theta$

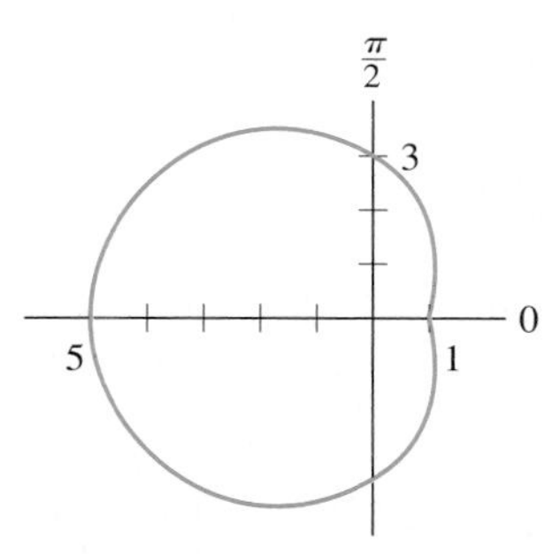

Graph A

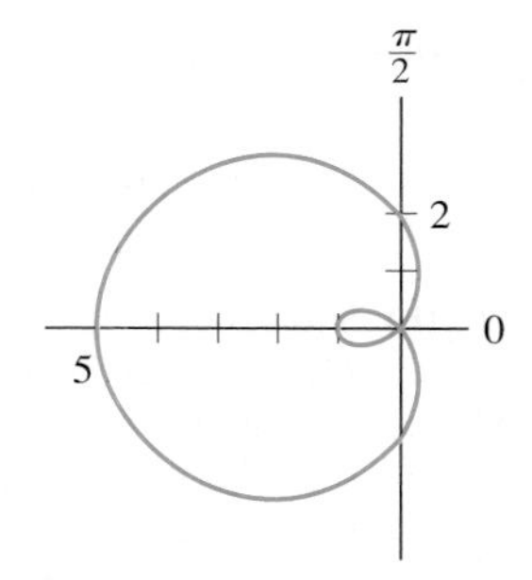

Graph B

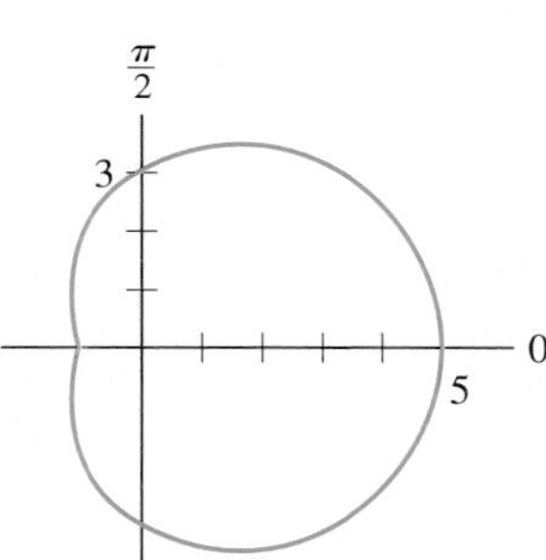

Graph C

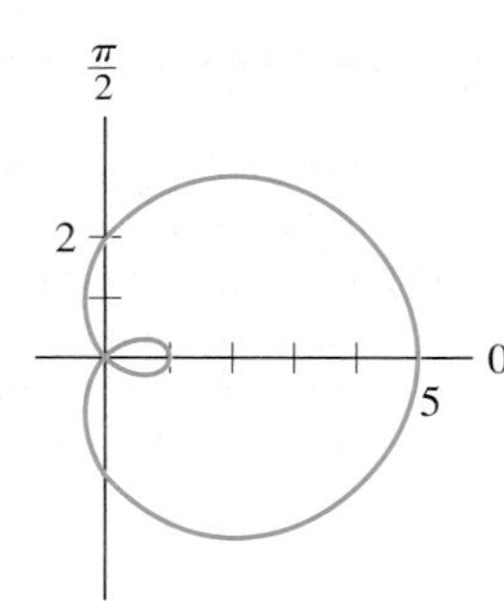

Graph D

Ⓖ **38.** The *limaçon of Pascal* is defined by the equation $r = a\cos\theta + b$. [The curve is named after Étienne Pascal (1588–1640), the father of Blaise Pascal. According to mathematics historian Howard Eves, however, the curve is misnamed; it appears earlier, in the writings of Albrecht Dürer (1471–1528).] When $a = b$, the curve is a cardioid; when $a = 0$ the curve is a circle. Graph the following limaçons.

(a) $r = 2\cos\theta + 1$

(b) $r = 2\cos\theta - 1$

(c) $r = \cos\theta + 2$

(d) $r = \cos\theta - 2$

(e) $r = 2\cos\theta + 2$

(f) $r = \cos\theta$

B

39. The figure on the next page shows a graph of the *logarithmic spiral* $\ln r = a\theta$ for $0 \le \theta \le 2\pi$. (In this equation, a denotes a positive constant.)

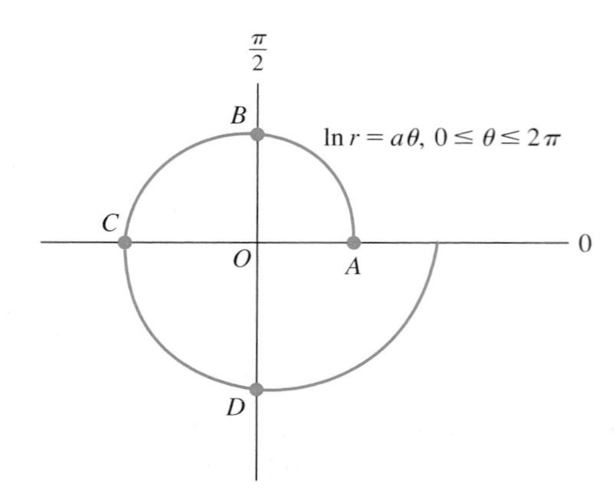

(a) Find the polar coordinates of the points A, B, C, and D, and then show that

$$\frac{OD}{OC} = \frac{OC}{OB} = \frac{OB}{OA} = e^{a\pi/2}$$

(b) Show that $\angle ABC$ and $\angle BCD$ are right angles. *Hint*: Use rectangular coordinates to compute slopes.

40. **(a)** Graph the polar curves $r = 2\cos\theta - 1$ and $r = 2\cos\theta + 1$. What do you observe?

(b) Part (a) shows that algebraically nonequivalent polar equations may have identical graphs. (This is another consequence of the fact that the polar coordinates of a point are not unique.) Show that both equations in part (a) can be written $(x^2 - 2x + y^2)^2 - x^2 - y^2 = 0$.

41. **(a)** Graph the polar curves $r = \cos(\theta/2)$ and $r = \sin(\theta/2)$. What do you observe?

(b) Part (a) shows that algebraically nonequivalent polar equations may have identical graphs. (See Exercise 40 for another example.) Show that both polar equations can be written in the rectangular form

$$(x^2 + y^2)[2(x^2 + y^2) - 1]^2 - x^2 = 0$$

42. Let k denote a positive constant, and let F_1 and F_2 denote the points with rectangular coordinates $(-k, 0)$ and $(k, 0)$, respectively. A curve known as the *lemniscate of Bernoulli* is defined as the set of points $P(x, y)$ such that $(F_1P) \times (F_2P) = k^2$.

(a) Show that the rectangular equation of the curve is $(x^2 + y^2)^2 = 2k^2(x^2 - y^2)$.

(b) Show that the polar equation is $r^2 = 2k^2\cos 2\theta$.

(c) Graph the equation $r^2 = 2k^2\cos 2\theta$.

Ⓖ **43.** Graph the *lemniscate of Bernoulli* with polar equation $r^2 = 4\cos 2\theta$. *Hint*: The equation is equivalent to the two equations $r = \pm 2\sqrt{\cos 2\theta}$. You'll find, however, that you need to consider only one of these.

Ⓖ **44.** Graph the *cissoid of Diocles*, $r = 10\sin\theta\tan\theta$. [The Greek mathematician Diocles (ca. 180 B.C.) used the geometric properties of this curve to solve the problem of duplicating the cube. For an explanation of the problem and historical background, see the text by Howard Eves, *An Introduction to the History of Mathematics,* 6th ed., pp. 109–127 (Philadelphia: Saunders College Publishing, 1990).]

Ⓖ **45.** The polar curve $r = \cos^4(\theta/4)$ has a property that is difficult to detect without the aid of a graphing utility. Graph the equation in the standard viewing rectangle. Note that the curve appears to have a simple inner loop to the left of the origin. Now zoom in on the origin. What do you observe?

In Exercises 46–48, graph the polar curves.

Ⓖ **46.** **(a)** $r = 2|\sin\theta|^{\sin\theta}$ **(b)** $r = 2|\sin\theta|^{\sin(\theta/2)}$

Ⓖ **47.** $r = (\sin\theta)/\theta, \quad 0 < \theta \le 4\pi$

Ⓖ **48.** Graph the polar curve

$$r = 1.5\sin\left(\frac{30\theta}{17} + \frac{\pi}{30}\right) + 0.5$$

Use a viewing rectangle extending from -2 to 2 in both the x- and y-directions. Let θ run from 0 to 34π.

Chapter 9 Summary of Principal Terms and Formulas

Terms or formulas	Page reference	Comments
1. $A = \frac{1}{2}ab\sin\theta$	636	The area of a triangle equals half the product of the lengths of two sides times the sine of the included angle.
2. $\frac{\sin A}{a} = \frac{\sin B}{b} = \frac{\sin C}{c}$ or $\frac{a}{\sin A} = \frac{b}{\sin B} = \frac{c}{\sin C}$	652	This is the *law of sines.* In words, it states that in any triangle the ratio of the sine of an angle to the length of its opposite side is constant. We use it when we know one angle and its opposite side and one other side or angle.

Terms or formulas	Page reference	Comments
3. $a^2 = b^2 + c^2 - 2bc \cos A$ $b^2 = a^2 + c^2 - 2ac \cos B$ $c^2 = a^2 + b^2 - 2ab \cos C$	654	This is the *law of cosines.* We can use it to calculate the third side of a triangle when we are given two sides and the included angle. The formula can also be used to calculate the angles of a triangle when we know the three sides, as in Example 5 on page 656.
4. Vector	667 678	Geometrically, a vector in the plane is a directed line segment. Vectors can be used to represent quantities that have both magnitude and direction, such as force and velocity. For examples, see Figures 1 and 2 in Section 9.3. Algebraically, a vector in the plane is an ordered pair of real numbers, denoted by $\langle a, b \rangle$; see Figure 3 in Section 9.4. The numbers a and b are called the *components* of the vector $\langle a, b \rangle$.
5. Vector equality	668 678	Geometrically, two vectors are said to be equal provided that they have the same length and the same direction. In terms of components, this means that two vectors are equal if and only if their corresponding components are equal.
6. Vector addition	668–670 679	Geometrically, two vectors can be added by using the parallelogram law, as in Figure 10 in Section 9.3. Algebraically, this is equivalent to the following componentwise formula for vector addition: $\langle u_1, u_2 \rangle + \langle v_1, v_2 \rangle = \langle u_1 + v_1, u_2 + v_2 \rangle$
7. $\lvert \mathbf{v} \rvert = \sqrt{v_1^2 + v_2^2}$	668 678	The expression on the left-hand side of the equation denotes the *length* or *magnitude* of the vector $\mathbf{v}$. The expression on the right-hand side of the equation tells us how to compute the length of $\mathbf{v}$ in terms of its respective x- and y-components v_1 and v_2.
8. $\overrightarrow{PQ} = \langle x_2 - x_1, y_2 - y_1 \rangle$	679	This formula gives the components of the vector $\overrightarrow{PQ}$, where P and Q are the points $P(x_1, y_1)$ and $Q(x_2, y_2)$; see Figure 5 in Section 9.4 for a derivation of this formula.
9. The zero vector	680	The zero vector, denoted by $\mathbf{0}$, is the vector $\langle 0, 0 \rangle$. The zero vector plays the same role in vector addition as does the real number zero in addition of real numbers. For any vector $\mathbf{v}$ we have $\mathbf{0} + \mathbf{v} = \mathbf{v} + \mathbf{0} = \mathbf{v}$
10. Scalar multiplication	680	For each real number k and each vector $\mathbf{v} = \langle x, y \rangle$ the vector $k\mathbf{v}$ is defined by the equation $k\mathbf{v} = \langle kx, ky \rangle$ Geometrically, the length of $k\mathbf{v}$ is $\lvert k \rvert$ times the length of $\mathbf{v}$. The operation that forms the vector $k\mathbf{v}$ from the scalar k and the vector $\mathbf{v}$ is called scalar multiplication.
11. The opposite of a vector	681	The opposite or negative of a vector $\mathbf{v}$, denoted $-\mathbf{v}$, is the vector defined by the equation $-\mathbf{v} = (-1)\mathbf{v}$

Terms or formulas	Page reference	Comments
12. Vector subtraction	681	The difference of two vectors **u** and **v**, denoted **u** − **v**, is the vector defined by the equation $\mathbf{u} - \mathbf{v} = \mathbf{u} + (-\mathbf{v})$
13. $\mathbf{i} = \langle 1, 0 \rangle$ $\mathbf{j} = \langle 0, 1 \rangle$	681–682	These two equations define the *unit vectors* **i** and **j**; see Figure 12 in Section 9.4. Any vector $\langle x, y \rangle$ can be expressed in terms of the unit vectors **i** and **j** as follows: $\langle x, y \rangle = x\mathbf{i} + y\mathbf{j}$
14. Parametric equations	690	Suppose the x- and y-coordinates of the points on a curve are expressed as functions of another variable t: $x = f(t)$ and $y = g(t)$. These equations are parametric equations of the curve, and t is called a parameter. An example: A pair of parametric equations for the ellipse $(x^2/a^2) + (y^2/b^2) = 1$ is $x = a\cos t$ and $y = b\sin t$.
15. $x = r\cos\theta$ $y = r\sin\theta$ $x^2 + y^2 = r^2$ $\tan\theta = y/x$	702	These formulas relate the rectangular coordinates of a point (x, y) to its polar coordinates (r, θ). See the boxed figure on page 702.
16. $d^2 = r_1^2 + r_2^2 - 2r_1r_2\cos(\theta_2 - \theta_1)$	704	This is the formula for the (square of the) distance d between the points with polar coordinates (r_1, θ_1) and (r_2, θ_2).
17. $r^2 + r_0^2 - 2rr_0\cos(\theta - \theta_0) = a^2$	705	This is a polar equation for the circle with center (r_0, θ_0) and radius a.
18. $r\cos(\theta - \alpha) = d$	706	This is a polar equation for a line not passing through the origin. In the equation, d and α are constants; d is the perpendicular distance from the origin to the line, and the (polar) point (d, α) is the foot of the perpendicular from the origin to the line.

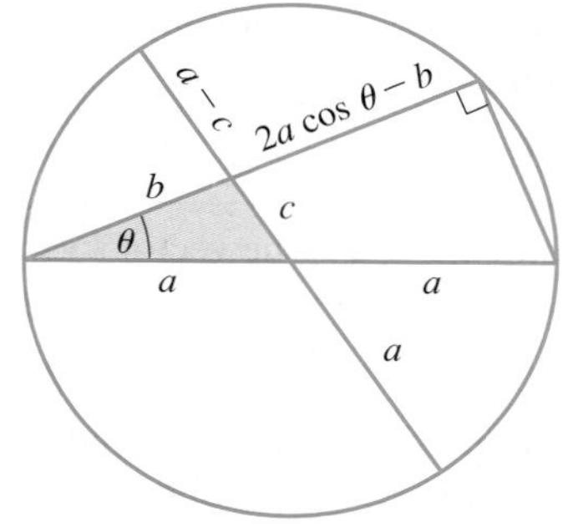

$(2a\cos\theta - b)b = (a - c)(a + c)$
$c^2 = a^2 + b^2 - 2ab\cos\theta$

Proof without words for the law of cosines, $\theta < 90°$ (by Professor Sidney H. Kung)

Writing Mathematics

1. A regular feature in *Mathematics Magazine* is the "Proof Without Words" section. The idea is to present a proof of a well-known theorem in such a way that the entire proof is "obvious" from a well-chosen picture and at most several equations. Of course, what is obvious to a mathematician might not be so obvious to her or his students (as you well know).

The following clever proof without words for the law of cosines was developed by Professor Sidney H. Kung of Jacksonville University. (It was published in *Mathematics Magazine,* vol. 63, no. 5, December 1990.) Study the proof (or discuss it with friends) until you see how it works. Then, in a paragraph or two, write out the details of the proof as if you were explaining it to a classmate. (To facilitate your explanation, you will probably need to label some of the points in the given figure.)

2. *Trisecting an angle using the spiral of Archimedes, $r = a\theta$*
Using *only* the ruler-and-compass constructions of elementary geometry, there is a well known method for bisecting an angle but no similar method for trisecting an angle. However, if we are given an accurate graph of the spiral of Archimedes, then the ruler and compass can indeed be used to trisect an angle. The method is described in the following paragraph. With a group of classmates, or on your own, study and analyze the method until you can show or explain why it is valid. (You'll need to think about what polar coordinates mean.) Then, strictly on your own, write a paragraph showing why the method works, that is, why $\angle AOE = \angle EOF = \angle FOB$.

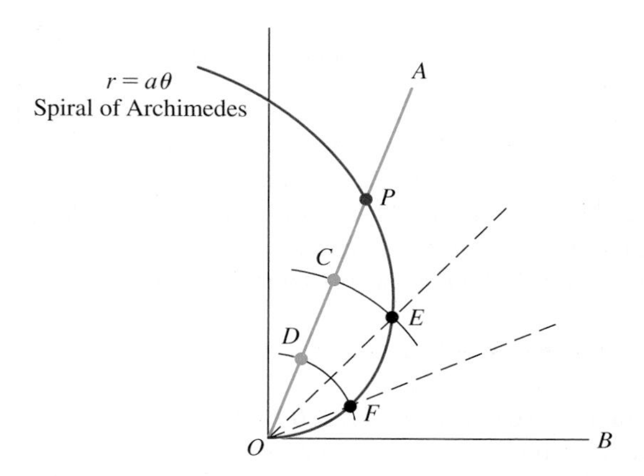

Let $\angle AOB$ be the angle to be trisected; place it in standard position, as shown. Let P be the point where the terminal side of $\angle AOB$ meets the spiral $r = a\theta$. Trisect the line segment $\overline{OP}$; let C and D be the trisection points. Now draw circular arcs with center O and radii OD and OC, as indicated in the figure. These arcs determine points F and E on the spiral, and it can be shown that $\angle AOE = \angle EOF = \angle FOB$. This is the required trisection.

Chapter 9 Review Exercises

1. Suppose θ is an acute angle and $\tan\theta + \cot\theta = 2$. Show that $\sin\theta + \cos\theta = \sqrt{2}$.
2. A 100-ft vertical antenna is on the roof of a building. From a point on the ground, the angles of elevation to the top and the bottom of the antenna are 51° and 37°, respectively. Find the height of the building.
3. In an isosceles triangle, the two base angles are each 35°, and the length of the base is 120 cm. Find the area of the triangle.
4. Find the perimeter and the area of a regular pentagon inscribed in a circle of radius 9 cm.
5. In triangle ABC, let h denote the length of the altitude from A to $\overline{BC}$. Show that $h = a/(\cot B + \cot C)$.
6. The length of each side of an equilateral triangle is $2a$. Show that the radius of the inscribed circle is $a/\sqrt{3}$ and the radius of the circumscribed circle is $2a/\sqrt{3}$.
7. From a helicopter h ft above the sea, the angle of depression to the pilot's horizon is θ. Show that $\cot\theta = R/\sqrt{2Rh + h^2}$, where R is the radius of the Earth.

In Exercises 8–14, the lengths of the three sides of a triangle are denoted a, b, and c; the angles opposite these sides are A, B, and C, respectively. In each exercise, use the given information to find the required quantities.

	GIVEN	FIND
8.	$b = 4, c = 5,$ $A = 150°$	area of $\triangle ABC$
9.	$A = 120°, b = 8,$ area $\triangle ABC = 12\sqrt{3}$	c
10.	$c = 4, a = 2, B = 90°$	$\sin^2 A + \cos^2 B$
11.	$B = 90°, 2a = b$	A
12.	$A = 30°, B = 120°,$ $b = 16$	a
13.	$a = 7, b = 8,$ $\sin C = 1/4$	area of $\triangle ABC$
14.	$a = b = 5,$ $\sin(C/2) = 9/10$	c and area of $\triangle ABC$

In Exercises 15–22 the lengths of the three sides of a triangle are denoted by a, b, and c; the angles opposite these sides are A, B, and C, respectively. In each exercise, use the given data to find the remaining sides and angles. Use a calculator, and round the final answers to one decimal place.

15. $\angle A = 40°, \angle B = 85°, c = 16$ cm

16. $\angle C = 84°, \angle B = 16°, a = 9$ cm

17. **(a)** $a = 8$ cm, $b = 9$ cm, $\angle A = 52°, \angle B < 90°$
(b) Use the data in part (a), but assume that $\angle B > 90°$.

18. **(a)** $a = 6.25$ cm, $b = 9.44$ cm, $\angle A = 12°, \angle B < 90°$
(b) Use the data in part (a), but assume that $\angle B > 90°$.

19. $a = 18$ cm, $b = 14$ cm, $\angle C = 24°$

20. $a = 32.16$ cm, $b = 50.12$ cm, $\angle C = 156°$

21. $a = 4$ cm, $b = 7$ cm, $c = 9$ cm

22. $a = 12.61$ cm, $b = 19.01$ cm, $c = 14.14$ cm

In Exercises 23–32, refer to the following figure. In each case, determine the indicated quantity. Use a calculator and round your result to two decimal places.

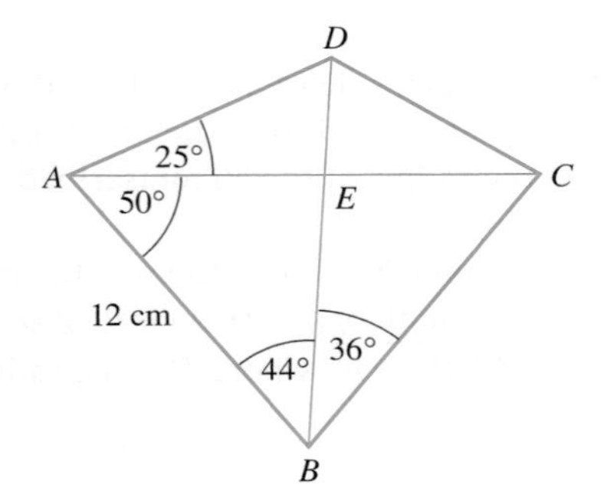

23. BE **24.** BC
25. area of $\triangle BCE$ **26.** BD
27. area of $\triangle ABD$ **28.** DE
29. CD **30.** AD
31. AC **32.** area of $\triangle DEC$

In Exercises 33 and 34, refer to the following figure, in which $\overline{BD}$ bisects $\angle ABC$.

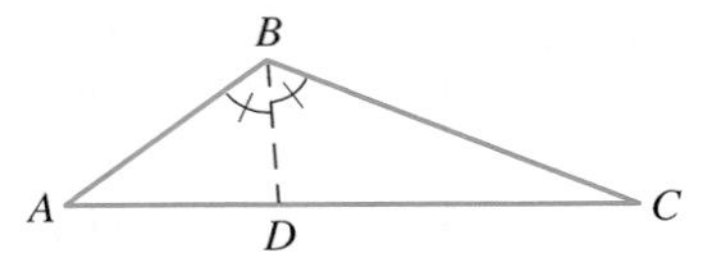

33. If $AB = 24$ cm, $BC = 40$ cm, and $AC = 56$ cm, find the length of the angle bisector $\overline{BD}$.

34. If $AB = 105$ cm, $BC = 120$ cm, and $AC = 195$ cm, find the length of the angle bisector $\overline{BD}$.

In Exercises 35 and 36, refer to the following figure, in which $AD = DC$.

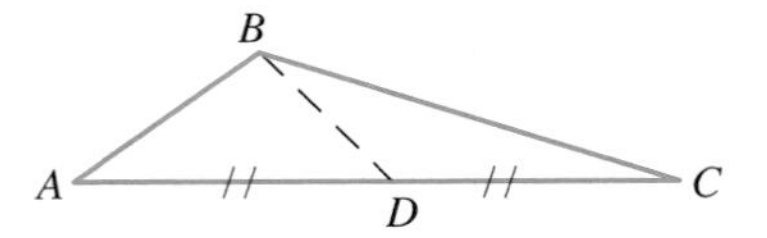

35. If $AB = 12$ cm, $BC = 26$ cm, and $AD = DC = 17$ cm, find BD.

36. If $AB = 16$ cm, $BC = 22$ cm, and $AD = DC = 17$ cm, find BD.

37. In $\triangle ABC$, $a = 4$, $b = 5$, and $c = 6$.
(a) Find $\cos A$ and $\cos C$.
(b) Using the results in part (a), show that angle C is twice angle A.

38. In $\triangle ABC$, suppose that angle C is twice angle A. Show that $ab = c^2 - a^2$.

In Exercises 39–44, refer to the framework shown in the accompanying figure. (Assume that the figure is drawn to scale.) In the figure, $\overline{AC}$ is parallel to $\overline{ED}$ and $BE = BD$. In each case, compute the indicated quantity. Round the final answers to two decimal places.

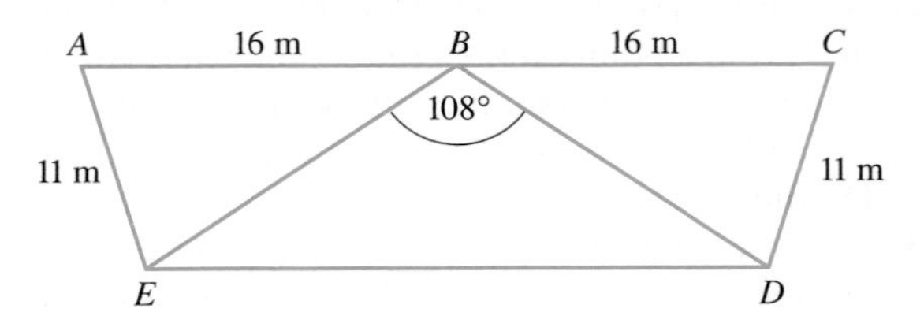

39. $\angle BED$ **40.** $\angle ABE$
41. $\angle AEB$ **42.** $\angle BAE$
43. BE **44.** DE

45. In this exercise you will verify some of the properties of the *pentagram star* shown in Figure A. Figure B shows how the pentagram is constructed. We start with a regular pentagon $ABCDE$ inscribed in a circle with center O. Drawing the diagonals of the pentagon then yields the pentagram star. As indicated in Figure B, the five intersection points (S, T, U, V, and W) of the diagonals form the vertices of a second, smaller regular pentagon. In this exercise you can assume the following facts (which can be established using elementary geometry):

- Each of the five triangles AUV, BVW, CWS, DST, and ETU is isosceles, with angles of 72°, 72°, and 36°.
- In a regular pentagon, each interior angle is 108°. (For example, in the large regular pentagon, $\angle AED = 108°$, and in the small pentagon, $\angle VUT = 108°$.)

Figure A Pentagram Star

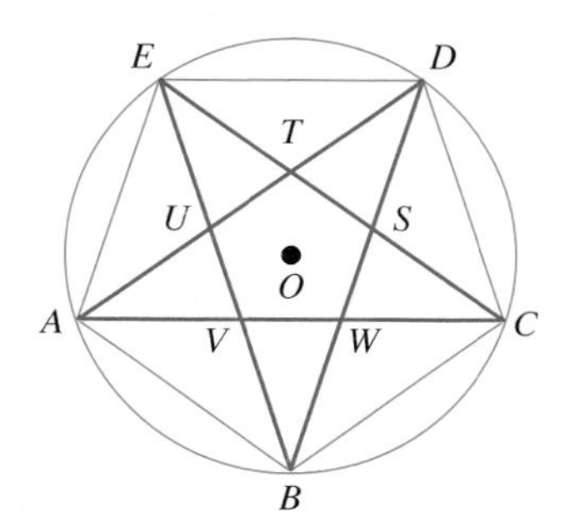

Figure B

(a) Assume (throughout this exercise) that the radius of the circle in Figure B is 1 unit. Show that $AB = 2 \sin 36°$. *Hints*: Why is $\angle AOB = 72°$? After answering, draw a perpendicular from O to $\overline{AB}$.

(b) Show that $AD = 2 \sin 72°$. *Hint*: Draw a perpendicular from O to $\overline{AD}$ and draw a radius from O to D. Then use the right triangle that is formed.

(c) Show $AV = \tan 36°$. *Hint*: Draw a perpendicular from V to $\overline{AB}$. [Recall that you found AB in part (a).]

(d) Show $VW = 2 \tan 36° \cos 72°$. *Hint*: From Figure B, note that $VW = AC - 2(AV)$. After substituting for AC and AV in this last expression, use the double-angle formulas.

(e) Show $AW = (\tan 36°)(1 + 2 \cos 72°)$. *Hint*: $AW = AV + VW$

(f) Use results from previous parts of this exercise to verify that $(AD/AB) = 2 \cos 36°$.

(g) For this last part of the exercise, assume as given the following two trigonometric values:

$$\sin 18° = (\sqrt{5} - 1)/4 \qquad \cos 36° = (\sqrt{5} + 1)/4$$

(The value for sin 18° was obtained in Exercise 68 in Section 8.2; the value for cos 36° can be derived from this by using the identity $\cos 2\theta = 1 - 2 \sin^2\theta$.) Verify each of the following statements by using results from previous parts of this exercise along with the values for sin 18° and cos 36°.

(i) $\dfrac{AV}{VW} = 2 \cos 36°$

(ii) $\dfrac{AW}{AV} = 2 \cos 36°$

(iii) $\dfrac{AC}{AW} = 2 \cos 36°$

Remark: The quantity $2 \cos 36° = (\sqrt{5} + 1)/2$ occurs with sufficient frequency in mathematics and its applications that a particular symbol is commonly used for it: ϕ (the Greek letter *phi*). For a thorough discussion of ϕ (or the *golden ratio,* as it is sometimes called), see any of the following books:

Pedoe, Dan, *Geometry and the Visual Arts* (New York: Dover Publications, Inc., 1976); Huntley, H. E., *The Divine Proportion* (New York: Dover Publications, Inc., 1970); Livio, Mario, *The Golden Ratio: The story of phi, the most astonishing number,* (New York: Broadway Books, 2003)

46. A pilot on a training flight is supposed to leave the airport and fly for 100 miles in the direction N40°E. Then he is supposed to turn around and fly directly back to the airport. By error, however, he makes the 100-mile return trip in the direction S25°W. How far is he now from the airport, and in what direction must he fly to get there? (Round the answers to one decimal place.)

47. The following figures show the same circle of diameter D inscribed first in an equilateral triangle and then in a regular hexagon. Let a denote the length of a side of the triangle, and let b denote the length of a side of the hexagon. Show that $ab = D^2$.

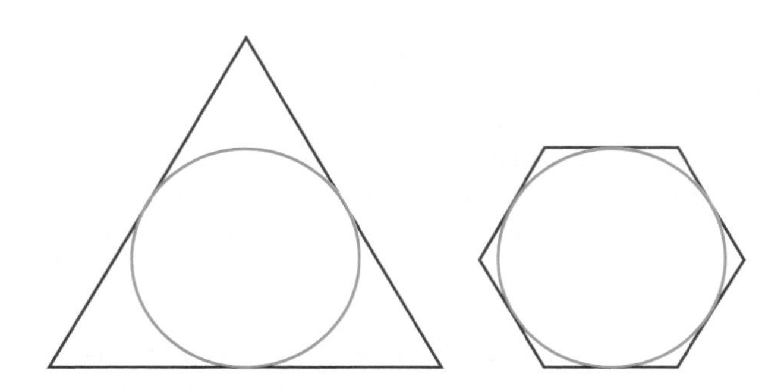

48. Suppose that $\triangle ABC$ is a right triangle with the right angle at C. Use the law of cosines to prove the following statements.

(a) The square of the distance from C to the midpoint of the hypotenuse is equal to one-fourth the square of the hypotenuse.

(b) The sum of the squares of the distances from C to the two points that trisect the hypotenuse is equal to five-ninths the square of the hypotenuse.

(c) Let P, Q, and R be points on the hypotenuse such that $AP = PQ = QR = RB$. Derive a result similar to your results in parts (a) and (b) for the sum of the squares of the distances from C to P, Q, and R.

49. The perimeter and the area of the triangle in the figure are 20 cm and $10\sqrt{3}$ cm^2, respectively. Find a and b. (Assume that $a < b$.)

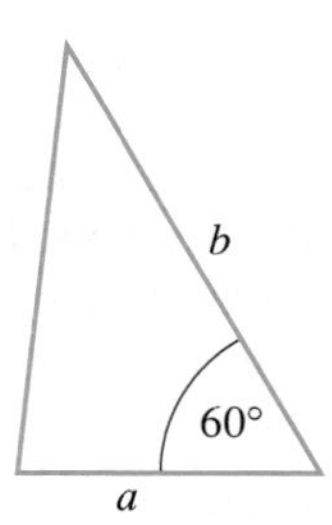

50. In $\triangle ABC$, suppose that $\angle B - \angle A = 90°$.

(a) Show that $c^2 = (b^2 - a^2)^2/(a^2 + b^2)$. *Hint*: Let P be the point on $\overline{AC}$ such that $\angle CBP = 90°$ and $\angle PBA = \angle A$. Show that $CP = (a^2 + b^2)/2b$.

(b) Show that $\dfrac{2}{c^2} = \dfrac{1}{(b+a)^2} + \dfrac{1}{(b-a)^2}$.

51. Two forces **F** and **G** act on an object, **G**, horizontally to the right with a magnitude of 15 N, and **F** vertically upward with a magnitude of 20 N. Determine the magnitude and direction of the resultant force. (Use a calculator to determine the angle between the horizontal and the resultant; round the result to one decimal place.)

52. Determine the resultant of the two forces in the accompanying figure. (Use a calculator, and round the values you obtain for the magnitude and direction to one decimal place.)

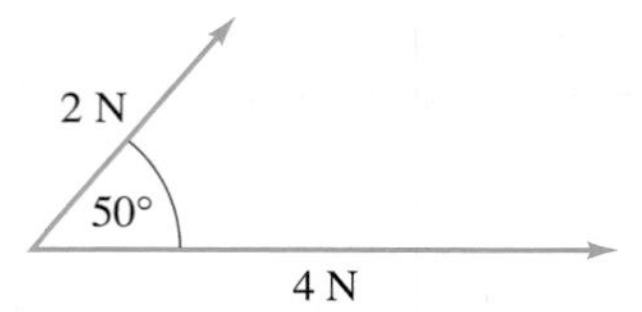

53. Determine the horizontal and vertical components of the velocity vector **v** in the following figure. (Use a calculator, and round your answers to one decimal place.)

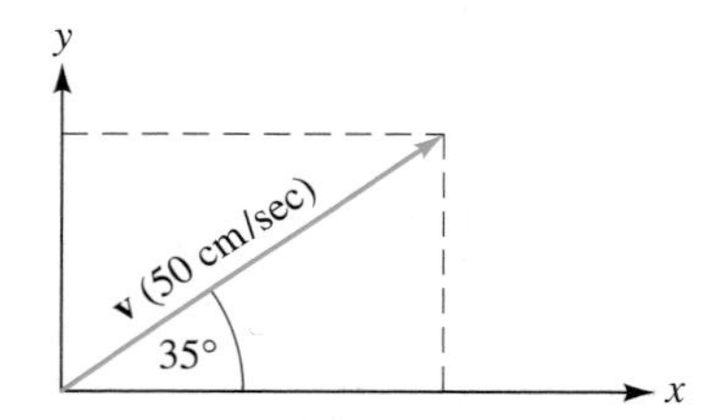

54. The heading and air speed of an airplane are 50° and 220 mph, respectively. If the wind is 60 mph from 140°, find the ground speed, the drift angle, and the course. (Use a calculator, and round your answers to one decimal place.)

55. A block rests on an inclined plane that makes an angle of 24° with the horizontal. The component of the weight parallel to the plane is 14.8 lb. Determine the weight of the block and the component of the weight perpendicular to the plane. (Use a calculator, and round your answers to one decimal place.)

56. Find the length of the vector $\langle 20, 99 \rangle$.

57. For which values of b will the vectors $\langle 2, 6 \rangle$ and $\langle -5, b \rangle$ have the same length?

58. Given the points $A(2, 6)$ and $B(-7, 4)$, find the components of the following vectors. Write your answers in the form $\langle x, y \rangle$.

(a) $\overrightarrow{AB}$ **(c)** $3\overrightarrow{AB}$

(b) $\overrightarrow{BA}$ **(d)** $\dfrac{1}{|\overrightarrow{AB}|}\overrightarrow{AB}$

In Exercises 59–70, compute each of the indicated quantities, given that the vectors **a**, **b**, **c**, *and* **d** *are defined as follows:*

$\mathbf{a} = \langle 3, 5\rangle \quad \mathbf{b} = \langle 7, 4\rangle \quad \mathbf{c} = \langle 2, -1\rangle \quad \mathbf{d} = \langle 0, 3\rangle$

59. $\mathbf{a} + \mathbf{b}$
60. $\mathbf{b} - \mathbf{d}$
61. $3\mathbf{c} + 2\mathbf{a}$
62. $|\mathbf{a}|$
63. $|\mathbf{b} + \mathbf{d}|^2 - |\mathbf{b} - \mathbf{d}|^2$
64. $\mathbf{a} + (\mathbf{b} + \mathbf{c})$
65. $(\mathbf{a} + \mathbf{b}) + \mathbf{c}$
66. $\mathbf{a} - (\mathbf{b} - \mathbf{c})$
67. $(\mathbf{a} - \mathbf{b}) - \mathbf{c}$
68. $|\mathbf{a} + \mathbf{b}|^2 + |\mathbf{a} - \mathbf{b}|^2 - 2|\mathbf{a}|^2 - 2|\mathbf{b}|^2$
69. $4\mathbf{c} + 2\mathbf{a} - 3\mathbf{b}$
70. $|4\mathbf{c} + 2\mathbf{a} - 3\mathbf{b}|$
71. Express the vector $\langle 7, -6\rangle$ in terms of the unit vectors **i** and **j**.
72. Express the vector $4\mathbf{i} - 6\mathbf{j}$ in the form $\langle a, b\rangle$.
73. Find a unit vector having the same direction as $\langle 6, 4\rangle$.
74. Find two unit vectors that are perpendicular to $\langle \cos\theta, \sin\theta\rangle$. Simplify the components as much as possible.

In Exercises 75 and 76, compute the distance between the points with the given polar coordinates. Use a calculator and round the final answers to two decimal places.

75. $(3, \frac{\pi}{12})$ and $(2, \frac{17\pi}{18})$
76. $(1, 1)$ and $(3, 2)$

In Exercises 77 and 78, determine a polar equation for the circle satisfying the given conditions.

77. The radius is 3 and the polar coordinates of the center are $(5, \frac{\pi}{6})$.
78. The radius is 1 and the polar coordinates of the center are $(-3, \frac{\pi}{4})$.

In Exercises 79 and 80, the polar equation of a line is given. In each case: **(a)** *specify the perpendicular distance from the origin to the line;* **(b)** *determine the polar coordinates of the points on the line corresponding to $\theta = 0$ and $\theta = \pi/2$;* **(c)** *specify the polar coordinates of the foot of the perpendicular from the origin to the line;* **(d)** *sketch the line; and* **(e)** *find a rectangular form for the equation of the line.*

79. $r\cos(\theta - \frac{\pi}{3}) = 3$
80. $r\cos(\theta - 1) = \sqrt{5}$

In Exercises 81–88, graph the polar equations.

81. **(a)** $r = 2 - 2\cos\theta$
(b) $r = 2 - 2\sin\theta$
82. **(a)** $r^2 = 4\cos 2\theta$
(b) $r^2 = 4\sin 2\theta$
83. **(a)** $r = 2\cos\theta - 1$
(b) $r = 2\sin\theta - 1$
84. **(a)** $r = 2\cos\theta$
(b) $r = 2\cos\theta + 1$
85. **(a)** $r = 4\sin 2\theta$
(b) $r = 4\cos 2\theta$
86. **(a)** $r = 4\sin 3\theta$
(b) $r = 4\cos 3\theta$
87. **(a)** $r = 1 + 2\sin(\theta/2)$
(b) $r = 1 - 2\cos(\theta/2)$
88. **(a)** $r = (1.5)^\theta \quad (\theta \geq 0)$
(b) $r = (1.5)^\theta \quad (\theta \leq 0)$

In Exercises 89–94, graph the curve (or line) determined by the parametric equations. Indicate the direction of travel along the curve as t increases. You may do this by plotting points or by eliminating the parameter (or both). Unless indicated, the domain is all real numbers.

89. $x = 3 - 5t,\ y = 1 + t$
90. $x = t - 1,\ y = 3t$
91. $x = 3\sin t,\ y = 6\cos t$
92. $x = 2\cos t,\ y = 2\sin t$
93. $x = 4\sec t,\ y = 3\tan t$ for $0 \leq t \leq 2\pi$, except $t = \pi/2$ and $t = 3\pi/2$
94. $x = 1 + 3\cos t,\ y = 2 + 4\sin t$
95. In the accompanying figure, P and Q are two points on the polar curve $r = (\sin\theta)/\theta$ such that $\angle AOP = \angle POQ$. Follow steps (a) through (f) to prove that $\overline{PQ} \perp \overline{OQ}$.

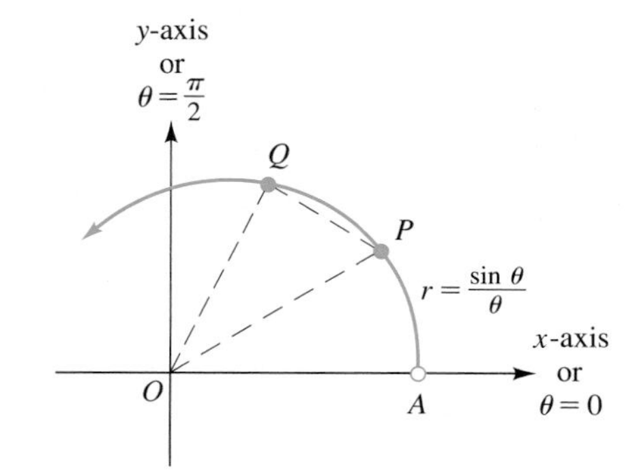

(a) Let $\alpha = \angle AOP = \angle POQ$. What are the polar coordinates of P and Q in terms of α?

(b) What are the rectangular coordinates of P and Q in terms of α?

(c) Using the coordinates determined in part (b), show that the slope of $\overline{OQ}$ is $\tan 2\alpha$.

(d) Using the coordinates determined in part (b), show that the slope of $\overline{PQ}$ is

$$\frac{\sin^2 2\alpha - 2\sin^2 \alpha}{\sin 2\alpha \cos 2\alpha - 2 \sin \alpha \cos \alpha}$$

(e) Show that the expression in part (d) can be simplified to $-\cot 2\alpha$. *Hint*: In the numerator, replace $\sin^2 2\alpha$ by $1 - \cos^2 2\alpha$ and $\sin^2 \alpha$ by $\frac{1 - \cos 2\alpha}{2}$.

(f) Use the results in parts (c) and (e) to explain why $\overline{PQ} \perp \overline{OQ}$.

96. In this exercise you will use the following figure to derive the *Mollweide formula*:

$$\frac{a+b}{c} = \frac{\cos \frac{1}{2}(A - B)}{\sin \frac{1}{2}C}$$

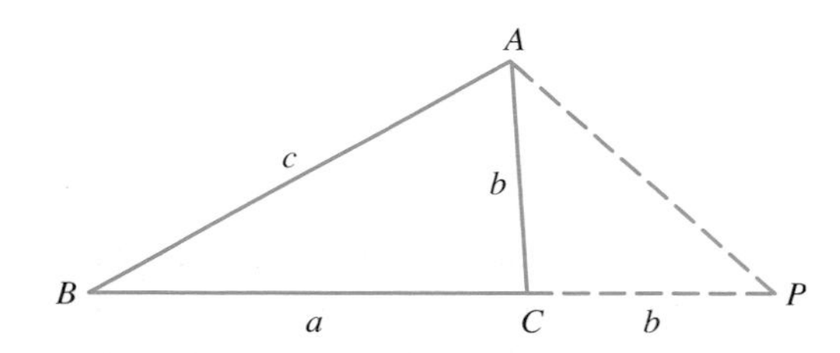

The figure is constructed as follows. Starting with $\triangle ABC$, extend side $\overline{BC}$ so that $CP = b$, as shown. Then draw line segment $\overline{AP}$.

(a) Show that $\angle APC = \angle PAC = \frac{1}{2}C$.

(b) Show that $\angle BAP = 90° + \frac{1}{2}(A - B)$. *Hint*: Start with the fact that $\angle BAP = A + \frac{1}{2}C$.

(c) Use the law of sines now in $\triangle ABP$ to show that

$$\frac{\sin[90° + \frac{1}{2}(A - B)]}{a + b} = \frac{\sin \frac{1}{2}C}{c}$$

(d) Use the result in part (c) to obtain the required identity.

97. In this exercise we use the law of sines to deduce the **law of tangents** for $\triangle ABC$:

$$\frac{a-b}{a+b} = \frac{\tan \frac{1}{2}(A - B)}{\tan \frac{1}{2}(A + B)}$$

This law was given by the Danish physician and mathematician Thomas Fink in his text *Geometria Rotundi*, published in Basel in 1583. (Our use of the terms "tangent" and "secant" is also due to Fink.)

(a) Suppose that a, b, x, and y are real numbers such that $\frac{a}{b} = \frac{x}{y}$. Verify that $\frac{a-b}{a+b} = \frac{x-y}{x+y}$.

(b) Use the law of sines and the result in part (a) to show that

$$\frac{a-b}{a+b} = \frac{\sin A - \sin B}{\sin A + \sin B}$$

(c) Use the result in part (a) and the sum-to-product formulas from Section 8.3 to complete the derivation of the law of tangents.

98. The following figure shows a sector of the unit circle $x^2 + y^2 = 1$. The central angle for the sector is α (a constant). The point $P(\cos\theta, \sin\theta)$ denotes an arbitrary point on the arc of the sector. From P, perpendiculars are drawn to the sides of the sector, meeting the sides at Q and R, as shown. Show that the distance from Q to R does not depend on θ.

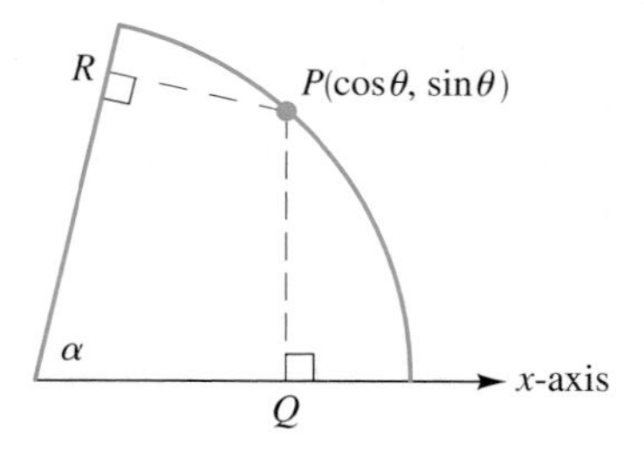

Chapter 9 Test

1. In triangle ABC, let $A = 120°$, $b = 5$ cm, and $c = 3$ cm. Find a.
2. The sides of a triangle are 2 cm, 3 cm, and 4 cm. Determine the cosine of the angle opposite the longest side. On the basis of your answer, explain whether or not the angle opposite the longest side is an acute angle.
3. Two of the angles in a triangle are 30° and 45°. If the side opposite the 45° angle is $20\sqrt{2}$ cm, find the side opposite the 30° angle.
4. A 10-ft ladder that is leaning against the side of a building makes an angle of 60° with the ground, as shown in the following figure. How far up the building does the ladder reach?

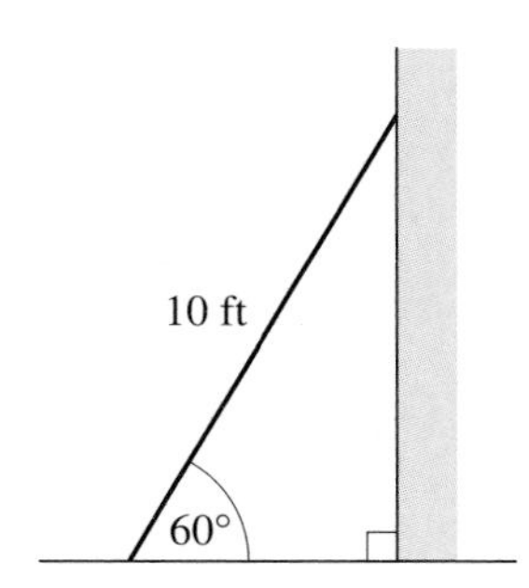

5. In $\triangle ABC$, $B = 90°$, $c = 2$, and $a = 5$. Find A; express the answer in degrees rounded to one decimal place.
6. Refer to the following figure. Express the area of $\triangle ABC$ as a function of θ.

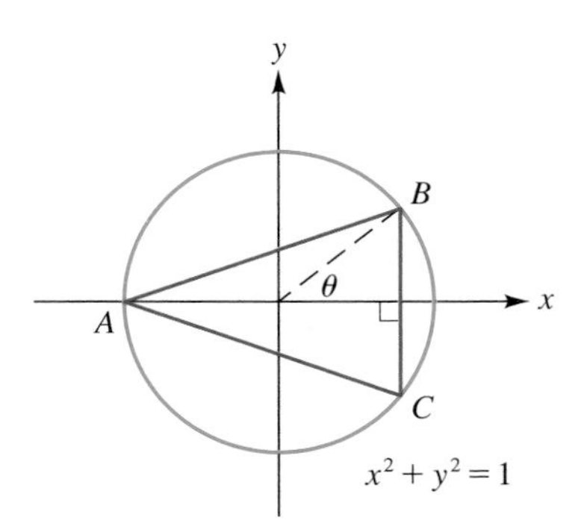

7. Each side of the square $STUV$ is 8 cm long. The point P lies on diagonal $\overline{SU}$ such that $SP = 2$ cm. Find the distance from P to V.
8. In $\triangle ABC$, $b = 5.8$ cm, $c = 3.2$ cm, and $A = 27°$. Find the remaining sides and angles of the triangle.
9. Two forces **F** and **G** act on an object. The force **G** acts horizontally with a magnitude of 2 N, and **F** acts vertically upward with a magnitude of 4 N.
   (a) Find the magnitude of the resultant.
   (b) Find $\tan\theta$, where θ is the angle between **G** and the resultant.
10. Two forces act on an object, as shown in the following figure.
    (a) Find the magnitude of the resultant. (Leave your answer in terms of radicals and the trigonometric functions.)
    (b) Find $\sin\theta$, where θ is the angle between the 12 N force and the resultant.

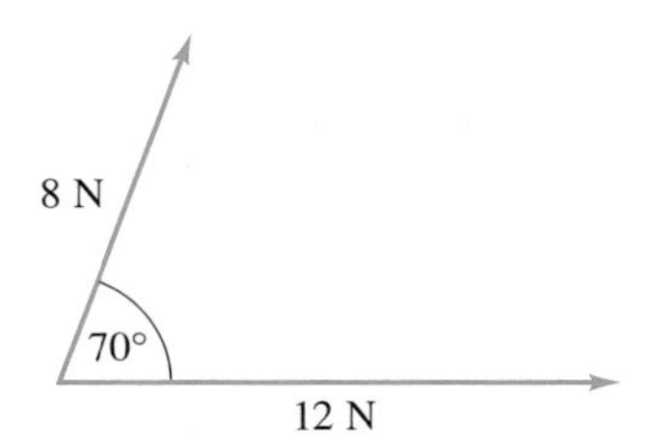

11. The heading and air speed of an airplane are 40° and 300 mph, respectively. If the wind is 50 mph from 130°, find the ground speed and the tangent of the drift angle. (Leave your answer in terms of radicals and the trigonometric functions.)
12. Let $\mathbf{A} = \langle 2, 4\rangle$, $\mathbf{B} = \langle 3, -1\rangle$, and $\mathbf{C} = \langle 4, -4\rangle$.
    (a) Find $2\mathbf{A} + 3\mathbf{B}$.
    (b) Find $|2\mathbf{A} + 3\mathbf{B}|$.
    (c) Express $\mathbf{C} - \mathbf{B}$ in terms of **i** and **j**.
13. Let P and Q be the points (4, 5) and (−7, 2), respectively. Find a unit vector having the same direction as $\overrightarrow{PQ}$.
14. Convert the polar equation $r^2 = \cos 2\theta$ to rectangular form.
15. Graph the equation $r = 2(1 - \cos\theta)$.
16. Given the parametric equations $x = 4\sin t$ and $y = 2\cos t$, eliminate the parameter t and sketch the graph.
17. Compute the distance between the two points with polar coordinates $(4, \frac{10\pi}{21})$ and $(1, \frac{\pi}{7})$.

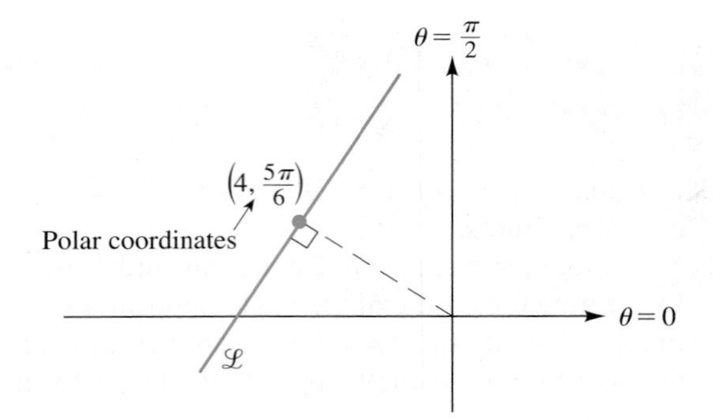

18. Find a polar equation for the circle with center (in polar coordinates) $(5, \frac{\pi}{2})$ and with radius 2. Does the (polar) point $(2, \frac{\pi}{6})$ lie on this circle?

19. Refer to the accompanying figure.

(a) Determine a polar equation for the line $\mathcal{L}$ in the figure.

(b) Use an addition formula to show that the polar equation of the line can be written $-r\sqrt{3}\cos\theta + r\sin\theta - 8 = 0$.

10 CHAPTER Systems of Equations

In this chapter we consider systems of equations. Roughly speaking, a system of equations is just a collection of equations with a common set of unknowns. In solving such systems, we try to find values for the unknowns that simultaneously satisfy each equation in the system. Matrices are introduced as a tool for solving systems of equations. In this chapter you'll see systems of equations and/or matrices used to:

- Determine conditions for market equilibrium (Example 6 in Section 10.1)
- Solve an airline scheduling problem (Example 6 in Section 10.2)
- Determine production requirements in various sectors of an economy using input-output analysis (projects following Sections 10.2 and 10.4)
- Analyze a communications network (project following Section 10.3)
- Encode and decode messages (Examples 4 and 5 in Section 10.4)

10.1 SYSTEMS OF TWO LINEAR EQUATIONS IN TWO UNKNOWNS

Many problems in a variety of disciplines give rise to . . . [linear] *systems. For example, in physics, in order to find the currents in an electrical circuit, . . . a system of linear equations must be solved. In chemistry, the balancing of chemical equations requires the solution of a system of linear equations. And in economics, the Leontief input-output model reduces problems concerning the production and consumption of goods to systems of linear equations.*—Leslie Hogben in *Elementary Linear Algebra* (Pacific Grove, Calif.: Brooks/Cole, 1999)

Both in theory and in applications it is often necessary to solve two equations in two unknowns. You may have been introduced to the idea of simultaneous equations in a previous course in algebra; however, to put matters on a firm foundation, we begin here with the basic definitions. By a **linear equation in two variables** we mean an equation of the form

$$ax + by = c$$

where a, b, and c are constants with a and b not both zero. The two variables needn't always be denoted by the letters x and y, of course; it is the *form* of the equation that matters. Table 1 (on the next page) displays some examples.

An ordered pair of numbers (x_0, y_0) is said to be a **solution of the linear equation** $ax + by = c$, provided that we obtain a true statement when we replace x and y in the equation by x_0 and y_0, respectively. For example, the ordered pair (3, 2) is a solution of the equation $x - y = 1$, since $3 - 2 = 1$. On the other hand, (2, 3) is not a solution of $x - y = 1$, since $2 - 3 \neq 1$.

Now consider a **system** of two linear equations in two unknowns:

$$\begin{cases} ax + by = c \\ dx + ey = f \end{cases}$$

An ordered pair that is a solution to both equations is called a **solution of the system.** Sometimes, to emphasize the fact that a solution must satisfy both equations, we refer to the system as a pair of **simultaneous equations.** A system that has at least one solution is said to be **consistent.** If there are no solutions, the system is **inconsistent.**

TABLE 1

Equations in two variables	Is it linear? Yes	No
$3x - 8y = 12$	✓	
$-s + 4t = 0$	✓	
$2x - 3y^2 = 1$		✓
$y = 4 - 2x$	✓	
$\frac{4}{u} + \frac{5}{v} = 3$		✓

EXAMPLE 1 Recognizing a solution of a system of equations

Consider the system

$$\begin{cases} x + y = 2 \\ 2x - 3y = 9 \end{cases}$$

(a) Is (1, 1) a solution of the system?
(b) Is (3, −1) a solution of the system?

SOLUTION

(a) Although (1, 1) is a solution of the first equation, it is not a solution of the system because it does not satisfy the second equation. (Check this for yourself.)

(b) (3, −1) satisfies the first equation:

$$3 + (-1) \stackrel{?}{=} 2 \qquad \text{True}$$

(3, −1) satisfies the second equation:

$$2(3) - 3(-1) \stackrel{?}{=} 9$$
$$6 + 3 \stackrel{?}{=} 9 \qquad \text{True}$$

Since (3, −1) satisfies both equations, it is a solution of the system.

We can gain an important perspective on systems of linear equations by looking at Example 1 in graphical terms. Table 2 shows how each of the statements in that example can be rephrased by using the geometric ideas with which we are already familiar.

In Example 1 we verified that (3, −1) is a solution of the system

$$\begin{cases} x + y = 2 \\ 2x - 3y = 9 \end{cases}$$

Are there any other solutions of this particular system? No: Figure 1 shows us that there are no other solutions, since (3, −1) is clearly the only point common to both

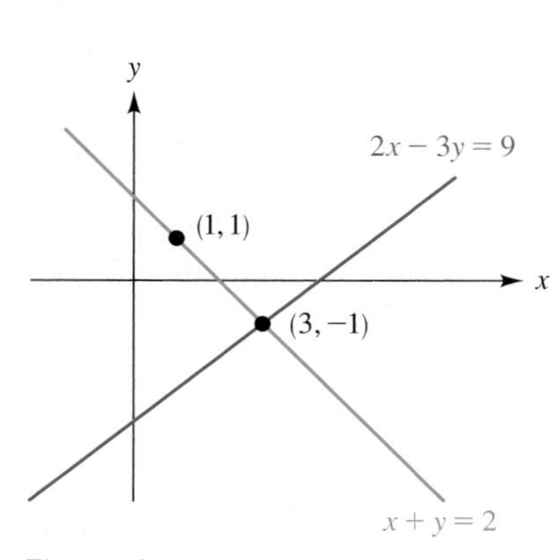

Figure 1

TABLE 2

Algebraic idea	Corresponding geometric idea
1. The ordered pair (1, 1) is a solution of the equation $x + y = 2$.	1. The point (1, 1) lies on the line $x + y = 2$. See Figure 1.
2. The ordered pair (1, 1) is not a solution of the equation $2x - 3y = 9$.	2. The point (1, 1) does not lie on the line $2x - 3y = 9$. See Figure 1.
3. The ordered pair (3, −1) is a solution of the system $\begin{cases} x + y = 2 \\ 2x - 3y = 9 \end{cases}$	3. The point (3, −1) lies on both of the lines $x + y = 2$ and $2x - 3y = 9$. See Figure 1.

lines. In a moment we'll look at two important methods for solving systems of linear equations in two unknowns. But even before we consider these methods, we can say something about the solutions of linear systems.

PROPERTY SUMMARY Possibilities for Solutions of Linear Systems

Given a system of two linear equations in two unknowns, exactly one of the following cases must occur.

CASE 1 The graphs of the two linear equations intersect in exactly one point. Thus there is exactly one solution to the system. See Figure 2.

CASE 2 The graphs of the two linear equations are parallel lines. Therefore the lines do not intersect, and the system has no solution. See Figure 3.

CASE 3 The two equations actually represent the same line. Thus, there are infinitely many points of intersection and correspondingly infinitely many solutions. See Figure 4.

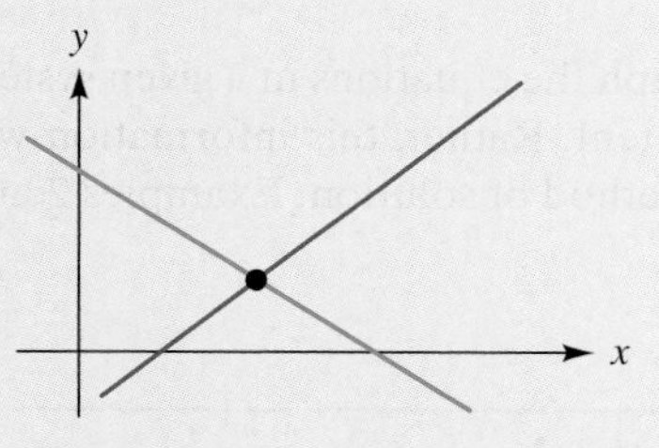

Figure 2
A consistent system with exactly one solution

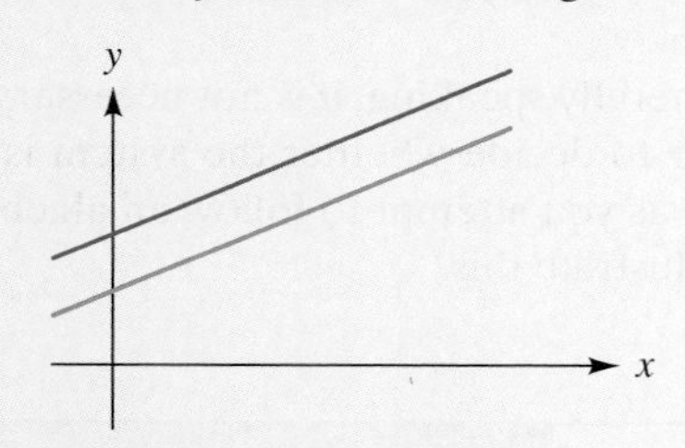

Figure 3
An inconsistent system has no solution

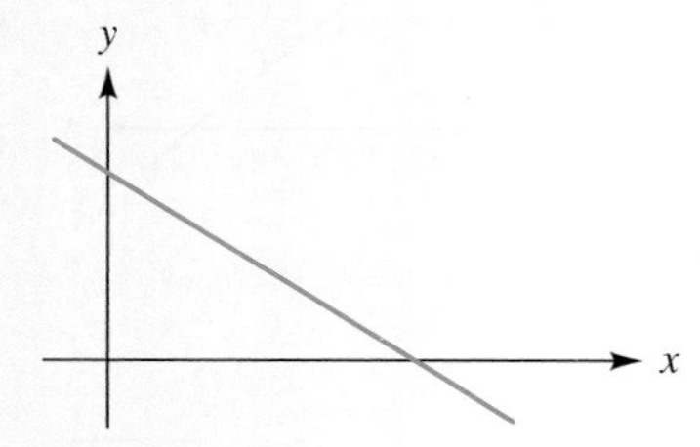

Figure 4
A consistent system with infinitely many solutions

We are going to review two methods from intermediate algebra for solving systems of two linear equations in two unknowns. These methods are the **substitution method** and the **addition–subtraction method.** We'll begin by demonstrating the substitution method. Consider the system

$$\begin{cases} 3x + 2y = 17 & (1) \\ 4x - 5y = -8 & (2) \end{cases}$$

We first choose one of the two equations and then use it to express one of the variables in terms of the other. In the case at hand, neither equation appears simpler than the other, so let's just start with the first equation and solve for x in terms of y. We have

$$\begin{aligned} 3x &= 17 - 2y \\ x &= \frac{1}{3}(17 - 2y) \qquad (3) \end{aligned}$$

Now we use equation (3) to substitute for x in the equation that we have not yet used, namely, equation (2). This yields

$$\begin{aligned} 4\left[\frac{1}{3}(17 - 2y)\right] - 5y &= -8 \\ 4(17 - 2y) - 15y &= -24 \qquad \text{multiplying by 3} \\ -23y &= -92 \\ y &= 4 \end{aligned}$$

The value $y = 4$ that we have just obtained can now be used in equation (3) to find x. Replacing y with 4 in equation (3) yields

$$x = \frac{1}{3}[17 - 2(4)] = \frac{1}{3}(9) = 3$$

We have now found that $x = 3$ and $y = 4$. According to our work on linear equations, each step that we used in the substitution method results in an **equivalent system,** that is, a system with the same solution set. So the original system of equations is equivalent to the system

$$\begin{cases} x = 3 \\ y = 4 \end{cases}$$

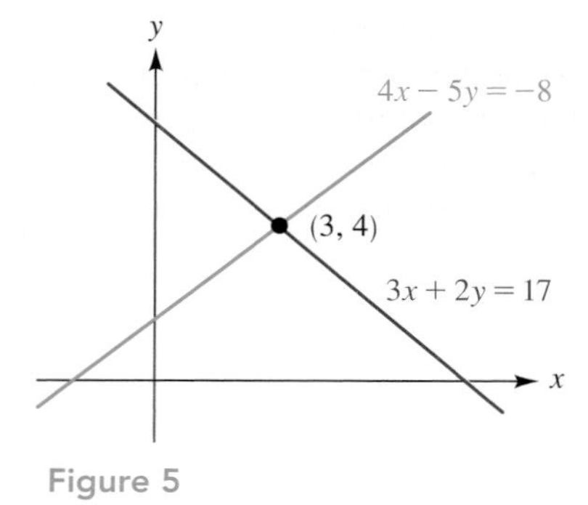

Figure 5

As you can easily check, this pair of values indeed satisfies both of the original equations. We write our solution as the ordered pair (3, 4). Figure 5 summarizes the situation. It shows that the system is consistent and that (3, 4) is the only solution.

Generally speaking, it is not necessary to graph the equations in a given system in order to decide whether the system is consistent. Rather, this information will emerge as you attempt to follow an algebraic method of solution. Examples 2 and 3 will illustrate this.

EXAMPLE 2 Using the substitution method

Solve the system

$$\begin{cases} \frac{3}{2}x - 3y = -9 \\ x - 2y = 4 \end{cases}$$

SOLUTION

We use the substitution method. Since it is easy to solve the second equation for x, we begin there:

$$\begin{aligned} x - 2y &= 4 \\ x &= 4 + 2y \end{aligned}$$

Now we substitute this result in the first equation of our system to obtain

$$\begin{aligned} \frac{3}{2}(4 + 2y) - 3y &= -9 \\ 6 + 3y - 3y &= -9 \\ 6 &= -9 \qquad \text{False} \end{aligned}$$

Since the substitution process results in equivalent systems and, in this case, leads to an obviously false statement, we conclude that the given system has no solution; that is, the system is inconsistent.

QUESTION What can you say about the graphs of the two given equations?

EXAMPLE 3 Another substitution example

Solve the system

$$\begin{cases} 3x + 4y = 12 \\ 2y = 6 - \dfrac{3}{2}x \end{cases}$$

SOLUTION

We use the method of substitution. Since it is easy to solve the second equation for y, we begin there:

$$2y = 6 - \frac{3}{2}x \qquad \text{and therefore} \qquad y = 3 - \frac{3}{4}x$$

Now we use this result to substitute for y in the first equation of the original system. The result is

$$\begin{aligned} 3x + 4\left(3 - \frac{3}{4}x\right) &= 12 \\ 3x + 12 - 3x &= 12 \\ 3x - 3x &= 12 - 12 \\ 0 &= 0 \qquad \text{Always true} \end{aligned}$$

This last identity imposes no restrictions on x. Graphically speaking, this says that our two lines intersect for every value of x. In other words, the two lines coincide. We could have foreseen this initially had we solved both equations for y. As you can verify, the result in both cases is

$$y = -\frac{3}{4}x + 3$$

Every point on this line yields a solution to our system of equations. In summary, then, our system is consistent and the solutions to the system have the form $(x, -\frac{3}{4}x + 3)$, where x can be any real number. For instance, when $x = 0$, we obtain the solution $(0, 3)$. When $x = 1$, we obtain the solution $(1, \frac{9}{4})$. The idea here is that for *each* value of x we obtain a solution; thus there are infinitely many solutions.

Now let's turn to the addition–subtraction method of solving systems of equations. By way of example, consider the system

$$\begin{cases} 2x + 3y = 5 \\ 4x - 3y = -1 \end{cases}$$

Notice that if we add these two equations, the result is an equation involving only the unknown x:

$$\begin{aligned} 6x &= 4 \\ x &= \frac{4}{6} = \frac{2}{3} \end{aligned}$$

There are now several ways in which the corresponding value of y can be obtained. As you can easily check, substituting the value $x = 2/3$ in either of the original equations leads to the result $y = 11/9$.

Another way to find y is by multiplying both sides of the first equation by -2. (You'll see why in a moment.) We display the work this way:

$$2x + 3y = 5 \xrightarrow{\text{Multiply by } -2} -4x - 6y = -10$$

$$4x - 3y = -1 \xrightarrow{\text{No change}} 4x - 3y = -1$$

Adding the last two equations then gives us $-9y = -11$, and therefore $y = 11/9$. As with the substitution method, the addition–subtraction method results in an equivalent system. The required solution is then $(2/3, 11/9)$.

In the previous example we were able to find x directly by adding the two equations. As the next example shows, it may be necessary first to multiply both sides of each equation by an appropriate constant.

EXAMPLE 4 Using the addition–subtraction method

Solve the system

$$\begin{cases} 5x - 3y = 4 \\ 2x + 4y = 1 \end{cases}$$

SOLUTION

To eliminate x, we could multiply the second equation by $5/2$ and then subtract the resulting equation from the first equation. However, to avoid working with fractions, we proceed as follows:

$$5x - 3y = 4 \xrightarrow{\text{Multiply by 2}} 10x - 6y = 8 \qquad (4)$$

$$2x + 4y = 1 \xrightarrow{\text{Multiply by 5}} 10x + 20y = 5 \qquad (5)$$

Subtracting equation (5) from equation (4) then yields

$$-26y = 3$$
$$y = -\frac{3}{26}$$

To find x, we return to the original system and work in a similar manner:

$$5x - 3y = 4 \xrightarrow{\text{Multiply by 4}} 20x - 12y = 16$$

$$2x + 4y = 1 \xrightarrow{\text{Multiply by 3}} 6x + 12y = 3$$

Upon adding the last two equations, we obtain

$$26x = 19$$
$$x = \frac{19}{26}$$

The solution of the given system of equations is therefore $(19/26, -3/26)$.

Alternatively, after solving for y, we can use the substitution method to solve for x.

We conclude this section with some problems that can be solved by using simultaneous equations.

EXAMPLE 5 Finding the equation of a parabola through given points

Determine the constants b and c so that the parabola $y = x^2 + bx + c$ passes through the points $(-3, 1)$ and $(1, -2)$.

SOLUTION

Since the point $(-3, 1)$ lies on the curve $y = x^2 + bx + c$, the coordinates must satisfy the equation. Thus we have

$$\begin{aligned} 1 &= (-3)^2 + b(-3) + c \\ -8 &= -3b + c \qquad (6) \end{aligned}$$

This gives us one equation in two unknowns. We need another equation involving b and c. Since the point $(1, -2)$ also lies on the graph of $y = x^2 + bx + c$, we must have

$$\begin{aligned} -2 &= 1^2 + b(1) + c \\ -3 &= b + c \qquad (7) \end{aligned}$$

Rewriting equations (6) and (7), we have the system

$$\begin{cases} -3b + c = -8 & (8) \\ \quad\ b + c = -3 & (9) \end{cases}$$

Subtracting equation (9) from (8) then yields

$$-4b = -5 \qquad \text{and therefore} \qquad b = \frac{5}{4}$$

One way to obtain the corresponding value of c is to replace b by $5/4$ in equation (9). This yields

$$\frac{5}{4} + c = -3$$

$$c = -3 - \frac{5}{4} = -\frac{17}{4}$$

The required values of b and c are therefore

$$b = \frac{5}{4} \qquad c = -\frac{17}{4}$$

(Exercise 39(b) at the end of this section asks you to check that the parabola with equation $y = x^2 + \frac{5}{4}x - \frac{17}{4}$ indeed passes through the given points.)

The next example involves an application in the economics of supply and demand. Figure 6 shows *supply* and *demand models* for a certain commodity. In the

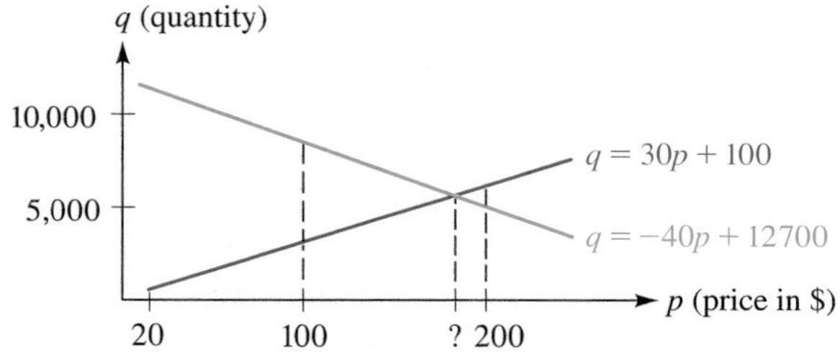

q (quantity)

10,000

5,000

$q = 30p + 100$

$q = -40p + 12700$

p (price in \$)

20 100 ? 200

Figure 6
Supply function:
$q = 30p + 100 \quad (p \geq 20)$
Demand function:
$q = -40p + 12{,}700 \quad (p \geq 20)$

figure, the **supply function** $q = 30p + 100$ $(p \geq 20)$ tells us the quantity q that the manufacturer will produce and supply (to the stores) when the selling price is p dollars per item. Note that this supply function is increasing: The higher the selling price, the more items the manufacturer will produce. The **demand function** in Figure 6, $q = -40p + 12{,}700$ $(p \geq 20)$, gives us the quantity q that can be sold when the selling price is p dollars per item. Note that the demand function is decreasing: The higher the price, the fewer people there are who want to buy, so the smaller the quantity demanded.

To see how these functions work, suppose that the selling price is $p = \$100$. Then we have

$$\text{Supply:} \quad q = 30p + 100 = 30(100) + 100 = 3100 \text{ items}$$
$$\text{Demand:} \quad q = -40p + 12{,}700 = -40(100) + 12{,}700 = 8700 \text{ items}$$

Thus, setting the market price at \$100 per item will create a shortage; customers will want to buy 8700 items at this price, but only 3100 items are available. We can see this qualitatively in Figure 6. For $p = \$100$, the graph of the (blue) demand function is above that of the (red) supply function. In other words, demand exceeds supply in this case. Figure 6 also shows that if the price is set at \$200, rather than \$100, then the demand is less than the supply. So there is a surplus in this case. The point where the graph of the supply function intersects the graph of the demand function is called the market **equilibrium point** for the given commodity. The values of p and q at this point are called the **equilibrium price** and the **equilibrium quantity,** respectively. At the equilibrium point, supply equals demand, so there is neither a surplus nor a shortage. In Figure 6 the equilibrium price is indicated with a question mark. We determine this price in Example 6.

EXAMPLE 6 Using supply and demand functions

Assume that the supply and demand functions for a commodity are as given in Figure 6.

$$\begin{cases} q = 30p + 100 \\ q = -40p + 12{,}700 \end{cases}$$

(As in Figure 6, we are assuming that p is in dollars and $p \geq 20$.) Find the equilibrium price and the corresponding equilibrium quantity.

SOLUTION

Use the first equation to substitute for q in the second equation. This yields

$$\begin{aligned} 30p + 100 &= -40p + 12{,}700 \\ 70p &= 12{,}600 \\ p &= \frac{12{,}600}{70} = \frac{1260}{7} = 180 \end{aligned}$$

The equilibrium price is therefore \$180. Notice that this is consistent with Figure 6, where the equilibrium price is between \$100 and \$200, closer to the latter than the former. For the equilibrium quantity we can substitute $p = 180$ in either equation of the given system. Using the first equation, we obtain $q = 30(180) + 100 = 5500$. The equilibrium quantity is 5500. This too is consistent with Figure 6. (Why?)

In the next example we solve a mixture problem using a system of two linear equations.

EXAMPLE 7 Solving a mixture problem

Suppose that a chemistry student can obtain two acid solutions from the stockroom. The first solution is 20% acid, and the second solution is 45% acid. (The percentages are by volume.) How many milliliters of each solution should the student mix together to obtain 100 ml of a 30% acid solution?

SOLUTION

We begin by assigning letters to denote the required quantities.

Let x denote the number of milliliters of the 20% solution to be used.

Let y denote the number of milliliters of the 45% solution to be used.

We summarize the data in Table 3. Since the final mixture must total 100 ml, we have the equation

$$x + y = 100 \tag{10}$$

This gives us one equation in two unknowns. However, we need a second equation.

TABLE 3

Type of solution	Number of ml	Percent of acid	Total acid (ml)
First solution (20% acid)	x	20	$(0.20)x$
Second solution (45% acid)	y	45	$(0.45)y$
Mixture	$x + y$	30	$(0.30)(x + y)$

Looking at the data in the right-hand column of Table 3, we can write

$$\underbrace{0.20x}_{\substack{\text{amount of acid}\\ \text{in } x \text{ ml of the}\\ 20\% \text{ solution}}} + \underbrace{0.45y}_{\substack{\text{amount of acid}\\ \text{in } y \text{ ml of the}\\ 45\% \text{ solution}}} = \underbrace{(0.30)(x + y)}_{\substack{\text{amount of acid in}\\ \text{the final mixture}}}$$

Thus

$$\begin{aligned} 0.20x + 0.45y &= 0.30(x + y) \\ 20x + 45y &= 30(x + y) \\ 4x + 9y &= 6(x + y) = 6x + 6y \\ -2x + 3y &= 0 \end{aligned} \tag{11}$$

Equations (10) and (11) can be solved by either the substitution method or the addition method. As Exercise 40 at the end of this section asks you to show, the results are

$$x = 60 \text{ ml} \quad \text{and} \quad y = 40 \text{ ml}$$

So we need to mix 60 milliliters of the 20% solution with 40 milliliters of the 40% solution.

EXERCISE SET 10.1

A

1. Which of the following are linear equations in two variables?
   (a) $3x + 3y = 10$
   (b) $2x + 4xy + 3y = 1$
   (c) $u - v = 1$
   (d) $x = 2y + 6$
2. Which of the following are linear equations in two variables?
   (a) $y = x$
   (b) $y = x^2$
   (c) $\frac{4}{x} - \frac{3}{y} = -1$
   (d) $2w + 8z = -4w + 3$
3. Is (5, 1) a solution of the following system?
$$\begin{cases} 2x - 8y = 2 \\ 3x + 7y = 22 \end{cases}$$
4. Is (14, −2) a solution of the following system?
$$\begin{cases} x + y = 12 \\ x - y = 4 \end{cases}$$
5. Is (0, −4) a solution of the following system?
$$\begin{cases} \frac{1}{6}x + \frac{1}{2}y = -2 \\ \frac{2}{3}x + \frac{3}{4}y = 2 \end{cases}$$
6. Is (12, −8) a solution of the system in Exercise 5?

G *In Exercises 7–10 you are given a system of two linear equations. By graphing the pair of equations, determine which one of the three cases described in Figures 2 through 4 (on page 731) applies. (You're not being asked to solve the system.)*

7. $\begin{cases} 3x + 7y = 10 \\ 6x - 3y = 1 \end{cases}$
8. $\begin{cases} y = \sqrt{3}(1 - 3x)/3 \\ \sqrt{3}y + 3x - 1 = 0 \end{cases}$
9. $\begin{cases} 5y = 10.5x - 25.5 \\ 21x = 50 + 10y \end{cases}$
10. $\begin{cases} 2y = x - 18 \\ y = 0.4x + 1 \end{cases}$

In Exercises 11–18, use the substitution method to find all solutions of each system.

11. $\begin{cases} 4x - y = 7 \\ -2x + 3y = 9 \end{cases}$
12. $\begin{cases} 3x - 2y = -19 \\ x + 4y = -4 \end{cases}$
13. $\begin{cases} 6x - 2y = -3 \\ 5x + 3y = 4 \end{cases}$
14. $\begin{cases} 4x + 2y = 3 \\ 10x + 4y = 1 \end{cases}$
15. $\begin{cases} \frac{3}{2}x - 5y = 1 \\ x + \frac{3}{4}y = -1 \end{cases}$
16. $\begin{cases} 13x - 8y = -3 \\ -7x + 2y = 0 \end{cases}$
17. $\begin{cases} 4x + 6y = 3 \\ -6x - 9y = -\frac{9}{2} \end{cases}$
18. $\begin{cases} -\frac{2}{5}x + \frac{1}{4}y = 3 \\ \frac{1}{4}x - \frac{2}{5}y = -3 \end{cases}$

In Exercises 19 and 20:

G **(a)** *Graph the pair of equations, and by zooming in on the intersection point, estimate the solution of the system (each value to the nearest one-tenth).*

(b) *Use the substitution method to determine the solution. Check that your answer is consistent with the graphical estimate in part (a).*

19. $\begin{cases} 0.02x - 0.03y = 1.06 \\ 0.75x + 0.50y = -0.01 \end{cases}$
20. $\begin{cases} \sqrt{2}x - \sqrt{3}y = \sqrt{3} \\ \sqrt{3}x - \sqrt{8}y = \sqrt{2} \end{cases}$

In Exercises 21–28, use the addition–subtraction method to find all solutions of each system of equations.

21. $\begin{cases} 5x + 6y = 4 \\ 2x - 3y = -3 \end{cases}$
22. $\begin{cases} -8x + y = -2 \\ 4x - 3y = 1 \end{cases}$
23. $\begin{cases} 4x + 13y = -5 \\ 2x - 54y = -1 \end{cases}$
24. $\begin{cases} 16x - 3y = 100 \\ 16x + 10y = 10 \end{cases}$
25. $\begin{cases} \frac{1}{4}x - \frac{1}{3}y = 4 \\ \frac{2}{7}x - \frac{1}{7}y = \frac{1}{10} \end{cases}$ *Suggestion for Exercise 25*: First clear both equations of fractions.
26. $\begin{cases} 2.1x - 3.5y = 1.2 \\ 1.4x + 2.6y = 1.1 \end{cases}$
27. $\begin{cases} 8x + 16y = 5 \\ 2x + 5y = \frac{5}{4} \end{cases}$
28. $\begin{cases} \sqrt{6}x - \sqrt{3}y = 3\sqrt{2} - \sqrt{3} \\ \sqrt{2}x - \sqrt{5}y = \sqrt{6} + \sqrt{5} \end{cases}$
29. Find b and c, given that the parabola $y = x^2 + bx + c$ passes through (0, 4) and (2, 14).
30. Determine the constants a and b, given that the parabola $y = ax^2 + bx + 1$ passes through (−1, 11) and (3, 1).
31. Determine the constants A and B, given that the line $Ax + By = 2$ passes through the points (−4, 5) and (7, −9).
32. **(a)** Determine constants a and b so that the graph of $y = ax^3 + b$ passes through the two points (2, 1) and (−2, −7).
    G **(b)** Using the values for a and b determined in part (a), graph the equation $y = ax^3 + b$ and see that it appears to pass through the two given points.

As background for Exercises 33 and 34, you need to have read the discussion on supply and demand preceding Example 6, as well as Example 6 itself. In each exercise, assume that the price p is in dollars.

33. Assume that the supply and demand functions for a commodity are as follows.
    Supply: $q = 200p$; Demand: $q = 9600 - 400p$; for $p \geq 0$.
    (a) If the price is set at \$6, will there be a shortage or a surplus of the commodity? What if the price is doubled?
    (b) If the price is \$20, will there be a shortage or a surplus?
    (c) Find the equilibrium price and the corresponding equilibrium quantity.
34. Assume that the supply and demand functions for a commodity are as follows.
    Supply: $q = 15p$; Demand: $q = 12{,}493 - 50p$; for $p \geq 0$.
    (a) If the price is set at \$50, will there be a shortage or a surplus of the commodity? What if the price is tripled?
    (b) Show that if the price is \$200, there will be a surplus. How many items will be left unsold?
    (c) Find the equilibrium price and the corresponding equilibrium quantity.
35. A student in a chemistry laboratory has access to two acid solutions. The first solution is 10% acid and the second is 35% acid. (The percentages are by volume.) How many

cubic centimeters of each should she mix together to obtain 200 cm^3 of a 25% acid solution?

36. One salt solution is 15% salt, and another is 20% salt. How many cubic centimeters of each solution must be mixed to obtain 50 cm^3 of a 16% salt solution?

37. A shopkeeper has two types of coffee beans on hand. One type sells for \$5.20/lb, the other for \$5.80/lb. How many pounds of each type must be mixed to produce 16 lb of a blend that sells for \$5.50/lb?

38. A certain alloy contains 10% tin and 30% copper. (The percentages are by weight.) How many pounds of tin and how many pounds of copper must be melted with 1000 lb of the given alloy to yield a new alloy containing 20% tin and 35% copper? *Hint:* Introduce variables for the weights of tin and copper to be added to the given alloy. Express the total weight of the new alloy in terms of these variables. The total weight of tin in the new alloy can be computed two ways, giving one equation. Computing the total weight of copper similarly gives a second equation.

39. In this exercise you'll check the result of Example 5, first visually, then algebraically.

Ⓖ **(a)** Graph the parabola $y = x^2 + \frac{5}{4}x - \frac{17}{4}$ in an appropriate viewing rectangle to see that it appears to pass through the two points $(-3, 1)$ and $(1, -2)$.

(b) Verify algebraically that the two points $(-3, 1)$ and $(1, -2)$ indeed lie on the parabola $y = x^2 + \frac{5}{4}x - \frac{17}{4}$.

40. Consider the following system from Example 7:

$$\begin{cases} x + y = 100 \\ -2x + 3y = 0 \end{cases}$$

(a) Solve this system using the method of substitution. (As was stated in Example 7, you should obtain $x = 60$, $y = 40$.)

(b) Solve the system using the addition–subtraction method.

B

41. Find constants a and b so that $(8, -7)$ is the solution of the system

$$\begin{cases} ax + by = 10 \\ bx + ay = -5 \end{cases}$$

42. (a) Sketch the triangular region in the first quadrant bounded by the lines $y = 5x$, $y = -3x + 6$, and the x-axis. One vertex of this triangle is the origin. Find the coordinates of the other two vertices. Then use your answers to compute the area of the triangle.

(b) More generally now, express the area of the shaded triangle in the following figure in terms of m, M, and b. Then use your result to check the answer you obtained in part (a) for that area.

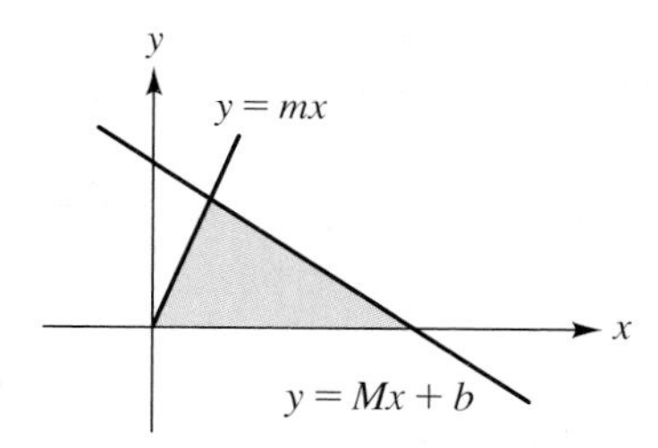

43. Find x and y in terms of a and b:

$$\begin{cases} \dfrac{x}{a} + \dfrac{y}{b} = 1 \\ \dfrac{x}{b} + \dfrac{y}{a} = 1 \end{cases}$$

Does your solution impose any conditions on a and b?

44. Solve the following system for x and y in terms of a and b, where $a \neq b$:

$$\begin{cases} ax + by = 1/a \\ b^2x + a^2y = 1 \end{cases}$$

45. Solve the following system for x and y in terms of a and b, where $a \neq b$:

$$\begin{cases} ax + a^2y = 1 \\ bx + b^2y = 1 \end{cases}$$

Does your solution impose any additional conditions on a and b?

46. Solve the following system for s and t:

$$\begin{cases} \dfrac{3}{s} - \dfrac{4}{t} = 2 \\ \dfrac{5}{s} + \dfrac{1}{t} = -3 \end{cases}$$

Hint: Make the substitutions $1/s = x$ and $1/t = y$ in order to obtain a system of two linear equations.

47. Solve the following system for s and t:

$$\begin{cases} \dfrac{1}{2s} - \dfrac{1}{2t} = -10 \\ \dfrac{2}{s} + \dfrac{3}{t} = 5 \end{cases}$$

(Use the hint in Exercise 46.)

48. Consider the following system: $\begin{cases} 2x^2 + 2y^2 = 55 \\ 4x^2 - 8y^2 = k \end{cases}$

(a) Assuming $k = 109$, solve the system. *Hint:* The substitutions $u = x^2$ and $v = y^2$ will give you a linear system.

(b) Follow part (a) using: $k = 110$; $k = 111$.

Ⓖ **(c)** Use a graphing utility to shed light on why the number of solutions is different for each of the values of k considered in parts (a) and (b).

In Exercises 49–55, find all solutions of the given systems. For Exercises 51–55, use a calculator to round the final answers to two decimal places.

49. $\begin{cases} \dfrac{2w-1}{3} + \dfrac{z+2}{4} = 4 \\ \dfrac{w+3}{2} - \dfrac{w-z}{3} = 3 \end{cases}$

50. $\dfrac{x-y}{2} = \dfrac{x+y}{3} = 1$

51. $\begin{cases} 2\ln x - 5\ln y = 11 \\ \ln x + \ln y = -5 \end{cases}$

52. $\begin{cases} 3\ln x + \ln y = 3 \\ 4\ln x - 6\ln y = -7 \end{cases}$

Hint: Let $u = \ln x$ and $v = \ln y$.

53. $\begin{cases} e^x - 3e^y = 2 \\ 3e^x + e^y = 16 \end{cases}$

54. $\begin{cases} e^x + 2e^y = 4 \\ \frac{1}{2}e^x - e^y = 0 \end{cases}$

Hint: Let $u = e^x$ and $v = e^y$.

55. $\begin{cases} 4\sqrt{x^2 - 3x} - 3\sqrt{y^2 + 6y} = -4 \\ \frac{1}{2}(\sqrt{x^2 - 3x} + \sqrt{y^2 + 6y}) = 3 \end{cases}$

Hint: Let $u = \sqrt{x^2 - 3x}$ and $v = \sqrt{y^2 + 6y}$.

56. The sum of two numbers is 64. Twice the larger number plus five times the smaller number is 20. Find the two numbers. (Let x denote the larger number and let y denote the smaller number.)

57. In a two-digit number, the sum of the digits is 14. Twice the tens digit exceeds the units digit by one. Find the number.

58. You have two brands of dietary supplements on your shelf. Among other ingredients, both contain protein and carbohydrates. The amounts of protein and carbohydrates in one unit of each supplement are given in the following table as percentages of the *recommended daily amount* (RDA). How many units of each supplement do you need in a day to obtain the RDA for both protein and carbohydrates?

	Protein (% of RDA in one unit)	Carbohydrates (% of RDA in one unit)
Supplement #1	8	12
Supplement #2	16	4

59. Solve the following system for x and y in terms of a and b, where a and b are nonzero and $a \neq b$.

$$\begin{cases} \dfrac{a}{bx} + \dfrac{b}{ay} = a + b \\ \dfrac{b}{x} + \dfrac{a}{y} = a^2 + b^2 \end{cases}$$

60. Solve for x and y in terms of a, b, c, d, e, and f:

$$\begin{cases} ax + by = c \\ dx + ey = f \end{cases}$$

(Assume that $ae - bd \neq 0$.)

C

61. (a) Given that the lines $7x + 5y = 4$, $x + ky = 3$, and $5x + y + k = 0$ are concurrent (pass through a common point), what are the possible values for k?

Ⓖ **(b)** Check that your answers are reasonable: For each value of k that you find, use a graphing utility to draw the three lines. Do they appear to be concurrent?

62. Solve the following system for x and y in terms of a and b, where $ab \neq -1$:

$$\begin{cases} \dfrac{x+y-1}{x-y+1} = a \\ \dfrac{y-x+1}{x-y+1} = ab \end{cases}$$

Answer: $x = (a+1)/(ab+1)$
$y = a(b+1)/(ab+1)$

PROJECT

GEOMETRY WORKBOOKS ON THE EULER LINE AND THE NINE-POINT CIRCLE

The exercises in this project form a unit introducing two remarkable geometric results. Workbook I uses a specific example to introduce the *Euler line* and the *nine-point circle* of a triangle. Workbook II develops the Euler line for the general case. The exercises are straightforward in that they involve only basic coordinate geometry and systems of two linear equations. However, a good deal of algebraic calculation is required in both workbooks, and just one error along the way will spoil the result at the end of that workbook. Thus this is a good occasion for group work, assuming that each person in a group takes an active role in checking results. For both workbooks, one group can be assigned to each exercise. Some exercises, however, require results from one or more previous exercises. For instance, in Workbook I, Exercises 4–7 depend upon the results of the previous exercises. So there is a need for some coordination here; the class as a whole needs to be certain that the results in Exercises 1–3 are correct before the groups assigned to Exercises 4–7 can begin their work. A similar situation occurs in several places in Workbook II.

WORKBOOK 1

For this workbook you are given $\triangle ABC$, with vertices as indicated in the following figure.

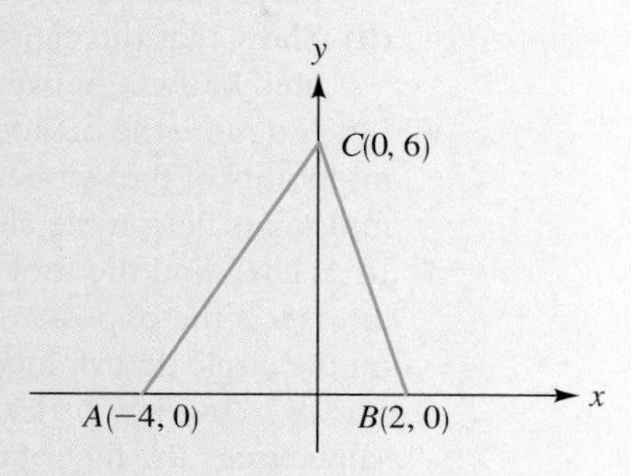

1. (a) A *median* of a triangle is a line segment drawn from a vertex to the midpoint of the opposite side. For $\triangle ABC$, find the equation of the line containing the median to side $\overline{BC}$. Next, find the equation of the line containing the median to side $\overline{AC}$. Now, find the point where these two medians intersect.

(b) A theorem from geometry states that in any triangle the three medians are concurrent (that is, intersect in a single point). Use this to check your last answer in part (a) as follows. Find the equation of the line containing the median to $\overline{AB}$. Then check that this median passes through the intersection point found in part (a).

Remark: For any triangle the point where the three medians intersect is called the *centroid* of the triangle. It can be shown that the centroid is the center of gravity or balance point of the triangle.

Ⓖ **(c)** Use a graphing utility to draw $\triangle ABC$ along with the three lines containing the medians. Check to see that the location of the centroid appears to be consistent with coordinates obtained in part (a).

2. An *altitude* of a triangle is a line segment drawn from a vertex perpendicular to the opposite side. A theorem from geometry states that in any triangle the three altitudes are concurrent. The point where they meet is called the *orthocenter* of the triangle. Follow Exercise 1 using the altitudes of $\triangle ABC$, rather than the medians.

3. A theorem from geometry states that in any triangle the perpendicular bisectors of the three sides are concurrent. The point where they intersect is called the *circumcenter* of the triangle. The circumcenter is the center of the circle that passes through the three vertices of the triangle. Follow Exercise 1 using the perpendicular bisectors of the sides of $\triangle ABC$ rather than the medians.

4. In Exercises 1–3 you found the coordinates of the centroid, the orthocenter, and the circumcenter of $\triangle ABC$.

(a) Show that these three points are collinear (that is, all lie on a straight line). *Hint*: Use slopes.

(b) Show that the distance from the orthocenter to the centroid is twice the distance from the centroid to the circumcenter.

Remark: It can be shown (Workbook II) that the results in parts (a) and (b) are valid for all triangles. The line containing the centroid, orthocenter, and circumcenter is called the *Euler line* of the triangle.

5. **(a)** Find the center and radius of the circle that passes through the midpoints of the sides of $\triangle ABC$. *Hint*: Find the center as the intersection of two perpendicular bisectors, as in Exercise 3.

(b) Show that the center lies on the Euler line of $\triangle ABC$ and that it is situated halfway between the orthocenter and circumcenter of $\triangle ABC$.

6. Let H denote the orthocenter of $\triangle ABC$ (determined in Exercise 2). Find the midpoints of the segments $\overline{HA}$, $\overline{HB}$, and $\overline{HC}$. Then check that each of these midpoints lies on the circle determined in Exercise 5.

7. In ΔABC, find the foot of each altitude (that is, the point where the altitude intersects the opposite side.) Then check that each of these three points lies on the circle determined in Exercise 5.

Remark: In view of Exercises 5–7, the following nine points all lie on the same circle: the feet of the three medians, the feet of the three altitudes, and the midpoints of the three segments drawn from the orthocenter to each vertex. It can be shown that this occurs no matter what triangle we begin with. The circle is called the *nine-point circle*.

Ⓖ **8.** Use a graphing utility to display $\triangle ABC$, the nine-point circle, and the nine special points that all lie on the circle.

WORKBOOK 2

For this workbook you are given an arbitrary triangle with vertices as indicated in the following figure. (All triangles can be represented in this fashion: The units on the y-axis are chosen so that the length of that altitude is 6; the constant a can be any negative real number, while b can be any positive real number. The factor 6 has been chosen so that the appearance of fractions is avoided in subsequent calculations.) For each exercise, supply the calculations to show that the given statements are correct.

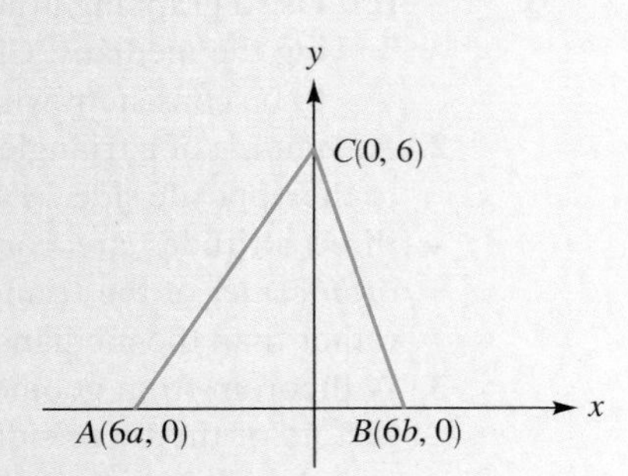

1. The equations of the lines containing the medians of ΔABC are $2x + (a + b)y = 6(a + b)$, $x - (a - 2b)y = 6b$, and $x - (b - 2a)y = 6a$.

2. The intersection point for each pair of medians of $\triangle ABC$ is $(2a + 2b, 2)$. This point is called the *centroid* of $\triangle ABC$, and it is usually denoted by the letter G.

3. The equations of the lines containing the altitudes of $\triangle ABC$ are $y = ax - 6ab$, $y = bx - 6ab$, and $x = 0$.

4. The intersection point for each pair of altitudes of $\triangle ABC$ is $(0, -6ab)$. This point is called the *orthocenter* of $\triangle ABC$, and it is usually denoted by the letter H.

5. The equations of the lines containing the perpendicular bisectors of the sides of $\triangle ABC$ are $x = 3a + 3b$, $bx - y = 3b^2 - 3$, and $ax - y = 3a^2 - 3$.
6. The intersection point for each pair of perpendicular bisectors is $(3a + 3b, 3ab + 3)$. This point is called the *circumcenter* of $\triangle ABC$, and it is usually denoted by the letter O.
7. The orthocenter H, the centroid G, and circumcenter O are collinear. The line through these three points is called the *Euler line* of $\triangle ABC$.
8. $HG = 2(GO)$. That is, the distance from the orthocenter to the centroid is twice the distance from the centroid to the circumcenter.

10.2 GAUSSIAN ELIMINATION

A method of solution is perfect if we can foresee from the start, and even prove, that following that method we shall obtain our aim.—Gottfried Wilhelm von Leibniz (1646–1716)

. . . the first electronic computer, the ABC, named after its designers Atanasoff and Berry, was built specifically to solve systems of 29 equations in 29 unknowns, a formidable task without the aid of a computer.
—Angela B. Shiflet in *Discrete Mathematics for Computer Science* (St. Paul: West Publishing Co., 1987)

In the previous section we solved systems of linear equations in two unknowns. In this section we introduce the technique known as Gaussian elimination for solving systems of linear equations in which there are more than two unknowns.*

As a first example, consider the following system of three linear equations in the three unknowns x, y, and z:

$$\begin{cases} 3x + 2y - z = -3 \\ \phantom{3x + {}} 5y - 2z = 2 \\ \phantom{3x + 5y - {}} 5z = 20 \end{cases}$$

This system is easy to solve by using the process of *back-substitution.* Dividing both sides of the third equation by 5 yields $z = 4$. Then, substituting $z = 4$ back into the second equation gives us

$$\begin{aligned} 5y - 2(4) &= 2 \\ 5y &= 10 \\ y &= 2 \end{aligned}$$

Finally, substituting the values $z = 4$ and $y = 2$ back into the first equation yields

$$\begin{aligned} 3x + 2(2) - 4 &= -3 \\ 3x &= -3 \\ x &= -1 \end{aligned}$$

We have now found that $x = -1$, $y = 2$, and $z = 4$. If you go back and check, you will find that these values indeed satisfy all three equations in the given system. Furthermore, the algebra we've just carried out shows that these are the only possible values for x, y, and z satisfying all three equations. We summarize by saying that the **ordered triple** $(-1, 2, 4)$ is the solution of the given system.

*The technique is named after Carl Friedrich Gauss (1777–1855). Early in the nineteenth century, Gauss used this technique (and introduced the method of least squares for minimizing errors) in analyzing the orbit of the asteroid Pallas. However, the essentials of Gaussian elimination were in existence long before Gauss's time. Indeed, a version of the method appears in the Chinese text *Chui-Chang Suan-Shu* ("Nine Chapters on the Mathematical Art"), written approximately two thousand years ago.

The system that we just considered was easy to solve (using back-substitution) because of the special form in which it was written. This form is called **upper-triangular form.** Although the following definition of upper-triangular form refers to systems with three unknowns, the same type of definition can be given for systems with any number of unknowns. Table 1 displays examples of systems in upper-triangular form.

Upper-Triangular Form (three variables)

A system of linear equations in x, y, and z is said to be in **upper-triangular form** provided that x appears in no equation after the first and y appears in no equation after the second. (It is possible that y may not even appear in the second equation.)

TABLE 1 Examples of Systems in Upper-Triangular Form

2 Unknowns: x, y	3 Unknowns: x, y, z	4 Unknowns: x, y, z, t
$\begin{cases} 3x + 5y = 7 \\ 8y = 5 \end{cases}$	$\begin{cases} 4x - 3y + 2z = -5 \\ 7y + z = 9 \\ -4z = 3 \end{cases}$	$\begin{cases} x - y + z - 4t = 1 \\ 3y - 2z + t = -1 \\ 3z - 5t = 4 \\ 6t = 7 \end{cases}$
	$\begin{cases} 15x - 2y + z = 1 \\ 3z = -8 \end{cases}$	$\begin{cases} 2x + y + 2z - t = -3 \\ 4z + 3t = 1 \\ 5t = 6 \end{cases}$
		$\begin{cases} 8x + 3y - z + t = 2 \\ 2y + z - 4t = 1 \end{cases}$

When we were solving linear systems of two equations with two unknowns in the previous section, we observed that there were three possibilities for the solution set: a unique solution, infinitely many solutions, and no solution. As the next three examples indicate, the situation is similar when dealing with larger systems.

EXAMPLE 1 Using back-substitution to solve a system

Find all solutions of the system

$$\begin{cases} x + y + 2z = 2 \\ 3y - 4z = -5 \\ 6z = 3 \end{cases}$$

SOLUTION

The system is in upper-triangular form, so we can use back-substitution. Dividing the third equation by 6 yields $z = 1/2$. Substituting this value for z back into the second equation then yields

$$3y - 4\left(\frac{1}{2}\right) = -5$$
$$3y = -3$$
$$y = -1$$

Now, substituting the values $z = 1/2$ and $y = -1$ back into the first equation, we obtain

$$x + (-1) + 2\left(\frac{1}{2}\right) = 2$$
$$x = 2$$

As you can easily check, the values $x = 2$, $y = -1$, and $z = 1/2$ indeed satisfy all three equations. Furthermore, the algebra we've just carried out shows that these are the only possible values for x, y, and z satisfying all three equations. We summarize by saying that the unique solution to our system is the ordered triple $(2, -1, \frac{1}{2})$.

EXAMPLE 2 A system with infinitely many solutions

Find all solutions of the system

$$\begin{cases} -2x + y + 3z = 6 \\ 2z = 10 \end{cases}$$

SOLUTION

Again, the system is in upper-triangular form, and we use back-substitution. Solving the second equation for z yields $z = 5$. Then replacing z by 5 in the first equation gives us

$$-2x + y + 3(5) = 6$$
$$-2x + y = -9$$
$$y = 2x - 9$$

At this point, we've made use of both equations in the given system. There is no third equation to provide additional restrictions on x, y, or z. We know from the previous section that the equation $y = 2x - 9$ has infinitely many solutions, all of the form

$$(x, 2x - 9) \qquad \text{where } x \text{ is a real number}$$

It follows, then, that there are infinitely many solutions to the given system and they may be written

$$(x, 2x - 9, 5) \qquad \text{where } x \text{ is a real number}$$

For instance, choosing in succession $x = 0$, $x = 1$, and $x = 2$ yields the solutions $(0, -9, 5)$, $(1, -7, 5)$, and $(2, -5, 5)$. (We remark in passing that any linear system in upper-triangular form in which the number of unknowns exceeds the number of equations will always have infinitely many solutions.)

EXAMPLE 3 A system with no solutions

Find all solutions of the system

$$\begin{cases} 4x - 7y + 3z = 1 \\ 3x + y - 2z = 4 \\ 4x - 7y + 3z = 6 \end{cases}$$

(Note that the system is not in upper-triangular form.)

SOLUTION

Look at the left-hand sides of the first and third equations: They are identical. Thus, if there were values for x, y, and z that satisfied both equations, it would follow that $1 = 6$, which is clearly impossible. We conclude that the given system has no solutions.

As Examples 1 and 2 have demonstrated, systems in upper-triangular form can be readily solved. In view of this, it would be useful to have a technique for converting a given system into an equivalent system in upper-triangular form. (Recall that an **equivalent system** means a system with exactly the same set of solutions as the original system.) **Gaussian elimination** is one such technique. We will demonstrate this technique in Examples 4, 5, 7, and 8. In using Gaussian elimination, we will rely on what are called the three **elementary operations,** listed in the box that follows. These are operations that, when performed on equations in a system, produce an equivalent system.

The Elementary Operations

1. Multiply both sides of an equation by a nonzero constant.
2. Interchange the order in which two equations of a system are listed.
3. To one equation add a multiple of another equation in the system.

EXAMPLE 4 Using Gaussian elimination to solve a system

Find all solutions of the system

$$\begin{cases} x + 2y + z = 3 \\ 2x + y + z = 16 \\ x + y + 2z = 9 \end{cases}$$

SOLUTION

First we want to eliminate x from the second and third equations. To eliminate x from the second equation, we add to it -2 times the first equation. The result is the equivalent system

$$\begin{cases} x + 2y + z = 3 \\ -3y - z = 10 \\ x + y + 2z = 9 \end{cases}$$

To eliminate x from the third equation, we add to it -1 times the first equation. The result is the equivalent system

$$\begin{cases} x + 2y + z = 3 \\ -3y - z = 10 \\ -y + z = 6 \end{cases}$$

Now to bring the system into upper-triangular form, we need to eliminate y from the third equation. We could do this by adding $-1/3$ times the second equation to the third equation. However, to avoid working with fractions as long as possible, we proceed instead to interchange the second and third equations to obtain the equivalent system

$$\begin{cases} x + 2y + z = 3 \\ \quad -y + z = 6 \\ \quad -3y - z = 10 \end{cases}$$

Now we add -3 times the second equation to the last equation to obtain the equivalent system

$$\begin{cases} x + 2y + \ \ z = 3 \\ \quad -y + \ \ z = 6 \\ \quad\quad\ \ - 4z = -8 \end{cases}$$

The system is now in upper-triangular form, and back-substitution yields, in turn, $z = 2$, $y = -4$, and $x = 9$. (Check this for yourself.) The required solution is therefore $(9, -4, 2)$.

In Table 2 we list some convenient abbreviations used in describing the elementary operations. Plan on using these abbreviations yourself; they'll make it simpler for you (and your instructor) to check your work. In Table 2 the notation E_i stands for the ith equation in a system. For instance, for the initial system in Example 4 the symbol E_1 denotes the first equation: $x + 2y + z = 3$.

TABLE 2 Abbreviations for the Elementary Operations

Abbreviation	Explanation
1. cE_i	Multiply both sides of the ith equation by c, for $c \neq 0$.
2. $E_i \leftrightarrow E_j$	Interchange the ith and jth equations.
3. $cE_i + E_j$	Add c times the ith equation to the jth equation.

EXAMPLE 5 Specifying the elementary operations in using Gaussian elimination

Solve the system

$$\begin{cases} 4x - 3y + 2z = 40 \\ 5x + 9y - 7z = 47 \\ 9x + 8y - 3z = 97 \end{cases}$$

SOLUTION

$$\begin{cases} 4x - 3y + 2z = 40 \\ 5x + 9y - 7z = 47 \\ 9x + 8y - 3z = 97 \end{cases} \xrightarrow{(-1)E_2 + E_1} \begin{cases} -x - 12y + 9z = -7 \\ 5x + \ \ 9y - 7z = 47 \\ 9x + \ \ 8y - 3z = 97 \end{cases}$$

adding -1 times the second equation to the first equation

$$\xrightarrow[9E_1 + E_3]{5E_1 + E_2} \begin{cases} -x - 12y + \ \ 9z = -7 \\ \quad -51y + 38z = 12 \\ \quad -100y + 78z = 34 \end{cases}$$

$$\xrightarrow{\frac{1}{2}E_3} \begin{cases} -x - 12y + 9z = -7 \\ -51y + 38z = 12 \\ -50y + 39z = 17 \end{cases}$$

$$\xrightarrow{(-1)E_3 + E_2} \begin{cases} -x - 12y + 9z = -7 \\ -y - z = -5 \\ -50y + 39z = 17 \end{cases}$$

to allow working with smaller but integral coefficients

$$\xrightarrow{-50E_2 + E_3} \begin{cases} -x - 12y + 9z = -7 \\ -y - z = -5 \\ 89z = 267 \end{cases}$$

The system is now in upper-triangular form. Solving the third equation, we obtain $z = 3$. Substituting this value back into the second equation yields $y = 2$. (Check this for yourself.) Finally, substituting $z = 3$ and $y = 2$ back into the first equation yields $x = 10$. (Again, check this for yourself.) The solution to the system is therefore (10, 2, 3).

EXAMPLE 6 Using Gaussian elimination to solve a scheduling problem

A charter tour company specializing in Hawaiian vacations has a total of 1328 reservations for an August departure from Los Angeles to Honolulu. The company has three different kinds of jets available, each with three classes of seats: economy, economy-upgrade, and first class. The available seating on each type of jet is given in the following table.

	Jet type I	Jet type II	Jet type III
Economy	120	80	60
Economy-upgrade	60	35	26
First class	20	45	16

Of the 1328 reservations, 760 are in economy class, 354 are in economy-upgrade, and the remaining 214 are in first class. How many jets of each type are needed to accommodate these passengers, given that the tour company would like to have all seats occupied on each plane? Or is that not possible?

SOLUTION

Let x be the required number of type I jets, y the number of type II jets, and z the number of type III jets. We start by considering economy class. According to the table, on each type I jet there are 120 economy seats available. Thus with x type I jets we have $120x$ economy seats available. Similarly, for type II jets there are $80y$ economy seats, and for type III jets there are $60z$ economy seats. Therefore we require that

$$\underbrace{120x + 80y + 60z}_{\text{the number of economy seats available}} = \underset{\text{the number of economy seats reserved}}{\overset{760}{\uparrow}}$$

That gives us one equation involving the three unknowns x, y, and z. In a similar fashion, working with economy-upgrade and then working with first class, we obtain the two additional equations $60x + 35y + 26z = 354$ and $20x + 45y + 16z = 214$. (Verify this for yourself.) In summary then, we have the following system of three linear equations:

$$\begin{cases} 120x + 80y + 60z = 760 \\ 60x + 35y + 26z = 354 \\ 20x + 45y + 16z = 214 \end{cases} \tag{1}$$

This system can be solved using Gaussian elimination, just as in Example 5 (although the arithmetic is a bit messier). Exercise 30(a) asks you to solve the system. The result turns out to be $x = 3$, $y = 2$, and $z = 4$. This tells us that the tour company should use 3 type I jets, 2 type II jets, and 4 type III jets in order to accommodate all of the passengers while filling all of the seats on each plane. If at least one of the values for x, y, or z in the solution had not been a whole number, then we would have needed to *round up* to the next whole number (why?), which would mean that at least one plane would be flying with one or more empty seats.

Each linear system that we considered in Examples 4–6 had exactly one solution. As the next two examples indicate, the method of Gaussian elimination also works in cases where the system has infinitely many solutions or no solution. Indeed, it's not necessary to know how many solutions there are (if any) at the start; rather, this information is revealed by the Gaussian elimination process.

EXAMPLE 7 A system with infinitely many solutions

Solve the system

$$\begin{cases} x + 2y + 4z = 0 \\ x + 3y + 9z = 0 \end{cases}$$

SOLUTION

This system is similar to the one in Example 2 in that there are fewer equations than there are unknowns. By subtracting the first equation from the second, we readily obtain an equivalent system in upper-triangular form:

$$\begin{cases} x + 2y + 4z = 0 \\ \qquad y + 5z = 0 \end{cases}$$

Although the system is now in upper-triangular form, notice that the second equation does not determine y or z uniquely; that is, there are infinitely many number pairs (y, z) satisfying the second equation. We can solve the second equation for y in terms of z; the result is $y = -5z$. Now we replace y with $-5z$ in the first equation to obtain

$$x + 2(-5z) + 4z = 0 \qquad \text{or} \qquad x = 6z$$

At this point we've used both of the equations in the system to express x and y in terms of z. Furthermore, there is no third equation in the system to provide additional restrictions on x, y, or z. We therefore conclude that the given system has infinitely many solutions. These solutions have the form

$$(6z, -5z, z) \qquad \text{where } z \text{ is any real number}$$

EXAMPLE 8 A system with no solutions

Solve the system

$$\begin{cases} x - 4y + z = 3 \\ 3x + 5y - 2z = -1 \\ 7x + 6y - 3z = 2 \end{cases}$$

SOLUTION

$$\begin{cases} x - 4y + z = 3 \\ 3x + 5y - 2z = -1 \\ 7x + 6y - 3z = 2 \end{cases} \xrightarrow[-7E_1+E_3]{-3E_1+E_2} \begin{cases} x - 4y + z = 3 \\ 17y - 5z = -10 \\ 34y - 10z = -19 \end{cases}$$

$$\xrightarrow{-2E_2+E_3} \begin{cases} x - 4y + z = 3 \\ 17y - 5z = -10 \\ 0 = 1 \end{cases}$$

From the third equation in this last system we conclude that this system, and consequently the original system, has no solution. (Reason: If there *were* values for x, y, and z satisfying the original system, then it would follow that $0 = 1$, which is clearly impossible.)

In the examples in this section we've seen the three possibilities regarding solutions of linear systems with more than two unknowns: a unique solution (as in Examples 1, 4, 5, and 6); infinitely many solutions (as in Examples 2 and 7); and no solution (as in Examples 3 and 8). Just as in the previous section, there are graphs to explain each case. But now, instead of lines in a two-dimensional coordinate system, the graphs involve *planes* in a three-dimensional coordinate system. We'll describe this in general, without going into detail.

In the familiar x-y coordinate system an equation of the form $ax + by = c$ represents a line. And, as you know, a solution (x_0, y_0) of a linear system with two unknowns is a point that lies on both of the given lines. Similarly, in three dimensions, an equation of the form $ax + by + cz = d$ represents a plane. A solution (x_0, y_0, z_0) of a linear system involving three unknowns can be interpreted as a point that lies on all of the given planes. Thus where we talked about intersecting lines in two dimensions, now we are talking about intersecting planes in three dimensions. Figure 1, for instance, shows a geometric interpretation for Example 7, in which there were two equations in three unknowns. Each equation describes a plane. As is indicated in Figure 1, these two planes intersect in a line. Each point on the line corresponds to a solution of the linear system. In Figure 2 we show some of the possibilities when there are three equations and three unknowns.

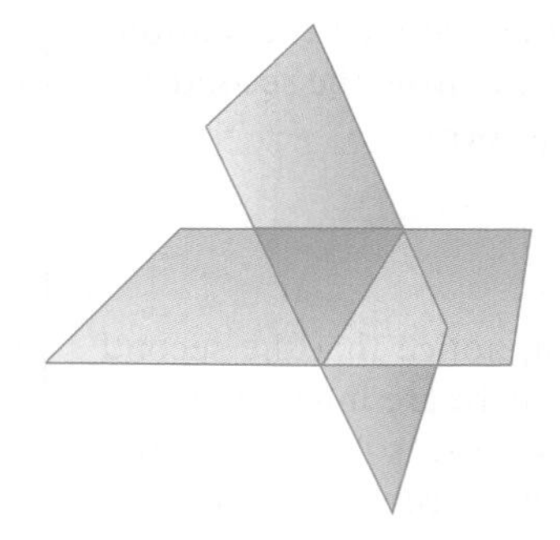

Figure 1
Geometric interpretation for Example 7

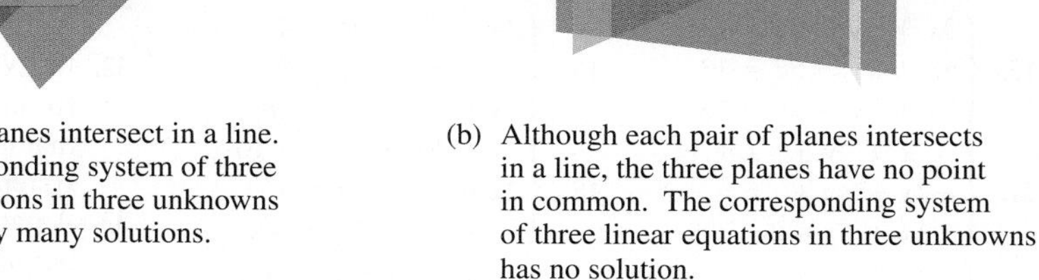

(a) The three planes intersect in a line. The corresponding system of three linear equations in three unknowns has infinitely many solutions.

(b) Although each pair of planes intersects in a line, the three planes have no point in common. The corresponding system of three linear equations in three unknowns has no solution.

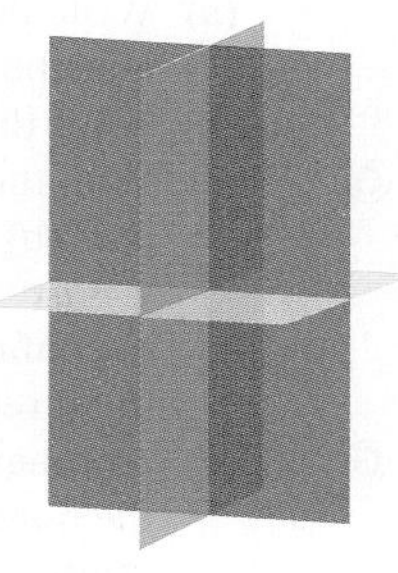

(c) There is exactly one point common to all three planes. The corresponding system of three linear equations in three unknowns has exactly one solution.

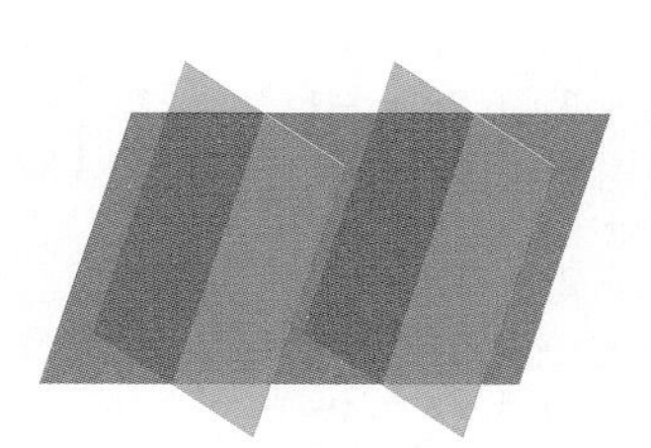

(d) Two parallel planes cut a third plane. There is no point common to all three planes. The corresponding system of three linear equations in three unknowns has no solution.

Figure 2
Some of the ways in which three planes can intersect (or fail to intersect)

EXERCISE SET 10.2 

A

In Exercises 1–10, the systems of linear equations are in upper-triangular form. Find all solutions of each system.

1. $$\begin{cases} 2x + y + z = -9 \\ 3y - 2z = -4 \\ 8z = -8 \end{cases}$$

2. $$\begin{cases} -3x + 7y + 2z = -19 \\ y + z = 1 \\ -2z = -2 \end{cases}$$

3. $$\begin{cases} 8x + 5y + 3z = 1 \\ 3y + 4z = 2 \\ 5z = 3 \end{cases}$$

4. $$\begin{cases} 2x + 7z = -4 \\ 5y - 3z = 6 \\ 6z = 18 \end{cases}$$

5. $$\begin{cases} -4x + 5y = 0 \\ 3y + 2z = 1 \\ 3z = -1 \end{cases}$$

6. $$\begin{cases} 3x - 2y + z = 4 \\ 3z = 9 \end{cases}$$

7. $$\begin{cases} -x + 8y + 3z = 0 \\ 2z = 0 \end{cases}$$

8. $$\begin{cases} -x + y + z + w = 9 \\ 2y - z - w = 9 \\ 3z + 2w = 1 \\ 11w = 22 \end{cases}$$

9. $$\begin{cases} 2x + 3y + z + w = -6 \\ y + 3z - 4w = 23 \\ 6z - 5w = 31 \\ -2w = 10 \end{cases}$$

10. $$\begin{cases} 7x - y - z + w = 3 \\ 2y - 3z - 4w = -2 \\ 3w = 6 \end{cases}$$

In Exercises 11–29, find all solutions of each system.

11. $\begin{cases} x + y + z = 12 \\ 2x - y - z = -1 \\ 3x + 2y + z = 22 \end{cases}$

12. $\begin{cases} A + B - C = -1 \\ 3A - B + 2C = 9 \\ 5A + 3B + 3C = 1 \end{cases}$

13. $\begin{cases} 2x - 3y + 2z = 4 \\ 4x + 2y + 3z = 7 \\ 5x + 4y + 2z = 7 \end{cases}$

14. $\begin{cases} x + 2z = 5 \\ y - 30z = -16 \\ x - 2y + 4z = 8 \end{cases}$

15. $\begin{cases} 3x + 3y - 2z = 13 \\ 6x + 2y - 5z = 13 \\ 7x + 5y - 3z = 26 \end{cases}$

16. $\begin{cases} 2x + 5y - 3z = 4 \\ 4x - 3y + 2z = 9 \\ 5x + 6y - 2z = 18 \end{cases}$

17. $\begin{cases} x + y + z = 1 \\ -2x + y + z = -2 \\ 3x + 6y + 6z = 5 \end{cases}$

18. $\begin{cases} 7x + 5y - 7z = -10 \\ 2x + y + z = 7 \\ x + y - 3z = -8 \end{cases}$

19. $\begin{cases} 2x - y + z = -1 \\ x + 3y - 2z = 2 \\ -5x + 6y - 5z = 5 \end{cases}$

20. $\begin{cases} -2x + 2y - z = 0 \\ 3x - 4y + z = 1 \\ 5x - 8y + z = 4 \end{cases}$

21. $\begin{cases} 2x - y + z = 4 \\ x + 3y + 2z = -1 \\ 7x + 5z = 11 \end{cases}$

22. $\begin{cases} 3x + y - z = 10 \\ 8x - y - 6z = -3 \\ 5x - 2y - 5z = 1 \end{cases}$

23. $\begin{cases} x + y + z + w = 4 \\ x - 2y - z - w = 3 \\ 2x - y + z - w = 2 \\ x - y + 2z - 2w = -7 \end{cases}$

24. $\begin{cases} x + y - 3z + 2w = 0 \\ -2x - 2y + 6z + w = -5 \\ -x + 3y + 3z + 3w = -5 \\ 2x + y - 3z - w = 4 \end{cases}$

25. $\begin{cases} 2x + 3y + 2z = 5 \\ x + 4y - 3z = 1 \end{cases}$

26. $\begin{cases} 4x - y - 3z = 2 \\ 6x + 5y - z = 0 \end{cases}$

27. $\begin{cases} x - 2y - 2z + 2w = -10 \\ 3x + 4y - z - 3w = 11 \\ -4x - 3y - 3z + 8w = -21 \end{cases}$

28. $\begin{cases} 2x + y + z + w = 1 \\ x + 3y - 3z - 3w = 0 \\ -3x - 4y + 2z + 2w = -1 \end{cases}$

29. $\begin{cases} 4x - 2y + 3z = -2 \\ 6y - 4z = 6 \end{cases}$

30. This exercise concerns Example 6 in the text.

(a) Solve system (1) on page 749. As stated in the text, you should obtain $x = 3$, $y = 2$, and $z = 4$.

Ⓖ **(b)** There are websites on the Internet that will solve a system of linear equations for you. They can be located by using any search engine to look for *equation solvers.* For instance, at the time of this writing, one such site is http://www.quickmath.com. Use one of these websites to solve system (1) on page 749.

31. Rework Example 6 in the text using the following new data. The number of reservations in economy, in economy-upgrade, and in first class are 830, 735, and 592, respectively. The seating capacity on each type of jet is as follows.

	Jet type I	Jet type II	Jet type III
Economy	40	80	70
Economy-upgrade	35	75	60
First class	30	44	56

32. In Exercise 31, suppose that the company had made an error in counting the number of economy reservations, and that it should be 825 rather than 830. Solve the resulting system and interpret the result.

33. A parabola $y = ax^2 + bx + c$ passes through the three points $(1, -2)$, $(-1, 0)$, and $(2, 3)$.

(a) Write down a system of three linear equations that must be satisfied by a, b, and c.

(b) Solve the system in part (a).

Ⓖ **(c)** With the values for a, b, and c that you found in part (b), use a graphing utility to draw the parabola $y = ax^2 + bx + c$. Find a viewing rectangle that seems to confirm that the parabola indeed passes through the three given points.

Ⓖ **(d)** Another way to check your result in part (b): Apply the quadratic regression option on a graphing utility after entering the three given data points $(1, -2)$, $(-1, 0)$, and $(2, 3)$. (This is a valid check because, in general, three noncollinear points determine a unique parabola.)

34. A curve $y = x^3 + Ax^2 + Bx + C$ passes through the three points $(1, -2)$, $(2, 3)$, and $(3, 20)$.

(a) Write down a system of three linear equations satisfied by A, B, and C.

(b) Solve the system in part (a).

Ⓖ **(c)** With the values for A, B, and C that you found in part (b), use a graphing utility to draw the curve $y = x^3 + Ax^2 + Bx + C$, checking to see that it appears to pass through the three given points.

35. A manufacturer of office chairs makes three models: Utility, Secretarial, and Managerial. Three materials common to the manufacturing process for all of the models are cloth, steel, and plastic. The amounts of these materials required for one chair in each category are specified in the following table. The company wants to use up its inventory of these materials because of upcoming design changes. How many of each model should the manufacturer build to deplete its current inventory consisting of 476 units of cloth, 440 units of steel, and 826 units of plastic?

	Utility	Secretarial	Managerial
Cloth	3	4	2
Steel	2	5	8
Plastic	6	4	1

36. The U.S. Food and Drug Administration lists the following RDI's (reference daily intakes) for the antioxidants vitamin C, vitamin E, and zinc.

Vitamin C: 60 mg Vitamin E: 30 mg Zinc: 15 mg

Remark on terminology: The Food and Drug Administration defines *RDI* as a weighted average of the recommended daily allowances for all segments of the U.S. population.

Suppose that you have three brands of dietary supplements on your shelf. Among other ingredients, all three contain the antioxidants mentioned above. The amounts of these antioxidants in each supplement are indicated in the following table. How many ounces of each supplement should you combine to obtain the RDI's for vitamin C, vitamin E, and zinc?

	Vitamin C (mg/oz.)	Vitamin E (mg/oz.)	Zinc (mg/oz.)
Supplement I	12	4	1
Supplement II	5	1.25	2.5
Supplement III	2	3	0.5

B

37. After reviewing records from previous years, the owner of a small company notices some trends in the data. One of the trends is that monthly revenues from July through February, although different from year to year, seem to rise and fall in a pattern similar to the following. The (mathematically inclined) owner observes that a portion of a parabola may be a good model here.

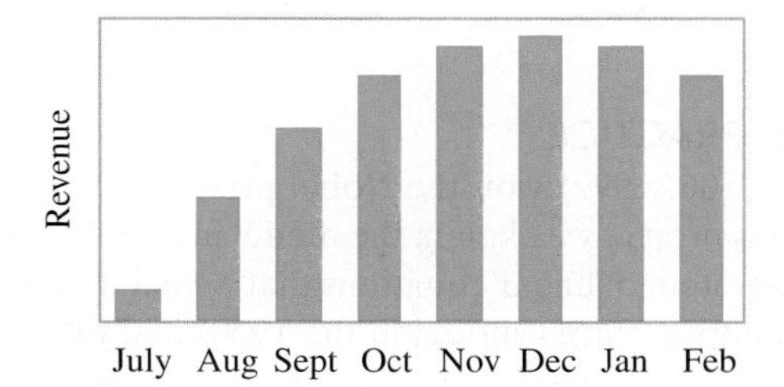

(a) This fiscal year, the monthly revenues for July, August, and September are as follows:

	Month		
	July	Aug.	Sept.
Revenue (units of $100,000)	1	3.7	5.8

Let $t = 0$ correspond to July, $t = 1$ to August, and so on. Let $R(t)$ be the revenue (in units of $100,000) for the month t. Find the quadratic function $R(t) = at^2 + bt + c$ whose graph passes through the three given data points (0, 1), (1, 3.7), and (2, 5.8).

(b) Use the vertex formula from Section 4.5 to determine the month for which the revenue is a maximum. Is your answer consistent with the bar graph shown at the start of this exercise? What is the corresponding maximum revenue?

38. Suppose that the height of an object as a function of time is given by $f(t) = at^2 + bt + c$, where t is time in seconds, $f(t)$ is the height in feet at time t, and a, b, and c are certain constants. If, after 1, 2, and 3 seconds, the corresponding heights are 184 ft, 136 ft, and 56 ft, respectively, find the time at which the object is at ground level (height = 0 ft).

39. Solve the following system for x, y, and z:

$$\begin{cases} e^x + e^y - 2e^z = 2a \\ e^x + 2e^y - 4e^z = 3a \\ \frac{1}{2}e^x - 3e^y + e^z = -5a \end{cases}$$

(Assume that $a > 0$.) *Hint*: Let $A = e^x$, $B = e^y$, and $C = e^z$. Solve the resulting linear system for A, B, and C.

40. Solve the following system for α, β, and γ.

$$\begin{cases} \ln \alpha - \ln \beta - \ln \gamma = 2 \\ 3 \ln \alpha + 5 \ln \beta - 2 \ln \gamma = 1 \\ 2 \ln \alpha - 4 \ln \beta + \ln \gamma = 2 \end{cases}$$

41. The following figure shows a rectangular metal plate. The temperature at each red point on the boundary is maintained at the constant value shown (in Fahrenheit degrees). The temperature is not regulated at the interior points (indicated by the blue dots). Under these conditions, over time, it can be shown that the temperature at each interior point is the average of the temperatures of the four surrounding points (that is, the points directly above, below, left, and right). For example, the temperature T_1 in the figure satisfies the condition $T_1 = (40 + 44 + T_2 + 41)/4$. Write down similar equations for T_2 and for T_3. Then solve the resulting system to determine the temperatures T_1, T_2, and T_3. *Remark*: Actually, the average-temperature condition that we've described is an approximation, and it holds only if the distances involved are sufficiently small.

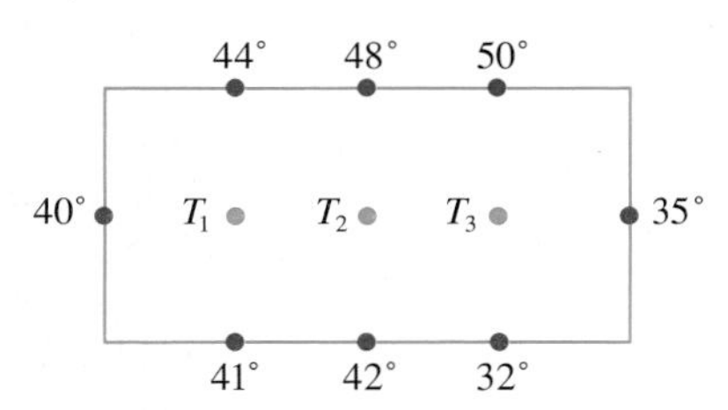

42. The following figure displays three circles that are mutually tangent. The line segments joining the centers have lengths a, b, and c, as shown. Let r_1, r_2, and r_3 denote the radii of the circles, as indicated in the figure. Show that

$$r_1 = \frac{a+c-b}{2} \qquad r_2 = \frac{a+b-c}{2} \qquad r_3 = \frac{b+c-a}{2}$$

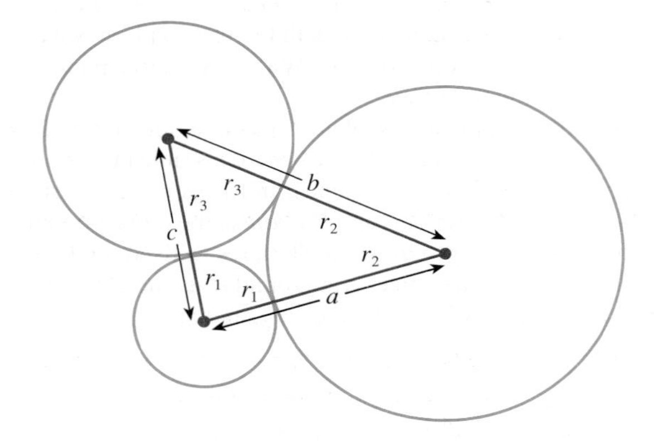

The following exercise appears in Algebra for Colleges and Schools *by H. S. Hall and S. R. Knight, revised by F. L. Sevenoak (New York: The Macmillan Co., 1906).*

43. A, B, and C are three towns forming a triangle. A man has to walk from one to the next, ride thence to the next, and drive thence to his starting point. He can walk, ride, and drive a mile in a, b, and c minutes, respectively. If he starts from B he takes $a + c - b$ hours, if he starts from C he takes $b + a - c$ hours, and if he starts from A he takes $c + b - a$ hours. Find the length of the circuit. [Assume that the circuit from A to B to C is counterclockwise.]

PROJECT THE LEONTIEF INPUT-OUTPUT MODEL

Extract of press release from the Royal Swedish Academy of Sciences, October 18, 1973: The Royal Swedish Academy of Sciences has awarded the 1973 year's Prize in Economic Science in Memory of Alfred Nobel to Professor Wassily Leontief for the development of the input-output method and for its application to important economic problems. Professor Leontief is the sole and unchallenged creator of the input-output technique. This important innovation has given to economic sciences an empirically-useful method to highlight the general interdependence in the production system of a society. In particular, the method provides tools for a systematic analysis of the complicated interindustry transactions in an economy.

PART I: EXPLANATION AND PRACTICE

Professor Wassily W. Leontief (1906–1999) won the Nobel prize in economics for his input-output model. In this project we explain the model in a simple case. As you'll see, the model yields a system of linear equations that we need to solve. Leontief began his work on the input-output model in the 1930s and often applied the model to the economy of the United States. You can see a readily available example of this at your local or college library in Leontief's article "The Structure of the U.S. Economy," which appears in the April 1965 issue of the magazine *Scientific American* (vol. 212, no. 4, pp. 25–35).

In the *Scientific American* article Leontief divided the American economy into 81 different industries or *sectors.* (For instance, 5 of the 81 sectors were coal mining, glass and glass products, primary iron and steel manufacturing, apparel, and aircraft and parts.) To keep things simple, let's suppose now that we have an economy with only three sectors: steel, coal, and electricity. Further, suppose that the sectors are interrelated, as given by the following *input-output* table, Table 1. We'll explain how to interpret the table in the next paragraph.

TABLE 1 An Input-Output Table for a Hypothetical Three-Sector Economy

	Outputs		
	Steel	Coal	Electricity
Inputs			
Steel	0.04	0.02	0.16
Coal	0.15	0	0.25
Electricity	0.14	0.10	0.04

The idea behind Table 1 is that each sector in the economy requires for its production process inputs from one or more of the other sectors. In the table, the first column of figures (beneath the output heading *Steel*) is interpreted as follows: The production of one unit of steel requires

0.04 unit of steel
0.15 unit of coal
0.14 unit of electricity

(While it's clear that coal and electricity would be required in the production of steel, you may be wondering why steel itself appears as one of the inputs. Think of the steel as being utilized in the building of equipment or factories for use by the steel industry.) The second and third columns of figures in the table are interpreted similarly. In particular, the second column of figures in Table 1 indicates that the production of 1 unit of coal requires 0.02 unit of steel, no units of coal, and 0.10 unit of electricity. The third column indicates that the production of 1 unit of electricity requires 0.16 unit of steel, 0.25 unit of coal, and 0.04 unit of electricity.

Table 1 can also be interpreted by reading across the rows, rather than down the columns. The first row of figures tells how much steel is required to produce 1 unit of output from each industry. In particular, to produce 1 unit of steel, 1 unit of coal, or 1 unit of electricity requires 0.04 unit of steel, 0.02 unit of steel, or 0.16 unit of steel, respectively. The second and third rows of figures are interpreted similarly.

Table 1 shows the demands that the three sectors place on one another for production. Now, outside of these three sectors, in other industries or in government, for example, there are, of course, additional demands for steel, coal, and electricity. These additional demands, from sources outside of the three given sectors, are referred to as *external demands.* By way of contrast, the demands in Table 1 are called *internal demands.* Let us suppose that the external demands are as given in Table 2. The problem to be solved then is as follows.

An Input-Output Problem

How many units should each of the three given sectors produce to satisfy both the internal and external demands on the economy?

TABLE 2

External demands*	
Steel	308 units
Coal	275 units
Electricity	830 units

*These are the demands for steel, coal, and electricity exclusive of the production requirements listed in Table 1.

To solve this input-output problem, let

x = the number of units of steel to be produced
y = the number of units of coal to be produced
z = the number of units of electricity to be produced

We'll use the information in Tables 1 and 2 to generate a system of three linear equations involving x, y, and z. First, since x represents the total number of units of steel to be produced, we have

$$\text{(internal demands for steel)} + \text{(external demands for steel)} = x$$
$$\text{(internal demands for steel)} + 308 = x \quad \text{using Table 2} \quad (1)$$

Regarding the internal demands for steel in equation (1), we'll make use of the first row of figures in Table 1 along with the following **proportionality assumption.** (For clarity, we state this for the case of steel output, but it's assumed for coal and electricity as well.) If the output of one unit of steel requires n units of a particular resource, then the output of x units steel requires $n \times x$ units of that resource. Now we use this proportionality assumption, the first row of figures in Table 1, and the definitions of x, y, and z, to write

$$\text{(internal demands for steel)} = 0.04x + 0.02y + 0.16z$$

Using this last equation to substitute in equation (1), we obtain

$$0.04x + 0.02y + 0.16z + 308 = x$$
$$-0.96x + 0.02y + 0.16z = -308 \quad (2)$$

Equation (2) is our first of three equations in three unknowns. A second equation can be generated in the same manner, but using coal, rather than steel, as the output under consideration. Again, similarly, a third equation can be obtained by starting with electricity rather than steel.

Working in groups or individually, obtain the remaining two equations and then solve the resulting system. As a check, the required answers are $x = 500$ units of steel, $y = 600$ units of coal, and $z = 1000$ units of electricity. *Suggestion*: Work on this in groups. Then after your group has obtained the correct answers and you are clear on the details, work individually on the next problem.

PART II: A PROBLEM TO DO AND THEN TO WRITE UP ON YOUR OWN

Consider a three-sector economy: plastics, energy, and transportation. The production of 1 unit of plastics requires 0.02 of unit plastics, 0.04 unit of energy, and 0.05 unit of transportation. The production of 1 unit of energy requires 0.03 unit of plastics, 0.08 unit of energy, and 0.01 unit of transportation. The production of 1 unit of transportation requires 0.01 unit of plastics, 0.02 unit of energy, and 0.06 unit of transportation. The external demands for plastics, energy, and transportation are 22.76 units, 36.68 units, and 43.45 units, respectively.

(a) Display the given data using tables similar to Tables 1 and 2.

(b) How many units should each sector produce in order to satisfy both the internal and external demands on the economy? *Hint for checking*: The final answers turn out to be integers.

10.3 MATRICES

Recall that in Gaussian Elimination, row operations are used to change the coefficient matrix to an upper triangular matrix. The solution is then found by back substitution, starting from the last equation in the reduced system. —Steven C. Althoen and Renate McLaughlin in "Gauss–Jordan Reduction: A Brief History," *American Mathematical Monthly,* vol. 94 (1987), pp. 130–142

Arthur Cayley (1821–1895) and James Joseph Sylvester (1814–1897), two English mathematicians, invented the matrix . . . in the 1850s. . . . The operations of addition and multiplication of matrices were later defined and the algebra of matrices was then developed. In 1925, Werner Heisenberg, a German physicist, used matrices in developing his theory of quantum mechanics, extending the role of matrices from algebra to the area of applied mathematics. —John K. Luedeman and Stanley M. Lukawecki in *Elementary Linear Algebra* (St. Paul: West Publishing Co., 1986)

As you saw in the previous section, there can be a good deal of bookkeeping involved in using Gaussian elimination to solve systems of equations. We can organize our work efficiently by using a **matrix** (pl.: **matrices**), which is simply a rectangular array of numbers enclosed in parentheses or brackets. Here are three examples:

$$\begin{pmatrix} 2 & 3 \\ -5 & 4 \end{pmatrix} \qquad \begin{pmatrix} -6 & 0 & 1 & \frac{1}{4} \\ \frac{2}{3} & 1 & 5 & 8 \end{pmatrix} \qquad \begin{pmatrix} \pi & 0 & 0 \\ 0 & 1 & 9 \\ -1 & -2 & 3 \\ -4 & 8 & 6 \end{pmatrix}$$

The particular numbers constituting a matrix are called its **entries** or **elements.** For instance, the entries in the matrix $\begin{pmatrix} 2 & 3 \\ -5 & 4 \end{pmatrix}$ are the four numbers 2, 3, -5, and 4. In this section the entries will always be real numbers. However, it is also possible to consider matrices in which some or all of the entries are nonreal complex numbers.

It is convenient to agree on a standard system for labeling the rows and columns of a matrix. The rows are numbered from top to bottom and the columns from left to right, as indicated in the following example:

$$\begin{matrix} & & \text{column 1} & \text{column 2} \\ & & \downarrow & \downarrow \\ \text{row 1} & \rightarrow & 5 & -3 \\ \text{row 2} & \rightarrow & 0 & 2 \\ \text{row 3} & \rightarrow & -1 & 16 \end{matrix}$$

We express the **size,** or **dimension,** of a matrix by specifying the number of rows followed by the number of columns. For instance, we would say that the matrix

$$\begin{pmatrix} 5 & -3 \\ 0 & 2 \\ -1 & 16 \end{pmatrix}$$

with three rows and two columns, is a 3×2 (read "3 by 2") matrix, *not* 2×3. The following example will help fix in your mind the terminology that we have introduced.

EXAMPLE 1 Matrix terminology

Consider the matrix

$$\begin{pmatrix} 1 & 3 & 5 \\ 7 & 9 & 11 \end{pmatrix}$$

(a) List the entries.
(b) What is the size of the matrix?
(c) Which element is in the second row, third column?

SOLUTION

(a) The entries are 1, 3, 5, 7, 9, and 11.
(b) Since there are two rows and three columns, this is a 2×3 matrix.
(c) To locate the element in the second row, third column, we imagine lines through the second row and the third column and see where they intersect:

$$\begin{pmatrix} 1 & 3 & 5 \\ 7 & 9 & 11 \end{pmatrix}$$

Thus the entry in row 2, column 3 is 11.

There is a natural way to use matrices to describe and solve systems of linear equations. Consider, for example, the following system of linear equations in **standard form** (with the x-, y-, and z-terms lined up on the left-hand side and the constant terms on the right-hand side of each equation):

$$\begin{cases} x + 2y - 3z = 4 \\ 3x \qquad\; + z = 5 \\ -x - 3y + 4z = 0 \end{cases}$$

The **coefficient matrix** of this system is the matrix

$$\begin{pmatrix} 1 & 2 & -3 \\ 3 & 0 & 1 \\ -1 & -3 & 4 \end{pmatrix}$$

As the name implies, the coefficient matrix of the system is the matrix whose entries are the coefficients of x, y, and z, written in the same relative positions as they appear in the system. Notice that a zero appears in the second row, second column of the matrix because the coefficient of y in the second equation is in fact zero. The **augmented matrix** of the system of equations considered here is

$$\left(\begin{array}{ccc|c} 1 & 2 & -3 & 4 \\ 3 & 0 & 1 & 5 \\ -1 & -3 & 4 & 0 \end{array}\right)$$

As you can see, the augmented matrix is formed by *augmenting* the coefficient matrix with the column of constant terms taken from the right-hand side of the given system of equations. The dashed line in the augmented matrix is used to visually separate the coefficient matrix and the right-hand sides of the equations of the linear system in standard form.

EXAMPLE 2 Specifying the coefficient matrix and augmented matrix

Write the coefficient matrix and the augmented matrix for the system

$$\begin{cases} 8x - 2y + z = 1 \\ 3x - 4z + y = 2 \\ 12y - 3z - 6 = 0 \end{cases}$$

SOLUTION

First we write the system in standard form, with the x-, y-, and z-terms lined up and the constant terms on the right. This yields

$$\begin{cases} 8x - 2y + z = 1 \\ 3x + y - 4z = 2 \\ 12y - 3z = 6 \end{cases}$$

The coefficient matrix is then

$$\begin{pmatrix} 8 & -2 & 1 \\ 3 & 1 & -4 \\ 0 & 12 & -3 \end{pmatrix}$$

and the augmented matrix is

$$\left(\begin{array}{ccc:c} 8 & -2 & 1 & 1 \\ 3 & 1 & -4 & 2 \\ 0 & 12 & -3 & 6 \end{array}\right)$$

In the previous section we used the three elementary operations in solving systems of linear equations. In Table 1 we express these operations in the language of matrices. The matrix operations are called the **elementary row operations.**

TABLE 1

Elementary operations for a system of linear equations	Corresponding elementary row operations for a matrix
1. Multiply both sides of an equation by a nonzero constant.	1′. Multiply each entry in a given row by a nonzero constant.
2. Interchange two equations.	2′. Interchange two rows.
3. To one equation, add a multiple of another equation.	3′. To one row, add a multiple of another row.

Table 2 displays an example of each elementary row operation. In the table, notice that the notation for describing these operations is essentially the same as that introduced in the previous section. For example, in the previous section, $10E_1$ indicated that the first equation in a system was multiplied by 10. Now, $10R_1$ indicates that each entry in the first *row* of a matrix is multiplied by 10.

TABLE 2

Examples of the elementary row operations	Comments
$\begin{pmatrix} 1 & 2 & 3 \\ 4 & 5 & 6 \\ 7 & 8 & 9 \end{pmatrix} \xrightarrow{10R_1} \begin{pmatrix} 10 & 20 & 30 \\ 4 & 5 & 6 \\ 7 & 8 & 9 \end{pmatrix}$	Multiply each entry in row 1 by 10.
$\begin{pmatrix} 1 & 2 & 3 \\ 4 & 5 & 6 \\ 7 & 8 & 9 \end{pmatrix} \xrightarrow{R_2 \leftrightarrow R_3} \begin{pmatrix} 1 & 2 & 3 \\ 7 & 8 & 9 \\ 4 & 5 & 6 \end{pmatrix}$	Interchange rows 2 and 3.
$\begin{pmatrix} 1 & 2 & 3 \\ 4 & 5 & 6 \\ 7 & 8 & 9 \end{pmatrix} \xrightarrow{-4R_1 + R_2} \begin{pmatrix} 1 & 2 & 3 \\ 0 & -3 & -6 \\ 7 & 8 & 9 \end{pmatrix}$	To each entry in row 2, add -4 times the corresponding entry in row 1.

We are now ready to use matrices to solve systems of equations. In the example that follows, we'll use the same system of equations used in Example 5 of the previous section. *Suggestion*: After reading the next example, carefully compare each step with the corresponding one taken in Example 5 of the previous section.

EXAMPLE 3 Using the elementary row operations to solve a system

Solve the system

$$\begin{cases} 4x - 3y + 2z = 40 \\ 5x + 9y - 7z = 47 \\ 9x + 8y - 3z = 97 \end{cases}$$

SOLUTION

$$\left(\begin{array}{rrr|r} 4 & -3 & 2 & 40 \\ 5 & 9 & -7 & 47 \\ 9 & 8 & -3 & 97 \end{array}\right) \xrightarrow{(-1)R_2 + R_1} \left(\begin{array}{rrr|r} -1 & -12 & 9 & -7 \\ 5 & 9 & -7 & 47 \\ 9 & 8 & -3 & 97 \end{array}\right)$$

$$\xrightarrow[9R_1 + R_3]{5R_1 + R_2} \left(\begin{array}{rrr|r} -1 & -12 & 9 & -7 \\ 0 & -51 & 38 & 12 \\ 0 & -100 & 78 & 34 \end{array}\right)$$

$$\xrightarrow{\frac{1}{2}R_3} \left(\begin{array}{rrr|r} -1 & -12 & 9 & -7 \\ 0 & -51 & 38 & 12 \\ 0 & -50 & 39 & 17 \end{array}\right)$$

$$\xrightarrow{(-1)R_3 + R_2} \left(\begin{array}{rrr|r} -1 & -12 & 9 & -7 \\ 0 & -1 & -1 & -5 \\ 0 & -50 & 39 & 17 \end{array}\right)$$

$$\xrightarrow{-50R_2 + R_3} \left(\begin{array}{rrr|r} -1 & -12 & 9 & -7 \\ 0 & -1 & -1 & -5 \\ 0 & 0 & 89 & 267 \end{array}\right)$$

This last augmented matrix represents a system of equations in upper-triangular form:

$$\begin{cases} -x - 12y + 9z = -7 \\ \quad\;\; -y - \;\; z = -5 \\ \qquad\qquad 89z = 267 \end{cases}$$

As you should now check for yourself, this yields the values $z = 3$, then $y = 2$, then $x = 10$. The solution of the original system is therefore $(10, 2, 3)$.

For the remainder of this section (and in part of the next) we will study matrices without referring to systems of equations. As motivation for this, we point out that matrices are essential tools in many fields of study. For example, a knowledge of matrices and their properties is needed for work in computer graphics. To begin, we need to say what it means for two matrices to be equal.

DEFINITION Equality of Matrices

Two matrices are equal provided that they have the same size (same number of rows, same number of columns) and the corresponding entries are equal.

EXAMPLES

$$\begin{pmatrix} 2 & 3 \\ 4 & 5 \end{pmatrix} = \begin{pmatrix} 2 & 3 \\ 4 & 5 \end{pmatrix}$$

$$\begin{pmatrix} 2 & 3 & 0 \\ 4 & 5 & 0 \end{pmatrix} \neq \begin{pmatrix} 2 & 3 \\ 4 & 5 \end{pmatrix}$$

$$\begin{pmatrix} 2 & 3 \\ 4 & 5 \end{pmatrix} \neq \begin{pmatrix} 2 & 3 \\ 5 & 4 \end{pmatrix}$$

Now we can define matrix addition and subtraction. These operations are defined only between matrices of the same size

DEFINITION Matrix Addition

To add (or subtract) two matrices of the same size, add (or subtract) the corresponding entries.

EXAMPLES

$$\begin{pmatrix} 2 & 3 \\ -1 & 4 \end{pmatrix} + \begin{pmatrix} 6 & 1 \\ 0 & -4 \end{pmatrix} = \begin{pmatrix} 2+6 & 3+1 \\ -1+0 & 4+(-4) \end{pmatrix} = \begin{pmatrix} 8 & 4 \\ -1 & 0 \end{pmatrix}$$

$$\begin{pmatrix} 2 & 3 \\ -1 & 4 \\ 9 & 10 \end{pmatrix} - \begin{pmatrix} 6 & 1 \\ 0 & -4 \\ 3 & 2 \end{pmatrix} = \begin{pmatrix} 2-6 & 3-1 \\ -1-0 & 4-(-4) \\ 9-3 & 10-2 \end{pmatrix} = \begin{pmatrix} -4 & 2 \\ -1 & 8 \\ 6 & 8 \end{pmatrix}$$

Many properties analogous to those of the real numbers are also valid for matrices. For instance, matrix addition is *commutative*:

$$A + B = B + A \qquad \text{where } A \text{ and } B \text{ are matrices of the same size}$$

Matrix addition is also *associative*:

$$A + (B + C) = (A + B) + C \qquad \text{where } A, B, \text{ and } C \text{ are matrices of the same size}$$

In the next example we verify these properties in two specific instances.

EXAMPLE 4 Adding matrices

Let $A = \begin{pmatrix} 1 & 2 \\ 3 & 4 \end{pmatrix}$, $B = \begin{pmatrix} 0 & -5 \\ 8 & -1 \end{pmatrix}$, and $C = \begin{pmatrix} 6 & 7 \\ 8 & 9 \end{pmatrix}$.

(a) Show that $A + C = C + A$.

(b) Show that $A + (B + C) = (A + B) + C$.

SOLUTION

(a) $A + C = \begin{pmatrix} 1 & 2 \\ 3 & 4 \end{pmatrix} + \begin{pmatrix} 6 & 7 \\ 8 & 9 \end{pmatrix} = \begin{pmatrix} 7 & 9 \\ 11 & 13 \end{pmatrix}$

$C + A = \begin{pmatrix} 6 & 7 \\ 8 & 9 \end{pmatrix} + \begin{pmatrix} 1 & 2 \\ 3 & 4 \end{pmatrix} = \begin{pmatrix} 7 & 9 \\ 11 & 13 \end{pmatrix}$

This shows that $A + C = C + A$, since both $A + C$ and $C + A$ represent the matrix $\begin{pmatrix} 7 & 9 \\ 11 & 13 \end{pmatrix}$.

(b) First we compute $A + (B + C)$:

$$\begin{aligned} A + (B + C) &= \begin{pmatrix} 1 & 2 \\ 3 & 4 \end{pmatrix} + \left[\begin{pmatrix} 0 & -5 \\ 8 & -1 \end{pmatrix} + \begin{pmatrix} 6 & 7 \\ 8 & 9 \end{pmatrix}\right] \\ &= \begin{pmatrix} 1 & 2 \\ 3 & 4 \end{pmatrix} + \begin{pmatrix} 6 & 2 \\ 16 & 8 \end{pmatrix} \\ &= \begin{pmatrix} 7 & 4 \\ 19 & 12 \end{pmatrix} \end{aligned}$$

Next we compute $(A + B) + C$:

$$\begin{aligned} (A + B) + C &= \left[\begin{pmatrix} 1 & 2 \\ 3 & 4 \end{pmatrix} + \begin{pmatrix} 0 & -5 \\ 8 & -1 \end{pmatrix}\right] + \begin{pmatrix} 6 & 7 \\ 8 & 9 \end{pmatrix} \\ &= \begin{pmatrix} 1 & -3 \\ 11 & 3 \end{pmatrix} + \begin{pmatrix} 6 & 7 \\ 8 & 9 \end{pmatrix} \\ &= \begin{pmatrix} 7 & 4 \\ 19 & 12 \end{pmatrix} \end{aligned}$$

We conclude from these calculations that $A + (B + C) = (A + B) + C$, since both sides of that equation represent the matrix $\begin{pmatrix} 7 & 4 \\ 19 & 12 \end{pmatrix}$.

We will now define an operation on matrices called **scalar multiplication.** First, the word *scalar* here just means *real number,* so we are talking about multiplying a matrix by a real number. (In more advanced work, nonreal complex scalars are also considered.)

DEFINITION **Scalar Multiplication**

To multiply a matrix by a scalar, multiply each entry in the matrix by that scalar.

EXAMPLES

$$2\begin{pmatrix} 5 & 9 & 0 \\ -1 & 2 & 3 \end{pmatrix} = \begin{pmatrix} 2\cdot 5 & 2\cdot 9 & 2\cdot 0 \\ 2\cdot(-1) & 2\cdot 2 & 2\cdot 3 \end{pmatrix} = \begin{pmatrix} 10 & 18 & 0 \\ -2 & 4 & 6 \end{pmatrix}$$

$$1\begin{pmatrix} 1 & 2 \\ 3 & 4 \end{pmatrix} = \begin{pmatrix} 1 & 2 \\ 3 & 4 \end{pmatrix}$$

There are two simple but useful properties of scalar multiplication that are worth noting at this point. We'll omit the proofs of these two properties; however, Example 5 does ask us to verify them for a particular case.

PROPERTY SUMMARY **Properties of Scalar Multiplication**

1. $c(kM) = (ck)M$, for all scalars c and k and any matrix M.
2. $c(M + N) = cM + cN$, where c is any scalar and M and N are any matrices of the same size.

EXAMPLE 5 **Demonstrating properties of scalar multiplication**

Let $c = 2$, $k = 3$, $M = \begin{pmatrix} 1 & 2 \\ 3 & 4 \end{pmatrix}$, and $N = \begin{pmatrix} 5 & 6 \\ 7 & 8 \end{pmatrix}$.

(a) Show that $c(kM) = (ck)M$.
(b) Show that $c(M + N) = cM + cN$.

SOLUTION

(a)
$$c(kM) = 2\left[3\begin{pmatrix} 1 & 2 \\ 3 & 4 \end{pmatrix}\right]$$
$$= 2\begin{pmatrix} 3 & 6 \\ 9 & 12 \end{pmatrix} = \begin{pmatrix} 6 & 12 \\ 18 & 24 \end{pmatrix}$$
$$(ck)M = (2\cdot 3)\begin{pmatrix} 1 & 2 \\ 3 & 4 \end{pmatrix}$$
$$= 6\begin{pmatrix} 1 & 2 \\ 3 & 4 \end{pmatrix} = \begin{pmatrix} 6 & 12 \\ 18 & 24 \end{pmatrix}$$

Thus $c(kM) = (ck)M$, since in both cases the result is $\begin{pmatrix} 6 & 12 \\ 18 & 24 \end{pmatrix}$.

(b)
$$c(M + N) = 2\left[\begin{pmatrix} 1 & 2 \\ 3 & 4 \end{pmatrix} + \begin{pmatrix} 5 & 6 \\ 7 & 8 \end{pmatrix}\right]$$
$$= 2\begin{pmatrix} 6 & 8 \\ 10 & 12 \end{pmatrix} = \begin{pmatrix} 12 & 16 \\ 20 & 24 \end{pmatrix}$$
$$cM + cN = 2\begin{pmatrix} 1 & 2 \\ 3 & 4 \end{pmatrix} + 2\begin{pmatrix} 5 & 6 \\ 7 & 8 \end{pmatrix}$$

$$= \begin{pmatrix} 2 & 4 \\ 6 & 8 \end{pmatrix} + \begin{pmatrix} 10 & 12 \\ 14 & 16 \end{pmatrix} = \begin{pmatrix} 12 & 16 \\ 20 & 24 \end{pmatrix}$$

Thus $c(M + N) = cM + cN$, since both sides equal $\begin{pmatrix} 12 & 16 \\ 20 & 24 \end{pmatrix}$.

A matrix with zeros for all of its entries plays the same role in matrix addition as does the number zero in ordinary addition of real numbers. For instance, in the case of 2×2 matrices, we have

$$\begin{pmatrix} a & b \\ c & d \end{pmatrix} + \begin{pmatrix} 0 & 0 \\ 0 & 0 \end{pmatrix} = \begin{pmatrix} a & b \\ c & d \end{pmatrix} \quad \text{and} \quad \begin{pmatrix} 0 & 0 \\ 0 & 0 \end{pmatrix} + \begin{pmatrix} a & b \\ c & d \end{pmatrix} = \begin{pmatrix} a & b \\ c & d \end{pmatrix}$$

for all real numbers a, b, c, and d. The **zero matrix** $\begin{pmatrix} 0 & 0 \\ 0 & 0 \end{pmatrix}$ is called the **additive identity** for 2×2 matrices. Similarly, any zero matrix is the additive identity for matrices of that size. It is sometimes convenient to denote an additive identity matrix by a boldface zero: $\mathbf{0}$. With this notation, we can write

$$A + \mathbf{0} = \mathbf{0} + A = A \qquad \text{for any matrix } A$$

For this matrix equation it is understood that the size of the matrix $\mathbf{0}$ is the same as the size of A. With this notation we also have

$$A - A = \mathbf{0} \qquad \text{for any matrix } A$$

Our last topic in this section is matrix multiplication. We will begin with the simplest case and then work up to the more general situation. By convention, a matrix with only one row is called a **row vector.** Examples of row vectors are

$$\begin{pmatrix} 2 & 13 \end{pmatrix}, \quad \begin{pmatrix} -1 & 4 & 3 \end{pmatrix}, \quad \text{and} \quad \begin{pmatrix} 0 & 0 & 0 & 1 \end{pmatrix}$$

Similarly, a matrix with only one column is c alled a **column vector.** Examples are

$$\begin{pmatrix} 2 \\ 13 \end{pmatrix}, \quad \begin{pmatrix} -1 \\ 4 \\ 3 \end{pmatrix}, \quad \text{and} \quad \begin{pmatrix} 0 \\ 0 \\ 0 \\ 1 \end{pmatrix}$$

The following definition tells us how to multiply a row vector and a column vector when they have the same number of entries.

DEFINITION **The Inner Product of a Row Vector and a Column Vector**

Let A be a row vector and B a column vector, and assume that the number of columns in A is the same as the number of rows in B. Then the **inner product** $A \cdot B$ is defined to be the number obtained by multiplying the corresponding entries and then adding the products.

EXAMPLES

$$\begin{pmatrix} 1 & 2 & 3 \end{pmatrix} \cdot \begin{pmatrix} 4 \\ 5 \\ 6 \end{pmatrix} = 1 \cdot 4 + 2 \cdot 5 + 3 \cdot 6 = 32$$

$$\begin{pmatrix} 1 & 2 \end{pmatrix} \cdot \begin{pmatrix} 4 \\ 5 \\ 6 \end{pmatrix} \quad \text{is not defined}$$

$$\begin{pmatrix} 1 & 2 & 3 \end{pmatrix} \cdot \begin{pmatrix} 4 \\ 5 \end{pmatrix} \quad \text{is not defined}$$

An important observation here is that the end result of taking the inner product is always just a number. The definition of matrix product that we now give depends on this observation.

DEFINITION **The Product of Two Matrices**

Let A and B be two matrices, and assume that the number of columns in A is the same as the number of rows in B. Then the **product matrix** AB is computed according to the following rule:

> The entry in the ith row and the jth column of AB is the inner product of the ith row of A with the jth column of B.

The matrix AB will have as many rows as A and as many columns as B.

As an example of matrix multiplication, we will compute the product AB, where $A = \begin{pmatrix} 1 & 2 \\ 3 & 4 \end{pmatrix}$ and $B = \begin{pmatrix} 5 & 6 & 0 \\ 7 & 8 & 1 \end{pmatrix}$. In other words, we will compute $\begin{pmatrix} 1 & 2 \\ 3 & 4 \end{pmatrix}\begin{pmatrix} 5 & 6 & 0 \\ 7 & 8 & 1 \end{pmatrix}$. Before we attempt to carry out the calculations of any matrix multiplication, however, we should check on two points.

1. Is the product defined? That is, does the number of columns in A equal the number of rows in B? In this case, yes; the common number is 2.
2. What is the size of the product? According to the definition, the product AB will have as many rows as A and as many columns as B. Thus the size of AB will be 2×3.

Schematically, then, the situation looks like this:

$$\begin{pmatrix} 1 & 2 \\ 3 & 4 \end{pmatrix}\begin{pmatrix} 5 & 6 & 0 \\ 7 & 8 & 1 \end{pmatrix} = \begin{pmatrix} ? & ? & ? \\ ? & ? & ? \end{pmatrix}$$

We have six positions to fill. The computations are presented in Table 3. Reading from the table, we see that our result is

$$AB = \begin{pmatrix} 1 & 2 \\ 3 & 4 \end{pmatrix}\begin{pmatrix} 5 & 6 & 0 \\ 7 & 8 & 1 \end{pmatrix} = \begin{pmatrix} 19 & 22 & 2 \\ 43 & 50 & 4 \end{pmatrix}$$

TABLE 3

Position	How to compute	Computation
row 1, column 1	inner product of row 1 and column 1 $\begin{pmatrix} 1 & 2 \\ 3 & 4 \end{pmatrix}\begin{pmatrix} 5 & 6 & 0 \\ 7 & 8 & 1 \end{pmatrix}$	$1 \cdot 5 + 2 \cdot 7 = 19$
row 1, column 2	inner product of row 1 and column 2 $\begin{pmatrix} 1 & 2 \\ 3 & 4 \end{pmatrix}\begin{pmatrix} 5 & 6 & 0 \\ 7 & 8 & 1 \end{pmatrix}$	$1 \cdot 6 + 2 \cdot 8 = 22$
row 1, column 3	inner product of row 1 and column 3 $\begin{pmatrix} 1 & 2 \\ 3 & 4 \end{pmatrix}\begin{pmatrix} 5 & 6 & 0 \\ 7 & 8 & 1 \end{pmatrix}$	$1 \cdot 0 + 2 \cdot 1 = 2$

(continues)

Position	How to compute	Computation
row 2, column 1	inner product of row 2 and column 1 $\begin{pmatrix}1 & 2\\3 & 4\end{pmatrix}\begin{pmatrix}5 & 6 & 0\\7 & 8 & 1\end{pmatrix}$	$3\cdot5+4\cdot7=43$
row 2, column 2	inner product of row 2 and column 2 $\begin{pmatrix}1 & 2\\3 & 4\end{pmatrix}\begin{pmatrix}5 & 6 & 0\\7 & 8 & 1\end{pmatrix}$	$3\cdot6+4\cdot8=50$
row 2, column 3	inner product of row 2 and column 3 $\begin{pmatrix}1 & 2\\3 & 4\end{pmatrix}\begin{pmatrix}5 & 6 & 0\\7 & 8 & 1\end{pmatrix}$	$3\cdot0+4\cdot1=4$

EXAMPLE 6 Computing matrix products

Let $A=\begin{pmatrix}1 & 2\\3 & 4\end{pmatrix}$ and $B=\begin{pmatrix}5 & 6\\7 & 8\end{pmatrix}$. By computing AB and then BA, show that $AB \neq BA$. This shows that, in general, matrix multiplication is not commutative.

SOLUTION

$$AB=\begin{pmatrix}1 & 2\\3 & 4\end{pmatrix}\begin{pmatrix}5 & 6\\7 & 8\end{pmatrix}=\begin{pmatrix}1\cdot5+2\cdot7 & 1\cdot6+2\cdot8\\3\cdot5+4\cdot7 & 3\cdot6+4\cdot8\end{pmatrix}=\begin{pmatrix}19 & 22\\43 & 50\end{pmatrix}$$

$$BA=\begin{pmatrix}5 & 6\\7 & 8\end{pmatrix}\begin{pmatrix}1 & 2\\3 & 4\end{pmatrix}=\begin{pmatrix}5\cdot1+6\cdot3 & 5\cdot2+6\cdot4\\7\cdot1+8\cdot3 & 7\cdot2+8\cdot4\end{pmatrix}=\begin{pmatrix}23 & 34\\31 & 46\end{pmatrix}$$

Comparing the two matrices AB and BA, we conclude that $AB \neq BA$.

Most graphing utilities perform matrix multiplication. This can be useful when the matrices are large or contain decimal entries. We'll briefly outline two examples here using the simple two-by-two matrices from Example 6. (For the details on these and for other types of graphing utilities, see your user's manual. Figure 1 shows the computation of the matrix product AB on a Texas Instruments TI-83+ graphing calculator. After specifying the dimensions of the two matrices and their elements in the matrix-edit menu, the matrices can be displayed as in Figure 1(a). The next screen, Figure 1(b), shows the matrix product AB.

Figure 2 shows the resulting calculation as carried out in a Microsoft *Excel* spreadsheet. The elements of the matrix A are typed into cells B1 through C2.

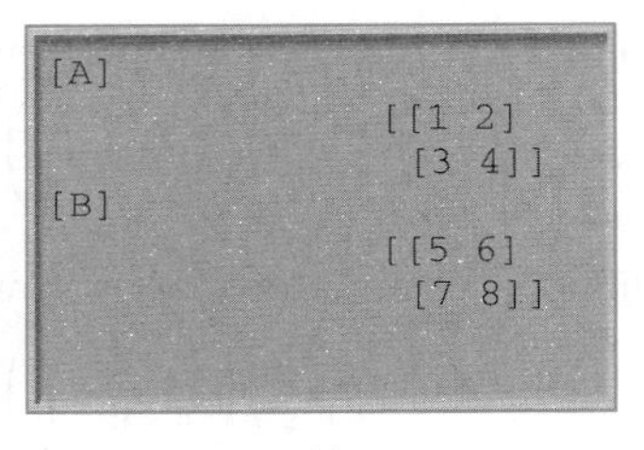

(a)

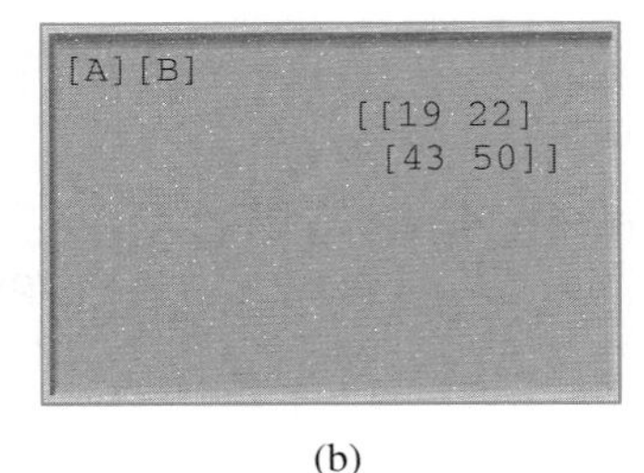

(b)

Figure 1
Computing the matrix product AB using a Texas Instruments TI-83+ graphing calculator

Likewise, matrix B is entered in cells B4 through C5. The product matrix is then obtained by means of the function MMULT, which stands for "matrix multiplication." Some additional details are given in the figure caption.

	A	B	C
1	A=	1	2
2		3	4
3			
4	B=	5	6
5		7	8
6			
7	AB=	19	22
8		43	50
9			

Figure 2
Computing the matrix product AB using Microsoft *Excel*. After typing in the entries of the two matrices A and B, the so-called array formula `{=MMULT(B1:C2,B4:C5)}` is used to compute and display AB. (See the documentation or help menu that came with *Excel*.)

EXERCISE SET 10.3

A

In Exercises 1–4, specify the size of each matrix.

1. (a) $\begin{pmatrix} -4 & 0 & 5 \\ 2 & 8 & -1 \end{pmatrix}$

(b) $\begin{pmatrix} 7 & 1 \\ 4 & -3 \\ 0 & 0 \end{pmatrix}$

2. (a) $\begin{pmatrix} 1 & 0 \\ 0 & -1 \end{pmatrix}$

(b) $\begin{pmatrix} 1 \\ 6 \\ 8 \\ 1 \end{pmatrix}$

3. $\begin{pmatrix} 1 & a & b & c \\ a & 1 & 0 & a \\ b & 0 & 1 & b \\ c & a & b & 1 \\ 0 & 0 & 0 & 1 \end{pmatrix}$

4. $\begin{pmatrix} -3 & 1 & 6 & 0 \end{pmatrix}$

In Exercises 5–8, write the coefficient matrix and the augmented matrix for each system.

5. $\begin{cases} 2x + 3y + 4z = 10 \\ 5x + 6y + 7z = 9 \\ 8x + 9y + 10z = 8 \end{cases}$

6. $\begin{cases} 5x - y + z = 0 \\ 4y + 2z = 1 \\ 3x + y + z = -1 \end{cases}$

7. $\begin{cases} x + z + w = -1 \\ x + y + 2w = 0 \\ y + z + w = 1 \\ 2x - y - z = 2 \end{cases}$

8. $\begin{cases} 8x - 8y = 5 \\ x - y + z = 1 \end{cases}$

In Exercises 9–22, use matrices to solve each system of equations.

9. $\begin{cases} x - y + 2z = 7 \\ 3x + 2y - z = -10 \\ -x + 3y + z = -2 \end{cases}$

10. $\begin{cases} 2x - 3y + 4z = 14 \\ 3x - 2y + 2z = 12 \\ 4x + 5y - 5z = 16 \end{cases}$

11. $\begin{cases} x + z = -2 \\ -3x + 2y = 17 \\ x - y - z = -9 \end{cases}$

12. $\begin{cases} 5x + y + 10z = 23 \\ 4x + 2y - 10z = 76 \\ 3x - 4y = 18 \end{cases}$

13. $\begin{cases} x + y + z = -4 \\ 2x - 3y + z = -1 \\ 4x + 2y - 3z = 33 \end{cases}$

14. $\begin{cases} 2x + 3y - 4z = 7 \\ x - y + z = -\frac{3}{2} \\ 6x - 5y - 2z = -7 \end{cases}$

15. $\begin{cases} 3x - 2y + 6z = 0 \\ x + 3y + 20z = 15 \\ 10x - 11y - 10z = -9 \end{cases}$

16. $\begin{cases} 3A - 3B + C = 4 \\ 6A + 9B - 3C = -7 \\ A - 2B - 2C = -3 \end{cases}$

17. $\begin{cases} 4x - 3y + 3z = 2 \\ 5x + y - 4z = 1 \\ 9x - 2y - z = 3 \end{cases}$

18. $\begin{cases} 6x + y - z = -1 \\ -3x + 2y + 2z = 2 \\ 5y + 3z = 1 \end{cases}$

19. $\begin{cases} x - y + z + w = 6 \\ x + y - z + w = 4 \\ x + y + z - w = -2 \\ -x + y + z + w = 0 \end{cases}$

20. $\begin{cases} x + 2y - z - 2w = 5 \\ 2x + y + 2z + w = -7 \\ -2x - y - 3z - 2w = 10 \\ z + w = -3 \end{cases}$

21. $\begin{cases} 15A + 14B + 26C = 1 \\ 18A + 17B + 32C = -1 \\ 21A + 20B + 38C = 0 \end{cases}$

22. $\begin{cases} A + B + C + D + E = -1 \\ 3A - 2B - 2C + 3D + 2E = 13 \\ 3C + 4D - 4E = 7 \\ 5A - 4B + E = 30 \\ C - 2E = 3 \end{cases}$

In Exercises 23–50, the matrices A, B, C, D, E, F, and G are defined as follows:

$$A = \begin{pmatrix} 2 & 3 \\ -1 & 4 \end{pmatrix} \qquad B = \begin{pmatrix} 1 & -1 \\ 3 & 0 \end{pmatrix} \qquad C = \begin{pmatrix} 1 & 0 \\ 0 & 1 \end{pmatrix}$$

$$D = \begin{pmatrix} -1 & 2 & 3 \\ 4 & 0 & 5 \end{pmatrix} \qquad E = \begin{pmatrix} 2 & 1 \\ 8 & -1 \\ 6 & 5 \end{pmatrix}$$

$$F = \begin{pmatrix} 5 & -1 \\ -4 & 0 \\ 2 & 3 \end{pmatrix} \qquad G = \begin{pmatrix} 0 & 0 \\ 0 & 0 \\ 0 & 0 \end{pmatrix}$$

In each exercise, carry out the indicated matrix operations if they are defined. If an operation is not defined, say so.

23. $A + B$ **24.** $A - B$ **25.** $2A + 2B$ **26.** $2(A + B)$
27. AB **28.** BA **29.** AC **30.** CA
31. $3D + E$ **32.** $E + F$ **33.** $2F - 3G$ **34.** DE
35. ED **36.** DF **37.** FD **38.** $A + D$
39. $G + A$ **40.** DG **41.** GD
42. $(A + B) + C$ **43.** $A + (B + C)$ **44.** CD
45. DC **46.** $5E - 3F$ **47.** $A^2\ (= AA)$
48. A^2A **49.** AA^2 **50.** C^2

51. Let

$$A = \begin{pmatrix} -1 & 3 & 4 \\ 3 & 2 & -3 \\ 9 & 1 & 6 \end{pmatrix} \qquad B = \begin{pmatrix} 7 & 0 & 1 \\ 0 & 0 & 3 \\ -1 & 2 & 4 \end{pmatrix}$$

$$C = \begin{pmatrix} 4 & 6 & 1 \\ 2 & 1 & 3 \\ -1 & -1 & 2 \end{pmatrix}$$

(a) Compute $A(B + C)$.
(b) Compute $AB + AC$.
(c) Compute $(AB)C$.
(d) Compute $A(BC)$.

Note: Parts (c) and (d) illustrate a specific example of the general property that matrix multiplication is associative.

Ⓖ *In Exercises 52–55, use a graphing utility to compute the matrix products.*

52. $\begin{pmatrix} 1.03 & 2.1 \\ -0.45 & 3.09 \end{pmatrix}\begin{pmatrix} 2.33 & 4.17 \\ 0 & -1.24 \end{pmatrix}$

53. $\begin{pmatrix} -32 & 14 \\ 27 & 9 \end{pmatrix}\begin{pmatrix} 83 & -19 \\ 13 & 41 \end{pmatrix}$

54. $\begin{pmatrix} 12 & -10 & 13 \\ 5 & 7 & 25 \\ -8 & 9 & 28 \end{pmatrix}\begin{pmatrix} -11 & 31 & 6 \\ 0 & 1 & -14 \\ 41 & 12 & -17 \end{pmatrix}$

55. $\begin{pmatrix} -6 & 9 & -5 & 1 \\ 9 & -1 & -5 & 2 \\ -5 & -5 & 9 & -3 \\ 1 & 2 & -3 & 1 \end{pmatrix}\begin{pmatrix} 0.5 & 1 & 1.5 & 2 \\ 1 & 2 & 3.5 & 5.5 \\ 1.5 & 3.5 & 7 & 12.5 \\ 2 & 5.5 & 12.5 & 25 \end{pmatrix}$

B

56. Let $A = \begin{pmatrix} 1 & 2 \\ 3 & 4 \end{pmatrix}$ and $B = \begin{pmatrix} 5 & 6 \\ 7 & 8 \end{pmatrix}$. Let A^2 and B^2 denote the matrix products AA and BB, respectively. Compute each of the following.
(a) $(A + B)(A + B)$
(b) $A^2 + 2AB + B^2$
(c) $A^2 + AB + BA + B^2$

57. Let $A = \begin{pmatrix} 3 & 5 \\ 7 & 9 \end{pmatrix}$ and $B = \begin{pmatrix} 2 & 4 \\ 6 & 8 \end{pmatrix}$. Compute each of the following.
(a) $A^2 - B^2$
(b) $(A - B)(A + B)$
(c) $(A + B)(A - B)$
(d) $A^2 + AB - BA - B^2$

58. Let

$$A = \begin{pmatrix} 1 & 0 \\ 0 & 1 \end{pmatrix} \qquad B = \begin{pmatrix} 1 & 0 \\ 0 & -1 \end{pmatrix}$$

$$C = \begin{pmatrix} -1 & 0 \\ 0 & 1 \end{pmatrix} \qquad D = \begin{pmatrix} -1 & 0 \\ 0 & -1 \end{pmatrix}$$

Complete the following multiplication table.

	A	B	C	D
A				
B			D	
C				
D				

Hint: In the second row, third column, D is the proper entry because (as you can check) $BC = D$.

59. In this exercise, let's agree to write the coordinates (x, y) of a point in the plane as the 2×1 matrix $\begin{pmatrix} x \\ y \end{pmatrix}$.

(a) Let $A = \begin{pmatrix} 1 & 0 \\ 0 & -1 \end{pmatrix}$ and $Z = \begin{pmatrix} x \\ y \end{pmatrix}$. Compute the matrix AZ. After computing AZ, observe that it represents the point obtained by reflecting $\begin{pmatrix} x \\ y \end{pmatrix}$ about the x-axis.

(b) Let $B = \begin{pmatrix} -1 & 0 \\ 0 & 1 \end{pmatrix}$ and $Z = \begin{pmatrix} x \\ y \end{pmatrix}$. Compute the matrix BZ. After computing BZ, observe that it represents the point obtained by reflecting $\begin{pmatrix} x \\ y \end{pmatrix}$ about the y-axis.

(c) Let A, B, and Z represent the matrices defined in parts (a) and (b). Compute the matrix $(AB)Z$, and then interpret it in terms of reflection about the axes.

60. In this exercise, we continue to explore some of the connections between matrices and geometry. As in Exercise 59, we will use 2×1 matrices to specify the coordinates of points in the plane. Let P, S, and T be the matrices defined as follows:

$$P = \begin{pmatrix} \cos x \\ \sin x \end{pmatrix} \qquad S = \begin{pmatrix} \cos\theta & -\sin\theta \\ \sin\theta & \cos\theta \end{pmatrix}$$

$$T = \begin{pmatrix} \cos\beta & -\sin\beta \\ \sin\beta & \cos\beta \end{pmatrix}$$

Notice that the point P lies on the unit circle.

(a) Compute the matrix SP. After computing SP, observe that it represents the point on the unit circle obtained by rotating P (about the origin) through an angle θ.

(b) Show that

$$ST = TS = \begin{pmatrix} \cos(\theta + \beta) & -\sin(\theta + \beta) \\ \sin(\theta + \beta) & \cos(\theta + \beta) \end{pmatrix}$$

(c) Compose $(ST)P$? What is the angle through which P is rotated?

61. A function f is defined as follows. The domain of f is the set of all 2×2 matrices (with real entries). If $A = \begin{pmatrix} a & b \\ c & d \end{pmatrix}$, then $f(A) = ad - bc$.

(a) Let $A = \begin{pmatrix} 1 & 2 \\ 3 & 4 \end{pmatrix}$, and $B = \begin{pmatrix} 3 & -1 \\ 5 & 8 \end{pmatrix}$. Compute $f(A), f(B)$, and $f(AB)$. Is it true, in this case, that $f(A) \cdot f(B) = f(AB)$?

(b) Let $A = \begin{pmatrix} a & b \\ c & d \end{pmatrix}$ and $B = \begin{pmatrix} e & f \\ g & h \end{pmatrix}$. Show that $f(A) \cdot f(B) = f(AB)$.

62. The **trace** of a 2×2 matrix $\begin{pmatrix} a & b \\ c & d \end{pmatrix}$ is defined by

$$\mathrm{tr}\begin{pmatrix} a & b \\ c & d \end{pmatrix} = a + d$$

(a) If $A = \begin{pmatrix} 1 & 2 \\ 3 & 4 \end{pmatrix}$ and $B = \begin{pmatrix} 5 & 6 \\ 7 & 8 \end{pmatrix}$, verify that $\mathrm{tr}(A + B) = \mathrm{tr}\,A + \mathrm{tr}\,B$.

(b) If $A = \begin{pmatrix} a & b \\ c & d \end{pmatrix}$ and $B = \begin{pmatrix} e & f \\ g & h \end{pmatrix}$, show that $\mathrm{tr}(A + B) = \mathrm{tr}\,A + \mathrm{tr}\,B$.

63. Let $A = \begin{pmatrix} a & b \\ c & d \end{pmatrix}$. The **transpose** of A is the matrix denoted by A^{T} and defined by $A^{\mathrm{T}} = \begin{pmatrix} a & c \\ b & d \end{pmatrix}$. In other words, A^{T} is obtained by switching the columns and rows of A. Show that the following equations hold for all 2×2 matrices A and B.

(a) $(A + B)^{\mathrm{T}} = A^{\mathrm{T}} + B^{\mathrm{T}}$ **(c)** $(AB)^{\mathrm{T}} = B^{\mathrm{T}}A^{\mathrm{T}}$
(b) $(A^{\mathrm{T}})^{\mathrm{T}} = A$

64. Find an example of two 2×2 matrices A and B for which $AB = \mathbf{0}$ but neither A nor B is $\mathbf{0}$.

PROJECT COMMUNICATIONS AND MATRICES

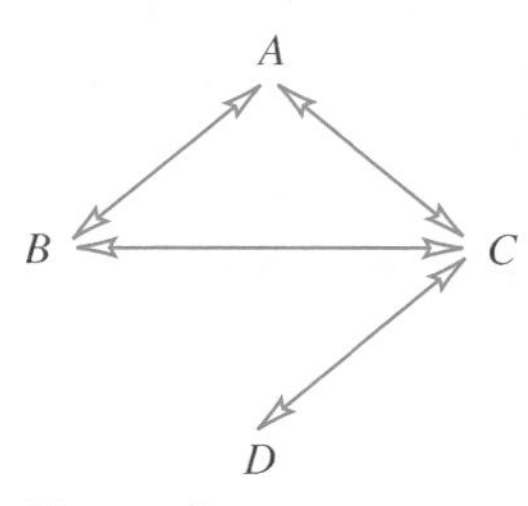

Figure 1
Communication links

$$\begin{array}{c} \\ A \\ B \\ C \\ D \end{array}\begin{array}{c} \begin{array}{cccc} A & B & C & D \end{array} \\ \begin{pmatrix} 0 & 1 & 1 & 0 \\ 1 & 0 & 1 & 0 \\ 1 & 1 & 0 & 1 \\ 0 & 0 & 1 & 0 \end{pmatrix} \end{array}$$

Figure 2
Communication matrix $\mathcal{M}$

Matrices and matrix products are used in communication theory. Suppose that four spies, A, B, C, and D, work for an intelligence agency under the following restrictions: Spies A, B, and C all can communicate directly with one another. Spy D, however, is super-secret and can communicate directly (back and forth) only with spy C. These communication links are summarized in Figure 1.

The communication links between the four spies can also be specified by means of the four-by-four *communication matrix* shown in Figure 2. We'll call this matrix $\mathcal{M}$.

In the communication matrix the rows and columns are associated with the spies as indicated by the red letters outside the matrix. An entry of 1 means that the two spies associated with that matrix location can communicate directly. For example, in the location row one, column two, there is a 1 because the corresponding spies A and B can communicate directly (according to Figure 1). In cases where two spies cannot communicate directly, 0 is used in the matrix. For instance, there is a 0 in the location row two, column four because the two spies B and D cannot communicate directly. There are four locations in the matrix that are not covered by what we've said so far. These are the locations corresponding to only one spy (that is, a spy and him/herself):

row 1, column 1: spy A row 3, column 3: spy C
row 2, column 2: spy B row 4, column 4: spy D

By definition, we use a 0 in those locations.

Looking at Figure 1, we see that spy D cannot send a message directly to spy A. However, spy D can get a message to A by using relays or *intermediaries.* For instance, using one intermediary, the message from D to A can be routed

$$D \rightarrow C \rightarrow A$$

Or using two intermediaries, the message can be routed

$$D \rightarrow C \rightarrow B \rightarrow A$$

One could even have three intermediaries (perhaps to confuse the enemy) with the route

$$D \rightarrow C \rightarrow B \rightarrow C \rightarrow A$$

We now display and explain a four-by-four matrix that shows how many ways messages can be sent from one spy to another using only one intermediary. See Figure 3.

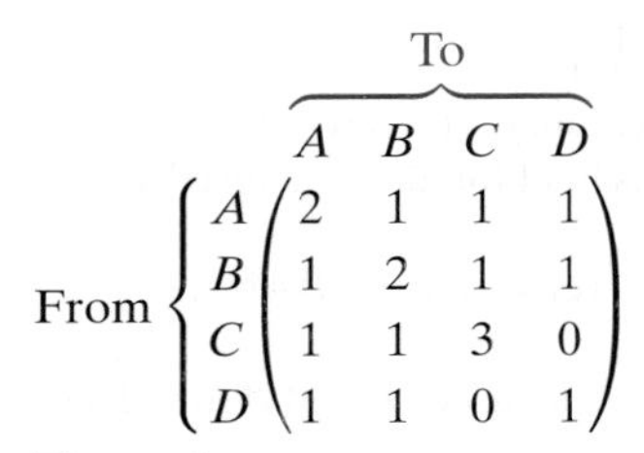

Figure 3
The one-intermediary matrix based on the spy network in Figure 1

As an example of how to interpret the entries in the one-intermediary matrix in Figure 3, consider the 1 that appears in the second row, first column. This tells us that there is exactly one way to send a message from B to A using an intermediary. You can check that this is the case by looking at Figure 1; the route is $B \rightarrow C \rightarrow A$. As another example, consider the 2 in the first row, first column. This tells us that there are two ways that A can send a message to A through one intermediary. As Figure 1 shows, the two routes are $A \rightarrow B \rightarrow A$ and $A \rightarrow C \rightarrow A$. (One possible reason why spy A might want to send a message to him/herself through an intermediary would be to test the integrity of that intermediary.) As a final example in interpreting the matrix in Figure 3, consider the 0 in the bottom row. This says that there is no way for D to send a message to C using exactly one intermediary. (Check this in Figure 1.)

There is a surprising connection between the communication matrix $\mathcal{M}$ in Figure 2 and the one-intermediary matrix in Figure 3. If you compute the matrix product of $\mathcal{M}$ with itself (this is abbreviated $\mathcal{M}^2$), you get the one-intermediary matrix in Figure 3. You should verify this now for yourself. That is, either by hand or with a graphing utility, compute $\mathcal{M}^2$ and check that the product is the same as the matrix in Figure 3. It can be shown that this situation holds in general.

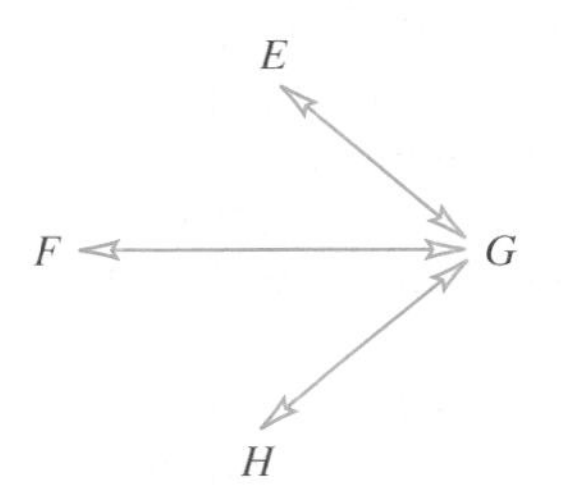

Figure 4

EXERCISES

1. Consider the communication links shown in Figure 4.

(a) Specify the communication matrix $\mathcal{M}$ for this system.

(b) By referring to Figure 4, work out the one-intermediary matrix for this system. Then check your answer by computing $\mathcal{M}^2$ and comparing the results.

2. Return to the spy network that we considered in Figure 1.

$$\begin{array}{c} \text{To} \\ \begin{array}{cc} & \begin{array}{cccc} A & B & C & D \end{array} \\ \text{From} \begin{array}{c} A \\ B \\ C \\ D \end{array} & \begin{pmatrix} 2 & ? & ? & ? \\ ? & ? & ? & ? \\ ? & ? & ? & ? \\ ? & 1 & 3 & ? \end{pmatrix} \end{array} \end{array}$$

Figure 5
Two-intermediary matrix for the spy network in Figure 1

(a) Work out the entries for the following *two-intermediary* matrix in Figure 5. As examples, three entries are already filled in. The 2 in the first row, first column indicates that there are two ways to send a message from A to A using two intermediaries. The routes are $A \rightarrow B \rightarrow C \rightarrow A$ and $A \rightarrow C \rightarrow B \rightarrow A$. As another example, the 1 in the fourth row, second column indicates that there is only one way to send a message from D to B using two intermediaries. As you can verify in Figure 1, the route is $D \rightarrow C \rightarrow A \rightarrow B$. Finally, consider the 3 in the fourth row, third column. This indicates that there are three ways to send a message from D to C

using two intermediaries. The three routes are $D \to C \to A \to C$, $D \to C \to B \to C$, and $D \to C \to D \to C$. In these routes from D to C, note that D and C themselves can be used as intermediaries.

(b) Let $\mathcal{M}$ be the communication matrix given in Figure 2. It can be shown that the matrix $\mathcal{M}^3$ $[=\mathcal{M}(\mathcal{M}^2)]$ equals the two-intermediary matrix for this system. Compute $\mathcal{M}^3$, either by hand or with a graphing utility, and then use the result to check the entries in your answer for part (a).

3. Consider a network of five spies, A, B, C, D, and E with the communication matrix $\mathcal{M}$ shown in Figure 6.

(a) Draw a diagram, of the type shown in Figure 1, showing the communication links.

(b) Using the diagram from part (a): How many one-intermediary routes can you find for sending a message from D to A? How many two-intermediary routes can you find for sending a message from D to A?

(c) Use a graphing utility to compute $\mathcal{M}^2$ and $\mathcal{M}^3$, and thereby check to see whether you missed any routes in part (b).

$$\begin{array}{c} \\ A \\ B \\ C \\ D \\ E \end{array}\begin{array}{c} \begin{array}{ccccc} A & B & C & D & E \end{array} \\ \begin{pmatrix} 0 & 1 & 1 & 1 & 1 \\ 1 & 0 & 1 & 0 & 0 \\ 1 & 1 & 0 & 1 & 0 \\ 1 & 0 & 1 & 0 & 1 \\ 1 & 0 & 0 & 1 & 0 \end{pmatrix} \end{array}$$

Figure 6
Communication matrix $\mathcal{M}$ for a network with five spies

Concluding Remarks

1. As you can see after working this exercise, for larger networks, it becomes preferable to work from the matrices rather than a diagram. For instance, what if there were, say, 50 spies, rather than only 4 or 5?
2. The mathematics in this project can be applied in many other areas. For instance, instead of communication between spies, the subject could be air travel between cities. Then, for example, a one-intermediary route from New York to San Francisco would correspond to a flight between those two cities with one stopover, for example, in Chicago.

10.4 THE INVERSE OF A SQUARE MATRIX

... when the chips are down we close the office door and compute with matrices like fury. —Paul Halmos (quoted here out of context), *Celebrating 50 Years of Mathematics,* J. H. Ewing et al. (ed.) (New York: Springer-Verlag, 1995)

We have noted Cayley's work in analytic geometry, especially in connection with the use of determinants; but [Arthur] *Cayley* [1821–1895] *also was one of the first men to study matrices, another instance of the British concern for form and structure in algebra.* —Carl B. Boyer in *A History of Mathematics,* 2nd ed., revised by Uta C. Merzback (New York: John Wiley and Sons, 1991)

A matrix that has the same number of rows and columns is called a **square matrix.** So, two examples of square matrices are

$$A = \begin{pmatrix} 1 & 2 \\ 3 & 4 \end{pmatrix} \quad \text{and} \quad B = \begin{pmatrix} -5 & 6 & 7 \\ \frac{1}{2} & 0 & 1 \\ 8 & 4 & -3 \end{pmatrix}$$

The matrix A is said to be a square matrix of **order two** (or, more simply, a 2×2 matrix); B is a square matrix of **order three** (that is, a 3×3 matrix). We will first present the concepts and techniques of this section in terms of square matrices of order two. After that, we'll show how the ideas carry over to larger square matrices.

We begin by defining a special matrix I_2:

$$I_2 = \begin{pmatrix} 1 & 0 \\ 0 & 1 \end{pmatrix}$$

The matrix I_2 plays the same role in the multiplication of 2×2 matrices as does the number 1 in the multiplication of real numbers. Specifically, I_2 has the following property: For every 2×2 matrix $A = \begin{pmatrix} a & b \\ c & d \end{pmatrix}$ we have (as you can easily verify)

$$\begin{pmatrix} a & b \\ c & d \end{pmatrix}\begin{pmatrix} 1 & 0 \\ 0 & 1 \end{pmatrix} = \begin{pmatrix} a & b \\ c & d \end{pmatrix} \quad \text{and} \quad \begin{pmatrix} 1 & 0 \\ 0 & 1 \end{pmatrix}\begin{pmatrix} a & b \\ c & d \end{pmatrix} = \begin{pmatrix} a & b \\ c & d \end{pmatrix}$$

In other words,

$$AI_2 = A \qquad \text{and} \qquad I_2A = A$$

The matrix I_2 is called the **multiplicative identity** or, more simply, the **identity matrix** for square matrices of order two.

If we have two real numbers a and b such that $ab = 1$, then we say that a and b are (multiplicative) inverses. In the box that follows, we apply this terminology to matrices. Note that in the case of inverse matrices, the multiplication *is* commutative. That is, AB and BA yield the same result (namely, I_2).

DEFINITION **The Inverse of a 2 × 2 Matrix**

Given a 2×2 matrix A, if there is a 2×2 matrix B such that

$$AB = I_2 \qquad \text{and} \qquad BA = I_2$$

then we say that A is **invertible** and B is an **inverse** of A.

A noninvertible matrix is also called a **singular** matrix, and an invertible matrix is also called a **nonsingular** matrix A. It can be shown that if a 2×2 matrix A is invertible then its inverse is unique. So we can talk about *the* inverse of A, denoted by A^{-1}.

EXAMPLE 1 **Finding a matrix inverse from first principles**

Find the inverse of $A = \begin{pmatrix} 1 & 2 \\ 3 & 4 \end{pmatrix}$.

SOLUTION

We need to find numbers a, b, c, and d such that the following two matrix equations are valid:

$$\begin{pmatrix} 1 & 2 \\ 3 & 4 \end{pmatrix}\begin{pmatrix} a & b \\ c & d \end{pmatrix} = \begin{pmatrix} 1 & 0 \\ 0 & 1 \end{pmatrix} \tag{1}$$

$$\begin{pmatrix} a & b \\ c & d \end{pmatrix}\begin{pmatrix} 1 & 2 \\ 3 & 4 \end{pmatrix} = \begin{pmatrix} 1 & 0 \\ 0 & 1 \end{pmatrix} \tag{2}$$

From equation (1) we have

$$\begin{pmatrix} a + 2c & b + 2d \\ 3a + 4c & 3b + 4d \end{pmatrix} = \begin{pmatrix} 1 & 0 \\ 0 & 1 \end{pmatrix}$$

For this last equation to be valid, the corresponding entries in the two matrices must be equal. (Why?) Consequently, we obtain four equations, two involving a and c and two involving b and d:

$$\begin{cases} a + 2c = 1 \\ 3a + 4c = 0 \end{cases} \qquad \begin{cases} b + 2d = 0 \\ 3b + 4d = 1 \end{cases} \tag{3}$$

The first system in (3) can be solved by using elementary row operations to transform the augmented matrix

$$\left(\begin{array}{cc|c} 1 & 2 & 1 \\ 3 & 4 & 0 \end{array}\right) \qquad \text{to} \qquad \left(\begin{array}{cc|c} 1 & 0 & -2 \\ 0 & 1 & \frac{3}{2} \end{array}\right)$$

in which case $a = -2$ and $c = 3/2$.

Similarly, the second system in (3) can be solved by using exactly the same elementary row operations to transform the augmented matrix

$$\left(\begin{array}{cc|c} 1 & 2 & 0 \\ 3 & 4 & 1 \end{array}\right) \quad \text{to} \quad \left(\begin{array}{cc|c} 1 & 0 & 1 \\ 0 & 1 & -\frac{1}{2} \end{array}\right)$$

in which case $b = 1$ and $d = -1/2$. Exercise 42(a) asks you to verify these claims. Furthermore (as you can check), these same values are obtained for a, b, c and d if we begin with matrix equation (2) rather than (1). Thus the required inverse matrix is

$$\begin{pmatrix} -2 & 1 \\ \frac{3}{2} & -\frac{1}{2} \end{pmatrix}$$

Exercise 42(b) at the end of this section asks you to carry out the matrix multiplication to confirm that we indeed have

$$\begin{aligned} \begin{pmatrix} 1 & 2 \\ 3 & 4 \end{pmatrix}\begin{pmatrix} -2 & 1 \\ \frac{3}{2} & -\frac{1}{2} \end{pmatrix} &= \begin{pmatrix} -2 & 1 \\ \frac{3}{2} & -\frac{1}{2} \end{pmatrix}\begin{pmatrix} 1 & 2 \\ 3 & 4 \end{pmatrix} \\ &= \begin{pmatrix} 1 & 0 \\ 0 & 1 \end{pmatrix} \end{aligned}$$

The work in Example 1 can be done more efficiently by augmenting the identity matrix I_2 to A to get

$$\left(\begin{array}{cc|cc} 1 & 2 & 1 & 0 \\ 3 & 4 & 0 & 1 \end{array}\right)$$

and using elementary row operations to transform this augmented matrix to

$$\left(\begin{array}{cc|cc} 1 & 0 & -2 & 1 \\ 0 & 1 & \frac{3}{2} & -\frac{1}{2} \end{array}\right)$$

In general, we use elementary row operations to transform

$$(A \mid I_2) \quad \text{to} \quad (I_2 \mid B)$$

If this is possible then A is invertible and $B = A^{-1}$. If this is not possible, then A is not invertible.

Let's apply this method to determine the inverse, if it exists, of the matrix

$$A = \begin{pmatrix} 6 & 2 \\ 13 & 4 \end{pmatrix}$$

The calculations run as follows:

$$\begin{aligned} \left(\begin{array}{cc|cc} 6 & 2 & 1 & 0 \\ 13 & 4 & 0 & 1 \end{array}\right) &\xrightarrow{-2R_1 + R_2} \left(\begin{array}{cc|cc} 6 & 2 & 1 & 0 \\ 1 & 0 & -2 & 1 \end{array}\right) \\ &\xrightarrow{R_1 \leftrightarrow R_2} \left(\begin{array}{cc|cc} 1 & 0 & -2 & 1 \\ 6 & 2 & 1 & 0 \end{array}\right) \\ &\xrightarrow{-6R_1 + R_2} \left(\begin{array}{cc|cc} 1 & 0 & -2 & 1 \\ 0 & 2 & 13 & -6 \end{array}\right) \\ &\xrightarrow{\frac{1}{2}R_2} \left(\begin{array}{cc|cc} 1 & 0 & -2 & 1 \\ 0 & 1 & \frac{13}{2} & -3 \end{array}\right) \end{aligned}$$

The inverse matrix can now be read off:

$$A^{-1} = \begin{pmatrix} -2 & 1 \\ \frac{13}{2} & -3 \end{pmatrix}$$

(You should verify for yourself that we indeed have $AA^{-1} = I_2$ and $A^{-1}A = I_2$.)

Inverse matrices can be used to solve certain systems of equations in which the number of unknowns is the same as the number of equations. Before explaining how this works, we describe how a system of equations can be written in matrix form. Consider the system

$$\begin{cases} x + 2y = 8 \\ 3x + 4y = 6 \end{cases} \tag{4}$$

As defined in the previous section, the coefficient matrix A for this system is

$$A = \begin{pmatrix} 1 & 2 \\ 3 & 4 \end{pmatrix}$$

Now we define two matrices X, the matrix of unknowns, and B, the matrix of right-hand sides:

$$X = \begin{pmatrix} x \\ y \end{pmatrix} \qquad B = \begin{pmatrix} 8 \\ 6 \end{pmatrix}$$

Then system (4) can be written as a single matrix equation:

$$AX = B \tag{5}$$

To see why this is so, we expand equation (5) to obtain

$$\begin{pmatrix} 1 & 2 \\ 3 & 4 \end{pmatrix}\begin{pmatrix} x \\ y \end{pmatrix} = \begin{pmatrix} 8 \\ 6 \end{pmatrix} \qquad \text{using the definitions of } A, X, \text{ and } B$$

$$\begin{pmatrix} x + 2y \\ 3x + 4y \end{pmatrix} = \begin{pmatrix} 8 \\ 6 \end{pmatrix} \qquad \text{carrying out the matrix multiplication}$$

By equating the corresponding entries of the matrices in this last equation, we obtain $x + 2y = 8$ and $3x + 4y = 6$, as given initially in system (4).

EXAMPLE 2 Using matrix algebra to solve a system

Use an inverse matrix to solve the system

$$\begin{cases} x + 2y = 8 \\ 3x + 4y = 6 \end{cases}$$

SOLUTION

As was explained just prior to this example, the matrix form for this system is

$$AX = B \tag{6}$$

where

$$A = \begin{pmatrix} 1 & 2 \\ 3 & 4 \end{pmatrix} \qquad X = \begin{pmatrix} x \\ y \end{pmatrix} \qquad B = \begin{pmatrix} 8 \\ 6 \end{pmatrix}$$

From Example 1 we know that

$$A^{-1} = \begin{pmatrix} -2 & 1 \\ \frac{3}{2} & -\frac{1}{2} \end{pmatrix}$$

Now we multiply both sides of equation (6) by A^{-1} to obtain

$$\begin{aligned} A^{-1}(AX) &= A^{-1}B \\ (A^{-1}A)X &= A^{-1}B \qquad \text{Matrix multiplication is associative. (See p. 768, Ex. 51.)} \\ I_2X &= A^{-1}B \\ X &= A^{-1}B \quad \text{(Why?)} \end{aligned}$$

Substituting the actual matrices into this last equation, we have

$$\begin{pmatrix} x \\ y \end{pmatrix} = \begin{pmatrix} -2 & 1 \\ \frac{3}{2} & -\frac{1}{2} \end{pmatrix} \begin{pmatrix} 8 \\ 6 \end{pmatrix} = \begin{pmatrix} -10 \\ 9 \end{pmatrix}$$

Therefore $x = -10$ and $y = 9$, as required.

All of the ideas we have discussed for square matrices of order two can be carried over directly to larger square matrices. It is easy to check that the following matrices, I_3 and I_4, are the multiplicative identities for 3×3 and 4×4 matrices, respectively:

$$I_3 = \begin{pmatrix} 1 & 0 & 0 \\ 0 & 1 & 0 \\ 0 & 0 & 1 \end{pmatrix} \qquad I_4 = \begin{pmatrix} 1 & 0 & 0 & 0 \\ 0 & 1 & 0 & 0 \\ 0 & 0 & 1 & 0 \\ 0 & 0 & 0 & 1 \end{pmatrix}$$

These identity matrices are described by saying that they have ones down the **main diagonal** and zeros everywhere else. (Larger identity matrices can be defined by following this same pattern.) Sometimes, when it is clear from the context or as a matter of convenience, we'll omit the subscript and denote the appropriately sized identity matrix simply by I. (This is done in the box that follows.)

PROPERTY SUMMARY The Inverse of a Square Matrix

1. Suppose that A and B are square matrices of the same size, and let I denote the identity matrix of that size. Then A and B are said to be **inverses** of one another provided

$$AB = I \quad \text{and} \quad BA = I$$

2. It can be shown that every square matrix has at most one inverse.
3. A square matrix that has an inverse is said to be **invertible** (or **nonsingular**). A square matrix that does not have an inverse is called **noninvertible** (or **singular**).
4. If the matrix A is **invertible,** then the inverse of A is denoted by A^{-1}. In this case, we have

$$AA^{-1} = I \quad \text{and} \quad A^{-1}A = I$$

EXAMPLE 3 Using elementary row operations to compute a matrix inverse

Let

$$A = \begin{pmatrix} 5 & 0 & 2 \\ 2 & 2 & 1 \\ -3 & 1 & -1 \end{pmatrix}$$

Use the elementary row operations to compute A^{-1}, if it exists.

SOLUTION

Following the method that we described for the 2×2 case, we first write down the augmented matrix $(A \mid I_3)$:

$$\left(\begin{array}{rrr|rrr} 5 & 0 & 2 & 1 & 0 & 0 \\ 2 & 2 & 1 & 0 & 1 & 0 \\ -3 & 1 & -1 & 0 & 0 & 1 \end{array}\right)$$

Now we carry out the elementary row operations, trying to obtain I_3 to the left of the dashed line. We have

$$\left(\begin{array}{rrr|rrr} 5 & 0 & 2 & 1 & 0 & 0 \\ 2 & 2 & 1 & 0 & 1 & 0 \\ -3 & 1 & -1 & 0 & 0 & 1 \end{array}\right) \xrightarrow{-2R_2+R_1} \left(\begin{array}{rrr|rrr} 1 & -4 & 0 & 1 & -2 & 0 \\ 2 & 2 & 1 & 0 & 1 & 0 \\ -3 & 1 & -1 & 0 & 0 & 1 \end{array}\right)$$

$$\xrightarrow[3R_1+R_3]{-2R_1+R_2} \left(\begin{array}{rrr|rrr} 1 & -4 & 0 & 1 & -2 & 0 \\ 0 & 10 & 1 & -2 & 5 & 0 \\ 0 & -11 & -1 & 3 & -6 & 1 \end{array}\right) \xrightarrow{1R_3+R_2} \left(\begin{array}{rrr|rrr} 1 & -4 & 0 & 1 & -2 & 0 \\ 0 & -1 & 0 & 1 & -1 & 1 \\ 0 & -11 & -1 & 3 & -6 & 1 \end{array}\right)$$

$$\xrightarrow{(-1)R_2} \left(\begin{array}{rrr|rrr} 1 & -4 & 0 & 1 & -2 & 0 \\ 0 & 1 & 0 & -1 & 1 & -1 \\ 0 & -11 & -1 & 3 & -6 & 1 \end{array}\right) \xrightarrow[11R_2+R_3]{4R_2+R_1} \left(\begin{array}{rrr|rrr} 1 & 0 & 0 & -3 & 2 & -4 \\ 0 & 1 & 0 & -1 & 1 & -1 \\ 0 & 0 & -1 & -8 & 5 & -10 \end{array}\right)$$

$$\xrightarrow{(-1)R_3} \left(\begin{array}{rrr|rrr} 1 & 0 & 0 & -3 & 2 & -4 \\ 0 & 1 & 0 & -1 & 1 & -1 \\ 0 & 0 & 1 & 8 & -5 & 10 \end{array}\right)$$

The required inverse is therefore

$$A^{-1} = \begin{pmatrix} -3 & 2 & -4 \\ -1 & 1 & -1 \\ 8 & -5 & 10 \end{pmatrix}$$

Graphing utilities can be used for matrix multiplication (as we saw in the previous section), they can also be used to compute matrix inverses. As in Section 10.3, we will briefly cite two examples, leaving the details for your user's manual. We'll use the matrix A given in Example 3. Figure 1 shows the two stages in computing the inverse on a Texas Instruments 83+ graphing calculator. After specifying the dimension and elements of the matrix A, it can be displayed as in Figure 1(a). Figure 1(b) then shows the result of computing A^{-1}. As you can see, it indeed agrees with the result in Example 3.

Figure 2 shows the resulting calculation as carried in a Microsoft *Excel* spreadsheet. The elements of the matrix A have been entered in cells B1 through D3. The inverse matrix is obtained by means of the function MINVERSE.

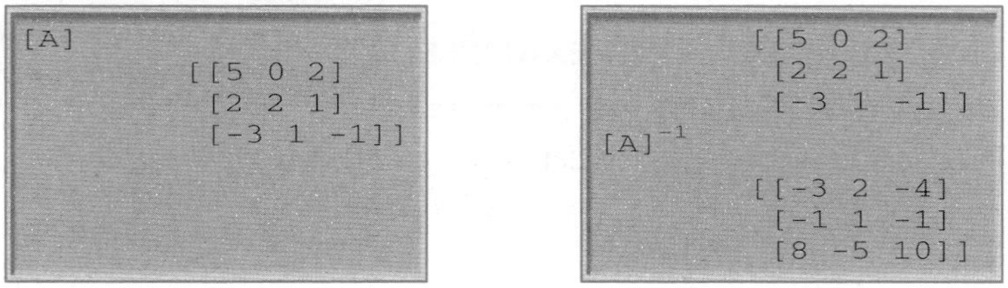

```
[A]
        [[5 0 2]
         [2 2 1]
         [-3 1 -1]]
```

(a)

```
        [[5 0 2]
         [2 2 1]
         [-3 1 -1]]
[A]^-1
        [[-3 2 -4]
         [-1 1 -1]
         [8 -5 10]]
```

(b)

Figure 1
Computing the inverse of the matrix A (from Example 3) on a Texas Instruments TI-83+ graphing calculator

	A	B	C	D
1		5		2
2	A=	2	2	1
3		–3	1	–1
4				
5		–3	2	–4
6	A^{-1} =	–1	1	–1
7		8	–5	10
8				

Figure 2
Computing the inverse of the matrix A (from Example 3) using Microsoft *Excel.* After typing in the entries for A in cells B1 through D3, the array formula `{=MINVERSE(B1:D3)}` is used to compute and display the inverse in cells B5 through D7.

We conclude this section with a description of how matrices can be used in sending and receiving coded messages. Any square matrix that has an inverse can be used to encode the message. Let's suppose that the *encoding matrix* is $\begin{pmatrix} 3 & 8 \\ 4 & 11 \end{pmatrix}$ and that the message to be encoded is *HIDE IT.* We begin by letting the letters of the alphabet correspond to the integers 1–26 in the natural way shown in Table 1. Additionally, as indicated in the table, the integer 0 is used either to indicate a space between words or as a placeholder at the end of the message.

TABLE 1

A	B	C	D	E	F	G	H	I	J	K	L	M
1	2	3	4	5	6	7	8	9	10	11	12	13
14	15	16	17	18	19	20	21	22	23	24	25	26
N	O	P	Q	R	S	T	U	V	W	X	Y	Z

Conventions regarding zero: The integer 0 is used either to indicate a space between words or as a placeholder at the end of the message.

Using the correspondences in Table 1, we have

H	I	D	E		I	T
8	9	4	5	0	9	20

At this point, the message has been converted to a list of seven integers: 8, 9, 4, 5, 0, 9, 20. From this list, we will create some 2×1 matrices. (If the encoding matrix had been a square matrix of order three, rather than order two, then we would create 3×1 matrices; you'll see this in Example 4.) To create the 2×1 matrices, we first pair up the integers in our list:

$$8, 9 \qquad 4, 5 \qquad 0, 9 \qquad 20, 0$$

The 0 at the end is inserted as a placeholder, because there was nothing to pair with 20. The four pairs are now written as one-column matrices:

$$\begin{pmatrix} 8 \\ 9 \end{pmatrix} \quad \begin{pmatrix} 4 \\ 5 \end{pmatrix} \quad \begin{pmatrix} 0 \\ 9 \end{pmatrix} \quad \begin{pmatrix} 20 \\ 0 \end{pmatrix}$$

The last step is to multiply each 2×1 matrix by the encoding matrix. As you should check for yourself, we obtain

$$\begin{pmatrix} 3 & 8 \\ 4 & 11 \end{pmatrix}\begin{pmatrix} 8 \\ 9 \end{pmatrix} = \begin{pmatrix} 96 \\ 131 \end{pmatrix}$$

$$\begin{pmatrix} 3 & 8 \\ 4 & 11 \end{pmatrix}\begin{pmatrix} 4 \\ 5 \end{pmatrix} = \begin{pmatrix} 52 \\ 71 \end{pmatrix}$$
$$\begin{pmatrix} 3 & 8 \\ 4 & 11 \end{pmatrix}\begin{pmatrix} 0 \\ 9 \end{pmatrix} = \begin{pmatrix} 72 \\ 99 \end{pmatrix}$$
$$\begin{pmatrix} 3 & 8 \\ 4 & 11 \end{pmatrix}\begin{pmatrix} 20 \\ 0 \end{pmatrix} = \begin{pmatrix} 60 \\ 80 \end{pmatrix}$$

The encoded message then consists of the eight integers that we have just calculated:

$$96, 131, 52, 71, 72, 99, 60, 80 \tag{7}$$

If you were to receive message (7) and you knew that the encoding matrix was $\begin{pmatrix} 3 & 8 \\ 4 & 11 \end{pmatrix}$, you could decode the message by undoing the previous steps as follows. Express message (7) as the four matrices

$$\begin{pmatrix} 96 \\ 131 \end{pmatrix} \quad \begin{pmatrix} 52 \\ 71 \end{pmatrix} \quad \begin{pmatrix} 72 \\ 99 \end{pmatrix} \quad \begin{pmatrix} 60 \\ 80 \end{pmatrix} \tag{8}$$

Next, compute $\begin{pmatrix} 3 & 8 \\ 4 & 11 \end{pmatrix}^{-1}$. As you can check, this turns out to be $\begin{pmatrix} 11 & -8 \\ -4 & 3 \end{pmatrix}$. Now multiply each of the matrices in (8) by $\begin{pmatrix} 11 & -8 \\ -4 & 3 \end{pmatrix}$. This yields

$$\begin{pmatrix} 11 & -8 \\ -4 & 3 \end{pmatrix}\begin{pmatrix} 96 \\ 131 \end{pmatrix} = \begin{pmatrix} 8 \\ 9 \end{pmatrix}$$
$$\begin{pmatrix} 11 & -8 \\ -4 & 3 \end{pmatrix}\begin{pmatrix} 52 \\ 71 \end{pmatrix} = \begin{pmatrix} 4 \\ 5 \end{pmatrix}$$
$$\begin{pmatrix} 11 & -8 \\ -4 & 3 \end{pmatrix}\begin{pmatrix} 72 \\ 99 \end{pmatrix} = \begin{pmatrix} 0 \\ 9 \end{pmatrix}$$
$$\begin{pmatrix} 11 & -8 \\ -4 & 3 \end{pmatrix}\begin{pmatrix} 60 \\ 80 \end{pmatrix} = \begin{pmatrix} 20 \\ 0 \end{pmatrix}$$

Use a calculator or graphing utility to check these results

This produces the list of eight integers 8, 9, 4, 5, 0, 9, 20, 0. Table 1 allows us to translate this list. As you can check, the result is the original message *HIDE IT.*

EXAMPLE 4 Encoding a message by means of a matrix

Encode the message *GO LATER* assuming that the encoding matrix is

$$\begin{pmatrix} 5 & 0 & 2 \\ 2 & 2 & 1 \\ -3 & 1 & -1 \end{pmatrix}$$

SOLUTION

Step 1 Convert the message to a list of integers. Using Table 1 and the convention regarding zero, the message *GO LATER* corresponds to the list 7, 15, 0, 12, 1, 20, 5, 18.

Step 2 Put the integers into one-column matrices. For a 3×3 encoding matrix we take the integers *three* at a time to obtain

$$\begin{pmatrix} 7 \\ 15 \\ 0 \end{pmatrix} \quad \begin{pmatrix} 12 \\ 1 \\ 20 \end{pmatrix} \quad \begin{pmatrix} 5 \\ 18 \\ 0 \end{pmatrix}$$

The zero in the matrix on the right is inserted as a placeholder.

Step 3 Multiply each matrix in Step 2 by the encoding matrix:

$$\begin{pmatrix} 5 & 0 & 2 \\ 2 & 2 & 1 \\ -3 & 1 & -1 \end{pmatrix} \begin{pmatrix} 7 \\ 15 \\ 0 \end{pmatrix} = \begin{pmatrix} 35 \\ 44 \\ -6 \end{pmatrix}$$

$$\begin{pmatrix} 5 & 0 & 2 \\ 2 & 2 & 1 \\ -3 & 1 & -1 \end{pmatrix} \begin{pmatrix} 12 \\ 1 \\ 20 \end{pmatrix} = \begin{pmatrix} 100 \\ 46 \\ -55 \end{pmatrix}$$

$$\begin{pmatrix} 5 & 0 & 2 \\ 2 & 2 & 1 \\ -3 & 1 & -1 \end{pmatrix} \begin{pmatrix} 5 \\ 18 \\ 0 \end{pmatrix} = \begin{pmatrix} 25 \\ 46 \\ 3 \end{pmatrix}$$

The encoded message is therefore 35, 44, −6, 100, 46, −55, 25, 46, 3.

EXAMPLE 5 Decoding a message using a matrix inverse

Check the result in Example 4 by decoding the message

$$35, 44, -6, 100, 46, -55, 25, 46, 3 \qquad (9)$$

The encoding matrix is $\begin{pmatrix} 5 & 0 & 2 \\ 2 & 2 & 1 \\ -3 & 1 & -1 \end{pmatrix}$.

SOLUTION

Step 1 Put the integers in (9) into one-column matrices. Since the encoding matrix is 3×3, each one-column matrix will have three rows. This produces

$$\begin{pmatrix} 35 \\ 44 \\ -6 \end{pmatrix} \quad \begin{pmatrix} 100 \\ 46 \\ -55 \end{pmatrix} \quad \begin{pmatrix} 25 \\ 46 \\ 3 \end{pmatrix}$$

Step 2 Compute the inverse matrix for the encoding matrix $\begin{pmatrix} 5 & 0 & 2 \\ 2 & 2 & 1 \\ -3 & 1 & -1 \end{pmatrix}$.

As we saw in Example 3, the inverse of this matrix is given by

$$\begin{pmatrix} 5 & 0 & 2 \\ 2 & 2 & 1 \\ -3 & 1 & -1 \end{pmatrix}^{-1} = \begin{pmatrix} -3 & 2 & -4 \\ -1 & 1 & -1 \\ 8 & -5 & 10 \end{pmatrix}$$

Step 3 Multiply each one-column matrix in Step 1 by the inverse matrix obtained in Step 2.

$$\begin{pmatrix} -3 & 2 & -4 \\ -1 & 1 & -1 \\ 8 & -5 & 10 \end{pmatrix}\begin{pmatrix} 35 \\ 44 \\ -6 \end{pmatrix} = \begin{pmatrix} 7 \\ 15 \\ 0 \end{pmatrix} \quad \begin{pmatrix} -3 & 2 & -4 \\ -1 & 1 & -1 \\ 8 & -5 & 10 \end{pmatrix}\begin{pmatrix} 100 \\ 46 \\ -55 \end{pmatrix} = \begin{pmatrix} 12 \\ 1 \\ 20 \end{pmatrix} \quad \begin{pmatrix} -3 & 2 & -4 \\ -1 & 1 & -1 \\ 8 & -5 & 10 \end{pmatrix}\begin{pmatrix} 25 \\ 46 \\ 3 \end{pmatrix} = \begin{pmatrix} 5 \\ 18 \\ 0 \end{pmatrix}$$

This gives us the list of nine integers

$$7, 15, 0, 12, 1, 20, 5, 18, 0$$

Table 1 can now be used to translate this list. As you can verify, the result is *GO LATER*.

EXERCISE SET 10.4

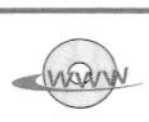

A

In Exercises 1–4 the matrices A, B, C, and D are defined as follows.

$$A = \begin{pmatrix} 4 & -1 \\ -5 & 2 \end{pmatrix} \qquad B = \begin{pmatrix} \frac{1}{2} & 5 \\ 3 & 1 \end{pmatrix}$$

$$C = \begin{pmatrix} 3 & 0 & -2 \\ 0 & 5 & 6 \\ 1 & 4 & -7 \end{pmatrix} \qquad D = \begin{pmatrix} 1 & 2 & 3 \\ 4 & 5 & 6 \\ 7 & 8 & 9 \end{pmatrix}$$

1. Compute AI_2 and I_2A to verify that $AI_2 = I_2A = A$.
2. Compute BI_2 and I_2B to verify that $BI_2 = I_2B = B$.
3. Compute CI_3 and I_3C to verify that $CI_3 = I_3C = C$.
4. Compute DI_3 and I_3D to verify that $DI_3 = I_3D = D$.

In Exercises 5–12, compute A^{-1}, if it exists, using the method of Example 1.

5. $A = \begin{pmatrix} 7 & 9 \\ 4 & 5 \end{pmatrix}$ **6.** $A = \begin{pmatrix} 3 & -8 \\ 2 & -5 \end{pmatrix}$

7. $A = \begin{pmatrix} -3 & 1 \\ 5 & 6 \end{pmatrix}$ **8.** $A = \begin{pmatrix} -4 & 0 \\ 9 & 3 \end{pmatrix}$

9. $A = \begin{pmatrix} -2 & 3 \\ -4 & 6 \end{pmatrix}$ **10.** $A = \begin{pmatrix} \frac{5}{3} & -2 \\ -\frac{2}{3} & 1 \end{pmatrix}$

11. $A = \begin{pmatrix} \frac{1}{3} & \frac{1}{3} \\ -\frac{1}{9} & \frac{2}{9} \end{pmatrix}$ **12.** $A = \begin{pmatrix} -3 & 7 \\ 12 & -28 \end{pmatrix}$

In Exercises 13–26, compute the inverse matrix, if it exists, using elementary row operations (as shown in Example 3).

13. $\begin{pmatrix} 2 & 1 \\ 3 & 2 \end{pmatrix}$ **14.** $\begin{pmatrix} -6 & 5 \\ 18 & -15 \end{pmatrix}$

15. $\begin{pmatrix} 0 & -11 \\ 1 & 6 \end{pmatrix}$ **16.** $\begin{pmatrix} -2 & 13 \\ -4 & 25 \end{pmatrix}$

17. $\begin{pmatrix} \frac{2}{3} & -\frac{1}{4} \\ -8 & 3 \end{pmatrix}$ **18.** $\begin{pmatrix} -\frac{2}{5} & \frac{1}{3} \\ -6 & 5 \end{pmatrix}$

19. $\begin{pmatrix} -5 & 4 & -3 \\ 10 & -7 & 6 \\ 8 & -6 & 5 \end{pmatrix}$ **20.** $\begin{pmatrix} 1 & 0 & -2 \\ -3 & -1 & 6 \\ 2 & 1 & -5 \end{pmatrix}$

21. $\begin{pmatrix} 1 & 2 & -1 \\ 0 & 3 & 0 \\ -4 & 0 & 5 \end{pmatrix}$ **22.** $\begin{pmatrix} 1 & -4 & -8 \\ 1 & 2 & 5 \\ 1 & 1 & 3 \end{pmatrix}$

23. $\begin{pmatrix} -7 & 5 & 3 \\ 3 & -2 & -2 \\ 3 & -2 & -1 \end{pmatrix}$ **24.** $\begin{pmatrix} 2 & -1 & -1 \\ 1 & 0 & -1 \\ -2 & 1 & 2 \end{pmatrix}$

25. $\begin{pmatrix} 1 & 2 & 3 \\ 4 & 5 & 6 \\ 7 & 8 & 9 \end{pmatrix}$ **26.** $\begin{pmatrix} 2 & 1 & 3 \\ 4 & 5 & -7 \\ 2 & 1 & 3 \end{pmatrix}$

Ⓖ *For Exercises 27–32 the matrices A, B, C, D, and E are defined as follows. In each exercise, use a graphing utility to carry out the indicated matrix operations. If an operation is undefined, state this.*

$$A = \begin{pmatrix} 2 & 3 \\ 4 & 5 \end{pmatrix} \qquad B = \begin{pmatrix} 3 & 2 & 6 \\ 1 & 1 & 2 \\ 2 & 2 & 5 \end{pmatrix} \qquad C = \begin{pmatrix} 9 & 4 & 4 \\ 2 & 2 & 1 \\ -3 & 1 & -1 \end{pmatrix}$$

$$D = \begin{pmatrix} -2 & 0 \\ 3 & 1 \\ 5 & 2 \end{pmatrix} \qquad E = \begin{pmatrix} 3 & -2 & -1 \\ 7 & -4 & 0 \end{pmatrix}$$

27. A^{-1} **28.** B^{-1} **29.** D^{-1}
30. $(CD)^{-1}$ **31.** $(DE)^{-1}$
32. $(BC)^{-1} - C^{-1}B^{-1}$ *Hint*: The order for carrying out the matrix operations is similar to that for real numbers. Thus the subtraction in this exercise is the last step.

33. If $A = \begin{pmatrix} 3 & 8 \\ 4 & 11 \end{pmatrix}$, then $A^{-1} = \begin{pmatrix} 11 & -8 \\ -4 & 3 \end{pmatrix}$. Use this fact and the method of Example 2 to solve the following systems.

(a) $\begin{cases} 3x + 8y = 5 \\ 4x + 11y = 7 \end{cases}$ **(b)** $\begin{cases} 3x + 8y = -12 \\ 4x + 11y = 0 \end{cases}$

34. If $A = \begin{pmatrix} 3 & -7 \\ 4 & -9 \end{pmatrix}$, then $A^{-1} = \begin{pmatrix} -9 & 7 \\ -4 & 3 \end{pmatrix}$. Use this fact and the method of Example 2 to solve the following systems.

(a) $\begin{cases} 3x - 7y = 30 \\ 4x - 9y = 39 \end{cases}$ **(b)** $\begin{cases} 3x - 7y = -45 \\ 4x - 9y = -71 \end{cases}$

35. The inverse of the matrix

$$A = \begin{pmatrix} 3 & 2 & 6 \\ 1 & 1 & 2 \\ 2 & 2 & 5 \end{pmatrix} \quad \text{is} \quad A^{-1} = \begin{pmatrix} 1 & 2 & -2 \\ -1 & 3 & 0 \\ 0 & -2 & 1 \end{pmatrix}$$

Use this fact to solve the following system.

$$\begin{cases} 3x + 2y + 6z = 28 \\ x + y + 2z = 9 \\ 2x + 2y + 5z = 22 \end{cases}$$

36. The inverse of the matrix

$$A = \begin{pmatrix} 1 & -1 & 1 \\ 2 & -3 & 2 \\ -4 & 6 & 1 \end{pmatrix} \quad \text{is} \quad A^{-1} = \begin{pmatrix} 3 & -\frac{7}{5} & -\frac{1}{5} \\ 2 & -1 & 0 \\ 0 & \frac{2}{5} & \frac{1}{5} \end{pmatrix}$$

Use this fact to solve the following system.

$$\begin{cases} x - y + z = 5 \\ 2x - 3y + 2z = -15 \\ -4x + 6y + z = 25 \end{cases}$$

G *In Exercises 37–41:*

(a) *Compute the inverse of the coefficient matrix for the system.*

(b) *Use the inverse matrix to solve the system. In cases in which the final answer involves decimals, round to three decimal places.*

37. $\begin{cases} x + 4y = 7 \\ 2x + 7y = 12 \end{cases}$

38. $\begin{cases} 8x - 5y = -13 \\ 3x + 4y = 48 \end{cases}$

39. $\begin{cases} x + 2y + 2z = 3 \\ 3x + y = -1 \\ x + y + z = 12 \end{cases}$

40. $\begin{cases} 5x - 2y - 2z = 15 \\ 3x + y = 4 \\ x + y + z = -4 \end{cases}$

41. $\begin{cases} 2x + 3y + z + w = 3 \\ 6x + 6y - 5z - 2w = 15 \\ x - y + z + \frac{1}{6}w = -3 \\ 4x + 9y + 3z + 2w = -3 \end{cases}$

42. (a) Solve the following two systems, and then check to see that your results agree with those given in Example 1.

$$\begin{cases} a + 2c = 1 \\ 3a + 4c = 0 \end{cases} \qquad \begin{cases} b + 2d = 0 \\ 3b + 4d = 1 \end{cases}$$

(b) At the end of Example 1, it is asserted that

$$\begin{pmatrix} 1 & 2 \\ 3 & 4 \end{pmatrix}\begin{pmatrix} -2 & 1 \\ \frac{3}{2} & -\frac{1}{2} \end{pmatrix} = \begin{pmatrix} -2 & 1 \\ \frac{3}{2} & -\frac{1}{2} \end{pmatrix}\begin{pmatrix} 1 & 2 \\ 3 & 4 \end{pmatrix} = \begin{pmatrix} 1 & 0 \\ 0 & 1 \end{pmatrix}$$

Carry out the indicated matrix multiplications to verify that these equations are valid.

43. Consider the matrix $\begin{pmatrix} 2 & 5 \\ 6 & 15 \end{pmatrix}$.

(a) Use the technique in Example 1 to show that the matrix does not have an inverse.

(b) Use the technique in Example 3 to show that the matrix does not have an inverse.

44. Follow Exercise 43 using $\begin{pmatrix} -5 & 2 \\ 20 & -8 \end{pmatrix}$.

G *In Exercises 45 and 46, encode each message using the matrix* $\begin{pmatrix} 9 & 8 \\ 8 & 7 \end{pmatrix}$. *For Exercises 47 and 48, encode using the matrix* $\begin{pmatrix} 1 & 1 & 0 \\ 0 & -1 & 2 \\ 1 & 0 & 1 \end{pmatrix}$.

45. TRY IT

46. DROP OFF

47. TURN NOW

48. WAIT

G *In Exercises 49 and 50, decode each message, assuming that the encoding matrix was* $\begin{pmatrix} 12 & 47 \\ -1 & -4 \end{pmatrix}$. *For Exercises 51 and 52, decode, assuming the encoding matrix was* $\begin{pmatrix} 2 & -1 & -1 \\ -2 & 1 & 2 \\ -1 & 1 & 1 \end{pmatrix}$.

49. 463, −39, 60, −5, 825, −70, 60, −5

50. 1038, −88, 274, −23, 379, −32, 156, −13

51. −3, 24, 22, −17, 17, 21

52. −7, 27, 21, −27, 52, 27, −4, 5, 4, 36, −36, −18

B

53. Use the elementary row operations (as in Example 3) to find the inverse of the following matrix.

$$\begin{pmatrix} 1 & 1 & 1 & 1 \\ 1 & 2 & 3 & 4 \\ 1 & 3 & 6 & 10 \\ 1 & 4 & 10 & 20 \end{pmatrix}$$

54. Let $A = \begin{pmatrix} 1 & -6 & 3 \\ 2 & -7 & 3 \\ 4 & -12 & 5 \end{pmatrix}$.

(a) Compute the matrix product AA. What do you observe?

(b) Use the result in part (a) to solve the following system.

$$\begin{cases} x - 6y + 3z = 19/2 \\ 2x - 7y + 3z = 11 \\ 4x - 12y + 5z = 19 \end{cases}$$

55. Let $A = \begin{pmatrix} 2 & 3 \\ 4 & 5 \end{pmatrix}$ and $B = \begin{pmatrix} 7 & 8 \\ 6 & 7 \end{pmatrix}$.

(a) Compute A^{-1}, B^{-1}, and $B^{-1}A^{-1}$.

(b) Compute $(AB)^{-1}$. What do you observe?

C

56. Let $A = \begin{pmatrix} a & b \\ c & d \end{pmatrix}$. Compute A^{-1}. (Assume that $ad - bc \neq 0$.)

57. (a) Use the result in Exercise 56 to find the inverse of the matrix $\begin{pmatrix} x & 1 + x \\ 1 - x & -x \end{pmatrix}$. What do you observe about the result?

(b) What's the inverse of the matrix $\begin{pmatrix} 11 & 12 \\ -10 & -11 \end{pmatrix}$? Of the matrix $\begin{pmatrix} \pi + 1 & \pi + 2 \\ -\pi & -1 - \pi \end{pmatrix}$?

PROJECT

THE LEONTIEF MODEL REVISITED

As background here, you need to be familiar with the project at the end of Section 10.2 on the Leontief input-output model. Additionally, you need to have studied Example 2 in the present section, where matrix methods are applied to solve a system of equations. If necessary, one or two groups can be assigned to work through or to review that material and then present it to the class at large. In this project, you'll use matrix methods to shorten the work in solving an input-output problem.

PART I: HOW TO SOLVE THE MATRIX EQUATION $AX + D = X$

In Example 2 of Section 10.4, we solved the matrix equation $AX = B$. Along the same lines, now we want to solve the matrix equation

$$AX + D = X \tag{1}$$

(We assume that A is a square matrix with an inverse and that each matrix has the appropriate numbers of rows and columns for the calculations to make sense.) We start by subtracting X from both sides of equation (1) to obtain

$$AX + D - X = \mathbf{0}$$ Recall from Section 10.3 that $\mathbf{0}$ denotes the zero matrix, in which every entry is the number 0.

$$AX - X = -D$$ subtracting D from both sides and using the property $\mathbf{0} - D = -D$

$$(A - I)X = -D$$ I denotes the identity matrix. We've used the fact that the distributive property is valid in matrix algebra just as it is in ordinary algebra.

To isolate X, assuming $A - I$ is invertible, multiply both sides of this last equation by $(A - I)^{-1}$. This yields

$$(A - I)^{-1}(A - I)X = (A - I)^{-1}(-D)$$
$$IX = (A - I)^{-1}(-D)$$
$$X = (A - I)^{-1}(-D) \qquad \text{using } IX = X \tag{2}$$

Equation (2) is the required solution. To see how this works in a simple case, suppose we want to solve equation (1), given that $A = \begin{pmatrix} 3 & 5 \\ 1 & 4 \end{pmatrix}$, $D = \begin{pmatrix} 10 \\ 20 \end{pmatrix}$, and $X = \begin{pmatrix} x \\ y \end{pmatrix}$. That is, the equation to be solved is

$$\begin{pmatrix} 3 & 5 \\ 1 & 4 \end{pmatrix}\begin{pmatrix} x \\ y \end{pmatrix} + \begin{pmatrix} 10 \\ 20 \end{pmatrix} = \begin{pmatrix} x \\ y \end{pmatrix} \tag{3}$$

Using equation (2), the solution is

$$X = (A - I)^{-1}(-D)$$
$$= \left[\begin{pmatrix} 3 & 5 \\ 1 & 4 \end{pmatrix} - \begin{pmatrix} 1 & 0 \\ 0 & 1 \end{pmatrix}\right]^{-1}\begin{pmatrix} -10 \\ -20 \end{pmatrix}$$
$$= \begin{pmatrix} 2 & 5 \\ 1 & 3 \end{pmatrix}^{-1}\begin{pmatrix} -10 \\ -20 \end{pmatrix} \qquad \text{carrying out the matrix subtraction}$$

$$= \begin{pmatrix} 3 & -5 \\ -1 & 2 \end{pmatrix}\begin{pmatrix} -10 \\ -20 \end{pmatrix} \quad \text{computing the inverse}$$

$$= \begin{pmatrix} 70 \\ -30 \end{pmatrix} \quad \text{carrying out the matrix multiplication}$$

Thus the solution of equation (3) is $X = \begin{pmatrix} 70 \\ -30 \end{pmatrix}$, and the values of the unknowns x and y are 70 and -30, respectively.

Practice Problems for Part I

1. Solve $AX + D = X$, given that $A = \begin{pmatrix} 11 & 9 \\ 6 & 5 \end{pmatrix}$, $D = \begin{pmatrix} 25 \\ 15 \end{pmatrix}$, and $X = \begin{pmatrix} x \\ y \end{pmatrix}$.
2. Solve $AX + D = X$, given that $A = \begin{pmatrix} 1 & -3 & 2 \\ -2 & 7 & -7 \\ 1 & -6 & 12 \end{pmatrix}$, $D = \begin{pmatrix} 20 \\ 50 \\ 70 \end{pmatrix}$, $X = \begin{pmatrix} x \\ y \\ z \end{pmatrix}$.

PART II: THE MATRIX EQUATION $AX + D = X$ APPLIED TO AN INPUT-OUTPUT PROBLEM

As our example, we'll use the same problem discussed in Part I of the project at the end of Section 10.2. For ease of reference, we restate the problem here. Given the input-output data in Table 1 and the external demands in Table 2, determine an appropriate production level for each sector so that both the internal and external demands on the economy are satisfied.

As in the solution in Section 10.2, we start by using x, y, and z to represent the required number of units of steel, coal, and electricity, respectively. Next, define three matrices, A, D, and X as follows:

$$A = \begin{pmatrix} 0.04 & 0.02 & 0.16 \\ 0.12 & 0 & 0.30 \\ 0.16 & 0.10 & 0.04 \end{pmatrix} \qquad D = \begin{pmatrix} 10 \\ 12 \\ 15 \end{pmatrix} \qquad X = \begin{pmatrix} x \\ y \\ z \end{pmatrix} \tag{4}$$

the input-output matrix, based on Table 1; the demand matrix based on Table 2; the matrix of unknowns

Your tasks now are to work through the following steps to solve this input-output problem using matrix methods.

1. Check that the matrix product AD is defined. What will be the dimension of the product AD?

TABLE 1 Input-Output Table (internal demands)

	Outputs		
	Steel	Coal	Electricity
Inputs			
Steel	0.04	0.02	0.16
Coal	0.12	0	0.30
Electricity	0.16	0.10	0.04

TABLE 2 External Demands

	External demands
Steel	10 units
Coal	12 units
Electricity	15 units

2. Explain why the matrix equation $AX + D = X$ holds. *Hint*: In the project for Section 10.2 look at the equation immediately preceding equation (2).
3. Use equation (2) from part I of this project to solve the equation $AX + D = X$. Make use of a graphing utility in carrying out the matrix operations, and round the answers for x, y, and z to two decimal places.

PART III: A SEVEN-SECTOR MODEL OF THE U.S. ECONOMY

The input-output examples above (and in the project for Section 10.2) involve two- or three-sector economies. By way of contrast, input-output tables for the U.S. economy (published by the U.S. Bureau of Economic Analysis) involve as many as 486 sectors. Table 3, which follows, is an *aggregated version* of one of those tables, in which the 486 sectors have been grouped into only 7 general categories or sectors. The names of the 7 sectors, abbreviated in Table 3, are spelled out in Table 4, which contains the external demands. Use a graphing utility to solve the input-output problem for this seven-sector economy. Round the final answers to the nearest integer. Then, given that each unit represents one million dollars, express the answers in dollars.

TABLE 3
Seven-Sector Input-Output Table for the 1967 U.S. Economy*

	Outputs						
	Ag	Min	Con	Mfg	T & T	Serv	Oth
Inputs							
Ag	.2939	.0000	.0025	.0516	.0009	.0081	.0203
Min	.0022	.0504	.0090	.0284	.0002	.0099	.0075
Con	.0096	.0229	.0003	.0042	.0085	.0277	.0916
Mfg	.1376	.0940	.3637	.3815	.0634	.0896	.1003
T & T	.0657	.0296	.1049	.0509	.0530	.0404	.0775
Serv	.0878	.1708	.0765	.0734	.1546	.1676	.1382
Oth	.0001	.0054	.0008	.0055	.0183	.0250	.0012

*Derived from more detailed tables published by the U.S. Bureau of Economic Analysis. Source for Tables 3 and 4: *Input-Output Analysis: Foundations and Extensions,* Ronald E. Miller and Peter D. Blair (Englewood Cliffs, N.J.: Prentice Hall, 1985)

TABLE 4 External Demands for 1967 U.S. Economy

External demands	
Agriculture	9300
Mining	1454
Construction	85,583
Manufacturing	278,797
Trade & transportation	141,468
Services	211,937
Others	2453

10.5 DETERMINANTS AND CRAMER'S RULE

The idea of the determinant . . . dates back essentially to Leibniz (1693), the Swiss mathematician Gabriel Cramer (1750), and Lagrange (1773); the name is due to Cauchy (1812). Y. Mikami has pointed out that the Japanese mathematician Seki Kōwa had the idea of a determinant sometime before 1683.—Dirk J. Struik in *A Concise History of Mathematics,* 4th ed. (New York: Dover Publications, 1987)

As you saw in the previous section, square matrices and their inverses can be used to solve certain systems of equations. In this section we will associate a number with each square matrix. This number is called the **determinant** of the matrix. As you'll see, this too has an application in solving systems of equations.

The determinant of a matrix A is denoted by det A or simply by replacing the parentheses of matrix notation with vertical lines. Thus, three examples of determinants are

$$\det\begin{pmatrix} 1 & 2 \\ 3 & 4 \end{pmatrix} \quad \text{or} \quad \begin{vmatrix} 1 & 2 \\ 3 & 4 \end{vmatrix}, \qquad \begin{vmatrix} 1 & 2 & 3 \\ 4 & 5 & 6 \\ 7 & 8 & 9 \end{vmatrix}, \qquad \begin{vmatrix} 3 & 7 & 8 & 9 \\ 5 & 6 & 4 & 3 \\ -9 & 9 & 0 & 1 \\ 1 & 3 & -2 & 1 \end{vmatrix}$$

A determinant with n rows and n columns is said to be an ***n*th-order determinant.** Therefore, the determinants we've just written are, respectively, second-, third-, and fourth-order determinants. As with matrices, we speak of the numbers in a determinant as its **entries.** We also number the rows and the columns of a determinant as we do with matrices. However, unlike matrices, each determinant has a numerical value. The value of a second-order determinant is defined as follows.

DEFINITION **2 × 2 Determinant**

$$\begin{vmatrix} a & b \\ c & d \end{vmatrix} = ad - bc$$

Table 1 illustrates how this definition is used to evaluate, or *expand,* a second-order determinant.

TABLE 1

Determinant	Value of determinant
$\begin{vmatrix} 3 & 7 \\ 5 & 10 \end{vmatrix}$	$\begin{vmatrix} 3 & 7 \\ 5 & 10 \end{vmatrix} = 3(10) - 7(5) = 30 - 35 = -5$
$\begin{vmatrix} 3 & -7 \\ 5 & 10 \end{vmatrix}$	$\begin{vmatrix} 3 & -7 \\ 5 & 10 \end{vmatrix} = 3(10) - (-7)(5) = 30 + 35 = 65$
$\begin{vmatrix} a & a^3 \\ 1 & a^2 \end{vmatrix}$	$\begin{vmatrix} a & a^3 \\ 1 & a^2 \end{vmatrix} = a(a^2) - a^3(1) = a^3 - a^3 = 0$

In general, the value of an nth-order determinant ($n > 2$) is defined in terms of certain determinants of order $n - 1$. For instance, the value of a third-order determinant is defined in terms of second-order determinants. We'll use the following example to introduce the necessary terminology here:

$$\begin{vmatrix} 8 & 3 & 5 \\ 2 & 4 & 6 \\ 9 & 1 & 7 \end{vmatrix}$$

Pick a given entry—for example, the entry 8 in the first row and first column—and imagine crossing out all entries occupying the same row and the same column as 8.

$$\begin{vmatrix} 8 & 3 & 5 \\ 2 & 4 & 6 \\ 9 & 1 & 7 \end{vmatrix}$$

Now we are left with the second-order determinant $\begin{vmatrix} 4 & 6 \\ 1 & 7 \end{vmatrix}$. This second-order determinant is called the **minor** of the entry 8. Similarly, to find the minor of the entry 6 in the original determinant, imagine crossing out all entries that occupy the same row and the same column as 6.

$$\begin{vmatrix} 8 & 3 & 5 \\ 2 & 4 & 6 \\ 9 & 1 & 7 \end{vmatrix}$$

We are left with the second-order determinant $\begin{vmatrix} 8 & 3 \\ 9 & 1 \end{vmatrix}$, which by definition is the minor of the entry 6. In the same manner, *the minor of any element is the determinant obtained by crossing out the entries occupying the same row and column as the given element.*

$$\begin{vmatrix} + & - & + \\ - & + & - \\ + & - & + \end{vmatrix}$$

Figure 1

Closely related to the minor of an entry in a determinant is the cofactor of that entry. The **cofactor** of an entry is defined as the minor multiplied by $+1$ or -1, according to the scheme displayed in Figure 1.

After looking at an example, we'll give a more formal rule for computing cofactors, one that will not rely on a figure and that will also apply to larger determinants.

EXAMPLE 1 Computing a minor and a cofactor

Consider the determinant

$$\begin{vmatrix} 1 & 2 & 3 \\ 4 & 5 & 6 \\ 7 & 8 & 9 \end{vmatrix}$$

Compute the minor and the cofactor of the entry 4.

SOLUTION

By definition, we have

$$\text{minor of } 4 = \begin{vmatrix} 2 & 3 \\ 8 & 9 \end{vmatrix} = 18 - 24 = -6$$

Thus the minor of the entry 4 is -6. To compute the cofactor of 4, we first notice that 4 is located in the second row and first column of the given determinant. On checking the corresponding position in Figure 1, we see a negative sign. Therefore we have, by definition,

$$\begin{aligned} \text{cofactor of } 4 &= (-1)(\text{minor of } 4) \\ &= (-1)(-6) = 6 \end{aligned}$$

The cofactor of 4 is therefore 6.

The following rule tells us how cofactors can be computed without relying on Figure 1. For reference we also restate the definition of a minor. (You should verify for yourself that this rule yields results that are consistent with Figure 1.)

DEFINITION Minors and Cofactors

The **minor** of an entry b in a determinant is the determinant formed by suppressing the entries in the row and in the column in which b appears.

Suppose that the entry b is in the ith row and the jth column. Then the **cofactor** of b is given by the expression

$$(-1)^{i+j}(\text{minor of } b)$$

We are now prepared to state the definition that tells us how to evaluate a third-order determinant. Actually, as you'll see later, the definition is quite general and may be applied to determinants of any size.

DEFINITION The Value of a Determinant

Multiply each entry in the first row of the determinant by its cofactor and then add the results. The value of the determinant is defined to be this sum.

To see how this definition is used, let's evaluate the determinant

$$\begin{vmatrix} 1 & 2 & 3 \\ 4 & 5 & 6 \\ 7 & 8 & 9 \end{vmatrix}$$

The definition tells us to multiply each entry in the first row by its cofactor and then add the results. Carrying out this procedure, we have

$$\begin{vmatrix} 1 & 2 & 3 \\ 4 & 5 & 6 \\ 7 & 8 & 9 \end{vmatrix} = 1\begin{vmatrix} 5 & 6 \\ 8 & 9 \end{vmatrix} - 2\begin{vmatrix} 4 & 6 \\ 7 & 9 \end{vmatrix} + 3\begin{vmatrix} 4 & 5 \\ 7 & 8 \end{vmatrix}$$

$$= 1(45 - 48) - 2(36 - 42) + 3(32 - 35)$$

$$= 0 \qquad \text{(Check the arithmetic!)}$$

So the value of this particular determinant is zero. The procedure that we've used here is referred to as **expanding the determinant along its first row.** The following theorem (stated here without proof) tells us that the value of a determinant can be obtained by expanding along any row or column; the result is the same in all cases.

THEOREM

Select any row or any column in a determinant and multiply each element in that row or column by its cofactor. Then add the results. The number obtained will be the value of the determinant. (In other words, the number that is obtained will be the same as that obtained by expanding the determinant along its first row.)

According to this theorem, we could have evaluated the determinant

$$\begin{vmatrix} 1 & 2 & 3 \\ 4 & 5 & 6 \\ 7 & 8 & 9 \end{vmatrix}$$

by expanding it along any row or any column. Let's expand it along the second column and check to see that the result agrees with the value we obtained earlier. Expanding along the second column, we have

$$\begin{aligned} \begin{vmatrix} 1 & 2 & 3 \\ 4 & 5 & 6 \\ 7 & 8 & 9 \end{vmatrix} &= -2\begin{vmatrix} 4 & 6 \\ 7 & 9 \end{vmatrix} + 5\begin{vmatrix} 1 & 3 \\ 7 & 9 \end{vmatrix} - 8\begin{vmatrix} 1 & 3 \\ 4 & 6 \end{vmatrix} \\ &= -2(36 - 42) + 5(9 - 21) - 8(6 - 12) \\ &= -2(-6) + 5(-12) - 8(-6) \\ &= 12 - 60 + 48 = 0 \qquad \text{as obtained previously} \end{aligned}$$

We would have obtained the same result had we chosen to begin with any other row or column. (Exercise 13 at the end of this section asks you to verify this.)

There are three basic rules that make it easier to evaluate determinants. These are summarized in the box that follows. (Suggestions for proving these can be found in the exercises.)

PROPERTY SUMMARY Rules for Manipulating Determinants

Examples

1. If each entry in a given row is multiplied by the constant k, then the value of the determinant is multiplied by k. This is also true for columns.

$$10\begin{vmatrix} 1 & 3 & 4 \\ 1 & 2 & 3 \\ 4 & 5 & 6 \end{vmatrix} = \begin{vmatrix} 10 & 30 & 40 \\ 1 & 2 & 3 \\ 4 & 5 & 6 \end{vmatrix}$$

$$k\begin{vmatrix} a & b & c \\ d & e & f \\ g & h & i \end{vmatrix} = \begin{vmatrix} a & kb & c \\ d & ke & f \\ g & kh & i \end{vmatrix}$$

2. If a multiple of one row is added to another row, the value of the determinant is not changed. This also applies to columns.

$$\begin{vmatrix} a & b & c \\ d & e & f \\ g & h & i \end{vmatrix} = \begin{vmatrix} a & b & c \\ d + ka & e + kb & f + kc \\ g & h & i \end{vmatrix}$$

3. If two rows are interchanged, then the value of the determinant is multiplied by -1. This also applies to columns.

$$\begin{vmatrix} 1 & 2 & 3 \\ 4 & 5 & 6 \\ a & b & c \end{vmatrix} = -\begin{vmatrix} 4 & 5 & 6 \\ 1 & 2 & 3 \\ a & b & c \end{vmatrix}$$

EXAMPLE 2 Evaluating a determinant

Evaluate the determinant $\begin{vmatrix} 15 & 14 & 26 \\ 18 & 17 & 32 \\ 21 & 20 & 42 \end{vmatrix}$.

SOLUTION

$$\begin{vmatrix} 15 & 14 & 26 \\ 18 & 17 & 32 \\ 21 & 20 & 42 \end{vmatrix} = (3 \times 2)\begin{vmatrix} 5 & 14 & 13 \\ 6 & 17 & 16 \\ 7 & 20 & 21 \end{vmatrix}$$ using Rule 1 to factor 3 from the first column and 2 from the third column

$$= 6\begin{vmatrix} 5 & 1 & 13 \\ 6 & 1 & 16 \\ 7 & -1 & 21 \end{vmatrix}$$ using Rule 2 to subtract the third column from the second column

$$= 6\begin{vmatrix} 12 & 0 & 34 \\ 13 & 0 & 37 \\ 7 & -1 & 21 \end{vmatrix}$$ using Rule 2 to add the third row to the first and second rows

$$= 6\left[-\left(-1\begin{vmatrix} 12 & 34 \\ 13 & 37 \end{vmatrix}\right)\right]$$ expanding the determinant along the second column

$$= 6\begin{vmatrix} 12 & 34 \\ 1 & 3 \end{vmatrix}$$ using Rule 2 to subtract the first row from the second row

$$= 6(36 - 34) = 12$$

The value of the given determinant is therefore 12. Notice the general strategy. We used Rules 1 and 2 until one column (or one row) contained two zeros. At that point, it is a simple matter to expand the determinant along that column (or row).

EXAMPLE 3 Applying the rules for manipulating a determinant

Show that

$$\begin{vmatrix} 1 & 1 & 1 \\ a & b & c \\ a^2 & b^2 & c^2 \end{vmatrix} = (b - a)(c - a)(c - b)$$

SOLUTION

$$\begin{vmatrix} 1 & 1 & 1 \\ a & b & c \\ a^2 & b^2 & c^2 \end{vmatrix} = \begin{vmatrix} 1 & 0 & 0 \\ a & b - a & c - a \\ a^2 & b^2 - a^2 & c^2 - a^2 \end{vmatrix}$$ using Rule 2 to subtract the first column from the second and third columns

$$= \begin{vmatrix} 1 & 0 & 0 \\ a & b - a & c - a \\ a^2 & (b - a)(b + a) & (c - a)(c + a) \end{vmatrix}$$

$$= (b - a)(c - a)\begin{vmatrix} 1 & 0 & 0 \\ a & 1 & 1 \\ a^2 & b + a & c + a \end{vmatrix}$$ using Rule 1 to factor $b - a$ from the second column and $c - a$ from the third column

$$= (b - a)(c - a)[(c + a) - (b + a)]$$ expanding along first row

$$= (b - a)(c - a)(c - b)$$

NOTE The determinant in Example 3 is the **Vandermonde determinant** of order three. Vandermonde determinants and variations of Vandermonde determinants are very useful in many applications. (See Exercises 32 and 34 at the end of this section.)

EXAMPLE 4 Two cases in which the value of a determinant is zero

Evaluate each determinant:

(a) $\begin{vmatrix} 9 & 4 & -3 \\ 13 & 17 & 5 \\ 9 & 4 & -3 \end{vmatrix}$ **(b)** $\begin{vmatrix} 2 & 7 & 3 \\ 15 & 40 & 79 \\ 20 & 70 & 30 \end{vmatrix}$

SOLUTION

(a) Notice that the first and third rows are identical. Thus we have

$$\begin{vmatrix} 9 & 4 & -3 \\ 13 & 17 & 5 \\ 9 & 4 & -3 \end{vmatrix} = \begin{vmatrix} 9 & 4 & -3 \\ 13 & 17 & 5 \\ 0 & 0 & 0 \end{vmatrix} \quad \text{subtracting row 1 from row 3}$$

$$= 0 \quad \text{expanding along the third row}$$

(b) Notice that the third row is a multiple of the first row. (A more precise, but less convenient, way to state this is: Each entry in the third row is a constant multiple of the corresponding entry in the first row.) So we have

$$\begin{vmatrix} 2 & 7 & 3 \\ 15 & 40 & 79 \\ 20 & 70 & 30 \end{vmatrix} = 10 \begin{vmatrix} 2 & 7 & 3 \\ 15 & 40 & 79 \\ 2 & 7 & 3 \end{vmatrix} = 10 \times 0 \quad \text{(Why?)}$$

$$= 0$$

The reasoning that we used in Example 4 can be generalized as indicated in the box that follows.

Two Instances When the Value of a Determinant Is Zero

1. If two rows are identical, the value of the determinant is zero. This also holds for two columns.
2. If one row is a multiple of another row, the value of the determinant is zero. This also holds for two columns.

Caution: The value of a determinant can still turn out to be zero even if neither of these conditions apply. See, for instance, the example on page 787.

The definition that we gave (on page 787) for third-order determinants can be extended to apply to fourth-order (or larger) determinants. Consider, for example, the fourth-order determinant given by

$$\begin{vmatrix} 2 & 7 & -1 & 9 \\ 4 & 0 & 3 & 6 \\ -8 & 5 & 1 & 3 \\ 11 & 2 & -6 & 1 \end{vmatrix}$$

By definition, we can evaluate this determinant by selecting the first row, multiplying each entry by its cofactor, and then adding the results. This yields

$$\begin{vmatrix} 2 & 7 & -1 & 9 \\ 4 & 0 & 3 & 6 \\ -8 & 5 & 1 & 3 \\ 11 & 2 & -6 & 1 \end{vmatrix}$$

$$= 2\begin{vmatrix} 0 & 3 & 6 \\ 5 & 1 & 3 \\ 2 & -6 & 1 \end{vmatrix} - 7\begin{vmatrix} 4 & 3 & 6 \\ -8 & 1 & 3 \\ 11 & -6 & 1 \end{vmatrix} + (-1)\begin{vmatrix} 4 & 0 & 6 \\ -8 & 5 & 3 \\ 11 & 2 & 1 \end{vmatrix} - 9\begin{vmatrix} 4 & 0 & 3 \\ -8 & 5 & 1 \\ 11 & 2 & -6 \end{vmatrix}$$

The problem is now reduced to evaluating four third-order determinants. As Exercise 31(a) asks you to check, the determinant is 174. If we had expanded along the second row rather than the first, then there would be one less third-order determinant to evaluate because of the zero in the second row. (In doing this, we would be relying on the theorem on page 787, which is valid for all determinants, not only three-by-three determinants.) Exercise 31(b) asks you to expand along the second row and again verify that the determinant is 174.

Alternatively, we can use the properties on page 788 to transform the determinant to an equivalent determinant containing more zero entries. Then, expanding by a row or column with many zeros requires less arithmetic.

As with matrix products and inverses, a graphing calculator can be used in computing determinants. We display two examples here using the four-by-four determinant given above. (For the details on these and for other types of graphing utilities, see your user's manual. Figure 2 shows the end results of the two steps in computing the determinant using a Texas Instruments TI-83+ graphing calculator. After the four-by-four matrix A corresponding to the determinant has been entered, the matrix can be displayed as in Figure 2(a). Figure 2(b) shows the viewing screen that then results after executing the command `det([A])`. In Figure 3 we've used Microsoft *Excel* to evaluate the determinant.

Determinants can be used to solve certain systems of linear equations in which there are as many unknowns as there are equations. In the box that follows, we state **Cramer's rule** for solving a system of three linear equations in three unknowns.*

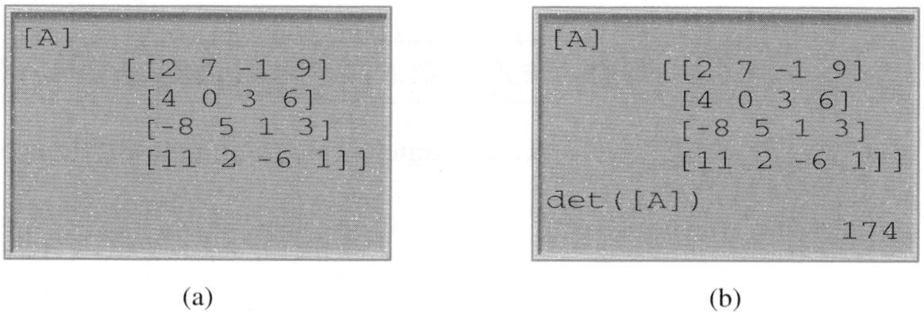

```
[A]
      [[2 7 -1 9]
       [4 0 3 6]
       [-8 5 1 3]
       [11 2 -6 1]]
```

(a)

```
[A]
      [[2 7 -1 9]
       [4 0 3 6]
       [-8 5 1 3]
       [11 2 -6 1]]
det([A])
                174
```

(b)

Figure 2
Evaluating a four-by-four determinant using a Texas Instruments TI-83+ graphing calculator (For details, see your user's manual.)

	A	B	C	D	E
1					
2		2	7	–1	9
3	A=	4	0	3	6
4		–8	5	1	3
5		11	2	–6	1
6					
7	det(A)=	174			

Figure 3
Computing a determinant using Microsoft *Excel*. After typing in the entries in cells B2 through E5, the formula `=MDETERM(B2:E5)` is used to compute the determinant. (See your user's manual.)

*The rule is named after one of its discoverers, the Swiss mathematician Gabriel Cramer (1704–1752).

A more general but entirely similar version of Cramer's rule holds for n equations in n unknowns.

Cramer's Rule

Consider the system

$$\begin{cases} a_1x + b_1y + c_1z = d_1 \\ a_2x + b_2y + c_2z = d_2 \\ a_3x + b_3y + c_3z = d_3 \end{cases}$$

Let the four determinants D, D_x, D_y, and D_z be defined as follows:

$$D = \begin{vmatrix} a_1 & b_1 & c_1 \\ a_2 & b_2 & c_2 \\ a_3 & b_3 & c_3 \end{vmatrix} \qquad D_x = \begin{vmatrix} d_1 & b_1 & c_1 \\ d_2 & b_2 & c_2 \\ d_3 & b_3 & c_3 \end{vmatrix}$$

$$D_y = \begin{vmatrix} a_1 & d_1 & c_1 \\ a_2 & d_2 & c_2 \\ a_3 & d_3 & c_3 \end{vmatrix} \qquad D_z = \begin{vmatrix} a_1 & b_1 & d_1 \\ a_2 & b_2 & d_2 \\ a_3 & b_3 & d_3 \end{vmatrix}$$

Then if $D \neq 0$, the system has a unique solution for x, y, and z given by

$$x = \frac{D_x}{D} \qquad y = \frac{D_y}{D} \qquad z = \frac{D_z}{D}$$

[If $D = 0$, the solutions (if any) can be found using Gaussian elimination.]

Notice that the determinant D in Cramer's rule is just the determinant of the coefficient matrix of the given system. If you replace the first column of D with the column of numbers on the right side of the given system, you obtain D_x. The determinants D_y and D_z are obtained similarly.

Before actually proving Cramer's rule, let's take a look at how it's applied.

EXAMPLE 5 Using Cramer's rule to solve a system

Use Cramer's rule to find all solutions of the following system of equations:

$$\begin{cases} 2x + 2y - 3z = -20 \\ x - 4y + z = 6 \\ 4x - y + 2z = -1 \end{cases}$$

SOLUTION

First we list the determinants D, D_x, D_y, and D_z:

$$D = \begin{vmatrix} 2 & 2 & -3 \\ 1 & -4 & 1 \\ 4 & -1 & 2 \end{vmatrix} \qquad D_x = \begin{vmatrix} -20 & 2 & -3 \\ 6 & -4 & 1 \\ -1 & -1 & 2 \end{vmatrix}$$

$$D_y = \begin{vmatrix} 2 & -20 & -3 \\ 1 & 6 & 1 \\ 4 & -1 & 2 \end{vmatrix} \qquad D_z = \begin{vmatrix} 2 & 2 & -20 \\ 1 & -4 & 6 \\ 4 & -1 & -1 \end{vmatrix}$$

The calculations for evaluating D begin as follows:

$$D = \begin{vmatrix} 2 & 2 & -3 \\ 1 & -4 & 1 \\ 4 & -1 & 2 \end{vmatrix} \underset{\uparrow}{=} \begin{vmatrix} 0 & 10 & -5 \\ 1 & -4 & 1 \\ 0 & 15 & -2 \end{vmatrix}$$

Subtract twice the second row from the first.
Subtract 4 times the second row from the third.

Since we now have two zeros in the first column, it is an easy matter to expand D along that column to obtain

$$D = -1\begin{vmatrix} 10 & -5 \\ 15 & -2 \end{vmatrix}$$
$$= -1[-20 - (-75)] = -1(55) = -55$$

The value of D is therefore -55. (Since this value is nonzero, Cramer's rule does apply.) As Exercise 37 at the end of this section asks you to verify, the values of the other three determinants are

$$D_x = 144 \qquad D_y = 61 \qquad D_z = -230$$

By Cramer's rule, then, the unique values of x, y, and z that satisfy the system are

$$x = \frac{D_x}{D} = \frac{144}{-55} = -\frac{144}{55} \qquad y = \frac{D_y}{D} = \frac{61}{-55} = -\frac{61}{55} \qquad z = \frac{D_z}{D} = \frac{-230}{-55} = \frac{46}{11}$$

One way we can prove Cramer's rule is to use Gaussian elimination to solve the system

$$\begin{cases} a_1x + b_1y + c_1z = d_1 \\ a_2x + b_2y + c_2z = d_2 \\ a_3x + b_3y + c_3z = d_3 \end{cases} \tag{1}$$

A much shorter and simpler proof, however, has been found by D. E. Whitford and M. S. Klamkin.* This is the proof we give here; it makes effective use of the rules employed in this section for manipulating determinants.

Consider the system of equations (1) and assume that $D \neq 0$. We will show that if x, y, and z satisfy the system, then in fact $x = D_x/D$, with similar equations giving y and z. (Exercise 68 at the end of this section then shows how to check that these values indeed satisfy the given system.) We have

$$D_x = \begin{vmatrix} d_1 & b_1 & c_1 \\ d_2 & b_2 & c_2 \\ d_3 & b_3 & c_3 \end{vmatrix} \qquad \text{by definition}$$

$$= \begin{vmatrix} (a_1x + b_1y + c_1z) & b_1 & c_1 \\ (a_2x + b_2y + c_2z) & b_2 & c_2 \\ (a_3x + b_3y + c_3z) & b_3 & c_3 \end{vmatrix} \qquad \text{using the equations in (1) to substitute for } d_1, d_2, \text{ and } d_3$$

*The proof was published in the *American Mathematical Monthly,* vol. 60 (1953), pp. 186–187.

$$= \begin{vmatrix} a_1x & b_1 & c_1 \\ a_2x & b_2 & c_2 \\ a_3x & b_3 & c_3 \end{vmatrix}$$ subtracting y times the second column as well as z times the third column from the first column

$$= x\begin{vmatrix} a_1 & b_1 & c_1 \\ a_2 & b_2 & c_2 \\ a_3 & b_3 & c_3 \end{vmatrix}$$ factoring x out of the first column

$$= xD$$ by definition

We now have $D_x = xD$, which is equivalent to $x = D_x/D$, as required. The formulas for y and z are obtained similarly.

EXERCISE SET 10.5

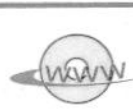

A

In Exercises 1–6, evaluate the determinants.

1. (a) $\begin{vmatrix} 2 & -17 \\ 1 & 6 \end{vmatrix}$ **(b)** $\begin{vmatrix} 1 & 6 \\ 2 & -17 \end{vmatrix}$

2. (a) $\begin{vmatrix} 1 & 0 \\ 0 & 1 \end{vmatrix}$ **(b)** $\begin{vmatrix} 0 & 1 \\ 0 & 1 \end{vmatrix}$

3. (a) $\begin{vmatrix} 5 & 7 \\ 500 & 700 \end{vmatrix}$ **(b)** $\begin{vmatrix} 5 & 500 \\ 7 & 700 \end{vmatrix}$

4. (a) $\begin{vmatrix} -8 & -3 \\ 4 & -5 \end{vmatrix}$ **(b)** $\begin{vmatrix} -3 & -8 \\ -5 & 4 \end{vmatrix}$

5. $\begin{vmatrix} \sqrt{2}-1 & \sqrt{2} \\ \sqrt{2} & \sqrt{2}+1 \end{vmatrix}$

6. $\begin{vmatrix} \sqrt{3}+\sqrt{2} & 1+\sqrt{5} \\ 1-\sqrt{5} & \sqrt{3}-\sqrt{2} \end{vmatrix}$

In Exercises 7–12, refer to the following determinant:

$$\begin{vmatrix} -6 & 3 & 8 \\ 5 & -4 & 1 \\ 10 & 9 & -10 \end{vmatrix}$$

7. Evaluate the minor of the entry 3.

8. Evaluate the cofactor of the entry 3.

9. Evaluate the minor of -10.

10. Evaluate the cofactor of -10.

11. (a) Multiply each entry in the first row by its minor and find the sum of the results.
(b) Multiply each entry in the first row by its cofactor and find the sum of the results.
(c) Which gives you the value of the determinant, part (a) or part (b)?

12. (a) Multiply each entry in the first column by its cofactor, and find the sum of the results.
(b) Follow the same instructions as in part (a), but use the second column.
(c) Follow the same instructions as in part (a), but use the third column.

13. Evaluate $\begin{vmatrix} 1 & 2 & 3 \\ 4 & 5 & 6 \\ 7 & 8 & 9 \end{vmatrix}$ by expanding it along
(a) the second row; **(b)** the third row; **(c)** the first column; **(d)** the third column.

In Exercises 14–24, evaluate the determinants.

14. $\begin{vmatrix} 1 & 2 & -1 \\ 2 & -1 & 1 \\ 4 & 0 & 2 \end{vmatrix}$ **15.** $\begin{vmatrix} 5 & 10 & 15 \\ 1 & 2 & 3 \\ -9 & 11 & 7 \end{vmatrix}$ **16.** $\begin{vmatrix} 8 & 4 & 2 \\ 3 & 9 & 3 \\ -2 & 8 & 6 \end{vmatrix}$

17. $\begin{vmatrix} 1 & 2 & -3 \\ 4 & 5 & -9 \\ 0 & 0 & 1 \end{vmatrix}$ **18.** $\begin{vmatrix} 9 & 9 & 12 \\ 4 & 4 & 6 \\ 7 & 7 & 5 \end{vmatrix}$ **19.** $\begin{vmatrix} 8 & 7 & 800 \\ 3 & 4 & 300 \\ 5 & 2 & 500 \end{vmatrix}$

20. $\begin{vmatrix} 12 & 21 & -4 \\ 0 & 0 & 0 \\ 73 & 82 & 14 \end{vmatrix}$ **21.** $\begin{vmatrix} 3 & 0 & 0 \\ 0 & 19 & 0 \\ 0 & 0 & 10 \end{vmatrix}$ **22.** $\begin{vmatrix} -6 & -8 & 18 \\ 25 & 12 & 15 \\ -9 & 4 & 13 \end{vmatrix}$

23. $\begin{vmatrix} 23 & 0 & 47 \\ -37 & 0 & 18 \\ 14 & 0 & 25 \end{vmatrix}$ **24.** $\begin{vmatrix} 16 & 0 & -64 \\ -8 & 15 & -12 \\ 30 & -20 & 10 \end{vmatrix}$

Ⓖ *In Exercises 25 and 26, use a graphing utility to evaluate the determinants. For Exercise 25, check your answer by referring to Example 2.*

25. $\begin{vmatrix} 15 & 14 & 26 \\ 18 & 17 & 32 \\ 21 & 20 & 42 \end{vmatrix}$ **26.** $\begin{vmatrix} 25 & 40 & 5 & 10 \\ 9 & 0 & 3 & 6 \\ -2 & 3 & 11 & -17 \\ -3 & 4 & 7 & 2 \end{vmatrix}$

27. Consider the two determinants

$$\begin{vmatrix} 1 & 2 & 3 \\ -7 & -4 & 5 \\ 9 & 2 & 6 \end{vmatrix} \quad \text{and} \quad \begin{vmatrix} 10 & 20 & 30 \\ -7 & -4 & 5 \\ 9 & 2 & 6 \end{vmatrix}$$

(a) According to Item 1 in the Property Summary box on page 788, how are the values of these determinants related?

(b) Evaluate each determinant to verify your answer in part (a).

28. Consider the two determinants

$$\begin{vmatrix} \sqrt{2} & \sqrt{5} & \sqrt{7} \\ -6 & 10 & 0 \\ 8 & -1 & -6 \end{vmatrix} \quad \text{and} \quad \begin{vmatrix} -6 & 10 & 0 \\ \sqrt{2} & \sqrt{5} & \sqrt{7} \\ 8 & -1 & -6 \end{vmatrix}$$

(a) According to Item 3 in the Property Summary box on page 788, how are the values of these determinants related?

Ⓖ (b) Evaluate each determinant to verify your answer in part (a).

In Exercises 29 and 30:

(a) *Evaluate the determinant as in Example 2.*

Ⓖ **(b)** *Use a graphing utility to evaluate the determinant.*

29. $\begin{vmatrix} 1 & -1 & 0 & 2 \\ 0 & 1 & -1 & 0 \\ 2 & 1 & 0 & -1 \\ -2 & 2 & 1 & 1 \end{vmatrix}$ **30.** $\begin{vmatrix} 3 & -2 & 3 & 4 \\ 1 & 4 & -3 & 2 \\ 6 & 3 & -6 & -3 \\ -1 & 0 & 1 & 5 \end{vmatrix}$

31. This exercise refers to the four-by-four determinant A on page 790.

(a) Evaluate the determinant by expanding along the first row. As stated in the text, you should obtain 174 as the value of the determinant.

(b) Evaluate the determinant by expanding along the second row.

32. Use the method illustrated in Example 3 to show that

$$\begin{vmatrix} 1 & 1 & 1 \\ a & b & c \\ a^3 & b^3 & c^3 \end{vmatrix} = (b-a)(c-a)(c^2-b^2+ac-ab)$$
$$= (b-a)(c-a)(c-b)(a+b+c)$$

33. Use the method shown in Example 3 to express the determinant $\begin{vmatrix} 1 & x & x^2 \\ 1 & y & y^2 \\ 1 & z & z^2 \end{vmatrix}$ as a product.

34. Show that

$$\begin{vmatrix} 1 & 1 & 1 \\ a^2 & b^2 & c^2 \\ a^3 & b^3 & c^3 \end{vmatrix} = (b-a)(c-a)(bc^2-b^2c+ac^2-ab^2)$$
$$= (b-a)(c-a)(c-b)(bc+ac+ab)$$

35. Simplify the determinant $\begin{vmatrix} 1 & 1 & 1 \\ 1 & 1+x & 1 \\ 1 & 1 & 1+y \end{vmatrix}$.

36. Use the method shown in Example 3 to express the following determinant as a product of three factors:

$$\begin{vmatrix} 1 & 1 & 1 \\ a & b & c \\ bc & ca & ab \end{vmatrix}$$

37. Verify the following statements (from Example 5).

(a) $\begin{vmatrix} -20 & 2 & -3 \\ 6 & -4 & 1 \\ -1 & -1 & 2 \end{vmatrix} = 144$

(b) $\begin{vmatrix} 2 & -20 & -3 \\ 1 & 6 & 1 \\ 4 & -1 & 2 \end{vmatrix} = 61$

(c) $\begin{vmatrix} 2 & 2 & -20 \\ 1 & -4 & 6 \\ 4 & -1 & -1 \end{vmatrix} = -230$

38. Consider the following system:

$$\begin{cases} x + y + 3z = 0 \\ x + 2y + 5z = 0 \\ x - 4y - 8z = 0 \end{cases}$$

(a) Without doing any calculations, find one obvious solution of this system.

(b) Calculate the determinant D.

(c) List all solutions of this system.

In Exercises 39–46, use Cramer's rule to solve those systems for which $D \neq 0$. In cases where $D = 0$, use Gaussian elimination or matrix methods.

39. $\begin{cases} 3x + 4y - z = 5 \\ x - 3y + 2z = 2 \\ 5x - 6z = -7 \end{cases}$ **40.** $\begin{cases} 3A - B - 4C = 3 \\ A + 2B - 3C = 9 \\ 2A - B + 2C = -8 \end{cases}$

41. $\begin{cases} 3x + 2y - z = -6 \\ 2x - 3y - 4z = -11 \\ x + y + z = 5 \end{cases}$ **42.** $\begin{cases} 5x - 3y - z = 16 \\ 2x + y - 3z = 5 \\ 3x - 2y + 2z = 5 \end{cases}$

43. $\begin{cases} 2x + 5y + 2z = 0 \\ 3x - y - 4z = 0 \\ x + 2y - 3z = 0 \end{cases}$ **44.** $\begin{cases} 4u + 3v - 2w = 14 \\ u + 2v - 3w = 6 \\ 2u - v + 4w = 2 \end{cases}$

45. $\begin{cases} 12x - 11z = 13 \\ 6x + 6y - 4z = 26 \\ 6x + 2y - 5z = 13 \end{cases}$ **46.** $\begin{cases} 3x + 4y + 2z = 1 \\ 4x + 6y + 2z = 7 \\ 2x + 3y + z = 11 \end{cases}$

Ⓖ *In Exercises 47 and 48, use Cramer's rule along with a graphing utility to solve the systems.*

47. $\begin{cases} x + y + z + w = -7 \\ x - y + z - w = -11 \\ 2x - 2y - 3z - 3w = 26 \\ 3x + 2y + z - w = -9 \end{cases}$

48. $\begin{cases} 2A - B - 3C + 2D = -2 \\ A - 2B + C - 3D = 4 \\ 3A - 4B + 2C - 4D = 12 \\ 2A + 3B - C - 2D = -4 \end{cases}$

B

49. Find all values of x for which

$$\begin{vmatrix} x-4 & 0 & 0 \\ 0 & x+4 & 0 \\ 0 & 0 & x+1 \end{vmatrix} = 0$$

50. Find all values of x for which

$$\begin{vmatrix} 1 & x & x^2 \\ 1 & 1 & 1 \\ 4 & 5 & 0 \end{vmatrix} = 0$$

51. By expanding the determinant $\begin{vmatrix} a & b & c \\ a & b & c \\ d & e & f \end{vmatrix}$ down the first column, show that its value is zero.

52. By expanding the determinant $\begin{vmatrix} ka & kb & kc \\ d & e & f \\ g & h & i \end{vmatrix}$ along its first row, show that it is equal to $k\begin{vmatrix} a & b & c \\ d & e & f \\ g & h & i \end{vmatrix}$.

53. Show that

$$\begin{vmatrix} a_1 + A_1 & b_1 & c_1 \\ a_2 + A_2 & b_2 & c_2 \\ a_3 + A_3 & b_3 & c_3 \end{vmatrix} = \begin{vmatrix} a_1 & b_1 & c_1 \\ a_2 & b_2 & c_2 \\ a_3 & b_3 & c_3 \end{vmatrix} + \begin{vmatrix} A_1 & b_1 & c_1 \\ A_2 & b_2 & c_2 \\ A_3 & b_3 & c_3 \end{vmatrix}$$

54. Show that

$$\begin{vmatrix} a_1 & b_1 & c_1 \\ a_2 & b_2 & c_2 \\ a_3 & b_3 & c_3 \end{vmatrix} = \begin{vmatrix} a_1 + kb_1 & b_1 & c_1 \\ a_2 + kb_2 & b_2 & c_2 \\ a_3 + kb_3 & b_3 & c_3 \end{vmatrix}$$

55. By expanding each determinant along a row or column, show that

$$\begin{vmatrix} a_1 & b_1 & c_1 \\ a_2 & b_2 & c_2 \\ a_3 & b_3 & c_3 \end{vmatrix} = -\begin{vmatrix} a_2 & b_2 & c_2 \\ a_1 & b_1 & c_1 \\ a_3 & b_3 & c_3 \end{vmatrix}$$

56. Solve for x in terms of a, b, and c:

$$\begin{vmatrix} a & a & x \\ c & c & c \\ b & x & b \end{vmatrix} = 0 \qquad (c \neq 0)$$

57. Show that

$$\begin{vmatrix} 1+a & 1 & 1 & 1 \\ 1 & 1+b & 1 & 1 \\ 1 & 1 & 1+c & 1 \\ 1 & 1 & 1 & 1+d \end{vmatrix} = abcd\left(\frac{1}{a} + \frac{1}{b} + \frac{1}{c} + \frac{1}{d} + 1\right)$$

58. Show that

$$\begin{vmatrix} 1 & a & a^2 \\ a^2 & 1 & a \\ a & a^2 & 1 \end{vmatrix} = (a^3 - 1)^2$$

59. Show that

$$\begin{vmatrix} 1 & bc & b+c \\ 1 & ca & c+a \\ 1 & ab & a+b \end{vmatrix} = (b-c)(c-a)(a-b)$$

60. Evaluate the determinant

$$\begin{vmatrix} a & b & c \\ a & a+b & a+b+c \\ a & 2a+b & 3a+2b+c \end{vmatrix}$$

61. Show that

$$\begin{vmatrix} 1 & a & a & a \\ 1 & b & a & a \\ 1 & a & b & a \\ 1 & a & a & b \end{vmatrix} = (b-a)^3$$

62. Show that

$$\begin{vmatrix} a & 1 & 1 & 1 \\ 1 & a & 1 & 1 \\ 1 & 1 & a & 1 \\ 1 & 1 & 1 & a \end{vmatrix} = (a-1)^3(a+3)$$

63. Solve the following system for x, y, and z:

$$\begin{cases} ax + by + cz = k \\ a^2x + b^2y + c^2z = k^2 \\ a^3x + b^3y + c^3z = k^3 \end{cases}$$

(Assume that the values of a, b, and c are all distinct and all nonzero.)

64. Show that the equation

$$\begin{vmatrix} x & y & 1 \\ x_1 & y_1 & 1 \\ 1 & m & 0 \end{vmatrix} = 0$$

represents a line that has slope m and passes through the point (x_1, y_1).

For Exercise 65, use the fact that the equation of a line passing through (x_1, y_1) and (x_2, y_2) can be written

$$\begin{vmatrix} x & y & 1 \\ x_1 & y_1 & 1 \\ x_2 & y_2 & 1 \end{vmatrix} = 0$$

65. Find the equation of the line passing through $(-3, -1)$ and $(2, 9)$. Write the answer in the form $y = mx + b$.

For Exercise 66, use the fact that the equation of a circle passing through (x_1, y_1), (x_2, y_2), and (x_3, y_3) can be written

$$\begin{vmatrix} x^2 + y^2 & x & y & 1 \\ x_1^2 + y_1^2 & x_1 & y_1 & 1 \\ x_2^2 + y_2^2 & x_2 & y_2 & 1 \\ x_3^2 + y_3^2 & x_3 & y_3 & 1 \end{vmatrix} = 0$$

66. Find the equation of the circle passing through $(7, 0)$, $(5, -6)$, and $(-1, -4)$. Write the answer in the form $(x-h)^2 + (y-k)^2 = r^2$.

C

67. Show that the area of the triangle in Figure A is $\frac{1}{2}\begin{vmatrix} a & b \\ c & d \end{vmatrix}$.

Hint: Figure B indicates how the required area can be found by using a rectangle and three right triangles.

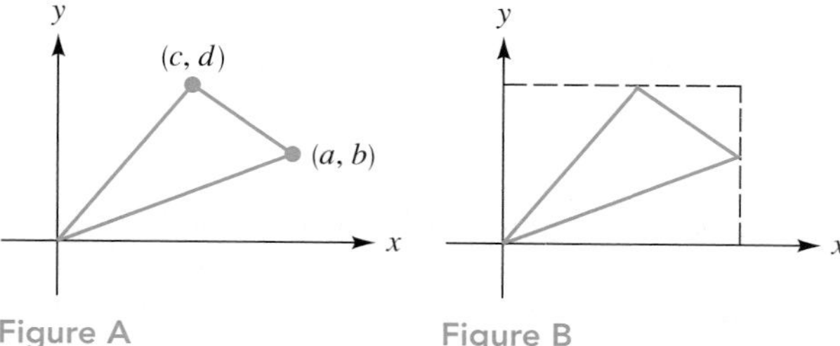

Figure A Figure B

68. This exercise completes the derivation of Cramer's rule given in the text. Using the same notation, and assuming $D \neq 0$, we need to show that the values $x = D_x/D$, $y = D_y/D$, and $z = D_z/D$ satisfy the equations in (1) on page 793. We will show that these values satisfy the first equation in (1), the verification for the other equations being entirely similar.

(a) Check that substituting the values $x = D_x/D$, $y = D_y/D$, and $z = D_z/D$ in the first equation of (1) yields an equation equivalent to

$$a_1D_x + b_1D_y + c_1D_z - d_1D = 0$$

(b) Show that the equation in part (a) can be written

$$a_1\begin{vmatrix} b_1 & c_1 & d_1 \\ b_2 & c_2 & d_2 \\ b_3 & c_3 & d_3 \end{vmatrix} - b_1\begin{vmatrix} a_1 & c_1 & d_1 \\ a_2 & c_2 & d_2 \\ a_3 & c_3 & d_3 \end{vmatrix} + c_1\begin{vmatrix} a_1 & b_1 & d_1 \\ a_2 & b_2 & d_2 \\ a_3 & b_3 & d_3 \end{vmatrix} - d_1\begin{vmatrix} a_1 & b_1 & c_1 \\ a_2 & b_2 & c_2 \\ a_3 & b_3 & c_3 \end{vmatrix} = 0$$

(c) Show that the equation in part (b) can be written

$$\begin{vmatrix} a_1 & b_1 & c_1 & d_1 \\ a_1 & b_1 & c_1 & d_1 \\ a_2 & b_2 & c_2 & d_2 \\ a_3 & b_3 & c_3 & d_3 \end{vmatrix} = 0$$

(d) Now explain why the equation in (c) indeed holds.

69. Let D denote the determinant of the matrix $\begin{pmatrix} a & b \\ c & d \end{pmatrix}$.

(a) Show that the inverse of this matrix is $\frac{1}{D}\begin{pmatrix} d & -b \\ -c & a \end{pmatrix}$. Assume that $ad - bc \neq 0$.

(b) Use the result in part (a) to find the inverse of the matrix $\begin{pmatrix} -6 & 7 \\ 1 & 9 \end{pmatrix}$.

70. Let $A = \begin{pmatrix} a & b \\ c & d \end{pmatrix}$ and $B = \begin{pmatrix} e & f \\ g & h \end{pmatrix}$. Is it true that $\det(AB) = (\det(A))(\det(B))$?

10.6 NONLINEAR SYSTEMS OF EQUATIONS

In the previous sections of this chapter we looked at several techniques for solving systems of linear equations. In the present section we consider **nonlinear systems** of equations, that is, systems in which at least one of the equations is not linear. There is no single technique that serves to solve all nonlinear systems. However, simple substitution will often suffice. The work in this section focuses on examples showing some of the more common approaches. In all of the examples and in the exercises, we will be concerned exclusively with solutions (x, y) in which both x and y are real numbers. In Example 1 the system consists of one linear equation and one quadratic equation. Such a system can always be solved by substitution.

EXAMPLE 1 Solving a simple nonlinear system by substitution

Find all solutions (x, y) of the following system, where x and y are real numbers:

$$\begin{cases} 2x + y = 1 \\ y = 4 - x^2 \end{cases}$$

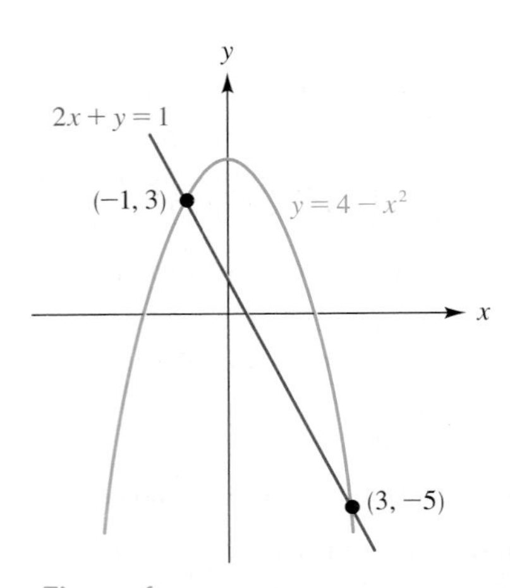

Figure 1

SOLUTION
We use the second equation to substitute for y in the first equation. This will yield an equation with only one unknown:

$$\begin{aligned} 2x + (4 - x^2) &= 1 \\ -x^2 + 2x + 3 &= 0 \\ x^2 - 2x - 3 &= 0 \\ (x - 3)(x + 1) &= 0 \end{aligned}$$

$$x - 3 = 0 \quad \Big| \quad x + 1 = 0$$
$$x = 3 \quad \Big| \quad x = -1$$

The values $x = 3$ and $x = -1$ can now be substituted back into either of the original equations. Substituting $x = 3$ in the equation $y = 4 - x^2$ yields $y = -5$. Similarly, substituting $x = -1$ in the equation $y = 4 - x^2$ gives us $y = 3$. We have now obtained two ordered pairs, $(3, -5)$ and $(-1, 3)$. As you can easily check, both of these are solutions of the given system. Figure 1 displays the graphical interpretation of this result. The line $2x + y = 1$ intersects the parabola $y = 4 - x^2$ at the points $(-1, 3)$ and $(3, -5)$.

EXAMPLE 2 Determining where two graphs intersect

Where do the graphs of the parabola $y = x^2$ and the circle $x^2 + y^2 = 1$ intersect? See Figure 2.

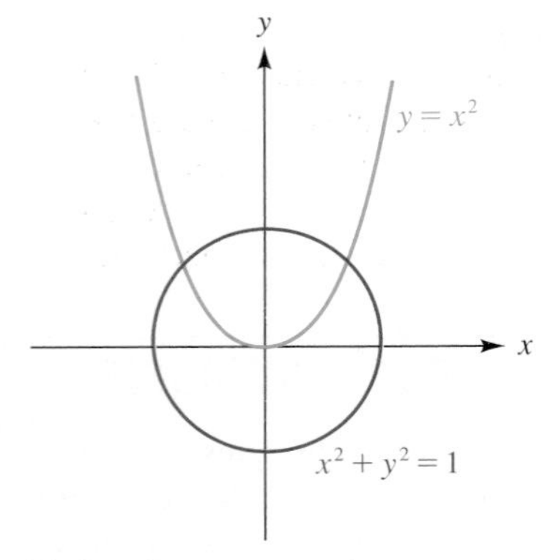

Figure 2

SOLUTION
The system that we wish to solve is

$$\begin{cases} y = x^2 & (1) \\ x^2 + y^2 = 1 & (2) \end{cases}$$

In view of equation (1) we can replace the x^2-term of equation (2) by y. Doing this yields

$$y + y^2 = 1$$

or

$$y^2 + y - 1 = 0$$

This last equation can be solved by using the quadratic formula with $a = 1$, $b = 1$, and $c = -1$. As you can check, the results are

$$y = \frac{-1 + \sqrt{5}}{2} \quad \text{and} \quad y = \frac{-1 - \sqrt{5}}{2}$$

However, from Figure 2 it is clear that the y-coordinate at each intersection point is positive. Therefore we discard the negative number $y = \frac{1}{2}(-1 - \sqrt{5})$ from further consideration in this context. Substituting the positive number $y = \frac{1}{2}(-1 + \sqrt{5})$ back in the equation $y = x^2$ then gives us

$$x^2 = (-1 + \sqrt{5})/2$$

Therefore

$$x = \pm\sqrt{(-1 + \sqrt{5})/2}$$

By choosing the positive square root, we obtain the x-coordinate for the intersection point in the first quadrant. That point is therefore

$$\left(\sqrt{(-1 + \sqrt{5})/2}, (-1 + \sqrt{5})/2\right) \approx (0.79, 0.62) \qquad \text{using a calculator}$$

Similarly, the negative square root yields the x-coordinate for the intersection point in the second quadrant. That point is

$$\left(-\sqrt{(-1 + \sqrt{5})/2}, (-1 + \sqrt{5})/2\right) \approx (-0.79, 0.62) \qquad \text{using a calculator}$$

We have now found the two intersection points, as required.

EXAMPLE 3 Using substitution and the quadratic formula to solve a system

Find all solutions (x, y) of the following system, where x and y are real numbers:

$$\begin{cases} xy = 1 & (3) \\ y = 3x + 1 & (4) \end{cases}$$

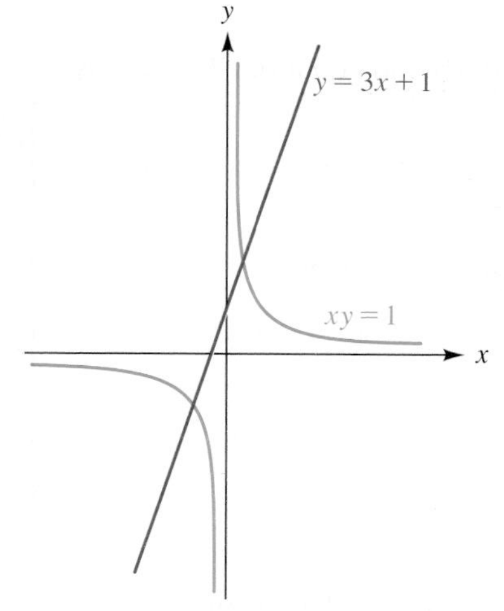

Figure 3

SOLUTION

Since these equations are easy to graph by hand, we do so, because that will tell us something about the required solutions. As Figure 3 indicates, there are two intersection points, one in the first quadrant, the other in the third quadrant. One way to begin now would be to solve equation (3) for one unknown in terms of the other. However, to avoid introducing fractions at the outset, let's use equation (4) to substitute for y in equation (3). This yields

$$x(3x + 1) = 1$$

or

$$3x^2 + x - 1 = 0$$

This last equation can be solved by using the quadratic formula. As you should verify, the solutions are

$$x = (-1 + \sqrt{13})/6 \qquad \text{and} \qquad x = (-1 - \sqrt{13})/6$$

The corresponding y-values can now be obtained by substituting for x in either of the given equations. We will substitute in equation (4). (Exercise 35 at the end of this section asks you to substitute in equation (3) as well and then to show that the very different-looking answers obtained in that way are in fact equal to those found here.) Substituting $x = (-1 + \sqrt{13})/6$ into the equation $y = 3x + 1$ gives us

$$y = 3\left(\frac{-1 + \sqrt{13}}{6}\right) + 1$$

$$= \frac{-1 + \sqrt{13}}{2} + \frac{2}{2} = \frac{1 + \sqrt{13}}{2}$$

Thus one of the intersection points is

$$((-1+\sqrt{13})/16),(1+\sqrt{13})/2)$$

Notice that this must be the first-quadrant point of intersection, since both coordinates are positive. The intersection point in the third quadrant is obtained in exactly the same manner. As you can check, substituting $x = -\frac{1}{6}(1+\sqrt{13})$ in the equation $y = 3x + 1$ yields $y = \frac{1}{2}(1-\sqrt{13})$. Thus the other intersection point is

$$(-(1+\sqrt{13})/6,(1-\sqrt{13})/2)$$

As Figure 3 indicates, there are no other solutions.

EXAMPLE 4 A system with no solution

Find all real numbers x and y that satisfy the system of equations

$$\begin{cases} y = \sqrt{x} \\ (x+2)^2 + y^2 = 1 \end{cases}$$

SOLUTION

We use the first equation to substitute for y in the second equation. This yields

$$\begin{aligned} (x+2)^2 + (\sqrt{x})^2 &= 1 \\ x^2 + 4x + 4 + x &= 1 \\ x^2 + 5x + 3 &= 0 \end{aligned}$$

Using the quadratic formula to solve this last equation, we have

$$x = \frac{-5 \pm \sqrt{5^2 - 4(1)(3)}}{2(1)} = \frac{-5 \pm \sqrt{13}}{2}$$

The two values of x are thus

$$\frac{-5+\sqrt{13}}{2} \quad \text{and} \quad \frac{-5-\sqrt{13}}{2}$$

However, notice that both of these quantities are negative. (The second is obviously negative; without using a calculator, can you explain why the first is negative?) Thus neither of these quantities is an appropriate x-input in the equation $y = \sqrt{x}$, since we are looking for y-values that are real numbers. We conclude from this that there are no pairs of real numbers x and y satisfying the given system. In geometric terms this means that the two graphs do not intersect; see Figure 4.

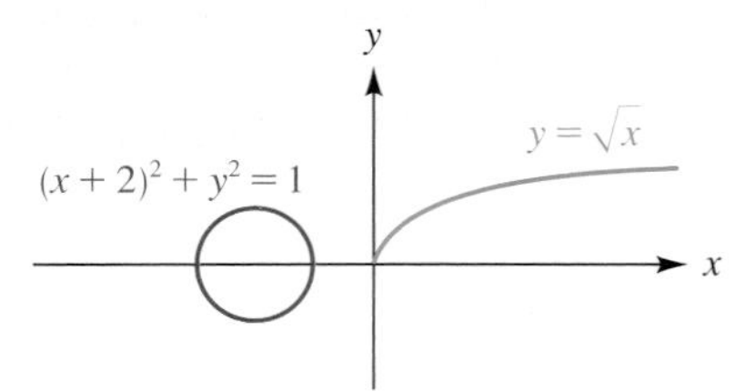

Figure 4

In the next example we look at a system that can be reduced to a linear system through appropriate substitutions.

EXAMPLE 5 Introducing a linear system to solve a nonlinear system

Solve the system

$$\begin{cases} \dfrac{2}{x^2} - \dfrac{3}{y^2} = -6 \\ \dfrac{3}{x^2} + \dfrac{4}{y^2} = 59 \end{cases}$$

SOLUTION

Let $u = 1/x^2$ and $v = 1/y^2$, so that the system becomes

$$\begin{cases} 2u - 3v = -6 \\ 3u + 4v = 59 \end{cases}$$

This is now a linear system. As you can verify using the methods of Section 10.1, the solution of this linear system is $u = 9$, $v = 8$. In view of the definitions of u and v, then, we have

$$\begin{aligned} \frac{1}{x^2} &= 9 \\ x^2 &= 1/9 \\ x &= \pm 1/3 \end{aligned} \qquad \Bigg| \qquad \begin{aligned} \frac{1}{y^2} &= 8 \\ y^2 &= 1/8 \\ y &= \pm 1/\sqrt{8} = \pm 1/(2\sqrt{2}) = \pm\sqrt{2}/4 \end{aligned}$$

This gives us four possible solutions for the original system:

$$\left(\frac{1}{3}, \frac{\sqrt{2}}{4}\right) \quad \left(\frac{1}{3}, -\frac{\sqrt{2}}{4}\right) \quad \left(-\frac{1}{3}, \frac{\sqrt{2}}{4}\right) \quad \left(-\frac{1}{3}, -\frac{\sqrt{2}}{4}\right)$$

As you can check, all four of these pairs satisfy the given system.

EXAMPLE 6 A system involving exponential functions

Determine all solutions (x, y) of the following system, where x and y are real numbers:

$$\begin{cases} y = 3^x & (5) \\ y = 3^{2x} - 2 & (6) \end{cases}$$

SOLUTION

We'll use the substitution method. First we rewrite equation (6) as

$$y = (3^x)^2 - 2 \tag{7}$$

Now, in view of equation (5), we can replace 3^x with y in equation (7) to obtain

$$\begin{aligned} y &= y^2 - 2 \\ 0 &= y^2 - y - 2 \\ 0 &= (y + 1)(y - 2) \end{aligned}$$

From this last equation we see that $y = -1$ or $y = 2$. With $y = -1$, equation (5) becomes $-1 = 3^x$, contrary to the fact that 3^x is positive for all real numbers x. Thus we discard the case in which $y = -1$. On the other hand, if $y = 2$, equation (5) becomes

$$2 = 3^x$$

Rewriting in logarithmic form, we obtain $x = \log_3 2$. Using the change of base formula (page 367), we can express x in terms of natural logarithms as

$$x = \frac{\ln 2}{\ln 3}$$

Alternatively, we can solve this exponential equation by taking the logarithm of both sides. Using natural logarithms, we have

$$\ln 2 = \ln 3^x = x \ln 3$$

and, consequently,

$$x = \frac{\ln 2}{\ln 3} \qquad \text{CAUTION: } \frac{\ln 2}{\ln 3} \neq \ln 2 - \ln 3$$

We've now found that $x = (\ln 2)/(\ln 3)$ and $y = 2$. Figure 5 displays a graphical interpretation of this result. (Using a calculator for the x-coordinate, we find that $x \approx 0.6$, which is consistent with Figure 5.)

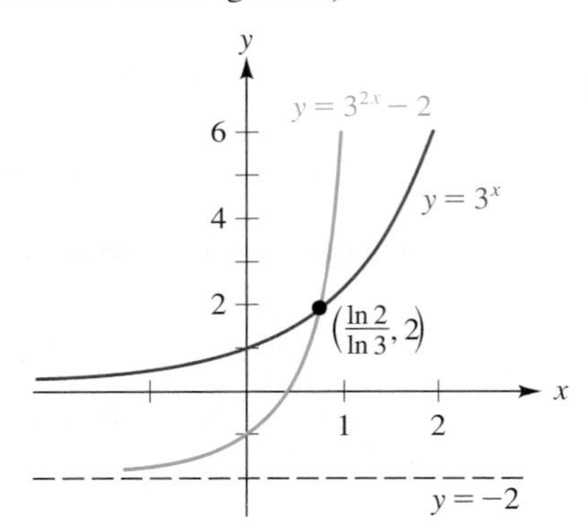

Figure 5

EXAMPLE 7 Using a graphing utility to approximate the solution of a nonlinear system

Use a graphing utility to approximate, to the nearest tenth, the solution(s) of the system

$$\begin{cases} y = 2 - x^2 \\ y = \ln x \end{cases}$$

SOLUTION

From work in previous chapters we know how to quickly sketch by hand the graphs of these two equations. If you do that, either on paper or in your head, you'll see that the given system has exactly one solution. As indicated by the viewing rectangle in Figure 6(a), the required value of x is between 1 and 2, and y is between 0 and 1. After some experimenting, we obtain the viewing rectangle shown in Figure 6(b), which indicates that, to the nearest tenth, the solution of the given system is (1.3, 0.3).

GRAPHICAL PERSPECTIVE

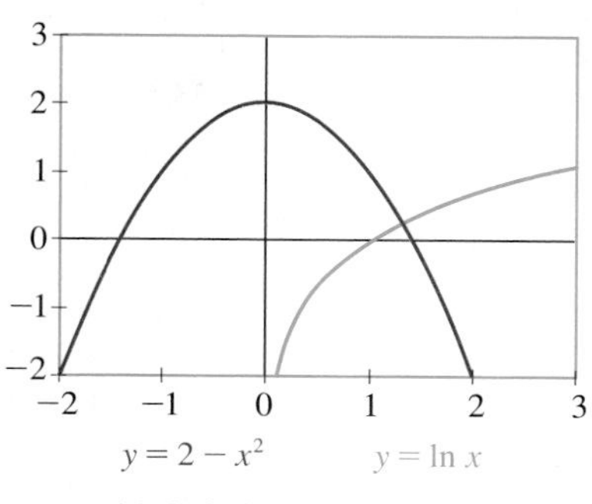

(a) [−2, 3, 1] by [−2, 3, 1]

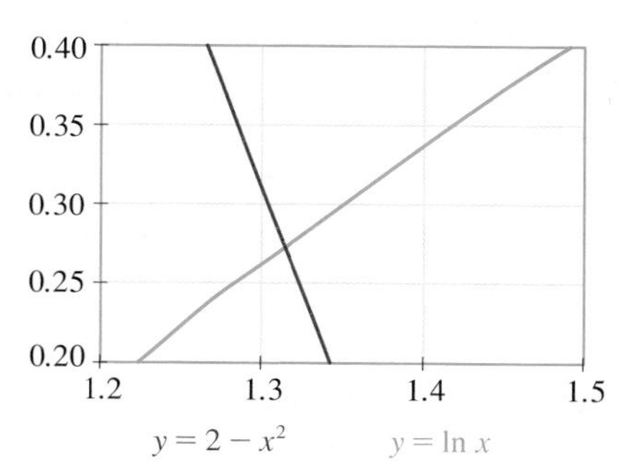

(b) [1.2, 1.5, 0.1] by [0.20, 0.40, 0.05]

Figure 6

EXERCISE SET 10.6 

A

In Exercises 1–22, find all solutions (x, y) of the given systems, where x and y are real numbers.

1. $\begin{cases} y = 3x \\ y = x^2 \end{cases}$

2. $\begin{cases} y = x + 3 \\ y = 9 - x^2 \end{cases}$

3. $\begin{cases} x^2 + y^2 = 25 \\ 24y = x^2 \end{cases}$

4. $\begin{cases} 3x + 4y = 12 \\ x^2 - y + 1 = 0 \end{cases}$

5. $\begin{cases} xy = 1 \\ y = -x^2 \end{cases}$

6. $\begin{cases} x + 2y = 0 \\ xy = -2 \end{cases}$

7. $\begin{cases} 2x^2 + y^2 = 17 \\ x^2 + 2y^2 = 22 \end{cases}$

8. $\begin{cases} x - 2y = 1 \\ y^2 - x^2 = 3 \end{cases}$

9. $\begin{cases} y = 1 - x^2 \\ y = x^2 - 1 \end{cases}$

10. $\begin{cases} xy = 4 \\ y = 4x \end{cases}$

11. $\begin{cases} xy = 4 \\ y = 4x + 1 \end{cases}$

12. $\begin{cases} \dfrac{2}{x^2} + \dfrac{5}{y^2} = 3 \\ \dfrac{3}{x^2} - \dfrac{2}{y^2} = 1 \end{cases}$

13. $\begin{cases} \dfrac{1}{x^2} - \dfrac{3}{y^2} = 14 \\ \dfrac{2}{x^2} + \dfrac{1}{y^2} = 35 \end{cases}$

14. $\begin{cases} y = -\sqrt{x} \\ (x - 3)^2 + y^2 = 4 \end{cases}$

15. $\begin{cases} y = -\sqrt{x - 1} \\ (x - 3)^2 + y^2 = 4 \end{cases}$

16. $\begin{cases} y = -\sqrt{x - 6} \\ (x - 3)^2 + y^2 = 4 \end{cases}$

17. $\begin{cases} y = 2^x \\ y = 2^{2x} - 12 \end{cases}$

18. $\begin{cases} y = e^{4x} \\ y = e^{2x} + 6 \end{cases}$

19. $\begin{cases} 2(\log_{10} x)^2 - (\log_{10} y)^2 = -1 \\ 4(\log_{10} x)^2 - 3(\log_{10} y)^2 = -11 \end{cases}$

20. $\begin{cases} y = \log_2(x + 1) \\ y = 5 - \log_2(x - 3) \end{cases}$

21. $\begin{cases} 2^x \cdot 3^y = 4 \\ x + y = 5 \end{cases}$

22. $\begin{cases} a^{2x} + a^{2y} = 10 \\ a^{x+y} = 4 \end{cases} \quad (a > 0)$

Hint: Use the substitutions $a^x = t$ and $a^y = u$.

In Exercises 23–34:

Ⓖ **(a)** *Use a graphing utility to approximate the solutions (x, y) of each system. Zoom in on the relevant intersection points until you are sure of the first two decimal places of each coordinate.*

(b) *In Exercises 23–28 only, also use an algebraic method of solution. Round the answers to three decimal places and check to see that your results are consistent with the graphical estimates obtained in part (a).*

23. $\begin{cases} y = x - 5 \\ y = -x^2 + 2 \end{cases}$

24. $\begin{cases} y = x^2 - 1 \\ y = -2x^4 + 3 \end{cases}$

25. $\begin{cases} y = \sqrt{x + 1} + 1 \\ 3x + 4y = 12 \end{cases}$

26. $\begin{cases} y = \sqrt{x} \\ y = 2x^3 \end{cases}$

27. $\begin{cases} y = 4^{2x} \\ y = 4^x + 3 \end{cases}$

28. $\begin{cases} y = \ln x \\ y = 1 + \ln(x - 5) \end{cases}$

29. $\begin{cases} y = e^{x/2} \\ y = x^2 \end{cases}$

30. $\begin{cases} x^2 - 2x + y = 0 \\ y = 3x^3 - x^2 - 10x \end{cases}$

31. $\begin{cases} y = \sqrt{x} - 1 \\ y = \ln x \end{cases}$
Hint: There are two solutions.

32. $\begin{cases} y = \sqrt[3]{x} \\ y = \ln x \end{cases}$
Hint: There are two solutions.

33. $\begin{cases} y = x^3 \\ y = (e^x + e^{-x})/2 \end{cases}$
Hint: There are two solutions.

34. $\begin{cases} y = x^3 \\ y = (e^x - e^{-x})/2 \end{cases}$
Hint: There are four solutions; use symmetry to cut the amount of work in half.

35. (a) Let $x = (-1 + \sqrt{13})/6$. Using this x-value, show that the equations $y = 3x + 1$ and $y = 1/x$ yield the same y-value. (This completes a detail mentioned in Example 3.)

(b) Solve the following system. (You should obtain $u = 9$ and $v = 8$, as stated in Example 5.)

$$\begin{cases} 2u - 3v = -6 \\ 3u + 4v = 59 \end{cases}$$

36. A sketch shows that the line $y = 100x$ intersects the parabola $y = x^2$ at the origin. Are there any other intersection points? If so, find them. If not, explain why not.

B

37. Solve the following system for x and y:

$$\begin{cases} ax + by = 2 \\ abxy = 1 \end{cases}$$

(Assume that neither a nor b is zero.)

38. Let a, b, and c be constants (with $a \neq 0$), and consider the system

$$\begin{cases} y = ax^2 + bx + c \\ y = k \end{cases}$$

For which value of k (in terms of a, b, and c) will the system have exactly one solution? What is that solution? What is the relationship between the solution you've found and the graph of $y = ax^2 + bx + c$?

39. Find all solutions of the system

$$\begin{cases} x^3 + y^3 = 3473 \\ x + y = 23 \end{cases}$$

40. Solve the following system for x, y, and z:

$$\begin{cases} yz = p^2 \\ zx = q^2 \\ xy = r^2 \end{cases}$$

(Assume that p, q, and r are positive constants.)

41. If the diagonal of a rectangle has length d and the perimeter of the rectangle is $2p$, express the lengths of the sides in terms of d and p.

42. Solve for x and y in terms of p, q, a, and b:

$$\begin{cases} q^{\ln x} = p^{\ln y} \\ (px)^{\ln a} = (qy)^{\ln b} \end{cases}$$

(Assume all constants and variables are positive.)

43. The accompanying figure shows the graph of a function $N = N_0e^{kt}$. In each case, determine the constants N_0 and k so that the graph passes through
(a) $(2, 3)$ and $(8, 24)$; **(b)** $(\frac{1}{2}, 1)$ and $(4, 10)$.

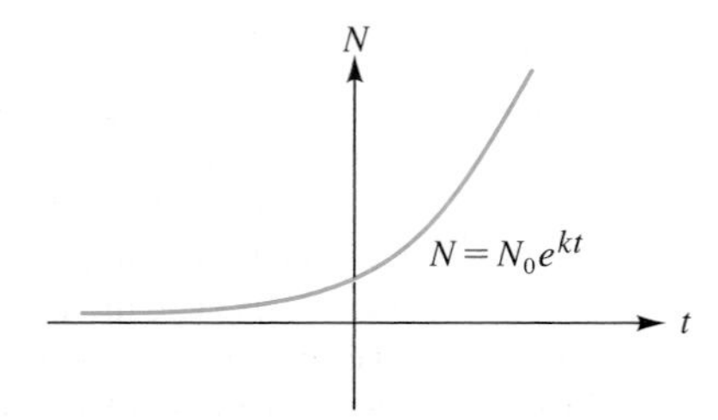

44. [As background for this exercise, first work Exercise 43(a).] Measurements (begun in 1958, by Charles D. Keeling of the Scripps Institution of Oceanography) show that the concentration of carbon dioxide in the atmosphere has increased exponentially over the years 1958–2000. In 1958, the average carbon dioxide concentration was $N = 315$ ppm (parts per million), and in 2000 it was $N = 369$ ppm.

(a) Use the data to determine the constants k and N_0 in the growth law $N = N_0e^{kt}$. *Hint:* The curve $N = N_0e^{kt}$ passes through the two points (1958, 315) and (2000, 369). Use this to obtain a system of two equations in two unknowns.

(b) Stephen H. Schneider in his article "The Changing Climate" [*Scientific American,* vol. 261 (1989), pp. 70–79] estimates that, under certain circumstances, the concentration of carbon dioxide in the atmosphere could reach 600 ppm by the year 2080. Use your results in part (a) to make a projection for the year 2080. Is your projection higher or lower than Dr. Schneider's?

45. Solve the following system for x, y, and z in terms of p, q, and r.

$$\begin{cases} x(x + y + z) = p^2 \\ y(x + y + z) = q^2 \\ z(x + y + z) = r^2 \end{cases}$$

(Assume that p, q, and r are nonzero.) *Hint:* Denote $x + y + z$ by w; then add the three equations.

46. (a) Find the points where the line $y = -2x - 2$ intersects the parabola $y = \frac{1}{2}x^2$.

(b) On the same set of axes, sketch the line $y = -2x - 2$ and the parabola $y = \frac{1}{2}x^2$. Be certain that your sketch is consistent with the results obtained in part (a).

47. If a right triangle has area 180 cm² and hypotenuse 41 cm, find the lengths of the two legs.

48. The sum of two numbers is 8, while their product is -128. What are the two numbers?

49. If a rectangle has perimeter 46 cm and area 60 cm², find the length and the width.

50. Find all right triangles for which the perimeter is 24 units and the area is 24 square units.

51. Solve the following system for x and y using the substitution method:

$$\begin{cases} x^2 + y^2 = 5 & (1) \\ xy = 2 & (2) \end{cases}$$

52. The substitution method in Exercise 51 leads to a quadratic equation. Here is an alternative approach to solving that system; this approach leads to linear equations. Multiply equation (2) by 2 and add the resulting equation to equation (1). Now take square roots to conclude that $x + y = \pm 3$. Next, multiply equation (2) by 2, and subtract the resulting equation from equation (1). Take square roots to conclude that $x - y = \pm 1$. You now have the following four linear systems, each of which can be solved (with almost no work) by the addition–subtraction method.

$$\begin{cases} x + y = 3 \\ x - y = 1 \end{cases} \qquad \begin{cases} x + y = 3 \\ x - y = -1 \end{cases}$$

$$\begin{cases} x + y = -3 \\ x - y = 1 \end{cases} \qquad \begin{cases} x + y = -3 \\ x - y = -1 \end{cases}$$

Solve these systems and compare your results with those obtained in Exercise 51.

53. Solve the following system using the method explained in Exercise 52.

$$\begin{cases} x^2 + y^2 = 7 \\ xy = 3 \end{cases}$$

54. Solve the following system using the substitution method:

$$\begin{cases} 3xy - 4x^2 = 2 \\ -5x^2 + 3y^2 = 7 \end{cases}$$

(Begin by solving the first equation for y.)

55. Here is an alternative approach for solving the system in Exercise 54. Let $y = mx$, where m is a constant to be determined. Replace y with mx in both equations of the system to obtain the following pair of equations:

$$x^2(3m - 4) = 2 \qquad (1)$$
$$x^2(-5 + 3m^2) = 7 \qquad (2)$$

Now divide equation (2) by equation (1). After clearing fractions and simplifying, you can write the resulting equation as $2m^2 - 7m + 6 = 0$. Solve this last equation by factoring. The values of m can then be used in equation (1) to determine values for x. In each case the corresponding y-values are determined by the equation $y = mx$.

C

56. Solve the following system for x and y:

$$\begin{cases} \dfrac{1}{x^2} + \dfrac{1}{xy} = \dfrac{1}{a^2} \\ \dfrac{1}{y^2} + \dfrac{1}{xy} = \dfrac{1}{b^2} \end{cases}$$

(Assume that a and b are positive.)

57. Solve the following system for x and y:

$$\begin{cases} x^4 = y^6 \\ \ln \dfrac{x}{y} = \dfrac{\ln x}{\ln y} \end{cases}$$

10.7 SYSTEMS OF INEQUALITIES

In the first section of this chapter we solved systems of equations in two unknowns. Now we will consider systems of inequalities in two unknowns. The techniques we develop are used in calculus when discussing functions of two variables and in business and economics in the study of linear programming.

Let a, b, and c denote real numbers, and assume that a and b are not both zero. Then all of the following are called **linear inequalities:**

$$ax + by + c < 0 \qquad ax + by + c > 0$$
$$ax + by + c \le 0 \qquad ax + by + c \ge 0$$

An ordered pair of numbers (x_0, y_0) is said to be a **solution** of a given inequality (linear or not) provided that we obtain a true statement on substituting x_0 and y_0 for x and y, respectively. For instance, $(-1, 1)$ is a solution of the inequality $2x + 3y - 6 < 0$, since substitution yields

$$2(-1) + 3(1) - 6 < 0$$
$$-5 < 0 \qquad \text{True}$$

Like a linear equation in two unknowns, a linear inequality has infinitely many solutions. For this reason, we often represent the solutions graphically. When we do this, we say that we are **graphing the inequality.**

For example, let's graph the inequality $y < 2x$. First we observe that the coordinates of the points on the line $y = 2x$ do not, by definition, satisfy this inequality. So it remains to consider points above the line and points below the line. We will in fact show that the required graph consists of all points that lie below the line. Take

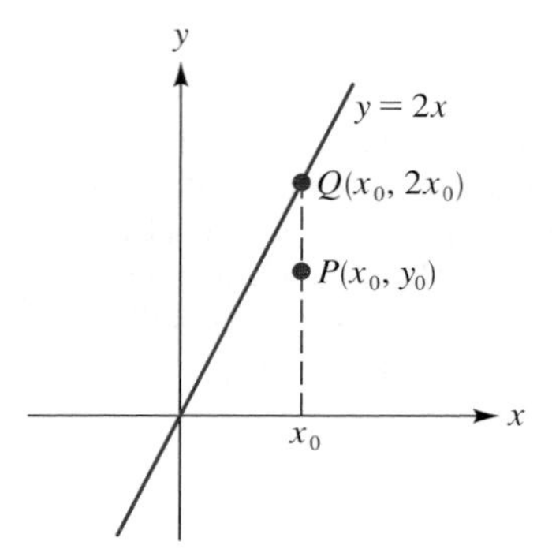

Figure 1

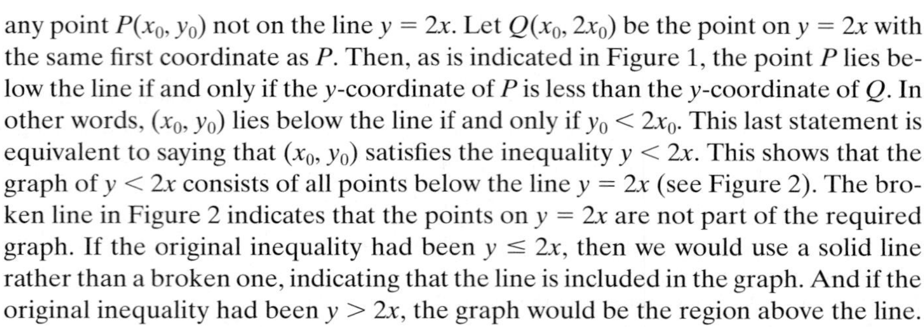

any point $P(x_0, y_0)$ not on the line $y = 2x$. Let $Q(x_0, 2x_0)$ be the point on $y = 2x$ with the same first coordinate as P. Then, as is indicated in Figure 1, the point P lies below the line if and only if the y-coordinate of P is less than the y-coordinate of Q. In other words, (x_0, y_0) lies below the line if and only if $y_0 < 2x_0$. This last statement is equivalent to saying that (x_0, y_0) satisfies the inequality $y < 2x$. This shows that the graph of $y < 2x$ consists of all points below the line $y = 2x$ (see Figure 2). The broken line in Figure 2 indicates that the points on $y = 2x$ are not part of the required graph. If the original inequality had been $y \leq 2x$, then we would use a solid line rather than a broken one, indicating that the line is included in the graph. And if the original inequality had been $y > 2x$, the graph would be the region above the line.

Just as the graph of $y < 2x$ is the region below the line $y = 2x$, it is true in general that the graph of $y < f(x)$ is the region below the graph of the function f. For example, Figures 3 and 4 display the graphs of $y < x^2$ and $y \geq x^2$.

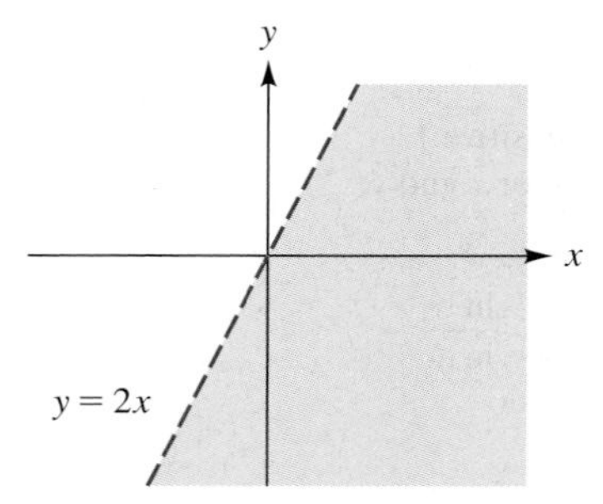

Figure 2
The graph of $y < 2x$

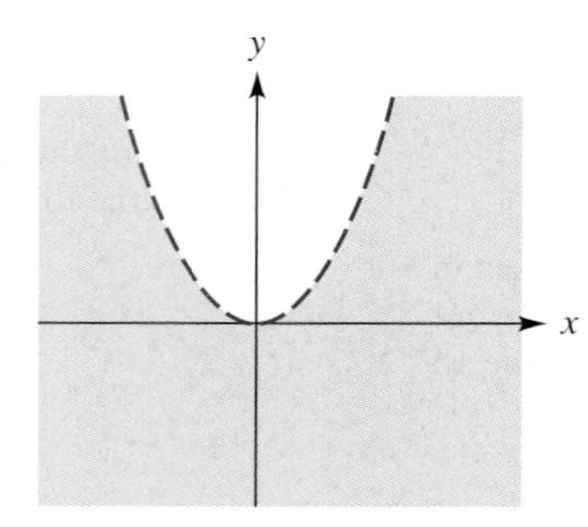

Figure 3
$y < x^2$

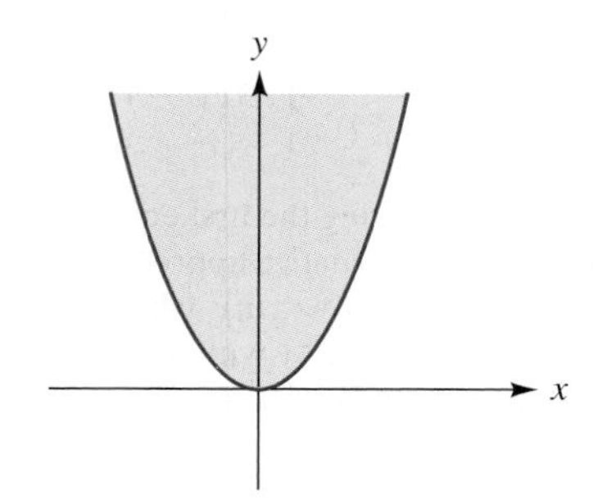

Figure 4
$y \geq x^2$

Example 1 summarizes the technique developed so far for graphing an inequality. Following this example, we will point out a useful alternative method.

EXAMPLE 1 Graphing an inequality

Graph the inequality $4x - 3y \leq 12$.

SOLUTION

The graph will include the line $4x - 3y = 12$ and either the region above the line or the region below it. To decide which region, we solve the inequality for y:

$$4x - 3y \leq 12$$
$$-3y \leq -4x + 12$$
$$y \geq \frac{4}{3}x - 4 \quad \text{Multiplying or dividing by a negative number reverses an inequality.}$$

This last inequality tells us that we want the region above the line, as well as the line itself. Figure 5 displays the required graph.

Figure 5
The graph of $4x - 3y \leq 12$

There is another method that we can use in Example 1 to determine which side of the line we want. This method involves a **test point.** We pick any convenient point that is not on the line $4x - 3y = 12$. Then we test to see whether this point satisfies the given inequality. For example, let's pick the point (0, 0). Substituting these coordinates in the inequality $4x - 3y \leq 12$ yields

$$4(0) - 3(0) \stackrel{?}{\leq} 12$$
$$0 \stackrel{?}{\leq} 12 \qquad \text{True}$$

We conclude from this that the required side of the line includes the point (0, 0). In agreement with Figure 5, we see this is the region above the line.

Next we discuss *systems* of inequalities in two unknowns. As with systems of equations, a **solution** of a system of inequalities is an ordered pair (x_0, y_0) that satisfies all of the inequalities in the system. As a first example, let's graph the points that satisfy the following nonlinear system.

$$\begin{cases} y - x^2 \geq 0 \\ x^2 + y^2 < 1 \end{cases}$$

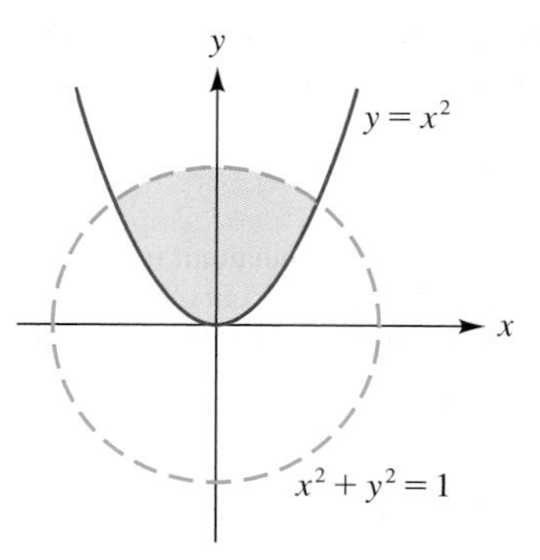

Figure 6

By writing the first inequality as $y \geq x^2$, we see that it describes the set of points on or above the parabola $y = x^2$. For the second inequality we must decide whether it describes the points inside or outside the circle $x^2 + y^2 = 1$. One way to do this is to choose (0, 0) as a test point. Substituting the values $x = 0$ and $y = 0$ in the inequality $x^2 + y^2 < 1$ yields the true statement $0^2 + 0^2 < 1$. Since (0, 0) lies within the circle and satisfies the inequality, we conclude that the inequality $x^2 + y^2 < 1$ describes the set of all points within the circle. Now we put our information together. We wish to graph the points that lie on or above the parabola $y = x^2$ but within the circle $x^2 + y^2 = 1$. This is the shaded region shown in Figure 6.

EXAMPLE 2 Graphing a system of inequalities

Graph the system

$$\begin{cases} -x + 3y \leq 12 \\ x + y \leq 8 \\ x \geq 0 \\ y \geq 0 \end{cases}$$

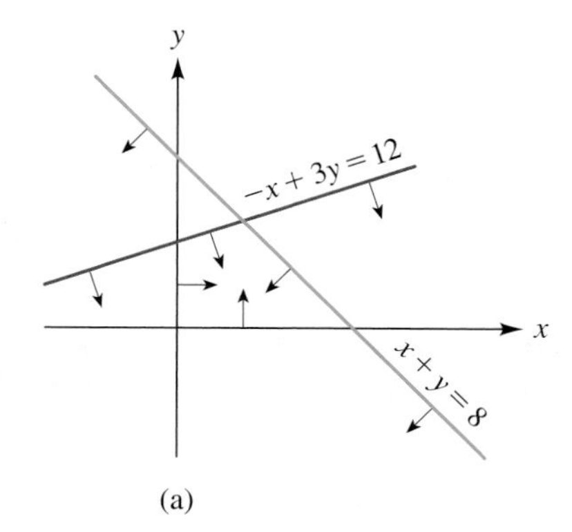

(a)

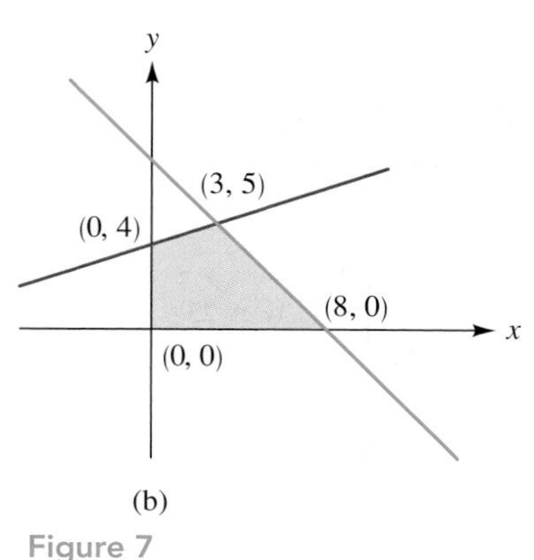

(b)

Figure 7

SOLUTION

Solving the first inequality for y, we have

$$3y \leq x + 12$$
$$y \leq \frac{1}{3}x + 4$$

Thus the first inequality is satisfied by the points on or below the line $-x + 3y = 12$. Similarly, by solving the second inequality for y, we see that it describes the set of points on or below the line $x + y = 8$. The third inequality, $x \geq 0$, describes the points on or to the right of the y-axis. Similarly, the fourth inequality, $y \geq 0$, describes the points on or above the x-axis. We summarize these statements in Figure 7(a). The arrows indicate which side of each line we wish to consider.

Finally, Figure 7(b) shows the graph of the given system. The coordinates of each point in the shaded region, including its boundary, satisfy all four of the given inequalities. The coordinates (3, 5) in Figure 7(b) were found by solving the system of linear equations $-x + 3y = 12$ and $x + y = 8$. (You'll need to carry out this work as part of Exercise 25.)

We can use Figure 7(b) to introduce some terminology that is useful in describing sets of points in the plane. A **vertex** of a region is a corner, or point, where two adjacent bounding sides meet. Thus the vertices of the shaded region in Figure 7(b) are the four points $(0, 0)$, $(8, 0)$, $(3, 5)$, and $(0, 4)$. The shaded region in Figure 7(b) is **convex.** This means that, given any two points in that region, the straight line segment joining these two points lies wholly within the region. Figure 8(a) displays another example of a convex set, while Figure 8(b) shows a set that is not convex. The shaded region in Figure 8(b) is also an example of a **bounded region.** By this we mean that the region can be wholly contained within some (sufficiently large) circle. Perhaps the simplest example of a region that is not bounded is the entire x-y plane itself. Figure 9 shows another example of an unbounded region.

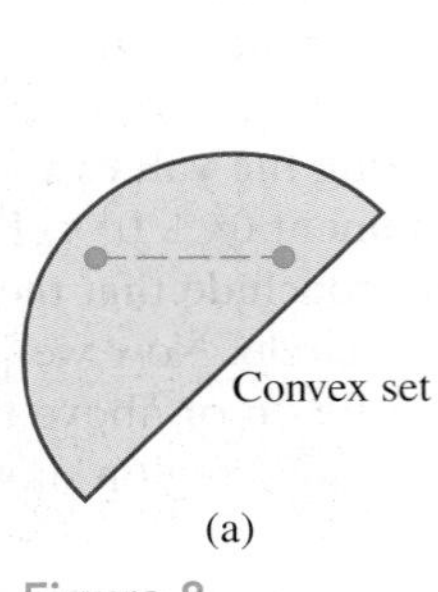
Convex set

(a)

Not a convex set

(b)

Figure 8

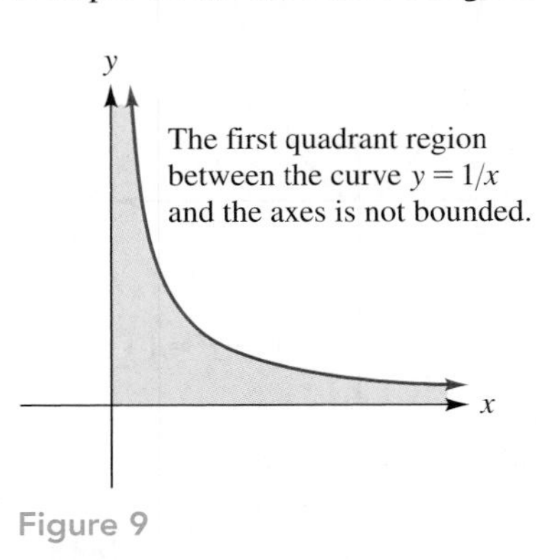
y
The first quadrant region
between the curve y = 1/x
and the axes is not bounded.
x

Figure 9

EXERCISE SET 10.7 

A

In Exercises 1 and 2, decide whether or not the ordered pairs are solutions of the given inequality.

1. $4x - 6y + 3 \geq 0$
(a) $(1, 2)$
(b) $(0, \frac{1}{2})$

2. $5x + 2y < 1$
(a) $(-1, 3)$
(b) $(0, 0)$

In Exercises 3–16, graph the given inequalities.

3. $2x - 3y > 6$
4. $2x - 3y < 6$
5. $2x - 3y \geq 6$
6. $2x - 3y \leq 6$
7. $x - y < 0$
8. $y \leq \frac{1}{2}x - 1$
9. $x \geq 1$
10. $y < 0$
11. $x > 0$
12. $y \leq \sqrt{x}$
13. $y > x^3 + 1$
14. $y \leq |x - 2|$
15. $x^2 + y^2 \geq 25$
16. $y \geq e^x - 1$

In Exercises 17–22, graph the systems of inequalities.

17. $\begin{cases} y \leq x^2 \\ x^2 + y^2 \leq 1 \end{cases}$

18. $\begin{cases} y < x \\ x^2 + y^2 < 1 \end{cases}$

19. $\begin{cases} y \geq 1 \\ y \leq |x| \end{cases}$

20. $\begin{cases} x \geq 0 \\ y \geq 0 \\ y < \sqrt{x} \\ x \leq 4 \end{cases}$

21. $\begin{cases} x \geq 0 \\ y \geq 0 \\ y \leq 1 - x^2 \end{cases}$

22. $\begin{cases} y < 2x \\ y > \frac{1}{2}x \end{cases}$

In Exercises 23–34, graph the systems of linear inequalities. In each case specify the vertices. Is the region convex? Is the region bounded?

23. $\begin{cases} y \leq x + 5 \\ y \leq -2x + 14 \\ x \geq 0 \\ y \geq 0 \end{cases}$

24. $\begin{cases} y \geq x + 5 \\ y \geq -2x + 14 \\ x \geq 0 \\ y \geq 0 \end{cases}$

25. $\begin{cases} -x + 3y \leq 12 \\ x + y \leq 8 \\ x \geq 0 \\ y \geq 0 \end{cases}$

26. $\begin{cases} y \geq 2x \\ y \geq -x + 6 \end{cases}$

27. $\begin{cases} 0 \leq 2x - y + 3 \\ x + 3y \leq 23 \\ 5x + y \leq 45 \end{cases}$

28. $\begin{cases} 0 \leq 2x - y + 3 \\ x + 3y \leq 23 \\ 5x + y \leq 45 \\ x \geq 0 \\ y \geq 0 \end{cases}$

29. $\begin{cases} 5x + 6y < 30 \\ y > 0 \end{cases}$

30. $\begin{cases} 5x + 6y < 30 \\ x > 0 \end{cases}$

31. $\begin{cases} 5x + 6y < 30 \\ x > 0 \\ y > 0 \end{cases}$

32. $\begin{cases} 2x + 3y \geq 6 \\ 2x + 3y \leq 12 \end{cases}$

33. $\begin{cases} x \geq 0 \\ y \geq 0 \\ 20 - x \geq 0 \\ 30 - y \geq 0 \\ x + y \leq 40 \\ x + y \geq 35 \end{cases}$

34. $\begin{cases} x \geq 0 \\ y \geq 0 \\ 3x - y + 1 \geq 0 \\ 0 \leq x - y + 3 \\ y \leq 5 \\ x \leq \frac{1}{3}(17 - y) \\ x \leq \frac{1}{2}(y + 8) \end{cases}$

B

35. Graph the following system of inequalities and specify the vertices.

$$\begin{cases} x \geq 0 \\ y \geq e^x \\ y \leq e^{-x} + 1 \end{cases}$$

A formula such as

$$f(x, y) = \sqrt{2x - y + 1} \qquad (1)$$

defines a **function of two variables.** *The inputs for such a function are ordered pairs (x, y) of real numbers. For example, using the ordered pair $(3, 5)$ as an input, we have*

$$f(3, 5) = \sqrt{2(3) - 5 + 1} = \sqrt{2}$$

So the input $(3, 5)$ yields an output of $\sqrt{2}$. We define the domain for this function just as we did in Chapter 3: The domain is the set of all inputs that yield real-number outputs. For instance, the ordered pair $(1, 4)$ is not in the domain of the function we have been discussing, because (as you should check for yourself) $f(1, 4) = \sqrt{-1}$, which is not a real number. We can determine the domain of the function in equation (1) by requiring that the quantity under the radical sign be nonnegative. Thus we require that $2x - y + 1 \geq 0$ and, consequently, $y \leq 2x + 1$. (Check this.) The following figure shows the graph of this inequality; the domain of our function is the set of ordered pairs making up the graph. In Exercises 36–41, follow a similar procedure and sketch the domain of the given function.

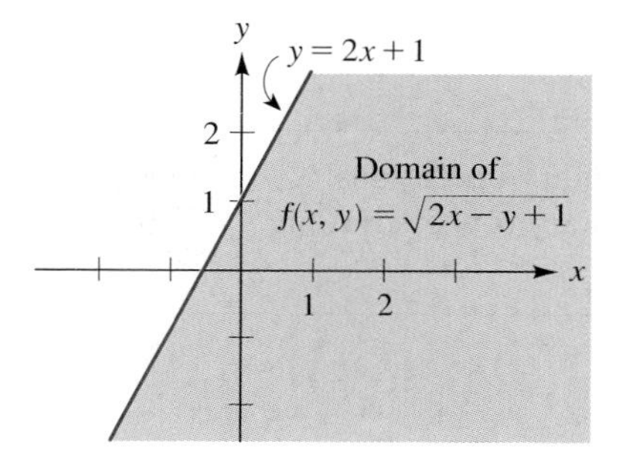

36. $f(x, y) = \sqrt{x + y + 2}$

37. $f(x, y) = \sqrt{x^2 + y^2 - 1}$

38. $g(x, y) = \sqrt{25 - x^2 - y^2}$

39. $g(x, y) = \ln(x^2 - y)$

40. $h(x, y) = \ln(xy)$

41. $h(x, y) = \sqrt{x} + \sqrt{y}$

Chapter 10 Summary of Principal Terms

Terms	Page reference	Comments
1. Linear equation in two variables	729	A linear equation in two variables is an equation of the form $ax + by = c$, where a, b, and c are constants, with both a and b not both zero, and x and y are variables or unknowns. Similarly, a linear equation in three variables is an equation of the form $ax + by + cz = d$, with a, b, and c not all zero
2. Solution of a linear equation	729	A solution of the linear equation $ax + by = c$ is an ordered pair of numbers (x_0, y_0) such that $ax_0 + by_0 = c$. Similarly, a solution of the linear equation $ax + by + cz = d$ is an ordered triple of numbers (x_0, y_0, z_0) such that $ax_0 + by_0 + cz_0 = d$.
3. Consistent system; inconsistent system	729	A system of equations is consistent if it has at least one solution; otherwise, the system is inconsistent. See Figures 2, 3, and 4 in Section 10.1 for a geometric interpretation of these terms.
4. Equivalent systems	732	Two systems of equations are equivalent if they have the same solution set.

Terms	Page reference	Comments
5. Upper-triangular form (three variables)	744	A system of linear equations in x, y, and z is said to be in upper-triangular form if x appears in no equation after the first and y appears in no equation after the second. This definition can be extended to include linear systems with any number of unknowns. See Table 1 in Section 10.2 for examples of systems that are in upper-triangular form. When a system is in upper-triangular form, it is a simple matter to obtain the solutions; see, for instance, Examples 1 and 2 in Section 10.2.
6. Gaussian elimination	746–748 758–760	This is a technique for converting a system of equations or its augmented matrix to upper-triangular form. See Examples 4 and 5 in Section 10.2 and Example 3 in Section 10.3.
7. Elementary operations, elementary row operations	746–747 759	These are operations that can be performed on an equation in a system without altering the solution set. See the boxes on pages 746, 747, and 759 for a list of these operations and notations.
8. Matrix	757	A matrix is a rectangular array of numbers, enclosed in parentheses or brackets. The numbers constituting the rectangular array are called the entries or the elements in the matrix. The size or dimension of a matrix is expressed by specifying the number of rows and the number of columns, in that order. For examples of this terminology, see Example 1 in Section 10.3.
9. Matrix equality	761	Two matrices are said to be equal provided that they are the same size and their corresponding entries are equal.
10. Matrix addition and scalar multiplication	761, 763	To add two matrices of the same size, add the corresponding entries. To multiply a matrix by a scalar, multiply each entry by that scalar.
11. Matrix multiplication	764, 765	Let A and B be two matrices. The matrix product AB is defined only when the number of columns in A is the same as the number of rows in B. In this case the matrix AB will have as many rows as A and as many columns as B. The entry in the ith row and jth column of AB is the number formed as follows: Multiply the corresponding entries in the ith row of A and the jth column of B, then add the results.
12. Square matrix	771	A matrix that has the same number of rows and columns is called a square matrix. An $n \times n$ square matrix is said to be an nth-order square matrix.
13. Multiplicative identity matrix	771, 772	For the set of $n \times n$ matrices the multiplicative identity matrix I_n is the $n \times n$ matrix with ones down the main diagonal (upper left corner to bottom right corner) and zeros everywhere else. For example, $I_2 = \begin{pmatrix} 1 & 0 \\ 0 & 1 \end{pmatrix}$ and $I_3 = \begin{pmatrix} 1 & 0 & 0 \\ 0 & 1 & 0 \\ 0 & 0 & 1 \end{pmatrix}$ For every $n \times n$ matrix A we have $AI_n = A$ and $I_nA = A$.

Terms	Page reference	Comments
14. Inverse matrix	772, 775	Given an $n \times n$ matrix A, the inverse matrix (if it exists) is denoted by A^{-1}. The defining equations for A^{-1} are $AA^{-1} = I$ and $A^{-1}A = I$. Here, I stands for the multiplicative identity matrix that is the same size as A. If a square matrix has an inverse, it is said to be invertible; if there is no inverse, then the matrix is noninvertible.
15. Minor	786, 787	The minor of an entry b in a determinant is the determinant formed by suppressing the entries in the row and column in which b appears.
16. Cofactor	786, 787	If an entry b appears in the ith row and the jth column of a determinant, then the cofactor of b is computed by multiplying the minor of b by the number $(-1)^{i+j}$.
17. Determinant	787	The value of the 2×2 determinant $\begin{vmatrix} a & b \\ c & d \end{vmatrix}$ is defined to be $ad - bc$. For larger determinants the value can be found as follows. Select any row or column and multiply each entry in that row or column by its cofactor, then add the results. The resulting sum is the value of the determinant. It can be shown that this value is independent of the particular row or column that is chosen.
18. Cramer's rule	792	This rule yields the solutions of certain systems of linear equations in terms of determinants. For a statement of the rule, see page 792. Example 5 on pages 792–793 shows how the rule is applied. The proof of Cramer's rule begins on page 793.
19. Linear inequality in two variables	805	A linear inequality in two variables is any one of the four types of inequalities that result when the equal sign in the equation $ax + by = c$ is replaced by one of the four symbols $<, \leq, >, \geq$. For any of these linear inequalities a solution is an ordered pair of numbers (x_0, y_0) with the property that a true statement is obtained when x and y (in the inequality) are replaced by x_0 and y_0, respectively.
20. Vertex	808	A vertex of a region in the x-y plane is a corner or point where two adjacent bounding sides meet.
21. Convex region	808	A region in the x-y plane is convex if, given any two points in the region, the line segment joining those points lies wholly within the region.
22. Bounded region	808	A region in the x-y plane is bounded if it can be completely contained within some (sufficiently large) circle.

Writing Mathematics

On your own or with a group of classmates, complete the following exercise. Then (strictly on your own) write out a detailed solution. This will involve a combination of English composition and algebra (much like the exposition in this textbook). At each stage, be sure to tell the reader (in complete sentences) where you are headed and why each of the main steps is necessary.

Three integers are said to form a **Pythagorean triple** if the square of the largest is equal to the sum of the squares of the other two. For example, 3, 4, and 5 form a Pythagorean triple because $3^2 + 4^2 = 5^2$. Do you know any other Pythagorean triples? This exercise shows how to develop expressions for generating an infinite number of Pythagorean triples.

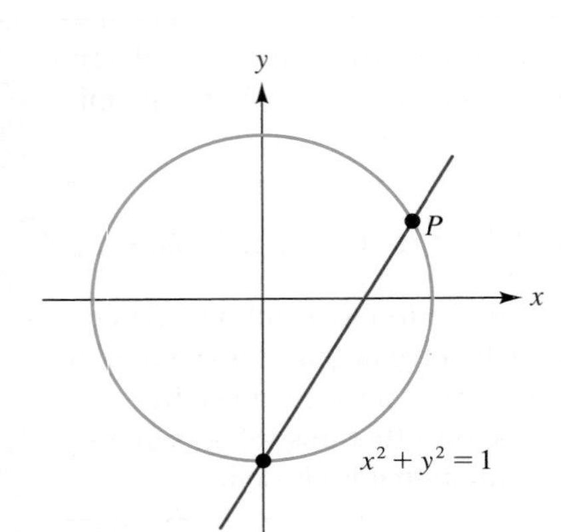

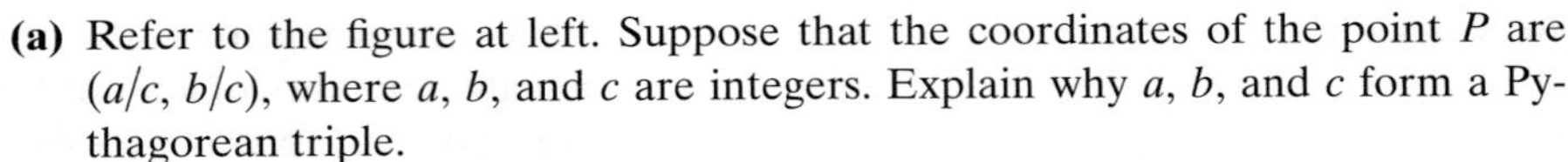

(a) Refer to the figure at left. Suppose that the coordinates of the point P are $(a/c, b/c)$, where a, b, and c are integers. Explain why a, b, and c form a Pythagorean triple.

(b) Let m denote the slope of the line passing through P and $(0, -1)$. Using the methods of Section 10.6, show that the coordinates of the point P are

$$\left(\frac{2m}{m^2 + 1}, \frac{m^2 - 1}{m^2 + 1}\right)$$

(c) Use the results in parts (a) and (b) to explain why the three numbers $2m$, $m^2 - 1$, and $m^2 + 1$ form a Pythagorean triple. Then complete the following table to obtain examples of Pythagorean triples.

m	$2m$	$m^2 - 1$	$m^2 + 1$
2			
4			
6			
8			

Chapter 10 Review Exercises

In Exercises 1–31, solve each system of equations. If there are no solutions in a particular case, say so. In cases in which there are literal (rather than numerical) coefficients, specify any restrictions that your solutions impose on those coefficients.

1. $\begin{cases} x + y = -2 \\ x - y = 8 \end{cases}$

2. $\begin{cases} x - y = 1 \\ x + y = 5 \end{cases}$

3. $\begin{cases} 2x + y = 2 \\ x + 2y = 7 \end{cases}$

4. $\begin{cases} 3x + 2y = 6 \\ 5x + 4y = 4 \end{cases}$

5. $\begin{cases} 7x + 2y = 9 \\ 4x + 5y = 63 \end{cases}$

6. $\begin{cases} \dfrac{x}{2} + \dfrac{y}{3} = 9 \\ \dfrac{x}{5} - \dfrac{y}{2} = -4 \end{cases}$

7. $\begin{cases} 2x - \dfrac{y}{2} = -8 \\ \dfrac{x}{3} + \dfrac{y}{8} = -1 \end{cases}$

8. $\begin{cases} 3x - 14y - 1 = 0 \\ -6x + 28y - 3 = 0 \end{cases}$

9. $\begin{cases} 3x + 5y - 1 = 0 \\ 9x - 10y - 8 = 0 \end{cases}$

10. $\begin{cases} 9x + 15y - 1 = 0 \\ 6x + 10y + 1 = 0 \end{cases}$

11. $\begin{cases} \dfrac{2}{3}x = -\dfrac{1}{2}y - 12 \\ \dfrac{x}{2} = y + 2 \end{cases}$

12. $\begin{cases} 0.1x + 0.2y = -5 \\ -0.2x - 0.5y = 13 \end{cases}$

13. $\begin{cases} \dfrac{1}{x} + \dfrac{1}{y} = -1 \\ \dfrac{2}{x} + \dfrac{5}{y} = -14 \end{cases}$

14. $\begin{cases} \dfrac{2}{x} + \dfrac{15}{y} = -9 \\ \dfrac{1}{x} + \dfrac{10}{y} = -2 \end{cases}$

15. $\begin{cases} ax + (1-a)y = 1 \\ (1-a)x + y = 0 \end{cases}$

16. $\begin{cases} ax - by - 1 = 0 \\ (a-1)x + by + 2 = 0 \end{cases}$

17. $\begin{cases} 2x - y = 3a^2 - 1 \\ 2y + x = 2 - a^2 \end{cases}$

18. $\begin{cases} 3ax + 2by = 3a^2 - ab + 2b^2 \\ 3bx + 2ay = 2a^2 + 5ab - 3b^2 \end{cases}$

19. $\begin{cases} 5x - y = 4a^2 - 6b^2 \\ 2x + 3y = 5a^2 + b^2 \end{cases}$

20. $\begin{cases} \dfrac{2b}{x} - \dfrac{3}{y} = 7ab \\ \dfrac{4a}{x} + \dfrac{5a}{by} = 3a^2 \end{cases}$

21. $\begin{cases} px - qy = q^2 \\ qx + py = p^2 \end{cases}$

22. $\begin{cases} x - y = \dfrac{a-b}{a+b} \\ x + y = 1 \end{cases}$

23. $\begin{cases} x + y + z = 9 \\ x - y - z = -5 \\ 2x + y - 2z = -1 \end{cases}$

24. $\begin{cases} x - 4y + 2z = 9 \\ 2x + y + z = 3 \\ 3x - 2y - 3z = -18 \end{cases}$

25. $\begin{cases} 4x - 4y + z = 4 \\ 2x + 3y + 3z = -8 \\ x + y + z = -3 \end{cases}$

26. $\begin{cases} x - 8y + z = 1 \\ 5x + 16y + 3z = 3 \\ 4x - 4y - 4z = -4 \end{cases}$

27. $\begin{cases} 4x + 2y - 3z = 15 \\ 2x + y + 3z = 3 \end{cases}$

28. $\begin{cases} 3x + 2y + 17z = 1 \\ x + 2y + 3z = 3 \end{cases}$

29. $\begin{cases} 9x + y + z = 0 \\ -3x + y - z = 0 \\ 3x - 5y + 3z = 0 \end{cases}$

30. $\begin{cases} ax + by - 2az = 4ab + 2b^2 \\ x + y + z = 4a + 2b \\ bx + ay + 4az = 5a^2 + b^2 \end{cases}$

31. $\begin{cases} x + y + z + w = 8 \\ 3x + 3y - z - w = 20 \\ 4x - y - z + 2w = 18 \\ 2x + 5y + 5z - 5w = 8 \end{cases}$

In Exercises 32–36, evaluate each of the determinants.

32. $\begin{vmatrix} \frac{1}{6} & 1 \\ 2 & 12 \end{vmatrix}$

33. $\begin{vmatrix} 2 & 6 & 4 \\ 6 & 18 & 24 \\ 15 & 5 & -10 \end{vmatrix}$

34. $\begin{vmatrix} 0 & 2 & 4 & 0 \\ 4 & 0 & 6 & 2 \\ 0 & 0 & 1 & 1 \\ 14 & 7 & 1 & 0 \end{vmatrix}$

35. $\begin{vmatrix} 1 & 0 & 0 & 0 \\ 0 & 2 & 0 & 0 \\ 0 & 0 & 3 & 0 \\ 0 & 0 & 0 & 4 \end{vmatrix}$

36. $\begin{vmatrix} 1 & a & b & c \\ 0 & 2 & d & e \\ 0 & 0 & 3 & f \\ 0 & 0 & 0 & 4 \end{vmatrix}$

37. Show that $\begin{vmatrix} a & b & c \\ b & c & a \\ c & a & b \end{vmatrix} = 3abc - a^3 - b^3 - c^3$.

38. Show that $\begin{vmatrix} 1 & 1 & 1 \\ 1 & 1+x & 1 \\ 1 & 1 & 1+x^2 \end{vmatrix} = x^3$.

39. Determine constants a and b so that the parabola $y = ax^2 + bx - 1$ passes through the points $(-2, 5)$ and $(2, 9)$.

40. Find two numbers whose sum and difference are 52 and 10, respectively.

41. This exercise appears in *Plane and Solid Analytic Geometry,* by W. F. Osgood and W. G. Graustein (New York: Macmillan, 1920): Let P be any point (a, a) of the line $x - y = 0$, other than the origin. Through P draw two lines, of arbitrary slopes m_1 and m_2, intersecting the x-axis in A_1 and A_2, and the y-axis in B_1 and B_2, respectively. Prove that the lines $\overline{A_1B_2}$ and $\overline{A_2B_1}$ will, in general, meet on the line $x + y = 0$.

42. Determine constants h, k, and r so that the circle $(x - h)^2 + (y - k)^2 = r^2$ passes through the three points $(0, 0)$, $(0, 1)$ and $(1, 0)$.

43. The vertices of a triangle are the points of intersection of the lines $y = x - 1$, $y = -x - 2$, and $y = 2x + 3$. Find the equation of the circle passing through these three intersection points.

44. The vertices of triangle ABC are $A(0, 0)$, $B(3, 0)$, and $C(0, 4)$.

(a) Find the center and the radius of the circle that passes though A, B, and C. This circle is called the **circumcircle** for triangle ABC.

(b) The figure on the following page shows the **inscribed circle** for triangle ABC; this is the circle that is tangent to all three sides of the triangle. Find the center and the radius of this circle using the following two facts.

(i) The center of the inscribed circle is the common intersection point of the three angle bisectors of the triangle.

(*continues*)

(ii) The line that bisects angle B has slope $-1/2$.

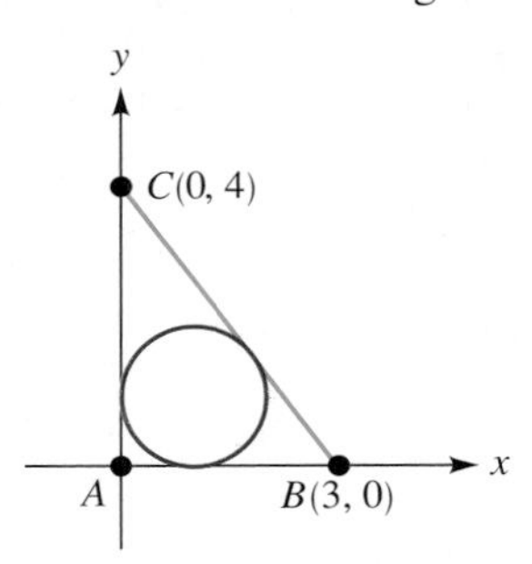

(c) Verify the following statement for triangle ABC. (The statement actually holds for any triangle; the result is known as *Euler's theorem.*) Let R and r denote the radii of the circles in parts (a) and (b), respectively. Then the distance d between the centers of the circles satisfies the equation

$$d^2 = R^2 - 2rR$$

(d) Verify the following statement for triangle ABC (which in fact holds for any triangle): The area of triangle ABC is equal to one-half the product of the perimeter and the radius of the inscribed circle.

(e) Verify the following statement for triangle ABC (which in fact holds for any triangle): The sum of the reciprocals of the lengths of the altitudes is equal to the reciprocal of the radius of the inscribed circle.

In Exercises 45–61, the matrices A, B, C, and D are defined as follows:

$$A = \begin{pmatrix} 3 & -2 \\ 1 & 5 \end{pmatrix} \qquad B = \begin{pmatrix} 2 & 1 \\ 1 & 8 \end{pmatrix}$$

$$C = \begin{pmatrix} -1 & 0 \\ 0 & -1 \end{pmatrix} \qquad D = \begin{pmatrix} 3 & -1 \\ 4 & 1 \\ -5 & 9 \end{pmatrix}$$

In each exercise, carry out the indicated matrix operations if they are defined. If an operation is not defined, say so.

45. $2A + 2B$ **46.** $2(A + B)$ **47.** $4B$
48. $B + 4$ **49.** AB **50.** BA
51. $AB - BA$ **52.** $B + D$ **53.** $B + C$
54. $A(B + C)$ **55.** $AB + AC$ **56.** $(B + C)A$
57. $BA + CA$ **58.** $(A + B) + C$ **59.** $A + (B + C)$
60. $(AB)C$ **61.** $A(BC)$

62. For a square matrix A, the notation A^2 means AA. Similarly, A^3 means AAA. If $A = \begin{pmatrix} 1 & 1 \\ 0 & 1 \end{pmatrix}$, verify that

$$A^2 = \begin{pmatrix} 1 & 2 \\ 0 & 1 \end{pmatrix} \quad \text{and} \quad A^3 = \begin{pmatrix} 1 & 3 \\ 0 & 1 \end{pmatrix}$$

63. Let A be the matrix $\begin{pmatrix} 0 & 0 & 0 \\ a & 0 & 0 \\ b & c & 0 \end{pmatrix}$. Compute A^2 and A^3.

For Exercises 64 and 65, in each case compute the inverse of the matrix in part (a), and then use that inverse to solve the system of equations in part (b).

64. (a) $\begin{pmatrix} 5 & -4 \\ 14 & -11 \end{pmatrix}$ **(b)** $\begin{cases} 5x - 4y = 2 \\ 14x - 11y = 5 \end{cases}$

65. (a) $\begin{pmatrix} 1 & -2 & 3 \\ 2 & -5 & 10 \\ -1 & 2 & -2 \end{pmatrix}$

(b) $\begin{cases} x - 2y + 3z = -2 \\ 2x - 5y + 10z = -3 \\ -x + 2y - 2z = 6 \end{cases}$

66. Find the inverse of the matrix $\begin{pmatrix} 1 & -1 & 0 & -3 \\ 5 & -2 & 3 & -11 \\ 2 & 3 & 2 & -3 \\ 4 & 5 & 5 & -5 \end{pmatrix}$.

In Exercises 67–74, compute D, D_x, D_y, D_z (and D_w where appropriate) for each system of equations. Use Cramer's rule to solve the systems in which $D \neq 0$. If $D = 0$, solve the system using Gaussian elimination or matrix methods.

67. $\begin{cases} 2x - y + z = 1 \\ 3x + 2y + 2z = 0 \\ x - 5y - 3z = -2 \end{cases}$ **68.** $\begin{cases} x + 2y - z = -1 \\ 2x - 3y + 3z = 3 \\ 2x + 3y + z = 1 \end{cases}$

69. $\begin{cases} x + 2y + 3z = -1 \\ 4x + 5y + 6z = 2 \\ 7x + 8y + 9z = -3 \end{cases}$ **70.** $\begin{cases} 3x + 2y - 2z = 0 \\ 2x + 3y - z = 0 \\ 8x + 7y - 5z = 0 \end{cases}$

71. $\begin{cases} 3x + 2y - 2z = 1 \\ 2x + 3y - z = -2 \\ 8x + 7y - 5z = 0 \end{cases}$

72. $\begin{cases} x + y + z + w = 5 \\ x - y - z + w = 3 \\ 2x + 3y + 3z + 2w = 21 \\ 4z - 3w = -7 \end{cases}$

73. $\begin{cases} 2x - y + z + 3w = 15 \\ x + 2y + 2w = 12 \\ 3y + 3z + 4w = 12 \\ -4x + y - 4z = -11 \end{cases}$

74. $\begin{cases} x + y + z = (a + b)^2 \\ \dfrac{bx}{a} + \dfrac{ay}{b} - z = 0 \\ x + y - z = (a - b)^2 \end{cases}$

In Exercises 75–86, find all solutions (x, y) for each system, where x and y are real numbers.

75. $\begin{cases} y = 6x \\ y = x^2 \end{cases}$

76. $\begin{cases} y = 4x \\ y = x^3 \end{cases}$

77. $\begin{cases} y = 9 - x^2 \\ y = x^2 - 9 \end{cases}$

78. $\begin{cases} x^3 - y = 0 \\ xy - 16 = 0 \end{cases}$

79. $\begin{cases} x^2 - y^2 = 9 \\ x^2 + y^2 = 16 \end{cases}$

80. $\begin{cases} 2x + 3y = 6 \\ y = \sqrt{x + 1} \end{cases}$

81. $\begin{cases} x^2 + y^2 = 1 \\ y = \sqrt{x} \end{cases}$

82. $\begin{cases} \dfrac{x}{11} + \dfrac{y}{12} = 2 \\ \dfrac{xy}{132} = 1 \end{cases}$

83. $\begin{cases} x^2 + 2xy + 3y^2 = 68 \\ 3x^2 - xy + y^2 = 18 \end{cases}$

84. $\begin{cases} \dfrac{x^2}{a^2} + \dfrac{y^2}{b^2} = 1 \\ \dfrac{x^2}{b^2} + \dfrac{y^2}{a^2} = 1 \end{cases} \quad (a > b > 0)$

85. $\begin{cases} 2(x - 3)^2 - (y + 1)^2 = -1 \\ -3(x - 3)^2 + 2(y + 1)^2 = 6 \end{cases}$

Hint: Let $u = x - 3$ and $v = y + 1$.

86. $\begin{cases} x^4 = y - 1 \\ y - 3x^2 + 1 = 0 \end{cases}$

Exercises 87–92 appear (in German) in an algebra text by Leonhard Euler, first published in 1770. The English versions given here are taken from the translated version, Elements of Algebra, 5th ed., *by Leonhard Euler (London: Longman, Orme, and Co., 1840). [This, in turn, has been reprinted by Springer-Verlag (New York, 1984).]*

87. Required two numbers, whose sum may be s, and their proportion as a to b.

Answer: $\dfrac{as}{a + b}$ and $\dfrac{bs}{a + b}$

88. The sum $2a$, and the sum of the squares $2b$, of two numbers being given; to find the numbers.

Answer: $a - \sqrt{b - a^2}$ and $a + \sqrt{b - a^2}$

89. To find three numbers, so that [the sum of] one-half of the first, one-third of the second, and one-quarter of the third, shall be equal to 62; one-third of the first, one-quarter of the second, and one-fifth of the third, equal to 47; and one-quarter of the first, one-fifth of the second, and one-sixth of the third, equal to 38.

Answer: 24, 60, 120

90. Required two numbers, whose product may be 105, and whose squares [when added] may together make 274.

91. Required two numbers, whose product may be m, and the sum of the squares n $(n \geq 2m)$.

92. Required two numbers such that their sum, their product, and the difference of their squares may all be equal.

In Exercises 93–96, graph each system of inequalities and specify whether the region is convex or bounded.

93. $\begin{cases} x^2 + y^2 \geq 1 \\ y - 4x \leq 0 \\ y - x \geq 0 \\ x \geq 0 \\ y \geq 0 \end{cases}$

94. $\begin{cases} x^2 + y^2 \leq 1 \\ x \geq 0 \\ y \geq 0 \end{cases}$

95. $\begin{cases} y - \sqrt{x} \leq 0 \\ y \geq 0 \\ x \geq 1 \\ x - 4 \leq 0 \end{cases}$

96. $\begin{cases} y - |x| \leq 0 \\ x + 1 \geq 0 \\ x - 1 \leq 0 \\ y + 1 \geq 0 \end{cases}$

Chapter 10 Test

1. Determine all solutions of the system

$$\begin{cases} 3x + 4y = 12 \\ y = x^2 + 2x + 3 \end{cases}$$

2. Find all solutions of the system

$$\begin{cases} x - 2y = 13 \\ 3x + 5y = -16 \end{cases}$$

3. (a) Find all solutions of the following system using Gaussian elimination:

$$\begin{cases} x + 4y - z = 0 \\ 3x + y + z = -1 \\ 4x - 4y + 5z = -7 \end{cases}$$

(b) Compute D, D_x, D_y, and D_z for the system in part (a). Then check your answer in part (a) using Cramer's rule.

4. Suppose that the matrices A and B are defined as follows:

$$A = \begin{pmatrix} 1 & -3 \\ 2 & -1 \end{pmatrix} \qquad B = \begin{pmatrix} 0 & 4 \\ 1 & 3 \end{pmatrix}$$

(a) Compute $2A - B$. **(b)** Compute BA.

5. Assume that the supply and the demand functions for a commodity are given by $q = 24p + 180$ and $q = -30p + 1314$, respectively, where p is in dollars. Find the equilibrium price and the corresponding equilibrium quantity.

6. Find all solutions of the system

$$\begin{cases} \dfrac{1}{2x} + \dfrac{1}{3y} = 10 \\ -\dfrac{5}{x} - \dfrac{4}{y} = -4 \end{cases}$$

7. Specify the coefficient matrix for the system

$$\begin{cases} x + y - z = -1 \\ 2x - y + 2z = 11 \\ x - 2y + z = 10 \end{cases}$$

Also specify the augmented matrix for this system.

8. Use matrix methods to find all solutions of the system displayed in the previous problem.

9. A student is taking four courses, Math, English, Chemistry, and Economics, and spends 40 hours per week studying. The following facts are known about how the student allocates study time each week. The combined amount of study time for English and Chemistry is the same as that for Math and Economics. The amount of time spent on Math is one-quarter of the amount spent on the other three courses combined. Finally, the combined amount of time for English and Economics is two-thirds as much as that for Math and Chemistry.

(a) Let w, x, y, and z denote the number of hours per week that the student studies Math, English, Chemistry, and Economics, respectively. Specify a system of four linear equations involving these four unknowns. Write the system in standard form.

(b) Write the coefficient matrix of the system in part (a).

Ⓖ **(c)** Use a graphing utility to find the inverse of the coefficient matrix. Use matrix methods to solve the system and thereby determine how many hours per week the student studies each subject.

10. Consider the determinant

$$\begin{vmatrix} 2 & 3 & -1 \\ 0 & 1 & 4 \\ 5 & -2 & 6 \end{vmatrix}$$

(a) What is the minor of the entry in the third row, second column?

(b) What is the cofactor of the entry in the third row, second column?

11. Evaluate the determinant

$$\begin{vmatrix} 4 & -5 & 0 \\ -8 & 10 & 7 \\ 16 & 20 & 14 \end{vmatrix}$$

12. Find all solutions of the system

$$\begin{cases} x^2 + y^2 = 15 \\ xy = 5 \end{cases}$$

13. Find the solutions of the system

$$\begin{cases} A + 2B + 3C = 1 \\ 2A - B - C = 2 \end{cases}$$

14. (a) Determine the inverse of the following matrix:

$$\begin{pmatrix} 10 & -2 & 5 \\ 6 & -1 & 4 \\ 1 & 0 & 1 \end{pmatrix}$$

(b) Use the inverse matrix determined in part (a) to solve the following system:

$$\begin{cases} 10u - 2v + 5w = -1 \\ 6u - v + 4w = -2 \\ u + w = 3 \end{cases}$$

15. Graph the inequality $5x - 6y \geq 30$.

16. Determine constants P and Q so that the parabola $y = Px^2 + Qx - 5$ passes through the two points $(-2, -1)$ and $(-1, -2)$.

17. Graph the inequality $(x - 2)^2 + y^2 > 1$. Is the solution set bounded? Is it convex?

18. Graph the following system of inequalities and specify the vertices:

$$\begin{cases} x \geq 0 \\ y \geq 0 \\ 2y - x \leq 14 \\ x + 3y \leq 36 \\ 9x + y \leq 99 \end{cases}$$

19. Given that the inverse of the matrix

$$\begin{pmatrix} 5 & 0 & 2 \\ 2 & 2 & 1 \\ -3 & 1 & -1 \end{pmatrix} \text{ is the matrix } \begin{pmatrix} -3 & 2 & -4 \\ -1 & 1 & -1 \\ 8 & -5 & 10 \end{pmatrix},$$

solve the following system of equations:

$$\begin{cases} 5 \ln x + 2 \ln z = 3 \\ 2 \ln x + 2 \ln y + \ln z = -1 \\ -3 \ln x + \ln y - \ln z = 2 \end{cases}$$

11 CHAPTER The Conic Sections

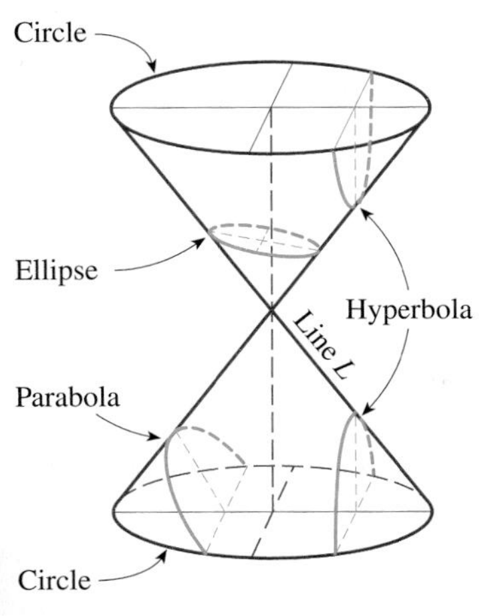

The Greeks knew the properties of the curves given by cutting a cone with a plane—the ellipse, the parabola and hyperbola. Kepler discovered by analysis of astronomical observations, and Newton proved mathematically . . . that the planets move in ellipses. The geometry of Ancient Greece thus became the cornerstone of modern astronomy.—John Lighton Synge (1897–1987)

In this chapter we study analytic geometry, with an emphasis on the *conic sections.* The **conic sections** (or **conics**) are the curves formed when planes intersect the surface of a right circular cone. As is indicated in the accompanying figure, these curves are the circle, the ellipse, the hyperbola, and the parabola. Although this is the context in which these curves were first identified by the ancient Greeks, in this chapter we'll follow a more algebraic approach involving the familiar *x-y* coordinate system. It can be shown that the two approaches are equivalent. If, after studying this chapter, you want to see the details, do a search on the World Wide Web for "Dandelin spheres." (Germinal Dandelin was a nineteenth-century Belgian/French mathematician who found a relatively simple way to see the equivalence of the two approaches.)

If not the first, certainly one of the first to discover and study the conics was the Greek mathematician Menaechmus (ca. 380–320 B.C.), who was a tutor of Alexander the Great. However, until the seventeenth century the conic sections were studied only as a portion of pure (as opposed to applied) mathematics. Then, in the seventeenth century, it was discovered that the conic sections were crucial in expressing some of the most important laws of nature. This is essentially the observation made in the opening quotation by the physicist J. L. Synge.

In this chapter you'll see examples where conics are used in:

- The design of a radio telescope (Example 6 in Section 11.2)
- The design of an arch (Exercise 41 in Section 11.2)
- Analyzing the orbit of a comet or planet (Example 7 in Section 11.4)
- Determining a location on earth without recourse to a global positioning system (Project for Section 11.5)

11.1 THE BASIC EQUATIONS

As background for the work in this chapter, you need to be familiar with the following results from Chapter 1.

1. The distance formula: $d = \sqrt{(x_2 - x_1)^2 + (y_2 - y_1)^2}$
2. The equation for a circle: $(x - h)^2 + (y - k)^2 = r^2$

3. The slope of a line: $m = \dfrac{y_2 - y_1}{x_2 - x_1}$
4. The point–slope formula: $y - y_1 = m(x - x_1)$
5. The slope–intercept formula: $y = mx + b$
6. The condition for two nonvertical lines to be parallel: $m_1 = m_2$
7. The condition for two nonvertical lines to be perpendicular: $m_1 m_2 = -1$
8. The midpoint formula: $(x_0, y_0) = \left(\dfrac{x_1 + x_2}{2}, \dfrac{y_1 + y_2}{2}\right)$

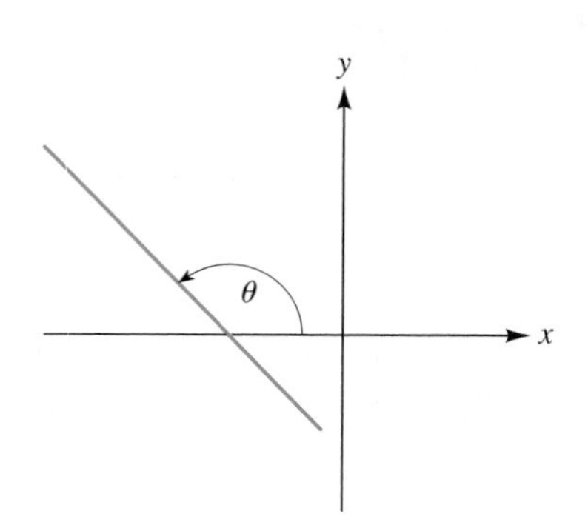

Figure 1
The angle of inclination

In this section we are going to develop two additional results to supplement those we just listed. The first of these results concerns the slope of a line. We begin by defining the **angle of inclination** (or simply the **inclination**) of a line to be the angle θ measured counterclockwise from the positive side of the x-axis to the line; see Figure 1. If θ denotes the angle of inclination, we always have $0° \le \theta < 180°$ if θ is measured in degrees and $0 \le \theta < \pi$ if θ is in radians. (Notice that when $\theta = 0°$, the line is horizontal.)

As you might suspect, there is a simple relationship between the angle of inclination and the slope of a line. We state this relationship in the following box.

PROPERTY SUMMARY Slope and the Angle of Inclination

A line's slope m and angle of inclination θ are related by the equation

$$m = \tan\theta$$

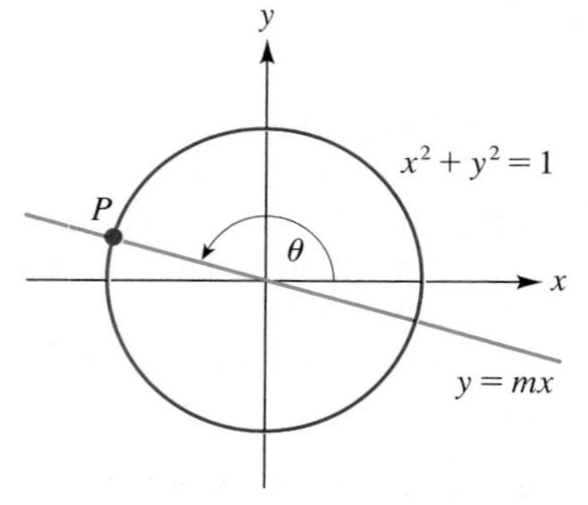

Figure 2

The formula $m = \tan\theta$ provides a useful connection between elementary coordinate geometry and trigonometry. To derive this formula, we can work with the line $y = mx$ rather than $y = mx + b$ (because these lines are parallel and therefore have equal angles of inclination). As is indicated in Figure 2, we let P denote the point where the line $y = mx$ intersects the unit circle. Then the coordinates of P are $(\cos\theta, \sin\theta)$, and we can compute the slope m using the point P and the origin:

$$m = \frac{y_2 - y_1}{x_2 - x_1} = \frac{\sin\theta - 0}{\cos\theta - 0} = \frac{\sin\theta}{\cos\theta} = \tan\theta \qquad \text{as required}$$

Note: $\cos\theta \ne 0$. Why?

EXAMPLE 1 Calculating an angle of inclination

Determine the acute angle θ between the x-axis and the line $y = 2x - 1$; see Figure 3. Express the answer in degrees, rounded to one decimal place.

SOLUTION

From the equation $y = 2x - 1$ we read directly that $m = 2$. Then, since $\tan\theta = m$, we conclude that

$$\tan\theta = 2$$

and therefore

$$\theta = \tan^{-1} 2$$
$$\theta \approx 63.4° \qquad \text{using a calculator set in the degree mode}$$

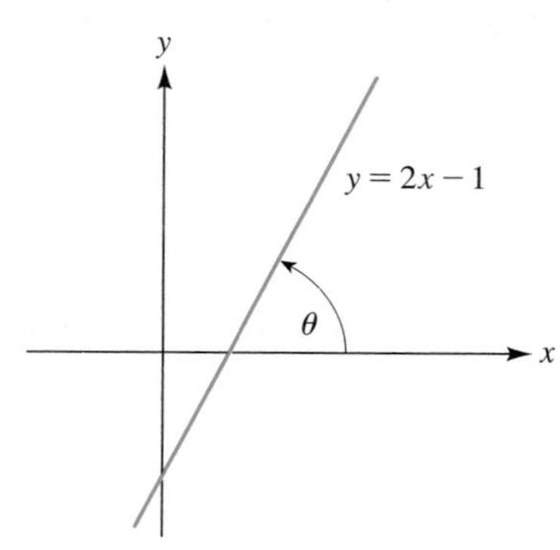

Figure 3

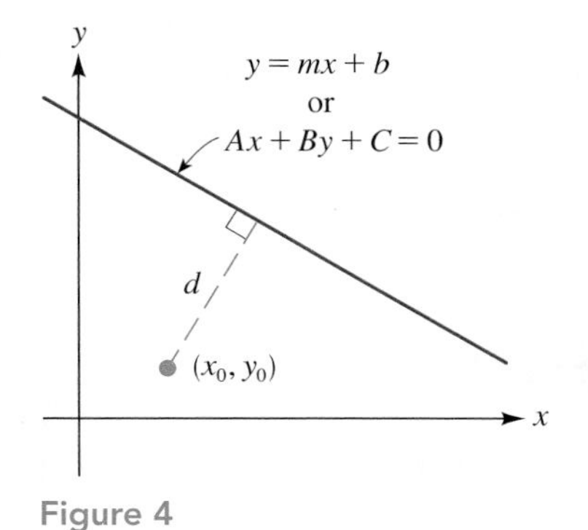

Figure 4

In Section 1.4 we reviewed the formula for the distance between two points. Now we consider the distance d from a point to a line. As is indicated in Figure 4, distance in this context means the *shortest* distance, which is the perpendicular distance. In the box that follows, we show two equivalent forms for this formula. Although the second form is more widely known, the first is just as useful and somewhat simpler to derive.

PROPERTY SUMMARY Distance from a Point to a Line

1. The distance d from the point (x_0, y_0) to the line $y = mx + b$ is given by

$$d = \frac{|mx_0 + b - y_0|}{\sqrt{1 + m^2}}$$

2. The distance d from the point (x_0, y_0) to the line $Ax + By + C = 0$ is given by

$$d = \frac{|Ax_0 + By_0 + C|}{\sqrt{A^2 + B^2}}$$

We will derive the first formula in the box with the aid of Figure 5. (Each dashed line in the figure is parallel to the x- or y-axis.)

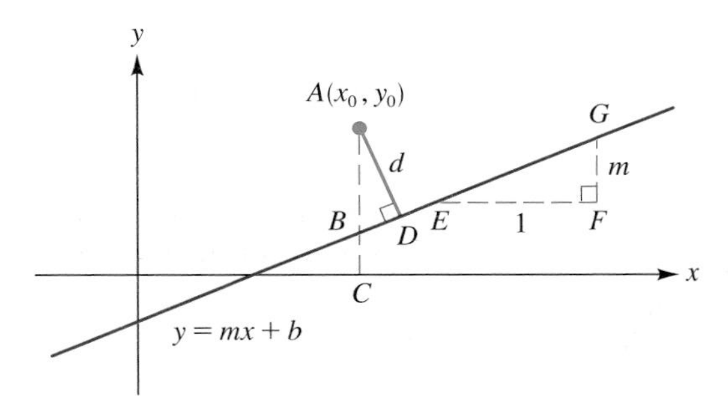

Figure 5

In Figure 5, $\triangle ABD$ is similar to $\triangle EGF$. (Exercise 44 asks you to verify this.) So we have

$$\frac{AB}{EG} = \frac{AD}{EF} \qquad \text{corresponding sides of similar triangles are proportional}$$

and therefore

$$\frac{AB}{\sqrt{1 + m^2}} = \frac{d}{1} \qquad \text{using the Pythagorean theorem and the fact that } EF = 1$$

or

$$d = \frac{AB}{\sqrt{1 + m^2}} \tag{1}$$

Next, from Figure 5 we have

$$\begin{aligned} AB &= AC - BC \\ &= y_0 - (mx_0 + b) \quad \text{(Why?)} \end{aligned}$$

Substituting into equation (1) yields

$$d = \frac{y_0 - (mx_0 + b)}{\sqrt{1 + m^2}}$$

For the general case (in which the point and line may not be situated as in Figure 5) we need to use the absolute value of the quantity in the numerator, to assure that AB and d are nonnegative. Why? We then have

$$d = \frac{|y_0 - (mx_0 + b)|}{\sqrt{1 + m^2}} = \frac{|mx_0 + b - y_0|}{\sqrt{1 + m^2}} \quad \text{as required}$$

EXAMPLE 2 The distance from a point to a line

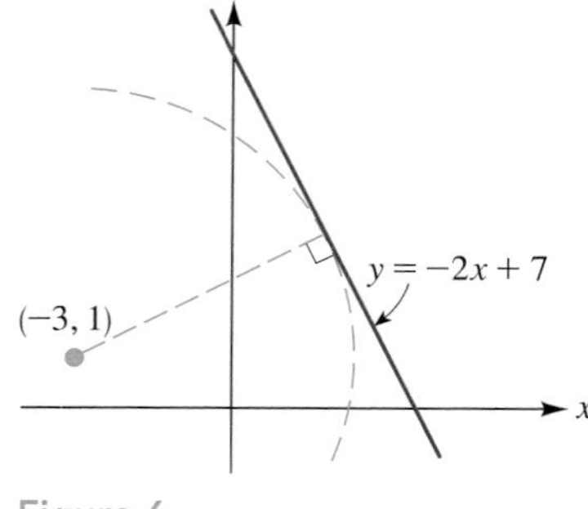

Figure 6

(a) Find the distance from the point $(-3, 1)$ to the line $y = -2x + 7$; see Figure 6.

(b) Find the equation of the circle that has center $(-3, 1)$ and that is tangent to the line $y = -2x + 7$.

SOLUTION

(a) To find the distance from the point $(-3, 1)$ to the line $y = -2x + 7$, we use the formula $d = |mx_0 + b - y_0|/\sqrt{1 + m^2}$ with $x_0 = -3$, $y_0 = 1$, $m = -2$, and $b = 7$. This yields

$$d = \frac{|(-2)(-3) + 7 - 1|}{\sqrt{1 + (-2)^2}} = \frac{12}{\sqrt{5}} = \frac{12\sqrt{5}}{5} \text{ units}$$

(b) The equation of a circle with center (h, k) and radius of length r is $(x - h)^2 + (y - k)^2 = r^2$. We are given here that (h, k) is $(-3, 1)$. Furthermore, the distance determined in part (a) is the length of the radius. (A theorem from geometry tells us that the radius drawn to the point of tangency is perpendicular to the tangent.) Thus the equation of the required circle is

$$[x - (-3)]^2 + (y - 1)^2 = \left(\frac{12\sqrt{5}}{5}\right)^2$$

or

$$(x + 3)^2 + (y - 1)^2 = \frac{144}{5}$$

EXAMPLE 3 Finding the area of a triangle using the formula for the distance from a point to a line

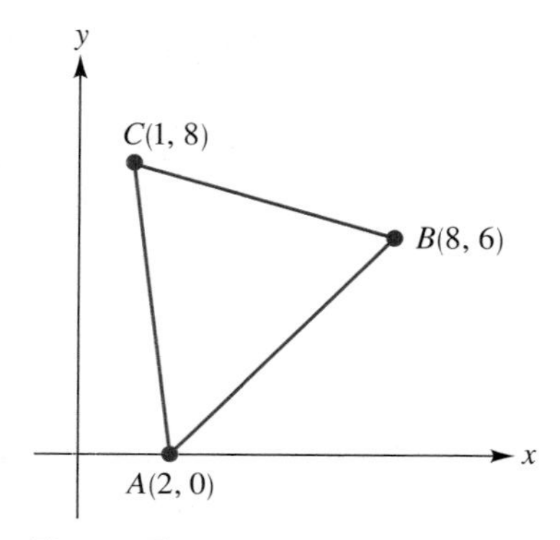

Figure 7

Find the area of triangle ABC in Figure 7.

SOLUTION

The area of any triangle is one-half the product of the base and the height. Let's view $\overline{AB}$ as the base. Then, using the formula for the distance between two points, we have

$$\begin{aligned} AB &= \sqrt{(x_2 - x_1)^2 + (y_2 - y_1)^2} \\ &= \sqrt{(8 - 2)^2 + (6 - 0)^2} \\ &= \sqrt{36 + 36} = \sqrt{2 \times 36} = 6\sqrt{2} \text{ units} \end{aligned}$$

With $\overline{AB}$ as the base, the height of the triangle is the perpendicular distance from C to $\overline{AB}$. To compute that distance, we first need the equation of the line through A and B. The slope of the line is

$$m = \frac{y_2 - y_1}{x_2 - x_1} = \frac{6 - 0}{8 - 2} = \frac{6}{6} = 1$$

Then, using the point (2, 0) and the slope $m = 1$, we have

$$\begin{aligned} y - y_1 &= m(x - x_1) \\ y - 0 &= 1(x - 2) \end{aligned}$$

and therefore

$$y = x - 2$$

We can now compute the height of the triangle by finding the perpendicular distance from $C(1, 8)$ to $y = x - 2$:

$$\begin{aligned} \text{height} &= \frac{|mx_0 + b - y_0|}{\sqrt{1 + m^2}} \\ &= \frac{|1(1) + (-2) - 8|}{\sqrt{1 + 1^2}} = \frac{|-9|}{\sqrt{2}} = \frac{9}{\sqrt{2}} \text{ units} \end{aligned}$$

Now we're ready to compute the area of the triangle, since we know the base and the height. We have

$$\begin{aligned} \text{area} &= \frac{1}{2}(6\sqrt{2})\frac{9}{\sqrt{2}} \\ &= 27 \text{ square units} \end{aligned}$$

EXAMPLE 4 The tangent line to a circle

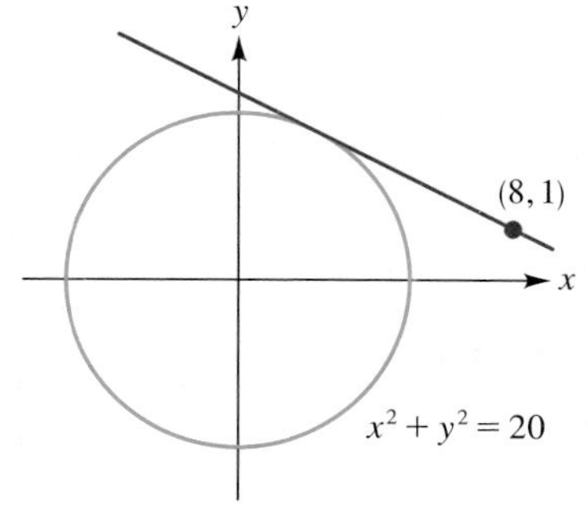

Figure 8

From the point (8, 1) a line is drawn tangent to the circle $x^2 + y^2 = 20$, as shown in Figure 8. Find the slope of this tangent line.

SOLUTION

This is a problem in which the direct approach is not the simplest. The direct approach would be first to determine the coordinates of the point of tangency. Then, using those coordinates along with (8, 1), the required slope could be computed. As it turns out, however, the coordinates of the point of tangency are not very easy to determine. In fact, one of the advantages of the following method is that those coordinates need not be found.

In Figure 8, let m denote the slope of the tangent line. Because the tangent line passes through (8, 1), we can write its equation

$$\begin{aligned} y - y_1 &= m(x - x_1) \\ y - 1 &= m(x - 8) \qquad \text{or} \qquad y = mx - 8m + 1 \end{aligned}$$

Now, since the distance from the origin to the tangent line is $\sqrt{20}$ units (the radius of the circle), we have

$$\begin{aligned}\sqrt{20} &= \frac{|mx_0 + b - y_0|}{\sqrt{1+m^2}}\\ &= \frac{|m(0) - 8m + 1 - 0|}{\sqrt{1+m^2}} \qquad \text{using } x_0 = 0 = y_0 \text{ and } b = -8m + 1\\ &= \frac{|-8m+1|}{\sqrt{1+m^2}}\end{aligned}$$

To solve this equation for m, we square both sides to obtain

$$20 = \frac{64m^2 - 16m + 1}{1 + m^2}$$

or

$$\begin{aligned}20(1 + m^2) &= 64m^2 - 16m + 1\\ 0 &= 44m^2 - 16m - 19\\ 0 &= (22m - 19)(2m + 1)\end{aligned}$$

From this we see that the two roots of the equation are 19/22 and $-1/2$. Because the slope of the tangent line specified in Figure 8 is negative, we choose the value $m = -1/2$; this is the required slope.

The last formula that we consider in this section is

$$d = \frac{|Ax_0 + By_0 + C|}{\sqrt{A^2 + B^2}}$$

As we pointed out earlier, this formula gives the distance from the point (x_0, y_0) to the line $Ax + By + C = 0$. To derive the formula, first note that the slope and the y-intercept of the line $Ax + By + C = 0$ are

$$m = -\frac{A}{B} \qquad \text{and} \qquad b = -\frac{C}{B}$$

So we have

$$d = \frac{|mx_0 + b - y_0|}{\sqrt{1 + m^2}} = \frac{|(-A/B)x_0 + (-C/B) - y_0|}{\sqrt{1 + (-A/B)^2}}$$

Now, to complete the derivation, we need to show that when this last expression is simplified, the result is $|Ax_0 + By_0 + C|/\sqrt{A^2 + B^2}$. Exercise 43 asks you to carry out the details.

EXAMPLE 5 Using the formula that we just derived

Use the formula $d = |Ax_0 + By_0 + C|/\sqrt{A^2 + B^2}$ to compute the distance from the point $(-3, 1)$ to the line $y = -2x + 7$.

NOTE In Example 2 we computed this quantity using the distance formula, $d = |mx_0 + b - y_0|/\sqrt{1 + m^2}$.

SOLUTION

First, we write the given equation $y = -2x + 7$ in the form $Ax + By + C = 0$:

$$2x + y - 7 = 0$$

From this we see that $A = 2$, $B = 1$, and $C = -7$. Thus we have

$$d = \frac{|Ax_0 + By_0 + C|}{\sqrt{A^2 + B^2}}$$
$$= \frac{|2(-3) + 1(1) + (-7)|}{\sqrt{2^2 + 1^2}} = \frac{12}{\sqrt{5}} = \frac{12\sqrt{5}}{5}$$

The required distance is therefore $12\sqrt{5}/5$ units, as we obtained previously in Example 2(a).

EXERCISE SET 11.1

A

Exercises 1–12 are review exercises. To solve these problems, you will need to utilize the formulas listed at the beginning of this section.

1. Find the distance between the points $(-5, -6)$ and $(3, -1)$.
2. Find an equation of the line that passes through $(2, -4)$ and is parallel to the line $3x - y = 1$. Write your answer in the form $y = mx + b$.
3. Find an equation of a line that is perpendicular to the line $4x - 5y - 20 = 0$ and has the same y-intercept as the line $x - y + 1 = 0$. Write your answer in the form $Ax + By + C = 0$.
4. Find an equation of the line passing through the points $(6, 3)$ and $(1, 0)$. Write your answer in the form $y = mx + b$.
5. Find an equation of the line that is the perpendicular bisector of the line segment joining the points $(2, 1)$ and $(6, 7)$. Write your answer in the form $Ax + By + C = 0$.
6. Find the area of the circle $(x - 12)^2 + (y + \sqrt{5})^2 = 49$.
7. Find the x- and y-intercepts of the circle with center $(1, 0)$ and radius 5.
8. Find an equation of the line that has a positive slope and is tangent to the circle $(x - 1)^2 + (y - 1)^2 = 4$ at one of its y-intercepts. Write your answer in the form $y = mx + b$.
9. Suppose that the coordinates of A, B, and C are $A(1, 2)$, $B(6, 1)$, and $C(7, 8)$. Find an equation of the line passing through C and through the midpoint of the line segment $\overline{AB}$. Write your answer in the form $ax + by + c = 0$.
10. Find an equation of the line passing through the point $(-4, 0)$ and through the point of intersection of the lines $2x - y + 1 = 0$ and $3x + y - 16 = 0$. Write your answer in the form $y = mx + b$.
11. Find the perimeter of $\triangle ABC$ in the following figure.

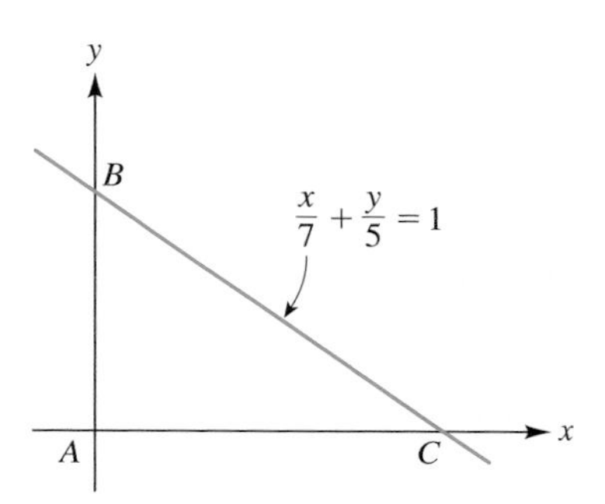

12. Find the sum of the x- and y-intercepts of the line

$$(\csc^2 \alpha)x + (\sec^2 \alpha)y = 1$$

(Assume that α is a constant.)

In Exercises 13–16, determine the angle of inclination of each line. Express the answer in both radians and degrees. In cases in which a calculator is necessary, round the answer to two decimal places.

13. $y = \sqrt{3}x + 4$
14. $x + \sqrt{3}y - 2 = 0$
15. **(a)** $y = 5x + 1$ **(b)** $y = -5x + 1$
16. **(a)** $3x - y - 3 = 0$ **(b)** $3x + y - 3 = 0$

In Exercises 17–20, find the distance from the point to the line using: **(a)** *the formula* $d = |mx_0 + b - y_0|/\sqrt{1 + m^2}$; and **(b)** *the formula* $d = |Ax_0 + By_0 + C|/\sqrt{A^2 + B^2}$.

17. $(1, 4)$; $y = x - 2$
18. $(-2, -3)$; $y = -4x + 1$
19. $(-3, 5)$; $4x + 5y + 6 = 0$
20. $(0, -3)$; $3x - 2y = 1$
21. **(a)** Find the equation of the circle that has center $(-2, -3)$ and is tangent to the line $2x + 3y = 6$.
    (b) Find the radius of the circle that has center $(1, 3)$ and is tangent to the line $y = \frac{1}{2}x + 5$.
22. Find the area of the triangle with vertices $(3, 1)$, $(-2, 7)$, and $(6, 2)$. *Hint*: Use the method shown in Example 3.
23. Find the area of the quadrilateral $ABCD$ with vertices $A(0, 0)$, $B(8, 2)$, $C(4, 7)$, and $D(1, 6)$. *Suggestion*: Draw a diagonal, and use the method shown in Example 3 for the two resulting triangles.

24. From the point $(7, -1)$, tangent lines are drawn to the circle $(x - 4)^2 + (y - 3)^2 = 4$. Find the slopes of these lines.

25. From the point $(0, -5)$, tangent lines are drawn to the circle $(x - 3)^2 + y^2 = 4$. Find the slope of each tangent.

26. Find the distance between the two parallel lines $y = 2x - 1$ and $y = 2x + 4$. *Hint*: Draw a sketch; then find the distance from the origin to each line.

27. Find the distance between the two parallel lines $3x + 4y = 12$ and $3x + 4y = 24$.

28. Find an equation of the line that passes through $(3, 2)$ and whose x- and y-intercepts are equal. (There are two answers.)

29. Find an equation of the line that passes through the point $(2, 6)$ in such a way that the segment of the line cut off between the axes is bisected by the point $(2, 6)$.

30. Find an equation of the line whose angle of inclination is 60° and whose distance from the origin is four units. (There are two answers.)

B

31. Find an equation of the angle bisector in the accompanying figure. *Hint*: Let (x, y) be a point on the angle bisector. Then (x, y) is equidistant from the two given lines. Or use angles of inclination.

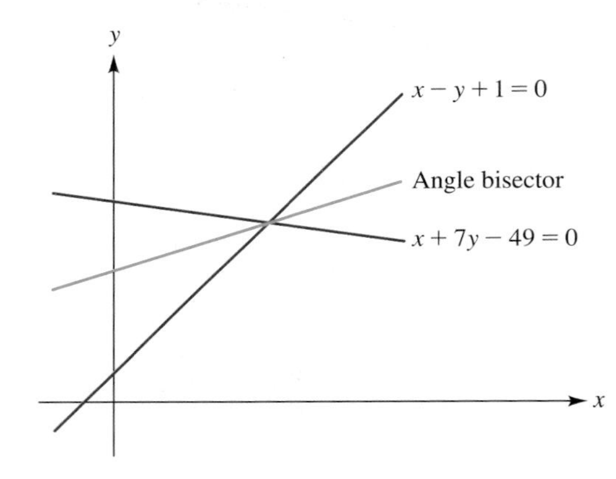

32. Find the center and the radius of the circle that passes through the points $(-2, 7)$, $(0, 1)$, and $(2, -1)$.

33. **(a)** Find the center and the radius of the circle passing through the points $A(-12, 1)$, $B(2, 1)$ and $C(0, 7)$.

(b) Let R denote the radius of the circle in part (a). In $\triangle ABC$, let a, b, and c be the lengths of the sides opposite angles A, B, and C, respectively. Show that the area of $\triangle ABC$ is equal to $abc/4R$.

34. (Continuation of Exercise 33.)

(a) Let H denote the point where the altitudes of $\triangle ABC$ intersect. Find the coordinates of H.

(b) Let d denote the distance from H to the center of the circle in Exercise 33(a). Show that

$$d^2 = 9R^2 - (a^2 + b^2 + c^2)$$

35. Suppose the line $x - 7y + 44 = 0$ intersects the circle $x^2 - 4x + y^2 - 6y = 12$ at points P and Q. Find the length of the chord $\overline{PQ}$.

36. The point $(1, -2)$ is the midpoint of a chord of the circle $x^2 - 4x + y^2 + 2y = 15$. Find the length of the chord.

37. Show that the product of the distances from the point $(0, c)$ to the lines $ax + y = 0$ and $x + by = 0$ is

$$\frac{|bc^2|}{\sqrt{a^2 + a^2b^2 + b^2 + 1}}$$

38. Suppose that the point (x_0, y_0) lies on the circle $x^2 + y^2 = a^2$. Show that the equation of the line tangent to the circle at (x_0, y_0) is $x_0x + y_0y = a^2$.

39. The vertices of $\triangle ABC$ are $A(0, 0)$, $B(8, 0)$, and $C(8, 6)$.

(a) Find the equations of the three lines that bisect the angles in $\triangle ABC$. *Hint*: Make use of the identity $\tan(\theta/2) = (\sin\theta)/(1 + \cos\theta)$.

(b) Find the points where each pair of angle bisectors intersect. What do you observe?

40. Show that the equations of the lines with slope m that are tangent to the circle $x^2 + y^2 = a^2$ are

$$y = mx + a\sqrt{1 + m^2} \quad \text{and} \quad y = mx - a\sqrt{1 + m^2}$$

41. The point (x, y) is equidistant from the point $(0, 1/4)$ and the line $y = -1/4$. Show that x and y satisfy the equation $y = x^2$.

42. The point (x_0, y_0) is equidistant from the line $x + 2y = 0$ and the point $(3, 1)$. Find (and simplify) an equation relating x_0 and y_0.

43. **(a)** Find the slope m and the y-intercept b of the line $Ax + By + C = 0$

(b) Use the formula $d = |mx_0 + b - y_0|/\sqrt{1 + m^2}$ to show that the distance from the point (x_0, y_0) to the line $Ax + By + C = 0$ is given by

$$d = \frac{|Ax_0 + By_0 + C|}{\sqrt{A^2 + B^2}}$$

44. Refer to Figure 5 on page 819. Show that $\triangle ABD$ is similar to $\triangle EGF$.

C

45. Show that the distance of the point (x_1, y_1) from the line passing through the two points (x_2, y_2) and (x_3, y_3) is given by $d = |D|/\sqrt{(x_2 - x_3)^2 + (y_2 - y_3)^2}$, where

$$D = \begin{vmatrix} x_1 & y_1 & 1 \\ x_2 & y_2 & 1 \\ x_3 & y_3 & 1 \end{vmatrix}$$

46. Let (a_1, b_1), (a_2, b_2), and (a_3, b_3) be three noncollinear points. Show that an equation of the circle passing through these three points is

$$\begin{vmatrix} x^2 + y^2 & x & y & 1 \\ a_1^2 + b_1^2 & a_1 & b_1 & 1 \\ a_2^2 + b_2^2 & a_2 & b_2 & 1 \\ a_3^2 + b_3^2 & a_3 & b_3 & 1 \end{vmatrix} = 0$$

47. Find an equation of the circle that passes through the points (6, 3) and (−4, −3) and that has its center on the line $y = 2x - 7$.

48. Let a be a positive number and suppose that the coordinates of points P and Q are $P\,(a\cos\theta, a\sin\theta)$ and $Q\,(a\cos\beta, a\sin\beta)$. Show that the distance from the origin to the line passing through P and Q is

$$a\left|\cos\left(\frac{\theta - \beta}{2}\right)\right|$$

49. Find an equation of a circle that has radius 5 and is tangent to the line $2x + 3y = 26$ at the point (4, 6). Write your answer in standard form. (There are two answers.)

50. Find an equation of the circle passing through (2, −1) and tangent to the line $y = 2x + 1$ at (1, 3). Write your answer in standard form.

51. For the last exercise in this section you will prove an interesting property of the curve $x^3 + y^3 = 6xy$. This curve, known as the *folium of Descartes,* is shown in Figure A. Figure B and the accompanying caption indicate the property that we will establish.

(a) Suppose that the slope of the line segment $\overline{OQ}$ is t. By solving the system of equations

$$\begin{cases} y = tx \\ x^3 + y^3 = 6xy \end{cases}$$

show that the coordinates of the point Q are $x = 6t/(1 + t^3)$ and $y = 6t^2/(1 + t^3)$.

(b) Show that the slope of the line joining the points P and Q is $(t^2 - t - 1)/(t^2 + t - 1)$.

(c) Suppose that the slope of the line segment $\overline{OR}$ is u. By repeating the procedure used in parts (a) and (b), you'll find that the slope of the line joining the points P and R is $(u^2 - u - 1)/(u^2 + u - 1)$. Now, since the points P, Q, and R are collinear (lie on the same line), it must be the case that

$$\frac{t^2 - t - 1}{t^2 + t - 1} = \frac{u^2 - u - 1}{u^2 + u - 1}$$

Working from this equation, show that $tu = -1$. This shows that $\angle ROQ$ is a right angle, as required.

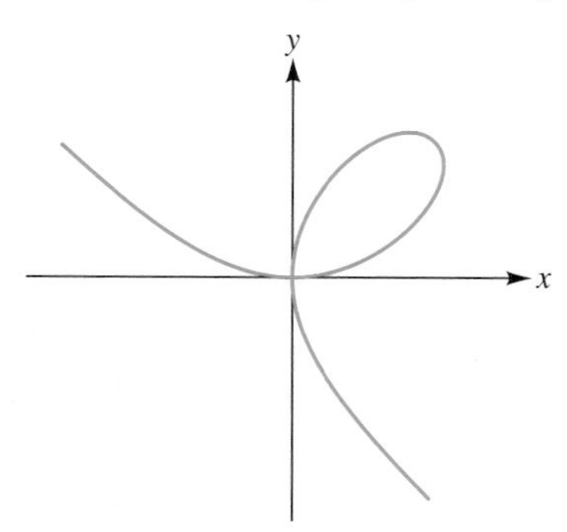

The folium of Descartes: $x^3 + y^3 = 6xy$

Figure A

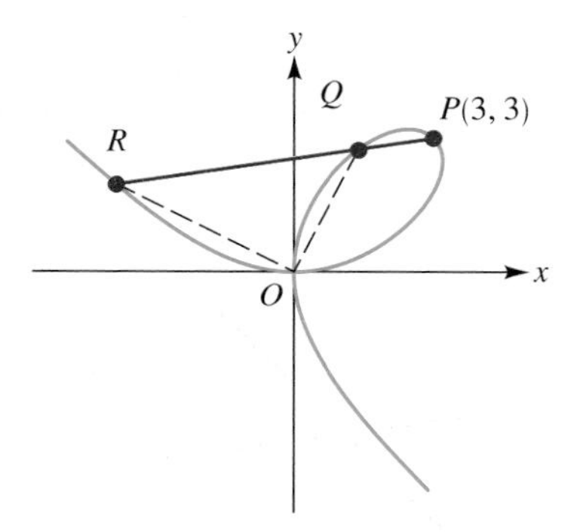

A property of the folium: Suppose that a line through the point $P(3, 3)$ meets the folium again at points Q and R. Then $\angle ROQ$ is a right angle.

Figure B

11.2 THE PARABOLA

In Section 4.2 we saw that the graph of a quadratic function $y = ax^2 + bx + c$ is a symmetric U-shaped curve called a *parabola.* In this section we give a more general definition of the parabola, a definition that emphasizes the geometric properties of the curve.

DEFINITION | **The Parabola**

A **parabola** is the set of all points in the plane equally distant from a fixed line and a fixed point not on the line. The fixed line is called the **directrix;** the fixed point is called the **focus.**

Let us initially suppose that the focus of the parabola is the point $(0, p)$ and the directrix is the line $y = -p$. We will assume throughout this section that p is positive. To understand the geometric context of the definition of the parabola, the special graph paper displayed in Figure 1(a) is useful. The common center of the concentric circles in Figure 1(a) is the focus $(0, p)$. Thus all the points on a given circle are at a fixed distance from the focus. The radii of the circles increase in increments of p units. Similarly, the dashed horizontal lines in the figure are drawn at intervals that are multiples of p units from the directrix $y = -p$. By considering the points where the circles intersect the horizontal lines, we can find a number of points that are equally distant from the focus $(0, p)$ and the directrix $y = -p$; see Figure 1(b).

Figure 1(b) shows that the points on the parabola are symmetric about a line, in this case the y-axis. Also, by studying the figure, you should be able to convince yourself that in this case there can be no points below the x-axis that satisfy the stated condition. However, to describe the required set of points completely and to show that our new definition is consistent with the old one, Figure 1(b) is inadequate. We need to bring algebraic methods to bear on the problem. Thus let d_1 denote the distance from the point $P(x, y)$ to the focus $(0, p)$, and let d_2 denote the distance from $P(x, y)$ to the directrix $y = -p$, as shown in Figure 2. The distance d_1 is then

$$d_1 = \sqrt{(x-0)^2 + (y-p)^2} = \sqrt{x^2 + y^2 - 2py + p^2}$$

The distance d_2 in Figure 2 is just the vertical distance between the points P and Q. Thus,

$$\begin{aligned} d_2 &= (y\text{-coordinate of } P) - (y\text{-coordinate of } Q) \\ &= y - (-p) = y + p \end{aligned}$$

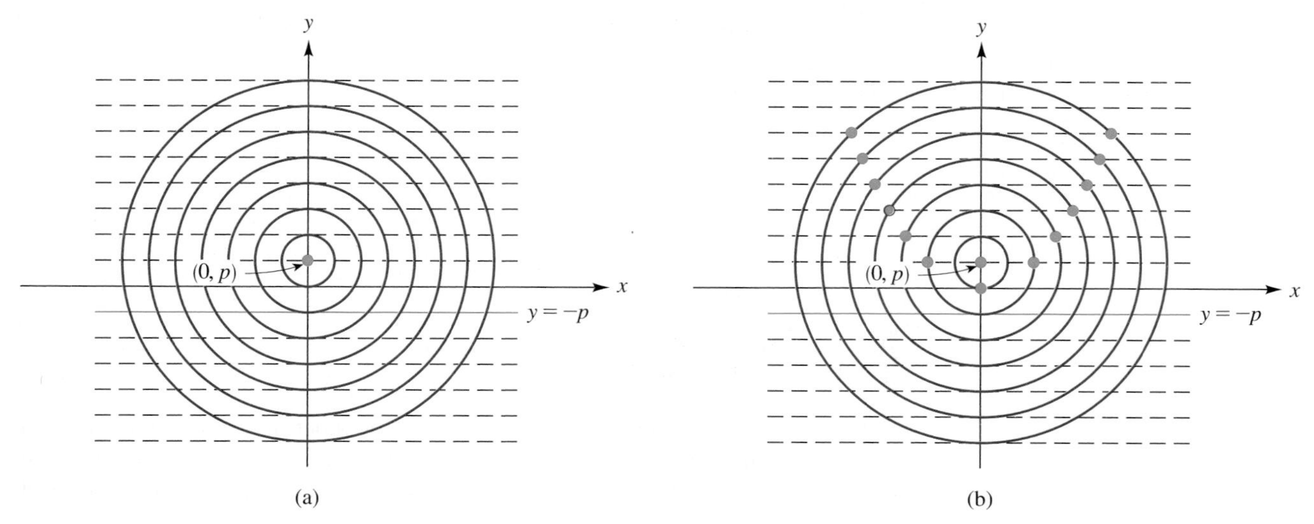

Figure 1

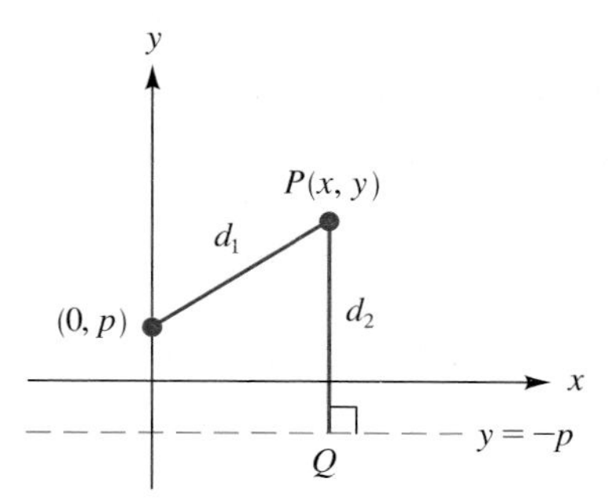

Figure 2

(Absolute value signs are unnecessary here because, as was noted earlier, P cannot lie below the x-axis.) The condition that P be equally distant from the focus $(0, p)$ and the directrix $y = -p$ can be expressed by the equation

$$d_1 = d_2$$

By using the expressions we've found for d_1 and d_2, this last equation becomes

$$\sqrt{x^2 + y^2 - 2py + p^2} = y + p$$

We can obtain a simpler but equivalent equation by squaring both sides. (Two nonnegative quantities are equal if and only if their squares are equal.) Thus

$$x^2 + y^2 - 2py + p^2 = y^2 + 2py + p^2$$

After like terms are combined, this equation becomes

$$x^2 = 4py \tag{1}$$

Conversely, it can be shown that if the pair of numbers x and y satisfy the preceding equation, then the point (x, y) is equidistant from the point $(0, p)$ and the line $y = -p$, that is, the point (x, y) lies on this parabola. So we call equation (1) the standard form for the equation of the parabola with focus $(0, p)$ and directrix $y = -p$. In the box that follows, we summarize the properties of the parabola $x^2 = 4py$. As Figure 3(b) indicates, the terminology we've introduced applies equally well to an arbitrary parabola for which the axis of symmetry is not necessarily parallel to one of the coordinate axes and the vertex is not necessarily the origin.

PROPERTY SUMMARY The Parabola

1. The parabola is the set of points equidistant from a fixed line called the **directrix** and a fixed point, not on the line, called the **focus.**
2. The **axis** of a parabola is the line drawn through the focus and perpendicular to the directrix.
3. The **vertex** of a parabola is the point where the parabola intersects its axis. The vertex is located halfway between the focus and the directrix. See Figure 3.

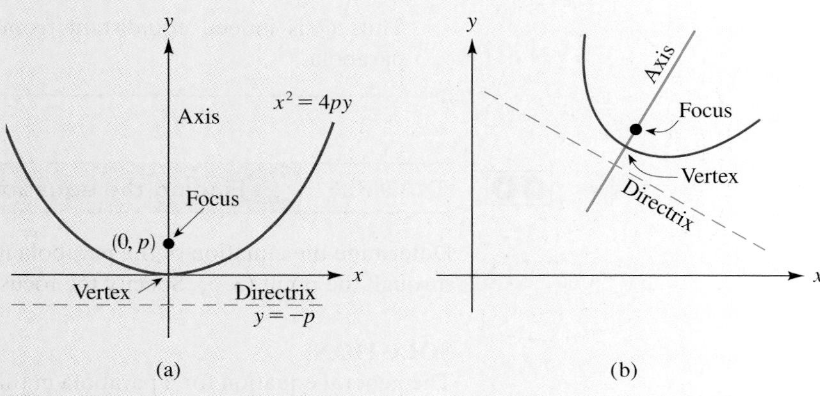

Figure 3

EXAMPLE 1 Analyzing a parabola

(a) Refer to Figure 4. Determine the focus and the directrix of the parabola $x^2 = 16y$.

(b) As is indicated in Figure 4, the point Q is on the parabola $x^2 = 16y$, and the x-coordinate of Q is -12. Find the y-coordinate of Q.

(c) Verify that the point Q is equidistant from the focus and the directrix of the parabola.

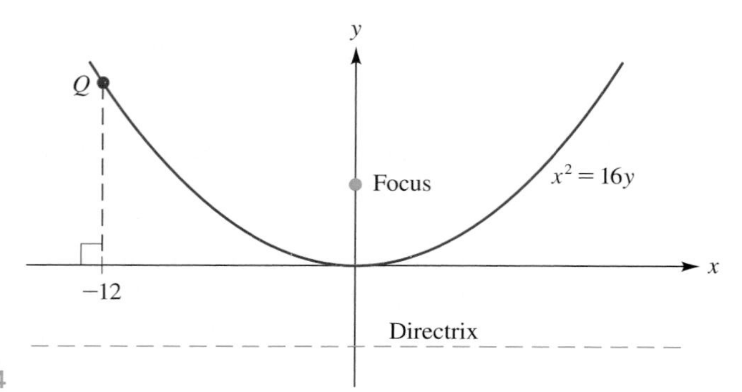

Figure 4

SOLUTION

(a) We know that for the basic parabola $x^2 = 4py$, the focus is the point $(0, p)$ and the directrix is the line $y = -p$. Comparing the given equation $x^2 = 16y$ with $x^2 = 4py$, we see that $4p = 16$ and therefore $p = 4$. Thus the focus of the parabola $x^2 = 16y$ is $(0, 4)$ and the directrix is $y = -4$.

(b) We are given that the x-coordinate of Q is -12. Substituting $x = -12$ in the given equation $x^2 = 16y$ yields $(-12)^2 = 16y$. Therefore $y = 144/16 = 9$, and the coordinates of Q are $(-12, 9)$.

(c) In part (b) we found that the y-coordinate of Q is 9. From part (a) we know that the equation of the directrix is $y = -4$. Thus the vertical distance from Q down to the directrix is $9 - (-4) = 13$. Next, to calculate the distance from $Q(-12, 9)$ to the focus $(0, 4)$, the distance formula yields

$$\begin{aligned} d &= \sqrt{(x_2 - x_1)^2 + (y_2 - y_1)^2} \\ &= \sqrt{(-12 - 0)^2 + (9 - 4)^2} = \sqrt{144 + 25} = \sqrt{169} = 13 \end{aligned}$$

Thus Q is indeed equidistant from the focus and the directrix of the parabola.

EXAMPLE 2 Finding the equation of a parabola

Determine the equation of the parabola in Figure 5, given that the curve passes through the point $(3, 5)$. Specify the focus and the directrix.

SOLUTION

The general equation for a parabola in this position is $x^2 = 4py$. Since the point $(3, 5)$ lies on the curve, its coordinates must satisfy the equation $x^2 = 4py$. Thus

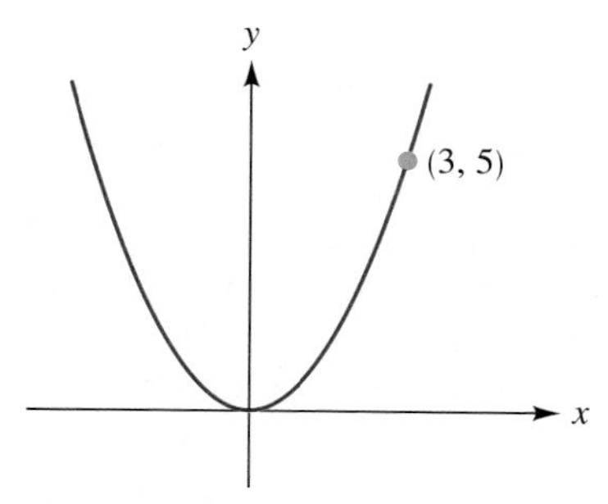

Figure 5

$$3^2 = 4p(5)$$
$$9 = 20p$$
$$\frac{9}{20} = p$$

With $p = 9/20$ the equation $x^2 = 4py$ becomes

$$x^2 = 4\left(\frac{9}{20}\right)y$$

or

$$x^2 = \left(\frac{9}{5}\right)y \qquad \text{as required}$$

Furthermore, since $p = 9/20$, the focus is $(0, 9/20)$, and the directrix is $y = -9/20$.

We have seen that the equation of a parabola with focus $(0, p)$ and directrix $y = -p$ is $x^2 = 4py$. By following the same method, we can obtain general equations for parabolas with other orientations. The basic results are summarized in the following box (see Figure 6).

PROPERTY SUMMARY Basic Equations for the Parabola

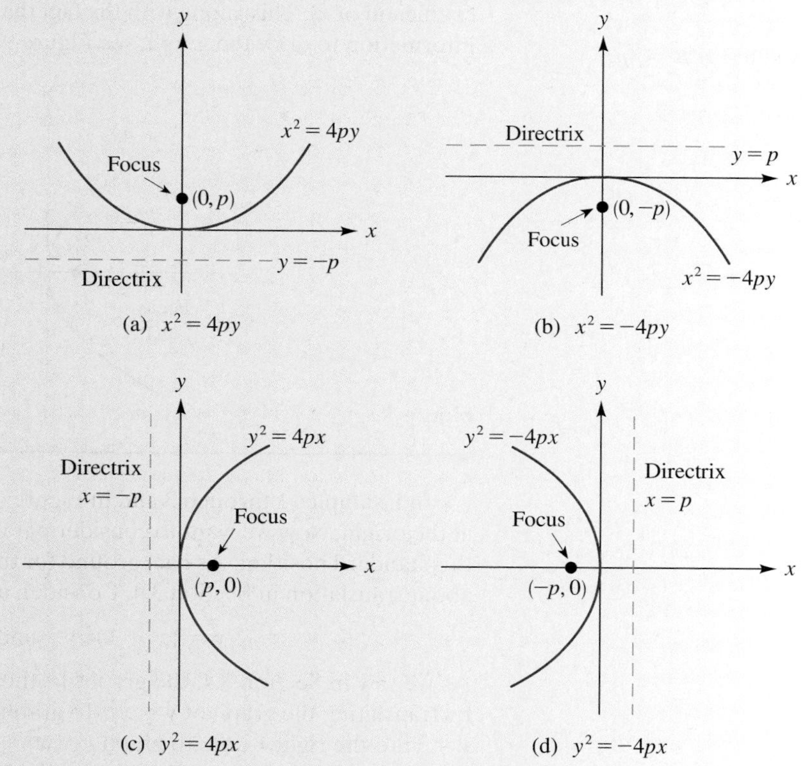

Directrix, Focus, $(0, p)$, $x^2 = 4py$, $y = -p$

(a) $x^2 = 4py$

Directrix, $y = p$, Focus, $(0, -p)$, $x^2 = -4py$

(b) $x^2 = -4py$

Directrix $x = -p$, Focus, $(p, 0)$, $y^2 = 4px$

(c) $y^2 = 4px$

Directrix $x = p$, Focus, $(-p, 0)$, $y^2 = -4px$

(d) $y^2 = -4px$

Figure 6

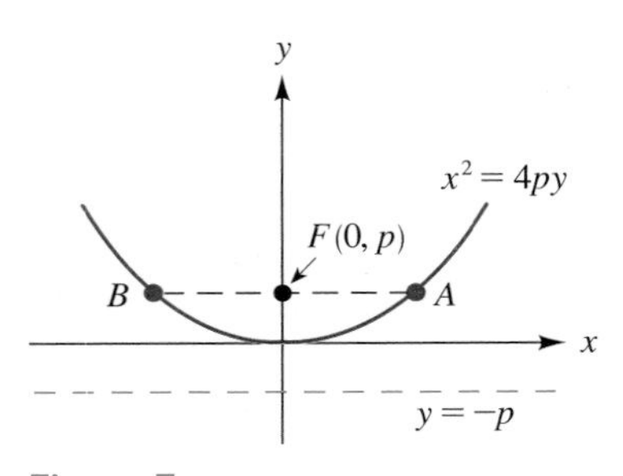

Figure 7
Focal width $= AB = 4p$

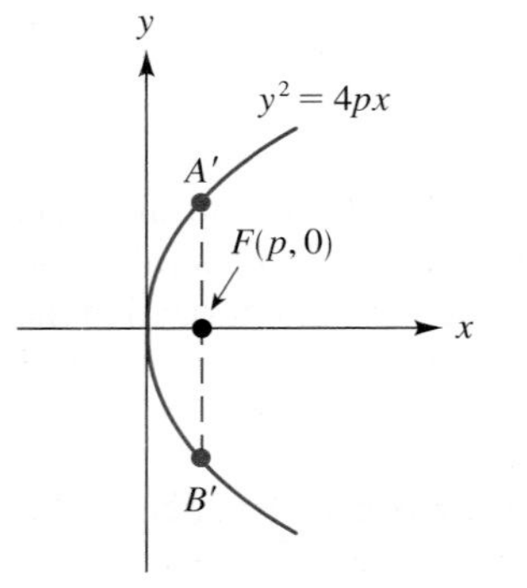

Figure 8
Focal width $= A'B' = 4p$

A **chord** of a parabola is a straight line segment joining any two points on the curve. If the chord passes through the focus, it is called a **focal chord.** For purposes of graphing, it is useful to know the length of the focal chord perpendicular to the axis of the parabola. This is the length of the horizontal line segment $\overline{AB}$ in Figure 7 and the vertical line segment $\overline{A'B'}$ in Figure 8. We will call this length the **focal width.**

In Figure 7 the distance from A to F is the same as the distance from A to the line $y = -p$. (Why?) But the distance from A to the line $y = -p$ is $2p$. Therefore, $AF = 2p$ and AB is twice this, or $4p$. We have shown that the focal width of the parabola is $4p$. In other words, given a parabola $x^2 = 4py$, the focal width is $4p$, the coefficient of y. In the same way, the length of the focal chord $\overline{A'B'}$ in Figure 8 is also $4p$, the coefficient of x in that case. In general, for the parabolas in Figure 6 the focal width, $4p$, is the absolute value of the coefficient of the x or y term.

EXAMPLE 3 Determining the focus and directrix of a parabola

Find the focus and the directrix of the parabola $y^2 = -4x$, and sketch the graph.

SOLUTION

Comparing the basic equation $y^2 = -4px$ [in Figure 6(d)] with the equation at hand, we see that $4p = 4$ and thus $p = 1$. The focus is therefore $(-1, 0)$, and the directrix is $x = 1$. The basic form of the graph will be as in Figure 6(d). For purposes of graphing, we note that the focal width is 4 (the absolute value of the coefficient of x). This, along with the fact that the vertex is $(0, 0)$, gives us enough information to draw the graph; see Figure 9.

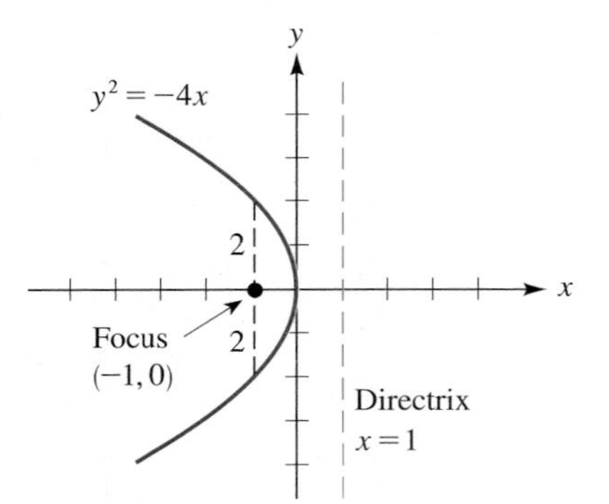

Figure 9

In Examples 1 through 3 and in Figure 6 the vertex of each parabola is located at the origin. Now we want to consider parabolas that are translated (shifted) from this standard position. As background for this, we review and generalize the results about translation in Section 3.4. Consider, as an example, the two equations

$$y = (x - 1)^2 \qquad \text{and} \qquad y = x^2 + 1$$

As we saw in Section 3.4, the graphs of these two equations can each be obtained by translating the graph of $y = x^2$. To graph $y = (x - 1)^2$, we translate the graph of $y = x^2$ to the right 1 unit; to graph $y = x^2 + 1$, we translate $y = x^2$ up 1 unit.

To see the underlying pattern here, let's rewrite the equation $y = x^2 + 1$ as $y - 1 = x^2$. Then the situation is as follows:

EQUATION	HOW GRAPH IS OBTAINED
$y = (x - 1)^2$	translation in positive x-direction
$y - 1 = x^2$	translation in positive y-direction

OBSERVATION *In the equation $y = x^2$, the effect of replacing x with $x - 1$ is to translate the graph one unit in the positive x-direction. Similarly, replacing y with $y - 1$ translates the graph one unit in the positive y-direction.*

As a second example before we generalize, consider the equations

$$y = (x + 1)^2 \quad \text{and} \quad y = x^2 - 1$$

The first equation involves a translation in the negative x-direction; the second involves a translation in the negative y-direction. As before, to see the underlying pattern, we rewrite the second equation as $y + 1 = x^2$. Now we have the following situation:

EQUATION	HOW GRAPH IS OBTAINED
$y = (x + 1)^2$	translation in negative x-direction
$y + 1 = x^2$	translation in negative y-direction

OBSERVATION *In the equation $y = x^2$ the effect of replacing x with $x + 1$ is to translate the graph one unit in the negative x-direction. Similarly, replacing y with $y + 1$ translates the graph one unit in the negative y-direction.*

Both of the examples that we have just considered are specific instances of the following basic result. (The result is valid whether or not the given equation and graph represent a function.)

PROPERTY SUMMARY Translation and Coordinates

Suppose we have an equation that determines a graph in the x-y plane, and let h and k denote positive numbers. Then, replacing x with $x - h$ or $x + h$, or replacing y with $y - k$ or $y + k$ has the following effects on the graph of the original equation.

REPLACEMENT	RESULTING TRANSLATION
1. x replaced with $x - h$	h units in the positive x-direction
2. y replaced with $y - k$	k units in the positive y-direction
3. x replaced with $x + h$	h units in the negative x-direction
4. y replaced with $y + k$	k units in the negative y-direction

EXAMPLE 4 A parabola with a vertex other than the origin

Graph the parabola $(y + 1)^2 = -4(x - 2)$. Specify the vertex, the focus, the directrix, and the axis of symmetry.

SOLUTION

The given equation is obtained from $y^2 = -4x$ (which we graphed in the previous example) by replacing x and y with $x - 2$ and $y + 1$, respectively. So the required graph is obtained by translating the parabola in Figure 9 to the right two units and down one unit. In particular, this means that the vertex moves from $(0, 0)$ to $(2, -1)$; the focus moves from $(-1, 0)$ to $(1, -1)$; the directrix

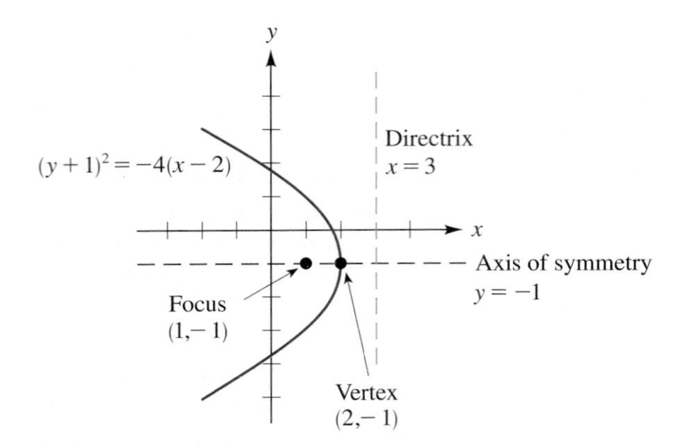

Figure 10

moves from $x = 1$ to $x = 3$; and the axis of symmetry moves from $y = 0$ (which is the x-axis) to $y = -1$. The required graph is shown in Figure 10.

EXAMPLE 5 Completing the square to analyze a parabola

Graph the parabola $2y^2 - 4y - x + 5 = 0$, and specify each of the following: vertex, focus, directrix, axis of symmetry, and focal width.

SOLUTION

Just as we did in Section 4.2, we use the technique of completing the square:

$$\begin{aligned} 2(y^2 - 2y) &= x - 5 \\ 2(y^2 - 2y + 1) &= x - 5 + 2 \qquad \text{adding 2 to both sides} \\ (y - 1)^2 &= \frac{1}{2}(x - 3) \end{aligned}$$

The graph of this last equation is obtained by translating the graph of $y^2 = \frac{1}{2}x$ "right 3, up 1." This moves the vertex from (0, 0) to (3, 1). Now, for $y^2 = \frac{1}{2}x$ the focus and directrix are determined by setting $4p = 1/2$. Therefore $p = 1/8$, and consequently, the focus and directrix of $y^2 = \frac{1}{2}x$ are (1/8, 0) and $x = -1/8$, respectively. Thus the focus of the translated curve is $(3\frac{1}{8}, 1)$, and the directrix is $x = 2\frac{7}{8}$. Figure 11(a) shows the graph of $y^2 = \frac{1}{2}x$, and Figure 11(b) shows the

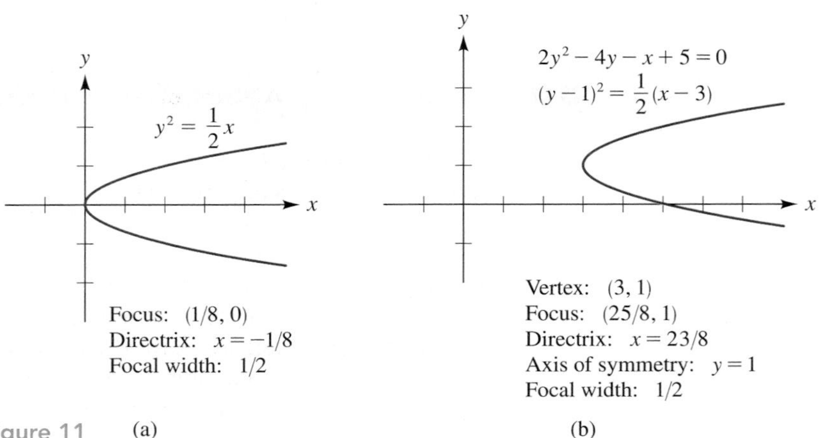

Figure 11 (a) (b)

translated graph. (You should check for yourself that the information accompanying Figure 11(b) is correct.)

Incoming light rays

Focus

Parabolic mirror

Figure 12

There are numerous applications of the parabola in the sciences. Many of these involve parabolic reflectors. Figure 12 shows a cross section of a parabolic mirror in a telescope. As is indicated in Figure 12, light rays coming in parallel to the axis of the parabola are reflected through the focus. Indeed, the word *focus* comes from a Latin word meaning "fireplace." [The mathematics (but not the physics) behind this focusing property is developed in Exercise 19 in Optional Section 11.3.] In addition to telescopes and radio telescopes, parabolic reflectors are used in communication systems such as satellite dishes for television, in surveillance systems, and in automobile headlights. To diagram the parabolic mirror of an automobile headlight, consider the light source to be at the focus and reverse the directions of the arrows on the light rays in Figure 12.

Geometrically, parabolic reflectors are described and designed as follows. We begin with a portion of a parabola and its axis of symmetry, as shown in Figure 13(a). By rotating the parabola about its axis, we obtain the bowl-shaped surface in Figure 13(b), known as a **paraboloid of revolution.** This is the surface used in a parabolic reflector.

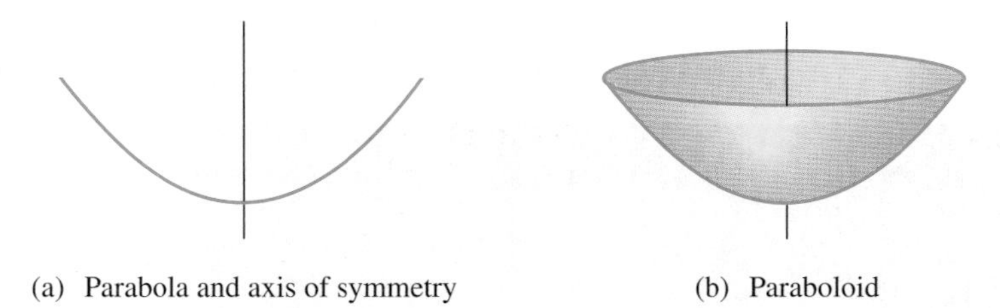

(a) Parabola and axis of symmetry (b) Paraboloid

Figure 13
Rotating a parabola about its axis of symmetry yields a bowl-shaped surface called a paraboloid of revolution.

As preparation for the next example, we define two terms that are used in describing optical systems: *focal length* and *focal ratio.* (Although we'll refer to a parabolic reflector in the definitions, the terms are often applied to other types of reflectors and lenses as well.)

DEFINITIONS Focal Length and Focal Ratio

Consider a parabolic reflector with diameter d and vertex-to-focus distance p.

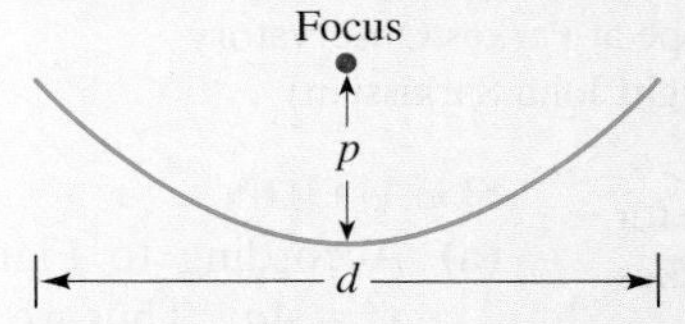

The **focal length** of the reflector is the distance p from the vertex to the focus. The **focal ratio** is the ratio of the focal length to the diameter; that is,

$$\text{focal ratio} = \frac{p}{d}$$

Thus, for example, if a parabolic reflector has a diameter of 2 ft and a focal length of 7 ft, then the focal ratio is $p/d = 7 \text{ ft}/2 \text{ ft} = 3.5$. *Remark*: In photography the focal ratio is referred to as the *f-stop* or *f-number,* and a focal ratio of 3.5 is often written as f/3.5.

In the next example we determine the focal length and focal ratio of a parabolic reflector in a particular radio telescope. The following quote provides some general background about the field of radio astronomy.

> *Since its earliest beginnings in 1932 radio astronomy has developed into one of the most important means of investigating the universe. An impressive confirmation of this statement is given in the fact that all observing astronomers who received a Nobel prize in physics were working in the field of radio astronomy. This happened in 1974, in 1978, and for the last time in 1993.*
>
> Max Planck Institute for Radio Astronomy, Bonn, Germany. Web page http://www.mpifr-bonn.mpg.de/index_e.html, accessed August 2004.

EXAMPLE 6 Finding the focal length and focal ratio of a parabolic reflector

Figure 14(a) shows the parabolic reflector dish of the radio telescope at Parkes Observatory, which is located approximately 220 miles west of Sydney, Australia. Figure 14(b) gives the specifications for the parabolic cross section of the dish.

(a) Find the focal length of the dish. Round the answer to two decimal places.

(b) Find the focal ratio. Round the answer to two decimal places.

(a) Radio telescope at Parkes Observatory (Photo copyright John Sarkissian)

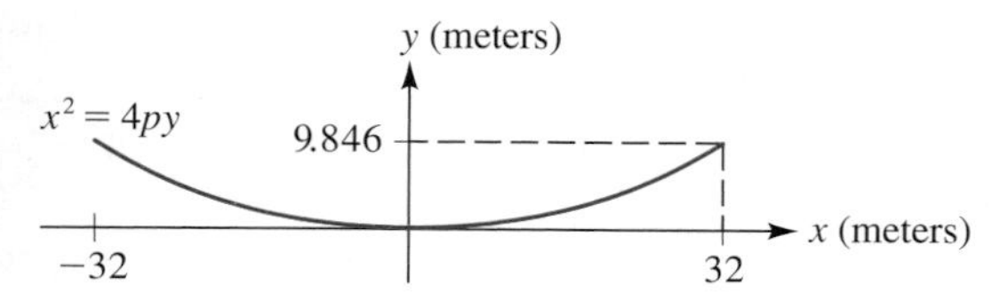

(b) Dimensions of parabolic cross section

Figure 14
Photograph and specifications for the radio telescope dish at Parkes Observatory

SOLUTION

(a) According to Figure 14(b), the point (32, 9.846) lies on the parabola $x^2 = 4py$. Thus we have $32^2 = 4p(9.846)$, and therefore

$$p = \frac{32^2}{4(9.846)}$$
$$\approx 26.00\text{m} \qquad \text{using a calculator}$$

Rounded to two decimal places, the focal length is 26.00 meters.

(b) From Figure 14(b) the diameter of the dish is 64 m ($= 32 + 32$). Using this value for the diameter and the expression for the focal length p obtained in part (a), we have

$$\text{focal ratio} = \frac{p}{d} = \frac{32^2/(4 \times 9.846)}{64}$$
$$\approx 0.41 \qquad \text{using a calculator}$$

Rounded to two decimal places, the focal ratio is 0.41.

EXERCISE SET 11.2 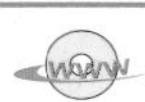

A

In Exercises 1–22, graph the parabolas. In each case, specify the focus, the directrix, and the focal width. For Exercises 13–22, also specify the vertex.

1. $x^2 = 4y$
2. $x^2 = 16y$
3. $y^2 = -8x$
4. $y^2 = 12x$
5. $x^2 = -20y$
6. $x^2 - y = 0$
7. $y^2 + 28x = 0$
8. $4y^2 + x = 0$
9. $x^2 = 6y$
10. $y^2 = -10x$
11. $4x^2 = 7y$
12. $3y^2 = 4x$
13. $y^2 - 6y - 4x + 17 = 0$
14. $y^2 + 2y + 8x + 17 = 0$
15. $x^2 - 8x - y + 18 = 0$
16. $x^2 + 6y + 18 = 0$
17. $y^2 + 2y - x + 1 = 0$
18. $2y^2 - x + 1 = 0$
19. $2x^2 - 12x - y + 18 = 0$
20. $y + \sqrt{2} = (x - 2\sqrt{2})^2$
21. $2x^2 - 16x - y + 33 = 0$
22. $\frac{1}{4}y^2 - y - x + 1 = 0$

23. The accompanying figure shows the specifications for a cross section of the parabolic reflector dish for the Lovell radio telescope, located at the Jodrell Bank Observatory in England.
(a) Find the focal length of the dish. Round the answer to one decimal place.
(b) Find the focal ratio. Round the answer to one decimal place.

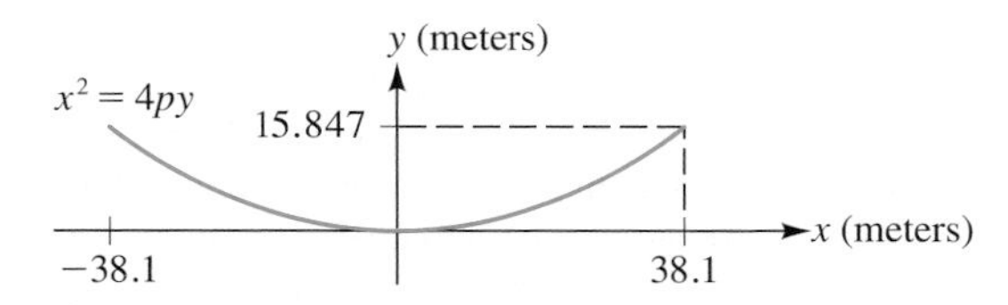

24. One of the largest radio telescopes in the world is the Effelsberg radio telescope, located 25 miles southwest of Bonn, Germany. Determine the diameter d of its parabolic reflector, given the specifications in the accompanying figure. (Assume that the units for x and y are meters.)

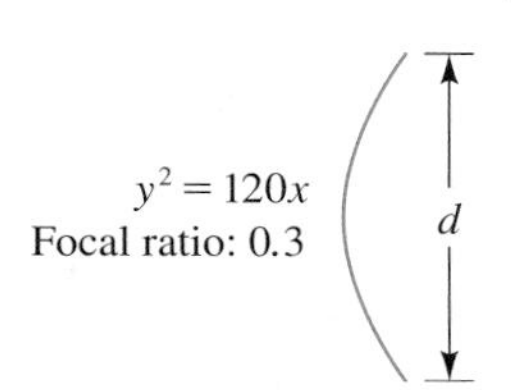

25. The figure shows the specifications for a parabolic arch. (The blue curve is a portion of a parabola.)
(a) Introduce a horizontal x-axis, passing through the vertex of the parabola, with the vertex corresponding to $x = 0$. Find an equation for the parabola of the form $x^2 = -4py$.
(b) Use the equation determined in part (a) to compute the height of the arch 5 feet away from the center (that is, at $x = 5$ and $x = -5$).

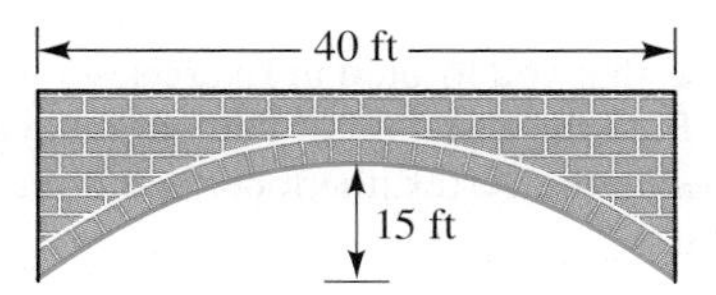

26. The figure shows a diagram for the main section of a suspension bridge. The shape of the curved cable can be closely approximated by a parabola. As shown in the figure, there are 17 vertical support wires. These, along with the two 10.6-m end posts, are equally spaced at 5-m intervals. At the center of the bridge, the vertical support wire (from the vertex of the parabola to the roadway) has a length of 0.6 m.
(a) In the figure, let the road correspond to the x-axis in an x-y coordinate system, with the vertex of the parabolic cable located at (0, 0.6). Find an equation of the form $x^2 = 4p(y - k)$ for the parabola.
(b) Find the lengths of the eight support wires on each side of the 0.6-m center wire.

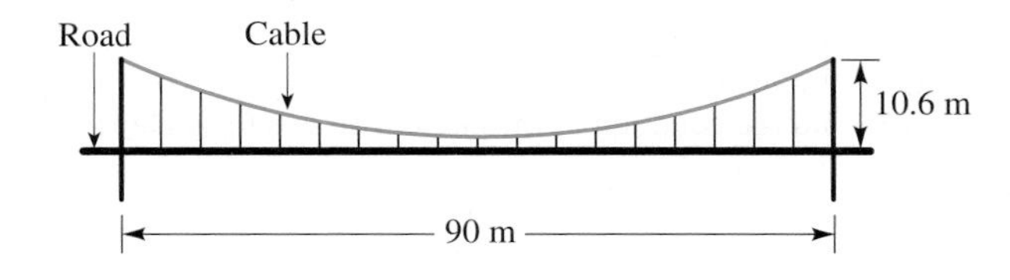

B

For Exercises 27 and 28, refer to Figure A.

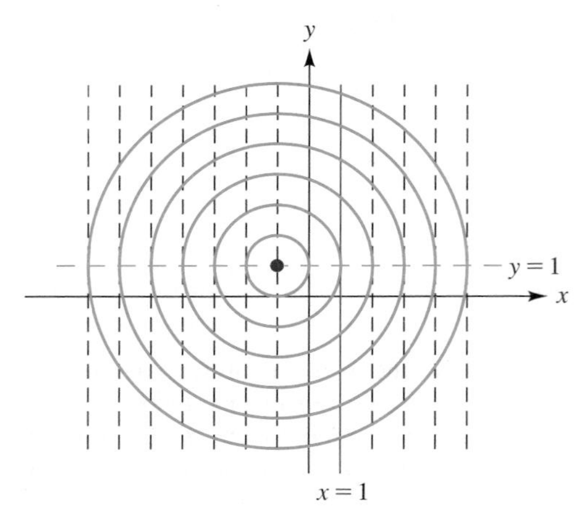

Figure A

27. Make a photocopy of Figure A. On your copy of Figure A, indicate (by drawing dots) 11 points that are equidistant from the point $(-1, 1)$ and the line $x = 1$. What is the equation of the line of symmetry for the set of dots?

28. The 11 dots that you located in Exercise 27 are part of a parabola. Find the equation of that parabola and sketch its graph. Specify the vertex, the focus and directrix, and the focal width.

For Exercises 29–33, find the equation of the parabola satisfying the given conditions. In each case, assume that the vertex is at the origin.

29. The focus is $(0, 3)$.

30. The directrix is $y - 8 = 0$.

31. The directrix is $x + 32 = 0$.

32. The focus lies on the y-axis, and the parabola passes through the point $(7, -10)$.

33. The parabola is symmetric about the x-axis, the x-coordinate of the focus is negative, and the length of the focal chord perpendicular to the x-axis is 9.

34. Let P denote the point $(8, 8)$ on the parabola $x^2 = 8y$, and let $\overline{PQ}$ be a focal chord.

- **(a)** Find the equation of the line through the point $(8, 8)$ and the focus.
- **(b)** Find the coordinates of Q.
- **(c)** Find the length of $\overline{PQ}$.
- **(d)** Find the equation of the circle with this focal chord as a diameter.
- **(e)** Show that the circle determined in part (d) intersects the directrix of the parabola in only one point. Conclude from this that the directrix is tangent to the circle. Draw a sketch of the situation.

35. The segments $\overline{AA'}$ and $\overline{BB'}$ are focal chords of the parabola $x^2 = 2y$. The coordinates of A and B are $(4, 8)$ and $(-2, 2)$, respectively.

- **(a)** Find the equation of the line through A and B'.
- **(b)** Find the equation of the line through B and A'.
- **(c)** Show that the two lines you have found intersect at a point on the directrix.

36. Let $\overline{PQ}$ be the horizontal focal chord of the parabola $x^2 = 8y$. Let R denote the point where the directrix of the parabola meets the y-axis. Show that $\overline{PR}$ is perpendicular to $\overline{QR}$.

37. Suppose $\overline{PQ}$ is a focal chord of the parabola $y = x^2$ and that the coordinates of P are $(2, 4)$.

- **(a)** Find the coordinates of Q.
- **(b)** Find the coordinates of M, the midpoint of $\overline{PQ}$.
- **(c)** A perpendicular is drawn from M to the y-axis, meeting the y-axis at S. Also, a line perpendicular to the focal chord is drawn through M, meeting the y-axis at T. Find ST, and verify that it is equal to one-half the focal width of the parabola.

38. Let F be the focus of the parabola $x^2 = 8y$, and let P denote the point on the parabola with coordinates $(8, 8)$. Let $\overline{PQ}$ be a focal chord. If V denotes the vertex of the parabola, verify that

$$PF \cdot FQ = VF \cdot PQ$$

39. In the following figure, ΔOAB is equilateral, and $\overline{AB}$ is parallel to the x-axis. Find the length of a side and the area of triangle OAB.

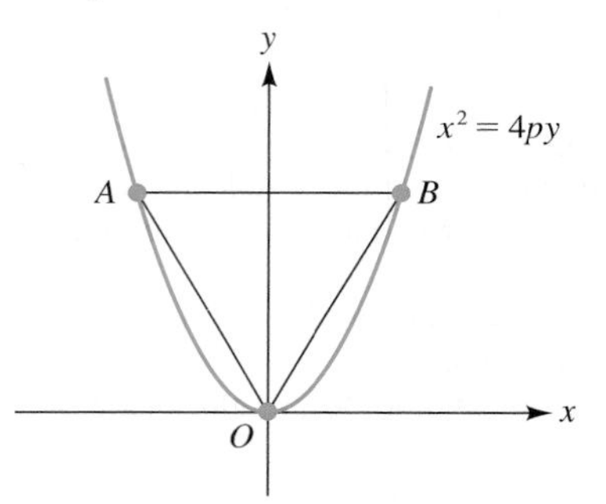

40. In designing an arch, architects and engineers sometimes use a parabolic arch rather than a semicircular arch. (One reason for this is that, in general, the parabolic arch can support more weight at the top than can the semicircular arch.) In the following figure, the blue arch is a semicircle of radius 1, centered at the origin. The red arch is a portion of a parabola. As is indicated in the figure, the two arches have the same base and the same height. Assume that the unit of distance for each axis is the meter.

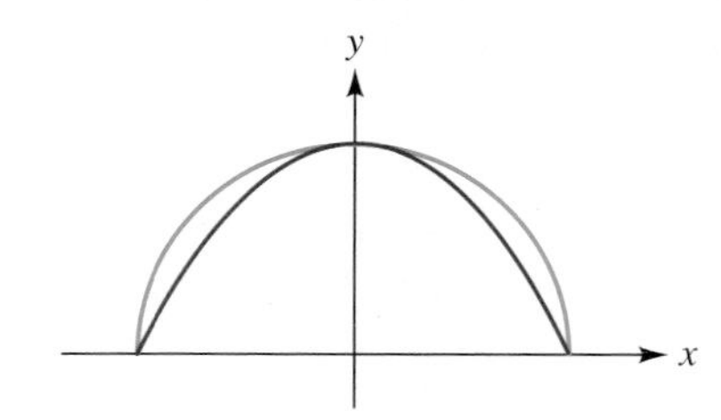

(a) Find the equation of the parabola in the figure.

(b) Using calculus, it can be shown that the area under this parabolic arch is $\frac{4}{3}$ m^2. Assuming this fact, show that the area beneath the parabolic arch is approximately 85% of the area beneath the semicircular arch.

(c) Using calculus, it can be shown that the length of this parabolic arch is $\sqrt{5} + \frac{1}{2}\ln(2 + \sqrt{5})$ meters. Assuming this fact, show that the length of the parabolic arch is approximately 94% of the length of the semicircular arch.

41. (a) The span of the parabolic arch in Figure B is 8 m. At a distance of 2 m from the center, the vertical clearance is 4.5 m. Find the height of the arch. *Hint*: Choose a convenient coordinate system in which the equation of the parabola will have the form $x^2 = -4p(y - k)$.

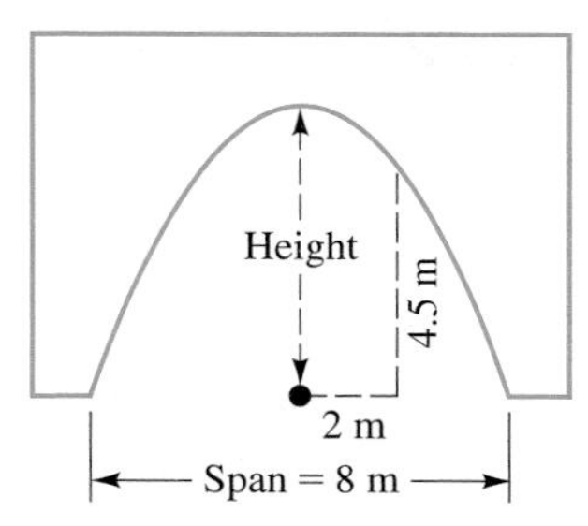

Figure B

(b) Figure C shows the same parabolic arch as in Figure B. Suppose that from a point P on the base of the arch, a laser beam is projected vertically upward. As indicated in the figure, the beam hits the arch at a point Q and is then reflected. Let R denote the point where the beam crosses the axis of the parabola. How far is it from R to the base of the arch? (That is, what is the perpendicular distance from R to the base?)

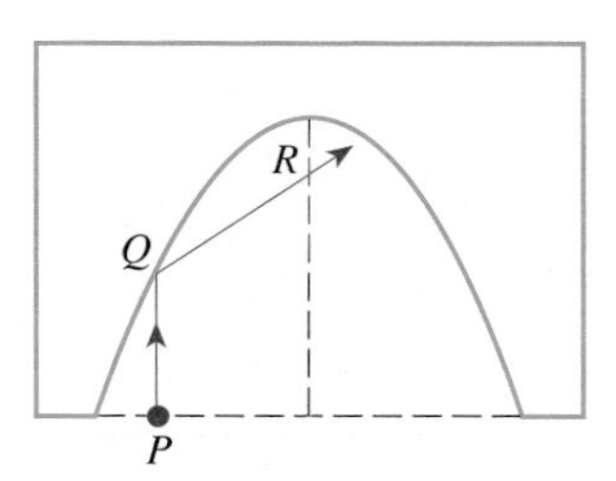

Figure C

C

42. If $\overline{PQ}$ is a focal chord of the parabola $x^2 = 4py$ and the coordinates of P are (x_0, y_0), show that the coordinates of Q are

$$\left(\frac{-4p^2}{x_0}, \frac{p^2}{y_0}\right)$$

43. If $\overline{PQ}$ is a focal chord of the parabola $y^2 = 4px$ and the coordinates of P are (x_0, y_0), show that the coordinates of Q are

$$\left(\frac{p^2}{x_0}, \frac{-4p^2}{y_0}\right)$$

44. Let F and V denote the focus and the vertex, respectively, of the parabola $x^2 = 4py$. If $\overline{PQ}$ is a focal chord of the parabola, show that

$$PF \cdot FQ = VF \cdot PQ$$

45. Let $\overline{PQ}$ be a focal chord of the parabola $y^2 = 4px$, and let M be the midpoint of $\overline{PQ}$. A perpendicular is drawn from M to the x-axis, meeting the x-axis at S. Also from M, a line segment is drawn that is perpendicular to $\overline{PQ}$ and that meets the x-axis at T. Show that the length of $\overline{ST}$ is one-half the focal width of the parabola.

46. Let $\overline{AB}$ be a chord (not necessarily a focal chord) of the parabola $y^2 = 4px$, and suppose that $\overline{AB}$ subtends a right angle at the vertex. (In other words, $\angle AOB = 90°$, where O is the origin in this case.) Find the x-intercept of the segment $\overline{AB}$. What is surprising about this result? *Hint*: Begin by writing the coordinates of A and B as $A(a^2/4p, a)$ and $B(b^2/4p, b)$.

47. Let $\overline{PQ}$ be a focal chord of the parabola $x^2 = 4py$. Complete the following steps to prove that the circle with $\overline{PQ}$ as a diameter is tangent to the directrix of the parabola. Let the coordinates of P be (x_0, y_0).

(a) Show that the coordinates of Q are

$$\left(\frac{-4p^2}{x_0}, \frac{p^2}{y_0}\right)$$

(b) Show that the midpoint of $\overline{PQ}$ is

$$\left(\frac{x_0^2 - 4p^2}{2x_0}, \frac{y_0^2 + p^2}{2y_0}\right)$$

(c) Show that the length of $\overline{PQ}$ is $(y_0 + p)^2/y_0$. *Suggestion*: This can be done using the formula for the distance between two points, but the following is simpler. Let F be the focus. Then $PQ = PF + FQ$. Now, both PF and FQ can be determined by using the definition of the parabola rather than the distance formula.

(d) Show that the distance from the center of the circle to the directrix equals the radius of the circle. How does this complete the proof?

48. P and Q are two points on the parabola $y^2 = 4px$, the coordinates of P are (a, b), and the slope of $\overline{PQ}$ is m. Find the y-coordinate of the midpoint of $\overline{PQ}$. *Hint for checking*: The final answer is independent of both a and b.

MINI PROJECT 1

A BRIDGE WITH A PARABOLIC ARCH

Figure A shows a bridge across a river. The arch of the bridge is a parabola, and the six vertical cables that help support the road are equally spaced at 4-m intervals. Figure B shows the parabolic arch in an x-y coordinate system, with the left end of the arch at the origin. As is indicated in Figure B, the length of the leftmost cable is 3.072 m. Determine an equation for the parabola of the form $(x - h)^2 = -4p(y - k)$. Then use the equation to determine the lengths of the other cables and also the maximum height of the arch above the road.

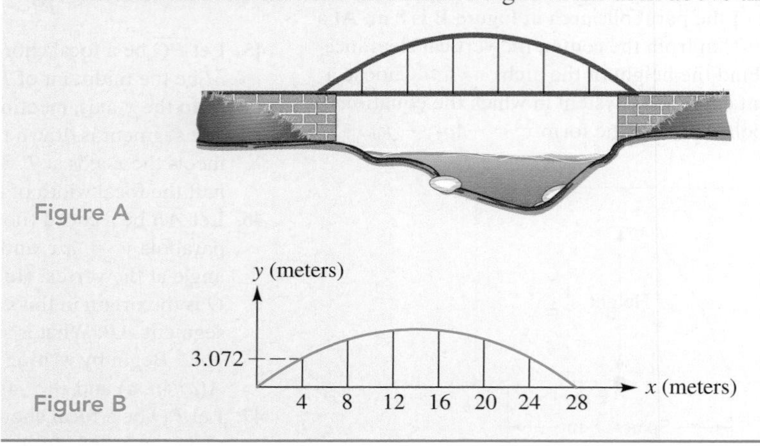

Figure A

Figure B

MINI PROJECT 2

CONSTRUCTING A PARABOLA

This hands-on mini project shows you how to construct a parabola. Follow the instructions, and then explain why the curve that you draw is indeed a parabola. In addition to a sheet of paper and a pen or marker, you'll need the following:

- A T-square and a drawing board or surface on which you can easily slide the T-square left and right in a straight horizontal path
- A piece of string that has the same length as the long bar of the T-square (see Figure C)
- Two thumb tacks

(Refer to Figure D.) Tack one end of the piece of string to a point on the drawing board. Call that point F. Tack the other end of the string to the top right corner of the T-square, as indicated in Figure D. Now, with the pen or marker, move the T-square horizontally subject to the following two constraints: The string

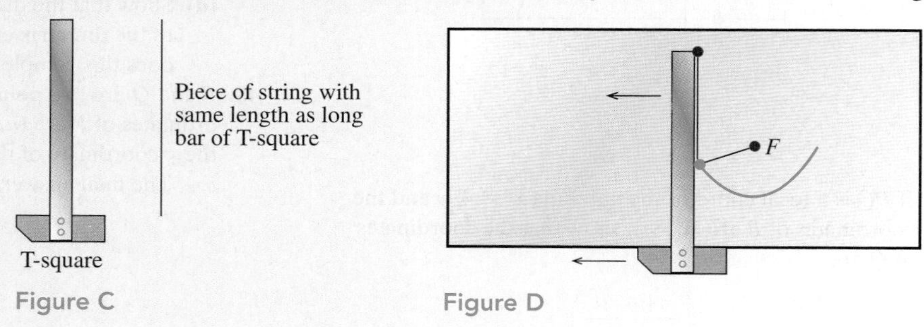

Figure C

Figure D

must be kept taut by the pen, and the pen must stay against the edge of the T-square. The curve traced out by the pen will be a portion of a parabola. (In practice, the piece of string that you begin with should actually be just slightly longer than the length of the T-square, since a slight amount of the string is taken up in the tacking.)

11.3 TANGENTS TO PARABOLAS (OPTIONAL SECTION)

And I dare say that this [the problem of finding tangents and normals to curves] *is not only the most useful and most general problem in geometry that I know, but even that I have ever desired to know.*—René Descartes, *La Géométrie* (1637).

Many of the more important properties of the parabola relate to the tangent to the curve. For instance, there is a close connection between the optical property illustrated in Figure 12 on page 833 and the tangents to the parabola. (See Exercise 19 at the end of this section for details.) In general, the techniques of calculus are required to deal with tangents to curves. However, for curves with equations and graphs as simple as the parabola, the methods of algebra are often adequate. The following discussion shows how we can determine the tangent to a parabola without using calculus.

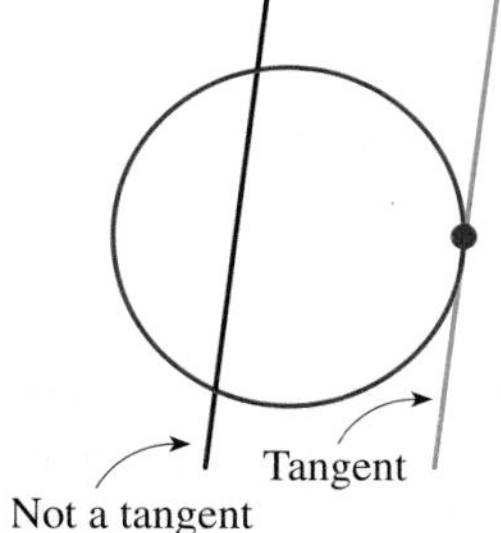

Figure 1

To begin with, we need a definition for the tangent to a parabola. As motivation, recall from geometry that a tangent line to a circle is defined to be a line that intersects the circle in exactly one point (see Figure 1). However, this definition is not quite adequate for the parabola. For instance, the y-axis intersects the parabola $y = x^2$ in exactly one point, but surely it is not tangent to the curve. With this in mind, we adopt the definition in the following box for the tangent to a parabola. Along with this definition we make the assumption that through each point P on a parabola, there is only one tangent line that can be drawn.

DEFINITION **Tangent to a Parabola**

Let P be a point on a parabola. Then a line through P is said to be tangent to the parabola at P provided that the line intersects the parabola only at P and the line is not parallel to the axis of the parabola.

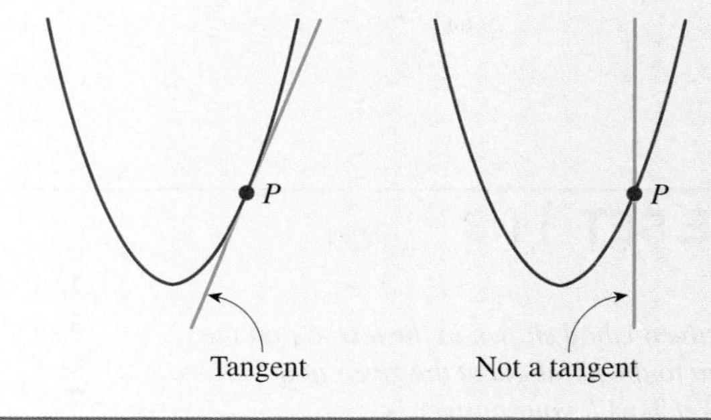

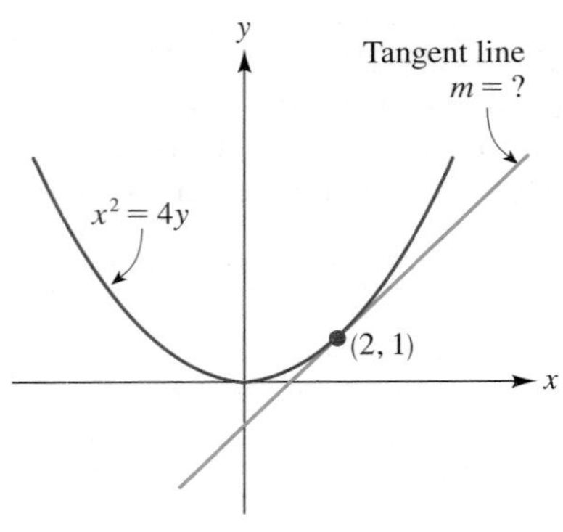

Figure 2

We'll demonstrate how to find tangents to parabolas by means of an example. The method used here can be used for any parabola. Suppose that we want to find the equation of the tangent to the parabola $x^2 = 4y$ at the point (2, 1) on the curve (see Figure 2). Let m denote the slope of the tangent line. Since the line must pass through the point (2, 1), its equation is

$$y - 1 = m(x - 2)$$

This is the tangent line, so it intersects the parabola in only one point, namely, (2, 1). Algebraically, this means that the ordered pair (2, 1) is the only solution of the following system of equations:

$$\begin{cases} y - 1 = m(x - 2) & (1) \\ x^2 = 4y & (2) \end{cases}$$

The strategy now is to solve this system of equations in terms of m. Then we'll reconcile the results with the fact that (2, 1) is known to be the unique solution. This will allow us to determine m. From equation (2) we have $y = x^2/4$. Then, substituting for y in equation (1), we obtain

$$\frac{x^2}{4} - 1 = m(x - 2)$$

$$x^2 - 4 = 4m(x - 2) \quad \text{multiplying both sides by 4}$$

$$(x - 2)(x + 2) = 4m(x - 2) \quad \text{factoring}$$

$$(x - 2)(x + 2) - 4m(x - 2) = 0 \quad \text{subtracting } 4m(x - 2) \text{ from both sides}$$

$$(x - 2)[(x + 2) - 4m] = 0 \quad \text{factoring out } x - 2$$

$$x - 2 = 0 \qquad x + 2 - 4m = 0$$

$$x = 2 \qquad x = 4m - 2$$

The first value, $x = 2$, yields no new information, since we knew from the start that this was the x-coordinate of the intersection point. However, the second value, $x = 4m - 2$, must also equal 2, since the system has but one solution. Thus we have

$$4m - 2 = 2$$

$$4m = 4$$

$$m = 1$$

Therefore the slope of the tangent line is 1. Substituting this value of m in equation (1) yields

$$y - 1 = 1(x - 2)$$

or

$$y = x - 1 \quad \text{as required}$$

EXERCISE SET 11.3 

A

In Exercises 1–8, use the method shown in the text to find the equation of the tangent to the parabola at the given point. In each case, include a sketch with your answer.

1. $x^2 = y$; (2, 4)
2. $x^2 = -2y$; (2, −2)
3. $x^2 = 8y$; (4, 2)
4. $x^2 = 12y$; (6, 3)
5. $x^2 = -y$; (−3, −9)
6. $x^2 = -6y$; $(\sqrt{6}, -1)$
7. $y^2 = 4x$; (1, 2)
8. $y^2 = -8x$; (−8, −8)

B

In Exercises 9–11, find the slope of the tangent to the curve at the indicated point. (Use the method shown in the text for parabolas.)

9. $y = \sqrt{x}$; $(4, 2)$

Hint: $x - 4 = (\sqrt{x} - 2)(\sqrt{x} + 2)$

10. $y = 1/x$; $(3, 1/3)$

11. $y = x^3$; $(2, 8)$

12. Consider the parabola $x^2 = 4y$, and let (x_0, y_0) denote a point on the parabola in the first quadrant.

(a) Find the y-intercept of the line tangent to the parabola at the point on the parabola where $y_0 = 1$.

(b) Repeat part (a) using $y_0 = 2$.

(c) Repeat part (a) using $y_0 = 3$.

(d) On the basis of your results in parts (a)–(c), make a conjecture about the y-intercept of the line that is tangent to the parabola at the point (x_0, y_0) on the curve. Verify your conjecture by computing this y-intercept.

C

13. Let (x_0, y_0) be a point on the parabola $x^2 = 4py$. Using the method explained in the text, show that the equation of the line tangent to the parabola at (x_0, y_0) is

$$y = \frac{x_0}{2p}x - y_0$$

Thus the y-intercept of the line tangent to $x^2 = 4py$ at (x_0, y_0) is just $-y_0$.

14. Let (x_0, y_0) be a point on the parabola $y^2 = 4px$. Show that the equation of the line tangent to the parabola at (x_0, y_0) is

$$y = \frac{2p}{y_0}x + \frac{y_0}{2}$$

Show that the x-intercept of this line is $-x_0$.

Exercises 15–20 contain results about tangents to parabolas. In working these problems, you'll find it convenient to use the facts developed in Exercises 13 and 14. Also assume, as given, the results about focal chords in Exercises 42 and 43 of Exercise Set 11.2. In some of the problems, reference is made to the normal line. *The* **normal line** *or* **normal** *to a parabola at the point (x_0, y_0) on the parabola is defined as the line through (x_0, y_0) that is perpendicular to the tangent at (x_0, y_0).*

15. Let $\overline{PQ}$ be a focal chord of the parabola $x^2 = 4py$.

(a) Show that the tangents to the parabola at P and Q are perpendicular to each other.

(b) Show that the tangents to the parabola at P and Q intersect at a point on the directrix.

(c) Let D be the intersection point of the tangents at P and Q. Show that the line segment from D to the focus is perpendicular to $\overline{PQ}$.

16. Let $P(x_0, y_0)$ be a point [other than $(0, 0)$] on the parabola $y^2 = 4px$. Let A be the point where the normal line to the parabola at P meets the axis of the parabola. Let B be the point where the line drawn from P perpendicular to the axis of the parabola meets the axis. Show that $AB = 2p$.

17. Let $P(x_0, y_0)$ be a point on the parabola $y^2 = 4px$. Let A be the point where the normal line to the parabola at P meets the axis of the parabola. Let F be the focus of the parabola. If a line is drawn from A perpendicular to $\overline{FP}$, meeting $\overline{FP}$ at Z, show that $ZP = 2p$.

18. Let $\overline{AB}$ be a chord (not necessarily a focal chord) of the parabola $x^2 = 4py$. Let M be the midpoint of the chord and let C be the point where the tangents at A and B intersect.

(a) Show that $\overline{MC}$ is parallel to the axis of the parabola.

(b) If D is the point where the parabola meets $\overline{MC}$, show that $CD = DM$.

19. In this exercise we prove the *reflection property* of parabolas. The following figure shows a line tangent to the parabola $y^2 = 4px$ at $P(x_0, y_0)$. The dashed line through H and P is parallel to the axis of the parabola. We wish to prove that $\alpha = \beta$. This is the reflection property of parabolas.

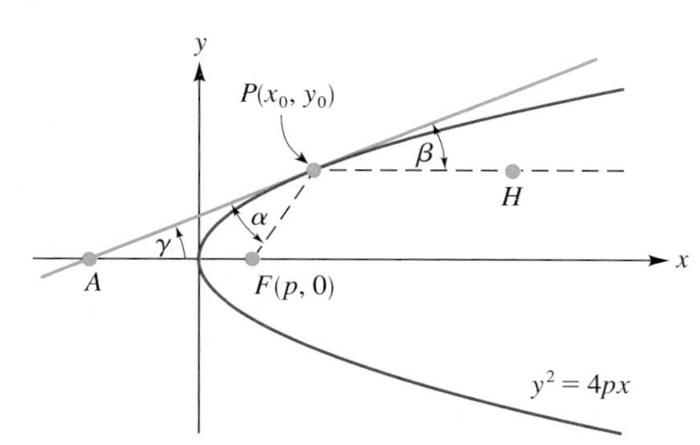

(a) Show that $FA = x_0 + p = FP$. Hint: Regarding FP, it is easier to rely on the definition of a parabola than on the distance formula.

(b) Why does $\alpha = \gamma$? Why does $\gamma = \beta$?

(c) Conclude that $\alpha = \beta$, as required.

20. The segment $\overline{AB}$ is a focal chord of the parabola $y^2 = 4x$ and the coordinates of A are $(4, 4)$. Normals drawn through A and B meet the parabola again at A' and B', respectively. Prove that $\overline{A'B'}$ is three times as long as $\overline{AB}$.

11.4 THE ELLIPSE

We have here apparently [in the work of Anthemius of Tralles (a sixth-century Greek architect and mathematician)] *the first mention of the construction of an ellipse by means of a string stretched tight round the foci.*—Sir Thomas Heath in *A History of Greek Mathematics,* Vol. II (Oxford: The Clarendon Press, 1921)

The true orbit of Mars was even less of a circle than the Earth's. It took almost two years for Kepler to realize that its orbital shape is that of an ellipse. An ellipse is the shape of a circle when viewed at an angle. —Phillip Flower in *Understanding the Universe* (St. Paul: West Publishing Co., 1990)

The heavenly motions are nothing but a continuous song for several voices, to be perceived by the intellect, not by the ear. —Johannes Kepler (1571–1630)

In this section we discuss the symmetric, oval-shaped curve known as the *ellipse.* As Kepler discovered and Newton later proved, this is the curve described by the planets in their motions around the sun.

DEFINITION The Ellipse

An **ellipse** is the set of all points in the plane, the sum of whose distances from two fixed points is constant. Each fixed point is called a **focus** (plural: **foci**).

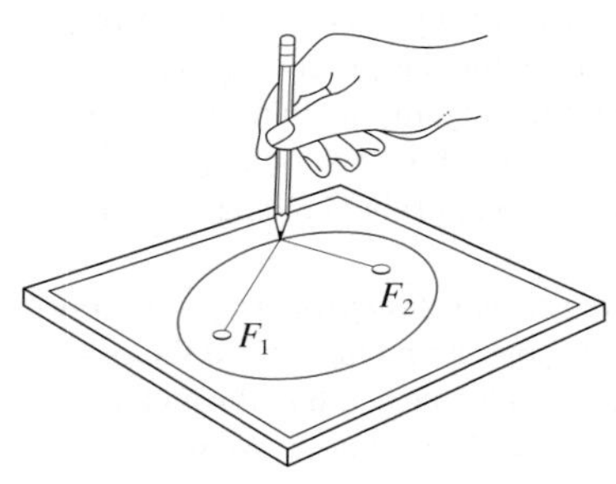

Figure 1

Subsequently, we will derive an equation describing the ellipse just as we found an equation for the parabola in Section 11.2. But first let's consider some rather immediate consequences of the definition. In fact, we can learn a great deal about the ellipse even before we derive its equation.

There is a simple mechanical method for constructing an ellipse that arises directly from the definition of the curve. Mark the given foci—say, F_1 and F_2—on a drawing board and insert thumbtacks at those points. Now take a piece of string that is longer than the distance from F_1 to F_2, and tie the ends of the string to the tacks. Next, pull the string taut with a pencil point, and touch the pencil point to the drawing board. Then if you move the pencil while keeping the string taut, the curve traced out will be an ellipse, as indicated in Figure 1. The reason the curve is an ellipse is that for each point on the curve, the sum of the distances from the foci is constant, the constant being the length of the string. By actually carrying out this construction for yourself several times, each time varying the distance between the foci or the length of the string, you can learn a great deal about the ellipse. For instance, when the distance between the foci is small in comparison to the length of the string, the ellipse begins to resemble a circle, as in Figure 2. On the other hand, when the distance between the foci is nearly equal to the length of the string, the ellipse becomes relatively flat, as in Figure 3.

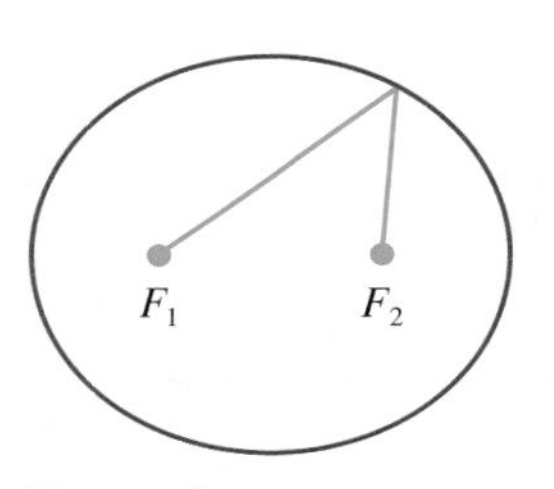

Figure 2
When the distance between the foci is small in comparison to the length of the string, the ellipse resembles a circle.

We now derive one of the standard forms for the equation of an ellipse. As is indicated in Figure 4, we assume that the foci are $F_1(-c, 0)$ and $F_2(c, 0)$ and that the sum of the distances from the foci to a point $P(x, y)$ on the ellipse is $2a$. Since the point P lies on the ellipse, we have

$$F_1P + F_2P = 2a$$

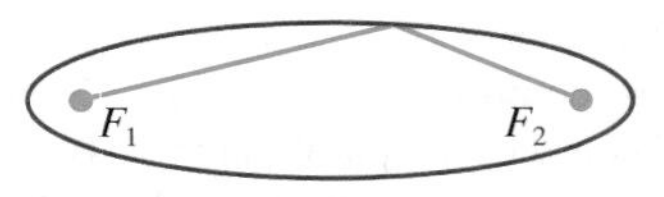

Figure 3
When the distance between the foci is nearly equal to the length of the string, the ellipse is relatively flat.

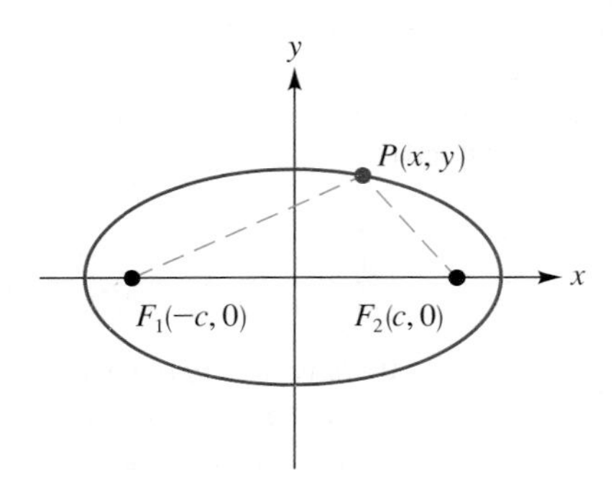

Figure 4
$F_1P + F_2P = 2a$

and therefore

$$\sqrt{(x+c)^2+(y-0)^2}+\sqrt{(x-c)^2+(y-0)^2}=2a$$

or

$$\sqrt{(x-c)^2+y^2}=2a-\sqrt{(x+c)^2+y^2} \qquad (1)$$

Now, by following a straightforward but lengthy process of squaring and simplifying (as outlined in detail in Exercise 44), we find that equation (1) becomes

$$(a^2-c^2)x^2+a^2y^2=a^2(a^2-c^2) \qquad (2)$$

To write equation (2) in a more symmetric form, we define the positive number b by the equation

$$b^2=a^2-c^2 \qquad (3)$$

NOTE For this definition to make sense, we need to know that the right-hand side of equation (3) is positive. (See Exercise 45 at the end of this section for details.)

Finally, using equation (3) to substitute for a^2-c^2 in equation (2), we obtain

$$b^2x^2+a^2y^2=a^2b^2$$

which can be written

$$\frac{x^2}{a^2}+\frac{y^2}{b^2}=1 \qquad (a>b) \qquad (4)$$

We have now shown that the coordinates of each point on the ellipse satisfy equation (4). Conversely, it can be shown that if x and y satisfy equation (4), then the sum of the distances from the point (x, y) to the points $(c, 0)$ and $(-c, 0)$ is $2a$; that is, the point (x, y) indeed lies on this ellipse (see Exercise 60). We refer to equation (4) as the **standard form** for the equation of an ellipse with foci $(-c, 0)$ and $(c, 0)$. For an ellipse in this form, it will always be the case that a is greater than b; this follows from equation (3).

For purposes of graphing, we want to know the intercepts of the ellipse. To find the x-intercepts, we set $y=0$ in equation (4) to obtain

$$\frac{x^2}{a^2}=1$$

$$x^2=a^2 \qquad \text{or} \qquad x=\pm a$$

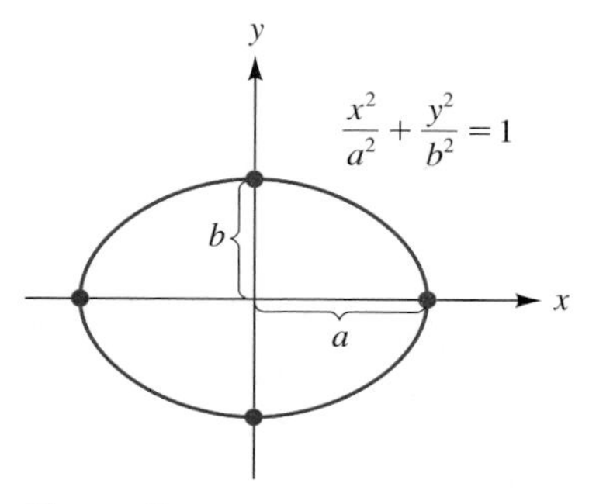

Figure 5
The intercepts of the ellipse $(x^2/a^2)+(y^2/b^2)=1$ are $x=\pm a$ and $y=\pm b$.

The x-intercepts are therefore a and $-a$. In a similar fashion you can check that the y-intercepts are b and $-b$. Also (according to the symmetry tests in Section 1.7), note that the graph of equation (4) must be symmetric about both coordinate axes. Figure 5 shows the graph of the ellipse $(x^2/a^2)+(y^2/b^2)=1$. (Calculator exercises at the end of this section will help to convince you that the general shape of the curve in Figure 5 is correct.)

In the box that follows, we define several terms that are useful in describing and analyzing the ellipse.

DEFINITION **Terminology for the Ellipse**

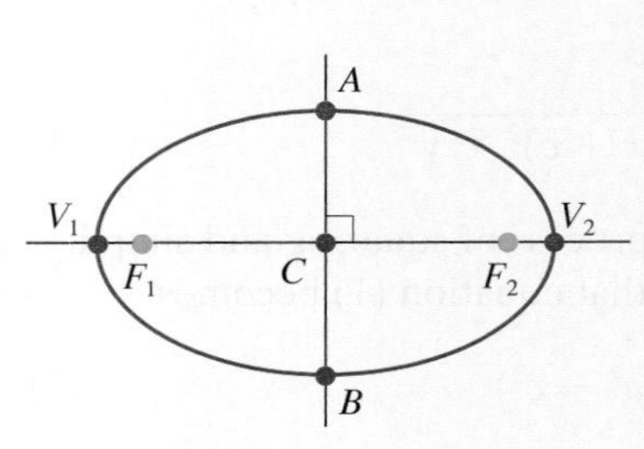

Figure 6

1. The **focal axis** is the line passing through the foci of the ellipse.
2. The **center** is the point midway between the foci. This is the point C in Figure 6.
3. The **vertices** (singular: **vertex**) are the two points where the focal axis meets the ellipse. These are the points V_1 and V_2 in Figure 6.
4. The **major axis** is the line segment joining the vertices. This is the line segment $\overline{V_1V_2}$ in Figure 6. Each of the (congruent) line segments $\overline{CV_1}$ and $\overline{CV_2}$ is referred to as a **semimajor axis.** The **minor axis** is the line segment through the center of the ellipse, perpendicular to the major axis, and with endpoints on the ellipse. In Figure 6 this is the line segment $\overline{AB}$. Each of the (congruent) line segments $\overline{CB}$ and $\overline{CA}$ is referred to as a **semiminor axis.**
5. The **eccentricity** is the ratio $\dfrac{c}{a} = \dfrac{\sqrt{a^2 - b^2}}{a}$.

The eccentricity (as defined in the box) provides a numerical measure of how much the ellipse deviates from being a circle. As Figure 7 indicates, the closer the eccentricity is to zero, the more the ellipse resembles a circle. In the other direction, as the eccentricity approaches 1, the ellipse becomes increasingly flat. As the eccentricity increases, the ellipse becomes increasingly flat. *Note*: The eccentricity is always between 0 and 1.

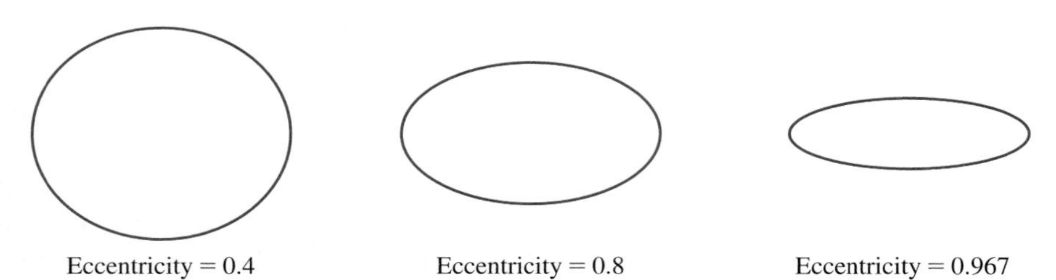

Figure 7
Eccentricity is a number between 0 and 1 that describes the shape of an ellipse. The narrowest, that is, the least circular, ellipse in this figure has the same proportions as the orbit of Halley's comet. By way of contrast, the eccentricity of the Earth's orbit is 0.0017; if an ellipse with this eccentricity were included in this figure, it would appear indistinguishable from a circle.

In the next box, we summarize our discussion up to this point. (We use the letter e to denote the eccentricity; this is the conventional choice, even though the same letter is used with a very different meaning in connection with exponential functions and logarithms.)

PROPERTY SUMMARY **The Ellipse $\dfrac{x^2}{a^2} + \dfrac{y^2}{b^2} = 1 \quad (a > b)$**

Foci: $(\pm c, 0)$, where $c^2 = a^2 - b^2$
Center: $(0, 0)$
Vertices: $(\pm a, 0)$
Length of major axis: $2a$
Length of semimajor axis: a
Length of minor axis: $2b$
Length of semiminor axis: b
Eccentricity: $e = \dfrac{c}{a} = \dfrac{\sqrt{a^2 - b^2}}{a}$

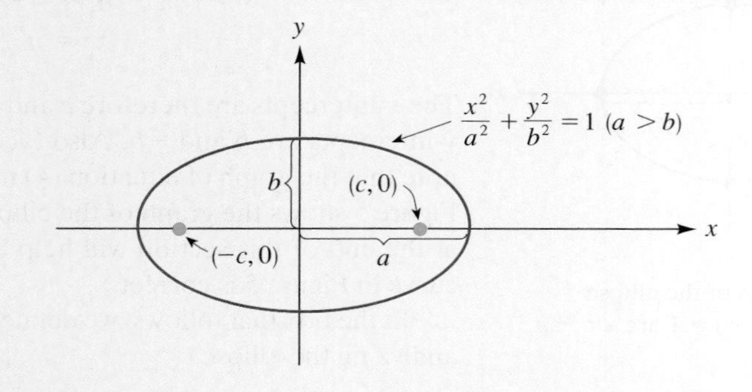

EXAMPLE 1 Analyzing an ellipse

Find the lengths of the major and minor axes of the ellipse $9x^2 + 16y^2 = 144$. Specify the coordinates of the foci and the eccentricity. Graph the ellipse.

SOLUTION

To convert the equation $9x^2 + 16y^2 = 144$ to standard form, we divide both sides by 144. This yields

$$\frac{9x^2}{144} + \frac{16y^2}{144} = \frac{144}{144}$$
$$\frac{x^2}{16} + \frac{y^2}{9} = 1$$
$$\frac{x^2}{4^2} + \frac{y^2}{3^2} = 1$$

This is the standard form. Comparing this equation with $(x^2/a^2) + (y^2/b^2) = 1$, we see that $a = 4$ and $b = 3$. Thus the major and minor axes are eight and six units, respectively. Next, to determine the foci, we use the equation $c^2 = a^2 - b^2$. We have

$$c^2 = 4^2 - 3^2 = 7 \qquad \text{and therefore} \qquad c = \sqrt{7}$$

(We choose the positive square root because $c > 0$.) It follows that the coordinates of the foci are $(-\sqrt{7}, 0)$ and $(\sqrt{7}, 0)$. We now calculate the eccentricity using the formula $e = c/a$. This yields $e = \sqrt{7}/4$. Figure 8 shows the graph along with the required information.

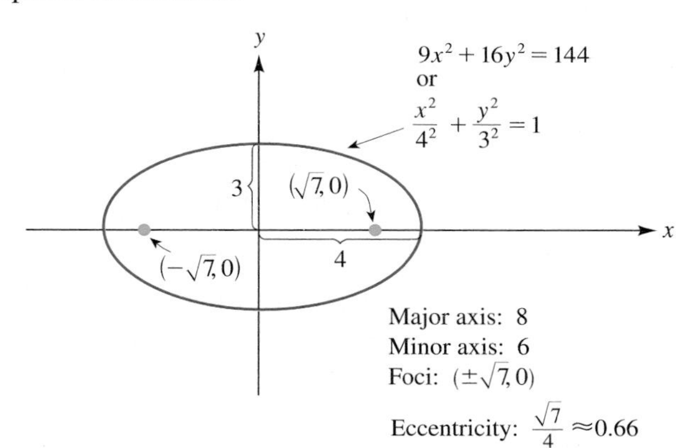

Figure 8

EXAMPLE 2 Finding the equation of an ellipse

The foci of an ellipse are $(-1, 0)$ and $(1, 0)$, and the eccentricity is $1/3$. Find the equation of the ellipse (in standard form), and specify the lengths of the major and minor axes.

SOLUTION

Since the foci are $(\pm 1, 0)$, we have $c = 1$. Using the equation $e = c/a$ with $e = 1/3$ and $c = 1$, we obtain

$$\frac{1}{3} = \frac{1}{a} \qquad \text{and therefore} \qquad a = 3$$

Recall now that b^2 is defined by the equation $b^2 = a^2 - c^2$. In view of this we have

$$b^2 = 3^2 - 1^2 = 8$$
$$b = \sqrt{8} = 2\sqrt{2}$$

The equation of the ellipse in standard form is therefore

$$\frac{x^2}{3^2} + \frac{y^2}{(2\sqrt{2})^2} = 1$$

Furthermore, since $a = 3$ and $b = 2\sqrt{2}$, the lengths of the major and minor axes are 6 and $4\sqrt{2}$ units, respectively.

EXAMPLE 3 Finding the distance between the foci of an ellipse

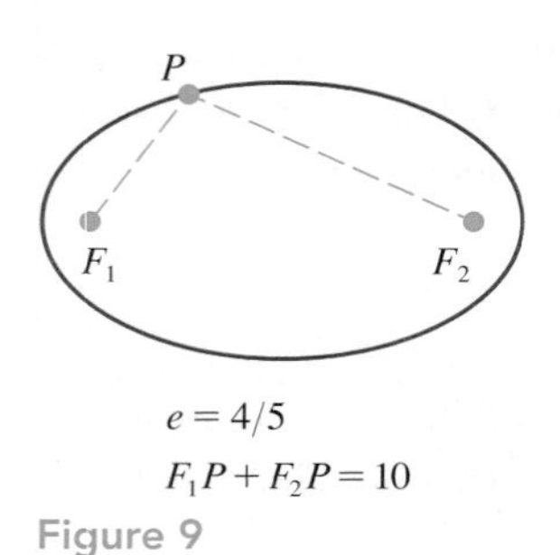

Figure 9

The eccentricity of an ellipse is 4/5, and the sum of the distances from a point P on the ellipse to the foci is 10 units. Compute the distance between the foci F_1 and F_2. See Figure 9.

SOLUTION

We are required to find the distance between the foci F_1 and F_2. By definition, this is the quantity $2c$. Since the sum of the distances from a point on the ellipse to the foci is 10 units, we have

$$2a = 10 \qquad \text{by definition of } 2a$$
$$a = 5$$

Now we substitute the values $a = 5$ and $e = 4/5$ in the formula $e = c/a$:

$$\frac{4}{5} = \frac{c}{5} \qquad \text{and therefore} \qquad c = 4$$

It now follows that the required distance $2c$ is 8 units.

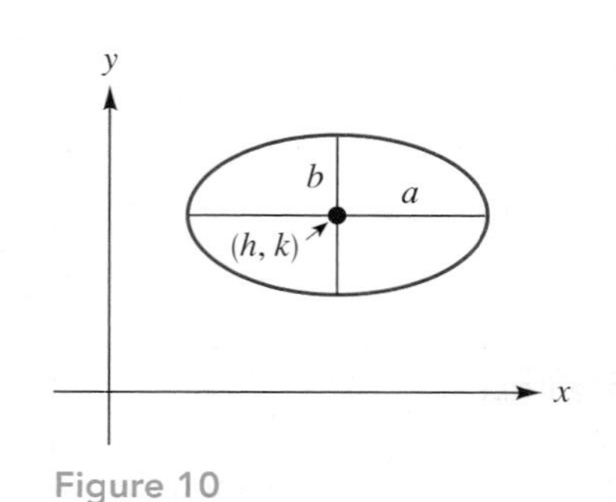

Figure 10

Suppose now that we translate the graph of $(x^2/a^2) + (y^2/b^2) = 1$ by h units in the x-direction and k units in the y-direction. Figure 10 shows the situation for h and k positive. Then the equation of the translated ellipse is

$$\frac{(x-h)^2}{a^2} + \frac{(y-k)^2}{b^2} = 1 \qquad (5)$$

Equation (5) is another **standard form** for the equation of an ellipse. As is indicated in Figure 10, the center of this ellipse is the point (h, k). In the next example we use the technique of completing the square to convert an equation for an ellipse to standard form. Once the equation is in standard form, the graph is readily obtained.

EXAMPLE 4 Completing the square to analyze an ellipse

Determine the center, foci, and eccentricity of the ellipse

$$4x^2 + 9y^2 - 8x - 54y + 49 = 0$$

Graph the ellipse.

SOLUTION

We will convert the given equation to standard form by using the technique of completing the square. We have

$$\begin{aligned} 4x^2 - 8x + 9y^2 - 54y &= -49 \\ 4(x^2 - 2x) + 9(y^2 - 6y) &= -49 \\ 4(x^2 - 2x + 1) + 9(y^2 - 6y + 9) &= -49 + 4(1) + 9(9) \\ 4(x - 1)^2 + 9(y - 3)^2 &= 36 \end{aligned}$$

Dividing this last equation by 36, we obtain

$$\frac{(x-1)^2}{9} + \frac{(y-3)^2}{4} = 1 \quad \text{or} \quad \frac{(x-1)^2}{3^2} + \frac{(y-3)^2}{2^2} = 1$$

This last equation represents an ellipse with center at (1, 3) and with $a = 3$ and $b = 2$. We can calculate c using the formula $c^2 = a^2 - b^2$:

$$c^2 = 3^2 - 2^2 = 5 \qquad \text{and therefore} \qquad c = \sqrt{5}$$

Since the center of this ellipse is (1, 3), the foci are therefore $(1 + \sqrt{5}, 3)$ and $(1 - \sqrt{5}, 3)$. Finally, the eccentricity is $c/a = \sqrt{5}/3$. Figure 11 shows the graph of this ellipse.

$b = 2$

$a = 3$

(1, 3)

$4x^2 + 9y^2 - 8x - 54y + 49 = 0$

or

$\frac{(x-1)^2}{3^2} + \frac{(y-3)^2}{2^2} = 1$

Figure 11

In developing the equation $(x^2/a^2) + (y^2/b^2) = 1$, we assumed that the foci of the ellipse were located on the x-axis at the points $(-c, 0)$ and $(c, 0)$. If instead the foci are located on the y-axis at the points $(0, c)$ and $(0, -c)$, then the same method we used in the previous case will show the equation of the ellipse to be

$$\frac{x^2}{b^2} + \frac{y^2}{a^2} = 1 \qquad (a > b)$$

where we still assume that $2a$ represents the sum of the distances from a point on the ellipse to the foci. In the box that follows, we summarize the situation for the ellipse with foci $(0, c)$ and $(0, -c)$ and constant sum of distances $2a$.

PROPERTY SUMMARY The Ellipse $\frac{x^2}{b^2} + \frac{y^2}{a^2} = 1 \quad (a > b)$

Foci: $(0, \pm c)$, where $c^2 = a^2 - b^2$
Center: (0, 0)
Vertices: $(0, \pm a)$
Length of major axis: $2a$
Length of minor axis: $2b$
Eccentricity: $e = \frac{c}{a} = \frac{\sqrt{a^2 - b^2}}{a}$

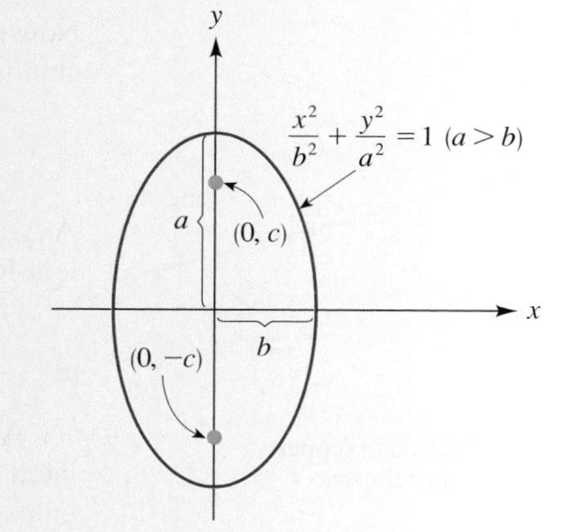

We now have two standard forms for the ellipse centered at the origin:

$$\text{Foci on the } x\text{-axis at } (\pm c, 0)\text{:} \quad \frac{x^2}{a^2} + \frac{y^2}{b^2} = 1 \qquad (a > b)$$

$$\text{Foci on the } y\text{-axis at } (0, \pm c)\text{:} \quad \frac{x^2}{b^2} + \frac{y^2}{a^2} = 1 \qquad (a > b)$$

Because a is greater than b in both standard forms, it is always easy to determine by inspection whether the foci lie on the x-axis or the y-axis. Consider, for instance, the equation $(x^2/5^2) + (y^2/6^2) = 1$. In this case, since $6 > 5$, we have $a = 6$. And since 6^2 appears under y^2, we conclude that the foci lie on the y-axis.

EXAMPLE 5 An ellipse with vertices on the y-axis

The point (5, 3) lies on an ellipse with vertices $(0, \pm 2\sqrt{21})$. Find the equation of the ellipse. Write the answer both in standard form and in the form $Ax^2 + By^2 = C$.

SOLUTION

Since the vertices are $(0, \pm 2\sqrt{21})$, the standard form for the equation in this case is $(x^2/b^2) + (y^2/a^2) = 1$. Furthermore, in view of the coordinates of the vertices, we have $a = 2\sqrt{21}$. Therefore,

$$\frac{x^2}{b^2} + \frac{y^2}{84} = 1$$

Now, since the point (5, 3) lies on the ellipse, its coordinates must satisfy this last equation. We thus have

$$\frac{5^2}{b^2} + \frac{3^2}{84} = 1$$

$$\frac{25}{b^2} + \frac{3}{28} = 1$$

$$\frac{25}{b^2} = \frac{25}{28}$$

From this last equation we see that $b^2 = 28$, and therefore

$$b = \sqrt{28} = 2\sqrt{7}$$

Now that we've determined a and b, we can write the equation of the ellipse in standard form. It is

$$\frac{x^2}{(2\sqrt{7})^2} + \frac{y^2}{(2\sqrt{21})^2} = 1$$

As you should verify for yourself, this equation can also be written in the equivalent form

$$3x^2 + y^2 = 84$$

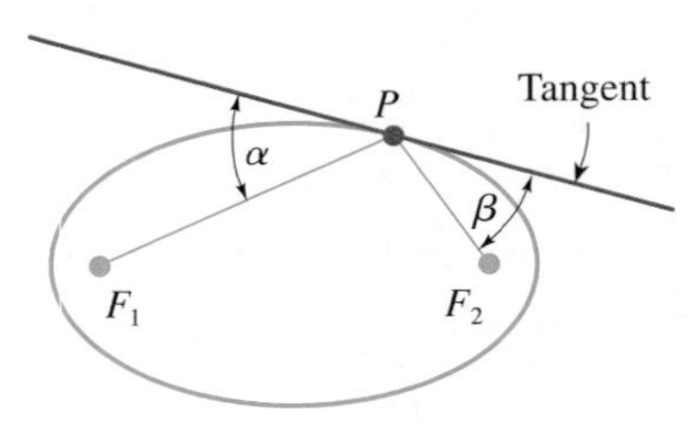

Reflection property of the ellipse: $\alpha = \beta$

Figure 12

As with the parabola, many of the interesting properties of the ellipse are related to tangent lines. We define a **tangent to an ellipse** as a line that intersects the ellipse in exactly one point. Figure 12 shows a line tangent to an ellipse at an arbitrary point P on the curve. Line segments $\overline{F_1P}$ and $\overline{F_2P}$ are drawn from the foci to the point of tangency. These two segments are called **focal radii.** One of the most basic properties of the ellipse is the **reflection property,** which says that the focal radii drawn to the point of tangency make equal angles with the tangent.

If $P(x_1, y_1)$ is a point on the ellipse $(x^2/a^2) + (y^2/b^2) = 1$, then the equation of the line tangent to the ellipse at $P(x_1, y_1)$ is

$$\frac{x_1 x}{a^2} + \frac{y_1 y}{b^2} = 1$$

This equation is easy to remember because it so closely resembles the equation of the ellipse itself. However, notice that the equation $(x_1 x/a^2) + (y_1 y/b^2) = 1$ is indeed linear, since x_1, y_1, a, and b all denote constants. This equation for the tangent to an ellipse can be derived by using the same technique we employed with the parabola in the previous section. (See Exercise 61 at the end of this section for the details.) Example 6 shows how this equation is used.

EXAMPLE 6 Determining the tangent to an ellipse

Find the equation of the line that is tangent to the ellipse $x^2 + 3y^2 = 57$ at the point $(3, 4)$ on the ellipse; see Figure 13. Write the answer in the form $y = mx + b$.

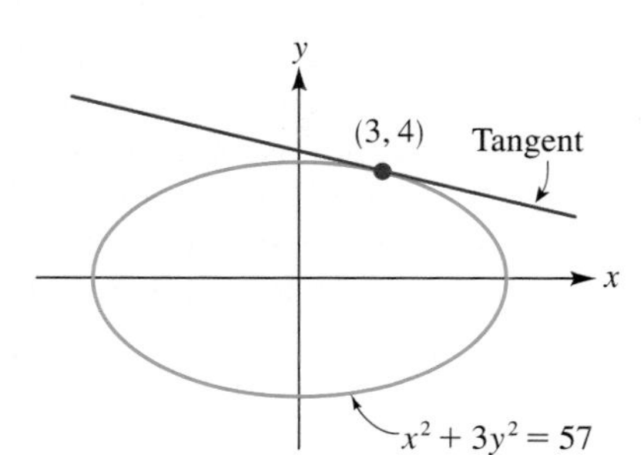

Figure 13

SOLUTION

We know that the equation of the line tangent to the ellipse at the point (x_1, y_1) is $(x_1 x/a^2) + (y_1 y/b^2) = 1$. We are given that $x_1 = 3$ and $y_1 = 4$. Thus we need to determine a^2 and b^2. To do that, we convert the equation $x^2 + 3y^2 = 57$ to standard form by dividing through by 57, which yields

$$\frac{x^2}{57} + \frac{y^2}{19} = 1$$

It follows that $a^2 = 57$ and $b^2 = 19$. The equation of the tangent line then becomes

$$\frac{3x}{57} + \frac{4y}{19} = 1$$

$$\frac{x}{19} + \frac{4y}{19} = 1$$

$$x + 4y = 19$$

Solving for y, we obtain

$$y = -\frac{1}{4}x + \frac{19}{4}$$

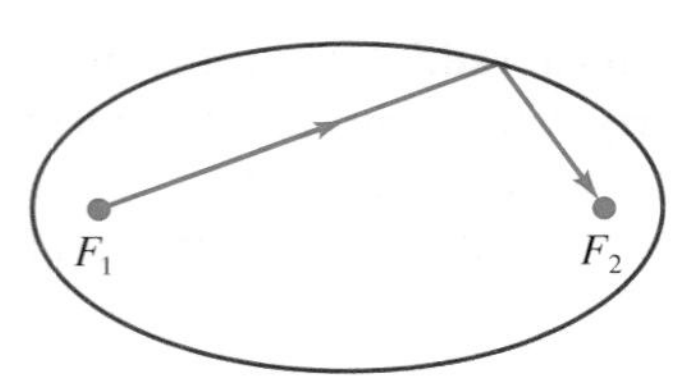

Figure 14
A consequence of the reflection property of the ellipse. A light ray or sound emitted from one focus is reflected through the other focus.

We conclude this section by mentioning a few applications of the ellipse. Some gears in machines are elliptical rather than circular. (In certain brands of racing bikes this is true of one of the gears in front.) As with the parabola, the reflection property of the ellipse has applications in optics and acoustics. As indicated in Figure 14, a light ray or sound emitted from one focus of an ellipse is always reflected through the other focus. This property is used in the design of "whispering galleries." In these rooms (with elliptical cross sections) a person standing at focus F_2 can hear a whisper from focus F_1 while others closer to F_1 might hear nothing. Statuary Hall in the Capitol building in Washington, D.C., is a whispering gallery. This idea is used in a modern medical device known as the *lithotripter,* in which high-energy sound waves are used to break up kidney stones. The patient is positioned with the kidney stone at one focus in an elliptical water bath while sound waves are emitted from the other focus.

An important application of the ellipse occurs in astronomy: For a planet or comet revolving around the Sun, the orbit is an ellipse with the Sun at one focus. (This fact was discovered empirically by Johannes Kepler in the early 1600s, and it was proved mathematically by Isaac Newton in the 1680s.) In this context, the vertices of the ellipse (that is, the endpoints of the major axis) have a special significance. We'll use Figure 15 to explain this. In Figure 15(a) the ellipse represents the orbit of a planet revolving about the Sun located at focus F. The red circle in the figure is drawn with center F and radius FV_2, where V_2 is the vertex closer to F. The figure demonstrates that the planet comes closest to the Sun (F) when it reaches the point V_2 in its orbit. (Why: Each point on the ellipse other than V_2 lies outside of the red circle and therefore is farther from F than is V_2.) The vertex V_2 in Figure 15(a) is referred to as the *perihelion* of the orbit. A similar argument, using Figure 15(b), shows that at the other vertex V_1, the planet reaches its farthest point from the Sun. This point is the *aphelion* of the orbit. For ease of reference we repeat this terminology in the box following Figure 15.

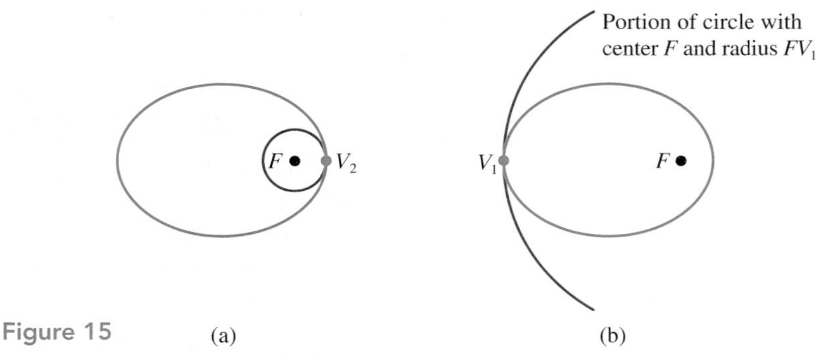

Figure 15

DEFINITION Perihelion and Aphelion

(Refer to the figure below.) The **perihelion** is the point of the orbit at which a planet (or comet) is closest to the sun. The **aphelion** is the point farthest from the Sun.

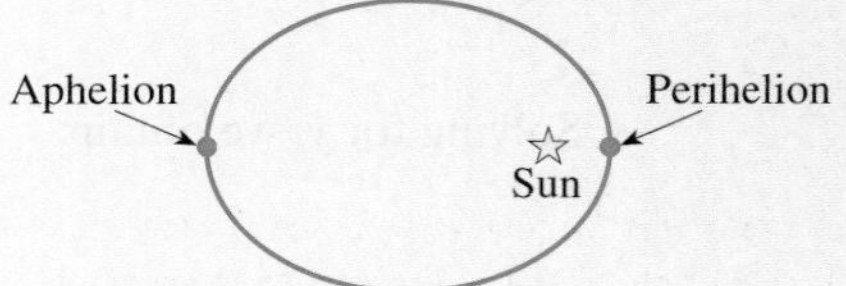

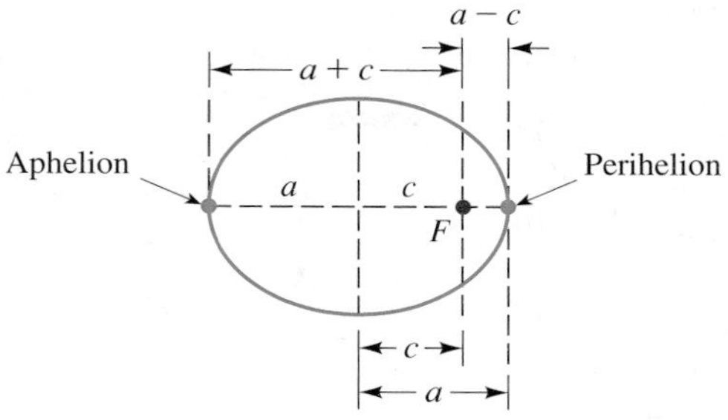

Figure 16
The distances from a planet or comet to the sun at perihelion and aphelion are $a - c$ and $a + c$, respectively.

As is indicated in Figure 16, at perihelion, the distance from a planet or comet to the Sun is given by the expression $a - c$, while at aphelion the distance is $a + c$. We use these expressions along with data for Halley's comet in Example 7.

EXAMPLE 7 Determining the distances to the Sun at perihelion and aphelion

Halley's comet follows an elliptical orbit around the Sun, reappearing in the skies above Earth once every 76.6 years. The semimajor axis of the ellipse is $a = 17.8$ astronomical units (AU). (One AU is 92,956,495 miles, the average distance of the Earth from the Sun.) The eccentricity of the ellipse is $e = 0.967$. Find the distance of the comet from the Sun at perihelion and at aphelion. That is, find the minimum and maximum distances of Halley's comet from the Sun.

SOLUTION

According to Figure 16, the distances of the comet from the Sun at perihelion and at aphelion are $a - c$ and $a + c$, respectively. Since we are given that $a = 17.8$ AU, we need to compute c. From the equation $e = c/a$ we have

$$\begin{aligned} c &= ae \\ &= (17.8)(0.967) && \text{using the values given for } a \text{ for } e \\ &= 17.2\,\text{AU} && \text{using a calculator and rounding to one decimal place} \end{aligned}$$

Using this value for c, we compute

$$a - c = 17.8\,\text{AU} - 17.2\,\text{AU} = 0.6\ \text{AU}$$

and

$$a + c = 17.8\,\text{AU} + 17.2\,\text{AU} = 35.0\ \text{AU}$$

In summary: At perihelion the distance from Halley's comet to the Sun is approximately 0.6 AU. This is the closest the comet ever gets to the Sun. At aphelion the distance from the comet to the Sun is approximately 35.0 AU. This is the farthest that the comet ever gets from the Sun.

EXERCISE SET 11.4

A

In Exercises 1–24, graph the ellipses. In each case, specify the lengths of the major and minor axes, the foci, and the eccentricity. For Exercises 13–24, also specify the center of the ellipse.

1. $4x^2 + 9y^2 = 36$

2. $4x^2 + 25y^2 = 100$

3. $x^2 + 16y^2 = 16$

4. $9x^2 + 25y^2 = 225$

5. $x^2 + 2y^2 = 2$

6. $2x^2 + 3y^2 = 3$

7. $16x^2 + 9y^2 = 144$

8. $25x^2 + y^2 = 25$

9. $15x^2 + 3y^2 = 5$

10. $9x^2 + y^2 = 4$

11. $2x^2 + y^2 = 4$

12. $36x^2 + 25y^2 = 400$

13. $\frac{(x-5)^2}{5^2} + \frac{(y+1)^2}{3^2} = 1$

14. $\frac{(x-1)^2}{2^2} + \frac{(y+4)^2}{3^2} = 1$

15. $\frac{(x-1)^2}{1^2} + \frac{(y-2)^2}{2^2} = 1$

16. $\frac{x^2}{4^2} + \frac{(y-3)^2}{2^2} = 1$

17. $\frac{(x+3)^2}{3^2} + \frac{y^2}{1^2} = 1$

18. $\frac{(x-2)^2}{2^2} + \frac{(y-2)^2}{2^2} = 1$

19. $3x^2 + 4y^2 - 6x + 16y + 7 = 0$

20. $16x^2 + 64x + 9y^2 - 54y + 1 = 0$

21. $5x^2 + 3y^2 - 40x - 36y + 188 = 0$

22. $x^2 + 16y^2 - 160y + 384 = 0$

23. $16x^2 + 25y^2 - 64x - 100y + 564 = 0$

24. $4x^2 + 4y^2 - 32x + 32y + 127 = 0$

In Exercises 25–32, find the equation of the ellipse satisfying the given conditions. Write the answer both in standard form and in the form $Ax^2 + By^2 = C$.

25. Foci $(\pm 3, 0)$; vertices $(\pm 5, 0)$

26. Foci $(0, \pm 1)$; vertices $(0, \pm 4)$

27. Vertices $(\pm 4, 0)$; eccentricity $1/4$

28. Foci $(0, \pm 2)$; endpoints of the minor axis $(\pm 5, 0)$

29. Foci $(0, \pm 2)$; endpoints of the major axis $(0, \pm 5)$

30. Endpoints of the major axis $(\pm 10, 0)$; endpoints of the minor axes $(0, \pm 4)$

31. Center at the origin; vertices on the x-axis; length of major axis twice the length of minor axis; $(1, \sqrt{2})$ lies on the ellipse

32. Eccentricity $3/5$; one endpoint of the minor axis $(-8, 0)$; center at the origin

33. Find the equation of the tangent to the ellipse $x^2 + 3y^2 = 76$ at each of the given points. Write your answers in the form $y = mx + b$.
(a) $(8, 2)$ **(b)** $(-7, 3)$ **(c)** $(1, -5)$

34. **(a)** Find the equation of the line tangent to the ellipse $x^2 + 3y^2 = 84$ at the point $(3, 5)$ on the ellipse. Write your answer in the form $y = mx + b$.
(b) Repeat part (a), but at the point $(-3, -5)$ on the ellipse.
(c) Are the lines determined in (a) and (b) parallel?

35. A line is drawn tangent to the ellipse $x^2 + 3y^2 = 52$ at the point $(2, 4)$ on the ellipse.
(a) Find the equation of this tangent line.
(b) Find the area of the first-quadrant triangle bounded by the axes and this tangent line.

36. Tangent lines are drawn to the ellipse $x^2 + 3y^2 = 12$ at the points $(3, -1)$ and $(-3, -1)$ on the ellipse.
(a) Find the equation of each tangent line. Write your answers in the form $y = mx + b$.
(b) Find the point where the tangent lines intersect.

In Exercises 37 and 38, you are given the eccentricity e and the length a of the semimajor axis for the orbits of the planets Pluto and Mars. Compute the distance of each planet from the Sun at perihelion and at aphelion (as in Example 7). For Pluto, round the final answers to two decimal places; for Mars, round to three decimal places.

37. Pluto: $e = 0.2484$; $a = 39.44$ AU

38. Mars: $e = 0.0934$; $a = 1.5237$ AU

39. As is the case for most asteroids in our solar system, the orbit of the asteroid Gaspra is located between the orbits of Mars and Jupiter. Given that its distances from the Sun at perihelion and aphelion are 2.132 AU and 2.288 AU, respectively, compute the length of the semimajor axis of the orbit and the eccentricity. Round both answers to two decimal places. *Remark*: Within the past decade the Galileo spacecraft (on a mission to Jupiter) took images of Gaspra, which can be viewed on the World Wide Web. For instance, at the time of this writing, one NASA website containing these images is http://www.jpl.nasa.gov/galileo. From this web page, use the search feature there to find "Gaspra."

40. The small asteroid Icarus was first observed in 1949 at Palomar Observatory in California. Given that the distance from Icarus to the Sun at aphelion is 1.969 AU and the eccentricity of the orbit is 0.8269, compute the semimajor axis of the orbit and the distance from Icarus to the Sun at perihelion. Round the answers to two decimal places. *Remark*: One reason for interest in the orbit of Icarus is that it crosses Earth's orbit. On June 14, 1968, there was a "close" approach in which Icarus came within approximately 4 million miles of Earth. According to the Jet Propulsion Laboratory, the next close approach will be June 16, 2015, at which time Icarus will come within approximately 5 million miles of the Earth.

B

41. Consider the equation $(x^2/3^2) + (y^2/2^2) = 1$.
(a) Use the symmetry tests from Section 1.7 to explain why the graph of this equation must be symmetric about both the x-axis and the y-axis.
(b) Show that solving the equation for y yields $y = \pm\frac{1}{3}\sqrt{36 - 4x^2}$.
(c) Let $y = \frac{1}{3}\sqrt{36 - 4x^2}$. Use the techniques of Section 2.4 to find the domain of this function. (You need to solve the inequality $36 - 4x^2 \geq 0$.) You should find that the domain is the closed interval $[-3, 3]$.
(d) Use a calculator to complete the following table. Then plot the resulting points and connect them with a smooth curve. This gives you a sketch of the first quadrant portion of the graph of $(x^2/3^2) + (y^2/2^2) = 1$.

x	0	0.5	1	1.5	2	2.5	3
$y = \frac{1}{3}\sqrt{36 - 4x^2}$	2						0

(e) Use your graph in part (d) along with the symmetry results in part (a) to sketch a complete graph of $(x^2/3^2) + (y^2/2^2) = 1$.

42. In this exercise we consider how the eccentricity e influences the graph of an ellipse $(x^2/a^2) + (y^2/b^2) = 1$.

(a) For simplicity, we suppose that $a = 1$ so that the equation of the ellipse is $x^2 + (y^2/b^2) = 1$. Solve this equation for y to obtain

$$y = \pm b\sqrt{1 - x^2} \qquad (1)$$

(b) Assuming that $a = 1$, show that b and e are related by the equation $b^2 = 1 - e^2$, from which it follows that $b = \pm\sqrt{1 - e^2}$. The positive root is appropriate here because $b > 0$. Thus, we have

$$b = \sqrt{1 - e^2} \qquad (2)$$

(c) Using equation (2) to substitute for b in equation (1) yields

$$y = \sqrt{1 - e^2}\sqrt{1 - x^2} \quad \text{or} \quad y = -\sqrt{1 - e^2}\sqrt{1 - x^2} \qquad (3)$$

This pair of equations represents an ellipse with semimajor axis 1 and eccentricity e. Using the value $e = 0.3$, graph equations (3) in the viewing rectangle $[-1, 1]$ by $[-1, 1]$. Use true proportions and, for comparison, add to your picture the circle with radius 1 and center $(0, 0)$. Note that the ellipse is nearly circular.

(d) Follow part (c) using $e = 0.017$. This is approximately the eccentricity for Earth's orbit around the Sun. How does the ellipse compare to the circle in this case?

(e) Follow part (c) using, in turn, $e = 0.4$, $e = 0.6$, $e = 0.8$, $e = 0.9$, $e = 0.99$, and $e = 0.999$. Then, in complete sentences, summarize what you've observed.

43. In the accompanying figure, the center of the circle and one focus of the ellipse coincide with the focus of the parabola $y^2 = 4x$. All three curves are symmetric about the x-axis and pass through the origin. The eccentricity of the ellipse is 3/4. Find the equations of the circle and the ellipse.

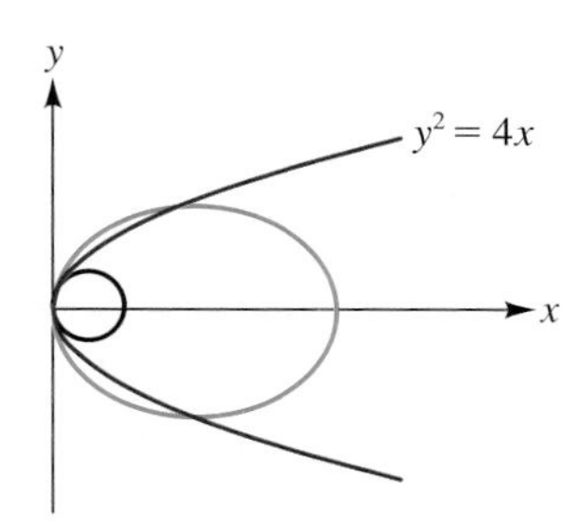

44. This exercise outlines the steps needed to complete the derivation of the equation $(x^2/a^2) + (y^2/b^2) = 1$.

(a) Square both sides of equation (1) on page 843. After simplifying, you should obtain

$$a\sqrt{(x + c)^2 + y^2} = a^2 + xc$$

(b) Square both sides of the equation in part (a). Show that the result can be written

$$a^2x^2 - x^2c^2 + a^2y^2 = a^4 - a^2c^2$$

(c) Verify that the equation in part (b) is equivalent to

$$(a^2 - c^2)x^2 + a^2y^2 = a^2(a^2 - c^2)$$

(d) Using the equation in part (c), replace the quantity $a^2 - c^2$ with b^2. Then show that the resulting equation can be rewritten $(x^2/a^2) + (y^2/b^2) = 1$, as required.

45. In the text we defined the positive number b^2 by the equation $b^2 = a^2 - c^2$. For this definition to make sense, we need to show that the quantity $a^2 - c^2$ is positive. This can be done as follows. First, recall that in any triangle, the sum of the lengths of two sides is always greater than the length of the third side. Now apply this fact to triangle F_1PF_2 in Figure 4 to show that $2a > 2c$. Conclude from this that $a^2 - c^2 > 0$, as required.

46. A line is drawn tangent to the ellipse $(x^2/a^2) + (y^2/b^2) = 1$ at the point (x_1, y_1) on the ellipse. Let P and Q denote the points where the tangent meets the y-and x-axes, respectively. Show that the midpoint of PQ is $(a^2/2x_1, b^2/2y_1)$.

47. A *normal* to an ellipse is a line drawn perpendicular to the tangent at the point of tangency. Show that the equation of the normal to the ellipse $(x^2/a^2) + (y^2/b^2) = 1$ at the point (x_1, y_1) can be written

$$a^2y_1x - b^2x_1y = (a^2 - b^2)x_1y_1$$

48. Let (x_1, y_1) be any point on the ellipse $(x^2/a^2) + (y^2/b^2) = 1$ $(a > b)$ other than one of the endpoints of the major or minor axis. Show that the normal at (x_1, y_1) does not pass through the origin. *Hint:* Find the y-intercept of the normal.

49. Find the points of intersection of the ellipses

$$\frac{x^2}{a^2} + \frac{y^2}{b^2} = 1 \quad \text{and} \quad \frac{x^2}{b^2} + \frac{y^2}{a^2} = 1 \quad (a > b)$$

Include a sketch with your answer.

50. Let F_1 and F_2 denote the foci of the ellipse $(x^2/a^2) + (y^2/b^2) = 1$ $(a > b)$. Suppose that P is one of the endpoints of the minor axis and angle F_1PF_2 is a right angle. Compute the eccentricity of the ellipse.

51. Let $P(x_1, y_1)$ be a point on the ellipse $(x^2/a^2) + (y^2/b^2) = 1$. Let N be the point where the normal through P meets the x-axis, and let F be the focus $(-c, 0)$. Show that $FN/FP = e$, where e denotes the eccentricity.

52. Let $P(x_1, y_1)$ be a point on the ellipse $b^2x^2 + a^2y^2 = a^2b^2$. Suppose that the tangent to the ellipse at P meets the y-axis at A and the x-axis at B. If $AP = PB$, what are x_1 and y_1 (in terms of a and b)?

53. **(a)** Verify that the points $A(5, 1)$, $B(4, -2)$, and $C(-1, 3)$ all lie on the ellipse $x^2 + 3y^2 = 28$.

(b) Find a point D on the ellipse such that $\overline{CD}$ is parallel to $\overline{AB}$.

(continues)

(c) If O denotes the center of the ellipse, show that the triangles OAC and OBD have equal areas. *Suggestion*: In computing the areas, the formula given at the end of Exercise 34 in Section 1.4 is useful.

54. Find the points of intersection of the parabola $y = x^2$ and the ellipse $b^2x^2 + a^2y^2 = a^2b^2$.

55. Recall that the two line segments joining a point on the ellipse to the foci are called *focal radii.* These are the segments $\overline{F_1P}$ and $\overline{F_2P}$ in the following figure.

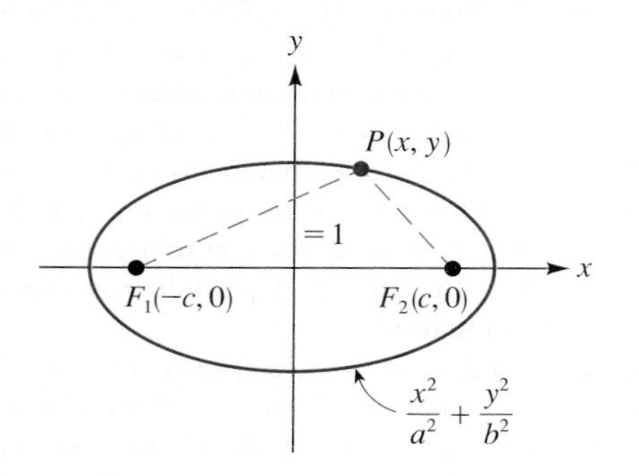

(a) Show that $F_1P = a + ex$. *Hint*: If you try to do this from scratch, it can involve a rather lengthy calculation. Begin instead with the equation $a\sqrt{(x + c)^2 + y^2} = a^2 + xc$ [from Exercise 44 (a)], and divide both sides by a.

(b) Show that $F_2P = a - ex$. *Hint*: Make use of the result in part (a), together with the fact that $F_1P + F_2P = 2a$, by definition.

56. Find the coordinates of a point P in the first quadrant on the ellipse $9x^2 + 25y^2 = 225$ such that $\angle F_2PF_1$ is a right angle.

57. The accompanying figure shows the two tangent lines drawn from the point (h, k) to the ellipse $b^2x^2 + a^2y^2 = a^2b^2$. Follow steps (a) through (d) to show that the equation of the line passing through the two points of tangency is $b^2hx + a^2ky = a^2b^2$.

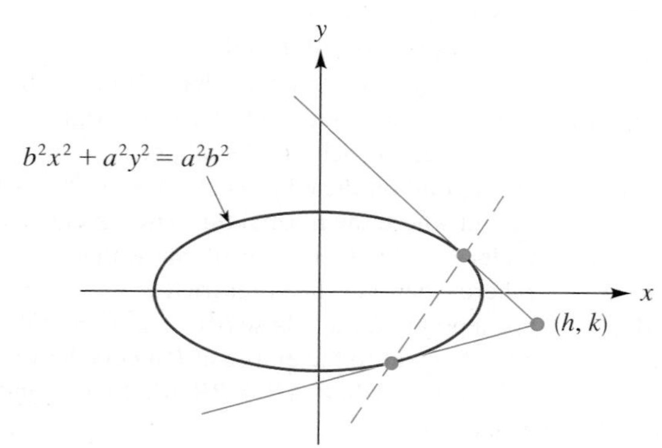

(a) Let (x_1, y_1) be one of the points of tangency. Check that the equation of the tangent line through this point is $b^2x_1x + a^2y_1y = a^2b^2$.

(b) Using the result in part (a), explain why $b^2x_1h + a^2y_1k = a^2b^2$.

(c) In a similar fashion, show that $b^2x_2h + a^2y_2k = a^2b^2$, where (x_2, y_2) is the other point of tangency.

(d) The equation $b^2hx + a^2ky = a^2b^2$ represents a line. Explain why this line must pass through the points (x_1, y_1) and (x_2, y_2).

58. (a) The *auxiliary circle* of an ellipse is defined to be the circle with diameter the same as the major axis of the ellipse. Determine the equation of the auxiliary circle for the ellipse $9x^2 + 25y^2 = 225$.

G **(b)** Graph the ellipse $9x^2 + 25y^2 = 225$ along with its auxiliary circle. (Use true proportions.)

59. As background for this exercise, you need to have worked Exercise 58. This exercise illustrates an interesting geometric result concerning an ellipse and its auxiliary circle.

G **(a)** Graph the ellipse $x^2 + 3y^2 = 12$ for $y \geq 0$.

(b) Find the values of a, b, and c for this ellipse.

G **(c)** Find the equation of the auxiliary circle for this ellipse. Add the graph of the top half of this circle to your picture from part (a).

(d) Verify (algebraically) that the point $P(3, 1)$ lies on the ellipse. Then find the equation of the tangent line to the ellipse at P.

G **(e)** Add the graph of the tangent from part (d) to the picture that you obtained in part (c).

G **(f)** In part (b) you determined the value of c. Find the equation of the line passing through $(-c, 0)$ and perpendicular to the tangent in part (d). Then add the graph of this line to your picture. If you've done things correctly, you should obtain a figure similar to the following one. This figure provides an example of this general result: The line through the focus and perpendicular to the tangent meets the tangent at a point on the auxiliary circle.

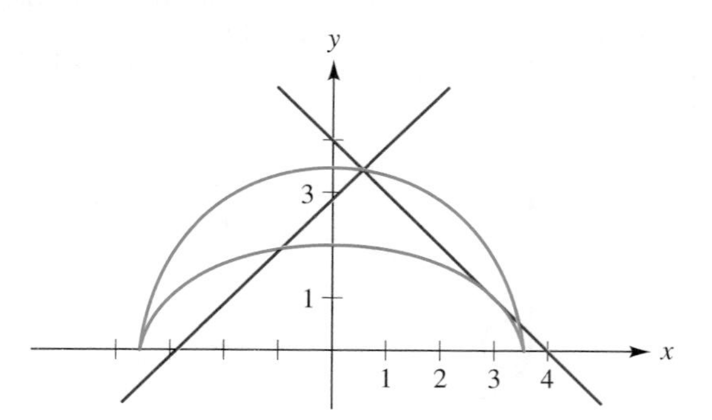

C

60. Complete the derivation of equation (4), the standard form for an equation of an ellipse with foci $F_1(-c, 0)$ and $F_2(c, 0)$. *Hint*: Let $b^2 = a^2 - c^2$. Show that if x and y satisfy equation (4), then the sum of the distances from the point (x, y) to the foci $(-c, 0)$ and $(c, 0)$ is

$$\left|\frac{a^2 + cx}{a}\right| + \left|\frac{a^2 - cx}{a}\right|$$

Simplify carefully using the fact $-a \le x \le a$ to show this sum is $2a$.

61. This exercise outlines the steps required to show that the equation of the tangent to the ellipse $(x^2/a^2) + (y^2/b^2) = 1$ at the point (x_1, y_1) on the ellipse is $(x_1x/a^2) + (y_1y/b^2) = 1$.

(a) Show that the equation $(x^2/a^2) + (y^2/b^2) = 1$ is equivalent to

$$b^2x^2 + a^2y^2 = a^2b^2 \qquad (1)$$

Conclude that (x_1, y_1) lies on the ellipse if and only if

$$b^2x_1^2 + a^2y_1^2 = a^2b^2 \qquad (2)$$

(b) Subtract equation (2) from equation (1) to show that

$$b^2(x^2 - x_1^2) + a^2(y^2 - y_1^2) = 0 \qquad (3)$$

Equation (3) is equivalent to equation (1) provided only that (x_1, y_1) lies on the ellipse. In the following steps, we will find the algebra much simpler if we use equation (3) to represent the ellipse, rather than the equivalent and perhaps more familiar equation (1).

(c) Let the equation of the line tangent to the ellipse at (x_1, y_1) be

$$y - y_1 = m(x - x_1)$$

Explain why the following system of equations must have exactly one solution, namely, (x_1, y_1):

$$\begin{cases} b^2(x^2 - x_1^2) + a^2(y^2 - y_1^2) = 0 & (4) \\ y - y_1 = m(x - x_1) & (5) \end{cases}$$

(d) Solve equation (5) for y, and then substitute for y in equation (4) to obtain

$$b^2(x^2 - x_1^2) + a^2m^2(x - x_1)^2 + 2a^2my_1(x - x_1) = 0 \qquad (6)$$

(e) Show that equation (6) can be written

$$(x - x_1)[b^2(x + x_1) + a^2m^2(x - x_1) + 2a^2my_1] = 0 \qquad (7)$$

(f) Equation (7) is a quadratic equation in x, but as was pointed out earlier, $x = x_1$ must be the only solution. (That is, $x = x_1$ is a double root.) Thus the factor in brackets must equal zero when x is replaced by x_1. Use this observation to show that

$$m = -\frac{b^2x_1}{a^2y_1}$$

This represents the slope of the line tangent to the ellipse at (x_1, y_1).

(g) Using this value for m, show that equation (5) becomes

$$b^2x_1x + a^2y_1y = b^2x_1^2 + a^2y_1^2 \qquad (8)$$

(h) Now use equation (2) to show that equation (8) can be written

$$\frac{x_1x}{a^2} + \frac{y_1y}{b^2} = 1$$

which is what we set out to show.

MINI PROJECT

THE CIRCUMFERENCE OF AN ELLIPSE

As you know, there is a simple expression for the circumference of a circle of radius a, namely, $2\pi a$. However, there is no similar type of elementary expression for the circumference of an ellipse. (The circumference of an ellipse can be computed to as many decimal places as required using the methods of calculus.) Nevertheless, there are some interesting elementary formulas that allow us to approximate the circumference of an ellipse quite closely. Four such formulas follow, along with the names of their discoverers and approximate dates of discovery. Each formula yields an approximate value for the circumference of the ellipse $(x^2/a^2) + (y^2/b^2) = 1$.

Discoverer	Date	Formula
Giuseppe Peano	1887	$C_1 = \pi\left[a + b + \frac{1}{2}(\sqrt{a} - \sqrt{b})^2\right]$
Scrinivasa Ramanujan	1914	$C_2 = \pi[3(a + b) - \sqrt{(a + 3b)(3a + b)}]$
Roger A. Johnson	1930	$C_3 = \frac{\pi}{2}[a + b + \sqrt{2(a^2 + b^2)}]$
Roger Maertens	2000	$C_4 = 4(a^y + b^y)^{1/y}$, where $y = \dfrac{\ln 2}{\ln(\pi/2)}$

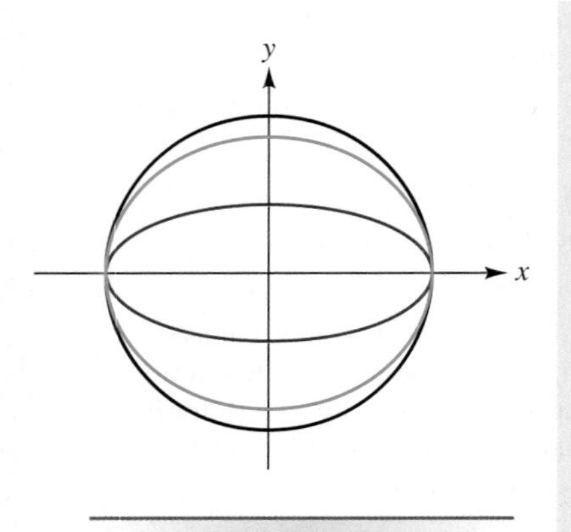

Approximation to circumference	Percentage error
C_1	
C_2	
C_3	
C_4	

(a) In the figure on the left, the red ellipse has an eccentricity of 0.9, and the blue ellipse an eccentricity of 0.5. The outer circle has a radius of 10, which is equal to the semimajor axis of each ellipse. As preparation for parts (b) and (c), determine the equation of each ellipse.

(b) For the red ellipse, use the approximation formulas given above to complete the table at left. Round the values of C_1, C_2, C_3, and C_4 to six decimal places. Round the percentage errors to two significant digits. In computing the percentage errors, use the fact that the actual circumference, rounded to six decimal places, is 23.433941. Which of the four approximations for the circumference of this ellipse is the best? Which is worst? How does the circumference of this ellipse compare to that of the black circle in the figure? That is, using the given six-place value for circumference, compute the ratio of the circumference of the ellipse to the circumference of the circle.

(c) Follow part (b) for the blue ellipse. The actual circumference here, rounded to six decimal places, is 29.349244. Compare your results for percentage errors to those in part (b), and summarize your observations (using complete sentences).

(d) What is the circumference of the circle in the given figure? What value does each approximation formula yield for the circumference of the circle?

11.5 THE HYPERBOLA

The ellipse is the general shape of any closed orbit. . . . It is also possible to have orbits that are not closed, whose shapes are represented by open curves. Even if the two bodies are not bound together by their mutual gravitational attraction, their gravitational attraction for each other still affects their relative motion. Then the general shape of their orbit is a hyperbola.—Theodore P. Snow in *The Dynamic Universe: An Introduction to Astronomy,* 4th ed. (St. Paul, Minn.: West Publishing Co., 1990)

In the previous section we defined an ellipse as the set of points P such that the sum of the distances from P to two fixed points is constant. By considering the difference instead of the sum, we are led to the definition of the hyperbola.

DEFINITION **The Hyperbola**

A **hyperbola** is the set of all points in the plane, the absolute value of the difference of whose distances from two fixed points is a positive constant. Each fixed point is called a **focus.**

As with the ellipse, we label the foci F_1 and F_2. Before obtaining an equation for the hyperbola, we can see the general features of the curve by using two sets of concentric circles, with centers F_1 and F_2, to locate points satisfying the definition of a hyperbola. In Figure 1 we've plotted a number of points P such that either $F_1P - F_2P = 3$ or $F_2P - F_1P = 3$. By joining these points, we obtain the graph of the hyperbola shown in Figure 1.

Unlike the parabola or the ellipse, the hyperbola is composed of two distinct parts, or **branches.** As you can check, the left branch in Figure 1 corresponds to the equation $F_2P - F_1P = 3$, while the right branch corresponds to the equation

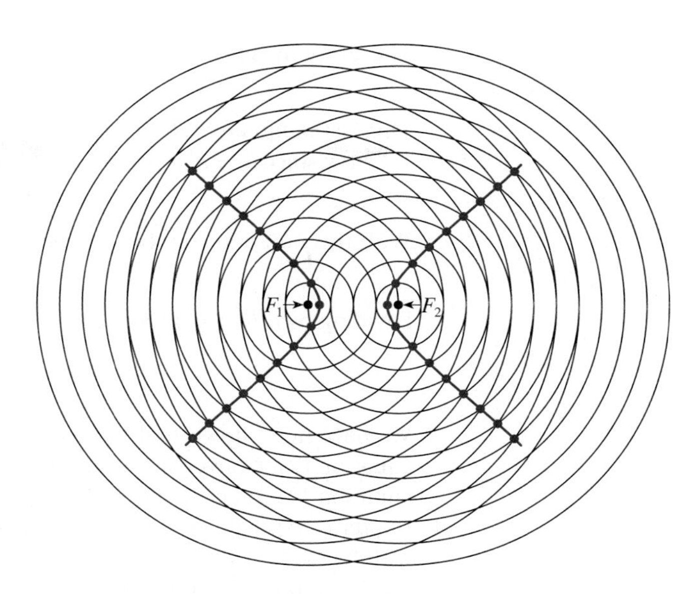

Figure 1

$F_1P - F_2P = 3$. Figure 1 also reveals that the hyperbola possesses two types of symmetry. First, it is symmetric about the line passing through the two foci F_1 and F_2; this line is referred to as the **focal axis** of the hyperbola. Second, the hyperbola is symmetric about the line that is the perpendicular bisector of the segment $\overline{F_1F_2}$.

To derive an equation for the hyperbola, let us initially assume that the foci are located at the points with foci $F_1(-c, 0)$ and $F_2(c, 0)$, as indicated in Figure 2. We will use $2a$ to denote the positive constant absolute value of the difference of distances referred to in the definition of the hyperbola. By definition, then, $P(x, y)$ lies on the hyperbola if and only if

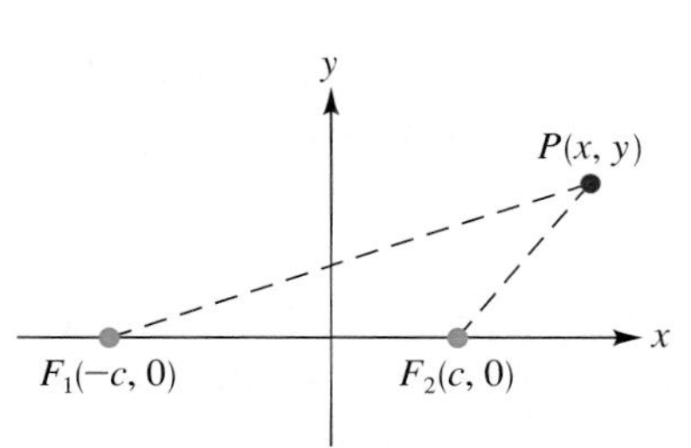

Figure 2

$$|F_1P - F_2P| = 2a$$

or, equivalently,

$$F_1P - F_2P = \pm 2a$$

If we use the formula for the distance between two points, this last equation becomes

$$\sqrt{(x + c)^2 + (y - 0)^2} - \sqrt{(x - c)^2 + (y - 0)^2} = \pm 2a$$

or

$$\sqrt{(x + c)^2 + y^2} - \sqrt{(x - c)^2 + y^2} = \pm 2a$$

We can simplify this equation by carrying out the same procedure that we used for the ellipse in the previous section. As Exercise 46 asks you to verify, the resulting equation is

$$(c^2 - a^2)x^2 - a^2y^2 = a^2(c^2 - a^2) \tag{1}$$

Before further simplifying equation (1), we point out that the quantity $c^2 - a^2$ [which appears twice in equation (1)] is positive. To see why this is so, refer back to Figure 2. In triangle F_1F_2P (as in any triangle) the length of any side is less than the sum of the lengths of the other two sides. Thus

$$F_1P < F_1F_2 + F_2P \qquad \text{and} \qquad F_2P < F_1F_2 + F_1P$$

and therefore

$$F_1P - F_2P < F_1F_2 \qquad \text{and} \qquad F_2P - F_1P < F_1F_2$$

These last two equations tell us that

$$|F_1P - F_2P| < F_1F_2$$

Therefore in view of the definitions of $2a$ and $2c$, we have

$$0 < 2a < 2c \qquad \text{or} \qquad 0 < a < c$$

This last inequality tells us that $c^2 - a^2$ is positive, as we wished to show.

Now, since $c^2 - a^2$ is positive, we can define the positive number b by the equation

$$b^2 = c^2 - a^2$$

With this notation equation (1) becomes

$$b^2x^2 - a^2y^2 = a^2b^2$$

Dividing by a^2b^2, we obtain

$$\frac{x^2}{a^2} - \frac{y^2}{b^2} = 1 \tag{2}$$

We have now shown that the coordinates of every point on the hyperbola satisfy equation (2). Conversely, it can be shown that if the coordinates of a point satisfy equation (2), then the point satisfies the original definition of the hyperbola (see Exercise 55). Equation (2) is the **standard form** for the equation of a hyperbola with foci $F_1(-c, 0)$ and $F_2(c, 0)$.

The intercepts of the hyperbola are readily obtained from equation (2). To find the x-intercepts, we set y equal to zero to obtain

$$\frac{x^2}{a^2} = 1$$

$$x^2 = a^2 \qquad \text{or} \qquad x = \pm a$$

Thus the hyperbola crosses the x-axis at the points $(-a, 0)$ and $(a, 0)$. On the other hand, the curve does not cross the y-axis, for if we set x equal to zero in equation (2), we obtain $-y^2/b^2 = 1$, or

$$y^2 = -b^2 \qquad (b > 0)$$

Since the square of any real number y is nonnegative, this last equation has no solution. Therefore the graph does not cross the y-axis. Finally, let us note that (according to the symmetry tests in Section 1.7) the graph of equation (2) must be symmetric about both coordinate axes.

Before graphing the hyperbola $(x^2/a^2) - (y^2/b^2) = 1$, we point out the important fact that the two lines $y = (b/a)x$ and $y = -(b/a)x$ are asymptotes for the curve. We can see why as follows. First we solve equation (2) for y:

$$\frac{x^2}{a^2} - \frac{y^2}{b^2} = 1$$

$$b^2x^2 - a^2y^2 = a^2b^2 \qquad \text{multiplying both sides by } a^2b^2$$

$$-a^2y^2 = a^2b^2 - b^2x^2$$

$$y^2 = \frac{b^2x^2 - a^2b^2}{a^2} = \frac{b^2(x^2 - a^2)}{a^2}$$

$$y = \pm\frac{b}{a}\sqrt{x^2 - a^2} \tag{3}$$

TABLE 1

x	$\sqrt{x^2 - 5^2}$
100	99.875
1000	999.987
10,000	9999.999

Now, as x grows arbitrarily large, the value of the quantity $\sqrt{x^2 - a^2}$ becomes closer and closer to x itself. Table 1 provides some empirical evidence for this statement in the case when $a = 5$. (A formal proof of the statement properly belongs to calculus.) In summary, then, we have the approximation $\sqrt{x^2 - a^2} \approx x$ as x grows arbitrarily large. So, in view of equation (3), we have

$$y = \pm\frac{b}{a}\sqrt{x^2 - a^2} \approx \pm\frac{b}{a}x \qquad \text{as } x \text{ grows arbitrarily large}$$

In other words, the two lines $y = \pm(b/a)x$ are **asymptotes** for the hyperbola.

A simple way to sketch the two asymptotes and then graph the hyperbola is as follows. First draw the *reference* rectangle with vertices (a, b), $(-a, b)$, $(-a, -b)$, and $(a, -b)$, as indicated in Figure 3(a). The slopes of the diagonals in this rectangle are b/a and $-b/a$. Thus by extending these diagonals as in Figure 3(b), we obtain the two asymptotes $y = \pm(b/a)x$. Now, since the x-intercepts of the hyperbola are a and $-a$, we can sketch the curve as shown in Figure 3(c). Also, note that c is the length of the segment from the origin to any corner of the box, for example, (a, b). Again, see Figure 3(c).

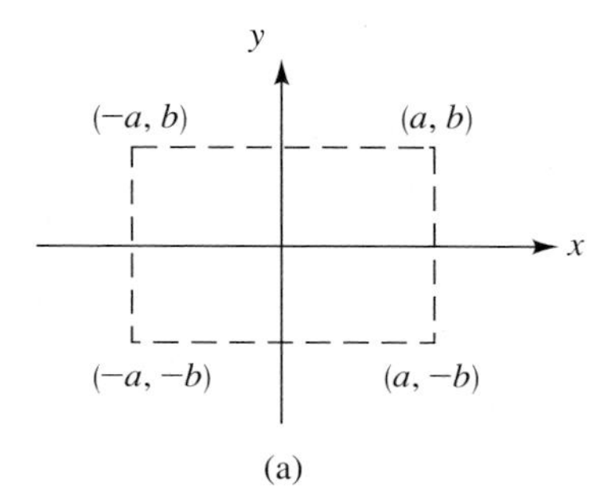

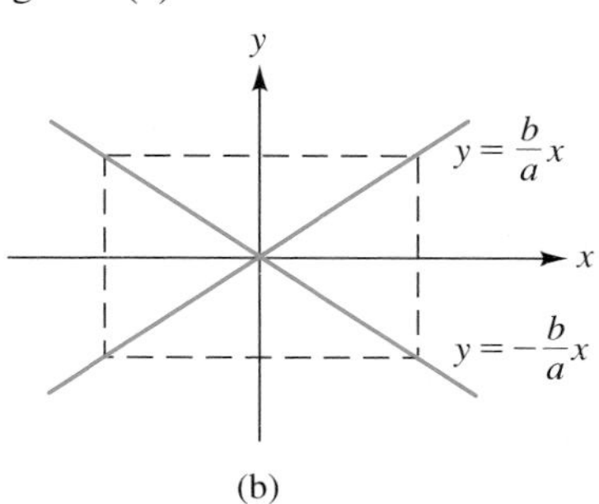

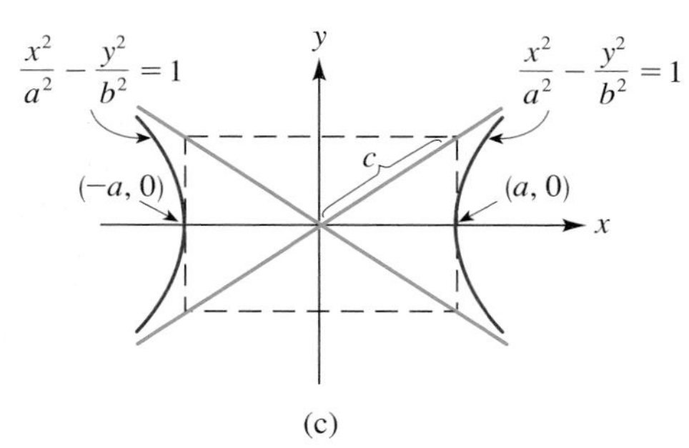

Figure 3
Steps in graphing the hyperbola $(x^2/a^2) - (y^2/b^2) = 1$ and its asymptotes

For the hyperbola in Figure 3(c) the two points $(\pm a, 0)$, where the curve meets the x-axis, are referred to as **vertices.** The midpoint of the line segment joining the two vertices is called the **center** of the hyperbola. (Equivalently, we can define the center as the point of intersection of the two asymptotes.) For the hyperbola in Figure 3(c) the center coincides with the origin. The line segment joining the vertices of a hyperbola is the **transverse axis** of the hyperbola. For reference, in the box that follows, we summarize our work on the hyperbola up to this point. Several new terms describing the hyperbola are also given in the box.

PROPERTY SUMMARY The Hyperbola $\frac{x^2}{a^2} - \frac{y^2}{b^2} = 1$

1. The **foci** are the points $F_1(-c, 0)$ and $F_2(c, 0)$. The hyperbola is the set of points P such that $|F_1P - F_2P| = 2a$. Given a and c, $b^2 = c^2 - a^2$.
2. The **focal axis** is the line passing through the foci.
3. The **vertices** are the points at which the hyperbola intersects its focal axis. In Figure 4 these are the two points $V_1(-a, 0)$ and $V_2(a, 0)$.
4. The **center** is the point on the focal axis midway between the foci. The center of the hyperbola in Figure 4 is the origin.
5. The **transverse axis** is the line segment joining the two vertices. In Figure 4 the length of the transverse axis $\overline{V_1V_2}$ is $2a$.
6. The **conjugate axis** is the line segment perpendicular to the transverse axis, passing through the center and extending a distance b on either side of the center. In Figure 4 this is the segment $\overline{AB}$.
7. The **eccentricity** e is defined by $e = c/a$, where $c^2 = a^2 + b^2$.

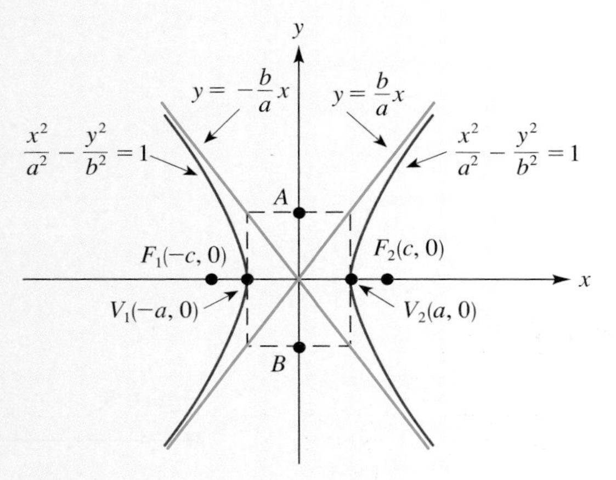

Figure 4

In general, if A, B, and C are positive numbers, then the graph of an equation of the form

$$Ax^2 - By^2 = C$$

will be a hyperbola of the type shown in Figure 4. Example 1 shows why this is so.

EXAMPLE 1 Analyzing a hyperbola

Graph the hyperbola $16x^2 - 9y^2 = 144$ after determining the following: vertices, foci, eccentricity, lengths of the transverse and conjugate axes, and asymptotes.

SOLUTION

First we convert the given equation to standard form by dividing both sides by 144. This yields

$$\frac{x^2}{9} - \frac{y^2}{16} = 1 \qquad \text{or} \qquad \frac{x^2}{3^2} - \frac{y^2}{4^2} = 1$$

By comparing this with the equation $(x^2/a^2) - (y^2/b^2) = 1$, we see that $a = 3$ and $b = 4$. Draw the reference rectangle. The value of c can be determined by using the equation $c^2 = a^2 + b^2$. We have

$$c^2 = 3^2 + 4^2 = 25 \qquad \text{and therefore} \qquad c = 5$$

Now that we know the values of a, b, and c, we can list the required information:

Vertices: $(\pm 3, 0)$ Length of transverse axis $(= 2a)$: 6

Foci: $(\pm 5, 0)$ Length of conjugate axis $(= 2b)$: 8

Eccentricity: $e = \frac{c}{a} = \frac{5}{3}$ Asymptotes: $y = \pm\frac{b}{a}x = \pm\frac{4}{3}x$

The graph of the hyperbola is shown in Figure 5.

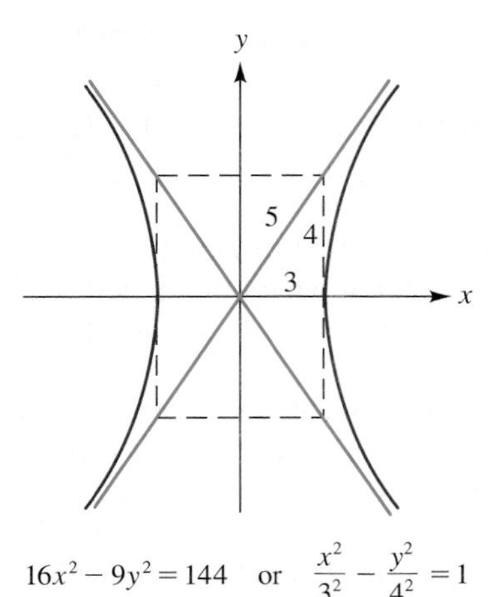

Figure 5

We can use the same method that we used to derive the equation for the hyperbola with foci $(\pm c, 0)$ when the foci are instead located on the y-axis at the points $(0, \pm c)$. The equation of the hyperbola in this case is

$$\frac{y^2}{a^2} - \frac{x^2}{b^2} = 1$$

We summarize the basic properties of this hyperbola in the following box.

PROPERTY SUMMARY The Hyperbola $\frac{y^2}{a^2} - \frac{x^2}{b^2} = 1$

Foci: $(0, \pm c)$, where $c^2 = a^2 + b^2$
Vertices: $(0, \pm a)$
Asymptotes: $y = \pm(a/b)x$
Length of transverse axis: $2a$
Length of conjugate axis: $2b$
Eccentricity: $e = c/a$

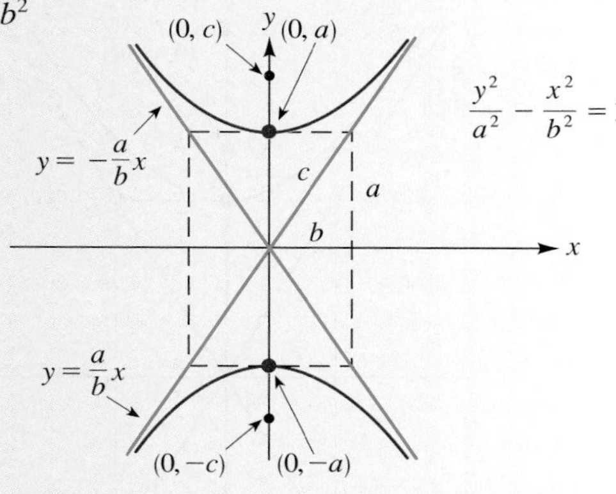

EXAMPLE 2 Completing the square to analyze and graph a hyperbola

Use the technique of completing the square to show that the graph of the following equation is a hyperbola:

$$9y^2 - 54y - 25x^2 + 200x - 544 = 0$$

Graph the hyperbola, and specify the center, the vertices, the foci, the length of the transverse axis, and the equations of the asymptotes.

SOLUTION

$$9(y^2 - 6y \qquad) - 25(x^2 - 8x \qquad) = 544 \qquad \text{factoring}$$

$$9(y^2 - 6y + 9) - 25(x^2 - 8x + 16) = 544 + 81 - 400 \qquad \text{completing the squares}$$

$$9(y-3)^2 - 25(x-4)^2 = 225$$

$$\frac{(y-3)^2}{5^2} - \frac{(x-4)^2}{3^2} = 1 \qquad \text{dividing by 225}$$

Now, the graph of this last equation is obtained by translating the graph of the equation

$$\frac{y^2}{5^2} - \frac{x^2}{3^2} = 1 \tag{4}$$

to the right four units and up three units. First we analyze the graph of equation (4). The general form of the graph is shown in the figure in the preceding Property Summary box. In this case we have $a = 5$, $b = 3$, and

$$c = \sqrt{a^2 + b^2} = \sqrt{5^2 + 3^2} = \sqrt{34} \; (\approx 5.8)$$

Consequently, the vertices [for equation (4)] are $(0, \pm 5)$, the foci are $(0, \pm\sqrt{34})$, and the asymptotes are $y = \pm\frac{5}{3}x$. Figure 6 shows the graph of this hyperbola. Finally, by translating the graph in Figure 6 to the right four units and up three units, we obtain the graph of the original equation, as shown in Figure 7. You should verify for yourself that the information accompanying Figure 7 is correct.

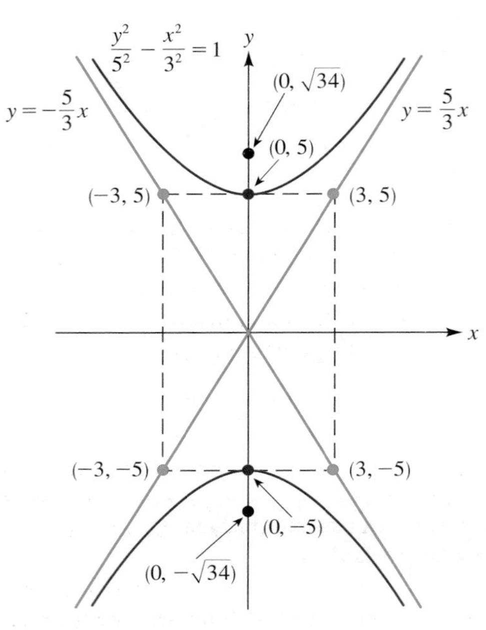

Figure 6

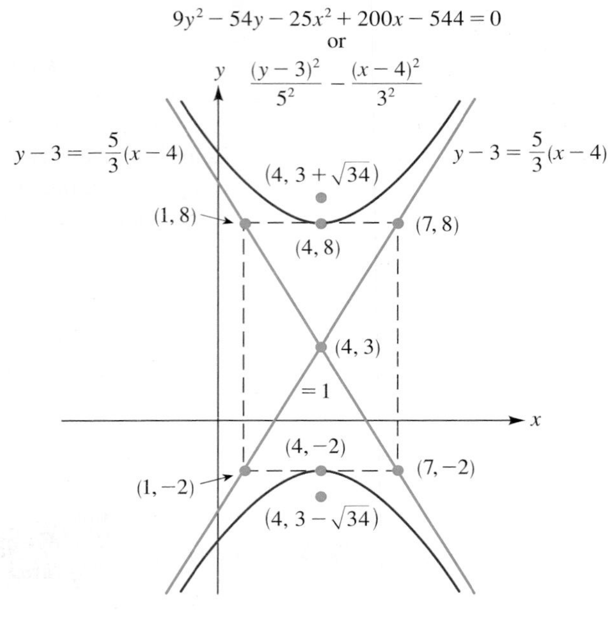

Figure 7

If you reread the example we have just completed, you'll see that it was not necessary to know in advance that the given equation represented a hyperbola. Rather, this fact emerged naturally after we completed the square. Indeed, completing the square is a useful technique for identifying the graph of any equation of the form

$$Ax^2 + Cy^2 + Dx + Ey + F = 0$$

EXAMPLE 3 Completing the square to identify a graph

Identify the graph of the equation

$$4x^2 - 32x - y^2 + 2y + 63 = 0$$

SOLUTION

As before, we complete the squares:

$$4(x^2 - 8x) - (y^2 - 2y) = -63$$
$$4(x^2 - 8x + 16) - (y^2 - 2y + 1) = -63 + 64 - 1$$
$$4(x - 4)^2 - (y - 1)^2 = 0$$

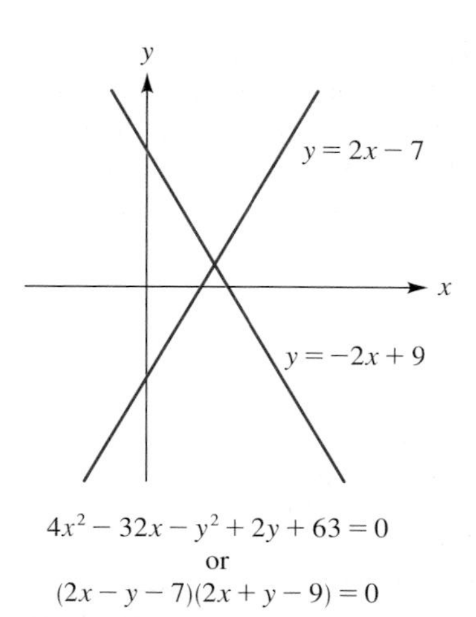

$4x^2 - 32x - y^2 + 2y + 63 = 0$
or
$(2x - y - 7)(2x + y - 9) = 0$

Figure 8

Since the right-hand side of this last equation is zero, dividing both sides by 4 will not bring the equation into one of the standard forms. Indeed, if we factor the left-hand side of the equation as a difference of two squares, we obtain

$$[2(x-4)-(y-1)][2(x-4)+(y-1)] = 0$$
$$(2x-y-7)(2x+y-9) = 0$$

$$2x - y - 7 = 0 \qquad\qquad 2x + y - 9 = 0$$
$$-y = -2x + 7 \qquad\qquad y = -2x + 9$$
$$y = 2x - 7$$

Thus the given equation is equivalent to the two linear equations $y = 2x - 7$ and $y = -2x + 9$. These two lines together constitute the graph. See Figure 8.

The two lines that we graphed in Figure 8 are actually the asymptotes for the hyperbola $4(x-4)^2 - (y-1)^2 = 1$. (Verify this for yourself.) For that reason the graph in Figure 8 is referred to as a **degenerate hyperbola.** There are other cases similar to this that can arise in graphing equations of the form $Ax^2 + Cy^2 + Dx + Ey + F = 0$. For instance, as you can check for yourself by completing the squares, the graph of the equation

$$x^2 - 2x + 4y^2 - 16y + 17 = 0$$

consists of the single point (1, 2). We refer to the graph in this case as a **degenerate ellipse.** Similarly, as you can check by completing the squares, the equation $x^2 - 2x + 4y^2 - 16y + 18 = 0$ has no graph; there are no points with coordinates that satisfy the equation.

We can obtain the equation of a tangent line to a hyperbola using the same ideas that were employed for the parabola and the ellipse in the previous sections. The result is this: The equation of the tangent to the hyperbola $(x^2/a^2) - (y^2/b^2) = 1$ at the point (x_1, y_1) on the curve is

$$\frac{x_1x}{a^2} - \frac{y_1y}{b^2} = 1$$

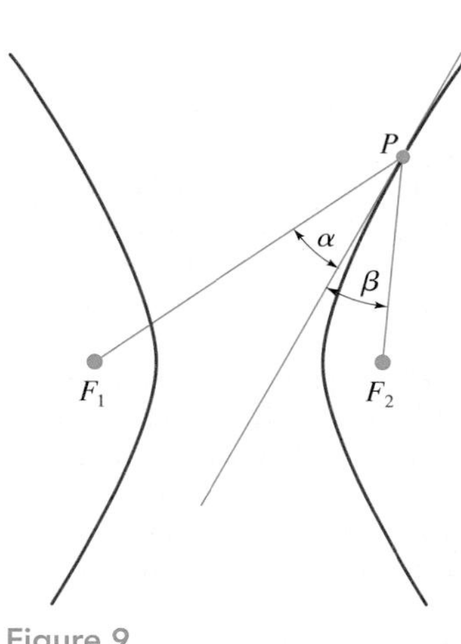

Figure 9
Reflection property of the hyperbola: The tangent at P bisects the angle formed by the focal radii drawn to P, so $\alpha = \beta$.

(See Exercise 56 at the end of this section for an outline of the derivation.) As with the parabola and the ellipse, many interesting properties of the hyperbola are related to the tangent lines. For instance, the hyperbola has a reflection property that is similar to the reflection properties of the parabola and the ellipse. To state this property, we first define a **focal radius** of a hyperbola as a line segment drawn from a focus to a point on the hyperbola. Then the **reflection property** of the hyperbola is that the tangent line bisects the angle formed by the focal radii drawn to the point of tangency, as indicated in Figure 9.

We conclude this section by listing several applications of the hyperbola. Some comets have hyperbolic orbits. Unlike Halley's comet, which has an elliptical orbit, these comets pass through the solar system once and never return. The Cassegrain telescope (invented by the Frenchman Sieur Cassegrain in 1672) uses both a hyperbolic mirror and a parabolic mirror. The Hubble Space Telescope, launched into orbit in April 1990, utilizes a Cassegrain telescope. The hyperbola is also used in some navigation systems. In the LORAN (LOng RAnge Navigation) system an airplane or a ship at a point P receives radio signals that are transmitted simultaneously from two locations, F_1 and F_2. The time difference between the two signals is converted to a difference in distances: $F_1P - F_2P$. This locates the ship along one branch of a hyperbola. Then, using data from a second set of signals, the ship is located along a second hyperbola. The intersection of the two hyperbolas then determines the location of the airplane or ship. The project at the end of this section presents a problem of this type.

EXERCISE SET 11.5 

A

For Exercises 1–24, graph the hyperbolas. In each case in which the hyperbola is nondegenerate, specify the following: vertices, foci, lengths of transverse and conjugate axes, eccentricity, and equations of the asymptotes. In Exercises 11–24, also specify the centers.

1. $x^2 - 4y^2 = 4$
2. $y^2 - x^2 = 1$
3. $y^2 - 4x^2 = 4$
4. $25x^2 - 9y^2 = 225$
5. $16x^2 - 25y^2 = 400$
6. $9x^2 - y^2 = 36$
7. $2y^2 - 3x^2 = 1$
8. $x^2 - y^2 = 9$
9. $4y^2 - 25x^2 = 100$
10. $x^2 - 3y^2 = 3$
11. $\dfrac{(x-5)^2}{5^2} - \dfrac{(y+1)^2}{3^2} = 1$
12. $\dfrac{(x-5)^2}{3^2} - \dfrac{(y+1)^2}{5^2} = 1$
13. $\dfrac{(y-2)^2}{2^2} - \dfrac{(x-1)^2}{1^2} = 1$
14. $\dfrac{(y-3)^2}{2^2} - \dfrac{x^2}{1^2} = 1$
15. $\dfrac{(x+3)^2}{4^2} - \dfrac{(y-4)^2}{4^2} = 1$
16. $\dfrac{(x+1)^2}{5^2} - \dfrac{(y-2)^2}{3^2} = 1$
17. $x^2 - y^2 + 2y - 5 = 0$
18. $16x^2 - 32x - 9y^2 + 90y - 353 = 0$
19. $x^2 - y^2 - 4x + 2y - 6 = 0$
20. $x^2 - 8x - y^2 + 8y - 25 = 0$
21. $y^2 - 25x^2 + 8y - 9 = 0$
22. $9y^2 - 18y - 4x^2 - 16x - 43 = 0$
23. $x^2 + 7x - y^2 - y + 12 = 0$
24. $9x^2 + 9x - 16y^2 + 4y + 2 = 0$

25. Let $P(x, y)$ be a point in the first quadrant on the hyperbola $(x^2/2^2) - (y^2/1^2) = 1$. Let Q be the point in the first quadrant with the same x-coordinate as P and lying on an asymptote to the hyperbola. Show that $PQ = (x - \sqrt{x^2 - 4})/2$.

26. The distance PQ in Exercise 25 represents the vertical distance between the hyperbola and the asymptote. Complete the following table to see numerical evidence that this separation distance approaches zero as x gets larger and larger. (Round each entry to one significant digit.)

x	10	50	100	500	1000	10,000
PQ						

In Exercises 27–36, determine the equation of the hyperbola satisfying the given conditions. Write each answer in the form $Ax^2 - By^2 = C$ or in the form $Ay^2 - Bx^2 = C$.

27. Foci $(\pm 4, 0)$; vertices $(\pm 1, 0)$

28. Foci $(0, \pm 5)$; vertices $(0, \pm 3)$

29. Asymptotes $y = \pm\frac{1}{2}x$; vertices $(\pm 2, 0)$

30. Asymptotes $y = \pm x$; foci $(0, \pm 1)$

31. Asymptotes $y = \pm\sqrt{10}x/5$; foci $(\pm\sqrt{7}, 0)$

32. Length of the transverse axis 6; eccentricity 4/3; center $(0, 0)$; focal axis horizontal

33. Vertices $(0, \pm 7)$; graph passes through the point $(1, 9)$

34. Eccentricity 2; foci $(\pm 1, 0)$

35. Length of the transverse axis 6; length of the conjugate axis 2; foci on the y-axis; center at the origin

36. Asymptotes $y = \pm 2x$; graph passes through $(1, \sqrt{3})$

37. Show that the two asymptotes of the hyperbola $x^2 - y^2 = 16$ are perpendicular to each other.

38. **(a)** Verify that the point $P(6, 4\sqrt{3})$ lies on the hyperbola $16x^2 - 9y^2 = 144$.
(b) In Example 1, we found that the foci of this hyperbola were $F_1(-5, 0)$ and $F_2(5, 0)$. Compute the lengths F_1P and F_2P, where P is the point $(6, 4\sqrt{3})$.
(c) Verify that $|F_1P - F_2P| = 2a$.

39. **(a)** Verify that the point $P(5, 6)$ lies on the hyperbola $5y^2 - 4x^2 = 80$.
(b) Find the foci.
(c) Compute the lengths of the line segments $\overline{F_1P}$ and $\overline{F_2P}$, where P is the point $(5, 6)$.
(d) Verify that $|F_1P - F_2P| = 2a$.

Ⓖ **40.** In this exercise you will look at the graph of the hyperbola $16x^2 - 9y^2 = 144$ from two perspectives.
(a) Solve the given equation for y, then graph the two resulting functions in the standard viewing rectangle.
(b) Determine the equations of the asymptotes. Add the graphs of the asymptotes to your picture from part (a).
(c) Looking at your picture from part (b), you can see that the hyperbola seems to be moving closer and closer to its asymptotes as $|x|$ gets large. To see more dramatic evidence of this, change the viewing rectangle so that both x and y extend from -100 to 100. At this scale, the hyperbola is virtually indistinguishable from its asymptotes.

Ⓖ **41.** In this exercise we graph the hyperbola

$$\frac{(y-3)^2}{5^2} - \frac{(x-4)^2}{3^2} = 1$$

(a) Solve the equation for y to obtain

$$y = 3 \pm 5\sqrt{1 + (x-4)^2/9}$$

(b) In the standard viewing rectangle, graph the two equations that you obtained in part (a). Then, for a better view, adjust the viewing rectangle so that both x and y extend from -20 to 20.

Ⓖ **42.** In this exercise we graph a hyperbola in which the axes of the curve are not parallel to the coordinate axes. The equation is $x^2 + 4xy - 2y^2 = 6$.

(a) Use the quadratic formula to solve the equation for y in terms of x. Show that the result can be written $y = x \pm \frac{1}{2}\sqrt{6x^2 - 12}$.

(b) Graph the two equations obtained in part (a). Use the standard viewing rectangle.

(c) It can be shown that the equations of the asymptotes are $y = (1 \pm 0.5\sqrt{6})x$. Add the graphs of these asymptotes to the picture that you obtained in part (b).

(d) Change the viewing rectangle so that both x and y extend from -50 to 50. What do you observe?

Ⓖ **43.** Use the technique indicated in Exercise 42 to graph the hyperbola $x^2 + xy - 2y^2 = 1$. Use a viewing rectangle that extends from -20 to 20 in both the x- and y-directions.

B

44. (a) Let e_1 denote the eccentricity of the hyperbola $x^2/4^2 - y^2/3^2 = 1$, and let e_2 denote the eccentricity of the hyperbola $x^2/3^2 - y^2/4^2 = 1$. Verify that $e_1^2 e_2^2 = e_1^2 + e_2^2$.

(b) Let e_1 and e_2 denote the eccentricities of the two hyperbolas $x^2/A^2 - y^2/B^2 = 1$ and $y^2/B^2 - x^2/A^2 = 1$, respectively. Verify that $e_1^2 e_2^2 = e_1^2 + e_2^2$.

45. (a) If the hyperbola $x^2/a^2 - y^2/b^2 = 1$ has perpendicular asymptotes, show that $a = b$. What is the eccentricity in this case?

(b) Show that the asymptotes of the hyperbola $x^2/a^2 - y^2/a^2 = 1$ are perpendicular. What is the eccentricity of this hyperbola?

46. Derive equation (1) on page 857 from the equation that precedes it.

47. Let $P(x, y)$ be a point on the right-hand branch of the hyperbola $x^2/a^2 - y^2/b^2 = 1$. As usual, let F_2 denote the focus located at $(c, 0)$. The following steps outline a proof of the fact that the length of the line segment $\overline{F_2P}$ in this case is given by $F_2P = xe - a$.

(a) Explain why

$$\sqrt{(x+c)^2 + y^2} - \sqrt{(x-c)^2 + y^2} = 2a$$

(b) In the preceding equation, add the quantity $\sqrt{(x-c)^2 + y^2}$ to both sides, and then square both sides. Show that the result can be written as

$$xc - a^2 = a\sqrt{(x-c)^2 + y^2}$$

or $$xc - a^2 = a(F_2P)$$

(c) Divide both sides of the preceding equation by a to show that $xe - a = F_2P$, as required.

48. Let $P(x, y)$ be a point on the right-hand branch of the hyperbola $x^2/a^2 - y^2/b^2 = 1$. As usual, let F_1 denote the focus located at $(-c, 0)$. Show that $F_1P = xe + a$. *Hint*: Use the result in Exercise 47 along with the fact that the right-hand branch is defined by the equation $F_1P - F_2P = 2a$.

49. Let P be a point on the right-hand branch of the hyperbola $x^2 - y^2 = k^2$. If d denotes the distance from P to the center of the hyperbola, show that

$$d^2 = (F_1P)(F_2P)$$

50. Let $P(x, y)$ be a point on the right-hand branch of the hyperbola $x^2/a^2 - y^2/b^2 = 1$. From P, a line segment is drawn perpendicular to the line $x = a/e$, meeting this line at D. If F_2 denotes (as usual) the focus located at $(c, 0)$, show that

$$\frac{F_2P}{PD} = e$$

The line $x = a/e$ is called the *directrix* of the hyperbola corresponding to the focus F_2. *Hint*: Use the expression for F_2P developed in Exercise 47.

51. By solving the system

$$\begin{cases} y = \frac{4}{3}x - 1 \\ 16x^2 - 9y^2 = 144 \end{cases}$$

show that the line and the hyperbola intersect in exactly one point. Draw a sketch of the situation. This demonstrates that a line that intersects a hyperbola in exactly one point does not have to be a tangent line.

In Exercises 52–54, find the equation of the line that is tangent to the hyperbola at the given point. Write your answer in the form $y = mx + b$.

52. $x^2 - 4y^2 = 16$; $(5, 3/2)$ **53.** $3x^2 - y^2 = 12$; $(4, 6)$

54. $16x^2 - 25y^2 = 400$; $(10, 4\sqrt{3})$

55. Complete the derivation of equation (2), the standard form for an equation of a hyperbola with foci $F_1(-c, 0)$ and $F_2(c, 0)$. *Hint*: Let $b^2 = c^2 - a^2$. Show that if x and y satisfy equation (2), then the difference of distances from the point (x, y) to the foci $(-c, 0)$ and $(c, 0)$ is

$$\sqrt{(x+c)^2 + y^2} - \sqrt{(x-c)^2 + y^2} = \left|\frac{a^2 + cx}{a}\right| - \left|\frac{a^2 - cx}{a}\right|$$

Simplify carefully in the cases $x \geq a$ and $x \leq -a$ to show this difference is $2a$ or $-2a$, respectively. Thus the absolute value of the difference of distances is $2a$.

C

56. We define a *tangent line to a hyperbola* as a line that is not parallel to an asymptote and that intersects the hyperbola in exactly one point. Show that the equation of the line tangent to the hyperbola $x^2/a^2 - y^2/b^2 = 1$ at the point (x_1, y_1) on the curve is

$$\frac{x_1x}{a^2} - \frac{y_1y}{b^2} = 1$$

Hint: Allow for signs, but follow exactly the same steps as were supplied in Exercise 61 of Exercise Set 11.4, where we found the tangent to the ellipse. You should find that the slope in the present case is $m = (b^2x_1/a^2y_1)$. Explain why this slope cannot equal the slope of an asymptote as long as (x_1, y_1) is on the hyperbola.

57. The *normal line* to a hyperbola at a point P on the hyperbola is the line through P that is perpendicular to the tangent at P. If the coordinates of P are (x_1, y_1), show that the equation of the normal line is

$$a^2y_1x + b^2x_1y = x_1y_1(a^2 + b^2)$$

58. Suppose the point $P(x_1, y_1)$ is on the hyperbola $x^2/a^2 - y^2/b^2 = 1$. A tangent is drawn at P, meeting the x-axis at A and the y-axis at B. Also, perpendicular lines are drawn from P to the x- and y-axes, meeting these axes at C and D, respectively. If O denotes the origin, show that: **(a)** $OA \cdot OC = a^2$; and **(b)** $OB \cdot OD = b^2$.

59. At the point P on the hyperbola $x^2/a^2 - y^2/b^2 = 1$, a tangent line is drawn that meets the lines $x = a$ and $x = -a$ at S and T, respectively. Show that the circle with $\overline{ST}$ as a diameter passes through the two foci of the hyperbola.

PROJECT USING HYPERBOLAS TO DETERMINE A LOCATION

An army base in the desert is located at the origin O of an x-y coordinate system. A soldier at the point P needs to determine his coordinates with respect to the army base, but he does not have a link to a global positioning system. He is, however, in radio contact with several small towns in the vicinity. As is indicated in Figure I, two of the towns are town A and town B. Town A is 6 miles due south of the army base; town B is 6 miles due north of the base. At an agreed-upon instant, both towns send out identifying radio signals. The soldier receives the signal from B slightly before the signal from A, so he knows that he is closer to B than to A. Furthermore, by measuring the time difference between receiving the two signals, the soldier is able to compute that he is 8 miles closer to town B than to town A. Explain why the point P must lie on a certain hyperbola, and then find the equation of the hyperbola.

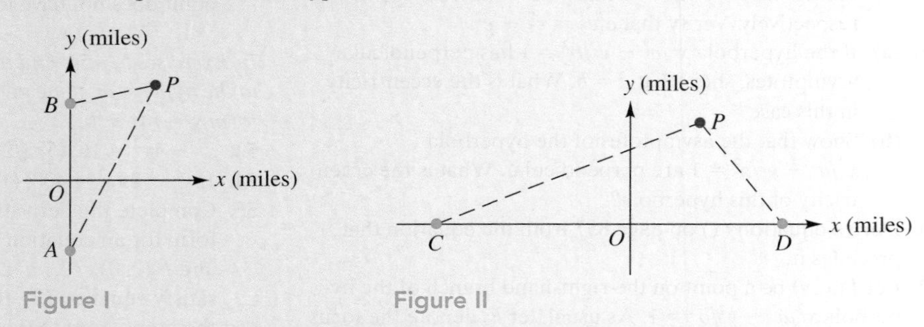

Figure I Figure II

Next, the soldier follows a similar procedure using the towns C and D shown in Figure II. Town C is 19 miles due west of the base; town D is 15 miles due east of the base. The soldier computes that he is 16 miles closer to town D than to town C. Using this information, explain why the point P must lie on a second hyperbola, and then determine its equation.

Use a graphing utility to draw the two hyperbolas and to estimate, to the nearest integers, the coordinates of the intersection point corresponding to the soldier's location P. Next, obtain more precise coordinates by using algebraic techniques (as in Section 10.6) to find the relevant intersection point of the two hyperbolas. At the end, use a calculator and round each coordinate to three decimal places.

11.6 THE FOCUS–DIRECTRIX PROPERTY OF CONICS

Here then [in the *Conics,* written by Appollonius of Perga (ca. 262–ca. 190 B.C.)] *we have the properties of the three curves expressed in the precise language of the Pythagorean application of areas, and the curves are named accordingly:* parabola *(παραβολη) where the rectangle is exactly* applied [equal], hyperbola *(υπερβολη) where it* exceeds, *and* ellipse *(ελλειψιζ) where it* falls short.—Sir Thomas Heath in *A History of Greek Mathematics,* Vol. II (Oxford: The Clarendon Press, 1921)

The most important contribution which Pappus [ca. 300] *made to our knowledge of the conics was his publication of the focus, directrix, eccentricity theorem.*—Julian Lowell Coolidge in *A History of the Conic Sections and Quadric Surfaces* (London: Oxford University Press, 1946)

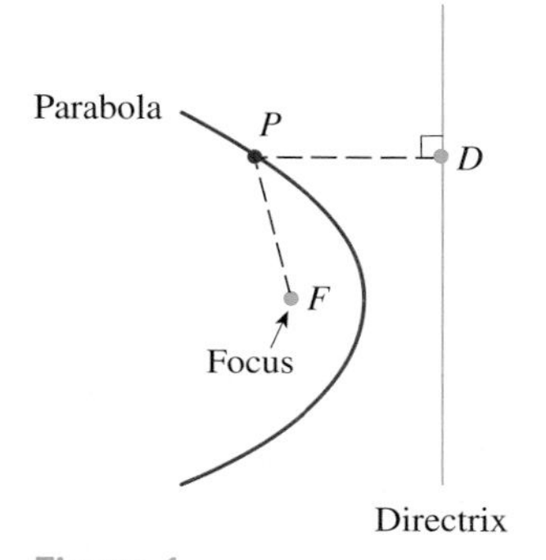

Figure 1

We begin by recalling the focus–directrix property that we used in defining the parabola. A point P is on a parabola if and only if the distance from P to the focus is equal to the distance from P to the directrix. For the parabola in Figure 1 this means that for each point P on the parabola, we have $FP = PD$ or, equivalently,

$$\frac{FP}{PD} = 1$$

The ellipse and the hyperbola also can be characterized by focus–directrix properties. We'll begin with the ellipse. To help us in subsequent computations, we need to know the lengths of the line segments $\overline{F_1P}$ and $\overline{F_2P}$ in Figure 2.

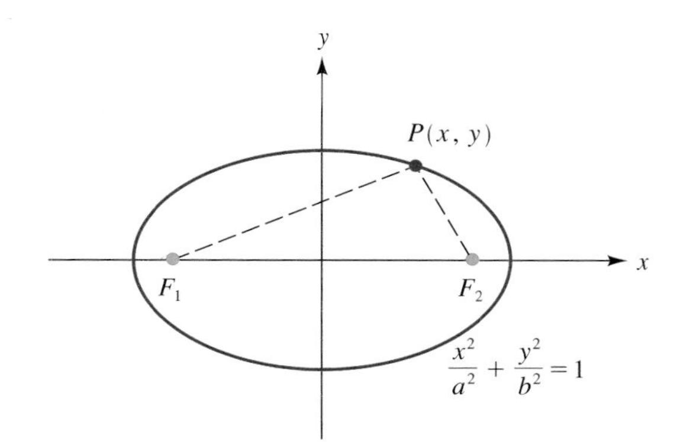

Figure 2
The segments drawn from a point on the ellipse to the foci are called focal radii. Here, $\overline{F_1P}$ and $\overline{F_2P}$ are the focal radii.

These line segments joining a point on the ellipse to the foci are called **focal radii.** In the box that follows, we give a formula for the length of each focal radius. In the formulas, e $(= c/a)$ denotes the eccentricity of the ellipse. (For the derivation of these formulas, see Exercise 15 at the end of this section.)

The Focal Radii of the Ellipse $\frac{x^2}{a^2} + \frac{y^2}{b^2} = 1$

As is indicated in Figure 2, let $P(x, y)$ be a point on the ellipse. Then the lengths of the focal radii are

$$F_1P = a + ex \qquad \text{and} \qquad F_2P = a - ex$$

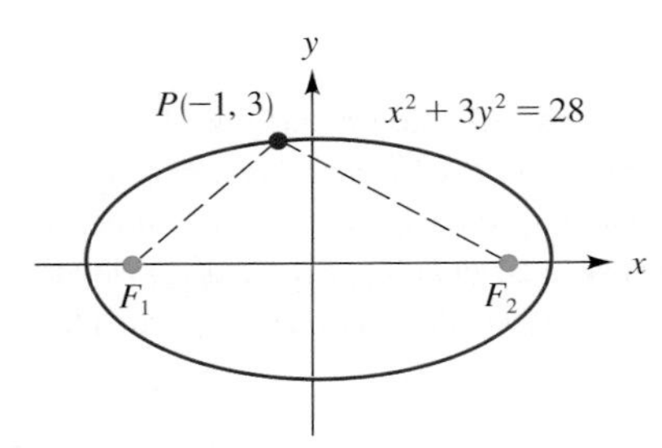

Figure 3

EXAMPLE 1 Computing the lengths of the focal radii of an ellipse

Figure 3 shows the ellipse $x^2 + 3y^2 = 28$. Compute the lengths of the focal radii drawn to the point $P(-1, 3)$.

SOLUTION

To apply the formulas in the box, we need to know the values of a and e for the given ellipse. (We already know that $x = -1$; that is given.) As you should verify for yourself using the techniques of Section 11.4, we have $a = 2\sqrt{7}$ and $e = \sqrt{6}/3$. Therefore

$$\begin{aligned} F_1P &= a + ex \\ &= 2\sqrt{7} + \left(\frac{\sqrt{6}}{3}\right)(-1) = \frac{6\sqrt{7} - \sqrt{6}}{3} \end{aligned}$$

and

$$\begin{aligned} F_2P &= a - ex \\ &= 2\sqrt{7} - \left(\frac{\sqrt{6}}{3}\right)(-1) = \frac{6\sqrt{7} + \sqrt{6}}{3} \end{aligned}$$

In Figure 4 we show the ellipse $(x^2/a^2) + (y^2/b^2) = 1$ and the vertical line $x = a/e$. The line $x = a/e$ is a **directrix** of the ellipse. In Figure 4, F_2P is the distance from the focus F_2 to a point P on the ellipse, and PD is the distance from P to the directrix $x = a/e$. We will prove the following remarkable *focus–directrix property of the ellipse:* For any point $P(x, y)$ on the ellipse, the ratio of F_2P to PD is equal to e, the eccentricity of the ellipse. To prove this, we proceed as follows:

$$\begin{aligned} \frac{F_2P}{PD} &= \frac{a - ex}{PD} && \text{using our formula for } F_2P \\ &= \frac{a - ex}{\frac{a}{e} - x} && \text{using Figure 4} \\ &= \frac{e(a - ex)}{a - ex} && \text{multiplying both numerator and denominator by } e \\ &= e \end{aligned}$$

So $F_2P/PD = e$, as we wished to show.

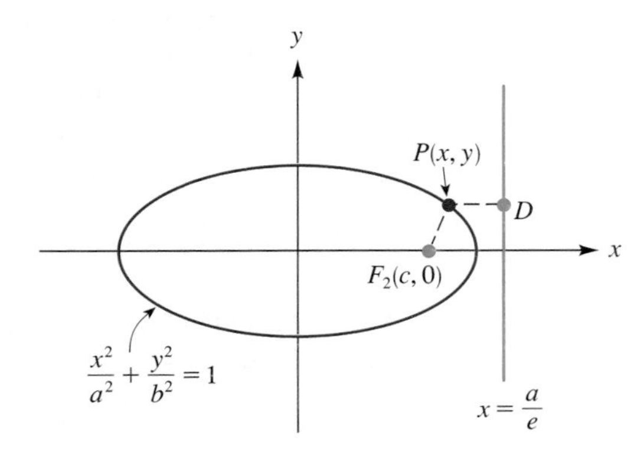

Figure 4

The directrix $x = a/e$ is associated with the focus $F_2(c, 0)$. What about the focus $F_2(-c, 0)$? From the symmetry of the ellipse it follows that the line $x = -a/e$ is the directrix associated with this focus. More specifically, referring to Figure 5, not only do we have $F_2P/PD = e$, we also have $F_1P/PE = e$.

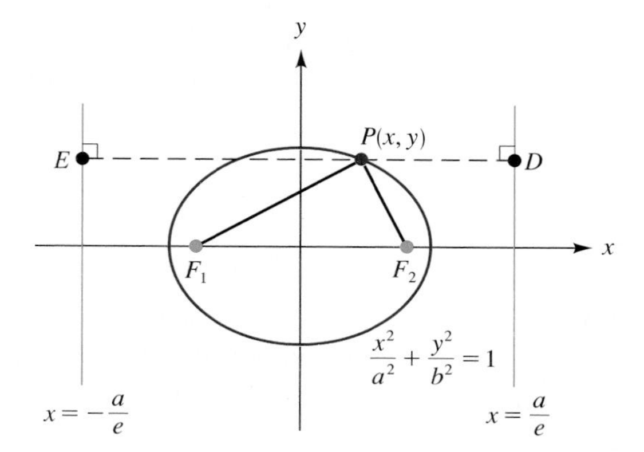

Figure 5
For every point P on the ellipse, we have $F_1P/PE = e$ and $F_2P/PD = e$.

We have seen that each point P on the ellipse in Figure 5 satisfies the following two equations:

$$\frac{F_1P}{PE} = e \tag{1}$$

$$\frac{F_2P}{PD} = e \tag{2}$$

Now, conversely, suppose that the point $P(x, y)$ in Figure 6 satisfies equations (1) and (2). We will show that $P(x, y)$ must, in fact, lie on the ellipse $(x^2/a^2) + (y^2/b^2) = 1$.*

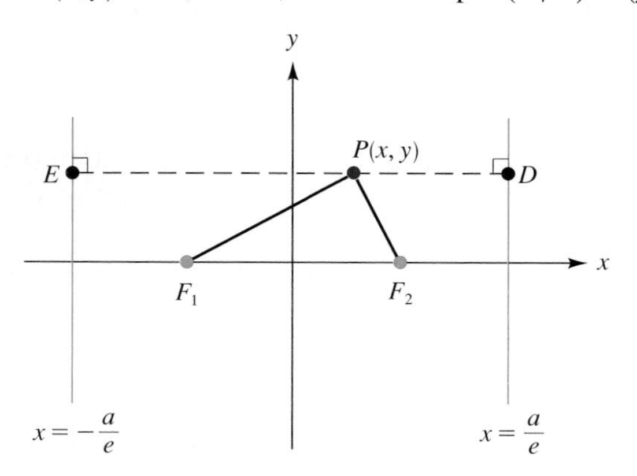

Figure 6

In Figure 6 the distance from E to D is $2a/e$. Therefore

$$PE + PD = \frac{2a}{e}$$

and, consequently,

$$\frac{F_1P}{e} + \frac{F_2P}{e} = \frac{2a}{e} \quad \text{using equations (1) and (2) to substitute for } PE \text{ and } PD$$

*The short proof given here was communicated to David Cohen by Professor Ray Redheffer. For a proof using equation (2) but not equation (1), see Exercise 16.

or

$$F_1P + F_2P = 2a \qquad \text{multiplying both sides by } e$$

Thus, by definition, P lies on the ellipse $(x^2/a^2) + (y^2/b^2) = 1$, as we wished to show.

In the box that follows, we summarize the focus–directrix properties of the ellipse $(x^2/a^2) + (y^2/b^2) = 1$. For reference we also include the original defining property of the ellipse from Section 11.4.

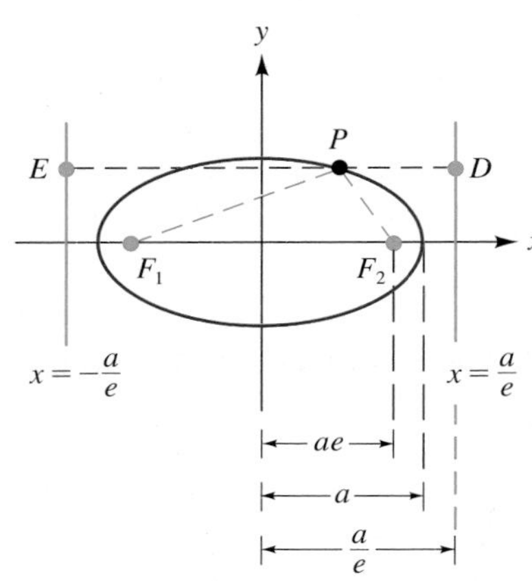

Figure 7
The ellipse $(x^2/a^2) + (y^2/b^2) = 1$ with foci $F_1(-c, 0)$ and $F_2(c, 0)$ and directrices $x = \pm a/e$

Focus and Focus–Directrix Properties of the Ellipse $\frac{x^2}{a^2} + \frac{y^2}{b^2} = 1$

Refer to Figure 7.

1. A point P is on the ellipse if and only if the sum of the distances from P to the foci $F_1(-c, 0)$ and $F_2(c, 0)$ is $2a$, where $c^2 = a^2 - b^2$.
2. A point P is on the ellipse if and only if

$$\frac{F_2P}{PD} = e$$

where F_2P is the distance from P to the focus F_2, PD is the distance from P to the directrix $x = a/e$, and e $(= c/a)$ is the eccentricity of the ellipse.

3. A point P is on the ellipse if and only if

$$\frac{F_1P}{PE} = e$$

where F_1P is the distance from P to the focus F_1 and PE is the distance from P to the directrix $x = -a/e$.

EXAMPLE 2 Using focus–directrix information to find the equation of an ellipse

The foci of an ellipse are $(\pm 3, 0)$, and the directrix corresponding to the focus $(3, 0)$ is $x = 5$. Find the equation of the ellipse. Write the answer in the form $Ax^2 + By^2 = C$.

SOLUTION

Using the given data (and referring to Figure 7), we have

$$ae = 3 \qquad \text{and} \qquad \frac{a}{e} = 5$$

From the first equation we obtain $e = 3/a$. Substituting this value of e in the second equation yields

$$\frac{a}{3/a} = 5 \qquad \text{and therefore} \qquad a^2 = 15$$

Now we can calculate b^2 by using the relation $b^2 = a^2 - c^2$ and the given information $c = 3$:

$$b^2 = a^2 - c^2 = 15 - 9 = 6$$

Substituting the values that we've obtained for a^2 and b^2 in the equation $(x^2/a^2) + (y^2/b^2) = 1$ yields

$$\frac{x^2}{15} + \frac{y^2}{6} = 1$$

or

$$2x^2 + 5y^2 = 30 \qquad \text{multiplying both sides by 30}$$

The focus–directrix property can be developed for the hyperbola in the same way that we have proceeded for the ellipse. In fact, the algebra is so similar that we shall omit the details here and simply summarize the results. The hyperbola $(x^2/a^2) - (y^2/b^2) = 1$ has two directrices: the vertical lines $x = a/e$ in Figure 8(a) and $x = -a/e$ in Figure 8(b). Notice that these equations are identical to those for the directrices of the ellipse. In the box following Figure 8 we summarize the focus and the focus–directrix properties of the hyperbola.

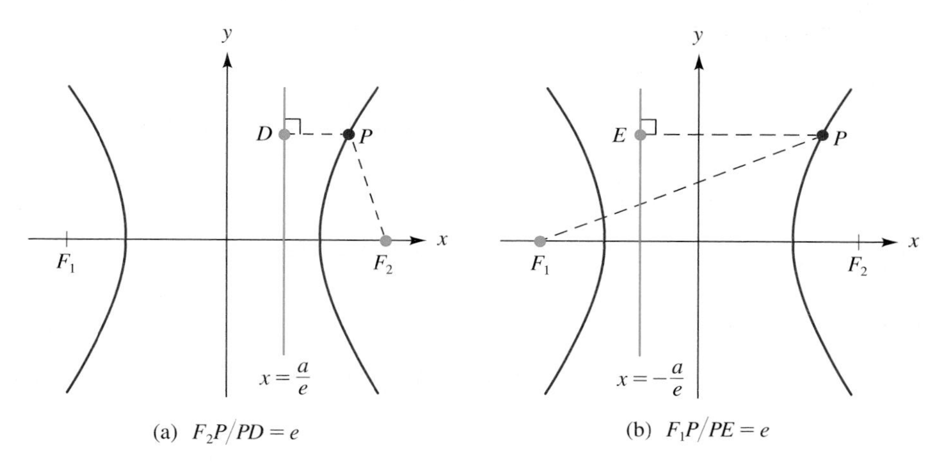

Figure 8
The hyperbola $(x^2/a^2) - (y^2/b^2) = 1$ has two directrices. The directrix $x = a/e$ corresponds to the focus $F_2(c, 0)$. The directrix $x = -a/e$ corresponds to the focus $F_1(-c, 0)$.

Focus and Focus–Directrix Properties of the Hyperbola $\frac{x^2}{a^2} - \frac{y^2}{b^2} = 1$

Refer to Figure 8.

1. A point P is on the hyperbola if and only if the absolute value of the difference of the distances from P to the foci $F_1(-c, 0)$ and $F_2(c, 0)$ is $2a$, where $c^2 = a^2 - b^2$.
2. A point P is on the hyperbola if and only if

$$\frac{F_2P}{PD} = e$$

where F_2P is the distance from P to the focus F_2, PD is the distance from P to the directrix $x = a/e$, and e $(= c/a)$ is the eccentricity of the hyperbola.

3. A point P is on the hyperbola if and only if

$$\frac{F_1P}{PE} = e$$

where F_1P is the distance from P to the focus F_1 and PE is the distance from P to the directrix $x = -a/e$.

EXAMPLE 3 An eccentricity computation

(a) Determine the foci, the eccentricity, and the directrices for the hyperbola $9x^2 - 16y^2 = 144$.
(b) Verify that the point $P(-5, 9/4)$ lies on this hyperbola.
(c) Compute F_2D and PD, and verify that

$$\frac{F_2P}{PD} = e$$

SOLUTION

(a) To convert the given equation to standard form, we divide both sides by 144. This yields

$$\frac{x^2}{4^2} - \frac{y^2}{3^2} = 1$$

So we have $a = 4$, $b = 3$, and

$$c = \sqrt{a^2 + b^2} = \sqrt{4^2 + 3^2} = 5$$

The foci are therefore $(\pm 5, 0)$, and the eccentricity is

$$e = \frac{c}{a} = \frac{5}{4}$$

The directrices are

$$x = \pm\frac{a}{e} = \pm\frac{4}{5/4} = \pm\frac{16}{5}$$

(b) Substituting the values $x = -5$ and $y = 9/4$ in the equation of the hyperbola yields

$$9(-5)^2 - 16\left(\frac{9}{4}\right)^2 = 144$$

or

$$225 - 81 = 144$$

Since this last equation is correct, we conclude that the point $P(-5, 9/4)$ indeed lies on the hyperbola. (See Figure 9.)

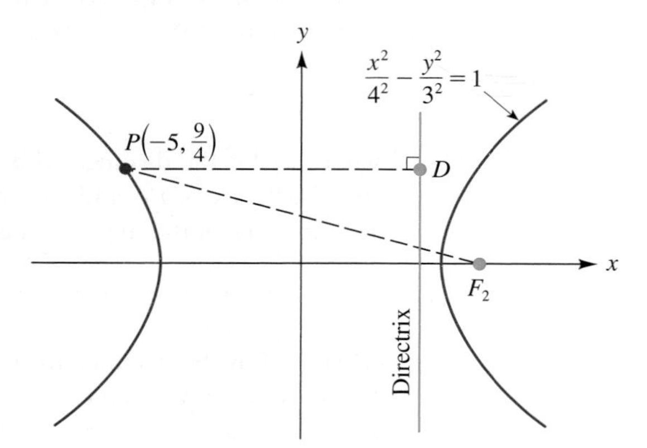

Figure 9

(c) From part (a) we know that the coordinates of the focus F_2 are (5, 0). So, using the distance formula, we have

$$F_2P = \sqrt{(5 - (-5))^2 + \left(0 - \frac{9}{4}\right)^2}$$
$$= \sqrt{100 + \frac{81}{16}} = \sqrt{\frac{1681}{16}} = \frac{41}{4}$$

Next, we use the fact from part (a) that the equation of the directrix in Figure 9 is $x = 16/5$. Thus in Figure 9,

$$PD = \frac{16}{5} - (-5) = \frac{41}{4}$$

Finally, we compute the ratio F_2P/PD:

$$\frac{F_2P}{PD} = \frac{41/4}{41/5} = \frac{5}{4}$$

This is the same number that we obtained for the eccentricity in part (a). So we have verified in this case that the ratio of F_2P to PD is equal to the eccentricity.

The focus–directrix property provides a unified approach to the parabola, the ellipse, and the hyperbola. For both the ellipse and the hyperbola we've seen that for any point P on the curve, we have

$$\frac{\text{distance from } P \text{ to a focus}}{\text{distance from } P \text{ to the corresponding directrix}} = e \qquad (3)$$

where e is a positive constant, the eccentricity of the curve. For the ellipse we have $0 < e < 1$, and for the hyperbola we have $e > 1$. With $e = 1$, equation (3) tells us that the distance from P to the focus equals the distance from P to the directrix, which is the defining condition for the parabola. So equation (3) also holds for the parabola. The following theorem summarizes these remarks.

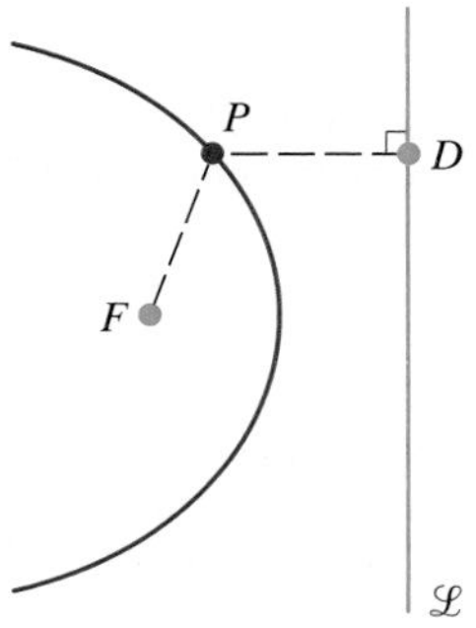

THEOREM The Focus–Directrix Property of Conics

Refer to Figure 10. Let $\mathcal{L}$ be a fixed line, F a fixed point, and e a positive constant. Consider the set of points P satisfying the condition $FP/PD = e$. (The point D is defined in the figure.) Then

(a) If $e = 1$, the set of points is a *parabola* with focus F and directrix $\mathcal{L}$.
(b) If $0 < e < 1$, the set of points is an *ellipse* with focus F, corresponding directrix $\mathcal{L}$, and eccentricity e.
(c) If $e > 1$, the set of points is a *hyperbola* with focus F, corresponding directrix $\mathcal{L}$, and eccentricity e.

Figure 10

NOTE In our development in this section we've always considered cases in which the directrix is vertical, for simplicity. However, the preceding theorem is valid for any orientation of the directrix $\mathcal{L}$.

EXERCISE SET 11.6

A

In Exercises 1–4 you are given an ellipse and a point P on the ellipse. Find F_1P and F_2P, the lengths of the focal radii.

1. $x^2 + 3y^2 = 76$; $P(-8, 2)$

2. $x^2 + 3y^2 = 57$; $P(3, -4)$

3. $(x^2/15^2) + (y^2/5^2) = 1$; $P(9, 4)$

4. $2x^2 + 3y^2 = 14$; $P(-1, -2)$

In Exercises 5–10, determine the foci, the eccentricity, and the directrices for each ellipse and hyperbola.

5. (a) $(x^2/4^2) + (y^2/3^2) = 1$
(b) $(x^2/4^2) - (y^2/3^2) = 1$

6. (a) $x^2 + 4y^2 = 1$
(b) $x^2 - 4y^2 = 1$

7. (a) $12x^2 + 13y^2 = 156$
(b) $12x^2 - 13y^2 = 156$

8. (a) $x^2 + 2y^2 = 2$
(b) $x^2 - 2y^2 = 2$

9. (a) $25x^2 + 36y^2 = 900$
(b) $25x^2 - 36y^2 = 900$

10. (a) $4x^2 + 25y^2 = 100$
(b) $4x^2 - 25y^2 = 100$

In Exercises 11 and 12, use the given information to find the equation of the ellipse. Write the answer in the form $Ax^2 + By^2 = C$.

11. The foci are $(\pm 1, 0)$ and the directrices are $x = \pm 4$.

12. The foci are $(\pm\sqrt{3}, 0)$ and the eccentricity is $2/3$.

In Exercises 13 and 14, use the given information to find the equation of the hyperbola. Write the answer in the form $Ax^2 - By^2 = C$.

13. The foci are $(\pm 2, 0)$, and the directrices are $x = \pm 1$.

14. The foci are $(\pm 3, 0)$, and the eccentricity is 2.

B

15. In this exercise we show that the focal radii of the ellipse $x^2/a^2 + y^2/b^2 = 1$ are $F_1P = a + ex$ and $F_2P = a - ex$. The method used here, which avoids the use of radicals, appears in the eighteenth-century text *Traité analytique des sections coniques* (Paris: 1707) by the Marquis de l'Hôpital (1661–1704). (For another method, one that does use radicals, see Exercise 55 in Section 11.4.) For convenience, let $d_1 = F_1P$ and $d_2 = F_2P$, as indicated in the accompanying figure.

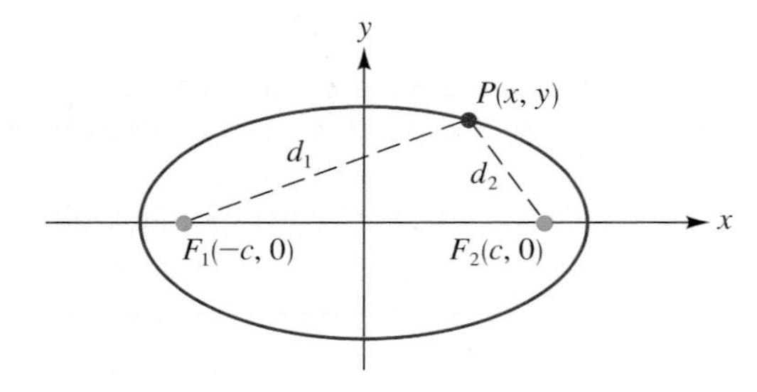

(a) Using the distance formula, verify that

$$d_1^2 = (x + c)^2 + y^2 \quad \text{and} \quad d_2^2 = (x - c)^2 + y^2$$

(b) Use the two equations in part (a) to show that $d_1^2 - d_2^2 = 4cx$.

(c) Explain why $d_1 + d_2 = 2a$.

(d) Factor the left-hand side of the equation in part (b) and substitute for one of the factors using the equation in part (c). Show that the result can be written $d_1 - d_2 = 2cx/a$.

(e) Add the equations in parts (c) and (d). Show that the resulting equation can be written $d_1 = a + ex$, as required.

(f) Use the equation in part (c) and the result in part (e) to show that $d_2 = a - ex$.

16. In this exercise we show that if the point $P(x, y)$ in the accompanying figure satisfies the condition

$$\frac{F_2P}{PD} = e \qquad \text{where } 0 < e < 1$$

then, in fact, the point $P(x, y)$ lies on the ellipse $x^2/a^2 + y^2/b^2 = 1$.

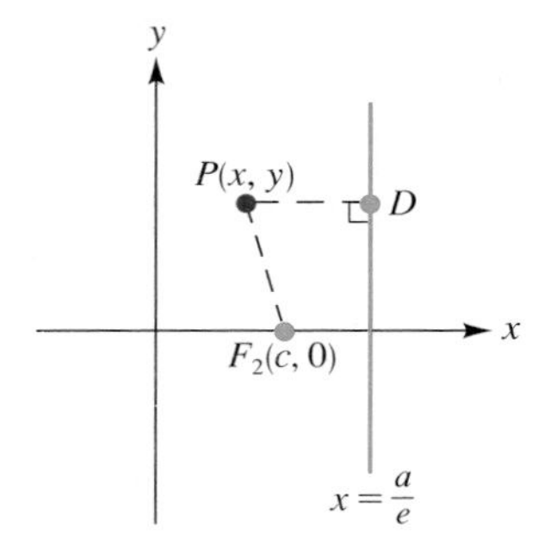

(a) From the given equation we have $(F_2P)^2 = e^2(PD)^2$. Use the distance formula and the figure to deduce from this equation that

$$(x - c)^2 + y^2 = e^2\left(\frac{a}{e} - x\right)^2$$

(b) In the equation in part (a), replace e with c/a. After carrying out the indicated operations and simplifying, show that the equation can be written

$$a^2x^2 - c^2x^2 + a^2y^2 = a^4 - a^2c^2$$

(c) The equation in part (b) is equivalent to $(a^2 - c^2)x^2 + a^2y^2 = a^2(a^2 - c^2)$. Now replace the quantity $a^2 - c^2$ by b^2 and show that the resulting equation can be written $x^2/a^2 + y^2/b^2 = 1$; thus P lies on the ellipse, as we wished to show.

(d) If $e > 1$, then let $b^2 = c^2 - a^2$ to show that the conic section is the hyperbola with equation

$$\frac{x^2}{a^2} - \frac{y^2}{b^2} = 1$$

11.7 THE CONICS IN POLAR COORDINATES

The use of polar coordinates permits a unified treatment of the conic sections, and it is the polar coordinate equations for these curves that are used in celestial mechanics. —Professor Bernice Kastner in her text *Space Mathematics,* published in 1985 by NASA (National Aeronautics and Space Administration)

As is indicated in the opening quotation, the polar coordinate equations of the parabola, ellipse, and hyperbola are useful in applications. We will develop these equations by using the focus–directrix property of the conics that we discussed in Section 11.6. Suppose that we have a conic with focus F, directrix $\mathcal{L}$, and eccentricity e. Then, as shown in Figure 1, we can set up a polar coordinate system in which the focus of the conic is the origin or pole and the directrix is perpendicular to the polar axis. In Figure 1 we've used d to denote the distance from the focus to the directrix.

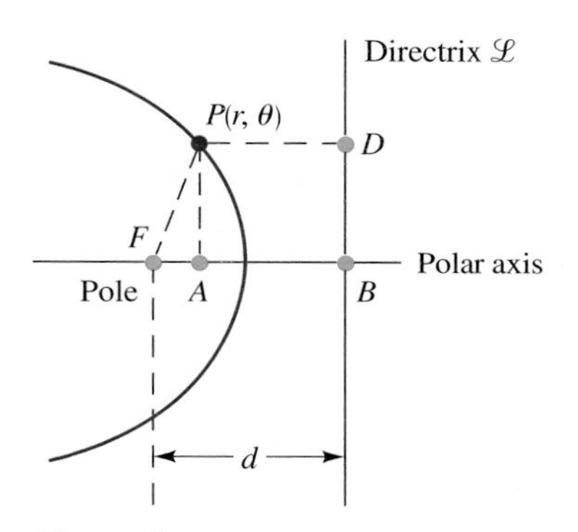

Figure 1
A conic with focus F and directrix $\mathcal{L}$

Using the focus–directrix property for the conic in Figure 1, we have

$$\frac{FP}{PD} = e \qquad \text{and therefore} \qquad \frac{r}{FB - FA} = e$$

In this last equation we can replace FB by d and FA by $r \cos \theta$. (Why?) This yields

$$\frac{r}{d - r \cos \theta} = e$$

Now, as Exercise 22 at the end of this section asks you to verify, when we solve this equation for r, we obtain

$$r = \frac{ed}{1 + e \cos \theta}$$

This is the polar form for the equation of the conic in Figure 1, in which the directrix is vertical and to the right of the focus. By the same technique we can obtain similar equations when the directrix is to the left of the focus and when the directrix is horizontal. We summarize the results in the following box. See Figures 2–5.

Polar Equations of the Conics

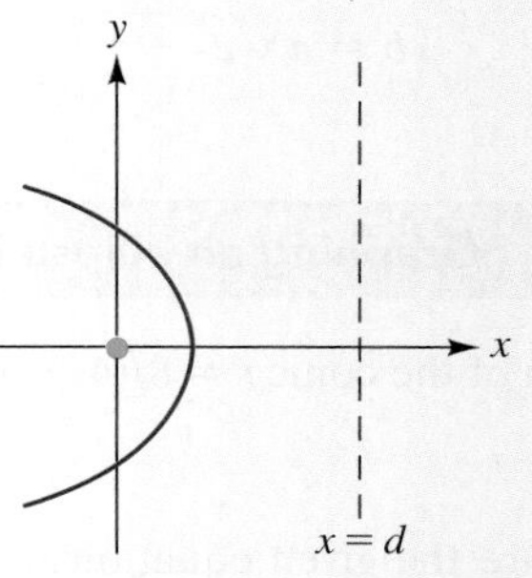

Figure 2

Polar equation: $r = \dfrac{ed}{1 + e \cos \theta}$

Focus: $(0, 0)$
Directrix: $x = d$

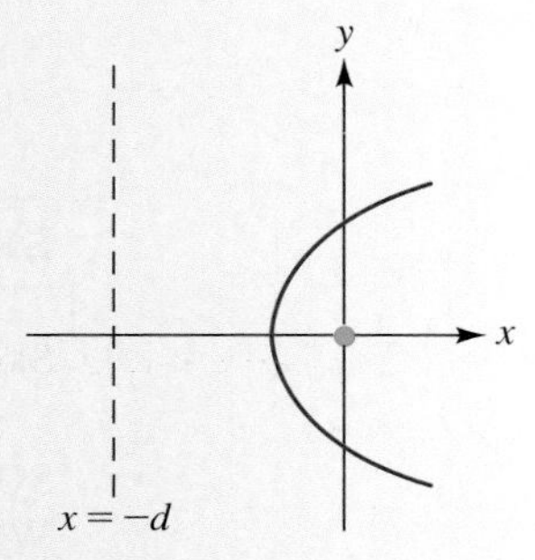

Figure 3

Polar equation: $r = \dfrac{ed}{1 - e \cos \theta}$

Focus: $(0, 0)$
Directrix: $x = -d$

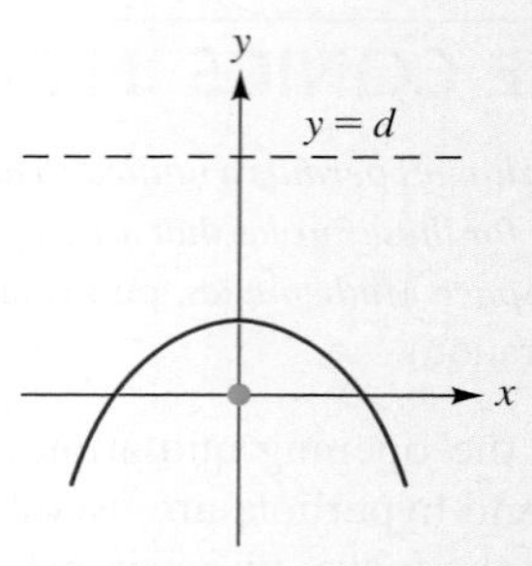

Figure 4

Polar equation: $r = \dfrac{ed}{1 + e \sin \theta}$

Focus: $(0, 0)$
Directrix: $y = d$

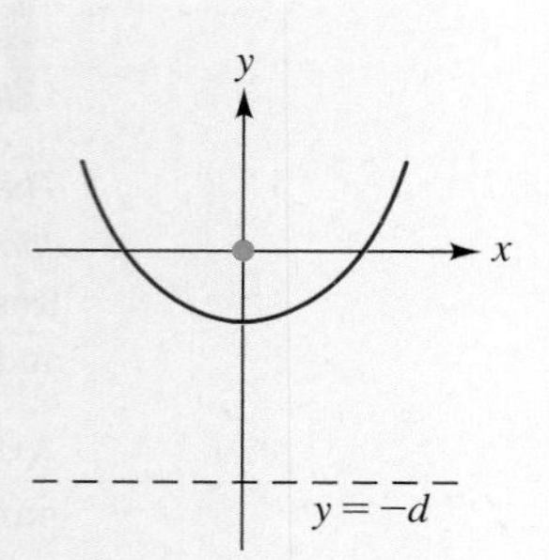

Figure 5

Polar equation: $r = \dfrac{ed}{1 - e \sin \theta}$

Focus: $(0, 0)$
Directrix: $y = -d$

Note: For each of the preceding illustrations we have given an equation of a conic in polar coordinates, while the graph, the coordinates of the focus, and the equation of the directrix are in a rectangular coordinate system. In giving a coordinate ordered pair for a point in a mixed representation of this kind, it is important to be clear about whether the coordinates are polar coordinates, (r, θ), or rectangular coordinates, (x, y).

For the example that follows and for the exercises at the end of this section it will be convenient to have formulas that express b in terms of e and a for the ellipse and the hyperbola. For the ellipse we have

$$b^2 = a^2 - c^2 = a^2 - (ae)^2 = a^2(1 - e^2)$$

and therefore

$$b = a\sqrt{1 - e^2} \qquad \text{for the ellipse} \tag{1}$$

Similarly, for the hyperbola we have

$$b^2 = c^2 - a^2 = (ae)^2 - a^2 = a^2(e^2 - 1)$$

and therefore

$$b = a\sqrt{e^2 - 1} \qquad \text{for the hyperbola} \tag{2}$$

EXAMPLE 1 Graphing an ellipse in polar coordinates

Sketch the graph of the conic $r = 8/(4 - 3 \cos \theta)$.

SOLUTION

When we compare the given equation with the four basic types shown in the box, we see that the appropriate standard equation is the one associated with Figure 3:

$$r = \frac{ed}{1 - e \cos \theta} \tag{3}$$

To write the given equation in this form (in which the first term in the denominator is 1), we divide both numerator and denominator on the right-hand side by 4. This yields

$$r = \frac{2}{1 - \frac{3}{4}\cos\theta} \tag{4}$$

Comparing equations (3) and (4), we see that

$$e = \frac{3}{4} \quad \text{and} \quad ed = 2$$

Therefore

$$d = \frac{2}{e} = \frac{2}{3/4} = \frac{8}{3}$$

The eccentricity e is $3/4$, which is less than 1, so the conic is an ellipse. From the result $d = 8/3$ (and the graph in Figure 3) we conclude that the directrix corresponding to the focus $(0, 0)$ is the vertical line $x = -8/3$. (Actually, as you'll see, we won't need this information about the directrix in drawing the graph.) Also from Figure 3 we know that the major axis of the ellipse lies along the polar or x-axis. Perhaps the simplest way to proceed now is to compute the value of r when $\theta = 0$, $\pi/2$, π, and $3\pi/2$. In Figure 6 we show the results of these computations and the four points that are determined.

$$r = \frac{8}{4 - 3\cos\theta}$$

θ	0	$\pi/2$	π	$3\pi/2$
r	8	2	8/7	2

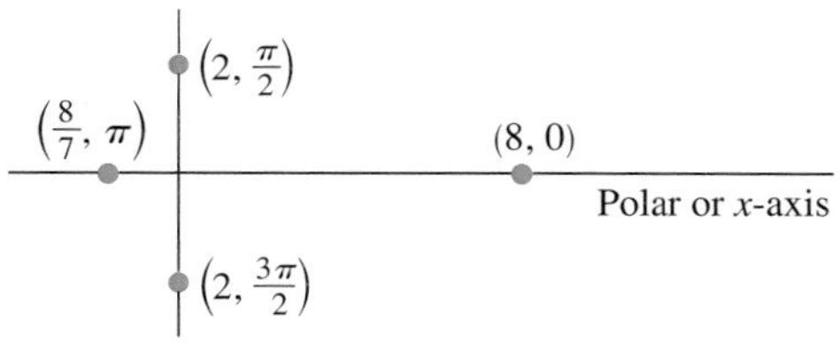

Figure 6
Four points on the ellipse $r = 8/(4 - 3\cos\theta)$

Since the major axis of this ellipse lies along the polar axis, the length of the major axis is

$$2a = 8 + \frac{8}{7} = \frac{64}{7} \quad \text{and therefore} \quad a = \frac{32}{7}$$

For the x-coordinate of the center of the ellipse we want the number (on the x-axis) that is halfway between $-8/7$ and 8. As you can check (by averaging the two numbers), this x-coordinate is $24/7$.

The last piece of information that we need for drawing an accurate sketch is the value of b for this ellipse. Using equation (1), we have

$$\begin{aligned} b &= a\sqrt{1 - e^2} \\ &= \frac{32}{7}\sqrt{1 - \left(\frac{3}{4}\right)^2} = \frac{32}{7}\sqrt{\frac{7}{16}} = \frac{8}{7}\sqrt{7} \end{aligned}$$

With a calculator we find that $b \approx 3.02$. We can now draw the graph, as shown in Figure 7.

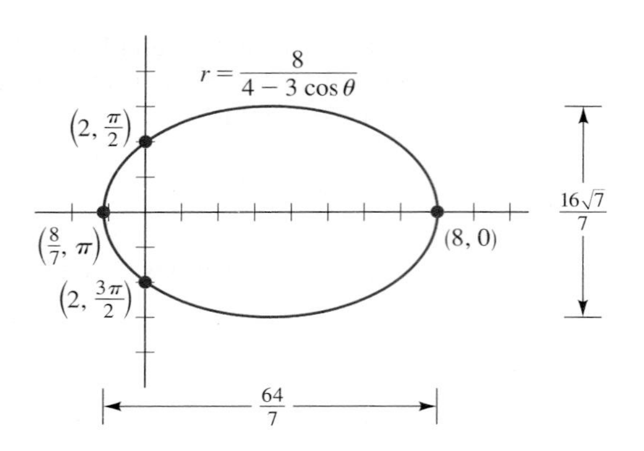

Figure 7

EXAMPLE 2 A polar equation for a parabola

Find the polar equation for the parabola in Figure 8. The focus of the parabola is (0, 0), and the (rectangular) equation of the directrix is $y = -9/4$.

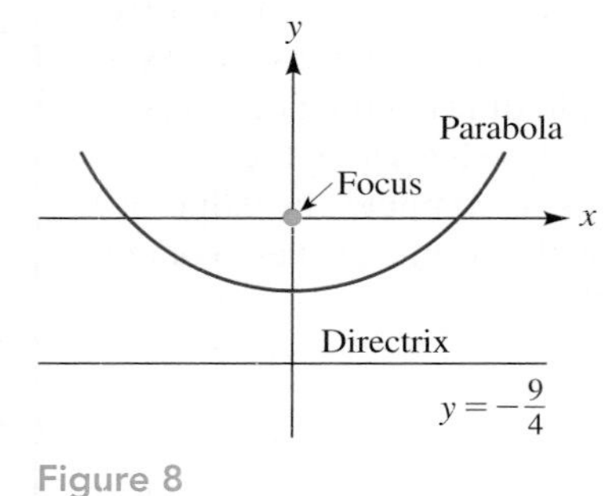

Figure 8

SOLUTION

The parabola in Figure 8 is a conic of the type shown in Figure 5, so the required equation must be of the form

$$r = \frac{ed}{1 - e\sin\theta}$$

For a parabola the eccentricity e is 1. From Figure 8 we see that the distance d from the focus to the directrix is 9/4. Using these values for e and d, the equation becomes

$$r = \frac{1(9/4)}{1 - (1)\sin\theta} = \frac{9/4}{1 - \sin\theta}$$

or

$$r = \frac{9}{4 - 4\sin\theta} \qquad \text{multiplying both numerator and denominator by 4}$$

EXAMPLE 3 Graphing a hyperbola in polar coordinates

Graph the conic $r = 9/(4 + 5\cos\theta)$.

SOLUTION

To put this equation in standard form, we divide both the numerator and the denominator of the fraction by 4. This yields

$$r = \frac{9/4}{1 + (5/4)\cos\theta}$$

Comparing this with the standard form

$$r = \frac{ed}{1 + e\cos\theta}$$

we see that the eccentricity e is 5/4, which is greater than 1, so the conic is a hyperbola. Figure 2 on page 875 shows us the general form for one branch of this hyperbola. From Figure 2 we see that the transverse axis of the hyperbola must lie along the polar or x-axis. Now, just as we did in Example 1, we compute the values of r corresponding to $\theta = 0$, $\pi/2$, π, and $3\pi/2$. Figure 9(a) shows the four points that are obtained. The two points $(1, 0)$ and $(-9, \pi)$ both lie on the transverse axis of the hyperbola, so they must be the vertices. This allows us to draw the rough sketch in Figure 9(b).

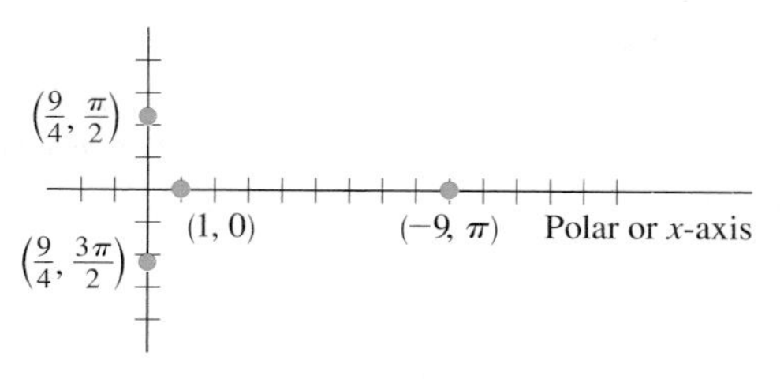

(a) The four points on the hyperbola corresponding to $\theta = 0$, $\pi/2$, π, and $3\pi/2$

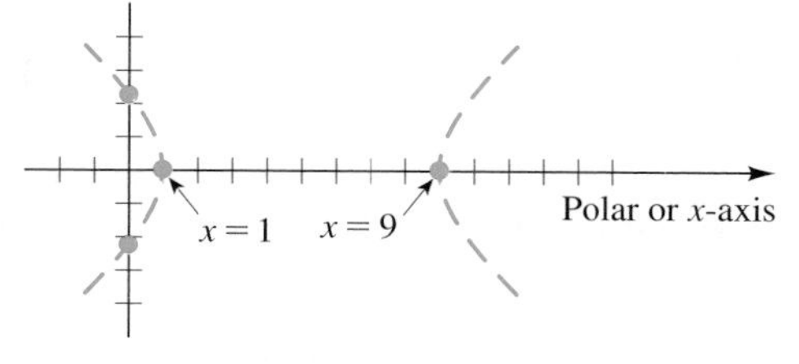

(b) A rough sketch of the hyperbola $r = 9/(4 + 5\cos\theta)$

Figure 9

For a more accurate drawing of the hyperbola we need to determine the values of a and b. Using Figure 9(b),

$$2a = 9 - 1 = 8 \quad \text{and therefore} \quad a = 4$$

The center of the hyperbola lies on the x-axis, halfway between the points with rectangular coordinates $(1, 0)$ and $(9, 0)$. Thus the center is $(5, 0)$, and consequently, $c = 5$. (We are using the fact that the origin is a focus.) Now that we know the values of a and c, we can calculate b:

$$b^2 = c^2 - a^2 = 5^2 - 4^2 = 9 \quad \text{and therefore} \quad b = 3$$

(Another way to calculate b is to use the formula on page 876: $b = a\sqrt{e^2 - 1}$). In summary, then, we have the following information to use in drawing the hyperbola in the x-y coordinate system: The rectangular coordinates of the vertices are $(1, 0)$ and $(9, 0)$; the foci are $(0, 0)$ and $(10, 0)$; the center is $(5, 0)$; and the values of a and b are 4 and 3, respectively, so the asymptotes are $y = \pm\frac{3}{4}(x - 5)$. This allows us to draw the graph as shown in Figure 10.

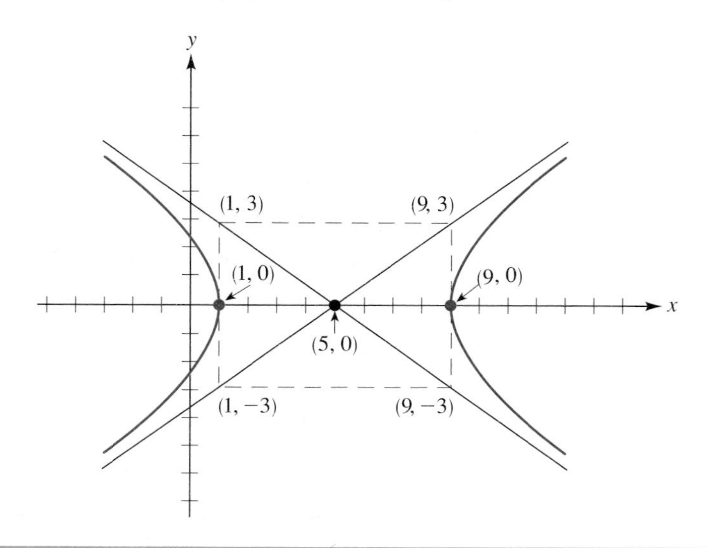

Figure 10
The hyperbola $r = 9/(4 + 5\cos\theta)$: The left-hand focus is located at the origin and the coordinates shown are x-y coordinates.

EXERCISE SET 11.7

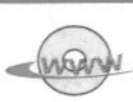

A

In Exercises 1 and 2, graph each ellipse. Specify the eccentricity, the center, and the endpoints of the major and minor axes.

1. (a) $r = \dfrac{6}{3 + 2\cos\theta}$ **(b)** $r = \dfrac{6}{3 - 2\cos\theta}$

2. (a) $r = \dfrac{12}{5 + 3\sin\theta}$ **(b)** $r = \dfrac{12}{5 - 3\sin\theta}$

In Exercises 3 and 4, graph each parabola. Specify the (rectangular) coordinates of the vertex and the equation of the directrix.

3. (a) $r = \dfrac{5}{2 + 2\cos\theta}$ **(b)** $r = \dfrac{5}{2 - 2\cos\theta}$

4. (a) $r = \dfrac{2}{1 + \sin\theta}$ **(b)** $r = \dfrac{2}{1 - \sin\theta}$

In Exercises 5 and 6, graph each hyperbola. Specify the eccentricity, the center, and the values of a, b, and c.

5. (a) $r = \dfrac{3}{2 + 4\cos\theta}$ **(b)** $r = \dfrac{3}{2 - 4\cos\theta}$

6. (a) $r = \dfrac{3}{3 + 4\sin\theta}$ **(b)** $r = \dfrac{3}{3 - 4\sin\theta}$

In Exercises 7–18, graph each conic section. If the conic is a parabola, specify (using rectangular coordinates) the vertex and the directrix. If the conic is an ellipse, specify the center, the eccentricity, and the lengths of the major and minor axes. If the conic is a hyperbola, specify the center, the eccentricity, and the lengths of the transverse and conjugate axes.

7. $r = \dfrac{24}{2 - 3\cos\theta}$

8. $r = \dfrac{16}{10 + 5\sin\theta}$

9. $r = \dfrac{8}{5 + 3\sin\theta}$

10. $r = \dfrac{9}{1 + 2\cos\theta}$

11. $r = \dfrac{12}{5 - 5\sin\theta}$

12. $r = \dfrac{5}{3 - 2\sin\theta}$

13. $r = \dfrac{12}{7 + 5\cos\theta}$

14. $r = \dfrac{5}{3 + 3\cos\theta}$

15. $r = \dfrac{4}{5 + 5\sin\theta}$

16. $r = \dfrac{14}{7 - 8\cos\theta}$

17. $r = \dfrac{9}{1 - 2\cos\theta}$

18. $r = \dfrac{2}{\sqrt{2} - \sqrt{2}\sin\theta}$

B

*For Exercises 19–21, refer to the following figure. The line segment $\overline{PQ}$ is called a **focal chord** of the conic.*

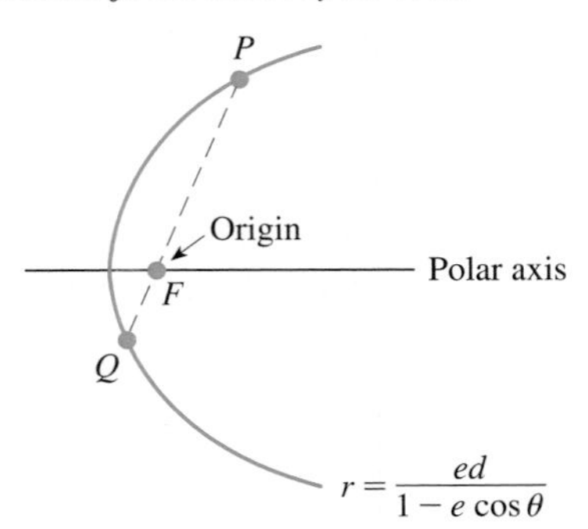

19. Show that $\dfrac{1}{FP} + \dfrac{1}{FQ} = \dfrac{2}{ed}$. What is remarkable about this result? *Hint*: Denote the polar coordinates of P by (r, θ), where $r = ed/(1 - e\cos\theta)$. Now, what are the polar coordinates of Q?

20. If the coordinates of P are (r, θ), show that

$$PQ = \frac{2ed}{1 - e^2\cos^2\theta}$$

Hint: $PQ = FP + FQ$

21. In the figure preceding Exercise 19, suppose that we draw a focal chord $\overline{AB}$ that is perpendicular to $\overline{PQ}$. Show that the sum $\dfrac{1}{PQ} + \dfrac{1}{AB}$ is a constant.

22. Solve the equation $r/(d - r\cos\theta) = e$ for r. [As stated in the text, you should obtain $r = ed/(1 + e\cos\theta)$.]

11.8 ROTATION OF AXES

In Section 11.5 we saw that the equation

$$Ax^2 + Cy^2 + Dx + Ey + F = 0$$

can represent, in general, one of three curves: a parabola, an ellipse (or circle), or a hyperbola. (We use the phrase "in general" here to allow for the so-called degenerate cases.) In the present section we will find that the second-degree equation

$$Ax^2 + Bxy + Cy^2 + Dx + Ey + F = 0 \qquad (1)$$

also represents, in general, one of these three curves. The difference now is that due to the xy-term in equation (1), the axes of the curves will no longer be parallel or perpendicular to the x- and y-axes. To study the curves defined by equation (1), it is useful first to introduce the technique known as *rotation of axes.*

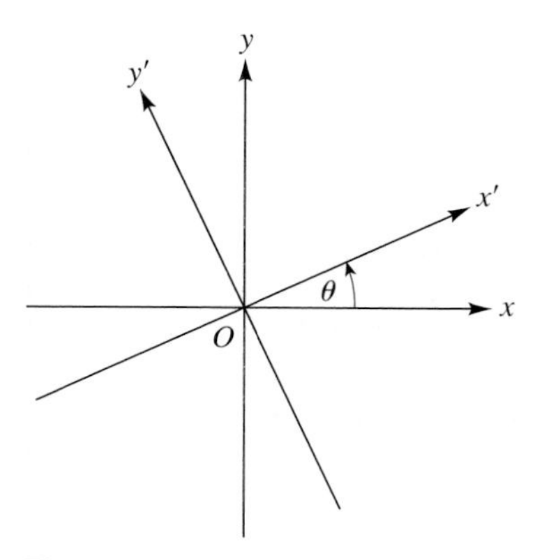

Figure 1

Suppose that the x- and y-axes are rotated through a positive angle θ to yield a new x'-y' coordinate system, as shown in Figure 1. This procedure is referred to as a **rotation of axes.** We wish to obtain formulas relating the old and new coordinates. Let P be a given point with coordinates (x, y) in the original coordinate system and (x', y') in the new coordinate system. In Figure 2, let r denote the distance OP and α the angle measured from the positive x-axis to $\overline{OP}$. From Figure 2 we have

$$\cos\alpha = \frac{\text{adjacent}}{\text{hypotenuse}} = \frac{x}{r} \quad \text{and} \quad \sin\alpha = \frac{\text{opposite}}{\text{hypotenuse}} = \frac{y}{r}$$

Thus

$$x = r\cos\alpha \quad \text{and} \quad y = r\sin\alpha \tag{2}$$

Again from Figure 2 we have

$$\cos(\alpha - \theta) = \frac{x'}{r} \quad \text{and} \quad \sin(\alpha - \theta) = \frac{y'}{r}$$

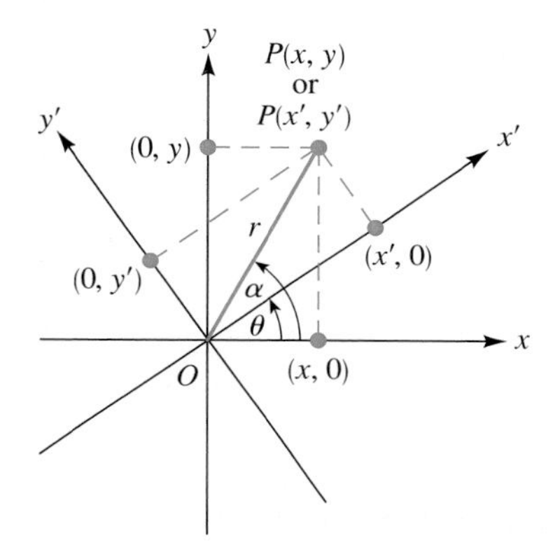

Figure 2

Thus

$$x' = r\cos(\alpha - \theta) \quad \text{and} \quad y' = r\sin(\alpha - \theta)$$

or

$$x' = r\cos\alpha\cos\theta + r\sin\alpha\sin\theta$$

and

$$y' = r\sin\alpha\cos\theta - r\cos\alpha\sin\theta$$

With the aid of equations (2) this last pair of equations can be rewritten as

$$x' = x\cos\theta + y\sin\theta \quad \text{and} \quad y' = -x\sin\theta + y\cos\theta$$

These two equations tell us how to express the new coordinates (x', y') in terms of the original coordinates (x, y) and the angle of rotation θ. On the other hand, it is also useful to express x and y in terms of x', y', and θ. This can be accomplished by treating the two equations we've just derived as a system of two equations in the unknowns x and y. As Exercise 43 at the end of this section asks you to verify, the results of solving this system for x and y are $x = x'\cos\theta - y'\sin\theta$ and $y = x'\sin\theta + y'\cos\theta$.

Formulas for the Rotation of Axes

$$\begin{cases} x = x'\cos\theta - y'\sin\theta \\ y = x'\sin\theta + y'\cos\theta \end{cases} \qquad \begin{cases} x' = x\cos\theta + y\sin\theta \\ y' = -x\sin\theta + y\cos\theta \end{cases}$$

EXAMPLE 1 Finding x-y coordinates from x'-y' coordinates

Suppose that the angle of rotation from the x-axis to the x'-axis is 45°. If the coordinates of a point P are $(2, 0)$ with respect to the x'-y' coordinate system, what are the coordinates of P with respect to the x-y system?

SOLUTION
Substitute the values $x' = 2$, $y' = 0$, and $\theta = 45°$ in the formulas

$$\begin{cases} x = x' \cos\theta - y' \sin\theta \\ y = x' \sin\theta + y' \cos\theta \end{cases}$$

This yields

$$x = 2\cos 45° - 0\sin 45° = 2\left(\frac{\sqrt{2}}{2}\right) = \sqrt{2}$$

$$y = 2\sin 45° + 0\cos 45° = 2\left(\frac{\sqrt{2}}{2}\right) = \sqrt{2}$$

Thus the coordinates of P in the x-y system are $(\sqrt{2}, \sqrt{2})$. See Figure 3.

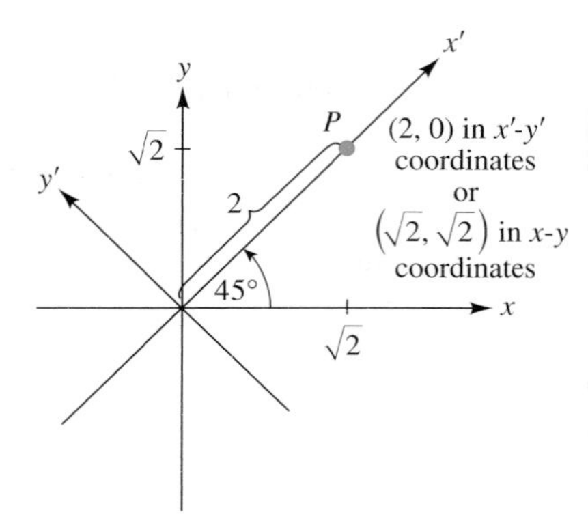

Figure 3

EXAMPLE 2 Transforming an x-y equation to x'-y' coordinates

Suppose that the angle of rotation from the x-axis to the x'-axis is 45°. Write the equation $xy = 1$ in terms of the x'-y' coordinate system, and then sketch the graph of this equation.

SOLUTION
With $\theta = 45°$ the rotation formulas for x and y become

$$\begin{cases} x = x' \cos 45° - y' \sin 45° \\ y = x' \sin 45° + y' \cos 45° \end{cases}$$

Thus we have

$$x = x'\left(\frac{\sqrt{2}}{2}\right) - y'\left(\frac{\sqrt{2}}{2}\right) = \frac{\sqrt{2}}{2}(x' - y')$$

and

$$y = x'\left(\frac{\sqrt{2}}{2}\right) + y'\left(\frac{\sqrt{2}}{2}\right) = \frac{\sqrt{2}}{2}(x' + y')$$

If we now substitute these expressions for x and y in the equation $xy = 1$, we obtain

$$\left[\frac{\sqrt{2}}{2}(x' - y')\right]\left[\frac{\sqrt{2}}{2}(x' + y')\right] = 1$$

$$\frac{1}{2}[(x')^2 - (y')^2] = 1$$

$$\frac{(x')^2}{(\sqrt{2})^2} - \frac{(y')^2}{(\sqrt{2})^2} = 1$$

This last equation represents a hyperbola in the x'-y' coordinate system. With respect to this x'-y' system, the hyperbola can be analyzed using the techniques developed in Section 11.5. The results (as you should verify) are as follows:

$$xy = 1 \quad \text{or} \quad \frac{(x')^2}{(\sqrt{2})^2} - \frac{(y')^2}{(\sqrt{2})^2} = 1 \qquad \begin{cases} \text{Focal axis:} & x'\text{-axis} \\ \text{Center:} & \text{origin} \\ \text{Vertices:} & (\pm\sqrt{2}, 0) \\ \text{Foci:} & (\pm 2, 0) \\ \text{Asymptotes:} & y' = \pm x' \end{cases}$$

These specifications are in terms of the x'-y' coordinate system.

As noted, the preceding specifications are in terms of the x'-y' coordinate system. However, since the original equation, $xy = 1$, is given in terms of the x-y system, we would like to express these specifications in terms of the same x-y coordinate system. This can be done by the method shown in Example 1. For the x-y equations of the focal axis and the asymptotes, use the second pair of equations on page 881. The results are as follows:

$$xy = 1 \quad \text{or} \quad \frac{(x')^2}{(\sqrt{2})^2} - \frac{(y')^2}{(\sqrt{2})^2} = 1 \qquad \begin{cases} \text{Focal axis:} & y = x \\ \text{Center:} & \text{origin} \\ \text{Vertices:} & (1, 1) \text{ and } (-1, -1) \\ \text{Foci:} & (\sqrt{2}, \sqrt{2}) \text{ and } (-\sqrt{2}, -\sqrt{2}) \\ \text{Asymptotes:} & x\text{- and } y\text{-axes} \end{cases}$$

These specifications are in terms of the x-y coordinate system.

Figure 4 displays the graph of this hyperbola.

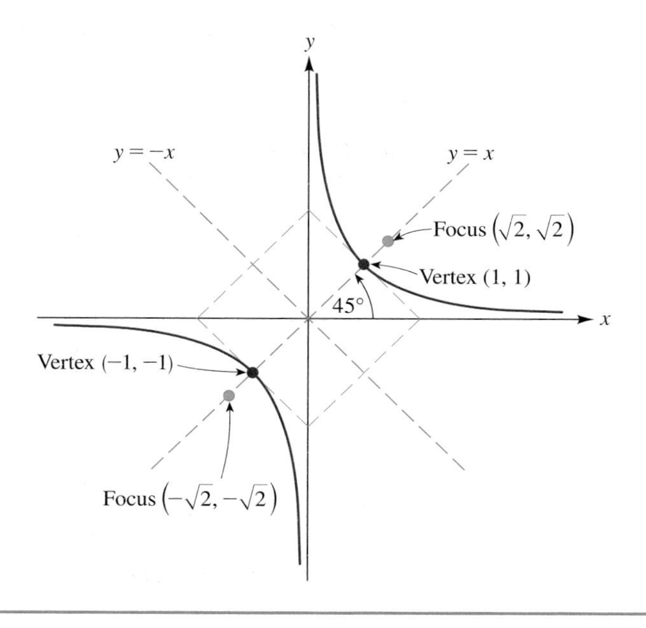

Figure 4
$xy = 1$

In Example 2 we saw that a rotation of 45° reduced the given equation to one of the standard forms with which we are already familiar. Now let us consider the situation in greater generality. We begin with the second-degree equation

$$Ax^2 + Bxy + Cy^2 + Dx + Ey + F = 0 \qquad (B \neq 0) \tag{3}$$

If we rotate the axes through an angle θ, equation (3) will, after some simplification, take on the form

$$A'(x')^2 + B'x'y' + C'(y')^2 + D'x' + E'y' + F' = 0 \tag{4}$$

for certain constants A', B', C', D', E', and F'. We wish to determine an angle of rotation θ for which $B' = 0$. The reason we want to do this is that if $B' = 0$, we will be able to analyze equation (4) using the techniques of the previous sections. We begin with the rotation formulas:

$$x = x' \cos\theta - y' \sin\theta$$
$$y = x' \sin\theta + y' \cos\theta$$

Substituting these expressions for x and y in equation (3) yields

$$\begin{aligned} A(x'\cos\theta - y'\sin\theta)^2 &+ B(x'\cos\theta - y'\sin\theta)(x'\sin\theta + y'\cos\theta) \\ &+ C(x'\sin\theta + y'\cos\theta)^2 + D(x'\cos\theta - y'\sin\theta) \\ &+ E(x'\sin\theta + y'\cos\theta) + F = 0 \end{aligned}$$

We can simplify this equation by performing the indicated operations and then collecting like terms. As Exercise 45 asks you to verify, the resulting equation is

$$A'(x')^2 + B'x'y' + C'(y')^2 + D'x' + E'y' + F' = 0 \tag{5}$$

where

$$\begin{aligned} A' &= A\cos^2\theta + B\sin\theta\cos\theta + C\sin^2\theta \\ B' &= 2(C - A)\sin\theta\cos\theta + B(\cos^2\theta - \sin^2\theta) \\ C' &= A\sin^2\theta - B\sin\theta\cos\theta + C\cos^2\theta \\ D' &= D\cos\theta + E\sin\theta \\ E' &= E\cos\theta - D\sin\theta \\ F' &= F \end{aligned}$$

Thus B' will be zero provided that

$$2(C - A)\sin\theta\cos\theta + B(\cos^2\theta - \sin^2\theta) = 0$$

By using the double-angle formulas, we can write this last equation as

$$(C - A)\sin 2\theta + B\cos 2\theta = 0$$

or

$$B\cos 2\theta = (A - C)\sin 2\theta$$

Now, dividing both sides by $B \sin 2\theta$, we obtain

$$\frac{\cos 2\theta}{\sin 2\theta} = \frac{A - C}{B}$$

or

$$\cot 2\theta = \frac{A - C}{B}$$

We have now shown that if θ satisfies the condition $\cot 2\theta = (A - C)/B$, then equation (5) will contain no $x'y'$-term. It can be shown that there is always a value of θ in the range $0° < \theta < 90°$ for which $\cot 2\theta = (A - C)/B$. (See Exercise 41.) In subsequent examples we will always choose θ in this range.

EXAMPLE 3 **Using rotation of axes to graph a quadratic equation with an *xy*-term**

Graph the equation $2x^2 + \sqrt{3}xy + y^2 = 2$.

SOLUTION

We first rotate the axes through an angle θ so that the new equation will contain no $x'y'$-term. We have $A = 2$, $B = \sqrt{3}$, and $C = 1$. To choose an appropriate value of θ, we require that

$$\cot 2\theta = \frac{A - C}{B} = \frac{2 - 1}{\sqrt{3}} = \frac{1}{\sqrt{3}}$$

Thus $\cot 2\theta = 1/\sqrt{3}$, from which we conclude that $2\theta = 60°$, or $\theta = 30°$. With this value of θ the rotation formulas become

$$x = x'\left(\frac{\sqrt{3}}{2}\right) - y'\left(\frac{1}{2}\right) \quad \text{and} \quad y = x'\left(\frac{1}{2}\right) + y'\left(\frac{\sqrt{3}}{2}\right)$$

Now we use these formulas to substitute for x and y in the given equation. This yields

$$2\left(\frac{x'\sqrt{3}}{2} - \frac{y'}{2}\right)^2 + \sqrt{3}\left(\frac{x'\sqrt{3}}{2} - \frac{y'}{2}\right)\left(\frac{x'}{2} + \frac{y'\sqrt{3}}{2}\right) + \left(\frac{x'}{2} + \frac{y'\sqrt{3}}{2}\right)^2 = 2.$$

After simplification this last equation becomes

$$(y')^2 + 5(x')^2 = 4$$

or

$$\frac{(y')^2}{2^2} + \frac{(x')^2}{(2/\sqrt{5})^2} = 1$$

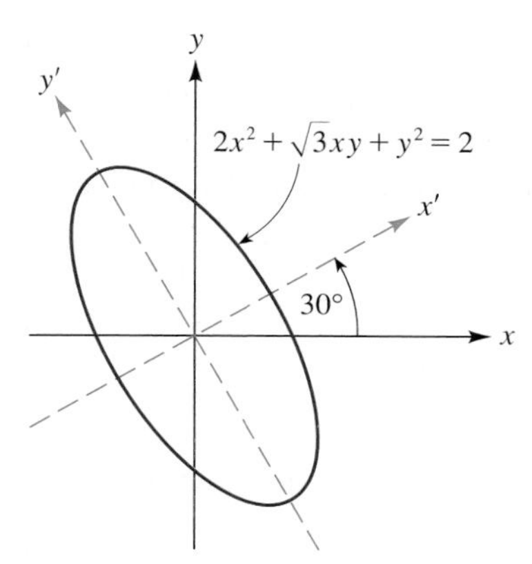

Figure 5

We recognize this as the equation of an ellipse in the x'-y' coordinate system. The focal axis is the y'-axis. The values of a and b are 2 and $2/\sqrt{5}$, respectively. The ellipse can now be sketched as in Figure 5.

In Example 3 we were able to determine the angle θ directly, since we recognized the quantity $1/\sqrt{3}$ as the value of $\cot 60°$. However, this is the exception rather than the rule. In most problems the value of $(A - C)/B$ is not so easily identified as the cotangent of one of the more familiar angles. The next examples demonstrate a technique that can be used in such cases. The technique relies on the following three trigonometric identities:

1. $\sec^2 \beta = 1 + \tan^2 \beta$
2. $\sin \theta = \sqrt{\dfrac{1 - \cos 2\theta}{2}}$
3. $\cos \theta = \sqrt{\dfrac{1 + \cos 2\theta}{2}}$

(2 and 3:) The positive square roots are appropriate, since $0° < \theta < 90°$.

The key step is to calculate $\cos 2\theta$.

EXAMPLE 4 **Using rotation of axes to graph a quadratic equation with an xy-term**

Graph the equation $x^2 + 4xy - 2y^2 = 6$.

SOLUTION

We have $A = 1$, $B = 4$, and $C = -2$. Therefore

$$\cot 2\theta = \frac{A - C}{B} = \frac{1 - (-2)}{4} = \frac{3}{4}$$

Since $\cot 2\theta = 3/4$, it follows that $\tan 2\theta = 4/3$. Therefore

$$\sec^2 2\theta = 1 + \left(\frac{4}{3}\right)^2 \quad \text{(Why?)}$$
$$= \frac{9}{9} + \frac{16}{9} = \frac{25}{9}$$

Thus

$$\sec 2\theta = \pm\frac{5}{3}$$

At this point, we need to decide whether the positive or negative sign is appropriate. Since we are assuming that $0° < \theta < 90°$, the angle 2θ must lie in either Quadrant I or Quadrant II. To decide which, we note that the value determined for $\cot 2\theta$ was positive. That rules out the possibility that 2θ might lie in Quadrant II. We conclude that in this case, $0° < 2\theta < 90°$. Therefore the sign of $\sec 2\theta$ must be positive, and we have

$$\sec 2\theta = \frac{5}{3} \quad \text{so} \quad \cos 2\theta = \frac{3}{5}$$

(*Note*: We could have found $\cos 2\theta$ by using right triangles rather than identities.)

The values of $\sin \theta$ and $\cos \theta$ can now be obtained as follows:

$$\sin \theta = \sqrt{\frac{1 - \cos 2\theta}{2}} = \sqrt{\frac{1 - \frac{3}{5}}{2}} = \frac{1}{\sqrt{5}} \quad \text{after simplifying}$$

$$\cos \theta = \sqrt{\frac{1 + \cos 2\theta}{2}} = \sqrt{\frac{1 + \frac{3}{5}}{2}} = \frac{2}{\sqrt{5}} \quad \text{after simplifying}$$

With these values for $\sin\theta$ and $\cos\theta$ the rotation formulas become

$$x = x'\left(\frac{2}{\sqrt{5}}\right) - y'\left(\frac{1}{\sqrt{5}}\right) = \frac{1}{\sqrt{5}}(2x' - y')$$

and

$$y = x'\left(\frac{1}{\sqrt{5}}\right) + y'\left(\frac{2}{\sqrt{5}}\right) = \frac{1}{\sqrt{5}}(x' + 2y')$$

We now substitute these expressions for x and y in the original equation, $x^2 + 4xy - 2y^2 = 6$. This yields

$$\left(\frac{2x' - y'}{\sqrt{5}}\right)^2 + 4\left(\frac{2x' - y'}{\sqrt{5}}\right)\left(\frac{x' + 2y'}{\sqrt{5}}\right) - 2\left(\frac{x' + 2y'}{\sqrt{5}}\right)^2 = 6$$

As Exercise 42 asks you to verify, this equation can be simplified to

$$2(x')^2 - 3(y')^2 = 6$$

or

$$\frac{(x')^2}{(\sqrt{3})^2} - \frac{(y')^2}{(\sqrt{2})^2} = 1$$

This last equation represents a hyperbola with center at the origin of the x'-y' coordinate system and with a focal axis that coincides with the x'-axis. We can sketch the hyperbola using the methods of Section 11.5, but it is first necessary to know the angle θ between the x- and x'-axes. We have

$$\sin\theta = \frac{1}{\sqrt{5}}$$

$$\theta = \sin^{-1}\left(\frac{1}{\sqrt{5}}\right)$$

$$\theta \approx 27^\circ \qquad \text{using a calculator set in the degree mode}$$

Figure 6 shows the required graph.

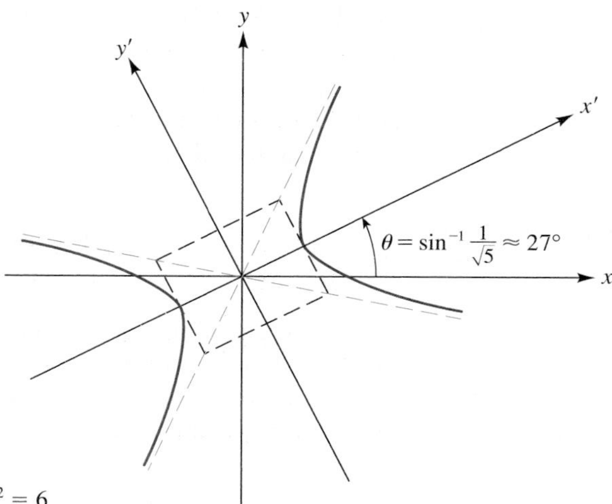

Figure 6
$x^2 + 4xy - 2y^2 = 6$

In Examples 3 and 4 we graphed equations of the form

$$Ax^2 + Bxy + Cy^2 + F = 0$$

The technique used in those examples is equally effective in graphing equations of the form $Ax^2 + Bxy + Cy^2 + Dx + Ey + F = 0$, in which the x- and y-terms are present. This is demonstrated in Example 5. Since the general technique employed in Example 5 is the same as in the previous examples, we will merely outline the procedure and the results in the solution, leaving the detailed calculations to Exercise 44 at the end of this section.

EXAMPLE 5 Graphing a quadratic equation with xy and linear terms

Graph the equation $16x^2 - 24xy + 9y^2 + 110x - 20y + 100 = 0$.

OUTLINE OF SOLUTION

$A = 16$, $B = -24$, and $C = 9$, so

$$\cot 2\theta = \frac{A - C}{B} = -\frac{7}{24}$$

Now, proceeding as in the last example, we find that $\cos 2\theta = -7/25$, $\cos\theta = 3/5$, and $\sin\theta = 4/5$. Thus the rotation formulas become

$$x = x'\left(\frac{3}{5}\right) - y'\left(\frac{4}{5}\right) = \frac{1}{5}(3x' - 4y')$$

and

$$y = x'\left(\frac{4}{5}\right) + y'\left(\frac{3}{5}\right) = \frac{1}{5}(4x' + 3y')$$

Next we substitute for x and y in the given equation. After straightforward but lengthy computations we obtain

$$(y')^2 + 2x' - 4y' + 4 = 0$$

We graphed equations of this form in Section 11.2 by completing the square. Using that technique here, we have

$$\begin{aligned}(y')^2 - 4y' &= -2x' - 4\\ (y')^2 - 4y' + 4 &= -2x' - 4 + 4\\ (y' - 2)^2 &= -2x'\end{aligned}$$

This is the equation of a parabola. With respect to the x'-y' system the vertex is $(0, 2)$, and the axis of the parabola is the line $y' = 2$. In terms of the x-y system the vertex is $(-\frac{8}{5}, \frac{6}{5})$, and the axis of the parabola is the line $-4x + 3y = 10$. Finally, the angle of rotation is $\theta = \sin^{-1}(\frac{4}{5}) \approx 53°$. The required graph is shown in Figure 7.

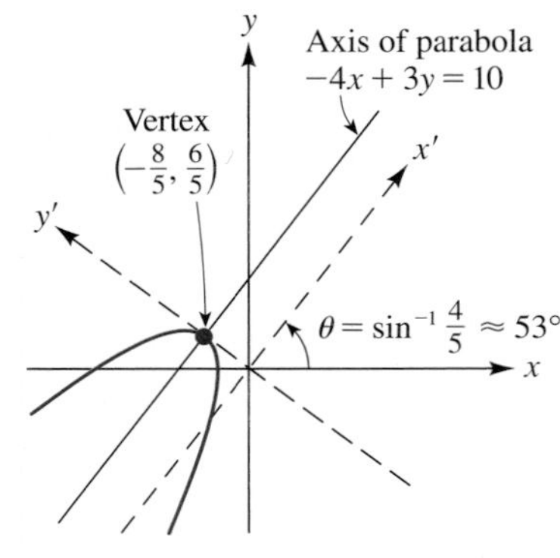

Figure 7
$16x^2 - 24xy + 9y^2 + 110x - 20y + 100 = 0$

EXERCISE SET 11.8

A

In Exercises 1–3 an angle of rotation is specified, followed by the coordinates of a point in the x′-y′ system. Find the coordinates of each point with respect to the x-y system.

1. $\theta = 30°; (x', y') = (\sqrt{3}, 2)$
2. $\theta = 60°; (x', y') = (-1, 1)$
3. $\theta = 45°; (x', y') = (\sqrt{2}, -\sqrt{2})$

In Exercises 4–6 an angle of rotation is specified, followed by the coordinates of a point in the x-y system. Find the coordinates of each point with respect to the x′-y′ system.

4. $\theta = 45°; (x, y) = (0, -2)$
5. $\theta = \sin^{-1}(\frac{5}{13}); (x, y) = (-3, 1)$
6. $\theta = 15°; (x, y) = (1, 0)$

In Exercises 7–14, find $\sin\theta$ *and* $\cos\theta$, *where* θ *is the (acute) angle of rotation that eliminates the x′y′-term.* Note: *You are not asked to graph the equation.*

7. $25x^2 - 24xy + 18y^2 + 1 = 0$
8. $x^2 + 24xy + 8y^2 - 8 = 0$
9. $x^2 - 24xy + 8y^2 - 8 = 0$
10. $220x^2 + 119xy + 100y^2 = 0$
11. $x^2 - 2\sqrt{3}xy - y^2 = 3$
12. $5x^2 + 12xy - 4 = 0$
13. $161xy - 240y^2 - 1 = 0$
14. $4x^2 - 5xy + 4y^2 + 2 = 0$
15. Suppose that the angle of rotation is 45°. Write the equation $2xy = 9$ in terms of the x'-y' coordinate system and then graph the equation.
16. Suppose that the angle of rotation is 45°. Write the equation $5x^2 - 6xy + 5y^2 + 16 = 0$ in terms of the x'-y' system.

In Exercises 17–40, graph the equations.

17. $7x^2 + 8xy + y^2 - 1 = 0$
18. $2x^2 - \sqrt{3}xy + y^2 - 20 = 0$
19. $x^2 + 4xy + 4y^2 = 1$
20. $x^2 + 4xy + 4y^2 = 0$
21. $9x^2 - 24xy + 16y^2 - 400x - 300y = 0$
22. $8x^2 + 12xy + 13y^2 = 34$
23. $4xy + 3y^2 + 4x + 6y = 1$
24. $x^2 - 2xy + y^2 + x - y = 0$
25. $3x^2 - 2xy + 3y^2 - 6\sqrt{2}x + 2\sqrt{2}y + 4 = 0$
26. $x^2 + 3xy + y^2 = 1$
27. $(x - y)^2 = 8(y - 6)$
28. $4x^2 - 4xy + y^2 - 4x + 2y + 1 = 0$
29. $3x^2 + 4xy + 6y^2 = 7$
30. $x^2 + 2\sqrt{3}xy + 3y^2 + 12\sqrt{3}x - 12y - 24 = 0$
31. $17x^2 - 12xy + 8y^2 - 80 = 0$
32. $7x^2 - 2\sqrt{3}xy + 5y^2 = 32$
33. $3xy - 4y^2 + 18 = 0$
34. $x^2 + y^2 = 2xy + 4x + 4y - 8$
35. $(x + y)^2 + 4\sqrt{2}(x - y) = 0$
36. $41x^2 - 24xy + 9y^2 = 3$
37. $3x^2 - \sqrt{15}xy + 2y^2 = 3$
38. $3x^2 + 10xy + 3y^2 - 2\sqrt{2}x + 2\sqrt{2}y - 10 = 0$
39. $3x^2 - 2xy + 3y^2 + 2 = 0$
40. $(x + y)(x + y + 1) = 2$

B

41. In transforming an equation of the form

$$Ax^2 + Bxy + Cy^2 + Dx + Ey + F = 0$$

to an x'-y' equation without an $x'y'$ term using rotation of axes, explain why there is always a value of θ in the interval $0° < \theta < 90°$ for which $\cot 2\theta = \frac{A - C}{B}$.

42. Simplify the equation:

$$\left(\frac{2x' - y'}{\sqrt{5}}\right)^2 + 4\left(\frac{2x' - y'}{\sqrt{5}}\right)\left(\frac{x' + 2y'}{\sqrt{5}}\right) - 2\left(\frac{x' + 2y'}{\sqrt{5}}\right)^2 = 6$$

Answer: $2(x')^2 - 3(y')^2 = 6$

43. Solve for x and y:

$$\begin{cases} (\cos\theta)x + (\sin\theta)y = x' \\ (-\sin\theta)x + (\cos\theta)y = y' \end{cases}$$

Answer: $x = x'\cos\theta - y'\sin\theta$
$y = x'\sin\theta + y'\cos\theta$

44. Refer to Example 5 in the text.
(a) Show that $\cos 2\theta = -7/25$.
(b) Show that $\cos\theta = 3/5$ and $\sin\theta = 4/5$.
(c) Make the substitutions $x = \frac{1}{5}(3x' - 4y')$ and $y = \frac{1}{5}(4x' + 3y')$ in the given equation $16x^2 - 24xy + 9y^2 + 110x - 20y + 100 = 0$ and show that the resulting equation simplifies to $(y')^2 + 2x' - 4y' + 4 = 0$.

45. Make the substitutions $x = x'\cos\theta - y'\sin\theta$ and $y = x'\sin\theta + y'\cos\theta$ in the equation $Ax^2 + Bxy + Cy^2 + Dx + Ey + F = 0$ and show that the result is

$$A'(x')^2 + B'x'y' + C'(y')^2 + D'x' + E'y' + F' = 0$$

where

$$\begin{aligned} A' &= A\cos^2\theta + B\sin\theta\cos\theta + C\sin^2\theta \\ B' &= 2(C - A)\sin\theta\cos\theta + B(\cos^2\theta - \sin^2\theta) \\ C' &= A\sin^2\theta - B\sin\theta\cos\theta + C\cos^2\theta \\ D' &= D\cos\theta + E\sin\theta \\ E' &= E\cos\theta - D\sin\theta \\ F' &= F \end{aligned}$$

46. (Refer to Exercise 45.) Show that $A + C = A' + C'$.

47. Complete the following steps to derive the equation $(B')^2 - 4A'C' = B^2 - 4AC$.

(a) Show that $A' - C' = (A - C)\cos 2\theta + B \sin 2\theta$.
(b) Show that $B' = B \cos 2\theta - (A - C)\sin 2\theta$.
(c) Square the equations in parts (a) and (b), then add the two resulting equations to show that $(A' - C')^2 + (B')^2 = (A - C)^2 + B^2$.
(d) Square the equation given in Exercise 46, then subtract the result from the equation in part (c). The result can be written $(B')^2 - 4A'C' = B^2 - 4AC$, as required.

48. Use Exercise 47 to prove the following theorem: The graph of $Ax^2 + Bxy + Cy^2 + Dx + Ey + F = 0$ is

an ellipse if $B^2 - 4AC < 0$
a parabola if $B^2 - 4AC = 0$
a hyperbola if $B^2 - 4AC > 0$

Chapter 11 Summary of Principal Terms and Formulas

Terms or formulas	Page reference	Comments
1. Conic sections	817	These are the curves that are formed when a plane intersects the surface of a right circular cone. As is indicated in the figure at the beginning of the chapter, these curves are the circle, the ellipse, the hyperbola, and the parabola.
2. Angle of inclination	818	The angle of inclination of a line is the angle between the x-axis and the line, measured counterclockwise from the positive side or positive direction of the x-axis to the line.
3. $m = \tan \theta$	818	The slope of a line is equal to the tangent of the angle of inclination.
4. $d = \dfrac{\lvert mx_0 + b - y_0 \rvert}{\sqrt{1 + m^2}}$	819	This is a formula for the (perpendicular) distance d from the point (x_0, y_0) to the line $y = mx + b$.
5. $d = \dfrac{\lvert Ax_0 + By_0 + C \rvert}{\sqrt{A^2 + B^2}}$	819, 822	This is a formula for the (perpendicular) distance d from the point (x_0, y_0) to the line $Ax + By + C = 0$.
6. Parabola	826	A parabola is the set of all points in the plane equally distant from a fixed line and a fixed point not on the line. The fixed line is called the *directrix,* and the fixed point is called the *focus.*
7. Axis (of a parabola)	827	This is the line that is drawn through the focus of the parabola, perpendicular to the directrix.
8. Vertex (of a parabola)	827	This is the point at which the parabola intersects its axis.
9. Focal chord (of a parabola)	830	A focal chord of a parabola is a line segment that passes through the focus and has endpoints on the parabola.
10. Focal width (of a parabola)	830	The focal width of a parabola is the length of the focal chord that is perpendicular to the axis of the parabola. For a given (positive) value of p the four parabolas $x^2 = 4py$, $x^2 = -4py$, $y^2 = 4px$, and $y^2 = -4px$ have the same focal width of $4p$.

Terms or formulas	Page reference	Comments
11. Tangent line (to a parabola)	839	A line that is not parallel to the axis of a parabola is tangent to the parabola provided that it intersects the parabola in exactly one point.
12. Ellipse	842	An ellipse is the set of all points in the plane, the sum of whose distances from two fixed points is constant. Each fixed point is called a *focus* of the ellipse.
13. Eccentricity (of an ellipse)	844	The eccentricity is a number that measures how much the ellipse deviates from being a circle. See Figure 7 in Section 11.4. The eccentricity e is defined by the formula $e = c/a$, where c and a are defined by the following conventions. The distance between the foci is denoted by $2c$. The sum of the distances from a point on the ellipse to the two foci is denoted by $2a$.
14. Focal axis (of an ellipse)	844	This is the line passing through the two foci.
15. Center (of an ellipse)	844	This is the midpoint of the line segment joining the foci.
16. Vertices (of an ellipse)	844	The two points at which an ellipse meets its focal axis are called the vertices of the ellipse.
17. Major axis (of an ellipse)	844	This is the line segment joining the two vertices of the ellipse.
18. Minor axis (of an ellipse)	844	This is the line segment through the center of the ellipse, perpendicular to the major axis, and with endpoints on the ellipse.
19. $\dfrac{x^2}{a^2} + \dfrac{y^2}{b^2} = 1 \quad (a > b)$	843	This is the standard form for the equation of an ellipse with foci $(\pm c, 0)$.
$\dfrac{x^2}{b^2} + \dfrac{y^2}{a^2} = 1 \quad (a > b)$	847	This is the standard form for the equation of an ellipse with foci $(0, \pm c)$.
20. Tangent line to an ellipse	848	A tangent to an ellipse is a line that intersects the ellipse in exactly one point. See Figure 12 in Section 11.4.
21. $\dfrac{x_1 x}{a^2} + \dfrac{y_1 y}{b^2} = 1$	849	This is an equation of the tangent line to the ellipse $(x^2/a^2) + (y^2/b^2) = 1$ at the point (x_1, y_1) on the ellipse.
22. Hyperbola	856	A hyperbola is the set of all points in the plane, the absolute value of the difference of whose distances from two fixed points is a positive constant. The two fixed points are the *foci,* and the line passing through the foci is the *focal axis.*
23. $\dfrac{x^2}{a^2} - \dfrac{y^2}{b^2} = 1$	858	This is the standard form for the equation of a hyperbola with foci $(\pm c, 0)$.

Terms or formulas	Page reference	Comments
23, *continued*		
$\dfrac{y^2}{a^2} - \dfrac{x^2}{b^2} = 1$	861	This is the standard form for the equation of a hyperbola with foci $(0, \pm c)$.
24. Asymptote	858, 859	A line is said to be an asymptote for a curve if the distance between the line and the curve approaches zero as we move farther and farther out along the line. The asymptotes for the hyperbola $(x^2/a^2) - (y^2/b^2) = 1$ are the two lines $y = \pm(b/a)x$. For the hyperbola $(y^2/a^2) - (x^2/b^2) = 1$, the asymptotes are $y = \pm(a/b)x$.
25. Focal axis (of a hyperbola)	859	This is the line passing through the foci.
26. Vertices (of a hyperbola)	859	The two points at which the hyperbola intersects its focal axis are called vertices.
27. Center (of a hyperbola)	859	This is the midpoint of the line segment joining the foci.
28. Transverse axis	859	This is the line segment joining the two vertices of a hyperbola.
29. Conjugate axis	859	This is the line segment perpendicular to the transverse axis of the hyperbola, passing through the center and extending a distance $b(=\sqrt{c^2 - a^2})$ on either side of the center.
30. Eccentricity (of a hyperbola)	859	For both of the standard forms for the hyperbola the eccentricity e is defined by $e = c/a$.
31. $\dfrac{x_1 x}{a^2} - \dfrac{y_1 y}{b^2} = 1$	863	This is an equation of the tangent line to the hyperbola $(x^2/a^2) - (y^2/b^2) = 1$ at the point (x_1, y_1) on the curve.
32. Focus–directrix property of conics	873	Refer to Figure 10 on page 873. Let $\mathscr{L}$ be a fixed line, F a fixed point, and e a positive constant. Consider the set of points P satisfying the condition $FP/PD = e$, where D is defined by Figure 10. Then: **(a)** If $e = 1$ the set of points is a *parabola* with focus F and directrix $\mathscr{L}$; **(b)** If $0 < e < 1$, the set of points is an *ellipse* with focus F, corresponding directrix $\mathscr{L}$ and eccentricity e; **(c)** If $e > 1$, the set of points is a *hyperbola* with focus F, corresponding directrix $\mathscr{L}$, and eccentricity e.
33. $r = \dfrac{ed}{1 \pm e\cos\theta}$ $r = \dfrac{ed}{1 \pm e\sin\theta}$	875, 876	These are polar equations of conics with the focus at the pole or origin, as indicated in Figures 2–5 on pages 875–876. The eccentricity is e, and d is the distance from the focus to the directrix.

Terms or formulas	Page reference	Comments
34. $\begin{cases} x = x' \cos\theta - y' \sin\theta \\ y = x' \sin\theta + y' \cos\theta \end{cases}$ $\begin{cases} x' = x\cos\theta + y\sin\theta \\ y' = -x\sin\theta + y\cos\theta \end{cases}$	881	These are the formulas that relate the coordinates of a point in the x-y system to the coordinates in the rotated x'-y' system. See Figures 1 and 2 in Section 11.8.
35. $\cot 2\theta = \dfrac{A - C}{B}$	885	This formula determines an angle of rotation θ so that when the equation $Ax^2 + Bxy + Cy^2 + Dx + Ey + F = 0 \quad (B \neq 0)$ is written in the rotated x'-y' system, the resulting equation does not contain an x'-y'-term. The graph can then be analyzed by means of the technique of completing the square.

Writing Mathematics

Write out your answers to Questions 1 and 2 in complete sentences. If you draw a diagram to accompany your writing or if you use equations, be sure that you clearly label any elements to which you refer.

1. Refer to Figure 1 on page 842. If the thumbtacks are 3 in. apart, what length of string should be used to produce an ellipse with eccentricity $2/3$?

2. Refer to Figure 1 on page 842 to explain each of the following.

(a) When the eccentricity of an ellipse is close to 1, the ellipse is very flat.

(b) When the eccentricity of an ellipse is close to 0, the ellipse resembles a circle.

3. Investigate the geometric significance of the eccentricity of a hyperbola by completing the three steps that follow. Then write a report telling what you have done, what patterns you have observed, and what relationship you have found between the eccentricity and the shape of the hyperbola.

(a) Use the definition of the eccentricity e to show, for the hyperbola $(x^2/a^2) - (y^2/b^2) = 1$, that $e = \sqrt{1 + (b/a)^2}$.
(This shows that the eccentricity is always greater than 1.)

(b) Compute the eccentricity for each of the following hyperbolas. (Use a calculator.)

(i) $(0.0201)x^2 - y^2 = 0.0201$

(ii) $3x^2 - y^2 = 3$

(iii) $8x^2 - y^2 = 8$

(iv) $15x^2 - y^2 = 15$

(v) $99x^2 - y^2 = 99$

(c) On the same set of axes, sketch the Quadrant I portion of each of the hyperbolas in part (b).

Chapter 11 Review Exercises

In Exercises 1–15, refer to the following figure and show that the given statements are correct.

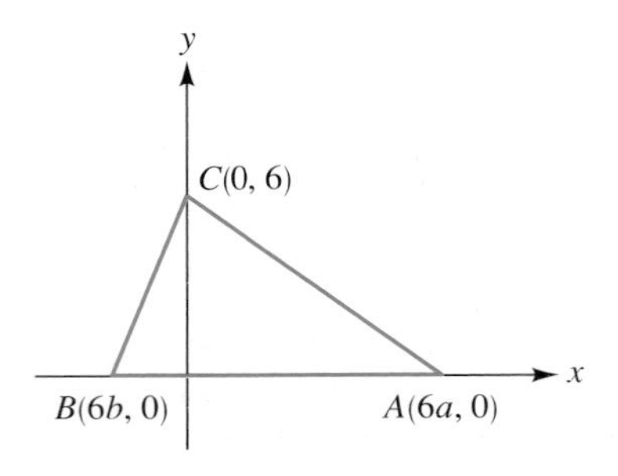

1. The equations of the lines forming the sides of $\triangle ABC$ are $x + by = 6b$, $x + ay = 6a$, and $y = 0$.
2. The equations of the lines forming the medians of $\triangle ABC$ are $2x + (a + b)y = 6(a + b)$, $x - (a - 2b)y = 6b$, and $x - (b - 2a)y = 6a$. (A *median* is a line segment drawn from a vertex of a triangle to the midpoint of the opposite side.)
3. Each pair of medians of $\triangle ABC$ intersect at the point $G(2a + 2b, 2)$. (The point G is called the *centroid* of $\triangle ABC$.)
4. The equations of the lines forming the altitudes of $\triangle ABC$ are $y = ax - 6ab$, $y = bx - 6ab$, and $x = 0$. (An *altitude* is a line segment drawn from a vertex to the opposite side, perpendicular to that side.)
5. Each pair of altitudes intersect at the point $H(0, -6ab)$. (The point H is the *orthocenter* of $\triangle ABC$.)
6. The equations of the perpendicular bisectors of the sides of $\triangle ABC$ are $x = 3a + 3b$, $bx - y = 3b^2 - 3$, and $ax - y = 3a^2 - 3$.
7. Each pair of perpendicular bisectors intersect at the point $O(3a + 3b, 3ab + 3)$. (The point O is called the *circumcenter* of $\triangle ABC$.)
8. The distance from the circumcenter O to each vertex is $3\sqrt{(a^2 + 1)(b^2 + 1)}$. (This distance, denoted by R, is the *circumradius* of $\triangle ABC$. Note that the circle with center O and radius R passes through the points A, B, and C.)
9. In $\triangle ABC$, let p, q, and r denote the lengths BC, AC, and AB, respectively. Then the area of $\triangle ABC$ is $pqr/4R$.
10. $AH^2 + BC^2 = 4(OA)^2$
11. $OH^2 = 9R^2 - (p^2 + q^2 + r^2)$ *Hint*: See Exercise 9.
12. $GH^2 = 4R^2 - \frac{4}{9}(p^2 + q^2 + r^2)$
13. $HA^2 + HB^2 + HC^2 = 12R^2 - (p^2 + q^2 + r^2)$
14. The points H, G, and O are collinear. (The line through these three points is the *Euler line* of $\triangle ABC$.)
15. $GH = 2(GO)$

In Exercises 16–18, find the angle of inclination for each line. Use a calculator to express your answers in degrees. (Round to one decimal place.)

16. $y = 4x - 3$ 17. $2x + 3y = 6$ 18. $y = 2x$
19. Find the distance from the point $(-1, -3)$ to the line $5x + 6y = 30$.
20. Find the distance from the point $(2, 1)$ to the line $y = \frac{1}{2}x + 4$.
21. The vertices of an equilateral triangle are $(\pm 6, 0)$ and $(0, 6\sqrt{3})$. Verify that the sum of the three distances from the point $(1, 2)$ to the sides of the triangle is equal to the height of the triangle. (It can be shown that, for any point inside an equilateral triangle, the sum of the distances to the sides is equal to the height of the triangle. This is *Viviani's theorem.*)
22. A tangent line is drawn from the point $(-12, -1)$ to the circle $x^2 + y^2 = 20$. Find the slope of this line, given that its y-intercept is positive.

In Exercises 23–26, find the equation of the parabola satisfying the given conditions. In each case, assume the vertex is $(0, 0)$.

23. **(a)** The focus is $(4, 0)$. **(b)** The focus is $(0, 4)$.
24. The focus lies on the x-axis, and the curve passes through the point $(3, 1)$.
25. The parabola is symmetric about the y-axis, the y-coordinate of the focus is positive, and the length of the focal chord perpendicular to the y-axis is 12.
26. The focus of the parabola is the center of the circle $x^2 - 8x + y^2 + 15 = 0$.

In Exercises 27–29, find the equation of the ellipse satisfying the given conditions. Write your answers in the form $Ax^2 + By^2 = C$.

27. Foci $(\pm 2, 0)$; endpoints of the major axis $(\pm 8, 0)$
28. Foci $(0, \pm 1)$; endpoints of the minor axis $(\pm 4, 0)$
29. Eccentricity $4/5$; one end of the minor axis at $(-6, 0)$; center at the origin
30. For any point P on a certain ellipse, the sum of the distances from $(1, 2)$ and $(-1, -2)$ is 12. Find the

equation of the ellipse. *Hint*: Use the distance formula and the definition of an ellipse.

31. Find an equation for the ellipse with foci $(\pm 2, 0)$ and directrices $x = \pm 5$.

In Exercises 32–35, find the equation of the hyperbola satisfying the given conditions. Write each answer in the form $Ax^2 - By^2 = C$ or in the form $Ay^2 - Bx^2 = C$.

32. Foci $(\pm 6, 0)$; vertices $(\pm 2, 0)$

33. Asymptotes $y = \pm 2x$; foci $(0, \pm 3)$

34. Eccentricity 4; foci $(\pm 3, 0)$

35. Length of the transverse axis 3; eccentricity 5/4; center $(0, 0)$; focal axis horizontal

36. Verify that the point $P(4, -2)$ lies on the hyperbola $3x^2 - 5y^2 = 28$, and compute the lengths F_1P and F_2P of the focal radii.

In Exercises 37–41, graph the parabolas, and in each case specify the vertex, the focus, the directrix, and the focal width.

37. $x^2 = 10y$

38. $x^2 = 5y$

39. $x^2 = -12(y - 3)$

40. $x^2 = -8(y + 1)$

41. $(y - 1)^2 = -4(x - 1)$

In Exercises 42–47, graph the ellipses, and in each case specify the center, the foci, the lengths of the major and minor axes, and the eccentricity.

42. $x^2 + 2y^2 = 4$

43. $4x^2 + 9y^2 = 144$

44. $49x^2 + 9y^2 = 441$

45. $9x^2 + y^2 = 9$

46. $\dfrac{(x - 1)^2}{5^2} + \dfrac{(y + 2)^2}{3^2} = 1$

47. $\dfrac{(x + 3)^2}{3^2} + \dfrac{y^2}{3^2} = 1$

In Exercises 48–53, graph the hyperbolas. In each case specify the center, the vertices, the foci, the equations of the asymptotes, and the eccentricity.

48. $x^2 - 2y^2 = 4$

49. $4x^2 - 9y^2 = 144$

50. $49y^2 - 9x^2 = 441$

51. $9y^2 - x^2 = 9$

52. $\dfrac{(x - 1)^2}{5^2} - \dfrac{(y + 2)^2}{3^2} = 1$

53. $\dfrac{(y + 3)^2}{3^2} - \dfrac{x^2}{3^2} = 1$

In Exercises 54–67, use the technique of completing the square to graph the given equation. If the graph is a parabola, specify the vertex, axis, focus, and directrix. If the graph is an ellipse, specify the center, foci, and lengths of the major and minor axes. If the graph is a hyperbola, specify the center, vertices, foci, and equations of the asymptotes. Finally, if the equation has no graph, say so.

54. $3x^2 + 4y^2 - 6x + 16y + 7 = 0$

55. $y^2 - 16x - 8y + 80 = 0$

56. $y^2 + 4x + 2y - 15 = 0$

57. $16x^2 + 64x + 9y^2 - 54y + 1 = 0$

58. $16x^2 - 32x - 9y^2 + 90y - 353 = 0$

59. $x^2 + 6x - 12y + 33 = 0$

60. $5x^2 + 3y^2 - 40x - 36y + 188 = 0$

61. $x^2 - y^2 - 4x + 2y - 6 = 0$

62. $9x^2 - 90x - 16y^2 + 32y + 209 = 0$

63. $x^2 + 2y - 12 = 0$

64. $y^2 - 25x^2 + 8y - 9 = 0$

65. $x^2 + 16y^2 - 160y + 384 = 0$

66. $16x^2 + 25y^2 - 64x - 100y + 564 = 0$

67. $16x^2 - 25y^2 - 64x + 100y - 36 = 0$

68. Let F_1 and F_2 denote the foci of the hyperbola $5x^2 - 4y^2 = 80$.

(a) Verify that the point P with coordinates $(6, 5)$ lies on the hyperbola.

(b) Compute the quantity $(F_1P - F_2P)^2$.

69. Show that the coordinates of the vertex of the parabola $Ax^2 + Dx + Ey + F = 0$ are given by

$$x = -\frac{D}{2A} \quad \text{and} \quad y = \frac{D^2 - 4AF}{4AE}$$

70. If the equation $Ax^2 + Cy^2 + Dx + Ey + F = 0$ represents an ellipse or a hyperbola, show that the center is the point $(-D/2A, -E/2C)$.

71. The figure shows the parabola $x^2 = 4py$ and a circle with center at the origin and diameter $3p$. If V and F denote the vertex and focus of the parabola, respectively, show that the common chord of the circle and parabola bisects the line segment $\overline{VF}$.

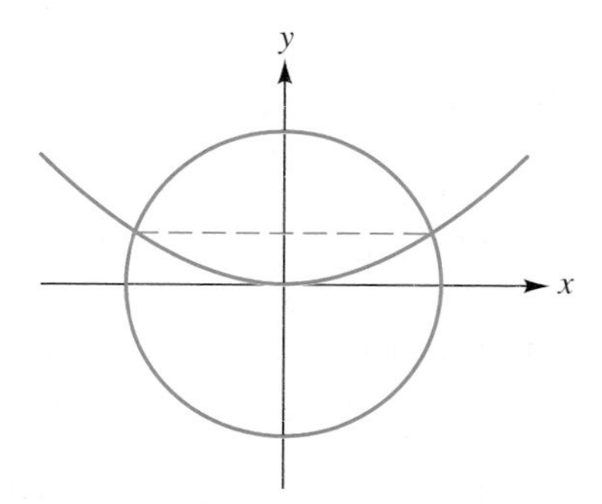

72. The figure on the following page shows an ellipse and a parabola. As is indicated in the figure, the curves are symmetric about the x-axis, and they both have an x-intercept of 5.

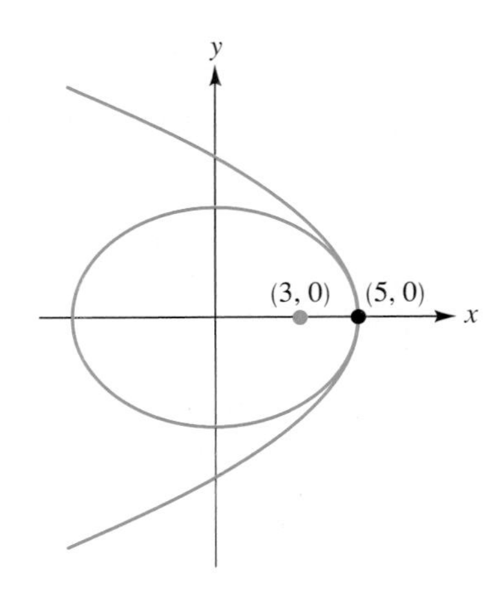

Find the equation of the ellipse and the parabola, given that the point (3, 0) is a focus for both curves.

73. Show that the equation of a line tangent to the circle $(x - h)^2 + (y - k)^2 = r^2$ at the point (a, b) on the circle is

$$(a - h)(x - h) + (b - k)(y - k) = r^2$$

74. The area A of an ellipse $(x^2/a^2) + (y^2/b^2) = 1$ is given by the formula $A = \pi ab$. Use this formula to compute the area of the ellipse $5x^2 + 6y^2 = 60$.

75. In the following figure, V is the vertex of the parabola $y = ax^2 + bx + c$.

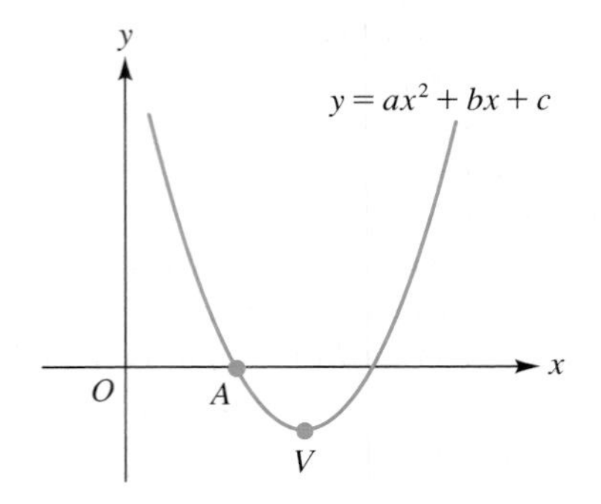

If r_1 and r_2 are the roots of the equation $ax^2 + bx + c = 0$, show that

$$VO^2 - VA^2 = r_1r_2$$

Chapter 11 Test

1. Find the focus and the directrix of the parabola $y^2 = -12x$, and sketch the graph.

2. Graph the hyperbola $x^2 - 4y^2 = 4$. Specify the foci and the asymptotes.

3. The distances from the planet Saturn to the Sun at aphelion and at perihelion are 9.5447 AU and 9.5329 AU, respectively. Compute the eccentricity of the orbit and the length of the semimajor axis. Round each answer to three decimal places.

4. (a) Determine an angle of rotation θ so that there is no $x'y'$-term present when the equation

$$x^2 + 2\sqrt{3}xy + 3y^2 - 12\sqrt{3}x + 12y = 0$$

is transformed to the x'-y' coordinate system.

(b) Graph the equation

$$x^2 + 2\sqrt{3}xy + 3y^2 - 12\sqrt{3}x + 12y = 0$$

5. Determine the angle of inclination for the line $y = (1/\sqrt{3})x - 4$.

6. The foci of an ellipse are $(0, \pm 2)$, and the eccentricity is 1/2. Determine the equation of the ellipse. Write your answer in standard form.

7. Tangents are drawn from the point $(-4, 0)$ to the circle $x^2 + y^2 = 1$. Find the slopes of the tangents.

8. The x-intercept of a line is 2, and its angle of inclination is 60°. Find the equation of the line. Write your answer in the form $Ax + By + C = 0$.

9. Determine the equation of the hyperbola with foci $(\pm 2, 0)$ and with asymptotes $y = \pm(1/\sqrt{3})x$. Write your answer in standard form.

10. Let F_1 and F_2 denote the foci of the hyperbola $5x^2 - 4y^2 = 80$.

(a) Verify that the point $P(6, 5)$ lies on the hyperbola.

(b) Compute the quantity $(F_1P - F_2P)^2$.

11. Graph the ellipse $4x^2 + 25y^2 = 100$. Specify the foci and the lengths of the major and minor axes.

12. Compute the distance from the point $(-1, 0)$ to the line $2x - y - 1 = 0$.

13. Graph the equation $16x^2 + y^2 - 64x + 2y + 65 = 0$.

14. Graph the equation $\dfrac{(x + 4)^2}{3^2} - \dfrac{(y - 4)^2}{1^2} = 1$.

15. Graph the equation $r = 9/(5 - 4\cos\theta)$. Which type of conic is this?

16. Graph the parabola $(x - 1)^2 = 8(y - 2)$. Specify the focal width and the vertex.

17. Consider the ellipse $(x^2/6^2) + (y^2/5^2) = 1$.

(a) What are the equations of the directrices?

(b) If P is a point on the ellipse in Quadrant I such that the x-coordinate of P is 3, compute F_1P and F_2P the lengths of the focal radii.

18. Determine the equation of the line that is tangent to the ellipse $x^2 + 3y^2 = 52$ at the point $(-2, 4)$. Write your answer in the form $y = mx + b$.

19. Find the equation of the line that is tangent to the parabola $x^2 = 2y$ at the point $(4, 8)$. Write your answer in the form $y = mx + b$.

20. The following figure shows the specifications for a cross section of a parabolic reflector. Determine the focal length and the focal ratio of the reflector.

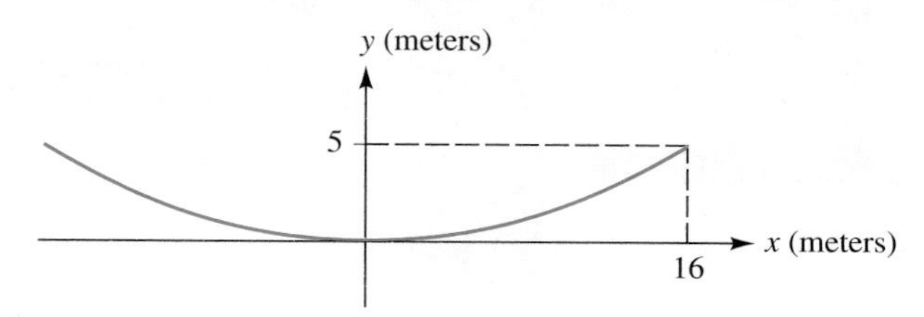

12 CHAPTER Roots of Polynomial Equations

It is necessary that I make some general statements concerning the nature of equations.—René Descartes (1596–1650) in *La Géométrie* (1637)

The polynomial can be derived using Hückel's molecular orbital theory. The roots of the polynomial represent the allowed energy levels of the pi electrons.—Professor Jun-ichi Aihara in his article "Why Aromatic Compounds Are Stable," *Scientific American,* vol. 266 (March 1992), p. 65

A number of problems in computer graphics reduce to finding approximate real roots of quartic and cubic equations in one unknown.—Professor Don Herbison-Evans in his article "Solving Quartics and Cubics for Graphics," http://linus.socs.uts.edu.au/~don/pubs/solving.html

As is indicated by the opening quotations, solving polynomial equations is an old subject with contemporary applications at the cutting edge of technology and science. This chapter continues the work we began in Sections 1.3 and 2.1 on solving polynomial equations. Here are some of the questions you should be able to answer after studying this chapter. See whether you can answer any of them now, on the basis of your previous algebra courses. Come back to this list again when you have finished the chapter, as a way of evaluating your reading.

- What is synthetic division, and how is it used to evaluate a polynomial and test for a root?
- In view of the quadratic formula, we know that every quadratic equation has two roots (although the two roots might be equal). In a similar way, does every cubic equation have three roots, and does every fourth-degree equation have four roots?
- Does every equation have a root?
- What does the fundamental theorem of algebra assert?
- The quadratic formula gives us the roots of a quadratic equation in terms of the coefficients. Is there a similar formula for cubic equations? What about a general formula for a polynomial equation of degree n?
- How can we determine whether an equation has rational roots?
- How can the table feature on a graphing utility be used to locate irrational roots? How can the ideas about iteration from Section 3.5 be used to locate irrational roots?
- Without a graphing utility and without solving the equation, can we tell whether the roots will be positive, negative, or complex?
- Is it always possible to factor a polynomial?

12.1 THE COMPLEX NUMBER SYSTEM

In the following I shall denote the expression $\sqrt{-1}$ *by the letter i so that* $ii = -1$.—Leonhard Euler in a paper presented to the Saint Petersburg Academy in 1777.

In elementary algebra, complex numbers appear as expressions $a + bi$, *where a and b are ordinary real numbers and* $i^2 = -1$. . . . *Complex numbers are manipulated by the usual rules of algebra, with the convention that* i^2 *is to be replaced by* -1 *whenever it occurs.*—Ralph Boas in *Invitation to Complex Analysis* (New York: Random House, 1987)

The preceding quotation from Professor Boas summarizes the basic approach we will follow in this section. Near the end of the section, after you've become accustomed to working with complex numbers, we'll present a formal list of some of the basic definitions and properties that can be used to develop the subject more rigorously, as is required in more advanced courses.

When we solve equations in this chapter, you'll see instances in which the real-number system proves to be inadequate. In particular, since the square of a real number is never negative, there is no real number x such that $x^2 = -1$. To overcome this inconvenience, mathematicians define the symbol i by the equation

$$\boxed{i^2 = -1}$$

For reasons that are more historical than mathematical, i is referred to as the **imaginary unit.** This name is unfortunate in a sense, because to an engineer or a mathematician, i is neither less "real" nor less tangible than any real number. Having said this, however, we do have to admit that i does not belong to the real-number system.

Algebraically, we operate with the symbol i as if it were any letter in a polynomial expression. However, when we see i^2, we must remember to replace it by -1. Here are four sample calculations involving i:

1. $3i + 2i = 5i$
2. $-2i^2 + 6i = -2(-1) + 6i = 2 + 6i$
3. $(-i)^2 = i^2 = -1$
4. $0i = 0$

An expression of the form $a + bi$, where a and b are real numbers, is called a **complex number.*** Four examples of complex numbers are:

$$2 + 3i$$

$$4 + (-5)i \qquad \text{(usually written } 4 - 5i\text{)}$$

$$1 - \sqrt{2}i \qquad \text{(also written } 1 - i\sqrt{2}\text{)}$$

$$\frac{1}{2} + \frac{3}{2}i \qquad \left(\text{also written } \frac{1 + 3i}{2}\right)$$

Given a complex number $a + bi$, we say that a is the **real part** of $a + bi$ and b is the **imaginary part** of $a + bi$. For example, the real part of $3 - 4i$ is 3, and the imaginary part is -4. Observe that both the real part and the imaginary part of a complex number are themselves real numbers.

*The term "complex number" is attributed to Carl Friedrich Gauss (1777–1855). The term "imaginary number" originated with René Descartes (1596–1650).

We define the notion of *equality* for complex numbers in terms of their real and imaginary parts. Two complex numbers are said to be **equal** if their corresponding real and imaginary parts are equal. We can write this definition symbolically as follows:

$$a + bi = c + di \quad \text{if and only if} \quad a = c \quad \text{and} \quad b = d$$

EXAMPLE 1 Equality of complex numbers

Determine the real numbers c and d such that $10 + 4i = 2c + di$.

SOLUTION

Equating the real parts of the two complex numbers gives us $2c = 10$, and therefore $c = 5$. Similarly, equating the imaginary parts yields $d = 4$. These are the required values for c and d.

As the next example indicates, addition, subtraction, and multiplication of complex numbers are carried out by using the usual rules of algebra, with the understanding (as was mentioned before) that i^2 is always to be replaced with -1. (We'll discuss division of complex numbers subsequently.)

EXAMPLE 2 Operations with complex numbers

Let $z = 2 + 5i$ and $w = 3 - 4i$. Compute each of the following:
(a) $w + z$; **(b)** $3z$; **(c)** $w - 3z$; **(d)** zw; **(e)** wz.

SOLUTION

(a) $w + z = (3 - 4i) + (2 + 5i) = (3 + 2) + (-4i + 5i) = 5 + i$
(b) $3z = 3(2 + 5i) = 6 + 15i$
(c) $w - 3z = (3 - 4i) - (6 + 15i) = (3 - 6) + (-4i - 15i) = -3 - 19i$
(d) $zw = (2 + 5i)(3 - 4i) = 6 - 8i + 15i - 20i^2$
$= 6 + 7i - 20(-1) = 26 + 7i$
(e) $wz = (3 - 4i)(2 + 5i) = 6 + 15i - 8i - 20i^2$
$= 6 + 7i - 20(-1) = 26 + 7i$

If you look over the result of part (a) in Example 2, you can see that the sum is obtained simply by adding the corresponding real and imaginary parts of the given numbers. Likewise, in part (c) the difference is obtained by subtracting the corresponding real and imaginary parts. In part (d), however, notice that the product is *not* obtained in a similar fashion. (Exercise 68 at the end of this section provides some perspective on this.) Finally, notice that the results in parts (d) and (e) are identical. In fact, it can be shown that for any two complex numbers z and w, we always have $zw = wz$. In other words, just as with real numbers, multiplication of complex numbers is *commutative.* Furthermore, along the same lines, it can be shown that all of the properties of real numbers listed in Appendix A.2 continue to hold for complex numbers. We'll return to this point at the end of this section and in the exercises.

As background for the discussion of division of complex numbers, we introduce the notion of a *complex conjugate,* or simply a *conjugate.**

DEFINITION **Complex Conjugate**

Let $z = a + bi$. The **complex conjugate** of z, denoted by $\bar{z}$, is defined by

$$\bar{z} = a - bi$$

EXAMPLES

If $z = 3 + 4i$, then $\bar{z} = 3 - 4i$.

If $w = 9 - 2i$, then $\bar{w} = 9 + 2i$.

EXAMPLE 3 **Computations involving conjugates**

(a) If $z = 6 - 3i$, compute $z\bar{z}$. **(b)** If $w = a + bi$, compute $w\bar{w}$.

SOLUTION

(a)
$$\begin{aligned} z\bar{z} &= (6 - 3i)(6 + 3i) \\ &= 36 + 18i - 18i - 9i^2 \\ &= 36 + 9 = 45 \end{aligned}$$

(b)
$$\begin{aligned} w\bar{w} &= (a + bi)(a - bi) \\ &= a^2 - abi + abi - b^2i^2 \\ &= a^2 + b^2 \end{aligned}$$

NOTE The result in part (b) shows that the product of a complex number and its conjugate is always a nonnegative real number.

Quotients of complex numbers are usually computed by using conjugates. Suppose, for example, that we wish to compute the quotient

$$\frac{5 - 2i}{3 + 4i}$$

and express the quotient in the form $a + bi$, where a and b are real numbers. To do this, we take the conjugate of the denominator, namely, $3 - 4i$, and then multiply the given fraction by $\frac{3 - 4i}{3 - 4i}$, which equals 1. This yields

$$\begin{aligned} \frac{5 - 2i}{3 + 4i} &= \frac{5 - 2i}{3 + 4i} \cdot \frac{3 - 4i}{3 - 4i} \\ &= \frac{15 - 26i + 8i^2}{9 - 16i^2} \\ &= \frac{7 - 26i}{25} && \text{since } 8i^2 = -8 \text{ and } -16i^2 = 16 \\ &= \frac{7}{25} - \frac{26}{25}i && \text{as required} \end{aligned}$$

*The term "conjugates" *(conjuguées)* was introduced by the nineteenth-century French mathematician A. L. Cauchy in his text *Cours d'Analyse Algébrique* (Paris: 1821).

In the box that follows, we summarize our procedure for computing quotients. The condition $w \neq 0$ means that w is any complex number other than $0 + 0i$.

Procedure for Computing Quotients

Let z and w be two complex numbers, $w \neq 0$. Then z/w is computed as follows:

$$\frac{z}{w} = \frac{z}{w} \cdot \frac{\overline{w}}{\overline{w}}$$

NOTE For any nonzero complex number w, $w\overline{w}$ is a nonzero real number. So multiplying $\frac{z}{w}$ by $\frac{\overline{w}}{\overline{w}}$ and "grouping" as $\frac{z\overline{w}}{w\overline{w}}$ gives us a real-number denominator.

EXAMPLE 4 Using conjugates to carry out division

Let $z = 3 + 4i$ and $w = 1 - 2i$. Compute each of the following quotients, and express your answer in the form $a + bi$, where a and b are real numbers.

(a) $\frac{1}{z}$ **(b)** $\frac{z}{w}$

SOLUTION

(a)
$$\begin{aligned}\frac{1}{z} &= \frac{1}{z} \cdot \frac{\overline{z}}{\overline{z}} = \frac{1}{3 + 4i} \cdot \frac{3 - 4i}{3 - 4i}\\ &= \frac{3 - 4i}{9 - 16i^2} = \frac{3 - 4i}{25}\\ &= \frac{3}{25} - \frac{4}{25}i\end{aligned}$$

(b)
$$\begin{aligned}\frac{z}{w} &= \frac{z}{w} \cdot \frac{\overline{w}}{\overline{w}} = \frac{3 + 4i}{1 - 2i} \cdot \frac{1 + 2i}{1 + 2i}\\ &= \frac{3 + 10i + 8i^2}{1 - 4i^2} = \frac{-5 + 10i}{5}\\ &= -1 + 2i\end{aligned}$$

We began this section by defining i with the equation $i^2 = -1$. This can be rewritten as

$$i = \sqrt{-1}$$

provided that we agree to certain conventions regarding principal square roots and negative numbers. In dealing with the principal square root of a negative real number, say, $\sqrt{-5}$, we shall write

$$\sqrt{-5} = \sqrt{(-1)(5)} = \sqrt{-1}\sqrt{5} = i\sqrt{5}$$

In other words, we are allowing the use of the rule $\sqrt{ab} = \sqrt{a}\sqrt{b}$ when a is -1 and b is a positive real number. However, the rule $\sqrt{ab} = \sqrt{a}\sqrt{b}$ *cannot* be used when both a and b are negative. If that were allowed, we could write

$$1 = (-1)(-1)$$

and then

$$\sqrt{1} = \sqrt{(-1)(-1)} = \sqrt{-1}\sqrt{-1} = (i)(i)$$

Consequently,

$$1 = i^2 \qquad \text{and therefore} \qquad 1 = -1$$

which is a contradiction. Again, the point here is that the rule $\sqrt{ab} = \sqrt{a}\sqrt{b}$ cannot be applied when both a and b are negative.

EXAMPLE 5 Calculations involving the imaginary unit *i*

Simplify:
(a) i^4; **(b)** i^{101}; **(c)** $\sqrt{-12} + \sqrt{-27}$; **(d)** $\sqrt{-9}\sqrt{-4}$.

SOLUTION

(a) We make use of the defining equation for i, which is $i^2 = -1$. Thus we have

$$i^4 = (i^2)^2 = (-1)^2 = 1$$

(The result, $i^4 = 1$, is worth remembering.)

(b) $i^{101} = i^{100}i = (i^4)^{25}i = 1^{25}i = i$

(c) $$\begin{aligned}\sqrt{-12} + \sqrt{-27} &= \sqrt{12}\sqrt{-1} + \sqrt{27}\sqrt{-1}\\ &= \sqrt{4}\sqrt{3}i + \sqrt{9}\sqrt{3}i\\ &= 2\sqrt{3}i + 3\sqrt{3}i = 5\sqrt{3}i\end{aligned}$$

(d) $\sqrt{-9}\sqrt{-4} = (3i)(2i) = 6i^2 = -6$

Note: $\sqrt{-9}\sqrt{-4} \neq \sqrt{36}$ (Why?)

We have developed the complex-number system in this section for the same reason that complex numbers were developed historically: They're needed to solve polynomial equations. Consider, for example, the quadratic equation

$$ax^2 + bx + c = 0$$

where a, b, and c are real numbers and $a \neq 0$. In Section 2.1 we noted that if the discriminant $b^2 - 4ac$ is negative, then the equation has no real roots. As the next example indicates, such equations do have two complex-number roots. Furthermore (assuming that the coefficients a, b, and c in the equation are real numbers), these roots always turn out to be complex conjugates.

EXAMPLE 6 A quadratic equation with complex roots

Solve the quadratic equation $x^2 - 4x + 6 = 0$.

SOLUTION

We use the quadratic formula with $a = 1$, $b = -4$, and $c = 6$:

$$\begin{aligned}x = \frac{-b \pm \sqrt{b^2 - 4ac}}{2a} &= \frac{4 \pm \sqrt{16 - 4(1)(6)}}{2(1)}\\ &= \frac{4 \pm \sqrt{-8}}{2} = \frac{4 \pm (2\sqrt{2})i}{2} \qquad \text{Check the algebra.}\\ &= 2 \pm \sqrt{2}i\end{aligned}$$

In summary, the two roots of the quadratic equation are the complex numbers $2 + \sqrt{2}i$ and $2 - \sqrt{2}i$. These numbers are complex conjugates.

NOTE When you write answers such as these on paper, be sure you make it clear that the symbol i is outside (not inside) the radical sign. To emphasize this distinction, sometimes it's helpful to write the roots in the form $2 + i\sqrt{2}$ and $2 - i\sqrt{2}$.

Near the beginning of this section we mentioned that we would eventually present a formal list of some of the basic definitions and properties that can be used to develop the subject more rigorously. These are given in the two boxes that follow. As you'll see in some of the exercises, these definitions and properties are indeed consistent with our work in this section and with the properties of real numbers listed in Appendix A.2.

DEFINITION **Addition, Subtraction, Multiplication, and Division for Complex Numbers**

Let $z = a + bi$ and $w = c + di$. Then $z + w$, $z - w$, and zw are defined as follows:

1. $z + w = (a + bi) + (c + di) = (a + c) + (b + d)i$
2. $z - w = (a + bi) - (c + di) = (a - c) + (b - d)i$
3. $zw = (a + bi)(c + di) = (ac - bd) + (ad + bc)i$

Furthermore, if $w \neq 0$, then z/w is defined as follows:

4. $\dfrac{z}{w} = \dfrac{a + bi}{c + di} \cdot \dfrac{c - di}{c - di} = \left(\dfrac{ac + bd}{c^2 + d^2}\right) + \left(\dfrac{bc - ad}{c^2 + d^2}\right)i$

PROPERTY SUMMARY **Properties of Complex Conjugates**

If $z = a + bi$, the complex conjugate, $\overline{z} = a - bi$, has the following properties.

1. $\overline{\overline{z}} = z$
2. $z = \overline{z}$ if and only if z is a real number
3. $\overline{z + w} = \overline{z} + \overline{w}$; $\overline{z - w} = \overline{z} - \overline{w}$
4. $\overline{zw} = \overline{z}\,\overline{w}$; $\dfrac{\overline{z}}{\overline{w}} = \overline{\left(\dfrac{z}{w}\right)}$
5. $(\overline{z})^n = \overline{z^n}$ for each natural number n

EXERCISE SET 12.1 

A

In any computation involving complex numbers, express your answer in the form $a + bi$, where a and b are real numbers. If a, or b, or both are zero, then simplify further.

1. Complete the table.

i^2	i^3	i^4	i^5	i^6	i^7	i^8
-1						

2. Simplify the following expression, and write the answer in the form $a + bi$.

$$1 + 3i - 5i^2 + 4 - 2i - i^3$$

For Exercises 3 and 4, specify the real and imaginary parts of each complex number.

3. (a) $4 + 5i$
(b) $4 - 5i$
(c) $\frac{1}{2} - i$
(d) $16i$

4. (a) $-2 + \sqrt{7}i$
(b) $1 + 5^{1/3}i$
(c) $-3i$
(d) 0

5. Determine the real numbers c and d such that

$$8 - 3i = 2c + di$$

6. Determine the real numbers a and b such that

$$27 - 64i = a^3 - b^3i$$

7. Simplify each of the following.
(a) $(5 - 6i) + (9 + 2i)$ **(b)** $(5 - 6i) - (9 + 2i)$

8. If $z = 1 + 4i$, compute $z - 10i$.

9. Compute each of the following.

(a) $(3 - 4i)(5 + i)$
(b) $(5 + i)(3 - 4i)$
(c) $\dfrac{3 - 4i}{5 + i}$
(d) $\dfrac{5 + i}{3 - 4i}$

10. Compute each of the following.

(a) $(2 + 7i)(2 - 7i)$ **(b)** $\dfrac{-1 + 3i}{2 + 7i}$ **(c)** $\dfrac{1}{2 + 7i}$ **(d)** $\dfrac{1}{2 + 7i} \cdot (-1 + 3i)$

In Exercises 11–36, evaluate each expression using the values $z = 2 + 3i$, $w = 9 - 4i$, and $w_1 = -7 - i$.

11. (a) $z + w$ **(b)** $\bar{z} + w$ **(c)** $z + \bar{z}$

12. (a) $\bar{z} + \bar{w}$ **(b)** $\overline{(z + w)}$ **(c)** $w - \bar{w}$

13. $(z + w) + w_1$ **14.** $z + (w + w_1)$ **15.** zw

16. wz **17.** $z\bar{z}$ **18.** $w\bar{w}$

19. $z(ww_1)$ **20.** $(zw)w_1$ **21.** $z(w + w_1)$

22. $zw + zw_1$ **23.** $z^2 - w^2$ **24.** $(z - w)(z + w)$

25. $(zw)^2$ **26.** z^2w^2 **27.** z^3

28. z^4 **29.** z/w **30.** w/z

31. $\bar{z}/\bar{w}$ **32.** $\overline{(z/w)}$ **33.** $z/\bar{z}$

34. $\bar{z}/z$ **35.** $(w - \bar{w})/(2i)$ **36.** $(w + \bar{w})/2$

For Exercises 37–40, compute each quotient.

37. $\dfrac{i}{5 + i}$ **38.** $\dfrac{1 - i\sqrt{3}}{1 + i\sqrt{3}}$ **39.** $\dfrac{1}{i}$ **40.** $\dfrac{i + i^2}{i^3 + i^4}$

In Exercises 41–44, use the fact that $i^4 = 1$ to simplify each expression [as in Example 5(b)].

41. i^{17} **42.** i^{36} **43.** i^{26} **44.** i^{83}

In Exercises 45–52, simplify each expression.

45. $\sqrt{-49} + \sqrt{-9} + \sqrt{-4}$

46. $\sqrt{-25} + i$

47. $\sqrt{-20} - 3\sqrt{-45} + \sqrt{-80}$

48. $\sqrt{-4}\sqrt{-4}$

49. $1 + \sqrt{-36}\sqrt{-36}$ **50.** $i - \sqrt{-100}$

51. $3\sqrt{-128} - 4\sqrt{-18}$ **52.** $64 + \sqrt{-64}\sqrt{-64}$

In Exercises 53–60:

(a) *Compute the discriminant of the quadratic and note that it is negative (and therefore the equation has no real-number roots).*

(b) *Use the quadratic formula to obtain the two complex-conjugate roots of each equation.*

53. $x^2 - x + 1 = 0$ **54.** $x^2 - 6x + 12 = 0$

55. $5z^2 + 2z + 2 = 0$ **56.** $-10z^2 + 4z - 2 = 0$

57. $2z^2 + 3z + 4 = 0$ **58.** $3z^2 - 7z + 5 = 0$

59. $\frac{1}{6}z^2 - \frac{1}{4}z + 1 = 0$ **60.** $\frac{1}{2}z^2 + 2z + \frac{9}{4} = 0$

61. (a) Evaluate the expression $x^2 - 4x + 6$ when $x = 2 - i\sqrt{2}$.

(b) How does your result in part (a) relate to Example 6 in this section?

62. (a) Find the roots of the quadratic equation $x^2 - 2x + 5 = 0$.

(b) Compute the product of the two roots that you obtained in part (a).

(c) Check your answer in part (b) by applying the theorem in the box on page 84.

63. Let $z = a + bi$ and $w = c + di$. Compute each of the following quantities, and then check that your results agree with the definitions in the first box on page 904.

(a) $z + w$ **(b)** $z - w$ **(c)** zw **(d)** $\dfrac{z}{w}$

B

64. Show that $\left(\dfrac{-1 + i\sqrt{3}}{2}\right)^2 + \left(\dfrac{-1 - i\sqrt{3}}{2}\right)^2 = -1.$

65. Let $z = \dfrac{-1 + i\sqrt{3}}{2}$ and $w = \dfrac{-1 - i\sqrt{3}}{2}$. Verify the following statements.

(a) $z^3 = 1$ and $w^3 = 1$

(b) $zw = 1$

(c) $z = w^2$ and $w = z^2$

(d) $(1 - z + z^2)(1 + z - z^2) = 4$

66. Let $z = a + bi$ and $w = c + di$.

(a) Show that $\bar{\bar{z}} = z$.

(b) Show that $\overline{(z + w)} = \bar{z} + \bar{w}$.

67. Show that the complex number 0 $(=0 + 0i)$ has the following properties.

(a) $0 + z = z$ and $z + 0 = z$, for all complex numbers z.
Hint: Let $z = a + bi$.

(b) $0 \cdot z = 0$ and $z \cdot 0 = 0$, for all complex numbers z.

68. This exercise indicates one of the reasons why multiplication of complex numbers is not carried out simply by multiplying the corresponding real and imaginary parts of the numbers. (Recall that addition and subtraction *are* carried out in this manner.) Suppose for the moment that we were to define multiplication in this seemingly less complicated way:

$$(a + bi)(c + di) = ac + (bd)i \qquad (*)$$

(a) Compute $(2 + 3i)(5 + 4i)$, assuming that multiplication is defined by (*).

(b) Still assuming that multiplication is defined by (*), find two complex numbers z and w such that $z \neq 0$, $w \neq 0$, but $zw = 0$ (where 0 denotes the complex number $0 + 0i$).

Now notice that the result in part (b) is contrary to our expectation or desire that the product of two nonzero numbers be nonzero, as is the case for real numbers. On the other hand, it can be shown that when multiplication is carried out as described in the text, then the product of two complex numbers is nonzero if and only if both factors are nonzero.

69. (a) Show that addition of complex numbers is commutative. That is, show that $z + w = w + z$ for all complex numbers z and w. *Hint*: Let $z = a + bi$ and $w = c + di$.

(b) Show that multiplication of complex numbers is commutative. That is, show that $zw = wz$ for all complex numbers z and w.

70. Let $z = a + bi$.
(a) Show that $(\bar{z})^2 = \overline{z^2}$.
(b) Show that $(\bar{z})^3 = \overline{z^3}$.

In Exercises 71–74, find all roots of each equation. Hints: First, factor by grouping. In Exercises 71 and 72 each equation has three roots; in Exercise 73 the equation has six roots; in Exercise 74 there are five roots.

71. $x^3 - 3x^2 + 4x - 12 = 0$
72. $2x^3 + 4x^2 + 3x + 6 = 0$
73. $x^6 - 9x^4 + 16x^2 - 144 = 0$
74. $x^5 + 4x^3 + 8x^2 + 32 = 0$

C

75. Let a and b be real numbers. Find the real and imaginary parts of the quantity

$$\frac{a + bi}{a - bi} + \frac{a - bi}{a + bi}$$

76. Find the real and imaginary parts of the quantity

$$\left(\frac{a + bi}{a - bi}\right)^2 - \left(\frac{a - bi}{a + bi}\right)^2$$

77. Find the real part of $\dfrac{(a + bi)^2}{a - bi} - \dfrac{(a - bi)^2}{a + bi}$.

78. **(a)** Let $\alpha = \left(\dfrac{\sqrt{a^2 + b^2} + a}{2}\right)^{1/2}$

and let $\beta = \left(\dfrac{\sqrt{a^2 + b^2} - a}{2}\right)^{1/2}$.

Show that the square of the complex number $\alpha + \beta i$ is $a + bi$.

(b) Use the result in part (a) to find a complex number z such that $z^2 = i$.

(c) Use the result in part (a) to find a complex number z such that $z^2 = -7 + 24i$.

One of the basic properties of real numbers is that if a product is equal to zero, then at least one of the factors is zero. Exercises 79 and 80 show that this property also holds for complex numbers. For Exercises 79 and 80, assume that $zw = 0$, where $z = a + bi$ and $w = c + di$.

79. If $a \neq 0$, prove that $w = 0$. (That is, prove that $c = d = 0$.)

80. If $b \neq 0$, prove that $w = 0$. (That is, prove that $c = d = 0$.)

MINI PROJECT — A GEOMETRIC INTERPRETATION OF COMPLEX ROOTS

As you know, if a quadratic equation $f(x) = x^2 + Bx + C = 0$ has two real roots, there is a geometric interpretation: The roots are the x-intercepts for the graph of f. Less well known is the fact that there is a geometric interpretation when the quadratic equation has nonreal complex roots. If these roots are $a \pm bi$, where a and b are real numbers, then the coordinates of the vertex of the graph of f are (a, b^2).

(a) Check that this is true in the case of the quadratic equation $f(x) = x^2 + 6x + 25 = 0$.

(b) Show that the result holds in general. That is, assume that the roots of $f(x) = x^2 + Bx + C = 0$ are $a \pm bi$, where a and b are real numbers with $b \neq 0$. Show that the vertex of the graph of f is (a, b^2). Hint: Use the theorem in the box on page 84 to express B and C in terms of a and b.

12.2 DIVISION OF POLYNOMIALS

Although the process of long division for polynomials is often taught in elementary algebra courses, it usually does not receive sufficient emphasis there. As with ordinary long division for numbers, the process is best learned by first watching someone do examples and then practicing on your own. The terms "quotient," "remainder," "divisor," and "dividend" will be used here in the same way they are used in ordinary division of numbers. For instance, when 7 is divided by 2, the quotient is 3 and the remainder is 1. We write this

$$\begin{array}{rll} & 3 & \text{quotient} \\ \text{divisor} \quad 2 & \overline{)7} & \text{dividend} \\ & \underline{6} & \\ & 1 & \text{remainder} \end{array}$$

or

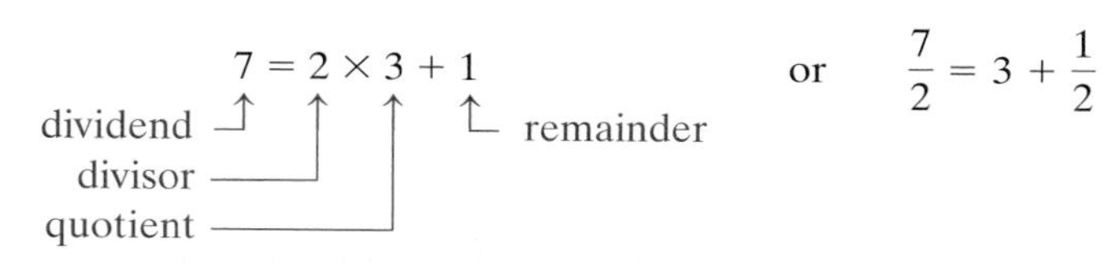

The process of long division for polynomials follows the same four-step cycle that is used in ordinary long division of numbers: Divide, multiply, subtract, bring down. As a first example, we divide $2x^2 - 7x + 8$ by $x - 2$. Notice that in setting up the division, we write both the dividend and the divisor in decreasing powers of x.

$$\begin{array}{r} 2x \quad - 3 \\ x - 2 \overline{)\, 2x^2 - 7x + 8} \\ \underline{2x^2 - 4x} \qquad \\ -3x + 8 \\ \underline{-3x + 6} \\ 2 \end{array}$$

1. Divide the first term of the dividend by the first term of the divisor: $\dfrac{2x^2}{x} = 2x$. The result becomes the first term of the quotient, as shown.
2. Multiply the divisor $x - 2$ by the term $2x$ obtained in the previous step. This yields the quantity $2x^2 - 4x$, which is written below the dividend, as shown.
3. From the quantity $2x^2 - 7x$ in the dividend, subtract the quantity $2x^2 - 4x$. This yields $-3x$.
4. Bring down the $+8$ in the dividend, as shown. The resulting quantity, $-3x + 8$, is now treated as the dividend and the entire process is repeated.

We've now found that when $2x^2 - 7x + 8$ is divided by $x - 2$, the quotient is $2x - 3$ and the remainder is 2. This is summarized by writing either

$$\frac{2x^2 - 7x + 8}{x - 2} = 2x - 3 + \frac{2}{x - 2} \tag{1}$$

or, after multiplying through by $x - 2$,

$$\underbrace{2x^2 - 7x + 8}_{\text{dividend}} = \underbrace{(x - 2)}_{\text{divisor}}\underbrace{(2x - 3)}_{\text{quotient}} + \underbrace{2}_{\text{remainder}} \tag{2}$$

There are two observations to be made here. First, notice that equation (2) is valid for all real numbers x, whereas equation (1) carries the implicit restriction that x may not equal 2. For this reason we often prefer to write our results in the form of equation (2). Second, notice that the degree of the remainder is less than the degree of the divisor. This is very similar to the situation with ordinary division of positive integers, in which the remainder is always less than the divisor.

As another example of the long division process, we divide $3x^4 - 2x^3 + 2$ by $x^2 - 1$. Notice in what follows that we have inserted the terms in the divisor and dividend that have coefficients of zero. These terms act as place holders.

$$
\begin{array}{r|l}
 & 3x^2 - 2x + 3 \\
\hline
x^2 + 0x - 1 & 3x^4 - 2x^3 + 0x^2 + 0x + 2 \\
 & \underline{3x^4 + 0x^3 - 3x^2} \\
 & \quad - 2x^3 + 3x^2 + 0x \\
 & \quad \underline{- 2x^3 + 0x^2 + 2x} \\
 & \qquad\qquad 3x^2 - 2x + 2 \\
 & \qquad\qquad \underline{3x^2 - 0x + 3} \\
 & \qquad\qquad\quad - 2x + 5
\end{array}
$$

We can write this result as

$$\frac{3x^4 - 2x^3 + 2}{x^2 - 1} = 3x^2 - 2x + 3 + \frac{-2x + 5}{x^2 - 1}$$

or, multiplying through by $x^2 - 1$,

$$\underbrace{3x^4 - 2x^3 + 2}_{\text{dividend}} = \underbrace{(x^2 - 1)}_{\text{divisor}}\underbrace{(3x^2 - 2x + 3)}_{\text{quotient}} + \underbrace{(-2x + 5)}_{\text{remainder}}$$

Notice that this last equation holds for all values of x, whereas the previous equation carries the restrictions that x may be neither 1 nor -1. Also, as in our previous example, observe that the degree of the remainder is less than that of the divisor.

There is a theorem, commonly referred to as the **division algorithm,** that summarizes rather nicely the key results of the long division process. We state the theorem here without proof.

The Division Algorithm

Let $p(x)$ and $d(x)$ be polynomials, and assume that $d(x)$ is not the zero polynomial. Then there are unique polynomials $q(x)$ and $R(x)$ such that

$$p(x) = d(x) \cdot q(x) + R(x)$$

where either $R(x)$ is the zero polynomial or the degree of $R(x)$ is less than the degree of $d(x)$.

The polynomials $p(x)$, $d(x)$, $q(x)$, and $R(x)$ are referred to as the *dividend, divisor, quotient,* and *remainder,* respectively. When $R(x) = 0$, we have $p(x) = d(x) \cdot q(x)$, and we say that $d(x)$ and $q(x)$ are **factors** of $p(x)$. Also, since $d(x)$ is not the zero polynomial, notice that the equation $p(x) = d(x) \cdot q(x) + R(x)$ implies that the degree of $q(x)$ is less than or equal to the degree of $p(x)$. (Why?)

EXAMPLE 1 Using long division to find a quotient and remainder

Let $p(x) = x^3 + 2x^2 - 4$ and $d(x) = x - 3$. Use the long division process to find the polynomials $q(x)$ and $R(x)$ such that

$$p(x) = d(x) \cdot q(x) + R(x)$$

where either $R(x) = 0$ or the degree of $R(x)$ is less than the degree of $d(x)$.

SOLUTION

After inserting the term $0x$ in the dividend $p(x)$, we use long division to divide $p(x)$ by $d(x)$:

$$\begin{array}{r|l} & x^2 + 5x + 15 \\ \hline x - 3 & x^3 + 2x^2 + 0x - 4 \\ & \underline{x^3 - 3x^2} \\ & \quad 5x^2 + 0x \\ & \quad \underline{5x^2 - 15x} \\ & \qquad 15x - 4 \\ & \qquad \underline{15x - 45} \\ & \qquad\qquad 41 \end{array}$$

We now have

$$\underbrace{x^3 + 2x^2 - 4}_{p(x)} = \underbrace{(x - 3)}_{d(x)}\underbrace{(x^2 + 5x + 15)}_{q(x)} + \underbrace{41}_{R(x)}$$

Thus $q(x) = x^2 + 5x + 15$ and $R(x) = 41$. Notice that the degree of $R(x)$ is less than the degree of $d(x)$.

The long division procedure for polynomials can be streamlined when the divisor is of the form $x - r$. This shortened version, known as **synthetic division,** will be useful in subsequent sections when we are solving polynomial equations.

We can explain the idea behind synthetic division by using the long division carried out in Example 1. The basic idea is that in the long division process, it is the *coefficients* of the various polynomials that carry all the necessary information. In our example, for instance, the quotient and remainder can be abbreviated by writing down a sequence of four numbers:

$$1 \quad 5 \quad 15 \quad 41$$

By studying the long division process, you will find that these numbers are obtained through the following four steps:

Step 1 Write down the first coefficient of the dividend. This will be the first coefficient of the quotient. *Result* $\boxed{1}$

Step 2 Multiply the 1 obtained in the previous step by the -3 in the divisor. Then subtract the result from the second coefficient of the dividend:

$$-3 \times 1 = -3 \qquad 2 - (-3) = 5 \qquad \textit{Result } \boxed{5}$$

Step 3 Multiply the 5 obtained in the previous step by the -3 in the divisor. Then subtract the result from the third coefficient of the dividend:

$$-3 \times 5 = -15 \qquad 0 - (-15) = 15 \qquad \textit{Result } \boxed{15}$$

Step 4 Multiply the 15 obtained in the previous step by the -3 in the divisor. Then subtract the result from the fourth coefficient of the dividend:

$$-3 \times 15 = -45 \qquad -4 - (-45) = 41 \qquad \textit{Result } \boxed{41}$$

A convenient format for setting up this process involves writing the constant term of the divisor and the coefficients of the dividend as follows:

$$\begin{array}{r|rrrr} -3 & 1 & 2 & 0 & -4 \\ \hline & & & & \end{array}$$

Now, using this format, let's again go through the four steps we have just described:

Step 1 Bring down the 1.

$$\begin{array}{r|rrrr} -3 & 1 & 2 & 0 & -4 \\ & & & & \\ \hline & 1 & & & \end{array}$$

Step 2 $-3 \times 1 = -3$
$2 - (-3) = 5$

$$\begin{array}{r|rrrr} -3 & 1 & 2 & 0 & -4 \\ & & -3 & & \\ \hline & 1 & 5 & & \end{array}$$

Step 3 $-3 \times 5 = -15$
$0 - (-15) = 15$

$$\begin{array}{r|rrrr} -3 & 1 & 2 & 0 & -4 \\ & & -3 & -15 & \\ \hline & 1 & 5 & 15 & \end{array}$$

Step 4 $-3 \times 15 = -45$
$-4 - (-45) = 41$

$$\begin{array}{r|rrrr} -3 & 1 & 2 & 0 & -4 \\ & & -3 & -15 & -45 \\ \hline & 1 & 5 & 15 & 41 \end{array}$$

We have now obtained the required sequence of coefficients, 1 5 15 41, but there is one further significant simplification that can be made. In Steps 2 through 4, if we use 3 instead of -3 in the initial format, we can *add* instead of *subtract* (why?), ideally minimizing arithmetic errors. (You will see the motivation for this in the next section when we discuss the remainder theorem.) We illustrate our *simplified* version of synthetic division in the following box. The method is applicable for any polynomial division in which the divisor has the form $x - r$.

To Divide $x^3 + 2x^2 - 4$ by $x - 3$ Using the Simplified Version of Synthetic Division

		COMMENTS
Format	$\begin{array}{r\|rrrr} 3 & 1 & 2 & 0 & -4 \\ \hline \end{array}$	Since the divisor is $x - 3$, the format begins with 3. The coefficients from the dividend are written in the order corresponding to decreasing powers of x. A zero coefficient is inserted as a place holder.
Procedure	$\begin{array}{r\|rrrr} 3 & 1 & 2 & 0 & -4 \\ & & 3 & 15 & 45 \\ \hline & 1 & 5 & 15 & 41 \end{array}$	**Step 1** Bring down the 1. **Step 2** $3 \times 1 = 3$; $2 + 3 = 5$ **Step 3** $3 \times 5 = 15$; $0 + 15 = 15$ **Step 4** $3 \times 15 = 45$; $-4 + 45 = 41$
Answer	Quotient: $x^2 + 5x + 15$ Remainder: 41	The degree of the first term in the quotient is one less than the degree of the first term of the dividend.

EXAMPLE 2 Synthetic division

Use synthetic division to divide $x^3 - 6x + 4$ by $x - 2$.

SOLUTION

$$\begin{array}{r|rrrr} 2 & 1 & 0 & -6 & 4 \\ & & 2 & 4 & -4 \\ \hline & 1 & 2 & -2 & 0 \end{array}$$

Looking at the third row of numbers in the synthetic division we've carried out, we see that the quotient is $x^2 + 2x - 2$ and the remainder is 0. In other words, both $x - 2$ and $x^2 + 2x - 2$ are factors of $x^3 - 6x + 4$, and we have

$$x^3 - 6x + 4 = (x - 2)(x^2 + 2x - 2)$$

EXAMPLE 3 Another example of synthetic division

Use synthetic division to divide $x^5 - a^5$ by $x - a$.

SOLUTION

$$\begin{array}{r|rrrrrr} a & 1 & 0 & 0 & 0 & 0 & -a^5 \\ & & a & a^2 & a^3 & a^4 & a^5 \\ \hline & 1 & a & a^2 & a^3 & a^4 & 0 \end{array}$$

As before, we read off the quotient and remainder from the third row of numbers. The quotient is

$$x^4 + ax^3 + a^2x^2 + a^3x + a^4$$

and the remainder is zero, so we have

$$\underbrace{x^5 - a^5}_{\text{dividend}} = \underbrace{(x - a)}_{\text{divisor}}\underbrace{(x^4 + ax^3 + a^2x^2 + a^3x + a^4)}_{\text{quotient}} \underbrace{+ 0}_{\text{remainder}}$$

or

$$x^5 - a^5 = (x - a)(x^4 + ax^3 + a^2x^2 + a^3x + a^4).$$

The last equation in Example 3 tells us how to factor $x^5 - a^5$. One factor is $x - a$. Notice the pattern in the second factor, $x^4 + ax^3 + a^2x^2 + a^3x + a^4$.

first term: a^0x^{5-1}

second term: a^1x^{5-2}

third term: a^2x^{5-3}

fourth term: a^3x^{5-4}

fifth term: a^4x^{5-5}

In the same way that we've found a factorization for $x^5 - a^5$, we can find a factorization for $x^n - a^n$ for any positive integer $n \geq 2$. In each case, the first factor is $x - a$, while the second factor follows the same pattern just described for $x^5 - a^5$. We state the general result in the box that follows. The result can be proved by using *mathematical induction* (discussed in a later chapter) or by using the *remainder theorem* (discussed in the next section).

Factorization of $x^n - a^n$

$$x^n - a^n = (x - a)(x^{n-1} + ax^{n-2} + a^2x^{n-3} + \cdots + a^{n-1})$$

In our development of synthetic division we assumed that the form of the divisor was $x - r$. The next example shows what to do when the form of the divisor is $x + r$.

EXAMPLE 4 Synthetic division when the divisor has the form $x + r$

Use synthetic division to divide $x^4 - 2x^3 + 5x^2 - 4x + 3$ by $x + 1$.

SOLUTION

We first need to write the divisor $x + 1$ in the form $x - r$. We have

$$x + 1 = x - (-1)$$

In other words, r is -1, and this is the value we use to set up the synthetic division format. The format then is

$$\begin{array}{r|rrrrr} -1 & 1 & -2 & 5 & -4 & 3 \\ & & & & & \\ \hline \end{array}$$

Now we carry out the synthetic division procedure:

$$\begin{array}{r|rrrrr} -1 & 1 & -2 & 5 & -4 & 3 \\ & & -1 & 3 & -8 & 12 \\ \hline & 1 & -3 & 8 & -12 & 15 \end{array}$$

The quotient is therefore $x^3 - 3x^2 + 8x - 12$, and the remainder is 15. We can summarize this result by writing

$$x^4 - 2x^3 + 5x^2 - 4x + 3 = (x + 1)(x^3 - 3x^2 + 8x - 12) + 15$$

(Notice that the degree of the remainder is less than the degree of the divisor, in agreement with the division algorithm.)

EXERCISE SET 12.2

A

In Exercises 1–20, use long division to find the quotients and the remainders. Also, write each answer in the form $p(x) = d(x) \cdot q(x) + R(x)$, as in equation (2) in the text.

1. $\dfrac{x^2 - 8x + 4}{x - 3}$

2. $\dfrac{x^3 - 4x^2 + x - 2}{x - 5}$

3. $\dfrac{x^2 - 6x - 2}{x + 5}$

4. $\dfrac{3x^2 + 4x - 1}{x - 1}$

5. $\dfrac{6x^3 - 2x + 3}{2x + 1}$

6. $\dfrac{x^4 - 4x^3 + 6x^2 - 4x + 1}{x - 1}$

7. $\dfrac{x^5 + 2}{x + 3}$

8. $\dfrac{4x^3 - x^2 + 8x - 1}{x^2 - x + 1}$

9. $\dfrac{x^6 - 64}{x - 2}$

10. $\dfrac{x^6 + 64}{x - 2}$

11. $\dfrac{5x^4 - 3x^2 + 2}{x^2 - 3x + 5}$

12. $\dfrac{8x^6 - 36x^4 + 54x^2 - 27}{2x^2 - 3}$

13. $\dfrac{3y^3 - 4y^2 - 3}{y^2 + 5y + 2}$

14. $\dfrac{4y^4 - y^3 + 2y - 1}{2y^2 - 3y - 4}$

15. $\dfrac{t^4 - 4t^3 + 4t^2 - 16}{t^2 - 2t + 4}$

16. $\dfrac{2t^5 - 6t^4 - t^2 + 2t + 3}{t^3 - 2}$

17. $\dfrac{z^5 - 1}{z - 1}$

18. $\dfrac{1 + z + z^2 + z^3}{1 + z + z^2}$

19. $\dfrac{ax^2 + bx + c}{x - r}$

20. $\dfrac{ax^3 + bx^2 + cx + d}{x - r}$

In Exercises 21–40, use synthetic division to find the quotients and remainders. Also, in each case, write the result of the division in the form $p(x) = d(x) \cdot q(x) + R(x)$, as in equation (2) in the text.

21. $\dfrac{x^2 - 6x - 2}{x - 5}$

22. $\dfrac{3x^2 + 4x - 1}{x - 1}$

23. $\dfrac{4x^2 - x - 5}{x + 1}$ **24.** $\dfrac{x^2 - 1}{x + 2}$

25. $\dfrac{6x^3 - 5x^2 + 2x + 1}{x - 4}$ **26.** $\dfrac{x^4 - 4x^3 + 6x^2 - 4x + 1}{x - 1}$

27. $\dfrac{x^3 - 1}{x - 2}$ **28.** $\dfrac{x^3 - 8}{x - 2}$

29. $\dfrac{x^5 - 1}{x + 2}$ **30.** $\dfrac{x^3 - 8x^2 - 1}{x + 3}$

31. $\dfrac{x^4 - 6x^3 + 2}{x + 4}$ **32.** $\dfrac{3x^3 - 2x^2 + x + 1}{x - \frac{1}{2}}$

33. $\dfrac{x^3 - 4x^2 - 3x + 6}{x - 10}$ **34.** $\dfrac{1 + 3x + 3x^2 + x^3}{x + 1}$

35. $\dfrac{x^3 - x^2}{x + 5}$

36. $\dfrac{5x^4 - 4x^3 + 3x^2 - 2x + 1}{x + \frac{1}{2}}$

37. $\dfrac{14 - 27x - 27x^2 + 54x^3}{x - \frac{2}{3}}$ **38.** $\dfrac{14 - 27x - 27x^2 + 54x^3}{x + \frac{2}{3}}$

39. $\dfrac{x^4 + 3x^2 + 12}{x - 3}$

40. (a) $\dfrac{x^4 - 16}{x - 2}$ **(b)** $\dfrac{x^4 + 16}{x + 2}$

In Exercises 41–44, each expression has the form $x^n - a^n$. Write each expression as a product of two factors (as in the box on page 911).

41. $x^5 - 32$ **42.** $y^6 - 1$ **43.** $z^4 - 81$ **44.** $x^7 - y^7$

In Exercises 45–48, use synthetic division to determine the quotient $q(x)$ and the remainder $R(x)$ in each case.

45. $\dfrac{6x^2 - 8x + 1}{3x - 4}$ *Hint*: Divide both numerator and denominator by 3. (Why?)

46. $\dfrac{4x^3 + 6x^2 - 6x - 5}{2x - 3}$ **47.** $\dfrac{6x^3 + 1}{2x + 1}$

48. $\dfrac{5x^3 - 3x^2 + 1}{3x + 1}$

49. When $x^3 + kx + 1$ is divided by $x + 1$, the remainder is -4. Find k.

50. (a) Show that when $x^3 + kx + 6$ is divided by $x + 3$, the remainder is $-21 - 3k$.

(b) Determine a value of k such that $x + 3$ will be a factor of $x^3 + kx + 6$.

51. When $x^2 + 2px - 3q^2$ is divided by $x - p$, the remainder is zero. Show that $p^2 = q^2$.

52. Given that $x - 3$ is a factor of $x^3 - 2x^2 - 4x + 3$, solve the equation $x^3 - 2x^2 - 4x + 3 = 0$.

The process of synthetic division applies equally well when some or all of the coefficients are nonreal complex numbers. In Exercises 53–56, use synthetic division to determine the quotient $q(x)$ and the remainder $R(x)$ in each case.

53. $\dfrac{x^2 - 4x + 1}{x - i}$ **54.** $\dfrac{x^3 - 2x^2 - 4}{x - 3i}$

55. $\dfrac{x^2 - 2x + 2}{x - (1 + i)}$ **56.** $\dfrac{x^3 - x^2 + 4x - 4}{x + 2i}$

57. Given that the identity $f(x) = d(x) \cdot q(x) + R(x)$ holds for the following polynomials, evaluate $f(\sqrt{3})$.
Hint (of sorts): There's an easy way and a tedious way.

$$f(x) = 2x^5 + 5x^4 - 8x^3 + 7x^2 - 9 \qquad d(x) = x^2 - 3$$
$$q(x) = 2x^3 + 5x^2 - 2x + 22 \qquad R(x) = -6x + 57$$

58. Given that the identity $f(t) = d(t) \cdot q(t) + R(t)$ holds for the following polynomials, evaluate $f(4)$.

$$f(t) = t^5 - 3t^4 + 2t^3 - 5t^2 + 6t - 7 \qquad d(t) = t - 4$$
$$q(t) = t^4 + t^3 + 6t^2 + 19t + 82 \qquad R(t) = 321$$

59. Find the remainder when $t^5 - 5a^4t + 4a^5$ is divided by $t - a$.

60. When $f(x)$ is divided by $(x - a)(x - b)$, the remainder is $Ax + B$. Apply the division algorithm to show that

$$A = \frac{f(a) - f(b)}{a - b} \quad \text{and} \quad B = \frac{bf(a) - af(b)}{b - a}$$

Descartes recommended [in his *La Géométrie* of 1637] *that all terms* [of an equation] *should be taken to one side and equated with zero. Though he was not the first to suggest this, he was the earliest writer to realize the advantage to be gained. He pointed out that a polynomial $f(x)$ was divisible by $(x - a)$ if and only if a was a root of $f(x)$.*—David M. Burton in *The History of Mathematics, An Introduction,* 2d ed. (Dubuque, Iowa: Wm. C. Brown Publishers, 1991)

12.3 THE REMAINDER THEOREM AND THE FACTOR THEOREM

The techniques for solving polynomial equations of degree 2 were discussed in Sections 1.3 and 2.1. Now we want to extend those ideas. Our focus in this section and in the remainder of the chapter is on solving polynomial equations of any degree, that is, equations of the form

$$f(x) = a_nx^n + a_{n-1}x^{n-1} + \cdots + a_1x + a_0 = 0 \tag{1}$$

Here, as in Chapter 1, a **root** or **solution** of equation (1) is a number r that when substituted for x leads to a true statement. Thus r is a root of equation (1) provided

$f(r) = 0$. For this reason we also refer to the number r in this case as a **zero** of the function f.

EXAMPLE 1 Checking for a zero or a root

(a) Is -3 a zero of the function f defined by $f(x) = x^4 + x^2 - 6$?
(b) Is $\sqrt{2}$ a root of the equation $x^4 + x^2 - 6 = 0$?

SOLUTION

(a) By definition, -3 will be a zero of f if $f(-3) = 0$. We have

$$f(-3) = (-3)^4 + (-3)^2 - 6 = 81 + 9 - 6 = 84 \neq 0$$

Thus -3 is not a zero of the function f.

(b) To check whether $\sqrt{2}$ is a root of the given equation, we have

$$\begin{aligned} (\sqrt{2})^4 + (\sqrt{2})^2 - 6 &\stackrel{?}{=} 0 \\ 4 + 2 - 6 &\stackrel{?}{=} 0 \\ 0 &\stackrel{?}{=} 0 \qquad \text{True} \end{aligned}$$

Thus $\sqrt{2}$ is a root of the equation $x^4 + x^2 - 6 = 0$.

There are cases in which a root of an equation is what we call a **repeated root.** Consider, for instance, the equation $x(x - 1)(x - 1) = 0$. We have

$$x(x - 1)(x - 1) = 0$$

$$x = 0 \quad \Big| \quad \begin{aligned} x - 1 &= 0 \\ x &= 1 \end{aligned} \quad \Big| \quad \begin{aligned} x - 1 &= 0 \\ x &= 1 \end{aligned}$$

The roots of the equation are therefore 0, 1, and 1. The repeated root here is $x = 1$. We say in this case that 1 is a **double root** or, equivalently, that 1 is a **root of multiplicity 2.** More generally, if a root is repeated k times, we call it a **root of multiplicity *k*.**

EXAMPLE 2 Specifying the multiplicity of a root

State the multiplicity of each root of the equation

$$(x - 4)^2(x - 5)^3 = 0$$

SOLUTION

We have $(x - 4)(x - 4)(x - 5)(x - 5)(x - 5) = 0$. By setting each factor equal to zero, we obtain the roots 4, 4, 5, 5, and 5. From this we see that the root 4 has multiplicity 2, while the root 5 has multiplicity 3. It is not really necessary to write out all the factors as we did here; the exponents of the factors in the original equation give us the required multiplicities.

There is a connection between repeated roots and graphs. If a polynomial equation $f(x) = 0$ has a repeated root r, then the graph of the function f is tangent to the x-axis at $x = r$. We saw examples demonstrating this in Section 4.6, although

GRAPHICAL PERSPECTIVE

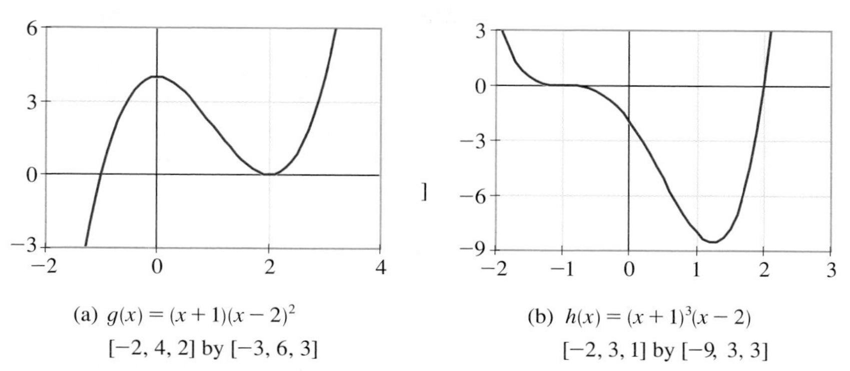

(a) $g(x) = (x + 1)(x - 2)^2$
$[-2, 4, 2]$ by $[-3, 6, 3]$

(b) $h(x) = (x + 1)^3(x - 2)$
$[-2, 3, 1]$ by $[-9, 3, 3]$

Figure 1
At a multiple root, the graph is tangent to the x-axis. In contrast, at a simple root [such as $x = 2$ in Figure 1(b)], the graph crosses the x-axis.

we had not defined the term *multiple root* at that stage. Figure 1 shows two examples of this connection. (For the examples in Section 4.6, see Figures 18 through 21 on pages 296–298.)

There are two simple but important theorems that will form the basis for much of our subsequent work with polynomials: the *remainder theorem* and the *factor theorem*. We begin with a statement of the remainder theorem.

The Remainder Theorem

When a polynomial $f(x)$ is divided by $x - r$, the remainder is $f(r)$.

Before turning to a proof of the remainder theorem, let's see what the theorem is saying in two particular cases. First, suppose that we divide the polynomial $f(x) = 2x^2 - 3x + 4$ by $x - 1$. Then according to the remainder theorem, the remainder in this case should be the number $f(1)$. Let's check:

$$\begin{array}{r|rrr} 1 & 2 & -3 & 4 \\ & & 2 & -1 \\ \hline & 2 & -1 & \textcircled{3} \end{array}$$

The remainder is 3, and $f(1) = 2(1)^2 - 3(1) + 4 = \textcircled{3}$.

As the calculations show, the remainder is indeed equal to $f(1)$. As a second example, let us divide the polynomial $g(x) = ax^2 + bx + c$ by $x - r$. According to the remainder theorem, the remainder should be $g(r)$. Again, let us check:

$$\begin{array}{r|lll} r & a & b & c \\ & & ar & ar^2 + br \\ \hline & a & ar + b & \boxed{ar^2 + br + c} \end{array}$$

The remainder is $ar^2 + br + c$, and $g(r) = \boxed{ar^2 + br + c}$.

The calculations show that the remainder is equal to $g(r)$, as we wished to check. Thus the remainder theorem holds for any quadratic polynomial $g(x) = ax^2 + bx + c$.

A general proof of the remainder theorem can easily be given along these same lines. The only drawback is that it becomes slightly cumbersome to carry out the synthetic division process when the dividend is

$$a_nx^n + a_{n-1}x^{n-1} + \cdots + a_1x + a_0$$

Instead, we base the proof of the remainder theorem on the division algorithm.

To prove the remainder theorem, we must show that when the polynomial $f(x)$ is divided by $x - r$, the remainder is $f(r)$. Now, according to the division algorithm, we can write

$$f(x) = (x - r) \cdot q(x) + R(x) \tag{2}$$

for unique polynomials $q(x)$ and $R(x)$. In this identity, either $R(x)$ is the zero polynomial or the degree of $R(x)$ is less than that of $x - r$. Since the degree of $x - r$ is 1, the degree of $R(x)$ must be zero. Thus in *either* case the remainder $R(x)$ is a constant. Denoting this constant by c, we can rewrite equation (2) as

$$f(x) = (x - r) \cdot q(x) + c$$

If we set $x = r$ in this identity, we obtain

$$f(r) = (r - r) \cdot q(r) + c = c$$

We have now shown that $f(r) = c$. But by definition, c is the remainder $R(x)$. Thus $f(r)$ is the remainder. This proves the remainder theorem.

EXAMPLE 3 Using the remainder theorem to evaluate a function and check for a factor

Let $f(x) = 2x^3 - 5x^2 + x - 6$.

(a) Use the remainder theorem to evaluate $f(3)$.

(b) Is $x - 3$ a factor of $f(x) = 2x^3 - 5x^2 + x - 6$?

SOLUTION

(a) According to the remainder theorem, $f(3)$ is the remainder when $f(x)$ is divided by $x - 3$. Using synthetic division, we have

$$\begin{array}{r|rrrr} 3 & 2 & -5 & 1 & -6 \\ & & 6 & 3 & 12 \\ \hline & 2 & 1 & 4 & 6 \end{array}$$

The remainder is 6, and therefore $f(3) = 6$.

(b) By definition, $x - 3$ is a factor of $f(x)$ if we obtain a zero remainder when $f(x)$ is divided by $x - 3$. But from our work in part (a) we know that the remainder is 6, not 0, so $x - 3$ is not a factor of $f(x)$.

EXAMPLE 4 Using the remainder theorem to check for a factor

In the previous section it was stated, but not proved, that $x - a$ is a factor of $f(x) = x^n - a^n$. Use the remainder theorem to prove this fact.

SOLUTION

We need to show that when $f(x) = x^n - a^n$ is divided by $x - a$, the remainder is zero. By the remainder theorem the remainder is equal to $f(a)$, which is easy to find:

$$f(a) = a^n - a^n = 0$$

Thus the remainder is zero, and $x - a$ is a factor of $x^n - a^n$, as required.

From our experience with quadratic equations we know that there is a close connection between factoring a quadratic polynomial $f(x)$ and solving the polynomial equation $f(x) = 0$. The factor theorem states this relationship between roots and factors in a precise form. Furthermore, the factor theorem tells us that this relationship holds for polynomials of all degrees, not just quadratics.

The Factor Theorem

Let $f(x)$ be a polynomial. If $f(r) = 0$, then $x - r$ is a factor of $f(x)$. Conversely, if $x - r$ is a factor of $f(x)$, then $f(r) = 0$.

In terms of roots, we can summarize the factor theorem by saying that r is a root of the equation $f(x) = 0$ if and only if $x - r$ is a factor of $f(x)$. To prove the factor theorem, let us begin by assuming that $f(r) = 0$. We want to show that $x - r$ is a factor of $f(x)$. Now, according to the remainder theorem, if $f(x)$ is divided by $x - r$, the remainder is $f(r)$. So we can write

$$f(x) = (x - r) \cdot q(x) + f(r) \qquad \text{for some polynomial } q(x)$$

But since $f(r)$ is zero, this equation becomes

$$f(x) = (x - r) \cdot q(x)$$

This last equation tells us that $x - r$ is a factor of $f(x)$, as we wished to prove.

Now, conversely, let us assume that $x - r$ is a factor of $f(x)$. We want to show that $f(r) = 0$. Since $x - r$ is a factor of $f(x)$, we can write

$$f(x) = (x - r) \cdot q(x) \qquad \text{for some polynomial } q(x)$$

If we now let $x = r$ in this last equation, we obtain

$$f(r) = (r - r) \cdot q(r) = 0$$

as we wished to show.

The example that follows indicates how the factor theorem can be used to solve equations.

EXAMPLE 5 Applying the factor theorem in solving an equation

Refer to Figure 2, which shows a graph of the function $y = x^3 - 5x + 2$. As is indicated by the graph, the equation

$$x^3 - 5x + 2 = 0 \qquad (3)$$

has (as least) three roots, one of which either is equal to 2 or is very close to 2. Confirm that $x = 2$ is indeed a root of equation (3), and then solve the equation.

GRAPHICAL PERSPECTIVE

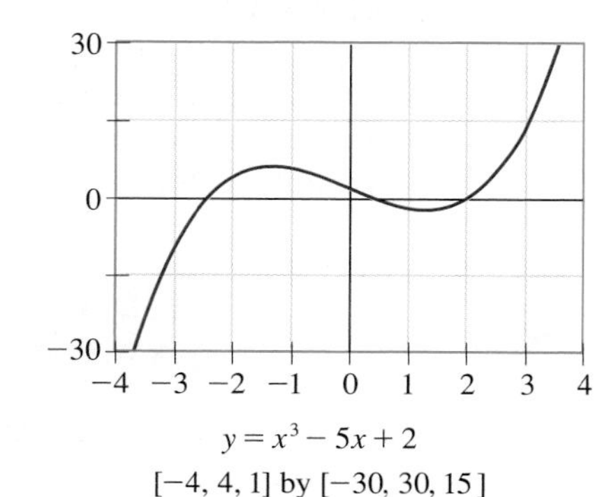

$y = x^3 - 5x + 2$
$[-4, 4, 1]$ by $[-30, 30, 15]$

Figure 2

SOLUTION

Replacing x by 2 in equation (3) yields $2^3 - 5(2) + 2 = 0$, which is true. Thus 2 is a root of the given equation. Now, since $x = 2$ is a root, the factor theorem tells us that $x - 2$ is a factor of $x^3 - 5x + 2$. In other words,

$$x^3 - 5x + 2 = (x - 2) \cdot q(x) \qquad \text{for some polynomial } q(x) \tag{4}$$

To isolate $q(x)$, we divide $x^3 - 5x + 2$ by $x - 2$. Using synthetic division, we have

$$\begin{array}{r|rrrr} 2 & 1 & 0 & -5 & 2 \\ & & 2 & 4 & -2 \\ \hline & 1 & 2 & -1 & 0 \end{array}$$

The quotient is $x^2 + 2x - 1$ with zero remainder. Thus equation (4) becomes

$$x^3 - 5x + 2 = (x - 2)(x^2 + 2x - 1)$$

With this identity the original equation becomes

$$(x - 2)(x^2 + 2x - 1) = 0$$

At this point, the problem is reduced to solving the linear equation $x - 2 = 0$ and the quadratic equation $x^2 + 2x - 1 = 0$. The linear equation yields $x = 2$, but we already know that 2 is a root. So if we are to find any additional roots, they must come from the quadratic equation $x^2 + 2x - 1 = 0$. As you should now verify for yourself, the quadratic formula yields $x = -1 \pm \sqrt{2}$. Thus we have three roots of the given cubic equation: 2, $-1 + \sqrt{2}$, and $-1 - \sqrt{2}$. As you'll see in the next section, a cubic equation can have at most three roots. So in this case we've determined all the roots; that is, we have solved the equation. As you can check, the calculator approximations for the two roots containing radicals are $-1 + \sqrt{2} \approx 0.41$ and $-1 - \sqrt{2} \approx -2.41$. Note that these values are consistent with the graph shown in Figure 2.

Before going on to other examples, let's take a moment to summarize the technique we used in Example 5. We want to solve a polynomial equation $f(x) = 0$, given that one root is $x = r$. Since r is a root, the factor theorem tells us that $x - r$ is a factor of $f(x)$. Then, with the aid of synthetic division, we obtain a factorization

$$f(x) = (x - r) \cdot q(x) \qquad \text{for some polynomial } q(x)$$

This gives rise to the two equations $x - r = 0$ and $q(x) = 0$. Since the first of these only reasserts that r is a root, we try to solve the second equation, $q(x) = 0$. We refer to the equation $q(x) = 0$ as the **reduced equation.** Example 5 showed you the idea behind this terminology: The degree of $q(x)$ is one less than that of $f(x)$. If, as in Example 5, the reduced equation happens to be a quadratic equation, then we can always determine the remaining roots by factoring or by the quadratic formula. In subsequent sections we will look at techniques that are helpful in cases in which $q(x)$ is not quadratic.

EXAMPLE 6 Determining a polynomial equation with prescribed roots

Solve the equation $x^4 + 2x^3 - 7x^2 - 20x - 12 = 0$, given that $x = 3$ and $x = -2$ are roots.

SOLUTION

We can check that $x = 3$ is a root as well as find the reduced equation by using synthetic division. We have

$$\begin{array}{r|rrrrr} 3 & 1 & 2 & -7 & -20 & -12 \\ & & 3 & 15 & 24 & 12 \\ \hline & 1 & 5 & 8 & 4 & 0 \end{array}$$

Thus our original equation is equivalent to

$$(x-3)(x^3+5x^2+8x+4)=0$$

Now, $x = -2$ is also a root of this equation. But $x = -2$ surely is not a root of the equation $x - 3 = 0$, so it must be a root of the reduced equation

$$x^3+5x^2+8x+4=0$$

Again, the factor theorem is applicable. Since $x = -2$ is a root of the reduced equation, $x + 2$ must be a factor of $x^3 + 5x^2 + 8x + 4$. Using synthetic division again, we have

$$\begin{array}{r|rrrr} -2 & 1 & 5 & 8 & 4 \\ & & -2 & -6 & -4 \\ \hline & 1 & 3 & 2 & 0 \end{array}$$

Thus our reduced equation can be written

$$(x+2)(x^2+3x+2)=0$$

This gives rise to a second reduced equation:

$$x^2+3x+2=0$$

In this case the roots are readily obtained by factoring. We have

$$\begin{aligned} x^2+3x+2&=0 \\ (x+2)(x+1)&=0 \end{aligned}$$

$$\begin{array}{r|r} x+2=0 & x+1=0 \\ x=-2 & x=-1 \end{array}$$

Now, of the two roots we've just found, -2 happens to be one of the roots that we were initially given in the statement of the problem. On the other hand, -1 is a distinct additional root. In summary, then, we have three distinct roots: 3, -2, and -1, where the root -2 has multiplicity 2. As you will see in the next section, a fourth-degree equation can have at most four (not necessarily distinct) roots. So we have found all the roots of the given equation.

EXAMPLE 7 Finding polynomial equations satisfying given conditions

In each case, find a polynomial equation $f(x) = 0$ satisfying the given conditions. If there is no such equation, say so.

(a) The numbers -1, 4, and 5 are roots.

(b) A factor of $f(x)$ is $x - 3$, and -4 is a root of multiplicity 2.

(c) The degree of f is 4, the number -5 is a root of multiplicity 3, and 6 is a root of multiplicity 2.

SOLUTION

(a) If $f(x)$ is any polynomial containing the factors $x + 1, x - 4$, and $x - 5$, then the equation $f(x) = 0$ will certainly be satisfied with $x = -1, x = 4$, or $x = 5$. The simplest polynomial equation in this case is therefore

$$(x + 1)(x - 4)(x - 5) = 0$$

This is a polynomial equation with the required roots. If required, we can carry out the multiplication on the left-hand side of the equation. As you can check, this yields $x^3 - 8x^2 + 11x + 20 = 0$.

(b) According to the factor theorem, since -4 is a root, $x + 4$ must be a factor of $f(x)$. In fact, since -4 is a root of multiplicity 2, the quantity $(x + 4)^2$ must be a factor of $f(x)$. The following equation therefore satisfies the given conditions:

$$(x - 3)(x + 4)^2 = 0$$

(c) We are given two roots: one with multiplicity 3, the other with multiplicity 2. Thus the degree of $f(x)$ must be at least $3 + 2 = 5$. (Why?) But this then contradicts the given condition that the degree of f should be 4. Consequently, there is no polynomial equation that satisfies the given conditions.

EXERCISE SET 12.3 

A

In Exercises 1–6, determine whether the given value for the variable is a root of the equation.

1. $12x - 8 = 112; x = 10$
2. $12x^2 - x - 20 = 0; x = 5/4$
3. $x^2 - 2x - 4 = 0; x = 1 - \sqrt{5}$
4. $1 - x + x^2 - x^3 = 0; x = -1$
5. $2x^2 - 3x + 1 = 0; x = 1/2$
6. $(x - 1)(x - 2)(x - 3) = 0; x = 4$

In Exercises 7–14, determine whether the given value is a zero of the function.

7. $f(x) = 3x - 2; x = 2/3$
8. $g(x) = 1 + x^2; x = -1$
9. $h(x) = 5x^3 - x^2 + 2x + 8; x = -1$
10. $F(x) = -2x^5 + 3x^4 + 8x^3; x = 0$
11. $f(t) = 1 + 2t + t^3 - t^5; t = 2$
12. $f(t) = 1 + 2t + t^3 - t^5; t = \sqrt{2}$
13. $f(x) = 2x^3 - 3x + 1$:
(a) $x = (\sqrt{3} - 1)/2$ **(b)** $x = (\sqrt{3} + 1)/2$
14. $g(x) = x^4 + 8x^3 + 9x^2 - 8x - 10$:
(a) $x = 1$ **(b)** $x = \sqrt{6} - 4$ **(c)** $x = \sqrt{6} + 4$

In Exercises 15 and 16, list the distinct roots of each equation. In the case of a repeated root, specify its multiplicity.

15. $(x - 1)(x - 2)^3(x - 3) = 0$
16. $x(x + 5)^4 = 0$

Ⓖ *In Exercises 17 and 18 you are given a polynomial equation $f(x) = 0$. Specify the multiplicity of each repeated root. Then use a graphing utility to visually verify that the graph of $y = f(x)$ is tangent to the x-axis at each repeated root.*

17. (a) $(x + 1)^2(x + 2) = 0$
(b) $(x + 1)(x + 2)^3 = 0$
(c) $(x + 1)^2(x + 2)^3 = 0$

18. (a) $x(x + 1)^2(x - 1) = 0$
(b) $x^2(x + 1)^2(x - 1) = 0$
(c) $x^3(x + 1)^2(x - 1) = 0$

In Exercises 19–24, use the remainder theorem (as in Example 3) to evaluate $f(x)$ for the given value of x.

19. $f(x) = 4x^3 - 6x^2 + x - 5; x = -3$
20. $f(x) = 2x^3 - x - 4; x = 4$
21. $f(x) = 6x^4 + 5x^3 - 8x^2 - 10x - 3; x = 1/2$
22. $f(x) = x^5 - x^4 - x^3 - x^2 - x - 1; x = -2$
23. $f(x) = x^2 + 3x - 4; x = -\sqrt{2}$
24. $f(x) = x^7 - 7x^6 + 5x^4 + 1; x = -3$
25. The graph in the accompanying figure indicates that the equation $2x^3 - x + 1 = 0$ has a root at or very near $x = -1$. Confirm that $x = -1$ is indeed a root, and then solve the equation, as in Example 5.

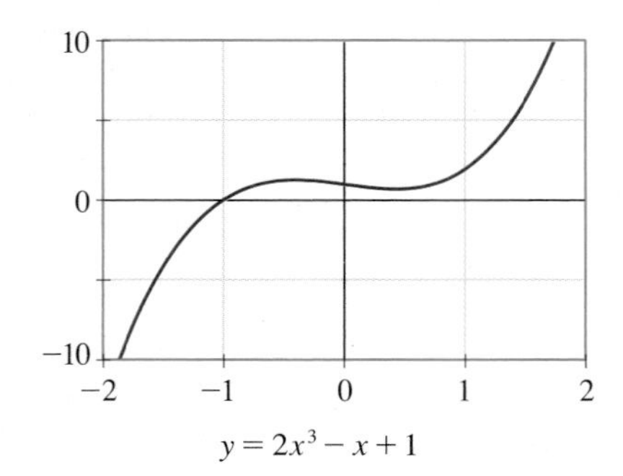

$y = 2x^3 - x + 1$

26. The graph in the accompanying figure indicates that the equation $x^3 - 14x - 8 = 0$ has a root at or very near $x = 4$.

Confirm that $x = 4$ is indeed a root of the equation, and then solve the equation, as in Example 5.

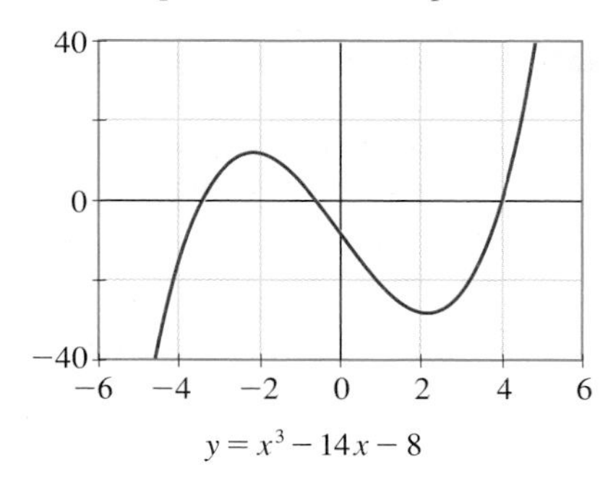

$y = x^3 - 14x - 8$

In Exercises 27–38 you are given a polynomial equation and one or more roots. Check that the given numbers really are roots, then solve each equation using the method shown in Examples 5 and 6. To help you decide whether you have found all the roots in each case, you may rely on the following theorem, discussed in the next section: A polynomial equation of degree n has at most n (not necessarily distinct) roots.

27. $x^3 - 4x^2 - 9x + 36 = 0$; -3 is a root.
28. $x^3 + 7x^2 + 11x + 5 = 0$; -1 is a root.
29. $x^3 + x^2 - 7x + 5 = 0$; 1 is a root.
30. $x^3 + 8x^2 - 3x - 24 = 0$; -8 is a root.
31. $3x^3 - 5x^2 - 16x + 12 = 0$; -2 is a root.
32. $2x^3 - 5x^2 - 46x + 24 = 0$; 6 is a root.
33. $2x^3 + x^2 - 5x - 3 = 0$; $-3/2$ is a root.
34. $6x^4 - 19x^3 - 25x^2 + 18x + 8 = 0$; 4 and $-1/3$ are roots.
35. $x^4 - 15x^3 + 75x^2 - 125x = 0$; 5 is a root.
36. $2x^3 + 5x^2 - 8x - 20 = 0$; 2 is a root.
37. $x^4 + 2x^3 - 23x^2 - 24x + 144 = 0$; -4 and 3 are roots.
38. $6x^5 + 5x^4 - 29x^3 - 25x^2 - 5x = 0$; $\sqrt{5}$ and $-1/3$ are roots.

Ⓖ *In Exercises 39–42, each polynomial equation $f(x) = 0$ has a root r that is a rational number. Use a graphing utility to zoom in on the x-intercepts of the graph of $y = f(x)$ until you think you know what r is. Then check algebraically that r is indeed a root. [You can check either by direct substitution or by using the remainder theorem and synthetic division to compute $f(r)$.] After you know the rational root, follow the procedure shown in Example 5 to determine the remaining roots. For roots that contain radicals, give both the exact expression for the answer and a calculator approximation rounded to two decimal places. Finally, check to see that the decimal values are consistent with the x-intercepts that you looked at initially.*

39. $x^3 + x^2 - 18x + 10 = 0$
40. $-2x^3 + 7x^2 - 2x - 6 = 0$
41. $6x^3 - 28x^2 + 19x - 2 = 0$
42. $5x^5 - x^4 - 15x^3 + 3x^2 - 15x + 3 = 0$

Ⓖ *In Exercises 43 and 44, each polynomial equation $f(x) = 0$ has at least two rational roots. Use a graphing utility to zoom in on the x-intercepts of the graph of $y = f(x)$ to see what the rational roots might be. For each suspected rational root, find out whether it is actually a root (either by using direct substitution or by using the remainder theorem and synthetic division). After you have determined the rational roots in this way, solve the equation as in Example 6.*

43. $4x^5 - 15x^4 + 8x^3 + 19x^2 - 12x - 4 = 0$
44. $x^4 - 10x^3 - 3.1x^2 + 28.9x + 21 = 0$

For Exercises 45 and 46, refer to the following tables for the functions f and g.

$f(t) = t^3 - 4t + 3$

t	f(t)
0.500	1.125000
0.750	0.421875
1.000	0.000000
1.250	−0.046875
1.500	0.375000

$g(t) = t^5 + 2t^4 + t^3 + 2t^2 - t - 2$

t	g(t)
−3.00	−89.0
−2.50	−22.15625
−2.25	−7.4228515625
−2.00	0.0
−1.75	2.8603515625

45. **(a)** What is the remainder when $f(t)$ is divided by $t - 1/2$? *Hint*: Don't calculate; use the remainder theorem.
(b) What is the remainder when $f(t)$ is divided by $t - 1.25$?
(c) Specify a linear factor of $f(t)$.
(d) Solve the equation $f(t) = 0$.

46. **(a)** What is the remainder when $g(t)$ is divided by $t + 3$?
(b) What is the remainder when $g(t)$ is divided by $t + 5/2$?
(c) Specify a linear factor of $g(t)$.
(d) Solve the equation $g(t) = 0$.

In Exercises 47–52, find a polynomial equation $f(x) = 0$ satisfying the given conditions. If no such equation is possible, state this.

47. Degree 3; the coefficient of x^3 is 1; three roots are 3, -4, and 5
48. Degree 3; the coefficient are integers; 1/2, 2/5, and $-3/4$ are roots
49. Degree 3; -1 is a root of multiplicity two; $x + 6$ is a factor of $f(x)$
50. Degree 3; 4 is a root of multiplicity two; -1 is a root of multiplicity two
51. Degree 4; 1/2 is a root of multiplicity three; $x^2 - 3x - 4$ is a factor of $f(x)$
52. Degree 4; the coefficients are integers; 1/2 is a root of multiplicity two; $2x^2 - 4x - 1$ is a factor of $f(x)$
53. In this exercise we verify that the remainder theorem is valid for the cubic polynomial $g(x) = ax^3 + bx^2 + cx + d$.
(a) Compute $g(r)$.
(b) Using synthetic division, divide $g(x)$ by $x - r$. Check that the remainder you obtain is the same as the answer in part (a).

Ⓖ **54.** **(a)** In the standard viewing rectangle, graph the four equations $y = x^n - 2^n$, $(n = 2, 3, 4, 5)$.
(b) Switch to a viewing rectangle that extends from -5 to 5 in the x-direction and from -40 to 20 in the

y-direction. Name one property that is shared by all four graphs.

(c) How does your answer in part (b) relate to the result in Example 4 of this section?

In Exercises 55 and 56, determine whether the given value is a zero of the function.

55. $Q(x) = ax^2 + bx + c; x = \dfrac{-b + \sqrt{b^2 - 4ac}}{2a}$

Hint: Look before you leap!

56. $f(x) = x^3 - 3x^2 + 3x - 3$:

(a) $x = \sqrt[3]{2} - 1$ **(b)** $x = \sqrt[3]{2} + 1$

57. Determine values for a and b such that $x - 1$ is a factor of both $x^3 + x^2 + ax + b$ and $x^3 - x^2 - ax + b$.

58. Determine a quadratic equation with the given roots:

(a) $a/b, -b/a$

(b) $-a + 2\sqrt{2b}, \ -a - 2\sqrt{2b}$

59. One root of the equation $x^2 + bx + 1 = 0$ is twice the other; find b. (There are two answers.)

60. Determine a value for a such that one root of the equation $ax^2 + x - 1 = 0$ is five times the other.

C

61. Solve the equation $x^3 - 12x + 16 = 0$, given that one of the roots has multiplicity two.

12.4 THE FUNDAMENTAL THEOREM OF ALGEBRA

One of the most intriguing problems was the question of the number of roots of an equation, which brought in negative and imaginary quantities, and led to the conclusion, in the work of Girard and Descartes, that an equation of degree n can have no more than n roots. The more precise statement, that an equation of degree n always has one root, and hence always has n roots (allowing for multiple roots), became known as the fundamental theorem of algebra. After several attempts by D'Alembert, Euler, and others, the proof was finally given by Gauss in 1799.—From *A Source Book in Mathematics, 1200–1800,* D. J. Struik, ed. (Princeton, N.J.: Princeton University Press, 1986)

Every equation of algebra has as many solutions as the exponent of the highest term indicates.—Albert Girard, 1629, as quoted by David M. Burton in *The History of Mathematics, An Introduction,* 2d ed. (Dubuque, Iowa: Wm. C. Brown Publishers, 1991)

. . . we arrive from all the investigations explained above at the rigorous proof of the theorem that every integral rational algebraic function of one variable can be decomposed into real factors of the first and second degree.—Carl Friedrich Gauss in his doctoral dissertation (1799)

Does every polynomial equation (of degree at least 1) have a root? Or, to put the question another way, is it possible to write a polynomial equation that has no solution? Certainly, if we consider only real roots, that is, roots that are real numbers, then it is easy to specify an equation with no real roots:

$$x^2 = -1$$

This equation has no real roots because the square of a real number is never negative. On the other hand, if we allow complex-number roots, then $x^2 = -1$ does indeed have a root. In fact, both i and $-i$ are roots in this case.

It turns out that the situation just described for the equation $x^2 = -1$ holds quite generally. That is, within the complex-number system, every polynomial equation of degree at least 1 has at least one root. This is the substance of a remarkable theorem that was first proved by the great mathematician Carl Friedrich Gauss in 1799. (Gauss was only 22 years old at the time.) Although there are many fundamental theorems in algebra, Gauss's result has come to be known as *the* **fundamental theorem of algebra.**

The Fundamental Theorem of Algebra

Every polynomial equation of the form

$$a_nx^n + a_{n-1}x^{n-1} + \cdots + a_1x + a_0 = 0 \qquad (n \geq 1, a_n \neq 0)$$

has at least one root within the complex number system. (This root may be a real number.)

Although most of this section deals with polynomials with real coefficients, Gauss's theorem still applies in cases in which some or all of the coefficients are nonreal complex numbers. The proof of Gauss's theorem is usually given in the post-calculus course called complex variables.

The fundamental theorem of algebra asserts that every polynomial equation of degree at least 1 has a root. There are two initial observations to be made here. First, notice that the theorem says nothing about actually finding the root. Second, notice that the theorem deals only with polynomial equations. Indeed, it is easy to specify a non-polynomial equation that does not have a root. Such an equation is $1/x = 0$. The expression on the left-hand side of this equation can never be zero because the numerator is 1.

EXAMPLE 1 Understanding what the fundamental theorem says

Which of the following equations has at least one root?

(a) $x^3 - 17x^2 + 6x - 1 = 0$

(b) $\sqrt{2}x^{47} - \pi x^{25} + \sqrt{3} = 0$

(c) $x^2 - 2ix + (3 + i) = 0$

SOLUTION

All three equations are polynomial equations, so according to the fundamental theorem of algebra, each equation has at least one root.

In Section 1.3 we used factoring as a tool for solving quadratic equations. The next theorem, a consequence of the fundamental theorem of algebra, tells us that (in principle, at least) any polynomial of degree n can be factored into a product of n linear factors. In proving the **linear factors theorem,** we will need to use the factor theorem. If you reread the proof of that theorem in the previous section, you will see that it makes no difference whether the number r appearing in the factor $x - r$ is a real number or a nonreal complex number. Thus the factor theorem is valid in either case.

The Linear Factors Theorem

Let $f(x) = a_nx^n + a_{n-1}x^{n-1} + \cdots + a_1x + a_0$, where $n \geq 1$ and $a_n \neq 0$. Then $f(x)$ can be expressed as a product of n linear factors:

$$f(x) = a_n(x - r_1)(x - r_2)\cdots(x - r_n)$$

(The complex numbers r_k that appear in these factors are not necessarily all distinct, and some or all of the r_k may be real numbers.)

Proof of the Linear Factors Theorem. According to the fundamental theorem of algebra, the equation $f(x) = 0$ has a root; let's call this root r_1. By the factor theorem, $x - r_1$ is a factor of $f(x)$, and we can write

$$f(x) = (x - r_1)\cdot Q_1(x)$$

for some polynomial $Q_1(x)$ that has degree $n - 1$ and leading coefficient a_n. If the degree of $Q_1(x)$ happens to be zero, we're done. On the other hand, if the degree of $Q_1(x)$ is at least 1, another application of the fundamental theorem of algebra followed by the factor theorem gives us

$$Q_1(x) = (x - r_2) \cdot Q_2(x)$$

where the degree of $Q_2(x)$ is $n - 2$ and the leading coefficient of $Q_2(x)$ is a_n. We now have

$$f(x) = (x - r_1)(x - r_2) \cdot Q_2(x)$$

We continue this process until the quotient is $Q_n(x) = a_n$. As a result, we obtain

$$\begin{aligned} f(x) &= (x - r_1)(x - r_2) \cdots (x - r_n)a_n \\ &= a_n(x - r_1)(x - r_2) \cdots (x - r_n) \end{aligned}$$

as we wished to show.

The linear factors theorem tells us that any polynomial can be expressed as a product of linear factors. The theorem gives us no information, however, as to how those factors can actually be obtained. The next example demonstrates a case in which the factors are readily obtainable; this is always the case with quadratic polynomials.

EXAMPLE 2 Writing quadratics as the products of linear factors

Express the following second-degree polynomials in the form $a_n(x - r_1)(x - r_2)$:
(a) $3x^2 - 5x - 2$; **(b)** $x^2 - 4x + 5$.

SOLUTION

(a) A factorization for $3x^2 - 5x - 2$ can be found by simple trial and error. We have

$$3x^2 - 5x - 2 = (3x + 1)(x - 2)$$

We now write the factor $3x + 1$ as $3(x + \frac{1}{3})$. This, in turn, can be written $3[x - (-\frac{1}{3})]$. The final factorization is then

$$3x^2 - 5x - 2 = 3\left[x - \left(-\frac{1}{3}\right)\right](x - 2)$$

(b) From the factor theorem, or from our more elementary work with quadratic equations, we know that if r_1 and r_2 are the roots of the equation $x^2 - 4x + 5 = 0$, then $x - r_1$ and $x - r_2$ are the factors of $x^2 - 4x + 5$. That is, $x^2 - 4x + 5 = (x - r_1)(x - r_2)$. The values for r_1 and r_2 in this case are readily obtained by using the quadratic formula. As you can check, the results are

$$r_1 = 2 + i \qquad \text{and} \qquad r_2 = 2 - i$$

The required factorization is therefore

$$x^2 - 4x + 5 = [x - (2 + i)][x - (2 - i)]$$

Using the linear factors theorem, we can show that every polynomial equation of degree $n \geq 1$ has exactly n roots. To help you follow the reasoning, we make two preliminary comments. First, we agree that a root of multiplicity k will be counted as k roots. For example, although the third-degree equation $(x-1)(x-4)^2 = 0$ has only two distinct roots, 1 and 4, it has three roots *if* we agree to count the repeated root 4 two times. The second preliminary comment concerns the *zero-product property,* which states that $pq = 0$ if and only if $p = 0$ or $q = 0$. When we stated this in Section 1.3, we were working within the real-number system. However, the property is also valid within the complex-number system. (See Exercises 79 and 80 in Section 12.1.) Now let's state and prove our theorem.

THEOREM

Every polynomial equation of degree $n \geq 1$ has exactly n roots, where a root of multiplicity k is counted k times.

Proof of the Theorem. Using the linear factors theorem, we can write the nth-degree polynomial equation $f(x) = a_n x^n + a_{n-1}x^{n-1} + \cdots + a_1 x + a_0 = 0$ as

$$f(x) = a_n(x - r_1)(x - r_2)\cdots(x - r_n) = 0 \qquad (a_n \neq 0) \tag{1}$$

By the factor theorem, each of the numbers $r_1, r_2, \ldots, r_n$ is a root. Some of these numbers may in fact be equal; in other words, we may have repeated roots in this list. In any case, if we agree to count a root of multiplicity k as k roots, then we obtain exactly n roots from the list $r_1, r_2, \ldots, r_n$. Furthermore, the equation $f(x) = 0$ can have no other roots, as we now show. Suppose that r is any number distinct from all the numbers $r_1, r_2, \ldots, r_n$. Replacing x with r in equation (1) yields

$$f(r) = a_n(r - r_1)(r - r_2)\cdots(r - r_n)$$

But the expression on the right-hand side of this last equation cannot be zero, because none of the factors is zero. Thus $f(r)$ is not zero, and so r is not a root. This completes the proof of the theorem.

EXAMPLE 3 Finding $f(x)$ given the roots and multiplicities of $f(x) = 0$

TABLE 1

Root	Multiplicity
3	2
−2	1
0	2

Find a polynomial $f(x)$ with leading coefficient 1 such that the equation $f(x) = 0$ has only those roots specified in Table 1. What is the degree of this polynomial?

SOLUTION

The expressions $(x-3)^2$, $x+2$, and $(x-0)^2$ all must appear as factors of $f(x)$, and furthermore, no other linear factor can appear. The form of $f(x)$ is therefore

$$f(x) = a_n(x-3)^2(x+2)(x-0)^2$$

Since the leading coefficient a_n is to be 1, we can rewrite this last equation as

$$f(x) = x^2(x-3)^2(x+2)$$

This is the required polynomial. The degree here is 5. This can be seen either by multiplying out the factors or by simply adding the multiplicities of the roots in Table 1. Figure 1 provides a graphical perspective on this example.

GRAPHICAL PERSPECTIVE

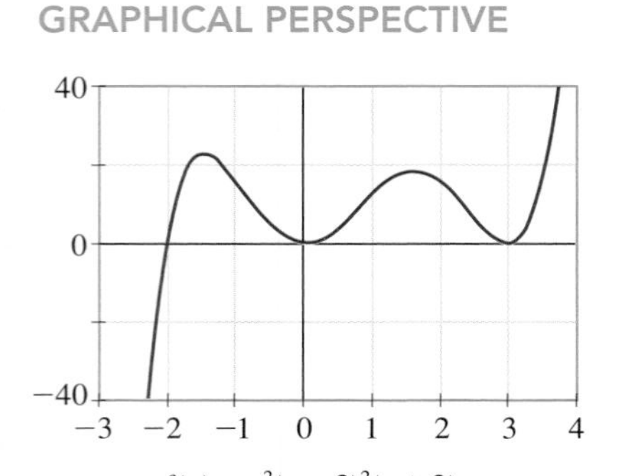

Figure 1
The roots of the equation $f(x) = x^2(x - 3)^2(x + 2) = 0$ are -2, 0, and 3. The roots 0 and 3 each have multiplicity 2, and correspondingly, the graph of f is tangent to the x-axis at those *x-intercepts.* The root -2 is a simple root, and the graph of f is not tangent to the x-axis at the intercept $x = -2$.

EXAMPLE 4 Using the linear factors theorem in determining a quadratic function

Find a quadratic function f that has zeros of 3 and 5 and a graph that passes through the point $(2, -9)$.

SOLUTION
The general form of a quadratic function with 3 and 5 as zeros is $f(x) = a_n(x - 3)(x - 5)$. Since the graph passes through $(2, -9)$, we have

$$-9 = a_n(2 - 3)(2 - 5) = a_n(3)$$
$$-3 = a_n$$

The required function is therefore $f(x) = -3(x - 3)(x - 5)$. If we wish, we can carry out the multiplication and rewrite this as

$$f(x) = -3x^2 + 24x - 45$$

Figure 2 summarizes this result.

GRAPHICAL PERSPECTIVE

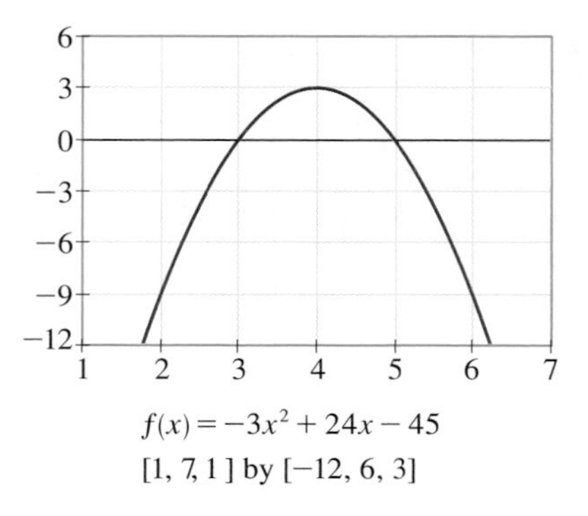

Figure 2
In Example 4 we determined that the function $f(x) = -3x^2 + 24x - 45$ has zeros of 3 and 5 and that the graph passes through the point $(2, -9)$.

The next example shows how we can use the factored form of a polynomial $f(x)$ to determine the relationships between the coefficients of the polynomial and the roots of the equation $f(x) = 0$.

EXAMPLE 5 Applying the linear factors theorem to relate roots and coefficients

Let r_1 and r_2 be the roots of the equation $x^2 + bx + c = 0$. Show that

$$r_1 r_2 = c \qquad \text{and} \qquad r_1 + r_2 = -b$$

REMARK We actually established these two results previously, in Section 2.1. If you look back, you'll see we did this by means of the quadratic formula. The advantage to the following technique is that it does not require a formula for the roots; thus it can be applied to polynomial equations of any degree. (We'll return to this idea after the example.)

SOLUTION

Since r_1 and r_2 are the roots of the equation $x^2 + bx + c = 0$, we have the identity

$$x^2 + bx + c = (x - r_1)(x - r_2)$$

After multiplying out the right-hand side, we can rewrite this identity as

$$x^2 + bx + c = x^2 - (r_1 + r_2)x + r_1 r_2$$

By equating coefficients, we readily obtain $r_1 r_2 = c$ and $r_1 + r_2 = -b$, as required.

The technique used in Example 5 can be used to obtain similar relationships between the roots and the coefficients of polynomial equations of any given degree. In Table 2, for instance, we show the relationships obtained in Example 5, along with the corresponding relationships that can be derived for a cubic equation. (Exercise 47 at the end of this section asks you to verify the results for the cubic equation.)

TABLE 2

Equation	Roots	Relationships between roots and coefficients
$x^2 + bx + c = 0$	r_1, r_2	$r_1 + r_2 = -b$ $r_1 r_2 = c$
$x^3 + bx^2 + cx + d = 0$	r_1, r_2, r_3	$r_1 + r_2 + r_3 = -b$ $r_1 r_2 + r_2 r_3 + r_3 r_1 = c$ $r_1 r_2 r_3 = -d$

We conclude this section with some remarks concerning the solving of polynomial equations by formulas. You know that the roots of the quadratic equation $ax^2 + bx + c = 0$ are given by the formula

$$x = \frac{-b \pm \sqrt{b^2 - 4ac}}{2a}$$

Are there similar formulas for the solutions of higher-degree equations? By "similar" we mean a formula involving the coefficients and radicals. To answer this question, we look at a bit of history. As early as 1700 B.C., Babylonian mathematicians were able to solve quadratic equations. This is clear from the study of the clay tablets with cuneiform numerals that archeologists have found. The ancient Greeks also were able to solve quadratic equations. Like the Babylonians, the Greeks worked without the aid of algebra as we know it. The mathematicians of ancient Greece used geometric constructions to solve equations. Of course, since all quantities were interpreted geometrically, negative roots were never considered.

The general quadratic formula was known to Islamic mathematicians sometime before A.D. 1000. For the next 500 years, mathematicians searched for, but did not discover, a formula to solve the general cubic equation. Indeed, in 1494, Luca Pacioli stated in his text *Summa di Arithmetica* that the general cubic equation could not be solved by the algebraic techniques then available. All of this was to change, however, within the next several decades.

Around 1515 the Italian mathematician Scipione del Ferro solved the cubic equation $x^3 + px + q = 0$ using algebraic techniques. This essentially constituted a solution of the seemingly more general equation $x^3 + bx^2 + cx + d = 0$. The reason for this is that if we make the substitution $x = y - b/3$ in the latter equation, the result is a cubic equation with no y^2-term. By 1540 the Italian mathematician Ludovico Ferrari had solved the general fourth-degree equation. Actually, at that time in Renaissance Italy there was considerable controversy as to exactly who discovered the various formulas first. Details of the dispute can be found in any text on the history of mathematics. But for our purposes here, the point is simply that by the middle of the sixteenth century, all polynomial equations of degree 4 or less could be solved by the formulas that had been discovered. The common feature of these formulas was that they involved the coefficients, the four basic operations of arithmetic, and various radicals. For example, a formula for a solution of the equation $x^3 + px + q = 0$ is as follows:

$$x = \sqrt[3]{\frac{-q}{2} + \sqrt{\frac{q^2}{4} + \frac{p^3}{27}}} + \sqrt[3]{\frac{-q}{2} - \sqrt{\frac{q^2}{4} + \frac{p^3}{27}}}$$

To get some idea of the practical difficulties inherent in computing with this formula, try using it to show that $x = -2$ is a root of the cubic equation $x^3 + 4x + 16 = 0$.

For more than 200 years after the cubic and quartic (fourth-degree) equations had been solved, mathematicians continued to search for a formula that would yield the solutions of the general fifth-degree equation. The first breakthrough, if it can be called that, occurred in 1770, when the French mathematician Joseph Louis Lagrange found a technique that served to unify and summarize all of the previous methods used for the equations of degrees 2, 3, and 4. However, Lagrange then showed that his technique could not work in the case of the general fifth-degree equation. Although we will not describe the details of Lagrange's work here, it is worth pointing out that he relied on the types of relationships between roots and coefficients that we looked at in Table 2 on page 927.

Finally, in 1826 the Norwegian mathematician Niels Henrik Abel proved that for the general polynomial equation of degree 5 or higher, there could be no formula yielding the solutions in terms of the coefficients and radicals. This is not to say that such equations do not possess solutions. In fact they must, as we saw earlier in this section. It is just that we cannot in every case express the solutions in terms of the coefficients and radicals. For example, it can be shown that the equation $x^5 - 6x + 3 = 0$ has a real root between 0 and 1, but this number cannot be expressed in terms of the coefficients and radicals. (We can, however, compute the root to as many decimal places as we wish, as you will see in the next section.) In 1831 the French mathematician Evariste Galois completed matters by giving conditions for determining exactly which polynomial equations can be solved in terms of coefficients and radicals.

NOTE Historical background on polynomial equations and the fundamental theorem of algebra is available at the History of Mathematics website maintained by the University of St. Andrews, in Scotland. The home page is http://www-groups.dcs.st-and.ac.uk/~history. Another website is http://alephO.clarku.edu:80/~djoyce/mathhis/subjects.html. Choose "History Topics Index." Also, there are a number of math history textbooks that discuss this material. To cite but three:

1. *A History of Mathematics: An Introduction,* 2nd ed., Victor. J. Katz (Boston: Addison-Wesley, 1998)
2. *History of Mathematics: An Introduction,* 3rd ed., David M. Burton (New York: McGraw Hill, 1997)
3. *Mathematics and Its History,* 2nd ed., John Stillwell (New York: Springer-Verlag, 2001)

EXERCISE SET 12.4

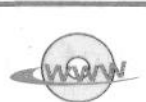

A

According to the fundamental theorem of algebra, which of the equations in Exercises 1 and 2 have at least one root?

1. (a) $x^5 - 14x^4 + 8x + 53 = 0$
(b) $4.17x^3 + 2.06x^2 + 0.01x + 1.23 = 0$
(c) $ix^2 + (2 + 3i)x - 17 = 0$
(d) $x^{2.1} + 3x^{0.3} + 1 = 0$

2. (a) $\sqrt{3}x^{17} + \sqrt{2}x^{13} + \sqrt{5} = 0$
(b) $17x^{\sqrt{3}} + 13x^{\sqrt{2}} + \sqrt{5} = 0$
(c) $1/(x^2 + 1) = 0$
(d) $2^{3x} - 2^x - 1 = 0$

Ⓖ *In each of Exercises 3–10 you are given a polynomial equation $f(x) = 0$. According to the fundamental theorem of algebra, each of these equations has at least one root. However, the fundamental theorem does not tell you whether the equation has any real-number roots. Use a graph to determine whether the equation has at least one real root.* Note: *You are not being asked to solve the equation.*

3. $x^2 - 3x + 2.26 = 0$
4. $x^2 - 2x - 290 = 0$
5. $x^3 - 3x^2 + 3 = 0$
6. $x^4 - 3x^2 + 3 = 0$
7. $x^4 + x^3 + x^2 + x + 1 = 0$
8. $x^5 + x^3 + x^2 + x + 1 = 0$
9. $0.2x^3 + 4.4x^2 - 109x - 1 = 0$
10. $x^4 - \sqrt{35}x^2 + 2.79\pi = 0$

In Exercises 11–18, express each polynomial in the form $a_n(x - r_1)(x - r_2) \cdots (x - r_n)$.

11. $x^2 - 2x - 3$
12. $x^3 - 2x^2 - 3x$
13. $4x^2 + 23x - 6$
14. $6x^2 + x - 12$
15. $x^2 - 5$
16. $x^2 + 5$
17. $x^5 - 7x^3 - 18x$
18. $x^3 + 2x^2 - 3x - 6$

In Exercises 19–24, find a polynomial $f(x)$ with leading coefficient 1 such that the equation $f(x) = 0$ has the given roots and no others. If the degree of $f(x)$ is 7 or more, express $f(x)$ in factored form; otherwise, express $f(x)$ in the form $a_nx^n + a_{n-1}x^{n-1} + \cdots + a_1x + a_0$.

19.

Root	1	−3
Multiplicity	2	1

20.

Root	0	4
Multiplicity	2	1

21.

Root	2	−2	$2i$	$-2i$
Multiplicity	1	1	1	1

22.

Root	$2 + i$	$2 - i$
Multiplicity	1	1

23.

Root	$\sqrt{3}$	$-\sqrt{3}$	$4i$	$-4i$
Multiplicity	2	2	1	1

24.

Root	5	1	$1 - i$	$1 + i$
Multiplicity	2	3	1	1

For Exercises 25–30:

(a) *Find a polynomial $f(x)$ with leading coefficient 1 such that the equation $f(x) = 0$ has the given roots and no others. If the degree of $f(x)$ is more than 4, leave $f(x)$ in factored form rather than multiplying it out.*

Ⓖ **(b)** *Use a graphing utility to check the following fact, mentioned in this section: If $x = r$ is a multiple root of $f(x) = 0$,* *then the graph of the function f is tangent to the x-axis at* $x = r$.

25.

Root	0	1	3
Multiplicity	2	1	1

26.

Root	0	1	3
Multiplicity	1	2	1

27.

Root	0	1	3
Multiplicity	1	2	2

28.

Root	0	1	3
Multiplicity	2	1	2

29.

Root	0	1	3
Multiplicity	2	1	3

30.

Root	0	1	3
Multiplicity	3	2	1

In Exercises 31–34, express the polynomial $f(x)$ *in the form* $a_nx^n + a_{n-1}x^{n-1} + \cdots + a_1x + a_0$.

31. Find a quadratic function that has zeros -4 and 9 and a graph that passes through the point $(3, 5)$.

32. **(a)** Find a quadratic function that has a maximum value of 2 and that has -2 and 4 as zeros.

Ⓖ **(b)** Use a graphing utility to check that your answer in part (a) appears to be correct.

33. **(a)** Find a third-degree polynomial function that has zeros -5, 2, and 3 and a graph that passes through the point $(0, 1)$.

Ⓖ **(b)** Use a graphing utility to check that your answer in part (a) appears to be correct.

34. Find a fourth-degree polynomial function that has zeros $\sqrt{2}, -\sqrt{2}, 1$, and -1 and a graph that passes through $(2, -20)$.

In Exercises 35–42, find a quadratic equation with the given roots. Write your answers in the form $Ax^2 + Bx + C = 0$. *Suggestion: Make use of Table 2.*

35. $r_1 = -i, r_2 = -\sqrt{3}$

36. $r_1 = 1 + i\sqrt{3}, r_2 = 1 - i\sqrt{3}$

37. $r_1 = 9, r_2 = -6$

38. $r_1 = 5, r_2 = 3/4$

39. $r_1 = 1 + \sqrt{5}, r_2 = 1 - \sqrt{5}$

40. $r_1 = 6 - 5i, r_2 = 6 + 5i$

41. $r_1 = a + \sqrt{b}, r_2 = a - \sqrt{b} \quad (b > 0)$

42. $r_1 = a + bi, r_2 = a - bi$

B

43. *Scipio Ferro of Bologna well-nigh thirty years ago discovered this rule and handed it on to Antonio Maria Fior of Venice, whose contest with Niccolò Tartaglia of Brescia gave Niccolò occasion to discover it. He [Tartaglia] gave it to me in response to my entreaties, though withholding the demonstration. Armed with this assistance, I sought out its demonstration in [various] forms.* —Girolamo Cardano, *Ars Magna* (Nuremberg, 1545)

The quotation is from the translation of *Ars Magna* by T. Richard Witmer (New York: Dover Publications, 1993).

In his book *Ars Magna* (*The Great Art*) the Renaissance mathematician Girolamo Cardano (1501–1576) gave the following formula for a root of the equation $x^3 + ax = b$:

$$x = \sqrt[3]{\frac{b}{2} + \sqrt{\frac{b^2}{4} + \frac{a^3}{27}}} - \sqrt[3]{\frac{-b}{2} + \sqrt{\frac{b^2}{4} + \frac{a^3}{27}}}$$

(a) Use this formula and your calculator to compute a root of the cubic equation $x^3 + 3x = 76$.

Ⓖ **(b)** Use a graph to check the answer in part (a). That is, graph the function $y = x^3 + 3x - 76$, and note the x-intercept. Also check the answer simply by substituting it in the equation $x^3 + 3x = 76$.

44. Consider the cubic equation $x^3 = 15x + 4$, which was solved by Rafael Bombelli in his text *L'Algebra parte maggiore del arithmetica* (Bologna: 1572). Bombelli was among the first to use complex numbers to obtain real-number solutions.

(a) Rewrite the equation in the form $x^3 - 15x = 4$, and apply Cardano's formula (given in the previous exercise). Show that this yields

$$x = \sqrt[3]{2 + 11i} - \sqrt[3]{-2 + 11i}$$

(b) Using paper and pencil (or a computer algebra system), show that

$$(2 + i)^3 = 2 + 11i \quad \text{and} \quad (-2 + i)^3 = -2 + 11i$$

(c) Use the results in part (b) to simplify the root given in part (a). You should obtain $x = 4$.

Ⓖ **(d)** Use a graph to check visually that $x = 4$ is a root of $x^3 = 15x + 4$. (That is, graph the function $y = x^3 - 15x - 4$ and note that $x = 4$ appears to be one of the x-intercepts.) Also, check, algebraically that the value $x = 4$ satisfies the equation.

(e) Given that $x = 4$ is a root of the equation $x^3 = 15x + 4$, use the techniques from Section 12.3 to determine the remaining two roots. Check your answers by using a graphing utility to determine the x-intercepts of the graph in part (d).

45. Express the polynomial $x^4 + 64$ as a product of four linear factors. *Hint*: Write $x^4 + 64 = (x^4 + 16x^2 + 64) - 16x^2$, then use the difference-of-squares factoring formula.

46. Suppose that p and q are positive integers with $p > q$. Find a quadratic equation with integer coefficients and roots $\sqrt{p}/(\sqrt{p} \pm \sqrt{p - q})$.

47. Let r_1, r_2, and r_3 be the roots of the equation $x^3 + bx^2 + cx + d = 0$. Use the method shown in Example 5 to verify the following relationships:

$$r_1 + r_2 + r_3 = -b$$
$$r_1r_2 + r_2r_3 + r_3r_1 = c$$
$$r_1r_2r_3 = -d$$

48. Let r_1, r_2, r_3, and r_4 be the roots of the equation $x^4 + bx^3 + cx^2 + dx + e = 0$. Use the method shown in Example 5 to prove the following facts:

$$r_1 + r_2 + r_3 + r_4 = -b$$
$$r_1r_2 + r_2r_3 + r_3r_4 + r_4r_1 + r_2r_4 + r_3r_1 = c$$
$$r_1r_2r_3 + r_2r_3r_4 + r_3r_4r_1 + r_4r_1r_2 = -d$$
$$r_1r_2r_3r_4 = e$$

49. Solve the equation $x^3 - 75x + 250 = 0$, given that two of the roots are equal. *Suggestion*: Use Table 2.

50. For this exercise, assume as given the following trigonometric identity:

$$\tan 3\theta = \frac{\tan^3 \theta - 3 \tan \theta}{3 \tan^2 \theta - 1}$$

(a) Use the given identity to show that the number tan 15° is a root of the cubic equation

$$x^3 - 3x^2 - 3x + 1 = 0 \qquad (1)$$

(b) Use a calculator to check that tan 15° indeed appears to be a root of equation (1).

(c) Factor (by grouping) the left-hand side of equation (1). Conclude that tan 15° is a root of the reduced equation

$$x^2 - 4x + 1 = 0 \qquad (2)$$

(d) The work in parts (a) and (c) shows that the number tan 15° is a root of equation (2). By following the same technique, show that the number tan 75° is also a root of equation (2).

(e) Use Table 2 on page 927 to evaluate each of the following quantities. Then use a calculator to check your answers.

(i) tan 15° + tan 75° **(ii)** tan 15° tan 75°

(f) Use the quadratic formula to solve equation (2). Which root is tan 15° and which is tan 75°?

51. For this exercise, assume as given the following two results which were derived in Exercises 58 and 59 of Exercise Set 8.2:

$$(\cos 72°)(\cos 144°) = -\frac{1}{4}$$

and

$$\cos 72° + \cos 144° = -\frac{1}{2}$$

(a) Find a quadratic equation, with integer coefficients, whose roots are the numbers cos 72° and cos 144°. *Hint*: Use the result in Example 5 or Table 2.

(b) Use the quadratic formula to solve the equation in part (a). Conclude that

$$\cos 72° = \frac{1}{4}(-1 + \sqrt{5})$$

and

$$\cos 144° = \frac{1}{4}(-1 - \sqrt{5})$$

(c) Usc a calculator to check the results in part (b).

52. (a) In the trigonometric identity $\cos 3\theta = 4 \cos^3 \theta - 3 \cos \theta$, make the substitution $\theta = 20°$ and conclude that the number cos 20° is a root of the equation $8x^3 - 6x - 1 = 0$.

(b) Follow the method in part (a) to show that each of the numbers cos 100° and cos 140° is a root of $8x^3 - 6x - 1 = 0$.

(c) Use Table 2 to explain why

$$(\cos 20°)(\cos 100°)(\cos 140°) = \frac{1}{8}$$

(d) Use the result in part (c) and the reference angle concept to show that

$$(\cos 20°)(\cos 40°)(\cos 80°) = \frac{1}{8}$$

(e) Use a calculator to check the result in part (d).

C

53. (a) Let r_1, r_2, r_3, and r_4 be four real roots of the equation $x^4 + ax^2 + bx + c = 0$. Show that

$$r_1 + r_2 + r_3 + r_4 = 0$$

Hint: Use the first formula in Exercise 48.

(b) Suppose a circle intersects the parabola $y = x^2$ in the points $(x_1, y_1), \ldots, (x_4, y_4)$. Show that

$$x_1 + x_2 + x_3 + x_4 = 0$$

Hint: Use the results in part (a).

54. (a) Let r_1, r_2, and r_3 be three distinct numbers that are roots of the equation $f(x) = Ax^2 + Bx + C = 0$. Show that $f(x) = 0$ for all values of x. *Hint*: You need to show that $A = B = C = 0$. First show that both A and B are zero as follows. If either A or B were nonzero, then the equation $f(x) = 0$ would be a polynomial equation of degree at most 2 with three distinct roots. Why is that impossible?

(b) Use the result in part (a) to prove the following identity:

$$\frac{a^2 - x^2}{(a-b)(a-c)} + \frac{b^2 - x^2}{(b-c)(b-a)} + \frac{c^2 - x^2}{(c-a)(c-b)} - 1 = 0$$

Hint: Let $f(x)$ denote the quadratic expression on the left-hand side of the equation. Compute $f(a)$, $f(b)$, and $f(c)$.

55. Prove that the following equation is an identity:

$$\frac{(x-a)(x-b)c^2}{(c-a)(c-b)} + \frac{(x-b)(x-c)a^2}{(a-b)(a-c)} + \frac{(x-c)(x-a)b^2}{(b-c)(b-a)} - x^2 = 0$$

Hint: Use the result in Exercises 54(a).

PROJECT

TWO METHODS FOR SOLVING CERTAIN CUBIC EQUATIONS

The first person known to have solved cubic equations algebraically was del Ferro but he told nobody of his achievement. On his deathbed, however, del Ferro passed on the secret to his (rather poor) student Fior. Fior began to boast that he was able to solve cubics and a challenge between him and Tartaglia was arranged in 1535.—From the MacTutor History of Mathematics website: http://www-groups.dcs.st-and.ac.uk/~history. The site is maintained by the University of St. Andrews, Scotland.

Stated in modern language, one of the problems that Fior challenged Tartaglia to solve in 1535 was essentially as follows.

A Restatement of One of Fior's Challenge Problems for del Ferro

A tree is 12 units high. At what height should it be cut so that the length of the portion left standing equals the cube root of the length of the portion cut off?

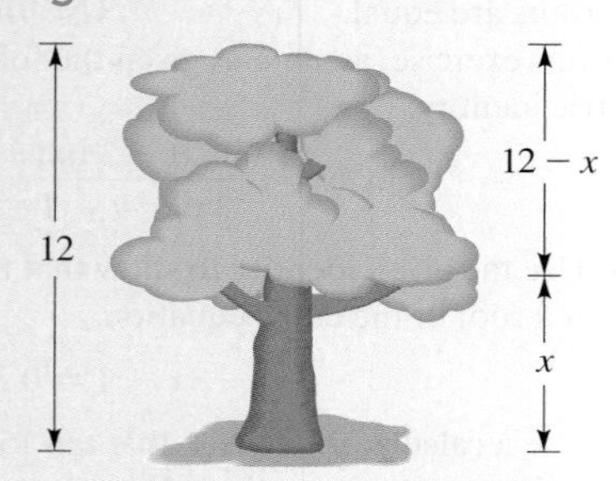

In this project you'll learn two different methods that can be used to determine the root of the cubic equation that solves Fior's tree-cutting problem. As is indicated in the figure above, we let x denote the length of the part of the tree left standing. The condition stated in the problem then gives us $x = \sqrt[3]{12 - x}$. After cubing both sides and rearranging, we have the cubic equation to be solved:

$$x^3 + x - 12 = 0 \qquad (1)$$

Method 1. To find a root of an equation of the form $x^3 + ax + b = 0$, when $a > 0$, make the substitution $x = y - \frac{a}{3y}$. After simplifying, this will produce a sixth-degree equation that is of *quadratic type* and that therefore can be solved as in Section 2.2. Apply this strategy to solve equation (1). *Hints*: At the beginning, you have $a = 1$, and so the substitution to make in equation (1) is $x = y - \frac{1}{3y}$. After simplifying and then solving the resulting equation of quadratic type, you should obtain $y = (6 \pm \sqrt{36 + \frac{1}{27}})^{1/3}$. Then, choosing the positive square root for the moment, you'll obtain the required value for x:

$$x = y - \frac{1}{3y} = (6 + \sqrt{36 + \tfrac{1}{27}})^{1/3} - \frac{1}{3(6 + \sqrt{36 + \frac{1}{27}})^{1/3}}$$

$$\approx 2.14404 \quad \text{using a calculator}$$

What value do you obtain for x if you use the negative square root, that is, $y = (6 - \sqrt{36 + \frac{1}{27}})^{1/3}$?

Method 2. This method uses function iteration. The explanation of iteration in Section 3.5 is a prerequisite for this discussion. It turns out that for certain functions f, a root of the equation $x = f(x)$ can be determined just by iterating the function f with an appropriate initial input. (Note that our tree-cutting equation was initially in the form $x = f(x)$; it was $x = \sqrt[3]{12 - x}$.)

We'll use Figures A and B to explain why iterating a function f can determine a root of the equation $x = f(x)$. Figure A shows the graph of a function f, along with the line $y = x$. Let (r, r) denote the point where the graph of $y = f(x)$ intersects the line $y = x$. (Why do the two coordinates of that point have to be the same?) The number r is then a root of the equation $x = f(x)$. We would like to be able to compute the value of r to as many decimal places as might be required. Figure B shows the iteration process for f using an initial input x_0. From the pattern you can see that the iterates approach the number r. Thus we can, in principle at least, determine the root r to as many decimal places as we want by carrying out the iteration process sufficiently many times.

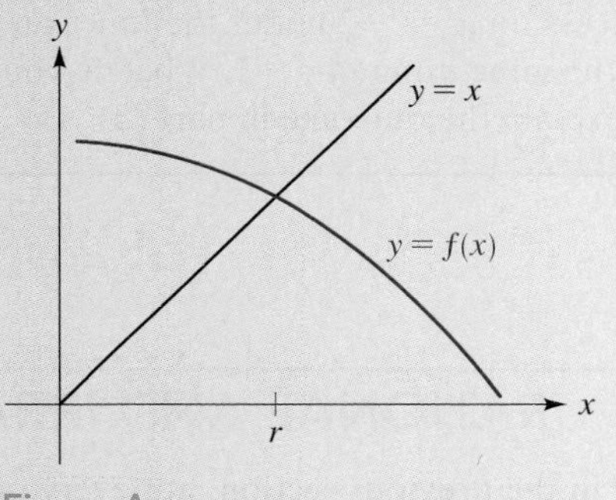

Figure A

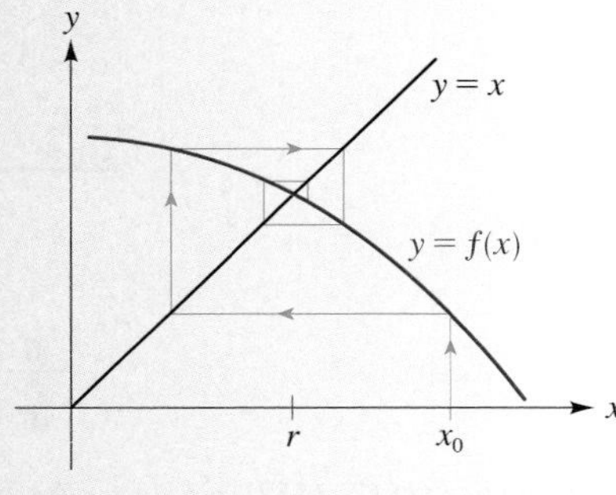

Figure B

Now let's see how this works in the case of Fior's tree-cutting equation $x = \sqrt[3]{12 - x}$. The function to be iterated is $f(x) = \sqrt[3]{12 - x}$. We need an initial input x_0. In general, one first makes an estimate of what the root might be and then uses that estimate or something close to it for x_0. But for the present function f, a remarkable fact is that any initial input x_0 will do. (A case in which the choice of the initial input is important is given in Exercise 1(c) at the end of this project.) Selecting (somewhat at random) the initial input $x_0 = 5$ now, we obtain the results shown in Table 1. You should check the entries in this table yourself and then set up similar tables using different initial inputs. In each case you will find that the iterates approach the required root, as determined in Method 1.

TABLE 1 The First Six Iterates of $x_0 = 5$ Under the Function $f(x) = \sqrt[3]{12 - x}$

x_1	1.91293 . . .
x_2	2.16066 . . .
x_3	2.14283 . . .
x_4	2.14412 . . .
x_5	2.14403 . . .
x_6	2.14404 . . .

The numbers are not rounded. The digits in red show the agreement with the value obtained using Method 1.

EXERCISES

1. Consider the cubic equation $g(x) = x^3 + 6x - 2 = 0$.

(a) Use a graphing utility to draw the graph of g. By zooming in on the x-intercept, approximate to the nearest hundredth the root of the given equation.

(b) Use Method 1 to obtain an exact expression for the root. Then use a calculator to check that the value is consistent with the result in part (a).

(c) From the equation $x^3 + 6x - 2 = 0$, we have $6x = 2 - x^3$, and consequently, $x = (2 - x^3)/6$. Use Method 2, with $f(x) = (2 - x^3)/6$, to determine a root of the equation $x = (2 - x^3)/6$. For the initial input x_0, use any number in the closed interval $[-2, 2]$. What happens if you take $x_0 = 3$?

(d) From the equation $x^3 + 6x - 2 = 0$, we have $x^3 = 2 - 6x$, and consequently, $x = (2 - 6x)^{1/3}$. Try using Method 2, with $f(x) = (2 - 6x)^{1/3}$, to determine a root. Experiment with various values for x_0.

2. (a) Use Method 1 to find a real root of the cubic equation $x^3 + 3x - 36 = 0$.

Answer: $x = (18 + 5\sqrt{13})^{1/3} - 1/(18 + 5\sqrt{13})^{1/3}$

(b) Use a calculator to evaluate the answer in part (a). Is the value shown on the calculator actually a root? Find all roots of the given equation.

3. (a) Method 1 will always yield a real root of the equation $x^3 + ax + b = 0$, when $a > 0$. In some cases it also works for $a < 0$. As an example, use Method 1 to find a root of the equation $x^3 - 3x + 4 = 0$. For the answer, give both the exact expression containing radicals and a calculator approximation rounded to four decimal places.

(b) The equation $x^3 - 3x + 4 = 0$ can be rewritten $x = (x^3 + 4)/3$. Try using Method 2, with $f(x) = (x^3 + 4)/3$, to determine a root of the equation.

(c) Let $f(x) = (x^3 + 4)/3$, and take $x_0 = -2$. Graphically display the iteration process for $x_0 = -2$ under the function f. (Go at least as far as x_4.) Then do the same using $x_0 = -1$. What do you observe? Use your observations to explain the outcome in part (b).

12.5 RATIONAL AND IRRATIONAL ROOTS

Jacques Peletier (1517–1582), a French man of letters, poet, and mathematician, had observed as early as 1558, that the root of an equation is a divisor of the last term. —Florian Cajori in *A History of Mathematics,* 4th ed. (New York: Chelsea Publishing Co., 1985)

As we saw in the previous section, not every polynomial equation has a real root. Furthermore, even if a polynomial equation does possess a real root, that root isn't necessarily a rational number. (The equation $x^2 = 2$ provides a simple example.) If a polynomial equation with integer coefficients does have a rational root, however, we can find that root by applying the **rational roots theorem,** which we now state.

The Rational Roots Theorem

Consider the polynomial equation

$$a_nx^n + a_{n-1}x^{n-1} + \cdots + a_1x + a_0 = 0 \qquad (n \geq 1, a_n \neq 0)$$

and suppose that all the coefficients are integers. Let p/q be a rational number, where p and q have no common factors other than ±1. If p/q is a root of the equation, then p is a factor of a_0 and q is a factor of a_n.

A proof of the rational roots theorem is outlined in Exercise 41 at the end of this section. For the moment, though, let's just see why the theorem is plausible. Suppose that the two rational numbers a/b and c/d are the roots of a certain quadratic equation. Then, from our experience with quadratics (or by the linear factors theorem) we know that the equation can be written in the form

$$k\left(x - \frac{a}{b}\right)\left(x - \frac{c}{d}\right) = 0 \qquad (1)$$

where k is a constant. Now, as Exercise 30 will ask you to check, if we carry out the multiplication and clear of fractions, equation (1) becomes

$$(kbd)x^2 - (kad + kbc)x + kac = 0 \tag{2}$$

Observe that a and c (the numerators of the two roots) are factors of the constant term kac in equation (2), just as the rational roots theorem asserts. Furthermore, b and d (the denominators of the roots) are factors of the coefficient of the x^2-term in equation (2), again as the theorem asserts.

The following example shows how the rational roots theorem can be used to solve a polynomial equation.

EXAMPLE 1 Applying the rational roots theorem

Find the rational roots (if any) of the equation $2x^3 - x^2 - 9x - 4 = 0$. Then solve the equation.

SOLUTION

First we list the factors of a_0, the factors of a_n, and the possibilities for rational roots:

factors of $a_0 = -4$: $p = \pm 1, \pm 2, \pm 4$

factors of $a_3 = 2$: $q = \pm 1, \pm 2$

possible rational roots: $\dfrac{p}{q} = \pm\dfrac{1}{1}, \pm\dfrac{1}{2}, \pm\dfrac{2}{1}, \pm\dfrac{4}{1}$

Now we can use synthetic division to test whether or not any of these possibilities is a root. (A zero remainder will tell us that we have a root.) As you can check, the first three possibilities (1, −1, and 1/2) are not roots. However, using −1/2, we have

$$\begin{array}{r|rrrr} -1/2 & 2 & -1 & -9 & -4 \\ & & -1 & 1 & 4 \\ \hline & 2 & -2 & -8 & 0 \end{array}$$

Thus $x = -1/2$ is a root. We could now continue to check the remaining possibilities in this same manner. At this point, however, it is simpler to consider the reduced equation $2x^2 - 2x - 8 = 0$, or $x^2 - x - 4 = 0$. Since this is a quadratic equation, it can be solved directly. We have

$$x = \frac{-(-1) \pm \sqrt{(-1)^2 - 4(1)(-4)}}{2(1)} = \frac{1 \pm \sqrt{17}}{2}$$

We have now determined three distinct roots. Since the degree of the original equation is 3, there can be no other roots. We conclude that $x = -1/2$ is the only rational root. The three roots of the equation are $-1/2$ and $(1 \pm \sqrt{17})/2$.

As Example 1 indicates, the number of possibilities for rational roots can be relatively large, even for rather simple equations. The next theorem that we develop allows us to reduce the number of possibilities. We say that a real number B is an **upper bound** for the roots of an equation if every real root is less than or equal to

B. Similarly, a real number b is a **lower bound** for the roots of an equation if every real root is greater than or equal to b. The following theorem tells us how synthetic division can be used in determining upper and lower bounds for roots.

The Upper and Lower Bound Theorem for Real Roots

Consider the polynomial equation

$$f(x) = a_n x^n + a_{n-1}x^{n-1} + \cdots + a_1 x + a_0 = 0$$

where all of the coefficients are real numbers and a_n is positive.

1. If we use synthetic division to divide $f(x)$ by $x - B$, where $B > 0$, and we obtain a third row containing no negative numbers, then B is an upper bound for the real roots of $f(x) = 0$.
2. If we use synthetic division to divide $f(x)$ by $x - b$, where $b < 0$, and we obtain a third row in which the numbers are alternately positive and negative, then b is a lower bound for the real roots of $f(x) = 0$. (In determining whether the signs alternate in the third row, zeros are counted as either positive or negative.)

CAUTION A number may fail the lower bound test but still be a lower bound. For an example, see Exercise 60.

We will prove the first part of this theorem. A proof of the second part can be developed along similar lines. To prove the first part of the theorem, we use the division algorithm to write

$$f(x) = (x - B) \cdot Q(x) + R \tag{3}$$

The remainder R here is a constant that may be zero. To show that B is an upper bound, we must show that any number greater than B is not a root. Toward this end, let p be a number that is greater than B. Note that p must be positive, since B is positive. Then with $x = p$, equation (3) becomes

$$f(p) = (p - B) \cdot Q(p) + R \tag{4}$$

We will now show that the right-hand side of equation (4) is positive. This will tell us that p is not a root. First, look at the factor $(p - B)$. This is positive, since p is greater than B. Next consider $Q(p)$. By hypothesis the coefficients of $Q(x)$ are all nonnegative. Furthermore, the leading coefficient of $Q(x)$ is a_n, which is positive. Since p is also positive, it follows that $Q(p)$ must be positive. Finally, the number R is nonnegative because, in the synthetic division of $f(x)$ by $x - B$, all the numbers in the third row are nonnegative. It now follows that the right-hand side of equation (4) is positive. Consequently, $f(p)$ is not zero, and p is not a root of the equation $f(x) = 0$. This is what we wished to show.

EXAMPLE 2 Solving an equation using the rational roots theorem and the upper and lower bound theorem

Determine the rational roots, or show that none exist, for the equation

$$\frac{1}{4}x^4 - \frac{3}{4}x^3 + \frac{17}{4}x^2 + 4x + 5 = 0$$

SOLUTION

We will use the rational roots theorem along with the upper and lower bound theorem. First of all, if we are to apply the rational roots theorem, then our equation must have integer coefficients. In view of this, we multiply both sides of the given equation by 4 to obtain

$$x^4 - 3x^3 + 17x^2 + 16x + 20 = 0$$

As in the previous example, we list the factors of a_0, the factors of a_n, and the possibilities for rational roots:

factors of $a_0 = 20$: $p = \pm1, \pm2, \pm4, \pm5, \pm10, \pm20$

factors of $a_4 = 1$: $q = \pm1$

possible rational roots: $\frac{p}{q} = \pm1, \pm2, \pm4, \pm5, \pm10, \pm20$

Our strategy here will be to first check for positive roots, beginning with 1 and working upward. The checks for $x = 1$, $x = 2$, and $x = 4$ are as follows:

1	1	−3	17	16	20
		1	−2	15	31
	1	−2	15	31	51
2	1	−3	17	16	20
		2	−2	30	92
	1	−1	15	46	112
4	1	−3	17	16	20
		4	4	84	400
	1	1	21	100	420

As you can see, none of the remainders here is zero. However, notice that in the division corresponding to $x = 4$, all the numbers that appear in the third row are nonnegative. It therefore follows that 4 is an upper bound for the roots of the given equation. In view of this, we needn't bother to check the remaining values $x = 5$, $x = 10$, and $x = 20$, since none of those can be roots. At this point we can conclude that the given equation has no positive rational roots.

Next we check for negative rational roots, beginning with -1 and working downward (if necessary). Checking $x = -1$, we have

−1	1	−3	17	16	20
		−1	4	−21	5
	1	−4	21	−5	25

Two conclusions can be drawn from this synthetic division. First, $x = -1$ is not a root of the equation. Second, -1 is a lower bound for the roots because the signs in the third row of the synthetic division alternate. This means that we needn't bother to check any of the numbers $-2, -4, -5, -15$, and -20; none of them can be roots, since they are all less than -1.

Let's summarize our results. We have shown that the given equation has no positive rational roots and no negative rational roots. Furthermore, by inspection we see that zero is not a root of the equation. Thus the given equation possesses no rational roots.

We conclude this section by demonstrating a method for approximating irrational roots. The method depends on the **location theorem.**

The Location Theorem

Let $f(x)$ be a polynomial, all of whose coefficients are real numbers. If a and b are real numbers such that $f(a)$ and $f(b)$ have opposite signs, then the equation $f(x) = 0$ has at least one real root between a and b.

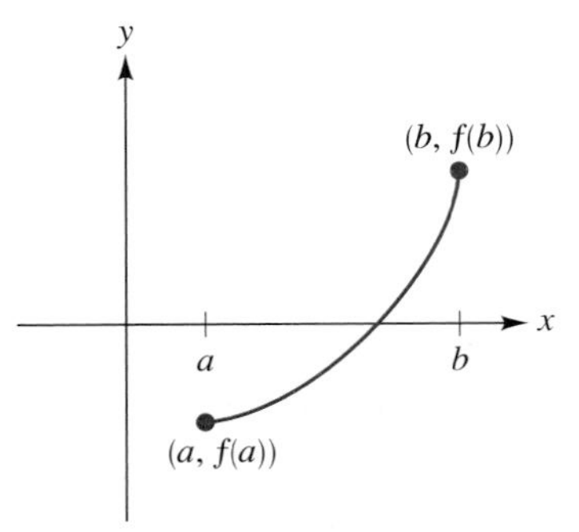

Figure 1

Figure 1 indicates why this theorem is plausible. If the point $(a, f(a))$ lies below the x-axis and $(b, f(b))$ lies above the x-axis, then it certainly seems that the graph of f must cross the x-axis at some point x_0 between a and b. At this intercept we have $f(x_0) = 0$; that is, x_0 is a root of the equation $f(x) = 0$. (The location theorem is a special case of the *intermediate value theorem,* which is usually discussed in calculus courses.)

Our technique for approximating (or "locating") irrational roots uses the **method of successive approximations.** We will demonstrate this method in Example 3.

EXAMPLE 3 Using the method of successive approximations to locate a root

As is indicated by both Table 1 and Figure 2, the equation

$$f(x) = x^5 - 3x^2 + 4x + 2 = 0$$

has a real root between -1 and 0. Use the method of successive approximations to locate this root between successive hundredths.

TABLE 1 $f(x) = x^5 - 3x^2 + 4x + 2$*

x	-3	-2	-1	0	1	2
$f(x)$	-280	-50	-6	2	4	30
			sign change			

*The sign change is a signal that the equation $x^5 - 3x^2 + 4x + 2 = 0$ has a real root between -1 and 0.

GRAPHICAL PERSPECTIVE

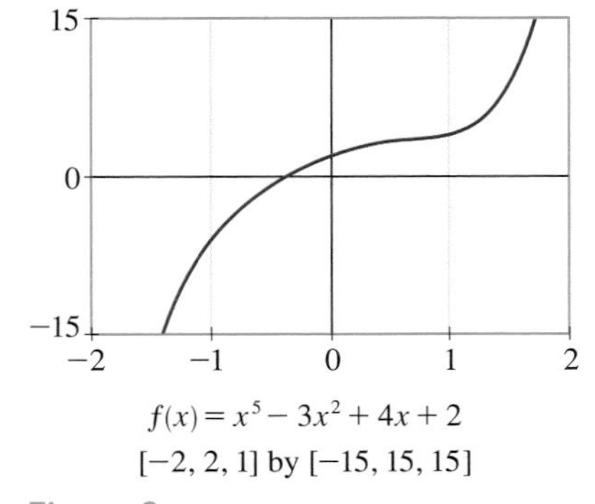

$f(x) = x^5 - 3x^2 + 4x + 2$
$[-2, 2, 1]$ by $[-15, 15, 15]$

Figure 2

SOLUTION

The table or the graph gives us the location of the root between the successive integers -1 and 0. The strategy now is to locate the root between successive tenths, and then between successive hundredths. In Table 2 we've used a graphing utility to generate values of $f(x)$ for inputs x running from -1 up to 0 in increments of 0.1 (For details on using a graphing utility to create a table of function values, refer to the user's manual for your graphing utility.) If a graphing utility is not available, the calculations can be carried out by using synthetic division, the remainder theorem, and an ordinary calculator.

TABLE 2 Decimal Values for $f(x)$ Rounded to Two Places*

x	−1.0	−0.9	−0.8	−0.7	−0.6	−0.5	−0.4	−0.3	−0.2	−0.1	0.0
$f(x)$	−6	−4.62	−3.45	−2.44	−1.56	−0.78	−0.09	0.53			
							sign change				

*Once a sign change is detected, it is not necessary to look at the remaining entries in the table (unless one is looking for a possible second root).

As is indicated in Table 2, there is a change in the sign of $f(x)$ as we go from $x = -0.4$ to $x = -0.3$. Thus, according to the location theorem, the equation $f(x) = 0$ has a root in the open interval $(-0.4, -0.3)$.

Next, we follow a similar procedure to locate the root between successive hundredths. This time, the inputs begin with $x = -0.40$ and they increase in increments of 0.01. See Table 3.

TABLE 3 Decimal Values for $f(x)$ Rounded to Two Places*

x	−0.40	−0.39	−0.38	−0.37	−0.36	−0.35	−0.34	−0.33	−0.32	−0.31	−0.30
$f(x)$	−0.09	−0.03	0.04								
		sign change									

*The sign change indicates the presence of a root in the open interval $(-0.39, -0.38)$.

In view of the location theorem, the sign change in Table 3 tells us that the given equation has a root between $x = -0.39$ and $x = -0.38$. So, we've located the root between successive hundredths, as required.

Remark: An alternative procedure for this example would be to begin immediately with a table of values with the inputs running from $x = -1$ to 0, in increments of 0.01. Which method uses fewer calculations?

EXERCISE SET 12.5

A

1. **(a)** State the rational roots theorem.
   (b) List the possibilities for the rational roots of the equation $x^7 - 144x^2 - 8x - 11 = 0$.
2. Use the rational roots theorem to list the possibilities for the rational roots of each equation.
   (a) $x^4 - 32x^3 + 40x^2 + 12x - 3 = 0$
   (b) $3x^4 - 32x^3 + 40x^2 + 12x - 1 = 0$
3. **(a)** Use the rational roots theorem to explain why the equation $x^5 - 14x - 5 = 0$ cannot have a rational root that is larger than 5. Is 5 a root?
   (b) The graph in the following figure indicates that 2 may be a root of the equation $x^5 - 14x - 5 = 0$. Use the rational roots theorem to explain why 2 cannot be a root. Also, show directly that 2 is not a root by substituting 2 in the given equation.

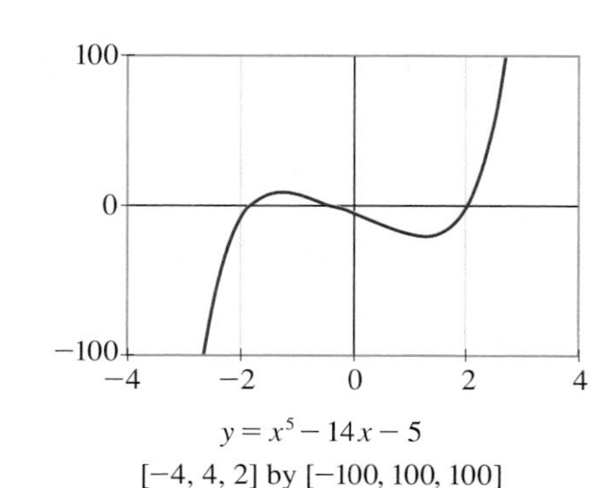

$y = x^5 - 14x - 5$
$[-4, 4, 2]$ by $[-100, 100, 100]$

Ⓖ **(c)** With a graphing utility, graph $y = x^5 - 14x - 5$; use a viewing rectangle that shows that 2 is not a root of the equation $x^5 - 14x - 5 = 0$.

4. **(a)** The graph in the following figure indicates that 4 may be a root of the equation $x^6 - 4x^5 + 8x^2 - 3x - 101 = 0$. Use the rational roots theorem to explain why 4 cannot be a root.

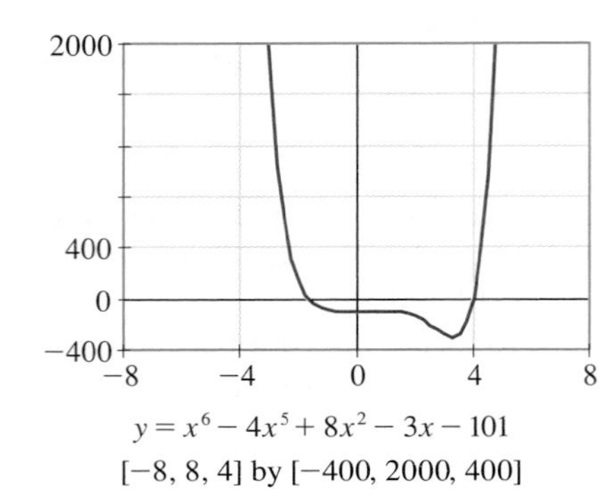

$y = x^6 - 4x^5 + 8x^2 - 3x - 101$
$[-8, 8, 4]$ by $[-400, 2000, 400]$

Ⓖ **(b)** With a graphing utility, graph $y = x^6 - 4x^5 + 8x^2 - 3x - 101$; use a viewing rectangle that shows that 4 is not a root of the equation $x^6 - 4x^5 + 8x^2 - 3x - 101 = 0$.

In Exercises 5–10, list the possibilities for rational roots.

5. $4x^3 - 9x^2 - 15x + 3 = 0$ **6.** $x^4 - x^3 + 10x^2 - 24 = 0$
7. $8x^5 - x^2 + 9 = 0$ **8.** $18x^4 - 10x^3 + x^2 - 4 = 0$
9. $\frac{2}{3}x^3 - x^2 - 5x + 2 = 0$
10. $\frac{1}{2}x^4 - 5x^3 + \frac{4}{3}x^2 + 8x - \frac{1}{3} = 0$

In Exercises 11–16, show that each equation has no rational roots.

11. $x^3 - 3x + 1 = 0$ **12.** $x^3 + 8x^2 - 1 = 0$
13. $x^3 + x^2 - x + 1 = 0$ **14.** $x^4 + 4x^3 + 4x^2 - 16 = 0$
15. $12x^4 - x^2 - 6 = 0$
16. $4x^5 - x^4 - x^3 - x^2 + x - 8 = 0$

For Exercises 17–27, find the rational roots of each equation, and then solve the equation. (Use the rational roots theorem and the upper and lower bound theorem, as in Example 2.)

17. $x^3 + 3x^2 - x - 3 = 0$ **18.** $2x^3 - 5x^2 - 3x + 9 = 0$
19. $4x^3 + x^2 - 20x - 5 = 0$ **20.** $3x^3 - 16x^2 + 17x - 4 = 0$
21. $9x^3 + 18x^2 + 11x + 2 = 0$ **22.** $4x^3 - 10x^2 - 25x + 4 = 0$
23. $x^4 + x^3 - 25x^2 - x + 24 = 0$
24. $10x^4 + 107x^3 + 301x^2 + 171x + 23 = 0$
25. $x^4 - 4x^3 + 6x^2 - 4x + 1 = 0$
26. $x^3 - \frac{5}{2}x^2 - 23x + 12 = 0$
27. $x^3 - \frac{17}{3}x^2 - \frac{10}{3}x + 8 = 0$

In Exercises 28 and 29, determine integral upper and lower bounds for the real roots of the equations. (Follow the method used within the solution of Example 2.)

28. **(a)** $x^3 + 2x^2 - 5x + 20 = 0$
(b) $x^5 - 3x^2 + 100 = 0$

29. **(a)** $5x^4 - 10x - 12 = 0$
(b) $3x^4 - 4x^3 + 5x^2 - 2x - 4 = 0$
(c) $2x^4 - 7x^3 - 5x^2 + 28x - 12 = 0$

30. Referring to equation (1) in this section, multiply out the left-hand side, then clear the equation of fractions. Check that your result agrees with equation (2).

In Exercises 31–36, each equation has exactly one positive root. In each case, locate the root between successive hundredths. For Exercises 31 and 32, you are given the successive integer bounds for the root. For the other exercises, determine the successive integer bounds by computing f(0), f(1), f(2), and so on, until you find a sign change.

31. $x^3 + x - 1 = 0$; root between 0 and 1
32. $x^3 - 2x - 5 = 0$; root between 2 and 3
33. $x^5 - 200 = 0$
34. $x^3 - 3x^2 + 3x - 26 = 0$
35. $x^3 - 8x^2 + 21x - 22 = 0$
36. $2x^4 - x^3 - 12x^2 - 16x - 8 = 0$

Ⓖ *In Exercises 37–40, each polynomial equation has exactly one negative root.*

(a) *Use a graphing utility to determine successive integer bounds for the root.*
(b) *Use the method of successive approximations to locate the root between successive thousandths. (Make use of the graphing utility to generate the required tables.)*

37. $x^3 + x^2 - 2x + 1 = 0$
38. $\dfrac{x^5}{10{,}000} - \dfrac{x^3}{50} + \dfrac{x}{1250} + \dfrac{1}{2000} = 0$
39. $x^3 + 2x^2 + 2x + 101 = 0$
40. $x^4 + 4x^3 - 6x^2 - 8x - 3 = 0$

B

41. This exercise outlines a proof of the rational roots theorem. At one point in the proof, we will need to rely on the following fact, which is proved in courses on number theory.

FACT FROM NUMBER THEORY Suppose that A, B, and C are integers and that A is a factor of the number BC. If A has no factor in common with C (other than ± 1), then A must be a factor of B.

(a) Let $A = 2$, $B = 8$, and $C = 5$. Verify that the fact from number theory is correct here.
(b) Let $A = 20$, $B = 8$, and $C = 5$. Note that A is a factor of BC, but A is not a factor of B. Why doesn't this contradict the fact from number theory?
(c) Now we're ready to prove the rational roots theorem. We begin with a polynomial equation with integer coefficients:

$$a_nx^n + a_{n-1}x^{n-1} + \cdots + a_1x + a_0 = 0 \qquad (n \geq 1, a_n \neq 0)$$

We assume that the rational number p/q is a root of the equation and that p and q have no common factors other than 1. Why is the following equation now true?

$$a_n\left(\frac{p}{q}\right)^n + a_{n-1}\left(\frac{p}{q}\right)^{n-1} + \cdots + a_1\left(\frac{p}{q}\right)a_0 = 0$$

(d) Show that the last equation in part (c) can be written

$$p(a_np^{n-1} + a_{n-1}qp^{n-2} + \cdots + a_1q^{n-1}) = -a_0q^n$$

Since p is a factor of the left-hand side of this last equation, p must also be a factor of the right-hand side. That is, p must be a factor of a_0q^n. But since p and q have no common factors, neither do p and q^n. Our fact from number theory now tells us that p must be a factor of a_0, as we wished to show. (The proof that q is a factor of a_n is carried out in a similar manner.)

42. The location theorem asserts that the polynomial equation $f(x) = 0$ has a root in the open interval (a, b) whenever $f(a)$ and $f(b)$ have unlike signs. If $f(a)$ and $f(b)$ have the same sign, can the equation $f(x) = 0$ have a root between a and b? *Hint*: Look at the graph of $f(x) = x^2 - 2x + 1$ with $a = 0$ and $b = 2$.

In Exercises 43–47, first graph the two functions. Then use the method of successive approximations to locate, between successive thousandths, the x-coordinate of the point where the graphs intersect. In Exercises 43 and 44, draw the graphs by hand. In Exercises 45–47, use a graphing utility to draw the graphs as well as to check your final answer. Finally, in Exercise 47, also check your answer by using an algebraic method to obtain the exact solution (as in Section 10.6).

43. $y = x^3 - 5;\ y = 2x - 3$ **44.** $y = x^3;\ y = 4 - x^2$

Ⓖ **45.** $y = e^{-x};\ y = \ln x$ *Remark*: The method of successive approximations is not restricted to polynomial functions.

Ⓖ **46.** $y = 20x^2 + 25x + 9;\ y = x^3$

Ⓖ **47.** $y = x^5 + 100;\ y = x^5 + \frac{1}{2}x^3$

48. In a note that appeared in *The Two-Year College Mathematics Journal* [vol. 12 (1981), pp. 334–336], Professors Warren Page and Leo Chosid explain how the process of testing for rational roots can be shortened. In essence, their result is as follows. Suppose that we have a polynomial with integer coefficients and we are testing for a possible root p/q. Then, if a noninteger is generated at any point in the synthetic division process, p/q cannot be a root of the polynomial. For example, suppose we want to know whether 4/3 is a root of $6x^4 - 10x^3 + 2x^2 - 9x + 8 = 0$. The first few steps of the synthetic division are as follows.

4/3	6	−10	2	−9	8
		8	−8/3		
	6	−2			

Since the noninteger −8/3 has been generated in the synthetic division process, the process can be stopped; 4/3 is not a root of the polynomial. Use this idea to shorten your work in testing to see whether the numbers 3/4, 1/8, and −3/2 are roots of the equation

$$8x^5 - 5x^4 + 3x^2 - 2x - 6 = 0$$

49. In a note that appeared in *The College Mathematics Journal* [vol. 20 (1989), pp. 139–141], Professor Don Redmond proved the following interesting result.

> Consider the polynomial equation $f(x) = a_nx^n + a_{n-1}x^{n-1} + \cdots + a_1x + a_0 = 0$, and suppose that the degree of $f(x)$ is at least 2 and that all of the coefficients are integers. If the three numbers a_0, a_n, and $f(1)$ are all odd, then the given equation has no rational roots.

Use this result to show that the following equations have no rational roots.

(a) $9x^5 - 8x^4 + 3x^2 - 2x + 27 = 0$
(b) $5x^5 + 5x^4 - 11x^2 - 3x - 25 = 0$

Ⓖ **50.** The following result is a particular case of a theorem proved by Professor David C. Kurtz in *The American Mathematical Monthly* [vol. 99 (1992), pp. 259–263].

> Suppose we have a cubic equation $a_3x^3 + a_2x^2 + a_1x + a_0 = 0$ in which all of the coefficients are positive real numbers. Furthermore, suppose that the following two inequalities hold.
>
> $$a_1^2 > 4a_0a_2 \qquad \text{and} \qquad a_2^2 > 4a_1a_3$$
>
> Then the cubic equation has three distinct real roots.

(a) Check that these inequalities are valid in the case of the equation $2x^3 + 8x^2 + 7x + 1 = 0$. This implies that the equation has three distinct real roots. Use a graphing utility to verify this and to estimate each root to the nearest one hundredth.

(b) Follow part (a) for the equation $3x^3 + 40x^2 + 100x + 30 = 0$.

(c) Use a graphing utility to demonstrate that the graph of $y = 6x^3 + 15x^2 + 11x + 2$ has three distinct x-intercepts. Thus, the equation $6x^3 + 15x^2 + 11x + 2 = 0$ has three distinct real roots. Now check that the condition $a_2^2 > 4a_1a_3$ fails to hold in this case. Explain why this does not contradict the result from Professor Kurtz stated above.

51. (a) Use a calculator to verify that the number tan 9° appears to be a root of the following equation:

$$x^4 - 4x^3 - 14x^2 - 4x + 1 = 0 \tag{1}$$

In parts (b) through (d) of this exercise, you will *prove* that tan 9° is indeed a root and that tan 9° is irrational.

(b) Use the trigonometric identity

$$\tan 5\theta = \frac{\tan^5\theta - 10\tan^3\theta + 5\tan\theta}{5\tan^4\theta - 10\tan^2\theta + 1}$$

to show that the number $x = \tan 9°$ is a root of the fifth-degree equation

$$x^5 - 5x^4 - 10x^3 + 10x^2 + 5x - 1 = 0 \tag{2}$$

(continues)

Hint: In the given trigonometric identity, substitute $\theta = 9°$.

(c) List the possibilities for the rational roots of equation (2). Then use synthetic division and the remainder theorem to show that there is only one rational root. What is the reduced equation in this case?

(d) Use your work in parts (b) and (c) to explain (in complete sentences) why the number tan 9° is an irrational root of equation (1).

52. As background for this exercise you need to have worked Exercise 51.

(a) Follow exactly the same method used in parts (b) through (d) of Exercise 51 to show that the number $-\tan 27°$ is an irrational root of equation (1) in Exercise 51.

(b) From Exercise 51 and part (a) of this exercise, we know that both of the numbers tan 9° and $-\tan 27°$ are roots of equation (1) in Exercise 51. By following the same method, it can also be shown that the numbers $-\tan 63°$ and $\tan 81°$ are roots of equation (1). Assuming this fact, along with the results in Exercise 48 on page 931, evaluate each of the following quantities, then use a calculator to check your results.

(i) $\tan 9° \tan 27° \tan 63° \tan 81°$

(ii) $\tan 9° - \tan 27° - \tan 63° + \tan 81°$

53. (a) Let $\theta = 2\pi/7$. Use the reference angle concept to explain why $\cos 3\theta = \cos 4\theta$, then use your calculator to confirm the result.

(b) For this portion of the exercise, assume as given the following two trigonometric identities:

$$\cos 3\theta = 4\cos^3\theta - 3\cos\theta$$
$$\cos 4\theta = 8\cos^4\theta - 8\cos^2\theta + 1$$

Use these identities and the result in part (a) to show that $\cos(2\pi/7)$ is a root of the equation

$$8x^4 - 4x^3 - 8x^2 + 3x + 1 = 0 \qquad (1)$$

(c) List the possibilities for the rational roots of equation (1). Then use synthetic division and the remainder theorem to show that there is only one rational root. Check that the reduced equation in this case is

$$8x^3 + 4x^2 - 4x - 1 = 0 \qquad (2)$$

(d) The work in parts (a) through (c) shows that the number $\cos(2\pi/7)$ is a root of equation (2). By following the same technique, it can be shown that the numbers $\cos(4\pi/7)$ and $\cos(6\pi/7)$ also are roots of equation (2). Use this fact, along with Table 2 in Section 12.4, to evaluate each of the following quantities. Then use a calculator to check your answers.

(i) $\cos\frac{2\pi}{7}\cos\frac{4\pi}{7}\cos\frac{6\pi}{7}$

(ii) $\cos\frac{2\pi}{7} + \cos\frac{4\pi}{7} + \cos\frac{6\pi}{7}$

C

In Exercises 54–58 you need to know that a ***prime number*** *is a positive integer greater than* 1 *with no factors other than itself and* 1. *Thus the first seven prime numbers are* 2, 3, 5, 7, 11, 13, *and* 17.

54. Find all prime numbers p for which the equation $x^2 + x - p = 0$ has a rational root.

55. Find all prime numbers p for which the equation $x^3 + x^2 + x - p = 0$ has at least one rational root. For each value of p that you find, find the corresponding *real* roots of the equation.

56. Consider the equation $x^2 + x - pq = 0$, where p and q are prime numbers. If this equation has rational roots, show that these roots must be -3 and 2. *Suggestion*: The possible rational roots are ± 1, $\pm p$, $\pm q$, and $\pm pq$. In each case, assume that the given number is a root, and see where that leads.

57. Consider the equation $x^3 + px - q = 0$, where p and q are prime numbers. Observe that there are only four possible rational roots here: $1, -1, q$, and $-q$.

(a) Show that if $x = 1$ is a root, then we must have $q = 3$ and $p = 2$. What are the remaining roots in this case?

(b) Show that none of the numbers -1, q, and $-q$ can be a root of the equation. *Hint*: For each case, assume the contrary, and deduce a contradiction.

58. If p and q are prime numbers, show that the equation $x^3 + px - pq = 0$ has no rational roots.

59. Find all integral values of b for which the equation $x^3 - b^2x^2 + 3bx - 4 = 0$ has a rational root.

60. Let $f(x) = x^3 + 3x^2 - x - 3$.

(a) Factor $f(x)$ by using the basic factoring techniques in Appendix B.4.

(b) Sketch the graph of $f(x) = x^3 + 3x^2 - x - 3$. Note that -3 is a lower bound for the roots.

(c) Show that the number -3 fails the lower bound test. This shows that a number may fail the lower bound test and yet be a lower bound. (We say that the lower bound test provides a *sufficient* but not a *necessary* condition for a lower bound.)

12.6 CONJUGATE ROOTS AND DESCARTES'S RULE OF SIGNS

An equation can have as many true [positive] *roots as it contains changes of sign, from plus to minus or from minus to plus; and as many false* [negative] *roots as the number of times two plus signs or two minus signs are found in succession.*—René Descartes (1637)

We now proceed to investigate a remarkable theorem, implicit in the work of [Thomas] *Harriot* [1560–1621] *but first used explicitly by Descartes (1637), which limits the number of positive or negative roots of an equation. . . .*

Remarkable as the Harriot-Descartes's Rule of Signs is, it still leaves uncertainty as to the exact number of real roots in an equation: it only gives an upper limit to them. The problem of finding an exact test . . . was finally solved in 1829 by [Jacques Charles François] *Sturm.*
—H. W. Turnbull in *Theory of Equations* (Edinburgh: Oliver and Boyd, 1939)

As you know from earlier work involving quadratic equations with real coefficients, when nonreal complex roots occur, they occur in conjugate pairs. For instance, as you can check by means of the quadratic formula, the roots of the equation $x^2 - 2x + 5 = 0$ are $1 + 2i$ and $1 - 2i$. The **conjugate roots theorem** tells us that the situation is the same for all polynomial equations with real coefficients.

The Conjugate Roots Theorem

Let $f(x)$ be a polynomial, all of whose coefficients are real numbers. Suppose that $a + bi$ is a root of the equation $f(x) = 0$, where a and b are real and $b \neq 0$. Then $a - bi$ is also a root of the equation.

To prove the conjugate roots theorem, we use four of the properties of complex conjugates listed in Section 12.1:

Property 1: $\overline{z_1}\,\overline{z_2} = \overline{z_1 z_2}$

Property 2: $(\overline{z})^m = \overline{z^m}$

Property 3: $\overline{r} = r$ for every real number r

Property 4: $\overline{z_1} + \overline{z_2} = \overline{z_1 + z_2}$

To prove the theorem, we begin with a polynomial with real coefficients:

$$f(x) = a_n x^n + a_{n-1}x^{n-1} + \cdots + a_1 x + a_0$$

We must show that if $z = a + bi$ is a root of $f(x) = 0$, then $\overline{z} = a - bi$ is also a root. We have

$$\begin{aligned}
f(\overline{z}) &= a_n(\overline{z})^n + a_{n-1}(\overline{z})^{n-1} + \cdots + a_1\overline{z} + a_0 \\
&= \overline{a_n}\,\overline{z^n} + \overline{a_{n-1}}\,\overline{z^{n-1}} + \cdots + \overline{a_1}\,\overline{z} + \overline{a_0} && \text{Properties 3 and 2} \\
&= \overline{a_n z^n} + \overline{a_{n-1}z^{n-1}} + \cdots + \overline{a_1 z} + \overline{a_0} && \text{Property 1} \\
&= \overline{a_n z^n + a_{n-1}z^{n-1} + \cdots + a_1 z + a_0} && \text{Property 4} \\
&= \overline{f(z)} = \overline{0} && f(z) = 0, \text{ since } z \text{ is a root} \\
&= 0 && \text{Property 3}
\end{aligned}$$

We have now shown that $f(\overline{z}) = 0$, given that $f(z) = 0$. Thus $\overline{z}$ is a root, as we wished to show.

Although the conjugate roots theorem concerns nonreal complex roots, it can nevertheless be used to obtain information about real roots, as the next two examples demonstrate.

EXAMPLE 1 Using the conjugate roots theorem in solving an equation

Solve the equation $f(x) = 2x^4 - 3x^3 + 12x^2 + 22x - 60 = 0$, given that one root is $1 + 3i$.

SOLUTION

Since all of the coefficients of $f(x)$ are real numbers, we know that the conjugate of $1 + 3i$ must also be a root. Thus $1 + 3i$ and $1 - 3i$ are roots, from which it follows that $[x - (1 + 3i)]$ and $[x - (1 - 3i)]$ are factors of $f(x)$. As you can check, the product of these two factors is $x^2 - 2x - 10$. Thus we must have

$$f(x) = (x^2 - 2x + 10) \cdot Q(x)$$

for some polynomial $Q(x)$. We compute $Q(x)$ using long division:

$$\begin{array}{r|l} & 2x^2 + x - 6 \\ \hline x^2 - 2x + 10 & 2x^4 - 3x^3 + 12x^2 + 22x - 60 \\ & \underline{2x^4 - 4x^3 + 20x^2} \\ & \qquad x^3 - 8x^2 + 22x \\ & \qquad \underline{x^3 - 2x^2 + 10x} \\ & \qquad\quad -6x^2 + 12x - 60 \\ & \qquad\quad \underline{-6x^2 + 12x - 60} \\ & \qquad\qquad\qquad\qquad 0 \end{array}$$

Thus $Q(x) = 2x^2 + x - 6$, and the original equation becomes

$$f(x) = (x^2 - 2x + 10)(2x^2 + x - 6) = 0$$

We can now find any additional roots by solving the equation $2x^2 + x - 6 = 0$. We have

$$\begin{aligned} 2x^2 + x - 6 &= 0 \\ (2x - 3)(x + 2) &= 0 \\ 2x - 3 = 0 \quad &\Big| \quad x + 2 = 0 \\ x = \frac{3}{2} \quad &\Big| \quad x = -2 \end{aligned}$$

We now have four distinct roots of the original equation: $1 + 3i$, $1 - 3i$, $3/2$, and -2. Since the degree of the equation is 4, there can be no other roots.

EXAMPLE 2 An analysis requiring the rational roots theorem and the conjugate roots theorem

Show that the equation $x^3 - 2x^2 + x - 1 = 0$ has at least one irrational root.

SOLUTION

We know that the equation has three roots. (Why?) The conjugate roots theorem tells us that complex roots of the equation come in conjugate pairs, so there are either zero or two complex roots. So there must be at least one real root. The rational root theorem tells us the only *possible* rational roots are $x = 1$ or -1, but neither 1 nor -1 is a root. (Check this.) So any real number root must be irrational. Thus the equation has at least one irrational root.

There is a theorem, similar to the conjugate roots theorem, that tells us about irrational roots of the form $a + b\sqrt{c}$. As background for this theorem, let's look at two preliminary examples. First, we consider the equation $x^2 - 2x - 5 = 0$. As you can check, the roots in this case are $1 + \sqrt{6}$ and $1 - \sqrt{6}$. However, it is not true in general that irrational roots such as these always occur in pairs. Consider as a second example the quadratic equation

$$(x + 2)(x - \sqrt{3}) = 0$$

or

$$x^2 + (2 - \sqrt{3})x - 2\sqrt{3} = 0$$

Here, one of the roots is $\sqrt{3}$, yet $-\sqrt{3}$ is not a root. This type of behavior can occur in polynomial equations in which not all of the coefficients are rational. On the other hand, when the coefficients are all rational, we do have the following theorem. (See Exercise 45 at the end of this section for a proof.)

THEOREM

Let $f(x)$ be a polynomial in which all the coefficients are rational. Suppose that $a + b\sqrt{c}$ is a root of the equation $f(x) = 0$, where a, b, and c are rational and $\sqrt{c}$ is irrational. Then $a - b\sqrt{c}$ is also a root of the equation.

EXAMPLE 3 Finding a quadratic equation given one irrational root

Find a quadratic equation with rational coefficients and a leading coefficient of 1 such that one of the roots is $r_1 = 4 + 5\sqrt{3}$.

SOLUTION

If one root is $r_1 = 4 + 5\sqrt{3}$, then the other is $r_2 = 4 - 5\sqrt{3}$. We denote the required equation by $x^2 + bx + c = 0$. Thus according to Table 2 in Section 12.4, we have

$$b = -(r_1 + r_2) = -[(4 + 5\sqrt{3}) + (4 - 5\sqrt{3})] = -8$$

and

$$c = r_1r_2 = (4 + 5\sqrt{3})(4 - 5\sqrt{3}) = 16 - 75 = -59$$

The required equation is therefore $x^2 - 8x - 59 = 0$. This answer can also be obtained without using the table. Since the roots are $4 \pm 5\sqrt{3}$, we can write the required equation as $[x - (4 + 5\sqrt{3})][x - (4 - 5\sqrt{3})] = 0$. As you can now check by multiplying out the two factors, this equation is equivalent to $x^2 - 8x - 59 = 0$, as obtained previously.

We conclude this section with a discussion of **Descartes's rule of signs.** This rule, published by Descartes in 1637, provides us with information about the types of roots an equation can have, even before we attempt to solve the equation. To state Descartes's rule of signs, we first explain what is meant by a variation in sign in a polynomial with real coefficients. Suppose that $f(x)$ is a polynomial with real coefficients, written in descending (or ascending) powers of x. For example, let $f(x) = 2x^3 - 4x^2 - 3x + 1$. Then we say that there is a **variation in sign** if two successive coefficients have opposite signs. In the case of $f(x) = 2x^3 - 4x^2 - 3x + 1$, there are two variations in sign, the first occurring as we go from 2 to -4 and the second occurring as we go from -3 to 1. In looking for variations in sign, we ignore terms with zero coefficients. (Contrast this with the Upper and Lower Bounds Theorem on real roots.) Table 1 shows a few more examples of how we count variations in sign.

TABLE 1

Polynomial	Number of variations in sign
$x^2 + 4x$	0
$-3x^5 + x^2 + 1$	1
$x^3 + 3x^2 - x + 6$	2

We now state Descartes's rule of signs and look at some examples. The proof of this theorem is rather lengthy, so we omit it.

Descartes's Rule of Signs

Let $f(x)$ be a polynomial, all of whose coefficients are real numbers, and consider the equation $f(x) = 0$. Then:

(a) The number of positive roots either is equal to the number of variations in sign of $f(x)$ or is less than that by an even integer.
(b) The number of negative roots either is equal to the number of variations in sign of $f(-x)$ or is less than that by an even integer.

EXAMPLE 4 Using Descartes's rule

Use Descartes's rule of signs to obtain information regarding the roots of the equation $x^3 + 8x + 5 = 0$.

SOLUTION
Let $f(x) = x^3 + 8x + 5$. Then, since there are no variations in sign for $f(x)$, we see from part (a) of Descartes's rule that the given equation has no positive roots. Next we compute $f(-x)$ to learn about the possibilities for negative roots: we have $f(-x) = -x^3 - 8x + 5$. So $f(-x)$ has one sign change, and consequently [by part (b) of Descartes's rule], the original equation has one negative root. Furthermore, notice that zero is not a root of the equation. Thus the equation has only one real root, a negative root. Since the equation has a total of three roots, we can conclude that we have one negative root and two nonreal complex roots. The two nonreal roots will be complex conjugates.

EXAMPLE 5 Another analysis using Descartes's rule

Use Descartes's rule to obtain information regarding the roots of the equation $x^4 + 3x^2 - 7x - 5 = 0$.

SOLUTION
Let $f(x) = x^4 + 3x^2 - 7x - 5$. Then $f(x)$ has one variation in sign, so according to part (a) of Descartes's rule, the equation has one positive root. That leaves three roots still to be accounted for, since the degree of the equation is 4. We have $f(-x) = x^4 + 3x^2 + 7x - 5$. Since $f(-x)$ has one sign change, we know from part (b) of Descartes's rule that the equation has one negative root. Noting now that zero is not a root, we conclude that the two remaining roots must be nonreal complex roots. In summary, then, the equation has one positive root, one negative root, and two nonreal complex (conjugate) roots.

EXAMPLE 6 A case in which Descartes's rule doesn't completely determine the nature of the roots

Use Descartes's rule to obtain information regarding the roots of the equation $f(x) = x^3 - x^2 + 3x + 2 = 0$.

SOLUTION

Since $f(x)$ has two variations in sign, the given equation has either two positive roots or no positive roots. To see how many negative roots are possible, we compute

$$\begin{aligned} f(-x) &= (-x)^3 - (-x)^2 + 3(-x) + 2 \\ &= -x^3 - x^2 - 3x + 2 \end{aligned}$$

Since $f(-x)$ has one variation in sign, we conclude from part (b) of Descartes's rule that the equation has exactly one negative root. In summary, then, there are two possibilities:

either: one negative root and two positive roots
or: one negative root and two nonreal complex roots

By using Descartes's rule in Examples 4 and 5, we were able to determine the exact numbers of positive and negative roots for the given equations. As Example 6 indicates, however, there are cases in which a direct application of Descartes's rule provides several distinct possibilities for the types of roots, rather than a single definitive result. (Exercises 55 and 56 in the Review Exercises for this chapter illustrate a technique that is sometimes useful in gaining additional information from Descartes's rule. In particular, Exercise 56 will show you that the equation in Example 6 has no positive roots.)

EXERCISE SET 12.6

A

In Exercises 1–16, an equation is given, followed by one or more roots of the equation. In each case, determine the remaining roots.

1. $x^2 - 14x + 53 = 0; x = 7 - 2i$
2. $x^2 - x - \frac{1535}{4} = 0; x = \frac{1}{2} + 8\sqrt{6}$
3. $x^3 - 13x^2 + 59x - 87 = 0; x = 5 + 2i$
4. $x^4 - 10x^3 + 30x^2 - 10x - 51 = 0; x = 4 + i$
5. $x^4 + 10x^3 + 38x^2 + 66x + 45 = 0; x = -2 + i$
6. $2x^3 + 11x^2 + 30x - 18 = 0; x = -3 - 3i$
7. $4x^3 - 47x^2 + 232x + 61 = 0; x = 6 - 5i$
8. $9x^4 + 18x^3 + 20x^2 - 32x - 64 = 0; x = -1 + \sqrt{3}i$
9. $4x^4 - 32x^3 + 81x^2 - 72x + 162 = 0; x = 4 + \sqrt{2}i$
10. $2x^4 - 17x^3 + 137x^2 - 57x - 65 = 0; x = 4 - 7i$
11. $x^4 - 22x^3 + 140x^2 - 128x - 416 = 0; x = 10 + 2i$
12. $4x^4 - 8x^3 + 24x^2 - 20x + 25 = 0; x = (1 + 3i)/2$
13. $15x^3 - 16x^2 + 9x - 2 = 0; x = (1 + \sqrt{2}i)/3$
14. $x^5 - 5x^4 + 30x^3 + 18x^2 + 92x - 136 = 0;$
    $x = -1 + i\sqrt{3}, x = 3 - 5i$
15. $x^7 - 3x^6 - 4x^5 + 30x^4 + 27x^3 - 13x^2 - 64x + 26 = 0;$
    $x = 3 - 2i, x = -1 + i, x = 1$
16. $x^6 - 2x^5 - 2x^4 + 2x^3 + 2x + 1 = 0; x = 1 + \sqrt{2}$

In Exercises 17–20, find a quadratic equation with rational coefficients, one of whose roots is the given number. Write your answer so that the coefficient of x^2 is 1. Use either of the methods shown in Example 3.

17. $r_1 = 1 + \sqrt{6}$
18. $r_1 = 2 - \sqrt{3}$
19. $r_1 = (2 + \sqrt{10})/3$
20. $r_1 = \frac{1}{2} + \frac{1}{4}\sqrt{5}$
21. Let $f(x) = 2x^4 - 3x^3 + 12x^2 + 22x - 60$.
    (a) Use Descartes's rule to verify that the equation $f(x) = 0$ has one negative root.
    (b) Use Descartes's rule to verify that the equation $f(x) = 0$ has either one or three positive roots.
    Ⓖ **(c)** Graph the equation $y = f(x)$. Use the graph to say which of the two cases in part (b) actually holds.
    Ⓖ **(d)** Use the graph to estimate the real roots of the equation $f(x) = 0$. Check that your answers are consistent with the values obtained in Example 1.
22. Let $f(x) = x^3 - 2x^2 + x - 1$.
    (a) Without using a graphing utility, explain in complete sentences why the equation $f(x) = 0$ must have either three real roots or only one real root. (If you get stuck, see Example 2 in the text.)
    Ⓖ **(b)** Use the graph of $y = f(x)$ to demonstrate that the equation $f(x) = 0$ has, in fact, only one real root.
    Ⓖ **(c)** According to Example 2, the root (or x-intercept) that you observed in part (b) is an irrational number. Use the graphing utility to obtain an approximation for this root.
23. Let $f(x) = x^3 + 8x + 5$.
    (a) Use Descartes's rule to explain in complete sentences why the equation $f(x) = 0$ has no positive roots and exactly one negative root. (If you need help, see Example 4 in the text.)

(continues)

Ⓖ **(b)** Use a graph to confirm the results in part (a). That is, graph $y = f(x)$ and note that there is but one x-intercept and it is negative.

Ⓖ **(c)** Use a graphing utility to compute the root of $x^3 + 8x + 5 = 0$. (Do this either by repeatedly zooming in on the x-intercept of the graph or by using a SOLVE key.)

(d) Use the general formula given on page 928 to compute the root of $x^3 + 8x + 5 = 0$. Check that the answer agrees with the value you obtained in part (c).

24. Let $f(x) = x^3 - x^2 + 3x + 2$.

(a) Use Descartes's rule to explain in complete sentences why the equation $f(x) = 0$ has

either: one negative root and two positive roots
or: one negative root and two nonreal complex roots

(If you need help, review Example 6 in the text.)

Ⓖ **(b)** Use a graph to determine which of the two possibilities in part (a) is actually the case.

Ⓖ **(c)** Use a graphing utility to compute the real root(s) of the equation $f(x) = 0$.

In Exercises 25–40, use Descartes's rule of signs to obtain information regarding the roots of the equations.

25. $x^3 + 5 = 0$

26. $x^4 + x^2 + 1 = 0$

27. $2x^5 + 3x + 4 = 0$

28. $x^3 + 8x - 3 = 0$

29. $5x^4 + 2x - 7 = 0$

30. $x^3 - 4x^2 + x - 1 = 0$

31. $x^3 - 4x^2 - x - 1 = 0$

32. $x^8 + 4x^6 + 3x^4 + 2x^2 + 5 = 0$

33. $3x^8 + x^6 - 2x^2 - 4 = 0$

34. $12x^4 - 5x^3 - 7x^2 - 4 = 0$

35. $x^9 - 2 = 0$

36. $x^9 + 2 = 0$

37. $x^8 - 2 = 0$

38. $x^8 + 2 = 0$

39. $x^6 + x^2 - x - 1 = 0$

40. $x^7 + x^2 - x - 1 = 0$

B

41. Consider the equation $x^4 + cx^2 + dx - e = 0$, where c, d, and e are positive. Show that the equation has one positive root, one negative root, and two nonreal complex roots.

42. Consider the equation $x^n - 1 = 0$.

(a) Show that the equation has $n - 2$ nonreal complex roots when n is even.

(b) How many nonreal complex roots are there when n is odd?

43. (a) Find the polynomial $f(x)$ of lowest degree, with integer coefficients and with leading coefficient 1, such that $\sqrt{3} + 2i$ is a root of the equation $f(x) = 0$.

(b) Find the other roots of the equation.

44. (a) Find a cubic polynomial $f(x)$ with integer coefficients and leading coefficient 1 such that $1 + \sqrt[3]{2}$ is a root of the equation $f(x) = 0$.

Ⓖ **(b)** Use a graphing utility to find out whether $1 - \sqrt[3]{2}$ is a root of the equation.

C

45. Let $f(x)$ be a polynomial, with rational coefficients. Suppose that $a + b\sqrt{c}$ is a root of $f(x) = 0$, where a, b, and c are rational and $\sqrt{c}$ is irrational. Complete the following steps to prove that $a - b\sqrt{c}$ is also a root of the equation $f(x) = 0$.

(a) If $b = 0$, we're done. Why?

(b) (From now on we'll assume that $b \neq 0$.) Let $d(x) = [x - (a + b\sqrt{c})][x - (a - b\sqrt{c})]$. Explain why $d(a + b\sqrt{c}) = 0$.

(c) Verify that $d(x) = (x - a)^2 - b^2c$. Thus $d(x)$ is a quadratic polynomial with rational coefficients.

(d) Now suppose that we use the long division process to divide the polynomial $f(x)$ by the quadratic polynomial $d(x)$. We'll obtain a quotient $Q(x)$ and a remainder. Since the degree of $d(x)$ is 2, our remainder will be of degree 1 or less. In other words, the general form of this remainder will be $Cx + D$. Furthermore, C and D will have to be rational, because all of the coefficients in $f(x)$ and in $d(x)$ are rational. In summary, we have the identity

$$f(x) = d(x) \cdot Q(x) + (Cx + D)$$

Now make the substitution $x = a + b\sqrt{c}$ in this identity, and conclude that $C = D = 0$. *Hint*: The sum of a rational number and an irrational number is not a rational number.

(e) Using the result in part (d), we have

$$f(x) = [x - (a + b\sqrt{c})][x - (a - b\sqrt{c})] \cdot Q(x)$$

Let $x = a - b\sqrt{c}$ in this last identity and conclude that $a - b\sqrt{c}$ is a root of $f(x) = 0$, as required.

46. (a) Find the polynomial $f(x)$ of lowest degree with integer coefficients and with leading coefficient 1, such that $\sqrt[3]{2} + \sqrt{2}$ is a root of the equation $f(x) = 0$.

Ⓖ **(b)** Use a graphing utility to find out whether any of the following three numbers seem to be a root of the equation that you determined in part (a):

$$-\sqrt[3]{2} + \sqrt{2} \qquad \sqrt[3]{2} - \sqrt{2} \qquad -\sqrt[3]{2} - \sqrt{2}$$

(c) For each number in part (b) that seems to be a root, carry out the necessary algebra to *prove* or *disprove* that it is a root.

12.7 INTRODUCTION TO PARTIAL FRACTIONS

In elementary Algebra, a group of fractions connected by the signs of addition and subtraction is reduced to a more simple form by being collected into one single fraction whose denominator is the lowest common denominator of the given fractions. But the converse process of separating a fraction into a group of simpler, or partial, *fractions is often required.*
—H. S. Hall and S. R. Knight, in *Higher Algebra* (London: Macmillan and Co., 1946). (This classic text was first published in 1887.)

In calculus there are times when it is helpful to express a given function in terms of simpler functions. In Section 3.5 we learned that composition of functions provided one way to do this. For example, if $R(x) = 1/(x^2 - 1)$, then, as you can easily check, $R(x)$ can be expressed as a composition of two simpler functions as follows:

$$R(x) = f(g(x)) \qquad \text{where } g(x) = x^2 - 1 \text{ and } f(x) = \frac{1}{x}$$

In this section we introduce another way that certain types of functions can be broken down into simpler functions. Instead of using composition of functions, now we use the sum and difference of functions. We can again use the example $R(x) = 1/(x^2 - 1)$. This time $R(x)$ can be expressed as a *difference* of two simpler expressions as follows:

$$\frac{1}{x^2 - 1} = \frac{1}{2(x - 1)} - \frac{1}{2(x + 1)}$$

You can check this result for yourself by using the least common denominator to combine the two fractions on the right-hand side of the equation.

Our basic goal in this and the next section is to be able to write a given fractional expression as a sum or difference of two or more simpler fractions. For instance, in Example 1 we will be given the fraction $\frac{2x - 3}{(x - 1)(x + 1)}$, and we will be asked to find constants A and B so that

$$\frac{2x - 3}{(x - 1)(x + 1)} = \frac{A}{x - 1} + \frac{B}{x + 1} \tag{1}$$

When A and B are determined, the right-hand side of equation (1) is called the **partial fraction decomposition** of the given fraction. The adjective *partial* is used because each denominator on the right-hand side of equation (1) is a *part* of the denominator on the other side of the equation.

One of the basic tools that can be used in finding partial fractions is supplied by the following theorem, which we state here without proof.

Equating-the-Coefficients Theorem

Suppose that $P(x)$ and $Q(x)$ are polynomials such that $P(x) = Q(x)$ for all x. Then the corresponding coefficients of the two polynomials are equal.

The next two statements supply examples of what this theorem is saying.

1. If $ax + b = 5x - 11$ for all x, then $a = 5$ and $b = -11$.
2. If $px^3 + qx^2 + rx + s = x^2 + 2$ for all x, then $p = 0$, $q = 1$, $r = 0$, and $s = 2$.

In Example 1 we apply the theorem to obtain a partial fraction decomposition.

EXAMPLE 1 A basic partial fractions example

Determine constants A and B so that the following equation is an identity:

$$\frac{2x - 3}{(x - 1)(x + 1)} = \frac{A}{x - 1} + \frac{B}{x + 1} \tag{2}$$

SOLUTION

First, to clear equation (2) of fractions, we multiply both sides by the least common denominator $(x - 1)(x + 1)$. This yields

$$\begin{aligned} 2x - 3 &= A(x + 1) + B(x - 1) \\ &= Ax + A + Bx - B \\ &= (A + B)x + (A - B) \end{aligned}$$

So we have

$$2x - 3 = (A + B)x + (A - B)$$

Now, since this last equation is supposed to be an identity, we can use our equating-the-coefficients theorem to obtain the two equations

$$\begin{cases} A + B = 2 \\ A - B = -3 \end{cases}$$

Adding these two equations gives us $2A = -1$, and therefore $A = -1/2$. Subtracting the two equations yields $2B = 5$, and consequently, $B = 5/2$. We've now found that $A = -1/2$ and $B = 5/2$, as required. The partial fraction decomposition is therefore

$$\begin{aligned} \frac{2x - 3}{(x - 1)(x + 1)} &= \frac{-1/2}{x - 1} + \frac{5/2}{x + 1} \\ &= \frac{-1}{2(x - 1)} + \frac{5}{2(x + 1)} \end{aligned}$$

You should check for yourself that combining the two fractions on the right-hand side of this last equation indeed yields $\dfrac{2x - 3}{(x - 1)(x + 1)}$.

In Example 1 we found the partial fraction decomposition by solving a system of two linear equations. There is a shortcut that is often useful in such problems. As before, we start by multiplying both sides of equation (2) by the common denominator $(x - 1)(x + 1)$ to obtain

$$2x - 3 = A(x + 1) + B(x - 1) \tag{3}$$

Now, equation (3) is an identity; in particular, it must hold when $x = 1$. (You're about to see why we've singled out $x = 1$.) Substituting $x = 1$ in equation (3) gives us

$$2(1) - 3 = A(1 + 1) + B(1 - 1)$$

or

$$-1 = 2A$$

and consequently,

$$A = -\frac{1}{2} \qquad \text{as was obtained previously}$$

The value for B is obtained similarly. Go back to equation (3), and this time let $x = -1$. This gives us

$$2(-1) - 3 = A(-1 + 1) + B(-1 - 1)$$

or

$$-5 = -2B \qquad \text{and therefore} \qquad B = \frac{5}{2}$$

Again, this agrees with the result obtained previously.

A graphing calculator can be used to provide a quick check on the result obtained in a partial fractions problem. For instance, to check Example 1, graph the two functions

$$y = \frac{2x - 3}{(x - 1)(x + 1)} \qquad \text{and} \qquad y = \frac{-1}{2(x - 1)} + \frac{5}{2(x + 1)}$$

in an appropriate viewing rectangle. If the graphs appear to be identical (as is the case in Figure 1), then it's unlikely that there is an error. (See, however, Exercise 29 and the mini project at the end of this section for cautionary examples.) On the other hand, if the graphs are not identical, there must be at least one algebraic error to find and correct in the partial fractions work.

GRAPHICAL PERSPECTIVE

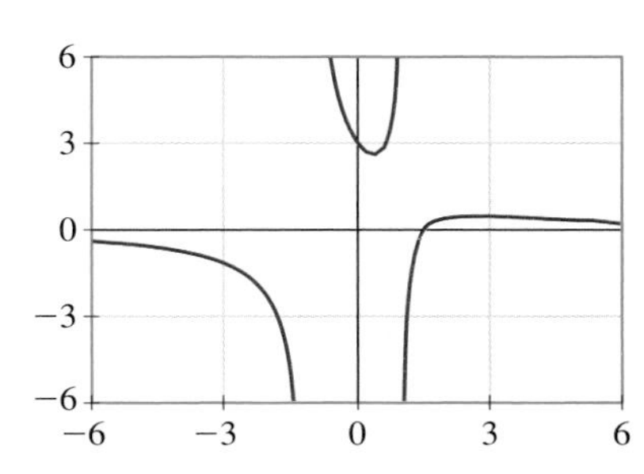

Figure 1
$y = \dfrac{2x - 3}{(x - 1)(x + 1)}$ and
$y = \dfrac{-1}{2(x - 1)} + \dfrac{5}{2(x + 1)}$;
[−6, 6, 3] by [−6, 6, 3]. The graphs appear to be identical.

EXAMPLE 2 Finding a partial fraction decomposition in two ways

Determine constants A and B so that the following equation is an identity:

$$\frac{x}{(x + 4)^2} = \frac{A}{x + 4} + \frac{B}{(x + 4)^2}$$

SOLUTION

We'll show two methods: first, the method that we used in Example 1, where we equated the coefficients; and second, the shortcut method explained after Example 1. For ease of reference we'll call this shortcut method the *convenient-values method.*

FIRST METHOD *(equating coefficients)* As in Example 1, we start by multiplying both sides of the given identity by the least common denominator. Here, the least common denominator is $(x + 4)^2$, and we obtain

$$x = A(x + 4) + B = Ax + 4A + B$$

That is,

$$x = Ax + (4A + B)$$

Now, by equating coefficients, we obtain the system of equations

$$\begin{cases} A = 1 \\ 4A + B = 0 \end{cases}$$

The first of these two equations gives us the value of A directly. Substituting $A = 1$ in the second equation then yields $4(1) + B = 0$, and therefore $B = -4$. In summary, $A = 1$, $B = -4$, and the partial fraction decomposition is

$$\frac{x}{(x+4)^2} = \frac{1}{x+4} - \frac{4}{(x+4)^2}$$

You should check for yourself that combining the two fractions on the right-hand side of this last equation indeed produces the fraction on the left-hand side of the equation.

SECOND METHOD *(convenient values)* As before, we start by multiplying both sides of the given identity by $(x + 4)^2$ to obtain

$$x = A(x+4) + B \qquad (4)$$

Letting $x = -4$ in identity (4) gives us

$$-4 = A(-4+4) + B$$

or $\quad B = -4 \quad$ as was obtained previously

With the value $B = -4$, identity (4) reads

$$x = A(x+4) - 4 \qquad (5)$$

At this point, substituting any value for x (other than -4, which we've already exploited) will produce the required value for A. For example, using $x = 0$ in equation (5) yields

$$0 = A(0+4) - 4 \quad \text{or} \quad -4A = -4$$

and therefore $\quad A = 1 \quad$ as was obtained previously

In the examples up to this point, the convenient-values method appears to be somewhat more efficient (that is, shorter) than equating the coefficients. (In Example 2, perhaps it's a toss-up.) In the next example we again show our two methods of solution. You can decide for yourself which method you prefer. In the convenient-values method you'll see that one of the values we choose to substitute in the identity is a complex number. For purposes of completeness we mention in passing the following theorem from post-calculus mathematics that justifies this technique.

> If p and q are polynomial functions such that $p(x) = q(x)$ for all real numbers x, then $p(z) = q(z)$ for all complex numbers z.

As we've just said, we're stating this result only for the purposes of completeness; you certainly don't need to memorize this theorem or know any more about it to work the exercises.

EXAMPLE 3 Another demonstration of the two methods

Determine real numbers A, B, and C so that the following equation is an identity:

$$\frac{7x^2 - 9x + 29}{(x-2)(x^2+9)} = \frac{A}{x-2} + \frac{Bx + C}{x^2+9}$$

SOLUTION

FIRST METHOD *(equating coefficients)* Multiplying both sides of the given identity by the least common denominator $(x - 2)(x^2 + 9)$ yields

$$\begin{aligned} 7x^2 - 9x + 29 &= A(x^2 + 9) + (Bx + C)(x - 2) \\ &= Ax^2 + 9A + Bx^2 - 2Bx + Cx - 2C \\ &= (A + B)x^2 + (-2B + C)x + (9A - 2C) \end{aligned}$$

Equating coefficients now gives us a system of three equations in three unknowns:

$$\begin{cases} A + B = 7 \\ -2B + C = -9 \\ 9A - 2C = 29 \end{cases}$$

Exercise 25(a) at the end of this section asks you to solve this system to obtain $A = 3$, $B = 4$, and $C = -1$. The partial fraction decomposition is therefore

$$\frac{7x^2 - 9x + 29}{(x - 2)(x^2 + 9)} = \frac{3}{x - 2} + \frac{4x - 1}{x^2 + 9}$$

SECOND METHOD *(convenient values)* Going back to the original identity and multiplying both sides by the least common denominator yields (as in the first method)

$$7x^2 - 9x + 29 = A(x^2 + 9) + (Bx + C)(x - 2) \tag{6}$$

Because of the factor $x - 2$ on the right-hand side of identity (6), we choose the convenient value $x = 2$. Substituting $x = 2$ in equation (6) gives us

$$7(2^2) - 9(2) + 29 = A(2^2 + 9) + 0$$

or

$$39 = 13A \qquad \text{and therefore} \qquad A = 3$$

Next, we want to choose another value for x that will make the factor $x^2 + 9$ zero. For example, we choose $x = 3i$. (The other root of the equation $x^2 + 9 = 0$, namely, $x = -3i$, would also be a suitable choice here.) Making the substitution $x = 3i$ (along with $x^2 = -9$) in identity (6) yields

$$\begin{aligned} 7(-9) - 9(3i) + 29 &= A(-9 + 9) + [B(3i) + C](3i - 2) \\ -34 - 27i &= -9B - 6Bi + 3Ci - 2C \\ &= (-9B - 2C) + (-6B + 3C)i \end{aligned}$$

Equating the real parts and equating the imaginary parts from both sides of this last equation gives the equations

$$-9B - 2C = -34 \tag{7}$$

and

$$-6B + 3C = -27$$

or

$$-2B + C = -9 \qquad \text{dividing by 3} \tag{8}$$

As Exercise 25(b) asks you to check, the solution to the system consisting of equations (7) and (8) is $B = 4$ and $C = -1$. In summary, then, we have $A = 3$, $B = 4$, and $C = -1$, as was obtained previously.

EXERCISE SET 12.7 

A

In Exercises 1–22, determine the constants (denoted by capital letters) so that each equation is an identity. For Exercises 1–6, do each problem in two ways: **(a)** *use the equating-the-coefficients theorem, as in Example 1; and* **(b)** *use the convenient-values method that was explained after Example 1. For the remainder of the exercises, use either method (or a combination).*

1. $\dfrac{7x-6}{(x-2)(x+2)} = \dfrac{A}{x-2} + \dfrac{B}{x+2}$

2. $\dfrac{5x+27}{(x+3)(x-3)} = \dfrac{A}{x+3} + \dfrac{B}{x-3}$

3. $\dfrac{6x-25}{(2x+5)(2x-5)} = \dfrac{A}{2x+5} + \dfrac{B}{2x-5}$

4. $\dfrac{x}{(4x+3)(4x-3)} = \dfrac{A}{4x+3} + \dfrac{B}{4x-3}$

5. $\dfrac{1}{(x+1)(3x-1)} = \dfrac{A}{x+1} + \dfrac{B}{3x-1}$

6. $\dfrac{1-x}{(4x+3)(2x-1)} = \dfrac{A}{4x+3} + \dfrac{B}{2x-1}$

7. $\dfrac{8x+3}{(x+3)^2} = \dfrac{A}{x+3} + \dfrac{B}{(x+3)^2}$

8. $\dfrac{7x}{(x-5)^2} = \dfrac{A}{x-5} + \dfrac{B}{(x-5)^2}$

9. $\dfrac{6-x}{(5x+4)^2} = \dfrac{A}{5x+4} + \dfrac{B}{(5x+4)^2}$

10. $\dfrac{30x-17}{(6x-1)^2} = \dfrac{A}{6x-1} + \dfrac{B}{(6x-1)^2}$

11. $\dfrac{3x^2+7x-2}{(x-1)(x^2+1)} = \dfrac{A}{x-1} + \dfrac{Bx+C}{x^2+1}$

12. $\dfrac{15x^2-35x+77}{(2x-5)(x^2+3)} = \dfrac{A}{2x-5} + \dfrac{Bx+C}{x^2+3}$

13. $\dfrac{x^2+1}{(x+1)(x^2+4)} = \dfrac{A}{x+1} + \dfrac{Bx+C}{x^2+4}$

14. $\dfrac{x-7}{x(x^2+2)} = \dfrac{A}{x} + \dfrac{Bx+C}{x^2+2}$

15. $\dfrac{1}{x(x^2-x+1)} = \dfrac{A}{x} + \dfrac{Bx+C}{x^2-x+1}$

16. $\dfrac{2x^2-11x-6}{(x+2)(x^2-2x+4)} = \dfrac{A}{x+2} + \dfrac{Bx+C}{x^2-2x+4}$

17. $\dfrac{3x^2-2}{(x-2)(x+1)(x-1)} = \dfrac{A}{x-2} + \dfrac{B}{x+1} + \dfrac{C}{x-1}$

18. $\dfrac{1}{(x-1)(x-2)(x-3)} = \dfrac{A}{x-1} + \dfrac{B}{x-2} + \dfrac{C}{x-3}$

19. $\dfrac{4x^2-47x+133}{(x-6)^3} = \dfrac{A}{x-6} + \dfrac{B}{(x-6)^2} + \dfrac{C}{(x-6)^3}$

20. $\dfrac{x^2+2x}{(x+1)^3} = \dfrac{A}{x+1} + \dfrac{B}{(x+1)^2} + \dfrac{C}{(x+1)^3}$

21. $\dfrac{x^2-2}{(x^2+2)^2} = \dfrac{Ax+B}{x^2+2} + \dfrac{Cx+D}{(x^2+2)^2}$

22. $\dfrac{x^3+2x^2}{(x^2+3)^2} = \dfrac{Ax+B}{x^2+3} + \dfrac{Cx+D}{(x^2+3)^2}$

Ⓖ **23. (a)** Find an appropriate viewing rectangle to demonstrate that the following purported partial fraction decomposition is incorrect:

$$\frac{2x+5}{(x-4)(x+3)} = \frac{13/7}{x-4} + \frac{2/7}{x+3}$$

Ⓖ **(b)** Follow part (a) using

$$\frac{2x+5}{(x-4)(x+3)} = \frac{13/7}{x-4} - \frac{1/7}{x+3}$$

(c) Determine the correct partial fraction decomposition, given that it has the general form

$$\frac{2x+5}{(x-4)(x+3)} = \frac{A}{x-4} + \frac{B}{x+3}$$

Ⓖ **24. (a)** Find an appropriate viewing rectangle to demonstrate that the following purported partial fraction decomposition is incorrect:

$$\frac{4}{x^2(x-5)} = \frac{-4/5}{x^2} + \frac{4/25}{x-5}$$

Ⓖ **(b)** Follow part (a) using

$$\frac{4}{x^2(x-5)} = \frac{-3/25}{x} + \frac{-2/5}{x^2} + \frac{6/25}{x-5}$$

(c) Determine the correct partial fraction decomposition, given that it has the general form

$$\frac{4}{x^2(x-5)} = \frac{A}{x} + \frac{B}{x^2} + \frac{C}{x-5}$$

25. (a) Solve the following system of equations. (As indicated in Example 3, you should obtain $A = 3$, $B = 4$, and $C = -1$.)

$$\begin{cases} A + B = 7 \\ -2B + C = -9 \\ 9A - 2C = 29 \end{cases}$$

(b) Solve the following system of equations. (As indicated in the text, you should obtain $B = 4$ and $C = -1$.)

$$\begin{cases} -9B - 2C = -34 \\ -2B + C = -9 \end{cases}$$

Exercises 26 provides practice using the convenient-values method with complex numbers. As background, you should review Example 3.

26. In this exercise you'll use the convenient-values method to determine real numbers A, B, and C such that the following equation is an identity:

$$\frac{x+1}{x(x^2+4)} = \frac{A}{x} + \frac{Bx+C}{x^2+4}$$

(a) Multiplying both sides of the given identity by the least common denominator yields

$$x + 1 = A(x^2 + 4) + (Bx + C)x \qquad (1)$$

Determine A by substituting $x = 0$ in identity (1).

(b) In identity (1), substitute $x = 2i$ and show that the resulting equation can be written

$$2i + 1 = -4B + 2iC \qquad (2)$$

(c) Determine B and C by equating real and imaginary parts in equation (2).

B

In Exercises 27 and 28, the equating-the-coefficients theorem is used as a tool to factor polynomials and thereby solve equations.

27. Consider the polynomial equation $f(x) = x^4 - x^3 + x^2 - x + 1 = 0$.

(a) Make use of the rational roots theorem in explaining why the equation has no rational roots.

Ⓖ **(b)** Use a graphing utility to graph the function $y = f(x)$, and conclude that the given polynomial equation has no real roots.

(c) Use the equating-the-coefficients theorem to find a factorization of $f(x)$ of the form

$$x^4 - x^3 + x^2 - x + 1 = (x^2 + bx + 1)(x^2 + cx + 1)$$

(d) Solve the equation $x^4 - x^3 + x^2 - x + 1 = 0$.

28. Let $f(x) = x^4 + 2x^3 + x^2 - 9$.

(a) Use a graphing utility to graph the function f. The graph indicates that the polynomial equation $f(x) = 0$ has two real roots. In this case, according to Section 12.4, how many complex roots must there be?

(b) Use the rational roots theorem to show that the equation $f(x) = 0$ had no rational roots. Why does this imply that the two real roots in part (a) must be irrational?

(c) Use the equating-the-coefficients theorem to find a factorization for $f(x)$ of the form $(x^2 + ax + b)(x^2 + ax - b)$.

(d) Use the factorization determined in part (c) to find the four roots of the equation $f(x) = 0$. For the two real roots, give both exact expressions and calculator approximations rounded to two decimal places. Check to see that these calculator values are consistent with the graph in part (a).

Exercise 29 provides an example in which an error in a partial fraction decomposition is not easily detected with a graphical approach. Indeed, this may be an example of a case in which, to check your partial fractions work, it's easier to repeat the algebra than to experiment with numerous viewing rectangles. Decide for yourself after completing the problem.

29. There is an error in the following partial fraction decomposition:

$$\frac{1}{(x+2)(x+5)(x-14)} = \frac{-1/48}{x+2} + \frac{1/57}{x+5} + \frac{1/305}{x-14}$$

Ⓖ **(a)** Let f and g denote the two functions defined by the expressions on the left side and the right side, respectively, in the above equation. Use a graphing utility to graph f and g, first in the standard viewing rectangle and then in the rectangle $[-15, 15, 5]$ by $[-0.02, 0.04, 0.02]$. In this latter rectangle, note that the graphs do *appear* to be identical. (People using a software graphing application and looking at the curves on a computer monitor may have a little advantage here over those drawing the graphs on a relatively small graphing calculator screen.)

Ⓖ **(b)** Find a viewing rectangle clearly demonstrating that the graphs of f and g are not identical.

(c) Find the correct partial fraction decomposition, given that the form is

$$\frac{1}{(x+2)(x+5)(x-14)} = \frac{A}{x+2} + \frac{B}{x+5} + \frac{C}{x-14}$$

MINI PROJECT CHECKING A PARTIAL FRACTION DECOMPOSITION

The partial fraction decomposition for the expression $\dfrac{3}{x(x+17)(x^2+13)}$ has the form

$$\frac{A}{x} + \frac{B}{x+17} + \frac{Cx+D}{x^2+13}$$

Suppose that a classmate of yours worked out the following values for the constants:

$$A = \frac{3}{221}, \quad B = -\frac{3}{5135}, \quad C = -\frac{51}{3925}, \quad D = -\frac{3}{302}$$

(a) Use a graphing utility to check this decomposition, as in the text, after Example 1. Unlike Example 1, however, not all of the constants given here are correct. Yet, initially, no difference in the two graphs is apparent. Can you find a viewing rectangle that clearly demonstrates that there must be an error in the decomposition? (It's certainly not the standard viewing rectangle.) Can you think of a simple nongraphical way to check for the presence of an error (short of rederiving the entire decomposition from scratch)?

(b) Determine correct values for the constants A, B, C, and D. If you have access to software or a graphing utility that computes partial fraction decompositions, use that. There are also interactive partial fraction calculators available on-line. Try using a search engine to locate "partial fraction expansion" or "partial fraction calculator." At the time of this writing, for example, one such calculator (developed by Professor Philip S. Crooke at Vanderbilt University) could be accessed from the web page http://MSS.math.vanderbilt.edu. Finally, if neither of these options works out for you, use the methods of this section to find the decomposition.

12.8 MORE ABOUT PARTIAL FRACTIONS

If the denominator of a rational function has two relatively prime factors, then this rational function can be expressed as the sum of two fractions whose denominators are equal to the two factors. —Leonhard Euler (1707–1783) in *Introductio in analysis infinitorum* (1748)

In the examples and exercises in the previous section you were told what the general form for each partial fraction should look like, and then the required constants were computed. In calculus, however, when a partial fraction decomposition is needed, you're usually not specifically told what the general form for the partial fractions should be. Rather, you're expected to know this. In this section we list the necessary guidelines. First, however, consider the following example indicating why these guidelines are needed.

Suppose, for instance, that we want to find the partial fraction decomposition for $1/[x(x^2 + 1)]$, and we guess that the general form is

$$\frac{1}{x(x^2 + 1)} = \frac{A}{x} + \frac{B}{x^2 + 1} \tag{1}$$

Multiplying both sides of this last equation by the least common denominator gives us

$$\begin{aligned} 1 &= A(x^2 + 1) + Bx \\ &= Ax^2 + A + Bx \\ &= Ax^2 + Bx + A \end{aligned}$$

Equating coefficients now, as we did in the previous section, yields $A = 0$, $B = 0$, and $A = 1$. The two equations $A = 0$ and $A = 1$ are contradictory. The only way out is to conclude that there is no partial fraction decomposition of the form shown in equation (1). This example demonstrates the need for guidelines for setting up partial fraction decompositions.

We begin with a few definitions. Suppose that we have two polynomials $p(x)$ and $q(x)$. Following the terminology in Section 4.7, we call an expression of the form $p(x)/q(x)$ a **rational expression.** A rational expression is said to be **proper** if the degree of the numerator is less than the degree of the denominator. So each of the following is a proper rational expression:

$$\frac{4x-5}{2x^2+3x-1} \qquad \frac{x^3-2x-5}{x^5-9x^3-4} \qquad \frac{x^2-5}{(x+1)(x+2)(x+3)}$$

An **improper** rational expression is one in which the degree of the numerator is greater than or equal to the degree of the denominator. Examples of improper rational expressions are $(x^2+1)/(x-1)$ and $(x^3-4)/(x^3+2x^2+1)$. All of the partial fraction guidelines that we are about to discuss pertain to proper rational expressions. (Near the end of this section we'll explain what to do for an improper rational expression.)

The final term that we need to define is *irreducible quadratic polynomial.* The definition is given in the box that follows.

DEFINITION Irreducible Quadratic Polynomial

Let $f(x)$ be a quadratic polynomial, all of whose coefficients are real. Then $f(x)$ is said to be **irreducible** provided that the equation $f(x) = 0$ has no real roots. This is equivalent to saying that $f(x)$ cannot be factored into the form $(ax+b)(cx+d)$, where a, b, c, and d are real numbers.

NOTE It follows that a quadratic polynomial ax^2+bx+c with real coefficients is irreducible if and only if its discriminant b^2-4ac is negative. As examples, note that x^2+5 is irreducible, but x^2-5 is not irreducible.

The first step in determining a partial fraction decomposition is to factor the denominator. In the previous section the denominators were already factored for you. If you glance back at the previous section, you'll see that the denominators contained one or more of the following types of factors and no others:

$$\begin{aligned} \text{Linear factors:}&\quad ax+b \\ \text{Powers of linear factors:}&\quad (ax+b)^n, \quad \text{integer } n \geq 2 \\ \text{Irreducible quadratic factors:}&\quad ax^2+bx+c \\ \text{Powers of irreducible quadratic factors:}&\quad (ax^2+bx+c)^n, \quad \text{integer } n \geq 2 \end{aligned}$$

A remarkable theorem tells us that, in fact, *every* polynomial with real coefficients can be factored (over the real numbers) using only the types of factors we've just listed. We state this theorem in the box that follows.

Linear and Quadratic Factors Theorem

Let $f(x)$ be a polynomial, all of whose coefficients are real numbers. Then $f(x)$ can be factored (over the real numbers) into a product of linear and/or irreducible quadratic factors.

At the end of this section we show how the linear and quadratic factors theorem follows from our earlier work in this chapter.

To see a familiar example, consider the cubic polynomial $x^3 - 1$. We know that

$$x^3 - 1 = (x - 1)(x^2 + x + 1)$$

So in this case there is one linear factor and one irreducible quadratic factor. (You should check for yourself, using the discriminant, that the quadratic factor is indeed irreducible.) However, not every polynomial, of course, is so easily factored. For instance, although the theorem tells us that the polynomial $x^3 + x + 1$ *can* be factored into linear and/or irreducible quadratic factors, it doesn't tell us how to find those factors. Indeed, that can be a very difficult job, quite beyond this course. For the examples and exercises in this section, you'll be able to find the factors using factoring techniques from Appendix B.4 or from this chapter.

For all of the guidelines that we give, we assume that we're starting with a proper rational expression $p(x)/q(x)$ and that $q(x)$ has been factored into linear and/or irreducible quadratic factors. We also assume that $p(x)$ and $q(x)$ have *no common factors;* that is, the fraction $p(x)/q(x)$ has been reduced to lowest terms. The first two guidelines tell us what to do when $q(x)$ contains a linear factor $ax + b$ or a power of this factor.

Guidelines for the Partial Fractions Setup: Linear Factors

(a) If the denominator contains a linear factor $ax + b$ and no higher power of this factor, then the partial fraction setup must contain a term $A_1/(ax + b)$, where A_1 is a constant to be determined.

(b) [This encompasses part (a).] More generally, for each factor in the denominator of the form $(ax + b)^n$, the partial fraction setup must contain the following sum of n fractions:

$$\frac{A_1}{ax + b} + \frac{A_2}{(ax + b)^2} + \cdots + \frac{A_n}{(ax + b)^n}$$

where $A_1, A_2, \ldots, A_n$ are constants to be determined.

EXAMPLE 1 Specifying the setup for a decomposition

Determine the form of the partial fraction decomposition:

(a) $\dfrac{5x + 1}{4x^2 - 9}$; **(b)** $\dfrac{x^2}{x^3 + 2x^2 - 5x - 10}$.

SOLUTION

(a) The denominator can be factored as a difference of squares:

$$4x^2 - 9 = (2x - 3)(2x + 3)$$

Thus the denominator contains two distinct linear factors. According to the first guideline in the box preceding this example, the form of the partial fraction decomposition is

$$\frac{5x + 1}{4x^2 - 9} = \frac{A_1}{2x - 3} + \frac{A_2}{2x + 3}$$

(b) The denominator can be factored by grouping (as in Appendix B.4):

$$\begin{aligned} x^3 + 2x^2 - 5x - 10 &= (x^3 - 5x) + (2x^2 - 10) \\ &= x(x^2 - 5) + 2(x^2 - 5) \\ &= (x^2 - 5)(x + 2) \end{aligned} \tag{2}$$

We're not finished with the factoring yet! If this were a factoring problem from Appendix B.4, equation (2) would indeed be our final result. That's because in Appendix B.4 we are restricted to factors with integer (or possibly rational) coefficients. But here the restriction is removed, and we can factor $x^2 - 5$ as a difference of squares:

$$x^2 - 5 = (x - \sqrt{5})(x + \sqrt{5})$$

Combining this last result with equation (2), our final factorization is

$$x^3 + 2x^2 - 5x - 10 = (x - \sqrt{5})(x + \sqrt{5})(x + 2)$$

Thus the denominator contains three distinct linear factors, and the form of the partial fraction decomposition is

$$\frac{x^2}{x^3 + 2x^2 - 5x - 10} = \frac{A_1}{x - \sqrt{5}} + \frac{A_2}{x + \sqrt{5}} + \frac{A_3}{x + 2}$$

EXAMPLE 2 Another setup with distinct linear factors in the denominator

Determine the form of the partial fraction decomposition for the following expression:

$$\frac{6x - 1}{x^3 + 2x^2 - 5x - 6}$$

SOLUTION

In Example 1(b) we factored a denominator that was quite similar to this one by grouping the terms. As you can check, however, that won't work here. The next strategy then is to apply the rational roots theorem to the polynomial equation

$$x^3 + 2x^2 - 5x - 6 = 0 \tag{3}$$

As Exercise 35 at the end of this section asks you to show, this leads to three rational roots: -1, 2, and -3. Consequently, we have a factorization into three distinct linear factors:

$$x^3 + 2x^2 - 5x - 6 = (x + 1)(x - 2)(x + 3)$$

The form of the partial fraction decomposition must therefore be

$$\frac{6x - 1}{x^3 + 2x^2 - 5x - 6} = \frac{A_1}{x + 1} + \frac{A_2}{x - 2} + \frac{A_3}{x + 3}$$

In Examples 1 and 2 none of the linear factors in the denominators are repeated. The next example shows cases in which there are repeated linear factors.

EXAMPLE 3 A setup with repeated linear factors in the denominator

Determine the form of the partial fraction decomposition for each of the following expressions:

(a) $\dfrac{5x-1}{x^2-6x+9}$; **(b)** $\dfrac{2x^4+x+1}{x^5+3x^4+3x^3+x^2}$.

SOLUTION

(a) The denominator is a perfect square of a linear factor:

$$x^2-6x+9=(x-3)^2$$

So according to the second guideline in the box preceding Example 1, the form of the partial fraction decomposition is

$$\frac{5x-1}{x^2-6x+9}=\frac{A_1}{x-3}+\frac{A_2}{(x-3)^2}$$

(b) The first step in factoring is to look for a common factor. (If you need to review factoring techniques, see Appendix B.4.) In this case there is one: It is x^2, and we have

$$x^5+3x^4+3x^3+x^2=x^2(x^3+3x^2+3x+1)$$

Now what? Regarding the second factor on the right-hand side of this last equation, we could try to factor it by grouping. As you can check, that does work. It's more direct, however, to observe that this factor is actually a perfect cube: $x^3+3x^2+3x+1=(x+1)^3$. Putting things together then, we have the final factorization

$$x^5+3x^4+3x^3+x^2=x^2(x+1)^3$$

In summary, there are two repeated linear factors: The factor x occurs twice, and the factor $x+1$ occurs three times. The form of the partial fraction decomposition is therefore

$$\frac{2x^4+x+1}{x^5+3x^4+3x^3+x^2}=\frac{A_1}{x}+\frac{A_2}{x^2}+\frac{B_1}{x+1}+\frac{B_2}{(x+1)^2}+\frac{B_3}{(x+1)^3}$$

The next two guidelines apply in cases in which the denominator contains one or more irreducible quadratic factors. After you read these guidelines, compare them to those that we listed previously for linear factors. You'll see that almost everything is the same. The only difference, in fact, is this: For the quadratic factors, the form of the numerator is $Ax+B$ rather than just A.

Guidelines for the Partial Fractions Setup: Irreducible Quadratic Factors

(a) If the denominator contains an irreducible quadratic factor $ax^2 + bx + c$ and no higher power of this factor, then the partial fractions setup must contain a term $\dfrac{A_1x + B_1}{ax^2 + bx + c}$, where A_1 and B_1 are constants to be determined.

(b) [This encompasses part (a).] More generally, for each factor in the denominator of the form $(ax^2 + bx + c)^n$, where $ax^2 + bx + c$ is irreducible, the partial fractions setup must contain the following sum of n fractions:

$$\frac{A_1x + B_1}{ax^2 + bx + c} + \frac{A_2x + B_2}{(ax^2 + bx + c)^2} + \cdots + \frac{A_nx + B_n}{(ax^2 + bx + c)^n}$$

where the A_i and B_i are constants to be determined.

EXAMPLE 4 A denominator with linear and irreducible quadratic factors

Determine the partial fraction decomposition for the following expression:

$$\frac{x^3 + 4x^2 + 1}{x^4 - 16}$$

SOLUTION

Our first job is to factor the denominator. Using difference-of-squares factoring, we have

$$\begin{aligned} x^4 - 16 &= (x^2 - 4)(x^2 + 4) \\ &= (x - 2)(x + 2)(x^2 + 4) \end{aligned}$$

So there are two linear factors and one irreducible quadratic factor, none of which is repeated. The form of the partial fraction decomposition is therefore

$$\frac{x^3 + 4x^2 + 1}{(x - 2)(x + 2)(x^2 + 4)} = \frac{A}{x - 2} + \frac{B}{x + 2} + \frac{Cx + D}{x^2 + 4}$$

(For ease of reading and writing, we are denoting the constants by A, B, C, and D instead of, say, A_1, A_2, B_1, and C_1, respectively.) Multiplying both sides of this equation by the common denominator gives

$$x^3 + 4x^2 + 1 = A(x + 2)(x^2 + 4) + B(x - 2)(x^2 + 4) + (Cx + D)(x - 2)(x + 2) \quad (4)$$

Substituting $x = 2$ in identity (4) yields

$$\begin{aligned} 2^3 + 4(2)^2 + 1 &= A(2 + 2)(2^2 + 4) + 0 + 0 \\ 25 &= 32A \\ A &= \frac{25}{32} \end{aligned}$$

Similarly, letting $x = -2$ in identity (4) gives us

$$(-2)^3 + 4(-2)^2 + 1 = 0 + B(-4)(8) + 0$$
$$9 = -32B$$
$$B = -\frac{9}{32}$$

At this point, we've found A and B, but we still need C and D. If we substitute the values we've just found for A and B in equation (4), we have

$$x^3 + 4x^2 + 1 = \frac{25}{32}(x + 2)(x^2 + 4) - \frac{9}{32}(x - 2)(x^2 + 4) + (Cx + D)(x - 2)(x + 2) \quad (5)$$

Observe now that in identity (5), letting $x = 0$ will yield an equation involving D alone. Exercise 36(a) at the end of this section asks you to follow through with the arithmetic and algebra to obtain $D = 15/8$. Finally, in identity (5), we replace D by $15/8$ and make the substitution $x = 1$. As Exercise 36(b) asks you to verify, the end result is $C = 1/2$. If we put everything together now, the required partial fraction decomposition is

$$\frac{x^3 + 4x^2 + 1}{x^4 - 16} = \frac{\frac{25}{32}}{x - 2} + \frac{-\frac{9}{32}}{x + 2} + \frac{\frac{1}{2}x + \frac{15}{8}}{x^2 + 4}$$
$$= \frac{25}{32(x - 2)} - \frac{9}{32(x + 2)} + \frac{4x + 15}{8(x^2 + 4)}$$

EXAMPLE 5 A denominator with a repeated irreducible quadratic factor

Determine the partial fraction decomposition for the following expression:

$$\frac{3x^3 - x^2 + 7x - 3}{x^4 + 6x^2 + 9}$$

SOLUTION

The denominator factors as a perfect square:

$$x^4 + 6x^2 + 9 = (x^2 + 3)^2$$

Since the repeated factor $x^2 + 3$ is irreducible, our guidelines tell us that the form of the partial fraction decomposition is

$$\frac{3x^3 - x^2 + 7x - 3}{(x^2 + 3)^2} = \frac{Ax + B}{x^2 + 3} + \frac{Cx + D}{(x^2 + 3)^2}$$

Multiplying both sides of this identity by $(x^2 + 3)^2$ gives us

$$3x^3 - x^2 + 7x - 3 = (Ax + B)(x^2 + 3) + (Cx + D) \quad (6)$$

At this point we can use either of the two techniques discussed in Section 12.7: the equating-the-coefficients technique or the convenient-values technique. As Exercise 37 at the end of this section asks you to show, the results are $A = 3$, $B = -1$, $C = -2$, and $D = 0$. The partial fraction decomposition then is

$$\frac{3x^3 - x^2 + 7x - 3}{x^4 + 6x^2 + 9} = \frac{3x - 1}{x^2 + 3} - \frac{2x}{(x^2 + 3)^2}$$

Each of the examples we've considered in this and the previous section involved proper rational expressions. For improper rational expressions we can first use long division to express the improper fraction in the general form

$$(\text{polynomial}) + (\text{proper rational expression})$$

Then the techniques that we've developed can be applied to the second term in this sum. For instance, suppose we're given the improper rational expression $(2x^3 + 4x^2 - 15x - 36)/(x^2 - 9)$. As you can check for yourself by using long division, we obtain

$$\frac{2x^3 + 4x^2 - 15x - 36}{x^2 - 9} = \underbrace{2x + 4}_{\text{polynomial}} + \underbrace{\frac{3x}{x^2 - 9}}_{\text{proper rational expression}}$$

At this stage, a partial fraction decomposition can be worked out for the proper rational expression $3x/(x^2 - 9)$. As Exercise 38(b) at the end of this section asks you to check, the result is

$$\frac{3x}{x^2 - 9} = \frac{3}{2(x - 3)} + \frac{3}{2(x + 3)}$$

So our final decomposition of the given improper rational fraction is

$$\frac{2x^3 + 4x^2 - 15x - 36}{x^2 - 9} = 2x + 4 + \frac{3}{2(x - 3)} + \frac{3}{2(x + 3)}$$

OPTIONAL NOTE The linear and quadratic factors theorem that we've discussed in this section is a direct consequence of two theorems that appeared earlier in the chapter: the linear factors theorem and the conjugate roots theorem. Here's why: If we start with a polynomial $f(x)$, the linear factors theorem says that we can decompose $f(x)$ into linear factors:

$$f(x) = a_n(x - c_1)(x - c_2) \cdots (x - c_n) \tag{7}$$

Now, it's possible that some, or even all, of these c_i may be nonreal complex numbers. For instance, suppose that c_1 is a nonreal complex number, with $c_1 = a + bi$. Then, assuming that all of the coefficients of $f(x)$ are real numbers, the conjugate roots theorem tells us that in one of the other factors, call it $x - c_2$ for simplicity, we must have $c_2 = a - bi$. Now look what happens when we compute the product of these two factors:

$$\begin{aligned}(x - c_1)(x - c_2) &= [x - (a + bi)][x - (a - bi)] \\ &= x^2 - (a - bi)x - (a + bi)x + (a + bi)(a - bi) \\ &= x^2 - 2ax + (a^2 + b^2) \qquad \text{(Check the algebra.)}\end{aligned}$$

As Exercise 45 asks you to check, this last quadratic polynomial is irreducible; that is, it has no real roots. In summary, the two linear factors $x - c_1$ and $x - c_2$ that contain nonreal complex numbers give rise to one irreducible quadratic factor with real coefficients. So, similarly, after pairing up any other linear factors in equation (7) that contain nonreal complex numbers, what are we left with? The right-hand side of equation (7) will contain only linear factors and/or irreducible quadratic factors, all with real coefficients, just as the linear and quadratic factors theorem asserts.

EXERCISE SET 12.8

A

In Exercises 1–4, determine whether the given quadratic polynomial is irreducible. [Recall from the text that a quadratic polynomial f(x) is irreducible if the equation f(x) = 0 has no real roots.]

1. **(a)** $x^2 - 16$
(b) $x^2 + 16$

2. **(a)** $x^2 + 17$
(b) $x^2 - 17$

3. **(a)** $x^2 + 3x - 4$
(b) $x^2 + 3x + 4$

4. **(a)** $24x^2 + x - 3$
(b) $x^2 + 24x + 144$

In Exercises 5–16: **(a)** *factor the denominator of the given rational expression;* **(b)** *determine the form of the partial fraction decomposition for the given rational expression; and* **(c)** *determine the values of the constants in the partial fraction decomposition that you gave in part (b). To help you in spotting errors, use the fact that in part (c), each of the required constants turns out to be an integer.*

5. $\dfrac{11x + 30}{x^2 - 100}$

6. $\dfrac{x + 18}{x^2 - 36}$

7. $\dfrac{8x - 2\sqrt{5}}{x^2 - 5}$

8. $\dfrac{2\sqrt{11}}{x^2 - 11}$

9. $\dfrac{7x + 39}{x^2 - x - 6}$

10. $\dfrac{19x - 15}{4x^2 - 5x}$

11. $\dfrac{3x^2 + 17x - 38}{x^3 - 3x^2 - 4x + 12}$

12. $\dfrac{16x^2 + 9x - 2}{3x^3 + x^2 - 2x}$

13. $\dfrac{5x^2 + 2x + 5}{x^3 + x^2 + x}$

14. $\dfrac{x^2 - 3x - 1}{x^3 - x^2 + 2x - 2}$

15. $\dfrac{2x^3 + 5x - 4}{x^4 + 2x^2 + 1}$

16. $\dfrac{11x^3 + 35x - 7}{x^4 + 6x^2 + 9}$

In Exercises 17–34, determine the partial fraction decomposition for each of the given rational expressions. Hint: *In Exercises 17, 18, and 26, use the rational roots theorem to factor the denominator.*

17. $\dfrac{x^2 + 2}{x^3 - 3x^2 - 16x - 12}$

18. $\dfrac{1}{x^3 + x^2 - 10x + 8}$

19. $\dfrac{5 - x}{6x^2 - 19x + 15}$

20. $\dfrac{2x}{32x^2 - 12x + 1}$

21. $\dfrac{2x + 1}{x^3 - 5x}$

22. $\dfrac{2x + 1}{x^3 + 5x}$

23. $\dfrac{x^3 + 2}{x^4 + 8x^2 + 16}$

24. $\dfrac{x^3 + 2}{x^4 - 8x^2 + 16}$

25. $\dfrac{x^3 + x - 3}{x^4 - 15x^3 + 75x^2 - 125x}$

26. $\dfrac{4x^2}{2x^3 - 5x^2 - 4x + 3}$

27. $\dfrac{1}{x^3 - 1}$

28. $\dfrac{x}{x^3 + 8}$

29. $\dfrac{7x^3 + 11x^2 - x - 2}{x^4 + 2x^3 + x^2}$

30. $\dfrac{4x - 5}{x^4 + 2x^3 + x^2 - 1}$ *Hint*: Review the factoring in Exercise 29.

31. $\dfrac{x^3 - 5}{x^4 - 81}$

32. $\dfrac{x + 1}{x^4 - 16}$

33. $\dfrac{1}{x^4 + x^3 + 2x^2 + x + 1}$ *Hint*: To factor the denominator, replace $2x^2$ with $x^2 + x^2$ and group as follows: $(x^4 + x^3 + x^2) + (x^2 + x + 1)$.

34. $\dfrac{x^3 + 4x^2 + 16x}{x^4 + 16}$ *Hint*: To factor the denominator, add and subtract the term $16x^2$.

35. Use the rational roots theorem and the remainder theorem to determine the roots of the equation $x^3 + 2x^2 - 5x - 6 = 0$. (This is to verify a statement made in Example 2.)

36. This exercise completes two details mentioned in Example 4.
(a) In identity (5) in the text, let $x = 0$ to obtain an equation involving D alone, then solve the equation. You should obtain $D = 15/8$.
(b) In identity (5) we replaced D by $15/8$ and made the substitution $x = 1$. Check that this leads to the result $C = 1/2$.

37. This exercise completes calculations mentioned in Example 5.
(a) Show that identity (6) in the text leads to the following system of equations:

$$\begin{cases} A = 3 \\ B = -1 \\ 3A + C = 7 \\ 3B + D = -3 \end{cases}$$

(b) From the system in part (a) we have $A = 3$ and $B = -1$, which agrees with the values given in Example 5. Now determine C and D, and check that your answers agree with the values given in Example 5.

38. This exercise completes the discussion of improper rational expressions in this section.
(a) Use long division to obtain the following result:

$$\frac{2x^3 + 4x^2 - 15x - 36}{x^2 - 9} = (2x + 4) + \frac{3x}{x^2 - 9}$$

(b) Find constants A and B such that $3x/(x^2 - 9) = A/(x - 3) + B/(x + 3)$. (According to the text, you should obtain $A = B = 3/2$.)

In Exercises 39–44, you are given an improper rational expression. First, use long division to rewrite the expression in the form

$$(polynomial) + (proper\ rational\ expression)$$

Next, obtain the partial fraction decomposition for the proper rational expression. Finally, rewrite the given improper rational expression in the form

$$(polynomial) + (partial\ fractions)$$

39. $\dfrac{6x^3 - 16x^2 - 13x + 25}{x^2 - 4x + 3}$

40. $\dfrac{2x^5 - 11x^4 - 4x^3 + 53x^2 - 24x - 5}{2x^3 + x^2 - 10x - 5}$

41. $\dfrac{x^5 - 10x^4 + 36x^3 - 55x^2 + 32x + 1}{x^4 - 6x^3 + 12x^2 - 8x}$

42. $\dfrac{x^6 + 3x^5 + 9x^3 + 26x^2 + 3x + 8}{x^3 + 8}$

43. $\dfrac{x^6 + 2x^5 + 5x^4 - x^2 - 2x - 4}{x^4 - 1}$

44. $\dfrac{2x^7 + 3x^6 - 2x^4 - 4x^3 + 2}{x^4 + 2x^3 + x^2 - 1}$
Hint: After the long division you can factor the denominator by writing it as $x^2(x^2 + 2x + 1) - 1^2$ and then using the difference-of-squares technique.

45. This exercise completes a detail mentioned in the text in the derivation of the linear and quadratic factors theorem. Let a and b be real numbers with $b \neq 0$. Show that the quadratic polynomial $x^2 - 2ax + a^2 + b^2$ is irreducible.

B

In Exercises 46–52, determine the partial fraction decomposition for each of the given expressions.

46. $\dfrac{1}{(x - a)(x - b)}$ $(a \neq b)$ **47.** $\dfrac{px + q}{(x - a)(x - b)}$ $(a \neq b)$

48. $\dfrac{1}{(x - a)(x + a)}$ $(a \neq 0)$ **49.** $\dfrac{px + q}{(x - a)(x + a)}$ $(a \neq 0)$

50. $\dfrac{x^2 + px + q}{(x - a)(x - b)(x - c)}$ (Assume that a, b, and c are all unequal.)

51. $\dfrac{1}{(1 - ax)(1 - bx)(1 - cx)}$ (Assume that a, b, and c all are nonzero and all unequal.)

52. (a) $\dfrac{x^2 - 1}{(x^2 + 1)^3}$ **(b)** $\dfrac{x^5 - 1}{(x^2 + 1)^3}$

C

53. Find the partial fraction decomposition: $\dfrac{1}{x^4 + 1}$.
Hint: To factor the denominator, add and subtract a term.

54. (a) Determine the general form for the partial fraction decomposition of $1/(x^4 - x^3 + x^2 - x + 1)$. (Note that you are not required to find the numerical values for each constant.) *Hint*: See Exercise 27 in Section 12.7.

(b) Determine the general form for the partial fraction decomposition of $1/(x^5 - 1)$.

MINI PROJECT AN UNUSUAL PARTIAL FRACTIONS PROBLEM

As background for this mini project, you need to have worked Exercise 27(c) in Exercise Set 12.7. That exercise indicates how the equating-the-coefficients theorem can be used to factor certain polynomials.

The guidelines in this section for setting up a partial fraction decomposition for a rational expression require us to first factor the denominator into linear and/or irreducible quadratic factors. Two students each used this approach in determining a partial fractions decomposition for

$$\frac{x}{x^4 - 7x^2 + 1}$$

Student #1 used the equating-the-coefficients theorem to find a factorization of $x^4 - 7x^2 + 1$ of the form

$$x^4 - 7x^2 + 1 = (x^2 + bx + 1)(x^2 - bx + 1)$$

After obtaining the factorization (and checking to see that each factor was irreducible), the student followed the methods of Sections 12.7 and 12.8 to obtain a partial fraction decomposition. Student #2 worked along these lines also but looked for a factorization of $x^4 - 7x^2 + 1$ having the form

$$x^4 - 7x^2 + 1 = (x^2 + Bx - 1)(x^2 - Bx - 1)$$

(a) Retrace each student's work to see what results were obtained. Do the two students end up with the same partial fraction decomposition? If not, who is correct (either, both, or neither)?

(b) Factoring has also been a key element in previous sections in which we solved polynomial equations. Consider the equation $x^4 - 7x^2 + 1 = 0$. Use a graphing utility to see that this equation has four distinct real roots. Then use the rational roots theorem to explain why none of the roots is a rational number.

(c) Solve the equation $x^4 - 7x^2 + 1 = 0$ using the factorization obtained by student #1. Leave the roots in radical form rather than obtaining calculator approximations. Next, solve the equation again, but this time use the factorization obtained by student #2. Do you obtain the same four roots?

(d) Using the values for b and B that you determined in part (a), graph the following four quadratic functions in the viewing rectangle $[-4, 4]$ by $[-3, 3]$. What do you observe regarding the x-intercepts, and what does this have to do with the last question in part (c)?

$$y_1 = x^2 + bx + 1 \qquad y_3 = x^2 + Bx - 1$$
$$y_2 = x^2 - bx + 1 \qquad y_4 = x^2 - Bx - 1$$

Chapter 12 Summary of Principal Terms and Notation

Terms or formulas	Page reference	Comments
1. $i^2 = -1$	899	This is the fundamental property of the number i. The quantity i is not a real number.
2. Complex numbers	899	A complex number is an expression of the form $a + bi$, where a and b are real numbers. With the exception of division, complex numbers are combined by using the usual rules of algebra, along with the convention $i^2 = -1$. Division is carried out using complex conjugates (defined in Item 3 below), as indicated in the first box on page 902.
3. If $z = c + di$, then $\bar{z} = c - di$.	901	$\bar{z}$ is called the complex conjugate of z.
4. The division algorithm	908	This is a theorem that summarizes the results of the long division process for polynomials. Suppose that $p(x)$ and $d(x)$ are polynomials and $d(x)$ is not the zero polynomial. Then according to the division algorithm, there are unique polynomials $q(x)$ and $R(x)$ such that $$p(x) = d(x) \cdot q(x) + R(x)$$ where either $R(x)$ is the zero polynomial or the degree of $R(x)$ is less than the degree of $d(x)$.

Terms or formulas	Page reference	Comments
5. Root, solution, zero	913, 914	A root, or solution, of a polynomial equation $f(x) = 0$ is a number r such that $f(r) = 0$. The root r is also called a zero of the function f.
6. The remainder theorem	915	The remainder theorem asserts that when a polynomial $f(x)$ is divided by $x - r$, the remainder is $f(r)$. Example 3 in Section 12.3 shows how this theorem can be used with synthetic division to evaluate a polynomial.
7. The factor theorem	917	The factor theorem makes two statements about a polynomial $f(x)$. First, if $f(r) = 0$, then $x - r$ is a factor of $f(x)$. Second, if $x - r$ is a factor of $f(x)$, then $f(r) = 0$.
8. The fundamental theorem of algebra	922	Let $f(x)$ be a polynomial of degree 1 or greater. The fundamental theorem of algebra asserts that the equation $f(x) = 0$ has at least one root among the complex numbers. (See Example 1 in Section 12.4.)
9. The linear factors theorem	923	Let $f(x) = a_nx^n + a_{n-1}x^{n-1} + \cdots + a_1x + a_0$, where $n \geq 1$ and $a_n \neq 0$. Then this theorem asserts that $f(x)$ can be expressed as a product of n linear factors: $$f(x) = a_n(x - r_1)(x - r_2)\cdots(x - r_n)$$ (The complex numbers $r_1, r_2, \ldots, r_n$ are not necessarily all distinct, and some or all of them may be real numbers.) On page 925 the linear factors theorem is used to prove that every polynomial equation of degree $n \geq 1$ has exactly n roots, where a root of multiplicity k is counted k times.
10. The rational roots theorem	934	Given a polynomial equation, this theorem tells us which rational numbers are candidates for roots of the equation. For a statement of the theorem, see page 934. A proof of the theorem is outlined in Exercise 41, Exercise Set 12.5. For an example of how the theorem is applied, see Example 1 in Section 12.5.
11. Upper bound for roots; lower bound for roots	935, 936	A real number B is an upper bound for the roots of an equation if every real root is less than or equal to B. Similarly, a real number b is a lower bound for the roots if every real root is greater than or equal to b.
12. The upper and lower bound theorem for real roots	936	This theorem tells how synthetic division can be used in determining upper and lower bounds for roots of equations. For the statement and proof of the theorem, see page 936. For a demonstration of how the theorem is applied, see Example 2 in Section 12.5.

Terms or formulas	Page reference	Comments
13. The location theorem	938	Let $f(x)$ be a polynomial with real coefficients. If a and b are real numbers such that $f(a)$ and $f(b)$ have opposite signs, then the equation $f(x) = 0$ has at least one root between a and b.
14. The conjugate roots theorem	943	Let $f(x)$ be a polynomial with real coefficients, and suppose that $a + bi$ is a root of the equation $f(x) = 0$, where a and b are real numbers and $b \neq 0$. Then this theorem asserts that $a - bi$ is also a root of the equation. (In other words, for polynomial equations in which all of the coefficients are real numbers, when complex non-real roots occur, they occur in conjugate pairs.) For illustrations of how this theorem is applied, see Examples 1 and 2 in Section 12.6.
15. Variation in sign	945	Suppose that $f(x)$ is a polynomial with real coefficients, written in descending or ascending powers of x. Then a variation in sign occurs whenever two successive coefficients have opposite signs. For examples, see Table 1 in Section 12.6.
16. Descartes's rule of signs	946	Let $f(x)$ be a polynomial, all of whose coefficients are real numbers, and consider the equation $f(x) = 0$. Then, according to Descartes's rule: **(a)** The number of positive roots either is equal to the number of variations in sign of $f(x)$ or is less than that by an even integer. **(b)** The number of negative roots either is equal to the number of variations in sign of $f(-x)$ or is less than that by an even integer. Examples 4–6 in Section 12.6 show how Descartes's rule is applied.
17. Proper rational expression	957	Let $p(x)$ and $q(x)$ be polynomials such that the degree of $p(x)$ is less than the degree of $q(x)$. Then the fraction $p(x)/q(x)$ is called a proper rational expression.
18. Irreducible quadratic polynomial	957	Let $ax^2 + bx + c$ be a quadratic polynomial, all of whose coefficients are real numbers. The polynomial is irreducible provided the equation $ax^2 + bx + c = 0$ has no real-number roots.
19. The linear and quadratic factors theorem	957	Let $f(x)$ be a polynomial, all of whose coefficients are real numbers. Then $f(x)$ can be factored (over the real numbers) into a product of linear and/or irreducible quadratic factors.
20. Partial fraction decomposition	949, 958, 961	A proper rational expression $p(x)/q(x)$ can be decomposed into a sum of simpler partial fractions. The denominators of the partial fractions are built from the linear and/or irreducible quadratic factors of the original denominator $q(x)$. For details, see the boxes on pages 958 and 961.

Writing Mathematics

Discuss each of the following statements with a classmate, and decide whether it is true or false. Then (on your own), write out the reason (or reasons) for each decision.

1. Every equation has a root.
2. Every polynomial equation of degree 4 has four distinct roots.
3. No cubic equation can have a root of multiplicity 4.
4. The degree of the polynomial $x(x - 1)(x - 2)(x - 3)$ is 3.
5. The degree of the polynomial $6(x + 1)^2(x - 5)^4$ is 6.
6. Every polynomial of degree n, where $n \geq 1$, can be written in the form $(x - r_1)(x - r_2) \cdots (x - r_n)$.
7. The sum of the roots of the polynomial equation $x^2 - px + q = 0$ is p.
8. The product of the roots of the polynomial equation $2x^3 - x^2 + 3x - 1 = 0$ is 1.
9. Although a polynomial equation of degree n may have n distinct roots, the fundamental theorem of algebra tells us how to find only one of the roots.
10. Every polynomial equation of degree $n \geq 1$ has at least one real root.
11. If all of the coefficients in a polynomial equation are real, then at least one of the roots must be real.
12. Every cubic equation with roots $\sqrt{5}$, $\sqrt{6}$, and $\sqrt{7}$ can be written in the form
$$a_3(x - \sqrt{5})(x - \sqrt{6})(x - \sqrt{7}) = 0$$
13. Every polynomial equation of degree ≥ 1 has at least one root.
14. According to the rational roots theorem, there are four possibilities for the rational roots of the equation $x^5 + 6x^2 - 2 = 0$.
15. According to the rational roots theorem, there are only two possibilities for the rational roots of the equation $\frac{1}{3}x^3 - 5x + 1 = 0$.
16. The sum of the roots of the equation $x^2 - 12x + 16 = 0$ is -12.
17. According to the location theorem, if $f(x)$ is a polynomial and $f(a)$ and $f(b)$ have the same sign, then the equation $f(x) = 0$ has at least one root between a and b.
18. According to Descartes's rule, the equation $x^7 - 4x + 3 = 0$ has two positive roots.

Chapter 12 Review Exercises

In Exercises 1 and 2 you are given polynomials $p(x)$ and $d(x)$. In each case, use synthetic division to determine polynomials $q(x)$ and $R(x)$ such that

$$p(x) = d(x) \cdot q(x) + R(x)$$

where either $R(x) = 0$ or the degree of $R(x)$ is less than the degree of $d(x)$.

1. $p(x) = x^4 + 3x^3 - x^2 - 5x + 1$; $d(x) = x + 2$
2. $p(x) = 4x^4 + 2x + 1$; $d(x) = x - 2$

In Exercises 3–8, use synthetic division to find the quotients and the remainders.

3. $\dfrac{x^4 - 2x^2 + 8}{x - 3}$
4. $\dfrac{x^3 - 1}{x - 2}$
5. $\dfrac{2x^3 - 5x^2 - 6x - 3}{x + 4}$
6. $\dfrac{x^2 + x - 3\sqrt{2}}{x - \sqrt{2}}$
7. $\dfrac{5x^2 - 19x - 4}{x + 0.2}$
8. $\dfrac{x^3 - 3a^2x^2 - 4a^4x + 9a^6}{x - a^2}$

In Exercises 9–16, use synthetic division and the remainder theorem to find the indicated function values. Use a calculator for Exercises 15 and 16.

9. $f(x) = x^5 - 10x + 4; f(10)$
10. $f(x) = x^4 + 2x^3 - x; f(-2)$
11. $f(x) = x^3 - 10x^2 + x - 1; f(1/10)$
12. $f(x) = x^4 - 2a^2x^2 + 3a^3x - a^4; f(-a)$
13. $f(x) = x^3 + 3x^2 + 3x + 1; f(a - 1)$
14. $f(x) = x^3 - 1; f(1.1)$
15. $f(x) = x^4 + 4x^3 - 6x^2 - 8x - 2;$
(a) $f(-0.3)$ (Round the result to two decimal places.)
(b) $f(-0.39)$ (Round the result to three decimal places.)
(c) $f(-0.394)$ (Round the result to five decimal places.)
16. $f(-4.907378)$, where f is the function in Exercise 15. (Round the result to three decimal places.)
17. Find a value for a such that 3 is a root of the equation $x^3 - 4x^2 - ax - 6 = 0$.
18. For which values of b will -1 be a root of the equation $x^3 + 2b^2x^2 + x - 48 = 0$?
19. For which values of a will $x - 1$ be a factor of the polynomial $a^2x^3 + 3ax^2 + 2$?
20. Use synthetic division to verify that $\sqrt{2} - 1$ is a root of the equation $x^6 + 14x^3 - 1 = 0$.
21. Let $f(x) = ax^3 + bx^2 + cx + d$ and suppose that r is a root of the equation $f(x) = 0$.
(a) Show that $r - h$ is a root of $f(x + h) = 0$.
(b) Show that $-r$ is a root of the equation $f(-x) = 0$.
(c) Show that kr is a root of the equation $f(x/k) = 0$.
22. Suppose that r is a root of the equation $a_2x^2 + a_1x + a_0 = 0$. Show that mr is a root of the quadratic equation $a_2x^2 + ma_1x + m^2a_0 = 0$.

In Exercises 23–28, list the possibilities for the rational roots of the equations.

23. $x^5 - 12x^3 + x - 18 = 0$
24. $x^5 - 12x^3 + x - 17 = 0$
25. $2x^4 - 125x^3 + 3x^2 - 8 = 0$
26. $\frac{3}{5}x^3 - 8x^2 - \frac{1}{2}x + \frac{3}{2} = 0$
27. $x^3 + x - p = 0$, where p is a prime number
28. $x^3 + x - pq = 0$, where both p and q are prime numbers

In Exercises 29–36, each equation has at least one rational root. Solve the equations. Suggestion: Use the upper and lower bound theorem to eliminate some of the possibilities for rational roots.

29. $2x^3 + x^2 - 7x - 6 = 0$
30. $x^3 + 6x^2 - 8x - 7 = 0$
31. $2x^3 - x^2 - 14x + 10 = 0$
32. $2x^3 + 12x^2 + 13x + 15 = 0$
33. $\frac{3}{2}x^3 + \frac{1}{2}x^2 + \frac{1}{2}x - 1 = 0$
34. $x^4 - 2x^3 - 13x^2 + 38x - 24 = 0$
35. $x^5 + x^4 - 14x^3 - 14x^2 + 49x + 49 = 0$
36. $8x^5 + 12x^4 + 14x^3 + 13x^2 + 6x + 1 = 0$
37. Solve the equation $x^3 - 9x^2 + 24x - 20 = 0$, using the fact that one of the roots has multiplicity 2.
38. One root of the equation $x^2 + kx + 2k = 0$ $(k \neq 0)$ is twice the other. Find k and find the roots of the equation.
39. State each of the following theorems.
(a) The division algorithm
(b) The remainder theorem
(c) The factor theorem
(d) The fundamental theorem of algebra
40. Find a quadratic equation with roots $a - \sqrt{a^2 - 1}$ and $a + \sqrt{a^2 - 1}$, where $a > 1$.

In Exercises 41–44, write each polynomial in the form

$$a_n(x - r_1)(x - r_2) \cdots (x - r_n)$$

41. $6x^2 + 7x - 20$
42. $x^2 + x - 1$
43. $x^4 - 4x^3 + 5x - 20$
44. $x^4 - 4x^2 - 5$

Each of Exercises 45–48 gives an equation, followed by one or more roots. Solve the equation.

45. $x^3 - 7x^2 + 25x - 39 = 0; x = 2 - 3i$
46. $x^3 + 6x^2 - 24x + 160 = 0; x = 2 + 2i\sqrt{3}$
47. $x^4 - 2x^3 - 4x^2 + 14x - 21 = 0; x = 1 + i\sqrt{2}$
48. $x^5 + x^4 - x^3 + x^2 + x - 1 = 0; x = (1 + i\sqrt{3})/2$, $x = (-1 - \sqrt{5})/2$

In Exercises 49–54, use Descartes's rule of signs to obtain information regarding the roots of the equations.

49. $x^3 + 8x - 7 = 0$
50. $3x^4 + x^2 + 4x - 2 = 0$
51. $x^3 + 3x + 1 = 0$
52. $2x^6 + 3x^2 + 6 = 0$
53. $x^4 - 10 = 0$
54. $x^4 + 5x^2 - x + 2 = 0$
55. Consider the equation $x^3 + x^2 + x + 1 = 0$.
(a) Use Descartes's rule to show that the equation has either one or three negative roots.
(b) Now show that the equation cannot have three negative roots. *Hint*: Multiply both sides of the equation by $x - 1$. Then simplify the left-hand side and reapply Descartes's rule to the new equation.
(c) Actually, the original equation can be solved using the basic factoring techniques discussed in Appendix B.4. Solve the equation in this manner.

56. Use Descartes's rule to show that the equation $x^3 - x^2 + 3x + 2 = 0$ has no positive roots. *Hint*: Multiply both sides of the equation by $x + 1$ and apply Descartes's rule to the resulting equation.

57. Let P be the point in the first quadrant where the curve $y = x^3$ intersects the circle $x^2 + y^2 = 1$. Locate the x-coordinate of P within successive hundredths.

58. Let P be the point in the first quadrant where the parabola $y = 4 - x^2$ intersects the curve $y = x^3$. Locate the x-coordinate of P within successive hundredths.

59. Consider the equation $x^3 - 36x - 84 = 0$.
- **(a)** Use Descartes's rule to check that this equation has exactly one positive root.
- **(b)** Use the upper and lower bound theorem to show that 7 is an upper bound for the positive root.
- **(c)** Using a calculator, locate the positive root within successive hundredths.

60. Consider the equation $x^3 - 3x + 1 = 0$.
- **(a)** Use Descartes's rule to check that this equation has exactly one negative root.
- **(b)** Use the upper and lower bound theorem to show that -2 is a lower bound for the negative root.
- **(c)** Using a calculator, locate the negative root within successive hundredths.

In Exercises 61–64, find polynomial equations that have integer coefficients and the given values as roots.

61. $4 - \sqrt{5}$

62. $a + b$ and $a - b$, where a and b are integers

63. $6 - 2i$ and $\sqrt{5}$

64. $\dfrac{5 + \sqrt{6}}{5 - \sqrt{6}}$ *Hint*: First rationalize the expression.

65. Find a fourth-degree polynomial equation with integer coefficients, such that $x = 1 + \sqrt{2} + \sqrt{3}$ is a root. *Hint*: Begin by writing the given relationship as $x - 1 = \sqrt{2} + \sqrt{3}$; then square both sides.

66. Find a cubic equation with integer coefficients, such that $x = 1 + \sqrt[3]{2}$ is a root. Is $1 - \sqrt[3]{2}$ also a root of the equation?

In Exercises 67–70, first determine the zeros of each function; then sketch the graph.

67. $y = x^3 - 2x^2 - 3x$

68. $y = x^4 + 3x^3 + 3x^2 + x$

69. $y = x^4 - 4x^2$

70. $y = x^3 + 6x^2 + 5x - 12$

For Exercises 71–78, carry out the indicated operations, and express your answer in the standard form $a + bi$.

71. $(3 - 2i)(3 + 2i) + (1 + 3i)^2$

72. $2i(1 + i)^2$

73. $(1 + i\sqrt{2})(1 - i\sqrt{2}) + (\sqrt{2} + i)(\sqrt{2} - i)$

74. $(2 + 3i)/(1 + i)$

75. $(3 - i\sqrt{3})/(3 + i\sqrt{3})$

76. $\dfrac{1 + i}{1 - i} + \dfrac{1 - i}{1 + i}$

77. $-\sqrt{-2}\sqrt{-9} + \sqrt{-8} - \sqrt{-72}$

78. $\dfrac{\sqrt{-4} - \sqrt{-3}\sqrt{-3}}{\sqrt{-100}}$

79. The **real part** of a complex number z is denoted by $\text{Re}(z)$. For instance, $\text{Re}(2 + 5i) = 2$. Show that for any complex number z, we have $\text{Re}(z) = \frac{1}{2}(z + \bar{z})$. *Hint*: Let $z = a + bi$.

80. The **imaginary part** of a complex number z is denoted by $\text{Im}(z)$. For instance, $\text{Im}(2 + 5i) = 5$. Show that for any complex number z, we have

$$\text{Im}(z) = \frac{1}{2i}(z - \bar{z})$$

81. The **absolute value** of the complex number $a + bi$ is defined by $|a + bi| = \sqrt{a^2 + b^2}$.
- **(a)** Compute $|6 + 2i|$ and $|6 - 2i|$.
- **(b)** As you know, the absolute value of the real number -3 is 3. Now write -3 in the form $-3 + 0i$ and compute its absolute value using the new definition. (The point here is to observe that the two results agree.)
- **(c)** Let $z = a + bi$. Show that $z\bar{z} = |z|^2$.

In Exercises 82–86, verify that the formulas are correct by carrying out the operations indicated on the left-hand side of the equations. (This list of formulas appears in A Treatise on Algebra, *by George Peacock, published in 1845.)*

82. $\dfrac{1}{a + bi} + \dfrac{1}{a - bi} = \dfrac{2a}{a^2 + b^2}$

83. $\dfrac{1}{a - bi} - \dfrac{1}{a + bi} = \dfrac{2bi}{a^2 + b^2}$

84. $\dfrac{a + bi}{a - bi} + \dfrac{a - bi}{a + bi} = \dfrac{2(a^2 - b^2)}{a^2 + b^2}$

85. $\dfrac{a + bi}{a - bi} - \dfrac{a - bi}{a + bi} = \dfrac{4abi}{a^2 + b^2}$

86. $\dfrac{a + bi}{c + di} + \dfrac{a - bi}{c - di} = \dfrac{2(ac + bd)}{c^2 + d^2}$

In Exercises 87–92, determine the partial fraction decomposition for each expression.

87. $\dfrac{2x - 1}{100 - x^2}$

88. $\dfrac{x}{x^2 - 20}$

89. $\dfrac{1}{x^3 + 2x^2 + x}$

90. $\dfrac{1}{x^4 + x^2}$

91. $\dfrac{x^3 + 2}{x^4 + 6x^2 + 9}$

92. $\dfrac{4x^3 - 3x^2 + 24x - 15}{x^4 + 11x^2 + 30}$

93. **(a)** Let $f(x) = x^4 - 2x^3 + x^2 - 1$. Use the equating-the-coefficients theorem to find a factorization for $f(x)$ of the form

$$f(x) = (x^2 + bx + 1)(x^2 + cx - 1)$$

(b) Find the partial fraction decomposition for $x^3/(x^4 - 2x^3 + x^2 - 1)$.

94. **(a)** Compute the product $(x^2 + rx + c)(x^2 - rx + c)$. [This can be useful in part (b).]

(b) If r is a root of the equation $x^2 + bx + c = 0$, show that r^2 is a root of $x^2 + (2c - b^2)x + c^2 = 0$.

Chapter 12 Test

1. Let $f(x) = 6x^4 - 5x^3 + 7x^2 - 2x - 2$. Use the remainder theorem and synthetic division to compute $f(1/2)$.

2. Solve the equation $x^3 + x^2 - 11x - 15 = 0$, given that one of the roots is -3.

3. List the possibilities for the rational roots of the equation $2x^5 - 4x^3 + x - 6 = 0$.

4. Find a quadratic function with zeros 1 and -8 and with a y-intercept of -24.

5. Use synthetic division to divide $4x^3 + x^2 - 8x + 3$ by $x + 1$.

6. **(a)** State the factor theorem.

(b) State the fundamental theorem of algebra.

(c) State the linear and quadratic factors theorem.

7. **(a)** The equation $x^3 - 2x^2 - 1 = 0$ has just one positive root. Use the upper and lower bound theorem to determine the smallest integer that is an upper bound for the root.

(b) Locate the root between successive tenths. (Use a calculator.)

8. Solve the equation $x^5 - 6x^4 + 11x^3 + 16x^2 - 50x + 52 = 0$, given that two of the roots are $1 + i$ and $3 - 2i$.

9. Let $p(x) = x^4 + 2x^3 - x + 6$ and $d(x) = x^2 + 1$. Find polynomials $q(x)$ and $R(x)$ such that $p(x) = d(x) \cdot q(x) + R(x)$.

10. Express the polynomial $2x^2 - 6x + 5$ in the factored form $a_2(x - r_1)(x - r_2)$.

11. Consider the equation $x^4 - x^3 + 24 = 0$.

(a) List the possibilities for rational roots.

(b) Use the upper and lower bound theorem to show that 2 is an upper bound for the roots.

(c) In view of parts (a) and (b), what possibilities now remain for positive rational roots?

(d) Which (if any) of the possibilities in part (c) are actually roots?

12. **(a)** Find the rational roots of the cubic equation $2x^3 - x^2 - x - 3 = 0$.

(b) Find all solutions of the equation in part (a).

13. Use Descartes's rule of signs to obtain information regarding the roots of the following equation: $3x^4 + x^2 - 5x - 1 = 0$.

14. Find a cubic polynomial $f(x)$ with integer coefficients, such that $1 - 3i$ and -2 are roots of the equation $f(x) = 0$.

15. Find a polynomial $f(x)$ with leading coefficient 1, such that the equation $f(x) = 0$ has the following roots and no other:

Root	Multiplicity
2	1
$3i$	3
$1 + \sqrt{2}$	2

Write your answer in the form $a_n(x - r_1)(x - r_2) \cdots (x - r_n)$.

16. Simplify the expression $(3 + 2i)(5 - 3i) + \sqrt{-3}$.

17. Write the expression $\dfrac{3 + i}{1 - 4i}$ in the form $a + bi$.

In Exercises 18–20, determine the partial fraction decomposition for each expression.

18. $\dfrac{3x - 1}{x^3 - 16x}$

19. $\dfrac{1}{x^3 - x^2 + 3x - 3}$

20. $\dfrac{4x^2 - 15x + 20}{x^3 - 4x^2 + 4x}$

13 CHAPTER Additional Topics in Algebra

A strong background in algebra is an important prerequisite for courses in calculus and in probability and statistics. In this final chapter we develop several additional topics that help to provide a foundation in those areas of study. We begin in Section 13.1 with the principle of mathematical induction. This gives us a framework for proving statements about the natural numbers. In Section 13.2 we discuss the binomial theorem, which is used to analyze and expand expressions of the form $(a + b)^n$. As you'll see, the proof of the binomial theorem uses mathematical induction. Section 13.3 introduces the related (but distinct) concepts of sequences and series. Then in the next two sections (Sections 13.4 and 13.5) we study arithmetic and geometric sequences and series. The last topic introduced in Section 13.5 concerns finding sums of infinite geometric series. In a sense, this is an appropriate topic for our last chapter, for it is closely related to the idea of a *limit,* which is the starting point for calculus. Finally, in the last section of this chapter (Section 13.6) we introduce DeMoivre's theorem.

... in the ordinary treatise on the elements of algebra, these [additional] *topics are either completely omitted or treated carelessly. For this reason, I am certain that the material I have gathered in this book is quite sufficient to remedy that defect.*— Leonhard Euler (1707–1783) in his classic text, *Introductio in analysis infinitorum* (1748)

13.1 MATHEMATICAL INDUCTION

Is mathematics an experimental science? The answer to this question is both yes and no, as the following example illustrates. Consider the problem of determining a formula for the sum of the first n odd natural numbers:

$$1 + 3 + 5 + \cdots + (2n - 1)$$

We begin by doing some calculations in the hope that this may shed some light on the problem. Table 1 shows the results of calculating the sum of the first n odd natural numbers for values of n ranging from 1 to 5. Upon inspecting the table, we observe that each sum in the right-hand column is the square of the corresponding entry in the left-hand column. For instance, for $n = 5$ we see that

$$\underbrace{1 + 3 + 5 + 7 + 9}_{\text{five terms}} = 5^2$$

TABLE 1

n	$1 + 3 + 5 + \cdots + (2n - 1)$
1	$1 = 1$
2	$1 + 3 = 4$
3	$1 + 3 + 5 = 9$
4	$1 + 3 + 5 + 7 = 16$
5	$1 + 3 + 5 + 7 + 9 = 25$

Now let's try the next case, in which $n = 6$, and see whether the pattern persists. That is, we want to know whether it is true that

$$\underbrace{1 + 3 + 5 + 7 + 9 + 11}_{\text{six terms}} = 6^2$$

As you can easily check, this last equation is true. Thus on the basis of the experimental (or empirical) evidence, we are led to the following conjecture:

Conjecture. The sum of the first n odd natural numbers is n^2. That is, $1 + 3 + 5 + \cdots + (2n - 1) = n^2$, for each natural number n.

At this point, the "law" we've discovered is indeed really only a conjecture. After all, we've checked it only for values of n ranging from 1 to 6. It is conceivable at this point (although we may feel it is unlikely) that the conjecture is false for certain values of n. For the conjecture to be useful, we must be able to prove that it holds without exception for *all* natural numbers n. In fact, we will subsequently prove that this conjecture is valid. But before explaining the method of proof to be used, let's look at one more example.

Again, let n denote a natural number. Then consider the following question: Which quantity is the larger, 2^n or $(n + 1)^2$? As before, we begin by doing some calculations. This is the experimental stage of our work. According to Table 2, the quantity $(n + 1)^2$ is larger than 2^n for each value of n up through $n = 5$. Thus we make the following conjecture:

TABLE 2

n	2^n	$(n+1)^2$
1	2	4
2	4	9
3	8	16
4	16	25
5	32	36

Conjecture. $(n + 1)^2 > 2^n$ for all natural numbers n.

Again, we note that this is only a conjecture at this point. Indeed, if we try the case in which $n = 6$, we find that the pattern does not persist. That is, when $n = 6$, we find that 2^n is 64, while $(n + 1)^2$ is only 49. So in this example the conjecture is not true in general; we have found a value of n for which it fails.

The preceding examples show that experimentation does have a place in mathematics, but we must be careful with the results. When experimentation leads to a conjecture, proof is required before the conjecture can be viewed as a valid law. For the remainder of this section we shall discuss one such method of proof: *mathematical induction.*

To state the principle of mathematical induction, we first introduce some notation. Suppose that for each natural number n we have a statement P_n to be proved. Consider, for instance, our first conjecture:

$$1 + 3 + 5 + \cdots + (2n - 1) = n^2$$

Denoting this statement by P_n, we have that

P_1 is the statement that $1 = 1^2$

P_2 is the statement that $1 + 3 = 2^2$

P_3 is the statement that $1 + 3 + 5 = 3^2$

With this notation we can now state the **principle of mathematical induction.**

Mathematical induction is not a method of discovery but a technique of proving rigorously what has already been discovered.
—David M. Burton in *The History of Mathematics, an Introduction* (Boston: Allyn and Bacon, 1985)

The Principle of Mathematical Induction

Suppose that for each natural number n, we have a statement P_n for which the following two conditions hold:

1. P_1 is true.
2. For each natural number k, if P_k is true, then P_{k+1} is true.

Then all of the statements are true; that is, P_n is true for all natural numbers n.

TABLE 3

Mathematical induction		Ladder analogy	
Hypotheses	1. P_1 is true. 2. If P_k is true, then P_{k+1} is true, for any k.	Hypotheses	1′. You can reach the first rung. 2′. If you are on the kth rung, you can reach the $(k + 1)$st rung, for any k.
Conclusion	3. P_n is true for all n.	Conclusion	3′. You can climb the entire ladder.

The idea behind mathematical induction is a simple one. Think of each statement P_n as the rung of a ladder to be climbed. Then we make the analogy shown in Table 3.

According to the principle of mathematical induction, we can prove that a statement or formula P_n is true for all n if we carry out the following two steps:

Step 1 Show that P_1 is true.

Step 2 Assume that P_k is true, and on the basis of this assumption, show that P_{k+1} is true.

In Step 2 the assumption that P_k is true is referred to as the **induction hypothesis.** (In computer science, Step 1 is sometimes referred to as the **initialization step.**) Now let's turn to some examples of proof by mathematical induction.

EXAMPLE 1 A basic induction proof

Use mathematical induction to prove that

$$1 + 3 + 5 + \cdots + (2n - 1) = n^2$$

for all natural numbers n.

SOLUTION

Let P_n denote the statement that $1 + 3 + 5 + \cdots + (2n - 1) = n^2$. Then we want to show that P_n is true for all natural numbers n.

Step 1 We must check that P_1 is true. But P_1 is just the statement that $1 = 1^2$, which is true.

Step 2 Assuming that P_k is true, we must show that P_{k+1} is true. Thus we assume that

$$1 + 3 + 5 + \cdots + (2k - 1) = k^2 \qquad (1)$$

That is the induction hypothesis. We must now show that

$$1 + 3 + 5 + \cdots + (2k - 1) + [2(k + 1) - 1] = (k + 1)^2 \qquad (2)$$

To derive equation (2) from equation (1), we add the quantity $[2(k + 1) - 1]$ to both sides of equation (1). (The motivation for this stems from the observation that the left-hand sides of equations (1) and (2) differ only by the quantity $[2(k + 1) - 1]$.) We obtain

$$
\begin{aligned}
1 + 3 + 5 + \cdots + (2k - 1) + [2(k + 1) - 1] &= k^2 + [2(k + 1) - 1] \\
&= k^2 + 2k + 1 \\
&= (k + 1)^2
\end{aligned}
$$

That is, $1 + 3 + 5 + \cdots + (2k + 1) + [2(k + 1) - 1] = (k + 1)^2$. So P_{k+1} is true.

Having now carried out Steps 1 and 2, we conclude by the principle of mathematical induction that P_n is true for all natural numbers n.

EXAMPLE 2 An induction proof requiring more algebra in Step 2

Use mathematical induction to prove that

$$2^3 + 4^3 + 6^3 + \cdots + (2n)^3 = 2n^2(n + 1)^2$$

for all natural numbers n.

SOLUTION
Let P_n denote the statement that

$$2^3 + 4^3 + 6^3 + \cdots + (2n)^3 = 2n^2(n + 1)^2$$

Then we want to show that P_n is true for all natural numbers n.

Step 1 We must check that P_1 is true, where P_1 is the statement that

$$2^3 = 2(1^2)(1 + 1)^2 \qquad \text{or} \qquad 8 = 8$$

Thus P_1 is true.

Step 2 Assuming that P_k is true, we must show that P_{k+1} is true. Thus we assume that

$$2^3 + 4^3 + 6^3 + \cdots + (2k)^3 = 2k^2(k + 1)^2 \tag{3}$$

We must now show that

$$2^3 + 4^3 + 6^3 + \cdots + (2k)^3 + [2(k + 1)]^3 = 2(k + 1)^2(k + 2)^2 \tag{4}$$

Adding $[2(k + 1)]^3$ to both sides of equation (3) yields

$$
\begin{aligned}
2^3 + 4^3 + 6^3 + \cdots + (2k)^3 + [2(k + 1)]^3 &= 2k^2(k + 1)^2 + [2(k + 1)]^3 \\
&= 2k^2(k + 1)^2 + 8(k + 1)^3 \\
&= 2(k + 1)^2[k^2 + 4(k + 1)] \\
&= 2(k + 1)^2(k^2 + 4k + 4) \\
&= 2(k + 1)^2(k + 2)^2
\end{aligned}
$$

We have now derived equation (4) from equation (3), as we wished to do. Having carried out Steps 1 and 2, we conclude by the principle of mathematical induction that P_n is true for all natural numbers n.

EXAMPLE 3 A case in which the statement to be proved is not a formula

As is indicated in Table 4, the number 3 is a factor of $2^{2n} - 1$ when $n = 1, 2, 3$, and 4. Use mathematical induction to show that 3 is a factor of $2^{2n} - 1$ for all natural numbers n.

TABLE 4

n	$2^{2n} - 1$
1	$3\ (= 3 \cdot 1)$
2	$15\ (= 3 \cdot 5)$
3	$63\ (= 3 \cdot 21)$
4	$255\ (= 3 \cdot 85)$

SOLUTION

Let P_n denote the statement that 3 is a factor of $2^{2n} - 1$. We want to show that P_n is true for all natural numbers n.

Step 1 We must check that P_1 is true. P_1 in this case is the statement that 3 is a factor of $2^{2(1)} - 1$, that is, 3 is a factor of 3, which is surely true.

Step 2 Assuming that P_k is true, we must show that P_{k+1} is true. Thus we assume that

$$3 \text{ is a factor of } 2^{2k} - 1 \tag{5}$$

and we must show that

$$3 \text{ is a factor of } 2^{2(k+1)} - 1$$

The strategy here will be to rewrite the expression $2^{2(k+1)} - 1$ in such a way that the induction hypothesis, statement (5), can be applied. We have

$$\begin{aligned} 2^{2(k+1)} - 1 &= 2^{2k+2} - 1 \\ &= 2^2 \cdot 2^{2k} - 1 \\ &= 4 \cdot 2^{2k} - 4 + 3 \\ 2^{2(k+1)} - 1 &= 4(2^{2k} - 1) + 3 \end{aligned} \tag{6}$$

Now look at the right-hand side of equation (6). By the induction hypothesis, 3 is a factor of $2^{2k} - 1$. Thus 3 is a factor of $4(2^{2k} - 1)$, from which it certainly follows that 3 is a factor of $4(2^{2k} - 1) + 3$. In summary, then, 3 is a factor of the right-hand side of equation (6). Consequently, 3 must be a factor of the left-hand side of equation (6), which is what we wished to show.

Having now completed Steps 1 and 2, we conclude by the principle of mathematical induction that P_n is true for all natural numbers n. In other words, 3 is a factor of $2^{2n} - 1$ for all natural numbers n.

There are instances in which a given statement P_n is false for certain initial values of n but true thereafter. An example of this is provided by the statement

$$2^n > (n + 1)^2$$

As you can easily check, this statement is false for $n = 1, 2, 3, 4$, and 5. But, as Example 4 shows, the statement is true for $n \geq 6$. In Example 4 we adapt the principle of mathematical induction by beginning in Step 1 with a consideration of P_6 rather than P_1.

EXAMPLE 4 A case in which the statement to be proved is an inequality

Use mathematical induction to prove that

$$2^n > (n + 1)^2 \qquad \text{for all natural numbers } n \geq 6$$

SOLUTION

Step 1 We must first check that P_6 is true. But P_6 is simply the assertion that

$$2^6 > (6 + 1)^2 \qquad \text{or} \qquad 64 > 49$$

Thus P_6 is true.

Step 2 Assuming that P_k is true, where $k \geq 6$, we must show that P_{k+1} is true. Thus we assume that

$$2^k > (k+1)^2 \qquad \text{where } k \geq 6 \tag{7}$$

We must show that

$$2^{k+1} > (k+2)^2$$

Multiplying both sides of inequality (7) by 2 gives us

$$2(2^k) > 2(k+1)^2 = 2k^2 + 4k + 2$$

This can be rewritten

$$2^{k+1} > k^2 + 4k + (k^2 + 2)$$

However, since $k \geq 6$, it is certainly true that

$$k^2 + 2 > 4$$

We therefore have

$$2^{k+1} > k^2 + 4k + 4 \qquad \text{or} \qquad 2^{k+1} > (k+2)^2$$

as we wished to show.

Having now completed Steps 1 and 2, we conclude that P_n is true for all natural numbers $n \geq 6$.

EXERCISE SET 13.1

A

In Exercises 1–18, use the principle of mathematical induction to show that the statements are true for all natural numbers.

1. $1 + 2 + 3 + \cdots + n = n(n+1)/2$
2. $2 + 4 + 6 + \cdots + 2n = n(n+1)$
3. $1 + 4 + 7 + \cdots + (3n-2) = n(3n-1)/2$
4. $5 + 9 + 13 + \cdots + (4n+1) = n(2n+3)$
5. $1^2 + 2^2 + 3^2 + \cdots + n^2 = n(n+1)(2n+1)/6$
6. $2^2 + 4^2 + 6^2 + \cdots + (2n)^2 = 2n(n+1)(2n+1)/3$
7. $1^2 + 3^2 + 5^2 + \cdots + (2n-1)^2 = n(2n-1)(2n+1)/3$
8. $2 + 2^2 + 2^3 + \cdots + 2^n = 2^{n+1} - 2$
9. $3 + 3^2 + 3^3 + \cdots + 3^n = (3^{n+1} - 3)/2$
10. $e^x + e^{2x} + e^{3x} + \cdots + e^{nx} = \dfrac{e^{(n+1)x} - e^x}{e^x - 1} \quad (x \neq 0)$
11. $1^3 + 2^3 + 3^3 + \cdots + n^3 = [n(n+1)/2]^2$
12. $2^3 + 4^3 + 6^3 + \cdots + (2n)^3 = 2n^2(n+1)^2$
13. $1^3 + 3^3 + 5^3 + \cdots + (2n-1)^3 = n^2(2n^2 - 1)$
14. $1 \cdot 2 + 3 \cdot 4 + 5 \cdot 6 + \cdots + (2n-1)(2n) = n(n+1)(4n-1)/3$
15. $1 \cdot 3 + 3 \cdot 5 + 5 \cdot 7 + \cdots + (2n-1)(2n+1) = n(4n^2 + 6n - 1)/3$
16. $\dfrac{1}{1 \times 3} + \dfrac{1}{2 \times 4} + \dfrac{1}{3 \times 5} + \cdots + \dfrac{1}{n(n+2)} = \dfrac{n(3n+5)}{4(n+1)(n+2)}$
17. $1 + \dfrac{3}{2} + \dfrac{5}{2^2} + \dfrac{7}{2^3} + \cdots + \dfrac{2n-1}{2^{n-1}} = 6 - \dfrac{2n+3}{2^{n-1}}$
18. $1 + 2 \cdot 2 + 3 \cdot 2^2 + 4 \cdot 2^3 + \cdots + n \cdot 2^{n-1} = (n-1)2^n + 1$
19. Show that $n \leq 2^{n-1}$ for all natural numbers n.
20. Show that 3 is a factor of $n^3 + 2n$ for all natural numbers n.
21. Show that $n^2 + 4 < (n+1)^2$ for all natural numbers $n \geq 2$.
22. Show that $n^3 > (n+1)^2$ for all natural numbers $n \geq 3$.

In Exercises 23–26, prove that the statement is true for all natural numbers in the specified range. Use a calculator to carry out Step 1.

23. $(1.5)^n > 2n;\ n \geq 7$
24. $(1.25)^n > n;\ n \geq 11$
25. $(1.1)^n > n;\ n \geq 39$
26. $(1.1)^n > 5n;\ n \geq 60$

B

27. Let $f(n) = \dfrac{1}{1 \cdot 2} + \dfrac{1}{2 \cdot 3} + \dfrac{1}{3 \cdot 4} + \cdots + \dfrac{1}{n(n+1)}$.

(a) Complete the following table.

n	1	2	3	4	5
$f(n)$					

(b) On the basis of the results in the table, what would you guess to be the value of $f(6)$? Compute $f(6)$ to see whether this is correct.

(c) Make a conjecture about the value of $f(n)$, and prove it using mathematical induction.

28. Let $f(n) = \frac{1}{1 \times 3} + \frac{1}{3 \times 5} + \frac{1}{5 \times 7} + \cdots + \frac{1}{(2n-1)(2n+1)}$.

(a) Complete the following table.

n	1	2	3	4
$f(n)$				

(b) On the basis of the results in the table, what would you guess to be the value of $f(5)$? Compute $f(5)$ to see whether your guess is correct.

(c) Make a conjecture about the value of $f(n)$, and prove it using mathematical induction.

29. Suppose that a function f satisfies the following conditions:

$$f(1) = 1$$
$$f(n) = f(n-1) + 2\sqrt{f(n-1)} + 1 \quad (n \geq 2)$$

(a) Complete the table.

n	1	2	3	4	5
$f(n)$					

(b) On the basis of the results in the table, what would you guess to be the value of $f(6)$? Compute $f(6)$ to see whether your guess is correct.

(c) Make a conjecture about the value of $f(n)$ when n is a natural number, and prove the conjecture using mathematical induction.

30. This exercise demonstrates the necessity of carrying out both Step 1 and Step 2 before considering an induction proof valid.

(a) Let P_n denote the statement that $n^2 + 1$ is even. Check that P_1 is true. Then give an example showing that P_n is not true for all n.

(b) Let Q_n denote the statement that $n^2 + n$ is odd. Show that Step 2 of an induction proof can be completed in this case, but not Step 1.

31. A *prime number* is a natural number that has no factors other than itself and 1. For technical reasons, 1 is not considered a prime. Thus, the list of the first seven primes looks like this: 2, 3, 5, 7, 11, 13, 17. Let P_n be the statement that $n^2 + n + 11$ is prime. Check that P_n is true for all values of n less than 10. Check that P_{10} is false.

32. Prove that if $x \neq 1$,

$$1 + 2x + 3x^2 + \cdots + nx^{n-1} = \frac{1 - x^n}{(1-x)^2} - \frac{nx^n}{1-x}$$

for all natural numbers n.

33. If $r \neq 1$, show that

$$1 + r + r^2 + \cdots + r^{n-1} = \frac{r^n - 1}{r - 1}$$

for all natural numbers n.

34. Use mathematical induction to show that

$$x^n - 1 = (x-1)(1 + x + x^2 + \cdots + x^{n-1})$$

for all natural numbers n.

35. Prove that 5 is a factor of $n^5 - n$ for all natural numbers $n \geq 2$.

36. Prove that 4 is a factor of $5^n + 3$ for all natural numbers n.

37. Prove that 5 is a factor of $2^{2n+1} + 3^{2n+1}$ for all nonnegative integers n.

38. Prove that 8 is a factor of $3^{2n} - 1$ for all natural numbers n.

39. Prove that 3 is a factor of $2^{n+1} + (-1)^n$ for all nonnegative integers n.

40. Prove that 6 is a factor of $n^3 + 3n^2 + 2n$ for all natural numbers n.

41. Use mathematical induction to show that $x - y$ is a factor of $x^n - y^n$ for all natural numbers n. *Suggestion for Step 2:* Verify and then use the fact that

$$x^{k+1} - y^{k+1} = x^k(x - y) + (x^k - y^k)y$$

In Exercises 42 and 43, use mathematical induction to prove that the formulas hold for all natural numbers n.

42. $\log_{10}(a_1 a_2 \ldots a_n) = \log_{10} a_1 + \log_{10} a_2 + \cdots + \log_{10} a_n$

43. $(1 + p)^n \geq 1 + np$, where $p > -1$

13.2 THE BINOMIAL THEOREM

[Blaise Pascal] *made numerous discoveries relating to this array and set them forth in his* Traité du triangle arithmétique, *published posthumously in 1665, and among these was essentially our present Binomial Theorem for positive integral exponents.*—David Eugene Smith in *History of Mathematics,* vol. II (New York: Ginn and Co., 1925)

A mathematician, like a painter or a poet, is a maker of patterns.—G. H. Hardy (1877–1947)

Recall from basic algebra that an expression that is the sum or difference of two terms is referred to as a *binomial expression.* Three examples of binomial expressions are

$$u + v, \quad 2x^2 - y^3, \quad \text{and} \quad 4m^3n + 5x^3y^4$$

Our goal in this section is to develop a general formula, known as the *binomial theorem,* for expanding any product of the form $(a + b)^n$, where n is a natural number.

We begin by looking for patterns in the expansion of $(a + b)^n$. To do this, let's list the expansions of $(a + b)^n$ for $n = 1, 2, 3, 4$, and 5. (Exercises 1 and 2 at the end of this section ask you to verify these results simply by repeated multiplication.)

$$(a + b)^1 = a + b$$
$$(a + b)^2 = a^2 + 2ab + b^2$$
$$(a + b)^3 = a^3 + 3a^2b + 3ab^2 + b^3$$
$$(a + b)^4 = a^4 + 4a^3b + 6a^2b^2 + 4ab^3 + b^4$$
$$(a + b)^5 = a^5 + 5a^4b + 10a^3b^2 + 10a^2b^3 + 5ab^4 + b^5$$

After surveying these results, we note the following patterns.

PROPERTY SUMMARY Patterns Observed in $(a + b)^n$ for $n = 1, 2, 3, 4, 5$

General Statement	Example
There are $n + 1$ terms.	There are 4 (=3 + 1) terms in the expansion of $(a + b)^3$.
The expansion begins with a^n and ends with b^n.	$(a + b)^3$ begins with a^3 and ends with b^3.
The sum of the exponents in each term is n.	The sum of the exponents in each term of $(a + b)^3$ is 3.
The exponents of a decrease by 1 from term to . term.	$(a + b)^3 = a^{\textcircled{3}} + 3a^{\textcircled{2}}\, b + 3a^{\textcircled{1}}\, b^2 + a^{\textcircled{0}}\, b^3$
The exponents of b increase by 1 from term to term.	$(a + b)^3 = a^3b^{\textcircled{0}} + 3a^2b^{\textcircled{1}} + 3ab^{\textcircled{2}} + b^{\textcircled{3}}$
When n is even, the coefficients are symmetric about the middle term.	The sequence of coefficients for $(a + b)^4$ is 1, 4, 6, 4, 1.
When n is odd, the coefficients are symmetric about the two middle terms.	The sequence of coefficients for $(a + b)^5$ is 1, 5, 10, 10, 5, 1.

古法七桒方圖

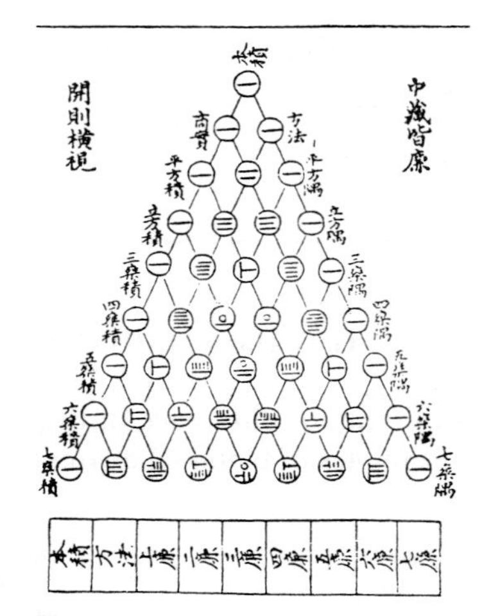

Figure 1
"Pascal's" triangle, by Chu-Shi-Kie, A.D. 1303

The patterns we have just observed for $(a + b)^n$ persist for all natural numbers n. This follows from the binomial theorem, which we prove at the end of this section. Thus, for example, the form of $(a + b)^6$ must be as follows. (Each question mark denotes a coefficient to be determined.)

$$(a + b)^6 = a^6 + ?a^5b + ?a^4b^2 + ?a^3b^3 + ?a^2b^4 + ?ab^5 + b^6$$

The problem now is to find the proper coefficient for each term. To do this, we need to discover additional patterns in the expansion of $(a + b)^n$.

We have already written out the expansions of $(a + b)^n$ for values of n ranging from 1 to 5. Let us now write only the coefficients appearing in those expansions. The resulting triangular array of numbers is known as **Pascal's triangle.*** For reasons of symmetry we begin with $(a + b)^0$ rather than $(a + b)^1$.

*The array is named after Blaise Pascal, a seventeenth-century French mathematician and philosopher. However, as Figure 1 indicates, the Pascal triangle was known to Chinese mathematicians centuries earlier.

$(a+b)^0$						1					
$(a+b)^1$					1		1				
$(a+b)^2$				1		2		1			
$(a+b)^3$			1		3		3		1		
$(a+b)^4$		1		4		6		4		1	
$(a+b)^5$	1		5		10		10		5		1

The key observation regarding Pascal's triangle is this: *Each entry in the array* (other than the 1's along the sides) *is the sum of the two numbers diagonally above it.* For instance, the 6 that appears in the fifth row is the sum of the two 3's diagonally above it. Using this observation, we can form as many additional rows as we please. The coefficients for $(a+b)^n$ will then appear in the $(n+1)$st row of the array.* For instance, to obtain the row corresponding to $(a+b)^6$, we have

sixth row, $(a+b)^5$: 1 ↘↙ 5 ↘↙ 10 ↘↙ 10 ↘↙ 5 ↘↙ 1

seventh row, $(a+b)^6$: 1 6 15 20 15 6 1

Thus the sequence of coefficients for $(a+b)^6$ is 1, 6, 15, 20, 15, 6, 1. This answers the question raised earlier about the expansion of $(a+b)^6$. We have

$$(a+b)^6 = a^6 + 6a^5b + 15a^4b^2 + 20a^3b^3 + 15a^2b^4 + 6ab^5 + b^6$$

For analytical work or for larger values of the exponent n, it is inefficient to rely on Pascal's triangle. For this reason we point out another pattern in the expansions of $(a+b)^n$.

In the expansion of $(a+b)^n$ the coefficient of any term after the first can be generated as follows: From the *previous* term, multiply the coefficient by the exponent of a and then divide by the number of that previous term.

To see how this method is used, let's compute the second, third, and fourth coefficients in the expansion of $(a+b)^6$. To compute the coefficient of the second term, we go back to the first term, which is a^6. We have

$$\text{coefficient of second term} = \frac{1 \cdot 6}{1} = 6$$

(1 = coefficient of first term; 6 = exponent of a in first term; denominator 1 = number of first term)

Thus the second term is $6a^5b$, and consequently, we have

$$\text{coefficient of third term} = \frac{6 \cdot 5}{2} = 15$$

(6 = coefficient of second term; 5 = exponent of a in second term; denominator 2 = number of second term)

---

*That these numbers actually are the appropriate coefficients follows from the binomial theorem, which is proved at the end of this section.

Continuing now with this method, you should check for yourself that the coefficient of the fourth term in the expansion of $(a + b)^6$ is 20.

NOTE We now know that the first four coefficients are 1, 6, 15, and 20. By symmetry it follows that the complete sequence of coefficients for this expansion is 1, 6, 15, 20, 15, 6, 1. No additional calculation of coefficients is necessary.

EXAMPLE 1 Expanding a binomial

Expand $(2x - y^2)^7$.

SOLUTION
First we write the expression of $(a + b)^7$ using the method explained just prior to this example, or using Pascal's triangle. As you should check for yourself, the expansion is

$$(a + b)^7 = a^7 + 7a^6b + 21a^5b^2 + 35a^4b^3 + 35a^3b^4 + 21a^2b^5 + 7ab^6 + b^7$$

Now we make the substitutions $a = 2x$ and $b = -y^2$. This yields

$$\begin{aligned}(2x - y^2)^7 = [2x + (-y^2)]^7 &= (2x)^7 + 7(2x)^6(-y^2) + 21(2x)^5(-y^2)^2 \\ &\quad + 35(2x)^4(-y^2)^3 + 35(2x)^3(-y^2)^4 \\ &\quad + 21(2x)^2(-y^2)^5 + 7(2x)(-y^2)^6 + (-y^2)^7 \\ &= 128x^7 - 448x^6y^2 + 672x^5y^4 - 560x^4y^6 + 280x^3y^8 \\ &\quad - 84x^2y^{10} + 14xy^{12} - y^{14}\end{aligned}$$

This is the required expansion. Notice how the signs alternate in the final answer; this is characteristic of expressions of the form $(a - b)^n$.

In preparation for the binomial theorem we introduce two notations that are used not only in connection with the binomial theorem, but also in many other areas of mathematics. The first of these notations is $n!$ (read "n factorial").

DEFINITION The Factorial Symbol

$n! = 1 \cdot 2 \cdot 3 \cdots n$ where n is a natural number

$0! = 1$

EXAMPLES

$3! = 1 \cdot 2 \cdot 3 = 6$

$$\frac{6!}{4!} = \frac{6 \cdot 5 \cdot 4 \cdot 3 \cdot 2 \cdot 1}{4 \cdot 3 \cdot 2 \cdot 1} = 6 \cdot 5 = 30$$

EXAMPLE 2 Practice with the factorial notation

Simplify the expression $\dfrac{(n + 1)!}{(n - 1)!}$.

SOLUTION

$$\frac{(n + 1)!}{(n - 1)!} = \frac{(n + 1) \cdot n \cdot (n - 1)!}{(n - 1)!} = (n + 1) \cdot n = n^2 + n$$

The second notation that we introduce in preparation for the binomial theorem is $\binom{n}{k}$. This notation is read "n choose k," because it can be shown that $\binom{n}{k}$ is equal to the number of different subsets of k elements from a set with n elements.

DEFINITION **The Binomial Coefficient $\binom{n}{k}$**

Let n and k be nonnegative integers, with $k \leq n$. Then the binomial coefficient $\binom{n}{k}$ is defined by

$$\binom{n}{k} = \frac{n!}{k!(n-k)!}$$

EXAMPLE

$$\binom{5}{2} = \frac{5!}{2!(5-2)!} = \frac{5!}{2!3!} = \frac{5 \cdot 4 \cdot 3 \cdot 2 \cdot 1}{(2 \cdot 1)(3 \cdot 2 \cdot 1)} = \frac{5 \cdot 4}{2 \cdot 1} = 10$$

The binomial coefficients are so named because they are indeed the coefficients in the expansion of $(a + b)^n$. More precisely, the relationship is this: The coefficients in the expansion of $(a + b)^n$ are the $n + 1$ numbers

$$\binom{n}{0}, \binom{n}{1}, \binom{n}{2}, \ldots, \binom{n}{n}$$

Subsequently, we will see why this statement is true. For now, however, let's look at an example. Consider the binomial coefficients $\binom{3}{0}$, $\binom{3}{1}$, $\binom{3}{2}$, and $\binom{3}{3}$. According to our statement, these four quantities should be the coefficients in the expansion of $(a + b)^3$. Let's check:

$$\binom{3}{0} = \frac{3!}{0!(3-0)!} = \frac{3!}{1(3!)} = 1$$

$$\binom{3}{1} = \frac{3!}{1!(3-1)!} = \frac{3 \cdot 2 \cdot 1}{1(2 \cdot 1)} = 3$$

$$\binom{3}{2} = \frac{3!}{2!(3-2)!} = \frac{3 \cdot 2 \cdot 1}{(2 \cdot 1)1} = 3$$

$$\binom{3}{3} = \frac{3!}{3!(3-3)!} = \frac{3!}{3!0!} = 1$$

The values of $\binom{3}{0}$, $\binom{3}{1}$, $\binom{3}{2}$, and $\binom{3}{3}$ are thus 1, 3, 3, and 1, respectively. But these last four numbers are indeed the coefficients in the expression of $(a + b)^3$, as we wished to check.

We are now in a position to state the binomial theorem, after which we will look at several applications. Finally, at the end of this section we will use mathematical induction to prove the theorem. In the statement of the theorem that follows, we are assuming that the exponent n is a natural number.

The Binomial Theorem

$$(a+b)^n = \binom{n}{0}a^n + \binom{n}{1}a^{n-1}b + \binom{n}{2}a^{n-2}b^2 + \cdots + \binom{n}{n-1}ab^{n-1} + \binom{n}{n}b^n$$

One of the uses of the binomial theorem is in identifying specific terms in an expansion without computing the entire expansion. This is particularly helpful when the exponent n is relatively large. Looking back at the statement of the binomial theorem, there are three observations we can make. First, the coefficient of the rth term is $\binom{n}{r-1}$. For instance, the coefficient of the third term is $\binom{n}{3-1} = \binom{n}{2}$. The second observation is that the exponent for a in the rth term is $n - (r-1)$. For instance, the exponent for a in the third term is $n - (3-1) = n - 2$. Finally, we observe that the exponent for b in the rth term is $r - 1$, the same quantity that appears in the lower position of the corresponding binomial coefficient. For instance, the exponent for b in the third term is $r - 1 = 3 - 1 = 2$. We summarize these three observations with the following statement.

The rth term in the expansion of $(a+b)^n$ is

$$\binom{n}{r-1}a^{n-r+1}b^{r-1}$$

EXAMPLE 3 Finding a term in a binomial expansion

Find the 15th term in the expansion of $\left(x^2 - \frac{1}{x}\right)^{18}$.

SOLUTION

Using the values $r = 15$, $n = 18$, $a = x^2$, and $b = -1/x$, we have

$$\begin{aligned}\binom{n}{r-1}a^{n-r+1}b^{r-1} &= \binom{18}{15-1}(x^2)^{18-15+1}\left(\frac{-1}{x}\right)^{15-1}\\ &= \binom{18}{14}x^8 \cdot \frac{1}{x^{14}}\\ &= \frac{18 \cdot 17 \cdot 16 \cdot 15 \cdot (14!)}{14!(4 \cdot 3 \cdot 2 \cdot 1)}x^{-6}\\ &= \frac{18 \cdot 17 \cdot 16 \cdot 15}{4 \cdot 3 \cdot 2 \cdot 1}x^{-6}\\ &= 3060x^{-6} \qquad \text{as required}\end{aligned}$$

EXAMPLE 4 Finding a coefficient in a binomial expansion

Find the coefficient of the term containing x^4 in the expansion of $(x + y^2)^{30}$.

SOLUTION

Again we use the fact that the rth term in the expansion of $(a + b)^n$ is $\binom{n}{r-1} a^{n-r+1}b^{r-1}$. In this case, n is 30, and x plays the role of a. The exponent for x is then $n - r + 1$ or $30 - r + 1$. To see when this exponent is 4, we write

$$30 - r + 1 = 4 \qquad \text{and therefore} \qquad r = 27$$

The required coefficient is therefore $\binom{30}{27-1}$. We then have

$$\binom{30}{26} = \frac{30!}{26!(30-26)!} = \frac{30 \cdot 29 \cdot 28 \cdot 27}{4 \cdot 3 \cdot 2 \cdot 1}$$

After carrying out the indicated arithmetic, we find that $\binom{30}{26} = 27{,}405$. This is the required coefficient.

EXAMPLE 5 Finding a coefficient in a binomial expansion

Find the coefficient of the term containing a^9 in the expansion of $(a + 2\sqrt{a})^{10}$.

SOLUTION

The rth term in this expansion is

$$\binom{10}{r-1} a^{10-r+1}(2\sqrt{a})^{r-1}$$

We can rewrite this as

$$\binom{10}{r-1} a^{10-r+1}(2^{r-1})(a^{1/2})^{r-1}$$

or

$$2^{r-1}\binom{10}{r-1} a^{10-r+1+(r-1)/2}$$

From this we see that the general form of the coefficient that we wish to find is $2^{r-1}\binom{10}{r-1}$. We now need to determine r when the exponent of a is 9. Thus we require that

$$10 - r + 1 + \frac{r-1}{2} = 9$$

$$-r + \frac{r-1}{2} = -2$$

$$-2r + r - 1 = -4 \qquad \text{multiplying both sides by 2}$$

$$r = 3$$

The required coefficient is now obtained by substituting $r = 3$ in the expression $2^{r-1}\binom{10}{r-1}$. Thus the required coefficient is

$$2^2\binom{10}{2} = \frac{4 \cdot 10!}{2!(10-2)!} = 2 \cdot 10 \cdot 9 = 180$$

Alternatively, we can write

$$(a + 2\sqrt{a})^{10} = (\sqrt{a})^{10}(\sqrt{a} + 2)^{10} = a^5(\sqrt{a} + 2)^{10}$$

Then the term containing a^4 is the term with $(\sqrt{a})^8$. So we want the coefficient of the term $a^5\binom{10}{2}(\sqrt{a})^8 \cdot 2^2$, that is, $2^2\binom{10}{2} = 180$.

There are three simple identities involving the binomial coefficients that will simplify our proof of the binomial theorem:

IDENTITY 1 $\binom{r}{0} = 1$ for all nonnegative integers r

IDENTITY 2 $\binom{r}{r} = 1$ for all nonnegative integers r

IDENTITY 3 $\binom{k}{r} + \binom{k}{r-1} = \binom{k+1}{r}$ for all natural numbers k and r with $r \le k$

All three of these identities can be proved directly from the definitions of the binomial coefficients, without the need for mathematical induction. The proofs of the first two are straightforward, and we omit them. Here is the proof of the third identity:

$$\begin{aligned}\binom{k}{r} + \binom{k}{r-1} &= \frac{k!}{r!(k-r)!} + \frac{k!}{(r-1)!(k-r+1)!} \\ &= \frac{k!}{r(r-1)!(k-r)!} + \frac{k!}{(r-1)!(k-r+1)(k-r)!}\end{aligned}$$

The common denominator on the right-hand side of the last equation is $r(r-1)!(k-r+1)(k-r)!$ Thus we have

$$\begin{aligned}\binom{k}{r} + \binom{k}{r-1} &= \frac{k!(k-r+1)}{r(r-1)!(k-r+1)(k-r)!} + \frac{k!r}{r(r-1)!(k-r+1)(k-r)!} \\ &= \frac{k!(k-r+1) + k!r}{r(r-1)!(k-r+1)(k-r)!} \\ &= \frac{k!(k-r+1+r)}{r(r-1)!(k-r+1)(k-r)!} = \frac{k!(k+1)}{r!(k-r+1)!} \\ &= \frac{(k+1)!}{r![(k+1)-r]!} \\ &= \binom{k+1}{r} \quad \text{as required}\end{aligned}$$

Taken together, the three identities show why the $(n+1)$st row of Pascal's triangle consists of the numbers $\binom{n}{0}, \binom{n}{1}, \binom{n}{2}, \ldots, \binom{n}{n}$. Identities 1 and 2 tell us that this row of numbers begins and ends with 1. Identity 3 is just a statement of the fact that each entry in the row, other than the initial and final 1, is generated by adding the two entries diagonally above it. For example, using Pascal's triangle for $(a + b)^4$ and $(a + b)^5$, we have

$$\begin{array}{ccccccccccc} & 1 & & 4 & & 6 & & 4 & & 1 & \\ 1 & & 5 & & 10 & & 10 & & 5 & & 1 \end{array} \qquad \text{or} \qquad \begin{array}{c} \text{Identity} \\ \binom{4}{2} + \binom{4}{3} = \binom{5}{3} \\ 6 \;+\; 4 \;=\; 10 \end{array}$$

$$6 + 4 = 10$$

We conclude this section by using mathematical induction to prove the binomial theorem. The statement P_n that we wish to prove for all natural numbers n is this:

$$(a + b)^n = \binom{n}{0}a^n + \binom{n}{1}a^{n-1}b + \cdots + \binom{n}{n-1}ab^{n-1} + \binom{n}{n}b^n$$

First, we check that P_1 is true. The statement P_1 asserts that

$$(a + b)^1 = \binom{1}{0}a^1 + \binom{1}{1}b$$

However, in view of Identities 1 and 2, this last equation becomes

$$(a + b)^1 = 1 \cdot a + 1 \cdot b$$

which is surely true. Now let's assume that P_k is true and, on the basis of this assumption, show that P_{k+1} is true. The statement P_k is

$$(a + b)^k = \binom{k}{0}a^k + \binom{k}{1}a^{k-1}b + \cdots + \binom{k}{k-1}ab^{k-1} + \binom{k}{k}b^k$$

Multiplying both sides of this equation by the quantity $(a + b)$ yields

$$\begin{aligned}
(a + b)^{k+1} &= (a + b)\left[\binom{k}{0}a^k + \binom{k}{1}a^{k-1}b + \cdots + \binom{k}{k-1}ab^{k-1} + \binom{k}{k}b^k\right] \\
&= a\left[\binom{k}{0}a^k + \binom{k}{1}a^{k-1}b + \cdots + \binom{k}{k-1}ab^{k-1} + \binom{k}{k}b^k\right] \\
&\quad + b\left[\binom{k}{0}a^k + \binom{k}{1}a^{k-1}b + \cdots + \binom{k}{k-1}ab^{k-1} + \binom{k}{k}b^k\right] \\
&= \binom{k}{0}a^{k+1} + \binom{k}{1}a^k b + \cdots + \binom{k}{k-1}a^2b^{k-1} + \binom{k}{k}ab^k \\
&\quad + \binom{k}{0}a^k b + \binom{k}{1}a^{k-1}b^2 + \cdots + \binom{k}{k-1}ab^k + \binom{k}{k}b^{k+1} \\
&= \binom{k}{0}a^{k+1} + \left[\binom{k}{1} + \binom{k}{0}\right]a^k b + \cdots + \left[\binom{k}{k} + \binom{k}{k-1}\right]ab^k + \binom{k}{k}b^{k+1}
\end{aligned}$$

We can now make some substitutions on the right-hand side of this last equation. The initial binomial coefficient $\binom{k}{0}$ can be replaced by $\binom{k+1}{0}$, because both are equal to 1, according to Identity 1. Similarly, the binomial coefficient $\binom{k}{k}$ appearing in the last term on the right-hand side of the equation can be replaced by

$\binom{k+1}{k+1}$, since both are equal to 1 according to Identity 2. Finally, we can use Identity 3 to simplify each of the sums in the brackets. We obtain

$$(a+b)^{k+1} = \binom{k+1}{0}a^{k+1} + \binom{k+1}{1}a^k b + \cdots + \binom{k+1}{k}ab^k + \binom{k+1}{k+1}b^{k+1}$$

This last equation is just the statement P_{k+1}; that is, we have derived P_{k+1} from P_k, as we wished to do. The induction proof is now complete.

EXERCISE SET 13.2 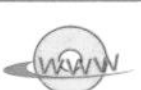

A

In Exercises 1 and 2, verify each statement directly, without using the techniques developed in this section.

1. (a) $(a+b)^2 = a^2 + 2ab + b^2$
(b) $(a+b)^3 = a^3 + 3a^2b + 3ab^2 + b^3$
Hint: $(a+b)^3 = (a+b)(a+b)^2$

2. (a) $(a+b)^4 = a^4 + 4a^3b + 6a^2b^2 + 4ab^3 + b^4$
Hint: Use the result in Exercise 1(b) and the fact that $(a+b)^4 = (a+b)(a+b)^3$.
(b) $(a+b)^5 = a^5 + 5a^4b + 10a^3b^2 + 10a^2b^3 + 5ab^4 + b^5$

In Exercises 3–12, evaluate or simplify each expression.

3. $5!$
4. (a) $3! + 2!$ **(b)** $(3+2)!$
5. $\binom{7}{3}\binom{3}{2}$
6. $\dfrac{20!}{18!}$
7. (a) $\binom{5}{3}$ **(b)** $\binom{5}{4}$
8. (a) $\binom{7}{7}$ **(b)** $\binom{7}{0}$
9. $\dfrac{(n+2)!}{n!}$
10. $\dfrac{n[(n-2)!]}{(n+1)!}$
11. $\binom{6}{4} + \binom{6}{3} - \binom{7}{4}$
12. $(3!)! + (3!)^2$

In Exercises 13–38, carry out the indicated expansions.

13. $(a+b)^9$
14. $(a-b)^9$
15. $(2A+B)^3$
16. $(1+2x)^6$
17. $(1-2x)^6$
18. $(3x^2-y)^5$
19. $(\sqrt{x}+\sqrt{y})^4$
20. $(\sqrt{x}-\sqrt{y})^4$
21. $(x^2+y^2)^5$
22. $(5A-B^2)^3$
23. $[1-(1/x)]^6$
24. $(3x+y^2)^4$
25. $(x/2-y/3)^3$
26. $(1-z^2)^7$
27. $(ab^2+c)^7$
28. $[x-(1/x)]^8$
29. $(x+\sqrt{2})^8$
30. $(4A-\frac{1}{2})^5$
31. $(\sqrt{2}-1)^3$
32. $(1+\sqrt{5})^4$
33. $(\sqrt{2}+\sqrt{3})^5$
34. $(\frac{1}{2}-2a)^6$
35. $(2\sqrt[3]{2}-\sqrt[3]{4})^3$
36. $(x+y+1)^4$
Suggestion: Rewrite the expression as $[(x+y)+1]^4$.
37. $(x^2-2x-1)^5$
Suggestion: Rewrite the expression as $[x^2-(2x+1)]^5$.
38. $[x^2-2x-(1/x)]^6$
39. Find the 15th term in the expansion of $(a+b)^{16}$.
40. Find the third term in the expansion of $(a-b)^{30}$.
41. Find the 100th term in the expansion of $(1+x)^{100}$.
42. Find the 23rd term in the expansion of $[x-(1/x^2)]^{25}$.
43. Find the coefficient of the term containing a^4 in the expansion of $(\sqrt{a}-\sqrt{x})^{10}$.
44. Find the coefficient of the term containing a^4 in the expansion of $(3a-5x)^{12}$.
45. Find the coefficient of the term containing y^8 in the expansion of $[(x/2)-4y]^9$.
46. Find the coefficient of the term containing x^6 in the expansion of $[x^2-(1/x)]^{12}$.
47. Find the coefficient of the term containing x^3 in the expansion of $(1-\sqrt{x})^8$.
48. Find the coefficient of the term containing a^8 in the expansion of $[a-(2/\sqrt{a})]^{14}$.
49. Find the term that does not contain A in the expansion of $[(1/A)+3A^2]^{12}$.
50. Find the coefficient of B^{-10} in the expansion of $[(B^2/2)-(3/B^3)]^{10}$.

B

51. Show that the coefficient of x^n in the expansion of $(1+x)^{2n}$ is $(2n)!/(n!)^2$.
52. Find n so that the coefficients of the 11th and 13th terms in $(1+x)^n$ are the same.
53. (a) Complete the following table.

k	0	1	2	3	4	5	6	7	8
$\binom{8}{k}$									

(b) Use the results in part (a) to verify that

$$\binom{8}{0} + \binom{8}{1} + \binom{8}{2} + \cdots + \binom{8}{8} = 2^8$$

(c) By taking $a = b = 1$ in the expansion of $(a+b)^n$, show that

$$\binom{n}{0} + \binom{n}{1} + \binom{n}{2} + \cdots + \binom{n}{n} = 2^n$$

C

54. Two real numbers A and B are defined by $A = \sqrt[99]{99!}$ and $B = \sqrt[100]{100!}$. Which number is larger, A or B?
Hint: Compare A^{9900} and B^{9900}.

55. This exercise outlines a proof of the identity

$$\binom{n}{0}^2 + \binom{n}{1}^2 + \binom{n}{2}^2 + \cdots + \binom{n}{n}^2 = \binom{2n}{n}$$

(a) Verify that

$$(1 + x)^n\left(1 + \frac{1}{x}\right)^n = \frac{(1 + x)^{2n}}{x^n} \qquad (1)$$

(This requires only basic algebra, not the binomial theorem.)

(b) Show that the coefficient of the term independent of x on the right side of equation (1) is $\binom{2n}{n}$.

(c) Use the binomial theorem to expand $(1 + x)^n$. Then show that the coefficient of the term independent of x on the left side of equation (1) is

$$\binom{n}{0}^2 + \binom{n}{1}^2 + \cdots + \binom{n}{n}^2$$

13.3 INTRODUCTION TO SEQUENCES AND SERIES

Students of calculus do not always understand that infinite series are primarily tools for the study of functions. —George F. Simmons in *Calculus with Analytic Geometry* (New York: McGraw-Hill, 1985)

This section and the next two sections in this chapter deal with numerical sequences. We will begin with a somewhat informal definition of this concept. Then, after looking at some examples and terminology, we'll point out how the function concept is involved in defining a sequence. A **numerical sequence** is an ordered list of numbers. Here are four examples:

Example A: $1, \sqrt{2}, 10$ Example C: $1, \frac{1}{2}, \frac{1}{4}, \frac{1}{8}, \ldots$

Example B: $2, 4, 6, 8, \ldots$ Example D: $1, 1, 1, 1, \ldots$

The individual entries in a numerical sequence are called the **terms** of the sequence. In this chapter the terms in each sequence will always be real numbers, so for convenience we will drop the adjective "numerical" and refer simply to *sequences.* (It is worth pointing out, however, that in more advanced courses, sequences are studied in which the individual terms are functions.) Any sequence that has only a finite number of terms is called a **finite sequence.** Thus the sequence in Example A is a finite sequence. On the other hand, Examples B, C, and D are examples of what we call **infinite sequences;** each contains infinitely many terms. As Example D indicates, it is not necessary that all the terms in a sequence be distinct. In this chapter, all the sequences we discuss will be infinite sequences.

In Examples B, C, and D the three dots are read "and so on." In using this notation, we are assuming that it is clear what the subsequent terms of the sequence are. Toward this end, we often specify a formula for the nth term in a sequence. Example B in this case would appear this way:

$$2, 4, 6, 8, \ldots, 2n, \ldots$$

A letter with subscripts is often used to denote the various terms in a sequence. For instance, if we denote the sequence in Example B by $a_1, a_2, a_3, \ldots$, then we have $a_1 = 2$, $a_2 = 4$, $a_3 = 6$, and, in general, $a_n = 2n$. Of course, there is nothing special about the letter a in this context; any other letter would do just as well.

EXAMPLE 1 Computing terms in a sequence

Consider the sequence $a_1, a_2, a_3, \ldots$, in which the nth term a_n is given by

$$a_n = \frac{n}{n+1}$$

Compute the first three terms of the sequence, as well as the 1000th term.

SOLUTION

To obtain the first term, we replace n by 1 in the given formula. This yields

$$a_1 = \frac{1}{1+1} = \frac{1}{2}$$

The other terms are similarly obtained. We have

$$a_2 = \frac{2}{2+1} = \frac{2}{3}$$
$$a_3 = \frac{3}{3+1} = \frac{3}{4}$$
$$a_{1000} = \frac{1000}{1000+1} = \frac{1000}{1001}$$

A word about notation: In Example 1 we denoted the sequence by $a_1, a_2, a_3, \ldots$. In this case the subscripts begin with 1 and run through the natural numbers. The next example shows a case in which the subscripts start with 0 rather than 1. (There is no new concept here; as we've said, it's just a matter of notation.)

EXAMPLE 2 Computing a sum of terms in a sequence

Consider the sequence $b_0, b_1, b_2, \ldots$, in which the general term is given by

$$b_n = (-10)^n \qquad \text{for } n \geq 0$$

Compute the sum of the first three terms of this sequence.

SOLUTION

We are asked to compute the sum $b_0 + b_1 + b_2$. Using the given formula for b_n, we have

$$b_0 = (-10)^0 = 1$$
$$b_1 = (-10)^1 = -10$$
$$b_2 = (-10)^2 = 100$$

Therefore $b_0 + b_1 + b_2 = 1 + (-10) + 100 = 91$.

In Examples 1 and 2 we were given a formula for the general term in each sequence. The next example shows a different way of specifying a sequence: Each term after the first (or after the first several) is defined in terms of preceding terms. This is an example of a **recursive definition.** Recursive definitions are particularly useful in computer programming.

EXAMPLE 3 **Computing terms in a sequence that is defined recursively**

Compute the first three terms of the sequence $b_1, b_2, b_3, \ldots$, which is defined recursively by

$$b_1 = 4$$
$$b_n = 2(b_{n-1} - 1) \qquad \text{for } n \geq 2$$

SOLUTION

We are given the first term: $b_1 = 4$. To find b_2, we replace n by 2 in the formula $b_n = 2(b_{n-1} - 1)$ to obtain

$$b_2 = 2(b_1 - 1) = 2(4 - 1) = 6$$

Thus $b_2 = 6$. Next we use this value of b_2 in the formula $b_n = 2(b_{n-1} - 1)$ to obtain b_3. Replacing n by 3 in this formula yields

$$b_3 = 2(b_2 - 1) = 2(6 - 1) = 10$$

We have now found the first three terms of the sequence: $b_1 = 4$, $b_2 = 6$, and $b_3 = 10$.

If you think about the central idea behind the concept of a sequence, you can see that a function is involved: For each input n, we have an output a_n. As the previous examples indicated, the inputs n may be the natural numbers $1, 2, 3, \ldots$, or they may be the nonnegative integers $0, 1, 2, 3, \ldots$. For these reasons the formal definition of a sequence is stated as follows.

DEFINITION **Sequence**

A sequence is a function whose domain is either the set of natural numbers or the set of nonnegative integers.

If we denote the function by f for the moment, then $f(1)$ denotes what we have been calling a_1. Similarly, $f(2) = a_2$, $f(3) = a_3$, and, in general, $f(n) = a_n$. So the sequence notation with subscripts is just another kind of function notation. Furthermore, since a sequence is a function, we can draw a graph, as indicated in the next example.

EXAMPLE 4 **Graphing a sequence**

Consider the sequence defined by

$$x_n = \frac{1}{2^n} \qquad \text{for } n \geq 0$$

Graph this sequence for $n = 0, 1, 2,$ and 3.

TABLE 1

n	0	1	2	3
x_n	1	1/2	1/4	1/8

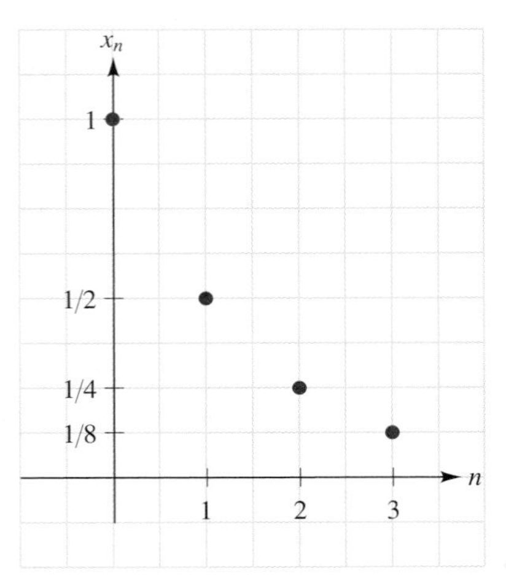

Figure 1
Graph of the sequence $x_n = \frac{1}{2^n}$ for $n = 0, 1, 2,$ and 3

SOLUTION
First we compute the required values of x_n, as shown in Table 1. Then, locating the inputs n on the horizontal axis and the outputs x_n on the vertical axis, we obtain the graph shown in Figure 1.

In part (b) of the next example we use a graphing utility to compute the terms of a recursive sequence and to obtain the graph. For details on using a graphing utility in this fashion, consult the user's manual for the graphing utility.

EXAMPLE 5 Using a recursive sequence to model population growth

Suppose that the following recursive sequence is a model for the size P_t of a population of rabbits on an island at the end of t years.

$$P_0 = 16$$
$$P_t = P_{t-1} + 0.0003P_{t-1}(3000 - P_{t-1}) \qquad (t \geq 1)$$

(a) Find $P_0, P_1, \ldots, P_5$. After all the computations are complete, round each answer to the nearest integer (because the projected numbers of rabbits must be integers). Then use the rounded values to sketch the graph of the population sequence for $t = 0, 1, \ldots, 5$. On the basis of the graph, give a general description of the population trend over this period.

(b) Use a graphing utility to graph the population sequence through the end of the 15th year and describe the population trend.

SOLUTION
(a) We are given that $P_0 = 16$. This is the initial population. For P_1, we substitute $t = 1$ in the equation

$$P_t = P_{t-1} + 0.0003P_{t-1}(3000 - P_{t-1}) \qquad (1)$$

This yields

$$\begin{aligned} P_1 &= P_0 + 0.0003P_0(3000 - P_0) \\ &= 16 + 0.0003(16)(3000 - 16) \qquad \text{using } P_0 = 16 \\ &= 30.3232 \qquad \text{using a calculator} \end{aligned}$$

Similarly, for P_2, we substitute $t = 2$ in equation (1) to obtain

$$\begin{aligned} P_2 &= P_1 + 0.0003P_1(3000 - P_1) \\ &= 30.3232 + 0.0003(30.3232)(3000 - 30.3232) \\ &\approx 57.33823106 \qquad \text{as displayed on calculator} \end{aligned}$$

The remaining calculations are carried out in the same way. The results are displayed in Table 2. The third row of the table shows the values rounded to the nearest integers. We interpret these rounded values as the population values predicted by the given recursive model. In Figure 2 we've drawn a scatter plot showing the size of the rabbit population at the end of years $t = 0$ through $t = 5$.

TABLE 2

t	0	1	2	3	4	5
P_t	16	30.3232	57.33823106	107.9563372	201.6206695	370.8840037
P_t (to nearest integer)	16	30	57	108	202	371

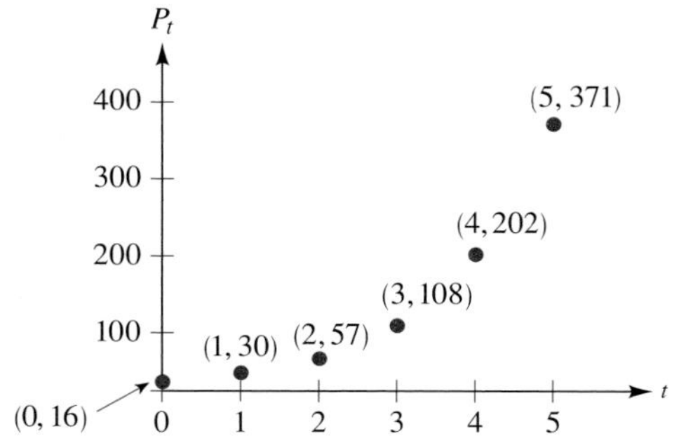

Figure 2
The size P_t of the rabbit population at the end of t years, $t = 0, 1, \ldots, 5$

In Figure 2, it appears that the size of the population increases slowly at first and then faster and faster. The picture suggests exponential growth; however, as you'll see in part (b), that model is inappropriate in the long run.

(b) Starting with the (unrounded) value of P_5 from Table 2 and applying equation (1), we use a graphing utility to compute P_6 through P_{15}. After rounding as in part (a), we obtain the population results displayed in Table 3. In Figure 3 we've used the graphing utility to draw a scatter plot using the data from both Tables 2 and 3.

In Figure 2 we observed that the size of the population increases slowly at first and then faster and faster (as in exponential growth). Now, with Figure 3, we see that the rapid growth continues only until the end of the ninth (or perhaps eighth) year. After the end of the ninth year, the size of the

TABLE 3 The Size of the Rabbit Population at the End of Years t = 6 through t = 15

t	6	7	8	9	10	11	12	13	14	15
P_t (to nearest integer)	663	1128	1762	2416	2839	2976	2997	3000	3000	3000

GRAPHICAL PERSPECTIVE

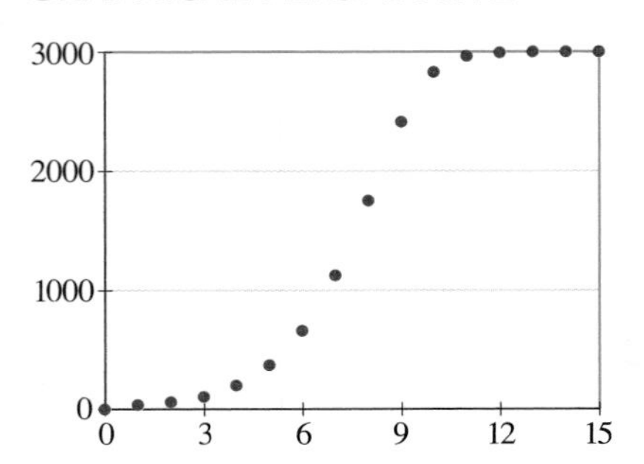

Figure 3
The size P_t of the rabbit population at the end of t years, $t = 0, 1, \ldots, 15$

population begins to level off and approach 3000. Indeed, as both Table 3 and Figure 3 indicated, the size of the population reaches and remains at 3000 from $t = 13$ onward. The population size of 3000 is referred to as the *equilibrium population.* Actually, up to this point, we have shown that the size of the population is 3000 only for $t = 13$, 14, and 15. We can use algebra to prove that once the population size reaches 3000, the recursive model projects that it will remain at that level for *all* subsequent years. In equation (1), assume that $P_{t-1} = 3000$. We then have

$$\begin{aligned} P_t &= 3000 + 0.0003(3000)(3000 - 3000) \\ &= 3000 + 0.0003(3000)(0) = 3000 \end{aligned}$$

This shows that if the size of the rabbit population is 3000 at the end of any year, then it will again be 3000 at the end of the next year.

We will often be interested in the sum of certain terms of a sequence. Consider, for example, the sequence

$$10, 20, 30, 40, \ldots$$

in which the nth term is $10n$. The sum of the first four terms in this sequence is

$$10 + 20 + 30 + 40 = 100$$

More generally, the sum of the first n terms in this sequence is indicated by

$$10 + 20 + 30 + \cdots + 10n$$

This expression is an example of a **finite series,** which simply means a sum of a finite number of terms.

We can indicate the sum of the first n terms of the sequence $a_1, a_2, a_3, \ldots$ by

$$a_1 + a_2 + a_3 + \cdots + a_n$$

Another way to indicate this sum uses what is called **sigma notation,** which we now introduce. The capital Greek letter sigma is written Σ. We define the notation $\sum_{k=1}^{n} a_k$ by the equation

$$\sum_{k=1}^{n} a_k = a_1 + a_2 + a_3 + \cdots + a_n$$

For example, $\sum_{k=1}^{3} a_k$ stands for the sum $a_1 + a_2 + a_3$, the idea in this case being to replace the subscript k successively by 1, 2, and 3 and then add the results.

For a more concrete example, let's evaluate $\sum_{k=1}^{4} k^2$. We have

$$\begin{aligned}\sum_{k=1}^{4} k^2 &= 1^2 + 2^2 + 3^3 + 4^2 \\ &= 1 + 4 + 9 + 16 = 30\end{aligned}$$

There is nothing special about the choice of the letter k in the expression $\sum_{k=1}^{4} k^2$. For instance, we could equally well write

$$\sum_{j=1}^{4} j^2 = 1^2 + 2^2 + 3^2 + 4^2 = 30$$

The letter k below the sigma in the expression $\sum_{k=1}^{4} k^2$ is called the **index of summation.** Similarly, the letter j appearing below the sigma in $\sum_{j=1}^{4} j^2$ is the index of summation in that case. As we have seen, the choice of the letter used for the index of summation has no effect on the value of the indicated sum. For this reason the index of summation is referred to as a *dummy variable.* The next two examples provide further practice with sigma notation.

EXAMPLE 6 Understanding sigma notation

Express each of the following sums without sigma notation.

(a) $\sum_{k=1}^{3} (3k - 2)^2$ **(b)** $\sum_{i=1}^{4} ix^{i-1}$ **(c)** $\sum_{j=1}^{5} (a_{j+1} - a_j)$

SOLUTION

(a) The notation $\sum_{k=1}^{3} (3k - 2)^2$ directs us to replace k successively by 1, 2, and 3 in the expression $(3k - 2)^2$ and then to add the results. We thus obtain

$$\sum_{k=1}^{3} (3k - 2)^2 = 1^2 + 4^2 + 7^2 = 66$$

(b) The notation $\sum_{i=1}^{4} ix^{i-1}$ directs us to replace i successively by 1, 2, 3, and 4 in the expression ix^{i-1} and then to add the results. We have

$$\sum_{i=1}^{4} ix^{i-1} = 1x^0 + 2x^1 + 3x^2 + 4x^3$$
$$= 1 + 2x + 3x^2 + 4x^3$$

(c) To expand $\sum_{j=1}^{5}(a_{j+1} - a_j)$, we replace j successively by 1, 2, 3, 4, and 5 in the expression $a_{j+1} - a_j$ and then add. We obtain

$$\sum_{j=1}^{5}(a_{j+1} - a_j) = (a_2 - a_1) + (a_3 - a_2) + (a_4 - a_3) + (a_5 - a_4) + (a_6 - a_5)$$
$$= a_2 - a_1 + a_3 - a_2 + a_4 - a_3 + a_5 - a_4 + a_6 - a_5$$

Combining like terms, we have

$$\sum_{j=1}^{5}(a_{j+1} - a_j) = a_6 - a_1$$

Sums such as $\sum_{j=1}^{5}(a_{j+1} - a_j)$ are known as **collapsing** or **telescoping sums.**

EXAMPLE 7 Using sigma notation

Use sigma notation to rewrite each sum.

(a) $\frac{x}{1!} + \frac{x^2}{2!} + \frac{x^3}{3!} + \cdots + \frac{x^{12}}{12!}$ **(b)** $\frac{x}{2!} + \frac{x^2}{3!} + \frac{x^3}{4!} + \cdots + \frac{x^n}{(n+1)!}$

SOLUTION

(a) Since the exponents on x run from 1 to 12, we choose a dummy variable, say k, running from 1 to 12. Also, we notice that if x^k is the numerator of a given term in the sum, then $k!$ is the corresponding denominator. Consequently, the sum can be written

$$\frac{x}{1!} + \frac{x^2}{2!} + \frac{x^3}{3!} + \cdots + \frac{x^{12}}{12!} = \sum_{k=1}^{12}\frac{x^k}{k!}$$

(b) Since the exponents on x run from 1 to n, we choose a dummy variable, say k, running from 1 to n. (Note that both of the letters n and x would be inappropriate here as dummy variables.) Also, we notice that if the numerator of a given term in the sum is x^k, then the corresponding denominator is $(k + 1)!$. Thus the given sum can be written

$$\frac{x}{2!} + \frac{x^2}{3!} + \frac{x^3}{4!} + \cdots + \frac{x^n}{(n+1)!} = \sum_{k=1}^{n}\frac{x^k}{(k+1)!}$$

EXERCISE SET 13.3

A

In Exercises 1–14, compute the first four terms in each sequence. For Exercises 1–10, assume that each sequence is defined for $n \geq 1$; in Exercises 11–14, assume that each sequence is defined for $n \geq 0$.

1. $a_n = n/(n + 1)$
2. $a_n = 1/n^2$
3. $b_n = (-1)^n$
4. $b_n = (n + 1)^2$
5. $c_n = 2^{-n}$
6. $c_n = (-1)^n(2^n)$
7. $x_n = 3n$
8. $x_n = (n - 1)^{n+1}$
9. $y_n = [1 + (1/n)]^n$
10. $y_n = [(-1)^n]\sqrt{n}$
11. $a_n = (n - 1)/(n + 1)$
12. $a_n = (-1)^n/(n + 2)$
13. $b_n = (-2)^{n+1}/(n + 1)^2$
14. $b_n = (n + 2)^{-n}$

In Exercises 15–22, the sequences are defined recursively. Compute the first five terms in each sequence.

15. $a_1 = 1; a_n = (1 + a_{n-1})^2, n \geq 2$

16. $a_1 = 2; a_n = \sqrt{a_{n-1}^2 + 1}, n \geq 2$

17. $a_1 = 2; a_2 = 2; a_n = a_{n-1}a_{n-2}, n \geq 3$

18. $F_1 = 1; F_2 = 1; F_n = F_{n-1} + F_{n-2}, n \geq 3$

19. $a_1 = 1; a_{n+1} = na_n, n \geq 1$

20. $a_1 = 1; a_2 = 2; a_n = a_{n-1}/a_{n-2}, n \geq 3$

21. $a_1 = 0; a_n = 2^{a_{n-1}}, n \geq 2$

22. $a_1 = 0; a_2 = 1; a_n = (a_{n-1} + a_{n-2})/2, n \geq 3$

In Exercises 23–32, sketch a graph showing the first five terms of the sequence. (A graphing utility is optional.)

23. $a_n = \frac{n}{2} + 1, n \geq 0$ **24.** $b_n = 16 - n^2, n \geq 0$

25. $c_n = 5/n, n \geq 1$ **26.** $d_n = (n - 1)!, n \geq 1$

27. $a_1 = 1; a_n = (a_{n-1})^2 - a_{n-1}, n \geq 2$

28. $a_0 = 3; a_n = (a_{n-1} + 1)/(a_{n-1} - 2), n \geq 1$

29. $b_0 = 2; b_n = (b_{n-1})^2 - 2b_{n-1} - 1, n \geq 1$

30. $b_1 = 0; b_n = b_{n-1} - n + 6, n \geq 2$

31. $A_0 = 0; A_n = \frac{A_{n-1} - 3}{A_{n-1} + 1}, n \geq 1$

32. $B_1 = 1; B_n = \frac{2B_{n-1} + 2}{-B_{n-1}}, n \geq 2$

In Exercises 33–38, find the sum of the first five terms of the sequence. (In Exercises 33–37, assume that the sequences are defined for $n \geq 1$.)

33. $a_n = 2^n$ **34.** $b_n = 2^{-n}$

35. $a_n = n^2 - n$ **36.** $b_n = (n - 1)!$

37. $a_n = (-1)^n/n!$

38. $a_1 = 1; a_n = \frac{1}{n - 1} - \frac{1}{n + 1}, n \geq 2$

39. Find the sum of the first five terms of the sequence that is defined recursively by $a_1 = 1; a_2 = 2; a_n = a_{n-1}^2 + a_{n-2}^2$, $n \geq 3$.

40. Find the sum of the first six terms of the sequence defined recursively by $a_1 = 1; a_2 = 2; a_{n+1} = a_n a_{n-1}, n \geq 2$.

In Exercises 41–52, express each of the sums without using sigma notation. Simplify your answers where possible.

41. $\sum_{k=1}^{3} (k - 1)$ **42.** $\sum_{k=1}^{5} k$

43. $\sum_{k=4}^{5} k^2$ **44.** $\sum_{k=2}^{6} (1 - 2k)$

45. $\sum_{n=1}^{3} x^n$ **46.** $\sum_{n=1}^{3} (n - 1)x^{n-2}$

47. $\sum_{n=1}^{4} \frac{1}{n}$ **48.** $\sum_{n=0}^{4} 3^n$

49. $\sum_{j=1}^{9} \log_{10}\left(\frac{j}{j + 1}\right)$ **50.** $\sum_{j=2}^{5} \log_{10} j$

51. $\sum_{j=1}^{6} \left(\frac{1}{j} - \frac{1}{j + 1}\right)$ **52.** $\sum_{j=1}^{5} (x^{j+1} - x^j)$

In Exercises 53–62, rewrite the sums using sigma notation.

53. $5 + 5^2 + 5^3 + 5^4$ **54.** $5 + 5^2 + 5^3 + \cdots + 5^n$

55. $x + x^2 + x^3 + x^4 + x^5 + x^6$

56. $x + 2x^2 + 3x^3 + 4x^4 + 5x^5 + 6x^6$

57. $\frac{1}{1} + \frac{1}{2} + \frac{1}{3} + \cdots + \frac{1}{12}$ **58.** $\frac{1}{1} + \frac{1}{2} + \frac{1}{3} + \cdots + \frac{1}{n}$

59. $2 - 2^2 + 2^3 - 2^4 + 2^5$

60. $\binom{10}{3} + \binom{10}{4} + \binom{10}{5} + \cdots + \binom{10}{10}$

61. $1 - 2 + 3 - 4 + 5$

62. $\frac{1}{2} - \frac{1}{4} + \frac{1}{8} - \frac{1}{16} + \frac{1}{32} - \frac{1}{64} + \frac{1}{128}$

B

63. A sequence is defined recursively as follows: $s_1 = 0.7$ and $s_n = (s_{n-1})^2$ for $n \geq 2$.

(a) Compute the first six terms of this sequence. (Use a calculator; for the answers, round to five decimal places.) What do you observe about the answers?

(b) Use a calculator to compute s_{10}. (Report the answer as shown on your calculator screen.)

(c) In view of your work in parts (a) and (b), what number do you think would be a very close approximation to s_{100}?

64. A sequence is defined recursively as follows:

$$x_1 = 1 \quad \text{and} \quad x_n = \frac{x_{n-1}}{1 + x_{n-1}} \quad \text{for } n > 1$$

(a) Complete the following table:

n	1	2	3	4
x_n				

(b) On the basis of the results in the table, make a guess about the value of x_5, then compute x_5 to see if your guess is correct.

(c) The sequence given at the start of this exercise is defined recursively. Make a conjecture about a simpler way to define this sequence, then use mathematical induction to prove that your conjecture is correct.

65. An important model that is used in population biology and ecology is the **Ricker model.** The Canadian biologist William E. Ricker introduced this model in his paper Stock and Recruitment (*Journal of the Fisheries Research Board of Canada,* **11** (1954) 559–623). For information on Ricker himself, see the web page

http://www.science.ca/scientists/scientistprofile.php?pID=17

The general form of the Ricker model that we will use here is defined by a recursive sequence of the form

P_0 = initial population at time $t = 0$

$P_t = rP_{t-1}e^{-kP_{t-1}}$ for $t \geq 1$, and where r and k are positive constants

G **(a)** Suppose that the initial size of a population is $P_0 = 300$ and that the size of the population at the end of year t

(continues)

is given by

$$P_t = 5P_{t-1}e^{-P_{t-1}/1000} \qquad (t \geq 1)$$

Use a graphing utility to compute the population sizes through the end of year $t = 5$. (As in Example 5, round the final answers to the nearest integers.) Then use the graphing utility to draw the population scatter plot for $t = 0, 1, \ldots, 5$. Describe in complete sentences how the size of the population changes over this period. Does the population seem to be approaching an equilibrium level?

Ⓖ **(b)** Using a graphing utility, compute the sizes of the population in part (a) through the end of the year $t = 20$, and draw the corresponding scatter plot. Note that the population seems to be approaching an equilibrium level of 1609 (or 1610).

(c) Determine the equilibrium population algebraically by solving the following equation for P_{t-1}. For the final answer, use a calculator and round to the nearest integer.

$$P_{t-1} = 5P_{t-1}e^{-P_{t-1}/1000}$$

Hint: Divide both sides of the equation by P_{t-1}, and then use the techniques of Chapter 5.

Ⓖ **66.** (The Ricker model continued) Suppose that the initial size of a population is $P_0 = 300$ and that the size of the population at the end of year t is given by

$$P_t = 10P_{t-1}e^{-P_{t-1}/1000} \qquad (t \geq 1)$$

(a) Use a graphing utility to compute the population sizes through the end of year $t = 5$. (As in Example 5, round the final answers to the nearest integers.) Then use the graphing utility to draw the population scatter plot for $t = 0, 1, \ldots, 5$. Give a general description (in complete sentences) of how the size of the population changes over this period.

(b) Use a graphing utility to compute the population sizes through the end of year $t = 20$, and draw the scatter plot. To help you see the pattern, use the option on your graphing utility that connects adjacent dots in a scatter plot with line segments. Describe the population trend that emerges over the period $t = 15$ to $t = 20$.

(c) For a clearer view of the long-term population behavior, use a graphing utility to compute the population sizes for the period $t = 25$ to $t = 35$, and draw the scatter plot. As in part (b), use the option on your graphing utility that connects adjacent dots with line segments. Summarize (in complete sentences) what you observe.

Exercises 67–76 involve the ***Fibonacci sequence,*** *which is defined recursively as follows:*

$$F_1 = 1; \qquad F_2 = 1; \qquad F_{n+2} = F_n + F_{n+1} \quad \text{for } n \geq 1$$

This sequence was first studied by the Italian mathematician and merchant Leonardo of Pisa (ca. 1170–1240), better known as Fibonacci ("son of Bonaccio").

67. (a) Complete the following table for the first ten terms of the Fibonacci sequence.

F_1	F_2	F_3	F_4	F_5	F_6	F_7	F_8	F_9	F_{10}
1	1								

(b) Given that $F_{20} = 6765$ and $F_{21} = 10946$, compute F_{22} and F_{19}.

(c) Given that $F_{29} = 514229$ and $F_{31} = 1346269$, compute F_{30}.

68. (a) Use the table in Exercise 67(a) to verify that $F_8 + F_9 + F_{10} = 2(F_8 + F_9)$.

(b) Show that $F_{100} + F_{101} + F_{102} = 2(F_{100} + F_{101})$.

(c) Show that $F_n + F_{n+1} + F_{n+2} = 2(F_n + F_{n+1})$ for all natural numbers $n \geq 3$. *Hint*: Mathematical induction is not needed.

(d) Does the identity in part (c) hold for either of the values $n = 1$ or $n = 2$?

In Exercises 69–71: **(a)** *Verify that the given equation holds for* $n = 1, n = 2,$ *and* $n = 3$; and **(b)** *use mathematical induction to show that the equation holds for all natural numbers.*

69. $F_1 + F_2 + F_3 + \cdots + F_n = F_{n+2} - 1$

70. $F_1^2 + F_2^2 + F_3^2 + \cdots + F_n^2 = F_nF_{n+1}$

71. $F_{n+1}^2 = F_nF_{n+2} + (-1)^n$ *Hint for part (b)*: Add $F_{k+1}F_{k+2}$ to both sides of the equation in the induction hypothesis. Then factor F_{k+1} from the left-hand side and factor F_{k+2} from the first two terms on the right-hand side.

72. Use mathematical induction to prove that $F_n \geq n$ for all natural numbers $n \geq 5$.

73. We've seen that the Fibonacci sequence is defined recursively; each term after the second is the sum of the previous two terms. There is, in fact, an explicit formula for the nth term of the Fibonacci sequence:

$$F_n = \frac{(1 + \sqrt{5})^n - (1 - \sqrt{5})^n}{2^n\sqrt{5}}$$

This formula was discovered by the French-English mathematician Abraham deMoivre more than 500 years after Fibonacci first introduced the sequence in 1202 in his book *Liber Abaci.* (For a proof of this formula, see Exercises 74 and 75.)

(a) Using algebra (and not your calculator), check that this formula gives the right answers for F_1 and F_2.

(b) Use this formula and your calculator to computer F_{24} and F_{25}.

(c) Use the formula and your calculator to compute F_{26}. Then check your answer by using the results in part (b).

74. (This result will be used in Exercise 75.) Suppose that x is a real number such that $x^2 = x + 1$. Use mathematical induction to prove that

$$x^n = F_nx + F_{n-1} \qquad \text{for } n \geq 2$$

Hint: You can carry out the induction proof without solving the given quadratic equation.

75. In this exercise we use the result in Exercise 74 to derive the following formula for the nth Fibonacci number:

$$F_n = \frac{(1 + \sqrt{5})^n - (1 - \sqrt{5})^n}{2^n\sqrt{5}} \qquad (1)$$

The clever method used here was discovered by Erwin Just; it appeared in *Mathematics Magazine,* vol. 44 (1971), p. 199.

Let α and β denote the roots of the quadratic equation $x^2 = x + 1$. Then, according to Exercise 74, for $n \geq 2$ we have

$$\alpha^n = F_n\alpha + F_{n-1} \qquad (2)$$

and

$$\beta^n = F_n\beta + F_{n-1} \qquad (3)$$

(a) Subtract equation (3) from equation (2) to show that

$$F_n = \frac{\alpha^n - \beta^n}{\alpha - \beta} \quad \text{for } n \geq 2 \qquad (4)$$

(b) Use the quadratic formula to show that the roots of the equation $x^2 = x + 1$ are given by $\alpha = (1 + \sqrt{5})/2$ and $\beta = (1 - \sqrt{5})/2$.

(c) In equation (4), substitute for α and β using the values obtained in part (b). Show that this leads to equation (1).

(d) The work in parts (a) through (c) shows that equation (1) holds for $n \geq 2$. Now complete the derivation by checking that equation (1) also holds for $n = 1$.

76. This exercise requires a knowledge of matrix multiplication. Use mathematical induction to show that

$$\begin{pmatrix} 1 & 1 \\ 1 & 0 \end{pmatrix}^n = \begin{pmatrix} F_{n+1} & F_n \\ F_n & F_{n-1} \end{pmatrix} \quad \text{for } n \geq 2$$

MINI PROJECT

PERSPECTIVE AND ALTERNATIVE SCENARIOS FOR EXAMPLE 5

As background for this mini project, someone in your group or in the class at large should review and summarize Example 5 for everyone else.

1. Refer to Figure 2 on page 993, which shows a scatter plot for the size of the rabbit population in Example 5 at the end of years $t = 0, 1, \ldots, 5$. In the text it was mentioned that the picture suggests exponential growth. Another possibility is quadratic growth. Explore this as follows. Using a graphing utility, enter the six data points specified in Figure 2 and create a scatter plot similar to the one shown in Figure 2. Then use the exponential regression option on the graphing utility to find the exponential function that best fits the data. Add the graph of the exponential function to your scatter plot. Next, add a quadratic regression curve to the picture. Which curve seems to fit the data points better?

2. In Example 5, suppose we make a change in the model so that the given recursive sequence is

$$\begin{aligned} P_0 &= 16 \\ P_t &= 0.0003P_{t-1}(3000 - P_{t-1}) \qquad (t \geq 1) \end{aligned}$$

Use a graphing utility to investigate the population trend or trends. On paper, in complete sentences, compare and/or contrast your results with those in Example 5.

3. In Example 5, instead of beginning with 16 rabbits, suppose that we start with the much larger initial population size $P_0 = 3200$. (Assume, however, that the recursion equation from Example 5 remains the same.) Use a graphing utility to investigate the population trend or trends. On paper, in complete sentences, compare and/or contrast your results with those in Example 5.

4. (a) Follow Exercise 3, assuming that the size of the initial population is $P_0 = 6330$. (Compare and/or contrast your results with those in Exercise 3, rather than Example 5.)

(b) Find out what happens if $P_0 = 6334$. In terms of the rabbit population, can you suggest a possible interpretation of the results?

13.4 ARITHMETIC SEQUENCES AND SERIES

One of the most natural ways to generate a sequence is to begin with a fixed number a and then repeatedly add a fixed constant d. This yields the sequence

$$a, a + d, a + 2d, a + 3d, \ldots \qquad (1)$$

Such a sequence is called an **arithmetic sequence** or **arithmetic progression.** Notice that the difference between any two consecutive terms is the constant d. We call d the **common difference.** Here are several examples of arithmetic sequences:

Example A: $1, 2, 3, \ldots$

Example B: $3, 7, 11, 15, \ldots$

Example C: $10, 5, 0, -5, \ldots$

In Example A the first term is $a = 1$, and the common difference is $d = 1$. For Example B we have $a = 3$. The value of d in this example is found by subtracting any term from the next term; thus $d = 4$. Finally, in Example C we have $a = 10$ and $d = -5$. Notice that when the common difference is negative, the terms of the sequence decrease.

There is a simple formula for the nth term in an arithmetic sequence. In arithmetic sequence (1), notice that

$$a_1 = a + 0d$$
$$a_2 = a + 1d$$
$$a_3 = a + 2d$$
$$a_4 = a + 3d$$

Following this pattern, it appears that the formula for a_n should be given by $a_n = a + (n - 1)d$. Indeed, this is the correct formula, and Exercise 32 at the end of this section asks you to verify it using mathematical induction.

The nth Term of an Arithmetic Sequence

The nth term of an arithmetic sequence $a, a + d, a + 2d, \ldots$ is given by

$$a_n = a + (n - 1)d$$

EXAMPLE 1 Finding a term in an arithmetic sequence

Determine the 100th term of the arithmetic sequence

$$7, 10, 13, 16, \ldots$$

SOLUTION

The first term is $a = 7$, and the common difference is $d = 3$. Substituting these values in the formula $a_n = a + (n - 1)d$ yields

$$a_n = 7 + (n - 1)3 = 3n + 4$$

To find the 100th term, we replace n by 100 in this last equation to obtain

$$a_{100} = 3(100) + 4 = 304 \qquad \text{as required}$$

EXAMPLE 2 Determining an arithmetic sequence from given data

Determine the arithmetic sequence in which the second term is -2 and the eighth term is 40.

SOLUTION

We are given that the second term is -2. Using this information in the formula $a_n = a + (n - 1)d$, we have

$$-2 = a + (2 - 1)d = a + d$$

This gives us one equation in two unknowns. We are also given that the eighth term is 40. Therefore

$$40 = a + (8 - 1)d = a + 7d$$

We now have a system of two equations in two unknowns:

$$\begin{cases} -2 = a + d \\ 40 = a + 7d \end{cases}$$

Subtracting the first equation from the second gives us

$$42 = 6d \qquad \text{or} \qquad 7 = d$$

To find a, we replace d by 7 in the first equation of the system. This yields

$$-2 = a + 7 \qquad \text{or} \qquad -9 = a$$

We have now determined the sequence, since we know that the first term is -9 and the common difference is 7. The first four terms of the sequence are -9, -2, 5, and 12.

Next we would like to derive a formula for the sum of the first n terms of an arithmetic sequence. Such a sum is referred to as an **arithmetic series.** If we use S_n to denote the required sum, we have

$$S_n = a + (a + d) + \cdots + [a + (n - 2)d] + [a + (n - 1)d] \tag{2}$$

Of course, we must obtain the same sum if we add the terms from right to left rather than left to right. That is, we must have

$$S_n = [a + (n - 1)d] + [a + (n - 2)d] + \cdots + (a + d) + a \tag{3}$$

Let's now add equations (2) and (3). Adding the left-hand sides is easy; we obtain $2S_n$. Now we add the corresponding terms on the right-hand sides. For the first term we have

$$\underset{\substack{\uparrow \\ \text{first term} \\ \text{in equation (2)}}}{a} \quad + \quad \underbrace{[a + (n - 1)d]}_{\substack{\text{first term} \\ \text{in equation (3)}}} = 2a + (n - 1)d$$

Next we add the second terms:

$$\underset{\substack{\uparrow \\ \text{second term} \\ \text{in equation (2)}}}{(a + d)} \quad + \quad \underbrace{[a + (n - 2)d]}_{\substack{\text{second term} \\ \text{in equation (3)}}} = 2a + d + (n - 2)d = 2a + (n - 1)d$$

Notice that the sum of the second terms is again $2a + (n - 1)d$, the same quantity we arrived at with the first terms. As you can check, this pattern continues all the way through to the last terms. For instance,

$$\underbrace{[a + (n - 1)d]}_{\substack{\text{last term}\\ \text{in equation (2)}}} + \underset{\substack{\uparrow\\ \text{last term}\\ \text{in equation (3)}}}{a} = 2a + (n - 1)d$$

We conclude from these observations that by adding the right-hand sides of equations (2) and (3), the quantity $2a + (n - 1)d$ is added a total of n times. Therefore

$$2S_n = n[2a + (n - 1)d]$$
$$S_n = \frac{n}{2}[2a + (n - 1)d]$$

This gives us the desired formula for the sum of the first n terms in an arithmetic sequence. There is an alternative form of this formula, which now follows rather quickly:

$$\begin{aligned} S_n &= \frac{n}{2}[2a + (n - 1)d] \\ &= \frac{n}{2}\{a + [a + (n - 1)d]\} \\ &= \frac{n}{2}(a + a_n) = n\left(\frac{a + a_n}{2}\right) \end{aligned}$$

This is consistent with our observation above in adding equations (2) and (3) term by term, namely, that $2S_n$ equals n times the sum of each "pair" of terms, for example, the first and last terms. So

$$2S_n = n(a + a_n) \qquad \text{or} \qquad S_n = n\left(\frac{a + a_n}{2}\right)$$

This last equation is easy to remember. It says that the sum of an arithmetic series is obtained by averaging the first and last terms and then multiplying this average by n, the number of terms. For reference we summarize both formulas in the following box.

Formulas for the Sum of an Arithmetic Series

$$S_n = \frac{n}{2}[2a + (n - 1)d]$$
$$S_n = n\left(\frac{a + a_n}{2}\right)$$

EXAMPLE 3 Finding a sum of terms in an arithmetic sequence

Find the sum of the first 30 terms of the arithmetic sequence 2, 6, 10, 14, ...

SOLUTION

We have $a = 2$, $d = 4$, and $n = 30$. Substituting these values in the formula $S_n = (n/2)[2a + (n - 1)d]$ then yields

$$S_{30} = \frac{30}{2}[2(2) + (30 - 1)4]$$
$$= 15[4 + 29(4)]$$
$$= 1800 \qquad \text{(Check the arithmetic.)}$$

EXAMPLE 4 Finding a sum of terms and the common difference of an arithmetic sequence

In a certain arithmetic sequence, the first term is 6, and the 40th term is 71. Find the sum of the first 40 terms and also the common difference for the sequence.

SOLUTION

We have $a = 6$ and $a_{40} = 71$. Using these values in the formula $S_n = n(a + a_n)/2$ yields

$$S_{40} = \frac{40}{2}(6 + 71) = 20(77) = 1540$$

The sum of the first 40 terms is thus 1540. The value of d can now be found by using the formula $a_n = a + (n - 1)d$. We have $a_{40} = a + (40 - 1)d$, and therefore

$$71 = 6 + 39d$$
$$39d = 65$$
$$d = \frac{65}{39} = \frac{5}{3}$$

The required value of d is $5/3$.

EXAMPLE 5 An arithmetic series defined using sigma notation

Show that $\sum_{k=1}^{50}(3k - 2)$ represents an arithmetic series, and compute the sum.

SOLUTION

There are two different ways to see that $\sum_{k=1}^{50}(3k - 2)$ is an arithmetic series. One way is simply to write out the first few terms and look at the pattern. We have

$$\sum_{k=1}^{50}(3k - 2) = 1 + 4 + 7 + 10 + \cdots + 148$$

From this it is clear that we are indeed summing the terms in an arithmetic sequence in which $d = 3$ and $a = 1$.

A more formal way to show that $\sum_{k=1}^{50}(3k - 2)$ represents an arithmetic series is to prove that the difference between successive terms in the indicated sum is a constant. Now, the form of a typical term in this sum is $3k - 2$. Thus the form of the next term must be $[3(k+1)-2]$. The difference between these terms is then

$$[3(k+1)-2]-(3k-2)=3k+3-2-3k+2=3$$

The difference therefore is constant, as we wished to show.

To evaluate $\sum_{k=1}^{50}(3k-2)$, we can use either of our two formulas for the sum of an arithmetic series. Using the formula $S_n=(n/2)[2a+(n-1)d]$, we obtain

$$S_{50}=\frac{50}{2}[2(1)+49(3)]=25(149)=3725$$

Thus the required sum is 3725. You should check for yourself that the same value is obtained by using the formula $S_n=n(a+a_n)/2$.

Note: There is another important way to compute this sum. See the project following Exercise Set 13.4.

EXERCISE SET 13.4 

A

1. Find the common difference d for each of the following arithmetic sequences.
(a) $1, 3, 5, 7, \ldots$
(b) $10, 6, 2, -2, \ldots$
(c) $2/3, 1, 4/3, 5/3, \ldots$
(d) $1, 1+\sqrt{2}, 1+2\sqrt{2}, 1+3\sqrt{2}, \ldots$

2. Which of the following are arithmetic sequences?
(a) $2, 4, 8, 16, \ldots$
(b) $5, 9, 13, 17, \ldots$
(c) $3, 11/5, 7/5, 3/5, \ldots$
(d) $-1, -1, -1, -1, \ldots$
(e) $-1, 1, -1, 1, \ldots$

In Exercises 3–8, find the indicated term in each sequence.

3. $10, 21, 32, 43, \ldots; a_{12}$
4. $7, 2, -3, -8, \ldots; a_{20}$
5. $6, 11, 16, 21, \ldots; a_{100}$
6. $2/5, 4/5, 6/5, 8/5, \ldots; a_{30}$
7. $-1, 0, 1, 2, \ldots; a_{1000}$
8. $42, 1, -40, -81, \ldots; a_{15}$

9. The fourth term in an arithmetic sequence is -6, and the 10th term is 5. Find the common difference and the first term.

10. The fifth term in an arithmetic sequence is $1/2$, and the 20th term is $7/8$. Find the first three terms of the sequence.

11. The 60th term in an arithmetic sequence is 105, and the common difference is 5. Find the first term.

12. Find the common difference in an arithmetic sequence in which $a_{10}-a_{20}=70$.

13. Find the common difference in an arithmetic sequence in which $a_{15}-a_7=-1$.

14. Find the sum of the first 16 terms in the sequence $2, 11, 20, 29, \ldots$.

15. Find the sum of the first 1000 terms in the sequence $1, 2, 3, 4, \ldots$.

16. Find the sum of the first 50 terms in an arithmetic series that has first term -8 and 50th term 139.

17. Find the sum: $\frac{\pi}{3}+\frac{2\pi}{3}+\pi+\frac{4\pi}{3}+\cdots+\frac{13\pi}{3}$.

18. Find the sum: $\frac{1}{e}+\frac{3}{e}+\frac{5}{e}+\cdots+\frac{21}{e}$.

19. Determine the first term of an arithmetic sequence in which the common difference is 5 and the sum of the first 38 terms is 3534.

20. The sum of the first 12 terms in an arithmetic sequence is 156. What is the sum of the first and 12th terms?

21. In a certain arithmetic sequence, the first term is 4, and the 16th term is -100. Find the sum of the first 16 terms and also the common difference for the sequence.

22. The fifth and 50th terms of an arithmetic sequence are 3 and 30, respectively. Find the sum of the first 10 terms.

23. The eighth term in an arithmetic sequence is 5, and the sum of the first 10 terms is 20. Find the common difference and the first term of the sequence.

In Exercises 24–26, find each sum.

24. $\sum_{i=1}^{10}(2i-1)=1+3+5+\cdots+19$

25. $\sum_{k=1}^{20}(4k+3)$

26. $\sum_{n=5}^{100}(2n-1)$

27. The sum of three consecutive terms in an arithmetic sequence is 30, and their product is 360. Find the three terms. *Suggestion*: Let x denote the *middle* term and d the common difference.

28. The sum of three consecutive terms in an arithmetic sequence is 21, and the sum of their squares is 197. Find the three terms.

29. The sum of three consecutive terms in an arithmetic sequence is 6, and the sum of their cubes is 132. Find the three terms.

30. In a certain arithmetic sequence, $a = -4$ and $d = 6$. If $S_n = 570$, find n.

31. Let $a_1 = 1/(1 + \sqrt{2})$, $a_2 = -1$, and $a_3 = 1/(1 - \sqrt{2})$.
(a) Show that $a_2 - a_1 = a_3 - a_2$.
(b) Find the sum of the first six terms in the arithmetic sequence

$$\frac{1}{1+\sqrt{2}}, \quad -1, \quad \frac{1}{1-\sqrt{2}}, \quad \ldots$$

32. Using mathematical induction, prove that the nth term of the sequence $a, a + d, a + 2d, \ldots$ is given by

$$a_n = a + (n - 1)d$$

B

33. Let b denote a positive constant. Find the sum of the first n terms in the arithmetic sequence

$$\frac{1}{1+\sqrt{b}}, \quad \frac{1}{1-b}, \quad \frac{1}{1-\sqrt{b}}, \quad \ldots$$

34. The sum of the first n terms in a certain arithmetic sequence is given by $S_n = 3n^2 - n$. Show that the rth term is given by $a_r = 6r - 4$.

35. Let $a_1, a_2, a_3, \ldots$ be an arithmetic sequence, and let S_k denote the sum of the first k terms. If $S_n/S_m = n^2/m^2$, show that

$$\frac{a_n}{a_m} = \frac{2n-1}{2m-1}$$

36. If the common difference in an arithmetic sequence is twice the first term, show that

$$\frac{S_n}{S_m} = \frac{n^2}{m^2}$$

37. The lengths of the sides of a right triangle form three consecutive terms in an arithmetic sequence. Show that the triangle is similar to the 3-4-5 right triangle.

38. Suppose that $1/a$, $1/b$, and $1/c$ are three consecutive terms in an arithmetic sequence. Show that

(a) $\dfrac{a}{c} = \dfrac{a-b}{b-c}$ **(b)** $b = \dfrac{2ac}{a+c}$

39. Suppose that a, b, and c are three positive numbers with $a > c > 2b$. If $1/a$, $1/b$, and $1/c$ are consecutive terms in an arithmetic sequence, show that

$$\ln(a + c) + \ln(a - 2b + c) = 2\ln(a - c)$$

PROJECT

MORE ON SUMS

In this project we develop some algebra for simplifying sums. First, here are four important properties of summation.

PROPERTIES OF SUMMATION

Let $a_1, a_2, \ldots, a_n, b_1, b_2, \ldots, b_n$, and c be real numbers. Then

1. $\displaystyle\sum_{k=1}^{n} (a_k + b_k) = \sum_{k=1}^{n} a_k + \sum_{k=1}^{n} b_k$
2. $\displaystyle\sum_{k=1}^{n} (a_k - b_k) = \sum_{k=1}^{n} a_k - \sum_{k=1}^{n} b_k$
3. $\displaystyle\sum_{k=1}^{n} ca_k = c\sum_{k=1}^{n} a_k$
4. $\displaystyle\sum_{k=1}^{n} c = nc$

PROOFS

1. To prove Property 1, we write out the sum term by term, then use the associative and commutative properties of addition to remove parentheses, then rearrange and regroup terms.

$$\sum_{k=1}^{n}(a_k + b_k) = (a_1 + b_1) + (a_2 + b_2) + \cdots + (a_n + b_n) \quad \text{definition of summation notation}$$

$$= a_1 + b_1 + a_2 + b_2 + \cdots + a_n + b_n \quad \text{remove parentheses}$$

$$= (a_1 + a_2 + \cdots + a_n) + (b_1 + b_2 + \cdots + b_n) \quad \text{commutative and associative properties of addition}$$

$$= \sum_{k=1}^{n} a_k + \sum_{k=1}^{n} b_k \quad \text{summation notation}$$

2. The proof of Property 2 is similar to that for Property 1. We leave it for you as Exercise 1.
3. Property 3 is a statement of the distributive property or, equivalently, factoring out a common factor from a sum of terms.

$$\sum_{k=1}^{n} ca_k = ca_1 + ca_2 + \cdots + ca_n \quad \text{summation notation}$$

$$= c(a_1 + a_2 + \cdots + a_n) \quad \text{factor out a common factor}$$

$$= c\sum_{k=1}^{n} a_k \quad \text{summation notation}$$

4. Property 4 is tricky. The summation notation says that there are n terms and each term is the same number c:

$$\sum_{k=1}^{n} c = \underbrace{c + c + \cdots + c}_{n \text{ terms}} = nc$$

EXAMPLE 1 Using properties of summation

Suppose $x_1, x_2, \ldots, x_{25}$ are numbers such that $\sum_{k=1}^{25} x_k = 30$ and $\sum_{k=1}^{25} x_k^2 = 100$. Calculate: $\sum_{k=1}^{25} (x_k + 3)^2$.

SOLUTION

$$\sum_{k=1}^{25}(x_k + 3)^2 = \sum_{k=1}^{25}(x_k^2 + 6x_k + 9)$$

$$= \sum_{k=1}^{25} x_k^2 + 6\sum_{k=1}^{25} x_k + \sum_{k=1}^{25} 9$$

$$= 100 + 6(30) + 25(9) = 505$$

Following, in summation notation, are three useful sums first stated in Exercises 1, 5, and 11 of Exercise Set 13.1.

USEFUL SUMS

1. $\sum_{k=1}^{n} k = \frac{n(n+1)}{2}$
2. $\sum_{k=1}^{n} k^2 = \frac{n(n+1)(2n+1)}{2}$
3. $\sum_{k=1}^{n} k^3 = \left[\frac{n(n+1)}{2}\right]^2$

As was proposed in the exercise for Section 13.1, these three summation formulas can be proven by using induction on the number n of terms in the sum. An alternative proof of Useful Sum 1 is sketched in Exercise 2.

Let's put the properties of summation together with the useful sums to simplify some sums.

EXAMPLE II Using properties of summation and useful sums

Simplify: $\sum_{k=1}^{50} (3k - 2)$. (This is Example 5 on page 1003.)

SOLUTION

$$\sum_{k=1}^{50} (3k - 2) = \sum_{k=1}^{50} 3k - \sum_{k=1}^{50} 2 \qquad \text{Summation Property 2}$$

$$= 3\sum_{k=1}^{50} k - 50(2) \qquad \text{Summation Properties 3 and 4}$$

$$= 3\left[\frac{50(50+1)}{2}\right] - 100 \qquad \text{Useful Sum 1}$$

$$= 3725$$

EXAMPLE III Using properties of summation and useful sums

Simplify: $\sum_{k=1}^{25} (2k - 7)^2$.

SOLUTION

$$\sum_{k=1}^{25} (2k - 7)^2 = \sum_{k=1}^{25} (4k^2 - 14k + 49)$$

$$= 4\sum_{k=1}^{25} k^2 - 14\sum_{k=1}^{25} k + \sum_{k=1}^{25} 49 \qquad \text{Why?}$$

$$= 4\left\{\frac{25(25+1)[2(25)+1]}{6}\right\} - 14\left[\frac{25(25+1)}{2}\right] + 25(49) \qquad \text{Useful Sums 1 and 2 and Summation Property 4}$$

$$= 18{,}775$$

EXAMPLE IV A generalization of Example III

Simplify: $\sum_{k=1}^{n}(2k-7)^2$.

SOLUTION

$$\sum_{k=1}^{n}(2k-7)^2 = 4\sum_{k=1}^{n}k^2 - 14\sum_{k=1}^{n}k + \sum_{k=1}^{n}49 \qquad \text{Why?}$$

$$= 4\left[\frac{n(n+1)(2n+1)}{6}\right] - 14\left[\frac{n(n+1)}{2}\right] + (49)n \qquad \text{Why?}$$

$$\vdots$$

$$= \frac{n}{3}(4n^2 - 15n + 128)$$

Note: With $n = 25$ we get $\frac{25}{3}[4(25)^2 - 15(25) + 128] = 18{,}775$, the result in Example III.

EXAMPLE V Using a variation of Useful Sum 1

Simplify: $\sum_{k=1}^{n-5}k$.

SOLUTION

By Useful Sum 1 with $n-5$ terms, we have

$$\sum_{k=1}^{n-5}k = \frac{(n-5)[(n-5)+1]}{2}$$

$$= \frac{(n-5)(n-4)}{2} = \frac{n^2-9n+20}{2}$$

EXAMPLE VI Using a variation of Useful Sum 1

Simplify: $\sum_{k=0}^{n-3}(k+2)$.

SOLUTION

$$\sum_{k=0}^{n-3}(k+2) = \sum_{k=0}^{n-3}k + \sum_{k=0}^{n-3}2$$

Now $\sum_{k=0}^{n-3}k = \sum_{k=1}^{n-3}k$. (Why?) And $\sum_{k=0}^{n-3}2$ contains $n-3+1 = n-2$ terms, all 2. (Why?) So

$$\sum_{k=0}^{n-3}(k+2) = \sum_{k=0}^{n-3}k + \sum_{k=0}^{n-3}2 = \sum_{k=1}^{n-3}k + (n-2)2$$

$$= \frac{(n-3)[(n-3)+1]}{2} + 2(n-2)$$

$$= \frac{(n-3)(n-2)}{2} + 2(n-2) = \frac{n-2}{2}[(n-3)+4]$$

$$= \frac{(n-2)(n+1)}{2} = \frac{n^2-n-2}{2}$$

EXAMPLE VII Shifting the index to compute a sum

Simplify: $\sum_{k=4}^{n} k$.

SOLUTION

We'll show two ways to simplify this sum. The first method is to apply Useful Sum 1 directly.

$$\sum_{k=4}^{n}k = \sum_{k=1}^{n}k - \sum_{k=1}^{3}k \quad \text{(Right?)}$$

$$= \frac{n(n+1)}{2} - \frac{3(3+1)}{2} = \frac{n^2+n-12}{2}$$

The second method is to apply Useful Sum 1 by "shifting the index." Substitute $i = k - 4$. Then as k goes from 4 to n, i goes from 0 to $n - 4$. So

$$\sum_{k=4}^{n}k = \sum_{i=0}^{n-4}(i+4) \qquad \text{substituting } k = i + 4$$

$$= \sum_{i=1}^{n-4}i + \sum_{i=0}^{n-4}4 \qquad \text{as in Example 6}$$

$$= \frac{(n-4)[(n-4)+1]}{2} + (n-3)4 = \frac{(n-3)}{2}[(n-4)+8]$$

$$= \frac{(n-3)(n+4)}{2} = \frac{n^2+n-12}{2}$$

EXERCISES

1. Derive Summation Property 2: $\sum_{k=1}^{n}(a_k - b_k) = \sum_{k=1}^{n}a_k - \sum_{k=1}^{n}b_k$.
2. Derive Useful Sum 1, $\sum_{k=1}^{n}k = \frac{n(n+1)}{2}$, by writing the summation twice, the second time with the terms in reverse order, as in the derivation of the formula for the sum of an arithmetic series, then adding.
3. Given, $\sum_{k=1}^{15}x_k = 30$ and $\sum_{k=1}^{15}x_k^2 = 50$, calculate $\sum_{k=1}^{15}(2x_k - 5)^2$.

4. Simplify the following sums.

(a) $\sum_{k=1}^{100}(4k+3)$ **(c)** $\sum_{k=1}^{n}(k^2+3k-5)$ **(e)** $\sum_{k=1}^{n}(5k-2)^2$

(b) $\sum_{k=1}^{30}(10-7k)$ **(d)** $\sum_{k=1}^{n}(k^3+2k)$ **(f)** $\sum_{k=1}^{n}(5k-2)^3$

5. Simplify the following sums. In part (c), use both methods of Example VII.

(a) $\sum_{k=1}^{n-3}5k$ **(b)** $\sum_{k=1}^{n-2}k^2$ **(c)** $\sum_{k=3}^{n}(3k-5)$

13.5 GEOMETRIC SEQUENCES AND SERIES

A **geometric sequence** or **geometric progression** is a sequence of the form

$$a, ar, ar^2, ar^3, \ldots \qquad \text{where } a \text{ and } r \text{ are nonzero constants}$$

As you can see, each term after the first in a geometric sequence is obtained by multiplying the previous term by r. The number r is called the **common ratio,** because the ratio of any term to the previous one is always r. For instance, the ratio of the fourth term to the third is $ar^3/ar^2 = r$. Here are two examples of geometric sequences:

$$1, \frac{1}{2}, \frac{1}{4}, \frac{1}{8}, \ldots$$

$$10, -100, 1000, -10{,}000, \ldots$$

In the first example we have $a = 1$ and $r = 1/2$; in the second example we have $a = 10$ and $r = -10$.

EXAMPLE 1 **Finding a term in a geometric sequence from given data**

In a certain geometric sequence, the first term is 2, the third term is 3, and the common ratio is negative. Find the second term.

SOLUTION

Let x denote the second term, so that the sequence begins

$$2, x, 3, \ldots$$

By definition the ratios $3/x$ and $x/2$ must be equal. Thus we have

$$\frac{3}{x} = \frac{x}{2}$$

$$x^2 = 6$$

$$x = \pm\sqrt{6}$$

Now, the second term must be negative, because the first term is positive and the common ratio is negative. Thus the second term is $x = -\sqrt{6}$.

TABLE 1

n	a_n
1	ar^0
2	ar^1
3	ar^2
4	ar^3
$\vdots$	$\vdots$

The formula for the nth term of a geometric sequence is easily deduced by considering Table 1. The table indicates that the exponent on r is 1 less than the value of n in each case. On the basis of this observation it appears that the nth term must be given by $a_n = ar^{n-1}$. Indeed, it can be shown by mathematical induction that this formula does hold for all natural numbers n. (Exercise 30 at the end of this section asks you to carry out the proof.) We summarize this result as follows:

nth Term of a Geometric Sequence

The nth term of the geometric sequence, $a, ar, ar^2, \ldots$ is given by

$$a_n = ar^{n-1}$$

EXAMPLE 2 Finding a term in a given geometric sequence

Find the seventh term in the geometric sequence 2, 6, 18, ...

SOLUTION

We can find the common ratio r by dividing the second term by the first. Thus $r = 3$. Now, using $a = 2$, $r = 3$, and $n = 7$ in the formula $a_n = ar^{n-1}$, we have

$$a_7 = 2(3)^6 = 2(729) = 1458$$

The seventh term of the sequence is therefore 1458.

Suppose that we begin with a geometric sequence $a, ar, ar^2, \ldots$, in which $r \neq 1$. If we add the first n terms and denote the sum by S_n, we have

$$S_n = a + ar + ar^2 + \cdots + ar^{n-2} + ar^{n-1} \tag{1}$$

This sum is called a **finite geometric series.** We would like to find a formula for S_n. To do this, we multiply equation (1) by r to obtain

$$rS_n = ar + ar^2 + ar^3 + \cdots + ar^{n-1} + ar^n \tag{2}$$

We now subtract equation (2) from equation (1). This yields (after combining like terms)

$$\begin{aligned} S_n - rS_n &= a - ar^n \\ S_n(1 - r) &= a(1 - r^n) \\ S_n &= \frac{a(1 - r^n)}{1 - r} \qquad (r \neq 1) \end{aligned}$$

This is the formula for the sum of a finite geometric series. We summarize this result in the box that follows.

Formula for the Sum of a Geometric Series

Let S_n denote the sum $a + ar + ar^2 + \cdots + ar^{n-1}$, and assume that $r \neq 1$. Then

$$S_n = \frac{a(1 - r^n)}{1 - r}$$

EXAMPLE 3 Computing the sum of a finite geometric series

Evaluate the sum

$$\frac{1}{2^1}+\frac{1}{2^2}+\frac{1}{2^3}+\cdots+\frac{1}{2^{10}}=\frac{1}{2}+\frac{1}{2}\left(\frac{1}{2}\right)+\frac{1}{2}\left(\frac{1}{2}\right)^2+\cdots+\frac{1}{2}\left(\frac{1}{2}\right)^9$$

SOLUTION

This is a finite geometric series with $a = 1/2$, $r = 1/2$, and $n = 10$. Using these values in the formula for S_n yields

$$S_{10}=\frac{\frac{1}{2}[1-(\frac{1}{2})^{10}]}{1-\frac{1}{2}}=\frac{\frac{1}{2}[1-\frac{1}{1024}]}{\frac{1}{2}}=1-\frac{1}{1024}=\frac{1023}{1024}$$

We would now like to give a meaning to certain expressions of the form

$$a + ar + ar^2 + \cdots$$

Such an expression is called an **infinite geometric series.** The three dots indicate (intuitively at least) that the additions are to be carried out indefinitely, without end. To see how to proceed here, let's start by looking at some examples involving finite geometric series. In particular, we'll consider the series

$$\frac{1}{2^1}+\frac{1}{2^2}+\frac{1}{2^3}+\cdots+\frac{1}{2^n}$$

for increasing values of n. The idea is to look for a pattern as n grows ever larger. Let $S_1=\frac{1}{2}$, $S_2=\frac{1}{2^1}+\frac{1}{2^2}$, $S_3=\frac{1}{2^1}+\frac{1}{2^2}+\frac{1}{2^3}$, and, in general,

$$S_n=\frac{1}{2^1}+\frac{1}{2^2}+\cdots+\frac{1}{2^n}$$

Then we can compute S_n for any given value of n by means of the formula for the sum of a finite geometric series. From Table 2, which displays the results of these calculations, it seems clear that as n grows larger and larger, the value of S_n grows ever closer to 1. More precisely (but leaving the details for calculus), it can be shown that the value of S_n can be made as close to 1 as we please, provided only that n is sufficiently large. For this reason we say that the *sum of the infinite geometric series* $\frac{1}{2^1}+\frac{1}{2^2}+\frac{1}{2^3}+\cdots$ is 1. That is,

$$\frac{1}{2^1}+\frac{1}{2^2}+\frac{1}{2^3}+\cdots=1$$

TABLE 2

n	$S_n=\frac{1}{2}+\frac{1}{2^2}+\cdots\frac{1}{2^n}$
1	0.5
2	0.75
5	0.96875
10	0.999023437 . . .
15	0.999969482 . . .
20	0.999999046 . . .
25	0.999999970 . . .

We can arrive at this last result another way. First we compute the sum of the *finite* geometric series $\frac{1}{2^1}+\frac{1}{2^2}+\cdots+\frac{1}{2^n}$. As you can check, the result is

$$S_n=1-\left(\frac{1}{2}\right)^n$$

Now, as n grows larger and larger, the value of $(1/2)^n$ gets closer and closer to zero. Thus as n grows ever larger, the value of S_n will get closer and closer to $1 - 0$, or 1.

Now let's repeat our reasoning to obtain a formula for the sum of the infinite geometric series

$$a + ar + ar^2 + \cdots \qquad \text{where } |r| < 1$$

First we consider the finite geometric series

$$a + ar + ar^2 + \cdots + ar^{n-1}$$

The sum S_n in this case is

$$S_n = \frac{a(1 - r^n)}{1 - r}$$

We want to know how S_n behaves as n grows ever larger. This is where the assumption $|r| < 1$ is crucial. Just as $(1/2)^n$ approaches zero as n grows larger and larger, so will r^n approach zero as n grows larger and larger. Thus as n grows ever larger, the sum S_n will more and more resemble

$$\frac{a(1 - 0)}{1 - r} = \frac{a}{1 - r}$$

For this reason we say that the sum of the infinite geometric series is $a/(1 - r)$. We will make free use of this result in the subsequent examples. A more rigorous development of infinite series properly belongs to calculus.

Formula for the Sum of an Infinite Geometric Series

Suppose that $|r| < 1$. Then the sum S of the infinite geometric series $a + ar + ar^2 + \cdots$ is given by

$$S = \frac{a}{1 - r}$$

EXAMPLE 4 Computing the sum of an infinite geometric series

Find the sum of the infinite geometric series $1 + \frac{2}{3} + \frac{4}{9} + \cdots$.

SOLUTION

In this case we have $a = 1$ and $r = 2/3$. Thus

$$S = \frac{a}{1 - r} = \frac{1}{1 - \frac{2}{3}} = \frac{1}{\frac{1}{3}} = 3$$

The sum of the series is 3.

EXAMPLE 5 Expressing a repeating decimal as a fraction

Find a fraction equivalent to the repeating decimal $0.2\overline{35}$.

SOLUTION

Let $S = 0.2\overline{35}$. Then we have

$$\begin{aligned} S &= 0.2353535\ldots \\ &= \frac{2}{10} + \frac{35}{1000} + \frac{35}{100{,}000} + \frac{35}{10{,}000{,}000} + \cdots \end{aligned}$$

Now, the expression following 2/10 on the right-hand side of this last equation is an infinite geometric series in which $a = 35/1000$ and $r = 1/100$. Thus

$$S = \frac{2}{10} + \frac{a}{1-r}$$
$$= \frac{2}{10} + \frac{\frac{35}{1000}}{1-\frac{1}{100}} = \frac{233}{990} \qquad \text{(Check the arithmetic!)}$$

The given decimal is therefore equivalent to 233/990.

EXERCISE SET 13.5

A

1. Find the second term in a geometric sequence in which the first term is 9, the third term is 4, and the common ratio is positive.

2. Find the fifth term in a geometric sequence in which the fourth term is 4, the sixth term is 6, and the common ratio is negative.

3. The product of the first three terms in a geometric sequence is 8000. If the first term is 4, find the second and third terms.

In Exercises 4–8, find the indicated term of the given geometric sequence.

4. $9, 81, 729, \ldots; a_7$ **5.** $-1, 1, -1, 1, \ldots; a_{100}$

6. $1/2, 1/4, 1/8, \ldots; a_9$ **7.** $2/3, 4/9, 8/27, \ldots; a_8$

8. $1, -\sqrt{2}, 2, \ldots; a_6$

9. Find the common ratio in a geometric sequence in which the first term is 1 and the seventh term is 4096.

10. Find the first term in a geometric sequence in which the common ratio is 4/3 and the tenth term is 16/9.

11. Find the sum of the first ten terms of the sequence $7, 14, 28, \ldots$.

12. Find the sum of the first five terms of the sequence $-1/2, 3/10, -9/50, \ldots$.

13. Find the sum: $1 + \sqrt{2} + 2 + \cdots + 32$.

14. Find the sum of the first 12 terms in the sequence $-4, -2, -1, \ldots$.

In Exercises 15–17, evaluate each sum.

15. $\sum_{k=1}^{6} \left(\frac{3}{2}\right)^k$ **16.** $\sum_{k=1}^{6} \left(\frac{2}{3}\right)^{k+1}$ **17.** $\sum_{k=2}^{6} \left(\frac{1}{10}\right)^k$

In Exercises 18–22, determine the sum of each infinite geometric series.

18. $\frac{1}{4} + \frac{1}{4^2} + \frac{1}{4^3} + \cdots$ **19.** $\frac{2}{3} - \frac{4}{9} + \frac{8}{27} - \cdots$

20. $\frac{9}{10} + \frac{9}{100} + \frac{9}{1000} + \cdots$ **21.** $1 + \frac{1}{1.01} + \frac{1}{(1.01)^2} + \cdots$

22. $-1 - \frac{1}{\sqrt{2}} - \frac{1}{2} - \cdots$

In Exercises 23–27, express each repeating decimal as a fraction.

23. $0.555\ldots$ **24.** $0.\overline{47}$ **25.** $0.1\overline{23}$

26. $0.050505\ldots$ **27.** $0.\overline{432}$

B

28. The lengths of the sides in a right triangle form three consecutive terms of a geometric sequence. Find the common ratio of the sequence. (There are two distinct answers.)

29. The product of three consecutive terms in a geometric sequence is -1000, and their sum is 15. Find the common ratio. (There are two answers.) *Suggestion:* Denote the terms by a/r, a, and ar.

30. Use mathematical induction to prove that the nth term of the geometric sequence $a, ar, ar^2, \ldots$ is ar^{n-1}.

31. Show that the sum of the following infinite geometric series is 3/2:

$$\frac{\sqrt{3}}{\sqrt{3}+1} + \frac{\sqrt{3}}{\sqrt{3}+3} + \cdots$$

32. Let A_1 denote the area of an equilateral triangle, each side of which is one unit long. A second equilateral triangle is formed by joining the midpoints of the sides of the first triangle. Let A_2 denote the area of this second triangle. This process is then repeated to form a third triangle with area A_3, and so on. Find the sum of the areas: $A_1 + A_2 + A_3 + \cdots$.

33. Let $a_1, a_2, a_3, \ldots$ be a geometric sequence such that $r \neq 1$. Let $S = a_1 + a_2 + a_3 + \cdots + a_n$, and let $T = \frac{1}{a_1} + \frac{1}{a_2} + \cdots + \frac{1}{a_n}$. Show that $\frac{S}{T} = a_1 a_n$.

34. Suppose that a, b, and c are three consecutive terms in a geometric sequence. Show that $\frac{1}{a+b}$, $\frac{1}{2b}$, and $\frac{1}{c+b}$ are three consecutive terms in an arithmetic sequence.

35. A ball is dropped from a height of 6 ft. Assuming that on each bounce, the ball rebounds to one-third of its previous height, find the total distance traveled by the ball.

13.6 DEMOIVRE'S THEOREM

Go to Mr. DeMoivre; he knows these things better than I do.—Isaac Newton [according to Boyer's *A History of Mathematics* (New York: John Wiley and Sons, Inc., 1968)]

In this section we explore one of the many important connections between trigonometry and the complex number system. We begin with an observation that was first made in the year 1797 by the Norwegian surveyor and mathematician Caspar Wessel. He realized, essentially, that the complex numbers can be visualized as points in the x-y plane by identifying the complex number $a + bi$ with the point (a, b). In this context we often refer to the x-y plane as the **complex plane,** and we refer to the complex numbers as points in this plane. In the complex plane the x-axis is also called the **real axis,** and the y-axis is also called **the imaginary axis.**

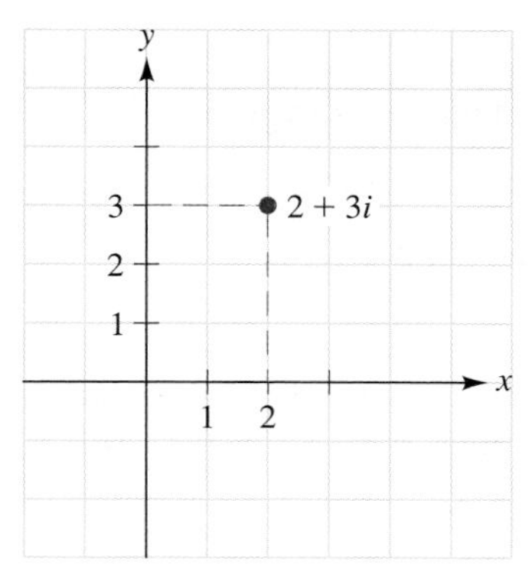

Figure 1

EXAMPLE 1 Graphing a complex number

Plot the point $2 + 3i$ in the complex plane.

SOLUTION

The complex number $2 + 3i$ is identified with the point $(2, 3)$. See Figure 1.

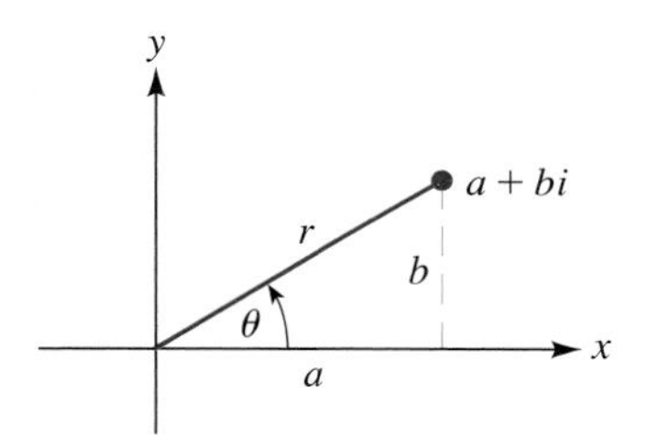

Figure 2

As Figure 2 indicates, the distance from the origin to the point $a + bi$ is denoted by r. We call the distance r the **modulus** of the complex number $a + bi$. The angle θ in Figure 2 (measured counterclockwise from the positive x-axis) is referred to as the **argument** of the complex number $a + bi$. (Using the terminology of Section 9.6, r and θ are the **polar coordinates** of the point $a + bi$.)

From Figure 2 we have the following three equations relating the quantities a, b, r, and θ. (Although Figure 2 shows $a + bi$ in the first quadrant, the equations remain valid for the other quadrants as well.)

$$r = \sqrt{a^2 + b^2} \tag{1}$$

$$a = r\cos\theta \tag{2}$$

$$b = r\sin\theta \tag{3}$$

If we have a complex number $z = a + bi$, we can use equations (2) and (3) to write

$$z = (r\cos\theta) + (r\sin\theta)i = r(\cos\theta + i\sin\theta)$$

That is,

$$z = r(\cos\theta + i\sin\theta) \tag{4}$$

The expression that appears on the right-hand side of equation (4) is called the **trigonometric** (or **polar**) **form** of the complex number z. In contrast, the expression $a + bi$ is referred to as the **rectangular form** of the complex number z.

EXAMPLE 2 Changing a complex number from trigonometric form to rectangular form

Express the complex number $z = 3(\cos\frac{\pi}{3} + i\sin\frac{\pi}{3})$ in rectangular form.

SOLUTION

$$\begin{aligned} z &= 3\left(\cos\frac{\pi}{3} + i\sin\frac{\pi}{3}\right) \\ &= 3\left(\frac{1}{2} + \frac{1}{2}\sqrt{3}i\right) = \frac{3}{2} + \frac{3}{2}\sqrt{3}i \end{aligned}$$

The rectangular form is therefore $\frac{3}{2} + \frac{3}{2}\sqrt{3}i$.

EXAMPLE 3 **Changing a complex number from rectangular form to trigonometric form**

Find the trigonometric form for the complex number $-\sqrt{2} + i\sqrt{2}$.

SOLUTION

We are asked to write the given number in the form $r(\cos\theta + i\sin\theta)$, so we need to find r and θ. Using equation (1) and the values $a = -\sqrt{2}$, $b = \sqrt{2}$, we have

$$r = \sqrt{(-\sqrt{2})^2 + (\sqrt{2})^2} = \sqrt{2 + 2} = \sqrt{4} = 2$$

Now that we know r, we can use equations (2) and (3) to determine θ. From equation (2) we obtain

$$\cos\theta = \frac{a}{r} = \frac{-\sqrt{2}}{2} \tag{5}$$

Similarly, equation (3) gives us

$$\sin\theta = \frac{b}{r} = \frac{\sqrt{2}}{2} \tag{6}$$

One angle that satisfies both of equations (5) and (6) is $\theta = 3\pi/4$. (There are other angles, and we'll return to this point in a moment.) In summary, then, we have $r = 2$ and $\theta = 3\pi/4$, so the required trigonometric form is

$$2\left(\cos\frac{3\pi}{4} + i\sin\frac{3\pi}{4}\right)$$

In Example 3 we noted that $\theta = 3\pi/4$ was only one angle satisfying the conditions $\cos\theta = -\sqrt{2}/2$ and $\sin\theta = \sqrt{2}/2$. Another such angle is $\frac{3\pi}{4} + 2\pi$. Indeed, any angle of the form $\frac{3\pi}{4} + 2\pi k$, where k is an integer, would do just as well. The significance of this is that θ, the argument of a complex number, is not uniquely determined. In Example 3 we followed a common convention in converting to trigonometric form: We picked θ in the interval $0 \le \theta < 2\pi$. Furthermore, although it won't cause us any difficulties in this section, you might also note that the argument θ is undefined for the complex number $0 + 0i$. (Why?)

We are now ready to derive a formula that will make it easy to multiply two complex numbers in trigonometric form. Suppose that the two complex numbers are

$$r(\cos\alpha + i\sin\alpha) \qquad \text{and} \qquad R(\cos\beta + i\sin\beta)$$

Then their product is

$$
\begin{aligned}
&rR[(\cos\alpha + i\sin\alpha)(\cos\beta + i\sin\beta)]\\
&\quad = rR[(\cos\alpha\cos\beta - \sin\alpha\sin\beta) + i(\sin\alpha\cos\beta + \cos\alpha\sin\beta)]\\
&\quad = rR[\cos(\alpha+\beta) + i\sin(\alpha+\beta)] \qquad \text{using the addition formulas from Section 8.1}
\end{aligned}
$$

Notice that the modulus of the product is rR, which is the product of the two original moduli. Also, the argument is $\alpha + \beta$, which is the *sum* of the two original arguments. So, to multiply two complex numbers, just multiply the moduli and add the arguments. There is a similar rule for obtaining the quotient of two complex numbers: Divide their moduli and subtract the arguments. These two rules are stated more precisely in the box that follows. (For a proof of the division rule, see Exercise 75 at the end of this section.)

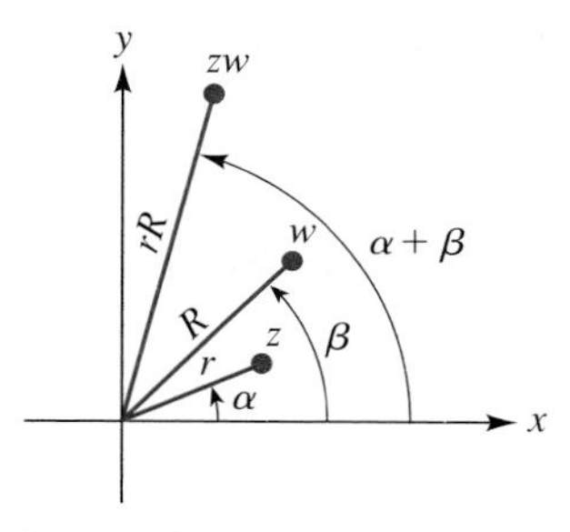

Figure 3

Formulas for Multiplication and Division of Complex Numbers in Trigonometric Form

Let $z = r(\cos\alpha + i\sin\alpha)$ and $w = R(\cos\beta + i\sin\beta)$; then

$$zw = rR[\cos(\alpha+\beta) + i\sin(\alpha+\beta)] \qquad \text{(See Figure 3.)}$$

Also, if $R \neq 0$, then

$$\frac{z}{w} = \frac{r}{R}[\cos(\alpha-\beta) + i\sin(\alpha-\beta)]$$

EXAMPLE 4 Multiplication and division of complex numbers in trigonometric form

Let $z = 8(\cos\frac{5\pi}{3} + i\sin\frac{5\pi}{3})$ and $w = 4(\cos\frac{2\pi}{3} + i\sin\frac{2\pi}{3})$. Compute **(a)** zw and **(b)** z/w. Express each answer in both trigonometric and rectangular form.

SOLUTION

(a)
$$
\begin{aligned}
zw &= (8)(4)[\cos(\tfrac{5\pi}{3} + \tfrac{2\pi}{3}) + i\sin(\tfrac{5\pi}{3} + \tfrac{2\pi}{3})]\\
&= 32(\cos\tfrac{7\pi}{3} + i\sin\tfrac{7\pi}{3}) = 32(\cos\tfrac{\pi}{3} + i\sin\tfrac{\pi}{3}) \qquad \text{trigonometric form}\\
&= 32(\tfrac{1}{2} + \tfrac{1}{2}\sqrt{3}i) = 16 + 16\sqrt{3}i \qquad \text{rectangular form}
\end{aligned}
$$

(b)
$$
\begin{aligned}
\frac{z}{w} &= \frac{8}{4}[\cos(\tfrac{5\pi}{3} - \tfrac{2\pi}{3}) + i\sin(\tfrac{5\pi}{3} - \tfrac{2\pi}{3})]\\
&= 2(\cos\pi + i\sin\pi) \qquad \text{trigonometric form}\\
&= 2(-1 + i\cdot 0) = -2 \qquad \text{rectangular form}
\end{aligned}
$$

EXAMPLE 5 Squaring a complex number

Compute z^2, where $z = r(\cos\theta + i\sin\theta)$.

SOLUTION

$$
\begin{aligned}
z^2 &= [r(\cos\theta + i\sin\theta)][r(\cos\theta + i\sin\theta)]\\
&= r^2[\cos(\theta+\theta) + i\sin(\theta+\theta)]\\
&= r^2(\cos 2\theta + i\sin 2\theta)
\end{aligned}
$$

The result in Example 5 is a particular case of an important theorem attributed to Abraham DeMoivre (1667–1754). In the box that follows, we state *DeMoivre's theorem.* (The theorem can be proved by using mathematical induction, which we discussed in Section 13.1.)

DeMoivre's Theorem

Let n be a natural number. Then

$$[r(\cos\theta + i\sin\theta)]^n = r^n(\cos n\theta + i\sin n\theta)$$

EXAMPLE 6 Using DeMoivre's theorem

Use DeMoivre's theorem to compute $(-\sqrt{2} + i\sqrt{2})^5$. Express your answer in rectangular form.

SOLUTION

In Example 3 we saw that the trigonometric form of $-\sqrt{2} + i\sqrt{2}$ is given by

$$-\sqrt{2} + i\sqrt{2} = 2\left(\cos\frac{3\pi}{4} + i\sin\frac{3\pi}{4}\right)$$

Therefore

$$\begin{aligned}(-\sqrt{2} + i\sqrt{2})^5 &= 2^5\left(\cos\frac{15\pi}{4} + i\sin\frac{15\pi}{4}\right)\\ &= 32\left[\frac{1}{2}\sqrt{2} + i\left(-\frac{1}{2}\sqrt{2}\right)\right]\\ &= 16\sqrt{2} - 16\sqrt{2}i\end{aligned}$$

The next two examples show how DeMoivre's theorem is used in computing roots. If n is a natural number and $z^n = w$, then we say that z is an ***n*th root** of w. The work in the examples also relies on the following observation about equality between nonzero complex numbers in trigonometric form: If $r(\cos\theta + i\sin\theta) = R(\cos A + i\sin A)$, then $r = R$ and $\theta = A + 2\pi k$, where k is an integer.

EXAMPLE 7 Finding cube roots of a complex number

Find the cube roots of $8i$.

SOLUTION

First we express $8i$ in trigonometric form. As can be seen from Figure 4,

$$8i = 8\left(\cos\frac{\pi}{2} + i\sin\frac{\pi}{2}\right)$$

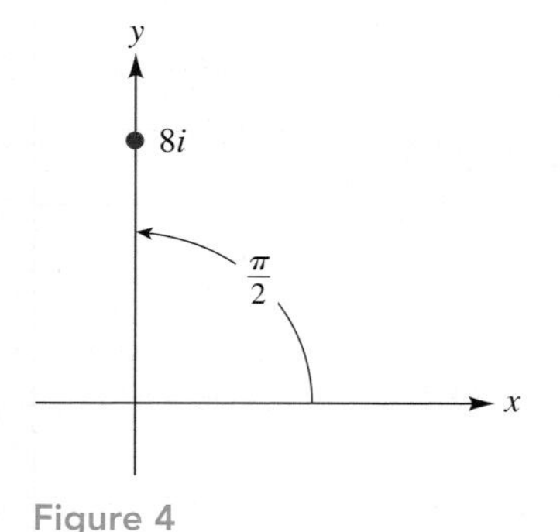

Figure 4

Now we let $z = r(\cos\theta + i\sin\theta)$ denote a cube root of $8i$. Then the equation $z^3 = 8i$ becomes

$$r^3(\cos 3\theta + i \sin 3\theta) = 8\left(\cos\frac{\pi}{2} + i \sin\frac{\pi}{2}\right) \tag{7}$$

From equation (7) we conclude that $r^3 = 8$ and, consequently, that $r = 2$. Also from equation (7) we have

$$3\theta = \frac{\pi}{2} + 2\pi k$$

or

$$\theta = \frac{\pi}{6} + \frac{2\pi k}{3} \qquad \text{dividing by 3} \tag{8}$$

If we let $k = 0$, equation (8) yields $\theta = \pi/6$. Thus one of the cube roots of $8i$ is

$$2\left(\cos\frac{\pi}{6} + i \sin\frac{\pi}{6}\right) = 2\left(\frac{1}{2}\sqrt{3} + i\cdot\frac{1}{2}\right) = \sqrt{3} + i$$

Next we let $k = 1$ in equation (8). As you can check, this yields $\theta = 5\pi/6$. So another cube root of $8i$ is

$$2\left(\cos\frac{5\pi}{6} + i \sin\frac{5\pi}{6}\right) = 2\left(-\frac{1}{2}\sqrt{3} + i\cdot\frac{1}{2}\right) = -\sqrt{3} + i$$

Similarly, using $k = 2$ in equation (8) yields $\theta = 3\pi/2$. (Verify this.) Consequently, a third cube root is

$$2\left(\cos\frac{3\pi}{2} + i \sin\frac{3\pi}{2}\right) = 2[0 + i(-1)] = -2i$$

We have now found three distinct cube roots of $8i$. If we were to continue the process, using $k = 3$, for example, we would find that no additional roots are obtained in this manner. We therefore conclude that there are exactly three cube roots of $8i$. We have plotted these cube roots in Figure 5. Notice that the points lie equally spaced on a circle of radius 2 centered at the origin of the complex plane.

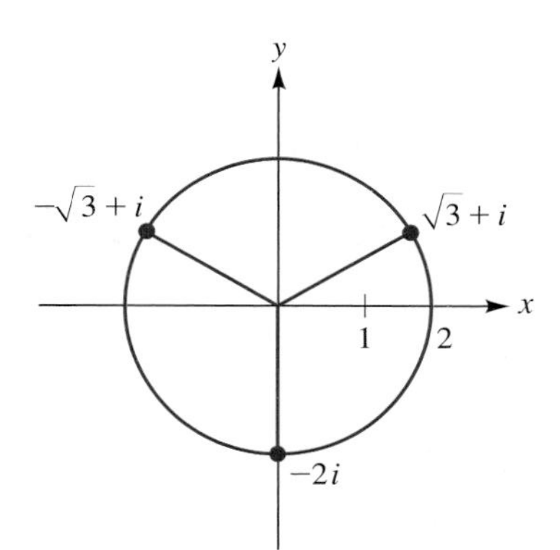

Figure 5

When we used DeMoivre's theorem in Example 7, we found that the number $8i$ had three distinct cube roots. It is true in general that any nonzero number $a + bi$ possesses exactly n distinct nth roots, equally spaced on a circle centered at the origin of the complex plane. This is true even if $b = 0$. In the next example we compute the five fifth roots of 2.

EXAMPLE 8 Finding fifth roots of a complex number

Compute the five fifth roots of 2.

SOLUTION

We will follow the procedure we used in Example 7. In trigonometric form, the number 2 becomes $2(\cos 0 + i \sin 0)$. Now we let $z = r(\cos\theta + i \sin\theta)$ denote a fifth root of 2. Then the equation $z^5 = 2$ becomes

$$r^5(\cos 5\theta + i \sin 5\theta) = 2(\cos 0 + i \sin 0) \tag{9}$$

From equation (9) we see that $r^5 = 2$, and consequently, $r = 2^{1/5}$. Also from equation (9) we have

$$5\theta = 0 + 2\pi k \quad \text{or} \quad \theta = \frac{2\pi k}{5}$$

Using the values $k = 0, 1, 2, 3$, and 4 in succession, we obtain the following results:

k	0	1	2	3	4
θ	0	$2\pi/5$	$4\pi/5$	$6\pi/5$	$8\pi/5$

The five fifth roots of 2 are therefore

$$z_1 = 2^{1/5}(\cos 0 + i \sin 0) = 2^{1/5} \qquad z_4 = 2^{1/5}\left(\cos\frac{6\pi}{5} + i\sin\frac{6\pi}{5}\right)$$

$$z_2 = 2^{1/5}\left(\cos\frac{2\pi}{5} + i\sin\frac{2\pi}{5}\right) \qquad z_5 = 2^{1/5}\left(\cos\frac{8\pi}{5} + i\sin\frac{8\pi}{5}\right)$$

$$z_3 = 2^{1/5}\left(\cos\frac{4\pi}{5} + i\sin\frac{4\pi}{5}\right)$$

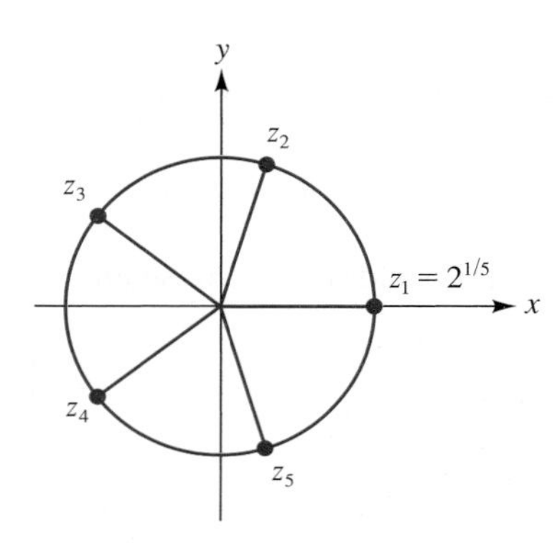

Figure 6

We have plotted these five fifth roots in Figure 6. Notice that the points are equally spaced on a circle of radius $2^{1/5}$ centered at the origin of the complex plane.

EXERCISE SET 13.6

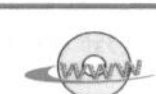

A

In Exercises 1–8, plot each point in the complex plane.

1. $4 + 2i$ **2.** $4 - 2i$ **3.** $-5 + i$
4. $-3 - 5i$ **5.** $1 - 4i$ **6.** i
7. $-i$ **8.** $1(= 1 + 0i)$

In Exercises 9–18, convert each complex number to rectangular form.

9. $2(\cos \frac{1}{4}\pi + i \sin \frac{1}{4}\pi)$ **10.** $6(\cos \frac{5}{3}\pi + i \sin \frac{5}{3}\pi)$
11. $4(\cos \frac{5}{6}\pi + i \sin \frac{5}{6}\pi)$ **12.** $3(\cos \frac{3}{2}\pi + i \sin \frac{3}{2}\pi)$
13. $\sqrt{2}(\cos 225° + i \sin 225°)$
14. $\frac{1}{2}(\cos 240° + i \sin 240°)$
15. $\sqrt{3}(\cos \frac{1}{2}\pi + i \sin \frac{1}{2}\pi)$ **16.** $5(\cos \pi + i \sin \pi)$
17. $4(\cos 75° + i \sin 75°)$ *Hint:* Use the addition formulas from Section 8.1 to evaluate cos 75° and sin 75°.
18. $2(\cos \frac{1}{8}\pi + i \sin \frac{1}{8}\pi)$ *Hint:* Use the half-angle formulas from Section 8.2 to evaluate $\cos(\pi/8)$ and $\sin(\pi/8)$.

In Exercises 19–28, convert from rectangular to trigonometric form. (In each case, choose an argument θ such that $0 \le \theta < 2\pi$.)

19. $\frac{1}{2}\sqrt{3} + \frac{1}{2}i$ **20.** $\sqrt{2} + \sqrt{2}i$
21. $-1 + \sqrt{3}i$ **22.** -4
23. $-2\sqrt{3} - 2i$ **24.** $-3\sqrt{2} - 3\sqrt{2}i$
25. $-6i$ **26.** $-4 - 4\sqrt{3}i$
27. $\frac{1}{4}\sqrt{3} - \frac{1}{4}i$ **28.** 16

In Exercises 29–54, carry out the indicated operations. Express your results in rectangular form for those cases in which the trigonometric functions are readily evaluated without tables or a calculator.

29. $2(\cos 22° + i \sin 22°) \times 3(\cos 38° + i \sin 38°)$
30. $4(\cos 5° + i \sin 5°) \times 6(\cos 130° + i \sin 130°)$
31. $\sqrt{2}(\cos \frac{1}{3}\pi + i \sin \frac{1}{3}\pi) \times \sqrt{2}(\cos \frac{4}{3}\pi + i \sin \frac{4}{3}\pi)$
32. $(\cos \frac{1}{5}\pi + i \sin \frac{1}{5}\pi) \times (\cos \frac{1}{20}\pi + i \sin \frac{1}{20}\pi)$
33. $3(\cos \frac{1}{7}\pi + i \sin \frac{1}{7}\pi) \times \sqrt{2}(\cos \frac{1}{7}\pi + i \sin \frac{1}{7}\pi)$
34. $\sqrt{3}(\cos 3° + i \sin 3°) \times \sqrt{3}(\cos 38° + i \sin 38°)$
35. $6(\cos 50° + i \sin 50°) \div 2(\cos 5° + i \sin 5°)$
36. $\sqrt{3}(\cos 140° + i \sin 140°) \div 3(\cos 5° + i \sin 5°)$
37. $2^{4/3}(\cos \frac{5}{12}\pi + i \sin \frac{5}{12}\pi) \div 2^{1/3}(\cos \frac{1}{4}\pi + i \sin \frac{1}{4}\pi)$
38. $\sqrt{6}(\cos \frac{16}{9}\pi + i \sin \frac{16}{9}\pi) \div \sqrt{2}(\cos \frac{1}{9}\pi + i \sin \frac{1}{9}\pi)$
39. $(\cos \frac{2}{5}\pi + i \sin \frac{2}{5}\pi) \div (\cos \frac{2}{5}\pi + i \sin \frac{2}{5}\pi)$
40. $(\cos \frac{2}{5}\pi + i \sin \frac{2}{5}\pi) \div (\cos \frac{1}{10}\pi + i \sin \frac{1}{10}\pi)$
41. $[3(\cos \frac{1}{3}\pi + i \sin \frac{1}{3}\pi)]^5$
42. $[\sqrt{2}(\cos \frac{5}{6}\pi + i \sin \frac{5}{6}\pi)]^4$
43. $[\frac{1}{2}(\cos \frac{1}{24}\pi + i \sin \frac{1}{24}\pi)]^6$

44. $[\sqrt{3}(\cos 70^\circ + i \sin 70^\circ)]^3$
45. $[2^{1/5}(\cos 63^\circ + i \sin 63^\circ)]^{10}$
46. $[2(\cos \frac{1}{5}\pi + i \sin \frac{1}{5}\pi)]^3$
47. $2(\cos 200^\circ + i \sin 200^\circ) \times \sqrt{2}(\cos 20^\circ + i \sin 20^\circ) \times \frac{1}{2}(\cos 5^\circ + i \sin 5^\circ)$
48. $\left[\dfrac{\cos(\pi/8) + i \sin(\pi/8)}{\cos(-\pi/8) + i \sin(-\pi/8)}\right]^5$
49. $[\frac{1}{2}(1 - \sqrt{3}i)]^5$ *Hint*: Convert to trigonometric form.
50. $(1 - i)^3$
51. $(-2 - 2i)^5$
52. $(-\frac{1}{2} + \frac{1}{2}\sqrt{3}i)^6$
53. $(-2\sqrt{3} - 2i)^4$
54. $(1 + i)^{16}$

In Exercises 55–61, use DeMoivre's theorem to find the indicated roots. Express the results in rectangular form.
55. Cube roots of $-27i$
56. Cube roots of 2
57. Eighth roots of 1
58. Square roots of i
59. Cube roots of 64
60. Square roots of $-\frac{1}{2} - \frac{1}{2}\sqrt{3}i$
61. Sixth roots of 729

Use a calculator to complete Exercises 62–65.
62. Compute $(9 + 9i)^6$.
63. Compute $(7 - 7i)^8$.
64. Compute the cube roots of $1 + 2i$. Express your answers in rectangular form, with the real and imaginary parts rounded to two decimal places.
65. Compute the fifth roots of i. Express your answers in rectangular form, with the real and imaginary parts rounded to two decimal places.

B

In Exercises 66–68, find the indicated roots. Express the results in rectangular form.
66. Find the fourth roots of i. *Hint*: Use the half-angle formulas from Section 8.2.
67. Find the fourth roots of $8 - 8\sqrt{3}i$. *Hint*: Use the addition formulas or the half-angle formulas.
68. Find the square roots of $7 + 24i$. *Hint*: You'll need to use the half-angle formulas from Section 8.2.
69. **(a)** Compute the three cube roots of 1.
(b) Let z_1, z_2, and z_3 denote the three cube roots of 1. Verify that $z_1 + z_2 + z_3 = 0$ and also that $z_1z_2 + z_2z_3 + z_3z_1 = 0$.
70. **(a)** Compute the four fourth roots of 1.
(b) Verify that the sum of these four fourth roots is 0.
71. Evaluate $(-\frac{1}{2} + \frac{1}{2}\sqrt{3}i)^5 + (-\frac{1}{2} - \frac{1}{2}\sqrt{3}i)^5$. *Hint*: Use DeMoivre's theorem.
72. Show that $(-\frac{1}{2} + \frac{1}{2}\sqrt{3}i)^6 + -\frac{1}{2} - \frac{1}{2}\sqrt{3}i)^6 = 2$.
73. Compute $(\cos\theta + i \sin\theta)(\cos\theta - i \sin\theta)$.
74. In the identity $(\cos\theta + i \sin\theta)^2 = \cos 2\theta + i \sin 2\theta$, carry out the multiplication on the left-hand side of the equation. Then equate the corresponding real parts and imaginary parts from each side of the equation that results. What do you obtain?
75. Show that

$$\frac{r(\cos\alpha + i \sin\alpha)}{R(\cos\beta + i \sin\beta)} = \frac{r}{R}[\cos(\alpha - \beta) + i \sin(\alpha - \beta)]$$

Suggestion: Begin with the quantity on the left side, and multiply it by $(\cos\beta - i \sin\beta)/(\cos\beta - i \sin\beta)$.
76. Show that

$$1 + \cos\theta + i \sin\theta = 2\cos\left(\frac{\theta}{2}\right)\left[\cos\left(\frac{\theta}{2}\right) + i \sin\left(\frac{\theta}{2}\right)\right]$$

77. If $z = r(\cos\theta + i \sin\theta)$ and z is not zero, show that

$$\frac{1}{z} = \frac{1}{r}(\cos\theta - i \sin\theta)$$

Hint: $1/z = [1(\cos 0 + i \sin 0)]/[r(\cos\theta + i \sin\theta)]$
78. **(a)** If $z = r(\cos\theta + i \sin\theta)$, z is not zero, and n is a positive integer, use the result of Exercise 77 and DeMoivre's theorem to show that $z^{-n} = r^{-n}[\cos(-n\theta) + i \sin(-n\theta)]$.
(b) If $z = r(\cos\theta + i \sin\theta)$and z is not zero, we define z^0 to be 1. Show that $z^0 = r^0(\cos 0 + i \sin 0)$.
(c) Finally, for any nonzero complex number $z = r(\cos\theta + i \sin\theta)$ and *any* integer n, obtain the following generalization of DeMoivre's theorem:

$$z^n = r^n[\cos(n\theta) + i \sin(n\theta)]$$

79. Show that $\dfrac{1 + \sin\theta + i \cos\theta}{1 + \sin\theta - i \cos\theta} = \sin\theta + i \cos\theta$.

Assume that $\theta \neq \frac{3}{2}\pi + 2\pi k$, where k is an integer. *Hint*: Work with the left-hand side; first "rationalize" the denominator by multiplying by the quantity

$$\frac{(1 + \sin\theta) + i \cos\theta}{(1 + \sin\theta) + i \cos\theta}$$

80. If $w + \dfrac{1}{w} = 2\cos\theta$, show that $w = \cos\theta \pm i \sin\theta$.

Hint: Use the quadratic formula to solve the given equation for w.

81. If $w + \dfrac{1}{w} = 2\cos\theta$, show that $w^n + \dfrac{1}{w^n} = 2\cos n\theta$.

Hint: Use the results in Exercises 80 and 77.

Chapter 13 Summary of Principal Terms and Formulas

Terms or formulas	Page reference	Comments
1. The principle of mathematical induction	974	This principle can be stated as follows. Suppose that for each natural number n we have a statement P_n. Suppose P_1 is true. Also suppose that P_{k+1} is true whenever P_k is true. Then according to the principle of mathematical induction, all of the statements are true; that is, P_n is true for all natural numbers n. The principle of mathematical induction has the status of an axiom; that is, we accept its validity without proof.
2. Pascal's triangle	980	Pascal's triangle refers to the triangular array of numbers displayed on page 981. Additional rows can be added to the triangle according to the following rule: Each entry in the array (other than the 1's along the sides) is the sum of the two numbers diagonally above it. The numbers in the nth row of Pascal's triangle are the coefficients of the terms in the expansion of $(a + b)^{n-1}$.
3. $n!$ (read: n factorial)	982	This denotes the product of the first n natural numbers. For example, $4! = (4)(3)(2)(1) = 24$.
4. $\binom{n}{k}$ (read: n choose k)	983	Let n and k be nonnegative integers with $k \leq n$. Then the *binomial coefficient* $\binom{n}{k}$ is defined by $$\binom{n}{k} = \frac{n!}{k!(n-k)!}$$
5. The binomial theorem	984	The binomial theorem is a formula that allows us to analyze and expand expressions of the form $(a + b)^n$. If we use *sigma notation,* then the statement of the binomial theorem that appears on page 984 can be abbreviated to read $$(a + b)^n = \sum_{k=0}^{n} \binom{n}{k} a^{n-k} b^k$$
6. Sequence	989, 991	In the context of the present chapter, a sequence is an ordered list of real numbers: $a_1, a_2, a_3, \ldots$. The numbers a_k are called the *terms* of the sequence. A sequence can also be defined as a *function* whose domain is the set of natural numbers or the set of nonnegative integers.
7. $\sum_{k=1}^{n} a_k$	995	This expression stands for the sum $a_1 + a_2 + a_3 + \cdots + a_n$. The letter k in the expression is referred to as the *index of summation.*

Terms or formulas	Page reference	Comments
8. Arithmetic sequence	1000	A sequence in which the successive terms differ by a constant is called an arithmetic sequence. The general form of an arithmetic sequence is $$a, a + d, a + 2d, a + 3d, \ldots$$ In this sequence, d is referred to as the *common difference.*
9. $a_n = a + (n - 1)d$	1000	This is the formula for the nth term of an arithmetic sequence.
10. $S_n = \frac{n}{2}[2a + (n - 1)d]$ $S_n = n\left(\frac{a + a_n}{2}\right)$	1002	These are the formulas for the sum of the first n terms of an arithmetic sequence.
11. Geometric sequence	1010	A sequence in which the ratio of successive terms is constant is called a geometric sequence. The general form of a geometric sequence is $$a, ar, ar^2, \ldots$$ The number r is referred to as the *common ratio.*
12. $a_n = ar^{n-1}$	1011	This is the formula for the nth term of a geometric sequence.
13. $S_n = \frac{a(1 - r^n)}{1 - r}$	1011	This is the formula for the sum of the first n terms of a geometric sequence.
14. $S = \frac{a}{1 - r}$	1013	This is the formula for the sum of the infinite geometric series $a + ar + ar^2 + \cdots$, where $\lvert r \rvert < 1$.
15. The complex plane	1015	Refer to Figure 2 on page 1015. A complex number $a + bi$ can be identified with the point (a, b) in the x-y plane. In this context the x-y plane is referred to as the complex plane. The distance r in the figure is the *modulus* of the complex number $a + bi$; the angle θ is an *argument* of $a + bi$.
16. $z = r(\cos\theta + \sin\theta)$	1015	The expression on the right-hand side is the *trigonometric form* of the complex number $z = a + bi$. (The expression $a + bi$ is the *rectangular form* of the complex number z.) The trigonometric form and the rectangular form are related by the following equations: $$r = \sqrt{a^2 + b^2} \qquad a = r\cos\theta \qquad b = r\sin\theta$$

Terms or formulas	Page reference	Comments
17. $zw = rR[\cos(\alpha + \beta) + i \sin(\alpha + \beta)]$ $\frac{z}{w} = \frac{r}{R}[\cos(\alpha - \beta) - i \sin(\alpha - \beta)]$	1017	These are the formulas for multiplication and division of complex numbers in trigonometric form $z = r(\cos \alpha + i \sin \alpha)$ and $w = R(\cos \beta + i \sin \beta)$.
18. DeMoivre's theorem	1018	Let n be a natural number and let $z = r(\cos \theta + i \sin \theta)$. Then DeMoivre's theorem states that $z^n = r^n(\cos n\theta + i \sin n\theta)$ For applications of this theorem, see Examples 6, 7, and 8 in Section 13.6.

Writing Mathematics

1. A sequence is defined recursively as follows: $a_0 = \pi$; $a_{n+1} = \dfrac{a_n}{a_n + 1}$ for $n \geq 0$. Compute a_0, a_1, and a_2. Describe (in a complete sentence or two) what pattern you observe. Make a conjecture about a general formula for a_n. Then use mathematical induction to prove that the formula is valid. (As in the examples in the text, use complete sentences in writing and explaining the proof.)

2. A sequence is defined recursively as follows.

$$\begin{aligned} a_1 &= 3/2 \\ a_{n+1} &= a_n^2 - 2a_n + 2 \qquad \text{for } n \geq 1 \end{aligned}$$

Use mathematical induction to show that $1 < a_n < 2$ for all natural numbers n.

3. Study the procedure that was used on page 1011 to obtain the formula for the sum of a finite geometric series. Work with classmates or your instructor to adapt the procedure to find the sum $1 + 2 \sin \theta + 3 \sin^2 \theta + \cdots + n \sin^{n-1} \theta$. Finally, on your own, write a complete summary of what you have done.

4. Study the following purported "proof" by mathematical induction, and explain the fallacy.

Let P_n denote the following statement: In any set of n numbers, the numbers are all equal.

Step 1 P_1 is clearly true.

Step 2 Assume that P_k is true. We must show that P_{k+1} is true. That is, given a set of $k + 1$ numbers $a_1, a_2, a_3, \ldots, a_{k+1}$, we must show that $a_1 = a_2 = a_3 = \cdots = a_{k+1}$. By the inductive hypothesis, however, we have $a_1 = a_2 = a_3 = \cdots = a_k$ and also $a_2 = a_3 = a_4 = \cdots = a_{k+1}$. These last two sets of equations imply that $a_1 = a_2 = a_3 = \cdots = a_k = a_{k+1}$, as we wished to show. Having now completed Steps 1 and 2, we conclude that P_n is true for every n.

Chapter 13 Review Exercises

In Exercises 1–10, use the principle of mathematical induction to show that the statements are true for all natural numbers.

1. $5 + 10 + 15 + \cdots + 5n = \frac{5}{2}n(n+1)$

2. $10 + 10^2 + 10^3 + \cdots + 10^n = \frac{10}{9}(10^n - 1)$

3. $1 \cdot 2 + 2 \cdot 3 + 3 \cdot 4 + \cdots + n(n+1) = \frac{1}{3}n(n+1)(n+2)$

4. $\dfrac{1}{2} + \dfrac{2}{2^2} + \dfrac{3}{2^3} + \cdots + \dfrac{n}{2^n} = 2 - \dfrac{2+n}{2^n}$

5. $1 + 3 \cdot 2 + 5 \cdot 2^2 + 7 \cdot 2^3 + \cdots + (2n-1) \cdot 2^{n-1} = 3 + (2n-3) \cdot 2^n$

6. $\dfrac{1}{1 \cdot 4} + \dfrac{1}{4 \cdot 7} + \dfrac{1}{7 \cdot 10} + \cdots + \dfrac{1}{(3n-2)(3n+1)} = \dfrac{n}{3n+1}$

7. $1 + 2^2 \cdot 2 + 3^2 \cdot 2^2 + 4^2 \cdot 2^3 + \cdots + n^2 \cdot 2^{n-1} = (n^2 - 2n + 3)2^n - 3$

8. 9 is a factor of $n^3 + (n+1)^3 + (n+2)^3$.

9. 3 is a factor of $7^n - 1$.

10. 8 is a factor of $9^n - 1$.

In Exercises 11–20, expand the given expressions.

11. $(3a + b^2)^4$

12. $(5a - 2b)^3$

13. $(x + \sqrt{x})^4$

14. $(1 - \sqrt{3})^6$

15. $(x^2 - 2y^2)^5$

16. $\left(\dfrac{1}{a} + \dfrac{2}{b}\right)^3$

17. $\left(1 + \dfrac{1}{x}\right)^5$

18. $\left(x^3 + \dfrac{1}{x^2}\right)^6$

19. $(a\sqrt{b} - b\sqrt{a})^4$

20. $(x^{-2} + y^{5/2})^8$

21. Find the fifth term in the expansion of $(3x + y^2)^5$.

22. Find the eighth term in the expansion of $(2x - y)^9$.

23. Find the coefficient of the term containing a^5 in the expansion of $(a - 2b)^7$.

24. Find the coefficient of the term containing b^8 in the expansion of $\left(2a - \dfrac{b}{3}\right)^{10}$.

25. Find the coefficient of the term containing x^3 in the expansion of $(1 + \sqrt{x})^8$.

26. Expand $(1 + \sqrt{x} + x)^6$. *Suggestion*: Rewrite the expression as $[(1 + \sqrt{x}) + x]^6$.

In Exercises 27–32, verify each assertion by computing the indicated binomial coefficients.

27. $\binom{2}{0}^2 + \binom{2}{1}^2 + \binom{2}{2}^2 = \binom{4}{2}$

28. $\binom{3}{0}^2 + \binom{3}{1}^2 + \binom{3}{2}^2 + \binom{3}{3}^2 = \binom{6}{3}$

29. $\binom{4}{0}^2 + \binom{4}{1}^2 + \binom{4}{2}^2 + \binom{4}{3}^2 + \binom{4}{4}^2 = \binom{8}{4}$

30. $\binom{2}{0} + \binom{2}{1} + \binom{2}{2} = 2^2$

31. $\binom{3}{0} + \binom{3}{1} + \binom{3}{2} + \binom{3}{3} = 2^3$

32. $\binom{4}{0} + \binom{4}{1} + \binom{4}{2} + \binom{4}{3} + \binom{4}{4} = 2^4$

In Exercises 33–38, compute the first four terms in each sequence. Also, in Exercises 33–35, graph the sequences for $n = 1, 2, 3,$ and 4.

33. $a_n = \dfrac{2n}{n+1}$

34. $a_n = \dfrac{3n-2}{3n+2}$

35. $a_n = (-1)^n\left(1 - \dfrac{1}{n+1}\right)$

36. $a_0 = 4;\ a_n = 2a_{n-1},\ n \geq 1$

37. $a_0 = -3;\ a_n = 4a_{n-1},\ n \geq 1$

38. $a_0 = 1;\ a_1 = 2;\ a_n = 3a_{n-1} + 2a_{n-2},\ n \geq 2$

In Exercises 39 and 40, evaluate each sum.

39. (a) $\displaystyle\sum_{k=1}^{3} (-1)^k(2k+1)$

(b) $\displaystyle\sum_{k=0}^{8} \left(\frac{1}{k+1} - \frac{1}{k+2}\right)$

40. (a) $\displaystyle\sum_{j=1}^{4} \frac{(-1)^j}{j}$

(b) $\displaystyle\sum_{n=1}^{5} \left(\frac{1}{n} - \frac{1}{n+1}\right)$

In Exercises 41 and 42, rewrite each sum using sigma notation.

41. $\dfrac{5}{3} + \dfrac{5}{3^2} + \dfrac{5}{3^3} + \dfrac{5}{3^4} + \dfrac{5}{3^5}$

42. $\dfrac{1}{2} - \dfrac{2}{2^2} + \dfrac{3}{2^3} - \dfrac{4}{2^4} + \dfrac{5}{2^5} - \dfrac{6}{2^6}$

In Exercises 43–46, find the indicated term in each sequence.

43. $5, 9, 13, 17, \ldots;\ a_{18}$

44. $5, 9/2, 4, 7/2, \ldots;\ a_{20}$

45. $10, 5, 5/2, 5/4, \ldots;\ a_{12}$

46. $\sqrt{2} + 1, 1, \sqrt{2} - 1, 3 - 2\sqrt{2}, \ldots;\ a_{10}$

47. Determine the sum of the first 12 terms of an arithmetic sequence in which the first term is 8 and the 12th term is 43/2.

48. Find the sum of the first 45 terms in the sequence $10, 29/3, 28/3, 9, \ldots$.

49. Find the sum of the first 10 terms in the sequence $7, 70, 700, \ldots$.

50. Find the sum of the first 12 terms in the sequence $1/3, -2/9, 4/27, -8/81, \ldots$.

51. In a certain geometric sequence, the third term is 4, and the fifth term is 10. Find the sixth term, given that the common ratio is negative.

52. For a certain infinite geometric series, the first term is 2, and the sum is 18/11. Find the common ratio r.

In Exercises 53–56, find the sum of each infinite geometric series.

53. $\frac{3}{5} + \frac{3}{25} + \frac{3}{125} + \cdots$

54. $\frac{7}{10} + \frac{7}{100} + \frac{7}{1000} + \cdots$

55. $\frac{1}{9} - \frac{1}{81} + \frac{1}{729} - \cdots$

56. $1 + \frac{1}{1 + \sqrt{2}} + \frac{1}{(1 + \sqrt{2})^2} + \cdots$

57. Find a fraction equivalent of $0.\overline{45}$.

58. Find a fraction equivalent to $0.2\overline{13}$.

In Exercises 59–62, verify each equation using the formula for the sum of an arithmetic series:

$$S_n = \frac{n}{2}[2a + (n - 1)d]$$

(The formulas given in Exercises 59–62 appear in Elements of Algebra *by Leonhard Euler, first published in 1770.)*

59. $1 + 2 + 3 + \cdots + n = n + n(n - 1)/2$

60. $1 + 3 + 5 + \cdots$ (to n terms) $= n + 2n(n - 1)/2$

61. $1 + 4 + 7 + \cdots$ (to n terms) $= n + 3n(n - 1)/2$

62. $1 + 5 + 9 + \cdots$ (to n terms) $= n + 4n(n - 1)/2$

63. In this exercise you will use the following (remarkably simple) formula for approximating sums of powers of integers:

$$1^k + 2^k + 3^k + \cdots + n^k \approx \frac{(n + \frac{1}{2})^{k+1}}{k + 1} \qquad (1)$$

[This formula appears in the article, "Sums of Powers of Integers" by B. L. Burrows and R. F. Talbot, published in the *American Mathematical Monthly,* vol. 91 (1984), p. 394.]

(a) Use formula (1) to estimate the sum $1^2 + 2^2 + 3^2 + \cdots + 50^2$. Round your answer to the nearest integer.

(b) Compute the exact value of the sum in part (a) using the formula $\sum_{k=1}^{n} k^2 = \frac{1}{6}n(n + 1)(2n + 1)$. (This formula can be proved using mathematical induction.) Then compute the percentage error for the approximation obtained in part (a). The percentage error is given by

$$\frac{|\text{actual value} - \text{approximate value}|}{\text{actual value}} \times 100$$

(c) Use formula (1) to estimate the sum $1^4 + 2^4 + 3^4 + \cdots + 200^4$. Round your answer to six significant digits.

(d) The following formula for the sum $1^4 + 2^4 + \cdots + n^4$ can be proved by using mathematical induction:

$$\sum_{k=1}^{n} k^4 = \frac{n(n + 1)(2n + 1)(3n^2 + 3n - 1)}{30}$$

Use this formula to compute the sum in part (c). Round your answer to six significant digits. Then use this result to compute the percentage error for the approximation in part (c).

64. According to *Stirling's formula* [named after James Stirling (1692–1770)], the quantity $n!$ can be approximated as follows:

$$n! \approx \sqrt{2\pi n}\left(\frac{n}{e}\right)^n$$

In this formula, e is the constant 2.718 . . . (discussed in Section 5.2). Use a calculator to complete the following table. Round your answers to five significant digits. As you will see, the numbers in the right-hand column approach 1 as n increases. This shows that in a certain sense the approximation improves as n increases.

n	$n!$	$\sqrt{2\pi n}(n/e)^n$	$\frac{n!}{\sqrt{2\pi n}(n/e)^n}$
10			
20			
30			
40			
50			
60			
65			

65. The sum of three consecutive terms in a geometric sequence is 13, and the sum of the reciprocals is 13/9. What are the possible values for the common ratio? (There are four answers.)

66. The nonzero numbers a, b, and c are consecutive terms in a geometric sequence, and $a + b + c = 70$. Furthermore, $4a$, $5b$, and $4c$ are consecutive terms in an arithmetic sequence. Find a, b, and c.

67. The nonzero numbers a, b, and c are consecutive terms in a geometric sequence, and a, $2b$, and c are consecutive terms in an arithmetic sequence. Show that the common ratio in the geometric sequence must be either $2 + \sqrt{3}$ or $2 - \sqrt{3}$.

68. If the numbers $\dfrac{1}{b+c}, \dfrac{1}{c+a}$, and $\dfrac{1}{a+b}$ are consecutive terms in an arithmetic sequence, show that a^2, b^2, and c^2 are also consecutive terms in an arithmetic sequence.

69. If a, b, and c are consecutive terms in a geometric sequence, prove that $\dfrac{1}{a+b}, \dfrac{1}{2b}$, and $\dfrac{1}{c+b}$ are consecutive terms in an arithmetic sequence.

70. **(a)** Find a value for x such that $3 + x$, $4 + x$, and $5 + x$ are consecutive terms in a geometric sequence.

(b) Given three numbers a, b, and c, find a value for x (in terms of a, b, and c) such that $a + x$, $b + x$, and $c + x$ are consecutive terms in a geometric sequence.

71. If $\ln(A + C) + \ln(A + C - 2B) = 2\ln(A - C)$, show that $1/A$, $1/B$, and $1/C$ are consecutive terms in an arithmetic sequence.

72. Let $a_1, a_2, a_3, \ldots$ be an arithmetic sequence with common difference d, and let $r(\neq 1)$ be a real number. In this exercise we develop a formula for the sum of the series

$$a_1 + ra_2 + r^2a_3 + \cdots + r^{n-1}a_n$$

The method that we use here is essentially the same as the method that was used in the text to derive the formula for the sum of a finite geometric series. So as background for this exercise, you should review the derivation on page 1011 for the sum of a geometric series.

(a) Let S denote the required sum. Show that

$$\begin{aligned} S - rS &= a_1 + rd + r^2d + \cdots + r^{n-1}d - r^na_n \\ &= a_1 + \frac{d(r - r^n)}{1 - r} - r^na_n \end{aligned}$$

(b) Show that $S = \dfrac{a_1 - r^na_n}{1 - r} + \dfrac{d(r - r^n)}{(1 - r)^2}$.

73. Use your calculator and the formula in Exercise 72(b) to find the sum of each of the following series.

(a) $1 + 2\cdot 2 + 2^2\cdot 3 + 2^3\cdot 4 + \cdots + 2^{13}\cdot 14$

(b) $2 + 4\cdot 5 + 4^2\cdot 8 + 4^3\cdot 11 + \cdots + 4^6\cdot 20$

(c) $3 - \dfrac{1}{2}\cdot 5 + \dfrac{1}{2^2}\cdot 7 - \dfrac{1}{2^3}\cdot 9 + \cdots +$ (to 10 terms)

In Exercises 74–77, convert each complex number to rectangular form.

74. $3(\cos\frac{1}{3}\pi + i\sin\frac{1}{3}\pi)$

75. $\cos\frac{1}{6}\pi + i\sin\frac{1}{6}\pi$

76. $2^{1/4}(\cos\frac{7}{4}\pi + i\sin\frac{7}{4}\pi)$

77. $5[\cos(-\frac{1}{4}\pi) + i\sin(-\frac{1}{4}\pi)]$

In Exercises 78–81, express the complex numbers in trigonometric form.

78. $\frac{1}{2}(1 + \sqrt{3}i)$

79. $3i$

80. $-3\sqrt{2} - 3\sqrt{2}i$

81. $2\sqrt{3} - 2i$

In Exercises 82–89, carry out the indicated operations. Express your results in rectangular form for those cases in which the trigonometric functions are readily evaluated without tables or a calculator.

82. $5(\cos\frac{1}{7}\pi + i\sin\frac{1}{7}\pi) \times 2(\cos\frac{3}{28}\pi + i\sin\frac{3}{28}\pi)$

83. $4(\cos\frac{1}{12}\pi + i\sin\frac{1}{12}\pi) \times 3(\cos\frac{1}{12}\pi + i\sin\frac{1}{12}\pi)$

84. $8(\cos\frac{1}{12}\pi + i\sin\frac{1}{12}\pi) \div 4(\cos\frac{1}{3}\pi + i\sin\frac{1}{3}\pi)$

85. $4(\cos 32^\circ + i\sin 32^\circ) \div 2^{1/2}(\cos 2^\circ + i\sin 2^\circ)$

86. $[3^{1/4}(\cos\frac{1}{36}\pi + i\sin\frac{1}{36}\pi)]^{12}$

87. $[2(\cos\frac{2}{15}\pi + i\sin\frac{2}{15}\pi)]^5$

88. $(\sqrt{3} + i)^{10}$

89. $(\sqrt{2} - \sqrt{2}i)^{15}$

In Exercises 90–93, use DeMoivre's theorem to find the indicated roots. Express your results in rectangular form.

90. Sixth roots of 1

91. Cube roots of $-64i$

92. Square roots of $\sqrt{2} - \sqrt{2}i$

93. Fourth roots of $1 + \sqrt{3}i$

94. Find the five fifth roots of $1 + i$. Use a calculator to express the roots in rectangular form. (Round each decimal to two places in the final answer.)

95. For our last exercise in the last chapter of this book, we take the following problem from the classic text by Isaac Todhunter, *Plane Trigonometry for the Use of Colleges and Schools,* 5th ed. (London: Macmillan & Co., 1874):

> If the cotangents of the angles of a triangle be in arithmetical progression, the squares of the sides will also be in arithmetical progression.

(continues)

Hint: Write the given information as an equation involving sines and cosines. Use the formulas on page 656 to express the cosines in terms of the lengths of the sides. For the sines, use the formulas in Exercise 56 in Section 9.2. [For a shorter proof, see page 68 in the January 1995 issue of *The College Mathematics Journal.* Although the solution given there applies to the converse of the present problem, it can easily be adjusted to fit this problem. (As is usual with all math journal articles, you'll find that there are algebra steps that you will need to fill in for yourself.)]

Chapter 13 Test

1. Use the principle of mathematical induction to show that the following formula is valid for all natural numbers n:

$$1^2 + 2^2 + 3^2 + \cdots + n^2 = \frac{n(n+1)(2n+1)}{6}$$

2. Express each of the following sums without using sigma notation, and then evaluate each sum.

(a) $\sum_{k=0}^{2} (10k - 1)$ **(b)** $\sum_{k=1}^{3} (-1)^k k^2$

3. (a) Write the formula for the sum S_n of a finite geometric series.

(b) Evaluate the sum $\frac{3}{2} + \frac{3^2}{2^2} + \frac{3^3}{2^3} + \cdots + \frac{3^{10}}{2^{10}}$.

4. (a) Determine the coefficient of the term containing a^3 in the expansion of $(a - 2b^3)^{11}$.

(b) Find the fifth term of the expansion in part (a).

5. Expand the expression $(3x^2 + y^3)^5$.

6. Determine the sum of the first 12 terms of an arithmetic sequence in which the first term is 8 and the 12th term is 43/2.

7. Find the sum of the following infinite geometric series:

$$\frac{7}{10} + \frac{7}{100} + \frac{7}{1000} + \cdots$$

8. A sequence is defined recursively as follows: $a_1 = 1$; $a_2 = 1$; and $a_n = (a_{n-1})^2 + a_{n-2}$ for $n \geq 3$. Determine the fourth and fifth terms in this sequence.

9. In a certain geometric sequence the third term is 4, and the fifth term is 10. Find the sixth term, given that the common ratio is negative.

10. What is the 20th term in the arithmetic sequence $-61, -46, -31, \ldots$?

11. Graph the first four terms of each sequence.

(a) $a_n = \frac{1}{2}(n^2 - n), n \geq 0$

(b) $b_1 = 1; b_n = (b_{n-1})^2 - nb_{n-1}, n \geq 2$

12. Find the rectangular form for the complex number

$$z = 2\left(\cos\frac{2}{3}\pi + i\sin\frac{2}{3}\pi\right)$$

13. Find the trigonometric form for the complex number $\sqrt{2} - \sqrt{2}i$.

14. Let $z = 3(\cos\frac{2\pi}{9} + i\sin\frac{2\pi}{9})$ and $w = 5(\cos\frac{\pi}{9} + i\sin\frac{\pi}{9})$. Compute the product zw, and express your answer in rectangular form.

15. Compute the cube roots of $64i$.

Appendix

A.1 SIGNIFICANT DIGITS

Many of the numbers that we use in scientific work and in daily life are approximations. In some cases the approximations arise because the numbers are obtained through measurements or experiments. Consider, for example, the following statement from an astronomy textbook:

The diameter of the Moon is 3476 km.

We interpret this statement as meaning that the actual diameter D is closer to 3476 km than it is to either 3475 km or 3477 km. In other words,

$$3475.5 \text{ km} \le D \le 3476.5 \text{ km}$$

The interval [3475.5, 3476.5] in this example provides information about the accuracy of the measurement. Another way to indicate accuracy in an approximation is by specifying the number of *significant digits* it contains. The measurement 3476 km has four significant digits. In general, the number of significant digits in a given number is found as follows.

Significant Digits

The number of significant digits in a given number is determined by counting the digits from left to right, beginning with the leftmost nonzero digit.

EXAMPLES

Number	Number of Significant Digits
1.43	3
0.52	2
0.05	1
4837	4
4837.0	5

Numbers obtained through measurements are not the only source of approximations in scientific work. For example, to five significant digits we have the following approximation for the irrational number π:

$$\pi \approx 3.1416$$

This statement tells us that π is closer to 3.1416 than it is to either 3.1415 or 3.1417. In other words,

$$3.14155 \le \pi \le 3.14165$$

Table 1 provides some additional examples of the ideas we've introduced.

TABLE 1

Number	Number of significant digits	Range of measurement
37	2	[36.5, 37.5]
37.0	3	[36.95, 37.05]
268.1	4	[268.05, 268.15]
1.036	4	[1.0355, 1.0365]
0.036	2	[0.0355, 0.0365]

There is an ambiguity involving zero that can arise in counting significant digits. Suppose that someone measures the width w of a rectangle and reports the result as 30 cm. How many significant digits are there? If the value 30 cm was obtained by measuring to the nearest 10 cm, then only the digit 3 is significant, and we can conclude only that the width w lies in the range 25 cm $\le w \le$ 35 cm. On the other hand, if the 30 cm was obtained by measuring to the nearest 1 cm, then both the digits 3 and 0 are significant, and we have 29.5 cm $\le w \le$ 30.5 cm.

By using **scientific notation,** we can avoid the type of ambiguity discussed in the previous paragraph. A number written in the form

$$b \times 10^n \qquad \text{where } 1 \le b < 10 \text{ and } n \text{ is an integer}$$

is said to be expressed in scientific notation. For the example in the previous paragraph, then, we would write

$$w = 3 \times 10^1 \text{ cm} \qquad \text{if the measurement is to the nearest 10 cm}$$

and

$$w = 3.0 \times 10^1 \text{ cm} \qquad \text{if the measurement is to the nearest 1 cm}$$

As the figures in Table 2 indicate, for a number $b \times 10^n$ in scientific notation the number of significant digits is just the number of digits in b. (This is one of the advantages in using scientific notation; the number of significant digits, and hence the accuracy of the measurement, is readily apparent.)

TABLE 2

Measurement	Number of significant digits	Range of measurement
Mass of the Earth:		
6×10^{27} g	1	[5.5×10^{27} g, 6.5×10^{27} g]
6.0×10^{27} g	2	[5.95×10^{27} g, 6.05×10^{27} g]
5.974×10^{27} g	4	[5.9735×10^{27} g, 5.9745×10^{27} g]
Mass of a proton:		
1.67×10^{-24} g	3	[1.665×10^{-24} g, 1.675×10^{-24} g]

Many of the numerical exercises in the text ask that you round the answers to a specified number of decimal places. Our rules for rounding are as follows.

Rules for Rounding a Number (with more than *n* Decimal Places) to *n* Decimal Places

1. If the digit in the $(n + 1)$st decimal place is greater than 5, increase the digit in the nth place by 1. If the digit in the $(n + 1)$st place is less than 5, leave the nth digit unchanged.
2. If the digit in the $(n + 1)$st decimal place is 5 and there is at least one nonzero digit to the right of this 5, increase the digit in the nth decimal place by 1.
3. If the digit in the $(n + 1)$st decimal place is 5 and there are no nonzero digits to the right of this 5, then increase the digit in the nth decimal place by 1 only if this results in an even digit.

The examples in Table 3 illustrate the use of these rules.

TABLE 3

Number	Rounded to one decimal place	Rounded to three decimal places
4.3742	4.4	4.374
2.0515	2.1	2.052
2.9925	3.0	2.992

These same rules can be adapted for rounding a result to a specified number of significant digits. As examples of this, we have

2347	rounded to two significant digits is $2300 = 2.3 \times 10^3$
2347	rounded to three significant digits is $2350 = 2.35 \times 10^3$
975	rounded to two significant digits is $980 = 9.8 \times 10^2$
0.985	rounded to two significant digits is $0.98 = 9.8 \times 10^{-1}$

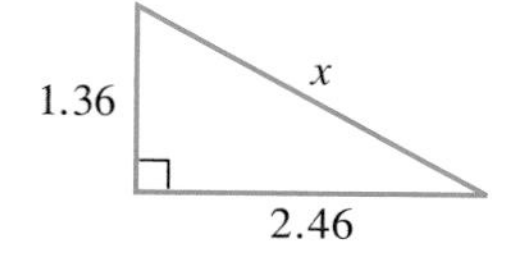

Figure 1

In calculator exercises that ask you to round your answers, it's important that you postpone rounding until the final calculation is carried out. For example, suppose that you are required to determine the hypotenuse x of the right triangle in Figure 1 to two significant digits. Using the Pythagorean theorem, we have

$$\begin{aligned} x &= \sqrt{(1.36)^2 + (2.46)^2} \\ &= 2.8 \quad \text{using a calculator and rounding the final result to two significant digits} \end{aligned}$$

On the other hand, if we first round each of the given lengths to two significant digits, we obtain

$$\begin{aligned} x &= \sqrt{(1.4)^2 + (2.5)^2} \\ &= 2.9 \quad \text{to two significant digits} \end{aligned}$$

This last result is inappropriate, and we can see why as follows. As Table 4 shows, the maximum possible values for the sides are 1.365 and 2.465, respectively.

TABLE 4

Number	Range of measurement
1.36	[1.355, 1.365]
2.46	[2.455, 2.465]

Thus the maximum possible value for the hypotenuse must be

$$\sqrt{(1.365)^2 + (2.465)^2} = 2.817\ldots \quad \text{calculator display}$$
$$= 2.8 \quad \text{to two significant digits}$$

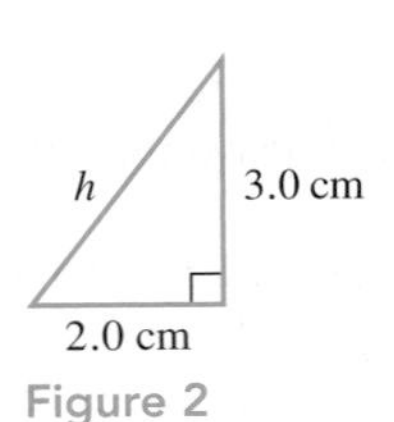

Figure 2

This shows that the value 2.9 is indeed inappropriate, as we stated previously.

An error that is often made by people working with calculators and approximations is to report a final answer with a greater degree of accuracy than the data warrant. Consider, for example, the right triangle in Figure 2. Using the Pythagorean theorem and a calculator with an eight-digit display, we obtain

$$h = 3.6055513 \text{ cm}$$

This value for h is inappropriate, because common sense tells us that the answer should be no more accurate than the data used to obtain that answer. In particular, since the given sides of the triangle apparently were measured only to the nearest tenth of a centimeter, we certainly should not expect any improvement in accuracy for the resulting value of the hypotenuse. An appropriate form for the value of h here would be $h = 3.6$ cm. In general, for calculator exercises in this text that do not specify a required number of decimal places or significant digits in the final results, you should use the following guidelines.

Guidelines for Computing with Approximations

1. *For adding and subtracting:* Round the final result so that it contains only as many decimal places as there are in the data with the fewest decimal places.
2. *For multiplying and dividing:* Round the final result so that it contains only as many significant digits as there are in the data with the fewest significant digits.
3. *For powers and roots:* In computing a power or a root of a real number b, round the result so that it contains as many significant digits as there are in b.

A.2 PROPERTIES OF THE REAL NUMBERS

I do not like × as a symbol for multiplication, as it is easily confounded with x; . . . often I simply relate two quantities by an interposed dot. . . . —G. W. Leibniz in a letter dated July 29, 1698

Today's familiar plus and minus signs were first used in 15th-century Germany as warehouse marks. They indicated when a container held something that weighed over or under a certain standard weight.
—Martin Gardner in "Mathematical Games" (*Scientific American,* June 1977)

In this section we will first list the basic properties for the real number system. Then we will use those properties to prove the familiar rules of algebra for working with signed numbers.

The set of real numbers is **closed** with respect to the operations of addition and multiplication. This means that when we add or multiply two real numbers, the result (that is, the **sum** or the **product**) is again a real number. Some of the other most basic properties and definitions for the real-number system are listed in the following box. In the box, the lowercase letters a, b, and c denote arbitrary real numbers.

PROPERTY SUMMARY Some Fundamental Properties of the Real Numbers

Commutative properties $a + b = b + a$ $ab = ba$

Associative properties $a + (b + c) = (a + b) + c$ $a(bc) = (ab)c$

Identity properties

1. There is a unique real number 0 (called **zero** or the **additive identity**) such that

$$a + 0 = a \quad \text{and} \quad 0 + a = a$$

2. There is a unique real number 1 (called **one** or the **multiplicative identity**) such that

$$a \cdot 1 = a \quad \text{and} \quad 1 \cdot a = a$$

Inverse properties

1. For each real number a there is a real number $-a$ (called the **additive inverse** of a or the **opposite** of a) such that

$$a + (-a) = 0 \quad \text{and} \quad (-a) + a = 0$$

2. For each real number $a \neq 0$ there is a real number denoted by $\frac{1}{a}$ (or $1/a$ or a^{-1}) and called the **multiplicative inverse** or **reciprocal** of a, such that

$$a \cdot \frac{1}{a} = 1 \quad \text{and} \quad \frac{1}{a} \cdot a = 1$$

Distributive properties $a(b + c) = ab + ac$ $(b + c)a = ba + ca$

On reading this list of properties for the first time, many students ask the natural question, "Why do we even bother to list such obvious properties?" One reason is that all the other laws of arithmetic and algebra (including the not-so-obvious ones) can be derived from our rather short list. For example, the rule $0 \cdot a = 0$ can be proved by using the distributive property, as can the rule that the product of two negative numbers is a positive number. We now state and prove those properties and several others.

THEOREM

Let a and b be real numbers. Then

(a) $a \cdot 0 = 0$
(b) $-a = (-1)a$
(c) $-(-a) = a$
(d) $a(-b) = -ab$
(e) $(-a)(-b) = ab$

Proof of Part (a)

$$a \cdot 0 = a \cdot (0 + 0) \quad \text{additive identity property}$$
$$a \cdot 0 = a \cdot 0 + a \cdot 0 \quad \text{distributive property}$$

Now since $a \cdot 0$ is a real number, it has an additive inverse, $-(a \cdot 0)$. Adding this to both sides of the last equation, we obtain

$$a \cdot 0 + [-(a \cdot 0)] = (a \cdot 0 + a \cdot 0) + [-(a \cdot 0)]$$
$$a \cdot 0 + [-(a \cdot 0)] = a \cdot 0 + \{a \cdot 0 + [-(a \cdot 0)]\} \quad \text{associative property of addition}$$
$$0 = a \cdot 0 + 0 \quad \text{additive inverse property}$$
$$0 = a \cdot 0 \quad \text{additive identity property}$$

Thus $a \cdot 0 = 0$, as we wished to show.

Proof of Part (b)

$$\begin{aligned} 0 &= 0 \cdot a && \text{using part (a) and the commutative of multiplication} \\ &= [1 + (-1)]a && \text{additive inverse property} \\ &= 1 \cdot a + (-1)a && \text{distributive property} \\ &= a + (-1)a && \text{multiplicative identity property} \end{aligned}$$

Now, by adding $-a$ to both sides of this last equation, we obtain

$$\begin{aligned} -a + 0 &= -a + [a + (-1)a] \\ -a &= (-a + a) + (-1)a && \text{additive identity property and associative property of addition} \\ -a &= 0 + (-1)a && \text{additive inverse property} \\ -a &= (-1)a && \text{additive identity property} \end{aligned}$$

This last equation asserts that $-a = (-1)a$, as we wished to show.

Proof of Part (c)

$$-(-a) + (-a) = 0 \qquad \text{additive inverse property}$$

By adding a to both sides of this last equation, we obtain

$$\begin{aligned} [-(-a) + (-a)] + a &= 0 + a \\ -(-a) + (-a + a) &= a && \text{associative property of addition and additive identity property} \\ -(-a) + 0 &= a && \text{additive inverse property} \\ -(-a) &= a && \text{additive identity property} \end{aligned}$$

This last equation states that $-(-a) = a$, as we wished to show.

Proof of Part (d)

$$\begin{aligned} a(-b) &= a[(-1)b] && \text{using part (b)} \\ &= [a(-1)]b && \text{associative property of multiplication} \\ &= [(-1)a]b && \text{commutative property of multiplication} \\ &= (-1)(ab) && \text{associative property of multiplication} \\ &= -(ab) && \text{using part (b)} \end{aligned}$$

Thus $a(-b) = -ab$, as we wished to show.

Proof of Part (e)

$$\begin{aligned} (-a)(-b) &= -[(-a)b] && \text{using part (d)} \\ &= -[b(-a)] && \text{commutative property of multiplication} \\ &= -[-(ba)] && \text{using part (d)} \\ &= ba && \text{using part (c)} \\ &= ab && \text{commutative property of multiplication} \end{aligned}$$

We've now shown that $(-a)(-b) = ab$, as required.

A.3 $\sqrt{2}$ IS IRRATIONAL

The proof is by reductio ad absurdum, *and* reductio ad absurdum, *which Euclid loved so much, is one of a mathematician's finest weapons.*
—G. H. Hardy (1877–1947)

We will use an *indirect proof* to show that the square root of 2 is an irrational number. The strategy is as follows:

1. We suppose that $\sqrt{2}$ is a rational number.
2. Using (1) and the usual rules of logic and algebra, we derive a contradiction.
3. On the basis of the contradiction in (2), we conclude that the supposition in (1) is untenable; that is, we conclude that $\sqrt{2}$ is irrational.

In carrying out the proof, we'll assume that the following three statements are known:

If x is an even natural number, then $x = 2k$ for some natural number k.

Any rational number can be written in the form a/b, where the integers a and b have no common integral factors other than ± 1. (In other words, any fraction can be reduced to lowest terms.)

If x is a natural number and x^2 is even, then x is even.

Our indirect proof now proceeds as follows. Suppose that $\sqrt{2}$ is a rational number. Then we can write

$$\sqrt{2} = \frac{a}{b} \qquad \text{where } a \text{ and } b \text{ are natural numbers with no common factor other than 1} \tag{1}$$

Since both sides of equation (1) are positive, we can square both sides to obtain the equation

$$2 = \frac{a^2}{b^2}$$

or

$$2b^2 = a^2 \tag{2}$$

Since the left-hand side of equation (2) is an even number, the right-hand side must also be even. But if a^2 is even, then a is even, and so

$$a = 2k \qquad \text{for some natural number } k$$

Using this last equation to substitute for a in equation (2), we have

$$2b^2 = (2k)^2 = 4k^2$$

or

$$b^2 = 2k^2$$

Hence (reasoning as before) b^2 is even, and therefore b is even. But then we have that both b and a are even, contrary to our hypothesis that b and a have no common factor other than 1. We conclude from this that equation (1) cannot hold; that is, there is no rational number a/b such that $\sqrt{2} = a/b$. Thus $\sqrt{2}$ is irrational, as we wished to prove.

B Appendix

B.1 REVIEW OF INTEGER EXPONENTS

I write a^{-1}, a^{-2}, a^{-3}, *etc., for* $\frac{1}{a}, \frac{1}{aa}, \frac{1}{aaa}$, *etc.*—Isaac Newton (June 13, 1676)

In basic algebra you learned the exponential notation a^n, where a is a real number and n is a natural number. Because that definition is basic to all that follows in this and the next two sections of this appendix, we repeat it here.

DEFINITION 1 Base and Exponent

Given a real number a and a natural number n, we define a^n by

$$a^n = \underbrace{a \cdot a \cdot a \cdots a}_{n \text{ factors}}$$

In the expression a^n the number a is the **base,** and n is the **exponent** or **power** to which the base is raised.

EXAMPLE 1 Using algebraic notation

Rewrite each expression using algebraic notation:
(a) x to the fourth power, plus five;
(b) the fourth power of the quantity x plus five;
(c) x plus y, to the fourth power;
(d) x plus y to the fourth power.

SOLUTION
(a) $x^4 + 5$ **(b)** $(x + 5)^4$ **(c)** $(x + y)^4$ **(d)** $x + y^4$

CAUTION Note that parts (c) and (d) of Example 1 differ only in the use of the comma. So for spoken purposes the idea in (c) would be more clearly conveyed by saying "the fourth power of the quantity x plus y." In the case of (d), if you read it aloud to another student or your instructor, chances are you'll be asked, "Do you mean $x + y^4$ or do you mean $(x + y)^4$?"

MORAL Algebra is a precise language; use it carefully. Learn to ask yourself whether what you've written will be interpreted in the manner you intended.

In basic algebra the four properties in the following box are developed for working with exponents that are natural numbers. Each of these properties is a direct consequence of the definition of a^n. For instance, according to the first property, we have $a^2a^3 = a^5$. To verify that this is indeed correct, we note that

$$a^2a^3 = (aa)(aaa) = a^5$$

PROPERTY SUMMARY Properties of Exponents

Property	Examples
1. $a^m a^n = a^{m+n}$	$a^5a^6 = a^{11}$; $(x+1)(x+1)^2 = (x+1)^3$
2. $(a^m)^n = a^{mn}$	$(2^3)^4 = 2^{12}$; $[(x+1)^2]^3 = (x+1)^6$
3. $\dfrac{a^m}{a^n} = \begin{cases} a^{m-n} & \text{if } m > n \\ \dfrac{1}{a^{n-m}} & \text{if } m < n \\ 1 & \text{if } m = n \end{cases}$	$\dfrac{a^6}{a^2} = a^4$; $\dfrac{a^2}{a^6} = \dfrac{1}{a^4}$; $\dfrac{a^5}{a^5} = 1$
4. $(ab)^m = a^m b^m$; $\left(\dfrac{a}{b}\right)^m = \dfrac{a^m}{b^m}$	$(2x^2)^3 = 2^3 \cdot (x^2)^3 = 8x^6$; $\left(\dfrac{x^2}{y^3}\right)^4 = \dfrac{x^8}{y^{12}}$

Of course, we assume that all of these expressions make sense; in particular, we assume that no denominator of a fraction is zero.

Now we want to extend our definition of a^n to allow for exponents that are integers but not necessarily natural numbers. We begin by defining a^0.

DEFINITION 2 Zero Exponent

For any nonzero real number a,

$$a^0 = 1$$

(0^0 is not defined.)

EXAMPLES

(a) $2^0 = 1$

(b) $(-\pi)^0 = 1$

(c) $\left(\dfrac{3}{1 + a^2 + b^2}\right)^0 = 1$

It's easy to see the motivation for defining a^0 to be 1. Assuming that the exponent zero is to have the same properties as exponents that are natural numbers, we can write

$$a^0a^n = a^{0+n}$$

That is,

$$a^0a^n = a^n$$

Now we divide both sides of this last equation by a^n to obtain $a^0 = 1$, which agrees with our definition.

Our next definition (see the box that follows) assigns a meaning to the expression a^{-n} when n is a natural number. Again, it's easy to see the motivation for this definition. We would like to have

$$a^na^{-n} = a^{n+(-n)} = a^0 = 1$$

That is,

$$a^na^{-n} = 1$$

Now we divide both sides of this last equation by a^n to obtain $a^{-n} = 1/a^n$, in agreement with Definition 3.

DEFINITION 3 Negative Exponent

For any nonzero real number a and natural number n,

$$a^{-n} = \frac{1}{a^n}$$

EXAMPLES

(a) $2^{-1} = \frac{1}{2^1} = \frac{1}{2}$

(b) $\left(\frac{1}{10}\right)^{-1} = \frac{1}{(\frac{1}{10})^1} = 10$

(c) $x^{-2} = \frac{1}{x^2}$

(d) $(a^2b)^{-3} = \frac{1}{(a^2b)^3} = \frac{1}{a^6b^3}$

(e) $\frac{1}{2^{-3}} = \frac{1}{1/2^3} = 2^3 = 8$

It can be shown that the four properties of exponents that we listed earlier continue to hold now for all integer exponents. We make use of this fact in the next three examples.

EXAMPLE 2 Using properties of exponents

Simplify the following expression. Write the answer in such a way that only positive exponents appear.

$$(a^2b^3)^2(a^5b)^{-1}$$

SOLUTION

FIRST METHOD

$$(a^2b^3)^2(a^5b)^{-1} = (a^2b^3)^2 \cdot \frac{1}{a^5b} = \frac{a^4b^6}{a^5b} = \frac{b^{6-1}}{a^{5-4}} = \frac{b^5}{a}$$

ALTERNATIVE METHOD

$$(a^2b^3)^2(a^5b)^{-1} = (a^4b^6)(a^{-5}b^{-1}) = a^{4-5}b^{6-1} = a^{-1}b^5 = \frac{b^5}{a} \quad \text{as obtained previously}$$

EXAMPLE 3 Using properties of exponents

Simplify the following expression, writing the answer so that negative exponents are not used.

$$\left(\frac{a^{-5}b^2c^0}{a^3b^{-1}}\right)^3$$

SOLUTION

We show two solutions. The first makes immediate use of the property $(a^m)^n = a^{mn}$. In the second solution we begin by working within the parentheses.

FIRST SOLUTION

$$\left(\frac{a^{-5}b^2c^0}{a^3b^{-1}}\right)^3 = \frac{a^{-15}b^6}{a^9b^{-3}} = \frac{b^{6-(-3)}}{a^{9-(-15)}} = \frac{b^9}{a^{24}}$$

ALTERNATIVE SOLUTION

$$\left(\frac{a^{-5}b^2c^0}{a^3b^{-1}}\right)^3 = \left(\frac{b^3}{a^8}\right)^3 = \frac{b^9}{a^{24}}$$

EXAMPLE 4 Simplifying with variable exponents

Simplify the following expressions, writing the answers so that negative exponents are not used. (Assume that p and q are natural numbers.)

(a) $\dfrac{a^{4p+q}}{a^{p-q}}$ **(b)** $(a^pb^q)^2(a^{3p}b^{-q})^{-1}$

SOLUTION

(a) $$\frac{a^{4p+q}}{a^{p-q}} = a^{4p+q-(p-q)} = a^{4p+q-p+q} = a^{3p+2q}$$

(b) $$(a^pb^q)^2(a^{3p}b^{-q})^{-1} = (a^{2p}b^{2q})(a^{-3p}b^q) = a^{-p}b^{3q} = \frac{b^{3q}}{a^p}$$

EXAMPLE 5 Simplifying numerical quantities involving exponents

Use the properties of exponents to compute the quantity $\dfrac{2^{10}\cdot 3^{13}}{27\cdot 6^{12}}$.

SOLUTION

$$\frac{2^{10}\cdot 3^{13}}{27\cdot 6^{12}} = \frac{2^{10}\cdot 3^{13}}{(3^3)(3\cdot 2)^{12}} = \frac{2^{10}\cdot 3^{13}}{3^3\cdot 3^{12}\cdot 2^{12}} = \frac{2^{10}\cdot 3^{13}}{3^{15}\cdot 2^{12}} = \frac{1}{(3^{15-13})(2^{12-10})} = \frac{1}{3^2\cdot 2^2} = \frac{1}{36}$$

As an application of some of the ideas in this section, we briefly discuss **scientific notation,** which is a convenient form for writing very large or very small numbers. Such numbers occur often in the sciences. For instance, the speed of light in a vacuum is

$$29{,}979{,}000{,}000 \text{ cm/sec}$$

As written, this number would be awkward to work with in calculating. In fact, the number as written cannot even be displayed on some hand-held calculators, because there are too many digits. To write the number 29,979,000,000 (or any positive number) in scientific notation, we express it as a number between 1 and 10, multiplied by an appropriate power of 10. That is, we write it in the form

$$a \times 10^n \qquad \text{where } 1 \leq a < 10 \text{ and } n \text{ is an integer}$$

According to this convention, the number 4.03×10^6 is in scientific notation, but the same quantity written as 40.3×10^5 is not in scientific notation. To convert a given number into scientific notation, we'll rely on the following two-step procedure.

To Express a Number Using Scientific Notation

1. First move the decimal point until it is to the immediate right of the first nonzero digit.
2. Then multiply by 10^n or 10^{-n}, depending on whether the decimal point was moved n places to the left or to the right, respectively.

For example, to express the number 29,979,000,000 in scientific notation, first we move the decimal point 10 places to the left so that it's located between the 2 and the 9, then we multiply by 10^{10}. The result is

$$29{,}979{,}000{,}000 = 2.9979000000 \times 10^{10}$$

or, more simply,

$$29{,}979{,}000{,}000 = 2.9979 \times 10^{10}$$

As additional examples we list the following numbers expressed in both ordinary and scientific notation. For practice, you should verify each conversion for yourself using our two-step procedure.

$$55{,}708 = 5.5708 \times 10^4$$
$$0.000099 = 9.9 \times 10^{-5}$$
$$0.0000002 = 2 \times 10^{-7}$$

All scientific calculators have keys for entering numbers in scientific notation. (Indeed, many calculators will convert numbers into scientific notation for you.) If you have questions about how your own calculator operates with respect to scientific notation, you should consult the user's manual. We'll indicate only one example here, using a Texas Instruments graphing calculator. The keystrokes on other brands are quite similar. Consider the following number expressed in scientific notation: 3.159×10^{20}. (This gargantuan number is the diameter, in miles, of a galaxy at the center of a distant star cluster known as Abell 2029.) The sequence of keystrokes for entering this number on the Texas Instruments calculator is

3.519 [EE] 20

EXERCISE SET B.1 

A

In Exercises 1–4, evaluate each expression using the given value of x.

1. $2x^3 - x + 4;\ x = -2$

2. $1 - x + 2x^2 - 3x^3;\ x = -1$

3. $\dfrac{1 - 2x^2}{1 + 2x^3};\ x = -\dfrac{1}{2}$

4. $\dfrac{1 - (x - 1)^2}{1 + (x - 1)^2};\ x = -1$

In Exercises 5–16, use the properties of exponents to simplify each expression.

5. **(a)** a^3a^{12}
(b) $(a + 1)^3(a + 1)^{12}$
(c) $(a + 1)^{12}(a + 1)^3$

6. **(a)** $(3^2)^3 - (2^3)^2$
(b) $(x^2)^3 - (x^3)^2$
(c) $(x^2)^a - (x^a)^2$

7. **(a)** yy^2y^8
(b) $(y + 1)(y + 1)^2(y + 1)^8$
(c) $[(y + 1)(y + 1)^8]^2$

8. **(a)** $\dfrac{x^{12}}{x^{10}}$
(b) $\dfrac{x^{10}}{x^{12}}$
(c) $\dfrac{2^{10}}{2^{12}}$

9. **(a)** $\dfrac{(x^2 + 3)^{10}}{(x^2 + 3)^9}$
(b) $\dfrac{(x^2 + 3)^9}{(x^2 + 3)^{10}}$
(c) $\dfrac{12^{10}}{12^9}$

10. **(a)** $(y^2y^3)^2$
(b) $y^2(y^3)^2$
(c) $2^m(2^n)^2$

11. **(a)** $\dfrac{t^{15}}{t^9}$
(b) $\dfrac{t^9}{t^{15}}$
(c) $\dfrac{(t^2 + 3)^{15}}{(t^2 + 3)^9}$

12. **(a)** $\dfrac{x^6}{x}$
(b) $\dfrac{x}{x^6}$
(c) $\dfrac{(x^2 + 3x - 2)^6}{x^2 + 3x - 2}$

13. **(a)** $\dfrac{x^6y^{15}}{x^2y^{20}}$
(b) $\dfrac{x^2y^{20}}{x^6y^{15}}$
(c) $\left(\dfrac{x^2y^{20}}{x^6y^{15}}\right)^2$

14. **(a)** $(x^2y^3z)^4$
(b) $2(x^2y^3z)^4$
(c) $(2x^2y^3z)^4$

15. **(a)** $4(x^3)^2$
(b) $(4x^3)^2$
(c) $\dfrac{(4x^2)^3}{(4x^3)^2}$

16. **(a)** $2(x - 1)^7 - (x - 1)^7$
(b) $[2(x - 1)]^7 - (x - 1)^7$
(c) $2(x - 1)^7 - [2(x - 1)]^7$

For Exercises 17–38, simplify each expression. Write the answers in such a way that negative exponents do not appear. In Exercises 35–38, assume that the letters p and q represent natural numbers.

17. **(a)** 64^0
(b) $(64^3)^0$
(c) $(64^0)^3$

18. **(a)** $2^0 + 3^0$
(b) $(2^0 + 3^0)^3$
(c) $(2^3 + 3^3)^0$

19. **(a)** $10^{-1} + 10^{-2}$
(b) $(10^{-1} + 10^{-2})^{-1}$
(c) $[(10^{-1})(10^{-2})]^{-1}$

20. **(a)** $4^{-2} + 4^{-1}$
(b) $[(\frac{1}{4})^{-1} + (\frac{1}{4})^{-2}]^{-1}$
(c) $[(\frac{1}{4})^{-1}(\frac{1}{4})^{-2}]^{-1}$

21. $(\frac{1}{3})^{-1} + (\frac{1}{4})^{-1}$

22. $(\frac{1}{3} + \frac{1}{4})^{-1}$

23. **(a)** $5^{-2} + 10^{-2}$
(b) $(5 + 10)^{-2}$

24. **(a)** $(\frac{1}{4})^{-2} + (\frac{3}{4})^{-2}$
(b) $(\frac{1}{4} + \frac{3}{4})^{-2}$

25. $(a^2bc^0)^{-3}$

26. $(a^3b)^3(a^2b^4)^{-1}$

27. $(a^{-2}b^{-1}c^3)^{-2}$

28. $(2^{-2} + 2^{-1} + 2^0)^{-2}$

29. $\left(\dfrac{x^3y^{-2}z}{xy^2z^{-3}}\right)^{-3}$

30. $\left(\dfrac{x^4y^{-8}z^2}{xy^2z^{-6}}\right)^2$

31. $\left(\dfrac{x^4y^{-8}z^2}{xy^2z^{-6}}\right)^{-2}$

32. $\left(\dfrac{a^3b^{-9}c^2}{a^5b^2c^{-4}}\right)^0$

33. $\dfrac{x^2}{y^{-3}} \div \dfrac{x^2}{y^3}$

34. $(2x^2)^{-3} - 2(x^2)^{-3}$

35. $\dfrac{b^{p+1}}{b^p}$

36. $\dfrac{b^{p+2q}}{(b^2)^q}$

37. $(x^p)(x^p)$

38. $(2^{p-q})(2^{q-p+1})$

In Exercises 39–42, use the properties of exponents in computing each quantity, as in Example 5. (The point here is to do as little arithmetic as possible.)

39. $\dfrac{2^8 \cdot 3^{15}}{9 \cdot 3^{10} \cdot 12}$

40. $\dfrac{2^{12} \cdot 5^{13}}{10^{12}}$

41. $\dfrac{24^5}{32 \cdot 12^4}$

42. $\left(\dfrac{144 \cdot 125}{2^3 \cdot 3^2}\right)^{-1}$

For Exercises 43–50, express each number in scientific notation.

43. The average distance (in miles) from Earth to the Sun:

92,900,000

44. The average distance (in miles) from the planet Pluto to the Sun:

3,666,000,000

45. The average orbital speed (in miles per hour) of Earth:

66,800

46. The average orbital speed (in miles per hour) of the planet Mercury:

107,300

47. The average distance (in miles) from the Sun to the nearest star:

25,000,000,000,000,000,000

48. The equatorial diameter (in miles) of
(a) Mercury: 3031
(b) Earth: 7927
(c) Jupiter: 88,733
(d) the Sun: 865,000

49. The time (in seconds) for light to travel

(a) one foot:

0.000000001

(b) across an atom:

0.000000000000000001

(c) across the nucleus of an atom:

0.000000000000000000000001

50. The mass (in grams) of

(a) a proton:

0.00000000000000000000000167

(b) an electron:

0.000000000000000000000000000911

B.2 REVIEW OF nTH ROOTS

Hindu mathematicians first recognized negative roots, and the two square roots of a positive number, . . . though they were suspicious also.—David Wells in *The Penguin Dictionary of Curious and Interesting Numbers* (Harmondsworth, Middlesex, England: Penguin Books, Ltd., 1986)

In this section we generalize the notion of square root that you studied in elementary algebra. The new idea is that of an nth root; the definition and some basic examples are given in the box that follows.

DEFINITION 1 ***n*th Roots**

Let n be a natural number. If a and b are real numbers and

$$a^n = b$$

then we say that a is an ***n*th root** of b. When $n = 2$ and when $n = 3$, we refer to the roots as **square roots** and **cube roots,** respectively.

EXAMPLES

Both 3 and -3 are square roots of 9 because $3^2 = 9$ and $(-3)^2 = 9$.

Both 2 and -2 are fourth roots of 16 because $2^4 = 16$ and $(-2)^4 = 16$.

2 is a cube root of 8 because $2^3 = 8$.

-3 is a fifth root of -243 because $(-3)^5 = -243$.

As the examples in the box suggest, square roots, fourth roots, and all *even* roots of positive numbers occur in pairs, one positive and one negative. In these cases we use the notation $\sqrt[n]{b}$ to denote the positive, or **principal,** nth root of b. As examples of this notation, we can write

$$\sqrt[4]{81} = 3 \qquad \text{(The principal fourth root of 81 is 3.)}$$

$$\sqrt[4]{81} \neq -3 \qquad \text{(The principal fourth root of 81 is not } -3.)$$

$$-\sqrt[4]{81} = -3 \qquad \text{(The negative of the principal fourth root of 81 is } -3.)$$

The symbol $\sqrt{}$ is called a **radical sign,** and the number within the radical sign is the **radicand.** The natural number n used in the notation $\sqrt[n]{}$ is called the **index** of the radical. For square roots, as you know from basic algebra, we suppress the index and simply write $\sqrt{}$ rather than $\sqrt[2]{}$. So, for example, $\sqrt{25} = 5$.

On the other hand, as we saw in the examples, cube roots, fifth roots, and in fact, all *odd* roots occur singly, not in pairs. In these cases, we again use the notation $\sqrt[n]{b}$ for the nth root. The definition and examples in the following box summarize our discussion up to this point.

DEFINITION 2 **Principal *n*th Root**

1. Let n be a natural number. If a and b are nonnegative real numbers, then

$$\sqrt[n]{b} = a \quad \text{if and only if} \quad b = a^n$$

The number a is the **principal *n*th root** of b.

2. If a and b are negative and n is an odd natural number, then

$$\sqrt[n]{b} = a \quad \text{if and only if} \quad b = a^n$$

EXAMPLES

$\sqrt{25} = 5$; $\sqrt{25} \neq -5$; $-\sqrt{25} = -5$

$\sqrt[4]{\frac{1}{16}} = \frac{1}{2}$; $\sqrt[3]{8} = 2$

$\sqrt[5]{-\frac{1}{32}} = -\frac{1}{2}$

There are five properties of nth roots that are frequently used in simplifying certain expressions. The first four are similar to the properties of square roots that are developed in elementary algebra. For reference we list these properties side by side in the following box. Property 5 is listed here only for the sake of completeness; we'll postpone discussing it until Appendix B.3.

PROPERTY SUMMARY **Properties of *n*th Roots**

Suppose that x and y are real numbers and that m and n are natural numbers. Then each of the following properties holds, provided only that the expressions on both sides of the equation are defined (and so represent real numbers).

	CORRESPONDING PROPERTIES FOR SQUARE ROOTS
1. $(\sqrt[n]{x})^n = x$	$(\sqrt{x})^2 = x$
2. $\sqrt[n]{xy} = \sqrt[n]{x}\sqrt[n]{y}$	$\sqrt{xy} = \sqrt{x}\sqrt{y}$
3. $\sqrt[n]{\frac{x}{y}} = \frac{\sqrt[n]{x}}{\sqrt[n]{y}}$	$\sqrt{\frac{x}{y}} = \frac{\sqrt{x}}{\sqrt{y}}$
4. n even: $\sqrt[n]{x^n} = \lvert x \rvert$	$\sqrt{x^2} = \lvert x \rvert$
n odd: $\sqrt[n]{x^n} = x$	
5. $\sqrt[m]{\sqrt[n]{x}} = \sqrt[mn]{x}$	

Our immediate use for these properties will be in simplifying expressions involving nth roots. In general, we try to factor the expression under the radical so that one factor is the largest perfect nth power that we can find. Then we apply Property 2 or 3. For instance, the expression $\sqrt{72}$ is simplified as follows:

$$\sqrt{72} = \sqrt{(36)(2)} = \sqrt{36}\sqrt{2} = 6\sqrt{2}$$

In this procedure we began by factoring 72 as (36)(2). Note that 36 is the largest factor of 72 that is a perfect square. If we were to begin instead with a different factorization, for example, $72 = (9)(8)$, we could still arrive at the same answer, but it would take longer. (Check this for yourself.) As another example, let's simplify $\sqrt[3]{40}$. First, what (if any) is the largest perfect-cube factor of 40? The first few perfect cubes are

$$1^3 = 1 \qquad 2^3 = 8 \qquad 3^3 = 27 \qquad 4^3 = 64$$

so we see that 8 is a perfect-cube factor of 40, and we write

$$\sqrt[3]{40} = \sqrt[3]{(8)(5)} = \sqrt[3]{8}\sqrt[3]{5} = 2\sqrt[3]{5}$$

EXAMPLE 1 Simplifying square roots

Simplify: **(a)** $\sqrt{12} + \sqrt{75}$; **(b)** $\sqrt{\dfrac{162}{49}}$.

SOLUTION

(a)
$$\begin{aligned}\sqrt{12} + \sqrt{75} &= \sqrt{(4)(3)} + \sqrt{(25)(3)} \\ &= \sqrt{4}\sqrt{3} + \sqrt{25}\sqrt{3} \\ &= 2\sqrt{3} + 5\sqrt{3} = (2 + 5)\sqrt{3} \\ &= 7\sqrt{3}\end{aligned}$$

(b)
$$\sqrt{\frac{162}{49}} = \frac{\sqrt{162}}{\sqrt{49}} = \frac{\sqrt{81}\sqrt{2}}{7} = \frac{9\sqrt{2}}{7}$$

EXAMPLE 2 Simplifying *n*th roots

Simplify: $\sqrt[3]{16} + \sqrt[3]{250} - \sqrt[3]{128}$.

SOLUTION

$$\begin{aligned}\sqrt[3]{16} + \sqrt[3]{250} - \sqrt[3]{128} &= \sqrt[3]{8}\sqrt[3]{2} + \sqrt[3]{125}\sqrt[3]{2} - \sqrt[3]{64}\sqrt[3]{2} \\ &= 2\sqrt[3]{2} + 5\sqrt[3]{2} - 4\sqrt[3]{2} \\ &= 3\sqrt[3]{2}\end{aligned}$$

EXAMPLE 3 Simplifying radicals containing variables

Simplify each of the following expressions by removing the largest possible perfect-square or perfect-cube factor from within the radical:

(a) $\sqrt{8x^2}$;

(b) $\sqrt{8x^2}$, where $x \geq 0$;

(c) $\sqrt{18a^7}$, where $a \geq 0$;

(d) $\sqrt[3]{16y^5}$.

SOLUTION

(a) $\sqrt{8x^2} = \sqrt{(4)(2)(x^2)} = \sqrt{4}\sqrt{2}\sqrt{x^2} = 2\sqrt{2}|x|$

(b)
$$\begin{aligned}\sqrt{8x^2} &= \sqrt{4}\sqrt{2}\sqrt{x^2} \\ &= 2\sqrt{2}x \qquad \sqrt{x^2} = x \text{ because } x \geq 0\end{aligned}$$

(c)
$$\begin{aligned}\sqrt{18a^7} &= \sqrt{(9a^6)(2a)} = \sqrt{9a^6}\sqrt{2a} \\ &= 3a^3\sqrt{2a}\end{aligned}$$

(d) $\sqrt[3]{16y^5} = \sqrt[3]{8y^3}\sqrt[3]{2y^2} = 2y\sqrt[3]{2y^2}$

EXAMPLE 4 Simplifying nth roots containing variables

Simplify each of the following, assuming that a, b, and c are positive:

(a) $\sqrt{8ab^2c^5}$; **(b)** $\sqrt[4]{\dfrac{32a^6b^5}{c^8}}$.

SOLUTION

(a) $$\sqrt{8ab^2c^5} = \sqrt{(4b^2c^4)(2ac)} = \sqrt{4b^2c^4}\sqrt{2ac} = 2bc^2\sqrt{2ac}$$

(b) $$\sqrt[4]{\frac{32a^6b^5}{c^8}} = \frac{\sqrt[4]{32a^6b^5}}{\sqrt[4]{c^8}} = \frac{\sqrt[4]{16a^4b^4}\sqrt[4]{2a^2b}}{c^2} = \frac{2ab\sqrt[4]{2a^2b}}{c^2}$$

In Examples 1 through 4 we used the definitions and properties of nth roots to simplify certain expressions. The box that follows shows some common errors to avoid in working with roots. Use the error box to test yourself: Cover up the columns labeled "Correction" and "Comment," and try to decide for yourself where the errors lie.

ERRORS TO AVOID

Error	Correction	Comment
$\sqrt[4]{16} = \pm 2$	$\sqrt[4]{16} = 2$	Although -2 is one of the fourth roots of 16, it is not the *principal* fourth root. The notation $\sqrt[4]{16}$ is reserved for the principal fourth root.
$\sqrt{25} = -5$	$\sqrt{25} = 5$	Although -5 is one of the square roots of 25, it is not the *principal* square root.
$\sqrt{a+b} = \sqrt{a} + \sqrt{b}$	The expression $\sqrt{a+b}$ cannot be simplified.	The properties of roots differ with respect to addition versus multiplication. For $\sqrt{ab}$ we do have the simplification $\sqrt{ab} = \sqrt{a}\sqrt{b}$ (assuming that a and b are nonnegative).
$\sqrt[3]{a+b} = \sqrt[3]{a} + \sqrt[3]{b}$	The expression $\sqrt[3]{a+b}$ cannot be simplified.	For $\sqrt[3]{ab}$ we do have $\sqrt[3]{ab} = \sqrt[3]{a}\sqrt[3]{b}$.

There are times when it is convenient to rewrite fractions involving radicals in alternative forms. Suppose, for example, that we want to rewrite the fraction $5/\sqrt{3}$ in an equivalent form that does not involve a radical in the denominator. This is called **rationalizing the denominator.** The procedure here is to multiply by 1 in this way:

$$\frac{5}{\sqrt{3}} = \frac{5}{\sqrt{3}} \cdot 1 = \frac{5}{\sqrt{3}} \cdot \frac{\sqrt{3}}{\sqrt{3}} = \frac{5\sqrt{3}}{3}$$

That is, $\dfrac{5}{\sqrt{3}} = \dfrac{5\sqrt{3}}{3}$, as required.

To rationalize a denominator of the form $a + \sqrt{b}$, we need to multiply the fraction not by $\sqrt{b}/\sqrt{b}$, but rather by 1 in the form $(a - \sqrt{b})/(a - \sqrt{b})$. To see why this is necessary, notice that

$$(a + \sqrt{b})\sqrt{b} = a\sqrt{b} + b$$

which still contains a radical, whereas

$$\begin{aligned}(a + \sqrt{b})(a - \sqrt{b}) &= a^2 - a\sqrt{b} + a\sqrt{b} - b \\ &= a^2 - b\end{aligned}$$

which is free of radicals. Note that we multiply $a + \sqrt{b}$ by $a - \sqrt{b}$ to obtain an expression that is free of square roots by taking advantage of the form of a difference of two squares. Similarly, to rationalize a denominator of the form $a - \sqrt{b}$, we multiply the fraction by $(a + \sqrt{b})/(a + \sqrt{b})$. (The quantities $a + \sqrt{b}$ and $a - \sqrt{b}$ are said to be **conjugates** of each other.) The next two examples make use of these ideas.

EXAMPLE 5 Rationalizing a square root denominator

Simplify: $\dfrac{1}{\sqrt{2}} - 3\sqrt{50}$.

SOLUTION

First, we rationalize the denominator in the fraction $1/\sqrt{2}$:

$$\frac{1}{\sqrt{2}} = \frac{1}{\sqrt{2}} \cdot 1 = \frac{1}{\sqrt{2}} \cdot \frac{\sqrt{2}}{\sqrt{2}} = \frac{\sqrt{2}}{2}$$

Next, we simplify the expression $3\sqrt{50}$:

$$3\sqrt{50} = 3\sqrt{(25)(2)} = 3\sqrt{25}\sqrt{2} = (3)(5)\sqrt{2} = 15\sqrt{2}$$

Now, putting things together, we have

$$\begin{aligned}\frac{1}{\sqrt{2}} - 3\sqrt{50} &= \frac{\sqrt{2}}{2} - 15\sqrt{2} = \frac{\sqrt{2}}{2} - \frac{30\sqrt{2}}{2} \\ &= \frac{\sqrt{2} - 30\sqrt{2}}{2} = \frac{(1 - 30)\sqrt{2}}{2} \\ &= \frac{-29\sqrt{2}}{2}\end{aligned}$$

EXAMPLE 6 Using the conjugate to rationalize a denominator

Rationalize the denominator in the expression $\dfrac{4}{2 + \sqrt{3}}$.

SOLUTION

We multiply by 1, writing it as $\dfrac{2 - \sqrt{3}}{2 - \sqrt{3}}$.

$$\frac{4}{2+\sqrt{3}}\cdot 1 = \frac{4}{2+\sqrt{3}}\cdot\frac{2-\sqrt{3}}{2-\sqrt{3}}$$
$$= \frac{4(2-\sqrt{3})}{4-(\sqrt{3})^2} = \frac{4(2-\sqrt{3})}{4-3} = \frac{8-4\sqrt{3}}{1}$$
$$= 8 - 4\sqrt{3}$$

Note: Check for yourself that multiplying the original fraction $\frac{2+\sqrt{3}}{2+\sqrt{3}}$ does *not* eliminate radicals in the denominator.

In the next example we are asked to rationalize the numerator rather than the denominator. This is useful at times in calculus.

EXAMPLE 7 Rationalizing a numerator using the conjugate

Rationalize the *numerator* of: $\frac{\sqrt{x}-\sqrt{3}}{x-3}$, where $x \geq 0, x \neq 3$.

SOLUTION

$$\frac{\sqrt{x}-\sqrt{3}}{x-3}\cdot 1 = \frac{\sqrt{x}-\sqrt{3}}{x-3}\cdot\frac{\sqrt{x}+\sqrt{3}}{\sqrt{x}+\sqrt{3}}$$
$$= \frac{(\sqrt{x})^2-(\sqrt{3})^2}{(x-3)(\sqrt{x}+\sqrt{3})} = \frac{x-3}{(x-3)(\sqrt{x}+\sqrt{3})}$$
$$= \frac{1}{\sqrt{x}+\sqrt{3}}$$

The strategy for rationalizing numerators or denominators involving nth roots is similar to that used for square roots. To rationalize a numerator or a denominator involving an nth root, we multiply the numerator or denominator by a factor that yields a product that itself is a perfect nth power. The next example displays two instances of this.

EXAMPLE 8 Rationalizing denominators containing nth roots

(a) Rationalize the denominator of: $\frac{6}{\sqrt[3]{7}}$.

(b) Rationalize the denominator of: $\frac{ab}{\sqrt[4]{a^2b^3}}$, where $a > 0, b > 0$.

SOLUTION

(a)
$$\frac{6}{\sqrt[3]{7}}\cdot 1 = \frac{6}{\sqrt[3]{7}}\cdot\frac{\sqrt[3]{7^2}}{\sqrt[3]{7^2}}$$
$$= \frac{6\sqrt[3]{7^2}}{\sqrt[3]{7^3}}$$
$$= \frac{6\sqrt[3]{49}}{7} \quad \text{as required}$$

(b) $$\frac{ab}{\sqrt[4]{a^2b^3}} \cdot 1 = \frac{ab}{\sqrt[4]{a^2b^3}} \cdot \frac{\sqrt[4]{a^2b}}{\sqrt[4]{a^2b}}$$
$$= \frac{ab\sqrt[4]{a^2b}}{\sqrt[4]{a^4b^4}} = \frac{ab\sqrt[4]{a^2b}}{ab} = \sqrt[4]{a^2b}$$

Finally, just as we used the difference of squares formula to motivate the method of rationalizing certain two-term denominators containing a square root term, we can use the difference or sum of cubes formula (see Appendix B.4) to rationalize a denominator involving a cube root term. In particular, in the sum of cubes formula $x^3 + y^3 = (x + y)(x^2 - xy + y^2)$, the expression $(x^2 - xy + y^2)$ is said to be the conjugate of the expression $x + y$.

EXAMPLE 9 Using the conjugate to rationalize a denominator involving a cube root

Rationalize the denominator in the expression $\dfrac{1}{\sqrt[3]{a} + \sqrt[3]{b}}$.

SOLUTION

If we let $x = \sqrt[3]{a}$ and $y = \sqrt[3]{b}$, the conjugate of the denominator $x + y = \sqrt[3]{a} + \sqrt[3]{b}$ is

$$x^2 - xy + y^2 = (\sqrt[3]{a})^2 - \sqrt[3]{a}\sqrt[3]{b} + (\sqrt[3]{b})^2 = \sqrt[3]{a^2} - \sqrt[3]{ab} + \sqrt[3]{b^2}$$

where, for example, $(\sqrt[3]{a})^2 = \sqrt[3]{a}\sqrt[3]{a} = \sqrt[3]{a^2}$. Then

$$\frac{1}{\sqrt[3]{a} + \sqrt[3]{b}} = \frac{1}{\sqrt[3]{a} + \sqrt[3]{b}} \cdot \frac{\sqrt[3]{a^2} - \sqrt[3]{ab} + \sqrt[3]{b^2}}{\sqrt[3]{a^2} - \sqrt[3]{ab} + \sqrt[3]{b^2}}$$
$$= \frac{\sqrt[3]{a^2} - \sqrt[3]{ab} + \sqrt[3]{b^2}}{(\sqrt[3]{a})^3 + (\sqrt[3]{b})^3} = \frac{\sqrt[3]{a^2} - \sqrt[3]{ab} + \sqrt[3]{b^2}}{a + b}$$

EXERCISE SET B.2 

A

In Exercises 1–8, determine whether each statement is TRUE *or* FALSE.

1. $\sqrt{81} = -9$
2. $\sqrt{256} = 16$
3. $-\sqrt{100} = -10$
4. $\sqrt{49} = -7$
5. $\sqrt{4/3} = 2/\sqrt{3}$
6. $\sqrt{10 + 6} = \sqrt{10} + \sqrt{6}$
7. $(\sqrt[3]{10})^3 = 10$
8. $\sqrt{(-5)^2} = -5$

For Exercises 9–18, evaluate each expression. If the expression is undefined (that is, does not represent a real number), say so.

9. **(a)** $\sqrt[3]{-64}$ **(b)** $\sqrt[4]{-64}$
10. **(a)** $\sqrt[5]{32}$ **(b)** $\sqrt[5]{-32}$
11. **(a)** $\sqrt[3]{8/125}$ **(b)** $\sqrt[3]{-8/125}$
12. **(a)** $\sqrt[3]{-1/1000}$ **(b)** $\sqrt[6]{-1/1000}$
13. **(a)** $\sqrt{-16}$ **(b)** $\sqrt[4]{-16}$
14. **(a)** $-\sqrt[4]{16}$ **(b)** $-\sqrt[4]{-16}$
15. **(a)** $\sqrt[4]{256/81}$ **(b)** $\sqrt[3]{-27/125}$
16. **(a)** $\sqrt[6]{64}$ **(b)** $\sqrt[6]{-64}$
17. **(a)** $\sqrt[5]{-32}$ **(b)** $-\sqrt[5]{-32}$
18. **(a)** $\sqrt[4]{(-10)^4}$ **(b)** $\sqrt[3]{(-10)^3}$

In Exercises 19–44, simplify each expression. Unless otherwise specified, assume that all letters in Exercises 35–44 represent positive numbers.

19. **(a)** $\sqrt{18}$ **(b)** $\sqrt[3]{54}$
20. **(a)** $\sqrt{150}$ **(b)** $\sqrt[3]{375}$
21. **(a)** $\sqrt{98}$ **(b)** $\sqrt[5]{-64}$
22. **(a)** $\sqrt{27}$ **(b)** $\sqrt[3]{-108}$

23. **(a)** $\sqrt{25/4}$
(b) $\sqrt[4]{16/625}$

24. **(a)** $\sqrt{225/49}$
(b) $\sqrt[5]{-256/243}$

25. **(a)** $\sqrt{2} + \sqrt{8}$
(b) $\sqrt[3]{2} + \sqrt[3]{16}$

26. **(a)** $4\sqrt{3} - 2\sqrt{27}$
(b) $2\sqrt[3]{81} + 3\sqrt[3]{24}$

27. **(a)** $4\sqrt{50} - 3\sqrt{128}$
(b) $\sqrt[4]{32} + \sqrt[4]{162}$

28. **(a)** $\sqrt{3} - \sqrt{12} + \sqrt{48}$
(b) $\sqrt[5]{-2} + \sqrt[5]{-64} - \sqrt[5]{486}$

29. **(a)** $\sqrt{0.09}$
(b) $\sqrt[3]{0.008}$

30. **(a)** $\sqrt[5]{-2} + \sqrt[5]{2}$
(b) $\sqrt{81/121} - \sqrt[3]{-8/1331}$

31. $4\sqrt{24} - 8\sqrt{54} + 2\sqrt{6}$

32. $\sqrt[3]{192} + \sqrt[3]{-81} + \sqrt{\sqrt[3]{9}}$

33. $\sqrt{\sqrt{64}}$

34. $\sqrt[3]{\sqrt{4096}}$

35. **(a)** $\sqrt{36x^2}$
(b) $\sqrt{36y^2}$, where $y < 0$

36. **(a)** $\sqrt{225x^4y^3}$
(b) $\sqrt[4]{16a^4}$, where $a < 0$

37. **(a)** $\sqrt{ab^2}\sqrt{a^2b}$
(b) $\sqrt{ab^3}\sqrt{a^3b}$

38. **(a)** $\sqrt[3]{125x^6}$
(b) $\sqrt[4]{64y^4}$, where $y < 0$

39. $\sqrt{72a^3b^4c^5}$

40. $\sqrt{(a+b)^5/(16a^2b^2)}$

41. $\sqrt[4]{16a^4b^5}$

42. $\sqrt[3]{8a^4b^6}$

43. $\sqrt[3]{16a^{12}b^2/c^9}$

44. $\sqrt[4]{ab^3}\sqrt[4]{a^3b}$

In Exercises 45–70, rationalize the denominators and simplify where possible. (Assume that all letters represent positive quantities.)

45. $4/\sqrt{7}$

46. $3/\sqrt{3}$

47. $1/\sqrt{8}$

48. $\sqrt{2}/\sqrt{3}$

49. $1/(1 + \sqrt{5})$

50. $\sqrt{2}/(1 - \sqrt{2})$

51. $(1 + \sqrt{3})/(1 - \sqrt{3})$

52. $(\sqrt{a} + \sqrt{b})/(\sqrt{a} - \sqrt{b})$

53. $(1/\sqrt{5}) + 4\sqrt{45}$

54. $(3/\sqrt{8}) - \sqrt{450}$

55. $1/\sqrt[3]{25}$

56. $4/\sqrt[3]{16}$

57. $3/\sqrt[4]{3}$

58. $\sqrt[3]{5}/\sqrt[3]{6}$

59. $3/\sqrt[5]{16a^4b^9}$

60. $3/\sqrt[4]{27a^5b^{11}}$

61. **(a)** $2/(\sqrt{3} + 1)$
(b) $2/(\sqrt{3} - 1)$

62. **(a)** $1/(2\sqrt{5} + 1)$
(b) $1/(2\sqrt{5} - 1)$

63. **(a)** $x/(\sqrt{x} - 2)$
(b) $x/(\sqrt{x} - y)$

64. **(a)** $\sqrt{x}/(5 + \sqrt{x})$
(b) $\sqrt{x}/(\sqrt{a} + \sqrt{x})$

65. **(a)** $(\sqrt{x} + \sqrt{2})/(\sqrt{x} - \sqrt{2})$
(b) $(\sqrt{x} + \sqrt{a})/(\sqrt{x} - \sqrt{a})$

66. **(a)** $(\sqrt{3} + 2\sqrt{y})/(\sqrt{3} - 2\sqrt{y})$
(b) $(\sqrt{x} - 2\sqrt{y})/(\sqrt{x} + 2\sqrt{y})$

67. $-2/(\sqrt{x+h} - \sqrt{x})$

68. $1/(\sqrt{x+h} - \sqrt{x-h})$

69. $1/(\sqrt[3]{a} - 1)$

70. $1/(\sqrt[3]{a} - \sqrt[3]{b})$

In Exercises 71–76, rationalize the numerator.

71. $(\sqrt{x} - \sqrt{5})/(x - 5)$

72. $(2\sqrt{y} + 3)/(4y - 9)$

73. $(\sqrt{2+h} + \sqrt{2})/h$

74. $(\sqrt{1+2h} - 1)/(2h)$

75. $(\sqrt{x+h} - \sqrt{x})/h$

76. $(\sqrt{x+2h} - \sqrt{x-2h})/h$

In Exercises 77–80, refer to the following table. The left-hand column of the table lists four errors to avoid in working with expressions containing radicals. In each case, give a numerical example showing that the expressions on each side of the equation are, in general, not equal. Use a calculator as necessary.

Errors to avoid	Numerical example showing that the formula is not, in general, valid
77. $\sqrt{a+b} = \sqrt{a} + \sqrt{b}$	
78. $\sqrt{x^2 + y^2} = x + y$	
79. $\sqrt[3]{u+v} = \sqrt[3]{u} + \sqrt[3]{v}$	
80. $\sqrt[3]{p^3 + q^3} = p + q$	

B

81. **(a)** Use a calculator to evaluate $\sqrt{8 - 2\sqrt{7}}$ and $\sqrt{7} - 1$. What do you observe?
(b) Prove that $\sqrt{8 - 2\sqrt{7}} = \sqrt{7} - 1$. *Hint*: In view of the definition of a principal square root, you need to check that $(\sqrt{7} - 1)^2 = 8 - 2\sqrt{7}$.

82. Use a calculator to provide empirical evidence indicating that both of the following equations may be correct:

$$\frac{2 - \sqrt{3}}{\sqrt{2} - \sqrt{2 - \sqrt{3}}} + \frac{2 + \sqrt{3}}{\sqrt{2} + \sqrt{2 + \sqrt{3}}} = \sqrt{2}$$

$$\sqrt{\sqrt{97.5 - (1/11)}} = \pi$$

The point of this exercise is to remind you that as useful as calculators may be, there is still the need for proofs in mathematics. In fact, it can be shown that the first equation is indeed correct, but the second is not.

B.3 REVIEW OF RATIONAL EXPONENTS

. . . *Nicole Oresme, a bishop in Normandy (about 1323–1382), first conceived the notion of fractional powers, afterwards rediscovered by* [Simon] *Stevin* [1548–1620] . . .—Florian Cajori in *A History of Elementary Mathematics* (New York: Macmillan Co., 1917)

If Descartes . . . had discarded the radical sign altogether and had introduced the notation for fractional as well as integral exponents, . . . it is conceivable that generations upon generations of pupils would have been saved the necessity of mastering the operations with two . . . notations when one alone (the exponential) would have answered all purposes.—Florian Cajori, *A History of Mathematical Notations,* Vol. 1 (La Salle, Ill.: The Open Court Publishing Co., 1928)

We can use the concept of an nth root to give a meaning to fractional exponents that is useful and, at the same time, consistent with our earlier work. First, by way of motivation, suppose that we want to assign a value to $5^{1/3}$. Assuming that the usual properties of exponents continue to apply here, we can write

$$(5^{1/3})^3 = 5^1 = 5$$

That is,

$$(5^{1/3})^3 = 5$$

or

$$5^{1/3} = \sqrt[3]{5}$$

By replacing 5 and 3 with b and n, respectively, we can see that we want to define $b^{1/n}$ to mean $\sqrt[n]{b}$. Also, by thinking of $b^{m/n}$ as $(b^{1/n})^m$, we see that the definition for $b^{m/n}$ ought to be $(\sqrt[n]{b})^m$. These definitions are formalized in the box that follows.

DEFINITION Rational Exponents

1. Let b denote a real number and n a natural number. We define $b^{1/n}$ by
$$b^{1/n} = \sqrt[n]{b}$$
(If n is even, we require that $b \geq 0$.)
2. Let m/n be a rational number reduced to lowest terms. Assume that n is positive and that $\sqrt[n]{b}$ exists. Then
$$b^{m/n} = (\sqrt[n]{b})^m$$
or, equivalently,
$$b^{m/n} = \sqrt[n]{b^m}$$

EXAMPLES

$4^{1/2} = \sqrt{4} = 2$

$(-8)^{1/3} = \sqrt[3]{-8} = -2$

$8^{2/3} = (\sqrt[3]{8})^2 = 2^2 = 4$

or, equivalently,

$8^{2/3} = \sqrt[3]{8^2} = \sqrt[3]{64} = 4$

It can be shown that the four properties of exponents that we listed in Appendix B.1 (on page A-9) continue to hold for rational exponents in general. In fact, we'll take this for granted rather than follow the lengthy argument needed for its verification. We will also assume that these properties apply to irrational exponents. For instance, we have

$$(2^{\sqrt{5}})^{\sqrt{5}} = 2^5 = 32$$

(The definition of irrational exponents is discussed in Section 5.1.) In the next three examples we display the basic techniques for working with rational exponents.

EXAMPLE 1 Evaluating rational exponents

Simplify each of the following quantities. Express the answers using positive exponents. If an expression does not represent a real number, say so.

(a) $49^{1/2}$ **(b)** $-49^{1/2}$ **(c)** $(-49)^{1/2}$ **(d)** $49^{-1/2}$

SOLUTION

(a) $49^{1/2} = \sqrt{49} = 7$

(b) $-49^{1/2} = -(49^{1/2}) = -\sqrt{49} = -7$

(c) The quantity $(-49)^{1/2}$ does not represent a real number because there is no real number x such that $x^2 = -49$.

(d) $49^{-1/2} = \sqrt{49^{-1}} = \sqrt{1/49} = \dfrac{\sqrt{1}}{\sqrt{49}} = \dfrac{1}{7}$

Alternatively, we have

$$49^{-1/2} = (49^{1/2})^{-1} = 7^{-1} = \frac{1}{7}$$

EXAMPLE 2 Simplifying expressions containing rational exponents

Simplify each of the following. Write the answers using positive exponents. (Assume that $a > 0$.)

(a) $(5a^{2/3})(4a^{3/4})$ **(b)** $\sqrt[5]{\dfrac{16a^{1/3}}{a^{1/4}}}$ **(c)** $(x^2 + 1)^{1/5}(x^2 + 1)^{4/5}$

SOLUTION

(a)
$$\begin{aligned}(5a^{2/3})(4a^{3/4}) &= 20a^{(2/3)+(3/4)} \\ &= 20a^{17/12} \qquad \text{because } \tfrac{2}{3} + \tfrac{3}{4} = \tfrac{17}{12}\end{aligned}$$

(b)
$$\begin{aligned}\sqrt[5]{\frac{16a^{1/3}}{a^{1/4}}} &= \left(\frac{16a^{1/3}}{a^{1/4}}\right)^{1/5} \\ &= (16a^{1/12})^{1/5} \qquad \text{because } \tfrac{1}{3} - \tfrac{1}{4} = \tfrac{1}{12} \\ &= 16^{1/5}a^{1/60}\end{aligned}$$

(c) $(x^2 + 1)^{1/5}(x^2 + 1)^{4/5} = (x^2 + 1)^1 = x^2 + 1$

EXAMPLE 3 Evaluating rational exponents

Simplify: **(a)** $32^{-2/5}$; **(b)** $(-8)^{4/3}$.

SOLUTION

(a) $32^{-2/5} = (\sqrt[5]{32})^{-2} = 2^{-2} = \dfrac{1}{2^2} = \dfrac{1}{4}$

Alternatively, we have

$$32^{-2/5} = (2^5)^{-2/5} = 2^{-2} = \frac{1}{2^2} = \frac{1}{4}$$

(b) $(-8)^{4/3} = (\sqrt[3]{-8})^4 = (-2)^4 = 16$

Alternatively, we can write

$$(-8)^{4/3} = [(-2)^3]^{4/3} = (-2)^4 = 16$$

Rational exponents can be used to simplify certain expressions containing radicals. For example, one of the properties of nth roots that we listed but did not discuss in Appendix B.2 is $\sqrt[m]{\sqrt[n]{x}} = \sqrt[mn]{x}$. Using exponents, it is easy to verify this property. We have

$$\sqrt[m]{\sqrt[n]{x}} = (x^{1/n})^{1/m}$$
$$= x^{1/mm} = \sqrt[mn]{x} \qquad \text{as we wished to show}$$

EXAMPLE 4 Using rational exponents to combine radicals

Consider the expression $\sqrt{x}\sqrt[3]{y^2}$, where x and y are positive.
(a) Rewrite the expression using rational exponents.
(b) Rewrite the expression using only one radical sign.

SOLUTION

(a) $\sqrt{x}\sqrt[3]{y^2} = x^{1/2}y^{2/3}$

(b) $\sqrt{x}\sqrt[3]{y^2} = x^{1/2}y^{2/3}$

$= x^{3/6}y^{4/6}$ rewriting the fractions using a common denominator

$= \sqrt[6]{x^3}\sqrt[6]{y^4} = \sqrt[6]{x^3y^4}$

EXAMPLE 5 Using rational exponents to simplify a radical expression

Rewrite the following expression using rational exponents (assume that x, y, and z are positive):

$$\sqrt{\frac{\sqrt[3]{x}\sqrt[4]{y^3}}{\sqrt[5]{z^4}}}$$

SOLUTION

$$\sqrt{\frac{\sqrt[3]{x}\sqrt[4]{y^3}}{\sqrt[5]{z^4}}} = \left(\frac{\sqrt[3]{x}\sqrt[4]{y^3}}{\sqrt[5]{z^4}}\right)^{1/2} = \left(\frac{x^{1/3}y^{3/4}}{z^{4/5}}\right)^{1/2}$$

$$= \frac{x^{1/6}y^{3/8}}{z^{2/5}} = x^{1/6}y^{3/8}z^{-2/5}$$

EXERCISE SET B.3

A

Exercises 1–14 are warm-up exercises to help you become familiar with the definition $b^{m/n} = (\sqrt[n]{b})^m = \sqrt[n]{b^m}$. In Exercises 1–8, write the expression in the two equivalent forms $(\sqrt[n]{b})^m$ and $\sqrt[n]{b^m}$. In Exercises 9–14, write the expression in the form $b^{m/n}$.

1. $a^{3/5}$ **2.** $x^{3/7}$ **3.** $5^{2/3}$
4. $10^{4/5}$ **5.** $(x^2 + 1)^{3/4}$ **6.** $(a + b)^{7/10}$
7. $2^{xy/3}$ **8.** $2^{(a+1)/b}$ **9.** $\sqrt[3]{p^2}$
10. $\sqrt[5]{R^3}$ **11.** $\sqrt[7]{(1 + u)^4}$ **12.** $\sqrt[6]{(1 + u)^5}$
13. $\sqrt[p]{(a^2 + b^2)^3}$ **14.** $\sqrt[3p]{(a^2 + b^2)^{2p}}$

In Exercises 15–50, evaluate or simplify each expression. Express the answers using positive exponents. If an expression is undefined (that is, does not represent a real number), say so. Assume that all letters represent positive numbers.

15. $16^{1/2}$ **16.** $100^{1/2}$ **17.** $(1/36)^{1/2}$
18. $0.09^{1/2}$ **19.** $(-16)^{1/2}$ **20.** $(-1)^{1/2}$
21. $625^{1/4}$ **22.** $(1/81)^{1/4}$ **23.** $8^{1/3}$
24. $0.001^{1/3}$ **25.** $8^{2/3}$ **26.** $64^{2/3}$
27. $(-32)^{1/5}$ **28.** $(-1/125)^{1/3}$ **29.** $(-1000)^{1/3}$
30. $(-243)^{1/5}$ **31.** $49^{-1/2}$ **32.** $121^{-1/2}$
33. $(-49)^{-1/2}$ **34.** $(-64)^{-1/3}$ **35.** $36^{-3/2}$

36. $(-0.001)^{-2/3}$ **37.** $125^{2/3}$ **38.** $125^{-2/3}$
39. $(-1)^{3/5}$ **40.** $27^{4/3} + 27^{-4/3} + 27^0$
41. $32^{4/5} - 32^{-4/5}$ **42.** $64^{1/2} + 64^{-1/2} - 64^{4/3}$
43. $(9/16)^{-5/2} - (1000/27)^{4/3}$ **44.** $(256^{-3/4})^{4/3}$
45. $(2a^{1/3})(3a^{1/4})$ **46.** $\sqrt[5]{3a^4}$
47. $\sqrt[4]{64a^{2/3}/a^{1/3}}$ **48.** $(x^2 + 1)^{2/3}(x^2 + 1)^{4/3}$
49. $\dfrac{(x^2 + 1)^{3/4}}{(x^2 + 1)^{-1/4}}$
50. $\dfrac{(2x^2 + 1)^{-6/5}(2x^2 + 1)^{6/5}(x^2 + 1)^{-1/5}}{(x^2 + 1)^{9/5}}$

For Exercises 51–58, follow Example 4 in the text to rewrite each expression in two ways: **(a)** *using rational exponents and* **(b)** *using only one radical sign. (Assume that x, y, and z are positive.)*

51. $\sqrt{3}\sqrt[3]{6}$ **52.** $\sqrt{5}\sqrt[3]{7}$
53. $\sqrt[3]{6}\sqrt[4]{2}$ **54.** $\sqrt[3]{2}\sqrt[5]{2}$
55. $\sqrt[3]{x^2}\sqrt[5]{y^4}$ **56.** $\sqrt{x}\sqrt[3]{y}\sqrt[4]{z}$
57. $\sqrt[4]{x^a}\sqrt[3]{x^b}\sqrt{x^{a/6}}$ **58.** $\sqrt[3]{27\sqrt{64x}}$

In Exercises 59–66, rewrite each expression using rational exponents rather than radicals. Assume that x, y, and z are positive.

59. $\sqrt[3]{(x + 1)^2}$ **60.** $(1/\sqrt{x}) + \sqrt{x}$
61. $(\sqrt[5]{x + y})^2$ **62.** $\sqrt{\sqrt{x}}$
63. $\sqrt[3]{\sqrt{x}} + \sqrt{\sqrt[3]{x}}$ **64.** $\sqrt[3]{\sqrt{2}}$
65. $\sqrt{\sqrt[3]{x}\sqrt[4]{y}}$ **66.** $\sqrt[5]{\sqrt{x}\sqrt[3]{y}/\sqrt[4]{z^2}}$
67. Use a calculator to determine which is larger, $9^{10/9}$ or $10^{9/10}$.
68. Use a calculator to determine which number is closer to 3: $13^{3/7}$ or $6560^{1/8}$.

In Exercises 69–74, refer to the following table. The left-hand column of the table lists six errors to avoid in working with expressions containing fractional exponents. In each case, give a numerical example showing that the expressions on each side of the equation are not equal. Use a calculator as necessary.

Errors to avoid	Numerical example showing that the formula is not, in general, valid
69. $(a + b)^{1/2} = a^{1/2} + b^{1/2}$	
70. $(x^2 + y^2)^{1/2} = x + y$	
71. $(u + v)^{1/3} = u^{1/3} + v^{1/3}$	
72. $(p^3 + q^3)^{1/3} = p + q$	
73. $x^{1/m} = \dfrac{1}{x^m}$	
74. $x^{-1/2} = \dfrac{1}{x^2}$	

B.4 REVIEW OF FACTORING

fac•tor *(fak′ter),* n. . . . **2.** Math. *one of two or more numbers, algebraic expressions, or the like, that when multiplied together produce a given product;* . . . v.t. **10.** Math. *to express (a mathematical quantity) as a product of two or more quantities of like kind, as* $30 = 2 \cdot 3 \cdot 5$, *or* $x^2 - y^2 = (x + y)(x - y)$.—*The Random House Dictionary of the English Language*, 2nd ed. (New York: Random House, 1987)

There are many cases in algebra in which the process of *factoring* simplifies the work at hand. To **factor** a polynomial means to write it as a product of two or more nonconstant polynomials. For instance, a factorization of $x^2 - 9$ is given by

$$x^2 - 9 = (x - 3)(x + 3)$$

In this case, $x - 3$ and $x + 3$ are the **factors** of $x^2 - 9$.

There is one convention that we need to agree on at the outset. If the polynomial or expression that we wish to factor contains only integer coefficients, then the factors (if any) should involve only integer coefficients. For example, according to this convention, we will not consider the following type of factorization in this section:

$$x^2 - 2 = (x - \sqrt{2})(x + \sqrt{2})$$

because it involves coefficients that are irrational numbers. (We should point out, however, that factorizations such as this are useful at times, particularly in calculus.) As it happens, $x^2 - 2$ is an example of a polynomial that cannot be factored using integer coefficients. In such a case we say that the polynomial is **irreducible over the integers.**

If the coefficients of a polynomial are rational numbers, then we do allow factors with rational coefficients. For instance, the factorization of $y^2 - \frac{1}{4}$ over the rational numbers is given by

$$y^2 - \frac{1}{4} = \left(y - \frac{1}{2}\right)\left(y + \frac{1}{2}\right)$$

We'll consider five techniques for factoring in this section. These techniques will be applied in the next two sections and throughout the text. In Table 1 we sum-

TABLE 1 Basic Factoring Techniques

Technique	Example or formula	Remark
Common factor	$3x^4 + 6x^3 - 12x^2 = 3x^2(x^2 + 2x - 4)$ $4(x^2 + 1) - x(x^2 + 1) = (x^2 + 1)(4 - x)$	In any factoring problem, the first step always is to look for the common factor of highest degree.
Difference of squares	$x^2 - a^2 = (x - a)(x + a)$	There is no corresponding formula for a sum of squares; $x^2 + a^2$ is irreducible over the integers.
Trial and error	$x^2 + 2x - 3 = (x + 3)(x - 1)$	In this example the only possibilities, or trials, are (a) $(x - 3)(x - 1)$ (c) $(x + 3)(x - 1)$ (b) $(x - 3)(x + 1)$ (d) $(x + 3)(x + 1)$ By inspection or by carrying out the indicated multiplications, we find that only case (c) checks.
Difference of cubes Sum of cubes	$x^3 - a^3 = (x - a)(x^2 + ax + a^2)$ $x^3 + a^3 = (x + a)(x^2 - ax + a^2)$	Verify these formulas for yourself by carrying out the multiplications. Then memorize the formulas.
Grouping	$x^3 - x^2 + x - 1 = (x^3 - x^2) + (x - 1)$ $= x^2(x - 1) + (x - 1) \cdot 1$ $= (x - 1)(x^2 + 1)$	This is actually an application of the common factor technique.

marize these techniques. Notice that three of the formulas in the table are just restatements of Special Products formulas. Remember that it is the form or pattern in the formula that is important, not the specific choice of letters.

The idea in factoring is to use one or more of these techniques until each of the factors obtained is irreducible. The examples that follow show how this works in practice.

EXAMPLE 1 Factoring using the difference-of-squares technique

Factor: **(a)** $x^2 - 49$; **(b)** $(2a - 3b)^2 - 49$.

SOLUTION

(a) $x^2 - 49 = x^2 - 7^2$

$= (x - 7)(x + 7)$ difference of squares

(b) Notice that the pattern or form is the same as in part (a): It's a difference of squares. We have

$$(2a - 3b)^2 - 49 = (2a - 3b)^2 - 7^2$$
$$= [(2a - 3b) - 7][(2a - 3b) + 7] \quad \text{difference of squares}$$
$$= (2a - 3b - 7)(2a - 3b + 7)$$

EXAMPLE 2 Factoring using common factor and difference of squares

Factor: **(a)** $2x^3 - 50x$; **(b)** $3x^5 - 3x$.

SOLUTION

(a) $2x^3 - 50x = 2x(x^2 - 25)$ common factor

$= 2x(x - 5)(x + 5)$ difference of squares

(b)
$$\begin{aligned} 3x^5 - 3x &= 3x(x^4 - 1) && \text{common factor} \\ &= 3x[(x^2)^2 - 1^2] \\ &= 3x(x^2 - 1)(x^2 + 1) && \text{difference of squares} \\ &= 3x(x - 1)(x + 1)(x^2 + 1) && \text{difference of squares, again} \end{aligned}$$

EXAMPLE 3 Factoring by trial and error

Factor: **(a)** $x^2 - 4x - 5$; **(b)** $(a + b)^2 - 4(a + b) - 5$.

SOLUTION

(a) $x^2 - 4x - 5 = (x - 5)(x + 1)$ trial and error

(b) Note that the form of this expression is

$$(\quad)^2 - 4(\quad) - 5$$

This is the same *form* as the expression in part (a), so we need only replace x with the quantity $a + b$ in the solution for part (a). This yields

$$\begin{aligned} (a + b)^2 - 4(a + b) - 5 &= [(a + b) - 5][(a + b) + 1] \\ &= (a + b - 5)(a + b + 1) \end{aligned}$$

EXAMPLE 4 Factoring by trial and error

Factor: $2z^4 + 9z^2 + 4$.

SOLUTION

We use trial and error to look for a factorization of the form $(2z^2 + ?)(z^2 + ?)$. There are three possibilities:

(i) $(2z^2 + 2)(z^2 + 2)$ **(ii)** $(2z^2 + 1)(z^2 + 4)$ **(iii)** $(2z^2 + 4)(z^2 + 1)$

Each of these yields the appropriate first term and last term, but (after checking) only possibility (ii) yields $9z^2$ for the middle term. The required factorization is then $2z^4 + 9z^2 + 4 = (2z^2 + 1)(z^2 + 4)$.

Question: Why didn't we consider any possibilities with subtraction signs in place of addition signs?

EXAMPLE 5 Two irreducible expressions

Factor: **(a)** $x^2 + 9$; **(b)** $x^2 + 2x + 3$.

SOLUTION

(a) The expression $x^2 + 9$ is irreducible over the integers. (This can be discovered by trial and error.) If the given expression had instead been $x^2 - 9$, then it could have been factored as a difference of squares. Sums of squares, however, cannot in general be factored over the integers.

(b) The expression $x^2 + 2x + 3$ is irreducible over the integers. (Check this for yourself by trial and error.)

EXAMPLE 6 Factoring using grouping and a special product

Factor: $x^2 - y^2 + 10x + 25$.

SOLUTION
Familiarity with the special product $(a + b)^2 = a^2 + 2ab + b^2$ suggests that we try grouping the terms this way:

$$(x^2 + 10x + 25) - y^2$$

Then we have

$$\begin{aligned} x^2 - y^2 + 10x + 25 &= (x^2 + 10x + 25) - y^2 \\ &= (x + 5)^2 - y^2 \\ &= [(x + 5) - y][(x + 5) + y] \qquad \text{difference of squares} \\ &= (x - y + 5)(x + y + 5) \end{aligned}$$

EXAMPLE 7 Factoring using grouping

Factor: $ax + ay^2 + bx + by^2$.

SOLUTION
We factor a from the first two terms and b from the second two to obtain

$$ax + ay^2 + bx + by^2 = a(x + y^2) + b(x + y^2)$$

We now recognize the quantity $x + y^2$ as a common expression that can be factored out. We then have

$$a(x + y^2) + b(x + y^2) = (x + y^2)(a + b)$$

The required factorization is therefore

$$ax + ay^2 + bx + by^2 = (x + y^2)(a + b)$$

Alternatively,

$$\begin{aligned} &ax + ay^2 + bx + by^2 \\ &\quad = (ax + bx) + (ay^2 + by^2) \\ &\quad = x(a + b) + y^2(a + b) \\ &\quad = (a + b)(x + y^2) \end{aligned}$$

EXAMPLE 8 Factoring sum and difference of cubes

Factor: **(a)** $t^3 - 125$; **(b)** $8 + (a - 2)^3$.

SOLUTION

(a) $$\begin{aligned} t^3 - 125 &= t^3 - 5^3 \\ &= (t - 5)(t^2 + 5t + 25) \qquad \text{difference of cubes} \end{aligned}$$

(b) $$\begin{aligned} 8 + (a - 2)^3 &= 2^3 + (a - 2)^3 \\ &= [2 + (a - 2)][2^2 - 2(a - 2) + (a - 2)^2] \qquad \text{sum of cubes} \end{aligned}$$

$$= a(4 - 2a + 4 + a^2 - 4a + 4)$$
$$= a(a^2 - 6a + 12)$$

As you can check now, the expression $a^2 - 6a + 12$ is irreducible over the integers. Therefore the required factorization is $a(a^2 - 6a + 12)$.

For some calculations (particularly in calculus) it's helpful to be able to factor an expression involving fractional exponents. For instance, suppose that we want to factor the expression

$$x(2x - 1)^{-1/2} + (2x - 1)^{3/2}$$

The common expression to factor out here is $(2x - 1)^{-1/2}$; the technique is to *choose the expression with the smaller exponent.*

EXAMPLE 9 Factoring with rational exponents

Factor: $x(2x - 1)^{-1/2} + (2x - 1)^{3/2}$.

SOLUTION

$$\begin{aligned} x(2x - 1)^{-1/2} + (2x - 1)^{3/2} &= (2x - 1)^{-1/2}[x + (2x - 1)^{3/2+1/2}] \\ &= (2x - 1)^{-1/2}[x + (2x - 1)^2] \\ &= (2x - 1)^{-1/2}(4x^2 - 3x + 1) \end{aligned}$$

We conclude this section with examples of four common errors to avoid in factoring. The first two errors in the box that follows are easy to detect; simply multiplying out the supposed factorizations shows that they do not check. The third error may result from a lack of familiarity with the basic factoring techniques listed on page A-26. The fourth error indicates a misunderstanding of what is required in factoring; the final quantity in a factorization must be expressed as a *product,* not a sum, of terms or expressions.

ERRORS TO AVOID

Error	Correction	Comment
$x^2 + 6x + 9 = (x - 3)^2$	$x^2 + 6x + 9 = (x + 3)^2$	Check the middle term.
$x^2 + 64 = (x + 8)(x + 8)$	$x^2 + 64$ is irreducible over the integers.	A sum of squares is, in general, irreducible over the integers. (A difference of squares, however, can always be factored.)
$x^3 + 64$ is irreducible over the integers	$x^3 + 64 = (x + 4)(x^2 - 4x + 16)$	Although a sum of squares is irreducible, a sum of cubes can be factored.
$x^2 - 2x + 3$ factors as $x(x - 2) + 3$	$x^2 - 2x + 3$ is irreducible over the integers.	The polynomial $x^2 - 2x + 3$ is the *sum* of the expression $x(x - 2)$ and the constant 3. By definition, however, the factored form of a polynomial must be a *product* (of two or more nonconstant polynomials).

EXERCISE SET B.4

A

In Exercises 1–66, factor each polynomial or expression. If a polynomial is irreducible, state this. (In Exercises 1–6 the factoring techniques are specified.)

1. (Common factor and difference of squares)
(a) $x^2 - 64$ **(b)** $7x^4 + 14x^2$ **(c)** $121z - z^3$ **(d)** $a^2b^2 - c^2$

2. (Common factor and difference of squares)
(a) $1 - t^4$ **(b)** $x^6 + x^5 + x^4$ **(c)** $u^2v^2 - 225$ **(d)** $81x^4 - x^2$

3. (Trial and error)
(a) $x^2 + 2x - 3$ **(b)** $x^2 - 2x - 3$ **(c)** $x^2 - 2x + 3$ **(d)** $-x^2 + 2x + 3$

4. (Trial and error)
(a) $2x^2 - 7x - 4$ **(b)** $2x^2 + 7x - 4$ **(c)** $2x^2 + 7x + 4$ **(d)** $-2x^2 - 7x + 4$

5. (Sum and difference of cubes)
(a) $x^3 + 1$ **(b)** $x^3 + 216$ **(c)** $1000 - 8x^6$ **(d)** $64a^3x^3 - 125$

6. (Grouping)
(a) $x^4 - 2x^3 + 3x - 6$
(b) $a^2x + bx - a^2z - bz$

7. (a) $144 - x^2$ **(b)** $144 + x^2$ **(c)** $144 - (y - 3)^2$

8. (a) $4a^2b^2 + 9c^2$ **(b)** $4a^2b^2 - 9c^2$ **(c)** $4a^2b^2 - 9(ab + c)^2$

9. (a) $h^3 - h^5$ **(b)** $100h^3 - h^5$ **(c)** $100(h + 1)^3 - (h + 1)^5$

10. (a) $x^4 - x^2$ **(b)** $3x^4 - 48x^2$ **(c)** $3(x + h)^4 - 48(x + h)^2$

11. (a) $x^2 - 13x + 40$ **(b)** $x^2 - 13x - 40$

12. (a) $x^2 + 10x + 16$ **(b)** $x^2 - 10x + 16$

13. (a) $x^2 + 5x - 36$ **(b)** $x^2 - 13x + 36$

14. (a) $x^2 - x + 6$ **(b)** $x^2 + x - 6$

15. (a) $3x^2 - 22x - 16$ **(b)** $3x^2 - x - 16$

16. (a) $x^2 - 4x + 1$ **(b)** $x^2 - 4x - 3$

17. (a) $6x^2 + 13x - 5$ **(b)** $6x^2 - x - 5$

18. (a) $16x^2 + 18x - 9$ **(b)** $16x^2 - 143x - 9$

19. (a) $t^4 + 2t^2 + 1$ **(b)** $t^4 - 2t^2 + 1$ **(c)** $t^4 - 2t^2 - 1$

20. (a) $t^4 - 9t^2 + 20$ **(b)** $t^4 - 19t^2 + 20$ **(c)** $t^4 - 19t^2 - 20$

21. (a) $4x^3 - 20x^2 - 25x$ **(b)** $4x^3 - 20x^2 + 25x$

22. (a) $x^3 + x^2 + x$ **(b)** $x^3 + 2x^2 + x$

23. (a) $ab - bc + a^2 - ac$
(b) $(u + v)x - xy + (u + v)^2 - (u + v)y$

24. (a) $3(x + 5)^3 + 2(x + 5)^2$
(b) $a(x + 5)^3 + b(x + 5)^2$

25. $x^2z^2 + xzt + xyz + yt$

26. $a^2t^2 + b^2t^2 - cb^2 - ca^2$

27. $a^4 - 4a^2b^2c^2 + 4b^4c^4$

28. $A^2 - B^2 + 16A + 64$

29. $A^2 + B^2$

30. $x^2 + 64$

31. $x^3 + 64$

32. $27 - (a - b)^3$

33. $(x + y)^3 - y^3$

34. $(a + b)^3 - 8c^3$

35. $x^3 - y^3 + x - y$

36. $8a^3 + 27b^3 + 2a + 3b$

37. (a) $p^4 - 1$ **(b)** $p^8 - 1$

38. (a) $p^4 + 4$ **(b)** $p^4 - 4$

39. $x^3 + 3x^2 + 3x + 1$

40. $-1 + 6x - 12x^2 + 8x^3$

41. $x^2 + 16y^2$

42. $4u^2 + 25v^2$

43. $\dfrac{25}{16} - c^2$

44. $\dfrac{81}{4} - y^2$

45. $z^4 - \dfrac{81}{16}$

46. $\dfrac{(a + b)^2}{4} - \dfrac{a^2b^2}{9}$

47. $\dfrac{125}{m^3n^3} - 1$

48. $\dfrac{x^3}{8} - \dfrac{512}{x^3}$

49. $\frac{1}{4}x^2 + xy + y^2$

50. $x^2 + x + 1$

51. $64(x - a)^3 - x + a$

52. $64(x - a)^4 - x + a$

53. $x^2 - a^2 + y^2 - 2xy$

54. $a^4 - (b + c)^4$

55. $21x^3 + 82x^2 - 39x$

56. $x^3a^2 - 8y^3a^2 - 4x^3b^2 + 32y^3b^2$

57. $12xy + 25 - 4x^2 - 9y^2$

58. $ax^2 + (1 + ab)xy + by^2$

59. $ax^2 + (a + b)x + b$

60. $(5a^2 - 11a + 10)^2 - (4a^2 - 15a + 6)^2$

61. $(x + 1)^{1/2} - (x + 1)^{3/2}$

62. $(x^2 + 1)^{3/2} + (x^2 + 1)^{7/2}$

63. $(x + 1)^{-1/2} - (x + 1)^{-3/2}$

64. $(x^2 + 1)^{-2/3} + (x^2 + 1)^{-5/3}$

65. $(2x + 3)^{1/2} - \frac{1}{3}(2x + 3)^{3/2}$

66. $(ax + b)^{-1/2} - \sqrt{ax + b}/b$

In Exercises 67 and 68, evaluate the given expressions using factoring techniques. (The point here is to do as little actual arithmetic as possible.)

67. (a) $100^2 - 99^2$ **(b)** $8^3 - 6^3$ **(c)** $1000^2 - 999^2$

68. (a) $10^3 - 9^3$ **(b)** $50^2 - 49^2$ **(c)** $\dfrac{15^3 - 10^3}{15^2 - 10^2}$

B

In Exercises 69–73, factor each expression.

69. $A^3 + B^3 + 3AB(A + B)$

70. $p^3 - q^3 - p(p^2 - q^2) + q(p - q)^2$

71. $2x(a^2 + x^2)^{-1/2} - x^3(a^2 + x^2)^{-3/2}$

72. $\frac{1}{2}(x - a)^{-1/2}(x + a)^{-1/2} - \frac{1}{2}(x + a)^{1/2}(x - a)^{-3/2}$

73. $y^4 - (p + q)y^3 + (p^2q + pq^2)y - p^2q^2$

74. (a) Factor $x^4 + 2x^2y^2 + y^4$.
(b) Factor $x^4 + x^2y^2 + y^4$. *Hint:* Add and subtract a term. [Keep part (a) in mind.]
(c) Factor $x^6 - y^6$ as a difference of squares.
(d) Factor $x^6 - y^6$ as a difference of cubes. [Use the result in part (b) to obtain the same answer as in part (c).]

B.5 REVIEW OF FRACTIONAL EXPRESSIONS

But if you do use a rule involving mechanical calculation, be patient, accurate, and systematically neat in the working. It is well known to mathematical teachers that quite half the failures in algebraic exercises arise from arithmetical inaccuracy and slovenly arrangement.— George Chrystal (1893)

The rules of basic arithmetic that you learned for working with simple fractions are also used for fractions involving algebraic expressions. In algebra, as in arithmetic, we say that a fraction is *reduced to lowest terms* or *simplified* when the numerator and denominator contain no common factors (other than 1 and −1). The factoring techniques that we reviewed in Appendix B.4 are used to reduce fractions. For example, to reduce the fraction $\frac{x^2-9}{x^2+3x}$, we write

$$\frac{x^2-9}{x^2+3x}=\frac{(x-3)\cancel{(x+3)}}{x\cancel{(x+3)}}=\frac{x-3}{x}$$

In the box that follows, we review two properties of negatives and fractions that will be useful in this section. Notice that Property 2 follows from Property 1 because

$$\frac{a-b}{b-a}=\frac{a-b}{-(a-b)}=-1$$

PROPERTY SUMMARY Negatives and Fractions

Property	Example
1. $b-a=-(a-b)$	$2-x=-(x-2)$
2. $\frac{a-b}{b-a}=-1$	$\frac{2x-5}{5-2x}=-1$

EXAMPLE 1 Using Property 2 to simplify a fractional expression

Simplify: $\frac{4-3x}{15x-20}$.

SOLUTION

$$\begin{aligned}\frac{4-3x}{15x-20}&=\frac{4-3x}{5(3x-4)}\\&=\frac{1}{5}\left(\frac{4-3x}{3x-4}\right)=-\frac{1}{5}\qquad\text{using Property 2}\end{aligned}$$

In Example 2 we display two more instances in which factoring is used to reduce a fraction. After that, Example 3 indicates how these skills are used to multiply and divide fractional expressions.

EXAMPLE 2 Using factoring to simplify a fractional expression

Simplify: **(a)** $\frac{x^3-8}{x^2-2x}$; **(b)** $\frac{x^2-6x+8}{a(x-2)+b(x-2)}$.

SOLUTION

(a) $$\frac{x^3-8}{x^2-2x}=\frac{(x-2)(x^2+2x+4)}{x(x-2)}=\frac{x^2+2x+4}{x}$$

(b) $$\frac{x^2-6x+8}{a(x-2)+b(x-2)}=\frac{(x-2)(x-4)}{(x-2)(a+b)}=\frac{x-4}{a+b}$$

EXAMPLE 3 Using factoring to simplify products and quotients

Carry out the indicated operations and simplify:

(a) $\dfrac{x^3}{2x^2+3x}\cdot\dfrac{12x+18}{4x^2-6x}$; **(b)** $\dfrac{2x^2-x-6}{x^2+x+1}\cdot\dfrac{x^3-1}{4-x^2}$;

(c) $\dfrac{x^2-144}{x^2-4}\div\dfrac{x+12}{x+2}$.

SOLUTION

(a) $$\frac{x^3}{2x^2+3x}\cdot\frac{12x+18}{4x^2-6x}=\frac{x^3}{x(2x+3)}\cdot\frac{6(2x+3)}{2x(2x-3)}$$
$$=\frac{6x^3}{2x^2(2x-3)}=\frac{3x}{2x-3}$$

(b) $$\frac{2x^2-x-6}{x^2+x+1}\cdot\frac{x^3-1}{4-x^2}=\frac{(2x+3)(x-2)}{(x^2+x+1)}\cdot\frac{(x-1)(x^2+x+1)}{(2-x)(2+x)}$$
$$=\frac{-(2x+3)(x-1)}{2+x}\quad\text{using the fact that } \frac{x-2}{2-x}=-1$$
$$=\frac{-2x^2-x+3}{2+x}$$

(c) $$\frac{x^2-144}{x^2-4}\div\frac{x+12}{x+2}=\frac{x^2-144}{x^2-4}\cdot\frac{x+2}{x+12}$$
$$=\frac{(x-12)(x+12)}{(x-2)(x+2)}\cdot\frac{x+2}{x+12}=\frac{x-12}{x-2}$$

As in arithmetic, to combine two fractions by addition or subtraction, the fractions must have a common denominator, that is, the denominators must be the same. The rules in this case are

$$\frac{a}{b}+\frac{c}{b}=\frac{a+c}{b}\quad\text{and}\quad\frac{a}{b}-\frac{c}{b}=\frac{a-c}{b}$$

For example, we have

$$\frac{4x-1}{x+1}-\frac{2x-1}{x+1}=\frac{4x-1-(2x-1)}{x+1}=\frac{4x-1-2x+1}{x+1}=\frac{2x}{x+1}$$

Fractions with unlike denominators are added or subtracted by first converting to a common denominator. For instance, to add $9/a$ and $10/a^2$, we write

$$\frac{9}{a}+\frac{10}{a^2}=\frac{9}{a}\cdot\frac{a}{a}+\frac{10}{a^2}$$
$$=\frac{9a}{a^2}+\frac{10}{a^2}=\frac{9a+10}{a^2}$$

Notice that the common denominator used was a^2. This is the **least common denominator.** In fact, other common denominators (such as a^3 or a^4) could be used here, but that would be less efficient. In general, the least common denominator for a given group of fractions is chosen as follows. Write down a product involving the irreducible factors from each denominator. The power of each factor should be equal to (but not greater than) the highest power of that factor appearing in any of the individual denominators. For example, the least common denominator for the two fractions $\frac{1}{(x+1)^2}$ and $\frac{1}{(x+1)(x+2)}$ is $(x+1)^2(x+2)$. In the example that follows, notice that the denominators must be in factored form before the least common denominator can be determined.

EXAMPLE 4 Using the least common denominator

Combine into a single fraction and simplify:

(a) $\frac{3}{4x} + \frac{7x}{10y^2} - 2$; **(b)** $\frac{x}{x^2-9} - \frac{1}{x+3}$; **(c)** $\frac{15}{x^2+x-6} + \frac{x+1}{2-x}$.

SOLUTION

(a) Denominators: $2^2 \cdot x;\ 2 \cdot 5 \cdot y^2;\ 1$

Least common denominator; $2^2 \cdot 5xy^2 = 20xy^2$

$$\begin{aligned}\frac{3}{4x} + \frac{7x}{10y^2} - \frac{2}{1} &= \frac{3}{4x}\cdot\frac{5y^2}{5y^2} + \frac{7x}{10y^2}\cdot\frac{2x}{2x} - \frac{2}{1}\cdot\frac{20xy^2}{20xy^2}\\ &= \frac{15y^2 + 14x^2 - 40xy^2}{20xy^2}\end{aligned}$$

Note: The final numerator is irreducible over the integers.

(b) Denominators: $x^2 - 9 = (x-3)(x+3);\ x+3$

Least common denominator: $(x-3)(x+3)$

$$\begin{aligned}\frac{x}{x^2-9} - \frac{1}{x+3} &= \frac{x}{(x-3)(x+3)} - \frac{1}{x+3}\\ &= \frac{x}{(x-3)(x+3)} - \frac{1}{x+3}\cdot\frac{x-3}{x-3}\\ &= \frac{x-(x-3)}{(x-3)(x+3)} = \frac{3}{(x-3)(x+3)}\end{aligned}$$

(c)
$$\begin{aligned}\frac{15}{x^2+x-6} + \frac{x+1}{2-x} &= \frac{15}{(x-2)(x+3)} - \frac{x+1}{x-2} \quad \text{using the fact that } 2-x = -(x-2)\\ &= \frac{15}{(x-2)(x+3)} - \frac{x+1}{x-2}\cdot\frac{x+3}{x+3}\\ &= \frac{15-(x^2+4x+3)}{(x-2)(x+3)} = \frac{-x^2-4x+12}{(x-2)(x+3)}\\ &= \frac{(x-2)(-x-6)}{(x-2)(x+3)} = \frac{-x-6}{x+3} = -\frac{x+6}{x+3}\end{aligned}$$

Notice in the last line that we were able to simplify the answer by factoring the numerator and reducing the fraction. (This is why we prefer to leave the least common denominator in factored form, rather than multiplying it out, in this type of problem.)

The next four examples illustrate using the least common denominator to simplify compound fractions, or complex fractions, which are fractions "within" fractions.

EXAMPLE 5 Simplifying a compound fraction

Simplify: $\dfrac{\dfrac{1}{3a} - \dfrac{1}{4b}}{\dfrac{5}{6a^2} + \dfrac{1}{b}}$.

SOLUTION

The least common denominator for the four individual denominators $3a, 4b, 6a^2$, and b is $12a^2b$. Multiplying the given expression by $12a^2b/12a^2b$, which equals 1, yields

$$\frac{12a^2b}{12a^2b} \cdot \frac{\dfrac{1}{3a} - \dfrac{1}{4b}}{\dfrac{5}{6a^2} + \dfrac{1}{b}} = \frac{4ab - 3a^2}{10b + 12a^2}$$

EXAMPLE 6 Simplifying a fraction containing negative exponents

Simplify: $(x^{-1} + y^{-1})^{-1}$. (The answer is *not* $x + y$.)

SOLUTION

After applying the definition of negative exponent to rewrite the given expression, we'll use the method shown in Example 5.

$$(x^{-1} + y^{-1})^{-1} = \left(\frac{1}{x} + \frac{1}{y}\right)^{-1} = \frac{1}{\dfrac{1}{x} + \dfrac{1}{y}}$$

$$= \frac{xy}{xy} \cdot \frac{1}{\dfrac{1}{x} + \dfrac{1}{y}} = \frac{xy}{y + x} \qquad \text{multiplying by } 1 = \frac{xy}{xy}$$

EXAMPLE 7 Simplifying a compound fraction

Simplify: $\dfrac{\dfrac{1}{x+h} - \dfrac{1}{x}}{h}$. (This type of expression occurs in calculus.)

SOLUTION

$$\frac{\dfrac{1}{x+h} - \dfrac{1}{x}}{h} = \frac{(x+h)x}{(x+h)x} \cdot \frac{\dfrac{1}{x+h} - \dfrac{1}{x}}{h} \qquad \text{multiplying by } 1 = \frac{(x+h)x}{(x+h)x}$$

$$= \frac{x - (x+h)}{(x+h)xh} = \frac{-h}{(x+h)xh} = -\frac{1}{x(x+h)}$$

EXAMPLE 8 Simplifying a compound fraction

Simplify: $\dfrac{x - \dfrac{1}{x^2}}{\dfrac{1}{x^2} - 1}$.

SOLUTION

$$\frac{x - \dfrac{1}{x^2}}{\dfrac{1}{x^2} - 1} = \frac{x^2}{x^2}\cdot\frac{x - \dfrac{1}{x^2}}{\dfrac{1}{x^2} - 1} \qquad \text{multiplying by } 1 = \frac{x^2}{x^2}$$

$$= \frac{x^3 - 1}{1 - x^2} = -\frac{x^3 - 1}{x^2 - 1}$$

$$= -\frac{(x - 1)(x^2 + x + 1)}{(x - 1)(x + 1)} = -\frac{x^2 + x + 1}{x + 1}$$

EXERCISE SET B.5

A

In Exercises 1–12, reduce the fractions to lowest terms.

1. $\dfrac{x^2 - 9}{x + 3}$

2. $\dfrac{25 - x^2}{x - 5}$

3. $\dfrac{x + 2}{x^4 - 16}$

4. $\dfrac{x^2 - x - 20}{2x^2 + 7x - 4}$

5. $\dfrac{x^2 + 2x + 4}{x^3 - 8}$

6. $\dfrac{a + b}{ax^2 + bx^2}$

7. $\dfrac{9ab - 12b^2}{6a^2 - 8ab}$

8. $\dfrac{a^3b^2 - 27b^5}{(ab - 3b^2)^2}$

9. $\dfrac{a^3 + a^2 + a + 1}{a^2 - 1}$

10. $\dfrac{(x - y)^2(a + b)}{(x^2 - y^2)(a^2 + 2ab + b^2)}$

11. $\dfrac{x^3 - y^3}{(x - y)^3}$

12. $\dfrac{x^4 - y^4}{(x^4y + x^2y^3 + x^3y^2 + xy^4)(x - y)^2}$

In Exercises 13–55, carry out the indicated operations and simplify where possible.

13. $\dfrac{2}{x - 2}\cdot\dfrac{x^2 - 4}{x + 2}$

14. $\dfrac{ax + 3}{2a + 1} \div \dfrac{a^2x^2 + 3ax}{4a^2 - 1}$

15. $\dfrac{x^2 - x - 2}{x^2 + x - 12}\cdot\dfrac{x^2 - 3x}{x^2 - 4x + 4}$

16. $(3t^2 + 4tx + x^2) \div \dfrac{3t^2 - 2tx - x^2}{t^2 - x^2}$

17. $\dfrac{x^3 + y^3}{x^2 - 4xy + 3y^2} \div \dfrac{(x + y)^3}{x^2 - 2xy - 3y^2}$

18. $\dfrac{a^2 - a - 42}{a^4 + 216a} \div \dfrac{a^2 - 49}{a^3 - 6a^2 + 36a}$

19. $\dfrac{4}{x} - \dfrac{2}{x^2}$

20. $\dfrac{1}{3x} + \dfrac{1}{5x^2} - \dfrac{1}{30x^3}$

21. $\dfrac{6}{a} - \dfrac{a}{6}$

22. $\dfrac{1}{a} + \dfrac{1}{b} + \dfrac{1}{c}$

23. $\dfrac{1}{x + 3} + \dfrac{3}{x + 2}$

24. $\dfrac{4}{x - 4} - \dfrac{4}{x + 1}$

25. $\dfrac{3x}{x - 2} - \dfrac{6}{x^2 - 4}$

26. $1 + \dfrac{1}{x} - \dfrac{1}{x^2}$

27. $\dfrac{a}{x - 1} + \dfrac{2ax}{(x - 1)^2} + \dfrac{3ax^2}{(x - 1)^3}$

28. $\dfrac{a^2 + 5a - 4}{a^2 - 16} - \dfrac{2a}{2a^2 + 8a}$

29. $\dfrac{4}{x - 5} - \dfrac{4}{5 - x}$

30. $\dfrac{x}{x + a} + \dfrac{a}{a - x}$

31. $\dfrac{a^2 + b^2}{a^2 - b^2} + \dfrac{a}{a + b} + \dfrac{b}{b - a}$

32. $\dfrac{3}{2x + 2} - \dfrac{5}{x^2 - 1} + \dfrac{1}{x + 1}$

33. $\dfrac{1}{x^2 + x - 20} - \dfrac{1}{x^2 - 8x + 16}$

34. $\dfrac{4}{6x^2 + 5x - 4} + \dfrac{1}{3x^2 + 4x} - \dfrac{1}{2x - 1}$

35. $\dfrac{2q + p}{2p^2 - 9pq - 5q^2} - \dfrac{p + q}{p^2 - 5pq}$

36. $\dfrac{1}{x - 1} + \dfrac{1}{x^2 - 1} + \dfrac{1}{x^3 - 1}$

37. $\dfrac{\frac{1}{x} + 1}{\frac{1}{x} - 1}$

38. $\dfrac{\frac{4}{a} - a}{\frac{2}{a} + 1}$

39. $\dfrac{\frac{1}{x} - \frac{1}{a}}{x - a}$

40. $\dfrac{\frac{1}{a} + \frac{1}{b}}{\frac{1}{a} - \frac{1}{b}}$

41. $\dfrac{a - \frac{1}{a}}{1 + \frac{1}{a}}$

42. $\dfrac{\frac{1}{x^2} - \frac{1}{y^2}}{\frac{1}{x} + \frac{1}{y}}$

43. $\dfrac{\frac{1}{2 + h} - \frac{1}{2}}{h}$

44. $\dfrac{\frac{3}{x^2 + h} - \frac{3}{x^2}}{h}$

45. $\dfrac{\frac{a}{x^2} + \frac{x}{a^2}}{a^2 - ax + x^2}$

46. $\dfrac{x + \frac{xy}{y - x}}{\frac{y^2}{x^2 - y^2} + 1}$

47. $(x^{-1} + 2)^{-1}$

48. $\dfrac{(x^{-2} + 2x)^{-1}}{x^2}$

49. $\left(\dfrac{1}{a^{-1}} + \dfrac{1}{a^{-2}}\right)^{-1}$

50. $(a^{-1} + a^{-2})^{-1}$

51. $x(x + y)^{-1} + y(x - y)^{-1}$

52. $x(2x - 2y)^{-1} + y(2y - 2x)^{-1}$

B

53. $\dfrac{\frac{a + b}{a - b} + \frac{a - b}{a + b}}{\frac{a - b}{a + b} - \frac{a + b}{a - b}} \cdot \dfrac{ab^3 - a^3b}{a^2 + b^2}$

54. $\dfrac{\sqrt{x + a}}{\sqrt{x - a}} - \dfrac{\sqrt{x - a}}{\sqrt{x + a}}$

C

55. $\dfrac{\left(a + \frac{1}{b}\right)^a \left(a - \frac{1}{b}\right)^b}{\left(b + \frac{1}{a}\right)^a \left(b - \frac{1}{a}\right)^b}$

56. Consider the three fractions

$$\frac{b - c}{1 + bc}, \quad \frac{c - a}{1 + ca}, \quad \text{and} \quad \frac{a - b}{1 + ab}$$

(a) If $a = 1$, $b = 2$, and $c = 3$, find the sum of the three fractions. Also compute their product. What do you observe?

(b) Show that the sum and the product of the three given fractions are, in fact, always equal.

Answers to Selected Exercises

CHAPTER 1

Exercise Set 1.1

1. (a) integer, rational number **(b)** rational number **3. (a)** natural number, integer, rational number **(b)** rational number **5. (a)** rational number **(b)** rational number **7.** irrational number **9.** irrational number

11.

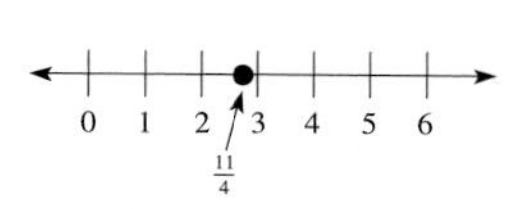

13.

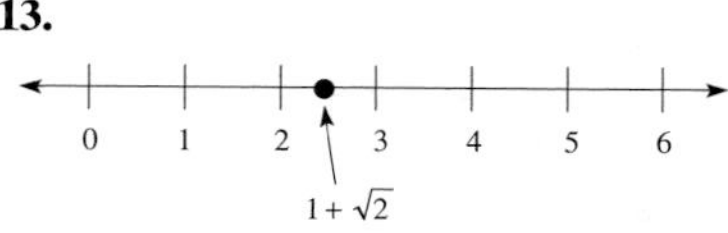

15.

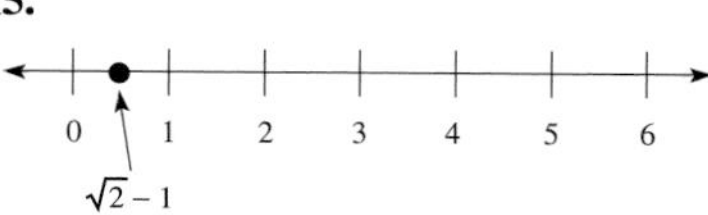

17.

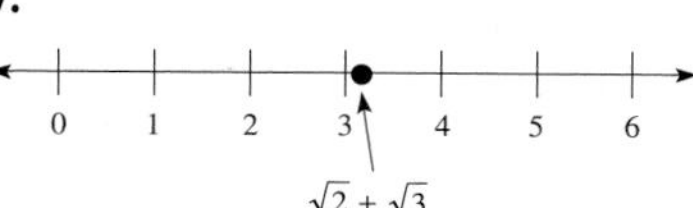

19.

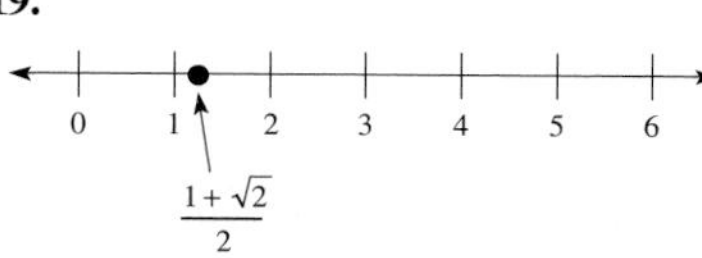

21.

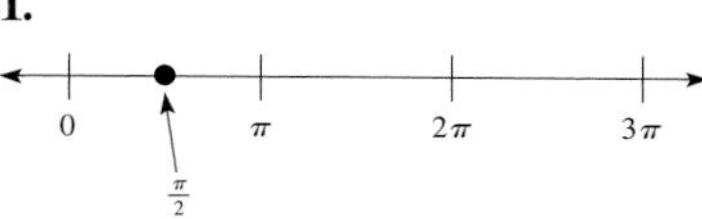

23.

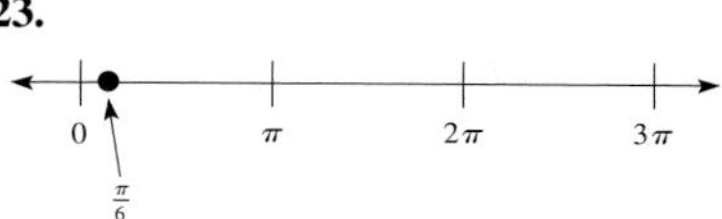

25.

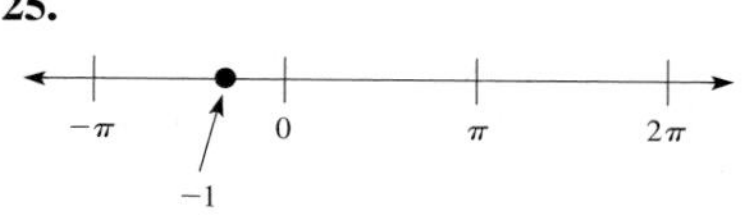

27.

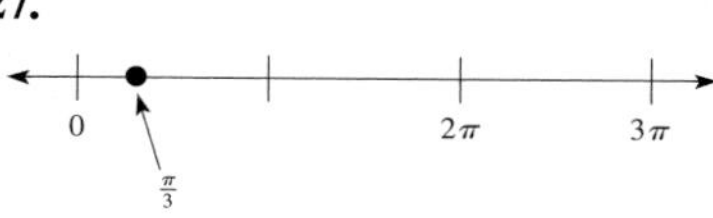

29.

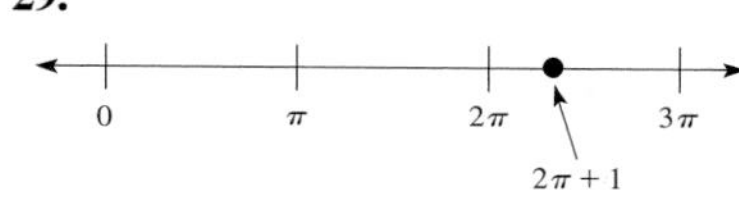

31. false **33.** true **35.** false **37.** false **39.** true

41. $2 < x < 5$

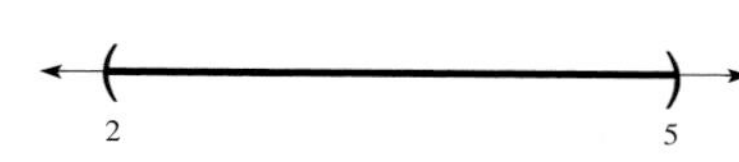

43. $1 \le x \le 4$

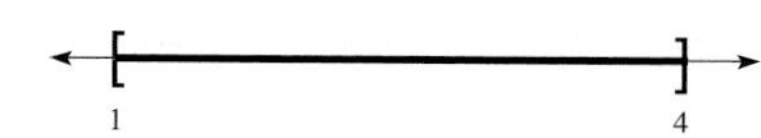

45. $0 \le x < 3$

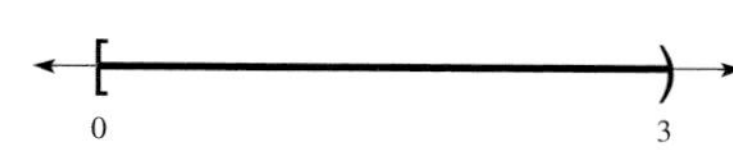

47. $-3 < x < \infty$

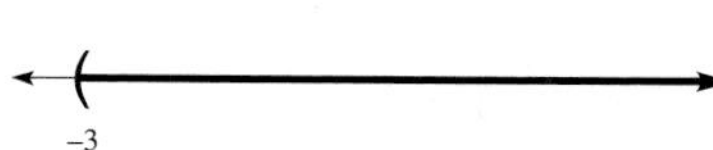

49. $-1 \le x < \infty$

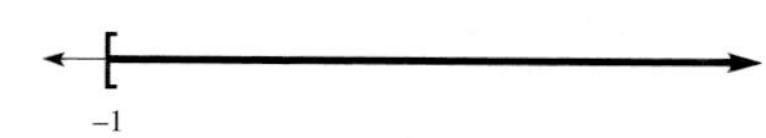

51. $-\infty < x < 1$

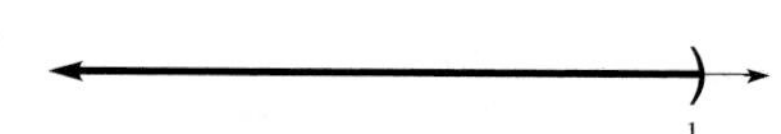

53. $-\infty < x \le \pi$

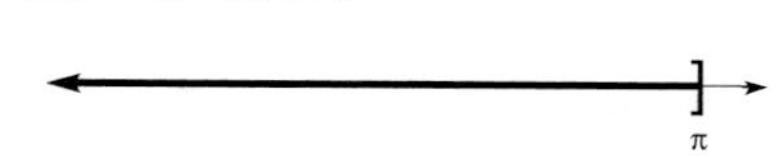

55. (a) one decimal place **(b)** two decimal places **(c)** six decimal places **(d)** nine decimal places **57. (a)** $a = \sqrt{2}, b = \sqrt{8}$ **(b)** $a = \sqrt{2}$, $b = \sqrt{3}$ **59. (a)** $2^{1/2} = \sqrt{2}$ **(b)** $(\sqrt{2})^2 = 2$

Exercise Set 1.2

1. 3 **3.** 6 **5.** 2 **7.** 0 **9.** 0 **11.** 17 **13.** 1 **15.** 0 **17.** 25 **19.** -1 **21.** 0 **23.** 3 **25.** $\sqrt{2}-2$ **27.** $x-3$ **29.** t^2+1 **31.** $\sqrt{3}+4$ **33.** $-2x+7$ **35.** 1 **37.** $3x+11$ **39.** $|x-1| = \frac{1}{2}$ **41.** $|x-1| \ge \frac{1}{2}$ **43.** $|y+4| < 1$ **45.** $|y| < 3$ **47.** $|x^2 - a^2| < M$

49.

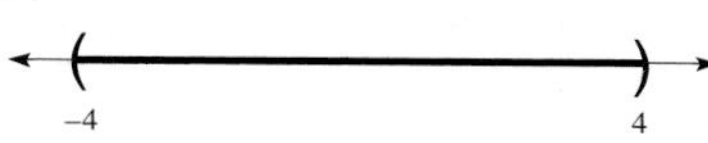

51.

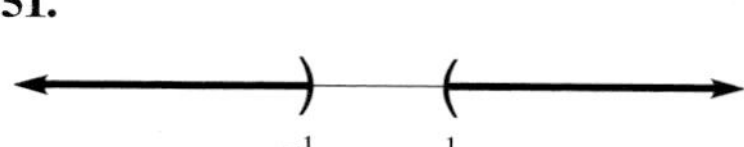

53.

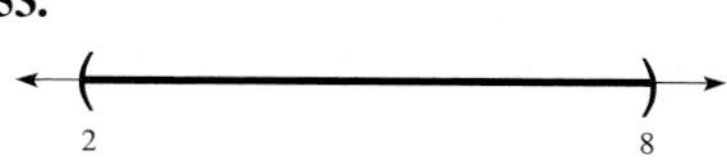

55.

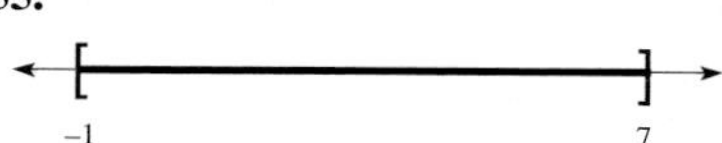

57.

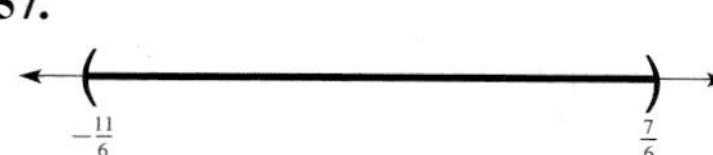

59.

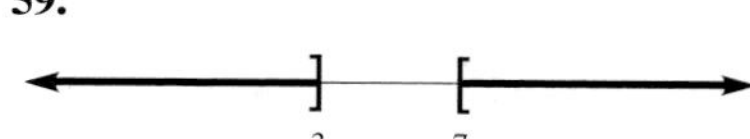

61. (a)

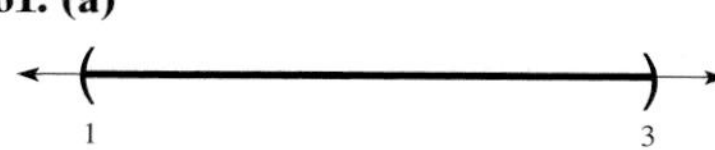

(b)

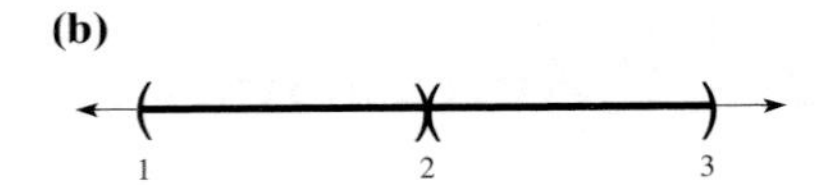

(c) The interval in part **(b)** does not include 2.

Exercise Set 1.3

1. is a solution **3.** not a solution **5.** is a solution **7.** $x = -1$ **9.** $m = 4$ **11.** $t = 14$ **13.** $y = -15$ **15.** $x = 3$ **17.** $x = 8$ **19.** no solution **21.** $x = -\frac{16}{9}$ **23. (a)** $x = -\frac{5}{11}$ **(b)** $x = -\frac{2}{11}$ **(c)** no solution **25.** $x = 3$ or $x = 2$ **27.** $t = -\frac{4}{3}$ or $t = -1$ **29.** $x = -8$ or $x = 5$ **31.** $x = -\frac{1}{3}$ or $x = 8$ **33.** $x = -\sqrt{5}$ **35.** $x = \dfrac{-3 \pm 3\sqrt{17}}{8} \approx -1.17, 1.92$ **37.** $x = \dfrac{-4 \pm \sqrt{10}}{3} \approx -2.39, -0.28$ **39.** $x = \sqrt{3} \pm \sqrt{2} \approx 0.32, 3.15$ **41.** $x = -\frac{5}{8}$ or $x = -\frac{1}{3}$ **43.** $y = \pm 5$ **45.** $x = \pm\sqrt[4]{5}$ **47. (a)** $x = -89$ or $x = -67$ **(b)** $y = -\frac{1}{6}$ or $y = \frac{13}{24}$ **49.** $x = \dfrac{b+1}{a}$ **51.** $x = 1$ **53.** $x = \dfrac{1}{a+b}$ **55.** $x = \dfrac{2ab}{b+a}$ **57.** $x = 1$ or $x = -1$ **59.** $x = \dfrac{a^2 + ab + b^2}{a - b}$ **61.** $x = b$ **63.** $x = \dfrac{a+b}{2}$ **65.** $h = \dfrac{S - 2\pi r^2}{2\pi r}$ **67.** $r = \dfrac{d}{1 - dt}$ **69.** $x = \frac{5}{2}$ or $x = -4$ **71.** $x = \frac{1}{3}$ or $x = \frac{1}{2}$ **73.** no solution

Exercise Set 1.4

1.

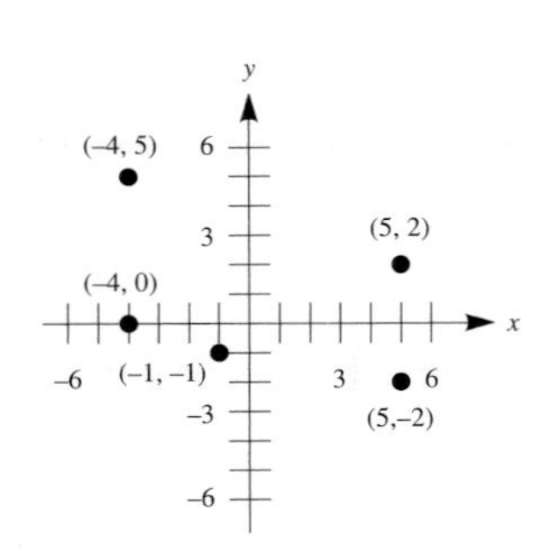

3. (a)

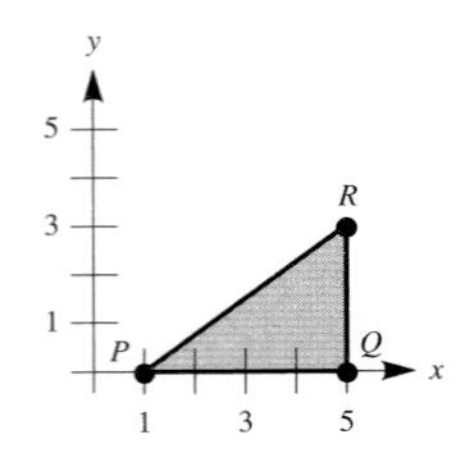

(b) 6 square units **5. (a)** 5 **(b)** 13 **7. (a)** 10 **(b)** 9 **9.** 4 **11. (a)** $(4, \frac{1}{2})$ is farther from the origin **(b)** $(-6, 7)$ is farther from the origin **13. (a)** is a right triangle **(b)** is a right triangle **(c)** not a right triangle **15.** area = 0; the three points are collinear **17. (a)** $(6, 5)$ **(b)** $(\frac{1}{2}, -\frac{3}{2})$ **(c)** $(1, -4)$ **19. (a)** $(1, 1)$ **(b)** $\sqrt{34}$ **21. (a)**

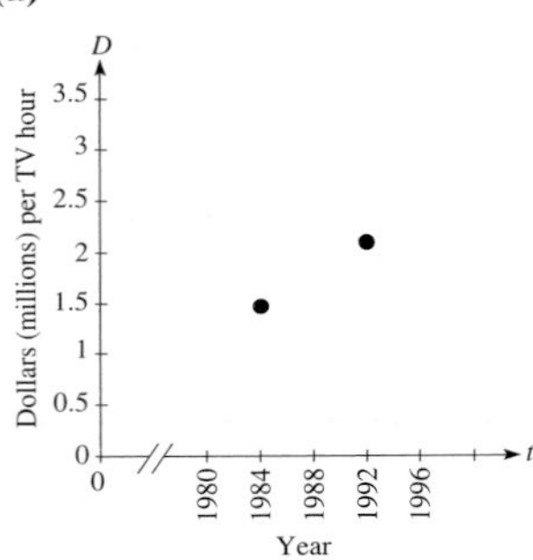

(b) \$1.8 million **(c)** 45.5% error **23. (a)**

t	1995	1997
n	14	30

22 million host computers
(b)

t	1985	1987
n	2	28

15 thousand host computers
(c) 1996: 0.83% error; 1986: 194.75% error
25. (a)

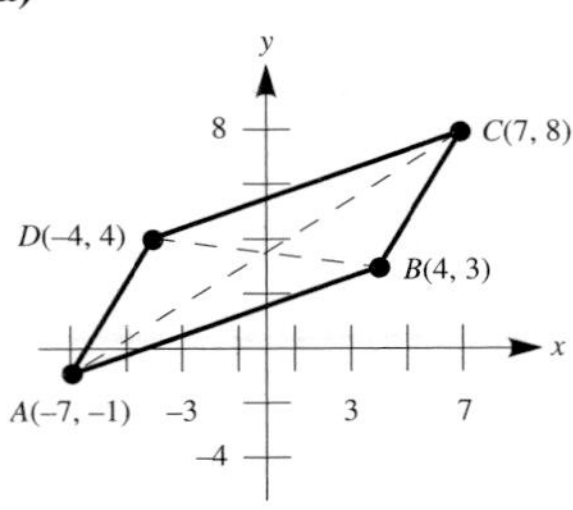

(b) both are $(0, \frac{7}{2})$ **(c)** The diagonals of a parallelogram bisect each other. **27. (a)** $a = \sqrt{2}$; $b = \sqrt{3}$; $c = \sqrt{4} = 2$; $d = \sqrt{5}$; $e = \sqrt{6}$; $f = \sqrt{7}$; $g = \sqrt{8} = 2\sqrt{2}$ **29. (b)** both are $\left(\dfrac{a+b}{2}, \dfrac{c}{2}\right)$

Exercise Set 1.5

1. does not lie on the graph **3.** does not lie on the graph **5.** lies on the graph **7. (a)** $y = \frac{2}{3}x + 1$;

x	-6	-3	0	3	6
y	-3	-1	1	3	5

(b)

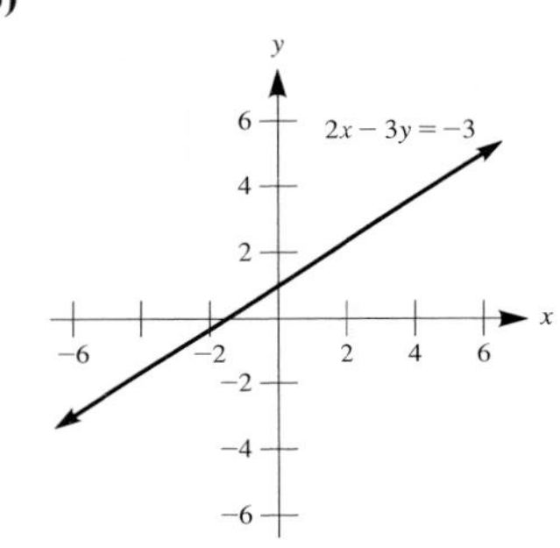

9. x-intercept: 4; y-intercept: 3

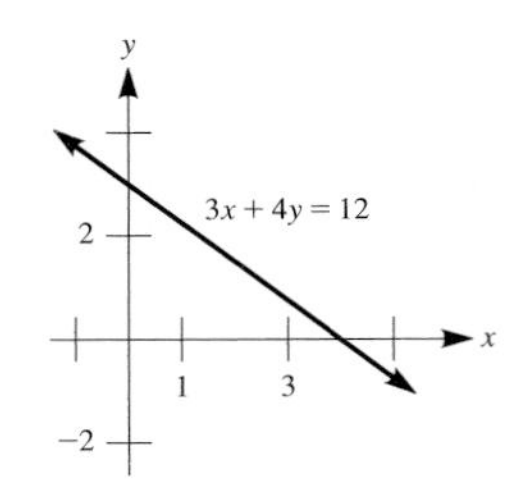

11. x-intercept: 2; y-intercept: -4

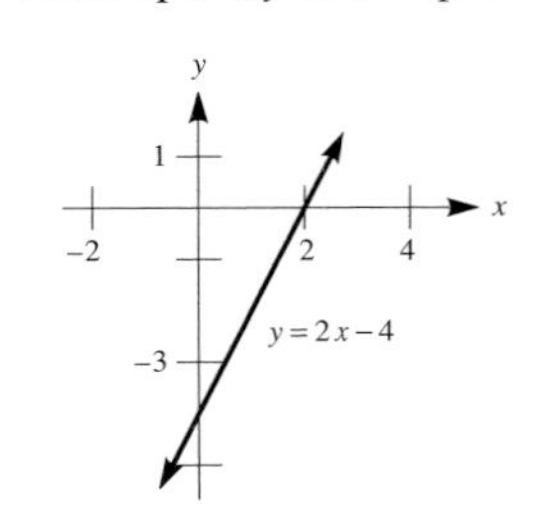

13. x-intercept: 1; y-intercept: 1

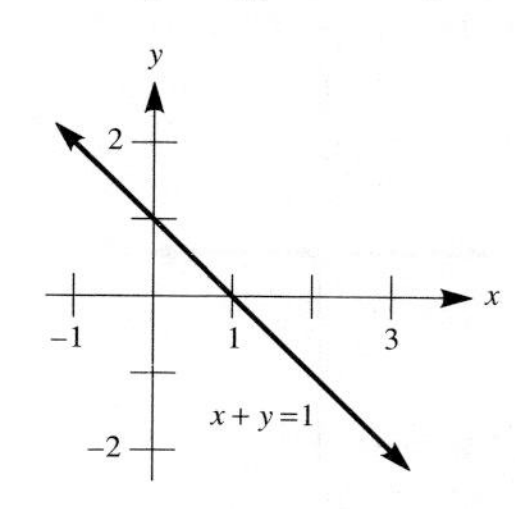

15. (a) Xmin = −3, Xmax = 3, Xscl = 1 and Ymin = −200, Ymax = 200, Yscl = 100; [−3, 3, 1] by [−200, 200, 100] **(b)** Xmin = −1, Xmax = 1.25, Xscl = 0.25 and Ymin = −0.2, Ymax = 0.1, Yscl = 0.1; [−1, 1.25, 0.25] by [−0.2, 0.1, 0.1] **17. (a)** x-intercepts: −2, −1; y-intercept: 2 **(b)** no x-intercepts; y-intercept: 3

19. (a) x-intercepts: $\frac{-1 \pm \sqrt{5}}{2}$; y-intercept: −1 **(b)** no x-intercepts; y-intercept: 1 **21.** x-intercepts: $-1 \pm 2\sqrt{3} \approx -4.46$, 2.46; y-intercept: 0 **23.** x-intercept: $-\frac{8}{3}$; y-intercepts: −2, 6 **25. (a)**

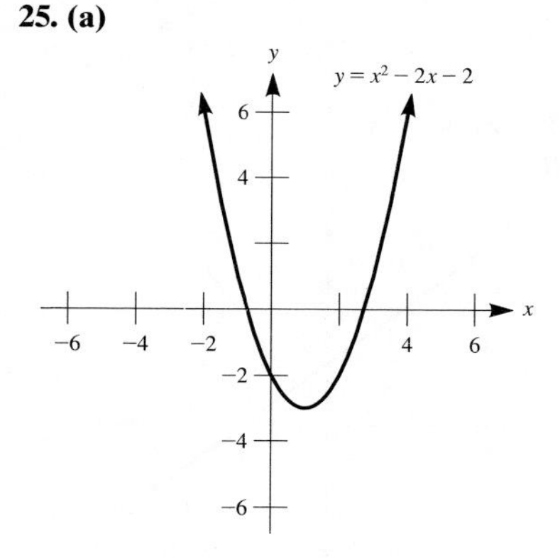

(b) −0.7, 2.7 **(c)** $1 \pm \sqrt{3} \approx -0.7, 2.7$

27. (a)

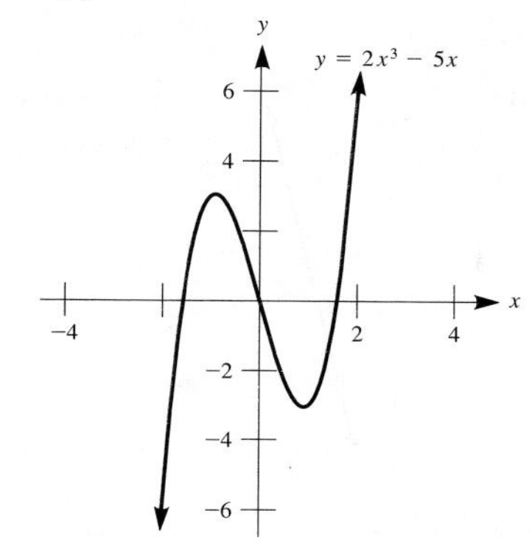

(b) −1.6, 0, 1.6 **(c)** $0, \pm\sqrt{5/2} \approx \pm 1.6$ **29. (a)**

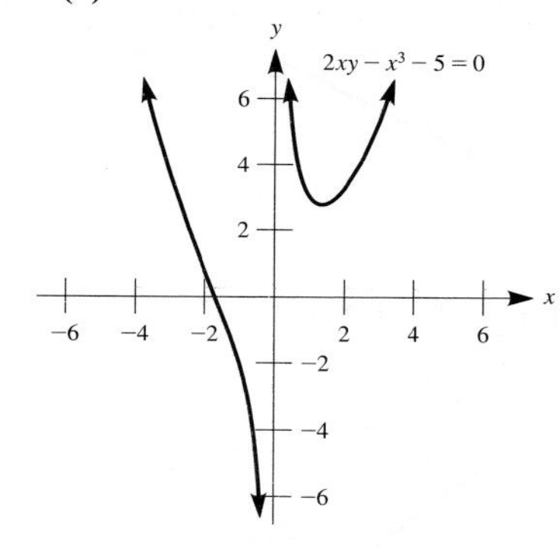

(b) −1.7 **(c)** $\sqrt[3]{-5} \approx -1.7$ **31.** x-intercepts: −1.879, 0.347, 1.532

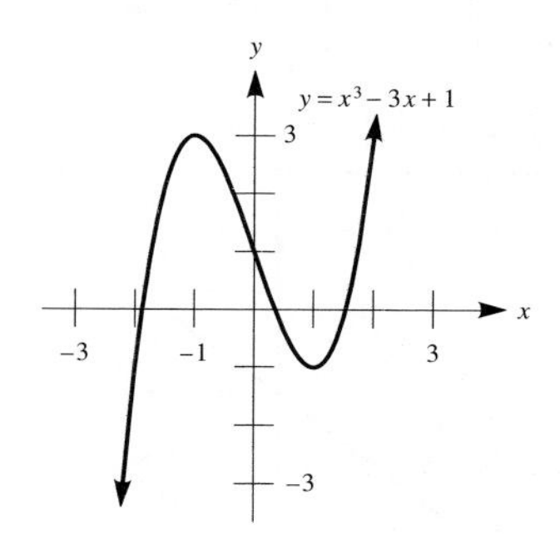

33. x-intercepts: −0.815, 0.875, 5.998

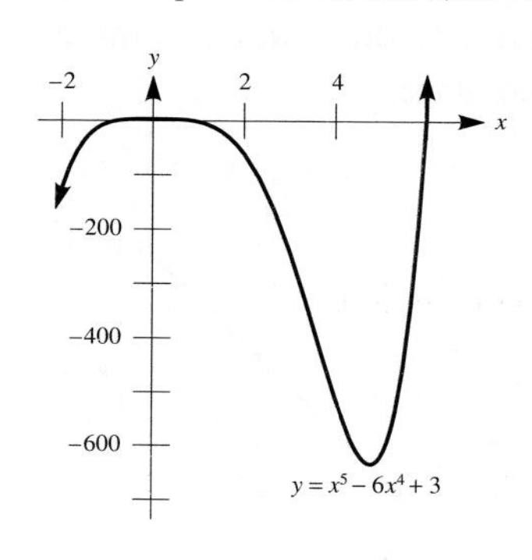

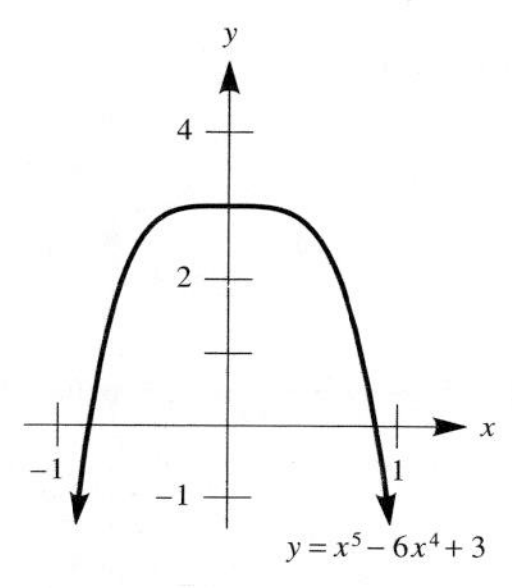

35. (a) The x-intercept is between −1 and 0.

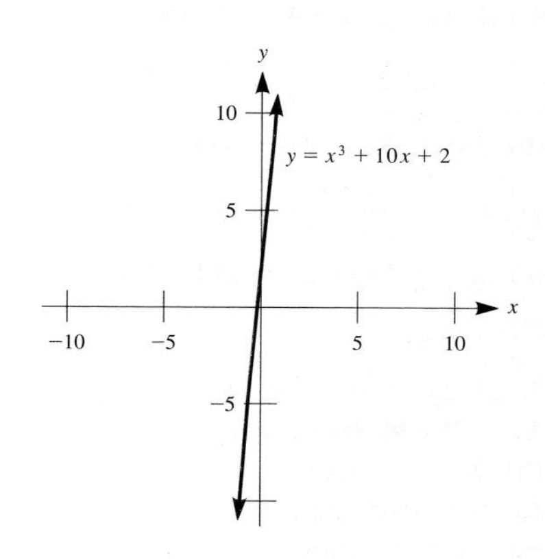

(c) [0, 10] **(d)** Start with a large viewing rectangle then zoom in near the x-intercepts. **37. (a)** −20°; this is the C-intercept of the graph. **(b)** $C = -17\frac{7}{9}$; this is the F-intercept of the graph. **39. (a)** $\sqrt{2} \approx 1.4$ **(b)** $\sqrt{3} \approx 1.7$ **(c)** $\sqrt{6} \approx 2.4$ **41. (a)** 500 bacteria **(b)** 1.5 hours **(c)** $t = 3.5$ hours **(d)** between $t = 3$ and $t = 4$ **43.** A: $x = \frac{2}{5}\sqrt{5} \approx 0.894$; B: $x = -\frac{1}{5}\sqrt{30} \approx -1.095$

Exercise Set 1.6

1. (a) -2 **(b)** 3 **(c)** $-\frac{7}{3}$ **(d)** 3

3. (a) slope = 1

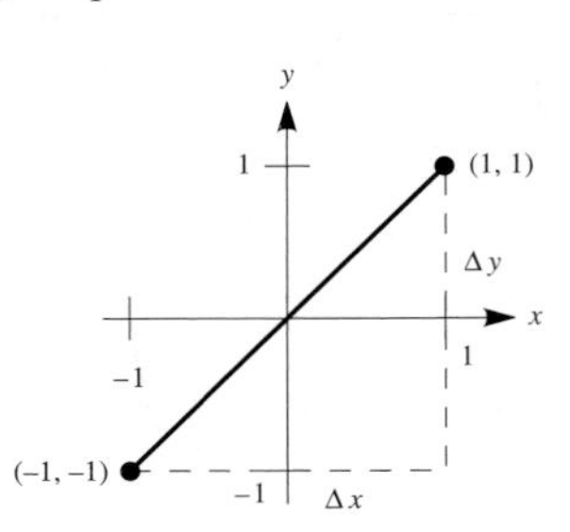

(b) slope = 0

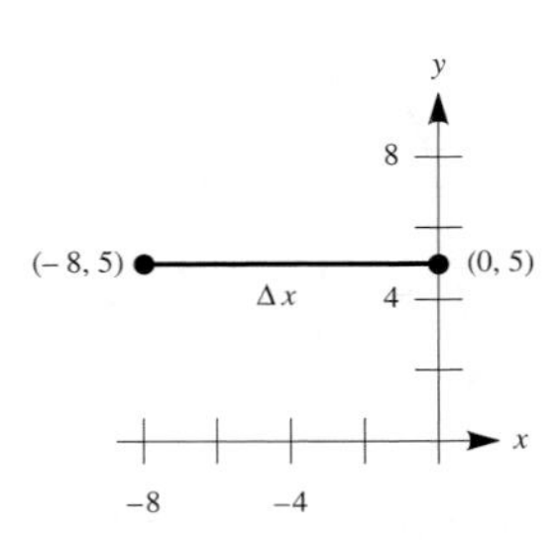

5. (a) $\Delta x = 5$ sec; $\Delta y = 15$ ft;
$\dfrac{\Delta y}{\Delta x} = 3$ ft/sec
(b) $\Delta x = 15$ sec; $\Delta y = 45$ ft;
$\dfrac{\Delta y}{\Delta x} = 3$ ft/sec
(c) $\Delta x = 25$ sec; $\Delta y = 75$ ft;
$\dfrac{\Delta y}{\Delta x} = 3$ ft/sec

7. (a) $\Delta x = 10$ years;
$\Delta y = 25{,}000{,}000$ people
(b) $\Delta x = 20$ years;
$\Delta y = 50{,}000{,}000$ people
(c) $\Delta x = 10$ years;
$\Delta y = 25{,}000{,}000$ people

9. $\Delta t = 3$ years;
$\Delta N = -0.119$ trillion cigarettes

11. $m_3 < m_2 < m_4 < m_1$

13. not collinear **15. (a)** $y = -5x - 9$
(b) $y = \frac{1}{3}x + \frac{4}{3}$ **17. (a)** $y = 2x$
(b) $y = -2x - 4$ **(c)** $y = \frac{1}{7}x - \frac{11}{7}$

19. (a) $x = -3$ **(b)** $y = 4$

21. $x = 0$ **23. (a)** $y = -4x + 7$
(b) $y = 2x + \frac{3}{2}$

25. (a) $y = 4x + 11$

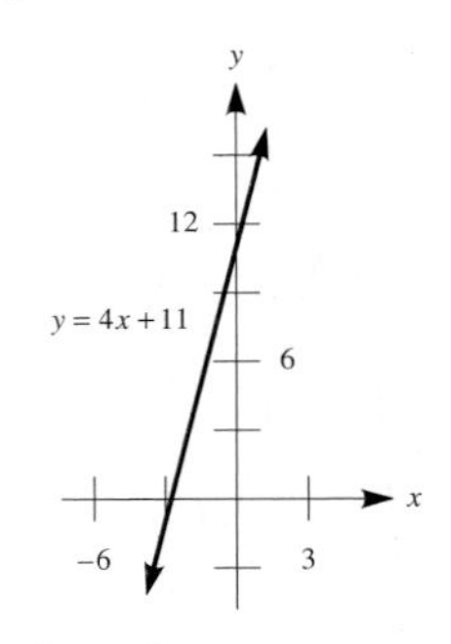

(b) $y = \frac{1}{2}x - \frac{5}{4}$

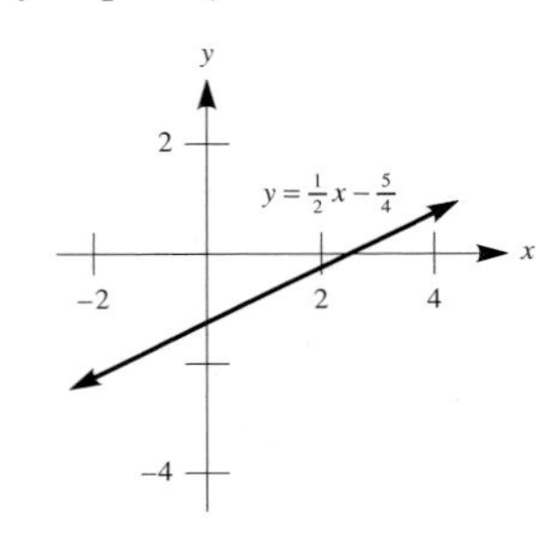

(c) $y = -\frac{5}{6}x + 5$

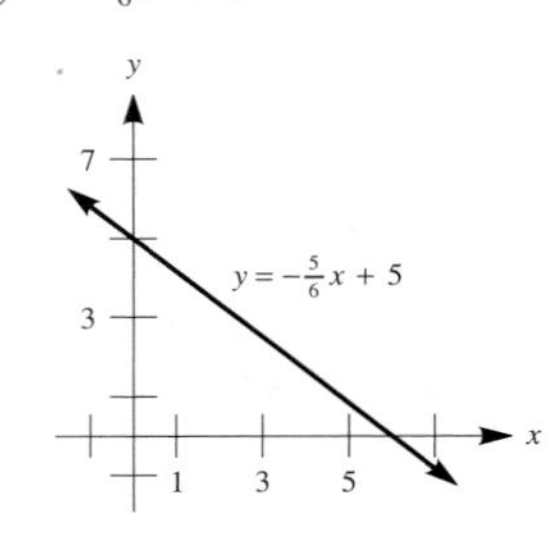

(d) $y = \frac{3}{4}x + \frac{3}{2}$

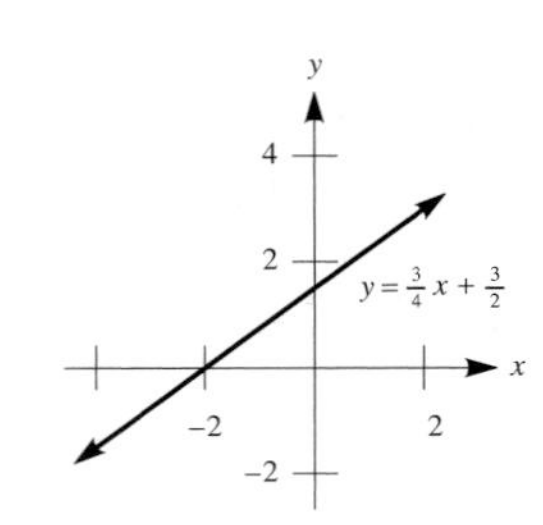

(e) $y = 4x - 2$

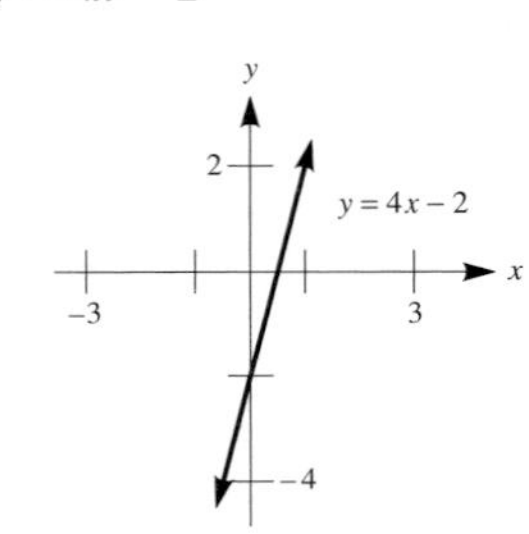

27. $y - 4 = 0$

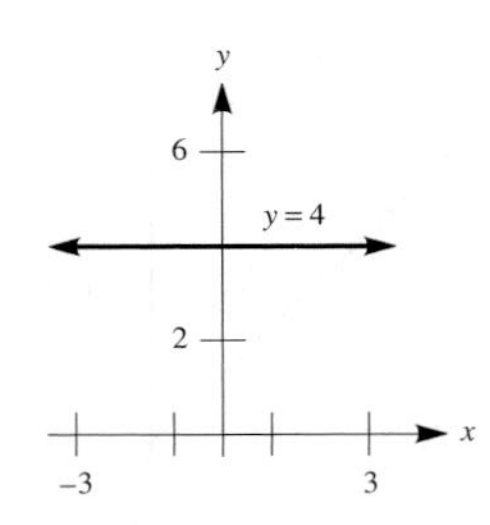

29. (a) x-intercept: 5; y-intercept: 3;
area $= \frac{15}{2}$; perimeter $= 8 + \sqrt{34}$
(b) x-intercept: 5; y-intercept: -3;
area $= \frac{15}{2}$; perimeter $= 8 + \sqrt{34}$

31. (a) neither **(b)** parallel
(c) perpendicular **(d)** perpendicular

33. $y = \frac{2}{5}x + \frac{12}{5}$; $2x - 5y + 12 = 0$

35. $y = -\frac{4}{3}x + \frac{16}{3}$; $4x + 3y - 16 = 0$

37. They appear parallel for each viewing rectangle.

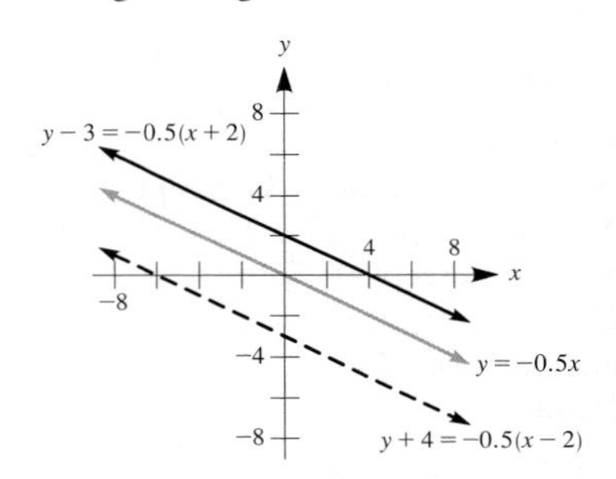

39. (a) $y = \frac{4}{3}x$
(b)

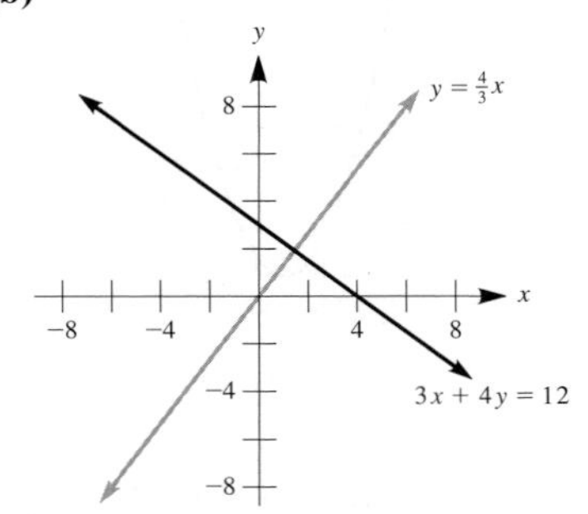

41. (a) $y = -\frac{2}{3}p + 410$ **(b)** 208 units
(c) \$183 **47.** 44.1 square units

49. (a)

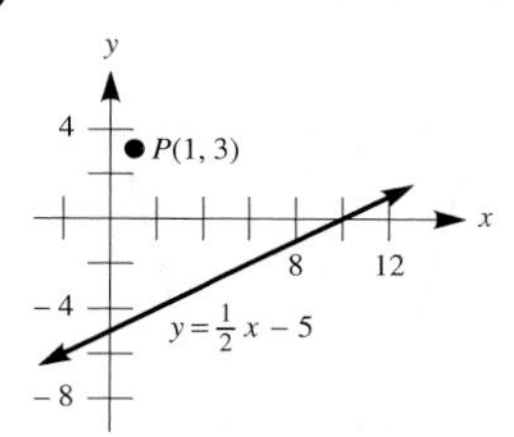

(b) $y = -2x + 5$ **(c)** $(4, -3)$
(d) $3\sqrt{5}$

51. (a)

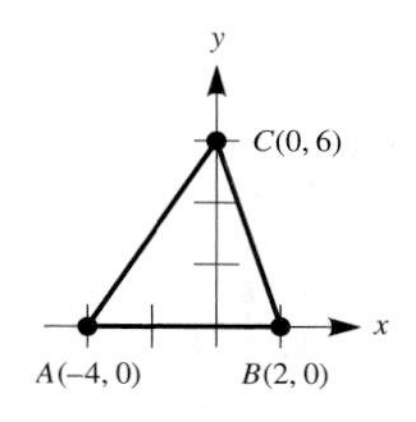

(b) point: $(-1, 2)$

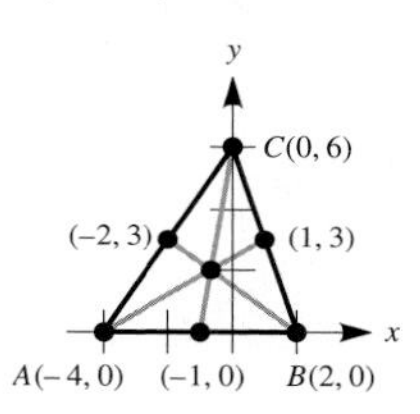

(c) $y = \frac{3}{5}x + \frac{12}{5}$; $y = -\frac{3}{4}x + \frac{3}{2}$; $(-\frac{2}{3}, 2)$

Exercise Set 1.7

1. (a)

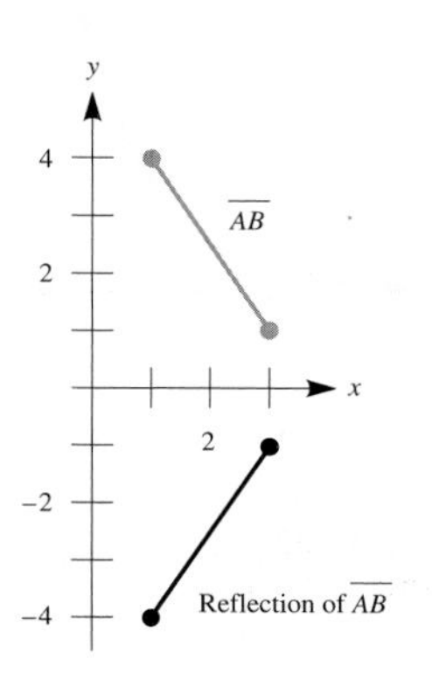

(b)

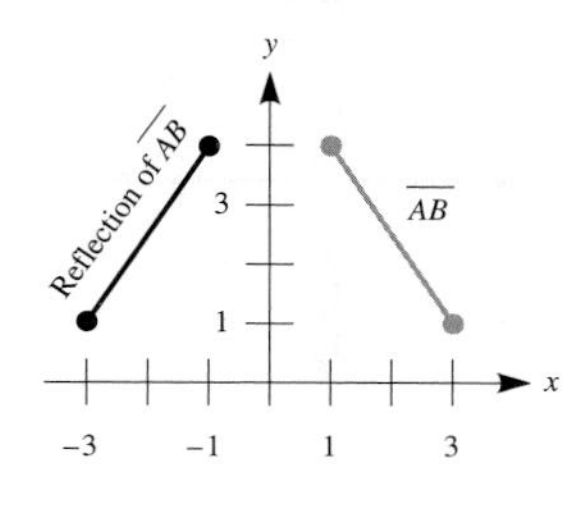

(c)

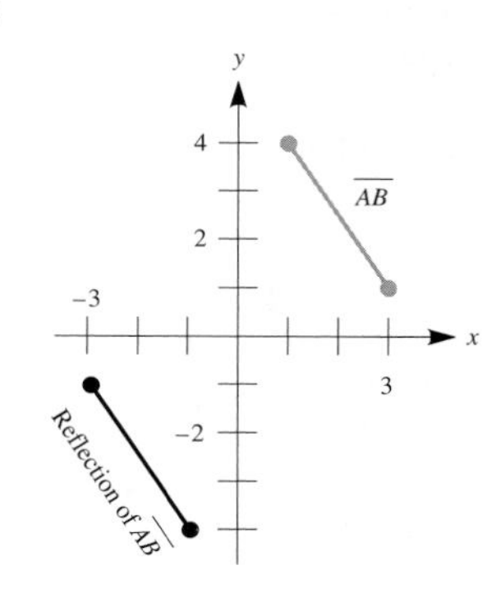

3. (a)

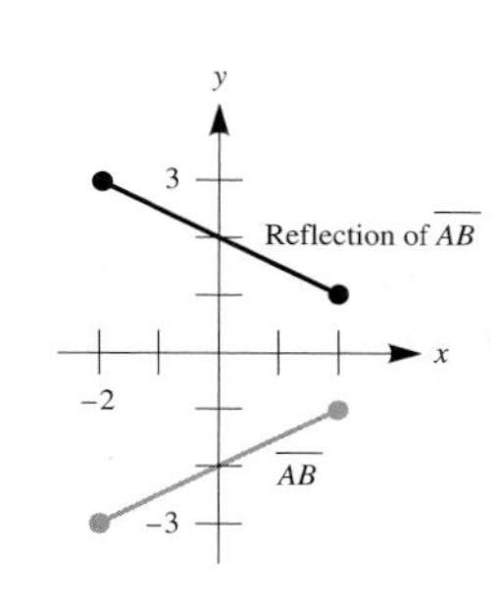

(b)

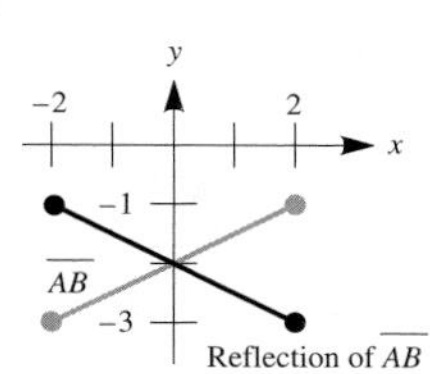

(c)

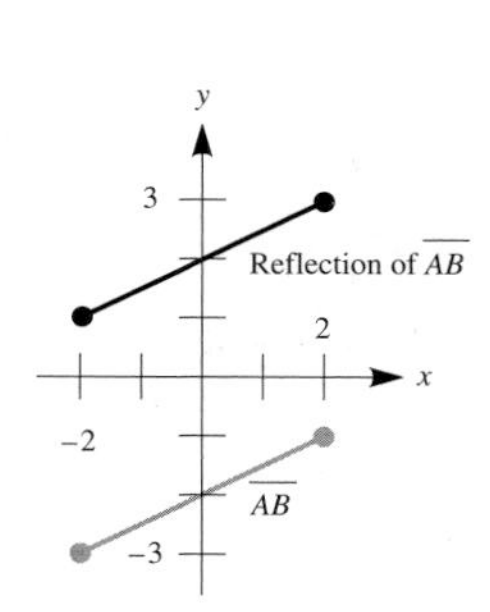

5. (a)

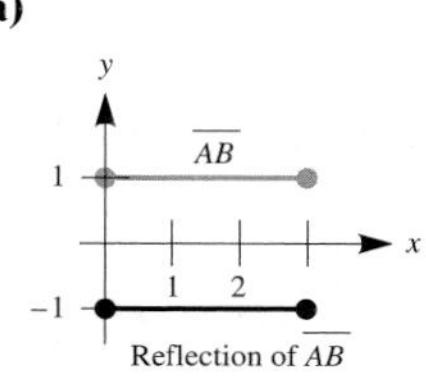

(b)

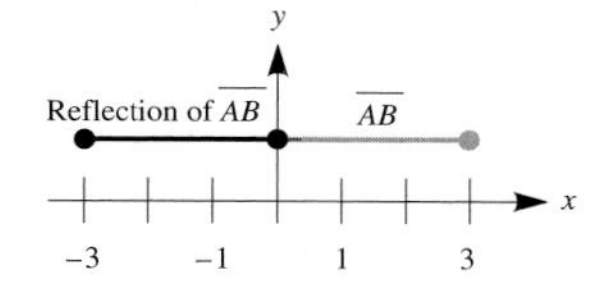

(c)

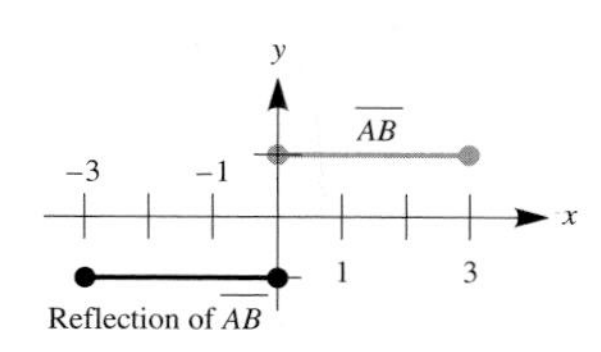

7. x-intercepts: 2 and -2; y-intercept: 4; symmetric about the y-axis

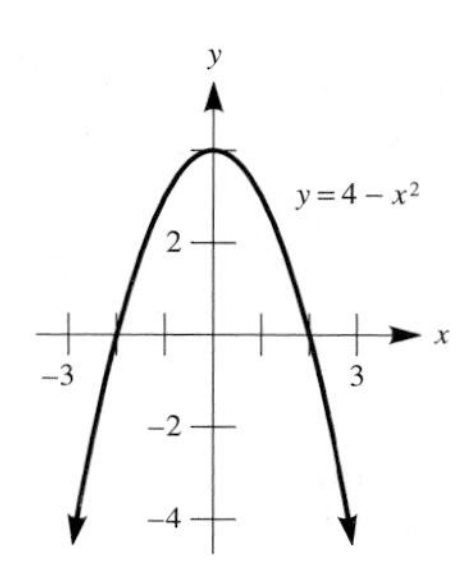

9. no x- or y-intercepts; symmetric about the origin

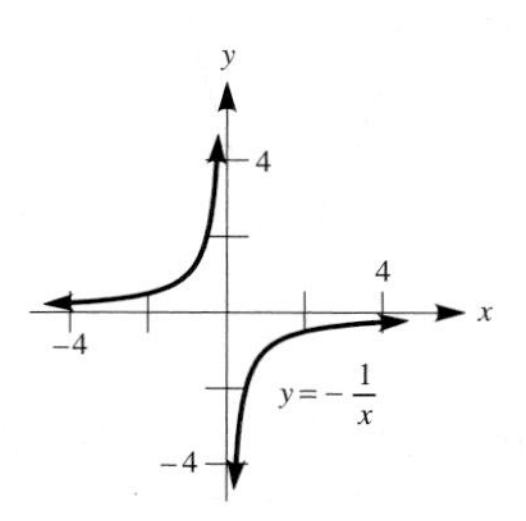

11. x- and y-intercepts: 0; symmetric about the y-axis

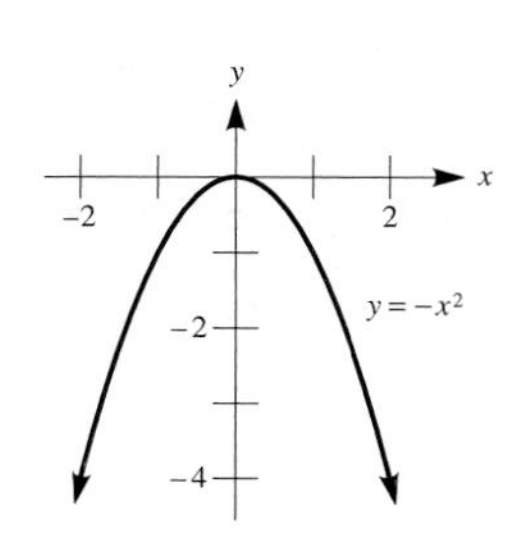

13. no x- or y-intercepts; symmetric about the origin

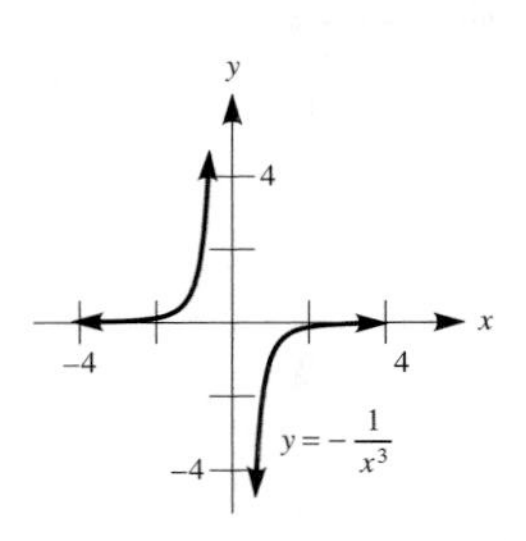

15. x- and y-intercepts: 0; symmetric about the y-axis

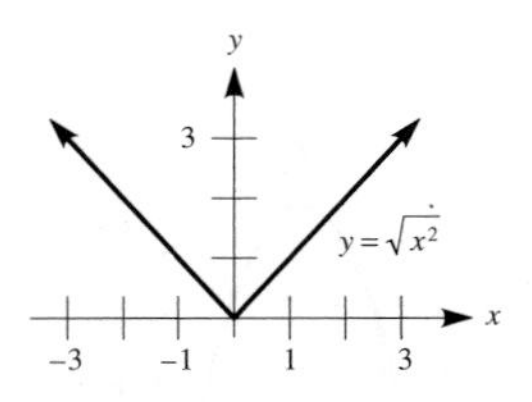

17. x- and y-intercepts: 1; no symmetry

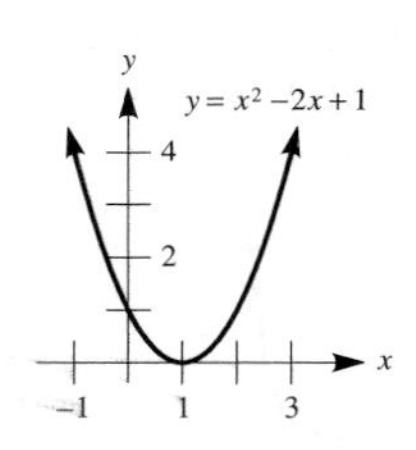

19. x-intercept: 2; no y-intercept; symmetric about the x-axis

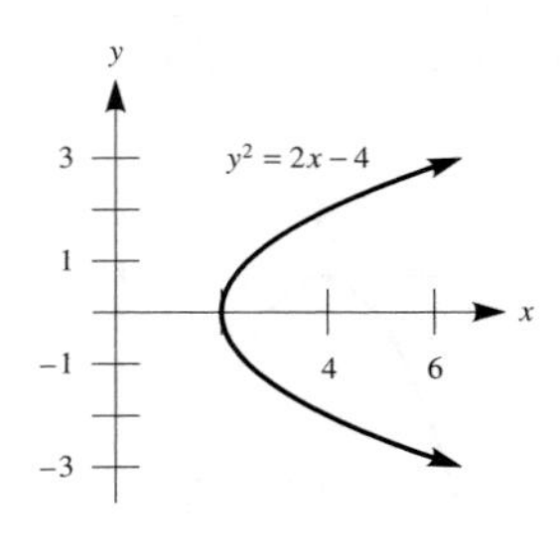

21. x-intercepts: $\frac{-1 \pm \sqrt{33}}{4} \approx -1.69, 1.19$; y-intercept: -4; no symmetry

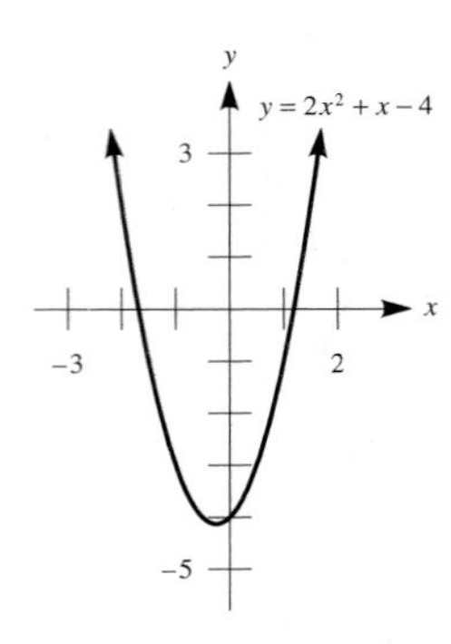

23. (a) x- and y-intercepts: 2; no symmetry

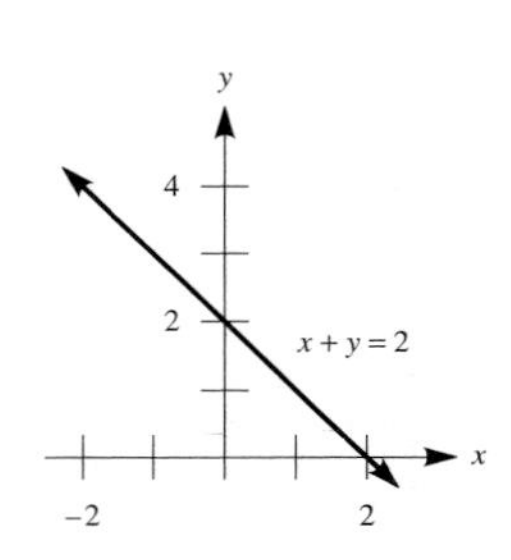

(b) x-intercepts: ± 2; y-intercept: 2; symmetric about the y-axis

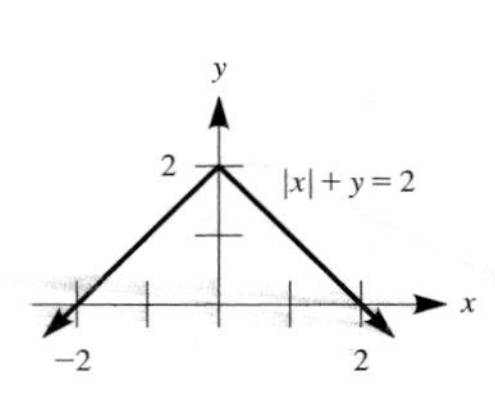

25. (a)

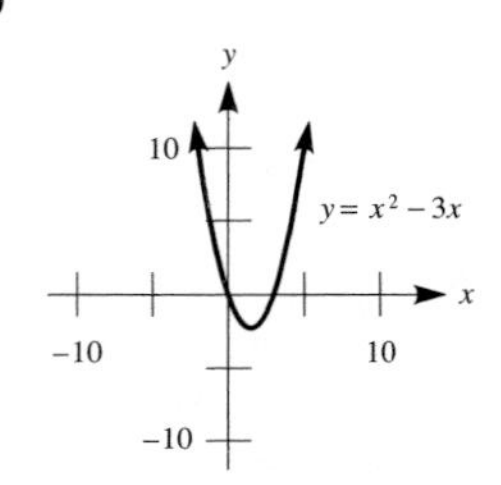

(b) no symmetry

(c)

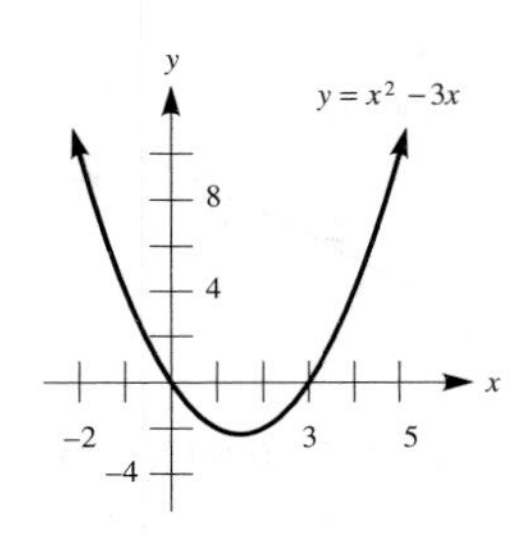

(d) no symmetry

27. (a)

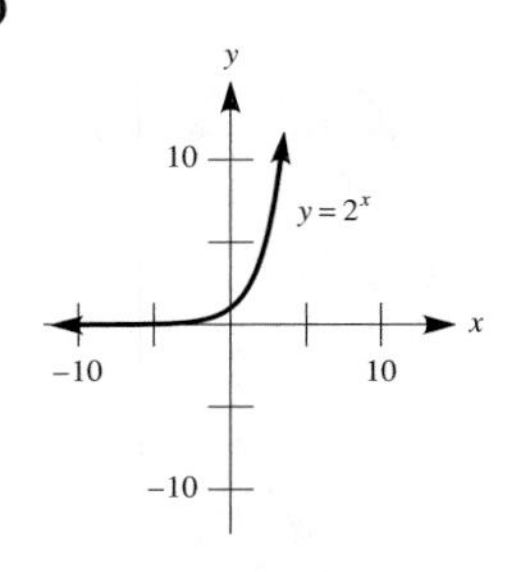

(b) no symmetry

(c)

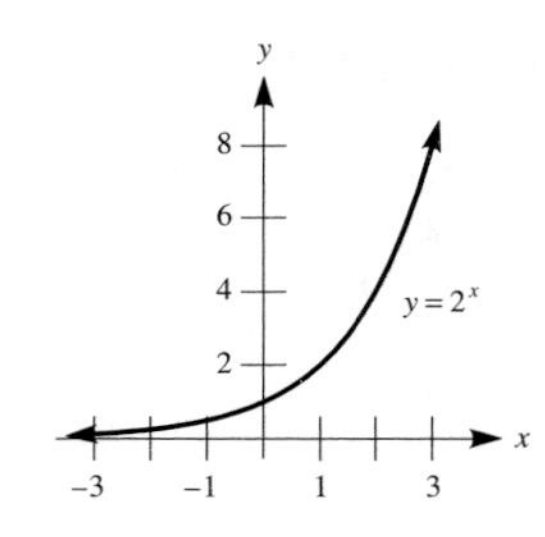

(d) no symmetry

29. (a)

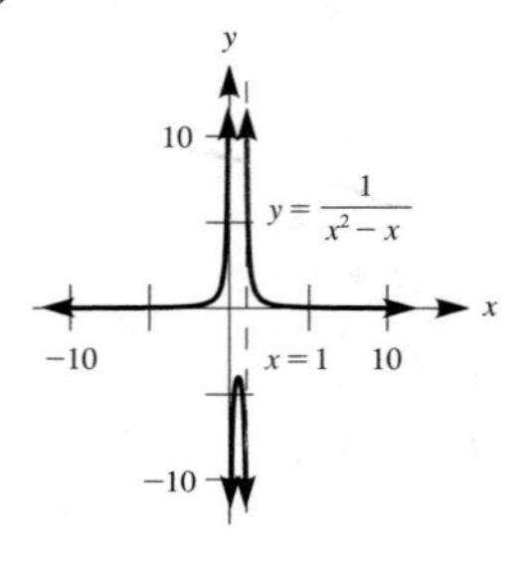

(b) no symmetry
(c)

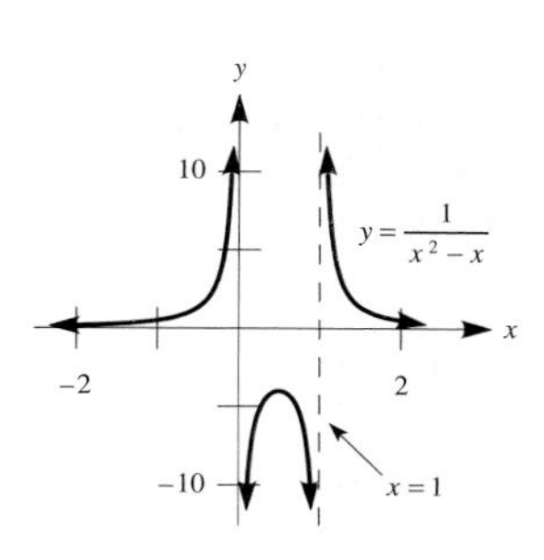

(d) no symmetry
31. (a)

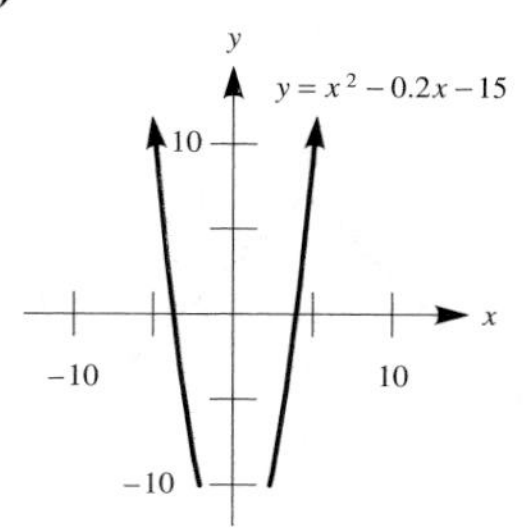

(b) y-axis symmetry
(c)

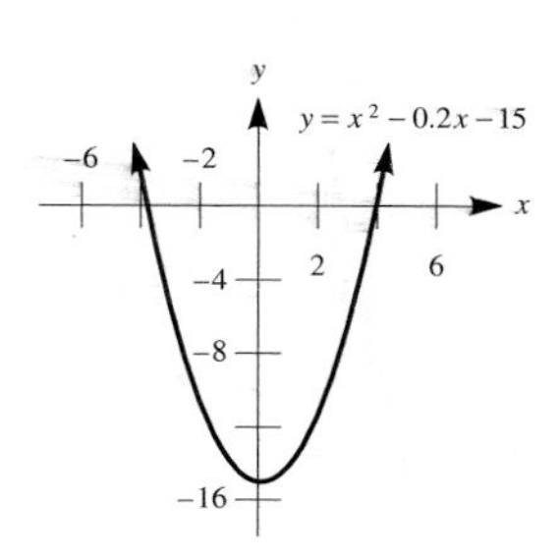

(d) no symmetry
33. (a)

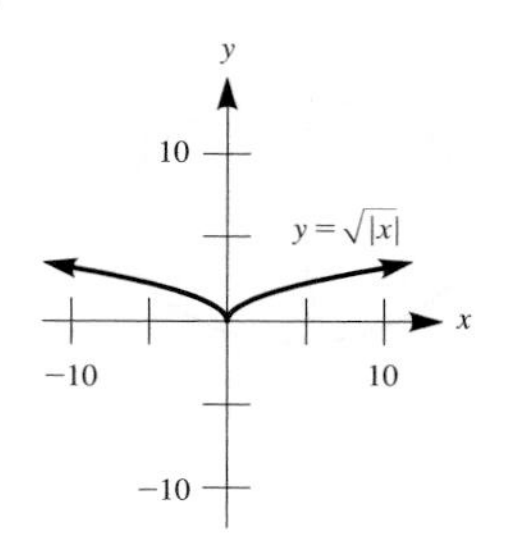

(b) y-axis symmetry
(c)

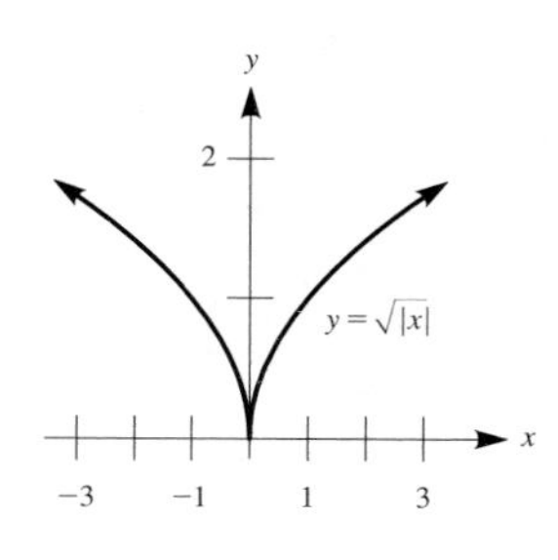

(d) y-axis symmetry
35. (a)

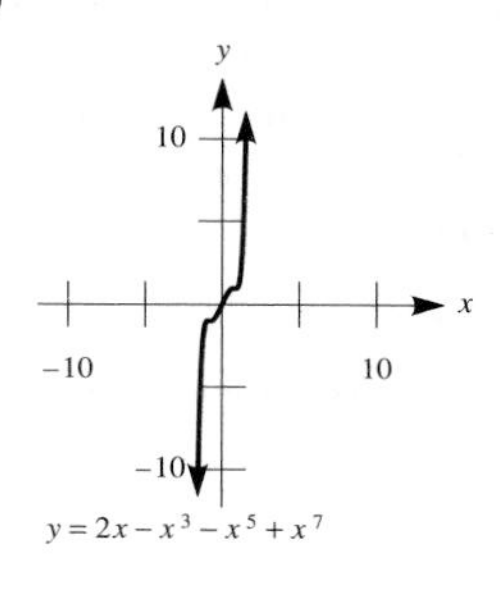

(b) origin symmetry
(c)

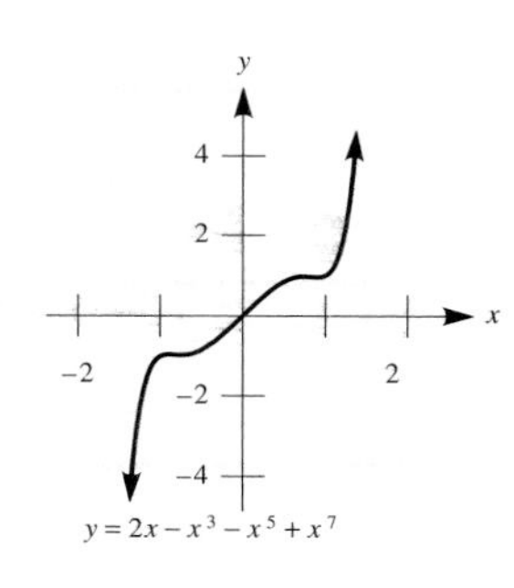

(d) origin symmetry
37. (a)

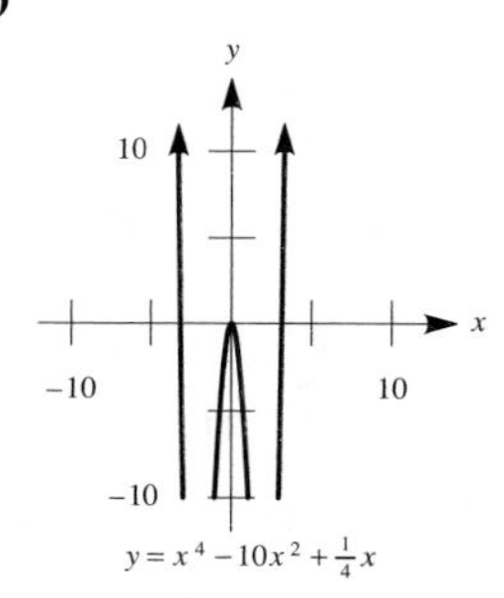

(b) y-axis symmetry
(c)

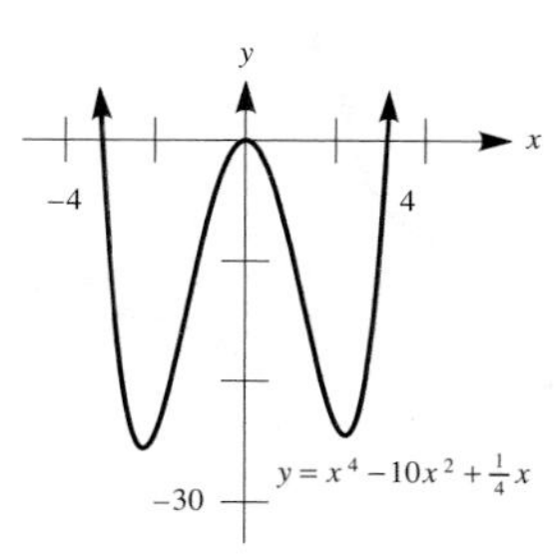

(d) no symmetry **39.** center: $(1, 5)$; radius: 13; $(6, -7)$ lies on the circle **41.** center: $(-8, 5)$; radius: $\sqrt{13}$; $(-5, 2)$ does not lie on the circle **43.** center: $(0, 0)$; radius: $\sqrt[4]{2}$; y-intercepts: $\pm\sqrt[4]{2}$ **45.** center: $(-4, 3)$; radius: 1; no y-intercepts **47.** center: $(-3, \frac{1}{3})$; radius: $\sqrt{2}$; no y-intercepts **49. (a)** center: $(2, -\frac{3}{2})$; radius: $\frac{13}{4}$
(b) x-intercepts: -0.88 and 4.88

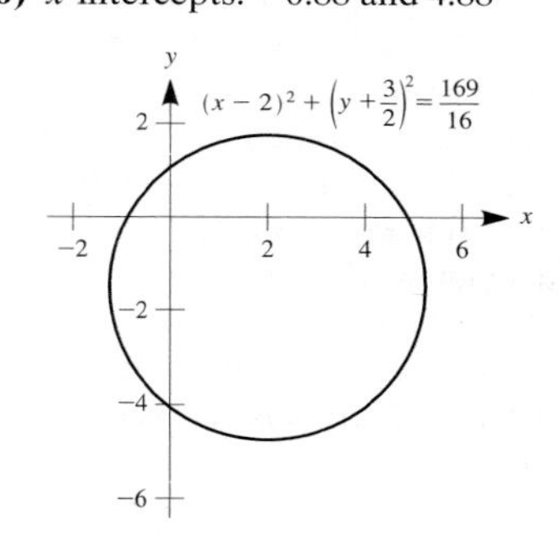

(c) $x = \dfrac{8 \pm \sqrt{133}}{4} \approx -0.88, 4.88$

51.

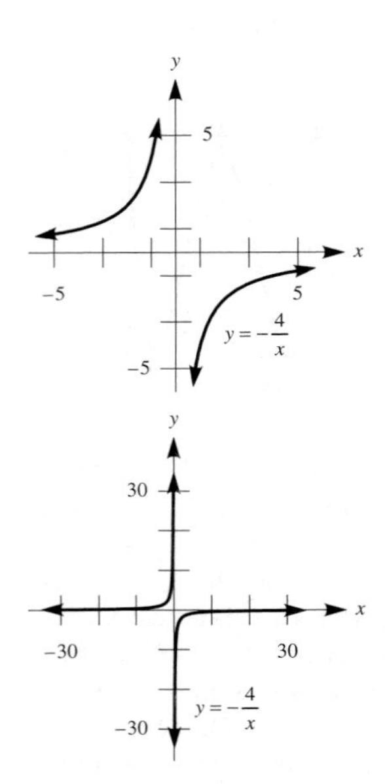

53. $(x - 3)^2 + (y - 2)^2 = 169$
55. $(x - 3)^2 + (y - 5)^2 = 9$
57. $y = 2 \pm \sqrt{19}$ **59. (a)** x-intercepts for each graph: $\frac{8}{3}$; y-intercept for $y = \frac{3}{4}x - 2$: -2; y-intercept for $y = |\frac{3}{4}x - 2|$: 2 **(b)** on the interval $[\frac{8}{3}, \infty)$ **(c)** The graph of $y = |\frac{3}{4}x - 2|$ can be obtained by reflecting $y = \frac{3}{4}x - 2$ about the x-axis for the interval $(-\infty, \frac{8}{3})$. For the interval $[\frac{8}{3}, \infty)$, no reflection is necessary.
61. (b)

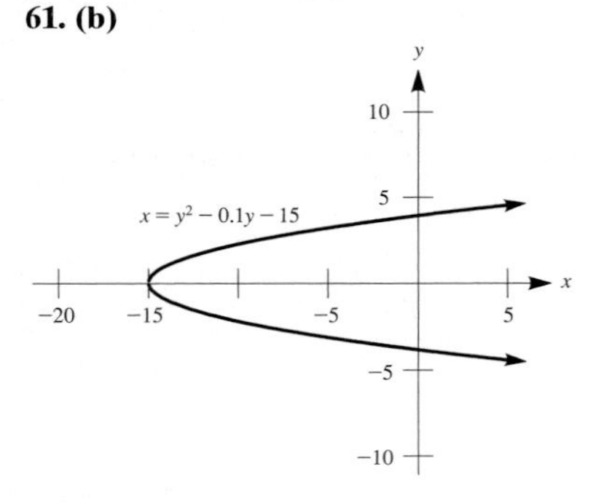

(c)

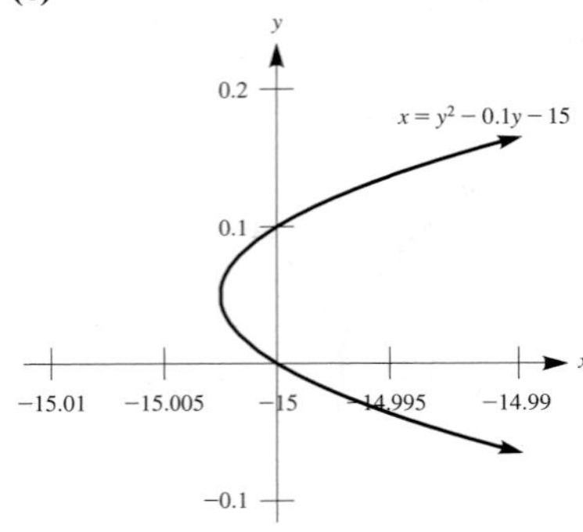

Chapter 1 Review Exercises

1. $|x - 6| = 2$ **3.** $|a - b| = 3$
5. $|x| > 10$ **7.** $\sqrt{6} - 2$
9. $x^4 + x^2 + 1$ **11. (a)** $-2x + 5$
(b) 1 **(c)** $2x - 5$
13. $3 < x < 5$

15. $-5 \le x < 0$

−5 0

17. $-1 \le x$

−1

19.

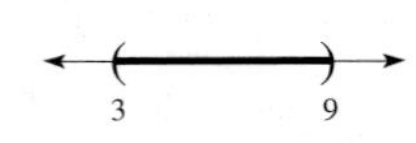

21.

−2 0

23. (a)

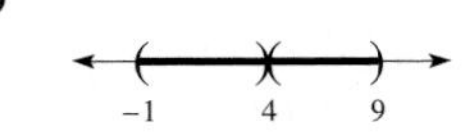

(b)

25. $x = \frac{1}{3}$ **27.** $t = -\frac{37}{11}$ **29.** $t = -\frac{11}{3}$
31. $y = 8$ **33.** $x = \frac{1}{3}, -\frac{1}{2}$
35. $x = -6, 4$ **37.** $x = 10, -1$
39. $t = \dfrac{-1 \pm \sqrt{3}}{2}$ **41.** $y = -2x - 6$
43. $y = \frac{1}{4}x - \frac{5}{2}$ **45.** $y = 2x + 8$
47. $y = -2$ **49.** $y = x + 1$
51. $9x - 4y + 14 = 0$ **53.** $2x + y = 0$
55. $3x - 4y = 0$ **57.** $x + y - 1 = 0$ and $x - 2y - 4 = 0$ **59.** y-axis symmetry **61.** y-axis symmetry
63. x-axis symmetry **65.** x-axis, y-axis, and origin symmetry
67. x-axis symmetry
69.

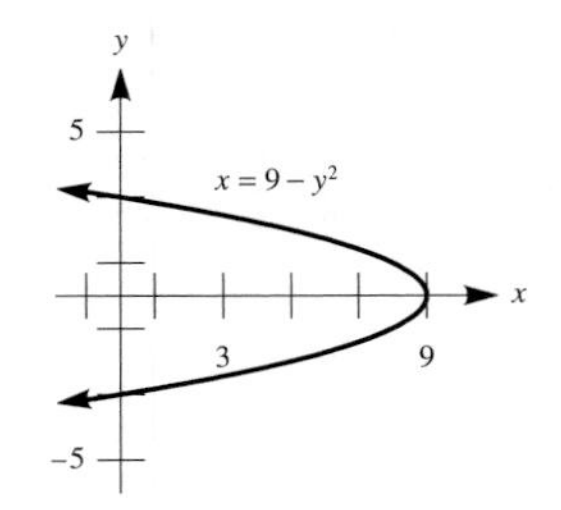

71.

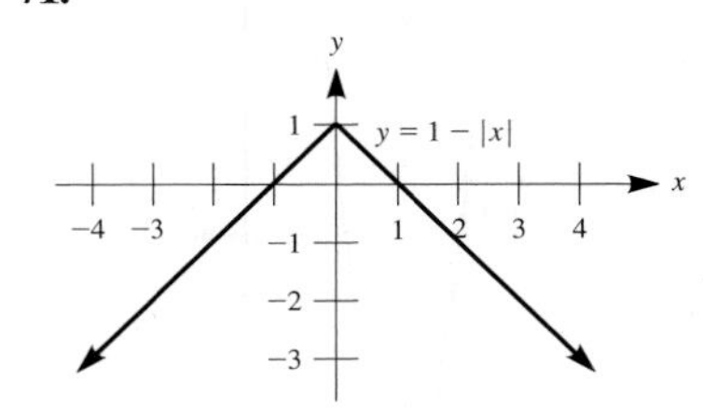

73.

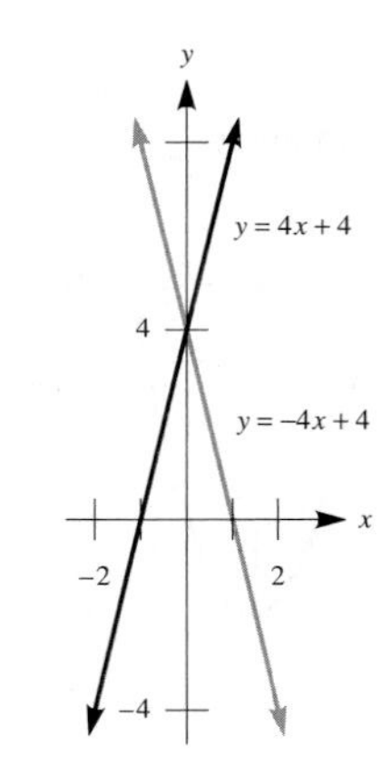

75. (a)

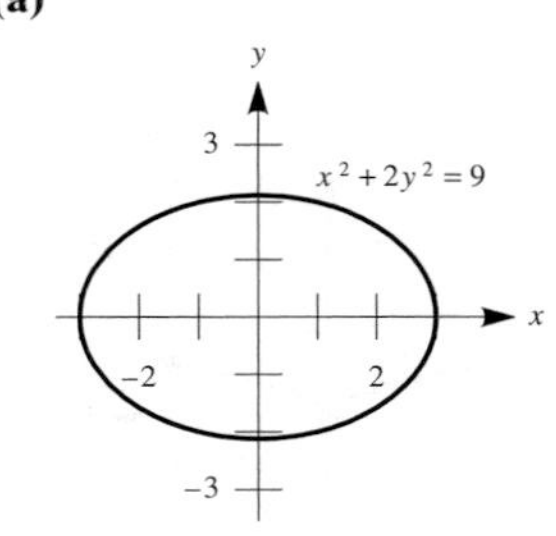

(b) intersect at $(-3, 0)$ and $(3, 0)$

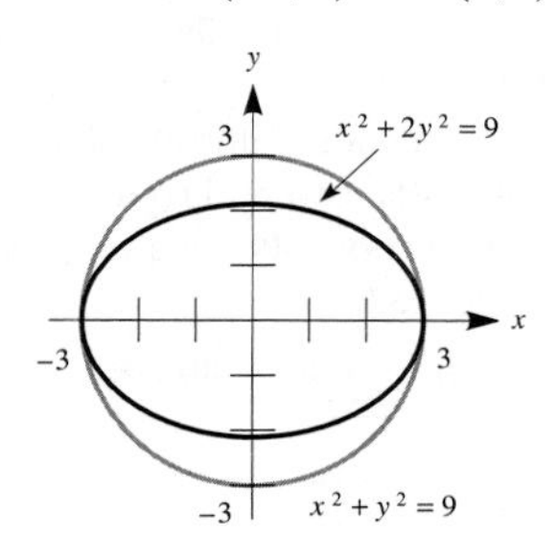

77. (a)

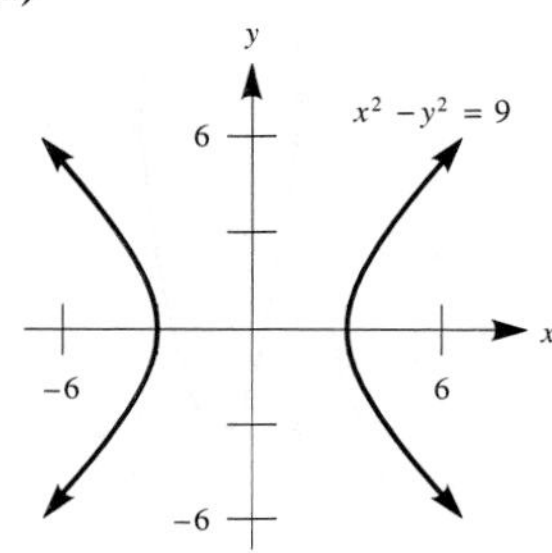

(b)

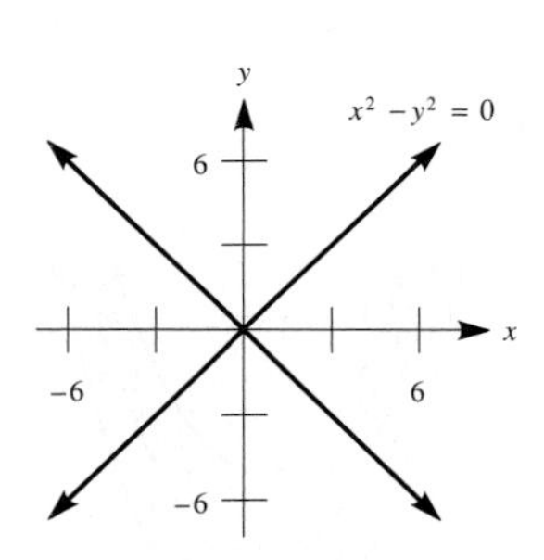

(c) no intersection points

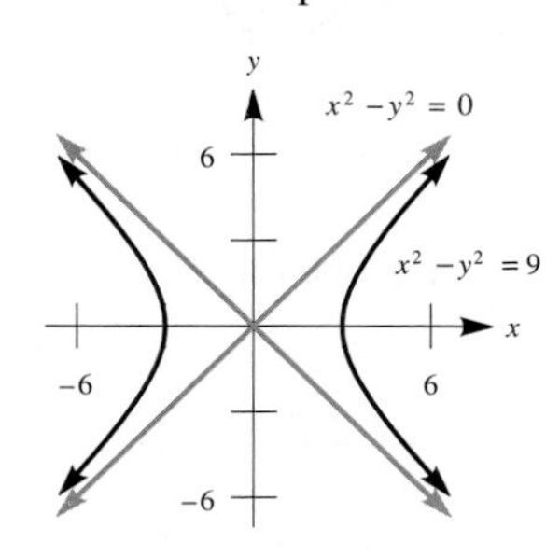

79.

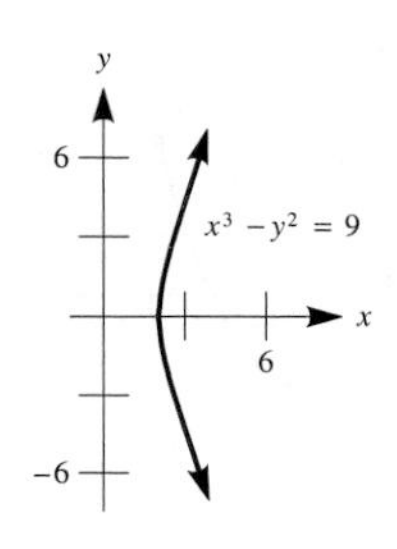

81. (a, b)

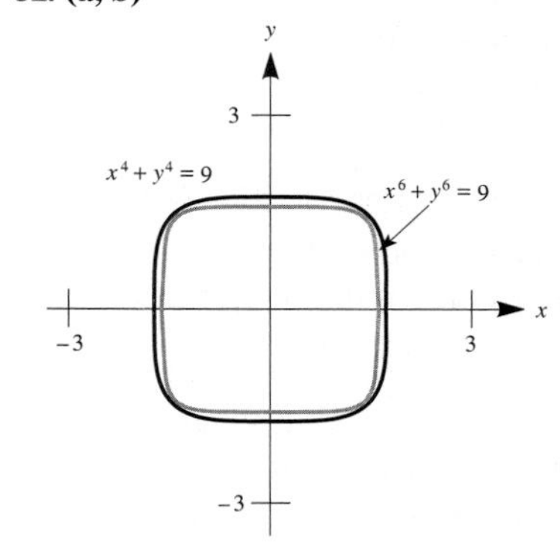

(c)

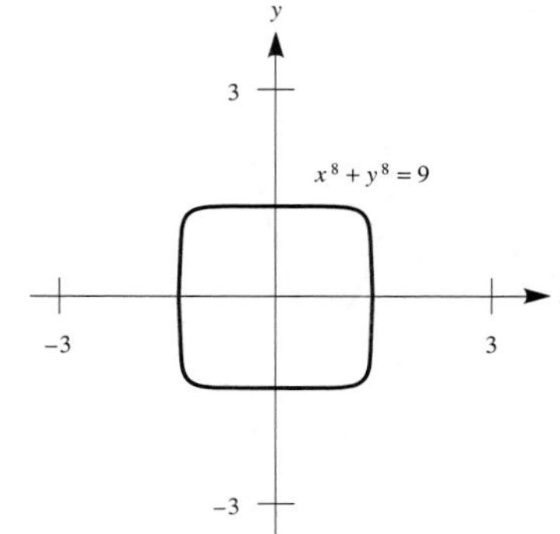

83. $MA = MB = MC = \sqrt{c^2 + b^2}$; all the same **85. (a)** $\sqrt{130}$ **(b)** $-\frac{11}{3}$ **(c)** $(\frac{7}{2}, -\frac{1}{2})$ **87.** $\frac{49}{6}$ square units
89. (a) $(\frac{5}{3}, \frac{11}{3})$; they all intersect at the same point.
(b) $\left(\dfrac{2a}{3} + \dfrac{2b}{3}, \dfrac{2c}{3}\right)$; they all intersect at the same point.
93. (a)

Reflection of $\overline{AB}$ $\overline{AB}$

(b)

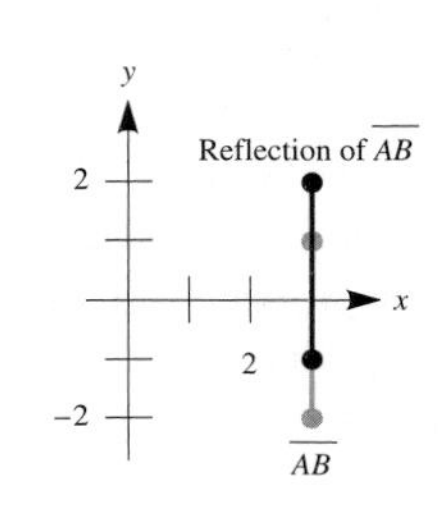

(c)

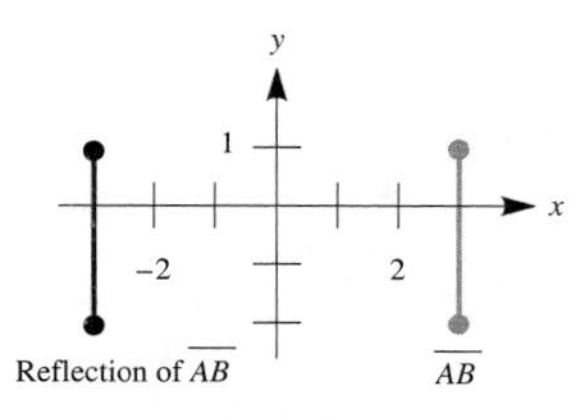

95. $13\sqrt{5}/5$ **97.** $17\sqrt{13}/13$
99. (a) $(x - 1)^2 + (y - 1)^2 = 1$
(b) $\overline{AT}$: $y = \frac{9}{8}x$; $\overline{BU}$: $y = -\frac{1}{4}x + 1$; $\overline{CS}$: $y = -3x + 3$ **(c)** $(\frac{8}{11}, \frac{9}{11})$; they intersect in the same point.

Chapter 1 Test

1. 1 **2. (a)** $(\frac{39}{10}, \frac{41}{10})$ **(b)** $[2, \infty)$
3. $x = 5$ **4.** $x = \frac{14}{11}, -1$
5. (a) $x = -5, 1$ **(b)** $x = -2 \pm \sqrt{5}$
6. $x = \dfrac{de - b}{a - ce}$ **7. (a)** $\Delta N = 777$
(b) $\dfrac{\Delta N}{\Delta t} \approx 194$ stations per year
(c) 11,759 stations
8. (a)

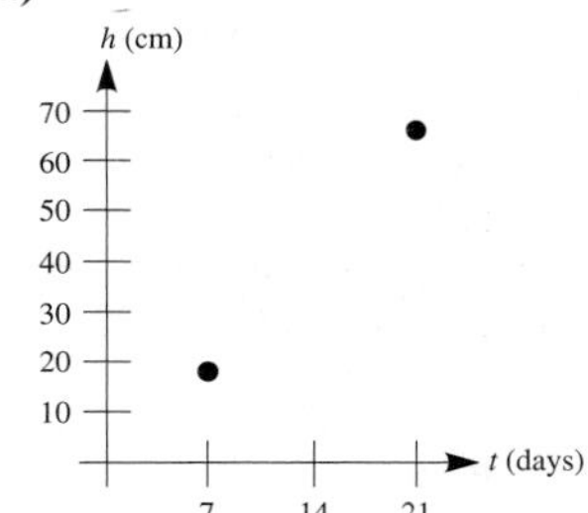

(b) 42.8 cm **(c)** 18% error
9. $y = 5x - 7$ **10.** $y = \frac{6}{5}x - \frac{17}{5}$
11. $(3, 9)$ **12. (a)** origin symmetry
(b) y-axis symmetry
(c) no symmetry
13. x-intercept: 5; y-intercept: -3

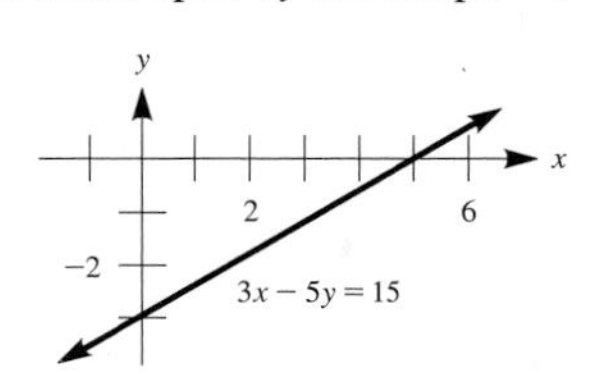

14. x-intercepts: $1 \pm \sqrt{5}$; y-intercepts: $-2 \pm 2\sqrt{2}$

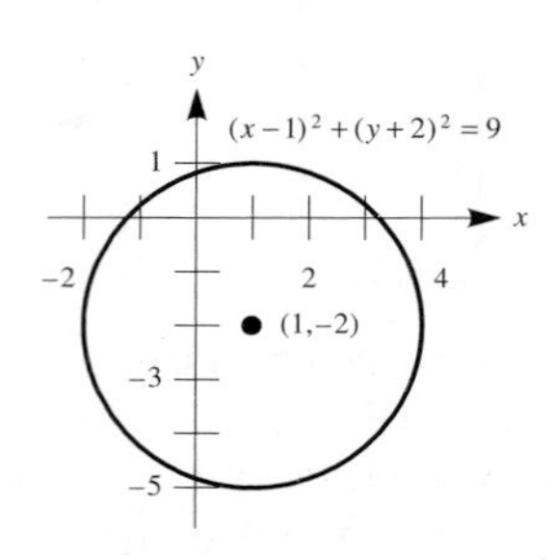

15. (a)

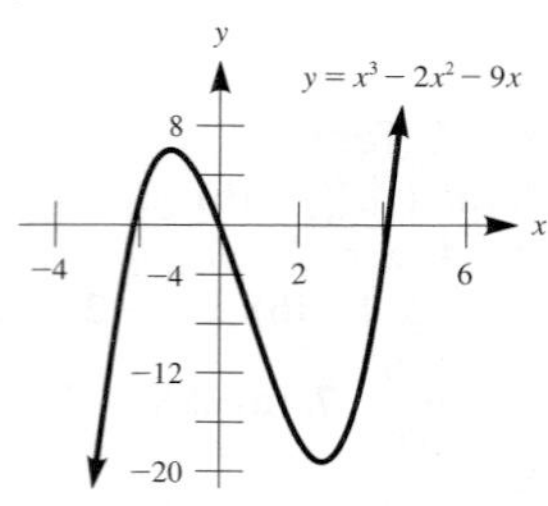

(b) x-intercepts: $-2.2, 0, 4.2$
(c) x-intercepts: 0, $1 \pm \sqrt{10} \approx -2.2, 4.2$ **16.** yes
17. (a)

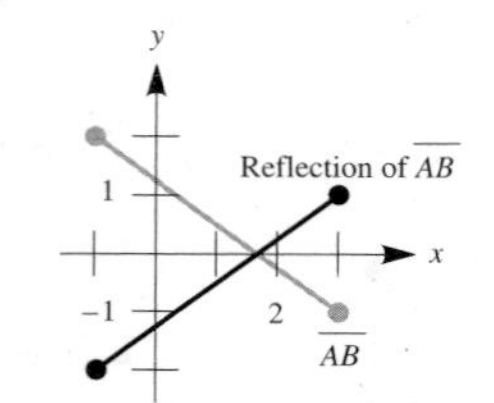

(b)

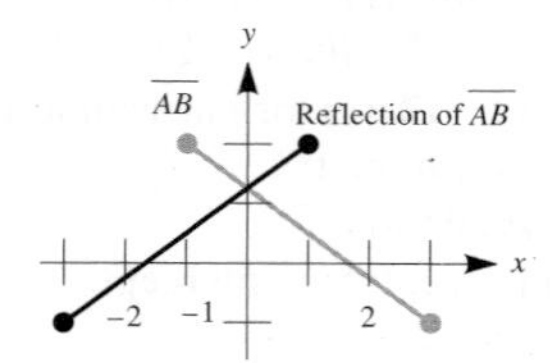

(c)

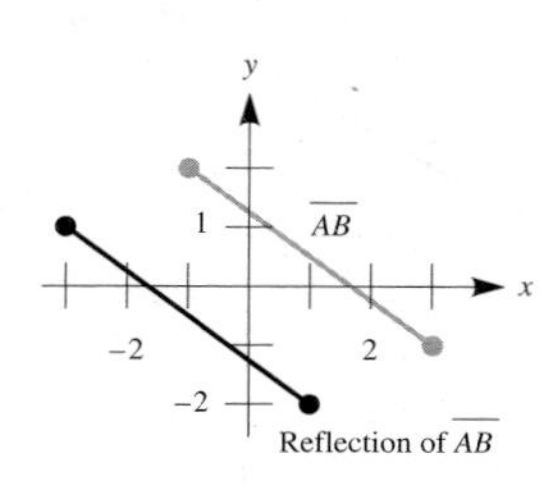

18. $\frac{121}{5}$ square units

CHAPTER 2

Exercise Set 2.1

1. $x = -4 \pm 3\sqrt{2} \approx -8.24, 0.24$
3. $x = -2 \pm \sqrt{3} \approx -3.73, -0.27$
5. $y = \dfrac{5 \pm \sqrt{41}}{4} \approx -0.35, 2.85$
7. no real roots **9.** $s = \frac{5}{2}$
11. $x = 4 \pm \sqrt{10} \approx 0.84, 7.16$
13. $x = \dfrac{1 \pm \sqrt{37}}{6} \approx -0.85, 1.18$
15. $y = -4 \pm \sqrt{15} \approx -7.87, -0.13$
17. no real roots **19. (a)** year 1917 **(b)** it is reasonable **21. (a)** year 1990; 3 years off **(b)** year 1997; 0 years off **23. (a)** x-intercepts: 0.268 and 3.732

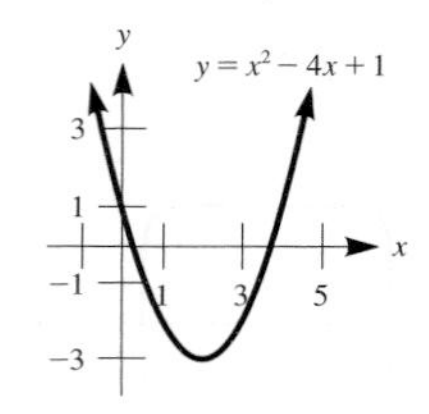

(b) $x = 2 \pm \sqrt{3} \approx 0.2679, 3.7321$
25. (a) x-intercepts: -16.367 and 0.367

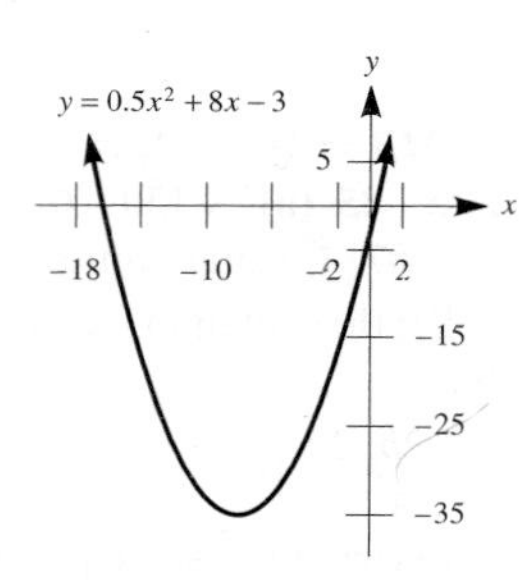

(b) $x = -8 \pm \sqrt{70} \approx -16.3666, 0.3666$
27. (a) x-intercept: -2.550

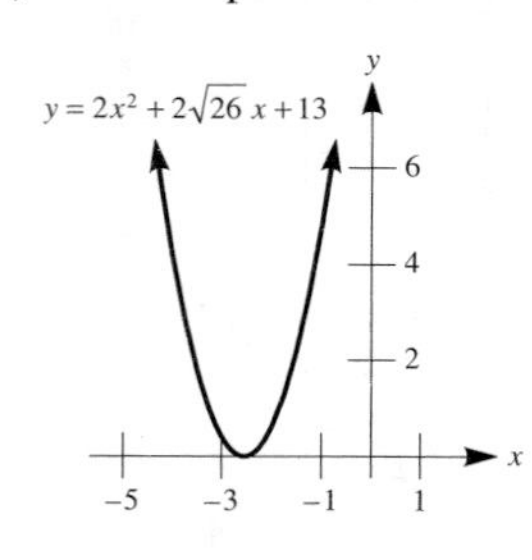

(b) $x = -\sqrt{26}/2 \approx -2.5495$
29. sum: -8; product: -20
31. sum: 7; product: $\frac{9}{4}$
33. $x^2 - 14x + 33 = 0$
35. $x^2 - 2x - 1 = 0$
37. $4x^2 - 8x - 1 = 0$ **39.** $\sqrt{2}$ and $\sqrt{5}$
41. (a) 6 seconds **(b)** 1 second (rising) 5 seconds (falling) **43.** two real roots **45.** two real roots
47. one real root **49.** two real roots
51. $k = 36$ **53.** $k = \pm 2\sqrt{5}$
55. $r = \dfrac{-h \pm \sqrt{h^2 + 40}}{2}$
57. $t = 0$ or $t = \dfrac{v_0}{16}$
59. (a)

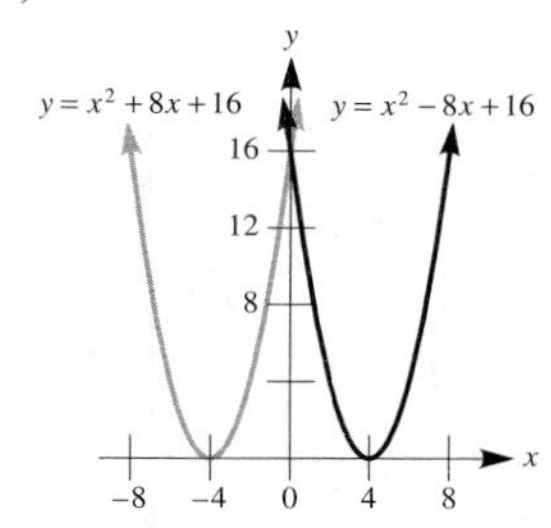

(b) -4 and 4; they are opposites
(c) $x = -4, 4$ **65.** $k = -\frac{10}{9}$ or $k = 2$
69. $x = 1$ or $x = \frac{1}{2}$
71. (a)

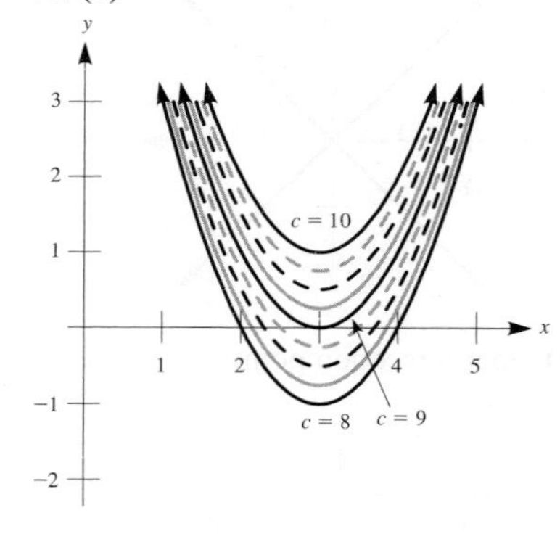

(b) $c = 9$ **(c)** one root: $x = 3$
(d) $c > 9$

Exercise Set 2.2

1. $x = 4, 6$ **3.** $x = -\frac{11}{2}, -\frac{13}{2}$
5. $x = -\frac{10}{3}, 5$ **7.** $x = 5$ **9.** $x = 0, \frac{4}{5}$
11. (a)

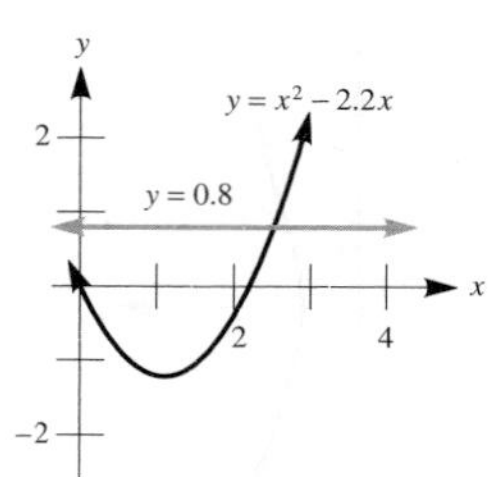

(b) $x \approx 2.518$

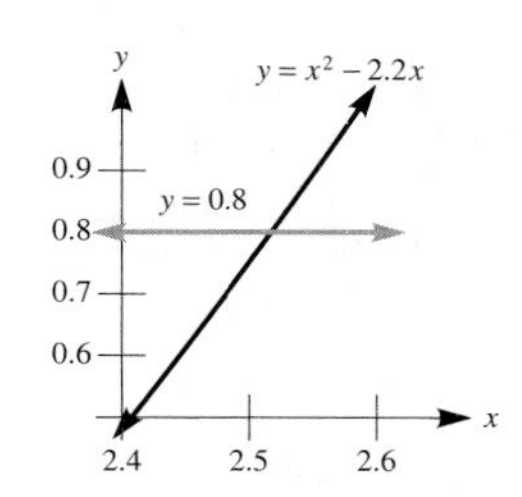

(c) $x = \dfrac{2.2 + \sqrt{8.04}}{2} \approx 2.5177$

13. (a)

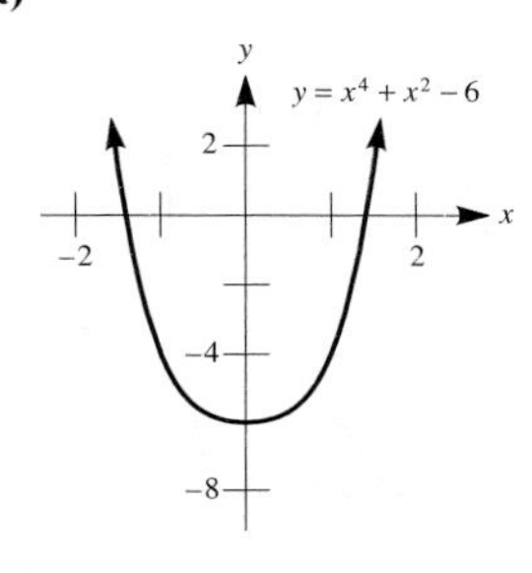

(b) $x \approx \pm 1.414$

15. $x = 0, 16$

17. $t = 5$ **19.** $x = 0, 2, -2$

21. $t = 0, -3, 1$ **23.** $x = 0, \frac{1}{4}, -6$

25. $x = \pm\sqrt{3}$ **27.** $y = -1, 1$

29. $t = \pm\sqrt{6}/3$

31. $x = 2 + \sqrt[3]{5} \approx 3.71$

33. $x = -4 - \sqrt[5]{16} \approx -5.74$

35. (a) $x = 3 \pm \sqrt[4]{30} \approx 5.34, 0.66$

(b) no solution

37. $x = 0, \pm\sqrt{6}, \pm 2$

39. $x = \pm\sqrt{\dfrac{-1 + \sqrt{5}}{2}}$

41. $x = \pm\sqrt{\dfrac{-3 + \sqrt{17}}{2}}$

43. $x = -2, 1$ **45.** $t = \frac{1}{3}, \frac{1}{4}$

47. $y = \frac{3}{5}, 4$ **49.** $x = \pm\frac{1}{3}$

51. $t = \pm 27$

53. $y = 1 + \sqrt[3]{7}$ **55.** $t = -4$

57. $x = \pm\frac{1}{27}, \pm 1$

59. $x = \pm\sqrt{1 + \sqrt{2}} \approx \pm 1.55$

61. $x = -1$ **63.** $x = 4$

65. $x = -6$ **67.** $x = -\frac{1}{2}$

69. $x = \pm 2, \pm 3$ **71.** $x = -4, -\frac{20}{9}$

73. no solutions **75.** $y = 2$

77. (a) (a) $x \approx -1.65, -1.51, 1.51, 1.65$

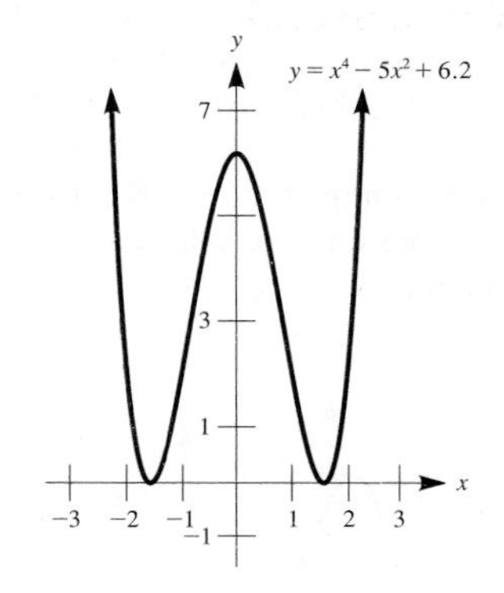

(b) $x = -\sqrt{\dfrac{5 - \sqrt{0.2}}{2}} \approx -1.51,$

$\sqrt{\dfrac{5 - \sqrt{0.2}}{2}} \approx 1.51,$

$-\sqrt{\dfrac{5 + \sqrt{0.2}}{2}} \approx -1.65,$

$\sqrt{\dfrac{5 + \sqrt{0.2}}{2}} \approx 1.65$

(b) (a) no real roots

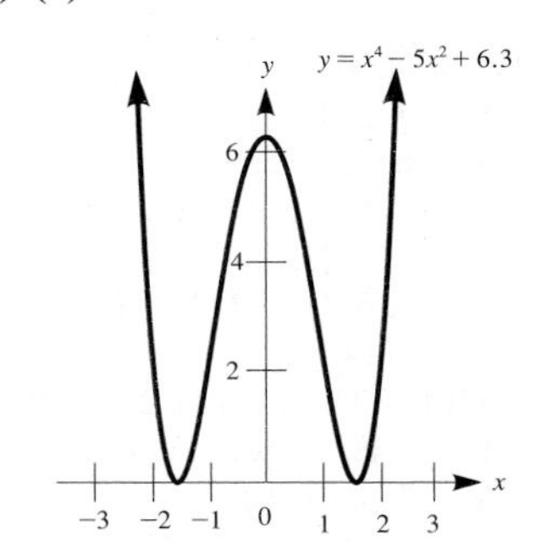

(b) no real roots

79. (a) $x \approx 0.49$ and 18.51

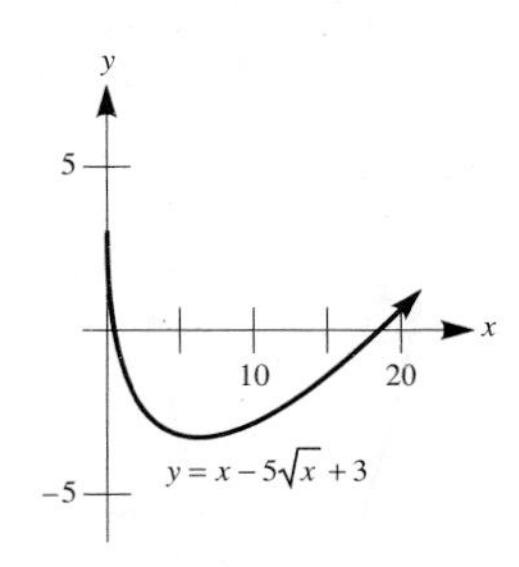

(b) $x = \dfrac{19 \pm 5\sqrt{13}}{2} \approx 0.4861, 18.5139$

81. (a) $x \approx 19.94$

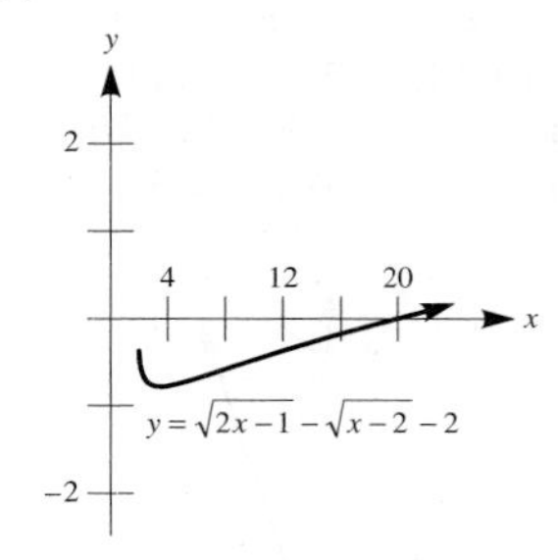

(b) $x = 11 + 4\sqrt{5} \approx 19.9443$

83. (a) $x \approx 2.78$

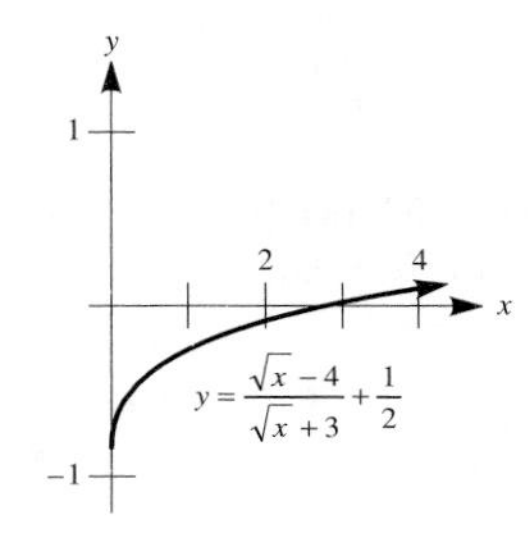

(b) $x = \frac{25}{9}$ **85. (a)** $x \approx -0.24$ and 4.24

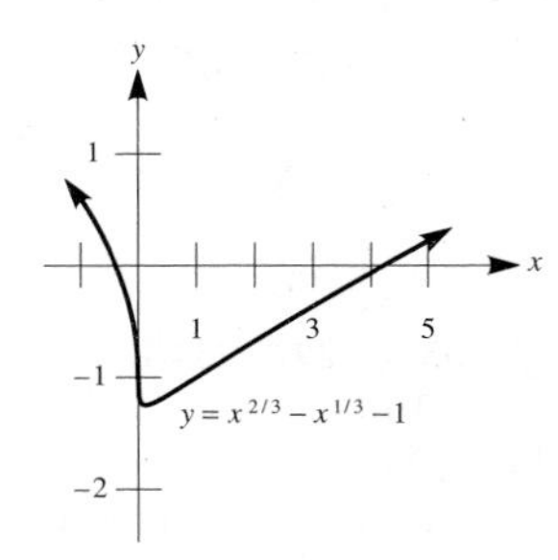

(b) $x = \left(\dfrac{1 \pm \sqrt{5}}{2}\right)^3 \approx -0.2361,$ 4.2361

89. $x = a + b$ **91.** $x = -\dfrac{a}{15}$

Exercise Set 2.3

1. $(-\infty, 9)$ **3.** $(-\infty, -\frac{9}{2})$ **5.** $(-\infty, \frac{5}{2})$

7. $(1, \infty)$ **9.** $[-\frac{1}{2}, 1]$

11. $(3.98, 3.998)$ **13.** $[-8, \frac{7}{5})$

15. (a) $[-\frac{1}{2}, \frac{1}{2}]$

(b) $(-\infty, -\frac{1}{2}) \cup (\frac{1}{2}, \infty)$

17. (a) $(-\infty, 0) \cup (0, \infty)$

(b) no solution **19. (a)** $(-\infty, 3)$

(b) $(1, 3)$ **(c)** $(-\infty, 1) \cup (3, \infty)$

21. (a) $[-4, \infty)$ **(b)** $[-4, 6]$

(c) $(-\infty, -4) \cup (6, \infty)$

23. (a) $(a - c, \infty)$ **(b)** $(a - c, a + c)$
(c) $(-\infty, a - c] \cup [a + c, \infty)$
25. $(-10, 14)$ **27.** $(-11, 1)$
29. (a) $(-2h, h)$ **(b)** $(h, -2h)$
31. (a) $[a, \infty)$ **(b)** $a \approx 0.3$

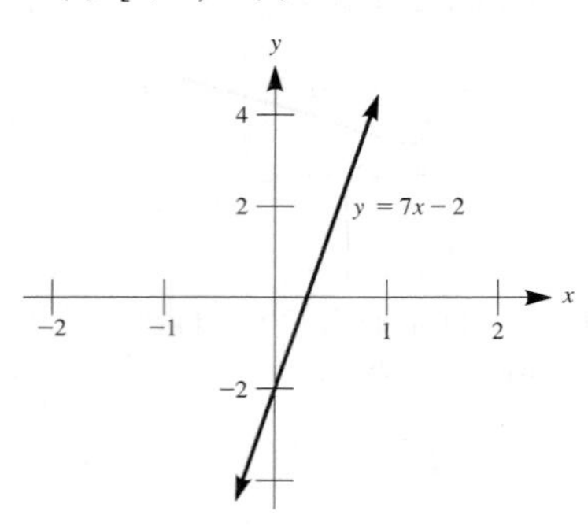

(c) $[\frac{2}{7}, \infty)$ **33. (a)** $[a, b]$
(b) $a \approx 0.1, b \approx 0.6$

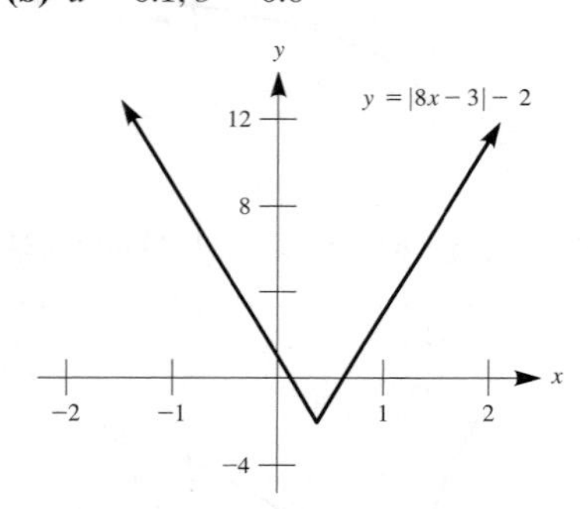

(c) $[\frac{1}{8}, \frac{5}{8}]$ **35.** $[-297°, 234°]$
37. $[-270°, 810°]$
39. $(-\infty, -\frac{7}{2}) \cup (\frac{5}{2}, \infty)$
43. (a) The positive slope tells us that the sulfur dioxide emissions are increasing.

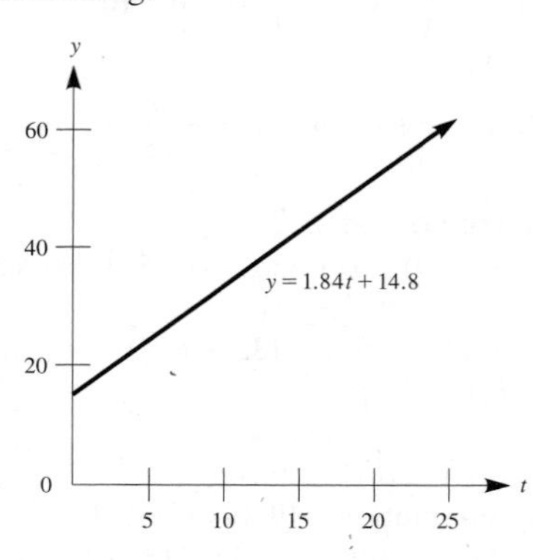

(b) year 2008

Exercise Set 2.4

1. (a) $(-\infty, -1] \cup [4, \infty)$ **(b)** $[-1, 4]$
3. (a) no solution **(b)** $(-\infty, \infty)$
5. (a) $[-1, 1] \cup [3, \infty)$
(b) $(-\infty, -1) \cup (1, 3)$ **7.** The curve is positive on the set $(-1, 0) \cup (3, \infty)$, which represents the solution set for the inequality.

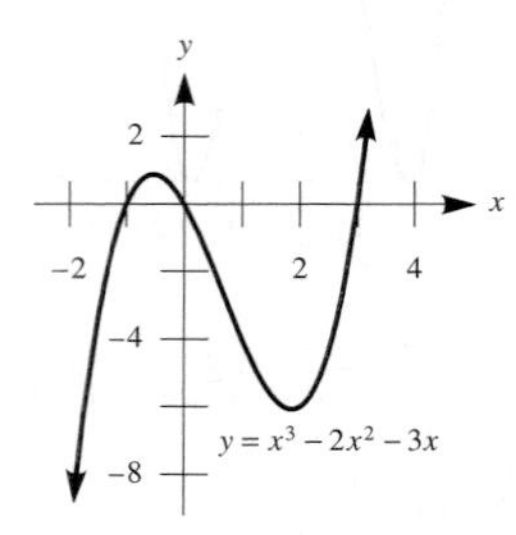

9. $(-3, 2)$ **11.** $(-\infty, 2) \cup (9, \infty)$
13. $(-\infty, 4] \cup [5, \infty)$
15. $(-\infty, -4] \cup [4, \infty)$
17. no solution
19. $(-7, -6) \cup (0, \infty)$ **21.** $(-\infty, \infty)$
23. $(-\infty, -\frac{3}{4}) \cup (-\frac{2}{3}, 0)$
25. $(-\infty, \frac{-1 - \sqrt{5}}{2}) \cup (\frac{-1 + \sqrt{5}}{2}, \infty)$
27. $[4 - \sqrt{14}, 4 + \sqrt{14}]$
29. $[-4, -3] \cup [1, \infty)$
31. $(-\infty, -6) \cup (-5, -4)$
33. $(-\infty, -\frac{1}{3}) \cup (\frac{1}{3}, 2) \cup (2, \infty)$
35. $(-\infty, -\frac{2}{3}) \cup \{3, -\frac{1}{2}\}$
37. $(-\infty, -\sqrt{5}] \cup [-2, 2] \cup [\sqrt{5}, \infty)$
39. $(-3, 3) \cup (4, \infty)$
41. $(-\sqrt{14}, -2\sqrt{2}) \cup (2\sqrt{2}, \sqrt{14})$
43. $(-1, 1]$ **45.** $(-\infty, \frac{3}{2}) \cup [2, \infty)$
47. $(-\infty, -1) \cup (0, 9)$
49. $(-\frac{7}{2}, -\frac{4}{3}) \cup (-1, 0) \cup (1, \infty)$
51. $(-\infty, -1)$ **53.** $(-1, 0)$
55. $[-1, 1) \cup (2, 4]$
57. $(-1, 2 - \sqrt{5}) \cup (1, 2 + \sqrt{5})$
59. $(-\infty, -1) \cup (1, 2) \cup (2, \infty)$
61. (a) $[0.697, 4.303]$

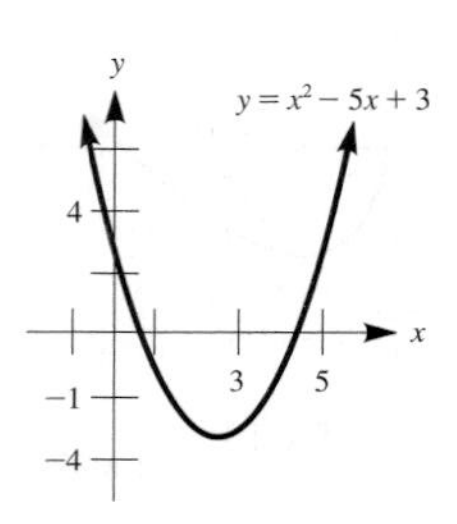

(b) $x = \dfrac{5 \pm \sqrt{13}}{2} \approx 0.697, 4.303$

63. (a) $(-0.329, 24.329)$

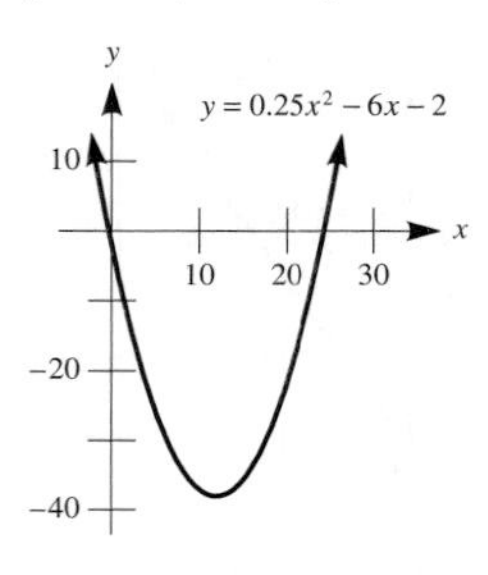

(b) $x = 12 \pm 2\sqrt{38} \approx -0.329, 24.329$
65. (a) $(-\infty, -1.554) \cup (1.554, \infty)$

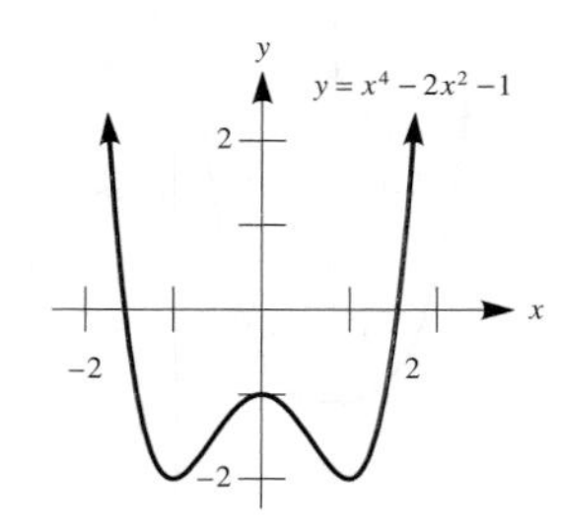

(b) $x = \pm\sqrt{1 + \sqrt{2}} \approx \pm 1.554$
67. (a) $[-2.236, 2.236]$

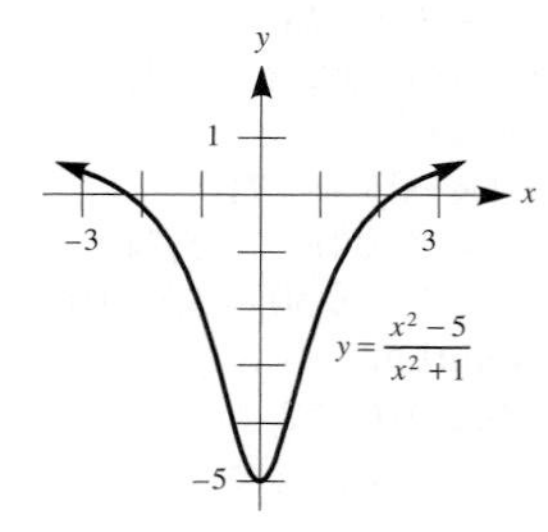

(b) $x = \pm\sqrt{5} \approx \pm 2.236$
69. (a) $(-2.236, 2.236)$

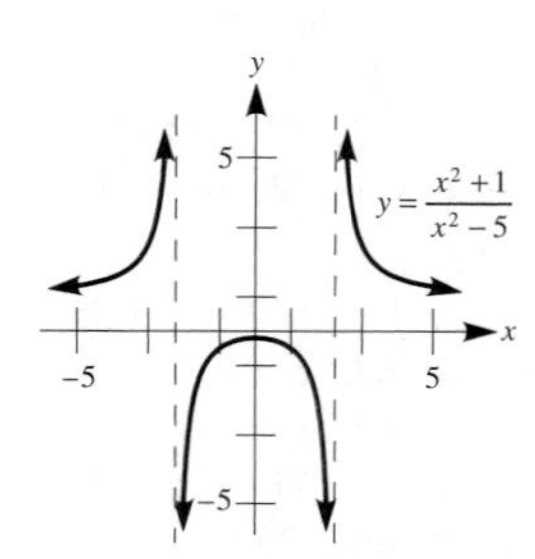

(b) $x = \pm\sqrt{5} \approx \pm 2.236$

71. $[-0.453, \infty)$

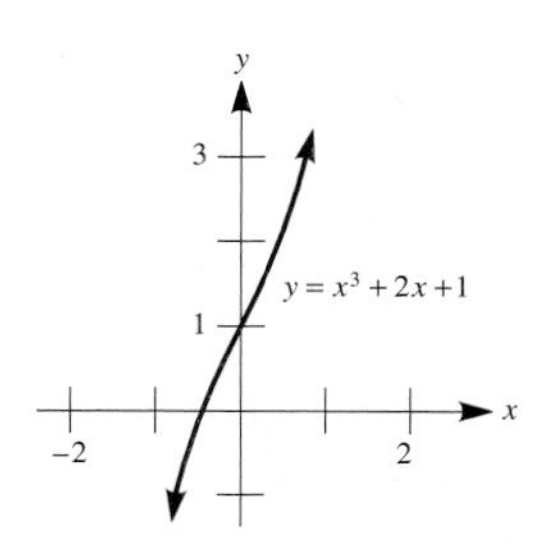

73. $(-\infty, 0.544) \cup (1, \infty)$

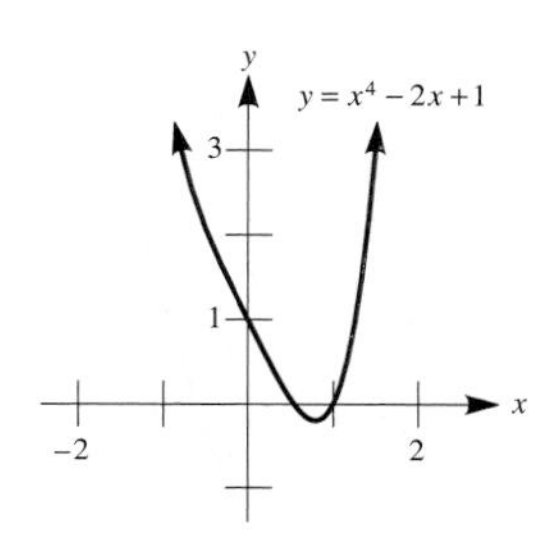

75. $(-3.079, 0) \cup (3.079, \infty)$

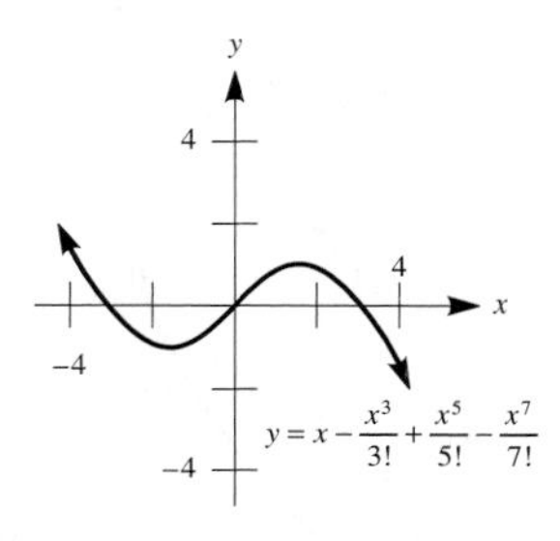

77. (a) April through September **(b)** January, October, November, December
79. $(-\infty, -2] \cup [2, \infty)$
81. $(-\infty, 0) \cup (\frac{1}{2}, \infty)$ **83.** $(\frac{1}{5}, \frac{4}{5})$
85. $c = 2$

Chapter 2 Review Exercises

1. true **3.** false **5.** false
7. true **9.** false **11.** false
13. $x = \frac{1}{3}, -\frac{1}{2}$ **15.** $x = -6, 4$
17. $x = -1, 10$ **19.** $x = \frac{2}{3}, 3$
21. $t = \dfrac{-1 \pm \sqrt{3}}{2}$ **23.** $x = -6, -8$
25. $x = 256, 6561$ **27.** $x = \frac{1}{4}$
29. $x = 10$ **31.** $x = 2$ **33.** $x = \dfrac{1}{2y}$
35. $x = \dfrac{a+b}{ab}, -\dfrac{2}{a}$
37. $x = -a$ or $x = -b$
39. $(-2, 1)$ **41.** $[-\frac{1}{2}, \frac{1}{2}]$
43. $(-\infty, -2] \cup [3, \infty)$ **45.** $(-8, 5)$
47. $(3 - \sqrt{10}, 3 + \sqrt{10})$
49. $(-5, -3) \cup (3, 5)$ **51.** $(-2, \infty)$
53. $(-\frac{5}{2}, 4)$
55. $\left(-\infty, -\sqrt{\dfrac{3+\sqrt{5}}{2}}\right) \cup \left(-\sqrt{\dfrac{3-\sqrt{5}}{2}}, 0\right) \cup \left(0, \sqrt{\dfrac{3-\sqrt{5}}{2}}\right) \cup \left(\sqrt{\dfrac{3+\sqrt{5}}{2}}, \infty\right)$
57. $(-\infty, 3 - 2\sqrt{2}] \cup [3 + 2\sqrt{2}, \infty)$
59. $x^2 + 7x + 12 = 0$ **61.** 7 and 12
63. $-1 + \sqrt{2}$ cm
65. (16 cm, ∞) **67.** 21, 22, 23
69. $\dfrac{-(a+b) + \sqrt{a^2 + 6ab + b^2}}{4}$
71. The ball thrown from 100 ft.
73. $x = \sqrt{2}/2$ **75.** $\dfrac{r}{1 + \sqrt{2}}$

Chapter 2 Test

1. (a) $x = -5, 1$ **(b)** $x = -2 \pm \sqrt{5}$
2. $x = \pm 2, \pm\sqrt{3}$ **3.** $x = -2, 2$
4. $x = -\frac{1}{3}, 1$ **5.** -4
6. $x^2 - 4x - 59 = 0$
7. $x = \pm\sqrt{\dfrac{3 + \sqrt{13}}{2}} \approx \pm 1.817$
8. (a) $x \approx -2.83, 2.83$

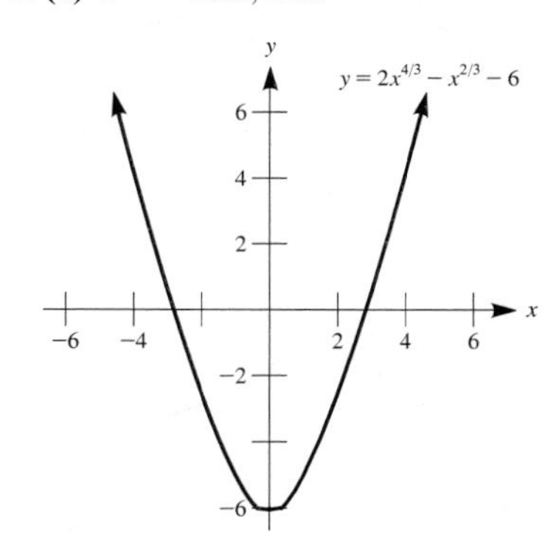

(b) $x = \pm 2\sqrt{2} \approx \pm 2.83$ **9.** $(-\infty, 3]$
10. $(\frac{27}{10}, \frac{31}{10})$ **11.** $[\frac{7}{3}, 3]$ **12.** $[-8, \infty)$
13. $(-2, \frac{-3-\sqrt{3}}{3}] \cup (-1, \frac{-3+\sqrt{3}}{3}] \cup (0, \infty)$ **14.** $[-\frac{3}{2}, \frac{3}{2}]$ **15.** $(-4, -7)$
16. $x = 6$ is a solution to the original equation. However, $x = -1$ results in the statement $\sqrt{4} = -2$, which is false. Thus $x = -1$ is an extraneous solution of the original equation, even though it is a solution to the subsequent equation $x^2 - 5x - 6 = 0$.

CHAPTER 3

Exercise Set 3.1

1. (a) $g(1975) = \$2.00$ **(b)** $g(1995) - g(1975) = \$2.25$; The minimum wage increased \$2.25 from 1975 to 1995. **3. (a)** range **(b)** $h(\text{Mars}) = 2$ **(c)** $h(\text{Neptune})$; Neptune has more moons than does Pluto. **5. (a)** f and g **(b)** range of f: $\{1, 2, 3\}$; range of g: $\{2, 3\}$ **7. (a)** g **(b)** range of g: $\{i, j\}$ **9. (a)** $(-\infty, \infty)$ **(b)** $(-\infty, \frac{1}{5}) \cup (\frac{1}{5}, \infty)$ **(c)** $(-\infty, \frac{1}{5}]$ **(d)** $(-\infty, \infty)$ **11. (a)** $(-\infty, \infty)$ **(b)** $(-\infty, -3) \cup (-3, 3) \cup (3, \infty)$ **(c)** $(-\infty, -3) \cup [3, \infty)$ **(d)** $(-\infty, \infty)$
13. (a) $(-\infty, \infty)$ **(b)** $(-\infty, 3) \cup (3, 5) \cup (5, \infty)$ **(c)** $(-\infty, 3] \cup [5, \infty)$ **(d)** $(-\infty, \infty)$
15. (a) $(-\infty, -3) \cup (-3, \infty)$ **(b)** $(-\infty, -3) \cup [2, \infty)$ **(c)** $(-\infty, -3) \cup (-3, \infty)$
17. domain: $(-\infty, \infty)$; range: $(-\infty, \infty)$
19. domain: $(-\infty, \infty)$; range: $(-\infty, \infty)$
21. domain: $(-\infty, 6) \cup (6, \infty)$; range: $(-\infty, \frac{4}{3}) \cup (\frac{4}{3}, \infty)$
23. (a) domain: $(-\infty, 5) \cup (5, \infty)$; range: $(-\infty, 1) \cup (1, \infty)$ **(b)** domain: $(-\infty, \sqrt[3]{5}) \cup (\sqrt[3]{5}, \infty)$; range: $(-\infty, 1) \cup (1, \infty)$
25. domain: $(-\infty, \infty)$; range: $[4, \infty)$
27. (a) $y = (x-3)^2$ **(b)** $y = x^2 - 3$ **(c)** $y = (3x)^2$ **(d)** $y = 3x^2$
29. (a) -1 **(b)** 1 **(c)** 5 **(d)** $-\frac{5}{4}$ **(e)** $z^2 - 3z + 1$ **(f)** $x^2 - x - 1$ **(g)** $a^2 - a - 1$ **(h)** $x^2 + 3x + 1$ **(i)** 1 **(j)** $4 - 3\sqrt{3}$ **(k)** $1 - \sqrt{2}$ **(l)** 2 **31. (a)** $12x^2$ **(b)** $6x^2$ **(c)** $3x^4$ **(d)** $9x^4$ **(e)** $\frac{3}{4}x^2$ **(f)** $\frac{3}{2}x^2$ **33. (a)** -3 **(b)** $-\frac{7}{18}$ **(c)** $-2x^2 - 4x - 1$ **(d)** $1 - 2x^2 - 4xh - 2h^2$ **35. (a)** 2 **(b)** 2 **(c)** 2 **37. (a)** $x = -2, 8$ **(b)** no real solutions **(c)** $x = 3$
39. 1041 tee shirts per month
41. first model: 124 sales; second model: 100 sales **43. (a)** $2x^2 + 5x + 2$ **(b)** $2x^2 - 11x + 14$ **(c)** $16x - 12$
45. $a = 1$ and $b = 2$ **47.** $a = 2$
49. (a) $f(a) = 0, f(2a) = \frac{1}{3}, f(3a) = \frac{1}{2}$: No **51.** $a = \frac{1}{2}$ and $b = -\frac{3}{2}$ **53.** z
55. $k = -1$ **57.** $c = 6$ **59.** 0

61. F is a function, since each person has exactly one birth mother. G is not a function, since a person can have more than one aunt or no aunt.
63. $f(8) = 4; f(10) = 4; f(50) = 15$
65. (a) $G(10) = 5; G(14) = 9$
(b) $G(100) = 9; G(750) = 0;$
$G(1000) = 9$

Exercise Set 3.2

1. $y = \sqrt{3} \approx 1.732$
3. $y = \sqrt{5}/5 \approx 0.447$ **5. (a)** yes
(b) no **(c)** no **(d)** yes
7. domain: $[-4, 2]$; range: $[-3, 3]$
9. domain: $[-3, 4]$; range: $[-2, 2]$
11. domain: $[-4, -1) \cup (-1, 4]$; range: $[-2, 3)$ **13.** domain: $[-4, 3]$; range: $\{2\}$ **15. (a)** 1 **(b)** -3 **(c)** no **(d)** $x = 2$ **(e)** -1 **17. (a)** positive
(b) $f(-2) = 4; f(1) = 1; f(2) = 2;$ $f(3) = 0$ **(c)** $f(2)$ **(d)** -3 **(e)** 3
(f) domain: $[-2, 4]$; range: $[-2, 4]$
19. (a) $g(-2)$ **(b)** 5 **(c)** $f(2) - g(2)$
(d) $x = -2$ or $x = 3$ **(e)** range of f
21. (a)

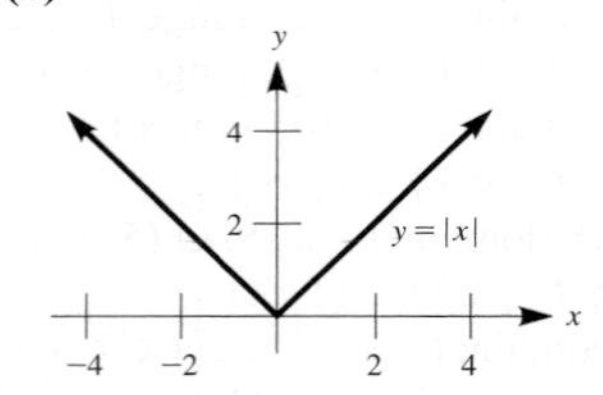

(b)

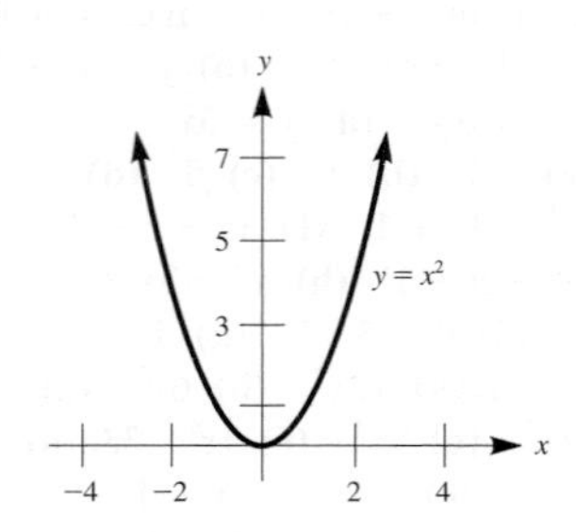

(c)

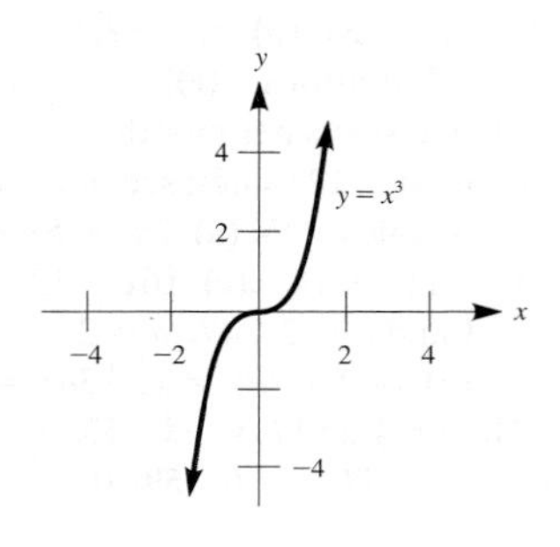

23.

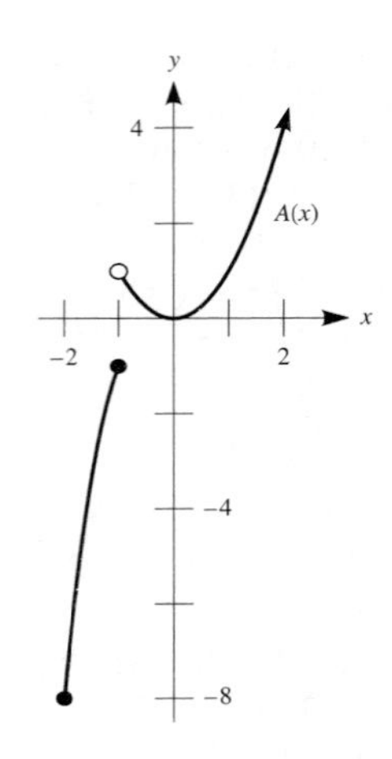

25.

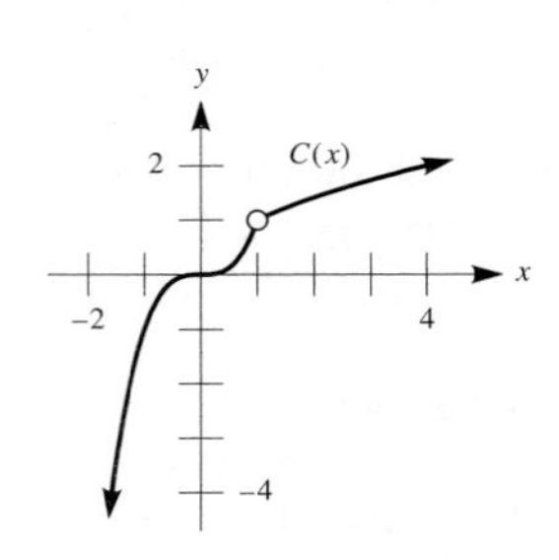

27. (a)

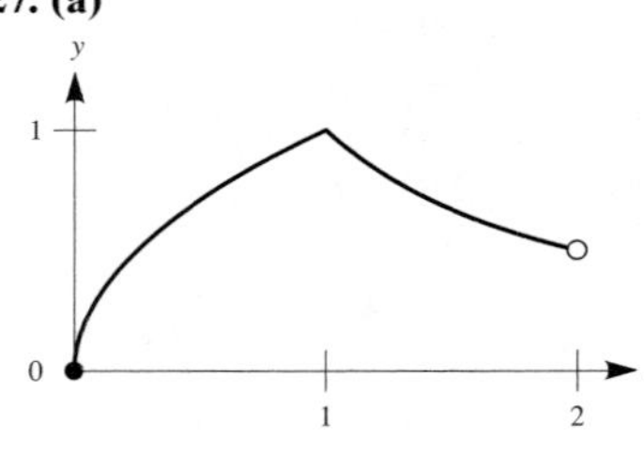

(b)

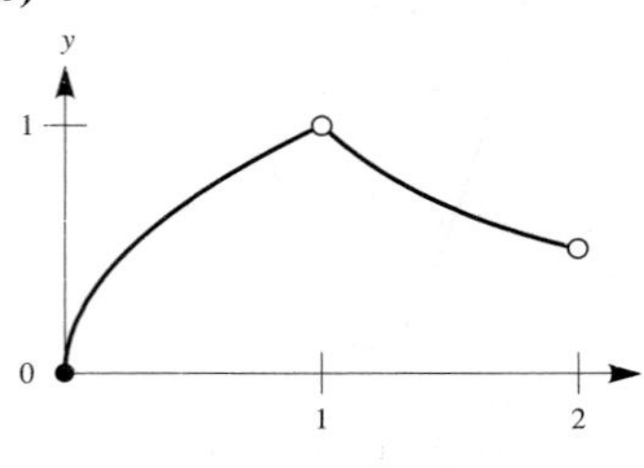

29. (a)

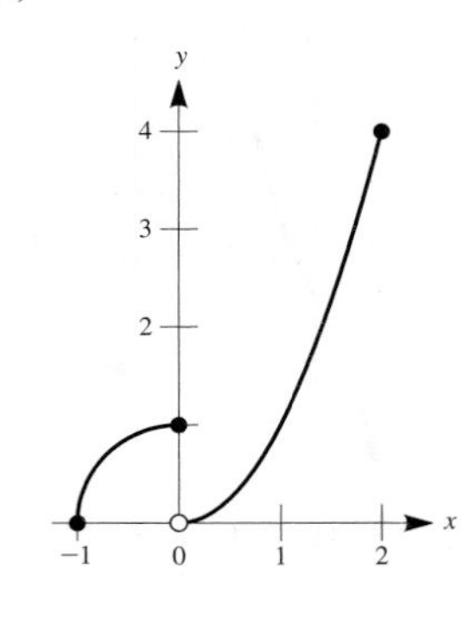

(b)

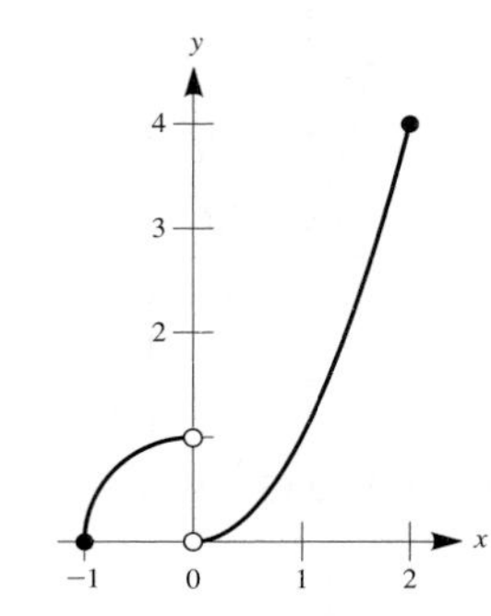

31. (a)

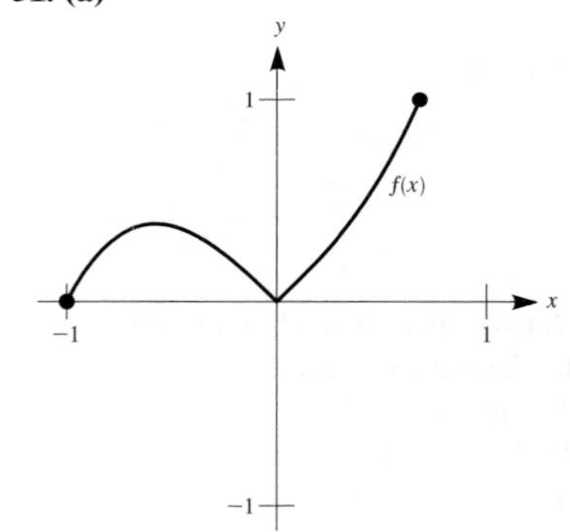

(b)

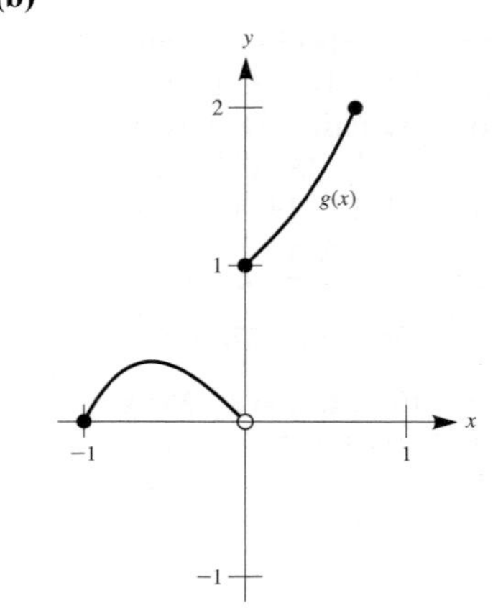

33. $P(4, \sqrt[3]{4}) \approx P(4, 1.587)$, $Q(\sqrt[3]{4}, \sqrt[3]{4}) \approx Q(1.587, 1.587)$, $R(\sqrt[3]{4}, \sqrt[9]{4}) \approx R(1.587, 1.167)$
35. $P(\sqrt{2}, -\sqrt{2}) \approx P(1.414, -1.414)$, $Q(-\sqrt{2}, -\sqrt{2}) \approx Q(-1.414, -1.414)$, $R(-\sqrt{2}, \sqrt{2}) \approx R(-1.414, 1.414)$
37. (a)

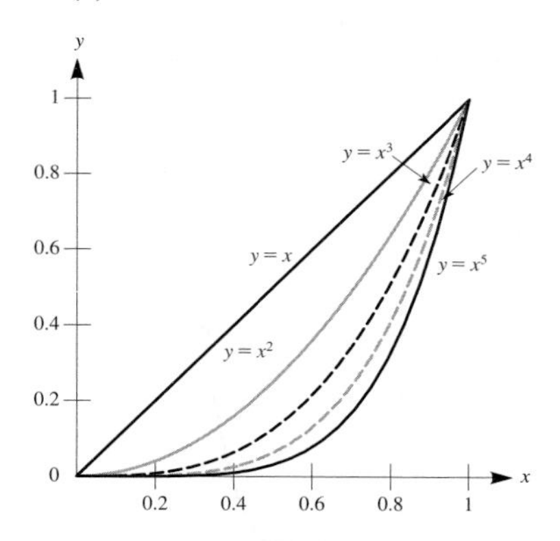

(b) As the power on x increases, the curve "hugs" the x-axis closer, before rising to the point $(1, 1)$.

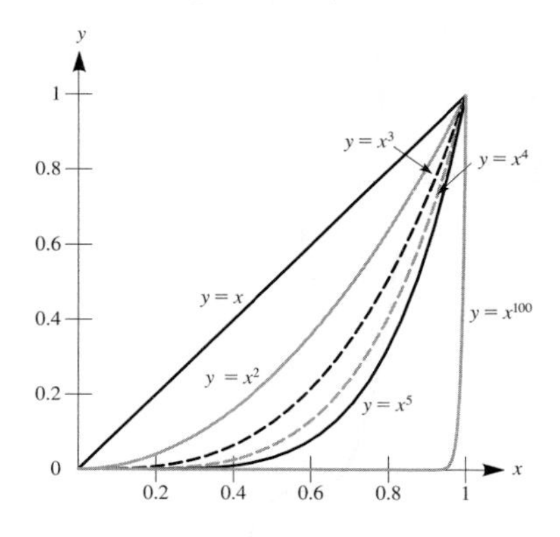

Exercise Set 3.3

1. (a) A turning point is a point where the function changes from increasing to decreasing or from decreasing to increasing. For example, $y = |x|$, $y = x^2$, and $y = \sqrt{1 - x^2}$ all have turning points at $x = 0$. **(b)** A maximum value is a y-value corresponding to the highest point on the graph. For example, $y = 1$ is a maximum value for the function $y = \sqrt{1 - x^2}$ occurring at $x = 0$. **(c)** A minimum value is a y-value corresponding to the lowest point on the graph. For example, $y = 0$ is a minimum value for the functions $y = |x|$ and $y = x^2$. **(d)** A function f is increasing on an interval if $f(x_1) < f(x_2)$ whenever $x_1 < x_2$ on the interval. For example, $y = |x|$ and $y = x^2$ are increasing on $[0, \infty)$, $y = x^3$ is increasing on $(-\infty, \infty)$, $y = \sqrt{x}$ is increasing on $[0, \infty)$, and $y = \sqrt{1 - x^2}$ is increasing on $[-1, 0]$. **(e)** A function f is decreasing on an interval if $f(x_1) > f(x_2)$ whenever $x_1 < x_2$ on the interval. For example, $y = |x|$ and $y = x^2$ are decreasing on $(-\infty, 0]$, $y = 1/x$ is decreasing on either $(-\infty, 0)$ or $(0, \infty)$, and $y = \sqrt{1 - x^2}$ is decreasing on $[0, 1]$.
3. (a) $[-1, 1]$ **(b)** 1 (occurring at $x = 1$) **(c)** -1 (occurring at $x = 3$) **(d)** $[0, 1]$ and $[3, 4]$ **(e)** $[1, 3]$
5. (a) $[-3, 0]$ **(b)** 0 (occurring at $x = 0$ and $x = 4$) **(c)** -3 (occurring at $x = 2$) **(d)** $[2, 4]$ **(e)** $[0, 2]$ **7.** turning points: $(-1.15, 1.08)$ and $(1.15, -5.08)$; increasing: $(-\infty, -1.15] \cup [1.15, \infty)$; decreasing: $[-1.15, 1.15]$ **9.** 10
11. 0 **13.** 2 **15. (a)** $\frac{1}{3}$ °C/min **(b)** $\frac{4}{3}$ °C/min **(c)** $\frac{1}{2}$ °C/min
17. (a) 1984–1990; 1978–1984: 1.8%/year; 1984–1990: 9.7%/year **(b)** 1978–1984: 1.7%/year; 1984–1990: 9.7%/year
19. (a) $\frac{1}{16} \approx 0.06$ million tons/year **(b)** 0.10 million tons/year
21. (a) $2x + 6$ **(b)** $2x + 2a$
23. (a) $4 + h$ **(b)** $2x + h$ **25. (a)** 8 **(b)** 8 **27. (a)** $x + a - 2$ **(b)** $2x + h - 2$ **29. (a)** $-\dfrac{1}{ax}$ **(b)** $-\dfrac{1}{x(x + h)}$
31. (a) $2x^2 + 2ax + 2a^2$ **(b)** $6x^2 + 6hx + 2h^2$
33. 48 feet/second
35. (a) $(64 + 16h)$ feet/second
(b)

h(seconds)	0.1	0.01	0.001	0.0001	0.00001
Average velocity $\Delta s/\Delta t$ on interval $[2, 2 + h]$	65.6	64.16	64.016	64.0016	64.00016

(c) 64 feet/second
37. $0 \le x \le 100$: $-\frac{3}{25}$ dollars/item; $300 \le x \le 400$: $-\frac{3}{200}$ dollars/item. As more items are produced, the price should decrease.
39.

Function	$\lvert x\rvert$	x^2	x^3
Domain	$(-\infty, \infty)$	$(-\infty, \infty)$	$(-\infty, \infty)$
Range	$[0, \infty)$	$[0, \infty)$	$(-\infty, \infty)$
Turning point	$(0, 0)$	$(0, 0)$	none
Maximum value	none	none	none
Minimum value	0	0	none
Interval(s) where increasing	$(0, \infty)$	$(0, \infty)$	$(-\infty, \infty)$
Interval(s) where decreasing	$(-\infty, 0)$	$(-\infty, 0)$	none

41. $b = 5$ **45.** turning point: $(-0.1875, -0.0016)$; increase: $[-0.1875, \infty)$; decrease: $(-\infty, -0.1875]$

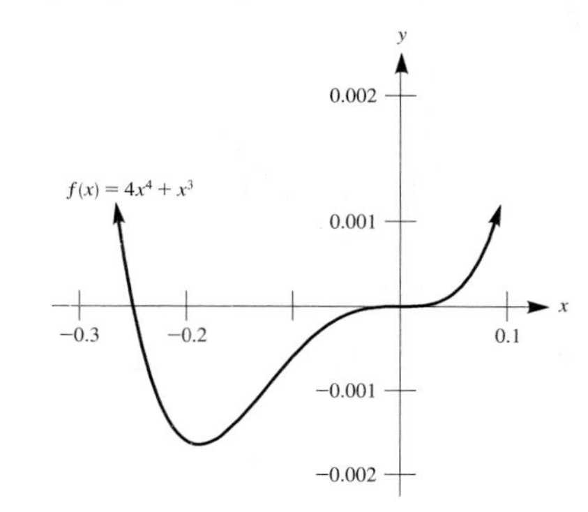

47. (a) They have the same graph.

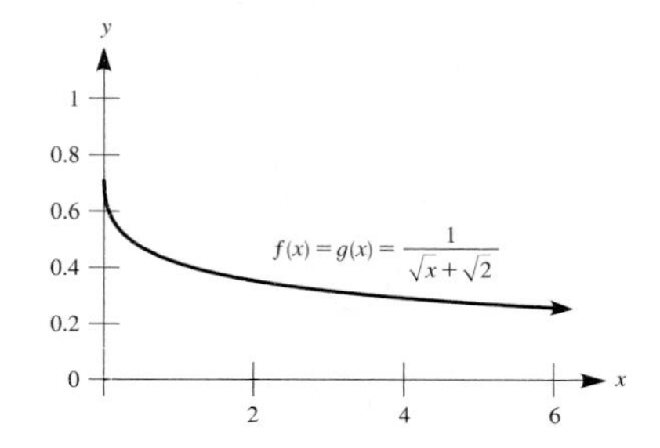

(b) $x = 2$ is not in the domain of $f(x)$ but is in the domain of $g(x)$, so the equation doesn't hold for $x = 2$.

Exercise Set 3.4

1. (a) C **(b)** F **(c)** I **(d)** A **(e)** J **(f)** K **(g)** D **(h)** B **(i)** E **(j)** H **(k)** G

3.

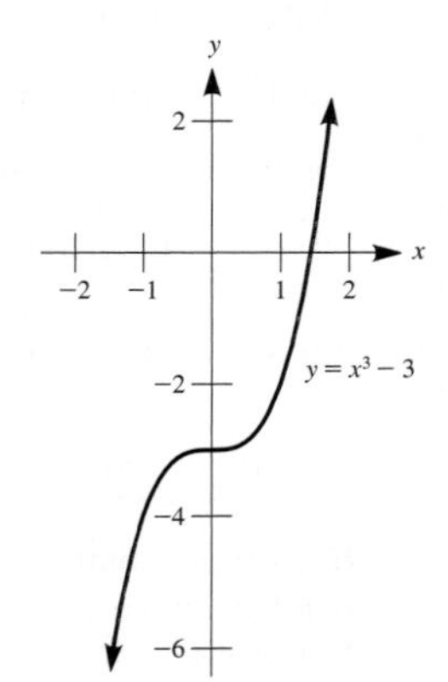

5.

$y = (x + 4)^2$

7.

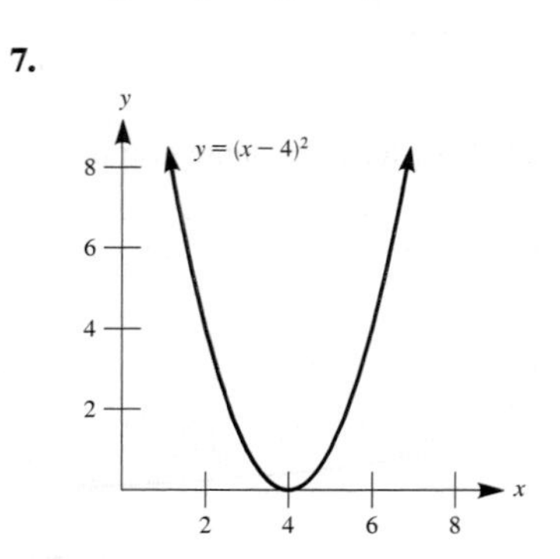

9.

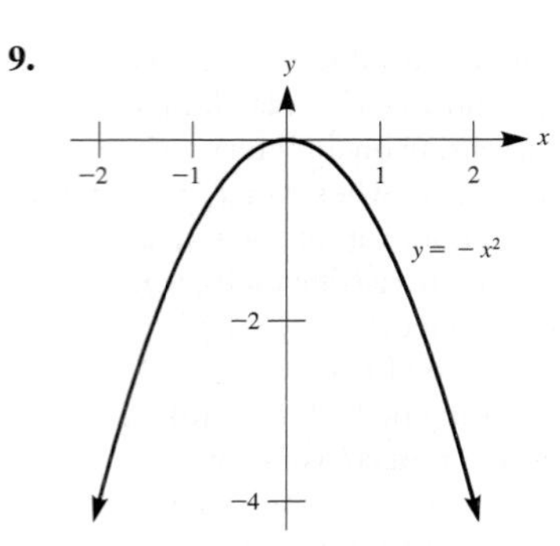

11.

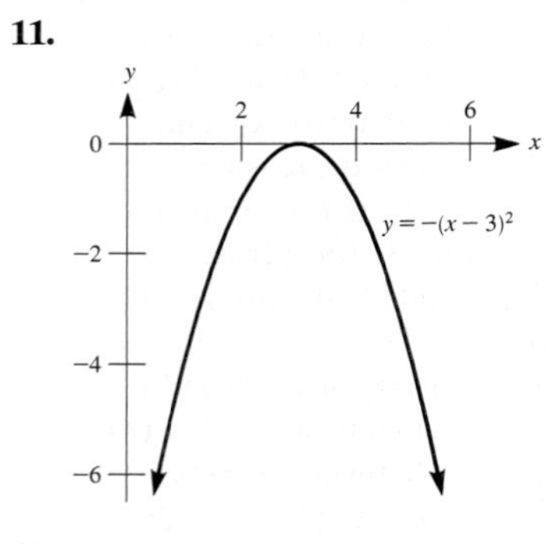

13.

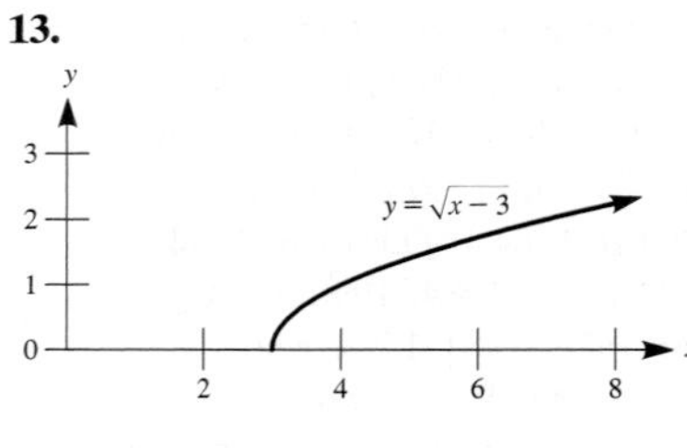

15.

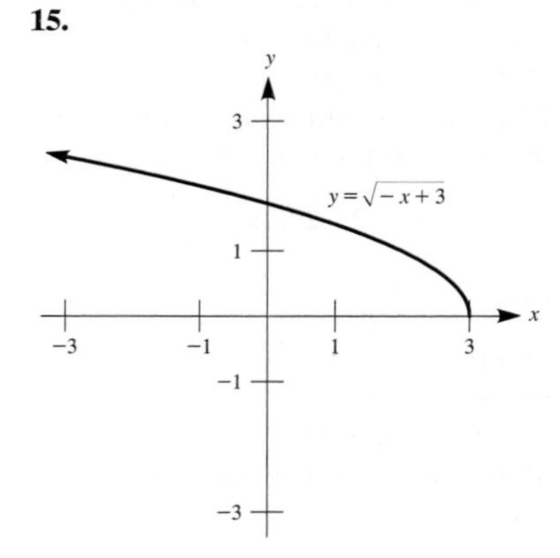

17.

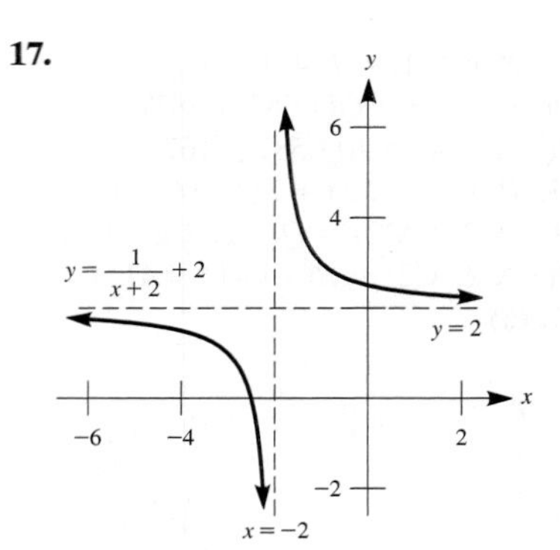

19.

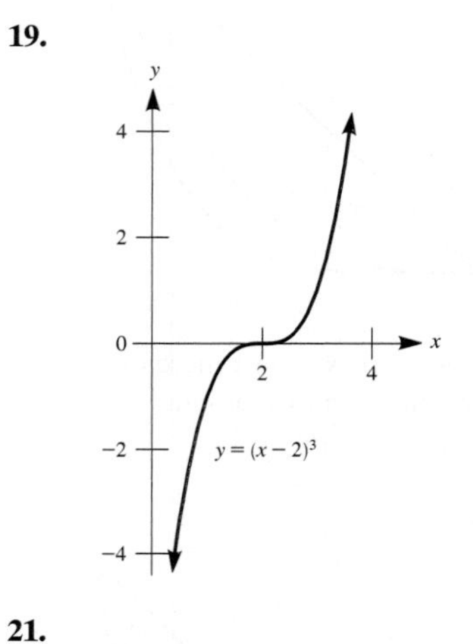

21.

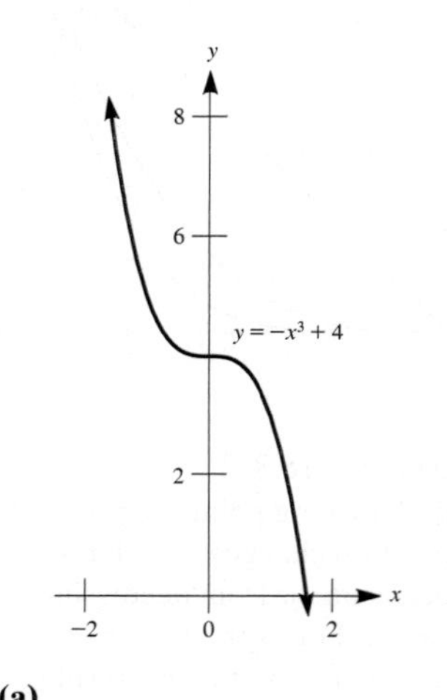

23. (a)

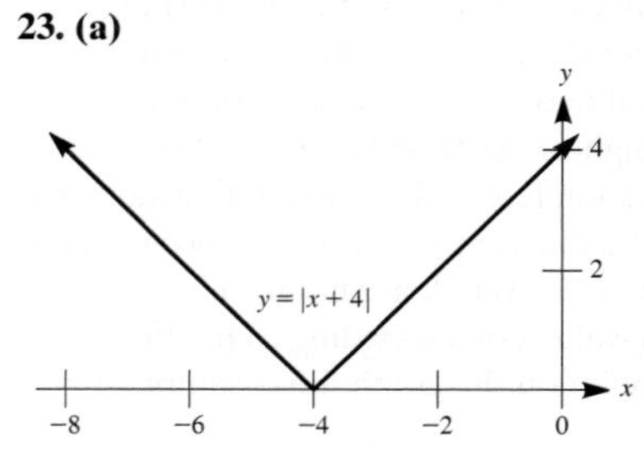

(b)

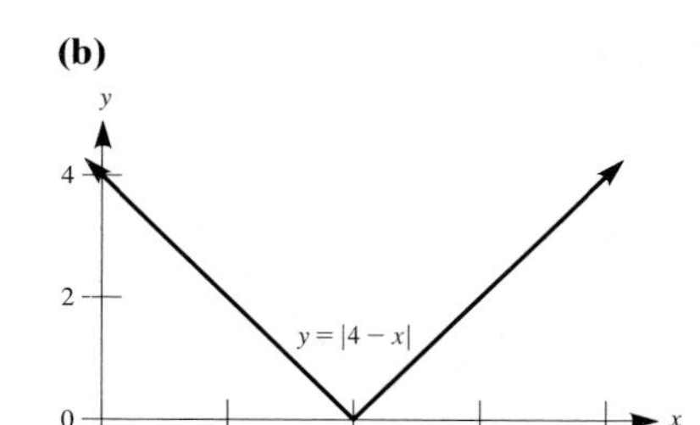

(c)

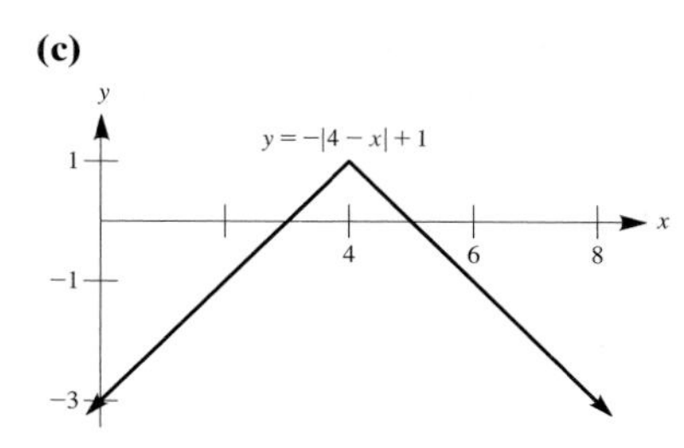

25.

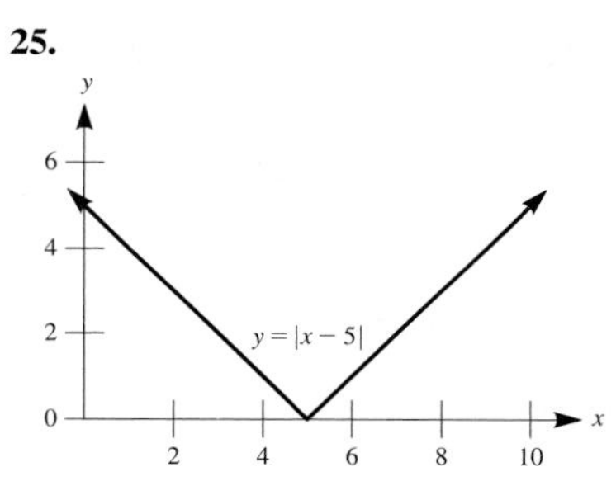

27.

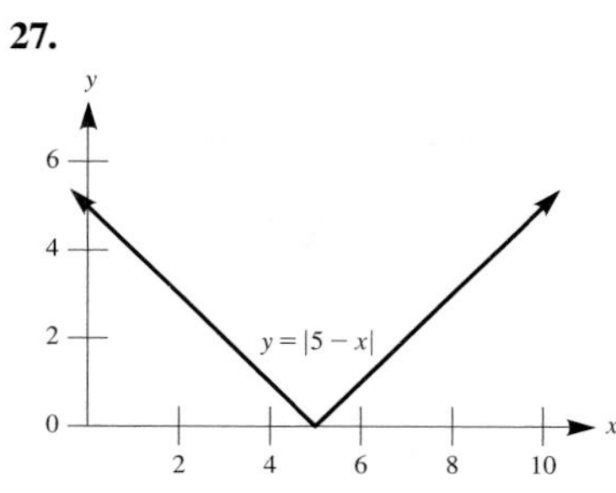

29.

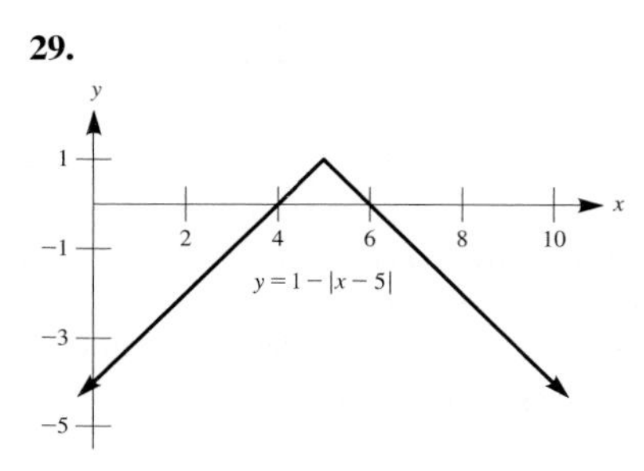

31.

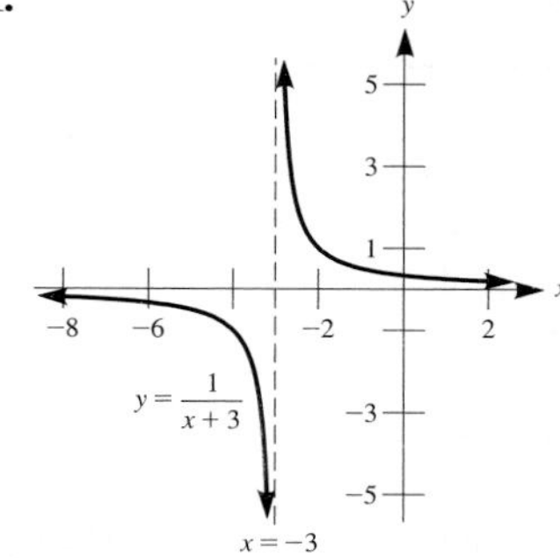

33.

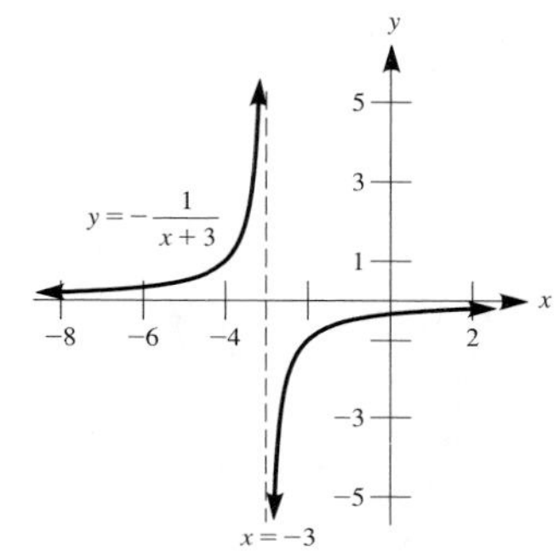

35.

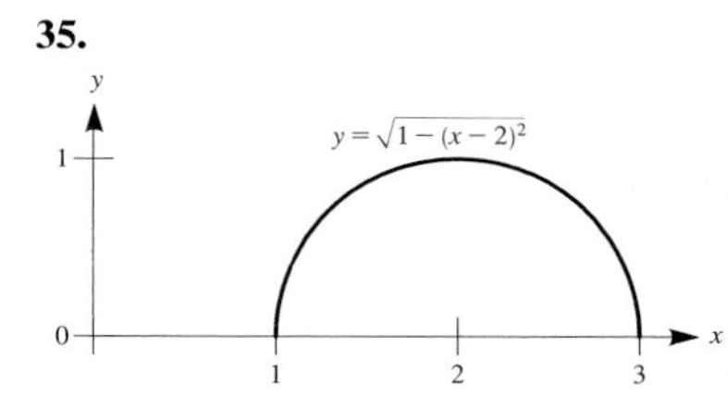

37.

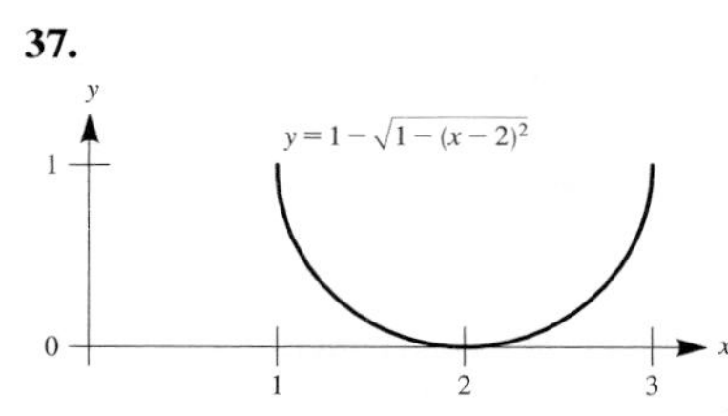

39.

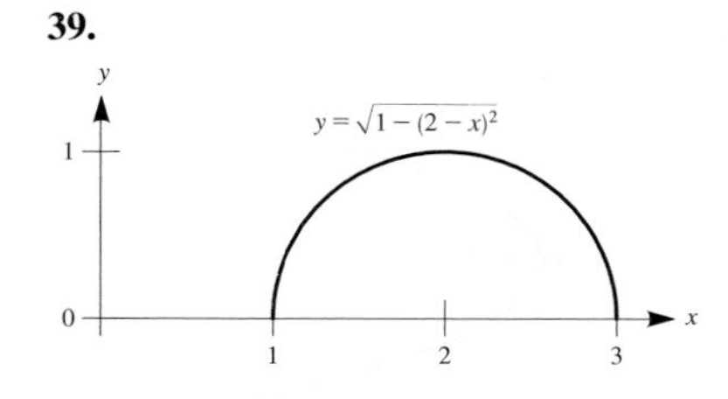

41. (a) $f(x)$ reflected across the y-axis

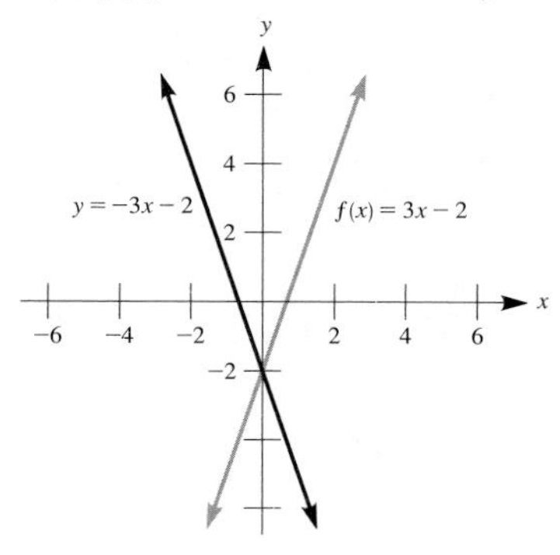

(b) $f(x)$ reflected across the x-axis

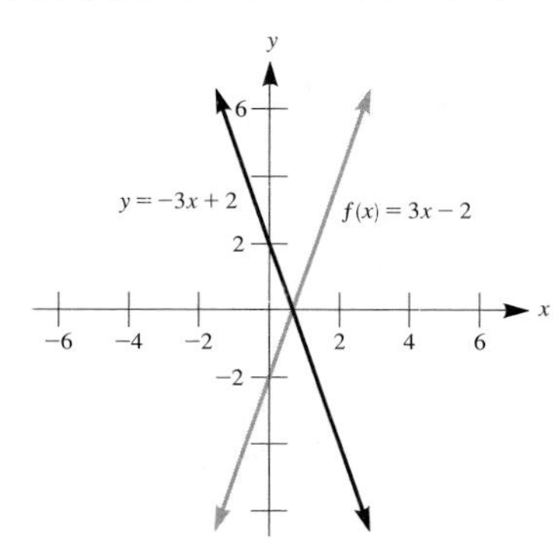

43. (a) $f(x)$ reflected across the y-axis

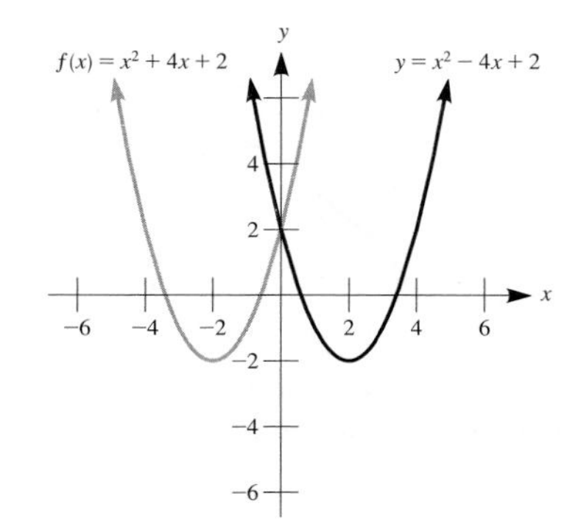

(b) $f(x)$ displaced down two units

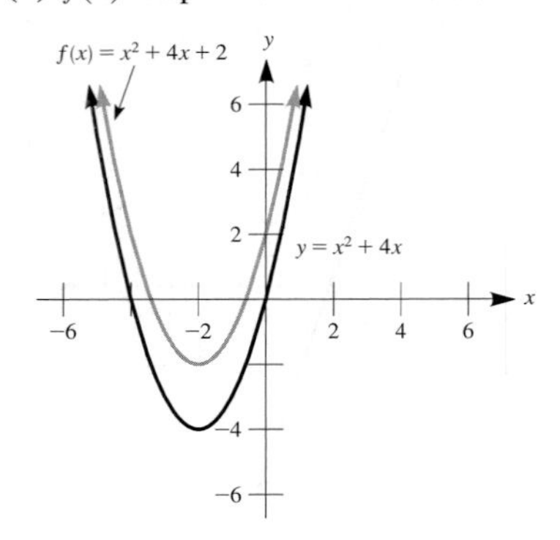

45. (a) $f(x)$ reflected across the y-axis

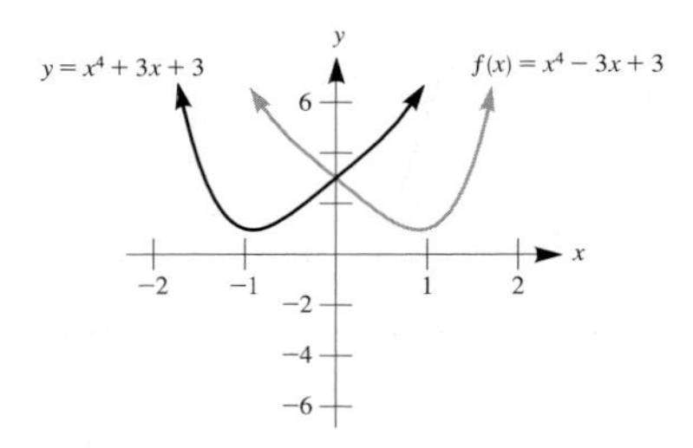

(b) $f(x)$ reflected across the x-axis

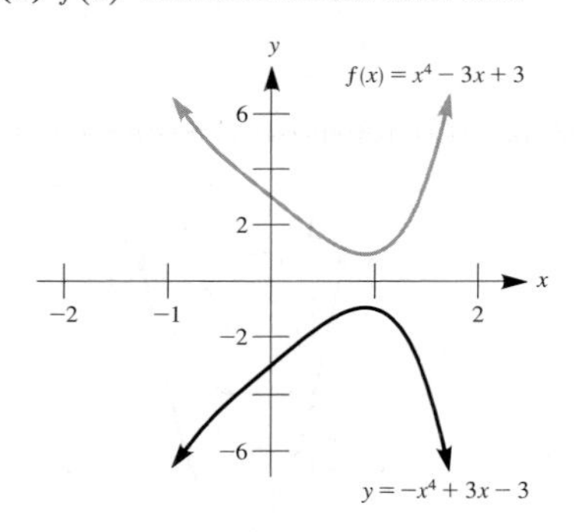

(c) $f(x)$ reflected across the x-axis, then displaced up three units

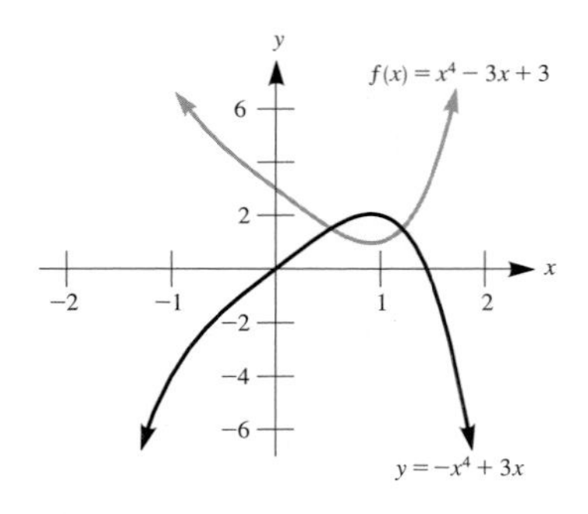

47. (a)

x	x^2	$x^2 - 1$	$x^2 + 1$
0	0	−1	1
±1	1	0	2
±2	4	3	5
±3	9	8	10

(b) $y = x^2 - 1$ is a vertical displacement down one unit from $y = x^2$, while $y = x^2 + 1$ is a vertical displacement up one unit (from $y = x^2$)

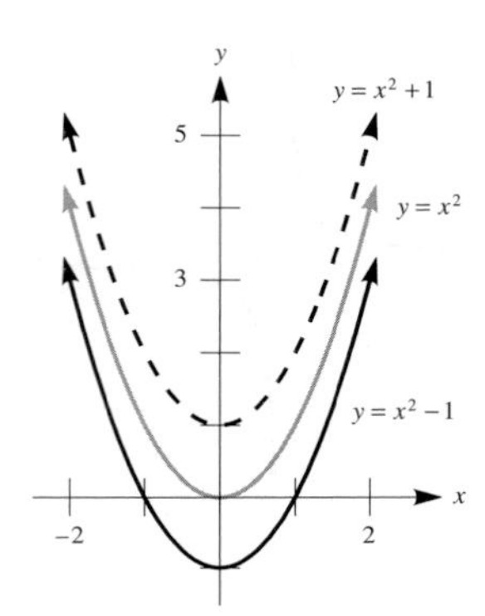

49. (a)

x	$\sqrt{x}$	$-\sqrt{x}$
0	0.0	0.0
1	1.0	−1.0
2	1.4	−1.4
3	1.7	−1.7
4	2.0	−2.0
5	2.2	−2.2

(b) $y = -\sqrt{x}$ is a reflection of $y = \sqrt{x}$ across the x-axis

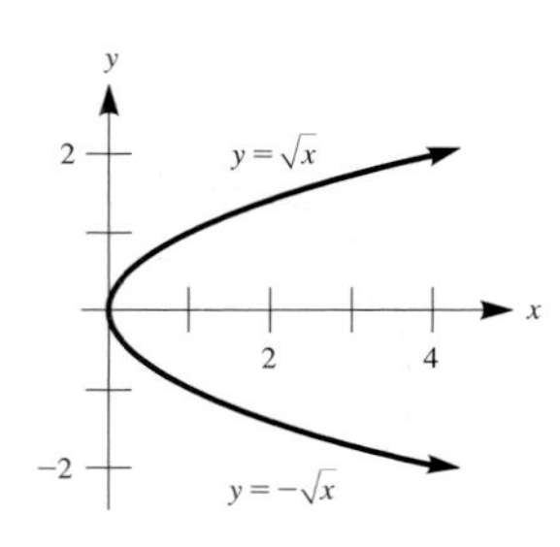

51. (a)

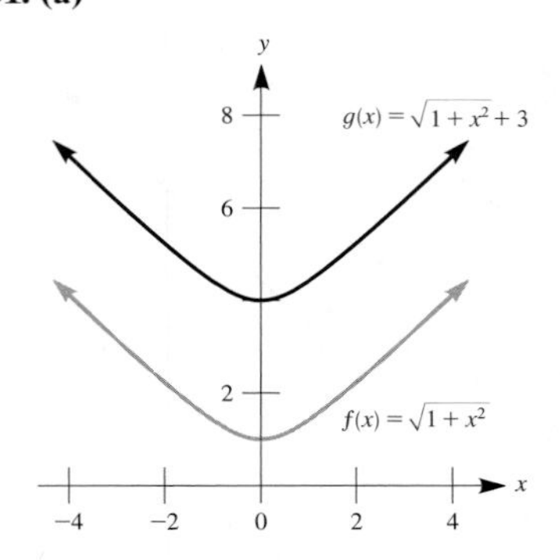

(b)

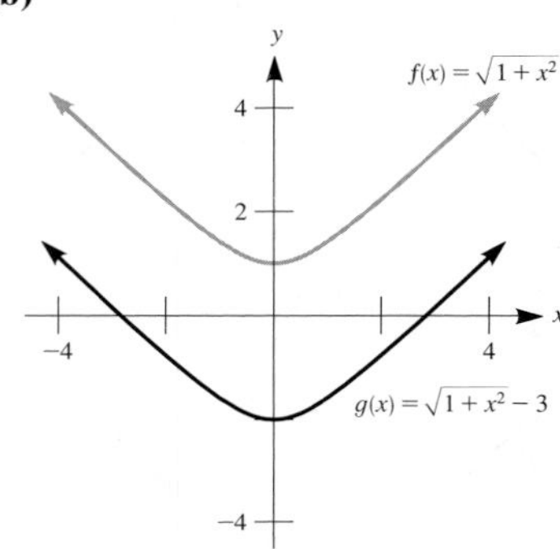

53. (a)

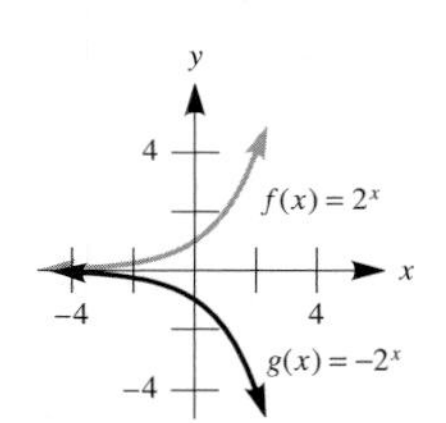

(b)

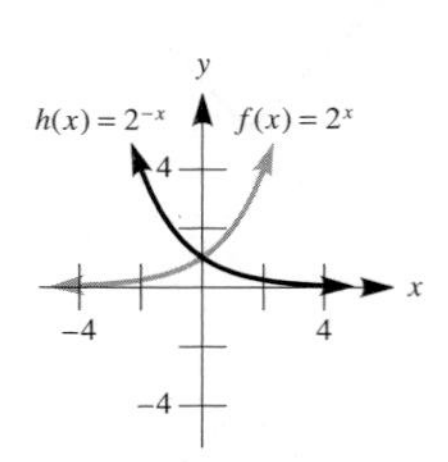

55. (a) $R(a + c, b + d)$
(b) $T(a + c, b + d)$
(c) The order of translations do not affect the final point.

57. (a)

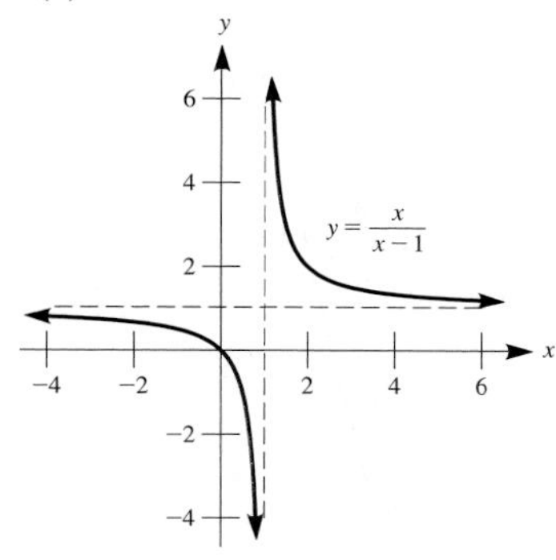

(c) Translate to the right one unit and one unit up. **59. (a)** $(-a, b+2)$
(b) $(-a, -b+2)$ **(c)** $(a+3, -b)$
(d) $(a-1, 1-b)$ **(e)** $(-a+1, b)$
(f) $(-a+1, -b+1)$

Exercise Set 3.5

1. (a) $x^2 - x - 7$
(b) $-x^2 + 5x + 5$ **(c)** 5
3. (a) $x^2 - 2x - 8$ **(b)** $-x^2 + 2x + 8$
5. (a) $4x - 2$ **(b)** $4x - 2$ **(c)** -4
7. (a) $\dfrac{-x^4 + 22x^2 - 4x - 80}{2x^3 - x^2 - 18x + 9}$
(b) $-\frac{80}{9}$ **9. (a)** $-6x - 14$ **(b)** -74
(c) $-6x - 7$ **(d)** -67
11. (a) $(f \circ g)(x) = 9x^2 - 3x - 6$, $(f \circ g)(-2) = 36$, $(g \circ f)(x) = -3x^2 + 9x + 14$, $(g \circ f)(-2) = -16$
(b) $(f \circ g)(x) = 2^{x^2+1}$, $(f \circ g)(-2) = 32$, $(g \circ f)(x) = 2^{2x} + 1$, $(g \circ f)(-2) = \frac{17}{16}$
(c) $(f \circ g)(x) = 3x^5 - 4x^2$, $(f \circ g)(-2) = -112$, $(g \circ f)(x) = 3x^5 - 4x^2$, $(g \circ f)(-2) = -112$
(d) $(f \circ g)(x) = x$, $(f \circ g)(-2) = -2$, $(g \circ f)(x) = x$, $(g \circ f)(-2) = -2$
13. (a) $\dfrac{-x+7}{6x}$ **(b)** $\dfrac{-t+7}{6t}$ **(c)** $\frac{5}{12}$
(d) $\dfrac{1-6x}{7}$ **(e)** $\dfrac{1-6y}{7}$ **(f)** $-\frac{11}{7}$
15. (a) $M(7) = \frac{13}{5}$, $M[M(7)] = 7$
(b) $(M \circ M)(x) = x$
(c) $(M \circ M)(7) = 7$
17. (a) $f[g(3)] = 1$ **(b)** $g[f(3)] = -3$
(c) $f[h(3)] = -1$ **(d)** $(h \circ g)(2) = 2$
(e) $h\{f[g(3)]\} = 2$
(f) $(g \circ f \circ h \circ f)(2) = -3$
19.

x	0	1	2	3	4
$(f \circ g)(x)$	1	3	2	undef.	2

x	−1	0	1	2	3	4
$(g \circ f)(x)$	0	0	3	4	2	undef.

21. (a) $f \circ g$ is the graph of f displaced 4 units to the left
(b)

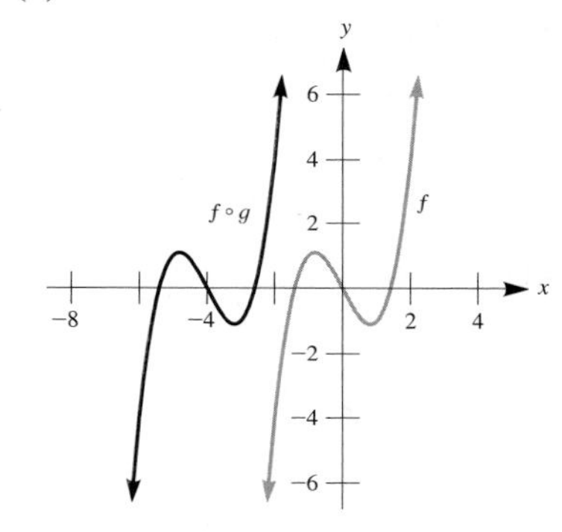

23. (a) domain: $[0, \infty)$; range: $[-3, \infty)$

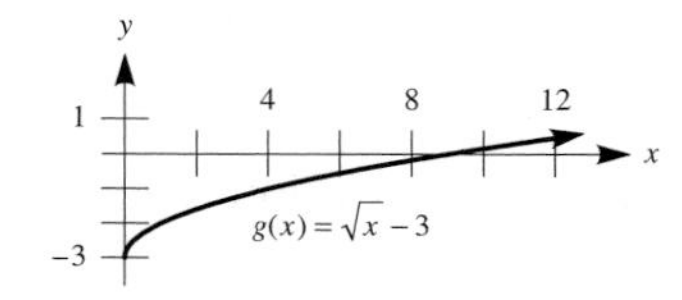

(b) domain: $(-\infty, \infty)$; range: $(-\infty, \infty)$

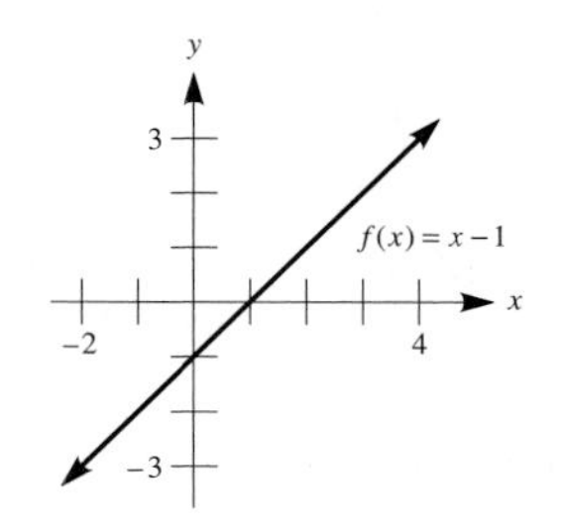

(c) $(f \circ g)(x) = \sqrt{x} - 4$; domain: $[0, \infty)$; range: $[-4, \infty)$

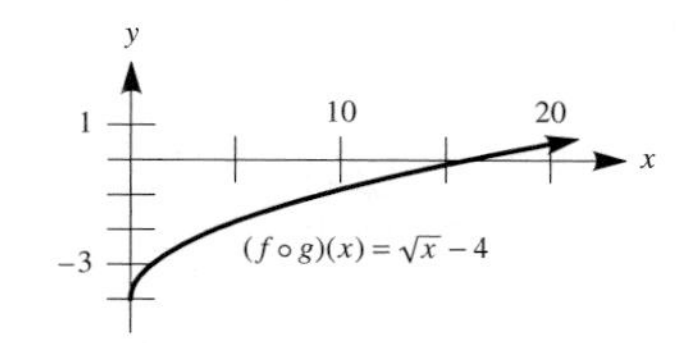

(d) $g[f(x)] = \sqrt{x-1} - 3$; domain: $[1, \infty)$
(e)

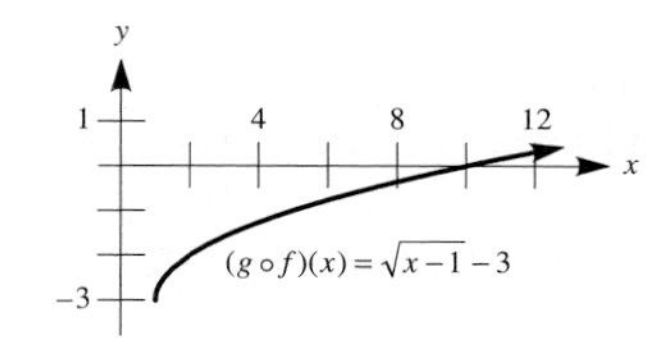

25. (a)

t	A	t	A
0	707	3	1402
0.5	737	3.5	1648
1	804	4	1940
1.5	903	4.5	2284
2	1034	5	2685
2.5	1199		

(b) 800 m² **(c)** initially: 707 m²; after three hours this area has doubled
(d) 0 to 2.5: 196.8 m²/hr; 2.5 to 5: 594.4 m²/hr; faster over the interval from $t = 2.5$ to $t = 5$
27. (a) $(C \circ f)(t) = 100 + 450t - 25t^2$
(b) \$1225 **(c)** \$1900; no
29. (a) $f(x) = \sqrt[3]{x}$, $g(x) = 3x + 4$; $F(x) = (f \circ g)(x)$ **(b)** $f(x) = |x|$, $g(x) = 2x - 3$; $G(x) = (f \circ g)(x)$
(c) $f(x) = x^5$, $g(x) = ax + b$; $H(x) = (f \circ g)(x)$ **(d)** $f(x) = \frac{1}{x}$, $g(x) = \sqrt{x}$; $T(x) = (f \circ g)(x)$
31. (a) $f(x) = (b \circ c)(x)$
(b) $g(x) = (a \circ d)(x)$
(c) $h(x) = (c \circ d)(x)$
(d) $K(x) = (c \circ b)(x)$
(e) $l(x) = (c \circ a)(x)$
(f) $m(x) = (a \circ c)(x)$

33. (b)

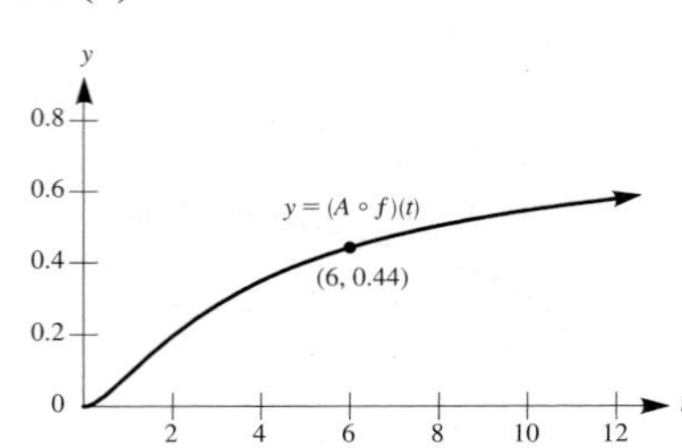

(c) 0.44; $(A \circ f)(6) = \dfrac{9\pi}{64} \approx 0.44$
35. (a) $x_1 = 2, x_2 = 4, x_3 = 8, x_4 = 16, x_5 = 32, x_6 = 64$
(b) all six iterates are 0 **(c)** $x_1 = -2, x_2 = -4, x_3 = -8, x_4 = -16, x_5 = -32, x_6 = -64$ **37. (a)** $x_1 = -3, x_2 = -5, x_3 = -9, x_4 = -17, x_5 = -33, x_6 = -65$ **(b)** all six iterates are -1
(c) $x_1 = 3, x_2 = 7, x_3 = 15, x_4 = 31, x_5 = 63, x_6 = 127$ **39. (a)** $x_1 = 0.81, x_2 \approx 0.656, x_3 \approx 0.430, x_4 \approx 0.185, x_5 \approx 0.034, x_6 \approx 0.001$
(b) all six iterates are 1
(c) $x_1 = 1.21, x_2 \approx 1.464, x_3 \approx 2.144, x_4 \approx 4.595, x_5 \approx 21.114, x_6 \approx 445.792$
41. $x_1 \approx 0.316, x_2 \approx 0.562, x_3 \approx 0.750, x_4 \approx 0.866$ **43. (a)** $f(1) = 4, f(2) = 1, f(3) = 10, f(4) = 2, f(5) = 16, f(6) = 3$
(b) $f(1) = 4, f(4) = 2, f(2) = 1$
(c) $f(3) = 10, f(10) = 5, f(5) = 16, f(16) = 8, f(8) = 4, f(4) = 2, f(2) = 1$
(d) $x_0 = 2$: 2, 1 $x_0 = 4$: 4, 2, 1
$x_0 = 5$: 16, 8, 4, 2, 1 $x_0 = 6$: 6, 3, 10, 5, 16, 8, 4, 2, 1 $x_0 = 7$: 22, 11, 34, 17, 52, 26, 13, 40, 20, 10, 5, 16, 8, 4, 2, 1
45. $f(x) = \dfrac{x + 6}{4}$
47. $a = -\frac{1}{2}$ and $b = \frac{1}{2}$
49. (a) $2x + 2a - 2$ **(b)** $4x + 4a - 4$
51. (a) $x_1 = 3, x_2 \approx 2.259259259, x_3 \approx 1.963308018, x_4 \approx 1.914212754, x_5 \approx 1.912932041, x_6 = \cdots = x_{10} \approx 1.912931183$
The iterates converge to 1.912931183.
(b) $\sqrt[3]{7} \approx 1.912931183$; they are the same **(c)** fifth iterate; sixth iterate

Exercise Set 3.6

1. (a) $h[k(x)] = x$ for every x in the domain of k. **(b)** $k[h(x)] = x$ for every x in the domain of h.

5. (a)

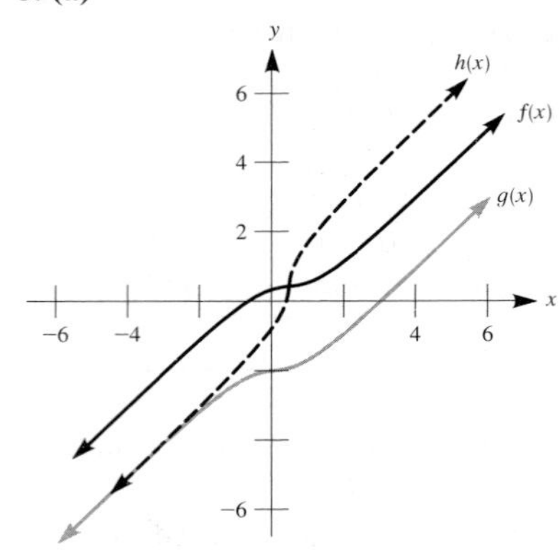

(b) f and h **7.** 7 **9. (a)** 4
(b) -1 **(c)** $\sqrt{2}$ **(d)** $t + 1$
11. (a) $f^{-1}(x) = \dfrac{x - 1}{2}$
(b) $f^{-1}(5) = 2, \dfrac{1}{f(5)} = \dfrac{1}{11}$; no
13. (a) $f^{-1}(x) = \dfrac{x + 1}{3}$
(c)

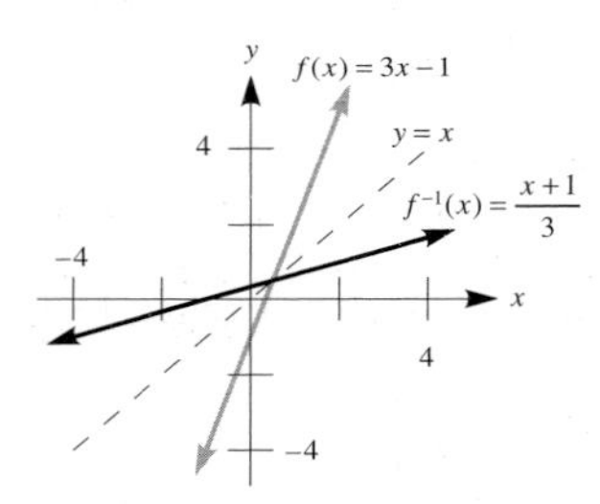

15. (a) $f^{-1}(x) = x^2 + 1$ for $x \geq 0$
(c)

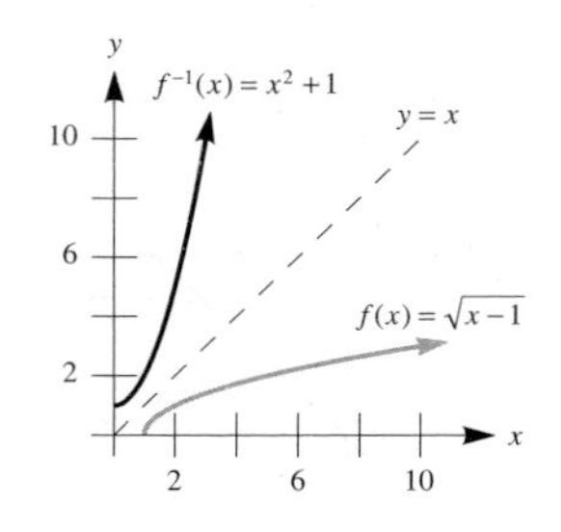

17. (a) domain: $(-\infty, 3) \cup (3, \infty)$; range: $(-\infty, 1) \cup (1, \infty)$
(b) $f^{-1}(x) = \dfrac{3x + 2}{x - 1}$
(c) domain: $(-\infty, 1) \cup (1, \infty)$; range: $(-\infty, 3) \cup (3, \infty)$; domain of f = range of f^{-1}, range of f = domain of f^{-1}

19. $f^{-1}(x) = \sqrt[3]{\dfrac{x - 1}{2}}$
21. $f[f^{-1}(x)] = f^{-1}[f(x)] = x$
23. (a)

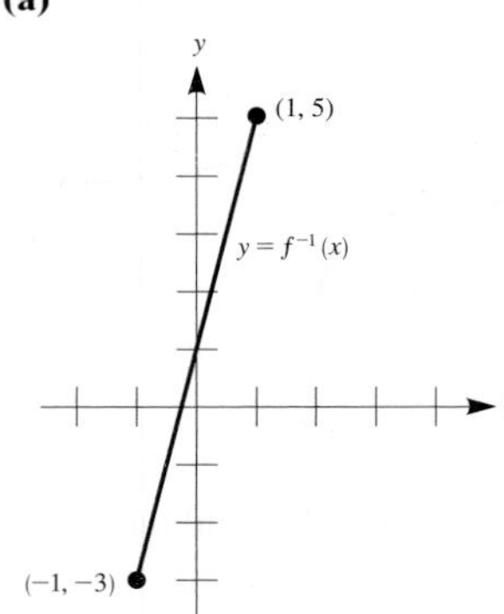

(b)

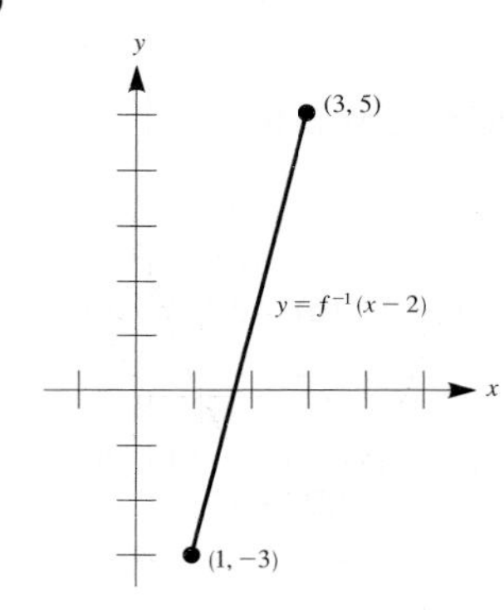

(c)

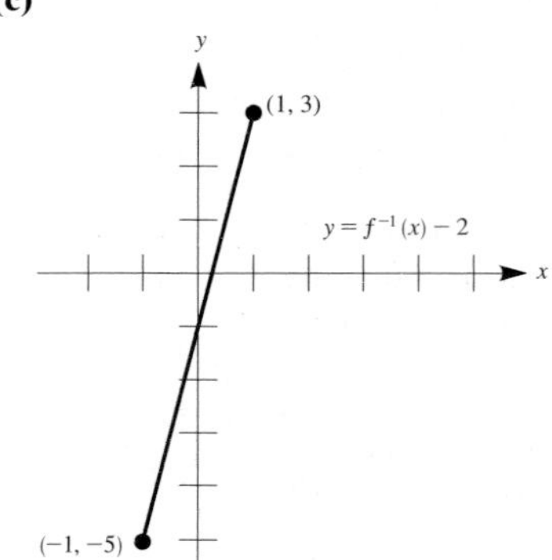

(d)

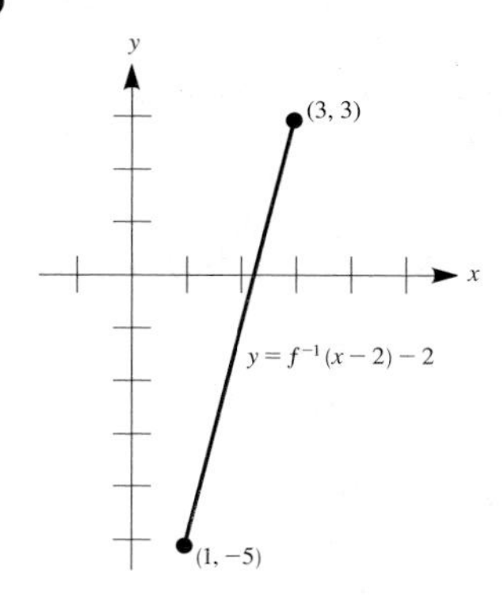

25. (a) $x = 7$ **(b)** $x = -3$
27. $t = \frac{9}{4}$ **29.** not one-to-one
31. not one-to-one **33.** not one-to-one **35.** not one-to-one

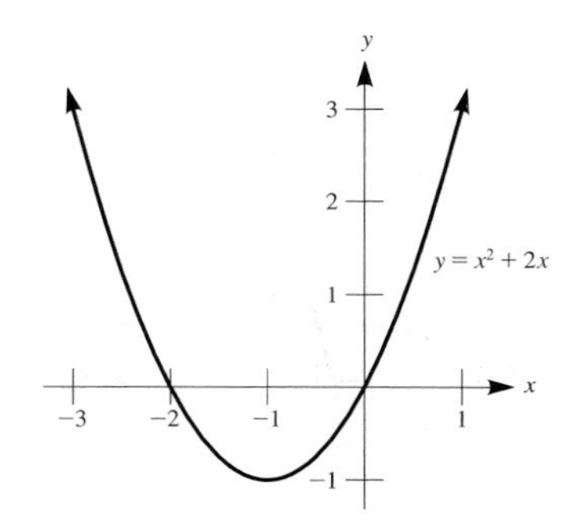

37. not one-to-one

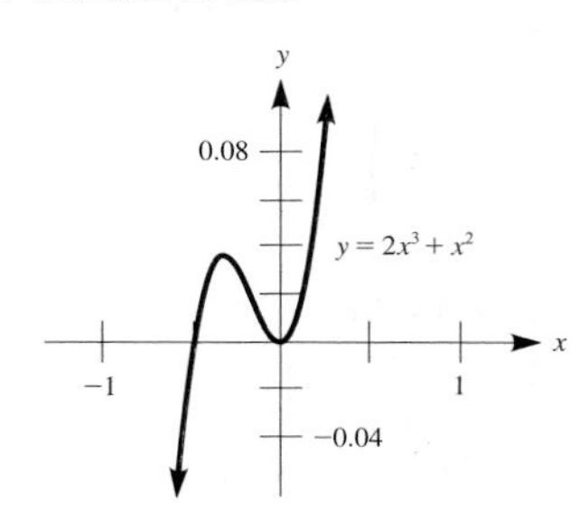

39. is one-to-one

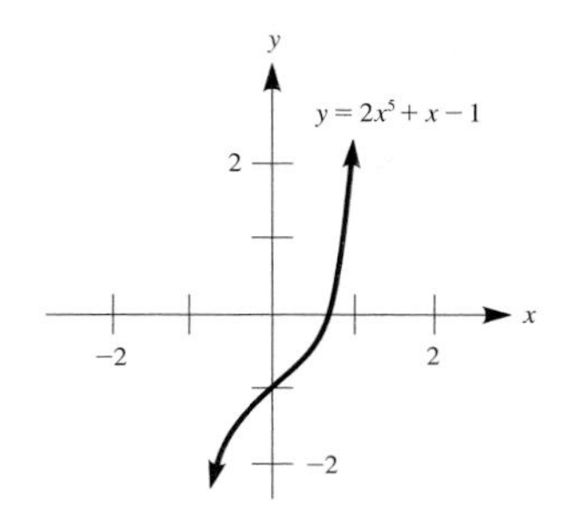

41. (a)

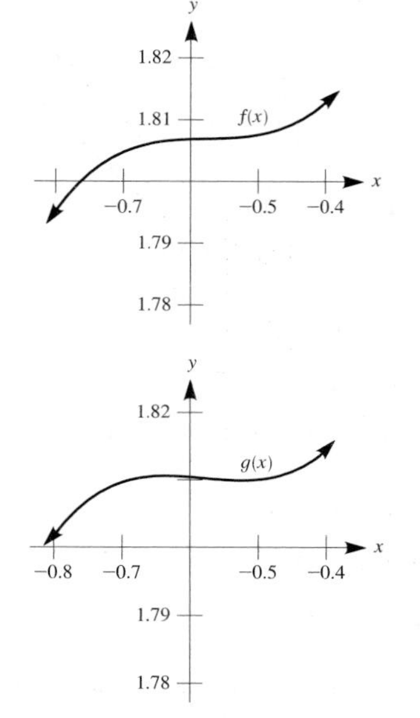

(b) f is one-to-one, g is not one-to-one
43. A: $(a, f(a))$; B: $(f(a), f(a))$; C: $(f(f(a)), f(a))$; D: $(f(a), f(f(a)))$
45. (a) $-\frac{1}{8}$ **(b)** $f^{-1}(x) = \dfrac{x+3}{x}$; -8; they are reciprocals
47. (a) E **(b)** C **(c)** L **(d)** A **(e)** J **(f)** G **(g)** B **(h)** M **(i)** K **(j)** D **(k)** I **(l)** H **(m)** N **(n)** F
49. (a) $-\dfrac{x+2}{x+1}$
(b) symmetry about the line $y = x$

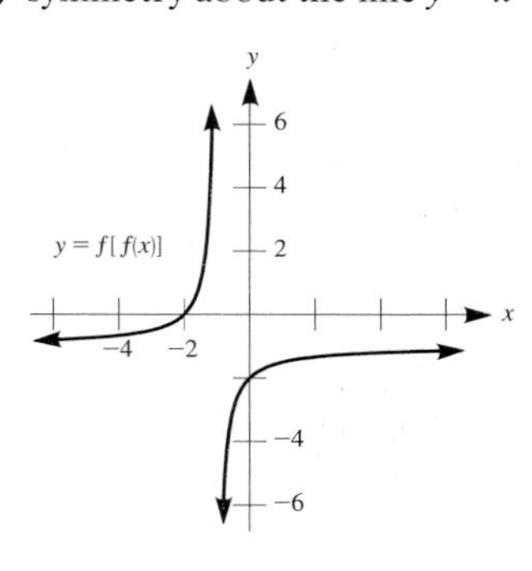

Chapter 3 Review Exercises

1. (a) $(-\infty, 3]$ **(b)** $(-\infty, \frac{1}{2}) \cup (\frac{1}{2}, \infty)$
3. yes **5. (a)** $-\dfrac{1}{ax}$
(b) $1 - 4x - 2h$
7. (a) $g^{-1}(x) = \dfrac{1}{3x+5}$

(b)

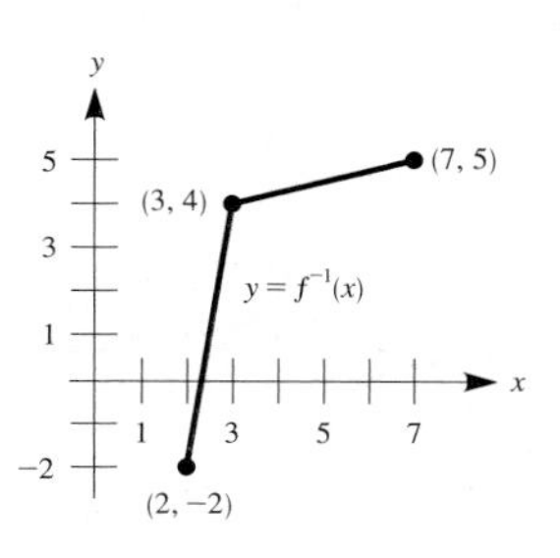

9. (a) x-intercepts: -5 and 1; y-intercept: -1

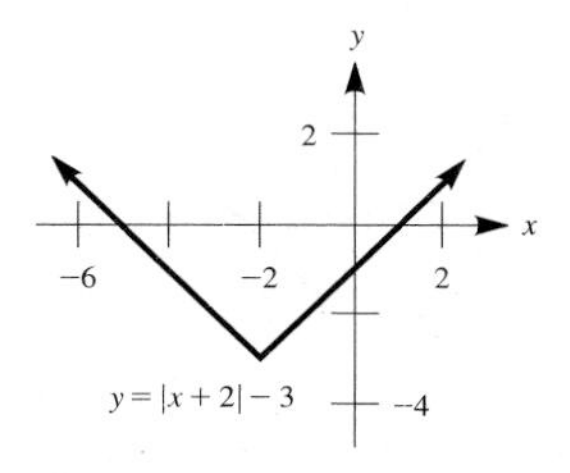

(b) x-intercept: -1; y-intercept: $-\frac{1}{2}$

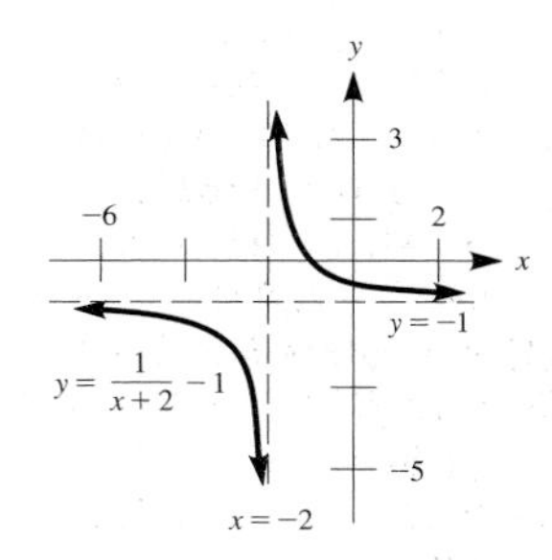

11. (a) 5 **(b)** $7 - 4\sqrt{2}$
13. $m(h) = 10 + h$
15.

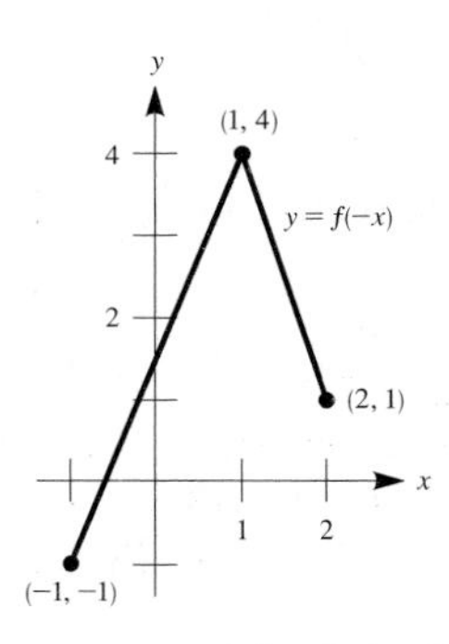

17. (a) 2.5, 3.25, 3.625
(b) A: $f(1) = 2.5$; B: $f(2.5) = 3.25$; C: $f(3.25) = 3.625$ **(c)** $y = 4$
(d) 3.813, 3.906, 3.953, 3.977, 3.988, 3.994, 3.997

19. no x-intercepts; y-intercept: 1

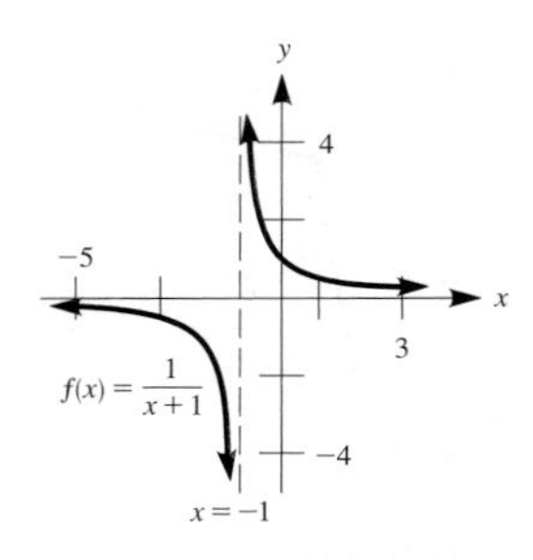

21. x-intercept: -3; y-intercept: 3

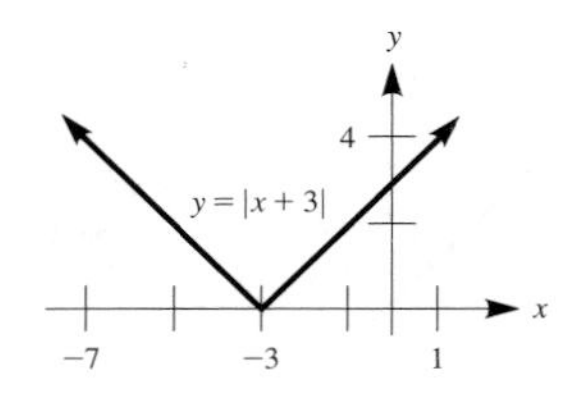

23. x-intercepts: ± 1; y-intercept: 1

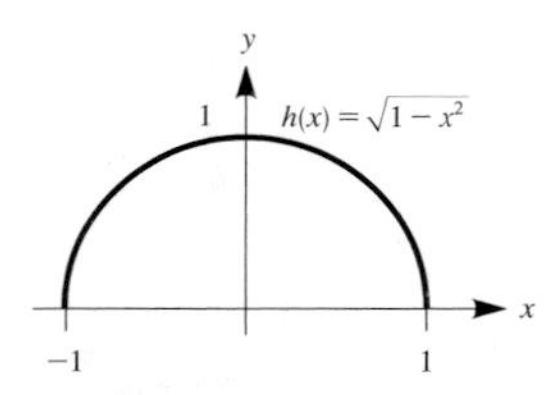

25. x-intercept: 0; y-intercept: 0

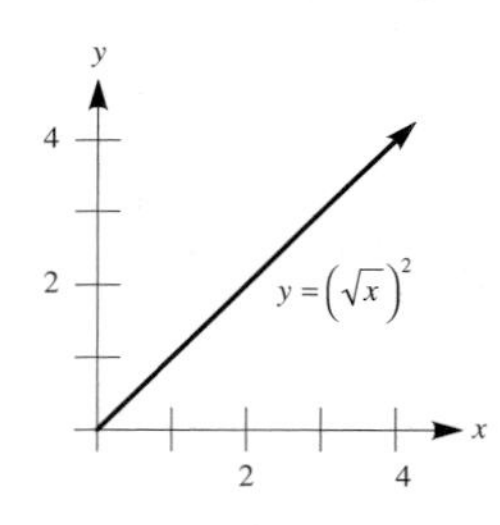

27. x-intercept: 1; no y-intercept

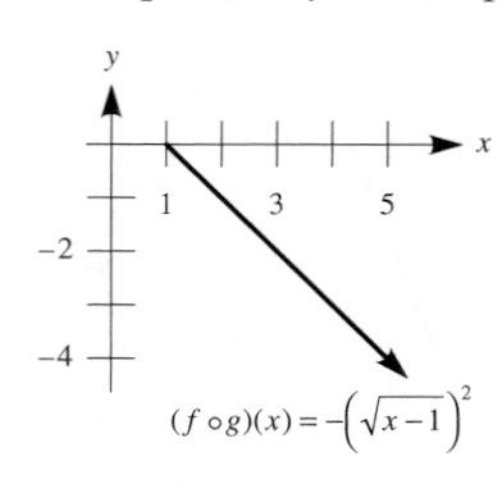

29. no x- or y-intercepts

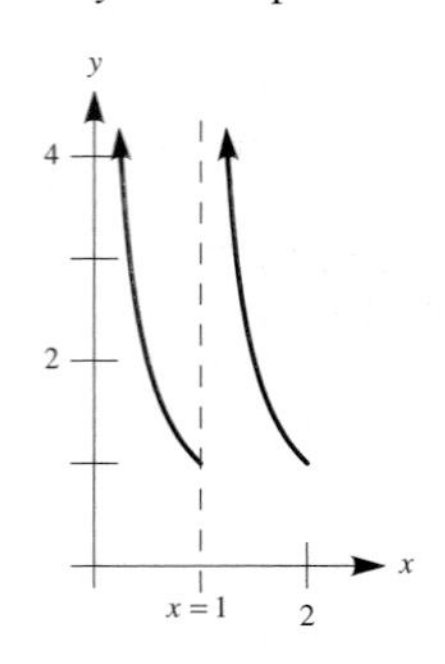

31. x-intercept: $\frac{1}{2}$; y-intercept: -1

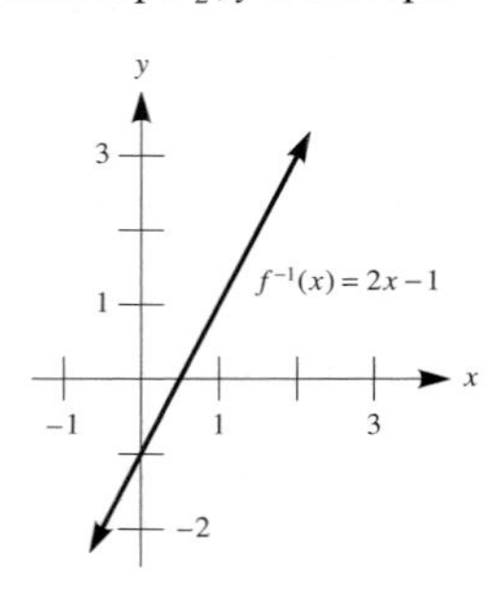

33. x-intercept: 0; y-intercept: 0

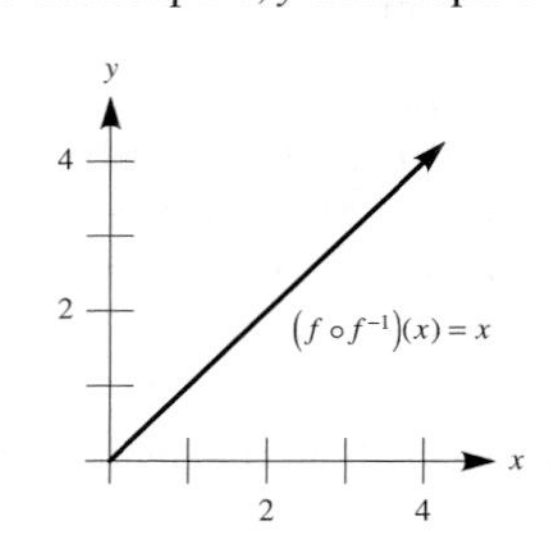

35. $(-\infty, -3) \cup (-3, 3) \cup (3, \infty)$
37. $(-\infty, 4]$ **39.** $(-\infty, -1] \cup [3, \infty)$
41. $(-\infty, \frac{1}{3}) \cup (\frac{1}{3}, \infty)$
43. $(-\infty, \frac{2}{5}) \cup (\frac{2}{5}, \infty)$
45. $a(x) = (f \circ g)(x)$
47. $c(x) = (G \circ g)(x)$
49. $A(x) = (g \circ f \circ G)(x)$
51. $C(x) = (g \circ G \circ G)(x)$ **53.** 12
55. $-\frac{9}{19}$ **57.** $t^2 + t$ **59.** $x^2 - 5x + 6$
61. $x^4 - x^2$ **63.** $-2x^3 + 3x^2 - x$
65. $4x^2 - 2x$ **67.** $-2x^2 + 2x + 1$
69. $\dfrac{2x + 2}{2x - 5}$ **71.** $2x + h - 1$
73. $F^{-1}(x) = \dfrac{4x + 3}{1 - x}$ **75.** x
77. $\dfrac{1 + x}{2}$ **79.** $\frac{22}{7}$ **81.** negative

83. -1 **85.** -1 **87.** $(0, -2)$ and $(5, 1)$ **89.** $[-6, 0]$ and $[5, 8]$
91. 0 at $x = 2$ **93.** no **95.** $x = 4$
97. (a) $x = 10$ **(b)** $x = 0$ **99. (a)** 5 **(b)** -3 **(c)** 4 **(d)** $\frac{1}{4}$
101. $(f \circ f)(10)$ **103.** $[0, 4]$ **105.** 5
107. $(1, 3) \cup (6, 10)$ **109.** $(4, 7)$

Chapter 3 Test

1. $(-\infty, -1] \cup [6, \infty)$
2. $(-\infty, \frac{2}{3}) \cup (\frac{2}{3}, \infty)$
3. (a) $2x^2 - 2x - 2$ **(b)** $2x^2 - 5x + 2$ **(c)** 54 **4.** $-\dfrac{2}{at}$
5. $4x + 2h - 5$ **6.** $g^{-1}(x) = -\dfrac{x}{6x + 4}$
7. $-\frac{8}{5}$ **8. (a)** x-intercepts: 4 and 2; y-intercept: -2

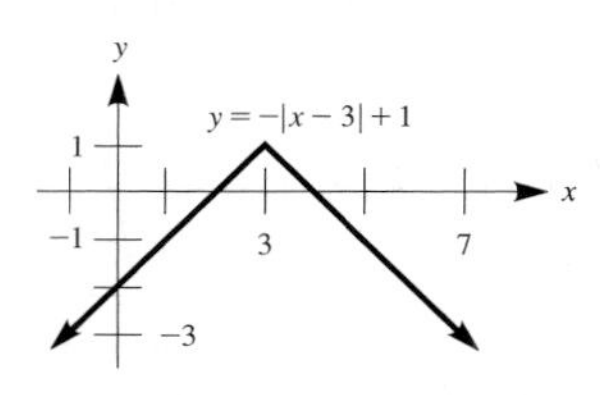

(b) x-intercept: $-\frac{5}{2}$; y-intercept: $-\frac{5}{3}$

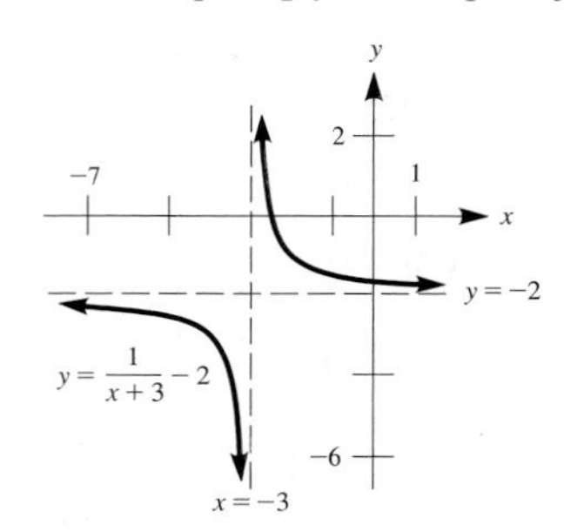

9. (a) $[-3, 1]$ **(b)** $(-2, 1)$ **(c)** -3 at $x = 2$ **(d)** 1 at $x = -2$ **(e)** $[-2, 2]$ **(f)** no **10. (a)** $\frac{23}{4}$ **(b)** $12 - 7\sqrt{3}$ **11.** domain: $(-\infty, -1) \cup (-1, \infty)$

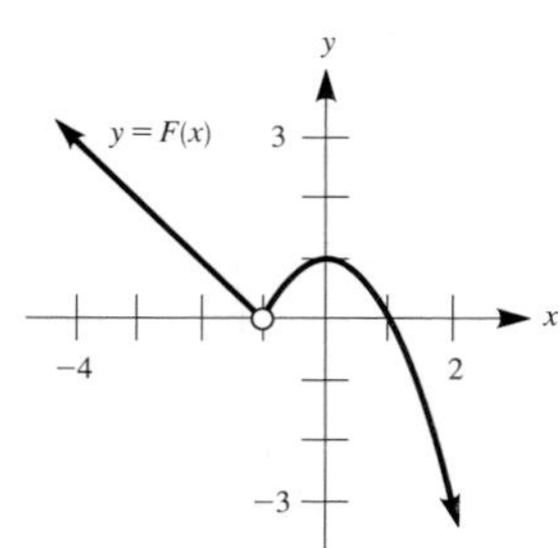

12. (a) 0.4, 0.8, 0.4, 0.8, 0.4, 0.8
(b) 0.4, 0.8, 0.4, 0.8, 0.4, 0.8
13.

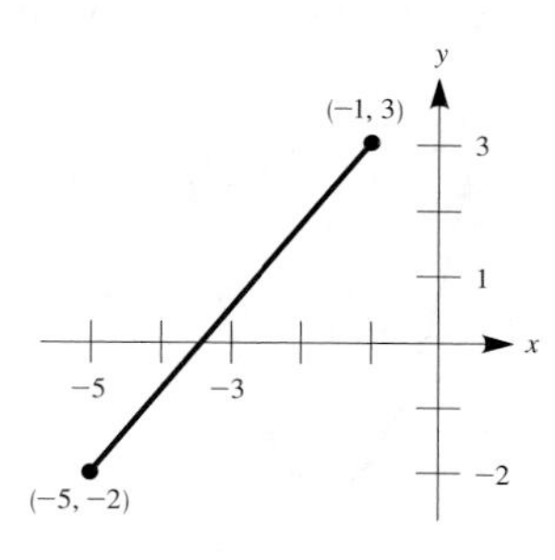

14. $t = 1$ **15.** $b = 10$

16. (a) $\dfrac{\Delta F}{\Delta t} = 0.16a - 0.8$

(b) −0.16 °C/hour

CHAPTER 4

Exercise Set 4.1

1. $f(x) = \frac{2}{3}x + \frac{2}{3}$ **3.** $g(x) = \sqrt{2}x$
5. $f(x) = x - \frac{7}{2}$ **7.** $f(x) = \frac{1}{2}x - 2$
9. yes **11.** $V(t) = -2375t + 20000$
13. (a) $V(t) = -12000t + 60000$
(b)

End of Year	Yearly Depreciation	Accumulated Depreciation	Value V
0	0	0	60,000
1	12,000	12,000	48,000
2	12,000	24,000	36,000
3	12,000	36,000	24,000
4	12,000	48,000	12,000
5	12,000	60,000	0

15. (a) \$530 **(b)** \$538 **(c)** \$8/fan
17. (a)

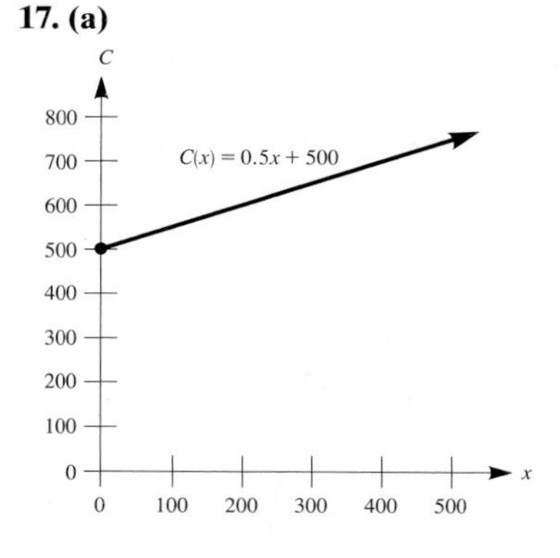

(b) \$575 **(c)** \$575.50; the cost of producing 151 units
19. (a) B is traveling faster
(b) A is farther to the right
(c) $t = 8$ sec
21. (a) $y = 362091.5x - 690{,}879{,}017.5$
(b) 33,304,000
(c) too low; 1.7% error
23. (a) $y = 3680x - 7{,}284{,}416$
(b) \$64,544 million

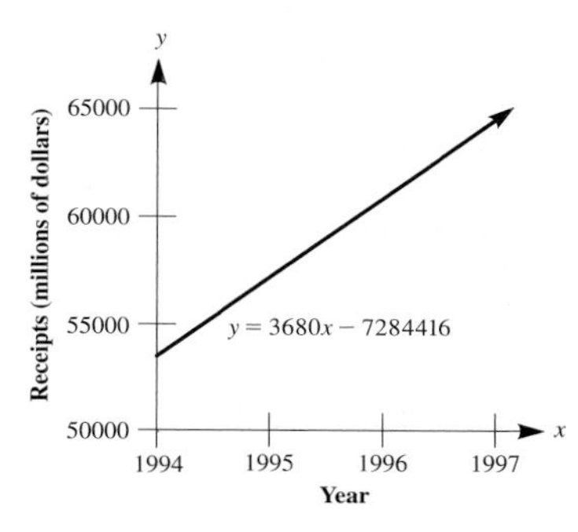

(c) 2.4% error
25. (a) $y = 4.92x - 35.55$
(b) 102.21 cm **(c)** too high; 4.2% error **(d)** 33.33 cm; too low; 8.3% error **(e)** 377.73 cm; too high; 48.4% error
27. (a)

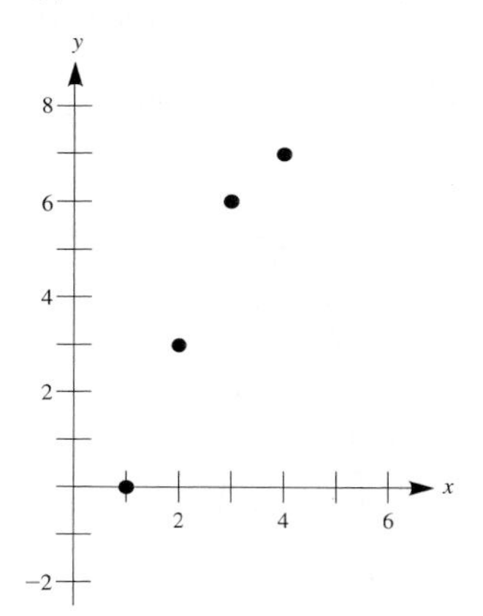

(b) slope: 2.5; y-intercept: −2

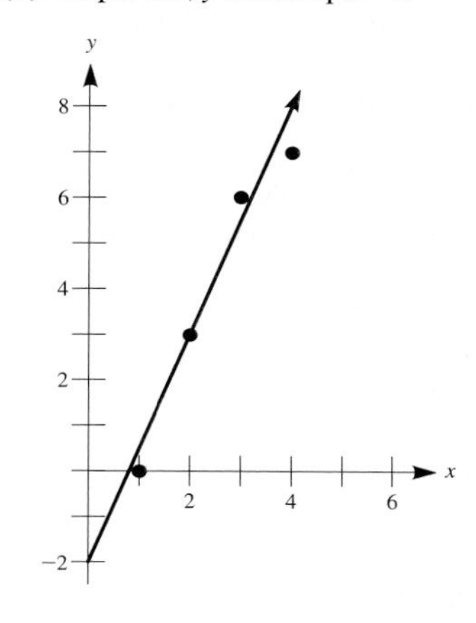

(c)

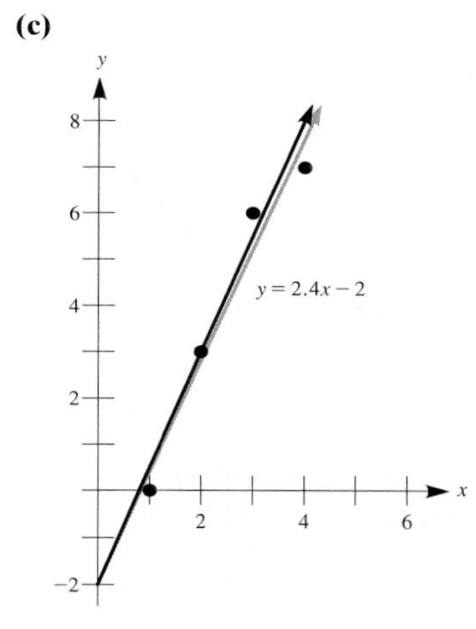

29. (a) 3,853,000; 0.8% error
(b) $f^{-1}(x) = \dfrac{x + 71238863.429}{37546.068}$
(c) year 2004
31. (a)

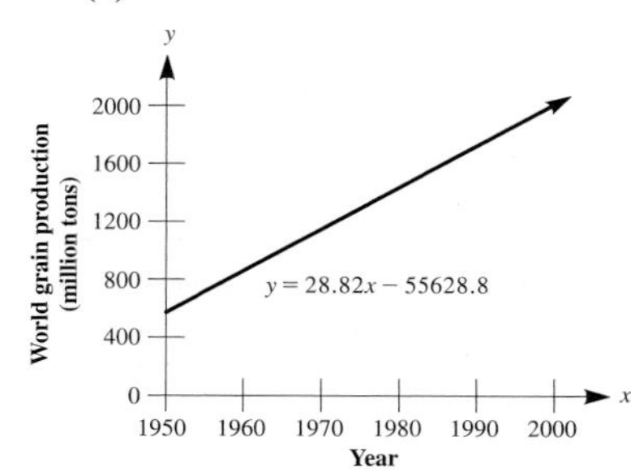

(b) 1809 million tons; 5.5% error
(c) 1954 million tons; 6.0% error
33. (a) 4,050,273 **(b)** no
35. (a) 3:57.6; too low; 0.17% error

(b) 4:11.3; too low; 1.61% error; This percent error is much larger, since 1911 is not in the data range.
37. (a) A **(b)** A^2 **(c)** AC
(d) AC **39. (a)** $\frac{1}{AC}$ **(b)** $\frac{1}{AC}$
43. $f(x) = 3x + 1$ **45.** $y = 2.4x - 2$
47. $y = 2.4x - 0.4$
49. $y = 0.084x + 37.241$
51. (a) $f(x) = \sqrt{2}x + (-1 + \sqrt{2})$ and $f(x) = -\sqrt{2}x + (-1 - \sqrt{2})$
(b) $f(x) = \sqrt[3]{2}x + (-1 + \sqrt[3]{2})$
53. $f(x) = x$

Exercise Set 4.2

1. (a)

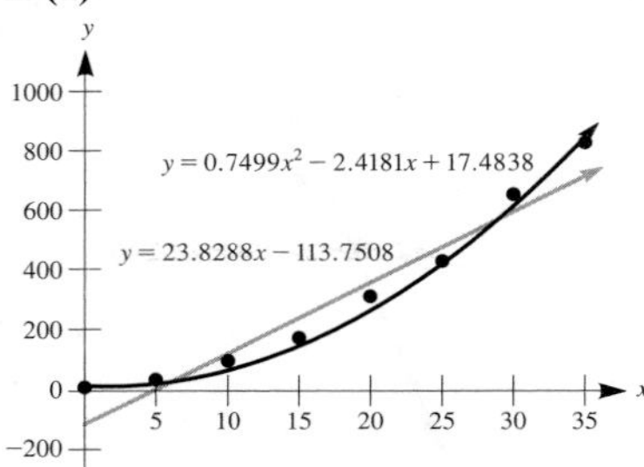

(b) 1990: linear ≈ 839.40 billion; quadratic ≈ 1314.05 billion
1998: linear ≈ 1030.03 billion; quadratic ≈ 1861.32 billion
(c) 1990: linear ≈ 26.8% error; quadratic ≈ 14.7% error.
1998: linear ≈ 35.0% error; quadratic ≈ 17.4% error.
The quadratic model produces the smaller percent error.

3.

	Projected U.S. Population (millions)	Actual U.S. Population (millions)	Projection Too High or Too Low?	% Error
1970	191.12	203.30	too low	5.99%
2000	258.35	275.60	too low	6.26%

5. vertex: $(-2, 0)$; axis of symmetry: $x = -2$; minimum value: 0; x-intercept: -2; y-intercept: 4

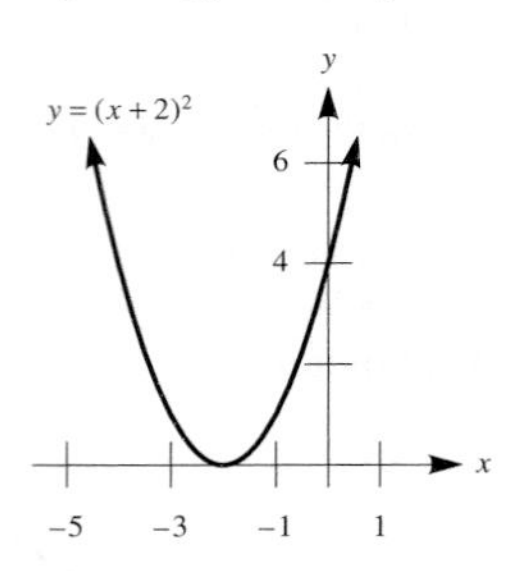

7. vertex: $(-2, 0)$; axis of symmetry: $x = -2$; minimum value: 0; x-intercept: -2; y-intercept: 8

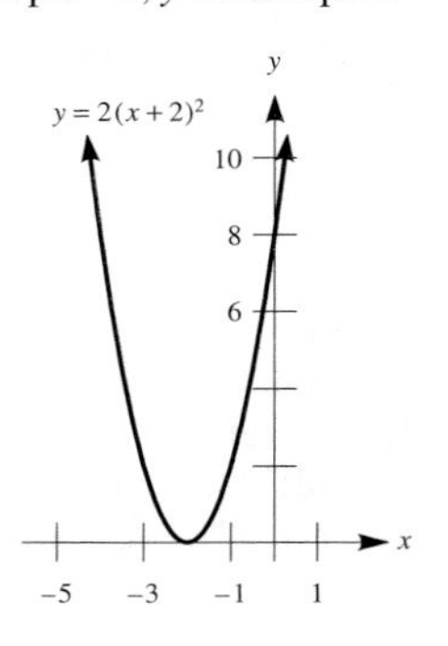

9. vertex: $(-2, 4)$; axis of symmetry: $x = -2$; maximum value: 4; x-intercepts: $-2 \pm \sqrt{2}$; y-intercept: -4

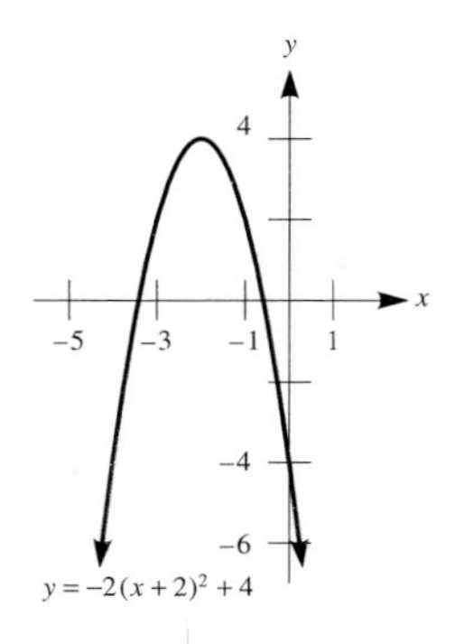

11. vertex: $(2, -4)$; axis of symmetry: $x = 2$; minimum value: -4; x-intercepts: 0 and 4; y-intercept: 0

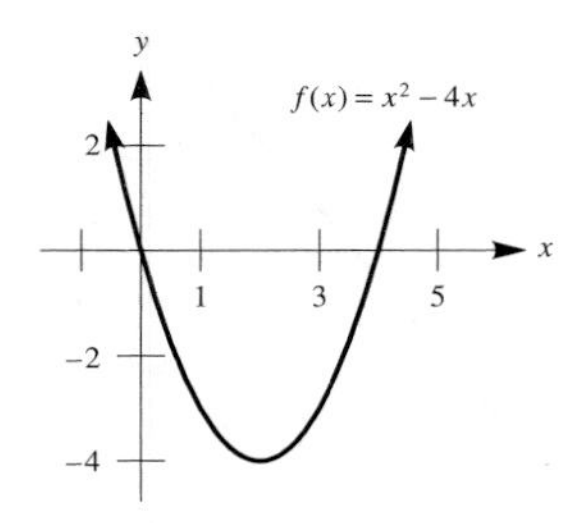

13. vertex: $(0, 1)$; axis of symmetry: $x = 0$; maximum value: 1; x-intercepts: ± 1; y-intercept: 1

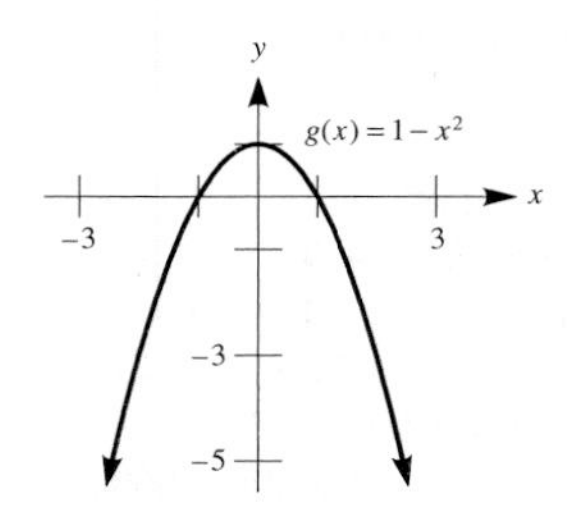

15. vertex: $(1, -4)$; axis of symmetry: $x = 1$; minimum value: -4; x-intercepts: 3 and -1; y-intercept: -3

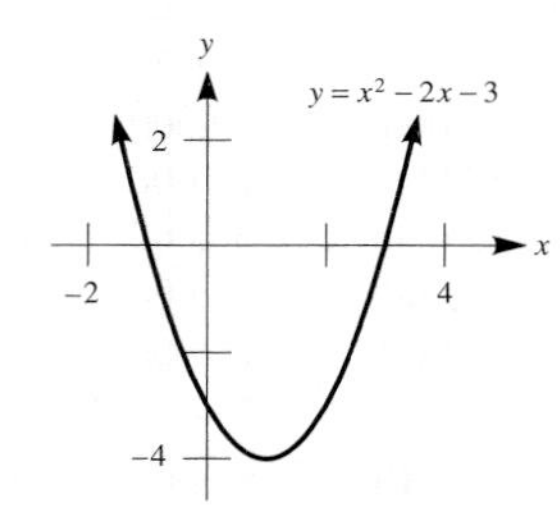

17. vertex: $(3, 11)$; axis of symmetry: $x = 3$; maximum value: 11; x-intercepts: $3 \pm \sqrt{11}$; y-intercept: 2

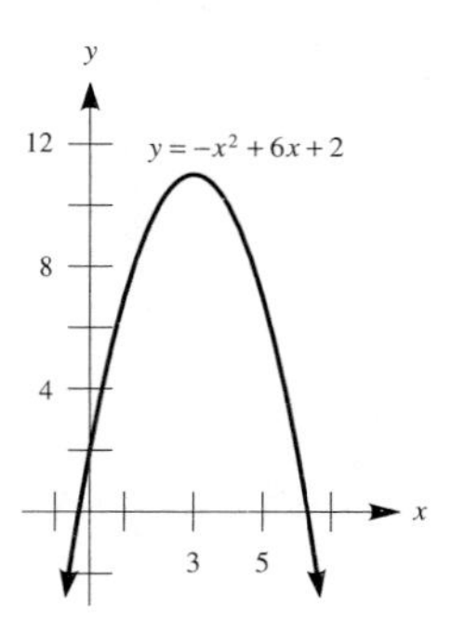

19. vertex: $(\frac{1}{6}, \frac{9}{4})$; axis of symmetry: $t = \frac{1}{6}$; maximum value: $\frac{9}{4}$; t-intercepts: $-\frac{1}{3}$ and $\frac{2}{3}$; s-intercept: 2

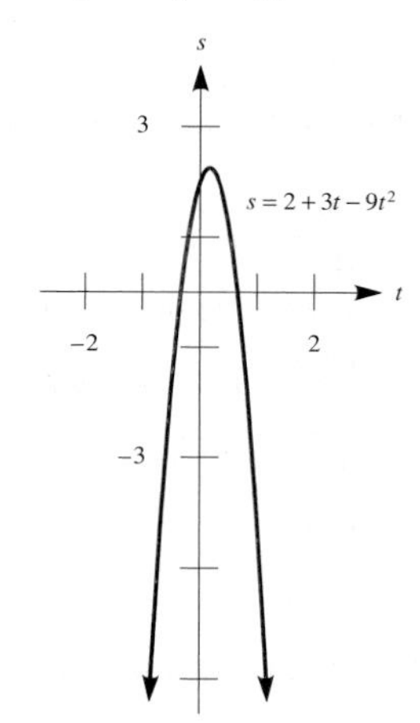

21. $x = 1$; minimum output
23. $x = \frac{3}{2}$; maximum output
25. $x = 0$; minimum output
27. minimum value: -13
29. maximum value: $\frac{25}{8}$
31. (a) minimum value
(b) minimum value: -4.0; $x_0 \approx 0.2$
(c) minimum value: $-\frac{97}{24}$; $x_0 = \frac{1}{12}$
33. (a) maximum value
(b) maximum value: -43.4; $t_0 \approx 2.2$
(c) maximum value: $-\frac{391}{9}$; $t_0 = \frac{20}{9}$
35. 5 units **37.** quadratic
39. neither **41.** linear
43. (a) minimum value $= 8$ at $x = 3$
(b) minimum value $= 4$ at $x = 3$
(c) minimum value $= 64$ at $x = \pm\sqrt{3}$
45. (a) maximum value $= 4$
(b) maximum value $= 2\sqrt[3]{2} \approx 2.5$
(c) maximum value $= 16$
47. neither **49.** quadratic
51. The graph of $y = 50x^2$ should be "narrower."

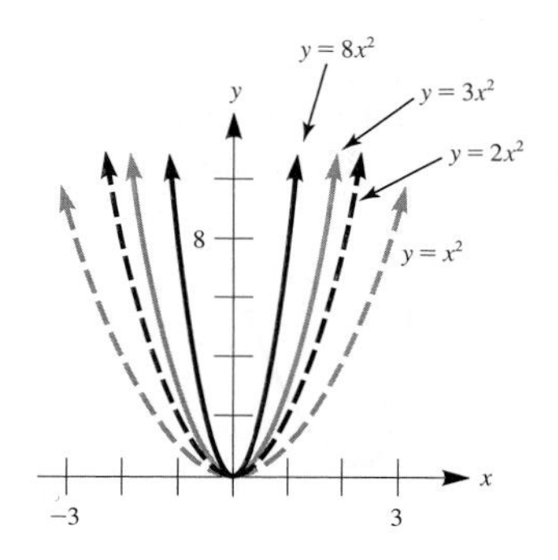

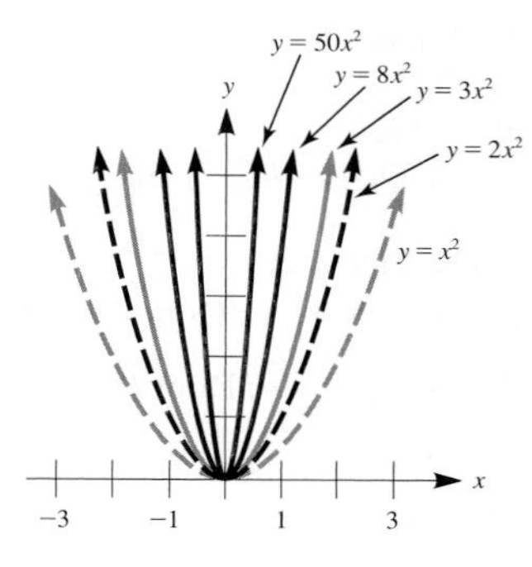

53. $x + a$ **55.** $x = \dfrac{a + b}{2}$
57. $y = -\frac{1}{2}(x - 2)^2 + 2$
59. $y = \frac{1}{4}(x - 3)^2 - 1$ **61.** $b = \pm 2$
63. (a)

x	a	$a + h$	$a + 2h$
$f(x)$	$ma + b$	$ma + mh + b$	$ma + 2mh + b$

(b) mh
65. (a) $y = 0.015089x^2 - 0.086075x + 3.364536$; 1989: 3.466 billion tons; 1996: 3.391 billion tons **(b)** 1989: 1.70% error; 1996: 1.08% error
(c) 3.642 billion tons; 9.40% error

Exercise Set 4.3

1. $x = 1$ **3.** no fixed points
5. $x = -1$ and $x = 5$ **7.** $t = 1$
9. $t = -3$ and $t = 4$ **11.** $x = 0$ and $x = \frac{4}{9}$ **13.** $u = -2$ and $u = 1$
15. $x = 10$ **17. (a)** one fixed point

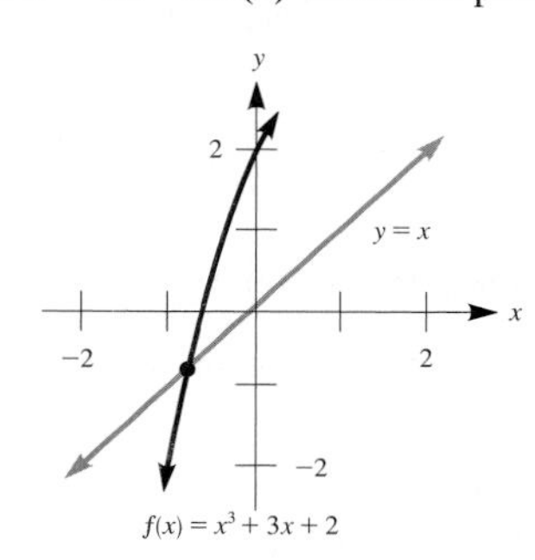

(b) $x \approx -0.771$

19. (a) three fixed points

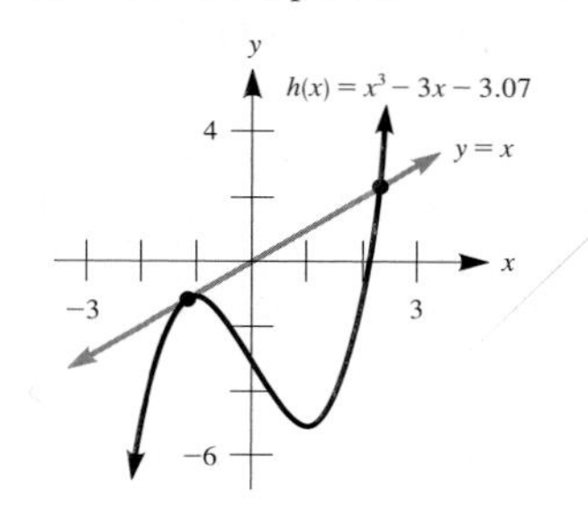

(b) $x \approx -1.206, -1.103, 2.309$
21. (a) two fixed points

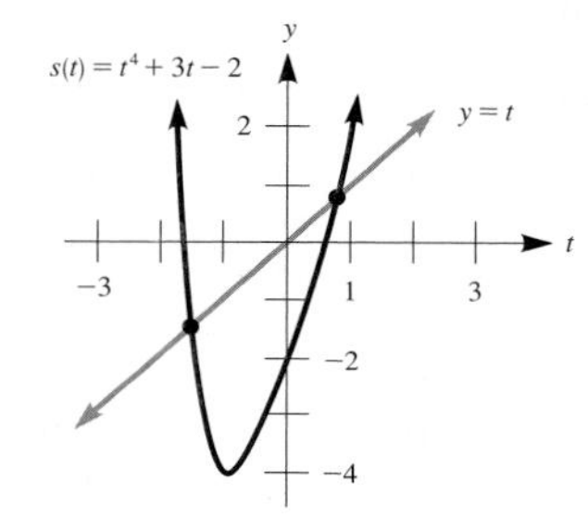

(b) $x \approx -1.495, 0.798$ **23. (b)** fourth iterate **(c)** twelfth iterate
25. (a)

	From graph	From calculator
x_1	1.7	1.72
x_2	0.8	0.796
x_3	1.4	1.443
x_4	1.0	0.990
x_5	1.3	1.307
x_6	1.1	1.085
x_7	1.2	1.240
x_8	1.1	1.132

(b) $x = \frac{20}{7} \approx 1.176$ **(c)** eighth iterate
27.

	From graph	From calculator
x_1	0.36	0.36
x_2	0.92	0.922
x_3	0.29	0.289
x_4	0.82	0.822
x_5	0.58	0.585
x_6	0.97	0.971
x_7	0.11	0.113
x_8	0.40	0.402
x_9	0.96	0.962

29. (a) $x_{21} \approx 0.6632, x_{22} \approx 0.6477, x_{23} \approx 0.6617, x_{24} \approx 0.6492, x_{25} \approx 0.6605$
(b)

n	20	21	22	23	24	25
Number of fish after n breeding seasons	323	332	324	331	325	330

(c) $x_1 = 0.0675, x_2 \approx 0.04721, x_3 \approx 0.03373, x_4 \approx 0.02445, x_5 \approx 0.01789$. The iterates are approaching 0, which is a fixed point.
31. (a)

n	x_n	Number of fish after n breeding seasons
0	0.1	50
1	0.279	140
2	0.6236	312
3	0.7276	364
4	0.6143	307
5	0.7345	367
6	0.6046	302
7	0.7411	371
8	0.5948	297
9	0.7471	374
10	0.5857	293

n	x_n	Number of fish after n breeding seasons
21	0.7633	382
22	0.5601	280
23	0.7638	382
24	0.5592	280
25	0.7641	382
26	0.5587	279

(b) $x = \frac{21}{31} \approx 0.6774$
(c) $a \approx 0.7646, b \approx 0.5580$
(d) 382 fish, 279 fish
33. (a) $x_0 = c$: $x_1 = d, x_2 = c, x_3 = d, x_4 = c, x_5 = d, x_6 = c$; $x_0 = d$: $x_1 = c, x_2 = d, x_3 = c, x_4 = d, x_5 = c, x_6 = d$; With an initial input of either c or d, the sequence of iterates alternates between c and d.
(b) 0.8, 0.4, 0.8, 0.4, 0.8, 0.4
(c) $\{0.4, 0.8\}$ **(d)** $x_1 = 0.8, x_2 = 0.4$
37. (a) $a = \frac{6}{7}, b = \frac{3}{7}$
(b) $P(\frac{3}{7}, \frac{3}{7}), Q(\frac{3}{7}, \frac{6}{7}), R(\frac{6}{7}, \frac{6}{7}), S(\frac{6}{7}, \frac{3}{7})$

Exercise Set 4.4

1. (a) $A(x) = 8x - x^2$; Dom $A = (0, 8)$
(b) $P(x) = 2x + \dfrac{170}{x}$; Dom $P = (0, \infty)$
3. (a) $D(x) = \sqrt{x^4 + 3x^2 + 1}$; Dom $D = (-\infty, \infty)$
(b) $m(x) = \dfrac{x^2 + 1}{x}$; Dom $m = (-\infty, 0) \cup (0, \infty)$
5. (a) $A(y) = \dfrac{\pi y^2}{4}$; Dom $A = (0, \infty)$
(b) $A(y) = \dfrac{\pi^2 y^2}{16}$; Dom $A = (0, \infty)$
7. (a) $P(x) = 16x - x^2$; Dom $P = (-\infty, \infty)$
(b) $S(x) = 2x^2 - 32x + 256$; Dom $S = (-\infty, \infty)$
(c) $D(x) = x^3 - (16 - x)^3$ or $D(x) = (16 - x)^3 - x^3$; Dom $D = (-\infty, \infty)$
(d) $A(x) = 8$; The average does not depend on what the two numbers are.
9. (a) Dom $P = [0, 20]$

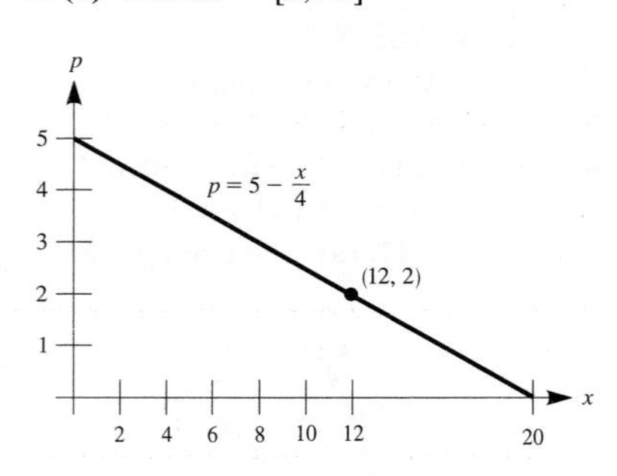

(b) 8 units; (8, 3) **(c)** \$2; (12, 2)
(d) $R(x) = 5x - \dfrac{x^2}{4}$; Dom $R = [0, 20]$

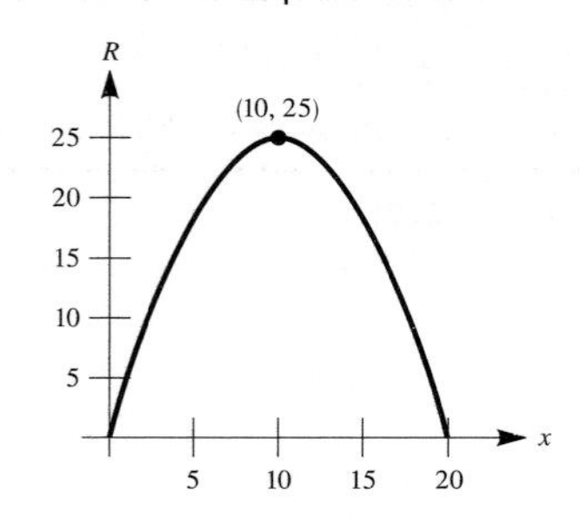

(e) $R(2) = \$8$; $R(8) = \$24$; $R(14) = \$21$
(f) $x = 10$; \$25; $p = \$2.50$
11. (a)

x	1	2	3	4	5	6	7
$P(x)$	17.88	19.49	20.83	21.86	22.49	22.58	21.75

(b) largest value: 22.58; $x = 6$
(c) 22.63
13. $A(x) = \dfrac{\sqrt{3}}{4}x^2$; Dom $A = [0, \infty)$
15. $R(V) = \sqrt[3]{V/2\pi}$; Dom $R = [0, \infty)$
17. $V(S) = \dfrac{S\sqrt{S\pi}}{6\pi}$; Dom $V = [0, \infty)$
19. $A(x) = \frac{1}{2}x\sqrt{400 - x^2}$; Dom $A = (0, 20)$
21. $AB(x) = \dfrac{(x + 4)\sqrt{x^2 + 25}}{x}$; Dom $AB = (0, \infty)$ **23. (a)** yes

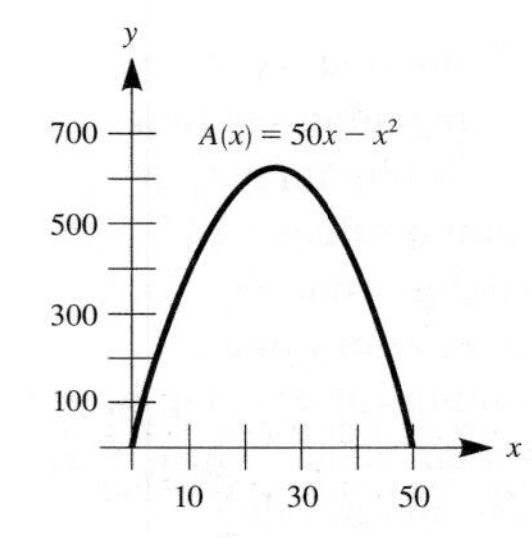

(b) one turning point
(c) maximum value = 625; no minimum value **25. (a)** no

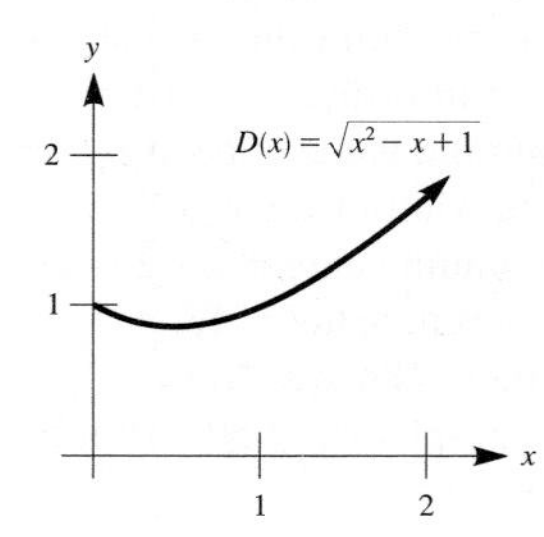

(b) one turning point **(c)** no maximum value; minimum value ≈ 0.87 **27. (a)** no

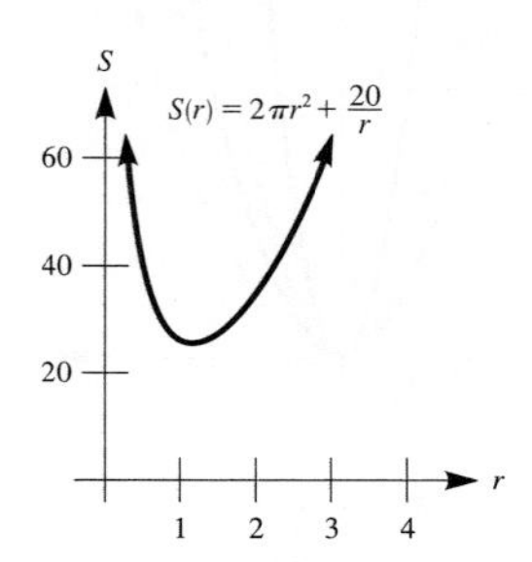

(b) one turning point **(c)** no maximum value; minimum value ≈ 25.69 **29. (a) (a)** no

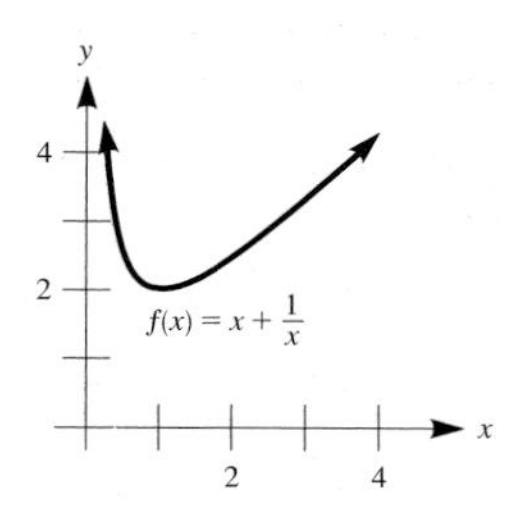

(b) one turning point **(c)** no maximum value; minimum value = 2
(b) (a) no

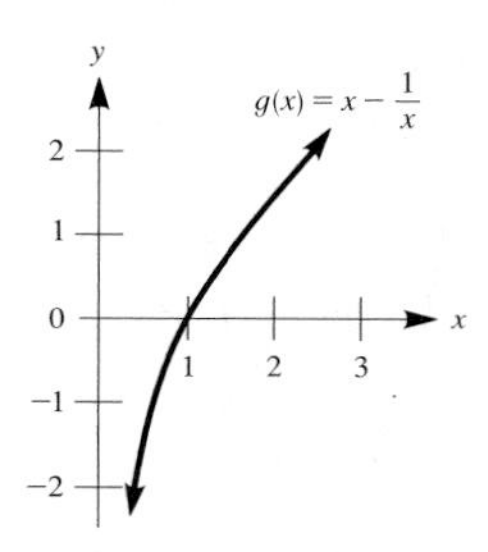

(b) no turning points **(c)** no maximum or minimum values

31. (a) $S(x) = x + \dfrac{\sqrt{11}}{x}$;
Dom $S = (0, \infty)$
minimum value ≈ 3.642

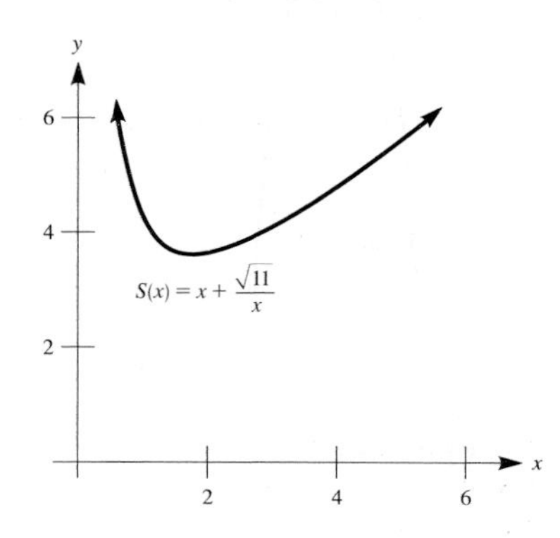

(b) $P(x) = -x^2 + \sqrt{11}x$;
Dom $P = (0, \sqrt{11})$
maximum value = 2.75

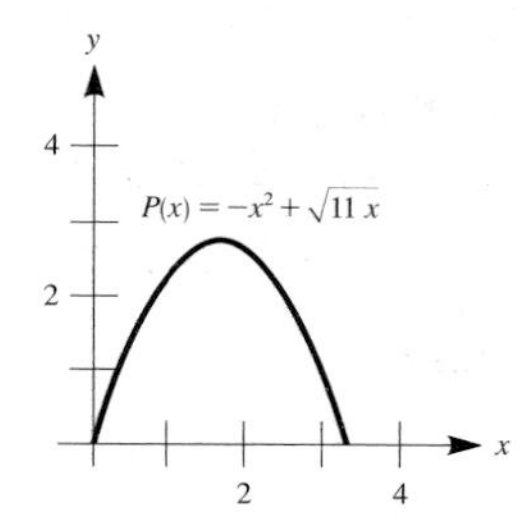

33. $A(x) = \frac{17}{144}x^2 - \frac{1}{3}x + \frac{1}{2}$;
Dom $A = (0, 3)$; yes

35. (a) $V(r) = \dfrac{\sqrt{3}}{3}\pi r^3$;
Dom $V = (0, \infty)$
(b) $S(r) = 2\pi r^2$; Dom $S = (0, \infty)$

37. (a) $r(h) = \dfrac{3h}{\sqrt{h^2 - 9}}$;
Dom $r = (3, \infty)$
(b) $h(r) = \dfrac{3r}{\sqrt{r^2 - 9}}$; Dom $h = (3, \infty)$

39. $A(x) = \dfrac{4x^2 + \pi(14 - x)^2}{16\pi}$;
Dom $A = (0, 14)$

41. $A(r) = \dfrac{r(1 - 4\pi r)}{4}$;
Dom $A = (0, \frac{1}{4\pi})$

43. $A(x) = \frac{\pi}{3}x^2$; Dom $A = (0, \infty)$

45. (a) $V(x) = 4x^3 - 28x^2 + 48x$;
Dom $V = (0, 3)$
(b)

x (in.)	0	0.5	1.0	1.5	2.0	2.5	3.0
volume (in.³)	0	17.5	24	22.5	16	7.5	0

(c) $x = 1.0$
(d)

x (in.)	0.8	0.9	1.0	1.1	1.2	1.3	1.4
volume (in.³)	22.5	23.4	24	24.2	24.2	23.9	23.3

(e) $x = 1.1$

47. (a) $A(r) = 32r - 2r^2 - \dfrac{\pi r^2}{2}$;
Dom $A = \left(0, \dfrac{32}{\pi + 2}\right)$
(b) downward; yes

49. (a) $y(s) = \dfrac{3s}{\sqrt{1 - s^2}}$;
Dom $y = (0, 1)$
(b) $s(y) = \dfrac{y}{\sqrt{y^2 + 9}}$; Dom $s = (0, \infty)$
(c) $z(s) = \dfrac{3}{\sqrt{1 - s^2}}$; Dom $z = (0, 1)$
(d) $s(z) = \dfrac{\sqrt{z^2 - 9}}{z}$; Dom $s = (3, \infty)$

51. (a) $m(a) = \dfrac{a^2 + 1}{a}$;
Dom $m = (-\infty, 0) \cup (0, \infty)$

53. $A(x) = \dfrac{8x - x^2}{4}$; Dom $A = (0, 8)$

55. $A(m) = \dfrac{2m^2 - 8m + 8}{m^2 - 4m}$;
Dom $m = (-\infty, 0)$

57. $A = \dfrac{(ma - b)^2}{-2m}$; Dom $A = (-\infty, 0)$

Exercise Set 4.5

1. $\frac{25}{4}$ **3.** $\frac{1}{2}$ **5.** $\frac{25}{4}$ m by $\frac{25}{4}$ m
7. 1250 in.² **9. (a)** 18 **(b)** $\frac{23}{4}$
(c) $\frac{47}{8}$ **(d)** $\frac{95}{16}$ **11. (a)** 16 feet; 12 feet
(b) 16 feet; 1 second
(c) $t = \frac{7}{4}$ sec or $t = \frac{1}{4}$ sec
13. $(\frac{7}{2}, \frac{2 + \sqrt{6}}{2})$; distance = $\sqrt{7}/2$
15. (a) $\frac{1}{2}$ **(b)** $\frac{1}{4}$ **17.** 125 ft by 250 ft
19. $x = 40$ **21.** $x = 60$;
maximum revenue = \$900; p = \$15
23. (a) $\frac{36}{13}$ **(b)** $6\sqrt{13}/13$; this is the square root.

25. (a)

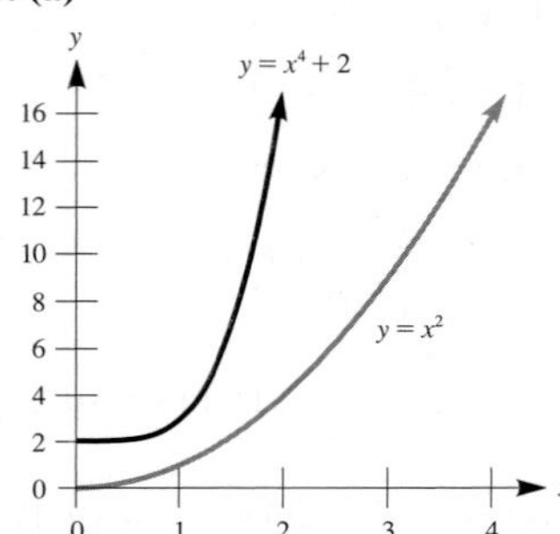

(b) 1.5 miles **(c)** $\frac{7}{4}$ miles **27. (a)** $\frac{225}{2}$

29. $x = \sqrt{2}/2$; area $= \frac{1}{2}$ **31.** $\frac{49}{12}$
35. 100 yd by 150 yd
39. (a) $p(x) = -2x + 500$
(b) maximum revenue = \$31,250; price = \$250 **41.** $t = \sqrt{3}$ or $t = -\sqrt{3}$ **43. (a)** $(x - \frac{1}{2})^2 \geq 0$, thus $(x - \frac{1}{2})^2 + \frac{3}{4} \geq \frac{3}{4} > 0$ **(b)** Since $D = -3 < 0$, there are no x-intercepts. Since the graph is a parabola opening up, and there are no x-intercepts, the y-values must be positive for all values of x.

45. (a) $A(x) = \dfrac{4 + \pi}{16\pi}x^2 - 2x + 16$;

Dom $A = (0, 16)$ **(b)** $x = \dfrac{16\pi}{4 + \pi}$

(c) $\dfrac{\pi}{4}$ **47. (a)** $A = \pi r^2 + \dfrac{1000}{r}$
(b) height ≈ 5.42 cm, radius ≈ 5.42 cm

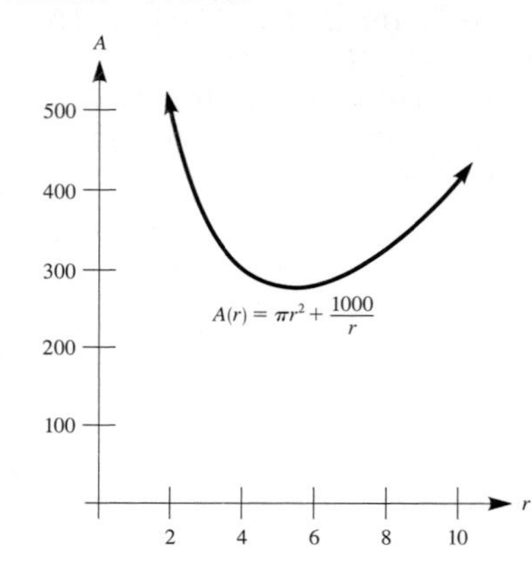

49. (a) $d = \sqrt{x^4 + x^2 - 6x + 9}$
(b) $(1, 1)$

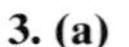

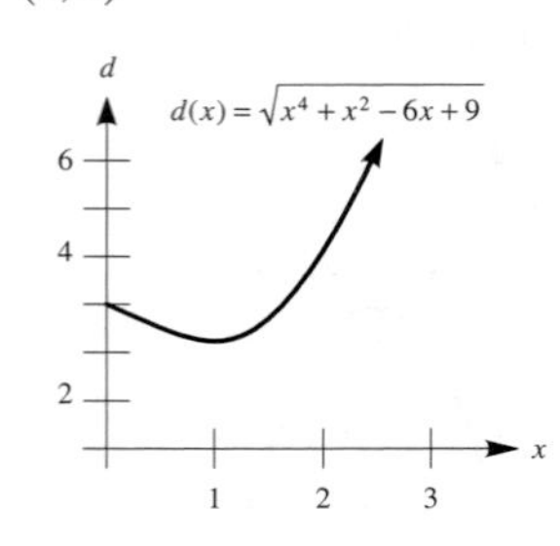

51. (a) $S = 2\sqrt{9 + y^2} + 6 - y$
(b) $P(0, 1.73)$

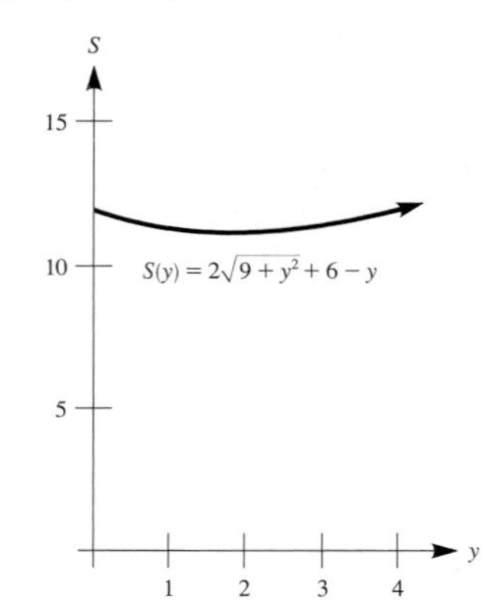

53. (a)

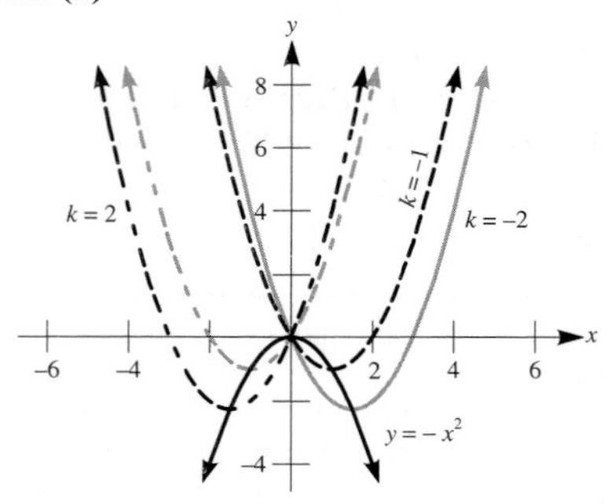

55. 2 **59.** 2

Exercise Set 4.6

1. (a) through (d)

3. (a)

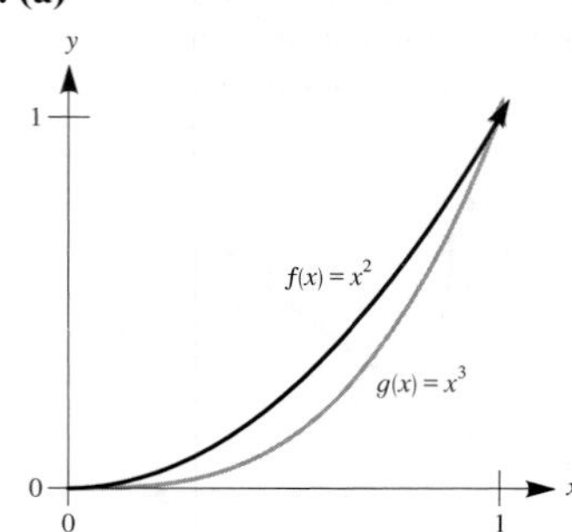

(b) $[0, 1]$: $\dfrac{\Delta f}{\Delta x} = 1, \dfrac{\Delta g}{\Delta x} = 1$; $[0, \frac{1}{2}]$: $\dfrac{\Delta f}{\Delta x} = \frac{1}{2}, \dfrac{\Delta g}{\Delta x} = \frac{1}{4}$ **5.** no x-intercepts; y-intercept: 5

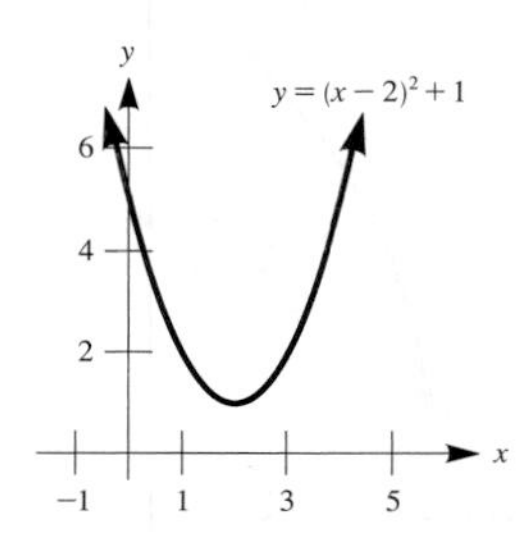

7. x-intercept: 1; y-intercept: -1

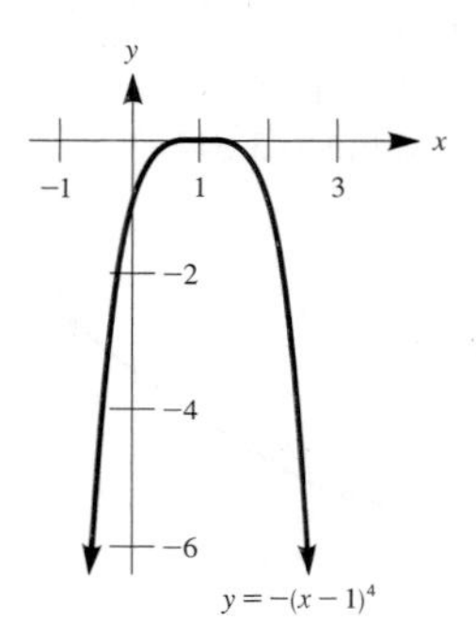

9. x-intercept: $4 + \sqrt[3]{2}$; y-intercept: -66

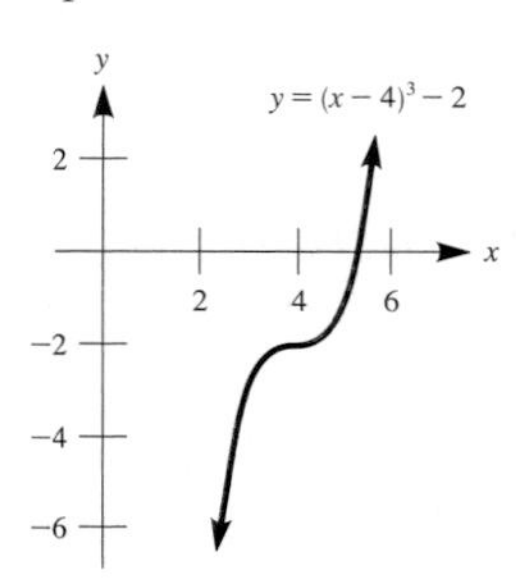

11. x-intercept: -5; y-intercept: -1250

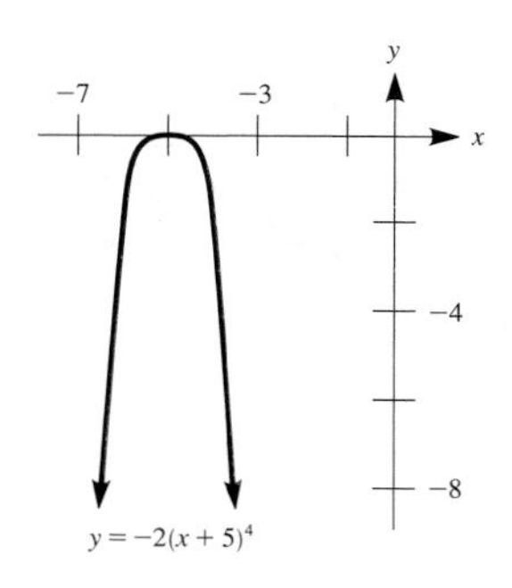

13. x-intercept: -1; y-intercept: $\frac{1}{2}$

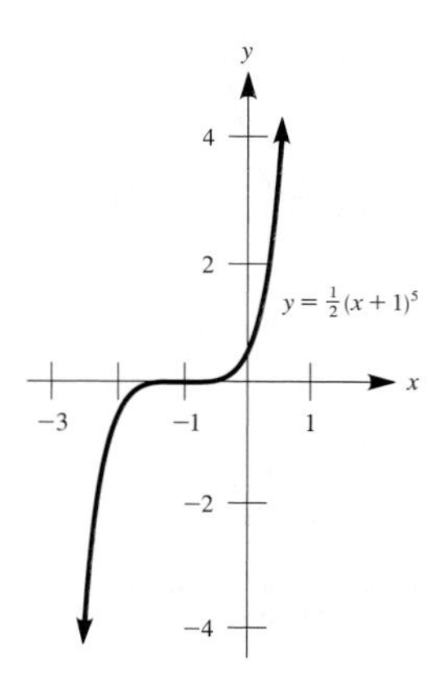

15. x- and y-intercepts: 0

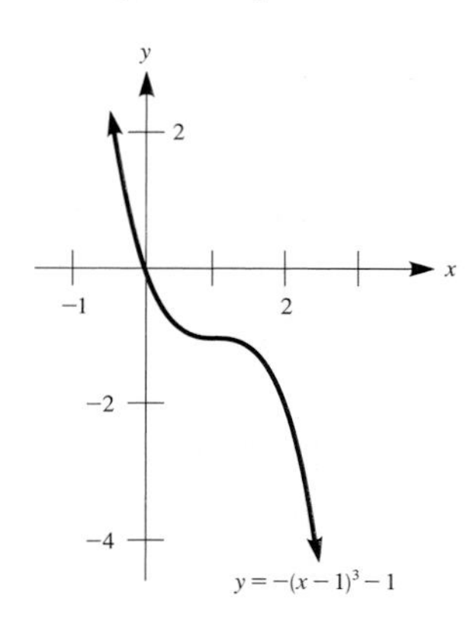

17. A polynomial function of degree 3 can have at most two turning points.
19. As $|x|$ gets very large, our function should be similar to $f(x) = a_3x^3$.
21. As $|x|$ gets very large with x negative, then the graph should resemble $2x^5$.
23. This graph has a corner.

25. (a)

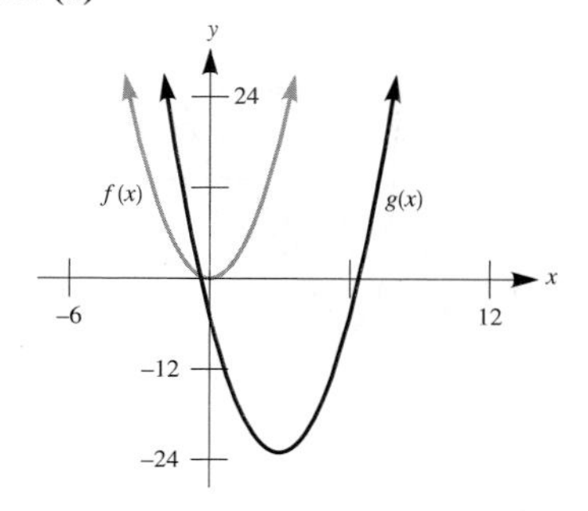

(b)

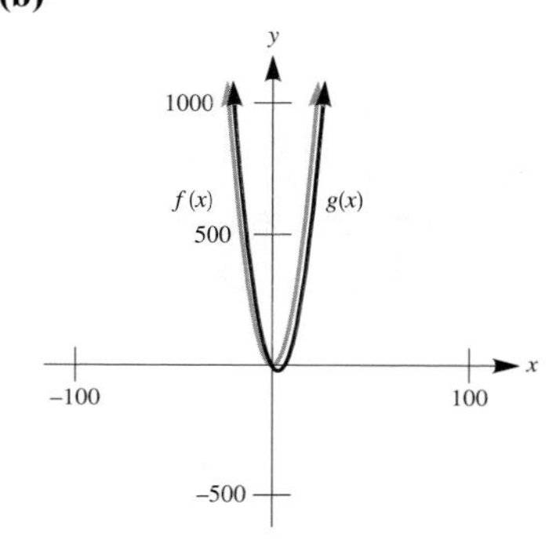

27. (a) x-intercepts: 2, 1, -1; y-intercept: 2

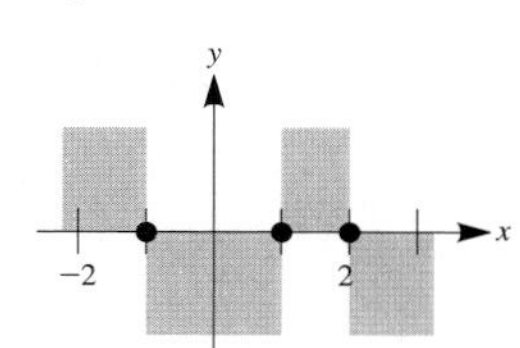

(b)

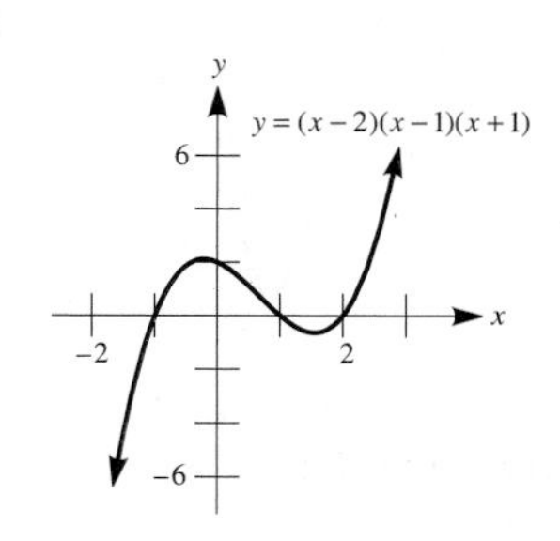

29. (a) x-intercepts: 0, 2, 1; y-intercept: 0

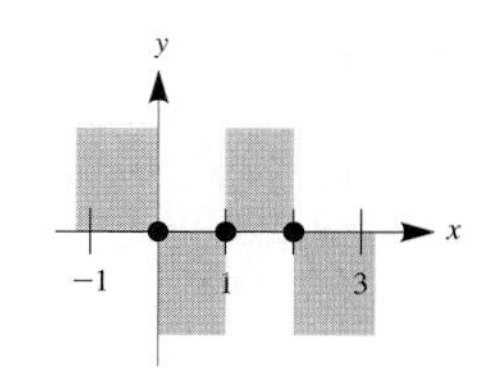

(b)

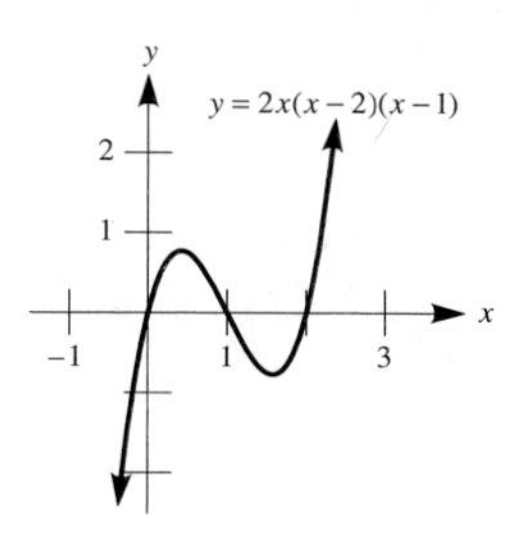

31. (a) x-intercepts: 0, 5, -1; y-intercept: 0

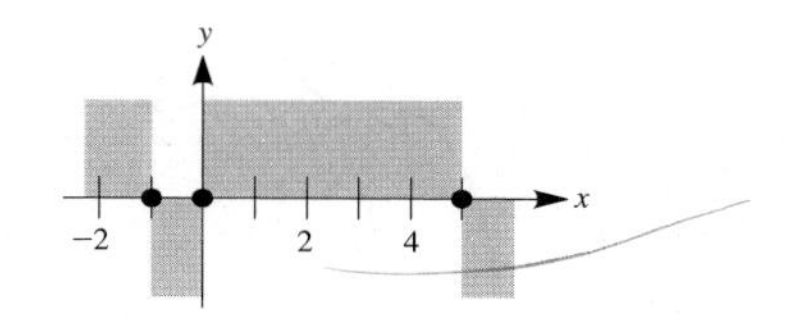

(b)

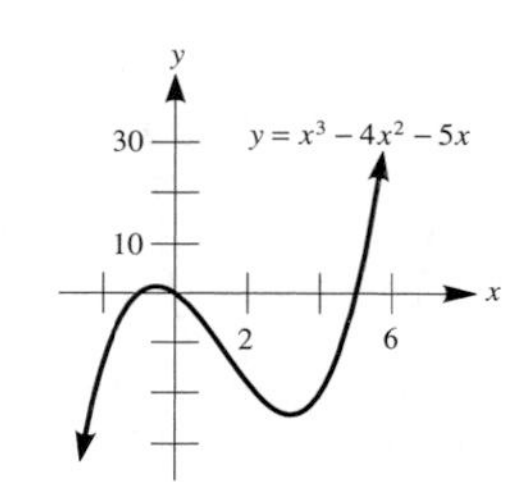

33. (a) x-intercepts: -3, -2, 2; y-intercept: -12

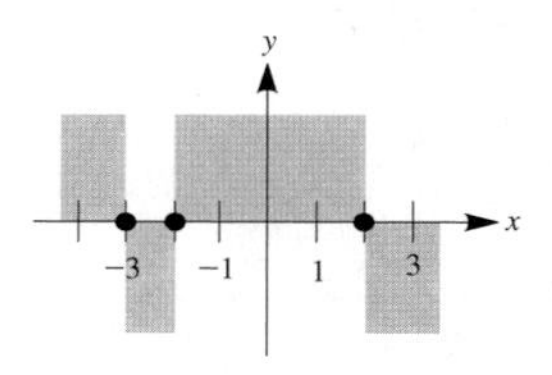

(b)

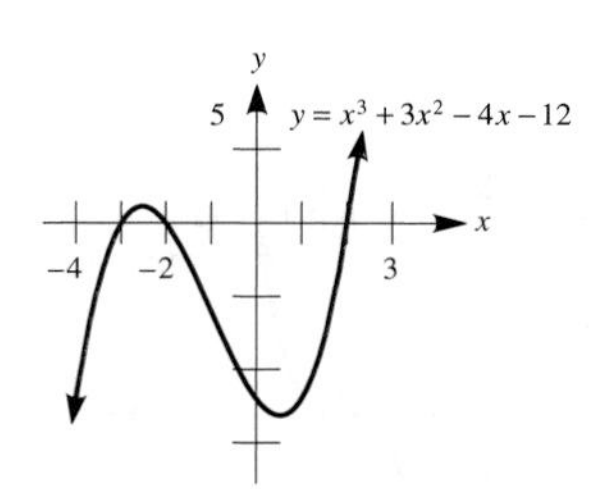

35. (a) x-intercepts: 0 and -2; y-intercept: 0

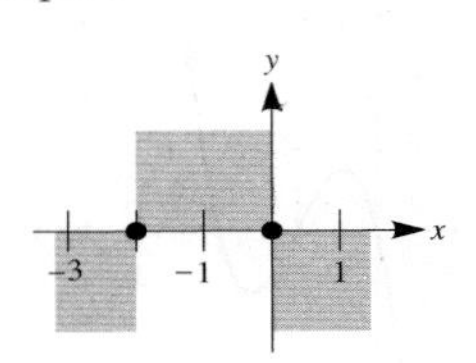

(b)

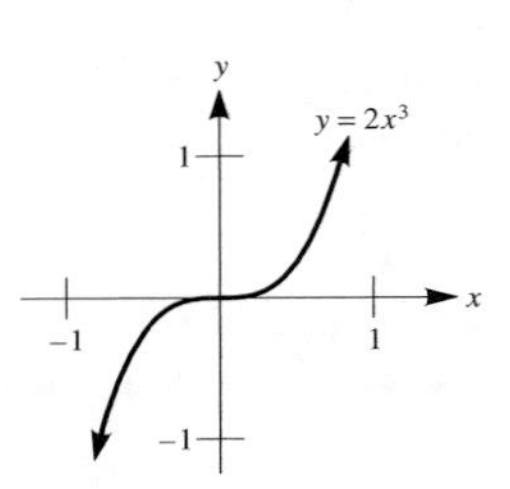

(c)

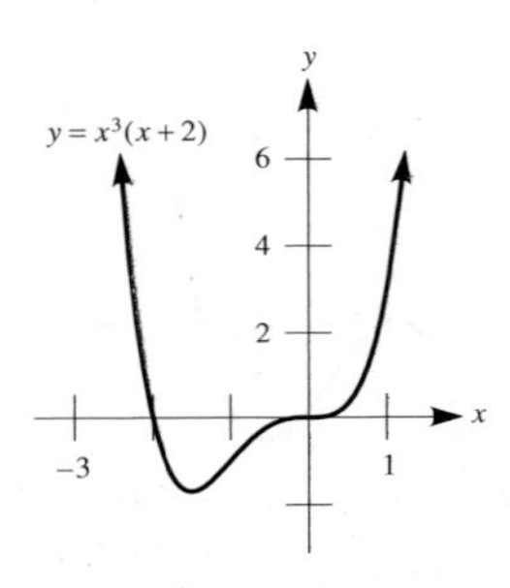

37. (a) x-intercepts: 1, 4; y-intercept: 128

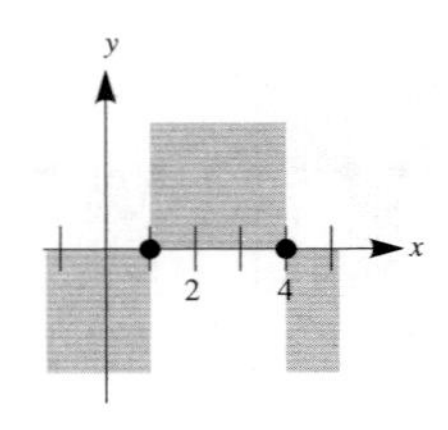

(b)

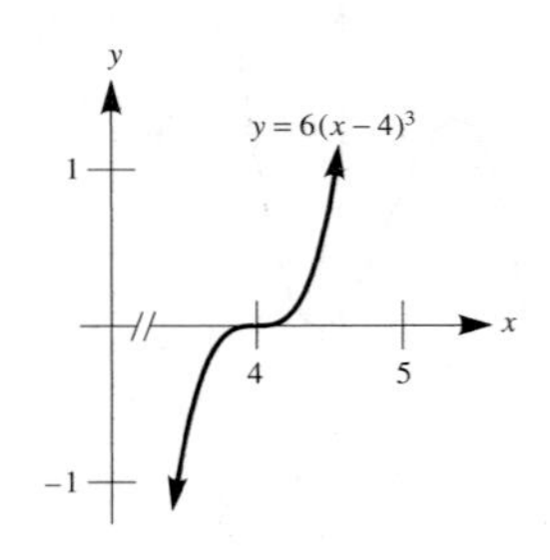

(c)

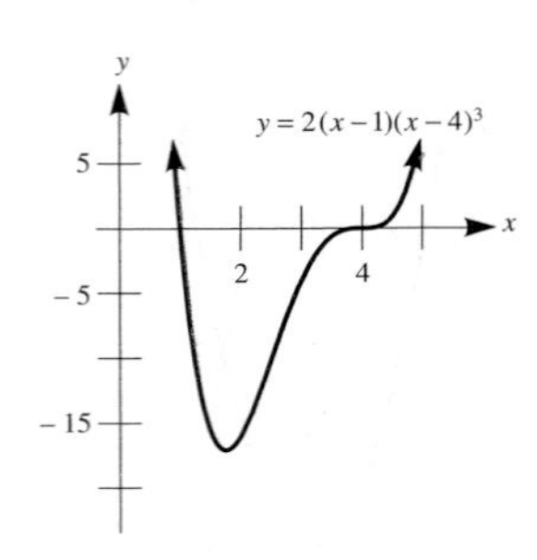

39. (a) x-intercepts: -1, 1, 3; y-intercept: 3

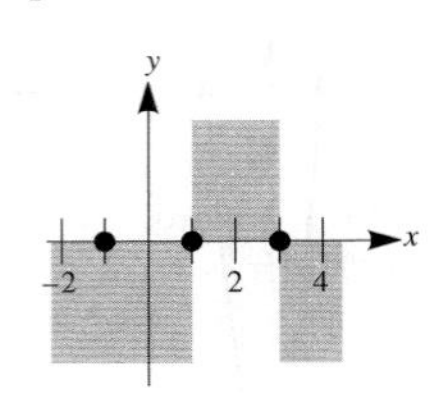

(b)

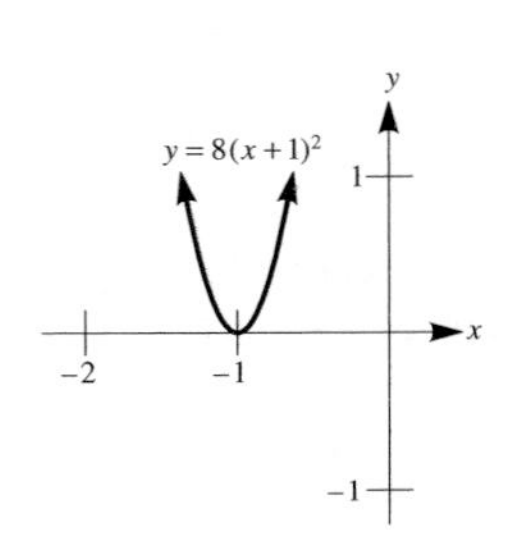

(c)

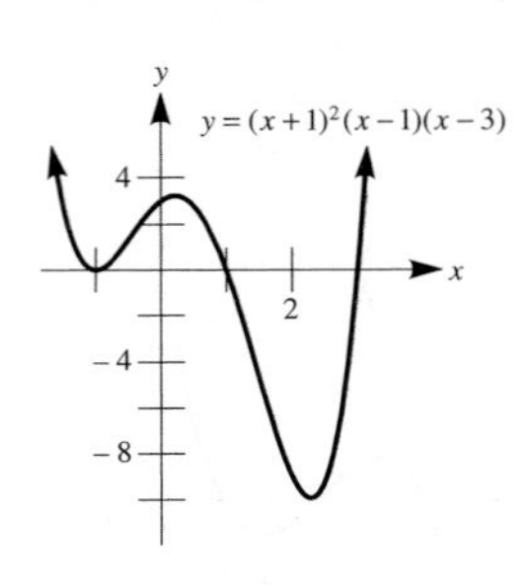

41. (a) x-intercepts: 0, 4, -2; y-intercept: 0

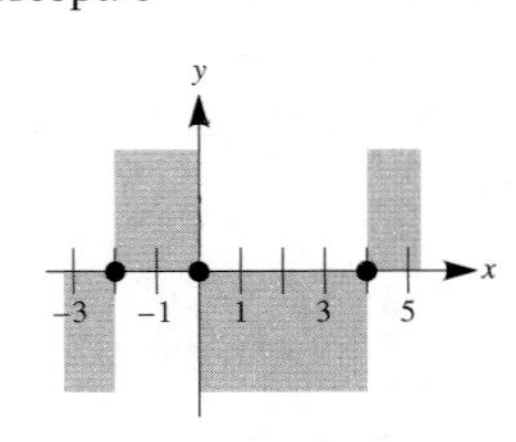

(b)

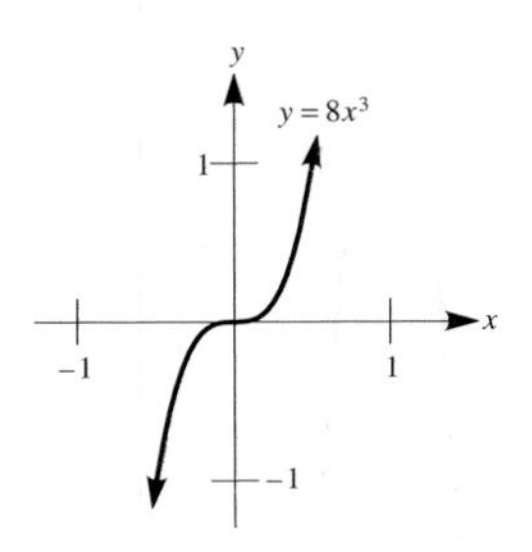

(c)

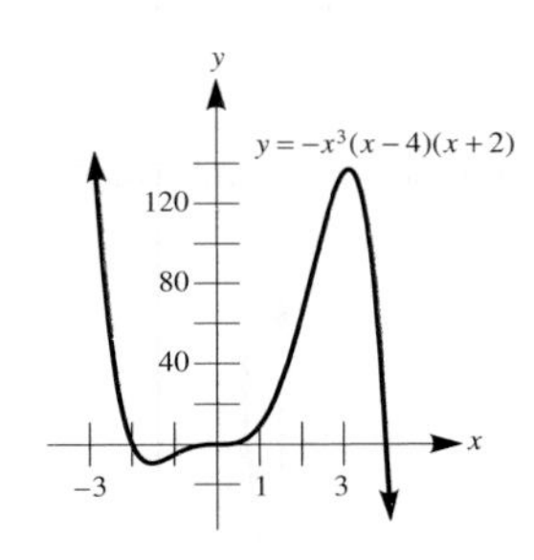

43. (a) x-intercepts: 0, 2, -2; y-intercept: 0

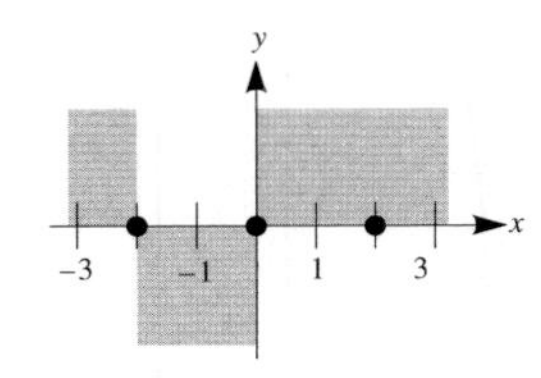

(b)

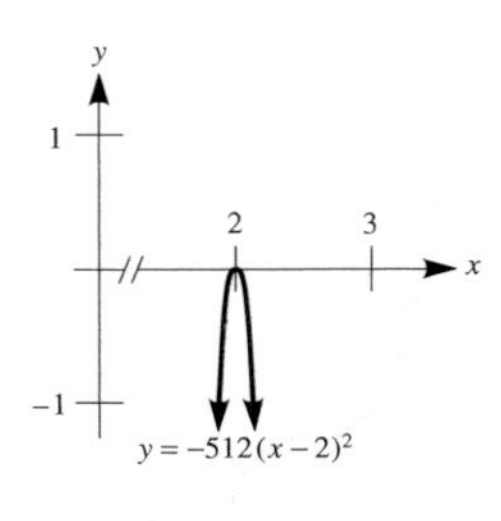

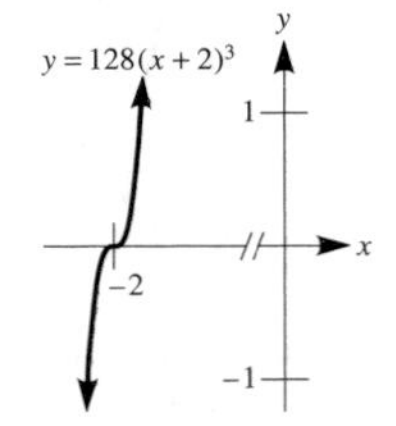

(c)

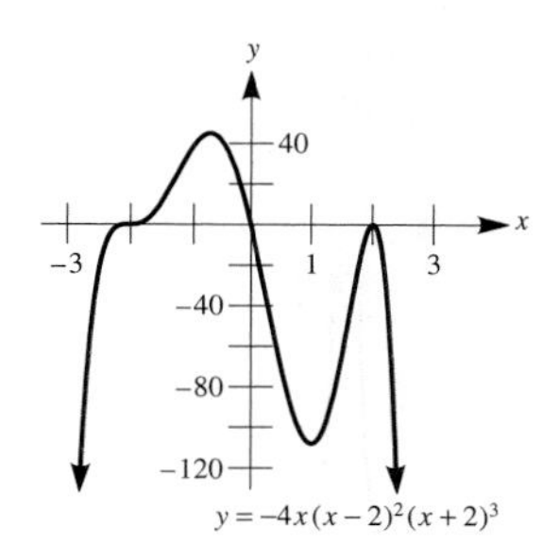

47. $x = 0, \dfrac{3 \pm \sqrt{29}}{2} \approx -1.193, 4.193$

49. $x = -6, \pm\sqrt{3} \approx \pm 1.732$

51. (a)

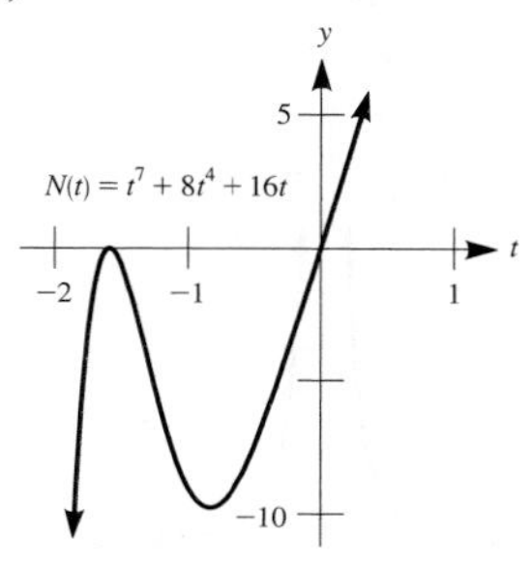

(b) t-intercepts: 0.0, −1.6
(c) $t = 0, -\sqrt[3]{4} \approx -1.587$
53. From left to right: $f(x) = x$, $g(x) = x^2$, $h(x) = x^3$, $F(x) = x^4$, $G(x) = x^5$, $H(x) = x^6$ **55.** $[0, 0.68)$
57. no such value **59. (a)** $(400, 1600)$
(b)

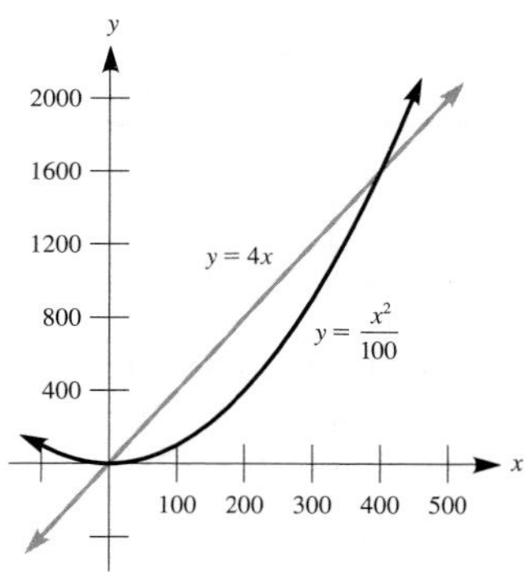

61. (a)

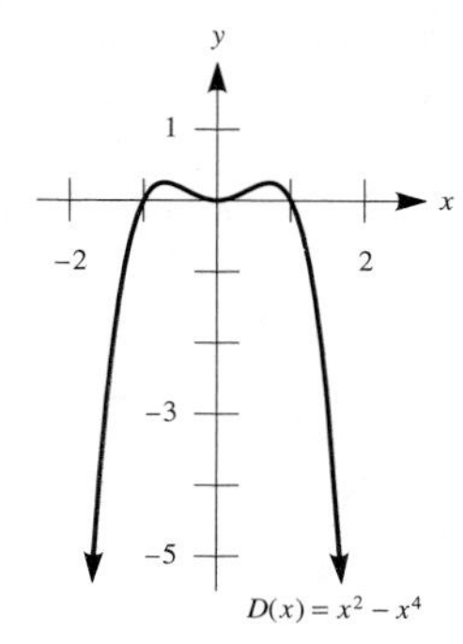

(b) $(\pm\frac{\sqrt{2}}{2}, \frac{1}{4})$, $(0, 0)$
(c) maximum vertical distance: $\frac{1}{4}$

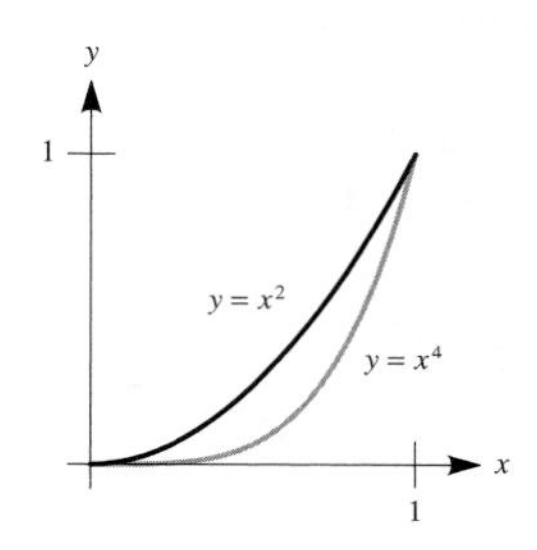

63. (a) $A(x) = x(1 - x^4) = x - x^5$
(b) maximum area: 0.53 square units

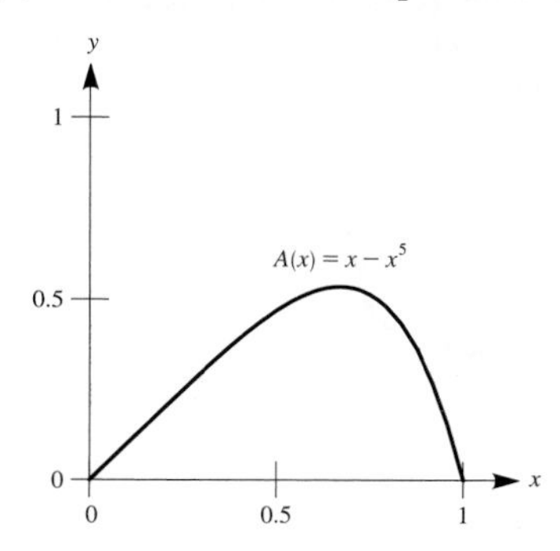

65. (b) From part **(a)**, if $|x| < 1$, then $x^n < x^m$, so the graph of $y = x^n$ lies below the graph of $y = x^m$. If $|x| > 1$, then $x^n > x^m$, so the graph of $y = x^n$ lies above the graph of $y = x^m$. These are valid, since m and n are even, thus ensuring that $x^m \geq 0$ and $x^n \geq 0$.
(c) If m and n are positive odd integers, from part **(a)** the graph of $y = x^m$ lies above the graph of $y = x^n$ for $0 < x < 1$, and it lies below the graph of $y = x^n$ for $x > 1$. If $-1 < x < 0$, however, $x^n > x^m$ (since x is negative), and so the graph of $y = x^n$ lies above the graph of $y = x^m$. If $x < 1$, then $x^n < x^m$, so the graph of $y = x^n$ lies below the graph of $y = x^m$.

Exercise Set 4.7

1. (a) domain: $(-\infty, 3) \cup (3, \infty)$; x-intercept: −5; y-intercept: $-\frac{5}{4}$; vertical asymptote: $x = 3$; horizontal asymptote: $y = \frac{3}{4}$
(b)

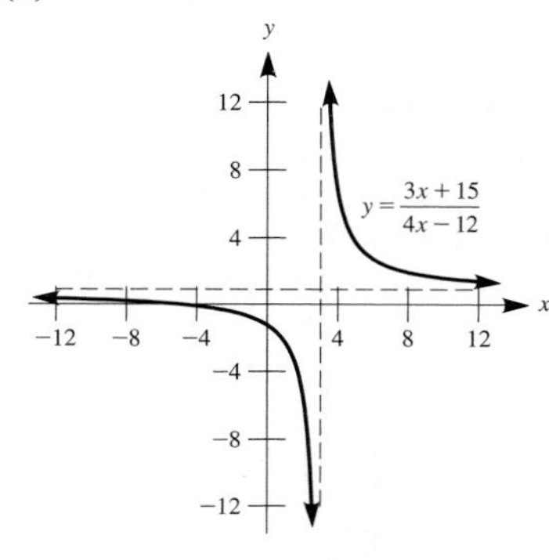

3. (a) domain: $(-\infty, 0) \cup (0, \infty)$; x-intercepts: $\frac{1}{3}, \frac{1}{2}$; no y-intercept; vertical asymptote: $x = 0$; horizontal asymptote: $y = 3$
(b)

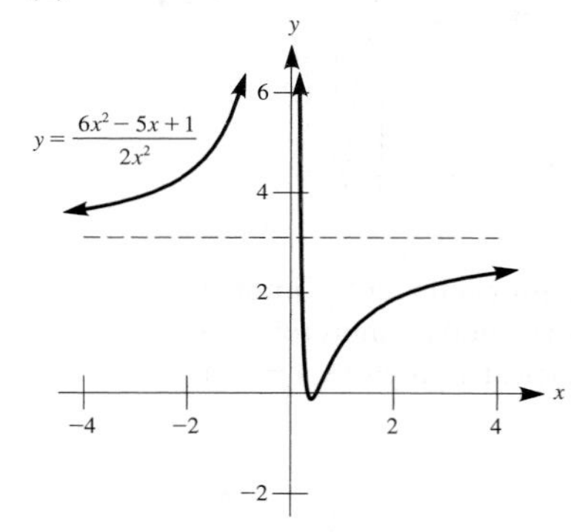

5. (a) domain: $(-\infty, -\frac{1}{2}) \cup (-\frac{1}{2}, \frac{1}{2}) \cup (\frac{1}{2}, \infty)$; x-intercepts: ±3; y-intercept: 9; vertical asymptotes: $x = \frac{1}{2}$ and $x = -\frac{1}{2}$; horizontal asymptote: $y = \frac{1}{4}$

(b)

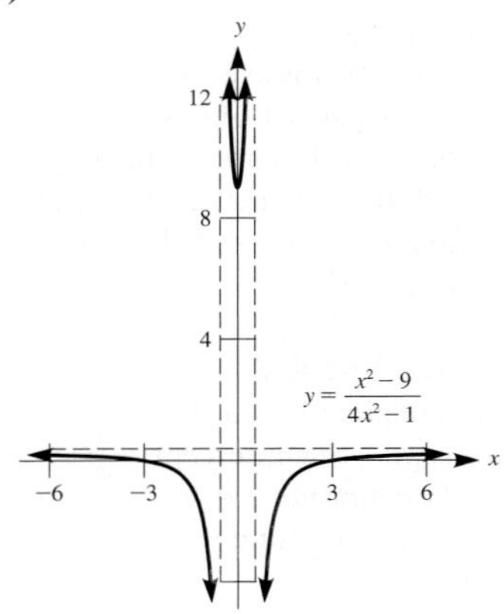

7. (a) domain: $(-\infty, -\frac{3}{2}) \cup (-\frac{3}{2}, 0) \cup (0, 1) \cup (1, \infty)$; x-intercepts: $-\frac{4}{3}$, 2; no y-intercept; vertical asymptotes: $x = -\frac{3}{2}$, $x = 0$, $x = 1$; horizontal asymptote: $y = 0$
(b)

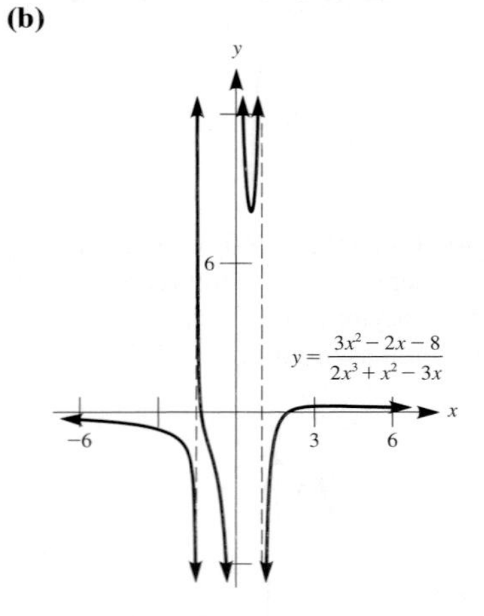

9. no x-intercept; y-intercept: $\frac{1}{4}$; horizontal asymptote: $y = 0$; vertical asymptote: $x = -4$

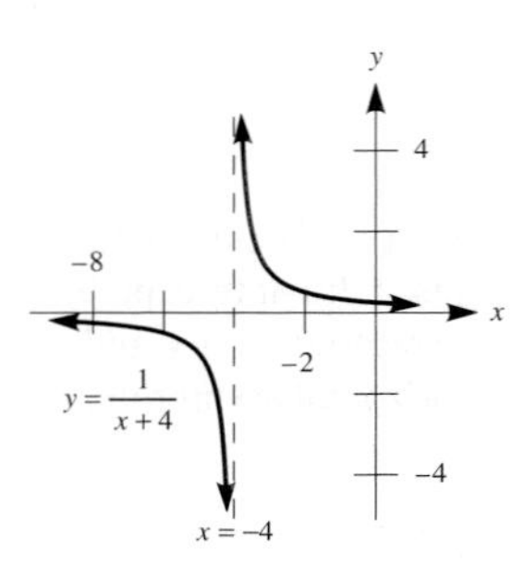

11. no x-intercept; y-intercept: $\frac{3}{2}$; horizontal asymptote: $y = 0$; vertical asymptote: $x = -2$

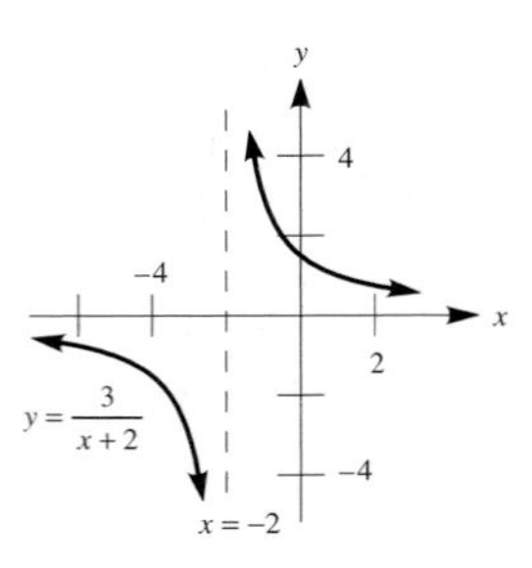

13. no x-intercept; y-intercept: $\frac{2}{3}$; horizontal asymptote: $y = 0$; vertical asymptote: $x = 3$

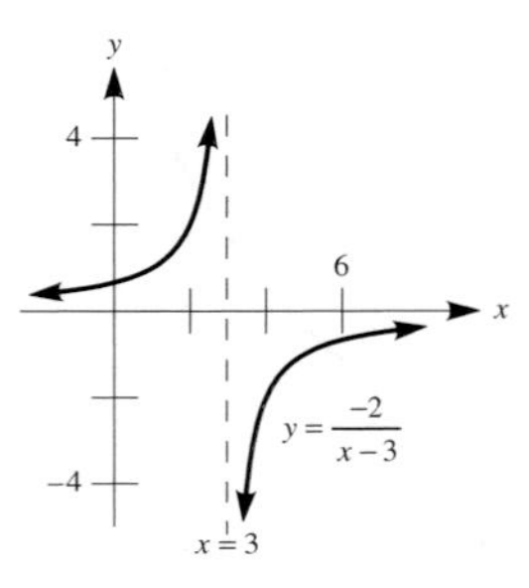

15. x-intercept: 3; y-intercept: 3; horizontal asymptote: $y = 1$; vertical asymptote: $x = 1$

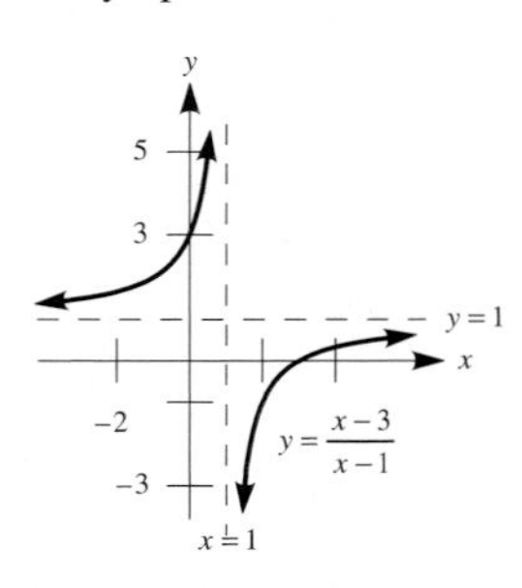

17. x-intercept: $\frac{1}{2}$; y-intercept: -2; horizontal asymptote: $y = 2$; vertical asymptote: $x = -\frac{1}{2}$

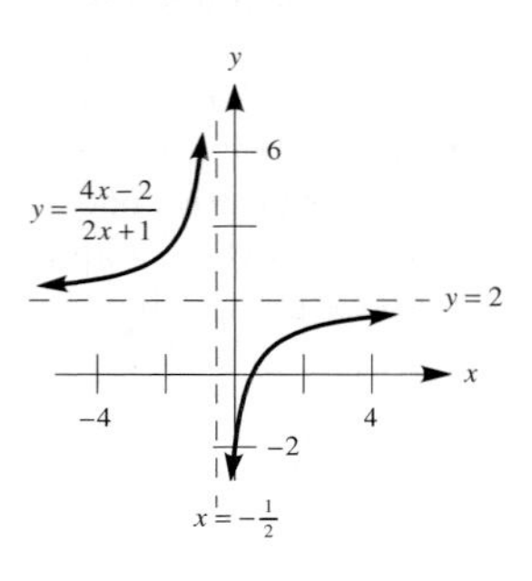

19. no x-intercept; y-intercept: $\frac{1}{4}$; horizontal asymptote: $y = 0$; vertical asymptote: $x = 2$

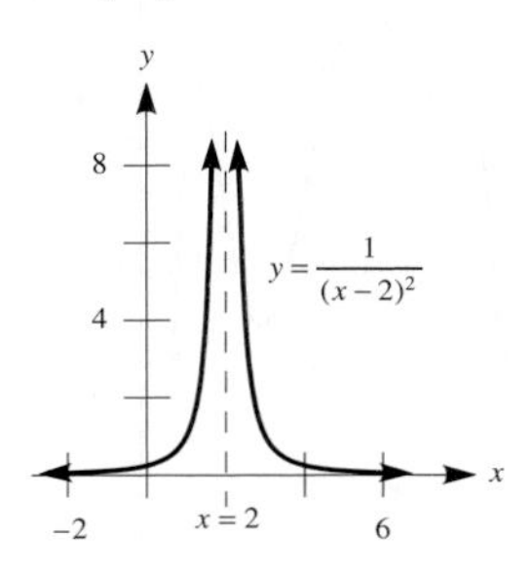

21. no x-intercept; y-intercept: 3; horizontal asymptote: $y = 0$; vertical asymptote: $x = -1$

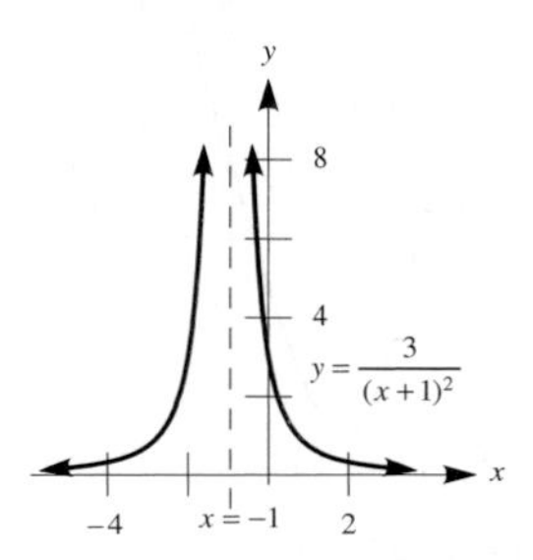

23. no x-intercept; y-intercept: $\frac{1}{8}$; horizontal asymptote: $y = 0$; vertical asymptote: $x = -2$

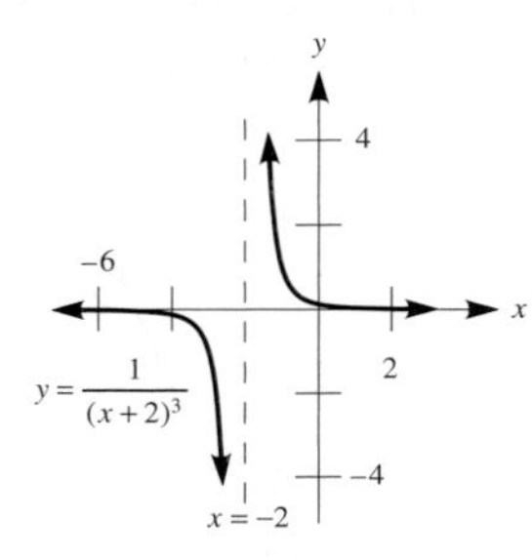

25. no x-intercept; y-intercept: $-\frac{4}{125}$; horizontal asymptote: $y = 0$; vertical asymptote: $x = -5$

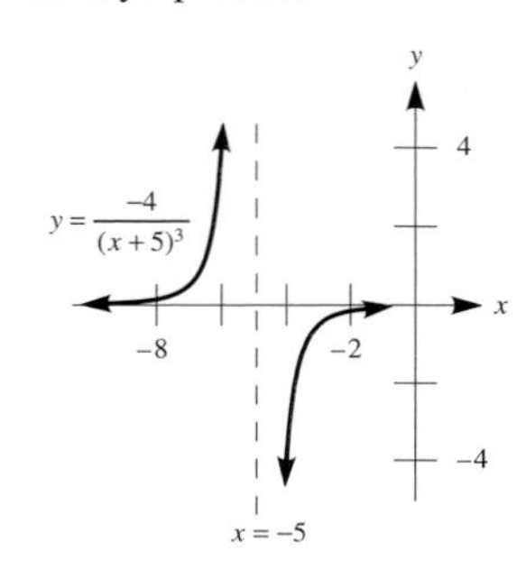

27. x- and y-intercepts: 0; horizontal asymptote: $y = 0$; vertical asymptotes: $x = -2, x = 2$

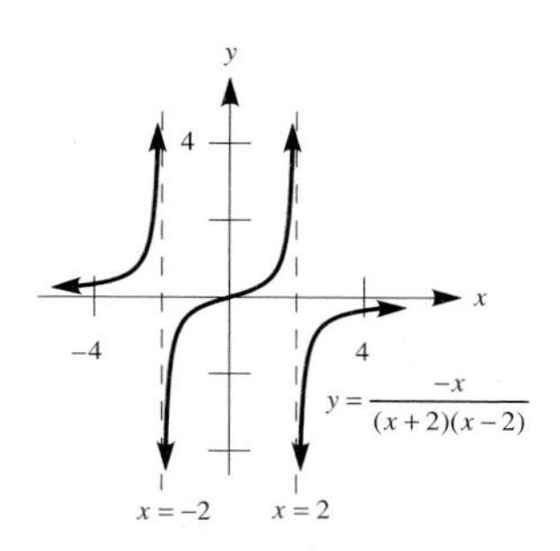

29. (a) x-intercept: 0; y-intercept: 0; horizontal asymptote: $y = 0$; vertical asymptotes: $x = 1, x = -3$

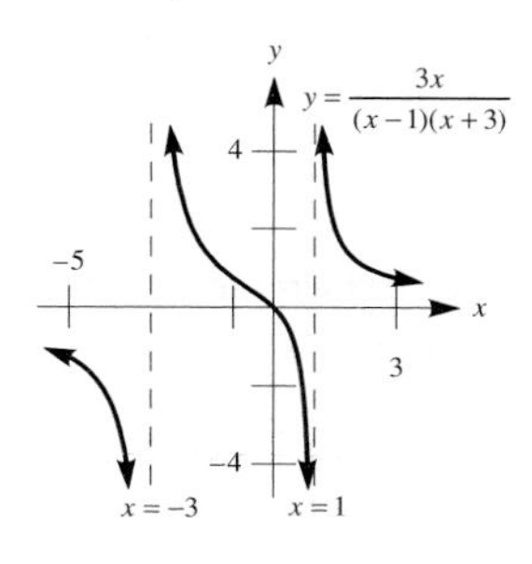

(b) x-intercept: 0; y-intercept: 0; horizontal asymptote: $y = 3$; vertical asymptotes: $x = 1, x = -3$

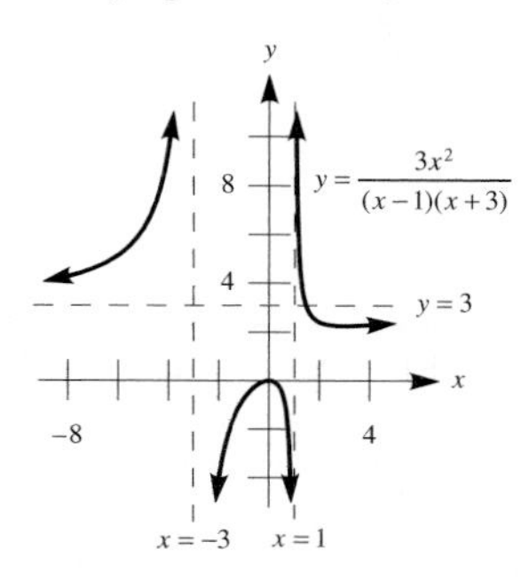

31. x-intercepts: $-\frac{5}{4}$, 1; y-intercept: 1; horizontal asymptote: $y = 2$; vertical asymptotes: $x = \frac{5}{2}, x = -1$

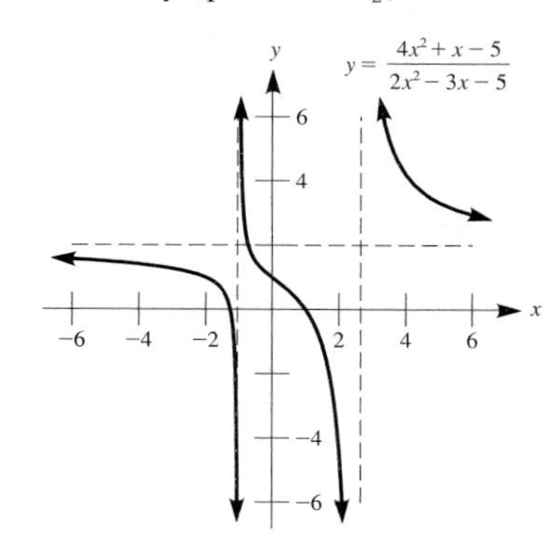

33. (a) x-intercepts: 2, 4; no y-intercept; horizontal asymptote: $y = 1$; vertical asymptotes: $x = 0, x = 1$

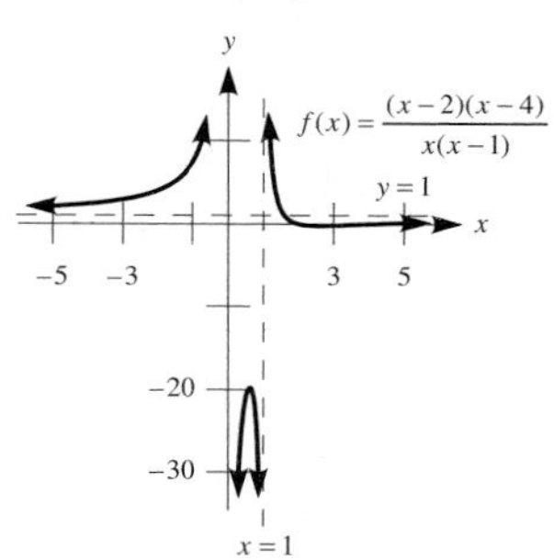

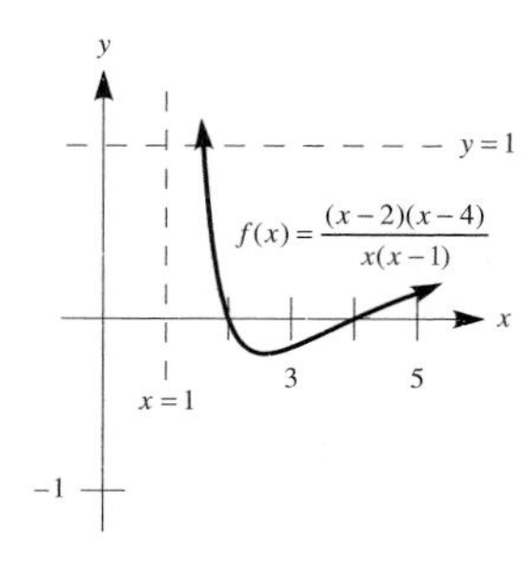

(b) x-intercepts: 2, 4; no y-intercept; horizontal asymptote: $y = 1$; vertical asymptotes: $x = 0, x = 3$

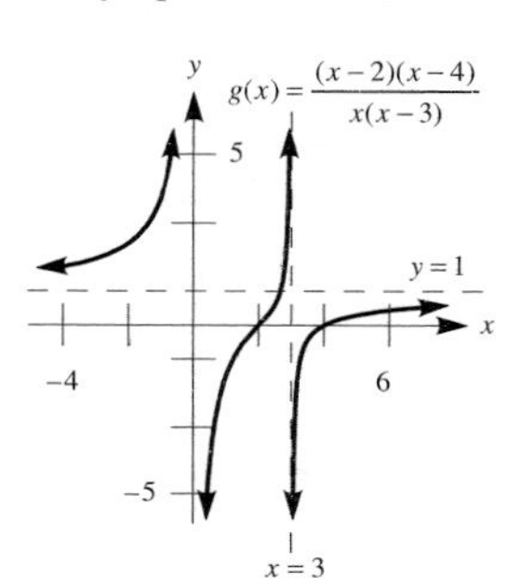

35. (a) 500 bacteria
(b) 2500 bacteria

37. (a) $A(x) = 521 - \dfrac{1500}{x} - 7x$
(b)

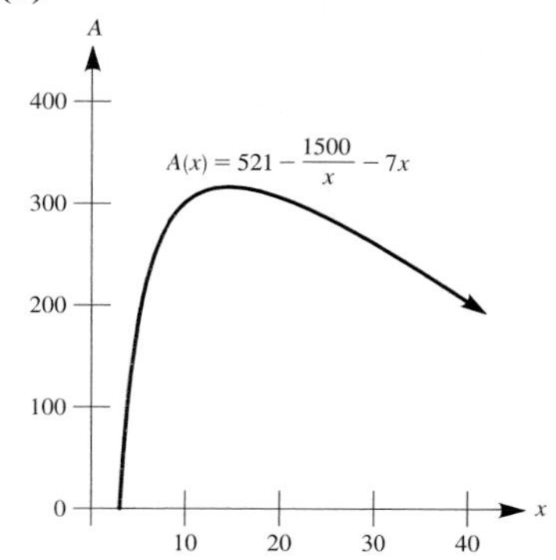

(c) width $\approx$ 14.6 inches, length $\approx$ 34.2 inches

39. crosses at $(\frac{11}{2}, 1)$

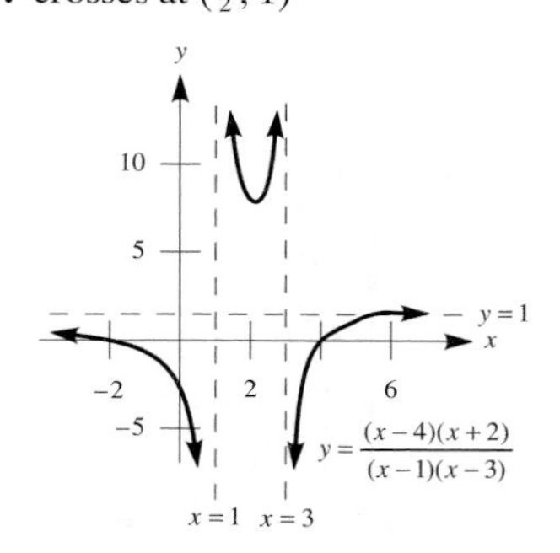

41. crosses at $(\frac{1}{3}, 1)$

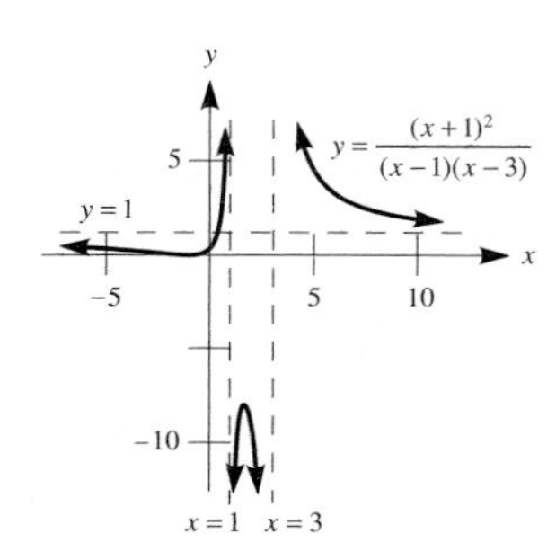

45. (a)

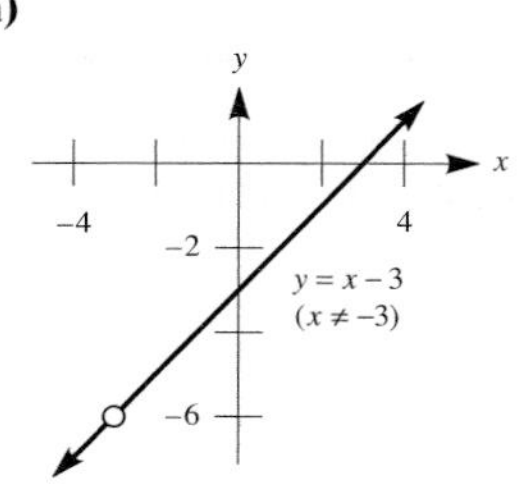

(b)

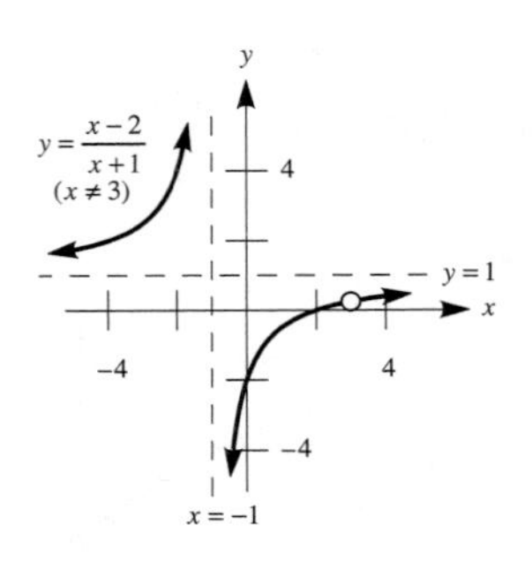

(c)

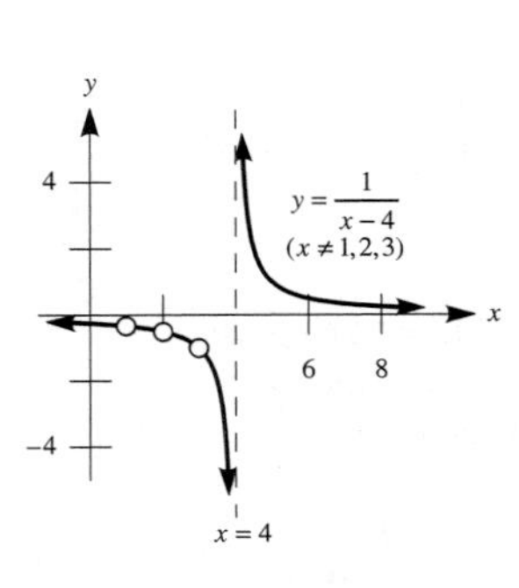

47. low point: $(-3, -\frac{1}{12})$

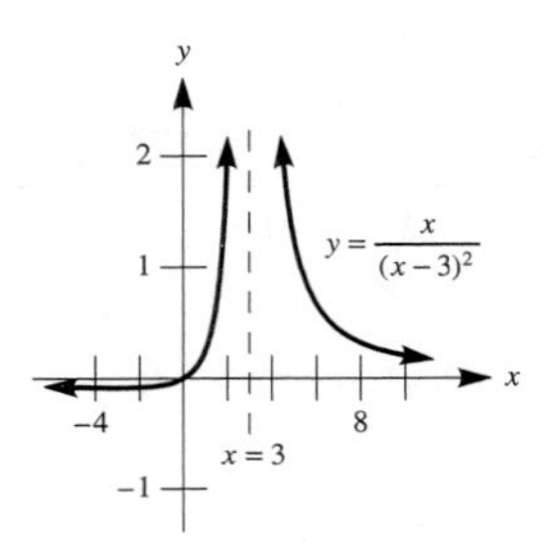

49. (b)

x	$x + 4$	$\frac{x^2 + x - 6}{x - 3}$
10	14	14.8571
100	104	104.0619
1000	1004	1004.0060
−10	−6	−6.4615
−100	−96	−96.0583
−1000	−996	−996.0600

(c) vertical asymptote: $x = 3$; x-intercepts: $-3, 2$; y-intercept: 2

(d)

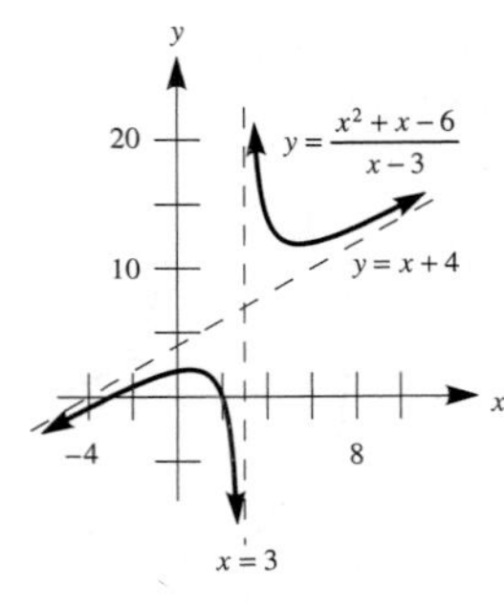

(e) $(3 + \sqrt{6}, 7 + 2\sqrt{6})$, $(3 - \sqrt{6}, 7 - 2\sqrt{6})$

51.

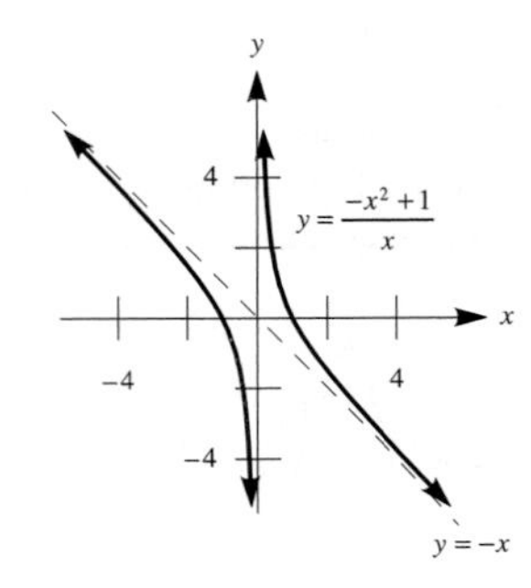

53. (a)

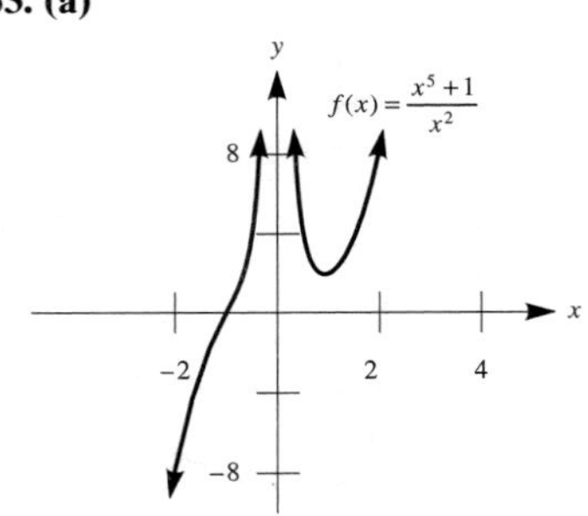

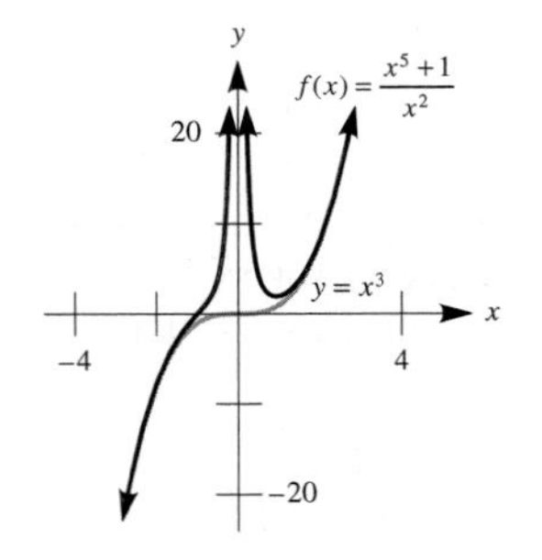

(b)

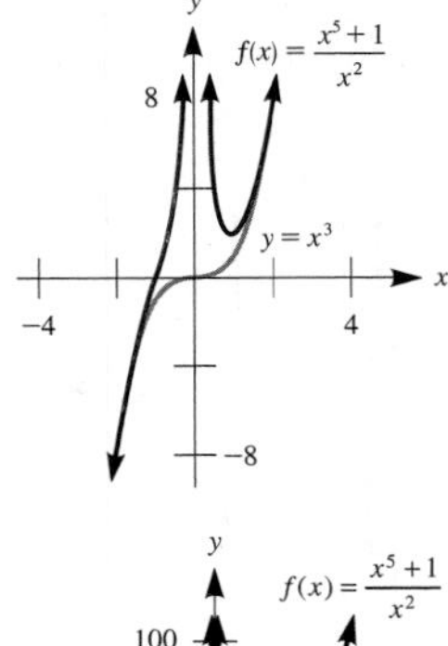

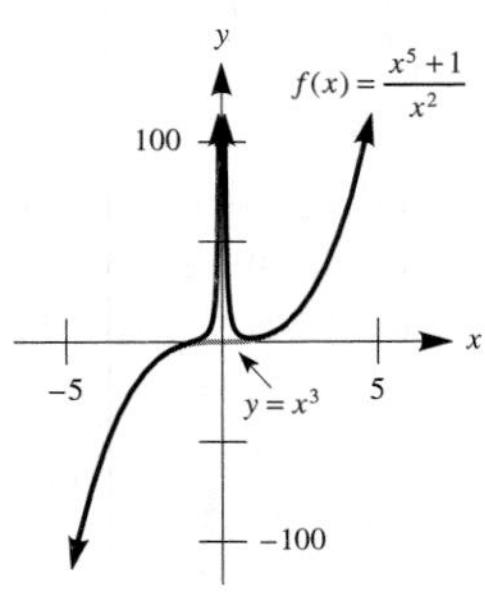

Note that as $|x|$ increases, the curve $f(x)$ approaches the curve $y = x^3$.

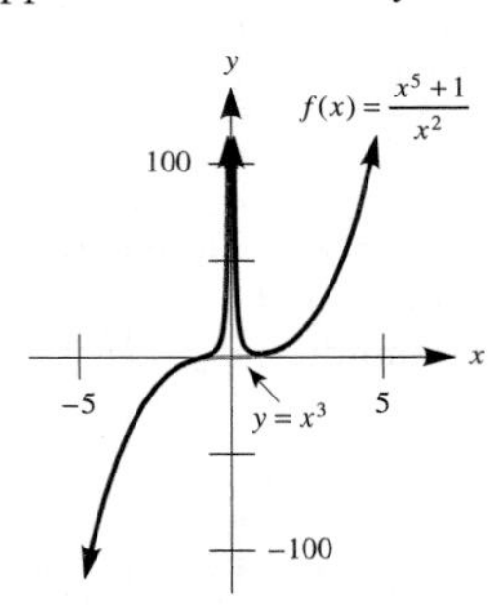

(c)

x	5	10	50	100	500
d	0.04	0.01	0.0004	0.0001	0.000004

x	−5	−10	−50	−100	−500
d	0.04	0.01	0.0004	0.0001	0.000004

(d) As $|x|$ increases, the quantity $1/x^2$ approaches 0, and thus $f(x) \approx x^3$ when $|x|$ gets very large.

Chapter 4 Review Exercises

1. $G(0) = -5$ **3.** $20,000

5.

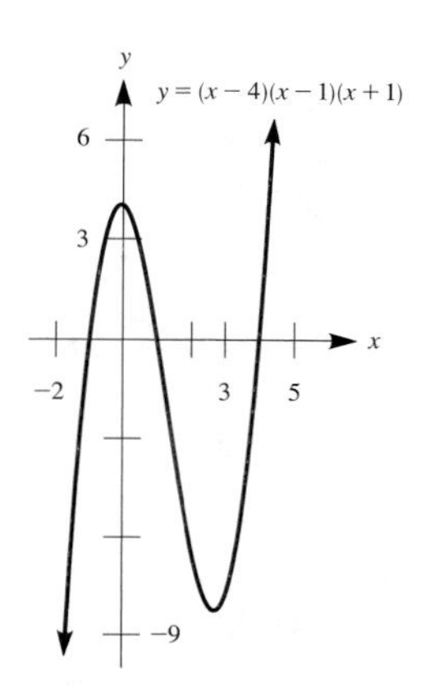

7. $V(t) = -180t + 1000$

9. x-intercept: $-\frac{5}{3}$; y-intercept: $\frac{5}{2}$; horizontal asymptote: $y = 3$; vertical asymptote: $x = -2$

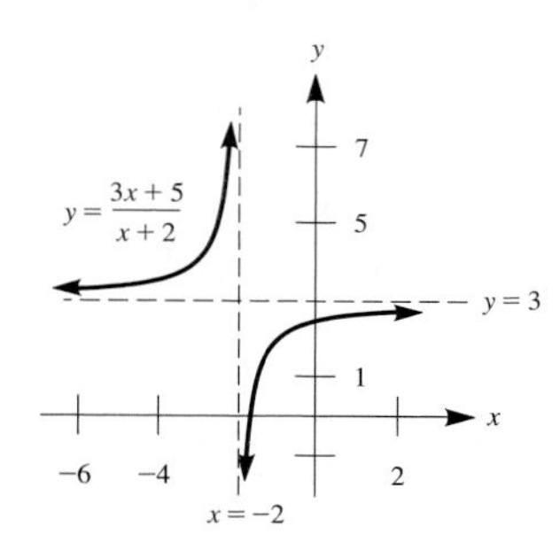

11.

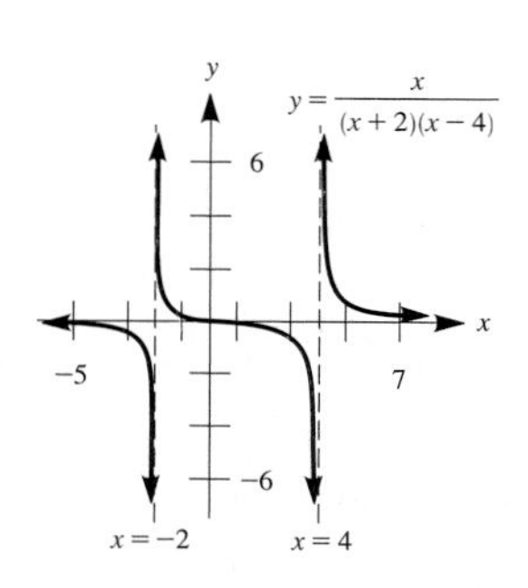

13. $P(w) = 2w + \dfrac{2\sqrt{144 - \pi^2 w^2}}{\pi}$

15. $f(x) = \frac{3}{8}x - \frac{5}{2}$

17. $f(x) = -\frac{3}{4}x + \frac{11}{4}$

19. vertex: $(-1, -4)$; x-intercepts: 1, -3; y-intercept: -3

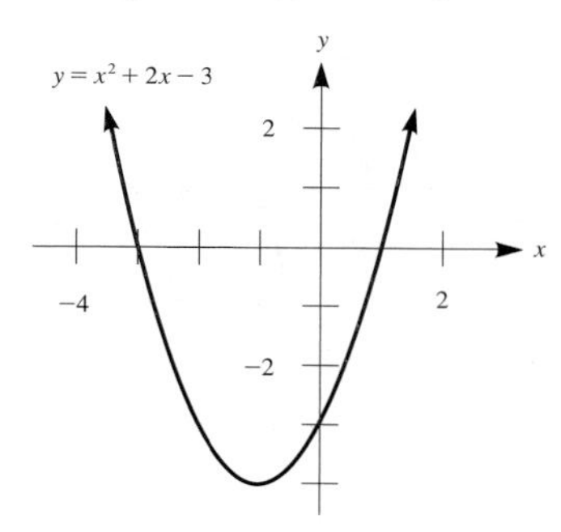

21. vertex: $(\frac{1}{2}, \frac{1}{2})$; no x-intercept; y-intercept: 1

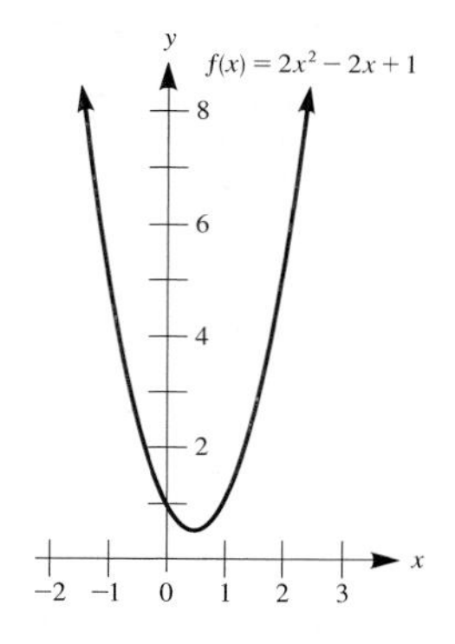

23. 5 **25. (a)** maximum height: $\dfrac{v_0^2}{32}$ ft; time: $t = \dfrac{v_0}{32}$ sec **(b)** $t = \dfrac{v_0}{16}$ sec

27. $b = \frac{17}{3}, -11$ **29.** 1 **31.** $\frac{225}{4}$ cm^2

33. $a = 1$ **35.** $x = \frac{432}{59}$ cm ≈ 7.32 cm

39. x-intercepts: -4, 2; y-intercept: -8

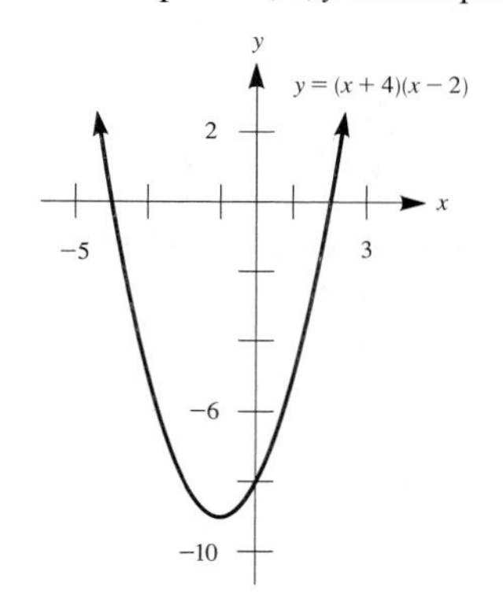

41. x-intercepts: -1, 0; y-intercept: 0

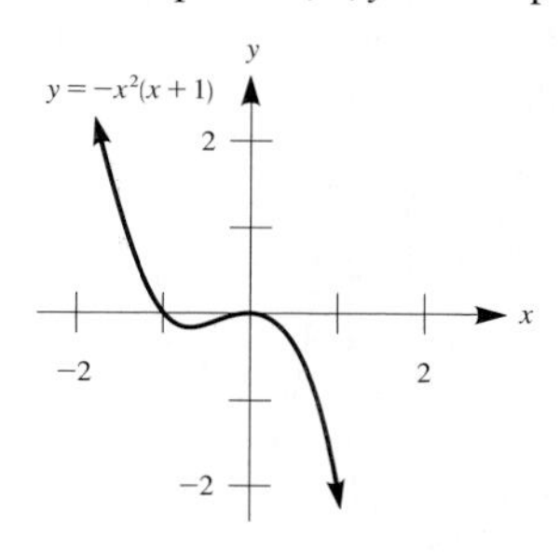

43. x-intercepts: 0, 2, -2; y-intercept: 0

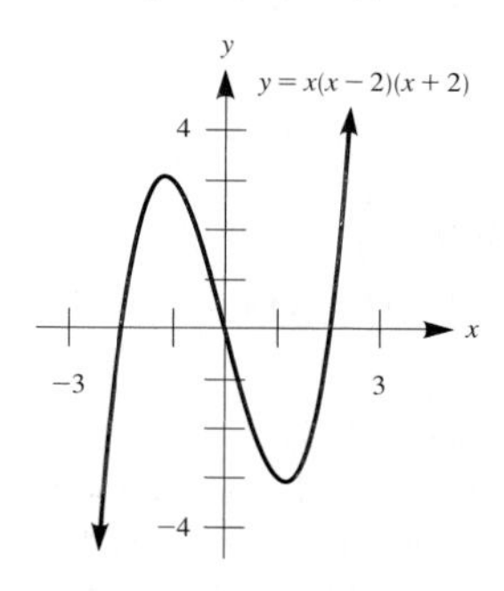

45. no x-intercept; y-intercept: -1; horizontal asymptote: $y = 0$; vertical asymptote: $x = 1$

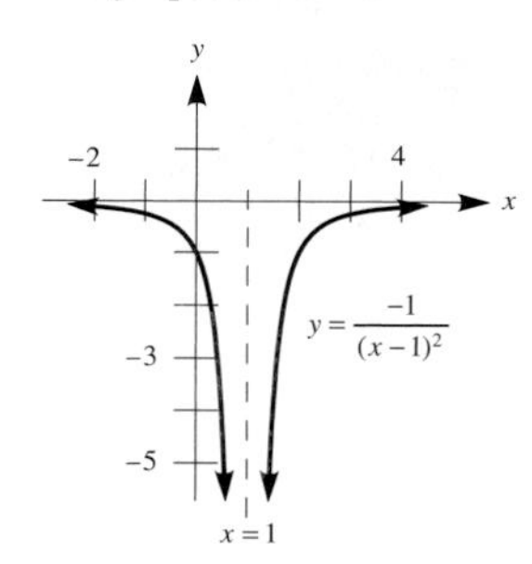

47. x-intercept: 2; y-intercept: $\frac{2}{3}$; horizontal asymptote: $y = 1$; vertical asymptote: $x = 3$

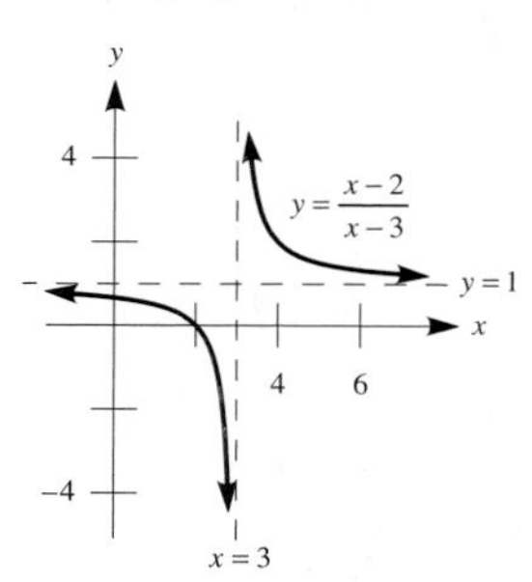

49. x-intercept: 1; y-intercept: $\frac{1}{4}$; horizontal asymptote: $y = 1$; vertical asymptote: $x = 2$

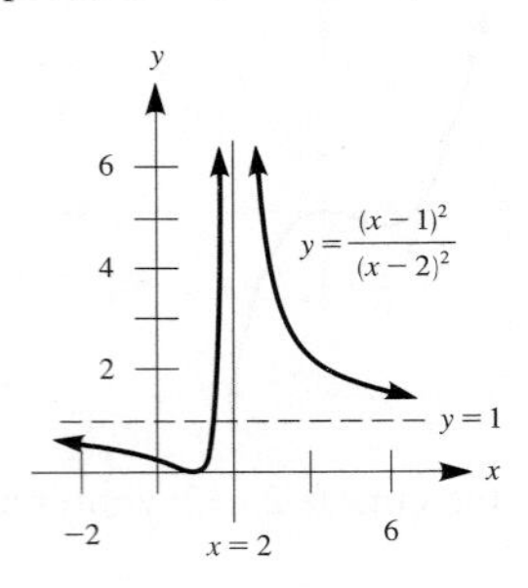

51. $k = 6$
53. $(-\infty, 4 - 2\sqrt{3}] \cup [4 + 2\sqrt{3}, \infty)$
55. $A(m) = \dfrac{m}{2}$
57. $A(x) = (1 - x)\sqrt{1 - x^2}$

Chapter 4 Test

1. $L(0) = -\frac{18}{7}$ **2. (a)** maximum value: 2; increasing: $(-\infty, 1)$
(b) $t = 0$ **3. (a)** x-intercepts: 3, −4; y-intercept: −48

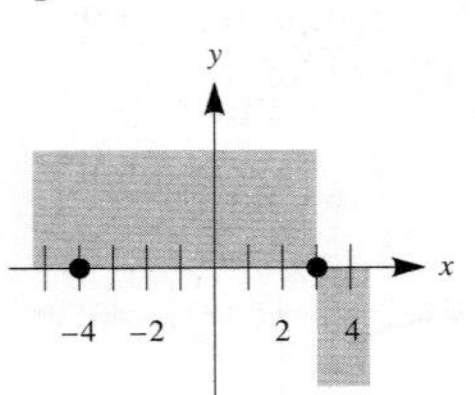

(b)

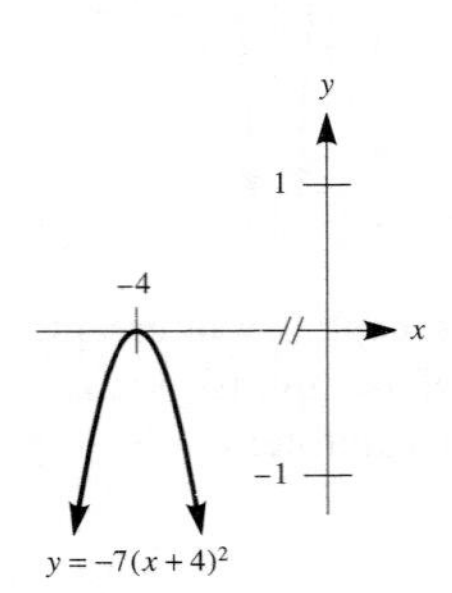

(c)

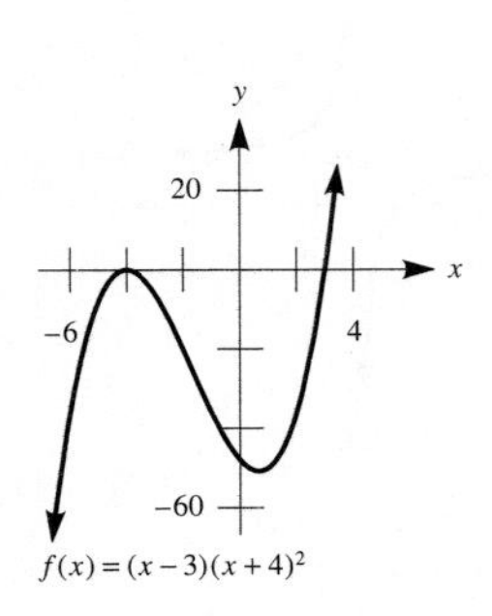

4. (a)

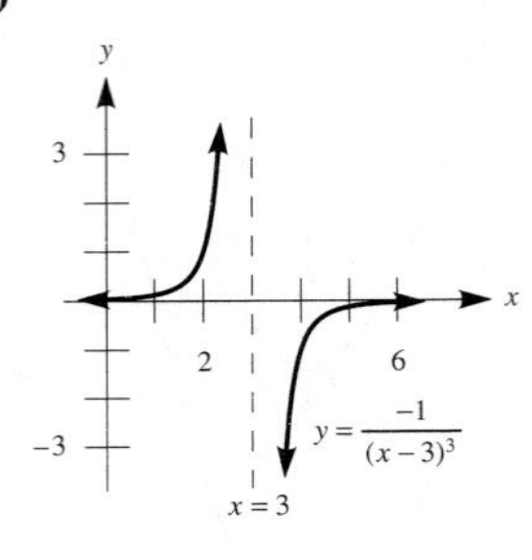

(b)

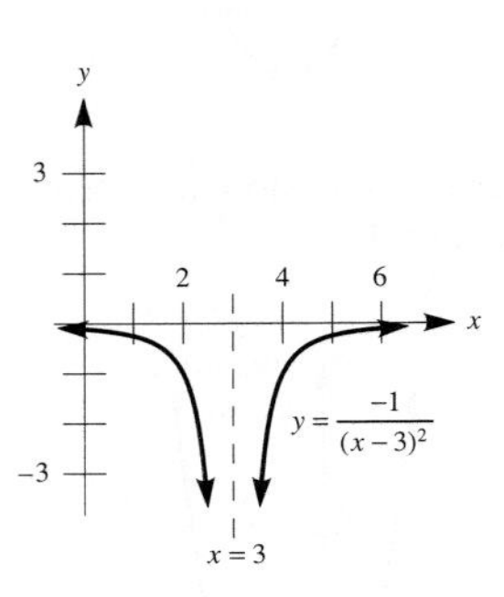

5. turning point: $(\frac{7}{2}, \frac{73}{4})$; x-intercepts: $\dfrac{7 \pm \sqrt{73}}{2}$; y-intercept: 6; axis of symmetry: $x = \frac{7}{2}$

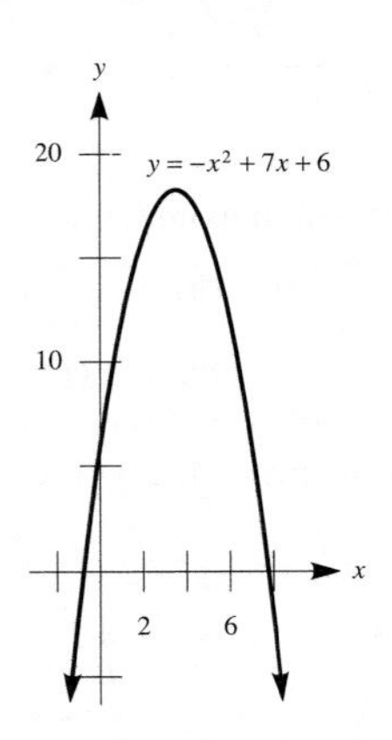

6. maximum revenue = \$9600; price = \$40/unit
7. $V(t) = -1325t + 14000$
8. x-intercept: 3; y-intercept: $-\frac{27}{2}$

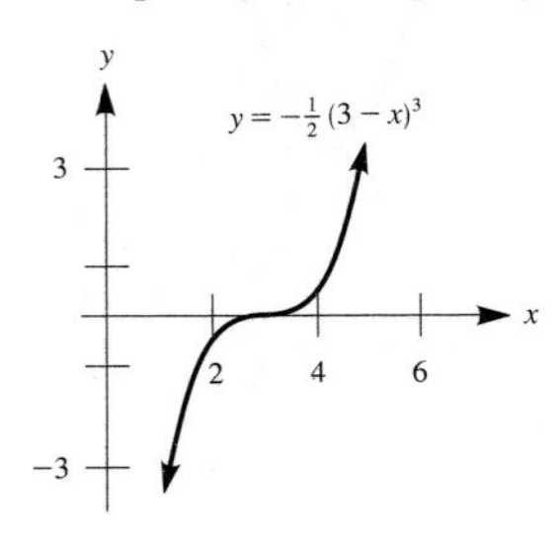

9. x-intercept: $\frac{3}{2}$; y-intercept: −3; vertical asymptote: $x = -1$; horizontal asymptote: $y = 2$

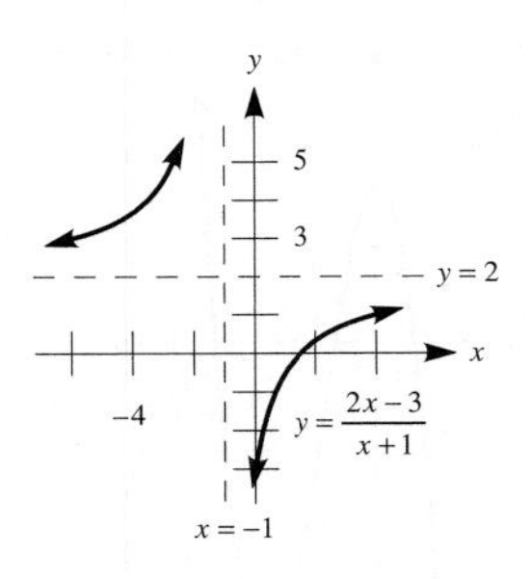

10. (a) $PQ(x) = \sqrt{10x^2 - 22x + 17}$
(b) $x = 1.1$ **11. (a)** vertical asymptotes: $x = 3$, $x = -3$; horizontal asymptote: $y = 1$
(b)

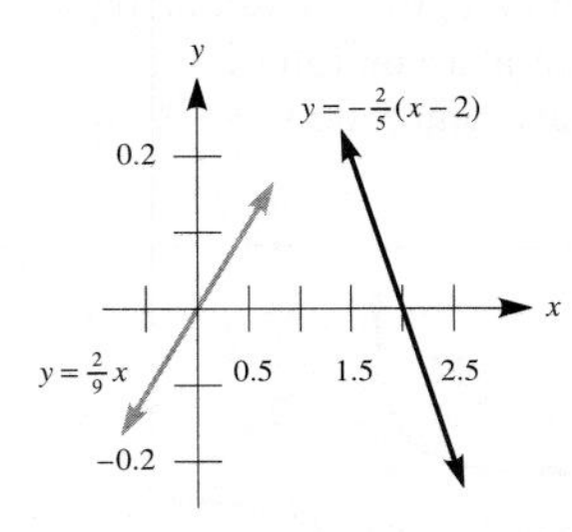

(c)

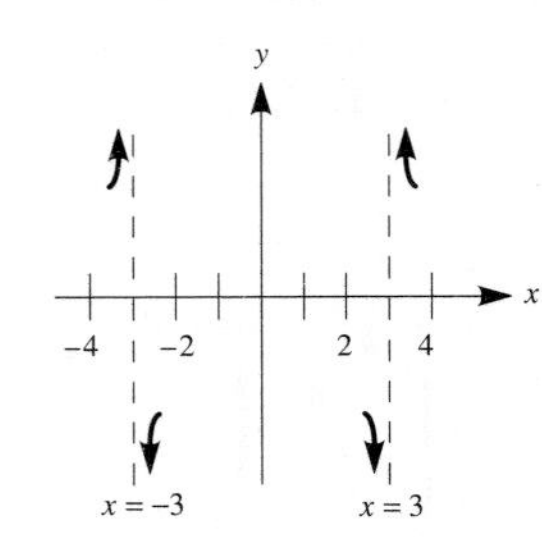

(d)

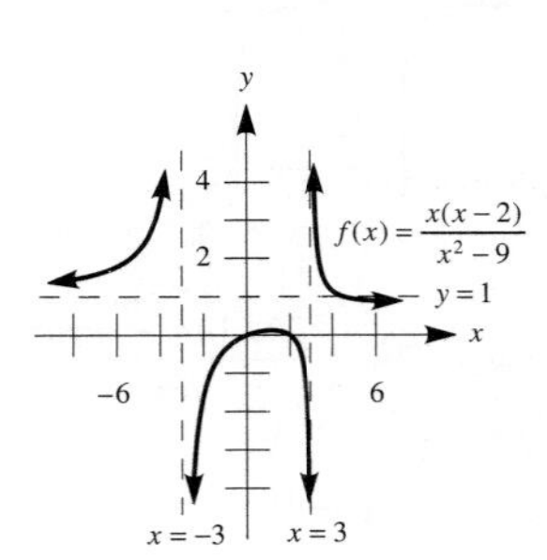

12. $A(w) = \frac{1}{2}w\sqrt{64 - w^2}$
13. (a) As $|x|$ increases in size for x positive, $-x^3$ increases in the negative direction, which the pictured function does not.

(b) This graph has four turning points, and a polynominal function with highest degree term $-x^3$ can have at most two turning points.
14. (a) slope ≈ -9 billion dollars/year

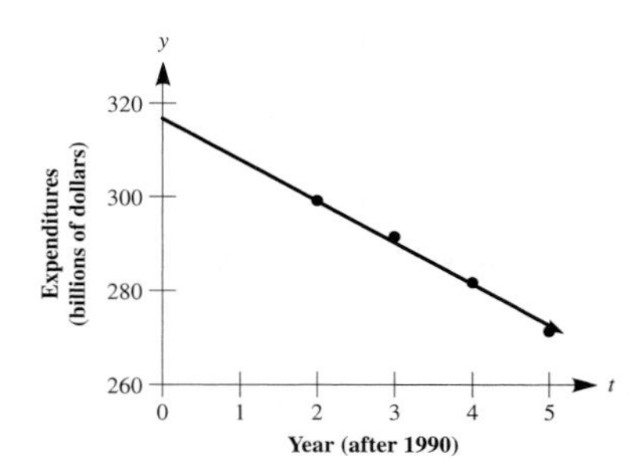

(b) $y = -8.84x + 316.74$; slope is close **(c)** 1996: 263.7 billion dollars; 1999: 237.18 billion dollars; 1996: 0.8% error: 1999: 14.3% error
15. (a) $x = 0, \frac{1}{2}$ **(b)** no fixed points
16. (a)

	x_1	x_2	x_3	x_4	x_5	x_6
From graph	0.56	0.28	0.52	0.33	0.49	0.36
From calculator	0.56	0.286	0.518	0.332	0.490	0.360

(b) $x = \dfrac{-5 - \sqrt{85}}{10} \approx -1.4220$, $x = \dfrac{-5 + \sqrt{85}}{10} \approx 0.4220$;
The iterates are approaching 0.4220.
(c) $x_1 = -0.4, x_2 = 0.44, x_3 = 0.4064$, $x_4 \approx 0.4348, x_5 \approx 0.4109, x_6 \approx 0.4311$. Yes, they are approaching the fixed point 0.4220.

CHAPTER 5

Exercise Set 5.1

1. (a) 10^9 **(b)** 10^{15} **3.** 125 **5.** 16 **7.** 8 **9.** $5^{\sqrt{2}}$ **11. (a)** $x = 3$ **(b)** $t = \frac{3}{2}$ **(c)** $y = \frac{1}{4}$ **(d)** $z = \frac{5}{2}$ **13.** $(-\infty, \infty)$ **15.** $(-\infty, \infty)$

17.

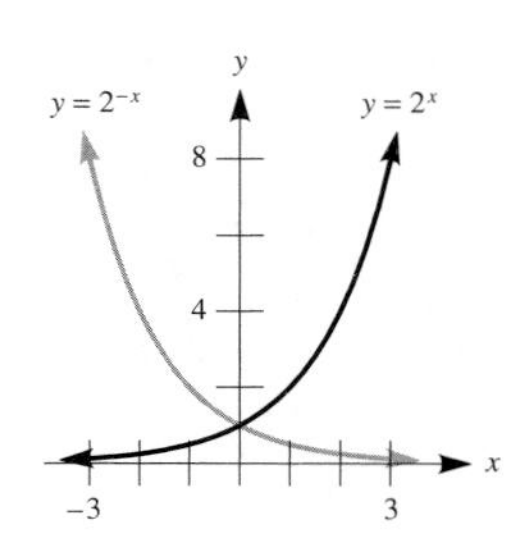

19.

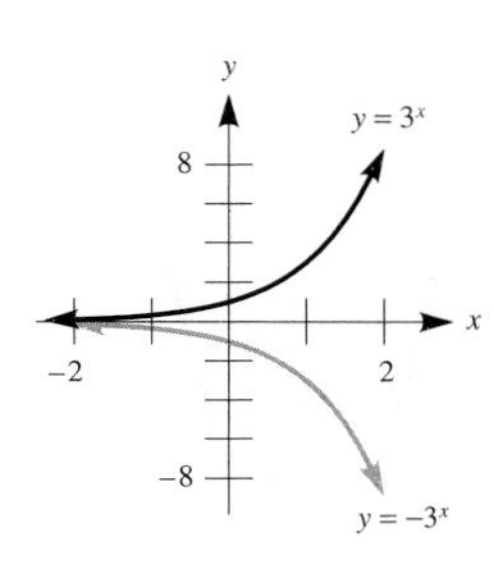

21.

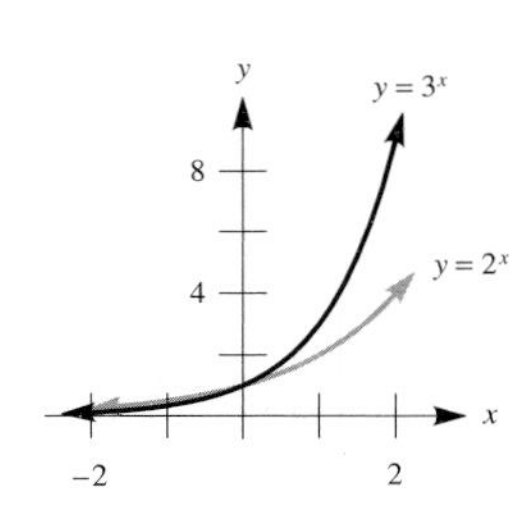

23.

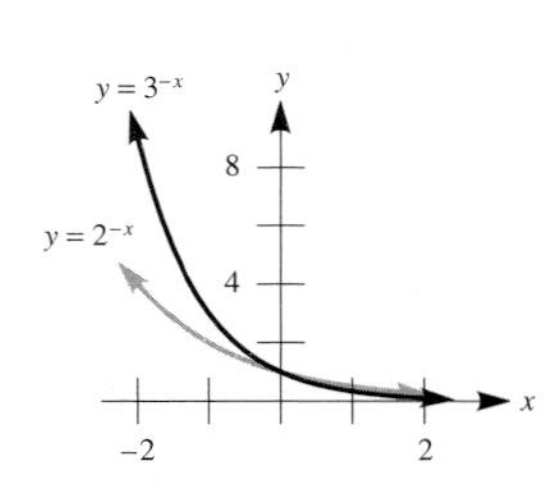

25. domain: $(-\infty, \infty)$; range: $(-\infty, 1)$; x- and y-intercepts: 0; asymptote: $y = 1$

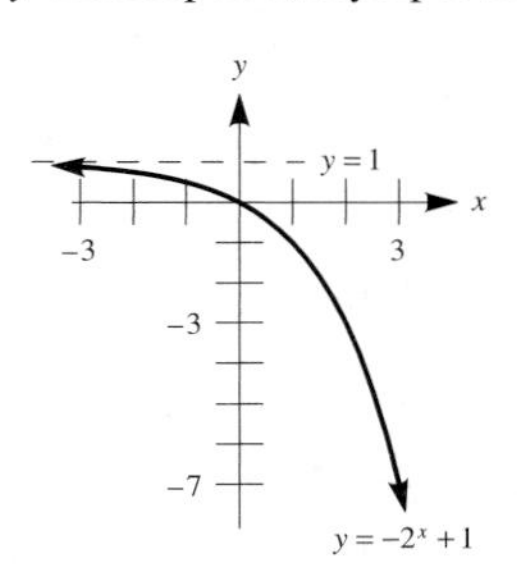

27. domain: $(-\infty, \infty)$; range: $(1, \infty)$; no x-intercept; y-intercept: 2; asymptote: $y = 1$

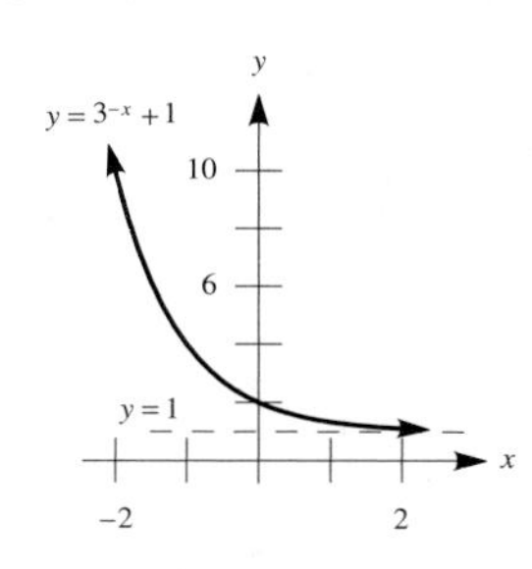

29. domain: $(-\infty, \infty)$; range: $(0, \infty)$; no x-intercept; y-intercept: $\frac{1}{2}$; asymptote: $y = 0$

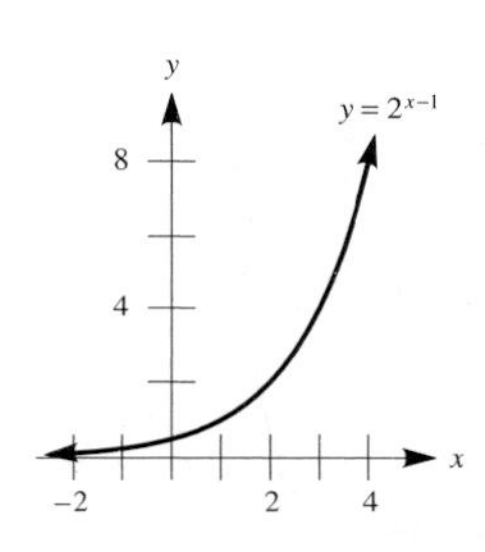

31. domain: $(-\infty, \infty)$; range: $(1, \infty)$; no x-intercept; y-intercept: 4; asymptote: $y = 1$

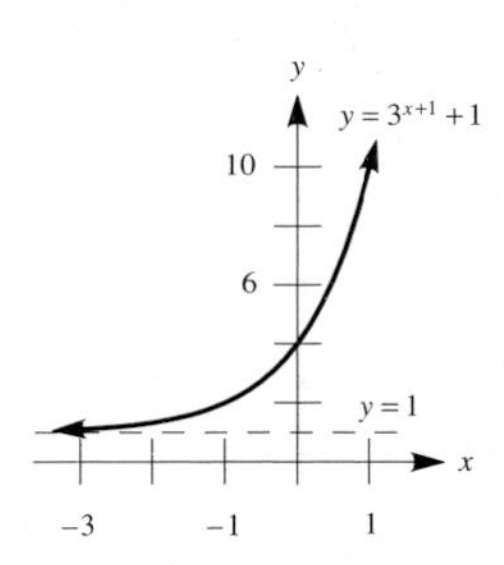

33. (a) no x-intercept; y-intercept: $-\frac{1}{9}$; asymptote: $y = 0$
(b)

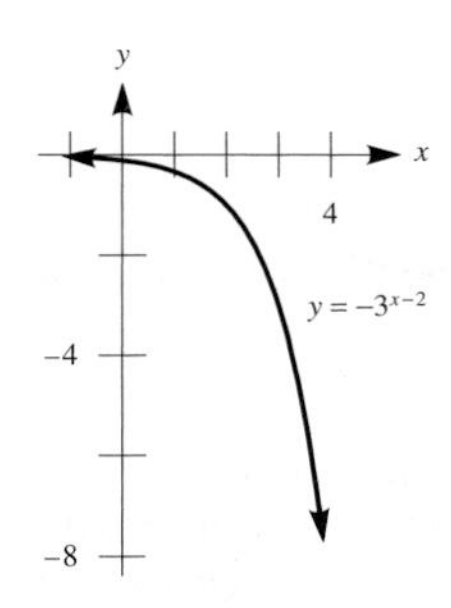

35. (a) x-intercept: -1; y-intercept: -3; asymptote: $y = -4$
(b)

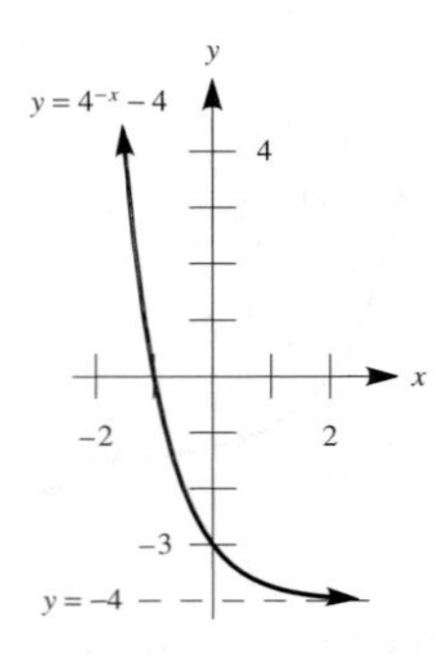

37. (a) no x-intercept; y-intercept: $\frac{1}{10}$; asymptote: $y = 0$
(b)

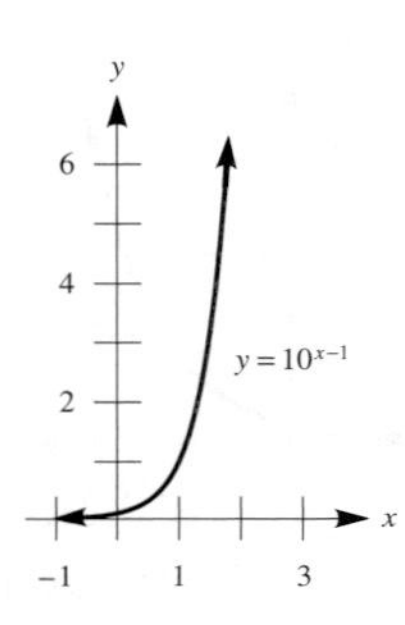

39. $x = -\frac{1}{3}$ **41.** $x = \frac{3}{2}, 1$

43. (a) $[0, 2]$: $\dfrac{\Delta f}{\Delta x} = 1.5, \dfrac{\Delta g}{\Delta x} = 1.5$;

$[2, 4]$: $\dfrac{\Delta f}{\Delta x} = 6, \dfrac{\Delta g}{\Delta x} = 6$

(b) $[4, 6]$: $\dfrac{\Delta f}{\Delta x} = 10.5, \dfrac{\Delta g}{\Delta x} = 24$;

$[6, 8]$: $\dfrac{\Delta f}{\Delta x} = 15, \dfrac{\Delta g}{\Delta x} = 96$

(c) $\dfrac{\Delta f}{\Delta x} = 24, \dfrac{\Delta g}{\Delta x} = 1536$

47. (a)

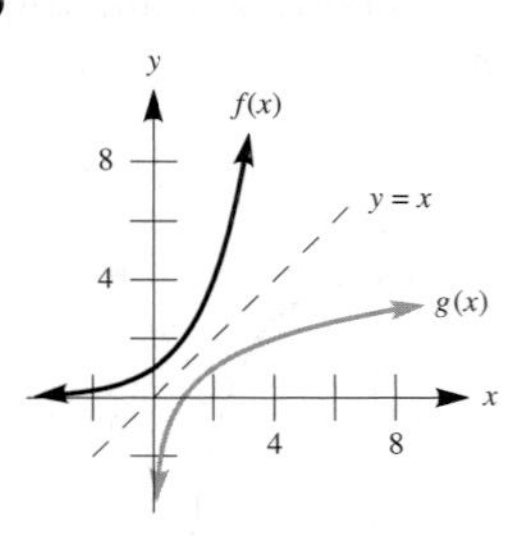

(b) domain: $(0, \infty)$; range: $(-\infty, \infty)$; x-intercept: 1; no y-intercept; asymptote: $x = 0$ **49. (a)** 1.4 **(b)** 1.41 **51. (a)** 1.5 **(b)** 1.52

53. (a) 1.7 **(b)** 1.73 **55. (a)** 1.6 **(b)** 1.62 **57.** $x \approx 0.3$ **59.** $x \approx 0.7$

61. (a)

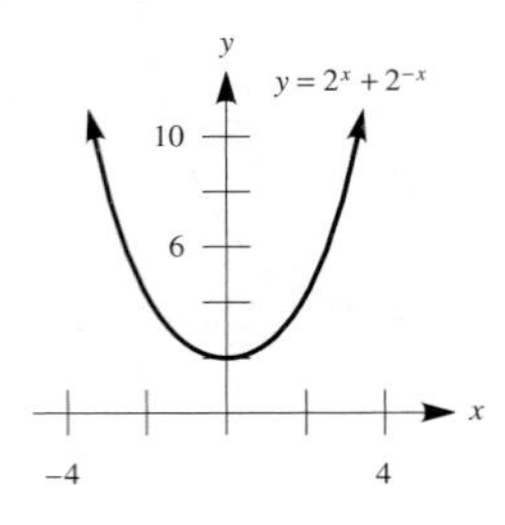

(b)

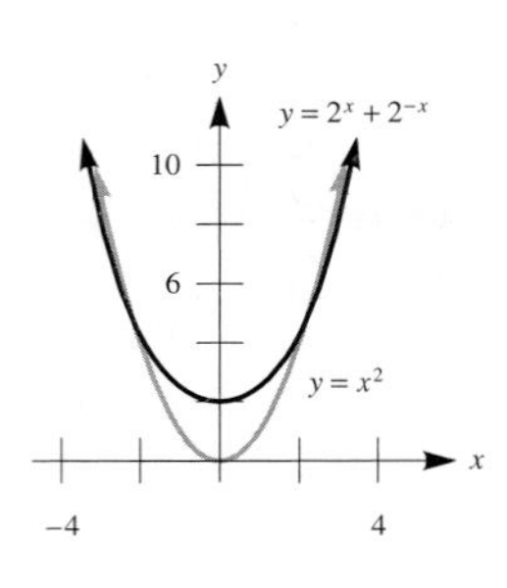

(c)

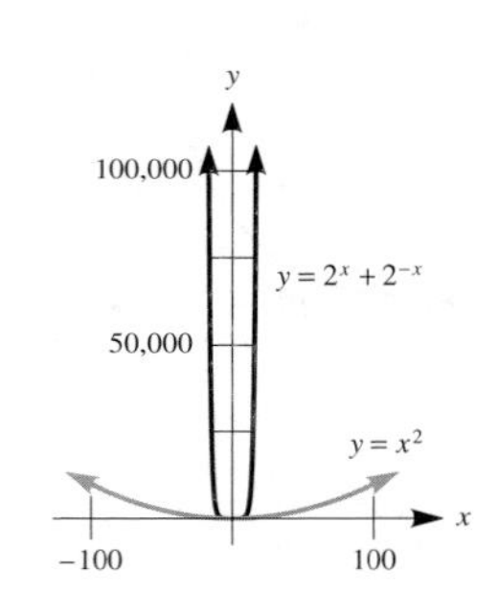

63. (a)

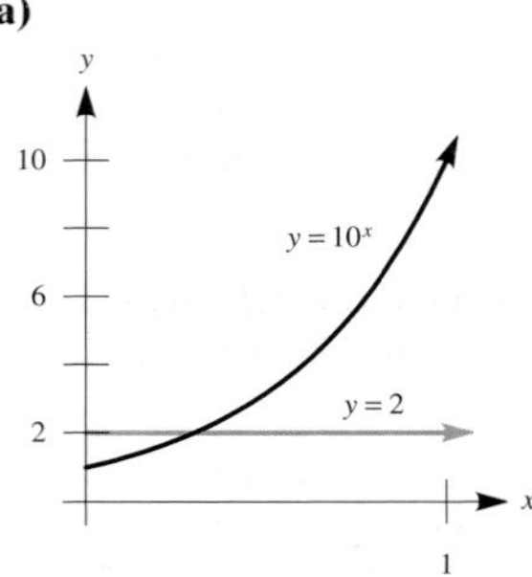

(b) $(0.3, 2)$ is the point of intersection

Exercise Set 5.2

1. false **3.** false **5.** true **7.** false
9. (a) two **(b)** two

11. domain: $(-\infty, \infty)$; range: $(0, \infty)$; no x-intercept; y-intercept: 1; asymptote: $y = 0$

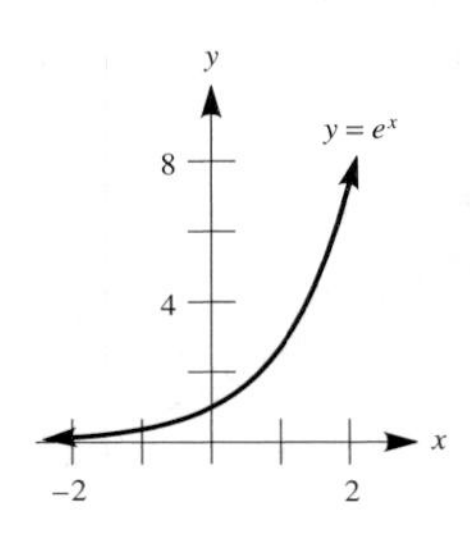

13. domain: $(-\infty, \infty)$; range: $(-\infty, 0)$; no x-intercept; y-intercept: -1; asymptote: $y = 0$

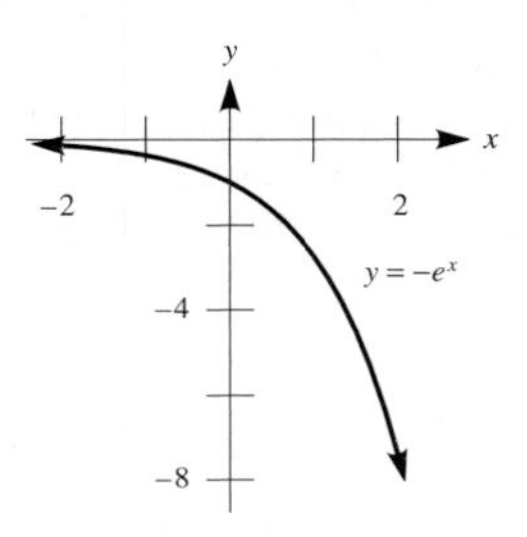

15. domain: $(-\infty, \infty)$; range: $(1, \infty)$; no x-intercept; y-intercept: 2; asymptote: $y = 1$

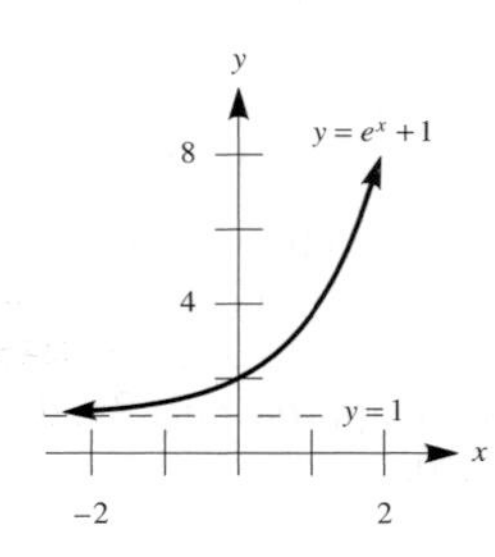

17. domain: $(-\infty, \infty)$; range: $(1, \infty)$; no x-intercept; y-intercept: $e + 1$; asymptote: $y = 1$

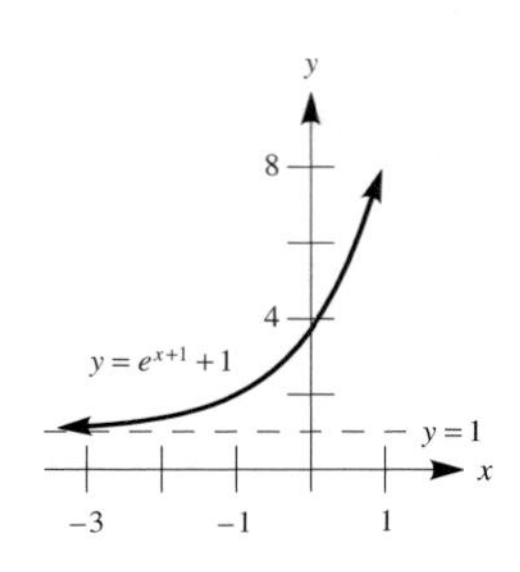

19. domain: $(-\infty, \infty)$; range: $(-\infty, e)$; x-intercept: 1; y-intercept: $e - 1$; asymptote: $y = e$

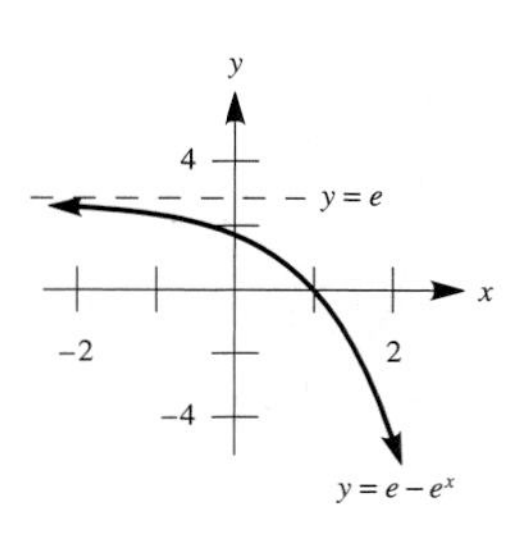

21. (a) $y = e^x$ displaced to the left two units

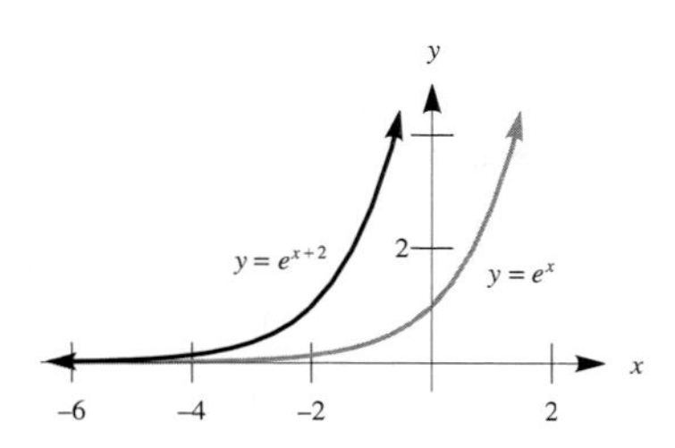

(b) $y = e^x$ displaced to the left two units, then reflected across the y-axis

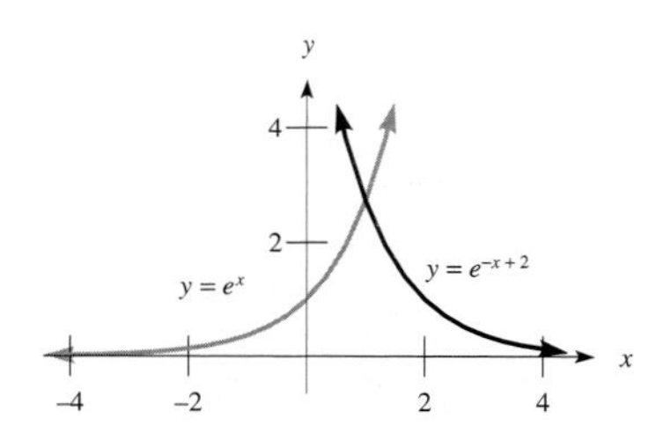

23. $\dfrac{\Delta f}{\Delta x} = 1, \dfrac{\Delta g}{\Delta x} = 1, \dfrac{\Delta h}{\Delta x} \approx 1.7$

25. (a)

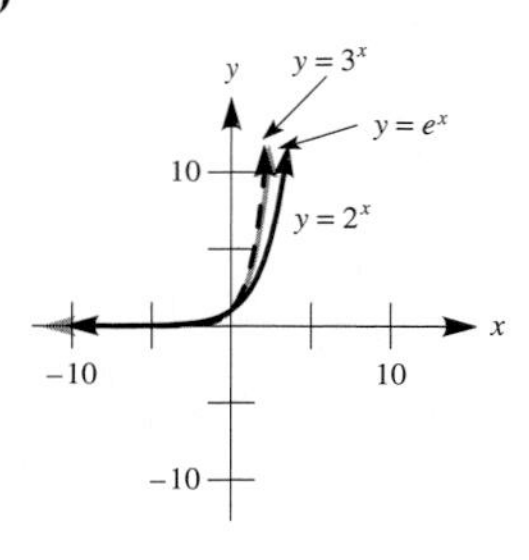

(b)

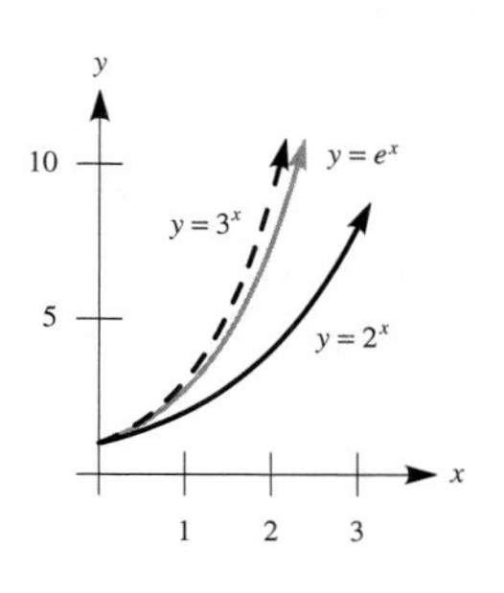

(c)

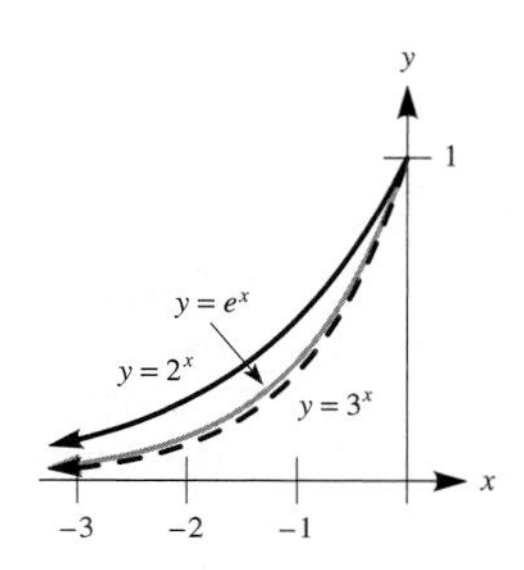

(d) For $x < 0, 2^{-x} < e^{-x} < 3^{-x}$ by (b), so $2^x > e^x > 3^x$ by taking reciprocals.

27. instantaneous rate of change $= 6$

Interval	$\Delta f/\Delta x$ for $f(x) = x^2$
[3, 3.1]	6.1
[3, 3.01]	6.01
[3, 3.001]	6.001
[3, 3.0001]	6.0001
[3, 3.00001]	6.00001
[2.9, 3]	5.9
[2.99, 3]	5.99
[2.999, 3]	5.999
[2.9999, 3]	5.9999
[2.99999, 3]	5.99999

29.

Interval	$\Delta f/\Delta x$ for $f(x) = e^x$
[1.9, 2]	7.031617
[1.99, 2]	7.352234
[1.999, 2]	7.385363
[1.9999, 2]	7.388687
[1.99999, 2]	7.389019
[1.999999, 2]	7.389053
[1.9999999, 2]	7.389056

31. (a) 255 bacteria per hour
(b) $t = 5$ hours: 148 bacteria per hour; $t = 5.5$ hours: 245 bacteria per hour

33. P: $e^{-1} \approx 0.37$; Q: 1; R: $e^{0.5} \approx 1.65$

35. A, D, E, G **37.** B, D, E, G
39. A, D, F, G, H **41.** A, D, F, G, H
43. graph: $e^{0.1} \approx 1.1$; calculator: $e^{0.1} \approx 1.105$ **45.** graph: $e^{-0.3} \approx 0.75$; calculator: $e^{-0.3} \approx 0.741$
47. graph: $e^{-1} \approx 0.35$; calculator: $e^{-1} \approx 0.368$

49. graph: $\dfrac{1}{\sqrt{e}} = e^{-0.5} \approx 0.6$;
calculator: $\dfrac{1}{\sqrt{e}} = e^{-0.5} \approx 0.607$

51. (a) $x \approx 0.4$ **(b)** $\ln 1.5 \approx 0.405$
53. (a) $x \approx 0.6$ **(b)** $\ln 1.8 \approx 0.588$
55. (a) $\cosh(0) = 1$, $\cosh(1) \approx 1.54$, $\cosh(-1) \approx 1.54$ **(b)** $(-\infty, \infty)$
(c) The graph is symmetric about the y-axis.
(d)

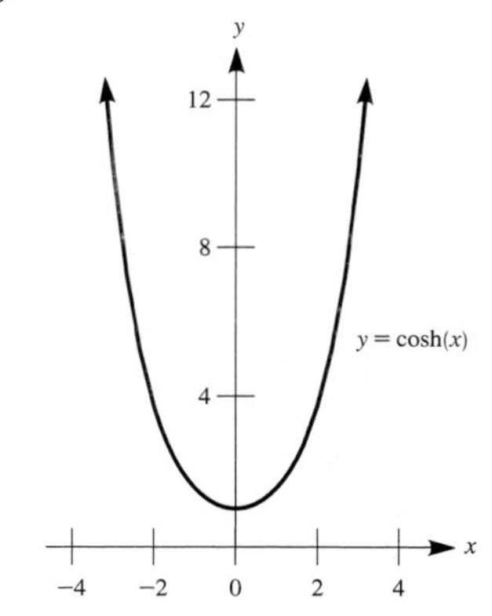

57. (a) $\sinh(0) = 0$, $\sinh(1) \approx 1.18$, $\sinh(-1) \approx -1.18$ **(b)** $(-\infty, \infty)$
(c) The graph is symmetric about the origin.
(d)

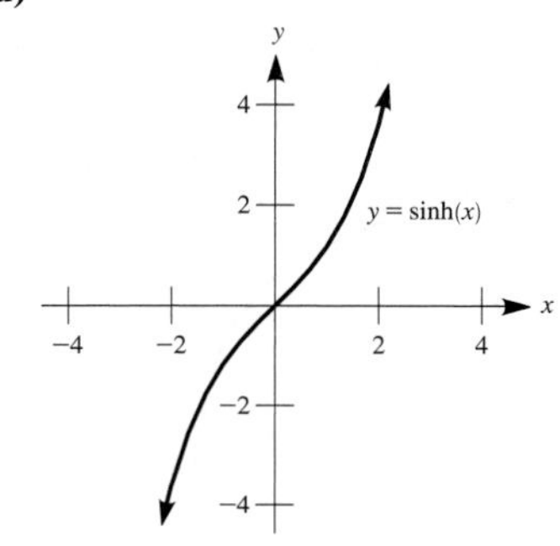

(e) three points

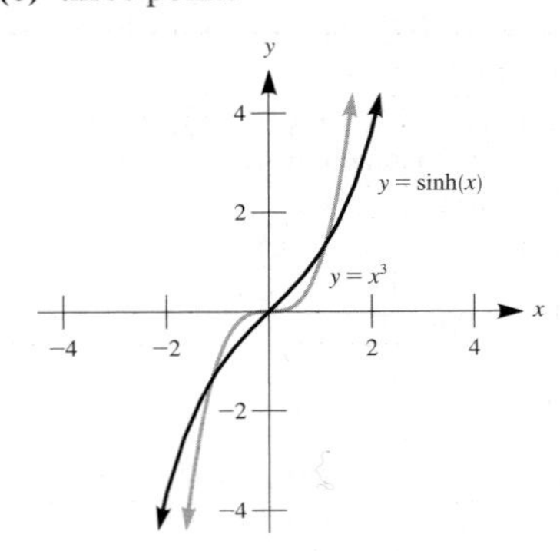

59. (a) e^{π} is larger **(b)** e^{π} is larger
61. (a)

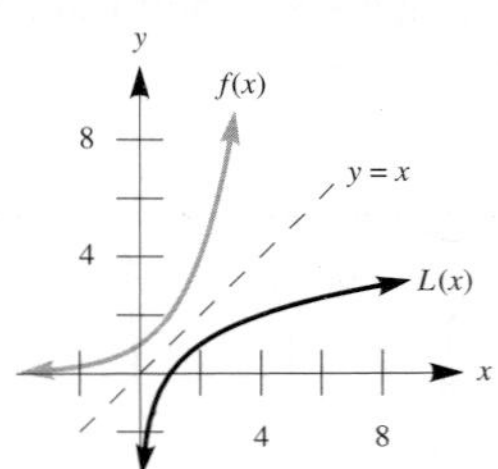

(b) domain: $(0, \infty)$; range: $(-\infty, \infty)$; x-intercept: 1; asymptote: $x = 0$
(c) (i) x-intercept: 1; asymptote: $x = 0$

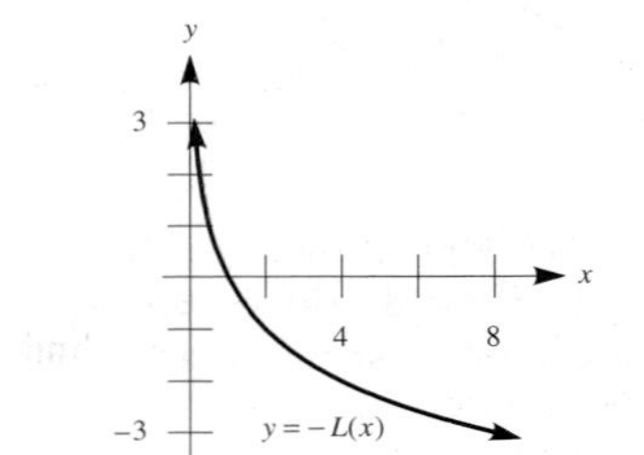

(ii) x-intercept: -1; asymptote: $x = 0$

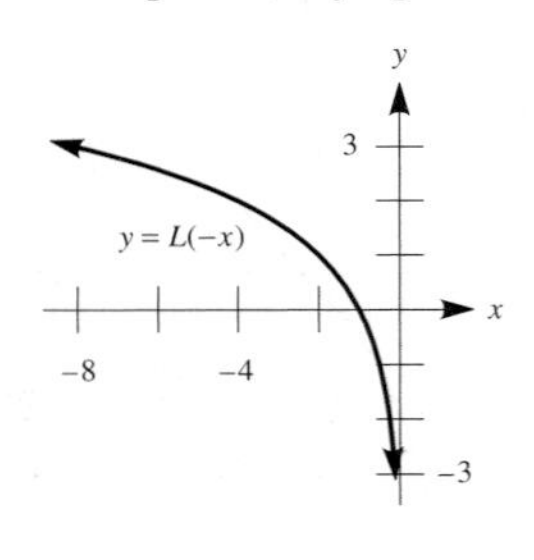

(iii) x-intercept: 2; asymptote: $x = 1$

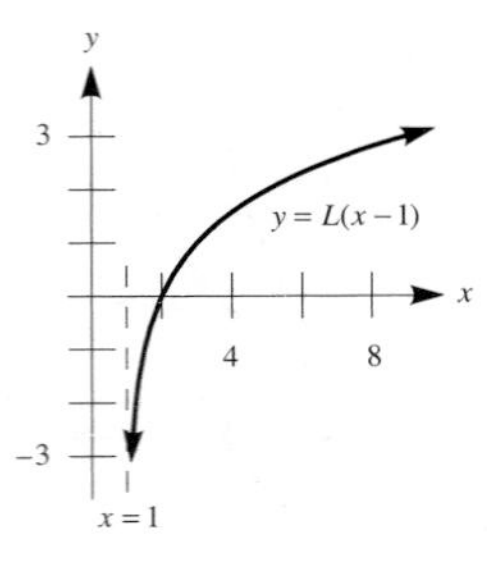

63. (a) $[\cosh(x)]^2 - 2[\sinh(x)]^2 = 1$ is not an identity; solution: $x = 0$

Exercise Set 5.3

1. (a) $f[g(x)] = x$ **(b)** $g[f(x)] = x$; f
3. (a) $g(x) = \log_2 x$ **(b)** $2^{\log_2 x} = x$; $\log_2(2^x) = x$ **(c)** $2^{\log_2 99} = 99$; $\log_2(2^{-\pi}) = -\pi$ **5. (a)** is one-to-one **(b)** not one-to-one **(c)** is one-to-one
7. $f[f^{-1}(6)] = 6$ **9. (a)** $\log_3 9 = 2$ **(b)** $\log_{10} 1000 = 3$ **(c)** $\log_7 343 = 3$ **(d)** $\log_2 \sqrt{2} = \frac{1}{2}$ **11. (a)** $2^5 = 32$ **(b)** $10^0 = 1$ **(c)** $e^{1/2} = \sqrt{e}$ **(d)** $e^{-1} = \dfrac{1}{e}$
13.

x	1	10	10^2	10^3	10^{-1}	10^{-2}	10^{-3}
$\log_{10}x$	0	1	2	3	−1	−2	−3

15. (a) $\log_5 30$ is larger **(b)** ln 17 is larger **17. (a)** $\frac{3}{2}$ **(b)** $-\frac{5}{2}$ **(c)** $\frac{3}{2}$
19. (a) $x = \frac{1}{16}$ **(b)** $x = e^{-2} \approx 0.14$
21. (a) $(0, \infty)$ **(b)** $(-\infty, \frac{3}{4})$ **(c)** $(-\infty, 0) \cup (0, \infty)$ **(d)** $(0, \infty)$ **(e)** $(-\infty, -5) \cup (5, \infty)$ **23.** A: (0, 1); B: (1, 0); C: (4, 2); D: (2, 4)
25. (a) domain: $(0, \infty)$; range: $(-\infty, \infty)$; x-intercept: 1; no y-intercept; asymptote: $x = 0$

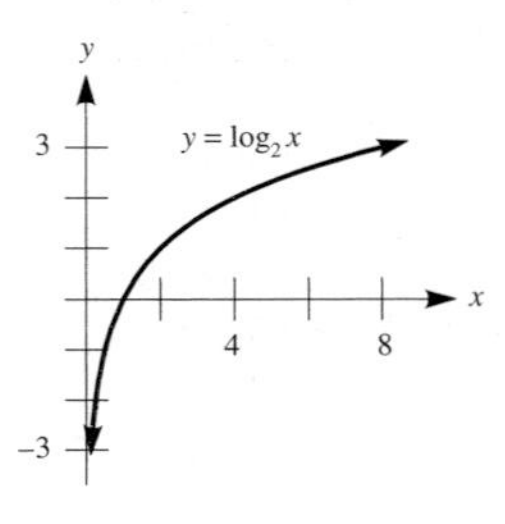

(b) domain: $(0, \infty)$; range: $(-\infty, \infty)$; x-intercept: 1; no y-intercept; asymptote: $x = 0$

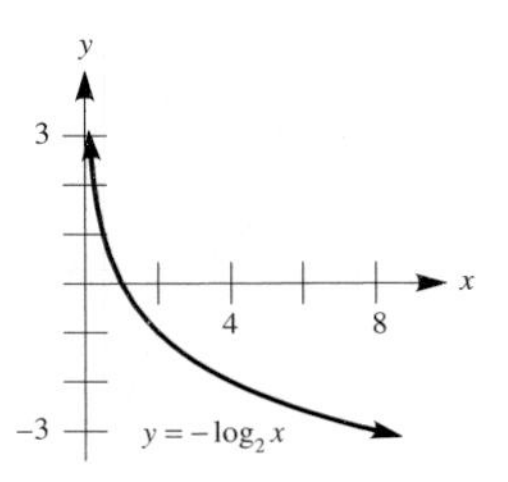

(c) domain: $(-\infty, 0)$; range: $(-\infty, \infty)$; x-intercept: -1; no y-intercept; asymptote: $x = 0$

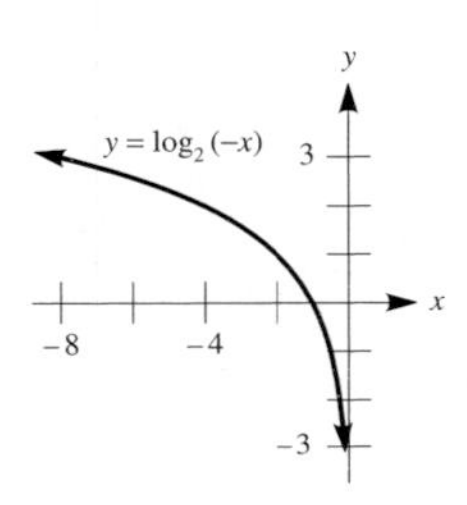

(d) domain: $(-\infty, 0)$; range: $(-\infty, \infty)$; x-intercept: -1; no y-intercept; asymptote: $x = 0$

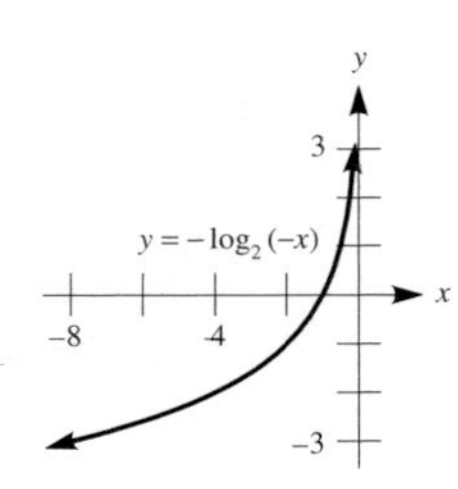

27. domain: $(2, \infty)$; range: $(-\infty, \infty)$; x-intercept: 5; no y-intercept; asymptote: $x = 2$

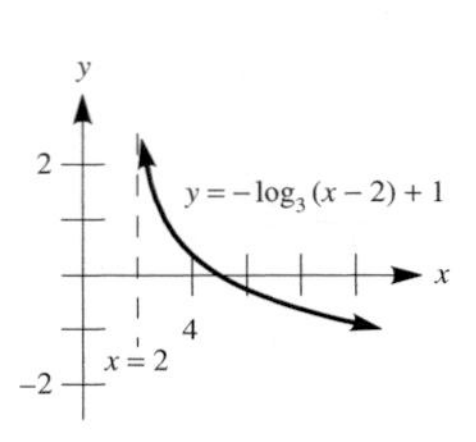

29. domain: $(-e, \infty)$; range: $(-\infty, \infty)$; x-intercept: $-e + 1$; y-intercept: 1; asymptote: $x = -e$

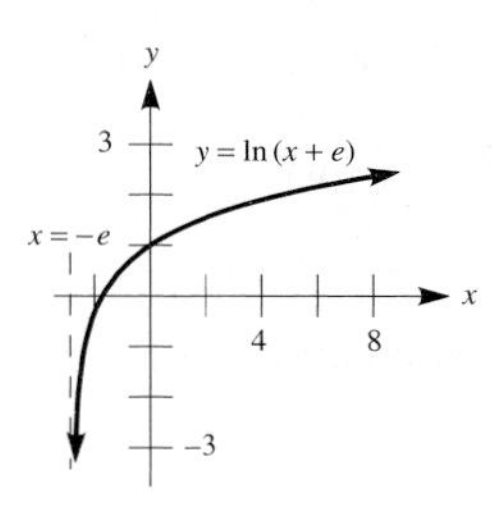

31. (a) 4 **(b)** -1 **(c)** $\frac{1}{2}$
33. (a) $x \approx 1.4$ **(b)** $x \approx 1.398$
35. (a) $x \approx \pm 1.3$ **(b)** $x \approx \pm 1.266$
37. (a) $t \approx -0.3$ **(b)** $t \approx -0.349$
39. (a) $t \approx -0.4$ **(b)** $t \approx -0.380$
41. (a) $(0, 1)$

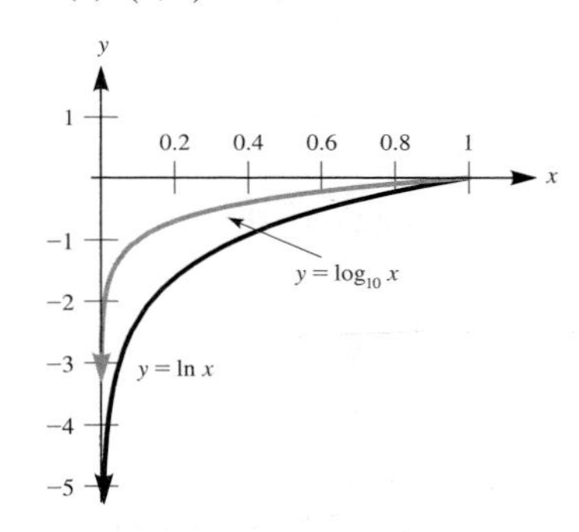

(b) $(1, \infty)$

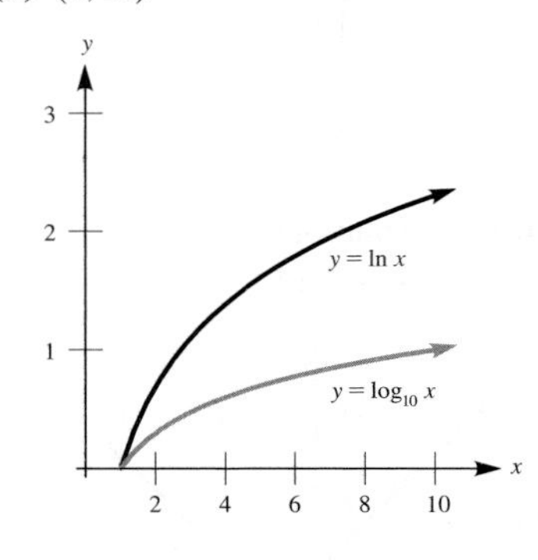

(c) $(0, \infty)$

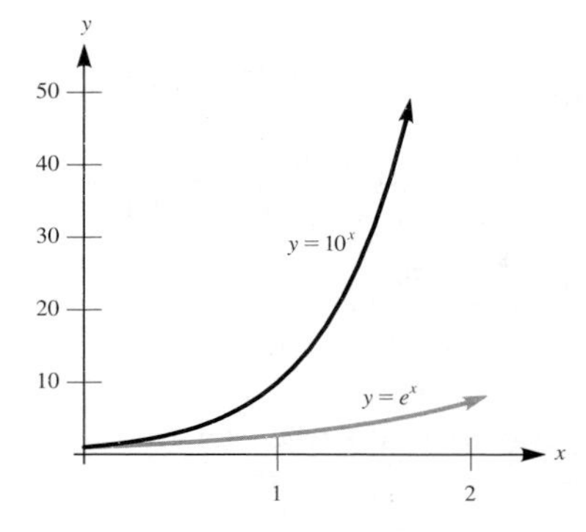

(d) $(-\infty, 0)$

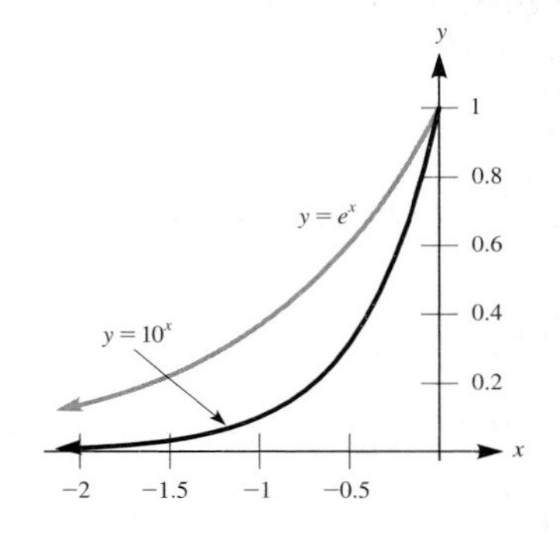

43. The San Salvador quake was 1.6 times stronger.
45. The second quake is 10^d times stronger. **47. (a)** $I = 10^{\beta/10} I_0$
(b) The power mower is 10^9 times more intense. **49. (a)** pH ≈ 3.5; acid
(b) pH $= 0$ **51.** $[H^+] = 10^{-5.9}$
53. D, E, H **55.** D, E, H
57. D, E, H **59.** D, E, H
61. $f^{-1}(x) = -1 + \ln x$; x-intercept: e; asymptote: $x = 0$

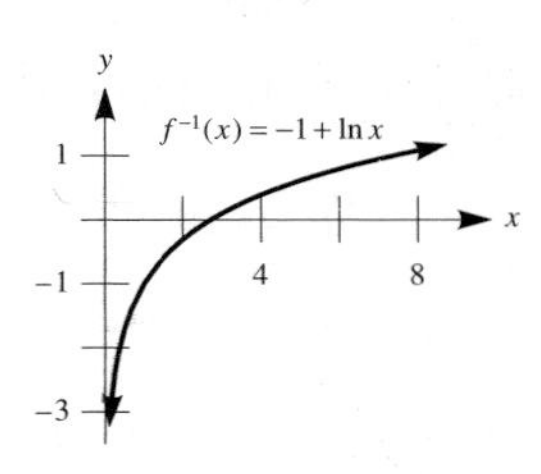

63. $x \approx 10^{30}$

65. (a)

Planet	x	y	ln x	ln y
Mercury	0.387	0.241	−0.95	−1.42
Venus	0.723	0.615	−0.32	−0.49
Earth	1.000	1.000	0.00	0.00
Mars	1.523	1.881	0.42	0.63
Jupiter	5.202	11.820	1.65	2.47

(b) $\ln y = 1.50 \ln x$
(d)

Planet	x	y (calculated)	y (observed)
Saturn	9.555	29.54	29.46
Uranus	19.22	84.26	84.01
Neptune	30.11	165.22	164.79
Pluto	39.44	247.69	248.50

67. (a) $y = -24.795459 + 7.453496 \ln x$

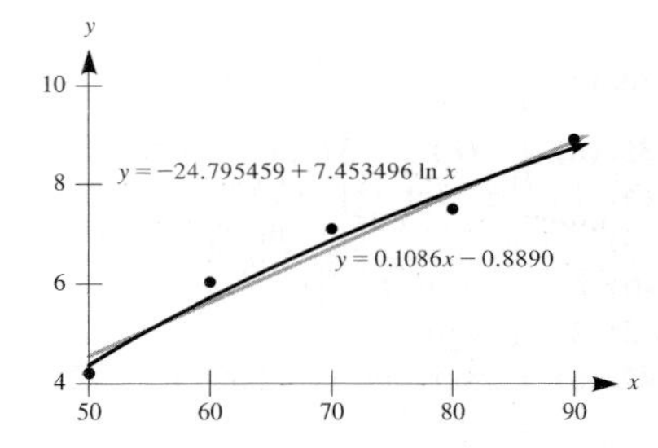

(b) $y = 0.1086x - 0.8890$
(c) logarithmic: 9.529 million; linear: 9.971 million **(d)** logarithmic model is closer; logarithmic: 0.11% error; linear: 4.75% error
69. (a) $(1, \infty)$ **(b)** $f^{-1}(x) = e^{e^x}$
71. (a)

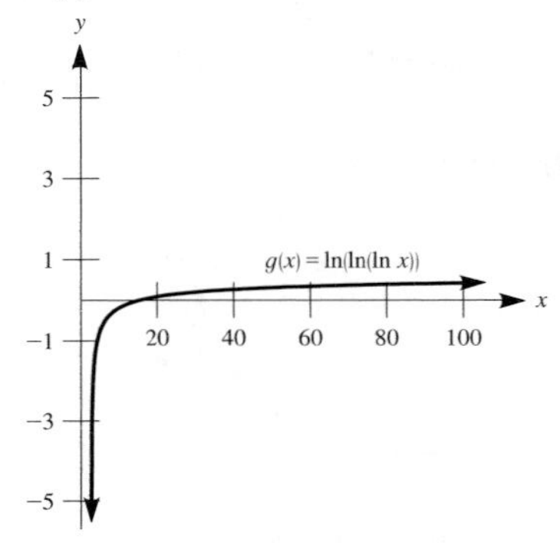

(b) $(-\infty, \infty)$ **(c)** no

Exercise Set 5.4

1. 1 **3.** $\frac{1}{2}$ **5.** 4 **7.** 0 **9.** 4
11. $\log_{10} 60$ **13.** $\log_5 20$ **15. (a)** $\ln 6$
(b) $\ln(\frac{3}{16384})$
17. $\log_b\left[\frac{4(1+x)^3}{(1-x)^{3/2}}\right]$
19. $\log_{10}\left[\frac{27\sqrt{x+1}}{(x^2+1)^6}\right]$
21. (a) $2\log_{10} x - \log_{10}(1+x^2)$
(b) $2\ln x - \frac{1}{2}\ln(1+x^2)$
23. (a) $\frac{1}{2}\log_{10}(3+x) + \frac{1}{2}\log_{10}(3-x)$
(b) $\frac{1}{2}\ln(2+x) + \frac{1}{2}\ln(2-x) - \ln(x-1) - \frac{3}{2}\ln(x+1)$
25. (a) $\frac{1}{2}\log_b x - \frac{1}{2}$
(b) $\ln(1+x^2) + \ln(1+x^4) + \ln(1+x^6)$ **27. (a)** $A+B$
(b) $-A-B$ **(c)** $3B$ **(d)** $-3B$
29. (a) $C-B$ **(b)** $B-C$
(c) $C-2B$ **(d)** $C-4A$
31. (a) $\frac{1}{B}$ **(b)** $\frac{A+C+1}{B}$
33. (a) $\frac{A}{B+1}$ **(b)** $\frac{B+C}{B+1}$
35. (a) 1 **(b)** 1 **37. (a)** $a+2b+3c$
(b) $1+\frac{1}{2}a$ **(c)** $\frac{1}{2}(1+a+b+c)$
(d) $1+a-\frac{1}{2}b-\frac{1}{2}c$ **39. (a)** $1+t$
(b) $u-t$ **(c)** $\frac{3}{2}t+\frac{1}{2}u-1$
(d) $2+t+\frac{1}{2}u$
41. (a) x-intercept: $x \approx 2.32$

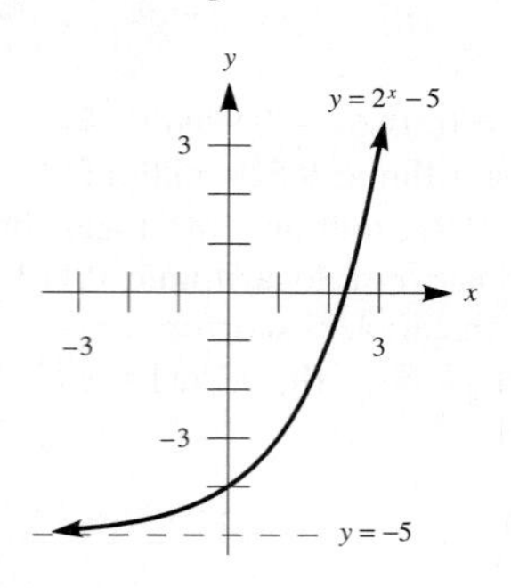

(b) x-intercept: $x \approx 4.64$

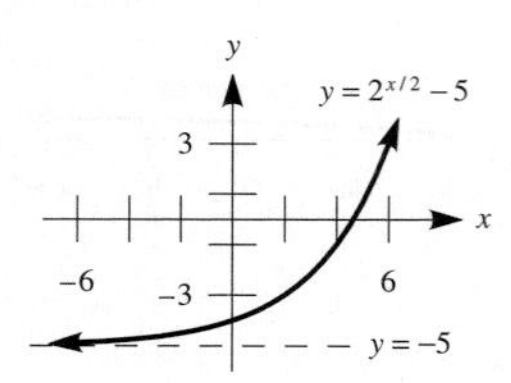

43. $x = \frac{\ln 5 - \ln 2 + 1}{2}$
45. $x = \frac{\ln 13}{\ln 2}$
47. $x = \frac{1}{\ln 10}$
49. $x = \pm\sqrt{1 + \frac{\log_{10} 12}{\log_{10} 3}} \approx \pm 1.806$
51. $\frac{\log_{10} 5}{\log_{10} 2}$ **53.** $\frac{\log_{10} 3}{\log_{10} e}$ **55.** $\frac{\log_{10} 2}{\log_{10} b}$
57. $\frac{\ln 6}{\ln 10}$ **59.** $\frac{1}{\ln 10}$
61. $\frac{\ln(\ln x) - \ln(\ln 10)}{\ln 10}$
63. (a) true **(b)** true **(c)** true
(d) false **(e)** true **(f)** false
(g) true **(h)** false **(i)** true
(j) false **(k)** false **(l)** true
(m) true **65. (f)** 2345.6
(g) 0.123456
67.

x	0.1	0.05	0.005	0.0005
$\ln(1+x)$	0.095310	0.048790	0.004987	0.000499

69. $a = 4,\ b = -\frac{\ln 243}{2.5}$
71. (a) $x = \frac{\log_{10} 3}{\log_{10} 2}$
(b) $\frac{\log_{10} 3}{\log_{10} 2} \approx 1.585,\ \frac{\ln 3}{\ln 2} \approx 1.585$
75. x^3
79. (c) $\log_\pi 2 + \frac{1}{\log_\pi 2}$ is larger

Exercise Set 5.5

1. $x = \frac{\ln 3}{2\ln 3 - \ln 5} \approx 1.869$
3. $x = e^{e^{1.5}} \approx 88.384$ **5.** $x = \pm 8$
7. $x = \frac{3-\sqrt{809}}{4} \approx -6.361$,
$x = \frac{3+\sqrt{809}}{4} \approx 7.861$
9. $x = \log_{10} 2 \approx 0.301$ **11. (a)** true for all $x > 0$
(b) $x = e^{-\sqrt{3}} \approx 0.177,\ x = e^{\sqrt{3}} \approx 5.652$
13. (a) true for all $x > 0$
(b) $x = 6^{1/5} \approx 1.431$
15. true for all $x > 0$
17. $x = \sqrt{3} \approx 1.732$ **19.** no real solutions **21.** no real solutions
23. $x = \frac{7}{3}$ **25. (a)** no real solutions
(b) $x = 0$ **(c)** $x = \ln 3 \approx 1.099$
(d) $x = \ln(1+\sqrt{5}) \approx 1.174$
27. $x = \ln\left(\frac{1+\sqrt{5}}{2}\right) \approx 0.481$
29. $x = \frac{3\ln 5}{5\ln 2 - \ln 3 - \ln 5} \approx 6.372$
31. $x = 2$ **33.** $x = \frac{1}{2}$ **35.** $x = 3$
37. $x = \frac{203}{99}$ **39.** $x = 3$ **41.** $x = 7$
43. (a) $x = \frac{10^y}{3(10^y) - 1}$
(b) $x = \frac{1-y}{2}$ for $y < \frac{1}{3}$
45. (a) one solution **(b)** $x \approx 0.57$

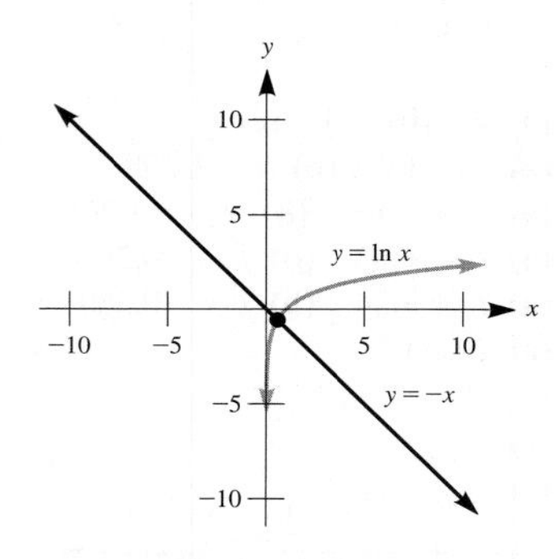

47. (a) one solution **(b)** $x \approx 1.49$

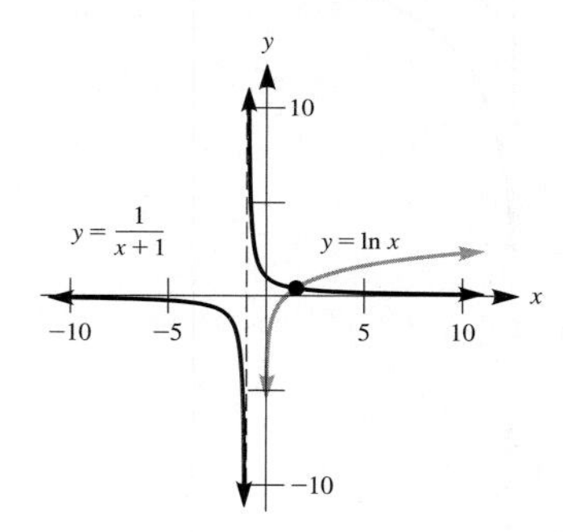

49. (a) two solutions
(b) $x \approx 1.56,\ x \approx 3.15$

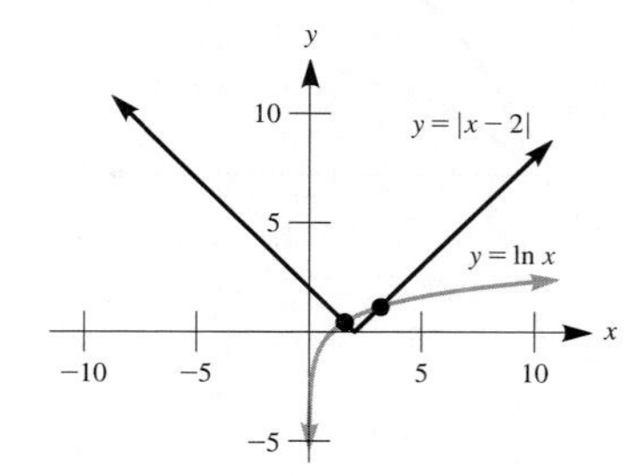

51. $x \le -1$ **53.** $x \ge \ln\frac{43}{4} \approx 2.375$
55. $x < \frac{2-e^2}{5} \approx -1.078$
57. $x \ge \ln 100 - 2 \approx 2.605$
59. all real numbers **61.** $-1 < x < \frac{1}{2}$
63. $x \le -1$ or $x \ge 5$ **65.** $0 < x < 1$

67. (a) $(4, \infty)$ **(b)** $4 < x \leq 7$
69. $0 < x < 3$ **71.** $x = \frac{1}{2}, 2$
73. The graphs are identical

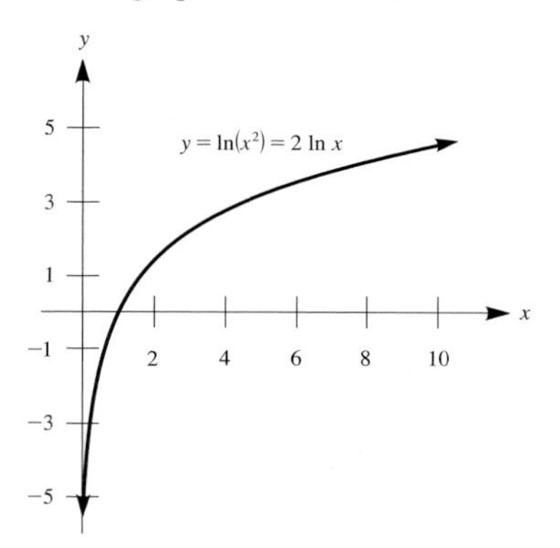

75. $x = e^2$ or $x = e^{-4/3}$
77. $(0, 1) \cup (1, \infty)$ **79.** $x = \beta^{-1/\alpha}$
81. $x = \dfrac{1}{k} \ln \dfrac{y}{A}$
83. $x = -\dfrac{1}{k} \ln\left(\dfrac{a - y}{by}\right)$
85. (a) $x = e^{e^2}$ **(b)** $x \approx 1618.178$
87. (a) $x = \dfrac{\ln 1.5}{\ln 3}$ **(b)** $x \approx 0.369$
89. Since $e > 0$, $2 - e < 2$, thus $\dfrac{2 - e}{3} < \dfrac{2}{3}$.
91. $(-3, -\sqrt{6}) \cup (\sqrt{6}, 3)$
93. $x = 1, x = 2$
95. $x = \dfrac{\ln(a - b)}{\ln(a + b)}$ **97.** $x = \dfrac{\pm b}{a + b}$

Exercise Set 5.6

1. \$1009.98 **3.** 8.45% **5.** \$767.27
7. (a) \$3869.68 **(b)** \$4006.39
9. 13 quarters **11.** \$3487.50
13. (a) 4.4 years **(b)** 4.43 years
15. 16 years **17.** \$2610.23
19. 5.83% **21.** 6% investment
23. (a) 14 years **(b)** 13.86 years
(c) 1.01% **25.** \$26.5 trillion
27. (a) 14 years
(b)

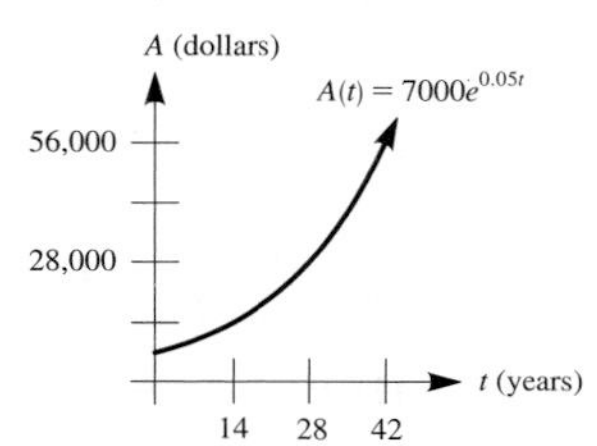

Exercise Set 5.7

1. (a) $k \approx 0.3209$ **(b)** 9951 bacteria
(c) 5.0 hours
3. $N_0 = 2000$; $k \approx 0.1769$
5. (a)

Region	World	More dev.	Less dev.
1995 population (billions)	5.702	1.169	4.533
Percent of population in 1995	100	20.5	79.5
Relative growth rate (percent per year)	1.5	0.2	1.9
Year 2000 population (billions)	6.146	1.181	4.985
Percent of world population in 2000	100	19.22	81.11

(b) world: 1.3% error; more developed: 0.3% error; less developed: 2.1% error
7. (a) Chad: $N(t) = 8.0e^{0.033t}$; United Kingdom: $N(t) = 59.8e^{0.001t}$
(b) 60 years **9. (a)** Niger: 15.8 million; Portugal: 10.2 million
(b) 13 years; Portugal: 10.1 million
11. (a) 6.332 billion **(b)** higher; 4% error **13. (a)** 116 years
(b) 43 years **(c)** 27 years
(d) 19 years
15. (a)

Year	1998 ($t = 10$)	2000 ($t = 12$)
Concentration of Carbon dioxide (ppm)	365.6	368.6

(b) 1998: too low, 0.3% error; 2000: too high, 0.1% error
17. (a) New York: $N(t) = 18.976e^{0.006t}$; Arizona: $N(t) = 5.131e^{0.04t}$
(b) year 2040

19. (a)

Region	1990 population (millions)	Growth rate (%)	2025 population
North America	275.2	0.7	351.6
Soviet Union	291.3	0.7	372.2
Europe	499.5	0.2	535.7
Nigeria	113.3	3.1	335.3

(b) 222.0 million **(c)** North America: 76.4 mil; Soviet Union: 80.9 mil; Europe: 36.2 mil; combined: 193.5 mil **(d)** Our results support this projection.
21. By the end of the year 2000 the population of Mexico exceeded 100 million people.
23. (a)

t (seconds)	0	550	1100	1650	2200
N (grams)	8	4	2	1	0.5

(b)

t (years)	0	4.9×10^9	9.8×10^9	14.7×10^9	19.6×10^9
N (grams)	10	5	2.5	1.25	0.625

25. 0.55 gram **27. (a)** 4.29 grams
(b) 79 hours
29. (a)

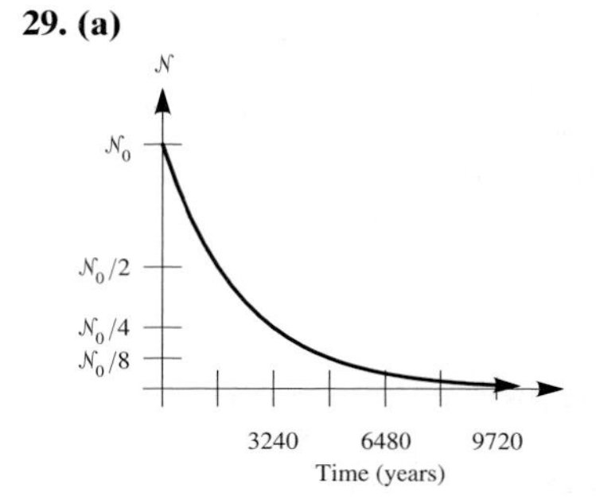

(b)

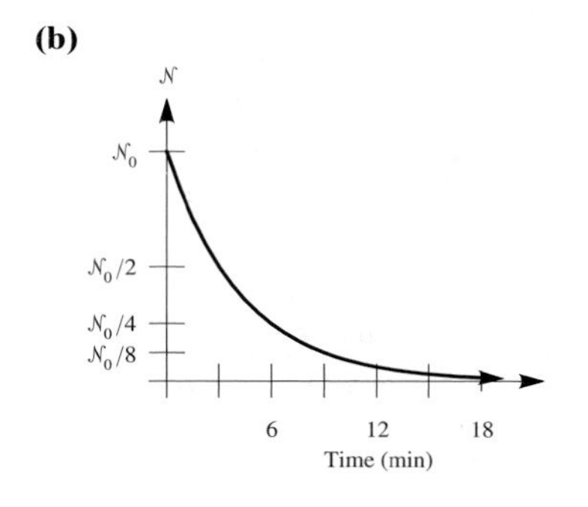

31. (a) 1.53 grams **(b)** 43 years
33. (a) 73,000 years **(b)** 10 half-lives
(c) 10 half-lives **35. (a)** $k \approx -0.0248$
(b) 279 years **(c)** 280 years
37. 0.02799997 ounces
39. (a) 8.5 billion; it is greater
(b) 0.934% per year
41. (b) $T \approx 32$ years; year 2004
(c) $y = 0.4606x + 24.0385$
(d) $T \approx 32$ years; year 2025
43. (a) $T \approx 34$ years; year 2033
(b) $T \approx 67$ years; year 2066
(c) $T \approx 52$ years; year 2051
47. 4.181 billion years old
49. 15,505 years old **51.** 8,000 years old; the site is older
53. years 10–5 B.C., which fit in the historical range
55. (a)

	$N(-1)$	$N(0)$	$N(1)$	$N(4)$	$N(5)$
From graph	0.25	0.5	1.0	3.5	3.75
From calculator	0.176	0.444	1.014	3.489	3.795

(b) $N(10) \approx 3.99854773$,
$N(15) \approx 3.99999021$,
$N(20) \approx 3.99999993$ **(c)** $t \approx 3$
(d) $t = \ln 24 \approx 3.178$
57. (a) $a = 4$ **(b)** $b = \ln \frac{8}{3} \approx 0.9808$
59. (a) $N(t) = \dfrac{162}{1 + 161e^{-0.50t}}$
(b) $N(4) \approx 7.1$, $N(8) \approx 41.0$, $N(12) \approx 115.8$, $N(16) \approx 153.7$; $N(4)$ and $N(8)$ are lower; $N(12)$ and $N(16)$
(c) 10 days 4 hours
61. (b) If k is close to 0, then $e^k - 1 \approx (k + 1) - 1 = k$.

Chapter 5 Review Exercises

1. $\log_5 126$ **3.** $\dfrac{4 \ln 1.5}{\ln 1.25}$ hours
5. f is not one-to-one

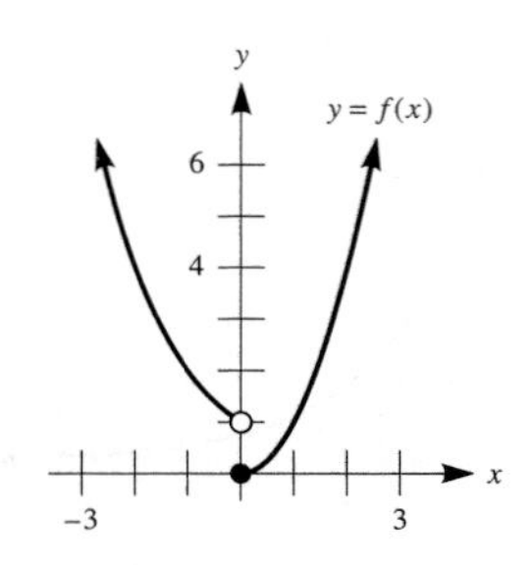

7. $x = \dfrac{e + 1}{e - 1}$ **9.** $y = e^x$: domain: $(-\infty, \infty)$, range: $(0, \infty)$; $y = \ln x$: domain: $(0, \infty)$, range: $(-\infty, \infty)$

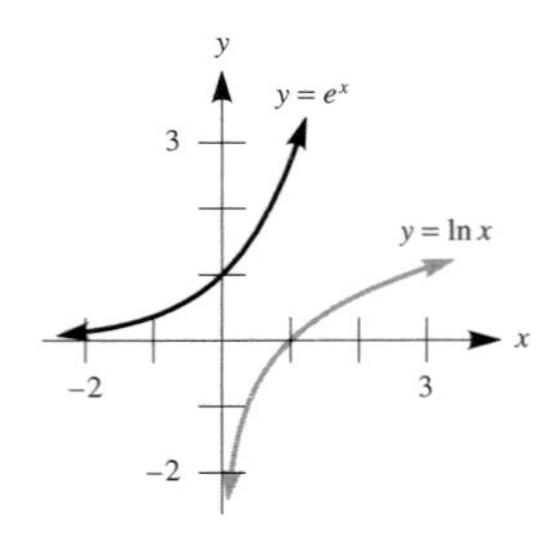

11. $-\frac{3}{2}$ **13.** $k \approx -0.05$
15. $x = 2 - \ln \frac{12}{5}$ **17.** year 2015
19. doubling time: 7 years

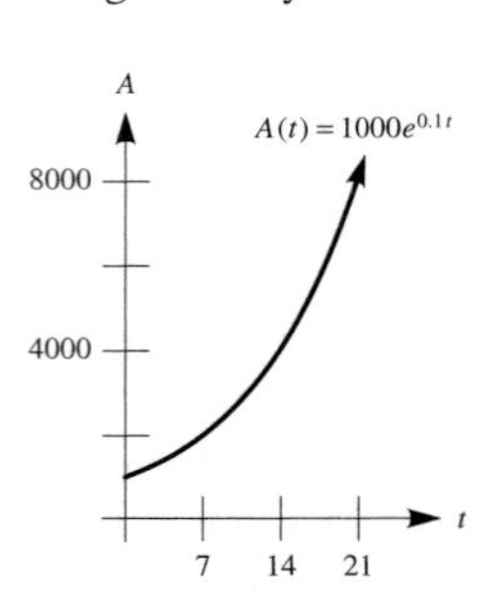

21. horizontal asymptote: $y = 0$; no vertical asymptote; no x-intercept; y-intercept: 1

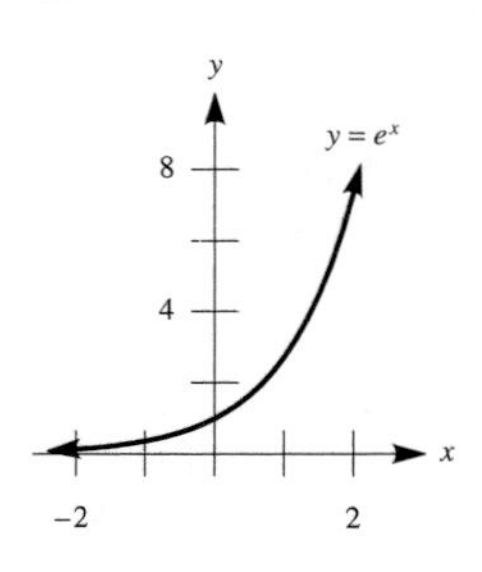

23. no horizontal asymptote; vertical asymptote: $x = 0$; x-intercept: 1; no y-intercept

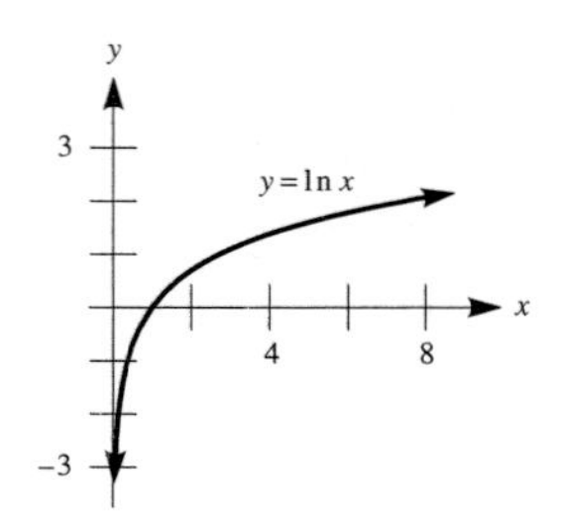

25. horizontal asymptote: $y = 1$; no vertical asymptote; no x-intercept; y-intercept: 3

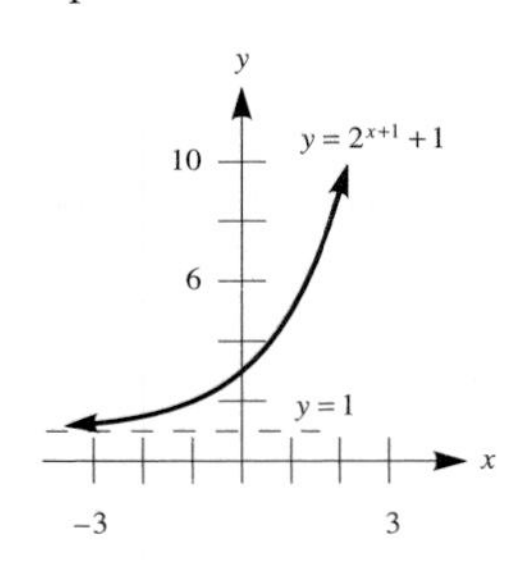

27. horizontal asymptote: $y = 0$; no vertical asymptote; no x-intercept; y-intercept: 1

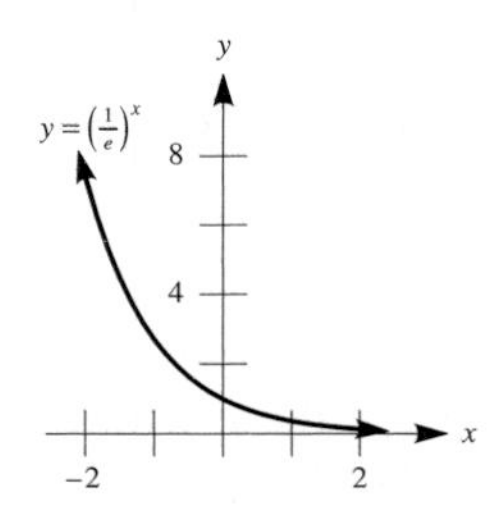

29. horizontal asymptote: $y = 1$; no vertical asymptote; no x-intercept; y-intercept: $e + 1$

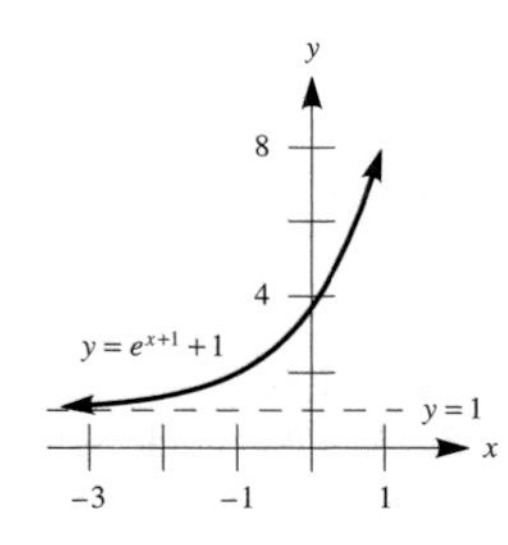

31. no asymptotes; x- and y-intercepts: 0

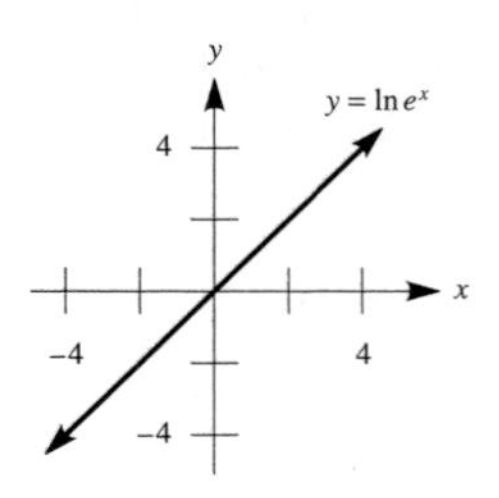

33. $x = 4$ **35.** $x = 3$ **37.** $x = \frac{2}{3}$
39. $x = \sqrt[3]{3}$ **41.** $x = \dfrac{1 - 2 \ln 3}{10}$

43. $x = \frac{200}{99}$ **45.** $x = 2$ **47.** all real numbers $x > 0$ **49.** $x = 1$ **51.** $\frac{1}{2}$
53. $\frac{1}{5}$ **55.** -1 **57.** 16 **59.** 4 **61.** 2
63. 2 **65.** $\frac{9}{14}$ **67.** $2a + 3b + \frac{c}{2}$
69. $8a + 4b$ **71.** between 2 and 3
73. between 2 and 3 **75.** -3 and -2
77. (a) third quadrant

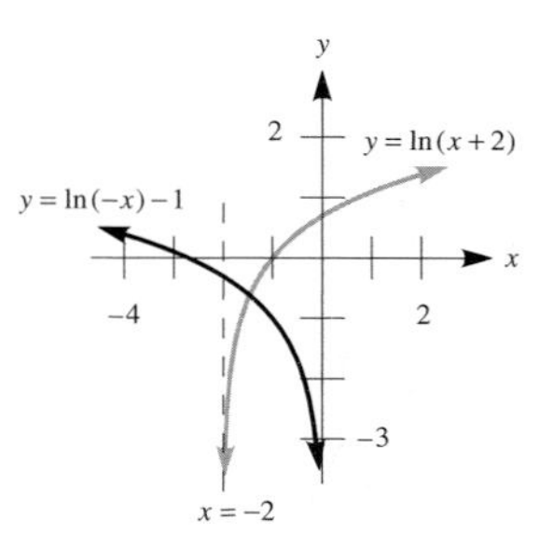

(b) $x = \dfrac{-2e}{e+1} \approx -1.46$ **79.** $k = \dfrac{\ln \frac{1}{2}}{T}$
81. 6.25% remaining
83. $\dfrac{d \ln \frac{c}{b}}{\ln \frac{1}{2}}$ days **85.** $\log_{10} 2$
87. $\ln 10$ **89.** $\ln x^a y^b$
91. $\frac{1}{2}\ln(x-3) + \frac{1}{2}\ln(x+4)$
93. $3\log_{10} x - \frac{1}{2}\log_{10}(1+x)$
95. $3\ln(1+2e) - 3\ln(1-2e)$
97. $\dfrac{\ln 2}{\ln(1 + \frac{R}{100})}$ years
99. 9.92% **101.** $\dfrac{100 \ln 2}{R}$ years
103. (a) $7\frac{3}{4}$ years
(b) $\dfrac{\ln n}{\ln(1 + \frac{R}{400})}$ years
105. (a) $(0, \infty)$ **(b)** $[1, \infty)$
107. $(0, e^2) \cup (e^2, \infty)$

Chapter 5 Test

1. domain: $(-\infty, \infty)$; range: $(-3, \infty)$; x-intercept: $-\dfrac{\ln 3}{\ln 2}$; y-intercept: -2; asymptote: $y = -3$

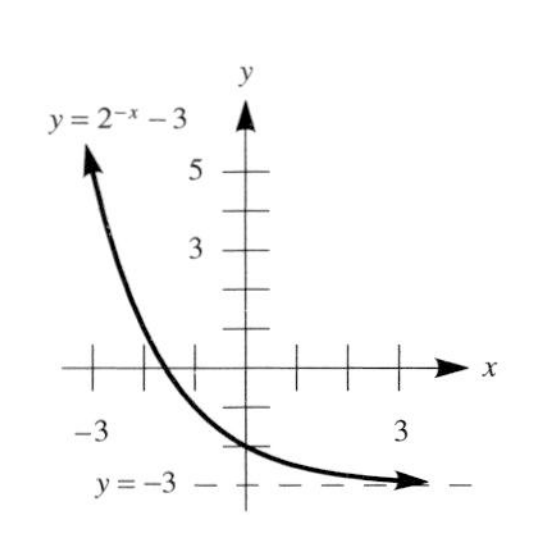

2. $\dfrac{\ln \frac{5}{3}}{\ln 1.033} \approx 16$ hours **3.** $\log_2 17$
4. $\dfrac{\ln 15}{\ln 2}$ **5.** 10^{12} **6.** 4 years
7. (a) $(0, \infty)$
(b)

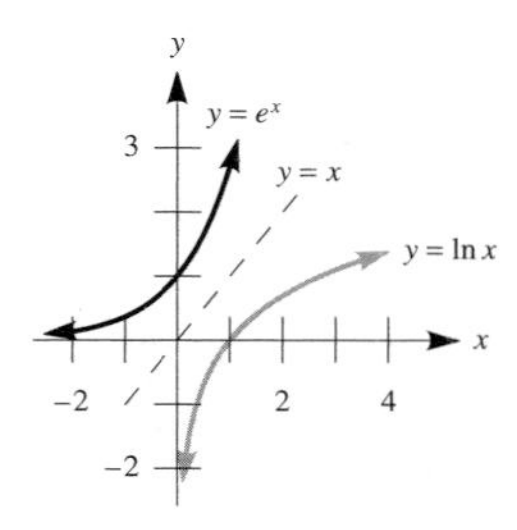

8. $3a - \frac{1}{2}b$ **9. (a)** $-\frac{1}{2}$ **(b)** -1
10. $x = 0, -2, 2$ **11. (a)** $k \approx -0.1733$
(b) 0.35 gram **12.** $\ln \dfrac{x^2}{(x^2+1)^{1/3}}$
13. no solution
14. $g^{-1}(x) = 10^x + 1$; range: $(1, \infty)$
15. (a) 12 years
(b)

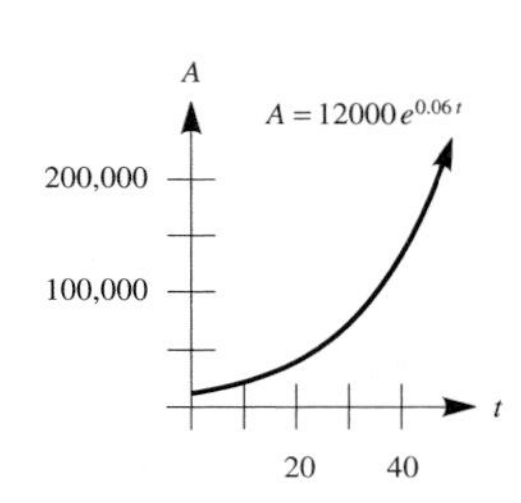

16. (a) $(0, \infty)$ **(b)** $x = 1, e^2$
17. (a) $\dfrac{1}{\ln \frac{4}{3}} \approx 3.48$ **(b)** 4
18. $x = 4$ **19.** $x = \ln \frac{3}{2} \approx 0.405$
20. (a) $x > \dfrac{\ln 1.6}{\ln 0.3} \approx -0.39$
(b) $(3, 4]$

CHAPTER 6

Exercise Set 6.1

1. $\theta = 2.5$ radians **3.** $\theta = 2$ radians
5. (a) $\pi/4 \approx 0.79$ radians
(b) $\pi/2 \approx 1.57$ radians
(c) $3\pi/4 \approx 2.36$ radians
7. (a) 0 radians **(b)** $2\pi \approx 6.28$ radians **(c)** $5\pi/2 \approx 7.85$ radians
9. (a) 15° **(b)** 30° **(c)** 45°
11. (a) 60° **(b)** 300° **(c)** 720°
13. (a) 114.59° **(b)** 171.89°
(c) 565.49° **15.** smaller
17. $30° = \pi/6$ radians; $45° = \pi/4$ radians; $60° = \pi/3$ radians; $120° = 2\pi/3$ radians; $135° = 3\pi/4$ radians; $150° = 5\pi/6$ radians
19. $s = 4\pi$ ft **21.** $s = \pi/2$ cm
23. (a) $A = 12\pi \text{ cm}^2 \approx 37.70 \text{ cm}^2$
(b) $A = 50\pi/9 \text{ m}^2 \approx 17.45 \text{ m}^2$
(c) $A = 72\pi/5 \text{ m}^2 \approx 45.24 \text{ m}^2$
(d) $A = 1.296\pi \text{ cm}^2 \approx 4.07 \text{ cm}^2$
25. $2\pi/5$ radians
27. (a) $\left(10 + \dfrac{5\pi}{6}\right)$ in. ≈ 12.62 in.
(b) $25\pi/12 \text{ in.}^2 \approx 6.54 \text{ in.}^2$
29. (a) 12π radians/sec
(b) 144π cm/sec **(c)** 72π cm/sec
31. (a) 6π radians/sec
(b) 150π cm/sec **(c)** 75π cm/sec
33. (a) $50\pi/3$ radians/sec
(b) 750π cm/sec **(c)** 375π cm/sec
35. (a) 0.000073 radians/sec
(b) 1040 mph **37.** 1 radian $\approx 57.30°$
39. $x = 0$ **41. (a)** 200π radians/min
(b) 2000π cm/min
(c) $\frac{1000}{3}\pi$ radians/min
(d) $\frac{500}{3}$ radians/min
43. 4930 mi **45.** 1470 mi
47. 2690 mi
51. (a) $P = \left(\dfrac{2\pi}{3} + 1\right)s$
(b) $A = \left(\dfrac{\pi}{3} - \dfrac{\sqrt{3}}{4}\right)s^2$
53. (a) $A_{ABC} = \left(\dfrac{\pi}{3} - \dfrac{\sqrt{3}}{4}\right)s^2$; $A_{ADF} = \left(\dfrac{\pi}{12} - \dfrac{\sqrt{3}}{16}\right)s^2$; $\dfrac{A_{ADF}}{A_{ABC}} = \dfrac{1}{4}$
(b) $A_{DBE} = \left(\dfrac{\pi}{8} - \dfrac{\sqrt{3}}{8}\right)s^2$
(c) $A_{DEF} = \left(\dfrac{\sqrt{3}}{4} - \dfrac{\pi}{8}\right)s^2$
(d) $A_{BEC} = \left(\dfrac{\pi}{12} - \dfrac{\sqrt{3}}{8}\right)s^2$
55. (a) $r(\theta) = \dfrac{12}{2 + \theta}$
(b) $A(\theta) = \dfrac{72\theta}{(2+\theta)^2}$; no
(c) $\theta(r) = \dfrac{12}{r} - 2$
(d) $A(r) = 6r - r^2$; yes
(e) maximum = 9 cm² when $r = 3$ cm, $\theta = 2$ radians

Exercise Set 6.2

1. (a) reference number: $\frac{\pi}{4}$

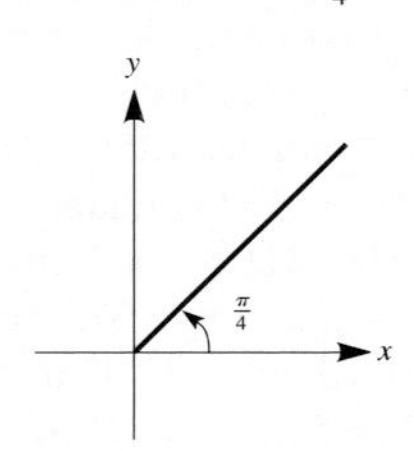

(b) reference number: $\frac{\pi}{4}$

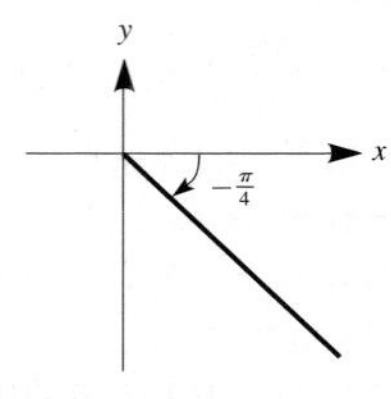

(c) reference number: $\frac{\pi}{4}$

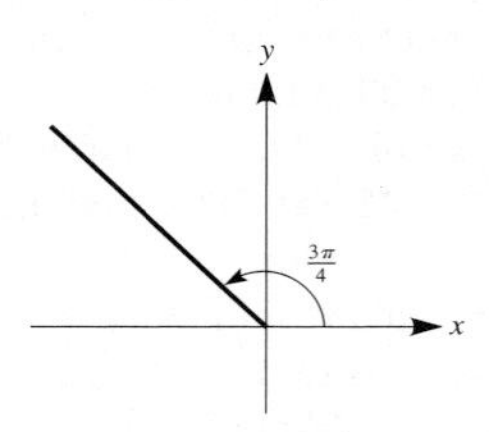

3. (a) reference number: $\frac{\pi}{3}$

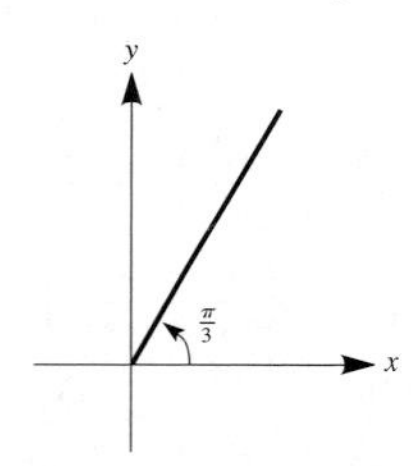

(b) reference number: $\frac{\pi}{3}$

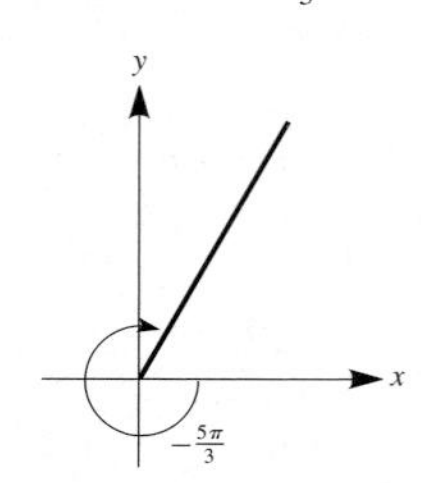

(c) reference number: $\frac{\pi}{3}$

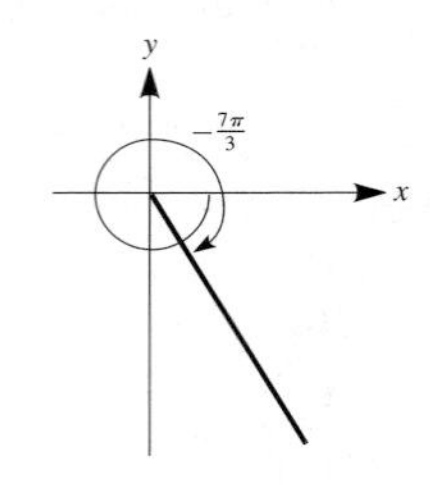

5. (a) reference angle: 30°

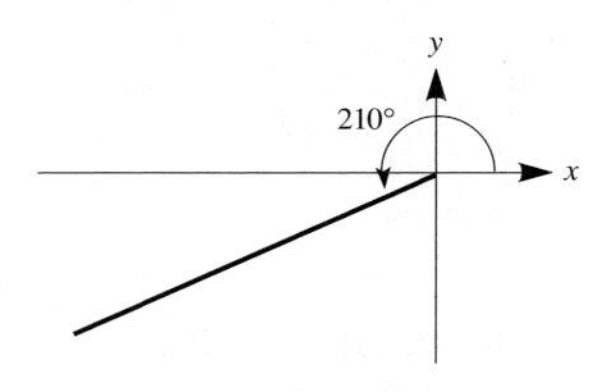

(b) reference angle: 30°

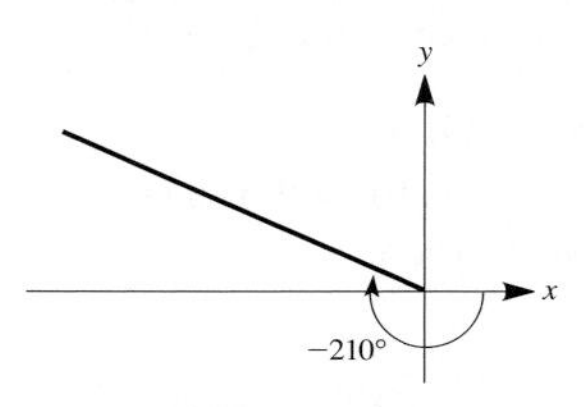

(c) reference angle: 30°

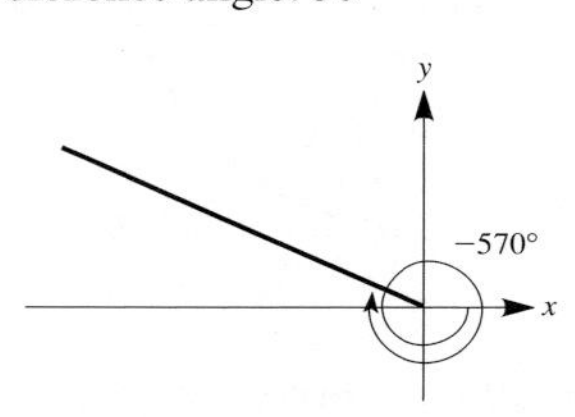

7. (a) reference angle: 60°

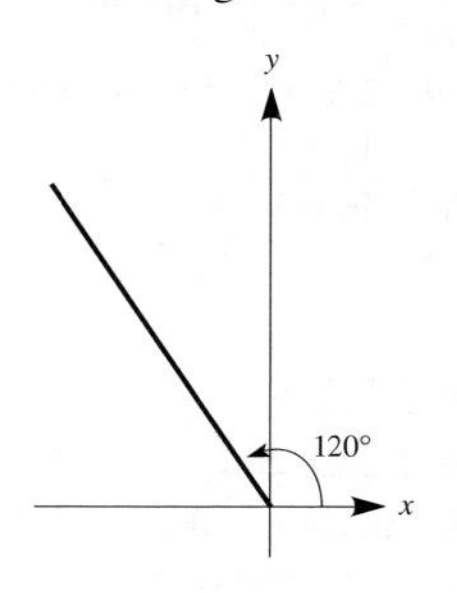

(b) reference angle: 60°

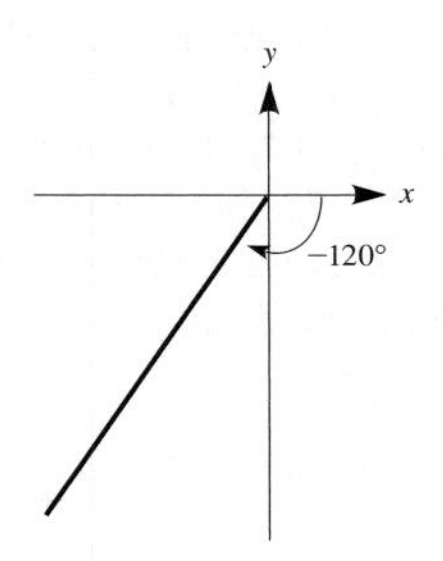

(c) reference angle: 60°

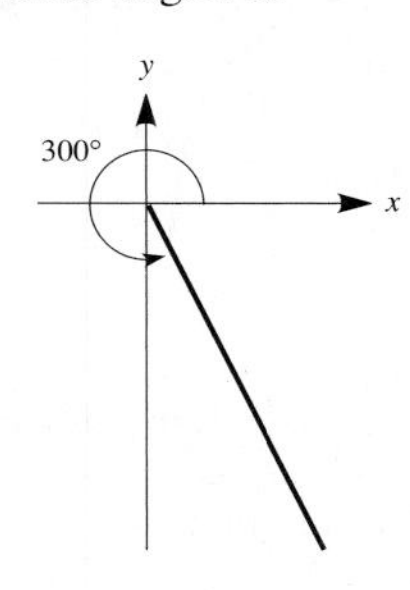

9. $\sin \pi = 0$, $\cos \pi = -1$, $\tan \pi = 0$, $\sec \pi = -1$, $\csc \pi$ is undefined, $\cot \pi$ is undefined **11.** $\sin(-2\pi) = 0$, $\cos(-2\pi) = 1$, $\tan(-2\pi) = 0$, $\sec(-2\pi) = 1$, $\csc(-2\pi)$ is undefined, $\cot(-2\pi)$ is undefined
13. $\sin(-3\pi/2) = 1$, $\cos(-3\pi/2) = 0$, $\tan(-3\pi/2)$ is undefined, $\sec(-3\pi/2)$ is undefined, $\csc(-3\pi/2) = 1$, $\cot(-3\pi/2) = 0$ **15.** $\sin 0 = 0$, $\cos 0 = 1$, $\tan 0 = 0$, $\sec 0 = 1$, $\csc 0$ is undefined, $\cot 0$ is undefined
17. $\sin 90° = 1$, $\cos 90° = 0$, $\tan 90°$ is undefined, $\sec 90°$ is undefined, $\csc 90° = 1$, $\cot 90° = 0$
19. $\sin(-270°) = 1$, $\cos(-270°) = 0$, $\tan(-270°)$ is undefined, $\sec(-270°)$ is undefined, $\csc(-270°) = 1$, $\cot(-270°) = 0$ **21.** $\sin 180° = 0$, $\cos 180° = -1$, $\tan 180° = 0$, $\sec 180° = -1$, $\csc 180°$ is undefined, $\cot 180°$ is undefined
23. (b) $\sin \theta = \frac{4}{5}$, $\cos \theta = -\frac{3}{5}$, $\tan \theta = -\frac{4}{3}$, $\sec \theta = -\frac{5}{3}$, $\csc \theta = \frac{5}{4}$, $\cot \theta = -\frac{3}{4}$

25.

θ	$\cos\theta$	$\sin\theta$	$\tan\theta$
0	1	0	0
$\pi/2$	0	1	undefined
π	−1	0	0
$3\pi/2$	0	−1	undefined
2π	1	0	0

θ	$\sec\theta$	$\csc\theta$	$\cot\theta$
0	1	undefined	undefined
$\pi/2$	undefined	1	0
π	−1	undefined	undefined
$3\pi/2$	undefined	−1	0
2π	1	undefined	undefined

27. (a) positive **(b)** negative **(c)** negative **29. (a)** positive **(b)** negative **31.** sin 2 is larger **33.** cos 2 is larger **35.** tan 4 is larger **37.** $0.5 < \cos 1 < 0.6$, $0.8 < \sin 1 < 0.9$; $\cos 1 \approx 0.54$, $\sin 1 \approx 0.84$ **39.** $0.5 < \cos(-1) < 0.6$, $-0.9 < \sin(-1) < -0.8$; $\cos(-1) \approx 0.54$, $\sin(-1) \approx -0.84$ **41.** $-0.7 < \cos 4 < -0.6$, $-0.8 < \sin 4 < -0.7$; $\cos 4 \approx -0.65$, $\sin 4 \approx -0.76$ **43.** $-0.7 < \cos(-4) < -0.6$, $0.7 < \sin(-4) < 0.8$; $\cos(-4) \approx -0.65$, $\sin(-4) \approx 0.76$ **45.** $\sin 10° \approx 0.2$, $\sin(-10°) \approx -0.2$; $\sin 10° \approx 0.17$, $\sin(-10°) \approx -0.17$ **47.** $\cos 80° \approx 0.2$, $\cos(-80°) \approx 0.2$; $\cos 80° \approx 0.17$, $\cos(-80°) \approx 0.17$ **49.** $\sin 120° \approx 0.9$, $\sin(-120°) \approx -0.9$; $\sin 120° \approx 0.87$, $\sin(-120°) \approx -0.87$ **51.** $\sin 150° = 0.5$, $\sin(-150°) = -0.5$; $\sin 150° = 0.5$, $\sin(-150°) = -0.5$ **53.** $\cos 220° \approx -0.8$, $\cos(-220°) \approx -0.8$; $\cos 220° \approx -0.77$, $\cos(-220°) \approx -0.77$ **55.** $\cos 310° \approx 0.6$, $\cos(-310°) \approx 0.6$; $\cos 310° \approx 0.64$, $\cos(-310°) \approx 0.64$ **57.** $0.8 < \sin(1 + 2\pi) < 0.9$; $\sin(1 + 2\pi) \approx 0.84$ **59.** $\sin\theta = 2\sqrt{2}/3$, $\cos\theta = \frac{1}{3}$, $\tan\theta = 2\sqrt{2}$, $\sec\theta = 3$, $\csc\theta = 3\sqrt{2}/4$, $\cot\theta = \sqrt{2}/4$ **61.** $\sin\theta = -\frac{4}{5}$, $\cos\theta = -\frac{3}{5}$, $\tan\theta = \frac{4}{3}$, $\sec\theta = -\frac{5}{3}$, $\csc\theta = -\frac{5}{4}$, $\cot\theta = \frac{3}{4}$ **63.** $\sin\theta = \frac{5}{13}$, $\cos\theta = -\frac{12}{13}$, $\tan\theta = -\frac{5}{12}$, $\sec\theta = -\frac{13}{12}$, $\csc\theta = \frac{13}{5}$, $\cot\theta = -\frac{12}{5}$ **65.** $\sin\theta = -\frac{3}{4}$, $\cos\theta = -\sqrt{7}/4$, $\tan\theta = -3\sqrt{7}/7$, $\sec\theta = -4\sqrt{7}/7$, $\csc\theta = -\frac{4}{3}$, $\cot\theta = -\sqrt{7}/3$ **67.** $\sin\theta = \sqrt{161}/15$, $\cos\theta = -\frac{8}{15}$, $\tan\theta = -\sqrt{161}/8$, $\sec\theta = -\frac{15}{8}$, $\csc\theta = 15\sqrt{161}/161$, $\cot\theta = -8\sqrt{161}/161$ **69.** $\sin\theta = -\frac{2}{9}$, $\cos\theta = \sqrt{77}/9$, $\tan\theta = -2\sqrt{77}/77$, $\sec\theta = 9\sqrt{77}/77$, $\csc\theta = -\frac{9}{2}$, $\cot\theta = -\sqrt{77}/2$ **71.** $\sin\theta = -\frac{24}{25}$, $\cos\theta = \frac{7}{25}$, $\tan\theta = -\frac{24}{7}$, $\sec\theta = \frac{25}{7}$, $\csc\theta = -\frac{25}{24}$, $\cot\theta = -\frac{7}{24}$ **73.** $\sin\theta = \sqrt{3}/2$, $\cos\theta = \frac{1}{2}$, $\tan\theta = \sqrt{3}$, $\sec\theta = 2$, $\csc\theta = 2\sqrt{3}/3$, $\cot\theta = \sqrt{3}/3$ **75.** $\sec 2.06 \approx -2.13$, $\csc 2.06 \approx 1.13$, $\cot 2.06 \approx -0.53$ **77.** $\sec 9 \approx -1.10$, $\csc 9 \approx 2.43$, $\cot 9 \approx -2.21$ **79.** $\sec(-0.55) \approx 1.17$, $\csc(-0.55) \approx -1.91$, $\cot(-0.55) \approx -1.63$ **81.** $\sec\frac{\pi}{6} \approx 1.15$, $\csc\frac{\pi}{6} = 2$, $\cot\frac{\pi}{6} \approx 1.73$ **83.** $\sec 1400 \approx 2.45$, $\csc 1400 \approx -1.10$, $\cot 1400 \approx -0.45$ **85.** $\sec 33° \approx 1.19$, $\csc 33° \approx 1.84$, $\cot 33° \approx 1.54$ **87.** $\sec(-125°) \approx -1.74$, $\csc(-125°) \approx -1.22$, $\cot(-125°) \approx 0.70$ **89.** $\sec 225° \approx -1.41$, $\csc 225° \approx -1.41$, $\cot 225° = 1$ **91. (a)** Since $\frac{\pi}{2}$ results in the point (0, 1) on the unit circle, $\sin\frac{\pi}{2} = 1$. Since π results in the point $(-1, 0)$ on the unit circle, $\sin\pi = 0$.

(b) $\sin\dfrac{\pi}{2} = 1$, $\dfrac{\sin\pi}{2} = 0$

Exercise Set 6.3

1. (a) reference angle: 70°

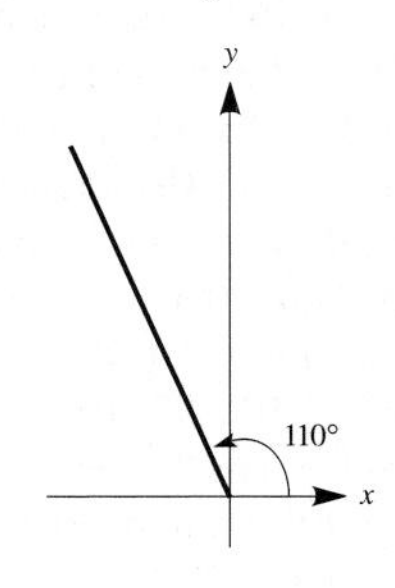

(b) reference angle: 60°

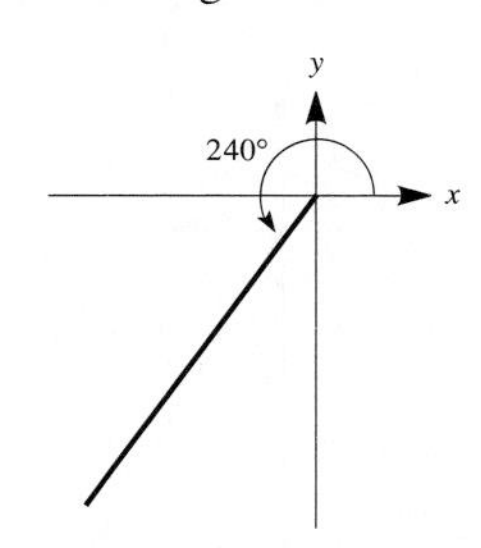

(c) reference angle: 60°

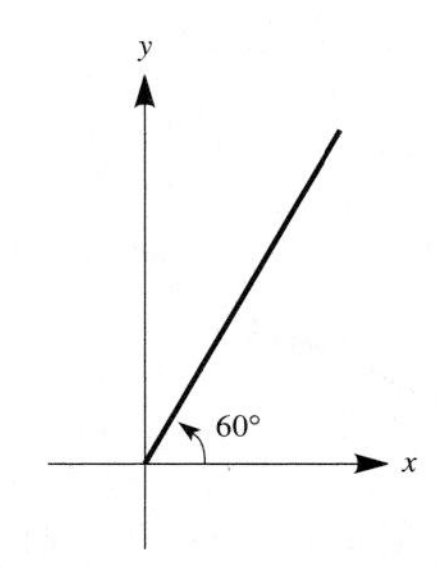

(d) reference angle: 60°

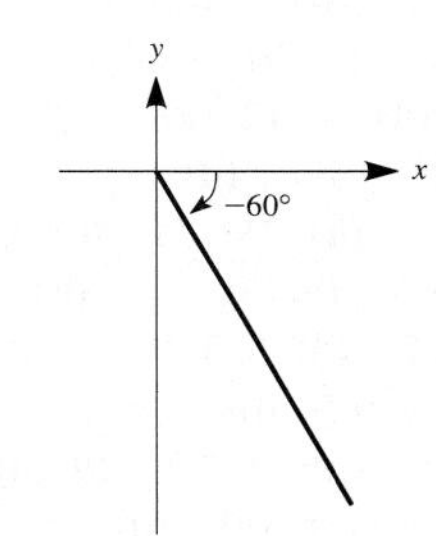

3. (a) reference number: $\frac{\pi}{4}$

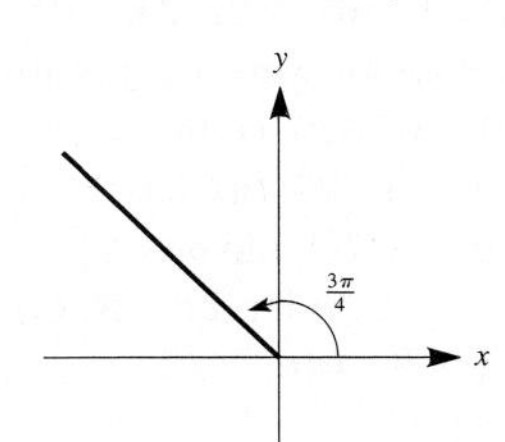

(b) reference number: $\frac{\pi}{6}$

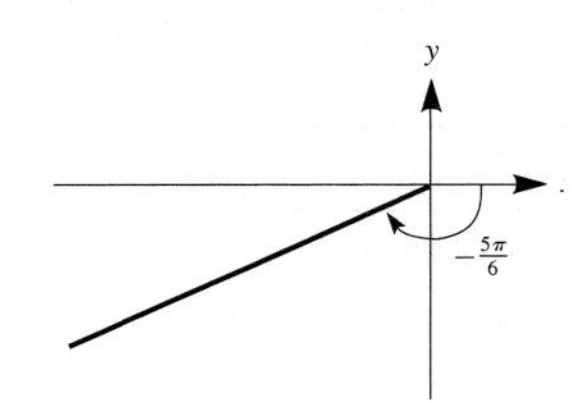

(c) reference number: $\frac{\pi}{3}$

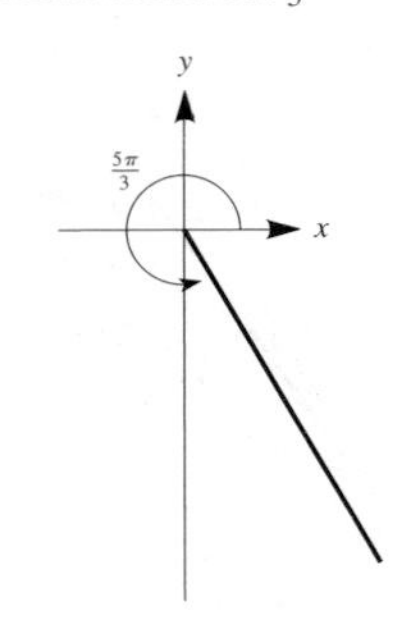

(d) reference number: $\frac{\pi}{6}$

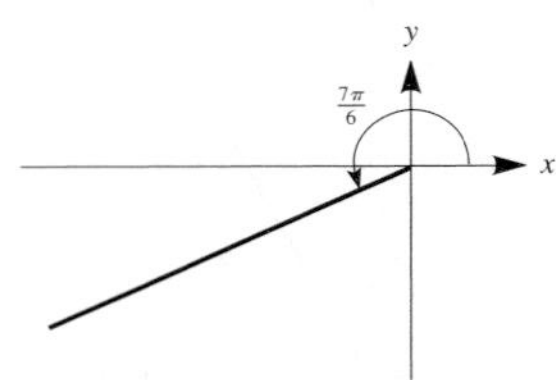

5. (a) B **(b)** A **(c)** D **(d)** A **(e)** B **(f)** C **7. (a)** $\frac{1}{2}$ **(b)** $\frac{1}{2}$ **(c)** $-\sqrt{3}/2$ **(d)** $\sqrt{3}/2$
9. (a) $-\sqrt{3}/2$ **(b)** $-\sqrt{3}/2$ **(c)** $-\frac{1}{2}$ **(d)** $\frac{1}{2}$ **11. (a)** $\sqrt{3}/2$ **(b)** $\sqrt{3}/2$ **(c)** $\frac{1}{2}$ **(d)** $-\frac{1}{2}$
13. (a) -2 **(b)** $2\sqrt{3}/3$ **(c)** $\sqrt{3}$ **(d)** $-\sqrt{3}/3$ **15. (a)** $-\frac{1}{2}$ **(b)** $-\frac{1}{2}$ **(c)** $-\sqrt{3}/2$ **(d)** $\sqrt{3}/2$
17. (a) $-\sqrt{2}/2$ **(b)** $-\sqrt{2}/2$ **(c)** $-\sqrt{2}/2$ **(d)** $\sqrt{2}/2$ **19. (a)** -2 **(b)** $2\sqrt{3}/3$ **(c)** $\sqrt{3}$ **(d)** $-\sqrt{3}/3$
21. (a) $-2\sqrt{3}/3$ **(b)** -2 **(c)** $-\sqrt{3}/3$ **(d)** $\sqrt{3}$ **23.** $-\frac{2\pi}{3}$, $\frac{2\pi}{3}$, $\frac{4\pi}{3}$ (Other answers are possible.)
25. $-30°$, $30°$, $330°$ (Other answers are possible.) **27. (a)** $\cos\theta = \frac{3}{4}$
(b) $\sin\theta = \sqrt{7}/4$, $\tan\theta = \sqrt{7}/3$
(c) $\cos(\beta - \frac{\pi}{2})$ is larger **29. (a)** 0
(b) 1 **(c)** 0 **(d)** $-\sqrt{2}$
(e) $\dfrac{4 - 2\sqrt{2}}{\pi}$ **(f)** $-2\sqrt{2}/\pi$
(g) $-\frac{3}{\pi}$ **(h)** 0
31. (a)

θ	$\sin\theta$	larger?
0.1	0.0998	θ
0.2	0.1987	θ
0.3	0.2955	θ
0.4	0.3894	θ
0.5	0.4794	θ

(b) In right ΔPQR the hypotenuse is $\overline{PR}$ while $\overline{PQ}$ is a leg, and thus $PQ < PR$. Assuming that the shortest path from point P to point R is a straight line, then $PR < \theta$. (Recall that θ is the arc length.) Thus $PQ < PR < \theta$.

33. (b) $Q\left(\dfrac{-b}{\sqrt{b^2 + a^2}}, \dfrac{a}{\sqrt{b^2 + a^2}}\right)$

(c) $P(a, b)$ lies on the unit circle $x^2 + y^2 = 1$.

Exercise Set 6.4

1. (a) $11SC$ **(b)** $11 \sin\theta\cos\theta$
3. (a) $-8C^3S$ **(b)** $-8\cos^3\theta\sin\theta$
5. (a) $1 + 2T + T^2$
(b) $1 + 2\tan\theta + \tan^2\theta$
7. (a) $T^2 + T - 6$
(b) $\tan^2\theta + \tan\theta - 6$
9. (a) -1 **(b)** -1
11. (a) $\dfrac{CS + 2}{S}$ **(b)** $\dfrac{\cos A \sin A + 2}{\sin A}$
13. (a) $(T - 1)(T + 9)$
(b) $(\tan\beta - 1)(\tan\beta + 9)$
15. (a) $(2C + 1)(2C - 1)$
(b) $(2\cos B + 1)(2\cos B - 1)$
17. (a) $3ST^2(3ST + 2)$
(b) $3\sec B\tan^2 B(3\sec B\tan B + 2)$
19. $\sin A + \cos A$ **21.** $\csc\theta$
23. $\cos^2 B$ **25.** $\cos A + 4$ **27.** $2\csc\theta$
29. 0 **31.** 1 **33.** $\cos\theta = -\frac{4}{5}$, $\tan\theta = \frac{3}{4}$ **35.** $\tan t = -\sqrt{39}/13$
37. $\csc\beta = \frac{17}{8}$, $\cot\beta = -\frac{15}{8}$
39. $\cos\theta = -2\sqrt{6}/5$, $\tan\theta = -\sqrt{6}/12$, $\sec\theta = -5\sqrt{6}/12$, $\csc\theta = 5$, $\cot\theta = -2\sqrt{6}$
41. $\sin\theta = -\frac{4}{5}$, $\tan\theta = \frac{4}{3}$, $\sec\theta = -\frac{5}{3}$, $\csc\theta = -\frac{5}{4}$, $\cot\theta = \frac{3}{4}$
43. $\sin\beta = 2\sqrt{2}/3$, $\cos\beta = \frac{1}{3}$, $\tan\beta = 2\sqrt{2}$, $\csc\beta = 3\sqrt{2}/4$, $\cot\beta = \sqrt{2}/4$ **67.** area $= \sin\theta$
69. (a) $m(\theta) = \tan\theta$ **(b) (i)** $-\sqrt{3}$ **(ii)** $\tan 1 \approx 1.6$ **71. (a)** not an identity ($\alpha = 30°$) **(b)** identity
79. (a) $(\cos\theta - \sin\theta)(1 + \cos\theta\sin\theta)$
85. (a) $\log_{10}(\sin^2 20°) = 2\log_{10}(\sin 20°) \approx -0.9319$
(b) $0° < \theta < 180°$ **87.** $(0°, 180°)$
89. (a) $\ln\sqrt{1 - \cos 20°} + \ln\sqrt{1 + \cos 20°} = \ln(\sin 20°) \approx -1.0729$ **(c)** $(0°, 180°)$

Exercise Set 6.5

1. (a) $\sin\theta = \frac{15}{17}$, $\cos\theta = \frac{8}{17}$, $\tan\theta = \frac{15}{8}$, $\cot\theta = \frac{8}{15}$, $\sec\theta = \frac{17}{8}$, $\csc\theta = \frac{17}{15}$
(b) $\sin\beta = \frac{8}{17}$, $\cos\beta = \frac{15}{17}$, $\tan\beta = \frac{8}{15}$, $\cot\beta = \frac{15}{8}$, $\sec\beta = \frac{17}{15}$, $\csc\beta = \frac{17}{8}$
3. (a) $\sin\theta = \sqrt{5}/5$, $\cos\theta = 2\sqrt{5}/5$, $\tan\theta = \frac{1}{2}$, $\cot\theta = 2$, $\sec\theta = \sqrt{5}/2$, $\csc\theta = \sqrt{5}$ **(b)** $\sin\beta = 2\sqrt{5}/5$, $\cos\beta = \sqrt{5}/5$, $\tan\beta = 2$, $\cot\beta = \frac{1}{2}$, $\sec\beta = \sqrt{5}$, $\csc\beta = \sqrt{5}/2$
5. (a) $\cos A = 3\sqrt{13}/13$, $\sin A = 2\sqrt{13}/13$, $\tan A = \frac{2}{3}$
(b) $\sec B = \sqrt{13}/2$, $\csc B = \sqrt{13}/3$, $\cot B = \frac{2}{3}$ **7.** $\sin B = \frac{12}{13}$, $\cos B = \frac{5}{13}$, $\tan B = \frac{12}{5}$, $\cot B = \frac{5}{12}$, $\sec B = \frac{13}{5}$, $\csc B = \frac{13}{12}$ **9. (a)** $\sin B = \frac{4}{5}$, $\cos A = \frac{4}{5}$ **(b)** $\sin A = \frac{3}{5}$, $\cos B = \frac{3}{5}$
(c) $(\tan A)(\tan B) = 1$
11. (a) $\cos A = \frac{24}{25}$, $\sin A = \frac{7}{25}$, $\tan A = \frac{7}{24}$ **(b)** $\cos B = \frac{7}{25}$, $\sin B = \frac{24}{25}$, $\tan B = \frac{24}{7}$ **(c)** $(\tan A)(\tan B) = 1$

25. (a) $\sin\theta = \dfrac{2x\sqrt{4x^2 + 9}}{4x^2 + 9}$, $\cos\theta = \dfrac{3\sqrt{4x^2 + 9}}{4x^2 + 9}$, $\tan\theta = \dfrac{2x}{3}$

(b) $\sin^2\theta = \dfrac{4x^2}{4x^2 + 9}$, $\cos^2\theta = \dfrac{9}{4x^2 + 9}$, $\tan^2\theta = \dfrac{4x^2}{9}$

(c) $\sin(90° - \theta) = \dfrac{3\sqrt{4x^2 + 9}}{4x^2 + 9}$, $\cos(90° - \theta) = \dfrac{2x\sqrt{4x^2 + 9}}{4x^2 + 9}$, $\tan(90° - \theta) = \dfrac{3}{2x}$

27. (a) $\sin\beta = \dfrac{\sqrt{16x^2 - 9}}{4x}$, $\cos\beta = \dfrac{1}{4x}$, $\tan\beta = \sqrt{16x^2 - 9}$

(b) $\csc\beta = \dfrac{4x\sqrt{16x^2 - 9}}{16x^2 - 9}$, $\sec\beta = 4x$, $\cot\beta = \dfrac{\sqrt{16x^2 - 9}}{16x^2 - 9}$

(c) $\sin(90° - \beta) = \dfrac{1}{4x}$, $\cos(90° - \beta) = \dfrac{\sqrt{16x^2 - 1}}{4x}$, $\tan(90° - \beta) = \dfrac{\sqrt{16x^2 - 1}}{16x^2 - 1}$

29. $\sin B = \sqrt{33}/7$, $\tan B = \sqrt{33}/4$, $\sec B = \frac{7}{4}$, $\csc B = 7\sqrt{33}/33$, $\cot B = 4\sqrt{33}/33$ **31.** $\cos\theta = \sqrt{13}/5$, $\tan\theta = 2\sqrt{39}/13$, $\sec\theta = 5\sqrt{13}/13$, $\csc\theta = 5\sqrt{3}/6$, $\cot\theta = \sqrt{39}/6$

33. $\sin A = \dfrac{2\sqrt{3} - \sqrt{6}}{6}$,

$\cos A = \dfrac{2\sqrt{3} + \sqrt{6}}{6}$,

$\sec A = 2\sqrt{3} - \sqrt{6}$,
$\csc A = 2\sqrt{3} + \sqrt{6}$, $\cot A = 3 + 2\sqrt{2}$

35. (a) $\cos\theta = \dfrac{\sqrt{4 - x^2}}{2}$,

$\tan\theta = \dfrac{x\sqrt{4 - x^2}}{4 - x^2}$, $\sec\theta = \dfrac{2\sqrt{4 - x^2}}{4 - x^2}$,

$\csc\theta = \dfrac{2}{x}$, $\cot\theta = \dfrac{\sqrt{4 - x^2}}{x}$

(b) Same as (a) except for change of sign for $\cos\theta$ and $\sec\theta$.

37. (a) $\sin\theta = \sqrt{1 - x^4}$,

$\tan\theta = \dfrac{\sqrt{1 - x^4}}{x^2}$, $\sec\theta = \dfrac{1}{x^2}$,

$\csc\theta = \dfrac{\sqrt{1 - x^4}}{1 - x^4}$, $\cot\theta = \dfrac{x^2\sqrt{1 - x^4}}{1 - x^4}$

(b) Same as (a) except for change of sign for $\sin\theta$, $\csc\theta$, $\tan\theta$, and $\cot\theta$.

39. false **41.** true **43.** true
45. true
47. (a) Since $RC < QB < PA$, $\sin 20° < \sin 40° < \sin 60°$.
(b) $\sin 20° \approx 0.3420$, $\sin 40° \approx 0.6428$, $\sin 60° \approx 0.8660$

57. $\tan\alpha = \dfrac{p}{q}\sqrt{\dfrac{1 - q^2}{p^2 - 1}}$,

$\tan\beta = \sqrt{\dfrac{1 - q^2}{p^2 - 1}}$

Chapter 6 Review Exercises

1.

θ	$\cos\theta$	$\sin\theta$	$\tan\theta$
0	1	0	0
$\pi/6$	$\sqrt{3}/2$	$1/2$	$\sqrt{3}/3$
$\pi/4$	$\sqrt{2}/2$	$\sqrt{2}/2$	1
$\pi/3$	$1/2$	$\sqrt{3}/2$	$\sqrt{3}$
$\pi/2$	0	1	undefined
$2\pi/3$	$-1/2$	$\sqrt{3}/2$	$-\sqrt{3}$
$3\pi/4$	$-\sqrt{2}/2$	$\sqrt{2}/2$	-1
$5\pi/6$	$-\sqrt{3}/2$	$1/2$	$-\sqrt{3}/3$
π	-1	0	0

θ	$\sec\theta$	$\csc\theta$	$\cot\theta$
0	1	undefined	undefined
$\pi/6$	$2\sqrt{3}/3$	2	$\sqrt{3}$
$\pi/4$	$\sqrt{2}$	$\sqrt{2}$	1
$\pi/3$	2	$2\sqrt{3}/3$	$\sqrt{3}/3$
$\pi/2$	undefined	1	0
$2\pi/3$	-2	$2\sqrt{3}/3$	$-\sqrt{3}/3$
$3\pi/4$	$-\sqrt{2}$	$\sqrt{2}$	-1
$5\pi/6$	$-2\sqrt{3}/3$	2	$-\sqrt{3}$
π	-1	undefined	undefined

3. $a = \frac{1}{2}$, $c = \sqrt{3}/2$ **5.** $b = \frac{35}{2}$
7. $\sin A = \sqrt{55}/8$, $\cot A = 3\sqrt{55}/55$

13. (b) $\sin\theta = \dfrac{\cos\alpha + \sin\alpha}{\sqrt{2}}$,

$\cos\theta = \dfrac{\cos\alpha - \sin\alpha}{\sqrt{2}}$

15. $\sin A \cos A$ **17.** $\sin^2 A$
19. $\cos A + \sin A$ **21.** $\sin A \cos A$
23. (a) $0.9 < \cos 6 < 1.0$; $\cos 6 \approx 0.96$
(b) $0.9 < \cos(-6) < 1.0$; $\cos(-6) \approx 0.96$
25. (a) $-0.8 < \cos 140° < -0.7$; $\cos 140° \approx -0.77$
(b) $-0.8 < \cos(-140°) < -0.7$; $\cos(-140°) \approx -0.77$
27. (a) $0.7 < \sin\frac{\pi}{4} < 0.8$; $\sin\frac{\pi}{4} \approx 0.71$
(b) $-0.8 < \sin(-\pi/4) < -0.7$; $\sin(-\pi/4) \approx -0.71$
29. (a) $-1.0 < \sin 250° < -0.9$; $\sin 250° \approx -0.94$
(b) $0.9 < \sin(-250°) < 1.0$; $\sin(-250°) \approx 0.94$
31. (a) $-0.7 < \cos 4 < -0.6$; $\cos 4 \approx -0.65$
(b) $-0.7 < \cos(-4) < -0.6$; $\cos(-4) \approx -0.65$
33. (a) $-0.7 < \cos(4 + 2\pi) < -0.6$, $\cos(4 + 2\pi) \approx -0.65$
(b) $-0.7 < \cos(-4 + 2\pi) < -0.6$, $\cos(-4 + 2\pi) \approx -0.65$
35. $\sqrt{1 - a^2}$ **37.** $\sqrt{1 - a^2}$
39. $-a$ **41.** $-a$ **43.** $-\sqrt{1 - a^2}$
45. $\sin\theta - \cos\theta$
65. $s = 2\pi$ cm, $A = 16\pi$ cm^2
67. $A = \frac{1}{2}$ cm^2 **69.** $r = \frac{20}{\pi}$ radians, $A = \frac{40}{\pi}$ cm^2 **71.** $r = 3$ cm, $\theta = 4$ radians **73.** $\frac{7\pi}{18}$ radians
75. $\sin\theta = \frac{2}{5}$, $\cos\theta = -\sqrt{21}/5$, $\tan\theta = -2\sqrt{21}/21$, $\cot\theta = -\sqrt{21}/2$, $\sec\theta = -5\sqrt{21}/21$, $\csc\theta = \frac{5}{2}$
77. $\sin\theta = -2\sqrt{6}/7$, $\cos\theta = -\frac{5}{7}$, $\tan\theta = 2\sqrt{6}/5$, $\cot\theta = 5\sqrt{6}/12$, $\sec\theta = -\frac{7}{5}$, $\csc\theta = -7\sqrt{6}/12$
79. $\sin\theta = \frac{8}{17}$, $\cos\theta = -\frac{15}{17}$, $\tan\theta = -\frac{8}{15}$, $\cot\theta = -\frac{15}{8}$, $\sec\theta = -\frac{17}{15}$, $\csc\theta = \frac{17}{8}$
81. $\sin\theta = -\frac{7}{25}$, $\cos\theta = \frac{24}{25}$, $\tan\theta = -\frac{7}{24}$, $\cot\theta = -\frac{24}{7}$, $\sec\theta = \frac{25}{24}$, $\csc\theta = -\frac{25}{7}$ **83.** $-\frac{5\pi}{6}, \frac{5\pi}{6}, \frac{7\pi}{6}$ (Other answers are possible.)
85. $-450°$, $-90°$, $270°$, $630°$ (Other answers are possible.)
87. $\sin\theta = -\frac{12}{13}$, $\tan\theta = \frac{12}{5}$
89. 0.9848 **91.** 0.1736

Chapter 6 Test

1. (a) -1 **(b)** -1 **(c)** 0
2. (a) $\dfrac{\sqrt{3}}{2}$ **(b)** $\dfrac{\sqrt{2}}{2}$ **(c)** 1
3. $\cos\theta = 2\sqrt{6}/5$, $\tan\theta = -\sqrt{6}/12$, $\cot\theta = -2\sqrt{6}$, $\sec\theta = 5\sqrt{6}/12$, $\csc\theta = -5$
4. $\sin\theta = \sqrt{31}/6$, $\tan\theta = -\sqrt{155}/5$, $\cot\theta = -\sqrt{155}/31$, $\sec\theta = -6\sqrt{5}/5$, $\csc\theta = 6\sqrt{31}/31$
5. $-\dfrac{\sqrt{3}}{2}$ **6.** -1 **7.** $\frac{1}{2}$ **8.** $-\sqrt{2}$
9. $\sin 2$ is larger **10. (a)** $\frac{11\pi}{12}$ radians **(b)** $\frac{540°}{\pi}$ **11.** $\frac{25\pi}{12}$ cm **12.** $\frac{125\pi}{24}$ cm^2
13. $-2\tan\theta - 1$
15. (a) 1200π radians/min
(b) 18000π cm/min
(c) 720π radians/min **(d)** 360 rpm

16. $\sin\theta = \dfrac{t\sqrt{1 + t^2}}{1 + t^2}$,

$\cos\theta = \dfrac{\sqrt{1 + t^2}}{1 + t^2}$, $\sec\theta = \sqrt{1 + t^2}$,

$\csc\theta = \dfrac{\sqrt{1 + t^2}}{t}$, $\cot\theta = \dfrac{1}{t}$

17. $\dfrac{\sqrt{17} - 1}{6}$

18. (a) $\tan(\angle CAB)$ is larger
(b) $\cos(\angle DAB)$ is larger

CHAPTER 7

Exercise Set 7.1

1. (a) $\sqrt{3}/2$ **(b)** $\sqrt{3}/2$ **(c)** $-\frac{1}{2}$ **(d)** $\frac{1}{2}$ **3. (a)** $\sqrt{3}/2$ **(b)** $\sqrt{3}/2$ **(c)** $\frac{1}{2}$ **(d)** $-\frac{1}{2}$ **5. (a)** $-\sqrt{2}/2$ **(b)** $-\sqrt{2}/2$ **(c)** $-\sqrt{2}/2$ **(d)** $\sqrt{2}/2$
7. (a) 2 **(b)** $2\sqrt{3}/3$ **(c)** $-\sqrt{3}$ **(d)** $\sqrt{3}/3$ **9. (a)** $t = \pi/2, 3\pi/2, 5\pi/2$, and $7\pi/2$ (Other answers are possible.)
(b) $t = -\pi/2, -3\pi/2, -5\pi/2$, and $-7\pi/2$ (Other answers are possible.)

11. (a) $\sin 2.06 \approx 0.88$; $\cos 2.06 \approx -0.47$; $\tan 2.06 \approx -1.88$; $\sec 2.06 \approx -2.13$; $\csc 2.06 \approx 1.13$; $\cot 2.06 \approx -0.53$
(b) $\sin(-2.06) \approx -0.88$; $\cos(-2.06) \approx -0.47$; $\tan(-2.06) \approx 1.88$; $\sec(-2.06) \approx -2.13$; $\csc(-2.06) \approx -1.13$; $\cot(-2.06) \approx 0.53$
13. (a) $\sin(\pi/6) \approx 0.50$; $\cos(\pi/6) \approx 0.87$; $\tan(\pi/6) \approx 1.73$; $\sec(\pi/6) \approx 1.15$; $\csc(\pi/6) \approx 2.00$; $\cot(\pi/6) \approx 0.58$
(b) $\sin(\frac{\pi}{6} + 2\pi) \approx 0.50$; $\cos(\frac{\pi}{6} + 2\pi) \approx 0.87$; $\tan(\frac{\pi}{6} + 2\pi) \approx 1.73$; $\sec(\frac{\pi}{6} + 2\pi) \approx 1.15$; $\csc(\frac{\pi}{6} + 2\pi) \approx 2.00$; $\cot(\frac{\pi}{6} + 2\pi) \approx 0.58$
25. $\cos t = -\frac{4}{5}$; $\tan t = \frac{3}{4}$
27. $\tan t = -\sqrt{39}/13$
29. $\sec \alpha = \frac{13}{5}$; $\cos \alpha = \frac{5}{13}$; $\sin \alpha = \frac{12}{13}$
35. $\dfrac{\sqrt{7}\cos\theta}{7}$ **37. (a)** $-\frac{2}{3}$ **(b)** $\frac{1}{4}$
(c) $\frac{1}{5}$ **(d)** $-\frac{1}{5}$ **39. (a)** $-\dfrac{1 + 2\sqrt{2}}{3}$
(b) 1 **41. (a)** $\sqrt{2}/2$ **(b)** $\sqrt{3}/2$
(c) 1 **43.** $\cos^2 t$ **45.** $\sin^2 \theta$ **55.** $\frac{13}{31}$
57. Assuming that the radius of the circle is 1, the coordinates of the point labeled (x, y) are $(\cos t, \sin t)$, and the coordinates of the point labeled $(-x, -y)$ are $(\cos(t + \pi), \sin(t + \pi))$. So $y = \sin t$ and $-y = \sin(t + \pi)$, from which it follows that $\sin(t + \pi) = -\sin t$. Similarly, $-x = \cos(t + \pi)$ and $x = \cos t$, from which it follows that $\cos(t + \pi) = -\cos t$. Since $t - \pi$ results in the same intersection point with the unit circle as $t + \pi$, identities (ii) and (iv) follow in a similar manner.
61. (a)

t	0.2	0.4	0.6	0.8
$f(t)$	219.07	50.53	19.70	9.55

t	1.0	1.2	1.4
$f(t)$	6.14	7.98	33.88

(b) The smallest output is 6.14, which occurs at $t = 1.0$.
(d) $\tan^2 t + 9\cot^2 t = (\tan t - 3\cot t)^2 + 6 \geq 0 + 6 \geq 6$

(e) The answer from part **(b)** is consistent with this result.
63. (a) Each side is 0. **(b)** Each side is $(2 - \sqrt{2})/2$. **(c)** no
65. (a)

t	$1 - \frac{1}{2}t^2$	$\cos\theta$
0.02	0.998	0.999800
0.05	0.99875	0.998750
0.1	0.995	0.995004
0.2	0.980	0.980067
0.3	0.955	0.955336

(b)

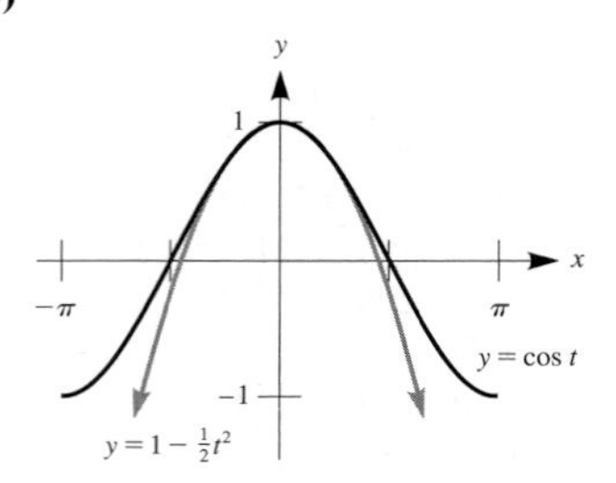

67. (a)

x	$\frac{1}{3}x^3 + x$	$\frac{2}{15}x^5 + \frac{1}{3}x^3 + x$	$\tan x$
0.1	0.100333	0.100335	0.100335
0.2	0.202667	0.202709	0.202710
0.3	0.309000	0.309324	0.309336
0.4	0.421333	0.422699	0.422793
0.5	0.541667	0.545833	0.546302

(b)

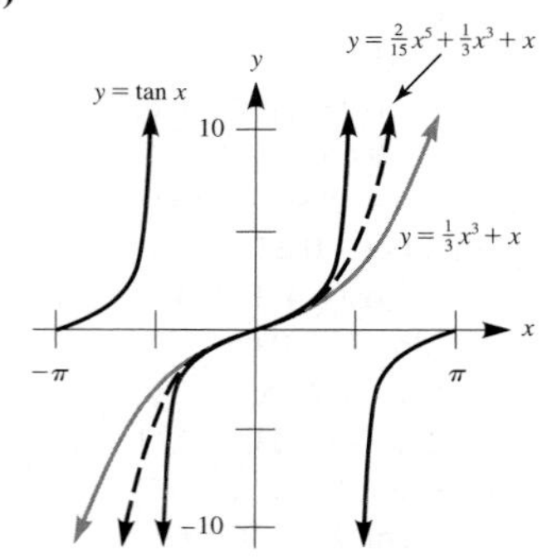

69. P: $\sin(\pi/12) \approx 0.259$; Q: $\frac{1}{2} = 0.5$; R: $\sqrt{2}/2 \approx 0.707$; S: $\sqrt{3}/2 \approx 0.866$; T: $\sin(5\pi/12) \approx 0.966$

Exercise Set 7.2

1. period: 2; amplitude: 1
3. period: 4; amplitude: 6
5. period: 4; amplitude: 2
7. period: 6; amplitude: $\frac{3}{2}$
9. $(-\frac{7\pi}{2}, 1) \approx (-10.996, 1)$
11. $(\frac{5\pi}{2}, 1) \approx (7.854, 1)$
13. $(-4\pi, 0) \approx (-12.566, 0)$
15. $(-3\pi, 0) \approx (-9.425, 0)$
17. $(-\pi, 0) \approx (-3.142, 0)$
19. increasing **21.** decreasing
23. $(\frac{9\pi}{2}, 0) \approx (14.137, 0)$
25. $(-4\pi, 1) \approx (-12.566, 1)$
27. $(\frac{\pi}{2}, 0) \approx (1.571, 0)$
29. $(4\pi, 1) \approx (12.566, 1)$
31. $(-\frac{5\pi}{2}, 0) \approx (-7.854, 0)$
33. decreasing **35.** increasing
37. (a) x-intercept: π

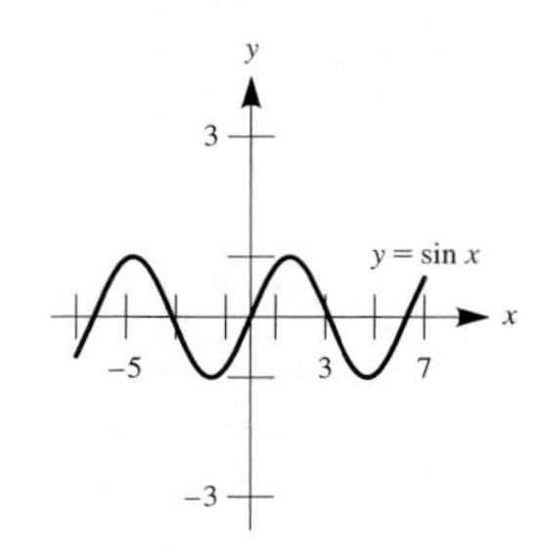

(b) four turning points; x-coordinate: $\frac{\pi}{2}$
(c) five turning points: x-coordinate: 2π

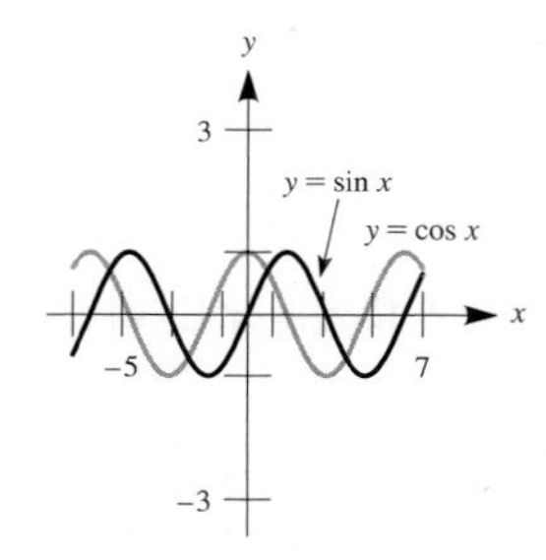

(d) $\frac{\pi}{2}$ units to the left
(e)

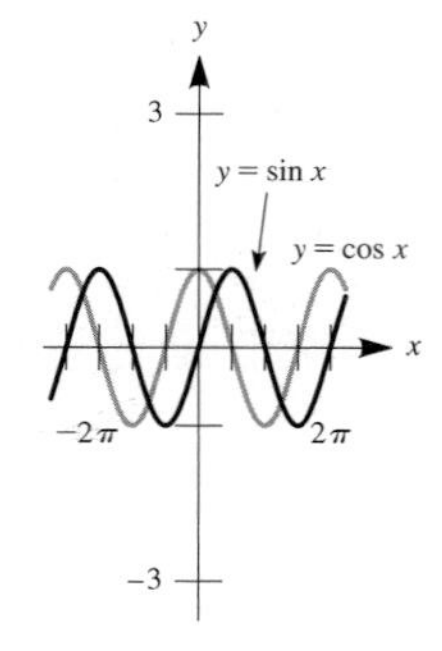

39. (a)

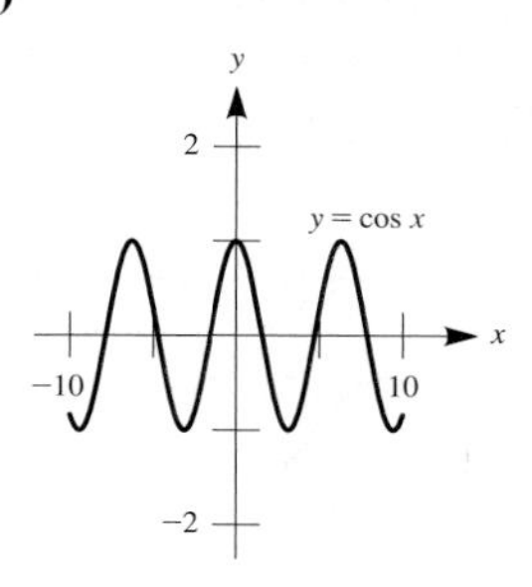

(b) In the vicinity of $x = 0$ the graphs of $y = \cos x$ and $y = 1 - 0.5x^2$ are very similar.

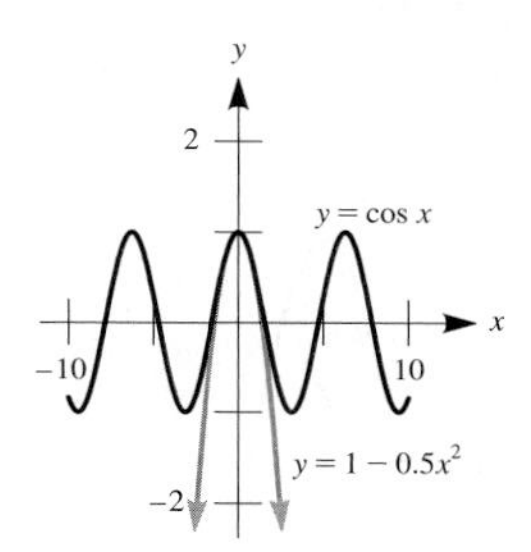

(c)

x	1	0.5	0.1
$\cos x$	0.54	0.8776	0.9950
$1 - 0.5x^2$	0.50	0.8750	0.9950

x	0.01	0.001
$\cos x$	0.99995	0.9999995
$1 - 0.5x^2$	0.99995	0.9999995

41. (a) 0.45 **(b)** 0.4510 **(c)** 5.8322 **(d)** 2.6906 and 3.5926 **43. (a)** 1.25 **(b)** 1.2661 **(c)** 5.0171 **(d)** 1.8755 and 4.4077 **45. (a)** 1.0 **(b)** 0.9884 **(c)** 5.2948 **(d)** 2.1532 and 4.1300 **47. (a)** 0.65 **(b)** 0.6435 **(c)** 2.4981 **(d)** 3.7851 and 5.6397 **49. (a)** 0.2 **(b)** 0.2014 **(c)** 2.9402 **(d)** 3.3430 and 6.0818 **51. (a)** 0.8 **(b)** 0.7754 **(c)** 2.3662 **(d)** 3.9170 and 5.5078

53. intersection points: $x \approx 2.498$ and $x \approx 3.785$

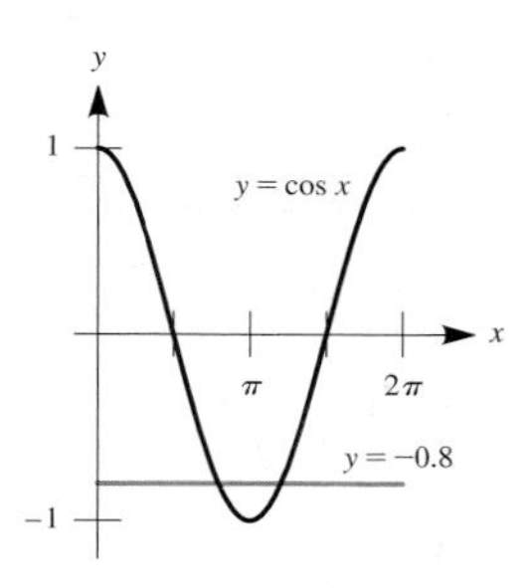

55. (a) $x \approx 0.7574$; $x \approx 2.3842$

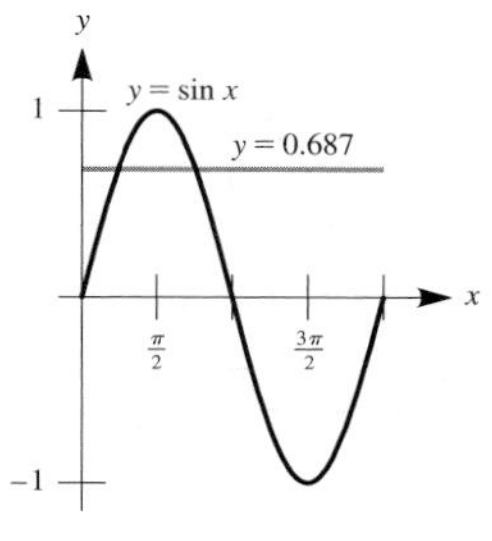

(b) $x \approx 3.8989$; $x \approx 5.5258$

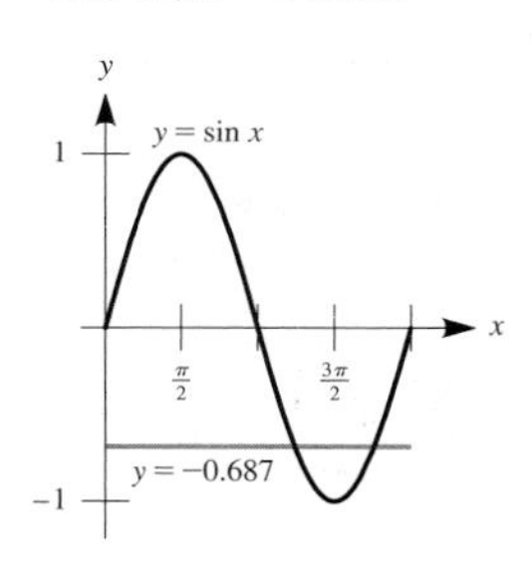

57. (a) 0 **(b)** 0 **59.** $(\frac{\pi}{2}, \pi)$

61. (a)

	x_1	x_2	x_3
From graph	1.0	0.55	0.85
From calculator	0.99500	0.54450	0.85539

	x_4	x_5	x_6
From graph	0.65	0.80	0.70
From calculator	0.65593	0.79248	0.70208

	x_7
From graph	0.75
From Calculator	0.76350

(b)

n	0	1	2	3
Number of fish after n breeding seasons	50	498	272	428

n	4	5	6	7
Number of fish after n breeding seasons	328	396	351	382

(c) 370 fish **63. (a)** $C(\cos\theta, \sin\theta)$

Exercise Set 7.3

1. (a) amplitude: 2; period: 2π; x-intercepts: 0, π, 2π; increasing: $(0, \frac{\pi}{2})$ and $(\frac{3\pi}{2}, 2\pi)$

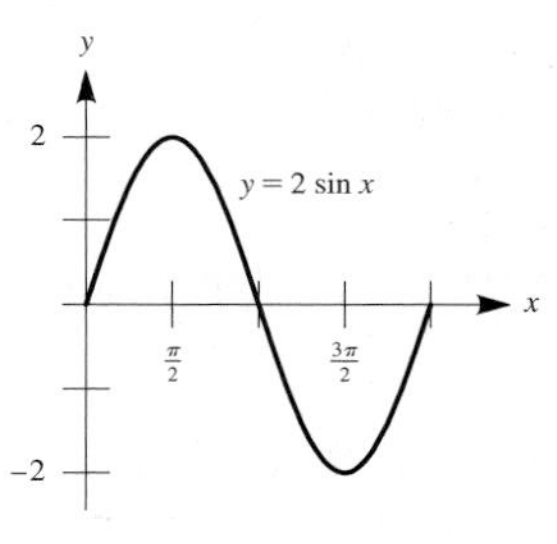

(b) amplitude: 1; period: π; x-intercepts: 0, $\frac{\pi}{2}$, π; increasing: $(\frac{\pi}{4}, \frac{3\pi}{4})$

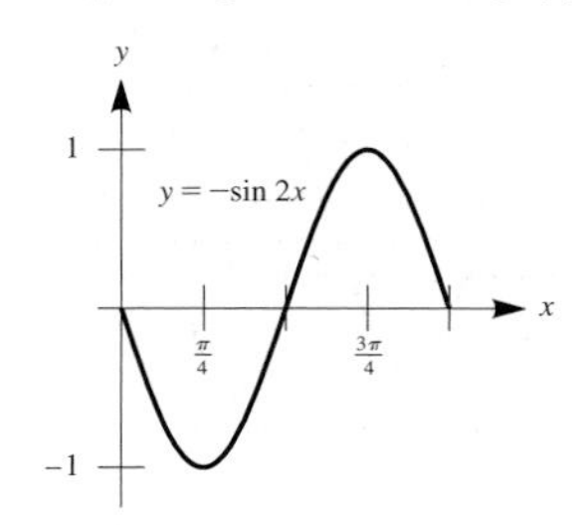

3. (a) amplitude: 1; period: π;
x-intercepts: $\frac{\pi}{4}, \frac{3\pi}{4}$; increasing: $(\frac{\pi}{2}, \pi)$

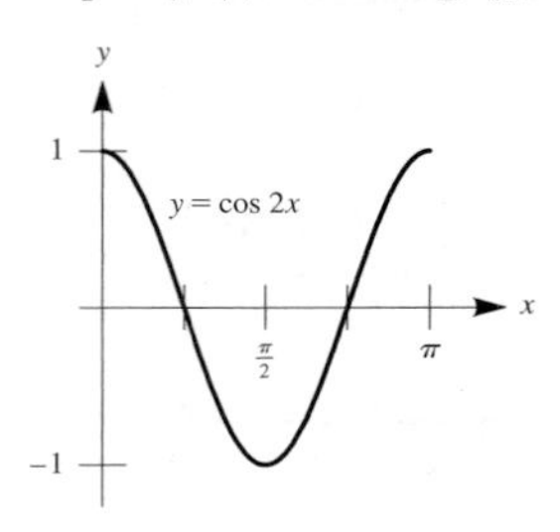

(b) amplitude: 2; period: π;
x-intercepts: $\frac{\pi}{4}, \frac{3\pi}{4}$; increasing: $(\frac{\pi}{2}, \pi)$

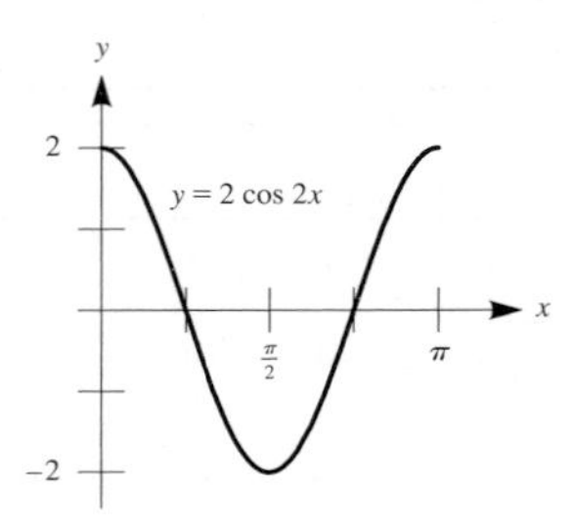

5. (a) amplitude: 3; period: 4;
x-intercepts: 0, 2, 4; increasing: (0, 1) and (3, 4)

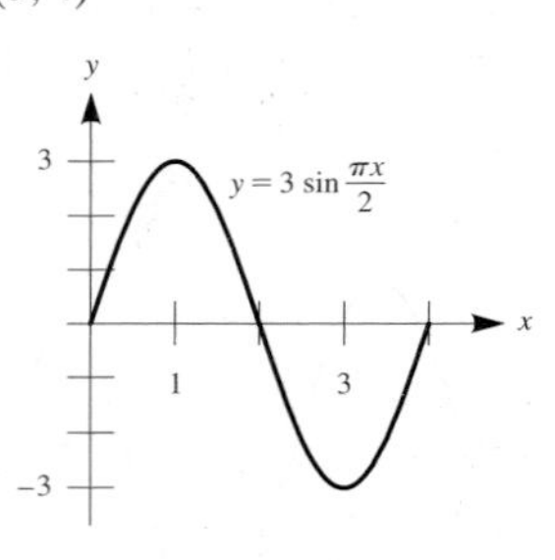

(b) amplitude: 3; period: 4;
x-intercepts: 0, 2, 4; increasing: (1, 3)

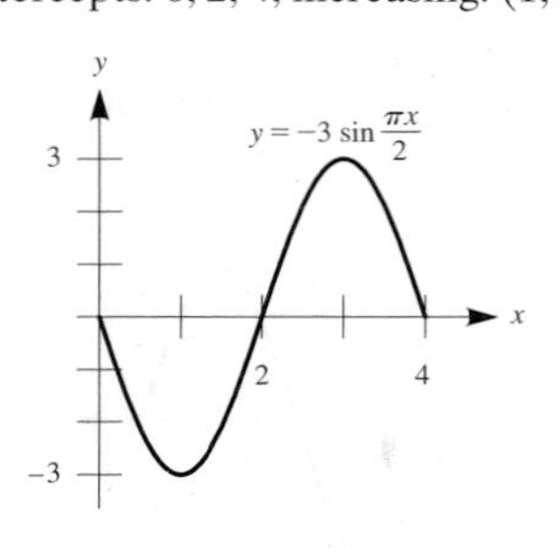

7. (a) amplitude: 1; period: 1;
x-intercepts: $\frac{1}{4}, \frac{3}{4}$; increasing: $(\frac{1}{2}, 1)$

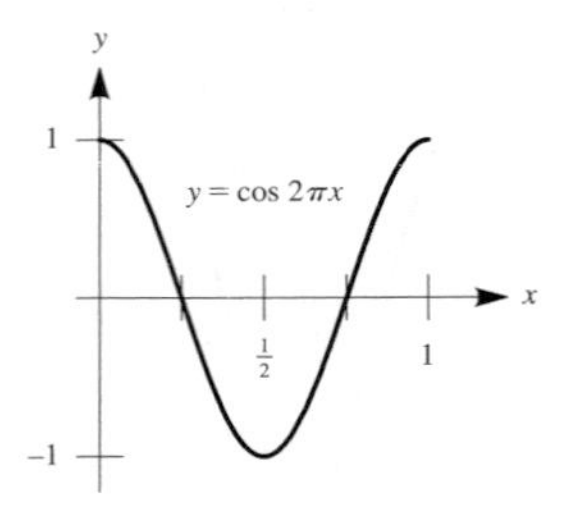

(b) amplitude: 4; period: 1;
x-intercepts: $\frac{1}{4}, \frac{3}{4}$; increasing: $(0, \frac{1}{2})$

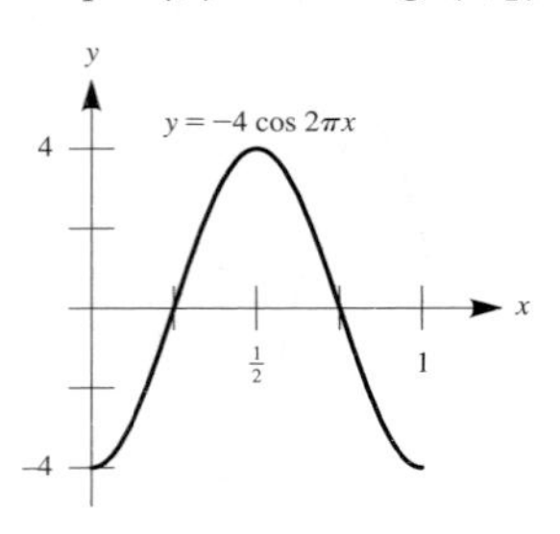

9. $y = \sin x$: amplitude $= 1$, period $= 2\pi$; $y = 2 \sin x$: amplitude $= 2$, period $= 2\pi$; $y = 3 \sin x$: amplitude $= 3$, period $= 2\pi$; $y = 4 \sin x$: amplitude $= 4$, period $= 2\pi$

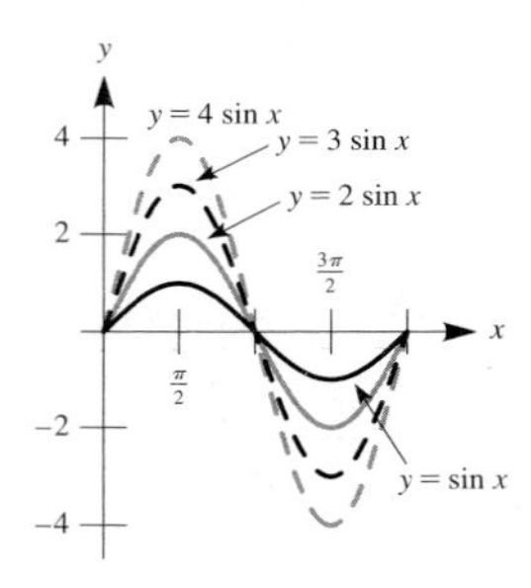

11. (a) $y = 2 \sin \pi x$: amplitude $= 2$, period $= 2$; $y = \sin 2\pi x$: amplitude $= 1$, period $= 1$
(b)

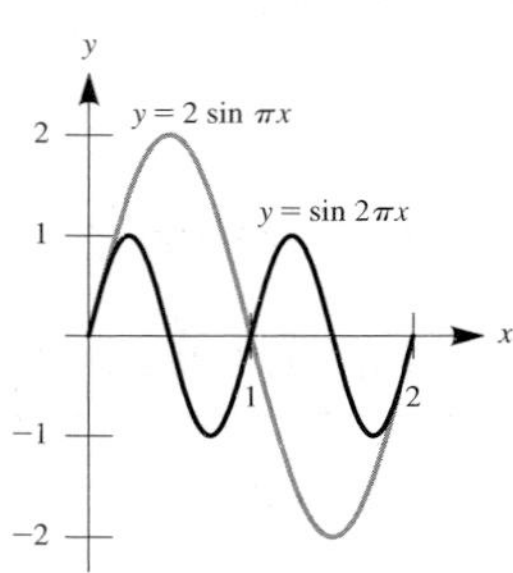

13. amplitude $= 1$; period $= 4\pi$; no x-intercepts; increasing on $(0, \pi)$ and $(3\pi, 4\pi)$

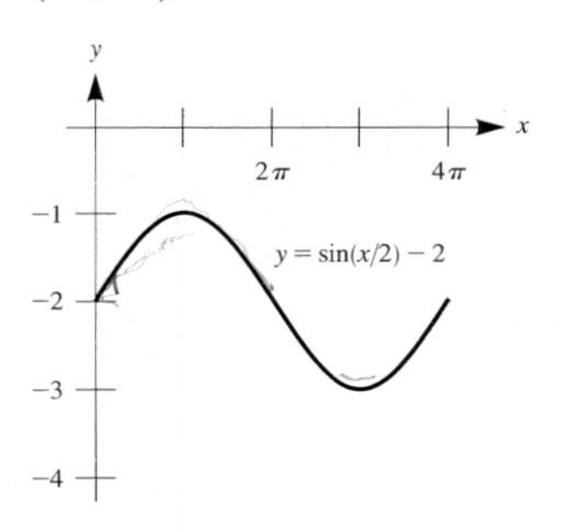

15. amplitude $= 2$; period $= \frac{2}{3}$;
x-intercept $= \frac{1}{3}$; increasing on $(0, \frac{1}{3})$

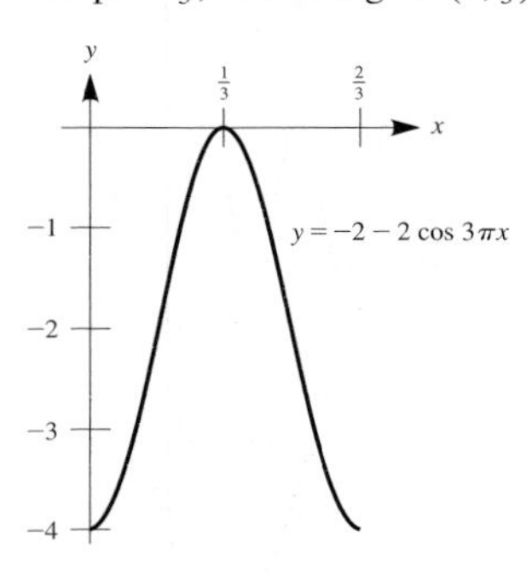

17. amplitude $= 1$; period $= 2\pi$;
phase shift $= -\frac{\pi}{3}$; x-intercepts $= \frac{\pi}{6}, \frac{7\pi}{6}$;
high points $= (-\frac{\pi}{3}, 1), (\frac{5\pi}{3}, 1)$;
low point $= (\frac{2\pi}{3}, -1)$

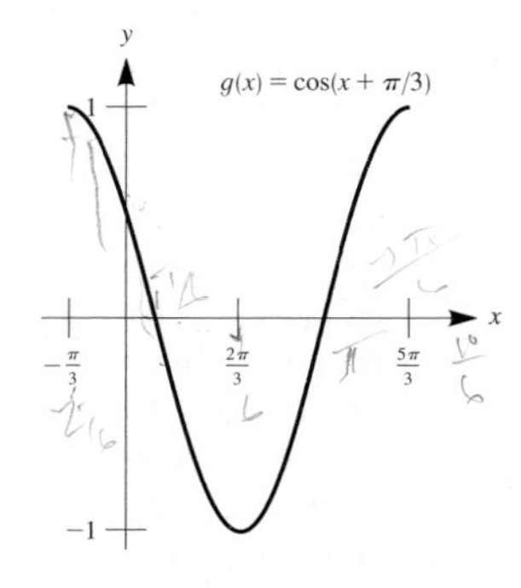

19. amplitude = 1; period = 2π; phase shift = -2; x-intercepts = $-2, \pi - 2, 2\pi - 2$; high point = $(\frac{3\pi}{2} - 2, 1)$; low point = $(\frac{\pi}{2} - 2, -1)$

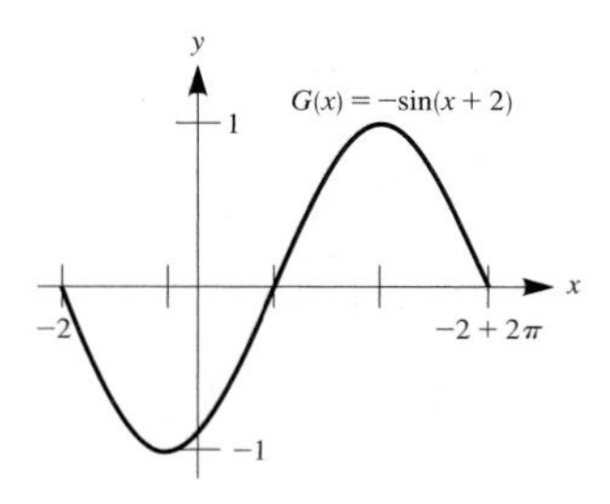

21. amplitude = 1; period = $\frac{2\pi}{3}$; phase shift = $-\frac{\pi}{6}$; x-intercepts = $-\frac{\pi}{6}, \frac{\pi}{6}, \frac{\pi}{2}$; high point = $(0, 1)$; low point = $(\frac{\pi}{3}, -1)$

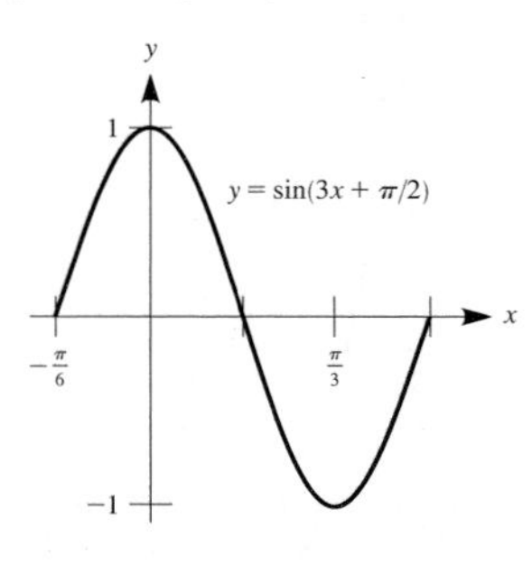

23. amplitude = 1; period = 2π; phase shift = $\frac{\pi}{2}$; x-intercepts = $\pi, 2\pi$; high points = $(\frac{\pi}{2}, 1), (\frac{5\pi}{2}, 1)$ low point = $(\frac{3\pi}{2}, -1)$

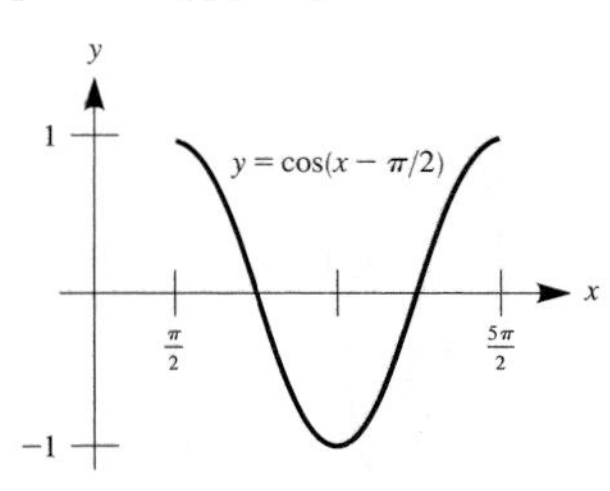

25. amplitude = 2; period = 2; phase shift = -1; x-intercepts = $-1, 0, 1$; high point = $(\frac{1}{2}, 2)$; low point = $(-\frac{1}{2}, -2)$

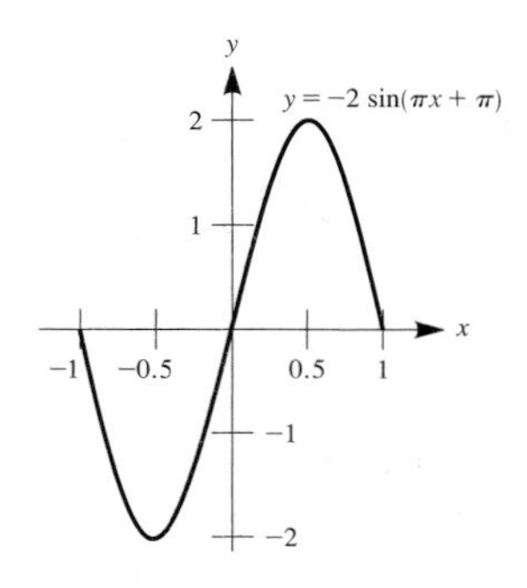

27. amplitude = 1; period = 2π; phase shift = -1; x-intercepts = $\frac{\pi}{2} - 1, \frac{3\pi}{2} - 1$; high points = $(-1, 1), (2\pi - 1, 1)$; low point = $(\pi - 1, -1)$

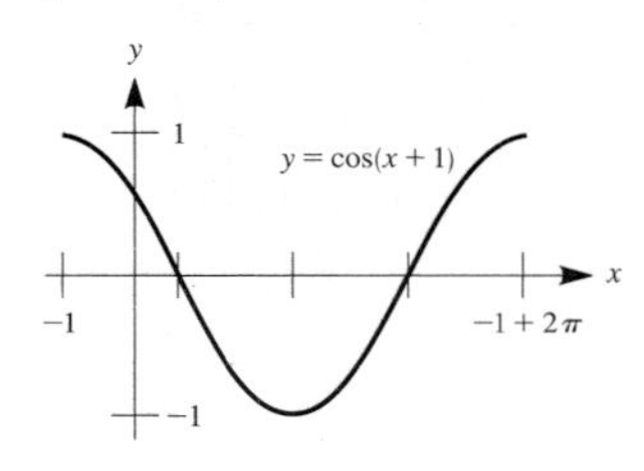

29. amplitude = 1; period = π; phase shift = $\frac{\pi}{6}$; x-intercept = $\frac{2\pi}{3}$; high points = $(\frac{\pi}{6}, 2), (\frac{7\pi}{6}, 2)$ low point = $(\frac{2\pi}{3}, 0)$

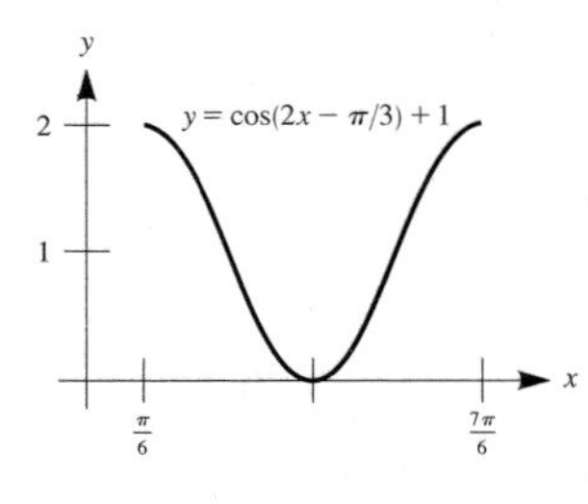

31. amplitude = 3; period = 3π; phase shift = $-\frac{\pi}{4}$; x-intercepts = $\frac{\pi}{2}, 2\pi$; high points = $(-\frac{\pi}{4}, 3), (\frac{11\pi}{4}, 3)$; low point = $(\frac{5\pi}{4}, -3)$

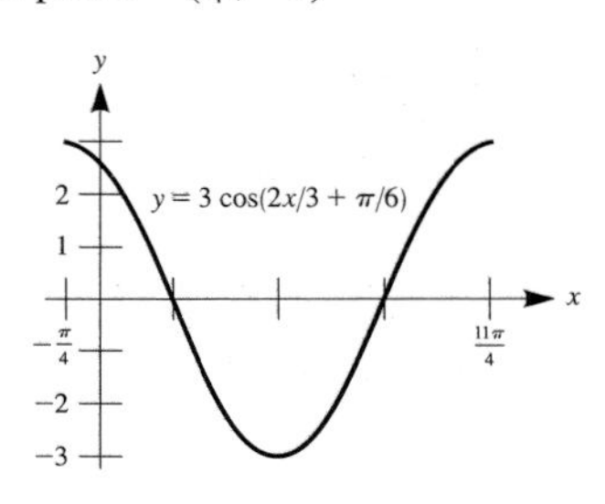

33. (a) amplitude: 2.5; period: $\frac{2}{3}$; phase shift: $-\frac{4}{3\pi}$
(b)

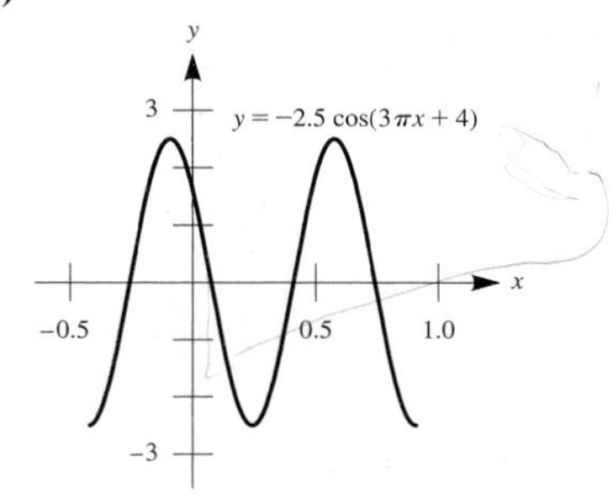

(c) high points: $(-0.09, 2.5)$, $(0.58, 2.5)$; low points: $(-0.42, -2.5)$, $(0.24, -2.5)$, $(0.91, -2.5)$
(d) high points: $(-\frac{4}{3\pi} + \frac{1}{3}, 2.5)$, $(-\frac{4}{3\pi} + 1, 2.5)$; low points: $(-\frac{4}{3\pi}, -2.5)$, $(-\frac{4}{3\pi} + \frac{2}{3}, -2.5)$, $(-\frac{4}{3\pi} + \frac{4}{3}, -2.5)$

35. (a) amplitude: 2.5; period: 6; phase shift: $-\frac{12}{\pi}$
(b)

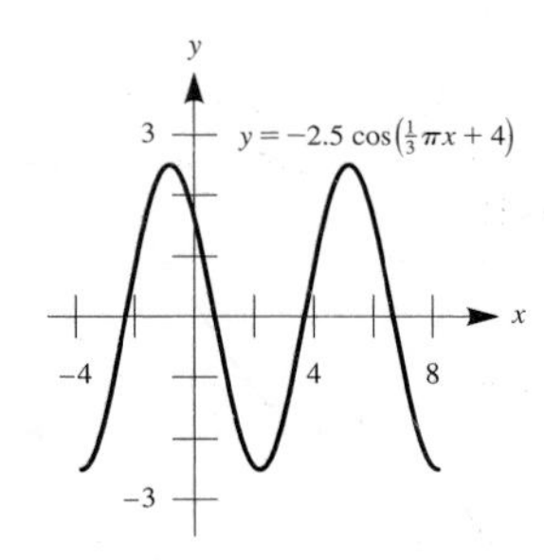

(c) high points: $(-0.82, 2.5)$, $(5.18, 2.5)$; low points: $(-3.82, -2.5)$, $(2.18, -2.5)$, $(8.18, -2.5)$
(d) high points: $(-\frac{12}{\pi} + 3, 2.5)$, $(-\frac{12}{\pi} + 9, 2.5)$; low points: $(-\frac{12}{\pi}, -2.5)$, $(-\frac{12}{\pi} + 6, -2.5)$, $(-\frac{12}{\pi} + 12, -2.5)$

37. (a) amplitude: 1; period: 4π; phase shift: -1.5
(b)

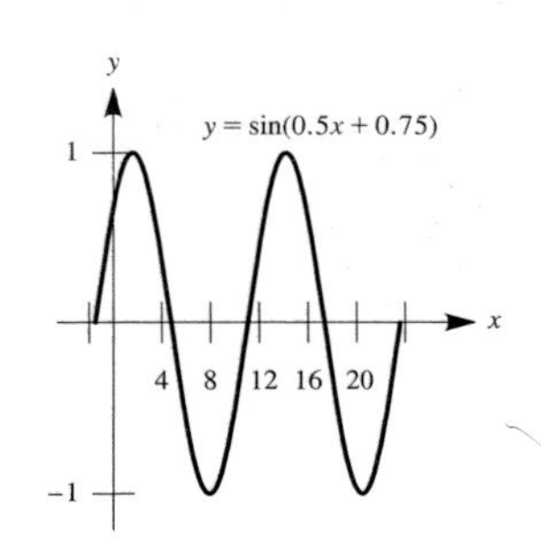

(c) high points: $(1.64, 1)$, $(14.21, 1)$; low points: $(7.92, -1)$, $(20.49, -1)$

(d) high points: $(-1.5 + \pi, 1)$, $(-1.5 + 5\pi, 1)$; low points: $(-1.5 + 3\pi, -1)$, $(-1.5 + 7\pi, -1)$
39. (a) amplitude: 0.02; period: 200; phase shift: 400
(b)

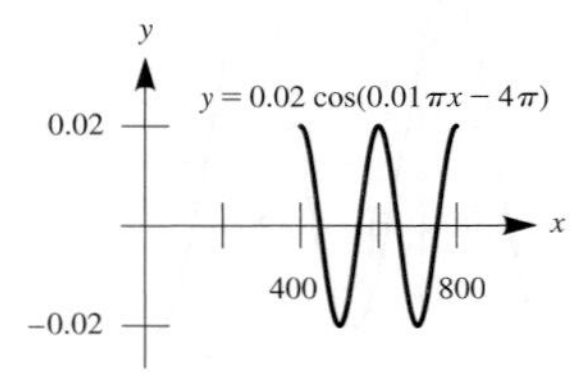

(c) high points: (400, 0.02), (600, 0.02), (800, 0.02); low points: (500, −0.02), (700, −0.02) **(d)** high points: (400, 0.02), (600, 0.02), (800, 0.02); low points: (500, −0.02), (700, −0.02)
41. $y = 1.5 \sin \frac{3}{2}x$ **43.** $y = \cos \frac{2\pi}{5}x$
45. $y = \pi \cos \frac{\pi}{4}x$
47. $y = 25.25 \sin(\frac{\pi}{6}t - \frac{2\pi}{3}) + 43.25$
49. $y = 6.3 \sin(\frac{\pi}{6}t + \frac{\pi}{4}) + 75.215$
51. $y = 25 \sin(\frac{\pi}{6}t - \frac{2\pi}{3}) + 90$
53.

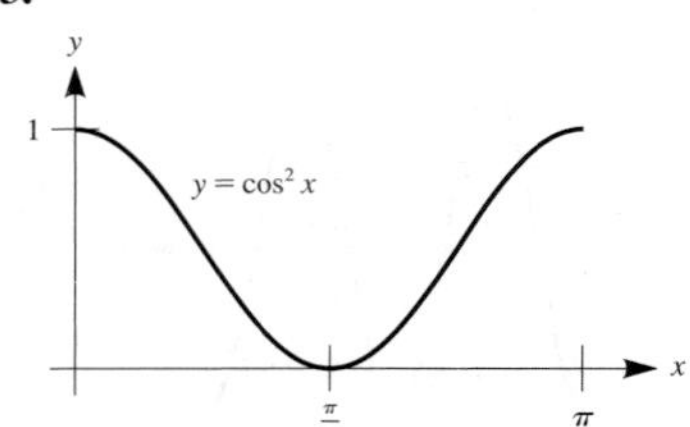

55.

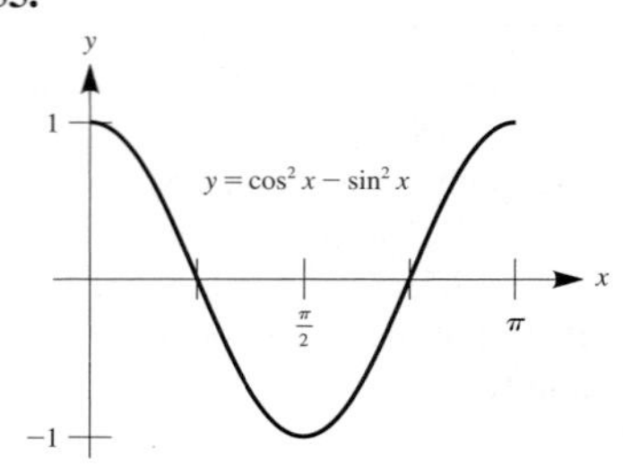

57. $G[F(x)]$ is the only one.

59. (a)

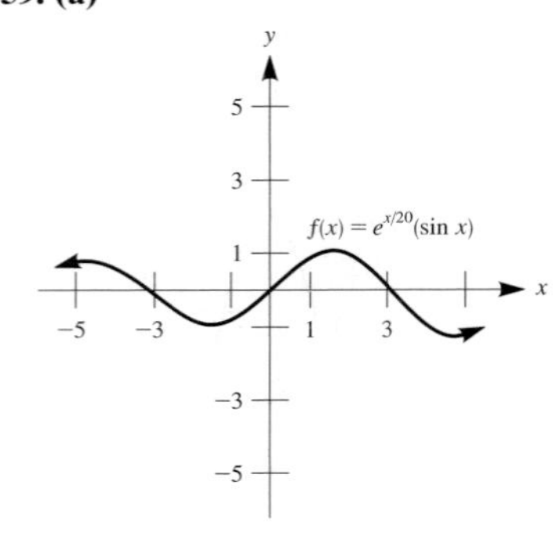

(b) The function is not periodic.

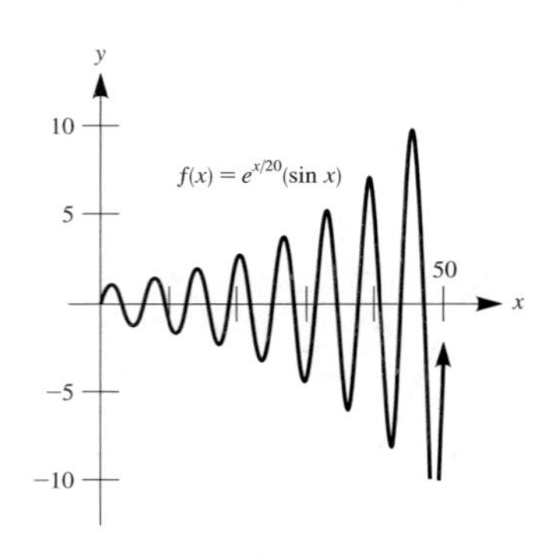

(c) The graph of $y = e^{x/20}$ touches near the top of each cycle of the curve, while the graph of $y = -e^{x/20}$ touches near the bottom of each cycle.

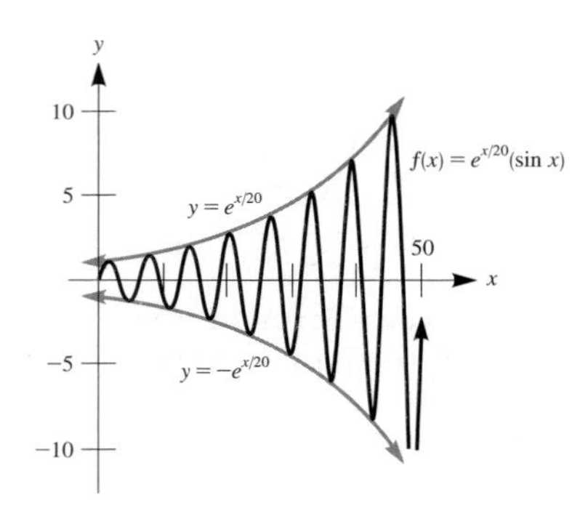

Exercise Set 7.4

1. (a) $t = 0$: $s = 4$ cm; $t = 0.5$: $s \approx 2.83$ cm; $t = 1$: $s = 0$ cm; $t = 2$: $s = -4$ cm **(b)** amplitude: 4 cm; period: 4 sec; frequency: $\frac{1}{4}$ cps

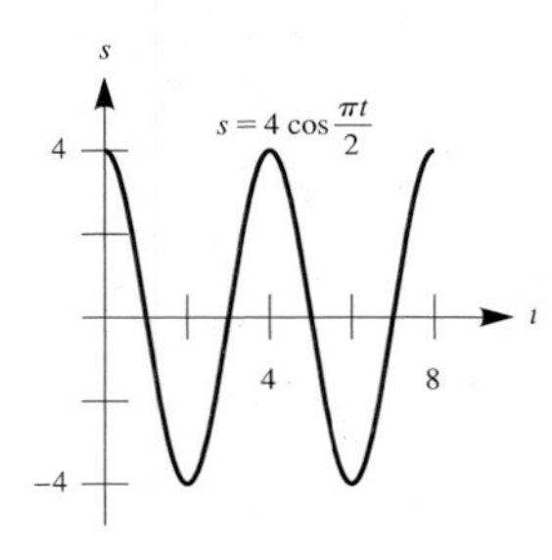

(c) $t = 0, t = 2, t = 4, t = 6$, and $t = 8$ sec **(d)** $t = 1, t = 3, t = 5$, and $t = 7$ sec **(e)** $2 < t < 4$ and $6 < t < 8$
3. (a) amplitude: 3 ft; period: 6 sec; frequency: $\frac{1}{6}$ cps

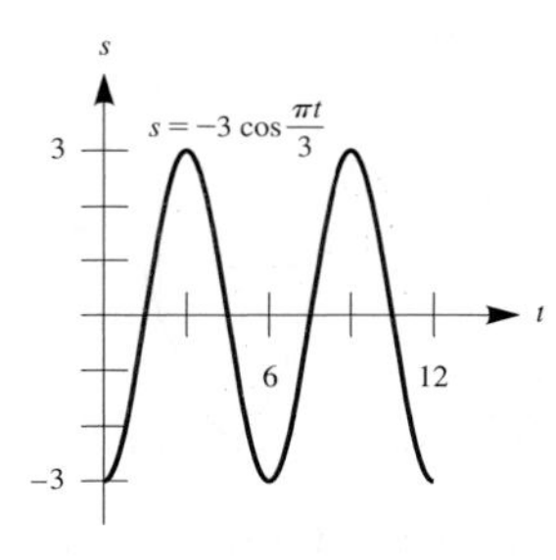

(b) $0 < t < 3$ and $6 < t < 9$
(c) $3 < t < 6$ and $9 < t < 12$
(d)

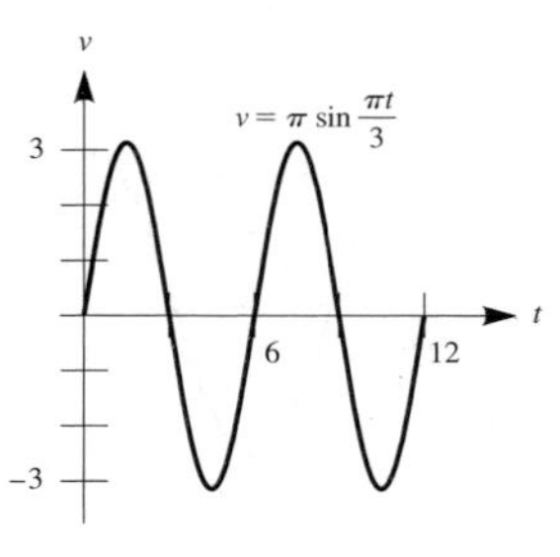

(e) $t = 0, t = 3, t = 6, t = 9$, and $t = 12$ sec; the mass (s-coordinate) is at −3, 3, −3, 3, and −3 ft, respectively. **(f)** $t = 1.5$ and $t = 7.5$ sec; the mass is at 0 ft. **(g)** $t = 4.5$ and $t = 10.5$ sec; the mass is at 0 ft.

(h)

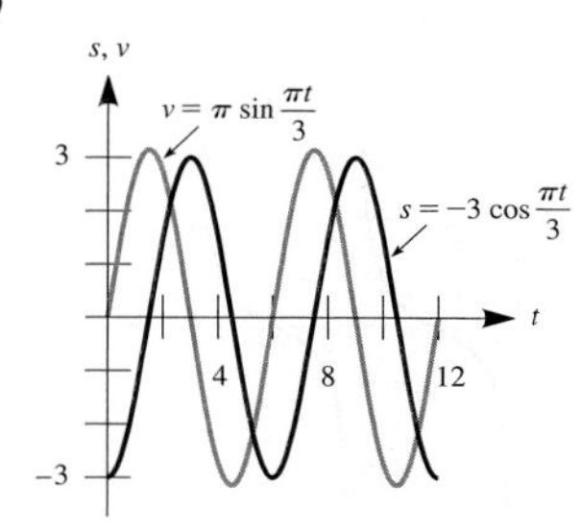

5. (a) amplitude: 170 volts; frequency: 60 cps
(b)

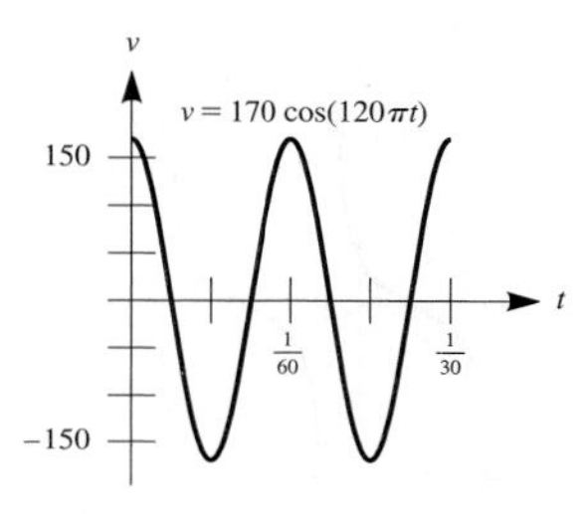

(c) $t = \frac{1}{60}$ sec and $t = \frac{1}{30}$ sec
7. (a)

t (sec)	0	1	2	3
θ (radians)	0	$\pi/3$	$2\pi/3$	π

t (sec)	4	5	6	7
θ (radians)	$4\pi/3$	$5\pi/3$	2π	$7\pi/3$

(b) $1, \frac{1}{2}, -\frac{1}{2}, -1, -\frac{1}{2}, \frac{1}{2}, 1$, and $\frac{1}{2}$
(c)

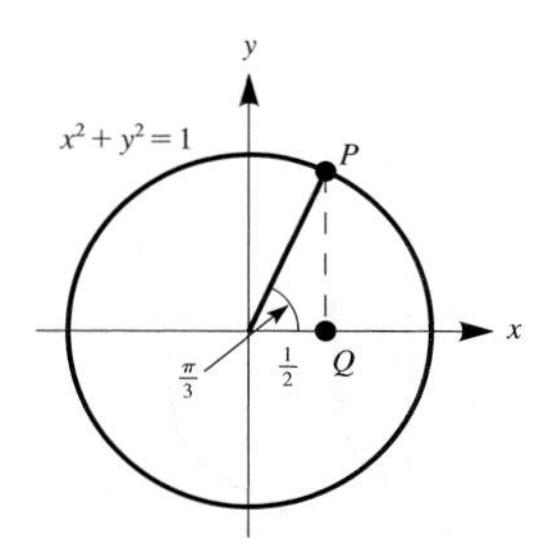

(d) $t = 2$ sec

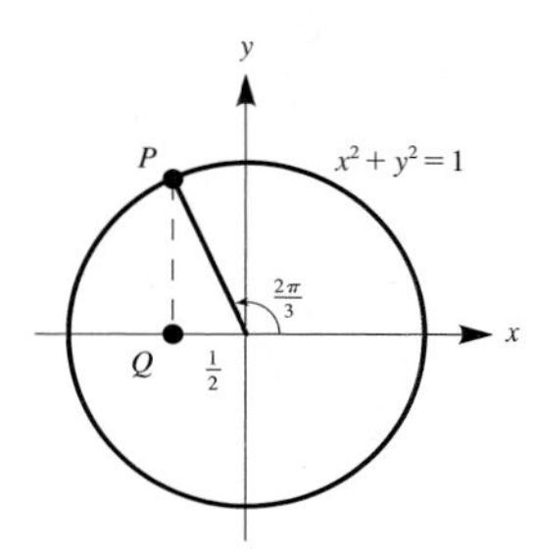

$t = 3$ sec

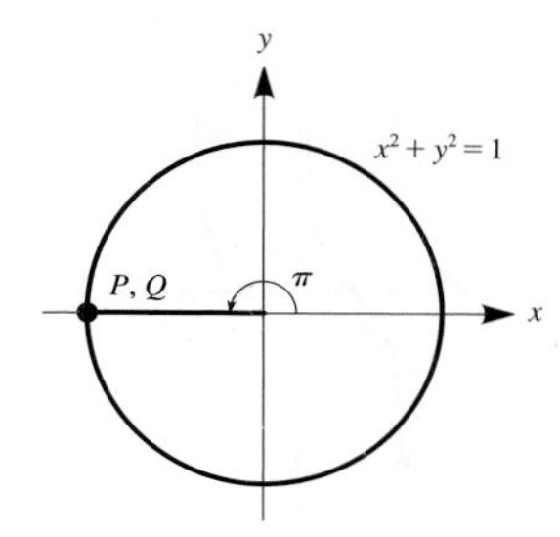

$t = 4$ sec

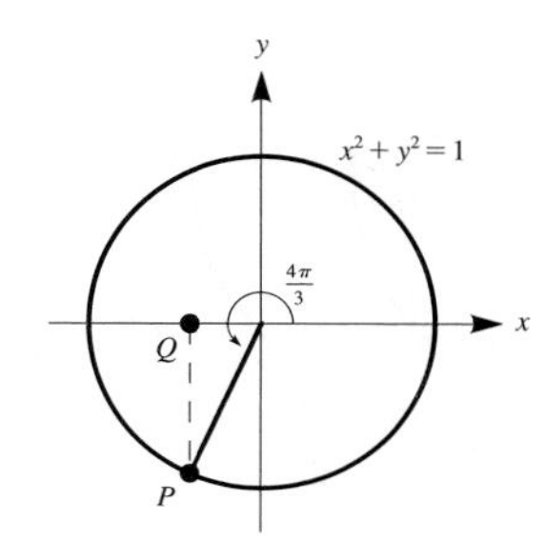

(e) The x-coordinate of Q is the same as the x-coordinate of P, which is $\cos \theta$.
(f) amplitude: 1; period: 6; frequency: $\frac{1}{6}$

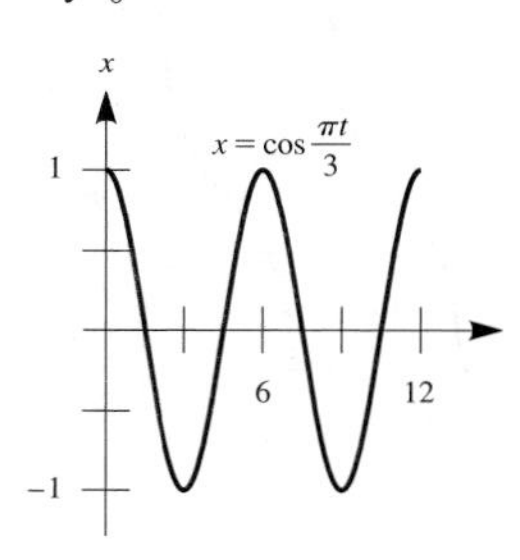

(g)

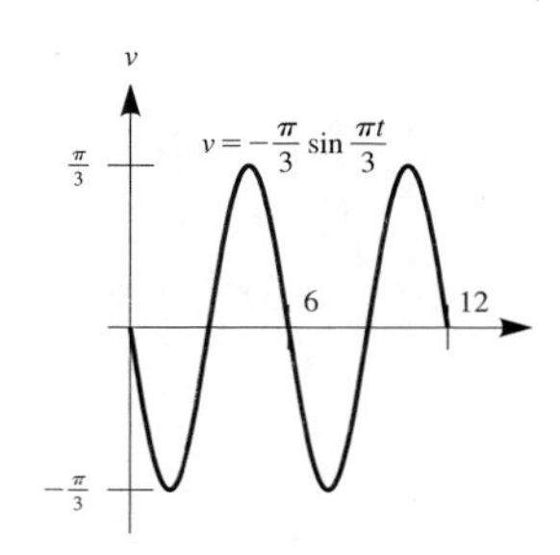

(h) $t = 0, 3, 6, 9$ and 12 sec; x-coordinates: $1, -1, 1, -1$ and 1, respectively **(i)** $t = 4.5$ sec and $t = 10.5$ sec; Q is located at the origin. **(j)** $t = 1.5$ sec and $t = 7.5$ sec; Q is located at the origin.

Exercise Set 7.5

1. (a) x-intercept: $-\frac{\pi}{4}$; y-intercept: 1; asymptotes: $x = -\frac{3\pi}{4}$ and $x = \frac{\pi}{4}$

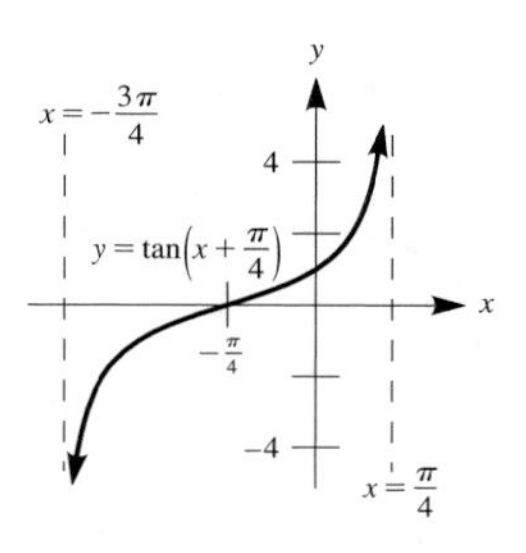

(b) x-intercept: $-\frac{\pi}{4}$; y-intercept: -1; asymptotes: $x = -\frac{3\pi}{4}$ and $x = \frac{\pi}{4}$

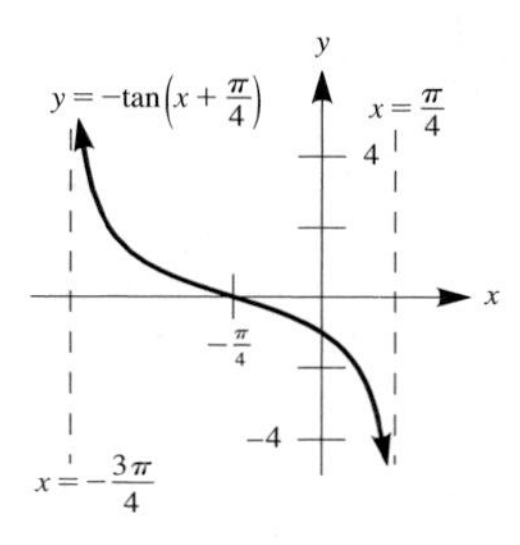

3. (a) x- and y-intercepts: 0; asymptotes: $x = -\frac{3\pi}{2}$ and $x = \frac{3\pi}{2}$

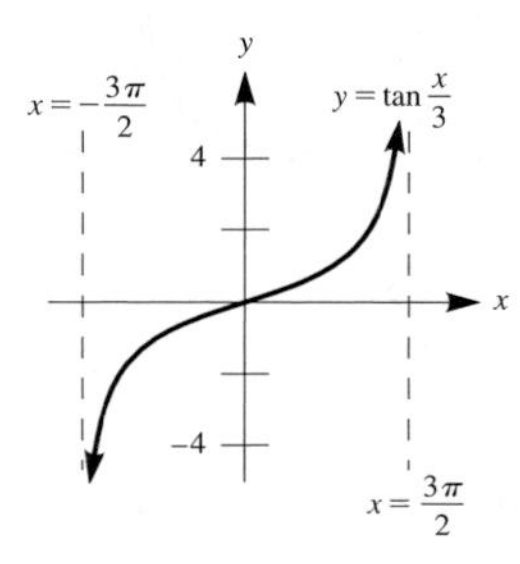

(b) x- and y-intercepts: 0; asymptotes: $x = -\frac{3\pi}{2}$ and $x = \frac{3\pi}{2}$

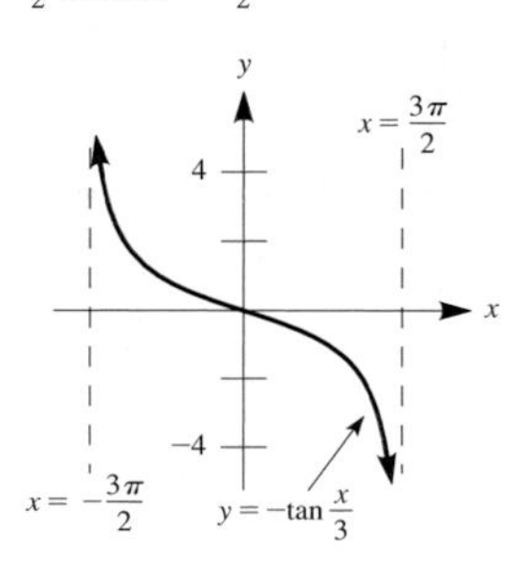

5. x- and y-intercepts: 0; asymptotes: $x = -1$ and $x = 1$

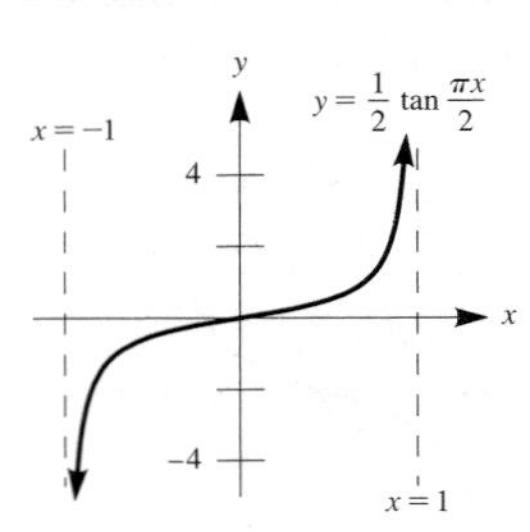

7. x-intercept: 1; no y-intercept; asymptotes: $x = 0$ and $x = 2$

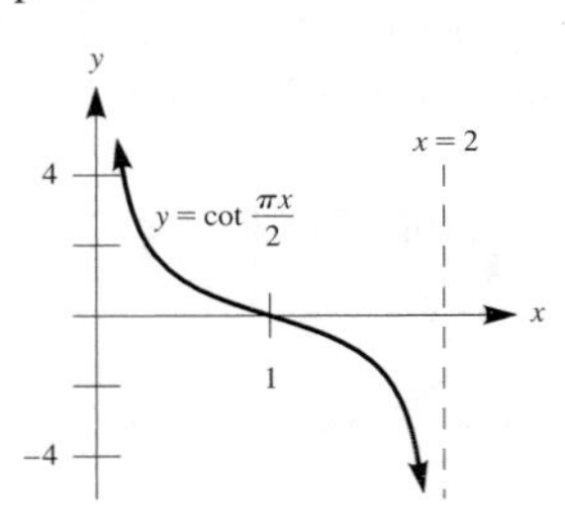

9. x-intercept: $\frac{3\pi}{4}$; y-intercept: 1; asymptotes: $x = \frac{\pi}{4}$ and $x = \frac{5\pi}{4}$

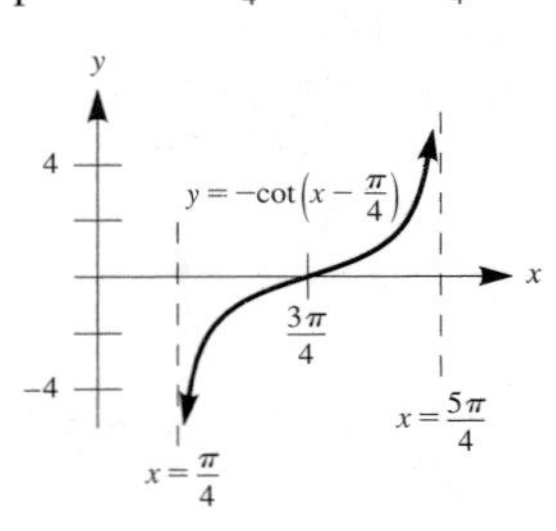

11. x-intercept: $\frac{\pi}{4}$; no y-intercept; asymptotes: $x = 0$ and $x = \frac{\pi}{2}$

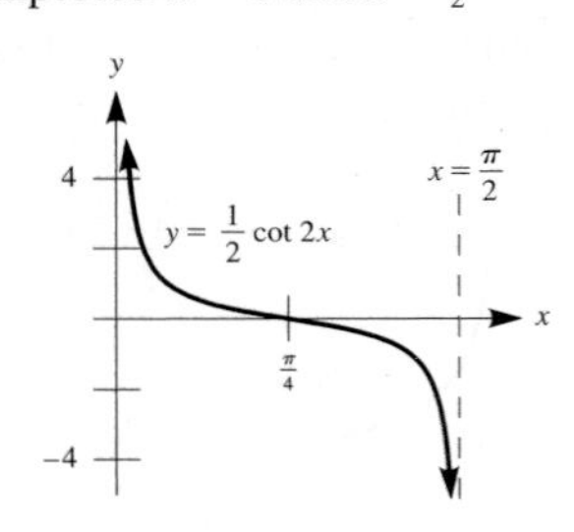

13. x-intercepts: $-\pi, 0, \pi$

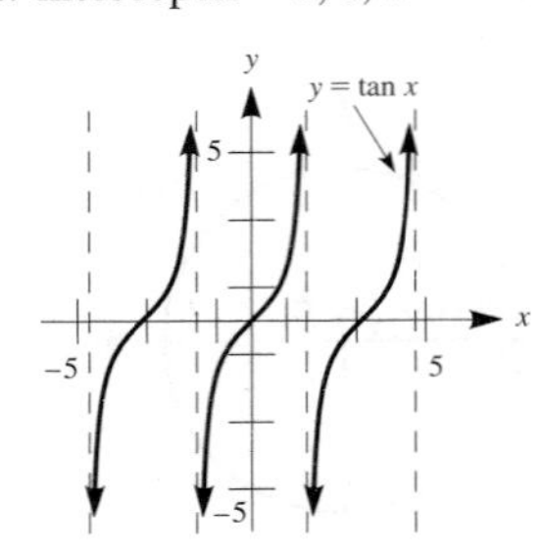

15. (a)

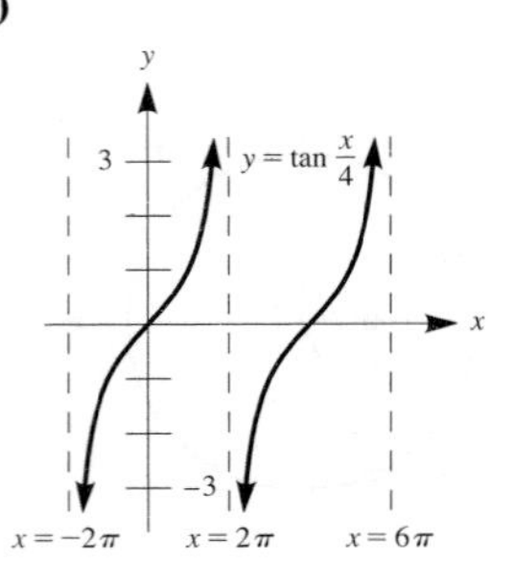

(b)

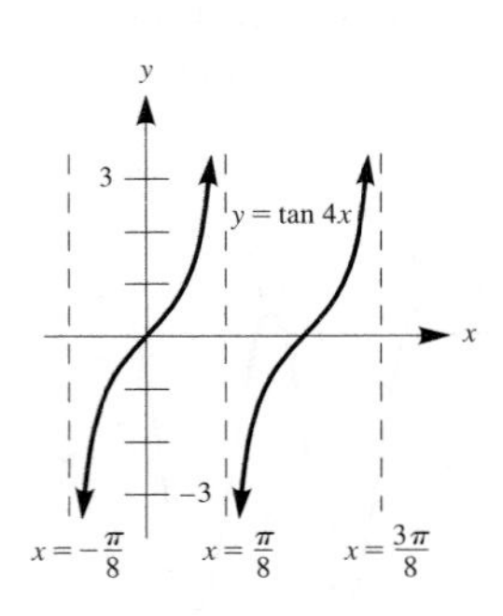

17. (a)

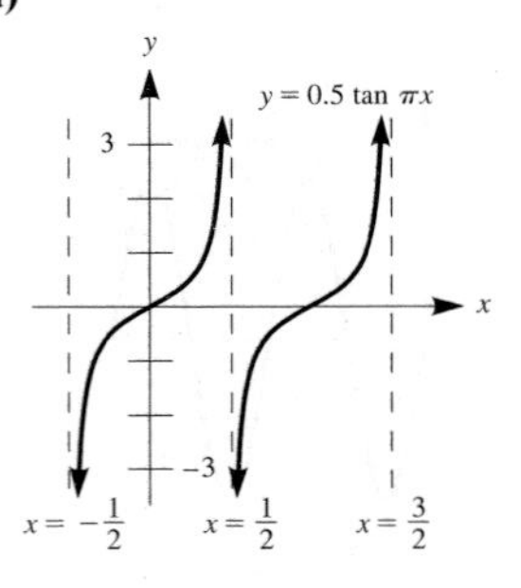

(b)

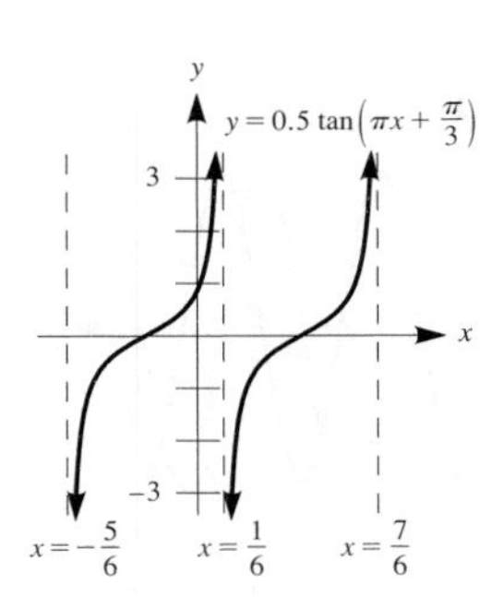

(c)

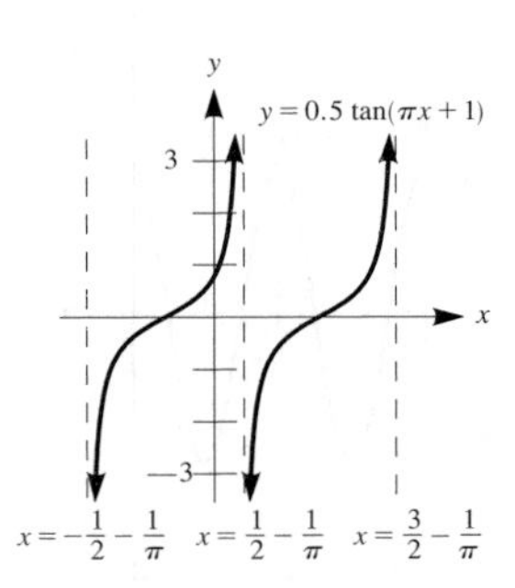

19. (a)

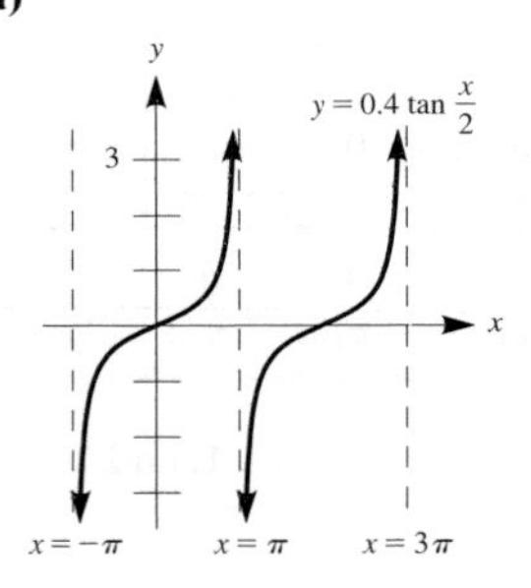

(b)

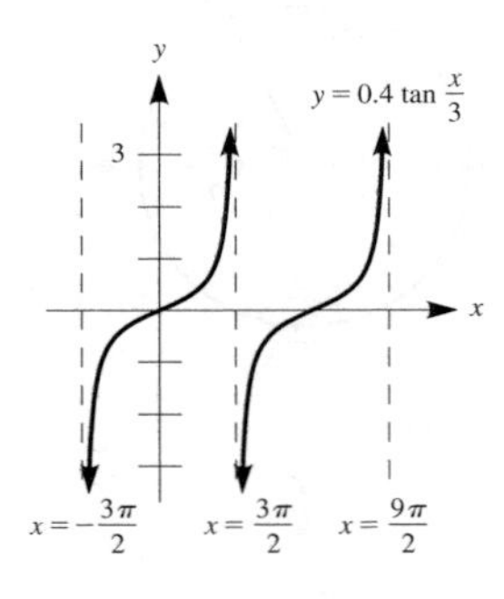

(c)

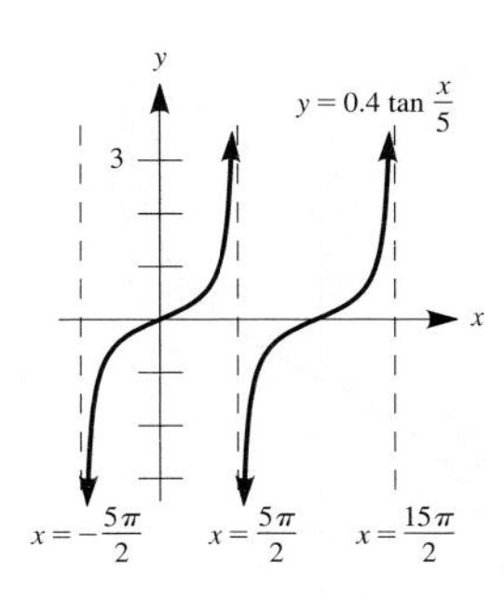

21. no x-intercept; y-intercept: $-\sqrt{2}$; asymptotes: $x = -\frac{3\pi}{4}$, $x = \frac{\pi}{4}$, and $x = \frac{5\pi}{4}$

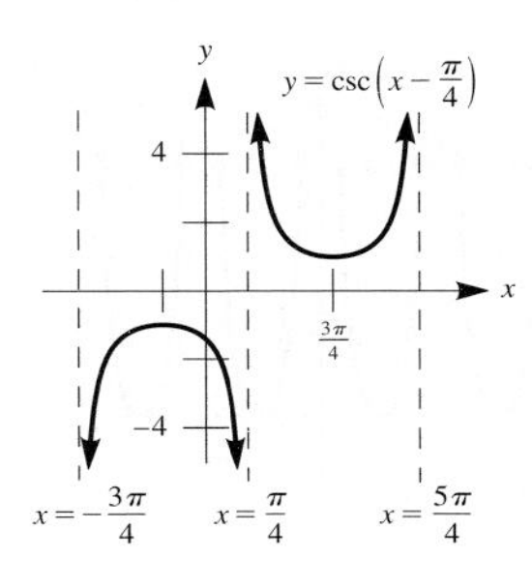

23. no x- or y-intercepts; asymptotes: $x = -2\pi$, $x = 0$, and $x = 2\pi$

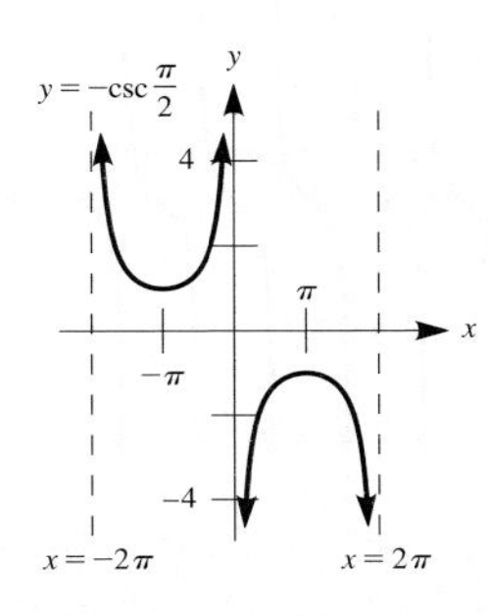

25. no x- or y-intercepts; asymptotes: $x = -1$, $x = 0$, and $x = 1$

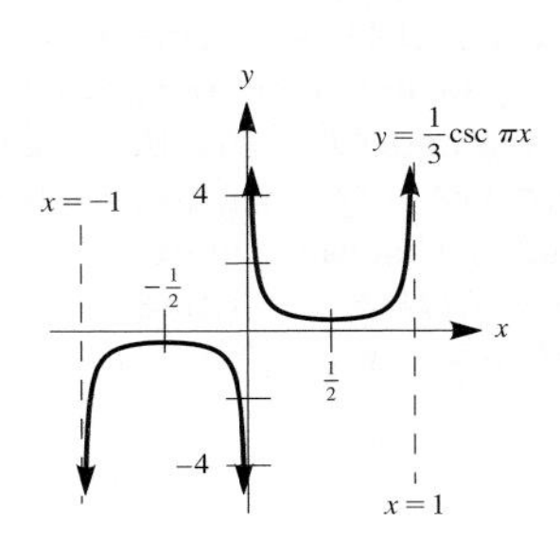

27. no x-intercept; y-intercept: -1; asymptotes: $x = -\frac{\pi}{2}$, $x = \frac{\pi}{2}$, and $x = \frac{3\pi}{2}$

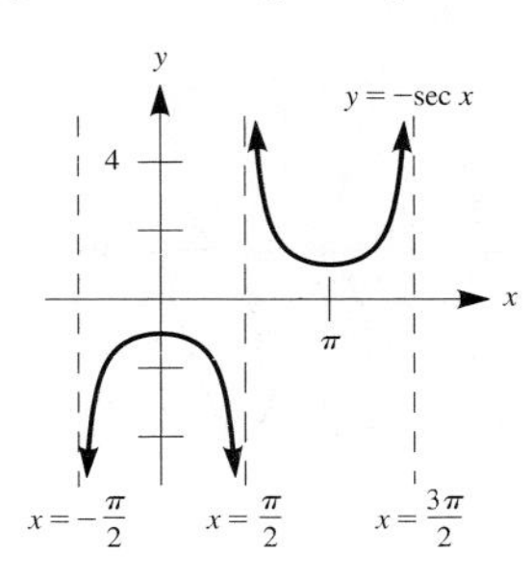

29. no x-intercept; y-intercept: -1; asymptotes: $x = \frac{\pi}{2}$, $x = \frac{3\pi}{2}$, and $x = \frac{5\pi}{2}$

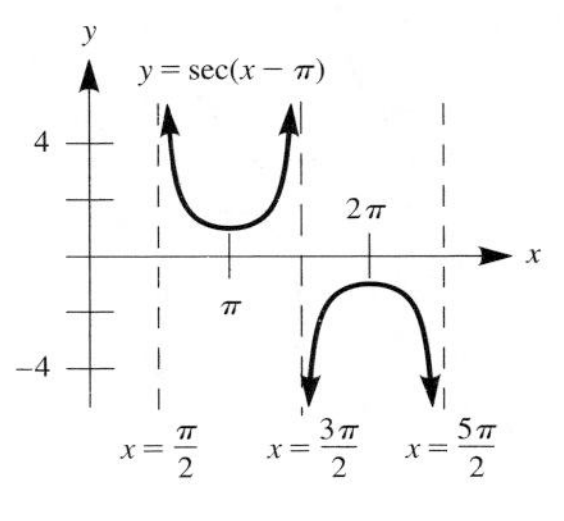

31. no x-intercept; y-intercept: 3; asymptotes: $x = -1$, $x = 1$, and $x = 3$

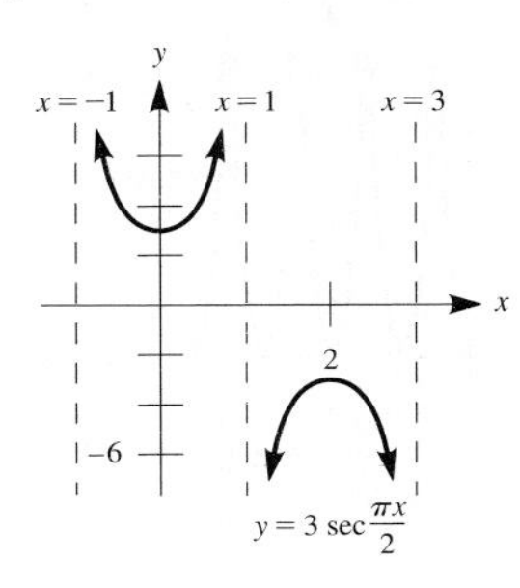

33. $\sin x = \csc x$ at odd multiples of $\frac{\pi}{2}$, such as $-\frac{3\pi}{2}$, $-\frac{\pi}{2}$, $\frac{\pi}{2}$, and $\frac{3\pi}{2}$; there are no points at which $\sin x = -\csc x$.

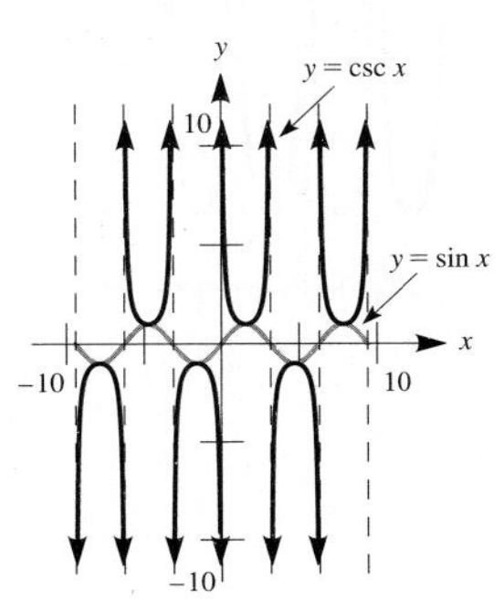

35. (a) x-intercepts: $-\frac{11}{18}$, $-\frac{5}{18}$, $\frac{1}{18}$, $\frac{7}{18}$ and $\frac{13}{18}$; y-intercept: -1; no asymptotes

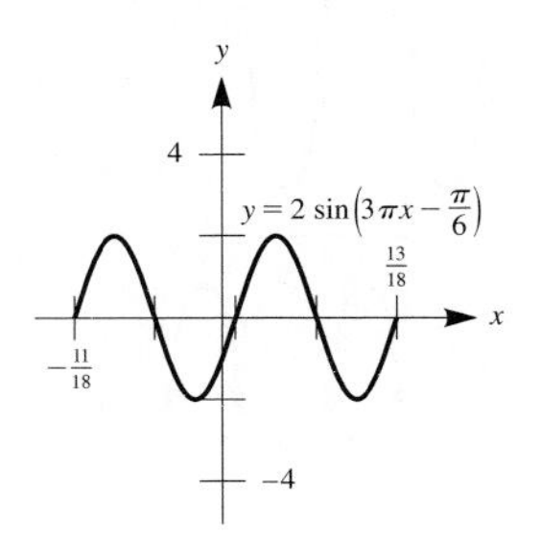

(b) no x-intercepts; y-intercept: -4; asymptotes: $x = -\frac{11}{18}$, $x = -\frac{5}{18}$, $x = \frac{1}{18}$, $x = \frac{7}{18}$, and $x = \frac{13}{18}$

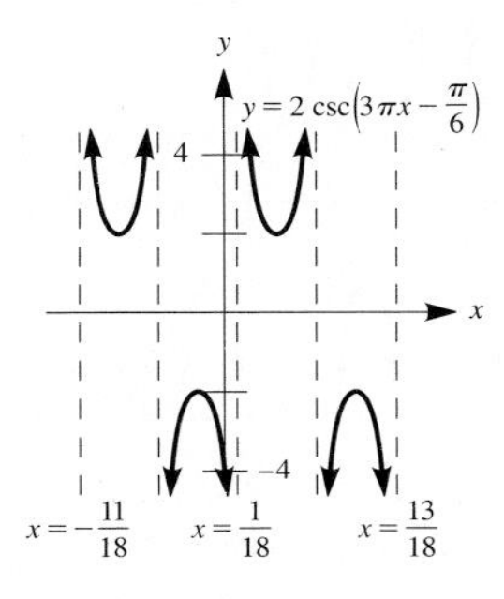

37. (a) x-intercepts: $-\frac{5}{8}$, $-\frac{1}{8}$, $\frac{3}{8}$, and $\frac{7}{8}$; y-intercept: $-3\sqrt{2}/2 \approx -2.12$; no asymptotes

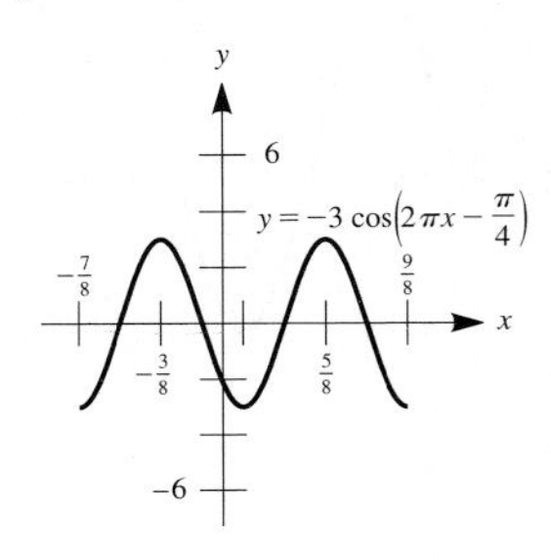

(b) no x-intercepts; y-intercept: $-3\sqrt{2}$; asymptotes: $x = -\frac{5}{8}$, $x = -\frac{1}{8}$, $x = \frac{3}{8}$, and $x = \frac{7}{8}$

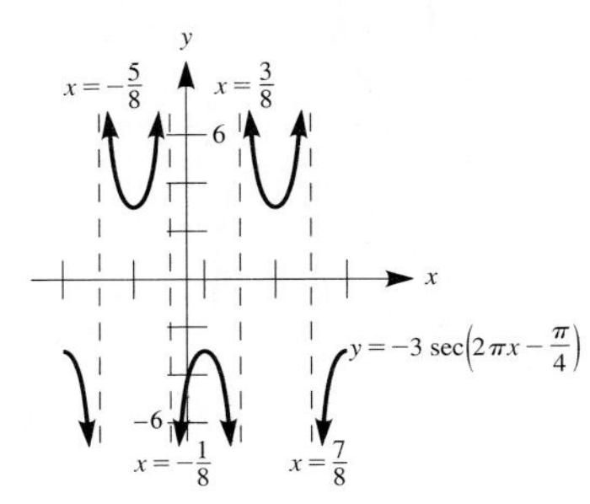

39.

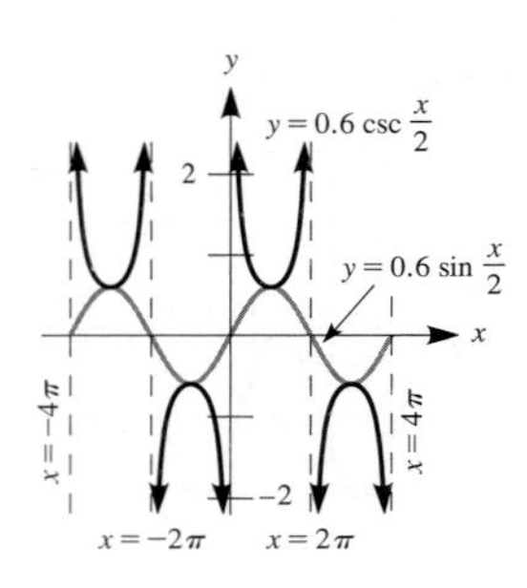

41.

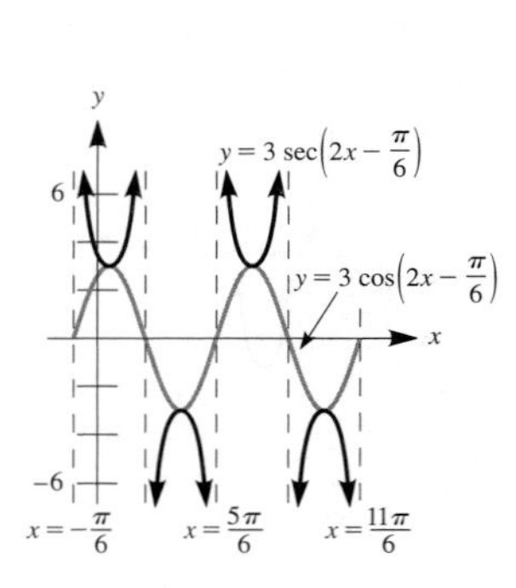

43. (a)

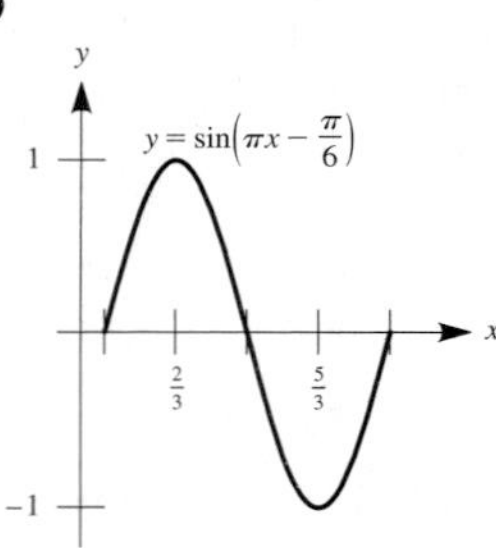

(b)

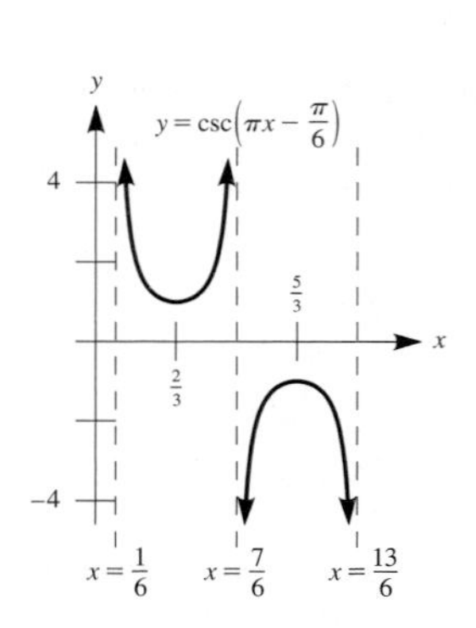

45. (a)

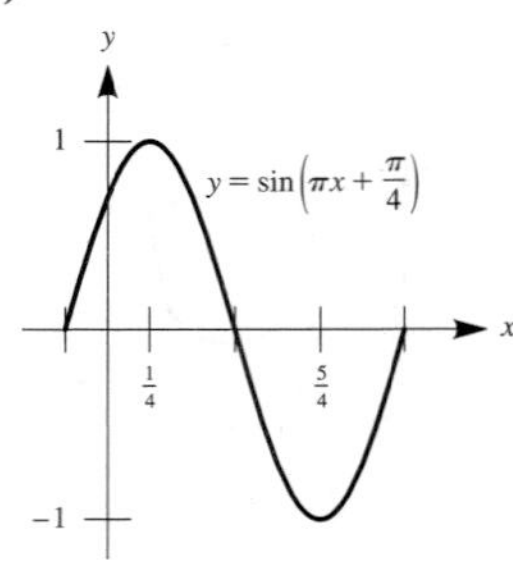

(b)

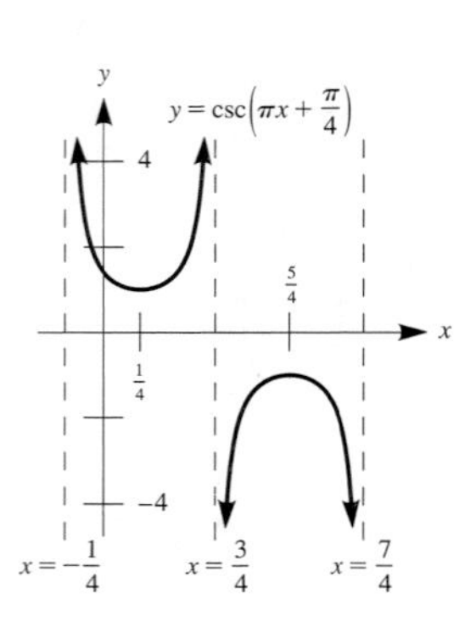

47.

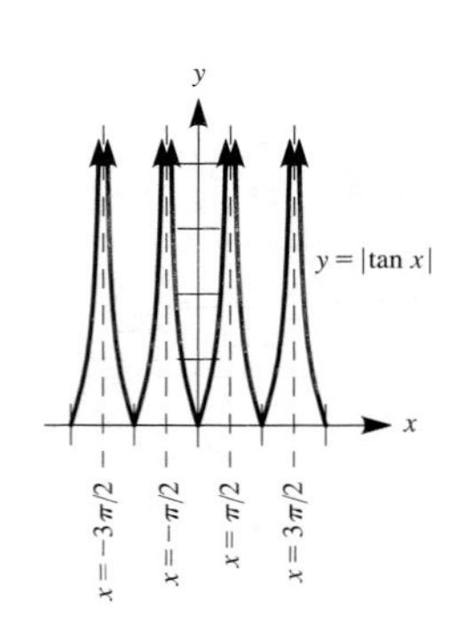

49.

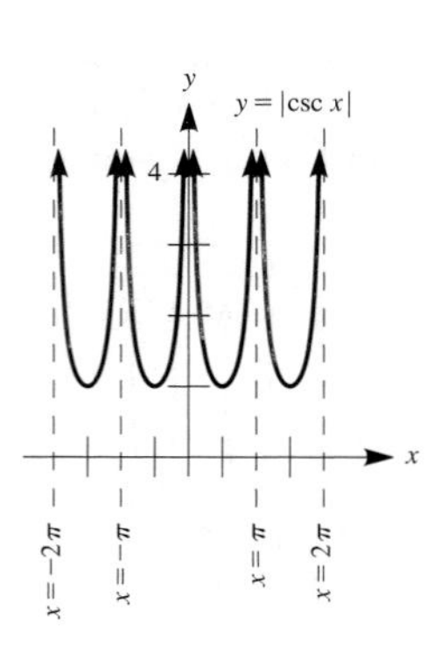

51.

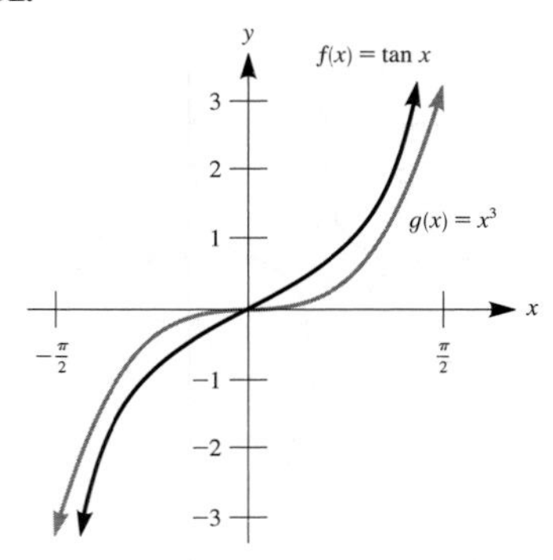

(a) $y = x^3$ **(b)** $y = \tan x$ **(c)** $y = x^3$

53. $\cot^2 x = \csc^2 x - 1$ is an identity.

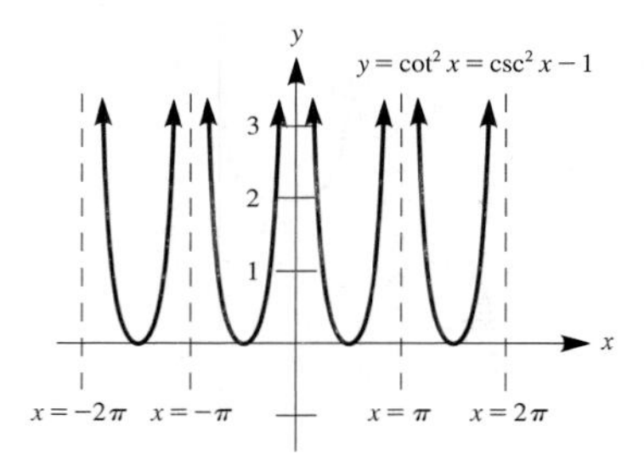

55. (a)

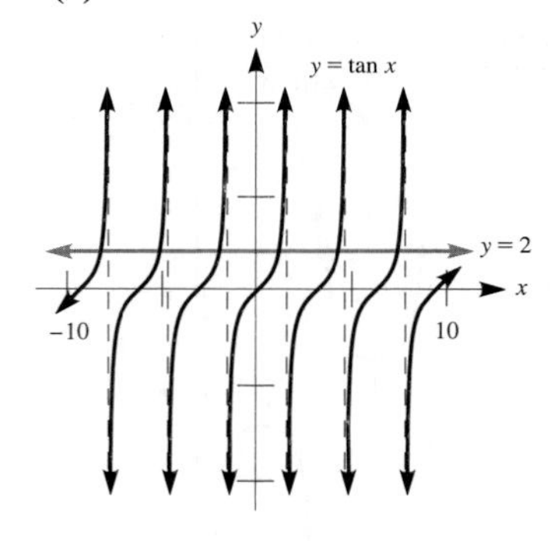

(b) $x \approx 1$ **(c)** $x \approx 1.1071$ **(d)** yes

57. (a) Since P and Q are both points on the unit circle, the coordinates are $P(\cos s, \sin s)$ and $Q(\cos(s - \frac{\pi}{2}), \sin(s - \frac{\pi}{2}))$.

(b) Since $\triangle OAP$ is congruent to $\triangle OBQ$ (labeling the third vertex B), $OA = OB$ and $AP = BQ$. Because the y-coordinate at Q is negative, we have concluded what was required.

59. (b) $\angle COD = \frac{\pi}{2} - s$, and $\angle AOD = \frac{\pi}{2} + s$

Chapter 7 Review Exercises

1. (a) $-\sqrt{3}/2$ **(b)** $-\sqrt{3}$ **3.** $2 \sin t$

5.

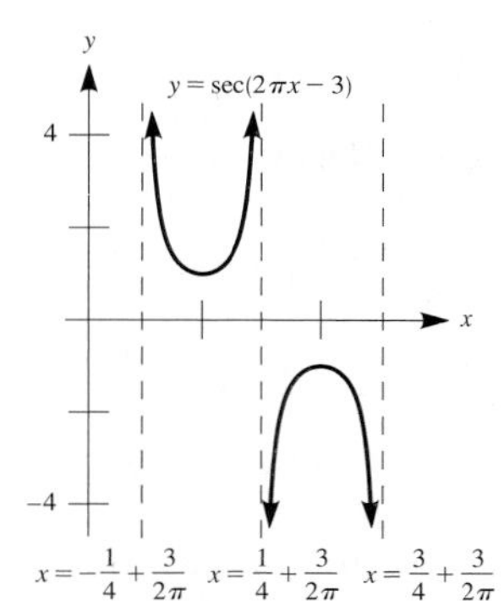

7. amplitude: 1; period: π; phase shift: $\frac{\pi}{2}$

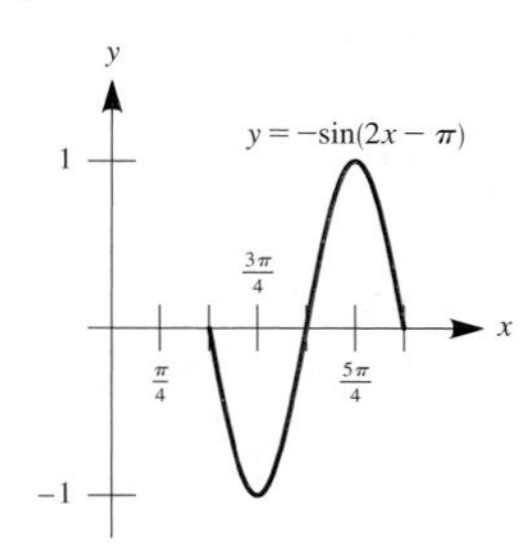

9. -1 **11.** $2\sqrt{3}/3$ **13.** $-\sqrt{3}/3$ **15.** $\frac{1}{2}$ **17.** 1 **19.** -2 **21.** 0.841 **23.** -1 **25.** 0.0123 **31.** $5 \cos \theta$ **33.** $10 \tan \theta$ **35.** $(\sqrt{5}/5)\cos \theta$ **37.** $-\frac{15}{8}$ **39.** $A(\theta) = \pi - \theta - \sin \theta$ **41.** $\frac{1}{2}(\pi - 2)\text{ cm}^2$ **43.** $y = 4 \sin x$ **45.** $y = -2 \cos 4x$ **47.** x-intercepts: $\frac{\pi}{8}$ and $\frac{3\pi}{8}$; high point: $(\frac{\pi}{4}, 3)$; low points: $(0, -3)$ and $(\frac{\pi}{2}, -3)$

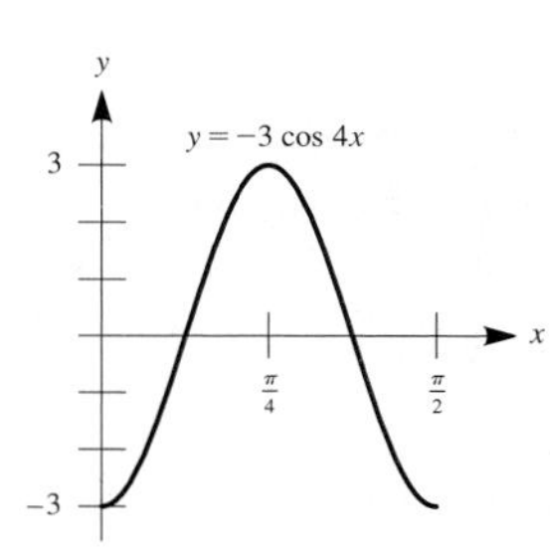

49. x-intercepts: $\frac{1}{2}, \frac{5}{2}$, and $\frac{9}{2}$; high point: $(\frac{3}{2}, 2)$; low point: $(\frac{7}{2}, -2)$

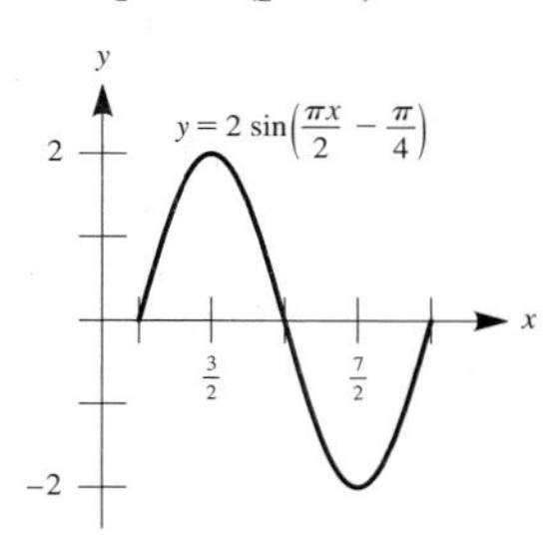

51. x-intercepts: $\frac{5}{2}$ and $\frac{11}{2}$; high points: $(1, 3)$, $(7, 3)$; low point: $(4, -3)$

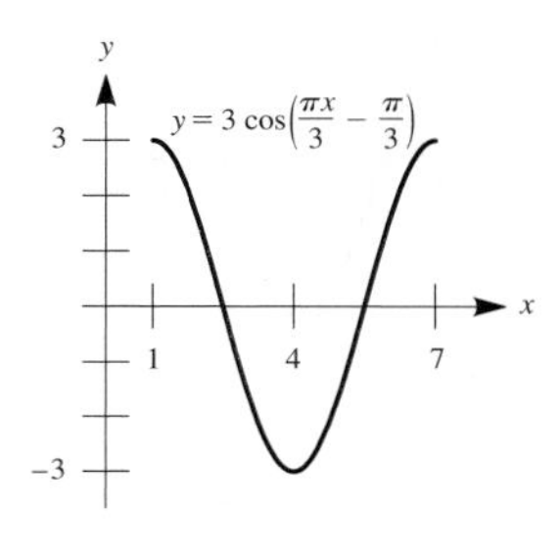

53. (a)

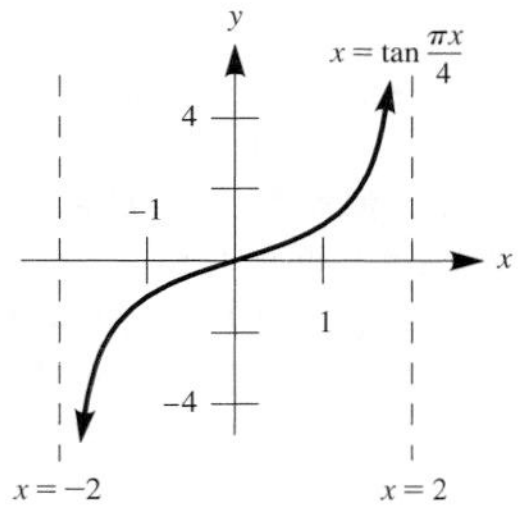

(b)

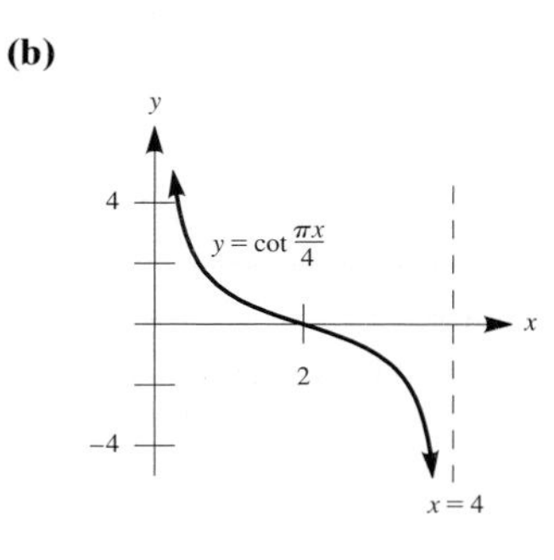

55. (a)

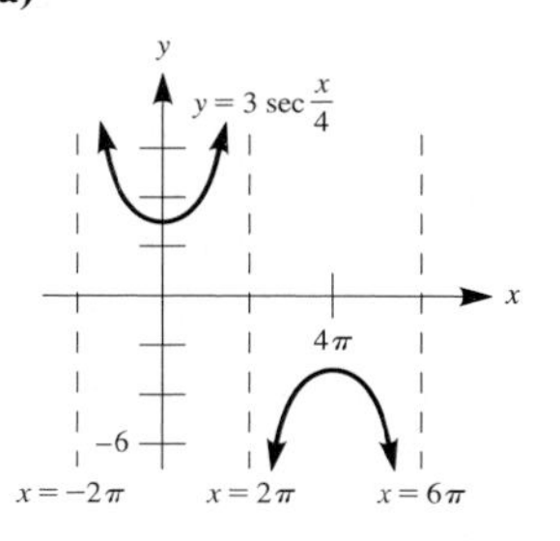

(b)

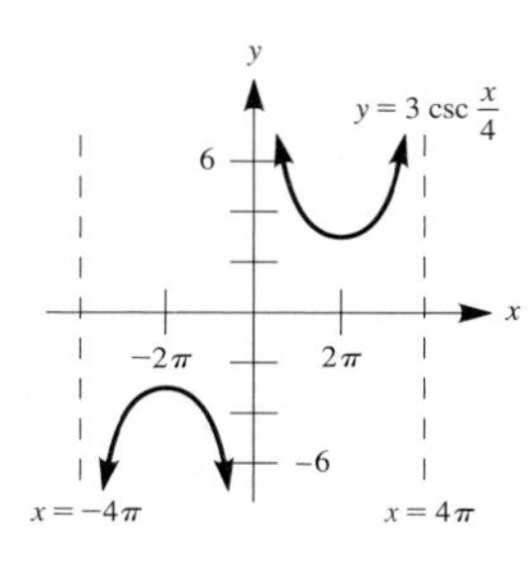

57. (a) amplitude: 2.5 cm; period: 16 sec; frequency: $\frac{1}{16}$ cps

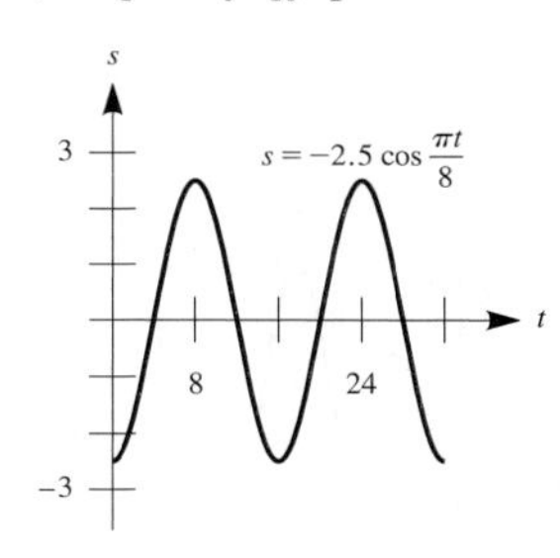

(b) $t = 0$, $t = 8$, $t = 16$, $t = 24$, and $t = 32$ sec **(c)** $t = 4$, $t = 12$, $t = 20$, and $t = 28$ sec

Chapter 7 Test

1. (a) $-\frac{1}{2}$ **(b)** -2 **(c)** 1 **2.** $\frac{1}{4} \csc u$

3.

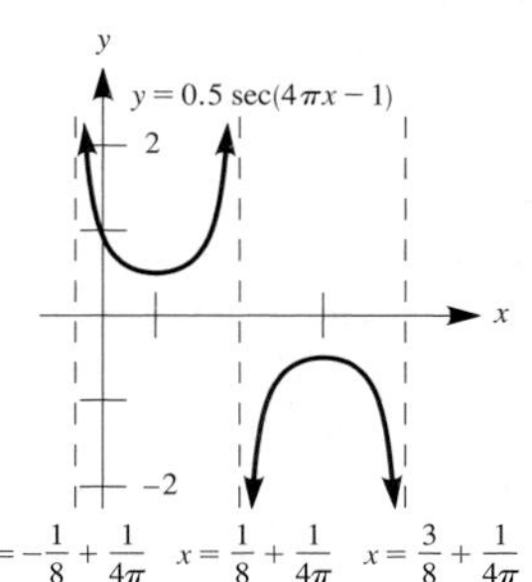

4. amplitude: 1; period: $\frac{2\pi}{3}$; phase shift: $\frac{\pi}{12}$

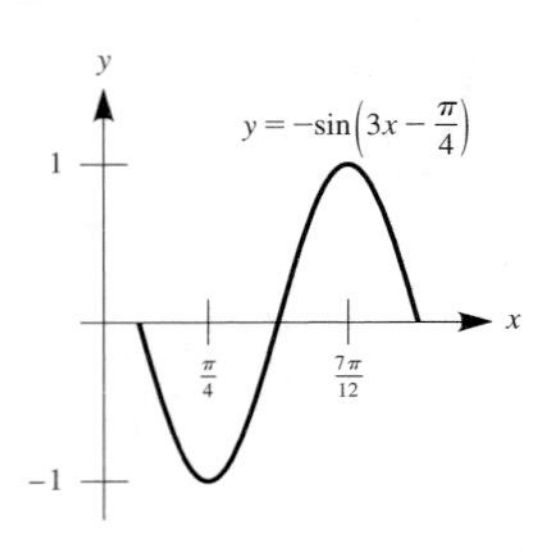

5.

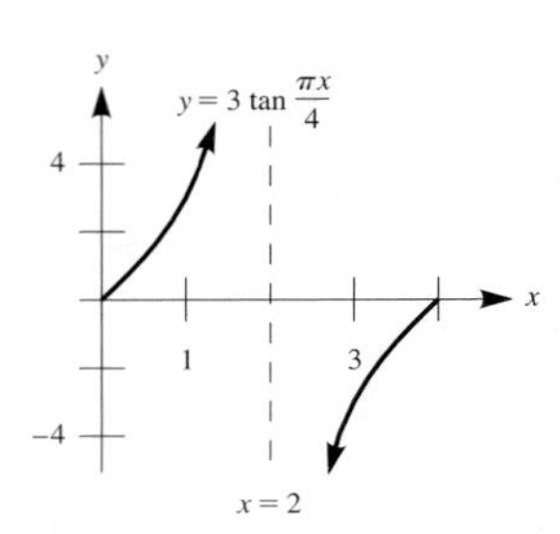

6. (a) 50π radians/sec **(b)** 250π cm/sec
8. (a)

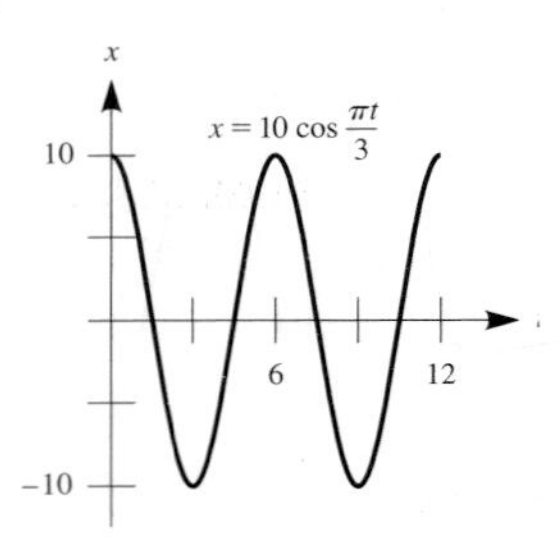

(b) passing through the origin: $t = 1.5$, $t = 4.5$, $t = 7.5$, and $t = 10.5$ sec; farthest from the origin: $t = 0$, $t = 3$, $t = 6$, $t = 9$, and $t = 12$ sec **9. (a)** 1 **(b)** 0 **(c)** 0 **10. (a)** 1.1 **(b)** 1.1198 **(c)** 2.0218
11. $y = 3.095\sin(\frac{\pi}{183}t - 1.296) + 12.087$

CHAPTER 8

Exercise Set 8.1

1. $\sin 3\theta$ **3.** $\sin 2\theta$ **5.** $\cos 5u$
7. $\sqrt{3}/2$ **9.** $\sin B$ **11.** $\cos\theta$
13. $-\cos\theta$ **15.** $\sin t$
17. $\dfrac{\sqrt{6} - \sqrt{2}}{4}$ **19.** $\dfrac{\sqrt{6} + \sqrt{2}}{4}$

21. $\sqrt{2}\sin s$ **23.** $\sqrt{3}\sin\theta$
25. (a) $-\frac{16}{65}$ **(b)** $\frac{63}{65}$ **27. (a)** $-\frac{44}{125}$
(b) $-\frac{4}{5}$ **29. (a)** $2\sqrt{6}/5$ **(b)** $4\sqrt{6}/25$
31. $\sin(\theta + \beta) = \dfrac{2\sqrt{39} - 3\sqrt{13}}{26}$;
$\cos(\beta - \theta) = \dfrac{2\sqrt{13} - 3\sqrt{39}}{26}$
37. $\tan(s + t) = -1$; $\tan(s - t) = -\frac{1}{7}$
39. $\tan(s + t) = \frac{5}{3}$; $\tan(s - t) = \frac{3}{5}$
41. $\tan 3t$ **43.** $\sqrt{3}$ **45.** $\tan x$
47. $-2 - \sqrt{3}$ **63. (b)** $\sqrt{a^2 + b^2}$
65. (b)

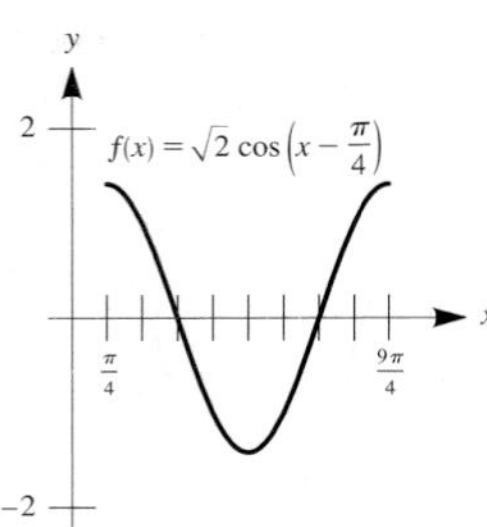

71. $\sqrt{3}/2$ **75. (a)** Using $\triangle ABH$, $\cos(\alpha + \beta) = AB/1$, so $\cos(\alpha + \beta) = AB$. **(b)** Using $\triangle ACF$, $\cos\alpha = AC/AF = AC/\cos\beta$, so $AC = \cos\alpha\cos\beta$. **(c)** Using $\triangle EFH$, $\sin(\angle EHF) = EF/HF$. But $\angle EHF = \alpha$, and $HF = \sin\beta$, so $\sin\alpha = EF/\sin\beta$, and thus $EF = \sin\alpha\sin\beta$. **(d)** From part (a), $\cos(\alpha + \beta) = AB = AC - BC$. But $AC = \cos\alpha\cos\beta$ from part (b), and $BC = EF = \sin\alpha\sin\beta$ from part (c), so $\cos(\alpha + \beta) = \cos\alpha\cos\beta - \sin\alpha\sin\beta$.
81. (a)

t	1	2	3	4
$f(t)$	1.5	1.5	1.5	1.5

(b) $f(t) = 1.5$.
85. (a) The two values are equal: $\tan A + \tan B + \tan C \approx -1.1918$; $\tan A \tan B \tan C \approx -1.1918$.
(b) $\tan\alpha + \tan\beta + \tan\gamma \approx -1.3764$; $\tan\alpha\tan\beta\tan\gamma \approx -1.3764$

Exercise Set 8.2

1. (a) $\frac{336}{625}$ **(b)** $-\frac{527}{625}$ **(c)** $-\frac{336}{527}$
3. (a) $-\frac{8}{17}$ **(b)** $-\frac{15}{17}$ **(c)** $\frac{8}{15}$
5. (a) $\frac{1}{2}$ **(b)** $\sqrt{3}/2$ **(c)** $\sqrt{3}/3$
7. (a) $2\sqrt{2}/3$ **(b)** $\frac{1}{3}$ **(c)** $2\sqrt{2}$

9. (a) $-3\sqrt{7}/8$ **(b)** $-\frac{1}{8}$
(c) $\dfrac{\sqrt{8 + 2\sqrt{7}}}{4}$ **(d)** $\dfrac{\sqrt{8 - 2\sqrt{7}}}{4}$
11. (a) $4\sqrt{2}/9$ **(b)** $-\frac{7}{9}$ **(c)** $\sqrt{6}/3$
(d) $-\sqrt{3}/3$ **13. (a)** $\dfrac{\sqrt{2 - \sqrt{3}}}{2}$
(b) $\dfrac{\sqrt{2 + \sqrt{3}}}{2}$ **(c)** $2 - \sqrt{3}$
15. (a) $\dfrac{\sqrt{2 + \sqrt{3}}}{2}$ **(b)** $-\dfrac{\sqrt{2 - \sqrt{3}}}{2}$
(c) $-2 - \sqrt{3}$ **17. (a)** $\frac{24}{25}$ **(b)** $\frac{7}{25}$
(c) $\frac{24}{7}$ **19. (a)** $\frac{24}{25}$ **(b)** $-\frac{7}{25}$ **(c)** $-\frac{24}{7}$
21. (a) $\sqrt{10}/10$ **(b)** $3\sqrt{10}/10$ **(c)** $\frac{1}{3}$
23. (a) $\sqrt{5}/5$ **(b)** $2\sqrt{5}/5$ **(c)** $\frac{1}{2}$
25. $\sin 2\theta = \dfrac{2x\sqrt{25 - x^2}}{25}$;
$\cos 2\theta = \dfrac{25 - 2x^2}{25}$
27. $\sin 2\theta = \dfrac{(x - 1)\sqrt{3 + 2x - x^2}}{2}$;
$\cos 2\theta = \dfrac{1 + 2x - x^2}{2}$
29. $\sin^4\theta = \dfrac{3 - 4\cos 2\theta + \cos 4\theta}{8}$
31. $\sin^4(\theta/2) = \dfrac{3 - 4\cos\theta + \cos 2\theta}{8}$
51. $\alpha + \beta = \pi/4$ **59. (a)** -0.5
(c) $\cos 108° \cos 36° = (-\cos 72°) \times (-\cos 144°) = \cos 72° \cos 144°$
63. (b) (i) This is true because of the double-angle formula for sine and the fact that $\cos 36° = \sin 54°$. **(ii)** This is true because of the double-angle formula for sine and the fact that $\cos 18° = \sin 72°$. **(iii)** Dividing each side by $4\sin 72°$ produces this result.

Exercise Set 8.3

1. $\frac{1}{2}\cos 50°$ **3.** $\frac{1}{2}\cos 80°$
5. $\frac{1}{2}\sin 10° + \frac{1}{2}$ **7.** $\frac{1}{2}\cos(3\pi/5) - \frac{1}{2}$
9. $\frac{1}{2}\cos(3\pi/7) + \frac{1}{2}$ **11.** $\dfrac{1}{2} + \dfrac{\sqrt{3}}{4}$
13. $\frac{1}{2}\cos x - \frac{1}{2}\cos 7x$
15. $\frac{1}{2}\sin\theta + \frac{1}{2}\sin 11\theta$
17. $\frac{1}{2}\sin\theta + \frac{1}{2}\sin 2\theta$
19. $\frac{1}{2}\cos 2y - \frac{1}{2}\cos 4x$
21. $\frac{1}{2}\sin(3t - s) + \frac{1}{2}\sin(t + s)$
23. $\sqrt{2}\cos 10°$ **25.** $-\sqrt{2}\sin(\pi/20)$
27. $-2\sin 4\theta\sin\theta$ **29.** $\cos 5°$
31. $\sin 2\theta$ **33.** $\sqrt{3}$
39. $4\cos\theta\cos 4\theta\cos 2\theta$

41. (a) $2\cos(\frac{x}{2} - \frac{\pi}{4})$
(b) amplitude: 2; period: 4π; phase shift: $\frac{\pi}{2}$

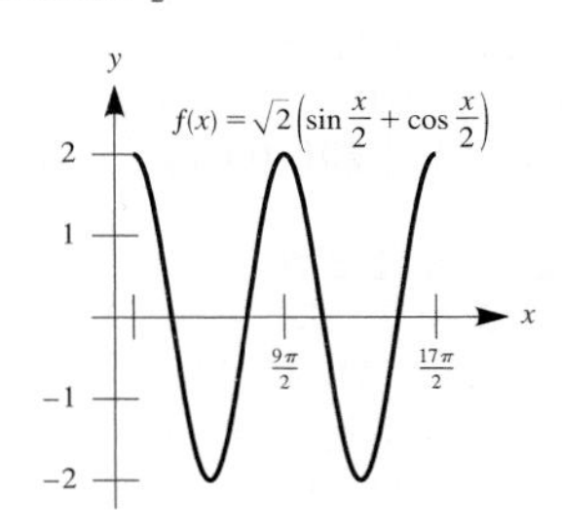

47. 1
51. (a)
$\cos 30° + \cos 70° + \cos 80° \approx 1.38$;
$\cos 40° + \cos 25° + \cos 115° \approx 1.25$;
$\cos 55° + \cos 55° + \cos 70° \approx 1.49$
(b) $\frac{3}{2}$ **(c) (i)** This is the sum-to-product formula for $\cos A + \cos B$.
(ii) This is true because $\cos[(A - B)/2] \leq 1$. **(iii)** This is true because $A + B = 180° - C$.
(iv) This is just division by 2.
(v) The identities used are $\cos(90° - \theta) = \sin\theta$ and $\cos\theta = 1 - 2\sin^2(\theta/2)$.
(vi) Multiplying this expression out shows they are equal. **(vii)** Since $2[\sin(C/2) - \frac{1}{2}]^2 \geq 0$, the expression is at most 3/2.

Exercise Set 8.4

1. yes **3.** no **5.** $\theta = (\pi/3) + 2\pi k$ or $\theta = (2\pi/3) + 2\pi k$, where k is any integer **7.** $\theta = (7\pi/6) + 2\pi k$ or $\theta = (11\pi/6) + 2\pi k$, where k is any integer **9.** $\theta = \pi + 2\pi k$, where k is any integer **11.** $\theta = (\pi/3) + \pi k$, where k is any integer
13. $x = \pi k$, where k is any integer
15. $\theta = (\pi/2) + \pi k$, $\theta = (2\pi/3) + 2\pi k$ or $\theta = (4\pi/3) + 2\pi k$, where k is any integer **17.** $t = \pi k$, where k is any integer **19.** $x = (\pi/6) + 2\pi k$, $x = (5\pi/6) + 2\pi k$ or $x = (3\pi/2) + 2\pi k$, where k is any integer **21.** $t = (\pi/4) + 2\pi k$ or $t = (3\pi/4) + 2\pi k$, where k is any integer

23. (a)

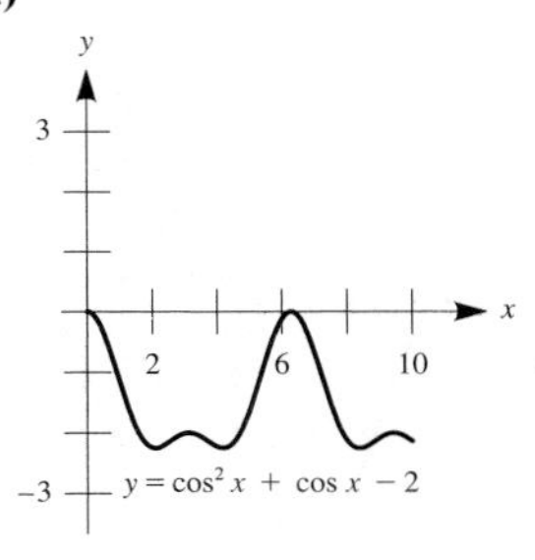

(b) $x \approx 6.28$ **(c)** $x = 2\pi$
25. $x \approx 1.39, 4.90$ **27.** $x \approx 0.46, 2.68$
29. $x \approx 1.41, 4.55$ **31.** $t \approx 1.37, 4.51$
33. $t \approx 1.11, 5.18$ **35.** $x \approx 1.25, 1.82, 4.39, 4.96$ **37.** $x \approx 0.85, 2.29$
39. $\theta \approx 14.5°, 165.5°$ **41.** $\theta \approx 53.1°, 126.9°, 233.1°, 306.9°$ **43.** $\theta \approx 128.2°, 231.8°$ **45.** $\theta = 0°, 120°, 240°$
47. $\theta = 75°, 105°, 195°, 225°, 315°, 345°$
49. $\theta = 90°, 270°$
51. $\theta = 30°, 120°, 210°, 300°$
53. $x \approx 0.898, 5.385$

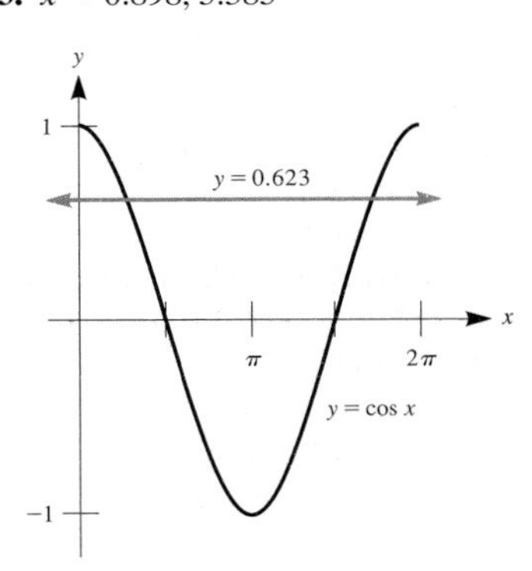

55. $x \approx 0.666, 2.475$

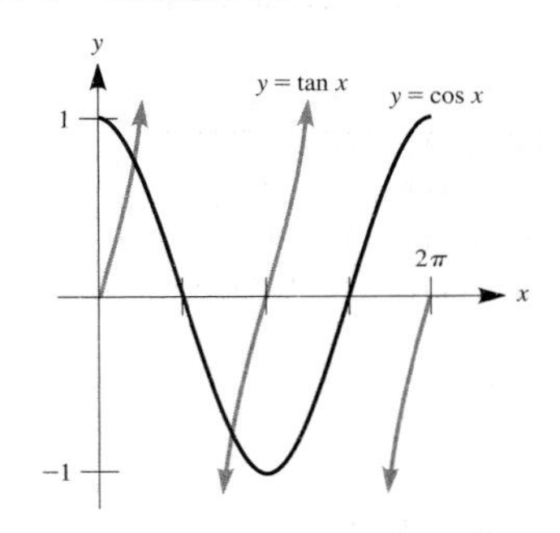

57. $x \approx 0.427, 2.715$

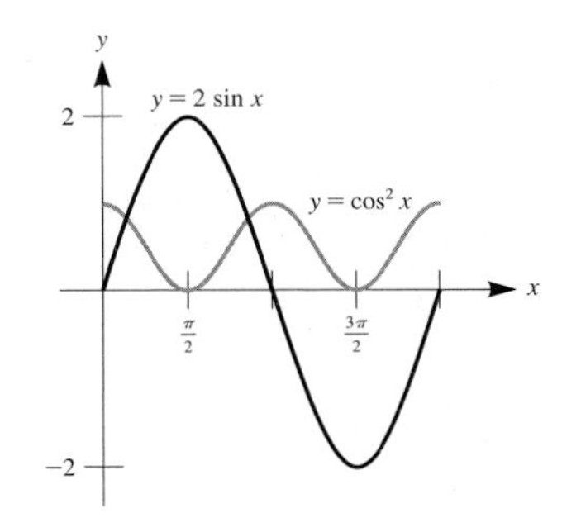

59. no solution

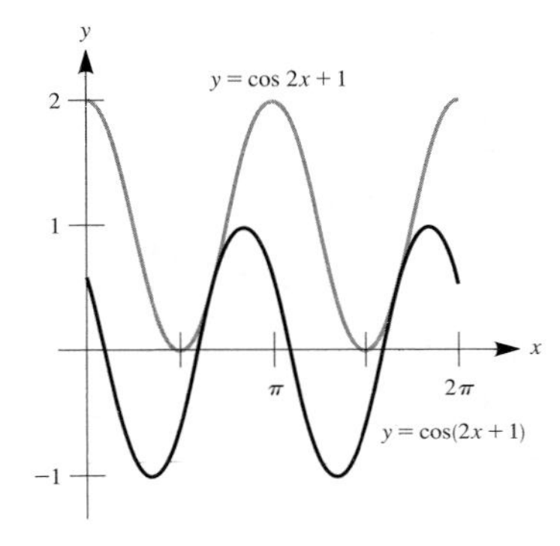

61. $x = 0, 1, x \approx 3.080, 4.080, 4.538, 5.538, 5.660$

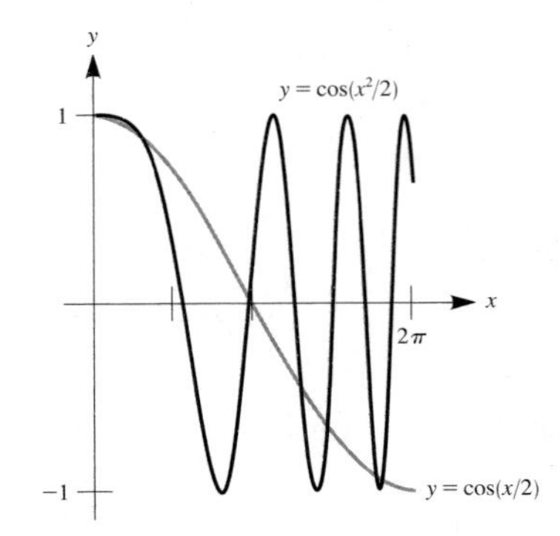

63. $x \approx 1.058, 3.739$

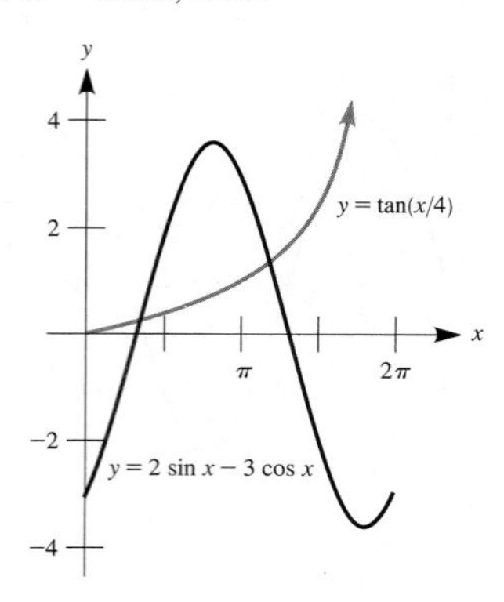

65. $x = 0, x \approx 0.695, 4.261$

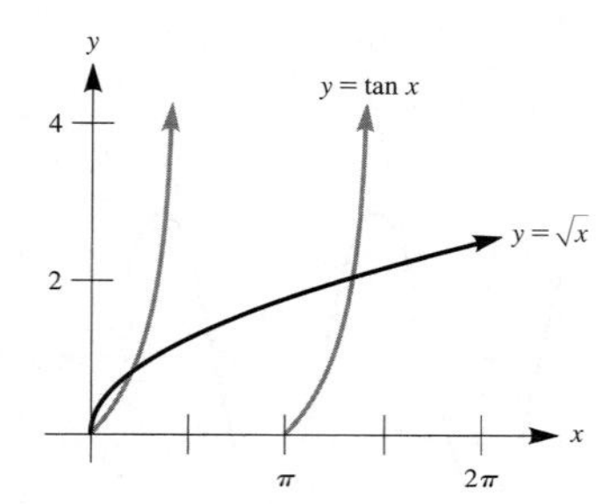

67. $x \approx 0.739, 3.881$

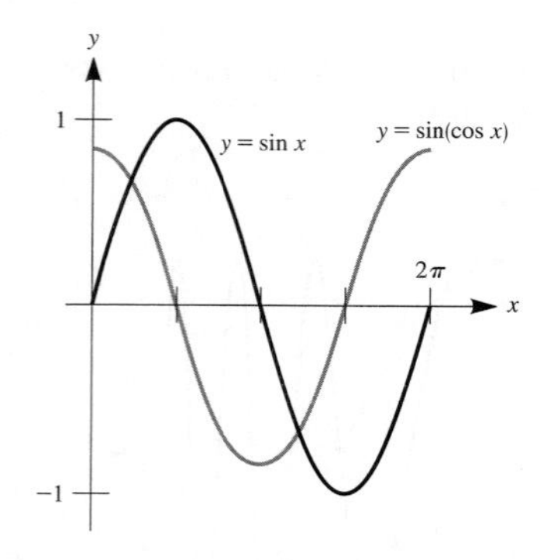

69. $x = 0, x \approx 4.493$

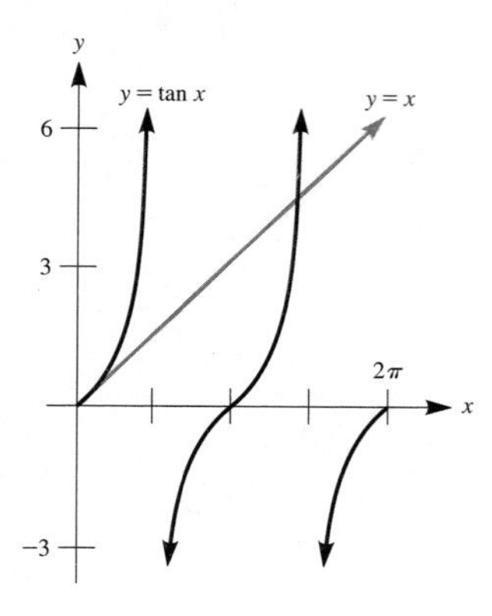

71. $x \approx 2.108, 5.746$

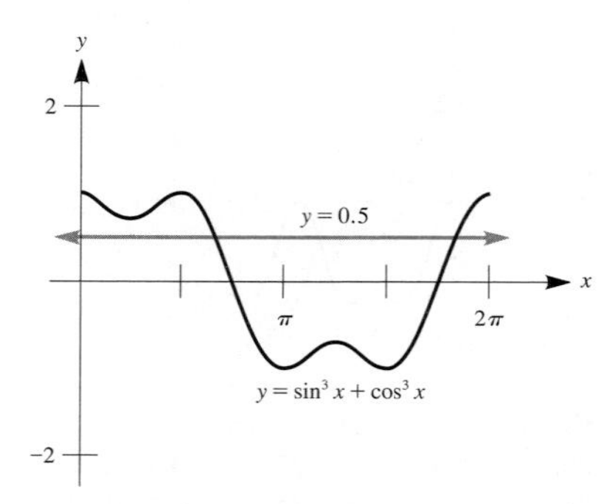

73. $x \approx 0.703$

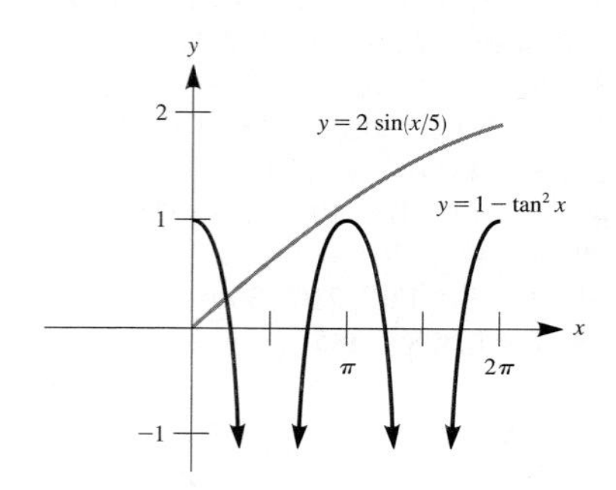

75. $x = 0$ or $x = \pi$
77. $x = 2\pi/3$ or $x = \pi$
79. $\theta = (3\pi/16) + k\pi/4$, where k is any integer **81.** $\theta \approx 60.45°$
85. (a) $x \approx 1000.173$
(b) $x \approx 1001.022$
87. P: 0.315; Q: 0.685; R: 1.315
89. (a) $\alpha = \frac{1}{2}\sin^{-1}\left(\frac{rg}{v_0^2}\right)$
(b) $\alpha = \frac{\pi}{12} = 15° \approx 0.26$ radians
(d) $\alpha = \frac{5\pi}{12} = 75° \approx 1.31$ radians
(e) $\alpha = \frac{\pi}{4} = 45°$ **91. (a)** $\alpha \approx 76.0°$
(b) $\alpha \approx 76.0°$ **93. (a)** (1.74, 0.99)
95. (a) $A(\theta) = \sin\theta(1 + \cos\theta)$
(b) $\theta \approx 49.78°$ **(c)** No, the maximum area is $1.30 < 0.42\pi$.
97. (a) $V(\theta) = \frac{1}{3}\pi \sin^2\theta \cos\theta$
(b)

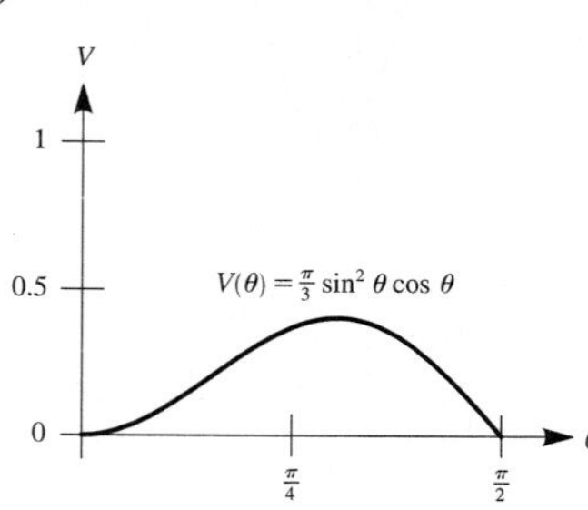

(c) $x \approx 0.905 \approx 51.8°$, $x \approx 1.005 \approx 57.6°$ **(d)** The maximum is 0.403, so the volume of the cone cannot equal 0.41 m^3.
99. (b) $\alpha \approx 30.0° \approx 0.52$ radians
(d) $\alpha \approx 75.0°$ **101. (c)** $r_{max} \approx 9.93$ ft

Exercise Set 8.5

1. $\pi/3$ **3.** $\pi/3$ **5.** $-\pi/6$ **7.** $\pi/4$
9. undefined **11.** $\frac{1}{4}$ **13.** $\frac{3}{4}$
15. $-\pi/7$ **17.** $\pi/2$ **19.** 0
21. (a) maximum value: 1.57 (when $x = 1$); minimum value: -1.57 (when $x = -1$)

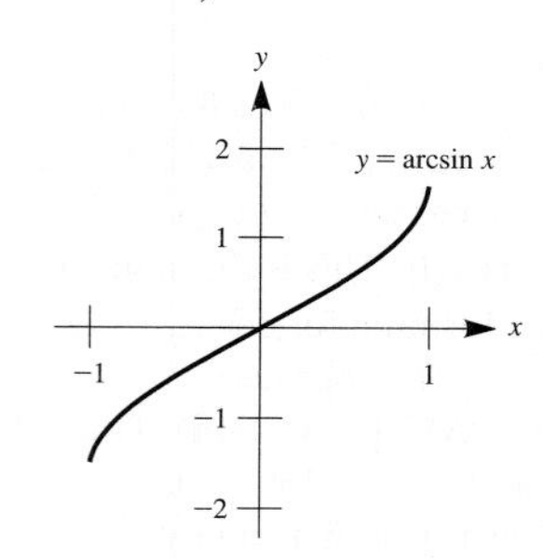

(b) maximum value: $\frac{\pi}{2}$ (when $x = 1$); minimum value: $-\frac{\pi}{2}$ (when $x = -1$)
23. (a)

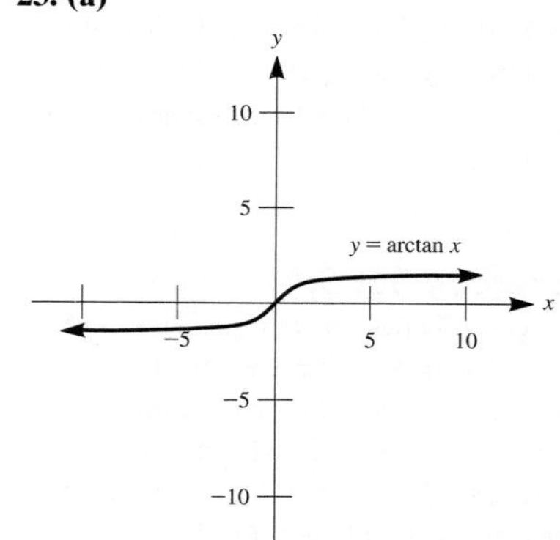

(b) $y = -\pi/2, y = \pi/2$

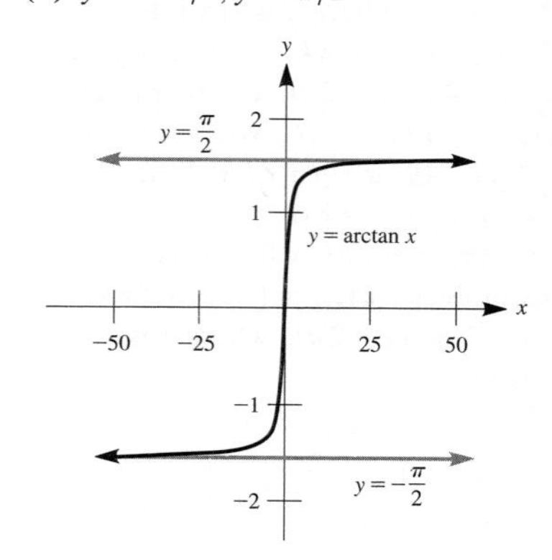

25. $3\sqrt{5}/7$ **27.** $-\sqrt{2}/2$ **29.** $\sqrt{5}/3$
31. $2\sqrt{2}/3$ **33.** $\frac{20}{21}$ **35.** -2
37. $\sqrt{3}/3$
39.

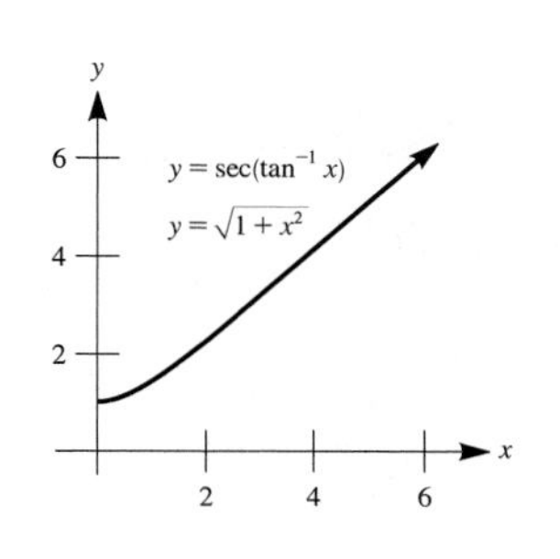

41.

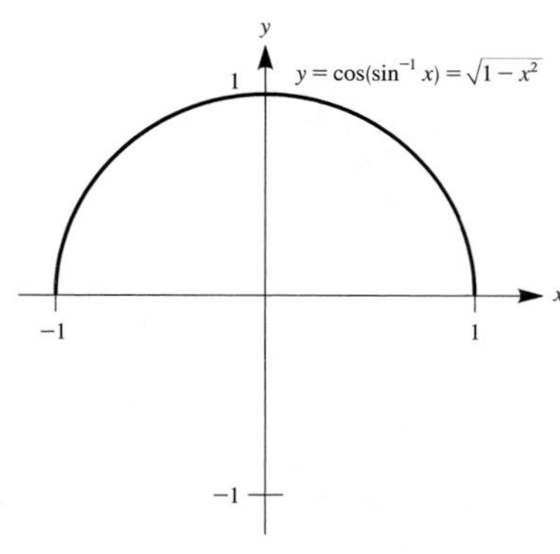

43. $\frac{\theta}{4} - \sin 2\theta =$

$\frac{1}{4}\sin^{-1}\left(\frac{3x}{2}\right) - \frac{3x\sqrt{4 - 9x^2}}{2}$

45. $\theta - \cos\theta =$

$\tan^{-1}\left(\frac{x-1}{2}\right) - \frac{2}{\sqrt{5 - 2x + x^2}}$

47. (a)

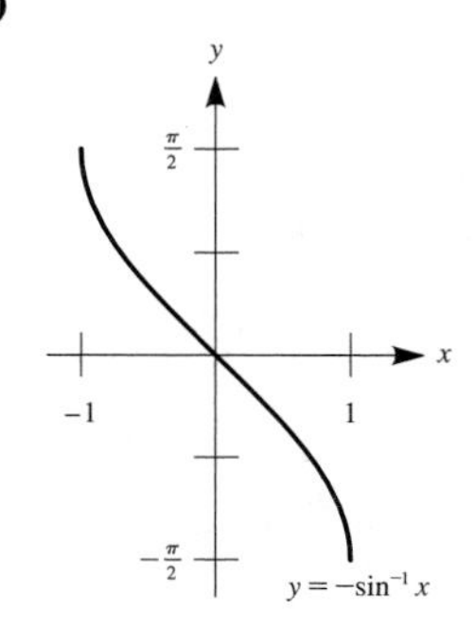

(b)

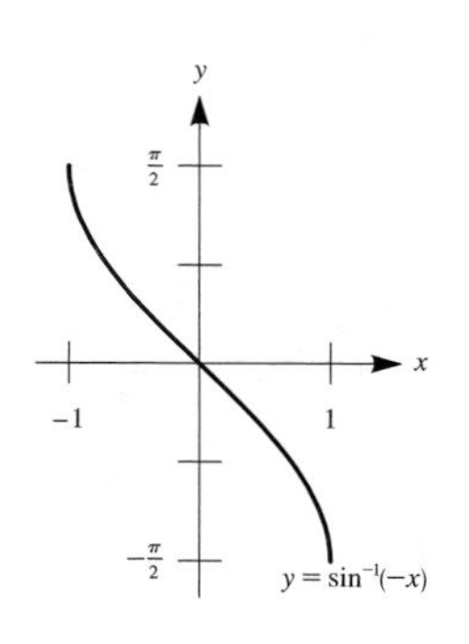

(c)

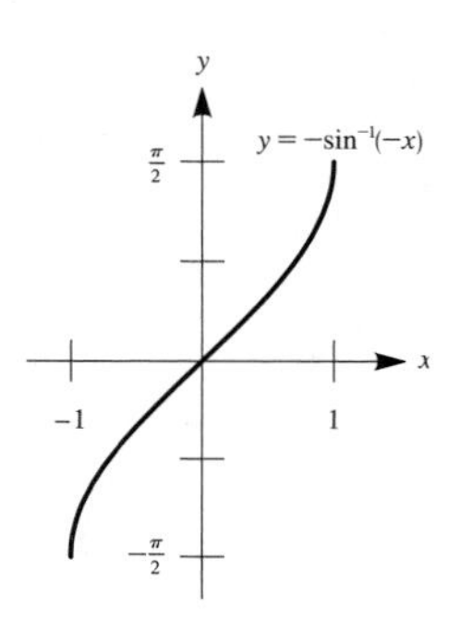

49. (a)

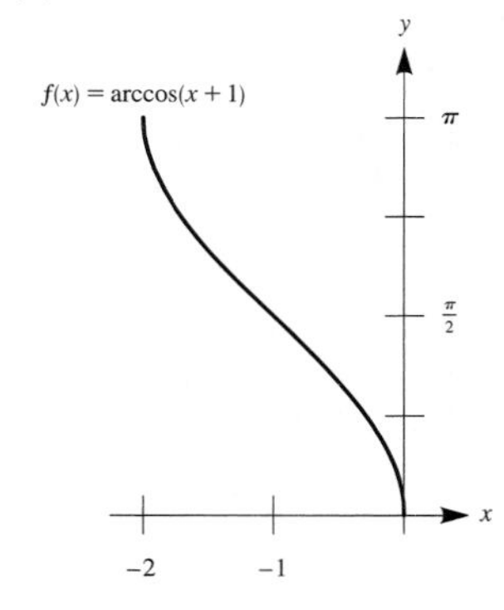

(b)

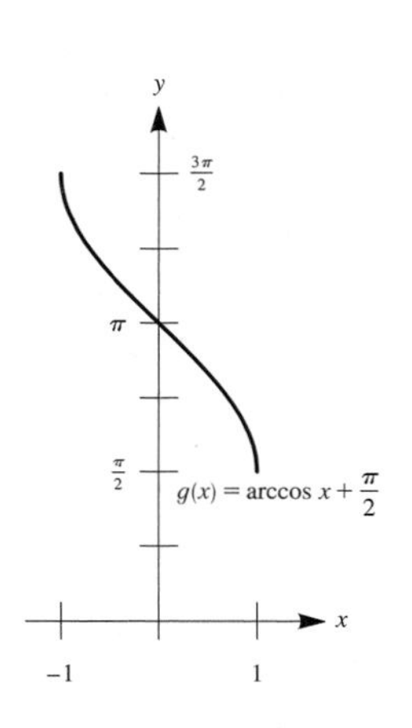

51. (a)

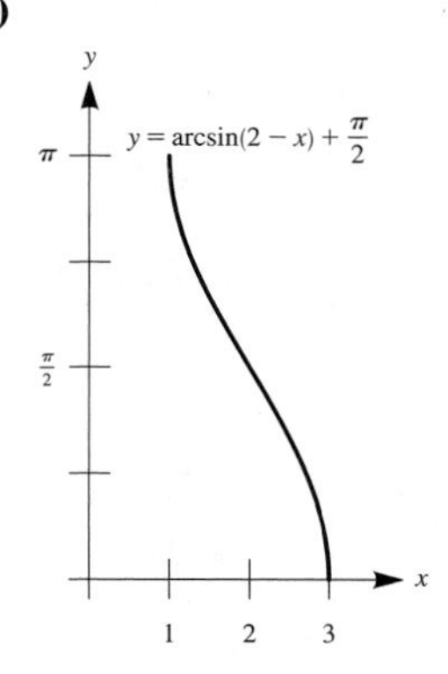

(b)

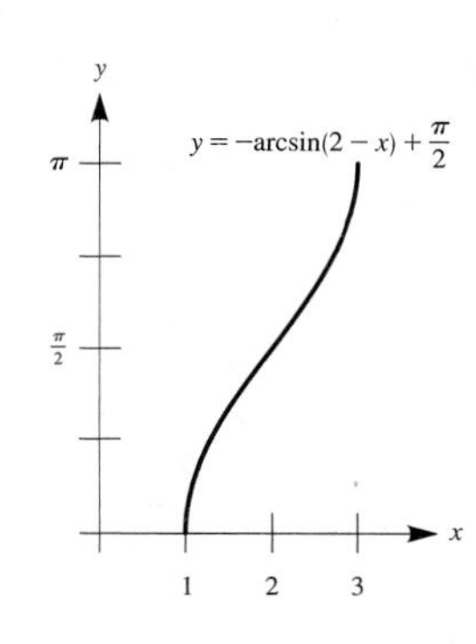

53. (a)

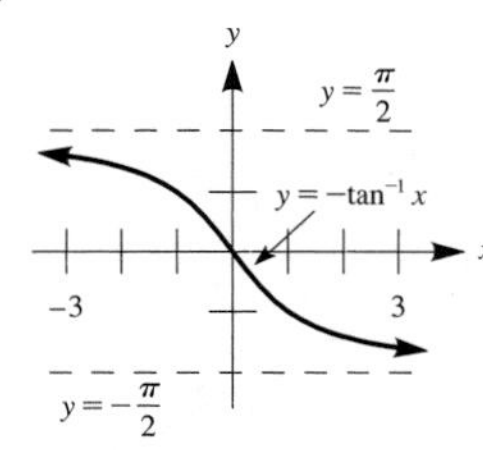

(b)

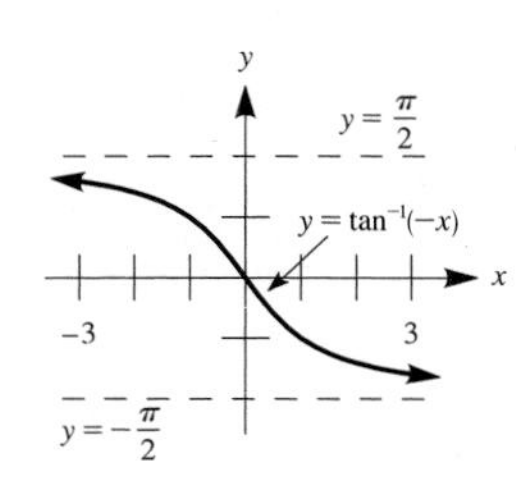

(c)

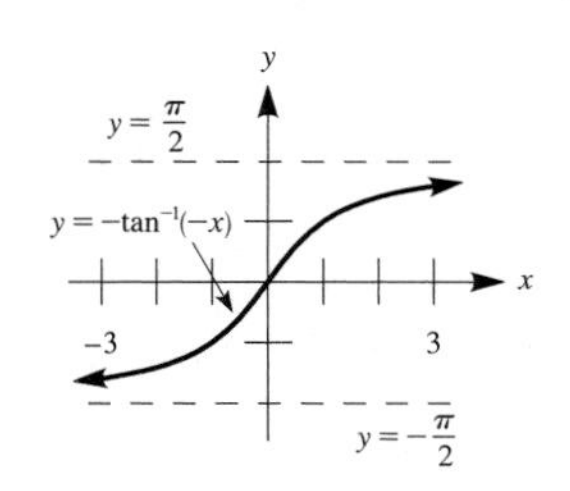

55.

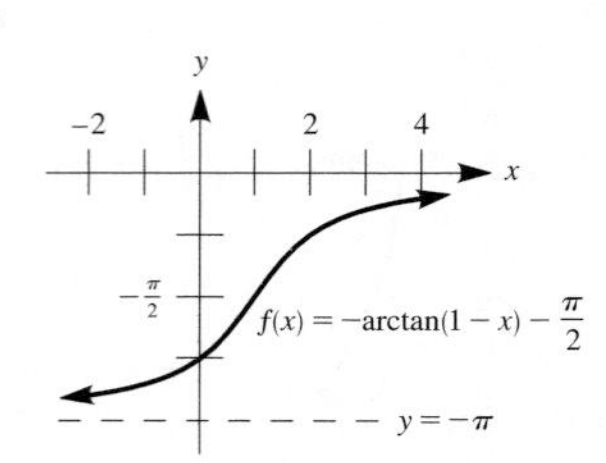

57. $\frac{8}{17}$ **59.** $31\sqrt{218}/1090$
61. (a) $-\frac{\pi}{2} \le \alpha \le \frac{\pi}{2}$ and $0 \le \beta \le \pi$, so $-\frac{\pi}{2} \le \alpha + \beta \le \frac{3\pi}{2}$ **63.** $t = \sqrt{2}/2$
67. $x = \pm\frac{1}{3}$ **69. (a)** $x = 0$ is not a solution, so $\cos^{-1} x > 0$. $\tan^{-1} x > 0$ only for $x > 0$. **(c)** (0.786, 0.666)
71. $x \approx 0.74$

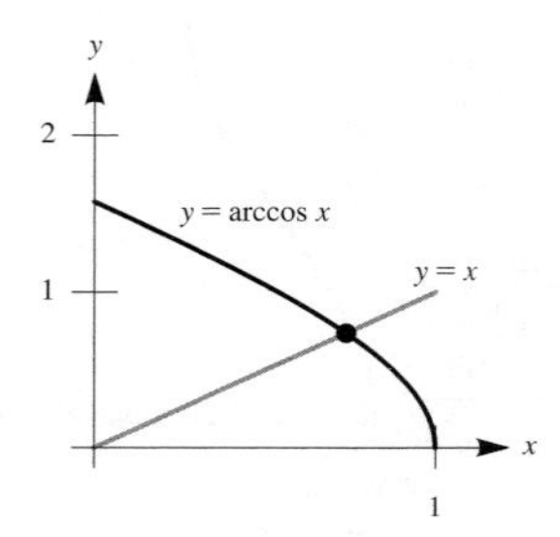

73. (a) $x \approx 0.96$

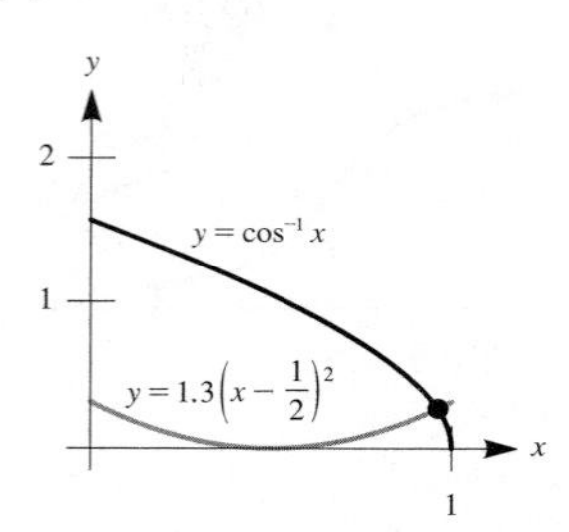

(b) $x \approx 0.96$

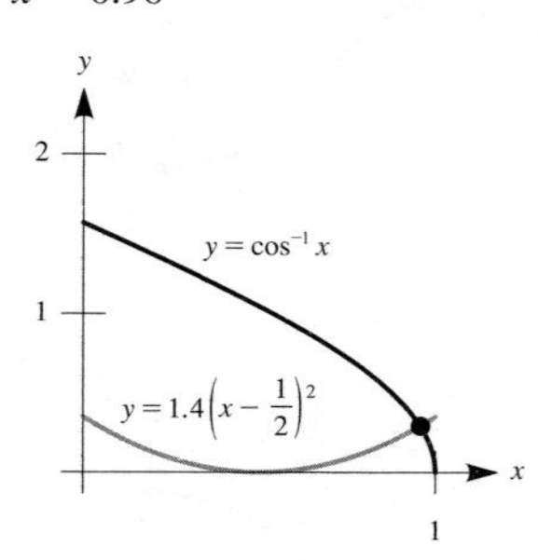

75. (a) $x \approx 0.24$

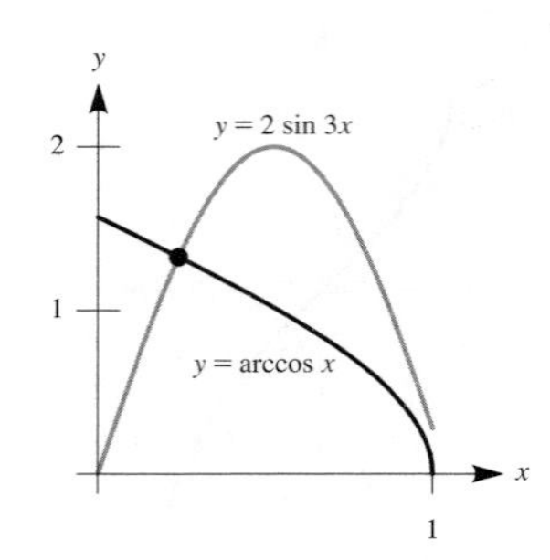

(b) $x \approx 0.19$ and $x \approx 0.68$

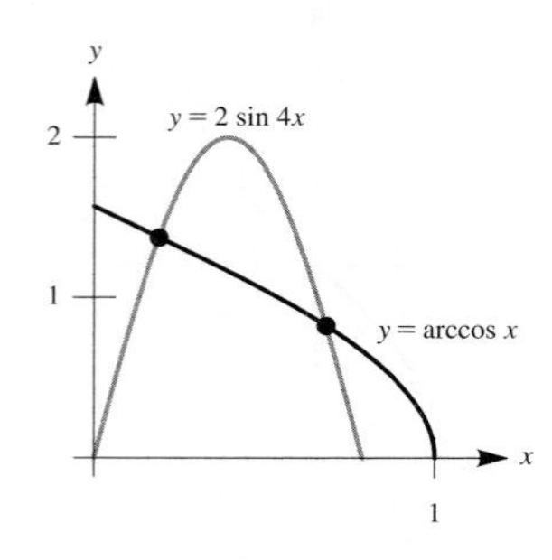

77. (a) $x \approx 0.56$

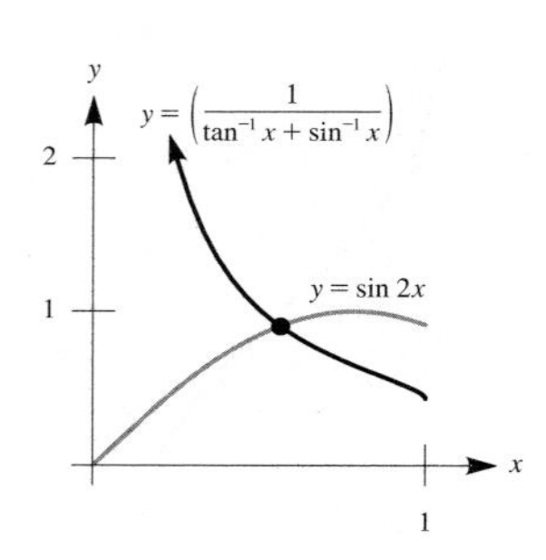

(b) $x \approx 0.51$ and $x \approx 0.84$

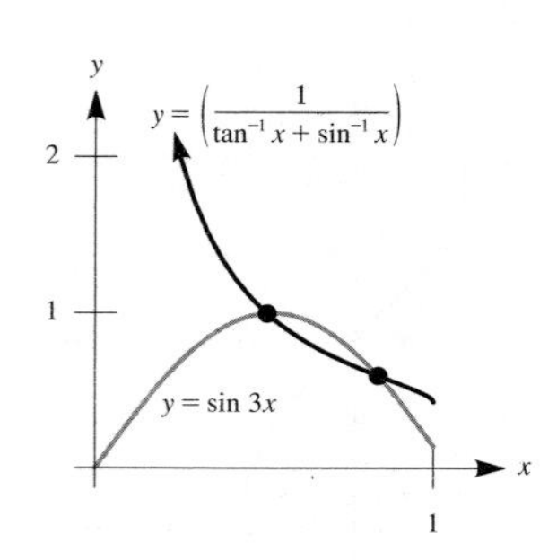

79. $x \approx 0.71$

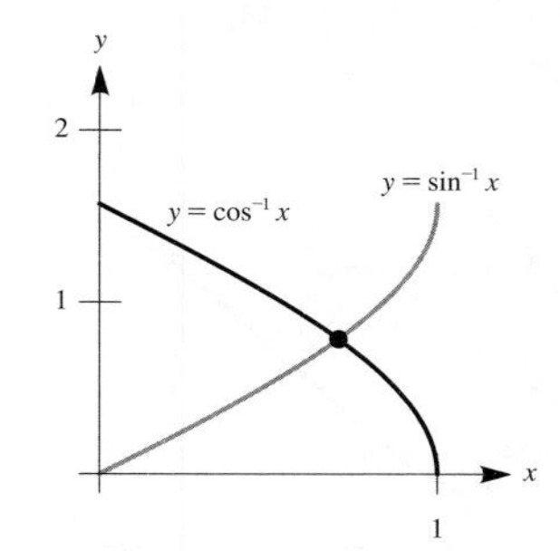

81. $x \approx 0.74$

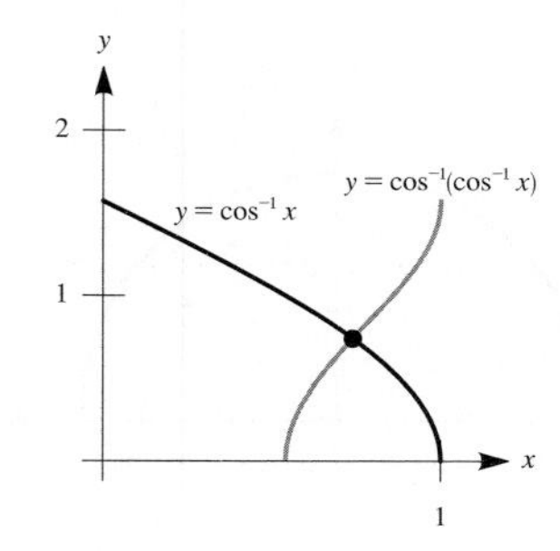

83. $x \approx 0.94$

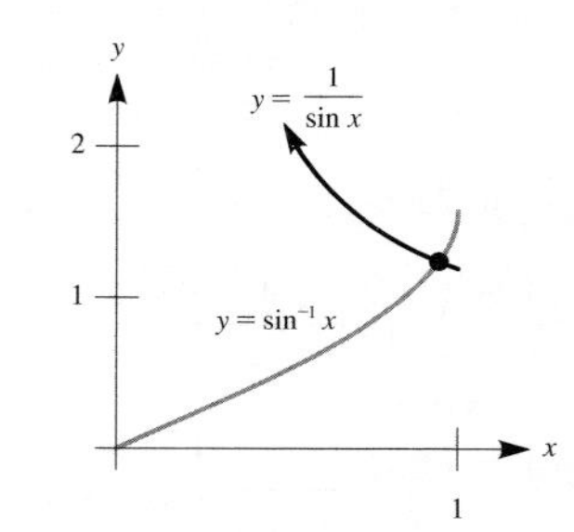

85. $x \approx 0.93$

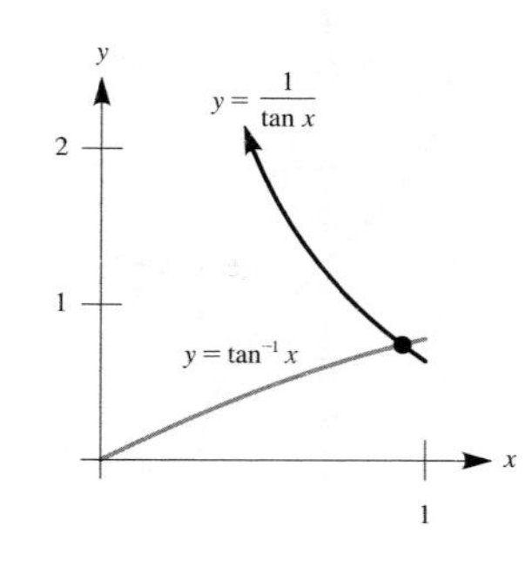

87. (b) The maximum value occurs when $x = 4$.

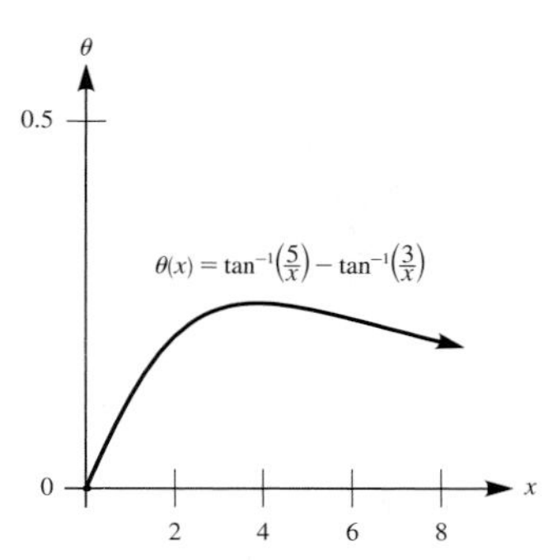

89. (b) $DE = \sqrt{2}, CE = \sqrt{2}, BE = 3\sqrt{2}, AB = 3\sqrt{10}$, and $BD = \sqrt{10}$

Chapter 8 Review Exercises

43. (a) $\pi/4$ **(b) (i)** $a \approx 14$ **(ii)** $a \approx 93$ **(iii)** $a \approx 1256$
45. $x \approx 1.34$ or 4.48 **47.** $x \approx 0.46$ or 2.68 **49.** $x = \pi/3, 2\pi/3, 4\pi/3$, or $5\pi/3$
51. $x = 0$ or $3\pi/2$ **53.** $x = 0, \pi, 7\pi/6$, or $11\pi/6$ **55.** $x = \pi/3, 2\pi/3, 4\pi/3$, or $5\pi/3$ **57.** $x = \pi/4, \pi/2, 3\pi/4$, or $5\pi/4$, $3\pi/2$, or $7\pi/4$ **59.** $x = \pi/6, \pi/3, 2\pi/3, 5\pi/6, 7\pi/6, 4\pi/3, 5\pi/3$, or $11\pi/6$
61. $x = 3\pi/4$ or $7\pi/4$ **63.** $\sqrt{3}/2$
65. $\pi/6$ **67.** $\pi/6$ **69.** $\pi/3$ **71.** $2\pi/3$
73. $\frac{2}{7}$ **75.** $2\pi - 5$ **77.** $-\sqrt{2}/2$
79. $3\sqrt{2}/2$ **81.** $\frac{17}{7}$ **83.** $-\frac{4}{3}$
85. $3\sqrt{10}/10$ **95. (a)** 0.0625

Chapter 8 Test

1. $-\cos\theta$ **2.** $-\frac{3}{5}$ **3.** $-\frac{3}{2}$
4. $x \approx 1.25, 4.39$
5. $x = 7\pi/6, 11\pi/6$ **6.** $\sqrt{5}/5$
7. $x = 30°$ **8.** $\dfrac{\sqrt{18 + 12\sqrt{2}}}{6}$
9. restricted sine function: domain $= [-\frac{\pi}{2}, \frac{\pi}{2}]$, range $= [-1, 1]$; $y = \sin^{-1} x$: domain $= [-1, 1]$, range $= [-\frac{\pi}{2}, \frac{\pi}{2}]$

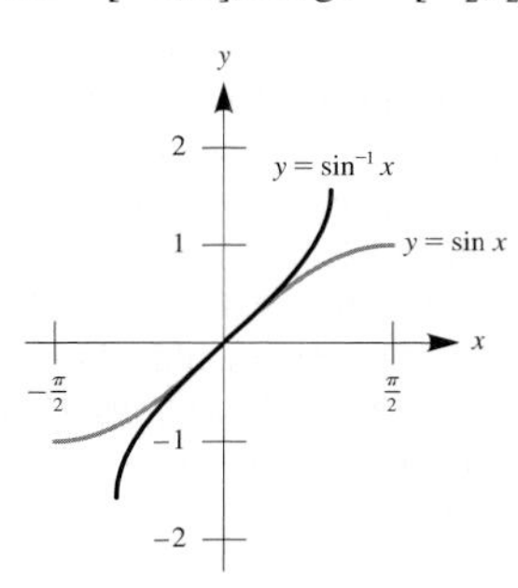

10. (a) $\pi/10$ **(b)** 0 **11.** $\sqrt{7}/4$
13. $\dfrac{\sqrt{3} + \sqrt{2}}{4}$ **14.** $\tan 4\theta$ **15.** x
16. domain: $(-\infty, \infty)$; range: $(-\frac{\pi}{2}, \frac{\pi}{2})$

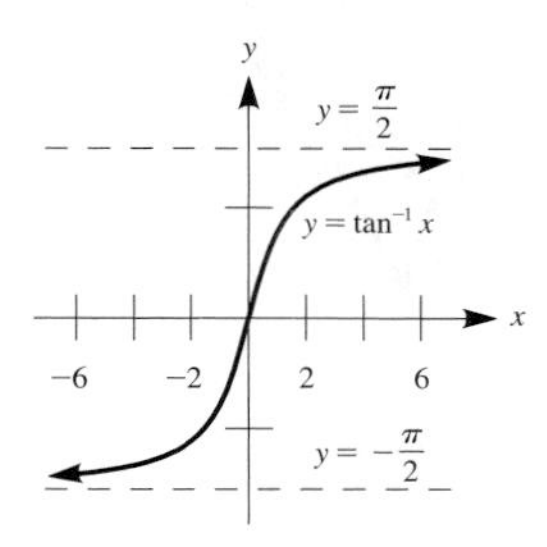

CHAPTER 9

Exercise Set 9.1

1. $BC = 30$ cm, $AC = 30\sqrt{3}$ cm
3. $AB = \dfrac{32\sqrt{3}}{3}$ cm, $BC = \dfrac{16\sqrt{3}}{3}$ cm
5. $AC \approx 11.5$ cm, $BC \approx 9.6$ cm
7. (a) 15.59 ft **(b)** 9 ft
9. (a) 22.6° **(b)** 22.6° **(c)** 22.6°
11. 34 million miles **13.** 141.1 m
15. (a) $h \approx 27.3$ ft **(b)** 906.9 ft^2
17. 1.5 in.^{2}
19. $\frac{7}{2}\sin(360°/7) \approx 2.736$ square units
21. $\pi - 2\sqrt{2} \approx 0.313$ square units
23. $\sqrt{3} \approx 1.732$ square units
25. $25.2 - 18 \sin 1.4 \approx 7.46$ cm^2
29. 10,660 ft **31.** 136 m
33. $BD = 18(\sqrt{3} - 1)$ cm
35. (a) $\angle BOA = 90° - \theta, \angle OAB = \theta, \angle BAP = 90° - \theta, \angle BPA = \theta$
(b) $AO = \sin\theta, AP = \cos\theta, OB = \sin^2\theta, BP = \cos^2\theta$
37. (a) $BC = \dfrac{5}{\sin\theta}$ **(b)** $AB = \dfrac{4}{\cos\theta}$
(c) $AC = 4\sec\theta + 5\csc\theta$
39. (a) $DE = \sin\theta$ **(b)** $OE = \cos\theta$
(c) $CF = \tan\theta$ **(d)** $OC = \sec\theta$
(e) $AB = \cot\theta$ **(f)** $OB = \csc\theta$
41. (b) 1080 miles
43. (d)

n	5	10	50	100
a_n	2.37764129	2.93892626	3.13333084	3.13952598

n	10^3	10^4	10^5
a_n	3.14157198	3.14159245	3.14159265

47. (a) $(\cos\theta, \sin\theta)$
49. (b) $\frac{\pi}{4}(1 - \cos\theta) - \frac{1}{2}(\theta - \sin\theta)$
(c) $\frac{\pi}{4}(1 + \cos\theta) - \frac{1}{2}(\pi - \theta - \sin\theta)$
(d) $\sin\theta$ **55. (b)** $\theta \approx 0.75$
(c) percentage error $\approx 1.5\%$, $\angle PCB \approx 48°$

Exercise Set 9.2

1. $4\sqrt{6}$ cm **3.** $20 \sin 50°$ cm
5. $a \approx 9.7$ cm; $c \approx 16.4$ cm
7. $A \approx 63.3°$; $C \approx 50.7°$; $c \approx 25.5$ cm
9. (a) 45° or 135° **(b)** 45°
(c) 14.5° or 165.5° **(d)** 131.8°
11. (a) $\sin B \approx 1.18 > 1$
13. (b) $\angle C = 105°$; $c \approx 1.93$
(c) $\angle C = 15°$; $c \approx 0.52$
(d) $A_1 \approx 0.68$; $A_2 \approx 0.18$
15. $a = \dfrac{2 \sin 110°}{\sin 20°}$ cm;
$b = \dfrac{2 \sin 50°}{\sin 20°}$ cm;
$c = \dfrac{2 \sin 50° \sin 70°}{\sin 20° \sin 95°}$ cm;
$d = \dfrac{2 \sin 50° \sin 15°}{\sin 20° \sin 95°}$ cm **17.** 160 ft
19. (a) $x = 7$ cm **(b)** $x = \sqrt{129}$ cm
21. (a) $x \approx 7.5$ cm **(b)** $x \approx 17.7$ cm
23. x is not the side opposite the 130° angle; $6^2 = x^2 + 3^2 - 2(x)(3)\cos 130°$
25. $\cos A = \frac{113}{140}$; $\cos B = \frac{29}{40}$; $\cos C = -\frac{5}{28}$
27. $A \approx 27.8°$; $B \approx 32.2°$; $C \approx 120°$
29. $A = 30°$; $B = 30°$; $C = 120°$
31. 5.9 units **33. (a)** $a \approx 4.2$ cm
(b) $C \approx 29.3°$ **(c)** $B \approx 110.7°$
35. approximately 31 miles
37. lighthouse A: 1.26 miles; lighthouse B: 1.60 miles
39. $D \approx 860{,}000$ miles
41. (b) $a = 5, b = 3, c = 7$ (using $m = 2$ and $n = 1$) **43.** 2:40 P.M.
47. (f)

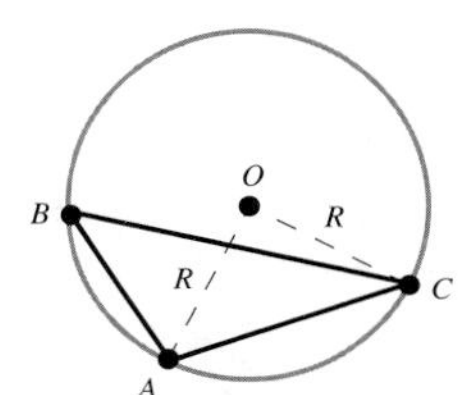

(g) Their circumscribed circles have the same radii.
55. $C = 45°$ or $135°$

Exercise Set 9.3

1. $|\overrightarrow{PQ}| = \sqrt{34}$

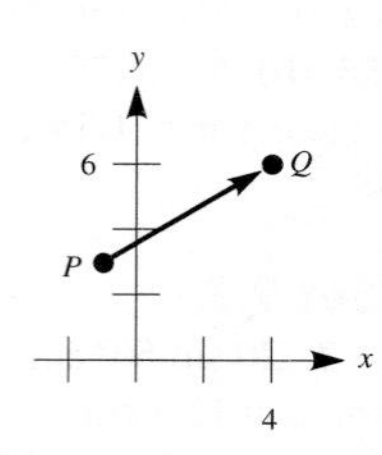

3. $|\overrightarrow{SQ}| = \sqrt{10}$

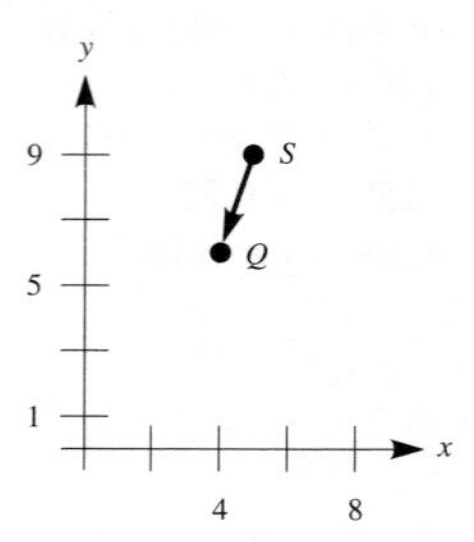

5. $|\overrightarrow{OP}| = \sqrt{10}$

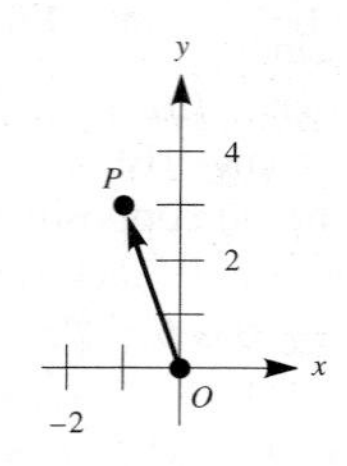

7. $|\overrightarrow{PQ} + \overrightarrow{QS}| = 6\sqrt{2}$

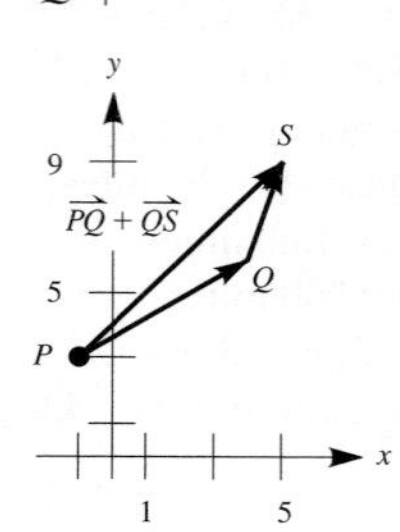

9. $|\overrightarrow{OP} + \overrightarrow{PQ}| = 2\sqrt{13}$

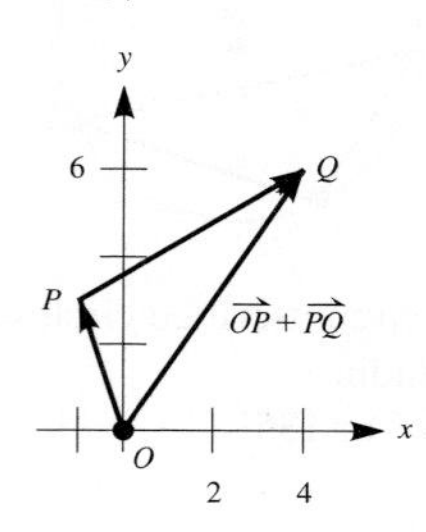

11. $|\overrightarrow{OS} + \overrightarrow{SQ} + \overrightarrow{QP}| = \sqrt{10}$

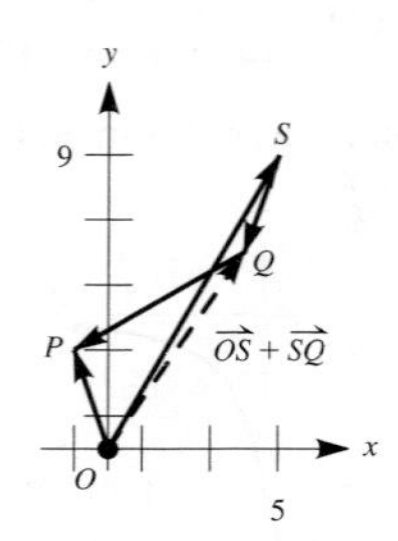

13. $|\overrightarrow{OP} + \overrightarrow{QS}| = 6$

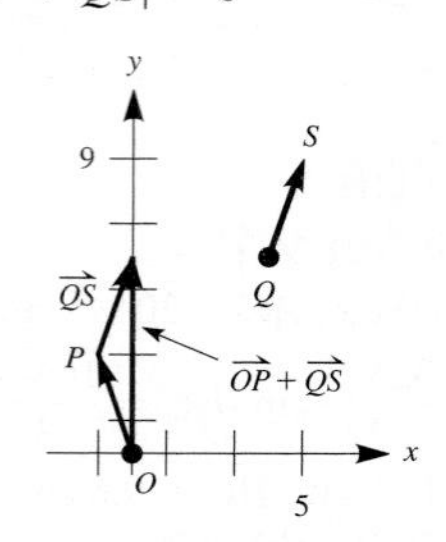

15. $|\overrightarrow{SR} + \overrightarrow{PO}| = 9$

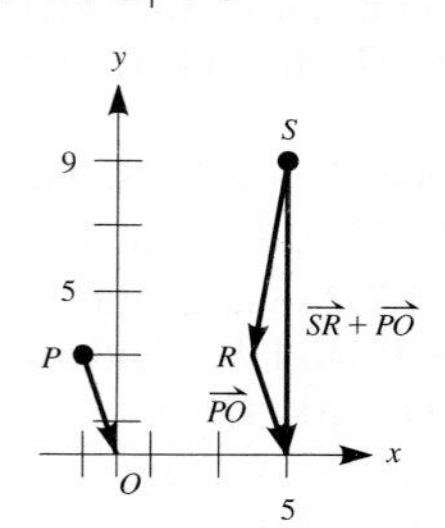

17. $|\overrightarrow{OP} + \overrightarrow{RQ}| = \sqrt{37}$

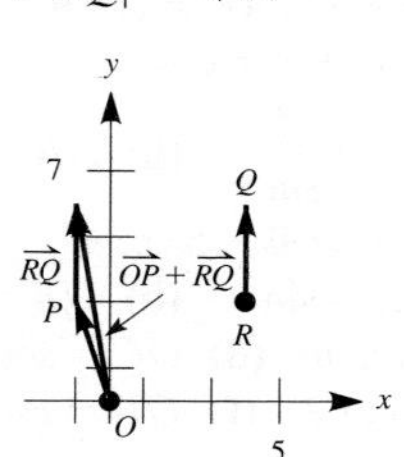

19. $|\overrightarrow{SQ} + \overrightarrow{RO}| = \sqrt{61}$

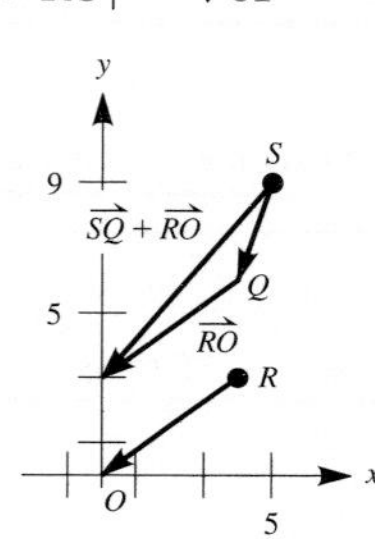

21. $|\overrightarrow{OP} + \overrightarrow{OR}| = 3\sqrt{5}$

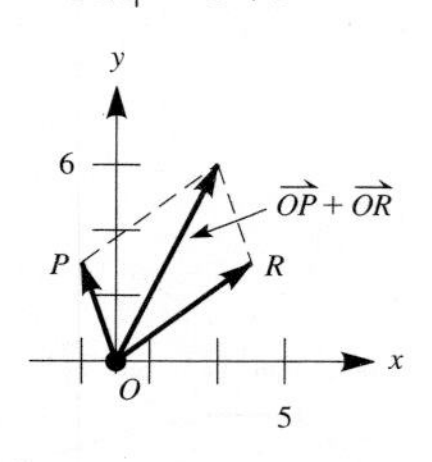

23. $|\overrightarrow{RP} + \overrightarrow{RS}| = 2\sqrt{13}$

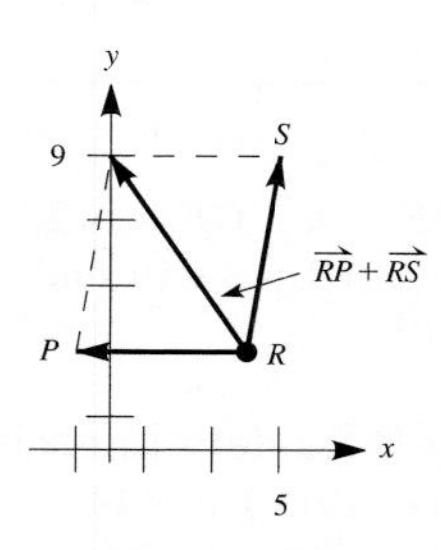

25. $|\overrightarrow{SO} + \overrightarrow{SQ}| = 6\sqrt{5}$

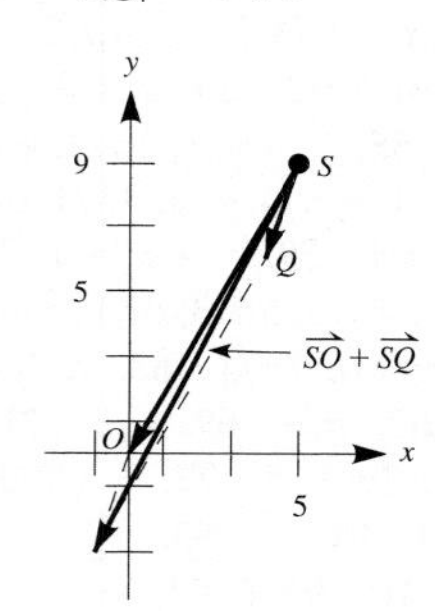

27. $|\mathbf{F} + \mathbf{G}| = \sqrt{41}$ N; $\theta \approx 38.7°$
29. $|\mathbf{F} + \mathbf{G}| = 9\sqrt{2}$ N; $\theta = 45°$
31. $|\mathbf{F} + \mathbf{G}| \approx 7.90$ N; $\theta \approx 24.1°$
33. $|\mathbf{F} + \mathbf{G}| \approx 6.92$ N; $\theta \approx 34.67°$
35. $|\mathbf{F} + \mathbf{G}| \approx 39.20$ N; $\theta \approx 21.46°$
37. $|\mathbf{F} + \mathbf{G}| \approx 38.96$ N; $\theta \approx 29.44°$
39. $V_x \approx 13.86$ cm/sec; $V_y = 8$ cm/sec
41. $F_x \approx 3.62$ N; $F_y \approx 13.52$ N
43. $V_x \approx -0.71$ cm/sec; $V_y \approx 0.71$ cm/sec
45. $F_x \approx -1.02$ N; $F_y \approx 0.72$ N
47. ground speed: 301.04 mph; drift angle: 4.76°; course: 25.24°
49. ground speed: 293.47 mph; drift angle: 8.82°; course: 91.18°
51. perpendicular: 9.83 lb; parallel: 6.88 lb **53.** perpendicular: 11.82 lb; parallel: 2.08 lb **55. (a)** initial point: $(-1, 2)$; terminal point: $(2, -3)$ **(b)** initial point: $(-1, 2)$; terminal point: $(2, -3)$

Exercise Set 9.4

1. length = 5

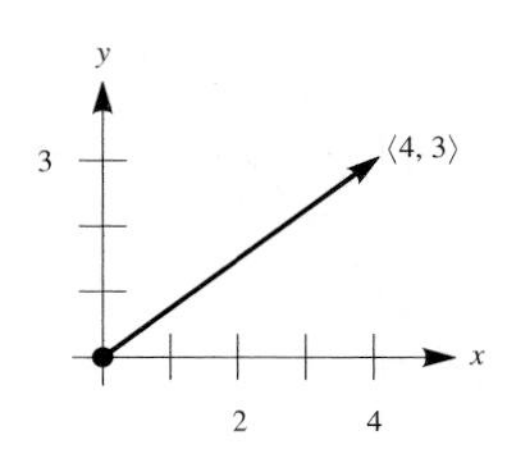

3. length = $2\sqrt{5}$

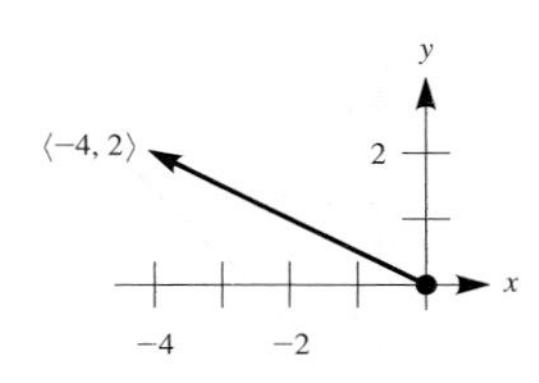

5. length = $\sqrt{13}/4$

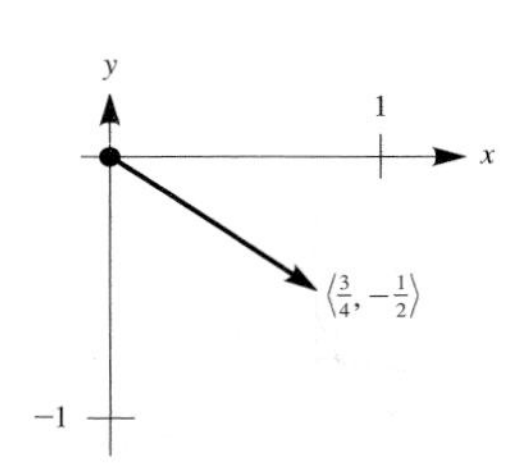

7. $\langle 1, 4\rangle$ **9.** $\langle -1, 1\rangle$ **11.** $\langle 8, -5\rangle$
13. $\langle 7, 7\rangle$ **15.** $\langle 24, 22\rangle$ **17.** $\sqrt{130}$
19. $2\sqrt{17} - \sqrt{13} - \sqrt{37}$
21. $\langle 13, 6\rangle$ **23.** $\langle 14, 21\rangle$
25. $\langle -3, -1\rangle$ **27.** $\langle 23, 12\rangle$
29. $\langle -9, 0\rangle$ **31.** -48 **33.** $3\mathbf{i} + 8\mathbf{j}$
35. $-8\mathbf{i} - 6\mathbf{j}$ **37.** $19\mathbf{i} + 23\mathbf{j}$
39. $\langle 1, 1\rangle$ **41.** $\langle 5, -4\rangle$
43. $\langle \frac{\sqrt{5}}{5}, \frac{2\sqrt{5}}{5}\rangle$
45. $\langle \frac{2\sqrt{5}}{5}, \frac{-\sqrt{5}}{5}\rangle$
47. $(8\sqrt{145}/145)\mathbf{i} - (9\sqrt{145}/145)\mathbf{j}$
49. $u_1 = \sqrt{3}/2;\ u_2 = \frac{1}{2}$
51. $u_1 = -\frac{1}{2};\ u_2 = \sqrt{3}/2$
53. $u_1 = -\sqrt{3}/2;\ u_2 = \frac{1}{2}$
61. (a) $\mathbf{u} \cdot \mathbf{v} = 8;\ \mathbf{v} \cdot \mathbf{u} = 8$
(b) $\mathbf{v} \cdot \mathbf{w} = -14;\ \mathbf{w} \cdot \mathbf{v} = -14$
63. (a) $\mathbf{v} \cdot \mathbf{v} = 25;\ |\mathbf{v}|^2 = 25$
(b) $\mathbf{w} \cdot \mathbf{w} = 29;\ |\mathbf{w}|^2 = 29$

65. $\cos\theta = 7/\sqrt{170};\ \theta \approx 57.53°$ or $\theta \approx 1.00$ radian

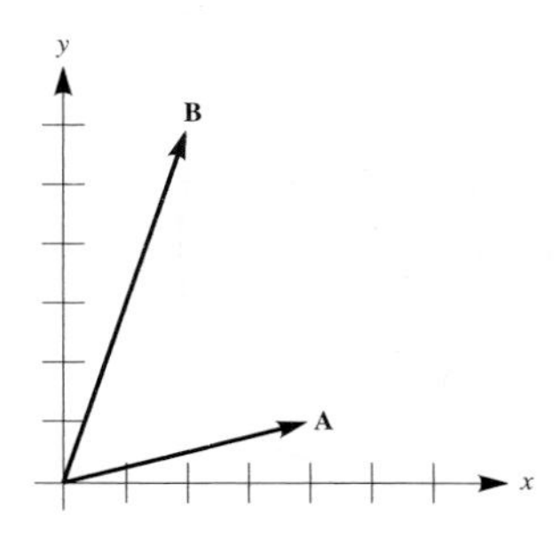

67. $\cos\theta = -57/\sqrt{3538};\ \theta \approx 163.39°$ or $\theta \approx 2.85$ radians

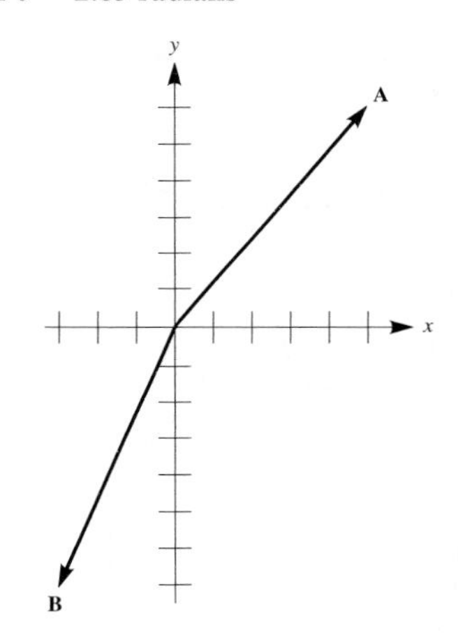

69. (a) $\cos\theta = -7/\sqrt{170};\ \theta \approx 122.47°$ or $\theta \approx 2.14$ radians

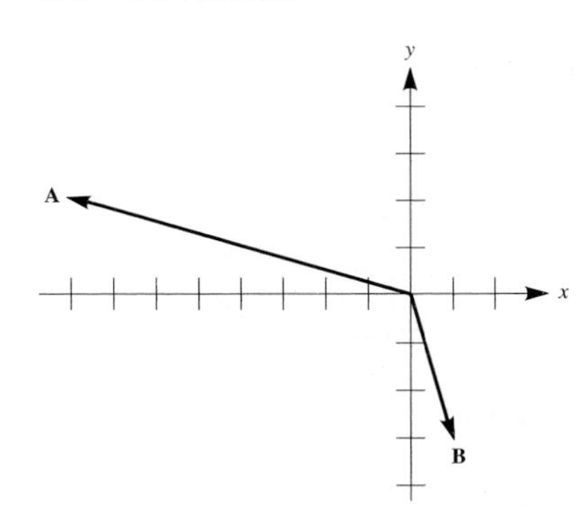

(b) $\cos\theta = 7/\sqrt{170};\ \theta \approx 57.53°$ or $\theta \approx 1.00$ radian

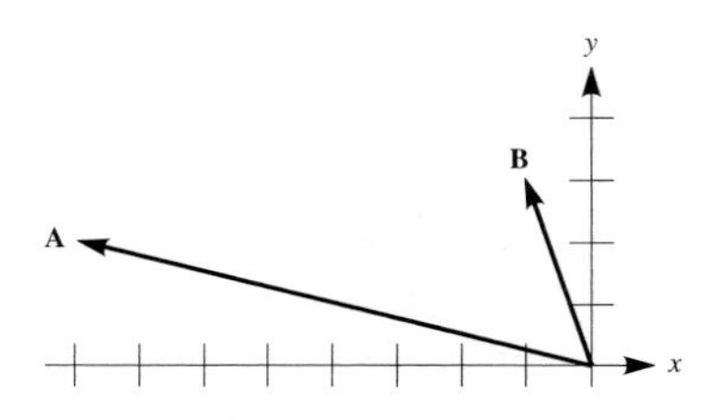

71. (a) $\cos\theta = 0$
(b) The vectors are perpendicular.
(c)

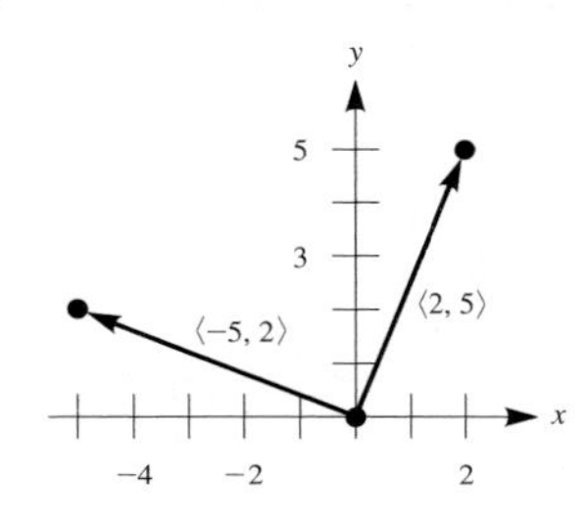

73. (a) $\mathbf{A} \cdot \mathbf{B} = 0$ implies $\cos\theta = 0$, also $0 \le \theta < \pi$, and thus $\theta = \pi/2$. So the vectors are perpendicular.
(b) If **A** and **B** are perpendicular, then $\cos\theta = 0$, so $\mathbf{A} \cdot \mathbf{B} = 0$.
75. $\langle \frac{5}{13}, \frac{12}{13}\rangle$ and $\langle -\frac{5}{13}, -\frac{12}{13}\rangle$

Exercise Set 9.5

1. $(2, 3)$ **3.** $(\frac{5\sqrt{3}}{2}, 1)$
5. $(\frac{3\sqrt{2}}{4}, \frac{3\sqrt{2}}{4})$
7. (a) $0 \le t \le 1$

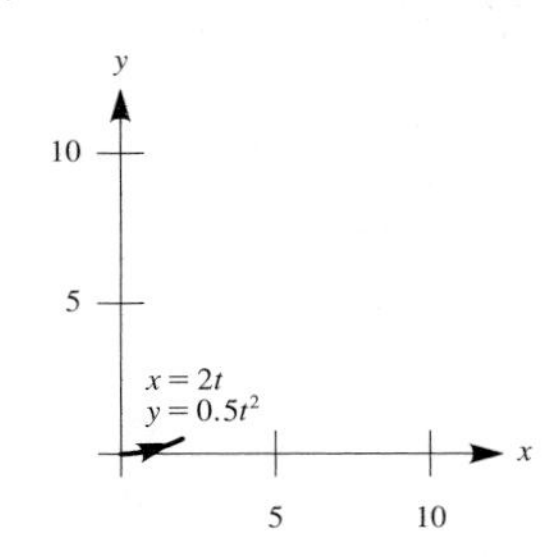

$0 \le t \le 3$

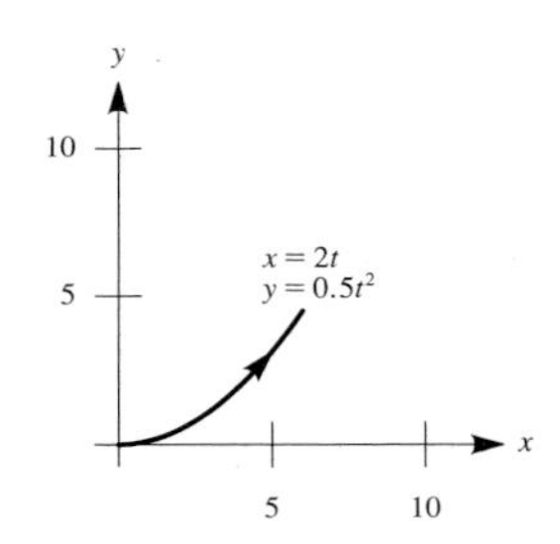

$0 \le t \le 4$

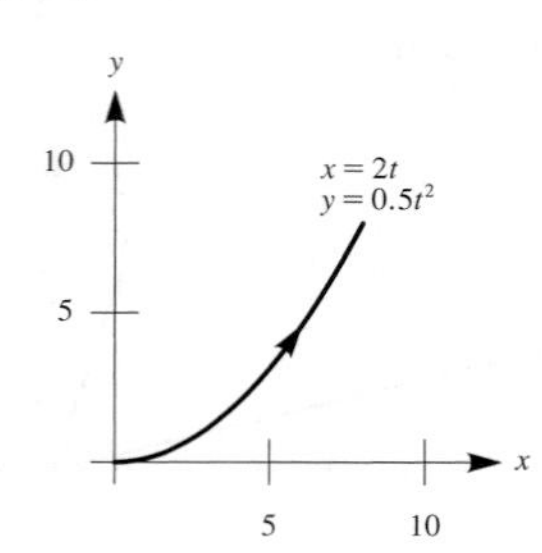

As the interval for t gets larger, the curve resembles a parabola.

(b) $-5 \le t \le 5$

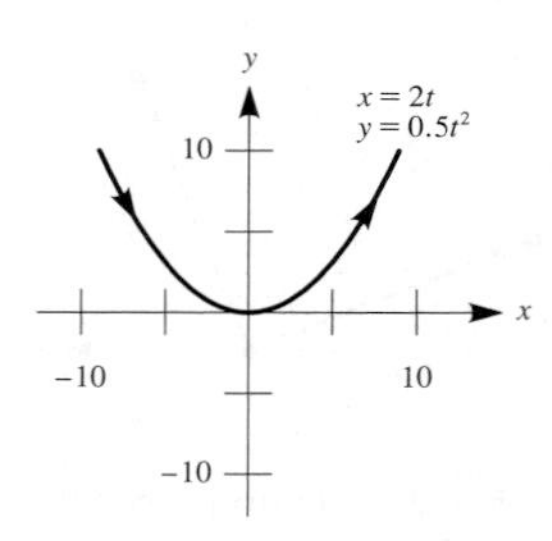

The restrictions on t in Figure 1(b) are $0 \le t \le 5$.

9.

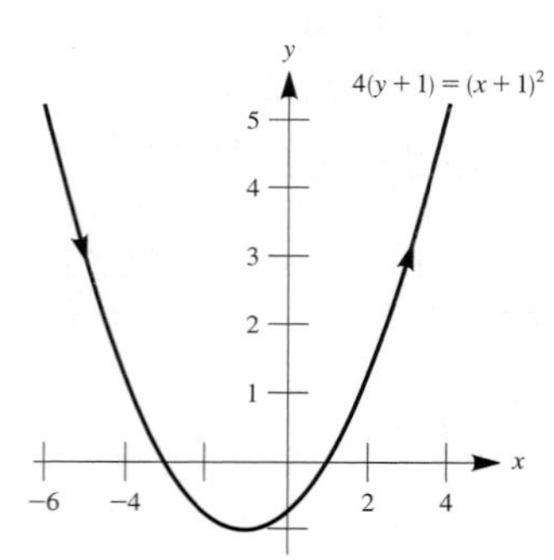

11.

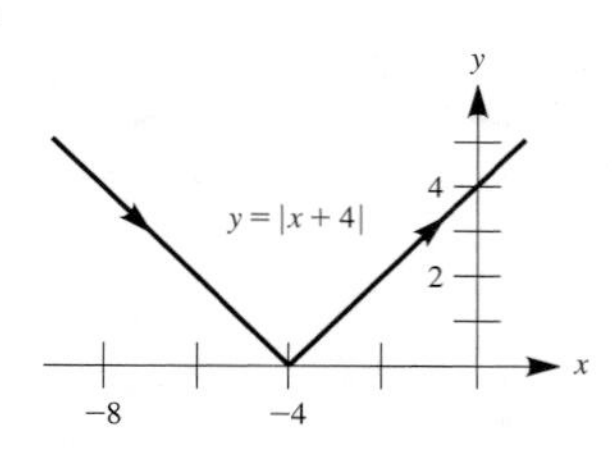

13.

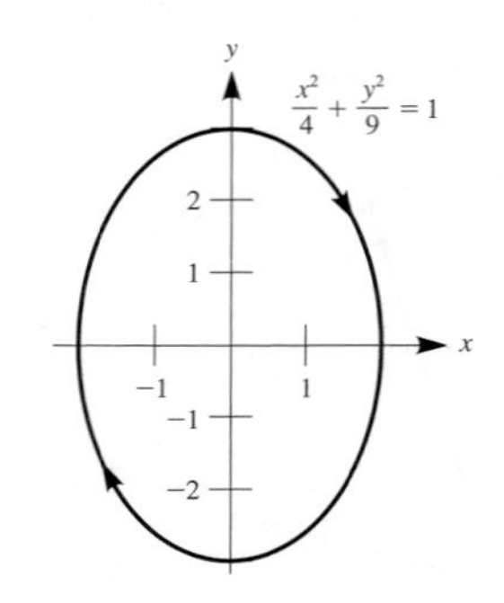

15.

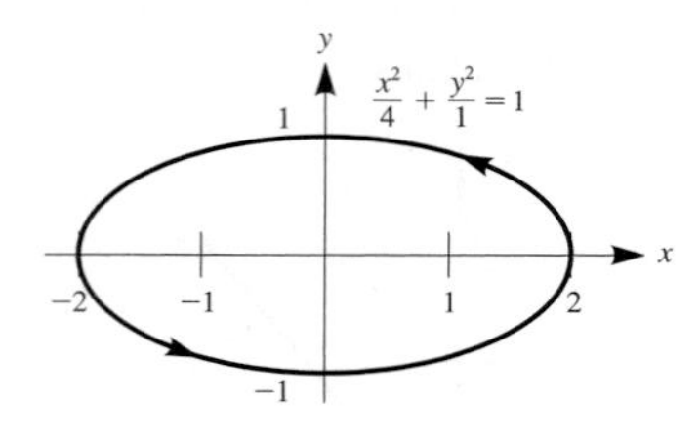

17. (a)

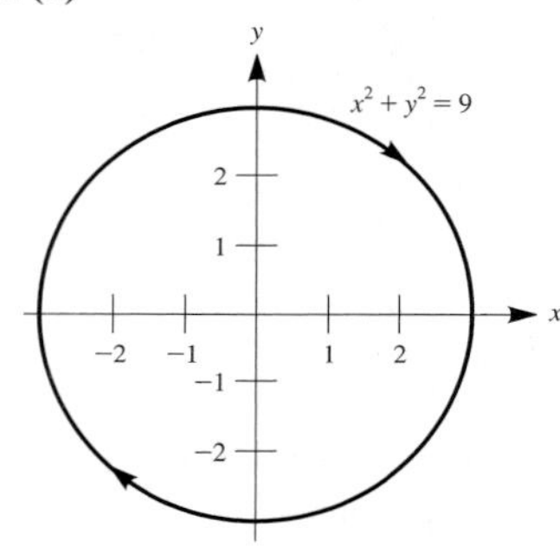

(b)

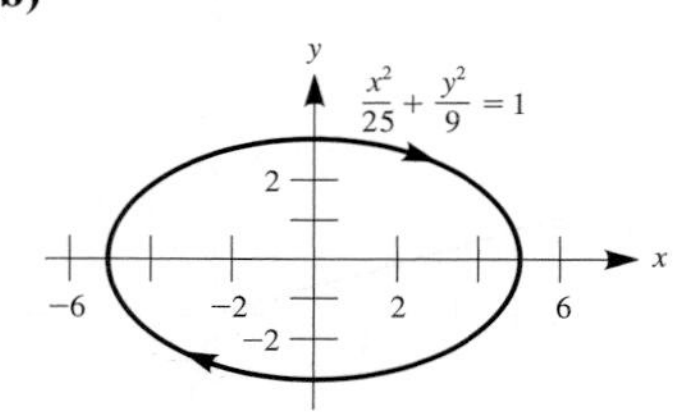

21. (a) $t = 1$: $x \approx 70.7$, $y \approx 59.7$
$t = 2$: $x \approx 141.4$, $y \approx 82.4$
$t = 3$: $x \approx 212.1$, $y \approx 73.1$
(b) 4.49 seconds; 317 feet

23. (a)

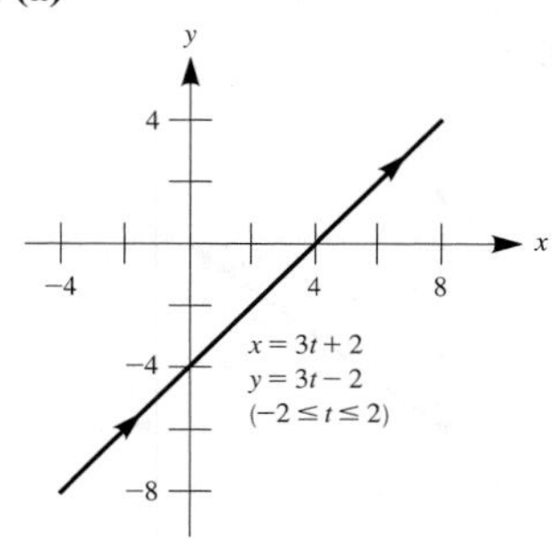

(b)

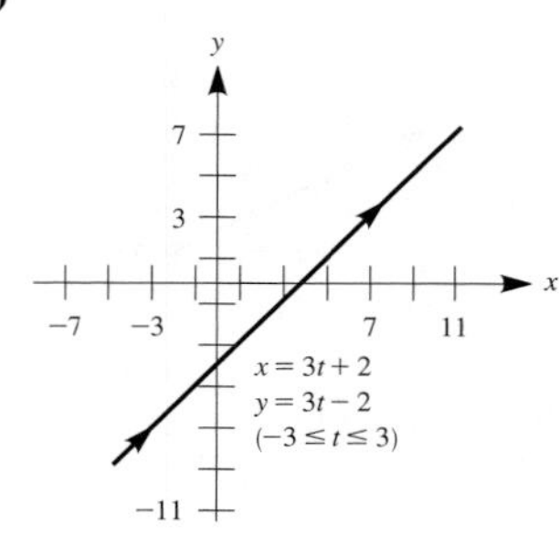

25.

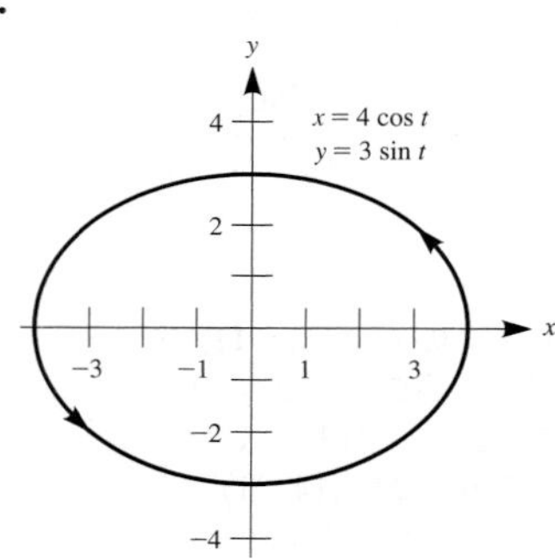

27.

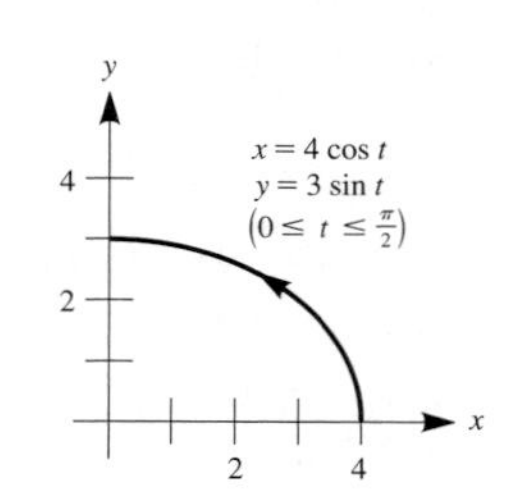

29.

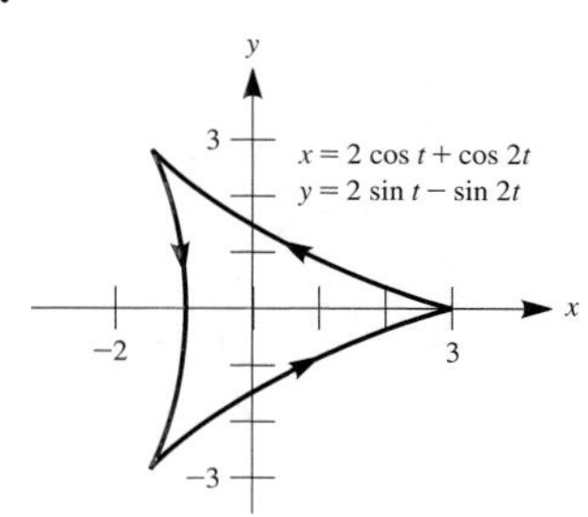

31.

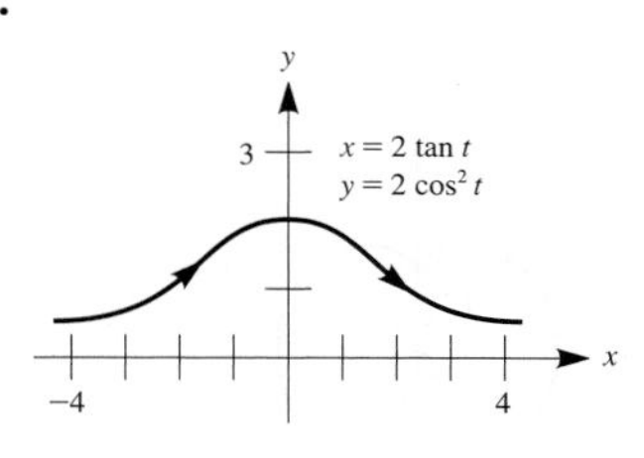

33.

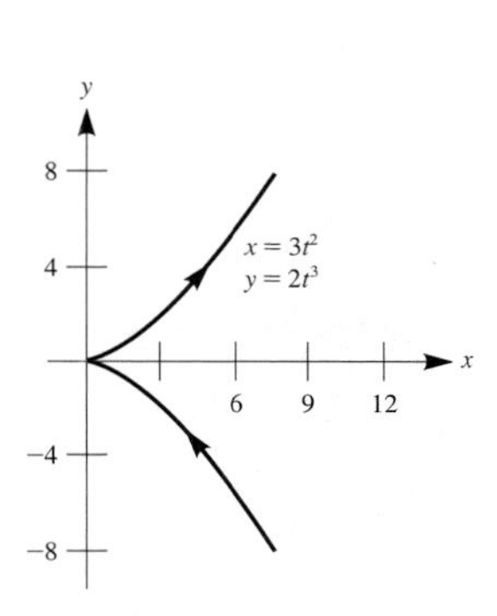

35.

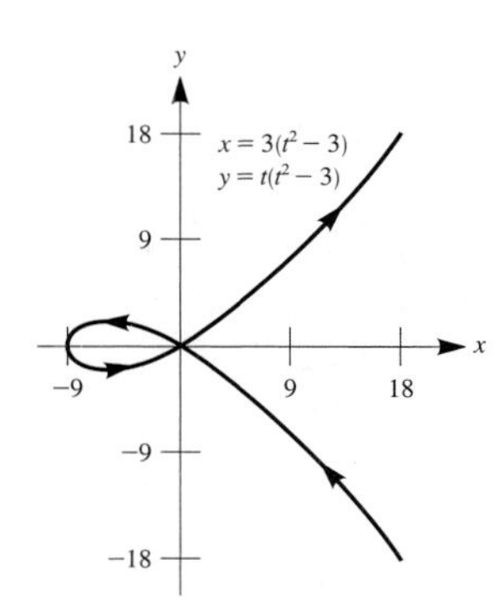

37.

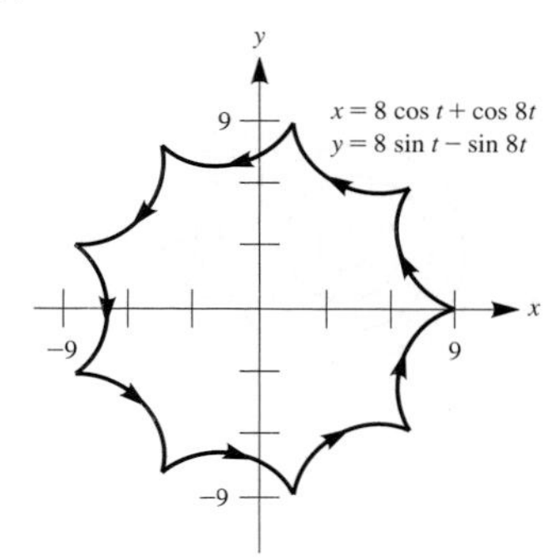

39.

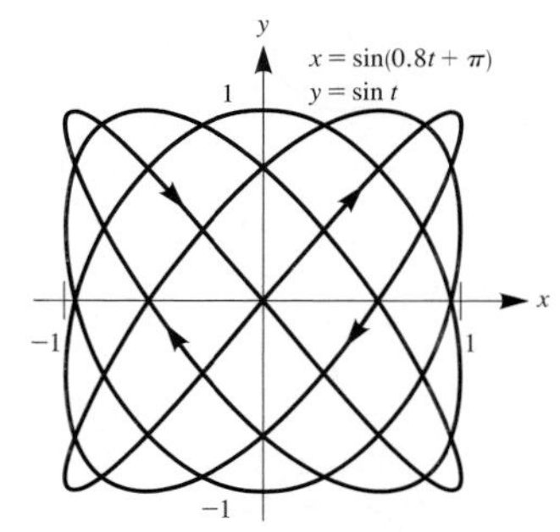

Exercise Set 9.6

1. (a) $(-\frac{3}{2}, \frac{3\sqrt{3}}{2})$
(b) $(2\sqrt{3}, -2)$

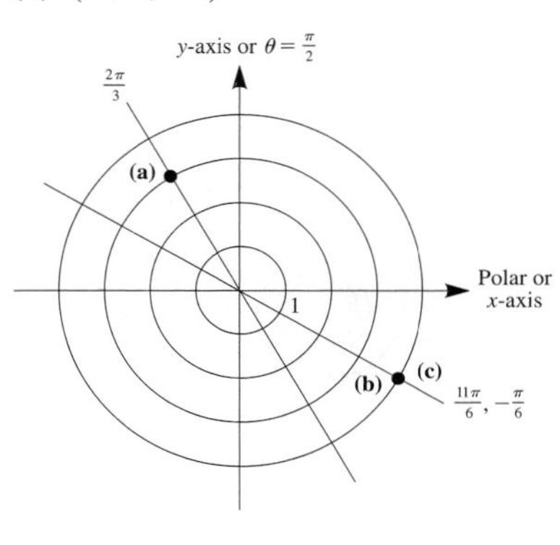

(c) $(2\sqrt{3}, -2)$

3. (a) $(0, 1)$
(b) $(0, 1)$

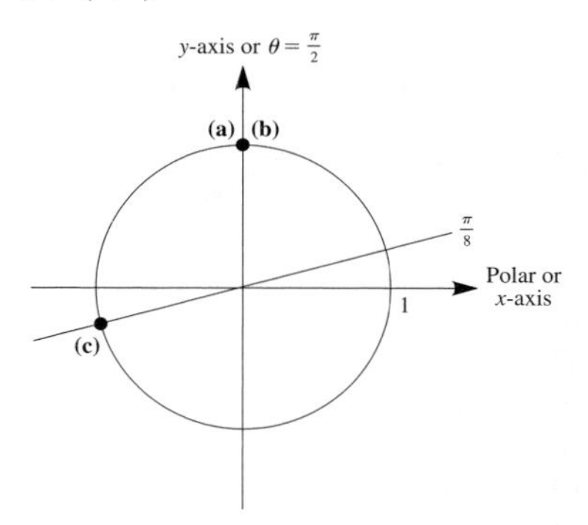

(c) $\left(-\frac{\sqrt{2+\sqrt{2}}}{2}, -\frac{\sqrt{2-\sqrt{2}}}{2}\right)$
5. $(\sqrt{2}, \frac{5\pi}{4})$ **7.** $(x-1)^2 + y^2 = 1$
9. $x^4 + x^2y^2 - y^2 = 0$
11. $(x^2 + y^2)^3 = 9(x^2 - y^2)^2$
13. $\frac{x^2}{4} + \frac{y^2}{8} = 1$ **15.** $y = -\sqrt{3}x + 4$
17. $r = \frac{2}{3\cos\theta - 4\sin\theta}$
19. $r = \tan^2\theta \sec\theta$ **21.** $r^2 = \csc 2\theta$
23. $r^2 = \frac{9}{9\cos^2\theta + \sin^2\theta}$
25. $A(\frac{8}{3}, \frac{\pi}{6}), B(\frac{8}{3}, \frac{5\pi}{6}), C(4, \pi), D(8, \frac{7\pi}{6})$
27. $A(1, \frac{\pi}{6}), B(1, \frac{5\pi}{6}), C(1, \frac{7\pi}{6}), D(-2, \frac{\pi}{2})$
29. $A(1, 0), B(1.14, \frac{\pi}{4}), C(1.48, \frac{3\pi}{4}),$
$D(1.69, \pi), E(1.92, \frac{5\pi}{4}), F(2.19, \frac{3\pi}{2}),$
$G(2.50, \frac{7\pi}{4}), H(2.85, 2\pi), I(3.25, \frac{9\pi}{4}),$
$J(4.22, \frac{11\pi}{4}), K(5.48, \frac{13\pi}{4})$
31. $2\sqrt{5}$ **33.** $\sqrt{21}$
35. (a) $r^2 - 8r\cos\theta = -12$
(b) $r^2 - 8r\cos(\theta - \frac{2\pi}{3}) = -12$
(c) $r = 2$ **37. (a)** $r = 2\cos(\theta - \frac{3\pi}{2})$
(b) $r = 2\cos(\theta - \frac{\pi}{4})$ **39. (a)** 2
(b) $(\frac{4\sqrt{3}}{3}, 0)$ and $(4, \frac{\pi}{2})$ **(c)** $(2, \frac{\pi}{6})$
(d)

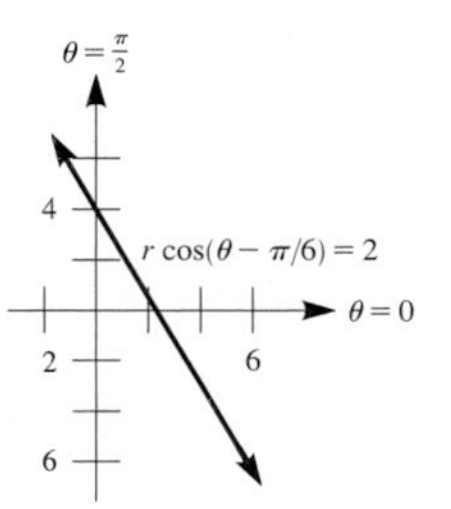

(e) $\sqrt{3}x + y = 4$
41. (a) 4 **(b)** $(-8, 0)$ and $(-\frac{8}{3}\sqrt{3}, \frac{\pi}{2})$
(c) $(4, -\frac{2\pi}{3})$

(d)

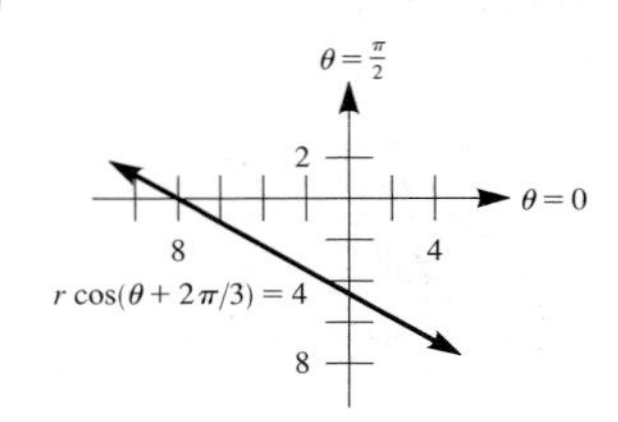

(e) $-x - \sqrt{3}y = 8$

43. (a) $(2, \frac{5\pi}{6})$

(b) $r\cos(\theta - \frac{5\pi}{6}) = 2$

(c) x-intercept: $-\frac{4\sqrt{3}}{3}$; y-intercept: 4

(d) $-\sqrt{3}x + y = 4$

45. (a)

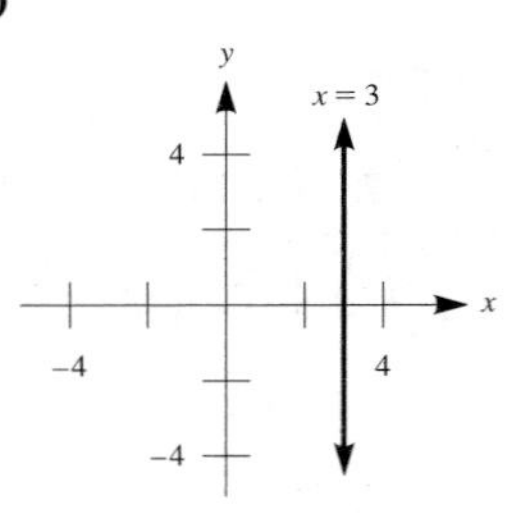

(b)

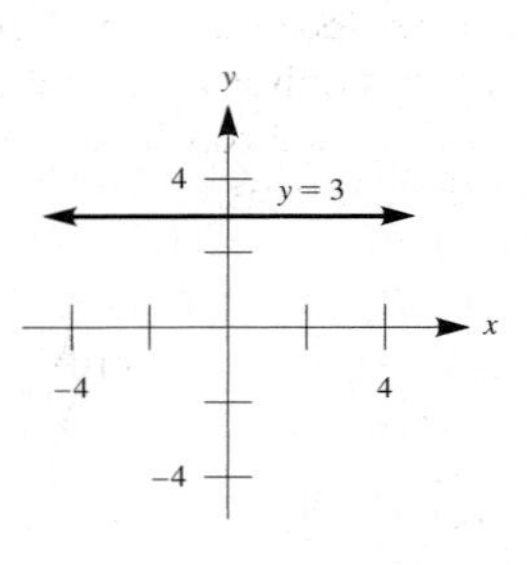

47. (a) $r\cos(\theta - \alpha) = d$

53. (a) $A(1, 0)$, $B(1, \frac{2\pi}{3})$, $C(1, \frac{4\pi}{3})$

Exercise Set 9.7

1. The graphs are four concentric circles with centers at the origin and radii 2, 4, 6, and 8.

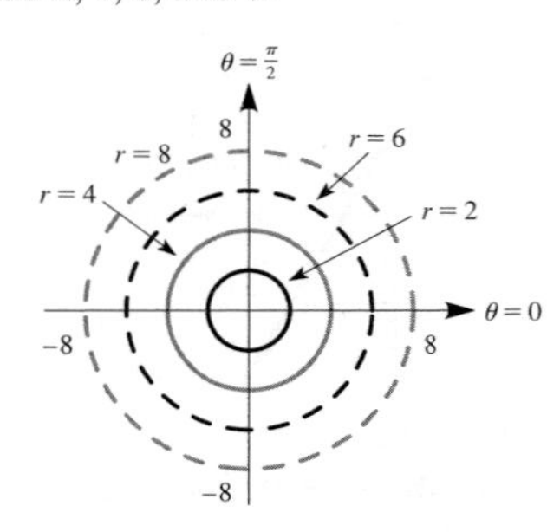

3.

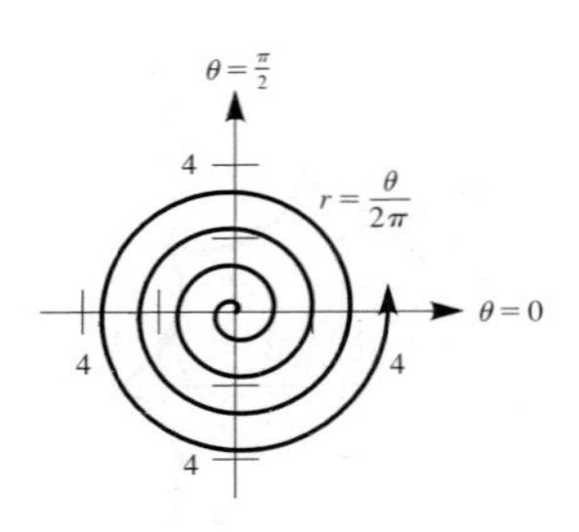

5.

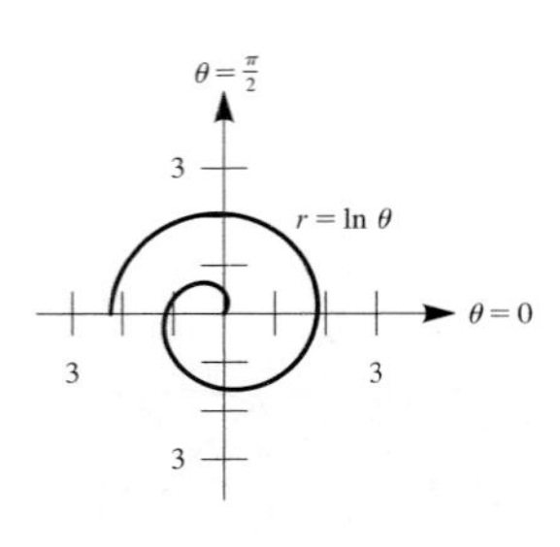

7.

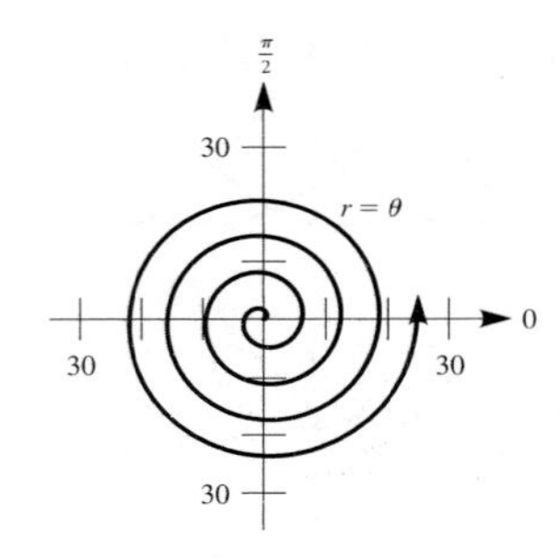

9.

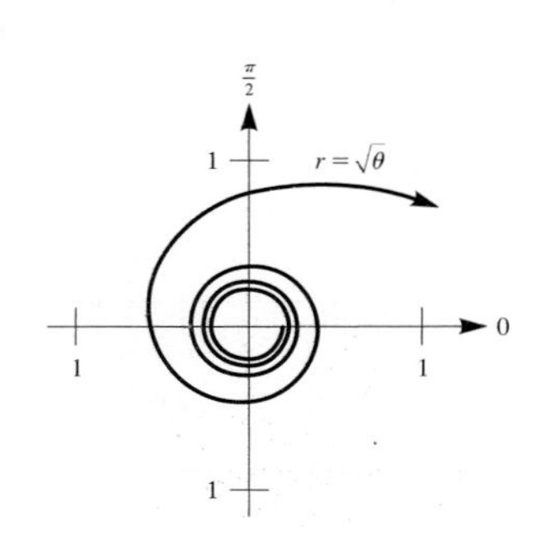

11.

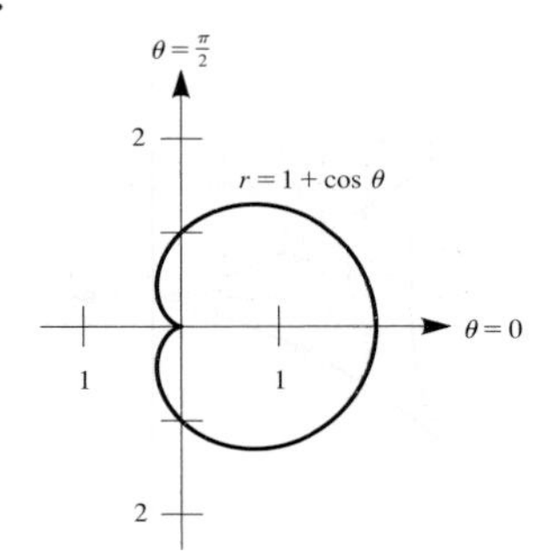

13.

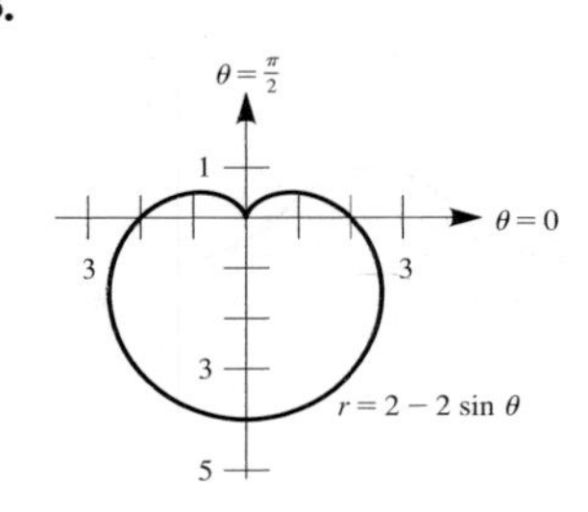

15.

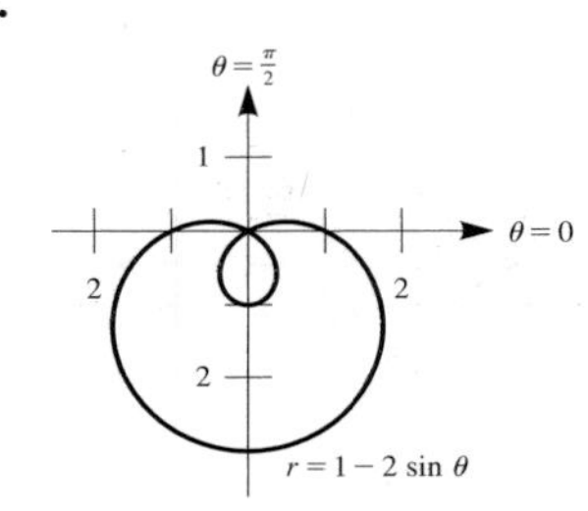

17.

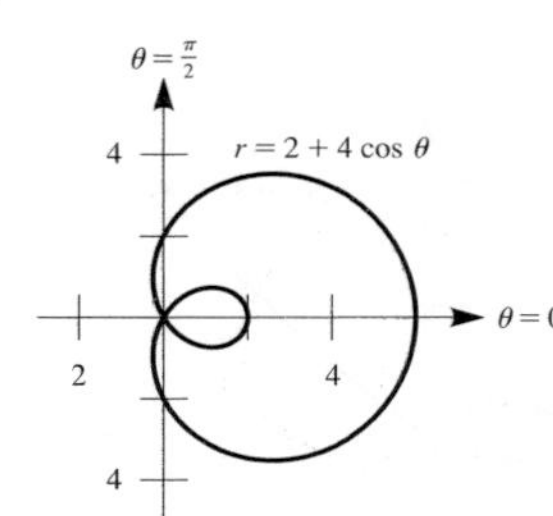

19.

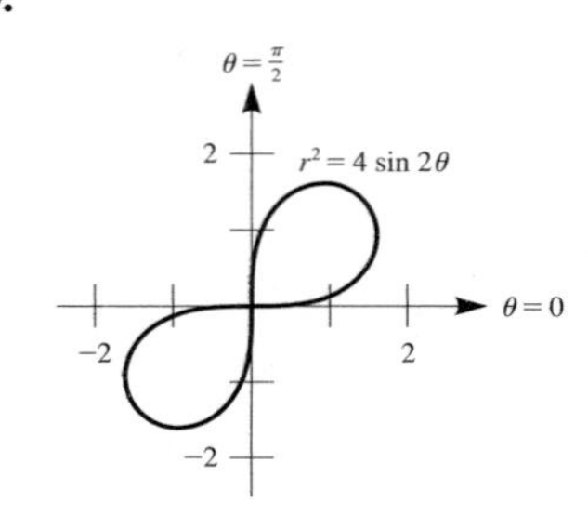

21.

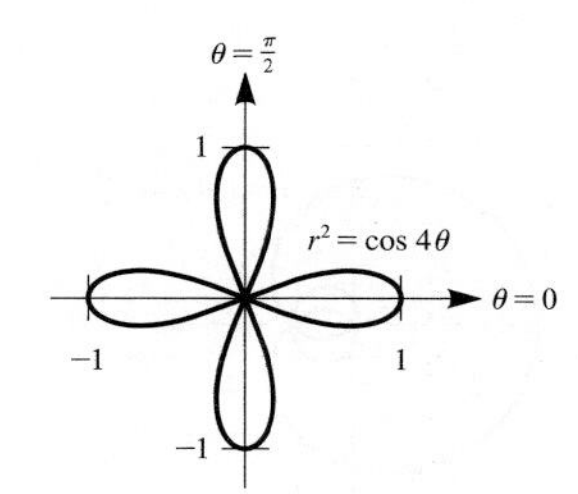

23.

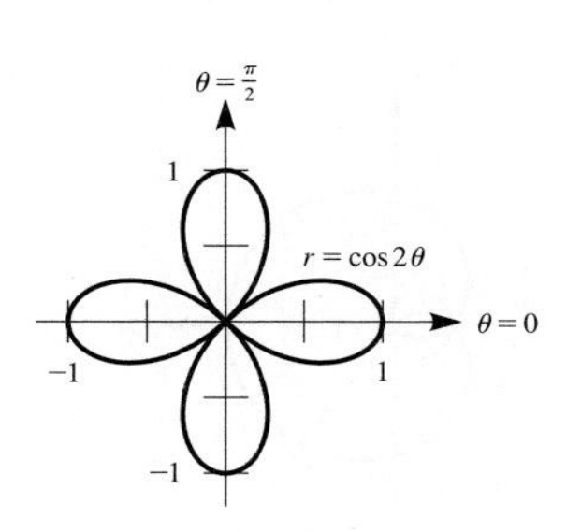

25.

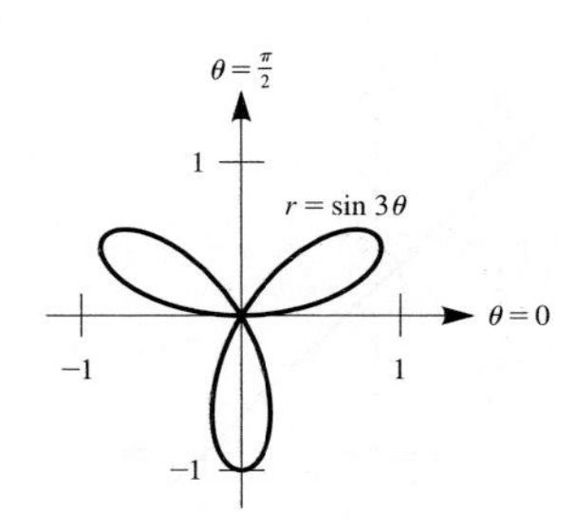

27.

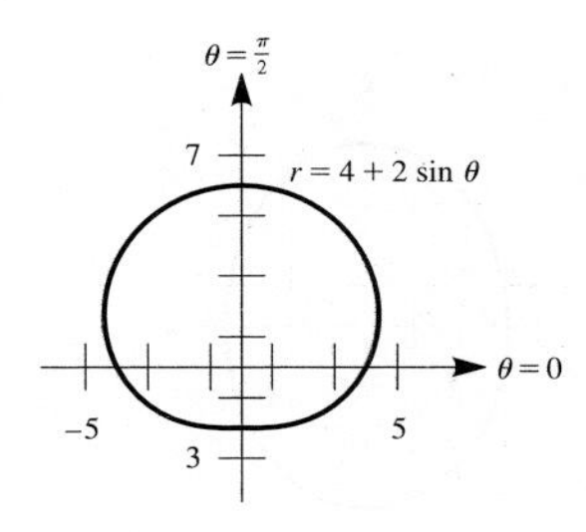

29.

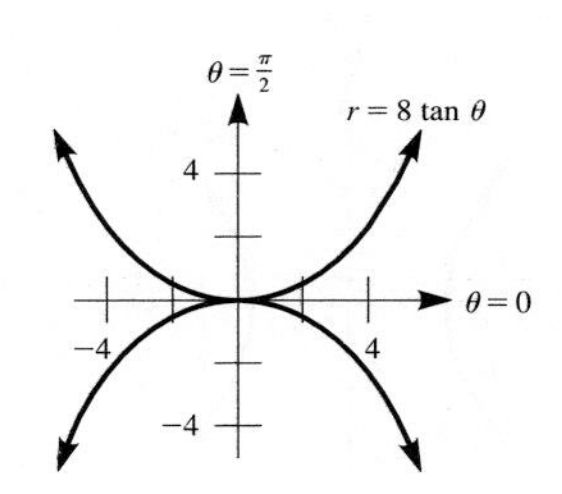

31.

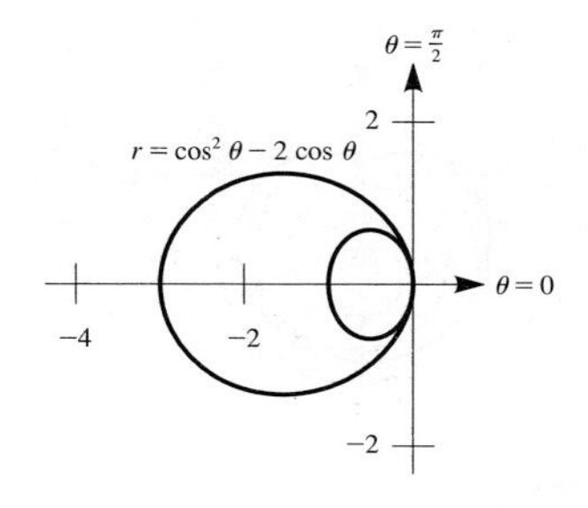

33.

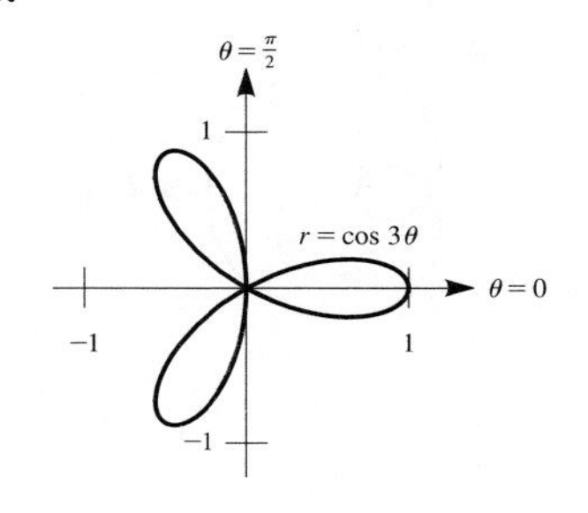

35. (a)

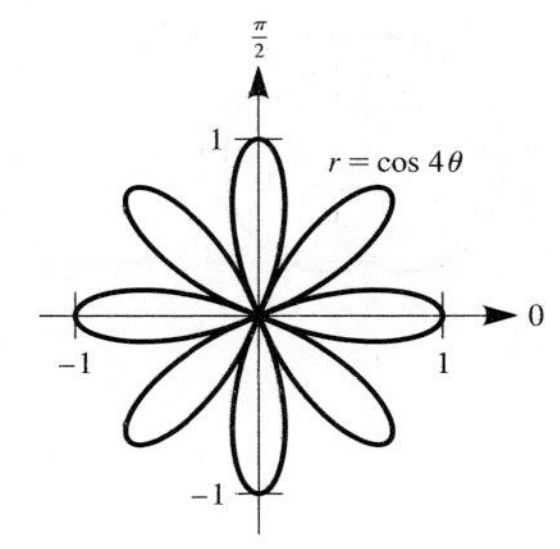

(b)

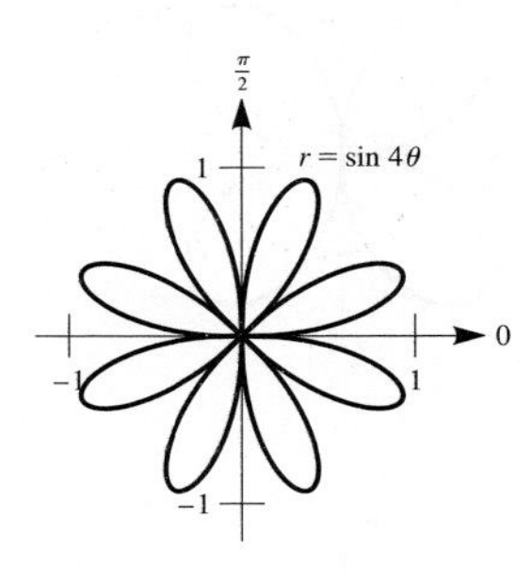

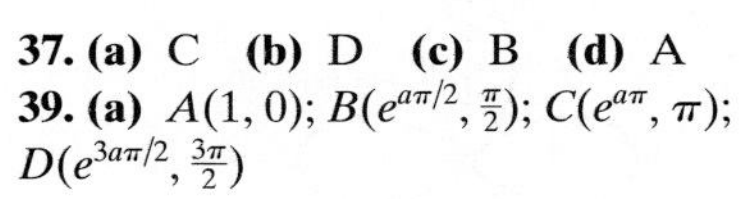

37. (a) C **(b)** D **(c)** B **(d)** A

39. (a) $A(1, 0)$; $B(e^{a\pi/2}, \frac{\pi}{2})$; $C(e^{a\pi}, \pi)$; $D(e^{3a\pi/2}, \frac{3\pi}{2})$

41. (a) The curves are identical.

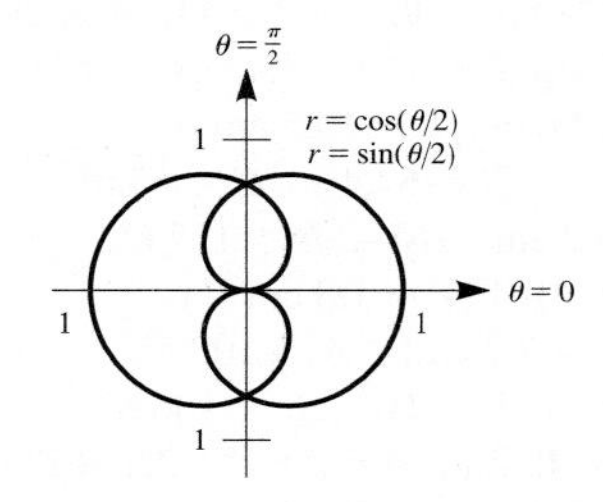

43.

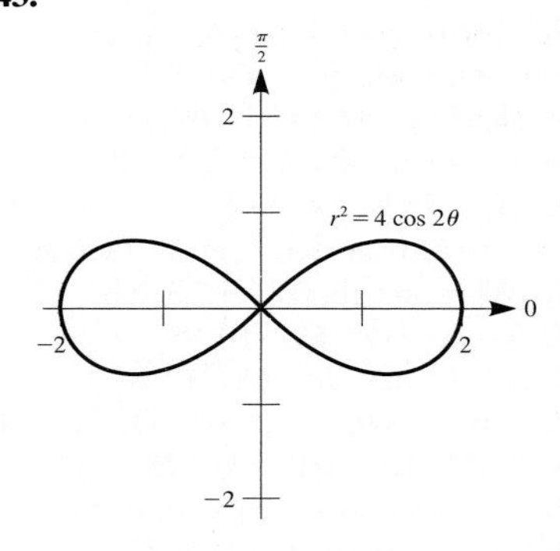

45. The inner loop near the origin is not simple but rather a cardioid-type shape that passes through both the first and fourth quadrants.

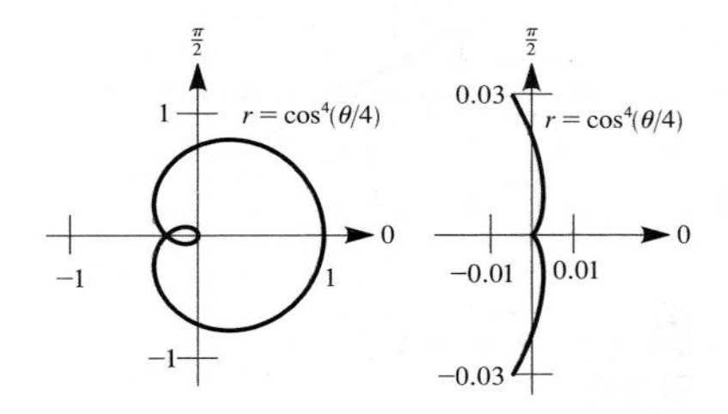

49.

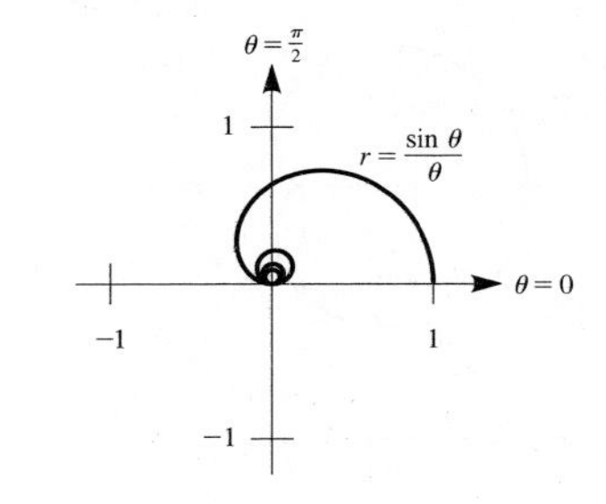

Chapter 9 Review Exercises

3. 2521 cm^2 **9.** $c = 6$ **11.** $A = 30°$ **13.** 7 square units **15.** $\angle C = 55°$, $a \approx 12.6$ cm, $b \approx 19.5$ cm **17. (a)** $\angle B \approx 62.4°$, $\angle C \approx 65.6°$, $c \approx 9.2$ cm **(b)** $\angle B \approx 117.6°$, $\angle C \approx 10.4°$, $c \approx 1.8$ cm **19.** $c \approx 7.7$ cm, $\angle A \approx 108.5°$, $\angle B \approx 47.5°$ **21.** $\angle C \approx 106.6°$, $\angle B \approx 48.2°$, $\angle A \approx 25.2°$ **23.** 9.21 cm **25.** 32.48 cm^2 **27.** 55.23 cm^2 **29.** 7.89 cm **31.** 15.43 cm **33.** 15 cm **35.** 11 cm **39. (a)** $\cos A = \frac{3}{4}$, $\cos C = \frac{1}{8}$ **41.** 36° **43.** 58.76° **45.** 18.65 m **47. (a)** 11.76 cm **(b)** 19.02 cm **49.** $a = 5, b = 8$ **51.** $|\mathbf{R}| = 25$ N, $\theta \approx 53.1°$ **53.** $v_x \approx 41.0$ cm/sec, $v_y \approx 28.7$ cm/sec **55.** $|\mathbf{W}| \approx 36.4$ lb, $|\mathbf{W}_p| \approx 33.2$ lb **57.** $b = \pm\sqrt{15}$ **59.** $\langle 10, 9\rangle$ **61.** $\langle 12, 7\rangle$ **63.** 48 **65.** $\langle 12, 8\rangle$ **67.** $\langle -6, 2\rangle$ **69.** $\langle -7, -6\rangle$ **71.** $7\mathbf{i} - 6\mathbf{j}$ **73.** $\langle 3\sqrt{13}/13, 2\sqrt{13}/13\rangle$ **75.** 4.89 **77.** $r^2 - 10r\cos(\theta - \frac{\pi}{6}) = -16$ **79. (a)** 3 **(b)** $(6, 0)$, $(2\sqrt{3}, \frac{\pi}{2})$ **(c)** $(3, \frac{\pi}{3})$
(d)

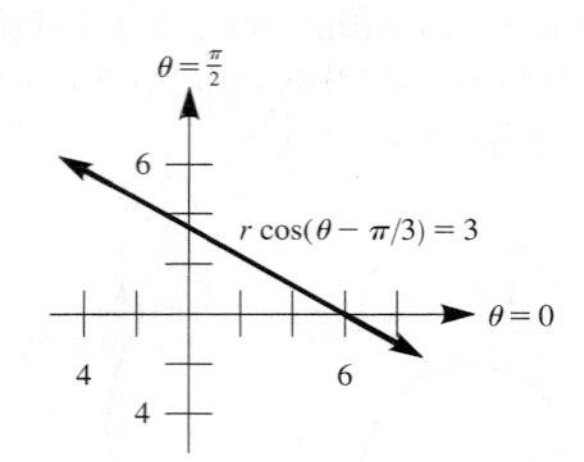

(e) $x + \sqrt{3}y = 6$
81. (a)

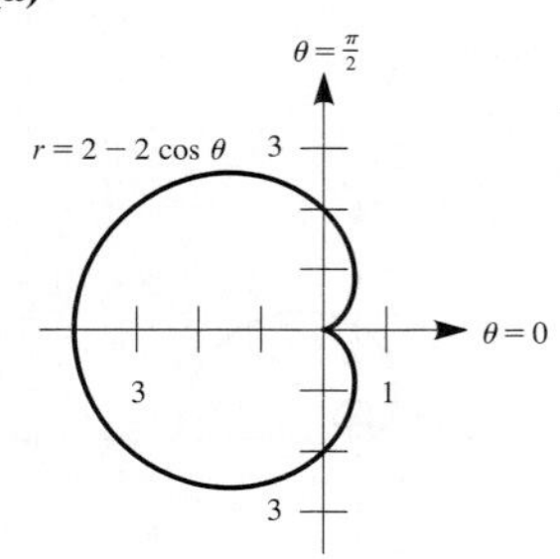

(b)

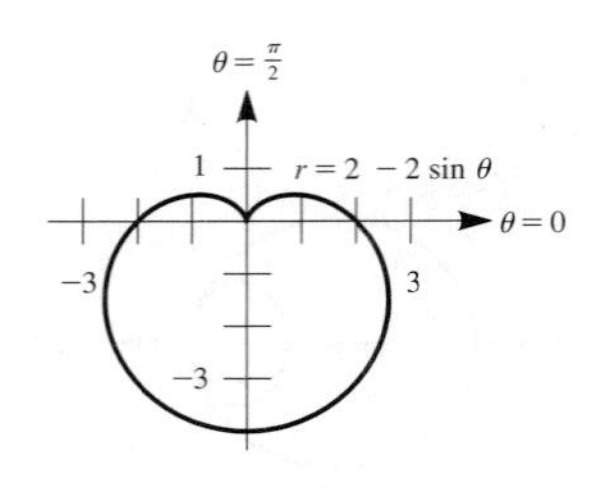

83. (a)

$r = 2\cos\theta - 1$

(b)

$r = 2\sin\theta - 1$

85. (a)

$r = 4\sin 2\theta$

(b)

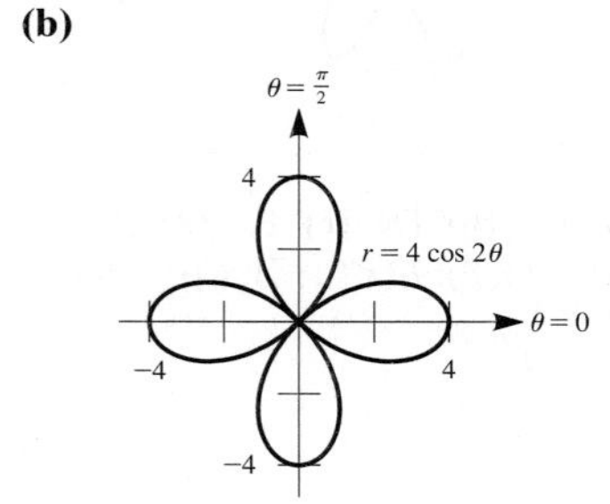

87. (a)

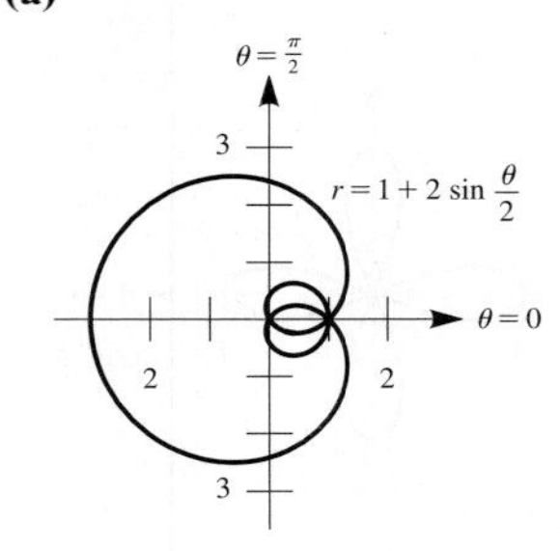

(b)

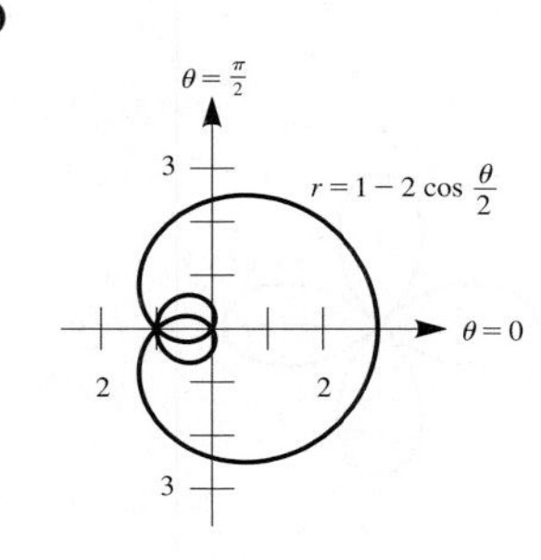

89.

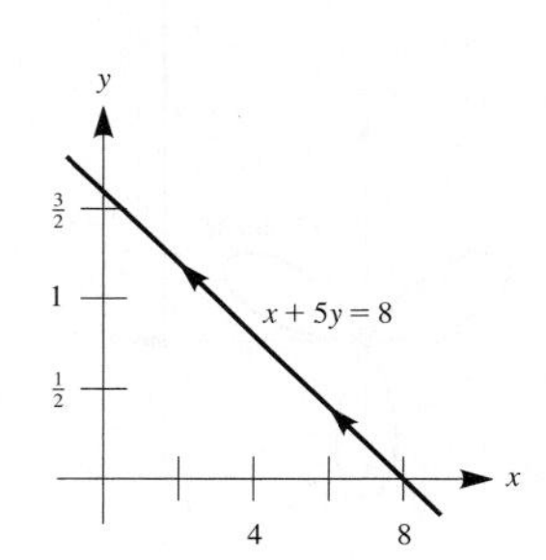

91.

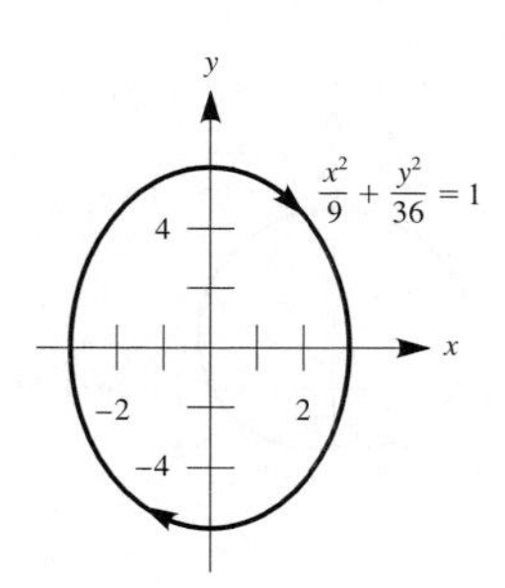

93.

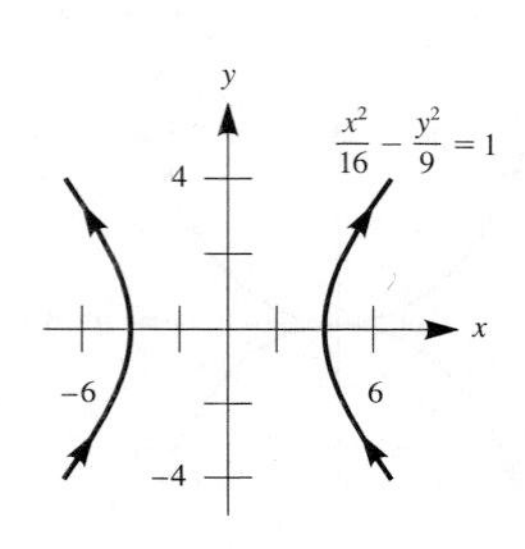

95. (a) $P\left(\frac{\sin\alpha}{\alpha}, \alpha\right), Q\left(\frac{\sin 2\alpha}{2\alpha}, 2\alpha\right)$

(b) $P\left(\frac{\sin\alpha\cos\alpha}{\alpha}, \frac{\sin^2\alpha}{\alpha}\right)$,

$Q\left(\frac{\sin 2\alpha\cos 2\alpha}{2\alpha}, \frac{\sin^2 2\alpha}{2\alpha}\right)$

(f) The product of the slopes of $\overline{PQ}$ and $\overline{OQ}$ is -1.

Chapter 9 Test

1. $a = 7$ cm **2.** If θ is the angle opposite the 4 cm side, then $\cos\theta = -\frac{1}{4} < 0$, so θ must be obtuse (not acute). **3.** 20 cm **4.** $5\sqrt{3}$ ft
5. $A \approx 68.2°$
6. $A(\theta) = \sin\theta(1 + \cos\theta)$
7. $2\sqrt{17 - 4\sqrt{2}}$ cm
8. $a \approx 3.3$ cm, $C \approx 26.2°$, $B \approx 126.8°$
9. (a) $2\sqrt{5}$ N **(b)** $\tan\theta = 2$
10. (a) $4\sqrt{13 - 12\cos 110°}$ N
(b) $\sin\theta = \dfrac{2\sin 110°}{\sqrt{13 - 12\cos 110°}}$
11. ground speed $= 50\sqrt{37}$ mph, $\tan\theta = \frac{1}{6}$
12. (a) $\langle 13, 5\rangle$ **(b)** $\sqrt{194}$ **(c)** $\mathbf{i} - 3\mathbf{j}$
13. $\langle \frac{-11\sqrt{130}}{130}, \frac{-3\sqrt{130}}{130}\rangle$
14. $(x^2 + y^2)^2 = x^2 - y^2$
15.

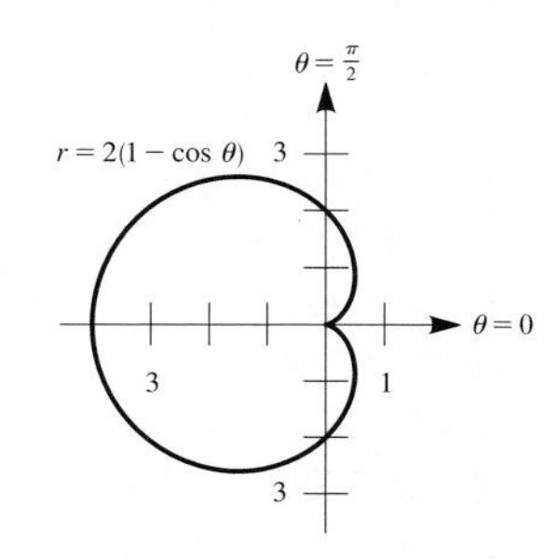

16.

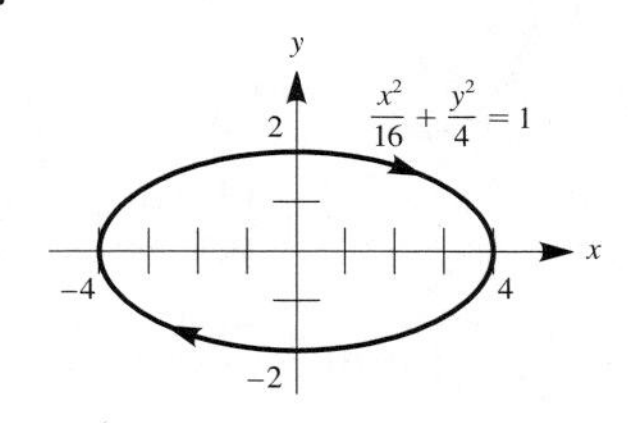

17. $\sqrt{13}$
18. $r^2 - 10r\cos(\theta - \frac{\pi}{2}) = -21$; no
19. (a) $r\cos(\theta - \frac{5\pi}{6}) = 4$

CHAPTER 10
Exercise Set 10.1

1. (a) yes **(b)** no **(c)** yes **(d)** yes **3.** yes **5.** no
7. consistent (one solution)

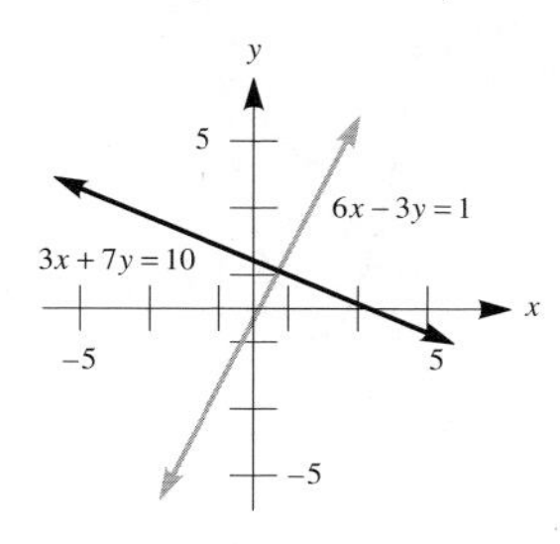

9. inconsistent (no solution)

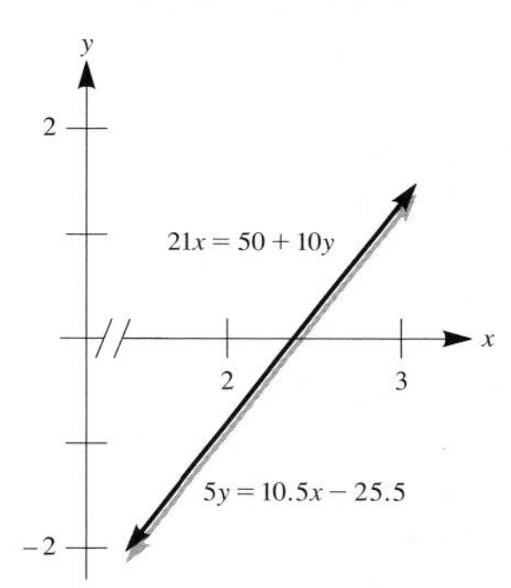

11. $(3, 5)$ **13.** $(-\frac{1}{28}, \frac{39}{28})$
15. $(-\frac{34}{49}, -\frac{20}{49})$ **17.** $\left(x, \frac{3 - 4x}{6}\right)$, $x =$ any real number
19. (a) $(16.30, -24.5)$

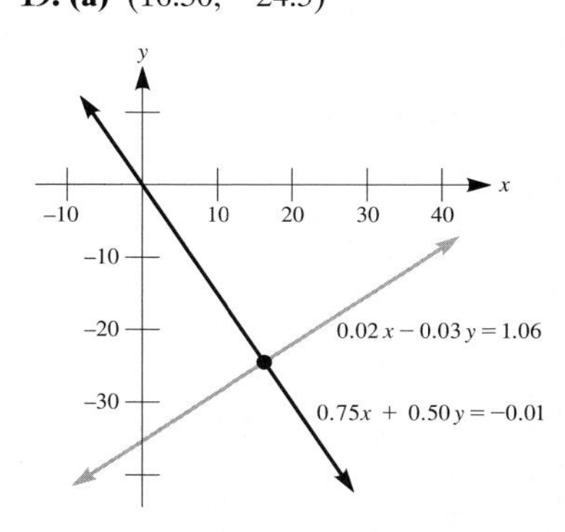

(b) $(\frac{5297}{325}, -\frac{7952}{325}) \approx (16.30, -24.47)$
21. $(-\frac{2}{9}, \frac{23}{27})$ **23.** $(-\frac{283}{242}, -\frac{3}{121})$
25. $(-\frac{226}{25}, -\frac{939}{50})$ **27.** $(\frac{5}{8}, 0)$
29. $y = x^2 + 3x + 4$
31. $A = -28$, $B = -22$
33. (a) $p = \$6$: shortage of 6000 items; $p = \$12$: shortage of 2400 items
(b) surplus of 2400 items
(c) equilibrium price: \$16; equilibrium quantity: 3200 items **35.** 80 cc of 10% solution, 120 cc of 35% solution
37. 8 pounds of \$5.20 coffee, 8 pounds of \$5.80 coffee
39. (a)

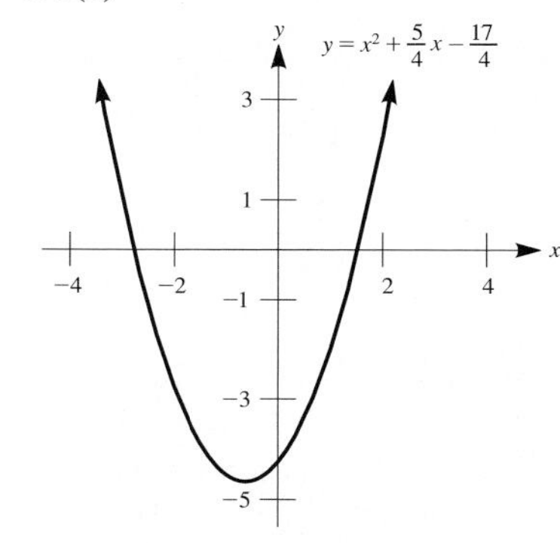

41. $a = 3$, $b = 2$
43. $\left(\frac{ab}{a + b}, \frac{ab}{a + b}\right)$; $a \neq \pm b$
45. $\left(\frac{a + b}{ab}, \frac{-1}{ab}\right)$; $a \neq 0$ and $b \neq 0$
47. $(-\frac{1}{11}, \frac{1}{9})$ **49.** $(5, 2)$
51. $(e^{-2}, e^{-3}) \approx (0.14, 0.05)$
53. $(\ln 5, 0) \approx (1.61, 0)$
55. $(4, -8), (4, 2), (-1, -8), (-1, 2)$
57. 59 **59.** $\left(\frac{1}{b}, \frac{1}{a}\right)$
61. (a) $k = -\frac{29}{7}$ and 2
(b) $k = -\frac{29}{7}$:

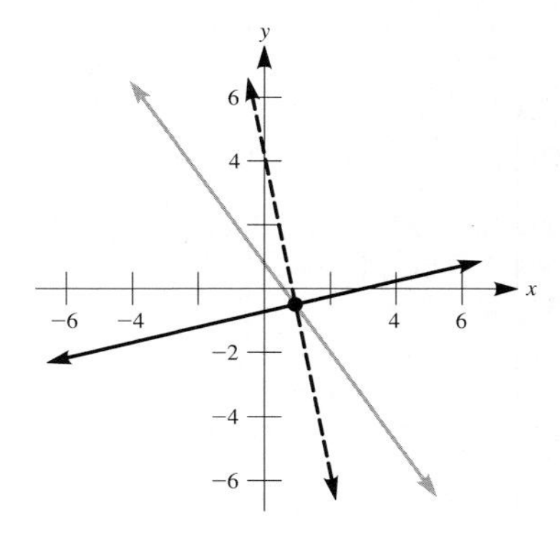

$k = 2$:

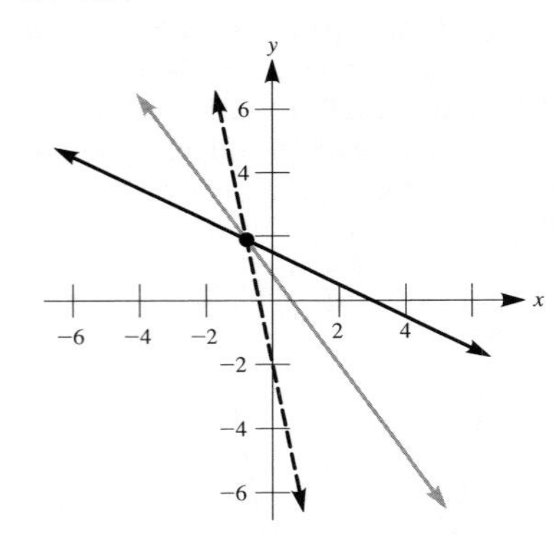

(c)

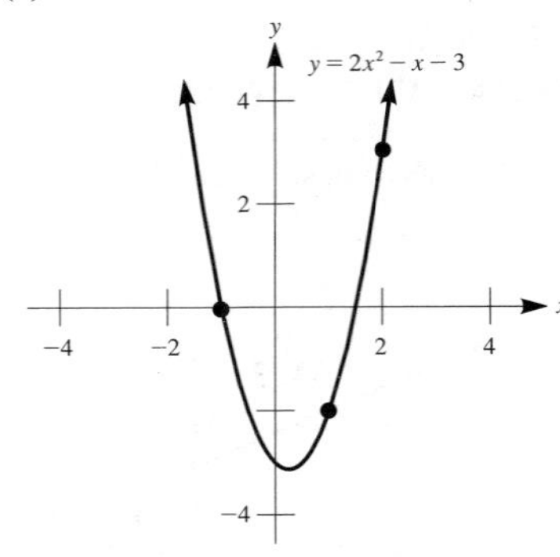

(d) $y = 2x^2 - x - 3$ **35.** 120 utility chairs, 24 secretarial chairs, 10 managerial chairs
37. (a) $R(t) = -0.3t^2 + 3t + 1$
(b) December; \$850,000
39. $x = \ln a, y = \ln 2a, z = \ln \frac{a}{2}$
41. $T_1 = 42, T_2 = 43, T_3 = 40$
43. 60 miles

Exercise Set 10.2

1. $(-3, -2, -1)$ **3.** $(-\frac{1}{60}, -\frac{2}{15}, \frac{3}{5})$
5. $(\frac{25}{36}, \frac{5}{9}, -\frac{1}{3})$ **7.** $(x, \frac{x}{8}, 0)$, x = any real number **9.** $(-1, 0, 1, -5)$
11. $(\frac{11}{3}, \frac{8}{3}, \frac{17}{3})$ **13.** $(1, 0, 1)$

15. $(2, 3, 1)$ **17.** inconsistent (no solution) **19.** $\left(\frac{z+1}{-7}, \frac{5(z+1)}{7}, z\right)$, z = any real number
21. $\left(\frac{11-5z}{7}, \frac{-3z-6}{7}, z\right)$, z = any real number **23.** $(4, 1, -3, 2)$
25. $\left(\frac{17-17z}{5}, \frac{8z-3}{5}, z\right)$, z = any real number
27. $\left(\frac{12+10w}{11}, \frac{146+19w}{55}, \frac{61w+159}{55}, w\right)$, w = any real number
29. $\left(-\frac{5z}{12}, \frac{2z+3}{3}, z\right)$, z = any real number **31.** 6 Type I jets, 3 Type II jets, 5 Type III jets
33. (a) $a + b + c = -2$
$a - b + c = 0$
$4a + 2b + c = 3$
(b) $a = 2, b = -1, c = -3$

Exercise Set 10.3

1. (a) two by three (2×3)
(b) three by two (3×2)
3. five by four (5×4)
5. coefficient matrix is
$\begin{pmatrix} 2 & 3 & 4 \\ 5 & 6 & 7 \\ 8 & 9 & 10 \end{pmatrix}$; augmented matrix:
$\begin{pmatrix} 2 & 3 & 4 & 10 \\ 5 & 6 & 7 & 9 \\ 8 & 9 & 10 & 8 \end{pmatrix}$

7. coefficient matrix:
$\begin{pmatrix} 1 & 0 & 1 & 1 \\ 1 & 1 & 0 & 2 \\ 0 & 1 & 1 & 1 \\ 2 & -1 & -1 & 0 \end{pmatrix}$;
augmented matrix:
$\begin{pmatrix} 1 & 0 & 1 & 1 & -1 \\ 1 & 1 & 0 & 2 & 0 \\ 0 & 1 & 1 & 1 & 1 \\ 2 & -1 & -1 & 0 & 2 \end{pmatrix}$
9. $(-1, -2, 3)$ **11.** $(-5, 1, 3)$
13. $(3, 0, -7)$ **15.** $(8, 9, -1)$
17. $\left(\frac{9z+5}{19}, \frac{31z-6}{19}, z\right)$, z = any real number **19.** $(2, -1, 0, 3)$
21. inconsistent (no solution)
23. $\begin{pmatrix} 3 & 2 \\ 2 & 4 \end{pmatrix}$ **25.** $\begin{pmatrix} 6 & 4 \\ 4 & 8 \end{pmatrix}$

27. $\begin{pmatrix} 11 & -2 \\ 11 & 1 \end{pmatrix}$ **29.** $\begin{pmatrix} 2 & 3 \\ -1 & 4 \end{pmatrix}$
31. not defined
33. $\begin{pmatrix} 10 & -2 \\ -8 & 0 \\ 4 & 6 \end{pmatrix}$
35. $\begin{pmatrix} 2 & 4 & 11 \\ -12 & 16 & 19 \\ 14 & 12 & 43 \end{pmatrix}$
37. $\begin{pmatrix} -9 & 10 & 10 \\ 4 & -8 & -12 \\ 10 & 4 & 21 \end{pmatrix}$
39. not defined
41. $\begin{pmatrix} 0 & 0 & 0 \\ 0 & 0 & 0 \\ 0 & 0 & 0 \end{pmatrix}$ **43.** $\begin{pmatrix} 4 & 2 \\ 2 & 5 \end{pmatrix}$
45. not defined
47. $\begin{pmatrix} 1 & 18 \\ -6 & 13 \end{pmatrix}$ **49.** $\begin{pmatrix} -16 & 75 \\ -25 & 34 \end{pmatrix}$
51. (a) $\begin{pmatrix} -13 & 1 & 40 \\ 43 & 17 & 0 \\ 89 & 61 & 60 \end{pmatrix}$
(b) $\begin{pmatrix} -13 & 1 & 40 \\ 43 & 17 & 0 \\ 89 & 61 & 60 \end{pmatrix}$
(c) $\begin{pmatrix} -52 & -82 & 61 \\ 87 & 141 & 0 \\ 216 & 318 & 165 \end{pmatrix}$
(d) $\begin{pmatrix} -52 & -82 & 61 \\ 87 & 141 & 0 \\ 216 & 318 & 165 \end{pmatrix}$
53. $\begin{pmatrix} -2474 & 1182 \\ 2358 & -144 \end{pmatrix}$
55. $\begin{pmatrix} 0.5 & 0 & 0 & 0 \\ 0 & 0.5 & 0 & 0 \\ 0 & 0 & 0.5 & 0 \\ 0 & 0 & 0 & 0.5 \end{pmatrix}$
57. (a) $\begin{pmatrix} 16 & 20 \\ 24 & 28 \end{pmatrix}$ **(b)** $\begin{pmatrix} 18 & 26 \\ 18 & 26 \end{pmatrix}$
(c) $\begin{pmatrix} 14 & 14 \\ 30 & 30 \end{pmatrix}$ **(d)** $\begin{pmatrix} 18 & 26 \\ 18 & 26 \end{pmatrix}$
59. (a) $\begin{pmatrix} x \\ -y \end{pmatrix}$ **(b)** $\begin{pmatrix} -x \\ y \end{pmatrix}$
(c) $\begin{pmatrix} -x \\ -y \end{pmatrix}$; this would represent a reflection about the origin.
61. (a) $f(A) = -2, f(B) = 29, f(AB) = -58$, yes

Exercise Set 10.4

5. $\begin{pmatrix} -5 & 9 \\ 4 & -7 \end{pmatrix}$ **7.** $\begin{pmatrix} -\frac{6}{23} & \frac{1}{23} \\ \frac{5}{23} & \frac{3}{23} \end{pmatrix}$

9. no A^{-1} exists **11.** $\begin{pmatrix} 2 & -3 \\ 1 & 3 \end{pmatrix}$

13. $\begin{pmatrix} 2 & -1 \\ -3 & 2 \end{pmatrix}$ **15.** $\begin{pmatrix} \frac{6}{11} & 1 \\ -\frac{1}{11} & 0 \end{pmatrix}$

17. does not exist

19. $\begin{pmatrix} -1 & 2 & -3 \\ 2 & 1 & 0 \\ 4 & -2 & 5 \end{pmatrix}$

21. $\begin{pmatrix} 5 & -\frac{10}{3} & 1 \\ 0 & \frac{1}{3} & 0 \\ 4 & -\frac{8}{3} & 1 \end{pmatrix}$ **23.** $\begin{pmatrix} 2 & 1 & 4 \\ 3 & 2 & 5 \\ 0 & -1 & 1 \end{pmatrix}$

25. does not exist

27. $A^{-1} = \begin{pmatrix} -2.5 & 1.5 \\ 2 & -1 \end{pmatrix}$

29. D^{-1} does not exist

31. $(DE)^{-1}$ **33. (a)** $x = -1$ and $y = 1$ **(b)** $x = -132$ and $y = 48$

35. $x = 2, y = -1, z = 4$

37. (a) $\begin{pmatrix} -7 & 4 \\ 2 & -1 \end{pmatrix}$ **(b)** $(-1, 2)$

39. (a) $\begin{pmatrix} -1 & 0 & 2 \\ 3 & 1 & -6 \\ -2 & -1 & 5 \end{pmatrix}$

(b) $(21, -64, 55)$

41. (a) $\begin{pmatrix} \frac{31}{108} & \frac{2}{27} & \frac{7}{18} & -\frac{11}{108} \\ -\frac{53}{162} & \frac{1}{81} & -\frac{5}{27} & \frac{31}{162} \\ -\frac{55}{54} & -\frac{1}{27} & \frac{5}{9} & \frac{23}{54} \\ \frac{131}{54} & -\frac{4}{27} & -\frac{7}{9} & -\frac{43}{54} \end{pmatrix}$

(b) $(\frac{10}{9}, -\frac{22}{27}, -\frac{59}{9}, \frac{88}{9})$

45. 324, 286, 225, 200, 241, 212

47. 41, 15, 38, 14, 28, 28, 38, −23, 15

49. SEE JOE **51.** STUDY

53. $\begin{pmatrix} 4 & -6 & 4 & -1 \\ -6 & 14 & -11 & 3 \\ 4 & -11 & 10 & -3 \\ -1 & 3 & -3 & 1 \end{pmatrix}$

55. (a) $A^{-1} = \begin{pmatrix} -\frac{5}{2} & \frac{3}{2} \\ 2 & -1 \end{pmatrix}$

$B^{-1} = \begin{pmatrix} 7 & -8 \\ -6 & 7 \end{pmatrix}$

$B^{-1}A^{-1} = \begin{pmatrix} -\frac{67}{2} & \frac{37}{2} \\ 29 & -16 \end{pmatrix}$

(b) $(AB)^{-1} = \begin{pmatrix} -\frac{67}{2} & \frac{37}{2} \\ 29 & -16 \end{pmatrix}$, $(AB)^{-1} = B^{-1}A^{-1}$

57. (a) $\begin{pmatrix} x & 1+x \\ 1-x & -x \end{pmatrix}$; The inverse is the same as the original matrix. **(b)** $\begin{pmatrix} 11 & 12 \\ -10 & -11 \end{pmatrix}$; $\begin{pmatrix} \pi+1 & \pi+2 \\ -\pi & -1-\pi \end{pmatrix}$

Exercise Set 10.5

1. (a) 29 **(b)** −29 **3. (a)** 0 **(b)** 0 **5.** −1 **7.** −60 **9.** 9

11. (a) 314 **(b)** 674 **(c)** part (b)

13. (a) 0 **(b)** 0 **(c)** 0 **(d)** 0

15. 0 **17.** −3 **19.** 0 **21.** 570

23. 0 **25.** 12 **27. (a)** The right-hand determinant should be 10 times the left-hand determinant.

(b) $\det A = 206$ and $\det B = 2060$

29. (a) 20 **(b)** 20 **31. (a)** 174 **(b)** 174 **33.** $(y-x)(z-x)(z-y)$

35. xy **39.** $(1, 1, 2)$

41. $(2, -3, -3)$ **43.** $(0, 0, 0)$

45. $(13 - \frac{11}{3}y, y, 13 - 4y)$, y = any real number **47.** $(1, 0, -10, 2)$

49. $x = 4, -4, -1$

63. $\left(\dfrac{k(k-b)(k-c)}{a(a-b)(a-c)}, \dfrac{k(k-a)(k-c)}{b(b-a)(b-c)}, \dfrac{k(k-a)(k-b)}{c(c-a)(c-b)}\right)$

65. $y = 2x + 5$ **69. (b)** $\begin{pmatrix} -\frac{9}{61} & \frac{7}{61} \\ \frac{1}{61} & \frac{6}{61} \end{pmatrix}$

Exercise Set 10.6

1. $(0, 0), (3, 9)$

3. $(2\sqrt{6}, 1), (-2\sqrt{6}, 1)$

5. $(-1, -1)$ **7.** $(2, 3), (2, -3), (-2, 3), (-2, -3)$ **9.** $(1, 0), (-1, 0)$

11. $\left(\dfrac{-1+\sqrt{65}}{8}, \dfrac{1+\sqrt{65}}{2}\right)$, $\left(\dfrac{-1-\sqrt{65}}{8}, \dfrac{1-\sqrt{65}}{2}\right)$

13. $(\frac{\sqrt{17}}{17}, 1), (\frac{\sqrt{17}}{17}, -1), (\frac{-\sqrt{17}}{17}, 1), (\frac{-\sqrt{17}}{17}, -1)$

15. $(1, 0), (4, -\sqrt{3})$ **17.** $(2, 4)$

19. $(100, 1000), (100, \frac{1}{1000}), (\frac{1}{100}, 1000), (\frac{1}{100}, \frac{1}{1000})$

21. $\left(\dfrac{2\ln 2 - 5\ln 3}{\ln 2 - \ln 3}, \dfrac{3\ln 2}{\ln 2 - \ln 3}\right)$

23. (a) $(2.19, -2.81), (-3.19, -8.19)$

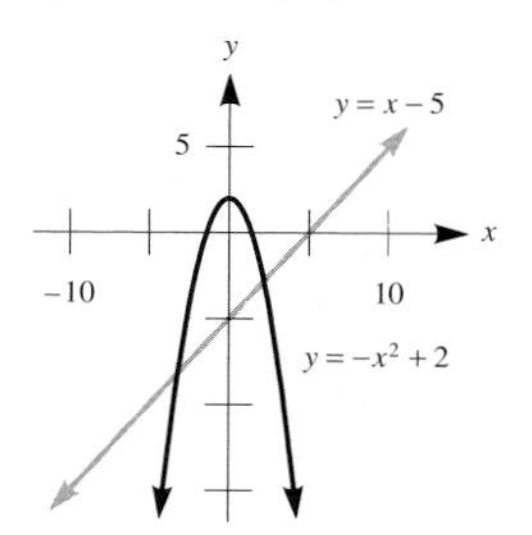

(b) $\left(\dfrac{-1+\sqrt{29}}{2}, \dfrac{-11+\sqrt{29}}{2}\right) \approx (2.193, -2.807)$, $\left(\dfrac{-1-\sqrt{29}}{2}, \dfrac{-11-\sqrt{29}}{2}\right) \approx (-3.193, -8.193)$

25. (a) $(0.85, 2.36)$

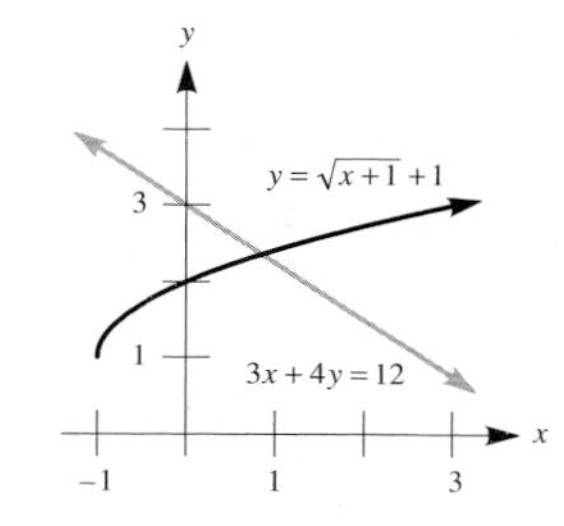

(b) $\left(\dfrac{32 - 4\sqrt{37}}{9}, \dfrac{1+\sqrt{37}}{3}\right) \approx (0.852, 2.361)$

27. (a) $(0.60, 5.30)$

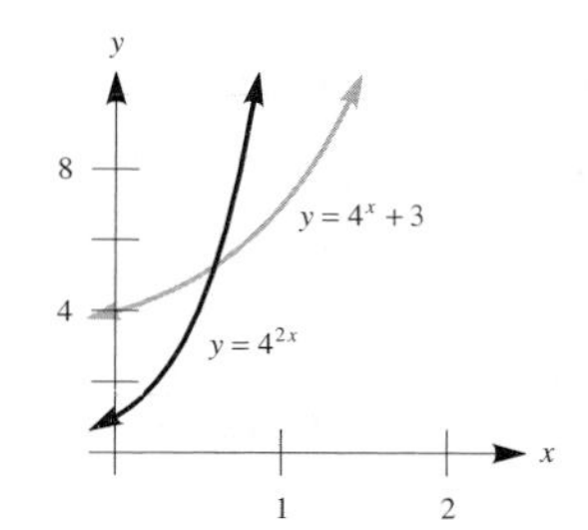

(b) $\left(\dfrac{\ln[(1+\sqrt{13})/2]}{\ln 4}, \dfrac{7+\sqrt{13}}{2}\right) \approx (0.602, 5.303)$

29. $(-0.82, 0.67), (1.43, 2.04)$

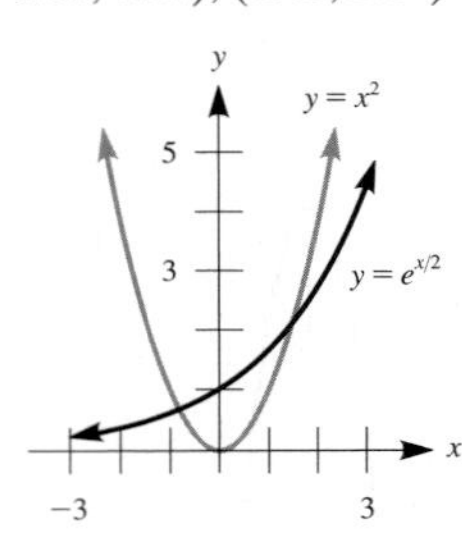

31. $(1, 0), (12.34, 2.51)$

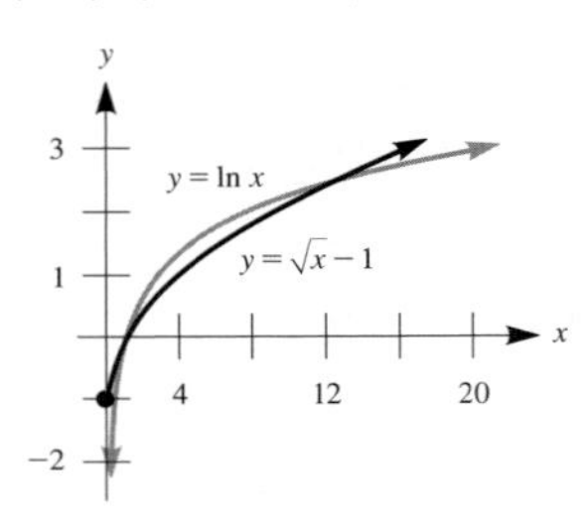

33. $(1.23, 1.86), (6.14, 230.95)$

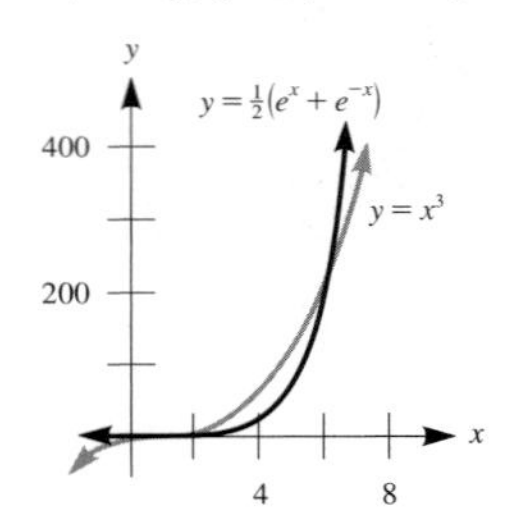

35. (b) $u = 9, v = 8$ **37.** $\left(\frac{1}{a}, \frac{1}{b}\right)$

39. $(9, 14), (14, 9)$

41. $\frac{p - \sqrt{2d^2 - p^2}}{2}$ by $\frac{p + \sqrt{2d^2 - p^2}}{2}$ **43. (a)** $N_0 = \frac{3}{2}$, $k = \frac{\ln 8}{6}$ **(b)** $N_0 = 10^{-1/7}$, $k = \frac{2}{7}\ln 10$ **45.** $\left(\frac{p^2}{A}, \frac{q^2}{A}, \frac{r^2}{A}\right)$ and $\left(\frac{-p^2}{A}, \frac{-q^2}{A}, \frac{-r^2}{A}\right)$, where $A = \sqrt{p^2 + q^2 + r^2}$

47. 9 cm and 40 cm

49. 3 cm by 20 cm

51. $(1, 2), (-1, -2), (2, 1), (-2, -1)$

53. $\left(\frac{1 + \sqrt{13}}{2}, \frac{-1 + \sqrt{13}}{2}\right)$, $\left(\frac{-1 + \sqrt{13}}{2}, \frac{1 + \sqrt{13}}{2}\right)$, $\left(\frac{1 - \sqrt{13}}{2}, \frac{-1 - \sqrt{13}}{2}\right)$, $\left(\frac{-1 - \sqrt{13}}{2}, \frac{1 - \sqrt{13}}{2}\right)$

55. $(2, 3), (-2, -3), (1, 2), (-1, -2)$

57. $(e^{9/2}, e^3)$

Exercise Set 10.7

1. (a) no **(b)** yes

3.

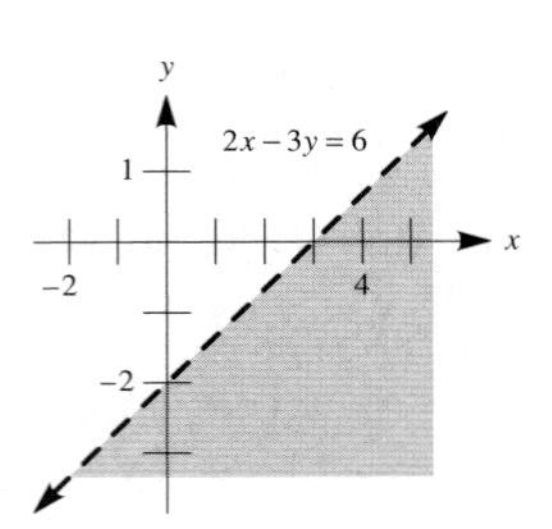

5.

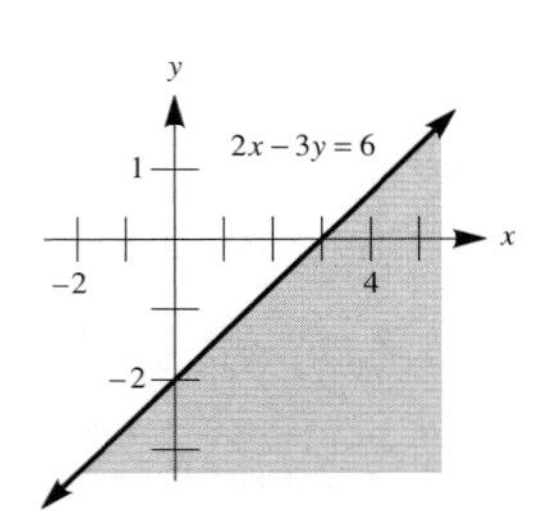

7.

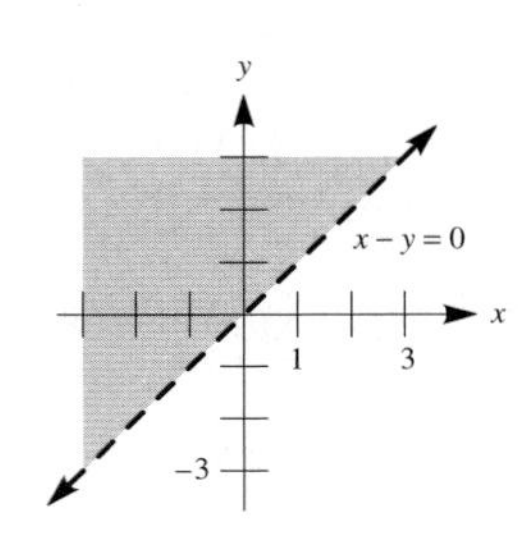

9.

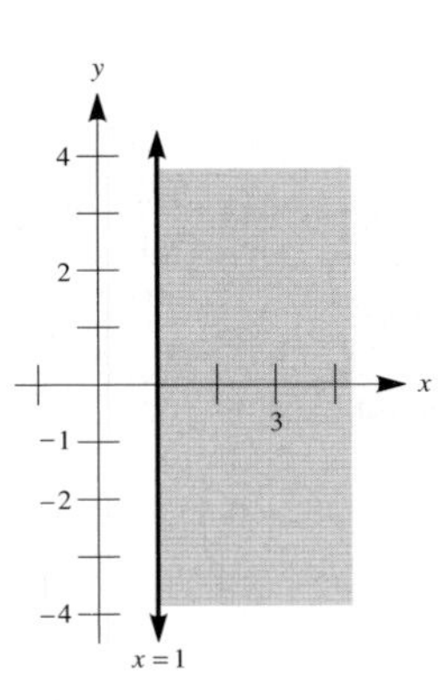

11.

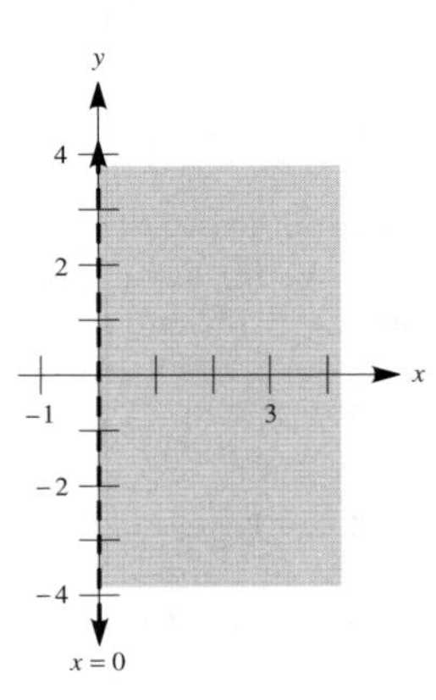

13.

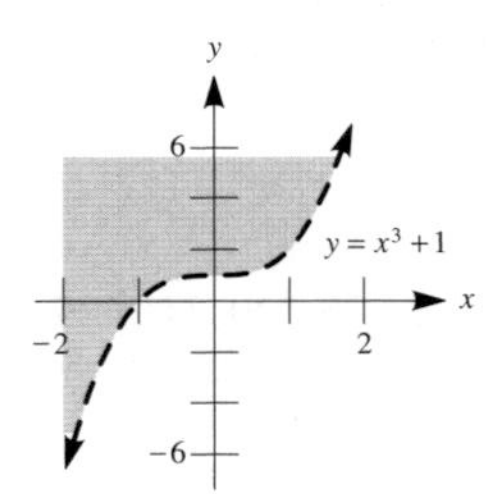

15.

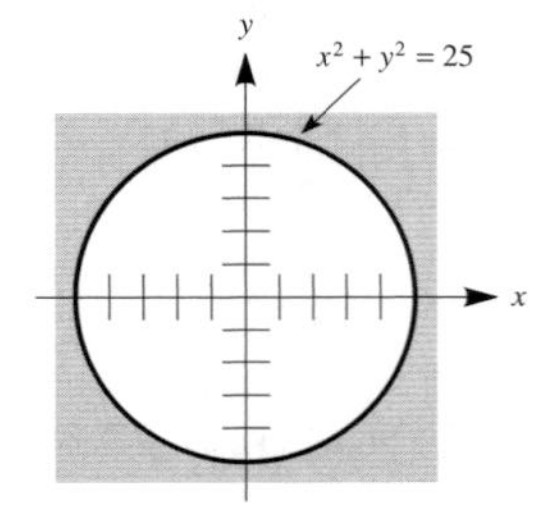

17.

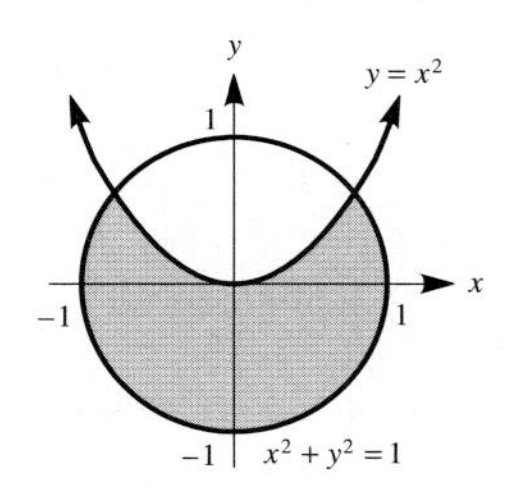

19.

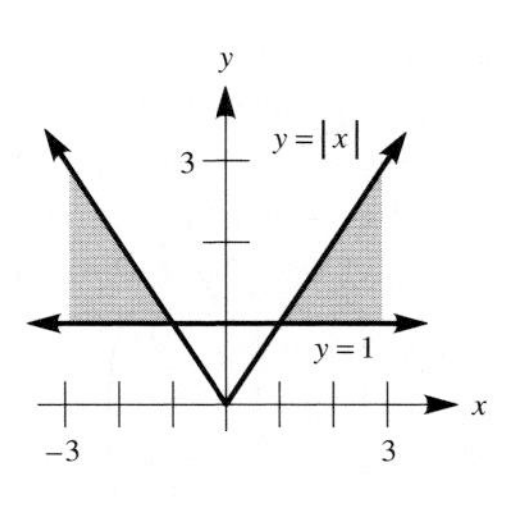

21.

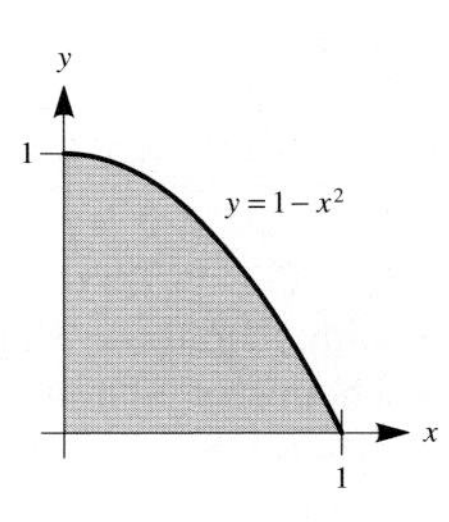

23. convex and bounded; vertices: (0, 0), (7, 0), (3, 8), (0, 5)

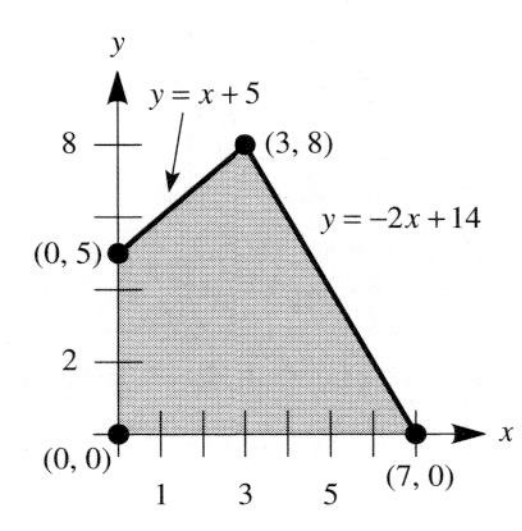

25. convex and bounded; vertices: (0, 0), (0, 4), (3, 5), (8, 0)

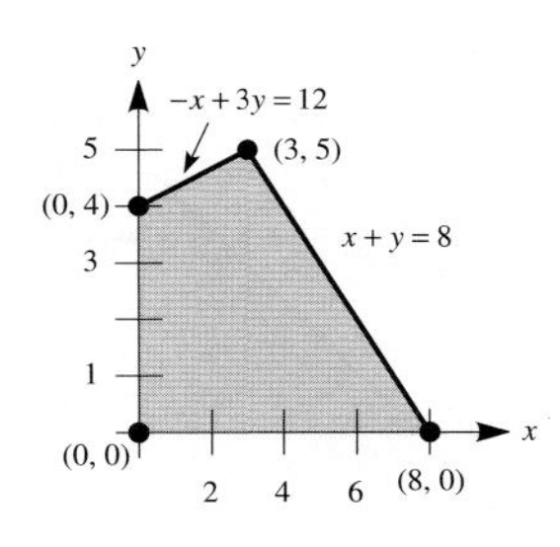

27. convex but not bounded; vertices: (2, 7), (8, 5)

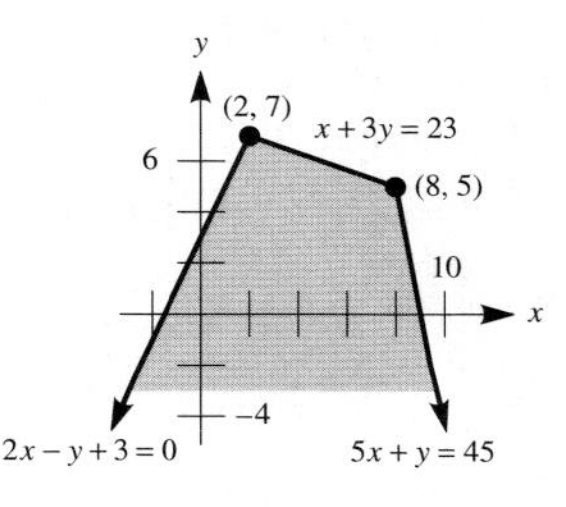

29. convex but not bounded; vertex: (6, 0)

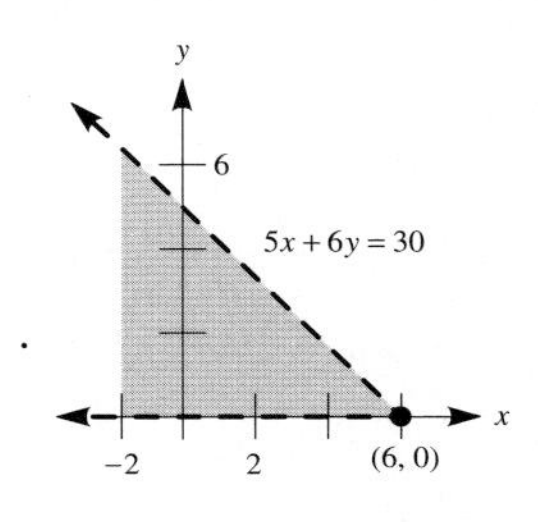

31. convex and bounded; vertices: (0, 0), (0, 5), (6, 0)

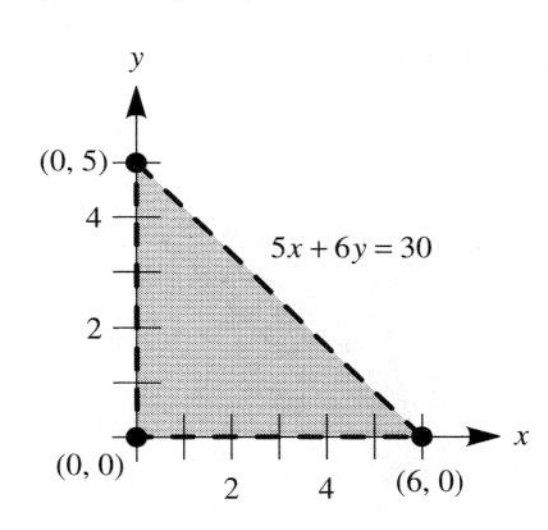

33. convex and bounded; vertices: (5, 30), (10, 30), (20, 15), (20, 20)

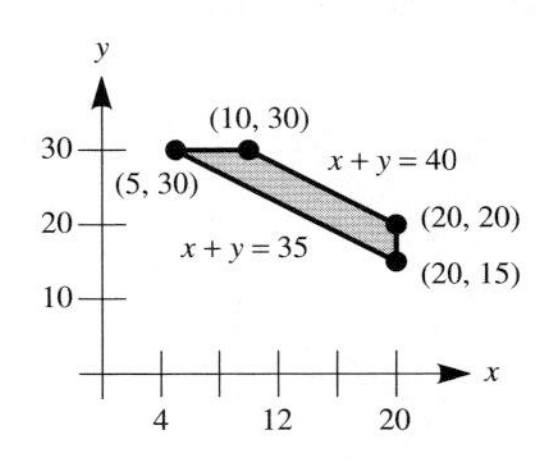

35. vertices: (0, 1), (0, 2), $\left(\ln\dfrac{1+\sqrt{5}}{2}, \dfrac{1+\sqrt{5}}{2}\right)$

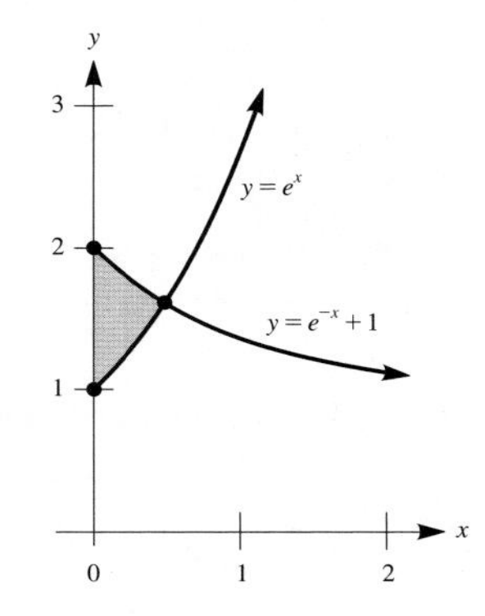

37.

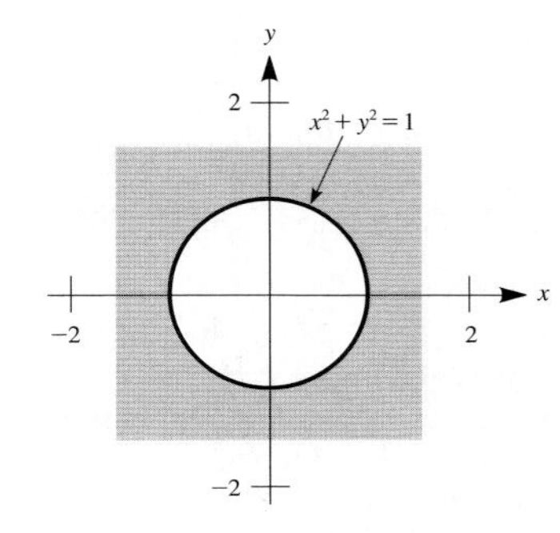

39.

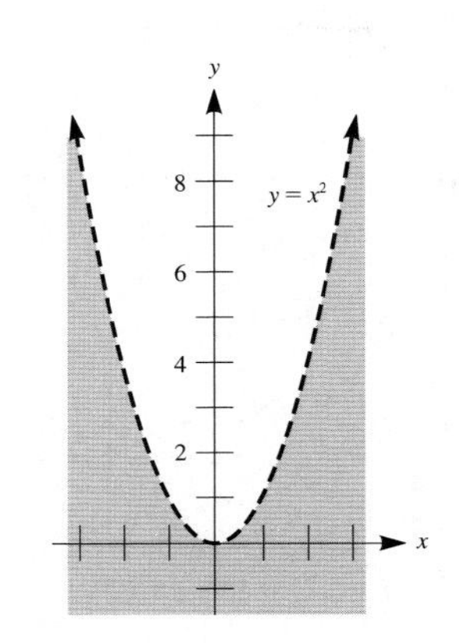

41.

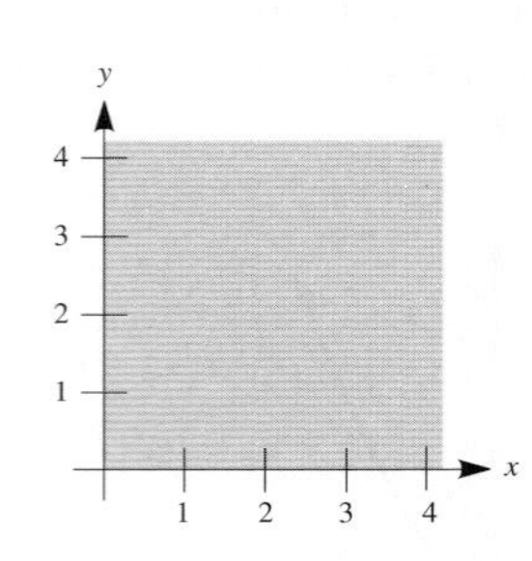

Chapter 10 Review Exercises

1. $(3, -5)$ **3.** $(-1, 4)$ **5.** $(-3, 15)$
7. $(-\frac{18}{5}, \frac{8}{5})$ **9.** $(\frac{2}{3}, -\frac{1}{5})$
11. $(-12, -8)$ **13.** $(\frac{1}{3}, -\frac{1}{4})$
15. $\left(\dfrac{-1}{a^2 - 3a + 1}, \dfrac{1 - a}{a^2 - 3a + 1}\right)$; $a \neq \dfrac{3 \pm \sqrt{5}}{2}$
17. $(a^2, 1 - a^2)$ **19.** $(a^2 - b^2, a^2 + b^2)$
21. $\left(\dfrac{pq(p + q)}{p^2 + q^2}, \dfrac{p^3 - q^3}{p^2 + q^2}\right)$; p and q not both zero **23.** $(2, 3, 4)$
25. $(-1, -2, 0)$ **27.** $(x, 6 - 2x, -1)$, x = any real number
29. $(-\frac{1}{6}z, \frac{1}{2}z, z)$, z = any real number
31. $(4, 3, -1, 2)$ **33.** 960
35. 24 **39.** $a = 2$ and $b = 1$
43. $\left(x + \dfrac{17}{6}\right)^2 + \left(y + \dfrac{8}{3}\right)^2 = \dfrac{245}{36}$
45. $\begin{pmatrix} 10 & -2 \\ 4 & 26 \end{pmatrix}$ **47.** $\begin{pmatrix} 8 & 4 \\ 4 & 32 \end{pmatrix}$
49. $\begin{pmatrix} 4 & -13 \\ 7 & 41 \end{pmatrix}$ **51.** $\begin{pmatrix} -3 & -14 \\ -4 & 3 \end{pmatrix}$
53. $\begin{pmatrix} 1 & 1 \\ 1 & 7 \end{pmatrix}$ **55.** $\begin{pmatrix} 1 & -11 \\ 6 & 36 \end{pmatrix}$
57. $\begin{pmatrix} 4 & 3 \\ 10 & 33 \end{pmatrix}$ **59.** $\begin{pmatrix} 4 & -1 \\ 2 & 12 \end{pmatrix}$
61. $\begin{pmatrix} -4 & 13 \\ -7 & -41 \end{pmatrix}$
63. $A^2 = \begin{pmatrix} 0 & 0 & 0 \\ 0 & 0 & 0 \\ ac & 0 & 0 \end{pmatrix}$; $A^3 = \begin{pmatrix} 0 & 0 & 0 \\ 0 & 0 & 0 \\ 0 & 0 & 0 \end{pmatrix}$
65. (a) $\begin{pmatrix} 10 & -2 & 5 \\ 6 & -1 & 4 \\ 1 & 0 & 1 \end{pmatrix}$
(b) $(16, 15, 4)$ **67.** $(-\frac{3}{5}, -\frac{13}{20}, \frac{31}{20})$
69. inconsistent (no solution)
71. $(-5 - 4y, y, -8 - 5y)$, y = any real number **73.** $(4, 1, -1, 3)$
75. $(0, 0)$ and $(6, 36)$
77. $(3, 0)$ and $(-3, 0)$
79. $(\frac{5\sqrt{2}}{2}, \frac{\sqrt{14}}{2})$, $(-\frac{5\sqrt{2}}{2}, \frac{\sqrt{14}}{2})$, $(\frac{5\sqrt{2}}{2}, -\frac{\sqrt{14}}{2})$, $(-\frac{5\sqrt{2}}{2}, -\frac{\sqrt{14}}{2})$
81. $\left(\dfrac{-1 + \sqrt{5}}{2}, \dfrac{\sqrt{-2 + 2\sqrt{5}}}{2}\right)$
83. $(\sqrt{2}, 3\sqrt{2})$, $(-\sqrt{2}, -3\sqrt{2})$, $(\frac{7\sqrt{22}}{33}, \frac{31\sqrt{22}}{33})$, $(\frac{-7\sqrt{22}}{33}, \frac{-31\sqrt{22}}{33})$
85. $(5, 2)$, $(5, -4)$, $(1, 2)$, $(1, -4)$
87. $\dfrac{as}{a + b}$ and $\dfrac{bs}{a + b}$
89. 24, 60, 120
91. $\dfrac{\sqrt{n + 2m} + \sqrt{n - 2m}}{2}$ and $\dfrac{\sqrt{n + 2m} - \sqrt{n - 2m}}{2}$, $\dfrac{\sqrt{n - 2m} - \sqrt{n + 2m}}{2}$, and $\dfrac{-\sqrt{n - 2m} - \sqrt{n + 2m}}{2}$
93. neither convex nor bounded

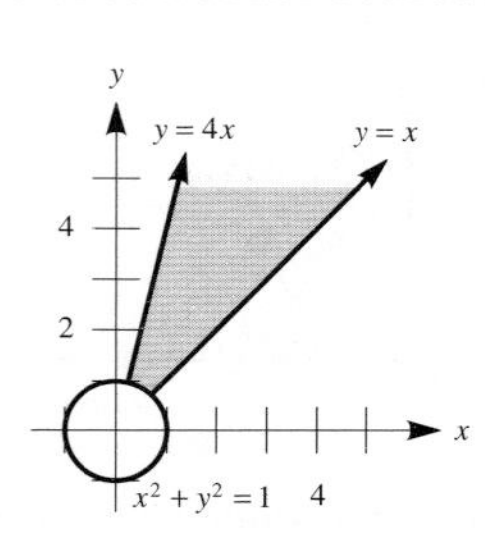

95. convex and bounded

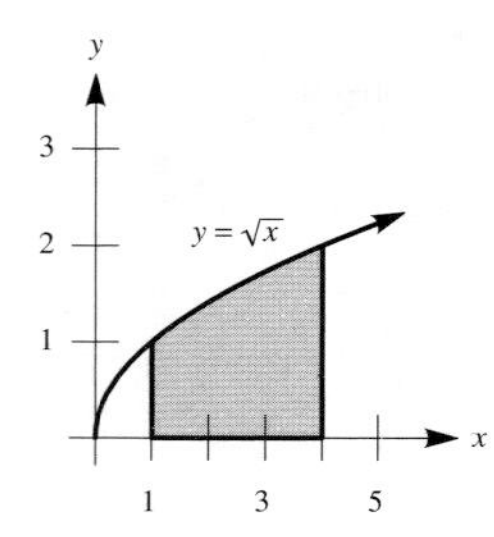

Chapter 10 Test

1. $(0, 3)$ and $(-\frac{11}{4}, \frac{81}{16})$ **2.** $(3, -5)$
3. (a) $(1, -1, -3)$ **(b)** $(1, -1, -3)$
4. (a) $\begin{pmatrix} 2 & -10 \\ 3 & -5 \end{pmatrix}$ **(b)** $\begin{pmatrix} 8 & -4 \\ 7 & -6 \end{pmatrix}$
5. equilibrium price = \$21; equilibrium quantity = 684 units
6. $(\frac{1}{116}, -\frac{1}{144})$
7. coefficient matrix: $\begin{pmatrix} 1 & 1 & -1 \\ 2 & -1 & 2 \\ 1 & -2 & 1 \end{pmatrix}$; augmented matrix: $\begin{pmatrix} 1 & 1 & -1 & -1 \\ 2 & -1 & 2 & 11 \\ 1 & -2 & 1 & 10 \end{pmatrix}$
8. $(3, -3, 1)$
9. (a)
$$\begin{aligned} w + x + y + z &= 40 \\ -w + x + y - z &= 0 \\ 4w - x - y - z &= 0 \\ -2w + 3x - 2y + 3z &= 0 \end{aligned}$$
(b) $\begin{pmatrix} 1 & 1 & 1 & 1 \\ -1 & 1 & 1 & -1 \\ 4 & -1 & -1 & -1 \\ -2 & 3 & -2 & 3 \end{pmatrix}$
(c) inverse matrix: $\begin{pmatrix} 0.2 & 0 & 0.2 & 0 \\ 0.1 & 0.5 & 0.2 & 0.2 \\ 0.4 & 0 & -0.2 & -0.2 \\ 0.3 & -0.5 & -0.2 & 0 \end{pmatrix}$
8 hours on math, 4 hours on English, 16 hours on chemistry, 12 hours on economics **10. (a)** 8
(b) -8 **11.** -1120
12. $\left(\dfrac{5 + \sqrt{5}}{2}, \dfrac{5 - \sqrt{5}}{2}\right)$, $\left(\dfrac{5 - \sqrt{5}}{2}, \dfrac{5 + \sqrt{5}}{2}\right)$, $\left(\dfrac{-5 + \sqrt{5}}{2}, \dfrac{-5 - \sqrt{5}}{2}\right)$, $\left(\dfrac{-5 - \sqrt{5}}{2}, \dfrac{-5 + \sqrt{5}}{2}\right)$
13. $(1 - \frac{1}{5}C, -\frac{7}{5}C, C)$, C = any real number
14. (a) $\begin{pmatrix} 1 & -2 & 3 \\ 2 & -5 & 10 \\ -1 & 2 & -2 \end{pmatrix}$
(b) $(12, 38, -9)$
15.

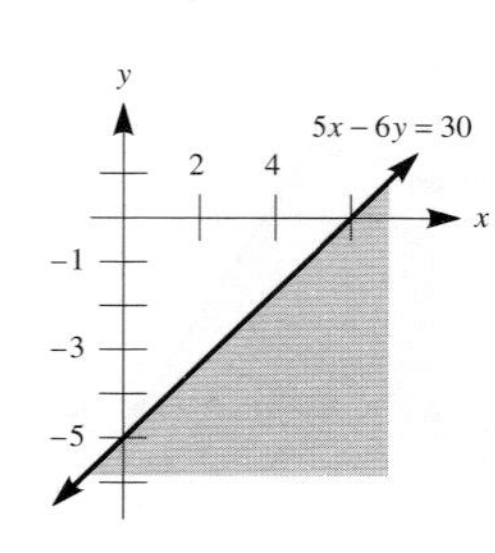

16. $P = -1$ and $Q = -4$
17. neither bounded nor convex

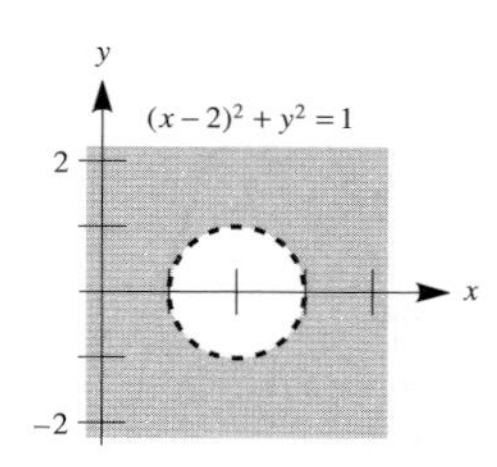

18. vertices: $(0, 0)$, $(0, 7)$, $(6, 10)$, $(\frac{261}{26}, \frac{225}{26})$, $(11, 0)$

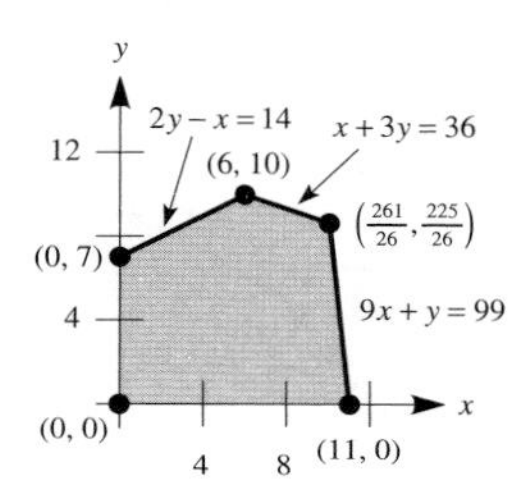

19. $(e^{-19}, e^{-6}, e^{49})$

CHAPTER 11

Exercise Set 11.1

1. $\sqrt{89}$ **3.** $5x + 4y - 4 = 0$
5. $2x + 3y - 20 = 0$
7. x-intercepts: 6, −4; y-intercepts: $\pm 2\sqrt{6}$ **9.** $13x - 7y - 35 = 0$
11. $12 + \sqrt{74}$ **13.** $\theta = \pi/3$ or 60°
15. (a) $\theta = 1.37$ or 78.69°
(b) $\theta = 1.77$ or 101.31°
17. (a) $5\sqrt{2}/2$ **(b)** $5\sqrt{2}/2$
19. (a) $19\sqrt{41}/41$ **(b)** $19\sqrt{41}/41$
21. (a) $(x + 2)^2 + (y + 3)^2 = \frac{361}{13}$
(b) $\sqrt{5}$ **23.** $\frac{65}{2}$ **25.** $\dfrac{15 \pm 2\sqrt{30}}{5}$
27. $\frac{12}{5}$ **29.** $y = -3x + 12$
31. $y = \frac{1}{3}x + \frac{9}{2}$
33. (a) center: $(-5, 2)$; radius: $5\sqrt{2}$
35. $PQ = 5\sqrt{2}$
39. (a) A: $y = \frac{1}{3}x$;
B: $y = -x + 8$; C: $y = 2x - 10$
(b) $(6, 2)$
43. (a) slope: $-\dfrac{A}{B}$; y-intercept: $-\dfrac{C}{B}$

47. $\left(x - \dfrac{26}{11}\right)^2 + \left(y + \dfrac{25}{11}\right)^2 = \dfrac{4964}{121}$

49. $\left(x - 4 - \dfrac{10\sqrt{13}}{13}\right)^2 + \left(y - 6 - \dfrac{15\sqrt{13}}{13}\right)^2 = 25$; $\left(x - 4 + \dfrac{10\sqrt{13}}{13}\right)^2 + \left(y - 6 + \dfrac{15\sqrt{13}}{13}\right)^2 = 25$

Exercise Set 11.2

1. focus: $(0, 1)$; directrix: $y = -1$; focal width: 4

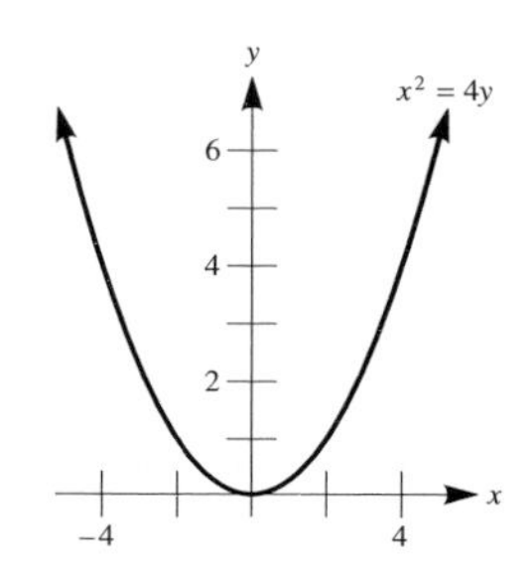

3. focus: $(-2, 0)$; directrix: $x = 2$; focal width: 8

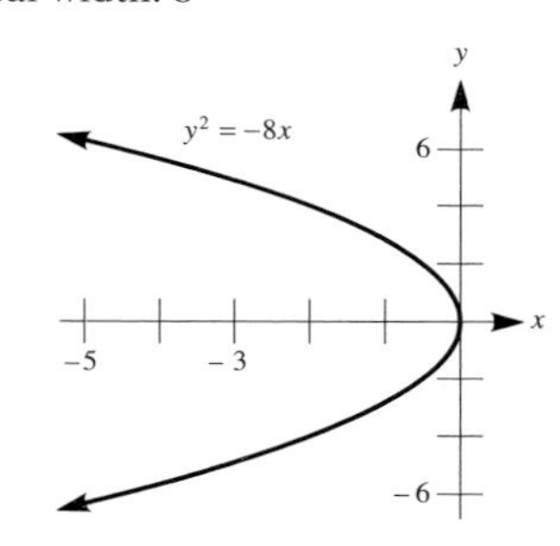

5. focus: $(0, -5)$; directrix: $y = 5$; focal width: 20

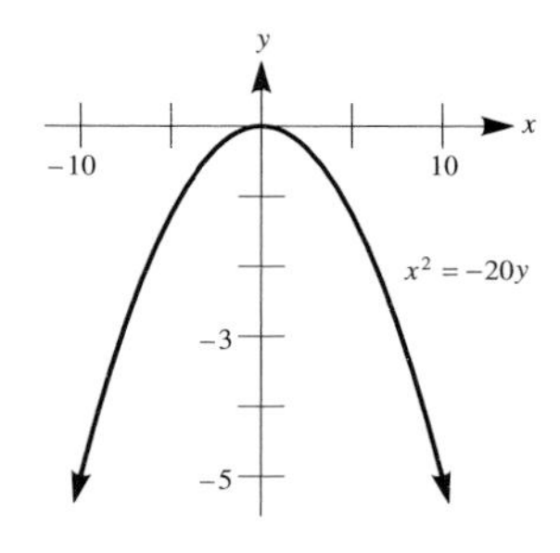

7. focus: $(-7, 0)$; directrix: $x = 7$; focal width: 28

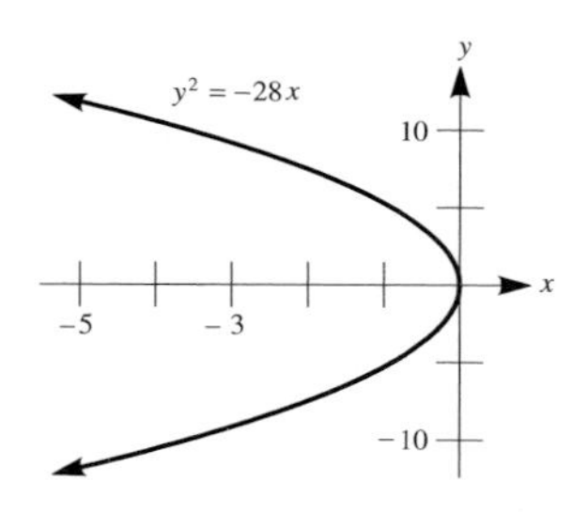

9. focus: $(0, \frac{3}{2})$; directrix: $y = -\frac{3}{2}$; focal width: 6

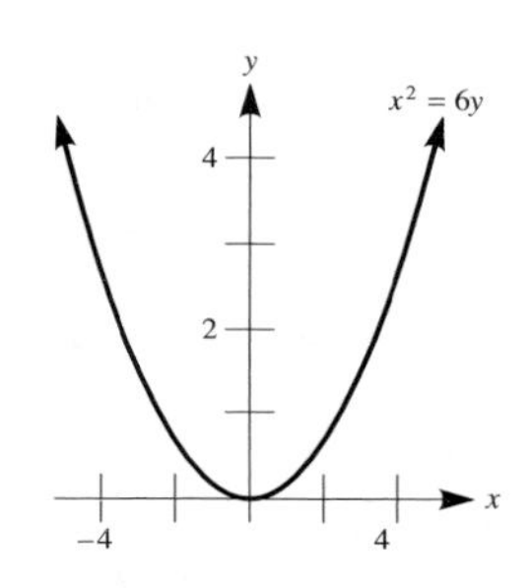

11. focus: $(0, \frac{7}{16})$; directrix: $y = -\frac{7}{16}$; focal width: $\frac{7}{4}$

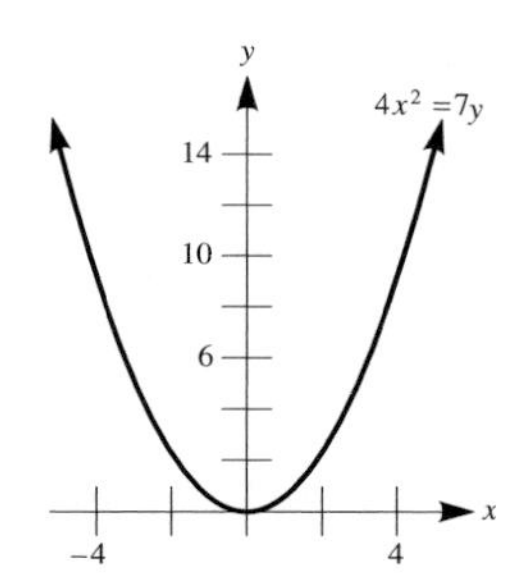

13. vertex: $(2, 3)$; focus: $(3, 3)$; directrix: $x = 1$; focal width: 4

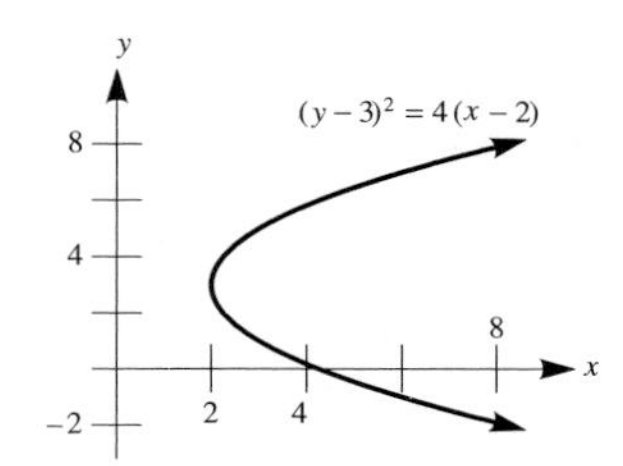

15. vertex: (4, 2); focus: $(4, \frac{9}{4})$; directrix: $y = \frac{7}{4}$; focal width: 1

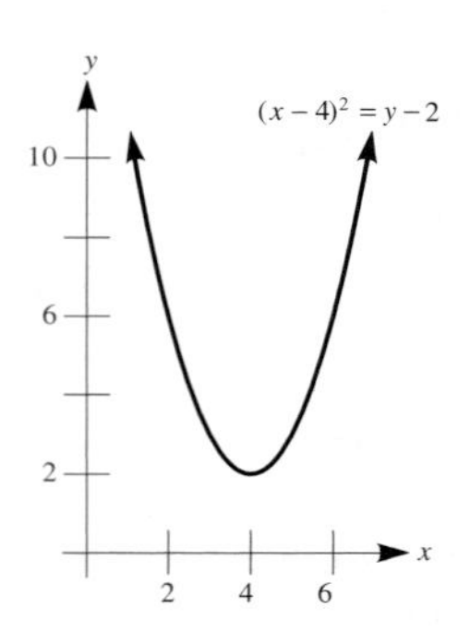

17. vertex: (0, −1); focus: $(\frac{1}{4}, -1)$; directrix: $x = -\frac{1}{4}$; focal width: 1

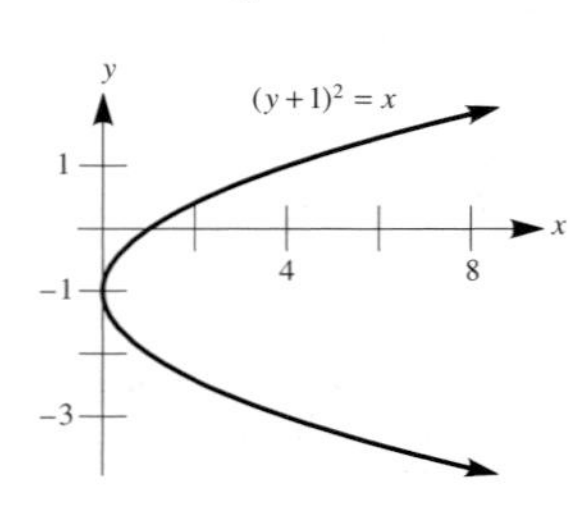

19. vertex: (3, 0); focus: $(3, \frac{1}{8})$; directrix: $y = -\frac{1}{8}$; focal width: $\frac{1}{2}$

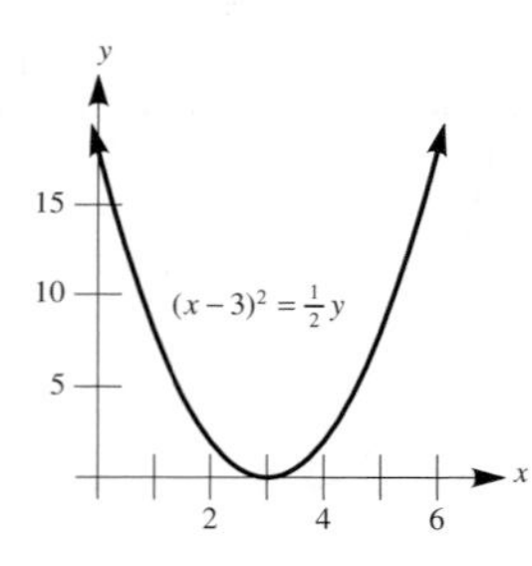

21. vertex: (4, 1); focus: $(4, \frac{9}{8})$; directrix: $y = \frac{7}{8}$; focal width: $\frac{1}{2}$

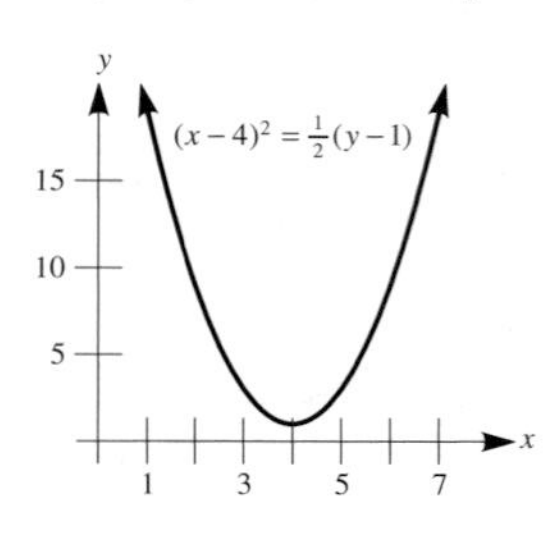

23. (a) 22.9 meters **(b)** 0.3
25. (a) $x^2 = -\frac{80}{3}y$ **(b)** $14\frac{1}{16}$ feet
27. line of symmetry: $y = 1$

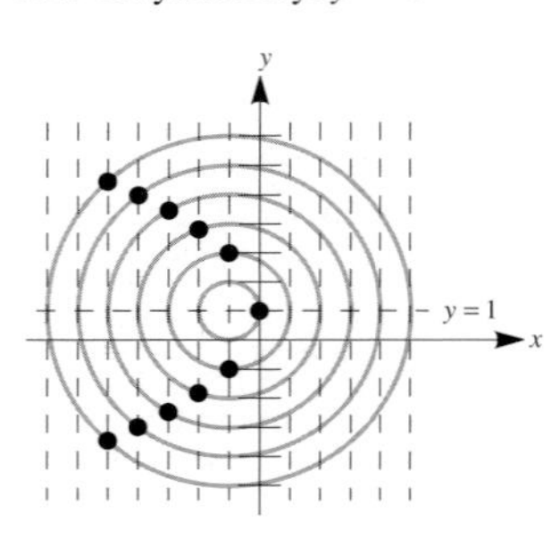

29. $x^2 = 12y$ **31.** $y^2 = 128x$ **33.** $y^2 = -9x$ **35. (a)** $y = \frac{9}{4}x - 1$ **(b)** $y = -\frac{9}{8}x - \frac{1}{4}$ **(c)** intersect at $(\frac{2}{9}, -\frac{1}{2})$ **37. (a)** $(-\frac{1}{8}, \frac{1}{64})$ **(b)** $(\frac{15}{16}, \frac{257}{128})$ **(c)** $\overline{ST} = \frac{1}{2}$
39. side $= 8\sqrt{3}p$ units, area $= 48\sqrt{3}p^2$ sq. units **41. (a)** 6 m **(b)** $5\frac{1}{3}$ m

Exercise Set 11.3

1. $y = 4x - 4$

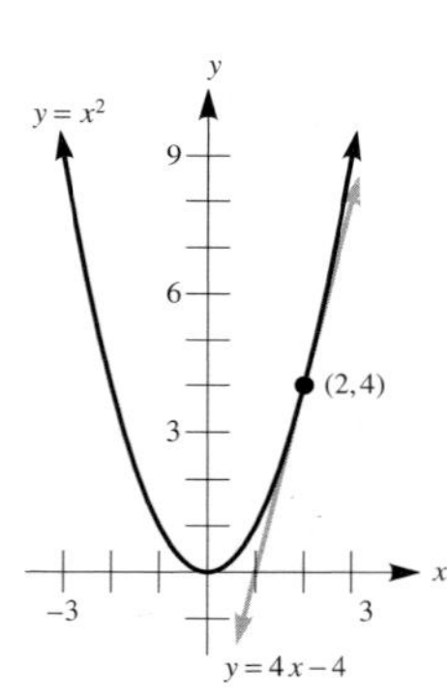

3. $y = x - 2$

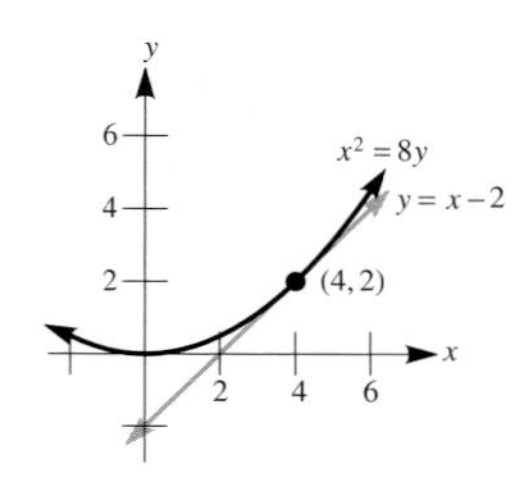

5. $y = 6x + 9$

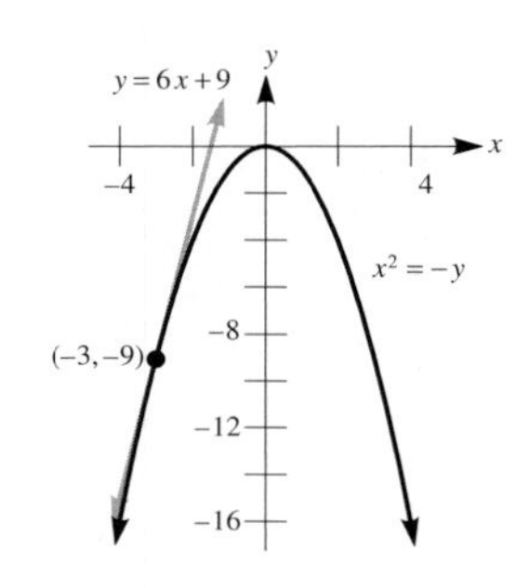

7. $y = x + 1$

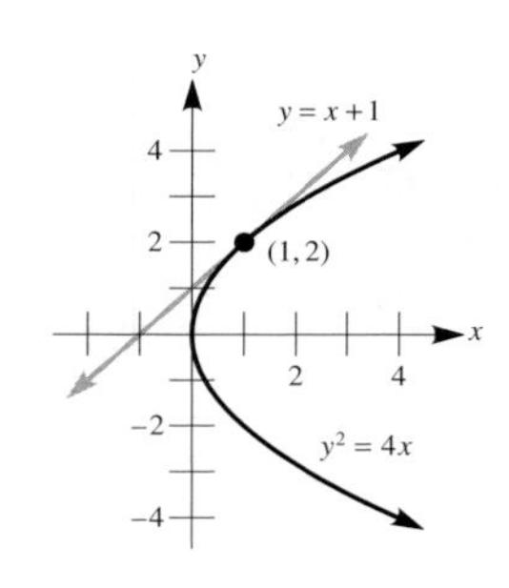

9. $m = \frac{1}{4}$ **11.** $m = 12$

Exercise Set 11.4

1. length of major axis: 6; length of minor axis: 4; foci: $(\pm\sqrt{5}, 0)$; eccentricity: $\sqrt{5}/3$

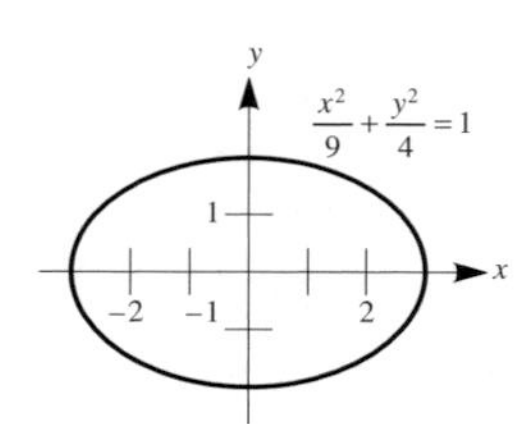

3. length of major axis: 8; length of minor axis: 2; foci: $(\pm\sqrt{15}, 0)$; eccentricity: $\sqrt{15}/4$

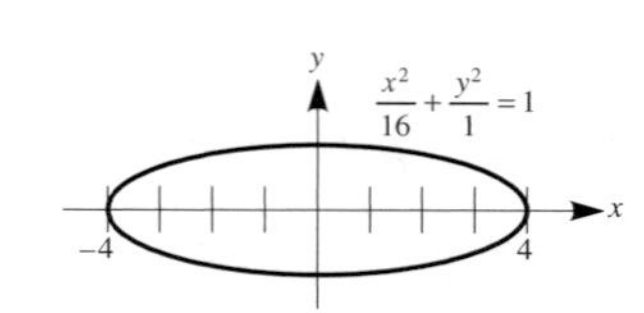

5. length of major axis: $2\sqrt{2}$; length of minor axis: 2; foci: $(\pm 1, 0)$; eccentricity: $\sqrt{2}/2$

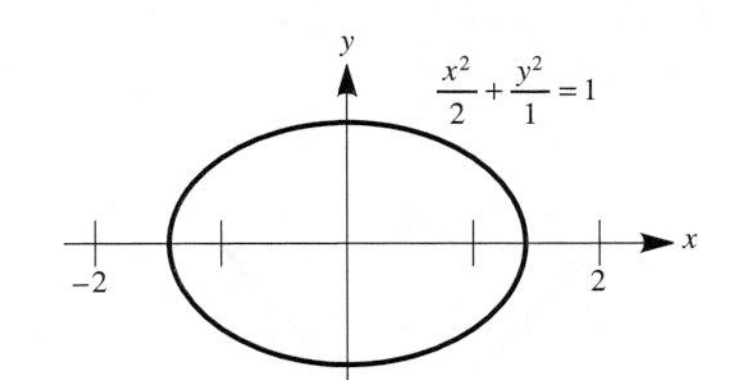

7. length of major axis: 8; length of minor axis: 6; foci: $(0, \pm\sqrt{7})$; eccentricity: $\sqrt{7}/4$

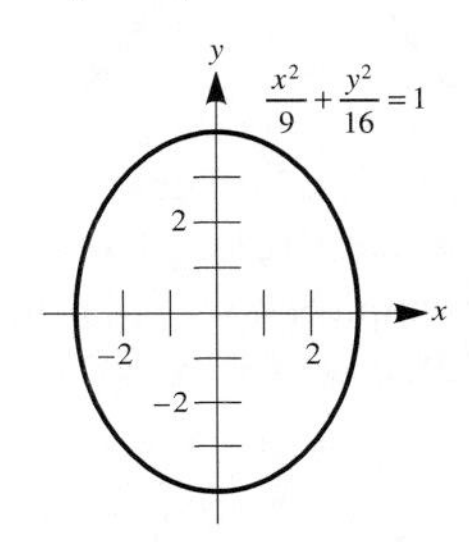

9. length of major axis: $2\sqrt{15}/3$; length of minor axis: $2\sqrt{3}/3$; foci: $(0, \pm\frac{2\sqrt{3}}{3})$; eccentricity: $2\sqrt{5}/5$

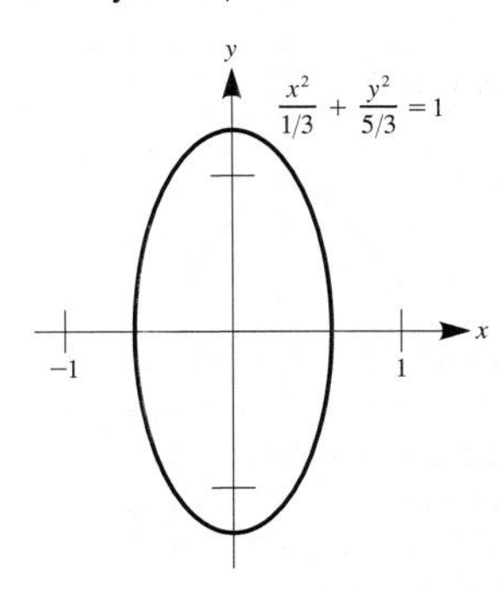

11. length of major axis: 4; length of minor axis: $2\sqrt{2}$; foci: $(0, \pm\sqrt{2})$; eccentricity: $\sqrt{2}/2$

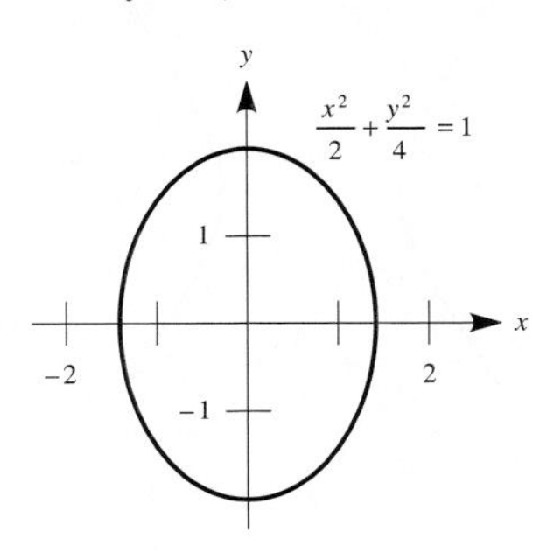

13. center: $(5, -1)$; length of major axis: 10; length of minor axis: 6; foci: $(9, -1)$ and $(1, -1)$; eccentricity: $\frac{4}{5}$

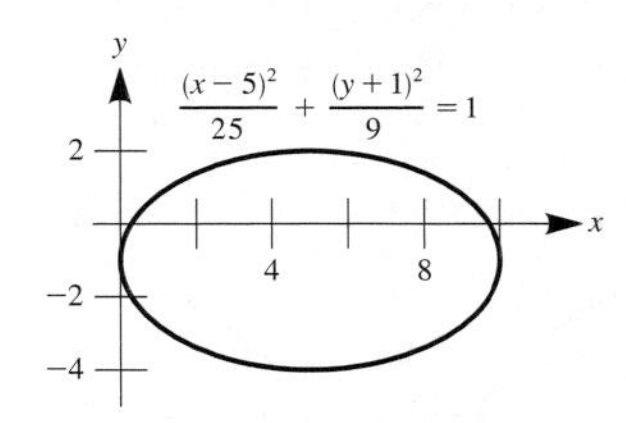

15. center: $(1, 2)$; length of major axis: 4; length of minor axis: 2; foci: $(1, 2 \pm \sqrt{3})$; eccentricity: $\sqrt{3}/2$

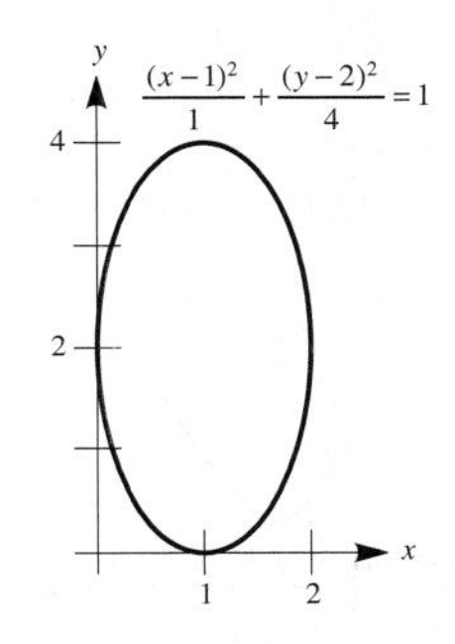

17. center: $(-3, 0)$; length of major axis: 6; length of minor axis: 2; foci: $(-3 \pm 2\sqrt{2}, 0)$; eccentricity: $2\sqrt{2}/3$

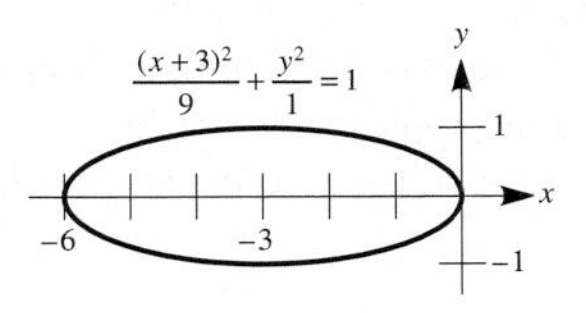

19. center: $(1, -2)$; length of major axis: 4; length of minor axis: $2\sqrt{3}$; foci: $(2, -2)$ and $(0, -2)$; eccentricity: $\frac{1}{2}$

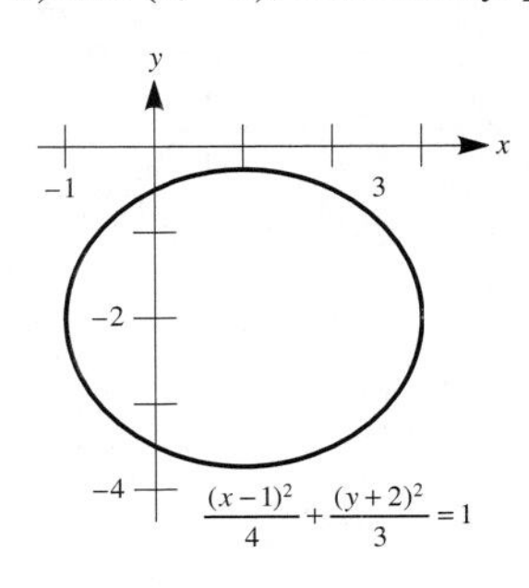

21. center: $(4, 6)$; degenerate ellipse

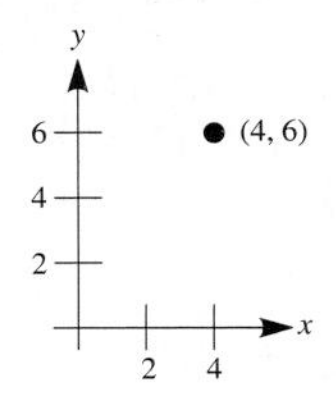

23. no graph

25. $\frac{x^2}{25} + \frac{y^2}{16} = 1$ or $16x^2 + 25y^2 = 400$

27. $\frac{x^2}{16} + \frac{y^2}{15} = 1$ or $15x^2 + 16y^2 = 240$

29. $\frac{x^2}{21} + \frac{y^2}{25} = 1$ or $25x^2 + 21y^2 = 525$

31. $\frac{x^2}{9} + \frac{y^2}{9/4} = 1$ or $x^2 + 4y^2 = 9$

33. (a) $y = -\frac{4}{3}x + \frac{38}{3}$
(b) $y = \frac{7}{9}x + \frac{76}{9}$ **(c)** $y = \frac{1}{15}x - \frac{76}{15}$

35. (a) $y = -6x + 26$ **(b)** $\frac{169}{3}$

37. perihelion: 29.64 AU; aphelion: 49.24 AU **39.** semimajor axis: 4.42 AU; eccentricity: 0.04

41. (c) $[-3, 3]$
(d)

x	0	0.5	1.0	1.5	2.0	2.5	3.0
y	±2	±1.97	±1.89	±1.73	±1.49	±1.11	0

(e)

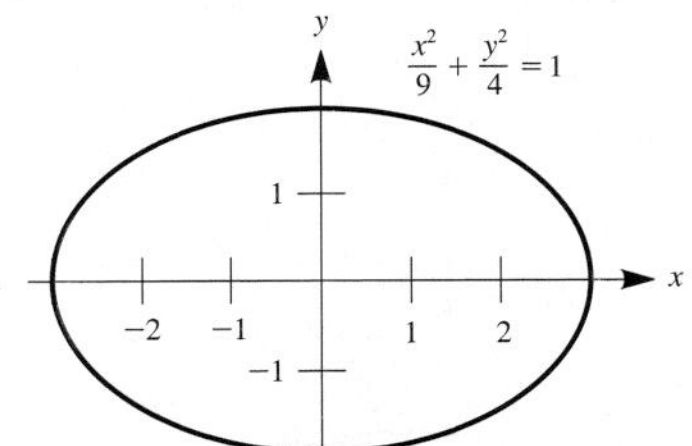

43. circle: $(x - 1)^2 + y^2 = 1$; ellipse: $\frac{(x-4)^2}{16} + \frac{y^2}{7} = 1$

49. intersection points: $\left(\frac{ab}{A}, \frac{ab}{A}\right)$, $\left(\frac{ab}{A}, -\frac{ab}{A}\right)$, $\left(-\frac{ab}{A}, \frac{ab}{A}\right)$, $\left(-\frac{ab}{A}, -\frac{ab}{A}\right)$, where $A = \sqrt{a^2 + b^2}$

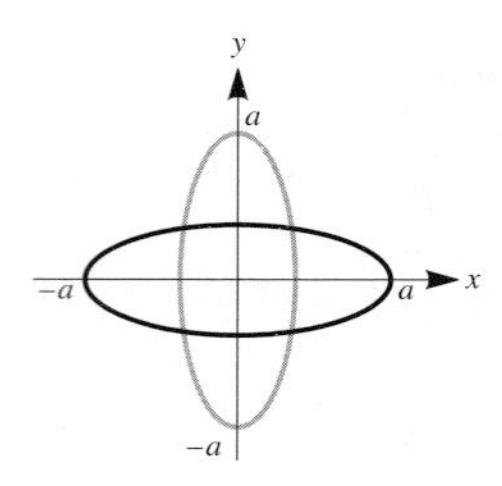

53. (b) $\left(-\frac{20}{7}, -\frac{18}{7}\right)$

59. (a)

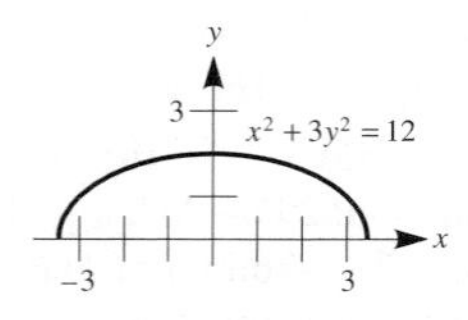

(b) $a = 2\sqrt{3}$, $b = 2$, $c = 2\sqrt{2}$

(c) auxiliary circle: $x^2 + y^2 = 12$

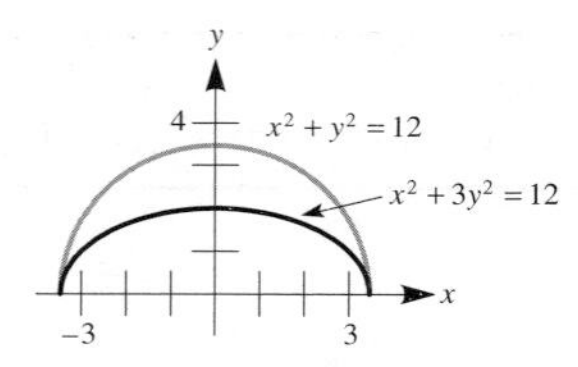

(d) $x + y = 4$

(e)

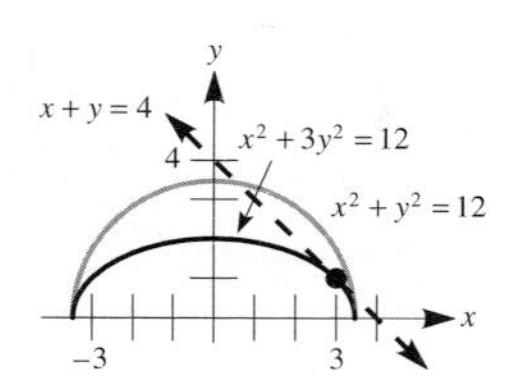

(f) $y = x + 2\sqrt{2}$

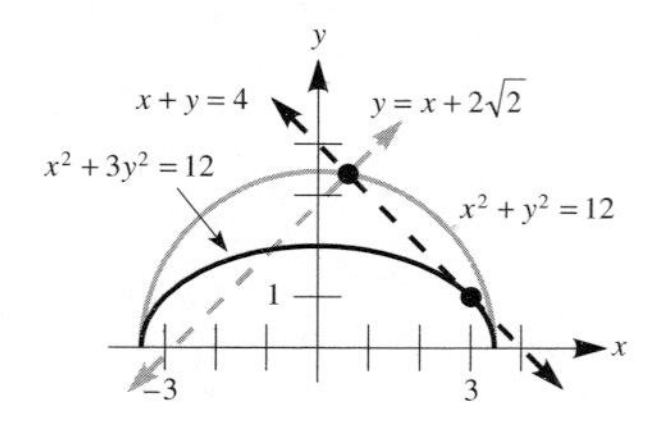

Exercise Set 11.5

1. vertices: $(\pm 2, 0)$; length of transverse axis: 4; length of conjugate axis: 2; asymptotes: $y = \pm\frac{1}{2}x$; foci: $(\pm\sqrt{5}, 0)$; eccentricity: $\sqrt{5}/2$

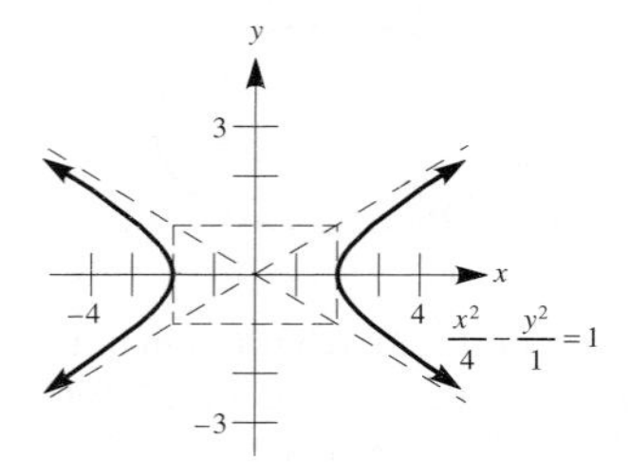

3. vertices: $(0, \pm 2)$; length of transverse axis: 4; length of conjugate axis: 2; asymptotes: $y = \pm 2x$; foci: $(0, \pm\sqrt{5})$; eccentricity: $\sqrt{5}/2$

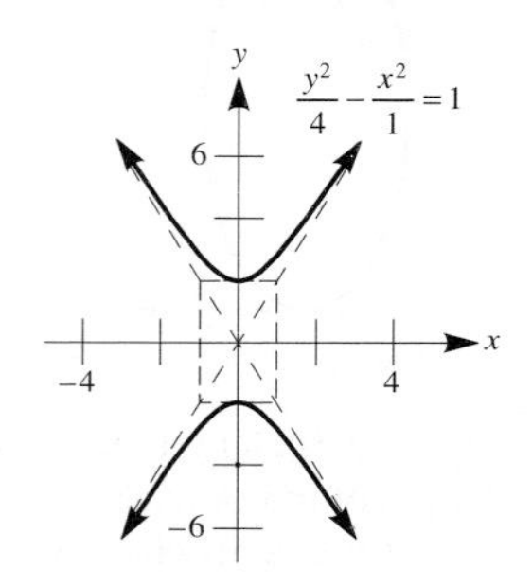

5. vertices: $(\pm 5, 0)$; length of transverse axis: 10; length of conjugate axis: 8; asymptotes: $y = \pm\frac{4}{5}x$; foci: $(\pm\sqrt{41}, 0)$; eccentricity: $\sqrt{41}/5$

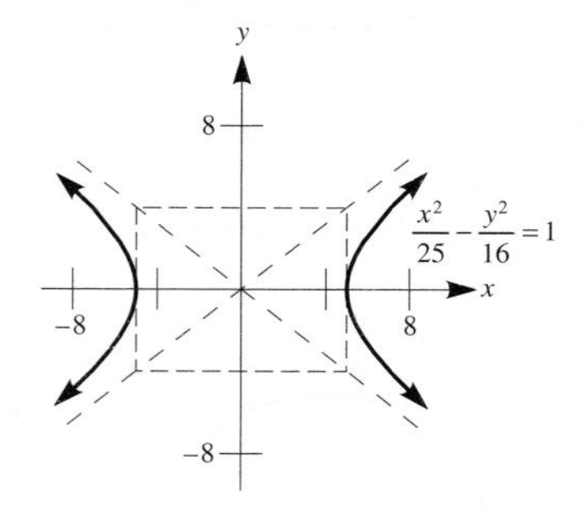

7. vertices: $(0, \pm\frac{\sqrt{2}}{2})$; length of transverse axis: $\sqrt{2}$; length of conjugate axis: $\frac{2\sqrt{3}}{3}$; asymptotes: $y = \pm\frac{\sqrt{6}}{2}x$; foci: $(0, \pm\frac{\sqrt{30}}{6})$; eccentricity: $\sqrt{15}/3$

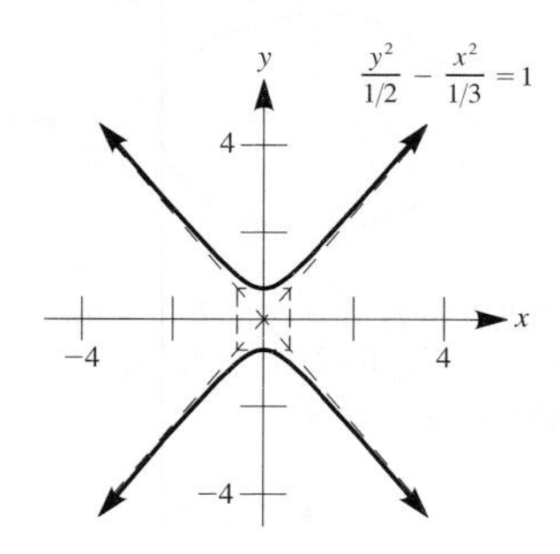

9. vertices: $(0, \pm 5)$; length of transverse axis: 10; length of conjugate axis: 4; asymptotes: $y = \pm\frac{5}{2}x$; foci: $(0, \pm\sqrt{29})$; eccentricity: $\sqrt{29}/5$

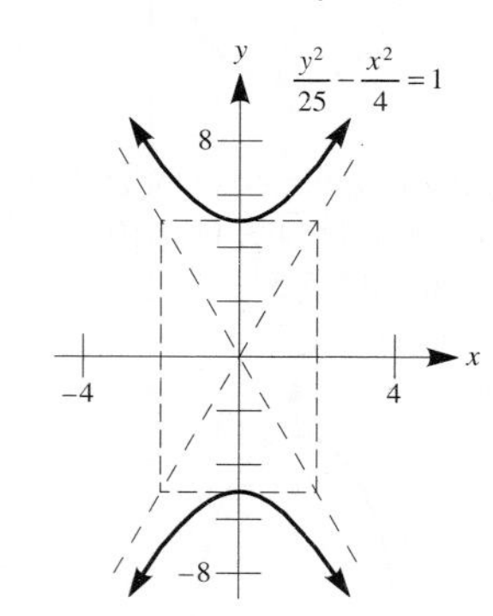

11. center: $(5, -1)$; vertices: $(10, -1)$ and $(0, -1)$; length of transverse axis: 10; length of conjugate axis: 6; asymptotes: $y = \frac{3}{5}x - 4$ and $y = -\frac{3}{5}x + 2$; foci: $(5 \pm \sqrt{34}, -1)$; eccentricity: $\sqrt{34}/5$

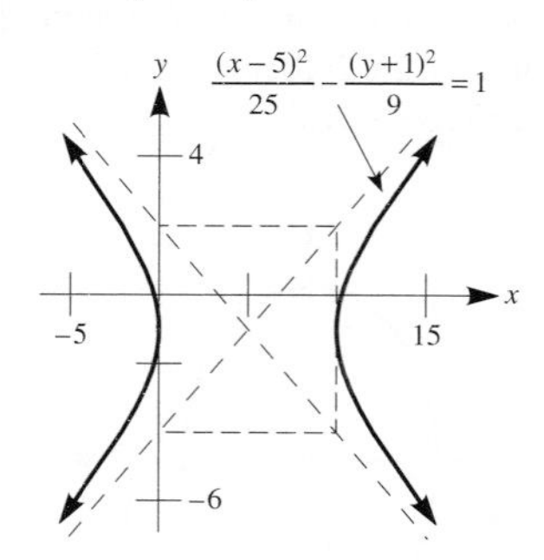

13. center: $(1, 2)$; vertices: $(1, 4)$ and $(1, 0)$; length of transverse axis: 4; length of conjugate axis: 2; asymptotes: $y = 2x$ and $y = -2x + 4$; foci: $(1, 2 \pm \sqrt{5})$; eccentricity: $\sqrt{5}/2$

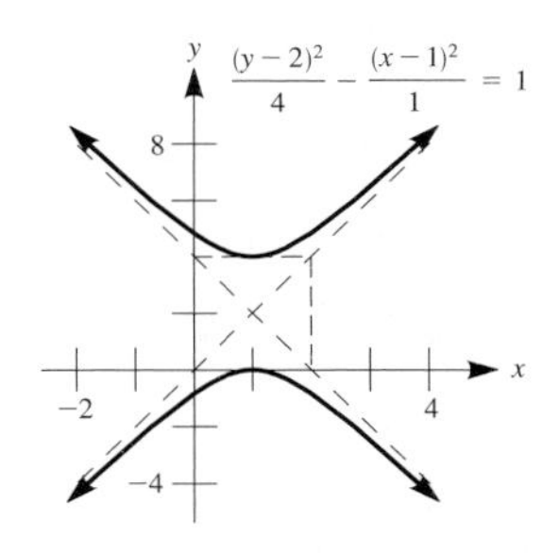

15. center: $(-3, 4)$; vertices: $(1, 4)$ and $(-7, 4)$; length of transverse axis: 8; length of conjugate axis: 8; asymptotes: $y = x + 7$ and $y = -x + 1$; foci: $(-3 \pm 4\sqrt{2}, 4)$; eccentricity: $\sqrt{2}$

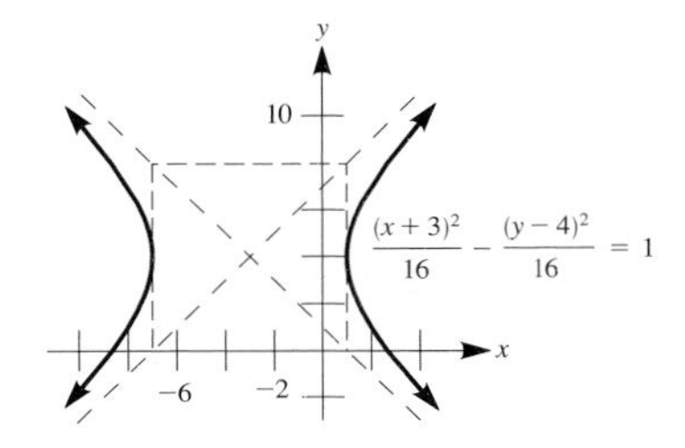

17. center: $(0, 1)$; vertices: $(2, 1)$ and $(-2, 1)$; length of transverse axis: 4; length of conjugate axis: 4; asymptotes: $y = x + 1$ and $y = -x + 1$; foci: $(\pm 2\sqrt{2}, 1)$; eccentricity: $\sqrt{2}$

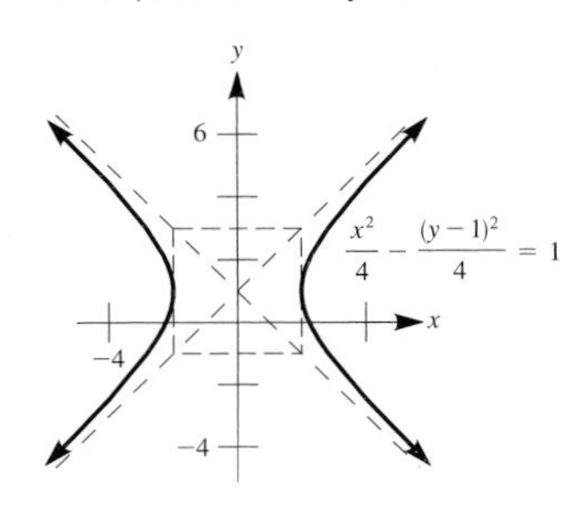

19. center: $(2, 1)$; vertices: $(5, 1)$ and $(-1, 1)$; length of transverse axis: 6; length of conjugate axis: 6; asymptotes: $y = x - 1$ and $y = -x + 3$; foci: $(2 \pm 3\sqrt{2}, 1)$; eccentricity: $\sqrt{2}$

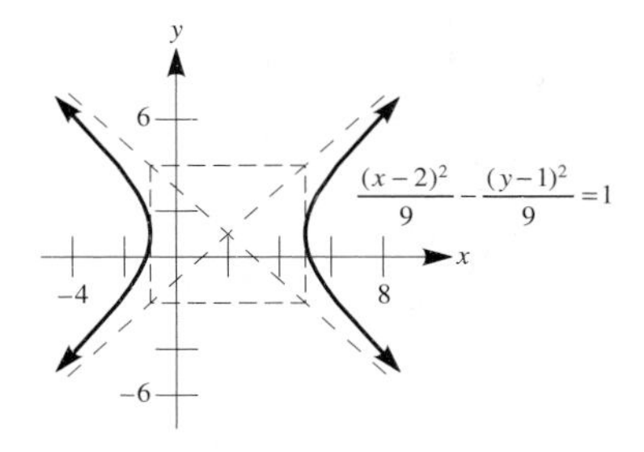

21. center: $(0, -4)$; vertices: $(0, 1)$ and $(0, -9)$; length of transverse axis: 10; length of conjugate axis: 2; asymptotes: $y = 5x - 4$ and $y = -5x - 4$; foci: $(0, -4 \pm \sqrt{26})$; eccentricity: $\sqrt{26}/5$

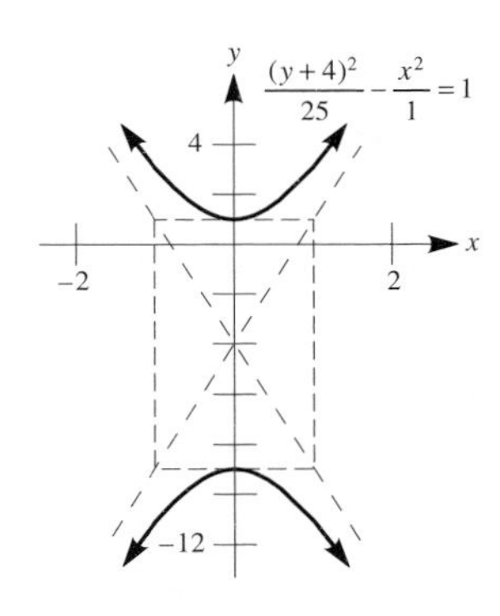

23. degenerate hyperbola

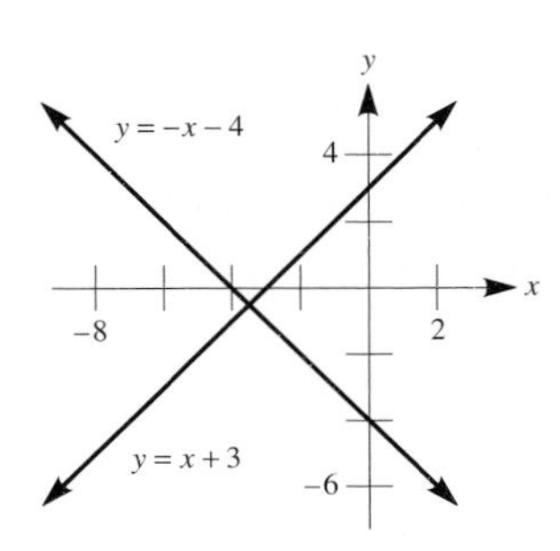

27. $15x^2 - y^2 = 15$ **29.** $x^2 - 4y^2 = 4$
31. $2x^2 - 5y^2 = 10$
33. $y^2 - 32x^2 = 49$ **35.** $y^2 - 9x^2 = 9$
39. (b) $(0, \pm 6)$ **(c)** $F_1P = 5$, $F_2P = 13$

41. (b)

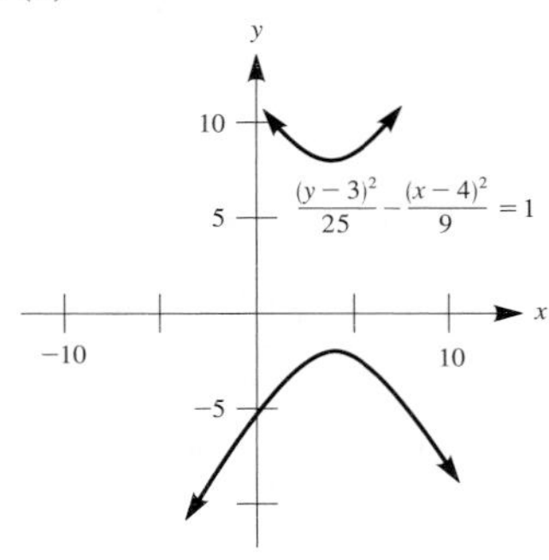

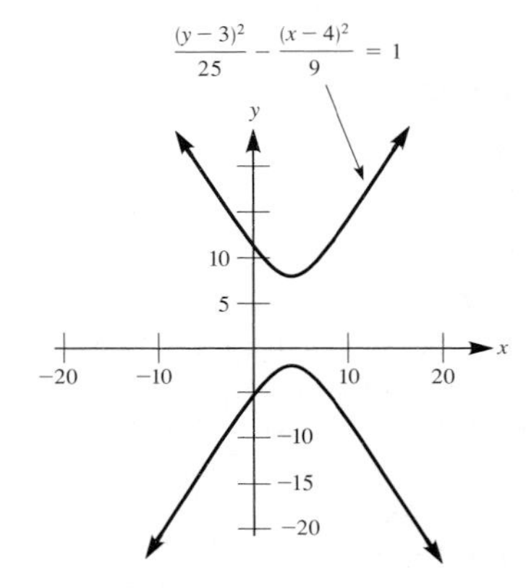

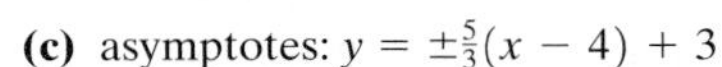
(c) asymptotes: $y = \pm\frac{5}{3}(x - 4) + 3$

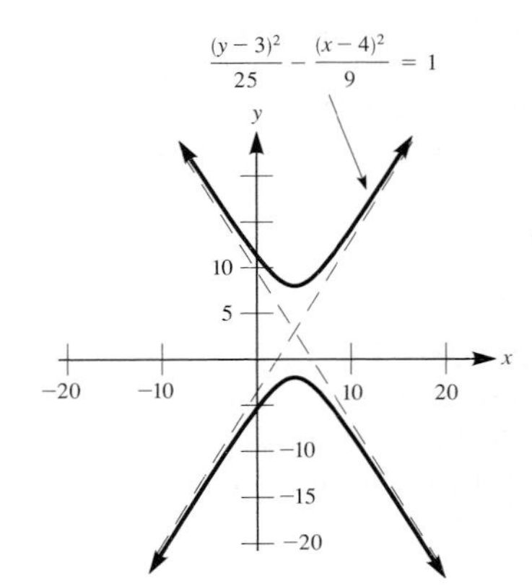

43.

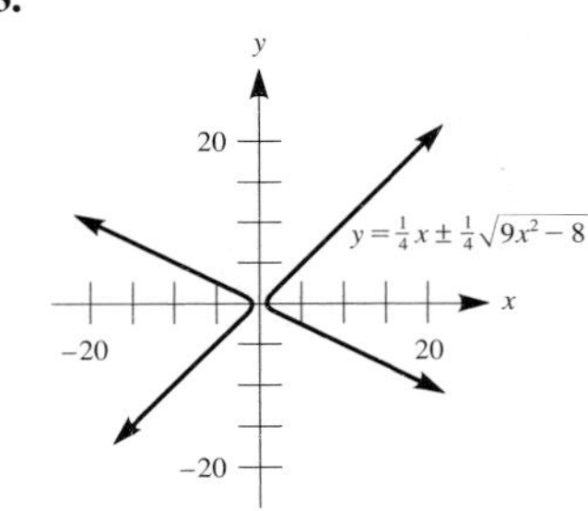

51. intersection point: $(\frac{51}{8}, \frac{15}{2})$

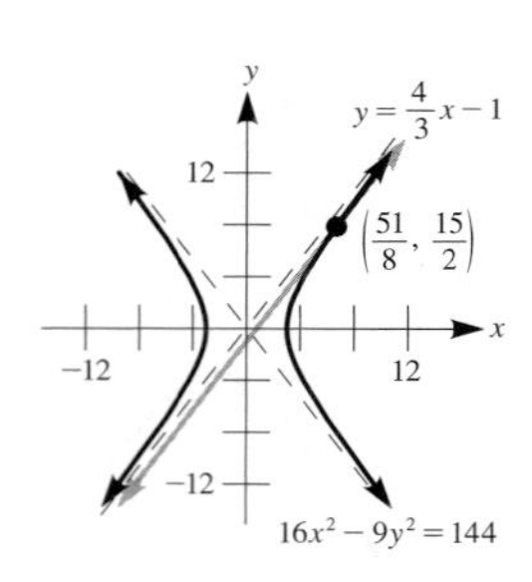

53. $y = 2x - 2$

Exercise Set 11.6

1. $F_1P = \dfrac{6\sqrt{19} - 8\sqrt{6}}{3}$, $F_2P = \dfrac{6\sqrt{19} + 8\sqrt{6}}{3}$

3. $F_1P = 15 + 6\sqrt{2}$, $F_2P = 15 - 6\sqrt{2}$

5. (a) foci: $(\pm\sqrt{7}, 0)$; eccentricity: $\sqrt{7}/4$; directrices: $x = \pm 16\sqrt{7}/7$
(b) foci: $(\pm 5, 0)$; eccentricity: $\frac{5}{4}$; directrices: $x = \pm\frac{16}{5}$ **7. (a)** foci: $(\pm 1, 0)$; eccentricity: $\sqrt{13}/13$; directrices: $x = \pm 13$ **(b)** foci: $(\pm 5, 0)$; eccentricity: $5\sqrt{13}/13$; directrices: $x = \pm\frac{13}{5}$ **9. (a)** foci: $(\pm\sqrt{11}, 0)$; eccentricity: $\sqrt{11}/6$; directrices: $x = \pm 36\sqrt{11}/11$
(b) foci: $(\pm\sqrt{61}, 0)$; eccentricity: $\sqrt{61}/6$; directrices: $x = \pm 36\sqrt{61}/61$

11. $3x^2 + 4y^2 = 12$

13. $x^2 - y^2 = 2$

Exercise Set 11.7

1. (a) eccentricity: $\frac{2}{3}$; center: $(-\frac{12}{5}, 0)$; endpoints of major axis: $(\frac{6}{5}, 0)$ and $(-6, 0)$; endpoints of minor axis: $(-\frac{12}{5}, \pm\frac{6\sqrt{5}}{5})$

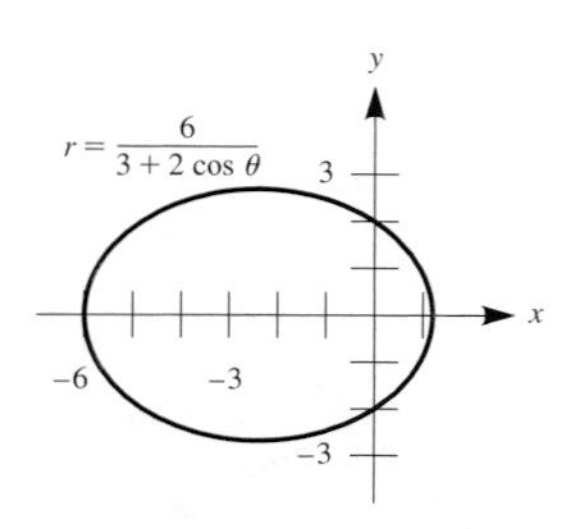

(b) eccentricity: $\frac{2}{3}$; center: $(\frac{12}{5}, 0)$; endpoints of major axis: $(6, 0)$ and $(-\frac{6}{5}, 0)$; endpoints of minor axis: $(\frac{12}{5}, \pm\frac{6\sqrt{5}}{5})$

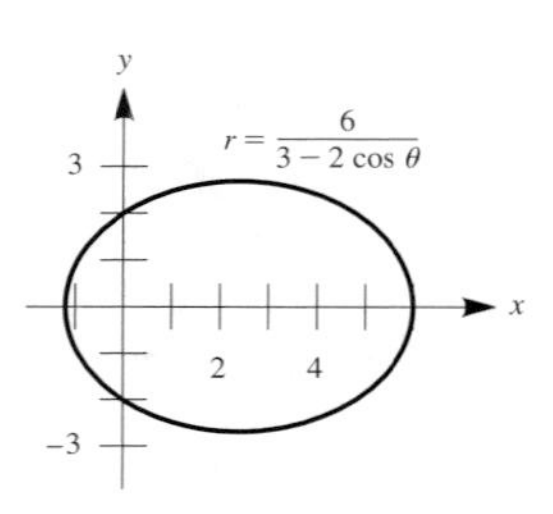

3. (a) vertex: $(\frac{5}{4}, 0)$; directrix: $x = \frac{5}{2}$

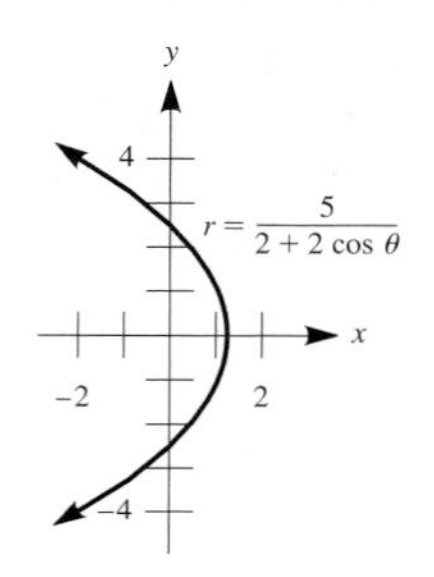

(b) vertex: $(-\frac{5}{4}, 0)$; directrix: $x = -\frac{5}{2}$

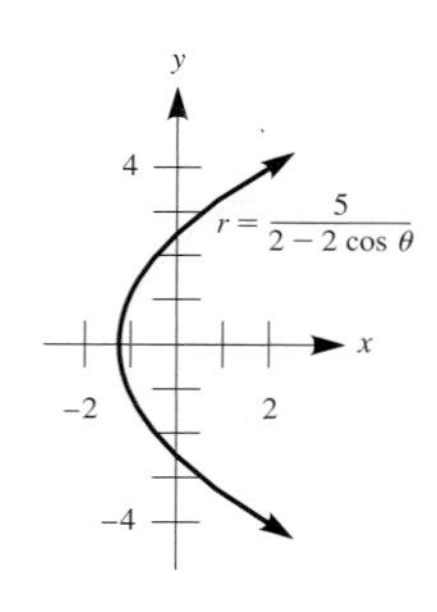

5. (a) eccentricity: 2; center: $(1, 0)$; $a = \frac{1}{2}$; $b = \frac{1}{2}\sqrt{3}$; $c = 1$

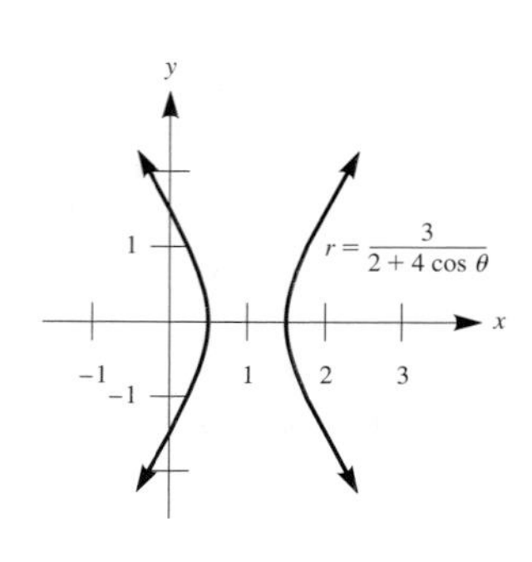

(b) eccentricity: 2; center: $(-1, 0)$; $a = \frac{1}{2}$; $b = \frac{1}{2}\sqrt{3}$; $c = 1$

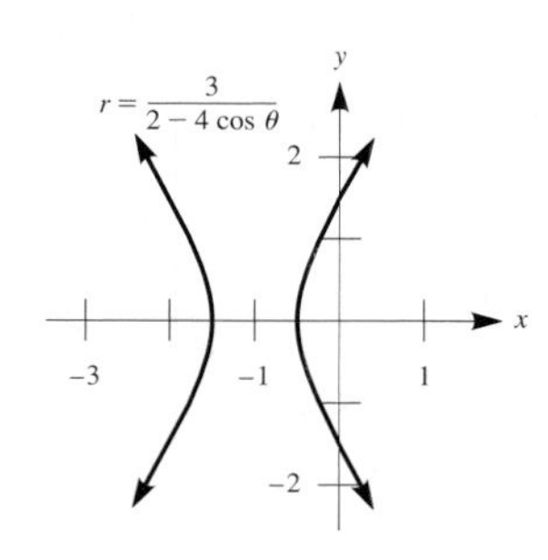

7. center: $(-\frac{72}{5}, 0)$; eccentricity: $\frac{3}{2}$; length of traverse axis: $\frac{96}{5}$; length of conjugate axis: $\frac{48}{5}\sqrt{5}$

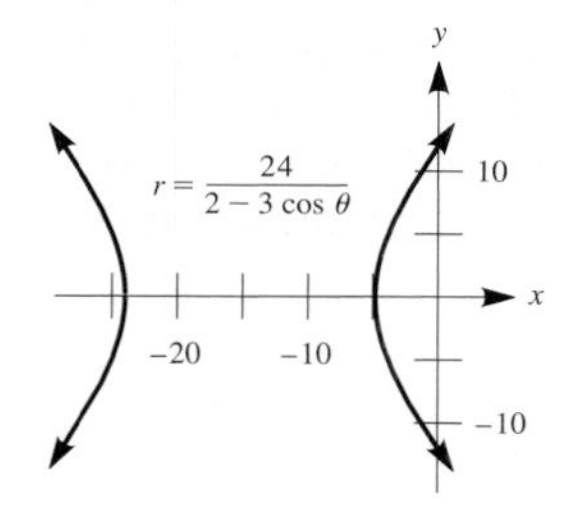

9. center: $(0, -\frac{3}{2})$; eccentricity: $\frac{3}{5}$; length of major axis: 5; length of minor axis: 4

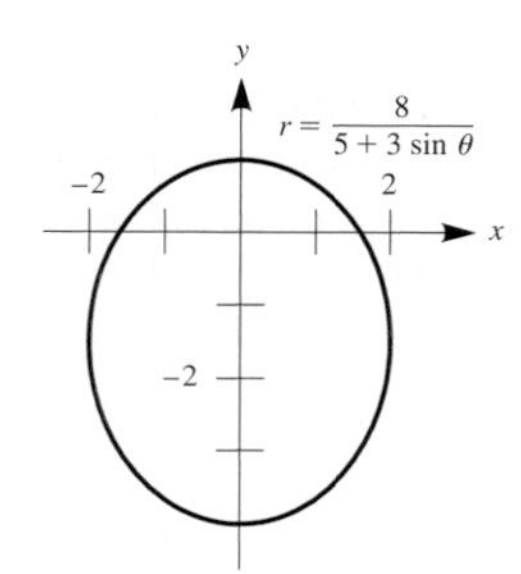

11. vertex: $(0, -\frac{6}{5})$; directrix: $y = -\frac{12}{5}$

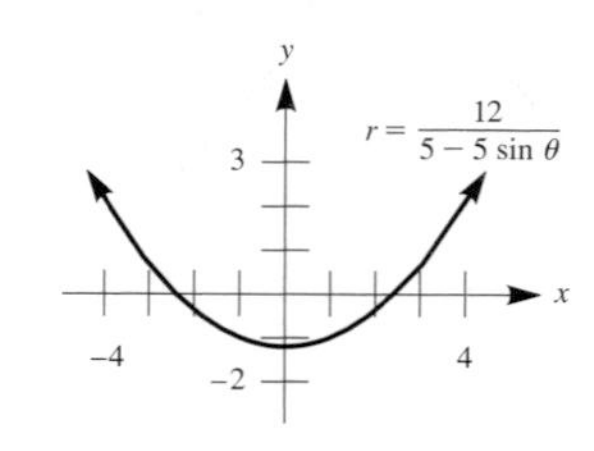

13. center: $(-\frac{5}{2}, 0)$; eccentricity: $\frac{5}{7}$; length of major axis: 7; length of minor axis: $2\sqrt{6}$

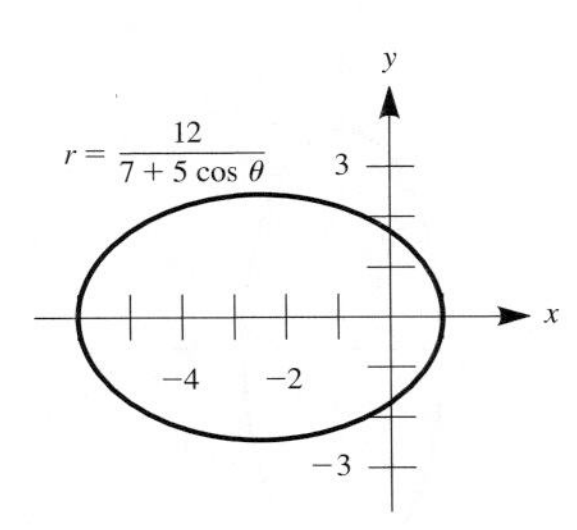

15. vertex: $(0, \frac{2}{5})$; directrix: $y = \frac{4}{5}$

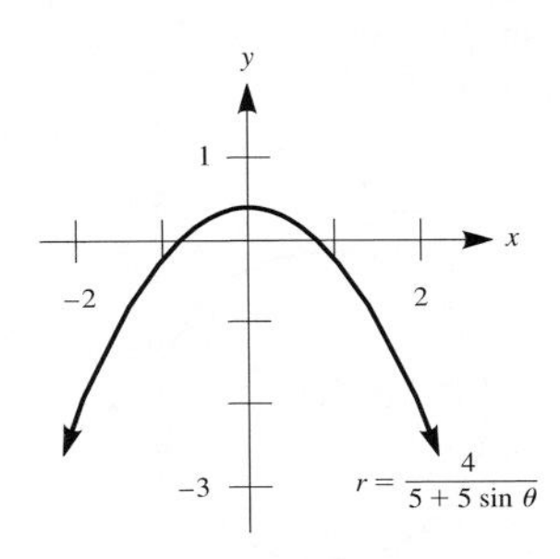

17. center: $(-6, 0)$; eccentricity: 2; length of transverse axis: 6; length of conjugate axis: $6\sqrt{3}$

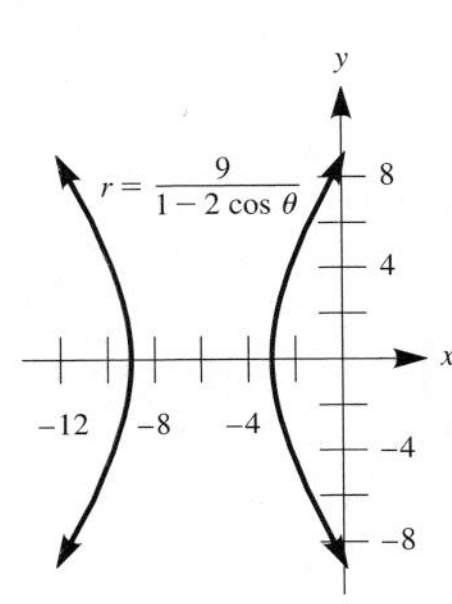

Exercise Set 11.8

1. $(\frac{1}{2}, \frac{3\sqrt{3}}{2})$ **3.** $(2, 0)$ **5.** $(-\frac{31}{13}, \frac{27}{13})$
7. $\sin\theta = \frac{4}{5}, \cos\theta = \frac{3}{5}$
9. $\sin\theta = \frac{3}{5}, \cos\theta = \frac{4}{5}$
11. $\sin\theta = \frac{\sqrt{3}}{2}, \cos\theta = \frac{1}{2}$
13. $\sin\theta = \frac{7}{34}\sqrt{2}, \cos\theta = \frac{23}{34}\sqrt{2}$

15. $(x')^2 - (y')^2 = 9$

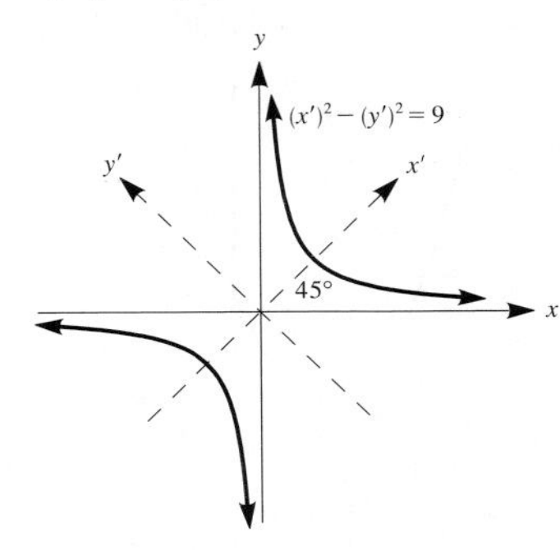

17.

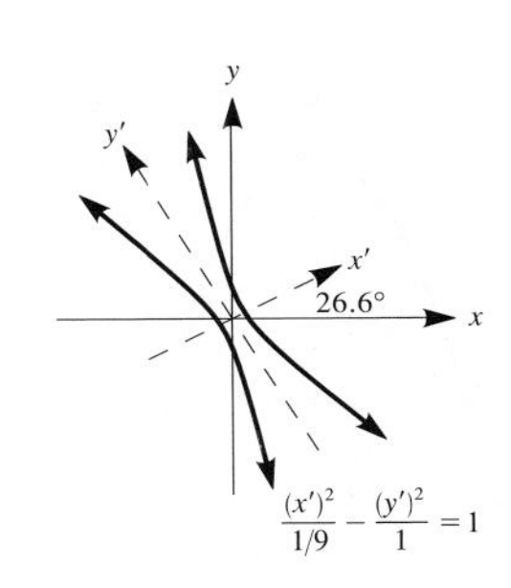

19.

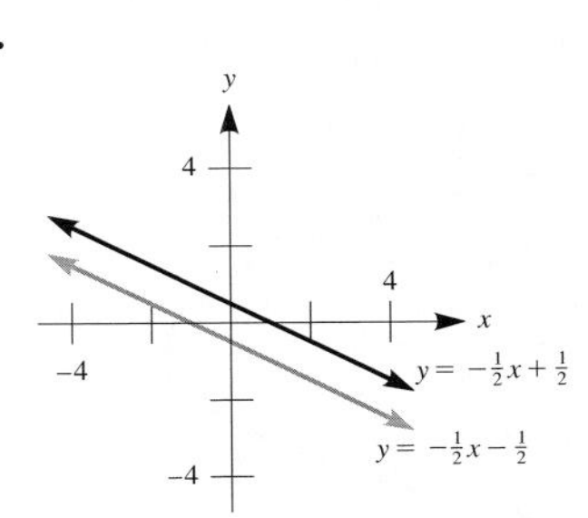

21.

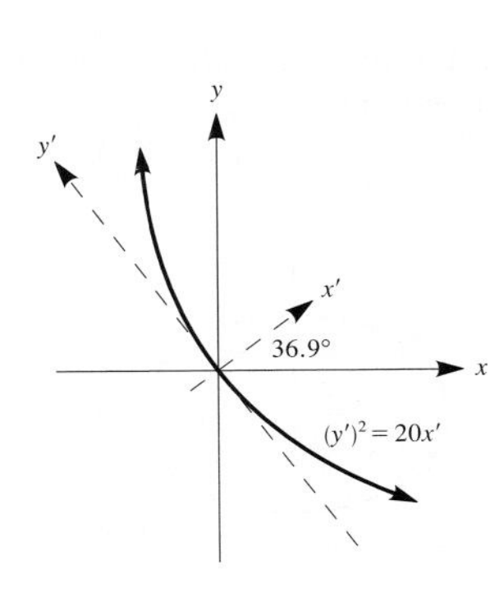

23.

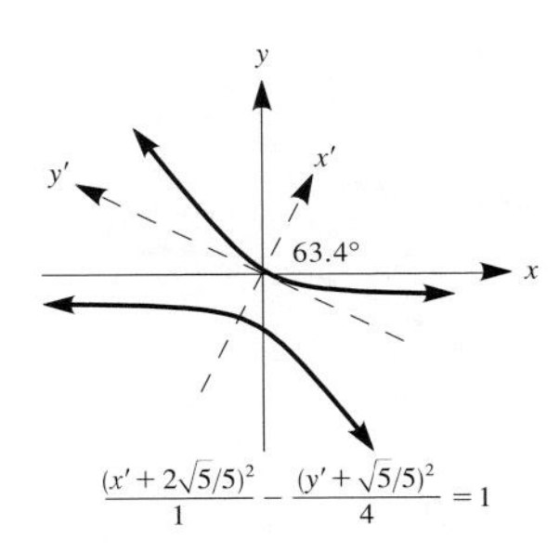

25.

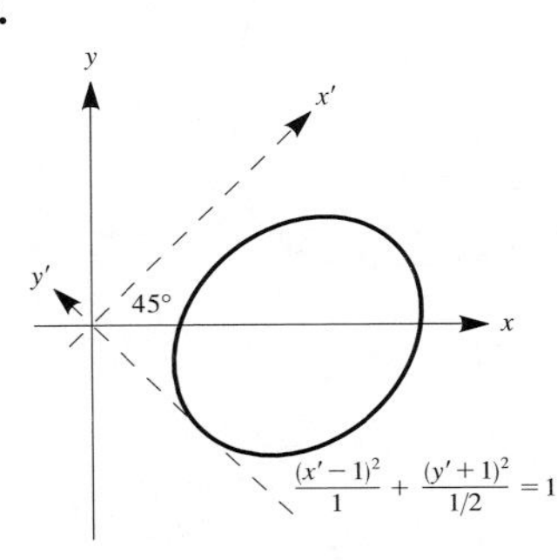

27.

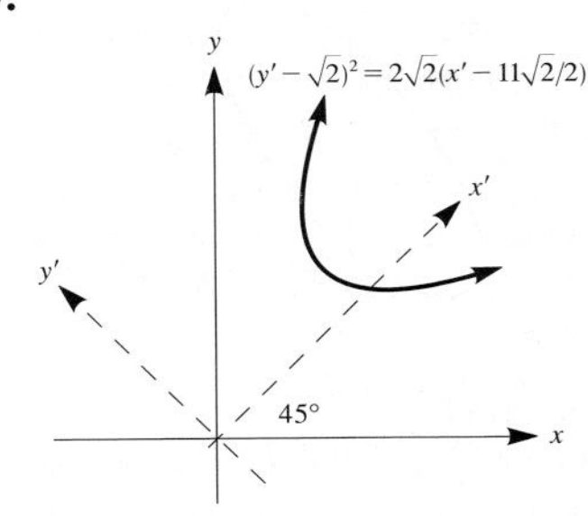

29.

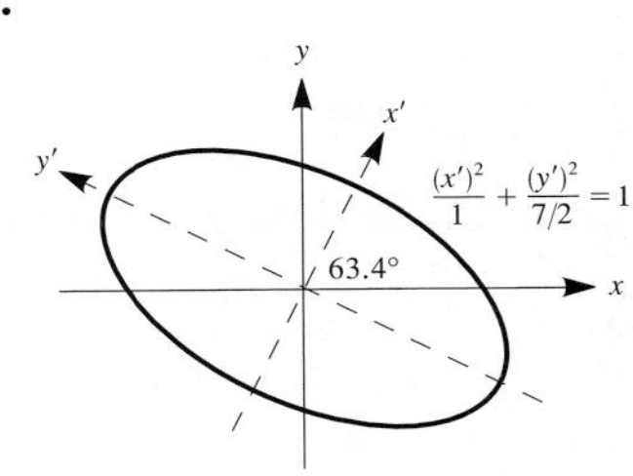

31.

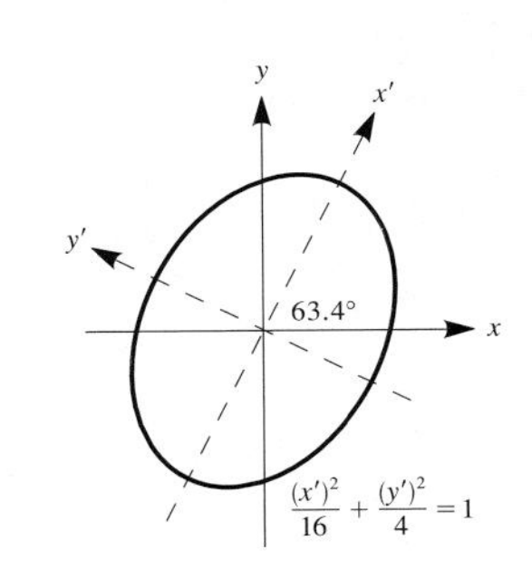

33.

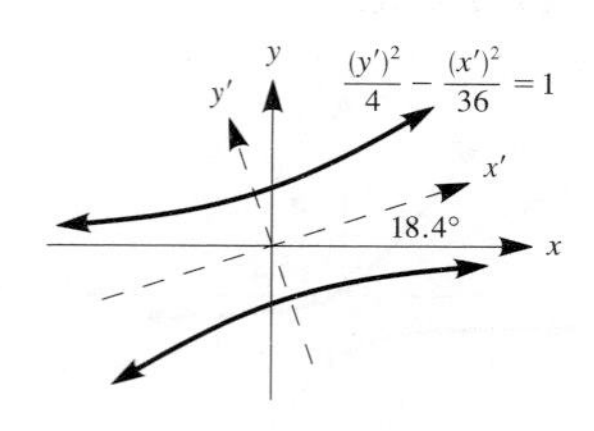

35.

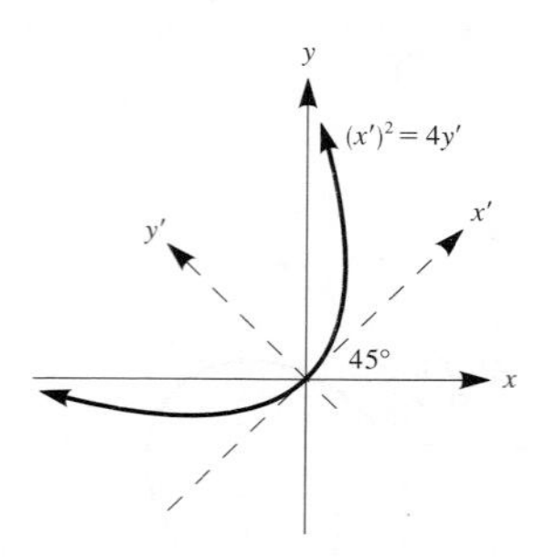

37.

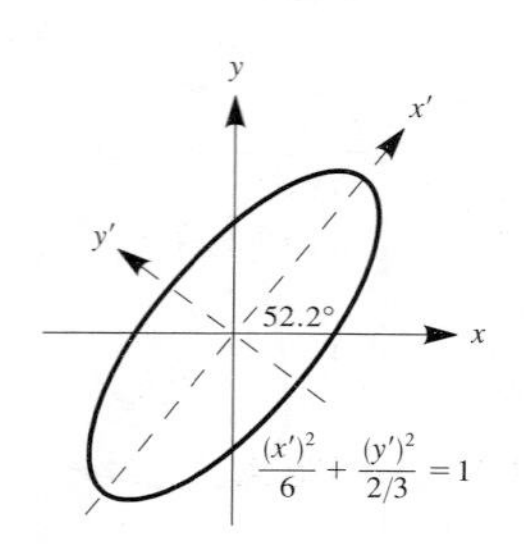

39. no graph
41. For $0° < \theta < 90°$, $0° < 2\theta < 180°$, which gives a full period of cot and Range cot $= (-\infty, \infty)$.
43. $x = x' \cos\theta - y' \sin\theta$; $y = x' \sin\theta + y' \cos\theta$

Chapter 11 Review Exercises

17. 146.3° **19.** $53\sqrt{61}/61$
23. (a) $y^2 = 16x$ **(b)** $x^2 = 16y$
25. $x^2 = 12y$ **27.** $15x^2 + 16y^2 = 960$
29. $25x^2 + 9y^2 = 900$
31. $3x^2 + 5y^2 = 30$
33. $5y^2 - 20x^2 = 36$
35. $36x^2 - 64y^2 = 81$

37. vertex: (0, 0); focus: $(0, \frac{5}{2})$; directrix: $y = -\frac{5}{2}$; focal width: 10

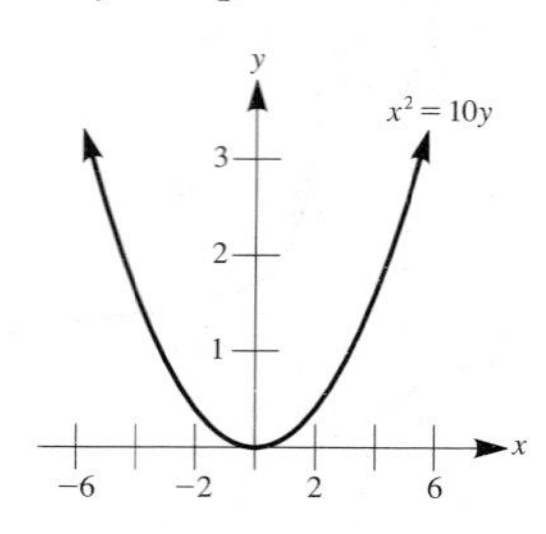

39. vertex: (0, 3); focus: (0, 0); directrix: $y = 6$; focal width: 12

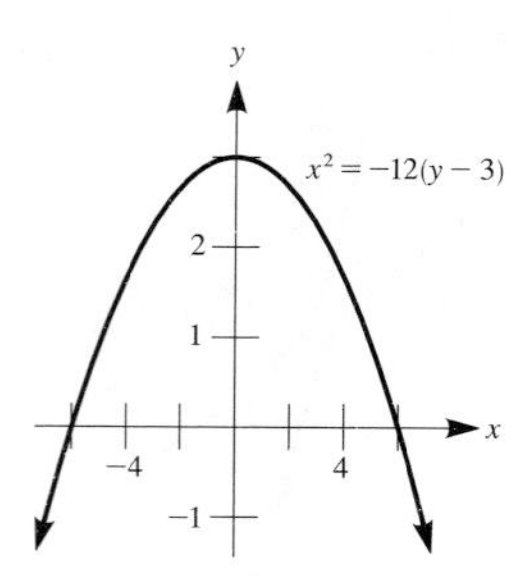

41. vertex: (1, 1); focus: (0, 1); directrix: $x = 2$; focal width: 4

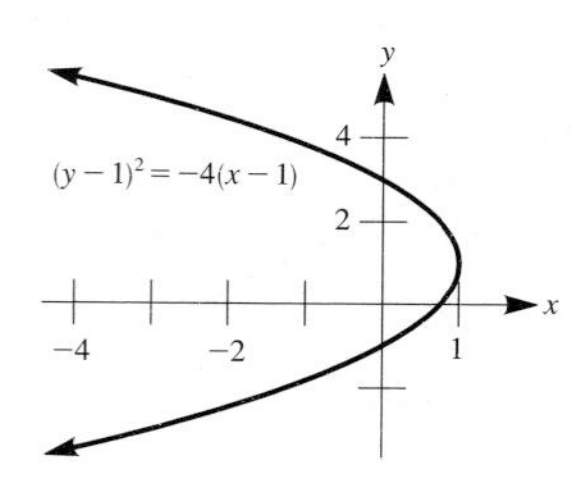

43. center: (0, 0); length of major axis: 12; length of minor axis: 8; foci: $(\pm 2\sqrt{5}, 0)$; eccentricity: $\sqrt{5}/3$

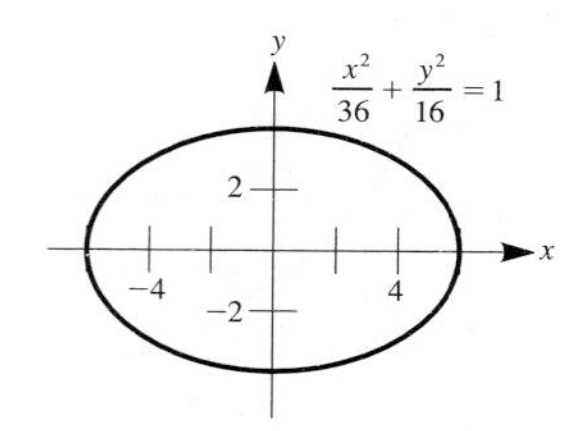

45. center: (0, 0); length of major axis: 6; length of minor axis: 2; foci: $(0, \pm 2\sqrt{2})$; eccentricity: $2\sqrt{2}/3$

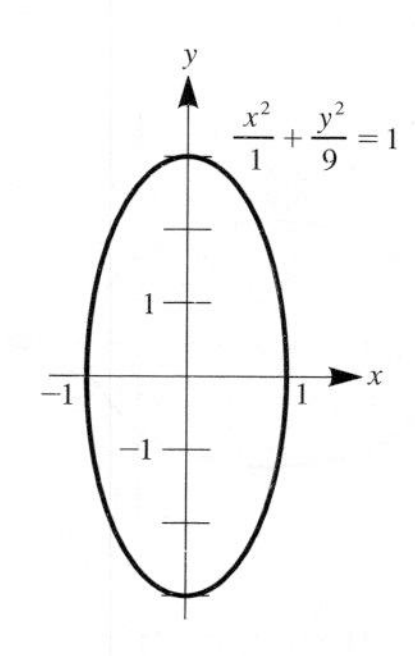

47. center: $(-3, 0)$; lengths of major and minor axes: 6; focus: $(-3, 0)$; eccentricity: 0

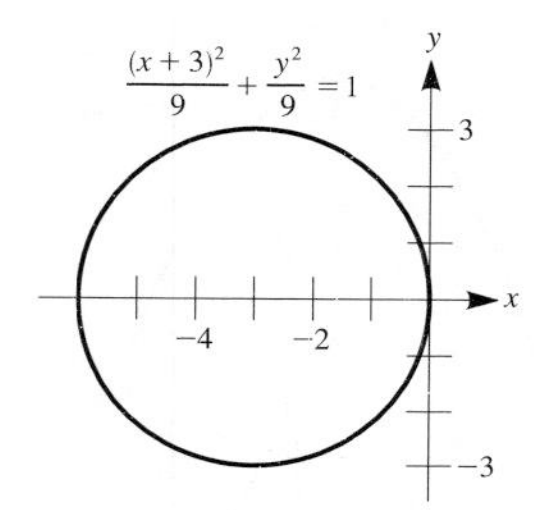

49. center: (0, 0); vertices: $(\pm 6, 0)$; asymptotes: $y = \pm\frac{2}{3}x$; foci: $(\pm 2\sqrt{13}, 0)$; eccentricity: $\sqrt{13}/3$

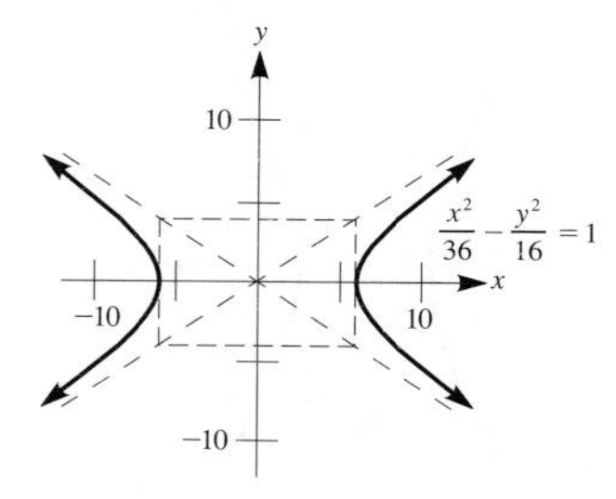

51. center: (0, 0); vertices: $(0, \pm 1)$; asymptotes: $y = \pm\frac{1}{3}x$; foci: $(0, \pm\sqrt{10})$; eccentricity: $\sqrt{10}$

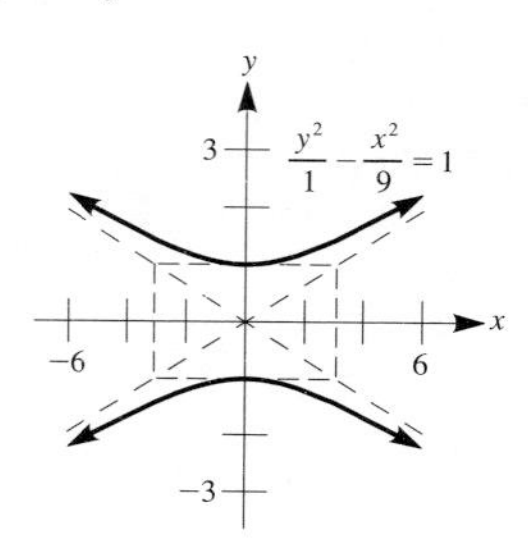

53. center: $(0, -3)$; vertices: $(0, 0)$ and $(0, -6)$; asymptotes: $y = x - 3$ and $y = -x - 3$; foci: $(0, -3 \pm 3\sqrt{2})$; eccentricity: $\sqrt{2}$

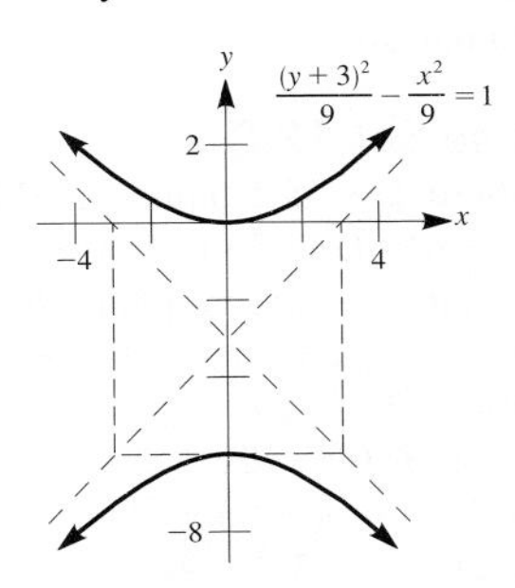

55. parabola: vertex: $(4, 4)$; axis of symmetry: $y = 4$; focus: $(8, 4)$; directrix: $x = 0$

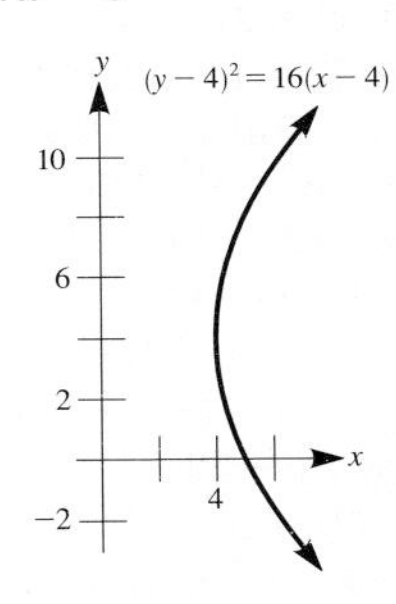

57. ellipse: center: $(-2, 3)$; length of major axis: 8; length of minor axis: 6; foci: $(-2, 3 \pm \sqrt{7})$

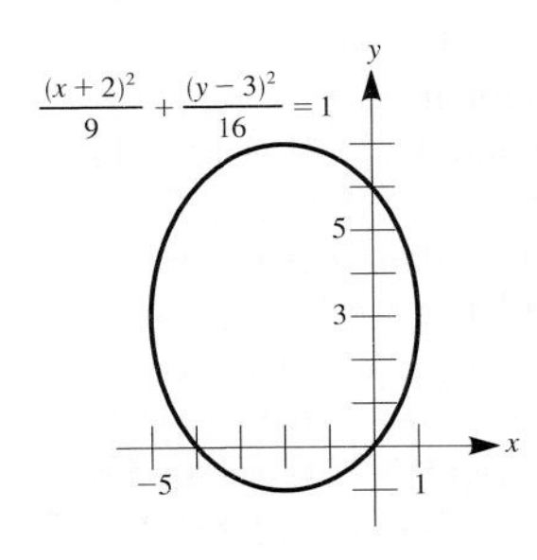

59. parabola: vertex: $(-3, 2)$; axis of symmetry: $x = -3$; focus: $(-3, 5)$; directrix: $y = -1$

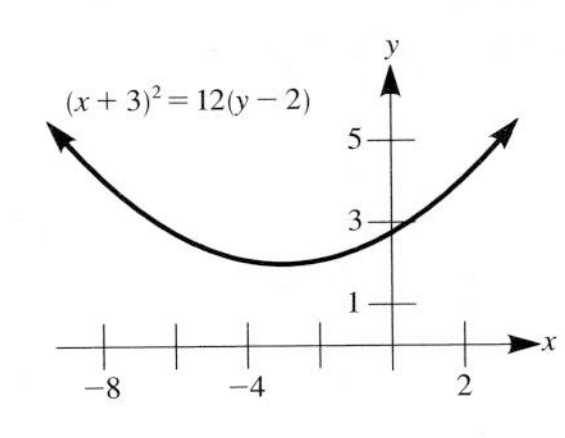

61. hyperbola: center: $(2, 1)$; vertices: $(5, 1)$ and $(-1, 1)$; asymptotes: $y = x - 1$ and $y = -x + 3$; foci: $(2 \pm 3\sqrt{2}, 1)$

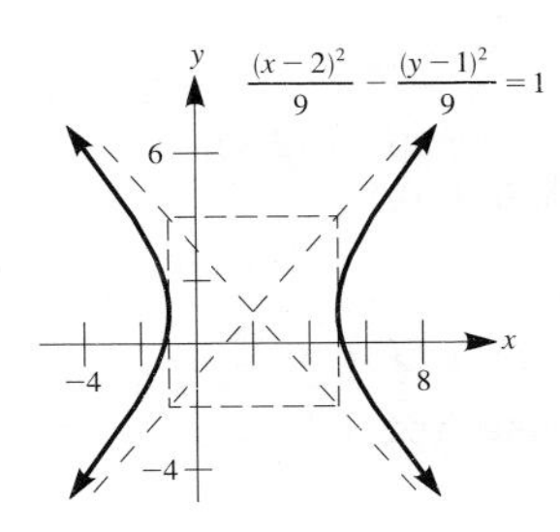

63. parabola: vertex: $(0, 6)$; axis of symmetry: $x = 0$; focus: $(0, \frac{11}{2})$; directrix: $y = \frac{13}{2}$

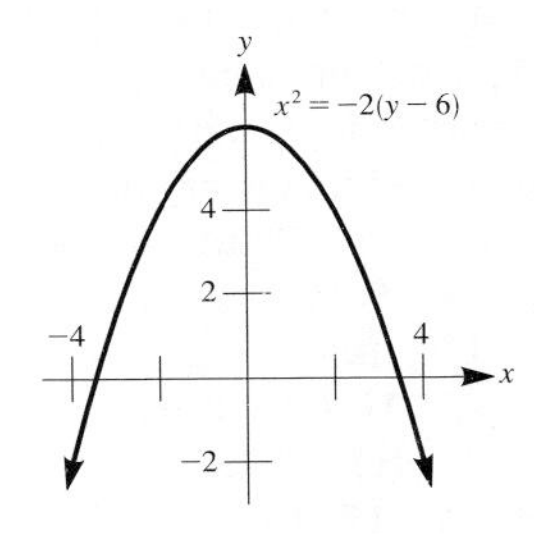

65. ellipse: center: $(0, 5)$; length of major axis: 8; length of minor axis: 2; foci: $(\pm\sqrt{15}, 5)$

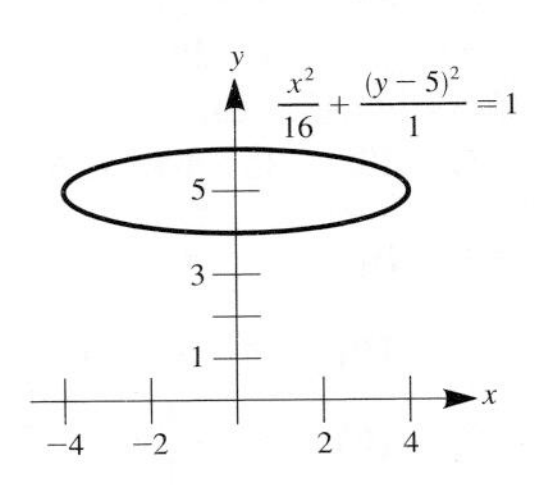

67. two lines: $4x - 5y = -2$ and $4x + 5y = 18$

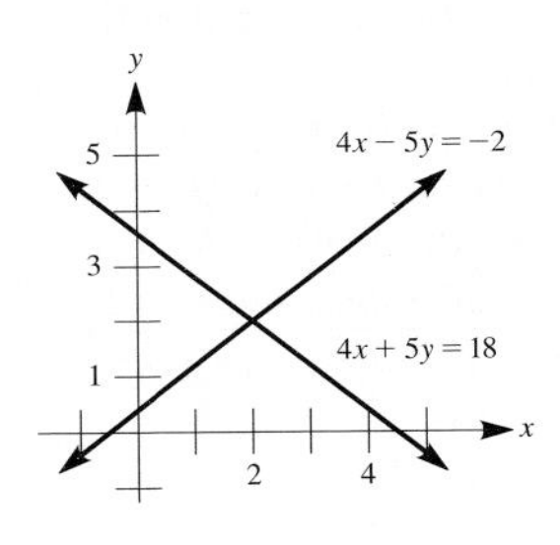

Chapter 11 Test

1. focus: $(-3, 0)$; directrix: $x = 3$

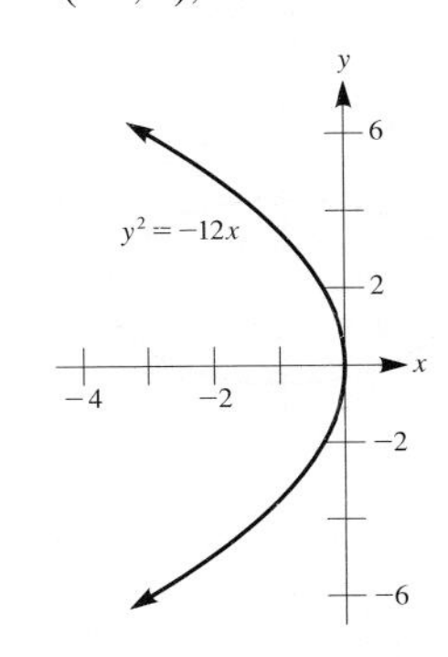

2. asymptotes: $y = \pm\frac{1}{2}x$; foci: $(\pm\sqrt{5}, 0)$

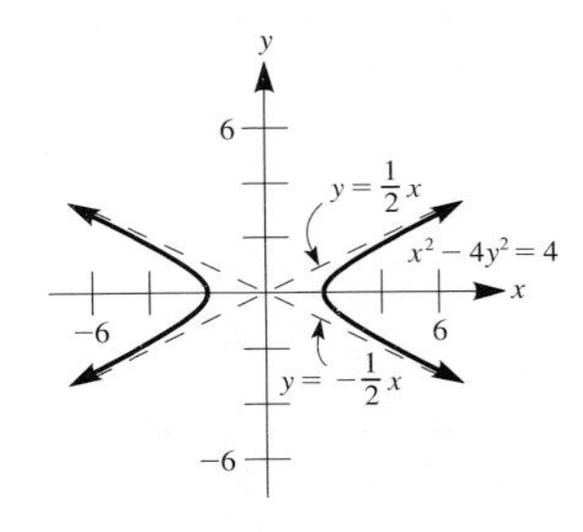

3. $e = 0.001$; 19.078 AU
4. (a) $\theta = 60°$
(b)

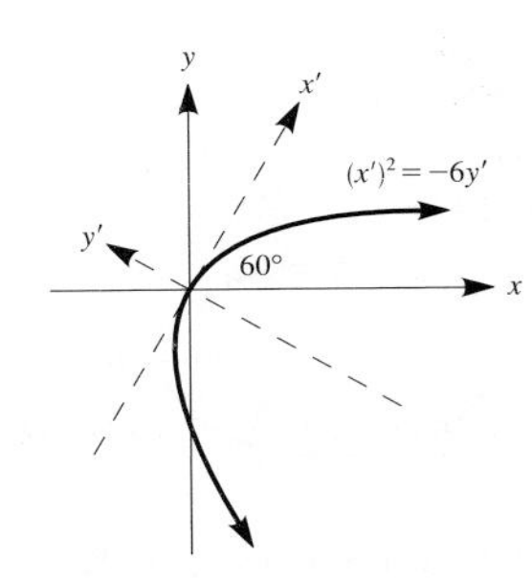

5. 30° **6.** $\frac{x^2}{12} + \frac{y^2}{16} = 1$
7. $m = \pm\sqrt{15}/15$
8. $\sqrt{3}x - y - 2\sqrt{3} = 0$
9. $\frac{x^2}{3} - \frac{y^2}{1} = 1$ **10. (b)** 64

11. length of major axis: 10; length of minor axis: 4; foci: $(\pm\sqrt{21}, 0)$

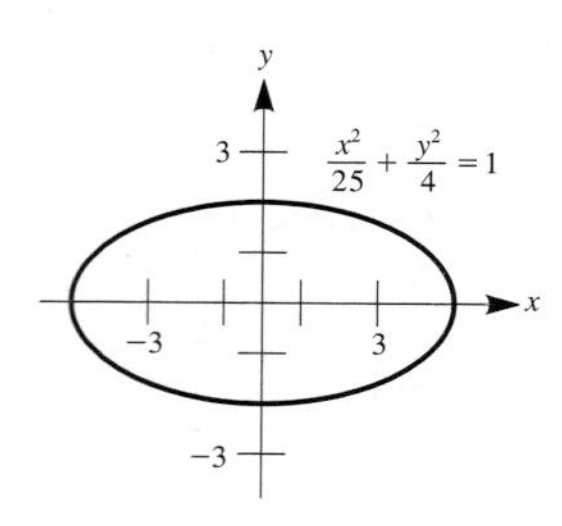

12. $3\sqrt{5}/5$

13.

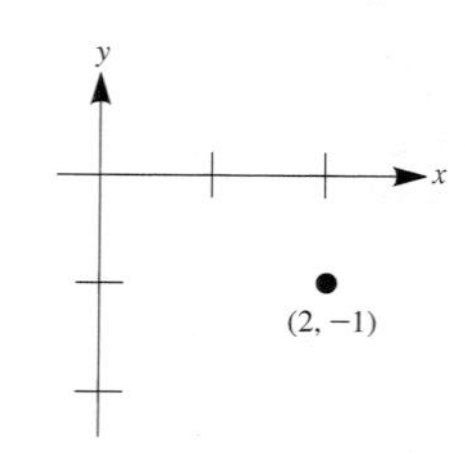

14.

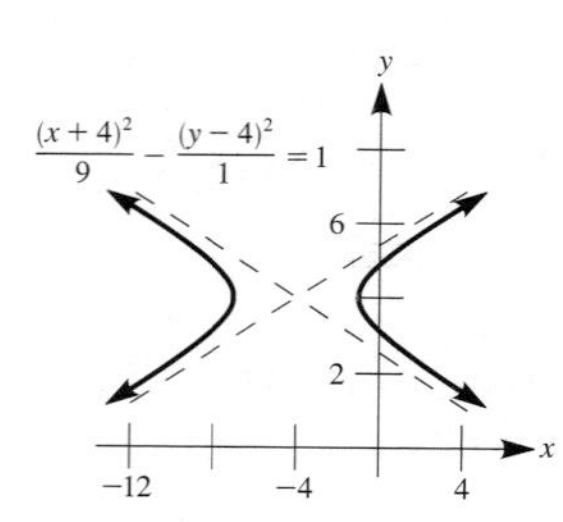

15. ellipse

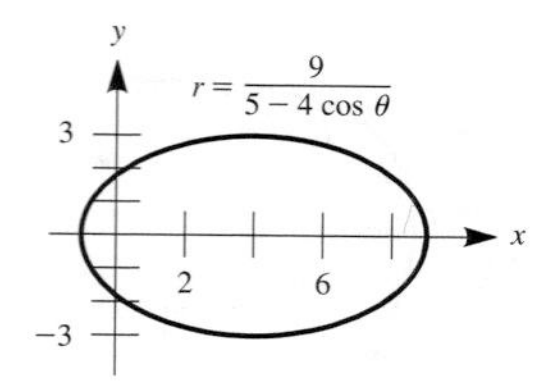

16. focal width: 8; vertex: (1, 2)

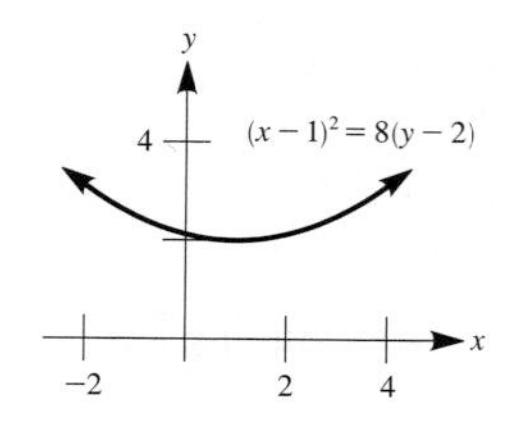

17. (a) $x = \pm 36\sqrt{11}/11$ **(b)** $F_1P = \dfrac{12 + \sqrt{11}}{2}$; $F_2P = \dfrac{12 - \sqrt{11}}{2}$

18. $y = \frac{1}{6}x + \frac{13}{3}$ **19.** $y = 4x - 8$

20. focal length = 12.8 m; focal ratio = 0.4

CHAPTER 12

Exercise Set 12.1

1.

i^2	i^3	i^4	i^5	i^6	i^7	i^8
-1	$-i$	1	i	-1	$-i$	1

3. (a) real part: 4; imaginary part: 5 **(b)** real part: 4; imaginary part: -5 **(c)** real part: $\frac{1}{2}$; imaginary part: -1 **(d)** real part: 0; imaginary part: 16

5. $c = 4, d = -3$ **7. (a)** $14 - 4i$ **(b)** $-4 - 8i$ **9. (a)** $19 - 17i$ **(b)** $19 - 17i$ **(c)** $\frac{11}{26} - \frac{23}{26}i$ **(d)** $\frac{11}{25} + \frac{23}{25}i$ **11. (a)** $11 - i$ **(b)** $11 - 7i$ **(c)** 4 **13.** $4 - 2i$

15. $30 + 19i$ **17.** 13

19. $-191 - 163i$ **21.** $19 - 4i$

23. $-70 + 84i$ **25.** $539 + 1140i$

27. $-46 + 9i$ **29.** $\frac{6}{97} + \frac{35}{97}i$

31. $\frac{6}{97} - \frac{35}{97}i$ **33.** $-\frac{5}{13} + \frac{12}{13}i$

35. -4 **37.** $\frac{1}{26} + \frac{5}{26}i$ **39.** $-i$

41. i **43.** -1 **45.** $12i$

47. $-3\sqrt{5}i$ **49.** -35 **51.** $12\sqrt{2}i$

53. (a) -3 **(b)** $x = \dfrac{1}{2} \pm \dfrac{\sqrt{3}}{2}i$

55. (a) -36 **(b)** $z = -\frac{1}{5} \pm \frac{3}{5}i$

57. (a) -23 **(b)** $z = -\dfrac{3}{4} \pm \dfrac{\sqrt{23}}{4}i$

59. (a) $-\frac{29}{48}$ **(b)** $z = \dfrac{3}{4} \pm \dfrac{\sqrt{87}}{4}i$

61. (a) 0 **(b)** This verifies the solution.

63. (a) $z + w = (a + c) + (b + d)i$ **(b)** $z - w = (a - c) + (b - d)i$ **(c)** $zw = (ac - bd) + (bc + ad)i$ **(d)** $\dfrac{z}{w} = \dfrac{ac + bd}{c^2 + d^2} + \dfrac{bc - ad}{c^2 + d^2}i$

71. $x = 3, \pm 2i$ **73.** $x = \pm 3, \sqrt{2} \pm \sqrt{2}i, -\sqrt{2} \pm \sqrt{2}i$ **75.** real part: $\dfrac{2a^2 - 2b^2}{a^2 + b^2}$; imaginary part: 0

77. real part: 0

Exercise Set 12.2

1. quotient: $x - 5$; remainder: -11; $x^2 - 8x + 4 = (x - 3)(x - 5) - 11$

3. quotient: $x - 11$; remainder: 53; $x^2 - 6x - 2 = (x + 5)(x - 11) + 53$

5. quotient: $3x^2 - \frac{3}{2}x - \frac{1}{4}$; remainder: $\frac{13}{4}$; $6x^3 - 2x + 3 = (2x + 1)(3x^2 - \frac{3}{2}x - \frac{1}{4}) + \frac{13}{4}$

7. quotient: $x^4 - 3x^3 + 9x^2 - 27x + 81$; remainder: -241; $x^5 + 2 = (x + 3)(x^4 - 3x^3 + 9x^2 - 27x + 81) - 241$

9. quotient: $x^5 + 2x^4 + 4x^3 + 8x^2 + 16x + 32$; remainder: 0; $x^6 - 64 = (x - 2)(x^5 + 2x^4 + 4x^3 + 8x^2 + 16x + 32) + 0$

11. quotient: $5x^2 + 15x + 17$; remainder: $-24x - 83$; $5x^4 - 3x^2 + 2 = (x^2 - 3x + 5) \times (5x^2 + 15x + 17) + (-24x - 83)$

13. quotient: $3y - 19$; remainder: $89y + 35$; $3y^3 - 4y^2 - 3 = (y^2 + 5y + 2) \times (3y - 19) + (89y + 35)$

15. quotient: $t^2 - 2t - 4$; remainder: 0; $t^4 - 4t^3 + 4t^2 - 16 = (t^2 - 2t + 4) \times (t^2 - 2t - 4) + 0$

17. quotient: $z^4 + z^3 + z^2 + z + 1$; remainder: 0; $z^5 - 1 = (z - 1) \times (z^4 + z^3 + z^2 + z + 1) + 0$

19. quotient: $ax + (b + ar)$; remainder: $c + r(b + ar) = ar^2 + br + c$; $ax^2 + bx + c = (x - r) \times (ax + (b + ar)) + (ar^2 + br + c)$

21. quotient: $x - 1$; remainder: -7; $x^2 - 6x - 2 = (x - 5)(x - 1) - 7$

23. quotient: $4x - 5$; remainder: 0; $4x^2 - x - 5 = (x + 1)(4x - 5) + 0$

25. quotient: $6x^2 + 19x + 78$; remainder: 313; $6x^3 - 5x^2 + 2x + 1 = (x - 4)(6x^2 + 19x + 78) + 313$

27. quotient: $x^2 + 2x + 4$; remainder: 7; $x^3 - 1 = (x - 2) \times (x^2 + 2x + 4) + 7$ **29.** quotient: $x^4 - 2x^3 + 4x^2 - 8x + 16$; remainder: -33; $x^5 - 1 = (x + 2) \times (x^4 - 2x^3 + 4x^2 - 8x + 16) - 33$
31. quotient: $x^3 - 10x^2 + 40x - 160$; remainder: 642; $x^4 - 6x^3 + 2 = (x + 4)(x^3 - 10x^2 + 40x - 160) + 642$ **33.** quotient: $x^2 + 6x + 57$; remainder: 576; $x^3 - 4x^2 - 3x + 6 = (x - 10) \times (x^2 + 6x + 57) + 576$
35. quotient: $x^2 - 6x + 30$; remainder: -150; $x^3 - x^2 = (x + 5)(x^2 - 6x + 30) - 150$
37. quotient: $54x^2 + 9x - 21$; remainder: 0; $54x^3 - 27x^2 - 27x + 14 = (x - \frac{2}{3})(54x^2 + 9x - 21) + 0$
39. quotient: $x^3 + 3x^2 + 12x + 36$; remainder: 120; $x^4 + 3x^2 + 12 = (x - 3)(x^3 + 3x^2 + 12x + 36) + 120$
41. $(x - 2)(x^4 + 2x^3 + 4x^2 + 8x + 16)$
43. $(z - 3)(z^3 + 3z^2 + 9z + 27)$
45. quotient: $2x$; remainder: 1
47. quotient: $3x^2 - \frac{3}{2}x + \frac{3}{4}$; remainder: $\frac{1}{4}$ **49.** $k = 4$
53. quotient: $x + (-4 + i)$; remainder: $-4i$ **55.** quotient: $x + (-1 + i)$; remainder: 0
57. $-6\sqrt{3} + 57$ **59.** 0

Exercise Set 12.3

1. yes **3.** yes **5.** yes **7.** is a zero
9. is a zero **11.** not a zero
13. (a) is a zero **(b)** not a zero
15. 1, 2 (multiplicity 3), 3
17. (a) multiplicity at -1: 2

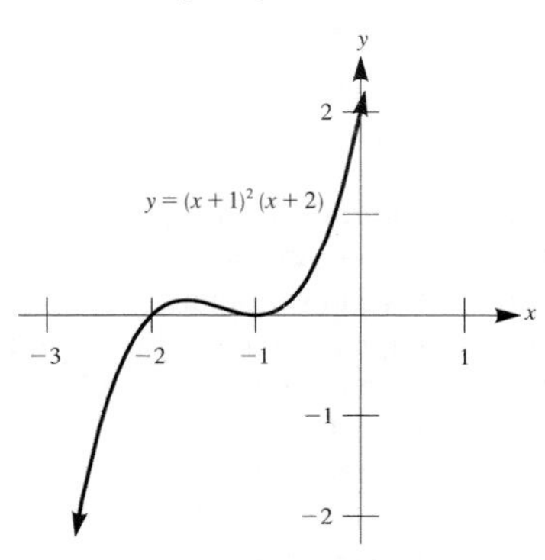

(b) multiplicity at -2: 3

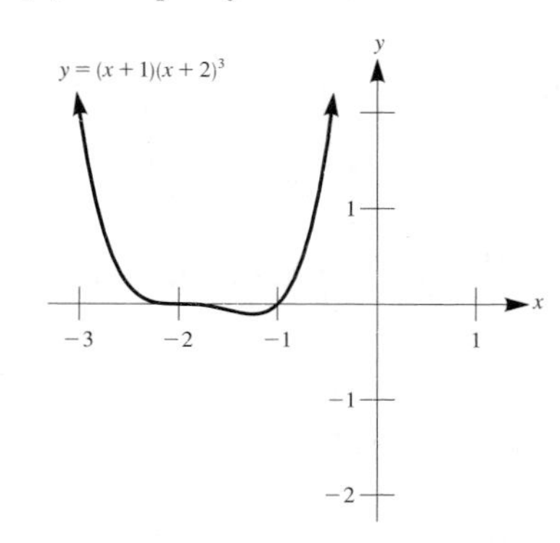

(c) multiplicity at -1: 2; multiplicity at -2: 3

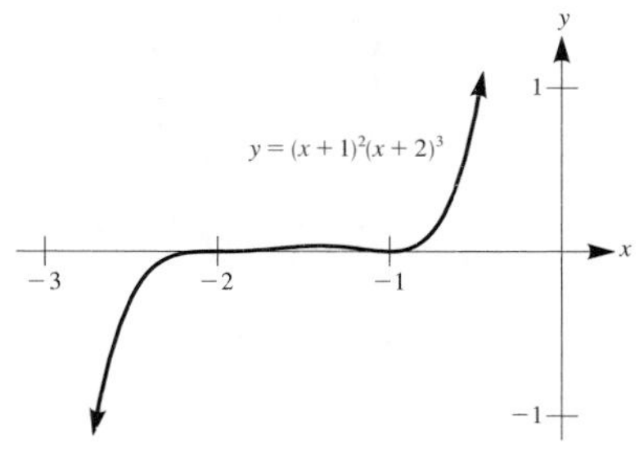

19. $f(-3) = -170$
21. $f(\frac{1}{2}) = -9$
23. $f(-\sqrt{2}) = -3\sqrt{2} - 2$
25. $x = -1, \dfrac{1 \pm i}{2}$ **27.** $x = \pm 3, 4$
29. $x = 1, -1 \pm \sqrt{6}$
31. $x = -2, \frac{2}{3}, 3$ **33.** $x = -\frac{3}{2}, \dfrac{1 \pm \sqrt{5}}{2}$
35. $x = 0, 5$ **37.** $x = -4, 3$
39. $x = -5, 2 - \sqrt{2} \approx 0.59$, $2 + \sqrt{2} \approx 3.41$

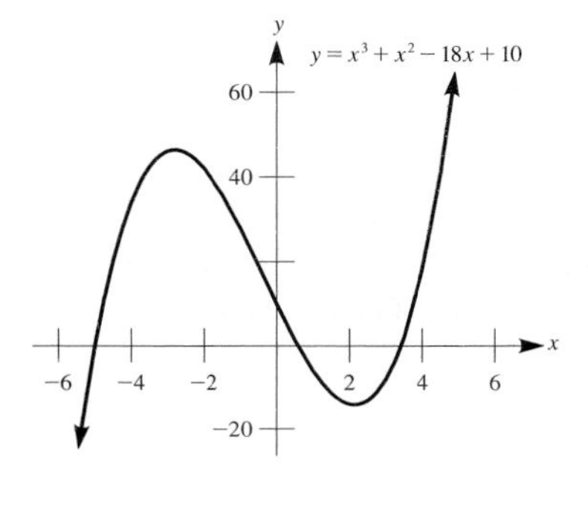

41. $x = \frac{2}{3}, \dfrac{4 - \sqrt{14}}{2} \approx 0.13$, $\dfrac{4 + \sqrt{14}}{2} \approx 3.87$

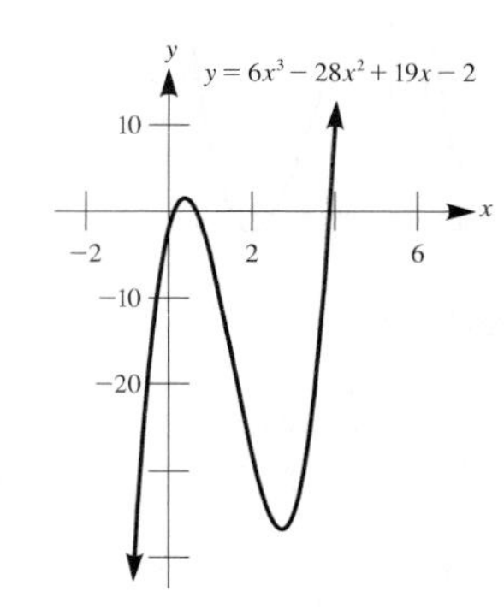

43. $x = 1, -1, -\frac{1}{4}, 2$

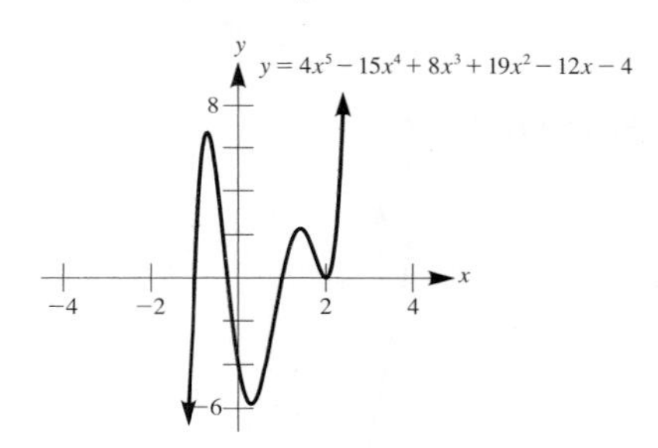

45. (a) 1.125 **(b)** -0.046875
(c) $t - 1$ **(d)** $t = 1, \dfrac{-1 \pm \sqrt{13}}{2}$
47. $x^3 - 4x^2 - 17x + 60 = 0$
49. $x^3 + 8x^2 + 13x + 6 = 0$
51. no such polynomial exists
53. (a) $g(r) = ar^3 + br^2 + cr + d$
(b) remainder: $ar^3 + br^2 + cr + d$
55. is a zero **57.** $a = -1, b = -1$
59. $b = \pm 3\sqrt{2}/2$
61. $x = 2$ (multiplicity 2), -4

Exercise Set 12.4

1. (a) yes **(b)** yes **(c)** yes
(d) no **3.** no real roots

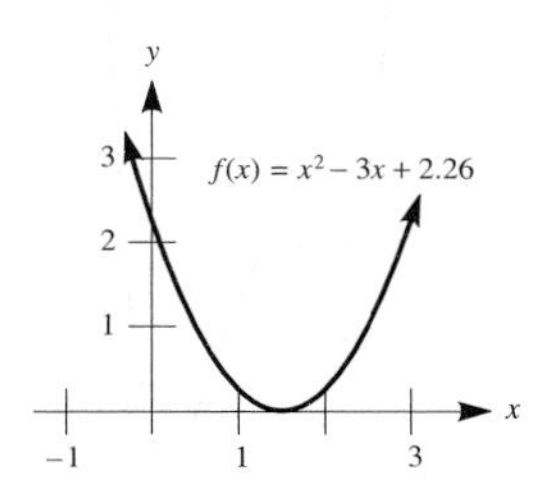

5. three real roots

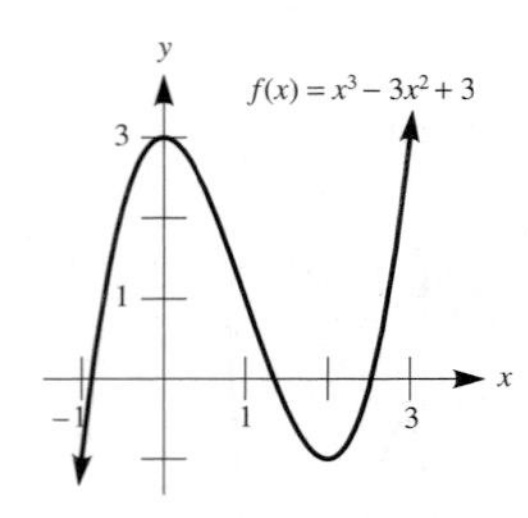

7. no real roots

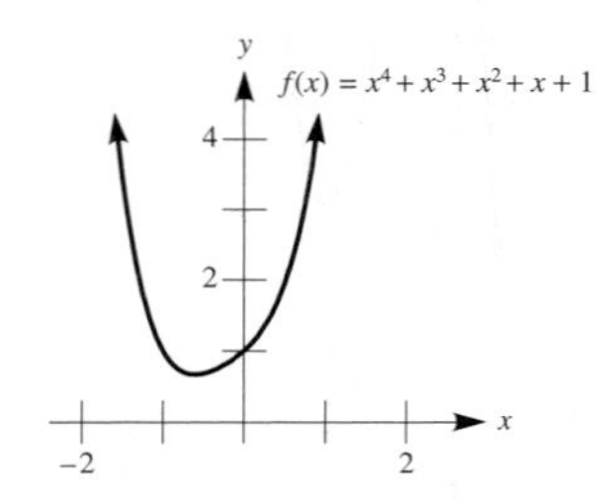

9. one real root

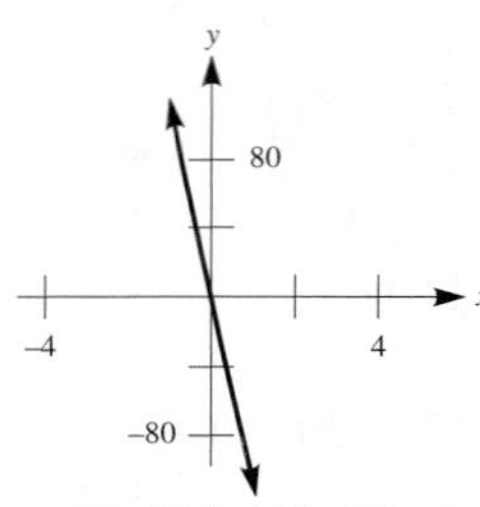

11. $[x - (-1)](x - 3)$
13. $4(x - \frac{1}{4})[x - (-6)]$
15. $[x - (-\sqrt{5})](x - \sqrt{5})$
17. $(x - 0)(x - (-3))(x - 3)$ $(x - (-\sqrt{2}i))(x - \sqrt{2}i)$
19. $f(x) = x^3 + x^2 - 5x + 3$
21. $f(x) = x^4 - 16$
23. $f(x) = x^6 + 10x^4 - 87x^2 + 144$
25. (a) $f(x) = x^4 - 4x^3 + 3x^2$
(b)

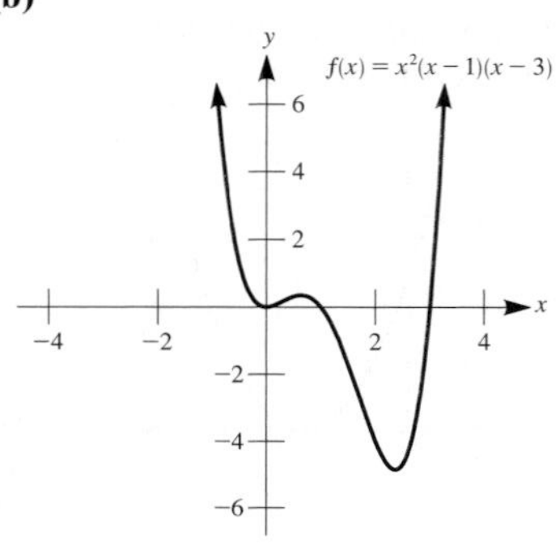

27. (a) $f(x) = x(x - 1)^2(x - 3)^2$

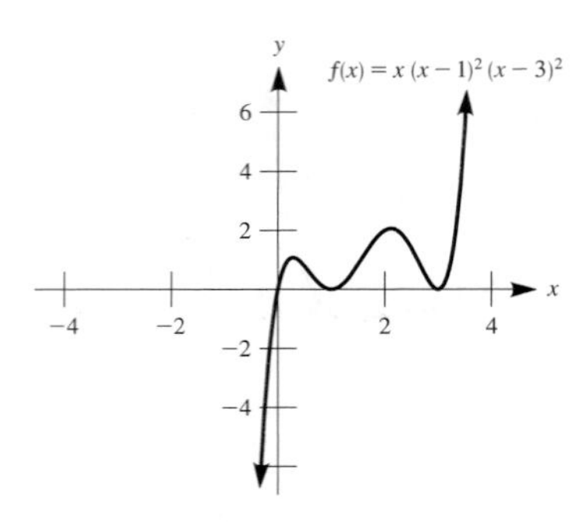

29. (a) $f(x) = x^2(x - 1)(x - 3)^3$
(b)

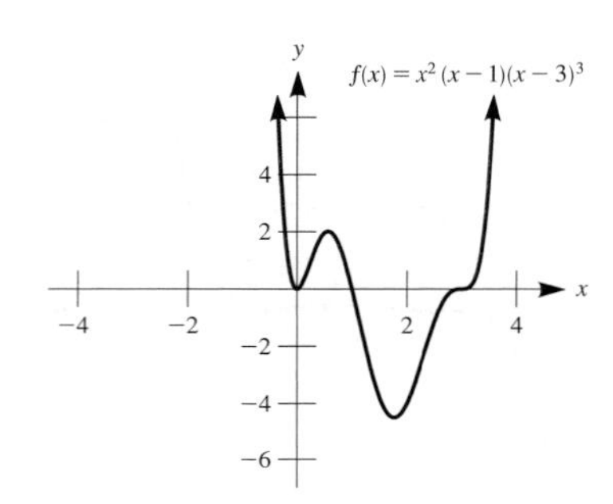

31. $f(x) = -\frac{5}{42}x^2 + \frac{25}{42}x + \frac{30}{7}$
33. (a) $f(x) = \frac{1}{30}x^3 - \frac{19}{30}x + 1$
(b)

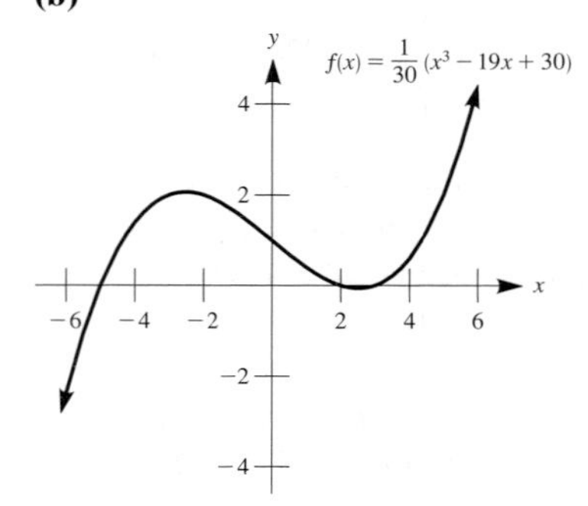

35. $x^2 + (i + \sqrt{3})x + i\sqrt{3} = 0$
37. $x^2 - 3x - 54 = 0$
39. $x^2 - 2x - 4 = 0$ **41.** $x^2 - 2ax + a^2 - b = 0$ **43. (a)** $x = 4$

(b)

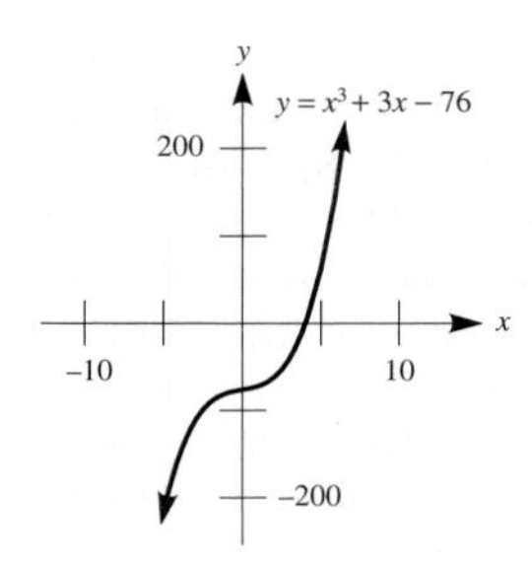

45. $(x + 2 - 2i)(x + 2 + 2i) \times (x - 2 - 2i)(x - 2 + 2i)$
49. $x = 5$ (multiplicity 2), -10
51. (a) $4x^2 + 2x - 1 = 0$

Exercise Set 12.5

1. (a) see text **(b)** $\pm 1, \pm 11$
3. (a) The possible rational roots are $\pm 1, \pm 5$, so there cannot be a rational root larger than 5. No, 5 is not a root.
(b) The possible rational roots are $\pm 1, \pm 5$, so 2 cannot be a root.
(c)

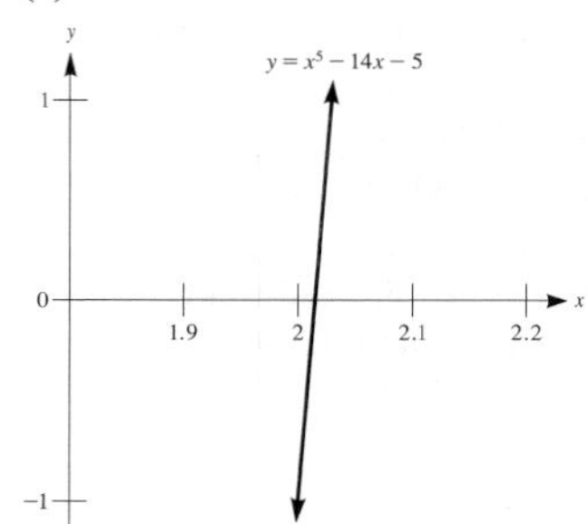

5. $\pm 1, \pm\frac{1}{2}, \pm\frac{1}{4}, \pm 3, \pm\frac{3}{2}, \pm\frac{3}{4}$
7. $\pm 1, \pm\frac{1}{2}, \pm\frac{1}{4}, \pm\frac{1}{8}, \pm 3, \pm\frac{3}{2}, \pm\frac{3}{4}, \pm\frac{3}{8}, \pm 9, \pm\frac{9}{2}, \pm\frac{9}{4}, \pm\frac{9}{8}$
9. $\pm 1, \pm\frac{1}{2}, \pm 2, \pm 3, \pm\frac{3}{2}, \pm 6$
17. $x = 1, -1, -3$ **19.** $x = -\frac{1}{4}, \pm\sqrt{5}$
21. $x = -1, -\frac{2}{3}, -\frac{1}{3}$
23. $x = 1, -1, \dfrac{-1 \pm \sqrt{97}}{2}$
25. $x = 1$ (multiplicity 4)
27. $x = 1, -\frac{4}{3}, 6$ **29. (a)** 2 is an upper bound, -1 is a lower bound.
(b) 2 is an upper bound, -1 is a lower bound. **(c)** 6 is an upper bound, -2 is a lower bound.
31. between 0.68 and 0.69
33. between 2.88 and 2.89
35. between 4.31 and 4.32
37. (a) between -3 and -2
(b) between -2.148 and -2.147

39. (a) between -6 and -5
(b) between -5.265 and -5.264
41. (b) A and C have a common factor of 5, so the result does not apply.
(c) $x = p/q$ is a root of the equation.
43. $x \approx 1.769$

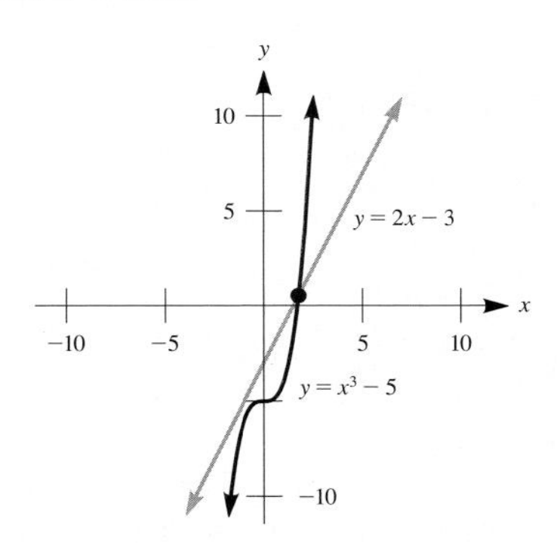

45. $x \approx 1.310$

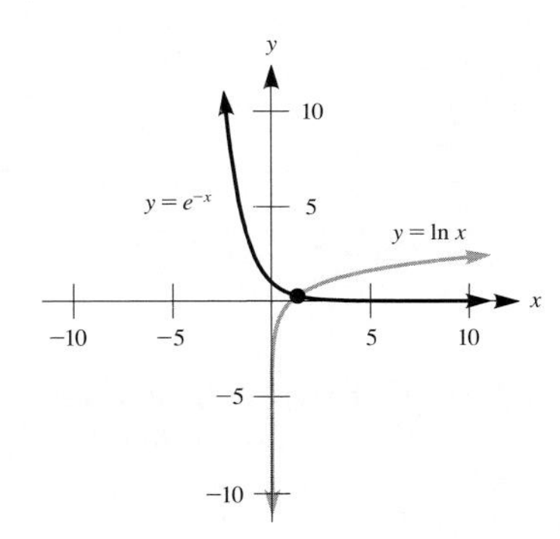

47. $x = \sqrt[3]{200} \approx 5.848$

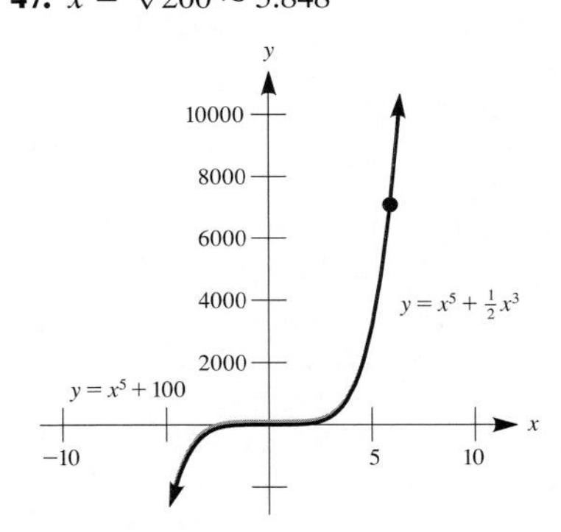

55. $p = 3, x = 1$
57. (a) $x = \dfrac{-1 \pm i\sqrt{11}}{2}$ **59.** $b = 2$

Exercise Set 12.6

1. $7 + 2i$ **3.** $5 - 2i, 3$ **5.** $-2 - i$, -3 (multiplicity 2) **7.** $6 + 5i, -\frac{1}{4}$
9. $4 - \sqrt{2}i, \pm\frac{3i}{2}$ **11.** $10 - 2i, 1 \pm \sqrt{5}$
13. $\dfrac{1 - i\sqrt{2}}{3}, \frac{2}{5}$ **15.** $3 + 2i, -1 - i$, $-1 \pm \sqrt{2}$ **17.** $x^2 - 2x - 5 = 0$
19. $x^2 - \frac{4}{3}x - \frac{2}{3} = 0$
21. (c) one positive real root

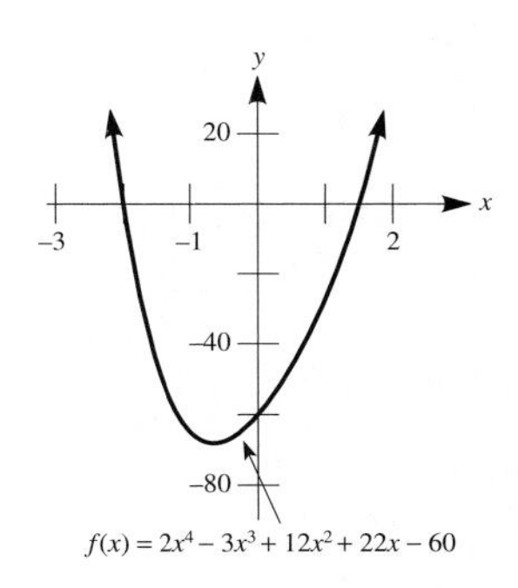

(d) $-2, \frac{3}{2}$ **23. (b)** one negative real root

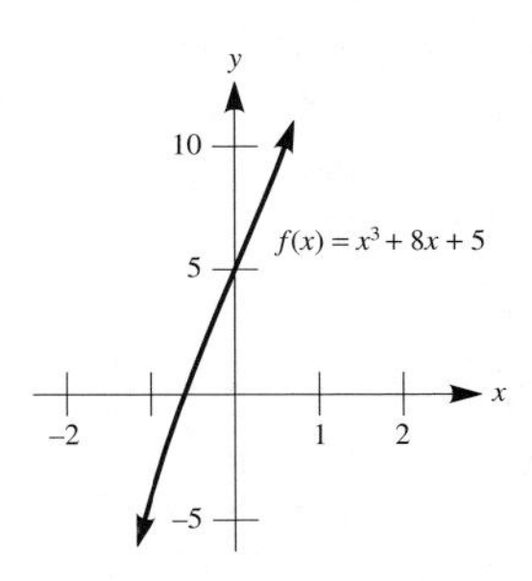

(c) -0.5982
(d) $x = \sqrt[3]{-2.5 + \sqrt{\frac{2723}{108}}} - \sqrt[3]{2.5 + \sqrt{\frac{2723}{108}}} \approx -0.5982$
25. 2 complex roots and 1 negative real root **27.** 4 complex roots and 1 negative real root **29.** 2 complex roots, 1 positive real root and 1 negative real root **31.** either 1 positive real root and 2 negative real roots, or 1 positive real root and 2 complex roots **33.** 1 positive real root, 1 negative real root and 6 complex roots **35.** 1 positive real root and 8 complex roots
37. 1 positive real root, 1 negative real root and 6 complex roots
39. 1 positive real root, 1 negative real root and 4 complex roots
41. 1 positive real root, 1 negative real root and 2 complex roots
43. (a) $f(x) = x^4 + 2x^2 + 49$
(b) $-\sqrt{3} \pm 2i, \sqrt{3} - 2i$

Exercise Set 12.7

1. (a) $A = 2, B = 5$ **(b)** $A = 2$, $B = 5$ **3. (a)** $A = 4, B = -1$
(b) $A = 4, B = -1$
5. (a) $A = -\frac{1}{4}, B = \frac{3}{4}$
(b) $A = -\frac{1}{4}, B = \frac{3}{4}$
7. $A = 8, B = -21$
9. $A = -\frac{1}{5}, B = \frac{34}{5}$
11. $A = 4, B = -1, C = 6$
13. $A = \frac{2}{5}, B = \frac{3}{5}, C = -\frac{3}{5}$
15. $A = 1, B = -1, C = 1$
17. $A = \frac{10}{3}, B = \frac{1}{6}, C = -\frac{1}{2}$
19. $A = 4, B = 1, C = -5$
21. $A = 0, B = 1, C = 0, D = -4$
23. (a)

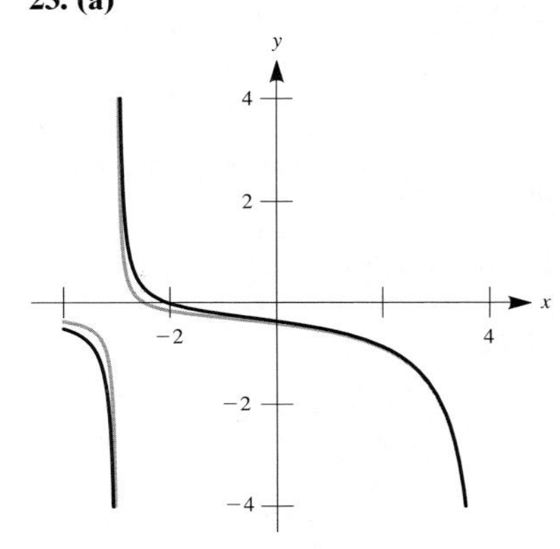

(b)

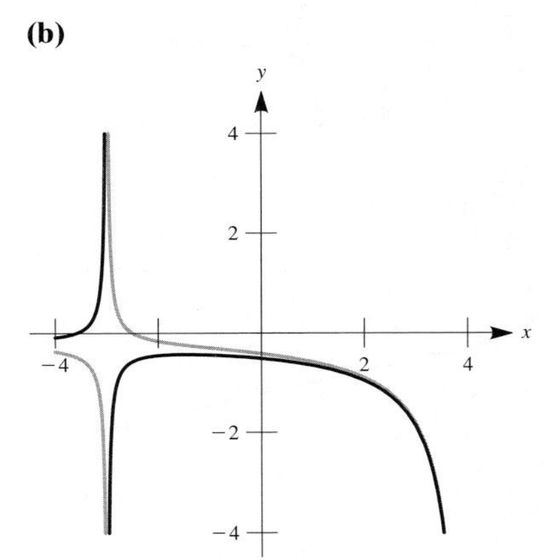

(c) $\dfrac{2x + 5}{(x - 4)(x + 3)} = \dfrac{13/7}{x - 4} + \dfrac{1/7}{x + 3}$
25. (a) $A = 3, B = 4, C = -1$
(b) $B = 4, C = -1$
27. (a) The possible rational roots are -1 and 1, neither of which are roots.

(b)

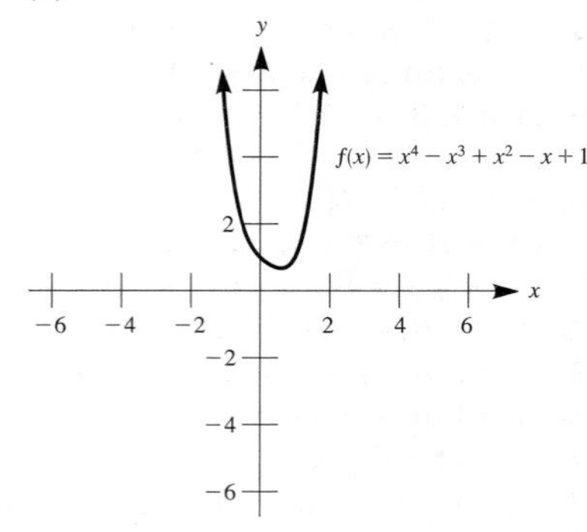

(c) $\left(x^2 + \frac{-1+\sqrt{5}}{2}x + 1\right) \times \left(x^2 + \frac{-1-\sqrt{5}}{2}x + 1\right)$

(d) $x = \frac{1-\sqrt{5} \pm i\sqrt{10+2\sqrt{5}}}{4}, \frac{1+\sqrt{5} \pm i\sqrt{10-2\sqrt{5}}}{4}$

29. (a)

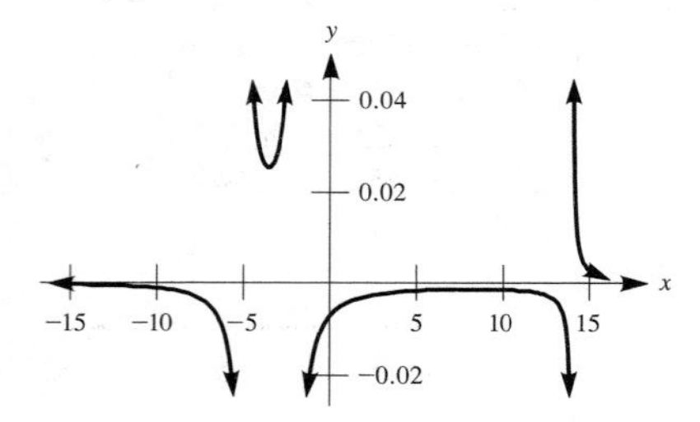

(b)

−14.005
−14
−0.00330
−0.00331

(c) $\frac{-1/48}{x+2} + \frac{1/57}{x+5} + \frac{1/304}{x-14}$

Exercise Set 12.8

1. (a) no **(b)** yes **3. (a)** no **(b)** yes **5. (a)** $(x+10)(x-10)$
(b) $\frac{A}{x+10} + \frac{B}{x-10}$
(c) $\frac{4}{x+10} + \frac{7}{x-10}$

7. (a) $(x+\sqrt{5})(x-\sqrt{5})$
(b) $\frac{A}{x+\sqrt{5}} + \frac{B}{x-\sqrt{5}}$
(c) $\frac{5}{x+\sqrt{5}} + \frac{3}{x-\sqrt{5}}$

9. (a) $(x-3)(x+2)$
(b) $\frac{A}{x+2} + \frac{B}{x-3}$
(c) $\frac{-5}{x+2} + \frac{12}{x-3}$

11. (a) $(x-3)(x-2)(x+2)$
(b) $\frac{A}{x-3} + \frac{B}{x-2} + \frac{C}{x+2}$
(c) $\frac{8}{x-3} + \frac{-2}{x-2} + \frac{-3}{x+2}$

13. (a) $x(x^2+x+1)$
(b) $\frac{A}{x} + \frac{Bx+C}{x^2+x+1}$
(c) $\frac{5}{x} + \frac{-3}{x^2+x+1}$

15. (a) $(x^2+1)^2$
(b) $\frac{Ax+B}{x^2+1} + \frac{Cx+D}{(x^2+1)^2}$
(c) $\frac{2x}{x^2+1} + \frac{3x-4}{(x^2+1)^2}$

17. $\frac{\frac{19}{28}}{x-6} + \frac{-\frac{3}{7}}{x+1} + \frac{\frac{3}{4}}{x+2}$

19. $\frac{10}{3x-5} + \frac{-7}{2x-3}$

21. $\frac{-1/5}{x} + \frac{(1-2\sqrt{5})/10}{x+\sqrt{5}} + \frac{(1+2\sqrt{5}/10}{x-\sqrt{5}}$

23. $\frac{x}{x^2+4} + \frac{-4x+2}{(x^2+4)^2}$

25. $\frac{\frac{3}{125}}{x} + \frac{\frac{122}{125}}{x-5} + \frac{\frac{253}{25}}{(x-5)^2} + \frac{\frac{127}{5}}{(x-5)^3}$

27. $\frac{\frac{1}{3}}{x-1} + \frac{-\frac{1}{3}x - \frac{2}{3}}{x^2+x+1}$

29. $\frac{3}{x} + \frac{-2}{x^2} + \frac{4}{x+1} + \frac{3}{(x+1)^2}$

31. $\frac{\frac{8}{27}}{x+3} + \frac{\frac{11}{54}}{x-3} + \frac{\frac{1}{2}x + \frac{5}{18}}{x^2+9}$

33. $\frac{-x}{x^2+1} + \frac{x+1}{x^2+x+1}$

35. $x = -3, -1, 2$

37. (b) $C = -2, D = 0$

39. $6x + 8 + \frac{2}{x-3} + \frac{-1}{x-1}$

41. $x - 4 + \frac{-\frac{1}{8}}{x} + \frac{\frac{1}{8}}{x-2} + \frac{\frac{3}{4}}{(x-2)^2} + \frac{\frac{5}{2}}{(x-2)^3}$

43. $x^2 + 2x + 5 + \frac{-\frac{1}{4}}{x+1} + \frac{\frac{1}{4}}{x-1} + \frac{-\frac{1}{2}}{x^2+1}$

47. $\frac{(pa+q)/(a-b)}{x-a} + \frac{(pb+q)/(b-a)}{x-b}$

49. $\frac{(pa+q)/2a}{x-a} + \frac{(pa-q)/2a}{x+a}$

51.
$$\frac{\frac{a^2}{(a-b)(a-c)}}{1-ax} + \frac{\frac{b^2}{(b-a)(b-c)}}{1-bx} + \frac{\frac{c^2}{(c-a)(c-b)}}{1-cx}$$

53. $\frac{\frac{\sqrt{2}}{4}x + \frac{1}{2}}{x^2 + \sqrt{2}x + 1} + \frac{-\frac{\sqrt{2}}{4}x + \frac{1}{2}}{x^2 - \sqrt{2}x + 1}$

Chapter 12 Review Exercises

1. $q(x) = x^3 + x^2 - 3x + 1, R(x) = -1$
3. quotient: $x^3 + 3x^2 + 7x + 21$; remainder: 71
5. quotient: $2x^2 - 13x + 46$; remainder: −187 **7.** quotient: $5x - 20$; remainder: 0
9. $f(10) = 99{,}904$
11. $f(\frac{1}{10}) = -\frac{999}{1000}$
13. $f(a-1) = a^3$
15. (a) $f(-0.3) \approx -0.24$
(b) $f(-0.39) \approx -0.007$
(c) $f(-0.394) \approx 0.00003$
17. $a = -5$ **19.** $a = -1, -2$
23. $\pm1, \pm2, \pm3, \pm6, \pm9, \pm18$
25. $\pm1, \pm\frac{1}{2}, \pm2, \pm4, \pm8$ **27.** $\pm p, \pm1$
29. $2, -\frac{3}{2}, -1$
31. $\frac{5}{2}, -1 \pm \sqrt{3}$ **33.** $\frac{2}{3}, \frac{-1 \pm i\sqrt{3}}{2}$
35. $-1, -\sqrt{7}$ (multiplicity 2), $\sqrt{7}$ (multiplicity 2)
37. 2 (multiplicity 2), 5
39. see text

41. $6(x - \frac{4}{3})[x - (-\frac{5}{2})]$
43. $(x - 4)(x - (-\sqrt[3]{5})) \times \left(x - \frac{\sqrt[3]{5} + i\sqrt{3\sqrt[3]{25}}}{2}\right) \times \left(x - \frac{\sqrt[3]{5} - i\sqrt{3\sqrt[3]{25}}}{2}\right)$
45. $2 \pm 3i, 3$ **47.** $1 \pm i\sqrt{2}, \pm\sqrt{7}$
49. 1 positive real root and 2 complex roots **51.** 1 negative real root and 2 complex roots **53.** 1 positive real root, 1 negative real root and 2 complex roots **55. (c)** $-1, \pm i$
57. between 0.82 and 0.83
59. (c) between 6.93 and 6.94
61. $x^2 - 8x + 11 = 0$
63. $x^4 - 12x^3 + 35x^2 + 60x - 200 = 0$
65. $x^4 - 4x^3 - 4x^2 + 16x - 8 = 0$
67. zeros: 0, 3, −1

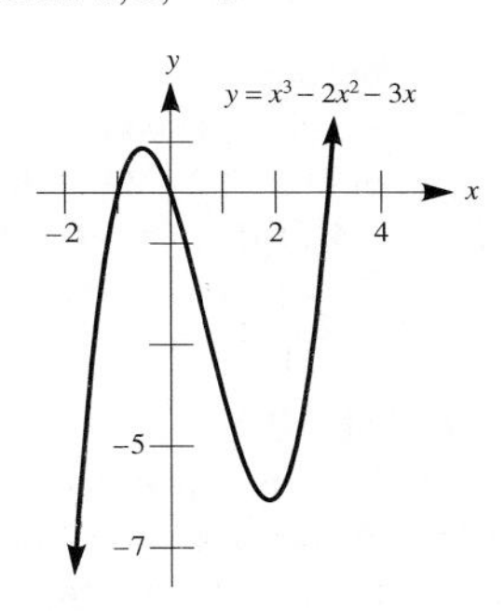

69. zeros: 0, −2, 2

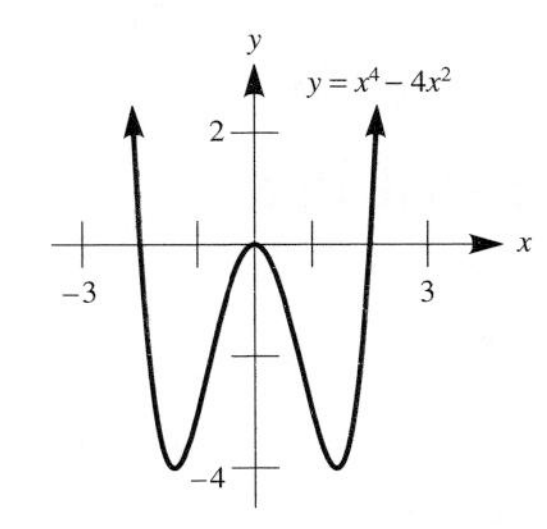

71. $5 + 6i$ **73.** 6 **75.** $\frac{1}{2} - \frac{\sqrt{3}}{2}i$
77. $3\sqrt{2} - 4\sqrt{2}i$ **81. (a)** $|6 + 2i| = 2\sqrt{10}$, $|6 - 2i| = 2\sqrt{10}$
(b) $|-3| = 3$
87. $\frac{-\frac{21}{20}}{10 + x} + \frac{\frac{19}{20}}{10 - x}$
89. $\frac{1}{x} + \frac{-1}{x + 1} + \frac{-1}{(x + 1)^2}$
91. $\frac{x}{x^2 + 3} + \frac{-3x + 2}{(x^2 + 3)^2}$
93. (a) $(x^2 - x + 1)(x^2 - x - 1)$
(b) $\frac{\frac{1}{2}}{x^2 - x + 1} + \frac{x + \frac{1}{2}}{x^2 - x - 1}$

Chapter 12 Test

1. $f(\frac{1}{2}) = -\frac{3}{2}$ **2.** $-3, 1 \pm \sqrt{6}$
3. $\pm 1, \pm\frac{1}{2}, \pm 2, \pm 3, \pm\frac{3}{2}, \pm 6$
4. $f(x) = 3x^2 + 21x - 24$
5. quotient: $4x^2 - 3x - 5$; remainder: 8 **6.** See text.
7. (a) 3 is an upper bound.
(b) between 2.2 and 2.3 **8.** $1 \pm i$, $3 \pm 2i, -2$ **9.** $q(x) = x^2 + 2x - 1$, $R(x) = -3x + 7$
10. $2[x - (\frac{3}{2} + \frac{1}{2}i)][x - (\frac{3}{2} - \frac{1}{2}i)]$
11. (a) $\pm 1, \pm 2, \pm 3, \pm 4, \pm 6, \pm 8, \pm 12, \pm 24$ **(c)** $x = 1$ **(d)** none
12. (a) $\frac{3}{2}$ **(b)** $\frac{3}{2}, \frac{-1 \pm i\sqrt{3}}{2}$
13. 1 positive real root, 1 negative real root, and 2 complex roots
14. $x^3 + 6x + 20 = 0$
15. $f(x) = (x - 2)(x - 3i)^3 [x - (1 + \sqrt{2})]^2$
16. $21 + (1 + \sqrt{3})i$ **17.** $-\frac{1}{17} + \frac{13}{17}i$
18. $\frac{\frac{1}{16}}{x} + \frac{-\frac{13}{32}}{x + 4} + \frac{\frac{11}{32}}{x - 4}$
19. $\frac{\frac{1}{4}}{x - 1} + \frac{-\frac{1}{4}x - \frac{1}{4}}{x^2 + 3}$
20. $\frac{5}{x} + \frac{-1}{x - 2} + \frac{3}{(x - 2)^2}$

CHAPTER 13

Exercise Set 13.1

27. (a)

n	1	2	3	4	5
$f(n)$	$\frac{1}{2}$	$\frac{2}{3}$	$\frac{3}{4}$	$\frac{4}{5}$	$\frac{5}{6}$

(b) $\frac{6}{7}$ **(c)** $f(n) = \frac{1}{1 \times 2} + \frac{1}{2 \times 3} + \cdots + \frac{1}{n(n + 1)} = \frac{n}{n + 1}$
29. (a)

n	1	2	3	4	5
$f(n)$	1	4	9	16	25

(b) 36 **(c)** $f(n) = n^2$

Exercise Set 13.2

3. 120 **5.** 105 **7. (a)** 10 **(b)** 5
9. $n^2 + 3n + 2$ **11.** 0
13. $a^9 + 9a^8b + 36a^7b^2 + 84a^6b^3 + 126a^5b^4 + 126a^4b^5 + 84a^3b^6 + 36a^2b^7 + 9ab^8 + b^9$
15. $8A^3 + 12A^2B + 6AB^2 + B^3$
17. $1 - 12x + 60x^2 - 160x^3 + 240x^4 - 192x^5 + 64x^6$
19. $x^2 + 4x\sqrt{xy} + 6xy + 4y\sqrt{xy} + y^2$
21. $x^{10} + 5x^8y^2 + 10x^6y^4 + 10x^4y^6 + 5x^2y^8 + y^{10}$
23. $1 - \frac{6}{x} + \frac{15}{x^2} - \frac{20}{x^3} + \frac{15}{x^4} - \frac{6}{x^5} + \frac{1}{x^6}$
25. $\frac{x^3}{8} - \frac{x^2y}{4} + \frac{xy^2}{6} - \frac{y^3}{27}$
27. $a^7b^{14} + 7a^6b^{12}c + 21a^5b^{10}c^2 + 35a^4b^8c^3 + 35a^3b^6c^4 + 21a^2b^4c^5 + 7ab^2c^6 + c^7$
29. $x^8 + 8\sqrt{2}x^7 + 56x^6 + 112\sqrt{2}x^5 + 280x^4 + 224\sqrt{2}x^3 + 224x^2 + 64\sqrt{2}x + 16$
31. $5\sqrt{2} - 7$ **33.** $89\sqrt{3} + 109\sqrt{2}$
35. $12 - 24\sqrt[3]{2} + 12\sqrt[3]{4}$
37. $x^{10} - 10x^9 + 35x^8 - 40x^7 - 30x^6 + 68x^5 + 30x^4 - 40x^3 - 35x^2 - 10x - 1$ **39.** $120a^2b^{14}$
41. $100x^{99}$ **43.** 45 **45.** 294912
47. 28 **49.** 40095
53. (a)

k	0	1	2	3	4	5	6	7	8
$\binom{8}{k}$	1	8	28	56	70	56	28	8	1

Exercise Set 13.3

1. $\frac{1}{2}, \frac{2}{3}, \frac{3}{4}, \frac{4}{5}$ **3.** $-1, 1, -1, 1$
5. $\frac{1}{2}, \frac{1}{4}, \frac{1}{8}, \frac{1}{16}$ **7.** 3, 6, 9, 12,
9. $2, \frac{9}{4}, \frac{64}{27}, \frac{625}{256}$ **11.** $-1, 0, \frac{1}{3}, \frac{1}{2}$
13. $-2, 1, -\frac{8}{9}, 1$
15. 1, 4, 25, 676, 458329
17. 2, 2, 4, 8, 32 **19.** 1, 1, 2, 6, 24
21. 0, 1, 2, 4, 16
23.

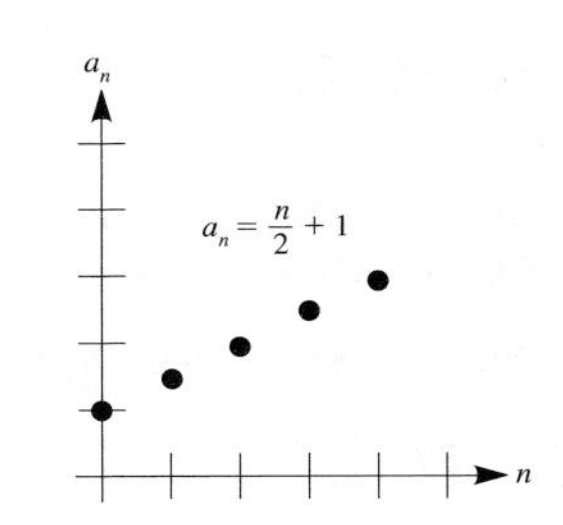

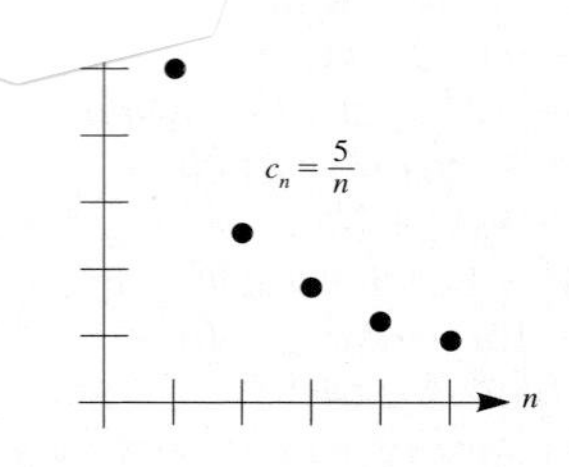

27.

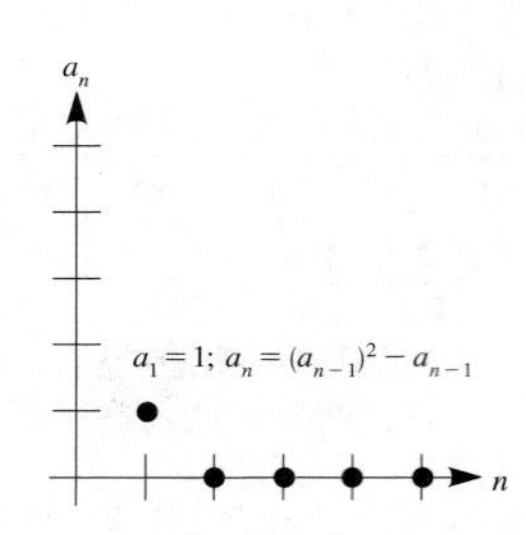

29.

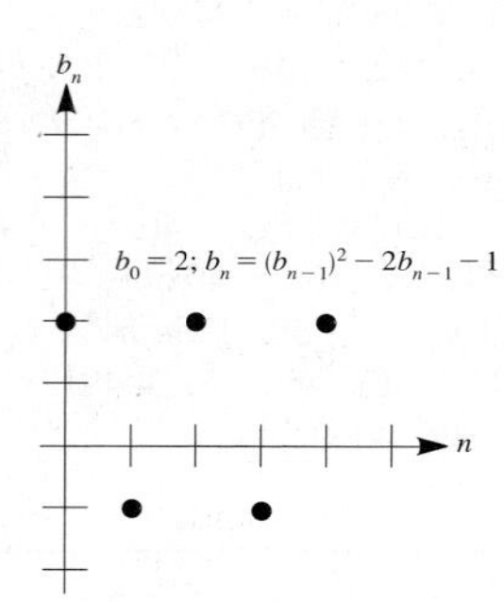

31.

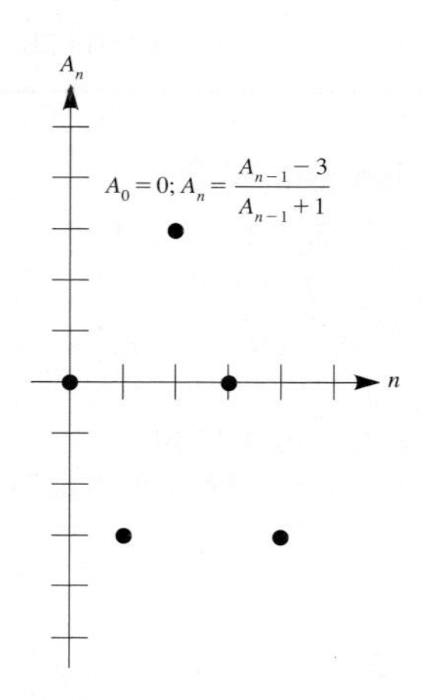

33. 62 **35.** 40 **37.** $-\frac{19}{30}$ **39.** 903 **41.** 3 **43.** 41 **45.** $x + x^2 + x^3$ **47.** $\frac{25}{12}$ **49.** -1 **51.** $\frac{6}{7}$ **53.** $\sum_{j=1}^{4} 5^j$ **55.** $\sum_{j=1}^{6} x^j$ **57.** $\sum_{k=1}^{12} \frac{1}{k}$

59. $\sum_{j=1}^{5} (-1)^{j+1} 2^j$ **61.** $\sum_{j=1}^{5} (-1)^{j+1} j$

63. (a) 0.7, 0.49, 0.2401, 0.05765, 0.00332, 0.00001; they approach 0. **(b)** $s_{10} \approx 4.90 \times 10^{-80}$ **(c)** 0

65. (a)

t	0	1	2	3	4	5
P_t	300	1111	1829	1468	1691	1559

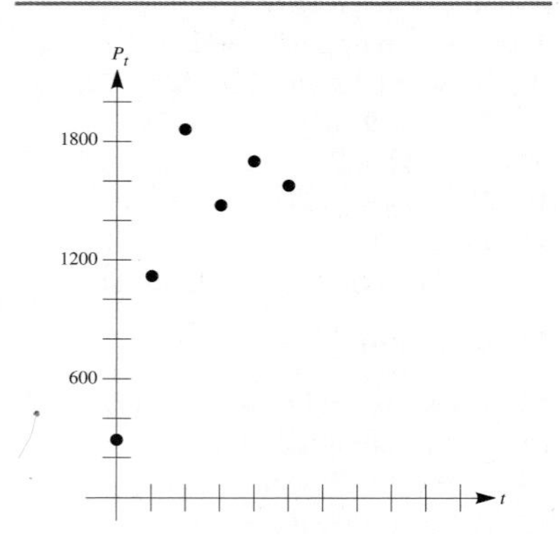

The population seems to be oscillating closer to a value near 1600.

(b)

t	P_t	t	P_t
0	300	11	1607
1	1111	12	1611
2	1829	13	1608
3	1468	14	1610
4	1691	15	1609
5	1559	16	1610
6	1640	17	1609
7	1591	18	1610
8	1621	19	1609
9	1602	20	1610
10	1614		

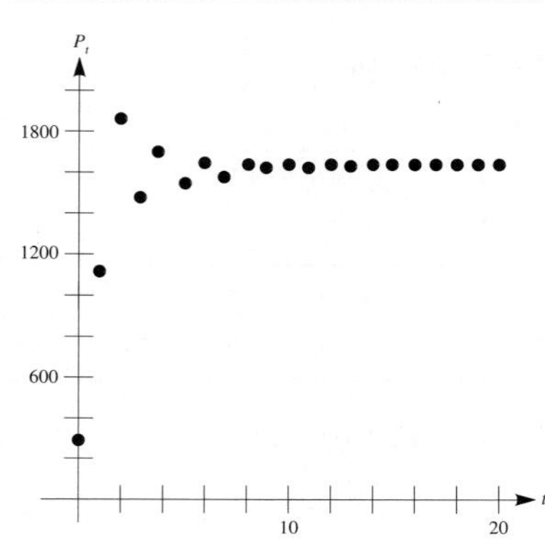

(c) $P_{t-1} \approx 1609$

67. (a)

F_1	F_2	F_3	F_4	F_5	F_6	F_7	F_8	F_9	F_{10}
1	1	2	3	5	8	13	21	34	55

(b) $F_{22} = 17{,}711$; $F_{19} = 4181$ **(c)** $F_{30} = 832{,}040$ **73. (a)** $F_1 = 1$; $F_2 = 1$ **(b)** $F_{24} = 46{,}368$; $F_{25} = 75{,}025$ **(c)** $F_{26} = 121{,}393$

Exercise Set 13.4

1. (a) 2 **(b)** -4 **(c)** $\frac{1}{3}$ **(d)** $\sqrt{2}$ **3.** $a_{12} = 131$ **5.** $a_{100} = 501$ **7.** $a_{1000} = 998$ **9.** $d = \frac{11}{6}$; $a = -\frac{23}{2}$ **11.** $a = -190$ **13.** $d = -\frac{1}{8}$ **15.** $S_{1000} = 500{,}500$ **17.** $S_{13} = \frac{91\pi}{3}$ **19.** $a = \frac{1}{2}$ **21.** $S_{16} = -768$; $d = -\frac{104}{15}$ **23.** $d = \frac{6}{5}$; $a = -\frac{17}{5}$ **25.** 900 **27.** 2, 10, 18 or 18, 10, 2 **29.** -1, 2, 5 or 5, 2, -1 **31. (b)** $S_6 = -6 - 9\sqrt{2}$ **33.** $S_n = \frac{n}{2(1-b)}[2 + (n-3)\sqrt{b}]$

Exercise Set 13.5

1. 6 **3.** 20; 100 **5.** $a_{100} = 1$ **7.** $a_8 = \frac{256}{6561}$ **9.** $r = \pm 4$ **11.** 7161 **13.** $63 + 31\sqrt{2}$ **15.** $\frac{1995}{64}$ **17.** $\frac{11111}{1000000} = 0.011111$ **19.** $\frac{2}{5}$ **21.** 101 **23.** $\frac{5}{9}$ **25.** $\frac{61}{495}$ **27.** $\frac{16}{37}$ **29.** $r = -\frac{1}{2}, -2$ **35.** 12 ft

Exercise Set 13.6

1.

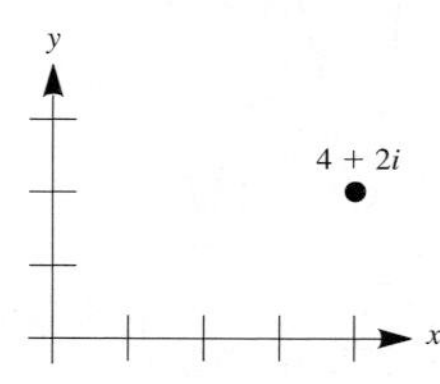

3.

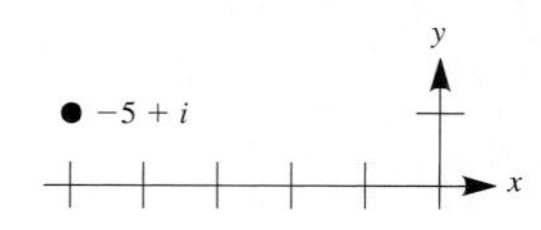

5.

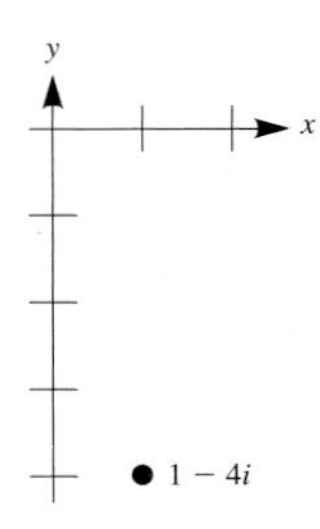

7.

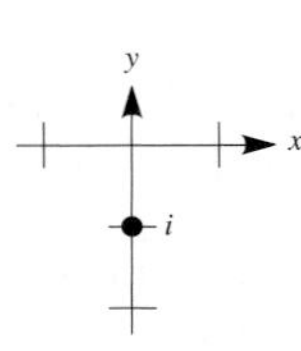

9. $\sqrt{2} + \sqrt{2}i$ **11.** $-2\sqrt{3} + 2i$
13. $-1 - i$ **15.** $\sqrt{3}i$
17. $(\sqrt{6} - \sqrt{2}) + (\sqrt{6} + \sqrt{2})i$
19. $\cos(\pi/6) + i\sin(\pi/6)$
21. $2[\cos(2\pi/3) + i\sin(2\pi/3)]$
23. $4[\cos(7\pi/6) + i\sin(7\pi/6)]$
25. $6[\cos(3\pi/2) + i\sin(3\pi/2)]$
27. $\frac{1}{2}[\cos(11\pi/6) + i\sin(11\pi/6)]$
29. $3 + 3\sqrt{3}i$ **31.** $1 - \sqrt{3}i$
33. $3\sqrt{2}\cos(2\pi/7) + [3\sqrt{2}\sin(2\pi/7)]i$
35. $\dfrac{3\sqrt{2}}{2} + \dfrac{3\sqrt{2}}{2}i$ **37.** $\sqrt{3} + i$
39. 1 **41.** $\dfrac{243}{2} - \dfrac{243\sqrt{3}}{2}i$
43. $\dfrac{\sqrt{2}}{128} + \dfrac{\sqrt{2}}{128}i$ **45.** $-4i$
47. $-1 - i$ **49.** $\dfrac{1}{2} + \dfrac{\sqrt{3}}{2}i$
51. $128 + 128i$ **53.** $-128 + 128\sqrt{3}i$
55. $3i, -\frac{3}{2}\sqrt{3} - \frac{3}{2}i, \frac{3}{2}\sqrt{3} - \frac{3}{2}i$
57. $1, \frac{1}{2}\sqrt{2} + \frac{1}{2}\sqrt{2}i, i, -\frac{1}{2}\sqrt{2} + \frac{1}{2}\sqrt{2}i, -1, -\frac{1}{2}\sqrt{2} - \frac{1}{2}\sqrt{2}i, -i, \frac{1}{2}\sqrt{2} - \frac{1}{2}\sqrt{2}i$
59. $4, -2 + 2\sqrt{3}i, -2 - 2\sqrt{3}i$
61. $3, \frac{3}{2} + \frac{3}{2}\sqrt{3}i, -\frac{3}{2} + \frac{3}{2}\sqrt{3}i, -3, -\frac{3}{2} - \frac{3}{2}\sqrt{3}i, \frac{3}{2} - \frac{3}{2}\sqrt{3}i$
63. 92,236,816
65. $0.95 + 0.31i, i, -0.95 + 0.31i, -0.59 - 0.81i, 0.59 - 0.81i$
67. $\frac{1}{2}(\sqrt{6} - \sqrt{2}) + \frac{1}{2}(\sqrt{6} + \sqrt{2})i, -\frac{1}{2}(\sqrt{6} + \sqrt{2}) + \frac{1}{2}(\sqrt{6} - \sqrt{2})i, \frac{1}{2}(\sqrt{2} - \sqrt{6}) - \frac{1}{2}(\sqrt{2} + \sqrt{6})i$, and $\frac{1}{2}(\sqrt{2} + \sqrt{6}) + \frac{1}{2}(\sqrt{2} - \sqrt{6})i$
69. (a) $1, -\frac{1}{2} + \frac{1}{2}\sqrt{3}i, -\frac{1}{2} - \frac{1}{2}\sqrt{3}i$
71. -1 **73.** 1

Chapter 13 Review Exercises

11. $81a^4 + 108a^3b^2 + 54a^2b^4 + 12ab^6 + b^8$
13. $x^4 + 4x^3\sqrt{x} + 6x^3 + 4x^2\sqrt{x} + x^2$
15. $x^{10} - 10x^8y^2 + 40x^6y^4 - 80x^4y^6 + 80x^2y^8 - 32y^{10}$
17. $1 + \dfrac{5}{x} + \dfrac{10}{x^2} + \dfrac{10}{x^3} + \dfrac{5}{x^4} + \dfrac{1}{x^5}$
19. $a^4b^2 - 4a^3b^2\sqrt{ab} + 6a^3b^3 - 4a^2b^3\sqrt{ab} + a^2b^4$ **21.** $15xy^8$
23. 84 **25.** 28 **27.** Each side is 6.
29. Each side is 70. **31.** Each side is 8. **33.** $1, \frac{4}{3}, \frac{3}{2}, \frac{8}{5}$

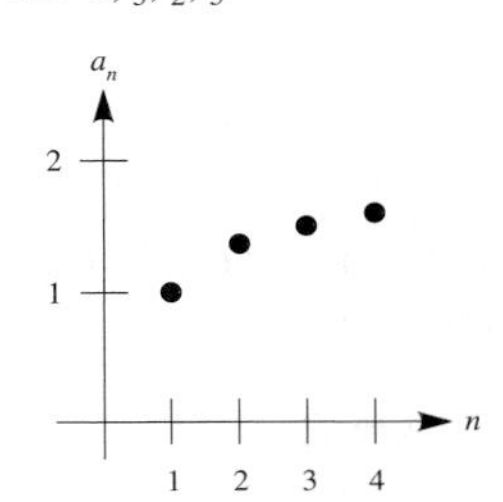

35. $-\frac{1}{2}, \frac{2}{3}, -\frac{3}{4}, \frac{4}{5}$

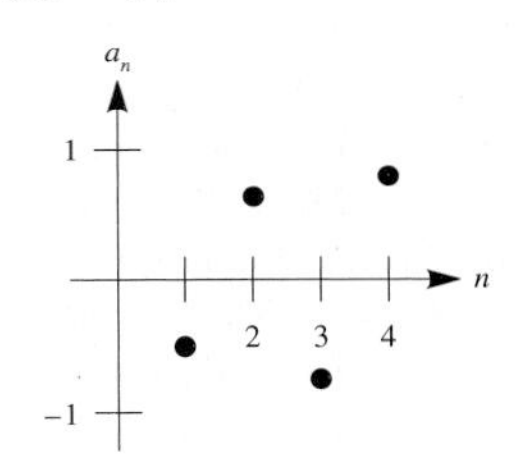

37. $-3, -12, -48, -192$
39. (a) -5 **(b)** $\frac{9}{10}$ **41.** $\displaystyle\sum_{k=1}^{5} \frac{5}{3k}$
43. $a_{18} = 73$ **45.** $a_{12} = \frac{5}{1024}$
47. $S_{12} = 177$
49. $S_{10} = 7{,}777{,}777{,}777$
51. $a_6 = -5\sqrt{10}$ **53.** $\frac{3}{4}$ **55.** $\frac{1}{10}$
57. $\frac{5}{11}$ **63. (a)** 42929 **(b)** 42925; percent error: 0.00932%
(c) 6.48040×10^{10}
(d) 6.48027×10^{10}; percent error; 2×10^{-3}%
65. $r = \frac{1}{3}, 3, \dfrac{-8 \pm \sqrt{55}}{3}$
73. (a) 212993 **(b)** 103766 **(c)** $\frac{789}{512}$
75. $\dfrac{\sqrt{3}}{2} + \dfrac{1}{2}i$ **77.** $\dfrac{5\sqrt{2}}{2} - \dfrac{5\sqrt{2}}{2}i$
79. $3(\cos\frac{\pi}{3} + i\sin\frac{\pi}{3})$
81. $4(\cos\frac{11\pi}{6} + i\sin\frac{11\pi}{6})$
83. $6\sqrt{3} + 6i$ **85.** $\sqrt{6} + \sqrt{2}i$
87. $-16 + 16\sqrt{3}i$
89. $16384\sqrt{2} + 16384\sqrt{2}i$
91. $4i, -2\sqrt{3} - 2i, 2\sqrt{3} - 2i$
93. $\dfrac{\sqrt{2\sqrt{2}} + \sqrt{6\sqrt{2}}}{4} + \dfrac{\sqrt{6\sqrt{2}} - \sqrt{2\sqrt{2}}}{4}i$, $\dfrac{\sqrt{2\sqrt{2}} - \sqrt{6\sqrt{2}}}{4} + \dfrac{\sqrt{6\sqrt{2}} + \sqrt{2\sqrt{2}}}{4}i$, $\dfrac{-\sqrt{6\sqrt{2}} - \sqrt{2\sqrt{2}}}{4} + \dfrac{\sqrt{2\sqrt{2}} - \sqrt{6\sqrt{2}}}{4}i$, $\dfrac{\sqrt{6\sqrt{2}} - \sqrt{2\sqrt{2}}}{4} - \dfrac{\sqrt{6\sqrt{2}} + \sqrt{2\sqrt{2}}}{4}i$

Chapter 13 Test

2. (a) 27 **(b)** -6
3. (a) $S_n = \dfrac{a(1 - r^n)}{1 - r}$ **(b)** $\frac{174075}{1024}$
4. (a) 42240 **(b)** $5280a^7b^{12}$
5. $243x^{10} + 405x^8y^3 + 270x^6y^6 + 90x^4y^9 + 15x^2y^{12} + y^{15}$
6. 177 **7.** $\frac{7}{9}$ **8.** 5 and 27
9. $a_6 = -5\sqrt{10}$ **10.** 224
11. (a)

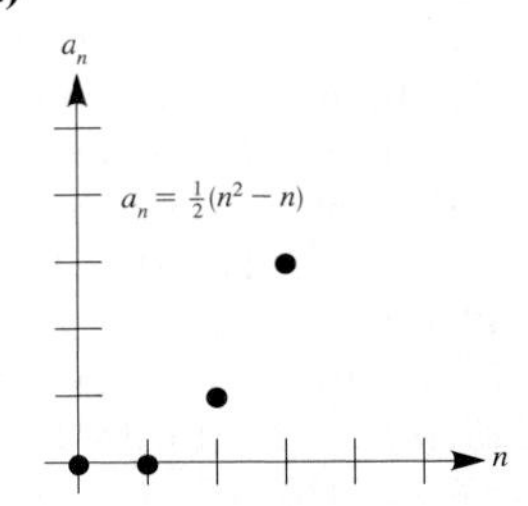

(b)

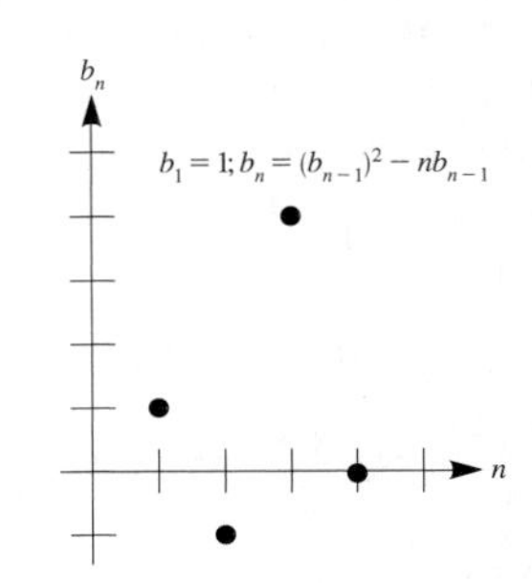

12. $-1 + \sqrt{3}i$
13. $2[\cos(7\pi/4) + i\sin(7\pi/4)]$
14. $\frac{15}{2} + \frac{15}{2}\sqrt{3}i$ **15.** $2\sqrt{3} + 2i$, $-2\sqrt{3} + 2i$, and $-4i$

APPENDIX B

Exercise Set B.1

1. -10 **3.** $\frac{2}{3}$ **5. (a)** a^{15} **(b)** $(a+1)^{15}$ **(c)** $(a+1)^{15}$ **7. (a)** y^{11} **(b)** $(y+1)^{11}$ **(c)** $(y+1)^{18}$ **9. (a)** $x^2 + 3$ **(b)** $\dfrac{1}{x^2+3}$ **(c)** 12 **11. (a)** t^6 **(b)** $\dfrac{1}{t^6}$ **(c)** $(t^2+3)^6$ **13. (a)** $\dfrac{x^4}{y^5}$ **(b)** $\dfrac{y^5}{x^4}$ **(c)** $\dfrac{y^{10}}{x^8}$ **15. (a)** $4x^6$ **(b)** $16x^6$ **(c)** 4 **17. (a)** 1 **(b)** 1 **(c)** 1 **19. (a)** $\frac{11}{100}$ **(b)** $\frac{100}{11}$ **(c)** 1000 **21.** 7 **23. (a)** $\frac{1}{20}$ **(b)** $\frac{1}{225}$ **25.** $\dfrac{1}{a^6b^3}$ **27.** $\dfrac{a^4b^2}{c^6}$ **29.** $\dfrac{y^{12}}{x^6z^{12}}$ **31.** $\dfrac{y^{20}}{x^6z^{16}}$ **33.** y^6 **35.** b **37.** x^{2p} **39.** 576 **41.** 12 **43.** 9.29×10^7 miles **45.** 6.68×10^4 mph **47.** 2.5×10^{19} miles **49. (a)** 1.0×10^{-9} seconds **(b)** 1.0×10^{-18} seconds **(c)** 1.0×10^{-24} seconds

Exercise Set B.2

1. false **3.** true **5.** true **7.** true **9. (a)** -4 **(b)** not a real number **11. (a)** $\frac{2}{5}$ **(b)** $-\frac{2}{5}$ **13. (a)** not a real number **(b)** not a real number **15. (a)** $\frac{4}{3}$ **(b)** $-\frac{3}{5}$ **17. (a)** -2 **(b)** 2 **19. (a)** $3\sqrt{2}$ **(b)** $3\sqrt[3]{2}$ **21. (a)** $7\sqrt{2}$ **(b)** $-2\sqrt[5]{2}$ **23. (a)** $\frac{5}{2}$ **(b)** $\frac{2}{5}$ **25. (a)** $3\sqrt{2}$ **(b)** $3\sqrt[3]{2}$ **27. (a)** $-4\sqrt{2}$ **(b)** $5\sqrt[4]{2}$ **29. (a)** 0.3 **(b)** 0.2 **31.** $-14\sqrt{6}$ **33.** $2\sqrt{2}$ **35. (a)** $6x$ **(b)** $-6y$ **37. (a)** $ab\sqrt{ab}$ **(b)** a^2b^2 **39.** $6ab^2c^2\sqrt{2ac}$ **41.** $2ab\sqrt[4]{b}$ **43.** $\dfrac{2a^4\sqrt[3]{2b^2}}{c^3}$ **45.** $\dfrac{4\sqrt{7}}{7}$ **47.** $\dfrac{\sqrt{2}}{4}$ **49.** $\dfrac{\sqrt{5}-1}{4}$ **51.** $-2-\sqrt{3}$ **53.** $\dfrac{61\sqrt{5}}{5}$ **55.** $\dfrac{\sqrt[3]{5}}{5}$ **57.** $\sqrt[4]{27}$ **59.** $\dfrac{3\sqrt[5]{2ab}}{2ab^2}$ **61. (a)** $\sqrt{3}-1$ **(b)** $\sqrt{3}+1$ **63. (a)** $\dfrac{x(\sqrt{x}+2)}{x-4}$ **(b)** $\dfrac{x(\sqrt{x}+y)}{x-y^2}$ **65. (a)** $\dfrac{x+2\sqrt{2x}+2}{x-2}$ **(b)** $\dfrac{x+2\sqrt{ax}+a}{x-a}$ **67.** $\dfrac{-2(\sqrt{x+h}+\sqrt{x})}{h}$ **69.** $\dfrac{\sqrt[3]{a^2}+\sqrt[3]{a}+1}{a-1}$ **71.** $\dfrac{1}{\sqrt{x}+\sqrt{5}}$ **73.** $\dfrac{1}{\sqrt{2+h}-\sqrt{2}}$ **75.** $\dfrac{1}{\sqrt{x+h}+\sqrt{x}}$ **77.** $a = 9, b = 16$ **79.** $u = 1, v = 8$ **79. (a)** 1.645751

Exercise Set B.3

1. $\sqrt[5]{a^3} = (\sqrt[5]{a})^3$ **3.** $\sqrt[3]{5^2} = (\sqrt[3]{5})^2$ **5.** $\sqrt[4]{(x^2+1)^3} = \sqrt[4]{(x^2+1)^3}$ **7.** $\sqrt[3]{2^{xy}} = (\sqrt[3]{2})^{xy}$ **9.** $p^{2/3}$ **11.** $(1+u)^{4/7}$ **13.** $(a^2+b^2)^{3/p}$ **15.** 4 **17.** $\frac{1}{6}$ **19.** not a real number **21.** 5 **23.** 2 **25.** 4 **27.** -2 **29.** -10 **31.** $\frac{1}{7}$ **33.** not a real number **35.** $\frac{1}{216}$ **37.** 25 **39.** -1 **41.** $\frac{255}{16}$ **43.** $-\frac{28976}{243}$ **45.** $6a^{7/12}$ **47.** $2^{3/2}a^{1/12}$ **49.** x^2+1 **51. (a)** $2^{1/3}3^{5/6}$ **(b)** $\sqrt[6]{972}$ **53. (a)** $2^{7/12}3^{1/3}$ **(b)** $\sqrt[3]{10368}$ **55. (a)** $x^{2/3}y^{4/5}$ **(b)** $\sqrt[15]{x^{10}y^{12}}$ **57. (a)** $x^{(a+b)/3}$ **(b)** $\sqrt[3]{x^{a+b}}$ **59.** $(x+1)^{2/3}$ **(b)** $(x+y)^{2/5}$ **63.** $2x^{1/6}$ **65.** $x^{1/6}y^{1/8}$ **67.** $9^{10/9}$ **69.** $a = 9, b = 16$ **71.** $u = 1, v = 8$ **73.** $x = 4, m = 2$

Exercise Set B.4

1. (a) $(x+8)(x-8)$ **(b)** $7x^2(x^2+2)$ **(c)** $z(11+z)(11-z)$ **(d)** $(ab+c)(ab-c)$ **3. (a)** $(x+3)(x-1)$ **(b)** $(x-3)(x+1)$ **(c)** irreducible **(d)** $(-x+3)(x+1)$ or $-(x-3)(x+1)$ **5. (a)** $(x+1)(x^2-x+1)$ **(b)** $(x+6)(x^2-6x+36)$ **(c)** $8(5-x^2)(25+5x^2+x^4)$ **(d)** $(4ax-5)(16a^2x^2+20ax+25)$ **7. (a)** $(12+x)(12-x)$ **(b)** irreducible **(c)** $(9+y)(15-y)$ **9. (a)** $h^3(1+h)(1-h)$ **(b)** $h^3(10+h)(10-h)$ **(c)** $(h+1)^3(11+h)(9-h)$ **11. (a)** $(x-8)(x-5)$ **(b)** irreducible **13. (a)** $(x+9)(x-4)$ **(b)** $(x-9)(x-4)$ **15. (a)** $(3x+2)(x-8)$ **(b)** irreducible **17. (a)** $(3x-1)(2x+5)$ **(b)** $(6x+5)(x-1)$ **19. (a)** $(t^2+1)^2$ **(b)** $(t+1)^2(t-1)^2$ **(c)** irreducible **21. (a)** $x(4x^2-20x-25)$ **(b)** $x(2x-5)^2$ **23. (a)** $(a-c)(b+a)$ **(b)** $(u+v-y)(x+u+v)$ **25.** $(xz+t)(xz+y)$ **27.** $(a^2-2b^2c^2)^2$ **29.** irreducible **31.** $(x+4)(x^2-4x+16)$ **33.** $x(x^2+3xy+3y^2)$ **35.** $(x-y)(x^2+xy+y^2+1)$ **37. (a)** $(p^2+1)(p+1)(p-1)$ **(b)** $(p^4+1)(p^2+1)(p+1)(p-1)$ **39.** $(x+1)^3$ **41.** irreducible **43.** $(\frac{5}{4}+c)(\frac{5}{4}-c)$ **45.** $(z^2+\frac{9}{4})(z+\frac{3}{2})(z-\frac{3}{2})$ **47.** $\left(\dfrac{5}{mn}-1\right)\left(\dfrac{25}{m^2n^2}+\dfrac{5}{mn}+1\right)$ **49.** $(\frac{1}{2}x+y)^2$ **51.** $(x-a)(8x-8a+1) \times (8x-8a-1)$ **53.** $(x-y-a)(x-y+a)$ **55.** $x(7x-3)(3x+13)$ **57.** $(5-2x+3y)(5+2x-3y)$ **59.** $(ax+b)(x+1)$ **61.** $-x(x+1)^{1/2}$ **63.** $x(x+1)^{-3/2}$ **65.** $-\frac{2}{3}x(2x+3)^{1/2}$ **67. (a)** 199 **(b)** 296 **(c)** 1999 **69.** $(A+B)^3$ **71.** $x(a^2+x^2)^{-3/2}(2a^2+x^2)$ **73.** $(y^2-pq)(y-p)(y-q)$

Exercise Set B.5

1. $x-3$ **3.** $\dfrac{1}{(x-2)(x^2+4)}$ **5.** $\dfrac{1}{x-2}$ **7.** $\dfrac{3b}{2a}$ **9.** $\dfrac{a^2+1}{a-1}$ **11.** $\dfrac{x^2+xy+y^2}{(x-y)^2}$ **13.** 2 **15.** $\dfrac{x^2+x}{(x+4)(x-2)}$

17. $\dfrac{x^2 - xy + y^2}{(x - y)(x + y)}$ **19.** $\dfrac{4x - 2}{x^2}$ **21.** $\dfrac{36 - a^2}{6a}$ **23.** $\dfrac{4x + 11}{(x + 3)(x + 2)}$ **25.** $\dfrac{3x^2 + 6x - 6}{(x - 2)(x + 2)}$ **27.** $\dfrac{6ax^2 - 4ax + a}{(x - 1)^3}$ **29.** $\dfrac{8}{x - 5}$ **31.** $\dfrac{2a}{a + b}$ **33.** $\dfrac{-9}{(x + 5)(x - 4)^2}$ **35.** $\dfrac{-p^2 - pq - q^2}{p(2p + q)(p - 5q)}$ **37.** $\dfrac{1 + x}{1 - x}$ **39.** $-\dfrac{1}{ax}$ **41.** $a - 1$ **43.** $-\dfrac{1}{4 + 2h}$ **45.** $\dfrac{a + x}{a^2x^2}$ **47.** $\dfrac{x}{1 + 2x}$ **49.** $\dfrac{1}{a + a^2}$ **51.** $\dfrac{x^2 + y^2}{x^2 - y^2}$ **53.** $\dfrac{a^2 - b^2}{2}$ **55.** $\left(\dfrac{a}{b}\right)^{a+b}$

Index

A

D

E

F

Q

R

S

T

U

V

W

X

Y

Z

CREDITS